W9-COS-037

PRINCIPLES OF GENETICS

D. Peter Snustad
University of Minnesota

Michael J. Simmons
University of Minnesota

John B. Jenkins
Swarthmore College

Epilogue by James F. Crow

John Wiley & Sons, Inc.
New York • Chichester • Brisbane • Toronto • Singapore • Weinheim

ACQUISITIONS EDITOR David Harris
SENIOR DEVELOPMENTAL EDITOR Barbara Heaney
SENIOR MARKETING MANAGER Catherine Faduska
SENIOR PRODUCTION EDITOR Katharine Rubin
DESIGNER Harry Nolan
MANUFACTURING MANAGER Dorothy Sinclair
PHOTO EDITOR Lisa Passmore
PHOTO RESEARCHER Ramón Rivera Moret
ILLUSTRATION EDITOR Edward Starr
DEVELOPMENTAL PROGRAM ASSISTANT Pui Szeto

Cover Photos *Background:* Philippe Plailly/Science Photo Library/Photo Researchers *From left to right, top to bottom:* Oliver Meches/Photo Researchers; M. Wurtz/Biozentrum, University of Basel/Science Photo Library/Photo Researchers; Biophoto Associates/Photo Researchers; Marc Chamberlain/Tony Stone Images; Philippe Plailly/Eurelios/Science Photo Library/Photo Researchers; David Scharf/Peter Arnold, Inc.; Courtesy of Lisa Passmore; Courtesy of Jeffrey L. Carpenter, Carolyn D. Silflow, and D. Peter Snustad, Department of Genetics & Cell Biology, University of Minnesota, St. Paul; NIBSC/Science Photo Library/Photo Researchers.

This book was set in 10/12 Palatino by Ruttle, Shaw & Wetherill, and printed and bound by Von Hoffmann Press. The cover was printed by Phoenix Color.

Recognizing the importance of preserving what has been written, it is a policy of John Wiley & Sons, Inc. to have books of enduring value published in the United States printed on acid-free paper, and we exert our best efforts to that end.

The paper in this book was manufactured by a mill whose forest management programs include sustained yield harvesting of its timberlands. Sustained yield harvesting principles ensure that the number of trees cut each year does not exceed the amount of new growth.

Library of Congress Cataloging in Publication Data:

Snustad, D. Peter.
Principles of genetics / D. Peter Snustad, Michael J. Simmons, John B. Jenkins.
p. cm.
Includes bibliographical references and index.
ISBN 0-471-31196-0 (alk. paper)
1. Genetics. I. Simmons, Michael J. II. Jenkins, John B. III. Title.
QH430.S68 1997
576.6—dc21 96-39102
CIP

Printed in the United States of America
10 9 8 7 6 5 4 3

About the Authors

D. Peter Snustad is a Professor in the Department of Genetics and Cell Biology at the University of Minnesota, Twin Cities. He received his B.S. degree in science specialization from the University of Minnesota and his M.S. and Ph.D. degrees in genetics from the University of California, Davis. During his 31 years as a member of the faculty at Minnesota, he has taught courses at all levels from general biology to advanced biochemical genetics. For 20 years, his research focused on bacteriophage T4 morphogenesis and the interaction between T4 and its host, *Escherichia coli*. For the past 11 years, his research group has studied the genetic control of the cytoskeleton in *Arabidopsis thaliana* and the glutamine synthetase gene family in *Zea mays*. He has served on the National Institutes of Health Molecular Cytology Study Section and as Program Chairperson for the Annual Meeting of the Genetics Society of America. His honors include the Morse-Amoco and Stanley Dagley Memorial teaching awards. A lifelong love of the Canadian wilderness has kept him in nearby Minnesota.

Michael J. Simmons is a Professor in the Department of Genetics and Cell Biology at the University of Minnesota, Twin Cities. He received his B.A. degree in biology from St. Vincent College in Latrobe, Pennsylvania, and his M.S. and Ph.D. degrees in genetics from the University of Wisconsin, Madison. Dr. Simmons has taught courses in general biology, genetics, population biology, population genetics, and molecular biology. His research activities are focused on the genetic significance of transposable genetic elements, especially those present in the genome of *Drosophila melanogaster*. He has served on advisory committees at the National Institutes of Health and is currently a member of the Editorial Board of *Genetics*, published by the Genetics Society of America. In 1986, Dr. Simmons received the Morse-Amoco teaching award from the University of Minnesota in recognition of his contributions to undergraduate education. One of his favorite activities, figure skating, is especially compatible with the Minnesota climate.

John B. Jenkins is a Professor in the Department of Biology at Swarthmore College, where he teaches genetics, human genetics, and evolution. He received his B.S. and M.S. degrees from Utah State University, where he studied with Eldon J. Gardner, and his Ph.D. from the University of California, Los Angeles, where he studied chemical mutagenesis with Elof Axel Carlson. Dr. Jenkins's research interests are in the areas of mutation, behavior genetics, cytogenetics, and Huntington's disease. He has authored textbooks on genetics and human genetics and has co-authored a collection of classic papers in genetics. He is a member of the Genetics Society of America, the American Society of Human Genetics, and other scientific societies. He has served on numerous committees and panels of the Genetics Society of America, American Society of Human Genetics, and the National Science Foundation. He has served on the Editorial Board of the Education Section of the *American Journal of Human Genetics* and the Board of Reviewers for the *Journal of Medical Genetics*. His love of fishing frequently takes him to the streams of Pennsylvania and the lakes of New Hampshire.

Dedications

To my son Eric, my best friend Judy, and my campfire buddies Jason and Oak. Thanks for sharing those special times.

D.P.S.

To my family, especially to Benjamin John.

M.J.S.

With much love to my extraordinary wife Dotti and to my wonderful children, John, Soraya, Jessica, and Charlotte, all of whom bring so much joy to my life.

J.B.J.

Preface

Although the science of genetics is only 130 years old, it has grown at an explosive rate during the last half of the twentieth century and now impacts virtually all aspects of our lives. What began with the simple and elegant experiments of Gregor Mendel has blossomed into a mature science that is the focus of research throughout the world. When we started teaching 30 years ago, genetics was considered by many biologists to be just one of several specialized disciplines in biology. Today, it is viewed as the core of biology. Genetic approaches to the dissection of biological processes have proven invaluable. The powerful tools of molecular genetics are now routinely used to investigate phenomena as diverse as photosynthesis, the immune response, memory, and evolution. In addition, numerous practical applications of genetics have been documented in fields such as medicine, agriculture, pharmaceutics, and forensics. Given the rapid growth of genetics during the last few decades, it follows that the teaching of genetics has also undergone major changes. *Principles of Genetics* is the culmination of our collective attempt to stay abreast of the exciting new developments in genetics without sacrificing rigor in the coverage of basic Mendelian genetics.

Our philosophy in teaching genetics—and in preparing this textbook—is that learning is best achieved by emphasizing basic principles, especially when these principles are introduced through the analysis of observations and experimental results. Not all genetics instructors will agree with all aspects of our approach to teaching genetics; however, we think that most of them will agree that an introductory course should not only convey concepts but also should demonstrate how genetics is done—how the results of genetic experiments lead to new knowledge.

Focus on the basic principles of genetics. Every week new and important discoveries in genetics are announced. Yet, all of these impressive accomplishments are rooted in basic principles that were elucidated by Mendel and his successors. This textbook strives to develop these principles carefully and thoroughly. We believe that an appreciation of current advances in genetics must be grounded on a foundation of basic principles. For example, the powerful tools of molecular genetics permit scientists to screen individuals for mutations that cause Huntington's disease (Chapter 20, Figure 20.11). However, only through an understanding of Mendelian genetics will students appreciate the 50 percent risk of Huntington's disease among the children of a parent with this disease (see *A Conversation with Nancy Wexler*).

Show how scientific concepts develop from observation and experimentation. No one should lose sight of the fact that genetics, like every science, is a human endeavor. What is known about genetics today is the result of considerable toil and effort. Thousands of geneticists have labored to discover the facts and develop the concepts that make up the science of genetics. Each discovery began with a rather tentative mix of ideas and observations. With careful scrutiny and experimental testing, these ideas and observations eventually gelled into something less tentative—a set of concepts supported by facts. Genetics is replete with examples of this process, many of which we present in this text to emphasize the central role of scientific experimentation in the evaluation of ideas and the importance of critical interpretation of experimental data in testing hypotheses. In addition, we have incorporated *Conversations* with eminent geneticists to emphasize the human component of the scientific process.

Incorporate human examples and show the relevance of genetics to societal issues. Experience has shown us that when we discuss human genetics, our students are more interested and pay closer attention. Because they are more attentive, students tend to comprehend complex concepts better when they are illustrated with human examples. For that reason, we have used human examples to illustrate genetic principles whenever possible. We have also included separate sections on the Human Genome Project, human gene mapping, human gene therapy, and human genetic disorders such as Huntington's disease, cystic fibrosis, and xeroderma pigmentosum. In addition, we have dedicated an entire chapter to behavior genetics (Chapter 26) in which we examine the effects of genes on traits such as intelligence, personality, and sexual orientation.

Genetics, perhaps more than any other science, has sparked numerous social, legal, and ethical debates, which is one reason why students find the subject so fascinating. Many of these controversial issues—genetic screening and its potential for misuse, DNA fingerprinting, genetic engineering, and gene therapy—are products of the Human Genome Project. Although some of these topics are controversial, we believe that it is important to involve students in an intelligent dialogue on these issues because society will increasingly be called upon to address such questions. Hopefully, this textbook will provide students with the background information needed to address these concerns in an informed manner.

Emphasize analytical thinking and problem solving. Genetics has always been a bit different from other disciplines in biology because of its heavy emphasis on analysis and problem solving. In this text, we have fleshed out the analytical nature of genetics in many ways—in the development of principles in classical genetics, in the discussion of experiments in molecular genetics, and in the presentation of calculations in population and quantitative genetics. We have provided special worked-out problems in the *Testing Your Knowledge* section at the end of each chapter to help students develop their analytical skills. For example, students usually master the concept of semiconservative DNA replication quite easily. However, when asked to properly package the parental and nascent DNA strands into chromatids and daughter chromosomes, they often are unable to do so. We have observed that when students work through *Testing Your Knowledge* problem 2 in Chapter 10, they develop an understanding of both semiconservative replication and the packaging of DNA into chromosomes.

AN ADAPTABLE ORGANIZATION

Our most difficult decisions in preparing this text related to organization and content. Should the text begin with Mendel and proceed roughly in chronological fashion? Should it begin with DNA and present classical transmission genetics afterward? Should population and quantitative genetics follow classical genetics, or should they be placed at the very end of the text? Should classical and molecular genetics be interwoven thoughout the text? Obviously, there is no single "correct" way to teach the basic principles of genetics. Thus our goal was to create a text that covers all the core topics and has an adaptable organization. We believe that this text can be adapted to a variety of class formats. For example, if an instructor prefers to discuss molecular genetics prior to classical genetics, she or he can cover Chapters 9–14 and then return to Chapters 3–8. Another goal was to keep the length of the text reasonable. As genetics has grown, so has the length of genetics textbooks. Without doubt, the most challenging choices that we made during the preparation of this text involved the omission of some of our favorite experiments, topics, and historical events. Genetics is a rich and diverse discipline, and we hope that we have made sensible decisions regarding what to include and exclude in this book.

The organization of the text is a mixture of traditional and unique. The organization is unique in that we have recognized the important contributions of viruses (Chapter 15) and bacteria (Chapter 16) to the basic concepts of genetics by covering each in a separate chapter. The genetic basis of cancer is covered in Chapter 22, which emphasizes the most important and recent research on this topic. It is also unique in dedicating an entire chapter (24) to immunogenetics, although this seems almost unavoidable given the impact of AIDS in our world. The organization is traditional in that we cover Mendelian genetics first (Chapters 3–8) and DNA, RNA, and proteins second (Chapters 9–14). The text contains 28 chapters and an epilogue. Chapters 1–2 provide an introduction to the science of genetics and the basic features of living organisms; Chapters 3–8 present the concepts of classical genetics; Chapters 9–14 present the core concepts of molecular genetics; Chapters 15–18 cover the genetics of viruses, bacteria, transposable elements and eukaryotic organelles; Chapters 19–24 contain more advanced topics in molecular genetics; and Chapters 25–28 contain topics on population, quantitative, evolutionary, and behavior genetics.

Finally, the epilogue, *Genetics: Yesterday, Today, and Tomorrow*, written by James F. Crow, an eminent geneticist and a leading contributor to genetics (and to the teaching of genetics), shares his personal views about the history and future prospects of this science. We do not believe that you can read his epilogue without feeling the excitement that has infected the people who have made genetics what it is. We thank Dr. Crow for providing a truly special ending to the text. His wit, charm, and wisdom are greatly appreciated; many thanks, Jim.

ART PROGRAM

Well-designed illustrations are an essential component of any science textbook. Stepped-out illustrations, showing each phase in a process, are an invaluable aid in communicating complex concepts. Thus we have worked very hard to make the illustrations in this text clear, attractive, and pedagogically effective. Approximately 700 figures and 200 photographs have been included to illustrate basic concepts, experimental procedures, and various genetic phenomena. Many complex figures are stepped-out and contain succinct labels to help students break down difficult processes into manageable parts. For example, in Chapter 12, the complex process of translation has been divided into chain initiation (Figure 12.15), chain elongation (Figure 12.17), and chain termination (Figure 12.19), with each component covered in a separate stepped-out illustration. In addition, in developing the illustrations, we have utilized the same color scheme throughout the text so that related items can be identified in different figures based on color.

CONVERSATIONS WITH GENETICISTS

During the writing of this book, one of our most enjoyable tasks was to have conversations with some truly remarkable scientists. The book contains nine conversations with prominent geneticists. We are indebted to these people for sharing their insights about science and education, and we hope that you will find these conversations as fascinating as we have. We thank:

Thomas J. Bouchard, Jr.	University of Minnesota, Twin Cities
Deborah and Brian Charlesworth	University of Chicago
James F. Crow	University of Wisconsin, Madison
Margaret G. Kidwell	University of Arizona
Edward B. Lewis	California Institute of Technology
Johng K. Lim	University of Wisconsin, Eau Claire
James V. Neel	University of Michigan
Mary Lou Pardue	Massachusetts Institute of Technology
Nancy Wexler	Hereditary Disease Foundation, Santa Monica, California

We are grateful to each of these individuals for their willingness to contribute to this book.

PEDAGOGY

This text includes special features designed to emphasize the relevance of the topics discussed, to facilitate the comprehension of important concepts, and to assist students in evaluating their grasp of these concepts. However, we have tried to make sure that these features do not interfere with the flow of the scientific content of the text. The features include:

- ***Sidelights (Technical, Historical, Human Genetics).*** Throughout the text, special topics are presented in separate *Sidelight* sections: *Technical Sidelights* describe important experimental techniques; *Historical Sidelights* provide insights into the history of genetics; and *Human Genetics Sidelights* examine important aspects of human genetics.
- ***Chapter-Opening Vignettes.*** Each chapter opens with a vignette or brief story—usually related to human genetics or historical developments—that emphasizes the relevance of the topics discussed in the chapter.
- ***Key points.*** These learning aids are in-text summaries that appear at the end of each major section of the text. They are designed to help students review for exams and focus on the major concepts covered in each section.
- ***Testing Your Knowledge.*** At the end of each chapter, we have provided worked-out problems to help students hone their analytical and problem-solving skills. The answers walk the students step by step through the solutions to the problems.
- ***Questions and Problems.*** Each chapter ends with 20 to 35 questions and problems. The range of questions will provide students with the opportunity to enhance their understanding of the concepts covered in the chapter and to further develop their analytical skills.
- ***Answers.*** Answers to the odd-numbered problems are given at the back of the text, and answers to all problems are included in the *Instructor's Manual and Test Questions* supplement prepared by Robert Ivarie of the University of Georgia.

SUPPLEMENTS

The Problems Workbook and Study Guide, by H. J. Price of Texas A & M University, is a hands-on workbook designed to improve problem solving skills and to reinforce terminology and concepts from the text. Features included are important concepts, terms, and names; additional problems for self-test; key figures from the text; thought challenging exercises, answers to problems, and approaches to problem solving.

Instructor's Resources on CD-ROM, designed as a lecture enhancer and database of instructor's materials, is highly functional and easy to use. This cross-platform CD-ROM contains the following components and is free to adopters of the text: (1) A database of all the illustrations from the text from which the instructor can create presentations, download to the desktop, and/or print to acetates (2) *Instructor's Manual and Test Questions*, written by Bob Ivarie of the University of Georgia, which contains sample syllabi, lecture outlines, key concepts, teaching strategies for difficult material, and approximately 40 test questions per chapter. The complete set of answers to problems in the text are included here as well. The material can be printed from the CD or exported to word-processing programs for creation of handouts or lecture notes. The *Instructor's Manual and Test Questions* will also be available in a print format. (3) The *Transparency Set* which contains full-color figures from the text. (4) *World Wide Web*. Materials relevant to *Principles of Genetics* which are available to faculty and students.

GenLink, a new electronic supplement that integrates the Table of Contents to the vast resources of the World Wide Web, provides relevant links to genetic sites for you and your students. Linked sites include the latest research findings, forums, and simulations. We invite you to make comments and suggestions at the GenLink homepage so that we can make this program an even more effective learning tool. Please visit GenLink at http://www.wiley.com/genlink.

Of Related Interest: Drlica, *Understanding DNA & Gene Cloning: A Guide For the Curious*, 3/e, 1997.

ACKNOWLEDGMENTS

This book has been greatly influenced by the genetics courses that we have taught over the last 30 years. We must, therefore, acknowledge the many contributions of our students at Swarthmore College and the University of Minnesota. We must also acknowledge the contributions of our teachers and of many colleagues who have contributed to our ongoing education. Their knowledge and wisdom are deeply appreciated.

Manuscript Reviewers

Our gratitude also goes out to a host of reviewers whose criticisms and comments helped to shape the content of this book. These generous critics gave their time and expertise to correct errors and suggest ways in which the book could be improved; we deeply appreciate all their efforts. Of course, any errors that remain are solely our own responsibility. In particular, we acknowledge the constructive comments of the following reviewers:

Faculty Reviewers

Robert Baker
University of Southern California

Anna Berkovitz
Purdue University

Rick Cavicchioli
University of New South Wales

Richard W. Cheney, Jr.
Christopher Newport University

James F. Crow
University of Wisconsin—Madison

Jerry Eberhart
Biology Consultant

Richard Gethman
University of Maryland—Baltimore County

Ben Golden
Kennesaw State College

Charles Green
Rowan College of New Jersey

Keith Hartberg
Bailey University

Richard B. Imberski
University of Maryland—College Park

Bob Ivarie
University of Georgia

Clint Magill
Texas A & M University

Sandra D. Michael
Binghamton University

Gregory Phillips
New Mexico State University

Ruth Phillips
University of Wisconsin

Jim Price
Texas A & M University

Susan Reimer
James Madison University

John Ringo
University of Maine

Dorothy Rosenthal
Science Education Consultant

Mark Sanders
University of California—Davis

John Schiefelbein
University of Michigan

Millard Susman
University of Wisconsin—Madison

Student Reviewers

Paul Bruinsma of Purdue University

Cameron Parry of New Mexico State University

Natalie Sanchez of New Mexico State University

Many people contributed to the development and production of this book. The project was initiated under the supervision of Sally Cheney, former biology editor at John Wiley & Sons. We are indebted to Sally for having confidence in our ability to produce a state-of-the-art genetics text. David Harris, current biology editor at John Wiley, guided the project to completion. David is responsible for many of the features that make this book unique, and we are especially grateful for his guidance and unbounded enthusiasm. We also thank Cathy Donovan, David's editorial program assistant, for her cheerful help throughout the project.

Many professionals at Wiley worked with us during the production of this text. Barbara Heaney, senior development editor, and Pui Szeto, developmental program assistant, kept the project on course. Their organizational skills and good judgment were indispensable for the completion of the project. We deeply appreciate their thoughtful input on a wide range of issues. Barbara Conover, developmental consultant, was responsible for editing our early drafts and for converting our ideas and crude diagrams into the stepped-out illustrations in the text. Barbara, thanks so much! The art program was designed and produced by Poole Visual Communication Group of Cincinnati, and we thank Anthony J. Poole, creative director, for his talent and patience. The art was polished by Wellington Studios. We also thank Edward Starr, Wiley art coordinator, who supervised the development of the art program. Lisa Passmore, photo editor, obtained all the photographs that we requested, and more; thank you, Lisa. We thank Harry Nolan, designer, for creating an eye-catching cover and for developing the text layout. Our gratitude also goes out to Katharine Rubin, production editor, and Betty Pessagno, copy editor, for correcting our mistakes and for putting all the pieces of this book into a common format. We thank Ethan Goodman, marketing manager, for his help in getting this textbook into the hands of prospective users; Bonnie Cabot, former supplements editor, for persuading Jim Price of Texas A & M University to develop a *Problems Workbook and Study Guide*, and Bob Ivarie of the University of Georgia to prepare the *Instructor's Manual and Test Questions*; great job, Bonnie. We also thank Jennifer Yee, current supplements editor, for guiding the supplements to completion.

Lastly, we would like to encourage everyone—students, teaching assistants, instructors, and other readers—to send any comments on the text, corrections, or suggestions for improvements to D. Peter Snustad at John Wiley & Sons, Inc., 605 Third Avenue, New York, NY, 10158-0012, or PSnustad@Wiley.Com.

BRIEF CONTENTS

CONTENTS

The Molecular Biology of Genes

A CONVERSATION WITH

JAMES NEEL

Dr. Neel's life can best be described as a long, varied, and inspirational journey. During his youth Dr. Neel faced adversity: his father died when he was 10, and with resources limited after the Great Depression, he had to work tirelessly to put himself through college. Dr. Neel has contributed enormously to our understanding of human genetics. In fact, some of the principles you will learn during this course are a result of his discoveries. Dr. Neel is currently Lee R. Dice Distinguished Professor Emeritus of Human Genetics and Professor Emeritus of Internal Medicine at the University of Michigan. He received his Ph.D. and M.D. degrees from the University of Rochester. Dr. Neel is also a prolific author. His most recent book, *Physician to the Gene Pool* (Wiley, 1994), reflects on his life in science and challenges some of today's genetic thinking.

Dr. Neel, it is an honor to have you reflect on some of your many achievements over a long and illustrious career. How did you enter the field of human genetics?

As a boy, I collected everything in sight—butterflies, beetles, and so on. When I registered for my first year in college at Wooster, it was a given that I would take a course in biology. I came to a chapter on genetics in the textbook. I was captivated by the Mendelian rules that explained the variation in nature. So, at my first opportunity at Wooster, I took a genetics course under a remarkable person, W.P. Spencer. In my senior year I did a special project on *Drosophila* population genetics that led to my first scientific paper several years later.

My graduate research at Rochester in the late 1930s was in *Drosophila* genetics. Although I was very happy working with *Drosophila,* there was a feeling in the air that maybe *Drosophila,* which had been so useful in the early days of genetics, had run its course as an experimental organism, and people were thinking about other organisms that might yield new genetic insights. This was about the time that genetic studies of bacteria and bacterial viruses were incubating. I found myself wondering if it was going to be possible to do some good genetics on humans. After I finished my degree at Rochester, I continued with a career in *Drosophila* genetics during the next several years, while I was teaching at Dartmouth College, but all the time this feeling was building that I'd like to do human genetics. I was absolutely convinced that if I was going to do genetic research on humans, I should get medical training, because I needed to know about the ills to which the organism I was going to study was subject. So, with that in mind, I found a way to get back to medical school at Rochester.

At Rochester, my mentor was Curt Stern. Stern and Spencer were quite sympathetic to my entry into human genetics, although both crossed their fingers, in the sense that human genetics was at the time not what it is today by any means. The Nazi misuse of genetics, in the name of eugenics, had created a cloud over the field. So there was a major risk involved, but I felt that if we proceeded properly it was the time for human genetics to come into being in this country.

You published important research on the effects of the radiation released by the atomic bombs dropped on Japan. How did you come to be involved in this research?

During the war, Rochester was one of the places doing intensified research on the biological effects of radiation, in connection with the development of the atomic bomb. Now we didn't know that during the war; all we knew was that there was a hush, hush research effort at Rochester. Spencer had come up to Rochester, to be involved with Stern in some of the research, but he never talked about it. After the war, it became clear that they had worked on extending the understanding of the genetic effects of radiation on *Drosophila* to much lower exposure levels than ever studied before.

I was either in medical school or a house officer during the war years. We medical students, of course, did not know how long the war would last, but we did expect to get called up in due time. Peace came before I was called up, but medical officers who had served during the war were being released, and we young fellows were being brought in to replace them. When I saw my military number coming up, I spoke to an acquaintance connected to the radiation research unit. I told him that as a junior officer, I'd probably end up in a field hospital taking care of venereal disease, respiratory infections, and diarrhea, that sort of thing. I asked if there was any thought being given to follow-up studies about the effects of the atomic bombs. One thing led to another, and, when I went on active duty in the Army Medical Corps, I was assigned as support personnel to two civilian VIPs who were going over to Japan to advise our National Academy of Sciences how they might proceed to establish a study of the late effects of exposure to the atomic bombs. This study was to have a genetic component, and I soon found myself deeply involved in how this should be designed. It was really, in retrospect, most unusual that the only Ph.D. geneticist in the Army at that time in the Medical Corps ended up in a genetics job.

You witnessed the devastation caused by the atomic bomb. How did that affect you?

I arrived in Japan about a year after the bombings, but still the devastation was extensive. All of us in the mission realized the full power of the new weaponry, and it was a pretty unsettling experience. It was clear that the world had entered into a new and difficult stage in human relationships. My experience in Japan was very sobering.

Did this experience change your thinking about relationships between science and society?

I wouldn't say it changed my thinking, but it certainly influenced it, particularly as to how difficult it was and is to transmit to society some of the implications of scientific developments. In this case, it was the implications of all-out nuclear war. How do you educate a population that has not seen at firsthand the devastation nuclear bombs create? One look is worth a thousand words. So, yes, my experience in Japan after the war influenced very much my thinking about how difficult it would be to educate the public concerning the risks of nuclear warfare.

What is an effective way to communicate with the community of genetic researchers and the public in general about issues in human genetic research?

It's not easy. Genetics has become so complex that specialists in one area have difficulty communicating with those in other areas, let alone the public. And when attempting to communicate with the public, usually through science reporters, misleading over-simplifications often creep in. Nevertheless, I think we must keep trying, since these days there are many developments in human genetics that really do relate to the detection and possible treatment of diseases.

Where are you currently going with your work?

First of all, I'm close to the end of a career, as you must appreciate. I am kind of a Father Methuselah. But I've been very fortunate in that exciting things kept happening since I retired. Right now I'm focusing on three interesting activities that have a future. One is getting a monitoring system set up for detecting genetic damage at the DNA level, whether by radiation or chemical pollutants. This involves some complex laboratory techniques but also developing some equally-complex computer algorithms to help us analyze the DNA preparations we use.

The second direction in which I am moving is tremendously exciting right now. You may remember that in my book, *Physician to the Gene Pool,* I describe how in the course of studies on a very remote American Indian tribe, the Yanomama, we observed that some of their white blood cells exhibited severe chromosomal damage. Now we believe we have discovered the cause of these cells and suspect the discovery may have some rather important implications for the whole field of cancer. There is increasing evidence that these strange cells are due to a virus called SV40. About eight precent of us test positive for infection with that virus, and the footprints of this virus are widespread through our tissues. Given how frequent chromosome damage is in cancer cells, we believe we may have stumbled upon one of the players in the complex process that leads to a cancer.

My third activity is following up the book, *Physican to the Gene Pool.* I'm very much in the pulpit these days. Geneticists should be getting more involved in the population and resources issues. Unless we can reconcile the size of the world's populations with its sustainable resources, there will in the next 50 years almost surely be a series of population "crashes". It is that stark. I may be wrong, but I think that humankind is in much bigger trouble than most geneticists, so preoccupied with molecular genetics, realize. Bringing population growth under control is a challenging issue that doesn't have the excitement that we associate with molecular genetics, such as the thrill of finding a new gene. But, in terms of importance, I can't think of a greater challenge than this issue offers. An equal challenge we must face is how we use the enormous mass of genetic information that is being generated. The ethical and practical issues involved in applying this knowledge are enormous, and solutions are not in sight.

A photo taken in 1881 of Charles Darwin at the age of 72. This was one of the last pictures ever taken of Darwin, who died a year later.

1

The Science of Genetics

CHAPTER OUTLINE

A Passion for Experimenting

If you happened to be out walking in the English countryside just beyond the small village of Downe in Kent on a summer evening 125 years ago, you might have come across a remarkable sight. In a greenhouse down a path about 100 yards from a large, elegant home, a tall man in his sixties was stooped over some small potted plants. Next to him sat a younger man playing a bassoon. The older man was Charles Darwin; the younger man was his son, Frank. They were conducting an experiment on insect-eating plants called sundews. Darwin wanted to know what caused the sundew to close its leaves when a fly lands on it and becomes trapped in its sticky liquid. He was exploring various factors that might trigger the leaves' closing. He had already experimented with numerous substances including sand, milk, urine, water, roast beef, and bits of hard-boiled egg. None of them worked. Now he was experimenting with sound, even though he did not think sound waves would trigger the plants' response. Darwin was conducting an experiment, methodically working through a list of potential stimulants.

Darwin never did find out what triggered the sundew's response, but he was not discouraged by this failure. He was content to leave the sundew problem for others to solve. Darwin in his sixties was a world-renowned scientist who had changed our whole understanding and perspective of the natural world. He remained intensely curious about nature, content to perform experiments that would bear fruit in the future.

Darwin was an explorer with a passion to uncover the mysteries of the natural world. In a way, that is a goal of this book. It asks questions; it encourages exploration; and it details discovery.

CLASSICAL AND MOLECULAR GENETICS

As sciences go, genetics is young, less than 100 years old. Yet for thousands of years people have sought answers to questions about inheritance. Scientifically sound answers became possible about 130 years ago. In 1865, an obscure Austrian monk, Gregor Johann Mendel reported his discoveries of the fundamental laws of inheritance. He suggested that every cell contained pairs of "factors" and that each pair determined a specific trait. The members of each pair segregated from each other during the process of sex-cell formation, so that a gamete contained one member of each pair. Furthermore, the segregation of each pair of factors was independent of the segregation of other pairs of factors. These deceptively simple ideas form the foundation of the modern science of genetics and are the core of classical genetics.

As profound and as far reaching as Mendel's observations and interpretations were, they were not recognized as such for 35 years. Part of the reason for this delay was the absence of an understanding of cell structure and of the process by which cells divide. By 1900, this was no longer the case. In 1900, with the description of cell structure and of cell division, there at last existed a solid cellular framework within which Mendel's principles could be properly interpreted. The year 1900 is important, for it marks the beginning of the modern era of genetics.

The spectacular unfolding of all the modern genetic concepts has taken less than a century. We have moved from the obscure units that Mendel called "factors," segregating and assorting independently of each other in the nuclei of sex-cells, to the identification of DNA (deoxyribonucleic acid) as the chemical basis of inheritance. We have moved from a vague understanding of the relationship between the units of inheritance and the physical appearance of an organism to an understanding of the elegantly organized sequence of events by which Mendel's factors, now called genes, express their encoded information in cells. We have moved from exclusive reliance on classical genetic analysis to the merger of classical genetic techniques with modern molecular techniques. Today, the recently developed molecular technology that allows us to manipulate genetic material has opened entirely new vistas to us, vistas of how genes work, how they are regulated, and how genetic defects can be detected, modified, or corrected.

This unfolding of genetic concepts is much more than the erection of a structure within which we can interpret various aspects of heredity. It is a model of scientific methodology, of human genius and creativity, and, above all, of human potential. The thousands of participants who have contributed and continue to contribute to our understanding of modern genetics reflect human qualities we should always treasure and support. As we explore the many aspects of genetics, we encounter individuals who discover or observe something very important about the natural world, but fail to interpret it correctly; others, who though incorrect in their thinking, are catalytic in the thinking of others who then go on to make new and important discoveries; and those few who integrate diverse and seemingly unrelated observations to arrive at unique and important insights that have a major influence on the thinking of others. Hence genetics is more than a conceptual science; it is a vital and dynamic science that touches all facets of our being.

GENETICS IN THE NEWS

Barely a week passes without the appearance of some item in the news media about a new genetic discovery or a new way that genetics applies to our daily lives (Figure 1.1). As we shall see in the next section, some of the most dramatic discoveries have been in the field of medical genetics. Geneticists now understand the metabolic basis for several hundred inherited disorders. Researchers have isolated mutant genes causing numerous inherited disorders, including cystic fibrosis, Duchenne muscular dystrophy, Huntington's disease, the fragile X syndrome, neurofibromatosis, Alzheimer's disease, familial hypercholesterolemia,

Early-onset Alzheimer's Gene is Isolated

Philadelphia Inquirer
18 August 1995

Genetic Flaw May Result in Obesity

Philadelphia Inquirer
10 August 1995

Biotechnology Trials -- and Tribulations

Vector I, the gene therapy drug, showed promise against cystic fibrosis, but it later caused inflammation. Vector II looked better, but other problems cropped up. Then came vector III ....

Philadelphia Inquirer
24 July 1995

Maps to Scientific Discovery

Scientists at Children's Hospital of Philadelphia are sorting through the bits and pieces of DNA on chromosome 22. What they find will bring clues to genetic illness -- and perhaps new treatments and cures.

Philadelphia Inquirer
20 February 1995

Scientists Make Aids Breakthrough

Additional receptor proteins provide gateways for HIV

Philadelphia Inquirer
20 June 1996

Gene Hunters Pursue Elusive and Complex Traits of the Mind

New York Times
31 October 1995

Mutated Gene Identified in Breast Cancer

Philadelphia Inquirer
15 September 1994

Figure 1.1 Genetics in the news; a sampling of some recent headlines.

adult polycystic kidney disease, and breast cancer, to name only a few.

Beyond medical applications, however, researchers study which aspects of our behavior and personality are controlled by our genetic constitution. This question has been debated for hundreds of years. Important recent studies centered at the University of Minnesota using identical twins raised apart show that many aspects of our personality and behavior are strongly influenced by genes (see conversation with Thomas Bouchard). Indeed, evidence is accumulating on the significant role genes may play in such behaviors as alcoholism and sexuality. Geneticists have known for some time that defective genes lie at the root of mental disorders such as schizophrenia and manic-depressive illness. Today, armed with powerful tools of molecular genetic analysis, researchers are poised to identify the genes that when mutated, cause these maladies.

Molecular genetic research has given us powerful tools to study genes. In 1985 Kary Mullis developed a technique called *polymerase chain reaction* (*PCR*) that allows researchers to take a minute sample of DNA, perhaps only a single cell's worth, and amplify it millions of times so that it can be analyzed. One application of this technique has been in forensic science, the use of scientific techniques in legal settings. A small sample of tissue at a crime scene, perhaps a single sperm, white blood cell, or hair follicle, can be analyzed in detail for its DNA content using PCR. A type of DNA analysis called *DNA fingerprinting* examines variation in the sequence of the subunits that compose DNA. A tissue sample not belonging to the victim, such as a hair follicle, a drop of blood, or a semen sample, can be analyzed and used to identify a primary suspect or to exclude a person as a suspect. A man was recently arrested in Camden, New Jersey for the 1985 murder of a young Florida girl. The man smoked an unusual brand of cigarette, a brand found at the murder scene. The investigating detective, who had stayed with this case over the years, suspected the man but could not arrest him based only on the cigarettes. The cigarette butts were saved, however, and cells were isolated from them. In 1985, DNA testing was unsophisticated. But in 1993, those cells were analyzed by more sophisticated DNA fingerprinting techniques, and a link was established between the sample and the suspect.

Genetic research may also be seen in the corner grocery store. To eat or not to eat genetically engineered food products was the question that many asked as the "flavr savr" tomato hit the supermarkets. This tomato was created by manipulating the tomato's genetic material. This new variety of tomato contains a gene that allows a ripe tomato to last longer and retain its flavor longer. There are those who openly fear food products that have been genetically manipulated and thus avoid them. Is the fear justified? Probably not, at least not in the case of the "flavr savr" tomato. However, other types of genetic manipulation of food products may well cause justifiable concern.

A list of recent news items can easily fill a book. Only a few have been mentioned here. One of the objectives of this book is to help you gain a deeper understanding for the science behind the headlines and in the process to think more critically about issues at the interface of science and society.

GENETICS AND MEDICINE

Nature is nowhere accustomed to more openly display her secret mysteries than in cases where she shows traces of her workings apart from the beaten path; nor is there any better way to advance the proper practice of medicine than to give our minds to the discovery of the usual law of Nature by careful investigation of cases of rarer forms. For it has been found, in almost all things, that what they contain of useful or applicable nature is hardly perceived unless we are deprived of them, or they become deranged in some way.

A letter written by William Harvey in 1657 (quoted by A. Garrod in 1928 in an article about rare maladies. Lancet, i, pp. 1055–60, 1928)

Modern genetics has had a profound impact on medicine. Although connections between certain diseases and inheritance were made centuries ago, one of the first and most important links between the newly discovered Mendelian principles and disease was published by Sir Archibald Garrod in 1902. The British physician was studying some rare metabolic disorders that ran in families and thus seemed to be inherited. One of these disorders was *alkaptonuria,* a metabolic disorder in which a substance called *alkapton* accumulates in cells and tissues and is excreted in the urine. Alkapton, known properly as *homogentisic acid,* is an intermediary metabolic product formed from the breakdown of the amino acid phenylalanine. Normally, homogentisic acid is converted into another compound (maleylacetoacetate), but in this case the enzyme that makes this conversion possible is defective. Homogentisic acid thus accumulates and is oxidized to a black product that accumulates in areas of the body rich in cartilage, turning ears, nose, joints, and other tissues black (Figure 1.2). The urine, too, turns black when exposed to oxygen. Although this is a relatively benign clinical condition, causing at its worst moderate arthritis, it persists throughout a patient's life.

Garrod, after discussions with William Bateson, one of the early champions of Mendelism, concluded that this familial disease was inherited according to Mendelian principles and that the Mendelian factors were controlling cellular chemistry. Garrod was the first to associate a gene with a gene product, in this case an enzymatic protein.

Garrod's insightful analysis of this rare inherited disease stands today as the basis of modern genetic medicine. Even though many inherited disorders are rare, understanding them provides insight as to how genes control normal cellular processes. For example, studies of alkaptonuria, where a normal metabolic process is disrupted, led to important discoveries of how the amino acid phenylalanine is normally metabolized. Such an application is the essence of William Harvey's statement at the opening of this section.

Since Garrod's time, researchers have made enormous strides in establishing links between defective genes and disease. We shall discuss a few of the major linkages here and come back to many of them throughout the text. The gene causing Huntington's disease, a fatal neurological disorder, was finally isolated in 1993 after an intense 10-year search. Not only was the gene isolated, but the mutation causing the disease joined a growing list of disease genes carrying an entirely new category of gene mutation called an expanded triplet repeat. The fragile X syndrome, the most common form of inherited mental retardation in humans, is also caused by the same type of mutation. The mutant genes causing both Duchenne muscular dystrophy and cystic fibrosis were isolated and their gene products identified. With the understanding of disease mechanisms comes the opportunity for treatment. Gene therapy, made possible only after a gene has been isolated, has been used to treat a devastating immune system disorder called severe combined immunodeficiency disease. There is hope that gene therapy may soon prove successful in the treatment of cystic fibrosis. Other diseases that are currently being treated in clinical trials of gene therapy are various types of cancer, Gaucher's disease, familial hypercholesterolemia, hemophilia, Fanconi's anemia, Hunter's syndrome, and acquired immune deficiency syndrome (AIDS). There is every reason to expect that gene therapy will eventually be successful in the treatment of these and other disorders.

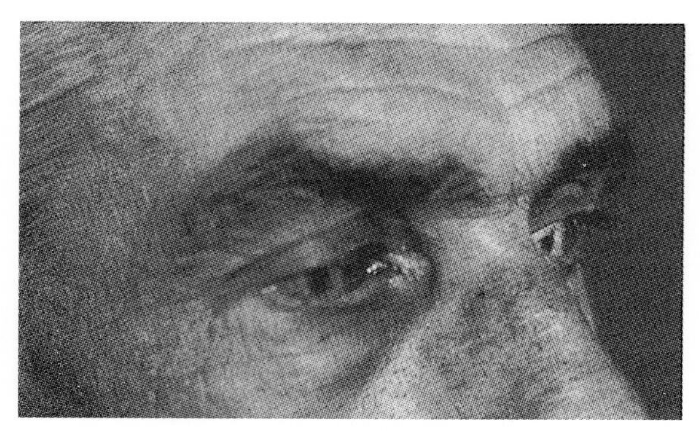

Figure 1.2 Alkaptonuria, a metabolic disorder. Notice the deposition of the brownish-black pigment, caused by oxidation of homogentisic acid, in the sclera of the eye.

Cancer is also a genetic disease. A few dozen genes controlling or influencing cell division and differentiation lie at the root of all cancers. When one or more of these genes becomes mutated or their regulatory mechanisms are somehow disrupted, the carefully regulated cell and all its descendants are transformed into unregulated cancerous cells.

Mutations that cause cancer are being identified and studied intensely. Recently, a gene (*BRCA1*) causing susceptibility to develop breast and ovarian cancer in certain families was isolated (Figure 1.3). A second breast cancer gene, *BRCA2*, was isolated in December 1995. They join a long list of genes that cause familial cancer, including Wilm's tumor, a kidney cancer in children, and Gardner syndrome, a type of colon cancer. As we learn more about how these cancer genes

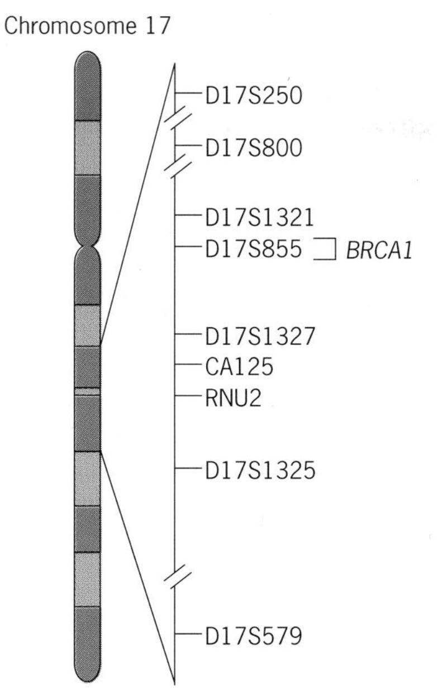

Figure 1.3 Schematic of human chromosome 17 showing the location of the breast cancer gene, *BRCA1*. Genes (*RNU2* and *CA125*) and molecular markers (everything that begins with D) in the vicinity of *BRCA1* are also shown.

function in normal and abnormal settings, we move that much closer to effective therapeutic treatments.

However, as has been the case for other genes identified by this strategy, it is a long way from isolating a gene to identifying the product it codes for and to finding out what the product does in the cell. Even with the knowledge of gene product function, designing a therapy is another problem of major proportion. For example, the gene for Huntington's disease was isolated in 1993 and its protein product, *huntingtin,* identified shortly thereafter. Yet as of this writing, the function of *huntingtin* remains largely speculative. Nancy Wexler, who was instrumental in the discovery of the Huntington's disease gene, shares her thoughts with us about the scientific, personal, and social aspects of Huntington's disease in a conversation at the beginning of Chapter 19. Once the function of the Huntington's disease gene is fully understood, it may still be a long time before any meaningful therapy for this devastating disease is established.

In contrast to Huntington's disease, we understand a great deal about the function of the cystic fibrosis gene product, called CFTR (cystic fibrosis transmembrane conductance regulator). This knowledge has opened the door to potentially successful gene therapy. In one strategy, a normal form of the *CFTR* gene is inserted into a common cold virus that has been genetically altered so that it cannot complete a life cycle. This engineered virus is then applied to the nasal passages of CF patients. The envelope that encapsulates the cold virus functions as the vehicle for transferring the normal *CFTR* gene into nasal epithelial cells. Once in the cell, the *CFTR* gene expresses a normal gene product, and the cells—previously crippled by the lack of CFTR—regain some of their normal function. However, this particular gene therapy treatment does not always work. Furthermore, it does not correct the pancreatic defect of this disease, nor does it correct CF-caused defects in other cells. The results, though preliminary, give us hope that gene therapy will soon provide a definitive treatment for the lung disease and other abnormalities in CF.

The isolation of the breast cancer genes *BRCA1* and *BRCA2* and their variant mutant forms is creating important social and ethical issues. For example, *BRCA1* is a very large gene with over a 100 different mutant forms. It is an expensive and labor-intensive job to screen women for all possible variants of this gene on a routine basis. To complicate matters further, some variants of *BRCA1* are benign and do not predispose women to breast cancer. As of this writing, it appears that the testing will be done primarily on women with a family history of breast cancer. Even in these cases, however, there is the risk of laboratory tests producing false positive or false negative results. In cases where a diagnostic test indicates a woman is carrying a mutant *BRCA1* or *BRCA2* gene that may predispose her to develop breast cancer, what treatment options can be offered? Is it always in the best interest of the woman to be told that she is carrying a mutation that predisposes her to develop breast cancer? Prophylactic mastectomy is a recommended treatment for carriers of mutant *BRCA* genes, but this treatment may not always be successful. Furthermore, since mutant *BRCA* genes also cause ovarian cancer, do genetic counselors suggest that women with *BRCA* mutations also have their ovaries removed? Many vexing problems are associated with the identification and isolation of breast cancer genes, problems that will have to be addressed.

GENETICS AND MODERN AGRICULTURE

In addition to its impact on medicine, modern genetics has had a tremendous impact on agriculture. One of the greatest achievements in modern agriculture was the application of Mendelian principles to the development of hybrid corn (Figure 1.4). Hybrid corn is produced by crossing different inbred strains, each of which has valuable properties, such as disease resistance, high protein or sugar content, and drought tolerance. During the period from 1940 to 1980, the average yield for corn increased by over 250 percent, in large part because of the development and introduc-

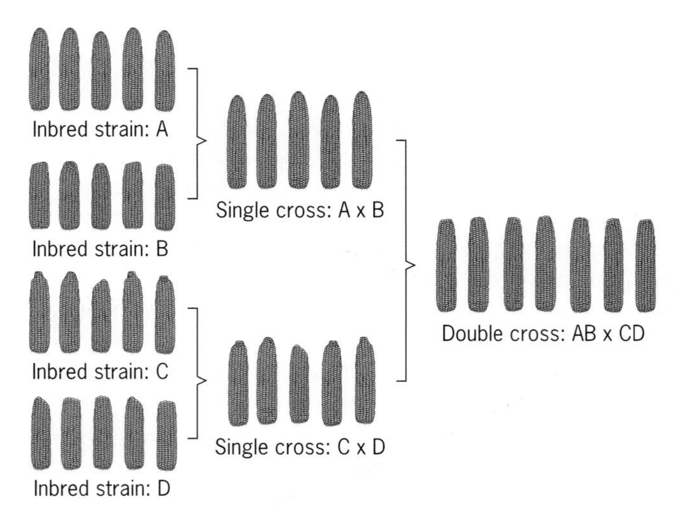

Figure 1.4 Ears of corn from four inbred strains. These strains are intercrossed to produce a double cross hybrid.

tion of hybrid corn varieties. In the United States, hybridization has resulted in dramatic yield increases in nearly all of the important food crops, such as barley, beans, oats, rice and wheat, though none as great as hybrid corn.

From the 1950s through the 1970s, Norman Borlaug and his team of researchers used classic genetic principles to develop new strains of wheat in Mexico that would do well under the stressed conditions often found in developing countries (Figure 1.5). His success in developing this wheat as well as other crop plants earned him a Nobel Peace Prize in 1970. His enormous contributions to agricultural productivity, and thus to nutrition, in developing countries helped launch the much-publicized "green revolution."

The modern tomato has benefited tremendously from the application of genetic principles. Genes that confer resistance to various pathogens, such as fungi and nematodes, have been bred into modern varieties. Most of these resistance genes came from tomato relatives growing in the wild. The growth pattern of the tomato plant has also been genetically altered so that the plant is bushier and more compact—characteristics that help keep the fruits off the ground. Breeders have developed a wide variety of morphological tomato types. These include round, medium-sized lemon boys; round, large beefsteaks; small, round cherry; pear-shaped orange; and others (Figure 1.6).

Selective breeding programs have produced chickens that are meatier, grow faster, are more disease resistant, and lay more eggs (Figure 1.7). Selective breeding programs have also produced cattle and pigs that are meatier, grow faster, are more efficient at converting feed into meat, and are better adapted to regional environments (Figure 1.8). Milk production per cow and butterfat content have also increased dramatically as a consequence of selective breeding.

One of the most effective techniques used to improve livestock quality is artificial insemination. A bull known to sire cows that produce large quantities of milk or offspring that are outstanding meat producers has his sperm collected at regular intervals and stored. This sperm is used to fertilize thousands of females all over the world.

The new genetic technology has had and continues to have a major impact on agriculture. For exam-

(a)

(b)

Figure 1.5 (*a*) Norman E. Borlaug and his prize-winning wheat at the International Maize and Wheat Improvement Center, Londres, Mexico. (*b*) Dwarf wheat growing in Mexico.

Figure 1.6 Some of the many varieties of tomatoes produced by hybridization and selection. Tomatoes are red, yellow, purple, orange, green, round, almond-shaped, and elongate.

ple, *Bacillus thuringiensis* is a bacterium containing a gene that encodes a protein toxic to many insects. Different subspecies of this bacteria produce toxins that kill different insects. One subspecies produces a toxin that kills the tobacco hornworm, an insect that is devastating to tomatoes. The gene for this toxin has been isolated from the bacterium and inserted into tomato plants. When tobacco hornworm larvae are allowed to feed on tomato plants carrying the *B. thuringiensis* toxin gene, the larvae all die within a few days. Tomato plants lacking this gene are unprotected and are quickly devoured by the larvae (Figure 1.9).

Inserting genes for resistance to insects or pathogens into crop plants is becoming a major weapon in the war against organisms that destroy so much of the world's food supply. Genetic technology is also being employed to improve the nutritional quality of crop plants such as corn, and to help plants synthesize their own nitrogen so that the need for externally applied fertilizers is reduced. There is no question that the application of genetic technology to agricultural problems has enormous potential.

While the formal application of molecular and classical genetic principles to agriculture has been going on now for less than 100 years, we should not lose sight of the fact that the human species has been using genetics in agriculture for centuries without formal knowledge of the laws of inheritance. Early humans carried out the first "genetic selection" experiments in wheat between 7000 and 10,000 years ago. Evidence indicates that almost all of our present food crops were domesticated during this early Neolithic period, coincident with the development of stone tools. The

Silver Seabright Bantam

Black Tailed White Japanese Bantam

Single Comb Light Brown Leghorn

Barred Plymouth Rock

Single Comb White Leghorn Hen

Figure 1.7 Some of the many varieties of chickens produced by hybridization and selection.

Yorkshire-Duroc Hybrid Piglets

Yorkshire Piglet

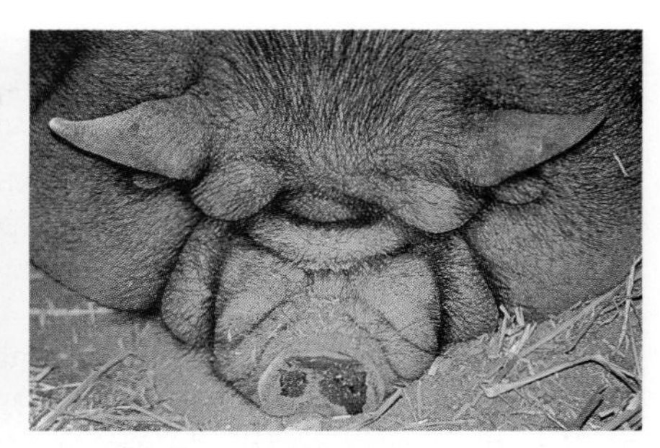

Vietnamese Potbellied Pig

Angus Cow

Polled Hereford Cow and Bull Calf

Shorthorn Cattle

Figure 1.8 Some of the many varieties of beef cattle and pigs produced by hybridization and selection.

Figure 1.9 Tomato plants engineered to be resistant to tobacco hornworm larvae. The plant on the right is expressing a gene from a bacterium that is toxic to the hornworm larvae.

largest, most vigorous individuals in the population were simply selected as parents for subsequent generations, a practice that is still the mainstay for modern plant and animal breeding.

GENETICS AND SOCIETY

One reason for so much media coverage of developments in genetics is that discoveries in genetics often have a direct impact on us as a society. Sometimes they hold promise for the cure of a fatal disease; sometimes they help create new food products; and sometimes they create complex, moral dilemmas for us.

A young couple has a daughter with cystic fibrosis. To have this devastating genetic disease, the child must inherit two copies of the defective gene. Thus, each of her parents had to carry one copy of the defective gene and pass that copy on to her. The mother became pregnant again, even though she and her husband knew there was a 25 percent risk they would produce another affected child. Their insurance company requested they obtain a prenatal diagnosis of the fetus to assess its cystic fibrosis status. The test results indicated that the fetus had cystic fibrosis. Representatives of the insurance company then requested that they terminate the pregnancy or risk losing insurance coverage. The couple refused to terminate the pregnancy, the insurance company terminated their insurance, and the couple sued. Do insurance companies have the right to cancel health insurance policies or insist on pregnancy termination when confronted with the birth a child with a medically expensive, nearly always fatal genetic disorder? The courts said no. However, such an issue does not end with a legal decision. In fact, families that are at risk for developing certain inherited diseases are being denied health insurance.

It is not fair, however, to argue that the insurance companies are insensitive and interested only in the bottom line. Genetic disorders such as cystic fibrosis are expensive to treat, often costing hundreds of thousands, even millions of dollars. Is it appropriate to ask the insurance company's policyholders to cover the entire costs for treatments? What rights or obligations do insurance companies have in the prevention of genetically defective births? What responsibilities do parents who are at risk have in the prevention of genetic disorders?

The **Human Genome Project,** an international project with the goal of mapping and sequencing all human genes (and other species too!), is adding to this health insurance–birth defect issue. The Human Genome Project continues to uncover information about the human genome, and insurance companies as well as employers would like to have access to that information. Information about the human genome is accumulating at a rapid pace. It is entirely possible that we will soon be able to make a number of important predictions about genetic health based on that information: for example, a person's likelihood to develop cancer, mental illness, or some other genetically based disorder. A complex problem that lies at the core of this issue is access to the information being gathered. Most of the human genome research is publicly funded. Should the public have unlimited access to the information? It is conceivable that a person could be denied employment or insurance based on DNA information.

The Human Genome Project also involves an important economic issue. It will cost more than $3 billion, a large percentage of the total federal research budget. A finite amount of money is available for research. Do we as a society want to commit such a large percentage of that money to the Human Genome Project at the expense of other worthwhile projects in ecology or evolution or mathematics that are not as human oriented or do not attract as much media attention? As you use this text to lay a genetic foundation for a thoughtful analysis of these and other issues, you will be in a stronger position to evaluate them.

THE MISUSE OF GENETICS

Genetics, perhaps more than any other science, has a great potential for both good and evil. Throughout this book we shall highlight how our knowledge of genetics has grown and how our lives have been made better by that knowledge. But we should bear in mind some unfortunate, even tragic misuses of genetics.

Darwin's theory of evolution by natural selection stated, among other things, that traits beneficial to a species replaced traits that were not beneficial. This seminal idea was quickly and inappropriately applied to the human species. Francis Galton, Charles Darwin's cousin, believed that many human mental and physical qualities were inherited and thus subject to the forces of selection. But Galton took this idea a step further. He suggested that the genetic constitution of the human species could be improved through the use of artificial selection, an idea he called **eugenics:** Parents expressing favorable traits would be encouraged to have larger families (*positive eugenics*), and parents with undesirable traits would be encouraged not to have children (*negative eugenics*). Traits that Galton thought were favorable included high intelligence, high levels of achievement, artistic creativity, and excellent health. Traits that he believed should be selected against included low intelligence, mental illness, criminal behavior, and alcoholism.

The eugenics movement gained strength in the United States during the early part of the twentieth century, especially negative eugenics. In 1907 Indiana passed laws that mandated sterilization of individuals who were "imbeciles, idiots, convicted rapists, and habitual criminals." By 1931 nearly half of the states carried such laws, and mandatory sterilization was extended to include such things as sexual perversion, drug addiction, alcoholism, and epilepsy. Implicit in these sterilization laws was the connection between inheritance and behavior. Sterilization of the mentally retarded may sound barbarous to us in the 1990s, but during the 1920s and 1930s sterilization procedures were advocated by some because mentally retarded people were poor parents and provided poor environments for their children. Because there were no good drug treatments enabling the mentally diseased to live normal lives and care for children, many more were institutionalized than are now. And sterilization was thought to be kinder than incarceration.

Another outgrowth of the eugenics movement during the 1920s were the immigration restriction laws. These laws, motivated largely by economics, favored "genetically desirable" ethnic groups (mainly northern European) and placed severe restrictions on the less "genetically desirable" groups (the Mediterranean area, Central Europe, China). These laws were founded on flawed, unsubstantiated "data" and on bigotry. Many of them remained on the books until the 1960s.

The eugenics movement took on its most twisted and perverted form in Nazi Germany. Between 1930 and 1945, millions of Jews, Gypsies, and others were systematically exterminated by the Hitler regime in an attempt to "cleanse" Germany of "inferior" genetic material.

In the United States, England, and elsewhere, most geneticists were appalled at the way the science of genetics, still so young, was being abused. There was no solid evidence that genetics played a role in most of the human traits deemed desirable or undesirable. People were taking anecdotal information and elevating it to the level of scientific fact (e.g., Mr. Smith

HISTORICAL SIDELIGHT

The Lysenko Affair

T. D. Lysenko had no real evidence to support his arguments that plant development and agricultural productivity could be improved by manipulating the environment, at least none that could be independently verified. However, he did attract a powerful supporter. Joseph Stalin was desperate to improve the Soviet Union's struggling agricultural program and saw in Lysenko's ideas an opportunity to do so in the context of Marxism. The idea that environmentally induced changes could produce specific, directed genetic changes was very appealing to Stalin. It was an idea perfectly compatible with Marxism: Proper social conditions would induce permanent, heritable changes in human behavior. Lysenko became a key part of Stalin's team and in 1940 was appointed director of the Institute of Genetics of the USSR Academy of Science. In this powerful position, Lysenko suppressed all genetic research that was based on Mendelism. He was ruthless in his opposition to all the developments in genetics taking place in the rest of the world. Soviet scientists who advocated Mendelism were swiftly identified and sent to prison, where many of them died.

Lysenko dominated Soviet genetics and agriculture until 1964. Soviet agriculture, based on the theory of acquired characteristics, was in shambles. Soviet genetic research was an international disgrace. Nikita Khrushchev replaced Stalin in the mid-1950s and continued to support Lysenko's policies, though with increasingly less enthusiasm than Stalin had. By the time Khrushchev was replaced, the bankruptcy of Lysenkoism was evident to all, and Lysenko was removed from his post and relegated to obscurity. Soviet science was then confronted with the daunting task of catching up with 27 years worth of genetic and agricultural research. As a point of reference, 1964—the year Lysenko's domination of agriculture and genetics ended—marked the eleventh anniversary of the discovery of the structure of DNA by Watson, Crick, and Wilkins.

Figure 1 T. D. Lysenko (far right) with Soviet Premier, Nikita Khrushchev (center right), Mikhail A. Suslov (far left), and a Soviet colleague, 1962.

was a wife and child abuser; his son was also a wife and child abuser; therefore, abusing one's wife and child is an inherited behavior). This perversion of genetics and society's willingness to accept it uncritically prompted many geneticists to avoid the study of human genetics for fear of being grouped together with the eugenicists. The field of human genetics suffered greatly because of these abuses.

Perhaps one of the most bizarre episodes in modern genetics, indeed in modern science, occurred in the Soviet Union. While the application of genetic principles to agriculture was resulting in dramatic increases in agricultural production in the United States in the period 1937 to 1964, agricultural production in the Soviet Union was stagnant. During these years, biology and agriculture in the Soviet Union were dominated by one person, T. D. Lysenko, a young plant breeder from the Ukraine. Lysenko argued that plant development and agricultural productivity could be vastly improved by manipulating the environment. Environmentally induced changes in plant growth and development, according to Lysenko, would be assimilated into the genetic material and passed on to the next generation. His ideas were similar to those commonly attributed to and formalized by Lamarck in the early nineteenth century. Lysenko's ideas were eventually discredited, but only after many years (See Historical Sidelight).

THE PRINCIPLES OF GENETICS: AN OVERVIEW

Genetics has had and continues to have a significant impact on our daily lives. At various times in this text, we shall discuss the complex personal and societal implications of some of the recent discoveries in genetics. Mainly, however, this book will address three major questions, though not necessarily in this order: What is the chemical nature of the genetic material? How is the genetic material transmitted? And what does it do? These questions form the core of modern genetics.

QUESTIONS AND PROBLEMS

1.1 Select one or two recent newspaper or magazine articles about genetics and write a brief report about them. Why are these articles interesting, and how do they relate to the issues raised in this chapter?

1.2 What social or economic considerations affect genetics research?

1.3 What three questions form the core of modern genetics?

BIBLIOGRAPHY

Borlaug, Norman E. 1983. Contributions of conventional plant breeding to feed production. *Science* 219:689–693.

Elias, Sherman, and George J. Annas. 1987. *Reproductive Genetics and the Law.* Year Book Medical Publishers, Chicago.

Garver, K. L., and B. Garver, 1991. Eugenics: Past, present, and future. *Am. J. Hum. Genet.* 49:1109–1118.

Horgan, J. 1993. Eugenics revisited. *Scient. Amer.* (June) 268:123–131.

Mange, Elaine J., and Arthur P. Mange. 1994. *Basic Human Genetics.* Sinauer, Sunderland, MA.

Medvedev, Z. A. 1969. *The Rise and Fall of T. D. Lysenko.* Columbia University Press, New York.

Morsy, M. A., et al. 1993. Progress toward human gene therapy. *J. Amer. Med. Assoc. 270:2338–2345.*

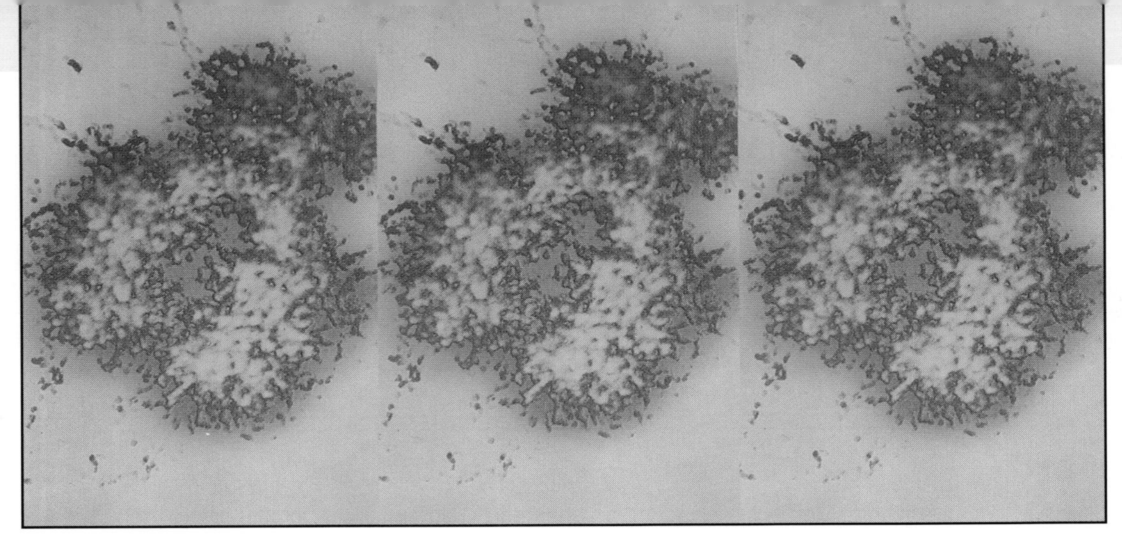

2

Chromosome 21. Individuals born with three copies of chromosome 21, instead of two copies, have Down Syndrome (color enhanced EM magnified at 10000X at 35mm).

Reproduction as the Basis of Heredity

CHAPTER OUTLINE

An Accident in Cell Division

After five years of marriage, Michael and Carole decided that they were ready to begin their family. Carole soon became pregnant, and she and Michael eagerly awaited the birth of their baby. It was a normal pregnancy and a normal delivery. When their baby boy was born, however, they knew immediately that something was wrong. The child was lethargic and had poor muscle tone. His head was short and very flat at the back; the eyes had a slanted appearance, and the irises had speckles around the edges; the nose had a low bridge and the tongue tended to protude. These and other abnormalities all pointed to one conclusion: The baby boy had Down syndrome. Examination of the baby's chromosomes showed that he had 47 instead of 46 chromosomes. The extra chromosome was the smallest chromosome, chromosome number 21. The diagnosis was confirmed.

The birth of a baby with Down syndrome is rare in young women, but happens about 2 percent of the time in 45-year-old women. Down syndrome illustrates several important genetic concepts. It emphasizes the central role of chromosomes in inheritance and the necessity for cell division processes to be error free. In this case, an error occurring in the cell division process that produced the mother's egg resulted in an extra chromosome 21 in that egg. After fertilization there were three copies of chromosome 21, two from the mother and one from the father. Thus, the child carried three doses of chromosome 21 genes instead of the normal two. This dosage imbalance caused the broad spectrum of abnormalities that are called Down syndrome. In addition, the boy will grow up to be seriously retarded and will have a 1500 percent increased risk for developing leukemia. If he survives to adulthood, he will almost certainly develop early-onset Alzheimer's disease. All of these problems develop because of the extra dose of genetic material he carries.

Chapter 1 has provided us with an introduction to the societal relevance of genetics and an overview of major questions that genetics addresses. In this chapter, as further preparation for our journey, we review the structure of cells and the behavior of chromosomes during cell division. We examine how sex cells, or gametes, are formed. We also look at the various strategies that have evolved in the natural world for transmitting the genetic material to the next generation, strategies known as life cycles. The cell is the basic unit of life, and the chromosome is the carrier of the genetic information. The behavior of chromosomes during cell division accounts for the principles of inheritance first described by Mendel.

THE CELL AS THE BASIC UNIT OF LIFE

All living things, with the exception of viruses, are made up of cells—small membrane-bounded structures filled with a variety of chemicals in an aqueous solution that interact to give life its basic properties. These properties include, among other things, the acquisition and use of energy, as well as the capacity to reproduce, to respond to the environment, to carry out a variety of controlled chemical reactions, and to maintain a relatively constant internal environment despite fluctuations in the external environment.

There are two types of cells in the living world, distinguishable by the type of internal structures they contain: prokaryotes and eukaryotes. Although both contain the same basic kinds of molecules, there are fundamental differences between them.

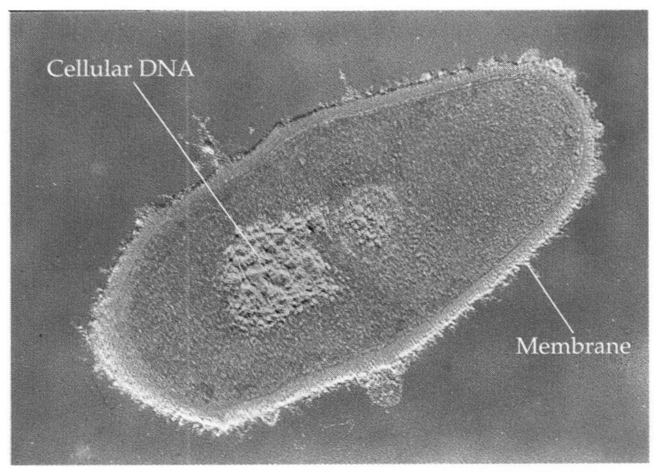

Figure 2.1 An electron micrograph of a longitudinal section through a bacterium, *Escherichia coli.* The cell's DNA is concentrated in a special region of the cytoplasm called the nucleoid.

The Prokaryotic Cell

The simpler of the two cell types is the prokaryotic cell (*pro* = before; *karyote* = nucleus). Prokaryotic cells include the various types of bacteria and their relatives, all members of the kingdom Monera (Figure 2.1). They possess a **cell wall** constructed of **peptidoglycan,** a substance unique to prokaryotes. Beneath the cell wall is a **plasma membrane** that encloses a **cytoplasm** containing DNA, RNA, proteins, and various small molecules. Under the highest magnification, the cytoplasm does not appear to have any distinct internal organization, although the DNA is sequestered in a poorly demarcated region of the cytoplasm called a **nucleoid.** The nucleoid is not surrounded by a membrane, nor are there any membrane-bounded cytoplasmic organelles in prokaryotes.

Bacteria are microscopic organisms that reproduce quickly by binary fission—simple, asexual division. A single cell in an environment with plentiful food resources divides about once every 20 minutes. At this rate, a single cell would produce 5 billion progeny cells in just under 11 hours, a number equal to the approximate human population of the earth! Bacteria remain the most abundant cell type on earth.

Bacteria may be simple in terms of their structure, but they are biochemically diverse. The ancestral prokaryote, a bacterium that appeared on earth over 3 billion years ago, evolved into a myriad of bacterial types that have adapted to and inhabit virtually all niches in the world, from the human gut to garden soil, from the ocean depths to mountain peaks, from swamps to hot acid springs. There are bacterial species that can utilize virtually any organic molecule as a food source, including some that use carbon from CO_2 and nitrogen from N_2 to make all of their necessary molecules. Bacteria are ancestral to all life forms on earth and have occupied this planet longer than any other life form.

The Eukaryotic Cell

The eukaryotic cell (*eu* = true; *karyote* = nucleus) is generally much larger than the prokaryotic cell and far more complex in terms of its structure and functions. The single most distinctive feature of a eukaryotic cell is a membrane-bound true **nucleus** that houses the genetic material. Between the nucleus and the plasma membrane is the cytoplasm, an aqueous matrix containing various membranous and nonmembranous organelles that perform a variety of complex metabolic functions. Sometimes a cell wall lies just exterior to the plasma membrane. The eukaryote cell wall is made of cellulose and other molecules, but *not* peptidoglycan. The vast majority of living species, including animals, plants, fungi, and protists, are eukaryotic. Let's look more closely at the eukaryotic cellular structure and functions (Figure 2.2).

Surrounding the cell is a plasma membrane constructed of a fluid phospholipid bilayer penetrated by a variety of proteins. The plasma membrane has two main functions:

1. It forms a protective barrier between the extracellular and intracellular matrices.
2. It regulates the flow of molecules into and out of the cell.

The membrane-bound proteins are commonly complexed with carbohydrates; thus they are **glycoproteins.** They usually interact with other molecules in the extracellular world. Those proteins and glycoproteins that receive signals from the external world and communicate with the cell's interior are called **receptors.** Receptors have particular importance to humans because mutations in the genes that code for them can lead to diseases such as cancer and cystic fibrosis.

A variety of organelles are distributed throughout the cytoplasm, some of which are surrounded by membranes and some not. Though cell structure and function vary widely from one cell type to another, all eukaryotic cells have the same basic types of organelles. The cell differences arise from the kinds of gene products being synthesized in the cells, the nature of their distribution in the cell, and the distribution and type of organelle systems in cells. We might think of organelles as compartments or structures in cells that carry on very specific types of metabolic activity, chemical processes by which energy is provided and used.

The **membrane-bound nucleus** is the largest structure in a eukaryotic cell (Figure 2.2). Inside the nucleus are the chromosomes, made up of DNA, RNA, and various proteins. Also inside the nucleus

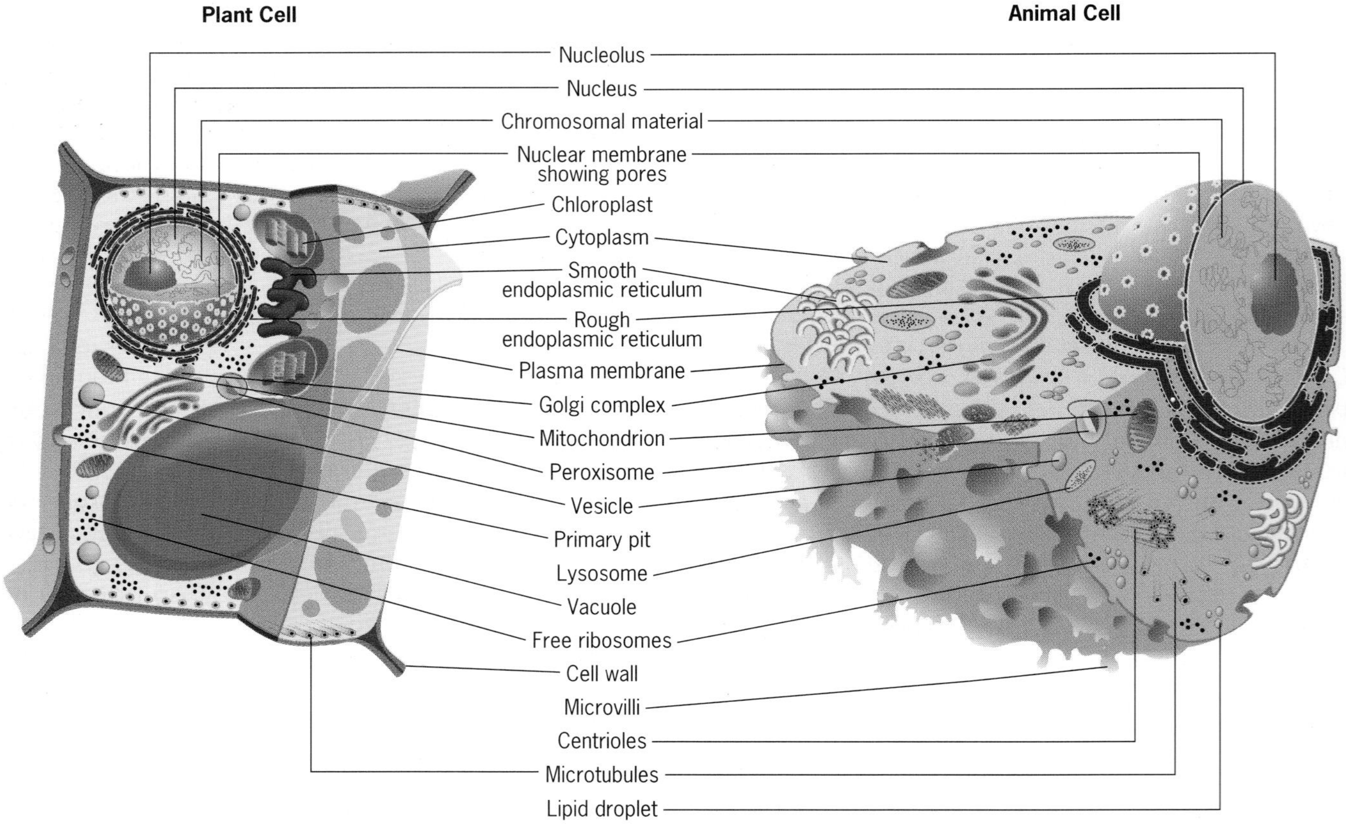

Figure 2.2 The generalized structure of eukaryotic cells. Both plant and animal cells have a plasma membrane, a membrane-bound nucleus, and cytoplasmic organelles. The two cell types differ in the types of organelles they contain and in the presence or absence of a cell wall. Plant cells have cell walls, animal cells do not.

and associated with the chromosomes are often found one or more **nucleoli** (sing. = nucleolus), structures that function in RNA synthesis. The chromosomes and nucleoli are contained in a nuclear matrix called the **nucleoplasm.** The nuclear membrane separating the cytoplasm from the nucleoplasm has numerous pores in it so that molecules can traverse it. The RNA synthesized in the nucleus is transported to the cytoplasm through the nuclear pores.

The cellular cytoplasm is filled with a vast interconnected network of tubular membranes called the **endoplasmic reticulum,** or **ER** (Figure 2.2). Some of the ER is rough in appearance because **ribosomes,** cytoplasmic structures made of RNA and protein involved in protein synthesis, are bound to it. Ribosomes are assembled in the nucleolus. Rough ER is the site of protein synthesis. Proteins synthesized here are transported across the ER membrane where they are directed to different cellular locations. ER that is not complexed with ribosomes is called smooth ER. This type of ER functions in the synthesis of certain types of hormones and in other cellular reactions.

The **Golgi complex** is an organized pile of flattened membranous sacs. It receives, stores, and often modifies proteins synthesized on the rough ER. For example, the insulin protein that regulates blood sugar levels is synthesized on rough ER as a large precursor protein called proinsulin that is subsequently cleaved to its functional form, insulin, in the Golgi complex. Inherited defects in this Golgi-based cleavage process lead to diabetes.The proteins stored in the Golgi complex also function in the synthesis of other important cellular molecules, such as the complex carbohydrates that make up the plant cell wall. Sometimes Golgi vesicles move to the plasma membrane, fuse with it, and release their contents to the external environment, a process called **exocytosis.**

Small vesicles called **lysosomes** pinch off from the Golgi complex, forming membranous sacs containing powerful digestive enzymes capable of degrading nearly any molecule inside the cell. The membrane surrounding the lysosome keeps these powerful enzymes harmlessly sequestered inside the cell. Mutations in genes coding for lysosomal enzymes cause such devastating diseases as Tay-Sachs disease. In the case of Tay-Sachs, a defective lysosomal enzyme is unable to properly degrade a ganglioside, a type of lipid molecule found in abundance in nerve cell membranes. Intermediate breakdown products accumulate in the lysosome to toxic levels and destroy brain cell functions, causing death before the age of three.

Similar to lysosomes are the **peroxisomes.** They are membrane-bounded vesicles containing a different battery of enzymes that use molecular oxygen to oxidize organic molecules, especially fatty acids. Some of the genetically encoded enzymes in peroxisomes generate dangerously reactive hydrogen peroxide (H_2O_2) during chemical reactions; other enzymes degrade hydrogen peroxide. Hydrogen peroxide is a powerful oxidizing agent.

In plants, membrane-bound, liquid-filled vesicles called **vacuoles** may occupy as much as 90 percent of a cell's total volume (Figure 2.2) . The plant vacuole may contain a variety of dissolved molecules such as salts, sugars, toxic waste, and pigments that account for the color of flowers and leaves. Vacuoles also function to maintain high internal water pressure in plant cells, which aids in the physical support of plant tissues.

Among the most prominent organelles in eukaryotic cells are the mitochondria and chloroplasts (Figure 2.3). Mitochondria are found in virtually all eukaryotic cells, while chloroplasts are found only in photosynthesizing plant cells. Animal and fungal cells do not contain chloroplasts.

Mitochondria and chloroplasts probably originated from small prokaryotic cells that were taken into a larger cell. Supporting this hypothesis is the fact that both of these organelles contain their own genetic material: small circular molecules of DNA, like those found in modern-day bacteria. They also synthesize many of their own proteins.

Mitochondria are cellular "power plants." They convert energy stored in energy-rich macromolecules (lipids and carbohydrates) into **adenosine triphosphate (ATP),** which is used to power virtually all the activities of the cell. As might be expected, the most energetically active cells have the largest number of mitochondria. The mitochondrion has a double membrane. Proteins embedded in the inner membrane function in energy transfer reactions. The inner membrane has a specific structure and organization so that reactions are carried out in a precise and orderly fashion. The inner membrane is folded into **cristae** that project into a **matrix**, thus increasing the surface area for these reactions.

Mutations in mitochondrial genes cause a number of disorders. For example, a type of eye disease called Leber's hereditary optic neuropathy (LHON) is caused by mutations in mitochondrial genes. Aging is associated with degenerative changes in mitochondrial DNA. Mutations in mitochondrial genes produce slow-growing colonies of the bread mold, *Neurospora.*

The **chloroplasts,** which are found only in plant cells, are the sites of **photosynthesis,** a complex process by which plants (and some bacteria) capture and use energy from the sun to drive the synthesis of complex organic molecules from carbon dioxide and water. Like mitochondria, chloroplasts have a double membrane system. The interior of the chloroplast contains a system of flattened membranous discs called **thylakoids.** Piles of thylakoids are located in the cen-

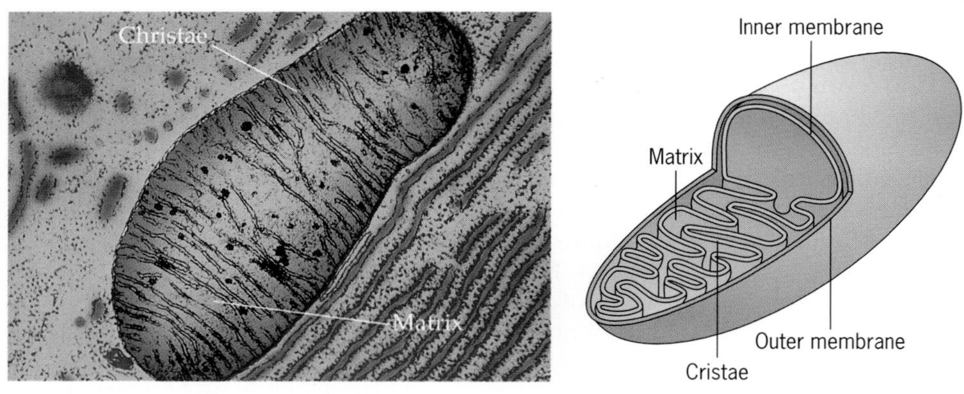

Figure 2.3*a* Electron micrograph and drawing of mitochondria showing the outer and inner membranes. The highly folded inner membrane forms the cristae which project into the matrix containing DNA, RNA, ribosomes, and various proteins.

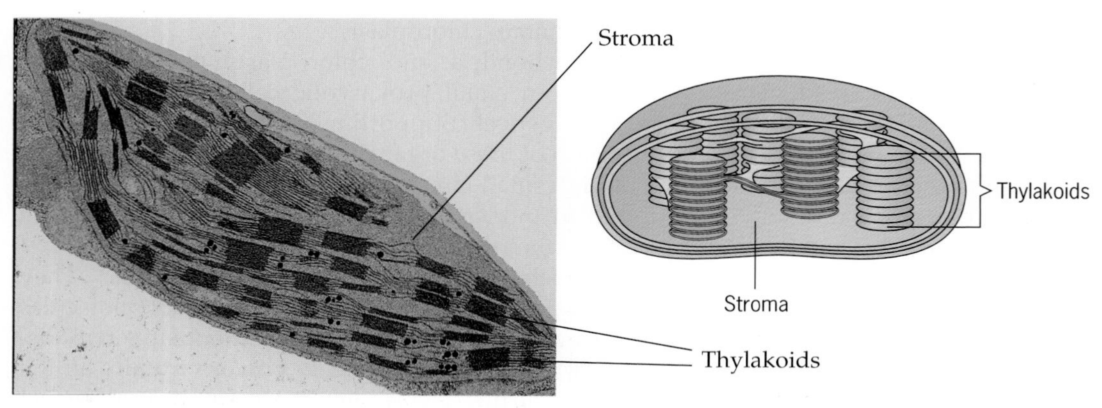

Figure 2.3*b* Electron micrograph and drawing of a chloroplast showing the inner and outer membranes as well as the stacks of membranous thylakoids contained in the central stroma. The stroma contains DNA, RNA, ribosomes, and various proteins.

tral region called the **stroma. Chlorophyll,** the green, sunlight-capturing pigment, is contained in the thylakoids.

Finally, all eukaryotic cells have a **cytoskeleton.** The cytoskeleton is a network of protein filaments, giving the cell its shape, its ability to move, and its ability to organize and position its organelles within its cytoplasm. Two of the most important filaments are the microfilaments and microtubules. The microfilaments are primarily involved in cell movement, like the flowing movements of the amoeba. Microtubules are the main structural elements in cilia and flagella—cell surface structures that propel cells through their environment.

Important inherited disorders are caused by defects in the cytoskeleton. For example, the gene causing Duchenne muscular dystrophy codes for a protein called *dystrophin,* which is a member of a family of cytoskeletal proteins. Although geneticists understand a great deal about dystrophin's structure and cellular location, its function remains elusive.

Key Points: **The prokaryotic cell has no membrane-bound nucleus. The prokaryote equivalent of the true nucleus is the nucleoid. The eukaryotic cell has a membrane-bound nucleus that houses the chromosomes. The eukaryotic cell has localized many of its metabolic processes into membrane-bound compartments or organelles, such as lysosomes, Golgi apparatus, mitochondria, and chloroplasts. Protein synthesis in eukaryotes takes place on ribosomes in the cytoplasm, in mitochondria, and in chloroplasts.**

THE CHROMOSOME: AN OVERVIEW

The key event in cell division is the proper distribution of genetic material to daughter cells. This genetic material is organized into structures called **chromosomes** (*chromo-* color; *soma-* body). One of the most important developments in the history of modern genetics was the research associating genes with chro-

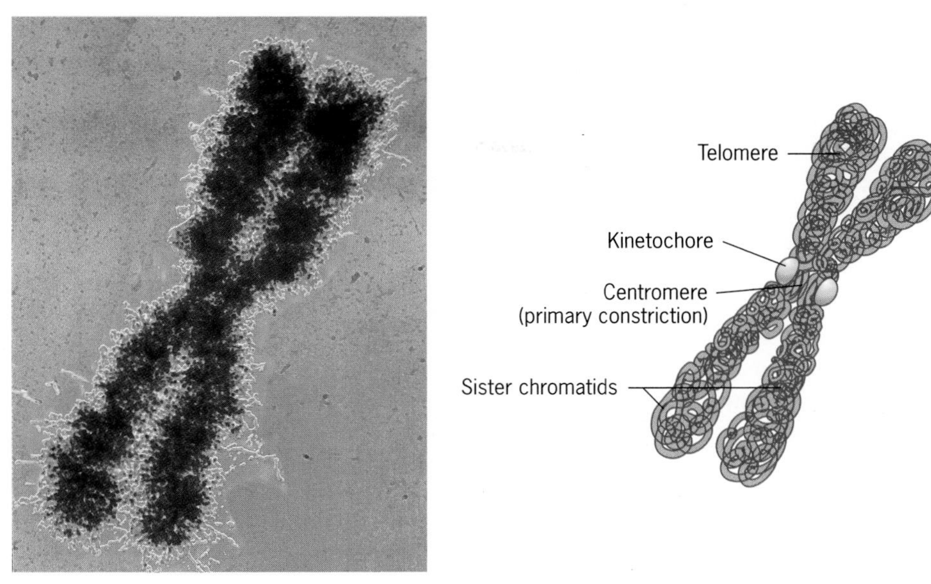

Figure 2.4 The structure of a mitotic chromosome.

mosomes, a topic that we will discuss in depth in Chapter 5. In prokaryotes, the chromosome is simply a molecule of double-stranded DNA. The eukaryote chromosome is considerably more complex. Each eukaryote chromosome is a single DNA molecule packaged in a protein matrix. Chromosomal DNA is compacted by coiling and supercoiling. If the DNA in a single human chromosome were uncoiled so that it was a perfectly linear molecule, it would measure between 2.5 and 5 centimeters in length. If all the DNA molecules in the nucleus of a human cell were uncoiled and lined up end to end, they would measure more than a meter in length. To imagine how this much DNA is packaged into chromosomes and then packed into a nucleus, we might visualize twisting a rubber band over and over: it coils and then forms supercoils as it compacts.

Chromosomes of eukaryotes and prokaryotes are involved in two main activities:

1. Those concerned with the transmission of genetic information from cell to cell and from generation to generation.
2. Those concerned with the orderly release of this information to control cellular function and development.

In a dividing cell, the chromosomal material condenses, becoming a relatively thick, dense, rodlike structure. In nondividing cells, the chromosome is extended and its structure is difficult to study.

The eukaryotic chromosome has key architectural landmarks at both the morphological and the molecular level. As seen under a light microscope, the chromosome has some distinct morphological features. The replicated chromosome is made up of two **sister chromatids** joined together by a **centromere** (Figure 2.4). The **kinetochore,** a protein structure at the centromere, functions in chromosome movement during cell division. The position of the centromere varies from the middle to the end of the chromosome. The ends of the chromosomes are called **telomeres.**

At the molecular level, several key **DNA sequence elements** are required to produce a stable chromosome (Figure 2.5). Various DNA sequences signal **replication origin,** the starting points for DNA

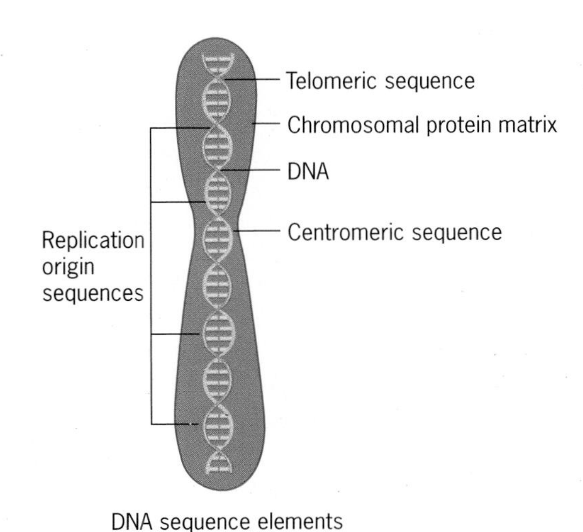

Figure 2.5 The DNA sequence elements required to produce a stable chromosome. There are several sequences where DNA replication begins. Telomeric DNA sequences are at the tip, and centromeric DNA sequences are in the centromere region.

replication. The centromeric DNA sequence has two functions: It binds the sister chromatids together after replication; and it is the region of attachment of the spindle apparatus that moves the chromatids to opposite poles of the cell during cell division. Finally, sequences at the ends of the chromosomes, called telomeric DNA sequences, are required for proper chromosome replication (Chapter 10).

Key Points: **The eukaryotic chromosome is a threadlike nuclear structure consisting mainly of a complex of DNA (the genetic material) and proteins. Chromosomes have key morphological and molecular features such as centromere, telomere, chromatid, and DNA sequences that designate replication origin.**

THE CELL CYCLE

In order for an organism to grow, whether it is single-celled or multicellular, three things must occur: (1) the cell mass must increase; (2) there must be a duplication of the genetic material; and (3) there must be a division process assuring that each daughter cell receives an equal and identical complement of the genetic material to ensure perpetuation of the cell line. In eukaryotes, these occurrences take place in an ordered progression of events during the cell's life span, or **cell cycle.** (See Technical Sidelight: Cell Cycle Checkpoints.)

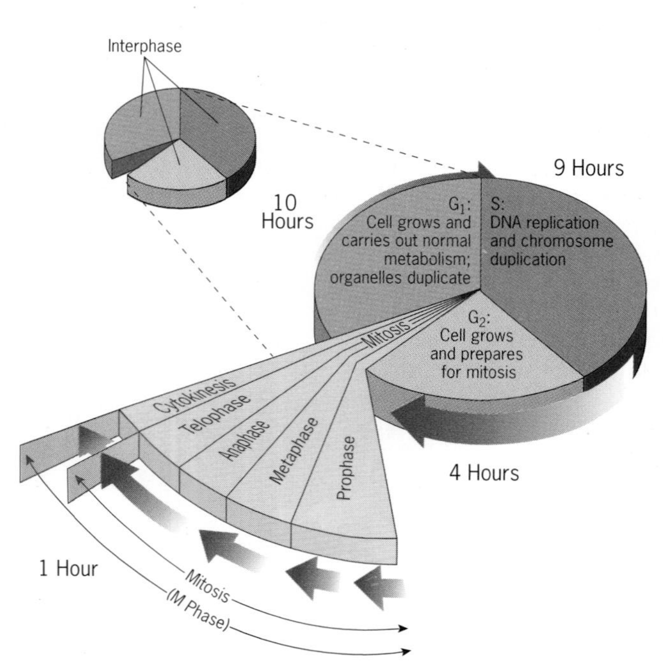

Figure 2.6 The cell cycle. Times indicated are for mammalian cells in tissue culture.

The initial event in the cell cycle (Figure 2.6) is the growth and increase in cell mass that occurs following cell division. This is called the **G_1 phase** (G = gap). For a mammalian cell taking 24 hours to complete the cell cycle, the G_1 phase might last about 10 hours. G_1 is devoted to the metabolic activities associated with cell growth and preparation for DNA replication. Following G_1, DNA replication occurs. This is the synthesis, or **S phase,** which lasts about nine hours. During the S phase, the genetic material of each and every chromosome is replicated. Thus, each chromosome consists of two sister chromatids. After completion of DNA replication, the cell enters another growth phase called the **G_2 phase**. During this post–DNA replication phase, which typically lasts about four hours, preparations for mitotic cell division are taking place. The **M phase,** or division phase, is the final part of the cell cycle. The division process takes about one hour to complete. The two sister chromatids separate from each other, one going to each of the two daughter cells. The two diploid daughter cells are genetically identical to each other and to the parent cell.

Key Points: **The cell cycle is a sequence of events involving the periodic replication of DNA and the segregation of this replicated DNA with cellular constituents to daughter cells. There are four phases of the cell cycle: G_1, S, G_2, and M or division. The two G phases are called "gaps"; the S phase is the period of DNA replication; and the division or M phase signals the actual division of the cell.**

CELL DIVISION: MITOSIS

The capacity of a cell to reproduce itself is perhaps the most fundamental property of life. There are two main processes by which eukaryotic cells divide: mitosis and meiosis. Mitotic cell division, which occurs in virtually all eukaryotic cells, is the process by which single cells reproduce themselves and multicellular organisms grow. The key feature of mitotic cell division is that the daughter cells are identical to each other and to the parent cell. The parent and daughter cells are **diploid (2n):** there are two copies of each type of chromosome. This is not true in meiotic cell division. Meiotic cell division is the basis of sexual reproduction in all higher plants and animals. It occurs during gamete formation in animals and spore formation in plants and fungi. The parent cell is diploid (2n), but following meiosis, the daughter cells (gametes or spores) are **haploid (n):** each gamete or spore has one member of each chromosome pair.

Mitosis, or mitotic cell division, was first described in 1879 by Walter Flemming. It consists of two interrelated processes: (1) mitosis, the division of the

TECHNICAL SIDELIGHT

Cell Cycle Checkpoints

Cycles in nature are ubiquitous. In the northern hemisphere, migratory birds fly south in the fall and north in the spring, stimulated to do so by the changing length of daylight hours, which in turn reflects the cycle of the earth's year-long journey around the sun. Deciduous trees sprout leaves in the spring and drop them in the fall as a response to these same light/dark cycles. The human female ovulates on a monthly cycle that is controlled by cycling hormone signals.

The cell cycle is also coordinated by external and internal signals, in this case chemical signals. The transition from each phase of the cycle, such as from G_1 to S or from G_2 to M, requires the integration of specific chemical signals and precise responses to those signals. If the signals are incorrectly sensed or if the cell is not properly prepared to respond, the cell could be transformed into a cancer cell.

The current view of cell cycle control is that the transitions between different cell cycle states (G_1, S, G_2, and M) are regulated at "checkpoints," also referred to as restriction points. Many checkpoints in turn are regulated by members of a family of proteins called cyclins and by members of a family of proteins that complex with the cyclins called cyclin-dependent kinases (CDK). The CDK proteins phosphorylate amino acids, and this regulates protein function. The cyclin-CDK complex is like a motor that drives the cell cycle. One of the most important cell cycle checkpoints, called *START,* is in mid G_1 (Figure 1). The cell receives both external and internal signals at this checkpoint which aid in the decision to drive the cell cycle into the S phase. This checkpoint is regulated in part by D-type cyclins and their main partner, CDK4. If the cell is driven past the *START* checkpoint by the cyclin-CDK complex, it becomes committed to another round of DNA replication. Inhibitory proteins with the capability of sensing problems in the late G_1 phase, such as low levels of nutrients or DNA damage, can put the brake on the cyclin-CDK motor and prevent the cell from entering the S phase.

In tumor cells, researchers have found that checkpoints are often deregulated. This deregulation is commonly due to defects in the regulation of the cyclin-CDK complexes caused by mutations in the genes that code for cyclins or CDK proteins or in genes that affect the activity of the cyclin-CDK complex. One consequence of a deregulated *START* checkpoint is an improper response to external signals, in which cell growth and cell division are no longer regulated in this cell line. Studies of different tumor lines have confirmed the aberrant nature of the cyclin-CDK complex at the *START* checkpoint.

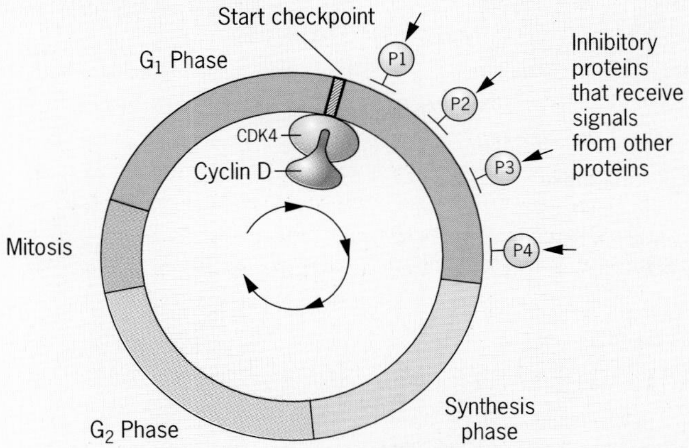

Figure 1 A schematic view of the START checkpoint in the mammalian cell cycle. Other Cyclin/CDK complexes monitor other checkpoints, but they are not shown in this figure.

A second consequence of a dysfunctional *START* checkpoint is that a cell with damaged DNA may move into the S phase. Normally, damaged DNA induces a signal to the cell for DNA repair. The damage-induced signal restricts the entry of the cell into the S phase. The cell that is arrested in the cell cycle does not move ahead until either the damage is repaired, or the cell is killed. But if the *START* checkpoint is unregulated, the cell with damaged DNA moves directly into the S phase so that damaged DNA is replicated. Unrepaired damage in the DNA can lead to further mutations, to tumors, and eventually to the formation of aggressive tumors.

Researchers continue to learn more about these checkpoints. In addition to the G_1 checkpoint (*START*) that detects damaged DNA, nutrient availability, and cell-cell contacts, G_2 checkpoints detect both DNA that has not been replicated, and aberrancies in the spindle apparatus that organizes and moves chromosomes during cell division. The function of checkpoints is to make cells wait for DNA damage to be repaired or to trigger the death of damaged cells in order to make certain that chromosomes will be replicated and segregated properly to give two daughter cells with the proper number and structure of chromosomes. If repair does not occur or if damaged cells are permitted to replicate, tumors commonly result.

nucleus (*mitos-* a thread, a term that refers to the threadlike appearance of the chromosomes); and (2) **cytokinesis,** the changes in the cytoplasm that include division of the cell proper. Figure 2.7 illustrates these processes as they occur in animal cells. The process is virtually identical in plants, with slight variations.

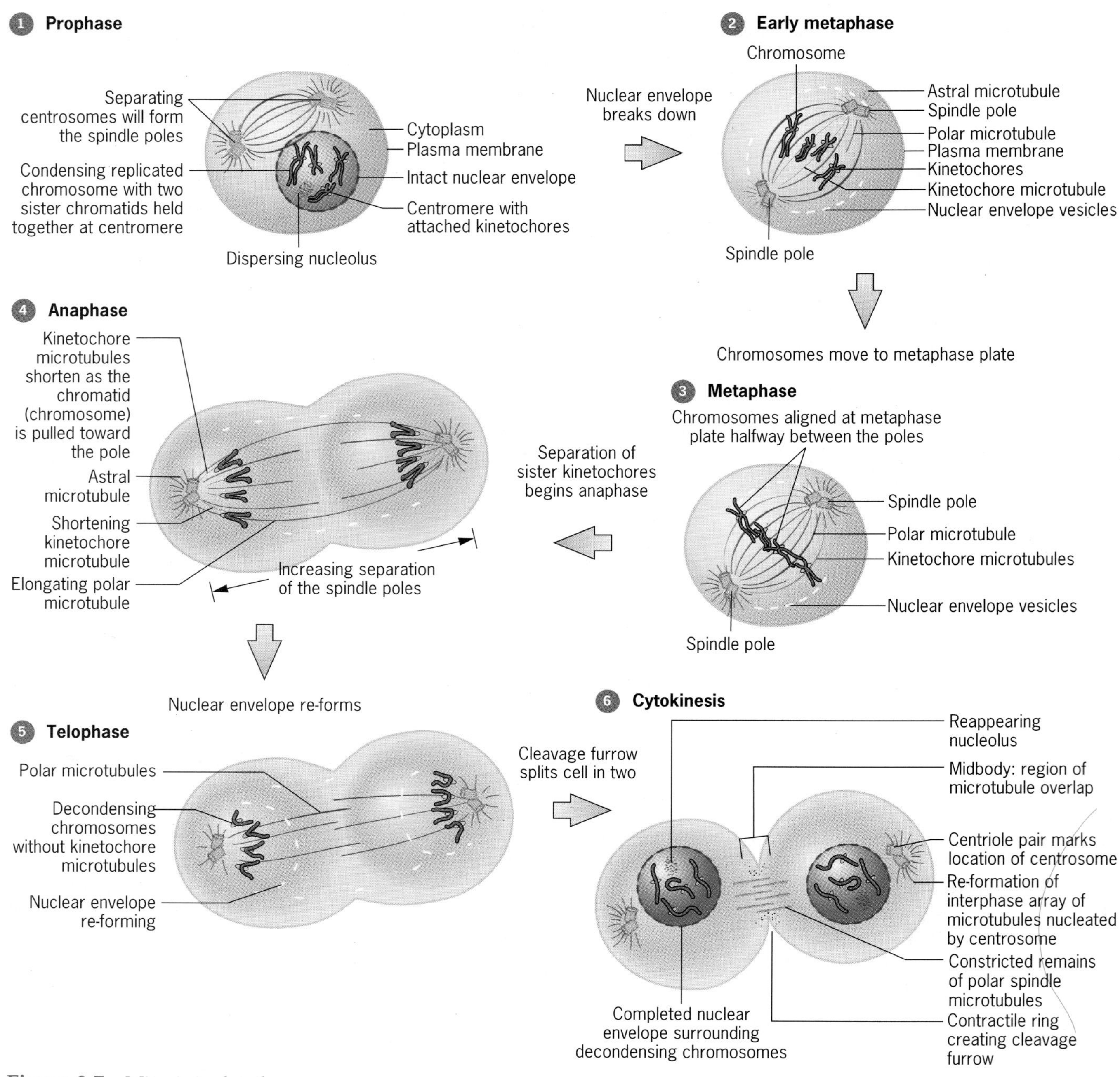

Figure 2.7 Mitosis in detail.

Mitosis is a continuous process, but for discussion purposes it is divided into five sequential phases: interphase (between divisions—consisting of G_1, S, and G_2 stages of the cell cycle), prophase, metaphase, anaphase, and telophase. Each phase is defined by the structure and behavior of the chromosomes. Prophase and telophase are usually long and involved, whereas metaphase and anaphase are commonly brief.

The first indications of approaching mitosis in animal cells are observed in the cytoplasm of the cell at **interphase.** The **centrosome** (Figure 2.8), a centrally located organelle, is the primary microtubule organizing center and acts as the spindle pole during mitosis. The microtubules, cytoskeletal structures that radiate from the centrosome, organize and coordinate the movement of chromosomes during mitotic cell division. The centrosome is duplicated by the cell during interphase so that each daughter cell eventually receives one. The centrosome in most animal cells has a pair of **centrioles** found in its center, and they, too, are

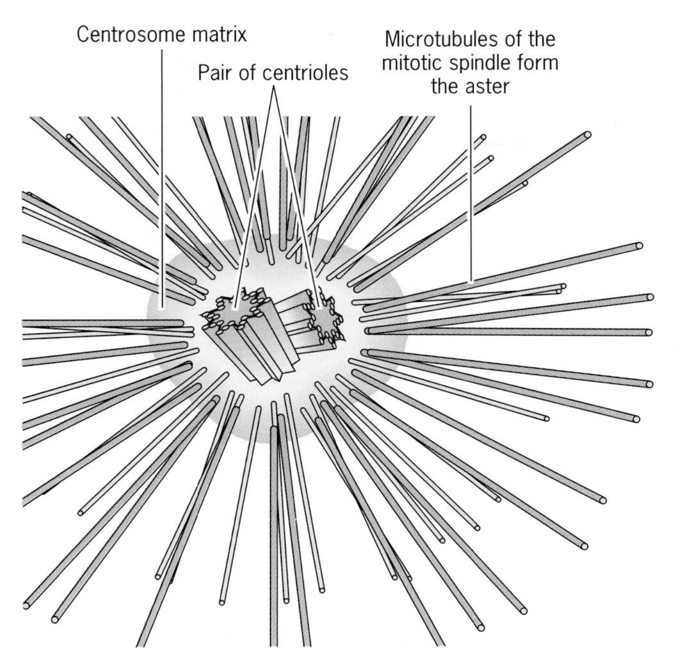

Figure 2.8 The centrosome.

replicated prior to division. As mitosis begins, the centrosome begins splitting into two, and the microtubules radiate out forming an **aster.**

The centrosome and its associated centrioles have perplexed and fascinated biologists for over a hundred years. How exactly do they replicate? What are they constructed of? What is their evolutionary history? Answers to these questions have so far eluded biologists.

In **early prophase** the two centrosomes separate and move toward opposite poles of the cell. The microtubules extending from them form the "rails" for chromosome movement. The chromosomes, which were replicated during the S phase of the cell cycle, are highly extended. They begin condensing through coiling and now appear as discrete entities.

By **late prophase** the chromosomes are highly condensed. The two chromatids of each chromosome are held together at the centromere, or primary constriction. The kinetochores complex with the centromere and function as the sites to which microtubules attach and guide chromosome movement. Centrosomes have now moved to opposite poles of the cell. A continuous microtubule network extends around the nuclear membrane and interconnects the poles. The nuclear membrane and nucleolus, the nuclear structure where ribosomes are assembled, become fragmented and dispersed in the cytoplasm. Microtubules now invade the nuclear region and become attached to each chromatid at the kinetochore in the centromeric region. Each of the two sister chromatids becomes attached to a different pole of the spindle, but the centromeres remain together.

At **metaphase** condensed sister chromatid pairs assume positions in the cell's center, or **equatorial plate,** between the two poles. Metaphase chromatids are tightly coiled and discrete, thus facilitating accurate chromosome counts and gross structural analysis. The diagnosis of disorders caused by the structural changes in chromosomes is commonly made on these metaphase chromosomes. Arms of sister chromatids are extended from the centromere region, but the chromatids are held together at the centromeres until the beginning of the next phase.

Anaphase begins when the centromeres separate and the sister chromatids of each chromosome disengage and move toward opposite poles of the cell. Movement is the result of microtubule shortening. After anaphase separation occurs, each chromatid has its own centromere and is now considered to be a chromosome. Anaphase chromosomes elongate somewhat by relaxation of the tight metaphase coiling and move to respective poles of the spindle. Mitosis ensures that each daughter cell has the same genetic information as the parent cell.

The last mitotic phase, **telophase,** occurs when chromosome movement is completed and the microtubules disassemble. By the completion of telophase, a nuclear membrane is reconstructed around each daughter nucleus, and the nucleolus begins to reappear. The chromosomes uncoil and become more extended. Mitosis is completed.

The cytoplasmic part of the cell now divides by cytokinesis (Figure 2.9). Animal cells, with their flexible outer layers, accomplish this by a midline constriction that eventually separates the two daughter cells. The surface around the equatorial region of the cell pushes in toward the center and pinches the cell into two parts. Plant cells, with their rigid cell walls, form a partition, or **cell plate**, between the daughter cells. After the cell plate is formed, walls of cellulose are deposited on either side.

As noted earlier, each cell division (nuclear and cytoplasmic phases) is a continuous process from the time a cell first shows evidence of beginning to divide until the two daughter cells are completely formed. Nuclear and cytoplasmic phases are distinct but coordinated processes. The entire process ordinarily requires a few hours to several days, with variation dependent on the type of organism and environmental conditions.

Key Points: **Mitosis, a mode of nuclear and cytoplasmic division, provides for the production of two daughter nuclei that contain identical chromosome sets and are genetically identical to each other and to the parent cell from which they arose. The division is**

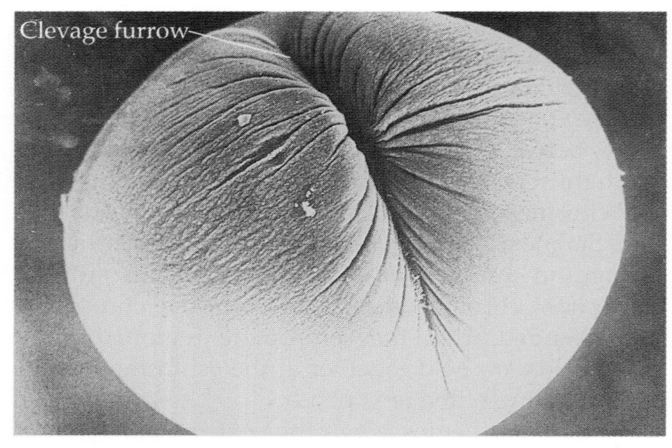

(a) Cytokinesis in a frog egg

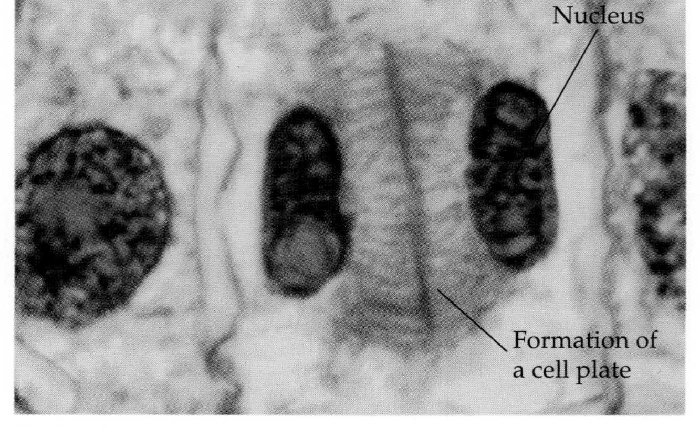

(b) Cytokinesis in an onion root cell

Figure 2.9 Cytokinesis in animal (*a*) and plant (*b*) cells. The animal cells constrict; the plant cells build a cell plate.

divided into five phases: interphase, prophase, metaphase, anaphase, and telophase. Cytokinesis, the division of the cytoplasm, follows telophase.

CELL DIVISION: MEIOSIS

In 1883, four years after mitosis was first described, Edouard van Beneden discovered that the egg of the roundworm, *Ascaris,* contains only half as many chromosomes as somatic cells (nonsex cells) and that the chromosome number characteristic of the species is reestablished at fertilization. He accurately described the process of **meiosis** (from the Greek meaning diminution), or meiotic cell division, whereby the diploid chromosome number ($2n$) is reduced by one-half to the haploid state (n) during gamete formation. But van Beneden erroneously concluded that all maternal chromosomes (those chromosomes contributed to the individual by the mother) go to one nucleus and all paternal chromosomes to the other. In fact, maternal and paternal chromosomes are shuffled and randomly distributed to daughter nuclei during meiosis. The behavior of chromosomes during meiosis was thus recorded, but no one at this time related meiosis to heredity.

In its broadest sense, meiosis is a lengthy process in which the chromosomal material replicates once and the cell divides twice, thus reducing the chromosome number by one-half. In pre-meiotic diploid cells, each chromosome has a pairing mate, or **homolog.** For each homologous pair, one member is contributed by the sperm (paternal chromosome) and the other contributed by the egg (maternal chromosome). At the conclusion of meiosis, each cell has one member of a homologous pair, either a maternal or a paternal chromosome, and is thus haploid.

In the absence of meiosis, the fusion of two diploid sex cells would produce a zygote with twice the number of chromosomes as the parents. In humans, where the diploid chromosome number is 46 (23 pairs), the fusion of two diploid cells would produce progeny with 92 chromosomes. The next generation would have 184 and the next 368. This type of chromosome doubling is not compatible with life. Meiosis ensures a constant chromosome number from generation to generation.

Meiosis has two successive divisions. The first, **reduction division** (meiosis I), is the more complex. During this division, homologous chromosomes pair, then segregate and move to different nuclei. During the second meiotic division, the **equational division** (meiosis II), the centromere splits and the sister chromatids separate into different nuclei. Thus, four haploid nuclei result from the meiotic division of one diploid nucleus. Figure 2.10 details the two divisions of meiosis and their stages.

The first stage of meiosis, a lengthy and complex *prophase I* is made up of five sequential substages. The first substage that can be recognized as distinct from interphase is *leptonema.* During this substage, the chromosomes first appear as distinct, threadlike nuclear objects. Chromosomes have been replicated by this substage (during interphase), but they appear as single, unreplicated structures.

During the next substage of prophase I, *zygonema,* homologous chromosomes pair tightly to form a four-stranded complex called a **tetrad.** The pairing process, which occurs continuously through zygonema at multiple points along the chromosome, is called **synapsis.** During this process the chromosomes continue to con-

Early prophase I

Replicated chromosomes become visible.

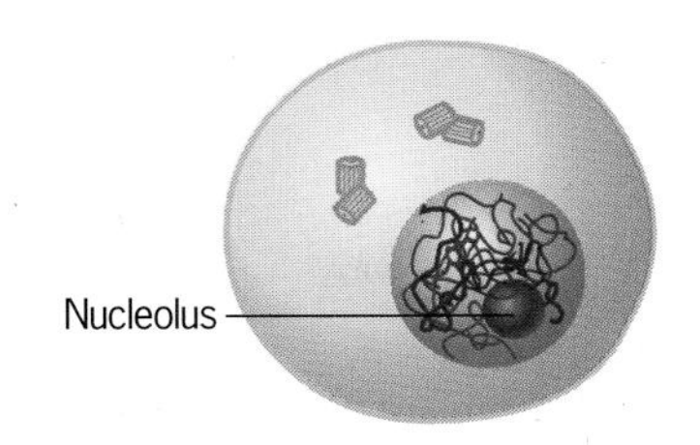

Middle prophase I

Homologous chromosomes shorten and thicken. The chromosomes synapse and crossing over occurs.

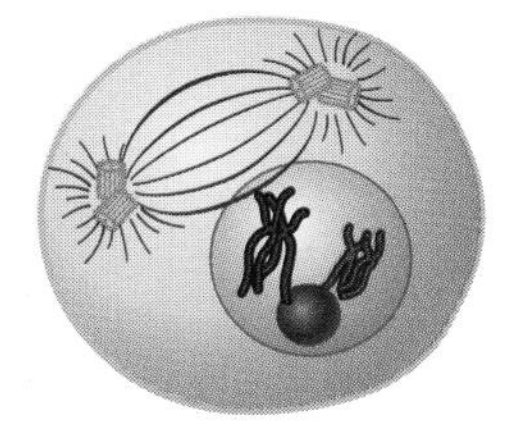

Late prophase I

Results of crossing over become visible as chiasmata. Nuclear membrane begins to disappear. Spindle apparatus begins to form. Nucleolus disperses.

Metaphase I

Assembly of spindle is completed. Each chromosome pair aligns across the metaphase plate of the spindle.

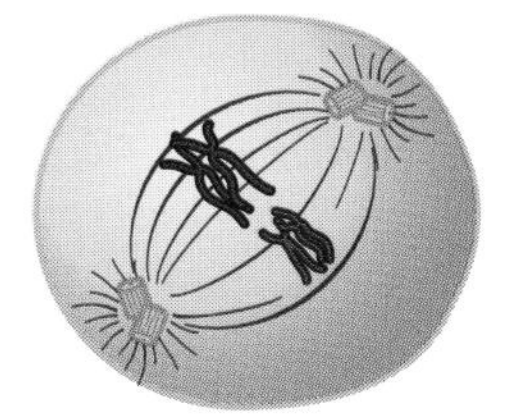

Anaphase I

Homologous chromosome pairs separate and migrate toward opposite poles.

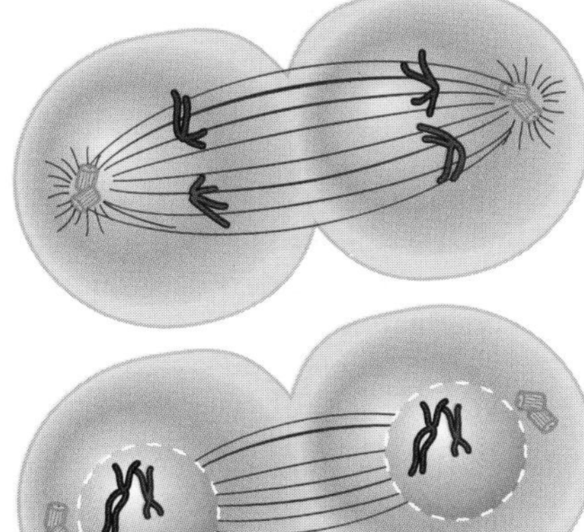

Telophase I

Chromosomes (each with two sister chromatids) complete migration to the poles, and new nuclear membranes may form.

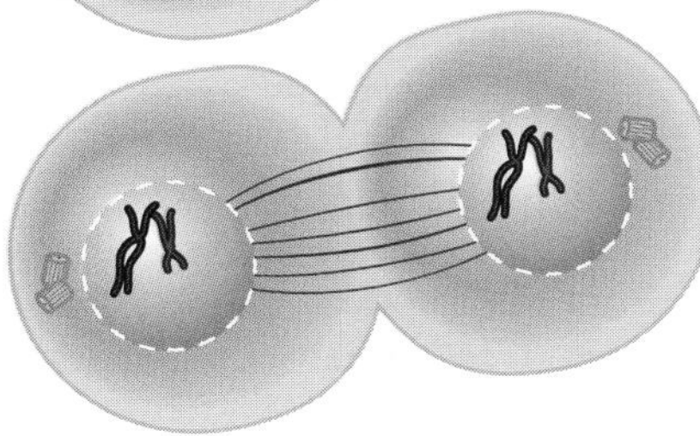

Cytokinesis

In most species, cytokinesis produces two daughter cells. Chromosomes do not replicate before meiosis II.

Prophase II

Chromosomes condense and move to metaphase plate.

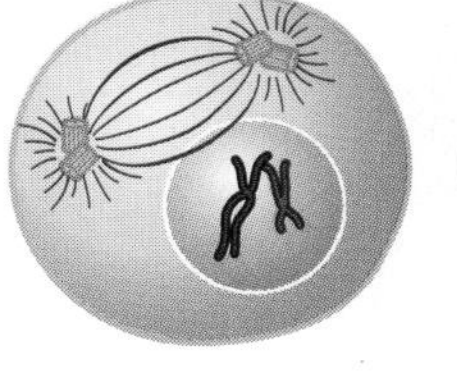
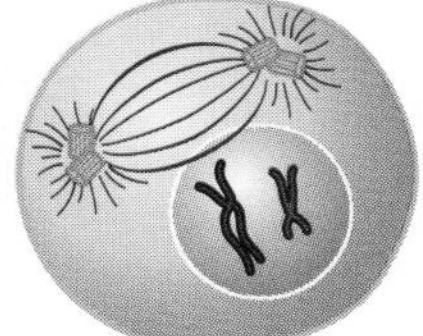

Metaphase II

Kinetochores attach to spindle fibers. Chromosomes line up on metaphase plate.

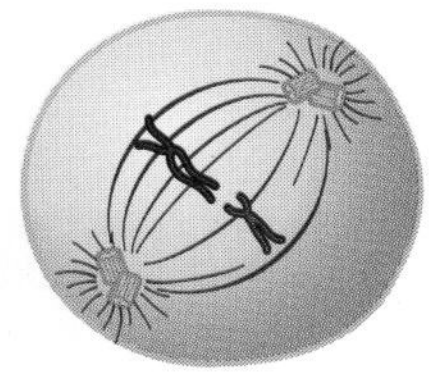

Anaphase II

Sister chromatids separate and move to opposite poles as separate chromosomes.

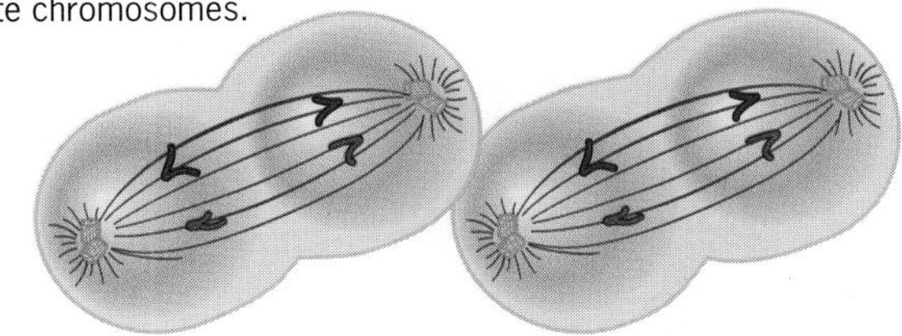

Telophase II

Nuclear membrane forms around chromosomes and chromosomes uncoil. Nucleolus reforms.

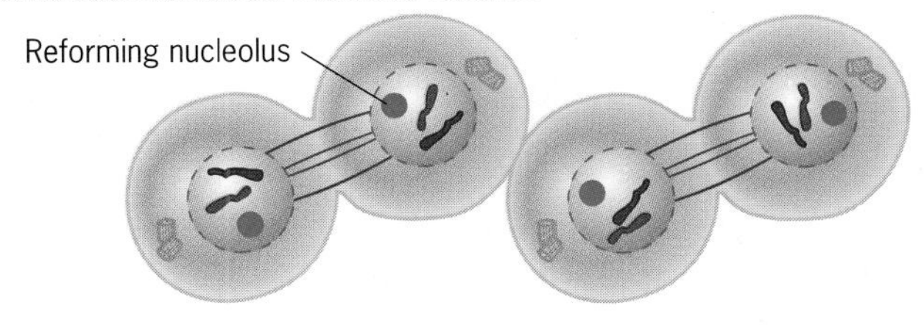

Four haploid cells form after cytokinesis.

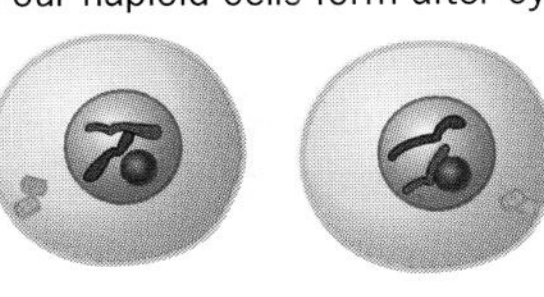
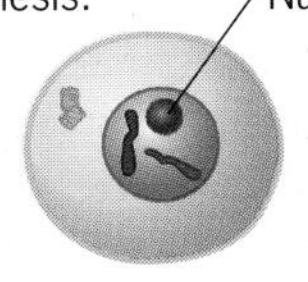

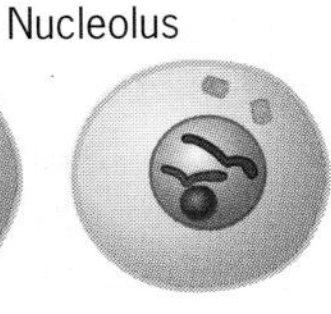

Figure 2.10 Meiosis in detail.

dense. However, they still appear to be single, unreplicated structures.

The third substage of prophase I is called *pachynema*. For the first time, the chromosomes can be seen as replicated structures—that is, as seen under the light microscope, each chromosome consists of two chromatids. The chromosomes continue to shorten and condense during pachytene. Nonsister chromatids within each homologous chromosome pair commonly exchange segments of genetic material with each other during early prophase, a process of genetic recombination called **crossing over.** Visual ev-

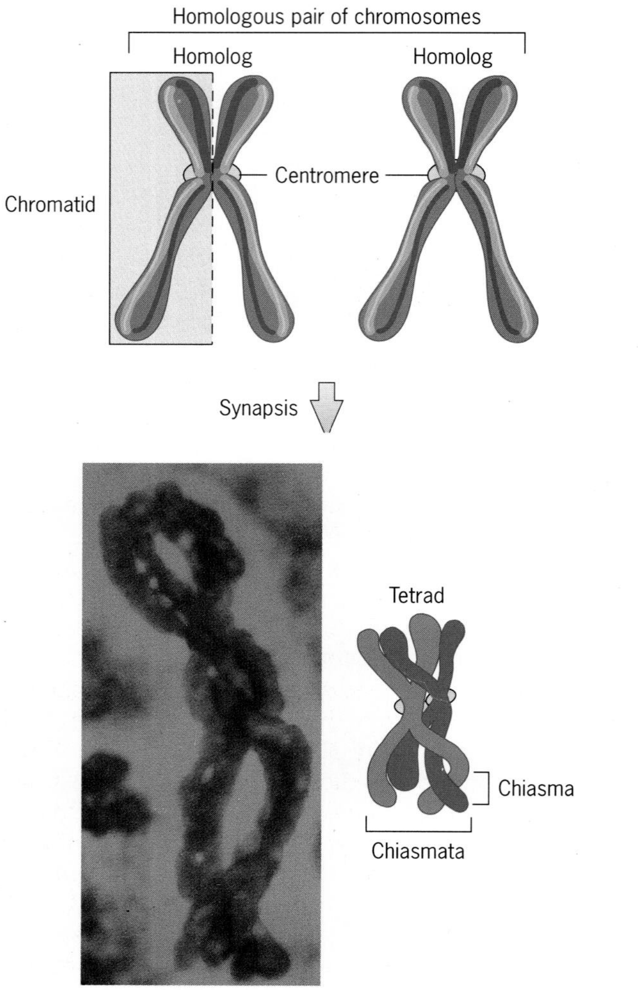

Figure 2.11 Chiasmata are cytologically visible X-shaped regions where crossing over has occurred in a synapsed pair of homologous chromosomes.

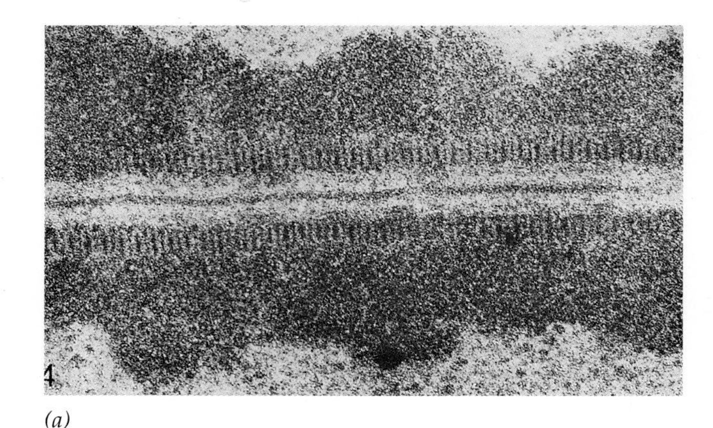

(a)

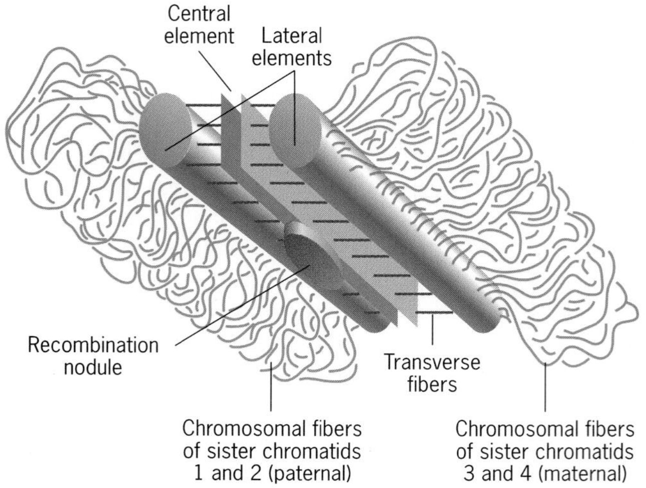

(b)

Figure 2.12 Crossing over and the synaptinemal complex. During prophase I of meiosis, homologous chromosomes synapse with one another, appearing as parallel fibers joined at the centromere region by the proteinaceous kinetochore (*a*). Close examinantion of the synapsed chromosomes reveals the ladderlike structure called the synaptinemal complex, which is diagrammed in (*b*). The recombination nodules are thought to contain the enzymatic machinery required for genetic recombination.

idence of crossing over are X-shaped structures called **chiasmata** (sing. = chiasma) (Figure 2.11). With the aid of an electron microscope, a ladderlike structure called the **synaptinemal complex** can be seen between the synapsed homologs, apparently binding the nonsister chromatids along their length and facilitating crossing over (Figure 2.12). The space in the middle of the ladder contains loops of DNA from each chromosome, and it appears likely that crossing over occurs between these DNA loops, coordinated by nodules of recombination enzymes. The genetic significance of crossing over, especially as it pertains to gene mapping, will be explored in more detail in Chapter 7.

During the fourth substage of prophase I, *diplonema,* homologous chromosomes seem to repel each other, or desynapse. However, the members of a pair remain attached at the chiasmata referred to earlier. The diplotene stage (diplotene is the adjective form of diplonema) in many animals is extremely prolonged. In human females, for example, the diplotene stage of prophase I is completed in the ovary during fetal life. About 400,000 immature eggs (oocytes) reach this stage, and then cell division stops. The eggs remain in this suspended state until the onset of puberty. After puberty and the onset of the menstrual cycle, one egg a month continues meiosis in the fallopian tube. Some eggs remain in this suspended meiotic state for several decades. This can have profound genetic consequences. Older eggs, held in this arrested state of meiosis, have an increasingly difficult time completing a normal meiosis because of changes that accumulate in the proteins controlling chromosome movement. The result is eggs with abnormal numbers of chromosomes.

During the last substage of prophase I, *diakinesis,* chromosome pairs continue to thicken and move toward the center of the nucleus. The nucleolus and nuclear membrane disperse, and microtubules emanating from the centrosomes attach to the kinetochore. Prophase I is now completed.

At *metaphase I* the chromosome pairs are highly condensed and arranged on opposite sides of the equatorial plate. Microtubules are attached to the kinetochore in the centromeric region. The chiasmata have moved to the ends of each chromosome pair.

During *anaphase I* the members of a homologous chromosome pair move toward opposite poles. The centromeres have not yet divided; thus each chromosome still consists of two chromatids.

Telophase I occurs when the chromosomes reach the opposite poles of the cell. Often this stage is accompanied by the re-formation of nuclear membranes around each haploid complement, cytokinesis, and then by a brief interphase (but an interphase in which the DNA is *not* replicated). Sometimes, however, membrane formation, cytokinesis, and interphase do not occur, and cells move directly into the second meiotic division.

Meiotic division II, the equational division, resembles a normal mitosis, except there is only one member of a chromosome pair in each nucleus. At *prophase II,* the chromosomes condense and move to the equatorial region. The centromere divides at *metaphase II,* and the former chromatids, now daughter chromosomes, move toward the opposite poles in *anaphase II.* When the chromosomes have reached the poles and stopped their movement *(telophase II),* nuclear membranes form around each set of haploid chromosomes, and cytokinesis follows. The result of meiotic divisions I and II is four haploid cells containing one member of each homologous chromosome pair.

Meiotic Abnormalities

On occasion, a problem occurs during meiosis such that homologous chromosomes fail to pair properly and separate from each other. This phenomenon is called **nondisjunction.** A result of nondisjunction, which can occur in either meiosis I or meiosis II, is that both members of a homologous chromosome pair (or both sister chromatids) may go to the same pole instead of segregating to opposite poles of the cell. Studies of sperm cells from normal fertile men indicate that almost 5 percent of these cells carry abnormal chromosome numbers; thus meiotic mistakes are surprisingly common.

The causes of nondisjunction are many and varied. A number of environmental agents cause nondisjunction. For example, heat and cold shock interfere with chromosome pairing in many species and lead to irregular meiosis. X rays produce meiotic abnormalities by disrupting either synapsis or the integrity of the spindle fiber apparatus. Chemicals, such as colchicine, an alkaloid commonly used to treat gout, causes the rapid disintegration of the spindle apparatus and thus disrupts normal cell division.

In humans, there is a striking relationship between advanced maternal age and nondisjunction. The frequency of Down syndrome, the disorder described in the opening of this chapter, is 4 per 10,000 births in 20-year-old women, but increases dramatically to about 200 per 10,000 in women who are 45 or older. It has been estimated that if women over the age of 35 refrained from having children, the incidence of children born with abnormal numbers of chromosomes would decrease by 35 to 50 percent. Perhaps the underlying cause of nondisjunction in older women is the deteriorating state of the egg, which has been held in suspended meiosis for 30 to 40 years or more. While maternal age is a major factor in meiotic mistakes, advanced paternal age is also important. It is estimated that about 20 percent of all children born with abnormal chromosome numbers are the result of paternal nondisjunction, most in older fathers.

Genes have also been implicated in meiotic mistakes, although it is difficult to identify and characterize specific genes. High frequencies of meiotic mistakes and children born with multiple meiotic abnormalities have been found in some families. These observations suggest that mutant genes in some families disrupt the meiotic process. Some of these genes may control or influence the structure of the synaptinemal complex or perhaps the attachment of microtubules at the centromeric region. We will consider the consequences of meiotic mistakes in detail in Chapter 6.

Key Points: **Meiosis, a fundamental process of cell division in sexually reproducing eukaryotes, involves three main events: the pairing of homologous chromosomes; the exchange of genetic material by crossing over; and the segregation of the members of a homologous pair of chromosomes into different daughter nuclei. Meiosis involves one round of DNA replication followed by two separate cell divisions. The result of meiosis in a diploid cell is four haploid cells. Meiotic nondisjunction is the failure of sister chromatids or homologous chromosomes to disjoin.**

THE EVOLUTIONARY SIGNIFICANCE OF MEIOSIS

The behavior of chromosomes during meiotic cell division accounts for the basic principles of Mendelian inheritance (Chapter 3). Mendel's principles of segregation and independent assortment describe perfectly the behavior of chromosomes during meiosis. But meiosis is important for other reasons too. Van Beneden, in his original description of meiosis, had suggested that at metaphase all the maternal chromosomes lined up on one side of the equatorial plate and all paternal chromosomes on the other. Such an arrangement would produce gametes carrying all maternal or all paternal chromosomes. This is not the case, however. Chromosomes align randomly along the equatorial plate so that the gametes carry a mixture of maternal and paternal chromosomes. For example, the human has 23 pairs of chromosomes. For each pair, the probability that a gamete will get a maternal chromosome of any pair is 1/2. The probability that a gamete will get *all* maternal chromosomes is $(1/2)^{23}$! Stated another way, there are 2^{23} possible combinations of chromosomes. When crossing over is factored into this equation, the number of possible different genetic combinations becomes astronomical. This is the true significance of meiosis and sexual reproduction: It is a mechanism for generating tremendous amounts of genetic variation in sexually reproducing populations. Populations with the greatest amount of genetic variation tend to be the most successful, in evolutionary terms.

Key Points: **The most important consequence of meiosis is that it produces an enormous amount of genetic variation through the production of different combinations of chromosomes.**

THE FORMATION AND UNION OF GAMETES

The completion of meiosis does not necessarily mean that gametes have formed. Meiosis produces four haploid cells. These cells must now mature into functional gametes, eggs and sperm. The meiotic process that produces haploid cells and the subsequent maturation of these cells into functioning gametes is called **gametogenesis.** Let's now look at the relationship between meiosis and male and female gamete formation.

Oogenesis: The Formation of the Egg

In the early stages of embryonic development, groups of cells become committed to differentiate into various cell types, such as liver cells, nerve cells, and muscle cells. One group of cells becomes committed to form the germ cell line, cells that will eventually develop into eggs if the embryo is female or sperm if the embryo is male. These are the only cells that undergo meiosis. These committed cells, primordial germ cells, increase in number through mitotic cell division. Later, they divide meiotically to produce mature sperm or eggs. The process of forming mature eggs is called **oogenesis.**

There are different stages of oogenesis, and while there is variation from species to species, the general stages of egg formation are similar (Figure 2.13). In this section we describe oogenesis in mammals. The primordial germ cells that migrate to the developing ovary during early embryogenesis become **oogonia** (sing. = oogonium). They multiply rapidly, undergoing several rounds of mitotic cell division, and eventually differentiate into **primary oocytes.**

The primary oocytes begin meiotic cell division. They complete the diplotene stage of prophase I and then division stops. They remain in this suspended prophase for a period of time that varies from species to species. During this period, the oocyte undergoes many changes that prepare it for the completion of meiosis and for fertilization. For example, it acquires a special coat that protects the developing egg from mechanical damage and in many cases acts as a barrier to sperm from other species. Just beneath the membrane, **cortical granules** develop that alter the egg coat so that only a single sperm fertilizes an egg. In addition, the primary oocyte accumulates large quantities of nutrients and other molecules that nourish the early embryo and coordinate and direct its early development.

With the attainment of sexual maturity, the next stage of oocyte development occurs, triggered by hormones. The oocyte completes meiosis I and two haploid nuclei are formed, each containing one member of each chromosome pair in a replicated state. But cytokinesis is very asymmetrical. One cell, called the **secondary oocyte,** gets virtually all the cytoplasm and is ancestral to the mature egg. The other cell, called a **polar body,** gets very little cytoplasm. Both of these cells, the secondary oocyte and the polar body, undergo meiosis II to produce four haploid nuclei. Again, cytokinesis is asymmetrical: Meiosis II in the secondary oocyte produces one large cell, the **ovum,** or egg, which has virtually all the cytoplasm, and a small polar body with very little cytoplasm. Thus, of the four meiotic products, only one forms the mature egg. The polar bodies, which are small with little cytoplasm to support their metabolism, eventually degenerate.

Oogenesis occurs in the ovaries of the female. Each primary oocyte is surrounded by a spherical cluster of cells in a cavity or sac called the primary or Graafian follicle. In response to hormone signals, the

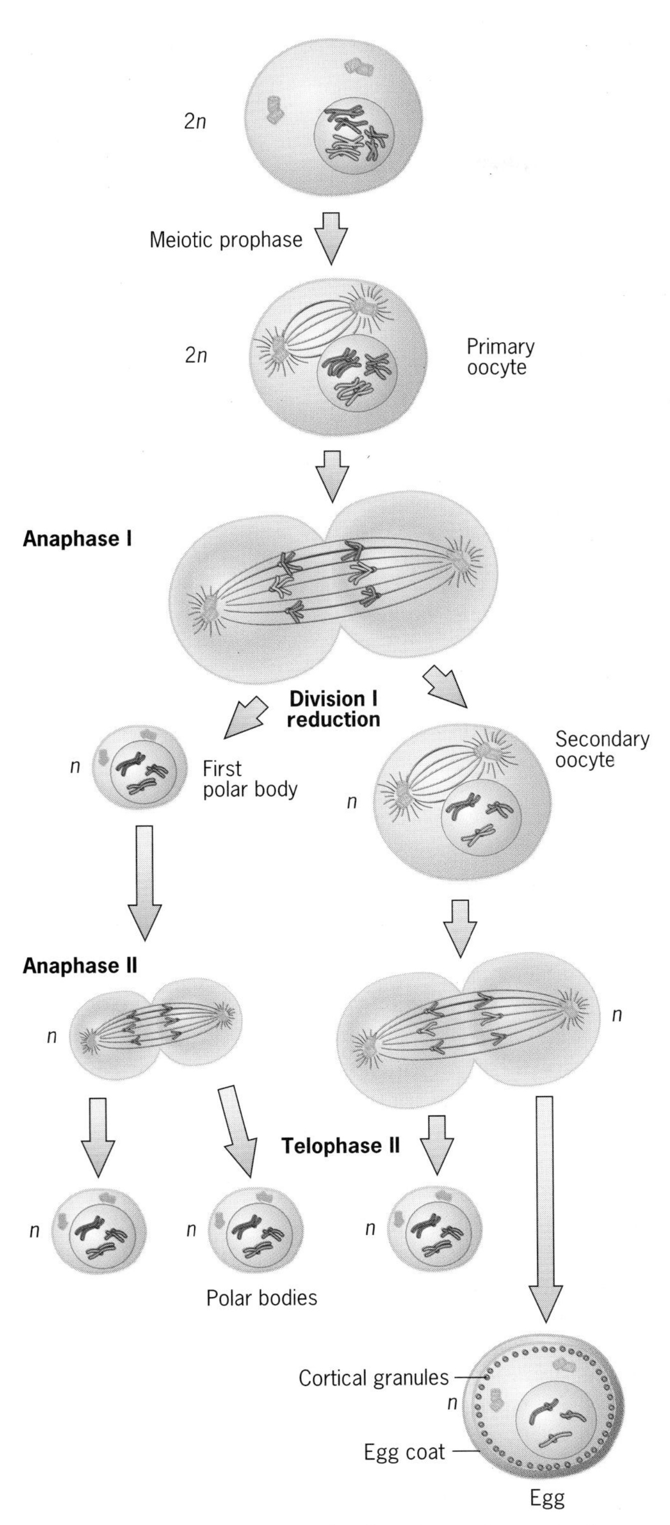

Figure 2.13 Oogenesis in the female results in the formation of one egg and three polar bodies.

primary oocyte completes meiosis I to become a secondary oocyte. The follicle then ruptures, releasing the secondary oocyte into the oviduct, where it begins the second meiotic division. In humans, the oviduct is called the Fallopian tube. The release of the secondary oocyte from the Graafian follicle is called **ovulation.**

As the secondary oocyte travels down the oviduct, it enters the second meiotic division. If it is fertilized, meiosis II is completed. If it is not fertilized, meiosis II is not completed and the egg degenerates. Upon fertilization, both male and female nuclei occupy the egg cytoplasm. The haploid nuclei fuse, restoring the diploid chromosome number. The **zygote,** or fertilized egg, passes down the oviduct into the uterus and attaches there. We discuss fertilization in more detail shortly.

Spermatogenesis: The Formation of Sperm

The egg is among the largest cells in an organism. The sperm is among the smallest. This size difference is related to the function of the two cell types. The egg is nonmotile and stores large quantities of material needed to support the growth and development of the early embryo. The sperm is highly motile and constructed to move through the female reproductive tract to fertilize the egg. In many respects, the sperm is simply a haploid nucleus with a tail. (See, however, Technical Sidelight: Giant Sperm.) There is neither an endoplasmic reticulum nor a Golgi apparatus, nor are there any ribosomes. The mitochondria coalesce around the tail and serve as an energy source for tail movement. The head of the sperm contains an acrosomal vesicle with several enzymes that help the sperm penetrate the egg.

Spermatogenesis begins in the human male at puberty. (Recall that in the human female, meiosis begins during early embryogenesis.) It takes place in the testes in coiled tubes called seminiferous tubules. The immature premeiotic germ cells, called **spermatogonia,** are located around the border of the seminiferous tubules. The spermatogonia proliferate mitotically. At various points, a subset of the spermatogonia stop dividing, enlarge, and differentiate into **primary spermatocytes.** Primary spermatocytes undergo meiosis I to produce **secondary spermatocytes** and then meiosis II to produce four haploid **spermatids.** The spermatids differentiate into mature **spermatozoa,** which pass into the lumen of the seminiferous tubule, and from there into the epididymis, a tube overlying the testis, where they undergo further maturation. Figure 2.14 summarizes the stages of spermatogenesis.

The developing sperm cells remain connected to each other by a common cytoplasm during the entire process. With the differentiation of spermatids into spermatozoa, the individual sperm are released. The common cytoplasm provides the developing spermatids with all the cytoplasmic products they require

TECHNICAL SIDELIGHT

Giant Sperm

The male fruit fly, *Drosophila melanogaster*, produces impressive sperm cells almost 2 mm in length. These sperm are almost 300 times larger than the sperm produced by human males. But *Drosophila melanogaster* sperm pale in comparison to the sperm cells of *Drosophila bifurca*, a distant relative of *D. melanogaster*. The sperm of *D. bifurca* are almost 6 cm in length, about 20 times larger than the flies that produce them! If a 6 foot tall human male produced sperm in similar proportions, the sperm cells would be over 40 yards in length!

The production of these enormous sperm cells creates a puzzle for evolutionary biologists. Males in general tend to produce vast numbers of sperm, invest little energy in each, and disseminate the sperm widely. *D. bifurca*, on the other hand, produces a greatly reduced number of sperm and is prudent in disseminating them. The large sperm require more energy to produce and to store than the smaller sperm. The testes of *D. bifurca*, for example, are 67 mm in length when they are laid out straight and account for about 11 percent of the male fly's dry weight. This compares with *D. melanogaster*'s testes which account for about 5 percent of the dry weight. The large sperm size also means that these giant sperm take longer to produce. This lengthens the time required for the male fly to reach sexual maturity, a time span he may not survive.

Are there any advantages to producing such long sperm? It may be that sperm with longer tails compete more effectively for eggs in the fertilization process, although there is some debate on this point. It may also be possible that the longer sperm contribute to some postfertilization function, such as facilitating the transfer of paternal mitochondria into the egg. Any advantage these large sperm confer must be weighed against the disadvantages: high energy costs, extended time required for males to reach sexual maturity during which they are exposed to the dangers of their environment, and drastically reduced number of progeny per mating. The selective value of the giant sperm is being investigated, but answers are elusive.

Pitnick, S., G. S. Spicer, and T. A. Markow. *Nature* 375:109 (1995).

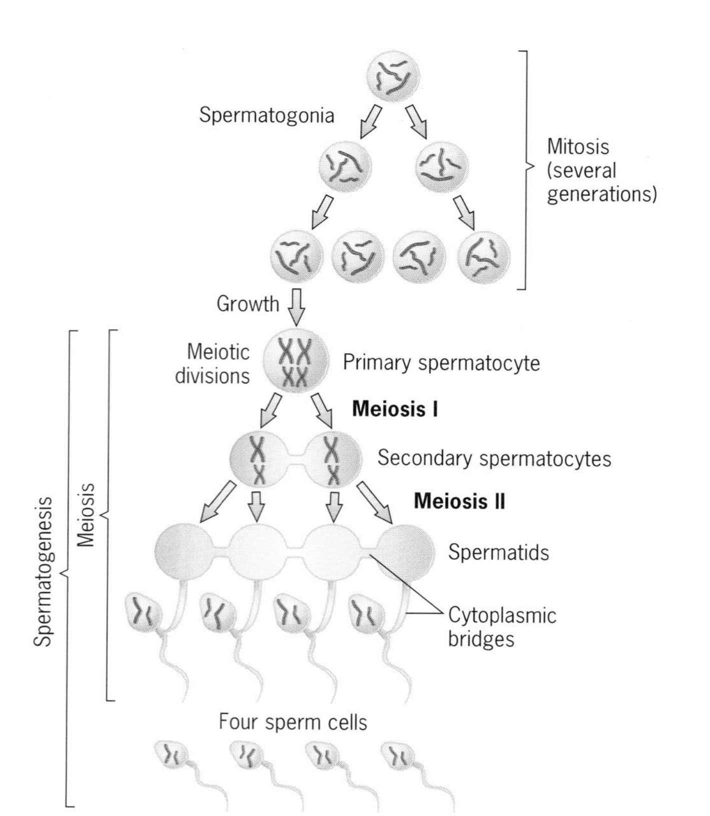

Figure 2.14 Spermatogenesis in the male results in the formation of four sperm cells. During the process, sperm remain connected via a common cytoplasm until the formation of mature sperm.

for differentiating into spermatozoa. In contrast to the mature egg that has sequestered virtually all the cytoplasm from the meiotic products, the sperm cells lose cytoplasm during spermatogenesis, producing spermatids that contain very little cytoplasm.

In mammals, **fertilization** occurs with the union of sperm and egg in the oviduct. The sperm head binds to the **zona pellucida,** or egg coat. The zona pellucida has two main functions: (1) it contains various glycoproteins that bind the sperm and serve as recognition signals for sperm of the same species; and (2) it functions as a barrier to sperm from other species. On binding, the contents of the sperm's acrosomal vesicle are released. This vesicle contains enzymes that aid the sperm in its penetration of the zona pellucida and facilitate the fusion of the sperm and egg membranes. After membrane fusion, the cortical granules just inside the egg membrane release a variety of molecules, including enzymes that alter the structure of zona pellucida in such a way that additional sperm are prevented from fusing with the egg membrane. This assures that only a single sperm can fertilize an egg. Once the sperm and egg nuclei fuse into a single diploid nucleus, the zygote begins development.

Gamete Formation in Plants

The major difference between the reproductive cycles in plants and animals is the mode of gamete formation. In animals, meiosis produces haploid cells that then mature into gametes. In plants, gametes rarely arise directly from meiosis. Meiosis takes place in plants, but the resultant haploid cells do not become gametes. The haploid products of meiosis in plants are called **spores.**

Before gametes form and fertilization occurs, haploid spores divide mitotically to produce a fully developed, multicellular haploid organism called a **gametophyte,** or gamete-producing plant. The gametophyte, which developed from the spore, produces gametes by mitosis. When two haploid gametes fuse, a diploid zygote is produced. The zygote divides mitotically and develops into a multicellular diploid organism called the **sporophyte,** or spore-producing plant. Later, specialized cells within the sporophyte, analogous to the germ cells in animals, undergo meiosis and produce haploid spores. The cycle then starts anew. Thus plants, unlike animals, exist as fully developed haploid and diploid organisms during their life cycle. Alternating sporophyte and gametophyte phases, referred to as the **alternation of generations,** vary in length and importance in different plants (Figure 2.15.)

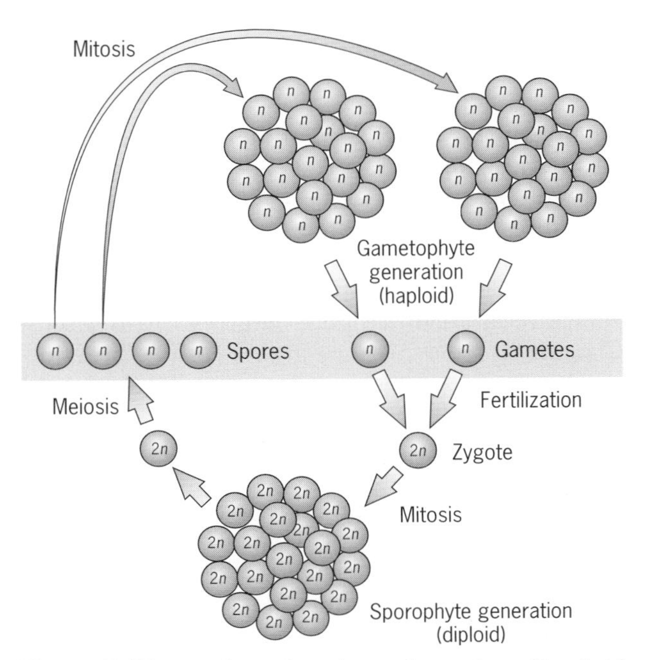

Figure 2.15 In plants there is an alternation of haploid and diploid generations.

Key Points: **Gametogenesis is the formation of male and female gametes or sex cells. Spermatogenesis and oogenesis are the formation of male and female gametes, respectively, by the processes of mitosis and meiosis. Gametes unite in the process of fertilization. In plants, meiosis in the sporophyte produces spores that develop into haploid gametophytes. Gametophytes produce haploid gametes that fuse to produce a diploid zygote, which divides mitotically to produce a diploid sporophyte.**

THE LIFE CYCLES OF SOME GENETICALLY IMPORTANT ORGANISMS

Meiosis generates haploid cells in which the parental genetic material has been reshuffled by independent assortment and crossing over. The fusion of haploid cells produces an almost infinite variety of new genetic combinations that the evolutionary processes can act upon. In general, the greater the amount of genetic variability in a population, the greater the chances for evolutionary success. Thus the life cycles of organisms present opportunities for the reshuffling of the genetic material to produce new genetic combinations. In this section, we look at the life cycles of a few sexually reproducing organisms of special genetic interest to see how they maximize their opportunities to generate new genetic recombinants. We discuss life cycle strategies of some nonsexually reproducing organisms, such as bacteria and viruses, in later chapters.

Neurospora crassa: The Simple Bread Mold

An organism that has made enormous contributions to our understanding of genetic concepts, especially to issues relating to recombination and the relationship between the gene and the gene product, is the salmon-colored bread mold, *Neurospora crassa* (Figure 2.16). Molds or fungi are predominantly terrestrial and haploid through most of their life cycle. Although some fungi are unicellular, most are multicellular and filamentous. *Neurospora* is a haploid filamentous fungus. The filaments, or hyphae, are divided into segments by perforated cross-walls called septae. The perforations allow cytoplasm to flow freely between the hyphal segments. *Neurospora* reproduces asexually by the formation of specialized multinucleated spores called **conidia** (sing.= conidium). The asexually pro-

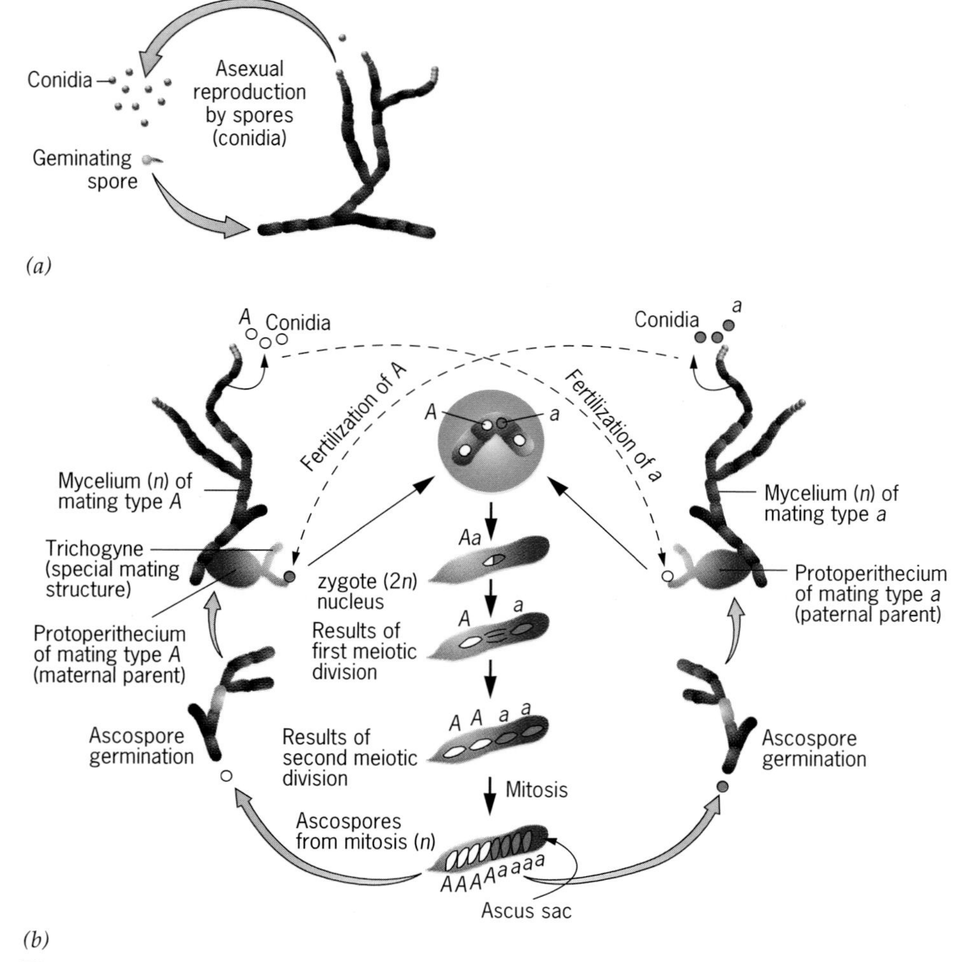

Figure 2.16 The life cycle of bread mold, *Neurospora crassa. Neurospora* reproduces (*a*) asexually or (*b*) sexually.

duced spores germinate into hyphal segments. Asexual reproduction is the predominant mode of reproduction in *Neurospora.*

Neurospora also has a sexual mode of reproduction. There are two mating types of *Neurospora, A* and *a*. Under certain conditions, a conidium from one mating type lands on a large, specialized hypha called a protoperithicium, or fruiting body, from an organism that is of the opposite mating type. The conidium fuses with a special mating structure, the trychogyne. The two nuclei now share a common cytoplasm, but they do not immediately fuse. Rather, each divides mitotically to form a multinucleated cell. Nuclei of opposite mating types fuse to form a diploid nucleus, the only diploid phase in the entire *Neurospora* life cycle. Each diploid nucleus becomes walled off in an **ascus** (plural- asci), a cell with a thick outer wall. Meiosis produces four haploid products, which in turn divide mitotically to produce eight meiotic products contained in a linear array. Each of the four meiotic products is represented twice. The eight haploid products mature into **ascospores** and are released when the ascus ruptures. Each ascospore gives rise to a haploid hyphal segment which continues to divide mitotically. The linearity of the ascospores has important genetic significance in gene mapping, which we will discuss in Chapter 8.

Corn (*Zea mays*)

Genetic studies using corn have led to many important discoveries. Several of the early Mendelian geneticists used corn as their experimental material to explore the basic laws of inheritance as well as variations of these laws. Barbara McClintock was awarded the Nobel Prize in 1983 for her pioneering research into genetic elements that move around the corn

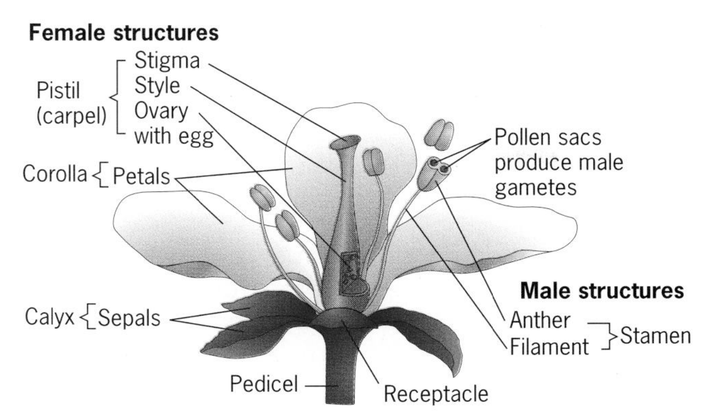

Figure 2.17 The structures of a typical flower, showing the male and female reproductive systems.

genome. These mobile genetic elements, called transposable genetic elements, have been discovered in many species and will be discussed in Chapter 17.

In *Neurospora* and in most lower plants, the haploid condition is the dominant phase of the life cycle. The diploid phase is fleeting, existing only after the fusion of two haploid spores, following which meiosis produces the haploid state once again. As plants evolved and became more complicated, the diploid phase of the life cycle became the dominant form. Thus we can see in the more ancient ferns and mosses a gradual lengthening of the diploid phase and a diminishing of the haploid phase.

In most flowering plants, including the pea plant that Mendel studied, the flower contains both male and female reproductive structures (Figure 2.17). In the corn plant, which represents the sporophyte generation, male and female reproductive structures are housed in separate flowers (Figure 2.18). The tassel at the top of the plant is an aggregation of **stamens** containing the diploid male spore mother cells, or **microspore mother cells**. The silk on the side of the plant is an aggregation of **pistils.** In the ovaries at the base of the pistils, we find the diploid female spore mother cells (**megaspore mother cells**).

In the stamens, microspore mother cells divide meiotically to produce four haploid microspores, the first cells of the gametophyte generation. Each microspore differentiates into a **pollen grain,** the male gametophyte. The haploid pollen nucleus divides mitotically, forming two nuclei. One becomes a tube nucleus and the other a generative nucleus. The generative nucleus divides mitotically again to produce two gamete nuclei. The pollen grain now has three haploid nuclei, two gamete nuclei, and one tube nucleus.

In the pistils, megaspore mother cells undergo meiosis to form four haploid nuclei. As in oogenesis, only one of the four meiotic products becomes functional and occupies the embryo sac or female gametophyte. The other three meiotic products degenerate. The remaining nucleus divides three times mitotically to produce eight identical haploid nuclei arranged linearly in the embryo sac. Two nuclei in the center fuse to produce a diploid endosperm nucleus. The three nuclei farthest away from the point in the sac where the pollen tube enters become the antipodal nuclei. The three closest to the pollen tube's point of entry form a gamete nucleus and two synergid nuclei.

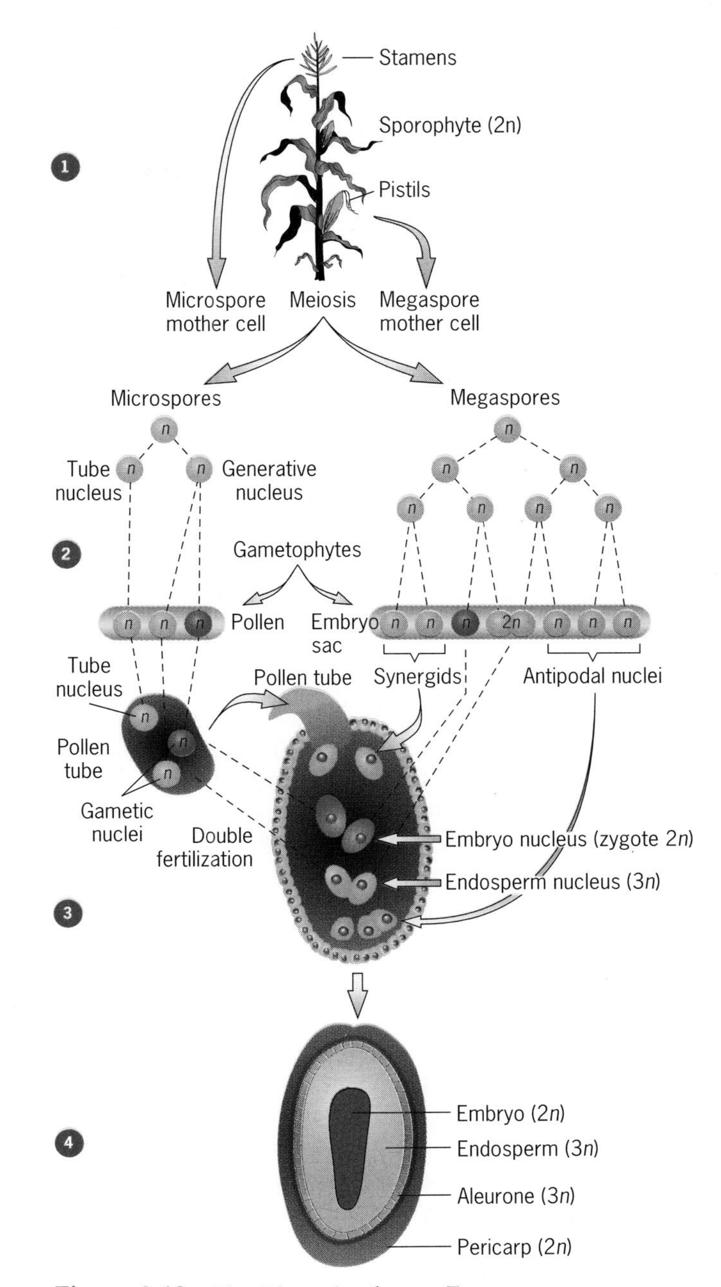

Figure 2.18 The life cycle of corn, *Zea mays*.

Pollen grains land on the stigma (silk) of the pistil and produce long **pollen tubes** that grow down the style. The development and growth of the tube is under the control and direction of the tube nucleus. The gamete nuclei enter the ovary (embryo sac). One gamete fertilizes the diploid endosperm nucleus, and the other gamete nucleus fertilizes the female gamete nucleus. The fusion of a male gamete nucleus with the diploid endosperm nucleus produces the triploid **endosperm,** a tissue with three copies of each chromosome that supplies nourishment to the developing embryo until it is able to photosynthesize. The fusion of the male and female gamete nuclei produces the diploid **embryo.** The synergid and antipodal nuclei degenerate after fertilization. It is not known if the synergid and antipodal nuclei have any function.

The outer layer of the triploid endosperm differentiates into the aleurone. The aleurone can accumulate pigments, giving corn kernels a range of colors, from the more familiar market varieties of white and yellow if *no pigment* is deposited in the aleurone (white and yellow are properties of the endosperm, visible through the colorless aleurone) to purples and reds if pigments are deposited. Nuclear genes control pigment deposition in the aleurone and the endosperm color. The pericarp, which originates from the maternal plant tissue, surrounds the developing kernel, which becomes the seed that germinates into the diploid sporophyte generation.

The Fruit Fly, *Drosophila melanogaster*

Perhaps no organism has proven so valuable to genetic research as the little fruit fly, *Drosophila melanogaster.* The development of the entire chromosome theory of inheritance, which established experimental links between Mendelian principles of inheritance and chromosomes, has been based almost entirely on research using *Drosophila.* Today, *Drosophila* is giving researchers valuable insights into the genetic control of developmental processes.

Drosophila is a small organism, about 2 mm in length, and can complete its life cycle in about 10 days at 25° C (Figure 2.19). The fertilized egg becomes an embryo that hatches into a small, segmented first instar **larva** that is very motile and that feeds rapidly. Specialized jaw hooks in the head region function like rakes as the larva burrows through the food material (rotting and fermenting fruit in the natural world; corn meal, agar, molasses, and yeast in the lab). As the larva grows, it eventually outgrows its chitinous cuticle, shedding it and growing a new one. This is the second instar stage of development. Further growth and development result in another cuticle molting to produce the large third instar larva. This stage appears about five days after fertilization.

Near the end of the third instar stage, the larva slows down and crawls out of the food onto a relatively dry place. The larval body shortens, and the cuticle thickens and accumulates pigment as it becomes the chitinous pupal case. Major developmental changes are now taking place as the **pupa** forms inside the pupal case. Adult structures, such as head, wings, and legs, developing in the pupal case, were actually present in the larva as small pre-adult clusters of determined tissue called **imaginal discs.** These discs are genetically committed to form specific adult struc-

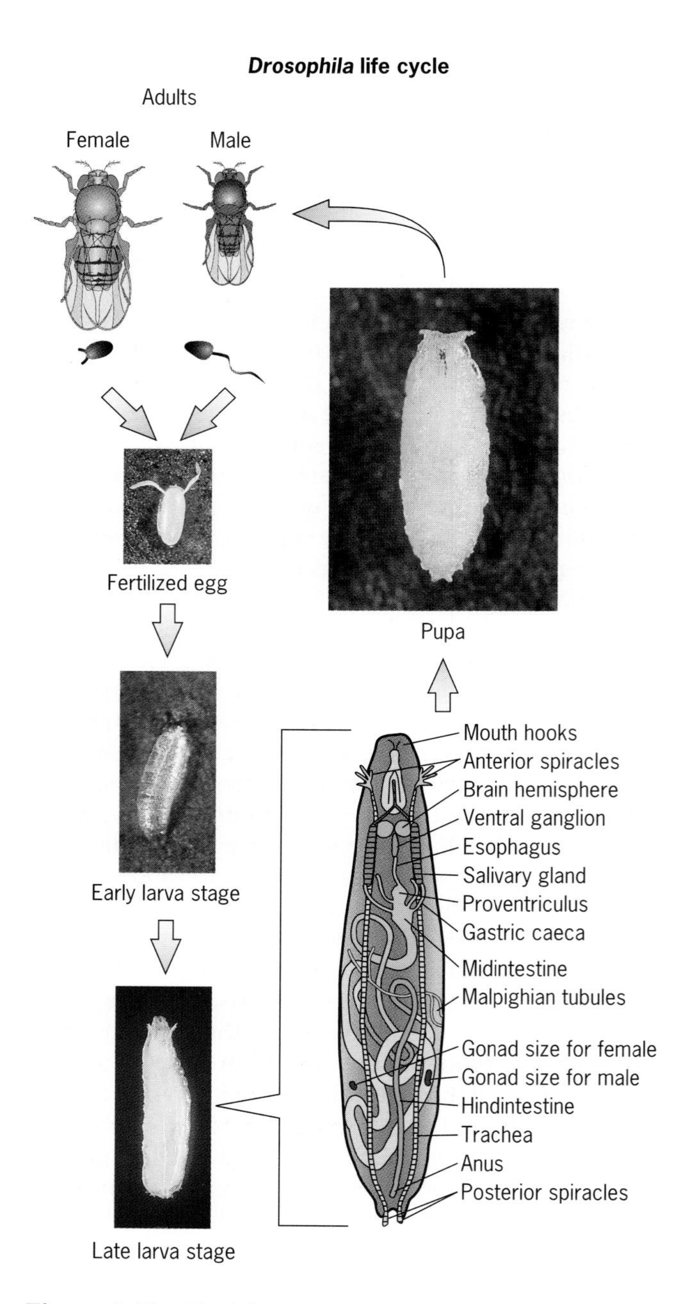

Figure 2.19 The life cycle of the fruit fly, *Drosophila melanogaster.*

tures, but they do not actually differentiate into those structures until the the pupal stage of development. Important research in developmental genetics has been carried out using imaginal discs. As adult structures develop, larval structures break down. About 10 days after fertilization, the adult emerges from the pupal case. If it is male, spermatogenesis produces mature sperm; if female, oogenesis produces mature eggs. Fertilization produces the zygote and the cycle begins anew.

Humans

Some of the most exciting genetic research today is being carried out in humans. In terms of its essential elements, the human life cycle is basically the same as that of the fruit fly: Haploid gametes (sperm and egg) unite in fertilization to produce a diploid zygote that develops into an embryo and, finally, a mature multicellular organism. Specialized cells in the gonad (testicle or ovary) of the mature organism produce gametes that unite in fertilization to produce a zygote, and the cycle begins anew.

As in most animals, the haploid phase of the human is reduced to a single-cell stage, sperm or egg. There are significant differences, however, between the larval instar developmental stages in *Drosophila* and human development. Six to eight days after fertilization, the human embryo completes its journey down the fallopian tube and implants into the uterus, rupturing uterine blood vessels in the process. These blood vessels become resealed, but the embryo remains bathed in a pool of maternal blood that is continuously replenished by fresh blood from maternal blood vessels. A **placenta,** which functions as the connection between mother and embryo, is constructed from embryonic and maternal membranes. It embeds in the uterine wall and is the site for the exchange of respiratory gases, nutrients, and waste matter. Connecting blood vessels in the umbilical cord link the embryonic circulatory system to the highly vascularized placenta.

During the first two months of development, all of the major organ systems appear in the embryo. Most of them are functioning by the end of this period. Beginning with the third month of development, the embryo is referred to as a **fetus.** During the last seven months of development, the fetus grows and refinements occur in its organ systems. The fetus is preparing for life outside the uterus.

From conception to birth, 266 days pass. These 266 days are divided into three periods, or trimesters. The most dramatic developmental changes occur during the first trimester (Figure 2.20). During this period, the single-celled zygote is transformed into a fetus with functioning organ systems that are only slightly less complex than those found in a full-term baby.

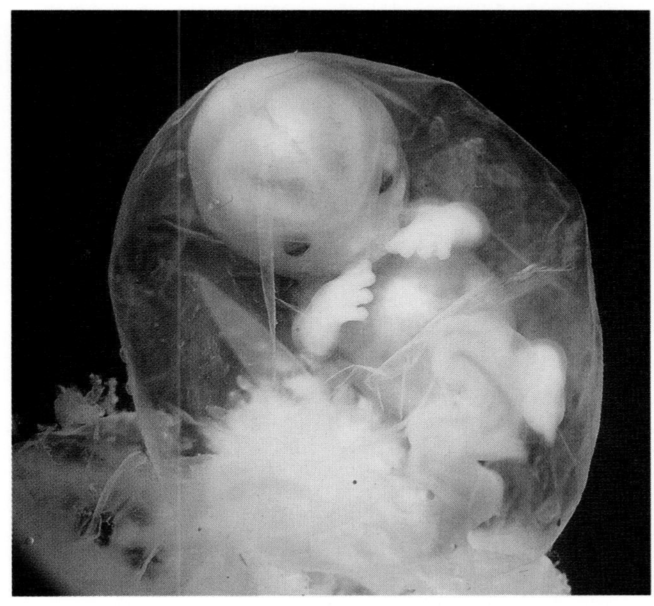

Figure 2.20 A two month human embryo. During this period, the single-celled zygote is transformed into a fetus with functioning organ systems that are only slightly less complex than those found in a full-term baby.

During the final trimester, the fetus prepares to live on its own. It increases in body weight 500 to 600 percent. The brain and peripheral nervous system mature at a rapid rate. As the end of the third trimester approaches, the fetus develops the ability to regulate its own temperature and to control its own breathing. Fat deposits form under the skin, giving the fetus a chubby appearance. Most fetuses change position in the uterus, becoming aligned for a head-first delivery through the birth canal.

One system does not mature until after birth. Human babies have no functioning immune system. During the last month of fetal life, antibodies from the mother cross the placental barrier and protect the baby from infectious diseases until its own immune system begins functioning. If the baby is breast-fed, maternal milk provides additional antibodies for the baby, augmenting its developing immune system.

Key Points: **The life cycle is the sequence of events from the individual's origin as a zygote to its ultimate death. It can also be described as the stages through which an organism passes between the production of gametes by one generation and the next. In many plants and some animals, the life cycle involves a regular alternation of generations between haploid and dipoid phases.**

TESTING YOUR KNOWLEDGE

1. Study the diagrams below of dividing cells. Is either of these cells haploid? What stages of mitosis are represented in these cells?

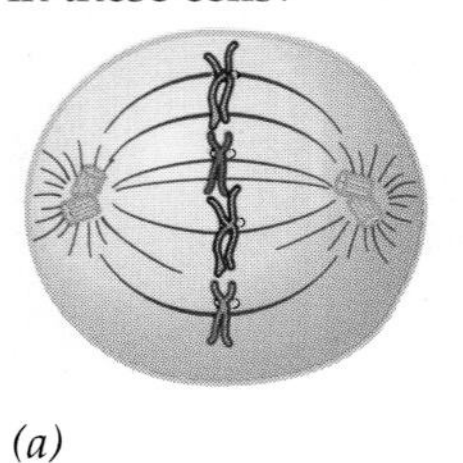

(a)

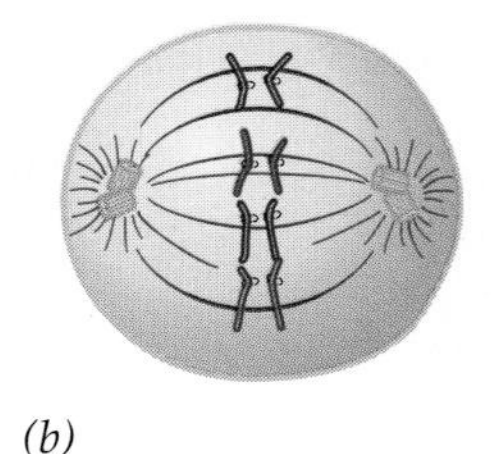

(b)

ANSWER

Careful examination of these two cells shows that they are both haploid because the four chromosomes are all different and in an unpaired state. Cell (*a*) is in metaphase because the chromosomes have replicated and lined up at the equatorial region, but the centromere has not yet divided. Cell (*b*) is in anaphase because the sister chromatids have separated and are beginning their journey to opposite poles of the cell.

2. What meiotic stages are represented by the following cells?

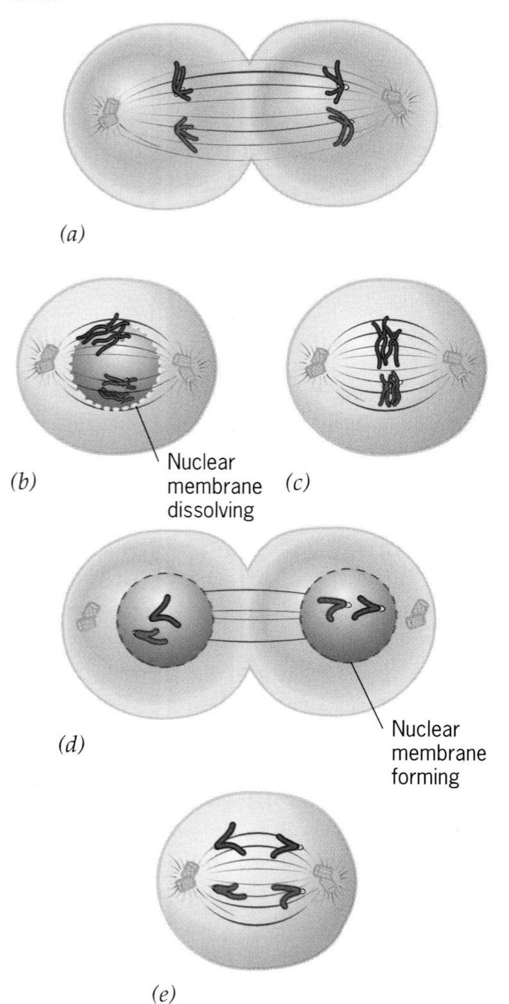

ANSWER

Cell (*a*) is in anaphase I because the pairs of replicated chromosomes are segregating from each other; the centromere has not divided. Cell (*b*) is in prophase I because the replicated chromosomes have paired but not yet aligned at the equatorial plate; the dissolving nuclear membrane is indicative of the approaching metaphase. Cell (*c*) is in metaphase I because the replicated chromosomes have paired at the equatorial plate and the nuclear membrane has dissolved. Cell (*d*) is in telophase II because the chromosomes in each of the haploid nuclei are unreplicated and a nuclear membrane is forming around them. Cell (*e*) is in anaphase II because the former sister chromatids are migrating to opposite poles of the cell as they approach telophase II.

QUESTIONS AND PROBLEMS

2.1 Mark the true statements with a (+) and the false with a zero (0). (a) Skin cells and gametes of the same animal contain the same number of chromosomes. (b) Any chromosome may pair with any other chromosome in the same cell in meiosis. (c) The gametes of an animal may contain more maternal chromosomes than its body cells contain. (d) Of 10 chromosomes in a mature sperm cell, 5 are always maternal. (e) Of 22 chromosomes in a primary oocyte, 15 may be paternal. (f) The homologous parts of two chromosomes lie opposite one another in paring. (g) A sperm has half as many postmitotic chromosomes as a spermatogonium of the same animal.

2.2 In each somatic cell of a particular animal species, there are 46 postmitotic chromosomes. How many should there be in a (a) mature egg, (b) first polar body, (c) sperm, (d) spermatid, (e) primary spermatocyte, (f) brain cell, (g) secondary oocyte, (h) spermatogonium?

2.3 If spermatogenesis is normal and all cells survive, how many sperm will result from (a) 50 primary spermatocytes and (b) 50 spermatids?

2.4 In humans, a type of myopia (an eye abnormality) is dependent on a gene (*M*). The normal gene is *m*. Represent diagrammatically (on the chromosomes) a cross between a woman with myopia who is heterozygous (*Mm*) and a normal man (*mm*). Show the kinds of gametes that each parent could produce and summarize the expected results from the cross.

2.5 Beginning with the myopic woman in Problem 2.4, diagram the oogenesis process that produces the egg involved in the production of an *Mm* child. Label all stages.

2.6 In what ways is cell division similar and different in animals and plants?

2.7 How does meiosis differ from mitosis? Consider differences in mechanism as well as end results.

2.8 How does gamete formation in higher plants differ from that in higher animals with reference to (a) gross mechanism and (b) chromosome mechanism?

2.9 How is double fertilization accomplished in plants, and what is the fate of the egg and the endosperm nuclei?

2.10 In humans, an abnormality of the large intestine called intestinal polyposis is dependent on gene *A*, and a neurological disorder called Huntington's disease is determined by gene *H*. The normal genes are *a* and *h*, respectively. A man who is *Aa hh* married a woman who is *aa Hh*. Assume the *A* and *H* are on nonhomologous chromosomes. Diagram the chromosomal constitution of the two parents and the gametes that each would produce. Which gametes would form a zygote that has an *A and* an *H*?

2.11 Beginning with the oogonium in the woman described in Problem 2.10, diagram the steps in the process of oogenesis necessary for formation of the egg that carries the chromomosome with the *H* gene. Label all stages.

2.12 Diagram completely the process of spermatogenesis involved in the production of the sperm in Problem 2.10 necessary for the production of a child carrying a chromosome with the *A* gene. Label all stages.

2.13 A man produces the following kinds of sperm in equal proportions: *AB, Ab, aB,* and *ab*. Diagram his chromosomal constitution.

2.14 Would greater genetic variability be expected among asexually reproducing organisms, self-fertilizing organisms, or bisexual organisms? Explain.

2.15 If biopsies were taken from follicle tissues of the human ovary at the following development periods, what stages in the process of oogenesis might be observed: (a) in the fifth month of intrauterine development, (b) at birth, (c) at 10 years of age, (d) at 17 years of age?

2.16 A cell has four pairs of homologous chromosomes (*Aa, Bb, Cc, and Dd*). How many different kinds of gametes can this cell produce?

2.17 A fruit fly has eight chromosomes in its somatic cells. How many of those chromosomes does it receive from its father?

2.18 If the garden pea has 14 chromosomes, how many different kinds of gametes can it form?

2.19 How is it possible for a species that is basically haploid to undergo meiosis?

2.20 Why is meiosis so evolutionarily important?

BIBLIOGRAPHY

ALBERTS, B., D. BRAY, J. LEWIS, M. RAFF, K. ROBERTS, AND J. D. WATSON. 1994. *Molecular Biology of the Cell, 3rd ed.* Garland Publishing, New York.

The Cell Cycle. 1991. *Cold Spring Harbor Symp. Quant. Biol.* 56.

GILBERT, S. F. 1997. *Developmental Biology, 5th ed.* Sinauer, Sunderland, MA.

KARP, GERALD, 1996. *Cell and Molecular Biology: Concepts and Experiments.* John Wiley and Sons, New York.

LODISH, H., D. BALTIMORE, A. BERK, S. L. ZIPURSKY, P. MATSUDAIRA, AND J. DARNELL. 1995. *Molecular Cell Biology, **3rd ed.*** W. H. Freeman, New York.

MIYAZAKI, W. Y., AND T. ORR-WEAVER. 1994. Sister chromatid cohesion in mitosis and meiosis. *Ann. Rev. Genet.* 28:167–188.

MURRAY, A., AND T. HUNT. 1993. *The Cell Cycle.* Oxford University Press, Oxford, UK.

THERMAN, EEVA, AND MILLARD SUSMAN. 1993. *Human Chromosome: Structure, Behavior, and Effects.* Springer-Verlag, New York.

A CONVERSATION WITH

JOHNG K. LIM

Johng Lim's life has certainly been varied and interesting. He was born and raised in S. Korea. He started studying Ancient Oriental History, but this study was interrupted by the Korean War. After the war he came to the United States, put himself through college, and received a BS/MS at the University of Minnesota in plant genetics. Johng completed his Ph.D. work in genetics at the University of Minnesota. Today he is recognized as one of the leading cytologists in the world and extensively researches the cytogenetics of *Drosophila*. He is currently a professor of biology at the University of Wisconsin–Eau Claire. Johng's avocation is horseback riding. As a youth, he won the Korean national equestrian championship and would have participated in the 1952 Olympic Games had it not been for the Korean War. In Wisconsin, he has had many enjoyable outings riding his horse "Saint."

How did you come to study science?

I was a college junior majoring in Ancient Oriental History when the Korean War started. After the war, I came to the United States and started as a college freshman again, this time with a major in agriculture. My initial educational plan was to study something useful in the U.S., like plant improvement through systematic breeding, and return home to help rebuild wartorn Korea. This was the way I came to study science.

How did you become interested in genetics?

I introduced myself to genetics by reading a textbook on my own. When I was a college sophomore, I found a copy of *Principles of Genetics* by Sinnot and Dunn on a library shelf. With great excitement, I read the entire book in a week. A few months later, I found another exciting genetics text—Sturtevant and Beadle's *An Introduction to Genetics*. After reading these books, I made sure that the fall quarter class schedule for my junior year included an introductory genetics course.

Your graduate studies involved field experiments with maize. Why did you switch the focus of your research to laboratory experiments with *Drosophila*?

One needs a farm or a field plot to do genetic experiments with maize. However, the college in Eau Claire, Wisconsin, where I found a teaching job, had no experimental plot to continue my maize experiments. Besides, the subject I wanted to study at that time was better suited to experimentation with *Drosophila*.

What technical demands have you faced in your scientific research?

The most challenging technical demand that I faced was to analyze *Drosophila* polytene chromosomes. In the early 1960s, there was a report that alkylating chemicals could induce cytologically detectable deletions in *Drosophila* chromosomes. I decided to investigate this claim by analyzing chromosomes that had been exposed to such chemicals. I learned to make salivary gland squashes in which the polytene chromosomes were very well-stretched. These preparations allowed me to check each chromosome for minute changes in structure, including deletions. I did this cytological analysis for almost five years and discovered—somewhat to my surprise—that I really enjoyed studying chromosomes. Thus began my lifelong interest in chromosome structure and function.

In addition to its technical side, science has an aesthetic element. Have you found a place for "art" in science?

Science is an objective art. You do not need to look hard for an aesthetic element in science. Whether it is a perfectly separated DNA fragment or an impeccably done Southern blot that shows a band you wanted to see at the end of a long experiment, there is a form, a proportion, a number, or a ratio that is absolutely pleasing. I suppose this is what James Watson meant when he said that Rosalind Franklin "saw the appeal of the base pairs and accepted the fact that the structure was too pretty not to be true."[1]

For a chromosome watcher like me, there is an ethereal quality to every chromosome. With their graceful curves, fancy geometric relationships, and puzzling spatial arrangements, polytene chromosomes have been both pleasing and tantalizing objects for me.

Is it important to understand the historical development of scientific concepts?

A person's life is a part of history. Understanding the development of scientific concepts helps one to sort out the relative importance of prevailing ideas. It also helps a person to chart his or her destiny. I feel that students in any discipline, not just biology, should read the biographies of key people and that they should study their discipline's history.

How does a scientist keep up with new discoveries and new techniques?

Techniques are essential tools in science. In fact, science cannot hope to advance without suitable techniques. However, we all know that these techniques have a way of changing, or being improved, or becoming obsolete. The most important single item in science is to be able to ask a significant question. If the question is important enough, you will be blessed with friends. Friends will help you understand what is going on in a particular field of science; they will be happy to share their technical knowledge with you.

You have been a teacher for more than 30 years—mainly of undergraduates—in a small public university. What has this experience meant to you?

I came to Eau Claire, Wisconsin—a town I hadn't heard of until I found this teaching job—with three things in mind: to teach, to study, and to live. Teaching a college course in genetics from 1963 to the present meant a great deal of reading and study. I am amused by the fact that none of the topics I now teach in my courses was included in the curriculum when I was a student. A small public university has been a wonderful place to teach and study during these years. The small size has meant relatively few academic hassles, and I have been able to do pretty much what I have thought best for the university, my students, and myself. Eau Claire is a delightful place to teach, to think, and, once in a while, to enjoy a quiet walk in the woods.

Are teaching and research complementary, or do they interfere with one another?

Teaching and research are self-complementary just as the two strands in the DNA double helix are complementary. Teaching in science is based on the results of research, and research in science is based on the concepts we teach. There should be no conflict between these two activities.

What advice do you have for students who are studying science?

Be versatile. The twenty-first century will demand each scientist to be more than a mere scientist. Science cannot dictate public policy, nor can it formulate practices. The great speed and phenomenal accomplishment in revealing the nature of the genetic material during the last half century are testimony to the remarkable ability and power of science. Yet, with all our godlike abilities to discover the secret of life—right down to the core properties of the DNA molecule—we do not know enough, we are not satisfied, and we are not happy. What are these creatures that we call humans, and why do they demand so much? Why are we so brilliant and yet so stupid and senseless to willingly destroy our planet? The coming century will demand scientists who are greater than Darwin, Mendel, Churchill, and Einstein all put together.

[1] Watson, J. 1980. *The Double Helix: A Personal Account of the Discovery of the Structure of DNA*, edited by G. S. Stent. W. W. Norton, New York.

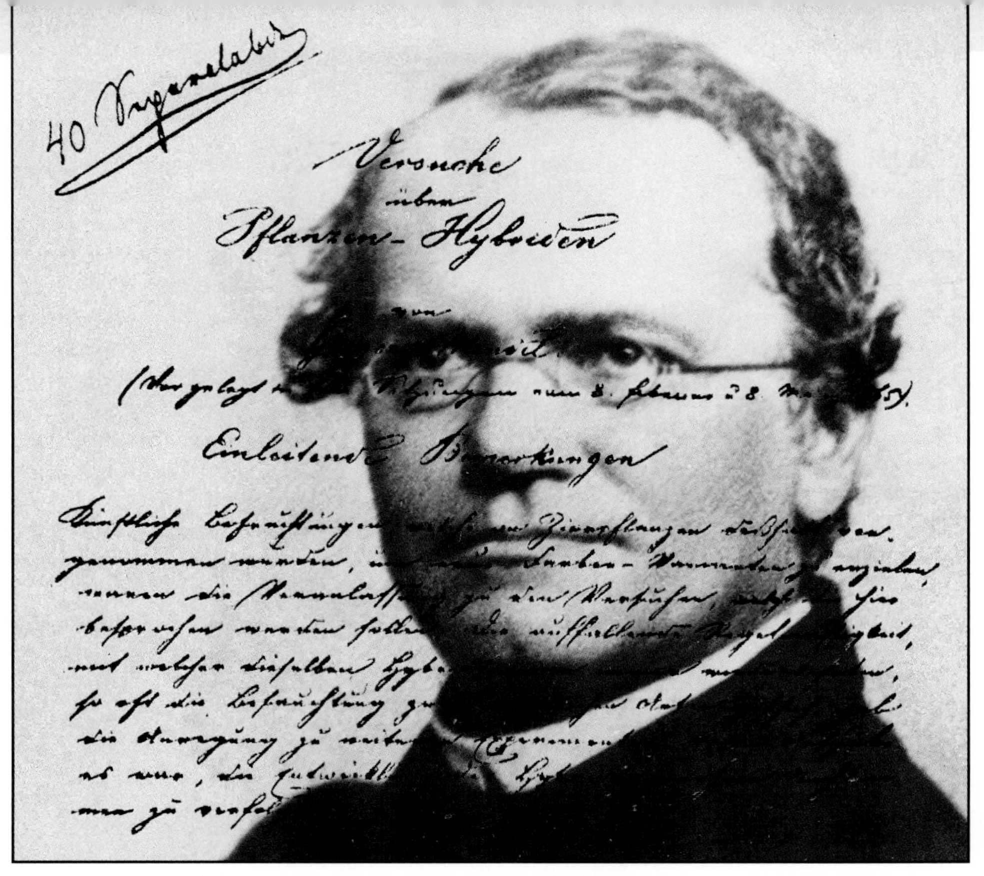

Front page of Mendel's original paper, with a portrait of Mendel.

3

Mendelism: The Basic Principles of Inheritance

CHAPTER OUTLINE

The Birth of Genetics: A Scientific Revolution

Science is a complex endeavor involving the careful observation of natural phenomena, reflective thinking about these phenomena, and the formulation of testable ideas about their causes and effects. Progress in science often depends on the work of a single insightful individual. Consider, for example, the effect that Nicolaus Copernicus had on astronomy, that Isaac Newton had on physics, or that Charles Darwin had on biology. Each of these individuals altered the course of their scientific disciplines by introducing radically new ideas. In effect, they began scientific revolutions.

In the middle of the nineteenth century, the Austrian monk Gregor Mendel, a contemporary of Darwin, laid the foundation for another revolution in biology, one that eventually produced an entirely new science—genetics. Mendel's ideas, published in 1866 under the title "Experiments in Plant Hybridization," endeavored to explain how the characteristics of organisms are inherited. Many people had attempted such an explanation previously but without much success. Indeed, Mendel commented on their failures in the opening paragraphs of his article:

To this object, numerous careful observers such as Kölreuter, Gärtner, Herbert, Lecoq, Wichura and others, have devoted a part of their lives with inexhaustible perseverance. . . . [However,] Those who survey the work in this department will arrive at the conviction that among all the numerous experiments made, not one has been carried out to such an extent and in such a way as to make it possible to determine the number of different forms under which the offspring of hybrids appear, or to arrange these forms with certainty according to their separate generations, or definitely to ascertain their statistical relations.

He then described his own efforts to elucidate the mechanism of heredity:

It requires indeed some courage to undertake a labor of such far-reaching extent; this appears, however, to be the only right way by which we can finally reach the solution of a question the importance of which cannot be overestimated in connection with the history of the evolution of organic forms.

The paper now presented records the results of such a detailed experiment. This experiment was practically confined to a small plant group, and is now, after eight years' pursuit, concluded in all essentials. Whether the plan upon which the separate experiments were conducted and carried out was the best suited to attain the desired end is left to the friendly decision of the reader.

This chapter describes Mendel's discoveries and their ramifications for the study of heredity. It also introduces the basic terms and techniques that are used in modern genetic analysis.

MENDEL'S STUDY OF HEREDITY

The life of Gregor Johann Mendel (1822–1884) spanned the middle of the nineteenth century. His parents were farmers in Moravia, then a part of the Hapsburg Empire in Central Europe. A rural upbringing taught him plant and animal husbandry and inspired an interest in nature. At the age of 21, Mendel left the farm and entered a Catholic monastery in the city of Brünn (today, Brno in the Czech Republic). In 1847 he was ordained a priest, adopting the clerical name Gregor. He subsequently taught at the local high school, taking time out between 1851 and 1853 to study at the University of Vienna. After returning to Brünn, he resumed his life as a teaching monk and began the genetic experiments that eventually made him famous.

Mendel performed experiments with several species of garden plants, and he even tried some experiments with honeybees. His greatest success, however, was with peas. He completed his experiments with peas in 1863, and spent the next two years analyzing and summarizing the data. In 1865, Mendel presented the results before the local Natural History Society, and the following year, he published a detailed report in the society's proceedings. Unfortunately, this paper languished in obscurity until 1900, when it was rediscovered by three botanists—Hugo de Vries in Holland, Carl Correns in Germany, and Eric von Tschermak-Seysenegg in Austria. As these men searched the scientific literature for data supporting their own theories of heredity, each found that Mendel had performed a detailed and careful analysis 35 years earlier. Mendel's ideas quickly gained acceptance, especially through the promotional efforts of a British biologist, William Bateson. This champion of Mendel's discoveries coined a new term to describe the study of heredity: **genetics,** from the Greek word meaning "to generate."

Mendel's Experimental Organism, the Garden Pea

One reason for Mendel's success is that he chose his experimental material astutely. The garden pea, *Pisum sativum*, is a **dicot**, a type of plant that sprouts two leaves, or cotyledons, from a germinating seed. Peas are easily grown in experimental gardens, or in pots in a greenhouse.

One peculiarity of pea reproduction is that the petals of the flower close down tightly, preventing pollen grains from entering or leaving. This enforces a system of self-fertilization, in which sperm and eggs from a particular flower unite with each other to produce seeds. As a result, individual pea strains are highly inbred, displaying little if any genetic variation from one generation to the next. Because of this uniformity, we say that such strains are **true-breeding**.

At the outset, Mendel obtained many different true-breeding varieties of peas, each distinguished by a particular characteristic. In one strain, for example, the plants were 6 to 7 feet high, whereas in another they measured only 9 to 18 inches. Another variety produced green seeds, and still another produced yellow seeds. Mendel took advantage of these contrasting traits to determine how the characteristics of pea plants are inherited.

Mendel's focus on these singular differences between pea strains allowed him to study the inheritance of one trait at a time—for example, plant height. Other biologists had attempted to follow the inheritance of many traits simultaneously, but because the results of such experiments were complex, they were never able to discover any fundamental principles about heredity. Mendel succeeded where these biologists had failed because he focused his attention on contrasting differences between plants that were otherwise the same—tall versus short, green seeds versus yellow seeds, and so forth. In addition, he kept careful records of the experiments that he performed.

Monohybrid Crosses: The Principles of Dominance and Segregation

In one experiment, Mendel **cross-fertilized**—or, simply, crossed—tall and dwarf pea plants to investigate how height was inherited (Figure 3.1). He carefully removed the anthers from one variety before its pollen had matured and then applied pollen from the other variety to the stigma, a sticky organ on top of the ovary. The seeds that resulted from these cross-fertilizations were sown the next year, yielding hybrids that were uniformly tall. Mendel obtained tall plants regardless of the way he performed the cross (tall male with dwarf female or dwarf male with tall female); thus the two reciprocal crosses gave the same results. Even more significantly, however, Mendel noted that the dwarf characteristic seemed to have disappeared in the progeny of the cross—all the hybrid plants were tall. To explore the hereditary makeup of

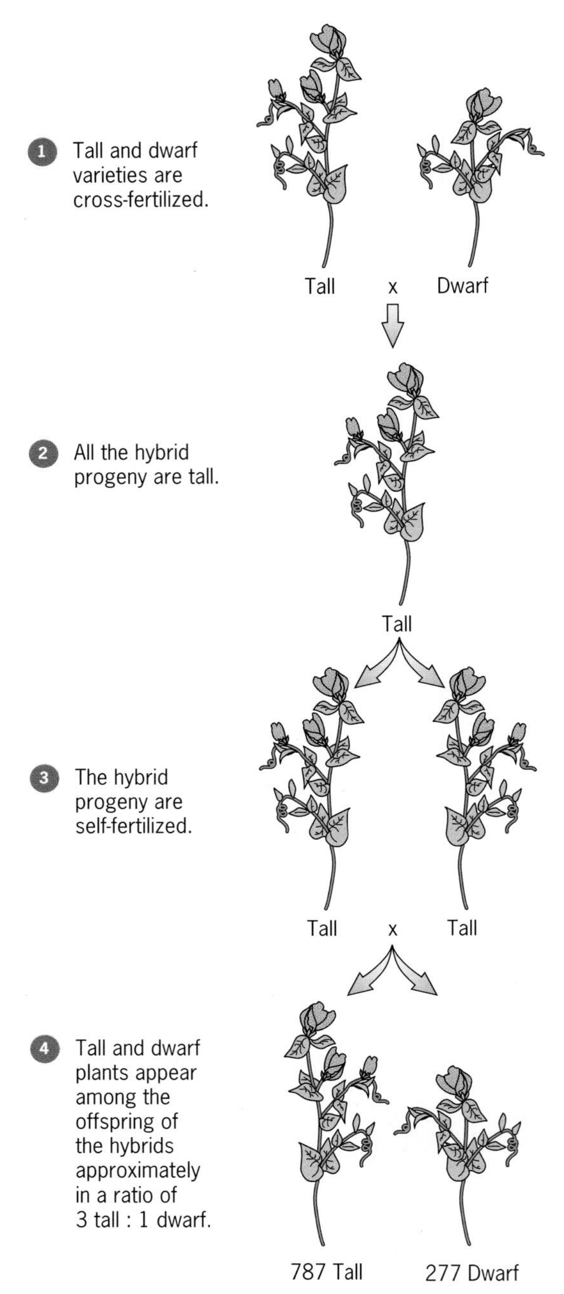

Figure 3.1 Mendel's crosses involving tall and dwarf varieties of peas.

these tall hybrids, Mendel allowed them to undergo self-fertilization—the natural course of events in peas. When he examined the progeny, he found that they consisted of both tall and dwarf plants. In fact, among 1064 progeny that Mendel cultivated in his garden, 787 were tall and 277 were dwarf—a ratio of approximately 3:1.

Mendel was struck by the reappearance of the dwarf characteristic. Clearly, the hybrids that he had made by crossing tall and dwarf varieties had the abil-

TABLE 3.1
Results of Mendel's Monohybrid Crosses

Parental Strains	*F_2 Progeny*	*Ratio*
Tall plants × dwarf plants	787 tall, 277 dwarf	2.84:1
Round seeds × wrinkled seeds	5474 round, 1850 wrinkled	2.96:1
Yellow seeds × green seeds	6022 yellow, 2001 green	3.01:1
Violet flowers × white flowers	705 violet, 224 white	3.15:1
Inflated pods × constricted pods	882 inflated, 299 constricted	2.95:1
Green pods × yellow pods	428 green, 152 yellow	2.82:1
Axial flowers × terminal flowers	651 axial, 207 terminal	3.14:1

ity to produce dwarf progeny even though they themselves were tall. Mendel inferred that these hybrids carried a latent genetic factor for dwarfness, one that was masked by the expression of another factor for tallness. He said that the latent factor was **recessive** and that the expressed factor was **dominant**. He also inferred that these recessive and dominant factors separated from each other when the hybrid plants reproduced. How else could he explain the reappearance of the dwarf characteristic in the next generation?

Mendel performed similar experiments to study the inheritance of six other traits: seed texture, seed color, pod shape, pod color, flower color, and flower position (Table 3.1). In each experiment—called a **monohybrid cross** because a single trait was being studied—Mendel observed that only one of the two contrasting characteristics appeared in the hybrids, and that when these hybrids were self-fertilized, they produced two types of progeny, each resembling one of the plants in the original crosses. Furthermore, he found that these progeny consistently appeared in a ratio of 3:1. Thus, each trait that Mendel studied seemed to be controlled by a heritable factor that existed in two forms, one dominant, the other recessive. These factors are now called **genes**, a word coined by the Danish plant breeder W. Johannsen in 1909; their dominant and recessive forms are called **alleles**—from the Greek word meaning "of one another."

The regular numerical relationships that Mendel observed in these crosses led him to another important conclusion: that genes come in pairs. Mendel proposed that each of the parental strains that he used in his experiments carried two identical copies of a gene—in modern terminology, they are diploid and **homozygous**. However, during the production of gametes, Mendel proposed that these two copies are reduced to one; that is, the gametes that emerge from meiosis carry a single copy of a gene—in modern terminology, they are haploid.

Mendel recognized that the diploid gene number would be restored when sperm and egg unite to form a zygote. Furthermore, he understood that if the pollen and egg came from genetically different plants—as they did in his crosses—the hybrid zygote would inherit two different alleles, one from the mother and one from the father. Such an offspring is said to be **heterozygous**. Mendel realized that the different alleles that are present in a heterozygote must coexist even though one is dominant and the other recessive, and that each of these alleles would have an equal chance of entering a gamete when the heterozygote reproduces. Furthermore, he realized that random fertilizations with a mixed population of gametes—half carrying the dominant allele and half carrying the recessive allele—would produce some zygotes in which both alleles were recessive. Thus, he could explain the reappearance of the recessive characteristic in the progeny of the hybrid plants.

Mendel used symbols to represent the hereditary factors that he postulated—a methodological breakthrough. With symbols, he could describe hereditary phenomena clearly and concisely, and he could analyze the results of crosses mathematically. He could even make predictions about the outcome of future crosses. Although the practice of using symbols to analyze genetic problems has been much refined since Mendel's time, the basic principles remain the same. The symbols stand for genes (or, more precisely, for their alleles), and they are manipulated according to the rules of inheritance that Mendel discovered. These manipulations are the essence of formal genetic analysis. As an introduction to this subject, let us consider the symbolic representation of the cross between tall and dwarf peas (Figure 3.2).

The two true-breeding varieties, tall and dwarf, are homozygous for different alleles of a gene controlling plant height. The allele for dwarfness, being recessive, is symbolized with a lower case letter, *d*; the allele for tallness, being dominant, is symbolized by the corresponding upper case letter, *D*. In genetics, the letter that is chosen to denote the alleles of a gene is usually taken from the word that describes the recessive trait (*d*, for dwarfness). Thus the tall and dwarf pea strains are symbolized by *DD* and *dd*, respectively. The allelic constitution of each strain is said to be its **genotype**. By contrast, the appearance of each

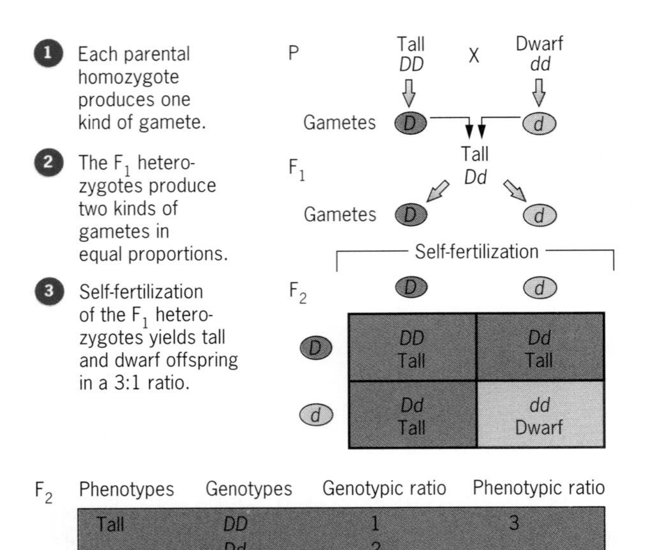

Figure 3.2 Symbolic representation of the cross between tall and dwarf peas and summary of phenotypic and genotypic results.

strain—that is, the tall or dwarf characteristic—is said to be its **phenotype**.

As the **parental** strains, the tall and dwarf pea plants form the **P** generation of the experiment. Their hybrid progeny are referred to as the first **filial** generation, or $\mathbf{F_1}$, from a Latin word meaning "offspring." Because each parent contributes equally to its offspring, the genotype of the F_1 plants must be *Dd;* that is, they are heterozygous for the alleles of the gene that control plant height. Their phenotype, however, is the same as that of the *DD* parental strain because *D* is dominant over *d*. During meiosis, these F_1 plants produce two kinds of gametes, *D* and *d*, in equal proportions. Neither allele is changed by having coexisted with the other in a heterozygous genotype; rather, they separate, or **segregate**, from each other during gamete formation. This process of allele segregation is perhaps the most important discovery that Mendel made. Upon self-fertilization, the two kinds of gametes produced by heterozygotes can unite in all possible ways. Thus they produce four kinds of zygotes: *DD* (the contribution of the egg is usually written first), *Dd*, *dD*, and *dd*. However, because of dominance, three of these genotypes have the same phenotype. Thus, in the next generation, called the $\mathbf{F_2}$, the plants are either tall or dwarf, in a ratio of 3:1.

Mendel took this analysis one step further. The F_2 plants were self-fertilized to produce an F_3. All the dwarf F_2 plants produced only dwarf offspring, demonstrating that they were homozygous for the *d* allele, but the tall F_2 plants comprised two categories. Approximately one-third of them produced only tall offspring, whereas the other two-thirds produced a mixture of tall and dwarf offspring. Mendel concluded—correctly—that the third that were true-breeding were *DD* homozygotes and that the two-thirds that were segregating were *Dd* heterozygotes. These proportions, 1/3 and 2/3, were exactly what his analysis predicted because, among the tall F_2 plants, the *DD* and *Dd* genotypes occur in a ratio of 1:2.

We summarize Mendel's analysis of this and other monohybrid crosses by stating the two key principles that he discovered:

1. **The Principle of Dominance:** *In a heterozygote, one allele may conceal the presence of another.* This principle is a statement about genetic function. Some alleles evidently control the phenotype even when they are present in a single copy. We consider the physiological explanation for this phenomenon in later chapters.
2. **The Principle of Segregation:** *In a heterozygote, two different alleles segregate from each other during the formation of gametes.* This principle is a statement about genetic transmission. An allele is transmitted faithfully to the next generation, even if it was present with a different allele in a heterozygote. The biological basis for this phenomenon is the pairing and subsequent separation of homologous chromosomes during meiosis, a process we discussed in Chapter 2. The experiments that led to this chromosome theory of heredity are discussed in Chapter 5.

Dihybrid Crosses: The Principle of Independent Assortment

Mendel also performed experiments with plants that differed in two traits (Figure 3.3). He crossed plants that produced yellow, round seeds with plants that

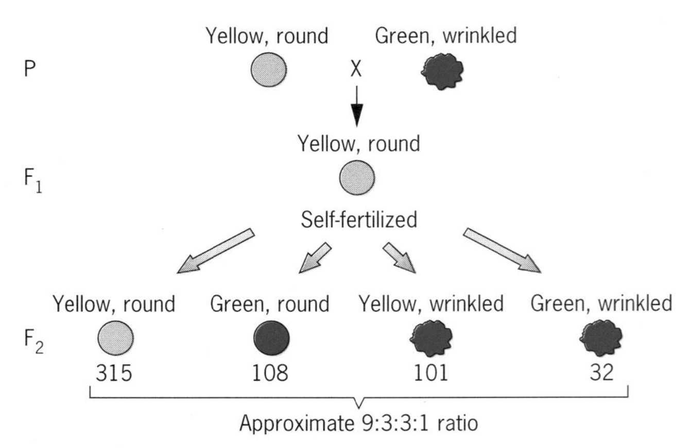

Figure 3.3 Mendel's crosses between peas that produced yellow, round seeds and peas that produced green, wrinkled seeds.

produced green, wrinkled seeds. The purpose of the experiment was to see if the two seed traits, color and texture, were inherited independently. Because the F_1 seeds were all yellow and round, the alleles for these two characteristics were dominant. Mendel grew plants from these seeds and allowed them to self-fertilize. He then classified the F_2 seeds and counted them by phenotype.

The four phenotypic classes in the F_2 represented all possible combinations of the color and texture traits. Two classes—yellow, round and green, wrinkled—resembled the parental strains. The other two—green, round and yellow, wrinkled—showed new combinations of traits. To Mendel, the numerical F_2 data, shown in Figure 3.3, suggested a simple explanation: each trait was controlled by a different gene segregating two alleles, and the two genes were inherited independently.

Let us analyze the results of this two-factor or **dihybrid cross,** using Mendel's methods. We denote each gene with a letter, using lower case for the recessive allele and upper case for the dominant (Figure 3.4). For the seed color gene, the two alleles are *g* (for green) and *G* (for yellow), and for the seed texture gene, they are *w* (for wrinkled) and *W* (for round). The parental strains, which were true-breeding, must have been doubly homozygous; the yellow, round plants were *GG WW* and the green, wrinkled plants were *gg ww*. Such two-gene genotypes are customarily written by separating pairs of alleles with a space.

The haploid gametes produced by a diploid plant contain one copy of each gene. Gametes from *GG WW* plants therefore contain one copy of the seed color gene (the *G* allele) and one copy of the seed texture gene (the *W* allele). Such gametes are symbolized by *G W*. By similar reasoning, the gametes from *gg ww* plants are written *g w*. Cross-fertilization of these two types of gametes produces F_1 hybrids that are doubly heterozygous, symbolized by *Gg Ww*, and their yellow, round phenotype indicates that the *G* and *W* alleles are dominant.

The Principle of Segregation predicts that the F_1 hybrids will produce four different gametic genotypes: (1) *G W*, (2) *G w*, (3) *g W*, and (4) *g w*. If each

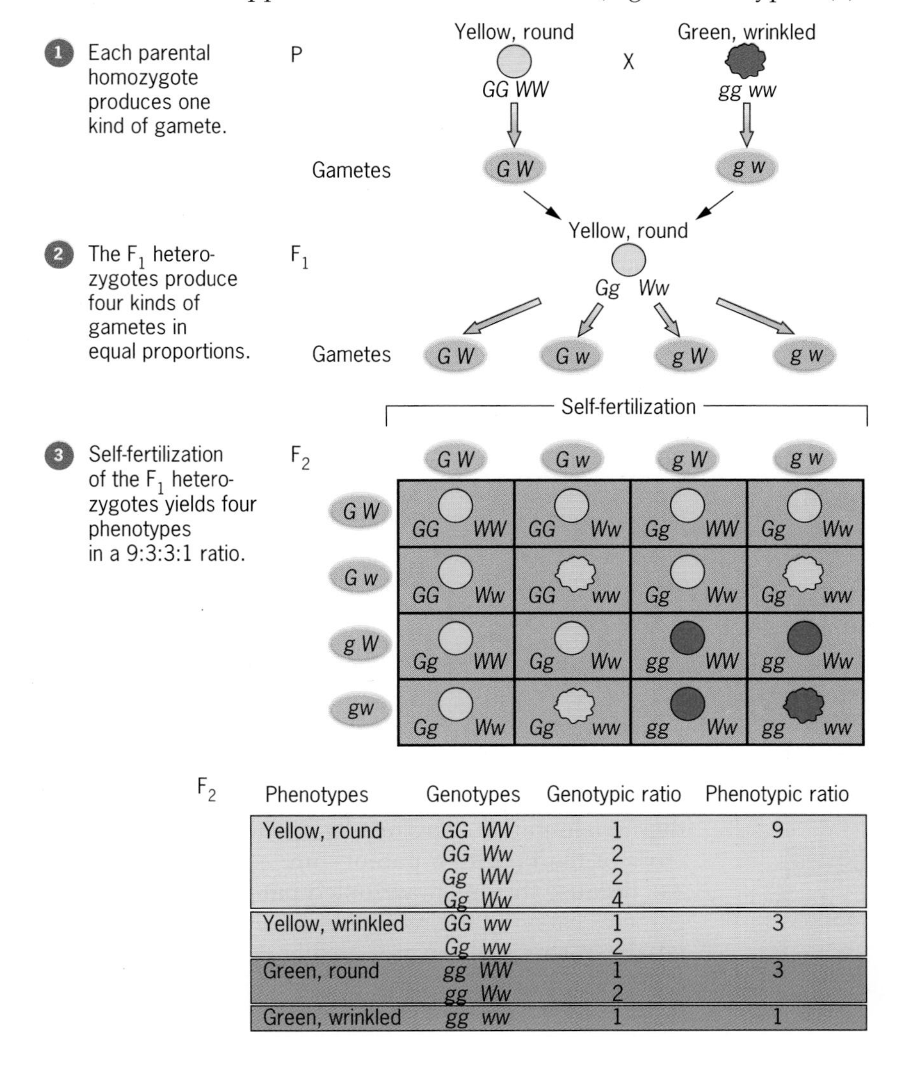

Phenotypes	Genotypes	Genotypic ratio	Phenotypic ratio
Yellow, round	GG WW	1	9
	GG Ww	2	
	Gg WW	2	
	Gg Ww	4	
Yellow, wrinkled	GG ww	1	3
	Gg ww	2	
Green, round	gg WW	1	3
	gg Ww	2	
Green, wrinkled	gg ww	1	1

Figure 3.4 Symbolic representation of the results of a cross between a variety of peas with yellow, round seeds and a variety with green, wrinkled seeds.

gene segregates its alleles independently, these four types will be equally frequent; that is, each will be 25 percent of the total. On this assumption, self-fertilization in the F_1 will produce an array of 16 equally frequent zygotic genotypes. The zygotic array is obtained by systematically combining the gametes, as shown in Figure 3.4. The phenotypes of these F_2 genotypes are then obtained by noting that *G* and *W* are the dominant alleles. Altogether, there are four distinguishable phenotypes, with relative frequencies indicated by the number of positions occupied in the array. For absolute frequencies, we divide each number by the total, 16:

yellow, round	9/16
yellow, wrinkled	3/16
green, round	3/16
green, wrinkled	1/16

This analysis is predicated on two assumptions: (1) that each gene segregates its alleles, and (2) that these segregations are independent of each other. The second assumption implies that there is no connection or linkage between the segregation events of the two genes. For example, a gamete that receives *W* through the segregation of the texture gene is just as likely to receive *G* as it is to receive *g* through the segregation of the color gene.

Do the experimental data fit with the predictions of our analysis? Figure 3.5 compares the predicted and observed frequencies of the four F_2 phenotypes in two ways—by proportions and by numerical frequencies. For the numerical frequencies, we calculate the predicted numbers by multiplying the predicted proportion by the total number of F_2 seeds examined. With either method, there is obvously good agreement between the observations and the predictions. Thus the assumptions on which we have built our analysis—independent segregation of the seed color and seed texture genes—are consistent with the observed data.

F_2 phenotypes	Observed Number	Observed Proportion	Expected Number	Expected Proportion
Yellow, round	315	0.567	313	0.563
Green, round	108	0.194	104	0.187
Yellow, wrinkled	101	0.182	104	0.187
Green, wrinkled	32	0.057	35	0.063
Total	**556**	**1.000**	**556**	**1.000**

Figure 3.5 Comparison of observed and expected results in the F_2 of Mendel's experiment involving the genes for seed color and seed texture in peas.

Mendel conducted similar experiments with other combinations of traits and in each case, observed that the genes segregated independently. The results of these experiments led him to a third principle:

3. **The Principle of Independent Assortment:** *The alleles of different genes segregate, or as we sometimes say, assort, independently of each other.* This principle is another rule of genetic transmission, based, as we will see in Chapter 5, on the behavior of different pairs of chromosomes during meiosis. However, not all genes abide by the Principle of Independent Assortment. In Chapters 7 and 8 we consider some important exceptions.

Key Points: **Mendel's experiments established three basic genetic principles: (1) Some alleles are dominant, others recessive. (2) During gamete formation, different alleles segregate from each other. (3) Different genes assort independently.**

APPLICATIONS OF MENDEL'S PRINCIPLES

If the genetic basis of a trait is known, Mendel's principles can be used to predict the outcome of crosses. There are three general procedures, two relying on the systematic enumeration of all the zygotic genotypes or phenotypes and one relying on mathematical insight.

The Punnett Square Method

For situations involving one or two genes, it is possible to write down all the gametes and combine them systematically to generate the array of zygotic genotypes. Once these have been obtained, the Principle of Dominance can be used to determine the associated phenotypes. This procedure, called the **Punnett Square method** after the British geneticist R. C. Punnett, is a straightforward way of predicting the outcome of crosses.

As an example, let us consider a cross between peas from the true-breeding green, wrinkled variety and the F_1 hybrids that Mendel obtained from his dihybrid cross; the green, wrinkled peas were used as the female parent. The results of this **backcross**—the cross of the F_1 with a parent—are analyzed in Figure 3.6. Because the green, wrinkled parent produces only gametes carrying the recessive alleles of both genes (*g w*), the outcome of the cross really depends on the gametes produced by the doubly heterozygous F_1 hybrids; such a cross is sometimes called a **testcross**. We

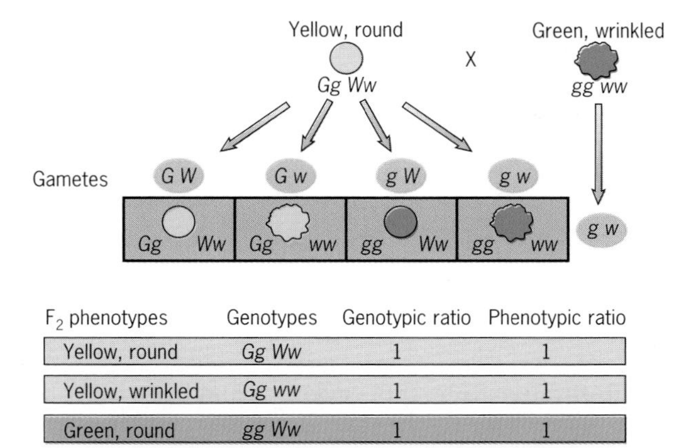

Figure 3.6 The Punnett Square method for predicting the results of a backcross involving two genes.

know that the F_1 plants produce four kinds of gametes. If we combine them with the single kind of gamete produced by the green, wrinkled variety, we obtain the zygotic genotypes. Using the Punnett Square (or, if you prefer here, rectangle), we can see that there are four genotypes, corresponding to the four types of sperm, and that these genotypes occur with equal frequencies.

We can also use the Punnett Square method to obtain the results of other kinds of crosses, provided that there are not too many different kinds of gametes. We have used it, for example, to analyze the zygotic output of self-fertilization with Mendel's yellow, round F_1 hybrids—a type of mating commonly called an **intercross**. However, in more complicated situations, like those involving more than two genes, the Punnett Square method is unwieldy.

The Forked-Line Method

Another procedure for predicting the outcome of a cross involving two or more genes is the **forked-line method**. However, instead of enumerating the progeny in a square (or rectangle) by combining the gametes systematically, we tally them in a diagram of branching lines.

As an example, let us consider an intercross between peas that are heterozygous for three independently assorting genes—one controlling plant height, one controlling seed color, and one controlling seed texture. This is a trihybrid cross—*Dd Gg Ww* × *Dd Gg Ww*—that can be partitioned into three monohybrid crosses—*Dd* x *Dd*, *Gg* × *Gg*, and *Ww* × *Ww*—because all the genes assort independently. For each gene, we expect the phenotypes to appear in a 3:1 ratio. Thus, for example, *Dd* x *Dd* will produce a ratio of 3 tall plants: 1 dwarf plant. Using the forked-line method (Figure 3.7), we can combine these separate ratios into an overall phenotypic ratio for the offspring of the cross.

The Probability Method

An alternative—and quicker—method to the Punnett Square and forked-line methods is based on the principle of **probability** (see Technical Sidelight: The Multiplicative and Additive Rules of Probability). Mendelian segregation is like a coin toss; when a heterozygote produces gametes, half contain one allele and half contain the other. If two segregating heterozygotes are crossed, their gametes are combined randomly, producing the zygotic genotypes (Figure 3.8). Let us suppose the cross is *Aa* × *Aa*. The chance that a zygote will be *AA* is simply the probability that

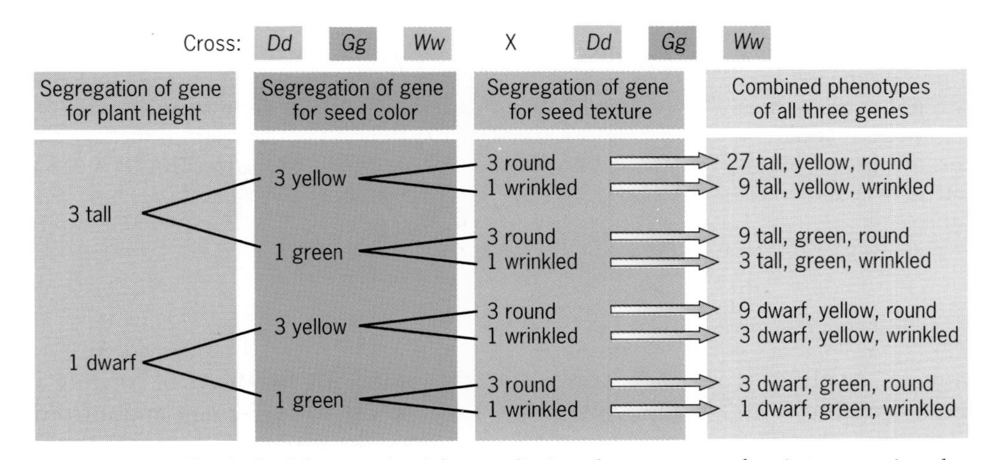

Figure 3.7 The forked-line method for predicting the outcome of an intercross involving three independently assorting genes in peas.

TECHNICAL SIDELIGHT

The Multiplicative and Additive Rules of Probability

Probability theory accounts for the frequency of events—for example, the chance of getting a head on a coin toss, drawing an ace from a deck of cards, obtaining a dominant homozygote from a mating between two heterozygotes. In each case, the event is the outcome of a process—tossing a coin, drawing a card, producing an offspring. To determine the probability of a particular event, we must consider all possible outcomes of the process. The collection of all events is called the *sample space*. For a coin toss, the sample space contains two events, head and tail; for drawing a card, it contains 52, one for each card, and for heterozygotes producing an offspring, it contains three, *GG*, *Gg*, and *gg*. *The probability of an event is the frequency of that event in the sample space.* For example, the probabilities associated with each of the progeny from a mating between two heterozygotes are 1/4 (for *GG*), 1/2 (for *Gg*), and 1/4 (for *gg*).

Two kinds of questions often arise in problems involving probabilities: (1) What is the probability that two events, A and B, will occur together? (2) What is the probability that at least one of two events, A or B, will occur at all? The first question specifies the joint occurrence of two events—A *and* B must occur together to satisfy this question. The second question is less stringent—if *either* A *or* B occurs, this question will be satisfied. A simple diagram can help to explain the different meanings of these two questions.

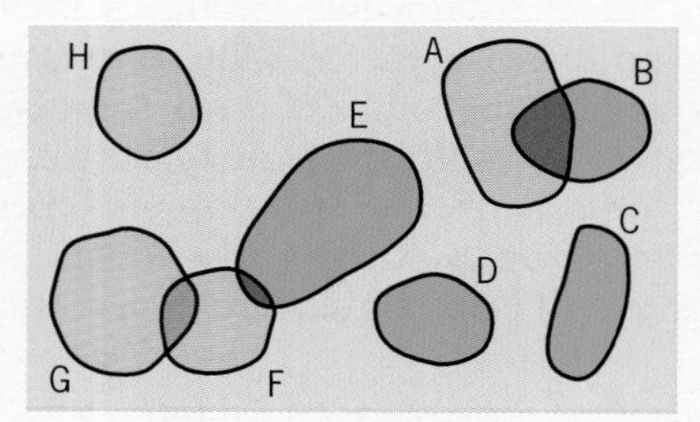

The shapes in the diagram represent events in the sample space, and the sizes of the shapes reflect their relative frequencies. Overlaps between shapes indicate the joint occurrence of two events. If the events do not overlap, then they can never occur together. The first question seeks the probability that both A and B will occur; this probability is represented by the size of the overlap between the two events. The second question seeks the probability that either A or B will occur; this probability is represented by the combined shapes of the two events, including, of course, the overlap between them.

The Multiplicative Rule: If the events A and B are independent, the probability that they occur together, denoted P(A and B), is P(A) × P(B).

Here P(A) and P(B) are the probabilties of the individual events. Note that independent does not mean that they do not overlap in the sample space. In fact, nonoverlapping, or disjoint, events are not independent, for if one occurs, then the other cannot. In probability theory, independent means that one event provides no information about the other. For example, if a card drawn from a deck turns out to be an ace, we have no clue about the card's suit. Thus drawing the ace of hearts represents the joint occurrence of two independent events—the card is an ace (A) and it is a heart (H). According to the Multiplicative Rule, P(A and H) = P(A) × P(H), and because P(A) = 4/52 and P(H) = 1/4, P(A and H) = (4/52) × (1/4) = 1/52.

The Additive Rule: If the events A and B are independent, the probability that at least one of them occurs, denoted P(A or B), is P(A) + P(B) − P(A) × P(B).

Here the term P(A) × P(B), which is the probability that A and B occur together, is subtracted from the sum of the probabilities, P(A) + P(B), because the straight sum includes this term twice. As an example, suppose we seek the probability that a card drawn from a deck is either an ace or a heart. According to the Additive Rule, P(A or H) = P(A) + P(H) − P(A) × P(H) = (4/52) + (1/4) − (4/52) × (1/4) = 16/52.

Of course, if the two events do not overlap in the sample space, the Additive Rule reduces to a simpler expression: P(A or B) = P(A) + P(B). For example, suppose we seek the probability that a card drawn from a deck is either an ace or a king (K). These two events do not overlap in the sample space; they are said to be mutually exclusive. Thus P(A or K) = P(A) + P(K) = (4/52) + 4(52) = 8/52.

---

each of the uniting gametes contains *A*, or (1/2) × (1/2) = (1/4), since the two gametes are produced independently. The chance for an *aa* homozygote is also 1/4. However, the chance for an *Aa* heterozygote is 1/2. The reason is that there are two ways of creating a heterozygote—*A* may come from the egg and *a* from the sperm, or vice versa. Because each of these events has a one-quarter chance of occurring, the total proba-

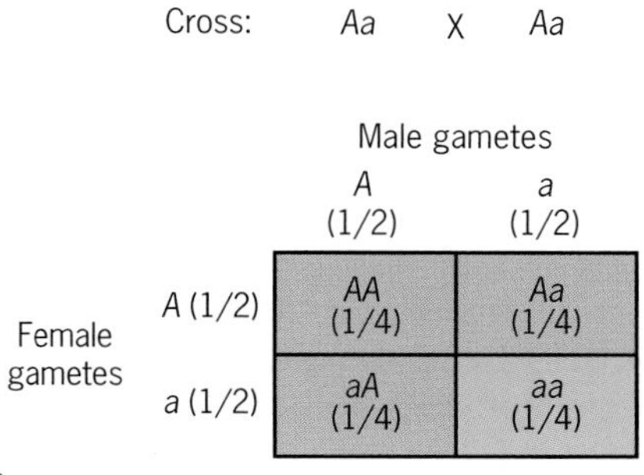

Progeny:

Genotype	Frequency	Phenotype	Frequency
AA	1/4	Dominant	3/4
Aa	1/2		
aa	1/4	Recessive	1/4

Figure 3.8 An intercross showing the probability method in the context of a Punnett Square. The frequency of each genotype from the cross is obtained from the frequencies in the Punnett Square, which are, in turn, obtained by multiplying the frequencies of the two types of gametes produced by the heterozygous parents.

bility that an offspring is heterozygous is (1/4) + (1/4) = (1/2). We therefore obtain the following **probability distribution** of the genotypes from the mating *Aa* × *Aa*:

AA	1/4
Aa	1/2
aa	1/4

We conclude that (1/4) + (1/2) = (3/4) of the progeny will have the dominant phenotype and 1/4 will have the recessive.

For such a simple situation, use of the probability method may seem unnecessary. However, in more complicated situations, it is clearly the most practical approach to predict the outcome of crosses. Consider, for example, a cross between plants heterozygous for four different genes, each assorting independently. What fraction of the progeny will be homozyous for all four recessive alleles? To answer this question, we consider the genes one at a time. For the first gene, the fraction of offspring that will be recessive homozygotes is 1/4, as it will be for the second, third, and fourth genes. Therefore, by the Principle of Independent Assortment, the fraction of offspring that will be quadruple recessive homozygotes is (1/4) × (1/4) × (1/4) × (1/4) = (1/256). Surely, using the probability method is a better approach than diagramming a Punnett Square with 256 entries!

Now let us consider a more difficult question. What fraction of the offspring will be homozygous for all four genes? Before computing any probabilities, we must first decide what genotypes satisfy the question. For each gene there are two types of homozygotes, the dominant and the recessive, and together they constitute half the progeny. The fraction of progeny that will be homozygous for all four genes will therefore be (1/2) × (1/2) × (1/2) × (1/2) = (1/16).

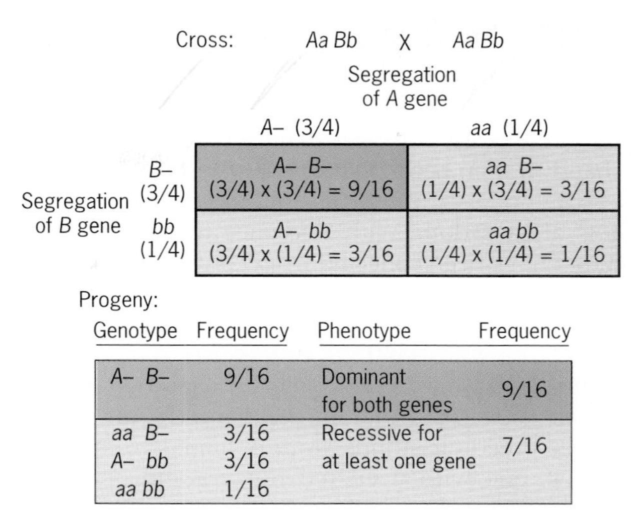

Genotype	Frequency	Phenotype	Frequency
A– B–	9/16	Dominant for both genes	9/16
aa B–	3/16	Recessive for at least one gene	7/16
A– bb	3/16		
aa bb	1/16		

Figure 3.9 Application of the probability method to an intercross involving two genes. In this cross, each gene segregates dominant and recessive phenotypes, with probabilities 3/4 and 1/4, respectively. Because the segregations occur independently, the frequencies of the combined phenotypes within the square are obtained by multiplying the marginal probabilities. The frequency of progeny showing the recessive phenotype for at least one of the genes is obtained by adding the frequencies in the relevant cells (yellow).

To see the full power of the probability method, we need to consider one more question. Suppose the cross is *Aa Bb* × *Aa Bb* and we want to know what fraction of the progeny will show the recessive phenotype for at least one gene (Figure 3.9). Clearly, three kinds of genotypes would satisfy this condition: (1) *A-; bb* (the dash stands for either *A* or *a*), (2) *aa; B-*, and (3) *aa; bb*. The answer to the question must therefore be the sum of the probabilities corresponding to each of these genotypes. The probability for *A- bb* is (3/4) × (1/4) = (3/16), that for *aa B-* is (1/4) × (3/4) = (3/16), and that for *aa bb* is (1/4) × (1/4) = (1/16). Adding these together, we find that the answer is 7/16.

Key Points: **The outcome of a cross can be predicted by the systematic enumeration of the genotypes in a Punnett Square. However, when more than two genes are involved, the forked-line or probability methods are used to predict the outcome of a cross.**

FORMULATING AND TESTING GENETIC HYPOTHESES

Sometimes the genetic basis of a trait is not known beforehand. It then becomes necessary to formulate and test an hypothesis about the inheritance of the trait. This is always done with reference to data, usually data that have been collected from controlled crosses.

There are no set rules for formulating a genetic hypothesis. Instead, we must rely on insight gained from the data and from previous experience. As an example, consider an experiment with the snapdragon, *Antirrhinum majus*, a popular garden plant (Figure 3.10*a*). Two true-breeding strains were obtained from nursery stock, one with deep red flowers and one with pure white flowers. These two strains were crossed (Figure 3.10*b*), yielding F_1 hybrids that had pink flowers. The F_1 were then intercrossed to produce three kinds of F_2 plants: red (62), pink (131), and white (57), with the actual numbers shown in parentheses. How might we explain the data?

We could postulate that flower color is controlled by a single gene with two alleles, *W* (for red) and *w* (for white), and that the flowers of *Ww* heterozygotes are pink because *W* is only partially dominant over *w*. According to this hypothesis, the P generation would be *WW* (red) × *ww* (white), producing F_1 hybrids that would be *Ww* (pink), which, when self-fertilized, would yield *WW* (red), *Ww* (pink), and *ww* (white) F_2 progeny in a 1:2:1 ratio. The actual numbers seem to bear this out, giving credence to the hypothesis.

The Chi-Square Test

We may ask whether data really do support a particular hypothesis. This question is critical, since the value of an hypothesis depends on its ability to explain the data. An hypothesis that does not fit needs to be modified or discarded altogether in favor of something better.

One procedure for testing the fit between the predictions of an hypothesis and actual data uses a statistic called **chi-square (χ^2)**. A statistic is a number calculated from data—for example, the mean of a set of examination scores. The χ^2 statistic allows a researcher to compare data, such as the numbers we get from a breeding experiment, with their predicted values. If the comparison is unfavorable, that is, the data are not in line with the predicted values, the χ^2 statistic will exceed a critical number and the genetic hypothesis will be rejected. If the χ^2 statistic is below this number, the hypothesis will be accepted. The χ^2 statistic therefore reduces hypothesis testing to a simple, objective procedure.

As an example, let us consider the snapdragon crosses described earlier. The F_2 data seemed to be consistent with the hypothesis that a single gene seg-

Figure 3.10 (*a*) Snapdragons, *Antirrhinum majus*. (*b*) Results of crosses between red and white varieties of snapdragons.

F_2 Phenotype	Observed number	Expected number
Red	62	(1/4) x 250 = 62.5
Pink	131	(1/2) x 250 = 125
White	57	(1/4) x 250 = 62.5
Total	**250**	**250**

Calculation of chi-square statistic to test for agreement between observed and expected numbers:

$$\chi^2 = \sum \frac{(\text{Observed} - \text{Expected})^2}{\text{Expected}}$$

$$= \frac{(62 - 62.5)^2}{62.5} + \frac{(131 - 125)^2}{125} + \frac{(57 - 62.5)^2}{62.5}$$

$$= 0.776$$

Figure 3.11 Comparison of the observed and expected results and calculation of χ^2 for an intercross with hybrid snapdragons.

regates two alleles. However, to evaluate this objectively we need to compare the data with their predicted values. Figure 3.11 illustrates the calculations.

The procedure is straightforward. For each phenotypic class, we compute the expected number of offspring by multiplying the Mendelian proportion and the total sample size. We then compute the difference between the observed and expected numbers and square these differences to eliminate the canceling effects of positive and negative values. After dividing each squared difference by the corresponding expected number of offspring, we sum all the terms and refer the resulting χ^2 statistic to a distribution of χ^2 values (Figure 3.12).

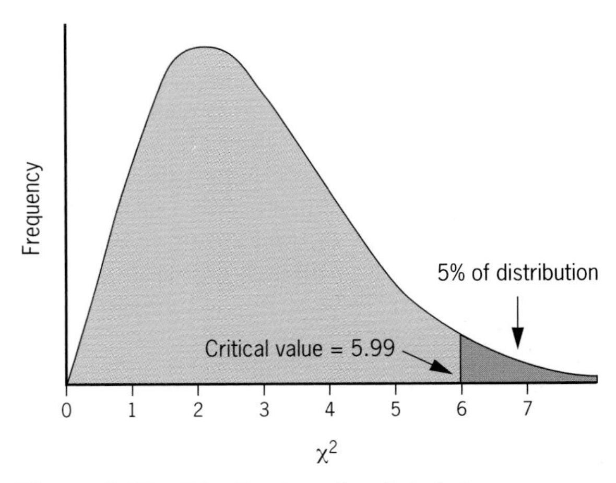

Figure 3.12 Distribution of a χ^2 statistic.

The distribution of χ^2 values, which is established by statistical theory, shows how often χ^2 will exceed a particular value solely by chance. Statisticians recommend focusing on the value that cuts off the upper 5 percent of the distribution. If the hypothesis is correct, the χ^2 statistic will exceed this **critical value** 5 percent of the time. However, if the hypothesis is incorrect, there is a much greater chance that χ^2 will exceed the critical value. An incorrect hypothesis is more likely to produce large differences between observations and expectations, thereby inflating the χ^2 statistic and pushing it to the right on the theoretical scale. It is customary to reject the hypothesis if the χ^2 statistic exceeds the critical value. Thus, if the hypothesis is actually true, there is a 5 percent chance of erroneously rejecting it.

To return to the example: The computed χ^2, 0.776, must be compared to the critical value from a theoretical distribution. It turns out, however, that there are many such distributions, and to select the appropriate one, we need to know something called the **degrees of freedom**. This is determined by subtracting one from the number of phenotypic classes; in this example, the number of degrees of freedom is $3 - 1 = 2$. We can now compare the computed χ^2 statistic with the critical value from the theoretical distribution with 2 degrees of freedom (see Table 3.2 for a list of critical values). The computed statistic, 0.776, is clearly less than the critical value, 5.991, so the hypothesis of one gene segregating two alleles is *not* rejected. We conclude that this hypothesis is an adequate explanation for the data.

A worked problem at the end of the chapter shows what happens when the χ^2 statistic is greater

TABLE 3.2
Table of Chi-Square (χ^2) 5% Critical Values[a]

Degrees of Freedom	*5% Critical Value*
1	3.841
2	5.991
3	7.815
4	9.488
5	11.070
6	12.592
7	14.067
8	15.507
9	16.919
10	18.307
15	24.996
20	31.410
25	37.652
30	43.773

[a]Selected entries from R. A. Fisher and Yates, 1943, *Statistical Tables for Biological, Agricultural and Medical Research.* Oliver and Boyd, London.

than the critical value. Other problems provide opportunities to use the χ^2 statistic.

Key Points: **The chi-square test is a simple way of evaluating whether the predictions of a genetic hypothesis agree with data from an experiment.**

MENDELIAN PRINCIPLES IN HUMAN GENETICS

The application of Mendelian principles to human genetics began soon after the rediscovery of Mendel's paper in 1900. However, because it is not possible to make controlled crosses with human beings, progress was obviously slow. The genetic analysis of human heredity depends on family records, which are often incomplete. In addition, human beings—unlike experimental organisms—do not produce many progeny, making it difficult to discern Mendelian ratios. Mistaken paternity is another problem in human genetics, introducing an element of confusion into the data. Time is also a factor, because some genetic conditions do not manifest themselves until a person reaches middle age. For all these reasons, human genetic analysis has been a difficult endeavor. Nonetheless, the drive to understand human heredity has been very strong, and today, despite all the obstacles, we have learned about thousands of human genes. Table 3.3 lists some of the conditions they control. We discuss many of these conditions in later chapters of this book.

TABLE 3.3
Inherited Conditions in Human Beings

Dominant Traits
Achondroplasia (dwarfism)
Brachydactyly (short fingers)
Congenital stationary night blindness
Ehler-Danlos syndrome (a connective tissue disorder)
Huntington's disease (a neurological disorder)
Marfan syndrome (tall, gangly stature)
Neurofibromatosis (tumorlike growths on the body)
Phenylthiocarbamide (PTC) tasting
Widow's peak
Woolly hair
Recessive Traits
Albinism (lack of pigment)
Alkaptonuria (a disorder of amino acid metabolism)
Ataxia telangiectasia (a neurological disorder)
Cystic fibrosis (a respiratory disorder)
Duchenne muscular dystrophy
Galactosemia (a disorder of carbohydrate metabolism)
Glycogen storage disease
Phenylketonuria (a disorder of amino acid metabolism)
Sickle-cell anemia (a hemoglobin disorder)
Tay-Sachs disease (a lipid storage disorder)

Pedigrees

Pedigrees are diagrams that show the relationships among the members of a family (Figure 3.13*a*). It is customary to represent males as squares and females as circles. A horizontal line connecting a circle and a square represents a mating. The offspring of the mating are shown beneath the mates, starting with the first born at the left and proceeding through the birth order to the right. Individuals that have a genetic condition are indicated by coloring or shading the symbols that represent them. The generations in a pedigree are usually denoted by Roman numerals, and particular individuals within a generation are referred to by Arabic numerals following the Roman numeral.

Traits caused by dominant alleles are the easiest to identify. Usually every individual who carries the dominant allele manifests the trait, making it possible to trace the transmission of the dominant allele through the pedigree (Figure 3.13*b*). Every affected individual is expected to have at least one affected parent, unless, of course, the dominant allele has just appeared in the family as a result of a new mutation. However, the frequency of new mutants is very low—on the order of one in a million; consequently, the spontaneous appearance of a dominant condition is an extremely rare event. Dominant traits that are associated with reduced viability or fertility never become frequent in a population. Thus most of the people who show such traits are heterozygous for the dominant allele. If their spouses do not have the trait, half their children should inherit the condition.

Recessive traits are not so easy to identify because they may occur in individuals whose parents are not affected. Sometimes several generations of pedigree data are needed to trace the transmission of a recessive allele (Figure 3.13*c*). Nevertheless, a large number of recessive traits have been observed in human beings—at last count, over 4000. Rare recessive traits are more likely to appear in a pedigree when spouses are related to each other—for example, when they are first cousins. This increased incidence occurs because relatives share alleles by virtue of their common ancestry. Siblings share one-half their alleles, half-siblings one-fourth their alleles, and first cousins, one-eighth their alleles. Thus, when such relatives mate, they have a greater chance of producing a child who is homozygous for a particular recessive allele than do unrelated parents. Many of the classical studies in human genetics have relied on the analysis of matings between relatives, principally first cousins.

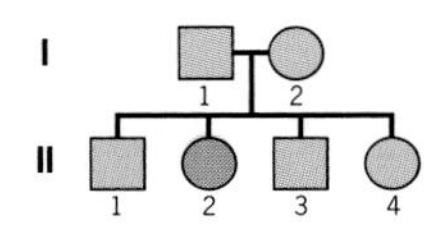

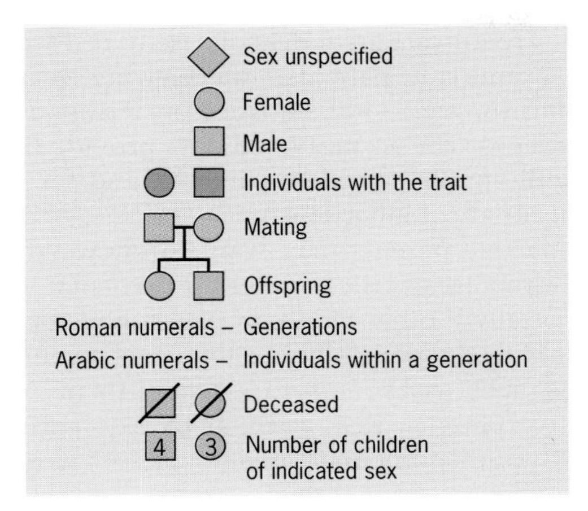

(*a*) Pedigree conventions

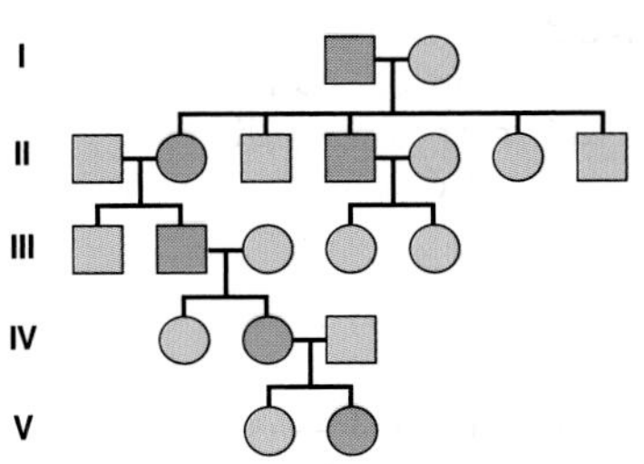

(*b*) Dominant trait

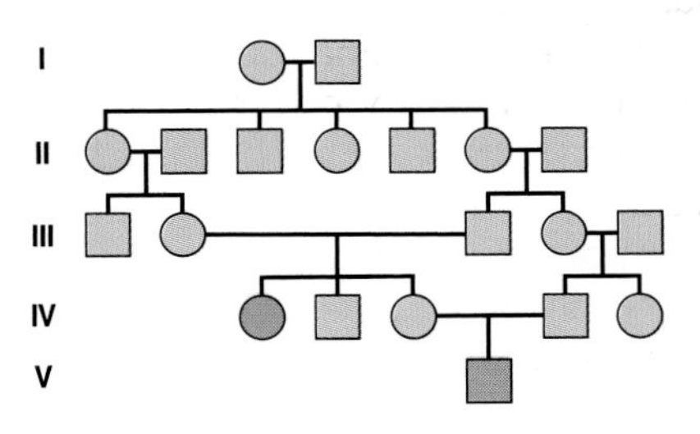

(*c*) Recessive trait

Figure 3.13 (*a*) Pedigree conventions. (*b*) Inheritance of a dominant trait. The trait appears in each generation. (*c*) Inheritance of a recessive trait. The two affected individuals are the offspring of relatives.

Mendelian Segregation in Human Families

In human beings, the number of children produced by a couple is typically small. Today in the United States, the average is around two. In developing countries, it is six to seven. Such numbers provide nothing close to the statistical power that Mendel had in his experiments with peas. Consequently, phenotypic ratios in human families often deviate significantly from their Mendelian expectations.

As an example, let us consider a couple who are each heterozygous for a recessive allele that, in homozygous condition, causes cystic fibrosis. If the couple were to have four children, would we expect exactly three to be normal and one to be affected by cystic fibrosis? The answer is no. Although this is clearly a possible outcome, it is not the only one. There are, in fact, five distinct possibilities:

1. Four normal, none affected.
2. Three normal, one affected.
3. Two normal, two affected.
4. One normal, three affected.
5. None normal, four affected.

Intuitively, the second outcome seems to be the most likely, since it conforms to Mendel's 3:1 ratio. We can calculate the probability of this outcome, and of each of the others, by using Mendel's principles and by treating each birth as an independent event (Figure 3.14).

For a particular birth, the chance that the child will be normal is 3/4. The probability that all four children will be normal is therefore $(3/4) \times (3/4) \times (3/4) \times (3/4) = (3/4)^4 = 81/256$. Similarly, the chance that a particular child will be affected is (1/4); thus the

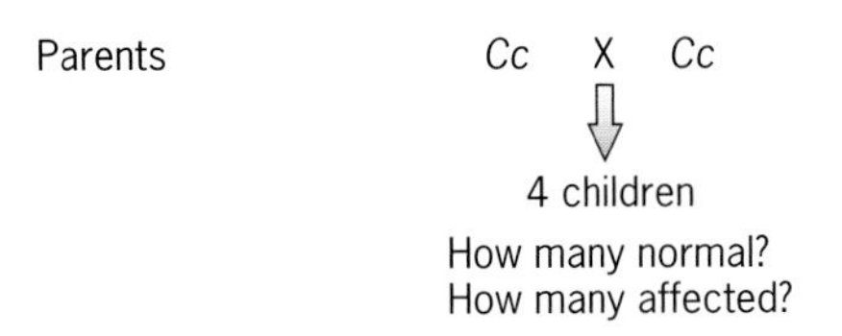

Number of children that are:

Normal	Affected	Probability
4	0	1 x (3/4) x (3/4) x (3/4) x (3/4) = 81/256
3	1	4 x (3/4) x (3/4) x (3/4) x (1/4) = 108/256
2	2	6 x (3/4) x (3/4) x (1/4) x (1/4) = 54/256
1	3	4 x (3/4) x (1/4) x (1/4) x (1/4) = 12/256
0	4	1 x (1/4) x (1/4) x (1/4) x (1/4) = 1/256

Probability distribution

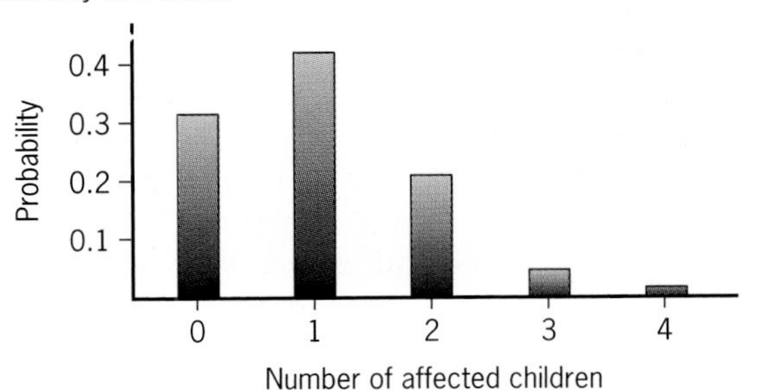

Figure 3.14 Probability distribution of families with four children segregating a recessive trait.

probability that all four will be affected is $(1/4)^4 = 1/256$. To find the probabilities for the three other outcomes, we need to recognize that each actually represents a collection of distinct events. The outcome of three normal children and one affected child, for instance, comprises four distinct events; if we let N symbolize a normal child and A an affected child, and if we write the children in their order of birth, we can represent these events as:

NNNA, NNAN, NANN, and ANNN

Because each has probability $(3/4)^3 \times (1/4)$, the total probability for three normal children and one affected, regardless of birth order, is $4 \times (3/4)^3 \times (1/4)$. The coefficient 4 is the number of ways in which three children could be normal and one could be affected in a family with four children. Similarly, the probability for two normal children and two affected is $6 \times (3/4)^2 \times (1/4)^2$, since in this case there are six distinct events. The probability for one normal child and three affected is $4 \times (3/4) \times (1/4)^3$, since in this case there are four distinct events. Figure 3.14 summarizes the calculations in the form of a probability distribution. As expected, three normal children and one affected child is the most probable outcome (probability 108/256). However, this outcome is not expected a majority of the time, because the other four outcomes have a combined probability of 148/256. The Technical Sidelight on page 55 generalizes this procedure to other situations in which the children fall into two possible phenotypic classes. Because there are only two classes, the probabilities associated with the outcomes are called **binomial probabilities**.

Genetic Counseling

The diagnosis of genetic conditions is often a difficult process. Typically, diagnoses are made by physicians who have been trained in genetics. The study of these conditions requires a great deal of careful research, including examining patients, interviewing relatives, and sifting through vital statistics on births, deaths, and marriages. The accumulated data provide the basis for defining the condition clinically and for determining its mode of inheritance.

Prospective parents may want to know whether their children are at risk to inherit a particular condition, especially if other family members have been affected. It is the responsibility of the genetic counselor to assess such risks and to explain them to the prospective parents. Risk assessment requires familiarity with probability and statistics, as well as a thorough knowledge of genetics.

As an example, let us consider a pedigree showing the inheritance of a rare form of **dwarfism** (Figure 3.15). The condition is manifested in all but the first generation, and every affected individual except the very first has an affected parent. This pattern strongly argues that the condition is caused by a dominant allele. The dwarfed woman in generation II probably represents a rare new mutation in the population.

The counseling issue arises in generation V. What is the chance that the dwarfed man will produce a dwarfed child? To answer this question we need to know the genotypes of the prospective parents. Because the condition is caused by a dominant allele, the mother must be homozygous for the normal (reces-

(a)

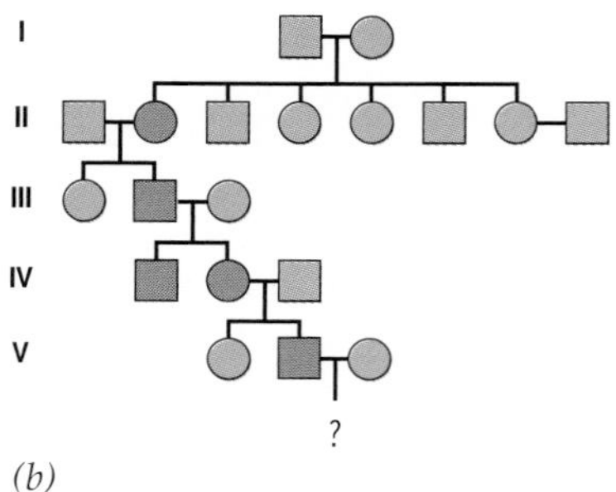

(b)

Figure 3.15 (*a*) Dwarfism illustrated by a family of seven siblings. The trait is caused by a dominant allele. (*b*) Pedigree showing the inheritance of dwarfism through five generations. The affected woman in generation II appears to be a new mutant because neither of her parents was affected.

TECHNICAL SIDELIGHT

Binomial Probabilities

The progeny of crosses sometimes segregate into two distinct classes—for example, male or female, healthy or diseased, normal or mutant, dominant phenotype or recessive phenotype. To be general, we can refer to these two kinds of progeny as P and Q, and note that for any individual offspring, the probability of being P is p and the probability of being Q is q. Because there are only two classes, $q = 1 - p$. Suppose that the total number of progeny is n and that each one is produced independently. We can calculate the **binomial probability** that exactly x of the progeny will fall into one class and y into the other:

$$\text{Probability of } x \text{ in class P and } y \text{ in class Q} = \left[\frac{(n!)}{(x!\,y!)}\right] p^x q^y.$$

The bracketed term contains three factorial functions ($n!$, $x!$, and $y!$), each of which is computed as a descending series of products. For example, $n! = n\,(n-1)\,(n-2)\,(n-3)...(3)\,(2)\,(1)$. If 0! is needed, it is defined as one. In the formula, the bracketed term enumerates the different ways, or orders, in which n offspring can be segregated so that x fall in the P class and y fall in the Q class. The other term, $p^x q^y$, gives the probability of obtaining a particular way or order. Because each of the orders is equally likely, multiplying this term by the bracketed term gives the probability of obtaining x progeny in the P class and y in the Q class, regardless of the order of occurrence.

If, for fixed values of n, p, and q, we systematically vary x and y, we can calculate a whole set of probabilities. This set constitutes a binomial probability distribution. With the distribution, we can answer questions such as "What is the probability that x will exceed a particular value?" or "What is the probability that x will lie between two particular values?"

For example, let us consider a family with six children. What is the probability that at least four will be girls? To answer this question, we note that for any given child, the probability that it will be a girl (p) is 1/2 and the probability that it will be a boy (q) is also 1/2. The probability that exactly four children in a family will be girls (and two will be boys) is therefore $[(6!)/(4!\,2!)](1/2)^4\,(1/2)^2 = 15/64$, which is one of the terms in the binomial distribution. However, the probability that at least four will be girls (and that no more than two will be boys) is the sum of three terms from this distribution:

Event	Binomial formula	Probability
4 girls and 2 boys:	$[(6!)/(4!\,2!)] \times (1/2)^4\,(1/2)^2 =$	15/64
5 girls and 1 boy:	$[(6!)/(5!\,1!)] \times (1/2)^5\,(1/2)^1 =$	6/64
6 girls and 0 boys:	$[(6!)/(6!\,0!)] \times (1/2)^6\,(1/2)^0 =$	1/64

Therefore, the answer is $(15/64) + (6/64) + (1/64) = 22/64$.

The binomial distribution also provides answers to other kinds of questions. For example, what is the probability that at least one but no more than four of the children will be girls? Here the answer is the sum of four terms:

Event	Binomial formula	Probability
1 girl and 5 boys	$[(6!)/(1!\,5!)] \times (1/2)^1\,(1/2)^5 =$	6/64
2 girls and 4 boys	$[(6!)/(2!\,4!)] \times (1/2)^2\,(1/2)^4 =$	15/64
3 girls and 3 boys	$[(6!)/(3!\,3!)] \times (1/2)^3\,(1/2)^3 =$	20/64
4 girls and 2 boys	$[(6!)/(4!\,2!)] \times (1/2)^4\,(1/2)^2 =$	15/64

Summing up, the answer is 56/64.

sive) allele (*dd*) and the father must be heterozygous for the dwarfism allele (*Dd*). The chance that the couple will have a dwarfed child is therefore 1/2.

As another example, consider the situation shown in Figure 3.16. A couple, denoted R and S, are concerned about the possibility that they will have a child (T) with **albinism**, an autosomal recessive condition characterized by a complete absence of melanin pigment in the skin, eyes, and hair. S, the prospective father, is an albino, and R, the prospective mother, has two albino sibilings. It would therefore seem that the child has some risk of being born albino.

This risk depends on two factors: (1) the probability that R is a heterozygous carrier of the albinism allele (*a*), and (2) the probability that she will transmit this allele to T if she actually is a carrier. S, who is obviously homozygous for the albinism allele, must transmit this allele to his offspring.

To determine the first probability, we need to consider the possible genotypes for R. One of these, that she is homozygous for the recessive allele (*aa*), is excluded because we know that she is not an albino. However, the other two genotypes, *AA* and *Aa*, remain distinct possibilities. To calculate the probabili-

(a)

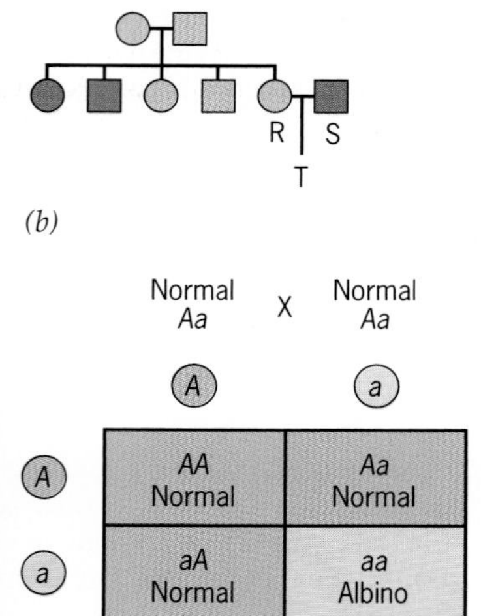

(c)

Figure 3.16 (*a*) Albinism in a family with six sons, two of whom are affected. (*b*) Pedigree showing the inheritance of albinism. (*c*) Punnett Square showing that among nonalbinos, the frequency of heterozygotes is 2/3.

ties associated with each of these, we note that both of R's parents must be heterozygotes, because they have had two albino children. The mating that produced R was therefore *Aa* × *Aa*, and from such a mating we would expect 2/3 of the *nonalbino* offspring to be *Aa* and 1/3 to be *AA*. Thus the probability that R is a heterozygous carrier of the albinism allele is 2/3. To determine the probability that she will transmit this allele to her child, we simply note that *a* will be present in half of her gametes.

In summary,

Risk that T is *aa*

= [Probability that R is *Aa*] × [Probability that R transmits *a*, assuming that R is *Aa*]

= (2/3) × (1/2) = (1/3)

The examples in Figures 3.15 and 3.16 illustrate simple counseling situations in which the risk can be determined precisely. Often the circumstances are much more complicated, making the task of risk assessment quite difficult. The genetic counselor's task is to analyze the pedigree information and determine the risk as precisely as possible.

Key Points: **Pedigrees are used to identify dominant and recessive traits in human families. The analysis of pedigrees allows genetic counselors to determine the probability that an individual will inherit a particular trait.**

TESTING YOUR KNOWLEDGE

1. Phenylketonuria, a metabolic disease in humans, is caused by a recessive allele, *k*. If two heterozygous carriers of the allele marry and plan a family of five children: (a) What is the chance that all their children will be normal? (b) What is the chance that four children will be normal and one affected with phenylketonuria? (c) What is the chance that at least three children will be normal? (d) What is the chance that the first child will be a normal girl? (e) What is the chance that a normal child will carry the recessive allele?

ANSWER

Before answering each of the questions, we must note that from a mating between two heterozygotes, the probability that a particular child will be normal is 3/4, and the probability that a particular child will be affected is 1/4. Furthermore, for any one child born, the chance that it will be a boy is 1/2 and the chance it will be a girl is 1/2.

(a) To calculate the chance that all five children will be

normal, we use the Multiplicative Rule. For each child, the chance that it will be normal is 3/4, and all five children are independent. Consequently, the probability of five normal children is $(3/4)^5 = 0.237$. This is the first term of the binomial probability distribution with $p = 3/4$ and $q = 1/4$.

(b) To calculate the chance that four children will be normal and one affected, we compute the second term of the binomial distribution:

$$\frac{5!}{4!\,1!} \times (3/4)^4 \times (1/4)^1 = 5 \times (81/1024) = 0.399$$

(c) To find the probability that at least three children will be normal, we must sum the first three terms of the binomial distribution:

Event	Binomial formula	Probability
5 normal, 0 affected	$[(5!)/(5!\,0!)] \times (3/4)^5\,(1/4)^0 =$	0.237
4 normal, 1 affected	$[(5!)/(4!\,1!)] \times (3/4)^4\,(1/4)^1 =$	0.399
3 normal, 2 affected	$[(5!)/(3!\,2!)] \times (3/4)^3\,(1/4)^2 =$	0.264
	Total	0.900

(d) To determine the probability that the first child will be a normal girl, we use the Multiplicative Rule: P(normal child and girl) = P(normal child) × P(girl) = (3/4) × (1/2) = 3/8.

(e) Among the normal children, there will be two genotypes, *KK* and *Kk*, in a ratio of 1:2. Consequently, the proportion of normal children that will be carriers is 2/3.

2. Mice from wild populations typically have gray-brown (or *agouti*) fur, but in one laboratory strain, some of the mice have yellow fur. A single yellow male is mated to several agouti females. Altogether, the matings produce 40 progeny, 22 with agouti fur and 18 with yellow fur. The agouti F_1 animals are then intercrossed with each other to produce an F_2, all of which are agouti. Similarly, the yellow F_1 animals are intercrossed with each other, but their F_2 progeny segregate into two classes; 30 are agouti and 54 are yellow. Subsequent crosses between yellow F_2 animals also segregate yellow and agouti progeny. What is the genetic basis of these coat color differences?

ANSWER

We note that the cross agouti × agouti produces only agouti animals, and that the cross yellow x yellow produces a mixture of yellow and agouti. Thus a reasonable hypothesis is that yellow fur is caused by a dominant allele, *Y*, and that agouti fur is caused by a recessive allele, *y*. According to this hypothesis, the agouti females used in the initial cross would be *yy*, and their yellow mate would be *Yy*. We hypothesize that the male was heterozygous because he produced approximately equal numbers of agouti and yellow F_1 offspring. Among these, the agouti animals should be *yy* and the yellow animals *Yy*. These genotypic assignments are borne out by the F_2 data, which show that the F_1 agouti mice have bred true and the F_1 yellow mice have segregated. However, the segregation ratio of yellow to agouti (54:30) seems to be out of line with the Mendelian expectation of 3:1. Is this lack of fit serious enough to warrant rejection of the hypothesis?

We can use the χ^2 procedure to test for disagreement between the data and the predictions of the hypothesis. According to the hypothesis, 3/4 of the F_2 progeny from the yellow × yellow intercross should be yellow and 1/4 should be agouti. Using these proportions, we can calculate the expected numbers of progeny in each class and then calculate a χ^2 statistic with $2 - 1 = 1$ degree of freedom.

F_2 phenotype	Obs	Exp	(Obs-Exp)²/Exp
yellow (*YY* and *Yy*)	54	(3/4) × 84 = 63	1.286
agouti (*yy*)	30	(1/4) × 84 = 21	5.762
Total	84	84	7.048

The χ^2 statistic (7.048) is much greater than the critical value (3.841) for a χ^2 distribution with 1 degree of freedom. Consequently, we reject the hypothesis that the coat colors are segregating in a 3:1 Mendelian fashion.

We obtain a clue to the lack of fit between the F_2 data and the predictions of the Mendelian hypothesis by noting that subsequent yellow x yellow crosses failed to establish a true-breeding yellow strain. One possible explanation is that all the yellow animals are *Yy* heterozygotes and that *YY* animals die. This is, in fact, why the yellow mice are underrepresented in the F_2 data. Examination of embryos within the uteruses of pregnant females reveals that about 1/4 of them are dead. The dead embryos must be genotypically *YY*. Thus a single copy of the *Y* allele produces a visible phenotypic effect (yellow fur), but two copies are lethal. Taking this lethality into account, we can modify the hypothesis and predict that 2/3 of the live-born F_2 progeny should be yellow (*Yy*) and 1/3 should be agouti (*yy*). We can then use the χ^2 procedure to test this modified hypothesis for consistency with the data.

F_2 phenotype	Obs	Exp	(Obs-Exp)²/Exp
yellow (*Yy*)	54	(2/3) × 84 = 56	0.071
agouti (*yy*)	30	(1/3) × 84 = 28	0.143
Total	84	84	0.214

This χ^2 statistic is clearly less than the critical value for a χ^2 distribution with 1 degree of freedom. Thus the data are in agreement with the predictions of the modified hypothesis.

QUESTIONS AND PROBLEMS

3.1 On the basis of Mendel's observations, predict the results from the following crosses with peas: (a) a tall (dominant and homozygous) variety crossed with a dwarf variety; (b) the progeny of (a) self-fertilized; (c) the progeny from (a) crossed with the original tall parent; (d) the progeny of (a) crossed with the original dwarf parent.

3.2 Mendel crossed pea plants that produced round seeds with those that produced wrinkled seeds and self-fertilized the progeny. In the F_2, he observed 5474 round seeds and 1850 wrinkled seeds. Using the letters *W* and *w* for the alleles controlling seed texture, diagram Mendel's crosses, showing the genotypes of the plants in each generation. Are Mendel's results consistent with the Principle of Segregation?

3.3 A geneticist crossed wild, gray-colored mice with white (albino) mice. All the progeny were gray. These progeny were intercrossed to produce an F_2, which consisted of 198 gray and 72 white mice. Propose an hypothesis to explain these results, diagram the crosses, and compare the results with the predictions of the hypothesis.

3.4 A woman has a rare abnormality of the eyelids called ptosis, which prevents her from opening her eyes completely. This condition is caused by a dominant allele, *P*. The woman's father had ptosis, but her mother had normal eyelids. Her father's mother had normal eyelids. (a) What are the genotypes of the woman, her father, and her mother? (b) What proportion of the woman's children will have ptosis if she marries a man with normal eyelids?

3.5 In pigeons, a dominant allele *C* causes a checkered pattern in the feathers; its recessive allele *c* produces a plain pattern. Feather coloration is controlled by an independently assorting gene; the dominant allele *B* produces red feathers, and the recessive allele *b* produces brown feathers. Birds from a true-breeding checkered, red variety are crossed to birds from a true-breeding plain, brown variety. (a) Predict the phenotype of their progeny. (b) If these progeny are intercrossed, what phenotypes will appear in the F_2, and in what proportions?

3.6 In mice, the allele *C* for colored fur is dominant over the allele *c* for white fur, and the allele *V* for normal behavior is dominant over the allele *v* for waltzing behavior, a form of discoordination. Give the genotypes of the parents in each of the following crosses: (a) colored, normal mice mated with white, normal mice produced 29 colored, normal and 10 colored, waltzing progeny; (b) colored, normal mice mated with colored, normal mice produced 38 colored, normal, 15 colored, waltzing, 11 white, normal, and 4 white, waltzing progeny; (c) colored, normal mice mated with white, waltzing mice produced 8 colored, normal, 7 colored, waltzing, 9 white, normal, and 6 white, waltzing progeny.

3.7 In rabbits, the dominant allele *B* causes black fur and the recessive allele *b* causes brown fur; for an independently assorting gene, the dominant allele *R* causes long fur and the recessive allele *r* (for *rex*) causes short fur. A homozygous rabbit with long, black fur is crossed with a rabbit with short, brown fur, and the offspring are intercrossed. In the F_2, what proportion of the rabbits with long, black fur will be homozygous for both genes?

3.8 In shorthorn cattle, the genotype *RR* causes a red coat, the genotype *rr* causes a white coat, and the genotype *Rr* causes a roan coat. A breeder has red, white, and roan cows and bulls. What phenotypes might be expected from the following matings, and in what proportions? (a) red × red; (b) red x roan; (c) red × white; (d) roan × roan; (e) roan × white; (f) white × white.

3.9 Albinism in humans is caused by a recessive allele *a*. From marriages between normally pigmented people known to be carriers (*Aa*) and albinos (*aa*), what proportion of the children would be expected to be albinos? What is the chance in a family of three children that one would be normal and two albino?

3.10 If both husband and wife are known to be carriers of the allele for albinism, what is the chance of the following combinations in a family of four children: (a) all four normal; (b) three normal and one albino; (c) two normal and two albino; (d) one normal and three albino?

3.11 In humans, cataracts in the eyes and fragility of the bones are caused by dominant alleles that assort independently. A man with cataracts and normal bones marries a woman without cataracts but with fragile bones. The man's father had normal eyes, and the woman's father had normal bones. What is the probability that the first child of this couple will (a) be free from both abnormalities; (b) have cataracts but not have fragile bones; (c) have fragile bones but not have cataracts; (d) have both cataracts and fragile bones?

3.12 Are the traits shown in the pedigrees below due to a dominant or a recessive allele?

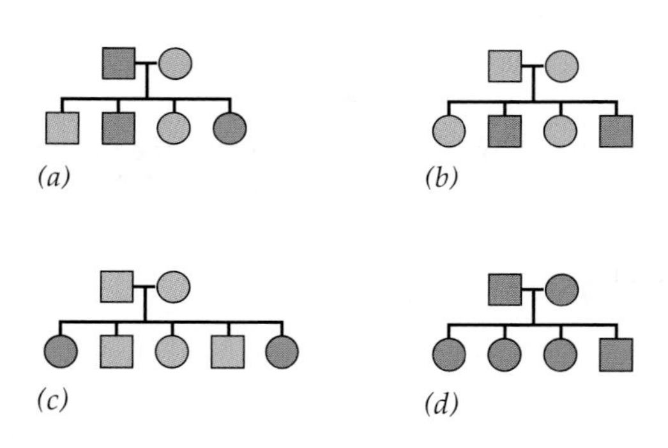

3.13 How many different kinds of F_1 gametes, F_2 genotypes, and F_2 phenotypes would be expected from the following crosses: (a) *AA* × *aa*; (b) *AA BB* × *aa bb*; (c) *AA BB CC* × *aa bb cc*? (d) What general formulas are suggested by these answers?

3.14 Mendel testcrossed pea plants grown from yellow, round F_1 seeds to plants grown from green, wrinkled seeds and obtained the following results: 31 yellow, round; 26 green, round; 27 yellow, wrinkled; and 26 green, wrinkled. Are these results consistent with the hypothesis that seed color and seed texture are controlled by independently assorting genes, each segregating two alleles?

3.15 If a man and a woman are heterozygous for a gene, and if they have three children, what is the chance that all three will also be heterozygous?

3.16 If four babies are born on a given day: (a) What is the chance that two will be boys and two girls? (b) What is the chance that all four will be girls? (c) What combination of boys and girls among four babies is most likely? (d) What is the chance that at least one baby will be a girl?

3.17 In a family of six children, what is the chance that at least three are girls?

3.18 The pedigree below shows the inheritance of a dominant trait. What is the chance that the offspring of the following matings will show the trait: (a) III-1 × III-3; (b) III-2 × III-4?

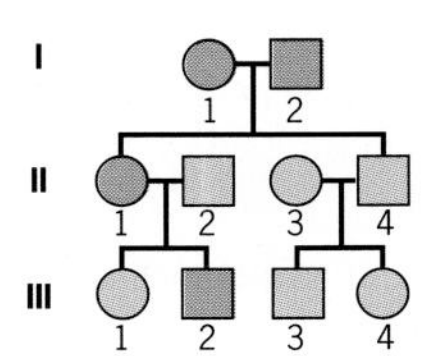

3.19 The pedigree below shows the inheritance of a recessive trait. Unless there is evidence to the contrary, assume that the individuals who have married into the family do not carry the recessive allele. What is the chance that the offspring of the following matings will show the trait: (a) III-1 × III-12; (b) III-4 × III-14; (c) III-6 × III-13; (d) IV-1 × IV-2?

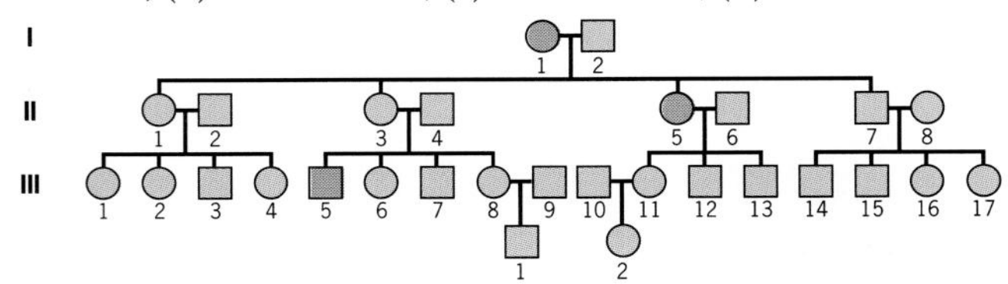

3.20 In the pedigrees below, determine whether the trait is more likely to be due to a dominant or a recessive allele.

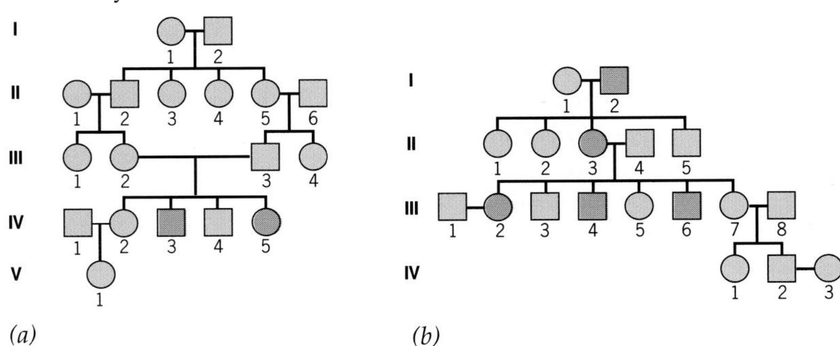

3.21 In pedigree (*b*) above, what is the chance that the couple III-1 and III-2 will have an affected child? What is the chance that the couple IV-2 and IV-3 will have an affected child?

3.22 The pedigree below shows the inheritance of a recessive trait. What is the chance that the couple III-3 and III-4 will have an affected child?

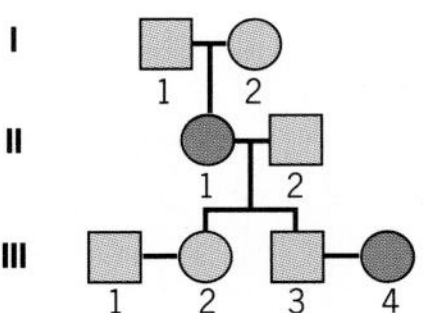

3.23 Peas heterozygous for three independently assorting genes were intercrossed. (a) What proportion of the offspring will be homozygous for all three recessive alleles? (b) What proportion of the offspring will be homozygous for all three genes? (c) What proportion of the offspring will be homozygous for one gene and heterozygous for the other two? (d) What proportion of the offspring will be homozygous for the recessive allele of at least one gene? (e) What proportion of the offspring will not be homozygous for any of the recessive alleles?

BIBLIOGRAPHY

Carlson, E. A. 1966. *The Gene: A Critical History*. W. B. Saunders, Philadelphia.

Dunn, L. C. 1965. *A Short History of Genetics*. McGraw-Hill, New York.

Iltis, H. 1932. *Life of Mendel*. (Translation by E. and C. Paul.) W. W. Norton, New York.

Peters, J. A., ed. 1959. *Classic Papers in Genetics*. Prentice-Hall, Engelwood Cliffs, NJ.

Stern, C. 1973. *Principles of Human Genetics*. W. H. Freeman, San Francisco.

Diverse species of plants growing in a garden. Experiments with many different plants extended Mendel's Principles of Dominance, Segregation, and Independent Assortment.

4

Extensions of Mendelism

CHAPTER OUTLINE

Genetics Grows Beyond Mendel's Monastery Garden

In 1902, enthused by what he read in Mendel's paper, the British biologist, William Bateson published an English translation of Mendel's German text and appended to it a brief account of what he called "Mendelism"—the Principles of Dominance, Segregation, and Independent Assortment. Later, in 1909, he published *Mendel's Principles of Heredity*, in which he summarized all the evidence then available to support Mendel's findings. This book was remarkable for two reasons. First, it examined the results of breeding experiments with many different plants and animals and in each case demonstrated that Mendel's principles applied. Second, it considered the implications of these experiments and raised questions about the fundamental nature of genes, or, as Bateson called them, "unit-characters." At the time Bateson's book was published, the word "gene" had not yet been invented.

Bateson's book played a crucial role in spreading the principles of Mendelism to the scientific world. Botanists, zoologists, naturalists, horticulturalists, and animal breeders all got the message in plain and simple language: Mendel's principles—tested by experiments with peas, beans, sunflowers, cotton, wheat, barley, tomatoes, maize, and assorted ornamental plants, as well as with cattle, sheep, cats, mice, rabbits, guinea pigs, chickens, pigeons, canaries, and moths—were universal. In the preface to his book, Bateson remarked that "The study of heredity thus becomes an organized branch of physiological science, already abundant in results, and in promise unsurpassed."

During the course of the twentieth century, the science that began so obscurely in Mendel's monastery garden, has grown to such an extent that today it encompasses much of biology. Bateson, who did so much to champion the principles of Mendelism, also gave this new science a name. In 1906, in an address to an International Congress of Botany, he remarked that "a new and well developed branch of Physiology has been created. To this study we may give the title *Genetics*." Two years later, Bateson's exertions on behalf of Mendelism were officially recognized: the University of Cambridge appointed him as professor of biology. In his inaugural lecture, "The Methods and Scope of Genetics," Bateson had some advice for all geneticists:

> *Treasure your exceptions! When there are none, the work gets so dull that no one cares to carry it further. Keep them always uncovered and in sight. Exceptions are like the rough brickwork of a growing building which tells that there is more to come and shows where the next construction is to be.*

In this chapter we consider some of the early extensions of Mendelian analysis: that a gene can exist in many different allelic states, that a particular gene can affect several different traits, and that a particular trait can be affected by several different genes. We also consider how genetic and environmental factors influence phenotypic variation.

ALLELIC VARIATION AND GENE FUNCTION

Mendel's experiments established that genes can exist in alternate forms. For each of the seven traits that he studied—seed color, seed texture, plant height, flower color, flower position, pod shape, and pod color—Mendel identified two alleles, one dominant, the other recessive. This discovery suggested a simple functional dichotomy between alleles, as if one allele did nothing and the other did everything to determine the phenotype. However, research early in the twentieth century demonstrated this to be an oversimplification. Genes can exist in more than two allelic states, and each allele can have a different effect on the phenotype.

Incomplete Dominance and Codominance

An allele is dominant if it has the same phenotypic effect in heterozygotes as in homozygotes—that is, the genotypes *Aa* and *AA* are phenotypically indistinguishable. Sometimes, however, a heterozygote has a phenotype different from that of either of its associated homozygotes. Flower color in the snapdragon, *Antirrhinum majus*, is an example. White and red varieties are homozygous for different alleles of a color-determining gene; when they are crossed, they produce heterozygotes that have pink flowers. The allele for red color (*W*) is therefore said to be **incompletely,** or **partially, dominant** over the allele for white color (*w*). The most likely explanation is that the intensity of pigmentation in this species depends on the amount of a product made by the color gene (Figure 4.1). If the *W* allele makes this product and the *w* allele does not, *WW* homozygotes will have twice as much of the product as *Ww* heterozygotes do, and therefore show deeper color. When the heterozygote's phenotype is

Phenotype	Genotype	Amount of gene products
Red	*WW*	2x
Pink	*Ww*	x
White	*ww*	0

Figure 4.1. Genetic basis of flower color in snapdragons. The allele *W* is incompletely dominant over *w*. Differences among the phenotypes could be due to differences in the amount of the product specified by the *W* allele.

Reactions with anti-sera		Blood type (antigen present)	Genotype
Anti-M serum	Anti-N serum		
		M	$L^M L^M$
		M N	$L^M L^N$
		N	$L^N L^N$

Figure 4.2. Detection of the M and N antigens on blood cells by agglutination with specific anti-sera. With the anti-M and anti-N sera, three blood types can be identified.

midway between the phenotypes of the two homozygotes, as it is here, the partially dominant allele is sometimes said to be **semidominant** (from the Latin word for "half"—thus half-dominant).

Another exception to the principle of simple dominance arises when a heterozygote shows characteristics found in each of the associated homozygotes. This occurs with human blood types, which are identified by testing for special cellular products called *antigens*. An antigen is detected by its ability to react with factors obtained from the serum portion of the blood. These factors, which are produced by the immune system, recognize antigens quite specifically. Thus, for example, one serum, called anti-M, recognizes only the M antigen on human blood cells; another serum, called anti-N, recognizes only the N antigen on these cells (Figure 4.2). When one of these sera detects its specific antigen in a blood-typing test, the cells clump together, causing a reaction called *agglutination*. Thus, by testing cells for agglutination with different sera, a medical technologist can identify which antigens are present and thereby determine the blood type.

The ability to produce the M and N antigens is determined by a gene with two alleles. One allele allows the M antigen to be produced; the other allows the N antigen to be produced. Homozygotes for the M allele produce only the M antigen, and homozygotes for the N allele produce only the N antigen; however, heterozygotes for these two alleles produce both kinds of antigens. Because the two alleles appear to contribute equally to the phenotype of the heterozygotes, they are said to be **codominant**. Codominance implies that there is an independence of allele function. Neither allele is dominant, or even partially dominant, over the other. It would therefore be inappropriate to distinguish the alleles by upper and lower case letters, as we have in all previous examples. Instead, codominant alleles are represented by superscripts on the symbol for the gene, which in this case is the letter L—a tribute to Karl Landsteiner, the discoverer of blood-typing. Thus the M allele is L^M and the N allele is L^N. Figure 4.2 shows the three possible genotypes formed by the L^M and L^N alleles, and their associated phenotypes.

Multiple Alleles

The Mendelian concept that genes exist in no more than two allelic states had to be modified when genes with three, four, or more alleles were discovered. A classic example of a gene with **multiple alleles** is the one that controls coat color in rabbits (Figure 4.3). The color-determining gene, denoted by the lower case letter c, has four alleles, three of which are distinguished by a superscript: c (*albino*), c^h (*himalayan*), c^{ch} (*chinchilla*), and c^+ (wild-type). In homozygous condition, each allele has a characteristic effect on the coat color: cc—white hairs over the entire body; c^hc^h—black hairs on the extremities, white hairs everywhere else; $c^{ch}c^{ch}$—white hair with black tips on the body; c^+c^+—colored hairs over the entire body.

Because most rabbits in wild populations are homozygous for the c^+ allele, this allele is called the **wild-type**. In genetics it is customary to represent wild-type alleles by a superscript plus sign after the letter for the gene. When the context is clear, the letter is sometimes omitted and only the plus sign is used; thus c^+ may be abbreviated simply as +.

The other alleles of the c gene are **mutants**—altered forms of the wild-type allele that must have arisen sometime during the evolution of the rabbit. The *himalayan* and *chinchilla* alleles are denoted by superscripts, but the *albino* allele is denoted simply by

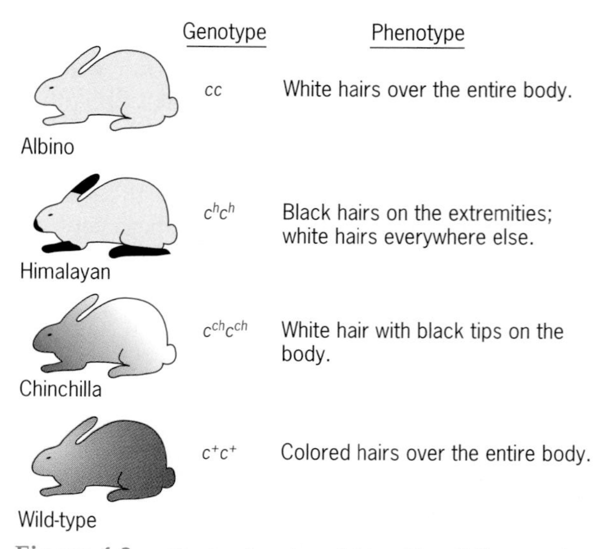

Figure 4.3. Coat colors in rabbits. The different phenotypes are caused by four different alleles of the c gene.

the letter *c* (for *colorless*, another word for the albino condition). This reflects another custom in genetics nomenclature: genes are often named for a mutant allele, usually the allele associated with the most abnormal phenotype. The convention of naming a gene for a mutant allele is usually consistent with the convention we saw in Chapter 3—that of naming genes for a recessive allele—because most mutant alleles are recessive. However, sometimes a mutant allele is dominant, in which case the gene is named after its associated phenotype. For example, there is a gene in mice that controls the length of the tail. The first mutant allele of this gene that was discovered caused a shortening of the tail in heterozygotes. This dominant mutant was therefore symbolized by *T*, for tail-less. All other alleles of this gene—and there are many—have been denoted by an upper or lower case letter, *T*, depending on whether they are dominant or recessive; different alleles are distinguished from each other by superscripts.

Another example of multiple alleles comes from the study of human blood types. The A, B, AB, and O blood types, like the M, N, and MN blood types, are identified by testing a blood sample with different sera. One serum detects the A antigen, another the B antigen. When only the A antigen is present on the cells, the blood is type A; when only the B antigen is present, the blood is type B. When both antigens are present, the blood is type AB, and when neither antigen is present, it is type O.

The gene responsible for producing the A and B antigens has three alleles: I^A, I^B, and I^O. The I^A allele specifies the production of the A antigen, and the I^B allele specifies the production of the B antigen. However, the I^O allele specifies nothing. Among the six possible genotypes, there are four distinguishable phenotypes—the A, B, AB, and O blood types (Table 4.1). In this system, the I^A and I^B alleles are codominant, since each is expressed equally in the $I^A I^B$ heterozygotes, and the I^O allele is recessive to both the I^A and I^B alleles. Because all three alleles are found at appreciable frequencies in human populations, the *I* gene is said to be **polymorphic**, from the Greek words for "having many forms." We consider the population and evolutionary significance of genetic polymorphisms in Chapters 27 and 28.

TABLE 4.1
Genotypes, Phenotypes, and Frequencies in the ABO Blood-Typing System

Genotype	*Blood Type*	*A Antigen Present*	*B Antigen Present*	*Frequency in U.S. White Population (%)*
I^AI^A or I^AI^O	A	+	−	41
I^BI^B or I^BI^O	B	−	+	11
I^AI^B	AB	+	+	4
I^OI^O	O	−	−	44

Allelic Series

The functional relationships among the members of a series of multiple alleles can be studied by making heterozygous combinations through crosses between homozygotes. For example, the four alleles of the *c* gene in rabbits can be combined with each other to make six different kinds of heterozygotes: cc^h, cc^{ch}, cc^+, c^hc^{ch}, c^hc^+, and $c^{ch}c^+$. These heterozygotes allow the dominance relations among the alleles to be studied (Figure 4.4). The wild-type allele is completely dominant over all the other alleles in the series; the *chinchilla* allele is partially dominant over the *himalayan* and *albino* alleles, and the *himalayan* allele is completely dominant over the *albino* allele. These dominance relations can be summarized as $c^+ > c^{ch} > c^h > c$. Notice that the dominance hierarchy parallels the effects that the alleles have on coat color. A plausible explanation is that the *c* gene controls a step in the formation of black pigment in the fur. The wild-type allele is fully functional in this process, producing colored hairs throughout the body. The *chinchilla* and *himalayan* alleles are only partially functional, producing some colored hairs, and the *albino* allele is not

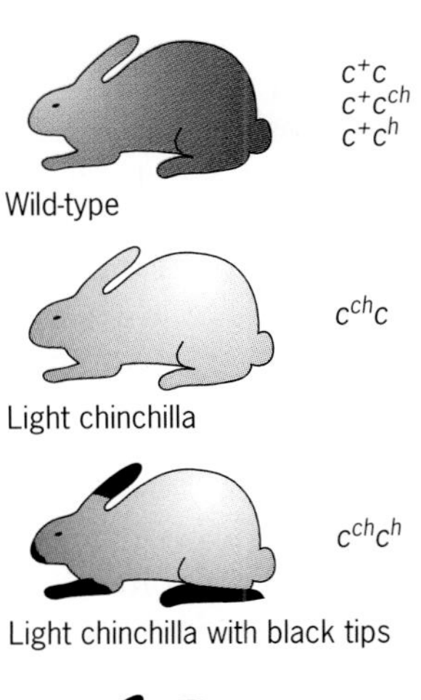

Figure 4.4. Phenotypes of different combinations of *c* alleles in rabbits. The alleles form a series, with the wild-type allele, c^+, dominant over all the other alleles and the null allele, *c* (albino), recessive to all the other alleles; one hypomorphic allele, c^{ch} (chinchilla), is dominant over the other, c^h (himalayan).

functional at all. Nonfunctional alleles are said to be **null** or **amorphic** (from the Greek words for "without form"); they are almost always completely recessive. Partially functional alleles are said to be **hypomorphic** (from the Greek words for "beneath form"); they are recessive to alleles that are more functional, including (usually) the wild-type allele. Later in this chapter we consider the biochemical basis for these differences.

New recessive mutation		Tester genotype		Hybrid phenotype	Conclusion
c^*c^*	X	$a\ a$	⇨	Wild-type	a and c^* not alleles
		$b\ b$	⇨	Wild-type	b and c^* not alleles
		$c\ c$	⇨	Mutant	c and c^* alleles
		$d\ d$	⇨	Wild-type	d and c^* not alleles

Figure 4.5. A general scheme to test recessive mutations for allelism. Two mutations are alleles if a hybrid that contains both of them has the mutant phenotype.

Testing Gene Mutations for Allelism

A mutant allele is created when an existing allele is changed to a new genetic state—a process called **mutation.** This event always involves a change in the physical composition of the gene (see Chapter 13) and sometimes produces an allele that has a detectable phenotypic effect. If, for example, the c^+ allele mutated to a null allele, a rabbit homozygous for this mutation would have the albino phenotype. However, it is not always possible to assign a new mutation to a gene on the basis of its phenotypic effect. For example, several genes determine coat color in rabbits, and a mutation in any one of them could reduce, alter, or abolish pigmentation in the hairs. Thus, if a new coat color appears in a population of rabbits, it is not immediately clear which gene has been mutated.

A simple test can be used to determine the allelic identity of a new mutation. As long as the new mutation is recessive, it can be identified by using crosses to combine it with recessive mutations of known genes (Figure 4.5). The hybrid progeny will show a mutant phenotype only if the new mutation and the tester mutation are alleles of the same gene. If they are alleles of different genes, the progeny will be wild-type. This test is based on the principle that allelic mutations impair the same genetic function. If two such mutations are combined, the organism should show a mutant phenotype, even if the mutations had an independent origin.

It is important to remember that this test applies only to recessive mutations. Dominant mutations cannot be tested in this way because they exert their effects regardless of what other mutations are present.

As an example, let us consider the analysis of two recessive mutations affecting eye color in the fruit fly, *Drosophila melanogaster* (Figure 4.6). This tiny organism has been investigated by geneticists for nearly a century, and many different mutations have been identified. Two independently isolated mutations, called *cinnabar* and *scarlet*, are phenotypically indistinguishable, each causing the eyes to be bright red. In wild-type flies, the eyes are dark red. We wish to know whether the *cinnabar* and *scarlet* mutations are alleles of a single color-determining gene, or if they are mutations in two different genes. To find the answer, we must cross the homozygous mutant strains with each other to produce hybrid progeny. If the hybrids have bright red eyes, we will conclude that *cinnabar* and *scarlet* are alleles of the same gene. If they have dark red eyes, we will conclude that they are mutations in different genes.

The hybrid progeny turn out to have dark red eyes; that is, they are wild-type rather than mutant.

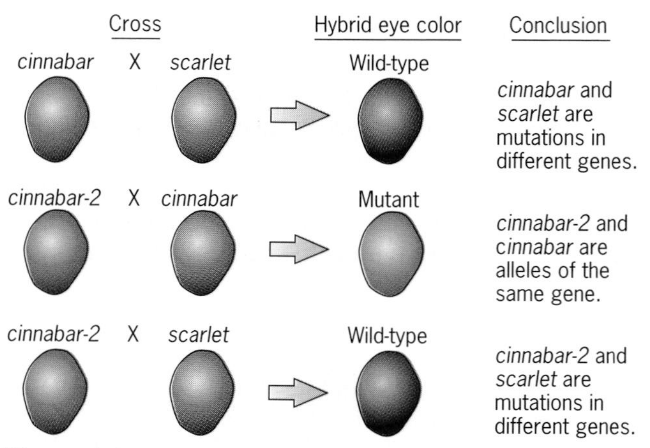

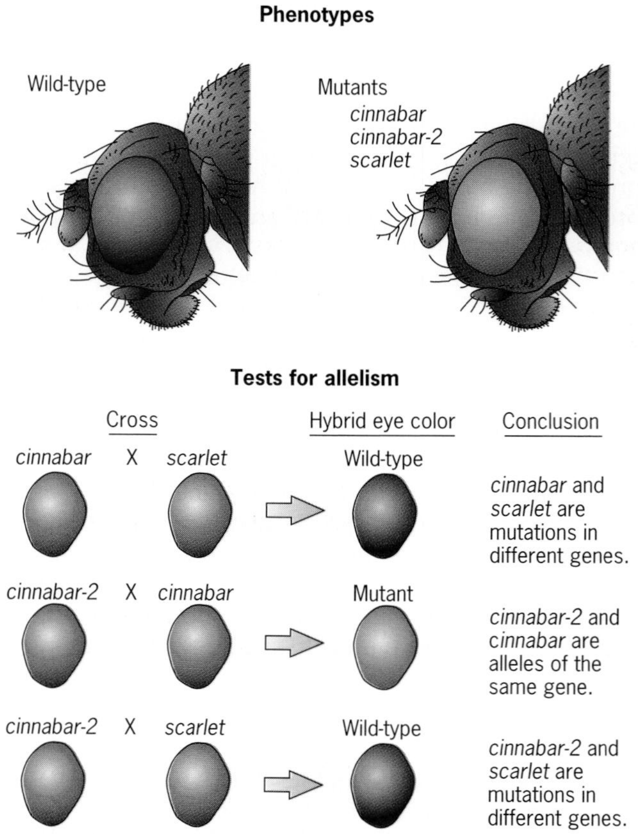

Figure 4.6. A test for allelism involving recessive eye color mutations in *Drosophila.* Three phenotypically identical mutations, *cinnabar*, *scarlet*, and *cinnabar-2*, are tested for allelism by making pairwise crosses between flies homozygous for different mutations. The phenotypes of the hybrids show that the *cinnabar* and *cinnabar-2* mutations are alleles of a single gene and that the *scarlet* mutation is not an allele of this gene.

Thus *cinnabar* and *scarlet* are not alleles of the same gene but, rather, mutations in two different genes, each apparently involved in the control of eye pigmentation. When we test a third mutation, called *cinnabar-2*, for allelism with the *cinnabar* and *scarlet* mutations, we find that the hybrid combination of *cinnabar-2* and *cinnabar* has the mutant phenotype (bright red eyes) and that the hybrid combination of *cinnabar-2* and *scarlet* has the wild phenotype (dark red eyes). These results tell us that the mutations *cinnabar* and *cinnabar-2* are alleles of one color-determining gene and that the *scarlet* mutation is an allele of another such gene. Research has revealed that more than a dozen genes are involved in the control of eye color in *Drosophila*.

The test to determine whether mutations are alleles of a particular gene is based on the phenotypic effect of combining the mutations in the same individual. If the hybrid combination is mutant, we conclude that the mutations are alleles; if it is wild-type, we conclude that they are not alleles. Chapter 14 discusses how this test—called the *complementation test* in modern terminology—enables geneticists to define the functions of individual genes.

Variation Among the Effects of Mutations

Genes are identified by mutations that alter the phenotype in some conspicuous way. For instance, a mutation may change the color or shape of the eyes, alter a behavior, or cause sterility or even death. The tremendous variation among the effects of individual mutations suggests that each organism carries many different kinds of genes, and that each of these can mutate in different ways. In nature, mutations provide the raw material for evolution (see Chapter 27).

Mutations that alter some aspect of morphology, such as seed texture or color, are called **visible mutations**. Most visible mutations are recessive, but a few are dominant. Geneticists have learned much about genes by analyzing the properties of these mutations. We will encounter many examples of this analysis throughout this textbook.

Mutations that limit reproduction are called **sterile mutations**. Some sterile mutations affect both sexes, but most affect either males *or* females. As with visible mutations, steriles can be either dominant or recessive. Some steriles completely prevent reproduction, whereas others only impair it slightly.

Mutations that interfere with necessary vital functions are called **lethal mutations**. Their phenotypic effect is death. We know that many—perhaps a majority of—genes are capable of mutating to the lethal state. Thus a large fraction of all genes are absolutely essential for life.

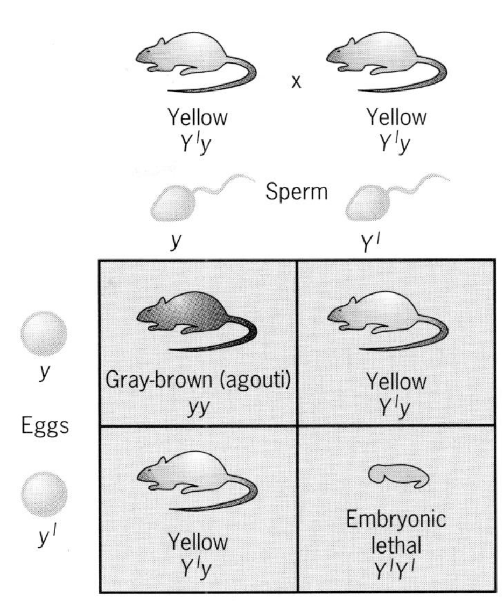

Figure 4.7. Y^l, the yellow-lethal mutation in mice: a dominant visible that is also a recessive lethal. A cross between carriers of this mutation produces yellow heterozygotes and gray-brown (agouti) homozygotes in a ratio of 2:1. The yellow homozygotes die as embryos.

Dominant lethals are lost one generation after they occur, but recessive lethals may linger a long time in a population because they can be hidden in heterozygous condition by a wild-type allele. Recessive lethal mutations are detected by observing unusual segregation ratios in the progeny of heterozygous carriers. An example is the *yellow-lethal* mutation, Y^l, in the mouse (Figure 4.7). This mutation is a dominant visible, causing the fur to be yellow instead of gray-brown (the wild-type color, also known as *agouti*). In addition, however, the Y^l mutation is a recessive lethal, killing Y^lY^l homozygotes early in their development. A cross between Y^ly heterozygotes produces two kinds of viable progeny, yellow (Y^ly) and gray-brown (yy), in a ratio of 2:1. Given a sufficiently large number of progeny, this ratio can easily be distinguished from the 3:1 ratio that would be obtained if Y^l were simply a dominant visible mutation.

Genes Function to Produce Polypeptides

The extensive variation revealed by mutations indicates that organisms contain many different genes, and that these genes can exist in multiple allelic states. However, it does not tell us how genes actually affect the phenotype. What is it about a gene that enables it to control a trait such as eye color, seed texture, or plant height?

The early geneticists had no answer to this question. However, today it is clear that most genes make

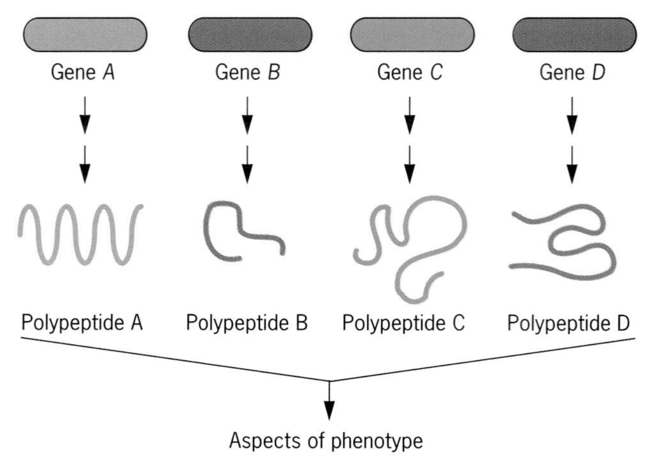

Figure 4.8. Relationship between genes and polypeptides. Each gene specifies a different polypeptide. These polypeptides then function to determine the organism's phenotype.

a product that subsequently affects the phenotype. This idea, which was discussed in Bateson's book and which was supported by the research of many scientists, including, most notably, the British physician Sir Archibald Garrod, was forcefully brought out in the middle of the twentieth century when George Beadle and Edward Tatum discovered that the products of genes are *polypeptides* (Figure 4.8).

Polypeptides are macromolecules built of a linear chain of *amino acids*. Every organism makes thousands of different polypeptides, each characterized by a specific amino acid sequence. These polypeptides are the fundamental constituents of *proteins*. Some proteins, called *enzymes*, function as catalysts in biochemical reactions; others form the structural components of cells; and still others are responsible for transporting substances within and between cells. Beadle and Tatum proposed that each gene is responsible for the synthesis of a particular polypeptide. When a gene is mutated, its polypeptide product either is not made or is altered in such a way that its role in the organism is changed. Mutations that eliminate or alter a polypeptide are often associated with a phenotypic effect. Whether this effect is dominant or recessive depends on the nature of the mutation. We consider the details of how genes produce polypeptides in Chapter 12. In that chapter, we also discuss the molecular basis of mutation.

Why Are Some Mutations Dominant and Others Recessive?

The discovery that genes produce polypeptides provides insight into the nature of dominant and recessive mutations. Dominant mutations have phenotypic effects in heterozygotes as well as in homozygotes, whereas recessive mutations have these effects only in homozygotes. What accounts for this striking difference in expression?

Recessive mutations occur when a gene loses its function, that is, when it no longer specifies a polypeptide or when it specifies a defective polypeptide (Figure 4.9). Recessive mutations are therefore typically **loss-of-function** alleles. In contrast, dominant mutations occur when a gene acquires a new function—that is, when it specifies a polypeptide that behaves in a new and different way. Dominant mutations are therefore typically **gain-of-function** alleles. According to this view, a recessive phenotype results from a partial or total loss of polypeptide function, whereas a dominant phenotype results from the gain of a new polypeptide function—a function that is expressed even in the presence of the polypeptide made by the wild-type allele. A recessive mutation that causes a total loss of polypeptide function is a null or amorphic allele, whereas one that causes a partial loss of function is a hypomorphic allele. A dominant mutation that creates a new polypeptide function is called a **neomorphic** allele, from the Greek words for "new form."

Why is a loss-of-function mutation recessive to a wild-type allele? In a heterozygote, such a mutation either will not produce a polypeptide or it will produce a defective polypeptide that has no effect on the phenotype (Figure 4.9). However, the wild-type allele will produce a fully functional polypeptide that will carry out its normal role in the organism. The phenotype of a mutant/wild heterozygote will therefore be the same, or essentially the same, as that of a wild-type homozygote. The only possible difference might be that the homozygote, with its two wild-type alleles, will have a more intense phenotype (for example, a deeper eye color) than the heterozygote, with its one such allele. In most instances, however, a single wild-type allele is sufficient to produce the wild phenotype.

We see this, for example, with the *cinnabar* gene in *Drosophila*. This gene specifies a polypeptide that functions as an enzyme in the synthesis of the brown pigment that is deposited in *Drosophila* eyes. Flies that are homozygous for a loss-of-function mutation in the *cinnabar* gene cannot produce this enzyme; consequently, they do not synthesize any brown pigment in their eyes. The phenotype of homozygous *cinnabar* mutants is bright red—the color of the mineral cinnabar, for which the gene is named. However, flies that are heterozygous for the *cinnabar* mutation and its wild-type allele have dark red eyes; that is, they are phenotypically identical to wild type. In these flies, the loss-of-function allele is recessive to the wild-type allele because the latter produces enough enzyme to synthesize brown eye pigment.

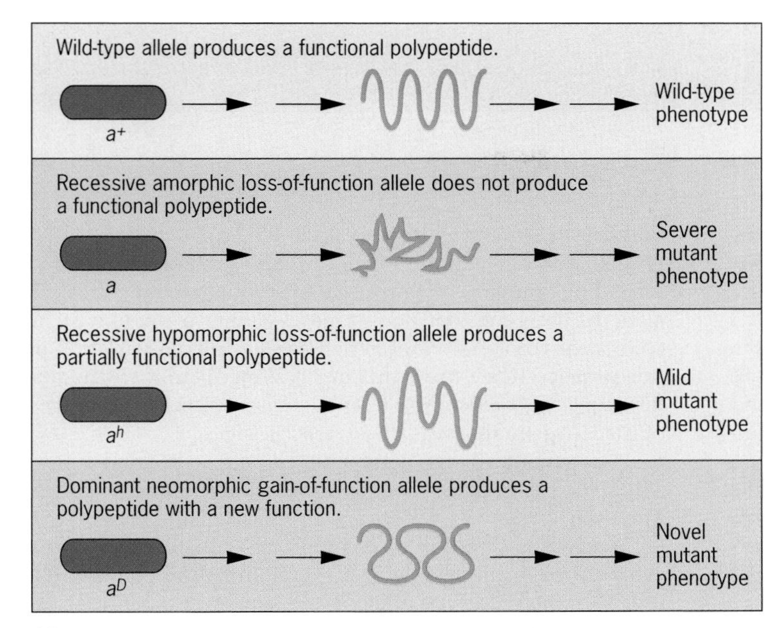

(a)

Genotype	Polypeptides present	Phenotype	Nature of mutant allele
$a^+ a$		Wild-type	Recessive
$a^+ a^h$		Wild-type	Recessive
$a^+ a^D$		Mutant	Dominant

(b)

Figure 4.9. Differences between recessive loss-of-function mutations and dominant gain-of-function mutations. (*a*) Polypeptide products of recessive and dominant mutations. (*b*) Phenotypes of heterozygotes carrying a wild-type allele and different types of mutant alleles.

Why is a gain-of-function mutation dominant to a wild-type allele? Such a mutation will produce a polypeptide with a novel function. In a mutant/wild heterozygote, this polypeptide can replace or interfere with the function of the normal polypeptide and thereby cause a mutant phenotype.

As an example, let us consider dominant mutations in the *T* gene of the mouse. We have already seen that in heterozygous condition, these mutations cause a shortening of the tail. In homozygous condition, they are lethal. This lethality is due to the complete failure of the trunk of the animal to develop. The *T* gene is therefore critical for the formation of a substantial portion of the mouse's body. The first dominant *T* mutation was discovered in 1927 by N. Dobrovolskaia-Zavadskaia, who worked at the Pasteur Institute in Paris. Many other dominant *T* mutations have since been identified, and recently several have been analyzed at the molecular level. From this analysis, mouse geneticists have learned that the wild-type *T* gene specifies a polypeptide of 436 amino acids.

Biochemical studies have indicated that this polypeptide is able to bind to DNA. It may therefore act as a master regulator of other genes whose products are needed for normal mouse development. Dominant mutations of the *T* gene result in the production of polypeptides that are slightly shorter than the wild-type polypeptide and also somewhat different in structure. The available evidence suggests that these mutant polypeptides counteract the function of the wild-type polypeptide, perhaps by interfering with its ability to bind DNA. This interference then alters the expression of the genes that are regulated by the T polypeptide, and thereby upsets the course of mouse embryological development. If only one dominant *T* mutation is present in the genotype, the animal is born with a short tail, or perhaps with no tail at all. If two dominant *T* mutations are present in the genotype, the animal dies as an embryo in the uterus.

In later chapters, we encounter other examples of dominant mutations. Some, like the *T* mutations in the mouse, are genuine gain-of-function mutations; however, others are not. We will see that the view presented here—that recessive alleles are typically loss-

HISTORICAL SIDELIGHT

Genetic Symbols: Evolution in a Dynamic Science

William Bateson improved on Mendel's genetic symbols by choosing gene symbols mnemonically. Thus, for example, he symbolized the dominant allele for tall pea plants as *T* and the recessive allele for short plants as *t*. Later, when it became customary to choose allele symbols based on the mutant trait, these alleles were represented by *D* (for tall) and *d* (for dwarf). This convention provided a simple and consistent notation in which the dominant and recessive alleles of a particular gene were represented by a *single* letter, and that letter was mnemonic for the trait influenced by the gene. Bateson also coined the words *genetics, alleleomorph* (which was later shortened to *allele*), *homozygote,* and *heterozygote,* and he introduced the practice of denoting the generations in a breeding scheme as P, F_1, F_2, and so forth.

The gene-naming system that Bateson developed worked well until the number of genes that had been identified exceeded the capacity of the English alphabet; thereupon it became necessary to use two or more letters to symbolize a gene. For example, there is a mutant allele in *Drosophila* that causes the eyes to be carmine instead of red. When this allele was discovered, it was given the symbol *cm* because the single letter *c* had already been used to represent a mutant allele that causes the wings to be curved instead of straight. Today, with thousands of genes identified, it is often necessary to use three or four letters, or combinations of letters and numbers, to symbolize genes. For example, mutations in the *cmp* gene of *Drosophila* cause the wings to be crumpled, and mutations in the *Sh1* and *Sh2* genes of maize cause the kernels to be shrunken.

The discovery of multiple alleles made genetic notation even more complicated; upper and lower case letters were no longer adequate to distinguish among alleles, so geneticists began to combine a basic gene symbol with an identification symbol. *Drosophila* geneticists were the first to apply this procedure. They made the identification symbol a superscript on the basic gene symbol. Usually, both the gene symbol and the superscript had some mnemonic significance. Thus, for example, cn^2 was used to symbolize the second cinnabar eye color allele that was discovered in *Drosophila*; and ey^D was used to symbolize a dominant allele that causes *Drosophila* to be eyeless. This convention was extended to other experimental animals, such as rabbits and mice. For example, c^{ch} was used to symbolize the chinchilla allele of the gene that determines whether a rabbit's fur is colored or colorless. Plant geneticists adopted a variation of this practice. They use hyphenated symbols to identify mutant alleles; for example, *sh2-6801* represents a mutant allele of the *Sh2* gene that was discovered in 1968.

As genetic nomenclature developed, it became necessary to use a special symbol to represent the wild-type allele. The early *Drosophila* geneticists proposed using a plus sign (+), sometimes written as a superscript on the basic gene symbol (for example, c^+). This simple notation conveys the idea that the wild-type allele is the standard, or normal, allele of the gene, and is widely used today. However, other gene-naming practices persist. Plant geneticists tend to use the gene symbol itself to represent the wild-type allele, but to make it stand out, they capitalize the first letter; thus *Sh2* is the wild-type allele of the second shrunken gene discovered in maize, whereas *sh2* is a mutant allele.

Genetic nomenclature has been further complicated by the discovery of genes through the polypeptides they specify. These discoveries have introduced gene symbols that are mnemonic for polypeptide gene products. For example, the human gene that specifies the polypeptide hypoxanthine-guanine phosphoribosyl transferase is symbolized by *HPRT*, and the plant gene that specifies the polypeptide alcohol dehydrogenase is symbolized by *Adh*. Whether upper case letters are used throughout the gene symbol or only for the first letter depends on the organism.

Today there are many specialized systems for symbolizing genes and alleles. Researchers who work with different organisms—*Drosophila,* mice, plants, and humans—each speak a slightly different language. Later, we will see that still other genetic dialects have been created to describe the genes of viruses, bacteria, and fungi. These different systems of nomenclature indicate that the symbols in genetics have evolved in response to new discoveries—visible evidence of growth in a dynamic, young science.

of-function mutations and that dominant alleles are typically gain-of-function mutations—is an oversimplification. The phenotypes associated with some dominant alleles, for example, actually result from a loss of gene function. A discussion of these atypical situations must, however, be postponed until after we have considered the molecular biology of genes. For a discussion of the conventions used in naming genes and mutations, see Historical Sidelight: Genetic Symbols: Evolution in a Dynamic Science.

Key Points: **Genes often have multiple alleles. Different alleles are created by the mutation of a wild-type allele. Mutant alleles may be dominant, recessive, incompletely dominant, or codominant. The allelism of different recessive mutations can be tested by com-**

bining them in the same individual; if the individual has a mutant phenotype, then the mutations are alleles; if it has a wild phenotype, then they are not alleles. The function of most genes is to produce a polypeptide. Recessive mutations can cause a partial or complete loss of polypeptide activity. Dominant mutations can endow a polypeptide with a new activity that may replace or interfere with the activity of the wild-type polypeptide.

GENE ACTION: FROM GENOTYPE TO PHENOTYPE

At the beginning of the twentieth century, geneticists had imprecise ideas about how genes evoke particular phenotypes. They knew nothing about the chemistry of gene structure or function, nor had they developed the techniques to study it. Everything that they proposed about the nature of gene action was inferred from the analysis of phenotypes. These analyses showed that genes do not act in isolation. Rather, they act in the context of an environment, and sometimes they act in concert with other genes. These analyses also showed that a particular gene can influence many different traits.

Interactions with the Environment

A gene must function in the context of both a biological and a physical environment. The factors in the physical environment are easier to study, for particular genotypes can be reared in the laboratory under controlled conditions, allowing an assessment of the effects of temperature, light, nutrition, and humidity.

Figure 4.10. Man with premature balding beginning on the crown of his head.

As an example, let us consider the *Drosophila* mutation known as *shibire*. At the normal culturing temperature, 25° C, *shibire* flies are viable and fertile, but extremely sensitive to a sudden shock. When a *shibire* culture is shaken, the flies—temporarily paralyzed—fall to the bottom of the culture. Indeed, *shibire* is the word for "paralysis" in Japanese. However, if a culture of *shibire* flies is placed at a slightly higher temperature, 29° C, all the flies fall to the bottom and die, even without a shock. Thus the phenotype of the *shibire* mutation is temperature-sensitive. At 25° C, the mutation is viable, but at 29° C it is lethal. A plausible explanation is that at 25° C the mutant gene makes a partially functional protein, but at 29° C, this protein is totally nonfunctional.

Environmental Effects on the Expression of Human Genes

Human genetic research provides an example of how the physical environment can influence a phenotype. **Phenylketonuria** (PKU) is a recessive disorder of amino acid metabolism. Infants homozygous for the mutant allele accumulate semitoxic substances in their brains; though not lethal, these substances can impair mental ability by affecting the brain's development. The harmful aspects of PKU are traceable to a particular amino acid, phenylalanine, which is ingested in the diet. Infants who are fed normal diets ingest enough phenylalanine to bring out the worst manifestations of the disease. However, infants who are fed low-phenylalanine diets usually mature without serious mental impairment. Because PKU can be diagnosed in newborn babies, the clinical impact of this disease can be reduced if infants that are PKU homozygotes are placed on a low-phenylalanine diet shortly after birth. This example illustrates how an environmental factor—diet—can be manipulated to modify a phenotype that would otherwise become a personal tragedy.

The biological environment can also influence the phenotypic expression of genes. **Pattern baldness** in humans is a well-known example (Figure 4.10). Here the relevant biological factor is gender. Premature pattern baldness is due to an allele that is expressed differently in the two sexes. In males, both homozygotes and heterozygotes for this allele develop bald patches, whereas in females, only the homozygotes show a tendency to become bald, and this is usually limited to general thinning of the hair. The expression of this allele is probably triggered by the male hormone, **testosterone**. Females produce much less of this hormone and are therefore seldom at risk to develop bald patches. The sex-influenced nature of pattern baldness shows that biological factors can control the expression of genes.

Penetrance and Expressivity

When individuals do not show a trait even though they have the appropriate genotype, the trait is said to exhibit **incomplete penetrance**. An example of incomplete penetrance in humans is **polydactyly**—the presence of extra fingers and toes (Figure 4.11*a*). This condition is due to a dominant mutation, *P*, that is manifested in some of its carriers. In the pedigree in Figure 4.11*b*, the individual labeled III-2 must be a carrier even though he does not have extra fingers or toes. The reason is that both his mother and three of his children were polydactylous—an indication of the transmission of the mutation through III-2. Incomplete penetrance can be a serious problem in pedigree analysis because it can lead to the incorrect assignment of genotypes.

The term **expressivity** is used if a trait is not manifested uniformly among the individuals that show it. The dominant *Lobe* eye mutation (Figure 4.12) in *Drosophila* is an example. The phenotype associated with this mutation is extremely variable. Some heterozygous flies have tiny compound eyes, whereas others have large, lobulated eyes; between these extremes, there is a full range of phenotypes. The *Lobe* mutation is therefore said to have **variable expressivity**.

Incomplete penetrance and variable expressivity indicate that the pathway between a genotype and its phenotypes is subject to considerable modulation. (See Human Genetics Sidelight: The Hapsburg Jaw.) Geneticists know that some of this modulation is due to environmental factors, but some is also due to factors in the genetic background. Clearcut evidence for such factors comes from breeding experiments that show that two or more genes can control a particular trait.

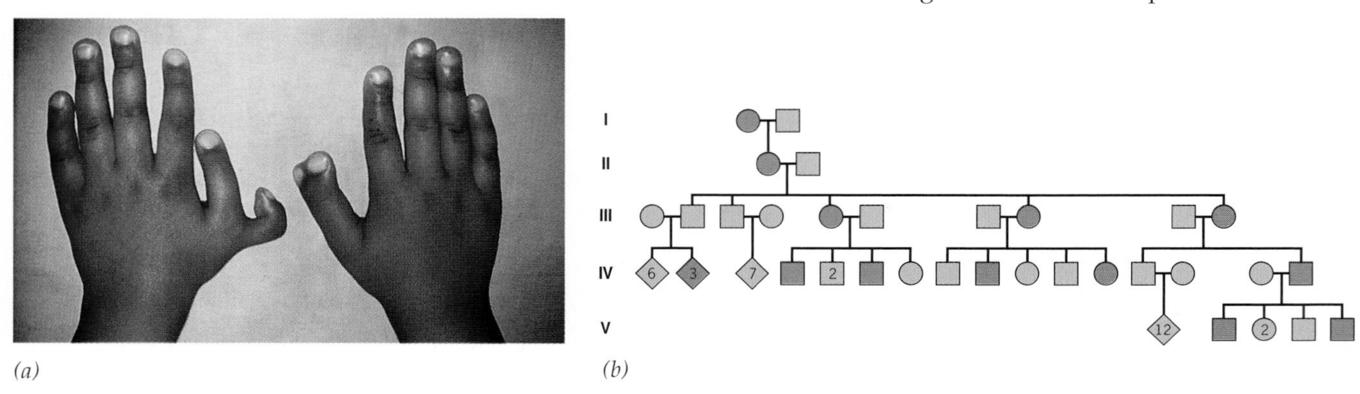

(*a*) (*b*)

Figure 4.11. Polydactyly in human beings. (*a*) Phenotype showing extra fingers. (*b*) Pedigree showing the inheritance of this dominant trait.

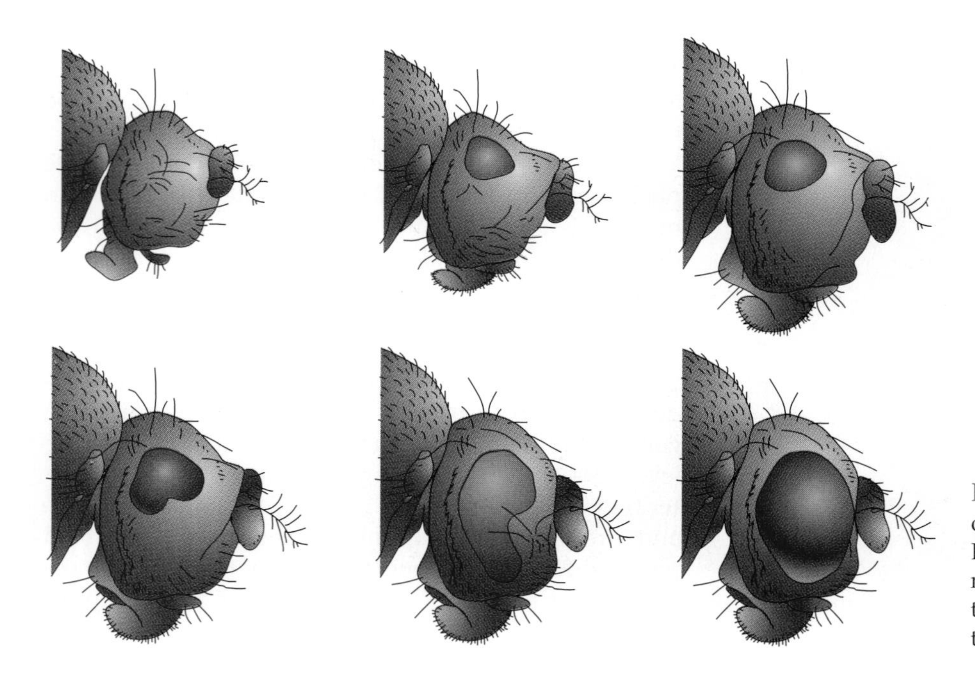

Figure 4.12. Variable expressivity of the *Lobe* mutation in *Drosophila*. Each fly is heterozygous for this dominant mutation; however, the phenotypes vary from complete absence of the eye to a nearly wild-type eye.

HUMAN GENETICS SIDELIGHT

The Hapsburg Jaw: A Dominant Trait with Incomplete Penetrance and Variable Expressivity in the European Nobility

From the Middle Ages to the beginning of the twentieth century, the House of Hapsburg ruled much of Europe. The members of this family included kings and queens, emperors and empresses. Although Hapsburg power was concentrated in Central Europe—especially in Austria—at various times it encompassed Spain, the Low Countries, and parts of Italy. The extension of Hapsburg power in Europe was due, in part, to the family's uncanny ability to arrange politically beneficial marriages. This skill was celebrated in a line of Latin verse: *Bella gerant allii, tu, felix Austria, nube!*—"Let others wage wars; you, happy Austria, marry!" Thus, through a foreign policy that was based on matrimonial vows, the Hapsburgs spread their genes among the European nobility. Marie Antoinette, the wife of the ill-fated French king Louis XVI, was, for example, a Hapsburg.

As the Hapsburgs spread their genes, so too did they spread the phenotypes associated with those genes. The most famous of these phenotypes is the Hapsburg jaw, a protrusion of the mandible often associated with difficulty in chewing. This condition is more properly known as **mandibular prognathism**. Its presence is recorded in the portraits of the Hapsburgs—in paintings, on coins, in sculpture, and more recently, in photographs.

The first documented case of prognathism among the Hapsburgs appears in portraits of the Emperor Maximilian I of Austria (Figure 1*a*), who lived from 1459 to 1519. Maximilian's son, Philip I, married Joanna ("The Mad"), the daughter of King Ferdinand and Queen Isabella of Spain. In this way, Philip I became king of Spain and established a Spanish line of Hapsburgs. Although it is not known whether Philip I manifested the Hapsburg jaw, his son, Charles V (1500–1558), clearly did (Figure 1*b*). Charles's severe prognathism caused his mouth to hang open, and he had difficulty chewing food. Charles grew up in Flanders, another Hapsburg dominion, and when he first came to Spain, it is said that a peasant shouted to him: "Your Majesty, shut your mouth; the flies in this country are very insolent."[1] In spite of Charles' condition, he proved to be an excellent leader. During his reign, Spain became the most powerful country in Europe.

When poor health forced Charles V to abdicate in 1556, his son, Philip II, became king. Portraits show that Philip II also had the Hapsburg jaw. It is not known whether Philip II's son, Philip III, manifested this trait. However, his son, Philip IV, clearly did. Diego Rodriguez de Silva y Velazquez (1599–1660)—regarded by some as Spain's greatest painter—produced a magnificent protrait of Philip IV when Philip was 18 years old (Figure 1*c*). The slightly empty expression and the sense of shyness in this portrait suggest that Philip IV was less than comfortable in the political realm. In fact, he turned the government of Spain over to professional politicians and focused his attention on the arts; Philip IV was a poet and a dramatist, and he esteemed Velazquez as one of his closest friends. Velazquez's portrait shows that like several of his forebears, Philip IV also had the Hapsburg jaw.

The Spanish line of Hapsburgs ended with the death of Philip IV's son, Carlos II, in 1700. Before his death, Carlos II designated Philip of Anjou as his successor. This Philip was from the House of Bourbon, the French royal family. However, because Bourbons had married Hapsburgs in previous generations, some of the Spanish Bourbons inherited the Hapsburg jaw. One example is Alfonso XIII (Figure 1*d*), who was king of Spain in the early part of the twentieth century.

Five hundred years of Hapsburg portraiture indicate that mandibular prognathism is caused by a dominant allele with incomplete penetrance and variable expressivity. This condition has been found to varying degrees in the portraits of both men and women. Although the mode of inheritance is typical for that of a dominant allele, it sometimes skips a generation, presumably because the dominant allele is not fully penetrant. The physiological basis of this allele's effects are currently unknown.

[1] Langdon-Davies, J., 1963, *Carlos, the King Who Would Not Die.* Prentice-Hall, Englewood Cliffs, NJ.

(*a*)

(*b*)

(*c*)

(*d*)

Figure 1 (*a*) Emperor Maximilian I of Austria, (*b*) King Charles V of Spain, (*c*) King Philip IV of Spain, and (*d*) King Alfonso XIII of Spain.

Gene Interactions

Some of the first evidence that a trait can be influenced by more than one gene was obtained by Bateson and Punnett from breeding experiments with chickens. Their work was carried out shortly after the rediscovery of Mendel's paper. Domestic breeds of chickens have different comb shapes (Figure 4.13): Wyandottes have "rose" combs, Brahmas have "pea" combs, and Leghorns have "single" combs. Crosses between Wyandottes and Brahmas produce chickens that have yet another type of comb called "walnut." Bateson and Punnett discovered that comb type is controlled by two independently assorting genes, *R* and *P*, each with two alleles (Figure 4.14). Wyandottes (with rose combs) have the genotype *RR pp*, and Brahmas (with pea combs) have the genotype *rr PP*. The F_1 hybrids between these two varieties are therefore *Rr Pp* and phenotypically they have walnut combs. If these hybrids are intercrossed with each other, all four types of combs appear in the progeny: 9/16 walnut (*R- P-*), 3/16 rose (*R- pp*), 3/16 pea (*rr P-*), and 1/16 single (*rr pp*). The Leghorn breed, which has the single-comb type, must therefore be homozygous for both of the recessive alleles.

The work of Bateson and Punnett clearly demonstrated that two independently assorting genes can control a trait. Different combinations of alleles from the two genes resulted in different phenotypes, presumably because of interactions between their products at the biochemical or cellular level.

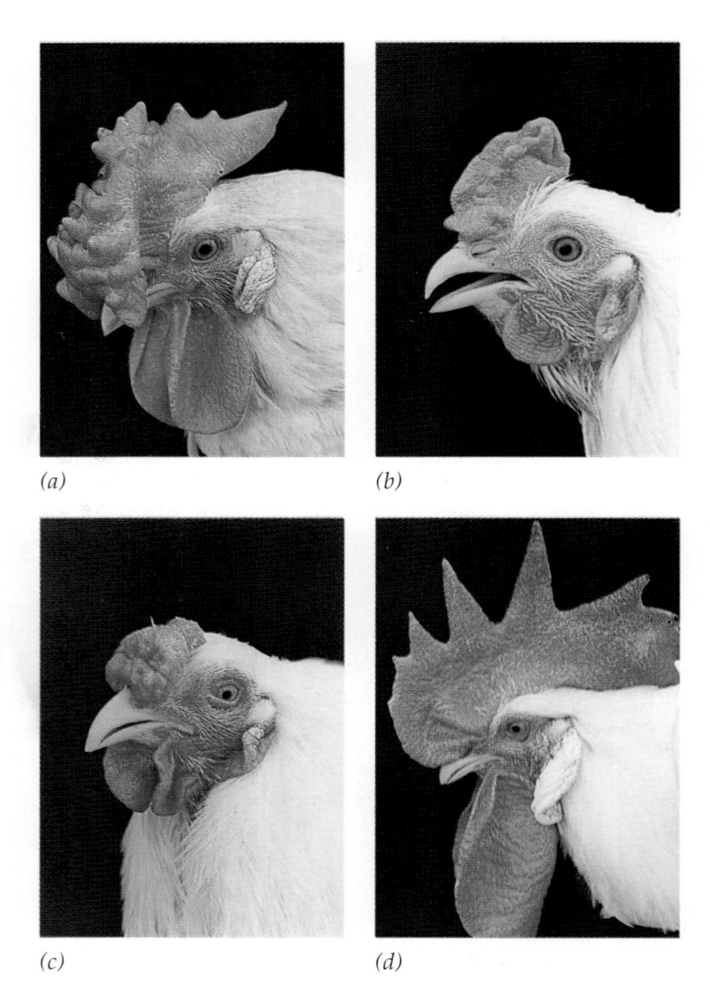

Figure 4.13. Comb shapes in chickens of different breeds. (*a*) rose, Wyandottes; (*b*) pea, Brahmas; (*c*) walnut, hybrid from cross between chickens with rose and pea combs; (*d*) single, Leghorns.

Epistasis

When two or more genes influence a trait, an allele of one of them may have an overriding effect on the phenotype. When an allele has such an overriding effect, it is said to be epistatic to the other genes that are involved; the term **epistasis** comes from a Greek word meaning to "stand above." For example, we know that eye pigmentation in *Drosophila* involves a large number of genes. If a fly is homozygous for a null allele in any one of these genes, the pigment-synthesizing pathway can be blocked, and an abnormal eye color will be produced. This allele essentially nullifies the

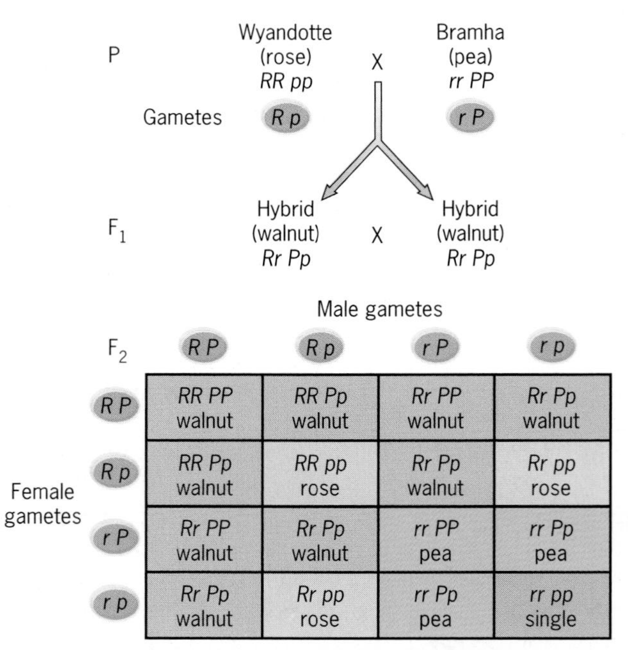

Figure 4.14. Bateson's and Punnett's experiment on comb shape in chickens. The intercross in the F_1 produces four phenotypes in a 9:3:3:1 ratio.

work of all the other genes, masking their contributions to the phenotype.

A mutant allele is said to be epistatic when it conceals the presence of another mutant allele in the genotype. We have already seen that a recessive mutation in the *cinnabar* gene of *Drosophila* causes the eyes of the fly to be bright red. A recessive mutation in a different gene causes the eyes to be bleach white. When both of these mutations are made homozygous in the same fly, the eye color is white. Thus the *white* mutation is epistatic to the *cinnabar* mutation.

What physiological mechanism makes the *white* mutation epistatic to the *cinnabar* mutation? For many years, the answer to this question was not known. However, recent molecular analysis has shown that the polypeptide product of the *white* gene transports pigment into the *Drosophila* eye. When this gene is mutated, the transporter polypeptide is not made and the eyes remain colorless, even though red pigment is synthesized in other tissues of the fly. A fly that is homozygous for the *cinnabar* and *white* mutations therefore has white eyes.

The analysis of epistatic relationships such as the one between *cinnabar* and *white* can suggest ways in which genes control a phenotype. A classic example of this analysis is again from the work of Bateson and Punnett, who studied the genetic control of flower color in the sweet pea, *Lathyrus odoratus* (Figure 4.15*a*). The flowers in this plant are either purple or white—purple if they contain anthocyanin pigment, and white if they do not. Bateson and Punnett crossed two different varieties with white flowers to obtain F_1 hybrids, which all had purple flowers. When these hybrids were intercrossed, Bateson and Punnett obtained a ratio of 9 purple: 7 white plants in the F_2. They explained the results by proposing that two independently assorting genes, *C* and *P*, are involved in anthocyanin synthesis, and that each gene has a recessive allele that abolishes pigment production (Figure 4.15*b*). Given this hypothesis, the parental varieties must have had complementary genotypes: *cc PP* and *CC pp*. When the two varieties were crossed, they produced *Cc Pp* double heterozygotes that had purple flowers. In this system, a dominant allele from each gene is necessary for the synthesis of anthocyanin pigment. In the F_2, 9/16 of the plants are *C- P-*, and have purple flowers; the remaining 7/16 are homozygous for at least one of the recessive alleles and have white flowers. Notice that the double recessive homozygotes, *cc pp*, are not phenotypically different from either of the single recessive homozygotes. Bateson and Punnett's work established that each of the recessive alleles is epistatic over the dominant allele of the other gene. A plausible explanation is that each dominant allele produces an enzyme that controls a step in the synthesis of anthocyanin from a biochemical precursor. If a dominant allele is not present, its step in the biosynthetic pathway is blocked and anthocyanin is not produced:

(*a*)

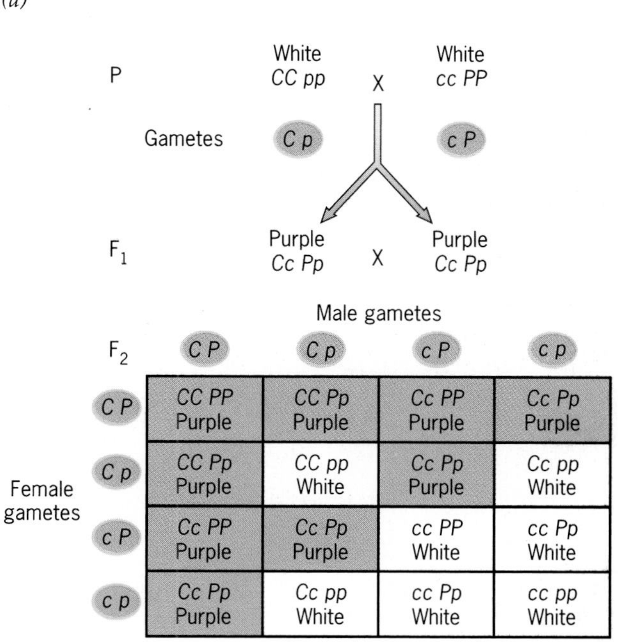

(*b*)

Figure 4.15. (*a*) Purple and white flowers of the sweet pea. (*b*) Bateson's and Punnett's experiment on flower color in sweet peas.

Gene		*C*		*P*	
	Precursor	→	Intermediate	→	Anthocyanin
Genotype					
C-; P-	+		+		+
cc; P-	+		−		−
C-; pp	+		+		−

Another classic study of epistasis was performed by George Shull using a weedy plant called the shepherd's purse, *Bursa bursa-pastoris* (Figure 4.16*a*). The seed capsules of this plant are either triangular or ovoid in shape. Ovoid capsules are produced only if a plant is homozygous for the recessive alleles of two genes—that is, if it has the genotype *aa bb*. If the dominant allele of either gene is present, the plant produces triangular capsules. The evidence for this conclusion comes from crosses between doubly heterozygous plants (Figure 4.16*b*). Such crosses produce progeny in a ratio of 15 triangular: 1 ovoid, indicating that the dominant allele of one gene is epistatic over the recessive allele of the other. The data suggest that capsule shape is determined by duplicate developmental pathways, either of which can produce a triangular capsule. One pathway involves the dominant allele of the *A* gene, and the other the dominant allele of the *B* gene. A precursor substance can be converted into a product that leads to a triangular seed capsule through either of these pathways. Only when both pathways are blocked by homozygous recessive alleles is the triangular phenotype suppressed and an ovoid capsule produced:

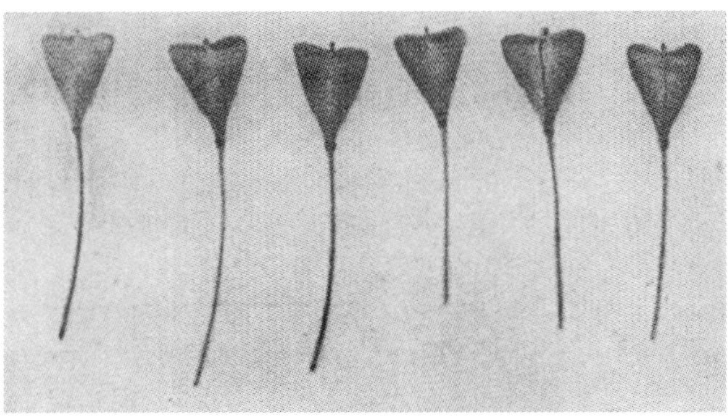

(*a*)

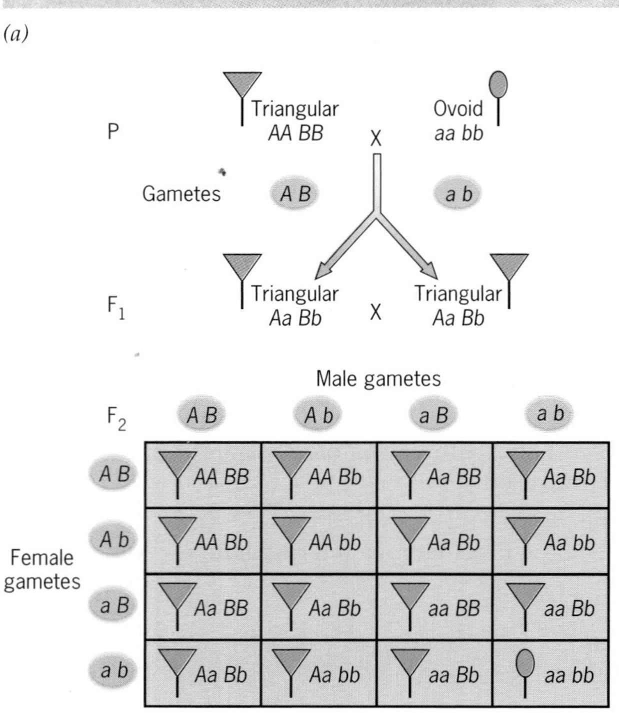

(*b*)

Figure 4.16. (*a*) Seed capsules of the shepherd's purse, *Bursa bursa-pastoris*. (*b*) Crosses showing duplicate gene control of seed capsule shape in the shepherd's purse.

Genotype

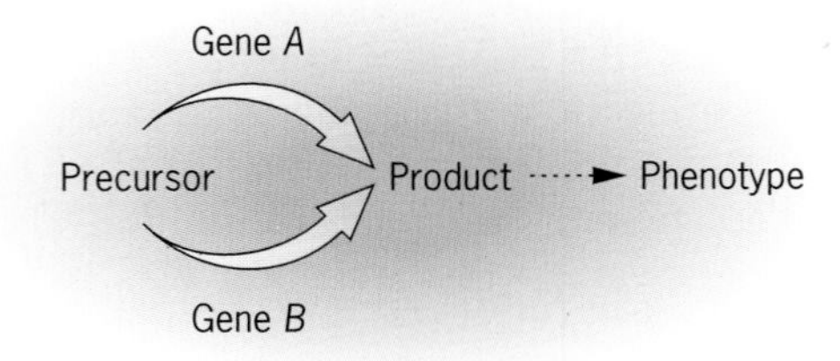

A-; B-	+	+	triangular
aa; B-	+	+	triangular
A-; bb	+	+	triangular
aa; bb	+	–	ovoid

These examples indicate that a particular phenotype is often the result of a process controlled by more than one gene. Each gene governs a step in a pathway that is part of the process. When a gene is mutated to a nonfunctional or partially functional state, the process can be disrupted, leading to a mutant phenotype. Much of modern genetic analysis is devoted to the analysis of pathways involved in important biological processes such as metabolism and development. Studying the epistatic relationships among genes can help to sort out the role that each gene plays in these processes.

Pleiotropy

Not only is it true that a phenotype can be influenced by many genes; it is also true that a gene can influence many phenotypes. When a gene affects many aspects of the phenotype, it is said to be **pleiotropic**, from the Greek words for "to take many turns." The gene for phenylketonuria in human beings is an example. The primary effect of recessive mutations in this gene is to cause toxic substances to accumulate in the brain, leading to mental impairment. However, these mutations also interfere with the synthesis of melanin pigment, lightening the color of the hair; individuals with PKU therefore frequently have light brown or blond hair. Biochemical tests also reveal that the blood and urine of PKU patients contain compounds that are rare or absent in normal individuals. This manifold of phenotypic effects is typical of most genes and probably results from interconnections between the biochemical and cellular pathways that the genes control.

The Genetic Basis of Continuous Phenotypic Variation

The rapid acceptance of the principles of Mendel was due to their ability to explain the inheritance of discrete traits such as flower color and seed shape. However, more complex traits such as height, weight, and agricultural yield seemed to defy Mendelian analysis. These complex traits vary more or less continuously in a population, making it difficult, if not impossible, to define distinct phenotypes. Attempts to study the inheritance of such traits began in the nineteenth century about the same time that Mendel was investigating inheritance in the garden pea. However, it was not until 1918 that continuous phenotypic variation could be understood in terms of Mendel's principles. The key paper was published by the British geneticist and statistician Ronald A. Fisher.

Fisher proposed that continuously varying traits are influenced by a large number of genes, each segregating different alleles. He further proposed that these traits are influenced by a host of environmental factors, each having a small effect. This combination of genetic and environmental factors creates a wide range of phenotypes distributed more or less continuously in a population of individuals. Thus, for example, the height of adult human males in the United States ranges from scarcely 3 feet to over 7 feet. However, very few individuals are found at the extremes. Most men in the U.S. population are between 5 and 6 feet tall (Figure 4.17). The distribution is approximately bell-shaped, with a peak corresponding to the mean height of 5 feet 9 inches. Fisher developed a theory to explain the inheritance of continuously varying traits such as height by extending Mendel's principles to a large number of genes. We discuss this theory, which has become the cornerstone of modern quantitative genetics, in Chapter 25.

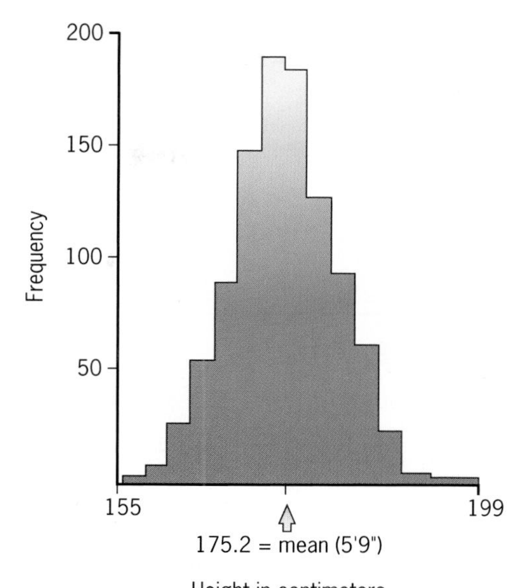

Figure 4.17. Distribution of height among 1000 male students at Harvard University early in the twentieth century.

Key Points: **Gene activity is affected by biological and physical factors in the environment. Two or more genes may determine a trait, and one of them may exert an overriding effect on it. A gene may affect many different phenotypes. Continuous variation in a trait can be explained by the combined action of many genes and environmental factors.**

TESTING YOUR KNOWLEDGE

A geneticist has obtained two true-breeding strains of mice, each homozygous for an independently discovered recessive mutation that prevents the formation of hair on the body. One mutant strain is called *naked*, and the other is called *hairless*. To determine whether the two mutations are alleles, the geneticist crosses *naked* and *hairless* mice with each other. All the offspring are phenotypically wild-type; that is, they have hairs all over their bodies. After intercrossing these F_1 mice, the geneticist observes 115 wild-type mice and 85 mutant mice in the F_2. Are the *naked* and *hairless* mutations alleles? How would you explain the segregation of wild-type and mutant mice in the F_2?

ANSWER

The *naked* and *hairless* mutations are not alleles because the F_1 hybrids are phenotypically wild-type. Thus *naked* and *hairless* are mutations of two different genes. To explain the phenotypic ratio in the F_2, let us first adopt symbols for these mutations and their dominant wild-type alleles:

n = *naked* mutation, N = wild-type allele

h = *hairless* mutation, H = wild-type allele

With these symbols, the genotypes of the true-breeding parental strains are $nn\ HH$ (*naked*) and $NN\ hh$ (*hairless*). The F_1 hybrids produced by crossing these strains are therefore $Nn\ Hh$. When these hybrids are intercrossed, we expect many different genotypes to appear in the offspring. However, each recessive allele, when homozygous, prevents the formation of hair on the body. Thus only mice that are genotypically N- H- will develop hair; all the others—homozygous nn or homozygous hh, or homozygous for both recessive alleles—will fail to develop body hair. We can predict the frequencies of the wild and mutant phenotypes if we assume that the naked and hairless genes assort independently. The frequency of mice that will be *N- H-* is $(3/4) \times (3/4) = 9/16 = 0.56$ (by the Multiplicative Rule of probability) and the frequency of mice that will be either *nn* or *hh* (or both) is $(1/4) + (1/4) - (1/4) \times (1/4) = 7/16 = 0.44$ (by the Additive Rule of probability). Thus, in a sample of 200 F_2 progeny, we expect $200 \times 0.56 = 112$ to be wild-type and $200 \times 0.44 = 88$ to be mutant. The observed frequencies of 115 wild-type and 85 mutant mice are close to these expected numbers, suggesting that the hypothesis of two independently assorting genes for body hair is, indeed, correct.

QUESTIONS AND PROBLEMS

4.1 What blood types could be observed in children born to a woman who has blood type M and a man who has blood type MN?

4.2 In rabbits, coloration of the fur depends on alleles of the gene *c*. From information given in the chapter, what phenotypes and proportions would be expected from the following crosses: (a) $c^+c^+ \times cc$; (b) $c^+c \times c^+c$; (c) $c^+c^h \times c^+ c^{ch}$; (d) $cc^{ch} \times cc$; (e) $c^+c^h \times c^+c$; (f) $c^hc \times cc$?

4.3 In mice, a series of five alleles determines fur color. In order of dominance, these alleles are: A^Y, yellow fur but homozygous lethal; A^L, agouti with light belly; A^+, agouti (wild-type); a^t, black and tan; and a, black. For each of the following crosses, give the coat color of the parents and the phenotypic ratios expected among the progeny: (a) A^YA^L x $A^Y A^L$; (b) $A^Ya \times A^La^t$:(c) $a^ta \times A^Ya$; (d) $A^La^t \times A^LA^L$; (e) $A^LA^L \times A^YA^+$ (f) $A^+a^t \times a^ta$; (g) $a^ta \times aa$; (h) $A^YA^L \times A^+a^t$; and (i) $A^Ya^L \times A^YA^+$.

4.4 In several plants, such as tobacco, primrose, and red clover, combinations of alleles in eggs and pollen have been found to influence the reproductive compatibility of the plants. Homozygous combinations, such as S^1S^1, do not develop because S^1 pollen is not effective on S^1- stigmas. However, S^1 pollen is effective on S^2S^3 stigmas. What progeny might be expected from the following crosses (seed parent written first): (a) $S^1S^2 \times S^2S^3$; (b) $S^1S^2 \times S^3S^4$; (c) $S^4S^5 \times S^4S^5$; and (d) $S^3S^4 \times S^5S^6$?

4.5 From information in the chapter about the ABO blood types, what phenotypes and ratios are expected from the following matings: (a) $I^AI^A \times I^BI^B$; (b) $I^AI^B \times I^OI^O$; (c) $I^AI^O \times I^BI^O$; and (d) $I^AI^O \times I^OI^O$?

4.6 A woman with type O blood gave birth to a baby, also with type O blood. The woman alleged that a man with type AB blood was the father of the baby. Is there any merit to her allegation?

4.7 Another woman with type AB blood gave birth to a baby with type B blood. Two different men claim to be the father. One has type A blood, the other type B blood. Can the genetic evidence decide in favor of either?

4.8 A woman who has blood type O and blood type M marries a man who has blood type AB and blood type MN. Assuming that the genes for the A-B-O and M-N blood-typing systems assort independently, what kinds of children could this couple have, and in what proportions?

4.9 A Japanese strain of mice has a peculiar uncoordinated gait called waltzing, which is due to a recessive allele, *v*. The dominant allele *V* causes mice to move in a coordinated fashion. A mouse geneticist has recently isolated another recessive mutation that causes uncoordinated movement. This mutation, called *tango*, could be an allele of the *waltzing* gene, or it could be a mutation in an entirely different gene. Propose a test to determine whether the *waltzing* and *tango* mutations are alleles, and if they are, propose symbols to denote them.

4.10 Congenital deafness in human beings is inherited as a recessive condition. In the pedigree below, two deaf individuals, each presumably homozygous for a recessive mutation, have married and produced four children with normal hearing. Propose an explanation.

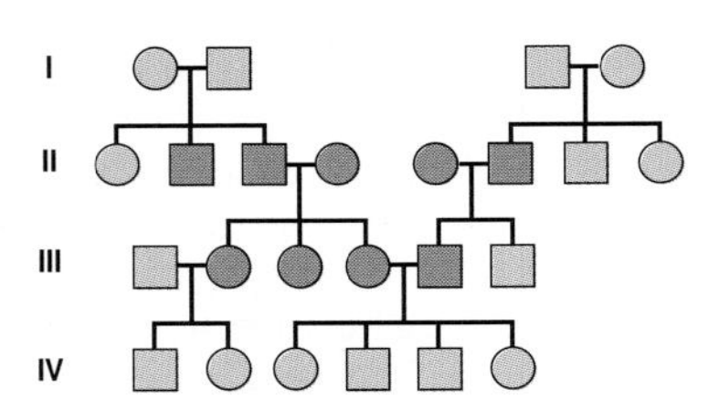

4.11 In the fruit fly, recessive mutations in either of two independently assorting genes, *brown* and *purple*, prevent the synthesis of red pigment in the eyes. Thus homozygotes for either of these mutations have brownish-purple eyes. However, heterozygotes for both of these mutations have dark red, that is, wild-type eyes. If such double heterozygotes are intercrossed, what kinds of progeny will be produced, and in what proportions?

4.12 The dominant mutation *Plum* in the fruit fly also causes brownish-purple eyes. Is it possible to determine whether *Plum* is an allele of the *brown* or *purple* genes?

4.13 From information given in the chapter, explain why mice with yellow coat color are not true-breeding.

4.14 A couple has four children. Neither the father nor the mother is bald; one of the two sons is bald, but neither of the daughters is bald. If one of the daughters marries a nonbald man, what is the chance that they will have a child who will become bald as an adult?

4.15 The pedigree below shows the inheritance of ataxia, a rare neurological disorder characterized by uncoordinated movements. Is ataxia caused by a dominant or a recessive allele? Explain.

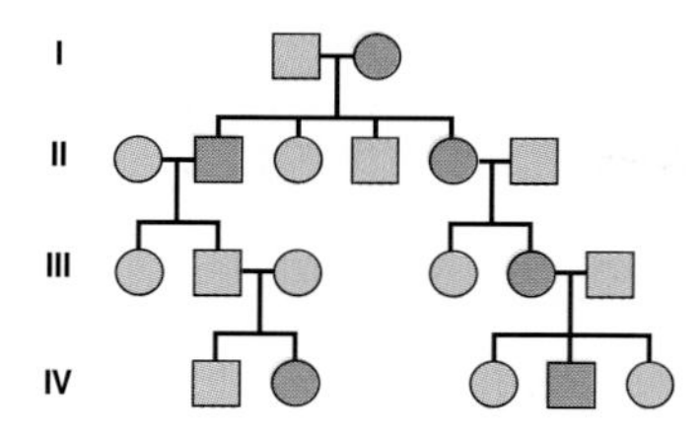

4.16 Chickens that carry both the alleles for rose comb (*R*) and pea comb (*P*) have walnut combs, whereas chickens that lack both of these alleles (that is, they are genotypically *rr pp*) have single combs. From the information about interactions between these two genes given in the chapter, determine the phenotypes and proportions expected from the following crosses: (a) *RR Pp* × *rr Pp*; (b) *rr PP* × *Rr Pp*; (c) *Rr Pp* × *Rr pp*; (d) *Rr pp* × *rr pp*.

4.17 Rose-comb chickens mated with walnut-comb chickens produced 15 walnut, 14 rose, 5 pea, and 6 single-comb chicks. Determine the genotypes of the parents.

4.18 Summer squash plants with the dominant allele *C* bear white fruit, whereas plants homozygous for the recessive allele *c* bear colored fruit. When the fruit is colored, the dominant allele *G* causes it to be yellow; in the absence of this allele (that is, with genotype *gg*), the fruit color is green. What are the F_2 phenotypes and proportions expected from intercrossing the progeny of *CC GG* and *cc gg* plants? Assume that the *C* and *G* genes assort independently.

4.19 The white Leghorn breed of chickens is homozygous for the dominant allele *C*, which produces colored feathers; however, this breed is also homozygous for the dominant allele *I* of an independently assorting gene that inhibits coloration of the feathers. Consequently, Leghorn chickens have white feathers. The white Wyandotte breed of chickens has neither the allele for color nor the inhibitor of color; it is therefore genotypically *cc ii*. What are the F_2 phenotypes and proportions expected from intercrossing the progeny of a white Leghorn hen and a white Wyandotte rooster?

4.20 Fruit flies homozygous for the recessive mutation *scarlet* have bright red eyes because they cannot synthesize brown pigment. Fruit flies homozygous for the recessive mutation *brown* have brownish-purple eyes because they cannot synthesize red pigment. Fruit flies homozygous for both of these mutations have white eyes because they cannot synthesize either type of pigment. The *brown* and *scarlet* mutations assort independently. If fruit flies that are heterozygous for both of these mutations are intercrossed, what kinds of progeny will they produce, and in what proportions?

4.21 Consider the following hypothetical scheme of determination of coat color in a mammal. Gene *A* controls the conversion of a white pigment P_0 into a gray pigment P_1; the dominant allele *A* produces the enzyme necessary for this conversion, and the recessive allele *a* produces an enzyme without biochemical activity. Gene *B* controls the conversion of the gray pigment P_1 into a black pigment P_2; the dominant allele *B* produces the active enzyme for this conversion, and the recessive allele *b* produces an enzyme without activity. The dominant allele *C* of a third gene produces a polypeptide that completely inhibits the activity of the enzyme produced by gene *A*; that is, it prevents the reaction $P_0 \rightarrow P_1$. Allele *c* of this gene produces a defective polypeptide that does not inhibit the reaction $P_0 \rightarrow P_1$. Genes *A*, *B*, and *C* assort independently, and no other genes are involved. In the F_2 of the cross *AA bb CC* × *aa BB cc*, what is the expected phenotypic segregation ratio?

4.22 What F_2 phenotypic segregation ratio would be expected for the cross described in the preceding problem if the dominant allele, *C*, of the third gene produced a product that completely inhibited the activity of the enzyme produced by gene *B*, that is, prevented the reaction $P_1 \rightarrow P_2$, rather than inhibiting the activity of the enzyme produced by gene *A*?

BIBLIOGRAPHY

BATESON, W. 1909. *Mendel's Principles of Heredity*. University Press, Cambridge, England.

CARLSON, E. A. 1966. *The Gene: A Critical History*. W. B. Saunders, Philadelphia.

DUNN, L. C. 1965. *A Short History of Genetics*. McGraw-Hill, New York.

HERRMANN, B. G. and A. KISPERT. 1994. The *T Genes* in embryogenesis. *Trends in Genet*. 10: 280–286.

HODGE, G. P. 1977. "A medical history of the Spanish Habsburgs." *J. Am. Med. Assn*. 238: 1169–1174.

PETERS, J. A. ED. 1959. *Classic Papers in Genetics*. Prentice-Hall, Englewood Cliffs, NJ.

STEWART, A. ED. 1994. Genetic nomenclature guide. *Trends in Genet*.

STURTEVANT, A. H. 1965. *A History of Genetics*. Harper and Row, New York.

5

Male and female Drosophila. *The male has white eyes because it carries a mutation on its X chromosome.*

The Chromosomal Basis of Mendelism

Sex, Chromosomes, and Genes

What causes organisms to develop as males or females? Why are there only two sexual phenotypes? Is the sex of an organism determined by its genes? These and related questions have intrigued geneticists since the rediscovery of Mendel's work at the beginning of the twentieth century.

The discovery that genes play a role in the determination of sex emerged from a fusion between two previously distinct scientific disciplines, genetics—the study of heredity—and cytology—the study of cells. Early in the twentieth century, these disciplines were brought together through a friendship between two remarkable American scientists, Thomas Hunt Morgan and Edmund Beecher Wilson. Morgan was the geneticist and Wilson the cytologist.

As the cytologist, Wilson was interested in the behavior of chromosomes. These structures would prove to be important for sex determination in many species, including our own. Wilson was one of the first to investigate differences in the chromosomes of the two sexes. Through careful study, he and his colleagues showed that these differences were confined to a special pair of chromosomes called sex chromosomes. Wilson found that the behavior of these chromosomes during meiosis could account for the inheritance of sex.

As the geneticist, Morgan was interested in the identification of genes. He focused his research on the fruit fly, *Drosophila melanogaster*, and rather quickly discovered a gene that gave different phenotypic ratios in males and females. Morgan hypothesized that this gene was located on one of the sex chromosomes, and one of his students, Calvin Bridges, eventually proved this hypothesis to be correct. Morgan's discovery that genes reside on chromosomes was a great achievement. The abstract genetic factors postulated by Mendel were finally localized on visible structures within cells. Geneticists could now explain the Principles of Segregation and Independent Assortment in terms of meiotic chromosome behavior.

The discovery that specific genes determine the sex of an organism came much later, only after another scientific discipline, molecular biology, had joined forces with genetics and cytology. The combined efforts of cytologists, geneticists, and molecular biologists identified specific sex-determining genes by studying rare individuals in which the sexual phenotype was inconsistent with the sex chromosomes that were present. Today, researchers in all three fields are earnestly trying to figure out how these genes control sexual development.

In this chapter we will see how cytological studies of differences between the sexes led to the localization of genes on chromosomes. We also discuss the genetic basis of sex determination and the ways in which the two sexes compensate for different numbers of sex chromosomes.

CHROMOSOMES

Chromosomes were discovered in the second half of the nineteenth century by a German cytologist, W. Waldeyer. Subsequent investigations with many different organisms established that chromosomes are characteristic of the nuclei of all cells. They are best seen by applying dyes to dividing cells; at this time, the material in a chromosome is packed into a small volume, giving it the appearance of a tightly organized cylinder. During the interphase between cell divisions, chromosomes are not so easily seen, even with the best of dyes. During this phase of the cell cycle they are much more loosely organized, forming thin threads that are distributed throughout the nucleus. Consequently, when dyes are applied, the whole nucleus is stained and individual chromosomes cannot be identified. This diffuse network of threads is called **chromatin**. Some regions of the chromatin stain more darkly than others, suggesting an underlying difference in organization. The light and dark regions, respectively called the **euchromatin** (eu—true chromatin) and the **heterochromatin** (heter—different chromatin), have different densities of chromosomal threads.

Chromosome Number

Within a species, the number of chromosomes is almost always an even multiple of a basic number. In human beings, for example, the basic number is 23; mature eggs and sperm have this number of chromosomes. Most other types of human cells have twice as many (46), although a few kinds, such as some liver cells, have four times (92) the basic number.

The **haploid**, or basic, chromosome number (**n**) defines a set of chromosomes called the **haploid genome**. Most somatic cells contain two of each of the chromosomes in this set and are therefore **diploid** (**2n**). Cells with four of each of the chromosomes are **tetraploid** (**4n**), those with eight of each are **octaploid** (**8n**), and so on. This ascending series, 1, 2, 4, 8, . . . , arises from successive rounds of chromosome replication.

The basic number of chromosomes varies among species. Chromosome number is unrelated to the size or biological complexity of an organism, with most species containing between 10 and 40 chromosomes in their genomes (Table 5.1). The muntjac, a tiny Asian deer, has only three chromosomes in its genome, whereas some species of ferns have many hundreds.

Sex Chromosomes

In some animal species—for example, grasshoppers—females have one more chromosome than males (Figure 5.1*a*). This extra chromosome, originally observed in certain bugs, is called the **X chromosome**. Females

TABLE 5.1
Chromosome Number in Different Organisms

Organism	*Haploid Chromosome Number*
Simple Eukaryotes	
Baker's yeast (*Saccharomyces cerevisiae*)	16
Bread mold (*Neurospora crassa*)	7
Unicellular green alga (*Chlamydomonas reinhardtii*)	17
Plants	
Maize (*Zea mays*)	10
Bread Wheat (*Triticum aestivum*)	21
Tomato (*Lycopersicon esculentum*)	12
Broad bean (*Vicia faba*)	6
Giant sequoia (*Sequoia sempivirens*)	11
Crucifer (*Arabidopsis thaliana*)	5
Invertebrate Animals	
Fruit fly (*Drosophila melanogaster*)	4
Mosquito (*Anopheles culicifacies*)	3
Starfish (*Asterias forbesi*)	18
Nematode (*Caenorhabditis elegans*)	6
Mussel (*Mytilus edulis*)	14
Vertebrate Animals	
Human being (*Homo sapiens*)	23
Chimpanzee (*Pan tryglodites*)	24
Cat (*Felis domesticus*)	36
Mouse (*Mus musculus*)	20
Chicken (*Gallus domesticus*)	39
Toad (*Xenopus laevis*)	17
Fish (*Esox lucius*)	25

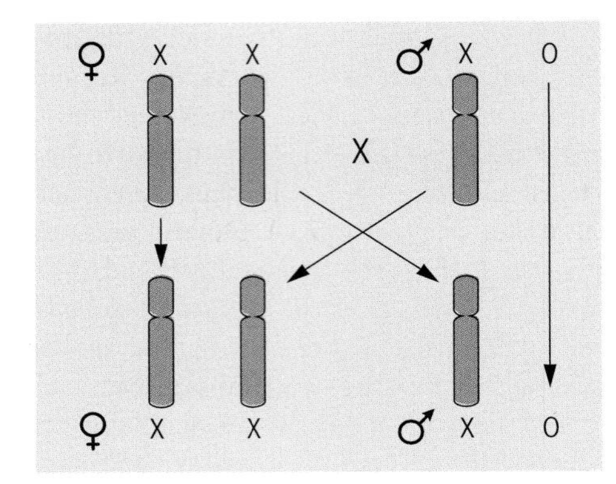

(*a*) Inheritance of sex chromosomes in animals with XX females and XO males.

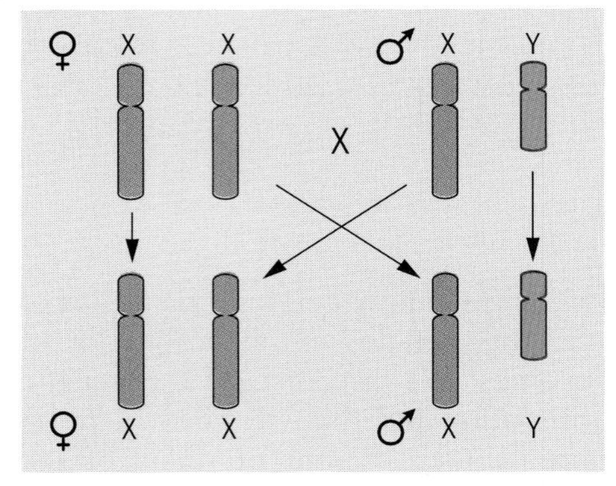

(*b*) Inheritance of sex chromosomes in animals with XX females and XY males.

Figure 5.1 Inheritance of sex chromosomes in animals.

of these species have two X chromosomes, and males have only one; thus females are cytologically XX and males are XO, where the "O" denotes the absence of a chromosome. During meiosis in the female, the two X chromosomes pair and then separate, producing eggs that contain a single X chromosome. During meiosis in the male, the solitary X chromosome moves independently of all the other chromosomes and is incorporated into half the sperm; the other half receive no X chromosome. Thus, when sperms and eggs unite, two kinds of zygotes are produced: XX, which develop into females, and XO, which develop into males. Because each of these types is equally likely, the reproductive mechanism preserves a 1:1 ratio of males to females in these species.

In many other animals, including human beings, males and females have the same number of chromosomes (Figure 5.1*b*). This numerical equality is due to the presence of a chromosome in the male, called the **Y chromosome**, which pairs with the X during meiosis. The Y chromosome is morphologically distinguishable from the X chromosome. In humans, for example, the Y is much shorter than the X, and its centromere is located closer to one of the ends (Figure 5.2). The material common to the human X and Y chromosomes is limited, consisting mainly of short terminal segments. During meiosis in the male, the X and Y chromosomes separate from each other, producing two kinds of sperm, X-bearing and Y-bearing; the frequencies of the two types are approximately equal. XX females produce only one kind of egg, which is X-bearing. If fertilization were to occur randomly, approximately half the zygotes would be XX and the other half would be XY, leading to a 1:1 sex ratio at conception. However,

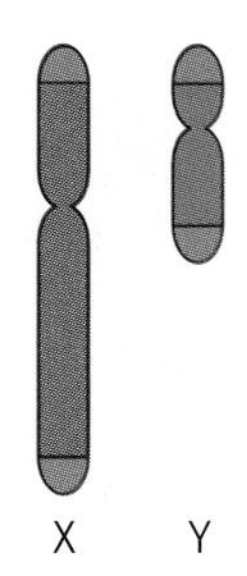

Figure 5.2 Human X and Y chromosomes. The terminal regions are common to both sex chromosomes.

in human beings, Y-bearing sperm have a fertilization advantage, and the zygotic sex ratio is actually about 1.3:1. During development, the excess of males is diminished by differential viability of XX and XY embryos, and at birth, males are only slightly more numerous than females (sex ratio 1.07:1). By the age of reproduction, the excess of males is essentially eliminated and the sex ratio is very close to 1:1.

The X and Y chromosomes are called **sex chromosomes**. All the other chromosomes in the genome are called **autosomes**. Sex chromosomes were discovered in the first few years of the twentieth century through the work of the American cytologists C. E. McClung, N. M. Stevens, W. S. Sutton, and E. B. Wilson. This discovery coincided closely with the emergence of Mendelism and stimulated research on the possible relationships between Mendel's principles and the meiotic behavior of chromosomes.

Key Points: **Chromosomes emerge from a diffuse network of chromatin fibers during cell division. Diploid somatic cells have twice as many chromosomes as haploid gametes. In some species, the X and Y sex chromosomes distinguish the cells of males and females—XY in males and XX in females. In other species, the female has two X chromosomes and the male has a single X and no Y.**

THE CHROMOSOME THEORY OF HEREDITY

By 1910 many biologists suspected that genes were situated on chromosomes, but they did not have definitive proof. Researchers needed to find a gene that could be unambiguously linked to a chromosome. This required that the gene be defined by a mutant allele and that the chromosome be morphologically distinguishable. Furthermore, the pattern of gene transmission had to reflect the chromosome's behavior during reproduction. All these requirements were fulfilled when the American biologist Thomas H. Morgan discovered a particular eye color mutation in the fruit fly, *Drosophila melanogaster*. Morgan began experimentation with this species of fly about 1909. It was ideally suited for genetics research because it reproduced quickly and prolifically and was inexpensive to rear in the laboratory. In addition, it had only four pairs of chromosomes, one being a pair of sex chromosomes—XX in the female and XY in the male. The X and Y chromosomes were morphologically distinguishable from each other and from each of the autosomes. Through careful experiments, Morgan was able to show that the eye color mutation was inherited along with the X chromosome, suggesting that a gene for eye color was physically situated on that chromosome. Later, his student, Calvin B. Bridges, obtained definitive proof for this Chromosome Theory of Heredity.

Experimental Evidence Linking the Inheritance of Genes to Chromosomes

Morgan's experiments commenced with his discovery of a mutant male fly that had white eyes instead of the red eyes of wild-type flies. When this male was crossed to wild-type females, all the progeny had red eyes, indicating that white was recessive to red. When these progeny were intercrossed with each other, Morgan observed a peculiar segregation pattern: all of the daughters, but only half of the sons, had red eyes; the other half of the sons had white eyes. This pattern suggested that the inheritance of eye color was linked to the sex chromosomes. Morgan proposed that a gene for eye color was present on the X chromosome, but not on the Y, and that the white and red phenotypes were due to two different alleles, a mutant allele denoted w and a wild-type allele denoted w^+.

Morgan's hypothesis is diagrammed in Figure 5.3. The wild-type females in the first cross are assumed to be homozygous for the w^+ allele. Their mate is assumed to carry the mutant w allele on its X chromo-

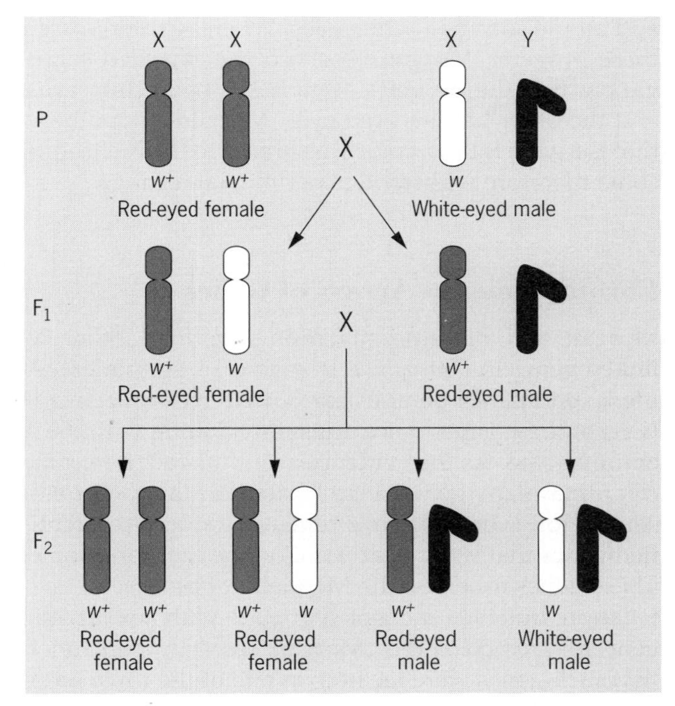

Figure 5.3 Morgan's experiment studying the inheritance of white eyes in *Drosophila*. The transmission of the mutant condition in association with sex suggested that the gene for eye color was present on the X chromosome but not on the Y chromosome.

some and neither of the alleles on its Y chromosome. An organism that has only one copy of a gene is called a **hemizygote**. Among the progeny from the cross, the sons inherit an X chromosome from their mother and a Y chromosome from their father; because the maternally inherited X carries the w^+ allele, these sons have red eyes. The daughters, in contrast, inherit an X chromosome from each parent—an X with w^+ from the mother and an X with w from the father. However, because w^+ is dominant to w, these heterozygous F_1 females also have red eyes.

When the F_1 males and females are intercrossed, four genotypic classes of progeny are produced, each representing a different combination of sex chromosomes. The XX flies, which are female, have red eyes because at least one w^+ allele is present. The XY flies, which are male, have either red or white eyes, depending on which X chromosome is inherited from the heterozygous F_1 females. Segregation of the w and w^+ alleles in these females is therefore the reason why half the F_2 males have white eyes.

Morgan carried out additional experiments to confirm the elements of his hypothesis. In one (Figure 5.4*a*), he crossed females assumed to be heterozygous for the eye color gene to mutant white males. As he expected, half the progeny of each sex had white eyes, and the other half had red eyes. In another experiment (Figure 5.4*b*), he crossed white-eyed females to red-eyed males. This time, all the daughters had red eyes, and all the sons had white eyes. When he intercrossed these progeny, Morgan observed the expected segregation: half the progeny of each sex had white eyes, and the other half had red eyes. Morgan's hypothesis that the gene for eye color was linked to the X chromosome therefore withstood experimental testing.

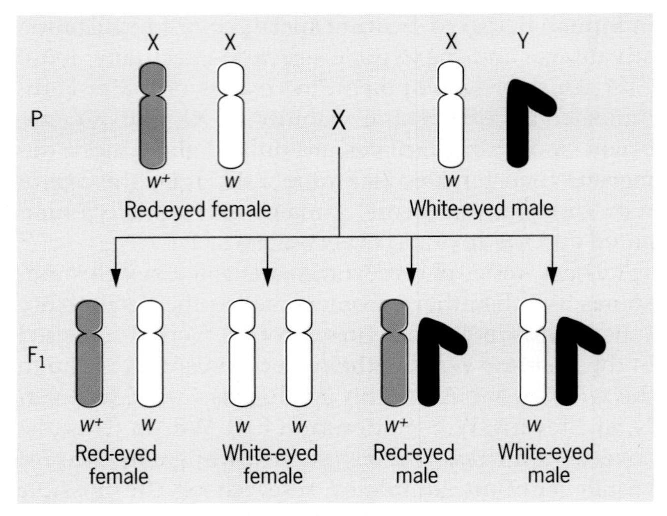

(*a*) Cross between a heterozygous female and a hemizygous mutant male.

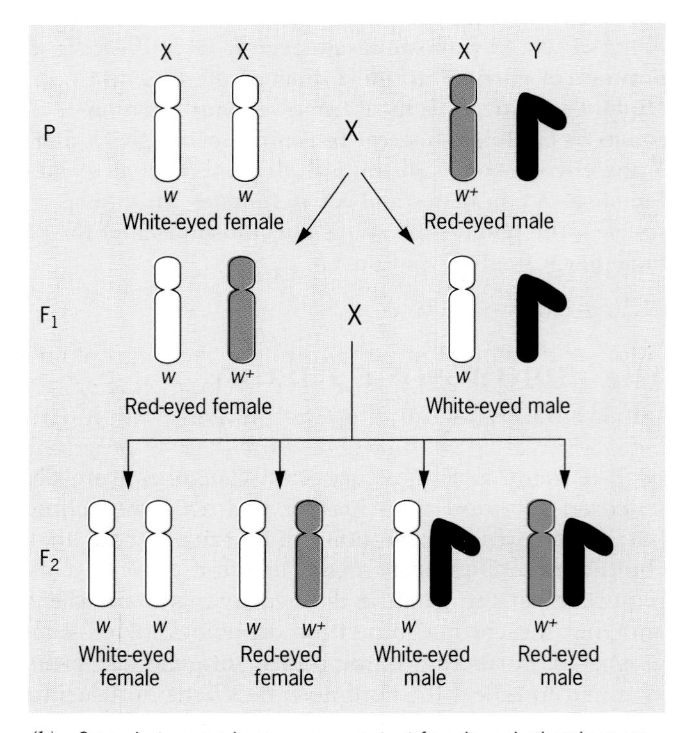

(*b*) Cross between a homozygous mutant female and a hemizygous wild-type male.

Figure 5.4 Experimental tests of Morgan's hypothesis that the gene for eye color in *Drosophila* is X-linked. In each experiment, eye color is inherited along with the X chromosome. Thus the results of these crosses supported Morgan's hypothesis that the gene for eye color is X-linked.

Chromosomes as Arrays of Genes

Morgan and his students soon identified other X-linked genes in *Drosophila*. In each case, simple breeding experiments demonstrated that recessive mutations of these genes were transmitted along with the X chromosome. As the evidence accumulated, it became clear that many genes were located on the X chromosome. However, Morgan's research group also identified genes that were clearly not on the X chromosome. These genes followed the Mendelian Principle of Segregation, but they did not segregate with sex, as the gene for eye color did. Morgan correctly concluded that such genes were located on one of the three autosomes in the *Drosophila* genome. Thus each *Drosophila* chromosome appeared to contain a different set of genes.

Morgan's laboratory then attempted to determine the relationships among the genes on a particular chromosome. They proceeded on the assumption that the genes were arranged in a linear array—an idea inspired by cytological evidence that the chromosome was a long, thin thread. In just a few years, Morgan's students were able to show that genes were indeed situated at different sites, or **loci** (from the Latin word for "place"; singular: **locus**), on a linear structure. This analysis, which we will discuss in Chapter 7, produced the world's first genetic maps—diagrams showing the positions of genes and the relative distances between them (Figure 5.5). Morgan's laboratory pioneered the methods for genetic mapmaking and laid the foundation for subsequent research on the physical structure of chromosomes. Eventually, the linearity of chromosomes was connected to the linear structure of DNA (see Chapter 9).

These early studies with *Drosophila*—primarily the work of Morgan and his laboratory (Historical Sidelight: *Drosphila*, T. H. Morgan, and "The Fly Room")—greatly strengthened the view that all genes were located on chromosomes and that Mendel's principles could be explained by the transmissional properties of chromosomes during reproduction. This idea, called the **Chromosome Theory of Heredity**, stands as one of the most important achievements in biology. Since its formulation in the early part of the twentieth century, the Chromosome Theory of Heredity has provided a unifying framework for all studies of inheritance.

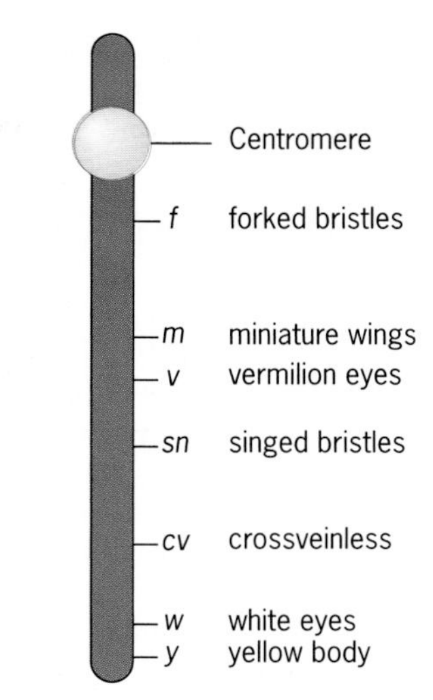

Figure 5.5 A map of genes on the X chromosome of *Drosophila*.

Nondisjunction as Proof of the Chromosome Theory

Morgan showed that a gene for eye color was on the X chromosome of *Drosophila* by correlating the inheritance of that gene with the transmission of the X chromosome during reproduction. However, as noted earlier, it was one of his students, C. B. Bridges, who secured the proof of the chromosome theory by showing that *exceptions* to the rules of inheritance could also be explained by chromosome behavior.

Bridges performed one of Morgan's experiments on a larger scale (Figure 5.6). He crossed white-eyed female *Drosophila* to red-eyed males and examined many F_1 progeny. Although nearly all the F_1 flies were either red-eyed females or white-eyed males, as expected, Bridges found a few exceptional flies—white-eyed females and red-eyed males. He crossed these exceptions to determine how they might have arisen. The exceptional males all proved to be sterile, but the exceptional females were fertile. When crossed to normal red-eyed males, they produced many progeny, including large numbers of white-eyed daughters and red-eyed sons. Thus the exceptional F_1 females, though rare in their own right, were prone to produce many exceptional progeny.

Bridges explained these results by proposing that the exceptional F_1 flies were the result of abnormal X chromosome behavior during meiosis in the females of the P generation. Ordinarily, the X chromosomes in these females should **disjoin**, or separate, from each other during meiosis. Occasionally, however, they might fail to separate, producing an egg with two X chromosomes, or an egg with no X chromosome at all. Fertilization of such abnormal eggs by normal sperm would produce zygotes with an abnormal number of

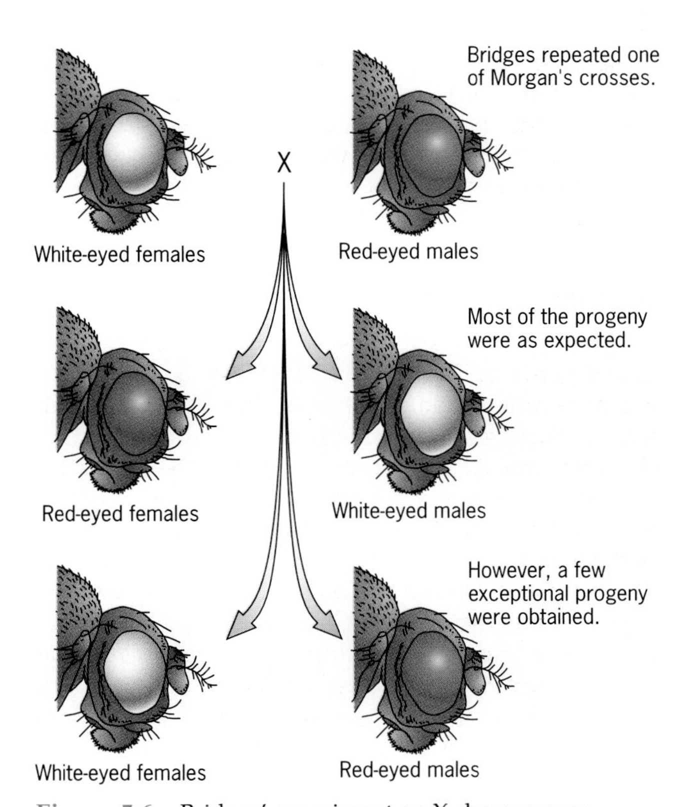

Figure 5.6 Bridges' experiment on X chromosome nondisjunction in *Drosophila*.

HISTORICAL SIDELIGHT

Drosophila, T. H. Morgan, and "The Fly Room"

Many people regard fruit flies as pests; they buzz around orchards and vineyards, especially in late summer, and make their way into houses when fruit is harvested. Because of this annoying behavior, geneticists often have difficulty explaining why these tiny creatures are among their favorite animals.

The genus *Drosophila* (from the Greek words meaning "lover of dew") comprises a large number of species, the most intensively studied being *D. melanogaster*. This species was described in the middle of the nineteenth century under the name *D. ampelophilia,* a name meaning "lover of grape vines." After being established as an experimental animal, *D. ampelophilia* was renamed *D. melanogaster,* which means "black belly."

C. W. Woodworth seems to have been the first person to culture *Drosophila* in the laboratory. From Woodworth, W. E. Castle, a professor of zoology at Harvard University, learned of the advantages of using this animal as an experimental organism. Castle recommended it to T. H. Morgan, who began to culture *Drosophila* in 1909.

From its intensive work with *Drosophila,* Morgan's laboratory at Columbia University became known as "The Fly Room" (see Figure 1). Numerous students worked in this laboratory from 1910 until 1926, the year Morgan moved his research to the California Institute of Technology. The most famous of Morgan's students were Calvin Bridges, Alfred Sturtevant, and Hermann Muller. Bridges provided the proof for the Chromosome Theory of Heredity, Sturtevant produced the world's first chromosome map, and Muller discovered that mutations could be induced by X-irradiation. These and other "*Drosophila* workers" were instrumental in developing much of classical genetic analysis. Indeed, William Bateson commented that "not even the most skeptical of readers can go through the *Drosophila* work unmoved by a sense of admiration for the zeal and penetration with which it has been conducted, and for the great extension of genetic knowledge to which it has led—greater far than has been made in any one line of work since Mendel's own experiments."

Figure 1 T. H. Morgan's laboratory at Columbia University.

In 1939 Morgan reminisced about the early days in the Fly Room:

> *It was not unusual for the six of us to carry on in this small room, the only space at our disposal. These were the days when bananas were used as fly food and in one corner of the room a bunch of bananas was generally on hand, —an adjunct to our researches which interested other members of the laboratory in a different way. As there were no incubators, a bookcase and a wallcase were rigged up with electric bulbs and a cheap thermostat, which behaved badly at times, with consequent loss of cultures. The use of milk bottles came into the program at an early date, but where they came from was not known, or at least not mentioned . . . . Our proximity to each other led to cooperation in everything that went on.*[1]

Later, Morgan's student, Alfred Sturtevant, noted that "There was a give-and-take atmosphere in the fly room. As each new result or new idea came along, it was discussed freely by the group."[2] The excitement of using *Drosophila* as an experimental organism, and the intellectual comraderie created by the close quarters of the Fly Room combined to stimulate research. The rapid isolation and analysis of new mutations turned *Drosophila* into the premier organism for experimental genetics.

Since the days of the Fly Room, research on *Drosophila* has grown into a worldwide endeavor. Hundreds of different laboratories are currently investigating the genes and chromosomes of this animal. In the United States, two large stock centers maintain cultures of mutant flies for distribution to interested researchers, and every year *Drosophila* workers gather for an international five-day meeting to discuss the results of their investigations; typically, more than a thousand people attend this annual meeting. In 1992, at the meeting in Philadelphia, a compendium of 80 years of *Drosophila* research was unveiled. This 1133-page volume, entiltled *The Genome of Drosophila melanogaster,*[3] describes more than 4000 genes. However, even as it went to press, the book was out of date. Information about *Drosophila* genes is now available from a regularly updated electronic database, called FlyBase, which is accessible on the Internet through servers such as Gopher (flybase.bio.indiana .edu) or through the World Wide Web (http://morgan.harvard.edu/). At the last major update, FlyBase listed 23,878 alleles of 8676 genes—a legacy that would certainly make T. H. Morgan happy.

[1]Morgan, T. H., 1939, "Personal recollections of Calvin B. Bridges." *Journal of Heredity* 30: 355.

[2]Sturtevant, A. H., 1965, *A History of Genetics*. Harper and Row: New York, p. 49.

[3]Lindsley, D. L., and G. G. Zimm, 1992, *The Genome of Drosophila melanogaster*. Academic Press: New York.

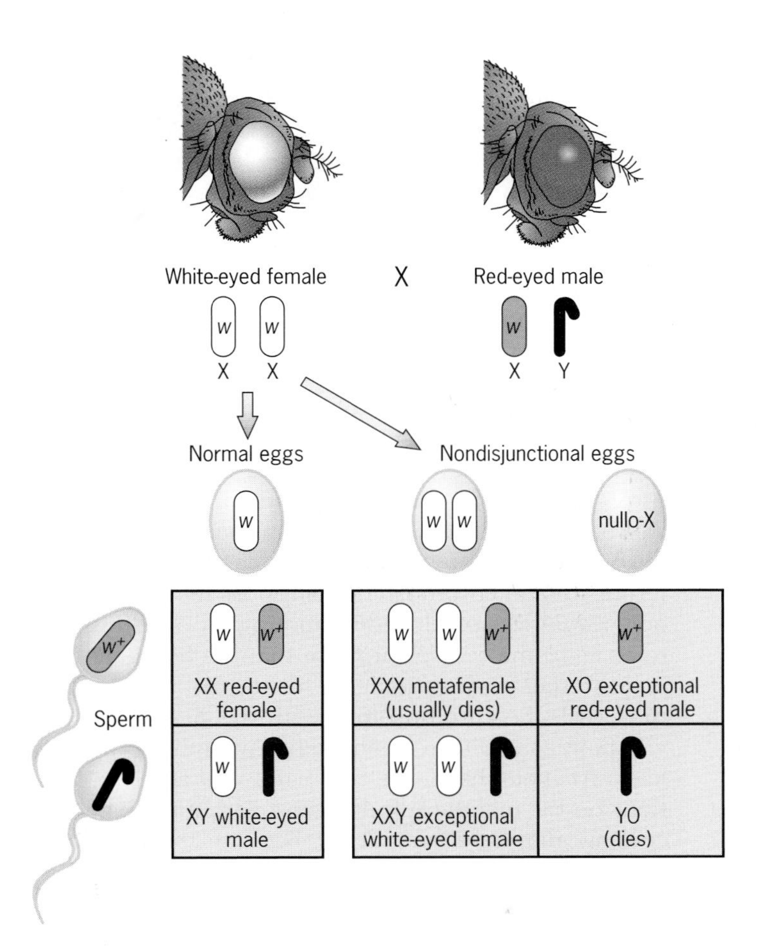

Figure 5.7 X chromosome nondisjunction is responsible for the exceptional progeny that appeared in Bridges' experiment. Nondisjunctional eggs that contain either two X chromosomes or no X chromosome unite with normal sperm that contain either an X chromosome or a Y chromosome to produce four types of zygotes. The XXY zygotes develop into white-eyed females, the XO zygotes develop into red-eyed, sterile males, and the XXX and YO zygotes die.

sex chromosomes. Figure 5.7 illustrates the possibilities.

If an egg with two X chromosomes (usually called a diplo-X egg; genotype X^wX^w) is fertilized by a Y-bearing sperm, the zygote will be X^wX^wY. Since each of the X chromosomes in this zygote carries a mutant w allele, the resulting fly will have white eyes. If an egg without an X chromosome (usually called a nullo-X egg) is fertilized by an X-bearing sperm (X^+), the zygote will be X^+O. (Once again, "O" denotes the absence of a chromosome.) Because the single X in this zygote carries a w^+ allele, it will develop into a red-eyed fly. Bridges inferred that XXY flies were female and that XO flies were male. The exceptional white-eyed females that he observed were therefore X^wX^wY, and the exceptional red-eyed males were X^+O. Bridges confirmed the chromosome constitutions of these exceptional flies by direct cytological observation. Because the XO animals were male, Bridges concluded that in *Drosophila* the Y chromosome actually has nothing to do with the determination of the sexual phenotype. However, because the XO males were always sterile, this chromosome must be important for male sexual function.

Bridges recognized that the fertilization of abnormal eggs by normal sperm could produce two additional kinds of zygotes: $X^wX^wX^+$, arising from the union of a diplo-X egg and an X-bearing sperm, and YO, arising from the union of a nullo-X egg and a Y-bearing sperm. The $X^wX^wX^+$ zygotes develop into females that are red-eyed, but weak and sickly. These "metafemales" can be distinguished from XX females by a syndrome of anatomical abnormalities, including ragged wings and etched abdomens. Generations of geneticists have inappropriately called them "superfemales"—a term coined by Bridges—even though there is nothing super about them. The YO zygotes turn out to be completely inviable, that is, they die. In *Drosophila*, as in most other organisms with sex chromosomes, at least one X chromosome is needed for viability.

Bridges' ability to explain the exceptional progeny that came from these crosses showed the power of the chromosome theory. Each of the exceptions was due to anomalous chromosome behavior during meiosis. Bridges called the anomaly **nondisjunction**, because it involved a failure of the chromosomes to disjoin during one of the meiotic divisions. This failure could result from faulty chromosome movement, inprecise or incomplete pairing, or centromere malfunction. From Bridges' data, it is impossible to specify the exact cause. However, Bridges did note that the exceptional

XXY females go on to produce a high frequency of exceptional progeny, presumably because their sex chromosomes can disjoin in different ways: the X chromosomes can disjoin from each other, or either X can disjoin from the Y. In the latter case, a diplo- or nullo-X egg is produced because the X that does not disjoin from the Y is free to move to either pole during the first meiotic division. When fertilized by normal sperm, these abnormal eggs will produce exceptional zygotes.

The Chromosomal Basis of Mendel's Principles of Segregation and Independent Assortment

Mendel established two principles of genetic transmission: (1) the alleles of a gene segregate from each other, and (2) the alleles of different genes assort independently. The finding that genes are located on chromosomes made it possible to explain these principles in terms of the meiotic behavior of chromosomes.

The Principle of Segregation (Figure 5.8). During the first meiotic division, homologous chromosomes pair. One of the homologs comes from the mother, the other from the father. If the mother was homozygous for an allele, *A*, of a gene on this chromosome, and the father was homozygous for a different allele, *a*, of the same gene, the offspring must be heterozygous, that is, *Aa*. In the anaphase of the first meiotic division, the paired chromosomes separate and move to opposite poles of the cell. One carries allele *A* and the other allele *a*. This physical separation of the two chromosomes segregates the alleles from each other; eventually, they will reside in different daughter cells. Mendel's Principle of Segregation is therefore based on the separation of homologous chromosomes during the anaphase of the first meiotic division.

The Principle of Independent Assortment (Figure 5.9). The Principle of Independent Assortment is also based on this anaphase separation. To understand the relationship, we need to consider genes on two different pairs of chromosomes. Suppose that a heterozygote *Aa Bb* was produced by mating an *AA BB* female to an *aa bb* male; also, suppose that the two genes are on different chromosomes. During the prophase of meiosis I, the chromosomes with the *A* and *a* alleles will pair, as will the chromosomes with the *B* and *b* alleles. At metaphase, the two pairs will take up positions on the meiotic spindle in preparation for the upcoming anaphase separation. Because there are two

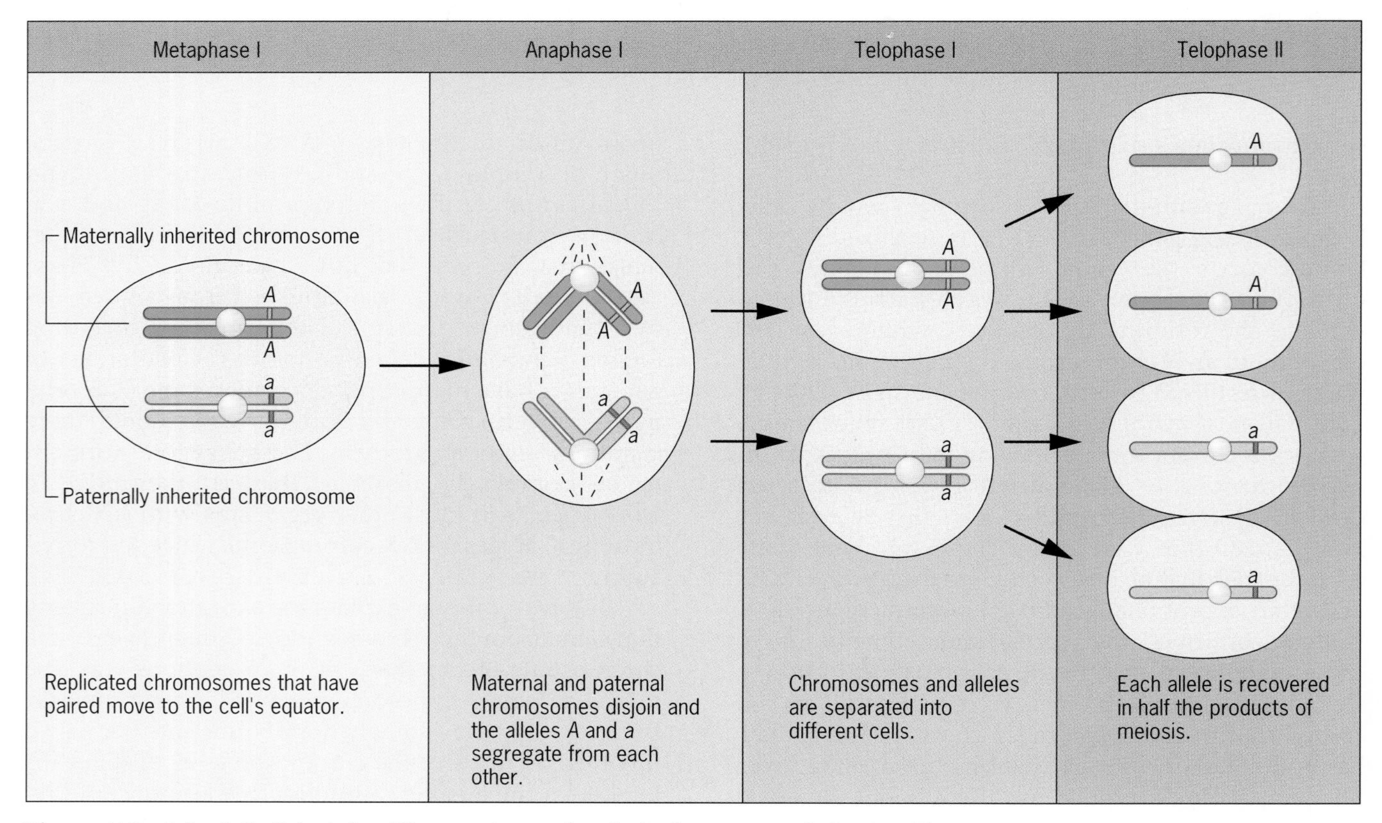

Figure 5.8 Mendel's Principle of Segregation and meiotic chromosome behavior. The segregation of alleles corresponds to the disjunction of paired chromosmes in the anaphase of the first meiotic division.

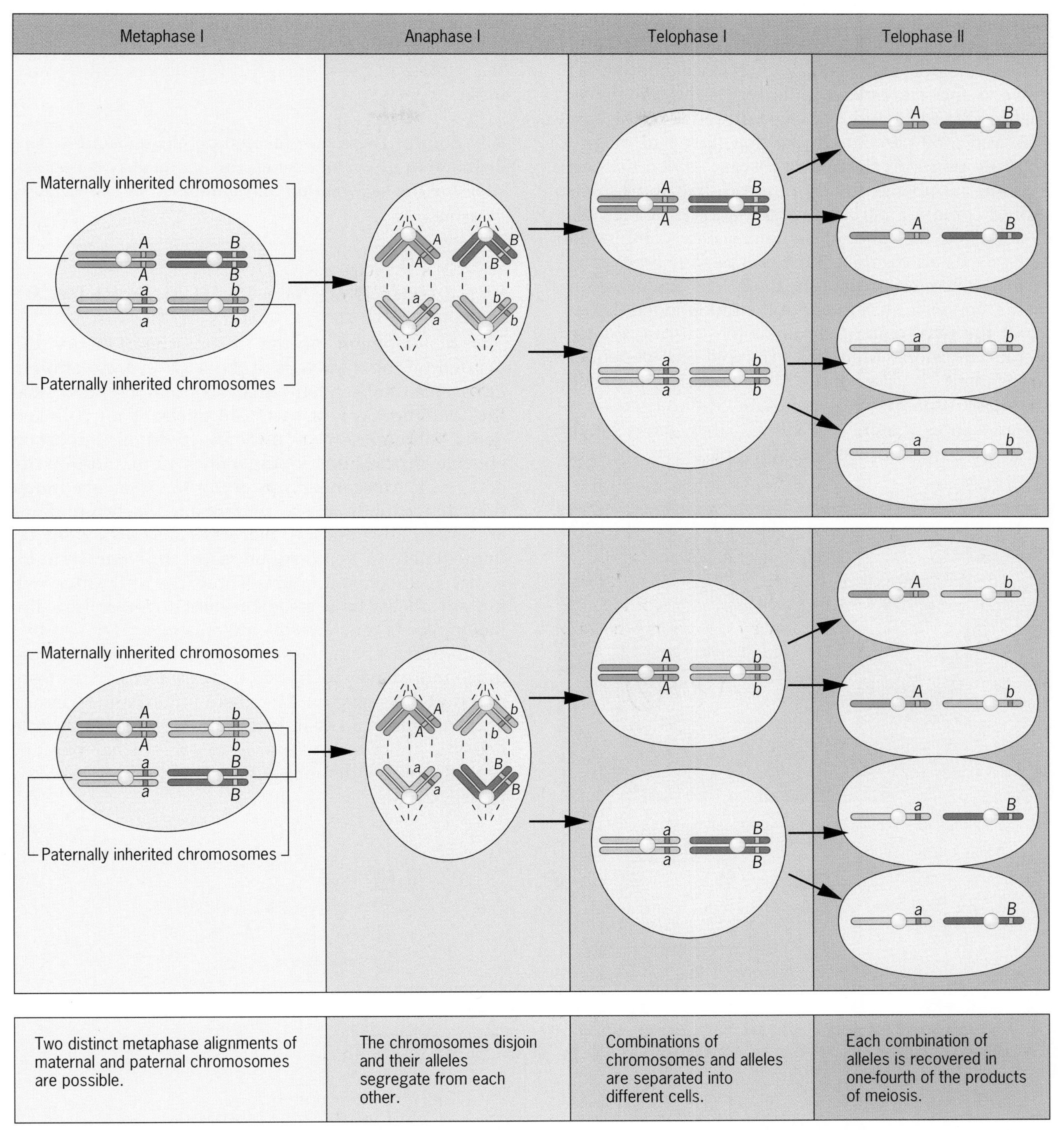

Figure 5.9 Mendel's Principle of Independent Assortment and meiotic chromosome behavior. Alleles on different pairs of chromosomes assort independently in the anaphase of the first meiotic division because maternally and paternally inherited chromosomes have aligned randomly on the cell's equator.

pairs of chromosomes, there are two distinguishable metaphase alignments:

$$\frac{A}{a} \quad \frac{B}{b} \quad \text{or} \quad \frac{A}{a} \quad \frac{b}{B}$$

Each of these is equally likely. Here the space separates different pairs of chromosomes, and the bar separates the homologous members of each pair; during anaphase, the alleles above each bar will move to one pole, and the alleles below it will move to the other.

When disjunction occurs, there is therefore a 50 percent chance that the *A* and *B* alleles will move together to the same pole and a 50 percent chance that they will move to opposite poles. Similarly, there is a 50 percent chance that the *a* and *b* alleles will move to the same pole and a 50 percent chance that they will move to opposite poles. At the end of meiosis, when the chromosome number is finally reduced, half the gametes should contain a parental combination of alleles (*A B* or *a b*), and half should contain a new combination (*A b* or *a B*). Altogether, there will be four types of gametes, each one-fourth of the total. This equality of gamete frequencies is a result of the independent behavior of the two pairs of chromosomes during the first meiotic division. Mendel's Principle of Independent Assortment is therefore a statement about the anaphase separation of genes on different pairs of chromosomes. Later, we will see that genes on the same pair of chromosomes do not assort independently. Instead, because they are physically linked to each other, they tend to travel together through meiosis, violating the Principle of Independent Assortment.

Key Points: **Genes are located on chromosomes. The disjunction of chromosomes during meiosis is responsible for the segregation and independent assortment of genes.**

SEX-LINKED GENES IN HUMAN BEINGS

The development of the chromosome theory depended on the discovery of the *white* eye mutation in *Drosophila*. Subsequent analysis demonstrated that this mutation was a recessive allele of an X-linked gene. Although some of us might credit this important episode in the history of genetics to extraordinarily good luck, Morgan's discovery of the *white* eye mutation was actually not so remarkable. Such mutations are among the easiest to detect because they show up immediately in hemizygous males. In contrast, autosomal recessive mutations show up only after two mutant alleles have been brought together in a homozygote—a much more unlikely event.

In human beings too, recessive X-linked traits are much more easily identified than are recessive autosomal traits. A male only needs to inherit one recessive allele to show an X-linked trait; however, a female needs to inherit two—one from each of her parents. Thus the preponderance of people who show X-linked traits are male.

(a)

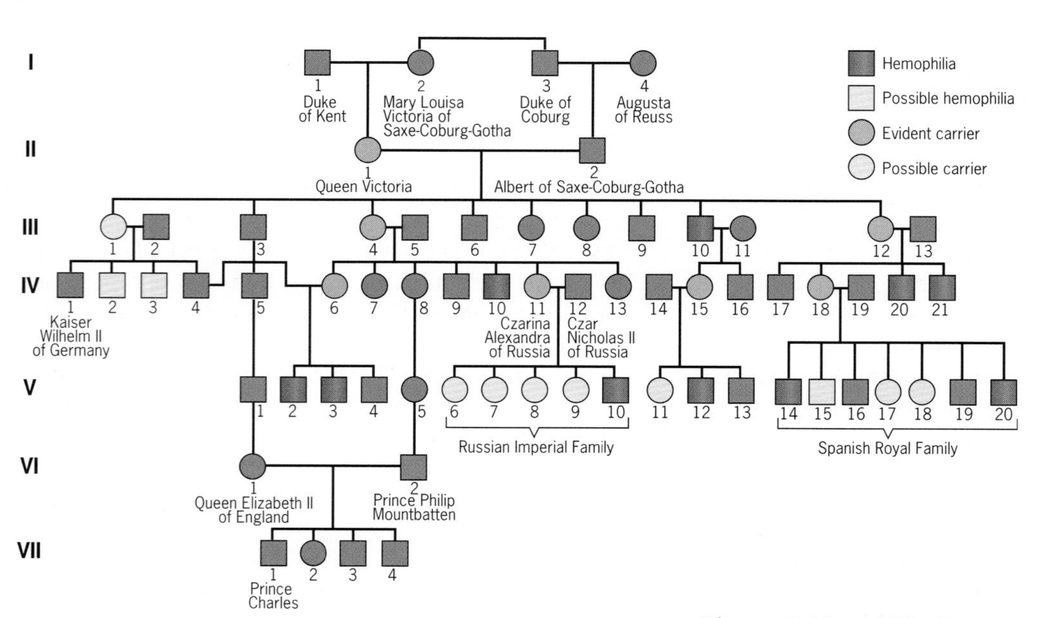

(b)

Figure 5.10 (*a*) The Russian imperial family of Czar Nicholas II. (*b*) Hemophilia in the royal families of Europe.

Hemophilia, an X-Linked Blood-Clotting Disorder

In human beings, **hemophilia** is one of the best known examples of an X-linked trait. People with this disease are unable to produce a factor needed for blood clotting; the cuts and wounds of hemophiliacs continue to bleed and, if not stopped by therapeutic treatment, can cause death. Nearly all the affected individuals in the population are male; only a few female hemophiliacs have ever been reported.

The most famous case of hemophilia occurred in the Russian imperial family at the beginning of the twentieth century (Figure 5.10). Czar Nicholas and Czarina Alexandra had four daughters and one son. The son, Alexis, suffered from hemophilia. The X-linked mutation responsible for Alexis' disease was transmitted to him by his mother, who was a heterozygous carrier. Czarina Alexandra was a granddaughter of Queen Victoria of Great Britain, who was also a carrier. Pedigree records show that Victoria transmitted the mutant allele to three of her nine children: Alice, who was Alexandra's mother, Beatrice, who had two sons with the disease, and Leopold, who had the disease himself. The allele that Victoria carried evidently arose as a new mutation in the germ cells of one of her parents, or in those of a more distant maternal ancestor. Some scholars have even argued that this mutation changed the course of history by contributing to the downfall of the Romanov dynasty in Russia and the subsequent rise of the Soviet state.

Color Blindness, an X-Linked Vision Disorder

In human beings, color perception is mediated by light-absorbing proteins in the specialized cone cells of the retina in the eye. Three such proteins have been identified—one to absorb blue light, one to absorb green light, and one to absorb red light. Color blindness may be caused by an abnormality in any of these receptor proteins. The classic type of color blindness, involving faulty perception of red and green light, follows an X-linked pattern of inheritance. About 5 to 10 percent of human males are red-green color blind; however, a much smaller fraction of females, less than 1 percent, have this disability, suggesting that the mutant alleles are recessive. Molecular studies have shown that there are actually two distinct genes for color perception on the X chromosome; one encodes the receptor for green light, and the other encodes the receptor for red light. Detailed analyses have demonstrated that these two receptors are structurally very similar, probably because the genes encoding them evolved from an ancient color-receptor gene. A third gene for color perception, the one encoding the receptor for blue light, is located on an autosome.

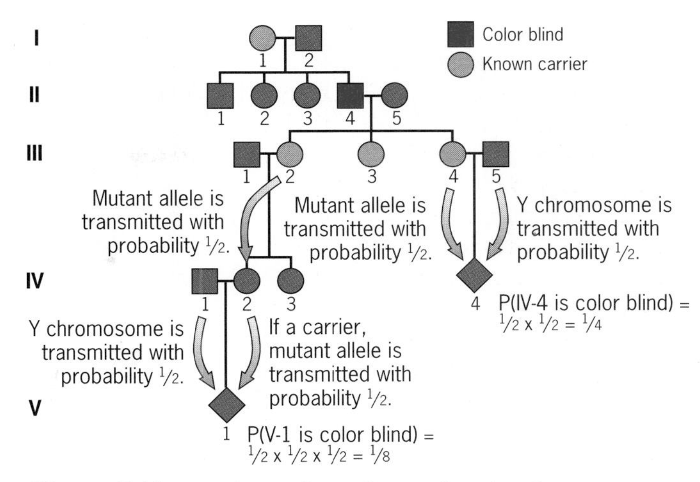

Figure 5.11 Analysis of a pedigree showing the segregation of X-linked color blindness.

In Figure 5.11 color blindness is used to illustrate the procedures for calculating the risk of inheriting a recessive X-linked condition. A heterozygous carrier, such as III-4 in the figure, has a 1/2 chance of transmitting the mutant allele to her children. However, the risk that a particular child will be color blind is only 1/4 since the child must be a male in order to manifest the trait. The female labeled IV-2 in the pedigree could be a carrier of the mutant allele for color blindness, because her mother was. This uncertainity about the genotype of IV-2 introduces another factor of 1/2 in the risk of having a color blind child; thus the risk for her child is $1/4 \times 1/2 = 1/8$.

The Fragile X Syndrome and Mental Retardation

In human beings, many cases of mental retardation appear to follow an X-linked pattern of inheritance. Most of these are associated with a cytological anomaly that is detectable in cells that have been cultured in the absence of certain nucleotides, the building blocks of DNA. This anomaly—a constriction near the tip of the long arm of the X chromosome—gives the impression that the tip is ready to detach from the rest of the chromosome (Figure 5.12*a*); hence the name fragile X chromosome. The clinical features of the fragile X syndrome vary considerably, making diagnosis difficult. Most patients show significant mental impairment, and some show facial and behavioral abnormalities; both males and females can be affected. Among children, the incidence of the fragile X syndrome is about one in 2000.

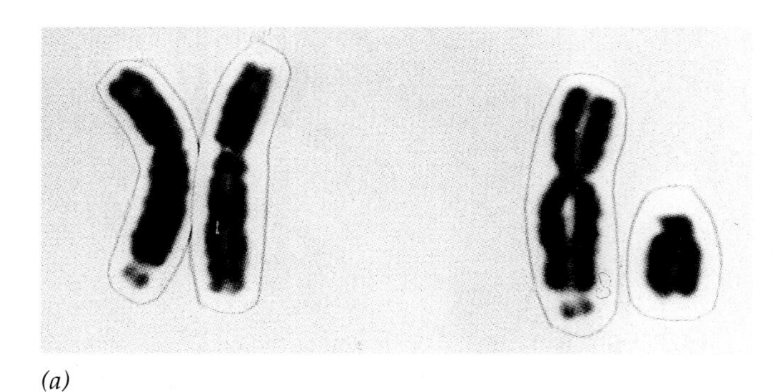

(a)

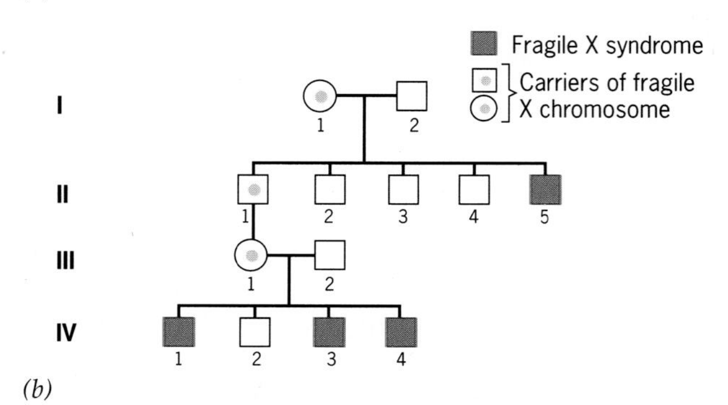

(b)

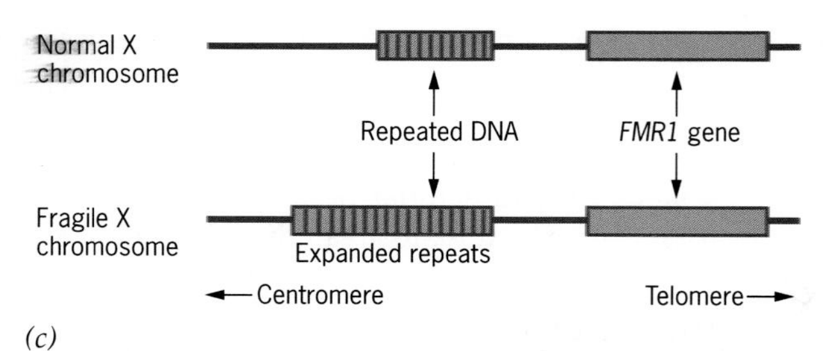

(c)

Figure 5.12 The fragile X chromosome. (*a*) A female (left) showing the fragile X and a normal X chromosome, and a male (right) showing the fragile X and a normal Y chromosome. (*b*) A pedigree showing the inheritance of the fragile X syndrome. The asymptomatic male II-1 is a carrier, indicating that the condition has incomplete penetrance. (*c*) Molecular basis of the fragile X syndrome. The mutation in the fragile X chromosome is due to an expansion of a region of repeats in the DNA flanking the *FMR1* gene. Chemical modification of the DNA around these repeats adversely affects the expression of the *FMR1* gene.

The fragile X syndrome has been described as an X-linked dominant disorder with incomplete penetrance. Affected females are heterozygous for the fragile X chromosome, and affected males are hemizygous for this chromosome. However, some carriers of the chromosome, both male and female, are asymptomatic for the disorder (Figure 5.12*b*). This lack of full penetrance complicates the analysis of pedigrees and makes genetic counseling difficult.

Molecular techniques have been used to isolate and analyze the fragile X site. It consists of a short repeating unit in the DNA adjacent to a gene of unknown function (Figure 5.12*c*). This repeating unit varies in length, being much larger in individuals who show the disorder. A combination of genetic and molecular analyses has demonstrated that the repeating unit tends to increase in size, probably as a result of faulty chromosome replication. This instability explains why individuals who do not show the disorder may have children who do; the repeating unit may expand in the mother's germ line and be passed on to the children, who will have it in all their somatic cells. Although the physiological details are not known, an expanded unit appears to be associated with chemical modification of the surrounding DNA. This modification has an adverse effect on the expression of nearby genes and is probably the actual cause of the fragile X syndrome.

Genes on the Human Y Chromosome

Very few genes have been localized to the human Y chromosome. This failure to find Y-linked genes is somewhat surprising, since a mutation in one of these genes should have an immediate phenotypic effect on the man who carries it. Furthermore, such a mutation should be passed on to all the man's sons, but to none of his daughters. A Y-linked gene should therefore be the easiest kind of gene to identify in conventional pedigree analysis. To date, however, only a few Y-linked genes have been discovered. One is responsible for the synthesis of a male-specific substance called the H-Y antigen, which is found on cell surfaces; another is involved in the production of a factor that is critical for the differentiation of the testes and the subsequent acquisition of male sexual characteristics. Advances in molecular genetics have provided new techniques to identify other Y-linked genes, but even with these, the view that the Y chromosome has fewer genes than any other human chromosome probably will not change.

Genes on Both the X and Y Chromosomes

Some genes are present on both the X and Y chromosomes, mostly near the ends of the short arms (see Figure 5.2). Alleles of these genes do not follow a distinct X- or Y-linked pattern of inheritance. Instead, they are transmitted from mothers and fathers to sons and daughters alike, mimicking the inheritance of an autosomal gene. Such genes are therefore called **pseudoautosomal genes**. In males, the regions that contain these genes seem to mediate pairing between the X and Y chromosomes.

Key Points: **Disorders caused by recessive X-linked mutations such as hemophilia and color blindness are more common in males than in females. In humans, the Y chromosome carries few genes. Some of these genes are also carried by the X chromosome.**

SEX CHROMOSOMES AND SEX DETERMINATION

In the animal kingdom, sex is perhaps the most conspicuous phenotype. Animals with distinct males and females are sexually dimorphic. Sometimes this dimorphism is established by environmental factors. In one species of turtles, for example, sex is determined by temperature. Eggs that have been incubated above 30° C hatch into females, whereas eggs that have been incubated at a lower temperature hatch into males. In many other species, sexual dimorphism is established by genetic factors, often involving a pair of sex chromosomes.

Sex Determination in Human Beings

The discovery that human females are XX and that human males are XY suggested that sex might be determined by the number of X chromosomes or by the presence or absence of a Y chromosome. As we now know, the second hypothesis is correct. In humans and other placental mammals, maleness is due to a dominant effect of the Y chromosome (Figure 5.13). The evidence for this fact comes from the study of individuals with an abnormal number of sex chromosomes. XO animals develop as females, and XXY animals develop as males. The dominant effect of the Y chromosome is manifested early in development, when it directs the primordial gonads to develop into testes. Once the testes have formed, they secrete testosterone, a hormone that stimulates the development of male secondary sexual characteristics.

Recent research has pinpointed the **testis-determining factor (TDF)** on the human Y chromosome. This factor appears to correspond to a gene called ***SRY*** **(for sex-determining region Y)**, which is located just outside the pseudoautosomal region in the chromo-

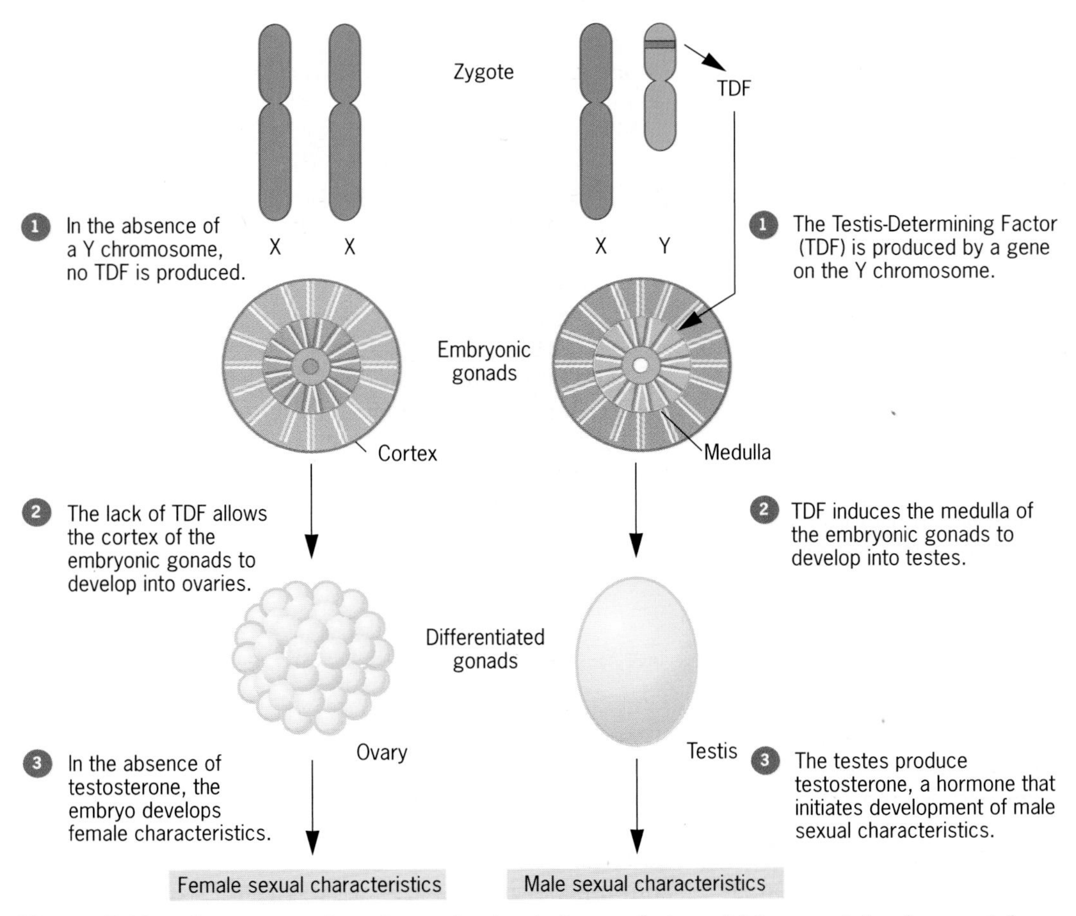

Figure 5.13 The process of sex determination in human beings. Male sexual development depends on the production of the testis-determining factor (TDF) by a gene on the Y chromosome. In the absence of this factor, the embryo develops as a female.

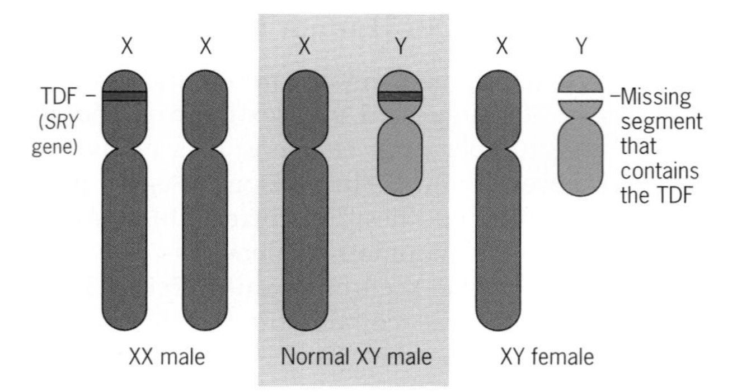

Figure 5.14 Evidence localizing the testis-determining factor (TDF) to the short arm of the Y chromosome in normal males. The TDF appears to correspond to the *SRY* gene. In XX males, a small region containing this gene has been inserted into one of the X chromosomes, and in XY females, it has been deleted from the Y chromosome.

some's short arm. The discovery of *SRY* was made possible by the identification of unusual individuals whose sex was inconsistent with their chromosome constitution—XX males and XY females (Figure 5.14). Some of the XX males were found to carry a small piece of the Y chromosome inserted into one of the X chromosomes. This piece evidently carried a gene responsible for maleness. Some of the XY females were found to carry an incomplete Y chromosome. The part of the Y chromosome that was missing corresponded to the piece that was present in the XX males; its absence in the XY females apparently prevented them from developing testes. These complementary lines of evidence showed that a particular segment of the Y chromosome was needed for male development. Molecular analyses subsequently identified the *SRY* gene in this male-determining segment. Additional research has shown that an *SRY* gene is present on the Y chromosome of the mouse, and that—like the human *SRY* gene—it specifies male development. Both human and mouse *SRY* genes encode a protein that can bind to DNA. However, the precise way in which these proteins trigger male development is not known. One hypothesis is that they regulate the activity of other genes that are directly involved in the process of testis formation.

After the testes have formed, testosterone secretion initiates the development of male sexual characteristics. Testosterone is a hormone that binds to receptors on many kinds of cells. Once bound, the hormone-receptor complex transmits a signal to the cell, instructing it in how to differentiate. The concerted differentiation of many types of cells leads to the development of distinctly male characteristics such as heavy musculature, beard, and deep voice. If the testosterone signaling system fails, these characteristics do not appear and the individual develops as a female. One reason for failure is an inability to make the testosterone receptor (Figure 5.15). XY individuals with this biochemical deficiency initially develop as males—testes are formed and testosterone is produced. However, the testosterone has no effect because it cannot reach its target cells to transmit the developmental signal. Individuals lacking the testosterone receptor therefore switch sexes during embryological development and acquire female sexual characteristics. They do not, however, develop ovaries and are therefore sterile. This syndrome, called **testicular feminization**, results from a mutation in an X-linked gene, *tfm*, which encodes the testosterone receptor. The *tfm* mutation is transmitted from mothers to sons (who are phenotypically female) in a typical X-linked pattern.

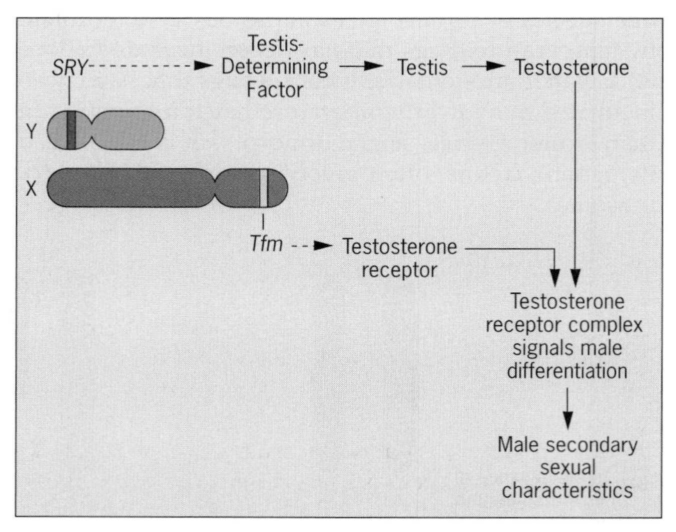

(*a*) Normal male with the wild-type *Tfm* gene.

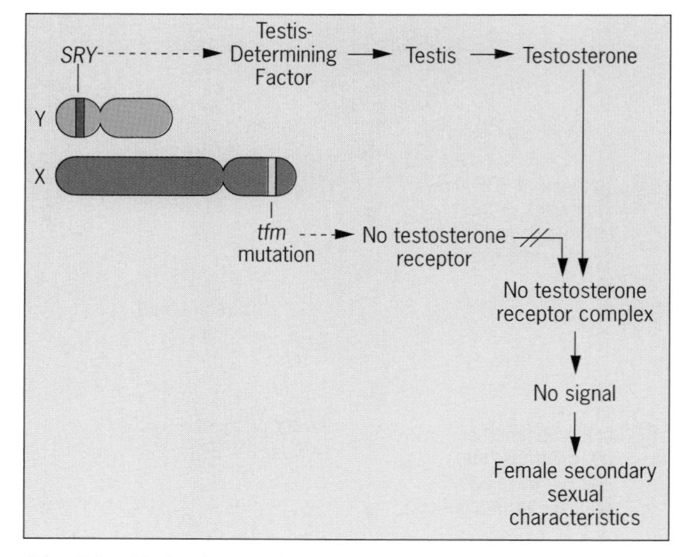

(*b*) Male with the tfm mutation and testicular feminization.

Figure 5.15 Testicular feminization, a condition caused by an X-linked mutation, *tfm*, that prevents the production of the testosterone receptor.

Sex Determination in *Drosophila*

The Y chromosome in *Drosophila*—unlike that in humans—plays no role in sex determination. Instead, the sex of the fly is determined by the ratio of X chromosomes to autosomes. This mechanism was first demonstrated by Bridges in 1921 through an analysis of flies with unusual chromosome constitutions.

Normal diploid flies have a pair of sex chromosomes, either XX or XY, and three pairs of autosomes, usually denoted AA; here, each A represents one haploid set of autosomes. Through genetic trickery, Bridges contrived flies with abnormal numbers of chromosomes (Table 5.2). He observed that whenever the ratio of X's to A's was 1.0 or greater, the fly was female, and whenever it was 0.5 or less, the fly was male. Flies with an X:A ratio between 0.5 and 1.0 developed characteristics of both sexes; thus Bridges called them intersexes. In none of these flies did the Y chromosome have any effect on the sexual phenotype. However, it was required for male fertility.

We now know that an X-linked gene called *Sex-lethal* (*Sxl*) is a key player in the *Drosophila* sex-determination system (Figure 5.16). An elaborate network of other X-linked genes, working with factors already present in the *Drosophila* egg, sets the level of *Sxl* activity in a zygote. If the X:A ratio is greater than or equal to one, the *Sxl* gene is activated and the zygote develops as a female; if it is less than or equal to 0.5, the *Sxl* gene is inactivated, and the zygote develops as a male. A ratio between 0.5 and 1.0 causes a mix of signals, and the zygote develops both male and female characteristics. This system has the remarkable ability to count chromosomes, compute the ratio of X's to A's, and then flip the *Sxl* switch either on or off. Inappropriate expression of *Sxl*—off in females or on in males—leads to embryonic death—hence the name of the gene, *Sex-lethal*. The molecular details of sex determination in *Drosophila*, and the nature of the lethality associated with inappropriate expression of the *Sxl* gene, are discussed in Chapters 22 and 23.

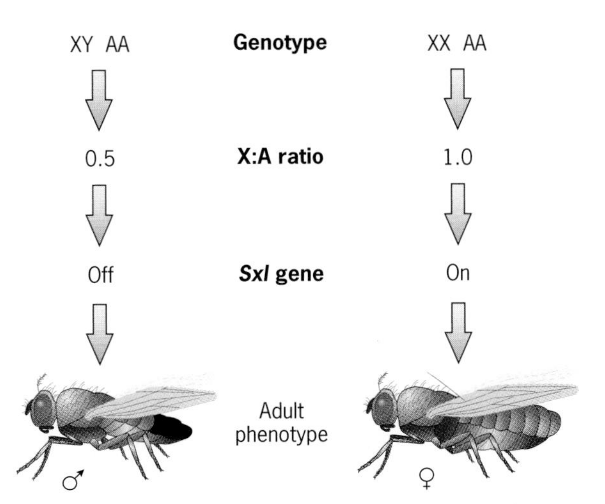

Figure 5.16 The *Sxl* sex determination switch in *Drosophila*. The *Sxl* gene is turned on in zygotes in which the X:A ratio is 1.0, and off in zygotes in which the X:A ratio is 0.5. Expression of the *Sxl* gene causes the zygote to develop into a female, whereas nonexpression of this gene causes the zygote to develop into a male.

TABLE 5.2
Ratio of X Chromosomes to Autosomes and the Corresponding Phenotype in *Drosophila*

X Chromosomes (X) and Sets of Autosomes (A)		*X:A Ratio*	*Phenotype*
1X	2A	0.5	Male
2X	2A	1.0	Female
3X	2A	1.5	Metafemale
4X	3A	1.33	Metafemale
4X	4A	1.0	Tetraploid female
3X	3A	1.0	Triploid female
3X	4A	0.75	Intersex
2X	3A	0.67	Intersex
2X	4A	0.5	Tetraploid male
1X	3A	0.33	Metamale

Sex Determination in Other Animals

In *Drosophila* and human beings, males produce two kinds of gametes, X-bearing and Y-bearing. For this reason, they are referred to as the **heterogametic** sex; in these species females are the **homogametic** sex. In birds, butterflies, and some reptiles, this situation is reversed (Figure 5.17). Males are homogametic (usually denoted ZZ) and females are heterogametic (ZW). However, little is known about the mechanism of sex determination in the Z-W sex chromosome system.

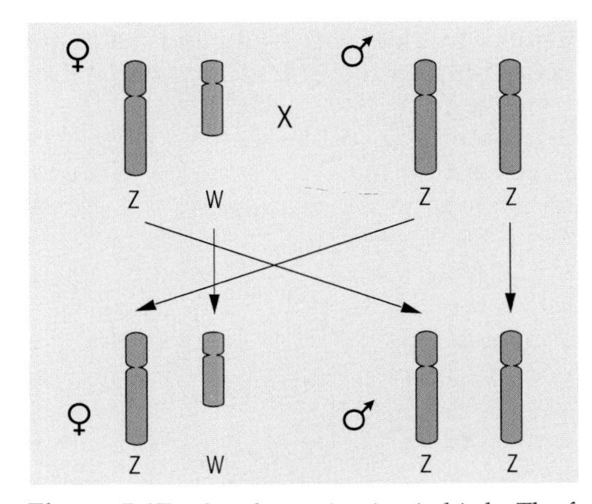

Figure 5.17 Sex determination in birds. The female is heterogametic (ZW) and the male is homogametic (ZZ). The sex of the offspring is determined by which of the sex chromosomes, Z or W, is transmitted by the female.

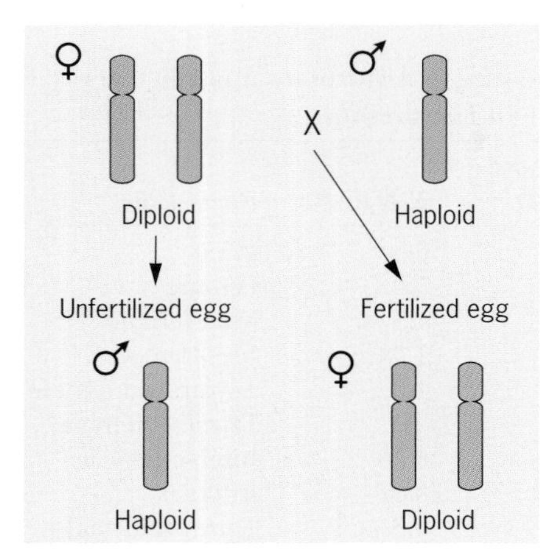

Figure 5.18 Sex determination in honeybees. Females, which are derived from fertilized eggs, are diploid, and males, which are derived from unfertilized eggs, are haploid.

In honeybees, sex is determined by ploidy (Figure 5.18). Diploid embryos, which develop from fertilized eggs, become females; haploid embryos, which develop from unfertilized eggs, become males. Whether or not a given female will mature into a reproductive form (queen) depends on how she was nourished as a larva. In this system, a queen can control the ratio of males to females by regulating the proportion of unfertilized eggs that she lays. Because this number is small, most of the progeny are female, albeit sterile, and serve as workers for the hive. In a haplo-diplo system of sex determination, eggs are produced through meiosis in the queen, and sperm are produced through mitosis in the male. This system ensures that fertilized eggs will have the diploid chromosome number and that unfertilized eggs will have the haploid number.

Some wasps also have a haplo-diplo method of sex determination. In these species diploid males are sometimes produced, but they are always sterile. Detailed genetic analysis in one species, *Bracon hebetor*, has indicated that the diploid males are homozygous for a sex-determining locus, called *X*; diploid females are always heterozygous for this locus. Evidently, the sex locus in *Bracon* has many alleles; crosses between unrelated males and females therefore almost always produce heterozygous diploid females. However, when the mates are related, there is an appreciable chance that their offspring will be homozygous for the sex locus, in which case they develop into sterile males.

Key Points: **In humans the testis-determining factor (TDF) on the Y chromosome causes an embryo to develop into a male. Without this factor, an embryo develops into a female. In *Drosophila*, sex is determined by the ratio of X chromosomes to sets of autosomes (X:A). For X:A ≤ 0.5 , the fly develops as a male, for X:A ≥ 1.0, it develops as a female, and for $0.5 <$ X:A < 1.0, it develops as an intersex.**

DOSAGE COMPENSATION OF X-LINKED GENES

Animal development is usually sensitive to an imbalance in the number of genes. Normally, each gene is present in two copies. Departures from this condition, either up or down, can cause abnormal phenotypes, and sometimes even death. It is therefore puzzling that so many species should have a sex-determination system based on females with two X chromosomes and males with only one. In these species, how is the numerical difference of X-linked genes accommodated? *A priori*, there are two possible mechanisms to compensate for this difference: (1) each X-linked gene works twice as hard in males as it does in females, or (2) one copy of each X-linked gene is inactivated in females. Extensive research has shown that both mecha-

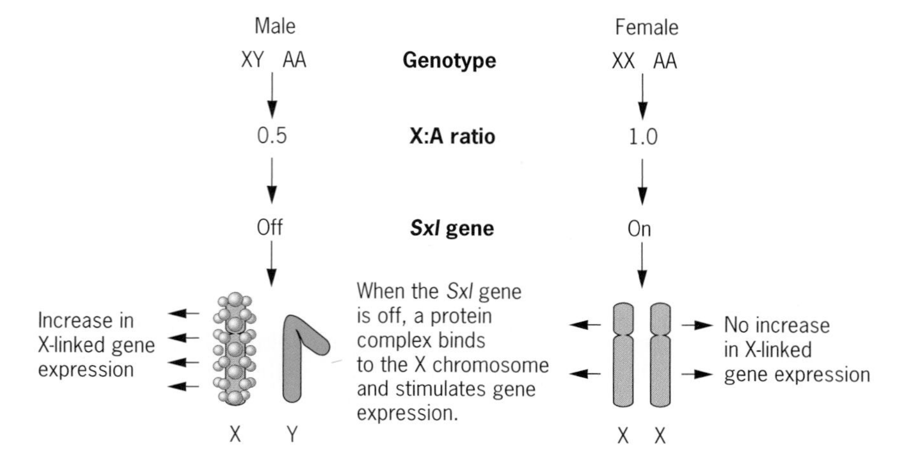

Figure 5.19 Dosage compensation in *Drosophila*. The single X chromosome in males is hyperactivated owing to a protein complex that binds to the chromosome. In females, where the *Sxl* gene is expressed, this complex does not bind to the X chromosomes, and these chromosomes are not hyperactivated.

nisms are utilized, the first in *Drosophila* and the second in mammals.

Hyperactivation of X-linked Genes in Male *Drosophila*

In *Drosophila*, dosage compensation of X-linked genes is achieved by an increase in the activity of these genes in males. This phenomenon, called **hyperactivation**, involves the *Sex-lethal* gene, which also plays a key role in sex determination (Figure 5.19). The *Sxl* gene is turned on in females and off in males. When the *Sxl* gene product is absent (as it is in males), a protein complex binds to many sites on the X chromosome and triggers a doubling of gene activity (see Chapter 22). When the *Sxl* gene product is present (as it is in females), this protein complex does not bind, and hyperactivation of X-linked genes does not occur. In this way, total X-linked gene activity in males and females is approximately equalized.

Inactivation of X-linked Genes in Female Mammals

In placental mammals, dosage compensation of X-linked genes is achieved by the **inactivation** of one of the female's X chromosomes (Figure 5.20). This mechanism was first proposed in 1961 by the British geneticist Mary Lyon, who inferred it from studies on mice. Subsequent research by Lyon and others has shown that the inactivation event occurs when the mouse embryo consists of a few thousand cells. At this time, each cell makes an independent decision to silence one of its X chromosomes. The chromosome to be inactivated is chosen at random; once chosen, however, it remains inactivated in all the descendants of that cell. Thus female mammals are **genetic mosaics** containing two types of cell clones; the maternally inherited X chromosome is inactivated in roughly half of these clones, and the paternally inherited X is inactivated in the other half. A female that is heterozygous for an X-linked gene is therefore able to show two different phenotypes.

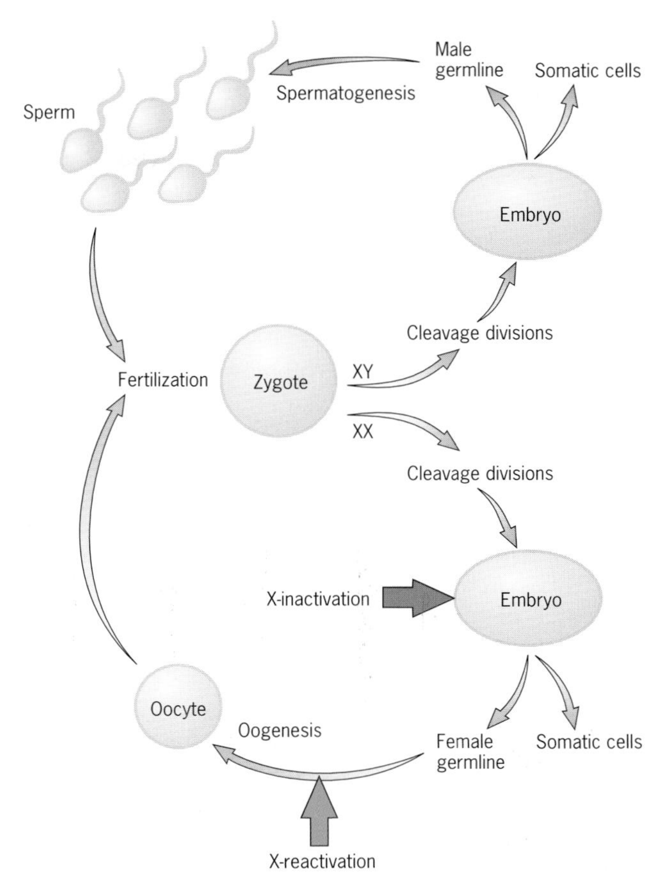

Figure 5.20 X chromosome inactivation in mammals. One of the X chromosomes in XX females is inactivated in each cell of the early embryo. In the germline, the inactivated X chromosomes are subsequently reactivated during oogenesis.

One of the best examples of this phenotypic mosacism comes from the study of fur coloration in cats and mice (Figure 5.21). In both of these species the X chromosome carries a gene for pigmentation of the fur. Females heterozygous for different alleles of this

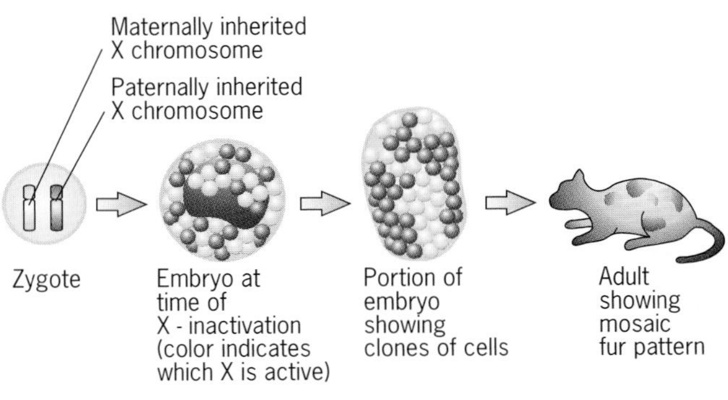

Figure 5.21 Color mosaics resulting from X chromosome inactivation in female mammals. (*a*) Formation of clones of cells in a cat embryo that produce different patches of fur in the adult. (*b*) A tortoise shell cat. This female is heterozgyous for an X-linked coat color gene. The orange and black patches are due to inactivation of different alleles in the pigment-producing cells of the body.

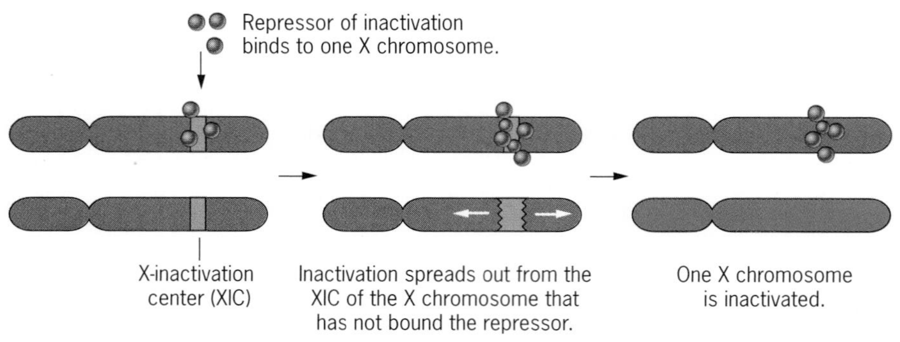

Figure 5.22 Mechanism of X chromosome inactivation.

gene show patches of light and dark fur. The light patches express one allele, and the dark patches express the other. In cats, where one allele produces black pigment and the other produces orange pigment, this patchy phenotype is called tortoise shell. Each patch of fur defines a clone of pigment-producing cells, or melanocytes, that were derived by mitosis from a precursor cell present at the time of X chromosome inactivation.

Many aspects of the mechanism for X chromosome inactivation are still a mystery. Genetic analyses have shown that in both humans and mice it begins at a site in the long arm of the X chromosome, and that it spreads in both directions to the chromosome's ends (Figure 5.22). The initiating site is called the **X-inactivation center (XIC)**. Curiously, this site is very close to a gene called *XIST* (X-inactive specific transcript), which is not silenced by the inactivation process. Ongoing research suggests that the product of the *XIST* locus may play a key role in this mechanism of dosage compensation (see Chapter 22).

An X chromosome that has been inactivated does not look or act like other chromosomes. Chemical analyses show that its DNA is modified by the addition of numerous methyl groups. In addition, it condenses into a darkly staining structure called a **Barr body** (Figure 5.23), after the Canadian geneticist Murray Barr, who first observed it. This structure becomes attached to the inner surface of the nuclear membrane, where it replicates out of step with the other chromosomes in the cell. The inactivated X chromosome remains in this altered state in all the somatic tissues. However, in the germ tissues it is reactivated, perhaps because two copies of some X-linked genes are needed for the successful completion of oogenesis.

Cytological studies have identified human beings with more than two X chromosomes (see Chapter 6). For the most part, these people are phenotypically normal females, apparently because all but one of their X chromosomes is inactivated. Often all the inactivated X's congeal into a single Barr body. These observations suggest that cells may have a limited amount of some factor needed to prevent X-inactivation. Once this factor has been used to keep one X chromosome active, all the others quietly succumb to the inactivation process.

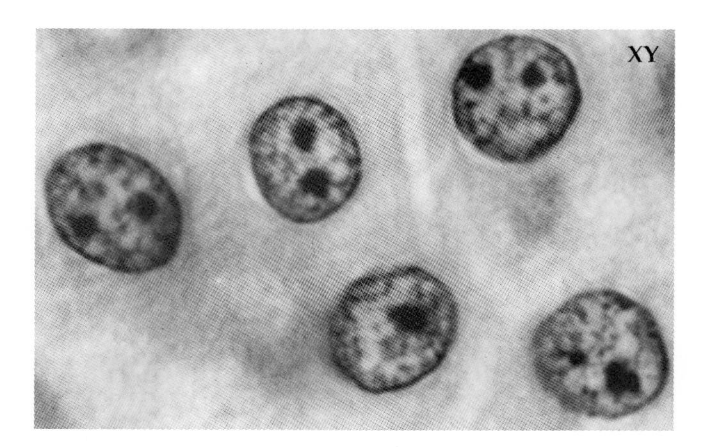

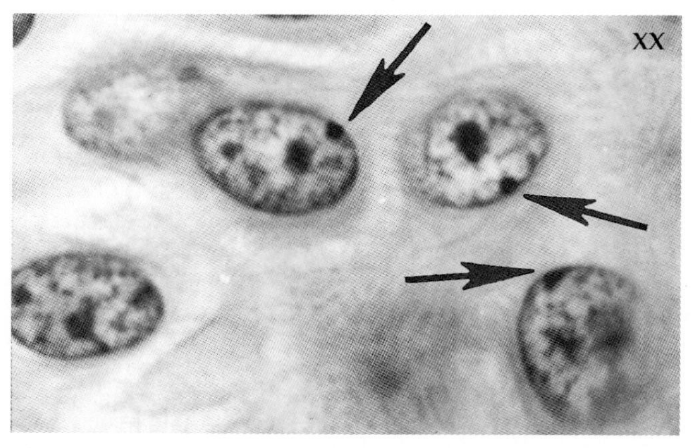

Figure 5.23 Barr bodies. Photograph of nuclei of normal human male (XY) and female (XX) cells. The Barr bodies (arrows) are visible only in the female nuclei.

Key Points: **In *Drosophila*, dosage compensation for X-linked genes is achieved by hyperactivating the single X chromosome in males. In mammals, dosage compensation is achieved by inactivating one of the two X chromosomes in females.**

TESTING YOUR KNOWLEDGE

1. The Lesch–Nyhan syndrome is a serious metabolic disorder affecting about one in 50,000 males in the population of the United States. A class of molecules called purines, which are biochemical precursors of DNA, accumulate in the nervous tissues and joints of people with the Lesch–Nyhan syndrome. This biochemical abnormality is caused by a deficiency for the enzyme hypoxanthine-guanine phosphoribosyltransferase (HPRT), which is encoded by a gene located on the X chromosome. Individuals deficient for this enzyme are unable to control their movements, and unwillingly engage in self-destructive behavior such as biting and scratching themselves. The males labeled IV-5 and IV-6 in the pedigree below have the Lesch-Nyhan syndrome. What are the risks that V-1 and V-2 will inherit this disorder?

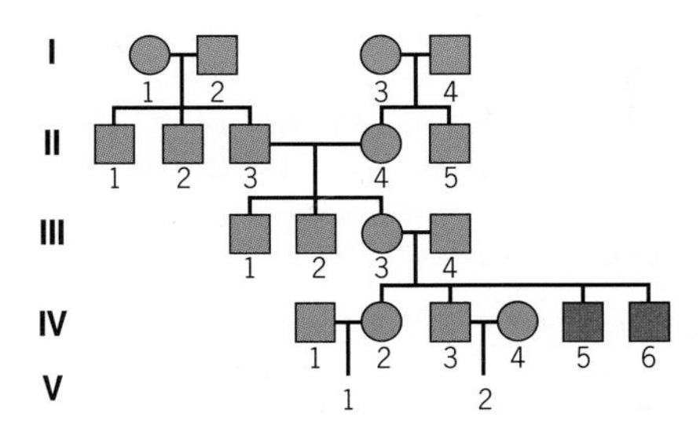

ANSWER

We know that III-3 must be a heterozygous carrier of the mutant allele (*h*) because two of her sons are affected. However, because she herself does not show the mutant phenotype, we know that her other X chromosome must carry the wild-type allele (*H*). Given that III-3 is genotypically *H/h*, there is a one-half chance that she passed the mutant allele to her daughter (IV-2). If she did, there is a one-half chance that IV-2 will transmit this allele to her child (V-1), and there is a one-half chance that this child will be a male. Thus the risk that V-1 will have the Lesch–Nyhan syndrome is (1/2) x (1/2) x (1/2) = 1/8. For V-2, the risk of inheriting the Lesch–Nyhan syndrome is essentially zero. This child's father (IV-3) is not a carrier, and even if he were, he would not transmit the mutant allele to a son. The child's mother comes from outside the family and is very unlikely to be a carrier because the trait is so rare in the general population. Thus V-2 has virtually no chance of suffering from the Lesch–Nyhan syndrome.

2. A geneticist crossed female *Drosophila* that had white eyes and ebony bodies to wild-type males, which had red eyes and gray bodies. Among the F_1, all the daughters had red eyes and gray bodies, and all the sons had white eyes and gray bodies. These flies were intercrossed to produce F_2 progeny, which were classified for eye and body color and then counted. Among 384 total progeny, the geneticist obtained the following results:

Combination of traits			
Eye color	**Body color**	**Males**	**Females**
white	ebony	20	21
white	gray	70	73
red	ebony	28	25
red	gray	76	71

How would you explain the inheritance of eye color and body color?

ANSWER

The results in the F_1 tell us that both mutant phenotypes are caused by recessive alleles. Furthermore, because the males and females have different eye color phenotypes, we know that the eye color gene is X-linked and that the body color gene is autosomal. In the F_2, the two genes assort independently, as we would expect for genes located on different chromosomes. Below, we show the genotypes of the different classes of flies in this experiment, using *w* for the white mutation and *e* for the ebony mutation; the wild-type alleles are denoted by plus signs. Following the convention of *Drosophila* geneticists, we write the sex chromosomes (X and Y) on the left and the autosomes on the right. A question mark in a genotype indicates that either the wild-type or mutant alleles could be present.

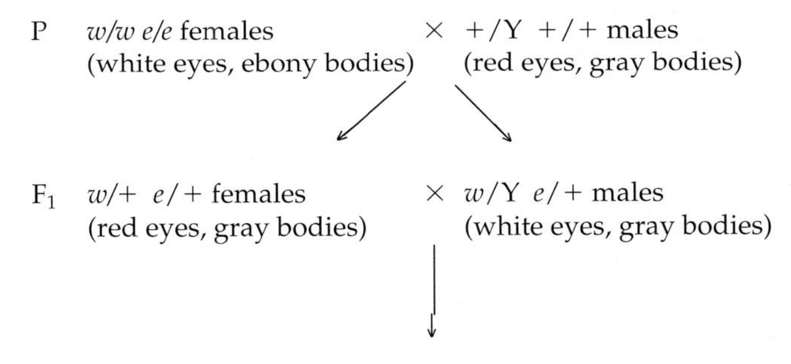

F_2

Phenotypes		Genotypes	
Eye color	Body color	Males	Females
white	ebony	*w*/Y *e/e*	*w/w e/e*
white	gray	*w*/Y +/?	*w/w* +/?
red	ebony	+/Y *e/e*	+/*w e/e*
red	gray	+/Y +/?	+/*w* +/?

QUESTIONS AND PROBLEMS

5.1 What are the genetic differences between male- and female-determining sperm in animals with heterogametic males?

5.2 A male with singed bristles appeared in a culture of *Drosophila*. How would you determine if this unusual phenotype was due to an X-linked mutation?

5.3 In grasshoppers, rosy body color is caused by a recessive mutation; the wild-type body color is green. If the gene for body color is on the X chromosome, what kind of progeny would be obtained from a mating between a homozygous rosy female and a hemizygous wild-type male? (In grasshoppers, females are XX and males are XO.)

5.4 In the mosquito *Anopheles culicifacies*, *golden* body (*go*) is a recessive X-linked mutation, and *brown* eyes (*bw*) is a recessive autosomal mutation. A homozygous XX female with golden body is mated to a homozygous XY male with brown eyes. Predict the phenotypes of their F_1 offspring. If the F_1 progeny are intercrossed, what kinds of progeny will appear in the F_2, and in what proportions?

5.5 What are the sexual phenotypes of the following genotypes in *Drosophila* : XX, XY, XXY, XXX, XO.

5.6 In human beings, a recessive X-linked mutation, *g*, causes green defective color vision; the wild-type allele, *G*, causes normal color vision. A man (a) and a woman (b), both with normal vision, have three children, all married to people with normal vision: a color-defective son (c), who has a daughter with normal vision (f); a daughter with normal vision (d), who has one color-defective son (g) and two normal sons (h); and a daughter with normal vision (e), who has six normal sons (i). Give the most likely genotypes for the individuals (a to i) in this family.

5.7 If a father and son both have defective color vision, is it likely that the son inherited the trait from his father?

5.8 A normal woman, whose father had hemophilia, marries a normal man. What is the chance that their first child will have hemophilia?

5.9 A man with X-linked color blindness marries a woman with no history of color blindness in her family. The daughter of this couple marries a normal man, and their daughter also marries a normal man. What is the chance that this last couple will have a child with color blindness? If this couple has already had a child with color blindness, what is the chance that their next child will be color blind?

5.10 A *Drosophila* female homozygous for a recessive X-linked mutation that causes vermilion eyes is mated to a wild-type male with red eyes. Among their progeny, all the sons have vermilion eyes, and nearly all the daughters have red eyes; however, a few daughters have vermilion eyes. Explain the origin of these vermilion-eyed daughters.

5.11 A *Drosophila* female heterozygous for the recessive X-linked mutation *w* (for white eyes) and its wild-type allele w^+ is mated to a wild-type male with red eyes. Among the sons, half have white eyes and half have red eyes. Among the daughters, nearly all have red eyes; however, a few have white eyes. Explain the origin of these white-eyed daughters.

5.12 Suppose that a mutation occurred in the *TDF* gene on the human Y chromosome, knocking out its ability to produce the testis-determining factor. Predict the phenotype of an individual who carried this mutation and a normal X chromosome.

5.13 A woman carries the testicular feminization mutation (*tfm*) on one of her X chromosomes; the other X carries the wild-type allele (*Tfm*). If the woman marries a normal man, what fraction of her children will be phenotypically female? Of these, what fraction will be fertile?

5.14 Would a human with two X chromosomes and a Y chromosome be male or female?

5.15 Predict the sex of *Drosophila* with the following chromosome compositions (A = haploid set of autosomes): (a) 4X 4A; (b) 3X 4A; (c) 2X 3A; (d) 1X 3A; (e) 2X 2A; (f) 1X 2A.

5.16 In *Drosophila*, the gene for bobbed bristles (recessive allele *bb*, bobbed bristles; wild-type allele +, normal bristles) is located on the X chromosome and on a homologous segment of the Y chromosome. Give the genotypes and phenotypes of the offspring from the following crosses: (a) $X^{bb}\,X^{bb} \times X^{bb}\,Y^{+}$; (b) $X^{bb}\,X^{bb} \times X^{+}\,Y^{bb}$; (c) $X^{+}\,X^{bb} \times X^{+}\,Y^{bb}$; (d) $X^{+}\,X^{bb} \times X^{bb}\,Y^{+}$.

5.17 In chickens, the absence of barred feathers is due to a recessive allele. A barred rooster was mated with a nonbarred hen, and all the offspring were barred. These F_1 chickens were intercrossed to produce F_2 progeny, among which all the males were barred; half the females were barred and half were nonbarred. Are these results consistent with the hypothesis that the gene for barred feathers is located on one of the sex chromosomes?

5.18 A *Drosophila* male carrying a recessive X-linked mutation for yellow body is mated to a homozygous wild-type female with gray body. The daughters of this mating all have uniformly gray bodies. Why aren't their bodies a mosaic of yellow and gray patches?

5.19 How many Barr bodies would be visible in the nuclei of human cells with the following chromosome compositions: (a) XY; (b) XX; (c) XXY; (d) XXX; (e) XXXX; (f) XYY?

5.20 Males in a certain species of deer have two nonhomologous X chromosomes, denoted X_1 and X_2, and a Y chromosome. Each X chromosome is about half as large as the Y chromosome, and its centromere is located near one of the ends; the centromere of the Y chromosome is located in the middle. Females in this species have two copies of each of the X chromosomes and lack a Y chromosome. How would you predict the X and Y chromosomes to pair and disjoin during spermatogenesis to produce equal numbers of male- and female-determining sperm?

BIBLIOGRAPHY

CLINE, T. W. 1993. "The *Drosophila* sex determination signal: how do flies count to two?" *Trends in Genet.* 9:385–390.

LYON, M. F. 1993. "Epigenetic inheritance in mammals." *Trends in Genet.* 9:123–128.

Wheat field in central North America.

6

Variation in Chromosome Number and Structure

Chromosomes, Agriculture, and Civilization

The cultivation of wheat originated some 10,000 years ago in the Middle East. Today, wheat is the principal food crop for more than a billion people. Wheat is grown in diverse environments, from Norway to Argentina. More than 17,000 varieties have been developed, each adapted to a different locality. The total wheat production of the world is hundreds of millions of metric tons annually, accounting for more than 20 percent of the food calories consumed by the entire human population. Wheat is clearly an important agricultural crop, and some would argue, a mainstay of civilization.

Modern cultivated wheat, *Triticum aestivum*, is a hybrid of at least three different species. Its progenitors were low-yielding grasses that grew in Syria, Iran, Iraq, and Turkey. Some of these grasses appear to have been cultivated by the ancient peoples of this region. Although we do not know the exact course of events, two of these species apparently interbred, producing a species that excelled as a crop plant. Through human cultivation, this hybrid species was selectively improved, and then it too interbred with a third species, yielding a triple-hybrid that was even better suited for agriculture. Modern wheat is descended from these triply hybrid plants.

What made the triple-hybrid wheats so superior to their ancestors? They had larger grains, they were more easily harvested, and they grew in a wider range of conditions. We now understand the chromosomal basis for these improvements. Triple-hybrid wheat contains the chromosomes of each of its progenitors. Genetically, it is an amalgamation of the genes of three different species.

Such chromosomal mixtures are common in the plant kingdom, especially among plants with an agricultural significance. Variations in chromosome number are also found in animals, including our own species, and some of these variations are associated with conspicuous phenotypes. In this chapter we consider different kinds of changes in chromosome number. We also consider several types of changes in chromosome structure.

CYTOLOGICAL TECHNIQUES

The nucleus of a cell is a repository of genetic information. The thousands of genes contained in each nucleus enable it to direct the synthesis of a large number of biochemical substances. The information for all this biological potential is organized into a set of chromosomes embedded in the nuclear matrix. Each species has a characteristic number of chromosomes, and each chromosome within the species has a characteristic structure. Geneticists can study these features—chromosome number and structure—by staining dividing cells with certain dyes and then examining them with a microscope. The analysis of stained chromosomes is the main activity of the discipline called **cytogenetics**.

Cytogenetics had its roots in the research of several nineteenth-century European biologists who discovered chromosomes and observed their behavior during mitosis, meiosis, and fertilization. This research blossomed during the twentieth century, as microscopes improved and better procedures for preparing and staining chromosomes were developed. The demonstration that genes reside on chromosomes boosted interest in this research and led to important studies on chromosome number and structure. Today, cytogenetics has significant applied aspects, especially in medicine, where it is used to determine whether disease conditions are associated with chromosome abnormalities.

Analysis of Mitotic Chromosomes

Researchers perform most cytological analyses on dividing cells, usually cells in the middle of mitosis. To enrich for cells at this stage, they have traditionally used rapidly growing tissues such as animal embryos and plant root tips. However, the development of cell-culturing techniques has made it possible to study chromosomes in other types of cells (Figure 6.1). For example, human white blood cells are collected from peripheral blood, separated from the nondividing red

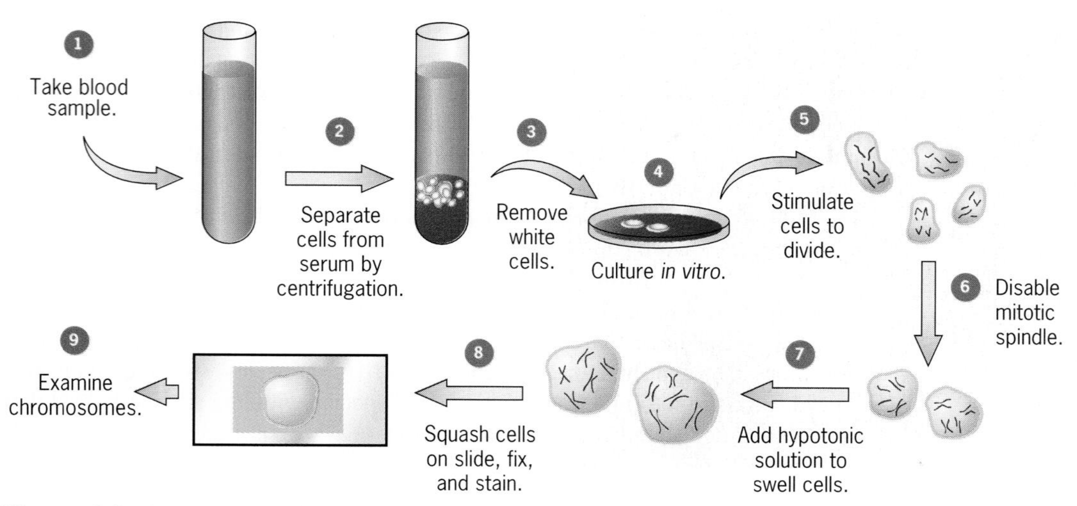

Figure 6.1 Preparation of cells for cytological analysis.

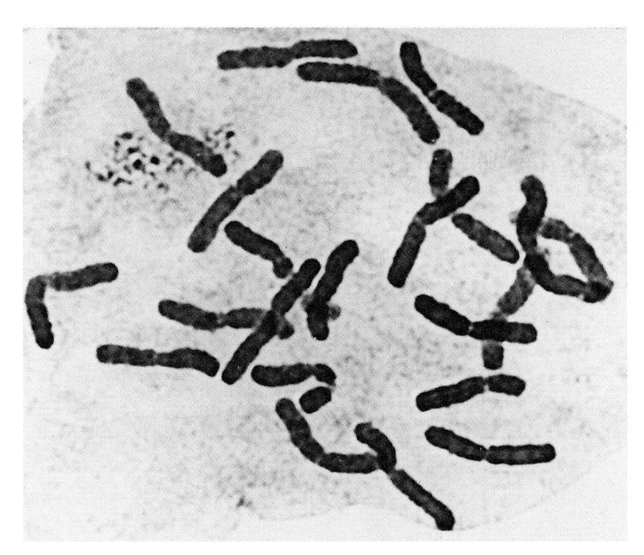

Figure 6.2 Chromosomes of the incense cedar (*Calocedrus decurrens*) stained with aceto-carmine.

blood cells, and put into culture. The white cells are then stimulated to divide by chemical treatment, and midway through division a sample of the cells is prepared for cytological analysis. The usual procedure is to treat the dividing cells with a chemical that disables the spindle apparatus. The effect of this interference is to trap the chromosomes in mitosis, when they are most easily seen. Mitotically arrested cells are then swollen by immersion in a hypotonic solution that causes the cells to take up water by osmosis. The contents of each cell are diluted by the additional water, so that when the cells are squashed on a microscope slide, the chromosomes are spread out in an uncluttered fashion. This technique greatly facilitates subsequent analysis, especially if the chromosome number is large. For many years it was erroneously thought that human cells contained 48 chromosomes. The correct number, 46, was determined only after the swelling technique was used to separate the chromosomes within individual mitotic cells.

Until the late 1960s and early 1970s, chromosome spreads were stained with Feulgen's reagent, a purple dye that reacts with the sugar molecules in DNA, or with aceto-carmine, a deep red dye (Figure 6.2). Although these stains are still used for routine cytological analyses, more detailed studies now use staining compounds that insert between the base pairs of DNA. A molecule with this insertional property is called an **intercalating agent**. Quinacrine, a chemical relative of the antimalarial drug quinine, is one such molecule. Chromosomes that have been stained with quinacrine show a characteristic pattern of bands. However, because quinacrine is a **fluorescent** compound, the bands appear only when the chromosomes are exposed to ultraviolet (UV) light. Ultraviolet irradiation causes some of the quinacrine molecules that have inserted into the chromosome to emit energy. As a result, parts of the chromosome shine brightly, while other parts remain dark. This differential fluorescence creates a pattern of bright and dark bands which is highly reproducible. Using these bands, cytogeneticists can identify particular chromosomes in a cell, and they can also analyze these chromosomes for structural abnormalities. This staining procedure is called **Q banding**, and the bands that it produces are called Q bands (Figure 6.3).

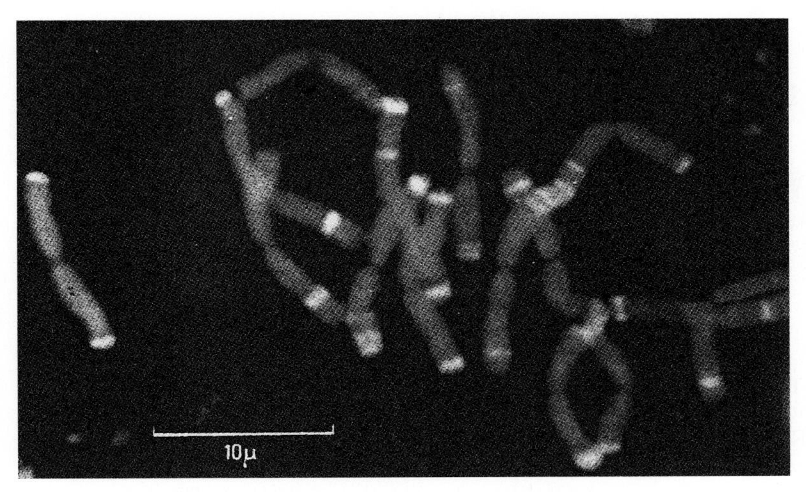

Figure 6.3 Metaphase chromosomes of the plant *Allium carinatum*, stained with quinacrine to reveal Q bands.

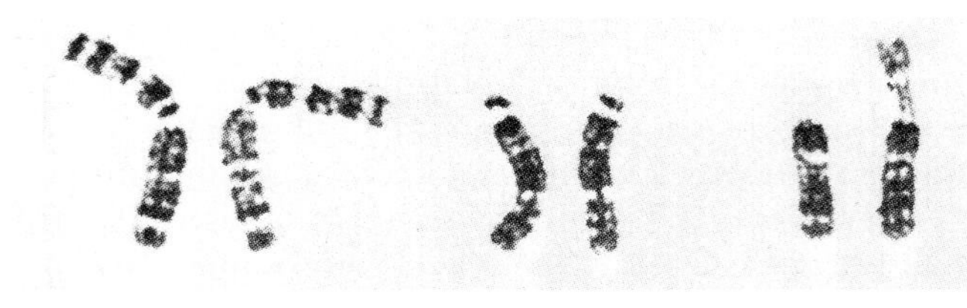

(a)

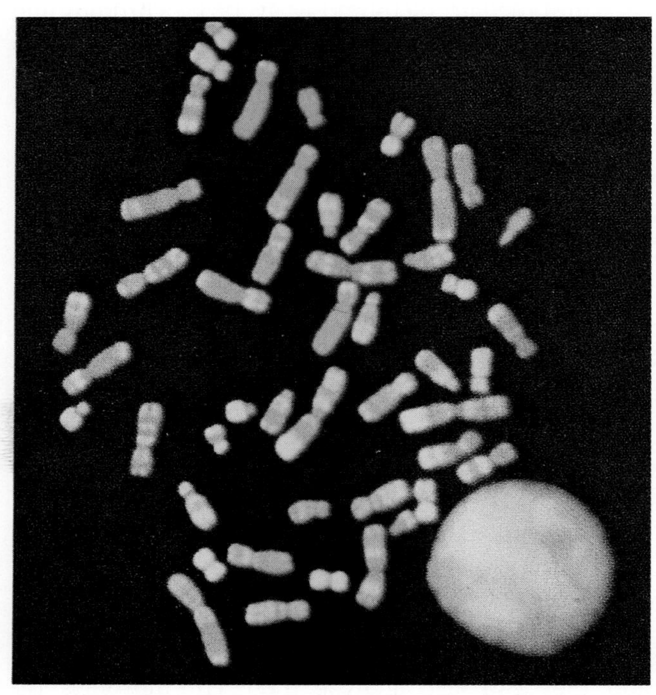

(b)

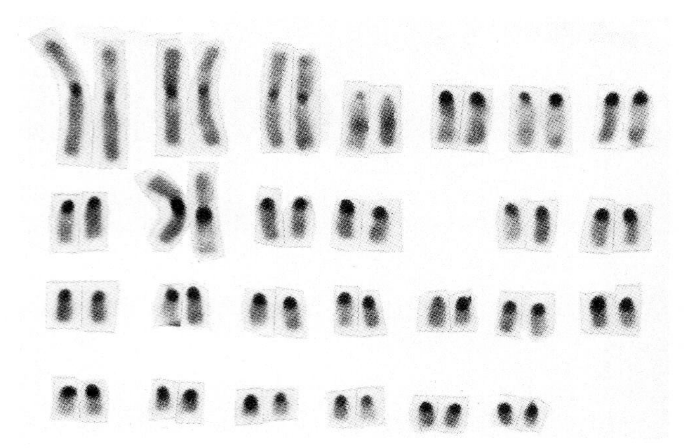

(c)

Figure 6.4 (*a*) Metaphase chromosomes of the Asian muntjak stained to show G banding. (*b*) Metaphase human chromosomes stained with acridine orange to show R banding. (*c*) Metaphase chromosomes of the domestic sheep stained to show C banding.

Excellent nonfluorescent staining techniques have also been developed. The most popular of these uses **Giemsa** stain, a mixture of dyes named after its inventor, Gustav Giemsa. Like quinacrine, Giemsa creates a reproducible pattern of bands on each chromosome. However, the nature of the banding pattern depends on how the chromosomes were prepared prior to staining. One procedure, called **G banding**, gives dark bands that correspond roughly to the bright bands obtained with quinacrine; another procedure, called **R banding**, gives the reverse pattern—dark bands that correspond roughly to light G bands. A third procedure, called **C banding**, stains the region around the centromere of each chromosome. These different banding techniques provide cytogeneticists with the means to analyze fine details of chromosome structure (Figure 6.4).

The Human Karyotype

Diploid human cells contain 46 chromosomes—44 autosomes and two sex chromosomes, which are XX in females and XY in males. At mitotic metaphase, each of the chromosomes can be recognized by its size, shape, and banding pattern. For cytological analysis, well-stained metaphase spreads are photographed, and then each of the chromosome images is cut out of the picture, matched with its partner, and arranged from largest to smallest on a chart (Figure 6.5). The largest autosome is number 1, and the smallest is number 21. (For historical reasons, the second smallest chromosome has been designated number 22.) The X chromosome is intermediate in size; the Y chromosome is about the same size as chromosome 22. This chart of chromosome cutouts is called a **karyotype** (from the Greek word meaning "kernel," a reference to the contents of the nucleus). A skilled researcher can use a karyotype to identify abnormalities in chromosome number and structure.

Before the banding techniques were available, it was difficult to distinguish one human chromosome from another. Cytogeneticists could only arrange the chromosomes into groups according to size, classifying the largest as group A, the next largest as group B, and so forth. Although they could recognize seven different groups, within these groups it was nearly impossible to identify a particular chromosome. Today—as a result of the banding techniques—we can routinely identify each of the 24 chromosomes. The banding techniques also make it possible to distinguish each arm of a chromosome. The short arm is denoted by the letter **p** (from the French word *petite*, meaning "small") and the long arm by the letter **q** (because it follows "p" in the alphabet). Thus, for example, a cytogeneticist can refer specifically to the short arm of chromosome 5 simply by writing "5p."

Analysis of Meiotic Chromosomes

Meiotic chromosomes are usually much more difficult to analyze than mitotic chromosomes. Meiosis occurs only in the germline, usually a minute portion of the total adult organism, and the cells that undergo meiosis are not easily cultured in the laboratory. Classical studies of meiotic chromosomes used plant material, especially maize and various species of lilies (Figure 6.6). Meiotic cells were obtained from these organisms by removing the male and female reproductive organs from flowers and then squashing samples of germ tissue. The meiotic chromosomes of many other organisms, including human beings, have also been studied, but sometimes only with great difficulty. In higher animals, such studies require surgery to obtain the necessary samples of germ tissue.

Cytogenetic Variation: An Overview

The phenotypes of many organisms are affected by changes in the number of chromosomes in their cells; sometimes even changes in part of a chromosome can be significant. These numerical changes are usually described as variations in the **ploidy** of the organism (from the Greek word meaning "fold," as in "twofold"). Organisms with complete, or normal, sets of chromosomes are said to be *euploid* (from the Greek words meaning "good" and "fold"). Organisms that carry extra sets of chromosomes are said to be *polyploid* (from the Greek words meaning "many" and "fold"), and the level of polyploidy is described by referring to a basic chromosome number, usually denoted n; n is defined as the number of chromosomes in a set. Thus diploids, with two chromosome sets, have $2n$ chromosomes; triploids, with three sets, have $3n$; tetraploids, with four sets, have $4n$; and so forth. Organisms in which a particular chromosome, or chromosome segment, is under- or overrepresented are said to be *aneuploid* (from the Greek words meaning "not," "good," and "fold"). These organisms therefore suffer from a specific genetic imbalance. The distinction between aneuploidy and polyploidy is that aneuploidy refers to a numerical change in part of the genome, usually just a single chromosome, whereas polyploidy refers to a numerical change in a whole set of chromosomes. Aneuploidy implies a genetic imbalance, but polyploidy does not.

Cytogeneticists have also catalogued various types of structural changes in the chromosomes of organisms. For example, a piece of one chromosome may be fused to another chromosome, or a segment within a chromosome may be inverted with respect to the rest of that chromosome. These structural changes are called rearrangements. Because rearrangements segregate irregularly during meiosis, they are often associated with aneuploidy. In the sections that follow, we consider all these cytogenetic variations—polyploidy, aneuploidy, and chromosome rearrangements.

Key Points: **Chromosomes are most easily analyzed in dividing cells. Among the various dyes used to stain chromosomes, quinacrine and Giemsa produce banding patterns that are useful in identifying individual chromosomes within a karyotype.**

POLYPLOIDY

Polyploidy, the presence of extra chromosome sets, is fairly common in plants but very rare in animals. One-half of all known plant genera contain polyploid species, and about two-thirds of all grasses are polyploids. Many of these species reproduce asexually. In animals, where reproduction is primarily by sexual means, polyploidy is rare, probably because it interferes with the sex-determination mechanism.

One general effect of polyploidy is that cell size is increased, presumably because there are more chromosomes in the nucleus. Often this increase in size is correlated with an overall increase in the size of the organism. Polyploid species tend to be larger and more robust than their diploid counterparts. These characteristics have a practical significance for human beings, who depend on many polyploid plant species for food. These species tend to produce larger seeds and fruits, and therefore provide greater yields in agriculture. Wheat, coffee, potatoes, bananas, strawberries, and cotton are all polyploid crop plants. Many ornamental garden plants, including roses, chrysanthemums, and tulips, are also polyploid (Figure 6.7).

Sterile Polyploids

In spite of their robust physical appearance, many polyploid species are sterile. Extra sets of chromosomes segregate irregularly in meiosis, leading to grossly unbalanced (that is, aneuploid) gametes. If such gametes unite in fertilization, the resulting zygotes almost always die. This inviability among the zygotes explains why many polyploid species are sterile.

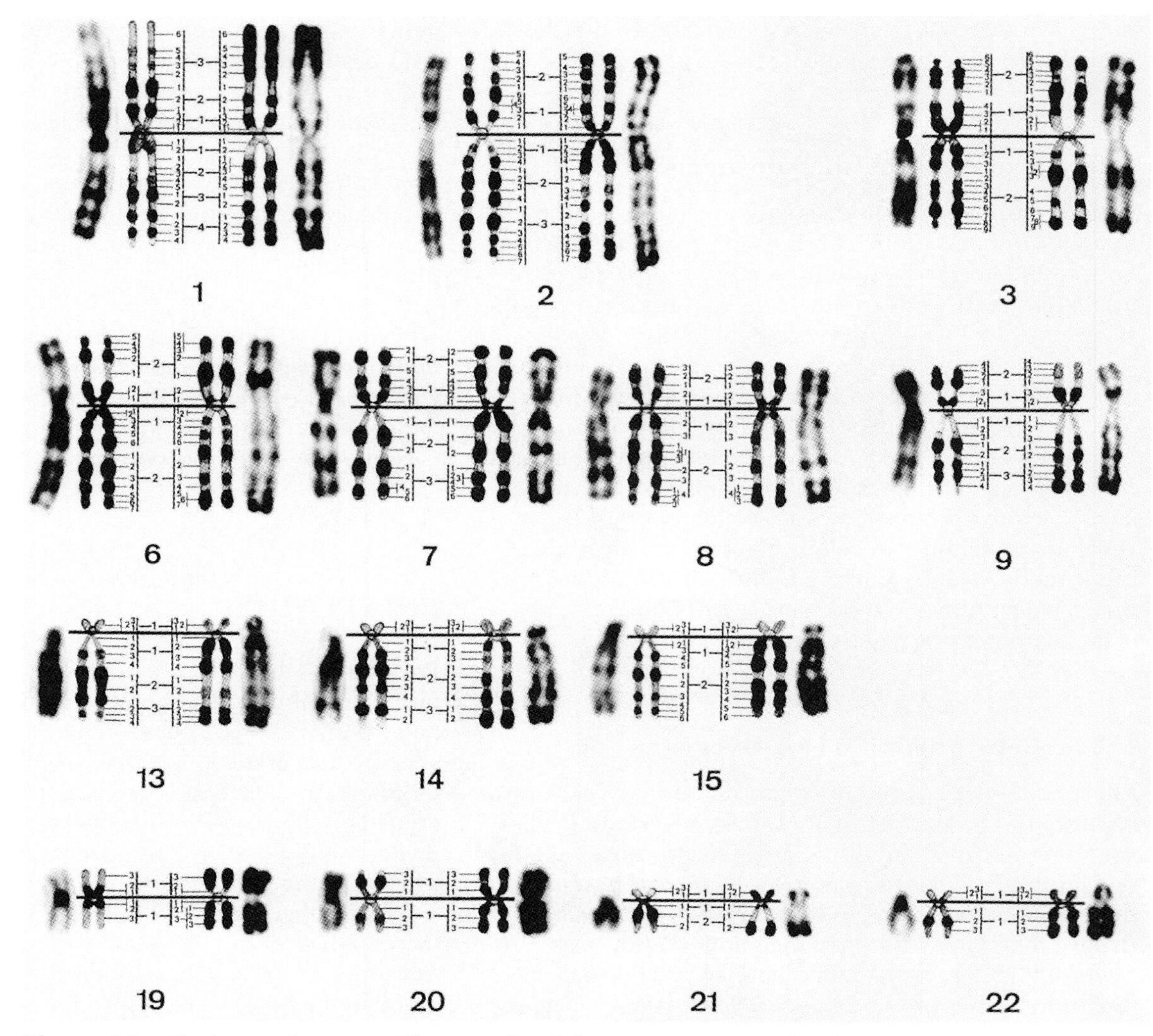

Figure 6.5 The human karyotpe. Photographs of chromosomes stained to show G banding (left) and R banding (right). Autosomes are numbered from 1 to 22. The short arm (p) and long arm (q) of each chromosome is subdivided into numbered regions, starting at the centromere.

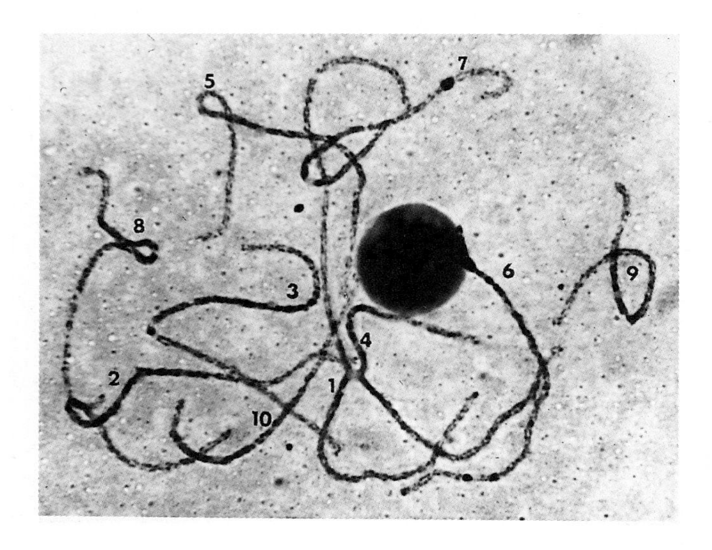

Figure 6.6 The 10 pairs of chromosomes in maize shown in prophase I of meiosis. The large nucleolus is associated with the end of chromosome 6.

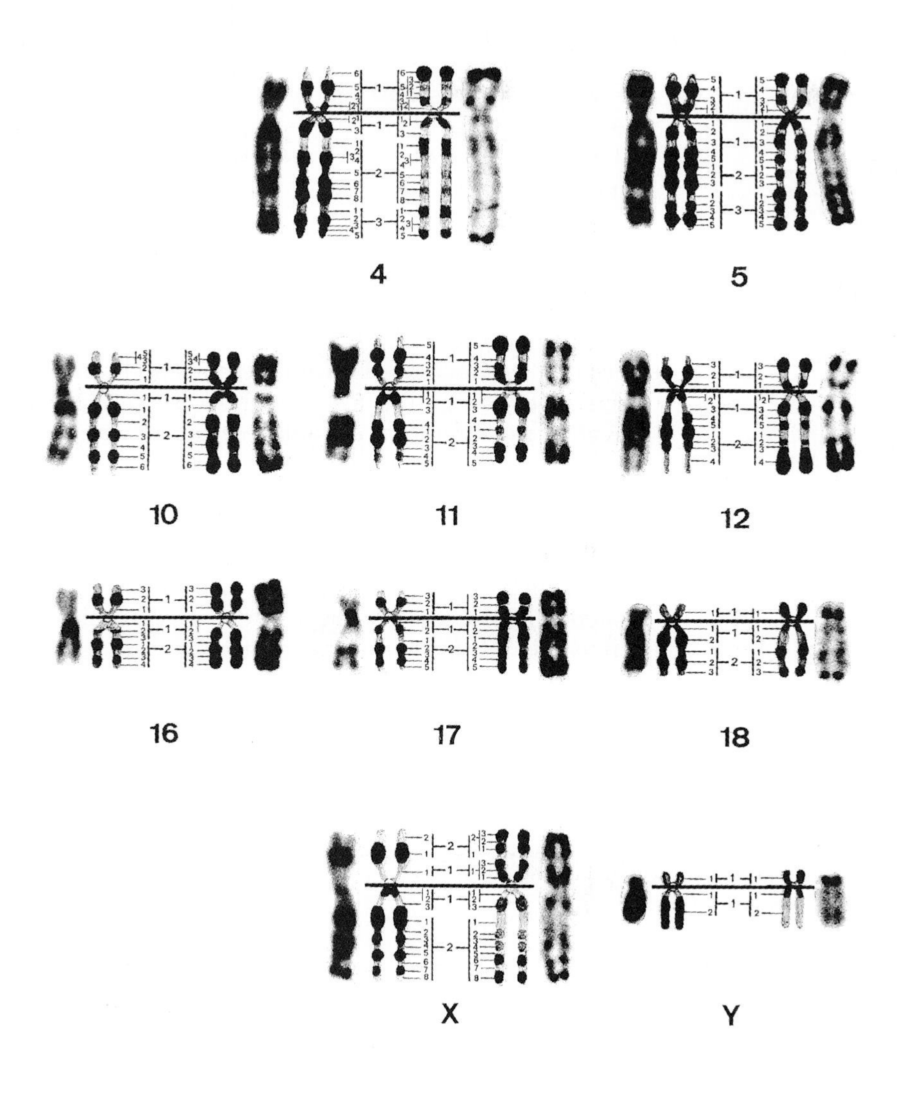

Figure 6.7 Polyploid plants with agricultural or horticultural significance.

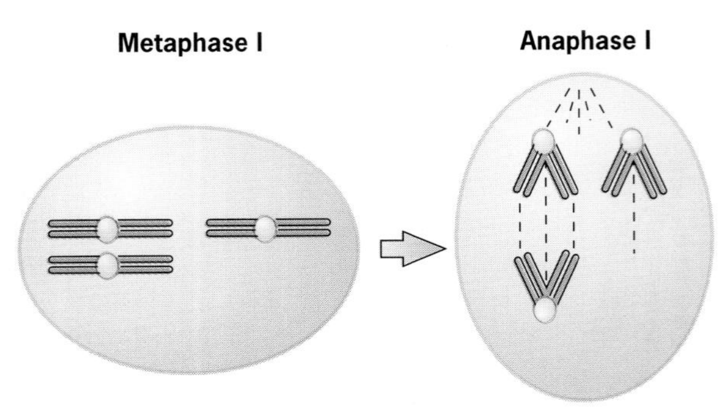

(*a*) Two of the three homologs synapse, leaving a univalent free to move to either pole during anaphase.

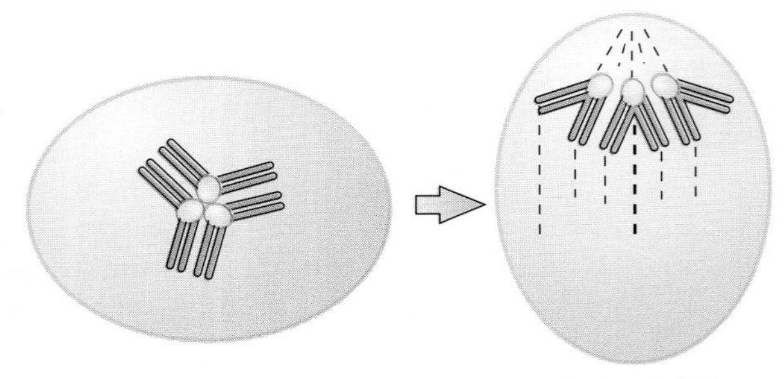

(*b*) All three homologs synapse, forming a trivalent, which may move as a unit to one pole during anaphase. However, other anaphase disjunctions are possible.

Figure 6.8 Meiosis in a triploid. (*a*) Univalent formation. (*b*) Trivalent formation.

As an example, let us consider a triploid species with three identical sets of n chromosomes. The total number of chromosomes is therefore $3n$. When meiosis occurs, each chromosome will try to pair with its homologs (Figure 6.8). One possibility is that two homologs will pair completely along their length, leaving the third without a partner; this solitary chromosome is called a **univalent**. Another possibility is that all three homologs will synapse, forming a **trivalent** in which each member is partially paired with each of the others. In either case, it is difficult to predict how the chromosomes will move during anaphase of the first meiotic division. The more likely event is that two of the homologs will move to one pole, and one homolog will move to the other, yielding gametes with one or two copies of the chromosome. However, all three homologs might move to one pole, producing gametes with zero or three copies of the chromosome. Because this segregational uncertainty applies to each trio of chromosomes in the cell, the total number of chromosomes in a gamete can vary from zero to $3n$.

Zygotes formed by fertilization with such gametes are almost certain to die; thus most triploids are completely sterile. In agriculture, this sterility is circumvented by propagating the species asexually. The many methods of asexual propagation include cultivation from cuttings (bananas), grafts (Winesap, Gravenstein, and Baldwin apples), and bulbs (tulips). In nature, polyploid plants can also reproduce asexually. One mechanism is **apomixis**, which involves a modified meiosis that produces unreduced eggs; these eggs then form seeds that germinate into new plants. A common plant that reproduces in this way is the dandelion, *Taraxacum officinale*, a highly successful polyploid weed.

Fertile Polyploids

The meiotic uncertainties that occur in triploids also occur in tetraploids with four identical chromosome sets. Such tetraploids are therefore also sterile. However, some tetraploids are able to produce viable progeny. Close examination shows that these species contain two distinct sets of chromosomes and that each set has been duplicated. Thus fertile tetraploids seem to have arisen by chromosome duplication in a hybrid that was produced by a cross of two different, but related, diploid species; most often these species have the same or very similar chromosome numbers. Figure 6.9 shows a plausible mechanism for the origin of such a tetraploid. Two diploids, denoted A and B, are crossed to produce a hybrid that receives one set of chromosomes from each of the parental species. Such a hybrid will probably be sterile because the A and B

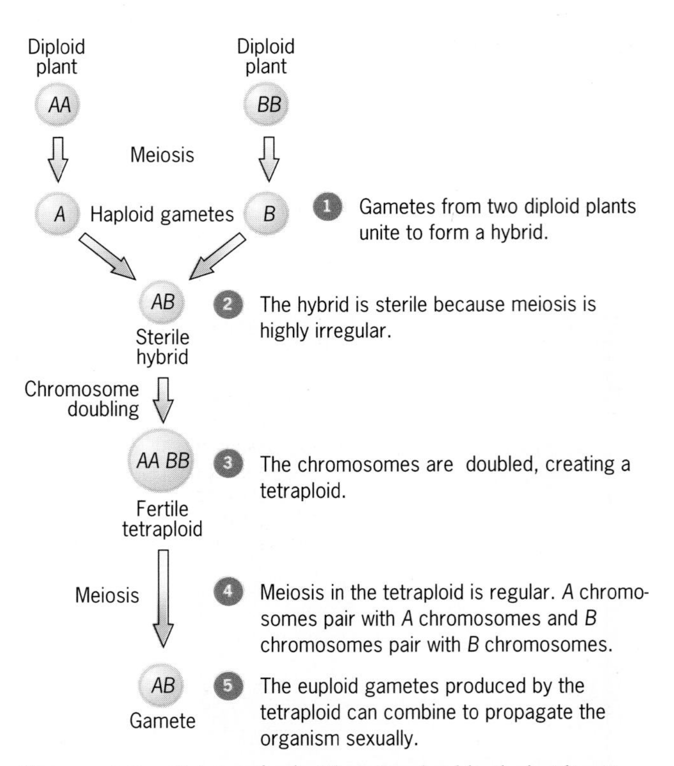

Figure 6.9 Origin of a fertile tetraploid by hybridization between two diploids and subsequent doubling of the chromosomes.

chromosomes cannot pair with each other. However, if the chromosomes in this hybrid are duplicated, meiosis will proceed in reasonably good order. Each of the A and B chromosomes will be able to pair with a perfectly homologous partner. Meiotic segregation can therefore produce gametes with a complete set of A and B chromosomes. In fertilization, these "diploid" gametes will unite to form tetraploid zygotes, which will survive because each of the parental sets of chromosomes will be balanced.

This scenario of hybridization between different but related species followed by chromosome doubling has evidently occurred many times during plant evolution. In some cases, the process has occurred repeatedly, generating complex polyploids with distinct chromosome sets. One of the best examples is modern bread wheat, *Triticum aestivum* (Figure 6.10). This important crop species is a hexaploid containing three different chromosome sets, each of which has been duplicated. There are seven chromosomes in each set, for a total of 21 in the gametes and 42 in the somatic cells. Modern wheat therefore seems to have been formed by two hybridization events. The first involved two diploid species that combined to form a tetraploid, and the second involved a combination between this tetraploid and another diploid, to produce a hexaploid. Cytogeneticists have identified primitive cereal plants in the Middle East that may have participated in this evolutionary process.

Because chromosomes from different species are less likely to interfere with each other's segregation during meiosis, polyploids arising from hybridizations between different species have a much greater chance of being fertile than do polyploids arising from the duplication of chromosomes in a single species. Polyploids created by hybridization between different species are called **allopolyploids**, (from the Greek prefix for "other"); in these polyploids, the contributing genomes are qualitatively different. Polyploids created by chromosome duplication within a species are called **autopolyploids**, (from the Greek prefix for "self"); in these polyploids, a single genome has been multiplied to create extra chromosome sets.

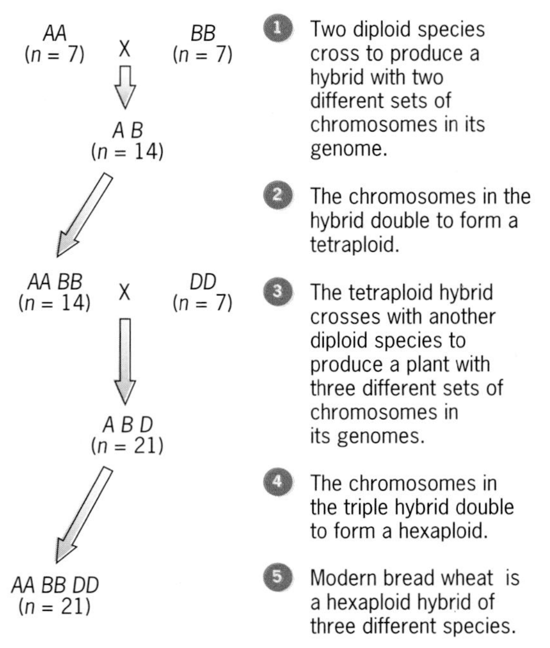

Figure 6.10 Origin of hexaploid wheat by sequential hybridization of different species. Each hybridization event is followed by doubling of the chromosomes.

Chromosome Doubling and the Origin of Polyploids

Where does the chromosome doubling process occur in the evolution of a polyploid plant species (Figure 6.11)? There are two possibilities: (1) in the meristematic cells that divide to produce a variety of plant tissues, or (2) in the reproductive cells as they go through meiosis.

In a diploid meristematic cell, the sister chromatids may fail to separate during mitosis. This failure of the normal mitotic mechanism will produce a tetraploid cell, which will divide to form a tetraploid clone of cells within an otherwise diploid organism. If this clone of tissue is detached from the plant and rooted, the tetraploid condition can be propagated in a new organism. Cuttings from this tetraploid plant can then be rooted to establish many additional tetraploid plants. If the original chromosome doubling occurs in a hybrid between two different species, the resulting tetraploids may be fertile, permitting subsequent sexual propagation. Allotetraploids of this sort are sometimes called **amphidiploids** (from the Greek words meaning "of two kinds"), in recognition of their status as the sum of two diploid species.

Chromosome doubling could also take place in a cell undergoing meiosis. The chromosomes might fail to separate in either of the meiotic divisions, leading to gametes with twice the normal chromosome number. If such a diploid gamete unites with a normal haploid gamete, a triploid zygote is produced. Alternatively, two diploid gametes might unite with each other, creating a tetraploid zygote.

Experimental Production of Polyploids

Numerous polyploids have been created in the laboratory, partly for scientific interest and partly for agricultural benefit. One of the earliest experiments involved hybrids between the radish, *Raphanus sativus*, and the cabbage, *Brassica oleracea*. Although these

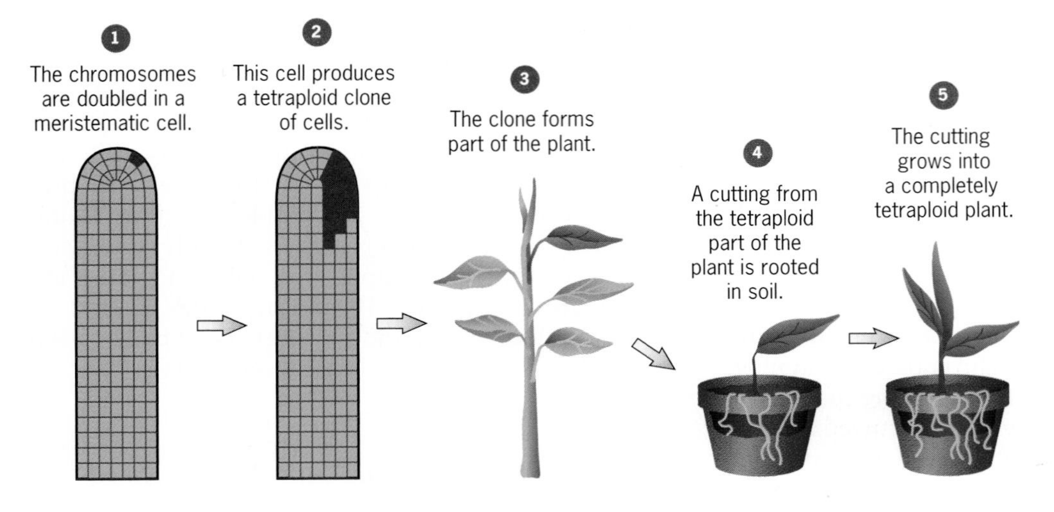

(*a*) Chromosome doubling in meristematic tissue.

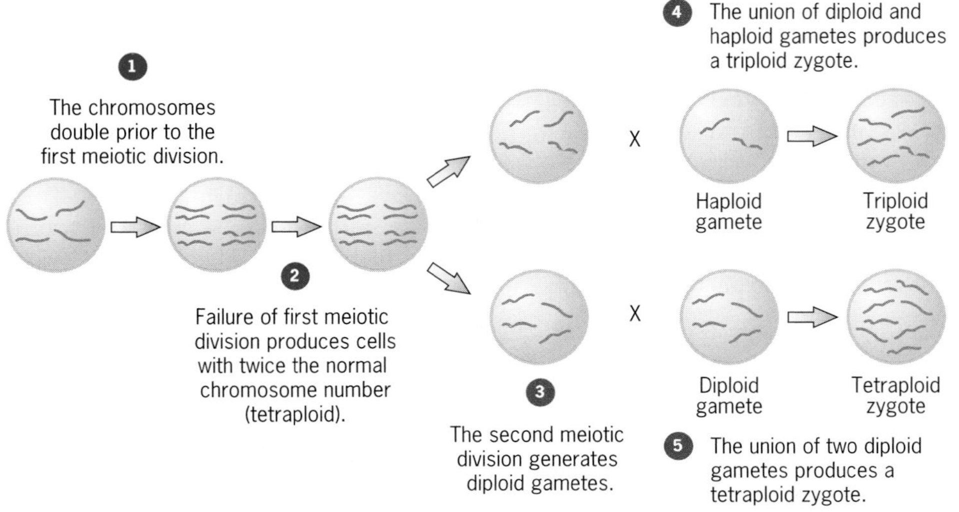

(*b*) Chromosome doubling during meiosis.

Figure 6.11 Chromosome doubling and the origin of polyploids.

plants belong to different genera, they can be crossed to make hybrids, which are usually sterile. Both species have nine pairs of chromosomes and the hybrids have one complete set of chromosomes from each parent, that is, 18 chromosomes altogether.

In the 1920s, the Russian cytologist G. D. Karpechenko found some unusual radish–cabbage hybrids that made unreduced gametes. Combining these gametes to form progeny that had 36 chromosomes—18 from each of the parental species—he established a strain of the new tetraploid and named it *Raphanobrassica*. With this work, Karpechenko demonstrated that cytogeneticists could create fertile interspecific hybrids in the laboratory and that they could combine traits from different species into a single polyploid plant. Unfortunately, his hope of creating a plant with the root of the radish and the foliage of the cabbage was not realized; both the foliage and the root of *Raphanobrassica* were inedible.

Polyploids have been induced experimentally in other plants, mainly by using the spindle-poisoning drug **colchicine**, an alkaloid extracted from the tissues of the autumn crocus, *Colchium autumnale*. The root tips are placed in a solution of colchicine to inhibit spindle formation, thereby arresting mitosis and doubling the chromosome number. When the treated roots are subsequently propagated, tetraploid plants can be obtained.

The feasibility of this procedure was demonstrated in the 1940s by J. O. Beasley, who studied hybrids between two species of cotton in the genus *Gossypium* (Figure 6.12). Old World cotton has 13 pairs of large chromosomes, whereas American cotton, which originated in Central or South America, has 13

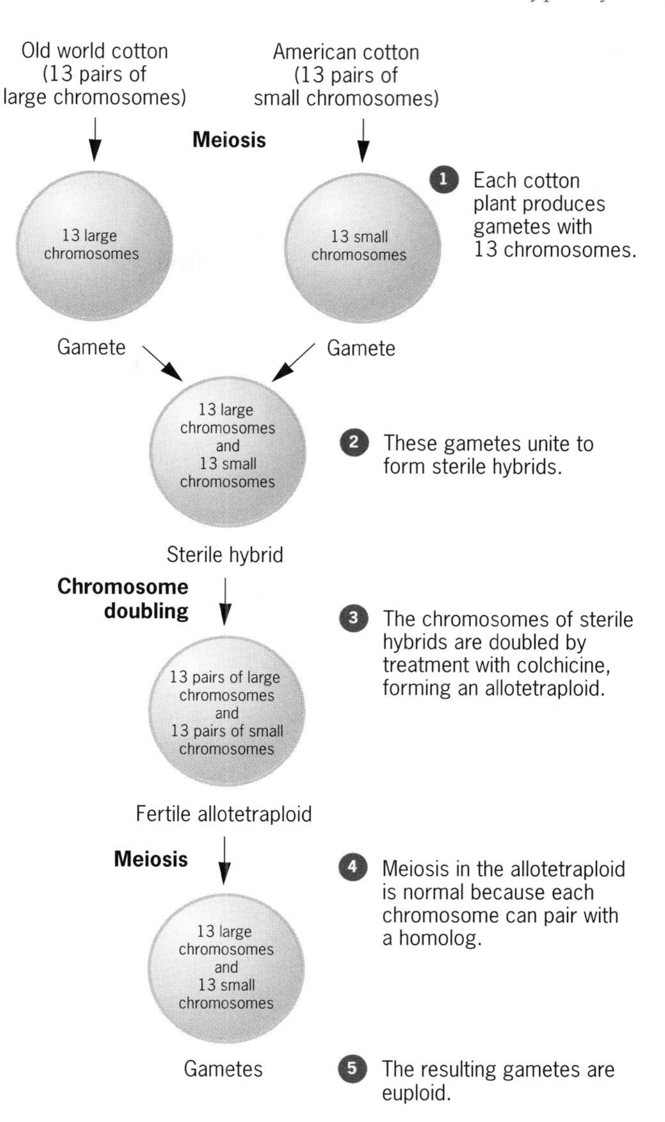

Figure 6.12 Experimental induction of polyploidy in cotton.

pairs of small chromosomes. New World cotton has 26 pairs, 13 large and 13 small. These numbers clearly suggest that New World cotton arose as an allotetraploid when the Old World and American species hybridized. Beasley showed that this hypothesis was correct by creating hybrids between Old World and American cotton and then doubling their chromosomes with colchicine. The resulting plants, which had 26 pairs of chromosomes, were fertile and resembled New World cotton. Thus Beasley duplicated in the laboratory the process by which polyploid cotton may have originated in nature.

Tissue-Specific Polyploidy and Polyteny

In some organisms, certain tissues become polyploid during development. This polyploidization is probably a response to the need for multiple copies of each chromosome and the genes it carries. The process that produces such polyploid cells, called **endomitosis**, involves chromosome duplication, followed by separation of the resulting sister chromatids. However, because there is no accompanying cell division, extra chromosome sets accumulate within a single nucleus. In human beings, for example, one round of endomitosis produces tetraploid cells in the liver and kidney.

Sometimes polyploidization occurs without the separation of sister chromatids. In these cases, the duplicated chromosomes pile up next to each other, forming a bundle of strands that are aligned in parallel. The resulting chromosomes are said to be **polytene**, from the Greek words meaning "many threads." The most spectacular examples of polytene chromosomes are found in the salivary glands of *Drosophila* larvae. Each chromosome undergoes about nine rounds of replication, producing a total of about 500 copies in each cell. All the copies pair tightly, forming a thick bundle of chromatin fibers. This bundle is so large that it can be seen under low magnification with a dissecting microscope. Differential coiling along the length of the bundle causes variation in the density of the chromatin. When dyes are applied to these chromosomes, the denser chromatin stains more deeply, creating a pattern of dark and light bands (Figure 6.13). This pattern is highly reproducible, permitting detailed analysis of chromosome structure.

The polytene chromosomes of *Drosophila* show two additional features:

1. *Homologous polytene chromosomes pair*. Ordinarily we think of pairing as a property of meiotic chromosomes; however, in many insect species the somatic chromosomes also pair—probably as a way of organizing the chromosomes within the nucleus. When *Drosophila* polytene chromosomes pair, the large chromatin bundles become even larger. Because this pairing is precise—point-for-point along the length of the chromosome—the two homologs

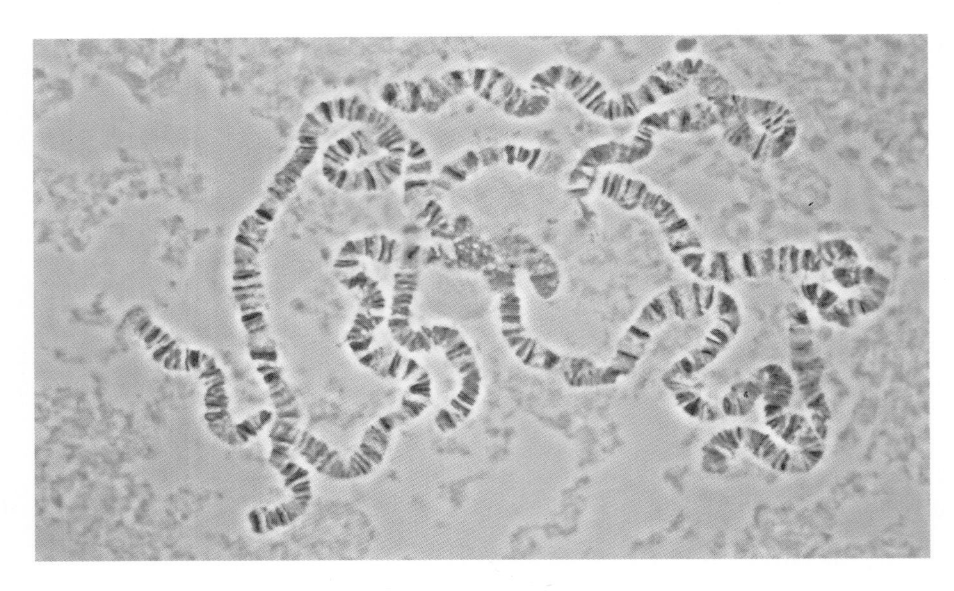

Figure 6.13 Polytene chromosomes of *Drosophila*.

come into perfect alignment. Thus the banding patterns of each are exactly in register, so much so that it is almost impossible to distinguish the individual members of a pair.

2. *All the centromeres of* Drosophila *polytene chromosomes congeal into a darkly staining body called the* ***chromocenter***. Material flanking the centromeres is also drawn into this dark mass. The result is that the chromosome arms seem to emanate out of the chromocenter. These arms, which are banded, consist of euchromatin, that portion of the chromosome that contains most of the genes; the chromocenter, which is uniformly stained, consists of heterochromatin, a gene-poor material that surrounds the centromere. The distinction between these different materials is that heterochromatin always stains darkly with dyes such as Feulgen and aceto-carmine.

Polytene chromosomes were discovered in the nineteenth century by the Italian cytologist E. G. Balbiani. However, it was not until the 1930s that Emil Heitz and Hans Bauer found them in garden midges, and Theophilus Painter found them in *Drosophila* larvae. Painter's discovery was extremely important because *Drosophila* had become the favorite organism for genetic analysis. After Painter, it was possible to combine sophisticated genetic work with detailed cytology to study the organization and structure of chromosomes. This combination of genetic and cytological analysis eventually culminated in maps showing the location of individual genes on these chromosomes. C. B. Bridges laid the foundation for this genetic cartography by publishing detailed drawings of the polytene chromosomes (Figure 6.14). Bridges arbitrarily divided each of the chromosomes into sections, which he numbered; each section was then divided into sub-

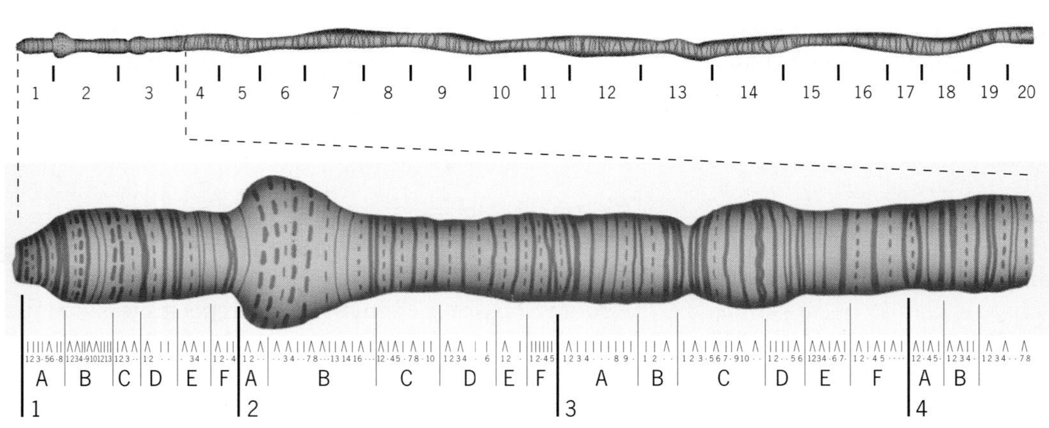

Figure 6.14 Bridges' polytene chromosome maps. *Top*, banding pattern of the polytene X chromosome. The chromosome is divided into 20 numbered sections. *Bottom*, detailed view of the left end of the polytene X chromosome showing Bridges' system for denoting individual bands.

sections, which were designated by the letters A-F. Within each subsection, Bridges enumerated all the dark bands, creating an alphanumeric directory of sites along the length of each chromosome. Bridges' alphanumeric system is still used today to describe the features of these remarkable chromosomes.

The polytene chromosomes of *Drosophila* are trapped in the interphase of the cell cycle. Thus, although most cytological analyses are performed on mitotic chromosomes, the most thorough and detailed analyses are actually performed on polytenized interphase chromosomes. Such chromosomes are found in many species within the insect order Diptera, including flies and mosquitoes. Unfortunately, human beings do not have polytene chromosomes; thus the high-resolution structural analysis that is possible for *Drosophila* is not possible for our own species.

Key Points: **Polyploidy involves the presence of extra sets of chromosomes. Many polyploids are sterile because their multiple sets of chromosomes segregate irregularly in meiosis. However, polyploids produced by chromosome doubling in interspecific hybrids may be fertile if their constituent genomes segregate independently. In some polyploid tissues, sister chromatids remain together, forming a large, polytene chromosome.**

ANEUPLOIDY

Aneuploidy describes a numerical change in part of the genome, usually a change in the dosage of a single chromosome. Individuals that have an extra chromosome, that are missing a chromosome, or that have a combination of these anomalies are aneuploid. This definition also includes pieces of chromosomes. Thus an individual in which a chromosome arm has been deleted is also considered to be aneuploid.

Aneuploidy was originally studied in plants, where it was shown that a chromosome imbalance usually has a phenotypic effect. The classic study was one by Albert Blakeslee and John Belling, who analyzed chromosome anomalies in Jimson weed, *Datura stramonium*. This diploid species has 12 pairs of chromosomes, for a somatic total of 24. Blakeslee collected plants with altered phenotypes and discovered that in some cases the phenotypes were inherited in an irregular way. These peculiar mutants were apparently caused by dominant factors that were transmitted primarily through the female. By examining the chromosomes of the mutant plants, Belling found that in every case an extra chromosome was present. Detailed analysis established that the extra chromosome was different in each mutant strain. Altogether there were 12 different mutants, each corresponding to a triplication of one of the *Datura* chromosomes (Figure 6.15). Such triplications are called **trisomies**. The transmissional irregularities of these mutants were due to anomalous chromosome behavior during meiosis.

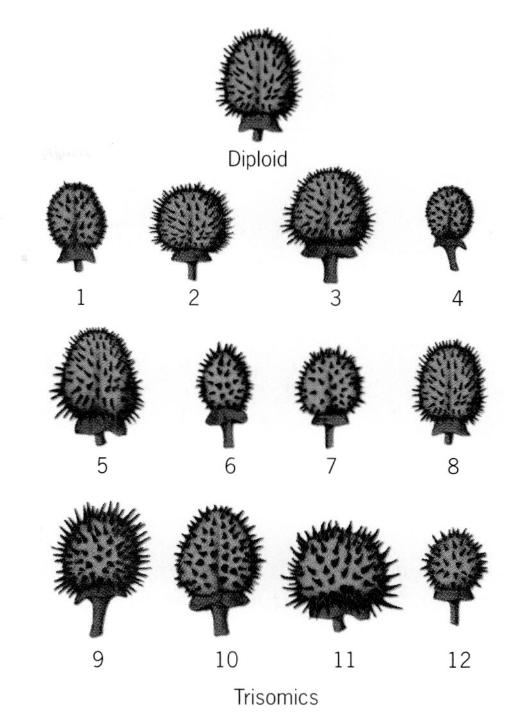

Figure 6.15 Seed capsules of normal and trisomic *Datura stamonium*. Each of the 12 trisomies is shown.

Belling also discovered the reason for the preferential transmission of the trisomic phenotypes through the female. During pollen tube growth, aneuploid pollen—in particular, pollen with $n + 1$ chromosomes—does not compete well with euploid pollen. Consequently, trisomic plants almost always inherit their extra chromosome from the female parent. Belling's work with *Datura* clearly demonstrated that each chromosome must be present in the proper dosage for normal growth and development.

Since Belling's work, aneuploids have been identified in many species, including our own. An organism in which a chromosome, or a piece of a chromosome, is underrepresented is referred to as a **hypoploid**, (from the Greek prefix for "under"). When a chromosome or chromosome segment is over-represented, the organism is said to be **hyperploid**, (from the Greek prefix for "over"). Each of these terms covers a wide range of abnormalities.

Trisomy in Human Beings

The best-known and one of the most common chromosome abnormality in humans is **Down syndrome**, a

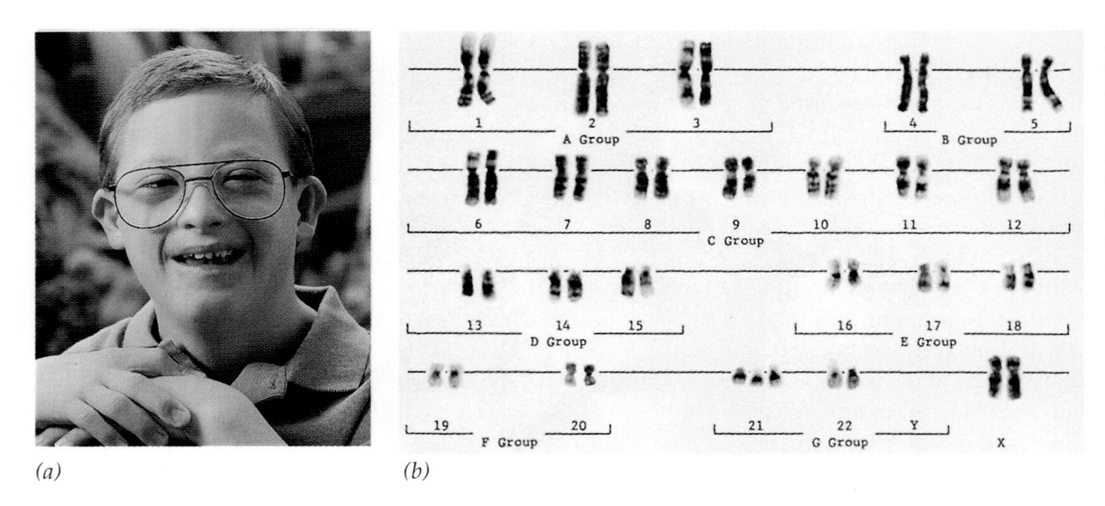

Figure 6.16 Down syndrome. (*a*) Facial features of a child with Down syndrome. (*b*) Karyotype of a child with Down syndrome, showing trisomy for chromosome 21 (47,XX, +21).

condition associated with an extra chromosome 21 (Figure 6.16*a*). This syndrome was first described in 1866 by a British physician, Langdon Down; however, its chromosomal basis was not clearly understood until 1959. People with Down syndrome are typically short in stature and loose-jointed, particularly in the ankles; they have broad skulls, wide nostrils, large tongues with a distinctive furrowing, and stubby hands with a crease on the palm. Impaired mental abilities require that they be given special training and care.

The extra chromosome 21 in Down syndrome is an example of a trisomy. Figure 6.16*b* shows the karyotype of a female Down patient. Altogether, there are 47 chromosomes, including two X chromosomes, plus the extra chromosome 21. The karyotype of this individual is therefore written **47, XX, +21**.

Trisomy 21 can be caused by chromosome nondisjunction in one of the meiotic cell divisions (Figure 6.17). The nondisjunction event can occur in either parent, but it seems to be more likely in females. In addition, the frequency of nondisjunction increases with maternal age. Thus, among mothers younger than 25 years old, the risk of having a child with Down syndrome is about 1 in 1500, whereas among mothers 40 years old, it is 1 in 100. This increased risk is due to factors that adversely affect meiotic chromosome behavior as a woman ages. In human females, meiosis begins in the fetus, but it is not completed until after the egg is fertilized. During the long time prior to fertilization, the meiotic cells are arrested in the prophase of the first division. In this suspended state, the chromosomes may become unpaired. The longer the time

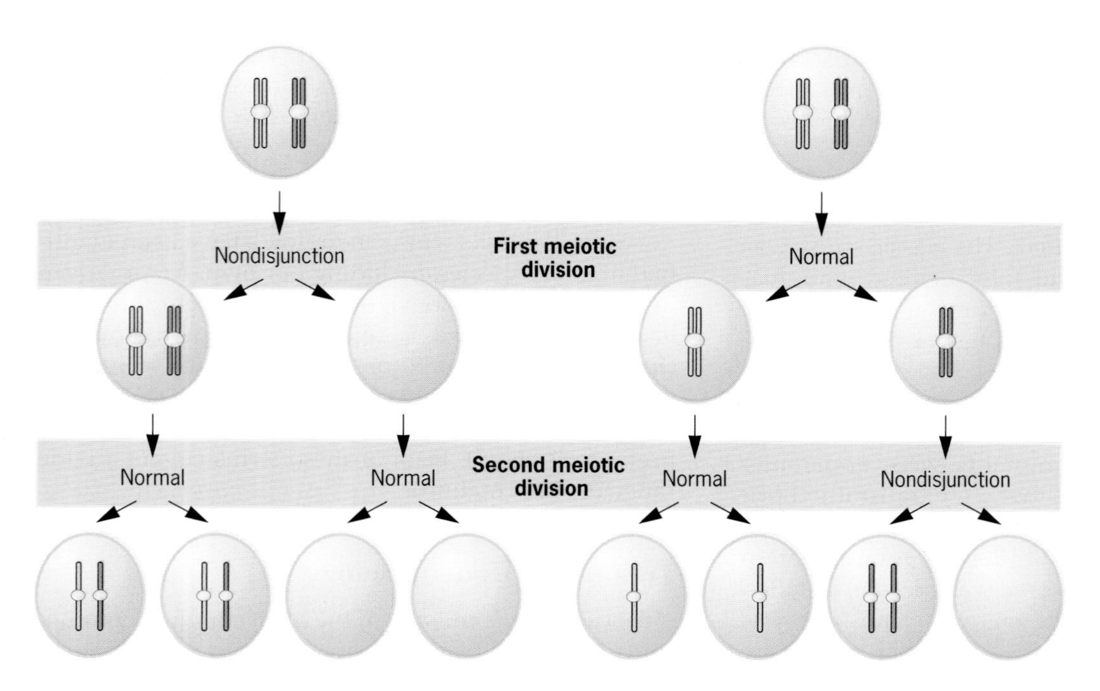

Figure 6.17 Meiotic nondisjunction and the origin of Down syndrome. Nondisjunction at meiosis I produces no normal gametes. Nondisjunction at meiosis II produces a gamete with two identical sister chromosomes, a gamete lacking chromosome 21, and two normal gametes.

in prophase, the greater the chance for unpairing and subsequent chromosome nondisjunction. Older females are therefore more likely than younger females to produce aneuploid eggs.

Trisomies for chromosomes 13 and 18 have also been reported. However, these are rare, and the affected individuals show serious phenotypic abnormalities and are short-lived, usually dying within the first few weeks after birth. Another viable trisomy that has been observed in human beings is the triplo-X karyotype, **47, XXX**. These individuals survive because two of the three X chromosomes are inactivated, effectively reducing the dosage of the X chromosome to the normal male level of one. Triplo-X individuals are female and are phenotypically normal, or nearly so; sometimes there is a slight mental impairment and reduced fertility.

The **47, XXY** karyotype is also a viable trisomy in human beings. These individuals have three sex chromosomes, two X's and one Y. Phenotypically, they are male, but they can show some female secondary sexual characteristics and are usually sterile. In 1942 H. F. Klinefelter described the abnormalities associated with this condition, now called **Klinefelter syndrome**; these include small testes, enlarged breasts, long limbs, knock knees, and underdeveloped body hair. The XXY karyotype can originate by fertilization of an exceptional XX egg with a Y-bearing sperm, or by fertilization of an X-bearing egg with an exceptional XY sperm. The XXY karyotype accounts for about three-fourths of all cases of Klinefelter syndrome. Other cases involve more complex karyotypes such as XXYY, XXXY, XXXYY, XXXXY, XXXXYY and XXXXXY. All individuals with Klinefelter syndrome have one or more Barr bodies in their cells, and those with more than two X chromosomes usually have some degree of mental impairment.

The **47, XYY** karyotype is another viable trisomy in human beings. These individuals are male, and except for a tendency to be taller than 46, XY men, they do not show a consistent syndrome of characteristics. A possible connection between the XYY karyotype and antisocial behavior has not been studied thoroughly enough to draw a firm conclusion.

All the other trisomies in human beings are embryonic lethals, demonstrating the importance of correct gene dosage. Unlike *Datura*, in which each of the possible trisomies is viable, human beings do not tolerate many types of chromosomal imbalance (see Table 6.1).

TABLE 6.1
Aneuploidy Resulting from Nondisjunction in Human Beings

Karyotype	*Chromosome Formula*	*Clinical Syndrome*	*Estimated Frequency At Birth*	*Phenotype*
47,+21	$2n+1$	Down	1/700	Short, broad hands with palmar crease, short stature, hyperflexibility of joints, mental retardation, broad head with round face, open mouth with large tongue, epicanthal fold.
47,+13	$2n+1$	Patau	1/20,000	Mental deficiency and deafness, minor muscle seizures, cleft lip and/or palate, cardiac anomalies, posterior heel prominence.
47,+18	$2n+1$	Edward	1/8000	Congenital malformation of many organs, low-set, malformed ears, receding mandible, small mouth and nose with general elfin appearance, mental deficiency, horseshoe or double kidney, short sternum, 90 percent die within first six months after birth.
45,X	$2n-1$	Turner	1/2500 female births	Female with retarded sexual development, usually sterile, short stature, webbing of skin in neck region, cardiovascular abnormalities, hearing impairment.
47,XXY 48,XXXY 48,XXYY 49,XXXXY 50,XXXXXY	$2n+1$ $2n+2$ $2n+2$ $2n+3$ $2n+4$	Klinefelter	1/500 male births	Male, subfertile with small testes, developed breasts, feminine-pitched voice, long limbs, knock knees.
47,XXX	$2n+1$	Triplo-X	1/700	Female with usually normal genitalia and limited fertility, slight mental retardation.

Monosomy

Monosomy occurs when one chromosome is missing in an otherwise diploid individual. In human beings, there is only one viable monosomy, the **45, X** karyotype. These individuals have a single X chromosome plus a diploid complement of autosomes. Phenotypically, they are female, but because their ovaries are rudimentary, they are almost always sterile. 45, X individuals are usually short in stature; they have webbed necks, hearing deficiencies, and significant cardiovascular abnormalities. Henry H. Turner first described the condition in 1938; thus it is now called **Turner syndrome**.

45, X individuals can originate from eggs or sperm that lack a sex chromosome or from the loss of a sex chromosome in mitosis sometime after fertilization (Figure 6.18). This latter possibility is supported by the finding that many Turner individuals are **somatic mosaics**. These people have two types of cells in their bodies; some are 45, X and others are 46, XX. This karyotypic mosaicism evidently arises when an X chromosome is lost during the development of a 46, XX zygote. All the descendants of the cell in which the loss occurred are 45, X. If the loss occurs early in development, an appreciable fraction of the body's cells will be aneuploid and the individual will show the features of Turner syndrome. If the loss occurs later, the aneuploid cell population will be smaller and the severity of the syndrome is likely to be reduced. (For a discussion of procedures used to detect aneuploidy in human fetuses, see Human Genetics Sidelight: Amniocentesis and Chorionic Biospy.)

XX/XO chromosome mosaics also occur in *Drosophila*, where they produce a curious phenotype. Because sex in this species is determined by the ratio of X chromosomes to autosomes, such flies are part female and part male. XX cells develop in the female direction, and XO cells develop in the male direction. Flies with both male and female structures are called **gynandromorphs** (from Greek words meaning "a combination of the two sexes)".

People with the 45, X karyotype do not have Barr bodies in their cells, indicating that the single X chromosome that is present is not inactivated. Why, then, should Turner patients, who have the same number of active X chromosomes as normal XX females, show any phenotypic abnormalities at all? The answer probably involves a small number of genes that remain active on both of the X chromosomes in normal 46, XX females. These noninactivated genes are apparently needed in double dose for proper growth and development. The finding that at least some of these special X-linked genes are also present on the Y chromosome would explain why XY males grow and develop normally. In addition, the X chromosome that has been

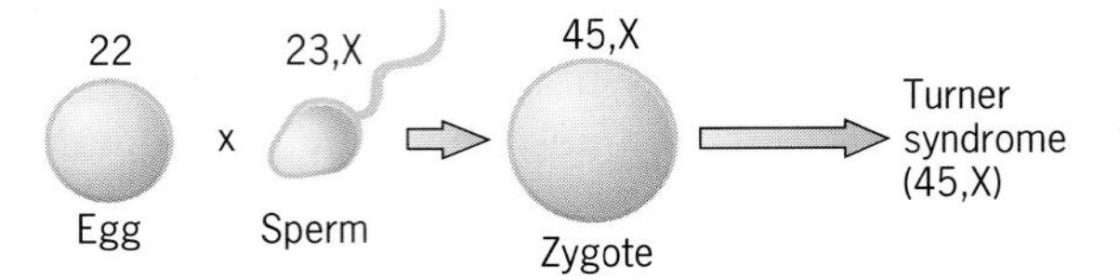

(*a*) Origin of monosomy at fertilization.

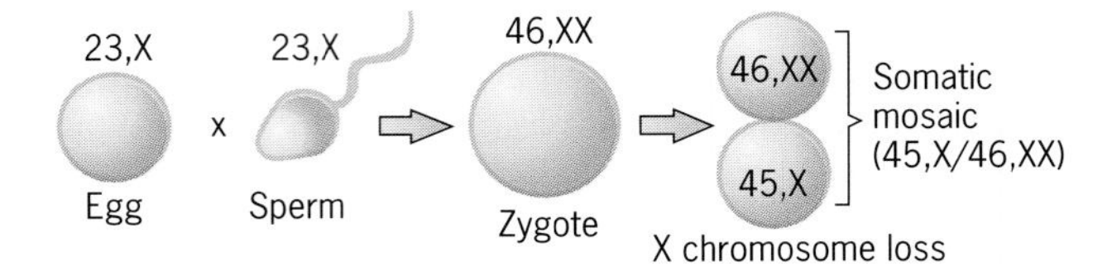

(*b*) Origin of monosomy in the cleavage division following fertilization.

Figure 6.18 Origin of the Turner syndrome karyotype.

inactivated in 46, XX females is reactivated during oogenesis, presumably because two copies of some X-linked genes are required for normal ovarian function. 45, X individuals, who have only one copy of these genes, cannot meet this quantitative requirement and are therefore sterile.

Curiously, the cognate of the XO Turner karyotype in the mouse is fully fertile and exhibits no anatomical abnormalities. This finding implies that the mouse homologs of the human genes that are involved in Turner syndrome need only be present in one copy for normal growth and development.

Deletions and Duplications of Chromosome Segments

A missing chromosome segment is referred to either as a **deletion** or as a **deficiency**. Large deletions can be detected cytologically, but small ones cannot. In a diploid organism, the deletion of a chromosome segment makes part of the genome hypoploid. This hypoploidy may be associated with a phenotypic effect, especially if the deletion is large. A classic example is the ***cri-du-chat*** **syndrome** (from the French words for "cry of the cat") in human beings (Figure 6.19). This condition is caused by a conspicuous deletion in the short arm of chromosome 5; about half the arm appears to be missing. Individuals heterozygous for this deletion and a normal chromosome have the karyotype 46 (5p-), where the term in parentheses indicates that part of the short arm (p) of one of the chromosomes 5 is missing. These individuals are severely impaired, mentally as well as physically; their plaintive, catlike crying gives the syndrome its name.

An extra chromosome segment is referred to as a **duplication**. The extra segment can be attached to one of the chromosomes, or it can exist as a new and sepa-

HUMAN GENETICS SIDELIGHT

Amniocentesis and Chorionic Biopsy: Procedures to Detect Aneuploidy in Human Fetuses

The Andersons, a couple living in Minneapolis, were expecting their first baby. Neither Donald nor Laura Anderson knew of any genetic abnormalities in their families, but because of Laura's age—38—they decided to have the fetus checked for aneuploidy.

Laura's physician performed a procedure called **amniocentesis**. A small amount of fluid was removed from the cavity surrounding the developing fetus by inserting a needle into Laura's abdomen (Figure 1). This cavity, called the amnionic sac, is enclosed by a membrane. To prevent discomfort during the procedure, Laura was given a local anesthetic. The needle was guided into position by following an ultrasound scan, and some of the amnionic fluid was drawn out. Because this fluid contains nucleated cells sloughed off from the fetus, it is possible to determine the fetus's karyotype (Figure 2). Usually the fetal cells are purified from the amnioic fluid by centrifugation, and then the cells are cultured for several days to a few weeks. Cytological analysis of these cells will reveal if the fetus is aneuploid. Additional tests may be performed on the fluid recovered from the amnionic sac to detect other sorts of abnormalities, including neural tube defects and some kinds of mutations. The results of all these tests may take up to three weeks. In Laura's case, no abnormalities of any sort were detected, and 20 weeks after the amniocentesis, she gave birth to a healthy baby girl.

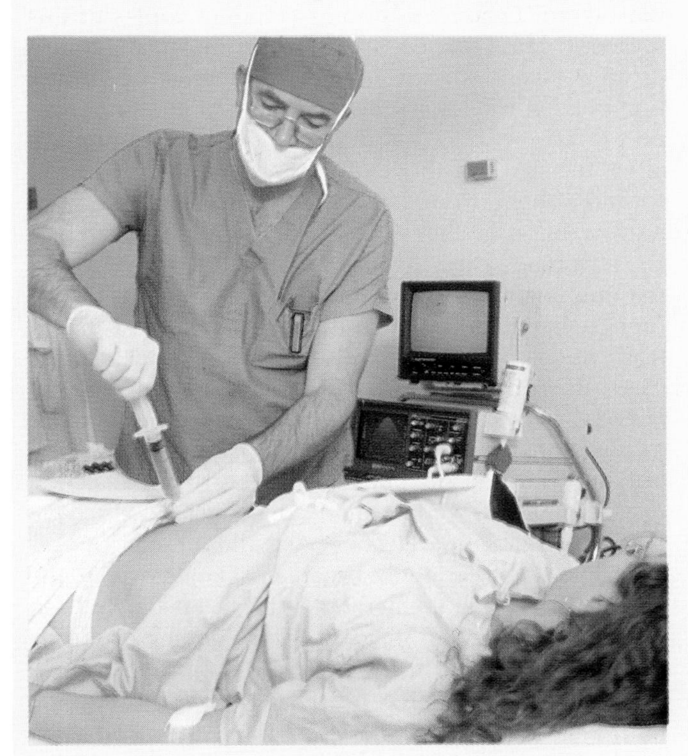

Figure 1 A physician taking a sample of fluid from the amniotic sac of a pregnant woman for prenatal diagnosis of a chromosomal or biochemical abnormality.

Chorionic biopsy provides another way of detecting chromosomal abnormalities in the fetus. The chorion is a fetal membrane that interdigitates with the uterine wall, eventually forming the placenta. The minute chorionic projections into the uterine tissue are called *villi* (singular, villus). At 10–11 weeks of gestation, before the placenta has developed, a sample of chorionic villi can be obtained by passing a hollow plastic tube into the uterus through the cervix. This tube can be guided by an ultrasound scan, and when it is in place, a tiny bit of material can be drawn up into the tube by aspiration. The recovered material usually consists of a mixture of maternal and fetal tissue. After separating these tissues by dissection, the fetal cells can be analyzed for chromosome abnormalities.

Chorionic biopsy can be performed earlier than amniocentesis (10–11 weeks gestation versus 14–16 weeks), but it is not as reliable. In addition, it seems to be associated with a slightly greater chance of miscarriage than amniocentesis, perhaps 2% to 3%. For these reasons, it tends to be used only in pregnancies where there is a strong reason to expect a genetic abnormality. In routine pregnancies, such as Laura Anderson's, amniocentesis is the preferred procedure.

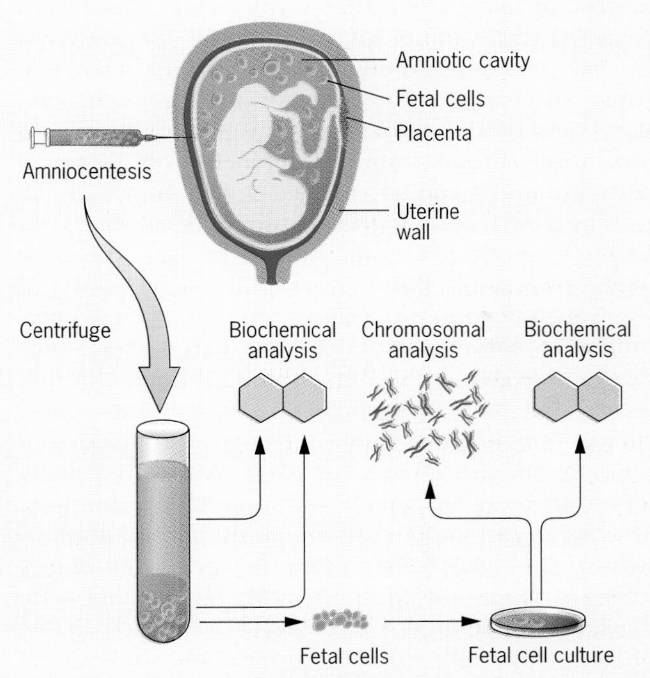

Figure 2 Amniocentesis and procedures for prenatal diagnosis of chromosomal and biochemical abnormalities.

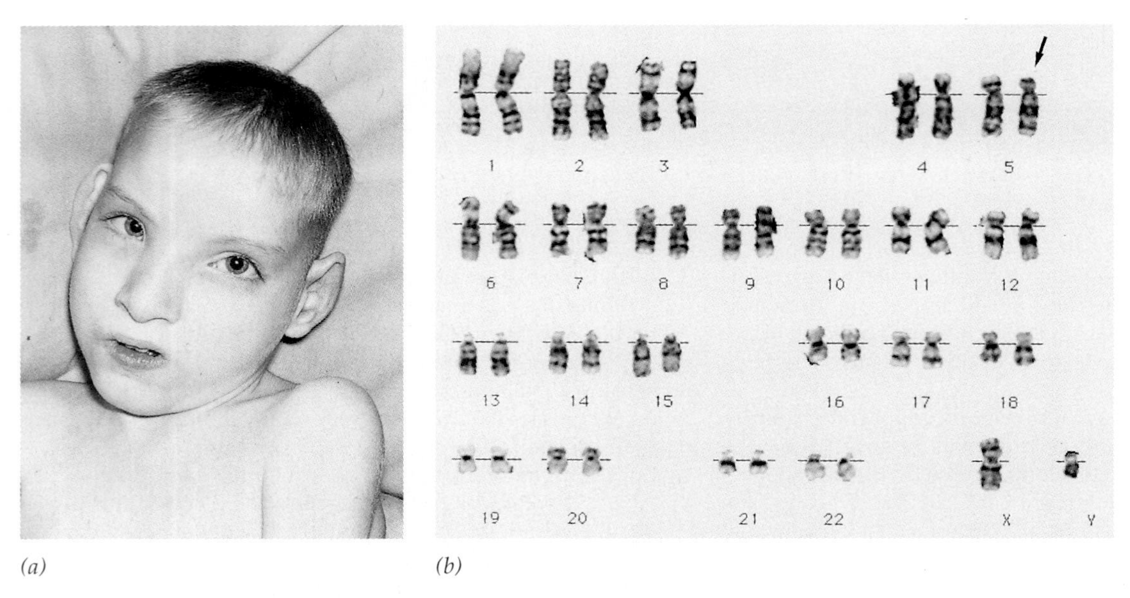

(a) (b)

Figure 6.19 *Cri-du-chat* syndrome. (*a*) Patient with *cri-du-chat* syndrome. (*b*) Karyotype of infant with *cri-du-chat* syndrome, 46, XY(5p-). There is a deletion in the short arm of chromosome 5 (arrow).

rate chromosome, that is, as a "free duplication." In either case, the effect is the same: The organism is hyperploid for part of its genome. As with deletions, this hyperploidy can be associated with a phenotypic effect. For example, in humans the long arm of chromosome 21 is sometimes found attached to the short arm of chromosome 14 (Figure 6.20). When this fusion chromosome is combined with normal chromosomes 14 and 21, the resulting individual will have two copies of nearly all the genes on these two chromosomes and will be phenotypically normal. However, if the fusion chromosome is combined with a normal chromosome 14 and two normal chromosomes 21, the resulting individual will be trisomic for all the genes on the long arm of chromosome 21 and will therefore have Down syndrome.

Deletions and duplications are two types of aberrations in chromosome structure. Large aberrations can be detected by examination of mitotic chromosomes that have been stained with banding agents such as quinacrine or Giemsa. However, small aberrations are difficult to detect in this way, and are usually identified by other genetic and molecular techniques. The best organism for studying deletions and duplications is *Drosophila*, where the polytene chromosomes afford an unparalleled opportunity for detailed cytological analysis. Figure 6.21*b* shows a deletion in one of two paired homologous chromosomes in a *Drosophila* salivary gland. Because the two chromosomes have separated slightly, we can see that a small region is missing in the lower one.

Duplicated segments can also be recognized in polytene chromosomes. Figure 6.21*c* shows a tandem duplication of a segment in the middle of the X chromosome of *Drosophila.* Because tandem copies of this segment pair with each other, the chromosome appears to have a knot in its middle. The *Bar* eye mutation in *Drosophila* is associated with a tandem duplication (Figure 6.22). This dominant X-linked mutation alters the size and shape of the compound eyes, transforming them from large, spherical structures into narrow bars. In the 1930s C. B. Bridges analyzed X chromosomes carrying the *Bar* mutation and found that the 16A region, which apparently contains a gene for eye shape, had been tandemly duplicated. Tandem triplications of 16A were also observed, and in these cases, the compound eye was extremely small—a phenotype referred to as double-bar. Many other tandem duplications have been found in *Drosophila,* where polytene chromosome analysis makes their detection relatively easy. Today, molecular techniques have made it possible to detect very small tandem duplications in a wide variety of organisms. For example, the genes that encode the hemoglobin proteins have been tandemly duplicated in mammals (Chapter 22). Gene duplications appear to be relatively common and provide a significant source of variation for evolution (see Chapter 28).

Key Points: **Aneuploidy involves the under- or over-representation of a chromosome or chromosome segment. In a trisomy, such as Down syndrome in hu-**

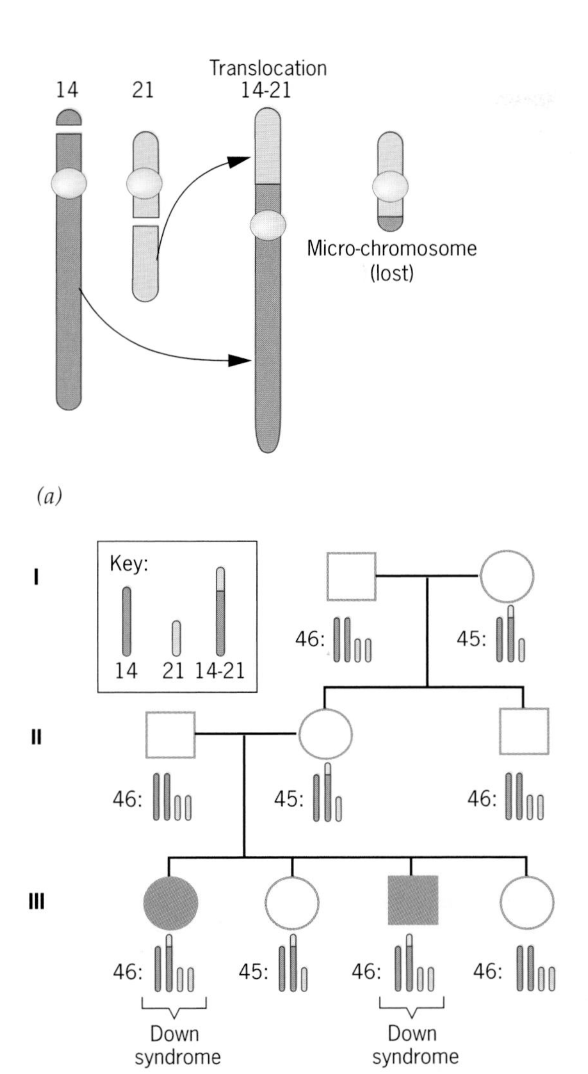

Figure 6.20 Translocation between chromosomes 14 and 21. (*a*) Origin of the translocation through breakage in the short arm of chromosome 14 and in the long arm of chromosome 21. (*b*) Transmission of the translocation in a pedigree. Two sibs in generation III have Down syndrome because they are trisomic for most of chromosome 21.

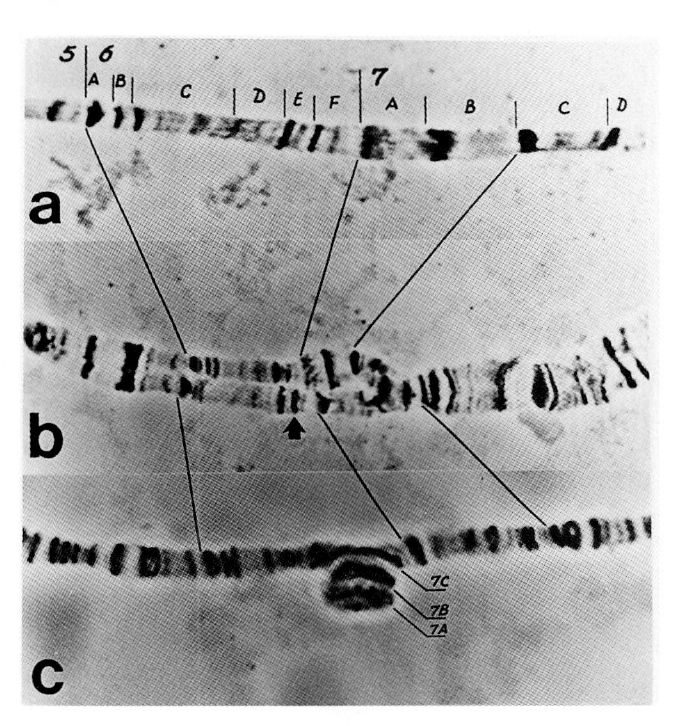

Figure 6.21 Polytene chromosomes showing (*a*) the normal structure of regions 6 and 7 in the middle of the *Drosophila* X chromosome, (*b*) a heterozygote with a deletion of region 6F-7C in one of the chromosomes (arrow), and (*c*) an X chromosome showing a reverse tandem duplication of region 6F-7C.

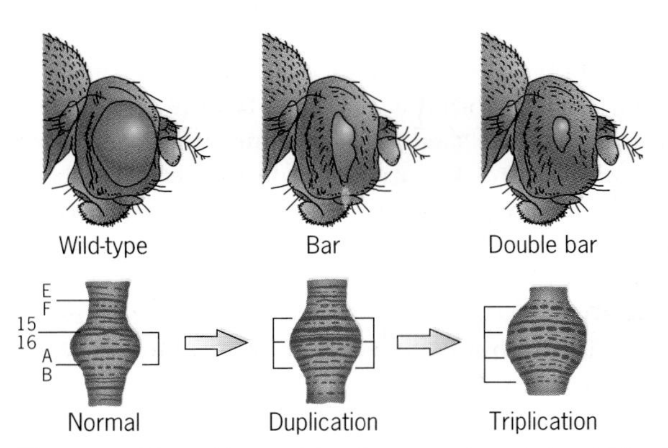

Figure 6.22 Effects of duplications for region 16A of the X chromosome on the size of the eyes in *Drosophila*.

man beings, a chromosome is overrepresented; in a monosomy, such as Turner syndrome, it is underrepresented. Deletions and duplications of particular chromosome segments also cause aneuploidy—hypoploidy in the case of a deletion and hyperploidy in the case of a duplication.

REARRANGEMENTS OF CHROMOSOME STRUCTURE: AN OVERVIEW

In nature there is considerable variation in the number and structure of chromosomes, even among closely related organisms. For example, *Drosophila melanogaster* has four pairs of chromosomes, including a pair of sex chromosomes, two pairs of large, metacentric autosomes (chromosomes with the centromere in the middle), and a pair of small, dotlike autosomes. *Drosophila virilis*, which is not too distantly related, has a pair of sex chromosomes, four pairs of acrocentric autosomes (chromosomes with the centromere near one end), and a pair of dotlike autosomes. Thus, even in the same genus, species can have different chromosome arrangements. These differences imply that over evolutionary time, segments of the genome are rearranged. In fact, the observation that chromosome rearrangements can be found as variants within a single species suggests that the genome is continuously being reshaped. These rearrangements may change the position of a segment within a chromosome, or they may bring together segments from different chromosomes. In either case, the order of the genes is altered. Cytogeneticists have identified many kinds of chromosome rearrangements, including the duplications and deficiencies discussed above. Here we consider two additional types, inversions, which involve a switch in the orientation of a segment within a chromosome, and translocations, which involve the fusion of segments from different chromosomes.

Inversions

Inversions occur when a chromosome segment is detached, flipped around 180°, and reattached to the rest of the chromosome; as a result, the order of the segment's genes is reversed (Figure 6.23). Such rearrangements can be induced in the laboratory by X-irradiation, which breaks chromosomes into pieces. Sometimes the pieces reattach, but in the process a segment gets turned around and an inversion occurs. There is also evidence that inversions are produced naturally through the activity of transposable elements, unusual DNA sequences capable of moving from one chromosomal position to another (Chapter 17). Sometimes, in the course of moving, these elements break a chromosome into pieces and the pieces reattach in an aberrant way, producing an inversion. Inversions may also be created by the reattachment of chromosome fragments generated by mechanical sheer, perhaps as a result of chromosome entanglement within the nucleus. No one really knows what fraction of naturally occurring inversions is caused by each of these mechanisms.

Cytogeneticists distinguish between two types of inversions based on whether or not the inverted segment includes the chromosome's centromere (Figure 6.24). **Pericentric** inversions include the centromere, whereas **paracentric** inversions do not. The significance is that a pericentric inversion may change the relative lengths of the two arms of the chromosome, whereas a paracentric inversion has no such effect. Thus, if an acrocentric chromosome acquires an inversion with a breakpoint in each of the chromosome's arms (that is, a pericentric inversion), it can be transformed into a metacentric chromosome. However, if an acrocentric chromosome acquires an inversion in which both of the breaks are in the chromosome's long arm (that is, a paracentric inversion), the morphology of the chromosome will not be changed. Hence, with the use of standard cytological methods, pericentric inversions are much easier to detect than paracentric inversions.

An individual in which one chromosome is inverted but its homolog is not is said to be an inversion heterozygote. During meiosis, the inverted and noninverted chromosomes pair point-for-point along their length. However, because of the inversion, the chromosomes must form a loop to allow for pairing in the region where their genes are in reversed order. Figure 6.25*a* shows this pairing configuration for two of the four chromatids in a tetrad; actually, only one of the chromatids is looped, and the other conforms around it. In practice, either the inverted or noninverted chromatid can form the loop to maximize pairing between them. However, near the ends of the inversion, the chromatids are stretched, and there is a tendency for

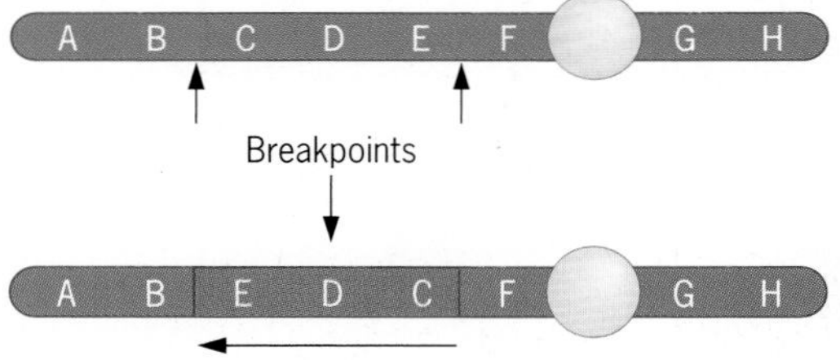

Figure 6.23 Structure of an inversion. The chromosome has been broken at two points, and the segment between them (containing regions C, D, and E) has been inverted.

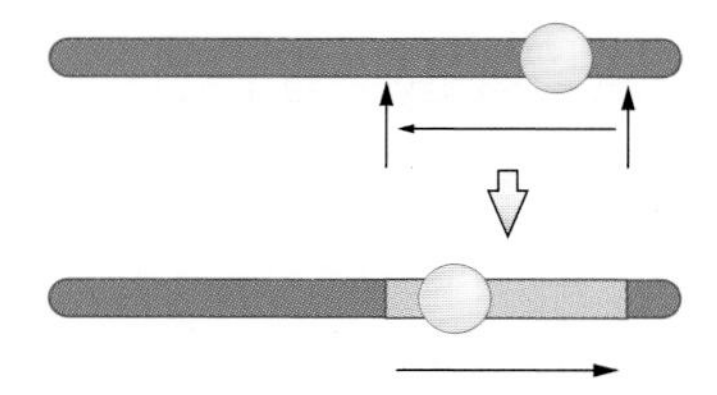

(*a*) Pericentric inversion–includes centromere

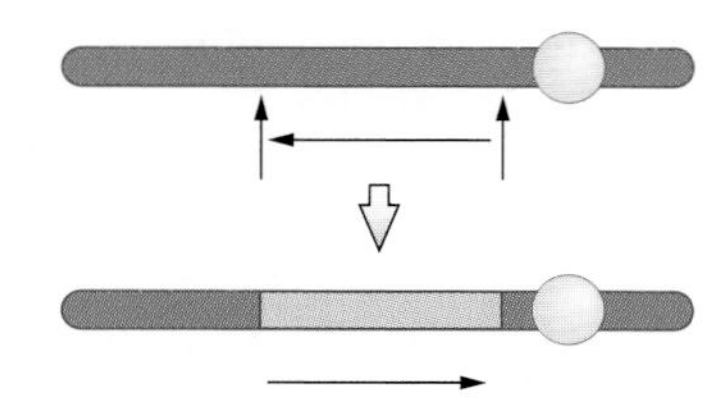

(*b*) Paracentric inversion–excludes centromere

Figure 6.24 Pericentric and paracentric inversions. A pericentric inversion changes the size of the chromosome arms because the centromere is included within the inversion.

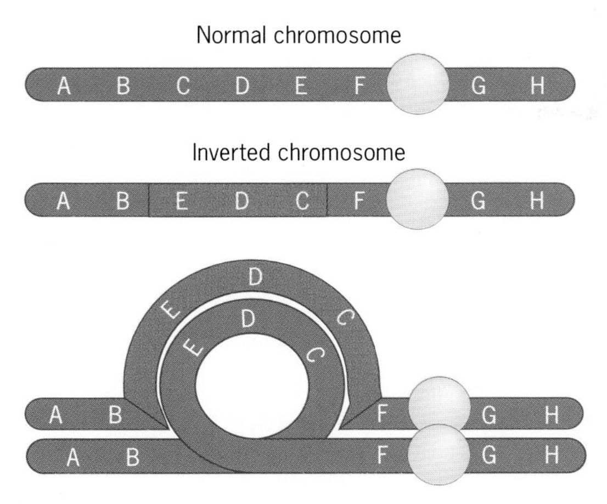

(*a*) Pairing between normal and inverted chromosomes.

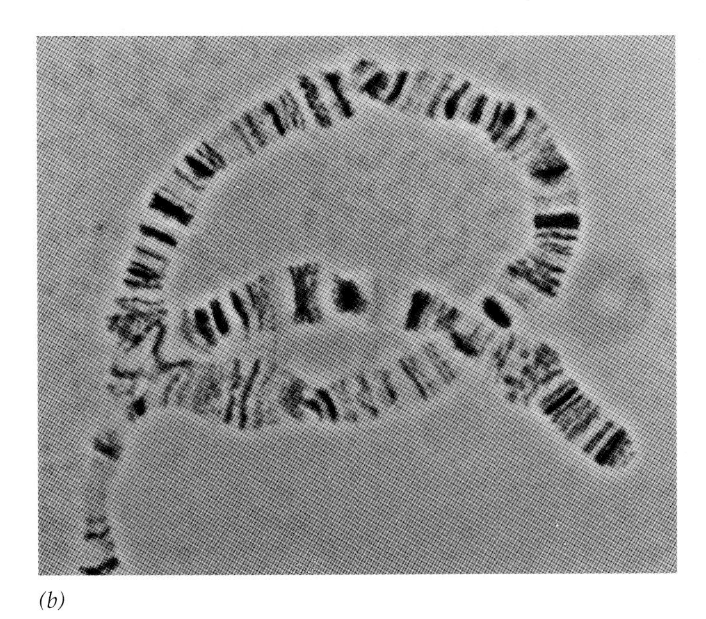

(*b*)

Figure 6.25 Pairing in inversion heterozygotes. (*a*) Diagram showing pairing between normal and inverted chromosomes. (*b*) Polytene chromosomes from an inversion heterozygote.

some de-synapsis. We consider the genetic consequences of inversion heterozygosity in Chapter 7.

In most organisms, meiotic inversion loops are difficult to observe clearly because the chromosomes are so small. Detailed analysis is practically impossible. However, the loop configuration is easily seen in the somatically paired polytene chromosomes of *Drosophila* (Figure 6.25*b*). Because of the banding pattern, the looped region can be clearly identified, and sometimes even the endpoints of the inversion can be precisely determined. Inversion loops are found only in heterozygotes. When inversion chromosomes are homozygous, they pair smoothly along their length without any loop formation.

Translocations

Translocations occur when a segment from one chromosome is detached and reattached to a different (that is, nonhomologous) chromosome. The genetic significance is that genes from one chromosome are transferred to another and their linkage relationships are altered. Translocations can be induced by X-irradiation, which can break two chromosomes simultaneously. Occasionally, the broken fragments are reattached in a novel way and segments from the different chromosomes are fused. In nature, other factors such as transposable elements and mechanical sheer may be involved in producing translocations; however, little is currently known about the relative importance of these factors.

When pieces of two nonhomologous chromosomes are interchanged without any net loss of genetic material, the event is referred to as a **reciprocal translocation**. Figure 6.26*a* shows a reciprocal translocation between the two large autosomes of *Drosophila*. These chromosomes have interchanged pieces of their right arms. During meiosis (and also in the polytene cells of the larval salivary glands), these translocated chromosomes would be expected to pair with their

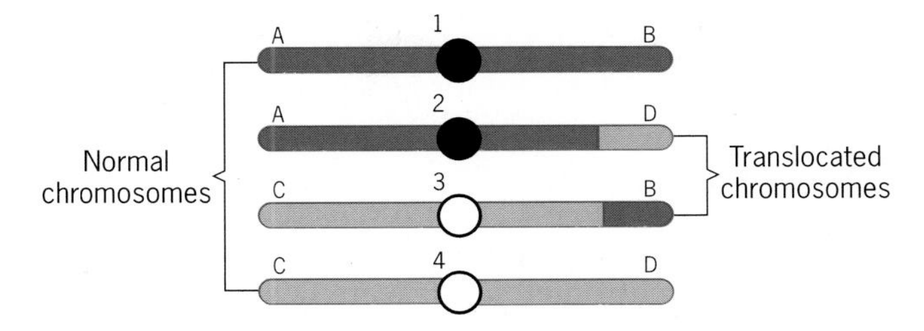

(*a*) Structure of chromosomes in translocation heterozygote.

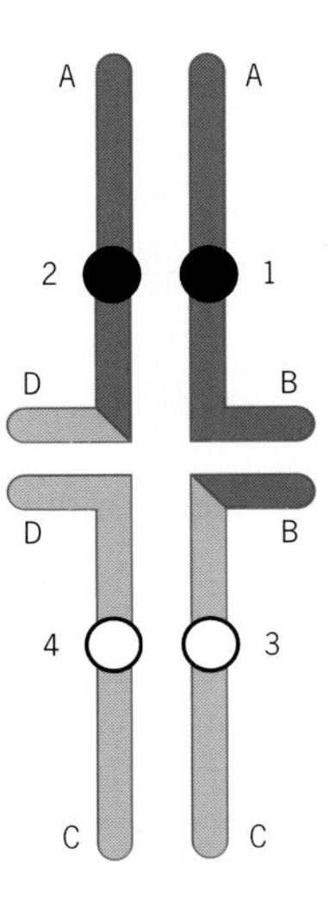

(*b*) Pairing of chromosomes in translocation heterozygote.

Figure 6.26 Structure and pairing behavior of a reciprocal translocation between chromosomes.

untranslocated homologs in a cruciform, or crosslike, pattern (Figure 6.26b). The two translocated chromosomes face each other opposite the center of the cross, and the two untranslocated chromosomes do likewise; to maximize pairing, the translocated and untranslocated chromosomes alternate with each other, forming the arms of the cross. This pairing configuration is diagnostic of a translocation heterozygote. Cells in which the translocated chromosomes are homozygous do not form a cruciform pattern. Instead, each of the translocated chromosomes pairs smoothly with its structurally identical partner.

Because cruciform pairing involves four centromeres, which may or may not be coordinately distributed to opposite poles in the first meiotic division, chromosome disjunction in translocation heterozygotes is a somewhat uncertain process, prone to produce aneuploid gametes. Altogether there are three possible disjunctional events, illustrated in Figure 6.27. This simplified figure shows only one of the two sister chromatids of each chromosome. In addition, each of the centromeres is labeled to keep track of chromosome movements; the two white centromeres are homologous (that is, derived from the same chromosome pair), as are the two black centromeres.

If centromeres 1 and 2 move to the same pole, forcing 3 and 4 to the opposite pole, all the resulting gametes will be aneuploid—because some chromosome segments will be deficient for genes, and others will be duplicated (Figure 6.27*a*). Similarly, if centromeres 2 and 4 move to one pole and 1 and 3 to the other, only aneuploid gametes will be produced (Figure 6.27*b*). Each of these cases is referred to as **adjacent disjunction** because centromeres that were next to each other in the cruciform pattern moved to the same pole. Another possibility is that centromeres 1 and 4 move to the same pole, forcing 2 and 3 to the opposite pole. This case, called **alternate disjunction**, produces only euploid gametes, although half of them will carry only translocated chromosomes (Figure 6.27*c*).

The production of aneuploid gametes by adjacent disjunction explains why translocation heterozygotes have reduced fertility. When such gametes fertilize a euploid gamete, the resulting zygote will be genetically unbalanced and therefore will be unlikely to survive. In plants, aneuploid gametes are themselves often inviable, especially on the male side, and fewer zygotes are produced. Translocation heterozygotes are therefore characterized by low fertility.

Sometimes the aneuploid gametes derived from adjacent disjunction do produce a euploid, viable zygote. Aneuploid gametes that have complementary deficiencies and duplications can unite in fertilization (Figure 6.28). A deficiency in one gamete is compensated for by a duplication in the other, and vice versa. The resulting zygote therefore has exactly two copies of all the chromosome segments. In this way, crosses between translocation heterozygotes can produce viable offspring. However, the number of offspring from such crosses is usually far fewer than it is from crosses between chromosomally normal individuals because the frequency of euploid zygotes is so low.

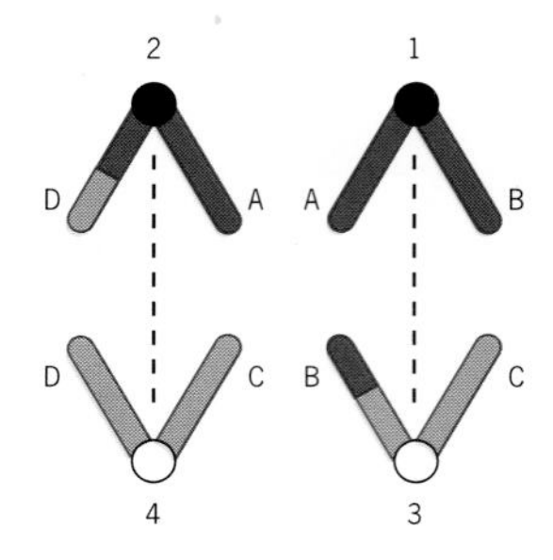

(*a*) Adjacent disjunction I.

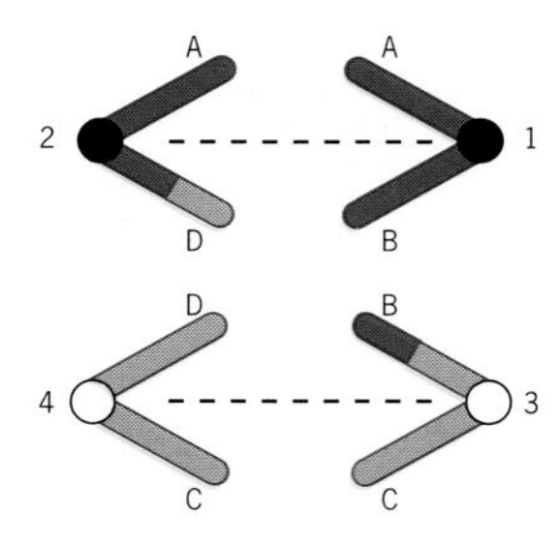

(*b*) Adjacent disjunction II.

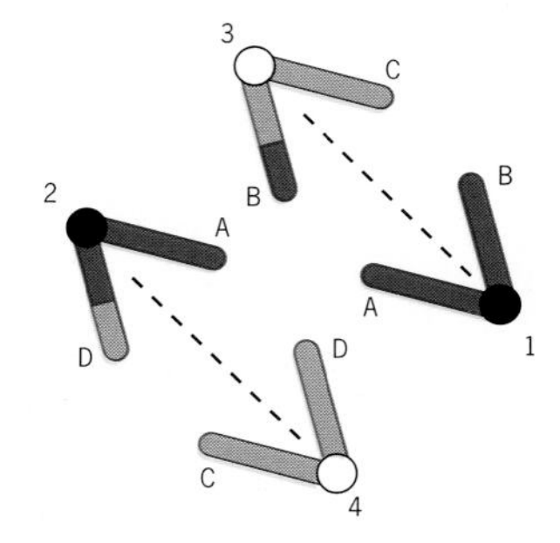

(*c*) Alternate disjunction.

Figure 6.27 Types of disjunction in a translocation heterozygote. (*a*) One form of adjacent disjunction in which homologous centromeres go to the same pole during anaphase. (*b*) Another form of adjacent disjunction in which homologous centromeres go to opposite poles during anaphase. (*c*) Alternate disjunction in which homologous centromeres go to opposite poles during anaphase.

In some species, such as the evening primroses in the genus *Oenothera*, the chromosomes are serially translocated with each other: chromsome 1 is translocated with chromosome 2, chromosome 2 with chromosome 3, chromosome 3 with chromosome 4, and so forth (see Technical Sidelight: *Oenothera*). During meiosis, this complicated situation is handled in a simple way: the chromosomes congregate to form a ring, with homologous segments synapsed along their lengths.

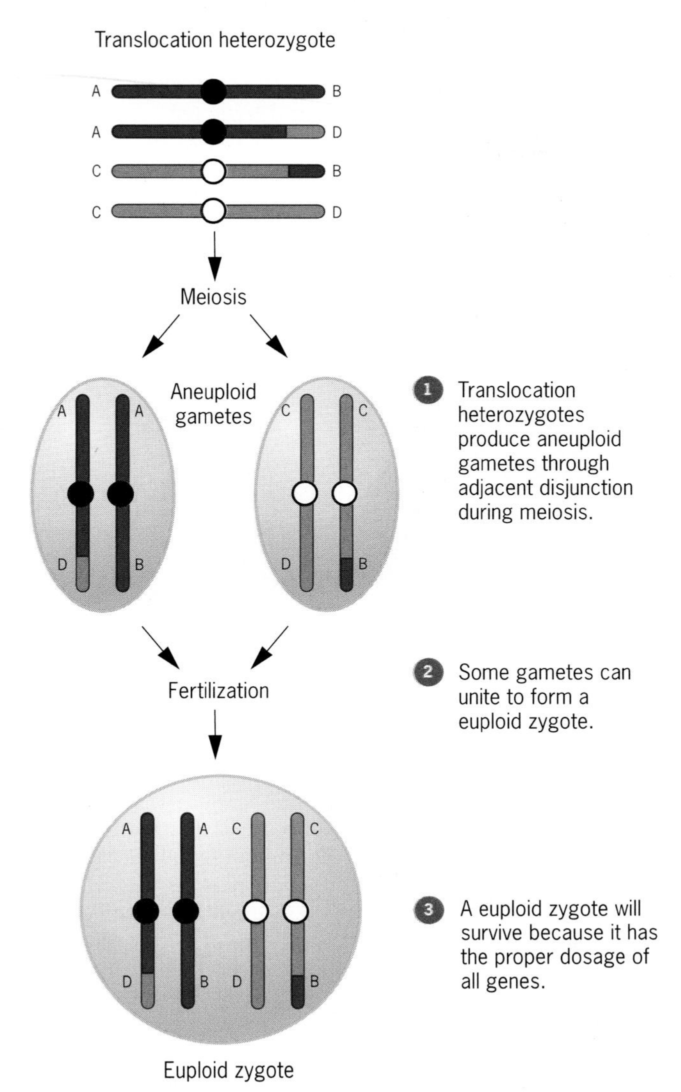

Figure 6.28 Euploid zygotes formed by the union of aneuploid gametes derived from a translocation heterozygote.

Then they disjoin in an alternate pattern, producing two types of euploid gametes. Self-fertilization with these gametes produces three kinds of zygotes; however, only the zygotes that are genotypically like their parent survive. Thus alternate disjunction during meiosis and selective survival among the offspring perpetuate the system of serially translocated chromosomes.

Compound Chromosomes and Robertsonian Translocations

Sometimes one chromosome fuses with its homolog, or two sister chromatids become attached to each other, forming a single genetic unit. A **compound chromosome** can exist stably in a cell as long as it has a single functional centromere; if there are two centromeres, each may move to a different pole during division, pulling the compound chromosome apart. A compound chromosome may also be formed by the union of homologous chromosome segments. For example, the right arms of the two second chromosomes in *Drosophila* might detach from their left arms and fuse at the centromere, creating a compound half-chromosome. Cytogeneticists sometimes call this structure an **isochromosome** (from the Greek prefix for "equal"), because its two arms are equivalent. Compound chromosomes differ from translocations in that they involve fusions of homologous chromosome segments. Translocations, by contrast, always involve fusions between nonhomologous chromosomes.

The first compound chromosome was discovered in 1922 by Lillian Morgan, the wife of T. H. Morgan. This compound was apparently formed by fusion of the two X chromosomes in *Drosophila*, creating double-X or **attached-X chromosomes**. The discovery occurred as a result of genetic experimentation rather than cytological analysis. Lillian Morgan crossed females homozygous for a recessive X-linked mutation to wild-type males. From such a cross, we would ordinarily expect all the daughters to be wild-type and all the sons to be mutant. However, Morgan observed just the opposite: all the daughters were mutant and all the sons were wild-type. Further work established that the X chromosomes in the mutant females had become attached to each other. Figure 6.29 illustrates the genetic significance of this attachment. The attached-X females produced two kinds of eggs, diplo-X and nullo-X, and their mates produced two kinds of sperm, X-bearing and Y-bearing. The union of these

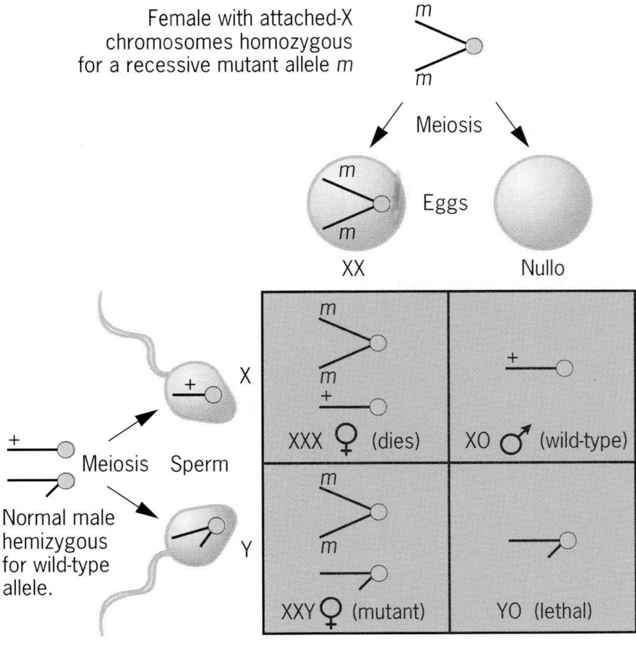

Figure 6.29 Results of a cross between a normal male and a female with attached-X chromosomes.

TECHNICAL SIDELIGHT

Oenothera: Chromosome Rings and Other Things

Hugo de Vries, one of the botanists who rediscovered Mendel's principles, was the first scientist to study the genetic properties of the evening primroses (Figure 1). In 1886 he discovered phenotypically different varieties of *Oenothera lamarckiana* growing in a waste field near Amsterdam. Intrigued by these differences, he brought specimens into his experimental garden, where he cultivated them under controlled conditions for several years. During this time, de Vries observed that the predominant form of *O. lamarckiana* occasionally produced offspring with unusual phenotypes. He gave each of these "sports" Latin names—*scintillans* (smooth, narrow leaves), *lata* (bubbly, broad leaves), *nanella* (dwarf), *gigas* (giant), and *rubrinervis* (red veins in the leaves)—and noted that each was heritable.

Figure 1 Photograph of one of the evening primroses in the genus *Oenothera*.

De Vries regarded these sports as new species of *Oenothera*. His extensive research led him to propose that "species are derived from other species by means of sudden small changes."[1] This view has come to be known as the mutation theory of evolution. Subsequent research has shown that De Vries reached the right conclusion for the wrong reason. The phenotypic changes that he observed were not caused by mutations in the true sense of the word; that is, they were not due to changes in genes. Rather, they were caused by chromosome aberrations—polyploidy, trisomy, and unusual segregations in complex translocation heterozygotes.

Most species of *Oenothera*, including *O. lamarckiana*, have 14 metacentric chromosomes in their somatic cells. During meiosis, these chromosomes associate in an unusual way. In *O. lamarckiana*, for example, two of the chromosomes pair, but the other 12 congregate to form a ring. In this ring, the arms of each chromosome appear to be paired with the arms of two adjacent chromosomes. This association suggested to the early cytologists—John Belling, in particular—that the chromosomes of *Oenothera* had been serially translocated. This hypothesis was subsquently verified by extensive cytological observations on a wide array of *Oenothera* species and on hybrids between them. This cytological analysis was mainly the work of the American botanist, Ralph Cleland. Each species of *Oenothera* proved to carry different translocation complexes—called Renner complexes after the German cytologist Oscar Renner, whose genetic analyses had first suggested their existence.

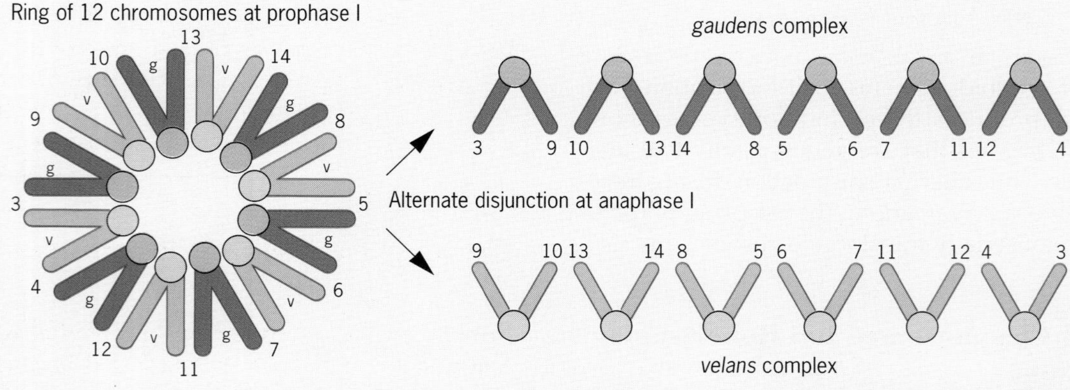

Figure 2 Formation of a ring of 12 chromosomes during prophase I in *O. lamarckiana,* and disjunction of the *gaudens* and *velans* complexes from that ring during anaphase I. The end of each chromosome arm is labeled with a number to indicate homology between the members of the *gaudens* and *velans* complexes. The chromosomes with arms 1 and 2 (not shown) form a bivalent during prophase I.

In *O. lamarckiana*, the ring of 12 chromosomes is produced by two sets of serial translocations. One set of translocated chromosomes constitutes the *gaudens* Renner complex; the other set constitutes the *velans* Renner complex. In the prophase of meiosis I, these two complexes associate to form a ring of 12 chromosomes (Figure 2). During anaphase I, the chromosomes disjoin alternately so that, at the conclusion of meiosis, two types of gametes are formed; one contains the *gaudens* complex, and the other the *velans* complex. Fertilization with these types of gametes produces three types of zygotes: *gaudens/gaudens*, *velans/velans*, and *gaudens/velans*; however, only the *gaudens/velans* zygotes survive. The other two genotypes die because they are homozygous for different recessive lethal mutations. This system therefore perpetuates the *gaudens/velans* genotype generation after generation.

The lethality of the *gaudens* and *velans* homozygotes explains why most of the offspring of *O. lamarckiana* are phenotypically like their parents; only the *gaudens/velans* heterozygotes survive, and they resemble their parents. At a frequency of about 1 percent, however, different offspring are produced. After all, it was these different phenotypes that attracted de Vries to the study of *Oenothera*. Some of the phenotypes that he observed were due to an exchange of alleles between the *gaudens* and *velans* chromosomes (see Chapter 7 for a discussion of the nature and significance of genetic exchange between chromosomes). For example, *nanella*, the dwarf type that he discovered, was caused by a reciprocal exchange between one chromosome in the *gaudens* complex and another in the *velans* complex. This exchange made a recessive allele for dwarfism homozygous in the otherwise heterozygous *gaudens/velans* genotype. Other phenotypes, such as *lata* and *scintillans*, were due to trisomy—undoubtedly caused by aberrant chromosome segregation from the meiotic ring, and the extremely rare *gigas* was due to polyploidy. Thus the remarkable mutability that de Vries observed in *O. lamarckiana* was actually the result of occasional slipups and breakdowns in a rather peculiar genetic system.

[1]de Vries, H., 1919, *Plant breeding*. Open Court Publishing Company, Chicago.

gametes in all possible ways produced two kinds of viable progeny: mutant XXY females, which inherit the attached-X chromosomes from their mothers and a Y chromosome from their fathers; and wild-type XO males, which inherit an X chromosome from their fathers and no sex chromosome from their mothers. Because the Y chromosome is needed for fertility, these XO males are sterile. Lillian Morgan was able to propagate the attached-X chromosomes by backcrossing XXY females to wild-type XY males from another stock. Because the sons of this cross inherited a Y chromosome from their mothers, they were fertile and could be crossed to their XXY sisters to establish a stock in which the attached-X chromosomes were permanently maintained in the female line.

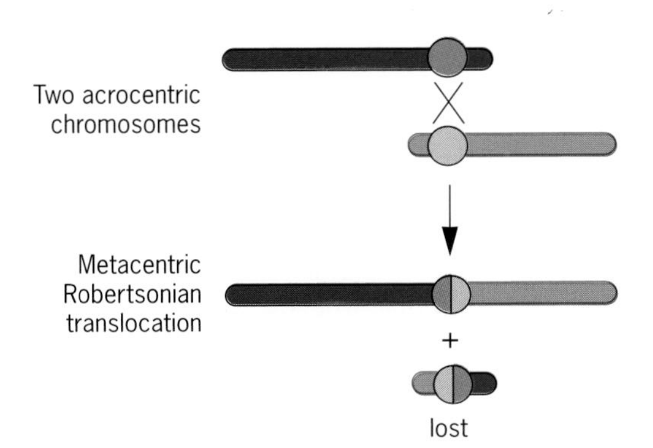

Figure 6.30 Formation of a metacentric Robertsonian translocation by exchange between two nonhomologous acrocentric chromosomes.

Nonhomologous chromosomes can also fuse at their centromeres, creating a structure called a **Robertsonian translocation** (Figure 6.30), named for the cytologist F. W. Robertson. For example, if two acrocentric chromosomes fuse, they will produce a metacentric chromosome; the tiny short arms of the participating chromosomes are simply lost in this process. Such chromosome fusions have apparently occurred quite often in the course of evolution.

Chromosomes can also fuse end-to-end to form a structure with two centromeres. If one of these is inactivated, the chromosome fusion will be stable. Such a fusion evidently occurred in the evolution of our own species. Human chromosome 2, which is a metacentric, has arms that correspond to two different acrocentric chromosomes in the genomes of the great apes. Detailed cytological analysis has shown that the ends of the short arms of these two chromosomes apparently fused to create human chromosome 2.

Phenotypic Effects of Chromosome Rearrangements

In homozygous condition, deletions that remove several genes are almost always lethal because some of the missing genes are likely to be essential for life. Duplications, in contrast, may be viable in homozygous condition, provided that they are not too large.

In heterozygous condition, deletions and duplications can affect the phenotype by altering the dosage of groups of genes. Usually, the larger the chromosome segment involved, the greater the phenotypic effect. In fact, aneuploidy for very large chromosome segments is typically lethal. However, sometimes even small deletions or duplications can have a lethal effect in heterozygous condition, indicating that the aneuploid region contains at least one gene with a strict requirement for proper dosage. In a diploid organism, a gene that is unable to sustain life when present in a single dose is said to be **haplo-lethal**, and a gene that kills the organism when it is duplicated is said to be **triplo-lethal**. Such genes have been identified in *Drosophila*, where researchers have analyzed thousands of different deficiencies and duplications scattered around the genome.

Inversions and translocations may also affect the phenotype. Sometimes the breakpoints of these rearrangements disrupt genes, rendering them mutant. When the rearrangements are then made homozygous, the mutant phenotype appears. In other cases, the breakpoints are not themselves disruptive, but the genes near them are put into a different chromosomal environment, where they may not function normally. A classic example involves the *white-mottled* (w^m) mutation of *Drosophila* (Figure 6.31). This mutation, discovered by Hermann J. Muller, is associated with an inversion on the X chromosome; one breakpoint is in the euchromatin near the *white* gene, and the other is in the distant heterochromatin flanking the centromere. Although the inversion does not physically disrupt the *white* gene, it adversely affects its function. Instead of having pigment evenly distributed throughout their eyes, flies with the inversion have a spotty distribution of pigment—a mosaic suggesting that the *white* gene is functioning properly in some eye cells but not in others. Muller reasoned that this effect was due to the juxtaposition of the *white* gene to centric heterochromatin, which tends to remain condensed during the cell cycle. In its new chromosomal environment, the *white* gene does not function normally. It is said to be influenced by a chromosomal **position effect**. Geneticists distinguish two types of position effects, stable and variegated, depending on how the phenotype is altered. With its spotty phenotype, the *white-mottled* mutation is clearly an example of variegated position effect. Such effects typically occur when euchromatic genes are juxtaposed near heterochromatin, which seems to exert a repressing effect on their function.

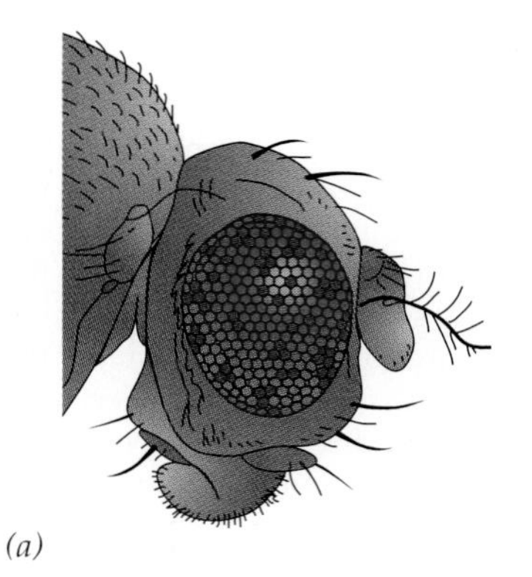

(*a*)

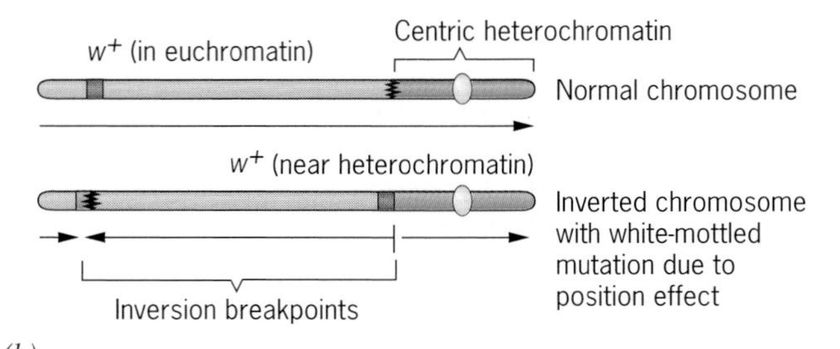

(*b*)

Figure 6.31 The *white-mottled* (w^m) mutation in *Drosophila*. (*a*) Mutant phenotype of the eye. (*b*) Origin of the mutation through an inversion that moves the wild-type white gene (w^+) near the centric heterochromatin, creating a position effect.

Key Points: **An inversion reverses the order of a segment within a single chromosome; a translocation interchanges segments between two nonhomologous chromosomes. During meiosis, the chromosomes in an inversion heterozygote pair by forming a loop; in a translocation heterozygote, they pair by forming a cross. Chromosome rearrangements may alter the expression of genes by moving them to new chromosomal positions.**

TESTING YOUR KNOWLEDGE

1. A *Drosophila* geneticist has obtained females that carry attached-X chromosomes homozygous for a recessive mutation (y) that causes the body to be yellow instead of gray. In one experiment, she crosses some of these females to ordinary wild-type males, and in another, she crosses these females to wild-type males that have their X and Y chromosomes attached to each other; that is, they carry a compound XY chromosome. Predict the phenotypes of the progeny from these two crosses and indicate which, if any, will be sterile.

ANSWER

To predict the phenotypes of the progeny, we need to know their genotypes. The easiest way to determine these genotypes is to diagram the kinds of zygotes produced by each cross.

First, we consider the cross between the yellow-bodied attached-X females and the ordinary wild-type males. The females produce two kinds of gametes, XX and nullo. The males also produce two kinds of gametes, X and Y. When these are combined in all possible ways, four types of zygotes are produced; however, only two types are viable. The XXY zygotes will develop into yellow-bodied females—like their mothers except that they carry a Y chromosome—and the XO zygotes will develop into gray-bodied males—like their fathers except that they lack a Y chromosome. The extra Y chromosome in the females will have no effect on fertility, but the missing Y chromosome in the males will cause them to be sterile.

		Eggs: $\widehat{X^yX^y}$	Eggs: 0
Sperm	X^+	$\widehat{X^yX^y}\,X^+$ (die)	X^+O gray males
Sperm	Y	$\widehat{X^yX^y}\,Y$ yellow females	YO (die)

Now we consider the cross between the yellow-bodied attached-X females and the males with a compound XY chromosome. Both sexes produce two kinds of gametes—for the females, the same as above, and for the males, either XY or nullo. When these are united in all possible ways, we find that two types of zygotes will be viable: yellow-bodied females with attached-X chromosomes and gray-bodied males with a compound XY chromosome. Both types of progeny will be fertile.

		Eggs: $\widehat{X^y\,X^y}$	Eggs: 0
Sperm	$\widehat{X^+Y}$	$\widehat{X^y\,X^y}\ \widehat{X^+Y}$ (die)	$\widehat{X^+Y}\ O$ gray males
Sperm	0	$\widehat{X^y\,X^y}\ O$ yellow females	O O (die)

2. A phenotypically normal man carries a translocated chromosome that contains the entire long arm of chromosome 14, part of the short arm of chromosome 14, and most of the long arm of chromosome 21:

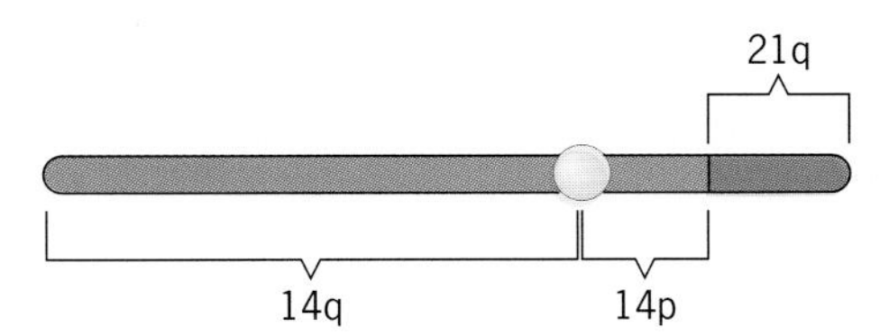

The man also carries a normal chromosome 14 and a normal chromosome 21. If he marries a cytologically (and phenotypically) normal woman, is there any chance that the couple will produce phenotypically abnormal children?

ANSWER

Yes, the couple could produce children with Down syndrome as a result of meiotic segregation in the cytologically abnormal man. During meiosis in this man, the translocated chromosome, T(14, 21), will synapse with the normal chromosomes 14 and 21, forming a trivalent. Disjunction from this trivalent will produce six different types of sperm, four of which are aneuploid.

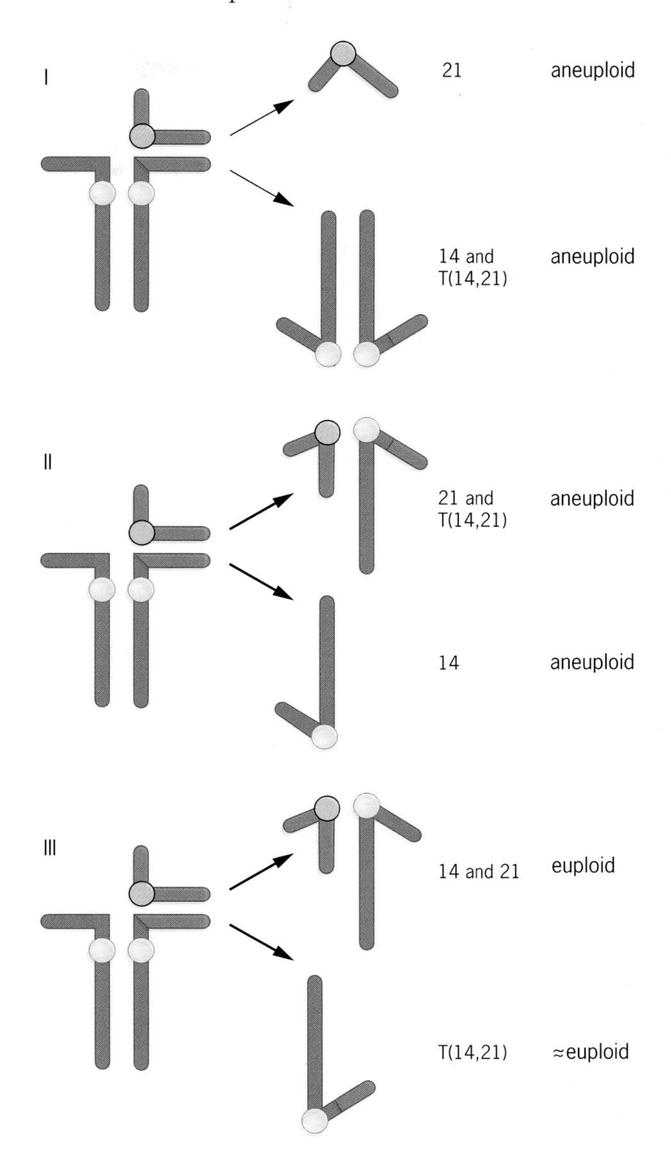

Fertilization of an egg containing one chromosome 14 and one chromosome 21 by any of the aneuploid sperm will produce an aneuploid zygote.

Disjunction	Sperm	Zygote	Outcome	
I	21	14, 21, 21	monosomy 14	dies
	14, T(14,21)	14, 14, T(14,21)	trisomy 14	dies
II	14	14, 14, 21	monosomy 21	dies
	T(14, 21), 21	14, T(14,21), 21, 21	trisomy 21	Down
III	14, 21	14, 14, 21, 21	euploid	normal
	T(14, 21)	14, T(14, 21)	≈ euploid	normal

Although trisomy or monosomy for chromosome 14 and monosomy for chromosome 21 are all lethal conditions, trisomy for chromosome 21 is not. Thus it is possible for the couple to give birth to a child with Down syndrome.

QUESTIONS AND PROBLEMS

6.1 In the human karyotype, the X chromosome is approximately the same size as seven of the autosomes (the so-called C group of chromosomes). What procedure could be used to distinguish the X chromosome from the other members of this group?

6.2 Distinguish between polyploidy and polyteny.

6.3 In human beings, a cytologically abnormal chromosome 22, called the "Philadelphia" chromosome because of the city in which it was discovered, is associated with chronic leukemia. This chromosome is missing part of its long arm. How would you denote the karyotype of an individual who had 46 chromosomes in his somatic cells, including one normal 22 and one Philadelphia chromosome?

6.4 Describe two ways in which polyploidy might occur in nature.

6.5 During meiosis, why do some tetraploids behave more regularly than triploids?

6.6 A plant species A, which has seven chromosomes in its gametes, was crossed with a related species B, which has nine. The hybrids were sterile, and microscopic observation of their pollen mother cells showed no chromosome pairing. A section from one of the hybrids that grew vigorously was propagated vegetatively, producing a plant with 32 chromosomes in its somatic cells. This plant was fertile. Explain.

6.7 A plant species X with $n = 5$ was crossed with a related species Y with $n = 7$. The F_1 hybrid produced only a few pollen grains, which were used to fertilize the ovules of species Y. A few plants were produced from this cross, and all had 19 chromosomes. Following self-fertilization, the F_1 hybrids produced a few F_2 plants, each with 24 chromosomes. These plants were phenotypically different from either of the original species and were highly fertile. Explain the sequence of events that produced these fertile F_2 hybrids.

6.8 Identify the sexual phenotypes of the following genotypes in human beings: XX, XY, XO, XXX, XXY, XYY.

6.9 If nondisjunction of chromosome 21 occurs in the division of a secondary oocyte in a human female, what is the chance that a mature egg derived from this division will receive two number 21 chromosomes?

6.10 A *Drosophila* female homozygous for a recessive X-linked mutation causing yellow body was crossed to a wild-type male. Among the progeny, one fly had sectors of yellow pigment in an otherwise gray body. These yellow sectors were distinctly male, whereas the gray areas were female. Explain the peculiar phenotype of this fly.

6.11 The *Drosophila* fourth chromosome is so small that flies monosomic or trisomic for it survive and are fertile. Several genes, including *eyeless* (*ey*), have been located on this chromosome. If a cytologically normal fly homozygous for a recessive eyeless mutation is crossed to a fly monosomic for a wild-type fourth chromosome, what kinds of progeny will be produced, and in what proportions?

6.12 A woman with X-linked color blindness and Turner syndrome had a color blind father and a normal mother. In which of her parents did nondisjunction of the sex chromosomes occur?

6.13 Although XYY men are phenotypically normal, would they be expected to produce more children with sex chromosome abnormalities than XY men? Explain.

6.14 What characteristics of *Drosophila* salivary gland chromosomes make them especially suitable for cytogenetic studies?

6.15 In a *Drosophila* salivary chromosome, the bands have a sequence of 1 2 3 4 5 6 7 8. The homolog with which this chromosome is synapsed has a sequence of 1 2 3 6 5 4 7 8. What kind of chromosome change has occurred? Draw the synapsed chromosomes.

6.16 Other chromosomes have sequences as follows: (a) 1 2 5 6 7 8; (b) 1 2 3 4 4 5 6 7 8; (c) 1 2 3 4 5 8 7 6. What kind of chromosome change is present in each? Illustrate how these chromosomes would pair with a chromosome whose sequence is 1 2 3 4 5 6 7 8.

6.17 One chromosome in a plant has the sequence A B C D E F, and another has the sequence M N O P Q R. A reciprocal translocation between these chromosomes produced the following arrangement: A B C P Q R on one chromosome and M N O D E F on the other. Illustrate how these translocated chromosomes would pair with their normal counterparts in a heterozygous individual during meiosis.

6.18 In *Drosophila*, the genes *bw* and *st* are located on chromosomes 2 and 3, respectively. Flies homozygous for *bw* mutations have brown eyes, flies homozygous for *st* mutations have scarlet eyes, and flies homozygous for *bw* and *st* mutations have white eyes. Doubly heterozygous males were mated individually to homozygous *bw; st* females. All but one of the matings produced four classes of progeny: wild-type, brown, scarlet and white-eyed flies. The single exception produced only wild-type and white-eyed progeny. Explain the nature of this exception.

6.19 A phenotypically normal boy has 45 chromosomes, but his sister, who has Down syndrome, has 46. Suggest an explanation for this paradox.

6.20 How do variegated position effects originate and how can they be explained?

6.21 A yellow-bodied *Drosophila* female with attached-X chromosomes was crossed to a white-eyed male. Both of the parental phenotypes are caused by X-linked recessive mutations. Predict the phenotypes of the progeny.

6.22 A man has attached chromosomes 21. If his wife is cytologically normal, what is the chance their first child will have Down syndrome?

6.23 Analysis of the polytene chromosomes of three populations of *Drosophila* has revealed three different banding sequences in a region of the second chromosome:

Population	Banding sequence
P1	1 2 3 4 5 6 7 8 9 10
P2	1 2 3 9 8 7 6 5 4 10
P3	1 2 3 9 8 5 6 7 4 10

Explain the evolutionary relationships among these populations.

6.24 The diagrams below show two pairs of chromosomes in the karyotypes of a man, a woman, and their child. The man and the woman are phenotypically normal, but the child (a boy) suffers from a syndrome of abnormalities, including poor motor control and severe mental impairment. What is the genetic basis of the child's abnormal phenotype? Is the child hyperploid or hypoploid for a segment in one of his chromosomes?

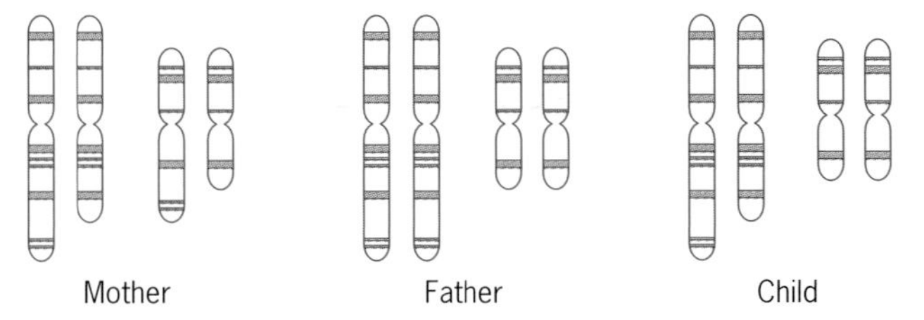

6.25 A male mouse that is heterozygous for a reciprocal translocation between the X chromosome and an autosome is crossed to a female mouse with a normal karyotype. The autosome involved in the translocation carries a gene responsible for coloration of the fur. The allele on the male's translocated autosome is wild-type, and the allele on its nontranslocated autosome is mutant; however, because the wild-type allele is dominant to the mutant allele, the male's fur is wild-type (dark in color). The female mouse has light color in her fur because she is homozygous for the mutant allele of the color-determining gene. When the offspring of the cross are examined, all the males have light fur and all the females have patches of light and dark fur. Explain these peculiar results.

6.26 In *Drosophila*, the autosomal genes *cinnabar* (*cn*) and *brown* (*bw*) control the production of brown and red eye pigments, respectively. Flies homozygous for *cinnabar* mutations have bright red eyes, flies homozygous for *brown* mutations have brown eyes, and flies homozygous for mutations in both of these genes have white eyes. A male homozygous for mutations in the *cn* and *bw* genes has bright red eyes because a small duplication that carries the wild-type allele of *bw* (bw^+) is attached to the Y chromosome. If this male is mated to a karyotypically normal female that is homozygous for the *cn* and *bw* mutations, what types of progeny will be produced?

6.27 Cytological examination of the sex chromosomes in a man has revealed that he carries an insertional translocation. A small segment has been deleted from the Y chromosome and inserted into the short arm of the X chromosome; this segment contains the gene responsible for male differentiation (*TDF*). If this man marries a karyotypically normal woman, what types of progeny will the couple produce?

BIBLIOGRAPHY

Bridges, C. B. 1935. "Salivary chromosome maps with a key to the banding of the chromosomes of *Drosophila melanogaster*." *J. Hered.* 26:60–64.

Burnham, C. R. 1962. *Discussions in Cytogenetics*. Burgess, Minneapolis.

DeGrouchy, J., and C. Turleau. 1984. *Clinical Atlas of Human Chromosomes*. Second edition. Wiley, New York.

Hsu, T. C. 1973. "The longitudinal differentiation of chromosomes." *Ann. Rev. Genet.* 7:153–176.

Simmonds, N. W., ed. 1976. *The Evolution of Crop Plants*. Longman, New York.

Stebbins, G. L. 1950. *Variation and Evolution in Plants*. Columbia University Press, New York.

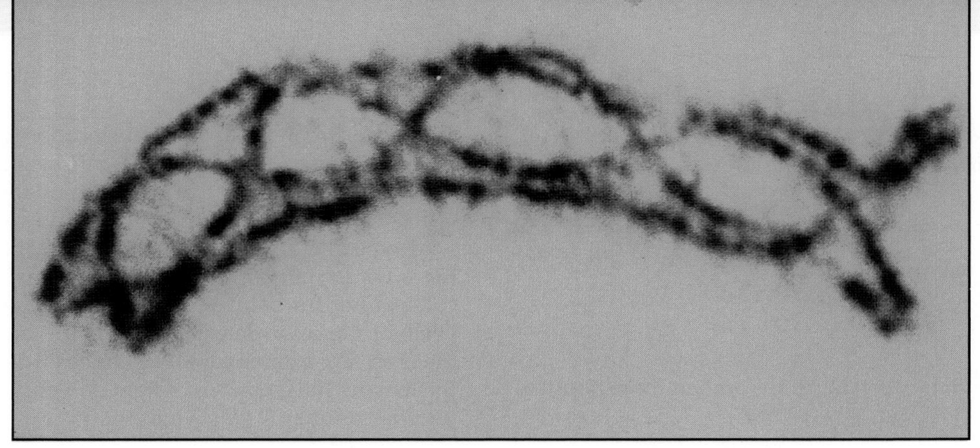

Chiasmata in the late prophase of the first meiotic divison. These cross-shaped figures are the result of exchanges between paired chromosomes.

7

Linkage, Crossing Over, and Chromosome Mapping in Eukaryotes

CHAPTER OUTLINE

The World's First Chromosome Map

The modern picture of chromosome organization emerged from a combination of genetic and cytological studies. T. H. Morgan laid the foundation for these studies when he demonstrated that the gene for *white* eyes in *Drosophila* was located on the X chromosome. Soon afterward, Morgan's students showed that other genes were X-linked, and eventually, they were able to locate each of these genes on a map of the chromosome. This map was a straight line, and each gene was situated at a particular point, or **locus**, on it (Figure 7.1). The structure of the map therefore implied that a chromosome was simply a linear array of genes. All subsequent research has supported this view. Genes are arranged in a linear order along the length of a chromosome. There are no branches or discontinuities. The basic geometric figure in genetics is a line, reflecting, as we now know, the linear structure of DNA.

The procedure for mapping chromosomes was invented by Alfred H. Sturtevant (Figure 7.2), an undergraduate working in Morgan's laboratory. One night in 1911 Sturtevant put aside his algebra homework in order to evaluate some experimental data. Before the sun rose the next day, he had constructed the world's first chromosome map.

The cytological appearance of chromosomes inspired Sturtevant to make his map a straight line. How was Sturtevant able to determine the map locations of individual genes? No microscope was powerful enough to see genes, nor was any measuring device accurate enough to obtain the distances between them. In fact, Sturtevant did not use any sophisticated instruments in his work. Instead, he relied completely on the analysis of data from experimental crosses with *Drosophila*.

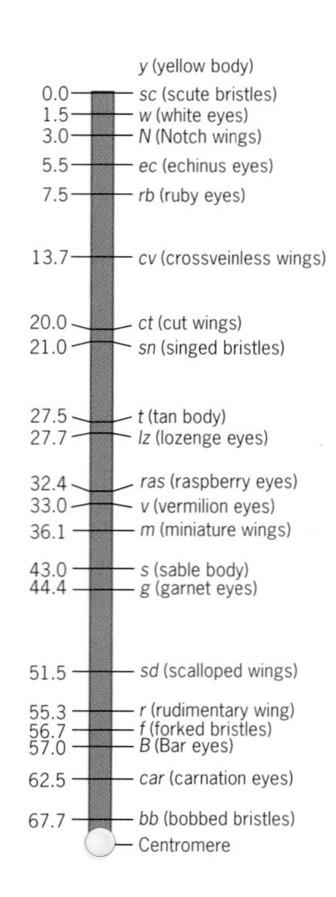

Figure 7.1 A map of genes on the X chromosome of *Drosophila melanogaster*.

Figure 7.2 Alfred H. Sturtevant.

LINKAGE, RECOMBINATION, AND CROSSING OVER

Sturtevant based his mapping procedure on the principle that genes on the same chromosome should be inherited together. Because such genes are physically attached to the same structure, they should travel as a unit through meiosis. This phenomenon is called **linkage**. The early geneticists were unsure about the nature of linkage, but some of them, including Morgan and his students, thought that genes were attached to one another much like beads on a string. Thus these researchers clearly had a linear model of chromosome organization in mind.

The early geneticists also knew that linkage was not absolute. Their experimental data demonstrated that genes on the same chromosome could be separated as they went through meiosis, and that new combinations of genes could be formed. However, this phenomenon, called **recombination**, was difficult to explain by simple genetic theory.

One hypothesis was that during meiosis, when homologous chromosomes paired, a physical exchange of material separated and recombined genes. This idea was inspired by the cytological observation that chromosomes could be seen in pairing configurations that suggested they had switched pieces with each other. At the switch points, the two homologs were crossed over, as if each had been broken and then reattached to its partner. A crossover point was called a **chiasma** (plural, **chiasmata**), from the Greek word meaning "cross." Geneticists began to use the term **crossing over** to describe the process that created the chiasmata—that is, the actual process of exchange between paired chromosomes. They considered recombination—the separation of linked genes and the formation of new gene combinations—to be a result of the physical event of crossing over.

Exceptions to the Mendelian Principle of Independent Assortment

The phenomena of linkage and recombination were first described by W. Bateson and R. C. Punnett shortly after the rediscovery of Mendel's work at the beginning of the twentieth century. Initially, this linkage was viewed as an exception to Mendel's Principle of Independent Assortment.

Some of the first evidence for linkage—and against independent assortment—came from experiments with sweet peas (Figure 7.3). Bateson and Punnett crossed varieties that differed in two traits, flower color and pollen length. They crossed plants with red flowers and long pollen grains to plants with white

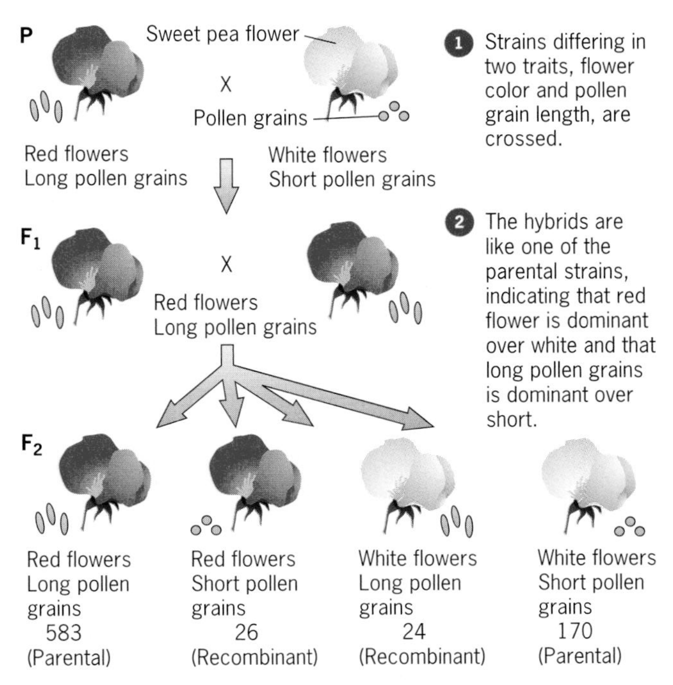

Figure 7.3 Bateson's and Punnett's experiment with sweet peas. The results in the F_2 indicate that the genes for flower color and pollen length do not assort independently.

flowers and short pollen grains. All the F_1 plants had red flowers and long pollen grains, thus indicating that the alleles for these two phenotypes were dominant. When the F_1 plants were self-fertilized, Bateson and Punnett observed a peculiar distribution of phenotypes among the offspring. Instead of the 9:3:3:1 ratio expected for two independently assorting genes, they obtained a ratio of 23.3:1:1:6.8. Obviously, the two parental classes were significantly overrepresented in the progeny. This departure from the expected Mendelian results was due to linkage between the gene for flower color and that for pollen length.

Although Bateson and Punnett devised a complicated explanation for their results, it turned out to be wrong. The correct explanation is that the genes for flower color and pollen length are located on the same chromosome; consequently, they tend to travel through meiosis together. This explanation is diagrammed in Figure 7.4. The alleles of the flower color gene are *R* (red) and *r* (white), and the alleles of the pollen length gene are *L* (long) and *l* (short); the *R* and *L* alleles are dominant. Because these two genes are linked, the parental combinations of alleles (*R* and *L*, and *r* and *l*) are more prevalent than the nonparental combinations (*R* and *l*, and *r* and *L*) in the gametes of the F_1 plants. Self-fertilization of the F_1 therefore produces a preponderance of parental phenotypes in the F_2. However, the genes for flower color and pollen length are not inextricably linked. Some nonparental progeny do appear, although at low frequency. Because these progeny indicate that the alleles of the two genes were recombined in the F_1, they are called **recombinants**.

Frequency of Recombination as a Measure of Linkage Intensity

We can use the frequency of recombination to measure the intensity of linkage between genes. Genes that are tightly linked seldom recombine, whereas genes that are loosely linked recombine often. The most direct way of estimating the frequency of recombination is to use a testcross. For example, the heterozygous F_1 sweet peas discussed above could have been crossed to homozygotes carrying both of the re-

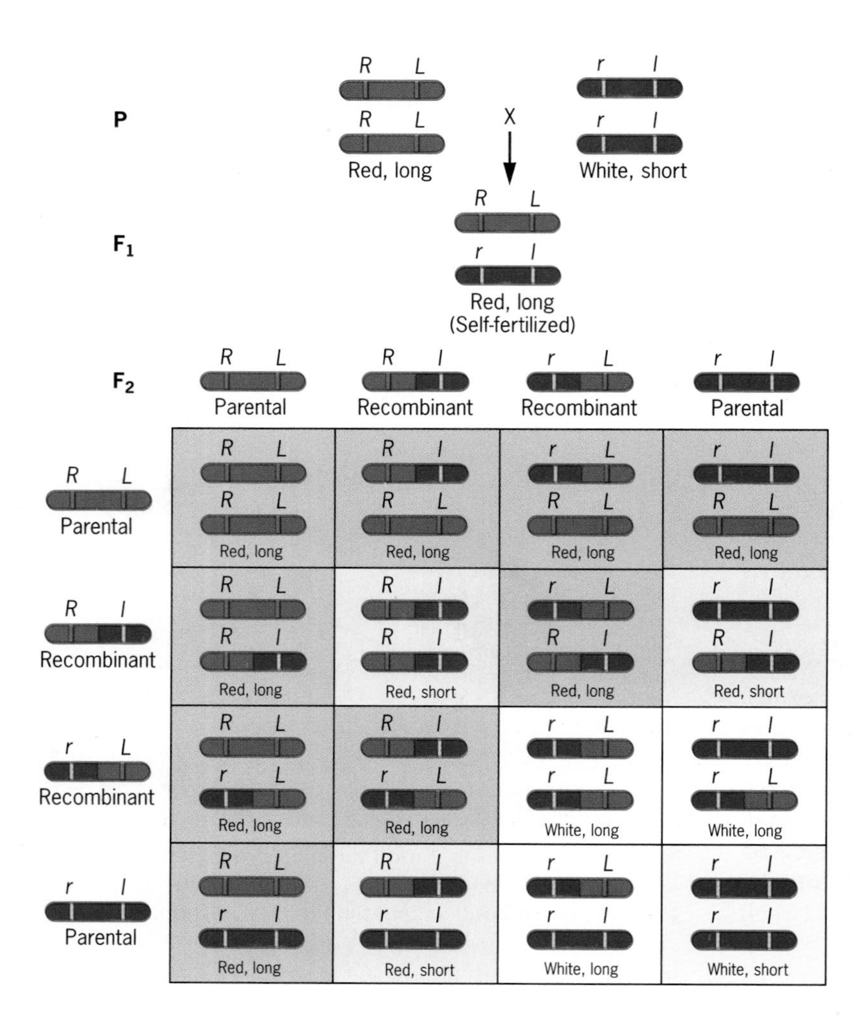

Figure 7.4 Genetic basis of the results of Bateson's and Punnett's experiment. Because the genes for flower color and pollen length are located on the same chromosome, they tend to travel through meiosis together.

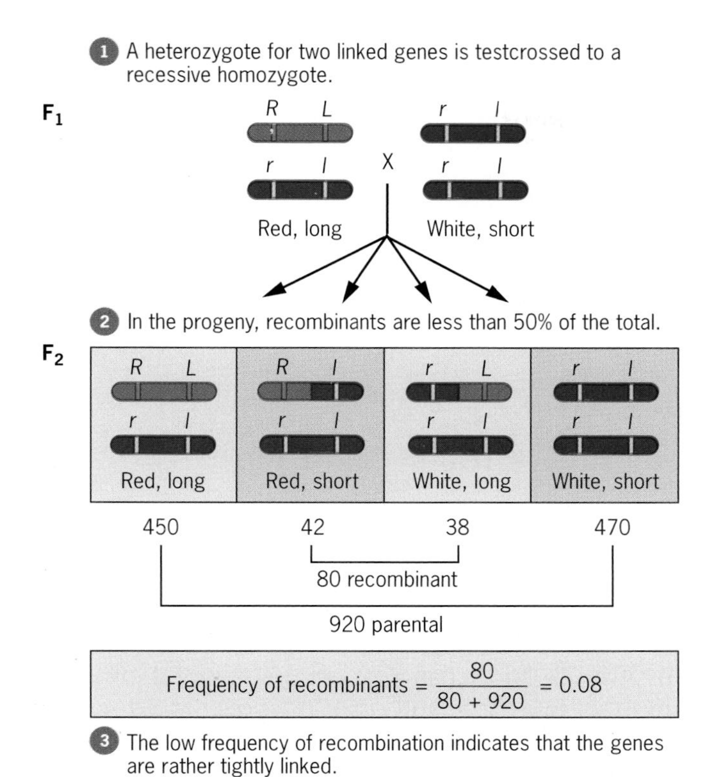

Figure 7.5 A testcross for linkage between genes in sweet peas. The vast excess of parental-type progeny in the F_2 indicates that the genes for flower color and pollen length are tightly linked.

cessive alleles. The progeny of this testcross would therefore reveal what kinds of gametes the F_1 plants produced, and in what proportions.

Figure 7.5 presents the analysis of testcross data. Among 1000 progeny scored, 920 resemble one or the other of the parental strains and the remaining 80 are recombinant. The frequency of recombinant gametes produced by the heterozygous F_1 plants is therefore $80/1000 = 0.08$. We can use this frequency as a measure of how often the flower color and pollen length genes recombine during meiosis. Notice that it represents the summation of two approximately equal frequencies, one for each of the two types of recombinant gametes.

For any two genes, the frequency of recombinant gametes never exceeds 50 percent. This upper limit is reached when genes are very far apart, perhaps at opposite ends of a chromosome. It is also reached when genes are on different chromosomes; 50 percent recombination is, in fact, what is meant when it is said that the genes assort independently. A frequency of recombination less than 50 percent implies that the genes are linked.

Crosses involving linked genes are usually diagrammed to show the **linkage phase**—the way in

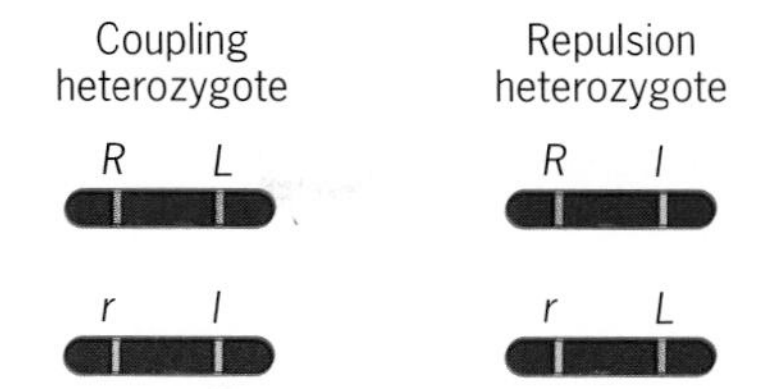

Figure 7.6 Coupling and repulsion linkage phases in double heterozygotes.

which the alleles are arranged in heterozygous individuals (Figure 7.6). In Bateson and Punnett's sweet pea experiment, the heterozygous F_1 plants received two dominant alleles, *R* and *L*, from one parent and two recessive alleles, *r* and *l*, from the other. Thus we write the genotype of these plants *R L*/*r l*, where the slash (/) separates alleles inherited from different parents. Another way of interpreting this symbolism is to say that the alleles on the left and right of the slash entered the genotype on different homologous chromosomes, one from each parent. Whenever the dominant alleles are all on one side of the slash, as in this example, the genotype has the *coupling* linkage phase. When the dominant and recessive alleles are split on both sides of the slash, as in *R l*/*r L*, the genotype has the *repulsion* linkage phase. These terms provide us with a way of distinguishing between the two kinds of double heterozygotes.

Crossing Over as the Physical Basis of Recombination

Recombinant gametes are produced as a result of crossing over between homologous chromosomes. This process involves a physical exchange between the chromosomes, as diagrammed in Figure 7.7. The exchange event occurs during the prophase of the first meiotic division, when duplicated chromosomes have paired. Although four homologous chromatids are present, forming what is called a **tetrad**, only two actually cross over at any one point. Each of these chromatids breaks at the site of the crossover, and the resulting pieces reattach to produce the recombinants. The other two chromatids are not recombinant at this site. Each crossover event therefore produces two recombinant chromatids among a total of four.

The evidence for crossing over between two chromatids within a tetrad comes from the study of certain fungi in the class Ascomycetes, the most important of which is baker's yeast, *Saccharomyces cerevisiae* (Figure 7.8). This single-celled haploid organism ordinarily reproduces asexually, but it also has a sexual phase that is initiated by the fusion of haploid cells of opposite mating types, denoted *a* and α *(alpha)*. This fusion cre-

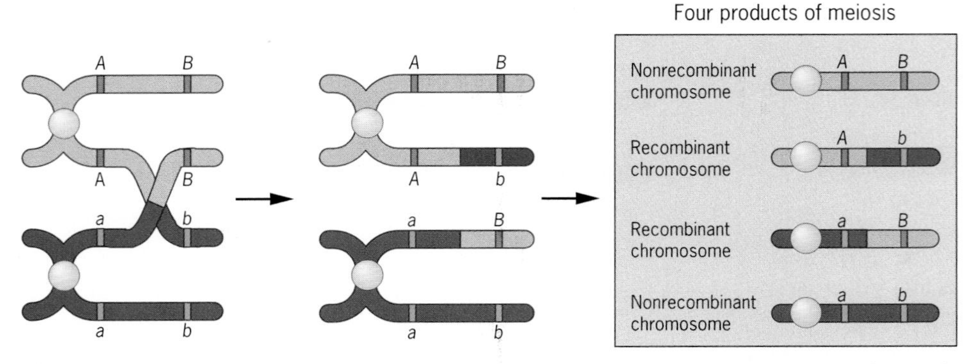

Figure 7.7 Crossing over as the basis of recombination between genes. An exchange between paired chromosomes during meiosis produces recombinant chromosomes at the end of meiosis.

ates a diploid cell, which promptly undergoes meiosis to produce four haploid cells, or ascospores, inside a small sac called the ascus. Each ascus therefore contains all four products of a single meiosis. By dissecting this sac, a researcher can isolate each product and place it in a culture dish to start a new yeast colony. Many mutations have been identified in *S. cerevisiae*, including some that alter colony shape and color. These mutations have been useful in the study of genetic recombination.

One important finding from studies with *S. cerevisiae* is that crossing over occurs *after* homologous chromosomes have duplicated. An ascus may contain two recombinant and two nonrecombinant ascospores, each of which has a different genotype (Figure 7.7). This observation can only be explained if crossing over occurs after chromosome duplication. If it occurred before duplication, an ascus could never contain more than two kinds of ascospores.

A second finding is that only two chromatids are involved in an exchange at any one point. However, the other two chromatids may cross over at a different point. Thus, there is a possibility for multiple exchanges in a tetrad of chromatids (Figure 7.9). There may, for example, be two, three, or even four separate exchanges—customarily called double, triple, or

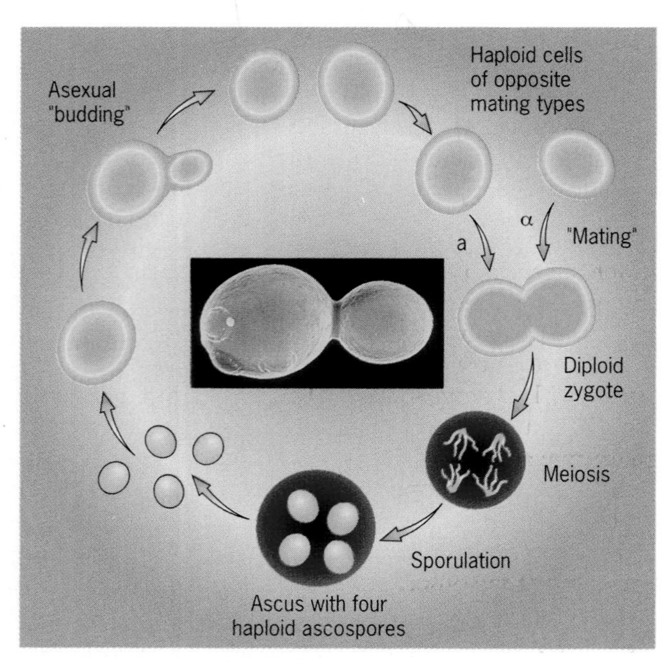

Figure 7.8 Life cycle of *Saccharomyces cerevisiae*.

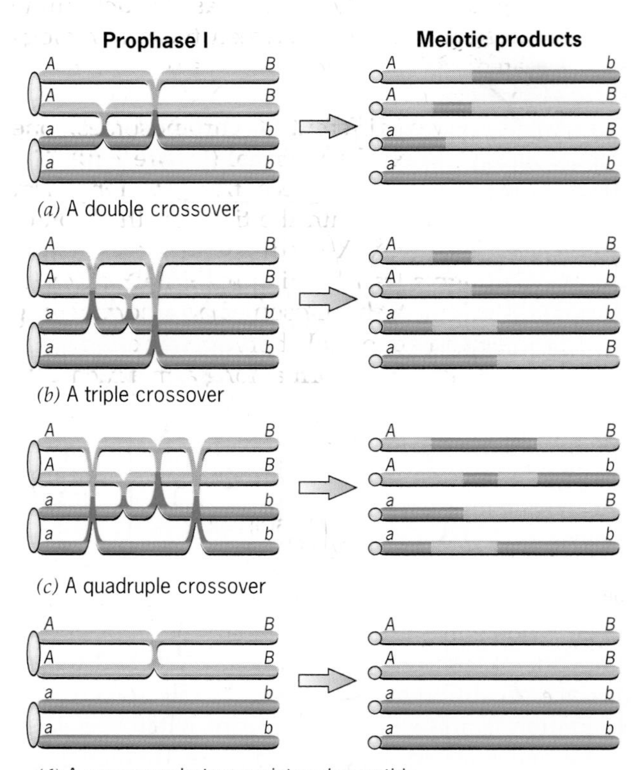

Figure 7.9 Consequences of multiple exchanges between chromosomes and exchange between sister chromatids during prophase I of meiosis.

quadruple crossovers. (We consider the genetic significance of these in a later section.) However, an exchange between sister chromatids cannot be detected because they are genetically identical.

What is responsible for the breakage of chromatids during crossing over? One hypothesis is that the chromatid fibers become entangled with each other and torsional stress causes them to break. Although such tangles may sometimes occur, it seems more likely that the breaks are induced by enzymes acting on the DNA within the chromatids. Enzymes are also responsible for repairing these breaks, that is, for reattaching chromatid fragments to each other. We consider the molecular details of this process in Chapter 13.

Evidence that Crossing Over Causes Recombination

In 1931 Curt Stern, Harriet Creighton, and Barbara McClintock obtained the first evidence that genetic recombination was associated with a material exchange between chromosomes. Stern worked with *Drosophila,* and Creighton and McClintock worked with maize. Each set of experiments depended on the availability of homologous chromosomes that were morphologically distinguishable. The goal was to determine whether physical exchange between these homologs was correlated with recombination between some of the genes they carried.

Stern used two different X chromosomes, one shorter and one longer than normal (Figure 7.10). The shorter X, X^s, had lost a piece from its long arm through a translocation with the tiny fourth chromosome, and the longer X, X^l, had gained a piece on its short arm through a translocation with the Y chromosome. Thus the long X chromosome had a normal long arm attached to an abnormal short arm, and the short X chromosome had an abnormal long arm attached to a normal short arm. The short X chromosome also carried mutant genes to allow genetic recombination to be detected. One of these was the dominant mutation for *Bar* eyes (symbol *B*), and the other was the recessive mutation for *carnation* eyes (symbol *car*). The long X chromosome carried the wild-type alleles of these two genes (B^+ and car^+).

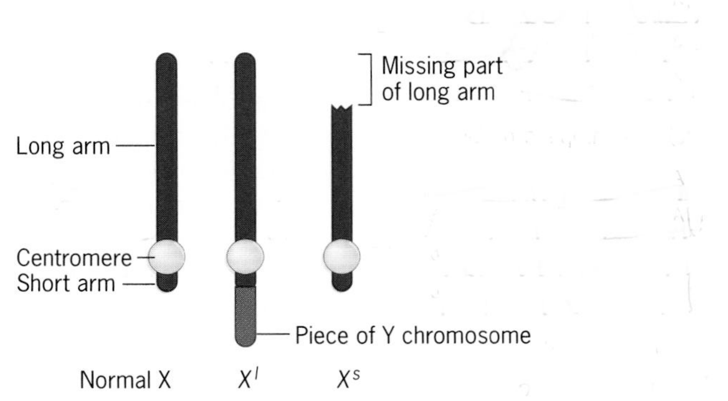

Figure 7.10 Structure of the X chromosomes used in Stern's experiment.

Stern crossed females that were heterozygous for the long and short X chromosomes to males that carried a structurally normal X marked with B^+ and *car.*

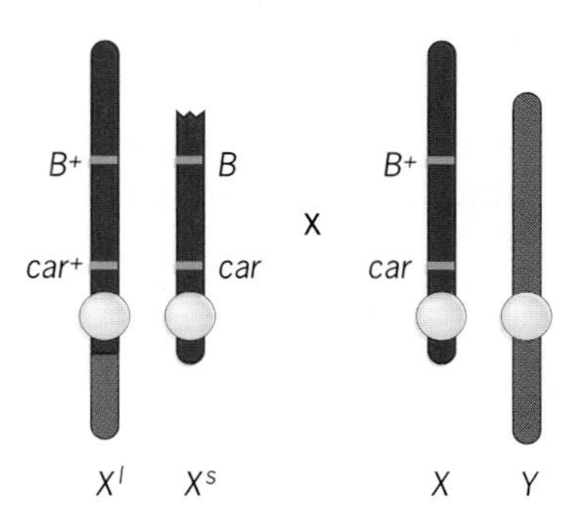

The critical question was whether or not the recombinant progeny from this cross carried an X chromosome that was produced by crossing over between the two abnormal X chromosomes. Stern showed that the answer was yes by demonstrating that the *B* car^+ recombinants carried an X chromosome in which both arms were abnormal:

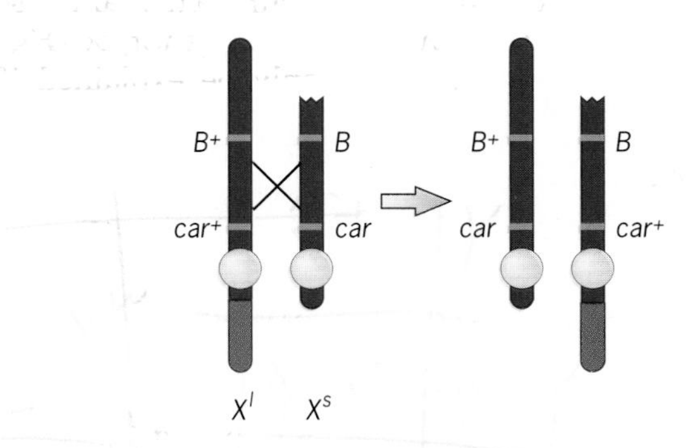

Such a chromosome could only have been produced by exchange between the two structurally abnormal X chromosomes that were used in the experiment.

This demonstration that recombination was correlated with chromosome exchange was independently confirmed by Creighton and McClintock in a study with maize. Two forms of chromosome 9 were available for analysis; one was normal, and the other had cytological aberrations at each end—a heterochromatic knob at one end and a piece of a different chromosome at the other. These two forms of chromosome 9 were also genetically marked to detect recombination. One marker gene controlled kernel color (*C*, colored; *c*, colorless), and the other controlled kernel texture (*Wx*, starchy; *wx*, waxy). Creighton and McClintock performed the following cross:

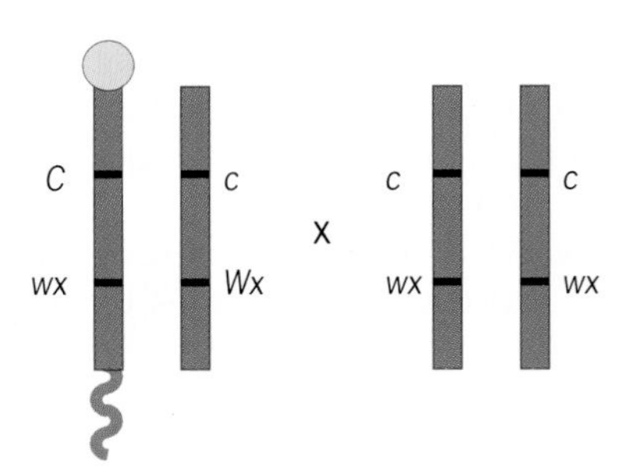

They then examined the recombinant progeny for evidence of exchange between the two different forms of chromosome 9. Their results showed that the *C Wx* and *c wx* recombinants carried a chromosome with only one of the abnormal cytological markers; the other abnormal marker had evidently been lost through an exchange with the normal chromosome 9 in the previous generation:

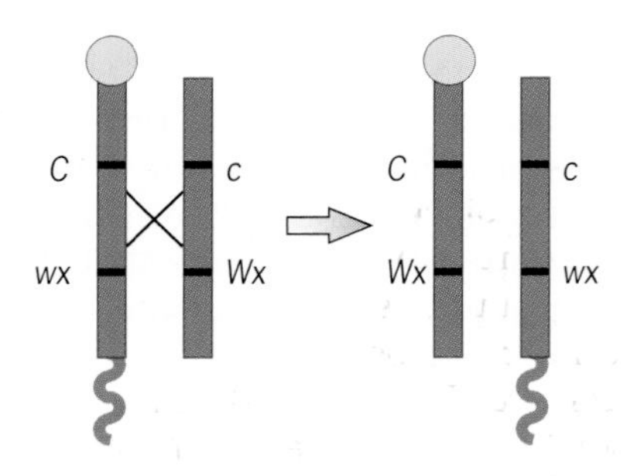

These findings strongly argued that recombination was caused by a physical exchange between paired chromosomes.

Chiasmata and the Time of Crossing Over

The cytological evidence for crossing over can be seen during late prophase of the first meiotic division when the chiasmata become clearly visible. At this time paired chromosomes repel each other slightly, maintaining close contact only at the centromere and at each chiasma (Figure 7.11). This partial separation makes it possible to count the chiasmata accurately. As we might expect, large chromosomes typically have more chiasmata than small chromosomes. Thus the number of chiasmata is roughly proportional to chromosome length.

The appearance of chiasmata late in the first meiotic prophase might imply that it is then that crossing over occurs. However, evidence from several different experiments suggests that it actually occurs earlier. Some of these experiments used heat shocks to alter the frequency of recombination. When the heat shocks were administered late in prophase, there was little effect, but when they were given earlier, the recombination frequency was changed. Thus the event that is responsible for recombination, namely, crossing over, occurs rather early in the meiotic prophase. Additional evidence comes from molecular studies on the time of DNA synthesis. Although almost all the DNA is synthesized during the interphase that precedes the onset of meiosis, a small amount is made during the first meiotic prophase. This limited DNA synthesis has been interpreted as part of a process to repair broken chromatids, which, as we have discussed, is thought to be associated with crossing over. Delicate timing experiments have shown that this DNA synthesis occurs in early to mid-prophase, but not later. The accumulated evidence therefore suggests that crossing over occurs in early to mid-prophase, long before the chiasmata can be seen.

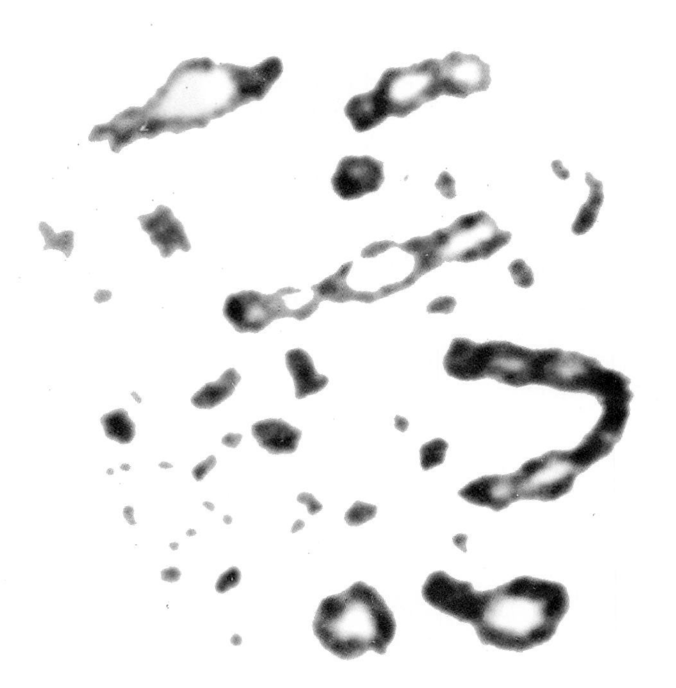

Figure 7.11 Meiotic prophase I chromosomes of the chicken showing multiple chiasmata in the large bivalents.

What, then, are chiasmata, and what do they mean? Most geneticists believe that the chiasmata are merely vestiges of the actual exchange process. Chromatids that have experienced an exchange probably remain entangled with each other during most of prophase. Eventually, these entanglements are resolved, and the chromatids are separated by the meiotic spindle apparatus to opposite poles of the cell. Therefore each chiasma probably represents an entanglement that was created by a crossover event earlier in prophase, probably in pachynema (Figure 7.12).

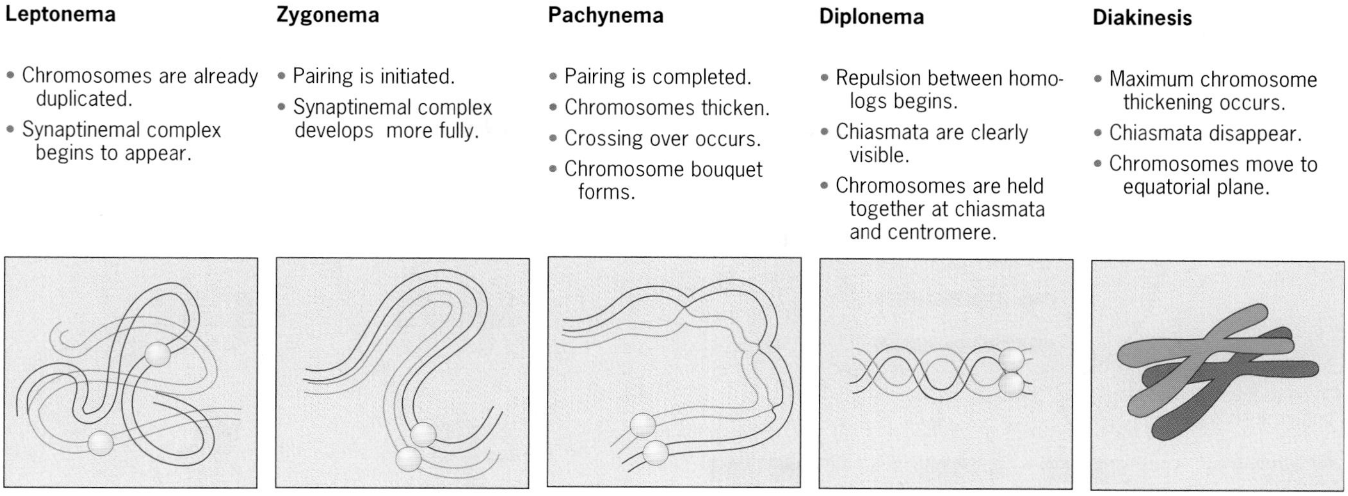

Leptonema

- Chromosomes are already duplicated.
- Synaptinemal complex begins to appear.

Zygonema

- Pairing is initiated.
- Synaptinemal complex develops more fully.

Pachynema

- Pairing is completed.
- Chromosomes thicken.
- Crossing over occurs.
- Chromosome bouquet forms.

Diplonema

- Repulsion between homologs begins.
- Chiasmata are clearly visible.
- Chromosomes are held together at chiasmata and centromere.

Diakinesis

- Maximum chromosome thickening occurs.
- Chiasmata disappear.
- Chromosomes move to equatorial plane.

Figure 7.12 Five stages of prophase of meiosis I. Crossing over occurs in pachynema. Chiasmata become visible in diplonema and disappear during diakinesis.

Key Points: **The frequency of recombination is a measure of linkage between genes on the same chromosome. Recombination is caused by a physical exchange between paired homologous chromosomes early in prophase of the first meiotic division, after the chromosomes have duplicated. At any one point, the process of exchange—crossing over—involves only two of the four chromatids in a meiotic tetrad. Later in prophase I, these crossovers become visible as chiasmata.**

CHROMOSOME MAPPING

Crossing over during the prophase of the first meiotic division has two observable outcomes:

1. Formation of chiasmata in late prophase.
2. Recombination between genes on opposite sides of the crossover point.

However, the second outcome can only be seen in the next generation, when the genes on the recombinant chromosomes are expressed.

Geneticists construct chromosome maps by counting the number of crossovers that occur during meiosis. However, because the actual crossover events cannot be seen, they cannot count them directly. Instead, they must estimate how many crossovers have taken place either by counting chiasmata or by counting recombinant chromosomes. Chiasmata are counted through cytological analysis, whereas recombinant chromosomes are counted through genetic analysis. Before we consider either of these procedures, we must define more clearly what we mean by distance on a chromosome map.

Crossing Over as a Measure of Genetic Distance

Sturtevant's fundamental insight was to estimate the distance between points on a chromosome by counting the number of crossovers between them. Points that are far apart should have more crossovers between them than points that are close together. However, the number of crossovers must be understood in a statistical sense. In any particular cell, the chance that a crossover will occur between two points may be low, but in a large population of cells, this crossover will probably occur several times simply because there are so many independent opportunities for it. Thus the quantity that we really need to measure is the average number of crossovers in a particular chromosome region. Genetic map distances are, in fact, based on such averages. This idea is sufficiently important to justify a formal definition: *The distance between two points on the genetic map of a chromosome is the average number of crossovers between them.*

One way for us to understand this definition is to consider 100 oogonia going through meiosis (Figure 7.13). In some cells, no crossovers will occur between sites A and B; in others, one, two, or more crossovers will occur between these loci. At the end of meiosis, there will be 100 gametes, each containing a chromosome with either zero, one, two, or more crossovers between A and B. We estimate the genetic map distance between these loci by calculating the average number of crossovers in this sample of chromosomes:

$$0 \times (15/100) + 1 \times (60/100) + 2 \times (15/100) + 3 \times (10/100) = 1.2$$

In practice, we cannot "see" each of the exchange points on the chromosomes coming out of meiosis. In-

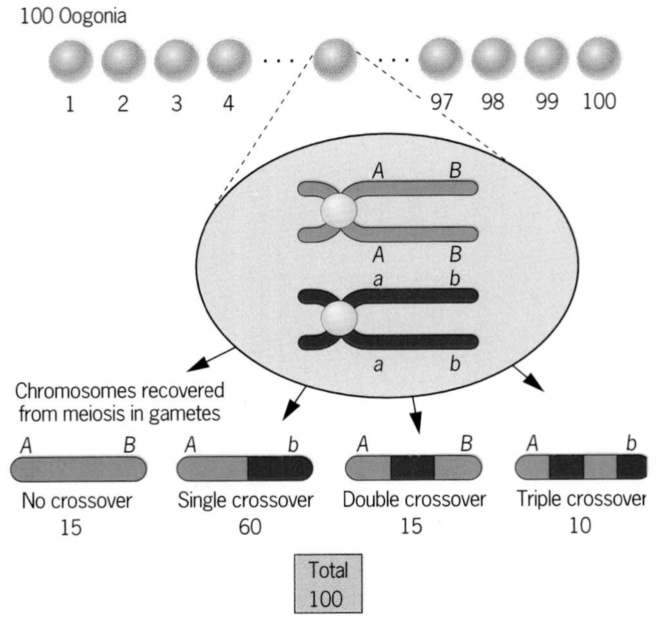

$$0 \times \left(\frac{15}{100}\right) + 1 \times \left(\frac{60}{100}\right) + 2 \times \left(\frac{15}{100}\right) + 3 \times \left(\frac{10}{100}\right) = 1.2$$

Figure 7.13 Calculating the average number of crossovers between genes on chromosomes recovered from meiosis.

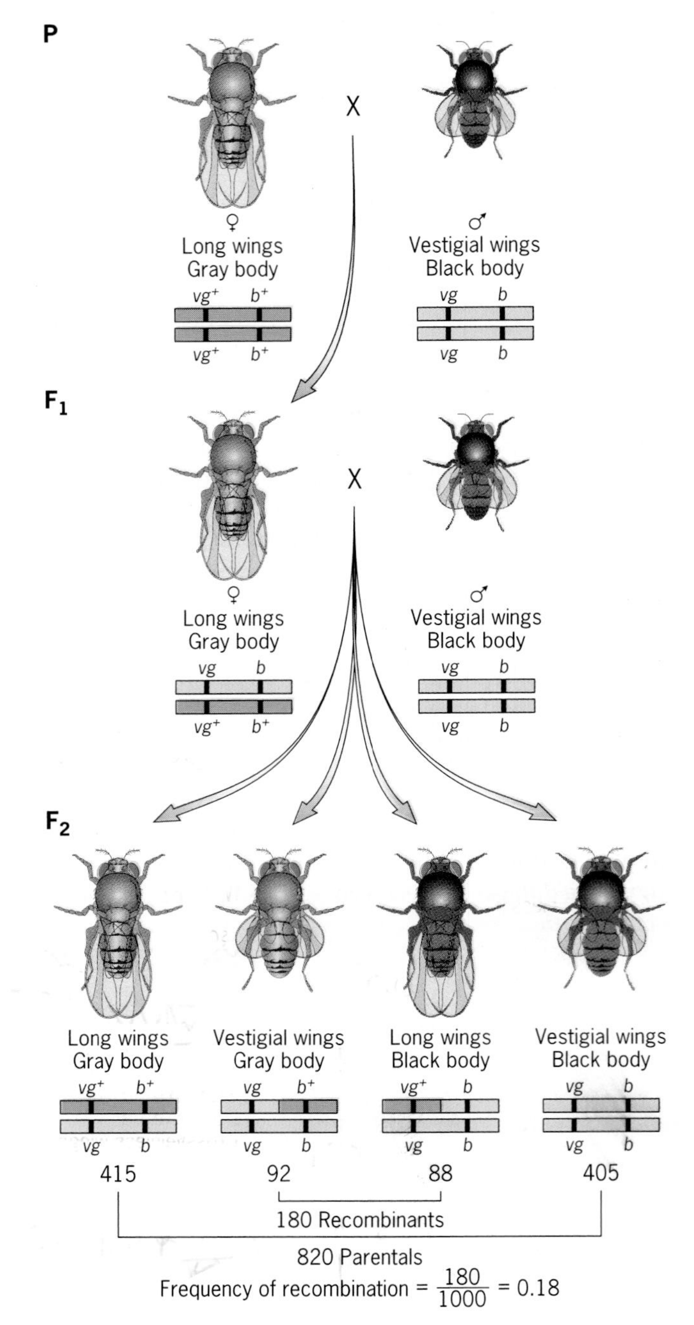

Figure 7.14 An experiment involving two linked genes, *vg* (*vestigial* wings) and *b* (*black* body), in *Drosophila*.

stead, we infer their existence by observing the recombination of the alleles that flank them. A chromosome in which alleles have recombined must have arisen by crossing over. Counting recombinant chromosomes therefore provides a way of counting crossover exchange points.

Recombination Mapping with a Two-Point Testcross

To illustrate the mapping procedure, let's consider the two-point testcross in Figure 7.14. Wild-type *Drosophila* females were mated to males homozygous for two autosomal mutations—*vestigial* (*vg*), which produces short wings, and *black* (*b*), which produces a black body. All the F_1 flies had long wings and gray bodies; thus the wild-type alleles (vg^+ and b^+) are dominant. The F_1 females were then testcrossed to vestigial, black males, and the F_2 progeny were sorted by phenotype and counted. As the data show, there were four phenotypic classes, two abundant and two rare. The abundant classes had the same phenotypes as the original parents, and the rare classes had recombinant phenotypes.

We know that the vestigial and black genes are linked because the recombinants are much fewer than 50 percent of the total progeny counted. These genes must therefore be on the same chromosome. To determine the distance between them, we must estimate the average number of crossovers in the gametes of the doubly heterozygous F_1 females. We can do this by calculating the frequency of recombinant F_2 flies and noting that each such fly inherited a chromosome that had crossed over once between *vg* and *b*. The average number of crossovers in the whole sample of progeny is therefore

nonrecombinants		recombinants		
$(0) \times 0.82$	$+$	$(1) \times 0.18$	$=$	0.18

In this expression, the number of crossovers for each class of flies is placed in parentheses; the other number is the frequency of that class. The nonrecombinant progeny obviously do not add any crossover chromosomes to the data, but we include them in the calculation to emphasize that we must calculate the average number of crossovers by using *all* the data, not just those from the recombinants.

This simple analysis indicates that on average, 18 out of 100 chromosomes recovered from meiosis had a crossover between *vg* and *b*. Thus *vg* and *b* are separated by 18 **units** on the genetic map. Sometimes geneticists call a map unit a **centiMorgan**, abbreviated cM, in honor of T. H. Morgan; 100 centiMorgans equal one Morgan (M). We can therefore say that *vg* and *b* are 18 cM (or 0.18 M) apart. Notice that the map distance is equal to the frequency of recombination, written as a percentage. (See a later section, however, for an important qualification of this statement.) The Human Genetics Sidelight shows how pedigrees can be analyzed for evidence of linkage between loci on a human chromosome.

Recombination Mapping with a Three-Point Testcross

We can also use the recombination mapping procedure with data from testcrosses involving more than two genes. Figure 7.15 illustrates an experiment by C. B. Bridges and T. M. Olbrycht, who crossed wild-type *Drosophila* males to females homozygous for three recessive X-linked mutations, *scute* (*sc*) bristles, *echinus* (*ec*) eyes, and *crossveinless* (*cv*) wings. They then intercrossed the F_1 progeny to produce F_2 flies, which they classified and counted. Because the F_1 males carried all three recessive mutations, this intercross was equivalent to a testcross.

The F_2 flies comprised eight phenotypically distinct classes, two of them parental and six recombinant. The parental classes were by far the most numerous. The less numerous recombinant classes each represented a different kind of crossover chromosome. To figure out which crossovers were involved in producing each type of recombinant, we must first

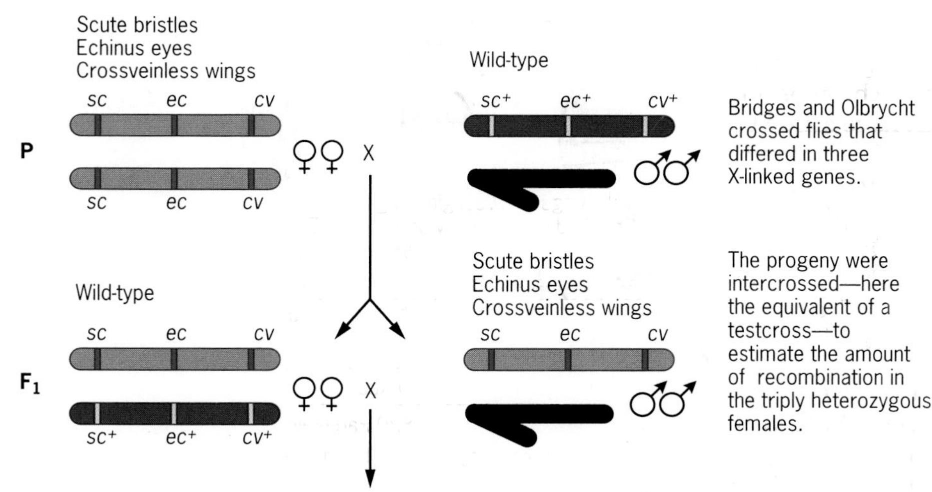

Class	Phenotype	Genotype of maternally inherited X chromosome			Number observed
1	Scute, echinus, crossveinless	*sc*	*ec*	*cv*	1158
2	Wild-type	*sc*+	*ec*+	*cv*+	1455
3	Scute	*sc*	*ec*+	*cv*+	163
4	Echinus, crossveinless	*sc*+	*ec*	*cv*	130
5	Scute, echinus	*sc*	*ec*	*cv*+	192
6	Crossveinless	*sc*+	*ec*+	*cv*	148
7	Scute, crossveinless	*sc*	*ec*+	*cv*	1
8	Echinus	*sc*+	*ec*	*cv*+	1
				Total:	3248

Figure 7.15 Bridges' and Olbrycht's three-point cross with the X-linked genes *sc* (*scute* bristles), *ec* (*echinus* eyes), and *cv* (*crossveinless* wings) in *Drosophila*.

HUMAN GENETICS SIDELIGHT

Evidence for Linkage Between the Genes for Hemophilia and Color Blindness

The detection and analysis of linkage in human beings has never been easy. A geneticist cannot arrange matings between men and women with particular genotypes in order to obtain linkage data. Rather, he or she must collect the data from pedigrees, which are often incomplete, and then analyze them for evidence of linkage. The procedures used in this analysis are considerably more complex than those used in the analysis of data from experimental organisms (see Chapter 8).

Some of the first evidence for linkage between human genes was obtained by studying the inheritance of red-green color blindness and hemophilia. Because both of these traits are caused by mutant genes on the X chromosome, they could be linked. However, if the genes controlling them were sufficiently far apart on the X chromosome, the traits would appear to be unlinked. Only pedigree analysis could distinguish between these two possibilities.

In the first half of the twentieth century, geneticists in England, the United States, Germany, and the Netherlands searched for linkage between color blindness and hemophilia. Their strategy was to study families in which both traits were segregating. Figure 1 shows three of several pedigrees that were discovered. To analyze these pedigrees, we note that both traits are caused by recessive mutations: *c* = color blindness, *h* = hemophilia; their normal alleles are denoted by the letters *C* and *H*, respectively.

Pedigree (*a*) shows four generations of a family described by Madlener in 1928. Because the great grandfather, I-1, has both color blindness and hemophilia, we know that his genotype is *c h*. His daughter, II-1, is phenotypically normal and must therefore carry the normal alleles, *C* and *H*. Moreover, because II-1 inherited *c* and *h* on the X chromosome transmitted by I-1, the two normal alleles must be carried by the other X chromosome—that is, the one she inherited from her mother. II-1's genotype is therefore *C H*/*c h*—that is, she is a coupling heterozygote for the two loci. III-2, the daughter of II-1, is also a coupling heterozygote. We infer that she has this genotype because her son has both color blindness and hemophilia (*c h*) and her father is phenotypically normal (*C H*). Evidently, III-2 inherited the *c h* chromosome from her mother.

To detect linkage between the two loci, we must examine the genotypes of the children produced by II-1 and III-2. Among the three sons of II-1, one is normal (*C H*) and two are

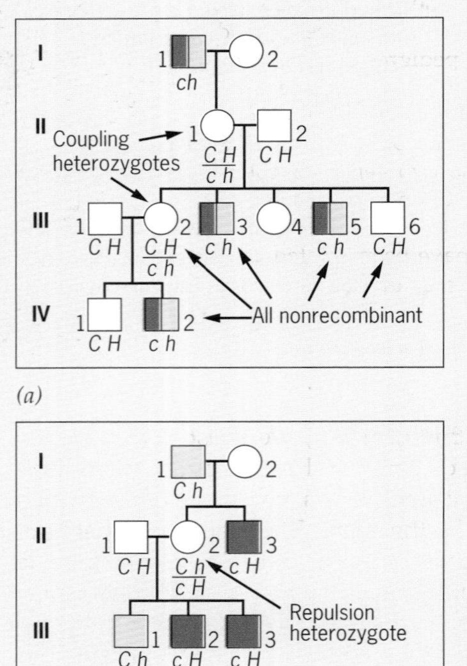

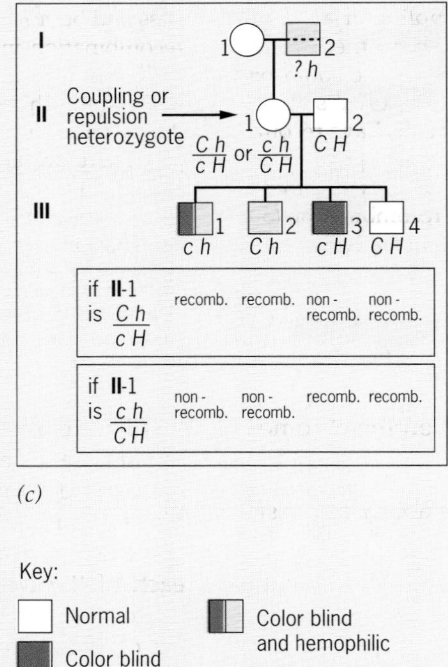

Figure 1 Pedigrees showing the inheritance of two X-linked traits, color blindness and hemophilia.

doubly mutant (*c h*). Thus each of the sons must have inherited a nonrecombinant X chromosome from II-1. As discussed above, II-1's first daughter (III-2) also inherited a nonrecombinant X chromosome; her second daughter (III-4) cannot be assigned a genotype. The other coupling heterozygote, III-2, produced two sons, one normal (*C H*) and one doubly mutant (*c h*). Clearly, both of these sons must also have inherited nonrecombinant X chromosomes from their mother. Thus we can ascertain the genotypes of six of the seven X chromosomes that were transmitted by II-1 and III-2, and we find that all six are nonrecombinant. This result argues rather persuasively that the color blindness and hemophilia loci are linked.

Pedigree (*b*), described by C. L. Birch in 1937, provides additional evidence for linkage. I-1 is genotypically *C h*. His wife has an uncertain genotype; however, because the couple has a color-blind son, we can infer that she carries the mutant allele *c*. The couple's phenotypically normal daughter (II-2) must carry both mutant alleles, *c* and *h*, because she has both color-blind and hemophilic sons. Moreover, her genotype must be *C h*/*c H* because she inherited a *C h* chromosome from her father. Thus II-2 is a repulsion heterozygote for the two loci. Among II-2's sons, we find that all three inherited nonrecombinant X chromosomes—an observation that suggests the two loci are linked.

Pedigree (*c*), described by B. Rath in 1938, provides evidence for recombination between the color blindness and hemophilia loci—the first ever observed. The key individual in this pedigree is II-1. This woman has four sons, each with a different genotype: one is color blind and hemophilic (*c h*), one is hemophilic (*C h*), one is color blind (*c H*), and one is normal (*C H*). Thus II-1 must carry both *c* and *h*. However, we cannot determine whether she is a coupling or a repulsion heterozygote because we do not know her father's complete genotype. (He died of hemophilia before he could be tested for color blindness.) All we can say is that II-1 is either *C H*/*c h* or *C h*/*c H*. In either case, half her sons have recombinant genotypes, and half have nonrecombinant genotypes. Thus we can infer that recombination occurs between these two X-linked loci. In fact, the frequency of recombination (50 percent) suggests that the loci are far apart on the X chromosome.

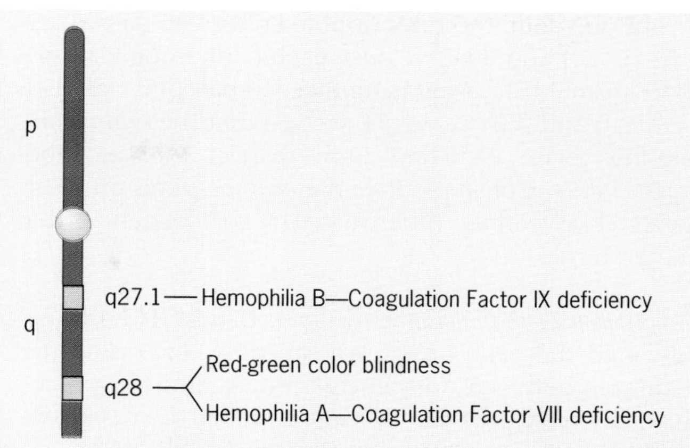

Figure 2 Map of the human X chromosome showing the location of the genes for hemophilia A and B and for red-green color blindness.

More detailed analyses of other pedigrees have shown that there are actually two forms of hemophilia, A and B, each involving the deficiency of a different blood clotting factor (VIII in A and IX in B). These two factors are produced by genes at different positions on the X chromosome (Figure 2). Factor VIII is encoded by a gene near the end of the chromosome's long arm, and Factor IX is encoded by a gene in the middle of the long arm. The color blindness gene is tightly linked to the Factor VIII gene but loosely linked to the Factor IX gene. These linkage relationships might explain why there is no recombination between color blindness and hemophilia in pedigrees (*a*) and (*b*), and 50 percent recombination in pedigree (*c*). The hemophilia observed in pedigrees (*a*) and (*b*) may have been caused by a Factor VIII deficiency, which is the more common type; in pedigree (*c*), it may have been caused by a Factor IX deficiency.

There are also two genes for color blindness on the X chromosome (see Chapter 5). However, these genes are so close together that they behave as one locus. In fact, the two genes appear to have been created by the duplication of a single gene sometime during the course of human evolution.

determine how the genes are ordered on the chromosome.

Determination of the Gene Order There are three possible gene orders:

(i)	*sc*	*ec*	*cv*
(ii)	*ec*	*sc*	*cv*
(iii)	*ec*	*cv*	*sc*

Other possibilities, such as *cv* - *ec* - *sc*, are the same as one of these because the left and right ends of the chromosome cannot be distinguished. Which of the orders is correct?

To answer this question, we must take a careful look at the six recombinant classes. Four of them must have come from a single crossover in one of the two regions delimited by the genes. The other two must have come from double crossing over—one exchange in each of the two regions. Because a double crossover switches the gene in the middle with respect to the genetic markers on either side of it, we have, in principle, a way of determining the gene order. Intuitively, we also know that a double crossover should occur much less frequently than a single crossover. Consequently, among the six recombinant classes, the two rare ones must represent the double crossover chromosomes.

In our data, the rare, double crossover classes are 7 (*sc ec*$^+$ *cv*) and 8 (*sc*$^+$ *ec cv*$^+$), each containing a single fly (Figure 7.15). Comparing these to parental classes 1 (*sc ec cv*) and 2 (*sc*$^+$ *ec*$^+$ *cv*$^+$), we see that the *echinus* allele has been switched with respect to *scute* and *crossveinless*. Consequently, the *echinus* gene must be located between the other two. The correct gene order is therefore (i) *sc - ec - cv*.

Calculation of the Distances Between Genes Having established the gene order, we can now determine the distances between adjacent genes. Again, the procedure is to compute the average number of crossovers in each chromosomal region (Figure 7.16).

We can obtain the length of the region between *sc* and *ec* by identifying the recombinant classes that involved a crossover between these genes. There are four such classes: 3 (*sc ec*$^+$ *cv*$^+$), 4 (*sc*$^+$ *ec cv*), 7 (*sc ec*$^+$ *cv*), and 8 (*sc*$^+$ *ec cv*$^+$). Classes 3 and 4 involved a single crossover between *sc* and *ec*, and classes 7 and 8 involved two crossovers, one between *sc* and *ec* and the other between *ec* and *cv*. We can therefore use the frequencies of these four classes to estimate the average number of crossovers between *sc* and *ec*:

$$\frac{\overset{\text{Class 3}}{163} + \overset{\text{Class 4}}{130} + \overset{\text{Class 7}}{1} + \overset{\text{Class 8}}{1}}{\text{Total}} = \frac{295}{3248} = 0.091$$

Thus, in every 100 chromosomes coming from meiosis in the F_1 females, 9.1 had a crossover between *sc* and *ec*. The distance between these genes is therefore 9.1 map units (or, if you prefer, 9.1 centiMorgans).

In a similar way, we can obtain the distance between *ec* and *cv*. Four recombinant classes involved a crossover in this region: 5 (*sc ec cv*$^+$), 6 (*sc*$^+$ *ec*$^+$ *cv*), 7, and 8. The double recombinants are also included here because one of their two crossovers was between *ec* and *cv*. The combined frequency of these four classes is:

$$\frac{\overset{\text{Class 5}}{192} + \overset{\text{Class 6}}{148} + \overset{\text{Class 7}}{1} + \overset{\text{Class 8}}{1}}{\text{Total}} = \frac{342}{3248} = 0.105$$

Consequently, *ec* and *cv* are 10.5 map units apart.

Combining the data for the two regions, we obtain the map

sc—9.1—*ec*—10.5—*cv*

Map distances computed in this way are *additive*. Thus we can estimate the distance between *sc* and *cv* by summing the lengths of the two map intervals between them:

$$9.1 \text{ cM} + 10.5 \text{ cM} = 19.6 \text{ cM}$$

We can easily verify this estimate by directly calculating the average number of crossovers between these genes:

Noncrossover classes	Single crossover classes	Double crossover classes
(1 and 2)	(3, 4, 5, and 6)	(7 and 8)

$$(0) \times 0.805 + (1) \times 0.195 + (2) \times 0.0006 = 0.196$$

Here the number of crossovers is given in parentheses, and its multiplier is the combined frequency of the classes with that many crossovers. In other words, each recombinant class contributes to the map distance according to the product of its frequency and the number of crossovers it represents.

Bridges and Olbrycht actually studied seven X-linked genes in their recombination experiment: *sc*, *ec*, *cv*, *ct* (*cut* wings), *v* (*vermilion* eyes), *g* (*garnet* eyes), and *f* (*forked* bristles). By calculating recombination frequencies between each pair of adjacent genes, they were able to construct a map of a large segment of the

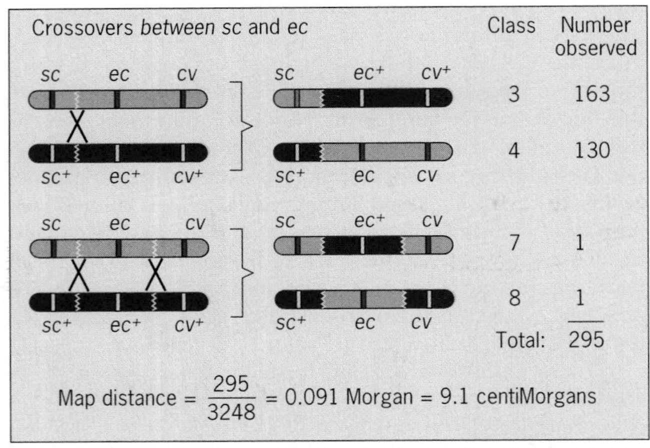

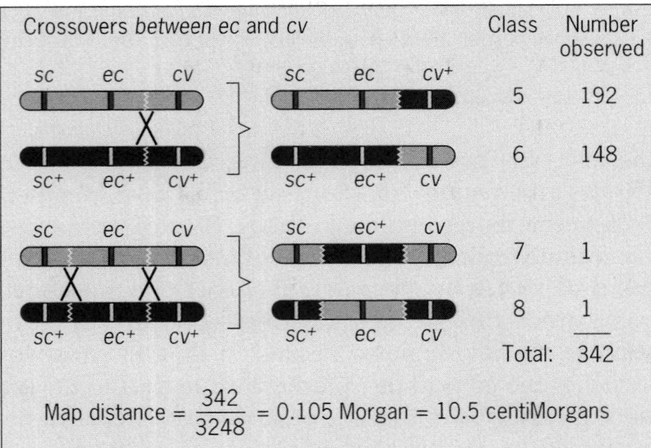

Figure 7.16 Calculation of genetic map distances from Bridges' and Olbrycht's data. The distance between each pair of genes is obtained by estimating the average number of crossovers.

X chromosome (Figure 7.17); *sc* was at one end, and *f* was at the other. Each of the seven genes that Bridges and Olbrycht studied was, in effect, a marker for a particular site on the X chromosome. Summing all the map intervals between these markers, Bridges and Olbrycht estimated the total length of the mapped segment to be 66.8 cM. Thus the average number of crossovers in this segment was 0.668.

Interference and the Coefficient of Coincidence A three-point cross has an important advantage over a two-point cross: it allows the detection of double crossovers, permitting us to ask if exchanges in adjacent regions are independent of each other. For example, does a crossover in the region between *sc* and *ec* (region I on the map of the X chromosome) occur independently of a crossover in the region between *ec* and *cv* (region II)? Or does one crossover inhibit the occurrence of another nearby?

To answer these questions, we must calculate the expected frequency of double crossovers, based on the idea of independence. We can do this by multiplying the crossover frequencies for two adjacent chromosome regions. For example, in region I on Bridges' and Olbrycht's map, the crossover frequency was (163 + 130 + 1 + 1)/3248 = 0.091, and in region II, it was (192 + 148 + 1 + 1)/3248 = 0.105. If we assume independence, the expected frequency of double crossovers in the interval between *sc* and *cv* would therefore be (0.091) × (0.105) = 0.0095. We can now compare this frequency with the observed frequency, which was 2/3248 = 0.0006. Clearly, double crossovers between *sc* and *cv* were much less frequent than expected. This result suggests that one crossover actually inhibited the occurrence of another nearby, a phenomenon called **interference**. The extent of the interference is customarily measured by the **coefficient of coincidence, c,** which is the ratio of the observed frequency of double crossovers to the expected frequency:

$$c = \frac{\text{observed frequency of double crossovers}}{\text{expected frequency of double crossovers}}$$

$$= \frac{0.0006}{0.0095} = 0.063$$

Because in this example the coefficient of coincidence is close to zero, its lowest possible value, interference was very strong. At the other extreme, a coefficient of coincidence equal to one would imply no interference at all; that is, it would imply that the crossovers occurred independently of each other.

Many studies have shown that interference is strong over map distances less than 20 cM; thus double crossovers seldom occur in short chromosomal regions. However, over long regions, interference weakens to the point that crossovers occur more or less independently. The strength of interference is therefore a function of map distance.

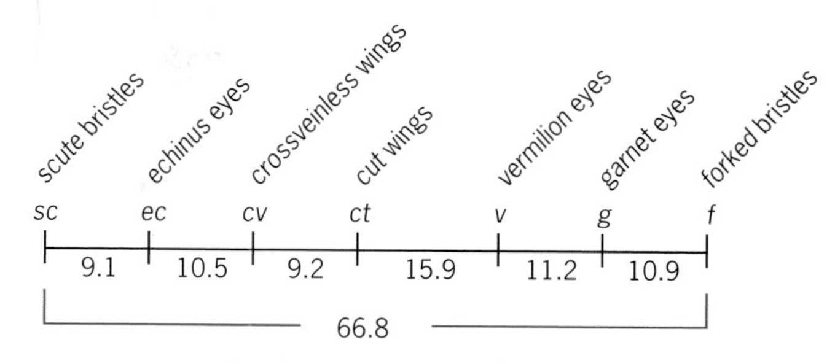

Figure 7.17 Bridges' and Olbrycht's map of seven X-linked genes in *Drosophila*. Distances are given in centiMorgans.

Recombination Frequency and Genetic Map Distance

In the preceding sections, we have considered how to construct chromosome maps from data on the recombination of genetic markers. These data allow us to infer where crossovers have occurred in a sample of chromosomes. By localizing and counting these crossovers, we can estimate the distances between genes and then place the genes on a chromosome map.

This method works well as long as the genes are fairly close together. However, when they are far apart, the frequency of recombination may not reflect the true map distance (Figure 7.18). As an example, let us consider the genes at the ends of Bridges' and Olbrycht's map of the X chromosome; *sc*, at the left end, was 66.8 cM away from *f*, at the right end. However, the frequency of recombination between *sc* and *f* was 50 percent—the maximum possible value. Using this frequency to estimate map distance, we would conclude that *sc* and *f* were 50 map units apart. Of course, the distance obtained by summing the lengths of the intervening regions on the map, 66.8 cM, is much greater.

This example shows that the *true* genetic distance, which depends on the average number of crossovers on a chromosome, may be much greater than the observed recombination frequency. Multiple crossovers may occur between widely separated genes, and some of these crossovers may not produce genetically recombinant chromosomes (Figure 7.19). To see this, let us assume that a single crossover occurs between two chromatids in a tetrad, causing recombination of the flanking genetic markers. If another crossover occurs between these same two chromatids, the flanking

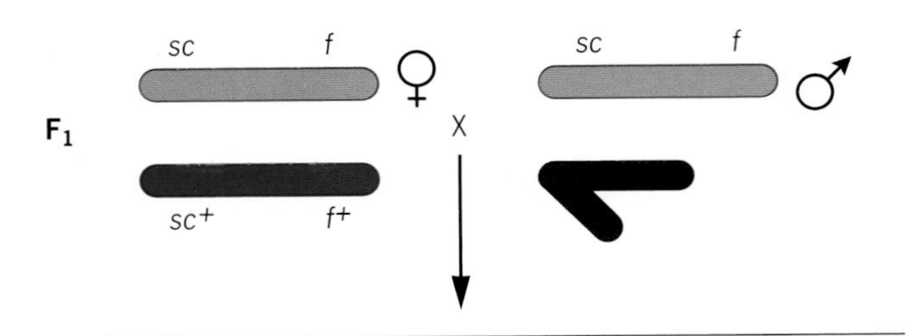

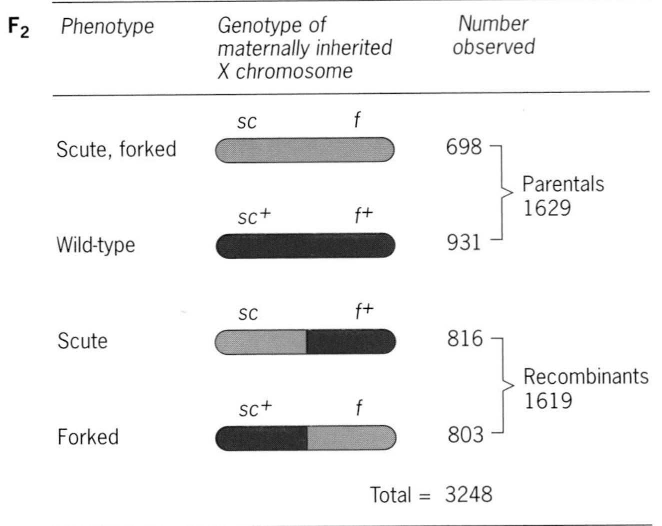

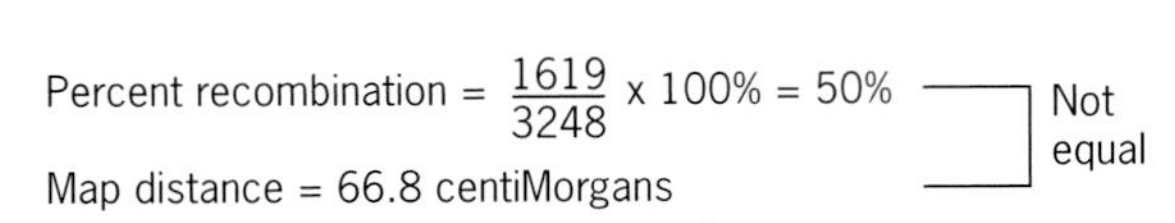

Figure 7.18 A discrepancy between map distance and percent recombination. The map distance between the genes *sc* and *f* is greater than the observed percent recombination between them.

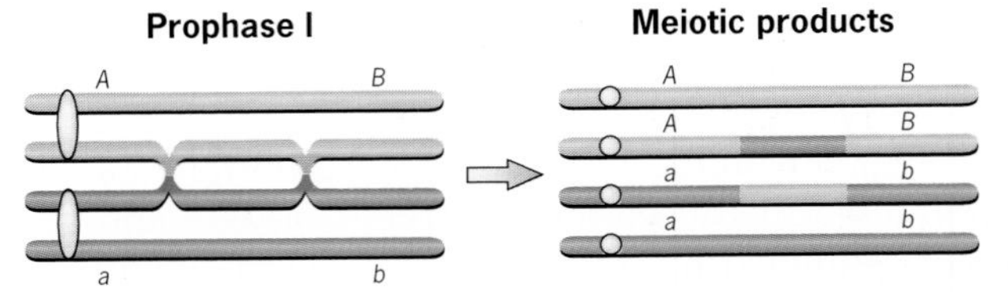

(*a*) Two-strand double crossovers produce all parental chromosomes.

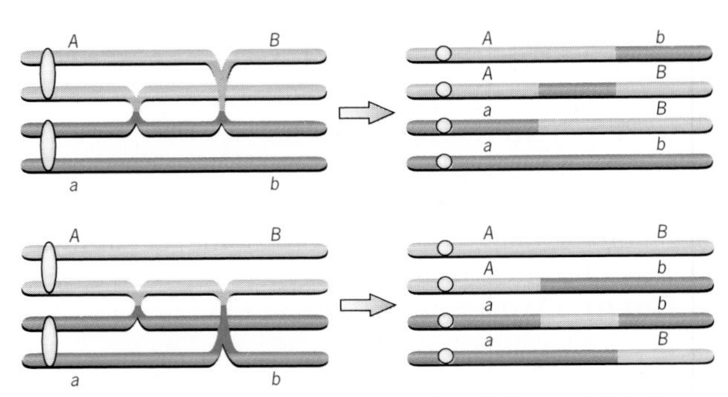

(*b*) Three-strand double crossovers produce half parental, half recombinant chromosomes.

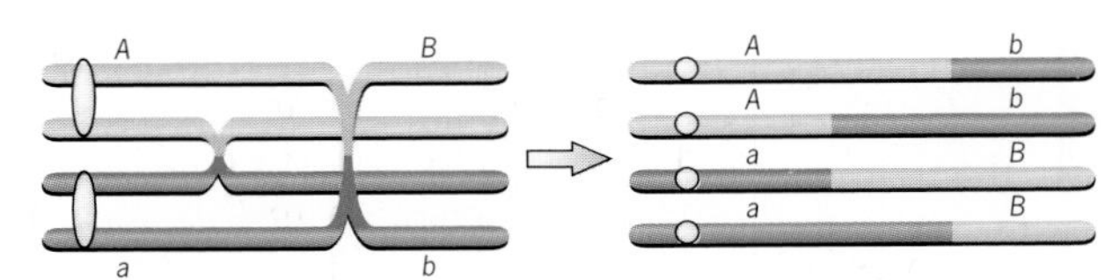

(*c*) Four-strand double crossovers produce all recombinant chromosomes.

Figure 7.19 Consequences of double crossing over between two loci. Two-strand double crossovers produce only nonrecombinant chromosomes, and four-strand double crossovers produce only recombinant chromosomes. Three-strand double crossovers produce half recombinant and half nonrecombinant chromosomes.

markers will be restored to their original configuration; the second crossover essentially cancels the effect of the first, converting the recombinant chromatids back into nonrecombinants. Thus, even though two crossovers have occurred in this tetrad, none of the chromatids that come from it will be recombinant for the flanking markers.

This second example shows that a double crossover may not contribute to the frequency of recombination even though it contributes to the average number of exchanges on a chromosome. A quadruple crossover would have the same effect. These and other multiple exchanges are responsible for the discrepancy between recombination frequency and genetic map distance. In practice, this discrepancy is small for distances less than 20–25 cM. Over such distances, interference is strong enough to suppress almost all multiple exchanges, and the recombination frequency is a good estimator of the true genetic distance. For values greater than 25 cM, these two quantities diverge, principally because multiple exchanges become much more likely. Figure 7.20 shows the mathematical relationship between recombination frequency and genetic map distance.

Chiasma Frequency and Genetic Map Distance

Recombination between genetic markers is one outcome of crossing over. The other is chiasma formation during prophase of the first meiotic division. Each chiasma is thought to represent the resolution of a crossover that occurred earlier in prophase. Thus, by counting chiasmata, we should be able to estimate the average number of crossovers occurring on a chromosome, which we can then use as an estimate of genetic map length.

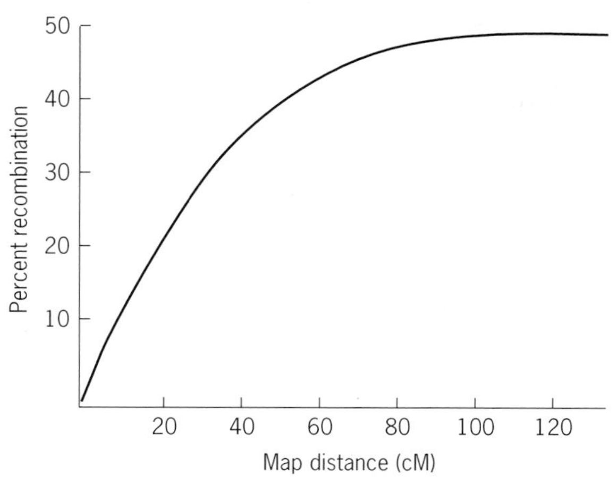

Figure 7.20 Relationship between frequency of recombination and genetic map distance. For values less than 25 cM, there is approximately a linear relationship between percent recombination and map distance; for values greater than 25 cM, the percent recombination underestimates the map distance.

Let's suppose, for example, that in a group of 100 cells going through meiosis, 5 cells show five chiasmata in a particular pair of chromosomes, 15 show four chiasmata in this pair, 15 show three, 30 show two, 25 show one, and 10 show none (Figure 7.21). Per cell, the average number of chiasmata in this pair of chromosomes is 2.15. This translates into an average of 1.07 chiasmata per chromatid because each chiasma affects only two of the four chromatids in the chromosome pair. Thus, at the completion of meiosis, the average number of crossovers per chromatid will be 1.07. This implies that the genetic length of this chromosome is 1.07 M, or 107 cM. If 2.15 chiasmata are equivalent to 107 cM, then one chiasma is equivalent to 107/2.15 = 50 cM. In general, we can conclude that each chiasma corresponds to 50 cM on the genetic map.

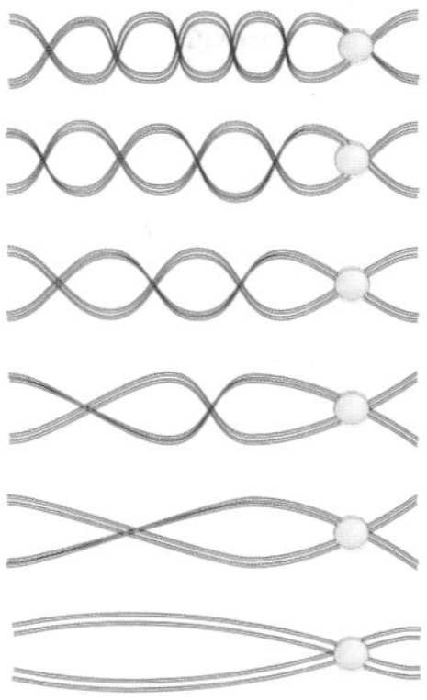

Meiotic chromosomes	*Number chiasmata* (A)	*Number of cells observed* (B)	*Product* (A × B)
	5	5	25
	4	15	60
	3	15	45
	2	30	60
	1	25	25
	0	10	0
	Total:	100	215

Average number of chiasmata per cell $= \frac{215}{100} = 2.15$

Average number of chiasmata per chromatid $= \frac{2.15}{2} = 1.07$

Genetic length of chromosome = 1.07 Morgans = 107 centiMorgans

Relationship between genetic length and average number of chiasmata $= \frac{107 \text{ cM}}{2.15 \text{ chiasmata}} = 50 \text{ cM/chiasma}$

Figure 7.21 Computing map length from chiasmata frequencies in meiosis. The average number of chiasmata per chromatid estimates the genetic length of the chromosome.

Genetic Distance and Physical Distance

The procedures for measuring genetic distance are based on the incidence of crossing over between paired chromosomes. Intuitively, we expect that long chromosomes should have more crossovers than short ones, and that this relationship will be reflected in the lengths of their genetic maps. For the most part, our assumption is true; however, within a chromosome some regions are more prone to crossing over than others. Thus distances on the genetic map do not correspond exactly to physical distances along the chromosome (Figure 7.22). Crossing over is less likely to occur near the ends of a chromosome and also around the centromere; consequently, these regions are condensed on the genetic map. Other regions, in which crossovers occur more frequently, are expanded.

Even though there is not a uniform relationship between genetic and physical distance, the genetic and physical maps of a chromosome are colinear; that is, particular sites have the same order. Recombination mapping therefore reveals the actual order of the genes along a chromosome. However, it does not tell us the actual physical distances between them.

Key Points: **The genetic maps of chromosomes are based on the average number of crossovers that occur during meiosis. Map distances are estimated by counting chiasmata in cytological preparations or by calculating the frequency of recombination between genes in experimental crosses. One chiasma is equivalent to a distance of 50 cM. Recombination frequencies less than 20 to 25 percent estimate map distance directly. However, frequencies greater than 20 to 25 percent usually underestimate map distance because some of the multiple crossovers that occur do not contribute to the recombination frequency.**

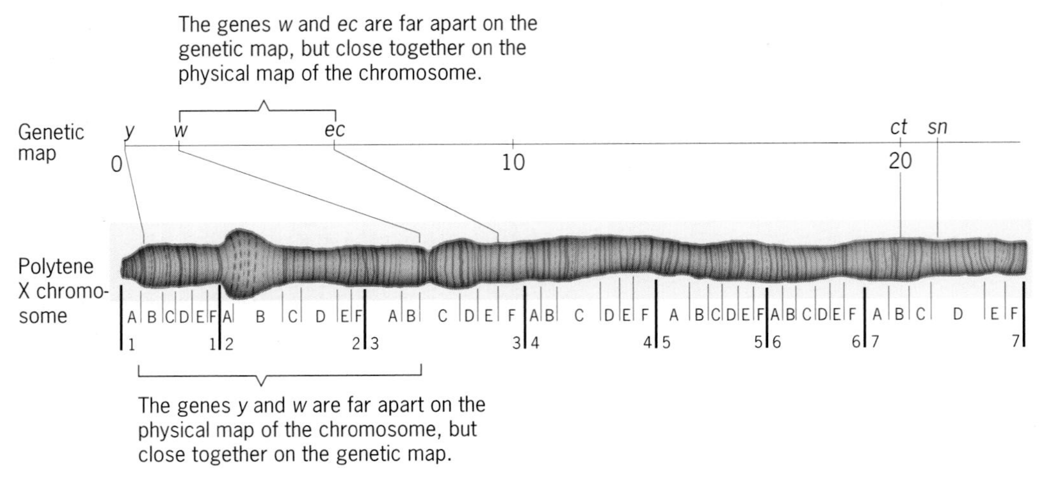

Figure 7.22 Left end of the polytene X chromosome of *Drosophila* and the corresponding portion of the genetic map.

RECOMBINATION AND EVOLUTION

Recombination is an essential feature of sexual reproduction. During meiosis, when chromosomes come together and cross over, there is an opportunity to create new combinations of alleles. Some of these may benefit the organism by enhancing survival or reproductive ability. Over time, such beneficial combinations would be expected to spread through a population and become standard features of the genetic makeup of the species. Meiotic recombination is therefore a way of shuffling genetic variation to potentiate evolutionary change.

Evolutionary Significance of Recombination

We can appreciate the evolutionary advantage of recombination by comparing two species, one capable of reproducing sexually and the other not. Let's suppose that a beneficial mutation has arisen in each species. Over time, we would expect these mutations to spread. Let's also suppose that while they are spreading, another beneficial mutation occurs in a nonmutant individual within each species. In the asexual organism, there is no possibility that this second mutation will be recombined with the first, but in the sexual organism, the two mutations can be recombined to produce a strain that is better than either of the single mutants by itself. This recombinant strain will be able to spread through the whole species population. In evolutionary terms, recombination can allow favorable alleles of different genes to come together in the same organism.

Suppression of Recombination by Inversions

The gene-shuffling effect of recombination can be thwarted by chromosome rearrangements. Crossing over is usually inhibited near the breakpoints of a rearrangement in heterozygous condition, probably because the rearrangement disrupts chromosome pairing. Many rearrangements are therefore associated with a reduction in the frequency of recombination. However, this effect is most pronounced in inversion heterozygotes because the inhibition of crossing over that occurs near the breakpoints of the inversion is compounded by the selective loss of chromosomes that have undergone crossing over within the inverted region.

To see this recombination-suppressing effect, we consider an inversion in the long arm of a chromosome (Figure 7.23). If a crossover occurs between inverted and noninverted chromatids within the tetrad, it will produce two recombinant chromatids; however, both of these chromatids are likely to be lost during or after meiosis. One of the chromatids lacks a centromere—it is an *acentric fragment*—and will therefore be unable to move to its proper place during anaphase of the first meiotic division. The other chromatid has two centromeres and will therefore be pulled in opposite directions, forming a *dicentric chromatid bridge.* Eventually, this bridge will break and split the chromatid into pieces. Even if the acentric and dicentric chromatids produced by crossing over within the inversion survive meiosis, they are not likely to form viable zygotes. Both of these chromatids are aneuploid—duplicate for some genes and deficient for others—and such aneuploidy is usually lethal. These

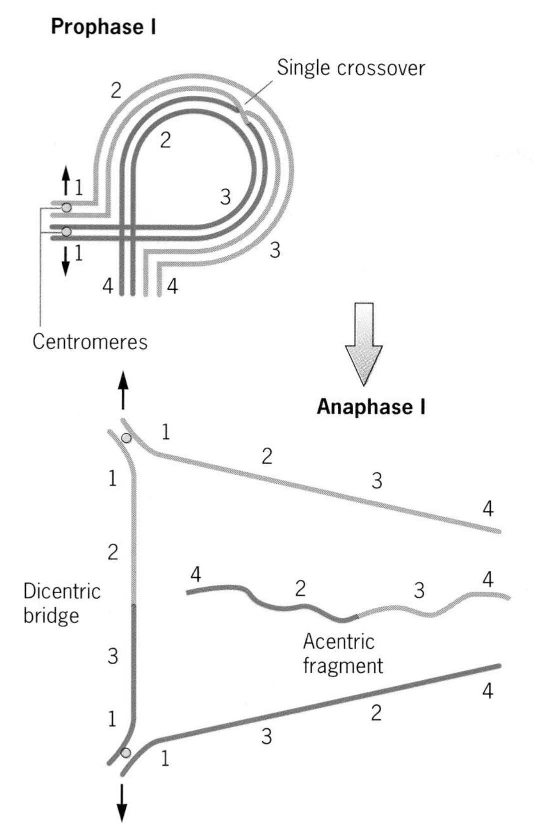

Figure 7.23 Suppression of recombination in an inversion heterozygote. The dicentric (1 2 3 1) and acentric (4 3 2 4) chromosomes formed from the crossover chromatids are aneuploid.

chromatids will therefore be eliminated by natural selection in the next generation. The net effect of this chromatid loss is to suppress recombination between inverted and noninverted chromosomes in heterozygotes.

Sometimes however, euploid products result from crossing over between inverted and noninverted chromatids, for example, when two crossovers occur within the inverted region (Figure 7.24). Both of the crossovers must involve the same two chromatids—a so-called two-strand double exchange. If they involve different chromatids, the products of the exchanges will be aneuploid.

Geneticists have exploited the recombination-suppressing properties of inversions to keep alleles of different genes together on the same chromosome. Let us assume, for example, that we have the recessive alleles *a, b, c, d,* and *e* on a chromosome that is structurally normal. If this chromosome is paired with another structurally normal chromosome that carries the corresponding wild-type alleles a^+, b^+, c^+, d^+, and e^+, the recessive and wild-type alleles will be scrambled by recombination. To prevent this scrambling, the chromosome with the recessive alleles can be paired with a wild-type chromosome that has an inversion. Unless double crossovers occur within the inverted region, this structural heterozygosity will suppress recombination. The multiply mutant chromosome can then be transmitted to the progeny as an intact genetic unit.

This recombination-suppressing technique has often been used in experiments with *Drosophila,* where the inverted chromosome usually carries a dominant mutation that permits it to be tracked through a whole series of crosses without cytological examination. Such marked inversion chromosomes are called **balancers** because they allow a mutant chromosome to be kept in heterozygous condition over the inversion.

Genetic Control of Recombination

It is not surprising that a process as important as recombination should be under genetic control. Studies with

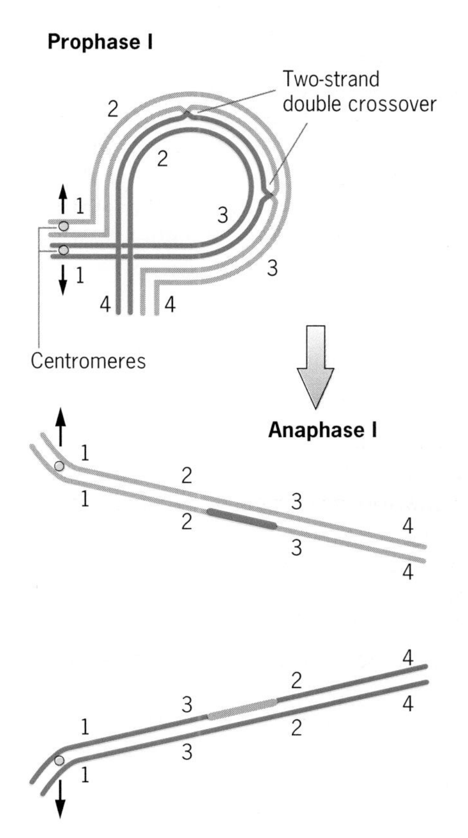

Figure 7.24 Double crossing over in an inversion heterozygote. None of the chromatids from a two-strand double crossover tetrad is aneuploid.

several organisms, including yeast and *Drosophila*, have demonstrated that recombination involves the products of many genes. Some of these gene products play a role in chromosome pairing, others catalyze the process of exchange, and still others help to rejoin broken chromatid segments. We shall consider some of these activities in greater detail in Chapter 13.

One curious phenomenon, which no one has yet explained, is that there is no crossing over in *Drosophila* males. In this regard, *Drosophila* is different from most species, including our own, where crossing over occurs in both sexes. In addition, we know that the amount of recombination varies among species. Perhaps the events that lead to recombination are themselves subject to evolutionary change.

Key Points: **Recombination can bring favorable mutations together in a population. Chromosome rearrangements, especially inversions, can suppress recombination.**

TESTING YOUR KNOWLEDGE

1. R. K. Sakai, K. Akhtar, and C. J. Dubash (1985, *J. of Heredity* 76: 140–141) reported data from a set of testcrosses with the mosquito *Anopheles culicifacies*, a vector for malaria in southern Asia. The data involved three mutations: *bw* (brown eyes), *c* (colorless eyes), and *Blk* (black body). In each cross, repulsion heterozygotes were mated to mosquitoes homozygous for the recessive alleles of the genes, and the progeny were scored as having either a parental or a recombinant genotype. Are any of the three genes studied in these crosses linked? If so, construct a map of the linkage relationships.

Cross	Repulsion Heterozygote	Progeny Parental	Progeny Recombinant	Pct. Recombinaton
1	*bw* +/+ *c*	850	503	37.2
2	*bw* +/+ *Blk*	750	237	24.0
3	*c* +/+ *Blk*	629	183	22.5

ANSWER

In each cross, the frequency of recombination is less than 50 percent, so all three loci are linked. To place them on a linkage map, we estimate the distances between each pair of genes from the observed recombination frequencies:

bw—-24.0—-*Blk*—-22.5—-*c*

37.2

Notice that the recombination frequency between *bw* and *c* (37.2 percent, from Cross 1) is substantially less than the actual distance between these genes (46.5). This shows that for widely separated genes, the recombination frequency underestimates the true map distance.

2. Singed bristles (*sn*), crossveinless wings (*cv*), and vermilion eye color (*v*) are due to recessive mutant alleles of three X-linked genes in *Drosophila melanogaster*. When a female heterozygous for each of the three genes was testcrossed with a singed, crossveinless, vermilion male, the following progeny were obtained:

Class	Phenotype	Number
1	singed, crossveinless, vermilion	3
2	crossveinless, vermilion	392
3	vermilion	34
4	crossveinless	61
5	singed, crossveinless	32
6	singed, vermilion	65
7	singed	410
8	wild-type	3
	Total =	1000

What is the correct order of these three genes on the X chromosome? What are the genetic map distances between *sn* and *cv*, *sn* and *v*, and *cv* and *v*? What is the coefficient of coincidence?

ANSWER

Before attempting to analyze these data, we must establish the genotype of the heterozygous female that produced the eight classes of offspring. We do this by identifying the two parental classes (2 and 7), which are the most numerous in the data. These classes tell us that the heterozygous female had the *cv* and *v* mutations on one of her X chromosomes and the *sn* mutation on the other. Her genotype was therefore (*cv* + *v*)/(+ *sn* +), with the parentheses indicating uncertainty about the gene order.

To determine the gene order, we must identify the double crossover classes among the six types of recombinant progeny. These are classes 1 and 8—the least numerous. They tell us that the *singed* gene is between *crossveinless* and *vermilion*. We can verify this by investigating the effect of a double crossover in a female with the genotype

$$\frac{cv\ +\ v}{+\ sn\ +}$$

Two exchanges in this genotype will produce gametes that are either *cv sn v* or + + +, which correspond to classes 1 and 8, the observed double crossovers. Thus, the proposed gene order—*cv sn v*—is correct.

Having established the gene order, we can now determine which recombinant classes represent crossovers between *cv* and *sn*, and which represent crossovers between *sn* and *v*.

Crossovers between *cv* and *sn*:

Class:	3		5		1		8		
Number:	34	+	32	+	3	+	3	=	72

Crossovers between *sn* and *v*:

Class:	4		6		1		8		
Number:	61	+	65	+	3	+	3	=	132

We determine the distances between these pairs of genes by calculating the average number of crossovers. Between *cv* and *sn*, the distance is 72/1000 = 7.2 cM, and between *sn* and *v* it is 132/1000 = 13.2 cM. We can estimate the distance between *cv* and *v* as the sum of these values: 7.2 + 13.2 = 20.4 cM. The linkage map of these three genes is therefore:

$$cv\text{—}7.2\text{—}sn\text{—}13.2\text{—}v$$

To calculate the coefficient of coincidence, we use the observed and expected frequencies of double crossovers:

$$c = \frac{\text{observed frequency of double crossovers}}{\text{expected frequency of double crossovers}} = \frac{(0.006)}{(0.072) \times (0.132)} = 0.63$$

which indicates only moderate interference.

3. A woman has two dominant traits, each caused by a mutation in a different gene: cataract (an eye abnormality), which she inherited from her father, and polydactyly (an extra finger), which she inherited from her mother. Her husband has neither trait. If the genes for these two traits are 15 cM apart on the same chromosome, what is the chance that the first child of this couple will have both cataract and polydactyly?

ANSWER

To calculate the chance that the child will have both traits, we first need to determine the linkage phase of the mutant alleles in the woman's genotype. Because she inherited the cataract mutation from her mother and the polydactyly mutation from her father, the mutant alleles must be on opposite chromosomes, that is, in the repulsion linkage phase:

$$\frac{C \quad +}{+ \quad P}$$

For a child to inherit both mutant alleles, the woman would have to produce an egg that carried a recombinant chromosome, *C P*. We can estimate the probability of this event from the distance between the two genes, 15 cM, which, because of interference, should be equivalent to 15 percent recombination. However, only half the recombinants will be *C P*. Thus the chance that the child will inherit both mutant alleles is (15/2) percent = 7.5 percent.

QUESTIONS AND PROBLEMS

7.1 From a cross between individuals with the genotypes *Cc Dd Ee* × *cc dd ee*, 1000 offspring were produced. The class that was *C- D- ee* included 351 individuals. Are the genes *c*, *d*, and *e* on the same or different chromosomes? Explain.

7.2 If *a* is linked to *b*, and *b* to *c*, and *c* to *d*, does it follow that a recombination experiment would detect linkage between *a* and *d*? Explain.

7.3 Two yeast strains differing in three linked genes were crossed: *A B C* x *a b c*. Among the tetrads that were analyzed, one contained the following spores: *A B C*, *A b C*, *a B c*, and *a b c*. How did this tetrad originate?

7.4 Mice have 19 autosomes in their genome, each about the same size. If two autosomal genes are chosen randomly, what is the chance that they will be on the same chromosome?

7.5 Mendel studied seven different traits in the pea plant. The pea has seven different chromosomes in its genome. What is the chance that seven different traits will all be due to genes on nonhomologous chromosomes?

7.6 If two loci are 10 cM apart, what proportion of the cells in prophase of the first meiotic division will contain a single crossover in the region between them?

7.7 Genes *a* and *b* are 20 cM apart. An $a^+ b^+/a^+ b^+$ individual was mated with an *a b*/ *a b* individual. (a) Diagram the cross and show the gametes produced by each parent and the genotype of the F_1. (b) What gametes can the F_1 produce, and in what proportions? (c) If the F_1 was crossed to *a b*/ *a b* individuals, what offspring would be expected, and in what proportions? (d) Is this an example of the coupling or repulsion linkage phase? (e) If the F_1 were intercrossed, what offspring would be expected, and in what proportions?

7.8 Answer questions (a)–(e) in the preceding problem under the assumption that the original cross was $a^+ b/a^+ b \times a\ b^+/a\ b^+$.

7.9 If the recombination frequency in the previous two problems were 40 percent instead of 20 percent, what change would occur in the proportions of gametes and testcross progeny?

7.10 A homozygous variety of maize with red leaves and normal seeds was crossed with another homozygous variety with green leaves and tassel seeds. The hybrids were then backcrossed to the green, tassel-seeded variety, and the following offspring were obtained: red, normal 124; red, tassel 126; green, normal 125; green, tassel 123. Are the genes for plant color and seed type linked? Explain.

7.11 A phenotypically wild-type female fruit fly that was heterozygous for genes controlling body color and wing length was crossed to a homozygous mutant male with black body (allele *b*) and vestigial wings (allele *vg*). The cross produced the following progeny: gray body, normal wings 126; gray body, vestigial wings 24; black body, normal wings 26; black body, vestigial wings 124. Do these data indicate linkage between the genes for body color and wing length? What is the frequency of recombination? Diagram the cross, showing the arrangement of the genetic markers on the chromosomes.

7.12 Another phenotypically wild-type female fruit fly heterozygous for the two genes mentioned in the previous problem was crossed to a homozygous black, vestigial male. The cross produced the following progeny: gray body, normal wings 23; gray body, vestigial wings 127; black body, normal wings 124; black body, vestigial wings 26. Do these data indicate linkage? What is the frequency of recombination? Diagram the cross, showing the arrangement of the genetic markers on the chromosomes.

7.13 In rabbits, the dominant allele *C* is required for colored fur; the recessive allele *c* makes the fur colorless (albino). In the presence of at least one *C* allele, another gene determines whether the fur is black (*B*, dominant) or brown (*b*, recessive). A homozygous strain of brown rabbits was crossed with a homozygous strain of albinos. The F_1 were then crossed to homozygous double recessive rabbits, yielding the following results: black 34; brown 66; albino 100. Are the genes *b* and *c* linked? What is the frequency of recombination? Diagram the crosses, showing the arrangement of the genetic markers on the chromosomes.

7.14 In tomatoes, tall vine (*D*) is dominant over dwarf (*d*), and spherical fruit shape (*P*) is dominant over pear shape (*p*). The genes for vine height and fruit shape are linked with 20 percent recombination between them. One tall plant (I) with spherical fruit was crossed with a dwarf, pear-fruited plant. The cross produced the following results: tall, spherical 81; dwarf, pear 79; tall, pear 22; dwarf spherical 17. Another tall plant with spherical fruit (II) was crossed with the dwarf, pear-fruited plant, and the following results were obtained: tall, pear 21; dwarf, spherical 18; tall, spherical 5; dwarf, pear 4. Diagram these two crosses, showing the genetic markers on the chromosomes. If the two tall plants with spherical fruit were crossed with each other, that is, I x II, what phenotypic classes would you expect from the cross, and in what proportions?

7.15 In *Drosophila*, genes *a* and *b* are located at positions 22.0 and 42.0 on chromosome 2, and genes *c* and *d* are located at positions 10.0 and 25.0 on chromosome 3. A fly homozygous for the wild-type alleles of these four genes was crossed with a fly homozygous for the recessive alleles, and the F_1 daughters were backcrossed to their quadruply recessive fathers. What offspring would you expect from this backcross, and in what proportions?

7.16 In *Drosophila*, the genes *sr* (*stripe* thorax) and *e* (*ebony* body) are located at 62 and 70 cM, respectively, from the left end of chromosome 3. A striped female homozygous for e^+ was mated with an ebony male homozygous for sr^+. All the offspring were phenotypically wild-type (gray body and unstriped). (a) What kind of gametes will be produced by the F_1 females, and in what proportions? (b) What kind of gametes will be produced by the F_1 males, and in what proportions? (c) If the F_1 females are mated with striped, ebony males, what offspring are expected, and in what proportions? (d) If the F_1 males and females are intercrossed, what offspring would you expect from this intercross, and in what proportions?

7.17 The *Drosophila* genes *vg* (*vestigial* wings) and *cn* (*cinnabar* eyes) are located at 67.0 and 57.0, respectively, on chromosome 2. A female from a homozygous strain of vestigial flies was crossed with a male from a homozygous strain of cinnabar flies. The F_1 hybrids were phenotypically wild-type (long wings and reddish-brown eyes). (a) How many different kinds of gametes could the F_1 females produce, and in what proportions? (b) If these females are mated with cinnabar, vestigial males, what kinds of progeny would you expect, and in what proportions?

7.18 In *Drosophila*, the genes *st* (*scarlet* eyes), *ss* (*spineless* bristles) and *e* (*ebony* body) are located on chromosome 3, with map positions as indicated:

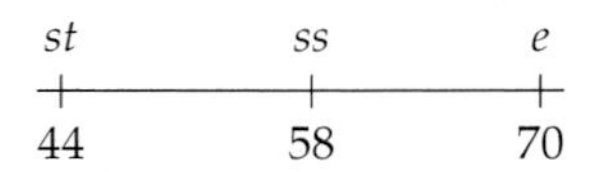

Each of these mutations is recessive to its wild-type allele (st^+, reddish-brown eyes; ss^+, smooth bristles; e^+, gray body). Phenotypically wild-type females with the genotype $st\ ss\ e^+/st^+\ ss^+\ e$ were crossed with triply recessive males. Predict the phenotypes of the progeny and the frequencies with which they will occur assuming (a) no interference and (b) complete interference.

7.19 In maize, the genes *Pl* for purple leaves (dominant over *pl* for green leaves), *sm* for salmon silk (recessive to *Sm* for yellow silk), and *py* for pigmy plant (recessive to *Py* for normal size plant) are on chromosome 6, with map positions as shown:

Pl	*sm*	*py*
45	55	65

Hybrids from the cross *Pl sm py/Pl sm py* x *pl Sm Py/ pl Sm Py* were testcrossed with *pl sm py/pl sm py* plants. Predict the phenotypes of the offspring and their frequencies assuming (a) no interference and (b) complete interference.

7.20 In maize, the genes *Tu*, *j2*, and *gl3* are located on chromosome 4 at map positions 101, 106, and 112, respectively. If plants homozygous for the recessive alleles of these genes are crossed with plants homozygous for the dominant alleles, and the F_1 plants are testcrossed to triply recessive plants, what genotypes would you expect, and in what proportions? Assume that interference is complete over this map interval.

7.21 A *Drosophila* geneticist made a cross between females homozygous for three X-linked recessive mutations (*y*, *yel-*

low-body; *ec*, *echinus* eye shape; *w*, *white* eye color and wild-type males. He then mated the F_1 females to triply mutant males and obtained the following results:

Females	Males	Number
+ + + / *y ec w*	+ + +	475
y ec w / *y ec w*	*y ec w*	469
y + + / *y ec w*	*y* + +	8
+ *ec w* / *y ec w*	+ *ec w*	7
y + *w* / *y ec w*	*y* + *w*	18
+ *ec* + / *y ec w*	+ *ec* +	23
+ + *w* / *y ec w*	+ + *w*	0
y ec + / *y ec w*	*y ec* +	0

Determine the order of the three loci *y*, *ec*, and *w*, and estimate the distances between them on the linkage map of the X chromosome.

7.22 A *Drosophila* geneticist crossed females homozygous for three X-linked mutations (*y*, *yellow* body; *B*, *bar* eye shape; *v*, *vermilion* eye color) to wild-type males. The F_1 females, which had gray bodies and bar eyes with reddish-brown pigment, were then crossed to *y* B^+ *v* males, yielding the following results:

Phenotype	Number
yellow, bar, vermilion wild-type	546
yellow bar, vermilion	244
yellow, vermilion bar	160
yellow, bar vermilion	25

Determine the order of these three loci on the X chromosome and estimate the distances between them.

7.23 Female *Drosophila* heterozygous for three recessive mutations *e* (*ebony* body), *st* (*scarlet* eyes), and *ss* (*spineless* bristles) were testcrossed, and the following progeny were obtained:

Phenotype	Number
wild-type	67
ebony	8
ebony, scarlet	68
ebony, spineless	347
ebony, scarlet, spineless	78
scarlet	368
scarlet, spineless	10
spineless	54

(a) What indicates that the genes are linked? (b) What was the genotype of the original heterozygous females? (c) What is the order of the genes? (d) What is the map distance between *e* and *st*? (e) Between *e* and *ss*? (f) What is the coefficient of coincidence? (g) Diagram the crosses in this experiment.

7.24 Consider a female *Drosophila* with the following X chromosome genotype:

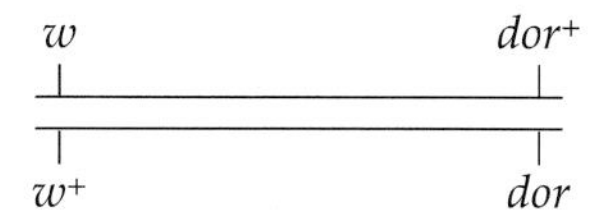

The recessive alleles *w* and *dor* cause mutant eye colors (white and deep orange, respectively). However, *w* is epistatic over *dor*; that is, the genotypes *w dor*/Y and *w dor*/*w dor* have white eyes. If there is 40 percent recombination between *w* and *dor*, what proportion of the sons from this heterozygous female will show a mutant phenotype? What proportion will have either red or deep orange eyes?

7.25 Assume that in *Drosophila* there are three genes *x*, *y*, and *z*, with each mutant allele recessive to the wild-type allele. A cross between females heterozygous for these three loci and wild-type males yielded the following progeny:

Females	+ + +	1010
Males	+ + +	39
	+ + *z*	430
	+ *y z*	32
	x + +	27
	x y +	441
	x y z	31
	Total =	2010

Using these data, construct a linkage map of the three genes and calculate the coefficient of coincidence.

7.26 A normal woman with a color blind father married a normal man and their first child, a boy, had hemophilia. Both color blindness and hemophilia are due to X-linked recessive mutations, and the relevant genes are separated by 10 cM. This couple plans to have a second child. What is the probability that it will have hemophilia? Color blindness? Both hemophilia and color blindness? Neither hemophilia nor color blindness?

7.27 In the nematode *Caenorhabditis elegans*, the linked genes *dpy* (*dumpy* body) and *unc* (*uncoordinated* behavior) recombine with a frequency *P*. If a repulsion heterozygote carrying recessive mutations in these genes is self-fertilized, what fraction of the offspring will be both dumpy and uncoordinated?

7.28 In the following testcross, genes *a* and *b* are 20 cM apart, and genes *b* and *c* are 10 cM apart: *a* + *c*/ + *b* + × *a b c*/*a b c*. If the coefficient of coincidence is 0.5 over this interval on the linkage map, how many triply homozygous recessive individuals are expected among 1000 progeny?

7.29 *Drosophila* females heterozgyous for three recessive mutations, *a*, *b*, and *c*, were crossed to males homozygous for all three mutations. The cross yielded the following results:

Phenotype	Number
+ + +	75
+ + *c*	348
+ *b* *c*	96
a + +	110
a *b* +	306
a *b* *c*	65

Construct a linkage map showing the correct order of these genes and estimate the distances between them.

7.30 A chromosome is 120 cM long. On the average, how many chiasmata will occur when it goes through meiosis?

7.31 The total map length of the *Drosophila* genome is about 270 cM. On the average, how many chiasmata will occur in an oocyte going through meiosis?

7.32 One of the metacentric chromosomes in a species of bean typically has one chiasma in each of its arms during meiosis. What is the genetic map length of this chromosome?

7.33 Two strains of maize, M1 and M2, are homozygous for four recessive mutations, *a*, *b*, *c*, and *d*, on one of the large chromosomes in the genome. Strain W1 is homozygous for the dominant alleles of these mutations. Hybrids produced by crossing M1 and W1 yield many different classes of recombinants, whereas hybrids produced by crossing M2 and W1 do not yield any recombinants at all. What is the difference between M1 and M2?

7.34 A *Drosophila* geneticist has identified a strain of flies with a large inversion in the left arm of chromosome 3. This inversion includes two mutations, *e* (*ebony* body) and *cd* (*cardinal* eyes), and is flanked by two other mutations, *sr* (*stripe* thorax) on the right and *ro* (*rough* eyes) on the left. The geneticist wishes to replace the *e* and *cd* mutations inside the inversion with their wild-type alleles; he plans to accomplish this by recombining the multiply mutant, inverted chromosome with a wild-type, inversion-free chromosome. What event is the geneticist counting on to achieve his objective? Explain.

BIBLIOGRAPHY

BRIDGES, C. B., AND T. M. OLBRYCHT. 1926. "The multiple stock 'XPLE' and its use." *Genetics* 11: 41–56.

CARLSON, E. A. 1966. *The Gene: A Critical History*. W. B. Saunders, Philadelphia.

CREIGHTON, H. B., AND B. MCCLINTOCK. 1931. "A correlation of cytological and genetical crossing over in *Zea mays*." *Proc. Natl. Acad. Sci. U.S.A.* 17: 492–497.

LINDSLEY, D. L., AND G. ZIMM. 1992. *The Genome of* Drosophila melanogaster. Academic Press, New York.

PETERS, J. A., ED. 1959. *Classic Papers in Genetics*. Prentice-Hall, Englewood Cliffs, N. J.

STURTEVANT, A. H., AND G. W. BEADLE. 1939. *An Introduction to Genetics*. W. B. Saunders, Philadelphia. (Republished in 1962 by Dover, New York.)

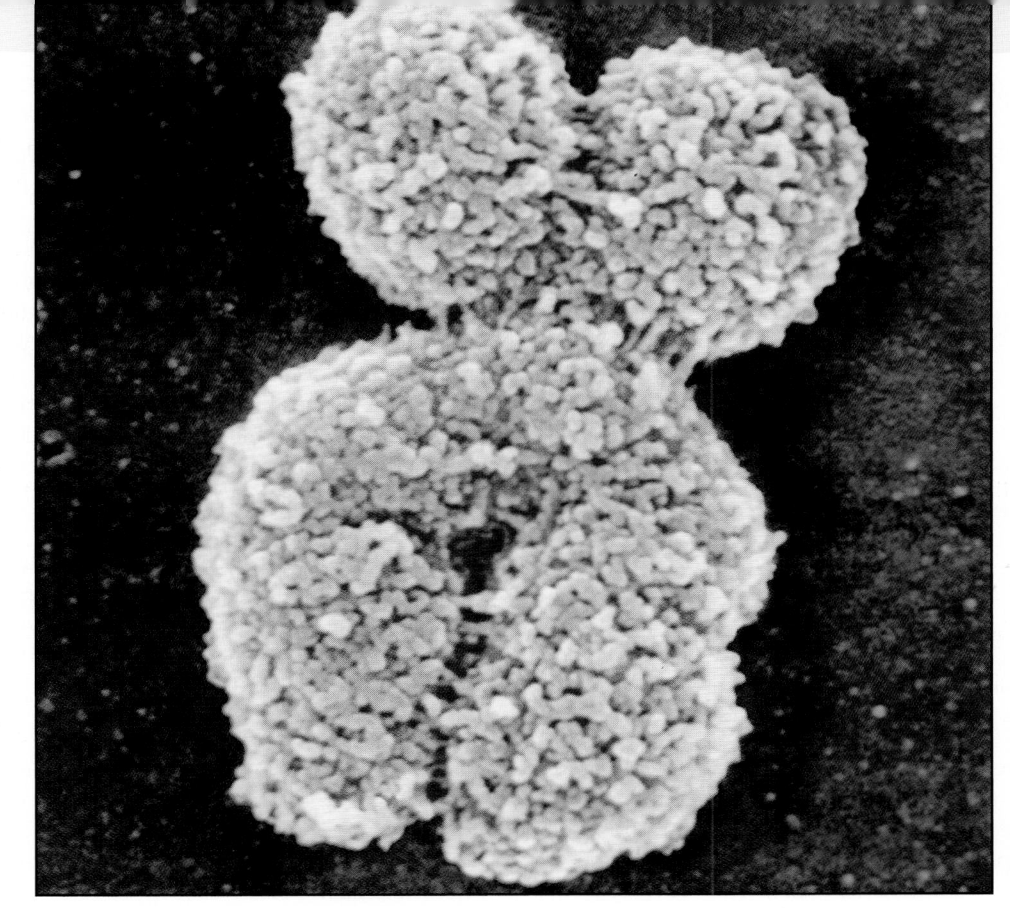

Highly magnified image of a replicated human chromosome.

8

Advanced Linkage Analysis

CHAPTER OUTLINE

Huntington's Disease: A Challenge in Gene Mapping

In 1872 a young boy accompanied his physician father as he made his rounds among the isolated villages of rural Long Island. As they passed through the countryside, they chanced upon a sight so mesmerizing that the young, impressionable boy never forgot it. Two women moved through the tall grass blanketing the clearing. Both were thin to the point of emaciation, and both were twisting, bowing, and grimacing. The father paused, spoke to the women, and left. It was a disturbing sight, but it made an enormous impact on the youngster.

The boy was George Huntington, and so fascinated was he by this encounter that he pondered it for many years. He wanted to understand this bizarre and terrifying malady. Young Huntington followed in his father's footsteps and entered into a career in medicine. His perseverance paid off, and the disease he observed in those two women today bears his name: **Huntington's disease**. Huntington concluded that the disease is inherited and that it involves a gradual deterioration of the central nervous system. This is a late-onset disease, with symptoms first appearing in approximately the third or fourth decade of life. Limb movement becomes spastic and uncontrolled. Mental functions deteriorate. Patients become bedridden and helpless. Death follows about 10 to 15 years after the symptoms make their first appearance. There is no cure, and there is little relief possible once the disease begins its course.

Though rare, this disease has fascinated researchers for decades. It was clearly an autosomal dominant disorder, but on which autosome was the gene? Since researchers did not know the biochemical basis of Huntington's disease, the strategy that evolved was first to locate and isolate the gene, and then determine what cell functions the gene controlled. That strategy is working. Researchers located the gene on chromosome 4 in 1983. Ten years later they isolated it, and shortly thereafter they determined the protein it encodes. The next step is to determine the function of the protein. The key advance in this attack on Huntington's disease was to map the gene.

In this chapter we consider some advanced techniques in gene mapping, including procedures to detect linkage in experimental organisms and to position genes with respect to chromosomal structures such as centromeres and bands. We also examine various strategies for mapping genes in human beings.

DETECTION OF LINKAGE IN EXPERIMENTAL ORGANISMS

During the hundred or so years that genetics has been a formal science, researchers have studied linkage in numerous organisms including peas, fruit flies, bread mold, mice, corn, yeast, mosquitoes, dogs, chickens, nematodes, tomatoes, and human beings. The most intensively studied organisms have been fungi—especially yeast—and fruit flies. In this section we consider some of the techniques that have been developed to detect linkage in these organisms.

Tetrad Analysis to Detect Linkage in Fungi

Fungi in the class Ascomycetes are unusual because all four products of a single meiosis are retained in the ascus sac. Geneticists have exploited this property to study the events of meiosis, including crossing over and recombination. Because these fungi are ordinarily haploid, each meiotic product, or ascospore, germinates into a new organism. Phenotypic analysis of these organisms can therefore provide information about the genotypes of the ascospores from which they were derived.

Ascospore formation in *Saccharomyces cerevisiae*, or baker's yeast, reflects the meiotic process (Figure 8.1). Each ascospore receives one of the chromatids that was present in a tetrad during prophase of the first meiotic division. The ascospores in an ascus therefore contain all four of the chromatids from a tetrad.

We detect linkage in yeast by analyzing tetrads of ascospores. A cross between two yeast strains with mutations on different chromosomes produces different classes of tetrads (Figure 8.2). We distinguish between alleles by using upper and lower case letters; the lower case letters represent mutants, and the upper case letters represent wild-type. The segregation and independent assortment of mutations on different chromosomes produce two main types of asci. In one type of ascus, two of the ascospores are genotypically like one of the parents and two are like the other parent. An ascus with this pattern is called a **parental ditype** ascus. In the other type, all the spores are recombinant; two of the spores show one genotype, and two show the reciprocal genotype. Neither of the two genotypes is like the parental genotypes; thus the ascus is called a **nonparental ditype** ascus. Because different chromosomes assort independently during the first meiotic division, a cross involving unlinked genes produces approximately equal frequencies of parental and nonparental ditype asci.

The segregation and independent assortment of unlinked genes can also produce a third type of ascus, called a **tetratype**. In this type there are four kinds of ascospores, one that is like each of the parents and two that are recombinant. Such asci occur whenever there is crossing over between one of the genes and its centromere (Figure 8.2). The existence of tetratype asci does not, however, break the rule that for unlinked

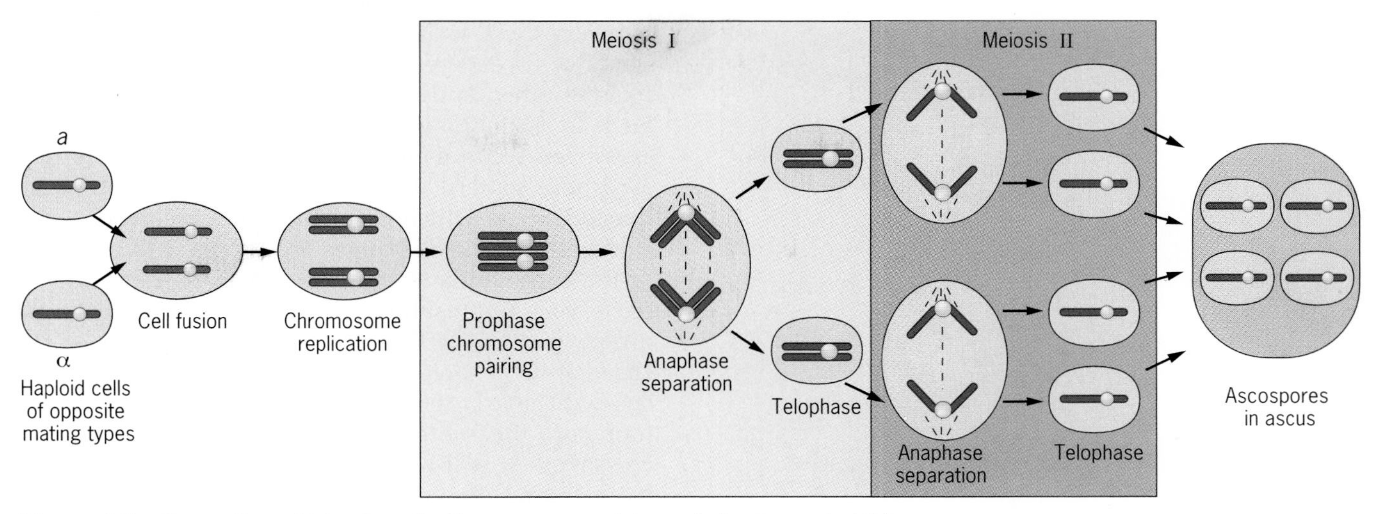

Figure 8.1 Formation of tetrads in *Saccharomyces cerevisiae*, or baker's yeast. All four products of meiosis are recovered in the ascus.

genes, parental and nonparental ditype asci occur with equal frequency.

A cross between two yeast strains with mutations on the same chromosome produces different results (Figure 8.3). If the mutations are tightly linked, we expect most of the asci to show the parental ditype pattern. The only exceptions arise from crossing over between the mutant loci. For example, a single crossover

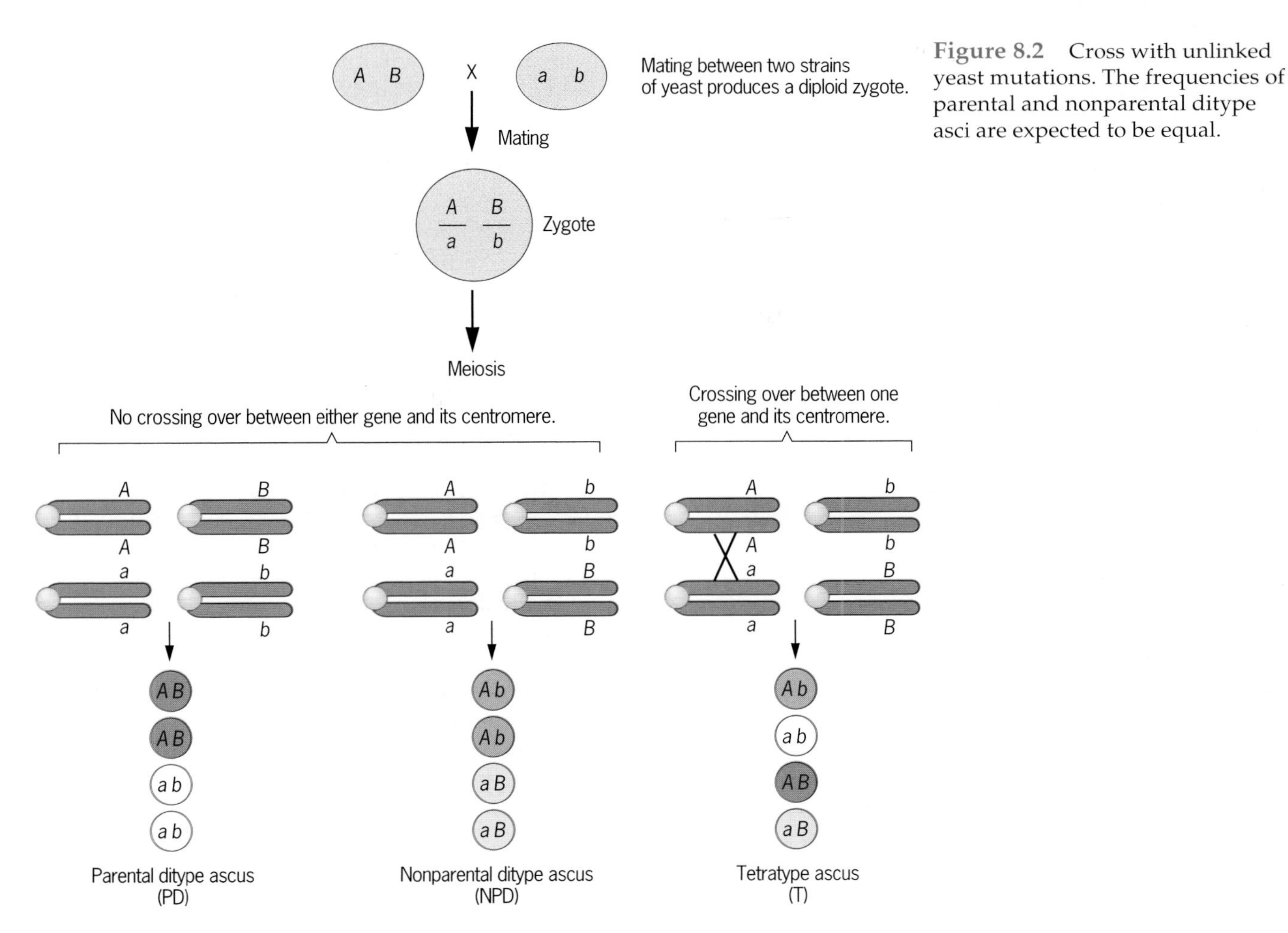

Figure 8.2 Cross with unlinked yeast mutations. The frequencies of parental and nonparental ditype asci are expected to be equal.

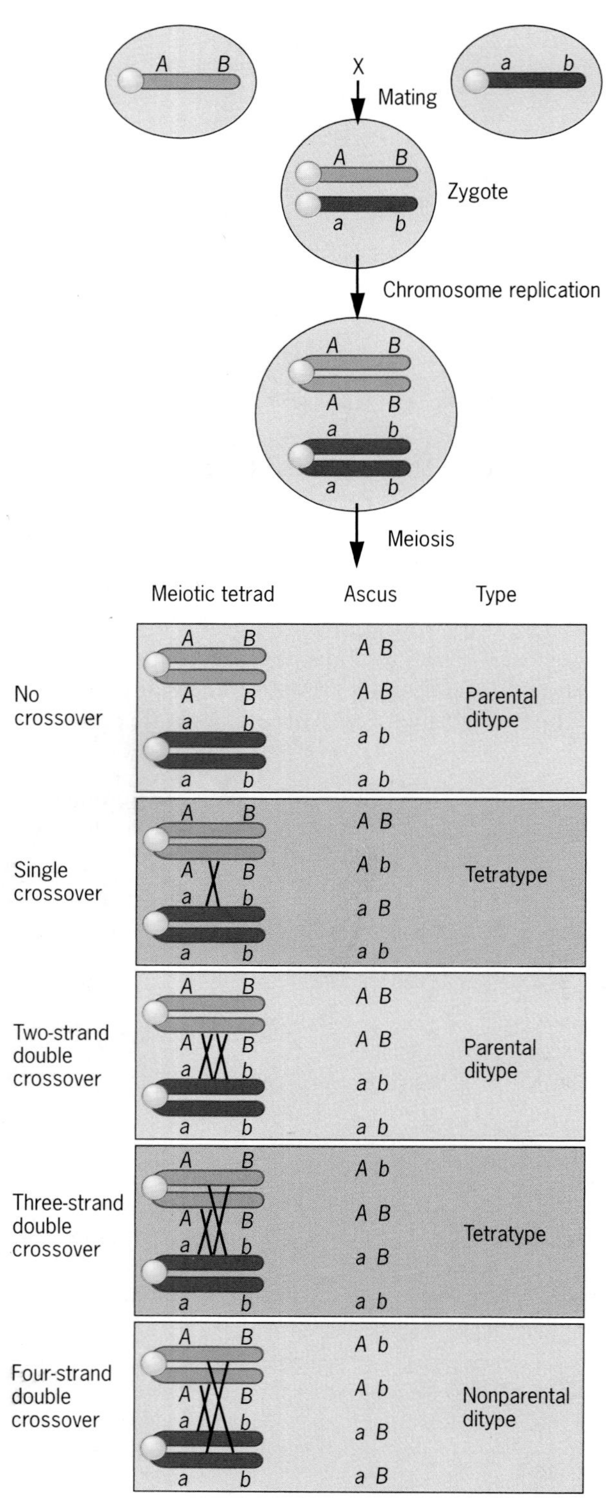

Figure 8.3 Cross with linked yeast mutations. The frequencies of parental and nonparental ditype asci are expected to be unequal. The nonparental ditype asci, which are produced by four-strand double crossovers, should be much less numerous than the parental ditype asci.

between these loci will produce an ascus with four kinds of ascospores: one spore like one parent, one like the other, and two with recombinant genotypes. Such an ascus shows the tetratype pattern. If a double crossover occurs between the two loci, the resulting ascospore pattern will be either ditype or tetratype, depending on which chromatids are involved in the exchanges. If the exchanges are confined to only two of the chromatids in the tetrad (a two-strand double crossover), the resulting ascus will have the parental ditype pattern. If the exchanges affect three of the chromatids (a three-strand double crossover), the ascus will have a tetratype pattern, and if they affect four of the chromatids (a four-strand double crossover), it will have a nonparental ditype pattern. Nonparental ditype asci can be produced only by a four-strand double crossover between the mutant loci. Even for loosely linked genes, this is a rare event. Consequently, a low frequency of nonparental ditype asci is considered strong evidence for genetic linkage.

Once we establish linkage, we can use the tetrad data to estimate the distance between the genes. A cross between two yeast strains illustrates how genetic distances are estimated (Figure 8.4). One strain carries two recessive mutations, *py* and *th*, which prevent growth unless the vitamins pyridoxine and thiamin are added to the medium; the *py* mutation causes the pyridoxine requirement, and the *th* mutation causes the thiamin requirement. The other strain, which can grow without any vitamin supplements, carries the wild-type alleles of these two mutations, *PY* and *TH*. Diploid cells formed by crossing the two haploid strains were induced to go through meiosis, and the ascospores from 100 of the resulting tetrads were carefully isolated and grown into independent yeast colonies. Cells from each colony were then separately tested for their ability to grow in the absence of pyridoxine or thiamin, thereby revealing the genotype of the spore from which they came. When all the tests were finished, each tetrad of spores could be classified into one of the three possible types.

The data indicate that the two genes are linked because nonparental ditype tetrads occur much less frequently than do parental ditype tetrads. To calculate the distance between the genes, we must estimate the average number of crossovers between them on a chromosome. A simple estimate is the frequency of recombination, which we obtain from the formula

$$\text{Recombination frequency} = [(1/2)\,\text{T} + \text{NPD}]/\text{total}$$

Here T and NPD are the observed numbers of tetratype and nonparental ditype tetrads. We multiply T by one-half because only half the chromatids in tetratype tetrads are genetically recombinant. However, all the chromatids in a nonparental ditype tetrad are recombinant; thus we count these tetrads fully in

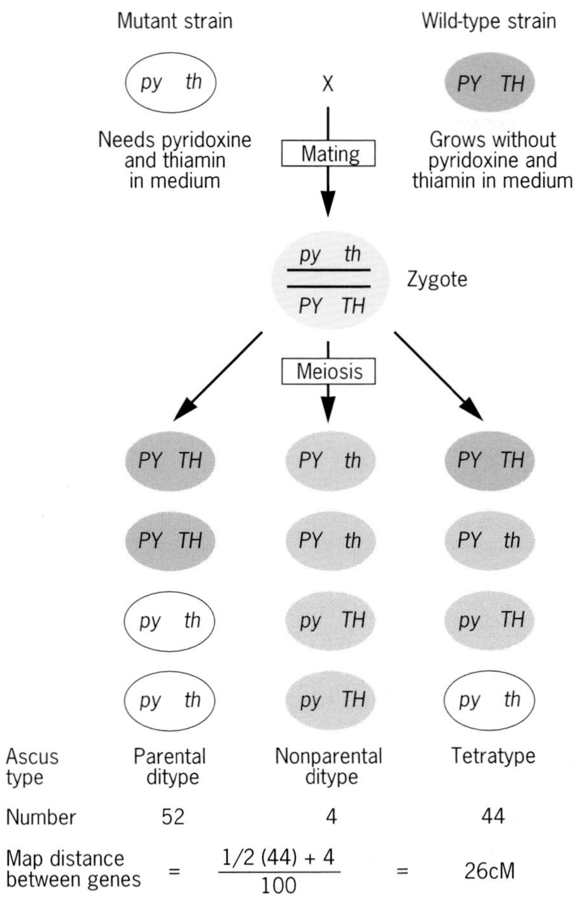

Figure 8.4 Tetrad analysis with two linked yeast mutations, *py* (pyridoxine requirement) and *th* (thiamin requirement).

the formula. We do not count the parental ditype tetrads because none of their chromatids is genetically recombinant.

Applying the formula, we find that

$$\text{Recombination frequency} = [(1/2)(44) + 4]/100 = 0.26 \text{ Morgan}$$

or a distance of 26 centiMorgans.

This mapping procedure usually underestimates the true distance between genes, albeit slightly. Two-strand double crossovers, which produce parental ditype tetrads, are not counted at all, and three-strand double crossovers, which produce tetratype tetrads, are not counted correctly. In these latter tetrads, the average number of exchanges per chromatid is 3/4 rather than 1/2; consequently, the coefficient 1/2 in the formula is too small. To compensate for these shortcomings, some researchers prefer to estimate the map distance between genes with the formula

$$\text{Map distance} = [(1/2)\,\text{T} + 3\,\text{NPD}]/\text{total}$$

The formula corrects for the failure to count the two- and three-strand double crossovers properly by inflating the contribution of the nonparental ditype tetrads. In this case,

$$\begin{aligned}\text{Map distance} &= [(1/2)(44) + (3)(4)]/100 \\ &= 0.34 \text{ Morgan} \\ &= 34 \text{ centiMorgans}\end{aligned}$$

Tetrad analysis is also possible for several other experimental organisms, including *Chlamydomonas reinhardii*, a unicellular green alga. In a later section, we consider a special application of tetrad analysis involving the Ascomycete *Neurospora crassa*. In this fungus, the ascospores are arranged in a way that reflects the progression of the chromosomes through the two meiotic divisions. These **ordered tetrads** permit mapping of the loci responsible for chromosome movement, that is, the centromeres. In yeast and *Chlamydomonas*, where the tetrads are unordered, it is also possible to map centromeres, but the statistical procedures for doing so are beyond the scope of this book.

Balancer Chromosome Technique to Assign a Gene to a Chromosome in *Drosophila*

Nearly a century of intensive work has made *Drosophila melanogaster* one of the best understood organisms for genetics research. Hundreds of genes have been mapped on its four chromosomes, and many genetically caused structural aberrations have been identified. The availability of these mutants and aberrations in *Drosophila* stock centers has greatly facilitated the isolation, identification, and localization of many new mutations in this organism.

One analytical procedure for mapping genes takes advantage of the recombination-suppressing effects of inversions (see Chapter 7). These aberrations are present in balancer chromosomes, which also carry dominant markers to allow them to be followed through a series of crosses. A new mutation can be localized to a chromosome on the basis of how it segregates in crosses with balancers.

The procedure involves crossing the new mutation—assumed to be recessive—to flies that carry the balancer chromosomes (Figure 8.5). In this example, the balancer second chromosome is marked with *Cy*, a dominant mutation for *Curly* wings, and the balancer third chromosome is marked with *Tb*, a dominant mutation for *Tubby* body. The homologs of these chromosomes also carry dominant markers—*Pm* for *Plum* eyes on the second chromosome and *Sb* for *Stubble* bristles on the third chromosome. However, because they lack extensive inversions, they are not good recombination-suppressors themselves. All four of the dominant markers on these chromosomes are also associated with recessive lethal effects; thus flies homozygous for any one of them cannot survive.

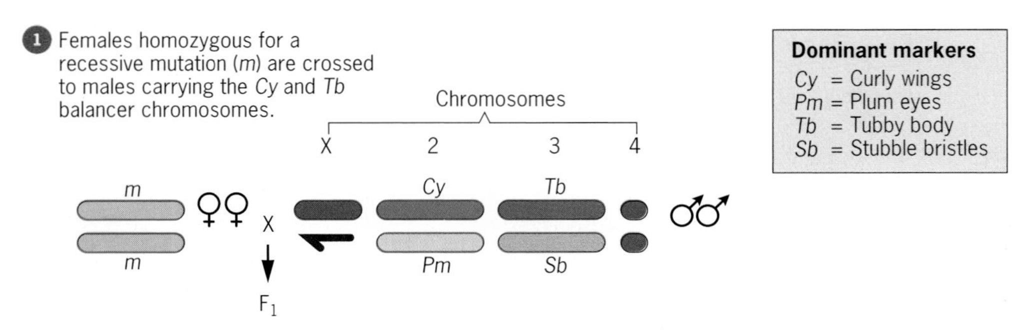

2 If the mutant is **X-linked,** the F_1 males will show the mutant phenotype but the F_1 females will not.

3 If the mutant is **autosomal**, none of the F_1 flies will show the mutant phenotype.
Balancer heterozygotes must be intercrossed to determine which autosome carries the mutation.

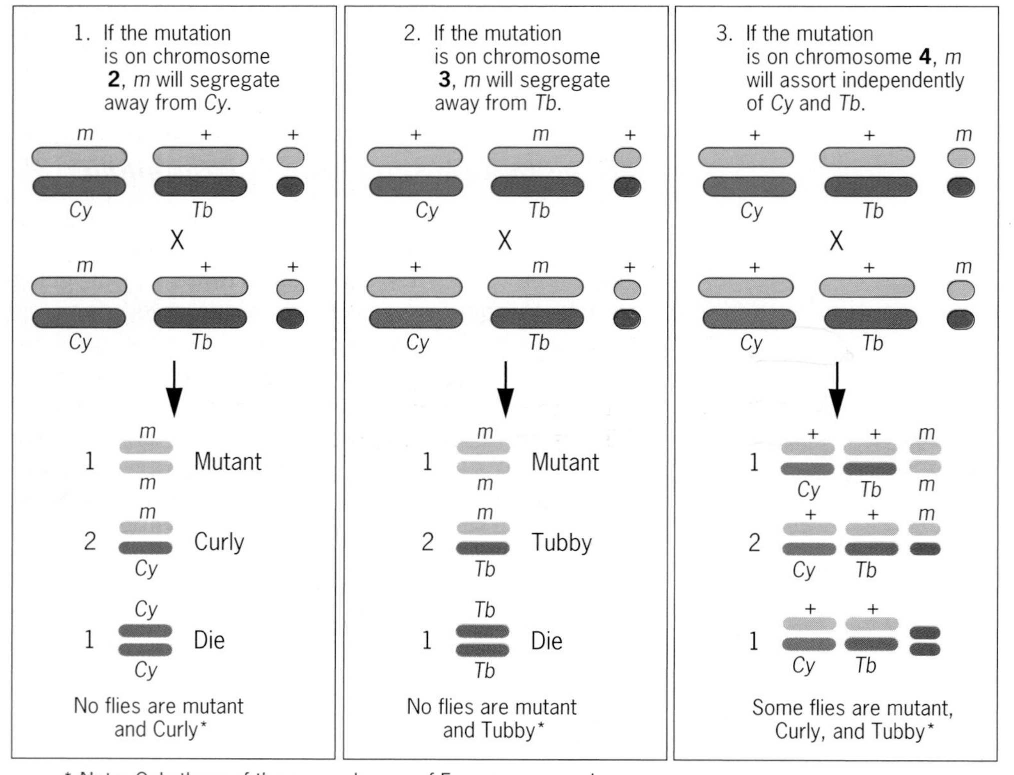

* Note: Only three of the many classes of F_2 progeny are shown.

Figure 8.5 Balancer chromosome technique to assign a gene marked by the recessive mutation *m* to a chromosome in *Drosophila*. The balancer chromosomes are marked by the dominant mutations *Cy* (*Curly* wings) and *Tb* (*Tubby* body), on chromosomes 2 and 3, respectively.

In the first cross designed to ascertain the chromosomal location of the new mutation, we use females homozygous for the new mutation. If all the sons from this cross show the mutant phenotype, we immediately know that the mutation is X-linked; otherwise, it must be autosomal. Should it turn out to be autosomal, we determine which autosome carries the mutation by intercrossing Curly, Tubby F_1 flies. A mutation on the second chromosome will segregate away from *Cy* because the *Cy* balancer suppresses recombination. The result is that none of the F_2 progeny will show the mutant and Curly phenotypes simultaneously. Segregation of a mutation on the third chromosome away from *Tb* prevents any F_2 flies from showing the mutant and Tubby phenotypes simultaneously. If the mutation is on the fourth chromosome, it will assort independently of both balancers and yield some flies that are simultaneously mutant, Curly, and Tubby. Analysis of the F_2 phentoypes can therefore reveal which of the four *Drosophila* chromosomes carries the new mutation. Testcrosses can then be designed to map its position precisely.

Key Points: **In organisms amenable to tetrad analysis, linkage is indicated when parental ditype tetrads are more frequent than nonparental ditype tetrads. In *Drosophila*, genes can be assigned to chromosomes by analyzing crosses involving specially marked balancer chromosomes.**

(*a*)

SPECIALIZED MAPPING TECHNIQUES

In some fungi, the ascospores are arranged in an ascus sac in a precise sequence. This enables us to map the centromere of the chomosome. In this section, we discuss how to map centromeres by analyzing ordered tetrads in the fungus *Neurospora crassa*. We also discuss procedures for aligning the genetic and cytological maps of *Drosophila* chromosomes.

Centromere Mapping with Ordered Tetrads in *Neurospora*

The bread mold *Neurospora crassa* (Figure 8.6) grows as a branching mass of haploid filaments called a *mycelium*. Cells from different *Neurospora* strains, if they belong to different mating types, can fuse to form a diploid cell. This diploid cell undergoes meiosis to produce four haploid nuclei within a narrow ascus. Constraints of space force these nuclei to separate in a linear fashion according to the way in which they were created on the meiotic spindles. Each of the four nuclei divides mitotically to produce two twin nuclei, which remain next to each other in the ascus. The end result is a set of eight nuclei representing the four meiotic products; each product is present twice as a result of the mitotic doubling that occurred after meiosis. These eight nuclei are then partitioned into separate cells, which develop into ascospores capable of germinating into new organisms. The genetic significance of this system is that the order of the ascospores in the ascus reflects the way in which the chromatids segregated during meiosis. Through careful dissection,

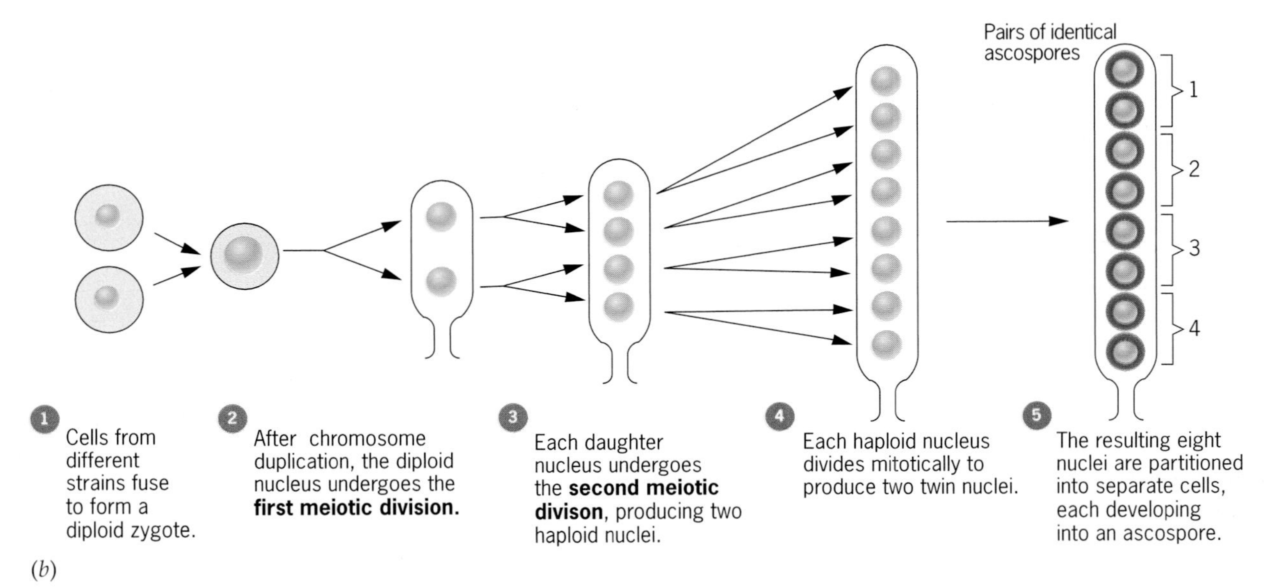

(*b*)

Figure 8.6 (*a*) Cultures of *Neurospora crassa*. (*b*) Production of ordered tetrads of ascospores in *Neurospora*.

each of the ascospores can be removed from the ascus and grown to determine its genotype.

The ordered array of ascospores makes it possible to map the loci responsible for chromosome movement during meiosis. These loci, or centromeres, are the points at which spindle fibers attach to the chromosomes. Although the centromeres have no phenotype themselves, we can map them by studying the segregation of nearby genes. Figure 8.7 illustrates the principles.

To map the centromere, let's assume that the *A* locus is a short distance away from the centromere, and that we can distinguish its two alleles, *A* and *a*, in haploid *Neurospora*. (They might, for example, produce different pigments in the ascospores; see Figure 8.8.) A heterozygous *Neurospora* cell going through meiosis can segregate these alleles in two different ways. One type of segregation produces an ascus in which the ascospores with the *A* allele are all at one end and those with the *a* allele are at the other; this is called the **first division segregation** pattern. The other type produces an alternating array of ascospores; this is called the **second division segregation** pattern.

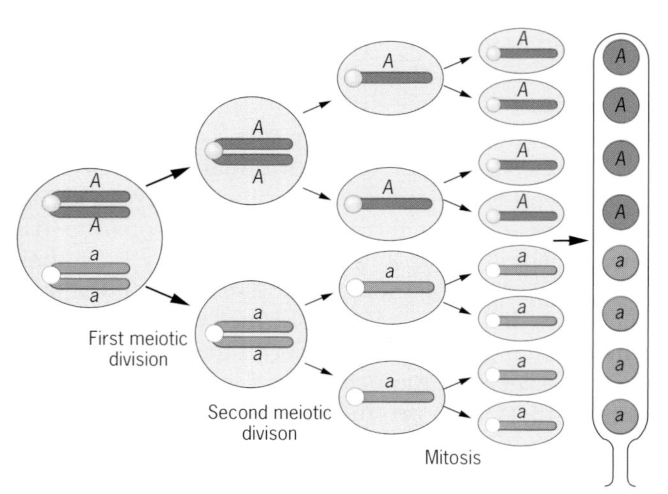

(*a*) No crossover between gene and centromere.

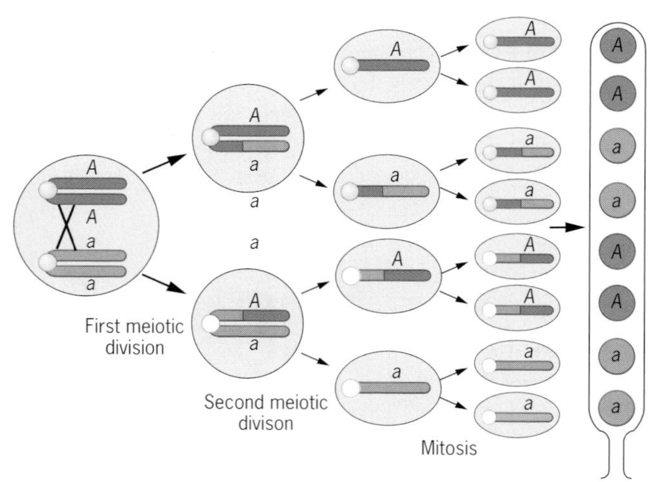

(*b*) Crossover between gene and centromere.

Figure 8.7 Segregation patterns in *Neurospora* asci. (*a*) First division segregation, resulting from the absence of a crossover between a gene and its centromere. (*b*) Second division segregation, resulting from a crossover between a gene and its centromere.

The genetic basis for these two patterns is simple. The first division segregation pattern occurs when the *A* and *a* alleles separate cleanly in the first meiotic division, that is, there is no crossing over. The sister chromatids with the *A* allele move to one pole, and the sister chromatids with the *a* allele move to the other. Because this positioning takes place within the narrow confines of the developing ascus, the alleles become "stacked" in a linear array. This array, which is preserved through the second meiotic division and the subsequent mitotic division, generates a total of eight ascospores. The four that carry *A* wind up at one end, and the four that carry *a* wind up at the other. The set of four that goes to the top of the ascus will depend on how the paired *A* and *a* alleles were oriented on the spindle of the first meiotic division.

The second division segregation pattern occurs when there is a crossover between the *A* locus and its centromere in the first meiotic division. Such a crossover puts both *A* and *a* into each of the daughter nuclei generated by that division. Then, in the second meiotic division, these alleles segregate from each other, creating an alternating pattern that persists through the ensuing mitotic division.

Because second division segregation patterns are produced by crossing over between the gene and its

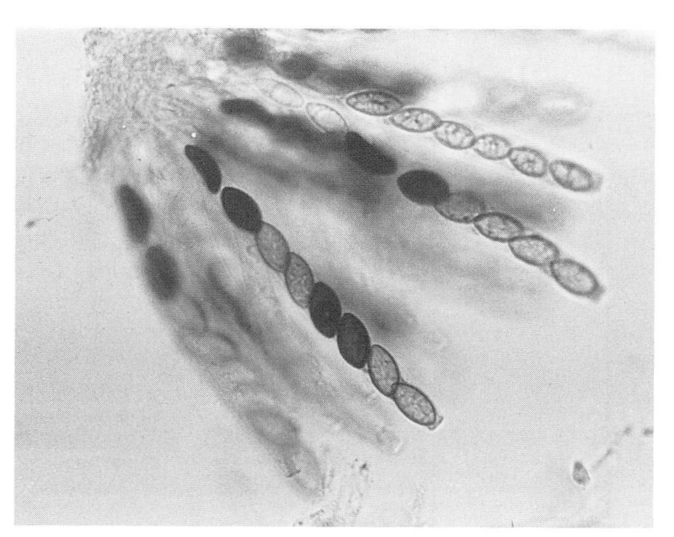

Figure 8.8 Photomicrograph showing the segregation of dark and light ascospores in *Neurospora*.

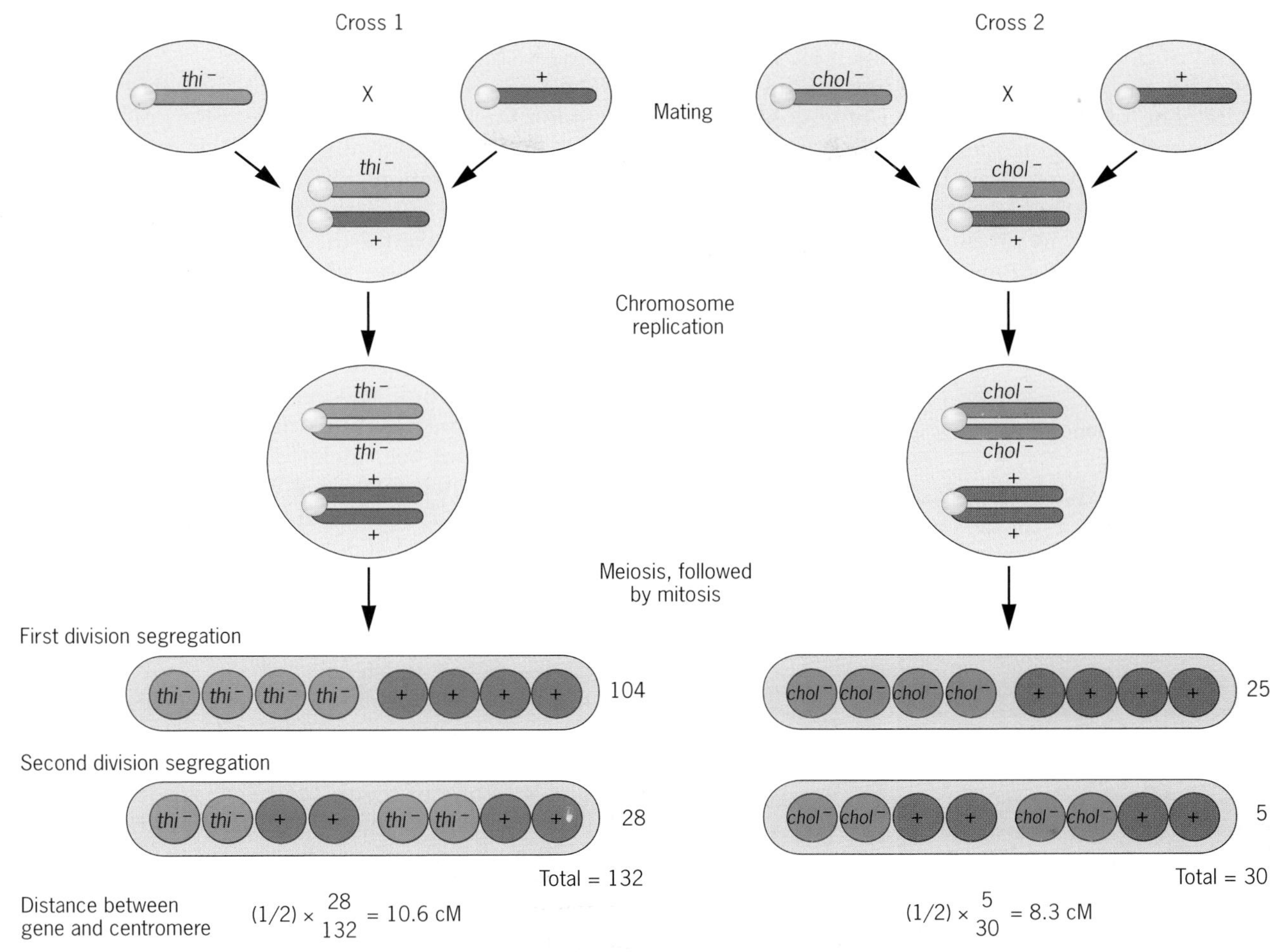

Figure 8.9 Centromere mapping in *Neurospora*. In Cross 1, the mutant strain, *thi*⁻, requires thiamin; in Cross 2, the mutant strain, *chol*⁻, requires choline.

centromere, their frequency allows us to estimate the distance between these two loci. The gene-to-centromere distance is simply (1/2) × frequency of second division segregation asci. The factor 1/2 enters in this expression because only half the chromatids in a second division segregation tetrad have crossed over between the gene and the centromere.

As an example of ordered tetrad analysis, let us consider data from two crosses performed with different strains of *Neurospora* by Mary Houlahan, George Beadle, and Hermione Calhoun (Figure 8.9). These researchers mapped a large number of *Neurospora* genes. In one experiment, they crossed a mutant strain (thi^-) that could not grow in the absence of the vitamin thiamin with a wild-type strain (thi^+); 132 asci were analyzed for first and second division segregation patterns. From the results, Houlahan, Beadle, and Calhoun inferred that the *thi* gene was (1/2)(28/132) = 10.6 cM from its centromere. In another experiment, they crossed a mutant strain ($chol^-$) that could not grow in the absence of choline with a wild-type strain ($chol^+$). From the 30 asci that were analyzed, they estimated the gene-to-centromere distance to be (1/2)(5/30) = 8.3 cM. Of course, these results do not tell us if the *thi* and *chol* genes are linked to the same centromere or to different ones. A cross between the two mutant strains would be needed to test for such linkage.

Cytogenetic Mapping with Deletions and Duplications in *Drosophila*

Recombination mapping allows us to determine the relative positions of genes by using the frequency of crossing over as a measure of distance. However, it does not allow us to localize genes with respect to cytological landmarks, such as bands, on chromosomes. This kind of localization requires a different procedure based on the phenotypic effects of combining a recessive mutation with a cytologically recognizable deletion or duplication.

This approach has been most thoroughly developed in *Drosophila* genetics, where the large, banded polytene chromosomes make it possible to define deletions and duplications accurately. As an example, let us consider the cytogenetic mapping of the X-linked *white* gene of *Drosophila*, a wild-type copy of which is required for pigmentation in the eyes. This gene is situated at map position 1.5 near one end of the X chromosome. But which of the two ends is it near, and how far is it, in physical terms, from that end? To answer these questions, we need to find the position of the *white* gene on the cytological map of the polytene X chromosome.

w/Df Genotype
Phenotype
w
white eyes
Df removes the w+ gene.
w
w+
red eyes
Df does not remove the w+ gene.

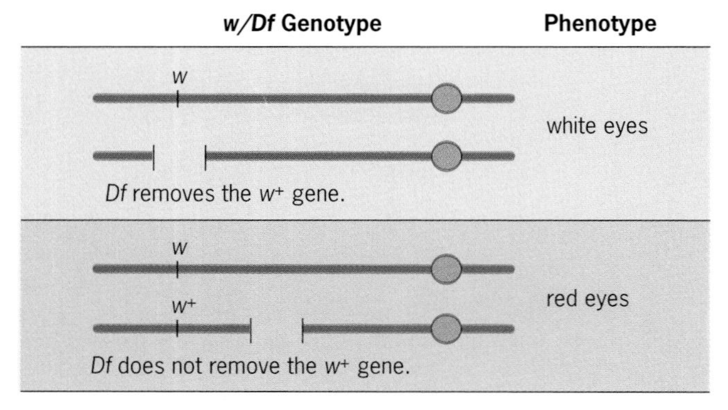

Figure 8.10 Principles of deletion mapping to localize a gene within a *Drosophila*. chromosome. The *white* gene on the X chromosome, defined by the recessive mutation *w* which causes white eyes, is used as an example.

The procedure is to produce flies that are heterozygous for a recessive null mutation of the *white* gene (*w*) and a cytologically defined deletion (or deficiency, usually symbolized *Df*) for part of the X chromosome (Figure 8.10). These *w/Df* heterozygotes provide a functional test for the location of *white* relative to the deficiency. If the *white* gene has been deleted from the *Df* chromosome, the *w/Df* heterozygotes will not be able to make eye pigment because they will not have a functional copy of the *white* gene on either of their X chromosomes. The eyes of the *w/Df* heterozygotes will therefore be white (mutant). If, however, the *white* gene has not been deleted from the *Df* chromosome, the *w/Df* heterozygotes will have a functional *white* gene somewhere on that chromosome, and their eyes will be red (wild type). By looking at the eyes of the *w/Df* heterozygotes, we can therefore determine whether or not a specific deficiency has deleted the *white* gene. If it has, *white* must be located within the boundaries of that deficiency.

Differerent X-chromosome deficiencies allowed researchers to locate the *white* gene (Figure 8.11). Each deficiency was combined with a recessive *white* mutation, but only one of them, $Df(1)w^{rJ1}$, produced white eyes. Thus we know that the *white* gene must be located within the chromosome segment that is missing in $Df(1)w^{rJ1}$, that is, somewhere between polytene

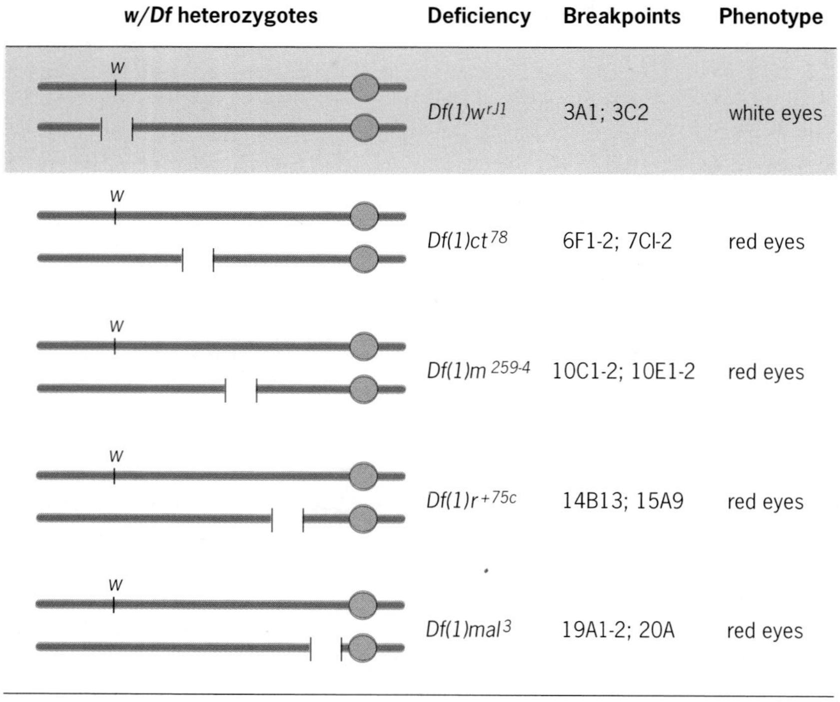

w/Df heterozygotes
Deficiency
Breakpoints
Phenotype
w
Df(1)wrJ1
3A1; 3C2
white eyes
w
Df(1)ct78
6F1-2; 7Cl-2
red eyes
w
Df(1)m259-4
10C1-2; 10E1-2
red eyes
w
Df(1)r+75c
14B13; 15A9
red eyes
w
Df(1)mal3
19A1-2; 20A
red eyes

The mutant eye color observed with $Df(1)w^{rJ1}$ indicates that the *white* gene is between the deficiency breakpoints in bands 3A1 and 3C2 on the X chromosome.

Figure 8.11 Localization of the *white* gene in the *Drosophila* X chromosome by deletion mapping.

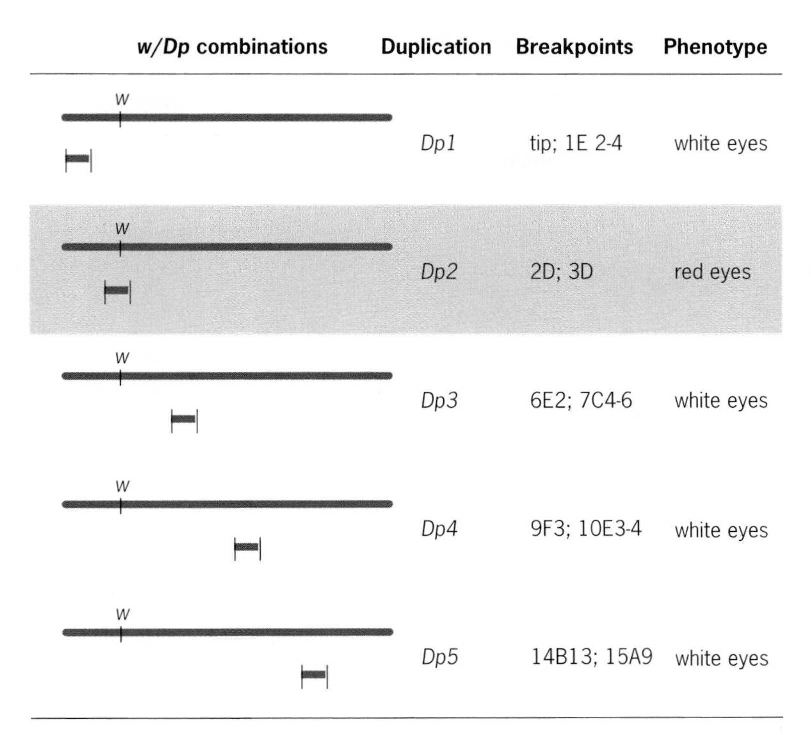

The wild-type eye color observed with *Dp2* indicates that the *white* gene is between the duplication breakpoints in regions 2D and 3D on the X chromosome.

Figure 8.12 Localization of the *white* gene in the *Drosophila* X chromosome by duplication mapping.

chromosome bands 3A1 to 3C2. With smaller deficiencies, the *white* gene has actually been localized to polytene chromosome band 3C2, near the right boundary of $Df(1)w^{rJ1}$.

Many other X-linked genes have been localized in this fashion. When we compare the resulting cytological map with the genetic map (see Figure 7.22), we see that the mapped sites are colinear but not identically spaced. Thus crossing over does not occur uniformly along the chromosome. Rather, it takes place more frequently in the middle than at the two ends.

We can also use duplications to determine the cytological locations of genes. The procedure is similar to that using deletions, except that we look for a duplication that masks the phenotype of a recessive mutation. Figure 8.12 shows an example utilizing duplications for small segments of the X chromosome. Only one of these duplications, *Dp2*, masked the *white* mutation; thus a wild-type copy of *white* must be present within it. This localizes the *white* gene somewhere between sections 2D and 3D on the polytene X chromosome, which is consistent with the results of the deletion tests discussed above.

Deletions and duplications have been extraordinarily useful in locating genes on the cytological maps of *Drosophila* chromosomes. The basic principle in **deletion mapping** is that a deletion that *uncovers* a recessive mutation must lack a wild-type copy of the mutant gene. This fact localizes that gene within the boundaries of the deletion. The basic principle in **duplication mapping** is that a duplication that *covers* a recessive mutation must contain a wild-type copy of the mutant gene. This fact localizes that gene within the boundaries of the duplication.

Burke Judd, Margaret Shen, and Thomas Kaufman used these principles to map all the genes in a small region near the left end of the *Drosophila* X chromosome (see Technical Sidelight: The Relationship Between Genes and Bands in the *Drosophila* Polytene X Chromosome). In fact, this work was so thorough that nearly all the genes could be localized to individual bands on the cytological map, suggesting a one gene–one band relationship. From subsequent work, we now know that this is an oversimplification; some bands in the polytene chromosomes clearly contain more than one gene. However, because the average number of genes per band is probably not much more than two, the total number of genes in the entire genome is somewhere between one and two times the total number of bands. The band number is around 5000; thus *Drosophila* probably has 5000 to 10,000 genes in its genome.

Key Points: **In organisms with ordered tetrads, the distance between a gene and its centromere is one-half the frequency of tetrads that show second division segregation. In *Drosophila*, genes can be localized on the polytene chromosome maps by testing**

TECHNICAL SIDELIGHT

The Relationship Between Genes and Bands in the *Drosophila* Polytene X Chromosome

Since their discovery by Theophilus Painter, the polytene chromosomes of *Drosophila* have afforded an unparalleled opportunity to study the fine details of chromosome organization. One idea that emerged from these studies is that polytene chromosome bands might correspond to individual genes. Over the years, this idea has been investigated by many researchers, but none have been as thorough as Burke Judd, Margaret Shen, and Thomas Kaufman. These researchers focused their attention on a small region near the left end of the X chromosome which was known to contain two genes that affect eye color, *zeste* (*z*) and *white* (*w*) (Figure 1*a*). Because these genes appeared to bracket the region, the segment between them has come to be called the *zeste-white* region. Cytological analysis had indicated that this region contained 15 polytene chromosome bands. Did this imply that it contained 15 genes? Judd, Shen, and Kaufman attempted to answer this question by collecting mutations in the *zeste-white* region and then mapping each of them cytologically.

The mutations were collected by screening for unusual phenotypes in flies that were heterozygous for an X chromosome that had been treated with a mutation-inducing agent, such as X rays, and an X chromosome that was missing the *zeste-white* region. The missing region was the result of a deletion called $Df(1)w^{rJ1}$. These flies therefore had the genotype

Mutagenized X chromosome: ——————*——————
Deletion X chromosome: ————┤ ├————
$Df(1)w^{rJ1}$

where the gap represents the deletion and the asterisk represents a newly induced recessive mutation.

If the induced mutation occurred in a gene in the *zeste-white* region, it would be expressed phenotypically because the $Df(1)w^{rJ1}$ chromosome has no wild-type gene to "cover" it. In the parlance of *Drosophila* genetics, we would say that the mutation was "uncovered" by the deletion. However, if the mutation occurred in a gene outside the *zeste-white* region, it would not be expressed because the $Df(1)w^{rJ1}$ chromosome would carry a wild-type copy of such a gene.

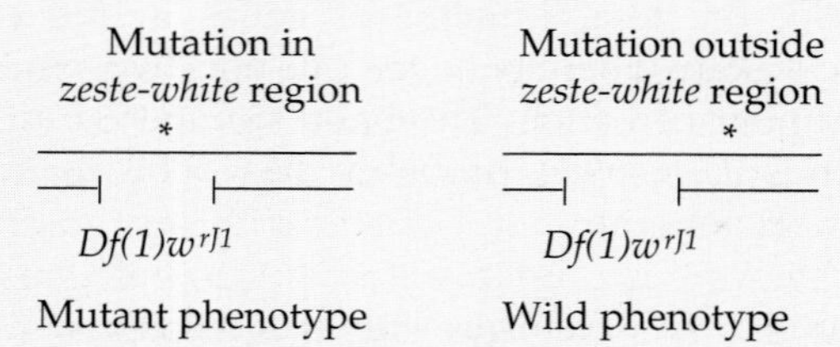

Judd, Shen, and Kaufman found many flies with a mutant phenotype, indicating that a mutation had occurred in the *zeste-white* region. Most of these mutations were reces-

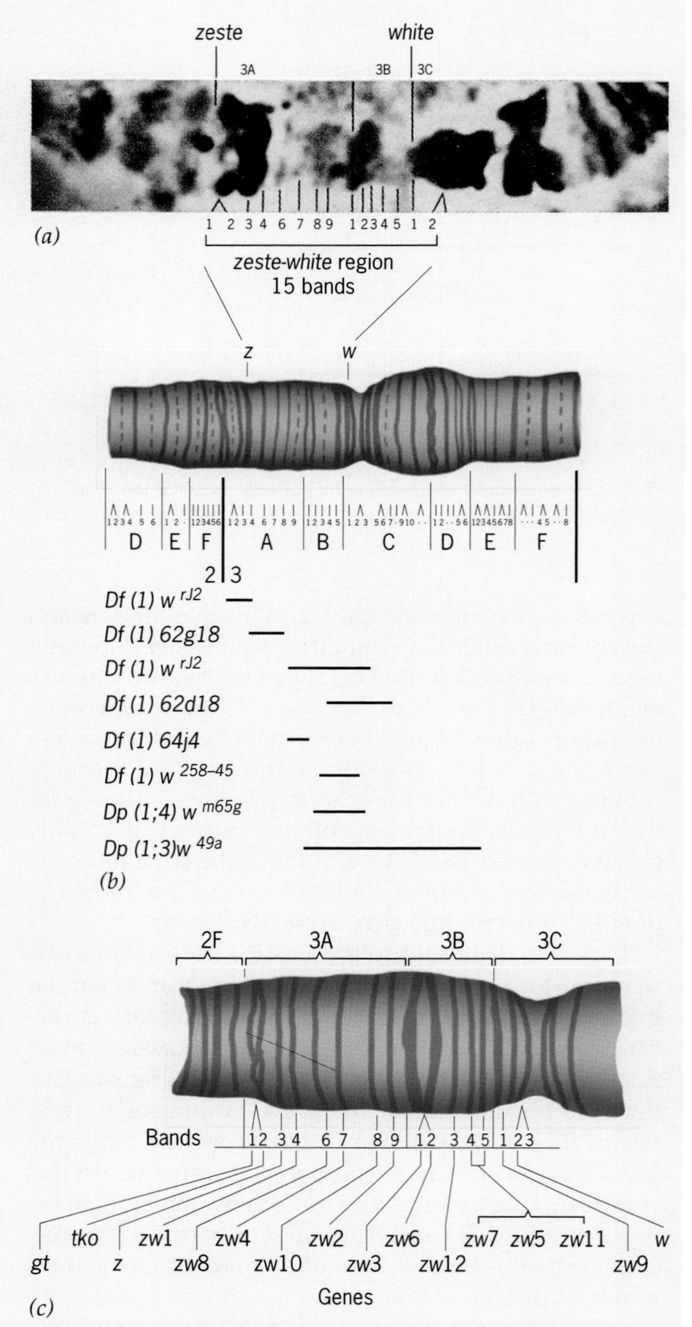

Figure 1 (*a*) Photomicrograph of the *zeste-white* region of the polytene X chromosome. (*b*) Some deletions and duplications that were used to localize mutations in the *zeste-white* region. (*c*) Cytogenetic map of the *zeste-white* region showing the locations of 16 different genes.

sive lethals, manifested by the failure of the */*Df(1)*w^{rJ1} flies to appear in the culture. Fortunately, these lethals could be saved for further analysis by putting them over a deletion-free, balancer X chromosome.

Judd, Shen, and Kaufman then tested each of the mutations they collected with a battery of cytologically defined deletions and duplications in and around the *zeste-white* region (Figure 1*b*). If a mutation was uncovered by a deletion or covered by a duplication, it could be localized within the deleted or duplicated segment. For example, let's consider tests with a mutation, *m*, and three different deletions, *Df(1)62g18*, *Df(1)64j4*, and *Df(1)62d18*:

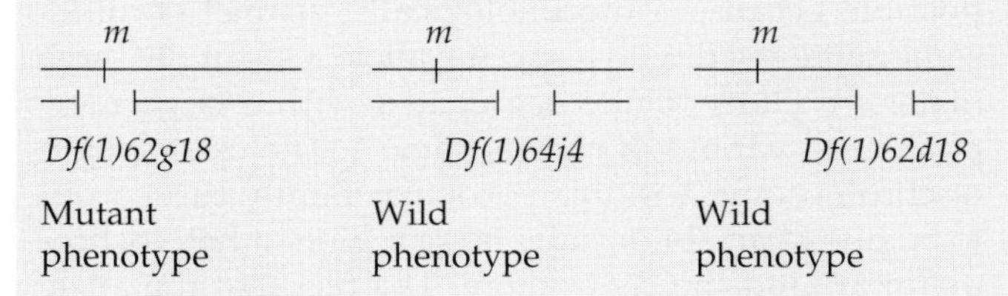

Df(1)62g18 uncovers the mutation, but the other two deletions do not. This tells us that the mutation resides in the region that is missing in *Df(1)62g18*, a segment that consists of four bands. By such analysis, Judd, Shen, and Kaufman were eventually able to localize most of the mutations they collected to individual bands. In the end, they were able to sort these mutations into 16 different genes, one more than the number of bands (Figure 1*c*). Thirteen of the genes each unambiguously corresponded to a single band. The other three, which were mapped to a pair of bands next to each other, could not be localized precisely because of technical limitations. Judd, Shen, and Kaufman therefore showed that the one gene–one band idea was *almost* correct.

Subsequent genetic studies have uncovered at least four more genes in the *zeste-white* region, bringing the number of known genes to 20. At the same time, electron microscopy has revealed that this region contains several more bands, perhaps a total of 23. The relationship between genes and bands is therefore still unclear. However, as Judd, Shen, and Kaufman suggested, it is probably not far from one to one.

recessive mutations against cytologically defined deletions and duplications. A deletion will reveal the phenotype of a recessive mutation located between its endpoints, whereas a duplication will conceal the mutant phenotype.

LINKAGE ANALYSIS IN HUMANS

Mapping human genes is critically important for a number of reasons. Knowledge of genes and their location gives us insight into the evolutionary relationship of humans to other primates and other vertebrate species. Equally important, mapping disease genes creates an opportunity for researchers to isolate the gene and understand how it causes a disease. In case after case, researchers have mapped a gene to a specific chromosome, and then proceeded to isolate the gene for further study. Duchenne muscular dystrophy, cystic fibrosis, Huntington's disease, and breast cancer are examples of the success of mapping disease genes to specific regions of specific chromosomes and subsequently isolating those genes.

Until recently, mapping genes in humans has been difficult. It was complicated by the fact that it is not possible to make the controlled crosses common with *Drosophila* or *Neurospora*. Although informative matings that help to locate genes do sometimes occur by chance, even in these instances, the number of offspring produced is usually too small to be of much use. With *Drosophila*, *Neurospora*, and other laboratory species, researchers set up controlled crosses and examine hundreds of offspring in a relatively short period of time. However, in spite of the limitations, there are several ways to analyze linkage in humans. The oldest and most direct way to ascertain linkage relationships in humans is the analysis of pedigrees. By itself, this is a cumbersome strategy. However, when pedigree analysis is coupled with modern molecular and somatic cell genetic techniques, researchers acquire important insights into the arrangement of genes on human chromosomes.

Detection of Linked Loci by Pedigree Analysis

Linkage relationships are most easily determined for X-linked genes because all genes that follow an X-linked inheritance pattern are obviously on the same chromosome (see Human Genetics Sidelight in Chapter 7). Determining linkage for autosomal genes is much more difficult because there are 22 pairs of autosomes. Sometimes, however, a pedigree following the inheritance of two autosomal traits shows clear evidence for linkage. For example, in 1955 J. H. Renwick and S. D. Lawler suggested that the gene for nail-patella syndrome (*NPS1*) is linked to the *ABO* blood group locus.

The nail-patella syndrome is a rare autosomal dominant trait that causes malformed nails and kneecaps, and often elbow and kidney abnormalities. A pedigree of this trait (Figure 8.13*a*) also records ABO blood group genotypes. The question we ask is whether the pedigree supports the hypothesis that the *ABO* locus is linked to the nail-patella locus, or if alleles at the two loci are assorting independently. If we examine the pedigree carefully, we note that, with one exception (II-8), each person with the nail-patella syndrome also has blood type B. This finding strongly suggests that the two loci are linked. A chi-square analysis of the results also supports the hypothesis that the two loci are linked.

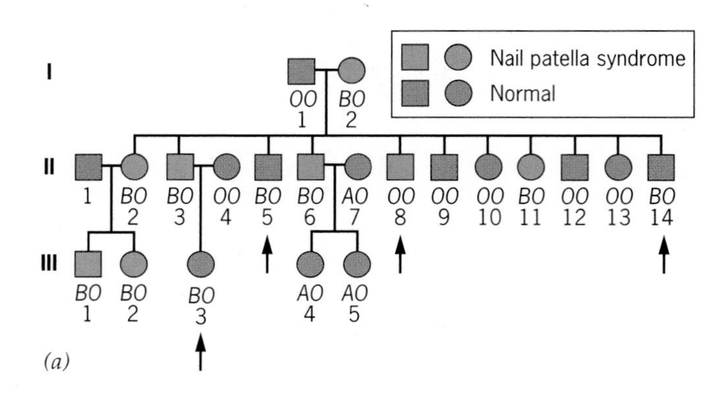

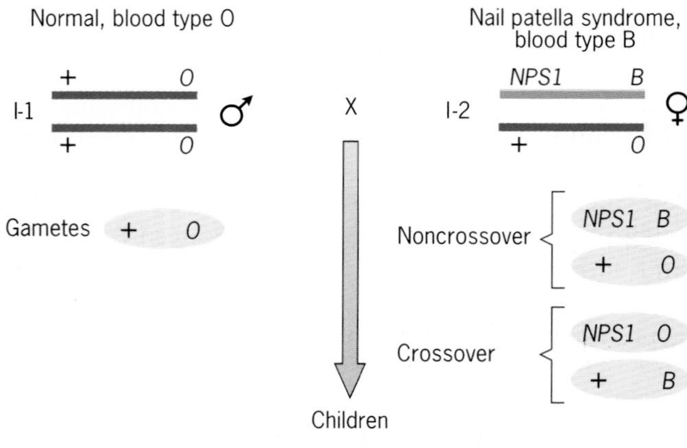

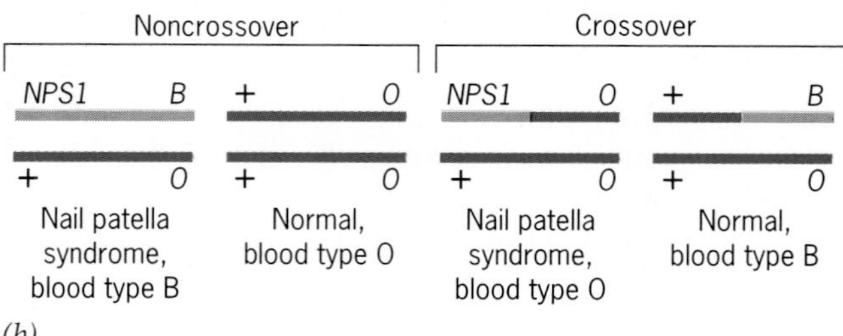

Figure 8.13 (*a*) A pedigree for the nail-patella syndrome (NPS1), with alleles of the ABO blood type locus noted. An arrow indicates an individual produced by a recombination event. (*b*) A genetic interpretation of the pedigree.

A genetic interpretation of the pedigree (Figure 8.13*b*) shows that individuals II-5, II-8, II-14, and III-3 (see arrows) resulted from crossing over in the grandmother (I-2) or the father (II-3). Of the 16 offspring in this pedigree, 4 show recombination by crossing over. Although the number of offspring is small, these data suggest that the two loci are linked, with about 25 map units (4/16) separating them. More recent studies have confirmed this linkage and refined the distance between them. The two loci are now known to be near the tip of chromosome 9, separated by about 10 map units.

Although the original study of *NPS1* and the *ABO* loci established linkage, it could not identify the specific autosome carrying the two loci. The first gene assigned to a specific autosome was the Duffy blood group locus (*FY*) in 1968. This discovery was made possible because a morphologically distinct chromosome segregated with a specific allele. One of the three major *FY* Duffy alleles segregated with a large morphological variant of chromosome 1. The extra length of chromosome 1 in this particular family turned out to be due to an abnormally large region of heterochromatin around the centromere. The fact that the allele and the morphologically distinct chromosome were always associated with each other suggested that the *FY* gene was on chromosome 1.

The Lod Score Method for Assessing Linkage It is often difficult to say with absolute certainty that two loci are linked on the same chromosome. We could have suggested, for example, that the *NPS1* and *ABO* loci assort independently, even though it appears more likely that they are linked. When we cannot determine linkage with absolute certainty, we must employ indirect or statistical methods to assess possible linkage relationships. These methods allow us to evaluate pedigrees where two or more traits are segregating from two perspectives:

1. The loci are linked by a certain number of map units.
2. The loci are unlinked.

By comparing the probability that two genes are linked by a certain number of map units with the probability of independent assortment, we can attain the most likely interpretation of the pedigree.

We can examine the possible linkage between two loci by using a technique known as the **lod score method**, an evaluative process developed in the 1950s by N. E. Morton. Using the lod score method (lod for log of odds), we can examine a pedigree by following the inheritance of alleles at two or more loci. The Huntington's disease (HD) pedigree in Figure 8.14*a* illustrates the technique. Each individual in the pedigree is classed according to his or her HD status (*H* =

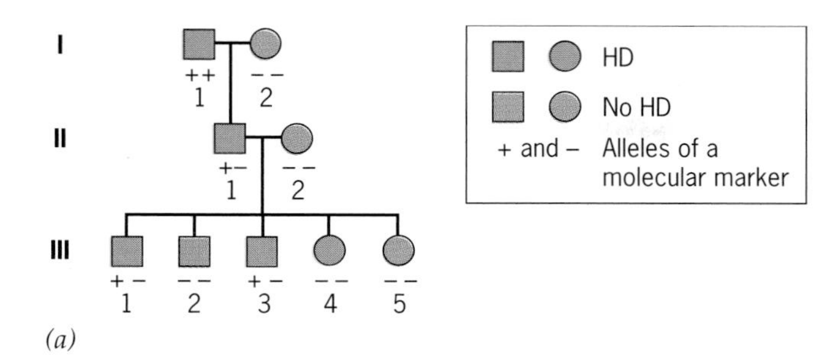

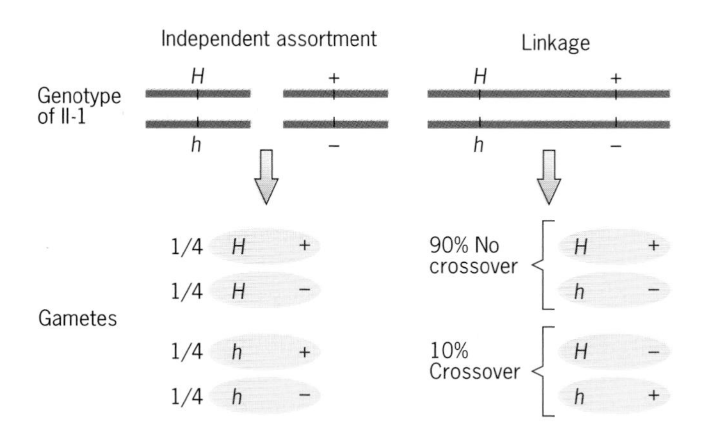

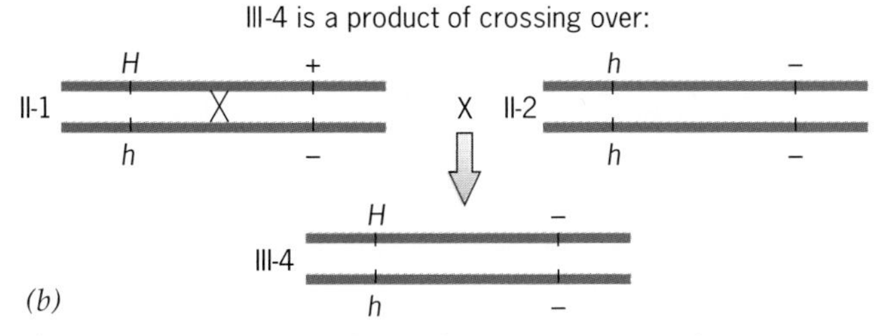

Figure 8.14 (*a*) A pedigree for Huntington's disease (HD), with alleles of a molecular marker shown. (*b*) Two interpretations of the pedigree. On the left is a model for independent assortment; on the right is a model for linkage with 10 percent crossing over. According to the linkage model, individual III-4 is a product of recombination.

Huntington's disease allele; h = the normal allele), and also for another marker, in this case, a molecular marker with two allelic forms: + and − (the "+" and "-" represent detectable DNA sequence variation). We then ask: What is the probability of obtaining this pedigree if the two loci are linked and separated by a specific amount of recombination (say 10 percent), compared with the probability that the two loci are assorting independently?

To answer this question, we set up a series of ratios, where the numerator is the probability of getting the pedigree for linked loci separated by "X" percent recombination ("X" map units) and the denominator is the probability of getting the pedigree if the two loci are independently assorting (50 percent recombination). First, we must determine the probability of the observed results given that the two loci are not linked. The father (II-1) is heterozygous at both loci (H/h and +/-); the mother (II-2) is homozygous at both loci (h/h and -/-). These parents thus represent a classic testcross. If the genes are not linked, the father will produce four classes of gametes in equal numbers and the mother one class:

Father: $1/4\,H, +$ $\quad 1/4\,H, -$ $\quad 1/4\,h, +$ $\quad 1/4\,h, -$

Mother: $h, -$

There are five children in the pedigree, and the probability of the genotype of each is 1/4. Thus the probability that these five children will be produced by these two parents is

$$(0.25)^5 = 0.000977$$

Next, we examine the same pedigree from the point of view that these two loci are linked and separated by 10 map units (10 percent recombination). We must first determine how the alleles are arranged on the two chromosomes. Individual I-1 is heterozygous for the HD alleles but homozygous for the molecular marker alleles (++). His wife, I-2, is homozygous for both alleles. They have a child (II-1) who later develops HD. II-1 is heterozygous for the two molecular alleles. I-2 had to pass a "-" and an "h" allele to her son. The father who passed HD to II-2 also had to pass the "+" allele to him. Thus the gene arrangement in II-2 must be

$$\frac{H \quad + \Leftarrow \text{from I-1}}{h \quad - \Leftarrow \text{from I-2}}$$

(Sometimes we are not able to determine the arrangement of the alleles on the chromosomes and must consider both possible arrangements; that is not the case here.) Given this gene arrangement, all the children except III-4 are nonrecombinant. III-4 is the product of crossing over between the two loci.

Since our hypothesis assumes that 10 map units separate the two loci, the two noncrossover gamete classes would account for 90 percent of the total, and the two crossover gamete classes for 10 percent. Each of the two noncrossover classes has a probability of 0.45, ($1/2 \times 0.90$), and each crossover class has a probability of 0.05, ($1/2 \times 0.10$). The probability of this pedigree, assuming 10 percent recombination, is therefore

$$(0.45)^4\,(0.05) = 0.002050$$

Our final step is to calculate the ratio. The probability of obtaining the pedigree given 10 percent recombination over the probability of obtaining the pedigree given independent assortment is

$$0.002050/0.000977 = 2.0985$$

TABLE 8.1
Lod Values for the Pedigree in Figure 8.14 Using Different Recombination Frequencies for *HD* and the Molecular Marker[a]

	Recombination Frequency					
	0.05	*0.10*	*0.20*	*0.30*	*0.40*	*0.50*
Odds (linked/ unlinked)	1.32	2.10	2.63	2.29	1.66	1.00
Log of odds (lod)	0.12	0.32	0.42	0.36	0.22	0.00

[a]The maximum lod value in this pedigree is obtained for a recombination frequency of about 20 percent.

If we calculate similarly for a variety of recombination values, we generate numbers like those seen in the first row of Table 8.1.

If the ratio of linked to unlinked is greater than 1.0—that is, if the probability of linkage exceeds the probability of independent assortment—there is an indication of linkage. If the ratio is less than 1.0, we have an indication of independent assortment. The maximum ratio we obtain points to the most likely explanation for the data. The data presented in Table 8.1 suggest that the most likely model for the two loci is one where they are linked by about 20 map units. Of course, this is only one family with five informative offspring. We would require additional families and more progeny before we can accept the linkage hypothesis.

This type of procedure can be cumbersome for the analysis of several pedigrees. However, we can simplify it by converting the ratio to a log value or lod score (remember that lod is a log of the odds). The log of 2.0985 is 0.32 (10 raised to the 0.32 power equals 2.0985). A series of lod values for different map distances is shown in Table 8.1. Lod values greater than 0 indicate linkage; negative lod scores indicate independent assortment. Again, the maximum lod value indicates the most likely model. For the pedigree shown, the lod score maximizes between 20 and 30 percent recombination; this indicates that the two loci in Figure 8-15 are most likely linked by 20 to 30 map units.

The value of using logs in these calculations is that we can simply sum the results from several individual pedigrees. If the sum of lod scores from several pedigrees totals 3.0 or more, we can conclude that the loci are linked. A lod score of 3 indicates that the data are 1000 times more likely under a model of linkage than under a model of independent assortment. A lod score of 2 is considered strong evidence for linkage.

The lod score method is a powerful tool for determining linkage relationships and can even indicate genetic heterogeneity—that is, more than one locus producing a particular trait (Figure 8.15). For example,

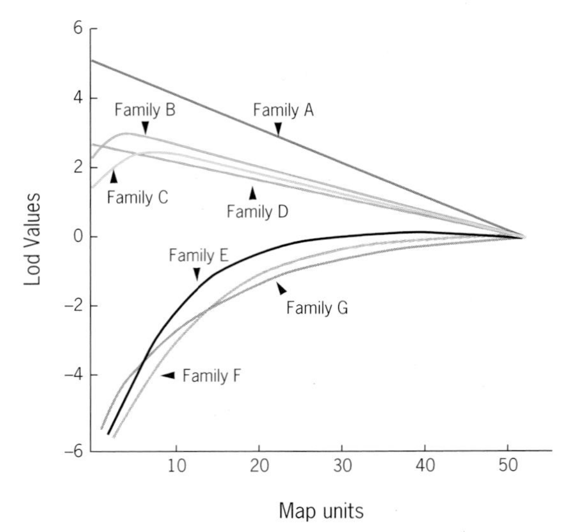

Figure 8.15 Plots of lod score values as a function of map units for seven families. The two loci being evaluated are the Rh factor and Elliptocytosis loci, both autosomal dominants. In families A, B, C, and D, the two genes are tightly linked with a maximum lod score value near 0 map units. In families E, F, and G, the two loci appear to be assorting independently with the lod score maximizing at 50 map units.

when studying seven families for the possible linkage of two loci (the *Rh* blood factor locus and elliptocytosis, or *EL*, a dominant red blood cell disorder), N. E. Morton in 1956 found that in four of the families the *Rh* and *EL* genes were tightly linked (with 0 map units the most likely model). In the other three families, the *Rh* and *EL* genes assorted independently. (Maximum lod scores were obtained when the map distance was set at 50; 50 percent recombination is the same as independent assortment.) Morton's findings suggest that there are two loci for elliptocytosis: one (*EL1*) is tightly linked to *Rh* on chromosome 1, and the other assorts independently of *Rh*. The second *EL* locus (*EL2*) was later discovered to be on chromosome 14.

By itself, the lod score method is of limited value because it does not provide information about specific autosomal locations. In conjunction with other techniques, however, it is a powerful tool for the construction of linkage maps.

Somatic Cell Genetics: An Alternative Approach to Gene Mapping

We can often tell from the analysis of pedigrees whether genes are sex-linked or autosomal. Through the application of direct or indirect methods, we can also usually ascertain whether or not particular genes are linked to each other. However, mapping genes concerns far more than these two factors. On *which* specific autosome is a particular gene located? *Where* on that chromosome is the gene?

In *Drosophila* and other experimental organisms, we are able to carry out carefully designed crosses that provide answers to these questions, as we have discussed in this and the previous chapter. Until recently, however, genetic analysis in humans has been difficult. Human gene mapping today relies on some very special techniques.

Cell Hybridization In the early 1960s, G. Barski and B. Ephrussi developed techniques for fusing different cells either from the same species or different species. The fused cells are called hybrids. These **cell hybridization** or **cell fusion** techniques have had a profound impact on human gene mapping. To form interspecific cell hybrids, human cells and cells from another species, usually a rodent, are mixed together in culture with a fusing agent, such as inactivated Sendai viruses or polyethylene glycol. The fusing agent first causes fusion of cell membranes followed by fusion of nuclear membranes (Figure 8.16). Then, as the hybrid cells divide, human chromosomes are randomly lost; the hybrids retain all the rodent chromosomes. Eventually, after several rounds of division, the hybrid stabilizes, retaining only one or a very few human chromosomes. We do not yet understand the mechanism that causes human chromosomes to be lost and rodent chromosomes to be retained.

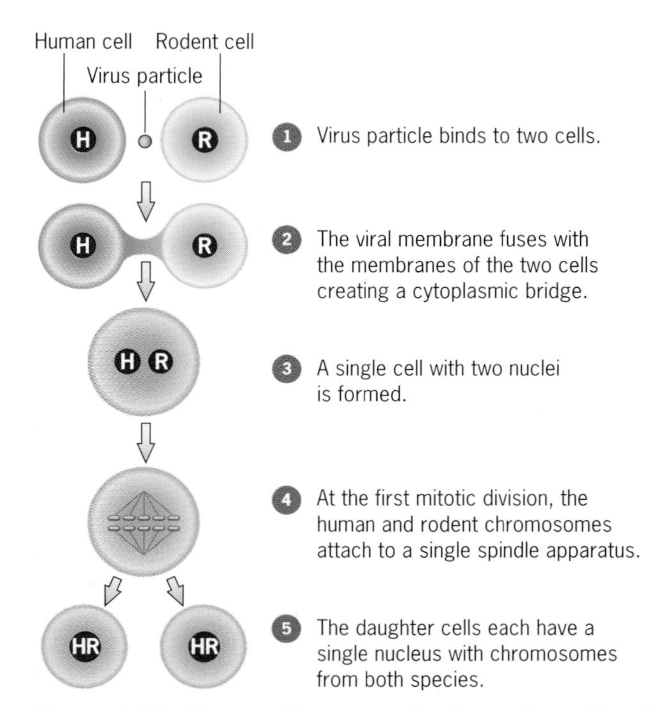

Figure 8.16 Fusion of human and rodent cells mediated by an inactivated Sendai virus.

Even under ideal research conditions, hybrid cells are not produced in large numbers. Some cells do not fuse, and some fusions involve same-species cells (mouse–mouse or human–human). To eliminate all but the true hybrids, the cells are placed on a medium that selects against all but the hybrid cells. The most commonly used selective medium is called the **HAT medium** (**h**ypoxanthine-**a**minopterin-**t**hymidine). In HAT medium, if one parental cell type is deficient for the enzyme thymidine kinase (TK^-) and the other parental cell type is deficient for the enzyme hypoxanthine phosphoribosyl transferase ($HPRT^-$), the parental cell types will not grow. Thus HAT medium selects against all cells that are *not* hybrid. The aminopterin in the medium blocks the synthesis of purines and thymidylate and forces the cells to use a "salvage pathway" requiring the use of the HPRT and TK enzymes (Figure 8.17*a*). Without functional copies of both enzymes, the parental cell types die, as do the mouse–mouse and human–human fused cells.

Hybrid cells that are $HPRT^+$ and TK^+ are able to use hypoxanthine and thymidine will survive on HAT because the medium contains both compounds (Figure 8.17*b*). The hybrid cells survive as long as they retain the human copy of the TK^+ allele. (The mouse chromosomes carry the $HPRT^+$ allele.) Once hybrid cells are selected, they are set up in clones or cell lines. Each cell line carries a sample of human chromosomes. The cell lines can be assayed for gene products, and the presence or absence of a specific gene product is correlated with the presence or absence of a specific human chromosome.

In 1971, using somatic cell hybridization, O. J. Miller and his colleagues assigned the first gene to an autosome. TK^- mouse cells ($HPRT^+$) were fused to $HPRT^-$ human cells (TK^+). After selection on HAT medium, they analyzed the surviving hybrid cells ($HPRT^+$ TK^+) for the human chromosomes that were retained. (Human and mouse chromosomes are cytologically distinct from each other.) One hybrid line retained only one human chromosome, 17, and still synthesized TK. Thus the *TK* gene must be located on chromosome 17. In a reciprocal experiment where $HPRT^-$ mouse cells were fused to TK^- human cells, the surviving hybrid cells all carried the human X chromosome. Thus it was confirmed that the *HPRT* gene was on the X chromosome. (Pedigree analysis had already strongly indicated X-linkage for this locus.)

Somatic cell hybrids can be used to map almost any human gene. The only requirement is that the hybrid cells express the human gene product and furthermore that the product be detectable and distinguishable from the rodent product being expressed in

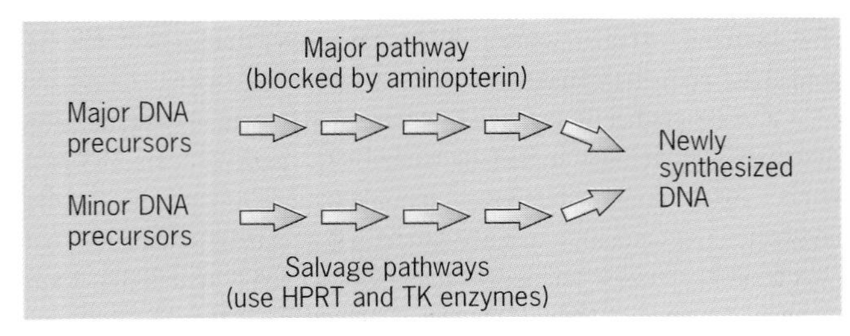

(a)

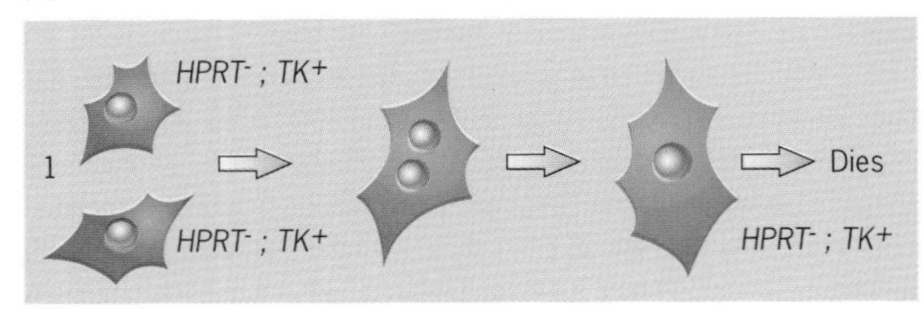

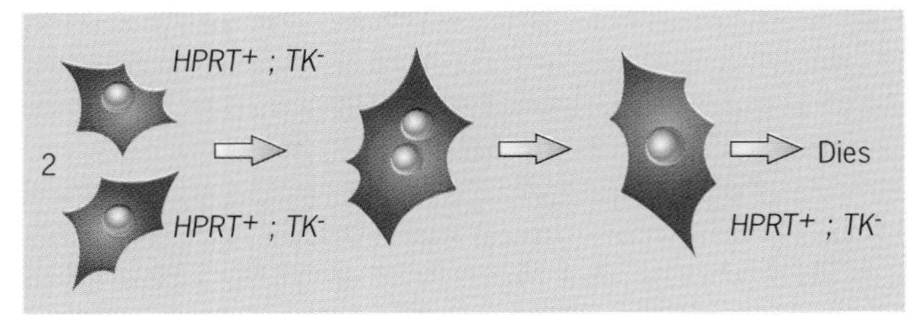

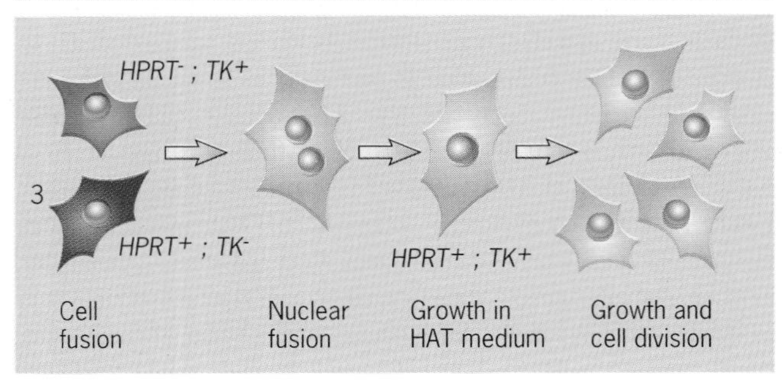

(b)

Figure 8.17 Selection of hybrid cells in HAT medium. Only cells that are *HPRT*⁺ and *TK*⁺ are able to grow in this medium. (*a*) Aminopterin in the HAT medium blocks the major DNA synthesis pathway. However, the medium supplies precursors (hypoxanthine and thymidine) that allow the wild-type cells to use a salvage pathway. (*b*) Parental cells and fused cells containing only one parental type do not survive on the HAT medium; the hybrid cells survive and divide.

the hybrids. For example, the gene for UMP kinase has been mapped to chromosome 1 using somatic cell hybridization techniques (Table 8.2). The enzyme is expressed and detectable in hybrid cells. In addition, mouse UMP kinase is distinct from human UMP kinase because of a few amino acid differences. These differences give the two forms of the enzyme different electrical properties that can be detected by **gel electrophoresis**, a technique that separates molecules in a gel in an electrical field. The human and mouse UMP kinases migrate to different regions of the gel, their location in the gel being determined by specific assays for UMP kinase.

TABLE 8.2

Assignment of the UMP-Kinase (*UMPK*) Gene to Chromosome 1 Using a Hybrid Cell Line Panel

Cell Line	*UMPK Activity*	*Human Chromosomes Present*[a]
WA-lla	−	2, 3, 4, 11, 12, 14, 16, X
JFA-14b	+	1, 2, 5, 8, 13, 16, 17, 18, X
WA-la	+	1, 2, 4, 7, 8, 10, 12, 13, 15, 16, 18, 19, 20, 21, 22, X
J-10-H-12	+	1, 3, 6, 7, 10, 11, 12, 14, 15, 16, 17, 18, 21, X
AIM-3a	−	2, 4, 5, 7, 8, 10, 11, 12, 13, 14, 17, 18, 20, X
AIM-8a	−	6, 7, 9, 11, 12, 13, 14, 15, 16, 17, 18, 19, 20, 21, 22, X
AIM-11a	−	2, 3, 7, 10, 12, X
AIM-23a	+	1, 2, 3, 4, 5, 7, 8, 10, 12, 14, 15, 16, 18, 21, 22, X

[a]Every time chromosome 1 is present, so is the enzyme; every time it is absent, so is the enzyme.

Until recently, somatic cell genetics could be used to map genes only if the hybrid cells expressed the genes. If a gene was not expressed in the hybrid cell, it could not be detected. After the development of molecular techniques, investigators could detect the genes themselves rather than their products. For example, the genes for α and β globin, the proteins that make up adult hemoglobin, are not expressed in hybrid cells. Nevertheless, we can detect the genes using DNA probes. We discuss these techniques in detail in Chapter 20.

Gene Mapping Using Chromosomal Rearrangements With somatic cell techniques, we can map a gene to a specific region of a chromosome if informative chromosome rearrangements are available. One type of rearrangement that is valuable in regionalizing a gene locus is a translocation (see Chapter 6). Let us consider an example in which hybrid cell lines were developed from a patient with a reciprocal translocation between the X chromosome and chromosome 14 (Figure 8.18). Most of the long arm of the X chromosome was translocated to the tip of chromosome 14, and a small piece of chromosome 14 was translocated to the X. Some of the hybrids that survived on HAT medium carried only the chromosome 14 with the long arm of the X attached. (The normal X, normal 14, and the X carrying the small piece of chromosome 14 were not present.) Since they survived on HAT they must be *HPRT*⁺, and since the *HPRT* gene is known to be X-linked, the *HPRT* gene must map to the long arm of the

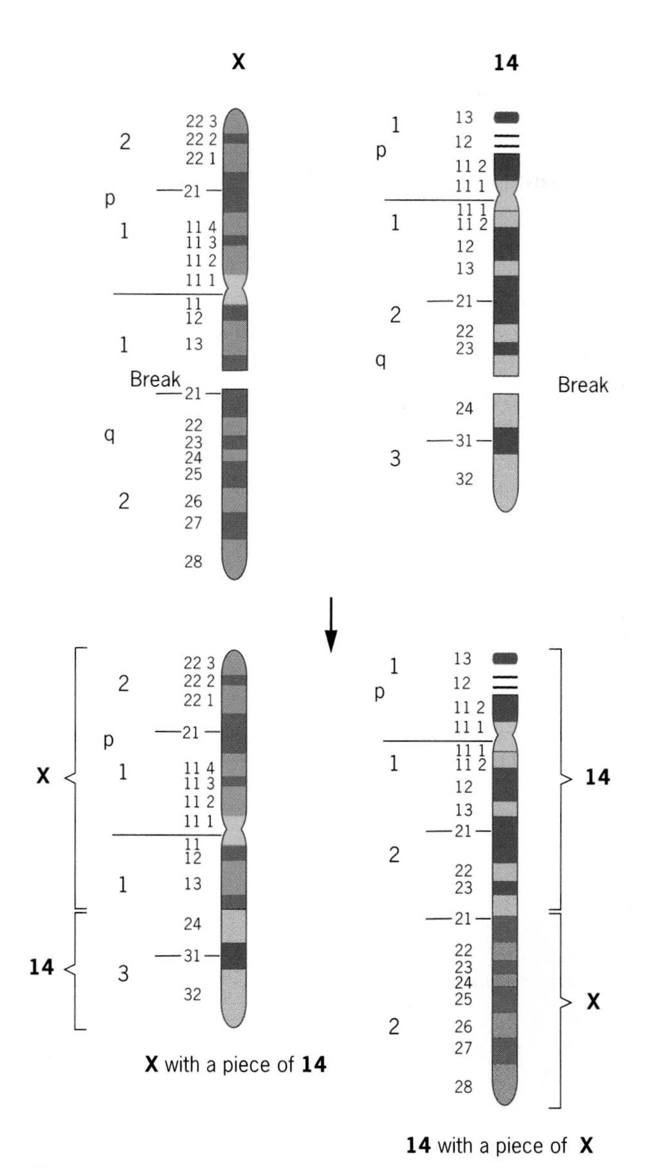

Figure 8.18 A reciprocal translocation between the X and chromosome 14. A translocation of this type established that the *HPRT* gene mapped to the long arm of the X chromosome.

X chromosome. Two other X-linked genes were also expressed in these hybrids: *PGK* (phosphoglycerate kinase) and *G6PD* (glucose-6-phosphate dehydrogenase). These loci must also map to the long arm of the X. In addition, this same cell line also expressed a human gene known only to be autosomal, *NP* (nucleoside phosphorylase). Since this hybrid line carried no other autosomes, the *NP* gene must be on the part of chromosome 14 remaining in the hybrid and not be located on the small piece of 14 that was translocated to the X.

By use of other X-autosome translocations, it was possible to establish the sequence of the X-linked genes *PGK*, *HPRT*, and *G6PD* on the q arm of the X (Figure 8.19). Because only translocated segments containing Xq13 expressed *PGK*, this locus must be close to the centromere. Only translocated segments carrying Xq26 expressed HPRT, and only segments carrying Xq28 expressed G6PD. Therefore the gene sequence must be centromere-*PGK-HPRT-G6PD*.

One of the most exciting stories in modern medical genetics is the discovery and eventual isolation of the X-linked gene causing **Duchenne muscular dystrophy (DMD)**. DMD is a fatal neuromuscular disease. Symptoms usually begin by age 6. The victim is chairbound by age 12 and dead by age 20. Mapping the *DMD* gene began with the discovery by R. Worton in 1984 that rare females with DMD carried an X-autosome balanced translocation. Usually DMD occurs only in males because it is X-linked recessive and males do not survive long enough to pass the defective allele on to their daughters. Why do these rare females develop DMD? One of a female's two X chromosomes is usually inactivated in cells shortly after development begins. This is a form of dosage compensation so that females who carry two X chromosomes produce about the same amount of X-linked gene products as males who have only one X. Because inactivation is a random process, in some cells the maternal X is inactivated and in other cells the paternal X is inactivated. However, in these rare DMD females with X-autosome translocations, inactivation is not random. The structurally intact, nontranslocated X is preferentially inactivated; thus the translocated X remains as the active chromosome (Figure 8.20). In all the DMD translocation females, the autosome involved in the translocation varies, but the breakpoint on the X is always in the same region: Xp21. The break

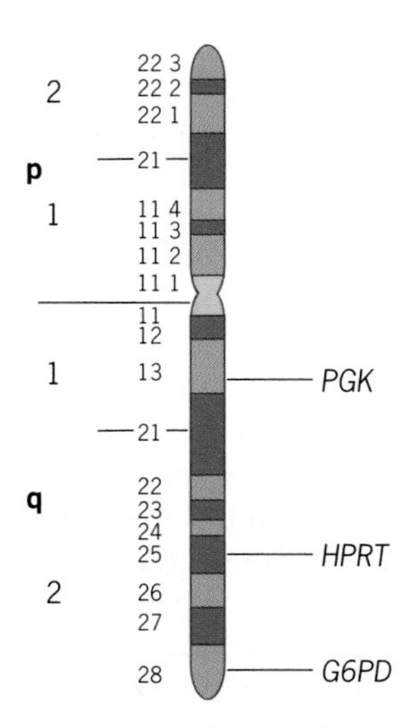

Figure 8.19 A schematic of the X chromosome showing the locations of the *PGK*, *HPRT*, and *G6PD* genes. These genes were physically mapped using translocations involving different X chromosome breakpoints.

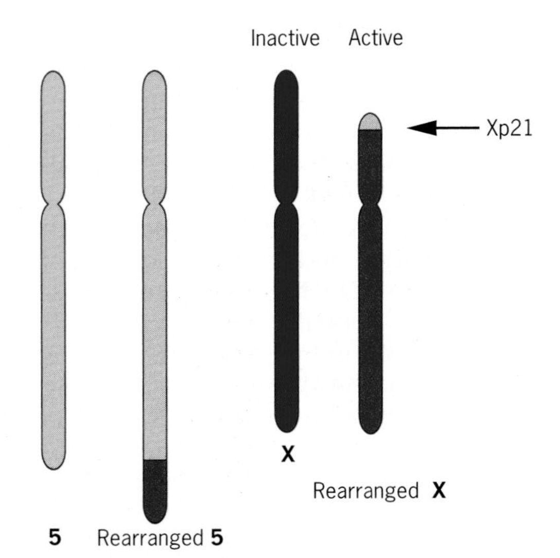

Figure 8.20 A reciprocal translocation between the X and chromosome 5 located the gene for Duchenne's muscular dystrophy to the p21 band of the X. The breakpoint at Xp21 broke the *DMD* gene. This female's normal X was inactivated, leaving her rearranged X with its disrupted *DMD* gene on the active chromosome. She expressed muscular dystrophy.

must have occurred in the *DMD* gene itself, thus destroying its function in these females. Because their normal X with its normal gene copy is inactivated, these females develop DMD.

Deletion Mapping Translocations are not the only type of chromosome abnormality that provides information about gene locations. Sometimes a segment of a chromosome and the genes it carries is simply deleted. If we are able to associate missing genes or gene products with a specific chromosome deletion, we can use this information to help us build a map. Let us consider the following case. A young boy with multiple rare congenital anomalies (Figure 8.21*a*) was suspected of carrying a chromosome abnormality. When his karyotype was prepared, it was discovered that he was missing the tip of chromosome 2 (breakpoint **A** in Figure 8.21*b*). Other studies already suggested that the gene for acid phosphatase (*ACP1*) mapped in this area of chromosome 2. Was the boy missing one of his two *ACP1* alleles? The *ACP1* gene has three distinct alleles: *A*, *B*, and *C*. Each allele encodes a form of *ACP1* that has slightly different electrical charges, and these differences are detectable using electrophoresis. Since the boy carried two alleles, *A* and *B*, we can conclude that the segment of the chromosome that was missing did not carry the *ACP1* locus. A second child also carrying a deletion of the tip of chromosome 2 was discovered (breakpoint **B** in Figure 8.21*b*), and this child carried only one allele for the *ACP1* gene. In this case, the deleted segment must have contained the *ACP1* locus. The breakpoints in both patients were different. Thus we can conclude that the *ACP1* gene lies between the two breakpoints.

A rare deletion event helped researchers confirm the location of the Duchenne muscular dystrophy locus (*DMD*) on the X chromosome. Recall that the *DMD* locus was known to be on the X chromosome because of the X-linked inheritance pattern, and the translocation DMD females suggested a location on the short arm of the X. In 1985 Ute Francke and her colleagues examined a young boy with several rare clinical conditions, all known to be X-linked recessive Mendelian traits: chronic granulomatous disease (CGD), an immune disorder associated with a deficiency of cytochrome b; the McLeod disorder, a red blood cell disorder; retinitis pigmentosa (RP), a vision disorder; and Duchenne muscular dystrophy. Careful analysis of his X chromosome revealed that he was missing a very small segment: Xp21. All four of these genes must be located in this segment of the X chromosome. Later that same year, Louis Kunkel and his colleagues in Boston and Ron Worton and his colleagues in Toronto, using DNA from this young boy

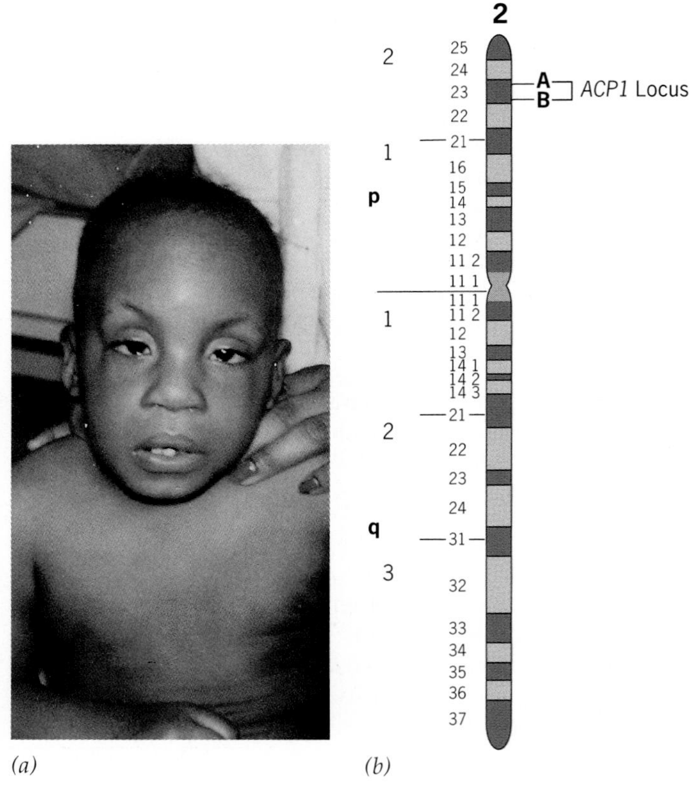

Figure 8.21 Genes can be located on chromosomes using deletions. (*a*) A young male who is missing the tip of the short arm of chromosome 2. (*b*) The breakpoints in two different patients with 2p deletions. The patient with the **A** breakpoint retains the *ACP1* gene; the patient with the **B** breakpoint has lost the *ACP1* gene. The *ACP1* gene must lie between the two breakpoints.

and from one of the translocation DMD females, isolated the *DMD* gene. We discuss this exciting achievement in detail in Chapter 20.

Duplication Mapping Chromosome rearrangements sometimes result in extra copies of genes. If we can determine how many copies of a gene are present and then correlate gene duplication events with a duplicated chromosome segment, we can assign a gene to that specific duplicated region of a chromosome.

We can see how **duplication mapping** works by looking at the methodology used to map the soluble (cytoplasmic) glutamic oxaloacetic transaminase (*GOTs*) gene.The *GOTs* gene was first assigned to the distal segment of the long arm of chromosome 10 by means of the somatic cell techniques we have already discussed (Figure 8.22*a*). Then, using hybrid cell lines with rearrangements that resulted in duplicated segments of chromosome 10, researchers were able to locate the gene to a specific region on the long arm of chromosome 10 (10q24).

Researchers used three different hybrid cell lines to map the *GOTs* locus. One cell line carried a translocation between chromosomes 10 and 17. A second cell line carried a translocation between chromosomes 10 and 21. The breakpoint on chromosome 10 was different in both cases (Figure 8.22*b*). The first cell line carried two copies of chromosome 10 and the small extra piece of chromosome 10 attached to chromosome 17; the second line carried two copies of chromosome 10 and the small extra piece of chromosome 10 attached to chromosome 21. A third cell line served as a control. It carried two copies each of chromosomes 10, 17, and 21. The researchers measured the GOTs enzyme activity in each line (Figure 8.23). The control cell line with

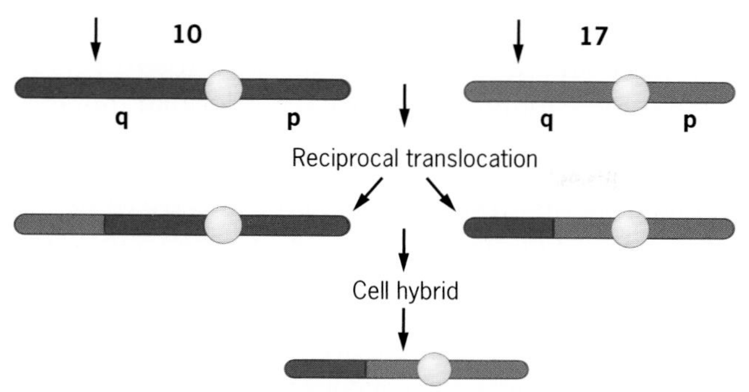

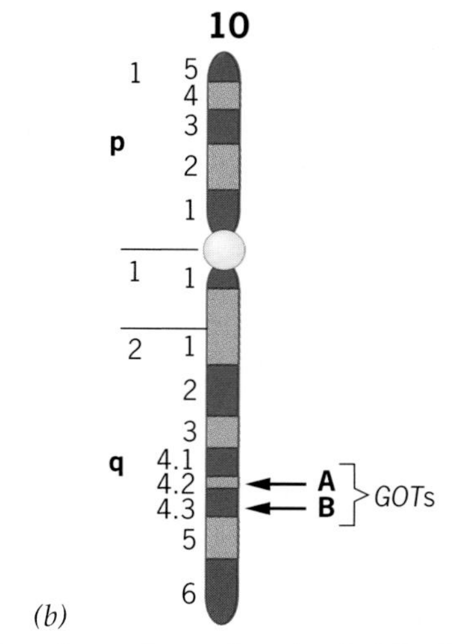

Figure 8.22 A reciprocal translocation between chromosomes 10 and 17. (*a*) Only one of the two translocation chromosomes carries the *GOTs* gene, indicating that this gene is near the tip of 10q. (*b*) Two different translocation breakpoints, **A** and **B**, in chromosome 10 were used in a duplication mapping experiment to locate the *GOTs* gene.

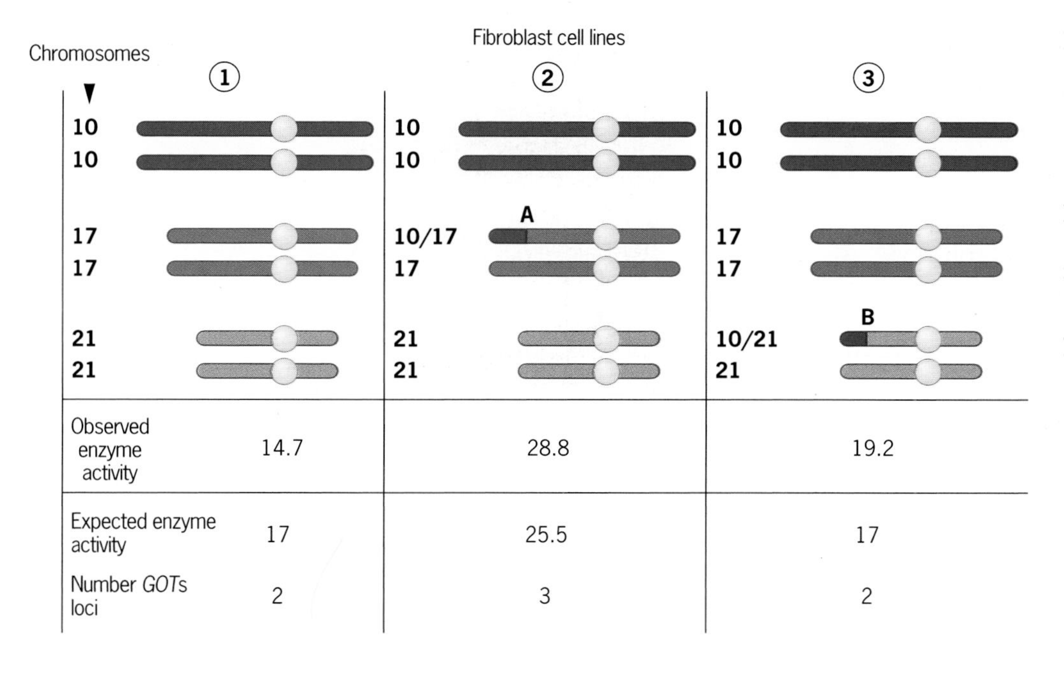

	1	2	3
Observed enzyme activity	14.7	28.8	19.2
Expected enzyme activity	17	25.5	17
Number *GOTs* loci	2	3	2

Figure 8.23 Duplication mapping. Three cell lines were used, each with a normal pair of chromosome 10s. Cell line 2 carrying translocational breakpoint **A** had three copies of the *GOTs* gene. Cell line 3 carrying translocation breakpoint **B** had two copies. The *GOTs* gene must lie between the two breakpoints.

two copies of the *GOTs* gene produced approximately 15 units of GOTs activity, and the cell line carrying the 10/17 translocation produced about 19 units of GOTs activity, findings that suggested that only two copies of the *GOTs* gene are present in the 10/17 translocation cell line. The translocated segment of chromosome 10 in this line must not carry a copy of the *GOTs* locus. The cell line carrying the 10/21 translocation produced about 30 units of GOTs activity. This is about 50 percent more than normal, an observation that suggested this cell line carried three copies of the *GOTs* locus. We can infer from these results that the *GOTs* gene lies between the two breakpoints in the 10q24 region.

The Human Gene Map

Somatic cell hybridization techniques have provided a major impetus for mapping human genes. However, more recent molecular techniques have greatly accelerated the process of human gene mapping. The much publicized **Human Genome Project** (**HUGO**) has as its primary goal the construction of maps for each human chromosome at increasingly finer resolutions. The basic strategy for this internationally coordinated project has two parts:

1. Dividing the chromosomes into segments that can be propagated and characterized.
2. Mapping these segments so that they correspond to their position on specific chromosomes.

Once this mapping is accomplished, the next step will be to sequence the DNA in each segment. The ultimate goal of the several research laboratories involved in HUGO is to identify the approximately 100,000 human genes and to sequence each one of them. This is an enormous undertaking if we consider that humans have about 3 billion DNA base pairs.

The rapidly developing map of the human genome describes the location and sequence of genes and molecular markers and the spacing between them on each chromosome. These maps are being constructed at several levels. The coarsest level of resolution is the genetic linkage map, which consists of the location and sequence of genes and DNA markers. More refined physical genetic maps describe the chemical characteristics of the DNA molecule itself at each chromosomal location.

At this time, approximately 5000 human genes have been identified, and about 50 percent of these have actually been mapped to specific chromosomal locations. There is clearly a very long way to go. The genes that have been identified are involved primarily in some kind of genetic disorder or disease state. We are just beginning to identify loci that influence normal, nondisease types of variation, such as height, weight, skin pigmentation, hair color, and disease resistance/susceptibility.

Key Points: **The lod score is a statistical method that tests genetic marker data in human families to determine whether two loci are linked. A gene can be located to a particular human chromosome by correlating its sequence or product with the human chromosomes that are retained in interspecific somatic cell hybrids. This location can be refined by using somatic cell hybrids that carry rearranged human chromosomes. The Human Genome Project is an international effort to map and sequence the entire human genome.**

TESTING YOUR KNOWLEDGE

1. Mary Houlahan, George Beadle, and Hermione Calhoun (1949. *Genetics* 34: 493–507) studied two mutant strains of *Neurospora crassa*. One strain (*pdx*) could not grow without pyridoxine and the other (*pan*) could not grow without pantothenic acid. Tetrads of ascospores from a cross between these strains were dissected and analyzed. From the results (shown below), determine the linkage relationships between the *pdx* and *pan* genes, and between each gene and its centromere. (A + represents the wild-type allele of a gene; thus, for example, *pdx* + = *pdx* pan^+ is the genotype of a spore that cannot grow without pyridoxine but that can grow without pantothenic acid.)

Tetrad classes

1	2	3	4	5	6
pdx +	*pdx pan*	*pdx* +	*pdx* +	*pdx* +	*pdx* +
pdx +	*pdx pan*	*pdx pan*	+ +	+ *pan*	+ *pan*
+ *pan*	+ +	+ +	*pdx pan*	*pdx* +	*pdx pan*
+ *pan*	+ +	+ *pan*	+ *pan*	+ *pan*	+ +

Number observed

15	1	17	1	13	2

ANSWER

Before analyzing the data, we note that the cross was between strains with the genotypes *pdx* + and + *pan*; these genotypes are indicative of parental-type ascospores in the progeny. Nonparental (that is, recombinant) ascospores are either *pdx pan* or + +. Using this nomenclature, we can classify each tetrad as parental ditype (PD), nonparental ditype (NPD), or tetratype (T). Furthermore, because *Neurospora* tetrads are ordered, we can classify them as having first (F) or second (S) division segregation patterns with respect to each marker:

Tetrad class	1	2	3	4	5	6
Number observed	15	1	17	1	13	2
Type	PD	NPD	T	T	PD	T
Segregation with respect to *pdx*	F	F	F	S	S	S
Segregation with respect to *pan*	F	F	S	F	S	S

The analysis of the data involves several steps. First, we note that the NPD tetrads (1) are much less numerous than the PD tetrads (15 + 13 = 28), indicating that *pdx* and *pan* are linked. Second, we use the standard mapping formula to estimate the distance between *pdx* and *pan*:

$$\frac{(1/2)\text{T} + \text{NPD}}{\text{total}} = \frac{(1/2)(17 + 1 + 2) + 1}{49} = 22.2 \text{ cM}$$

Third, we estimate the distance between *pdx* and its centromere by calculating the frequency of second division segregation patterns:

$$\frac{(1/2)\text{S}}{\text{total}} = \frac{(1/2)(1 + 13 + 2)}{49} = 16 \text{ cM}$$

Similarly, we estimate the distance between *pan* and its centromere:

$$\frac{(1/2)\text{S}}{\text{total}} = \frac{(1/2)(17 + 13 + 2)}{49} = 32.7 \text{ cM}$$

The combined results tell us that *pdx* and *pan* are on the same side of the centromere, and that *pan* is farther away. We summarize our analysis in a linkage map:

centromere—16 cM—*pdx*—22.2 cM—*pan*

Notice that the frequency of second division segregation patterns underestimates the distance between *pan* and the centromere because it overlooks a few double exchanges. We can obtain a better estimate by summing the distances between the centromere and *pdx* and between *pdx* and *pan*: 16 + 22.2 = 38.2 cM.

2. The pedigree below is for an X-linked recessive trait, ocular albinism (OA). Beneath some members of this family is information about a molecular marker located about 5 centiMorgans from the gene. The marker has two alleles, *1* and *2*.

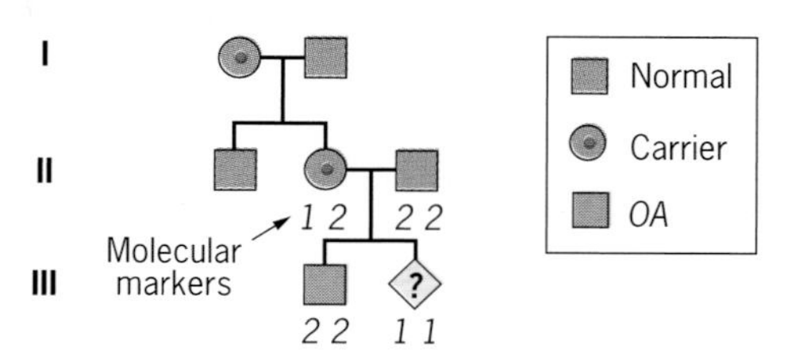

(a) Based on the information in this pedigree, what is the most likely arrangement of the alleles on the mother's (II-2) X chromosomes?

(b) The mother wishes to know if her fetus (III-2) inherited the trait. What is the most likely diagnosis of the fetus based on data for the molecular marker and the knowledge that the fetus is a male?

(c) At a later point, person I-2, the maternal grandfather, comes in for testing and is found to carry molecular marker allele *2*. How does this information affect your answer to (a)?

(d) Based on this additional information, how do you explain the affected male (III-1)?

(e) What further studies are needed to confirm the diagnosis of III-1?

(f) Does this additional information alter your diagnosis of the fetus (III-2)? Explain.

ANSWER

(a) The mother passed the *OA* allele and marker *2* to her affected son. Since the two points are about 5 cM apart, there is a 95 percent chance she is:

—— *2* —— *OA* —
—— *1* —— + —

(b) The fetus received marker *1* from his mother. There is a 95 percent chance she has the genotype diagrammed in (a), and a 5 percent chance she has the *1* allele linked to *OA*. If she is as shown in (a), there is a 95 percent chance she will pass *1* and + to the fetus and a 5 percent chance she will pass *1* and *OA* to the fetus. If she has the *1* allele linked to *OA*, then there is a 95 percent chance she will pass *1* and *OA* to the fetus. Thus, in total, there is a $(0.95 \times 0.05) + (0.05 \times 0.95) = 9.5$ percent chance that the fetus will be affected.

(c) Since the grandfather was —*2*— + —, the mother must be:

—— *1* —— *OA* —
—— *2* —— + —

(d) The affected male is —— 2 —— *OA* —. He must have received a recombinant X chromosome from the mother.

(e) By looking at other markers to the left and right of the *OA* gene in the mother and her offspring, we can confirm a crossover event.

(f) Since we now know with certainty what the arrangement of the alleles is in the mother, there is a 95 percent chance that the fetus (III-2) will have *OA* and a 5 percent chance he will not.

QUESTIONS AND PROBLEMS

8.1 In yeast, the *ad* mutation requires adenine for growth, and the *in* mutation requires inositol for growth. The wild-type alleles of these two mutations are *AD* and *IN*. C. C. Lindegren analyzed 48 tetrads from the cross *AD IN* × *ad in*; among these tetrads, 22 were parental ditype, 3 were nonparental ditype, and 23 were tetratype. Are the *AD* and *IN* genes linked? If so, what is the distance between them?

8.2 *Drosophila* females homozygous for a recessive mutation causing purple eyes (*pr*) were crossed to *Cy/Pm; Tb/Sb* males (see Figure 8.5 for an explanation of these genetic symbols). None of the F_1 flies had purple eyes, but when Curly, Tubby F_1 flies were intercrossed, they produced some purple offspring, and none of these had a Tubby body. Which chromosome carries the *pr* gene?

8.3 A geneticist obtained the following ordered tetrad data from a cross with *Neurospora:*

Top of ascus	Spore pairs		Bottom of ascus	Number of tetrads
(1,2)	(3,4)	(5,6)	(7,8)	
A	*A*	*a*	*a*	61
a	*a*	*A*	*A*	55
a	*A*	*a*	*A*	40
A	*a*	*A*	*a*	44
				Total = 200

What is the distance between the *A* locus and its centromere?

8.4 The following tetrad data were obtained from the cross *A B* × *a b* in *Neurospora*:

Top of ascus	Spore pairs		Bottom of ascus	Number of tetrads
(1,2)	(3,4)	(5,6)	(7,8)	
AB	*AB*	*ab*	*ab*	1766
AB	*aB*	*Ab*	*ab*	220
AB	*Ab*	*aB*	*ab*	14
				Total = 2000

Use these data to construct a linkage map of the two genes and their centromere.

8.5 In *Neurospora*, the mutations *arg*, *thi*, and *leu* block the synthesis of arginine, thiamine, and leucine, respectively. The following tetrad data come from two crosses with these mutations:

Cross	Spore pairs				Number of tetrads
	(1,2)	(3,4)	(5,6)	(7,8)	
arg × *thi*	*arg*+	*arg*+	+*thi*	+*thi*	46
	arg thi	*arg thi*	++	++	56
					Total = 100
arg × *leu*	*arg*+	*arg*+	+*leu*	+*leu*	155
	arg+	*arg leu*	++	+*leu*	44
	arg+	+*leu*	*arg* +	+*leu*	1
					Total = 200

Using these data, construct the linkage map(s) for these genes. Show the position of the centromere(s).

8.6 The order of three genes and the centromere on one *Neurospora* chromosome is

centromere *x* *y* *z*

A cross between + + + and *x y z* produced one ascus with the following ordered array of ascospores (only one member of each spore pair is shown):

(+ + *z*) (+ *y z*) (*x* + +) (*x y* +)

(a) Is this ascus most likely the result of a meiotic event in which 0, 1, 2, or 3 crossovers occurred?
(b) In what interval(s) did the crossover(s) most likely occur?
(c) If double or triple crossovers were involved, were they two-strand, three-strand, or four-strand multiple crossovers?

8.7 *Drosophila* females with X chromosomes attached at the centromere carried the recessive mutations *y* (*yellow* body) and *sn* (*singed* bristles) on one of their X chromosomes and the wild-type alleles of these genes on the other. When these females were crossed to wild-type males, they produced mostly wild-type progeny. However, a few of their daughters had yellow bodies, and fewer still had yellow bodies and singed bristles. How would you explain the occurrence of these exceptional females? Can you infer anything about the location of *y* and *sn* with respect to the centromere?

8.8 A *Drosophila* second chromosome that carried a recessive lethal mutation was balanced over the *Cy* chromosome. *l/Cy* females were systematically crossed to males carrying different second chromosome deletions (all homozygous lethal) balanced over the *Cy* chromosome. Each cross was scored for the presence or absence of non-Curly progeny.

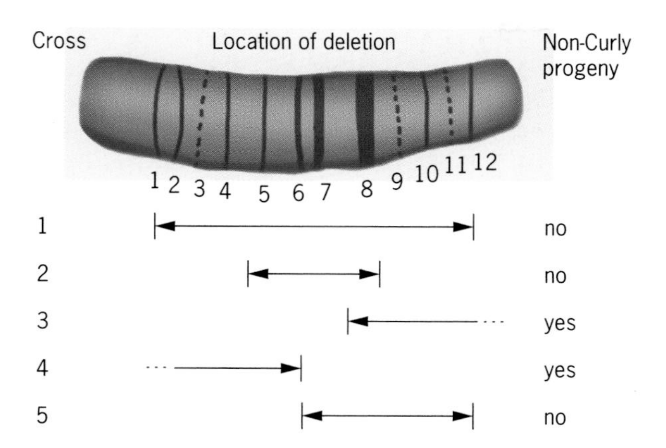

In which band is the lethal mutation located?

8.9 A woman gives birth to six sons. Two are normal, three are both color blind and hemophilic, and one is just color blind. These are sex-linked traits. Draw the most likely linkage relationship of the genes on the two X chromosomes.

8.10 A woman gives birth to seven sons. Three are hemophilic, three are color blind, and one is normal. Draw the most likely linkage relationship of the genes on the two X chromosomes.

8.11 A woman with the genetic constitution

A ———— *b*
a ———— *B*

marries a man with the genetic constitution

a ———— *b*
a ———— *b*

A and *B* are dominant to *a* and *b*, and the genes are 20 map units apart. What classes of offspring would this couple produce, and in what approximate proportions?

8.12 A woman is heterozygous for three X-linked genes: *Aa*, *Bb*, and *Cc*. She has seven sons with the following genotypes:

2 *ABc*
3 *abC*
1 *ABC*
1 *aBc*

Draw the mother's two X chromosomes with the appropriate alleles on each. Which of the sons are the result of recombination?

8.13 A human cell line carries a gene that codes for enzyme E. Mouse cells do not carry this gene. Human–mouse hybrids are formed, and the following stable lines are studied for the human chromosomes present and the presence or absence of enzyme E.

Hybrid Cell Population	Human Chromosomes Present	E Activity Present
1	21, 18, 16, 7, 5, 2	Yes
2	18, 16, 8, 3	Yes
3	21, 18, 1	Yes
4	23, 19, 17, 15, 14, 11 9, 8, 3, 1	No
5	18, 6, 2	Yes
6	21, 5	No

Using these data, determine where the gene for E is located.

8.14 The following translocation occurred between the X and chromosome 14:

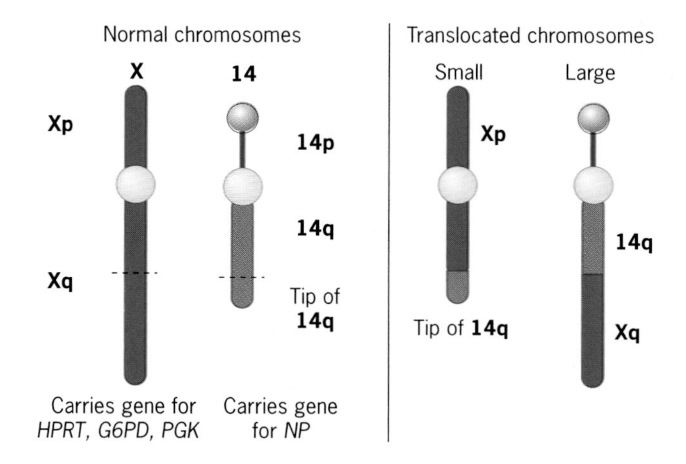

A cell line carrying this translocation was fused with a mouse cell line, and the following hybrid cell lines were isolated:

Translocated chromosomes in hybrid cell lines

Enzymes contained in hybrid cell lines		Both	Large one	Small one	Neither
	G6PD	+	+	–	–
	HPRT	+	+	–	–
	PGK	+	+	–	–
	NP	+	+	–	–

You know that the genes for *G6PD*, *HPRT*, and *PGK* are X-linked and that the gene for *NP* is autosomal. What else can you deduce from these data?

8.15 A child was born with the karyotype shown below:

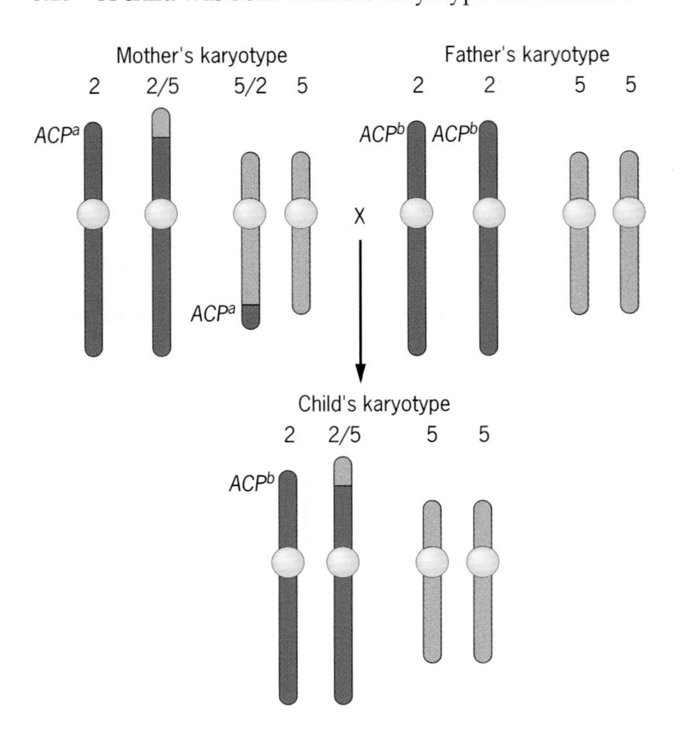

The mother carried a reciprocal translocation and was homozygous for an acid phosphatase allele (ACP^a). The father's karyotype was normal, and he was homozygous for a different *ACP* allele, ACP^b. The *ACP* alleles are codominant, but the child expressed only the ACP^b allele. What can you conclude about the *ACP* gene from this case?

8.16 There is a heritable fragile site on the long arm of chromosome 16. Somatic cell hybrids show that the *HPA* gene (a polypeptide of the haptoglobin protein) is also on chromosome 16. Among the offspring produced by persons heterozygous for both the fragile site and the *HPA* gene, about 8 percent were recombinant. What do you conclude from this?

8.17 The X-linked genes *A* and *B* show 10 percent recombination. Genes *A* and *C* show 16 percent recombination. Construct the gene maps that are compatible with these data.

8.18 Which of the models suggested in Problem 8.17 would be most likely if *B* and *C* showed about 6 percent recombination?

8.19 Fujibayashi and his colleagues (*Am. J. Hum. Genet.* 37:741, 1985) mapped the *SAP2* gene (sphingolipid activator protein-2) using somatic cell hybrids. Their results are presented in the accompanying table. On which chromosome is the human *SAP2* gene?

Hybrid	*Human Chromosomes Present*	*SAP-2*
CP3-1	4, 5, 11, 12, 14, 16, 17, 18, 19, 20, 21, X	−
CP4-1	4, 5, 8, 11, 14, 22, X	−
CP5-1	1, 5, 8, 9, 12, 14, 15, 17, 19, 21, 22	−
CP6-1	4, 12, 14, 17, 18, 19, 21, 22	−
CP12.3.2	2, 8, 9, 11, 12, 21, 22, X	−
CP12-7	2, 4, 8, 9, 10, 11, 12, 13, 21, 22, X	+
CP14-1	4, 5, 17, 21, 22, X	−
CP15-1	4, 5, 11, 12, 15, 16, 17	−
CP16-1	5, 14, 17, 19, 20, 21, 22	−
CP17-1	1, 4, 5, 12, 17	−
CP18-1	1, 8, 11, 14, 15, 17, 18, 19	−
CP20-1	2, 5, 9, 17, 21	−
CP26-1	1, 4, 5, 6, 7, 9, 10, 11, 12, 13, 17, 22	+
CP28-1	1, 4, 5, 8, 9, 18, 19, 20	−
153-E4A-1	4, 5, 8, 9, 11, 13, 15	−
822-19B3	1, 2, 3, 4, 5, 6, 10, 11, 12, 14, 15, 19, 21	+
Q72-18	4, 5, 6, 8, 9, 10, 11, 12 , 14, 15, 16, 19, 20, 21, 22, X	+
640-53B	3, 4, 9, 10, 13, 14, 15, 21, 22	+
640-34	4, 6, 9, 10, 12, 18, 19, 21	+
Q21-10A	1, 3, 4, 5, 6, 11, 12, 14, 15, 19, 20, 21, X	−
Q21-3A	1, 3, 4, 5, 6, 10, 11, 14, 15, 19, 20, 21	+
Q68-20-1A	3, 5, 10, 11, 12, 14, 18, 19, 21, X	+

8.20 Kondo and Hamaguchi (*Am. J. Hum. Genet.* 37:1106, 1985) evaluated linkage between two genes, *LCP1* (lymphocyte cytosol polypeptide) and *ESD* (esterase D). They looked at several pedigrees, one of which is as follows:

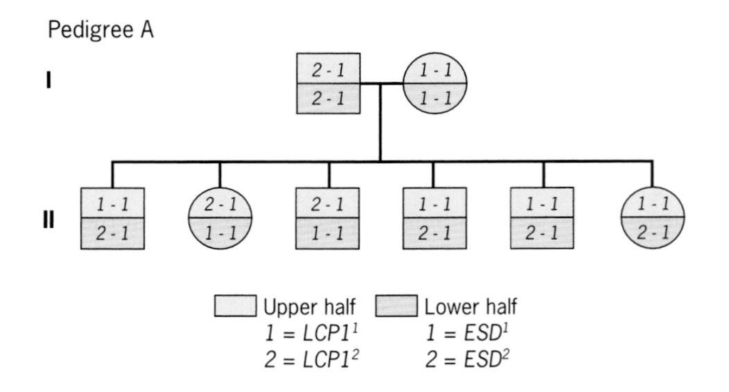

Lod Scores for Linkage of *LCP1* and *ESD*

		Frequency of Recombination					
Family	*No. of Children*	*0*	*0.05*	*0.10*	*0.20*	*0.30*	*0.40*
A	6	1.505	1.371	1.231	0.924	0.578	0.211
B	2	0.301	0.258	0.215	0.134	0.064	0.017
C	2	0.301	0.258	0.215	0.134	0.064	0.017
D	7	2.114	1.951	1.787	1.429	1.023	0.554
Sum of lod scores		4.221	3.838	3.448	2.621	1.729	0.799

(a) What is the probability that the two genes are assorting independently?
(b) What is the probability that the two genes are linked by 0 map units?
(c) What is the lod score for this pedigree?
(d) Data from the accompanying table sum the lod scores for linkage of *LCP1* and *ESD* for four pedigrees. Based on these data, what do you conclude about the linkage of *LCP1* and *ESD*?

BIBLIOGRAPHY

ASHBURNER, M. 1989. *Drosophila: A Laboratory Handbook.* Cold Spring Harbor Laboratory.

HOULAHAN, M., G. BEADLE, AND H. CALHOUN. 1949. "Linkage studies with biochemical mutants of *Neurospora crassa.*" *Genetics* 34:493–507.

JUDD, B. H., M. W. SHEN, AND T. C. KAUFMAN. 1972. "The anatomy and function of a segment of the *X* chromosome of *Drosophila melanogaster.*" *Genetics* 71:139–156.

KAO, F., C. JONES, AND T. T. PUCK. 1976. "Genetics of somatic mammalian cells: genetic, immunologic, and biochemical analysis with Chinese hamster cell hybrids containing selected human chromosomes. "*Proc. Natl. Acad. Sci. USA* 73: 193–197.

LINDEGREN, C. C. 1949. *The Yeast Cell, Its Genetics and Cytology.* Educational Publishers, St. Louis.

MORTON, N. E. 1955. "Sequential tests for the detection of linkage." *Amer. J. Hum. Genet.* 7: 277–318.

RUDDLE, F. H., AND R. S. KUCHERLAPATI. 1974. "Hybrid cells and human genes." *Sci. Amer.* 231(1): 36–44.

THOMPSON, M. W., R. R. MCINNES, AND H. F. WILLARD. 1991. *Genetics in Medicine.* 5th ed. W. B. Saunders, Philadelphia.

A CONVERSATION WITH
MARY LOU PARDUE

What inspired you to become a biologist?

I've always been interested in plants and animals, and in how things are put together. I've always liked knowing how everything works.

Along the way, did you have mentors who inspired you?

Well, I've been very lucky in the people I've worked for along the way. Interestingly enough, none of them have been women. Women mentors were few and far away. Of course, I was told about people like Barbara McClintock and Lottie Auerbach, but they seemed to be a very special set of people. I got to know them later in my career, and they proved to be as special and wonderful as I had been told. The men I've worked for have been very good and very helpful. That is especially true of Joe Gall, who was my Ph.D. mentor at Yale University. He has been a very important influence in my career.

Your Ph.D. thesis is one of the most famous. How did you do it? How did it feel to have invented a technique that is now so widely used?

My thesis project was the development of the *in situ* hybridization technique. It presented the first application of a technique for localizing DNA sequences within chromosomes. Joe Gall always said that it was too tough a problem to put a student on, and I always said it was too tough a problem for a *smart* student to take on. Well, I couldn't help trying it, and when it worked, I forgot about an alternative thesis project and went on with this tough one. It was a great relief to find out that it actually works to answer so many questions.

Mary Lou Pardue is a leader in molecular genetics whose research findings have had a dramatic impact on our knowledge of genetics. She received a B.S. degree from William and Mary College in 1955, an M.S. from the University of Tennessee in 1959, and a Ph.D. from Yale University in 1970. Her pioneering work on the molecular organization of chromosomes has promoted our understanding of genetics. In 1983 she served as president of the Genetics Society of America and the American Society for Cell Biology. Currently, Dr. Pardue is a member of the National Academy of Sciences and the Boris Magasanik Professor of biology at MIT.

When you did this pioneering work, the experimental materials were very precious. Did this cause you to think long and hard about which experiments you would do?

Oh, you better believe it did. Not only did I worry about wasting carefully collected materials. I also worried about how much time I was going to spend preparing the DNA and fractionating it to get the satellite DNA, which was the component I wanted to study. And then I had to collect the fractions and transcribe the satellite component using RNA polymerase, which I had to isolate for that purpose. In those days we couldn't buy kits from a scientific supply company to help with the work. What we order a kit for today would have been a semester's work in the 1960s. Yes, I did a lot of thinking in those days about the kinds of experiments I wanted to do.

Currently, you are studying chromosome ends—telomeres—in *Drosophila?*

Yes. When you look at a metaphase chromosome, telomeres appear to be the least interesting part of a chromosome, but it is beginning to look like they do a lot more than we ever suspected. Viruses have developed different ways of keeping the ends of their chromosomes from shortening, but eukaryotes have evolved a totally different mechanism, one that involves an RNA/enzyme complex called telomerase. What intrigues me about *Drosophila* telomeres is that at first they look so different from other eukaryotic telomeres because they have no telomerase but instead contain retroposons, a special class of transposable genetic elements. However, retroposons and the reverse transcriptase that they encode are really very similar to telomerase, which copies part of a guide RNA molecule onto the chromosome's end. What's happening at *Drosophila* telomeres is not so different from what happens at more typical telomeres. Both the telomerase-type telomere and the retroposon-type telomere are RNA-templated extensions of the chromosome. I think the two mechanisms of forming telomeres evolved from the same ancestral process.

Your discovery of the retroposon mechanism for lengthening *Drosophila* telomeres was serendipitous, wasn't it?

Yes, it came out of wanting to know what DNA sequences are located in heterochromatin that aren't just simple satellite sequences. I was looking for heterochromatic sequences that weren't just randomly repeated sequences, and I found a sequence that was not a typical satellite sequence, one that occurs only at chromosome ends (heterochromatic regions). It seemed that this sequence might have an interesting role, and indeed, it does.

Over your career as a student and as a scientist, what were the greatest challenges you faced?

This is a hard question to answer because in my day I hadn't really seen women in the jobs I wanted, so I didn't drive myself for a career when I started out. I just did what I felt was fun. There weren't so many challenges because I was just enjoying myself. It wasn't a process of striving for a career. It was a process of doing what was most interesting each moment, and that turned into a career.

What do you view as some of the most challenging and exciting problems in genetics?

I think there are two. One is to understand how the genome evolved. I think so much of what we need to know about the genome depends on understanding its history. It was not planned but developed historically. Another challenge is understanding how to read the sequence of the genome once we get it. I think we understand how to read protein sequences now, but I'm not sure that we understand how to read a lot of important genetic information that's not coding for proteins.

Do you have any advice for students who are interested in the study of genetics?

I have undergraduates doing research in the lab, and from time to time, I teach an undergraduate course in developmental biology so I get asked this question often. I believe that a person should do what he or she enjoys, because the future is hard to plan out. I don't think that I have any special advice except that biology is an awfully interesting field to be in, and one should do it if one enjoys it. I've been a research scientist at a particularly interesting point in the history of biology, and it's been a good time.

In the last 40 years, what would you identify as the most significant discoveries in genetics?

The idea that genes can move around—that elements can transpose in the genome, and that there can be splicing of RNA, and that RNA molecules can do many things that we never suspected they could. I think these are important discoveries because they showed us that the genome is more fluid than we ever expected it to be, and because they introduced us to several levels at which evolution and regulation can occur.

9

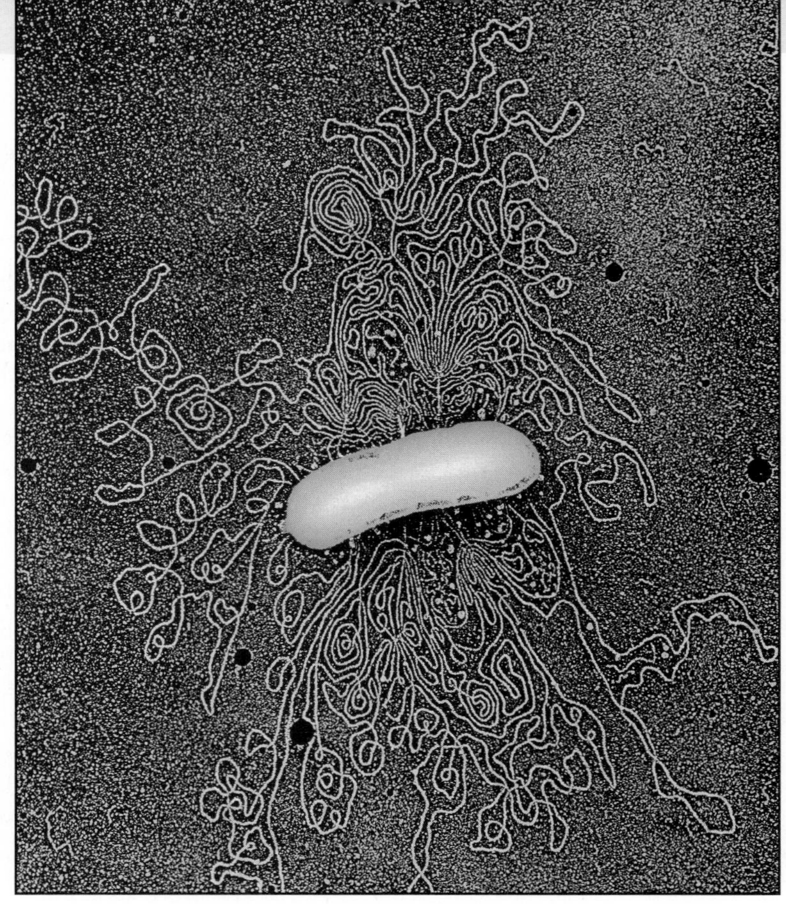

Electron micrograph of a ruptured E. coli *cell with much of the DNA extruded.*

DNA and the Molecular Structure of Chromosomes

CHAPTER OUTLINE

Discovery of Nuclein

In 1868, Johann Friedrich Miescher, a young Swiss medical student, became fascinated with an acidic substance that he isolated from pus cells obtained from bandages used to dress human wounds. To obtain larger quantities of pus for his investigations, Miescher began purchasing used bandages from a surgical clinic in Tübingen, Germany. He first separated the pus cells from the bandages and associated debris, and then treated the cells with pepsin, a proteolytic enzyme that he isolated from the stomachs of pigs. After the pepsin treatment, he recovered an acidic substance that he called "nuclein." Miescher's nuclein was unusual in that it contained large amounts of both nitrogen and phosphorus, two elements known at the time to coexist only in certain types of fat. Miescher wrote a paper describing his discovery of nuclein in human pus cells and submitted it for publication in 1869. However, the editor of the journal to which the paper was sent was skeptical of the results and decided to repeat the experiments himself. As a result, Miescher's paper describing nuclein was not published until 1871, two entire years after its submission.

At the time, the importance of the substance that Miescher called nuclein could not have been anticipated. The existence of polynucleotide chains, the key component of the acidic material in Miescher's nuclein, was not documented until the 1940s. The role of nucleic acids in storing and transmitting genetic information was not established until 1944, and the double-helix structure of DNA was not discovered until 1953. Even in 1953, many geneticists were reluctant to accept the idea that nucleic acids, rather than proteins, carried the genetic information because nucleic acids exhibited less structural variability than proteins.

FUNCTIONS OF THE GENETIC MATERIAL

In 1865, Mendel showed that genes transmitted genetic information, and in the first few decades of the twentieth century, their patterns of transmission from generation to generation were studied extensively. Although these classical genetic studies provided little insight into the molecular nature of genes, they did demonstrate that, whatever its chemical nature, the genetic material must perform three essential functions.

1. The genotypic function, **replication.** The genetic material must store genetic information and accurately transmit that information from parents to offspring, generation after generation.
2. The phenotypic function, **gene expression**. The genetic material must control the development of the phenotype of the organism. That is, the genetic material must dictate the growth and differentiation of the organism from the single-celled zygote to the mature adult.
3. The evolutionary function, **mutation**. The genetic material must undergo change so that organisms can adapt to modifications in the environment. Without such changes, evolution could not occur.

Other early genetic studies established a precise correlation between the patterns of transmission of genes and the behavior of chromosomes during sexual reproduction, providing strong evidence that genes are usually located on chromosomes. Thus further attempts to discover the chemical basis of heredity focused on the molecules present in chromosomes.

Chromosomes are composed of two types of large organic molecules (macromolecules) called **proteins** and **nucleic acids.** The nucleic acids are of two types: **deoxyribonucleic acid (DNA)** and **ribonucleic acid (RNA)**. During the 1940s and early 1950s, the results of some elegant experiments clearly established that the genetic information is stored in nucleic acids, not in proteins. These experiments are discussed in the following section of this chapter. In most organisms, the genetic information is encoded in the structure of DNA. In some small viruses, RNA carries the genetic information; these viruses contain no DNA.

Key Points: **The genetic material must provide the genotypic, phenotypic, and evolutionary functions of replication, gene expression, and mutation, respectively.**

PROOF THAT GENETIC INFORMATION IS STORED IN DNA

Several lines of indirect evidence suggested that DNA harbors the genetic information of living organisms.

1. Most of the DNA of cells is located in the chromosomes, whereas RNA and proteins are also abundant in the cytoplasm.
2. A precise correlation exists between the amount of DNA per cell and the number of sets of chromosomes per cell. For example, most somatic cells of diploid organisms contain twice the amount of DNA as the haploid germ cells (gametes) of the same species.
3. The molecular composition of the DNA is the same (with rare exceptions) in all the different cells of an organism, whereas the composition of both RNA and proteins is highly variable from one cell type to another.
4. DNA is more stable than RNA or proteins, which are synthesized and degraded quite rapidly in living organisms. Since the genetic material must store and transmit information from parents to offspring, we might expect it to be stable, like DNA.

Although these correlations strongly suggest that DNA is the genetic material, they by no means prove it.

TABLE 9.1
Characteristics of *Diplococcus pneumoniae* Strains When Grown on Blood Agar Medium

	Colony Morphology				*Reaction with Antiserum Prepared Against*	
Type	*Appearance*	*Size*	*Capsule*	*Virulence*	*Type IIS*	*Type IIIS*
IIR[a]	Rough	Small	Absent	Avirulent	None	None
IIS	Smooth	Large	Present	Virulent	Agglutination	None
IIIR[a]	Rough	Small	Absent	Avirulent	None	None
IIIS	Smooth	Large	Present	Virulent	None	Agglutination

[a]Although Type R cells are nonencapsulated, they carry genes that would direct the synthesis of a specific kind (antigenic Type II or III) of capsule if the block in capsule formation were not present. When Type R cells mutate back to encapsulated Type S cells, the capsule Type (II or III) is determined by these genes. Thus, R cells derived from Type IIS cells are designed Type IIR. When these Type IIR cells mutate back to encapsulated Type S cells, the capsules are of Type II.

Discovery of Transformation in Bacteria

Bacterial transformation, a type of recombination occurring in bacteria, was discovered by Frederick Griffith in 1928. Although his experiments on *Diplococcus pneumoniae* (pneumococcus) provided no evidence that DNA was involved in the transformation process, they set the stage for later discoveries.

Pneumococci, like all other living organisms, exhibit genetic variability that can be recognized by the existence of different phenotypes (Table 9.1). The two phenotypic characteristics of importance in Griffith's demonstration of transformation are (1) the presence or absence of a surrounding polysaccharide (complex sugar polymer) capsule, and (2) the type of capsule—that is, the specific molecular composition of the polysaccharides present in the capsule. When grown on blood agar medium in petri dishes, pneumococci with capsules form large, smooth colonies and are thus designated Type S. Encapsulated pneumococci are virulent (pathogenic), causing pneumonia in mammals such as mice and humans. The virulent Type S pneumococci mutate to an avirulent (nonpathogenic) form that has no polysaccharide capsule at a frequency of about once per 10^7 cells. When grown on blood agar medium, such nonencapsulated, avirulent pneumococci produce small, rough-surfaced colonies and are thus designated Type R (Table 9.1). The polysaccharide capsule is required for virulence because it protects the bacterial cell from destruction by white blood cells. When a capsule is present, it may be of several different antigenic types (Type I, II, III, and so forth), depending on the specific molecular composition of the polysaccharides and, of course, ultimately on the genotype of the cell.

The different capsule types can be identified immunologically. If Type II cells are injected into the bloodstream of rabbits, the immune system of the rabbits will produce antibodies that react specifically with Type II cells. Such Type II antibodies will aggluti-

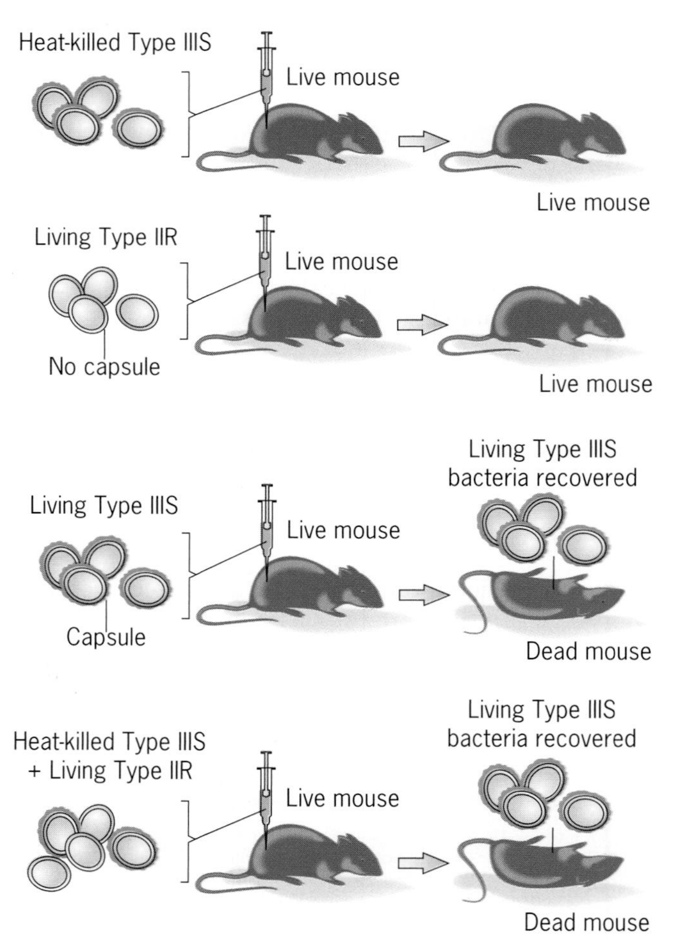

Figure 9.1 Griffith's discovery of transformation in *Diplococcus* pneumoniae.

nate Type II pneumococci but not Type I or Type III pneumococci.

Griffith's unexpected discovery was that if he injected heat-killed Type IIIS pneumococci (virulent when alive) plus live Type IIR pneumococci (avirulent) into mice, many of the mice succumbed to pneumonia, and live Type IIIS cells were recovered from the carcasses (Figure 9.1). When mice were injected with heat-killed Type IIIS pneumococci alone (Figure 9.1, top), none of the mice died. The observed virulence was therefore not due to a few Type IIIS cells that survived the heat treatment. The live pathogenic pneumococci recovered from the carcasses had Type III polysaccharide capsules. This result is important because nonencapsulated Type R cells can mutate back to encapsulated Type S cells. However, when such a mutation occurs in a Type IIR cell, the resulting cell will become Type IIS, not Type IIIS. Thus the transformation of avirulent Type IIR cells to virulent Type IIIS cells cannot be explained by mutation. Instead, some component of the dead Type IIIS cells (the "transforming principle") must have converted living Type IIR cells to Type IIIS.

Subsequent experiments showed that the phenomenon described by Griffith, now called **transformation**, was not mediated in any way by a living host. The same phenomenon occurred in the test tube when live Type IIR cells were grown in the presence of dead Type IIIS cells or extracts of Type IIIS cells. Since Griffith's experiments demonstrated that the Type IIIS phenotype of the transformed cells was passed on to progeny cells—that is, was due to a permanent inherited change in the genotype of the cells—the demonstration of transformation set the stage for determining the chemical basis of heredity in pneumococcus.

Proof That DNA Mediates Transformation

The first direct evidence showing that the genetic material is DNA rather than protein or RNA was published by Oswald Avery, Colin MacLeod, and Maclyn McCarty in 1944. They showed that the component of the cell responsible for transformation (the "transforming principle") in *Diplococcus pneumoniae* is DNA. They demonstrated that if highly purified DNA from

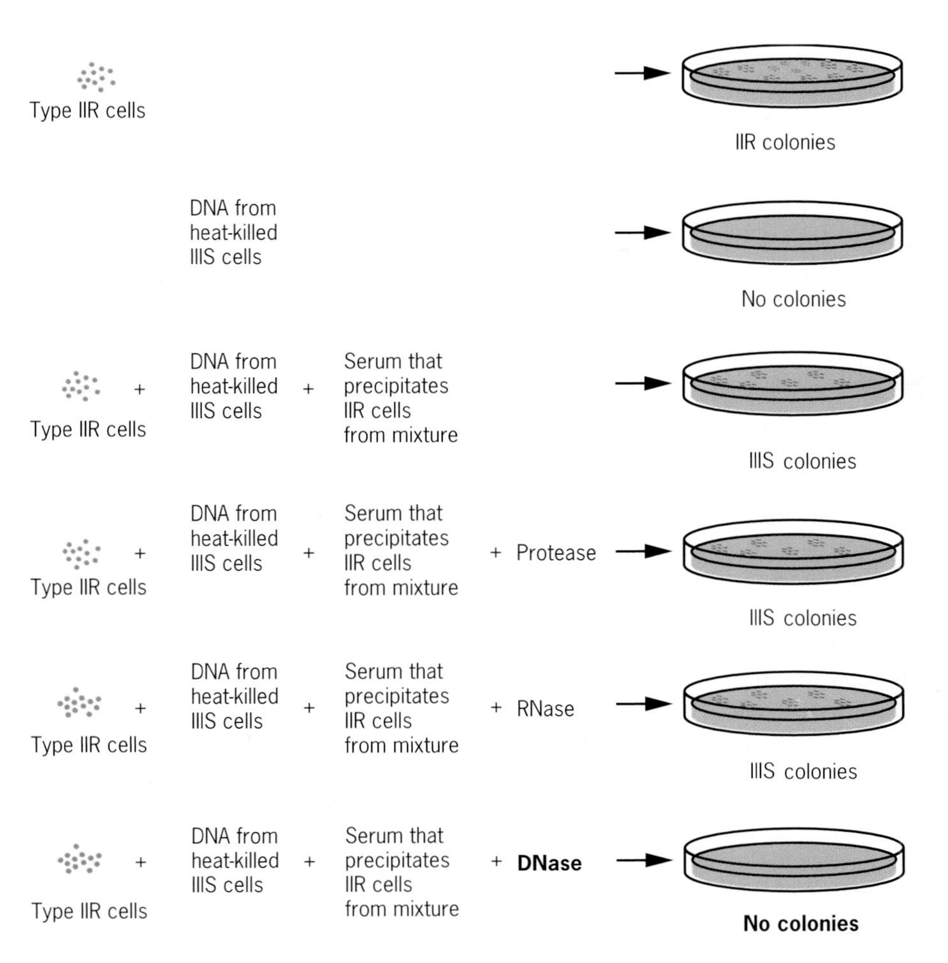

Figure 9.2 Avery, MacLeod, and McCarty's proof that the "transforming principle" is DNA.

Type IIIS pneumococci was present with Type IIR pneumococci, some of the pneumococci were transformed to Type IIIS (Figure 9.2). But how could they be sure that the DNA was really pure? Proving the complete purity of any macromolecular substance is extremely difficult. Maybe the DNA preparation contained a few molecules of protein, and these contaminating proteins were responsible for the observed transformation. The most definitive experiments in Avery, MacLeod, and McCarty's proof that DNA was the transforming principle involved the use of enzymes that degrade DNA, RNA, or protein. In separate experiments, highly purified DNA from Type IIIS cells was treated with the enzymes (1) **deoxyribonuclease (DNase)**, which degrades DNA, (2) **ribonuclease (RNase)**, which degrades RNA, or (3) **proteases**, which degrade proteins; the DNA was then tested for its ability to transform Type IIR cells to Type IIIS. Only DNase treatment had any effect on the transforming activity of the DNA preparation—it eliminated all transforming activity (Figure 9.2).

Although the molecular mechanism by which transformation occurs remained unknown for many years, the results of Avery and coworkers clearly established that the genetic information in pneumococcus is present in DNA. Geneticists now know that the segment of DNA in the chromosome of pneumococcus that carries the genetic information specifying the synthesis of a Type III capsule is physically inserted into the chromosome of the Type IIR recipient cell during the transformation process.

Proof That DNA Carries the Genetic Information in Bacteriophage T2

Additional evidence demonstrating that DNA is the genetic material was published in 1952 by Alfred Hershey (1969 Nobel Prize winner) and Martha Chase. The results of their experiments showed that the genetic information of a particular bacterial virus (bacteriophage T2) was present in DNA. Their results had a major impact on scientists' acceptance of DNA as the genetic material. This impact was the result of the simplicity of the Hershey–Chase experiment.

Viruses are the smallest living organisms; they are living at least in the sense that their reproduction is controlled by genetic information stored in nucleic acids via the same processes as in cellular organisms (Chapter 15). However, viruses are acellular parasites that can reproduce only in appropriate host cells. Their reproduction is totally dependent on the metabolic machinery (ribosomes, energy-generating systems, and other components) of the host. Viruses have been extremely useful in the study of many genetic processes because of their simple structure and chemical composition (many contain only proteins and nucleic acids) and their very rapid reproduction (15 to 20 minutes for some bacterial viruses under optimal conditions).

Bacteriophage T2, which infects the common colon bacillus *Escherichia coli*, is composed of about 50 percent DNA and about 50 percent protein (Figure 9.3). Experiments prior to 1952 had shown that all bacteriophage T2 reproduction takes place within *E. coli* cells. Therefore, when Hershey and Chase showed that the DNA of the virus particle entered the cell, whereas most of the protein of the virus remained adsorbed to the outside of the cell, the implication was that the genetic information necessary for viral reproduction was present in DNA. The basis for the Hershey–Chase experiment is that DNA contains phosphorus but no sulfur, whereas proteins contain sulfur but virtually no phosphorus. Thus Hershey and Chase were able to label specifically either (1) the phage DNA by growth in a medium containing the radioactive isotope of phosphorus, ^{32}P, in place of the normal isotope, ^{31}P; or (2) the phage protein coats by growth in a medium containing radioactive sulfur, ^{35}S, in place of the normal isotope, ^{32}S (Figure 9.3).

When T2 phage particles labeled with ^{35}S were mixed with *E. coli* cells for a few minutes and the phage-infected cells were then subjected to shearing forces in a Waring blender, most of the radioactivity (and thus the proteins) could be removed from the cells without affecting progeny phage production. When T2 particles in which the DNA was labeled with ^{32}P were used, however, essentially all the radioactivity was found inside the cells; that is, the DNA was not subject to removal by shearing in a blender. The sheared-off phage coats were separated from the infected cells by low-speed centrifugation, which pellets (sediments) cells while leaving phage particles suspended. These results indicated that the DNA of the virus enters the host cell, whereas the protein coat remains outside the cell. Since progeny viruses are produced inside the cell, Hershey and Chase's results indicated that the genetic information directing the synthesis of both the DNA molecules and the protein coats of the progeny viruses must be present in the parental DNA. Moreover, the progeny particles were shown to contain some of the ^{32}P, but none of the ^{35}S of the parental phage.

There was one flaw in Hershey and Chase's proof that the genetic material of phage T2 is DNA. A significant amount of ^{35}S (and thus protein) was found to be injected into the host cells with the DNA. Thus it could be argued that this small fraction of the phage proteins contained the genetic information. More recently, scientists have developed procedures by which

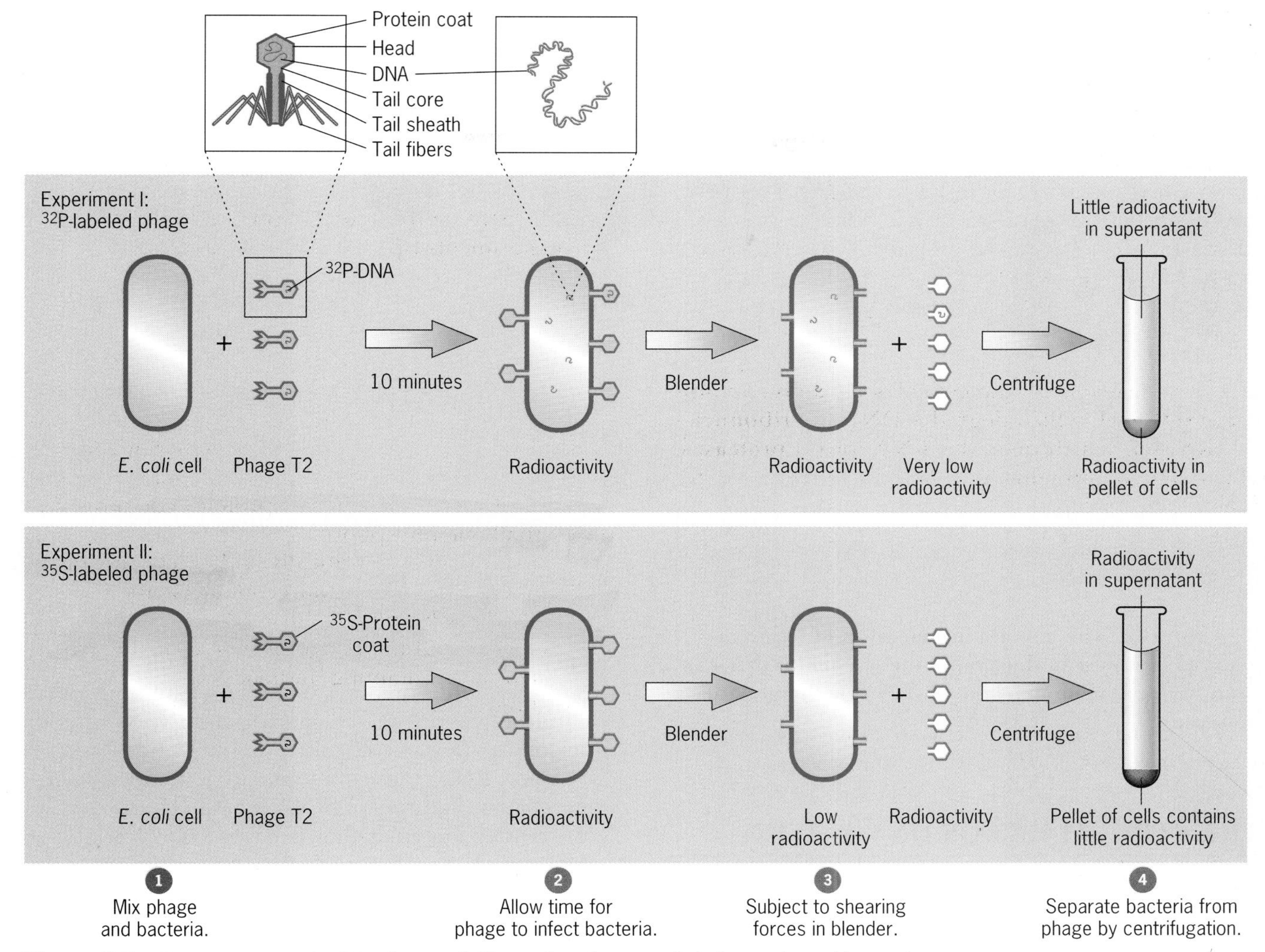

Figure 9.3 Demonstration by Hershey and Chase that the genetic information of bacteriophage T2 resides in its DNA.

protoplasts (cells with the walls removed) of *E. coli* can be infected with pure phage DNA. Normal infective progeny phage are produced in these experiments, called **transfection** experiments, proving that the genetic material of such bacterial viruses is DNA.

Proof That RNA Stores the Genetic Information in Some Viruses

As more and more viruses were identified and studied, it became apparent that many of them contain RNA and proteins, but no DNA. In all cases studied to date, it is clear that these RNA viruses—like all other organisms—store their genetic information in nucleic acids rather than in proteins, although in these viruses the nucleic acid is RNA. One of the first experiments that established RNA as the genetic material in RNA viruses was the so-called reconstitution experiment of Heinz Fraenkel-Conrat and coworkers, published in 1957. Their simple, but definitive, experiment was done with tobacco mosaic virus (TMV), a small virus composed of a single molecule of RNA encapsulated in a protein coat. Different strains of TMV can be identified on the basis of differences in the chemical composition of their protein coats.

Fraenkel-Conrat and colleagues treated TMV particles of two different strains with chemicals that dissociate the protein coats of the viruses from the RNA molecules and then separated the proteins from the RNA. They then mixed the proteins from one strain with the RNA molecules from the other strain under conditions that result in the reconstitution of complete, infective viruses composed of proteins from one strain and RNA from the other strain. When tobacco leaves were infected with these reconstituted mixed viruses, the progeny viruses were always phenotypically and genotypically identical to the parent strain

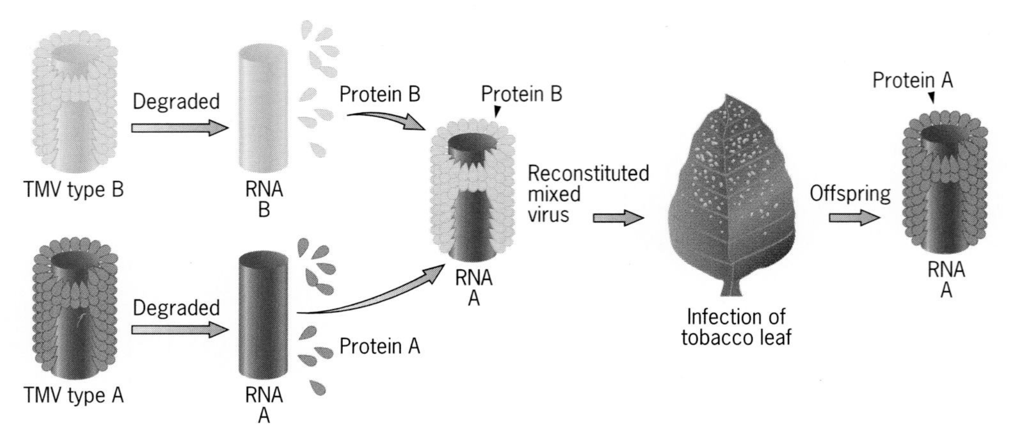

Figure 9.4 Demonstration that the genetic material of tobacco mosaic virus (TMV) is RNA, not protein. TMV contains no DNA; it is composed of just RNA and protein.

from which the RNA had been obtained (Figure 9.4). Thus the genetic information of TMV is stored in RNA, not in protein.

Key Points: **The genetic information of living organisms is stored in deoxyribonucleic acid (DNA). In some viruses, ribonucleic acid (RNA) is the genetic material.**

THE STRUCTURES OF DNA AND RNA

The genetic information of all living organisms, except the RNA viruses, is stored in DNA. What is the structure of DNA, and in what form is the genetic information stored? What features of the structure of DNA facilitate the accurate transmission of genetic information from generation to generation? The answers to these questions, defining the nature of the genetic code and the role of the complementary strands of the DNA double helix in the transmission of genetic information, are without doubt two of the most important facets of our understanding of the nature of life.

Nature of the Chemical Subunits in DNA and RNA

Nucleic acids, the major components of Miescher's nuclein, are macromolecules composed of repeating subunits called **nucleotides**. Each nucleotide is composed of (1) a phosphate group, (2) a five-carbon sugar (or pentose), and (3) a cyclic nitrogen-containing compound called a base (Figure 9.5). In DNA, the sugar is 2-deoxyribose (thus the name deoxyribonucleic acid); in RNA, the sugar is ribose (thus ribonucleic acid). Four different bases commonly are found in DNA: **adenine (A)**, **guanine (G)**, **thymine (T)**, and **cytosine (C)**. RNA also usually contains adenine, guanine, and cytosine but has a different base, **uracil (U)**, in place of thymine. Adenine and guanine are double-ring bases called **purines**; cytosine, thymine, and uracil are single-ring bases called **pyrimidines**. Both DNA and RNA, therefore, contain four different subunits or nucleotides, two purine nucleotides, and two pyrimidine nucleotides (Figure 9.6). In polynucleotides such as DNA and RNA, these subunits are joined together in long chains (Figure 9.7). RNA usually exists as a single-stranded polymer that is composed of a long sequence of nucleotides. DNA has one additional—and very important—level of organization; DNA is usually a double-stranded molecule.

DNA Structure: The Double Helix

The correct structure of DNA was first deduced in 1953 by James Watson and Francis Crick (Figure 9.8). Their double-helix model of DNA structure was based on two major kinds of evidence.

1. When Erwin Chargaff and colleagues analyzed the composition of DNA from many different organisms, they found that the concentration of thymine was always equal to the concentration of adenine and the concentration of cytosine was always equal to the concentration of guanine. Their results strongly suggested that thymine and adenine as well as cytosine and guanine were present in DNA in some fixed interrelationship. Their data also showed that the total concentration of pyrimidines (thymine plus cytosine) was always equal to the total concentration of purines (adenine plus guanine; see Table 9.2). In contrast, the [thymine + adenine]/[cytosine + guanine] ratio varied widely in DNAs of different species.

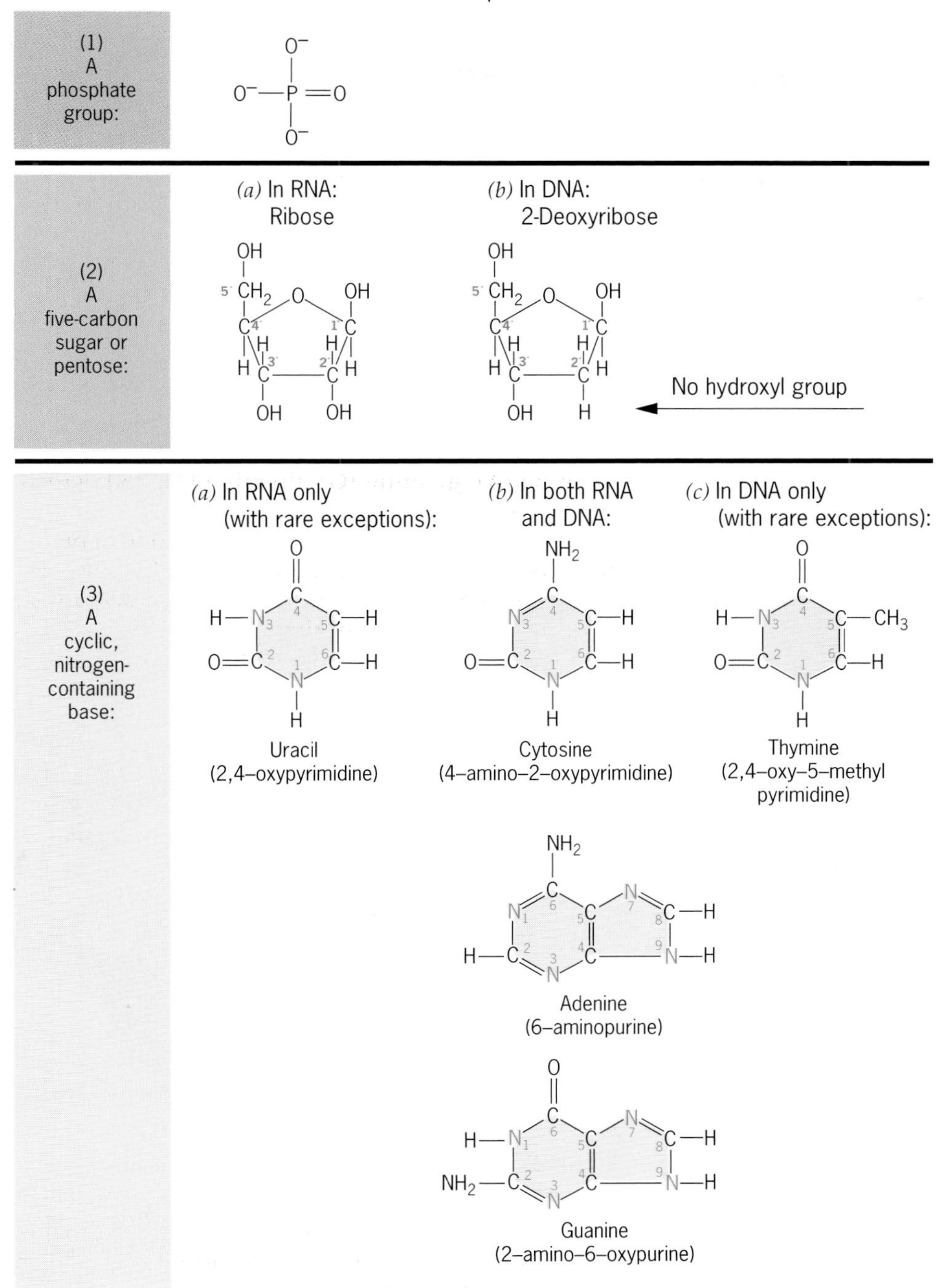

Figure 9.5 Structural components of nucleic acids.

2. When X rays are focused through crystals of purified molecules, the rays are deflected by the atoms of the molecules in specific patterns, called diffraction patterns, which provide information about the organization of the components of the molecules. These **X-ray diffraction patterns** can be recorded on X-ray-sensitive film just as patterns of light can be recorded with a camera and light-sensitive film. Watson and Crick used X-ray crystallographic data on DNA structure (Figure 9.9) provided by Maurice Wilkins, Rosalind Franklin, and their coworkers. These data indicated that DNA was a highly or-

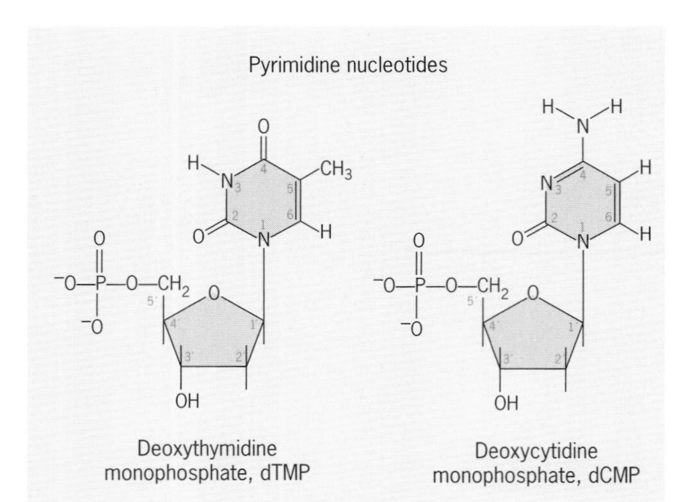

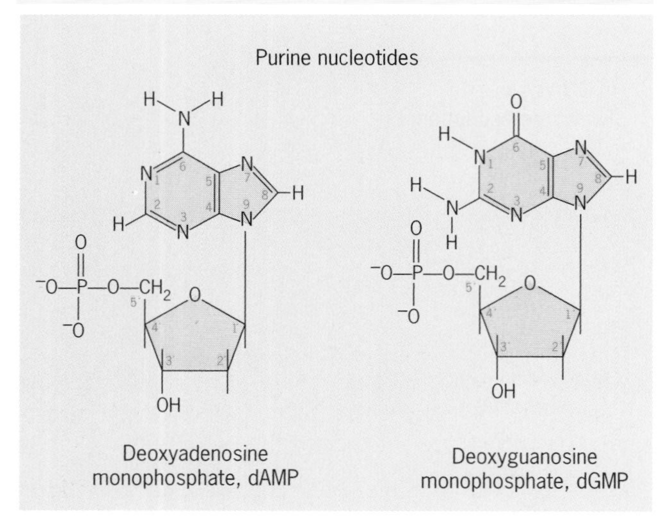

Figure 9.6 Structures of the four common deoxribonucleotides present in DNA.

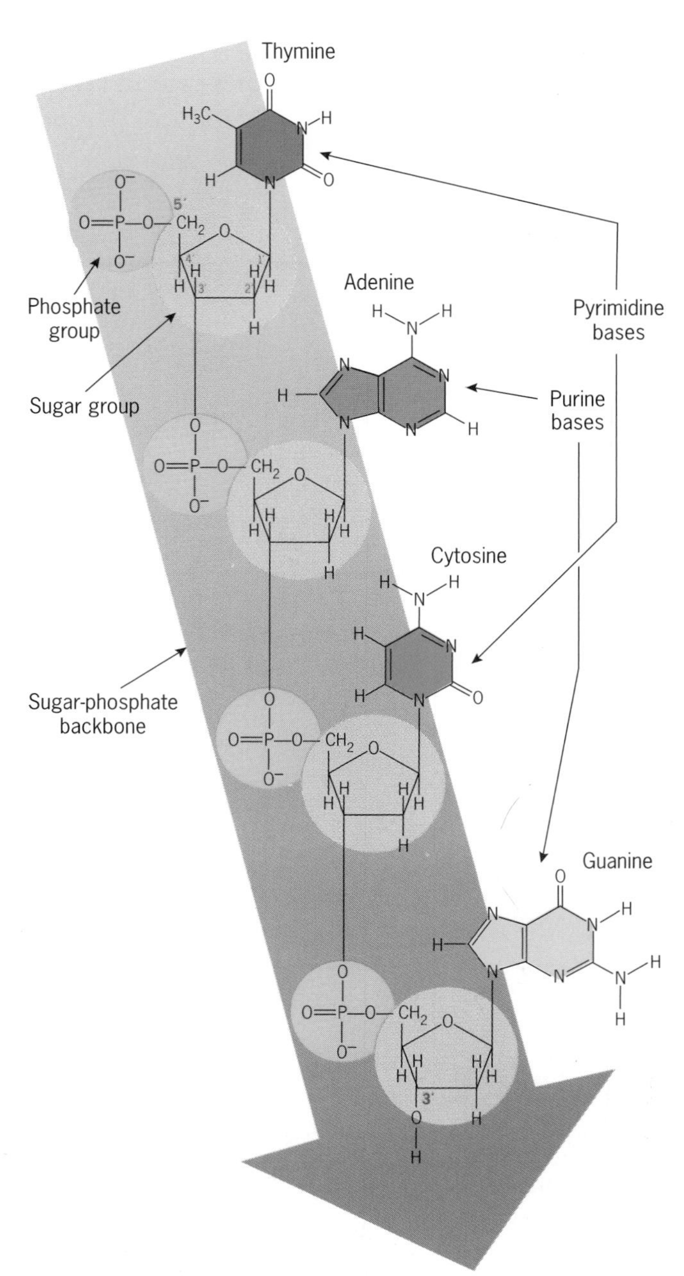

Figure 9.7 Formation of a polynucleotide chain by joining nucleotides with phosphodiester linkages. Note that the polynucleotide has a 5' to 3' chemical polarity (top to bottom); each phosphodiester linkage joins the 5' carbon of 2'-deoxyribose in one nucleotide to the 3' carbon of 2'-deoxyribose in the adjacent nucleotide, resulting in the chemical polarity of the polynucleotide chain.

Figure 9.8 James D. Watson (left) and Francis H. C. Crick.

TABLE 9.2
Base Composition of DNA from Various Organisms

Species	% *Adenine*	% *Guanine*	% *Cytosine*	% *Thymine*	*Molar Ratios* $\frac{A+G}{T+C}$	*Molar Ratios* $\frac{A+T}{G+C}$
I. **Viruses**						
Bacteriophage λ	26.0	23.8	24.3	25.8	0.99	1.08
Bacteriophage T2	32.6	18.1	16.6	32.6	1.03	1.88
Herpes simplex	13.8	37.7	35.6	12.8	1.06	0.36
Vaccinia	31.5	18.0	19.0	31.5	0.98	1.70
II. **Bacteria**						
Escherichia coli	26.0	24.9	25.2	23.9	1.04	1.00
Micrococcus lysodeikticus	14.4	37.3	34.6	13.7	1.07	0.39
Ramibacterium ramosum	35.1	14.9	15.2	34.8	1.00	2.32
III. **Fungi**						
Neurospora crassa	23.0	27.1	26.6	23.3	1.00	0.86
Aspergillus niger	25.0	25.1	25.0	24.9	1.00	1.00
Saccharomyces cerevisiae	31.7	18.3	17.4	32.6	1.00	1.80
IV. **Higher Eukaryotes**						
Zea mays (corn)	25.6	24.5	24.6	25.3	1.00	1.04
Nicotiana tabacum (tobacco)	29.3	23.5	16.5	30.7	1.12	1.50
Drosophila melanogaster	30.7	19.6	20.2	29.4	1.01	1.51
Homo sapiens (human)	30.2	19.9	19.6	30.3	1.01	1.53

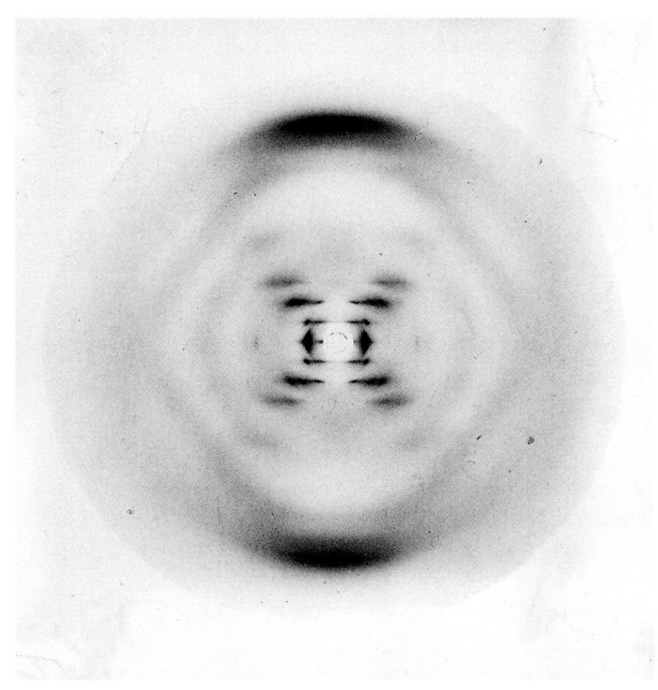

Figure 9.9 Photograph of the X-ray diffraction pattern obtained with DNA. The central cross-shaped pattern indicates that the DNA molecule has a helical structure, and the dark bands at the top and bottom indicate that the bases are stacked perpendicular to the axis of the molecule with a periodicity of 0.34 nm.

dered, two-stranded structure with repeating substructures spaced every 0.34 nanometer (1 nm = 10^{-9} meter) along the axis of the molecule.

On the basis of Chargaff's chemical data, Wilkins' and Franklin's X-ray diffraction data, and inferences from model building, Watson and Crick proposed that DNA exists as a right-handed **double helix** in which the two polynucleotide chains are coiled about one another in a spiral (Figure 9.10). Each polynucleotide chain consists of a sequence of nucleotides linked together by phosphodiester bonds, joining adjacent deoxyribose moieties (Table 9.3). The two polynucleotide strands are held together in their helical configuration by hydrogen bonding (Table 9.3) between bases in opposing strands; the resulting base pairs are stacked between the two chains perpendicular to the axis of the molecule like the steps of a spiral staircase (Figure 9.10). The base-pairing is specific: adenine is always paired with thymine, and guanine is always paired with cytosine. Thus all base pairs consist of one purine and one pyrimidine. The specificity of base-pairing results from the hydrogen-bonding capacities of the bases in their normal configurations (Figure 9.11). In their common structural configurations, adenine and thymine form two hydrogen bonds, and guanine and cytosine form three hydrogen bonds. Hydrogen bonding is not possible between cytosine and adenine or

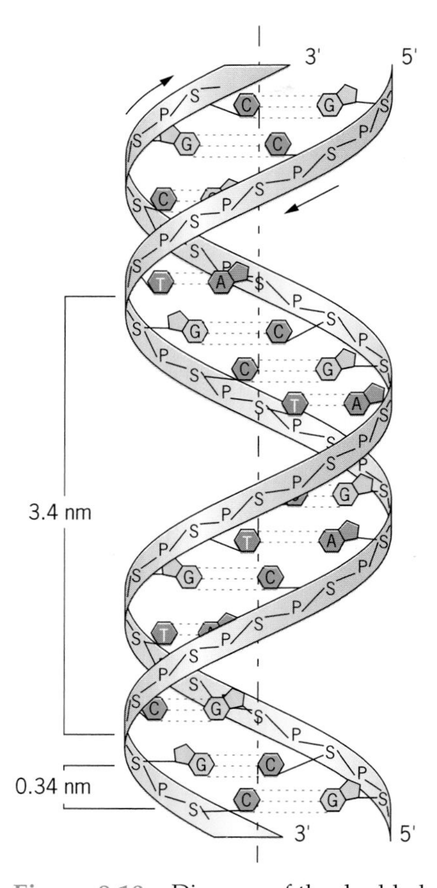

Figure 9.10 Diagram of the double-helix structure of DNA.

thymine and guanine when they exist in their common structural states.

Once the sequence of bases in one strand of a DNA double helix is known, the sequence of bases in the other strand is also known because of the specific base-pairing. The two strands of a DNA double helix are thus said to be complementary. *This property, the* ***complementarity*** *of the two strands of the double helix, makes DNA uniquely suited to store and transmit genetic information from generation to generation.*

The base pairs in DNA are stacked about 0.34 nm apart, with 10 base pairs per turn (360°) of the double helix (Figure 9.10). The sugar-phosphate backbones of the two complementary strands are **antiparallel;** that is, they have **opposite chemical polarity** (Figure 9.11). Unidirectionally along a DNA double helix, the phosphodiester bonds in one strand go from a 3' carbon of one nucleotide to a 5' carbon of the adjacent nucleotide, whereas those in the complementary strand go from a 5' carbon to a 3' carbon. This opposite polarity of the complementary strands of a DNA double helix plays an important role in DNA replication, transcription, and recombination.

The stability of DNA double helices results in part from the large number of hydrogen bonds between the base pairs (even though each hydrogen bond by itself is weak, much weaker than a covalent bond) and in part from the hydrophobic bonding (or stacking forces) between the stacked base pairs (Table 9.3). The

TABLE 9.3

Chemical Bonds Important in DNA Structure

(a) Covalent bonds

Strong chemical bonds formed by sharing of electrons between atoms.

(1) In bases and sugars

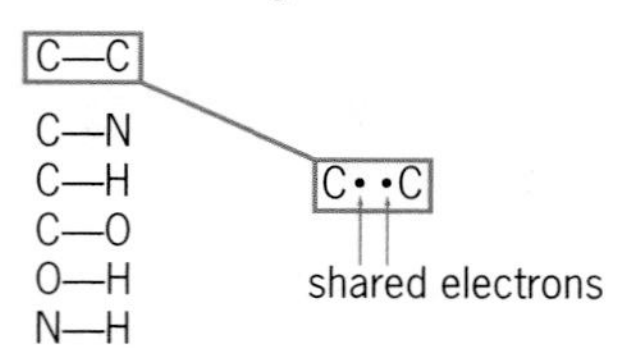

(2) In phosphodiester linkages

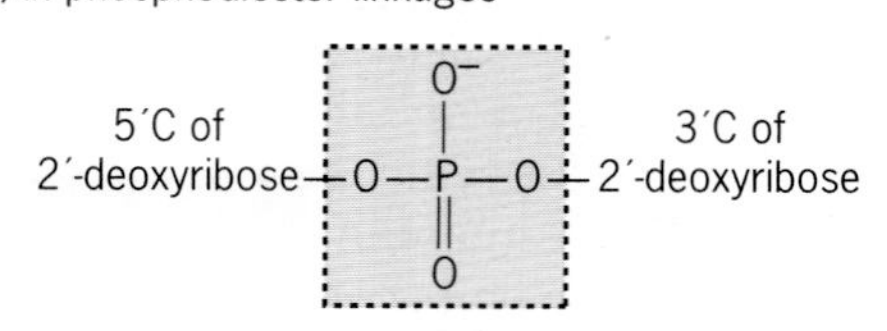

(b) Hydrogen bonds

A weak bond between an electronegative atom and a hydrogen atom (electropositive) that is covalently linked to a second electronegative atom.

N — H• • • • • O —

N — H• • • • • N

(c) Hydrophobic "bonds"

The association of nonpolar groups with each other when present in aqueous solutions because of their insolubility in water.

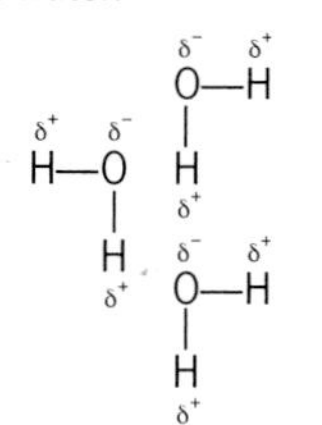

Water molecules are very polar (δ^- O and δ^+ Hs)
Compounds that are similarly polar are very soluble in water ("hydrophilic").
Compounds that are non-polar (no charged groups) are very insoluble in water ("hydrophobic").

The stacked base pairs provide a hydrophobic core.

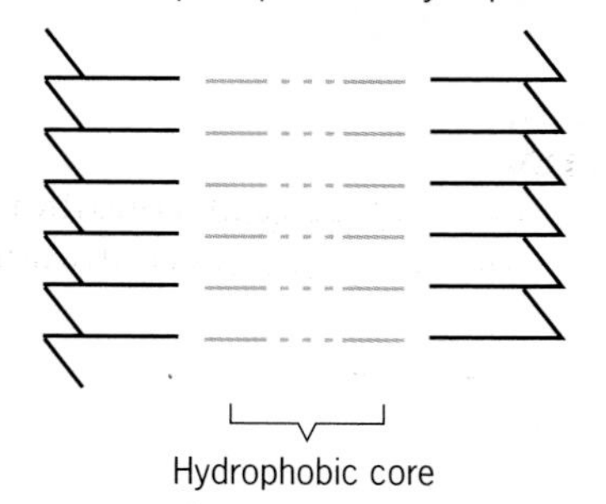

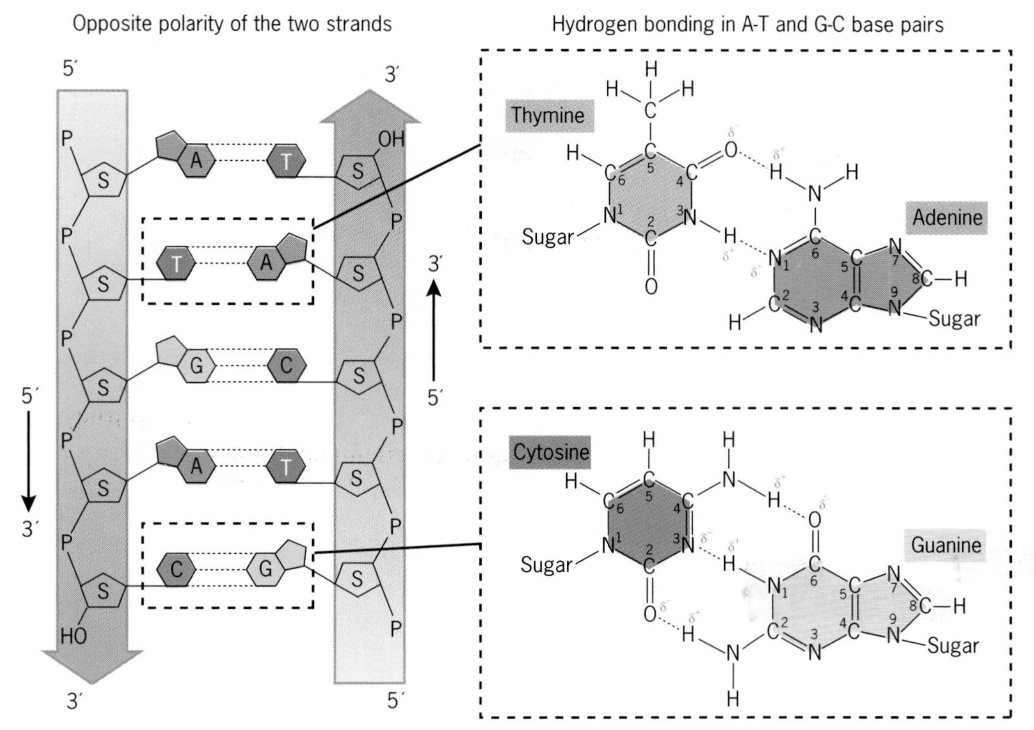

Figure 9.11 Diagram of a DNA double-helix, illustrating the opposite chemical polarity (see Fig. 9.7) of the two strands and the hydrogen bonding between thymine (T) and adenine (A) and between cytosine (C) and guanine (G). The base-pairing in DNA, T with A and C with G, is governed by the hydrogen-bonding potential of the bases.

stacked nature of the base pairs is best illustrated with a space-filling diagram of DNA structure (Figure 9.12). The planar sides of the base pairs are relatively nonpolar and thus tend to be hydrophobic (water insoluble). This hydrophobic core of stacked base pairs contributes considerable stability to DNA molecules present in the aqueous protoplasms of living cells. The space-filling drawing also shows that the two grooves of a DNA double helix are not identical; one, the major groove, is much wider than the other, the minor groove.

DNA Structure: Alternate Forms of the Double Helix

The Watson–Crick double-helix structure just described is called **B-DNA**. B-DNA is the conformation that DNA takes under physiological conditions (in aqueous solutions containing low concentrations of salts). The vast majority, if not all, of the DNA molecules present in the aqueous protoplasms of living cells exist in the B conformation. However, DNA is not a static, invariant molecule. To the contrary, DNA molecules exhibit a considerable amount of conformational flexibility.

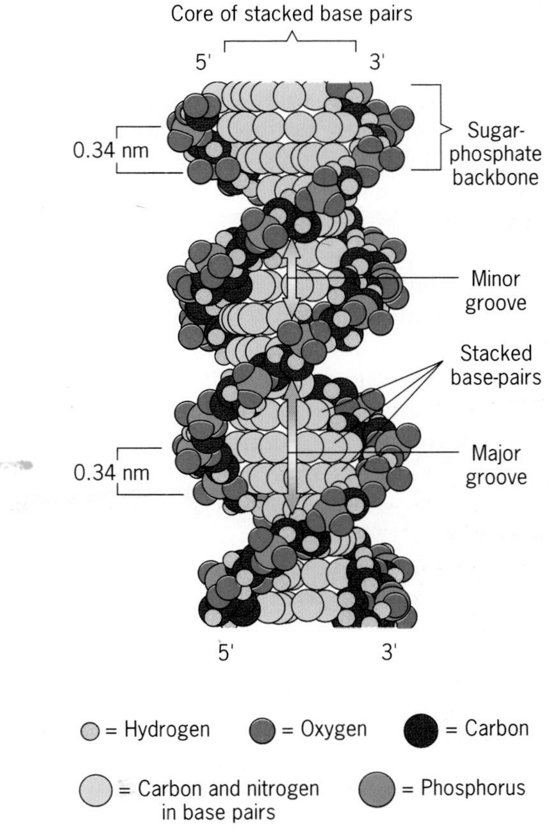

Figure 9.12 Space-filling diagram of a DNA double-helix.

TABLE 9.4
Alternate Forms of DNA

Helix Form	*Helix Direction*	*Base Pairs per Turn*	*Helix Diameter*
A	Right-handed	11	2.3 nm
B	Right-handed	10	1.9 nm
Z	Left-handed	12	1.8 nm

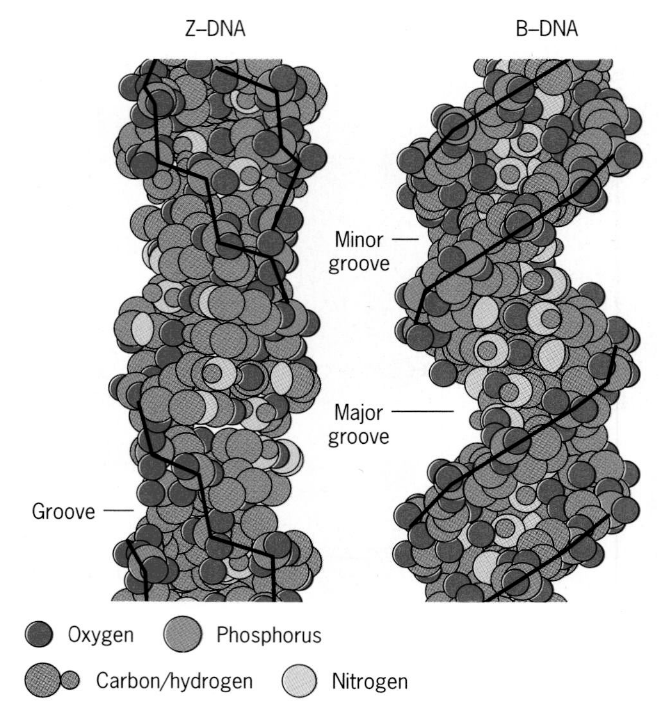

Figure 9.13 Comparison of the structures of Z-DNA and B-DNA. Note the left-handed winding and the zig-zagged paths of the sugar-phosphate backbones (solid lines) of the Z-DNA double-helix in contrast to the smooth paths and right-handed winding of the backbones in the B-DNA helix.

The structures of DNA molecules change as a function of their environment. The exact conformation of a given DNA molecule or segment of DNA molecule will depend on the nature of the molecules with which it is interacting. In fact, intracellular B-DNA appears to have an average of 10.4 nucleotide pairs per turn, rather than precisely 10 as shown in Figure 9.10. In high concentrations of salts or in a partially dehydrated state, DNA exists as **A-DNA**, which is a right-handed helix like B-DNA, but with 11 nucleotide pairs per turn (Table 9.4). A-DNA is a shorter, thicker double helix with a diameter of 0.23 nm. It has a narrow, deep major groove and a broad, shallow minor groove. DNA molecules almost certainly never exist as A-DNA *in vivo*. However, the A-DNA conformation is important because DNA-RNA heteroduplexes (double helices containing a DNA strand base-paired with a complementary RNA strand) or RNA-RNA duplexes exist in a very similar structure *in vivo*. Additional right-handed helical forms of DNA, designated C, D, and E, have been shown to exist under specific environmental conditions (dehydration, etc.). However, since these forms are not believed to exist in living cells, they will not be discussed here.

Certain DNA sequences have been shown to exist in a left-handed, double-helical form called **Z-DNA** (Z for the zigzagged path of the sugar-phosphate backbones of the structure). Z-DNA was discovered by X-ray diffraction analysis of crystals formed by DNA oligomers containing alternating G:C and C:G base pairs. Z-DNA occurs in double helices that are G:C-rich and contain alternating purine and pyrimidine residues. In addition to its unique left-handed helical structure, Z-DNA differs from B-DNA in having 12 base pairs per turn, a diameter of 0.18 nm, and a single deep groove (Table 9.4; Figure 9.13). Whether Z-DNA exists in living cells is still uncertain.

DNA Structure: Negative Supercoils *In Vivo*

All the functional DNA molecules present in living cells display one other very important level of organization—they are supercoiled. **Supercoils** are introduced into a DNA molecule when one or both strands are cleaved and when the complementary strands at one end are rotated or twisted around each other with the other end held fixed in space—and thus not allowed to spin. This supercoiling causes a DNA molecule to collapse into a tightly coiled structure similar to a coiled telephone cord or twisted rubber band (Figure 9.14, lower right). Supercoils are introduced into and removed from DNA molecules by enzymes that play essential roles in DNA replication (Chapter 10) and other processes.

Supercoiling occurs only in DNA molecules with fixed ends, ends that are not free to rotate. Obviously, the ends of the circular DNA molecules (Figure 9.14) present in most prokaryotic chromosomes and in the chromosomes of eukaryotic organelles such as mitochondria are fixed. The large linear DNA molecules present in eukaryotic chromosomes are also fixed by their attachment at intervals and at the ends to non-DNA components of the chromosomes. These attachments allow enzymes to introduce supercoils into the linear DNA molecules present in eukaryotic chromosomes, just as they are incorporated into the circular DNA molecules present in most prokaryotic chromosomes.

We can perhaps visualize supercoiling most easily by considering a circular DNA molecule. If we cleave

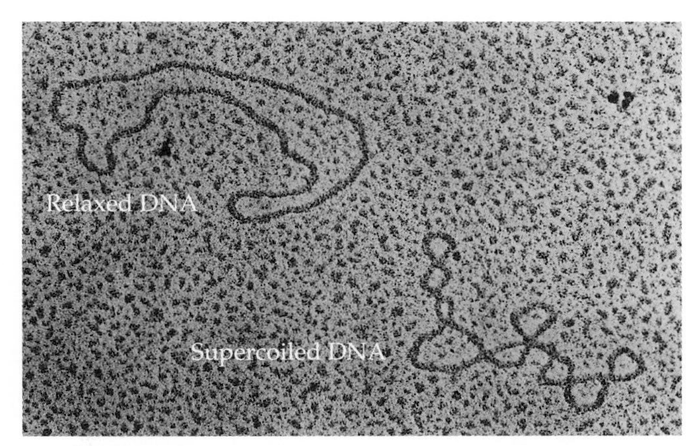

Figure 9.14 Comparison of the relaxed and negatively supercoiled structures of DNA. The relaxed structure is B-DNA with 10.4 base pairs per turn of the helix. The negatively supercoiled structure results when B-DNA is underwound, with less than one turn of the helix for every 10.4 base pairs.

one strand of a covalently closed, circular double helix of DNA, and rotate the end of one strand a complete turn (360°) around the complementary strand while holding the other end fixed, we will introduce one supercoil into the molecule (Figure 9.15). If we rotate the free end in the same direction as the DNA double helix is wound (right-handed), a positive supercoil (overwound DNA) will be produced. If we rotate the free end in the opposite direction (left-handed), a negative supercoil (underwound DNA) will result. Although this is the simplest way to define supercoiling in DNA, it is not the mechanism by which supercoils are produced in DNA *in vivo*. That mechanism is discussed in Chapter 10.

The DNA molecules of all organisms, from the smallest viruses to the largest eukaryotes, exhibit **negative supercoiling** (secondary coiling produced by left-handed rotations in DNA resulting in underwound double helices) *in vivo*, and many of the biological functions of chromosomes can be carried out only when the participating DNA molecules are negatively supercoiled. Considerable evidence indicates that negative supercoiling is involved in replication (Chapter 10), recombination, gene expression, and the regulation of gene expression. Supercoils are introduced into DNA by enzymes called **topoisomerases**, enzymes that catalyze changes in the topology of DNA molecules. The topoisomerase **DNA gyrase**, which catalyzes the formation of negative supercoils in DNA during its replication, has been isolated from several species, both prokaryotic and eukaryotic. The DNA gyrase of *E. coli* is inhibited by novobiocin and nalidixic acid, two potent inhibitors of DNA synthesis in bacteria; this result indicates that DNA gyrase activity is required for DNA replication. Similar amounts of negative supercoiling exist in the DNA molecules present in bacterial chromosomes and eukaryotic chromosomes.

Key Points: Nucleic acids are of two types: deoxyribonucleic acid and ribonucleic acid. DNA usually exists as a double helix, with the two strands held together by hydrogen bonds between the complementary bases: adenine paired with thymine and guanine paired with cytosine. The complementarity of the two strands of a double helix makes DNA uniquely suited to store and transmit genetic information from generation to generation. The two strands of a DNA double helix have opposite chemical polarity. RNA usually exists as a single-stranded molecule containing uracil instead of thymine. The DNA molecules present in chromosomes are negatively supercoiled.

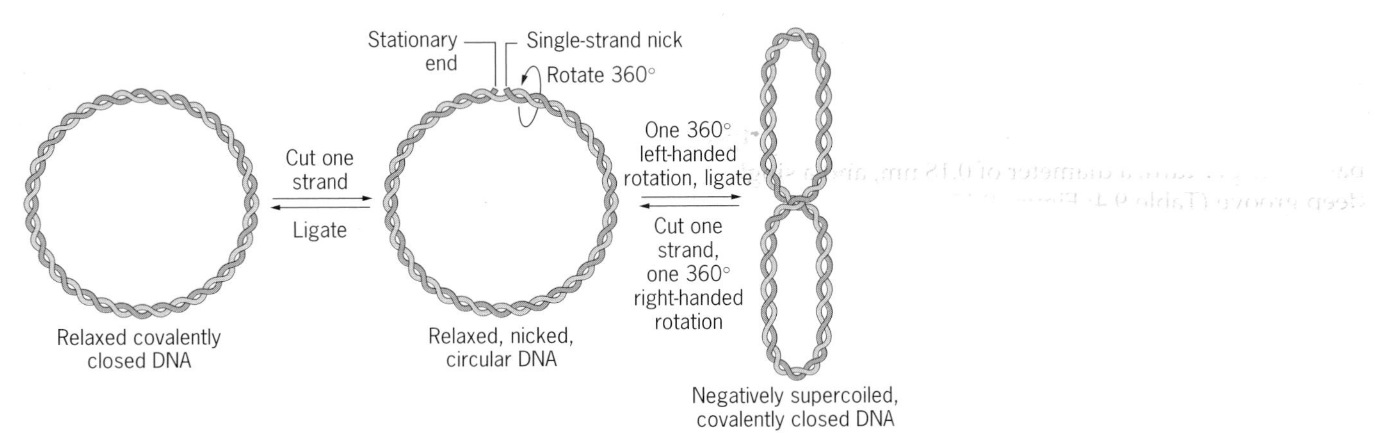

Figure 9.15 A visual definition of negatively supercoiled DNA. Although the structure of DNA supercoils is most clearly illustrated by the mechanism shown here, DNA supercoils are produced by a different mechanism *in vivo* (see Chapter 10).

CHROMOSOME STRUCTURE IN PROKARYOTES

Much of the information about the structure of DNA has come from studies of prokaryotes, primarily because they are less complex, both genetically and biochemically, than eukaryotes. Prokaryotes are monoploid (*mono* = one); they have only one set of genes (one copy of the genome). In most viruses and prokaryotes, the single set of genes is stored in a single chromosome, which in turn contains a single molecule of nucleic acid (either RNA or DNA).

The smallest known RNA viruses have only three genes, and the complete nucleotide sequences of the genomes of many viruses are known. For example, the single RNA molecule in the genome of bacteriophage MS2 consists of 3569 nucleotides and contains 4 genes. The smallest known DNA viruses have only 9 to 11 genes. Again, the complete nucleotide sequences are known in several cases. For example, the genome of bacteriophage ϕX174 is a single DNA molecule 5386 nucleotides in length that harbors 11 genes. The largest DNA viruses, like bacteriophage T2 and the animal pox viruses, contain about 150 genes. Bacteria like *E. coli* have 2500 to 3500 genes, most of which are present in a single molecule of DNA.

Until recently, prokaryotic chromosomes were often characterized as "naked molecules of DNA," in contrast to eukaryotic chromosomes with their associated proteins and complex morphology. This misconception resulted in part because (1) most of the published pictures of prokaryotic "chromosomes" were electron micrographs of isolated DNA molecules, not metabolically active or functional chromosomes, and (2) most of the published photographs of eukaryotic chromosomes were of highly condensed meiotic or mitotic chromosomes—again, metabolically inactive chromosomal states. Functional prokaryotic chromosomes, or nucleoids (nucleoids rather than nuclei because they are not enclosed in a nuclear membrane), are now known to bear little resemblance to the isolated viral and bacterial DNA molecules seen in electron micrographs, just as the metabolically active interphase chromosomes of eukaryotes have little morphological resemblance to mitotic or meiotic metaphase chromosomes.

The contour length of the circular DNA molecule present in the chromosome of the bacterium *Escherichia coli* is about 1100 μm. Since an *E. coli* cell has a diameter of only 1 to 2 μm, the large DNA molecule present in each bacterium must exist in a highly condensed (folded or coiled) configuration. When *E. coli* chromosomes are isolated by gentle procedures in the absence of ionic detergents (commonly used to lyse cells) and are kept in the presence of a high concentration of cations such as polyamines (small basic or positively charged proteins) or 1 M salt to neutralize the negatively charged phosphate groups of DNA, the chromosomes remain in a highly condensed state comparable in size to the nucleoid *in vivo*. This structure, called the **folded genome**, is the functional state of a bacterial chromosome. Though smaller, the functional intracellular chromosomes of bacterial viruses are very similar to the folded genomes of bacteria.

Within the folded genome, the large DNA molecule in an *E. coli* chromosome is organized into 50 to 100 **domains** or loops, each of which is independently negatively supercoiled (Figure 9.16). Both RNA and

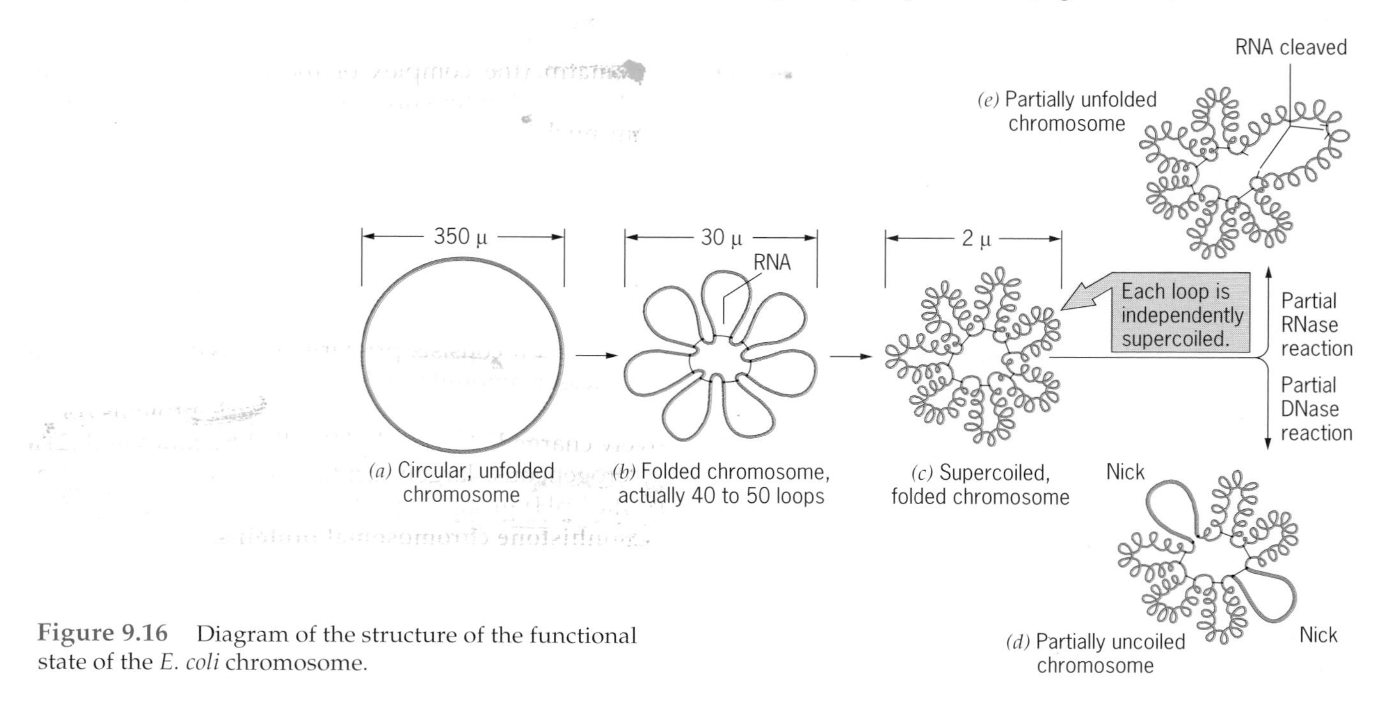

Figure 9.16 Diagram of the structure of the functional state of the *E. coli* chromosome.

protein are components of the folded genome, which can be partially relaxed by treatment with either deoxyribonuclease (DNase) or ribonuclease (RNase). Since each domain of the chromosome is independently supercoiled, the introduction of single-strand "nicks" in DNA by treatment of the chromosomes with a DNase that cleaves DNA at internal sites will relax the DNA only in the nicked domains, and all unnicked loops will remain supercoiled. Destruction of the RNA connectors by RNase will unfold the folded genome partially by eliminating the segregation of the DNA molecule into 50 to 100 loops. However, RNase treatment will not affect the supercoiling of the domains of the chromosome.

Key Points: **The DNA molecules in prokaryotic chromosomes are segregated into negatively supercoiled domains. Bacterial chromosomes contain circular molecules of DNA segregated into about 50 such domains.**

CHROMOSOME STRUCTURE IN EUKARYOTES

Eukaryotic genomes contain levels of complexity which are not encountered in prokaryotes. In contrast to prokaryotes, most eukaryotes are diploid, having two complete sets of genes, one from each parent. As we discussed in Chapter 6, some flowering plants are polyploid; that is, they carry several copies of the genome. Moreover, although eukaryotes have only about 2 to 25 times as many genes as *E. coli*, they have orders of magnitude more DNA (Figure 9.17). One of the fascinating questions currently being investigated is, what are the functions of this excess DNA that does not seem to carry genes, at least not genes encoding proteins or RNA molecules?

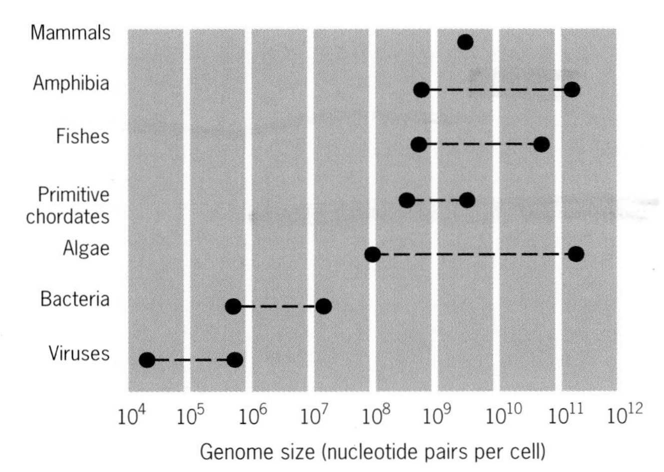

Figure 9.17 Increased genome size in organisms with increased developmental complexity.

Not only do most eukaryotes contain many times the amount of DNA of prokaryotes, but also this DNA is packaged in several chromosomes, and each chromosome is present in two (diploids) or more (polyploids) copies. Recall that the chromosome of *E. coli* has a contour length of 1100 μm, or about 1 mm. Now consider that the haploid chromosome complement, or genome, of the human contains about 1000 mm of DNA (or about 2000 mm per diploid cell). Moreover, this meter of DNA is subdivided among 23 chromosomes of variable size and shape, each chromosome containing from 15 to 85 mm of DNA. Until recently, geneticists had little information as to how this DNA was arranged in the chromosomes. Is there one molecule of DNA per chromosome as in prokaryotes, or are there many? If many, how are the molecules arranged relative to each other? How does the 85 mm (85,000 μm) of DNA in the largest human chromosome get condensed into a mitotic metaphase structure that is about 0.5 μm in diameter and 10 μm long? What are the structures of the metabolically active interphase chromosomes? We consider the answers to some of these questions in the following sections.

Chemical Composition of Eukaryotic Chromosomes

Interphase chromosomes are usually not visible with the light microscope. Moreover, electron microscopy of thin sections cut through eukaryotic nuclei has provided essentially no information about their structure. Recently, however, chemical analysis, electron microscopy, and X-ray diffraction studies on isolated **chromatin** (the complex of the DNA, chromosomal proteins, and other chromosome constituents isolated from nuclei) have provided a solid framework for a rapidly emerging picture of chromosome structure in eukaryotes.

When chromatin is isolated from interphase nuclei, the individual chromosomes are not recognizable. Instead, one observes an irregular aggregate of nucleoprotein. Chemical analysis of isolated chromatin shows that it consists primarily of DNA and proteins with lesser amounts of RNA (Figure 9.18). The proteins are of two major classes: (1) basic proteins (positively charged at neutral pH) called **histones** and (2) a heterogeneous, largely acidic (negatively charged at neutral pH) group of proteins collectively referred to as **nonhistone chromosomal proteins**.

Histones play a major structural role in chromatin. They are present in the chromatin of all higher eukaryotes in amounts equivalent to the amounts of DNA. This relationship suggests that an interaction occurs

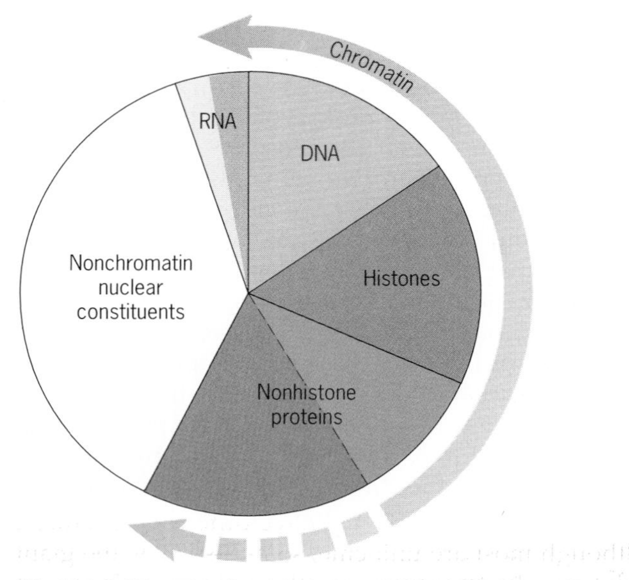

Figure 9.18 The chemical composition of chromatin as a function of the total nuclear content.

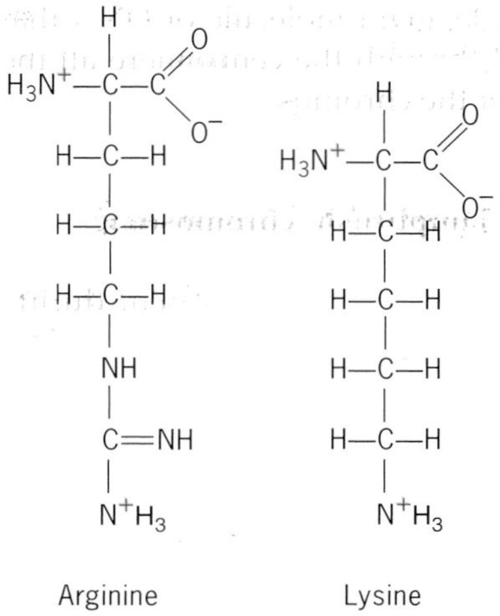

Arginine: $H_3N^+{-}CH(COO^-){-}CH_2{-}CH_2{-}CH_2{-}NH{-}C(=NH){-}N^+H_3$

Lysine: $H_3N^+{-}CH(COO^-){-}CH_2{-}CH_2{-}CH_2{-}CH_2{-}N^+H_3$

Figure 9.19 Structures of the amino acids arginine and lysine (at pH 7), which together account for 20 to 30 percent of the amino acid residues in histones.

between histones and DNA that is conserved in eukaryotes. The histones of all plants and animals consist of five classes of proteins. These five major histone types, called **H1, H2a, H2b, H3,** and **H4**, are present in almost all cell types. A few exceptions exist, most notably some sperm, where the histones are replaced by another class of small basic proteins called **protamines**.

The five histone types are present in molar ratios of approximately 1 H1: 2 H2a: 2 H2b: 2 H3: 2 H4. They are specifically complexed with DNA to produce the basic structural subunits of chromatin, small (approximately 11 nm in diameter by 6 nm high) ellipsoidal beads called **nucleosomes**. The histones have been highly conserved during evolution—four of the five types of histone are similar in all eukaryotes.

Most of the 20 amino acids in proteins are neutral in charge; that is, they have no charge at pH 7. However, a few are basic and a few are acidic. The histones are basic because they contain 20 to 30 percent arginine and lysine, two positively charged amino acids (Figure 9.19). The exposed $-NH_3^+$ groups of arginine and lysine allow histones to act as polycations. The positively charged side groups on histones are important in their interaction with DNA, which is polyanionic because of the negatively charged phosphate groups.

The remarkable constancy of histones H2a, H2b, H3, and H4 in all cell types of an organism and even between widely divergent species is consistent with the idea that they are important in chromatin structure (DNA packaging) and are only nonspecifically involved in the regulation of gene expression.

In contrast, the nonhistone protein fraction of chromatin consists of a large number of heterogeneous proteins. Moreover, the composition of the nonhistone chromosomal protein fraction varies widely among different cell types of the same organism. Thus the nonhistone chromosomal proteins probably do not play central roles in the packaging of DNA into chromosomes. Instead, they are likely candidates for roles in regulating the expression of specific genes or sets of genes.

One Large DNA Molecule per Chromosome

A typical eukaryotic chromosome contains from 1 to 20 cm (10^4 to 2×10^5 μm) of DNA. During metaphase of meiosis and mitosis, this DNA is packaged in a chromosome with a length of only 1 to 10 μm. How is all of this DNA condensed into the compact chromosomes that are present during mitosis and meiosis? Do many DNA molecules run parallel throughout the chromosome—the **multineme** or "multistrand" model—or is there just one DNA double helix extending from one end of the chromosome to the other—the **unineme** or "single-strand" model? (Note that strand here refers to the DNA double helix, not the individual polynucleotide chains of DNA.) Are there many DNA molecules joined end-to-end or arranged in some other fashion in the chromosome, or does one giant, continuous molecule of DNA extend from one end to the other in a highly coiled and folded form? The evidence supporting the unineme model of chromosome structure is now conclusive. In addition, solid evidence supports the concept of chromosome-size DNA molecules. That is, each chromosome appears to contain a single, giant molecule of DNA that extends from one end through the centromere all the way to the other end of the chromosome.

Some of the strongest evidence supporting the unineme model of chromosome structure has come from studies of the **lampbrush chromosomes** (so-named because they resemble the brushes used to clean the mantles of kerosene lamps) present during prophase I of oogenesis in many vertebrates, particularly amphibians. Lampbrush chromosomes are up to 800 μm long; thus they provide favorable material for cytological studies. The homologous chromosomes are paired, and each has duplicated to produce two chromatids at the lampbrush stage. Each lampbrush chromosome contains a central axial region, where the two chromatids are highly condensed, and numerous pairs of lateral loops (Figure 9.20). The loops are transcriptionally active regions of single chromatids. The integrity of both the central axis and the lateral loops depends on DNA. Treatment with DNase fragments both the axis and the loops. Treatment with RNase or proteases removes surrounding matrix material, but does not destroy the continuity of either the axis or the loops. Electron microscopy of RNase- and protease-treated lampbrush chromosomes reveals a central filament of about 2 nm in diameter in the lateral loops. Since each loop is a segment of one chromatid, and since the diameter of a DNA double helix is 1.9 nm, these lampbrush chromosomes must be unineme structures (Figure 9.20). This conclusion is supported by studies on the kinetics of nuclease digestion of lampbrush chromosomes. That is, the kinetics observed are those expected if the central filament of each loop is one DNA double helix. The axial region then contains two DNA molecules, one from each of the two tightly paired chromatids.

Lampbrush chromosomes are germ-line chromosomes. Thus their structure is particularly relevant to an understanding of genetic phenomena. Chromosomes of somatic cells may have different structures. Although most are unineme, some—such as the giant polytene chromosomes in the salivary glands of *Drosophila*—are known to be multineme structures.

The question of whether the unineme chromosomes of eukaryotes contain a single large molecule of DNA or many smaller molecules linked end-to-end

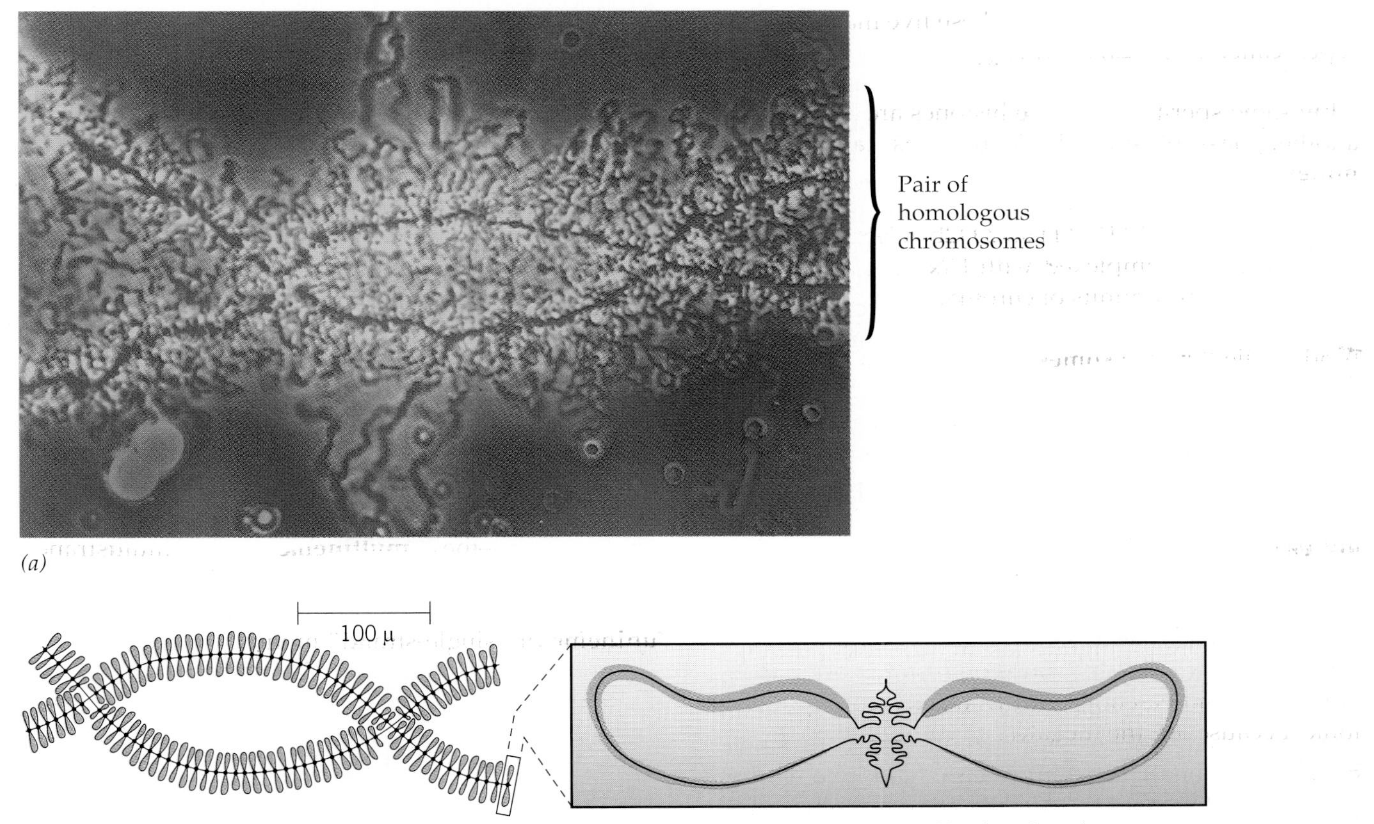

(*b*) Diagram of the chromosome pair shown above.

(*c*) Diagram of the two chromatids of one homolog.

Figure 9.20 Phase contrast micrograph (*a*) and diagram (*b*) of a pair of lampbrush chromosomes in an oocyte of the newt, *Triturus viridescens.* The structures of axial and loop regions of the two chromatids of a single lampbrush chromosome are shown in (*c*). The central element in each chromatid (both axial regions and lateral loops) is a single molecule of DNA. The matrix is primarily nascent RNA that is synthesized on the extended DNA in the loop regions.

›roven difficult to answer with rigorous experi-... tal evidence. A centimeter-long molecule of DNA has a length-to-width (diameter) ratio of 5 million to 1. Such a structure is extremely sensitive to shearing. If such a DNA molecule is in solution in a test tube, the slightest vibration will break the molecule into many fragments. For this and other reasons, accurate estimates of the sizes of eukaryotic DNAs cannot be obtained with the procedures used to analyze prokaryotic DNA molecules. However, by modifying old techniques and developing some new ones and applying these to the problem, scientists have obtained solid evidence indicating that each eukaryotic chromosome, no matter how large, contains one giant DNA double helix.

Some lower eukaryotic organisms, such as the mold *Neurospora crassa* and the yeast *Saccharomyces cerevisiae*, have relatively small chromosomes. In the case of these organisms, a procedure called **pulsed-field gel electrophoresis** has been used to demonstrate that each chromosome contains a single molecule of DNA. The technique of gel electrophoresis is a powerful tool for separating macromolecules such as proteins and nucleic acids based on their size and charge (see Chapter 20). A semisolid gel (usually polyacrylamide or agarose) provides an inert matrix with pores in a given size range through which the macromolecules migrate when placed in an electric field. Positively charged molecules migrate toward the cathode (the negative electrode), and negatively charged molecules move toward the anode (the positive electrode). Proteins may be either positively or negatively charged, depending on their amino acid composition. Nucleic acids are negatively charged with one phosphate group per nucleotide. Thus nucleic acids have an approximately constant charge per unit of mass and would all migrate at the same rate in the absence of sieving. However, polyacrylamide gels have relatively small pores, and agarose gels have somewhat larger pores. These gels act as molecular sieves such that small molecules migrate faster than larger molecules with the same charge per unit of mass. As a result, the rate of migration of a nucleic acid during gel electrophoresis is almost exclusively a function of its size. Sometimes, conformation is a factor; for example, supercoiled DNAs migrate faster than relaxed molecules of the same size.

Pulsed-field gel electrophoresis differs from standard gel electrophoresis in that instead of a single (one-dimensional), constant electric field, two electric fields offset by about 90° are applied across the gel in an alternating or pulsed manner. In standard gel electrophoresis, the DNA molecules pass through the gel in an end-first or snakelike fashion. In pulsed-field gel electrophoresis, the application of intermittent and alternating electric fields requires the molecules to reorient themselves before continuing to migrate through the gel. Larger molecules take longer to undergo these reorientation events and move more slowly. As a result, pulsed-field gel electrophoresis yields superior separation of very large DNA molecules. When this technique was used to separate intact DNA molecules from the fungi *N. crassa* and *S. cerevisiae*, the results showed that the number of different sized DNA molecules was equal to the number of nonhomologous chromosomes in these species (Figure 9.21).

Unfortunately, the very large DNA molecules present in the chromosomes of higher eukaryotes such as *Drosophila* and humans cannot be separated even by pulsed-field gel electrophoresis. Researchers have used additional approaches in attempts to demonstrate that the large chromosomes of higher animals and plants each contain one molecule of DNA. Autoradiography and viscoelastometry are two approaches that have yielded important results.

Autoradiography is a method for detecting and localizing radioactive isotopes in cytological preparations or macromolecules by exposure to a photographic emulsion that is sensitive to low-energy radiation. The emulsion contains silver halides that produce tiny black spots—often called silver grains—when they are exposed to the charged particles emit-

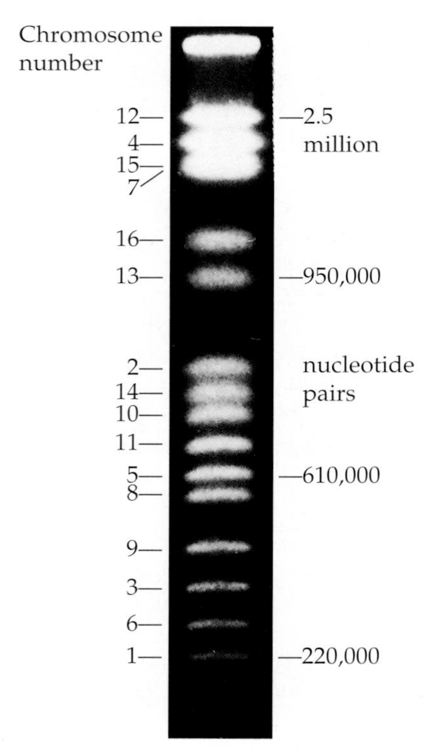

Figure 9.21 Separation of the chromosome-size DNA molecules of the yeast *Saccharomyces cerevisiae* by pulse-field agarose gel electrophoresis. The large DNA molecules present in 16 of the 17 yeast chromosomes can be resolved by this procedure.

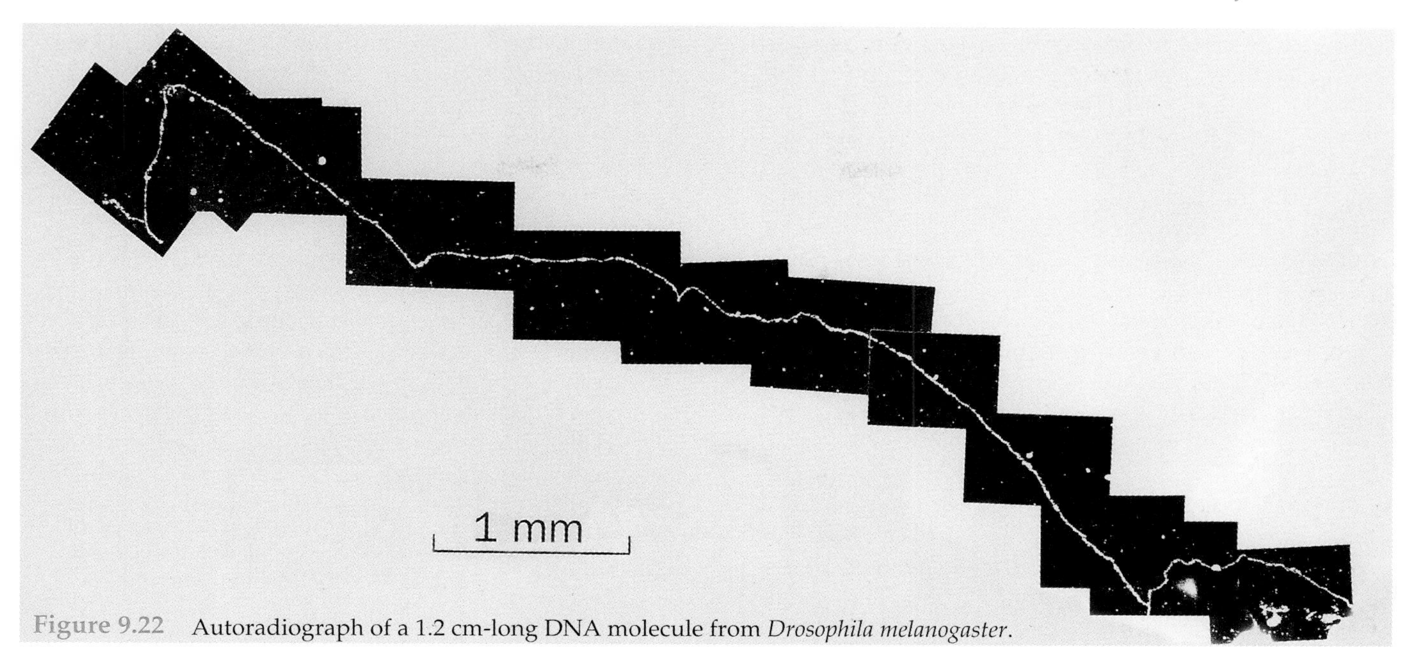

Figure 9.22 Autoradiograph of a 1.2 cm-long DNA molecule from *Drosophila melanogaster*.

ted during the decay of radioactive isotopes. Autoradiography permits a researcher to prepare an image of the localization of radioactivity in macromolecules, cells, or tissues just as photography permits us to make a picture of what we see. The difference is that the film used for autoradiography is sensitive to radioactivity, whereas the film we use in a camera is sensitive to visible light. Autoradiography is particularly useful in studying DNA metabolism because DNA can be specifically labeled by growing cells on [^{3}H]thymidine, the tritiated deoxyribonucleoside of thymine. Thymidine is incorporated almost exclusively into DNA; it is not present in any other major component of the cell.

Ruth Kavenoff, Lynn Klotz, and Bruno Zimm grew *Drosophila* cells in culture medium containing [^{3}H]thymidine for 24 hours, lysed the cells gently so as not to break the chromosomal DNA molecules, and carefully collected the DNA molecules on protein-coated glass slides. They then covered the slides with emulsion sensitive to β-particles (the low-energy electrons emitted during decay of tritium) and stored them in the dark for a period of time to allow sufficient radioactive decays. The greatest challenge to Kavenoff and her coworkers was to spread out the molecules with no tangles or overlaps on the slides so that the entire length of a molecule would be visible. Their best autoradiographs showed DNA molecules with contour lengths of up to 1.2 cm (Figure 9.22). DNA molecules of this length would have a mass of about 3×10^{10} daltons (one dalton is the mass of one hydrogen atom) and would contain about two-thirds of the DNA known to be present in the largest chromosomes of *D. melanogaster*. Thus these results provide support for the concept of chromosome-size DNA molecules in *Drosophila*.

Kavenoff and colleagues also used a technique called **viscoelastometry**, a procedure for analyzing the viscosity of molecules in solution, to determine the sizes of the DNA molecules in the largest *Drosophila* chromosomes. They measured the sizes of the largest DNA molecules present in *Drosophila* cells growing in suspension cultures. Kavenoff and coworkers' viscoelastometric data indicate that the largest DNA molecules in *Drosophila* have a mass of 4.1×10^{10} daltons. Since the largest chromosomes of *Drosophila* have been shown to contain about 4.3×10^{10} daltons of DNA (total, whether one molecule or many) by direct biochemical analysis, the viscoelastometric estimate of the size of the largest DNA molecules in *Drosophila* nuclei correlates almost exactly with the total amount of DNA present in the largest chromosome.

These and other results have provided convincing evidence that each eukaryotic chromosome contains one long double helix of DNA extending from one end of the chromosome through the centromere all the way to the other end. However, as we will discuss in the following section, this giant DNA molecule is highly condensed (coiled and folded) within the chromosome.

Three Levels of DNA Packaging in Eukaryotic Chromosomes

The largest chromosome in the human genome contains about 85 mm (85,000 μm, or 8.5×10^7 nm) of DNA that is believed to exist as one giant molecule.

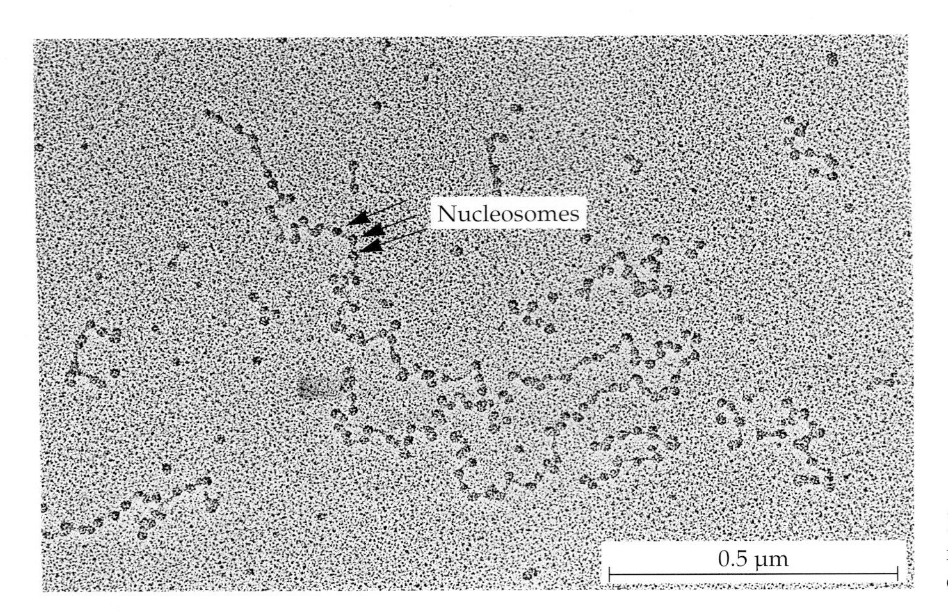

Figure 9.23 Electron micrograph of rat liver chromatin showing the beads-on-a-string nucleosome substructure.

This DNA molecule somehow gets packaged into a metaphase structure that is about 0.5 μm in diameter and about 10 μm in length—a condensation of almost 10^4-fold in length from the naked DNA molecule to the metaphase chromosome. How does this condensation occur? What components of the chromosomes are involved in the packaging processes? Are DNA molecules packaged in different chromosomes in different ways, or is there a universal packaging scheme? Are there different levels of packaging? Clearly, meiotic and mitotic chromosomes are more extensively condensed than are interphase chromosomes. What additional levels of condensation occur in these special structures that are designed to assure the proper segregation of the genetic material during cell divisions? Are DNA sequences of genes that are being expressed packaged differently than those of genes that are not being expressed? Let us investigate some of the evidence that establishes the existence of three different levels of packaging of DNA into chromosomes.

When isolated chromatin is examined by electron microscopy, it is found to consist of a series of ellipsoidal beads (about 11 nm in diameter and 6 nm high) joined by thin threads (Figure 9.23). Further evidence for a regular, periodic packaging of DNA has come from studies on the digestion of chromatin with various nucleases. These studies have shown that segments of DNA 146 nucleotide pairs in length were somehow protected from degradation by nucleases. Moreover, partial digestion of chromatin with these nucleases yielded fragments of DNA in a set of discrete sizes that were integral multiples of the smallest size fragment. These results are nicely explained if chromatin has a repeating structure, supposedly the bead seen by electron microscopy (Figure 9.23), within which the DNA is packaged in a nuclease-resistant form (Figure 9.24). This "bead" or chromatin subunit is called the **nucleosome**. According to the present concept of chromatin structure, the **linkers**, or inter-bead threads of DNA, are susceptible to nuclease attack.

After partial digestion of the DNA in chromatin with an endonuclease (an enzyme that cleaves DNA internally), DNA approximately 200 nucleotide pairs in length is associated with each nucleosome (produced by a cleavage in each linker region). After extensive nuclease digestion, a 146-nucleotide-pair-long

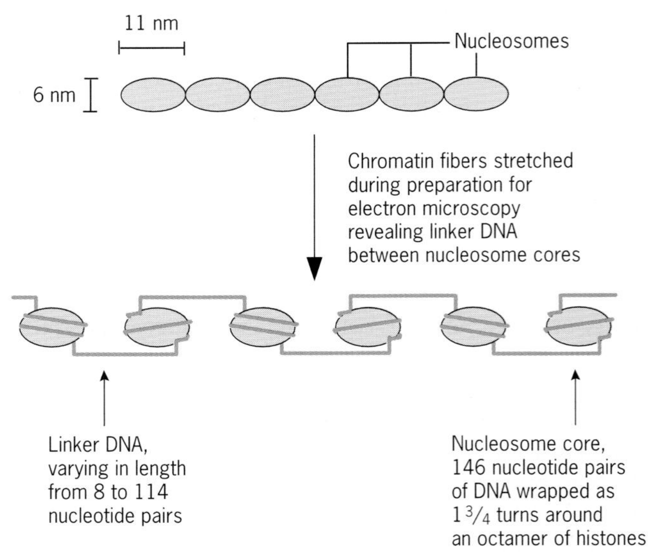

Figure 9.24 Diagram of the nucleosome structure of chromatin.

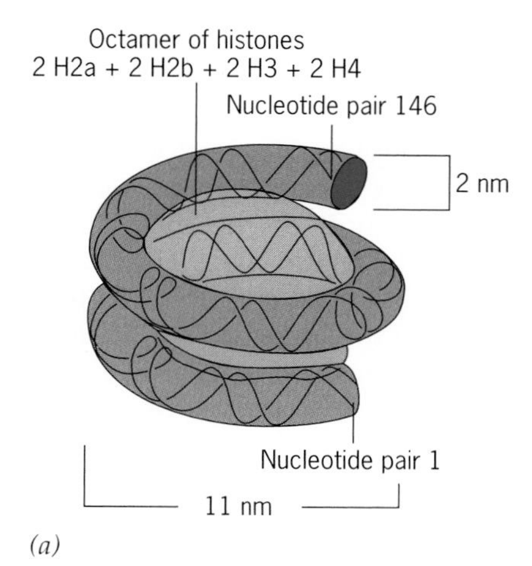

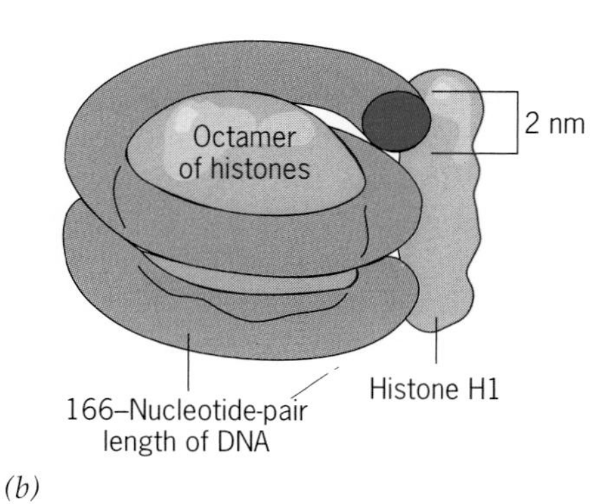

Figure 9.25 Diagrams of the macromolecular structure of the nucleosome core (*a*) and the complete nucleosome (*b*).

segment of DNA remains present in each nucleosome. This nuclease-resistant structure is called the **nucleosome core**. Its structure—essentially invariant in all eukaryotes—consists of a 146-nucleotide-pair length of DNA and two molecules each of histones H2a, H2b, H3, and H4. The histones protect the segment of DNA in the nucleosome core from cleavage by endonucleases. Physical studies (X-ray diffraction and similar analyses) of nucleosome-core crystals have shown that the DNA is wound as $1\frac{3}{4}$ turns of a superhelix around the outside of the histone octamer (Figure 9.25*a*).

The complete chromatin subunit consists of the nucleosome core, the linker DNA, an average of one molecule of histone H1, and the associated nonhistone chromosomal proteins. However, it has not been firmly established that histone H1 is evenly distributed, one molecule per nucleosome or linker, in chromatin. The size of the linker DNA varies from species to species and from one cell type to another. Linkers as short as eight nucleotide pairs and as long as 114 nucleotide pairs have been reported. Evidence suggests that the complete nucleosome (as opposed to the nucleosome core) contains one molecule of histone H1, which stabilizes two full turns of DNA superhelix (a 166-nucleotide-pair length of DNA) on the surface of the histone octamer (Figure 9.25*b*).

The basic structural component of eukaryotic chromatin clearly is the nucleosome. But are the structures of all nucleosomes the same? What role(s), if any, does nucleosome structure play in gene expression and the regulation of gene expression? The structure of nucleosomes in genetically active regions of chromatin is known to differ from that of nucleosomes in genetically inactive regions. But what are the details of this structure–function relationship? Present and future studies on the fine structure of nucleosomes will undoubtedly prove informative with regard to these and other questions.

Electron micrographs of isolated metaphase chromosomes show masses of tightly coiled or folded lumpy fibers (Figure 9.26). These **chromatin fibers** have an average diameter of 30 nm. When the structures seen by light and electron microscopy during earlier stages of meiosis are compared, it becomes clear that the light microscope simply permits one to

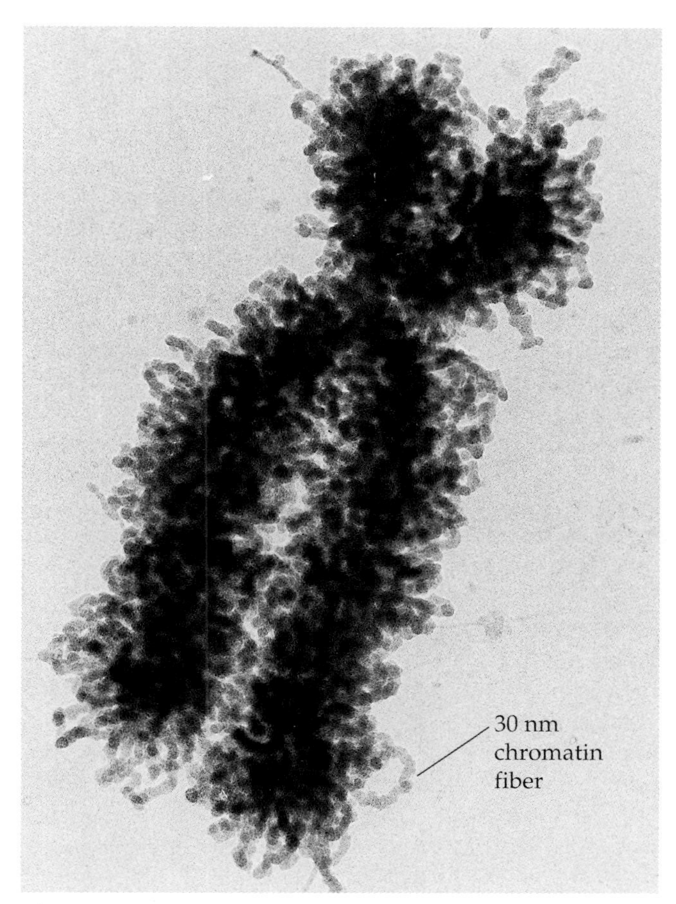

Figure 9.26 Electron micrograph of a human metaphase chromosome showing the presence of 30-nm chromatin fibers. The available evidence indicates that each chromatid contains one large, highly coiled or folded 30-nm fiber.

see those regions where these 30-nm fibers are tightly packed or condensed.

What is the substructure of the 30-nm fiber seen in mitotic and meiotic chromosomes? Although biologists do not have a definitive answer to this question, they do know that the DNA is wound as a supercoil about a histone octamer to yield the roughly 10 nm in diameter nucleosome. *In vivo*, the nucleosomes are probably in direct juxtaposition with each other without detectable linker regions; if so, they will form a 10-nm nucleosome fiber (see Figure 9.24, top). If this 10-nm fiber, in turn, is wound in a higher-order supercoil (a solenoid), a 30-nm fiber can be generated. Although scientists still do not understand all the details of the structure of this 30-nm chromatin fiber, there is good evidence that it represents a solenoidlike structure such as that shown in Figure 9.27.

Metaphase chromosomes contain the maximum degree of condensation observed in normal eukaryotic chromosomes. Clearly, the role of these highly condensed chromosomes is to organize and package the giant DNA molecules of eukaryotic chromosomes into structures that will facilitate their segregation to daughter nuclei without the DNA molecules of different chromosomes becoming entangled and, as a result, being broken during the anaphase separation of the daughter chromosomes. As we noted in the preceding section, the basic structural unit of the metaphase chromosome is the 30-nm chromatin fiber. However, the next obvious question is, how are these 30-nm fibers further condensed into the observed metaphase structure? Unfortunately, there is still no clear answer to this question. There is evidence that the gross structure of metaphase chromosomes is not dependent on histones. Electron micrographs of isolated metaphase chromosomes from which the histones have been removed reveal a **scaffold**, or central core, which is surrounded by a huge pool or halo of DNA (Figure 9.28). This chromosome scaffold must be composed of nonhistone chromosomal proteins. Note the absence of any apparent ends of DNA molecules in the micrograph shown in Figure 9.28; this finding again supports the concept of one giant DNA molecule per chromosome.

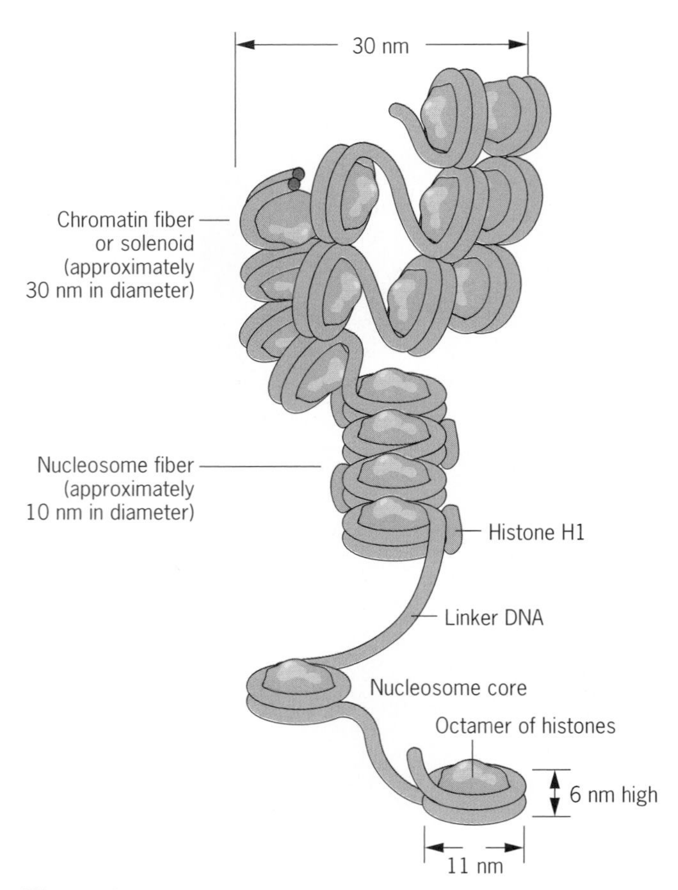

Figure 9.27 Diagram of the solenoid model of the 30-nm chromatin fiber. Histone H1 appears to stabilize the 10-nm nucleosome fiber and contribute to the formation of the 30-nm chromatin fiber.

In summary, at least three levels of condensation are required to package the 10^3 to 10^5 μm of DNA in a eukaryotic chromosome into a metaphase structure a few microns long. (1) The first level of condensation involves packaging DNA as a supercoil into nucleosomes, to produce the 10-nm-diameter interphase chromatin fiber. This clearly involves an octamer of histone molecules, two each of histones H2a, H2b, H3, and H4. (2) The second level of condensation involves an additional folding or supercoiling of the 10-nm nucleosome fiber, to produce the 30-nm chromatin fiber characteristic of mitotic and meiotic chromosomes. Histone H1 is involved in this supercoiling of the 10-nm nucleosome fiber to produce the 30-nm chromatin fiber. (3) Finally, nonhistone chromosomal proteins form a scaffold that is involved in condensing the 30-nm chromatin fiber into the tightly packed metaphase chromosomes. This third level of condensation appears to involve the separation of segments of the giant DNA molecules present in eukaryotic chromosomes into independently supercoiled domains or loops (Figure 9.28). The mechanism by which this third level of condensation occurs is not known.

Centromeres and Telomeres

As we discussed in Chapter 2, the two homologous chromosomes (each containing two sister chromatids) of each chromosome pair separate to opposite poles of the meiotic spindle during anaphase I of meiosis. Similarly, during anaphase II of meiosis and the single anaphase of mitosis, the sister chromatids of each chromosome move to opposite spindle poles and be-

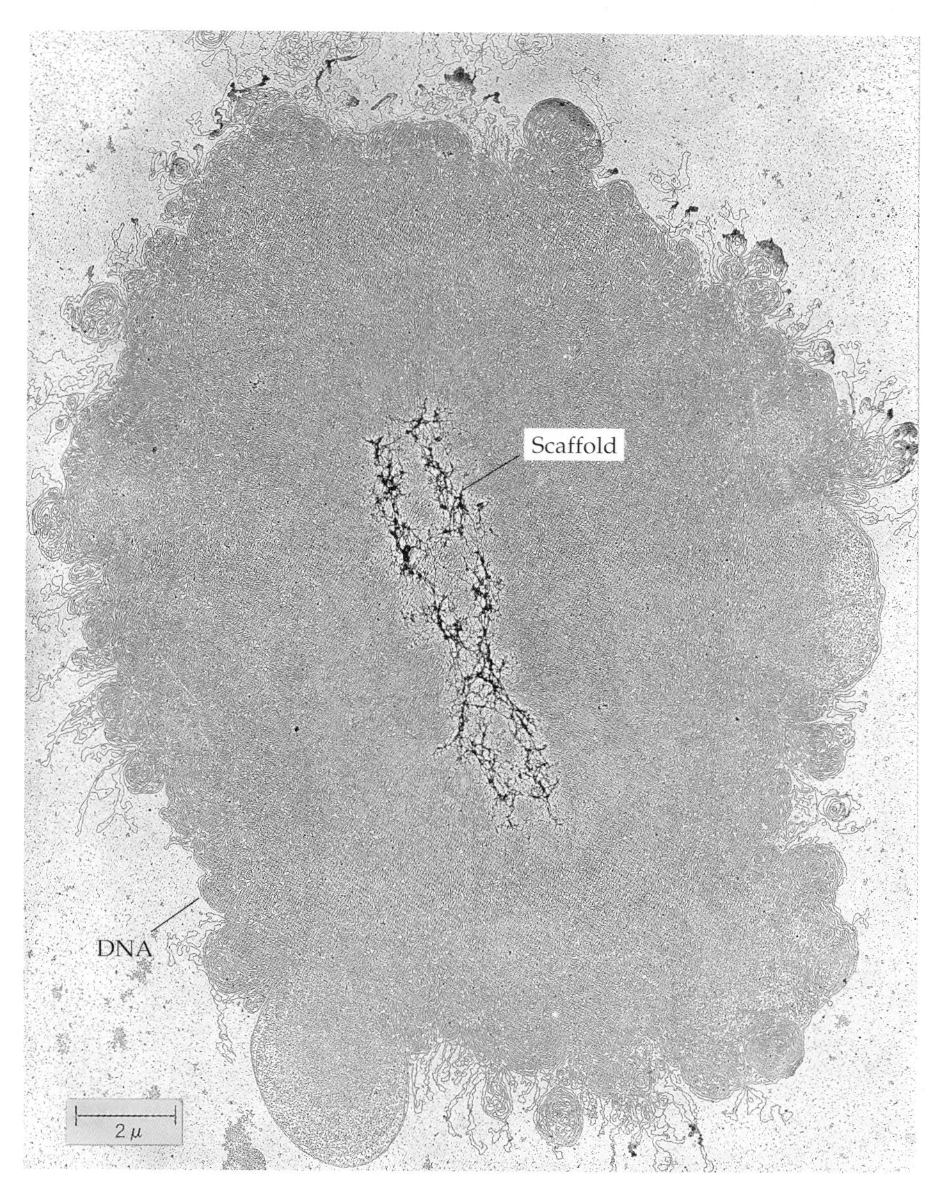

Figure 9.28 Electron micrograph of a human metaphase chromosome from which the histones have been removed. A huge pool of DNA surrounds a central "scaffold" composed of nonhistone chromosomal proteins. Note that the scaffold has roughly the same shape as the metaphase chromosome prior to the removal of the histones.

come daughter chromosomes. These anaphase movements depend on the attachment of spindle microtubules to specific regions of the chromosomes, the centromeres. Because all centromeres perform the same basic function, it is not surprising that the centromeres of different chromosomes of a species contain similar structural components.

The centromeres of metaphase chromosomes can usually be recognized as constricted regions (see Figure 9.26) where the chromosomes do not appear to have duplicated (Figure 9.29). In fact, the production of two functional centromeres from one parental centromere is a key step in the transition from metaphase to anaphase, and a functional centromere must be present on each daughter chromosome to avoid the deleterious effects of nondisjunction. Acentric chromoso-

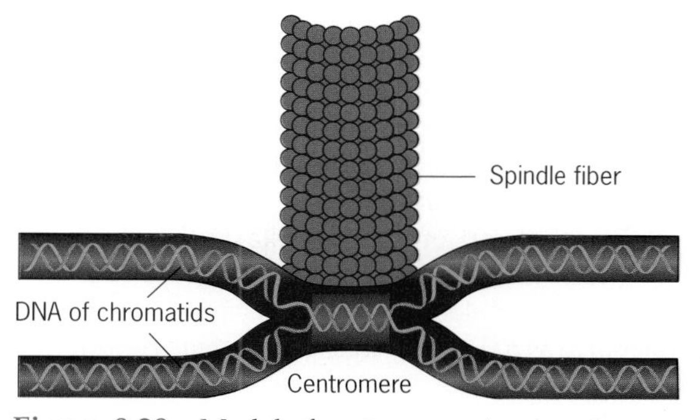

Figure 9.29 Model of centromere structure in a metaphase chromosome. The spindle fibers, which attach to centromeres, are reponsible for the separation of homologous chromosomes during anaphase I of meiosis and progeny chromosomes (derived from chromatids) during anaphase II of meiosis and anaphase of mitosis (Chapter 2).

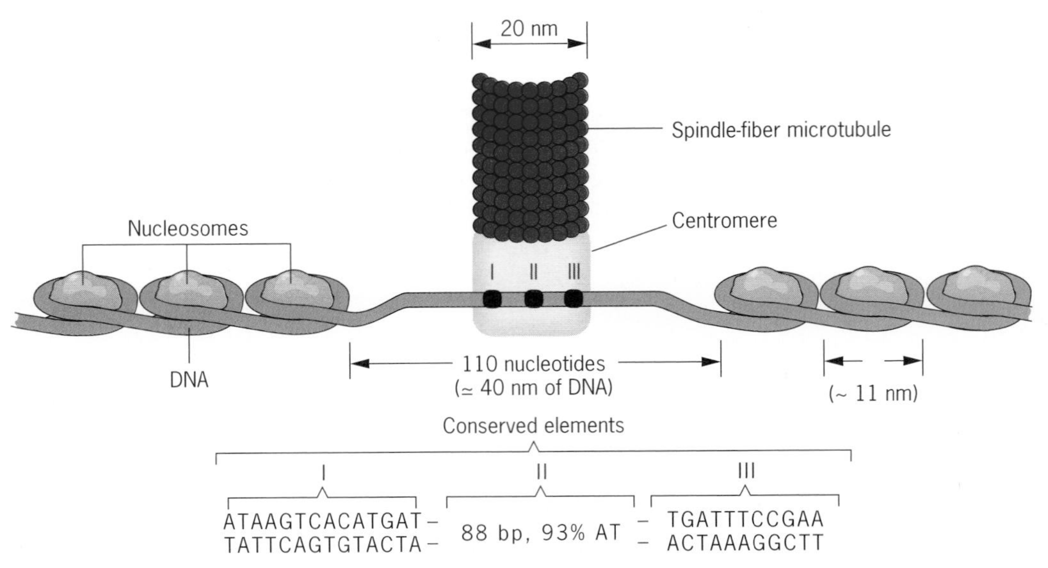

Figure 9.30 Diagram of the conserved structure of the centromeres in *Saccharomyces cerevisiae* (top) and the sequence of the CEN region of chromosome 3 of this species (bottom).

mal fragments are usually lost during mitotic and meiotic divisions.

The centromeres (*CEN* regions) of most of the chromosomes of baker's yeast, *S. cerevisiae*, have been isolated and characterized. The *CEN* regions of different chromosomes are interchangeable: replacing the *CEN* region of one chromosome with the *CEN* region of another chromosome has no detectable effect on the host cell or its capacity to undergo a normal cell division. Molecular studies have shown that a functional *S. cerevisiae* centromere is 110 to 120 nucleotide pairs in length and has three essential regions (Figure 9.30). Regions I and III are short, conserved boundary sequences, and region II is an A:T-rich (>90 percent A:T) central segment about 90 nucleotide pairs long. The length and A:T-rich nature of region II are probably more important than its actual nucleotide sequence, whereas regions I and III contain specific sequences that serve as binding sites for proteins thought to be involved in spindle-fiber attachment. The precise macromolecular structure of the centromere spindle-fiber attachment complex has not yet been determined for any eukaryotic chromosome.

The **telomeres** or ends of eukaryotic chromosomes have been known to have unique properties for several decades. In a classical study of maize chromosomes, Barbara McClintock (see Historical Sidelight: Barbara McClintock, the Discoverer of Transposable Elements in Chapter 17) demonstrated that new ends of broken chromosomes are sticky and tend to fuse with each other. In contrast, the natural ends of normal (unbroken) chromosomes are stable and show no tendency to fuse with other broken or native ends. McClintock's results indicated that telomeres must have special structures different from the ends produced by breakage of chromosomes.

Another reason for postulating that telomeres have unique structures is that the known mechanisms of replication of linear DNA molecules do not permit duplication of both strands of DNA all the way to the ends of the molecules (Chapter 10). Thus telomeres must have unique structures that facilitate their replication, or there must be some special replication enzyme that resolves this enigma. Whatever their structure, telomeres must provide at least three important functions. They must (1) prevent deoxyribonucleases from degrading the ends of the linear DNA molecules, (2) prevent fusion of the ends with other DNA molecules, and (3) facilitate replication of the ends of the linear DNA molecules without loss of material.

During the last decade, telomeres have been isolated and characterized from several species. In almost all cases, the telomeres have been shown to have unique structures that include short nucleotide sequences present as tandemly repeated units. Although these sequences are somewhat variable in different species, the basic repeat unit in all species studied to date has the pattern $5'\text{-}T_{1\text{-}4}A_{0\text{-}1}G_{1\text{-}8}\text{-}3'$. For example, the repeat sequence in humans is TTAGGG, that of the protozoan *Tetrahymena thermophila* is TTGGGG, and that of the plant *Arabidopsis thaliana* is TTTAGGG. The number of copies of this basic repeat unit in telomeres varies from species to species, from chromosome to chromosome within a species, and even on the same chromosome at different stages of the life cycle. The significance of variation in telomere length and the factors that control this variation are still being studied.

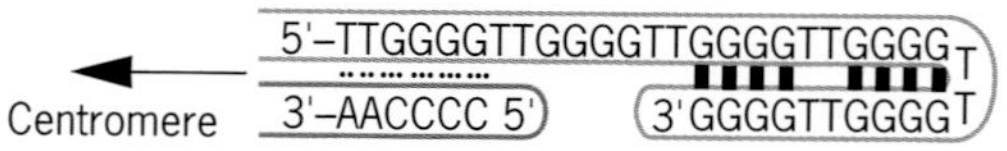

5'-TTGGGGTTGGGGTTGGGGTTGGGG T
Centromere 3'-AACCCC 5' 3' GGGGTTGGGG T

Figure 9.31 Proposed hairpin model of telomere structure in *Tetrahymena thermophila* based on pairing between methylated guanine residues within the 3' overhang regions at the DNA terminus.

The telomeres characterized to date terminate with a single-stranded region of the DNA strand with the 3' end (a so-called 3' overhang). Terminal bases of this single-stranded end exhibit unique patterns of methylation (covalently attached methyl groups) that probably contribute to the formation of a unique "hairpin" or folded structure at the very tip of the telomeric DNA (Figure 9.31). Additional repetitive DNA sequences are present adjacent to the telomere; these are referred to as telomere-associated sequences. Although the actual three-dimensional structures of telomeres are still uncertain, it is clear that the telomeres do possess special features that give them their unique properties. In addition, it is known that telomere sequences are added to chromosomes by a special enzyme called telomere transferase or telomerase (Chapter 10). Lastly, specific telomere-binding proteins have been identified in several species, and these proteins are thought to cap the telomeric DNA and provide further stability to the ends of chromosomes.

Key Points: **Each eukaryotic chromosome contains one giant molecule of DNA packaged into 10-nm ellipsoidal beads called nucleosomes. In the condensed chromosomes present during meiosis and mitosis, the 10-nm nucleosome fibers are further coiled into chromatin fibers about 30 nm in diameter. At metaphase, these 30-nm fibers are organized into domains by scaffolds composed of nonhistone chromosomal proteins. The spindle-fiber-attachment regions (centromeres) and ends (telomeres) of chromosomes have unique structures that facilitate their respective functions.**

EUKARYOTIC GENOMES: REPEATED DNA SEQUENCES

The chromosomes of prokaryotes almost exclusively contain DNA molecules with unique (nonrepeated) base-pair sequences. That is, each gene (with a few exceptions) is present only once in the genome. If the DNA molecules in prokaryotic chromosomes are broken into many short fragments, each fragment will contain a different sequence of base pairs. The chromosomes of eukaryotes are much more complex in this respect. Certain base sequences are repeated many times in the haploid chromosome complement, sometimes as many as a million times. DNA containing such repeated sequences, called **repetitive DNA**, often represents a major component (20 to 50 percent) of the eukaryotic genome.

The first evidence for repetitive DNA came from centrifugation studies of eukaryotic DNA. When the DNA of a prokaryote, such as *E. coli*, is isolated, fragmented, and centrifuged at high speeds for long periods of time in a 6 M cesium chloride (CsCl) solution, the DNA will form a single band in the centrifuge tube at the position where its density is equal to the density of the CsCl solution (Chapter 10). For *E. coli*, this band will form at a position where the CsCl density is equal to the density of DNA containing about 50 percent A:T and 50 percent G:C base pairs. DNA density increases with increasing G:C content. The extra hydrogen bond in a G:C base pair is believed to result in a tighter association between the bases and thus a higher density than for A:T base pairs. The centrifugation of DNAs from eukaryotes to equilibrium conditions in such CsCl solutions usually reveals the presence of one large mainband of DNA and one to several small bands. These small bands of DNA are called **satellite bands**, and the DNAs in these bands are often referred to as **satellite DNAs**. For example, the genome of *Drosophila virilis*, a distant relative of *Drosophila melanogaster*, contains three distinct satellite DNAs, each composed of a repeating sequence of seven base pairs. Other satellite DNAs in eukaryotes have long repetitive sequences.

The chromosomal locations of several satellite DNAs have been determined, and these repetitive DNA sequences are usually localized in regions flanking the centromeres or adjacent to telomeres of chromosomes. Satellite DNA sequences usually are not expressed; that is, they do not encode RNA or protein gene products. A repetitive DNA sequence will be identified as satellite DNA only if the sequence has a base composition sufficiently different from that of main-band DNA to produce a distinct band during density-gradient centrifugation. Therefore, many repetitive DNA sequences cannot be identified by this procedure.

Detection of Repeated Sequences: DNA Renaturation Kinetics

A more complete picture of the frequency and complexity of repetitive DNA sequences in eukaryotes has resulted from studies of DNA renaturation rates. The two strands of a DNA double helix are held together by a large number of relatively weak hydrogen bonds between complementary bases. When DNA molecules in aqueous solution are heated to near 100°C, these bonds are broken and the complementary strands of DNA separate. This process is called **denaturation**. If the complementary single strands of DNA are cooled slowly under the right conditions, the complementary base sequences will find each other and will re-form base-paired double helices. This re-formation of double helices from the complementary single strands of DNA is called **renaturation**. Analyses of the kinetics of renaturation of DNAs of eukaryotic organisms have yielded a wealth of information about the types of repetitive DNA sequences present in the genomes of eukaryotes.

Let's consider a long DNA molecule with no repeated sequences (for example, a DNA molecule from a prokaryotic chromosome). If such a molecule is sheared into fragments of a particular length, say 400 nucleotide pairs, is denatured, and is then allowed to renature under appropriate conditions, the rate of renaturation will depend on (1) the concentration of DNA in solution and (2) the complexity of the DNA, that is, the number of different 400 base-pair fragments. These two factors will determine the probability that any two complementary single strands of DNA will undergo a collision that can lead to renaturation. The effect of DNA concentration is quite obvious, because the higher the concentration of single strands in the solution, the greater the chance of a collision between any two of them.

The complexity of DNA is the total size of the nonrepeated nucleotide-pair sequences in the genome. In a double helix with poly A in one strand and poly T in the other stand, the complementary strands of every fragment will be the same. Thus all collisions between DNA single strands derived from complementary strands will lead to renaturation. As the complexity of the DNA increases, the proportion of the random collisions between single strands of DNA that are between complementary single strands will decrease and the rate of renaturation will decrease accordingly. Consider genomes that contain 4, 400, 4,000, and 40,000 different genes. In a solution containing 100 μg of DNA per ml, the concentration of any one gene will be 25, 0.25, 0.025, and 0.0025 μg/ml, respectively, for the four genomes. As a result, random collisions between the complementary strands derived from any one gene will decrease in frequency as the complexity of the genome (the number of different genes) increases. This effect of genome complexity on renaturation kinetics is illustrated in Figure 9.32, which shows the very rapid renaturation of poly A:T and the very slow renaturation of the nonrepeated sequences in the genome of *Bos taurus* (domestic cattle).

Mathematical analyses (see Technical Sidelight: DNA Renaturation Kinetics) of the kinetics of renaturation of eukaryotic DNAs have been extremely informative. The results of DNA renaturation kinetics experiments have demonstrated that eukaryotic genomes contain many different kinds of repetitive DNA sequences. The DNA sequences present in eukaryotes are commonly divided into three classes:

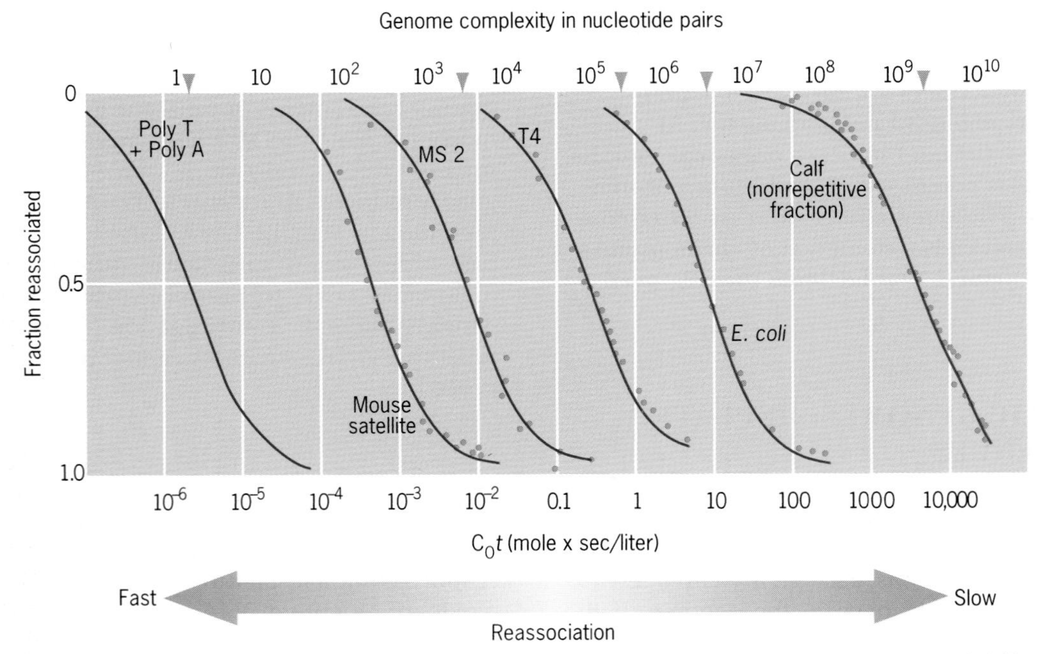

Figure 9.32 Illustration of the effect of genome complexity on DNA renaturation rates. In poly T/poly A duplexes, all fragments produced by shearing will have the same complementary strands and renaturation will occur rapidly (left). In complex genomes with lots of different sequences, for example, the nonrepeated sequences of calf DNA (right), renaturation will occur less rapidly.

TECHNICAL SIDELIGHT

DNA Renaturation Kinetics

Definitive evidence for the presence of repeated DNA sequences in eukaryotic genomes has been obtained by analyzing the kinetics of DNA renaturation. Sequences that are present in multiple copies per genome will renature more rapidly than sequences that are present only once. Indeed, as we discuss below, the rate at which a specific sequence renatures is inversely proportional to its copy number—the higher the copy number, the less time required for renaturation.

Consider a particular 400 base-pair fragment composed of two complementary strands, *a* and *a'*.

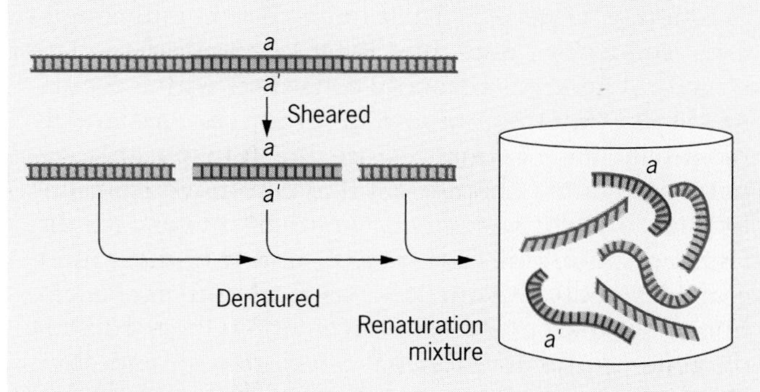

Reassociation of *a* and *a'* will require a specific collision between these two single strands. Collisions between *a* or *a'* and any other single strand will not lead to hybridization. For a given concentration, the larger the DNA molecule, and thus the more nonidentical 400 base-pair fragments, the slower the reassociation reaction will be, because a smaller proportion of the random collisions will be between complementary single strands such as *a* and *a'*.

Note that every reassociation event, like *a* with *a'*, will require a collision between two single strands that are present in the renaturation mixture in equal concentration. Because the reaction requires the interaction of two equally frequent molecules, the rate of renaturation will be a function of the square of the concentration of single strands (so-called second-order or bimolecular reaction kinetics), or

$$\frac{-dC}{dt} = kC^2$$

$$\left(\text{or } \frac{-dC}{C^2} = kdt, \text{ rearranged for integration}\right)$$

where

C = the concentration of single-stranded DNA in moles of nucleotides per liter

t = time in seconds

k = a second-order rate constant in liters per mole seconds

Literally, this equation states that the change (decrease) in concentration of single-stranded DNA ($-dC$) with time (dt) is equal to the proportionality constant (k) times the square of the concentration of single-stranded DNA.

Integration of the preceding equation from the initial conditions ($t = 0$ seconds and $C = C_o$, where C_o equals the concentration of single-stranded DNA at $t = 0$) yields

$$\frac{1}{C} - \frac{1}{C_o} = kt$$

or, rearranged,

$$\frac{C}{C_o} = \frac{1}{1 + kC_ot}$$

This equation states that the fraction of input single-stranded DNA remaining in a renaturation reaction mixture (C/C_o) at any given time (t) is a function of the initial concentration (C_o) times elapsed time (t), or C_ot. It is thus convenient to present data on hybridization kinetics in a plot of C/C_o versus C_ot. These so-called **C_ot** (pronounced "caught") **curves** (see Figures 9.32 and 9.33) have provided a great deal of information about the types of repetitive DNA in eukaryotic genomes.

Consider a DNA molecule containing a 400 base-pair sequence that is repeated (present twice).

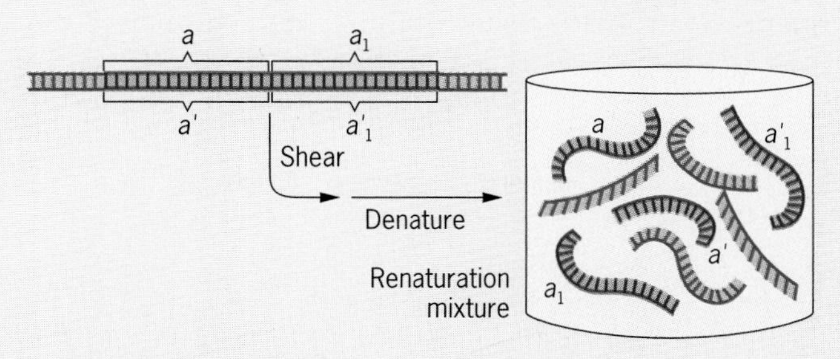

Now a and a_1 are identical single strands, and a' and a_1' are identical complementary strands. Reassociation of each repeated sequence will take only half as long as in the previous example, where each fragment contained a unique base sequence. Now reassociation will result from a collision of a with either a' or a_1' (also for a_1 and either a' or a_1'). Thus the time required for reassociation of a particular DNA sequence is inversely proportional to the number of times that sequence is present in the genome. Clearly, highly repetitive DNA sequences will renature very rapidly.

The proportion of the DNA that has renatured at any time can be quantitated in several ways. One method is simply to treat samples taken at various times with a nuclease that is specific for single-stranded nucleic acids. The DNA remaining in the samples after digestion (all double-stranded) can then be quantitated by direct chemical analysis.

1. **Unique** or **single-copy DNA sequences**—1 to 10 copies per genome.
2. **Moderately repetitive DNA sequences**—10 to 10^5 copies per genome.
3. **Highly repetitive DNA sequences**—more than 10^5 copies per genome.

Figure 9.33 shows a DNA renaturation curve demonstrating the presence of these three classes of DNA sequences in the human genome. The proportions of these three classes of DNA sequences vary from species to species. Single-copy sequences make up about 40 to 70 percent of the genome in most plants and animals. The moderately repetitive class of DNA sequences appears to be very heterogeneous in most eukaryotes; it contains many different sequences with different degrees of reiteration. For example, in *Drosophila melanogaster*, 12 percent of the DNA contains moderately repetitive sequences with an average reiteration frequency of 70. The highly repetitive DNAs of eukaryotes frequently contain both satellite and nonsatellite DNA sequences.

When the sequence organization of eukaryotic genomes is studied in greater detail—by combining the techniques of density-gradient centrifugation, hybridization kinetic analysis, biochemical analyses, electron microscopy, and cloning and sequencing of substantial segments of chromosomal DNA—a pattern emerges. Much of the genome consists of middle-repetitive sequences interspersed with single-copy sequences. In toads, sea urchins, and humans, the sequences are quite short. The middle-repetitive sequences average 300 nucleotide pairs in length; the single-copy sequences are about 800 to 1200 nucleotide pairs long. *Drosophila melanogaster* DNA also exhibits interspersion of middle-repetitive and single-copy sequences, but the sequences are much longer (5,600 and 13,000 nucleotide pairs, respectively).

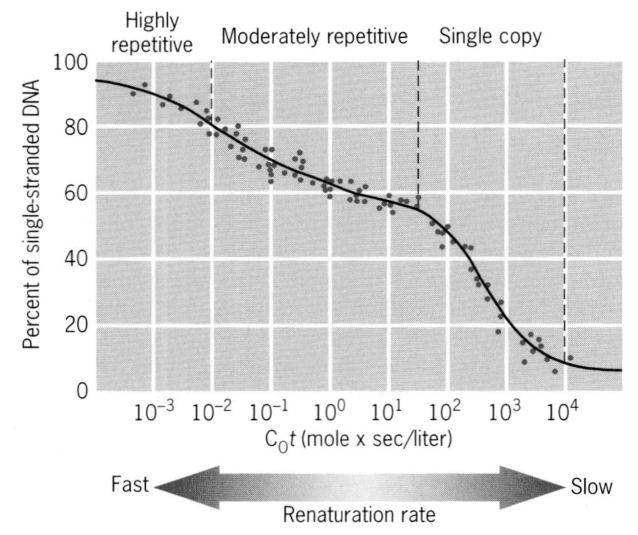

Figure 9.33 The renaturation kinetics of human DNA. Sequences that are highly repetitive (>10^5 copies per genome) renature rapidly (left); moderately repeated sequences (from 10 to 10^5 copies per genome) renature at intermediate rates; and single copy sequences (one to ten copies per genome) renature slowly.

What are the functions of the different kinds of DNA sequences in eukaryotes? Most of the structural genes (genes coding for proteins or RNA molecules) are single-copy sequences. The genes for histones, rRNA molecules, and ribosomal proteins, all gene products that are present in the cell in large quantities, are exceptions; these are redundant (present in moderately repetitive DNA) genes. Many geneticists have postulated that most other moderately repetitive sequences may be involved in the regulation of gene expression (Chapter 22). Their interspersion with single-copy sequences, and thus their location adjacent to structural genes, is certainly consistent with a regulatory role. Another important group of moderately repetitive DNA sequences are the **transposable genetic elements** (Chapter 17) that can move from one location in a chromosome to another or even to a different chromosome. The transposability of these moderately repetitive sequences has led to much speculation about their possible involvement in the regulation of gene expression during development and their roles in the evolution of eukaryotic genomes. These transposable elements are responsible for a surprisingly large number of the naturally occurring mutations in eukaryotic organisms.

The function(s) of highly repetitive DNA—most, if not all, of which is located in genetically inactive heterochromatic regions of chromosomes—is completely unknown. Postulated functions for highly repetitive DNA include (1) structural or organizational roles in chromosomes; (2) involvement in chromosome pairing during meiosis; (3) involvement in crossing over or recombination (Chapter 13); (4) protection of important structural genes, like histone, rRNA, or ribosomal protein genes; (5) a repository of unessential DNA sequences for use in the future evolution of the species; and (6) no function at all—just junk DNA that is carried along by the processes of replication and segregation of chromosomes. The validity of any of these postulated roles remains to be established.

Repeated Sequences in the Human Genome

In humans, as in other eukaryotes, much of the dispersed moderately repetitive DNA consists of families of transposable elements. Two important groups of moderately repetitive transposable elements are called the **LINE** (**L**ong **I**nterspersed **N**uclear **E**lement) and the **SINE** (**S**hort **I**nterspersed **N**uclear **E**lement) families.

LINE transposable elements have an average length of 6500 nucleotide pairs and are dispersed throughout the human genome. The best known member of the human LINE family of sequences is the transposable element *L1* (Chapter 17), which is present in 20,000 to 50,000 copies per genome.

SINE elements are only 150 to 300 base pairs in length. The most extensively studied SINE sequences are the **Alu sequences** in humans. (The designation Alu is derived from the name of an enzyme that cleaves DNA at a specific sequence located within the Alu elements.) There are approximately 500,000 Alu sequences, each about 300 nucleotide pairs long, in the haploid human genome. They are related, but not identical, in nucleotide sequence and are widely dispersed on all the human chromosomes. Collectively, they constitute about 5 percent of the human genome, with an Alu sequence present on average once every 5000 nucleotide pairs. Despite speculation that the Alu elements might contain sequences involved in the initiation of DNA replication and the regulation of gene expression, their biological function(s) is still subject to much speculation.

As is observed with other transposable genetic elements (Chapter 17), the transposition of an Alu sequence into genes and DNA rearrangements caused by recombination between two Alu sequences can produce mutant phenotypes. Mutant alleles that cause **acholinesterasemia** (a deficiency of the enzyme cholinesterase resulting in transient paralysis upon exposure to muscle relaxants) and **neurofibromatosis** (tumors of nerve tissues just under the skin and the central nervous system) have been shown to result from the insertion of Alu sequences into genes. In addition, numerous cases of **hypercholesterolemia** (excessive amounts of cholesterol in the blood due to a deficiency of cholesterol receptors on liver cells) and **thalassemia** (hemoglobin deficiencies) have been shown to result from gene rearrangements produced by recombination between closely linked Alu sequences. These and other results have clearly demonstrated that Alu sequences have played, and continue to play, a major role in the evolution of the human genome.

Key Points: **Eukaryotic genomes contain repeated DNA sequences, with some sequences present a million times or more. Eukaryotic DNA sequences are commonly grouped into three classes: (1) unique or single-copy sequences present in one to a few copies per genome, (2) moderately repetitive sequences present in 10 to 10^5 copies, and (3) highly repetitive sequences present in over 10^5 copies per genome. Two families of moderately repetitive transposable elements make up 10 percent of the human genome and play an important role in its evolution.**

TESTING YOUR KNOWLEDGE

1. The red alga *Polyides rotundus* stores its genetic information in double-stranded DNA. When DNA was extracted from *P. rotundus* cells and analyzed, 32 percent of the bases were found to be guanine residues. From this information, can you determine what percentage of the bases in this DNA were thymine residues? If so, what percentage? If not, why not?

ANSWER

The two strands of a DNA double helix are complementary to each other, with guanine (G) in one strand always paired with cytosine (C) in the other strand and, similarly, adenine (A) always paired with thymine (T). Therefore, the concentrations of G and C are always equal, as are the concentrations of A and T. If 32 percent of the bases in double-stranded DNA are G residues, then another 32 percent are C residues. Together, G and C comprise 64 percent of the bases in *P. rotundus* DNA; thus, 36 percent of the bases are A's and T's. Since the concentration of A must equal the concentration of T, 18 percent (36% × 1/2) of the bases must be T residues.

2. The *E. coli* virus ΦX174 stores its genetic information in single-stranded DNA. When DNA was extracted from ΦX174 virus particles and analyzed, 21 percent of the bases were found to be G residues. From this information, can you determine what percentage of the bases in this DNA were thymine residues? If so, what percentage? If not, why not?

ANSWER

No! The A=T and G=C relationships occur only in double-stranded DNA molecules, because of their complementary strands. Since base-pairing doesn't occur or occurs only as limited intrastrand pairing in single-stranded nucleic acids, you cannot determine the percentage of any of the other three bases from the G content of the DNA.

3. If each G_1-stage human chromosome contains a single molecule of DNA, how many DNA molecules would be present in the chromosomes of the nucleus of (a) a human egg, (b) a human sperm, (c) a human diploid somatic cell in stage G_1, (d) a human diploid somatic cell in stage G_2, (e) a human primary oocyte?

ANSWER

A normal human haploid cell contains 23 chromosomes, and a normal human diploid cell contains 46 chromosomes, or 23 pairs of homologs. If prereplication chromosomes contain a single DNA molecule, postreplication chromosomes will contain two DNA molecules, one in each of the two chromatids. Thus normal human eggs and sperm contain 23 chromosomal DNA molecules; diploid somatic cells contain 46 and 92 chromosomal DNA molecules at stages G_1 and G_2, respectively; and a primary oocyte contains 92 such DNA molecules.

QUESTIONS AND PROBLEMS

9.1 (a) How did the transformation experiments of Griffith differ from those of Avery and his associates? (b) What was the significant contribution of each? (c) Why was Griffith's work not evidence for DNA as the genetic material, whereas the experiments of Avery and coworkers provided direct proof that DNA carried the genetic information?

9.2 A cell-free extract is prepared from Type IIIS pneumococcal cells. What effect will treatment of his extract with (a) protease, (b) RNase, and (c) DNase have on its subsequent capacity to transform recipient Type IIR cells to Type IIIS? Why?

9.3 How could it be demonstrated that the mixing of heat-killed Type III pneumococcus with live Type II resulted in a transfer of genetic material from Type III to Type II rather than a restoration of viability to Type III by Type II?

9.4. What is the macromolecular composition of a bacterial virus or bacteriophage such as phage T2?

9.5 What chemical properties do DNA and proteins possess that allow researchers to label specifically one or the other of these macromolecules with a radioactive isotope?

9.6 (a) What was the objective of the experiment carried out by Hershey and Chase? (b) How was the objective accomplished? (c) What is the significance of this experiment?

9.7 (a) What background material did Watson and Crick have available for developing a model of DNA? (b) What was their contribution to the building of the model?

9.8 (a) Why was a double helix chosen as the basic structure for the model of DNA proposed by Watson and Crick? (b) Why were hydrogen bonds placed in the model to connect the bases?

9.9 (a) If a virus particle contains double-stranded DNA with 200,000 base pairs, how many nucleotides would be present? (b) How many complete spirals would occur on each strand? (c) How many atoms of phosphorus would be present? (d) What would be the length of the DNA configuration in the virus?

9.10 If one strand of DNA in the Watson–Crick double helix has a base sequence of 5'-GTCATGAC-3', what is the base sequence of the complementary strand?

9.11 What are the differences between DNA and RNA?

9.12 DNA was extracted from cells of *Staphylococcus afermentans* and analyzed for base composition. It was found that 37 percent of the bases are cytosine. With this information, is it possible to predict what percentage of the bases are adenine? If so, what percentage? If not, why not?

9.13 RNA was extracted from TMV (tobacco mosaic virus) particles and found to contain 20 percent cytosine (20 percent of the bases were cytosine). With this information, is it possible to predict what percentage of the bases in TMV are adenine? If so, what percentage? If not, why not?

9.14 Indicate whether each of the following statements about the structure of DNA is true or false. (Each letter is used to refer to the concentration of that base in DNA.) (a) $A + T = G + C$. (b) $A = G$; $C = T$. (c) $A/T = C/G$. (d) $T/A = C/G$. (e) $A + G = C + T$. (f) $G/C = 1$. (g) $A = T$ within each single strand. (h) Hydrogen bonding provides stability to the double helix in aqueous cytoplasms. (i) Hydrophobic bonding provides stability to the double helix in aqueous cytoplasms. (j) When separated, the two strands of a double helix are identical. (k) Once the base sequence of one strand of a DNA double helix is known, the base sequence of the second strand can be deduced. (l) The structure of a DNA double helix is invariant. (m) Each nucleotide pair contains two phosphate groups, two deoxyribose molecules, and two bases.

9.15 The available evidence indicates that each eukaryotic chromosome (excluding polytene chromosomes) contains a single giant molecule of DNA. What different levels of organization of this DNA molecule are apparent in chromosomes of eukaryotes at various times during the cell cycle?

9.16 A diploid nucleus of *Drosophila melanogaster* contains about 2×10^8 nucleotide pairs. Assume (1) that all the nuclear DNA is packaged in nucleosomes and (2) that an average internucleosome linker size is 60 nucleotide pairs. How many nucleosomes would be present in a diploid nucleus of *D. melanogaster*? How many molecules of histone H2a, H2b, H3, and H4 would be required?

9.17 The satellite DNAs of *Drosophila virilis* can be isolated, essentially free of mainband DNA, by density-gradient centrifugation. If these satellite DNAs are sheared into approximately 40-nucleotide-pair-long fragments and are analyzed in denaturation–renaturation experiments, how would you expect their hybridization kinetics to compare with the renaturation kinetics observed using similarly sheared mainband DNA under the same conditions? Why?

9.18 Experimental evidence indicates that most highly repetitive DNA sequences in the chromosomes of eukaryotes do not produce any RNA or protein products. What does this indicate about the function of highly repetitive DNA?

9.19 (a) Are (1) single-copy DNA sequences, (2) moderately repetitive DNA sequences, (3) highly repetitive DNA sequences more prevalent in euchromatin or heterochromatin?

(b) Are most (1) single-copy DNA sequences, (2) highly repetitive DNA sequences expressed? (c) What are the presumed functions of most (1) single-copy DNA sequences, (2) highly repetitive DNA sequences? (d) What roles are the moderately repetitive DNA sequences believed to play in higher eukaryotes?

9.20 Are eukaryotic chromosomes metabolically most active during prophase, metaphase, anaphase, telophase, or interphase?

9.21 (a) What functions do (1) centromeres, (2) telomeres provide? (b) Do telomeres have any unique structural features? (c) What is the function of telomerase? (d) When chromosomes are broken by exposure to high-energy radiation such as X rays, the resulting broken ends exhibit a pronounced tendency to stick to each other and fuse. Why might this occur?

9.22 Of what special interest is the biophysical technique of viscoelastometry to geneticists?

9.23 How many DNA molecules are present in (a) the axial regions, (b) the lateral loops of lampbrush chromosomes?

9.24 Are the scaffolds of eukaryotic chromosomes composed of histone or nonhistone chromosomal proteins? How has this been determined experimentally?

9.25 (a) Which class of chromosomal proteins, histones or nonhistones, is the more highly conserved in different eukaryotic species? Why might this be expected? (b) If one compares the histone and nonhistone chromosomal proteins of chromatin isolated from different tissues or cell types of a given eukaryotic organism, which class of proteins will exhibit the greater heterogeneity? Why are both classes of proteins not expected to be equally homogeneous in chromosomes from different tissues or cell types?

9.26 During DNA renaturation experiments, why is the rate of renaturation ($-dC/dt$) proportional to the square of the concentration of single strands (kC^2) rather than simply proportional to the concentration of single strands (kC)?

9.27 Studies of the renaturation kinetics of DNA from the human genome reveal that at $t = 0$ (the earliest time point at which the proportion of double-stranded DNA can be measured in standard renaturation experiments) approximately 6 percent of the DNA is already double stranded (see Figure 9.33). Assuming that all of the DNA was initially denatured in these experiments, how can this result (6 percent renatured at $t = 0$) be explained?

9.28 (a) If the haploid human genome contains 3×10^9 nucleotide pairs and the average molecular weight of a nucleotide pair is 660, how many copies of the human genome are present, on average, in 1 μg of human DNA? (b) What is the weight of one copy of the human genome? (c) If the haploid genome of the small plant *Arabidopsis thaliana* contains 7.7×10^7 nucleotide pairs, how many copies of the *A. thaliana* genome are present, on average, in 1 μg of *A. thaliana* DNA? (d) What is the weight of one copy of the *A. thaliana* genome? (e) Of what importance are calculations of the above type to geneticists?

BIBLIOGRAPHY

AVERY, O. T., C. M. MACLEOD, AND M. MCCARTY. 1944. "Studies on the chemical nature of the substance inducing transformation in pneumococcal types." J. EXPL. MED. 79:137–158.

BRITTEN, R. J., AND D. E. KOHNE. 1970. "Repeated segments of DNA." *Sci. Amer.* 222(4):24–31.

HERSHEY, A. D., AND M. CHASE. 1952. "Independent functions of viral protein and nucleic acid in growth of bacteriophage." *J. Gen. Physiol.* 36:39–56. (Reprinted in G. S. Stent. 1960. *Papers on Bacterial Viruses*, 2nd ed. Little, Brown, Boston.)

KORNBERG, R. D., AND A. KLUG. 1981. "The nucleosome." *Sci. Amer.* 244(2):52–64.

LEWIN, B. 1994. *Genes V*. Cell Press, Cambridge, MA/Oxford University Press, Oxford.

WATSON, J. D. 1968. *The Double Helix*. Atheneum, New York.

WATSON, J. D., AND F.H.C. CRICK. 1953. "A structure for deoxyribose nucleic acid." *Nature* 171:737–738.

WORCEL, A., AND E. BURGI. 1972. "On the structure of the folded chromosome of *Escherichia coli*." *J. Mol. Biol.* 71:127–147.

10

Seven pairs of identical twins. Although the twins were given no instructions regarding how to pose for the photograph, note the similar posture, hand placement, and facial expressions of both members of each pair of twins.

Replication of DNA and Chromosomes

CHAPTER OUTLINE

Monozygotic Twins: Are They Identical?

From the day of their birth, through childhood, adolescence, and adulthood, Merry and Sherry have been mistaken for one another. When they are apart, Merry is called Sherry about half of the time, and Sherry is misidentified as Merry with equal frequency. When they are together, their friends distinguish them by the clothes they wear rather than by differences in phenotypic traits. Even their parents have trouble distinguishing them. Merry and Sherry are monozygotic ("identical") twins; they both developed from a single fertilized egg. At an early cleavage stage, the embryo split into two cell masses, and both groups of cells developed into complete embryos. Both embryos developed normally through the embryonic and fetal stages, and on April 7, 1955, one newborn was named Merry, the other Sherry.

People often explain the nearly identical phenotypes of monozygotic twins like Merry and Sherry by stating that "they contain the same genes." Of course, that is not true. Like anything else, a gene can only exist in one place at any given time. To be accurate, the statement should be that identical twins contain progeny replicas of the same parental genes. But this simple colloquialism suggests that most people do, indeed, believe that the progeny replicas of a gene actually are identical. If the human genome contains 50,000 to 100,000 genes, are the progeny replicas of all these genes exactly the same in identical twins?

A human life emerges from a single cell, a tiny sphere about 0.1 mm in diameter. That cell gives rise to hundreds of billions of cells during fetal development. At maturity, a human of average size contains about 65 trillion (65,000,

000,000,000) cells. With some exceptions, each of these 65 trillion cells contains a progeny replica of each of the 50,000 to 100,000 genes. Are the progeny replicas of all of these genes identical in all 65 trillion cells? If so, the process by which the genes are duplicated must be extremely accurate. Moreover, the cells of the body are not static; in some tissues, old cells are continuously being replaced by new cells. For example, in a healthy individual, the bone marrow cells produce about 2 million red blood cells per minute. Although not all of the progeny replicas of genes in the human body are identical, the process by which these genes are duplicated is very accurate. The human haploid genome contains about 3×10^9 nucleotide pairs of DNA, all of which must be duplicated during each cell division. In this chapter, we examine how DNA replicates and focus on the mechanisms that ensure the fidelity of this process.

BASIC FEATURES OF DNA REPLICATION *IN VIVO*

In humans, the synthesis of a new strand of DNA occurs at the rate of about 3000 nucleotides per minute. In bacteria, about 30,000 nucleotides are added to a nascent DNA chain per minute. Clearly, the cellular machinery reponsible for DNA replication must work very fast, but, even more importantly, it must work with great precision. Indeed, the fidelity of DNA replication is amazing, with an average of only one mistake per billion nucleotides incorporated. Thus the majority of the genes of identical twins are indeed identical, but some will have changed owing to replication errors and other types of mutations (Chapter 13). Most of the key features of the mechanism by which the rapid and accurate replication of DNA occurs are now known, although many molecular details remain to be elucidated.

The synthesis of DNA, like the synthesis of RNA (Chapter 11) and proteins (Chapter 12), involves three steps: (1) chain initiation, (2) chain extension or elongation, and (3) chain termination. In this and the following two chapters, we examine the mechanisms by which cells carry out each of the three steps in the synthesis of these important macromolecules. First, however, we consider some key features of DNA replication.

Semiconservative Replication

When Watson and Crick deduced the double-helix structure of DNA with its complementary base-pairing, they immediately recognized that the base-pairing specificity could provide the basis for a simple mechanism for DNA duplication. If the two complementary strands of a double helix separated, by breaking the hydrogen bonds of each base pair, each parental strand could direct the synthesis of a new complementary strand because of the specific base-pairing requirements (Figure 10.1). That is, each parental strand could function as a template, a single strand of DNA that specifies the nucleotide sequence of a new complementary strand. Adenine, for example, in the parent strand would serve as a template via its hydrogen-bonding potential for the incorporation of thymine in the nascent complementary strand. This mechanism of DNA replication is called **semiconservative replication**, because each of the complementary strands of the parental double helix is conserved (or the double helix is "half-conserved") during the process.

Hypothetically, DNA replication could occur by any of three different mechanisms (Figure 10.2). In addition to (1) the semiconservative mechanism proposed by Watson and Crick, replication could occur by either (2) a conservative mechanism, with the parental double helix being conserved and directing the synthesis of a new progeny double helix, or (3) a dispersive mechanism, with segments of parental and progeny strands interspersed as a result of the synthesis and rejoining of short segments of DNA. The first critical test of Watson and Crick's proposal that DNA replicates by a semiconservative mechanism was performed by Matthew Meselson and Franklin Stahl in 1958. Their results showed that the chromosome (now known to contain a single double helix of DNA) of the common colon bacillus *E. coli*, a prokaryote, replicates semiconservatively.

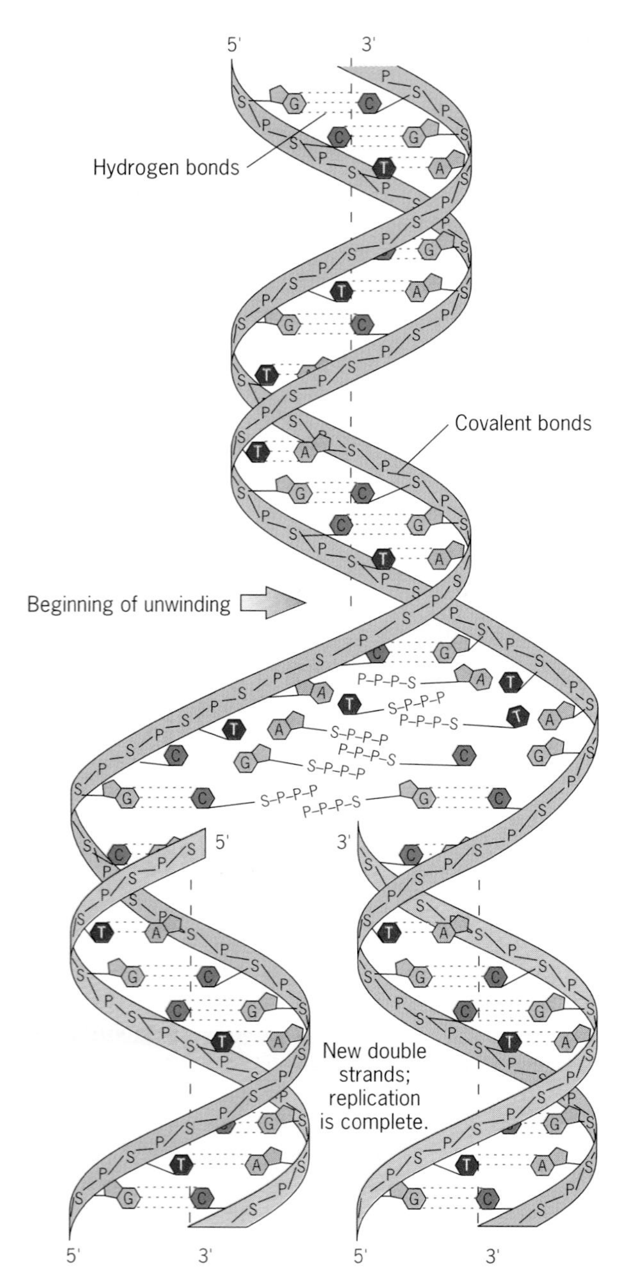

Figure 10.1 Semiconservative model of DNA replication based on complementary base pairing as proposed by Watson and Crick. Note that each of the parental strands is conserved and serves as a template for the synthesis of a new complementary strand. That is, the base sequence in each progeny strand is determined by the hydrogen-bonding potentials of the bases in the parental strand.

Meselson and Stahl grew *E. coli* cells for many generations in a medium in which the heavy isotope of nitrogen, ^{15}N, had been substituted for the normal, light isotope, ^{14}N. The purine and pyrimidine bases in DNA contain nitrogen; thus the DNA of cells grown on medium containing ^{15}N will have a greater density (weight per unit volume) than the DNA of cells grown on medium containing ^{14}N. Since molecules of different densities can be separated by a procedure called **equilibrium density-gradient centrifugation** (See Technical Sidelight: Cesium Chloride Equilibrium Density-Gradient Centrifugation . . .), Meselson and Stahl were able to distinguish between the three possible modes of DNA replication by following the changes in the density of DNA of cells grown on ^{15}N medium and then transferred to ^{14}N medium for various periods of time—so-called density-transfer experiments.

The density of most DNAs is about the same as the density of concentrated solutions of heavy salts such as cesium chloride (CsCl). For example, the density of 6 M CsCl is about 1.7 g/cm^3. *Escherichia coli* DNA containing ^{14}N has a density of 1.710 g/cm^3. Substitution of ^{15}N for ^{14}N increases the density of *E. coli* DNA to 1.724 g/cm^3. When a 6 M CsCl solution is centrifuged at very high speeds for long periods of time, an equilibrium density gradient is formed (see Technical Sidelight). If DNA is present in such a gradient, it will move to a position where the density of the CsCl solution is equal to its own density.

Meselson and Stahl took cells that had been growing in medium containing ^{15}N for several generations (and thus contained "heavy" DNA), washed them to remove the medium containing ^{15}N, and transferred them to medium containing ^{14}N. After the cells were allowed to grow in the presence of ^{14}N for varying periods of time, the DNAs were extracted and analyzed in CsCl equilibrium density gradients. The results of their experiment (Figure 10.3) are consistent only with semiconservative replication, excluding both conservative and dispersive models of DNA synthesis. All the DNA isolated from cells after one generation of growth in medium containing ^{14}N had a density halfway between the densities of "heavy" DNA and "light" DNA. This intermediate density is usually referred to as "hybrid" density. After two generations of growth in medium containing ^{14}N, half of the DNA was of hybrid density and half was light. These results are precisely those predicted by the Watson and Crick semiconservative mode of replication (Figure 10.1). One generation of semiconservative replication of a parental double helix containing ^{15}N in medium containing only ^{14}N would produce two progeny double helices, both of which had ^{15}N in one strand (the "old" strand) and ^{14}N in the other strand (the "new" strand). Such molecules would be of hybrid density.

Conservative replication would not produce any DNA molecules with hybrid density; after one generation of conservative replication of heavy DNA in light medium, half of the DNA would still be heavy and the other half would be light. If replication were disper-

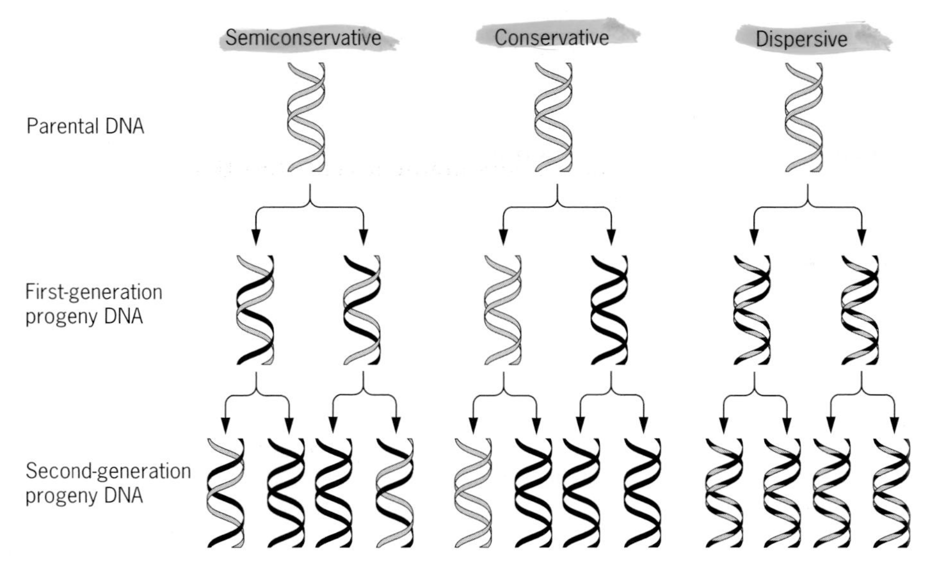

Figure 10.2 Three possible modes of DNA replication: (1) semiconservative, in which each strand of the parental double helix is conserved and directs the synthesis of a new complementary progeny strand; (2) conservative, in which the parental double helix is conserved and directs the synthesis of a new progeny double helix; and (3) dispersive, in which segments of each parental strand are conserved and direct the synthesis of new complementary strand segments that are subsequently joined to produce new progeny strands.

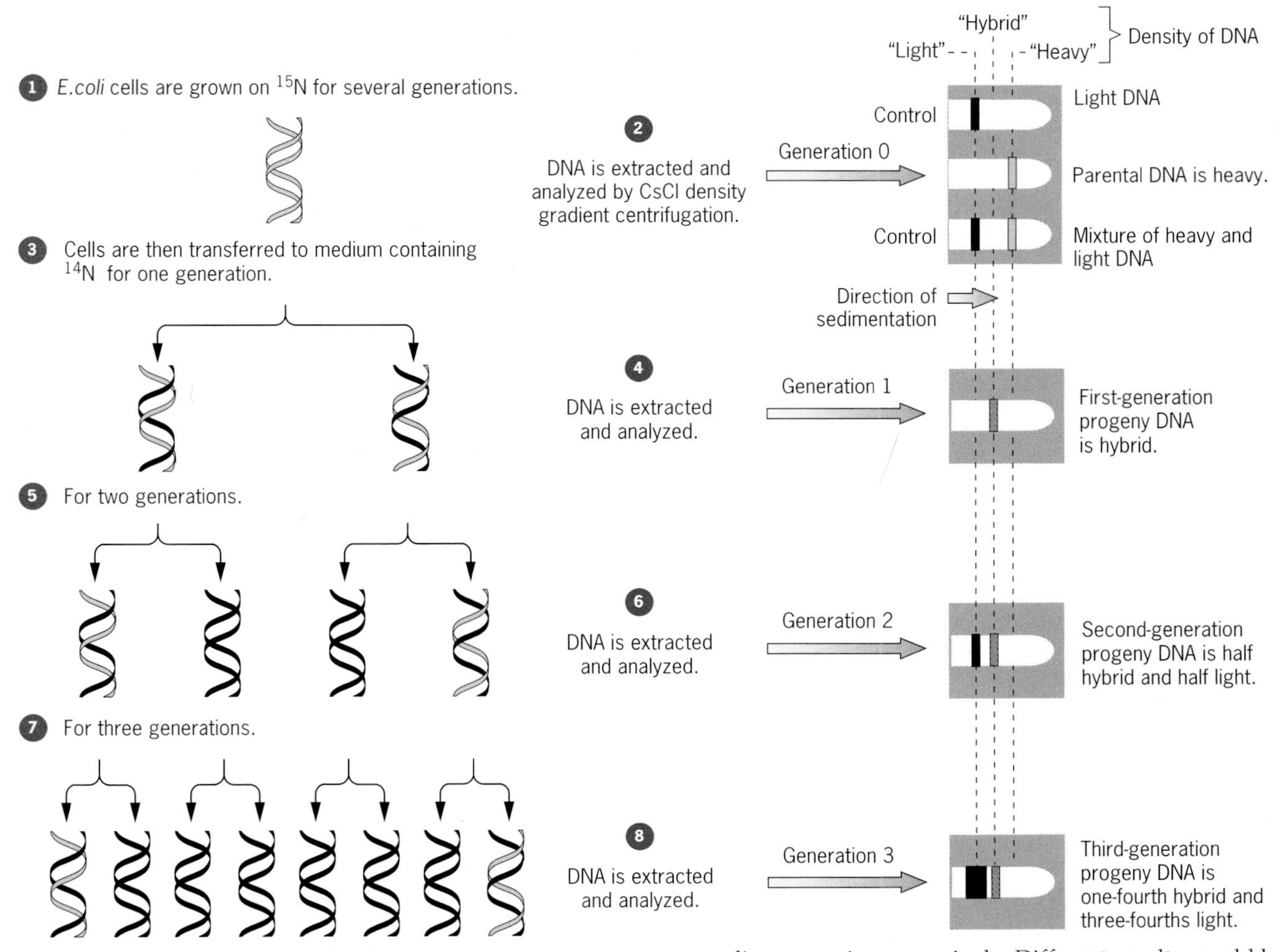

Figure 10.3 Meselson and Stahl's experiment demonstrating that DNA replicates by a semiconservative mechanism in *E. coli*. The diagram shows that the results of their experiment are those expected if the *E. coli* chromosome replicates semiconservatively. Different results would have been obtained if DNA replication in *E. coli* were either conservative or dispersive (see Fig. 10.2).

TECHNICAL SIDELIGHT

Cesium Chloride Equilibrium Density-Gradient Centrifugation and Sucrose Velocity Density-Gradient Centrifugation: Two Important and Distinct Tools

Although cesium chloride (CsCl) and sucrose density gradients are both used to study nucleic acids, these two techniques are used for totally different purposes.

CSCL EQUILIBRIUM DENSITY-GRADIENT CENTRIFUGATION

The CsCl density gradients that are used to analyze nucleic acids are **equilibrium density gradients**. They are used to separate nucleic acid molecules based on their densities, which determine the positions at which the molecules will band in the linear density gradient produced in the centrifuge tube at equilibrium conditions.

When a heavy salt solution such as 6 M CsCl is centrifuged at very high speeds (30,000 to 50,000 revolutions per minute) for 48 to 72 hours, an equilibrium density gradient is formed (see diagram *a*). The centrifugal force caused by spinning the solution at high speeds sediments the salt toward the bottom of the tube. Diffusion, on the other hand, results in movement of salt molecules back toward the top (low salt concentration) of the tube. After a sufficient period of high-speed centrifugation, an equilibrium between sedimentation and diffusion is reached, at which time a linear gradient of increasing density exists from the top to the bottom of the tube.

The density of 6 M CsCl is about 1.7 g/cm^3; at equilibrium, the linear density gradient produced by high-speed centrifugation of a 6 M CsCl solution will range from about 1.65 g/cm^3 at the top of the tube to about 1.75 g/cm^3 at the bottom of the tube. The densities of most naturally occurring nucleic acids fall within this range. If DNA is present in such a gradient, it will move to a position where the density of the salt solution is equal to its own density. Thus, if a mixture of *E. coli* DNA containing the heavy isotope of nitrogen, ^{15}N, and *E. coli* DNA containing the normal light nitrogen isotope, ^{14}N, is subjected to CsCl equilibrium density-gradient centrifugation, the DNA molecules will separate into two "bands," one consisting of "heavy" (^{15}N-containing) DNA and the other consisting of "light" (^{14}N-containing) DNA.

SUCROSE VELOCITY DENSITY-GRADIENT CENTRIFUGATION

The sucrose density gradients that are used to analyze nucleic acids are **velocity density gradients**. They are used to estimate the sizes of nucleic acid molecules based on their rates or velocities of movement through a preformed sucrose density gradient when exposed to a centrifugal force.

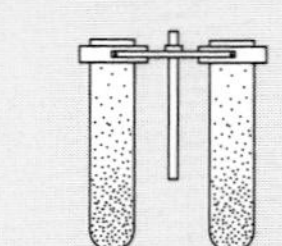

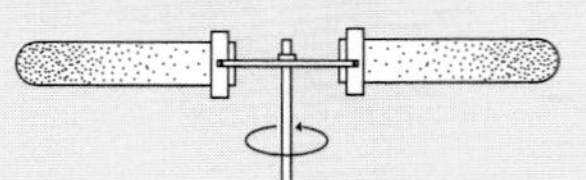

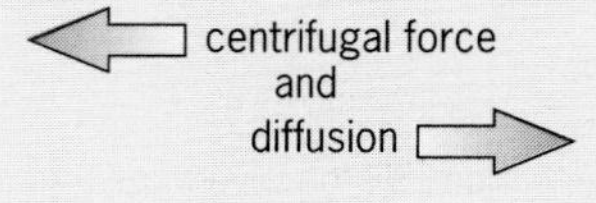

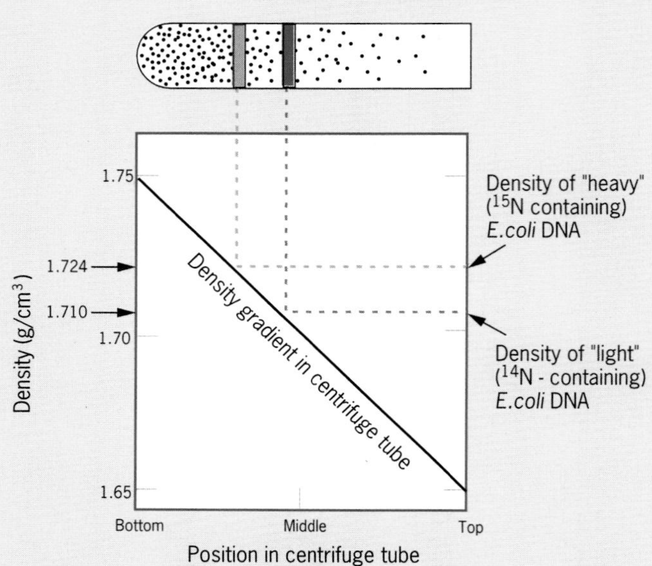

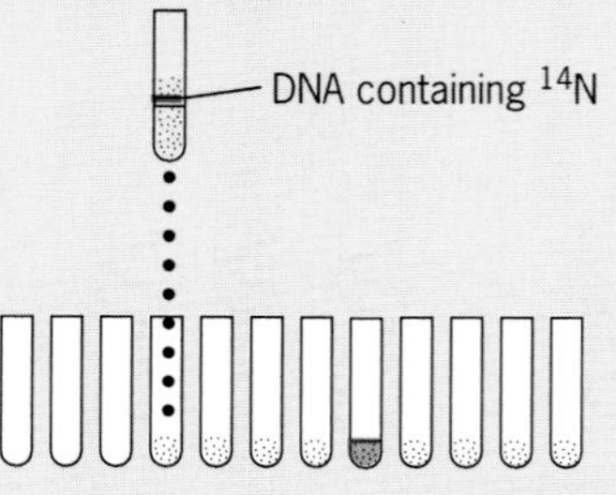

(*a*) CsCl Equilibrium Density Gradient.

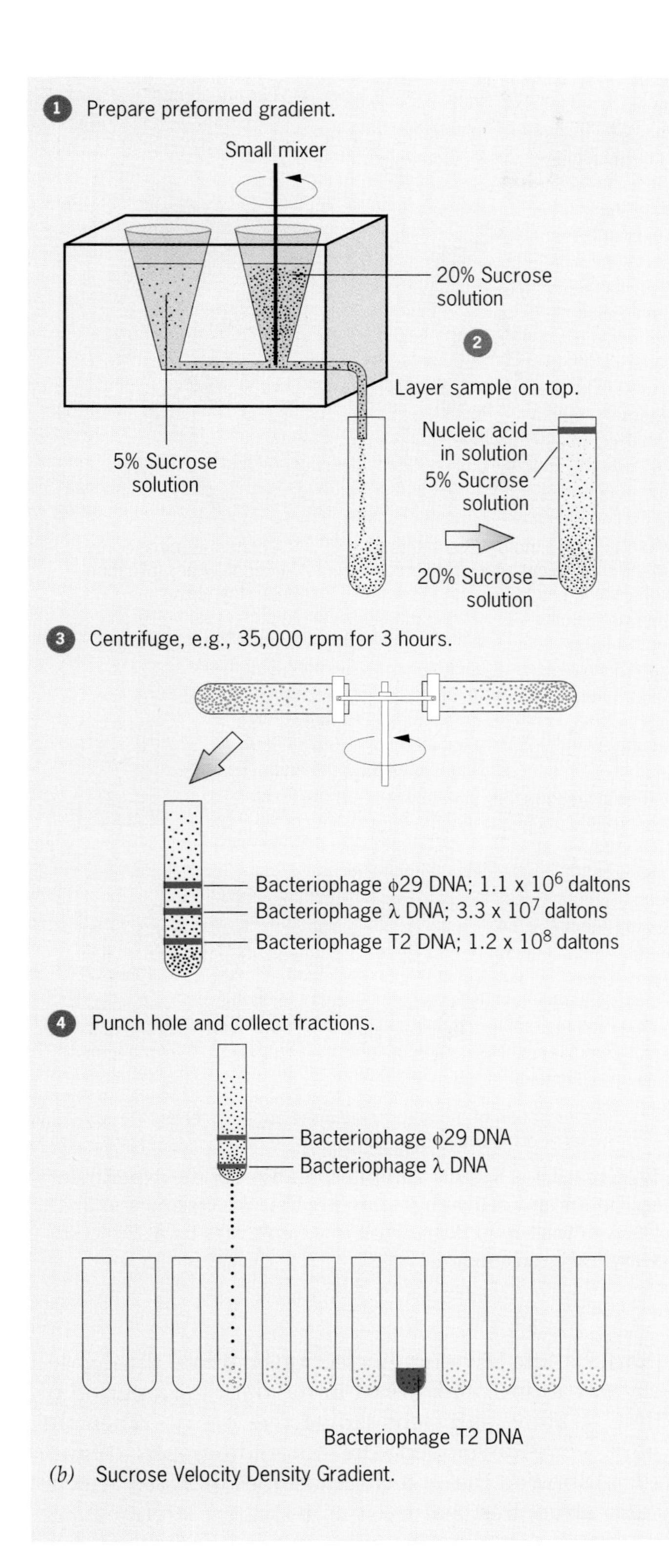

(*b*) Sucrose Velocity Density Gradient.

A gradient maker is used to prepare a preformed density gradient in a centrifuge tube (see diagram *b*). The gradient maker contains two chambers for sucrose solutions. One chamber is filled with a sucrose solution of the concentration desired at the bottom of the centrifuge tube; the other is filled with a sucrose solution of the concentration desired at the top of the tube. As the tube is filled, the high concentration sucrose solution is progressively diluted with the low concentration solution, producing a linear gradient in the tube.

Although any range of sucrose concentrations can be used, 5 to 20 percent and 10 to 40 percent sucrose gradients are widely used to separate nucleic acids. The solution containing the molecules to be analyzed is layered on the top of the sucrose gradient prior to centrifugation. The density of this solution must be lower than the density of the sucrose at the top of the centrifuge tube so that it will stay on the surface of the sucrose gradient. The tube is then placed in a swinging bucket rotor and centrifuged at high speed for one to a few hours. The centrifugal force causes the macromolecules to migrate through the gradient, with larger molecules moving faster than smaller molecules with the same shape. After centrifugation, a small hole is punched through the bottom of the centrifuge tube, and the resulting drops of solution are collected in test tubes for subsequent analysis. The collected fractions represent successive layers of the sucrose gradient and any molecules therein at the time that centrifugation was terminated.

The velocity at which a macromolecule will move through such a gradient is determined primarily by its molecular weight and shape. A molecule's rate of movement through a solution of lower density during sedimentation is measured by its sedimentation coefficient (s), which equals velocity/centrifugal force. Most macromolecules have s values between 10^{-13} and 10^{-11} seconds. Thus a sedimentation coefficient of 10^{-13} seconds has been designated one Svedberg (S) unit. It was named in honor of The Svedberg, the scientist who invented the ultracentrifuge. The Svedberg unit is the commonly used unit of sedimentation velocity in dealing with macromolecules or macromolecular complexes. For example, nucleic acids are sometimes referred to as 5S or 10S molecules, and ribosomal subunits of prokaryotes are frequently called 30S and 50S subunits.

sive, Meselson and Stahl would have observed a shift of the DNA from heavy toward light in each generation (that is, "half heavy" or hybrid after one generation, "quarter heavy" after two generations, and so forth). These possibilities are clearly inconsistent with the results of Meselson and Stahl's experiment.

The semiconservative replication of eukaryotic chromosomes was first demonstrated in 1957 by the results of experiments carried out by J. Herbert Taylor, Philip Woods, and Walter Hughes on root-tip cells of the broad bean, *Vicia faba*. Taylor and colleagues labeled *Vicia faba* chromosomes by growing root tips for eight hours (less than one cell generation) in medium containing radioactive [^{3}H]thymidine. The root tips were then removed from the radioactive medium, washed, and transferred to nonradioactive medium containing the alkaloid colchicine. As we discussed in Chapter 6, colchicine binds to microtubules and prevents the formation of functional spindle fibers. As a result, daughter chromosomes do not undergo their normal anaphase separation. Thus the number of chromosomes per nucleus will double once per cell cycle in the presence of colchicine. This doubling of the chromosome number each cell generation allowed Taylor and his colleagues to determine how many DNA duplications each cell had undergone subsequent to the incorporation of radioactive thymidine. At the first metaphase in colchicine (c-metaphase), nuclei will contain 12 pairs of chromatids (still joined at the centromeres). At the second c-metaphase, nuclei will contain 24 pairs, and so on.

When the distribution of radioactivity in the *Vicia faba* chromosomes was examined by autoradiography (Chapter 9), both chromatids of each pair were similarly labeled at the first c-metaphase (Figure 10.4*a*). However, at the second c-metaphase, only one of the chromatids of each pair was radioactive (Figure 10.4*b*). These are precisely the results expected if DNA replication is semiconservative, given one DNA molecule per chromosome (Figure 10.4*c*). In 1957, Taylor and his colleagues were able to conclude that chromosomal DNA in *Vicia faba* segregated in a semiconservative manner during each cell division. The conclusion that the double helix replicated semiconservatively in the broad bean had to await subsequent evidence indicating that each chromosome contains a single molecule of DNA. Analogous experiments have subsequently been carried out with several other eukaryotes, and, in all cases, the results indicate that replication is semiconservative.

Autoradiographs of *Vicia faba* chromosomes

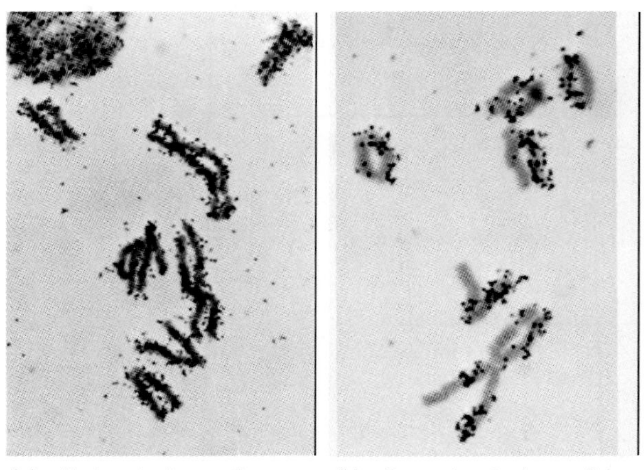

(*a*) First metaphase after replication in ^{3}H-thymidine. (*b*) Second metaphase after an additional replication in ^{1}H-thymidine.

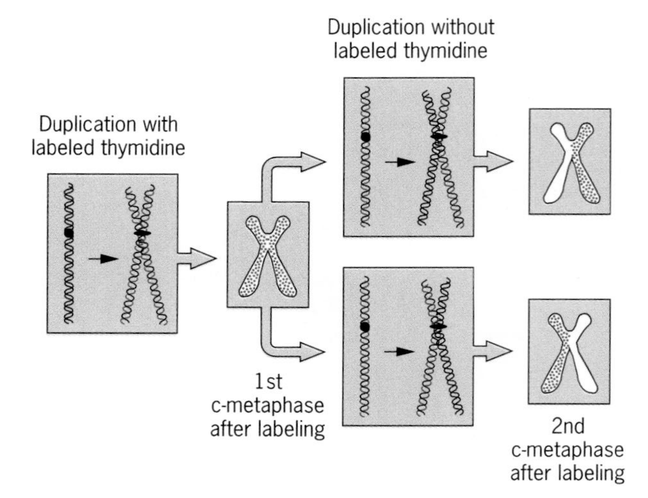

(*c*) Interpretation of the autoradiographs above in terms of semiconservative replication.

Figure 10.4 Results and interpretation of the experiment by Taylor and colleagues showing that the chromosomal DNA of the broad bean, *Vicia faba*, replicates by a semiconservative mechanism.

Visualization of Replication Forks by Autoradiography

The gross structure of replicating bacterial chromosomes was first determined by John Cairns in 1963, again by means of autoradiography. Cairns grew *E. coli* cells in medium containing [^{3}H]thymidine for varying periods of time, lysed the cells gently so as not to break the chromosomes (long DNA molecules are shear-sensitive), and carefully collected the chromosomes on membrane filters. These filters were affixed to glass slides, coated with emulsion sensitive to β-particles (the low-energy electrons emitted during decay of tritium), and stored in the dark for a period of time to allow sufficient radioactive decay. When the films were developed, the autoradiographs (Figure 10.5*a*) showed that the chromosomes of *E. coli* are circular structures that exist as θ-shaped intermediates during replication. The autoradiographs further indicated that the unwinding of the two complementary parental strands (which is necessary for their separation) and their semiconservative replication occur simultaneously or are closely coupled. Since the parental double helix must rotate 360° to unwind each gyre of the helix, some kind of "swivel" must exist. Geneticists now know that the required swivel is a transient single-strand break (cleavage of one phosphodiester bond in one strand of the double helix) produced by the action of enzymes called topoisomerases.

Cairns' interpretation of the autoradiographs was that semiconservative replication of the *E. coli* chromosome started at a specific site, which he called the *origin*, and proceeded sequentially and unidirectionally around the circular structure (Figure 10.5*b*). Subsequent evidence has shown his original interpretation to be incorrect on one point: replication of the *E. coli* chromosome actually proceeds bidirectionally, not unidirectionally. Each Y-shaped structure is a **replication fork**, and the two replication forks move in opposite directions sequentially around the circular chromosome (Figure 10.5*c*).

Unique Origins of Replication

Cairns' results established the existence of a site of initiation or **origin** of replication on the circular chromosome of *E. coli*, but provided no hint as to whether the origin was a unique site or occurred at randomly located sites in a population of replicating chromosomes. In bacterial and viral chromosomes, there is usually one unique origin per chromosome, and this single origin controls the replication of the entire chromosome. In the large chromosomes of eukaryotes, multiple origins collectively control the replication of the giant DNA molecule present in each chromosome. Current evidence indicates that these multiple replication origins in eukaryotic chromosomes also occur at specific sites. Each origin controls the replication of a unit of DNA called a *replicon*; thus most prokaryotic chromosomes contain a single replicon, whereas eukaryotic chromosomes usually contain many replicons.

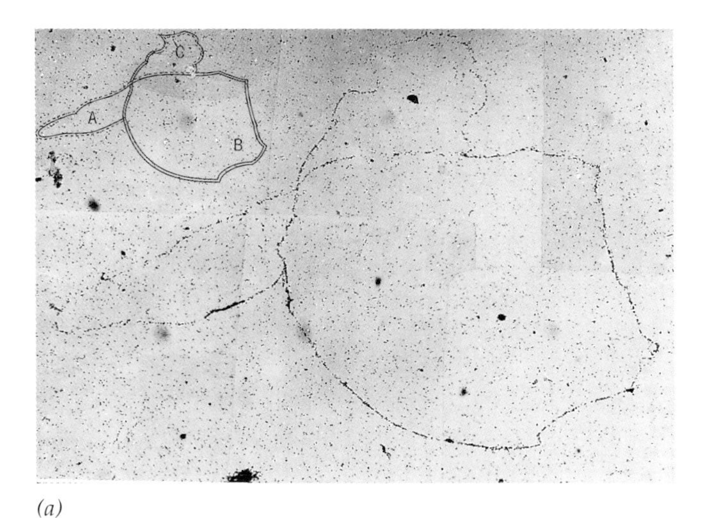

(*a*)

Swivel

Replication fork

(*b*) Original interpretation: unidirectional replication.

Origin

Replication fork

Replication fork

(*c*) Correct interpretation: bidirectional replication.

Figure 10.5 Visualization of the replication of the *E. coli* chromosome by autoradiography. (*a*) One of Cairns' autoradiographs of a θ-shaped replicating chromosome from a cell that had been grown for two generations in the presence of [^{3}H]thymidine, with his interpretative diagram shown at the upper left. Radioactive strands of DNA are shown as solid lines and nonradioactive strands as dashed lines. Loops A and B have completed a second replication in [^{3}H]thymidine; section C remains to be replicated the second time. The two possible interpretations of Cairns' results are shown in (*b*) and (*c*). Cairns originally interpreted his results in terms of unidirectional replication (*b*). DNA replication in *E. coli* was subsequently shown to be bidirectional (*c*).

The single origin of replication, called ***OriC***, in the *E. coli* chromosome has been characterized in considerable detail. *OriC* is 245 nucleotide pairs long and contains two different conserved repeat sequences (Figure 10.6). One 13-bp sequence is present as three tandem repeats. These three repeats are rich in A:T base pairs, facilitating the formation of a localized region of strand separation referred to as the **replication bubble**. Recall that A:T base pairs are held together by only two hydrogen bonds as opposed to three in G:C base pairs (Chapter 9). Thus the two strands of A:T-rich regions of DNA come apart more easily, that is, with the input of less energy. The formation of a localized zone of denaturation is an essential first step in the replication of all double-stranded DNAs. Another conserved component of *OriC* is a 9-bp sequence that is repeated four times and is interspersed with other sequences. These four sequences are binding sites for a protein that plays a key role in the formation of the replication bubble. We discuss additional details of the process of initiation of DNA synthesis at origins and the proteins that are involved later in this chapter.

The multiple origins of replication in eukaryotic chromosomes also appear to be specific DNA sequences. In the yeast *Saccharomyces cerevisiae*, segments of chromosomal DNA that allow a fragment of circularized DNA to replicate as an independent unit (autonomously), that is, as an extrachromosomal self-replicating unit, have been identified and characterized. These sequences are called **ARS** (for **A**utonomously **R**eplicating **S**equences) **elements**. Their frequency in the yeast genome corresponds well with the number of origins of replication, and some have been shown experimentally to function as origins. ARS elements are about 50 base pairs in length and include a core 11-bp A:T-rich sequence,

ATTTATPuTTTA
TAAATAPyAAAT

(where Pu is either of the two purines and Py is either of the two pyrimidines) and additional imperfect copies of this sequence. The ability of ARS elements to function as origins of replication is abolished by base-pair changes within this conserved core sequence. Although the molecular details remain to be worked out, the A:T-rich cores of ARS elements presumably are the sites of the initial strand separations leading to DNA replication.

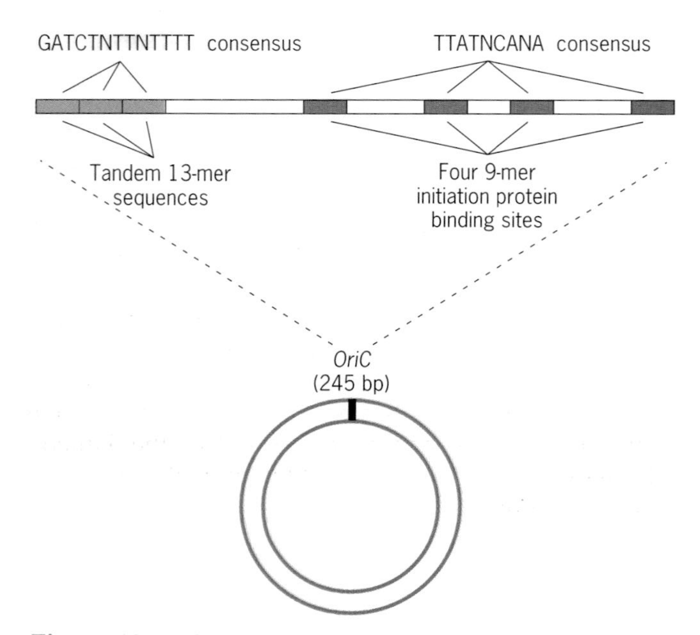

Figure 10.6 Structure of *OriC*, the single origin of replication in the *E. coli* chromosome.

Bidirectional Replication

Cairns interpreted his autoradiograms of replicating *E. coli* chromosomes based on the prevailing view that replication started at a single origin and that a single replication fork moved unidirectionally around the circular chromosome (Figure 10.5*b*). However, his results can be explained equally well by the formation of two forks at the single origin, with the two forks moving in opposite directions or bidirectionally around the circular chromosome (Figure 10.5*c*). We now have definitive evidence showing that replication in *E. coli* and many other prokaryotes proceeds bidirectionally from a unique origin.

Bidirectional replication was first convincingly demonstrated by experiments with some of the small bacterial viruses that infect *E. coli*. Bacteriophage lambda (phage λ) contains a single linear molecule of DNA only 17.5 μm long. The phage λ chromosome is somewhat unique in that it has a single-stranded region, 12 base pairs long, at the 5' end of each complementary strand (Figure 10.7). These single-stranded ends, called "cohesive" or "sticky" ends, are complementary to each other. The cohesive ends of a lambda chromosome can thus base-pair to form a hydrogen-bonded circular structure. One of the first events to occur after a lambda chromosome is injected into a host cell is its conversion to a covalently closed circular molecule (Figure 10.7). This conversion from the hydrogen-bonded circular form to the covalently closed circular form is catalyzed by *DNA ligase*, an important enzyme that seals single-stranded breaks in DNA double helices. DNA ligase is required in all organisms for DNA replication, DNA repair, and recombination between DNA molecules. Like the *E. coli* chro-

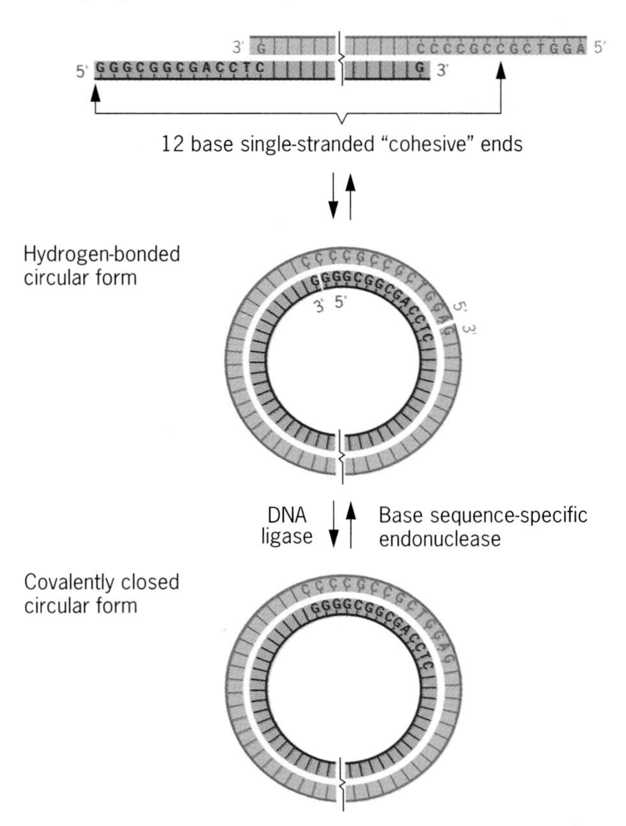

Figure 10.7 Interconversion of the linear lambda chromosome with its complementary cohesive ends, the hydrogen-bonded circular lambda chromosome, and the covalently closed circular lambda chromosome. The linear form of the chromosome appears to be an adaptation to facilitate its injection from the phage head through the small opening in the phage tail into the host cell during infection. Prior to replicating in the host cell, the chromosome is converted to the covalently closed circular form. Only the ends of the chromosome of the mature phage are shown; the jagged vertical line indicates that the central portion of the chromosome is not shown. The entire lambda chromosome is about 4.8×10^4 nucleotide pairs long.

mosome, the lambda chromosome replicates in its circular form via θ-shaped intermediates.

The feature of the lambda chromosome that facilitated the demonstration of bidirectional replication is its differentiation into regions containing high concentrations of adenine and thymine (A:T-rich regions) and regions with large amounts of guanine and cytosine (G:C-rich regions). In particular, it contains a few segments with high A:T content (A:T-rich clusters). Maria Schnös and Ross Inman used these A:T-rich clusters as physical markers to demonstrate, by means of a technique called denaturation mapping, that replication of the lambda chromosome is initiated at a unique origin and proceeds bidirectionally rather than unidirectionally.

When DNA molecules are exposed to high temperature (100°C) or high pH (11.4), the hydrogen and hydrophobic bonds that hold the complementary strands together in the double-helix configuration are broken, and the two strands separate—a process called denaturation. Because A:T base pairs are held together by only two hydrogen bonds, compared with three hydrogen bonds in G:C base pairs, A:T-rich molecules denature more easily (at lower pH or temperature) than G:C-rich molecules. When lambda chromosomes are exposed to pH 11.05 for 10 minutes under the appropriate conditions, the A:T-rich clusters denature to form denaturation bubbles, which are detectable by electron microscopy, whereas the G:C-rich regions remain in the duplex state (Figure 10.8). These denaturation bubbles can be used as physical markers whether the lambda chromosome is in its mature linear form, its circular form, or its θ-shaped replicative intermediates. By examining the positions of the branch points (Y-shaped structures) relative to the positions of the denaturation bubbles in a large number of θ-shaped replicative intermediates, Schnös and Inman demonstrated that both branch points are replication forks that move in opposite directions around the circular chromosome. Figure 10.9 shows the results expected in Schnös and Inman's experiment if replication is (*a*) unidirectional or (*b*) bidirectional. The results clearly demonstrated that replication of the lambda chromosome is bidirectional.

Bidirectional replication from a fixed origin has also been demonstrated for several organisms with chromosomes that replicate as linear structures. Replication of the chromosome of phage T7, another small bacteriophage, begins at a unique site near one end to form an "eye" structure (Figure 10.10*a*) and then proceeds bidirectionally until one fork reaches the nearest end. Replication of the Y-shaped structure (Figure 10.10*b*) continues until the second fork reaches the other end of the molecule, producing two progeny chromosomes.

Replication of chromosomal DNA in eukaryotes is also bidirectional in those cases where it has been investigated. However, bidirectional replication is not universal. The chromosome of coliphage P2, which replicates as a θ-shaped structure like the lambda chromosome, replicates unidirectionally from a unique origin.

Key Points: **DNA replication is semiconservative. As the two complementary strands of a parental double helix unwind and separate, each serves as a template for the synthesis of a new complementary strand. The**

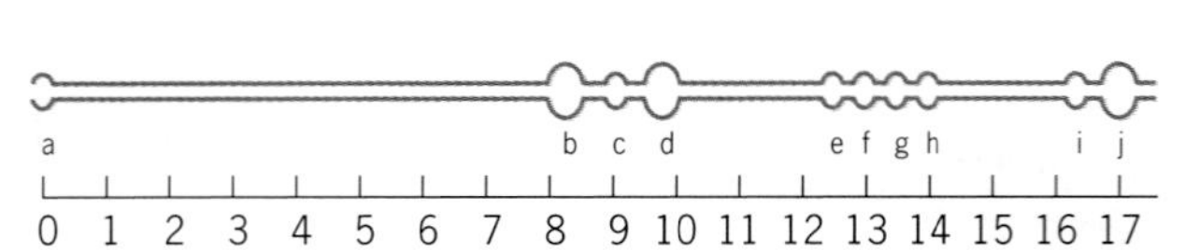

(*a*) Diagram showing the locations of AT-rich denaturation sites in the linear λ chromosome.

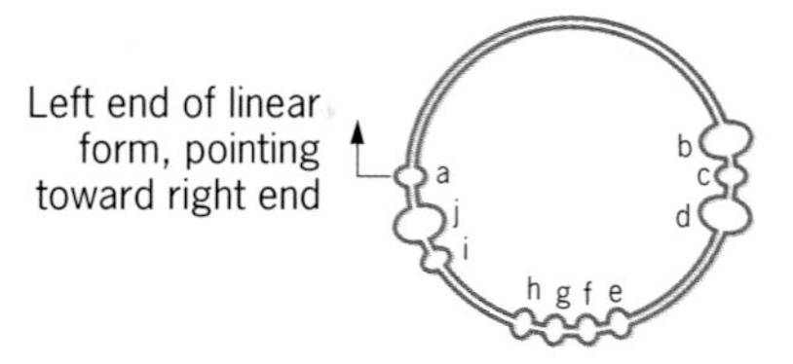

(*b*) Diagram showing the positions of AT-rich denaturation sites in the circular form of the λ chromosome.

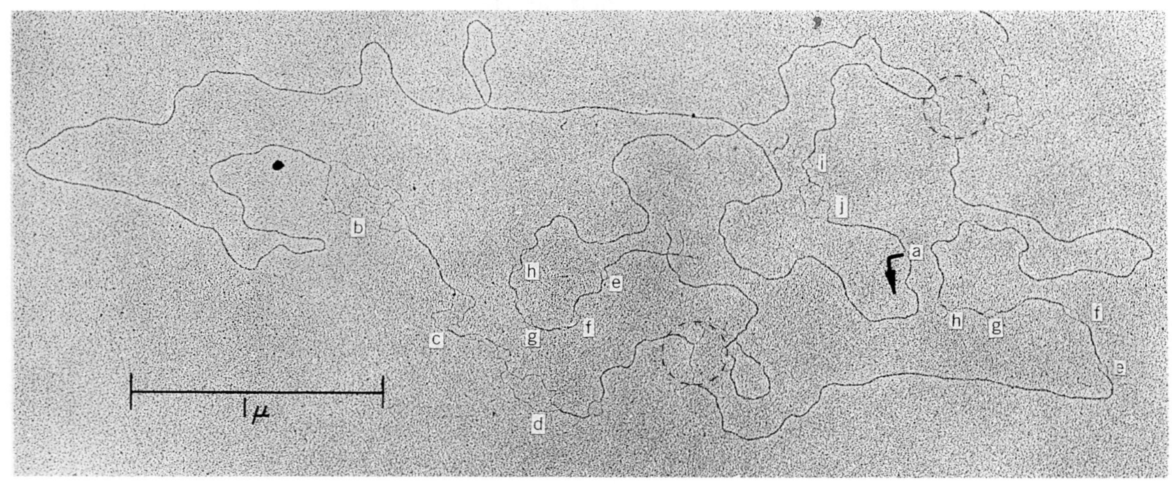

(*c*) Electron micrograph showing the locations of the AT-rich denaturation bubbles in a θ -shaped replicating λ chromosome.

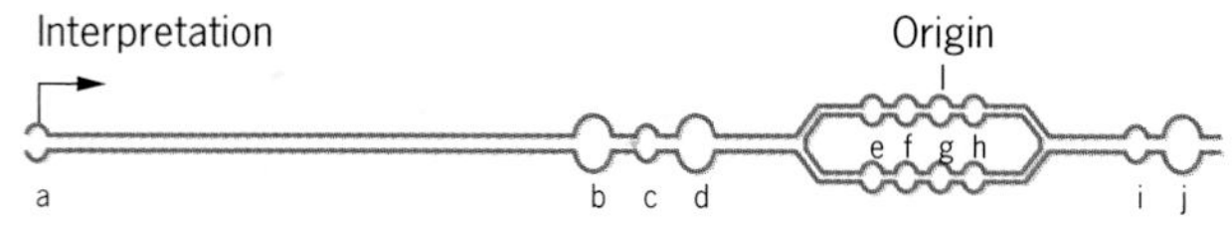

(*d*) Diagram in linear form of the λ replicative intermediate in (*c*) showing the positions of denaturation bubbles relative to the replication forks.

Figure 10.8 Illustration of the use of AT-rich denaturation sites as physical markers to prove that the phage λ chromosome replicates bidirectionally rather than unidirectionally.

hydrogen-bonding potentials of the bases in the template strands specify complementary base sequences in the newly synthesized strands. Replication is initiated at unique origins and usually proceeds bidirectionally from each origin.

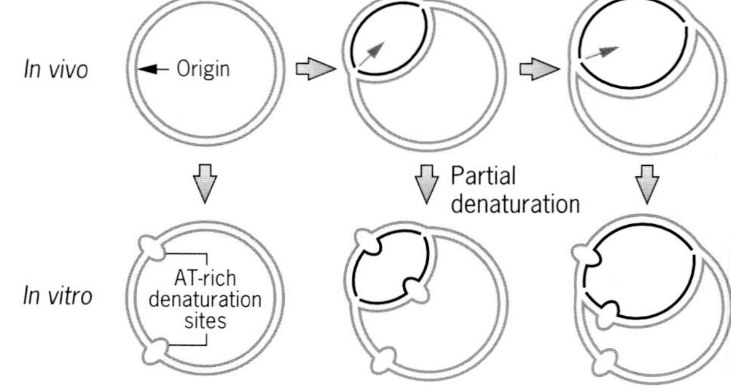

(*a*) Unidirectional replication.

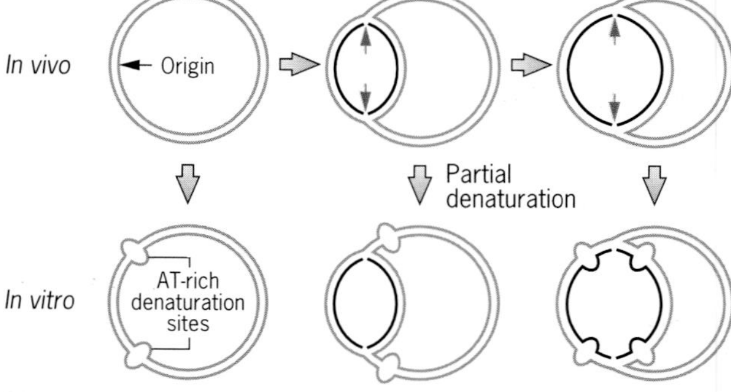

(*b*) Bidirectional replication.

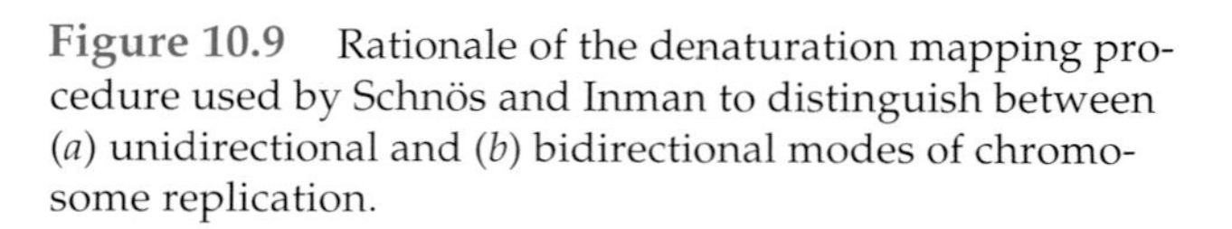

Figure 10.9 Rationale of the denaturation mapping procedure used by Schnös and Inman to distinguish between (*a*) unidirectional and (*b*) bidirectional modes of chromosome replication.

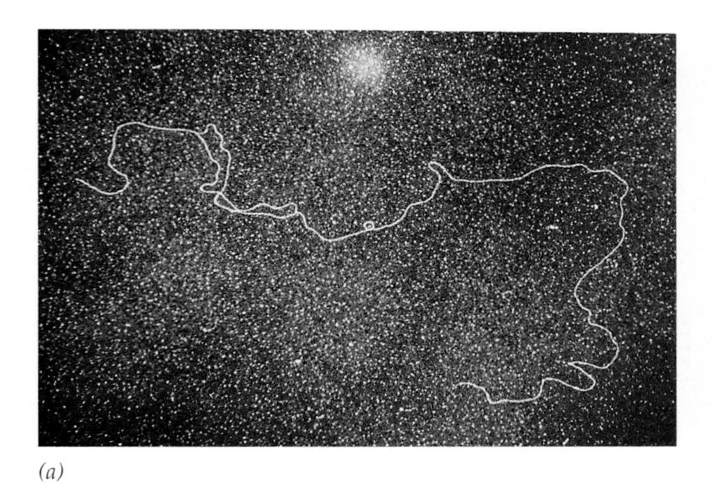

(*a*)

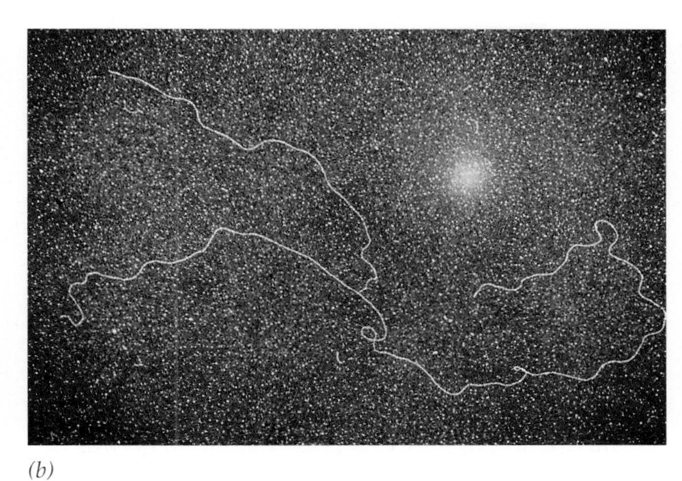

(*b*)

Figure 10.10 Electron micrographs of replicating bacteriophage T7 chromosomes. The phage T7 chromosome, unlike the *E. coli* and phage λ chromosomes, replicates as a linear structure. Its origin of replication is located 17 percent of the length of the chromosome from one end (the left end of the chromosomes shown). The chromosome in (*a*) illustrates the "eye" form (–O–) characteristic of early stages in the replication of linear DNA molecules. Replication proceeds bidirectionally from the origin until the fork moving in a leftward direction reaches the left end of the molecule, yielding a Y-shaped structure such as that shown in (*b*).

DNA POLYMERASES AND DNA SYNTHESIS *IN VITRO*

Much has been learned about the molecular mechanisms involved in biological processes by disrupting cells, separating the various organelles, macromolecules, and other components, and then reconstituting systems in the test tube, so-called *in vitro* systems that are capable of carrying out particular metabolic events. Such *in vitro* systems can be dissected biochemically much more easily than *in vivo* systems. Clearly, the information obtained from studies on *in vitro* systems has been invaluable. However, we should never assume that a phenomenon demonstrated *in vitro* occurs *in vivo*. Such an extrapolation should be made only when independent evidence from *in vivo* studies validates the *in vitro* studies.

Discovery of DNA Polymerase I in *Escherichia coli*

The *in vitro* synthesis of DNA was first accomplished by Arthur Kornberg and his co-workers in 1957. Kornberg, who received a Nobel Prize in 1959 for this work, isolated an enzyme from *E. coli* that catalyzes the covalent addition of nucleotides to preexisting DNA chains. Initially called DNA polymerase or "Kornberg's enzyme," it is now known as **DNA polymerase I**. The enzyme requires the 5'-triphosphates of each of the four deoxyribonucleosides—deoxyadenosine triphosphate (dATP), deoxythymidine triphosphate (dTTP), deoxyguanosine triphosphate (dGTP), and deoxycytidine triphosphate (dCTP)—and is active only in the presence of Mg^{2+} ions and preexisting DNA. This DNA must provide two essential components, one serving a primer function and the other a template function (Figure 10.11).

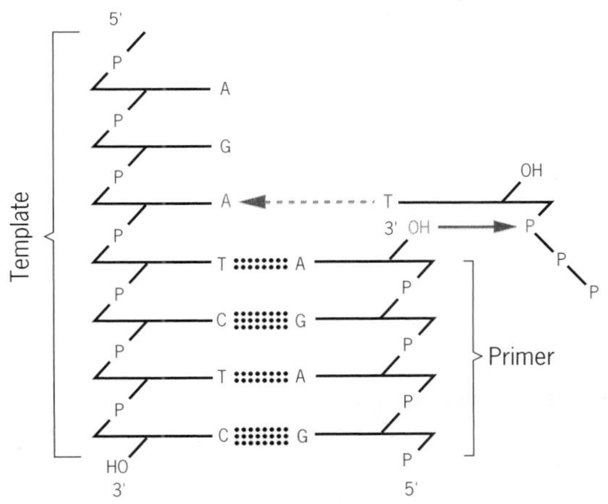

Figure 10.11 Template and primer requirements of DNA polymerases. All DNA polymerases require a primer strand (shown on the right) with a free 3'-hydroxyl. The primer strand is covalently extended by the addition of nucleotides (such as dTMP, derived from the incoming precursor dTTP shown). In addition, DNA polymerases require a template strand (shown on the left), which determines the base sequence of the strand being synthesized. The new strand will be complementary to the template strand.

1. The **primer DNA** provides a terminus with a free 3'-OH to which nucleotides are added during DNA synthesis. DNA polymerase I cannot initiate the synthesis of DNA chains *de novo.* It has an absolute requirement for a free 3'-hydroxyl on a preexisting DNA chain. DNA polymerase I catalyzes the formation of a phosphodiester bridge between the 3'-OH at the end of the primer DNA chain and the 5'-phosphate of the incoming deoxyribonucleotide.
2. The **template DNA** provides the nucleotide sequence that specifies the complementary sequence of the nascent DNA chain. DNA polymerase I requires a DNA template whose base sequence dictates, by its base-pairing potential, the synthesis of a complementary base sequence in the strand being synthesized.

The reaction catalyzed by DNA polymerase I is a nucleophilic attack by the 3'-OH at the terminus of the primer strand on the nucleotidyl or interior phosphorus atom of the nucleoside triphosphate precursor with the elimination of pyrophosphate. This reaction mechanism explains the absolute requirement of DNA polymerase I for a free 3'-OH group on the primer DNA strand that is being covalently extended and dictates that *the direction of synthesis is always 5′ → 3′* (Figure 10.12).

DNA polymerase I is a single polypeptide with a molecular weight of 109,000 encoded by a gene called *polA*. However, subsequent research has shown that DNA polymerase I is not the true "DNA replicase" in *E. coli*. It does not catalyze the semiconservative replication of the *E. coli* chromosome; that function is performed by another enzyme. Nevertheless, DNA polymerase I does perform important functions in the *E. coli* cell, including playing a key role in chromosome replication and a central role in repairing damaged DNA. To understand how DNA polymerase I performs these functions, we must get to know this enzyme better.

In addition to the polymerase activity illustrated in Figure 10.12, DNA polymerase I has two other enzymatic activities, both exonuclease activities. A **nuclease** is an enzyme that degrades nucleic acids. An **exonuclease** degrades nucleic acids starting at one or both ends, whereas an **endonuclease** cleaves nucleic acids at internal sites. DNA polymerase I contains both **5' → 3' exonuclease activity**, which cuts back DNA strands starting at 5' termini, and **3' → 5' exonuclease activity**, which cleaves off mononucleotides from the 3' termini of DNA strands. The 5' → 3' exonuclease activity of DNA polymerase I usually excises small oligomers containing up to 10 nucleotides. Thus DNA polymerase I has three different enzymatic activities: (1) a 5' → 3' polymerase activity, (2) a 5' → 3' exonuclease activity, and (3) a 3' → 5' exonuclease activity. These three activities are illustrated in Figure 10.13.

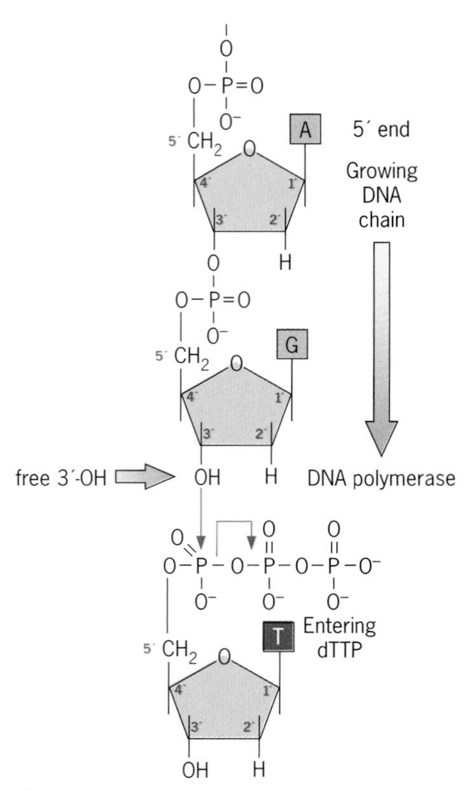

Figure 10.12 Mechanism of action of DNA polymerase I: covalent extension of a DNA primer strand in the 5' → 3' direction. The existing chain terminates at the 3' end with the nucleotide deoxyguanylate (or deoxyguanosine-5'-phosphate). The diagram shows the DNA polymerase-catalyzed addition of deoxythymidine monophosphate (from the precursor deoxythymidine triphosphate, dTTP) to the 3' end of the chain with the release of pyrophosphate (P_2O_7).

The first evidence that DNA polymerase I was not the true DNA replicase was published in 1969 by Paula DeLucia and John Cairns. They reported that DNA replication occurred in an *E. coli* strain lacking this enzyme owing to a mutation in the *polA* gene. Their results demonstrated that DNA replication in *E. coli* does not require DNA polymerase I activity, at least not the 5' → 3' polymerase activity of the enzyme. However, DeLucia and Cairns also discovered that this *polA1* mutant was extremely sensitive to ultraviolet light (UV). All three enzymatic activities of DNA polymerase I play important roles in the cell. The major function of DNA polymerase I in *E. coli* is to repair defects in DNA, such as those induced by UV (Chapter 13). However, as we will see later in this chapter, the 5' → 3' exonuclease activity of DNA polymerase I is also involved at one stage of chromosome replication.

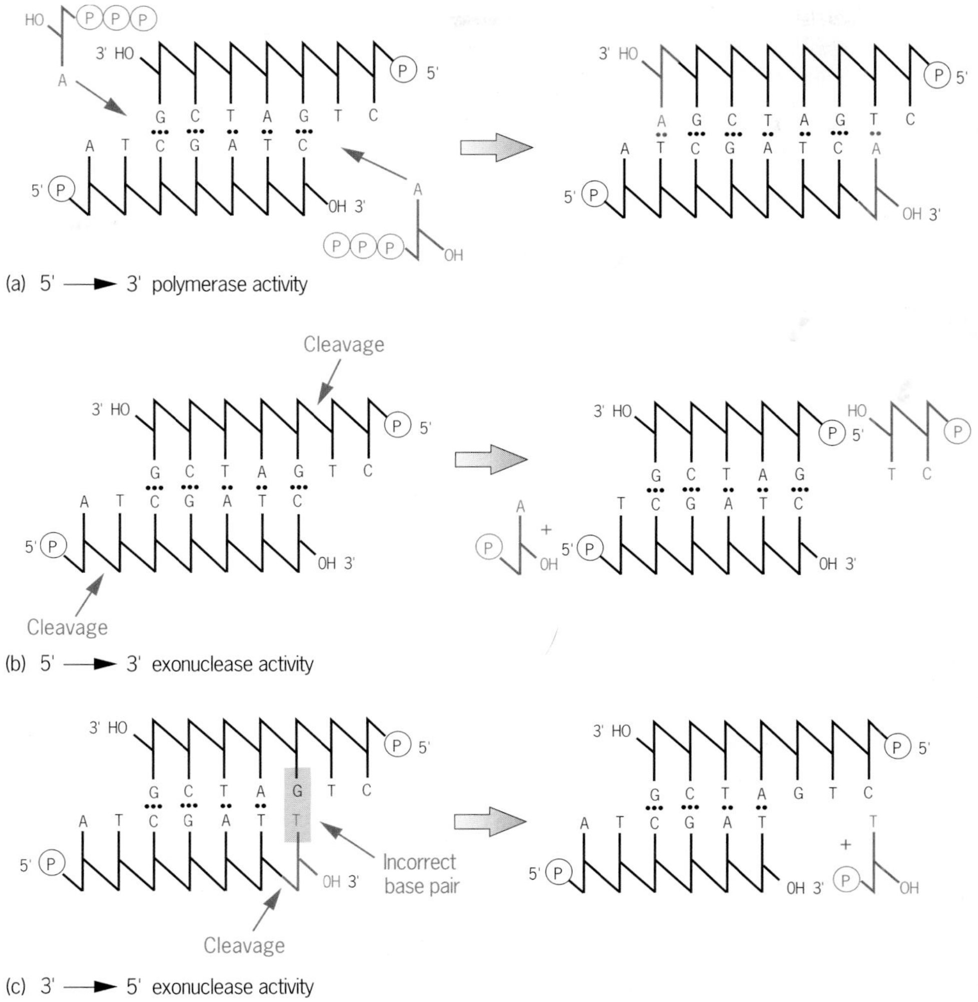

(a) 5' → 3' polymerase activity

(b) 5' → 3' exonuclease activity

(c) 3' → 5' exonuclease activity

Figure 10.13 The three activities of DNA polymerase I of *E. coli*. As is discussed in the text, all three activities play important roles in the replication of the *E. coli* chromosome.

Multiple DNA Polymerases

If DNA polymerase I does not catalyze the semiconservative replication of the *E. coli* chromosome, another polymerase must carry out this function. In fact, there are two other DNA polymerases, **DNA polymerase II** and **DNA polymerase III**, in *E. coli*. Like DNA polymerase I, DNA polymerase II is a DNA repair enzyme; but it represents a small proportion of the polymerase activity in an *E. coli* cell. DNA polymerase II is a single polypeptide with 5' → 3' polymerase and 3' → 5' exonuclease activities. However, it has no 5' → 3' exonuclease activity. In contrast to DNA polymerases I and II, DNA polymerase III is a complex enzyme composed of many different subunits. Like DNA polymerase II, DNA polymerase III has 5' → 3' polymerase and 3' → 5' exonuclease activities, but no 5' → 3' exonuclease activity.

Eukaryotic organisms are even more complex—with up to five different DNA polymerases having been identified. The five DNA polymerases characterized in mammals have been named α (sometimes I), β, γ, δ (sometimes III), and ϵ (sometimes II). Two of the DNA polymerases (α and δ) work together to carry out the semiconservative replication of nuclear DNA. DNA polymerase γ is responsible for the replication of DNA in mitochondria, and DNA polymerases β and ϵ are nuclear DNA repair enzymes. Some of the eukaryotic DNA polymerases lack the 3' → 5' exonuclease activity that is present in prokaryotic DNA polymerases.

All of the DNA polymerases studied to date, prokaryotic and eukaryotic, catalyze the same basic reaction: a nucleophilic attack by the free 3'-OH at the primer strand terminus on the nucleotidyl phosphorus of the nucleoside triphosphate precursor. Thus, all

DNA polymerases have an absolute requirement for a free 3'-hydroxyl group on a preexisting primer strand; none of these DNA polymerases can initiate new DNA chains *de novo*, and all DNA synthesis occurs in the 5' → 3' direction.

DNA Polymerase III: The Replicase in *Escherichia coli*

Evidence indicating that DNA polymerase III is the true DNA replicase responsible for the semiconservative replication of DNA in *E. coli* was first provided by the isolation and characterization of a mutant strain with a mutation in a gene called *polC*, now renamed *dnaE*. This mutant strain produced active DNA polymerase III when grown at 25°C, but totally inactive polymerase III when grown at 43°C. When *dnaE* mutant cells growing at 25°C were shifted to 43°C, DNA replication stopped, indicating that the product of the *dnaE* gene is required for DNA synthesis The *dnaE* gene was subsequently shown to encode the catalytic α subunit, the subunit with 5' → 3' polymerase activity, of DNA polymerase III.

DNA polymerase III is a multimeric enzyme (an enzyme with many subunits) with a molecular mass of about 900,000 daltons in its complete or **holoenzyme form**. The minimal core that has catalytic activity *in vitro* contains three subunits: α (the *dnaE* gene product), ϵ (the *dnaQ* product), and θ (gene unknown). The addition of the τ subunit (the *dnaX* product) results in dimerization of the catalytic core and increased activity. The catalytic core synthesizes rather short DNA strands because of its tendency to fall off the DNA template. In order to synthesize the long DNA molecules present in chromosomes, this frequent dissociation of the polymerase from the template must be eliminated. The β subunit (the *dnaN* gene product) of DNA polymerase III forms a dimeric clamp that keeps the polymerase from falling off the template DNA (Figure 10.14). The β-dimer forms a ring that encircles the replicating DNA molecule and allows DNA polymerase III to slide along the DNA while remaining tethered to it. The DNA polymerase III holoenzyme, which is responsible for the synthesis of both nascent DNA strands at a replication fork, contains at least 20 polypeptides. The structural complexity of the DNA polymerase III holoenzyme is illustrated in Figure 10.15; the diagram shows 16 of the best-characterized polypeptides encoded by seven different genes. We will consider the functions of some of the subunits in subsequent sections of this chapter.

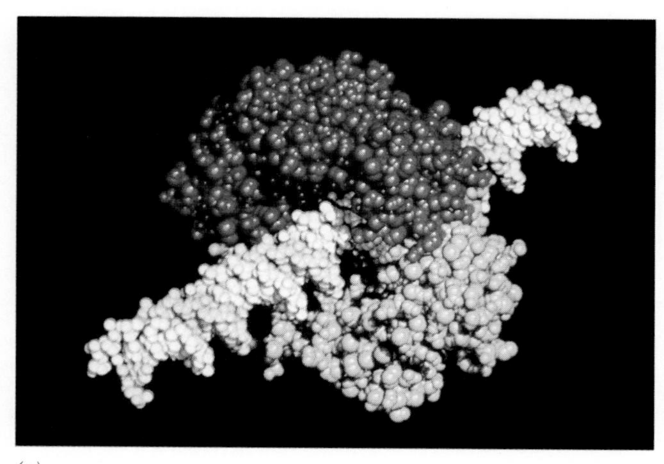

(a)

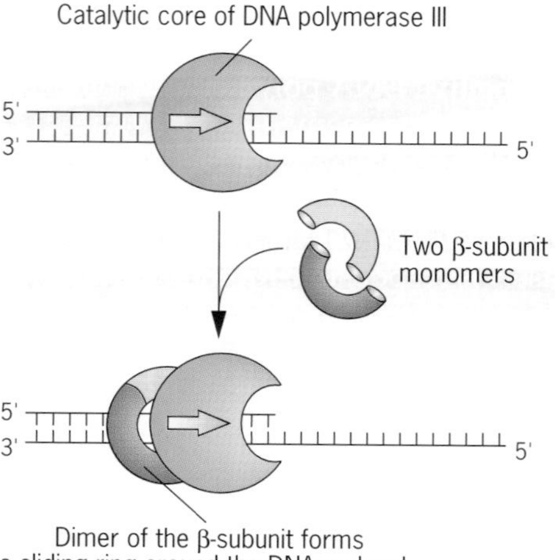

(b)

Figure 10.14 The β subunit of DNA polymerase III clamps the enzyme to the DNA molecule.

Proofreading Activities of DNA Polymerases

As we discussed earlier, the fidelity of DNA replication is amazing—with only about one error in every billion base-pair duplications. This high fidelity is necessary to keep the mutation load at a tolerable level, especially in large genomes such as those of mam-

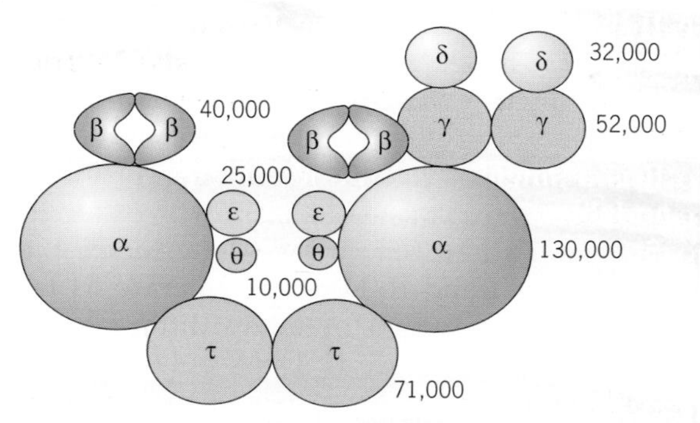

Figure 10.15 Structure of the *E. coli* DNA polymerase III holoenzyme.

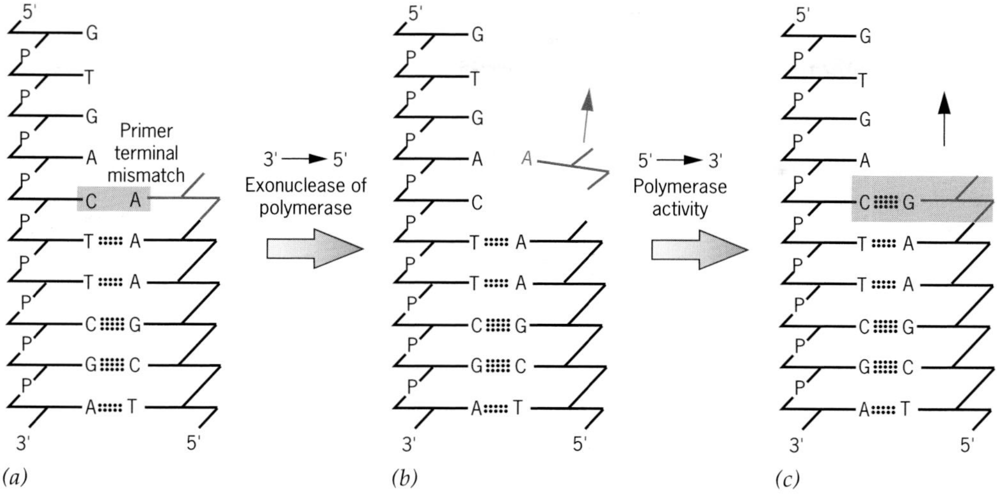

Figure 10.16 Proofreading by the 3' → 5' exonuclease activity of DNA polymerases during DNA replication. If DNA polymerase is presented with a template and primer containing a 3' primer terminal mismatch (*a*), it will not catalyze covalent extension (polymerization). Instead, the 3' → 5' exonuclease activity, an integral part of many DNA polymerases, will cleave off the mismatched terminal nucleotide (*b*). Then, presented with a correctly base-paired primer terminus, DNA polymerase will catalyze 5' → 3' covalent extension of the primer strand (*c*).

mals, which contain 3×10^9 nucleotide pairs. Without the high fidelity of DNA replication, the monozygotic twins discussed at the beginning of this chapter would be less similar in phenotype. Indeed, based on the dynamic structures of the four nucleotides in DNA, the observed fidelity of DNA replication is much higher than expected. The thermodynamic changes in nucleotides that allow the formation of hydrogen-bonded base pairs other than A:T and G:C predict error rates of 10^{-5} to 10^{-4}, or one error per 10,000 to 100,000 incorporated nucleotides. The predicted error rate of 10,000 times the observed error rate raises the question of how this high fidelity of DNA replication can be achieved.

Living organisms have solved the potential problem of insufficient fidelity during DNA replication by evolving a mechanism for **proofreading** the nascent DNA chain as it is being synthesized. The proofreading process involves scanning the termini of nascent DNA chains for errors and correcting them. This process is carried out by the 3' → 5' exonuclease activities of DNA polymerases. When a template-primer DNA has a terminal mismatch (an unpaired or incorrectly paired base or sequence of bases at the 3' end of the primer), the 3' → 5' exonuclease activity of the DNA polymerase clips off the unpaired base or bases (Figure 10.16). When an appropriately base-paired terminus is produced, the 5' → 3' polymerase activity of the enzyme begins resynthesis by adding nucleotides to the 3' end of the primer strand. In monomeric enzymes like DNA polymerase I of *E. coli*, this activity is built in. In multimeric enzymes, the 3' → 5' proofreading exonuclease activity is often present on a separate subunit. In the case of DNA polymerase III of *E. coli*, this proofreading function is carried out by the ϵ subunit. In eukaryotes, DNA polymerases γ, δ, and ϵ contain 3' → 5' proofreading exonuclease activities, but polymerases α and β lack this activity. Given the importance of proofreading, we might speculate that accessory proteins must carry out the proofreading function for DNA polymerases α and β.

Key Points: **DNA synthesis is catalyzed by enzymes called DNA polymerases. All DNA polymerases have an absolute requirement for a primer strand, which is extended, and a template strand, which is copied. All DNA polymerases have an absolute requirement for a free 3'-OH on the primer strand, and all DNA synthesis occurs in the 5' to 3' direction. The 3' → 5' exonuclease activities of DNA polymerases proofread nascent strands as they are synthesized, removing any mispaired nucleotides at the 3' termini of primer strands.**

THE COMPLEX REPLICATION APPARATUS

The results of studies of DNA replication by autoradiography and electron microscopy indicate that the two progeny strands being synthesized at each replicating fork are being extended in the same overall direction at the macromolecular level. Because the complementary strands of a double helix have opposite polarity, synthesis is occurring at the 5' end of one strand (or 3' → 5') and the 3' end of the other strand (5' → 3'). However, as we have previously discussed, all known polymerases have an absolute requirement for a free 3'-hydroxyl; they only carry out 5' → 3' synthesis. These apparently contradictory results created an interesting paradox. For many years biochemists searched for new polymerases that could catalyze 3' → 5' synthesis. No such polymerase was ever found. Instead, experimental evidence has shown that all DNA synthesis occurs in the 5' → 3' direction.

Clearly, the mechanism of DNA replication must be more complex than researchers originally thought (see Figure 10.1). In addition, given the absolute requirement of DNA polymerase for a free 3'-OH on the primer strand, this enzyme cannot begin the synthesis of a new strand *de novo*. How is the synthesis of a new DNA strand initiated? Since the two parental strands of DNA must be unwound, we have to deal with the need for a swivel or axis of rotation, especially for circular DNA molecules like that present in the *E. coli* chromosome. Finally, how does the localized zone of strand separation or replication bubble form at the origin? These considerations and others indicate that DNA replication is more complicated than scientists thought when the semiconservative mechanism of replication was proposed by Watson and Crick in 1953.

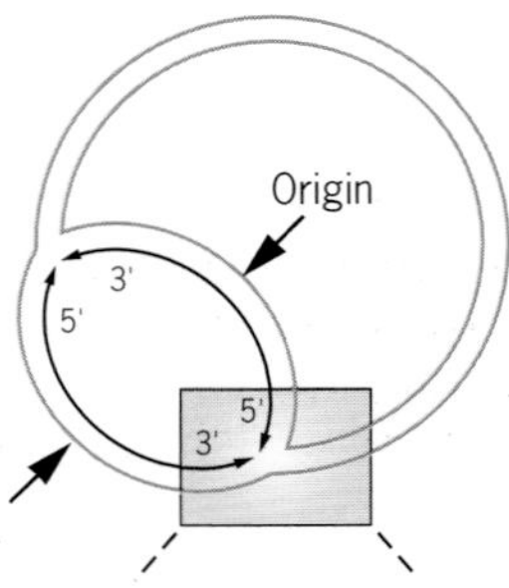

(*a*) Relatively low-resolution techniques such as autoradiography and electron microscopy show that at the macromolecular level both nascent DNA chains are extended in the same overall direction at each replication fork.

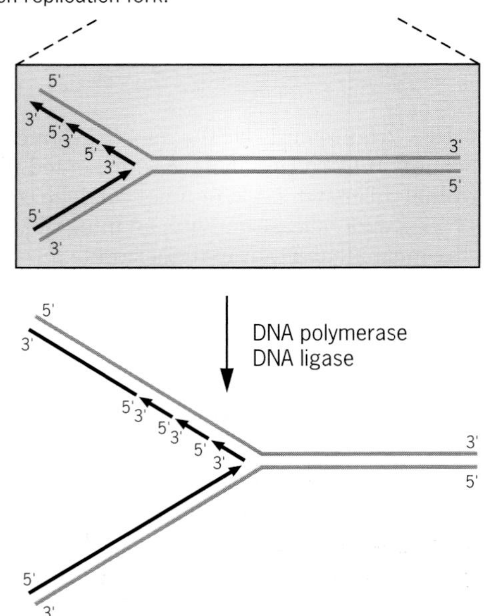

(*b*) High-resolution biochemical techniques such as pulse-labeling and density gradient analysis show that replication of the lagging strand is discontinuous—short fragments are synthesized in the 5' → 3' direction and subsequently joined by DNA ligase.

Figure 10.17 Continuous synthesis of the leading strand and discontinuous synthesis of the lagging strand at a DNA replication fork.

Continuous Synthesis of One Strand; Discontinuous Synthesis of the Other Strand

As discussed in the preceding section, the two nascent DNA strands being synthesized at each replicating fork are being extended in the same direction at the macromolecular level. Because the complementary strands of a DNA double helix have opposite chemical polarity, one strand is being extended in an overall 5' → 3' direction and the other strand is being extended in an overall 3' → 5' direction (Figure 10.17, top). But DNA polymerases can only catalyze synthesis in the 5' → 3' direction. The resolution of this paradox occurred with the demonstration that the synthesis of one strand of DNA is **continuous**, whereas synthesis of the other strand is **discontinuous**. At the molecular level, synthesis of the complementary strands of DNA is occurring in opposite physical directions (Figure 10.17, bottom), but both new strands are extended in the same 5' → 3' chemical direction. The synthesis of the strand being extended in the overall 5' → 3' direction, called the **leading strand**, is **continuous**. The strand being extended in the overall 3' → 5' direction, called the **lagging strand**, grows by the synthesis of short fragments (synthesized 5' → 3') and the subsequent covalent joining of these short fragments. Thus the synthesis of the lagging strand occurs by a discontinuous mechanism.

The first evidence for this discontinuous mode of DNA replication came from studies in which intermediates in DNA synthesis were radioactively labeled by growing *E. coli* cells and bacteriophage T4-infected *E. coli* cells for very short periods of time in medium containing [^{3}H]thymidine (pulse-labeling experiments). The labeled DNAs were then isolated, denatured, and characterized by measuring their velocity of sedimentation through sucrose gradients during high-speed centrifugation (see Technical Sidelight: Cesium Chloride . . .). When *E. coli* cells were pulse-labeled for 15 seconds, for example, much of the label was found in small fragments of DNA, 1000 to 2000 nucleotides long. These small fragments of DNA have been named Okazaki fragments after Reiji and Tuneko Okazaki, the scientists who discovered them in the late 1960s. In eukaryotes, the Okazaki fragments are only 100 to 200 nucleotides in length. When longer pulse-labeling periods are used, more of the label is recovered in large DNA molecules, presumably the size of *E. coli* or phage T4 chromosomes. If cells are pulse-labeled with ^{3}H-thymidine for a short period and then are transferred to nonradioactive medium for an extended period of growth (pulse-chase experiments), the labeled thymidine is present in chromosome-size DNA molecules. The results of these pulse-chase experiments are important because they indicate that the Okazaki fragments are true intermediates in DNA replication and not some type of metabolic byproduct.

Covalent Closure of Nicks in DNA by DNA Ligase

If the lagging strand of DNA is synthesized discontinuously as described in the preceding section, a mechanism is needed to link the Okazaki fragments together to produce the large DNA strands present in mature chromosomes. This mechanism is provided by the enzyme **DNA ligase.** DNA ligase catalyzes the covalent closure of nicks (missing phosphodiester linkages; no missing bases) in DNA molecules by using energy from nicotinamide adenine dinucleotide (NAD) or adenosine triphosphate (ATP). The *E. coli* DNA ligase uses NAD as a cofactor, but some DNA ligases use ATP. The reaction catalyzed by DNA ligase is shown in Figure 10.18. First, AMP of the ligase-AMP intermediate forms a phosphoester linkage with the 5'-phosphate at the nick, and then a nucleophilic attack by the 3'-OH at the nick on the DNA-proximal phosphorus atom produces a phosphodiester linkage between the adjacent nucleotides at the site of the nick. DNA ligase has no activity at breaks in DNA where one or more nucleotides are missing—so-called gaps. Gaps can be sealed only by the combined action of a DNA polymerase and DNA ligase. DNA ligase plays an essential role not only in DNA replication, but also in DNA repair and in the pathways for recombination, as we will see in Chapter 13.

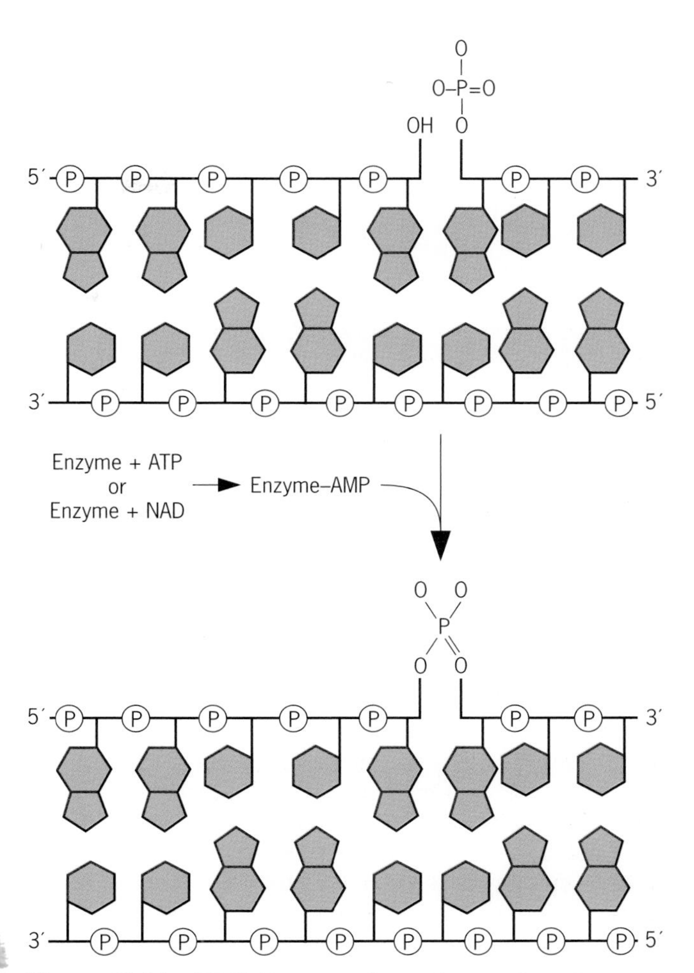

Figure 10.18 DNA ligase catalyzes the covalent closure of nicks in DNA. The energy required to form the ester linkage is provided by either adenosine triphosphate (ATP) or nicotinamide-adenine dinucleotide (NAD), depending on the species.

Initiation of DNA Chains with RNA Primers

As discussed earlier, all known DNA polymerases have absolute requirements for a free 3'-OH on a DNA primer strand and an appropriate DNA template strand for activity. No known DNA polymerase can initiate the synthesis of a new strand of DNA. Thus some special mechanism must exist to initiate or prime new DNA chains. Whereas the continuous synthesis of the leading strand requires the priming function only at the origin of replication, a priming event is required to initiate each Okazaki fragment during the discontinuous synthesis of the lagging strand. RNA polymerase, a complex enzyme that catalyzes the synthesis of RNA molecules from DNA templates, has long been known to be capable of initiating the synthesis of new RNA chains at specific sites on the DNA. When this occurs, an RNA-DNA hybrid is formed in which the nascent RNA is hydrogen-bonded to the DNA template. Since DNA polymerases are capable of extending polynucleotide chains containing an RNA primer with a free 3'-OH, scientists began testing the idea that DNA synthesis is initiated by RNA primers, and their results proved that this idea is correct.

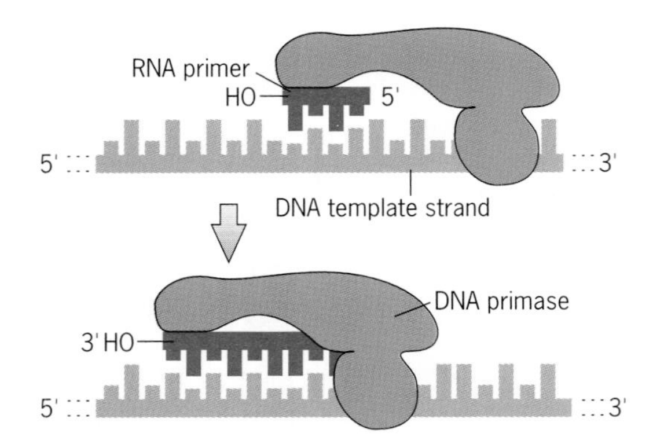

Figure 10.19 The initiation of DNA strands with RNA primers. The enzyme DNA primase catalyzes the synthesis of short (10 to 60 nucleotides long) RNA strands that are complementary to the template strands. DNA polymerase III then uses the free 3'-hydroxyls of the RNA primers to extend the chains by the addition of deoxyribonucleotides (see Figure 10.20).

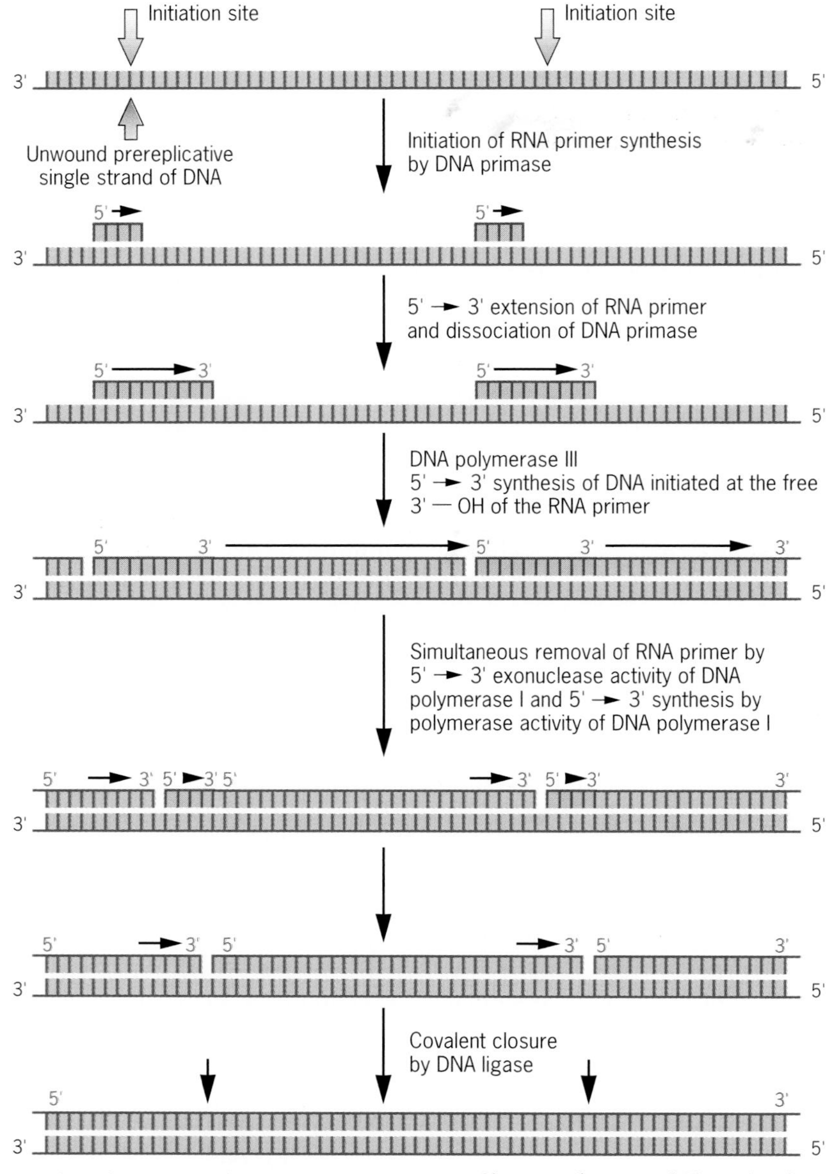

Figure 10.20 Synthesis and replacement of RNA primers during replication of the lagging strand of DNA. A short RNA strand is synthesized to provide a 3'-OH primer for DNA synthesis (see Figure 10.19). The RNA primer is subsequently removed and replaced with DNA by the dual 5' → 3' exonuclease and 5' → 3' polymerase activities built into DNA polymerase I. DNA ligase then covalently closes the nascent DNA chain, catalyzing the formation of phosphodiester linkages between adjacent 3'-hydroxyls and 5'-phosphates (see Figure 10.18).

Subsequent research has shown that each new DNA chain is initiated by a short **RNA primer** synthesized by **DNA primase** (Figure 10.19). The *E. coli* DNA primase is the product of the *dnaG* gene. In prokaryotes, these RNA primers are 10 to 60 nucleotides long, whereas in eukaryotes they are shorter, only about 10 nucleotides long. The RNA primers provide the free 3'-OHs required for covalent extension of polynucleotide chains by DNA polymerases. In *E. coli*, deoxyribonucleotides are added to the RNA primers by DNA polymerase III, either continuously on the leading strand or discontinuously by the synthesis of Okazaki fragments on the lagging stand. DNA polymerase III terminates an Okazaki fragment when it bumps into the RNA primer of the preceding Okazaki fragment.

The RNA primers subsequently are excised and replaced with DNA chains. This step is accomplished by DNA polymerase I in *E. coli*. Recall that of the three DNA polymerases in *E. coli*, only DNA polymerase I possesses 5' → 3' exonuclease activity. The 5' → 3' exonuclease activity of DNA polymerase I excises the RNA primer, and, at the same time, the 5' → 3' polymerase activity of the enzyme replaces the RNA with a DNA chain by using the adjacent Okazaki fragment with its free 3'-OH as a primer. As we might expect based on this mechanism of primer replacement, *E.*

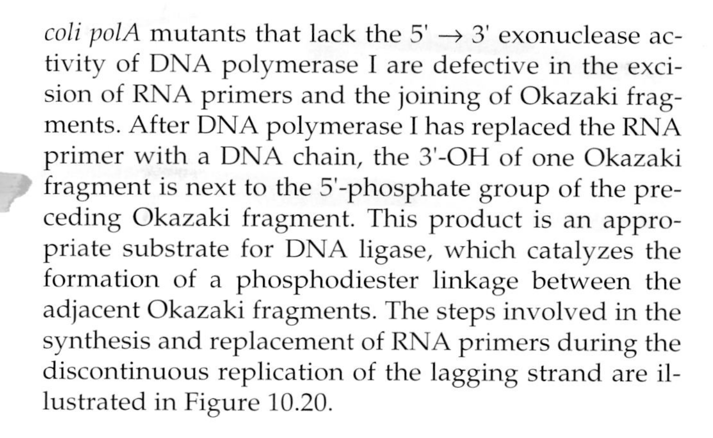

coli polA mutants that lack the 5' → 3' exonuclease activity of DNA polymerase I are defective in the excision of RNA primers and the joining of Okazaki fragments. After DNA polymerase I has replaced the RNA primer with a DNA chain, the 3'-OH of one Okazaki fragment is next to the 5'-phosphate group of the preceding Okazaki fragment. This product is an appropriate substrate for DNA ligase, which catalyzes the formation of a phosphodiester linkage between the adjacent Okazaki fragments. The steps involved in the synthesis and replacement of RNA primers during the discontinuous replication of the lagging strand are illustrated in Figure 10.20.

Unwinding DNA with Helicases, DNA-Binding Proteins, and Topoisomerases

Semiconservative replication requires that the two strands of a parental DNA molecule be separated during the synthesis of new complementary strands. Since a DNA double helix is a plectonemic coil with two strands that cannot be separated without untwisting them turn by turn, DNA replication requires an unwinding mechanism. Given that each gyre, or turn, is about 10 nucleotide pairs long, a DNA molecule must be rotated 360° once for each 10 replicated base pairs. In *E. coli*, DNA replicates at a rate of about 30,000 nucleotides per minute. Thus a replicating DNA molecule must spin at 3000 revolutions per minute to facilitate the unwinding of the parental DNA strands. The unwinding process (Figure 10.21*a*) is catalyzed by enzymes called **DNA helicases**. The major DNA helicase in *E. coli* is the product of the

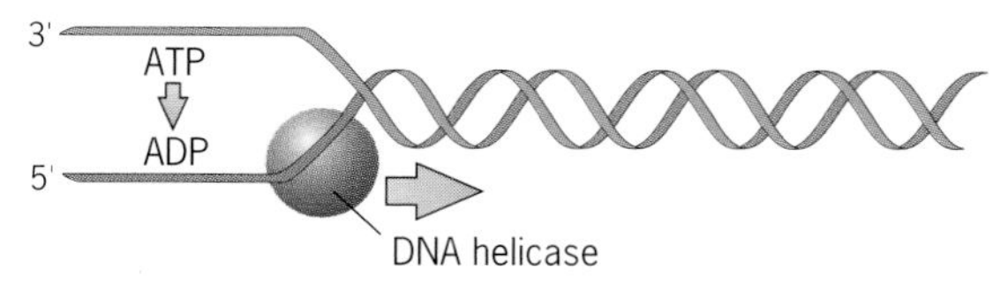

(*a*) DNA helicase catalyzes the unwinding of the parental double helix.

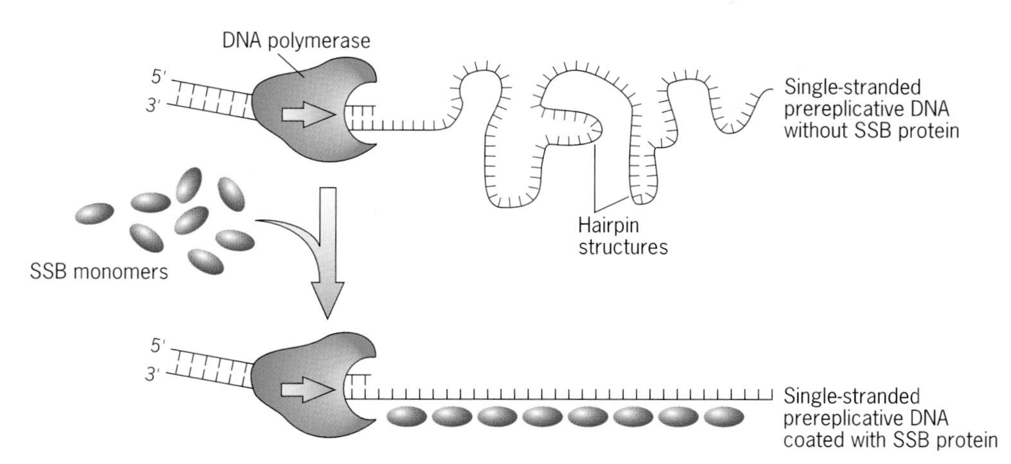

(*b*) Single-stranded DNA-binding (SSB) proteins keep the unwound strands in an extended form for replication.

Figure 10.21 The formation of functional template DNA requires (*a*) DNA helicase, which unwinds the parental double helix, and (*b*) single-strand DNA-binding (SSB) protein, which keeps the unwound DNA strands in an extended form. In the absence of SSB protein, DNA single strands can form hairpin structures by intrastrand base-pairing (*b*, top), and the hairpin structures will retard or arrest DNA synthesis.

dnaB gene. DNA helicases unwind DNA molecules using energy derived from ATP.

Once the DNA strands are unwound by DNA helicase, they must be kept in an extended single-stranded form for replication. They are maintained in this state by a coating of **single-strand DNA-binding proteins** (SSB proteins) (Figure 10.21*b*). The binding of SSB proteins to single-stranded DNA is cooperative; that is, the binding of the first SSB monomer stimulates the binding of additional monomers at contiguous sites on the DNA chain. Because of the cooperativity of SSB protein binding, an entire single-stranded region of DNA is rapidly coated with SSB protein. Without the SSB protein coating, the complementary strands could renature or form intrastrand hairpin structures by hydrogen bonding between short segments of complementary or partially complementary nucleotide sequences. Such hairpin structures are known to impede the activity of DNA polymerases. In *E. coli*, the SSB protein is encoded by the *ssb* gene.

Recall that the *E. coli* chromosome contains a circular molecule of DNA. With the *E. coli* DNA spinning at 3000 revolutions per minute to allow the unwinding of the parental strands during replication (Figure 10.22), what provides the swivel or axis of rotation that prevents the DNA from becoming tangled (positively supercoiled) ahead of the replication fork? The required axes of rotation during the replication of circular DNA molecules are provided by enzymes called **DNA topoisomerases**. The topoisomerases catalyze transient breaks in DNA molecules but use covalent linkages to themselves to hold on to the cleaved molecules. The topoisomerases are of two types: (1) DNA topoisomerase I enzymes produce temporary single-strand breaks or nicks in DNA, and (2) DNA topoisomerase II enzymes produce transient double-strand breaks in DNA. An important result of this difference is that topoisomerase I activities remove supercoils from DNA one at a time, whereas topoisomerase II enzymes remove and introduce supercoils two at a time.

The transient single-strand break produced by the activity of topoisomerase I provides an axis of rotation that allows the segments of DNA on opposite sides of the break to spin independently, with the phosphodiester bond in the intact strand serving as a swivel (Figure 10.23*a*). Thus, during DNA replication, only a short segment of DNA in front of the replication fork needs to spin—the segment up to the closest transient nick by topoisomerase I. Topoisomerase I enzymes are energy efficient. They conserve the energy of the cleaved phosphodiester linkages by storing it in covalent linkages between themselves and the phosphate groups at the cleavage sites; they then reuse this energy to reseal the breaks. Some topoisomerase I enzymes can relax both negatively and positively supercoiled DNAs; others, such as the *E. coli topA* gene product, can act only on negatively supercoiled DNAs. The *E. coli* topoisomerase I removes one nega-

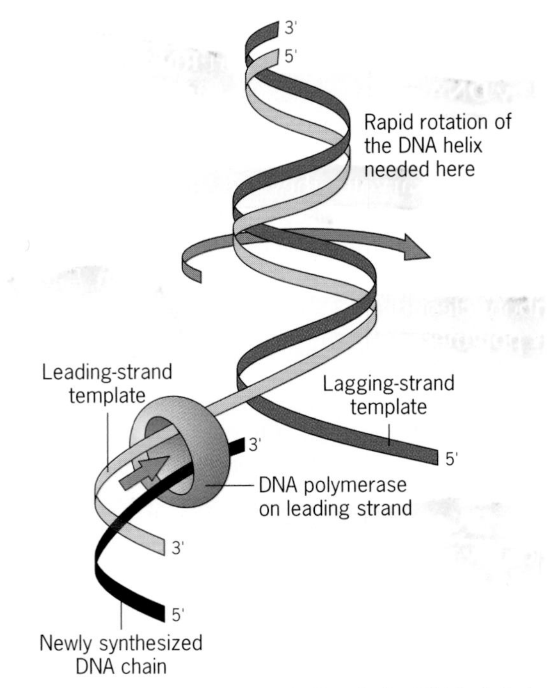

(*a*) To unwind the template strands in *E. coli*, the DNA helix in front of the replication fork must spin at 3000 rpm.

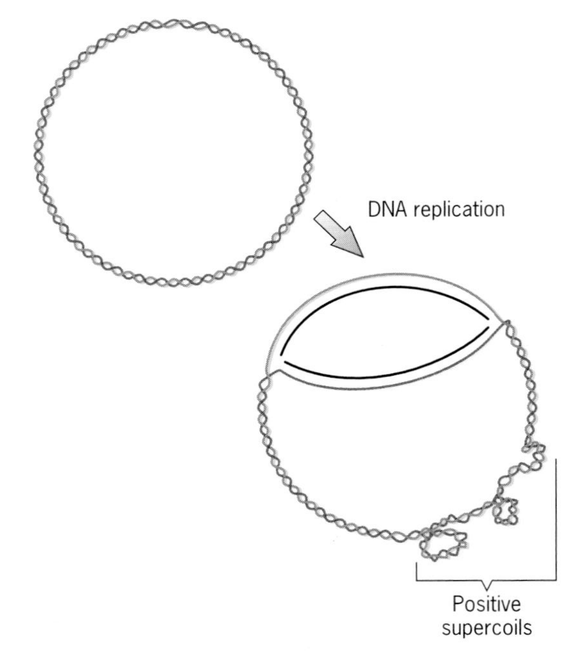

(*b*) Without a swivel or axis of rotation, the unwinding process would produce positive supercoils in front of the replication forks.

Figure 10.22 A swivel or axis of rotation is required during the replication of circular molecules of DNA like those in the *E. coli* or phage λ chromosomes.

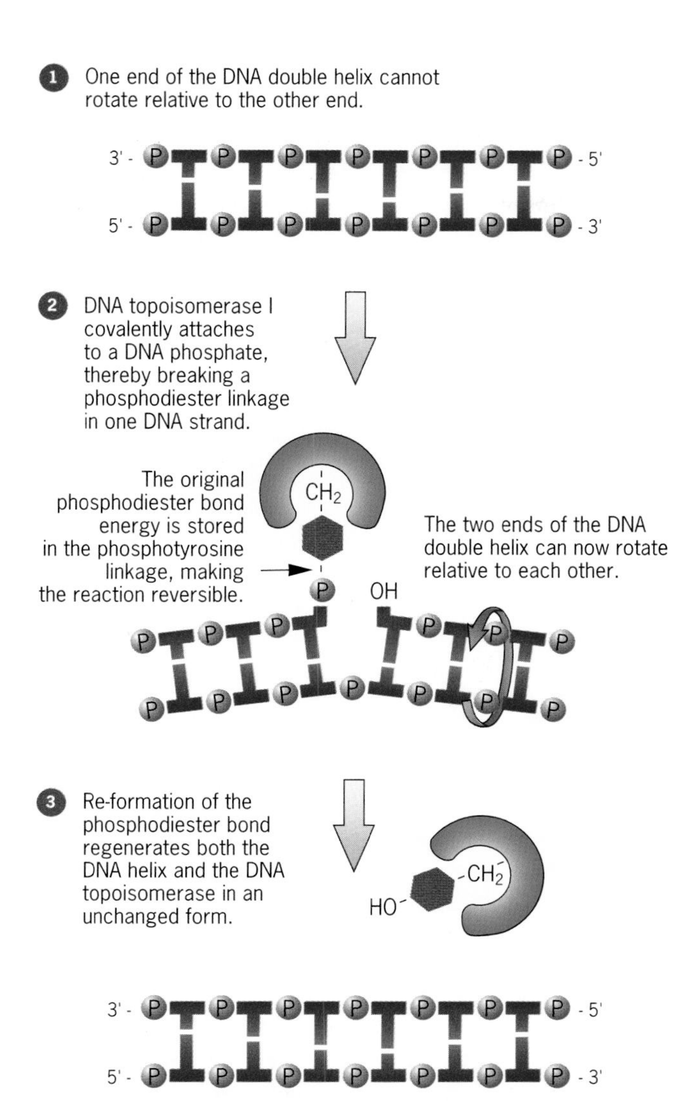

(*a*) DNA topoisomerase I provides transient single-strand breaks that serve as axes of rotation or swivels in DNA molecules.

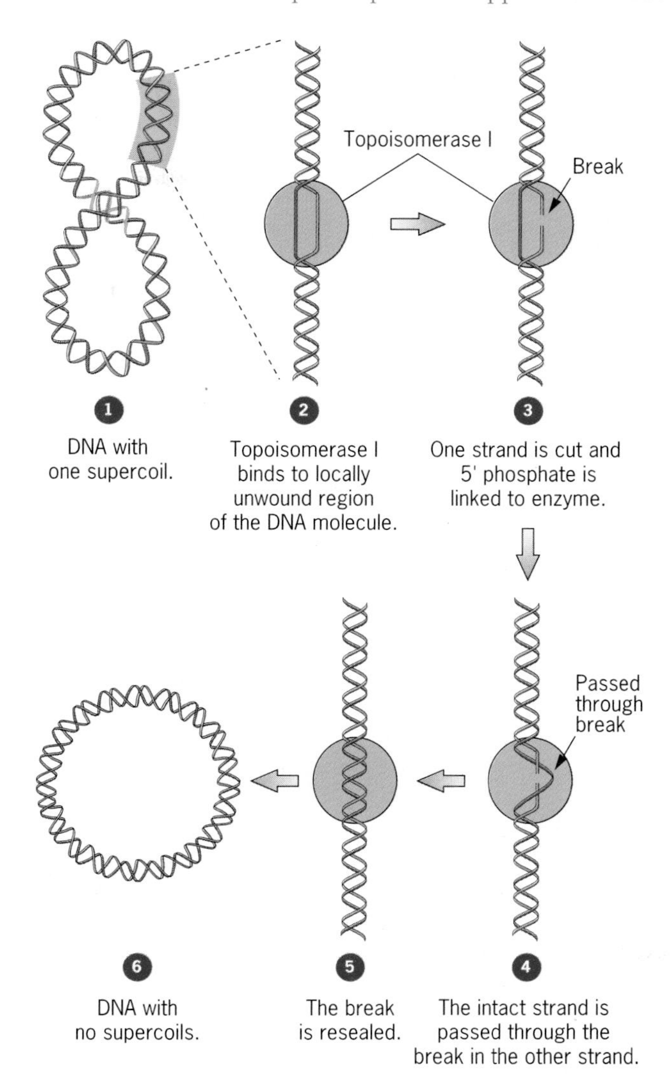

(*b*) DNA topoisomerase I removes supercoils from DNA one at a time

Figure 10.23 DNA topoisomerase I produces transient single-strand breaks in DNA (*a*) that act as axes of rotation or swivels during DNA replication. The reactions catalyzed by DNA topoisomerase I remove supercoils from DNA one at a time (*b*), whereas those catalyzed by DNA topoisomerase II remove supercoils two at a time (see Figure 10.24).

tive supercoil from DNA by breaking one strand, passing the intact strand through the break, and resealing the transient break (Figure 10.23*b*).

DNA topoisomerase II enzymes induce transient double-strand breaks and add or remove supercoils two at a time by an energy (ATP)-requiring mechanism. They carry out this process by cutting both strands of DNA, holding on to the ends at the cleavage site via covalent bonds, passing the intact double helix through the cut, and resealing the break (Figure 10.24*a*). In addition to relaxing supercoiled DNA and introducing negative supercoils into DNA, topoisomerase II enzymes can separate interlocking circular molecules of DNA (Figure 10.24*b*). Given the high concentration of DNA in nuclei and nucleoids and the large sizes of the DNA molecules involved, we can appreciate the important role of type II topoisomerases in keeping these DNA molecules from becoming irreversibly entangled.

The best-characterized type II topoisomerase is an enzyme named **DNA gyrase** in *E. coli*. DNA gyrase is a tetramer with two α subunits encoded by the *gyrA* gene (originally *nalA*, for nalidixic acid) and two β subunits specified by the *gyrB* gene (formerly *cou*, for coumermycin). Nalidixic acid and coumermycin are antibiotics that block DNA replication in *E. coli* by inhibiting the activity of DNA gyrase. Nalidixic acid and coumermycin inhibit DNA synthesis by binding to the

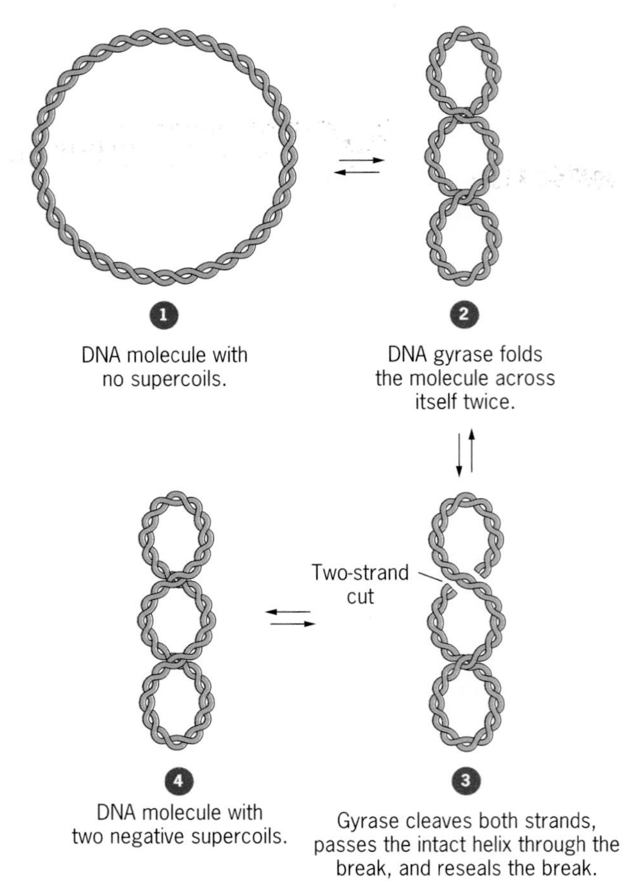

(a) DNA gyrase, a DNA topoisomerase II, introduces and removes negative supercoils two at a time.

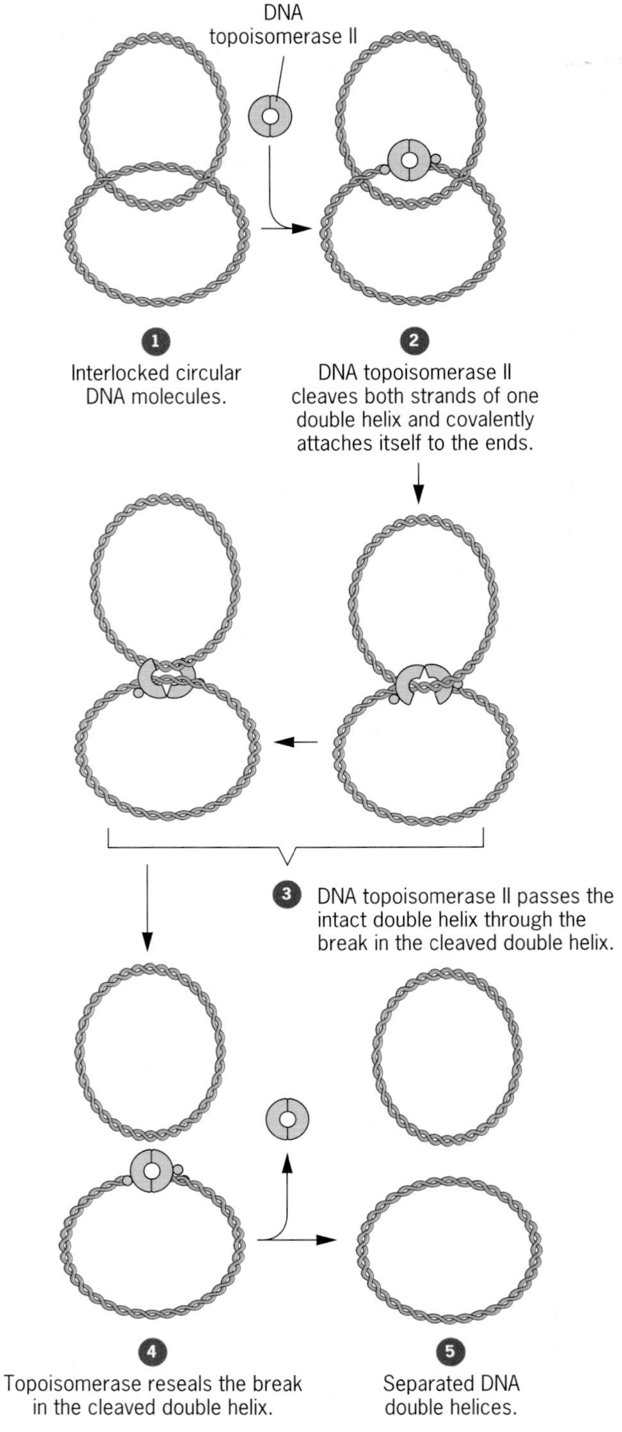

(b) Separation of interlocked circular DNA molecules by DNA topoisomerase II.

Figure 10.24 Activities of DNA topoisomerase II enzymes. (*a*) Mechanism of action of DNA gyrase, an *E. coli* DNA topoisomerase II required for DNA replication. (*b*) Separation of interlocked circular DNA molecules by DNA topoisomerase II.

α and β subunits, respectively, of DNA gyrase. Thus DNA gyrase activity is required for DNA replication to occur in *E. coli*.

Recall that chromosomal DNA is negatively supercoiled in *E. coli* (Chapter 9). The negative supercoils in bacterial chromosomes are introduced by DNA gyrase, with energy supplied by ATP. This activity of DNA gyrase provides another solution to the unwinding problem. Instead of creating positive supercoils ahead of the replication fork by unwinding the complementary strands of relaxed DNA, replication may produce relaxed DNA ahead of the fork by unwinding negatively supercoiled DNA. Because superhelical tension is reduced during unwinding—that is, strand separation is energetically favored—the negative supercoiling behind the fork may drive the unwinding process. If so, this nicely explains why DNA gyrase activity is required for DNA replication in bacteria.

The Replication Apparatus: Prepriming Proteins, Primosomes, and Replisomes

The basic features of DNA replication and most of the important components of the replication apparatus have been introduced in the preceding sections of this chapter. Now we will put the pieces together and look at the coordinated activities of these components during the replication of chromosomal DNA. To do so, we will consider the sequence of events that occur during the replication of the circular molecule of DNA in the *E. coli* chromosome. Figure 10.25 shows the most important components of the replication apparatus in *E. coli* and indicates where they are located on a replication fork.

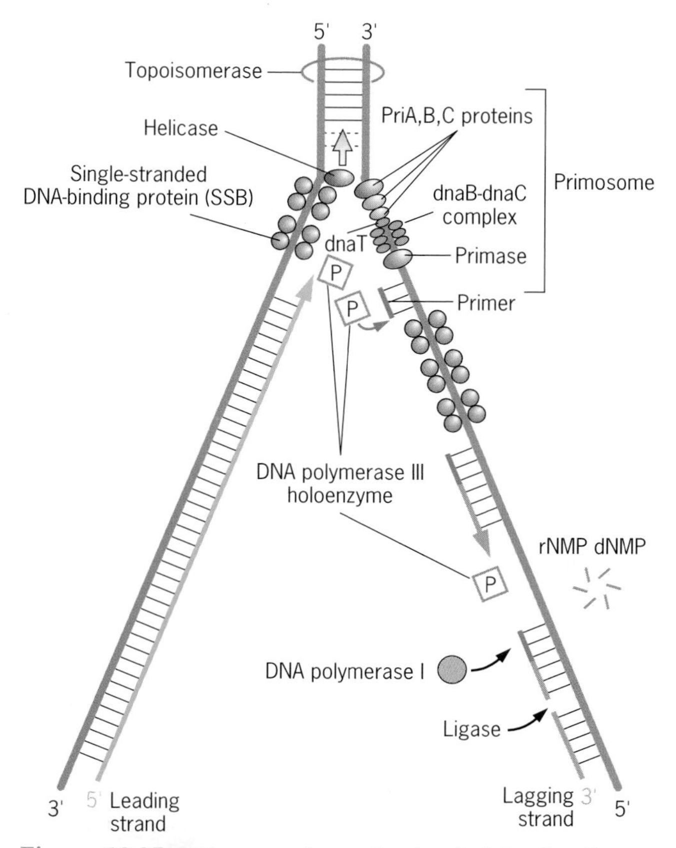

Figure 10.25 Diagram of a replication fork in *E. coli* showing the major components of the replication apparatus.

The replication of the *E. coli* chromosome begins at ***OriC***, the unique sequence at which replication is initiated, with the formation of a localized region of strand separation called the **replication bubble**. This replication bubble is formed by the interaction of **prepriming proteins** with *OricC* (Figure 10.26). The first step in prepriming appears to be the binding of four mole-

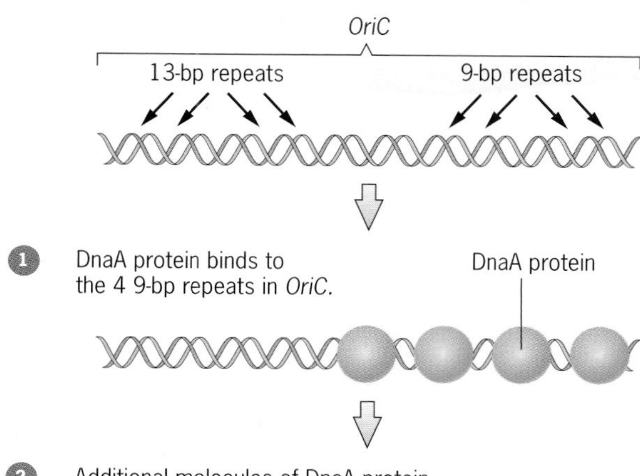

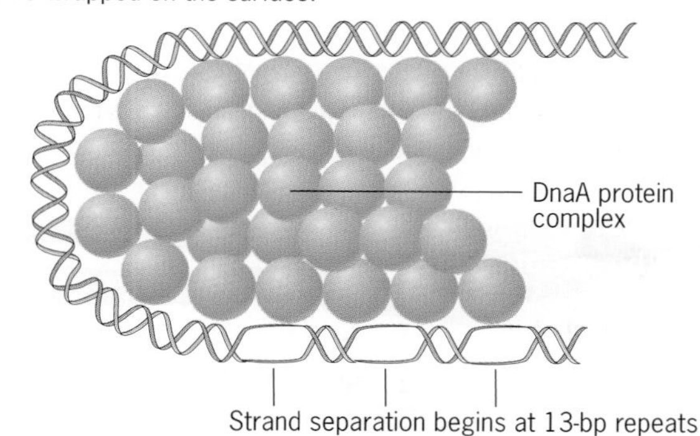

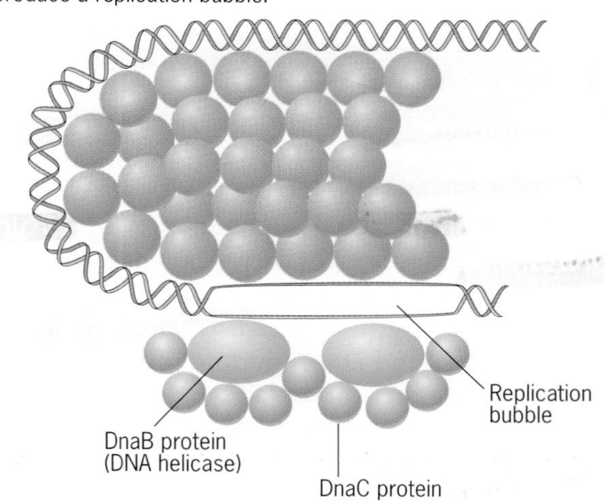

Figure 10.26 Prepriming of DNA replication at *OriC* in the *E. coli* chromosome.

cules of the *dnaA* gene product—**DnaA protein**—to the four 9-base-pair (bp) repeats in *OriC*. Next, DnaA proteins bind cooperatively to form a core of 20 to 40 polypeptides with *OriC* DNA wound on the surface of the protein complex. Strand separation begins within the three tandem 13-bp repeats in *OriC* and spreads until the replication bubble is created. A complex of DnaB protein (the hexameric DNA helicase) and **DnaC protein** (six molecules also) joins the initiation complex and contributes to the formation of two bidirectional replication forks. The DnaT protein also is present in the prepriming protein complex, but its function is unknown. Other proteins that are found associated with the initiation complex at *OriC* are DnaJ protein, DnaK protein, PriA protein, PriB protein, PriC protein, DNA-binding protein HU, DNA gyrase, and single-strand DNA-binding (SSB) protein. However, in some cases, their functional involvement in the prepriming process has not been established; in other cases, they are known to be involved, but their roles are unknown. The DnaA protein appears to be largely responsible for the localized strand separation at *OriC* during the initiation process.

Once a replication fork has formed, the synthesis of new DNA strands is initiated by RNA primers synthesized by DNA primase. A single RNA primer is sufficient for the continuous replication of the leading strand, but the discontinuous replication of the lagging strand requires an RNA primer to start the synthesis of each Okazaki fragment. The initiation of Okazaki fragments on the lagging strand is carried out by the **primosome**, a protein complex containing DNA primase and DNA helicase. The primosome moves along a DNA molecule, powered by energy from ATP; DNA helicase unwinds the double helix, and DNA primase synthesizes the RNA primers for successive Okazaki fragments. The RNA primers are covalently extended with deoxyribonucleotides by DNA polymerase III. DNA topoisomerases provide transient breaks in DNA that serve as swivels for DNA unwinding and keep the DNA untangled. Single-strand DNA-binding protein coats the unwound prereplicative DNA and keeps it in an extended state for DNA polymerase III. The RNA primers are replaced with DNA by DNA polymerase I, and the single-strand breaks left by polymerase are sealed by DNA ligase. The DNA is then condensed into the nucleoid or folded genome of *E. coli*, in part through negative supercoiling introduced by DNA gyrase. All of these enzymes and DNA-binding proteins function in concert at each replication fork.

As a replication fork moves along a parental double helix, two DNA strands (the leading strand and the lagging strand) are replicated in the highly coordinated series of reactions described above. The complete replication apparatus moving along the DNA

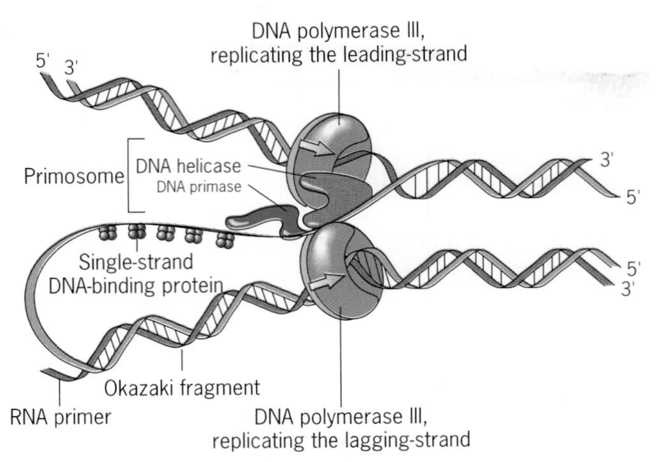

Figure 10.27 Diagram of the *E. coli* replisome, showing the two catalytic cores of DNA polymerase III replicating the leading and lagging strands and the primosome unwinding the parental double helix and initiating the synthesis of new chains with RNA primers. The entire replisome moves along the parental double helix, with each component performing its respective function in a concerted manner.

molecule at a replication fork is called the **replisome** (Figure 10.27). The replisome contains the DNA polymerase III holoenzyme; one catalytic core replicates the leading strand, the second catalytic core replicates the lagging strand, and the primosome unwinds the parental DNA molecule and synthesizes the RNA primers needed for the discontinuous synthesis of the lagging strand. In order for the two catalytic cores of the polymerase III holoenzyme to synthesize both the nascent leading and lagging strands, the lagging strand is thought to form a loop from the primosome to the second catalytic core of DNA polymerase III (Figure 10.27).

At the beginning of this chapter, we noted the striking fidelity of DNA replication. Now that we have examined the cellular machinery responsible for DNA replication in living organisms, this fidelity seems less amazing. A very sophisticated apparatus, with built-in safeguards against malfunctions, has evolved to assure that the genetic information of *E. coli* is transmitted accurately from generation to generation.

Rolling-Circle Replication

In the preceding sections of this chapter, we have considered θ-shaped, eye-shaped, and Y-shaped replicating DNAs. We will now examine another important type of DNA replication called **rolling-circle replication**. Rolling-circle replication is used (1) by many viruses to duplicate their genomes (Chapter 15), (2) to

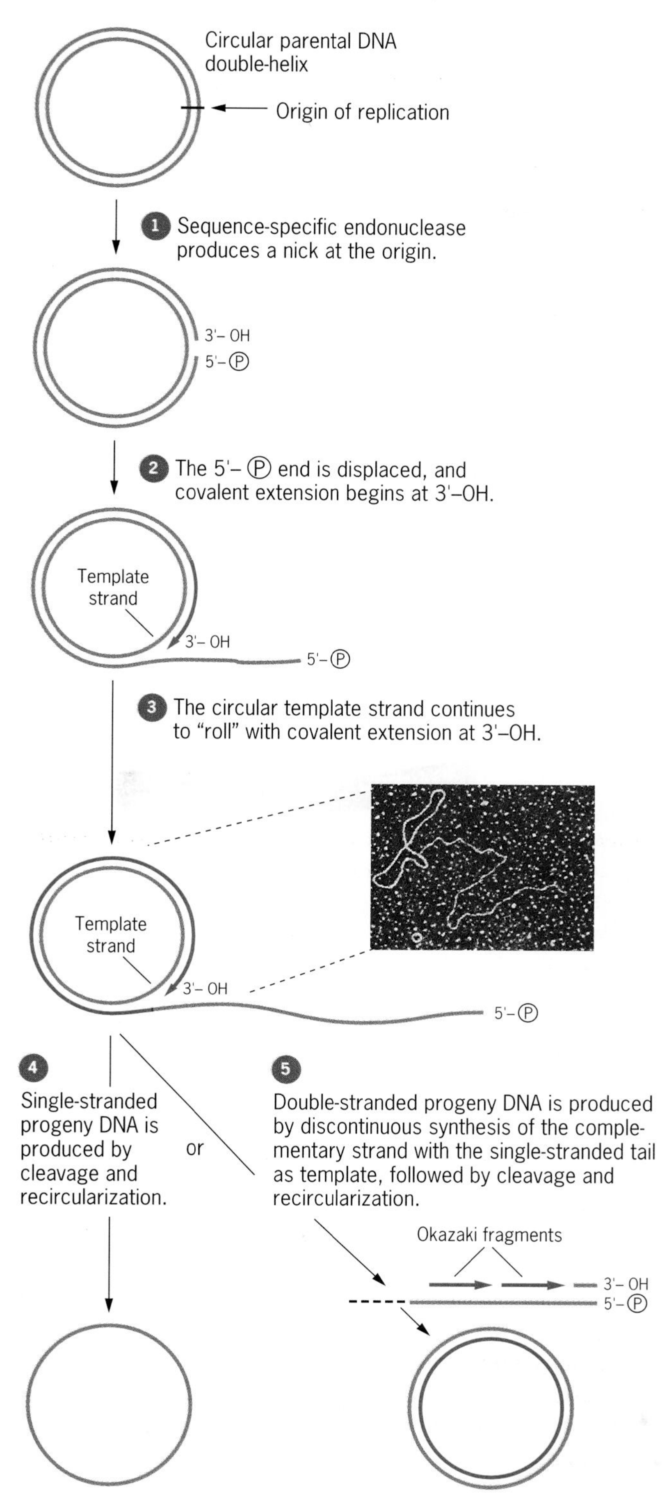

Figure 10.28 The rolling-circle mechanism of DNA replication. The inset shows an electron micrograph of a bacteriophage ΦX174 DNA molecule replicating by the rolling-circle mechanism. A single-stranded tail extends from a double-stranded, circular replicative DNA.

transfer DNA from donor cells to recipient cells during one type of genetic exchange in bacteria (Chapter 16), and (3) to amplify extrachromosomal DNAs carrying clusters of ribosomal RNA genes during oogenesis in amphibians (Chapter 22).

As the name implies, rolling-circle replication is a mechanism for replicating circular DNA molecules. The unique aspect of rolling-circle replication is that one parental circular DNA strand remains intact and rolls (thus the name rolling-circle) or spins while serving as a template for the synthesis of a new complementary strand (Figure 10.28). Replication is initiated when a sequence-specific endonuclease cleaves one strand at the origin, producing 3'-OH and 5'-phosphate termini. The 5' terminus is displaced from the circle as the intact template strand turns about its axis. Covalent extension occurs at the 3'-OH of the cleaved strand. Since the circular template DNA may turn 360° many times, with the synthesis of one complete or unit-length DNA strand during each turn, rolling-circle replication generates single-stranded tails longer than the contour length of the circular chromosome (Figure 10.28). Rolling-circle replication can produce either single-stranded or double-stranded progeny DNAs. Circular single-stranded progeny molecules are produced by site-specific cleavage of the single-stranded tails at the origins of replication and recircularization of the resulting unit-length molecules. To produce double-stranded progeny molecules, the single-stranded tails are used as templates for the discontinuous synthesis of complementary strands prior to cleavage and circularization. The enzymes involved in rolling-circle replication and the reactions catalyzed by these enzymes are basically the same as those responsible for DNA replication involving θ-type intermediates.

Key Points: **DNA replication is complex, requiring the participation of a large number of proteins. DNA synthesis is continuous on the progeny strand that is being extended in the overall 5' → 3' direction, but is discontinuous on the strand growing in the overall 3' → 5' direction. New DNA chains are initiated by short RNA primers synthesized by DNA primase. The enzymes and DNA-binding proteins involved in replication assemble into a replisome at each replication fork and act in concert as the fork moves along the parental DNA molecule.**

UNIQUE ASPECTS OF EUKARYOTIC CHROMOSOME REPLICATION

Most of the information about DNA replication has resulted from studies of *E. coli* and some of its viruses. Less information is available about DNA replication in eukaryotic organisms. However, enough information is available to conclude that most aspects of DNA replication are basically the same in both prokaryotes and eukaryotes, including humans. RNA primers and Okazaki fragments are shorter in eukaryotes than in prokaryotes, but the leading and lagging strands replicate by continuous and discontinuous mechanisms, respectively, in eukaryotes just as in prokaryotes. On the other hand, a few aspects of eukaryotic DNA replication are unique to these structurally more complex species. For example, DNA synthesis takes place within a small portion of the cell cycle in eukaryotes, not continuously as in prokaryotes. The giant DNA molecules present in eukaryotic chromosomes would take much too long to replicate if each chromosome contained a single origin. Thus eukaryotic chromosomes contain multiple origins of replication. Rather than using two catalytic complexes of one DNA polymerase to replicate the leading and lagging strands at each replication fork, eukaryotic organisms utilize two different polymerases.

As we discussed in Chapter 9, eukaryotic DNA is sequestered in histone-containing structures called nucleosomes. Do these nucleosomes impede the movement of replication forks? If not, how does a replisome move past a nucleosome? Is the nucleosome completely or partially disassembled, or does the fork somehow slide past the nucleosome as the replisome duplicates the DNA molecule while it is still present on the surface of the nucleosome? Lastly, eukaryotic chromosomes contain linear DNA molecules, and the discontinuous replication of the ends of linear DNA molecules creates a special problem. We will address these aspects of chromatin replication in eukaryotes in the final sections of this chapter.

The Cell Cycle

When bacteria are growing on rich media, DNA replication occurs nonstop throughout the cell cycle. However, in eukaryotes, DNA replication is restricted to the S phase (for synthesis; Chapter 2). Recall that a normal eukaryotic cell cycle consists of G_1 phase (immediately following the completion of mitosis; G for gap), S phase, G_2 phase (preparation for mitosis), and M phase (mitosis). In rapidly dividing embryonic cells, G_1 and G_2 are very short or nonexistent. In all cells, decisions to continue on through the cell cycle occur at two points: (1) entry into S phase and (2) entry into mitosis (see Chapter 2 for details).

Multiple Replicons per Chromosome

The giant DNA molecules in the largest chromosomes of *Drosophila melanogaster* contain about 6.5×10^7 nucleotide pairs. The rate of DNA replication in *Drosophila* is about 2600 nucleotide pairs per minute at 25°C. A single replication fork would therefore take about 17.5 days to replicate one of these giant DNA molecules. With two replication forks moving bidirectionally from a central origin, such a DNA molecule could be replicated in just over 8.5 days. Given that the chromosomes of *Drosophila* embryos replicate within 3 to 4 minutes and the nuclei divide once every 9 to 10 minutes during the early cleavage divisions, it is clear that each giant DNA molecule must contain many origins of replication. Indeed, the complete replication of the DNA of the largest *Drosophila* chromosomes within 3.5 minutes would require over 7000 replication forks distributed at equal intervals along the molecules. Thus multiple origins of replication are required to allow the very large DNA molecules in eukaryotic chromosomes to replicate within the observed cell division times.

The first evidence for multiple origins in eukaryotic chromosomes resulted from pulse-labeling experiments with Chinese hamster cells growing in culture. In 1968, when Joel Huberman and Arthur Riggs pulse-labeled cells with [^{3}H]thymidine for a few minutes, extracted the DNA, and performed autoradiographic analysis of the labeled DNA, they observed tandem arrays of exposed silver grains. The simplest interpretation of their results is that individual macromolecules of DNA contain multiple origins of replication (Figure 10.29*a*). When the pulse-labeling period was followed by a short interval of growth in nonradioactive medium (pulse-chase experiments), the tandem arrays contained central regions of high-grain density with tails of decreasing grain density at *both* ends (Figure 10.29*b*). This result indicates that replication in eukaryotes is bidirectional just as it is in most prokaryotes. The tails of decreasing grain density result from the gradual dilution of the intracellular pools of [^{3}H]thymidine by [^{1}H]thymidine as the replication forks move bidirectionally from central origins toward replication termini (Figure 10.29*c*).

A segment of DNA whose replication is under the control of one origin and two termini is called a **replicon**. In prokaryotes, the entire chromosome is usually one replicon. The existence of multiple replicons per eukaryotic chromosome has been verified directly by electron microscopy (Figure 10.30). Clearly, the number of replicons per chromosome is not fixed throughout the growth and development of a multicellular eukaryote. Replication is initiated at more sites during the very rapid cell divisions of embryogenesis than during later stages of development. Unfortunately, geneticists don't know what factors determine which origins are operational at any given time or in a particular type of cell.

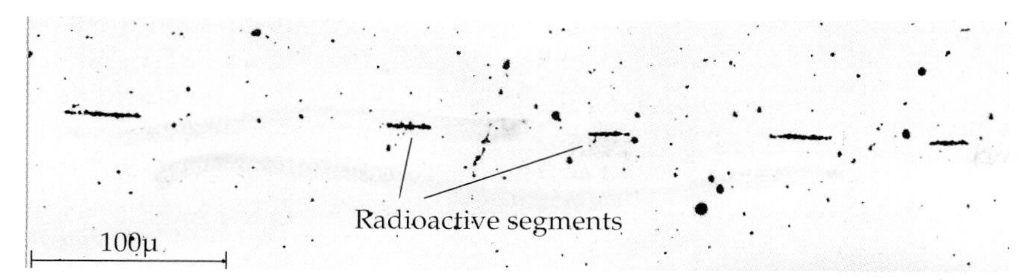

(a) Autoradiograph of a portion of a DNA molecule from a Chinese hamster cell that had been pulse-labeled with ^{3}H-thymidine.

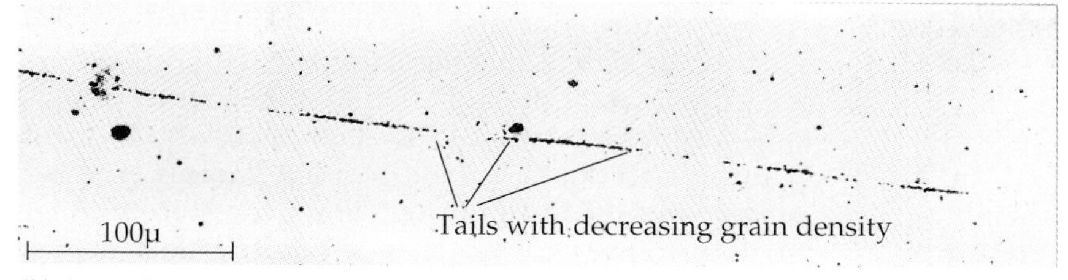

(b) Autoradiograph of a segment of a DNA molecule from a Chinese hamster cell that was pulse-labeled with ^{3}H-thymidine and then transferred to nonradioactive medium for an additional growth period.

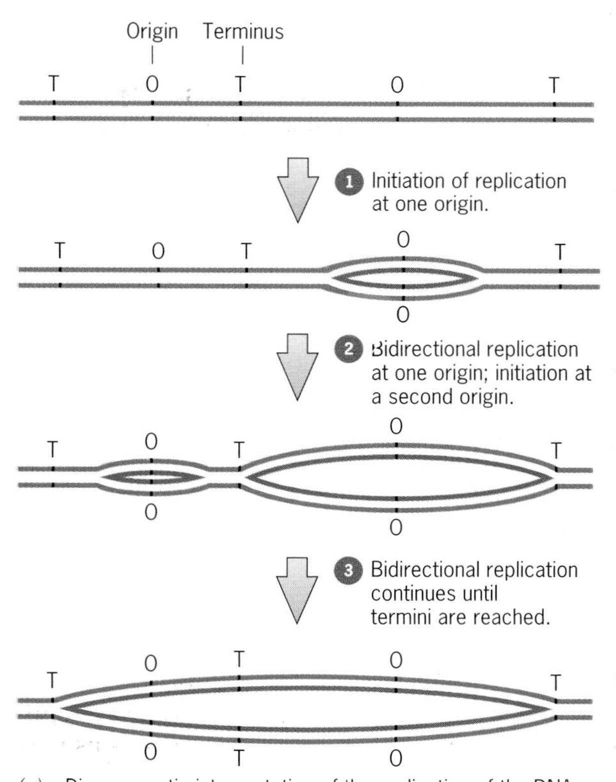

(c) Diagrammatic interpretation of the replication of the DNA molecules visualized above.

Figure 10.29 Evidence for bidirectional replication of the multiple replicons in the giant DNA molecules of eukaryotes. The tandem arrays of radioactivity in (*a*) indicate that replication occurs at multiple origins, and tails with decreasing grain density observed in (*b*) indicate that replication occurs bidirectionally from each origin (*c*).

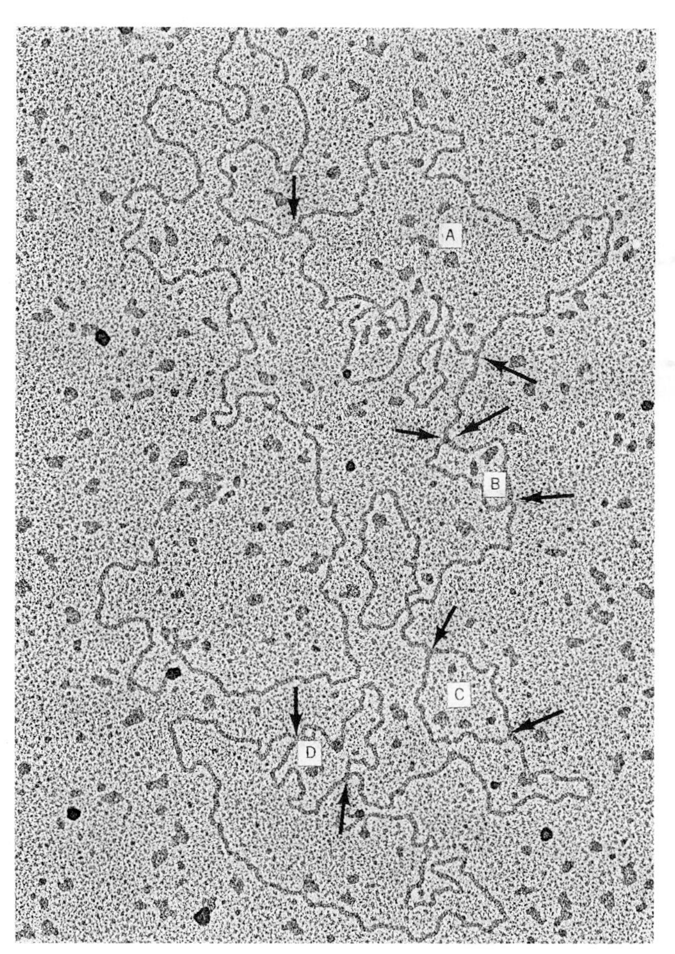

Figure 10.30 Electron micrograph of a DNA molecule in *D. melanogaster* showing multiple sites of replication. Four eye-shaped (–O–) replication structures (labeled A-D) are present in the segment of the DNA molecule shown. The arrows show the positions of replication forks.

Two DNA Polymerases at a Single Replication Fork

Given the complexity of the replisome in the simple bacterium *E. coli* (see Figures 10.25 and 10.27), it seems likely that the replication apparatus is even more complex in eukaryotes. Although knowledge of the structure of the replicative machinery in eukaryotes is still somewhat limited, many features of DNA replication are the same in both eukaryotes and prokaryotes. One difference is that two different DNA polymerases function at each replication fork in eukaryotes.

As in the case of prokaryotes, much of the information about DNA synthesis in eukaryotes has come from the development and dissection of *in vitro* DNA replication systems. Studies of the replication of DNA viruses of eukaryotes have proven informative, and, of these viruses, Simian virus 40 (SV40) has proven particularly useful. The replication of SV40 is carried out almost entirely by the host cell's replication apparatus. Only one viral protein, the so-called T antigen, is required for the replication of the SV40 chromosome. Two of the host cell's DNA polymerases, α and δ, are required to replicate the SV40 DNA molecule *in vitro* (Figure 10.31). DNA polymerase α appears to be the most similar to DNA polymerase III of *E. coli*. **DNA polymerase α** is multimeric in structure, and two of its subunits interact to provide DNA primase activity. Thus polymerase α contains the enzymatic activities required to carry out the discontinuous replication of the lagging strand. **DNA polymerase δ** lacks primase activity, and evidence suggests that it catalyzes the replication of the leading strand. As in prokaryotes, a helicase is required to unwind the double helix ahead of the replication fork.

Several other proteins have been identified as replication factors (for example, RF-A, RF-B, RF-C, and so forth); their functions remain to be elucidated. A protein called PCNA (proliferating cell nuclear antigen) appears to be a cofactor of polymerase δ; its presence results in a much increased rate of replication. Other accessory proteins undoubtedly will be identified in the future. Given the utility of the *in vitro* DNA replication systems now available, we can anticipate a more detailed picture of the eukaryotic replisome in the near future. However, the real challenges are (1) to identify the signals that regulate DNA replication during growth and differentiation in multicellular organisms such as humans, (2) to determine which signal pathways are malfunctioning in the various types of cancer, and (3) to figure out how to restore normal function to the various types of cancer cells (Chapter 22).

Duplication of Nucleosomes at Replication Forks

As we discussed in Chapter 9, the DNA in eukaryotic interphase chromosomes is packaged in approximately 10-nm beads called nucleosomes. Each nucleosome contains 166 nucleotide pairs of DNA wound in two turns around an octamer of histone molecules. Given the size of nucleosomes and the large size of DNA replisomes, it seems unlikely that a replication fork can move past an intact nucleosome. Yet, electron micrographs of replicating chromatin in *Drosophila* clearly show nucleosomes with apparently normal structure and spacing on both sides of replication forks (Figure 10.32*a*); that is, nucleosomes appear to

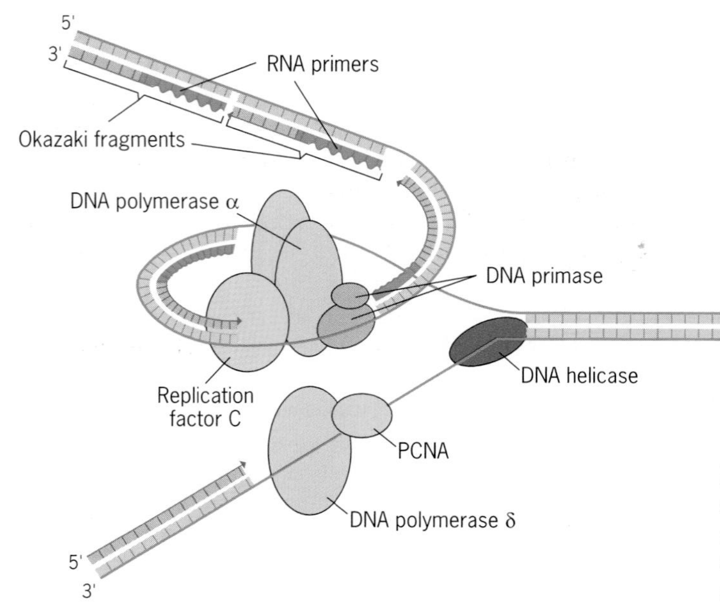

Figure 10.31 Two DNA polymerases, α and δ, function at the replication fork in eukaryotes. DNA polymerase δ replicates the leading strand, and DNA polymerase α replicates the lagging strand. PCNA = proliferating cell nuclear antigen.

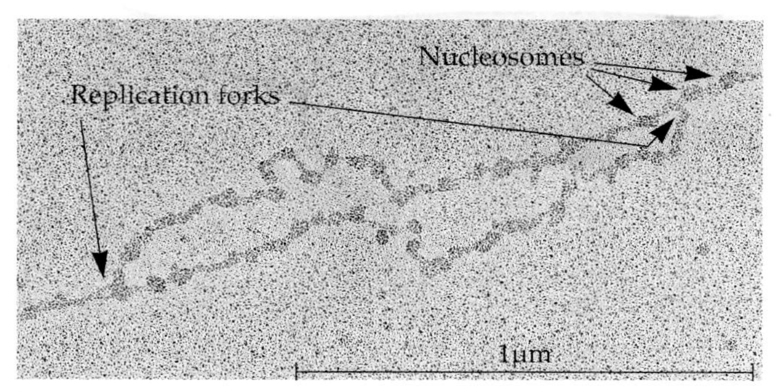

(*a*) Electron micrograph showing nucleosomes on both sides of each of two replication forks in *Drosophila*.

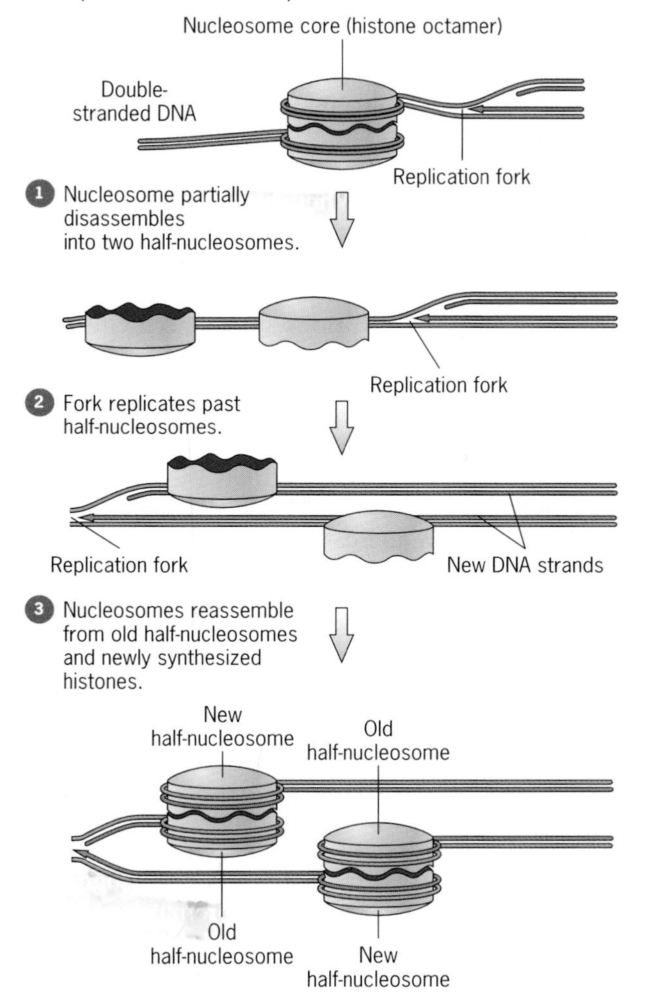

(*b*) Model of the movement of replisomes past nucleosomes.

Figure 10.32 Movement of a replication fork past nucleosomes. (*a*) Electron micrograph showing nucleosomes on both sides of two replication forks in *Drosophila*. Recall that DNA replication is bidirectional in eukaryotes; thus each branch point is a replication fork. (*b*) One model for the movement of replication forks past nucleosomes.

have the same structure and spacing immediately behind a replication fork (postreplicative DNA) as they do in front of a replication fork (prereplicative DNA). Thus various models have been proposed for transient alterations in nucleosome structure that would allow replisomes to synthesize DNA as they move past nucleosomes. In one popular model, the nucleosome splits into two half-nucleosomes while the replisome moves past, and the halves then reassemble into intact nucleosomes behind the replication fork (Figure 10.32*b*).

Regardless of the mechanism by which the replication fork moves past nucleosomes, DNA replication and nucleosome assembly are tightly coupled in eukaryotes. Since the mass of the histones in nucleosomes is equivalent to that of the DNA, large quantities of histones must be synthesized during each cell generation in order for the nucleosomes to duplicate. Although histone synthesis occurs throughout the cell cycle, there is a burst of histone biosynthesis during S phase that generates enough histones for chromatin duplication. When density-transfer experiments were performed to examine the mode of nucleosome duplication, the nucleosomes on both progeny DNA molecules were found to contain both old or prereplicative histone complexes and new or postreplicative complexes. Thus, at the protein level, nucleosome duplication appears to occur by a dispersive mechanism.

Telomerase: Replication of Chromosome Termini

We discussed the unique structures of telomeres, or chromosome ends, in Chapter 9. An early reason for thinking that telomeres must have special structures was that DNA polymerases cannot replicate the terminal DNA segment of the lagging strand of a linear chromosome. At the end of the DNA molecule being replicated discontinuously, there would be no DNA strand to provide a free 3'-OH (primer) for polymerization of deoxyribonucleotides after the RNA primer of the terminal Okazaki fragment has been excised (Figure 10.33*a*). This requires that either (1) the telomere must have a unique structure that facilitates its replication or (2) there must be a special enzyme that resolves this enigma of replicating the terminus of the lagging strand. Indeed, evidence has shown that both are correct. The special structure of telomeres provides a neat mechanism for the addition of telomeres by an RNA-containing enzyme called **telomerase**.

The telomere of *Tetrahymena thermophila*, which contains the tandemly repeated sequence GGGTTG, will be used to illustrate how telomerase adds ends to chromosomes. Telomerase recognizes the G-rich telomere sequence on the 3' overhang and extends it 5' → 3' one repeat unit at a time. The unique feature of telomerase is that it contains a built-in RNA strand template (Figure 10.33*b*). After several telomere repeat

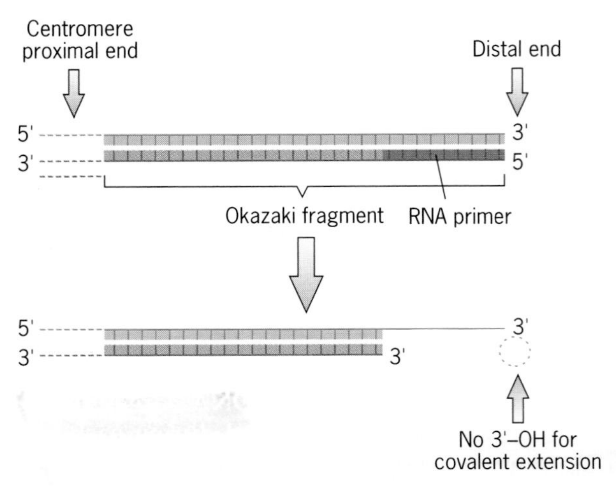

(*a*) The telomere lagging-strand primer problem.

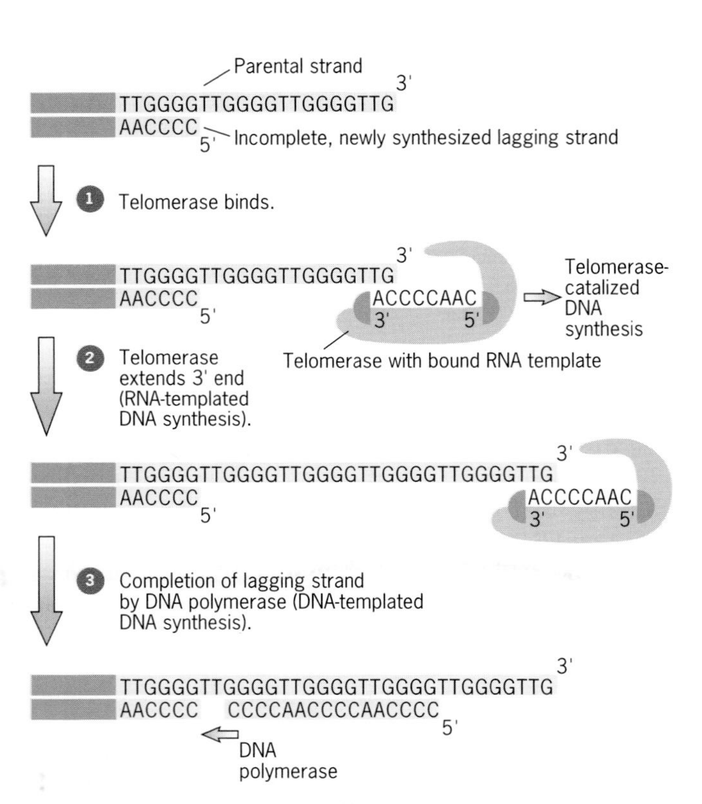

(*b*) Telomerase resolves the terminal primer problem.

Figure 10.33 Replication of chromosome telomeres. (*a*) Because of the requirement for a free 3'-OH at the end of the primer strand, DNA polymerases cannot replace the RNA primer that initiates DNA synthesis at the terminus of the lagging strand. (*b*) These termini of chromosomes are replicated by a special enzyme called telomerase, which prevents the ends of chromosomes from becoming shorter during each replication. The nucleotide sequence at the terminus of the lagging strand is specified by a short RNA molecule present as an essential component of telomerase. The telomere sequence shown is that of *Tetrahymena*.

units are added by telomerase, DNA polymerase catalyzes the synthesis of the complementary strand. Without telomerase activity, linear chromosomes would become progressively shorter. If the resulting terminal deletions extended into an essential gene or genes, this chromosome shortening would be lethal.

Telomere Length and Aging in Humans

As in other eukaryotes, telomerase is required to replicate the termini of human chromosomes. Without telomerase activity, the human chromosomes would become shorter during each replication. Eventually, essential genes located near the ends of chromosomes would be lost, and the cells that contain the shortened chromosomes would die.

Unlike germ-line cells, most human somatic cells lack telomerase activity. When human somatic cells are grown in culture, they divide only a limited number of times (usually only 20 to 70 cell generations) before senescence and death occur. When telomere lengths are measured in various somatic cell cultures, a correlation is observed between telomere length and the number of cell divisions preceding senescence and death. Cells with longer telomeres survive longer—go through more cell divisions—than cells with shorter telomeres. As would be expected in the absence of telomerase activity, telomere length decreases as the age of the cell culture increases. Occasionally, somatic cells are observed to acquire the ability to proliferate in culture indefinitely, and these immortal cells have been shown to contain telomerase activity, unlike their progenitors. Since the one common feature of all cancers is uncontrolled cell division or immortality, scientists have proposed that one way to combat human cancers would be to inhibit the telomerase activity in cancer cells.

Further evidence of a relationship between telomere length and aging in humans has come from studies of individuals with disorders called **progerias**, inherited diseases characterized by premature aging. In the most severe form of progeria, Hutchinson–Gilford syndrome (Figure 10.34), senescence—wrinkles, baldness, and other symptoms of aging—begins immediately after birth and death usually occurs in the teens. In a less severe form, Werner syndrome, senescence begins in the teenage years, with death usually occurring in the 40s. Consistent with the hypothesis that decreasing telomere length contributes to the aging process, the somatic cells of individuals with progeria have short telomeres and exhibit decreased proliferative capacity when grown in culture.

At present, the relationship between telomere length and cell senescence is entirely correlative. There is no direct evidence indicating that telomere

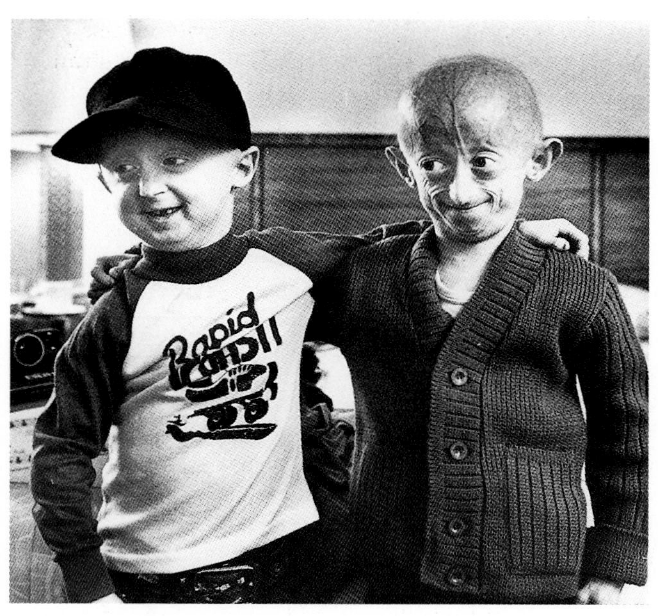

Figure 10.34 Photograph of children with Hutchinson–Gilford syndrome, showing the premature aging characteristic of this disorder. The children are eight (right) and nine years old, respectively.

shortening causes aging. Nevertheless, the correlation is striking, and the hypothesis that telomere shortening contributes to the aging process in humans warrants further study.

Key Points: **DNA replication is basically the same in both prokaryotes and eukaryotes, including humans. Replication of the giant DNA molecules in eukaryotic chromosomes occurs bidirectionally from multiple origins. Two DNA polymerases (α and δ) are present at each replication fork. Telomeres, the unique sequences at the ends of chromosomes, are added to chromosomes by a special RNA-containing enzyme called telomerase.**

TESTING YOUR KNOWLEDGE

1. *Escherichia coli* cells were grown for many generations in a medium in which the only available nitrogen was the heavy isotope ^{15}N. The cells were then collected by centrifugation, washed with a buffer, and transferred to a medium containing ^{14}N (the normal light nitrogen isotope). After two generations of growth in the ^{14}N-containing medium, the cells were transferred back to ^{15}N-containing medium for one final generation of growth. After this final generation of growth in the presence of ^{15}N, the cells were collected by centrifugation. The DNA was then extracted from these cells and analyzed by CsCl equilibrium density-gradient centrifugation. How would you expect the DNA from these cells to be distributed in the gradient?

ANSWER

Meselson and Stahl demonstrated that DNA replication in *E. coli* is semiconservative. Their control experiments showed that DNA double helices with (1) ^{14}N in both strands, (2) ^{14}N in one strand and ^{15}N in the other strand, and (3) ^{15}N in both strands separated into three distinct bands in the gradient, called (1) the light band, (2) the hybrid band, and (3) the heavy band, respectively. If you start with a DNA double helix with ^{15}N in both strands, and replicate it semiconservatively for two generations in the presence of ^{14}N and then for one generation in the presence of ^{15}N, you will end up with eight DNA molecules, two with ^{15}N in both strands and six with ^{14}N in one strand and ^{15}N in the other strand, as shown below. Therefore, 75 percent (6/8) of the DNA will appear in the hybrid band, and 25 percent (2/8) will appear in the heavy band.

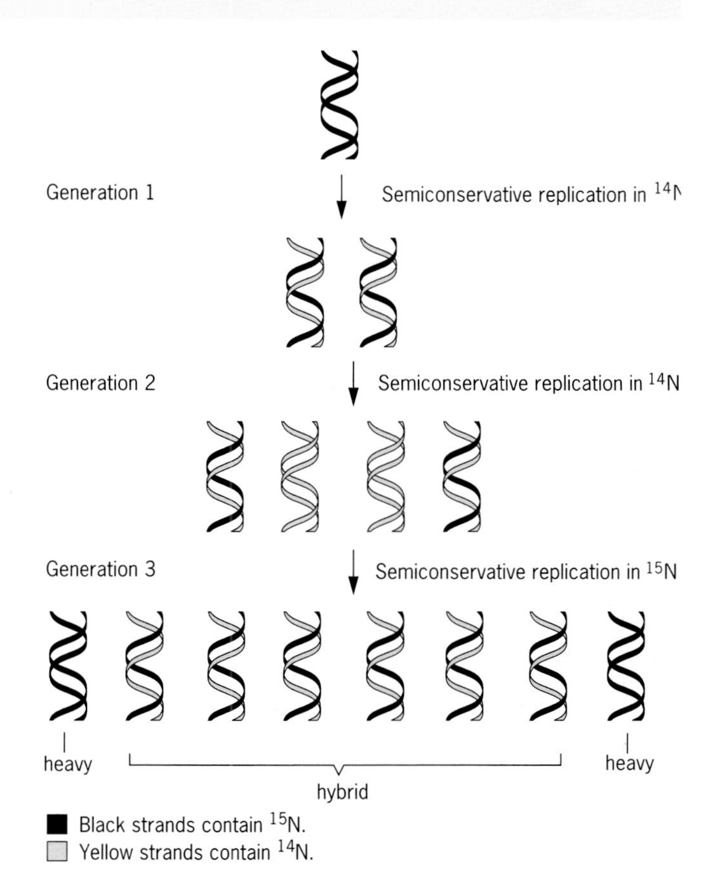

2. *Haplopappus gracilis* is a diploid plant with only two pairs of chromosomes. Assume that a G_1-stage cell of this plant, not previously exposed to radioactivity of any kind, was growing in culture medium containing ^{3}H-thymidine. After one generation of growth in this medium, the two progeny cells were transferred to medium containing ^{1}H-thymidine and colchicine. They were allowed to grow in this medium for about one and a half generations. Then, the chromosomes from each cell were spread on a microscope slide, stained, photographed, and exposed to an emulsion sensitive to low-energy radiation. One of the daughter cells exhibited a metaphase plate with eight chromosomes, each with two daughter chromatids. Draw this metaphase plate showing the predicted distribution of radioactivity on the autoradiograph.

ANSWER

Each chromosome contains one giant double helix that replicates semiconservatively. At the start of the experiment, both strands of this double helix contained nonradioactive (^{1}H) thymidine. Since all four chromosomes will go through the same processes of replication, you need to follow only one of them and then multiply your result by four. Since colchicine was not present during the first replication, the chromosome number per cell will remain unchanged, but the newly synthesized strands will be radioactive (contain ^{3}H-thymidine). Because the second and third replications occurred in the presence of ^{1}H-thymidine, strands synthesized during these replications will be nonradioactive. However, colchicine was present; thus the chromosome number will double during each duplication. These replications are as follows, with radioactive strands of DNA shown in red and nonradioactive strands in blue.

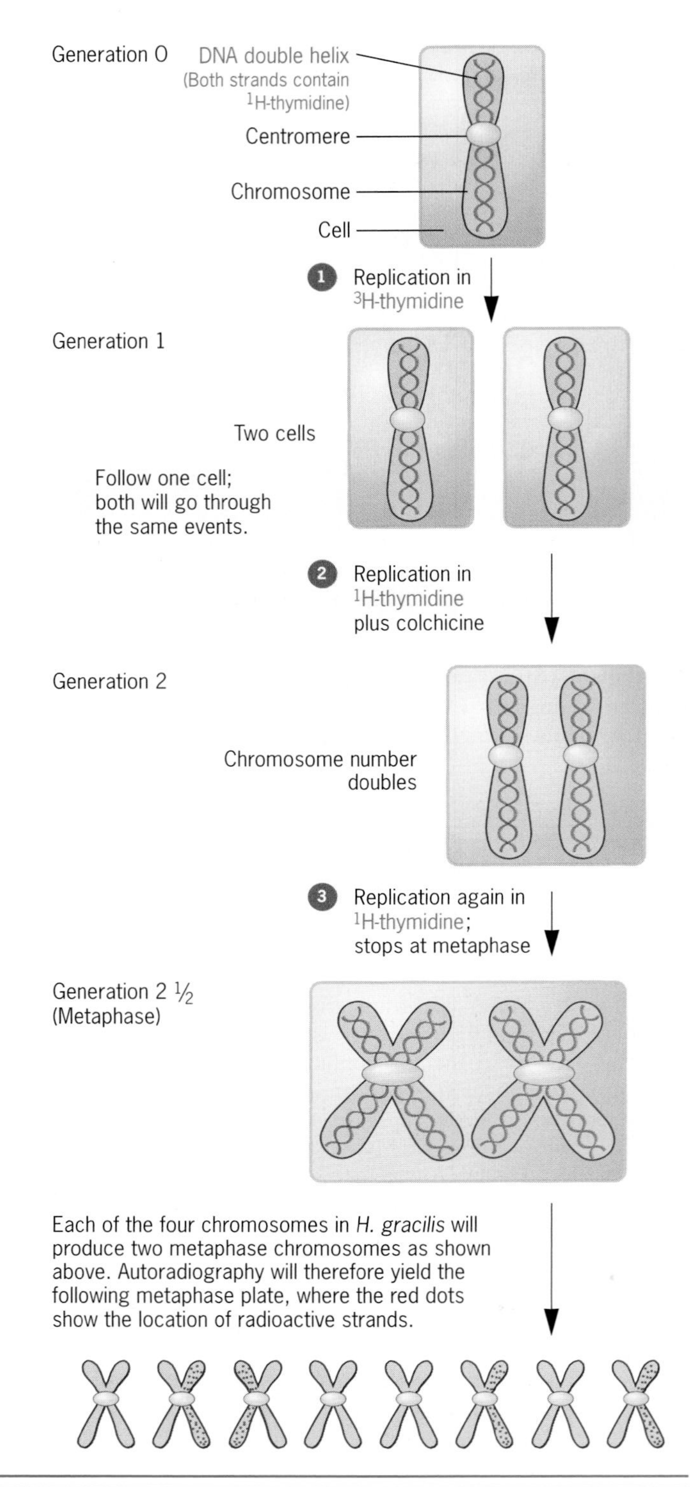

QUESTIONS AND PROBLEMS

10.1 *Escherichia coli* cells are grown for many generations in a medium in which the only available nitrogen is the heavy isotope ^{15}N. They are then transferred to a medium containing ^{14}N as the only source of nitrogen. (a) What distribution of ^{15}N and ^{14}N would be expected in the DNA molecules of cells that had grown for one generation in the ^{14}N-containing medium assuming that DNA replication was (i) conservative, (ii) semiconservative, or (iii) dispersive? (b) What distribution would be expected after two generations of growth in the ^{14}N-containing medium assuming (i) conservative, (ii) semiconservative, or (iii) dispersive replication?

10.2 A culture of bacteria is grown for many generations in a medium in which the only available nitrogen is the heavy isotope (^{15}N). The culture is then switched to a medium containing only ^{14}N for one generation of growth; it is then returned to a ^{15}N-containing medium for one final generation of growth. If the DNA from these bacteria is isolated and centrifuged to equilibrium in a CsCl density gradient, how would you predict the DNA to band in the gradient?

10.3 DNA polymerase I of *E. coli* is a single polypeptide of molecular weight 109,000. (a) What enzymatic activities other than polymerase activity does this polypeptide possess? (b) What are the *in vivo* functions of these activities? (c) Are these activities of major importance to an *E. coli* cell? Why?

10.4 A DNA template plus primer with the structure

3' Ⓟ — TGCGAATTAGCGACAT — Ⓟ 5'
5' Ⓟ — ATCGGTACGACGCTTAAC — OH 3'

(where Ⓟ= a phosphate group) is placed in an *in vitro* DNA synthesis system (Mg^{2+}, an excess of the four deoxyribonucleoside triphosphates, etc.) containing a mutant form of *E. coli* DNA polymerase I that lacks 5' → 3' exonuclease activity. The 5' → 3' polymerase and 3' → 5' exonuclease activities of this aberrant enzyme are identical to those of normal *E. coli* DNA polymerase I. It simply has no 5' → 3' exonuclease activity. (a) What will be the structure of the final product? (b) What will be the first step in the reaction sequence?

10.5 How might continuous and discontinuous modes of DNA replication be distinguished experimentally?

10.6 Identify the proteins that are involved in DNA replication in *E. coli* and list their known or putative functions.

10.7 The Boston straggler is an imaginary plant with a diploid chromosome number of 4. Boston straggler cells are easily grown in suspended cell cultures. [^{3}H]Thymidine was added to the culture medium in which a G_1-stage cell of this plant was growing. After one cell generation of growth in [^{3}H]thymidine-containing medium, colchicine was added to the culture medium. The medium now contained both [^{3}H]thymidine and colchicine. After two "generations" of growth in [^{3}H]thymidine-containing medium (the second "generation" occurring in the presence of colchicine as well), the two progeny cells (each now containing eight chromosomes) were transferred to culture medium containing nonradioactive thymidine ([^{1}H]thymidine) plus colchicine. Note that a "generation" in the presence of colchicine consists of a normal cell cycle's chromosomal duplication, but no cell division. The two progeny cells were allowed to continue to grow, proceeding through the "cell cycle," until each cell contained a set of metaphase chromosomes that looked like the following.

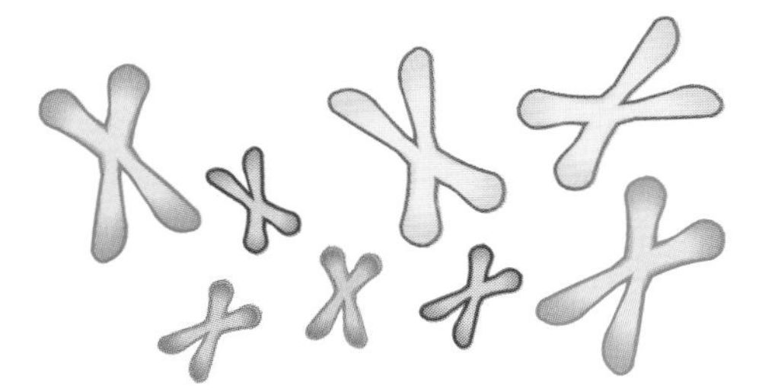

If autoradiography were carried out on these metaphase chromosomes (four large plus four small), what pattern of radioactivity (as indicated by silver grains on the autoradiograph) would be expected? (Assume no recombination between DNA molecules.)

10.8 Suppose that the experiment described in Problem 10.7 as carried out again, except this time replacing the [^{3}H]thymidine with nonradioactive thymidine at the same time that the colchicine was added (after one cell generation of growth in [^{3}H]thymidine-containing medium). The cells were then maintained in colchicine plus nonradioactive thymidine until the metaphase shown in Problem 10.7 occurred. What would the autoradiographs of these chromosomes look like?

10.9 Suppose that the DNA of cells (growing in a cell culture) in a eukaryotic species was labeled for a short period of time by the addition of [^{3}H]thymidine to the medium. Next assume that the label was removed and the cells were resuspended in nonradioactive medium. After a short period of growth in nonradioactive medium, the DNA was extracted from these cells, diluted, gently layered on filters, and autoradiographed. If autoradiographs of the type

were observed, what would this indicate about the nature of DNA replication in these cells? Why?

10.10 Are eukaryotic chromosomes metabolically most active during prophase, metaphase, anaphase, telophase, or interphase?

10.11 Five distinct DNA polymerases: α, β, γ, δ, and ϵ have been characterized in mammals. What are the intracellular locations and functions of these polymerases?

10.12 The *E. coli* chromosome contains approximately 4×10^6 nucleotide pairs and replicates as a single bidirectional replicon in approximately 40 minutes under a wide variety of growth conditions. The largest chromosome of *D. melanogaster* contains about 6×10^7 nucleotide pairs. (a) If this chromosome contains one giant molecule of DNA that replicates bidirectionally from a single origin located precisely in the middle of the DNA molecule, how long would it take to replicate the entire chromosome if replication in *Drosophila* occurred at the same rate as replication in *E. coli*? (b) Actually, replication rates are slower in eukaryotes than in prokaryotes. If each replication bubble grows at a rate of 5000 nucleotide pairs per minute in *Drosophila* and 100,000 nucleotide pairs per minute in *E. coli*, how long will it take to replicate the largest *Drosophila* chromosome if it contains a single bidirectional replicon as described in (a) above? (c) During the early cleavage divisions in *Drosophila* embryos, the nuclei divide every 9 to 10 minutes. Based on your calculations in (a) and (b) above, what do these rapid nuclear divisions indicate about the number of replicons per chromosome in *Drosophila*?

10.13 What experimental techniques can be used to separate DNA molecules of mass 3×10^7 daltons isolated from lambda bacteriophage and DNA molecules of mass 1.3×10^8 daltons isolated from T2 bacteriophage?

10.14 The bacteriophage lambda chromosome has several AT-rich segments that denature when exposed to pH 11.05 for 10 minutes. After such partial denaturation, the linear packaged form of the lambda DNA molecule has the structure shown in Figure 10.8*a*. Following its injection into an *E. coli* cell, the lambda DNA molecule is converted to a covalently closed circular molecule by hydrogen bonding between its complementary single-stranded termini and the action of DNA ligase. It then replicates as a θ-shaped structure. The entire lambda chromosome is 17.5 μm long. It has a unique origin of replication located 14.3 μm from the left end of the linear form shown in Figure 10.8*a*. Draw the structure that would be observed by electron microscopy after both (1) replication of an approximately 6 μm-long segment of the lambda chromosomal DNA molecule (*in vivo*) and (2) exposure of this partially replicated DNA molecule to pH 11.05 for 10 minutes (*in vitro*), (a) *if* replication had proceeded bidirectionally from the origin, and (b) *if* replication had proceeded unidirectionally from the origin.

10.15 What enzyme activity catalyzes each of the following steps in the semiconservative replication of DNA in prokaryotes: (a) the formation of negative supercoils in progeny DNA molecules, (b) the synthesis of RNA primers, (c) the removal of RNA primers, (d) the covalent extension of DNA chains at the 3'-OH termini of primer strands, and (e) proofreading of the nucleotides at the 3'-OH termini of DNA primer strands?

10.16 Why must each of the giant DNA molecules in eukaryotic chromosomes contain multiple origins of replication?

10.17 In *E. coli*, polA mutants have been isolated that produce a defective gene product with little or no 5' → 3' polymerase activity, but normal 5' → 3' exonuclease activity. However, no *polA* mutant has been identified that is completely deficient in the 5' → 3' exonuclease activity, while retaining 5' → 3' polymerase activity, of DNA polymerase I. How can these results be explained?

10.18 Other *polA* mutants of *E. coli* lack the 3' → 5' exonuclease activity of DNA polymerase I. Will the rate of DNA synthesis be altered in these mutants? What effect(s) will these *polA* mutations have on the phenotype of the organism?

10.19 Many of the origins of replication that have been characterized contain A:T-rich core sequences. Are these A:T-rich cores of any functional significance? If so, what?

10.20 (a) Why isn't DNA primase activity required to initiate rolling-circle replication? (b) DNA primase is required for the discontinuous synthesis of the lagging strand, which occurs on the single-stranded tail of the rolling circle. Why?

10.21 DNA polymerase I is needed to remove RNA primers during chromosome replication in *E. coli*. However, DNA polymerase III is the true replicase in *E. coli*. Why doesn't DNA polymerase III remove the RNA primers?

10.22 In *E. coli*, three different proteins are required to unwind the parental double helix and keep the unwound strands in an extended template form. What are these proteins, and what are their respective functions?

10.23 How similar are the structures of DNA polymerase I and DNA polymerase III in *E. coli*? What is the structure of the DNA polymerase III holoenzyme? What is the function of the *dnaN* gene product in *E. coli*?

10.24 The *dnaA* gene product of *E. coli* is required for the initiation of DNA synthesis at *OriC*. What is its function? How do we know that the DnaA protein is essential to the initiation process?

10.25 What is a primosome, and what are its functions? What essential enzymes are present in the primosome? What are the major components of the *E. coli* replisome? How can geneticists determine whether these components are required for DNA replication?

10.26 The chromosomal DNA of eukaryotes is packaged into nucleosomes during the S phase of the cell cycle. What obstacles do the size and complexity of both the replisome and the nucleosome present during the semiconservative replication of eukaryotic DNA? How might these obstacles be overcome?

10.27 Design an experiment with *E. coli* that will allow you to distinguish between (1) continuous synthesis of one DNA strand and discontinuous synthesis of the other DNA strand and (2) discontinuous synthesis of both progeny DNA strands.

10.28 Two mutant strains of *E. coli* each have a temperature-sensitive mutation in a gene that encodes a product required for chromosome duplication. Both strains replicate their DNA and divide normally at 25°C, but are unable to replicate their DNA or divide at 42°C. When cells of one strain are shifted from growth at 25°C to growth at 42°C, DNA synthesis stops immediately. When cells of the other strain are subjected to the same temperature shift, DNA synthesis continues, albeit at a decreasing rate, for about a half hour. What can you conclude about the functions of the products of these two genes?

10.29 In what ways does chromosomal DNA replication in eukaryotes differ from DNA replication in prokaryotes?

10.30 (*a*) The chromosome of the bacterium *Salmonella typhimurium* contains about 4×10^6 nucleotide pairs. Approximately how many Okazaki fragments are produced during one complete replication of the *S. typhimurium* chromosome? (b) The largest chromosome of *D. melanogaster* contains approximately 6×10^7 nucleotide pairs. About how many Okazaki fragments are produced during the replication of this chromosome?

10.31 In the yeast *S. cerevisiae*, haploid cells carrying a mutation called *est1* (for *e*ver-*s*horter *t*elomeres) lose distal telomere sequences during each cell division. Predict the ultimate phenotypic effect of this mutation on the progeny of these cells.

BIBLIOGRAPHY

ALBERTS, B., D. BRAY, J. LEWIS, M. RAFF, K. ROBERTS, AND J. D. WATSON. 1994. *Molecular Biology of the Cell*, 3rd ed. Garland, New York.

CAIRNS, J. 1966. "The bacterial chromosome." *Sci. Amer.* 214(1):36–44.

DE LANGE, T. 1994 "Activation of telomerase in a human tumor." *Proc. Natl. Acad. Sci. U.S.A.* 91:2882–2885.

HUBERMAN, J. A., AND A. D. RIGGS. 1968. "On the mechanism of DNA replication in mammalian chromosomes." *J. Mol. Biol.* 32:327–341.

KORNBERG, A., AND T. A. BAKER. 1992. *DNA Replication*, 2nd ed., Freeman, San Francisco.

LEWIN, B. 1994. *Genes V*. Cell Press, Cambridge, MA/Oxford University Press, Oxford.

MESELSON, M. S., AND F. W. STAHL. 1958. "The replication of DNA in *Escherichia coli*." *Proc. Natl. Acad. Sci. U. S.A.* 44:671–682.

TAYLOR, J. H., P. S. WOODS, AND W. L. HUGHES. 1957. "The organization and duplication of chromosomes as revealed by autoradiographic studies using tritium-labeled thymidine." *Proc. Natl. Acad. Sci., U.S.A.* 43:122–128.

11

Spliceosome processing a gene transcript.

Transcription and RNA Processing

Storage and Transmission of Information with Simple Codes

We live in the age of the computer. It has an impact on virtually all aspects of our lives, from driving to work to watching spaceships land on the moon. These electronic wizards can store, retrieve, and analyze data with lightning-like speed. The "brain" of the computer is a small chip of silicon, the microprocessor, which contains a sophisticated and integrated array of electronic circuits capable of responding almost instantaneously to coded bursts of electrical energy. In carrying out its amazing feats, the computer uses a binary code, a language based on 0's and 1's. Thus the alphabet used by computers is like that of the Morse code (dots and dashes) used in telegraphy. Both consist of only two symbols—in marked contrast to the 26 letters of the English alphabet. Obviously, if the computer can perform its wizardry with a binary alphabet, vast amounts of information can be stored and retrieved without using complex codes or lengthy alphabets. In this and the following chapter, we examine (1) how the genetic information of living creatures is written in an alphabet with just four letters, the four base pairs in DNA, and (2) how this genetic information is expressed during the growth and development of an organism. We shall see that RNA plays a key role in the process of gene expression.

THE GENETIC CONTROL OF METABOLISM: AN OVERVIEW

The ability of living organisms to grow and reproduce depends on a vast number of chemical reactions. Organisms must synthesize the many different molecules of which they are composed and degrade other molecules to obtain energy for growth. Plants have the ability to use energy from sunlight to synthesize macromolecules, but animals must obtain energy from the food they eat. This energy is obtained by breaking large molecules into smaller molecules and converting the energy derived from this process to stored chemical energy. All of the reactions that occur in living organisms collectively are called *metabolism*, and an organic molecule that is synthesized or degraded is called a *metabolite*. In the next four chapters, we examine a detailed picture of the genetic control of metabolism and present evidence documenting the most important features of this picture. We begin with a brief overview of the genetic control of metabolism, which should help us integrate the various pieces of the picture as each component is added.

The recognition of reactions specific to living creatures dates back to at least the early 1800s, when scientists observed the digestion of meat by secretions from the stomachs of animals and the conversion of starch to sugar by extracts of plant tissues. In the 1850s, Louis Pasteur studied the conversion of sugar into alcohol by "ferments" in yeasts and concluded that the fermentation process required the intact yeast cell. Pasteur's ferments were later called *enzymes*, and, in 1897, Eduard Buchner first extracted these enzymes from yeast cells, allowing their activities to be studied *in vitro*.

A catalyst is a substance that causes a reaction to occur without itself being modified in the process. Enzymes catalyze metabolic reactions by binding tightly to specific molecules, called *substrates*, and holding them in close juxtaposition so that the energy required for the reaction, the *activation energy*, is reduced. The net result is that covalent bonds can be rearranged at biological temperatures, the body temperatures of living organisms.

Although cell lysates were used extensively to study enzyme-catalyzed reactions throughout the first quarter of the twentieth century, the chemical nature of enzymes remained open to debate. Then, in 1926, James Sumner extracted the enzyme urease from jack beans, purified it to crystalline form, and demonstrated that it consisted entirely of a type of macromolecule called *protein*. Proteins were known to contain constituents called ***amino acids*** in the early 1900s, but it was not until the 1950s, when the structure of the small protein insulin was determined, that the complexity of each protein was appreciated. Amino acids are small organic molecules with a range of chemical structures, but all contain an amino ($-NH_2$) group and a carboxyl ($-COOH$) group. The 20 different amino acids present in proteins provide them with great structural diversity. A protein may contain a single chain of amino acids, called a *polypeptide*, or two or more chains (Chapter 12).

Thousands of enzymes have now been purified. Most are proteins with catalytic activity, but some are more complex, with additional components. Enzymes exhibit striking specificity—each enzyme catalyzes one reaction or a few very similar reactions. They range in size from about 12,000 to over 1 million daltons. The specificity of each enzyme results from its unique sequence of the 20 amino acids. Indeed, it was this essentially unlimited variability of protein structure that led many early geneticists to believe that proteins, not nucleic acids, stored genetic information.

The amino acid sequence of each polypeptide is controlled by one gene, and each polypeptide has either a structural, regulatory, or an enzymatic role in the cell. Moreover, each enzyme catalyzes a specific metabolic reaction. However, the picture gets more complex at this point, because metabolic processes seldom involve just one enzyme-catalyzed reaction. Instead, the synthesis or degradation of a particular metabolite usually occurs by a series of enzyme-catalyzed reactions. The complete set of enzyme-catalyzed reactions required to synthesize or degrade a given metabolite is called a **metabolic pathway**. For example, consider the biosynthesis of arginine, one of the 20 amino acids present in most proteins. In *E. coli*, the synthesis of arginine occurs in eight steps, each catalyzed by a specific enzyme (Figure 11.1). The eight enzymes are composed of nine gene products; one enzyme, ornithine carbamoyltransferase, contains two different gene products (ArgF and ArgI).

In general, each step in a metabolic pathway requires the activity of an enzyme, and each enzyme contains one or more polypeptides. Since each polypeptide is specified by one gene, each of the steps in a metabolic pathway will require the product of at least one gene. A generalized picture of the genetic control of metabolic pathways is presented in Figure 11.2.

Whereas many genes specify polypeptide products that are components of enzymes, other genes encode polypeptides with structural or regulatory roles. Some structural and regulatory proteins are components of cell membranes; others are present in the cytoskeleton, an intracellular network of fibers that controls cell shape and participates in the intracellular transport of macromolecules, organelles, and other macromolecular assemblages. Although most genes encode proteins, the final products of some genes are

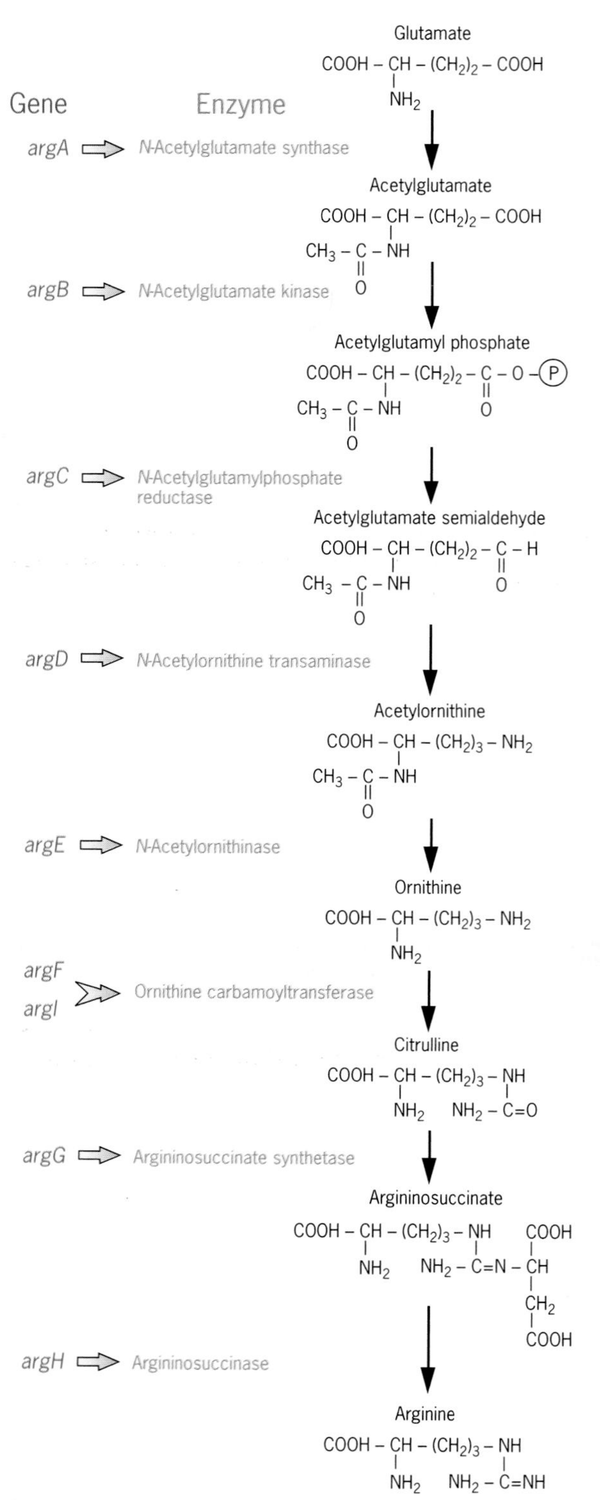

Figure 11.1 The genetic control of the arginine biosynthetic pathway in *E. coli*. Nine genes encode eight enzymes that catalyze the eight steps in this metabolic pathway.

RNA molecules. Several of these RNA molecules play essential roles in protein synthesis; we consider their functions in detail in Chapter 12. Since genes control the structures of RNAs and proteins, we should next ask how sequences of nucleotide pairs in DNA molecules specify the sequences of nucleotides in RNA and amino acids in protein molecules.

Key Points: **Metabolism occurs by sequences of enzyme-catalyzed reactions, with each enzyme specified by one or more genes. Therefore, each step in a metabolic pathway is under genetic control.**

TRANSFER OF GENETIC INFORMATION: THE CENTRAL DOGMA

According to the central dogma of molecular biology, genetic information flows (1) from DNA to DNA during its transmission from generation to generation and (2) from DNA to protein during its phenotypic expression in an organism (Figure 11.3). The transfer of genetic information from DNA to protein involves two steps: (1) **transcription**, the transfer of the genetic information from DNA to RNA, and (2) **translation,** the transfer of information from RNA to protein. In addition, genetic information flows from RNA to DNA during the conversion of the genomes of RNA tumor viruses to their DNA proviral forms (Chapter 15). Thus the transfer of genetic information from DNA to RNA is sometimes reversible, whereas the transfer of information from RNA to protein is always irreversible.

Transcription and Translation

As we discussed above, the expression of genetic information occurs in two steps: transcription and translation (Figure 11.3). During transcription, one strand of DNA of a gene is used as a template to synthesize a complementary strand of RNA, called the gene **transcript**. For example, in Figure 11.3, the DNA strand containing the nucleotide sequence AAA is used as a template to produce the complementary sequence UUU in the RNA transcript. During translation, the sequence of nucleotides in the RNA transcript is converted into the sequence of amino acids in the polypeptide gene product. This conversion is governed by the **genetic code**, the specification of amino acids by nucleotide triplets called **codons** in the gene transcript. For example, the UUU triplet in the RNA transcript shown in Figure 11.3 specifies the amino acid phenylalanine (Phe) in the polypeptide gene product. Translation takes place on intricate macromolecular machines called **ribosomes**, which are composed of three to five RNA molecules and 50 to 90 different proteins. However, the process of translation

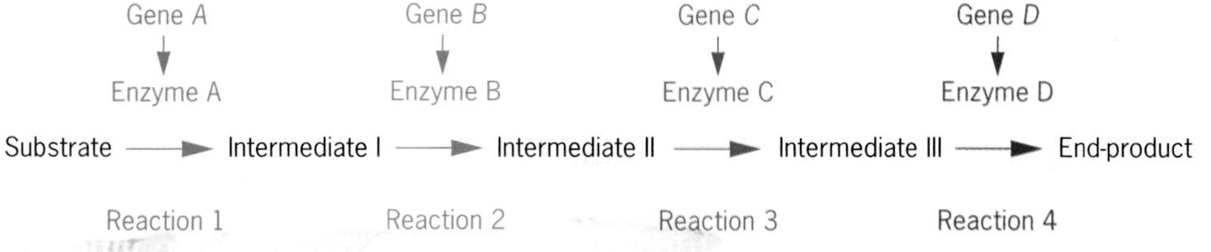

Figure 11.2 The genetic control of metabolism. The metabolic pathway shown contains four steps; some pathways are shorter and others are much longer. In addition, some enzymes contain the products of two or more different genes.

also requires the participation of many other macromolecules. This chapter focuses on transcription; translation is the subject of Chapter 12.

The RNA molecules that are translated on ribosomes are called **messenger RNAs (mRNAs)**. In prokaryotes, the product of transcription, the **primary transcript**, usually is equivalent to the mRNA molecule (Figure 11.4*a*). In eukaryotes, primary transcripts often must be processed by the excision of specific sequences and the modification of both termini before they can be translated (Figure 11.4*b*). Thus, in eukaryotes, primary transcripts usually are precursors to mRNAs and, as such, are called **pre-mRNAs**. Most of the nuclear genes in higher eukaryotes and some in lower eukaryotes contain noncoding sequences called *introns* that separate the coding sequences or *exons* of these genes. The entire sequences of these *split genes* are transcribed into pre-mRNAs, and the noncoding intron sequences are subsequently removed by *splicing reactions* carried out on macromolecular structures called *spliceosomes*.

Four Types of RNA Molecules

Four different classes of RNA molecules play essential roles in gene expression. We have already discussed messenger RNAs, the intermediaries that carry genetic

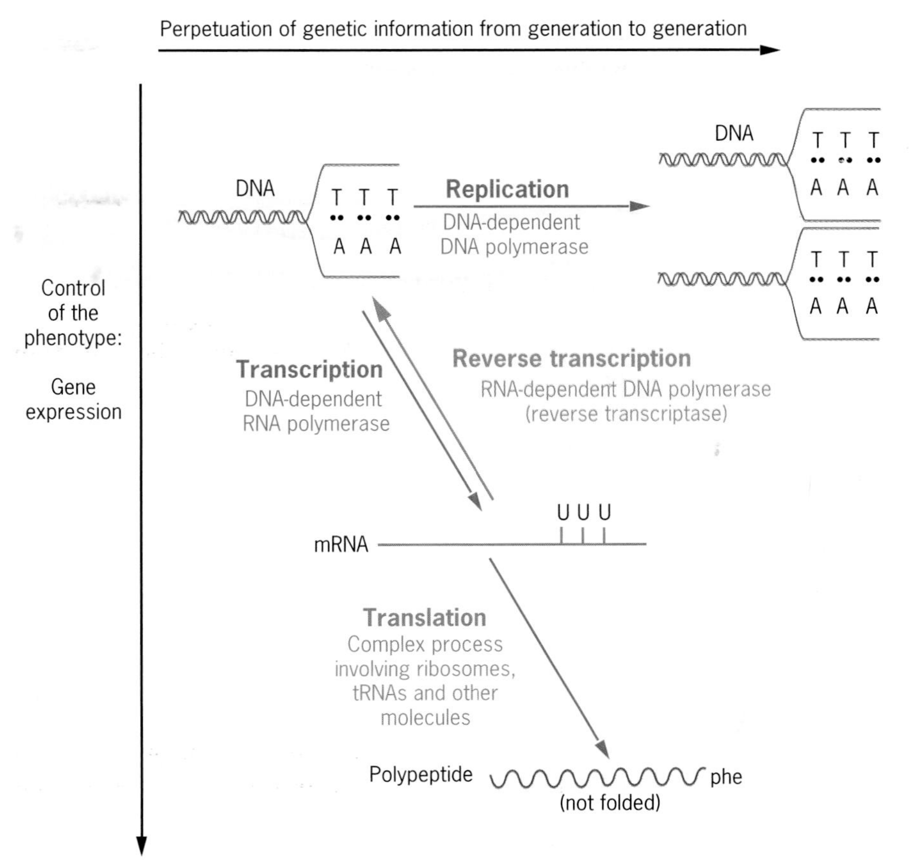

Figure 11.3 The flow of genetic information according to the central dogma of molecular biology. Replication, transcription, and translation occur in all organisms; reverse transcription occurs only in cells infected with certain RNA viruses.

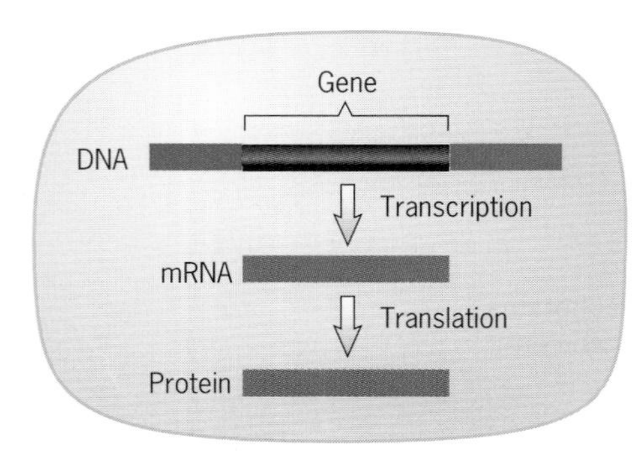

(*a*) Prokaryotes.

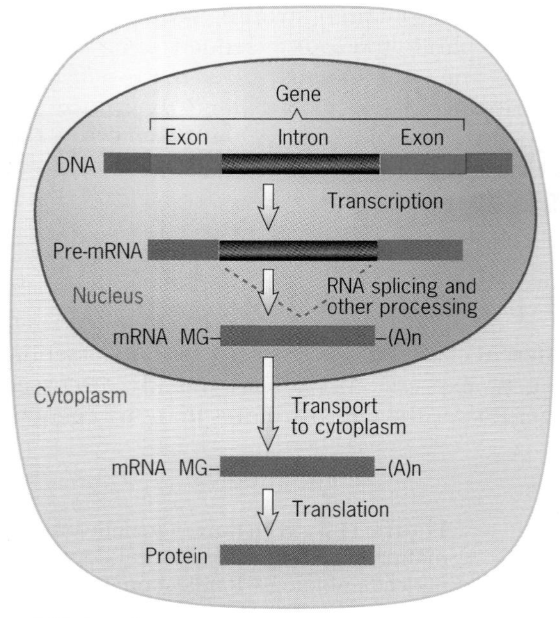

(*b*) Eukaryotes.

Figure 11.4 Protein synthesis involves two steps: (1) transcription and (2) translation in both prokaryotes (*a*) and eukaryotes (*b*). In addition, in eukaryotes, the primary transcripts or pre-mRNAs often must be processed by the excision of introns and the addition of 5' 7-methyl guanosine caps (MG) and 3' poly(A) tails $[(A)]_n$.

information from DNA to the ribosomes where proteins are synthesized. **Transfer RNAs (tRNAs)** are small RNA molecules that function as adaptors between amino acids and the codons in mRNA during translation. **Ribosomal RNAs (rRNAs)** are structural components of the ribosomes, the intricate machines that translate nucleotide sequences of mRNAs into amino acid sequences of polypeptides. **Small nuclear RNAs (snRNAs)** are structural components of spliceosomes, the nuclear structures that excise introns from nuclear genes. The roles of mRNAs and snRNAs are discussed in this chapter. Because we examine the structures and functions of tRNAs and rRNAs in detail in Chapter 12, the discussion here will be brief.

There are from one to four tRNAs for each of the 20 amino acids specified by the genetic code. The tRNA molecules are from 70 to 90 nucleotides long and fold into cloverleaf-shaped structures by intramolecular base-pairing. In its folded form, each tRNA has a specific amino acid attached at one end and a triplet nucleotide sequence called the **anticodon** at the other end. The recognition of nucleotide sequences in mRNAs by tRNAs is mediated by base-pairing between the anticodon in the tRNAs and the codon in the mRNA.

Ribosomal RNAs are essential components of ribosomes. The ribosomal RNAs interact with over 50 different ribosomal proteins to produce the complex three-dimensional structure of the ribosome. Prokaryotic ribosomes contain three rRNAs, about 120, 1540, and 2900 nucleotides in length, and about 50 proteins. Prokaryotic ribosomal RNA molecules are designated 5S, 16S, and 23S rRNAs based on their rates of sedimentation during sucrose gradient centrifugation (see Technical Sidelight in Chapter 10). Eukaryotic ribosomes are more complex than those of prokaryotes; they contain four rRNA molecules and about 80 different proteins. The four eukaryotic rRNAs, designated 5S, 5.8S, 18S, and 28S rRNAs, are approximately 120, 160, 1900, and 4700 nucleotides in length, respectively.

Five different small nuclear RNAs are present in spliceosomes. As the name implies, they are small in size, ranging from 100 to 215 nucleotides long in mammals. Unlike the other RNAs discussed here, snRNAs never leave the nucleus. Instead, they interact with about 40 nuclear proteins to form spliceosomes, and they play key roles in the excision of noncoding sequences from the transcripts of nuclear genes.

All four types of RNA—mRNA, tRNA, rRNA, and snRNA—are produced by transcription. Unlike mRNAs, which specify polypeptides, the final products of tRNA, rRNA, and snRNA genes are RNA molecules. Transfer RNA, ribosomal RNA, and snRNA molecules are not translated. Figure 11.5 shows an overview of protein synthesis in eukaryotes, emphasizing the transcriptional origin and functions of the four types of RNA molecules. The process is similar in prokaryotes. However, in prokaryotes, the DNA is not separated from the ribosomes by a nuclear envelope. In addition, prokaryotic genes seldom contain noncoding sequences that are removed during RNA transcript processing.

Key Points: **The central dogma of molecular biology is that genetic information flows from DNA to DNA during chromosome replication and from DNA to protein during gene expression. The flow of information from DNA to protein occurs in two steps: (1) transcription, DNA to RNA, and (2) translation, RNA to protein. Transcription involves the synthesis of an**

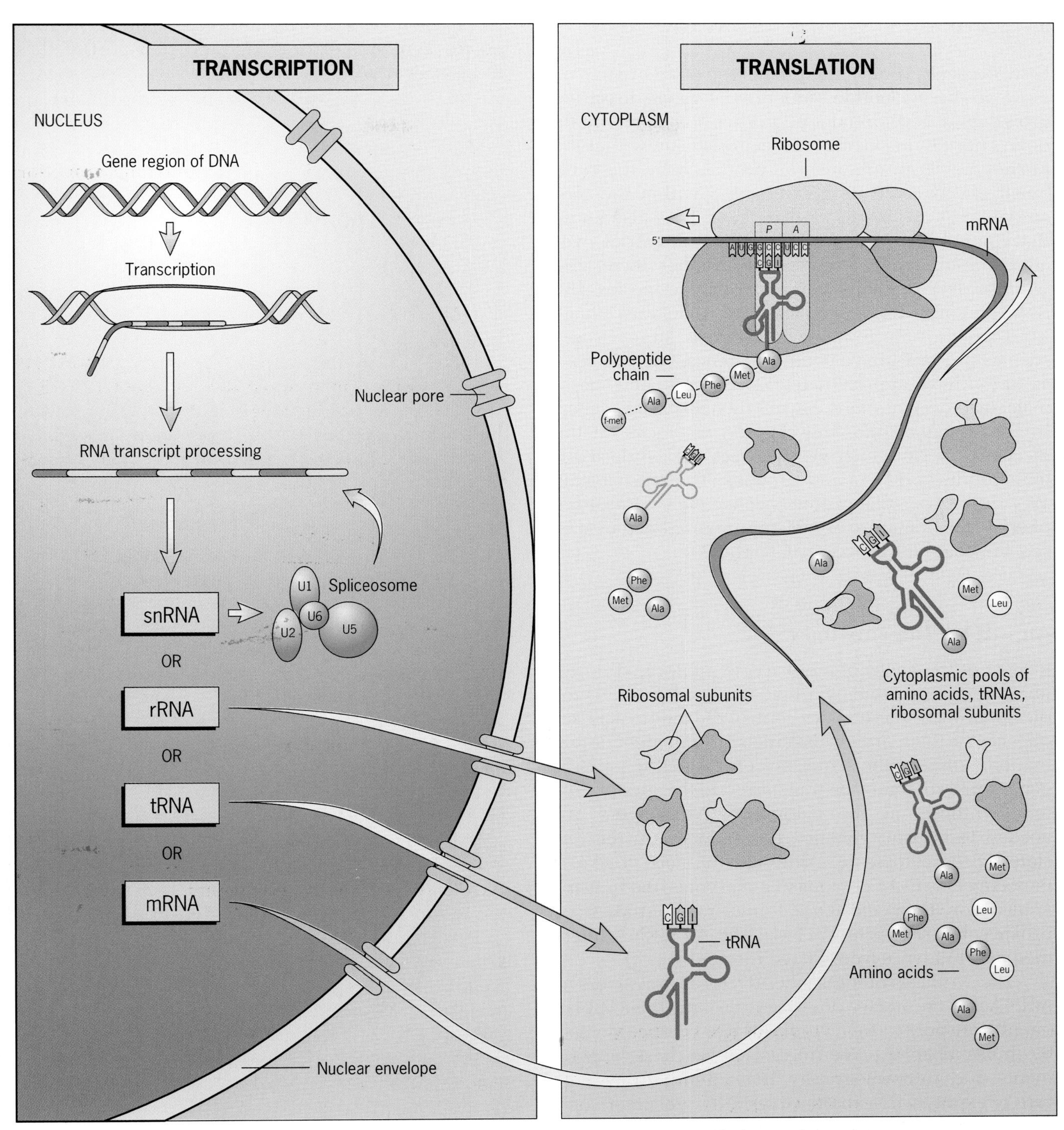

(*a*) Transcription and RNA processing occur in the nucleus.

(*b*) Translation occurs in the cytoplasm.

Figure 11.5 An overview of protein synthesis, emphasizing the transcriptional origin of snRNA, tRNA, rRNA, and mRNA, the splicing function of snRNA, and the translational roles of tRNA, rRNA, mRNA, and ribosomes.

RNA transcript complementary to one strand of DNA of a gene. Translation is the conversion of information stored in the sequence of nucleotides in the RNA transcript into the sequence of amino acids in the polypeptide gene product, according to the specifications of the genetic code.

THE PROCESS OF GENE EXPRESSION

How do genes control the phenotype of an organism? How do the nucleotide sequences of genes direct the growth and development of a cell, a tissue, an organ, or an entire living creature? Geneticists know that the phenotype of an organism is produced by the combined effects of all its genes, acting within the constraints imposed by the environment. They also know that the number of genes in an organism varies over an enormous range, with gene number increasing with the developmental complexity of the species. The RNA genomes of the smallest viruses such as phage MS2 contain only four genes, whereas large viruses such as phage T4 have about 200 genes. Bacteria such as *E. coli* have approximately 2000 genes, and mammals, including humans, have 50,000 to 100,000 genes. In this and the following chapter, we focus on the mechanisms by which genes direct the synthesis of their products, RNAs and proteins. The mechanisms by which these gene products collectively control the phenotypes of mature organisms are discussed in subsequent chapters, especially Chapter 23.

An mRNA Intermediary

If most of the genes of a eukaryote are located in the nucleus, and if proteins are synthesized in the cytoplasm, how do these genes control the amino acid sequences of their protein products? The genetic information stored in the sequences of nucleotide pairs in genes must somehow be transferred to the sites of protein synthesis in the cytoplasm. Messengers are needed to transfer genetic information from the nucleus to the cytoplasm. Although the need for such messengers is most obvious in eukaryotes, the first evidence for their existence came from studies of prokaryotes (see Technical Sidelight: An mRNA Intermediary: Evidence from Phage-Infected *E. coli*).

The synthesis of messenger RNAs or precursors to mRNAs in the nuclei of eukaryotes and their subsequent transport to the cytoplasm can be documented by pulse-labeling experiments, pulse-chase experiments, and autoradiography. If a cell growing in culture is exposed to a radioactive RNA precursor such as ^{3}H-uridine for a few minutes, and the intracellular location of the incorporated radioactivity is determined by autoradiography, the labeled RNA is present almost exclusively in the nucleus (Figure 11.6*a*). However, if the short exposure to ^{3}H-uridine is followed by a period of growth in nonradioactive medium, most of the incorporated radioactivity is present in the cytoplasm (Figure 11.6*b*). Thus the unstable RNA intermediaries are synthesized in the nucleus and are transported to the cytoplasm, where they direct the synthesis of proteins. In this chapter, we focus on the synthesis, processing, and transport of these messenger RNA molecules.

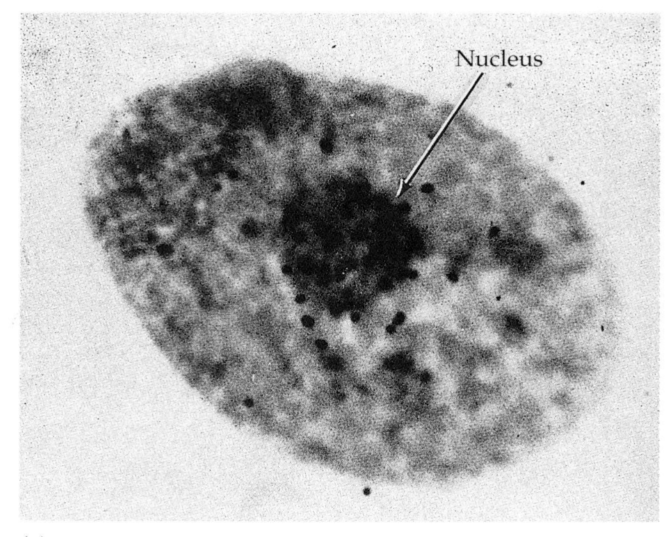

(*a*)

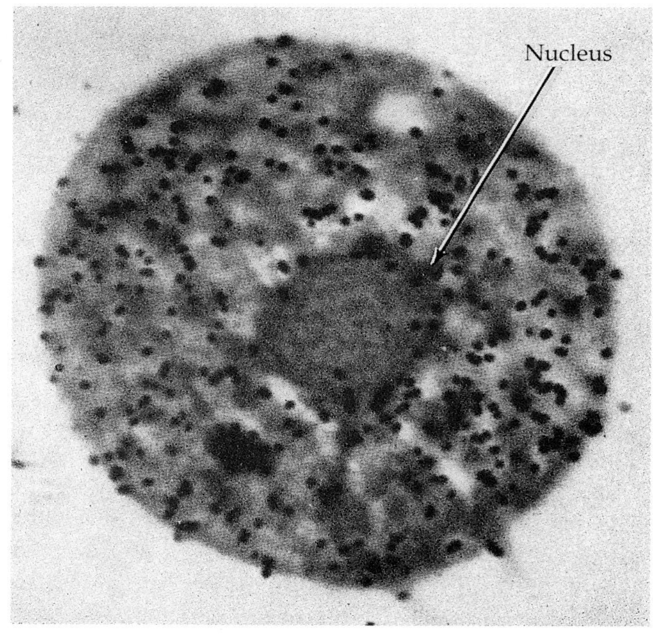

(*b*)

Figure 11.6 Autoradiographs demonstrating (1) the synthesis of RNA in the nucleus and (2) its subsequent transport to the cytoplasm. Each autoradiograph is superimposed on a photograph of a thin section of the cell. The black dots represent silver granules in the autoradiographic emulsion that have reacted with electrons emitted by the decay of ^{3}H atoms. (*a*) A *Tetrahymena* cell labeled with ^{3}H-cytidine for 15 minutes. (*b*) A *Tetrahymena* cell that was grown on nonradioactive medium for 88 minutes after exposure to ^{3}H-cytidine for 12 minutes.

TECHNICAL SIDELIGHT

An mRNA Intermediary: Evidence from Phage-Infected *E. coli*

The first evidence for the existence of an RNA intermediary in protein synthesis came from studies by Elliot Volkin and Lawrence Astrachan on bacteria infected with bacterial viruses. Their results, published in 1956, suggested that the synthesis of viral proteins in infected bacteria involved unstable RNA molecules specified by viral DNA. Volkin and Astrachan observed a burst of RNA synthesis after infecting *E. coli* cells with bacteriophage T2. By labeling RNA with the radioactive isotope ^{32}P, they demonstrated that the newly synthesized RNA molecules were unstable, turning over with half-lives of only a few minutes. In addition, they showed that the nucleotide composition of the unstable RNAs was similar to the composition of T2 DNA and unlike that of *E. coli* DNA. Their results were soon extended by studies in other laboratories.

In 1961, Sol Spiegelman and coworkers reported that the unstable RNAs synthesized in phage T4-infected cells could form RNA–DNA duplexes with denatured T4 DNA, but not with denatured *E. coli* DNA. They pulse-labeled bacteria with ^{3}H-uridine at various times after infection with T4 phage, isolated total RNA from these cells, and determined whether the radioactive RNA molecules hybridized with *E. coli* DNA or phage T4 DNA. Their experiment is diagrammed as follows.

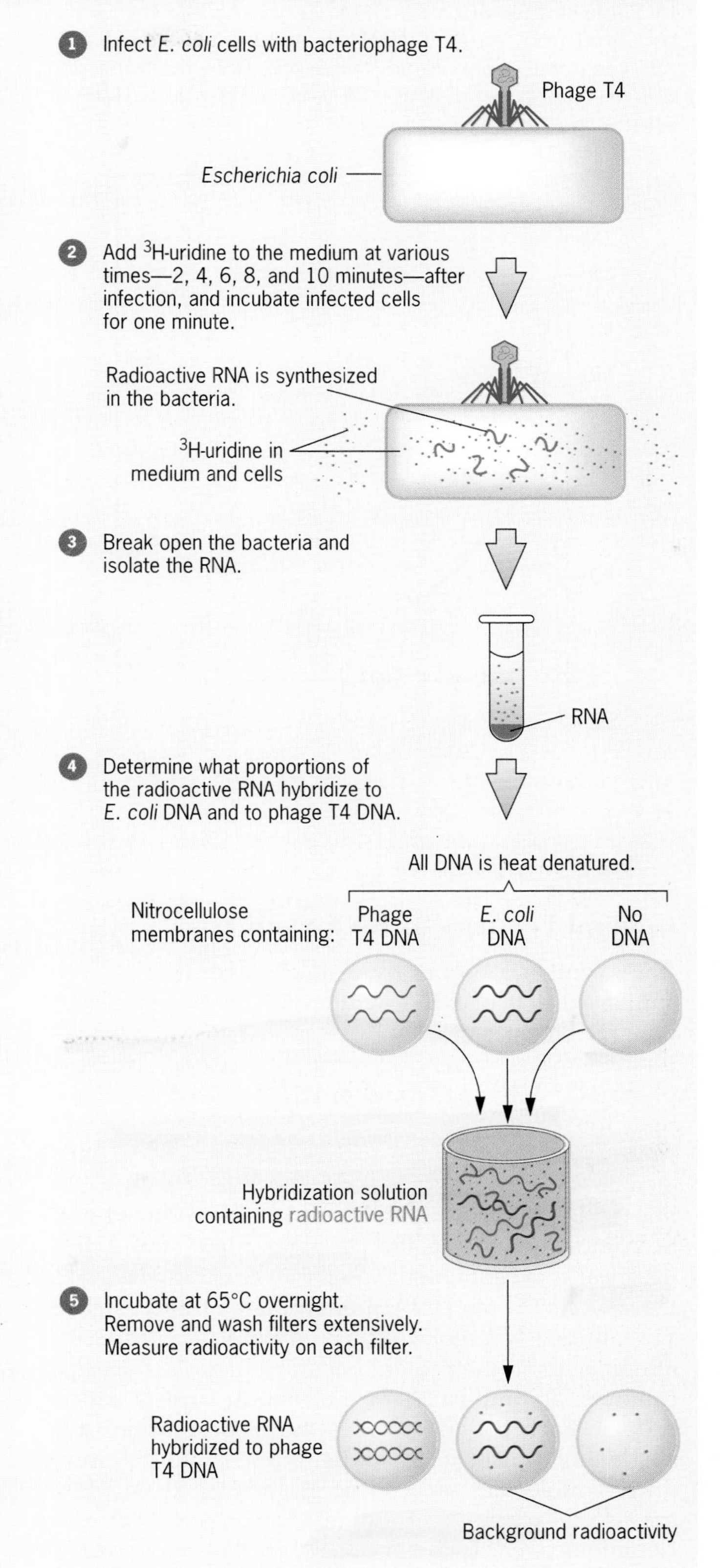

(*Box continues on next page.*)

Their results demonstrated that most of the short-lived RNA molecules synthesized after infection were complementary to single strands of phage T4 DNA and noncomplementary to single strands of *E. coli* DNA, indicating that they were produced from phage T4 DNA templates, not from *E. coli* DNA templates. This rapid switchover from the transcription of *E. coli* genes to phage T4 genes in infected bacteria is illustrated as follows.

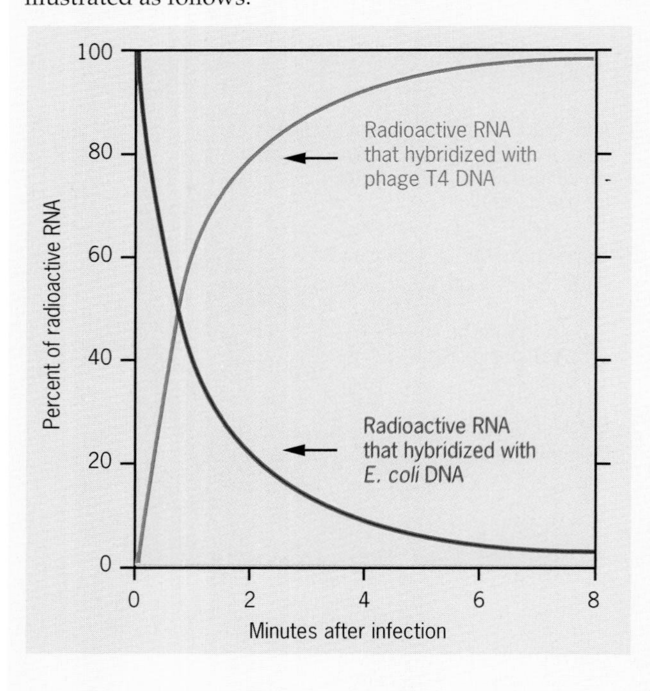

In the same year that Spiegelman and colleagues published their results, Sydney Brenner, François Jacob, and Matthew Meselson demonstrated that phage T4 proteins were synthesized on *E. coli* ribosomes. Thus the amino acid sequences of T4 proteins were not controlled by components of the ribosomes. Instead, the ribosomes provided the workbenches on which protein synthesis occurred, but did not provide the specifications for individual proteins. These results strengthened the idea, first formally proposed by François Jacob and Jacques Monod in 1961, that unstable RNA molecules carried the specifications for the amino acid sequences of individual gene products from the genes to the ribosomes. Subsequent research firmly established the role of these unstable RNAs, now called messenger RNAs or mRNAs, in the transfer of genetic information from genes to the sites of protein synthesis in the cytoplasm.

General Features of RNA Synthesis

RNA synthesis occurs by a mechanism that is very similar to that of DNA synthesis (Chapter 9) except that (1) the precursors are **ribonucleoside triphosphates** rather than deoxyribonucleoside triphosphates, (2) only one strand of DNA is used as a template for the synthesis of a complementary RNA chain in any given region, and (3) RNA chains can be initiated *de novo*, without any requirement for a preexisting primer strand. The RNA molecule produced will be complementary to the DNA **template strand** and identical, except that uridine residues replace thymidines, to the DNA **nontemplate strand** (Figure 11.7). If the RNA molecule is an mRNA, it will specify amino acids in the protein gene product. This specification is accomplished by nucleotide triplets called codons. Each codon specifies the incorporation of a single amino acid in a polypeptide chain. Therefore, mRNA molecules are coding strands of RNA. They are also called **sense strands** of RNA because their nucleotide sequences "make sense" in that they specify sequences of amino acids in the protein gene products.

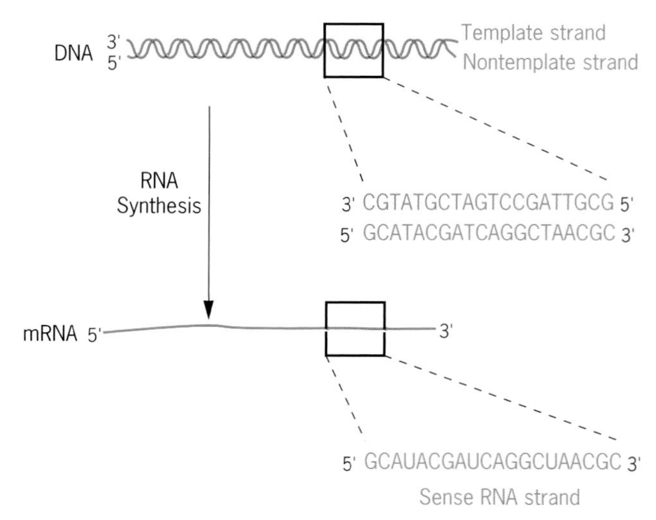

Figure 11.7 RNA synthesis utilizes only one DNA strand of a gene as template.

An RNA molecule that is complementary to an mRNA is referred to as **antisense RNA**. This terminology is sometimes extended to the two strands of DNA. However, usage of the terms *sense* and *antisense* to denote DNA strands has been inconsistent. Thus we will use *template strand* and *nontemplate strand* to refer to the transcribed and nontranscribed strands, respectively, of a gene.

The synthesis of RNA chains, like DNA chains, occurs in the 5′ → 3′ direction, with the addition of ribonucleotides to the 3′-hydroxyl group at the end of the chain (Figure 11.8). The reaction involves a nucleophilic attack by the 3′-OH on the nucleotidyl (interior)

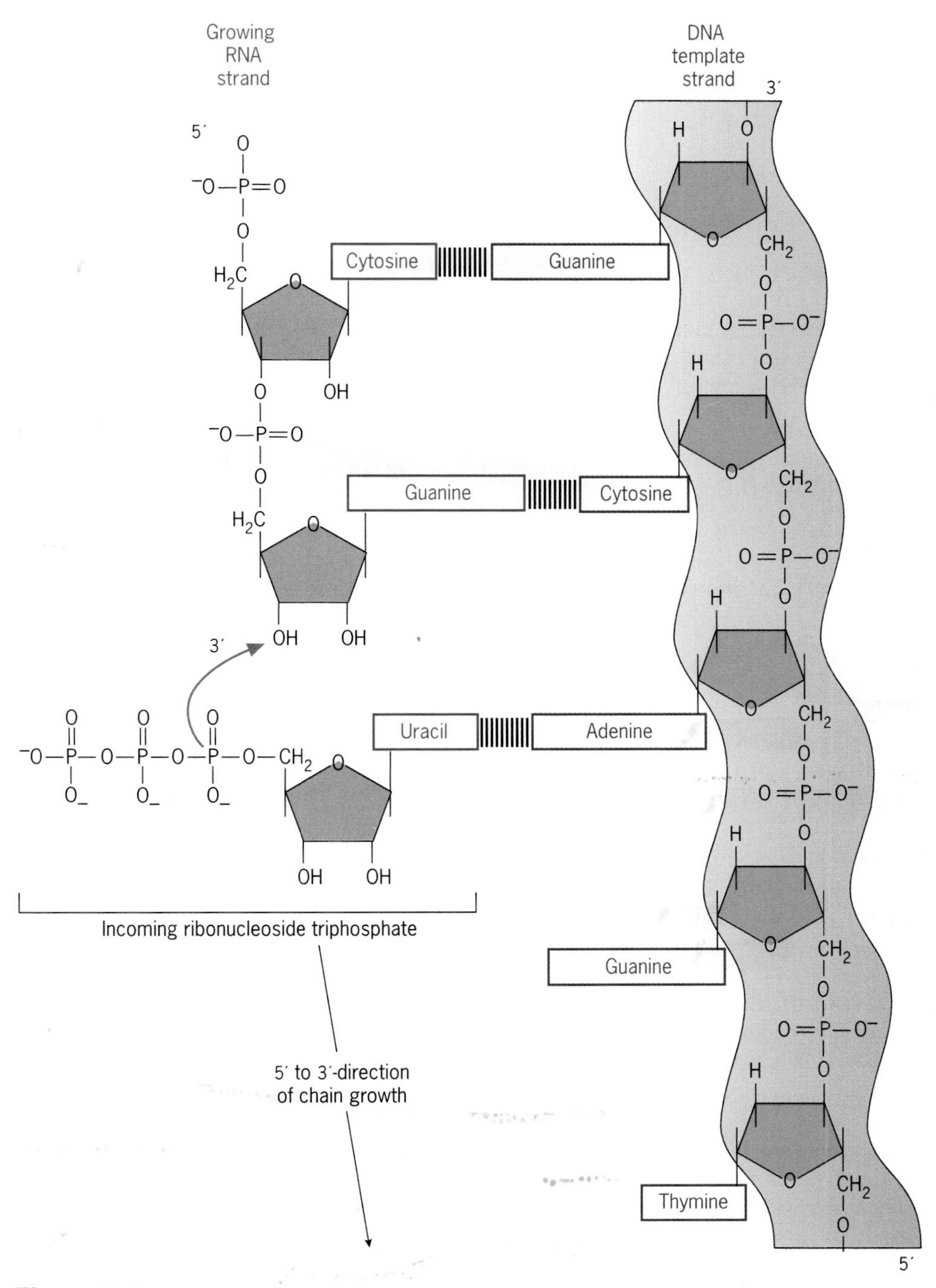

Figure 11.8 The RNA chain elongation reaction catalyzed by RNA polymerase.

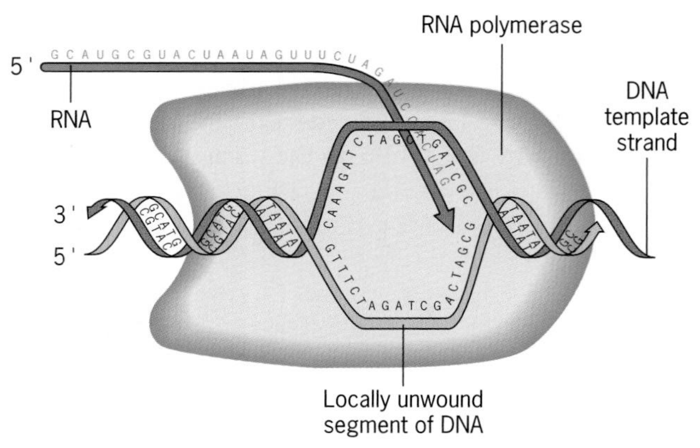

Figure 11.9 RNA synthesis occurs within a locally unwound segment of DNA. This transcription bubble allows a few nucleotides in the template strand to base-pair with the growing end of the RNA chain. The unwinding and rewinding of the DNA molecule are catalyzed by RNA polymerase.

phosphorus atom of the ribonucleoside triphosphate precursor with the elimination of pyrophosphate, just as in DNA synthesis. This reaction is catalyzed by enzymes called **RNA polymerases**. The overall reaction is as follows:

$$n(\text{RTP}) \xrightarrow[\text{RNA polymerase}]{\text{DNA template}} (\text{RMP})_n + n\,(\text{PP})$$

where n is the number of moles of ribonucleotide triphosphate (RTP) consumed, ribonucleotide monophosphate (RMP) incorporated into RNA, and pyrophosphate (PP) produced.

RNA polymerases initiate transcription at specific nucleotide sequences called **promoters**, which are different in prokaryotes and eukaryotes. A single RNA polymerase carries out all transcription in most prokaryotes, whereas three different RNA polymerases are present in eukaryotes, with each polymerase responsible for the synthesis of a distinct class of RNAs. RNA synthesis takes place within a locally unwound segment of DNA, sometimes called a **transcription bubble**, which is produced by RNA polymerase (Figure 11.9). The nucleotide sequence of an RNA molecule is complementary to that of its DNA template strand, and RNA synthesis is governed by the same base-pairing rules as DNA synthesis, but uracil replaces thymine.

Key Points: **In eukaryotes, genes are present in the nucleus, whereas polypeptides are synthesized in the cytoplasm. Messenger RNA molecules function as intermediaries that carry genetic information from DNA to the ribosomes, where proteins are synthesized. RNA synthesis, catalyzed by RNA polymerases, is similar to DNA synthesis in many respects. However, RNA synthesis occurs within a localized region of strand separation, and only one strand of DNA functions as a template for RNA synthesis.**

TRANSCRIPTION IN PROKARYOTES

The basic features of transcription are the same in both prokaryotes and eukaryotes, but many of the details—such as the promoter sequences—are different. The RNA polymerase of *E. coli* has been studied in great detail. It catalyzes all RNA synthesis in this species. There are about 7000 copies of RNA polymerase in an *E. coli* cell, and the majority of them are engaged in RNA synthesis.

A segment of DNA that is transcribed to produce one RNA molecule is called a **transcription unit**. Transcription units may be equivalent to individual genes, or they may include several contiguous genes. Large transcripts that carry the coding sequences of several genes are common in bacteria. The process of transcription can be divided into three stages: (1) initiation of a new RNA chain, (2) elongation of the chain, and (3) termination of transcription and release of the nascent RNA molecule (Figure 11.10).

When discussing transcription, biologists often use the terms *upstream* and *downstream* to refer to regions located toward the 5′ end or the 3′ end, respectively, of the transcript from some site in the mRNA molecule. These terms are based on the fact that RNA synthesis always occurs in the 5′ to 3′ direction. Upstream and downstream regions of genes are the DNA sequences specifying the corresponding 5′ and 3′ segments of their transcripts relative to a specific reference point.

RNA Polymerase: A Complex Enzyme

The RNA polymerases that catalyze transcription are complex, multimeric proteins. The *E. coli* RNA polymerase has a molecular weight of about 480,000 and consists of five polypeptides. Two of these are identical; thus the enzyme contains four distinct polypeptides. The complete RNA polymerase molecule, the **holoenzyme**, has the composition $\alpha_2\beta\beta'\sigma$. The α subunits are involved in the assembly of the **tetrameric core** ($\alpha_2\beta\beta'$) of RNA polymerase. The β subunit contains the ribonucleoside triphosphate binding site, and the β' subunit harbors the DNA template binding region.

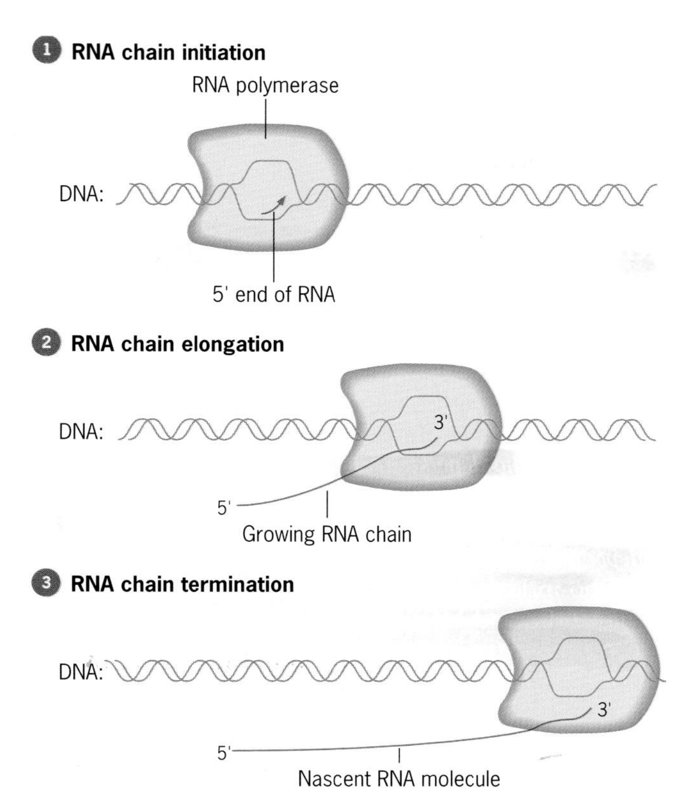

Figure 11.10 The three stages of transcription: initiation, elongation, and termination.

One subunit, the **sigma (σ) factor**, is involved only in the initiation of transcription; it plays no role in chain elongation. After RNA chain initiation has occurred, the σ-factor is released, and chain elongation (see Figure 11.8) is catalyzed by the core enzyme ($\alpha_2 \beta \beta'$). The function of sigma is to recognize and bind RNA polymerase to the transcription initiation or **promoter sites** in DNA. The core enzyme (with no σ) will catalyze RNA synthesis from DNA templates *in vitro*, but, in so doing, it will initiate RNA chains at random sites on both strands of DNA. In contrast, the holoenzyme (σ present) initiates RNA chains *in vitro* only at sites used *in vivo*.

Initiation of RNA Chains

Initiation of RNA chains involves three steps: (1) binding of the RNA polymerase holoenzyme to a promoter region in DNA, (2) the localized unwinding of the two strands of DNA by RNA polymerase, providing a template strand free to base-pair with incoming ribonucleotides, and (3) the formation of phosphodiester bonds between the first few ribonucleotides in the nascent RNA chain. The holoenzyme remains bound at the promoter region during the synthesis of the first eight or nine bonds, then the sigma factor is released, and the core enzyme begins the elongation phase of RNA synthesis. During initiation, short chains of two to nine ribonucleotides are synthesized and released. This abortive synthesis stops once chains of ten or more ribonucleotides have been synthesized and RNA polymerase has begun to move downstream from the promoter.

As mentioned earlier, the sigma subunit of RNA polymerase mediates its binding to promoters in DNA. Hundreds of *E. coli* promoters have been sequenced and found to have surprisingly little in common. Two short sequences within these promoters are sufficiently conserved to be recognized, but even these are seldom identical in two different promoters. The midpoints of the two conserved sequences occur at about 10 and 35 nucleotide pairs, respectively, before the transcription-initiation site (Figure 11.11). Thus they are called the **-10 sequence** and the **-35 sequence**, respectively. Although these sequences vary slightly from gene to gene, some nucleotides are highly conserved. The nucleotide sequences that are present in such conserved genetic elements most often are called **consensus sequences**. The -10 consensus sequence in the nontemplate strand is TATAAT; the -35 consensus sequence is TTGACA. The sigma subunit initially recognizes and binds to the -35 sequence; thus this sequence is sometimes called the **recognition sequence**. The AT-rich -10 sequence facilitates the localized unwinding of DNA, which is an essential prerequisite to the synthesis of a new RNA chain. The distance between the -35 and -10 sequences is highly conserved in *E. coli* promoters, never being less than 15 or more than 20 nucleotide pairs in length. In addition, the first or 5′ base in *E. coli* RNAs is usually (>90%) a purine. Except for these short conserved sequences, different *E. coli* promoters have little in common.

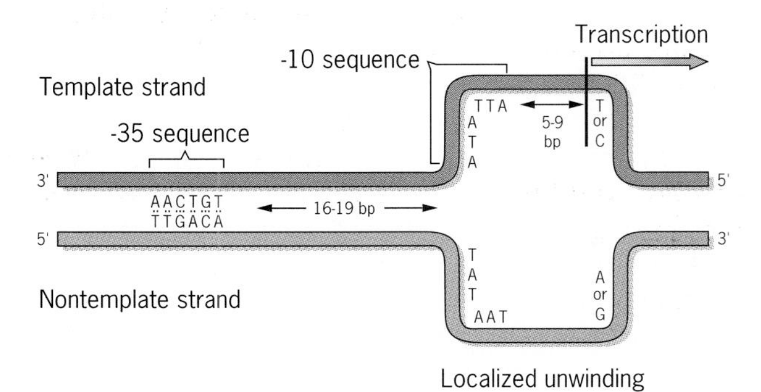

Figure 11.11 Structure of a typical promoter in *E. coli*. RNA polymerase binds to the –35 sequence of the promoter and initiates unwinding of the DNA strands at the AT-rich –10 sequence. Transcription begins within the transcription bubble at a site five to nine base pairs beyond the –10 sequence.

Elongation of RNA Chains

Elongation of RNA chains is catalyzed by the RNA polymerase core enzyme, after the release of the σ subunit. The covalent extension of RNA chains (see Figure 11.8) takes place within the transcription bubble, a locally unwound segment of DNA. The RNA polymerase molecule contains both DNA unwinding and DNA rewinding activities. RNA polymerase continuously unwinds the DNA double helix ahead of the polymerization site and rewinds the complementary DNA strands behind the polymerization site as it moves along the double helix (Figure 11.12). In *E. coli*, the average length of a transcription bubble is 18 nucleotide pairs, and about 40 ribonucleotides are incorporated into the growing RNA chain per second. The nascent RNA chain is displaced from the DNA template strand as RNA polymerase moves along the DNA molecule. The region of transient base-pairing between the growing chain and the DNA template strand is very short, perhaps only three base pairs in length. The stability of the transcription complex is maintained primarily by the binding of the DNA and the growing RNA chain to RNA polymerase, rather than by the base-pairing between the template strand of DNA and the nascent RNA.

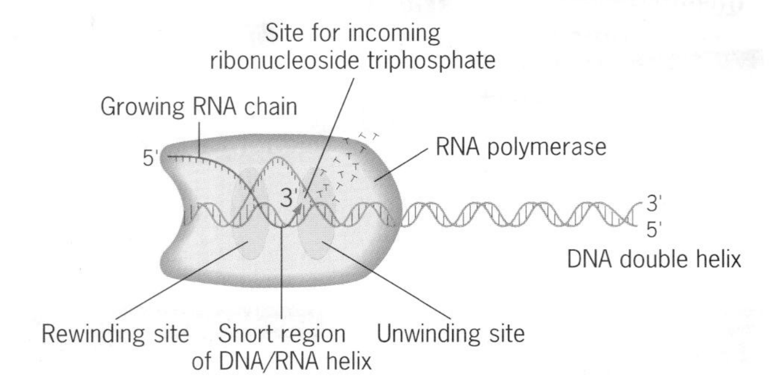

(*a*) RNA polymerase is bound to DNA and is covalently extending the RNA chain.

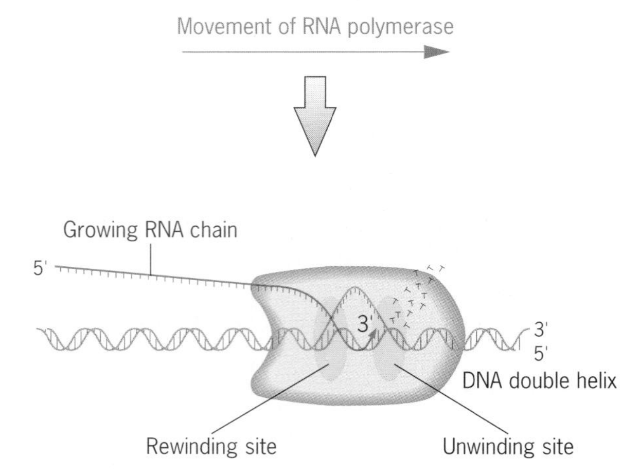

(*b*) RNA polymerase has moved downstream from its position in (*a*) processively extending the nascent RNA chain.

Figure 11.12 Elongation of an RNA chain catalyzed by RNA polymerase in *E. coli*.

Termination of RNA Chains

Termination of RNA chains occurs when RNA polymerase encounters a **termination signal**. When this occurs, the transcription complex dissociates, releasing the nascent RNA molecule. There are two types of transcription terminators in *E. coli*. One type results in termination only in the presence of a protein called *rho* (ρ); therefore, such termination sequences are called *rho-dependent terminators*. The other type results in the termination of transcription without the involvement of rho; such sequences are called *rho-independent terminators*.

Rho-independent terminators contain a G:C-rich region followed by six or more A:T base pairs, with the A's present in the template strand (Figure 11.13, top). The nucleotide sequence of the G:C-rich region is such that regions of the single-stranded RNA can base-pair and form hairpin-like structures (Figure 11.13, bottom). The RNA hairpin structures form immediately after the synthesis of the participating regions of the RNA chain and retard the movement of RNA polymerase molecules along the DNA, causing pauses in chain extension. Since A:U base-pairing is weak, requiring less energy to separate the bases than any of the other standard base pairs, the run of U's after the hairpin region is thought to facilitate the release of the newly synthesized RNA chains from the DNA template when the hairpin structure causes RNA polymerase to pause at this site.

The mechanism by which rho-dependent termination of transcription occurs is still uncertain. Rho-dependent termination sequences are 50 to 90 base pairs long, rich in C residues and largely devoid of G's. Beyond that, different rho-dependent termination signals have little in common. The rho protein binds to the growing RNA chain and moves 5′ to 3′ along the RNA, seeming to pursue the RNA polymerase molecule catalyzing the synthesis of the chain. When RNA polymerase slows down or pauses at the rho-dependent termination sequence, rho catches up with the polymerase and pulls the nascent RNA chain from the transcription bubble.

Concurrent Transcription, Translation, and mRNA Degradation

In prokaryotes, the translation and degradation of an mRNA molecule often begin before its synthesis (transcription) is complete. Since mRNA molecules are synthesized, translated, and degraded in the 5′ to 3′

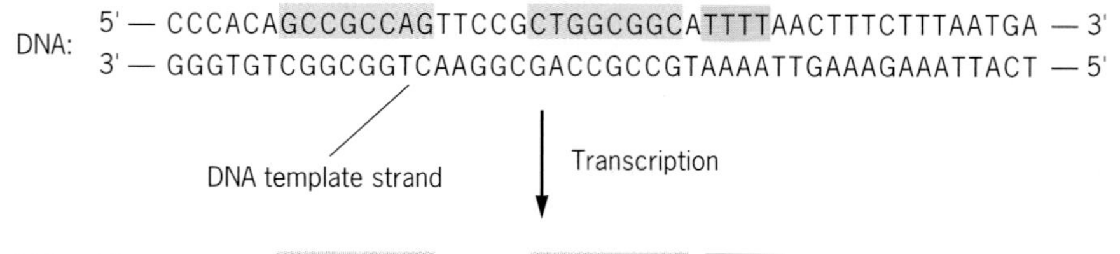

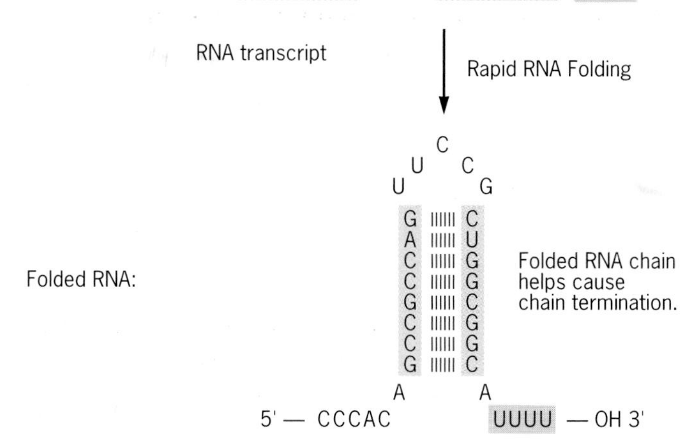

Figure 11.13 Structure of a rho-independent transcription terminator. Rho-independent terminator sequences contain a G:C-rich region followed by at least six A:T base pairs. Transcription of such a terminator sequence produces an RNA chain with G:C-rich segments that base-pair with each other immediately after synthesis to form a hairpin-like structure. This structure retards the movement of RNA polymerase along the DNA molecule, which results in the termination of transcription in the adjacent A:T tract and the release of the nascent RNA chain.

direction, all three processes can occur simultaneously on the same RNA molecule. In prokaryotes, the polypeptide synthesizing machinery is not separated by a nuclear envelope from the site of mRNA synthesis. Therefore, once the 5′ end of an mRNA has been synthesized, it can immediately be used as a template for polypeptide synthesis. Indeed, transcription and translation often are tightly coupled in prokaryotes. Oscar Miller, Barbara Hamkalo, and colleagues developed electron microscopy techniques that allowed them to directly visualize this coupling between transcription and translation in bacteria. One of their photographs showing the coupled transcription of a gene and translation of its mRNA product in *E. coli* is reproduced in Figure 11.14.

Key Points: **RNA synthesis occurs in three stages: (1) initiation, (2) elongation, and (3) termination. The RNA polymerases that catalyze transcription are complex, multimeric proteins. Covalent extension of RNA chains occurs within transcription bubbles, locally unwound segments of DNA. Chain elongation stops when RNA polymerase encounters transcription–termination signals in DNA. Transcription, translation, and degradation of a specific mRNA molecule often occur simultaneously in prokaryotes.**

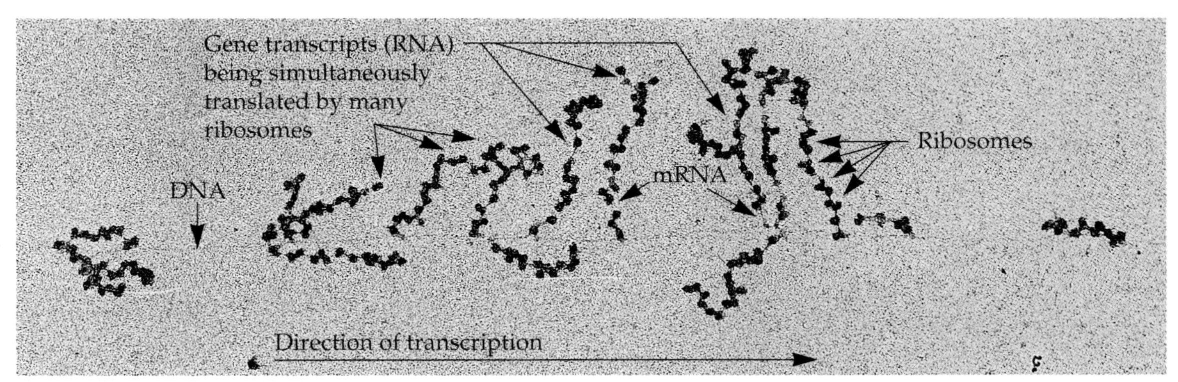

Figure 11.14 Electron micrograph prepared by Oscar Miller and Barbara Hamkalo showing the coupled transcription and translation of a gene in *E. coli*. DNA, mRNAs, and the ribosomes translating individual mRNA molecules are visible. The nascent polypeptide chains being synthesized on the ribosomes are not visible as they fold into their three-dimensional configuration during synthesis.

TRANSCRIPTION AND RNA PROCESSING IN EUKARYOTES

Although the overall process of RNA synthesis is similar in prokaryotes and eukaryotes, the process is considerably more complex in eukaryotes. In eukaryotes, RNA is synthesized in the nucleus, and RNAs that encode proteins must be transported to the cytoplasm for translation on ribosomes. Prokaryotic mRNAs often contain the coding regions of two or more genes; such mRNAs are said to be multigenic. On the other hand, many of the eukaryotic transcripts that have been characterized contain the coding region of a single gene (are monogenic). However, recent evidence indicates that up to one-fourth of the transcription units in the small worm *Caenorhabditis elegans* may be multigenic. Clearly, eukaryotic mRNAs may be either monogenic or multigenic. Three different RNA polymerases are present in eukaryotes; each enzyme catalyzes the transcription of a specific class of genes. Moreover, in eukaryotes, the majority of the primary transcripts of genes that encode polypeptides undergo three major modifications prior to their transport to the cytoplasm for translation (Figure 11.15).

1. 7-Methyl guanosine caps are added to the 5′ ends of the primary transcripts.
2. Poly(A) tails are added to the 3′ ends of the transcripts, which are generated by cleavage rather than termination of chain extension.
3. When present, noncoding intron sequences are spliced out of transcripts.

The **5′ cap** on most eukaryotic mRNAs is a 7-methyl guanosine residue joined to the initial nucleoside of

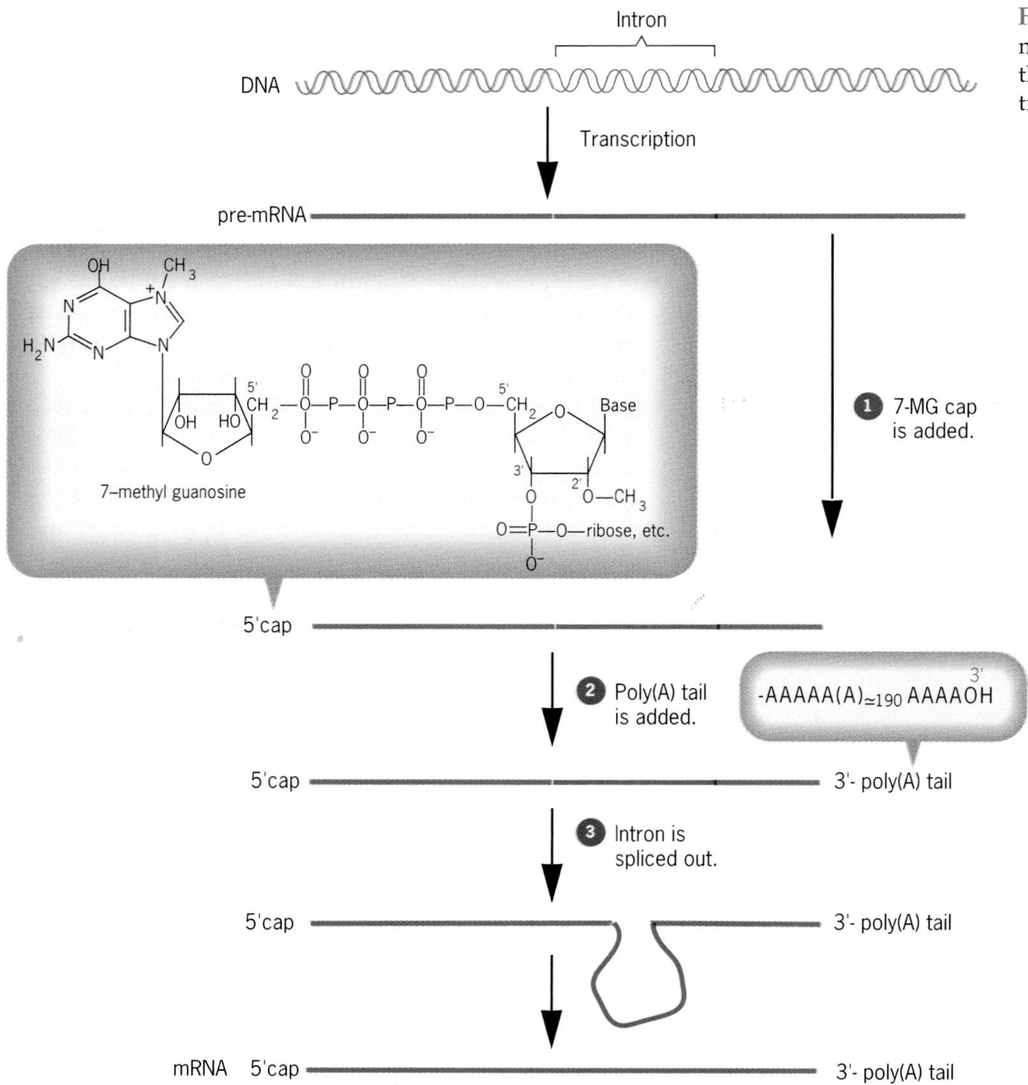

Figure 11.15 In eukaryotes, most gene transcripts undergo three different types of post-transcriptional processing.

the transcript by a 5′-5′ phosphate linkage. The **3′ poly(A) tail** is a polyadenosine tract 20 to 200 nucleotides long.

In eukaryotes, the population of primary transcripts in a nucleus is called **heterogeneous nuclear RNA (hnRNA)** because of the large variation in the sizes of the RNA molecules present. Major portions of these hnRNAs are noncoding intron sequences, which are excised from the primary transcripts and degraded in the nucleus. Thus much of the hnRNA actually consists of pre-mRNA molecules undergoing various processing events before leaving the nucleus. Also, in eukaryotes, RNA transcripts are coated with RNA-binding proteins during or immediately after their synthesis. These proteins protect gene transcripts from degradation by ribonucleases, enzymes that degrade RNA molecules, during processing and transport to the cytoplasm. The average half-life of a gene transcript in eukaryotes is about five hours, in contrast to an average half-life of less than five minutes in *E. coli*. This enhanced stability of gene transcripts in eukaryotes is provided, at least in part, by their presence in complexes with RNA-binding proteins.

Three RNA Polymerases/Three Sets of Genes

Whereas a single RNA polymerase catalyzes all transcription in *E. coli*, eukaryotes ranging in complexity from the single-celled yeasts to humans contain three different RNA polymerases. All three eukaryotic enzymes, designated **RNA polymerases I, II,** and **III**, are more complex, with 10 or more subunits, than the *E. coli* RNA polymerase. Moreover, unlike the *E. coli* enzyme, all three eukaryotic RNA polymerases require the assistance of other proteins called **transcription factors** in order to initiate the synthesis of RNA chains.

The key features of the three eukaryotic RNA polymerases are summarized in Table 11.1. RNA polymerase I is located in the nucleolus, a distinct region of the nucleus where rRNAs are synthesized and combined with ribosomal proteins. RNA polymerase I catalyzes the synthesis of all ribosomal RNAs except the small 5S rRNA. RNA polymerase II transcribes nuclear genes that encode proteins and perhaps other genes specifying hnRNAs. RNA polymerase III catalyzes the synthesis of the transfer RNA molecules, the 5S rRNA molecules, and small nuclear RNAs.

The three RNA polymerases exhibit very different sensitivities to inhibition by α-amanitin, a metabolic poison produced by the mushroom *Amanita phalloides*. Whereas RNA polymerase I activity is insensitive to α-amanitin, RNA polymerase II activity is inhibited completely by low concentrations of α-amanitin, and RNA polymerase III exhibits an intermediate level of sensitivity to this drug. Thus α-amanitin can be used to determine which RNA polymerase catalyzes the transcription of a particular gene.

Initiation of RNA Chains

Unlike their prokaryotic counterparts, eukaryotic RNA polymerases cannot initiate transcription by themselves. All three eukaryotic RNA polymerases require the assistance of protein transcription factors to start the synthesis of an RNA chain. Indeed, these transcription factors must bind to a promoter region in DNA and form an appropriate initiation complex before RNA polymerase will bind and initiate transcription. Different promoters and transcription factors are utilized by the RNA polymerases I, II, and III. In this section, we focus on the initiation of pre-mRNA synthesis by RNA polymerase II, which transcribes the vast majority of eukaryotic genes.

In all cases, the initiation of transcription involves the formation of a locally unwound segment of DNA, providing a DNA strand that is free to function as a template for the synthesis of a complementary strand of RNA (Figure 11.9). The formation of the locally unwound segment of DNA required to initiate transcription involves the interaction of several transcription factors with specific ***cis*-acting sequences** in the promoter for the transcription unit. The promoters recognized by RNA polymerase II consist of short conserved elements, or modules, located upstream from the transcription startpoint (Figure 11.16). The conserved element closest to the transcription start site

TABLE 11.1
Characteristics of the Three RNA Polymerases of Eukaryotes

Enzyme	*Location*	*Products*	*Sensitivity to α-Amanitin*
RNA polymerase I	Nucleolus	Ribosomal RNAs, excluding 5S rRNA	No sensitivity
RNA polymerase II	Nucleus	Nuclear Pre-mRNAs	Complete sensitivity
RNA polymerase III	Nucleus	tRNAs, 5S rRNA, and other small nuclear RNAs	Intermediate sensitivity

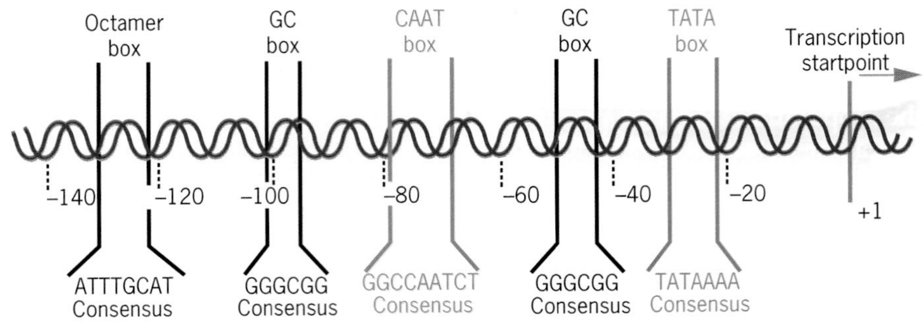

Figure 11.16 Structure of a promoter recognized by RNA polymerase II. The TATA and CAAT boxes are located at about the same positions in the promoters of most nuclear genes encoding proteins. The GC and octamer boxes may be present or absent; when present, they occur at many different locations, either singly or in multiple copies. The sequences shown here are the consensus sequences for each of the promoter elements. The conserved promoter elements are shown at their locations in the mouse thymidine kinase gene.

(position +1) is called the **TATA box**; it has the consensus sequence TATAAAA (reading 5′ to 3′ on the nontemplate strand) and is centered at about position –30. The TATA box plays an important role in positioning the transcription startpoint. The second conserved element is called the **CAAT box**; it usually occurs near position –80 and has the consensus sequence GGCCAATCT. Two other conserved elements, the **GC box**, consensus GGGCGG, and the **octamer box**, consensus ATTTGCAT, often are present in RNA polymerase II promoters; but they may occur at different positions and in one or multiple copies. The CAAT, GC, and octamer sequences influence the efficiency of a promoter in initiating transcription. All active promoters contain at least one of these elements, but all three elements are not present in every promoter.

The initiation of transcription by RNA polymerase II requires the assistance of several **basal transcription factors**. Still other transcription factors modulate the efficiency of initiation (Chapter 22). The basal transcription factors must interact with promoters in the correct sequence to initiate transcription effectively (Figure 11.17). Each basal transcription factor is denoted **TFIIX (Transcription Factor for polymerase II,** where **X is a letter identifying the individual factor).** The positions at which these transcription factors bind to the promoter DNA have been established in part by determining which segments of DNA are protected from degradation by nucleases when a specific factor has been added to the initiation complex.

TFIID is the first to interact with the promoter; it contains a TATA-binding protein (TBP) and several small TBP-associated proteins (Figure 11.17). Next, TFIIA joins the complex, followed by TFIIB. TFIIF first associates with RNA polymerase II, and then TFIIF and RNA polymerase II join the transcription initiation complex together. TFIIF contains two subunits, one of which has DNA-unwinding activity. Thus TFIIF probably catalyzes the localized unwinding of the DNA double helix required to initiate transcription. TFIIE then joins the initiation complex, binding to the DNA downstream from the transcription startpoint. Two other factors, TFIIH and TFIIJ, join the complex after TFIIE, but their locations in the complex are unknown. TFIIH contains an enzyme activity that phosphorylates RNA polymerase II. One hypothesis is that the phosphorylation of RNA polymerase II by TFIIF releases the enzyme from the initiation complex at the promoter so that it can move downstream and mediate RNA chain extension. Additional experiments must be done to test the validity of this hypothesis, but we do know that RNA polymerase must be disengaged from the initiation complex before it can catalyze processive chain elongation.

Similar transcription factor/promoter complexes are required to initiate transcription by RNA polymerases I and III. Polymerases I and III use some of the same transcription factors as polymerase II and some factors specific to each enzyme. The promoters of genes transcribed by polymerases I and III are quite different from the promoters utilized by RNA polymerase II. However, these promoters often contain some of the same *cis*-acting elements as the promoters recognized by RNA polymerase II. Interestingly, the promoters of some of the genes transcribed by RNA polymerase III are located within the transcription units, downstream from the transcription startpoints, rather than upstream as in units transcribed by RNA polymerase II.

RNA Chain Elongation and the Addition of 5′ Methyl Guanosine Caps

Once eukaryotic RNA polymerases have been released from their initiation complexes, they catalyze RNA chain elongation by the same mechanism as the

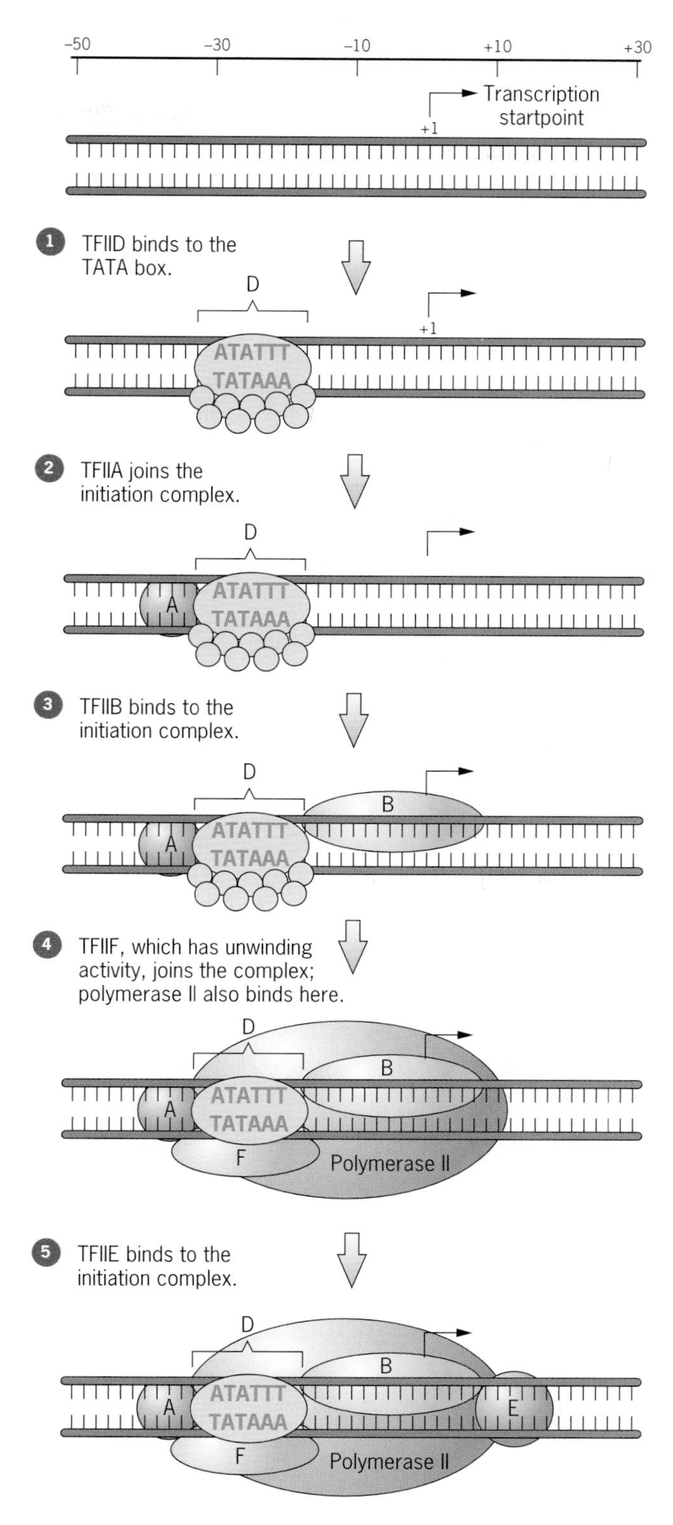

Figure 11.17 The initiation of transcription by RNA polymerase II requires the formation of a basal transcription initiation complex at the promoter region. The assembly of this complex begins when TFIID, which contains the TATA-binding protein (TBP), binds to the TATA box. The other transcription factors and RNA polymerase II join the complex in the sequence shown.

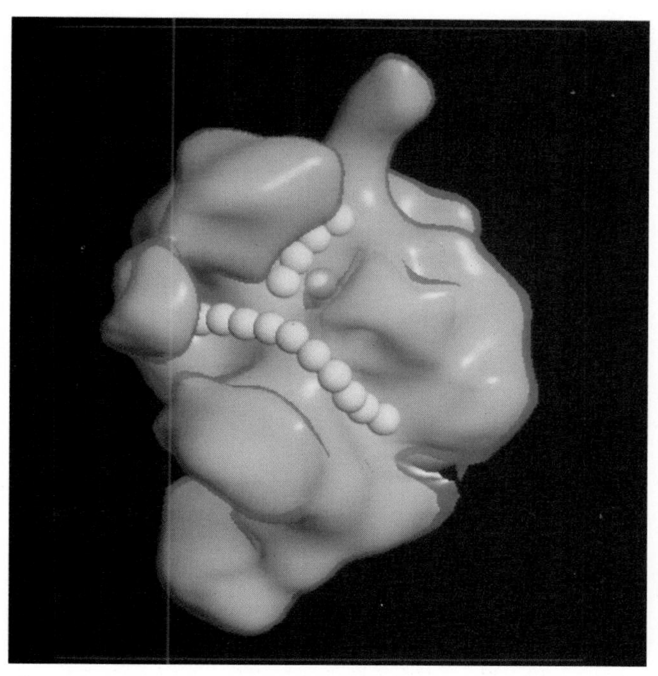

Figure 11.18 Model of the structure of RNA polymerase II in the yeast *S. cerevisiae*. Surface grooves on the enzyme are thought to be binding sites for DNA and the growing RNA chain. One groove about 2.5 nm wide and 1 nm deep is the putative DNA (the ten-beads-long chain) binding site; another groove about 1.5 nm wide and 2 nm deep could bind the growing RNA chain (the four-beads-long chain).

RNA polymerases of prokaryotes (Figures 11.8 and 11.9). Other details of the elongation process in eukaryotes are still uncertain. The exact structures of eukaryotic RNA polymerases are still unknown, but it is known that they are more complex than their prokaryotic counterparts. RNA polymerase II of the yeast *S. cerevisiae* contains 13 or more polypeptides, at least 10 of which are different gene products. Structural analysis of this enzyme reveals the presence of surface grooves thought to be DNA and RNA binding sites (Figure 11.18).

Early in the elongation process, the 5' ends of eukaryotic pre-mRNAs are modified by the addition of 7-methyl guanosine (7-MG) caps. These 7-MG caps are added when the growing RNA chains are only about 30 nucleotides long (Figure 11.19). The 7-MG cap contains an unusual 5'-5' triphosphate linkage (see Figure 11.15) and two or more methyl groups. These 5' caps are added post-transcriptionally by the biosynthetic pathway shown in Figure 11.19. The 7-MG caps are recognized by protein factors involved in the initiation of translation (Chapter 12) and also help protect the growing RNA chains from degradation by nucleases.

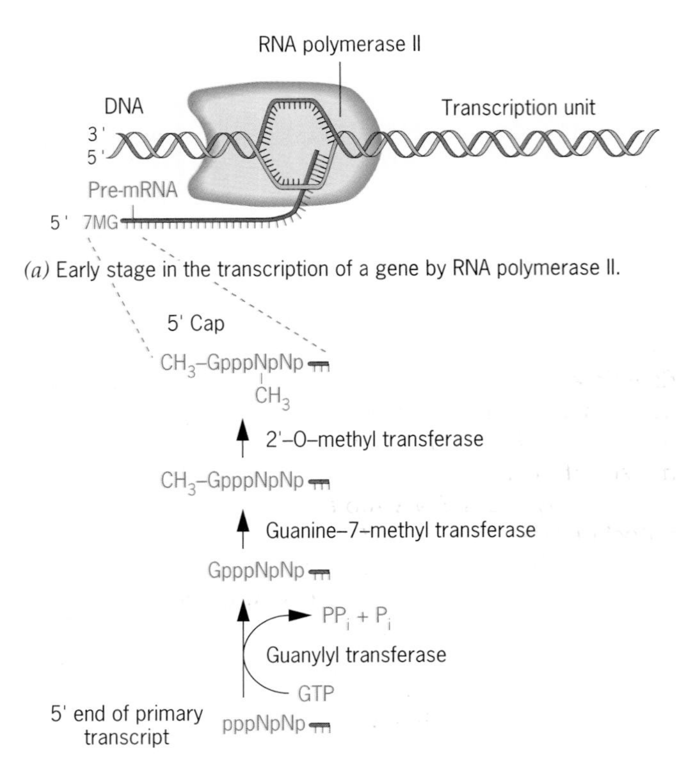

(*a*) Early stage in the transcription of a gene by RNA polymerase II.

(*b*) Pathway of biosynthesis of the 7MG cap.

Figure 11.19 7-Methyl guanosine (7MG) caps are added to the 5' ends of pre-mRNAs shortly after the elongation process begins.

Termination by Chain Cleavage and the Addition of 3' Poly(A) Tails

The 3' ends of RNA transcripts synthesized by RNA polymerase II are produced by endonucleolytic cleavage of the primary transcripts rather than by the termination of transcription (Figure 11.20). The actual transcription termination events often occur at multiple sites that are located 1000 to 2000 nucleotides downstream from the site that will become the 3' end of the mature transcript. That is, transcription proceeds beyond the site that will become the 3' terminus, and the distal segment is removed by endonucleolytic cleavage. The cleavage event that produces the 3' end of a transcript usually occurs at a site 11 to 30 nucleotides downstream from a conserved sequence, consensus AAUAAA, located near the end of the transcription unit. After cleavage, the enzyme **poly(A) polymerase** adds poly(A) tails, tracts of adenosine monophosphate residues about 200 nucleotides long, to the 3' ends of the transcripts (Figure 11.20). The addition of poly(A) tails to eukaryotic mRNAs is called **polyadenylation**.

The formation of poly(A) tails on transcripts requires a **specificity component** that recognizes the AAUAAA sequences in RNAs, binds to them, and directs both the cleavage and the polyadenylation reactions. The specificity component, the endonuclease, and the poly(A) polymerase are present in a multimeric complex that catalyzes both cleavage and polyadenylation in tightly coupled reactions. The poly(A) tails of eukaryotic mRNAs enhance their stability and play an important role in their transport from the nucleus to the cytoplasm.

In contrast to RNA polymerase II, both RNA polymerase I and III respond to discrete termination signals. RNA polymerase I terminates transcription in response to an 18-nucleotide-long sequence that is recognized by an associated terminator protein. RNA polymerase III reponds to a termination signal that is similar to the rho-independent terminator in *E. coli* (Figure 11.13). However, many details of the mechanism of termination of polymerase III-catalyzed transcription are still unknown.

RNA Editing: Altering the Information Content of mRNA Molecules

According to the central dogma of molecular biology, genetic information flows from DNA to RNA to protein during gene expression. Normally, the genetic in-

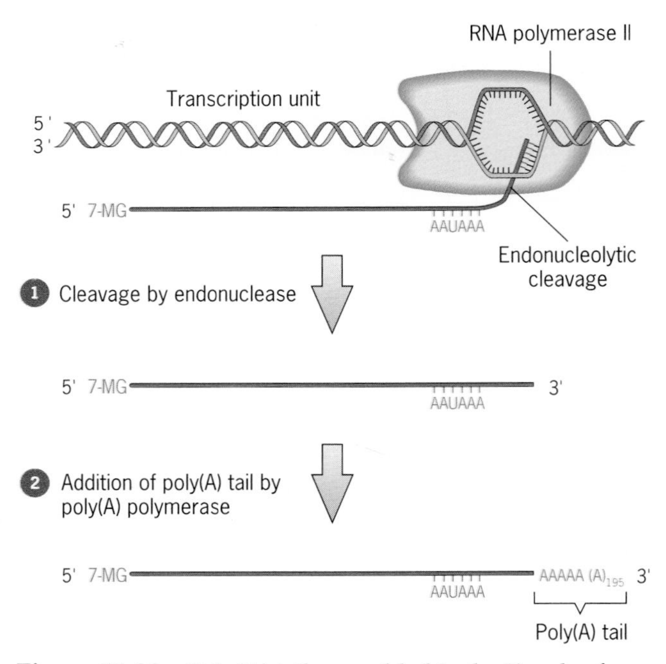

Figure 11.20 Poly(A) tails are added to the 3' ends of transcripts by the enzyme poly(A) polymerase. The 3' end substrates for poly(A) polymerase are produced by endonucleolytic cleavage of the transcript downstream from a polyadenylation signal, which has the consensus sequence AAUAAA.

formation is not altered in the mRNA intermediary. However, the discovery of **RNA editing** has shown that exceptions do occur. RNA editing processes alter the information content of gene transcripts in two ways: (1) by changing the structures of individual bases, and (2) by inserting or deleting uridine monophosphate residues.

The first type of RNA editing, which results in the substitution of one base for another base, is rare. This type of editing was discovered in studies of the apolipoprotein-B (*apo-B*) genes and mRNAs in rabbits and humans. Apolipoproteins are blood proteins that transport certain types of fat molecules in the circulatory system. In the liver, the *apo-B* mRNA encodes a large protein 4563 amino acids long. In the intestine, the *apo-B* mRNA directs the synthesis of a protein only 2153 amino acids long. Here, a C residue in the pre-mRNA is converted to a U, generating an internal UAA translation–termination codon, which results in the truncated apolipoprotein (Figure 11.21). UAA is one of three codons that terminate polypeptide chains during translation. If a UAA codon is produced within the coding region of an mRNA, it will prematurely terminate the polypeptide during translation, yielding an incomplete gene product. The C → U conversion is catalyzed by a sequence-specific RNA-binding protein with an activity that removes amino groups from cytosine residues. A similar example of RNA editing has been documented for an mRNA specifying a protein (the glutamate receptor) present in rat brain cells. More extensive mRNA editing of the C → U type occurs in the mitochondria of plants, where most of the gene transcripts are edited to some degree. Mitochondria have their own DNA genomes and protein-synthesizing machinery (Chapter 18). In some transcripts present in plant mitochondria, most of the C's are converted to U residues.

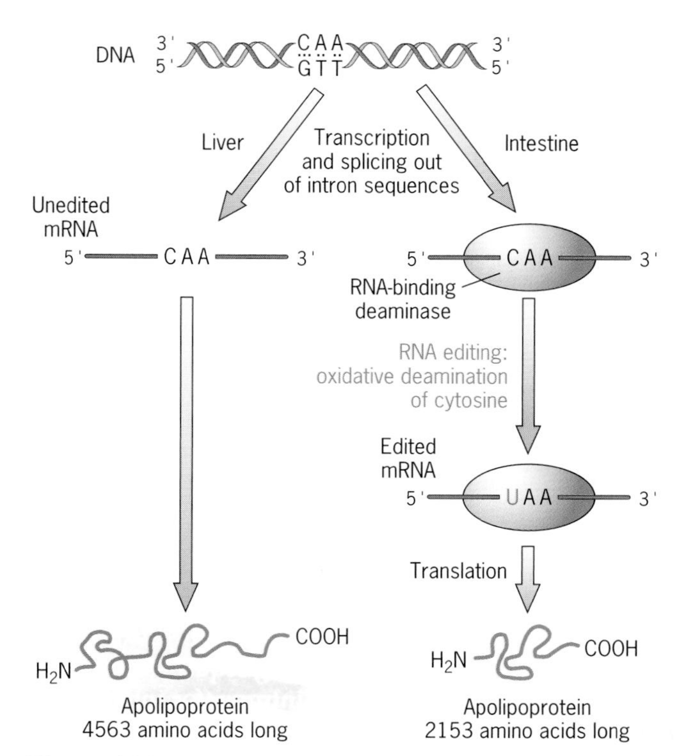

Figure 11.21 Editing of the apolipoprotein-B mRNA in the intestines of mammals.

A second, more complex type of RNA editing occurs in the mitochondria of trypanosomes (a group of flagellated protozoa that cause sleeping sickness in humans). In this case, uridine monophosphate residues are inserted (occasionally deleted) into gene transcripts, causing major changes in the polypeptides specified by the mRNA molecules. The trypanosome mitochondrial type of editing is illustrated in Figure 11.22, which shows the sequences of the DNA coding strand, the pre-edited RNA transcript, and the edited mRNA of the cytochrome *b* gene of *Leishmania tarentolae*. (Cytochrome b is a protein involved in the production of chemical energy.) This RNA editing process is mediated by **guide RNAs** transcribed from distinct mitochondrial genes. The guide RNAs contain sequences that are partially complementary to the pre-mRNAs to be edited. Pairing between the guide RNAs and the pre-mRNAs results in gaps with unpaired A residues in the guide RNAs. The guide RNAs serve as templates for editing, as U's are inserted in the gaps in pre-mRNA molecules opposite the A's in the guide RNAs (Figure 11.22).

In some cases, two or more different guide RNAs participate in the editing of a single pre-mRNA. The guide RNAs not only provide the templates for the editing process, but also donate the uridine monophosphates that are inserted into the pre-mRNAs. The guide RNAs have oligo(U) 3' tails, 5 to 24 nucleotides long. These U residues are inserted into pre-mRNAs during the editing process by a complex two-step mechanism (Figure 11.23). In the first step, the free 3'-OH on the guide RNA attacks and breaks an internal ester bond in the pre-mRNA, forming a covalent linkage with the 3' portion of the cleaved pre-mRNA. In the second step, the free 3'-OH on the 5' portion of the cleaved pre-mRNA attacks an ester linkage between two U residues at the terminus of the oligo(U) tail of the guide RNA, re-forming the pre-mRNA molecule, but now with a U residue inserted at the site of the initial attack.

Why do these RNA editing processes occur? Why are the final nucleotide sequences of these mRNAs not specified by the sequences of the mitochondrial genes as they are in most nuclear genes? As yet, answers to these interesting questions are purely speculative. Trypanosomes are primitive single-celled eukaryotes that diverged from other eukaryotes early in evolu-

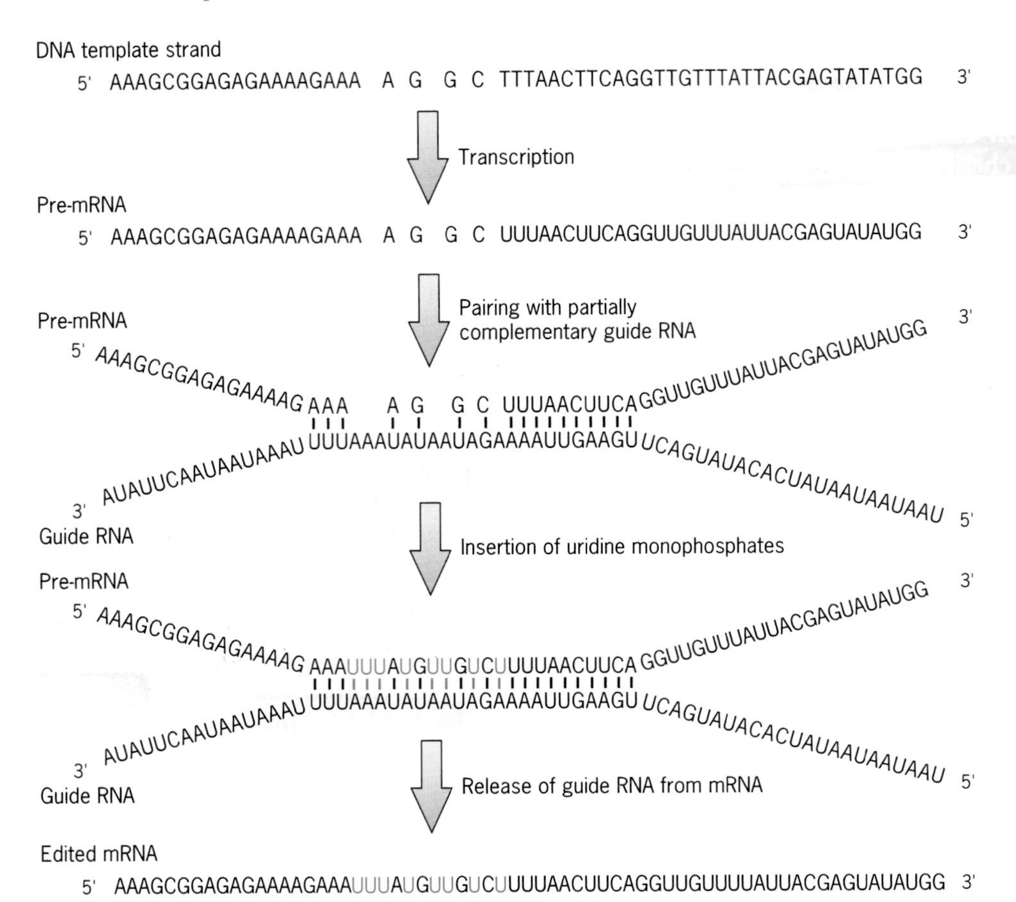

Figure 11.22 Editing of the mitochondrial cytochrome b pre-mRNA in the trypanosome *Leishmania tarentolae*. The uridine monophosphate residues (U) that are inserted in the gaps in the pre-mRNA during the editing process are highlighted by orange shading. Base-pairing between the pre-mRNA and the guide RNA is represented by vertical lines between the two molecules.

tion. Some evolutionists have speculated that RNA editing was common in ancient cells, where many reactions are thought to have been catalyzed by RNA molecules instead of proteins. Another view is that RNA editing is a primitive mechanism for altering patterns of gene expression. For whatever reason, RNA editing plays a major role in the expresson of genes in the mitochondria of trypanosomes and plants.

Key Points: **Three different RNA polymerases are present in eukaryotes, and each polymerase transcribes a distinct set of genes. Eukaryotic gene transcripts usually undergo three major modifications: (1) the addition of 7-methyl guanosine caps to 5' termini, (2) the addition of poly(A) tails to 3' ends, and (3) the excision of noncoding intron sequences. The information content of some eukaryotic transcripts is altered by RNA editing, which changes the nucleotide sequences of transcripts prior to their translation.**

INTERRUPTED GENES IN EUKARYOTES: EXONS AND INTRONS

Most of the well-characterized genes of prokaryotes consist of continuous sequences of nucleotide pairs, which specify colinear sequences of amino acids in the polypeptide gene products. However, in 1977, molecular analyses of three eukaryotic genes yielded a major surprise. Studies of mouse and rabbit β-globin (one of two different proteins in hemoglobin) genes and the chicken ovalbumin (an egg storage protein) gene revealed that they contain noncoding sequences intervening between coding sequences. Soon thereafter, noncoding sequences were shown to occur within many genes in a number of eukaryotic species. The coding sequences of these interrupted genes are called **exons** (for **ex**pressed sequences); the noncoding sequences between the coding sequences are called **introns** (for **int**e**r**vening sequences). Geneticists now know that introns interrupt most, but not all, eukaryotic genes.

1 Cleavage of an ester bond in pre-mRNA and formation of an ester bond between the guide RNA and the 3' portion of the pre-mRNA.

Pre-mRNA

5' O—P—O 3'

attack U—OH U U 5' Guide RNA

5' OH

U— 3' U U 5' Guide RNA

2 Cleavage of the ester bond between the two terminal U's of the guide RNA, reforming the pre- mRNA, but with a U inserted at the site of the initial cleavage in Step 1.

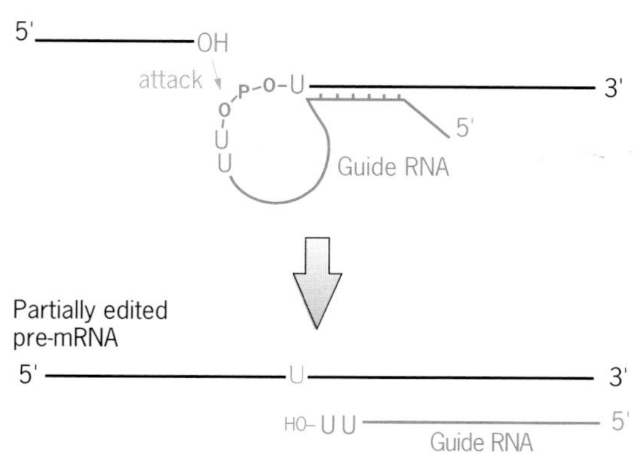

Figure 11.23 The two-step mechanism by which uridine monophosphate residues are inserted into pre-mRNA molecules during their editing in mitochondria of trypanosomes.

Early Evidence for Noncoding Sequences in Eukaryotic Genes

In the mid-1970s, researchers began to isolate or "clone" and characterize individual genes (Chapter 19). Comparisons of the structures of eukaryotic genes and the mRNA molecules produced by these genes yielded a totally unexpected result. The genes were shown to contain nucleotide sequences that were not present in the mRNA products of these genes. In October through December 1977, evidence for interrupted genes was reported from three laboratories. Philip Leder and colleagues reported evidence for a nucleotide sequence in a mouse β-globin gene that was not present in the corresponding β-globin mRNA molecule. One month later, Pierre Chambon and co-workers published evidence for noncoding sequences in the chicken ovalbumin gene, and, the following month, Alan Jeffries and Richard Flavell reported the presence of two introns in a rabbit β-globin gene. These initial reports were rapidly followed by numerous publications from these and other laboratories documenting the presence of introns in a large number of eukaryotic genes.

Some of the earliest evidence for introns in mammalian β-globin genes resulted from the visualization of genomic DNA-mRNA hybrids by electron microscopy. DNA-RNA duplexes are more stable than DNA double helices. Thus, if DNA double helices are partially denatured and are incubated with homologous RNA molecules under the appropriate conditions, the RNA strands will hybridize with the complementary DNA strands, displacing the equivalent DNA strands (Figure 11.24). The resulting DNA-RNA hybrid structures will contain single-stranded regions of DNA called **R-loops**, where RNA molecules have displaced DNA strands to form DNA-RNA duplex regions. These R-loop structures can be visualized directly by electron microscopy. Thus R-loop hybridiza-

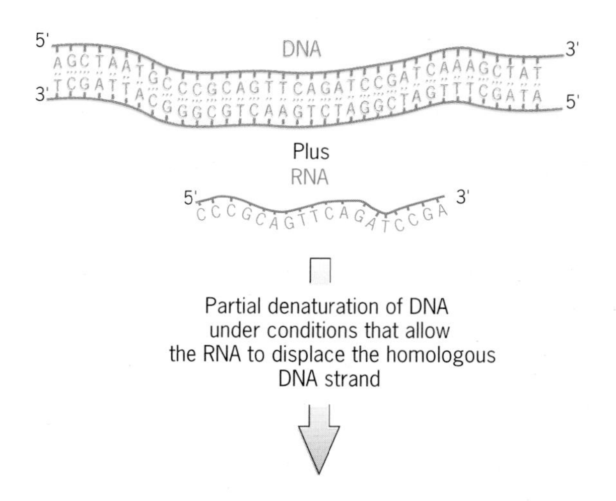

The RNA pairs with the complementary strand of DNA forming a DNA–RNA duplex, leaving a single-stranded region of DNA called an R–loop.

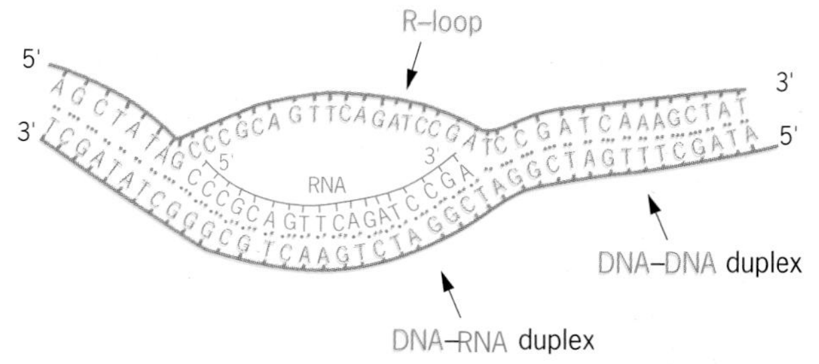

Figure 11.24 The technique of R-loop hybridization.

tion and electron microscopy provide a powerful tool that has been used extensively in studies of eukaryotic gene structure.

When Philip Leder and colleagues hybridized purified mouse β-globin mRNA to a DNA molecule that contained the mouse β-globin gene, they observed two R-loops separated by a loop of double-stranded DNA (Figure 11.25*a*). Their result demonstrated the presence of a sequence of nucleotide pairs in the middle of the β-globin gene that is not present in β-globin mRNA and, therefore, does not code for amino acids in the β-globin polypeptide. When Leder and coworkers repeated the R-loop experiments using purified β-globin gene transcripts isolated from nuclei and believed to be primary gene transcripts or pre-mRNA molecules, in place of cytoplasmic β-globin mRNA, they observed only one R-loop (Figure 11.25*b*). This result indicated that the primary transcript contains the complete structural gene sequence, including both exons and introns. Together, the R-loop results obtained with cytoplasmic mRNA and nuclear pre-mRNA demonstrate that the intron sequence is excised and the exons sequences are spliced together during processing events that convert the primary transcript to the mature mRNA.

Leder and colleagues quickly verified their interpretation of the R-loop results by sequencing a segment of the mouse β-globin gene that spanned the junction between this large intron and the exon preceding it. When they used the established codon assignments to compare the nucleotide sequence of the mouse β-globin gene with the known amino acid sequence of mouse β-globin, the presence of the noncoding sequence was clear (Figure 11.26). If there were no intron at this position in the mouse β-globin gene, the nucleotide sequence predicts that amino acid residues 105 through 110 would be Val-Ser-Leu-Met-Gly-Thr. However, the amino acid sequence of mouse β-globin has been determined experimentally, and amino acids 105 through 110 are Leu-Leu-Gly-Asn-Met-Ile. In a later paper, Leder and coworkers sequenced the rest of the mouse β-globin gene and showed that amino

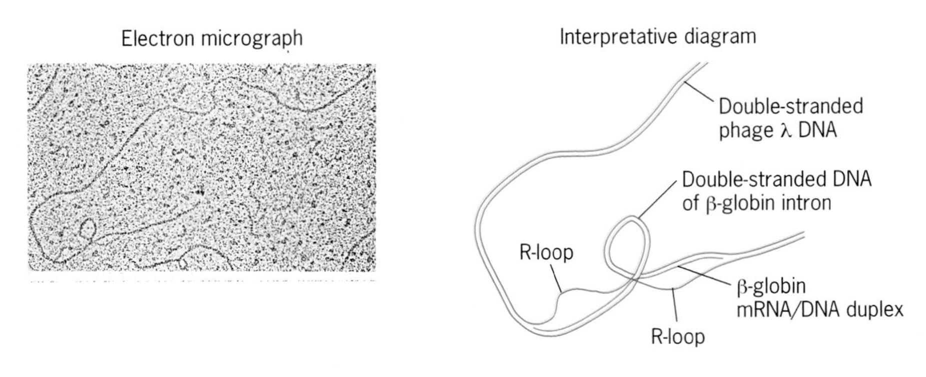

(*a*) R-loops formed by β-globin mRNA.

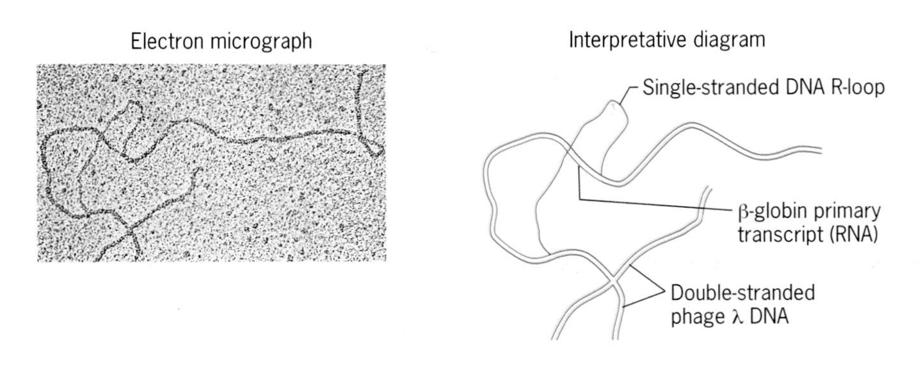

(*b*) R-loop formed by β-globin primary transcript (pre-mRNA).

Figure 11.25 R-loop evidence for an intron in the mouse β-globin gene. (*a*) When mouse β-globin genes and mRNAs were hybridized under R-loop conditions, two R-loops were observed in the resulting DNA-RNA hybrids. (*b*) When primary transcripts or pre-mRNAs of mouse β-globin genes were used in the R-loop experiments, a single R-loop was observed. These results demonstrate that the intron sequence is present in the primary transcript, but is removed during the processing of the primary transcript to produce the mature mRNA.

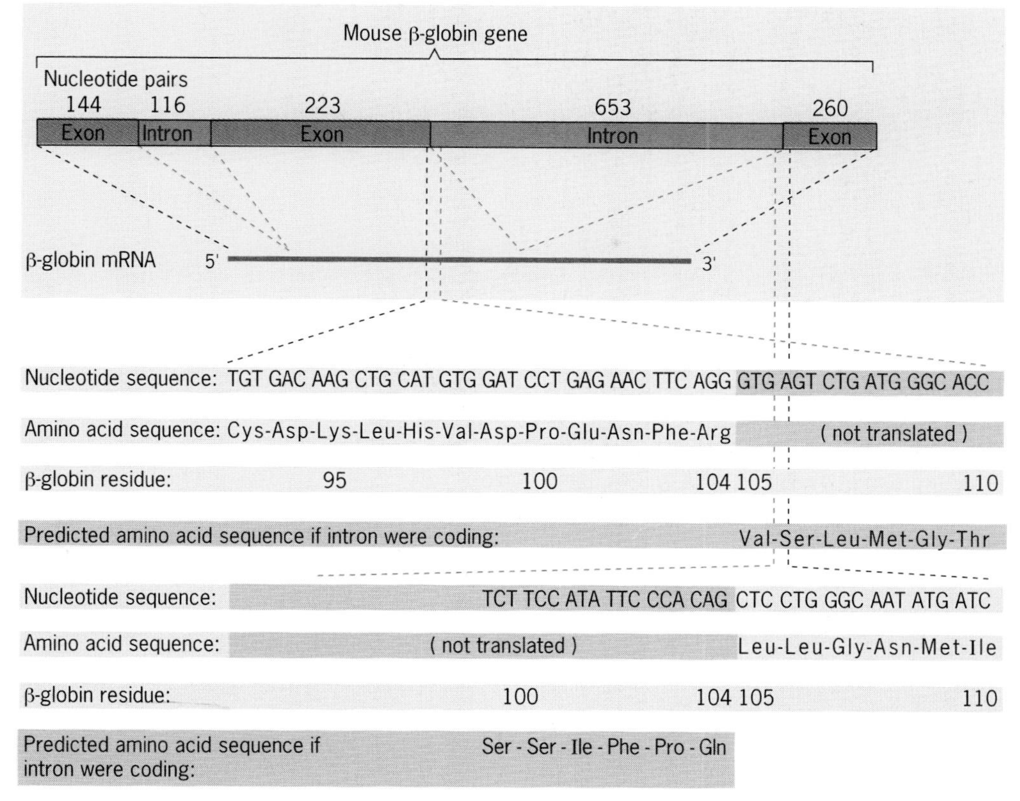

Figure 11.26 Structure of the mouse major β-globin gene. The three exons (coding sequences) are shown in purple, and the two introns (noncoding sequences) are shown in red. The relationship of the exon sequences in the gene to mRNA sequences is indicated by the dashed diagonal lines. The nucleotide sequences of the nontranscribed DNA strand are shown for two exon–intron junctions. These nucleotide sequences are the same as the mRNA sequences, but T replaces U.

acids 105 through 110 are encoded by the first six nucleotide-pair triplets in the exon distal to this intron. The large β-globin gene intron characterized in these early studies is 653 nucleotide pairs long. Shortly after the discovery of the large intron, Leder and colleagues found that the mouse β-globin gene contains a second, smaller intron, 116 nucleotide pairs long (Figure 11.26).

Studies by Tom Maniatis, Richard Flavell, and many others have shown that the structure of the human β-globin gene is very similar to that of the mouse β-globin gene; both contain two introns that separate three exons. Each of the two human α-globin genes also contains two introns. In fact, the human genes encoding the embryonic α-like (ζ) and β-like (ϵ) globin chains, the fetal β-like (γ) globin chain, and the adult minor β-like (δ) globin chain, all contain two introns and three exons. Moreover, all of the human genes encoding α-like globins have the introns at the same positions (separating codon 31 from 32 and codon 99 from 100), as do all the human genes encoding β-like globins (separating codon 30 from 31 and codon 104 from 105). Thus intron positions have been highly conserved in the human globin genes.

While Philip Leder, Richard Flavell, and their coworkers were demonstrating the presence of two introns in mammalian β-globin genes, Pierre Chambon, Bert W. O'Malley, and their colleagues were performing similar experiments on the chicken ovalbumin gene. When the transcribed strand of the chicken ovalbumin gene was hybridized to the ovalbumin mRNA and visualized by electron microscopy, seven single-stranded DNA loops were observed (Figure 11.27), suggesting that the ovalbumin gene contains seven introns. Subsequent experiments, including direct comparisons of the sequences of the gene and the mRNA, showed conclusively that the chicken ovalbumin gene contains seven introns separating eight exons. In the case of the ovalbumin gene, one intron is present in the 5′ transcribed, but untranslated region of the gene. Thus introns are not restricted to the protein coding regions of genes.

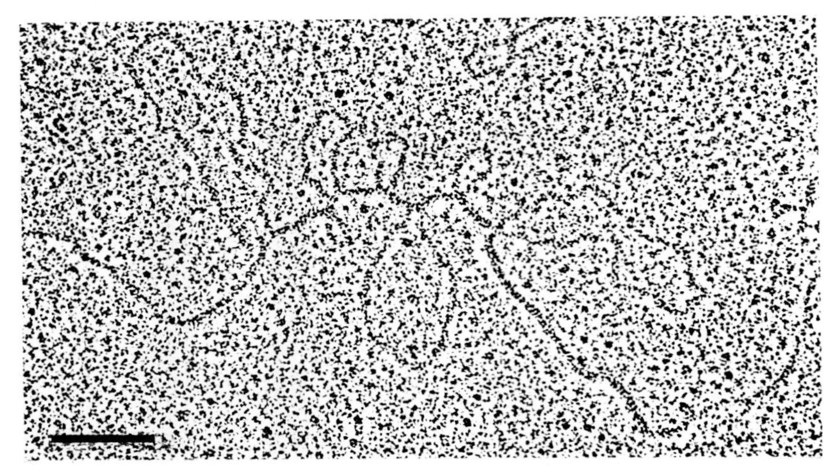

(*a*) Electron micrograph of ovalbumin DNA–mRNA heteroduplex.

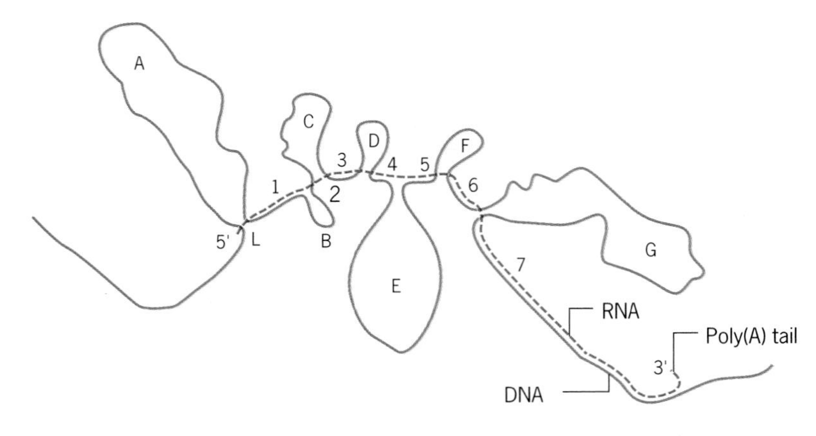

(*b*) Interpretative diagram of the DNA–mRNA heteroduplex.

Figure 11.27 Evidence for seven introns in the chicken ovalbumin gene. The mRNA molecule is represented by a red line; the DNA strand is shown as a green line. Intron sequences are single-stranded DNA loops labeled A through G.

Some Very Large Eukaryotic Genes

Subsequent to the pioneering studies on the mammalian globin genes and the chicken ovalbumin gene, noncoding introns have been demonstrated in a large number of eukaryotic genes. In fact, interrupted genes are much more common than uninterrupted genes in higher animals and plants. For example, the rat serum albumin gene contains 13 introns, and the *Xenopus laevis* gene that encodes vitellogenin A2 (which ends up as egg yolk protein) contains 33 introns. The chicken 1α2 collagen gene contains at least 50 introns, with many of the introns being only 45 or 54 nucleotide pairs long. This gene spans 37,000 nucleotide pairs, but gives rise to an mRNA molecule only about 4600 nucleotides long. Other genes contain relatively few introns, but some of the introns are very large. For example, the *Ultrabithorax* (*Ubx*) gene of *Drosophila* contains an intron that is approximately 70,000 nucleotide pairs in length. The largest gene characterized to date is the human *DMD* gene, which causes Duchenne muscular dystrophy when rendered nonfunctional by mutation (Chapter 20). The *DMD* gene spans 2.5 million nucleotide pairs and contains 78 introns.

The first evidence for introns in plant genes was the demonstration that the gene that encodes phaseolin (a major storage protein) in the French bean contains three introns. Subsequent studies have shown that most of the genes of higher plants—like those of higher animals—contain multiple introns.

The only structural features that seem to be shared by different introns are the dinucleotide sequences at their ends. The primary transcripts of genes almost always begin introns with the sequence GU (5′) and end them with the sequence AG (3′). These consensus sequences at intron–exon junctions are important in the mechanism by which introns are spliced out of primary transcripts.

Although introns are present in most genes of higher animals and plants, they are not essential because not all such genes contain introns. The sea urchin histone genes and four *Drosophila* heat shock genes were among the first animal genes shown to lack introns. We now know that many genes of higher animals and plants lack introns.

Introns: Biological Significance?

At present, scientists know relatively little about the biological significance of the exon–intron structure of eukaryotic genes. Introns are highly variable in size, ranging from a few nucleotide pairs to thousands of nucleotide–pairs in length. This fact has led to speculation that introns may play a role in regulating gene expression. The transcription of genes with large introns will take longer than the transcription of genes with small introns or no introns. Thus the presence of large introns in genes would be expected to decrease the rate of transcript accumulation in a cell. The fact that introns accumulate new mutations much more rapidly than exons indicates that the specific nucleotide-pair sequences of introns, excluding the ends, are not very important.

In some cases, the different exons of genes encode different functional domains of the protein gene products. This is most apparent in the case of the genes encoding heavy and light antibody chains (Chapter 24). In the case of the mammalian globin genes, the middle exon encodes the heme-binding domain of the protein. There has been considerable speculation that the exon-intron structure of eukaryotic genes has resulted from the evolution of new genes by the fusion of uninterrupted (single exon) ancestral genes. If this hypothesis is correct, introns may merely be relics of the evolutionary process.

Alternatively, introns may provide a selective advantage by increasing the rate at which coding sequences in different exons of a gene can reassort by recombination, thus speeding up the process of evolution. In the case of the mitochondrial gene of yeast encoding cytochrome b, the introns contain exons of genes encoding enzymes involved in processing the primary transcript of the gene. Thus different introns may indeed play different roles, and many introns may have no biological significance. Since many eukaryotic genes contain no introns, these noncoding regions clearly are not required for normal gene expression.

Key Points: **Most, but not all, eukaryotic genes are split into coding sequences called exons and noncoding sequences called introns. Some genes contain very large introns; others harbor large numbers of small introns. The biological significance of introns is still open to debate.**

REMOVAL OF INTRON SEQUENCES BY RNA SPLICING

Most nuclear genes that encode proteins in multicellular eukaryotes contain introns. Fewer, but still many, of the genes of unicellular eukaryotes such as the yeasts contain noncoding introns. Rare genes of a few viruses of prokaryotes and of an archebacterium (a primitive bacterium) also contain introns. In the case of these "split" genes, with coding sequences interrupted by noncoding sequences, the primary transcript contains the entire sequence of the gene and the noncoding sequences are spliced out during RNA processing (see Figure 11.15).

For genes that encode proteins, the splicing mechanism must be precise; it must join exon sequences with accuracy to the single nucleotide to assure that codons in exons distal to introns are read correctly (Figure 11.28). Accuracy to this degree would seem to require precise splicing signals, presumably nucleotide sequences within introns and at the exon-intron junctions. However, in the primary transcripts of nuclear genes, the only completely conserved se-

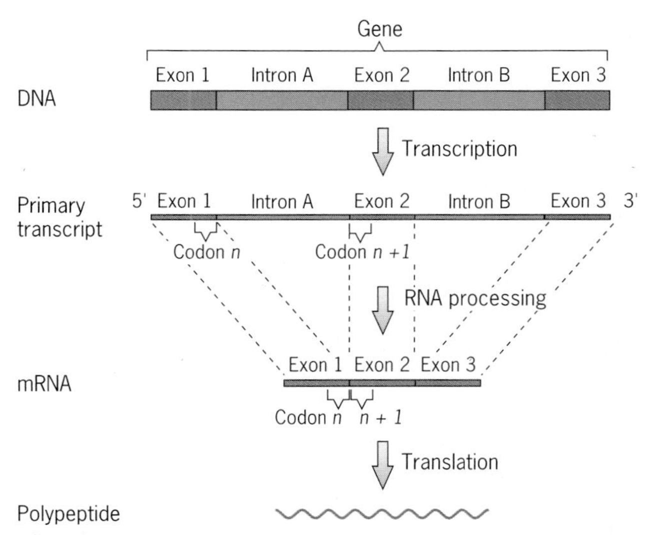

Figure 11.28 The excision of intron sequences from primary transcripts by RNA splicing. The splicing mechanism must be accurate to the single nucleotide to assure that codons in downstream exons are translated correctly to produce the right amino acid sequence in the polypeptide product.

quences of different introns are the dinucleotide sequences at the ends of introns, namely,

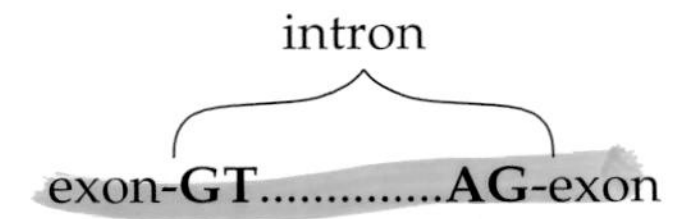

The sequences shown here are for the DNA nontemplate strand (equivalent to the RNA transcript, but with T rather than U). In addition, there are short consensus sequences at the exon–intron junctions. For nuclear genes, the consensus junctions are

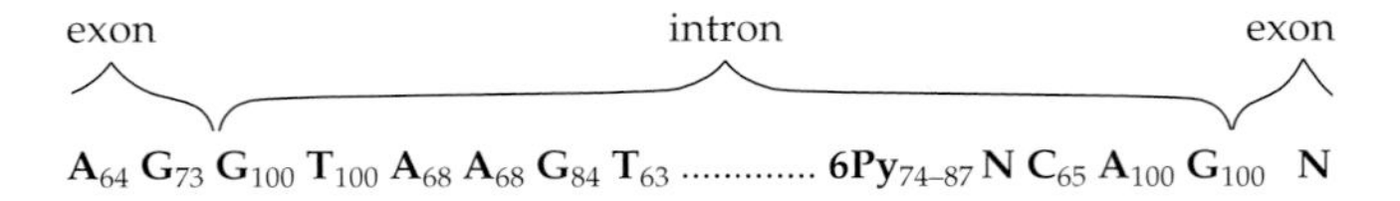

The numerical subscripts indicate the percentage frequencies of the consensus bases at each position; thus a 100 subscript indicates that a base is always present at that position. **N** indicates that any of the four standard nucleotides may be present at the indicated position. The exon–intron junctions are different for tRNA genes and structural genes in mitochondria and chloroplasts, which utilize different RNA splicing mechanisms (see the following section). There is only one short conserved sequence, the **TACTAAC box**, located about 30 nucleotides upstream from the 3' splice site of introns in nuclear genes, and it is rather poorly conserved. The TACTAAC box does exhibit a strong preference for either a purine or a pyrimidine at each site as follows:

$$\mathbf{Py_{80}\ N\ Py_{80}\ Py_{87}\ Pu_{75}\ A_{100}\ Py_{95}}$$

The adenine residue at position six in the TACTAAC box is completely conserved and is known to play a key role in the splicing reaction. With the exception of the terminal dinucleotides and the TACTAAC box, the intron sequences of nuclear genes are highly divergent, apparently random sequences. The introns of genes of mitochondria and chloroplasts also contain conserved sequences, but they are different from those of nuclear genes.

The highy conserved nature of the 5' and 3' splice sites and the TACTAAC box indicates that they play an important role in the process of gene expression. Direct evidence for their importance has been provided by mutations at these sites that cause mutant phenotypes in many different eukaryotes. Indeed, such mutations are sometimes responsible for inherited diseases in humans, such as hemoglobin disorders.

The discovery of noncoding introns in genes stimulated intense interest in the mechanism(s) by which intron sequences are removed during gene expression. The early demonstration that the intron sequences in eukaryotic genes were transcribed along with the exon sequences focused research on the processing of primary gene transcripts. Just as *in vitro* transcription and translation systems were instrumental in elucidating those processes, the key to deciphering RNA splicing events was the development of *in vitro* splicing systems. By using these systems, researchers have shown that there are three totally distinct types of intron excision from RNA transcripts, presented here in the order of increasing complexity, not in the order of importance.

1. The introns of tRNA precursors are excised by precise endonucleolytic cleavage and ligation reactions catalyzed by special splicing endonuclease and ligase activities.
2. The introns of some rRNA precursors are removed autocatalytically in a unique reaction mediated by the RNA molecule itself. (No protein enzymatic activity is involved.)
3. The introns of nuclear pre-mRNA (hnRNA) transcripts are spliced out in two-step reactions carried out by complex ribonucleoprotein particles called spliceosomes.

Many genes contain large numbers of introns. For example, the chicken 1α2 collagen (a structural protein in cartilage) gene contains over 50 introns, which leads to the question of the order in which multiple introns are removed. For certain genes that have been studied, the introns are excised in a preferred, but not fixed, order. Other introns have been found to undergo alternative pathways of splicing leading to mRNAs that produce families of related proteins. An intron in the mitochondrial cytochrome *b* gene of yeast includes part of the coding sequence for a protein, an RNA maturase, which is responsible for excising the second intron from the transcript of that gene. This discovery suggests interesting mechanisms for regulating the expression of genes at the level of intron processing. Clearly, variations in the use, structure, and excision of intron sequences occur in various oganisms, and novel intron structures may be discovered in the future. We will consider the three major mechanisms of intron excision in the following three sections.

tRNA Precursor Splicing: Unique Nuclease and Ligase Activities

The tRNA precursor splicing reaction has been worked out in detail in the yeast *Saccharomyces cerevisiae*. Both *in vitro* splicing systems and temperature-sensitive splicing mutants have been used in dissecting the

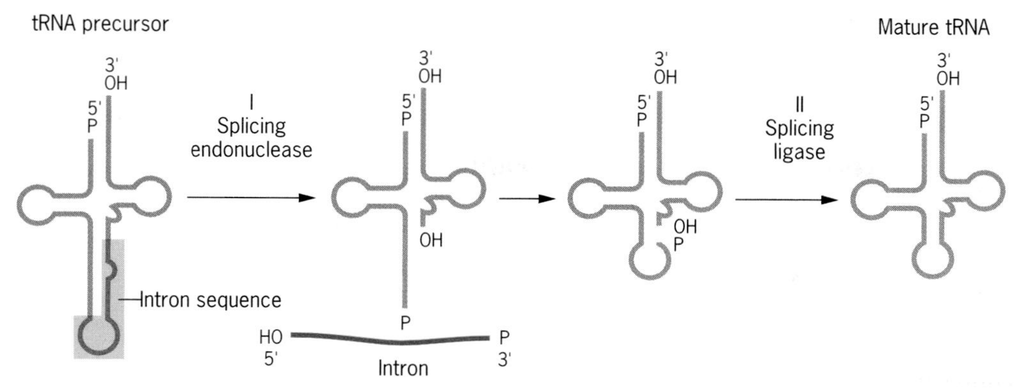

Figure 11.29 The excision of introns from tRNA precursors is a two-stage process: (I) endonucleolytic cleavages at the ends of the intron, and (II) ligation of the newly formed ends of the two exons. The ligation process actually involves three distinct reactions, all catalyzed by the splicing ligase. This unique enzyme contains phosphodiesterase, kinase, and ligase activities.

tRNA splicing mechanism in *S. cerevisiae*. The excision of introns from yeast tRNA precursors occurs in two stages (Figure 11.29). In stage I, a nuclear membrane-bound **splicing endonuclease** makes two cuts precisely at the ends of the intron. Then, in stage II, a **splicing ligase** joins the two halves of the tRNA to produce the mature form of the tRNA molecule. The specificity for these reactions resides in conserved three-dimensional structural features of the tRNA precursors, not in the nucleotide sequences per se.

Cleavage of the tRNA precursor yields 5'-OH termini and 2'-3' cyclic phosphate groups at the 3' termini. The stage II ligation process actually involves four separate reactions. (1) The first reaction is the addition of a phosphate group to the 5'-OH terminus; this reaction requires kinase activity and a phosphate donor (ATP). (Kinases are enzymes that add phosphate groups to molecules.) (2) Then, the 5' phosphate group is activated by the transfer of an AMP group to the terminus from an AMP-ligase intermediate. (3) The 2'-3' cyclic phosphate is opened by a cyclic phosphodiesterase that produces a 2′ phosphate and a free 3' hydroxyl. (4) The final ligation reaction occurs via a nucleophilic attack of the free 3'-OH on the interior 5' phosphate with the release of AMP. All four of these reactions are catalyzed by the splicing ligase. Finally, the 2' phosphate group (remaining from the 2'-3' cyclic phosphate produced by the original cleavage reaction) is removed by a phosphatase activity to yield the mature tRNA molecule.

The overall two-stage mode of tRNA intron excision appears to occur in other organisms as well. In fact, the mechanism may involve the same reactions in plants. However, in mammals, the reactions are not the same. Splicing still occurs in two stages, but the ligation reaction directly joins the 2'-3' cyclic phosphate terminus to the 5'-OH terminus. The details of this process of tRNA precursor splicing in mammalian cells are not yet as clearly established as they are in yeast.

Autocatalytic Splicing

As we reviewed in the first section of this chapter, the general theme in biology is that metabolism occurs via sequences of enzyme-catalyzed reactions. These all-important enzymes are generally proteins, sometimes single polypeptides and sometimes complex heteromultimers. Occasionally enzymes require nonprotein cofactors to perform their functions. When covalent bonds are being altered, it is usually assumed that the reaction is being catalyzed by an enzyme. Thus the discovery in *Tetrahymena thermophila* that the intron in the rRNA precursor was excised without the involvement of any protein was quite surprising to most biologists. However, it is now clearly established that the splicing activity that excises the intron from this rRNA precursor is intrinsic to the RNA molecule itself. Moreover, such **self-splicing** or **autocatalytic activity** has been shown to occur in rRNA precursors of several lower eukaryotes and in a large number of rRNA, tRNA, and mRNA precursors in mitochondria and chloroplasts of many different species. In the case of many of these introns, the self-splicing mechanism is the same as or very similar to that utilized by the *Tetrahymena* rRNA precursors (see Figure 11.30). For others, the self-splicing mechanism is similar to the splicing mechanism observed with nuclear mRNA precursors, but without the involvement of the spliceosome (see Figure 11.31).

The autocatalytic excision of the intron in the *Tetrahymena* rRNA precursor and certain other introns requires no external energy source and no protein. In-

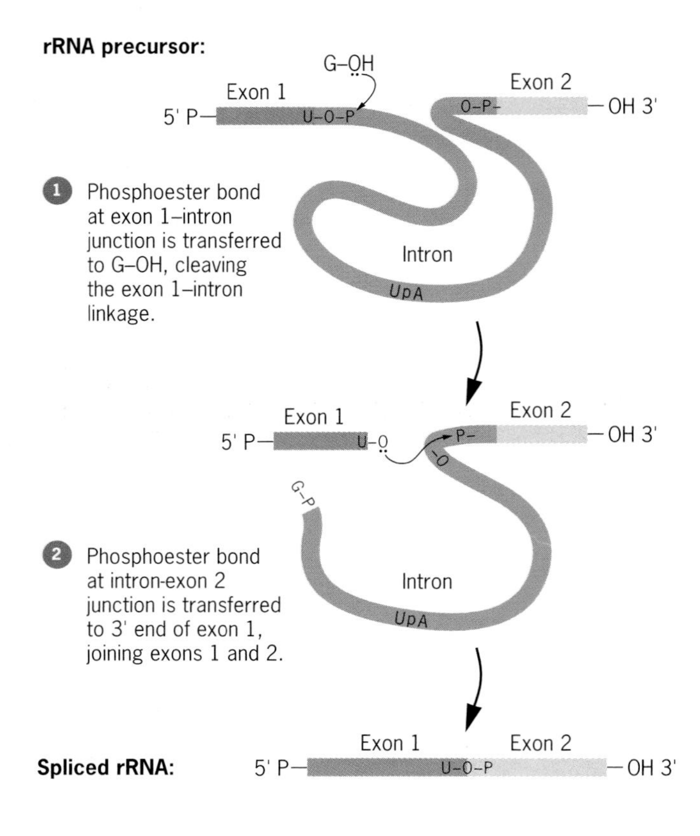

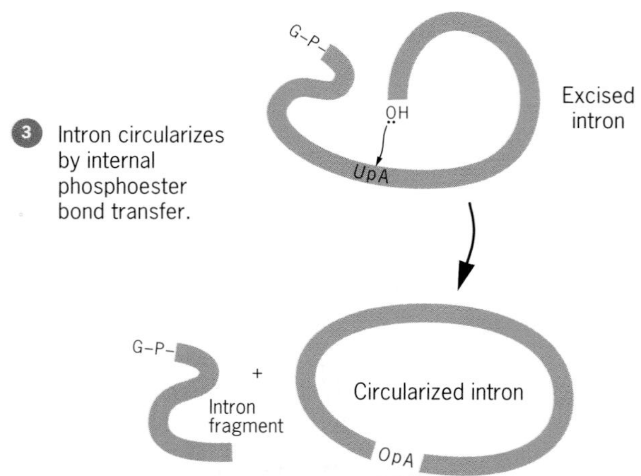

Figure 11.30 Diagram of the mechanism of self-splicing of the *Tetrahymena thermophila* rRNA precursor and the subsequent circularization of the excised intron.

stead, the splicing mechanism involves a series of phosphoester bond transfers, with no bonds lost or gained in the process. The reaction requires a guanine nucleoside or nucleotide with a free 3'-OH group (GTP, GDP, GMP, or guanosine all work) as a cofactor plus a monovalent cation and a divalent cation. The requirement for the G-3'-OH is absolute; no other base can be substituted in the nucleoside or nucleotide cofactor. The intron is excised by means of two phosphoester bond transfers, and the excised intron can subsequently circularize by means of another phosphoester bond transfer. These reactions are diagrammed in Figure 11.30.

The autocatalytic circularization of the excised intron suggests that the self-splicing of these rRNA precursors resides primarily, if not entirely, within the intron structure itself. Presumably, the autocatalytic activity is dependent on the secondary structure of the intron or at least the secondary structure of the RNA precursor molecule. The secondary structures of these self-splicing RNAs must bring the reactive groups into close juxtaposition to allow the phosphoester bond transfers to occur. Since the self-splicing phosphoester bond transfers are potentially reversible reactions, rapid degradation of the excised introns or export of the spliced rRNAs to the cytoplasm may drive splicing in the forward direction.

Note that the autocatalytic splicing reactions are intramolecular in nature and thus are not dependent on concentration. Moreover, the RNA precursors are capable of forming an active center in which the guanosine-3'-OH cofactor binds. The autocatalytic splicing of these rRNA precursors demonstrates that catalytic sites are not restricted to proteins; however, there is no *trans* catalytic activity as for enzymes, only *cis* catalytic activity.

Pre-mRNA Splicing: snRNAs, snRNPs, and the Spliceosome

The introns in nuclear pre-mRNAs are excised in two steps like the introns in yeast tRNA precursors and *Tetrahymena* rRNA precursors that were discussed in the preceding two sections. However, the introns are not excised by simple splicing nucleases and ligases or autocatalytically. Instead, nuclear pre-mRNA splicing is carried out by complex RNA/protein structures called **spliceosomes** (Figure 11.31). These structures are in many ways like small ribosomes. They contain a set of small RNA molecules called snRNAs (small nuclear RNAs) and a set of proteins that are still not completely defined. The two steps in nuclear pre-mRNA splicing are known (Figure 11.32); however, some of the details of the splicing process are still uncertain.

Five snRNAs, called U1, U2, U4, U5, and U6, are involved in nuclear pre-mRNA splicing as components of the spliceosome. (snRNA U3 is localized in the nucleolus and probably is involved in the formation of ribosomes.) In mammals, these snRNAs range in size from 100 nucleotides (U6) to 215 nucleotides (U3). Some of the snRNAs in the yeast *S. cerevisiae* are much larger. These snRNAs do not exist as free RNA molecules. Instead, they are present in small nuclear RNA-protein complexes called **snRNPs** (small nuclear ribonucleoproteins). Spliceosomes are assembled from

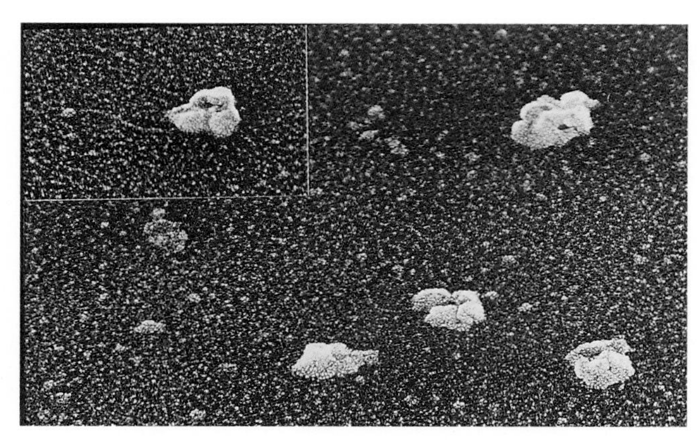

Figure 11.31 Electron microscope photographs of purified spliceosomes. Note the striking substructure of the particles, which have dimensions of 40 to 60 nanometers. The inset (top left) shows a particle with a thin filament and a smaller particle at the end of the filament.

four different snRNPs and protein splicing factors during the splicing process. Characterization of snRNPs has been facilitated by the discovery that some patients with a disease called **systemic lupus erythematosus** produce antibodies that react with snRNP proteins. These antibodies are called autoantibodies because they react with the patient's own proteins; normally, the human immune system will produce only antibodies that react with foreign proteins (Chapter 24). The autoantibodies from patients with systemic lupus erythematosus can be used to precipitate snRNPs; thus they greatly facilitate the purification of snRNPs for structural and functional studies.

Each of the snRNAs U1, U2, and U5 is present by itself in a specific snRNP particle. snRNAs U4 and U6 are present together in a fourth snRNP; U4 and U6 snRNAs contain two regions of intermolecular complementarity that probably are base-paired in the U4/U6 snRNP. Each of the four types of snRNP particles contains a subset of seven well-characterized snRNP proteins plus one or more proteins unique to the particular type of snRNP particle. All four snRNP complexes are present in the isolated spliceosomes shown in Figure 11.31. However, the exact protein composition of intact spliceosomes still is not established.

The first step in nuclear pre-mRNA splicing involves cleavage at the 5' intron splice site (↓GT-intron) and the formation of an intramolecular phosphodiester linkage between the 5' carbon of the G at the cleavage site and the 2' carbon of a conserved A residue near the 3' end of the intron. This step occurs on complete spliceosomes (Figure 11.32) and requires the hydrolysis of ATP. Evidence indicates that the U1 snRNP must bind at the 5' splice site prior to the initial cleavage reaction. The recognition of the cleavage site at the 5' end of the intron probably involves base-pairing between the consensus sequence at this site and a complementary sequence near the 5' terminus of snRNA U1. However, the specificity of the binding of at least some of the snRNPs to intron consensus sequences involves both the snRNAs and specific snRNP proteins. Thus the base-pairing between the intron 5' consensus sequence and the complementary sequence in the snRNA U1 may provide only part of the specificity for the functional binding of the U1 snRNP to the pre-mRNA molecule.

The second snRNP to be added to the splicing complex appears to be the U2 snRNP; it binds at the

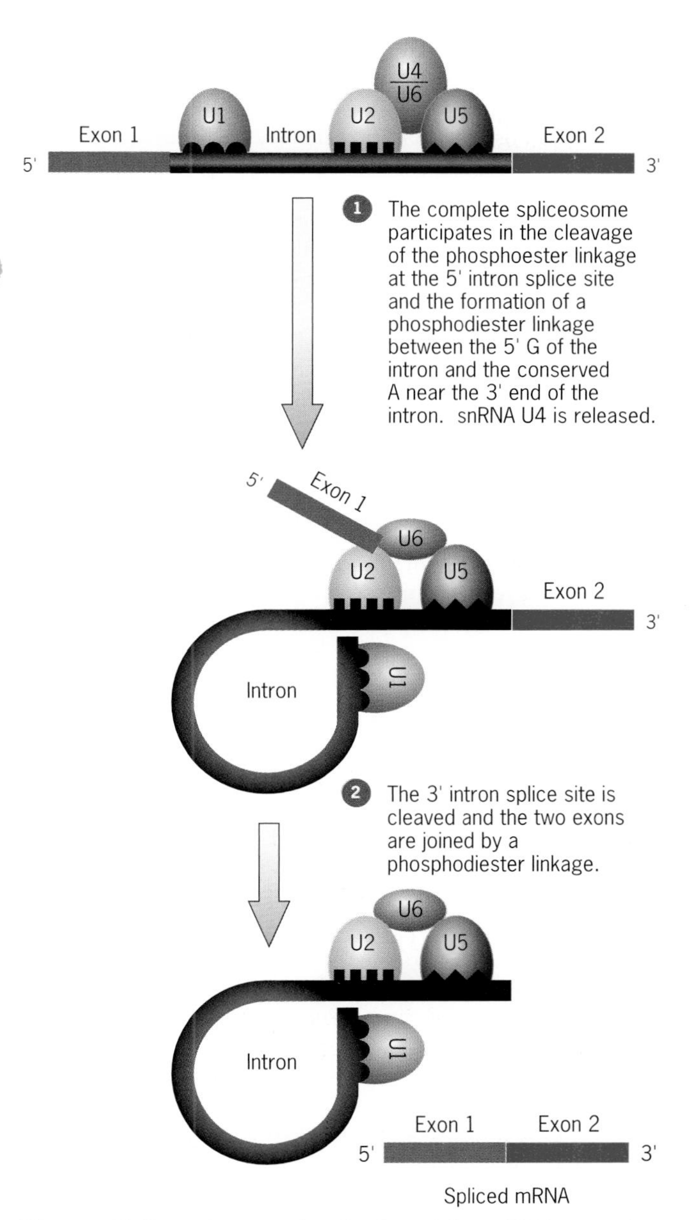

Figure 11.32 The postulated roles of the snRNA-containing snRNPs in nuclear pre-mRNA splicing.

consensus sequence that contains the conserved A residue that forms the branch point in the lariat structure of the spliced intron. Thereafter, the U5 snRNP binds at the 3' splice site, and the U4/U6 snRNP is added to the complex to yield the complete spliceosome (Figures 11.31 and 11.32). When the 5' intron splice site is cleaved in step 1, the U4 snRNA is released from the spliceosome. In step 2 of the splicing reaction, the 3' splice site of the intron is cleaved, and the two exons are joined by a normal 5' to 3' phosphodiester linkage (Figure 11.32). The spliced mRNA is now ready for export to the cytoplasm and translation on ribosomes.

Key Points: **Noncoding intron sequences are excised from RNA transcripts in the nucleus prior to their transport to the cytoplasm. Introns in tRNA precursors are removed by the concerted action of a splicing endonuclease and ligase, whereas introns in some rRNA precursors are spliced out autocatalytically—with no protein involved. The introns in nuclear pre-mRNAs are excised on complex ribonucleoprotein structures called spliceosomes. This excision process must be precise, with accuracy to the nucleotide level, to assure that codons in exons distal to introns are read correctly during translation.**

TESTING YOUR KNOWLEDGE

1. Certain medically important human proteins such as insulin and growth hormone are now being produced in bacteria. By using the tools of genetic engineering, DNA sequences encoding these proteins have been introduced into bacteria. You wish to introduce a human gene into *E. coli* and have that gene produce large amounts of the human gene product in the bacterial cells. Assuming that the human gene of interest can be isolated and introduced into *E. coli*, what problems might you encounter in attempting to achieve your goal?

ANSWER

The promoter sequences that are required to initiate transcription are very different in mammals and bacteria. Therefore, your gene will not be expressed in *E. coli* unless you first fuse its coding region to a bacterial promoter. In addition, your human gene probably will contain introns. Since *E. coli* cells do not contain spliceosomes or equivalent machinery with which to excise introns from RNA transcripts, your human gene will not be expressed correctly if it contains introns. As you can see, expressing eukaryotic genes in prokaryotic cells is not a trivial task.

2. A human β-globin gene has been purified and inserted into a linear bacteriophage lambda chromosome, producing the following DNA molecule.

λ DNA Exon 1 Intron 1 Exon 2 Intron 2 Exon 3 λ DNA

If this DNA molecule is hybridized to human β-globin mRNA using conditions that favor DNA:RNA duplexes over DNA:DNA duplexes (R-loop mapping conditions) and the product is visualized by electron microscopy, what nucleic acid structure would you expect to see?

ANSWER

The primary transcript of this human β-globin gene will contain both introns and all three exons. However, prior to its export to the cytoplasm, the intron sequences will be spliced out of the transcript. Thus the mature mRNA molecule will contain the three exon sequences spliced together with no intron sequences present. Under R-loop conditions, the mRNA will anneal with the complementary strand of DNA, displacing the equivalent DNA strand. However, since the mRNA contains no intron sequences, the introns will remain as regions of double-stranded DNA as shown in the following diagram.

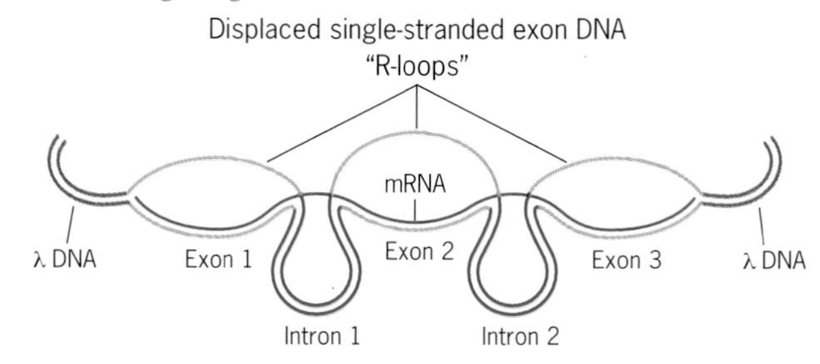

QUESTIONS AND PROBLEMS

11.1 Distinguish between DNA and RNA (a) chemically, (b) functionally, and (c) by location in the cell.

11.2 What bases in the mRNA transcript would represent the following DNA template sequence: 5'-TGCAGACA-3'?

11.3 What bases in the transcribed strand of DNA would give rise to the following mRNA base sequence: 5'-CUGAU-3'?

11.4 On the basis of what evidence was the messenger RNA hypothesis established?

11.5 At what locations in the cell does protein synthesis occur?

11.6 What different types of RNA molecules are present in prokaryotic cells? In eukaryotic cells? What roles do these different classes of RNA molecules play in the cell?

11.7 Many eukaryotic genes contain noncoding introns that separate the coding sequences or exons of these genes. At what stage during the expression of these mosaic genes are the noncoding intron sequences removed?

11.8 For several decades, the dogma in biology has been that molecular reactions in living cells are catalyzed by enzymes composed of polypeptides. We now know that the introns of some precursor RNA molecules such as the rRNA precursors in *Tetrahymena* are removed autocatalytically ("self-spliced") with no involvement of any protein. What does the demonstration of autocatalytic splicing indicate about the dogma that biological reactions are always catalyzed by proteinaceous enzymes?

11.9 What role(s) do spliceosomes play in pathways of gene expression? What is their macromolecular structure?

11.10 What components of the introns of nuclear genes that encode proteins in higher eukaryotes are conserved and required for the correct excision of intron sequences from primary transcripts by spliceosomes?

11.11 (a) Which of the following nuclear pre-mRNA nucleotide sequences potentially contains an intron?

(1) 5'—UGACCAUGGCGCUAACACUGCCAAUUGGCA-AUACUGACCUGAUAGCAUCAGCCAA—3'

(2) 5'—UAGUCUCAUCUGUCCAUUGACUUCGAAACU-GAAUCGUAACUCCUACGUCUAUGGA—3'

(3) 5'—UAGCUGUUUGUCAUGACUGACUGGUCACUA-UCGUACUAACCUGUCAUGCAAUGUC—3'

(4) 5'—UAGCAGUUCUGUCGCCUCGUGGUGCUGCUG-GCCCUUCGUCGCUCGGGCUUAGCUA—3'

(5) 5'—UAGGUUCGCAUUGACGUACUUCUGAGACUA-CUAACUACUAACGCAUCGAGUCUCAA—3'

(b) One of the five pre-mRNAs shown in (a) may undergo RNA splicing to excise an intron sequence. What mRNA nucleotide sequence would be expected to result from this splicing event?

11.12 What is the function of the introns in eukaryotic genes?

11.13 A particular gene is inserted into the phage lambda chromosome and is shown to contain three introns. (*a*) The primary transcript of this gene is purified from isolated nuclei. When this primary transcript is hybridized under R-loop conditions with the recombinant lambda chromosome carrying the gene, what will the R-loop structure(s) look like? Label your diagram. (*b*) The mRNA produced from the primary transcript of this gene is then isolated from cytoplasmic polyribosomes and similarly examined by the R-loop hybridization procedure using the recombinant lambda chromosome carrying the gene. Diagram what the R-loop structure(s) will look like when the cytoplasmic mRNA is used. Again, label the components of your diagram.

11.14 A segment of DNA in *E. coli* has the following sequence of nucleotide pairs:

```
3'-ATGCTACTGCTATTCGCTGTATCG-5'
   | | | | | | | | | | | | | | | | | | | | | | | |
5'-TACGATGACGATAAGCGACATAGC-3'
```

When this segment of DNA is transcribed by RNA polymerase, what will be the sequence of nucleotides in the RNA transcript if the promoter is on the left?

11.15 A segment of DNA in *E. coli* has the following sequence of nucleotide pairs:

```
3'-ATATTACTGCAATGGGCTGTATCG-
   | | | | | | | | | | | | | | | | | | | | | | | |
5'-TATAATGACGTTACCCGACATAGC-

ATGCTACTGCTATTCGCTGTATCG-5'
| | | | | | | | | | | | | | | | | | | | | | | |
TACGATGACGATAAGCGACATAGC-3'
```

When this segment of DNA is transcribed by RNA polymerase, what will be the sequence of nucleotides in the RNA transcript?

11.16 A segment of DNA in *E. coli* has the following sequence of nucleotide pairs:

```
3'-AACTGTACGTGCTACCTTGCTGATATTACT-
   | | | | | | | | | | | | | | | | | | | | | | | | | | | | | |
5'-TTGACATGCACGATGGAACGACTATAATGA-

GCAATGGGCTGTATCGATGCTACTGCTAT-5'
| | | | | | | | | | | | | | | | | | | | | | | | | | | | |
CGTTACCCGACATAGCTACGATGACGATA-3'
```

When this segment of DNA is transcribed by RNA polymerase, what will be the sequence of nucleotides in the RNA transcript?

11.17 A segment of human DNA has the following sequence of nucleotide pairs:

```
3'-ATATTTACGTGCTACCTTGCTGATAGGACT-
   | | | | | | | | | | | | | | | | | | | | | | | | | | | | | |
5'-TATAAATGCACGATGGAACGACTATCCTGA-

GCAATGGGCTGTATCGATGCTACTGCTAT-5'
| | | | | | | | | | | | | | | | | | | | | | | | | | | | |
CGTTACCCGACATAGCTACGATGACGATA-3'
```

When this segment of DNA is transcribed by RNA polymerase, what will be the sequence of nucleotides in the RNA transcript?

11.18 The genome of a human must store a tremendous amount of information using the four nucleotide pairs present in DNA. What does the Morse code and the language of computers tell us about the feasibility of storing large amounts of information using an alphabet composed of just four letters?

11.19 The biosynthesis of metabolite X occurs via six steps catalyzed by six different enzymes. What is the minimal number of genes required for the genetic control of this metabolic pathway? Might more genes be involved? Why?

11.20 What is the central dogma of molecular genetics? What impact did the discovery of RNA tumor viruses have on the central dogma?

11.21 What do the processes of DNA synthesis, RNA synthesis, and polypeptide synthesis have in common?

11.22 What are the two stages of gene expression? Where do they occur in the cell?

11.23 Compare the structures of primary transcripts with those of mRNAs in prokaryotes and eukaryotes. On average, in which group of organisms do they differ the most?

11.24 What four types of RNA molecules participate in the process of gene expression? What are the functions of each type of RNA? Which types of RNA perform their function(s) in (a) the nucleus and (b) the cytoplasm?

11.25 Two eukaryotic genes encode two different polypeptides, each of which is 335 amino acids long. One gene contains a single exon; the other gene contains an intron 41,324 nucleotide pairs long. Which gene would you expect to be transcribed in the least amount of time? Why? When the mRNAs specified by these genes are translated, which mRNA would you expect to be translated in the least time? Why?

11.26 Why was the need for an RNA intermediary in protein synthesis most obvious in eukaryotes? How did researchers first demonstrate that RNA synthesis occurred in the nucleus and that protein synthesis occurred in the cytoplasm?

11.27 Design an experiment to demonstrate that RNA transcripts are synthesized in the nucleus of eukaryotes and are subsequently transported to the cytoplasm.

11.28 What was the first evidence for an unstable mRNA intermediary in prokaryotes? How was this initial evidence substantiated?

11.29 DNA and RNA synthesis occur by very similar mechanisms. In what ways do they differ?

11.30 Total RNA was isolated from human cells growing in culture. This RNA was mixed with nontemplate strands (single strands) of the human gene encoding the enzyme thymidine kinase, and the RNA–DNA mixture was incubated for 12 hours under renaturation conditions. Would you expect any RNA–DNA duplexes to be formed during the incubation? If so, why? If not, why not? The same experiment was then performed using the template strand of the thymidine kinase gene. Would you expect any RNA–DNA duplexes to be formed in this second experiment? If so, why? If not, why not?

11.31 Two preparations of RNA polymerase from *E. coli* are used in separate experiments to catalyze RNA synthesis *in vitro* using a purified fragment of DNA carrying the *argH* gene (Figure 11.1) as template DNA. One preparation catalyzes the synthesis of RNA chains that are highly heterogeneous in size. The other preparation catalyzes the synthesis of RNA chains that are all the same length. What is the most likely difference in the composition of the RNA polymerases in the two preparations?

11.32 Transcription and translation are coupled in prokaryotes. Why is this not the case in eukaryotes?

11.33 You are studying the expression of a recently identified gene in *Drosophila*, and you have shown that this gene is expressed in *Drosophila* cells growing in culture. How can you determine whether this gene is transcribed by RNA polymerase I, RNA polymerase II, or RNA polymerase III?

11.34 What two elements are almost always present in the promoters of eukaryotic genes that are transcribed by RNA polymerase II? Where are these elements located relative to the transcription start site? What are their functions?

11.35 In what ways are most eukaryotic gene transcripts modified? What are the functions of these post-transcriptional modifications?

11.36 How does RNA editing contribute to protein diversity in eukaryotes? What role(s) do guide RNAs play in RNA editing?

11.37 How do the mechanisms by which the introns of tRNA precursors, *Tetrahymena* rRNA precursors, and nuclear pre-mRNAs are excised differ? In which process are snRNAs involved? What role(s) do these snRNAs play?

11.38 What role did the human disease systemic lupus erythematosus play in the characterization of human snRNPs?

11.39 A mutation in an essential human gene changes the 5′ splice site of a large intron from GT to CC. Predict the phenotype of an individual homozygous for this mutation.

11.40 Total RNA was isolated from nuclei of human cells growing in culture. This RNA was mixed with a purified, denatured DNA fragment that carried a large intron of a housekeeping gene (a gene expressed in essentially all cells), and the RNA–DNA mixture was incubated for 12 hours under renaturation conditions. Would you expect any RNA–DNA duplexes to be formed during the incubation? If so, why? If not, why not? The same experiment was then performed using total cytoplasmic RNA from these cells. Would you expect any RNA–DNA duplexes to be formed in this second experiment? If so, why? If not, why not?

BIBLIOGRAPHY

BLUM, B., N. R. STURM, A. M. SIMPSON, AND L. SIMPSON. 1991. "Chimeric gRNA-mRNA molecules with oligo(U) tails covalently linked at sites of RNA editing suggest that U addition occurs by transesterification." *Cell* 65: 543-550.

CHAMBON, P. 1981. "Split genes." *Sci. Amer.* 244(5):60–71.

CONAWAY, R. C., AND J. W. CONAWAY. 1993. "General initiation factors for RNA polymerase II." *Annu. Rev. Biochem.* 62:161–190.

DAS, A. 1993. "Control of transcription termination by RNA-binding proteins." *Annu. Rev. Biochem.* 62:893–930.

GUTHRIE, C., AND B. PATTERSON. 1988. "Spliceosomal snRNAs." *Annu. Rev. Genet.* 22:387–419.

REED, R., J. GRIFFITH, AND T. MANIATIS. 1988. "Purification and visualization of native spliceosomes." *Cell* 53:949–961.

SHARP, P. 1987. "Splicing of messenger RNA precursors." *Science* 238:729–730.

STEITZ, J. A. 1988. "Snurps." *Sci. Amer.* 258(6):56–63.

TILGHMAN, S. M., D. C. TIEMEIER, J. G. SEIDMAN, B. M. PETERLIN, M. SULLIVAN, J. V. MAIZEL, AND P. LEDER. 1978. "Intervening sequence of DNA identified in the stuctural portion of a mouse β-globin gene." *Proc. Natl. Acad. Sci. USA* 75:725–729.

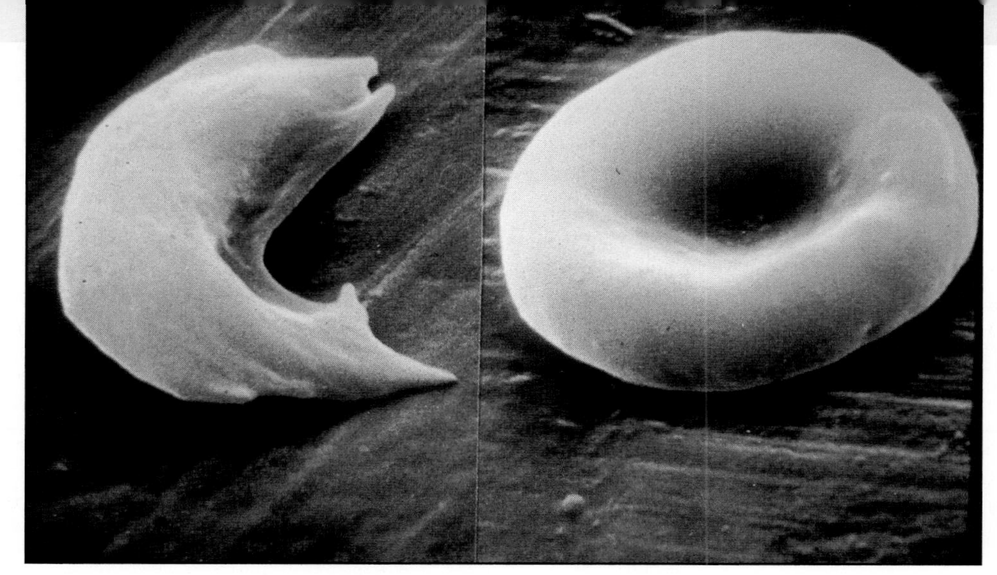

Scanning electron micrograph of sickle-shaped (left) and normal red blood cells.

12

Translation and the Genetic Code

CHAPTER OUTLINE

Sickle-Cell Anemia: Devastating Effects of a Single Base-Pair Change

In 1904, James Herrick, a respected Chicago physician, and Ernest Irons, a medical intern working under Herrick's supervision, examined the blood cells of one of their patients. They noticed that many of the red blood cells of the young man were thin and elongated, in striking contrast to the round, donut-like red cells of their other patients. Herrick and Irons had examined the blood cells of many other patients but had never seen red cells like these. They obtained fresh blood samples and repeated their microscopic examinations several times, always with the same result. The blood of this patient always contained cells shaped like the sickles that farmers used to harvest grain at that time.

The patient was a 20-year-old college student who was experiencing periods of weakness and dizziness. In many respects, the patient seemed normal, both physically and mentally. His major problem was fatigue. However, a physical exam showed an enlarged heart and enlarged lymph nodes. His heart always seemed to be working too hard, even when he was resting. Blood tests showed that the patient was anemic; the hemoglobin content of his blood was about half the normal level. Hemoglobin is the complex protein that carries oxygen from the lungs to other tissues. Herrick and Irons were puzzled by the clinical symptoms and the abnormal red blood cells in this patient. Herrick charted this patient's symptoms for six years before publishing his observations in 1910. In his paper, Herrick emphasized the chronic nature of the anemia and the presence of the sickle-shaped red cells. In 1916, at age 32, the patient died from severe anemia and kidney damage.

James Herrick was the first to publish a description of sickle-cell anemia, the first inherited human disease to be understood at the molecular level. In 1949, Linus Pauling and coworkers documented a difference between the hemoglobin of healthy individuals and those with sickle-cell anemia. Hemoglobin contains four polypeptides, two α-globin chains and two β-globin chains, and an iron-containing heme group. In 1957, Vernon Ingram and colleagues demonstrated that the sixth amino acid of the β-chain of sickle-cell hemoglobin was glutamic acid, whereas valine was present at this position in normal adult human hemoglobin. This single amino acid change in a single polypeptide chain is responsible for all the symptoms of sickle-cell anemia. Otherwise, the hemoglobin molecules in sickled and normal red blood cells are identical.

How does the genetic information of an organism, stored in the sequence of nucleotide pairs in DNA, control the phenotype of the organism? How does a nucleotide-pair change—like the mutation that causes sickle-cell anemia—in a gene alter the structure of a protein, the emissary through which the gene acts? In Chapter 11, we discussed the transfer of genetic information stored in the sequences of nucleotide pairs in DNA to the sequences of nucleotides in mRNA molecules, which carry that information from the nucleus to the sites of protein synthesis in the cytoplasm. The transfer of information from DNA to RNA, transcription, and RNA processing occur in the nucleus. In this chapter, we examine the process by which genetic information stored in sequences of nucleotides in mRNAs is used to specify the sequences of amino acids in polypeptide gene products. This process, **translation**, takes place in the cytoplasm on complex workbenches called ribosomes and requires the participation of many macromolecules.

PROTEIN STRUCTURE

Collectively, the proteins constitute about 15 percent of the wet weight of cells. Water molecules account for 70 percent of the total weight of living cells. With the exception of water, proteins are by far the most prevalent component of living organisms in terms of total mass. Not only are proteins major components in terms of cell mass, but they also play many roles vital to the lives of all cells. Before discussing the synthesis of proteins, we need to become more familiar with their structure.

Polypeptides: Twenty Different Amino Acid Subunits

As we know from Chapter 11, proteins are composed of polypeptides, and every polypeptide is encoded by a gene. Each polypeptide consists of a long sequence of amino acids linked together by covalent bonds. Twenty different amino acids are present in most proteins. Occasionally, one or more of the amino acids are chemically modified after a polypeptide is synthesized, yielding a novel amino acid in the mature protein. The structures of the 20 common amino acids are shown in Figure 12.1. All the amino acids except proline contain a **free amino group** and a **free carboxyl group**.

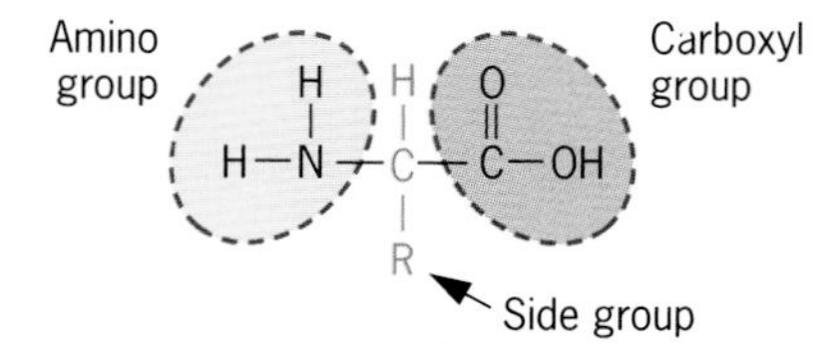

The amino acids differ from each other by the **side groups** (designated **R** for radical) that are present. The highly varied side groups provide the structural diversity of proteins. These side chains are of four types: (1) hydrophobic or nonpolar groups, (2) hydrophilic or polar chains, (3) acidic or negatively charged groups, and (4) basic or positively charged chains (Figure 12.1). The chemical diversity of the side groups of the amino acids is responsible for the enormous structural and functional versatility of proteins.

A **peptide** is a compound composed of two or more amino acids. Polypeptides are long sequences of amino acids, ranging in length from 51 amino acids in

1. Hydrophobic or nonpolar side groups

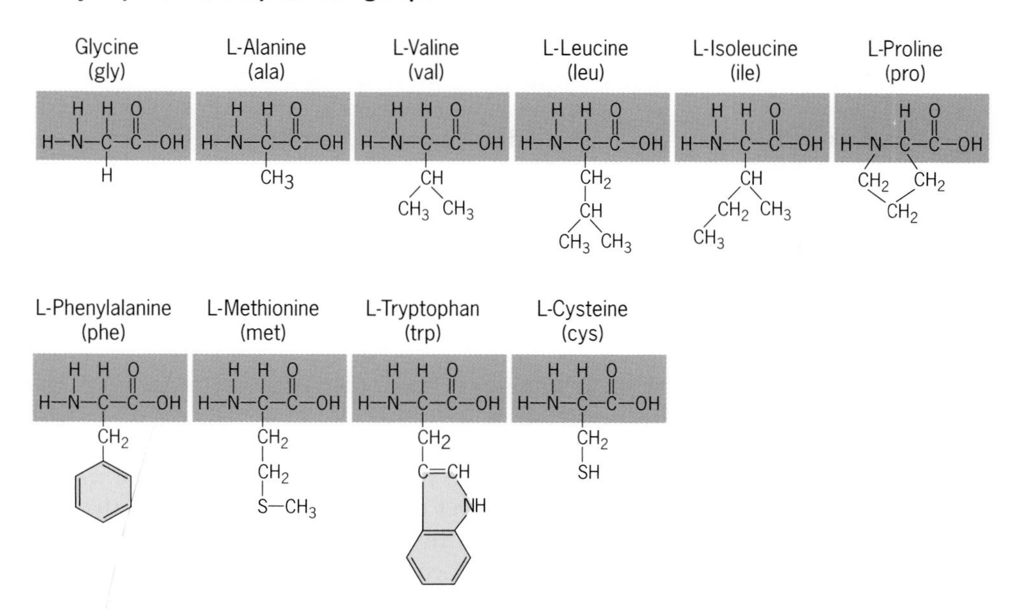

2. Hydrophilic or polar side groups

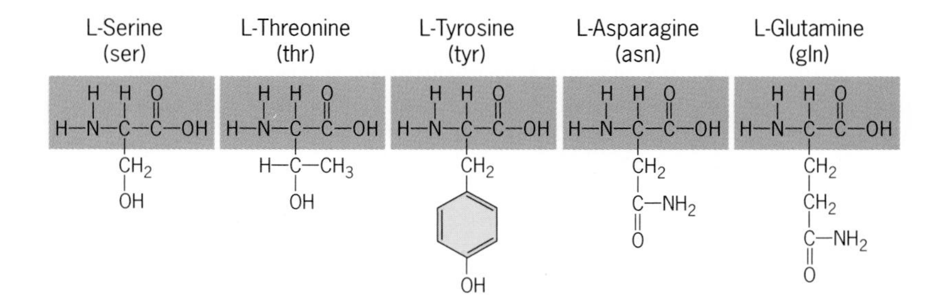

3. Acidic side groups

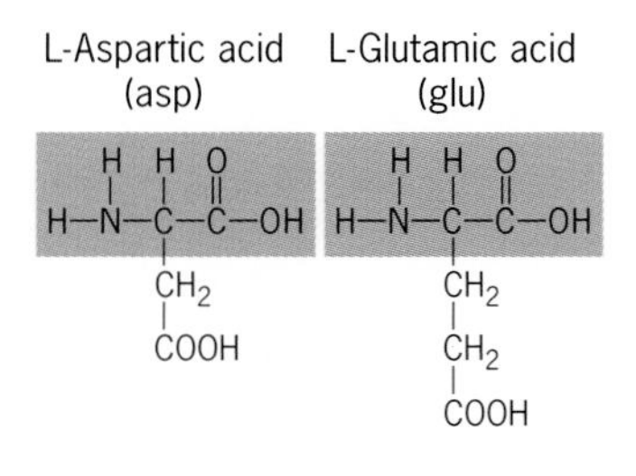

4. Basic side groups

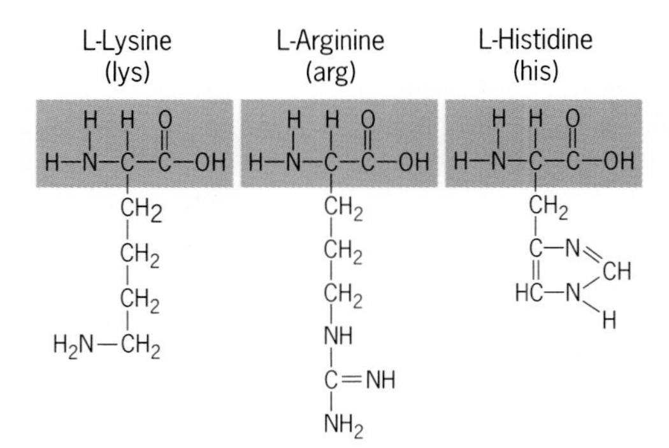

Figure 12.1 Structures of the 20 amino acids commonly found in proteins. The amino and carboxyl groups, which participate in peptide bond formation during protein synthesis, are shown in the shaded areas. The side groups, which are different for each amino acid, are shown below the shaded areas. The standard three-letter abbreviations are shown in parentheses.

insulin to over 1000 amino acids in the silk protein fibroin. Given the 20 different amino acids commonly found in polypeptides, the number of different polypeptides that are possible is truly enormous. For example, the number of different amino acid sequences that can occur in a polypeptide containing 100 amino acids long is 20^{100}! Since 20^{100} is too large to comprehend, let's consider a short peptide. There are 1.28 billion (20^7) different amino acid sequences possible in a peptide seven amino acids long. The amino

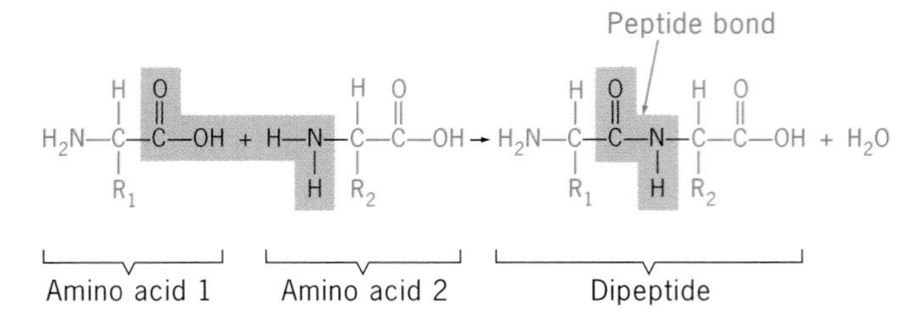

Figure 12.2 The formation of a peptide bond between two amino acids by the removal of water. Each peptide bond connects the amino group of one amino acid and the carboxyl group of the adjacent amino acid.

acids in polypeptides are covalently joined by linkages called **peptide bonds**. Each peptide bond is formed by a reaction between the amino group of one amino acid and the carboxyl group of a second amino acid with the elimination of a water molecule (Figure 12.2).

Proteins: Complex Three-Dimensional Structures

Four different levels of organization—primary, secondary, tertiary, and quaternary—are distinguished in the complex three-dimensional structures of proteins. The **primary structure** of a polypeptide is its amino acid sequence, which is specified by the nucleotide sequence of a gene. The **secondary structure** of a polypeptide refers to the spatial interrelationships of the amino acids in segments of the polypeptide. The **tertiary structure** of a polypeptide refers to its overall folding in three-dimensional space, and **quaternary structure** refers to the association of two or more polypeptides in a multimeric protein. Hemoglobin provides an excellent example of the complexity of proteins, exhibiting all four levels of structural organization (Figure 12.3).

Most polypeptides will fold spontaneously into specific conformations dictated by their primary structures. If denatured (unfolded) by treatment with appropriate solvents, most proteins will re-form their original conformations when the denaturing agent is removed. Thus, in most cases, all of the information required for shape determination resides in the primary structure of the protein.

The two most common types of secondary structure in proteins are **α helices** and **β sheets** (Figure 12.4). Both structures are maintained by hydrogen bonding between peptide bonds located in close proximity to one another. The α helix is a rigid cylinder in which every peptide bond is hydrogen bonded to another nearby peptide bond (Figure 12.4*a*). A β sheet occurs when a polypeptide folds back onto itself, sometimes repeatedly, and the parallel segments are held in place by hydrogen bonding between neighboring peptide bonds (Figure 12.4*b*).

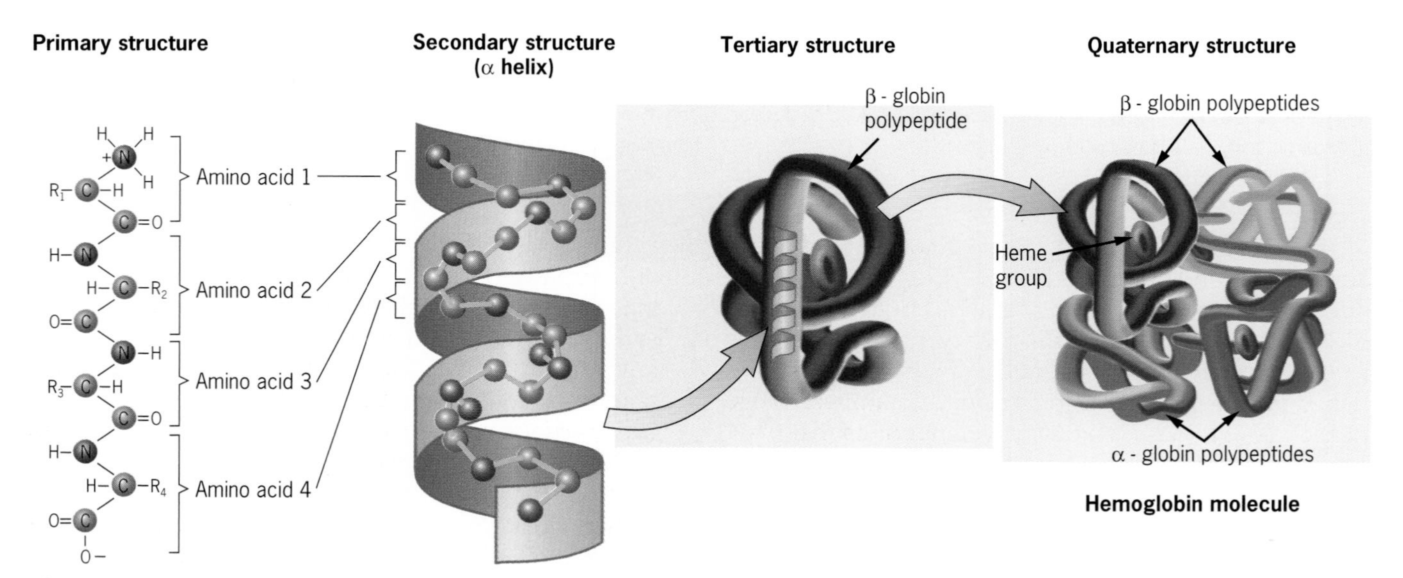

Figure 12.3 The four levels of organization in proteins—(1) primary, (2) secondary, (3) tertiary, and (4) quaternary structure—are illustrated using human hemoglobin as an example.

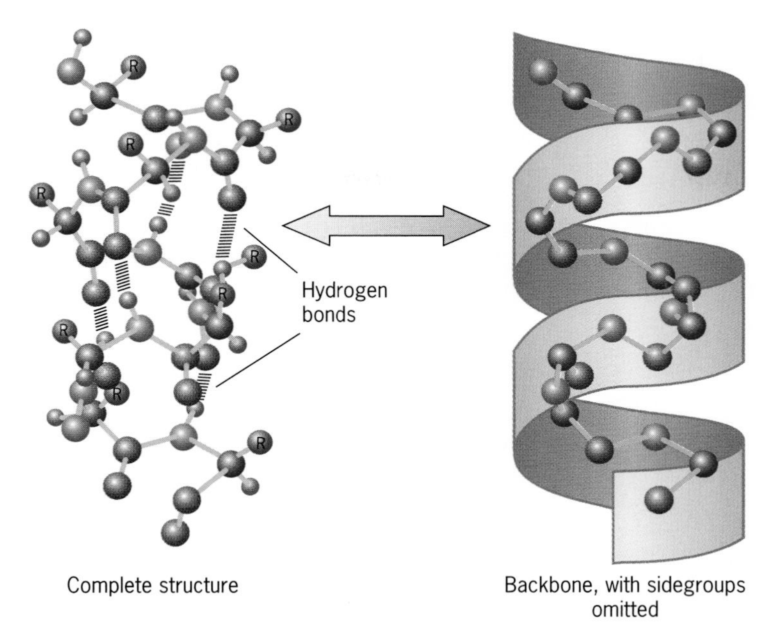

(a) Structure of an α-helix region of a polypeptide.

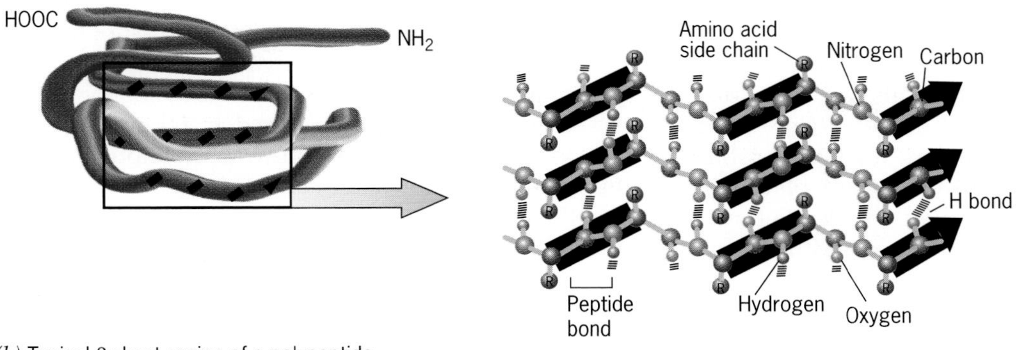

(b) Typical β-sheet region of a polypeptide.

Figure 12.4 Secondary structure in proteins. The α helix shown is from myoglobin, the oxygen-transporting protein in vertebrate muscle; the β sheet illustrated here is from a human immunoglobulin (antibody) molecule.

Whereas the spatial organization of adjacent amino acids and segments of a polypeptide determine its secondary structure, the overall folding of the complete polypeptide defines its tertiary structure or **conformation**. The tertiary structure of a protein is maintained primarily by a large number of relatively weak noncovalent bonds. The only covalent bonds that play a significant role in protein conformation are disulfide (S—S) bridges that form between appropriately positioned cysteine residues (Figure 12.5). However, four different types of noncovalent bonds are involved: (1) ionic bonds, (2) hydrogen bonds, (3) hydrophobic interactions, and (4) Van der Waals interactions (Figure 12.5).

Ionic bonds occur between amino acid side chains with opposite charges—for example, the side groups of lysine and glutamic acid (see Figure 12.1). **Ionic bonds** are strong forces under some conditions, but they are relatively weak interactions in the aqueous interiors of living cells because the polar water molecules partially neutralize or shield the charged groups. **Hydrogen bonds** are weak interactions between electronegative atoms (atoms with a partial negative charge) and hydrogen atoms (which are electropositive) that are linked to other electronegative atoms. **Hydrophobic interactions** are associations of nonpolar groups with each other when present in aqueous solutions because of their insolubility in water. Hy-

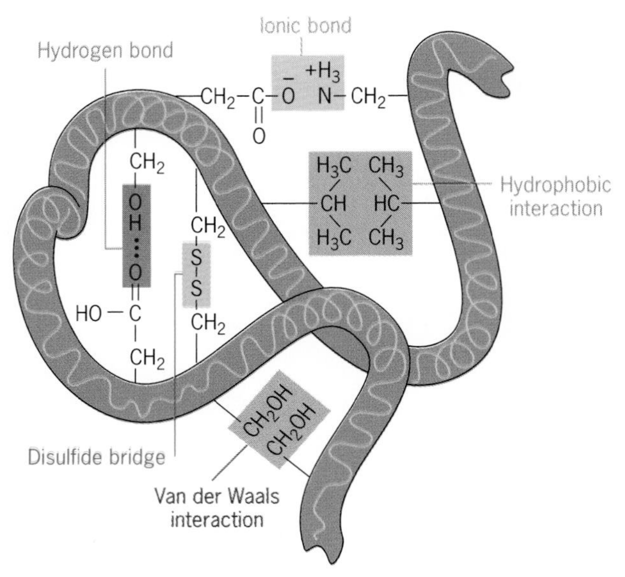

Figure 12.5 The five types of molecular interactions that determine the tertiary structure or three-dimensional conformation of a polypeptide. The disulfide bridge is a covalent bond; all other interactions are noncovalent.

drogen bonds and hydrophobic interactions play important roles in DNA structure; thus we have discussed them in some detail in Chapter 9 (see Table 9.3). **Van der Waals interactions** are weak attractions that occur between atoms when they are placed in close proximity to one another. Van der Waals forces are very weak, with about one-thousandth of the strength of a covalent bond, but they play an important role in maintaining the conformations of closely aligned regions of macromolecules.

Quaternary structure exists only in proteins that contain more than one polypeptide. Hemoglobin provides a good illustration of quaternary structure, being a tetrameric molecule composed of two α-globin chains and two β-globin chains, plus four iron-containing heme groups (see Figure 12.3). Since the secondary, tertiary, and quaternary structures of proteins usually are determined by the primary structure(s) of the polypeptide(s) involved, we will focus in the rest of this chapter on the mechanisms by which genes control the primary structures of polypeptides.

Key Points: **Most genes exert their effect(s) on the phenotype of an organism through proteins, which are large macromolecules composed of polypeptides. Each polypeptide is a chain-like polymer assembled from different amino acids. The amino acid sequence of each polypeptide is specified by the nucleotide sequence of a gene. The vast functional diversity of proteins results in part from their complex three-dimensional structures.**

PROTEIN SYNTHESIS: TRANSLATION

The process by which the genetic information stored in the sequence of nucleotides in an mRNA is translated, according to the specifications of the genetic code, into the sequence of amino acids in the polypeptide gene product is complex, requiring the functions of a large number of macromolecules. These include (1) over 50 polypeptides and from 3 to 5 RNA molecules present in each ribosome (the exact composition varies from species to species), (2) at least 20 amino acid-activating enzymes, (3) from 40 to 60 different tRNA molecules, and (4) numerous soluble proteins involved in polypeptide chain initiation, elongation, and termination. Because many of these macromolecules, particularly the components of the ribosome, are present in large quantities in each cell, the translation system makes up a major portion of the metabolic machinery of each cell.

Overview of Protein Synthesis

Before focusing on the details of the translation process, we should preview the process of protein synthesis in its entirety. An overview of protein synthesis, illustrating its complexity and the major macromolecules involved, is presented in Figure 12.6. The first step in gene expression, transcription, involves the transfer of information stored in genes to messenger RNA (mRNA) intermediaries, which carry that information to the sites of polypeptide synthesis in the cytoplasm. Transcription is discussed in detail in Chapter 11. The second step, translation, involves the transfer of the information in mRNA molecules into the sequences of amino acids in polypeptide gene products.

Translation occurs on ribosomes, which are complex macromolecular structures located in the cytoplasm. Translation involves three types of RNA, all of which are transcribed from DNA templates (chromosomal genes). In addition to mRNAs, three to five RNA molecules (rRNA molecules) are present as part of the structure of each ribosome, and 40 to 60 small RNA molecules (tRNA molecules) function as adaptors by mediating the incorporation of the proper amino acids into polypeptides in response to specific nucleotide sequences in mRNAs. The amino acids are attached to the correct tRNA molecules by a set of activating enzymes called **aminoacyl-tRNA synthetases**.

The nucleotide sequence of an mRNA molecule is translated into the appropriate amino acid sequence according to the dictations of the genetic code. Each amino acid is specified by one or more codons, and each codon contains three nucleotides. Of the 64 possible nucleotide triplets, 61 specify amino acids and 3

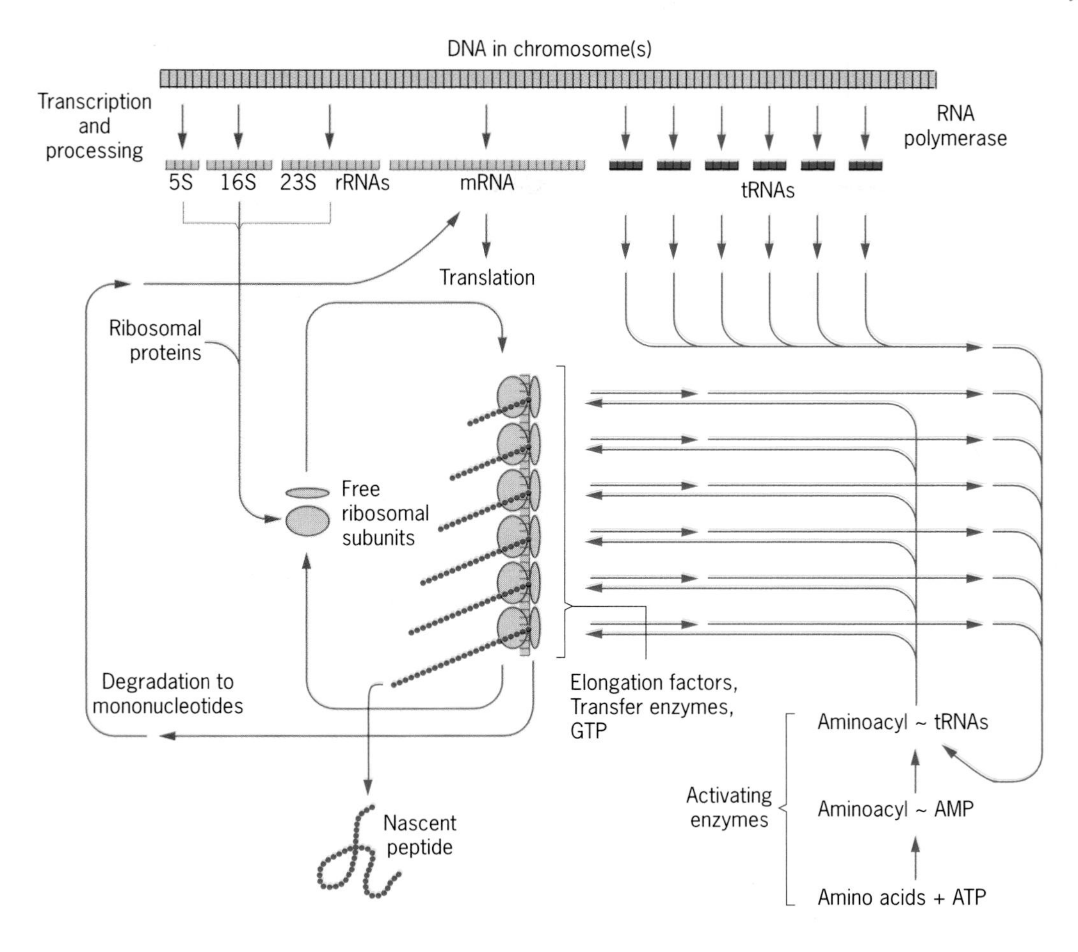

Figure 12.6 Overview of protein synthesis. The sizes of the rRNA molecules shown are correct for bacteria; larger rRNAs are present in eukaryotes. For simplicity, all RNA species have been transcribed from contiguous segments of a single DNA molecule. In reality, the various RNAs are transcripts of genes located at different positions on from one to many chromosomes.

specify polypeptide chain termination. The tRNA molecules contain nucleotide triplets called anticodons, which base-pair with the codons in mRNA during the translation process.

The ribosomes may be thought of as workbenches, complete with machines and tools needed to make a polypeptide. They are nonspecific in the sense that they can synthesize any polypeptide (any amino acid sequence) encoded by a particular mRNA molecule. Each mRNA molecule is simultaneously translated by several ribosomes, resulting in the formation of polyribosomes. Given this brief overview of protein synthesis, we will now examine some of the more important components of the translation machinery more closely.

Components Required for Protein Synthesis: Ribosomes and Transfer RNAs

Living cells devote more energy to the synthesis of proteins than to any other aspect of metabolism. About one-third of the total dry mass of most cells consists of molecules that participate directly in the biosynthesis of proteins. In *E. coli,* the approximately 200,000 ribosomes account for 25 percent of the dry weight of each cell. This commitment of a major proportion of the metabolic machinery of cells to the process of protein synthesis documents its importance in the life forms that exist on our planet.

When the sites of protein synthesis were labeled by growing cells for short intervals in the presence of radioactive amino acids and were visualized by autoradiography, the results showed that proteins are synthesized on the ribosomes. In prokaryotes, ribosomes are distributed throughout cells; in eukaryotes, they are located in the cytoplasm, frequently on the extensive intracellular membrane network of the endoplasmic reticulum.

Ribosomes are approximately half protein and half RNA (Figure 12.7). They are composed of two subunits, one large and one small, which dissociate when the translation of an mRNA molecule is completed and reassociate during the initiation of translation. Ribosome sizes are most frequently expressed in terms of their rates of sedimentation during centrifugation, in Svedberg units (see Technical Sidelight in Chapter 10). The *E. coli* ribosome, like those of other prokaryotes, has a molecular weight of 2.5×10^6, a size of 70S, and dimensions of about 20 nm $\times$ 25 nm.

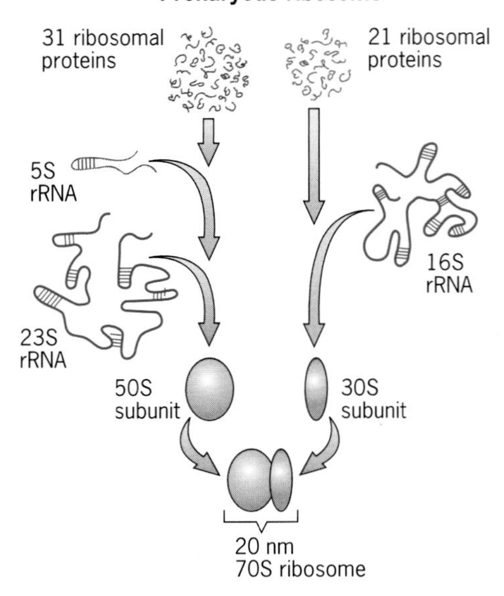

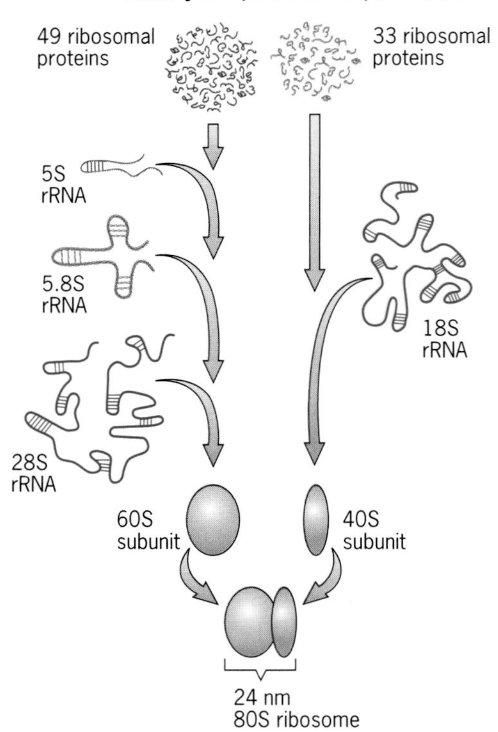

Figure 12.7 Macromolecular composition of prokaryotic and eukaryotic ribosomes.

The ribosomes of eukaryotes are larger (usually about 80S); however, size varies from species to species. The ribosomes present in the mitochondria and chloroplasts of eukaryotic cells are smaller (usually about 60S).

Although the size and macromolecular composition of ribosomes vary, the overall three-dimensional structure of the ribosome is basically the same in all organisms. In *E. coli*, the small (30S) ribosomal subunit contains a 16S (molecular weight about 6×10^5) RNA molecule plus 21 different polypeptides, and the large (50S) subunit contains two RNA molecules (5S, molecular weight about 4×10^4, and 23S, molecular weight about 1.2×10^6) plus 31 polypeptides. In mammalian ribosomes, the small subunit contains an 18S RNA molecule plus 33 polypeptides, and the large subunit contains three RNA molecules of size 5S, 5.8S, and 28S plus 49 polypeptides. In organelles, the corresponding rRNA sizes are 5S, 13S, and 21S.

Masayasu Nomura and his colleagues were able to disassemble the 30S ribosomal subunit of *E. coli* into the individual macromolecules and then reconstitute functional 30S subunits from the components. In this way, they studied the functions of individual rRNA and ribosomal protein molecules.

The ribosomal RNA molecules, like mRNA molecules, are transcribed from a DNA template. In eukaryotes, rRNA synthesis occurs in the nucleolus (Figure 12.8) and is catalyzed by RNA polymerase I. The nucleolus is a highly specialized component of the nucleus devoted exclusively to the synthesis of rRNAs and their assembly into ribosomes. The ribosomal RNA genes are present in tandemly duplicated arrays separated by intergenic spacer regions. The transcription of these tandem sets of rRNA genes can be visualized directly by electron microscopy (Figure 12.9).

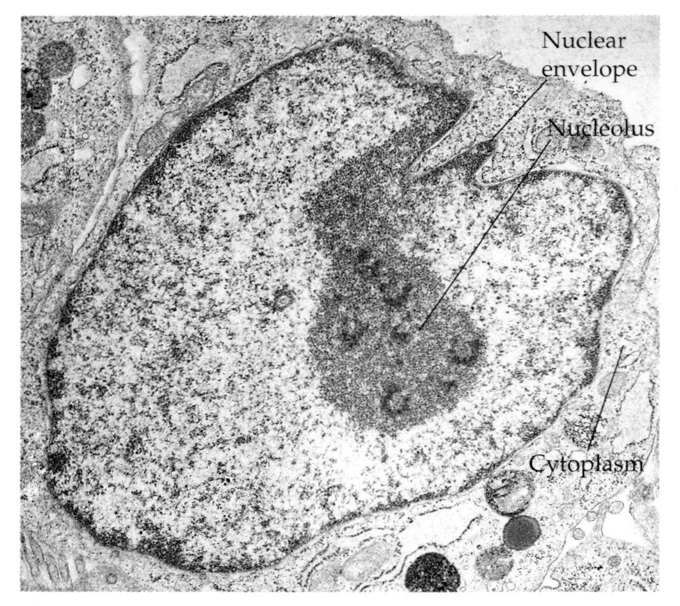

Figure 12.8 Electron micrograph of a human fibroblast, showing the nucleolus, the site of rRNA synthesis, and ribosome assembly.

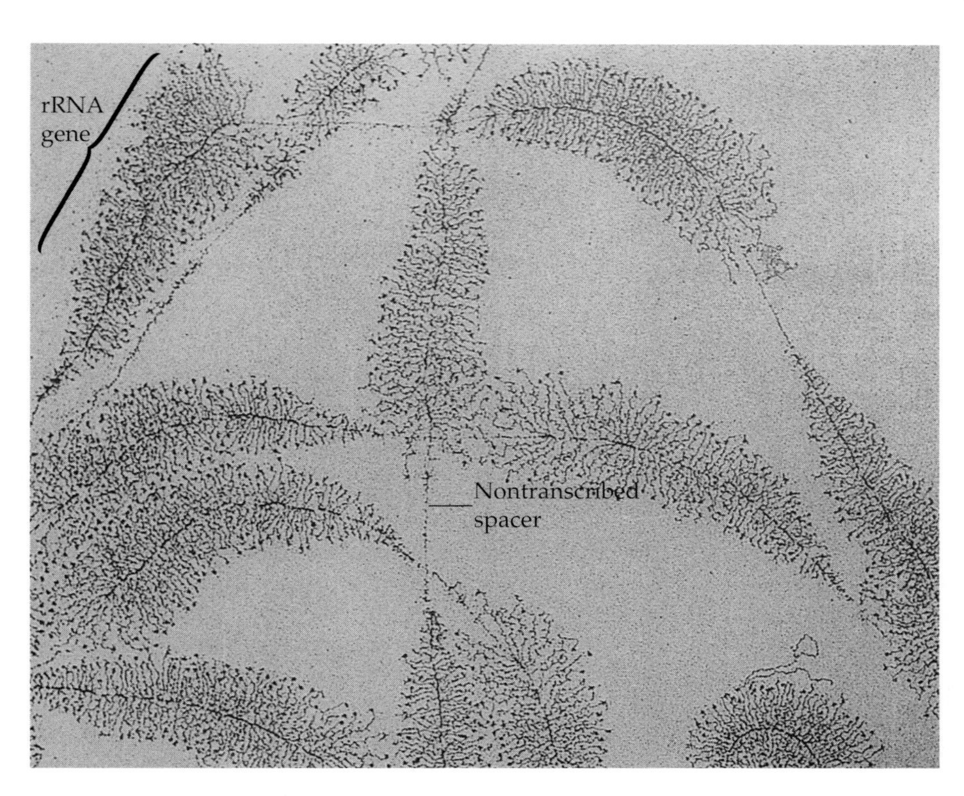

Figure 12.9 Electron micrograph showing the transcription of tandemly repeated rRNA genes in the nucleolus of the newt *Triturus viridescens*. A gradient of fibrils of increasing length is observed for each rRNA gene, and nontranscribed spacer regions separate the genes.

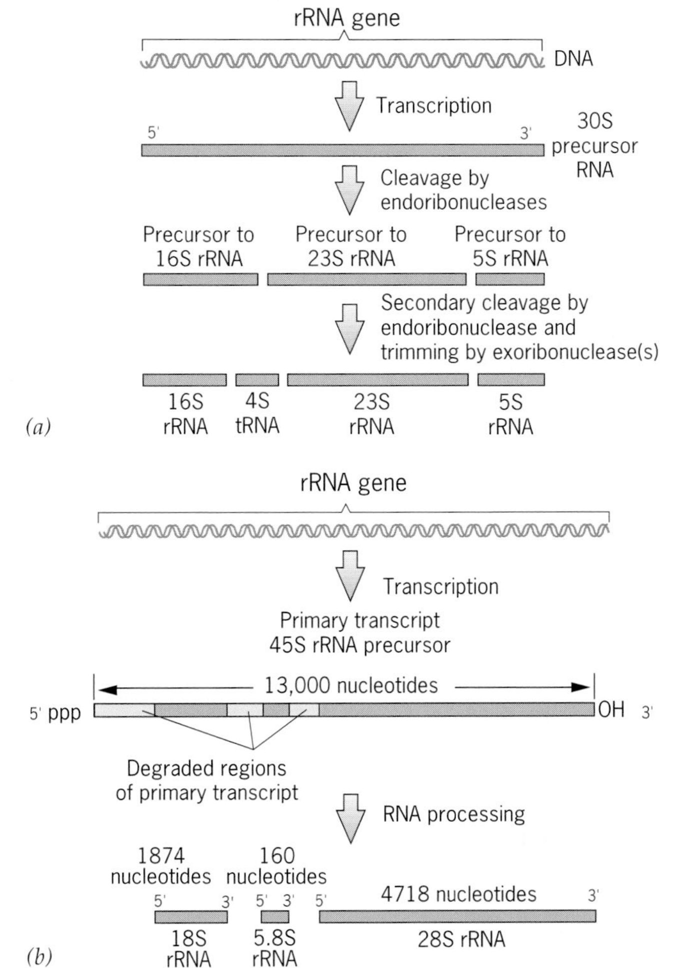

Figure 12.10 Synthesis and processing of (*a*) the 30S rRNA precursor in *E. coli* and (*b*) the 45S rRNA precursor in mammals.

The transcription of the rRNA genes produces precursors that are much larger than the RNA molecules found in ribosomes. These rRNA precursors undergo post-transcriptional processing to produce the mature rRNA molecules. In *E. coli*, the rRNA gene transcript is a 30S precursor, which undergoes endonucleolytic cleavages to produce the 5S, 16S, and 23S rRNAs plus one 4S transfer RNA molecule (Figure 12.10*a*). In mammals, the 5.8S, 18S, and 28S rRNAs are cleaved from a 45S precursor (Figure 12.10*b*), whereas the 5S rRNA is produced by post-transcriptional processing of a separate gene transcript. In addition to the post-transcriptional cleavages of rRNA precursors, many of the nucleotides in rRNAs are post-transcriptionally methylated. The methylation is thought to protect rRNA molecules from degradation by ribonucleases.

Multiple copies of the genes for rRNA are present in the genomes of all organisms that have been studied to date. This redundancy of rRNA genes is not surprising considering the large number of ribosomes present per cell. In *E. coli*, seven rRNA (*rrn*) genes are distributed among three distinct sites on the chromosome. In eukaryotes, the rRNA genes are present in hundreds to thousands of copies. The 5.8S-18S-28S rRNA genes of eukaryotes are present in tandem arrays in the nucleolar organizer regions of the chromosomes. In some eukaryotes, such as maize, there is a single pair of nucleolar organizers (on chromosome 6 in maize). In *Drosophila* and the South African clawed toad, *Xenopus laevis*, the sex chromosomes carry the nucleolar organizers. Humans have five pairs of nucleolar organizers located on the short arms of chromosomes 13, 14, 15, 21, and 22. The 5S rRNA genes in eukaryotes are not located in the nucleolar organizer regions. Instead, they are distributed over several chromosomes. However, the 5S rRNA genes are highly redundant, just as are the 5.8S-18S-28S rRNA genes.

The nucleolar organizer region in *Xenopus laevis* contains about 500 copies of the 5.8S-18S-28S rRNA gene. The estimates of rRNA gene redundancy are similar for other animals. Plants exhibit a greater variation in rRNA gene redundancy, with several thousand copies present in some genomes. Intraspecies variation in the amount of rRNA gene redundancy has also been documented in several species.

Although the ribosomes provide many of the components required for protein synthesis, and the specifications for each polypeptide are encoded in an mRNA molecule, the translation of a coded mRNA message into a sequence of amino acids in a polypeptide requires one additional class of RNA molecules, the transfer RNA (tRNA) molecules. Chemical considerations suggested that direct interactions between the amino acids and the nucleotide triplets or codons in mRNA were unlikely. Thus, in 1958, Francis Crick proposed that some kind of an adaptor molecule must mediate the specification of amino acids by codons in mRNAs during protein synthesis. The adapter molecules were soon identified by other researchers and shown to be small (4S, 70–90 nucleotides long) RNA molecules. These molecules, first called soluble RNA (sRNA) molecules and subsequently transfer RNA (tRNA) molecules, contain a triplet nucleotide sequence, the anticodon, which is complementary to and base-pairs with the codon sequence in mRNA during translation. There are from one to four tRNAs for each of the 20 amino acids.

The amino acids are attached to the tRNAs by high-energy (very reactive) bonds between the carboxyl groups of the amino acids and the 3'-hydroxyl termini of the tRNAs. The tRNAs are activated or charged with amino acids in a two-step process, with both reactions catalyzed by the same enzyme, aminoacyl-tRNA synthetase. There is at least one aminoacyl-tRNA synthetase for each of the 20 amino acids. The first step in aminoacyl-tRNA synthesis involves the activation of the amino acid using energy from adenosine triphosphate (ATP):

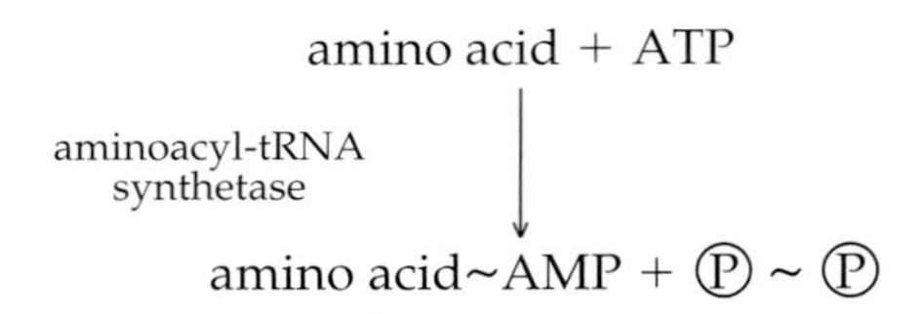

The amino acid~AMP intermediate is not normally released from the enzyme before undergoing the second step in aminoacyl-tRNA synthesis, namely, the reaction with the appropriate tRNA:

$$\text{amino acid} \sim \text{AMP} + \text{tRNA} \xrightarrow{\text{aminoacyl-tRNA synthetase}} \text{amino acid} \sim \text{tRNA} + \text{AMP}$$

The aminoacyl~tRNAs are the substrates for polypeptide synthesis on ribosomes, with each activated tRNA recognizing the correct mRNA codon and presenting the amino acid in a steric configuration (three-dimensional structure) that facilitates peptide bond formation.

The tRNAs are transcribed from genes. As in the case of rRNAs, the tRNAs are transcribed in the form of larger precursor molecules that undergo post-transcriptional processing (cleavage, trimming, methylation, and so forth). The mature tRNA molecules contain several nucleosides that are not present in the primary tRNA gene transcripts. These unusual nucleosides, such as inosine, pseudouridine, dihydrouridine, 1-methylguanosine, and several others, are produced by post-transcriptional, enzyme-catalyzed modifications of the four nucleosides incorporated into RNA during transcription.

Because of their small size (70 to 80 nucleotides long), tRNAs have been more amenable to structural analysis than the other, larger molecules of RNA involved in protein synthesis. The complete nucleotide sequence and proposed cloverleaf structure of the alanine tRNA of yeast (Figure 12.11) was published by Robert W. Holley and colleagues in 1965; Holley shared the 1968 Nobel Prize in physiology and medicine for this work. Since then, many tRNAs have been sequenced, and the yeast alanine tRNA gene has been synthesized *in vitro* from mononucleotides by H. Ghobind Khorana and coworkers. The three-dimensional structure of the phenylalanine tRNA of yeast was determined by X-ray diffraction studies in 1974 (Figure 12.12). The anticodon of each tRNA occurs within a loop (nonhydrogen-bonded region) near the middle of the molecule.

It should be apparent that tRNA molecules must contain a great deal of specificity despite their small size. Not only must they (1) have the correct anticodon sequences, so as to respond to the right codons, but

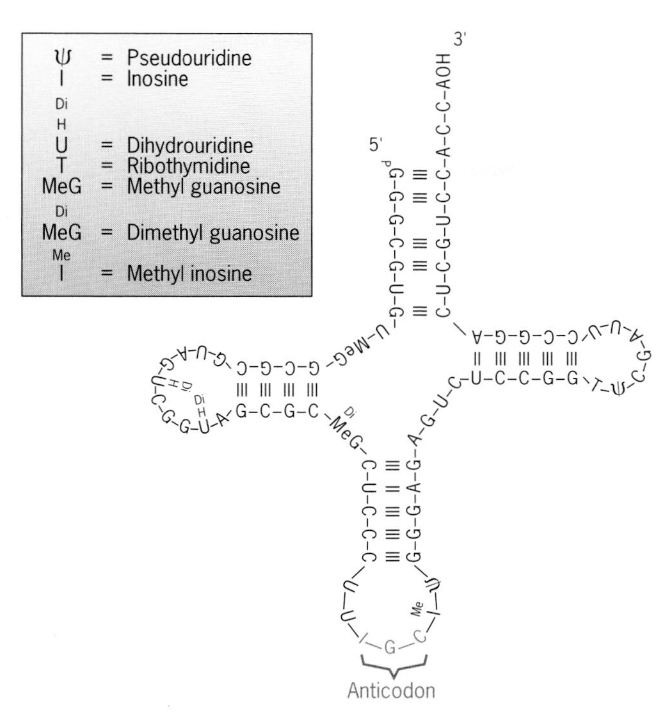

Figure 12.11 Nucleotide sequence and cloverleaf configuration of the alanine tRNA of *S. cerevisiae*. The names of the modified nucleosides present in the tRNA are shown in the inset.

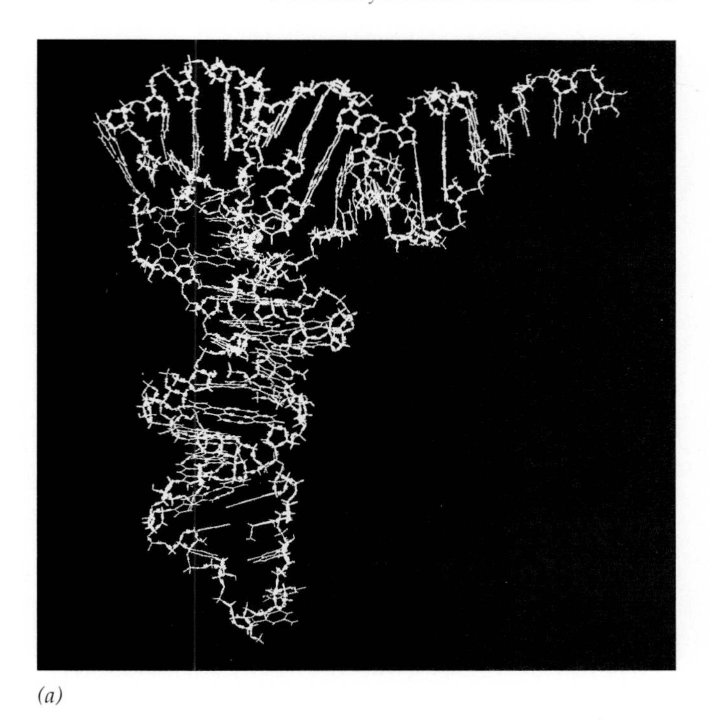

(a)

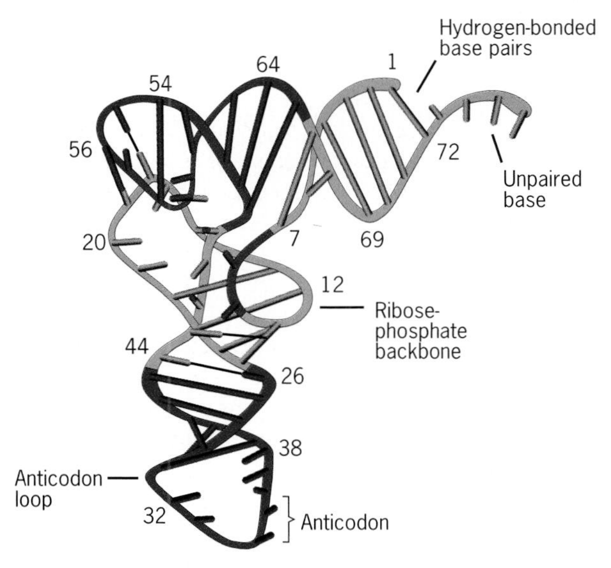

(b)

Figure 12.12 Photograph (a) and interpretative drawing (b) of a molecular model of the yeast phenylalanine tRNA based on X-ray diffraction data.

they also (2) must be recognized by the correct aminoacyl-tRNA synthetases, so that they are activated with the correct amino acids, and (3) bind to the appropriate sites on the ribosomes to carry out their adaptor functions.

François Chapeville and Günter von Ehrenstein and colleagues have proven, by means of a simple and direct experiment (Figure 12.13), that the specificity for codon recognition resides in the tRNA portion of an aminoacyl-tRNA, rather than in the amino acid. They treated cysteyl-tRNA^{Cys} (the cysteine tRNA activated with cysteine) with a strongly reducing nickel powder (Raney nickel), which converted (reduced) the cysteine to alanine while still attached to the cysteine tRNA. When this hybrid aminoacyl-tRNA, alanyl-tRNA^{Cys}, was used in *in vitro* protein-synthesizing systems, alanine was incorporated into polypeptides at positions normally occupied by cysteine. Thus tRNAs really do function as the adaptor molecules that Crick proposed must mediate the interaction between the codons in mRNAs and the amino acids that the codons specify during the translation process.

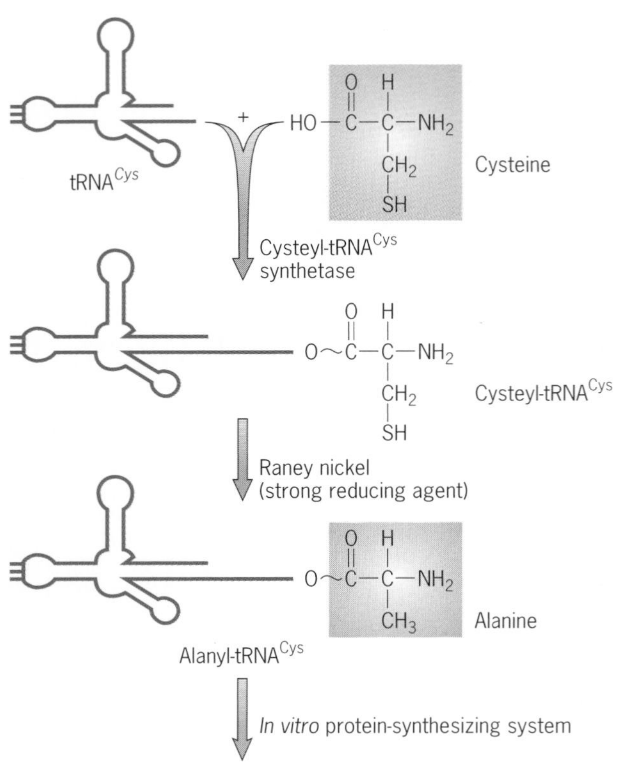

Experiment I:
Using poly-UG (UGUGU etc., repeating UG copolymer) as an artificial messenger RNA. Alanine attached to tRNA^{Cys} was incorporated, despite the fact that the alanine codons are GCU, GCC, GCA, and GCG. UGU = cysteine codon!

Experiment II:
Using the hemoglobin-synthesizing rabbit reticulocyte system containing native hemoglobin mRNAs. Demonstrated that alanine from alanyl-tRNA^{Cys} was incorporated into positions in the rabbit globin chains normally occupied by cysteine.

Figure 12.13 Proof that the codon-recognizing specificity of an aminoacyl-tRNA complex resides in the tRNA rather than in the amino acid.

Three tRNA binding sites are present on each ribosome (Figure 12.14). The *A* or **aminoacyl site** binds the incoming aminoacyl-tRNA, the tRNA carrying the next amino acid to be added to the growing polypeptide chain. The ***P*** **or peptidyl site** binds the tRNA to which the growing polypeptide is attached. The ***E*** or **exit site** binds the departing uncharged tRNA. The specificity for aminoacyl-tRNA binding in these sites is provided by the mRNA codons that make up part of the binding sites. As the ribosome moves along an mRNA (or as the mRNA is shuttled across the ribosome), the specificity for the aminoacyl-tRNA binding in the *A*, *P*, and *E* sites changes as different mRNA codons move into register in the binding sites. The ribosomal binding sites by themselves (minus mRNA) are thus capable of binding any aminoacyl-tRNA.

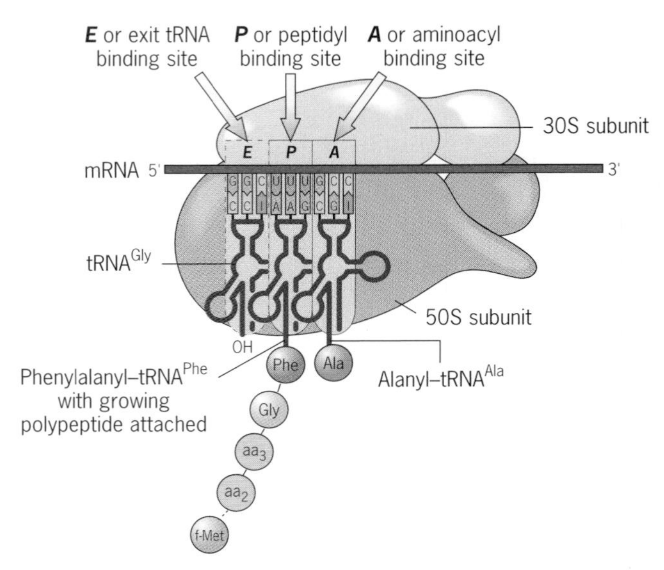

Figure 12.14 Each ribosome-mRNA complex contains three aminoacyl-tRNA binding sites. The *A* or aminoacyl-tRNA site is occupied by an alanyl-tRNA^{Ala} complex. The *P* or peptidyl site is occupied by a phenylalanyl-tRNA^{Phe} complex, with the growing polypeptide chain covalently linked to the phenylalnine tRNA. The *E* or exit site is occupied by the tRNA^{Gly} prior to its release from the ribosome.

Translation: The Synthesis of Polypeptides Using mRNA Templates

We now have reviewed all the major components of the protein-synthesizing system. The mRNA molecules provide the specifications for the amino acid sequences of the polypeptide gene products. The ribosomes provide many of the macromolecular components required for the translation process. The tRNAs provide the adaptor molecules needed to incorporate amino acids into polypeptides in response to codons in mRNAs. In addition, several soluble proteins participate in the process. The translation of the sequence of nucleotides in an mRNA molecule into the sequence of amino acids in its polypeptide product can be divided into three stages: (1) polypeptide chain initiation, (2) chain elongation, and (3) chain termination.

The **initiation** of translation includes all events that precede the formation of a peptide bond between the first two amino acids of the new polypeptide chain. Although several aspects of the initiation process are the same in prokaryotes and eukaryotes, some are different. Thus first we will examine the initiation of polypeptide chains in *E. coli*, and then we

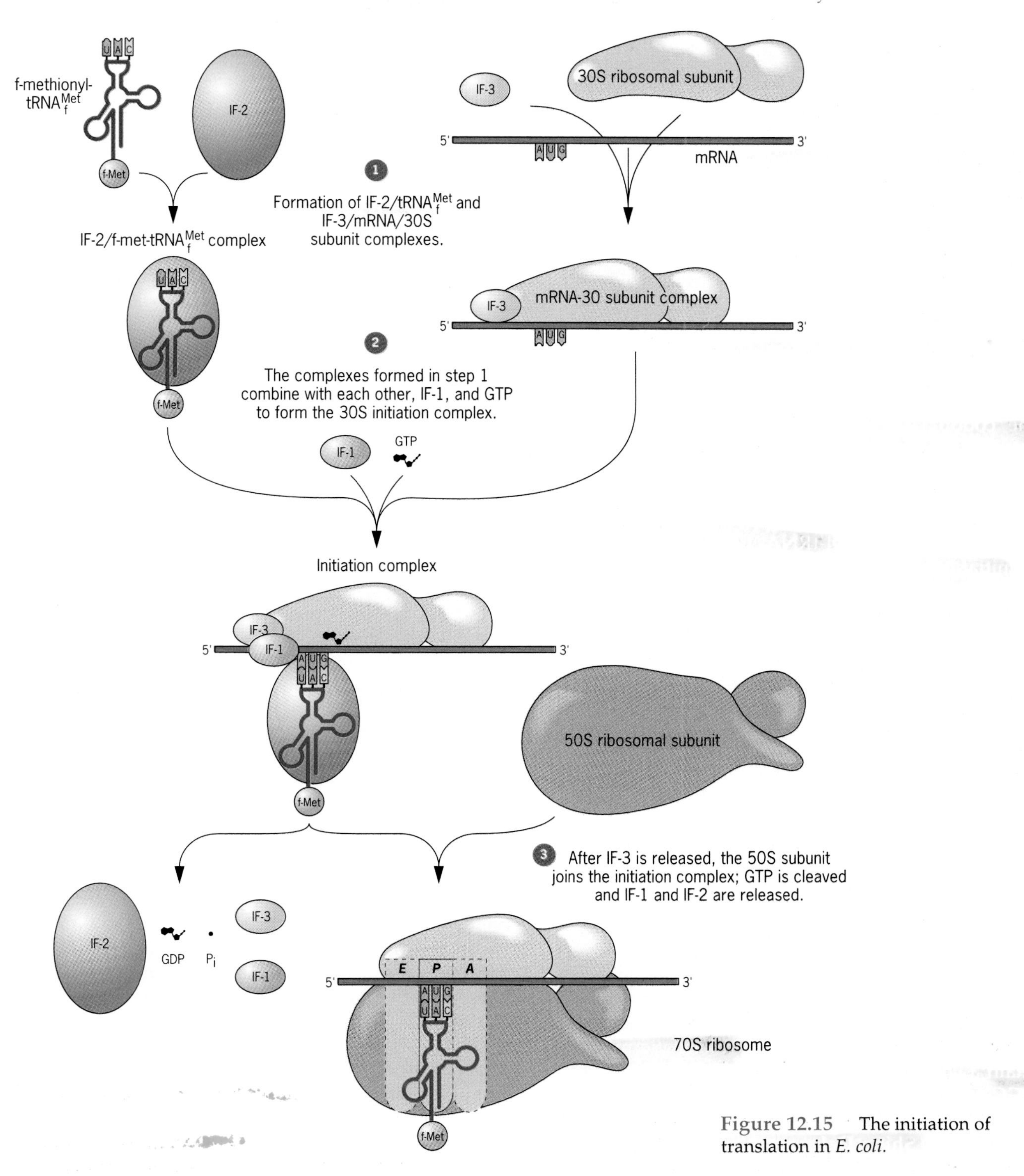

Figure 12.15 The initiation of translation in *E. coli.*

will look at the unique aspects of translational initiation in eukaryotes.

In *E. coli,* the initiation process involves the 30S subunit of the ribosome, a special initiator tRNA, an mRNA molecule, three soluble protein **initiation factors: IF-1, IF-2,** and **IF-3,** and one molecule of GTP (Figure 12.15). Translation occurs on 70S ribosomes, but the ribosomes dissociate into their 30S and 50S subunits each time they complete the synthesis of a polypeptide chain. In the first stage of the initiation of translation, a free 30S subunit interacts with an mRNA molecule and the initiation factors. The 50S subunit joins the complex to form the 70S ribosome in the final step of the initiation process.

The synthesis of polypeptides is initiated by a special tRNA, designated **$tRNA_f^{Met}$**. This means that all polypeptides begin with methionine during synthesis. The amino-terminal methionine is subsequently cleaved from many polypeptides. Thus functional proteins need not have an amino-terminal methionine. The methionine on the initiator $tRNA_f^{Met}$ has the amino group blocked with a formyl ($-\overset{\overset{\displaystyle O}{\|}}{C}-H$) group. A distinct methionine tRNA, **$tRNA^{Met}$**, responds to internal methionine codons. Both methionine tRNAs have the same anticodon, and both respond to the same codon (AUG) for methionine. However, the formylated amino group on methionyl-$tRNA_f^{Met}$ prevents the formation of a peptide bond between the amino group and the carboxyl group of the amino acid at the end of the growing polypeptide chain. In addition, only methionyl-$tRNA_f^{Met}$ interacts with protein initiation factor IF-2 to begin the initiation process (Figure 12.15). Thus only methionyl-tRNA binds to the ribosome in response to AUG initiation codons in mRNAs, leaving methionyl-$tRNA^{Met}$ to bind in response to internal AUG codons. Methionyl-$tRNA_f^{Met}$ also binds to ribosomes in response to an alternate initiator codon, GUG (a valine codon when present at internal positions), that occurs in a some mRNA molecules.

Polypeptide chain initiation begins with the formation of two complexes: (1) one contains initiation factor IF-2 and methionyl-$tRNA_f^{Met}$, and (2) the other contains an mRNA molecule, a 30S ribosomal subunit and initiation factor IF-3 (Figure 12.15). The 30S subunit/mRNA complex will form only in the presence of IF-3; thus IF-3 controls the ability of the 30S subunit to begin the initiation process. The formation of the 30S subunit/mRNA complex depends in part on base-pairing between a nucleotide sequence near the 3' end of the 16S rRNA and a sequence near the 5' end of the mRNA molecule (Figure 12.16). Prokaryotic mRNAs contain a conserved polypurine tract, consensus AGGAGG, located about seven nucleotides upstream from the AUG initiation codon. This conserved hexamer, called the **Shine-Dalgarno sequence** after the scientists who discovered it, is complementary to a sequence near the 5' terminus of the 16S ribosomal RNA. When the Shine-Dalgarno sequences of mRNAs are modified so that they can no longer base-pair with the 16S rRNA, the modified mRNAs are either not translated or are translated very inefficiently, indicating that this base-pairing plays an important role in translation.

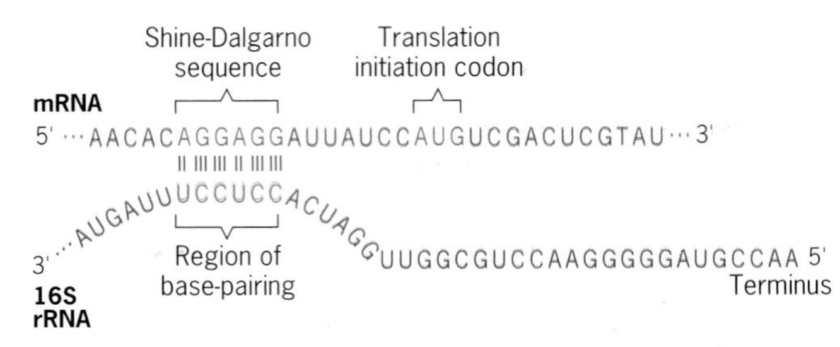

Figure 12.16 Base-pairing between the Shine-Dalgarno sequence in mRNA and a complementary sequence near the 5' terminus of the 16S rRNA is involved in the formation of the mRNA/30S ribosomal subunit initiation complex.

The IF-2/methionyl-$tRNA_f^{Met}$ complex and the mRNA/30S subunit/IF-3 complex then combine with each other and with initiation factor IF-1 and one molecule of GTP to form the complete 30S initiation complex. The final step in the initiation of translation is the addition of the 50S subunit to the 30S initiation complex to produce the complete 70S ribosome. Initiation factor IF-3 must be released from the complex before the 50S subunit can join the complex; IF-3 and the 50S subunit are never found to be associated with the 30S subunit at the same time. The addition of the 50S subunit requires energy from GTP and the release of initiation factors IF-1 and IF-2. The results of early experiments suggested that one molecule of GTP was required for polypeptide chain initiation. More recent data indicate that two molecules of GTP are required to form the initiation complex; however, there is still not complete agreement as to whether one or two molecules of GTP are required for initiation.

The addition of the 50S ribosomal subunit positions the initiator tRNA, methionyl-$tRNA_f^{Met}$, in the peptidyl (*P*) site with the anticodon of the tRNA aligned with the AUG initiation codon of the mRNA. Methionyl-$tRNA_f^{Met}$ is the only aminoacyl-tRNA that can enter the *P* site directly, without first passing through the aminoacyl (*A*) site. With the initiator AUG positioned in the *P* site, the second codon in the mRNA is in register with the *A* site, dictating the aminoacyl-tRNA binding specificity at that site and setting the stage for the second phase in polypeptide synthesis, chain elongation.

The initiation of translation is more complex in eukaryotes, involving several soluble initiation factors. Nevertheless, the overall process is similar except for two features. (1) The amino group of the methionine on the initiator tRNA is not formylated as in prokaryotes. (2) The initiation complex forms at the 5' terminus of the mRNA, not at the Shine-Dalgarno/AUG translation start site as in *E. coli*. In eukaryotes, the initiation complex scans the mRNA, starting at the 5' end, searching for an AUG translation-initiation codon. Thus, in eukaryotes, translation usually begins at the AUG closest to the 5' terminus of the mRNA molecule, although the efficiency with which a given AUG is used to initiate translation depends on the contiguous nucleotide sequence.

Like prokaryotes, eukaryotes contain a special initiator tRNA, **tRNA$_i^{Met}$**, but the amino group of the methionyl-tRNA$_i^{Met}$ is not formylated. The initiator methionyl-tRNA$_i^{Met}$ interacts with a soluble initiation factor and enters the *P* site directly during the initiation process, just as in *E. coli*.

In eukaryotes, a cap-binding protein (CBP) binds to the 7-methyl guanosine cap at the 5' terminus of the mRNA. Then, other initiation factors bind to the CBP-mRNA complex, followed by the small (40S) subunit of the ribosome. The entire initiation complex moves 5' → 3' along the mRNA molecule, searching for an AUG codon. When an AUG triplet is found, the initiation factors dissociate from the complex, and the large (60S) subunit binds to the methionyl-tRNA/mRNA/40S subunit complex, forming the complete (80S) ribosome. The 80S ribosome/mRNA/tRNA complex is ready to begin the second phase of translation, chain elongation.

The process of polypeptide chain **elongation** is basically the same in both prokaryotes and eukaryotes. The addition of each amino acid to the growing polypeptide occurs in three steps: (1) binding of an aminoacyl-tRNA to the *A* site of the ribosome, (2) transfer of the growing polypeptide chain from the tRNA in the *P* site to the tRNA in the *A* site by the formation of a new peptide bond, and (3) translocation of the ribosome along the mRNA to position the next codon in the *A* site (Figure 12.17). During step 3, the nascent polypeptide-tRNA and the uncharged tRNA are translocated from the *A* and *P* sites to the *P* and *E* sites, respectively. These three steps are repeated in a cyclic manner throughout the elongation process. The soluble factors involved in chain elongation in *E. coli* are described here. Similar factors participate in chain elongation in eukaryotes.

In the first step, an aminoacyl-tRNA enters and becomes bound to the *A* site of the ribosome, with the specificity provided by the mRNA codon in register with the *A* site (Figure 12.17). The three nucleotides in the anticodon of the incoming aminoacyl-tRNA must pair with the nucleotides of the mRNA codon present at the *A* site. This step requires **elongation factor Tu** carrying a molecule of GTP (**EF-TuGTP**). The GTP is required for aminoacyl-tRNA binding at the *A* site, but is not cleaved until the peptide bond is formed. After the cleavage of GTP, EF-TuGDP is released from the ribosome. EF-TuGDP is inactive and will not bind to aminoacyl-tRNAs. EF-TuGDP is converted to the active EF-TuGTP form by **elongation factor Ts** (**EF-Ts**), which hydrolyzes one molecule of GTP in the process. EF-Tu interacts with all of the aminoacyl-tRNAs except methionyl-tRNA.

The second step in chain elongation is the formation of a peptide bond between the amino group of the aminoacyl-tRNA in the *A* site and the carboxyl terminus of the growing polypeptide chain attached to the tRNA in the *P* site. This uncouples the growing chain from the tRNA in the *P* site and covalently joins the chain to the tRNA in the *A* site (Figure 12.17). This key reaction is catalyzed by **peptidyl transferase**, an enzymatic activity built into the 50S subunit of the ribosome. We should note that the peptidyl transferase activity resides in the 23S rRNA molecule rather than in a ribosomal protein. Peptide bond formation requires the hydrolysis of the molecule of GTP brought to the ribosome by EF-Tu in step 1.

During the third step in chain elongation, the peptidyl-tRNA present in the *A* site of the ribosome is translocated to the *P* site, and the uncharged tRNA in the *P* site is translocated to the *E* site, as the ribosome moves three nucleotides toward the 3' end of the mRNA molecule. The translocation step requires GTP and **elongation factor G (EF-G)**. The ribosome undergoes changes in conformation during the translocation process, suggesting that it may shuttle along the mRNA molecule. The energy for the movement of the ribosome is provided by the hydrolysis of GTP. The translocation of the peptidyl-tRNA from the *A* site to the *P* site leaves the *A* site unoccupied and the ribosome ready to begin the next cycle of chain elongation.

The elongation of one eukaryotic polypeptide, the silk protein fibroin, can be visualized with the electron microscope by using techniques developed by Oscar Miller, Barbara Hamkalo, and colleagues. Most proteins fold up on the surface of the ribosome during their synthesis. However, fibroin remains extended from the surface of the ribosome under the conditions used by Miller and coworkers. As a result, nascent polypeptide chains of increasing length can be seen attached to the ribosomes as they are scanned from the 5' end of the mRNA to the 3' end (Figure 12.18). Fibroin is a large protein with a mass of over 200,000 daltons; it is synthesized on large polyribosomes containing 50 to 80 ribosomes.

Polypeptide chain elongation proceeds rapidly. In *E. coli*, all three steps required to add one amino acid to the growing polypeptide chain occur in about 0.05 second. Thus the synthesis of a polypeptide containing 300 amino acids takes only about 15 seconds. Given its complexity, the accuracy and efficiency of the translational apparatus are indeed amazing.

Polypeptide chain elongation undergoes **termination** when any of three **chain-termination codons**

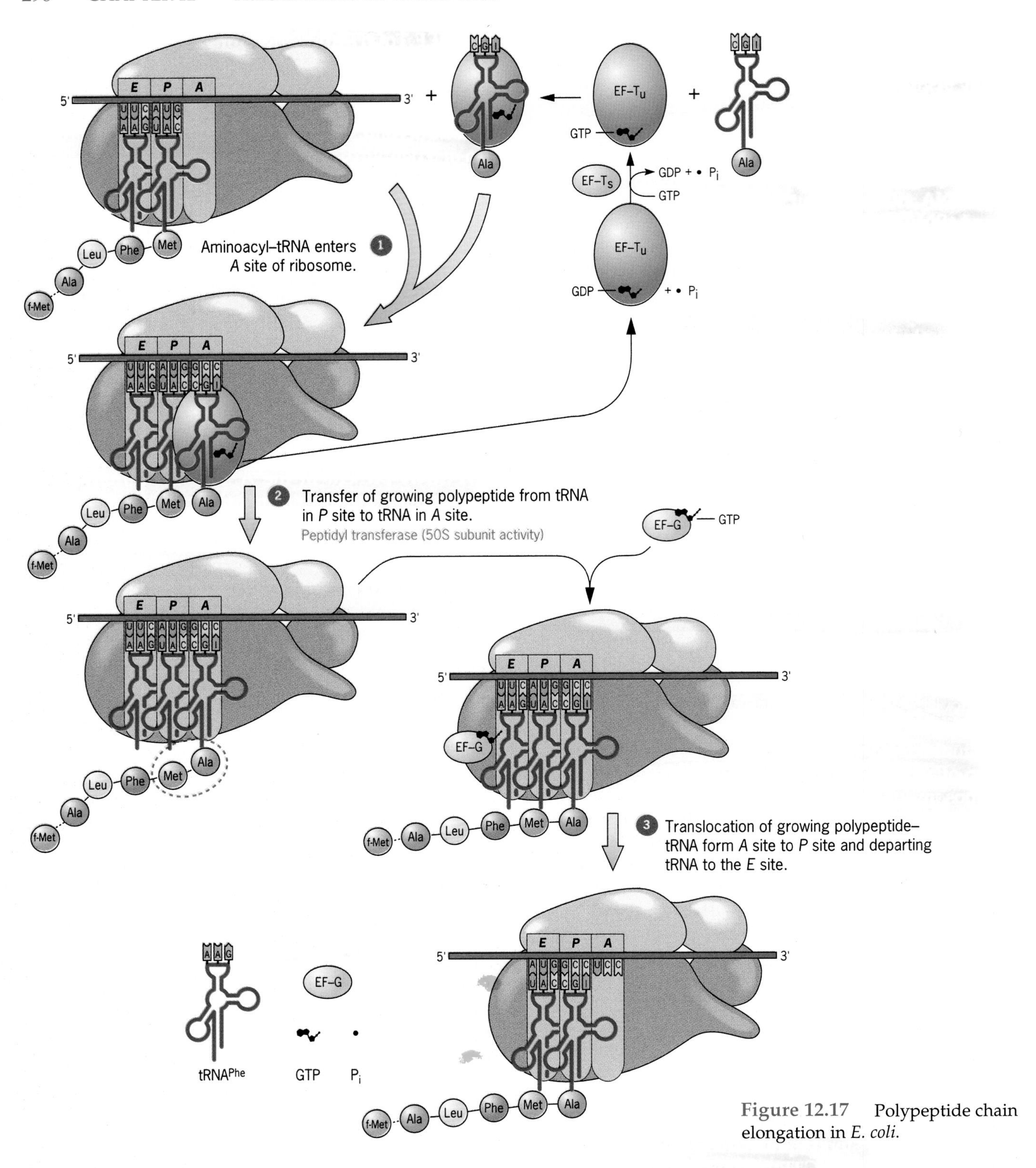

Figure 12.17 Polypeptide chain elongation in *E. coli.*

(UAA, UAG, or UGA) enters the *A* site on the ribosome (Figure 12.19). These three stop codons are recognized by soluble proteins called **release factors (RFs)**. In *E. coli*, there are two release factors, RF-1 and RF-2. RF-1 recognizes termination codons UAA and UAG; RF-2 recognizes UAA and UGA. In eukaryotes, a single release factor (**eRF**) recognizes all three termination codons. The presence of a release factor in the *A* site alters the activity of peptidyl transferase such that it adds a water molecule to the carboxyl terminus of the nascent polypeptide. This releases the polypeptide from the tRNA molecule in the *P* site and triggers the

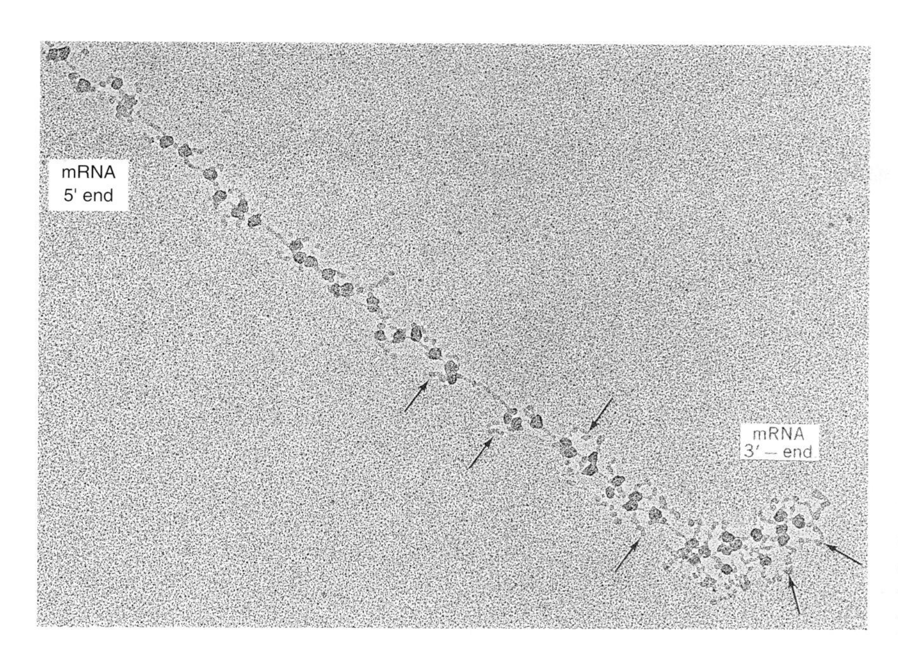

Figure 12.18 Visualization of the elongation of fibroin polypeptides in the posterior silk gland of the silkworm *Bombyx mori*. The arrows point to growing fibroin polypeptides. Note their increasing length as one approaches the 3' end of the mRNA molecule.

translocation of the free tRNA to the *E* site. Termination is completed by the release of the mRNA molecule from the ribosome and the dissociation of the ribosome into its subunits. The ribosomal subunits are then ready to initiate another round of protein synthesis, as previously described.

Key Points: **Genetic information carried in the sequence of nucleotides in an mRNA molecule is translated into a sequence of amino acids in a polypeptide gene product by intricate macromolecular machines called ribosomes. The translation process is complex, requiring the participation of many different RNA and protein molecules. Transfer RNA molecules serve as adaptors, mediating the interaction between amino acids and codons in mRNA. The process of translation involves the initiation, elongation, and termination of polypeptide chains and is governed by the specifications of the genetic code.**

THE GENETIC CODE

As it became evident that genes controlled the structure of polypeptides, attention focused on how the sequence of the four different nucleotides in DNA could control the sequence of the 20 amino acids present in proteins. With the discovery of the mRNA intermediary, the question became one of how the sequence of the four bases present in mRNA molecules could specify the amino acid sequence of a polypeptide. What is the nature of the genetic code relating mRNA base sequences to amino acid sequences? Clearly, the symbols or letters used in the code must be the bases; but what comprises a codon, the unit or word specifying one amino acid or, actually, one aminoacyl-tRNA?

Properties of the Genetic Code: An Overview

The main features of the genetic code were worked out during the 1960s. Cracking the code was one of the most exciting eras in the history of science, with new information reported almost daily. By the mid-1960s, the genetic code was largely solved. Before focusing on specific features of the code, let us consider its most important properties.

1. *The genetic code is composed of nucleotide triplets.* Three nucleotides in mRNA specify one amino acid in the polypeptide product; thus each codon contains three nucleotides.
2. *The genetic code is nonoverlapping.* Each nucleotide in mRNA belongs to just one codon except in rare cases where genes overlap.
3. *The genetic code is comma-free.* There are no commas or other forms of punctuation within the coding regions of mRNA molecules. During translation, the codons are read consecutively.
4. *The genetic code is degenerate.* All but two of the amino acids are specified by more than one codon.
5. *The genetic code is ordered.* Multiple codons for a given amino acid and codons for amino acids with similar chemical properties are closely related, usually differing by a single nucleotide.
6. *The genetic code contains start and stop codons.* Specific

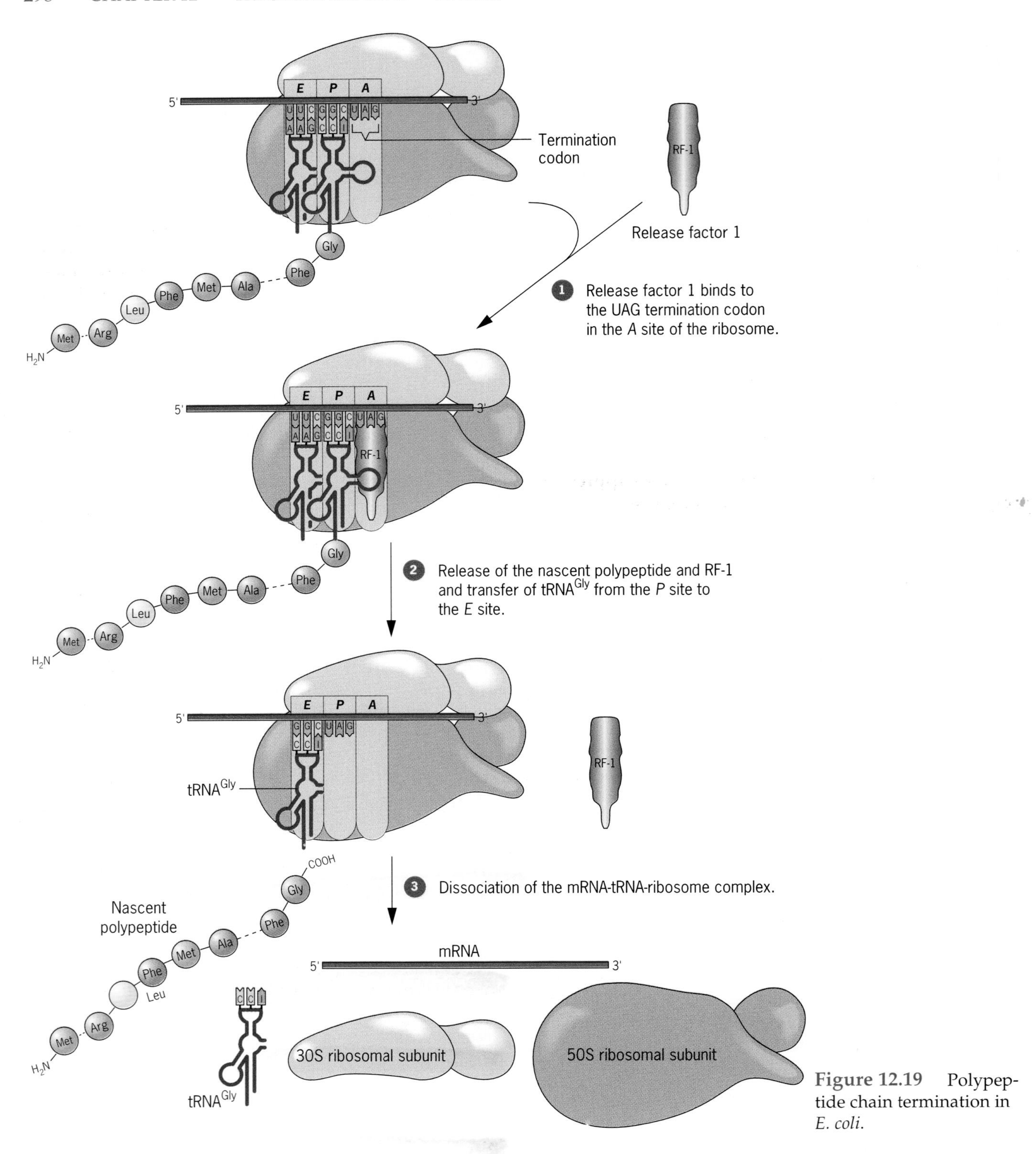

Figure 12.19 Polypeptide chain termination in *E. coli*.

codons are used to initiate and to terminate polypeptide chains.

7. *The genetic code is nearly universal.* With minor exceptions, the codons have the same meaning in all living organisms, from viruses to humans.

Three Nucleotides per Codon

Twenty different amino acids are incorporated into polypeptides during translation. Thus at least 20 different codons must be formed with the four bases available in mRNA. Two bases per codon would result in only 4^2 or 16 possible codons—clearly not enough. Three bases per codon yield 4^3 or 64 possible codons—an apparent excess.

In 1961, Francis Crick and colleagues published the first strong evidence in support of a **triplet code** (three nucleotides per codon). Crick and coworkers carried out a genetic analysis of mutations induced at the *r*II locus of bacteriophage T4 by the chemical proflavin. Proflavin is a mutagenic agent that causes single base-pair additions and deletions (Chapter 13). Phage T4 *r*II mutants are unable to grow in cells of *E. coli* strain K12, but grow like wild-type phage in cells of *E. coli* strain B (Chapter 15). Wild-type T4 grows equally well on either strain. Crick and coworkers isolated proflavin-induced revertants of a proflavin-induced mutation. These revertants were shown to result from the occurrence of additional mutations at nearby sites rather than reversion of the original mutation. Second-site mutations that restore the wild-type phenotype in a mutant organism are called **suppressor mutations** because they cancel, or suppress, the effect(s) of the original mutation.

Crick and colleagues reasoned that if the original mutation was a single base-pair addition or deletion, then the suppressor mutations must be single base-pair deletions or additions, respectively, occurring at a site or sites near the original mutation. If sequential nucleotide triplets in an mRNA specify amino acids, then every nucleotide sequence can be recognized or read during translation in three different ways. For example, the sequence AAAGGGCCCTTT can be read (1) AAA, GGG, CCC, TTT, (2) A, AAG, GGC, CCT, TT, or (3) AA, AGG, GCC, CTT, T. The **reading frame** of an mRNA is the series of nucleotide triplets that are read (positioned in the *A* site of the ribosome) during translation. A single base-pair addition or deletion will alter the reading frame of the gene and mRNA for that portion of the gene distal to the mutation. This effect is illustrated in Figure 12.20*a*. The suppressor mutations were then isolated as single mutants by screening progeny of backcrosses to wild-type. Like the original mutation, the suppressor mutations were found to produce mutant phenotypes. Crick and colleagues next isolated proflavin-induced suppressor mutations of the original suppressor mutations, and so on.

Crick and colleagues then classified all the isolated mutations into two groups, plus (+) and minus (−) (for additions and deletions, although they had no

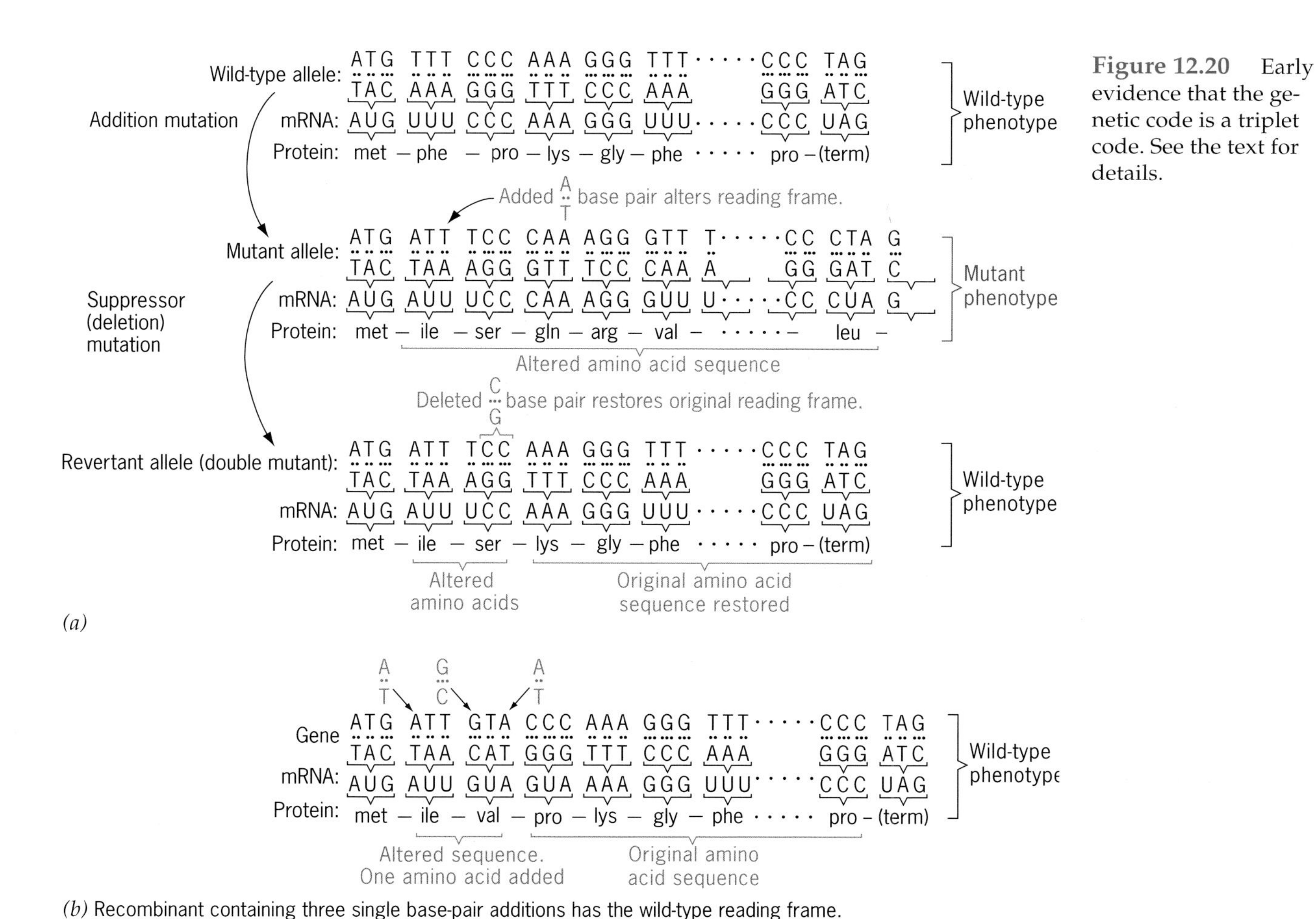

(*b*) Recombinant containing three single base-pair additions has the wild-type reading frame.

Figure 12.20 Early evidence that the genetic code is a triplet code. See the text for details.

idea which group was which), based on the reasoning that a (+) mutation would suppress a (−) mutation, but not another (+) mutation, and vice versa (Figure 12.20). Then, Crick and coworkers constructed recombinants that carried various combinations of the (+) and the (−) mutations. Like the single mutants, recombinants with two (+) mutations or two (−) mutations always had the mutant phenotype. The critical result was that recombinants with three (+) mutations (Figure 12.20*b*) or three (−) mutations often exhibited the wild-type phenotype. This indicated that the addition of three base pairs or the deletion of three base pairs left the distal portion of the gene with the wild-type reading frame. This result would be expected only if each codon contained three nucleotides.

Evidence from *in vitro* translation studies soon supported the results of Crick and colleagues and firmly established the triplet nature of the code. Some of the more important results follow: (1) Trinucleotides were sufficient to stimulate specific binding of aminoacyl-tRNAs to ribosomes. For example, 5'-UUC-3' stimulated the binding of phenylalanyl-$tRNA^{Phe}$ to ribosomes. (2) Chemically synthesized mRNA molecules that contained repeating dinucleotide sequences directed the synthesis of copolymers (large chainlike molecules composed of two different subunits) with alternating amino acid sequences. For example, when $poly(UG)_n$ was used as an artificial mRNA in an *in vitro* translation system, the repeating copolymer $(cys\text{-}val)_m$ was synthesized. (The subscripts n and m refer to the number of nucleotides and amino acids in the respective polymers.) (3) In contrast, mRNAs with repeating trinucleotide sequences directed the synthesis of a mixture of three homopolymers (initiation being at random on such mRNAs in the *in vitro* systems). For example, $poly(UUG)_n$ directed the synthesis of a mixture of polyleucine, polycysteine, and polyvaline. These results are consistent only with a triplet code, with its three different reading frames. When $poly(UUG)_n$ is translated in reading frame 1, UUG, UUG, polyleucine is produced, whereas translation in reading frame 2, UGU, UGU, yields polycysteine, and translation in reading frame 3, GUU, GUU, produces polyvaline. Ultimately, the triplet nature of the code was definitively established by comparing the nucleotide sequences of genes and mRNAs with the amino acid sequences of their polypeptide products.

Deciphering the Code

The cracking of the genetic code in the 1960s took several years and involved intense competition between many different research laboratories. New information accumulated rapidly but sometimes was inconsistent with earlier data. Indeed, cracking the code proved to be a major challenge.

Deciphering the genetic code required scientists to obtain answers to several questions. (1) Which codons specify each of the 20 amino acids? (2) How many of the 64 possible triplet codons are utilized? (3) How is the code punctuated? (4) Do the codons have the same meaning in viruses, bacteria, plants, and animals? The answers to these questions were obtained primarily from the results of two types of experiments, both of which were performed with cell-free systems. The first type of experiment involved translating artificial mRNA molecules *in vitro* and determining which of the 20 amino acids were incorporated into proteins. In the second type of experiment, ribosomes were activated with mini-mRNAs just three nucleotides long. Then, researchers determined which aminoacyl-tRNAs were stimulated to bind to ribosomes activated with each of the trinucleotide messages.

The first major breakthrough came in 1961 when Marshall Nirenberg (1968 Nobel Prize recipient) and J. Heinrich Matthaei demonstrated that synthetic RNA molecules could be used as artificial mRNAs to direct *in vitro* protein synthesis. When ribosomes, aminoacyl-tRNAs, and the soluble factors required for translation are purified free of natural mRNAs, these components can be combined *in vitro* and stimulated to synthesize polypeptides by the addition of chemically synthesized RNA molecules. If these synthetic mRNA molecules are of known nucleotide content, the amino acid composition of the resulting polypeptides can be used to deduce which codons specify which amino acids.

The first codon assignment (UUU for phenylalanine) was made when Nirenberg and Matthaei demonstrated that polyuridylic acid [$poly(U) = (U)_n$] directed the synthesis of polyphenylalanine [$(phenylalanine)_m$]. Nirenberg and Matthaei used radioactively labeled phenylalanine (^{14}C-phenylalanine) as a substrate in an *in vitro* translation system with poly(U), poly(C), and poly(A) as artificial messenger RNAs. The ^{14}C-phenylalanine was incorporated into polyphenylalanine only when poly(U) was used as the mRNA (Table 12.1). When each of 17 other labeled amino acids was used in the poly(U)-stimulated translation system, no radioactivity was incorporated into polypeptides. Given the triplet nature of the genetic code, Nirenberg and Matthaei concluded that UUU must be a codon for phenylalanine. Shortly thereafter, poly(C) and poly(A) were shown to encode polylysine and polyproline, respectively, allowing Nirenberg and colleagues to assign codon CCC to proline and AAA to lysine. Poly(G) does not function as an mRNA in the *in vitro* translation systems because of base-pairing between the guanine residues, resulting in the formation of complex three-stranded structures.

TABLE 12.1
Incorporation of ^{14}C-phenylalanine into Polyphenylalanine in an *In Vitro* Translation System Activated with Synthetic RNA Homopolymers[a]

Synthetic mRNA	*Radioactivity Incorporated (counts per minute)*
None	44
Poly (U)	39,800
Poly (A)	50
Poly (C)	38
Poly (I)[b]	57

[a]Data are from Nirenberg and Matthaei, *Proc. Natl. Acad. Sci. USA* 47:1588–1602, 1961.

[b]Poly (I) is polyinosinic acid, which contains the purine hypoxanthine. Hypoxanthine is like guanine in that it base-pairs with cytosine.

Researchers next began using synthetic mRNAs that contained two or more different nucleotides—first in random order and later in known order—to probe the nature of the code. The results of some of these experiments are discussed in the Technical Sidelight: Cracking the Genetic Code. By using RNAs with both random and fixed nucleotide sequences as mRNAs in *in vitro* translation systems, researchers were able to establish the meaning—which trinucleotide sequences specified which amino acids—of most of the 64 possible codons.

Additional information on the nature of the genetic code was obtained by assaying the binding of aminoacyl-tRNAs to ribosomes activated with small RNA oligomers. When an *in vitro* translation system is activated with poly(U), only one aminoacyl-tRNA, phenylalanyl-tRNA^{Phe}, binds to the ribosome. This specificity exists because the mRNA codon is part of the binding site, and binding involves base-pairing between the codon and the anticodon of the tRNA. In 1964, Nirenberg and Philip Leder developed an assay for aminoacyl-tRNA binding to ribosomes activated with **trinucleotides,** mini-mRNAs only three nucleotides long. Nirenberg and Leder synthesized trinucleotides of known sequence and tested their ability to stimulate the binding of specific aminoacyl-tRNAs to ribosomes. They assayed the ability of each trinucleotide to serve as a mini-mRNA by using labeled amino acids to detect the formation of trinucleotide-aminoacyl-tRNA-ribosome complexes (Figure 12.21). For example, the trinucleotides 5'-UUU-3' and 5'-UUC-3' both stimulated the binding of phenylalanyl-tRNA^{Phe} to ribosomes, indicating that UUU and UUC are both phenylalanine codons.

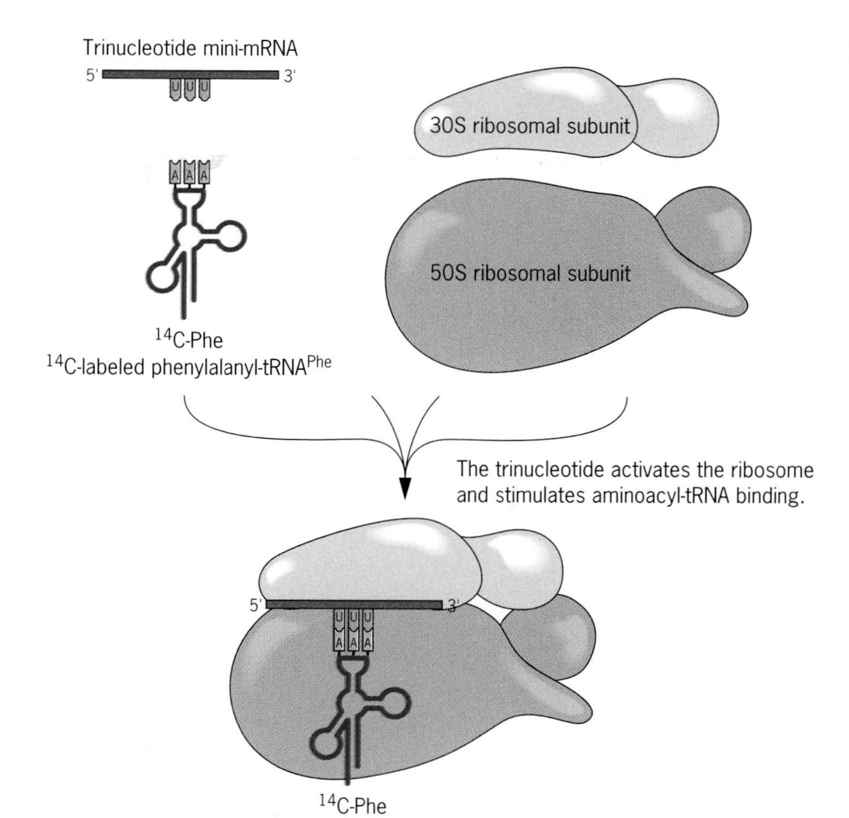

Figure 12.21 Stimulation of aminoacyl-tRNA binding to ribosomes by synthetic trinucleotide mini-mRNAs. The results of these trinucleotide-activated ribosome binding assays helped scientists crack the genetic code.

TECHNICAL SIDELIGHT

Cracking the Genetic Code: Synthetic mRNAs

Marshall Nirenberg and J. Heinrich Matthaei's 1961 demonstration that poly(U) directed the synthesis of polyphenylalanine in an in vitro translation system created excitement about the possibility of using such artificial mRNAs to probe the nature of the genetic code and stimulated a flurry of activity in several laboratories. Researchers in the laboratories of Nirenberg and Severo Ochoa (1959 Nobel Prize recipient) quickly extended the use of artificial mRNAs to synthetic copolymers with random sequences. For example, Nirenberg and co-workers synthesized a random copolymer containing approximately equal amounts of adenine and cytosine and used it as an artificial mRNA in their in vitro translation system. A random AC copolymer composed of equal amounts of A and C will contain 12.5 percent ($1/2 \times 1/2 \times 1/2 = 1/8$) of each of the eight possible codons: AAA, AAC, ACA, CAA, CCA, CAC, ACC, and CCC. Nirenberg and colleagues observed that poly(AC) directs the incorporation of six amino acids—asparagine, glutamine, histidine, lysine, proline, and threonine—into polypeptides. Since they already knew that AAA and CCC were lysine and proline codons, their results indicated that codons composed of two A's plus one C and two C's plus one A specified asparagine, glutamine, histidine, and threonine. Approximately equal amounts of asparagine, glutamine, histidine, and lysine were incorporated, but threonine and proline were incorporated in about double the amount of each of the other four amino acids. This result indicated that random AC copolymers might contain twice as many threonine and proline codons as asparagine codons. Indeed, there are two threonine codons (ACC and ACA) and two proline codons (CCC and CCA), but only one asparagine codon (AAC), in random AC copolymers.

The use of random copolymers as synthetic mRNAs provided lots of information about the nucleotide composition of codons but relatively little information about the specific nucleotide sequences of codons. By varying the nucleotide composition of random copolymers, Nirenberg, Ochoa, and their colleagues could alter the relative frequencies of the eight codons and look for correlations with the relative frequencies of the amino acids in the polypeptides synthesized in response to the copolymers. For example, Ochoa and coworkers synthesized random copolymers containing A and C in 5:1 and 1:5 ratios. When these copolymers were used as mRNAs in vitro, the same six amino acids were incorporated as in Nirenberg's experiments with a 1A:1C random copolymer, but their relative frequencies in the polypeptide products were very different. The results of Ochoa and colleagues are summarized in the following table.

A.
Predicted Codon Frequencies in Random Copolymers

		Frequency of Codon in Random Copolymer of Composition			
		5A:1C		1A:5C	
Codon Composition	*Possible Codons*	*Each Codon*	*Total*	*Each Codon*	*Total*
3A	AAA	$(5/6)^3 = 57.9\%$	57.9%	$(1/6)^3 = 0.4\%$	0.4%
2A + 1C	AAC, ACA, CAA	$(5/6)^2(1/6) = 11.6\% \times 3 =$	34.8%	$(1/6)^2(5/6) = 2.3\% \times 3 =$	6.9%
1A + 2C	CCA, CAC, ACC	$(5/6)(1/6)^2 = 2.3\% \times 3 =$	6.9%	$(1/6)(5/6)^2 = 11.6\% \times 3 =$	34.8%
3C	CCC	$(1/6)^3 = 0.4\%$	0.4%	$(5/6)^3 = 57.9\%$	57.9%

B.
Amino Acids Incorporated into Proteins In Vitro

	Proportion of Total Amino Acid Incorporation with Copolymer of Composition	
Amino Acid	5A:1C	1A:5C
Lysine	51.0%	1.2%
Glutamine	12.3%	4.6%
Asparagine	12.7%	4.4%
Threonine	13.8%	13.6%
Histidine	6.4%	15.4%
Proline	3.8%	60.8%

[a]Data are from Speyer et al. *Cold Spring Harbor Symp. Quant. Biol.* 28:559–567, 1963.

With the 1A:5C copolymer, proline and lysine represented 60.8 percent and 1.2 percent, respectively, of the amino acid residues incorporated, in agreement with the earlier demonstration that CCC and AAA were proline and lysine codons. The values were reversed when the 5A:1C copolymer was used; 51 percent and 3.8 percent of the incorporated residues were lysine and proline, respectively.

Ochoa's results indicated that the glutamine and asparagine codons contain 2 As and 1 C, whereas the histidine codon in poly(AC) contains 1 A and 2 C's. Approximately the same amount of threonine was incorporated in response to both the 5A:1C and the 1A:5C copolymers. At first, the threonine incoporation data did not seem logical. However, recall that Nirenberg's results with the 1A:1C copolymer indicated that poly(AC) contains two threonine codons. Taken together with Nirenberg's results, Ochoa's results with the 5A:1C and 1A:5C copolymers suggested that one of the threonine codons contains 1 A and 2 C's and the other contains 2 A's and 1 C. Also, given that one proline codon is CCC, the combined results of the two studies indicated that the second proline codon in poly(AC) probably contains 1 A and 2 C's. In the 5A:1C copolymer, 0.4 percent of the codons will be CCC and 6.9 percent will contain 1 A and 2 Cs (2.3 percent of each ACC, CAC, and CCA). If one of the 1 A and 2 C codons specifies proline, the expected frequency of proline in the 5A:1C product will be 2.3 percent (ACC or CAC or CCA) plus 0.4 percent (CCC) or 2.7 percent. This value agrees well with the experimental value of 3.8 percent.

A major breakthrough in deciphering the genetic code occurred when H. Ghobind Khorana (1968 Nobel Prize recipient) and colleagues developed a procedure by which copolymers with known repeating di-, tri- and tetranucleotide sequences could be synthesized. Because the codon sequences in these repeating copolymers were fixed, their use as synthetic mRNAs yielded results that were more easily interpreted than those obtained with random copolymers. For example, an RNA molecule with a repeating UG dinucleotide sequence directed the synthesis of polypeptides containing alternating cysteine and valine residues.

5'-UGU GUG UGU GUG UGU GUG-3'

H_2N—Cys—Val—Cys—Val—Cys—Val—COOH

Because the triplets UGU and GUG alternate in $poly(UG)_n$, these two codons must specify cysteine and valine, but the result doesn't tell us which codon specifies which amino acid. In contrast, a repeating UUG trinucleotide polymer directed the synthesis of a mixture of polyleucine, polycysteine, and polyvaline.

Synthetic mRNA

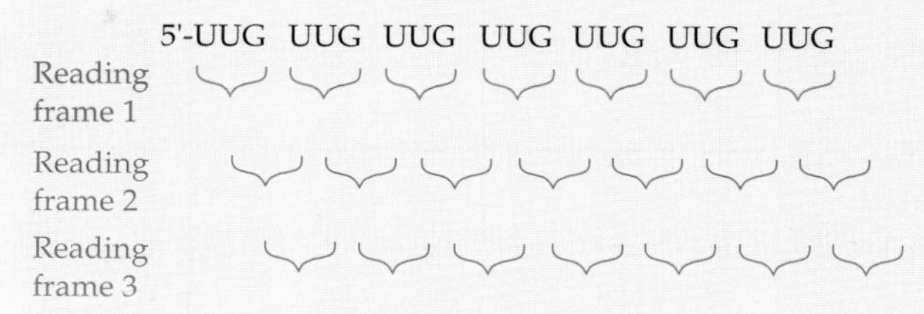

Polypeptide Products

Polyleucine Leu—Leu—Leu—Leu—Leu—Leu—Leu
Polycysteine Cys—Cys—Cys—Cys—Cys—Cys—Cys
Polyvaline Val—Val—Val—Val—Val—Val—Val

Because the initiation of translation occurs at random in these *in vitro* systems, some ribosomes will translate these polymers as UUG, UUG, UUG, and so on, while others will translate them as UGU, UGU, UGU. . . , and still others as GUU, GUU, GUU. . . . Thus these three codons must specify leucine, cysteine, and valine. Because the $poly(UG)_n$ product was a cysteine-valine copolymer, one of the new codons in $poly(UUG)_n$, either UUG or GUU, must be a leucine codon. By analyzing the amino acids incorporated in response to many different repeating di-, tri-, and tetranucleotide copolymers, Khorana and coworkers were able to assign many of the codons to specific amino acids. For example, UGU specifies cysteine, UUG leucine, and GUU valine.

By using synthetic RNAs of varied composition and sequence to activate *in vitro* translation systems, Nirenberg, Ochoa, and Khorana, their coworkers, and other researchers were able to determine the amino acids specified by most of the 64 possible trinucleotides. Their results were soon supported by the results of trinucleotide-binding experiments (see text) and were subsequently verified by direct comparisons of the nucleotide sequences of genes and the amino acid sequences of their polypeptide products.

This simple binding assay allowed Nirenberg, Leder, and others to test the ability of each of the 64 possible trinucleotides to activate ribosomes so that specific aminoacyl-tRNA would be bound. Not all of the trinucleotides worked; in some cases, the level of binding was very low, yielding ambiguous results. However, in most instances, a given trinucleotide caused one specific aminoacyl-tRNA to be bound to the activated ribosomes. By combining the results of trinucleotide binding assays and *in vitro* translation experiments performed with synthetic mRNAs, Nirenberg, Severo Ochoa, Khorana, and others were able to decipher the meaning of all 64 triplet codons (Table 12.2). These codon assignments are now firmly established, supported by definitive data from both *in vitro* and *in vivo* studies.

Initiation and Termination Codons

The genetic code also provides for punctuation of genetic information at the level of translation. In both prokaryotes and eukaryotes, the codon AUG is used to initiate polypeptide chains (Table 12.2). In rare in-

TABLE 12.2
The Genetic Code[a]

First (5′) letter	U	C	A	G	Third (3′) letter
U	UUU Phe	UCU Ser	UAU Tyr	UGU Cys	U
U	UUC Phe	UCC Ser	UAC Tyr	UGC Cys	C
U	UUA Leu	UCA Ser	UAA *Ochre* (terminator)	UGA *Opal* (terminator)	A
U	UUG Leu	UCG Ser	UAG *Amber* (terminator)	UGG Trp	G
C	CUU Leu	CCU Pro	CAU His	CGU Arg	U
C	CUC Leu	CCC Pro	CAC His	CGC Arg	C
C	CUA Leu	CCA Pro	CAA Gln	CGA Arg	A
C	CUG Leu	CCG Pro	CAG Gln	CGG Arg	G
A	AUU Ileu	ACU Thr	AAU Asn	AGU Ser	U
A	AUC Ileu	ACC Thr	AAC Asn	AGC Ser	C
A	AUA Ileu	ACA Thr	AAA Lys	AGA Arg	A
A	AUG *Met* (initiator)	ACG Thr	AAG Lys	AGG Arg	G
G	GUU Val	GCU Ala	GAU Asp	GGU Gly	U
G	GUC Val	GCC Ala	GAC Asp	GGC Gly	C
G	GUA Val	GCA Ala	GAA Glu	GGA Gly	A
G	GUG Val	GCG Ala	GAG Glu	GGG Gly	G

(light shading) = Polypeptide chain initiation codon
(dark shading) = Polypeptide chain termination codon

[a]Each triplet nucleotide sequence or codon refers to the nucleotide sequence in **mRNA** (not DNA) that specifies the incorporation of the indicated amino acid or polypeptide chain termination.

stances, GUG is used as an initiation codon. In both cases, the initiation codon is recognized by an initiator tRNA, tRNA$_f^{Met}$ in prokaryotes and tRNA$_i^{Met}$ in eukaryotes. In prokaryotes, an AUG codon must follow an appropriate nucleotide sequence in the 5' nontranslated segment of the mRNA molecule in order to serve as translation initiation codon. In eukaryotes, the codon must be the first AUG encountered by the ribosome as it scans from the 5' end of the mRNA molecule. At internal positions, AUG is recognized by tRNA^{Met}, and GUG is recognized by a valine tRNA.

Three codons—UAA, UAG, and UGA—specify polypeptide chain termination (Table 12.2). These codons are recognized by protein release factors, rather than by tRNAs. Prokaryotes contain two release factors, RF-1 and RF-2. RF-1 terminates polypeptides in response to codons UAA and UAG, whereas RF-2 causes termination at UAA and UGA codons. Eukaryotes contain a single release factor that recognizes all three termination codons.

A Degenerate and Ordered Code

All the amino acids except methionine and tryptophan are specified by more than one codon (Table 12.2). Three amino acids—leucine, serine, and arginine—are each specified by six different codons. Isoleucine has three codons. The other amino acids each have either two or four codons. The occurrence of more than one codon per amino acid is called **degeneracy** (although the usual connotations of the term are hardly appropriate). The degeneracy in the genetic code is not at random; instead, it is highly ordered. In most cases, the multiple codons specifying a given amino acid differ by only one base, the third or 3' base of the codon. The degeneracy is primarily of two types. (1) Partial degeneracy occurs when the third base may be either of the two pyrimidines (U or C) or, alternatively, either of the two purines (A or G). With partial degeneracy, changing the third base from a purine to a pyrimidine, or vice versa, will change the amino acid specified by the codon. (2) In the case of complete degeneracy, any of the four bases may be present at the third position in the codon, and the codon will still specify the same amino acid. For example, valine is encoded by GUU, GUC, GUA, and GUG (Table 12.2).

Scientists have speculated that the **order** in the genetic code has evolved as a way of minimizing mutational lethality. Many base substitutions at the third position of codons do not change the amino acid specified by the codon. Moreover, amino acids with similar chemical properties (such as leucine, isoleucine, and valine) have codons that differ from each other by only one base. Thus many single base-pair substitutions will result in the substitution of one amino acid for another amino acid with very similar chemical properties (for example, valine for isoleucine). In most cases, conservative substitutions of this type will yield active gene products, which minimizes the effects of mutations.

A Nearly Universal Code

Vast quantities of information are now available from *in vitro* studies, from amino acid replacements due to mutations, and from correlated nucleic acid and polypeptide sequencing, which allow a comparison of the meaning of the 64 codons in different species. These data all indicate that the genetic code is nearly **universal**; that is, the codons have the same meaning, with minor exceptions, in all species.

The most important exceptions to the universality of the code occur in mitochondria of mammals, yeast, and several other species. Mitochondria have their own chromosomes and protein synthesizing machinery (Chapter 18). Although the mitochondrial and cytoplasmic systems are similar, there are some differences. In the mitochondria of humans and other mammals, (1) UGA specifies tryptophan rather than chain termination (2) AUA is a methionine codon, not an isoleucine codon, and (3) AGA and AGG are chain-termination codons rather than arginine codons. The other 60 codons have the same meaning in mammalian mitochondria as in nuclear mRNAs (Table 12.2). There are also rare differences in codon meaning in the mitochondria of other species and in nuclear transcripts of some protozoa. However, since these exceptions are rare, the genetic code should be considered nearly universal.

Key Points: **Each of the 20 amino acids in proteins is specified by one or more nucleotide triplets in mRNA. Of the 64 possible triplets, given the four bases in mRNA, 61 specifiy amino acids and 3 signal chain termination. The code is nonoverlapping, with each nucleotide part of a single codon, degenerate, with most amino acids specified by two or four codons, and ordered, with similar amino acids specified by related codons. The genetic code is nearly universal; with minor exceptions, the 64 triplets have the same meaning in all organisms.**

CODON-tRNA INTERACTIONS

The translation of a sequence of nucleotides in mRNA into the correct sequence of amino acids in the polypeptide product requires the accurate recognition of codons by aminoacyl-tRNAs. Because of the degeneracy of the genetic code, either there must be several different tRNAs that recognize the different codons specifying a given amino acid or the anticodon of a given tRNA must be able to base-pair with several different codons. Actually, both of these occur. Several tRNAs exist for certain amino acids, and some tRNAs recognize more than one codon.

Recognition of Codons by tRNAs: The Wobble Hypothesis

The hydrogen bonding between the bases in the anticodons of tRNAs and the codons of mRNAs appears to follow strict base-pairing rules only for the first two bases of the codon. The base-pairing involving the third base of the codon is apparently less stringent, allowing what Crick has called **wobble** at this site. On the basis of molecular distances and steric (three-dimensional structure) considerations, Crick proposed that wobble would allow several types, but not all types, of base-pairing at the third codon base in the codon–anticodon interaction. His proposal has since been strongly supported by experimental data. Table 12.3 shows the base-pairing predicted by Crick's wobble hypothesis.

The **wobble hypothesis** predicted the existence of at least two tRNAs for each amino acid with codons that exhibit complete degeneracy, and this has proven to be true. The wobble hypothesis also predicted the

TABLE 12.3
Base-pairing between the 5' Base of the Anticodon of tRNAs and the 3' Base of Codons of mRNAs according to the Wobble Hypothesis

Base in Anticodon	*Base in Codon*
G	U or C
C	G
A	U
U	A or G
I	A, U, or C

occurrence of three tRNAs for the six serine codons. Three serine tRNAs have been characterized: (1) $tRNA^{Ser1}$ (anticodon AGG) binds to codons UCU and UCC, (2) $tRNA^{Ser2}$ (anticodon AGU) binds to codons UCA and UCG, and (3) $tRNA^{Ser3}$ (anticodon UCG) binds to codons AGU and AGC. These specificities were verified by the trinucleotide-stimulated binding of purified aminoacyl-tRNAs to ribosomes *in vitro*.

Finally, several tRNAs contain the base inosine, which is a ribonucleoside made from the purine hypoxanthine. Inosine is produced by a post-transcriptional modification of adenosine. Crick's wobble hypothesis predicted that when inosine is present at the 5' end of an anticodon (the wobble position), it would base-pair with adenine, uracil, or cytosine in the codon. In fact, purified alanyl-tRNA containing inosine (I) at the 5' position of the anticodon (see Figure 12.12) binds to ribosomes activated with GCU, GCC, or GCA trinucleotides (Figure 12.22). The same result has been obtained with other purified tRNAs with inosine at the 5′ position of the anticodon. Thus Crick's wobble hypothesis nicely explains the relationships between tRNAs and codons given the degenerate, but ordered, genetic code.

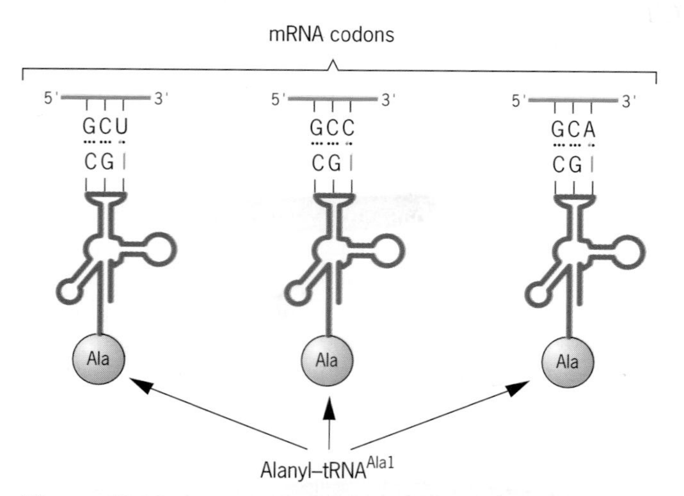

Figure 12.22 Base-pairing between the anticodon of alanyl-$tRNA^{Ala1}$ and mRNA codons GCU, GCC, and GCA according to Crick's wobble hypothesis. Trinucleotide-activated ribosome binding assays have shown that alanyl-$tRNA^{Ala1}$ does indeed base-pair with all three codons.

Suppressor Mutations That Produce tRNAs with Altered Codon Recognition

Even if we exclude the mitochondria, the genetic code is not absolutely universal. Minor variations in codon recognition and translation are well documented. In *E. coli* and yeast, for example, some mutations in tRNA genes alter the anticodons and thus the codons recognized by the mutant tRNAs. These mutations were initially detected as suppressor mutations, nucleotide substitutions that suppressed the effects of other mutations. The suppressor mutations were subsequently shown to occur in tRNA genes. Many of these suppressor mutations changed the anticodons of the altered tRNAs.

The best-known examples of suppressor mutations that alter tRNA specificity are the suppressors of mutations that produce UAG chain-termination triplets within the coding sequences of genes that specify polypeptide products. Such mutations, called *amber* mutations, result in the synthesis of truncated polypeptides. Mutations that produce chain-termination triplets within genes have come to be known as **nonsense mutations**, in contrast to **missense mutations**, which change a triplet so that it specifies a different amino acid. Genes that contain missense mutations encode complete polypeptides, but with amino acid substitutions in the polypeptide gene products. Nonsense mutations result in truncated polypeptides, with the lengths of the chains depending on the positions of the mutations within the gene. Nonsense mutations frequently result from single base-pair substitutions as illustrated in Figure 12.23*a*. The polypeptide fragments produced from genes containing nonsense mutations often are completely nonfunctional.

Suppression of nonsense mutations has been shown to result from mutations in tRNA genes that cause the mutant tRNAs to recognize the nonsense (UAG, UAA, or UGA) codons, albeit with varying efficiencies. These mutant tRNAs are referred to as suppressor tRNAs. When the *amber* (UAG) suppressor tRNA produced by the *amber su3* mutation in *E. coli* was sequenced, it was found to have an altered anticodon. This particular *amber* suppressor mutation occurs in the $tRNA^{Tyr2}$ gene (one of two tyrosine tRNA genes in *E. coli*). The anticodon of the wild-type (nonsuppressor) $tRNA^{Tyr2}$ was shown to be 5'-G'UA-3' (where G' is a derivative of guanine). The anticodon of the mutant (suppressor) $tRNA^{Tyr2}$ is 5'-CUA-3'. Because of the single-base substitution, the anticodon of the suppressor $tRNA^{Tyr2}$ base-pairs with the 5'-UAG-3'

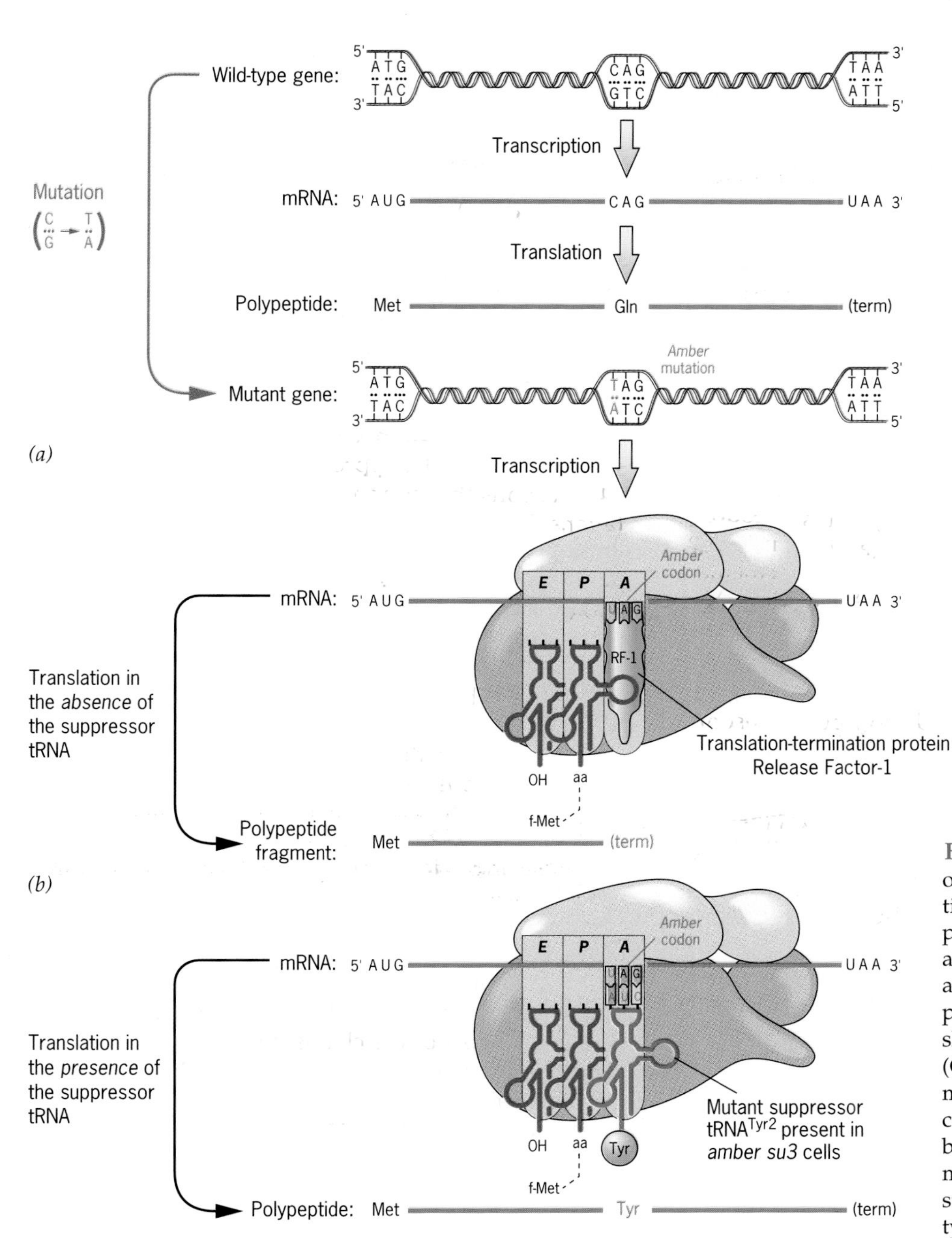

Figure 12.23 (*a*) The formation of an *amber* (UAG) chain-termination mutation. (*b*) Its effect on the polypeptide gene product in the absence of a suppressor tRNA, and (*c*) in the presence of a suppressor tRNA. The *amber* mutation shown changes a CAG glutamine (Gln) codon to a UAG chain-termination codon. The polypeptide containing the tyrosine inserted by the suppressor tRNA may or may not be functional; however, suppression of the mutant phenotype will occur only when the polypeptide is functional

amber codon (recall that base-pairing always involves strands of opposite polarity); that is,

5'-UAG-3' (codon)
3'-AUC-5' (anticodon)

Thus suppressor tRNAs allow complete polypeptides to be synthesized from mRNAs containing nonsense codons (Figure 12.23). Such polypeptides will be functional if the amino acid inserted by the suppressor tRNA does not significantly alter the protein's chemical properties.

Key Points: **The wobble hypothesis explains how a single tRNA can respond to two or more degenerate codons. According to this hypothesis, the pairing between the third base of a codon and the first base of an anticodon is less stringent than normal, permitting wobble at this site. Some suppressor mutations alter the anticodons of tRNAs so that the mutant tRNAs recognize chain-termination codons and insert amino acids in response to their presence in mRNA molecules.**

IN VIVO EVIDENCE CONFIRMS THE NATURE OF THE GENETIC CODE

The codon assignments shown in Table 12.2 were initially based on the results of *in vitro* translation and trinucleotide-stimulated aminoacyl-tRNA binding studies. This raised an obvious question. Are the assignments based on *in vitro* experiments valid *in vivo?* Several lines of evidence now indicate that these codon assignments are correct for protein synthesis *in vivo* for most, if not all, species. When the amino acid substitutions that result from mutations induced by chemical mutagens with specific effects (Chapter 13) are determined by amino acid sequencing, the substitutions are almost always consistent with the codon assignments given in Table 12.2 and the known effect of the mutagen.

More convincingly, when the nucleotide sequences of genes or mRNAs are determined and compared with the amino acid sequences of the polypeptides encoded by those genes or mRNAs, the observed correlations are those predicted from the codon assignments shown in Table 12.2. This was first demonstrated for bacteriophage MS2 by Walter Fiers and colleagues in 1972. Fiers and coworkers determined the amino acid sequence of the phage MS2 coat protein and the nucleotide sequence of the MS2 gene that encodes the coat protein. Phage MS2 stores its genetic information in single-stranded RNA; its chromosome is equivalent to an mRNA molecule in organisms with DNA genomes. Fiers and colleagues then compared the nucleotide sequence of the coat protein gene with the amino acid sequence of the coat polypeptide. The amino acid sequence of the coat protein was exactly the sequence predicted from the nucleotide sequence of the coat protein gene and the codon assignments shown in Table 12.2.

In the last two decades, similar comparisons of the nucleotide sequences of genes and the amino acid sequences of their polypeptide products have clearly established that the codon assignments shown in Table 12.2 are indeed valid *in vivo*. When the nucleotide sequence of the normal human β-globin gene was determined, the sequence predicted the sequence of the 146 amino acids in the human β-globin polypeptide, including the valine at position six. When the sequence of the sickle-cell allele was determined, it predicted the presence of glutamic acid at position six. Indeed, scientists now understand the molecular basis of all the symptoms of sickle-cell anemia described by Herrick and Irons in 1910.

Key Point: **Comparisons of the nucleotide sequences of genes with the amino acid sequences of their polypeptide products verify the codon assignments deduced from *in vitro* studies.**

TESTING YOUR KNOWLEDGE

1. The average mass of the 20 common amino acids is about 137 daltons. Estimate the approximate length of an mRNA molecule that encodes a polypeptide with a mass of 65,760 daltons. Assume that the polypeptide contains equal amounts of all 20 amino acids.

ANSWER

Based on this assumption, the polypeptide would contain about 480 amino acids (65,760 daltons/137 daltons per amino acid). Since each codon contains three nucleotides, the coding region of the mRNA would have to be 1440 nucleotides long (480 amino acids × 3 nucleotides per amino acid).

2. What sequence of nucleotide pairs in a gene in *Drosophila* will encode the amino acid sequence methionine-tryptophan (reading from the amino terminus to the carboxyl terminus)?

ANSWER

The codons for methionine and tryptophan are AUG and UGG, respectively. Thus the nucleotide sequence in the mRNA specifying the dipeptide sequence methionine-tryptophan must be 5'-AUG-UGG-3'. The template DNA strand must be complementary and antiparallel to the mRNA sequence (3'-TAC-ACC-5'), and the other strand of DNA must be complementary to the template strand. Therefore, the sequence of base pairs in the gene must be

5'-ATG-TGG-3'
3'-TAC-ACC-5'

3. The antibiotic streptomycin kills sensitive *E. coli* by inhibiting the binding of $tRNA_f^{Met}$ to the *P* site of the ribosome and by causing misreading of codons in mRNA. In sensitive bacteria, streptomycin is bound by protein S12 in the 30S subunit of the ribosome. Resistance to streptomycin can result from a mutation in the gene-encoding protein S12 so that the altered protein will no longer bind the antibiotic. In 1964, Luigi Gorini and Eva Kataja isolated mutants of *E. coli* that grew on minimal medium supplemented with either the amino acid arginine or streptomycin. That is, in the absence of streptomycin, the mutants behaved like typical arginine-requiring bacteria. However, in the absence of arginine, they were streptomycin-dependent conditional-lethal mutants. That is, they grew in the presence of streptomycin but not in the absence of streptomycin. How can Gorini and Kataja's results be explained?

ANSWER

The streptomycin-dependent conditional-lethal mutants isolated by Gorini and Kataja contained missense mutations in genes encoding arginine biosynthetic enzymes. If arginine was present in the medium, these enzymes were unessential. However, these enzymes were required for growth in the absence of arginine (one of the 20 amino acids required for protein synthesis).

Streptomycin causes misreading of mRNA codons in bacteria. This misreading allowed the codons that contained the missense mutations to be translated ambiguously—with the wrong amino acids incorporated—when the antibiotic was present. When streptomycin was present in the mutant bacteria, an amino acid occasionally would be inserted (at the site of the mutation) that resulted in an active enzyme, which, in turn, allowed the cells to grow, albeit slowly. In the absence of streptomycin, no misreading occurred, and all of the mutant polypeptides were inactive.

QUESTIONS AND PROBLEMS

12.1 In a general way, describe the molecular organization of proteins and distinguish proteins from DNA, chemically and functionally. Why is the synthesis of proteins of particular interest to geneticists?

12.2 At what locations in the cell does protein synthesis occur?

12.3 Characterize ribosomes in general as to size, location, function, and macromolecular composition.

12.4 (a) Where in the cells of higher organisms do ribosomes originate? (b) Where in the cells are ribosomes most active in protein synthesis?

12.5 Identify three different types of RNA that are involved in translation and list the characteristics and functions of each.

12.6 (a) How is messenger RNA related to polysome formation? (b) How does rRNA differ from mRNA and tRNA in specificity? (c) How does the tRNA molecule differ from that of DNA and mRNA in size and helical arrangement?

12.7 Outline the process of aminoacyl-tRNA formation.

12.8 What types of experimental evidence were used to decipher the genetic code?

12.9 In what sense and to what extent is the genetic code (a) degenerate, (b) ordered, and (c) universal?

12.10 Draw an analogy between the processes of transcription and translation and the process of building a house.

12.11 The thymine analog 5-bromouracil is a chemical mutagen that induces single base-pair substitutions in DNA called transitions (substitutions of one purine for another purine and one pyrimidine for another pyrimidine). Using the known nature of the genetic code (Table 12.2), which of the following amino acid substitutions should you expect to be induced by 5-bromouracil with the highest frequency: (a) Met → Val; (b) Met → Leu; (c) Lys → Thr; (d) Lys → Gln; (e) Pro → Arg; or (f) Pro → Gln? Why?

12.12 Using the information given in Problem 12.11, would you expect 5-bromouracil to induce a higher frequency of His → Arg or His → Pro substitutions? Why?

12.13 How is translation (a) initiated and (b) terminated?

12.14 If the average molecular mass of an amino acid is assumed to be 100 daltons, about how many nucleotides will be present in an mRNA coding sequence specifying a single polypeptide with a molecular mass of 27,000 daltons?

12.15 Of what significance is the wobble hypothesis?

12.16 The bases A, G, U, C, I (inosine) all occur at the 5' positions of anticodons in tRNAs. (a) Which base can pair with three different bases at the 3' positions of codons in mRNA? (b) What is the minimum number of tRNAs required to recognize all codons of amino acids specified by codons with complete degeneracy?

12.17 (a) Why is the genetic code a triplet code instead of a singlet or doublet code? (b) How many different amino acids are specified by the genetic code? (c) How many different amino acid sequences are possible in a polypeptide 146 amino acids long?

12.18 What are the basic differences between translation in prokaryotes and eukaryotes?

12.19 What is the function of each of the following components of the protein-synthesizing apparatus: (a) aminoacyl-tRNA synthetase, (b) release factor 1, (c) peptidyl transferase, (d) initiation factors, (e) elongation factor G?

12.20 An *E. coli* gene has been isolated and shown to be 68 nm long. What is the maximum number of amino acids that this gene could encode?

12.21 (a) What is the difference between a nonsense mutation and a missense mutation? (b) Are nonsense or missense mutations more frequent in living organisms? (c) Why?

12.22 The human α-globin chain is 141 amino acids long. How many nucleotides in mRNA are required to encode human α-globin?

12.23 (a) What are the functions of the *A* and *P* aminoacyl-tRNA binding sites on the ribosome? (b) Why are at least two aminoacyl-tRNA binding sites required on each ribosome?

12.24 (a) In what ways does the order in the genetic code minimize mutational lethality? (b) Why do base-pair changes that cause the substitution of a leucine for a valine in the polypeptide gene product seldom produce a mutant phenotype?

12.25 (a) What is the function of the Shine-Dalgarno sequence in prokaryotic mRNAs? (b) What effect does the deletion of the Shine-Dalgarno sequence from a mRNA have on its translation?

12.26 (a) In what ways are ribosomes and spliceosomes similar? (b) In what ways are they different?

12.27 If you were to (1) purify cysteine transfer RNA and charge it with labeled cysteine (that is, activate it by attaching ^{3}H-labeled cysteine), (2) use Raney nickel (a highly reducing nickel powder) to convert the cysteine to alanine still attached to the cysteyl-specific transfer RNA, and (3) place the alanine-charged cysteyl transfer RNA into an *in vitro* protein-synthesizing system activated with poly UG templates that normally stimulate the incorporation of cysteine, but not alanine, into polypeptide chains (that is, when you use the normal alanine-charged alanyl and cysteine-charged cysteyl transfer RNAs), what result would you expect?

12.28 A partial (5' subterminal) nucleotide sequence of a prokaryotic mRNA is as follows:

5'-.....AGGAGGCUCGAACAUGUCAAUAUGCUUGUUC-
CAAUCGUUAGCUGCGCAGGACCGUCCCGGA......3'.

When this mRNA is translated, what amino acid sequence will be specified by this portion of the mRNA?

12.29 The 5' terminus of a human mRNA has the following sequence:

5' cap-GAAGAGACAAGGTCAUGGCCAUAUGCUUGUU
CCAAUCGUUAGCUGCGCAGGAUCGCCCUGGG ......3'.

When this mRNA is translated, what amino acid sequence will be specified by this portion of the mRNA?

12.30 Alan Garen extensively studied a particular nonsense (chain-termination) mutation in the alkaline phosphatase gene of *E. coli.* This mutation resulted in the termination of the alkaline phosphatase polypeptide chain at a position where the amino acid tryptophan occurred in the wild-type polypeptide. Garen induced revertants (in this case, mutations altering the same codon) of this mutant with chemical mutagens that induced single base-pair substitutions and sequenced the polypeptides in the revertants. Seven different types of revertants were found, each with a different amino acid at the tryptophan position of the wild-type polypeptide (termination position of the mutant polypeptide fragment). The amino acids present at this position in the various revertants included tryptophan, serine, tyrosine, leucine, glutamic acid, glutamine, and lysine. Was the nonsense mutation studied by Garen an *amber* (UAG), an *ochre* (UAA), or an *opal* (UGA) nonsense mutation? Explain the basis of your deduction.

BIBLIOGRAPHY

Cox, R. A., and H. R. V. Arnstein. 1995. "Translation of RNA to protein." In *Molecular Biology and Biotechnology* (Meyers, R. A., ed.), pp. 914–922. VCH Publishers, New York.

Crick, F. H. C. 1962. "The genetic code." *Sci. Amer.* 207(4):66–77.

Crick, F. H. C. 1966. "The genetic code: III." *Sci. Amer.* 215(4):55–62.

Crick, F. H. C. 1966. "Codon-anticodon pairing: The wobble hypothesis." *J. Mol. Biol.* 19:548–555.

The Genetic Code. 1966. *Cold Spring Harbor Symp. Quant. Biol.*, Vol. 31. Cold Spring Harbor Laboratory Press, Cold Spring Harbor, NY.

Lake, J. A. 1981. "The ribosome." *Sci. Amer.* 245(2):84–97.

The Mechanism of Protein Synthesis. 1970. *Cold Spring Harbor Symp. Quant. Biol.*, Vol. 34. Cold Spring Harbor Laboratory Press, Cold Spring Harbor, NY.

Nirenberg, M. W. 1963. "The genetic code: II." *Sci. Amer.* 208(3):80–94.

Nomura, M. 1984. "The control of ribosome synthesis." *Sci. Amer.* 250(1):102–114.

Yanofsky, C. 1967. "Gene structure and protein structure." *Sci. Amer.* 216(5):80–94.

Scanning electron micrograph of Tetraptera, *a mutant in* Drosophila *that has four wings instead of two.*

13

Mutation, DNA Repair, and Recombination

CHAPTER OUTLINE

Xeroderma Pigmentosum: Defective Repair of Damaged DNA in Humans

The sun shone brightly on a mid-summer day—a perfect day for most children to spend at the beach. All of Nathan's friends were dressed in shorts or swimsuits. As Nathan prepared to join his friends, he pulled on full-length sweat pants and a long-sleeved shirt. Then he put on a wide-brimmed hat and applied a thick layer of sunscreen to his hands and face. Whereas his friends enjoy sunshine and the tan that results, Nathan lives in constant fear of the effects of sunlight. Nathan was born with the inherited disorder xeroderma pigmentosum, an autosomal recessive trait that affects about one of 250,000 children. Nathan's skin cells are extremely sensitive to ultraviolet radiation—the high-energy rays of sunlight. Ultraviolet light causes chemical changes in the DNA in Nathan's skin cells, changes that lead not only to intense freckling but also to skin cancer (Figure 13.1).

Nathan's friends gave little thought to playing in the sun; sunburn was their only major concern. Their skin cells contain enzymes that correct the changes in DNA resulting from exposure to ultraviolet light. However, Nathan's skin cells

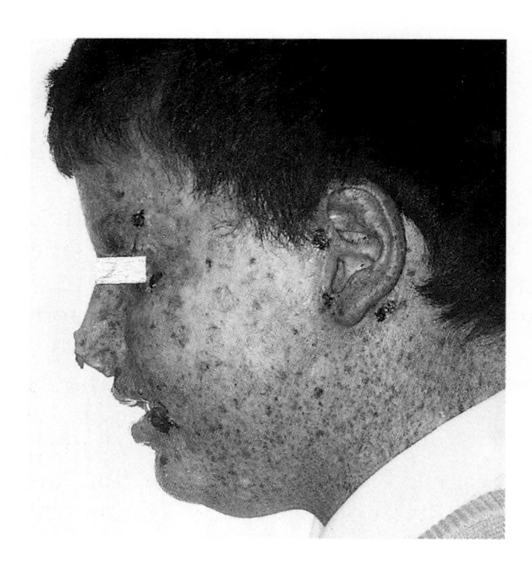

are lacking one of the enzymes required to repair ultraviolet light-induced alterations in the structure of DNA. Xeroderma pigmentosum can result from inherited defects in any of nine different human genes. Moreover, other inherited disorders are known to result from the failure to repair DNA damaged by other physical and chemical agents. The life-threatening consequences of these inherited defects in the DNA repair enzymes dramatically emphasize their importance.

Figure 13.1 Phenotypic effects of the inherited disease xeroderma pigmentosum. Individuals with this malignant disease develop extensive skin tumors after exposure to sunlight.

Given the key role that DNA plays in the growth and development of living organisms, the evolution of mechanisms to protect its integrity would seem inevitable. Indeed, as we discuss below, living cells contain numerous enzymes that constantly scan DNA to search for damaged or incorrectly paired nucleotides. When detected, these defects are corrected by a small army of **DNA repair enzymes**, each designed to combat a particular type of damage. In some cases, incorrect nucleotide pairs are produced during recombination between homologous DNA molecules. The repair of these mismatched base pairs results in a type of nonreciprocal recombination called **gene conversion**, a process during which one allele of a gene is changed or converted to another allele. In this chapter, we examine the types of changes that occur in DNA, the processes by which these alterations are corrected, and the related processes of recombination between homologous DNA molecules.

MUTATION: SOURCE OF THE GENETIC VARIABILITY REQUIRED FOR EVOLUTION

We know from preceding chapters that inheritance is based on genes that are transmitted from parents to offspring during reproduction and that the genes store genetic information encoded in the sequences of nucleotide pairs in DNA or nucleotides in RNA. We have examined how this genetic information is accurately duplicated during the semiconservative replication of DNA. This accurate replication was shown to depend in part on proofreading activities built into the DNA polymerases that catalyze DNA synthesis. Thus mechanisms have evolved to facilitate the faithful transmission of genetic information from generation to generation. Nevertheless, mistakes in the genetic material do occur. Such heritable changes in the genetic material are called mutations.

The term **mutation** refers to both (1) the change in the genetic material and (2) the process by which the change occurs. An organism that exhibits a novel phenotype resulting from a mutation is called a **mutant**. Used in its broad historical sense, mutation refers to any sudden, heritable change in the genotype of an organism. However, geneticists must be careful to distinguish changes in the genotype, and thus in the phenotype, of an organism that result from recombination events that produce new combinations of preexisting genetic variation from changes caused by new mutations. Both events sometimes give rise to new phenotypes at very low frequencies. Mutational changes in the genotype of an organism include changes in chromosome number and structure (Chapter 6), as well as changes in the structures of individual genes. Today, the term *mutation* often is used in a narrow sense to refer only to changes occurring within genes. In this chapter, we explore the process of mutation as defined in the narrow sense. Mutations that involve changes at specific sites in a gene are referred to as **point mutations**. They include the substitution of one base pair for another or the insertion or deletion of one or a few nucleotide pairs at a specific site in a gene.

Mutation is the ultimate source of all genetic variation; it provides the raw material for evolution. Recombination mechanisms rearrange genetic variability into new combinations, and natural or artificial selection preserves the combinations best adapted to the existing environmental conditions or desired by the plant or animal breeder. Without mutation, all genes would exist in only one form. Alleles would not exist, and classical genetic analysis would not be possible. Most important, organisms would not be able to evolve and adapt to environmental changes. Some level of mutation is essential to provide new genetic variability and allow organisms to adapt to new environments. At the same time, if mutations occurred too frequently, they would disrupt the faithful transfer of genetic information from generation to generation. As we would expect, the rate of mutation is under genetic control, and mechanisms have evolved that regulate the level of mutation that occurs under various environmental conditions.

Key Points: **Mutations are heritable changes in the genetic material that provide the raw material for evolution.**

MUTATION: BASIC FEATURES OF THE PROCESS

Mutations occur in all genes of all living organisms. These mutations provide new genetic variability that allows organisms to adapt to environmental changes. Thus mutations have been, and continue to be, essential to the evolutionary process. Before we discuss specific phenotypic effects of mutations and the mechanisms by which various types of mutation occur, we will consider some of the basic features of this important process.

Mutation: Somatic or Germinal

A mutation may occur in any cell and at any stage in the development of a multicellular organism. The immediate effects of the mutation and its ability to produce a phenotypic change are determined by its dominance, the type of cell in which it occurs, and the time at which it takes place during the life cycle of the organism. In higher animals, the germ-line cells that give rise to the gametes separate from other cell lineages early in development (Chapter 2). All nongerm-line cells are somatic cells. **Germinal mutations** are those that occur in germ-line cells, whereas **somatic mutations** occur in somatic cells.

If a mutation occurs in a somatic cell, the resulting mutant phenotype will occur only in the descendants of that cell. The mutation will not be transmitted through the gametes to the progeny. The Delicious apple and the navel orange are examples of mutant phenotypes that resulted from mutations occurring in somatic cells. The fruit trees in which the original mutations occurred were somatic mosaics. The changes that give these two fruits their desirable qualities are believed to have resulted from spontaneous mutations occurring in somatic cells. In each case, the cell carrying the mutant gene reproduced, eventually producing an entire branch that had the characteristics of the mutant type. Fortunately, vegetative propagation was feasible for both the Delicious apple and the navel orange, and today numerous progeny from grafts and buds have perpetuated the original mutations.

If dominant mutations occur in germ-line cells, their effects may be expressed immediately in progeny. If the mutations are recessive, their effects are often obscured in diploids. Germinal mutations may occur at any stage in the reproductive cycle of the organism, but they are more common during some stages than others. If the mutation arises in a gamete, only a single member of the progeny is likely to have the mutant gene. If a mutation occurs in a primordial germ-line cell of the testis or ovary, several gametes may receive the mutant gene, enhancing its potential for perpetuation. Thus the dominance of a mutant allele and the stage in the reproductive cycle at which a mutation occurs are major factors in determining the likelihood that the mutant allele will be manifest in an organism in a population.

The earliest recorded dominant germinal mutation in domestic animals was that observed by Seth Wright in 1791 on his farm by the Charles River in Dover, Massachusetts. Among his flock of sheep, Wright noticed a peculiar male lamb with unusually short legs (Figure 13.2). It occurred to him that it

(*a*)

(*b*)

Figure 13.2 Effect of a dominant mutation. (*a*) Short-legged sheep of the Ancon breed; (*b*) sheep with normal length legs.

would be an advantage to have a whole flock of these short-legged sheep, which could not jump over the low stone fences in his New England neighborhood. Wright used the new short-legged ram to breed his 15 ewes in the next season. Two of the 15 lambs had short legs. Short-legged sheep were then bred together, and a line was developed in which the new trait was expressed in all individuals.

Mutation: Spontaneous or Induced

When a new mutation—such as the one that produced Wright's short-legged sheep—occurs, is it caused by some agent in the environment or does it result from an inherent process in living organisms? **Spontaneous mutations** are those that occur without a known cause. They may be truly spontaneous, resulting from a low level of inherent metabolic errors, or they may actually be caused by unknown agents present in the environment. **Induced mutations** are those resulting from exposure of organisms to physical and chemical agents that cause changes in DNA (or RNA in some viruses). Such agents are called **mutagens**; they include ionizing irradiation, ultraviolet light, and a wide variety of chemicals.

Operationally, it is impossible to prove that a particular mutation occurred spontaneously or was induced by a mutagenic agent. Researchers have only begun to assess the potential mutagenicity of both human-made and naturally occurring chemicals.

Geneticists cannot distinguish between spontaneous and induced mutations when considering individual mutations. They must restrict such distinctions to the population level. If the mutation rate is increased a hundredfold by treatment of a population with a mutagen, an average of 99 of every 100 mutations present in the population will have been induced by the mutagen. Researchers can thus make valid comparisons between spontaneous and induced mutations statistically by comparing populations exposed to a mutagenic agent with control populations that have not been exposed to the mutagen.

Spontaneous mutations occur infrequently, although the observed frequencies vary from gene to gene and from organism to organism. Measurements of spontaneous mutation frequencies for various genes of phage and bacteria range from about 10^{-8} to 10^{-10} detectable mutations per nucleotide pair per generation. For eukaryotes, estimates of mutation rates range from about 10^{-7} to 10^{-9} detectable mutations per nucleotide pair per generation (considering only those genes for which extensive data are available). In comparing mutation rates per nucleotide with mutation rates per gene, the average gene is usually assumed to be 1000 nucleotide pairs in length. Thus the mutation rate per gene varies from about 10^{-4} to 10^{-7} per generation.

Treatment with mutagenic agents can increase mutation frequencies by orders of magnitude. The mutation frequency per gene in bacteria and viruses can be increased to over 1 percent by treatment with potent chemical mutagens. That is, over 1 percent of the genes of the treated organisms will contain a mutation, or, stated differently, over 1 percent of the phage or bacteria in the population will have a mutation in a given gene.

Mutation: Usually a Random, Nonadaptive Process

The rats in many cities are no longer affected by the anticoagulants that have traditionally been used as rodent poisons. Many cockroach populations are insensitive to Chlordane, the poison used to control them in the 1950s. Housefly populations often exhibit high levels of resistance to many insecticides. More and more pathogenic microorganisms are becoming resistant to antibiotics such as penicillin and streptomycin developed to control them. The introduction of these pesticides and antibiotics by humans produced new environments for these organisms. They responded to the imposed environmental changes by evolving to forms resistant to these chemicals. Mutations producing resistance to these pesticides and antibiotics occurred, and the mutant organisms were at a large selective advantage in environments where the agents were present. The sensitive organisms were killed, and the mutants multiplied to produce new resistant populations. Many such cases of evolution via mutation and natural selection are well documented.

These examples raise a basic question about the nature of mutation. Is mutation a purely random event with the environmental stress merely preserving preexisting mutations? Or is mutation directed by the environmental stress? For example, if you cut off the tails of mice for many generations, will you eventually produce a strain of tailless mice? Despite the beliefs of Jean Lamarck and his follower Trofim Lysenko, whom we discussed in Chapter 1, the answer is no; the mice will continue to be born with tails.

Today, it is hard to understand how Lysenko could have sold his belief in Lamarckism—the inheritance of acquired traits—to those in power in the Soviet Union in 1937 through 1964. However, disproving Lamarckism was not an easy task, especially in the case of microorganisms, where even small cultures often contain billions of organisms. As an example, let us consider a population of bacteria such as *E. coli* growing in a streptomycin-free environment. When exposed to streptomycin, most of the bacteria will be killed by the antibiotic. However, if the population is large enough, it will soon give rise to a streptomycin-resistant culture in which all the cells are resistant to the antibiotic. Does streptomycin simply select rare, randomly occurring mutants that preexist in the population, or do all of the cells have some low probability of developing resistance in response to the presence of streptomycin? How can geneticists distinguish between these two possibilities? They can detect resistance to streptomycin only by treating the culture with the antibiotic. How, then, can they determine whether resistant bacteria are present prior to exposure to streptomycin or are induced by the antibiotic?

In 1952, Joshua and Esther Lederberg developed an important new technique called **replica plating**. This technique allowed them to demonstrate the presence of antibiotic-resistant mutants in bacterial cultures prior to their exposure to the antibiotic (Figure 13.3). They first diluted the bacterial cultures, spread the bacteria on the surface of semisolid nutrient agar medium in petri dishes, and incubated the plates until each bacterium had produced a visible colony on the surface of the agar. They next inverted each plate and pressed it onto sterile velvet placed over a wood block. Some of the cells from each colony stuck to the velvet. They then gently pressed a sterile plate of nutrient agar medium containing streptomycin onto the velvet. They repeated this replica-plating procedure with many plates, each containing about 200 bacterial colonies. After they incubated the selective plates (those containing streptomycin) overnight, rare streptomycin-resistant colonies had formed. The Lederbergs subsequently tested the colonies on the nonselective plates (those not containing streptomycin) for their ability to grow on medium containing streptomycin. Their results were definitive. The colonies that grew on the selective replica plates almost always contained streptomycin-resistant cells, whereas those that did not grow on the selective medium seldom contained any resistant cells (Figure 13.3). By using their replica-plating technique, the Lederbergs demonstrated the existence of streptomycin-resistant mutants in a population of bacteria prior to their exposure to the antibiotic. Their results and the results of many other experiments have shown that environmental stress does not direct or cause genetic changes as Lysenko believed; it simply selects rare preexisting mutations that result in phenotypes that are better adapted to the new environment.

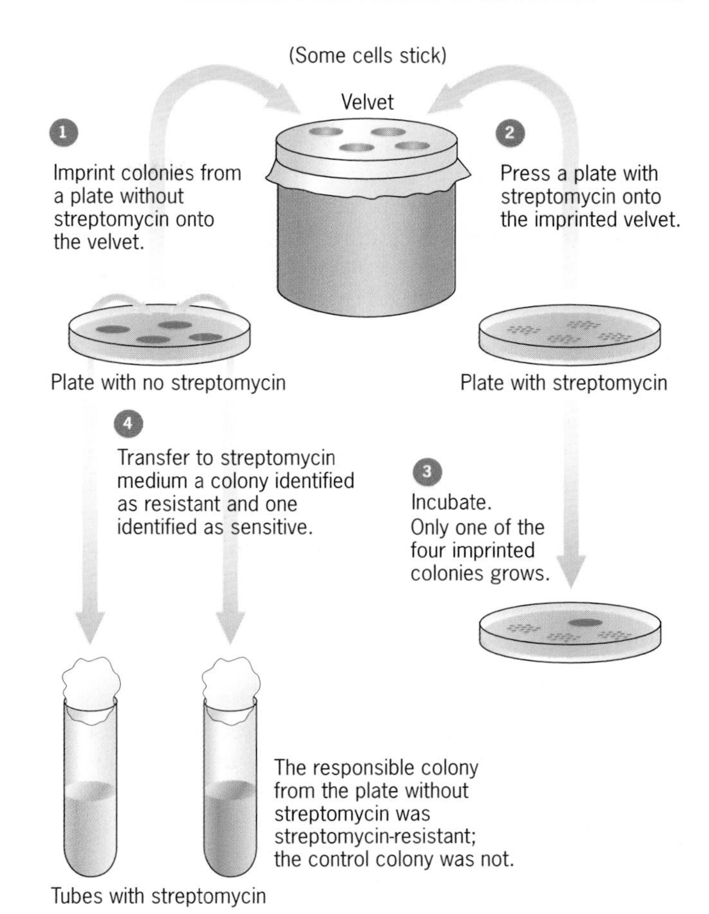

Figure 13.3 Joshua and Esther Lederberg's use of replica plating to demonstrate the random or nondirected nature of mutation. For simplicity, only four colonies are shown on each plate, and only two are tested for streptomycin resistance in step 4. Actually, each plate contains about 200 colonies, and many plates must be used to find an adequate number of mutant colonies.

Mutation: A Reversible Process

As we discussed above, a mutation in a wild-type gene can produce a mutant allele that results in an abnormal phenotype. However, the mutant allele can also mutate back to a form that restores the wild-type phenotype. That is, mutation is a reversible process.

The mutation of a wild-type gene to a form that results in a mutant phenotype is referred to as **forward mutation**. However, sometimes the designation of the wild-type and mutant phenotypes is quite arbitrary. They may simply represent two different, but normal, phenotypes. For example, geneticists consider the alleles for brown and blue eye color in humans both to be wild-type. However, in a population composed almost entirely of brown-eyed individuals, the allele for blue eyes might be thought of as a mutant allele.

When a second mutation restores the original phenotype lost because of an earlier mutation, the process is called **reversion** or **reverse mutation**. Reversion may occur in two different ways: (1) by **back mutation**, a second mutation at the same site in the gene as the original mutation, restoring the wild-type nucleotide sequence, or (2) by the occurrence of a **sup-**

pressor mutation, a second mutation at a different location in the genome, which compensates for the effects of the first mutation (Figure 13.4). Back mutation restores the original wild-type nucleotide sequence of the gene, whereas a suppressor mutation does not. Suppressor mutations may occur at distinct sites in the same gene as the original mutation or in different genes, even on different chromosomes.

Some mutations revert primarily by back mutation, whereas others do so almost exclusively through the occurrence of suppressor mutations. Thus, in genetic studies, researchers often must distinguish between these two possibilities by backcrossing the phenotypic revertant with the original wild-type organism. If the wild-type phenotype is restored by a suppressor mutation, the original mutation will still be present and can be separated from the suppressor mutation by recombination (Figure 13.4). If the wild-type phenotype is restored by back mutation, all of the progeny of the backcross will be wild-type.

Key Points: **Mutations occur in both germ-line and somatic cells, but only germ-line mutations are transmitted to progeny. Mutations may occur spontaneously or be induced by mutagenic agents in the environment. Mutation usually is a nonadaptive process, with an environmental stress simply selecting organisms with preexisting, randomly occurring mutations. Restoration of the wild-type phenotype in a mutant organism may occur by either back mutation or the occurrence of a suppressor mutation.**

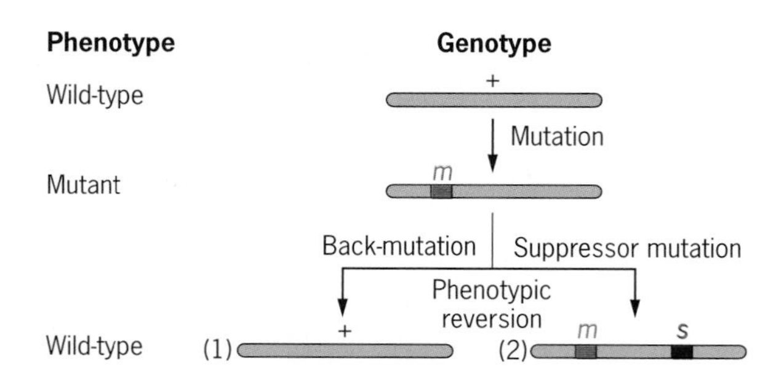

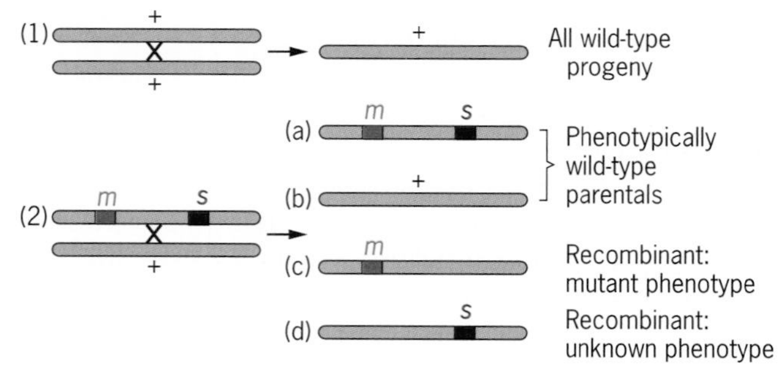

Figure 13.4 Restoration of the original wild-type phenotype of an organism may occur by (1) back mutation or (2) suppressor mutation. Some mutants can revert to the wild-type phenotype by both mechanisms. Revertants of the two types can be distinguished by backcrosses to the original wild-type. If back mutation has occurred, all backcross progeny will be wild-type. If a suppressor mutation is responsible, some of the backcross progeny will have the mutant phenotype (2c).

MUTATION: PHENOTYPIC EFFECTS

Mutations must result in some detectable phenotypic change to be recognized. The effects of mutations on phenotype range from alterations so minor that they can be detected only by special genetic or biochemical techniques to gross modifications of morphology to lethals. A gene is a sequence of nucleotide pairs that usually encodes a specific polypeptide. Any mutation occurring within a given gene will thus produce a new allele of that gene. Genes containing mutations with small effects that can be recognized only by special techniques are called **isoalleles**. Other mutations produce **null alleles** that result in totally nonfunctional gene products. If mutations of the latter type occur in genes that are required for the growth of the organism, individuals that are homozygous for the mutation will not survive. Such mutations are called **recessive lethals**.

Mutations can be either recessive or dominant. In monoploid organisms such as viruses and bacteria, both recessive and dominant mutations can be recognized by their effect on the phenotype of the organism in which they occur. In diploid organisms such as fruit flies and humans, recessive mutations will alter the phenotype only when present in the homozygous condition. Thus, in diploids, most recessive mutations will not be recognized at the time of their occurrence because they will be present in the heterozygous state. Sex-linked recessive mutations are an exception; they will be expressed in the hemizygous state in the heterogametic sex (males in humans and fruit flies; females in birds). Sex-linked recessive lethal mutations will alter the sex ratio because hemizygous individuals that carry the lethal will not survive (Figure 13.5).

Mutations with Phenotypic Effects: Usually Deleterious and Recessive

Most of the thousands of mutations that have been identified and studied by geneticists are deleterious and recessive. This result is to be expected if we consider what is known about the genetic control of metabolism and the techniques available for identifying

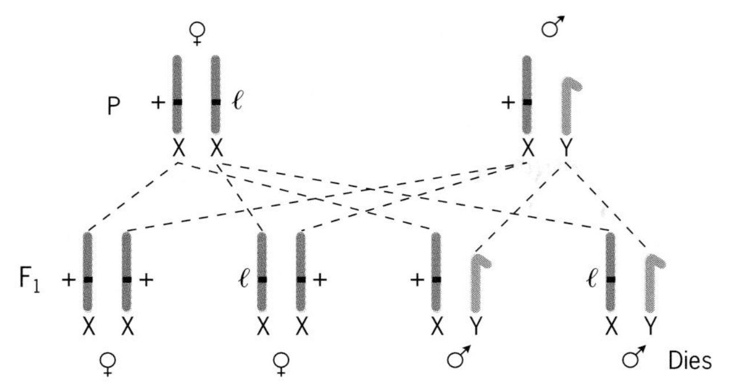

Figure 13.5 Alteration of the sex ratio by an X-linked recessive lethal mutation. Females heterozygous for an X-linked recessive lethal will produce female and male progeny in a 2:1 ratio.

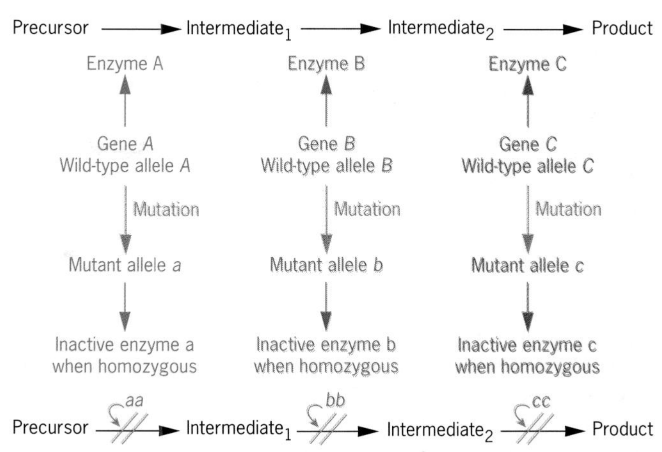

Figure 13.6 Recessive mutant alleles often result in blocks in metabolic pathways. The pathways can be only a few steps long, as diagrammed here, or many steps long. The wild-type allele of each gene usually encodes a functional enzyme that catalyzes the appropriate reaction. Most mutations that occur in wild-type genes result in altered forms of the enzyme with reduced or no activity. In the homozygous state, mutant alleles that produce inactive products cause metabolic blocks (⇸) owing to the lack of the required enzyme activity.

mutations. As we discussed in Chapter 11, metabolism occurs by sequences of chemical reactions, with each step catalyzed by a specific enzyme encoded by one or more genes. Mutations in these genes frequently produce blocks in metabolic pathways (Figure 13.6). These blocks occur because alterations in the base-pair sequences of genes often cause changes in the amino acid sequences of polypeptides (Figure 13.7), which may result in nonfunctional products (Figure 13.6). Indeed, this is the most commonly observed effect of easily detected mutations. Given a wild-type allele encoding an active enzyme and mutant alleles encoding less active or totally inactive enzymes, it is apparent why most of the observed mutations would be recessive. If a cell contains both active and inactive forms of a given enzyme, the active form usually will catalyze the reaction in question. Therefore, the allele specifying the active product usually will be dominant, and the allele encoding the inactive product will be recessive (Chapter 4).

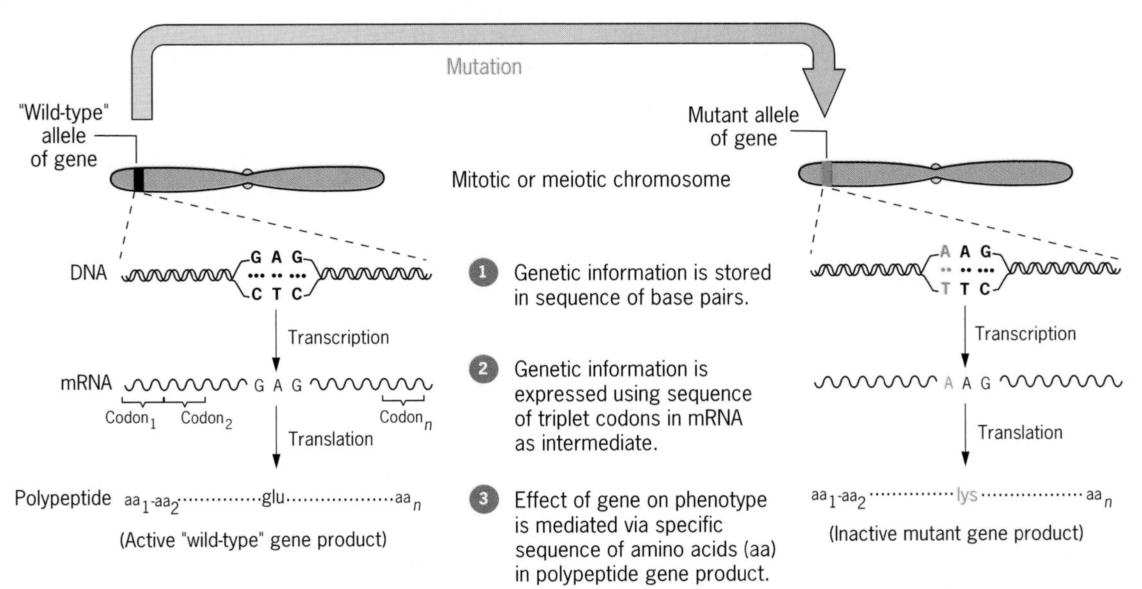

Figure 13.7 Overview of the mutation process and the expression of wild-type and mutant alleles. Mutations alter the sequences of nucleotide pairs in genes, which, in turn, cause changes in the amino acid sequences of the polypeptides encoded by these genes. A G:C base pair (top, left) has mutated to an A:T base pair (top, right). This mutation changes one mRNA codon from GAG to AAG and one amino acid in the polypeptide product from glutamic acid (glu) to lysine (lys). Such changes often yield nonfunctional gene products.

Because of the degeneracy and order in the genetic code (Chapter 12), many mutations have no effect on the phenotype of the organism; they are called **neutral mutations**. But why should most mutations with phenotypically recognizable effects result in decreased gene-product activity or no gene-product activity? A wild-type allele of a gene encoding a wild-type enzyme or structural protein will have been selected for optimal activity during the course of evolution. Thus mutations, which cause random changes in the highly adapted amino acid sequences, usually will produce less active or totally inactive products. You can make an analogy with any complex, carefully engineered machine such as a watch or an automobile. If you randomly modify an essential component, the machine seldom performs as well as it did prior to the change. This view of mutation and the interaction between mutant and wild-type alleles fits with the observation that most mutations with recognizable phenotypic effects are recessive and deleterious.

Effects of Mutations in Human Globin Genes

Mutant human hemoglobins provide good illustrations of the deleterious effects of mutation. Hemoglobin is the oxygen-transporting macromolecule present in the red blood cells of chordate animals. It serves the essential function of transporting oxygen from the lungs to all of the various tissues of the body. Each hemoglobin molecule is composed of four polypeptide chains, two of each of the two different polypeptides, plus a heme group linked to each. The major form of hemoglobin in adults (hemoglobin A) contains two identical **alpha (α) chains** and two identical **beta (β) chains**. Each α polypeptide consists of a specific sequence of 141 amino acids, whereas each β chain is 146 amino acids long. Because of similarities in their amino acid sequences, all the globin chains (and, thus, their structural genes) are believed to have evolved from a common progenitor.

Many different variants of adult hemoglobin have been identified in human populations, and several of them have severe phenotypic effects. Many of the variants were initially detected by their altered electrophoretic behavior (movement in an electric field due to charge differences). The hemoglobin variants provide an excellent illustration of the effects of mutation on the structures and functions of gene products and, ultimately, on the phenotypes of the affected individuals. In Chapter 12, we briefly discussed the traumatic consequences of one hemoglobin variant, sickle-cell hemoglobin. Sickle-cell hemoglobin (hemoglobin S) contains an altered β chain. Individuals homozygous ($Hb^s{}_\beta/Hb^s{}_\beta$) for the hemoglobin S allele develop severe hemolytic anemia, which is often fatal. Hemoglobin S molecules precipitate when deoxygenated, forming crystalloid aggregates that distort the morphology of red blood cells. The resulting sickle-shaped cells clog small blood vessels and cut off oxygen transport to various tissues.

When the amino acid sequences of the β chains of hemoglobin A and hemoglobin S were determined and compared, hemoglobin S was found to differ from hemoglobin A at only one position. The sixth amino acid from the amino terminus of the β chain of hemoglobin A is glutamic acid (a negatively charged amino acid). The β chain of hemoglobin S contains valine (no charge at neutral pH) at that position. The α chains of hemoglobin A and hemoglobin S are identical. Thus the change of a single amino acid in one polypeptide can have severe effects on the phenotype. A large number of similar effects of mutation on protein structure, and ultimately on the phenotype, are now well documented.

In the case of hemoglobin S, the substitution of valine for glutamic acid at the sixth position in the β chain allows a new bond to form, which changes the conformation of the protein and leads to aggregation of hemoglobin molecules. This change results in the grossly abnormal shape of the red blood cells. The mutational change in the $Hb^A{}_\beta$ gene that gave rise to $Hb^s{}_\beta$ was a substitution of an A:T base pair for a T:A base pair, with a T in the transcribed strand in the first case and an A in the transcribed strand in the second case (Figure 13.8). This T:A → A:T base-pair substitution was first predicted from protein sequence data and the known codon assignments, and was later verified by sequencing the $Hb^A{}_\beta$ and $Hb^S{}_\beta$ genes.

Over 100 hemoglobin variants with amino acid changes in the β chain are known. Most of them differ from the normal β chain of hemoglobin A by a single amino acid substitution. A few examples are illustrated in Figure 13.9. Numerous variants of the α polypeptide also have been identified.

The hemoglobin examples show that mutation is a process in which changes in gene structure, often changes in one or a few base pairs, cause changes in the amino acid sequences of the polypeptide gene products. These alterations in protein structure, in turn, cause changes in the phenotype that are recognized as mutant.

Mutation in Humans: Blocks in Metabolic Pathways

In Chapter 11, we discussed the genetic control of metabolic pathways, with each step in a pathway catalyzed by an enzyme encoded by one or more genes. When mutations occur in such genes, they often cause metabolic blocks (Figure 13.6) that lead to abnormal phenotypes. This picture of the genetic control of me-

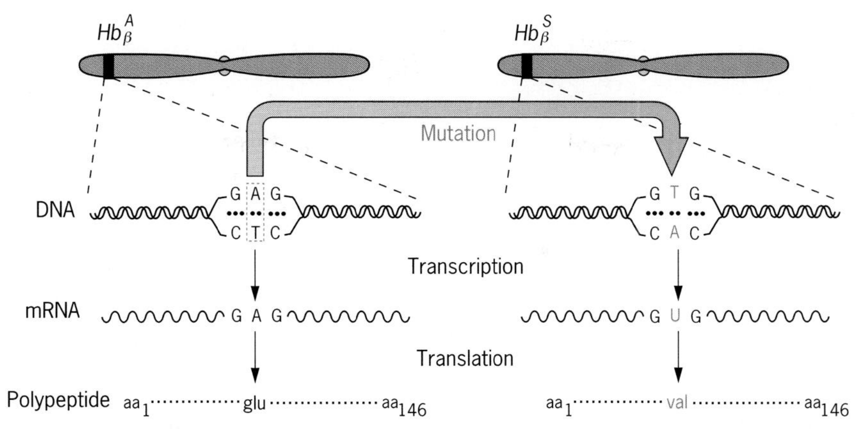

Figure 13.8 The mutational origin of sickle-cell hemoglobin (hemoglobin S). The mutation that produced the $Hb^S{}_\beta$ allele from the $Hb^A{}_\beta$ gene was the substitution of a $\frac{T}{A}$ base pair for an $\frac{A}{T}$ base pair, in which the bottom strand of DNA is the transcribed strand. The result is the substitution of the amino acid valine in the β chain of hemoglobin S for the glutamic acid residue present at the same position in hemoglobin A.

tabolism is valid for all living organisms, including humans (see the Human Genetics Sidelight).

We can illustrate the effects of mutations on human metabolism by considering virtually any metabolic pathway. However, the metabolism of the aromatic amino acids phenylalanine and tyrosine provides an especially good example because some of the early studies of mutations in humans revealed blocks in this pathway (Figure 13.10). Phenylalanine and tyrosine are essential amino acids required for protein synthesis; they are not synthesized *de novo* in humans as they are in microorganisms. Thus both amino acids must be obtained from dietary proteins.

The best-known inherited defect in phenylalanine-tyrosine metabolism is phenylketonuria, which is caused by the absence of phenylalanine hydroxylase,

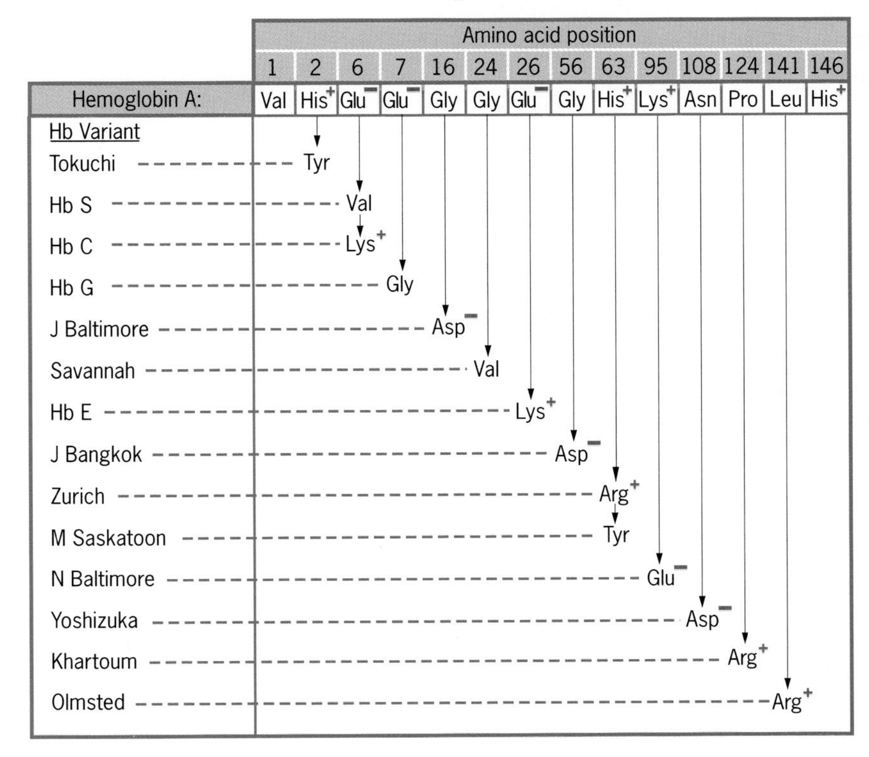

Figure 13.9 A few of the amino acid substitutions that have been documented in human β-globin chains. All the examples shown resulted from single base-pair substitutions in the $Hb^A{}_\beta$ gene (analogous to that shown for hemoglobin S in Figure 13.8). Note that many of the amino acid substitutions involve changes in charge, even though only 5 of the 20 common amino acids have a net charge at neutral pH. Substitutions involving changes in charge alter protein conformation more often than do other amino acid substitutions. Thus such substitutions are more likely to cause changes in protein function.

HUMAN GENETICS SIDELIGHT

Tay-Sachs Disease, A Childhood Tragedy

Of all the inherited human disorders, perhaps none is more tragic than Tay-Sachs disease. Infants homozygous for the mutant gene that causes Tay-Sachs disease are normal at birth. However, within a few months, they become hypersensitive to loud noises and develop a cherry-red spot on the retina of the eye. These early symptoms of the disease often go undetected by parents and physicians. At six months to one year after birth, Tay-Sachs children begin to undergo progressive neurological degeneration that rapidly leads to mental retardation, blindness, deafness, and general loss of control of body functions. By two years of age, they are usually totally paralyzed and develop chronic respiratory infections. Death commonly occurs at three to four years of age.

Although the molecular defect responsible for Tay-Sachs disease is known, there is no effective treatment of the disorder. The only positive aspect of Tay-Sachs disease is that it is rare in most populations. However, this is of little comfort to the Ashkenazi Jewish people of Central Europe and their descendants. Tay-Sachs disease occurs in about 1 of 3600 of their children, and about 1 of 30 adults in these Jewish populations carries the mutant gene in the heterozygous state. If two individuals from these populations marry, the chance that both will carry the mutant gene is about 1 in 1000 (0.033 × 0.033); if both are carriers, on average, one-fourth of their children will be homozygous for the mutant gene and develop Tay-Sachs disease.

The mutation that causes Tay-Sachs disease is located in the *HEXA* gene, which encodes the enzyme hexosaminidase A. This enzyme acts on a complex lipid called ganglioside GM2, cleaving it into a smaller ganglioside (GM3) and *N*-acetyl-*D*-galactosamine, as shown in the following.

N-acetylgalactosamine-β-1,4,-galactose-β-1,4-glucose-β-1,1-ceramide

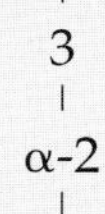

N-acetylneuramic acid

Ganglioside GM_2

Hexosaminidase A ↓ Tay-Sachs disease

N-acetylgalactosamine + Galactose-β-1,4-glucose-β-1,1-ceramide

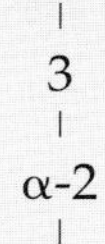

N-acetylneuramic acid

Ganglioside GM_3

The function of ganglioside GM2 is to coat nerve cells, insulating them from events occurring in neighboring cells and thus speeding up the transmission of nerve impulses. In the absence of the enzyme, ganglioside GM2 accumulates and literally smothers nerve cells. This buildup of complex lipids on neurons blocks their action, leading to deterioration of the nervous system and eventually paralysis.

Although Tay-Sachs disease was described by Warren Tay in 1881 and the biochemical basis has been known for over 20 years, there is still no effective treatment of this tragic disorder. Whereas some inherited disorders can be treated by enzyme therapy—by supplying the missing enzyme to patients, this approach is not feasible with Tay-Sachs disease because the enzyme will not penetrate the barrier separating brain cells from the circulatory system. Moreover, somatic-cell gene therapy—providing functional copies of the defective gene to somatic cells (Chapter 20)—is not possible at present because there is no established procedure for introducing genes into neurons. Indeed, scientists still don't know which nerve cells are responsible for the neurological degeneration that occurs in children with the disease.

Tay-Sachs disease can be detected prenatally by amniocentesis (Chapter 6), and this procedure has been used extensively to diagnose the disorder. Recently, a sensitive DNA test has been developed that allows scientists to detect the mutant gene that causes Tay-Sachs disease in DNA isolated from a single cell (Chapter 20). This DNA test has been used to screen eight-cell pre-embryos produced by *in vitro* fertilization for the Tay-Sachs mutation. One cell is used for the DNA test, and the other seven cells retain the capacity to develop into a normal embryo when implanted into the uterus of the mother. Only embryos that test normal—those not homozygous for the deadly Tay-Sachs gene—are implanted. This procedure allows parents who are both carriers of the mutant gene to have children without worrying about the birth of a child with Tay-Sachs disease.

the enzyme that converts phenylalanine to tyrosine. Newborns with phenylketonuria, an autosomal recessive disease, develop severe mental retardation if not placed on a diet low in phenylalanine. (This disorder is discussed in detail in Chapter 26.) The first inherited disorder in phenylalanine-tyrosine metabolism to be studied in humans was alkaptonuria, which is caused by autosomal recessive mutations that inactivate the

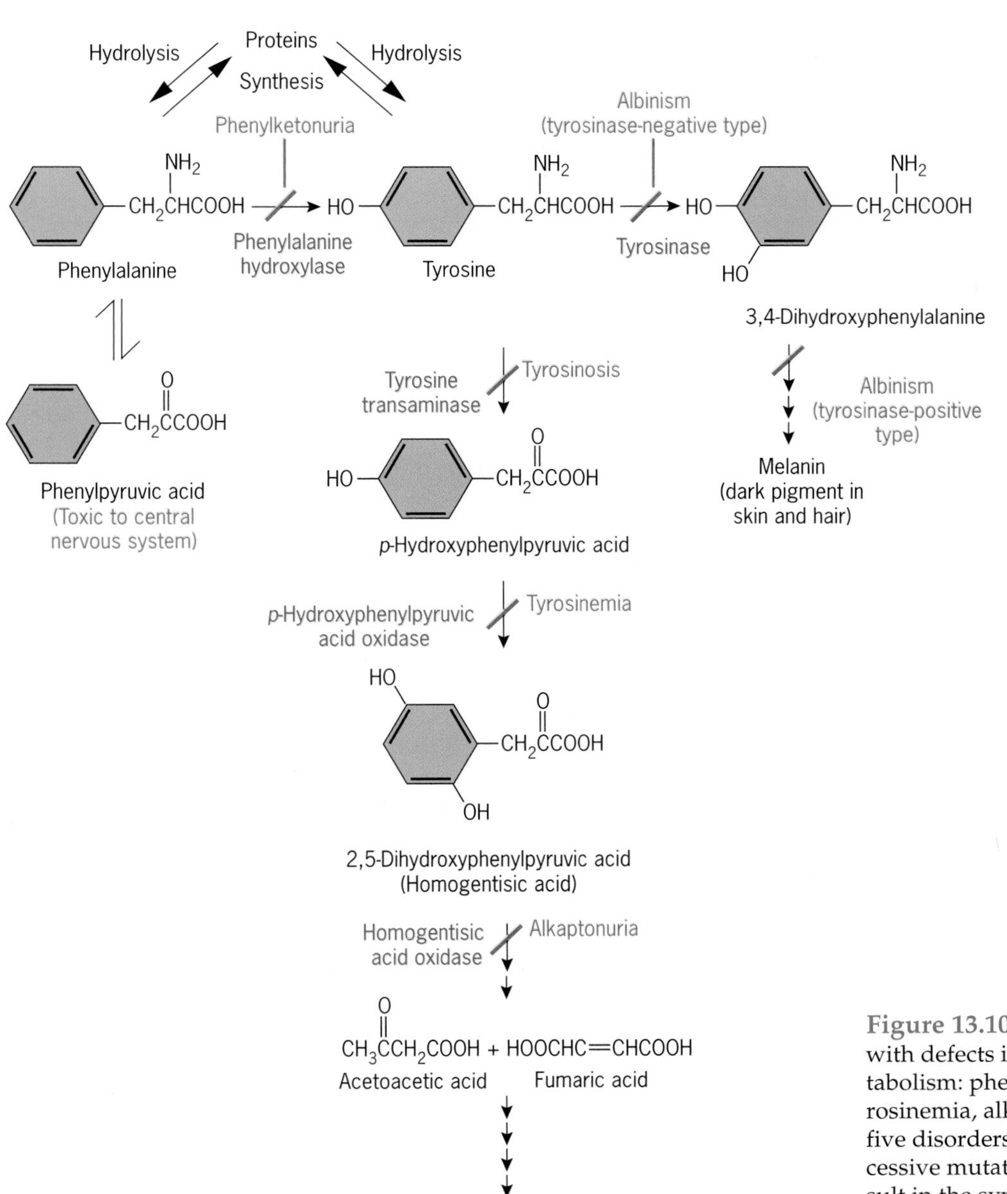

Figure 13.10 Inherited human disorders with defects in phenylalanine-tyrosine metabolism: phenylketonuria, tyrosinosis, tyrosinemia, alkaptonuria, and albinism. All five disorders are caused by autosomal recessive mutations. The mutations, which result in the synthesis of inactive enzymes, block phenylalanine-tyrosine metabolism at the steps indicated.

enzyme homogentisic acid oxidase. Alkaptonuria, which played an important role in the evolution of the concept of the gene, is discussed in Chapter 14.

Two other inherited disorders are caused by mutations in genes encoding enzymes required for the catabolism of tyrosine; both are inherited as autosomal recessives. Tyrosinosis and tyrosinemia result from the lack of the enzymes tyrosine transaminase and *p*-hydroxyphenylpyruvic acid oxidase, respectively (Figure 13.10). Tyrosinosis is very rare; only a few cases have been studied. Individuals with tyrosinosis show pronounced increases in tyrosine levels in their blood and urine and have various congenital abnormalities. Individuals with tyrosinemia have elevated levels of both tyrosine and *p*-hydroxyphenylpyruvic acid in their blood and urine. Most newborns with tyrosinemia die within six months after birth because of liver failure.

Albinism, the absence of pigmentation in the skin, hair, and eyes, results from a mutational block in the conversion of tyrosine to the dark pigment melanin. One type of albinism is caused by the absence of tyrosinase, the enzyme that catalyzes the first step in the synthesis of melanin from tyrosine. Other types of albinism result from blocks in subsequent steps in the conversion of tyrosine to melanin. Albinism is inherited as an autosomal recessive trait; heterozygotes usually have normal levels of pigmentation.

Thus studies of a single metabolic pathway, phenylalanine-tyrosine metabolism, have revealed five different inherited disorders, all caused by mutations in genes that control steps in this pathway. The same picture of the genetic control of metabolism can be obtained by examining essentially any other metabolic pathway in humans.

Conditional Lethal Mutations: Powerful Tools for Genetic Studies

Of all the mutations—from isoalleles to lethals, **conditional lethal mutations** are the most useful for genetic studies. These are mutations that are (1) lethal in one environment, the **restrictive conditions**, but are (2) viable in a second environment, the **permissive conditions**. Conditional lethal mutations allow geneticists to identify and study mutations in essential genes that result in complete loss of gene-product activity even in haploid organisms. Mutants carrying conditional lethals can be propagated under permissive conditions, and information about the functions of the gene products can be inferred by studying the consequences of their absence under the restrictive conditions. Conditional lethal mutations have been used to investigate a vast array of biological processes from development to photosynthesis. They have also been used to construct detailed genetic maps of chromosomes (Chapters 15 and 16).

The three major classes of mutants with conditional lethal phenotypes are (1) auxotrophic mutants, (2) temperature-sensitive mutants and (3) suppressor-sensitive mutants. **Auxotrophs** are mutants that are unable to synthesize an essential metabolite (amino acid, purine, pyrimidine, vitamin, and so forth) that is synthesized by wild-type or *prototrophic* organisms of the same species. The auxotrophs will grow and reproduce when the metabolite is supplied in the medium (the permissive conditions); they will not grow when the essential metabolite is absent (the restrictive conditions). **Temperature-sensitive mutants** will grow at one temperature but not at another. Most temperature-sensitive mutants are heat-sensitive; however, some are cold-sensitive. The temperature sensitivity usually results from increased heat or cold lability of the mutant gene product, for example, an enzyme that is active at low temperature but partially or totally inactive at higher temperatures. Occasionally, only the synthesis of the gene product is sensitive to temperature, and, once synthesized, the mutant gene product may be as stable as the wild-type gene product. **Suppressor-sensitive mutants** are viable when a second genetic factor, a suppressor, is present, but they are nonviable in the absence of the suppressor. The suppressor gene may correct or compensate for the defect in phenotype that is caused by the suppressor-sensitive mutation, or it may render the gene product altered by the mutation nonessential. We have discussed one class of suppressor-sensitive mutations, the *amber* mutations, in Chapter 12.

Now, let's briefly consider how conditional lethal mutations can be used to investigate biological processes—to dissect biological processes into their individual parts or steps. Let's begin with a simple biosynthetic pathway:

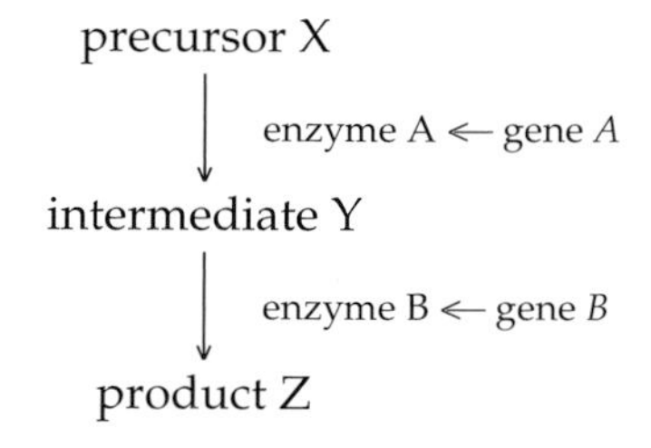

Intermediate Y is produced from precursor X by the action of enzyme A, the product of gene *A*, but intermediate Y may be rapidly converted to product Z by enzyme B, the product of gene *B*. If so, intermediate Y may be present in minute quantities and be difficult to isolate and characterize. However, in a mutant organism that has a mutation in gene *B*, resulting in the syn-

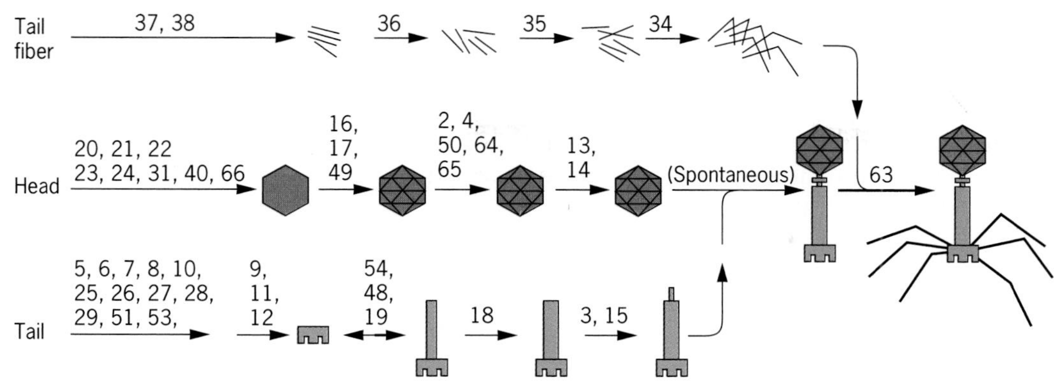

Figure 13.11 Abbreviated map of morphogenesis in bacteriophage T4. The pathway of morphogenesis is based on the work of Edgar, Wood, King, and colleagues. The head, the tail, and the tail fibers are produced via separate branches of the pathway and are then joined in the final stages of morphogenesis. The numbers identify the T4 genes whose products are required at each step in the pathway. The sequences of early steps in head and tail formation are known, but are omitted here to keep the diagram concise.

thesis of either an inactive form of enzyme B or no enzyme B, intermediate Y will often accumulate to much higher concentrations, facilitating its isolation and characterization. Similarly, a mutation in gene *A* may aid in the identification of precursor X.

In this way, the sequence of steps in a given metabolic pathway can usually be determined. Morphogenesis in living organisms occurs in part by the sequential addition of proteins to macromolecular structures to produce the final three-dimensional conformations. Again, the sequence of protein additions can often be determined by isolating and studying mutant organisms with defects in the genes encoding the proteins involved. Because an appropriate mutation will eliminate the activity of a single polypeptide, mutations provide a powerful tool with which to dissect biological processes—to break the processes down into individual steps.

The resolving power of mutational dissection of biological processes has been elegantly documented by the research of Robert Edgar, Jonathan King, William Wood, and colleagues, who worked out the complete pathway of morphogenesis of bacteriophage T4. This complex process involves the products of about 50 of the roughly 200 genes in the T4 genome. Each gene encodes a structural protein of the virus or an enzyme that catalyzes one or more steps in the morphogenetic pathway. By (1) isolating mutant strains of phage T4 with temperature-sensitive and suppressor-sensitive conditional lethal mutations in each of the approximately 50 genes, and (2) analyzing the structures that accumulate when these mutant strains are grown under the restrictive conditions by electron microscopy and biochemical techniques, Edgar, King, Wood, and coworkers have established the complete pathway of phage T4 morphogenesis (Figure 13.11).

Many other biological processes also have been successfully dissected by mutational studies. Examples include the photosynthetic electron transport chains in plants and pathways of nitrogen fixation in bacteria. Currently, mutational dissection is yielding new insights into the processes of differentiation and development in higher plants and animals (Chapter 23). Seymour Benzer and colleagues are using mutations to dissect behavior and learning in *Drosophila*. In principle, scientists should be able to use mutations to dissect any process that is under genetic control. Every gene can mutate to a nonfunctional state. Thus mutational dissection of biological processes is limited only by the ingenuity of researchers in identifying mutations of the desired types.

Key Points: **The effects of mutations on the phenotypes of living organisms range from minor to lethal changes. Most mutations exert their effects on the phenotype by altering the amino acid sequences of polypeptides, the primary gene products. The mutant polypeptides, in turn, cause blocks in metabolic pathways. Conditional lethal mutations provide powerful tools with which to dissect biological processes such as morphogenesis.**

MULLER'S DEMONSTRATION THAT X RAYS ARE MUTAGENIC

In 1927, Hermann J. Muller first demonstrated that mutation could be induced by an external factor.

Muller showed that X-ray treatment markedly increased the frequency of sex-linked recessive lethal mutations in *D. melanogaster*. X rays are a form of electromagnetic radiation with shorter wavelengths and higher energy than visible light; we discuss their effects on DNA later in this chapter. Muller's unambiguous demonstration of the mutagenicity of X rays became possible by his development of a technique facilitating the simple and accurate identification of lethal mutations on the X chromosome of *Drosophila*. This technique, called the ***CℓB* method**, involves the use of females heterozygous for a normal X chromosome and an X chromosome—the ***CℓB* chromosome**—specifically constructed for Muller's experiment.

The *CℓB* chromosome has three essential components. (1) The crossover (*C*) suppressor refers to the presence of a long inversion (an inverted segment of the chromosome) that renders nonviable the products of crossing over between the *CℓB* chromosome and the structurally normal X chromosome in heterozygous females. The inversion does not prevent crossing over between the two chromosomes, but causes progeny carrying recombinant X chromosomes produced by crossing over between the *CℓB* chromosome and the normal X chromosome to abort because of the presence of duplications and deficiencies (repeated and missing sets of genes; see Chapter 6). The inversion is required in Muller's experiment to assure that the markers on the *CℓB* chromosome stay together through meiosis. (2) The *ℓ* refers to a recessive lethal mutation on the *CℓB* chromosome. Homozygous females and hemizygous males carrying this X-linked lethal mutation are nonviable. (3) The mutant gene *B* produces the bar eye phenotype, which is a narrow, bar-shaped eye. Because it is partially dominant, the *B* mutation allows females heterozygous for the *CℓB* chromosome to be readily identified. Both the recessive lethal (*ℓ*) and the bar eye mutation (*B*) are located within the inverted segment of the *CℓB* chromosome.

Cross I: Females heterozygous for the *CℓB* chromosome are mated with irradiated males.

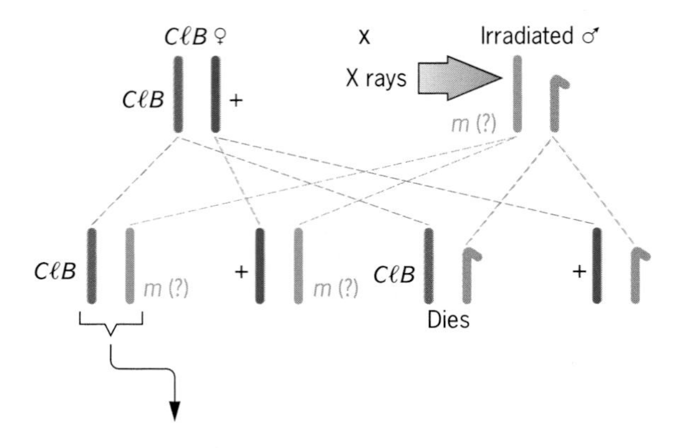

Cross II: *CℓB* female progeny of cross I are mated with wild-type males.

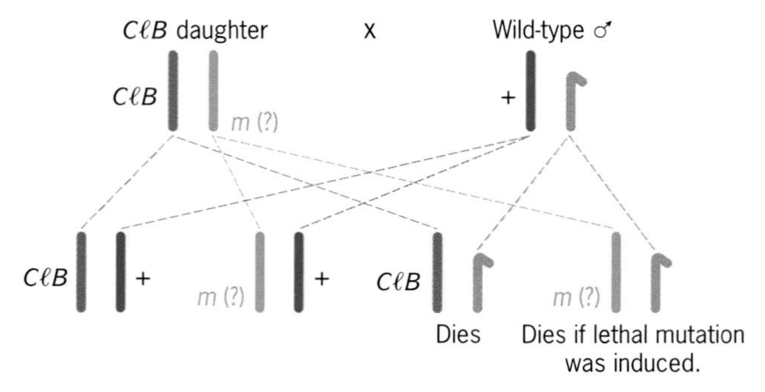

Figure 13.12 The *CℓB* technique used by Muller to detect sex-linked recessive lethal mutations in *Drosophila*. The mating shown in cross II will produce only female progeny if an X-linked recessive lethal is present on the irradiated X chromosome. One-third of the progeny produced from cross II will be males if there is no recessive lethal on the irradiated X chromosome. Thus scoring for lethal mutations simply involves screening the progeny of cross II for the presence or absence of males.

Given females heterozygous for the *CℓB* chromosome, Muller's experiment was quite simple (Figure 13.12). Male flies were irradiated and mated with *CℓB* females. The bar-eyed daughters of this mating will all carry the *CℓB* chromosome of the female parent and the irradiated X chromosome of the male parent. Because the entire population of reproductive cells of the males was irradiated, each bar-eyed daughter carries a potentially mutated X chromosome. The bar-eyed daughters were then mated individually (in separate cultures) with wild-type males. If the irradiated X chromosome carried by a bar-eyed daughter contains a sex-linked lethal, all of the progeny of the mating will be female. Because males are hemizygous for the X chromosome, those receiving the *CℓB* chromosome will die because of the recessive lethal (*ℓ*) that it carries. Those receiving the irradiated X chromosome will also die if a recessive lethal has been induced on it. Matings of bar-eyed daughters carrying an irradiated X chromosome in which no lethal mutation has been induced with wild-type males will produce female and male progeny in a ratio of 2:1 (only the males with the *CℓB* chromosome will die). With the use of the *CℓB* technique, scoring for the presence of recessive sex-linked lethals is unambiguous and error free—simply scoring for the presence or absence of male progeny. By this procedure, Muller was able to demonstrate a 150-fold increase in the frequency of X-linked lethals after treatment of male flies with X rays.

After Muller's pioneering experiments with *Drosophila*, other researchers soon demonstrated that X rays are mutagenic to other organisms, including plants, animals, and microbes. Moreover, other types of high-energy electromagnetic radiation and many chemicals were soon shown to be potent mutagens. We examine the mechanisms by which some of these agents induce mutations in the following sections of this chapter.

Key Points: Hermann J. Muller first demonstrated that an external agent can be mutagenic when, in 1927, he showed that X rays induce recessive lethals on the X chromosome of *Drosophila*.

THE MOLECULAR BASIS OF MUTATION

When Watson and Crick described the double-helix structure of DNA and proposed its semiconservative replication based on specific base-pairing to account for the accurate transmission of genetic information from generation to generation, they also proposed a mechanism to explain spontaneous mutation. Watson and Crick pointed out that the structures of the bases in DNA are not static. Hydrogen atoms can move from one position in a purine or pyrimidine to another position, for example, from an amino group to a ring nitrogen. Such chemical fluctuations are called **tautomeric shifts**. Although tautomeric shifts are rare, they may be of considerable importance in DNA metabolism because some alter the pairing potential of the bases. The nucleotide structures that we discussed in Chapter 9 are the common, more stable forms, in which adenine always pairs with thymine and guanine always pairs with cytosine. The more stable keto forms of thymine and guanine and the amino forms of adenine and cytosine may infrequently undergo tautomeric shifts to less stable enol and imino forms, respectively (Figure 13.13). The bases would be expected to exist in their less stable tautomeric forms for only short periods of time. However, if a base existed in the rare form at the moment that it was being replicated or being incorporated into a nascent DNA chain, a mutation would result. When the bases are present in their rare imino or enol states, they can form adenine-cytosine and guanine-thymine base pairs (Figure 13.14*a*). The net effect of such an event, and the subsequent replication required to segregate the mismatched base pair, is an A:T to G:C or a G:C to A:T base-pair substitution (Figure 13.14*b*).

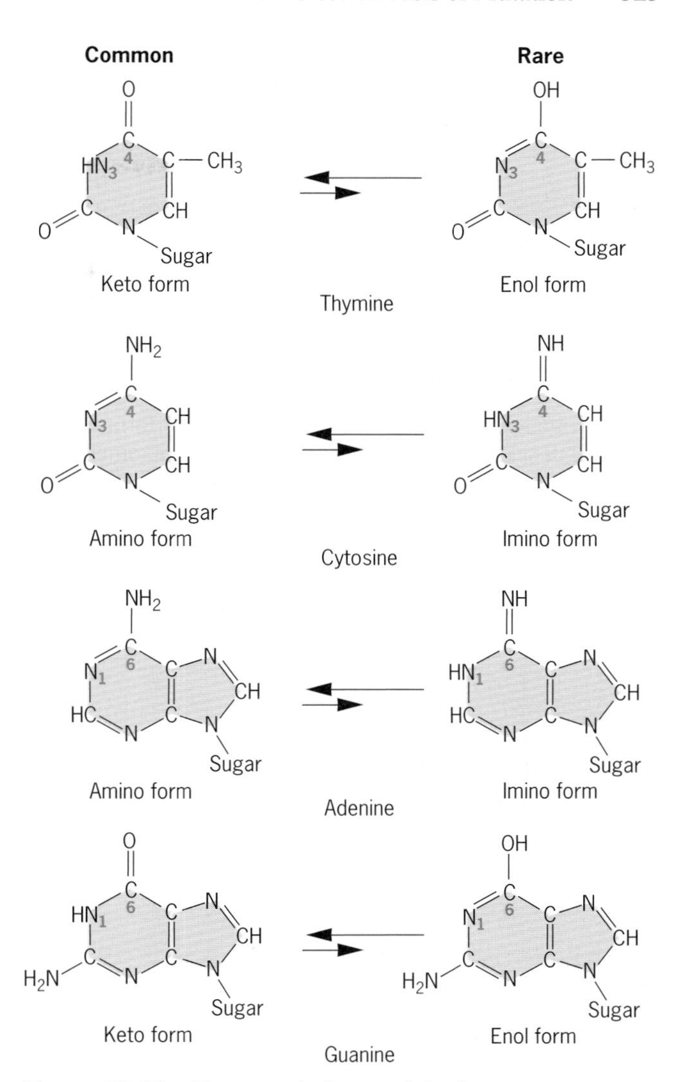

Figure 13.13 Tautomeric forms of the four common bases in DNA. The shifts of hydrogen atoms between the number 3 and number 4 positions of the pyrimidines and between the number 1 and number 6 positions of the purines change their base-pairing potential.

Mutations resulting from tautomeric shifts in the bases of DNA involve the replacement of a purine in one strand of DNA with the other purine and the replacement of a pyrimidine in the complementary strand with the other pyrimidine. Such base-pair substitutions are called **transitions**. Base-pair substitutions involving the replacement of a purine with a pyrimidine and vice versa are called **transversions**. Four different transitions and eight different transversions are possible (Figure 13.15*a*). A third type of point mutation involves the addition or deletion of one or a few base pairs. Base-pair additions and deletions are collectively referred to as **frameshift mutations** because they alter the reading frame of all base-pair triplets (specifying codons in mRNA and amino acids in the polypeptide gene product) in the gene that are distal to the site at which the mutation occurs (Figure 13.15*b*).

All three types of point mutations—transitions, transversions, and frameshift mutations—are present among spontaneously occurring mutations. A surpris-

Rare A:C base pair

Cytosine
(rare imino form)

Adenine
(common amino form)

Rare G:T base pair

Thymine
(common keto form)

Guanine
(rare enol form)

(*a*) Hydrogen-bonded A:C and G:T base pairs that form when cytosine and guanine are in their rare imino and enol tautomeric forms.

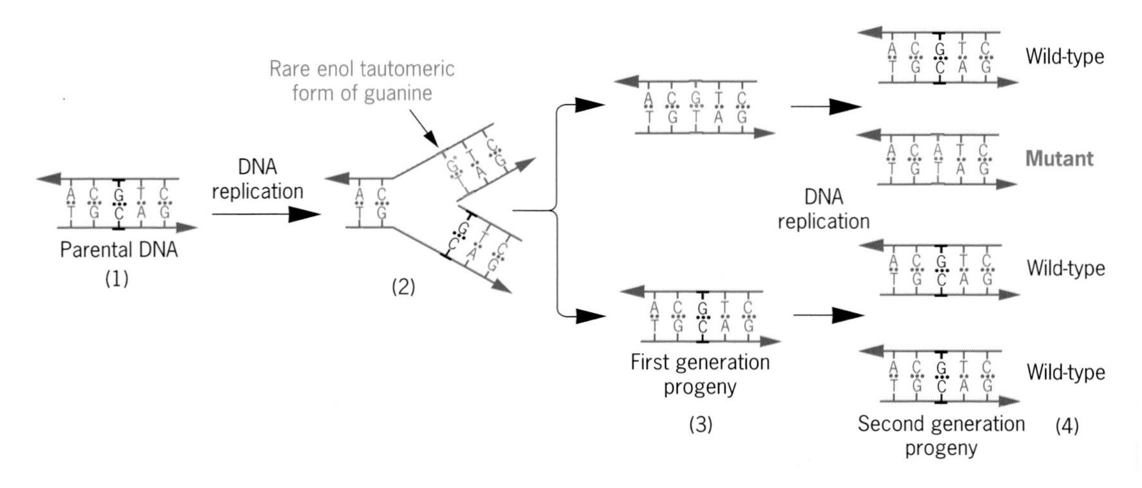

(*b*) Mechanism by which tautomeric shifts in the bases in DNA cause mutations.

Figure 13.14 The effects of tautomeric shifts in the nucleotides in DNA on (*a*) base pairing and (*b*) mutation. Rare A:C and G:T base pairs like those shown in (*a*) also form when thymine and adenine are in their rare enol and imino forms, respectively. (*b*) A guanine (1) undergoes a tautomeric shift to its rare enol form (G^*) at the time of replication (2). In its enol form, guanine pairs with thymine (2). During the subsequent replication (3 to 4), the guanine shifts back to its more stable keto form. The thymine incorporated opposite the enol form of guanine (2) directs the incorporation of adenine during the next replication (3 to 4). The net result is a G:C to A:T base-pair substitution.

ingly large proportion of the spontaneous mutations that have been studied in prokaryotes are found to be single base-pair additions and deletions rather than base-pair substitutions.

Although much remains to be learned about the causes, molecular mechanisms, and frequency of spontaneously occurring mutations, three major factors are (1) the accuracy of the DNA replication machinery, (2) the efficiency of the mechanisms that have evolved for the repair of damaged DNA, and (3) the degree of exposure to mutagenic agents present in the environment. Perturbations of the DNA replication apparatus or DNA repair systems have been shown to cause large increases in mutation rates.

Mutations Induced by Chemicals

Muller's discovery of the mutagenic effects of X rays in 1927 provided a method by which researchers could induce large numbers of mutations. However, because X rays have many effects on living tissues, X-ray-induced mutations provided little information about the molecular mechanisms that are involved. The discovery of chemical mutagens with known effects on DNA led to a better understanding of mutation at the molecular level. Many chemicals are mutagenic; our discussion here focuses on a few chemicals that have specific or potent mutagenic effects (Table 13.1).

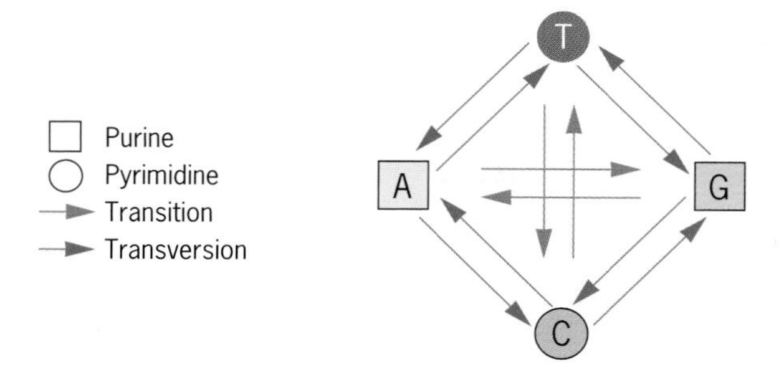

(*a*) Twelve different base substitutions can occur in DNA.

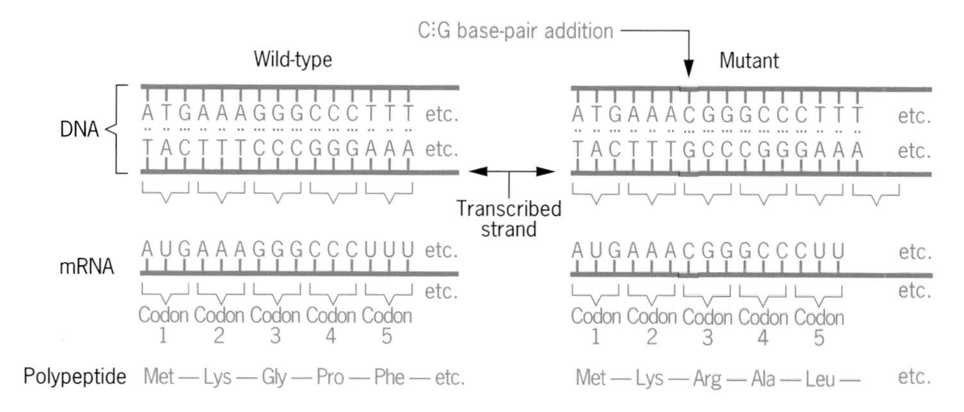

(*b*) Insertions or deletions of one or two base pairs alter the reading frame of the gene distal to the site of the mutation.

Figure 13.15 Types of point mutations that occur in DNA: (a) base substitutions and (b) frameshift mutations. (*a*) The base substitutions include four transitions (purine for purine and pyrimidine for pyrimidine; green arrows) and eight transversions (purine for pyrimidine and pyrimidine for purine; blue arrows). (*b*) A mutant gene (top, right) was produced by the insertion of a C:G base pair between the sixth and seventh base pairs of the wild-type gene (top, left). This insertion alters the reading frame of that portion of the gene distal to the mutation, relative to the direction of transcription and translation (left to right, as diagrammed). The shift in reading frame, in turn, changes all of the codons in the mRNA and all of the amino acids in the polypeptide specified by base-pair triplets distal to the mutation.

Mustard gas (sulfur mustard) was the first chemical shown to be mutagenic. Charlotte Auerbach and her associates discovered the mutagenic effects of mustard gas and related compounds during World War II. However, because of the potential use of mustard gas in chemical warfare, the British government placed their results on the classified list. Thus Auerbach and coworkers could neither publish their results nor discuss them with other geneticists. The compounds that they studied are examples of a large class of chemical mutagens that transfer alkyl (CH_3-, CH_3CH_2-, and so forth) groups to the bases in DNA; thus they are called **alkylating agents**.

Chemical mutagens can be divided into two groups: (1) those that are mutagenic only to replicating DNA, such as base analogs—purines, and pyrimidines with structures similar to the normal bases of DNA, and (2) those that are mutagenic to both replicating and nonreplicating DNA, such as the alkylating agents and nitrous acid. The base analogs must be incorporated into DNA chains in the place of normal bases during replication to exert their mutagenic effects. The first group of mutagens also includes the acridine dyes, which intercalate into DNA and increase the probability of mistakes during replication.

The mutagenic **base analogs** have structures similar to the normal bases and are incorporated into DNA during replication. However, their structures are sufficiently different from the normal bases in DNA that they increase the frequency of mispairing and thus mutation during replication. The two most commonly used base analogs are 5-bromouracil and 2-aminopurine. The pyrimidine 5-bromouracil is a thymine analog; the bromine at the 5 position is similar in several

TABLE 13.1
Some of the More Potent Chemical Mutagens

Chemical Name	*Common Name or Abbreviation*	*Structure*
I. Alkylating agents		
Di-(2-chloroethyl) sulfide	Mustard gas or sulfur mustard	$Cl{-}CH_2{-}CH_2{-}S{-}CH_2{-}CH_2{-}Cl$
Di-(2-chloroethyl) methylamine	Nitrogen mustard	$Cl{-}CH_2{-}CH_2{-}N(CH_3){-}CH_2{-}CH_2{-}Cl$
Ethylmethane sulfonate	EMS	$CH_3{-}CH_2{-}O{-}SO_2{-}CH_3$
Ethylethane sulfonate	EES	$CH_3{-}CH_2{-}O{-}SO_2{-}CH_2{-}CH_3$
N-Methyl-*N′*-nitro-*N*-nitrosoguanidine	NTG	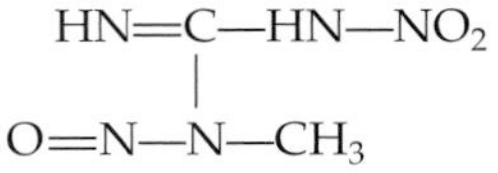 $HN{=}C{-}HN{-}NO_2$ with $O{=}N{-}N{-}CH_3$ attached to C
II. Base analogs		
5-Bromouracil	5-BU	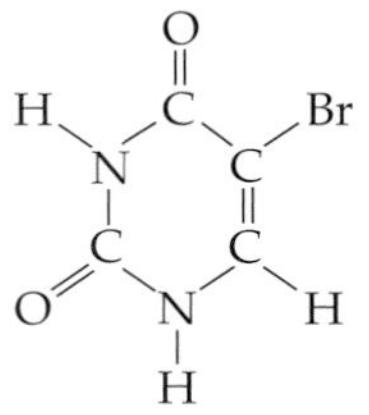 Pyrimidine ring: H–N, C=O, C–Br, C–H, N–H, C=O
2-Aminopurine	2-AP	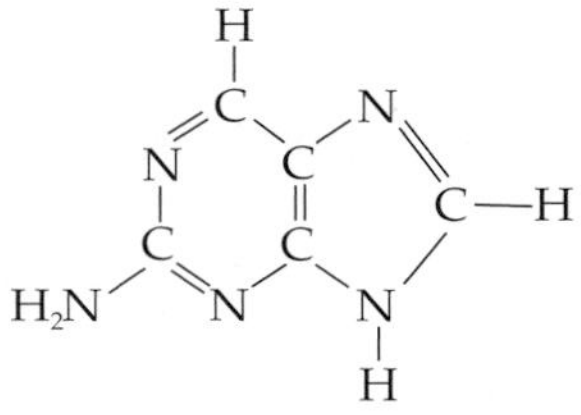 Purine ring: H–C, N, C, N, C–H, H_2N–C, N, C, N–H
III. Acridines		
2,8-Diamino acridine	Proflavin	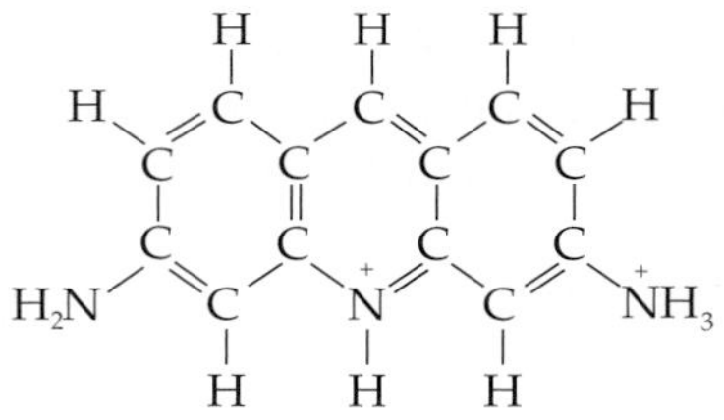 Acridine ring with H atoms, H_2N, $\overset{+}{N}$–H, $\overset{+}{N}H_3$
IV. Deaminating agents		
Nitrous acid	—	HNO_2
V. Miscellaneous		
Hydroxylamine	HA	NH_2OH

respects to the methyl ($-CH_3$) group at the 5 position in thymine. However, the bromine at this position changes the charge distribution and increases the freguency of tautomeric shifts (Figure 13.13). In its more stable keto form, 5-bromouracil pairs with adenine. After a tautomeric shift to its enol form, 5-bromouracil pairs with guanine (Figure 13.16). The mutagenic effect of 5-bromouracil is the same as that predicated for tautomeric shifts in normal bases (Figure 13.14*b*), namely, transitions.

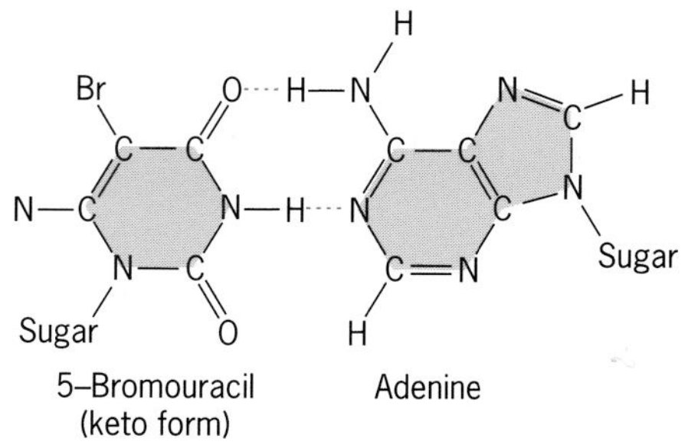

5–Bromouracil (keto form) Adenine

(*a*) 5–Bromouracil: adenine base pair.

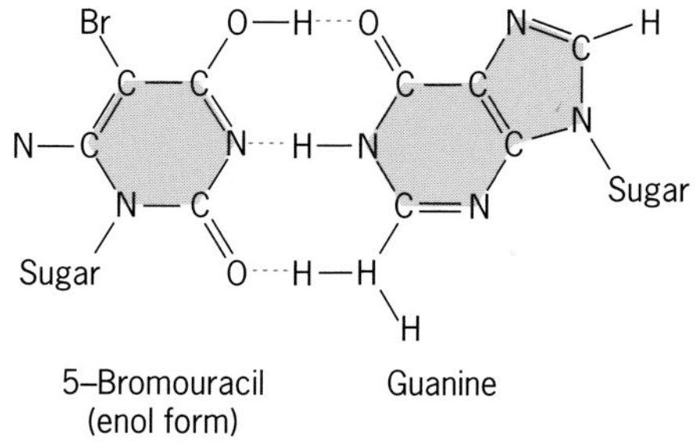

5–Bromouracil (enol form) Guanine

(*b*) 5–Bromouracil: guanine base pair.

Figure 13.16 Base-pairing between 5-bromouracil and (*a*) adenine or (*b*) guanine.

If 5-bromouracil is present in its less frequent enol form as a nucleoside triphosphate at the time of its incorporation into a nascent strand of DNA, it will be incorporated opposite guanine in the template strand and cause a G:C → A:T transition (Figure 13.17*a*). If, however, 5-bromouracil is incorporated in its more frequent keto form opposite adenine (in place of thymine) and undergoes a tautomeric shift to its enol form during a subsequent replication, it will cause an A:T → G:C transition (Figure 13.17*b*). Thus 5-bromouracil induces transitions in both directions, A:T ↔ G:C. An important consequence of the bidirectionality of 5-bromouracil-induced transitions is that mutations originally induced with this thymine analog can also be induced to mutate back to the wild-type with 5-bromouracil. 2-Aminopurine acts in a similar manner but is incorporated in place of adenine or guanine.

Nitrous acid (HNO_2) is a potent mutagen that acts on either replicating or nonreplicating DNA. Nitrous acid causes oxidative deamination of the amino groups in adenine, guanine, and cytosine. This reaction converts the amino groups to keto groups and changes the hydrogen-bonding potential of the modified bases (Figure 13.18). Adenine is deaminated to hypoxanthine, which base-pairs with cytosine rather than thymine. Cytosine is converted to uracil, which base-pairs with adenine instead of guanine. Deamination of guanine produces xanthine, but xanthine—just like guanine—base-pairs with cytosine. Thus the deamination of guanine is not mutagenic. Because the deamination of adenine results in A:T → G:C transitions, and the deamination of cytosine produces G:C → A:T transitions, nitrous acid induces transitions in both directions, A:T ↔ G:C. As a result, nitrous acid-induced mutations also are induced to mutate back to wild-type by nitrous acid.

The **acridine dyes** such as proflavin (Table 13.1), acridine orange, and a whole series of related compounds are potent mutagens that induce frameshift mutations (Figure 13.15*b*). The positively charged acridines intercalate, or sandwich themselves, be-

Effect of enol form of 5-bromouracil during:

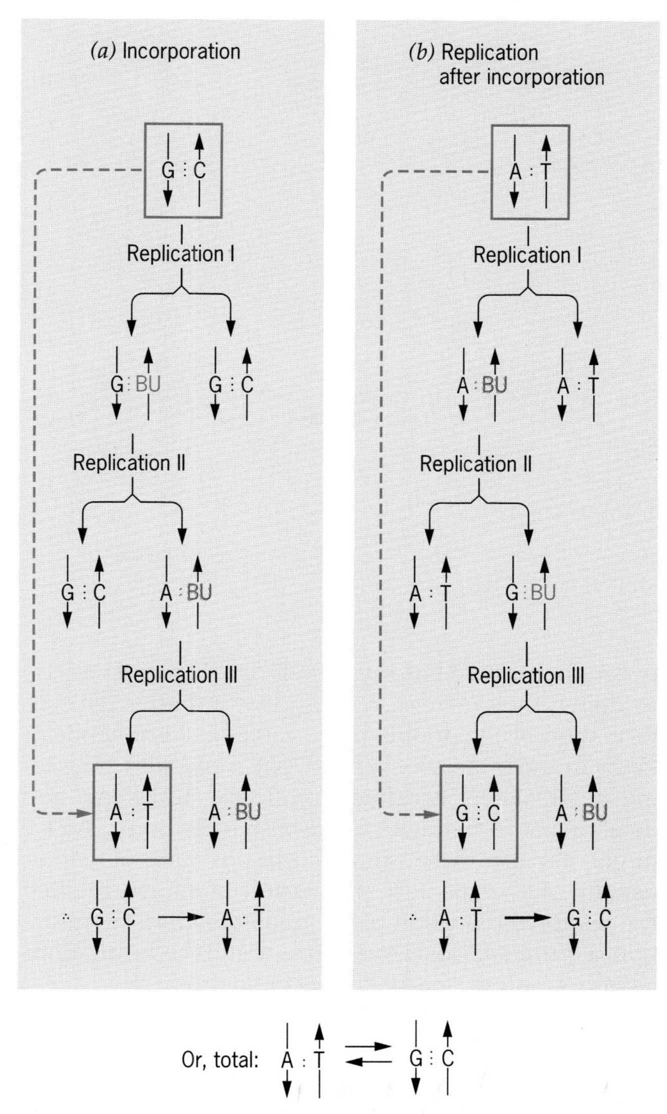

Figure 13.17 The mutagenic effects of 5-bromouracil. (*a*) When 5-bromouracil (BU) is present in its less frequent enol form (orange) at the time of incorporation into DNA, it induces G:C → A:T transitions. (*b*) When 5-bromouracil is incorporated into DNA in its more common keto form (blue) and shifts to its enol form during a subsequent replication, it induces A:T → G:C transitions. Thus 5-bromouracil can induce transitions in both directions, A:T ↔ G:C.

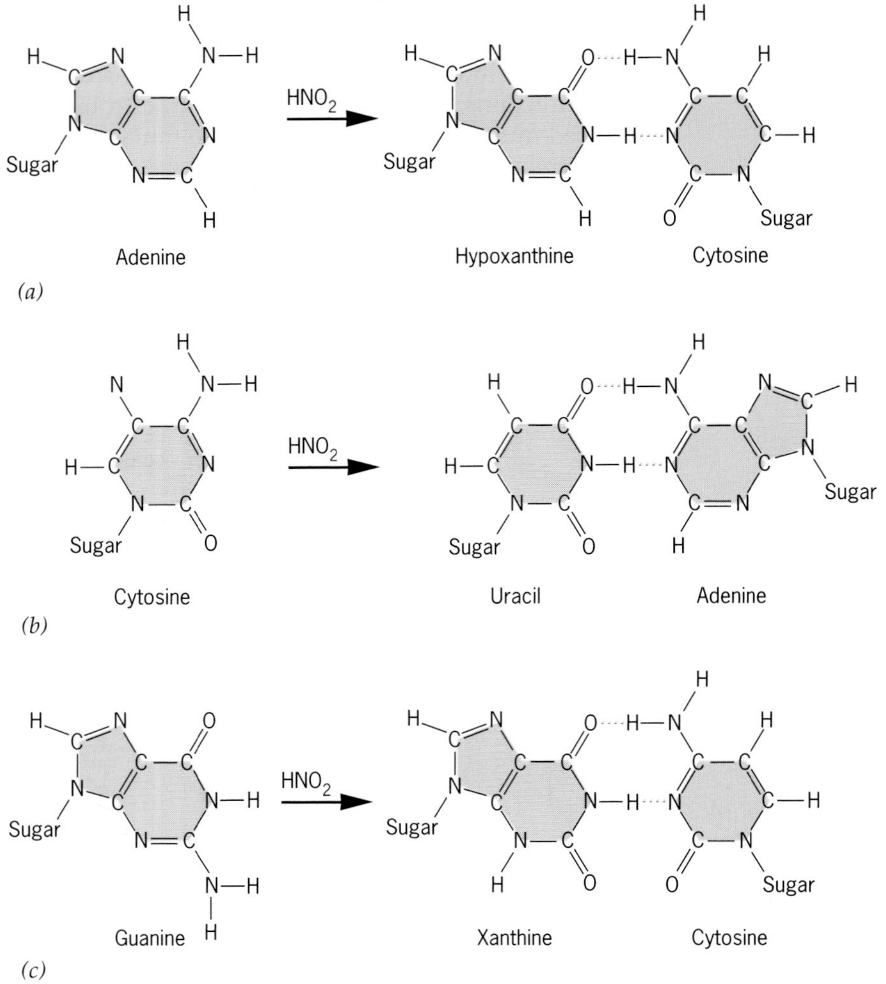

Adenine → HNO_2 → Hypoxanthine Cytosine

(*a*)

Cytosine → HNO_2 → Uracil Adenine

(*b*)

Guanine → HNO_2 → Xanthine Cytosine

(*c*)

Figure 13.18 Nitrous acid induces mutations by oxidative deamination of the bases in DNA. Nitrous acid converts (*a*) adenine to hypoxanthine, causing A:T → G:C transitions; (*b*) cytosine to uracil, causing G:C → A:T transitions; and (*c*) guanine to xanthine, which is not mutagenic. Together, the effects of nitrous acid on adenine and cytosine explain its ability to induce transitions in both directions, A:T ↔ G:C.

tween the stacked base pairs in DNA (Figure 13.19). In so doing, they increase the rigidity and alter the conformation of the double helix, causing slight bends or kinks in the molecule. When DNA molecules containing intercalated acridines replicate, additions and deletions of from one to a few base pairs occur. As we might expect, these small additions and deletions, usually of a single base pair, result in altered reading frames for the portion of the gene distal to the mutation (Figure 13.15*b*). Thus acridine-induced mutations usually result in nonfunctional gene products.

Alkylating agents are chemicals that donate alkyl groups to other molecules. They include the nitrogen and sulfur mustards, methyl and ethyl methanesulfonate (MMS and EMS), and nitrosoguanidine (NTG; Table 13.1)—chemicals that have multiple effects on DNA. One mechanism of mutagenesis by alkylating agents involves the transfer of methyl or ethyl groups to the bases resulting in altered base-pairing potentials. For example, EMS causes ethylation of the bases in DNA at the 7-*N* and the 6-*O* positions. When 7-ethylguanine is produced, it base-pairs with thymine (Figure 13.20*a*) to cause G:C → A:T transitions. Other base alkylation products activate error-prone DNA repair processes that introduce transitions, transversions, and frameshift mutations during the repair

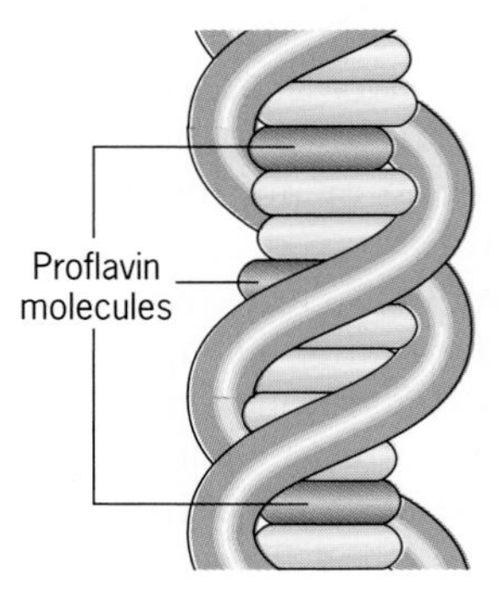

Figure 13.19 Intercalation of proflavin into the DNA double helix. X-ray diffraction studies have shown that these positively charged acridine dyes become sandwiched between the stacked base pairs.

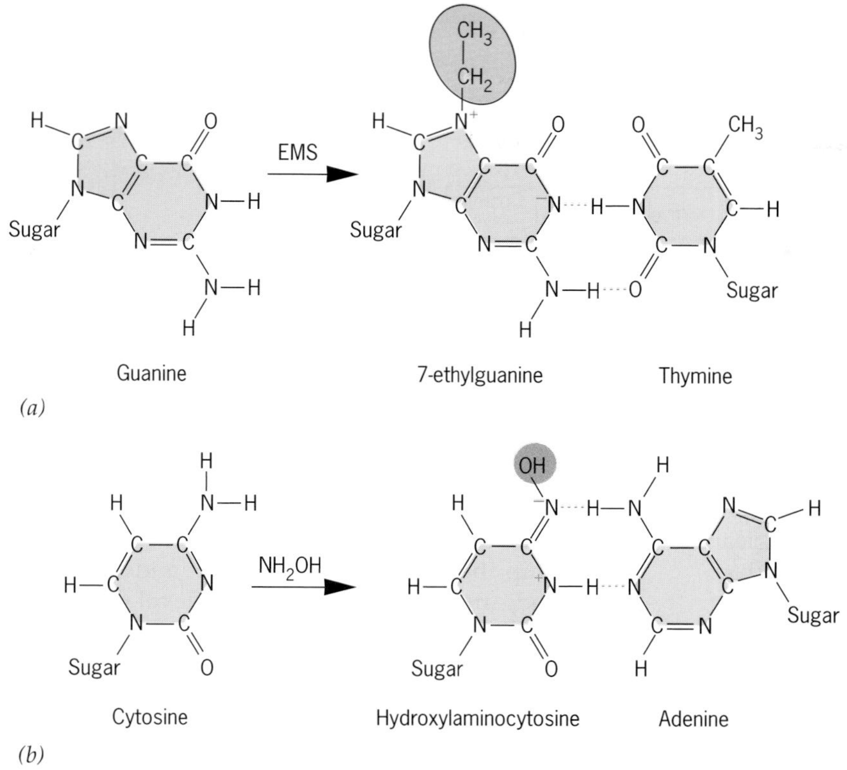

Figure 13.20 Proposed mutagenic effect of (*a*) the alkylating agent ethyl methanesulfonate (EMS) and (*b*) the hydroxylating agent hydroxylamine (NH_2OH). Alkylating agents such as EMS also induce mutations by other mechanisms.

process. Some alkylating agents, particularly difunctional alkylating agents (those with two reactive alkyl groups), cross-link DNA strands or molecules and induce chromosome breaks, which result in various kinds of chromosomal aberrations (Chapter 6). Alkylating agents as a class therefore exhibit less specific mutagenic effects than do base analogs, nitrous acid, or acridines. Alkylating agents induce all types of mutations, including transitions, transversions, frameshifts, and even chromosome aberrations, with relative frequencies that depend on the reactivity of the agent involved.

Nitrosoguanidine (NTG), one of the most potent chemical mutagens known, induces clusters of closely linked mutations in segments of chromosomes that are replicated during the mutagenic treatment. Mutants isolated after NTG treatment often carry multiple, closely linked mutations, making them less useful for genetic studies.

In contrast to most alkylating agents, the **hydroxylating agent** hydroxylamine, (NH_2OH) has a specific mutagenic effect. It induces only G:C → A:T transitions. When DNA is treated with hydroxylamine, the amino group of cytosine is hydroxylated (Figure 13.20*b*). The resulting hydroxylaminocytosine base-pairs with adenine, leading to G:C → A:T transitions. Because of its specificity, hydroxylamine has been very useful in classifying transition mutations. Mutations that are induced to revert by nitrous acid or base analogs, and therefore were caused by transitions, can be divided into two classes on the basis of their revertibility with hydroxylamine. (1) Those with an A:T base pair at the mutant site will not be induced to revert by hydroxylamine. (2) Those with a G:C base pair at the mutant site will be induced to revert by hydroxylamine. Thus hydroxylamine can be used to determine whether a particular mutation was an A:T → G:C or a G:C → A:T transition.

Mutations Induced by Radiation

The portion of the electromagnetic spectrum (Figure 13.21) with wavelengths shorter and of higher energy than visible light is subdivided into **ionizing radiation** (X rays, gamma rays, and cosmic rays) and **nonionizing radiation** (ultraviolet light). Ionizing radiations are of high energy and are useful for medical diagnosis because they penetrate living tissues for substantial distances. In the process, these high-energy rays collide with atoms and cause the release of electrons, leaving positively charged free radicals or ions. The ions, in turn, collide with other molecules and cause the release of additional electrons. The result is that a cone of ions is formed along the track of each high-energy ray as it passes through living tissues. This process of ionization is induced by machine-produced X rays, protons, and neutrons, as well as by the alpha,

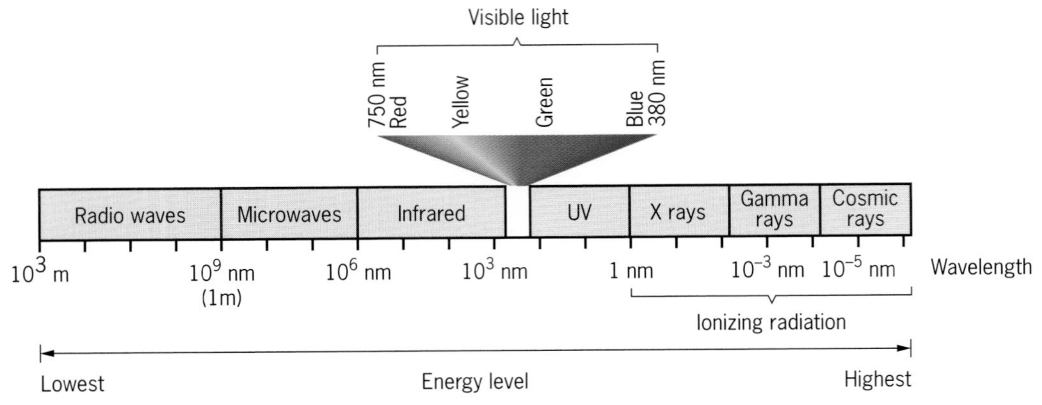

Figure 13.21 The electromagnetic spectrum.

beta, and gamma rays released by radioactive isotopes such as ^{32}P, ^{35}S, and the uranium 238 used in nuclear reactors. Ultraviolet rays, having lower energy than ionizing radiations, penetrate only the surface layer of cells in higher plants and animals and do not cause ionizations. Ultraviolet rays dissipate their energy to atoms that they encounter, raising the electrons in the outer orbitals to higher energy levels, a state referred to as **excitation**. Molecules containing atoms in either ionic forms or excited states are chemically more reactive than those containing atoms in their normal stable states. The increased reactivity of atoms present in DNA molecules is responsible for the mutagenicity of ionizing radiation and ultraviolet light.

X rays and other forms of ionizing radiation are quantitated in **roentgen** (**r**) units, which are measures of the number of ionizations per unit volume under a standard set of conditions. Specifically, one roentgen unit is a quantity of ionizing radiation that produces 2.083×10^9 ion pairs in one cubic centimeter of air at 0°C and a pressure of 760 mm of mercury. Note that the dosage of irradiation in roentgen units does not involve a time scale. The same dosage may be obtained by a low intensity of irradiation over a long period of time or a high intensity of irradiation for a short period of time. This point is important because in most studies the frequency of induced point mutations is directly proportional to the dosage of irradiation (Figure 13.22). For example, X-irradiation of *Drosophila* sperm causes an approximately 3 percent increase in mutation rate for each 1000 r increase in irradiation dosage. This linear relationship shows that the induction of mutations by X rays exhibits single-hit kinetics, which means that each mutation results from a single ionization event. That is, every ionization has a fixed probability of inducing a mutation under a standard set of conditions.

What is a safe level of irradiation? The development and use of the atomic bomb and the more recent accidents at nuclear power plants have generated concern about exposure to ionizing radiations. The linear relationship between mutation rate and radiation dosage indicates that there is no safe level of irradiation. Rather, the results indicate that the higher the dosage of irradiation, the higher the mutation rate, and the lower the dosage, the lower the mutation frequency. Even very low levels of irradiation have certain low, but real, probabilities of inducing mutations. However, the probabilities are not the same for all species.

In *Drosophila* sperm, chronic irradiation (low levels of irradiation over long periods of time) is as effective in inducing mutations as acute irradiation (the same total dosage of irradiation administered at high intensity for short periods of time). However, in mice, chronic irradiation results in fewer mutations than the same dosage of acute irradiation. Moreover, when

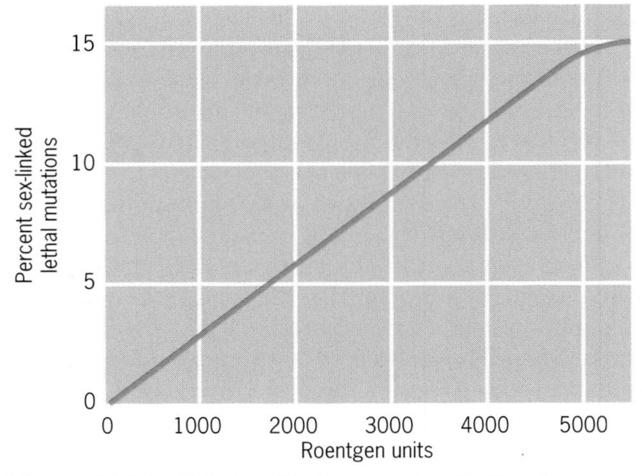

Figure 13.22 Relationship between irradiation dosage and mutation frequency in *Drosophila*.

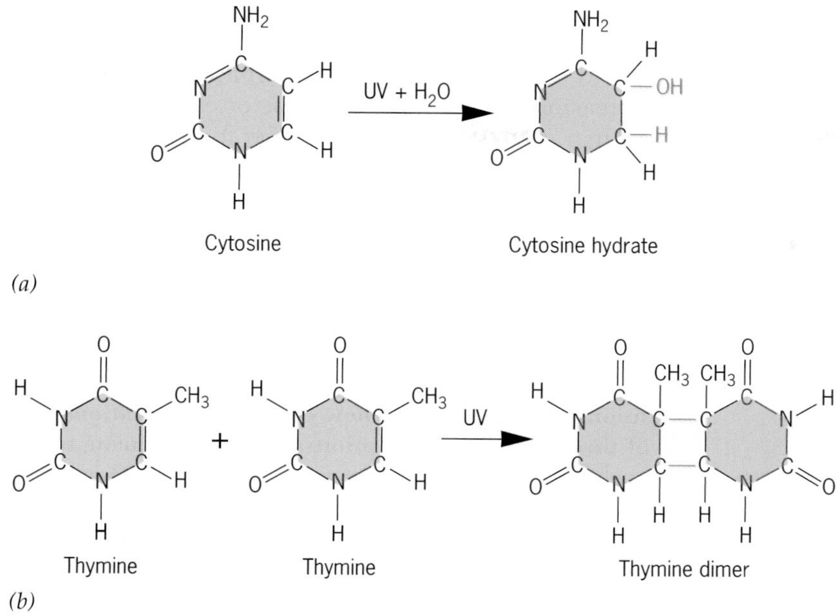

UV + H_2O

Cytosine

Cytosine hydrate

(a)

UV

Thymine

Thymine

Thymine dimer

(b)

Figure 13.23 Pyrimidine photoproducts of UV irradiation. (*a*) Hydrolysis of cytosine to a hydrate form that may cause mispairing of bases during replication. (*b*) Cross-linking of adjacent thymine molecules to form thymine dimers, which block DNA replication.

mice are treated with intermittent doses of irradiation, the mutation frequency is slightly lower than when they are treated with the same total amount of irradiation in a continuous dose. The differential response of fruit flies and mammals to chronic irradiation is thought to result from differences in the efficiency with which these species repair irradiation-induced damage in DNA. Repair mechanisms may exist in the spermatogonia and oocytes of mammals that do not function in *Drosophila* sperm. Nevertheless, we should emphasize that all of these irradiation treatments are mutagenic, albeit to different degrees, in both *Drosophila* and mammals.

Ionizing radiation also induces gross changes in chromosome structure, including deletions, duplications, inversions, and translocations (Chapter 6). These chromosome aberrations result from radiation-induced breaks in chromosomes. Because these aberrations require two chromosomal breaks, they exhibit two-hit kinetics rather than the single-hit kinetics observed for point mutations.

Ultraviolet (UV) radiation does not possess sufficient energy to induce ionizations. However, it is readily absorbed by many organic molecules such as the purines and pyrimidines in DNA, which then enter a more reactive or excited state. UV rays penetrate tissue only slightly. Thus, in multicellular organisms, only the epidermal layer of cells usually is exposed to the effects of UV. However, ultraviolet light is a potent mutagen for unicellular organisms. The maximum absorption of UV by DNA is at a wavelength of 254 nm. Maximum mutagenicity also occurs at 254 nm, suggesting that the UV-induced mutation process is mediated directly by the absorption of UV by purines and pyrimidines. *In vitro* studies show that the pyrimidines absorb strongly at 254 nm and, as a result, become very reactive. Two major products of UV absorption by pyrimidines are pyrimidine hydrates and pyrimidine dimers (Figure 13.23). Several lines of evidence indicate that thymine dimerization is probably the major mutagenic effect of UV. Thymine dimers cause mutations in two ways. (1) Dimers perturb the structure of DNA double helices and interfere with accurate DNA replication. (2) Errors occur during the cellular processes that repair defects in DNA, such as UV-induced thymine dimers (see the section, DNA Repair Mechanisms later in this chapter).

Mutations Induced by Transposable Genetic Elements

Living organisms contain remarkable DNA elements that can jump from one site in the genome to another site. These **transposons**, or transposable genetic elements, are the subject of Chapter 17. The insertion of one of these transposons into a gene will often render the gene nonfunctional (Figure 13.24). If the gene encodes an important product, a mutant phenotype is likely to result. Indeed, geneticists now know that many of the classical mutants of maize, *Drosophila*, *E. coli*, and other organisms were caused by the insertion of transposable genetic elements into important genes (see Chapter 17, especially Figure 17.2).

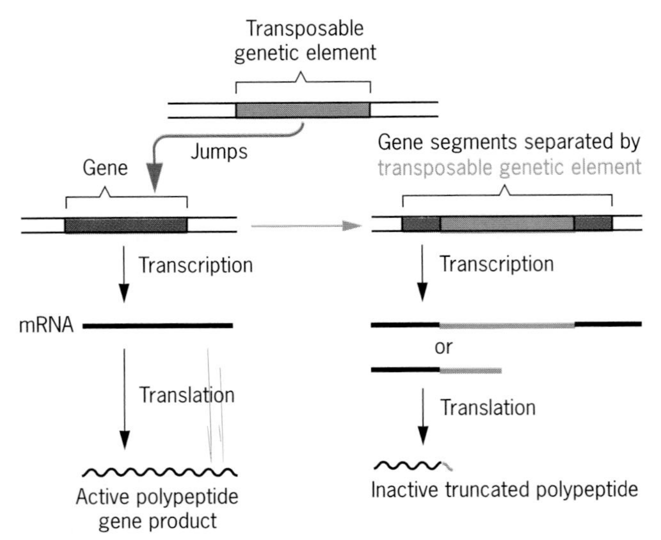

Figure 13.24 Mechanism of transposon-induced mutation. The insertion of a transposable genetic element (red) into a wild-type gene (left) will usually render the gene nonfunctional (right). A truncated gene product usually results from transcription- or translation-termination signals, or both located within the transposon.

Expanding Trinucleotide Repeats and Inherited Human Diseases

All of the types of mutations discussed in the preceding sections of this chapter occur in humans. In addition, another type of mutation occurs that, to date, is unique to humans. Tandemly repeated sequences of from one to six nucleotides pairs are known as **simple tandem repeats**. Such repeats are dispersed throughout the human genome. Repeats of three nucleotide pairs, **trinucleotide repeats**, can increase in copy number and cause inherited diseases in humans. Several trinucleotides have been shown to undergo such increases in copy number. Expanded CGG trinucleotide repeats on the X chromosome are responsible for fragile X syndrome, the most common form of inherited mental retardation in humans (Chapter 5). Normal X chromosomes contain from 6 to about 50 copies of the CGG repeat at the *FRAXA* site. Mutant X chromosomes contain up to 1000 copies of the tandem CGG repeat at this site.

CAG and CTG trinucleotide repeats are involved in several inherited neurological diseases including Huntington disease (discussed in detail in Chapter 20), myotonic dystrophy, Kennedy disease, dentatorubral pallidoluysian atrophy, Machado Joseph disease, and spinocerebellar ataxia. In all of these neurological disorders, the severity of the disease is correlated with trinucleotide copy number—the higher the copy number, the more severe the disease symptoms. In addition, the expanded trinucleotides associated with these diseases are unstable in somatic cells and between generations. This instability gives rise to the phenomenon of **anticipation**, which is the increasing severity of the disease or earlier age of onset that occurs in successive generations as the trinucleotide copy number increases. The mechanism of trinucleotide expansion is unknown, and, so far, this mutagenic process has been documented only in humans.

Key Points: **Mutations are induced by chemicals, ionizing irradiation, ultraviolet light, and endogenous transposable genetic elements that jump from one position in the genome to new sites. Point mutations are of three types: (1) transitions, purine for purine and pyrimidine for pyrimidine substitutions, (2) transversions, purine for pyrimidine and pyrimidine for purine substitutions, and (3) frameshift mutations, additions or deletions of one or two nucleotide pairs, which alter the reading frame of the gene distal to the site of the mutation. Several inherited human diseases are caused by expanded trinucleotide repeats.**

SCREENING CHEMICALS FOR MUTAGENICITY: THE AMES TEST

Mutagenic agents are also **carcinogens**; that is, they induce cancers. The one characteristic that the hundreds of types of cancer have in common is that the malignant cells continue to divide after cell division would have stopped in normal cells. Of course, cell division, like all other biological processes, is under genetic control. Specific genes encode products that regulate cell division in response to intracellular, intercellular, and environmental signals. When these genes mutate to nonfunctional states, uncontrolled cell division sometimes results. Clearly, we wish to avoid being exposed to mutagenic and carcinogenic agents. However, our technological society is often dependent on the extensive use of chemicals in both industry and agriculture. Hundreds of new chemicals are being produced each year, and the mutagenicity and carcinogenicity of these chemicals need to be evaluated before their use becomes widespread.

Traditionally, the carcinogenicity of chemicals has been tested on rodents, usually newborn mice. These studies involved feeding or injecting the substance being tested and subsequently examining the animals for tumors. Mutagenicity tests have been done in a similar fashion. However, because mutation is a low-frequency event and because maintaining large populations of mice is an expensive undertaking, the tests have been relatively insensitive; that is, low levels of mutagenicity would not be detected.

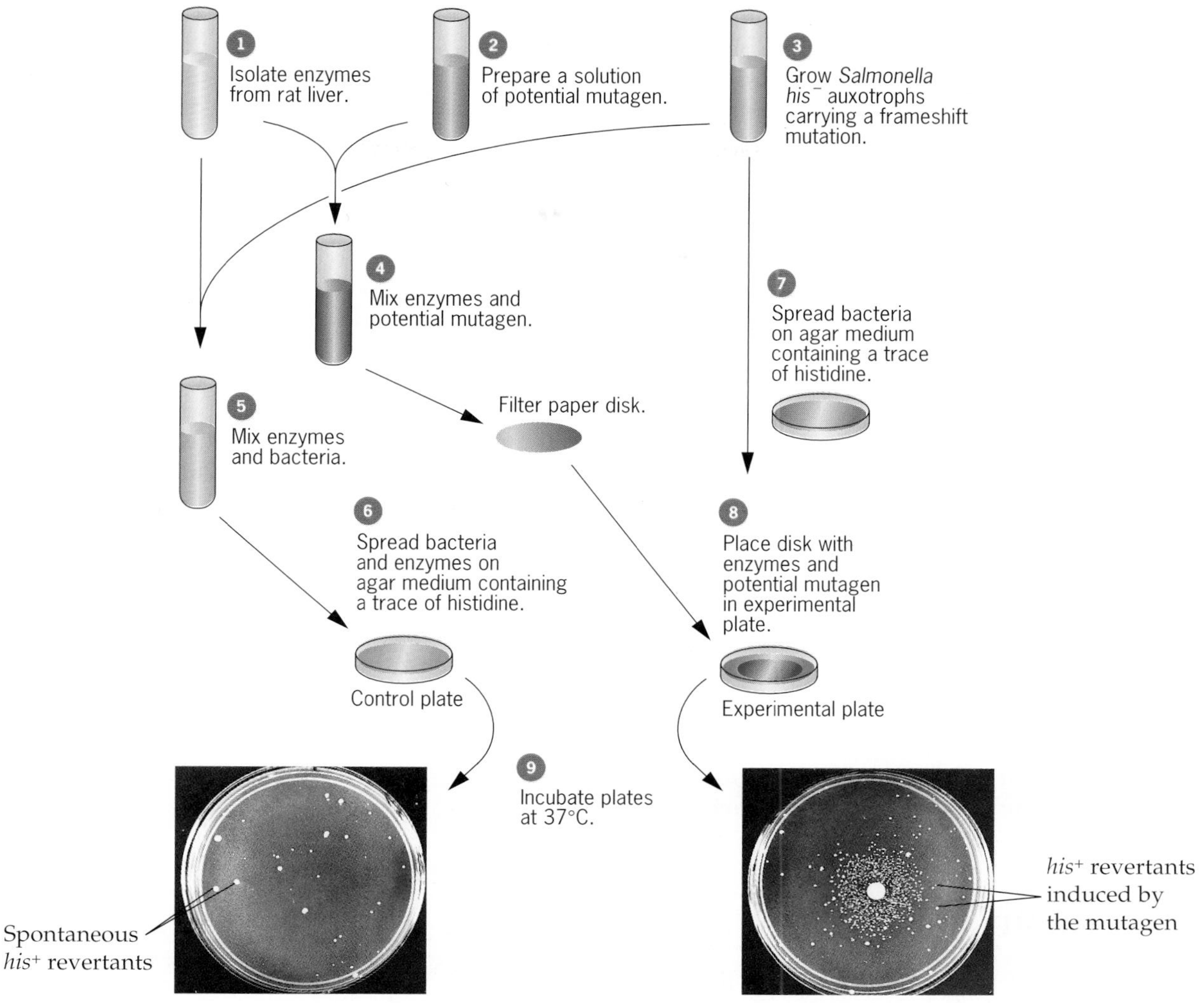

Figure 13.25 The Ames test for mutagenicity. The medium in each petri dish contains a trace of histidine and a known number of *his*⁻ cells of a specific *Salmonella typhimurium* "tester strain" harboring a frameshift mutation. The control plate shown on the left provides an estimate of the frequency of spontaneous reversion of this particular tester strain. The experimental plate on the right shows the frequency of reversion induced by the potential mutagen, in this case, the carcinogen 2-aminofluorene.

Bruce Ames and his associates developed sensitive techniques that allow the mutagenicity of large numbers of chemicals to be tested quickly at relatively low cost. Ames and coworkers constructed auxotrophic strains of the bacterium *Salmonella typhimurium* carrying various types of mutations—transitions, transversions, and frameshifts—in genes required for the biosynthesis of the amino acid histidine. They monitored the reversion of these auxotrophic mutants to prototrophy by placing a known number of mutant bacteria on medium lacking histidine and scoring the number of colonies produced by prototrophic revertants. Because some chemicals are mutagenic only to replicating DNA, they added a small amount of histidine—enough to allow a few cell divisions, but not the formation of visible colonies—to the medium. They measured the mutagenicity of a chemical by comparing the frequency of reversion in its presence with the spontaneous reversion frequency (Figure 13.25). They assessed its ability to induce different types of mutations by using a set of tester strains that carry different types of mutations—one strain with a transition, one with a frameshift mutation, and so forth.

Over a period of several years during which they tested hundreds of different chemicals, Ames and his colleagues observed a greater than 90 percent correlation between the mutagenicity and the carcinogenicity of the substances tested. Initially, they found several potent carcinogens to be nonmutagenic to the tester strains. Subsequently, they discovered that many of these carcinogens are metabolized to strongly mutagenic derivatives in eukaryotic cells. Thus Ames and his associates added a rat liver microsomal fraction to their assay systems in an attempt to detect the mutagenicity of metabolic derivatives of the substances being tested. Coupling of the rat liver microsomal activation system to the microbial mutagenicity tests expanded the utility of the system considerably. For example nitrates are not themselves mutagenic or carcinogenic. However, in eukaryotic cells, nitrates are converted to nitrosamines, which are highly mutagenic and carcinogenic. Ames's mutagenicity tests demonstrated the presence of frameshift mutagens in several components of chemically fractionated cigarette smoke condensates. In some cases, activation by the liver microsomal preparation was required for mutagenicity; in other cases, activation was not required. The Ames test provides a rapid, inexpensive, and sensitive procedure for testing the mutagenicity of chemicals. Since mutagenic chemicals are also carcinogens, the Ames test can be used to prevent the widespread use of mutagenic and carcinogenic chemicals.

Key Point: **By using *Salmonella* tester strains that carry various types of mutations in genes encoding histidine biosynthetic enzymes, Bruce Ames and coworkers developed an inexpensive and sensitive method for detecting the mutagenicity of chemicals.**

DNA REPAIR MECHANISMS

The multiplicity of repair mechanisms that have evolved in organisms ranging from bacteria to humans emphatically documents the importance of keeping mutation, both somatic and germ-line, at a tolerable level. For example, *E. coli* cells possess at least five distinct mechanisms for the repair of defects in DNA: (1) light-dependent repair or photoreactivation, (2) excision repair, (3) mismatch repair, (4) postreplication repair, and (5) the error-prone repair system. Moreover, there are at least two different types of excision repair, and the excision repair pathways can be initiated by several different enzymes, each acting on a specific kind of damage in DNA. Mammals seem to possess all of the repair mechanisms found in *E. coli* except photoreactivation. Since most mammalian cells do not have access to light, photoreactivation would be of relatively little value to them. Humans and other mammals undoubtedly possess some DNA repair mechanisms that are not present in bacteria, but many details of these repair processes require further documentation.

The importance of DNA repair pathways to human health is clear. Inherited disorders such as xeroderma pigmentosum, which was discussed at the beginning of this chapter, vividly document the serious consequences of defects in DNA repair. We discuss these inherited human diseases with defects in DNA repair in a subsequent section of this chapter.

Light-Dependent Repair

Light-dependent repair or **photoreactivation** of DNA in bacteria is carried out by a light-activated enzyme called **DNA photolyase**. When DNA is exposed to ultraviolet light, thymine dimers are produced by covalent cross-linkages between adjacent thymine residues (Figure 13.23). DNA photolyase binds to thymine dimers in DNA and uses light energy to cleave the covalent cross-links (Figure 13.26). Photolyase will bind

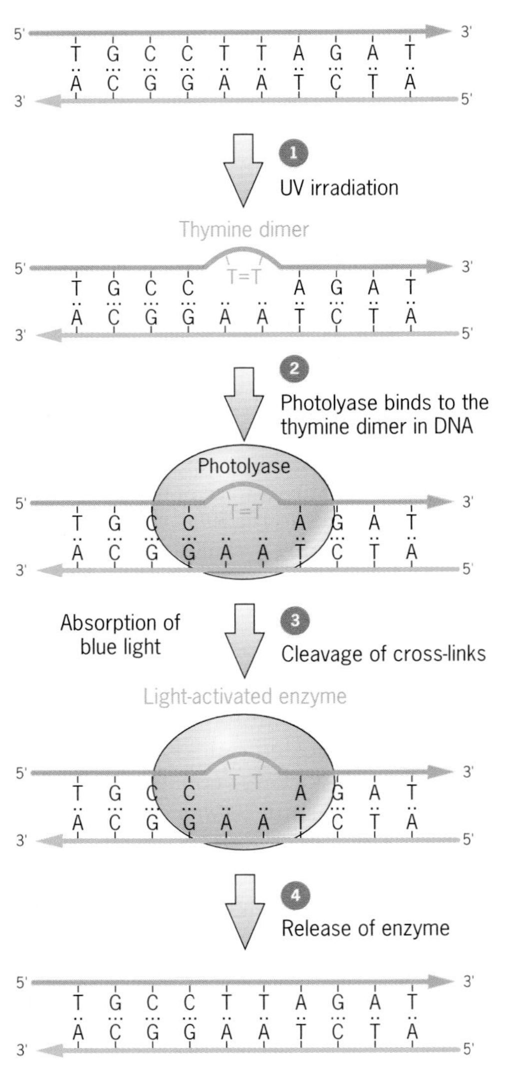

Figure 13.26 Cleavage of thymine dimer cross-links by light-activated photolyase. The arrows indicate the opposite polarity of the complementary strands of DNA.

to thymine dimers in DNA in the dark, but it cannot catalyze cleavage of the bonds joining the thymine moieties without energy derived from visible light, specifically light within the blue region of the spectrum. Photolyase also splits cytosine dimers and cytosine-thymine dimers. Thus, when ultraviolet light is used to induce mutations in bacteria, the irradiated cells are grown in the dark for a few generations to maximize the mutation frequency.

Excision Repair

Excision repair of damaged DNA involves at least three steps. In step 1, a DNA repair endonuclease or endonuclease-containing enzyme complex recognizes, binds to, and excises the damaged base or bases in DNA. In step 2, a DNA polymerase fills in the gap by using the undamaged complementary strand of DNA as template. In step 3, the enzyme DNA ligase seals the break left by DNA polymerase to complete the repair process. There are two major types of excision repair: **base excision repair** systems remove abnormal or chemically modified bases from DNA, whereas **nucleotide excision repair** pathways remove larger defects like thymine dimers. Both excision pathways are operative in the dark, and both occur by very similar mechanisms in *E. coli* and humans.

Base excision repair (Figure 13.27) can be initiated by any of a group of enzymes called DNA glycosylases that recognize abnormal bases in DNA. Each glycosylase recognizes a specific type of altered base, such as deaminated bases, oxidized bases, and so on. The glycosylases cleave the glycosidic bond between the abnormal base and 2-deoxyribose, creating apurinic or apyrimidinic sites (AP sites) with missing bases. These AP sites are recognized by AP endonucleases, which act together with phosphodiesterases to excise the sugar-phosphate groups at sites where no base is present. DNA polymerase then replaces the missing nucleotide according to the specifications of the complementary strand, and DNA ligase seals the nick.

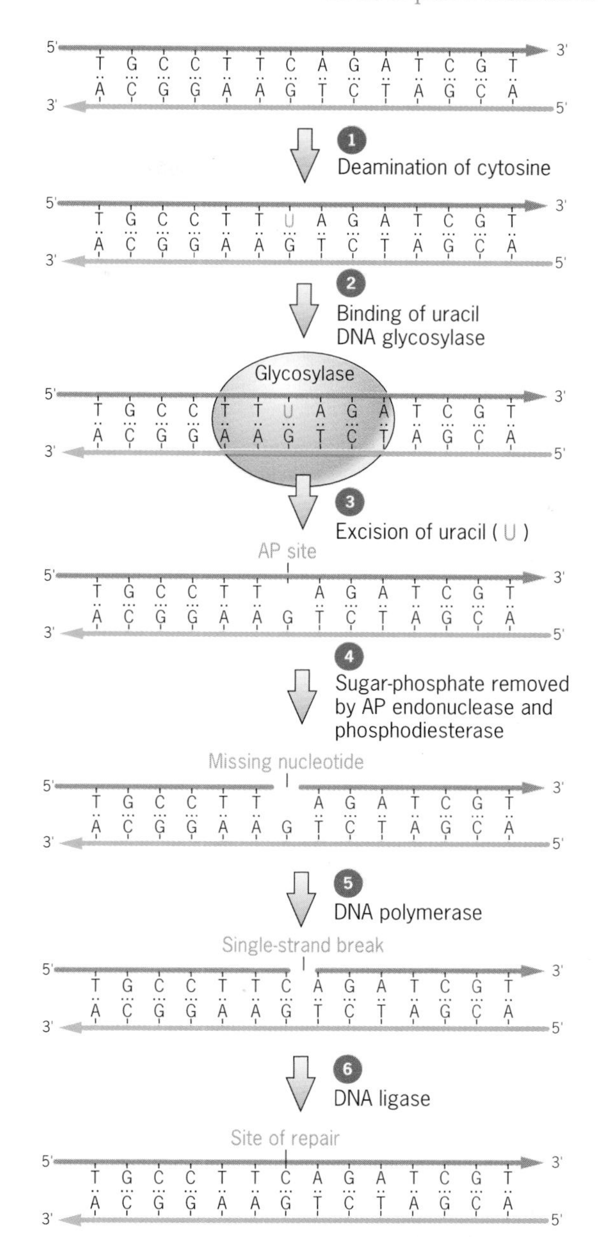

Figure 13.27 Repair of DNA by the base excision pathway. Base excision repair may be initiated by any one of several different DNA glycosylases. In the example shown, uracil DNA glycosylase starts the repair process.

Nucleotide excision repair removes larger lesions like thymine dimers and bases with bulky side-groups from DNA. In nucleotide excision repair, a unique excision nuclease activity produces cuts on either side of the damaged nucleotide(s) and excises an oligonucleotide containing the damaged base(s). This nuclease is called an **excinuclease** to distinguish it from the endonucleases and exonucleases that play other roles in DNA metabolism.

The *E. coli* nucleotide excision repair pathway is shown in Figure 13.28. In *E. coli*, excinuclease activity requires the products of three genes, *uvrA*, *uvrB*, and *uvrC* (designated *uvr* for **UV** repair). A trimeric protein containing two UvrA polypeptides and one UvrB polypeptide recognizes the defect in DNA, binds to it, and uses energy from ATP to bend the DNA at the damaged site. The UvrA dimer is then released, and the UvrC protein binds to the UvrB-DNA complex. The UvrB protein cleaves the fifth phosphodiester bond from the damaged nucleotide(s) on the 3' side, and the UvrC protein hydrolyzes the eighth phosphodiester linkage from the damage on the 5' side. The *uvrD* gene product, DNA helicase II, releases the ex-

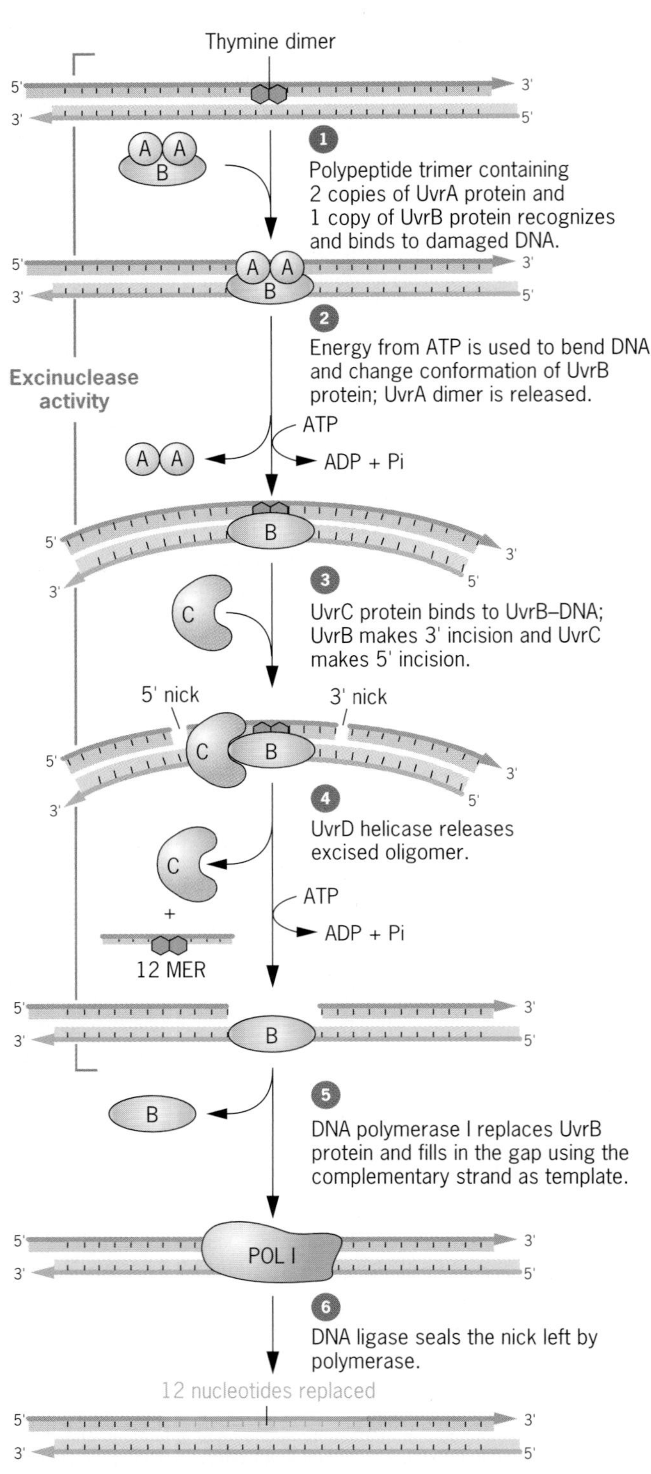

Figure 13.28 Repair of DNA by the nucleotide excision pathway in *E. coli.* The excinuclease (excision nuclease) activity requires the products of three genes—*UvrA, UvrB,* and *UvrC.* Nucleotide excision occurs by a similar pathway in humans, except that many more proteins are involved and a 29-nucleotide-long oligomer is excised.

cised dodecamer. In the last two steps of the pathway, DNA polymerase I fills in the gap, and DNA ligase seals the remaining nick in the DNA molecule.

Nucleotide excision repair in humans occurs through a pathway similar to the one in *E. coli,* but it involves about four times as many proteins. In humans, the excinuclease activity requires at least 17 polypeptides. Protein XPA (for **x**eroderma **p**igmentosum protein **A**) recognizes and binds to the damaged nucleotide(s) in DNA. It then recruits the other proteins required for excinuclease activity. In humans, the excised oligomer is 29 nucleotides long rather than the 12-mer removed in *E. coli.* The gap is filled in by either DNA polymerase δ or ϵ in humans, and DNA ligase completes the job.

Mismatch Repair

In Chapter 10, we examined the mechanism by which the 3' → 5' exonuclease activity built into DNA polymerases proofreads DNA strands during their synthesis, removing any mismatched nucleotides at the 3' termini of growing strands. The **mismatch repair** pathway provides a backup to this replicative proofreading by correcting mismatched nucleotides remaining in DNA after replication. Mismatches often involve the normal four bases in DNA. For example, a T may be mispaired with a G. Because both T and G are normal components of DNA, mismatch repair systems need some way to determine whether the T or the G is the correct base at a given site. The repair system makes this distinction by identifying the template strand, which contains the original nucleotide sequence, and the newly synthesized strand, which contains the misincorporated base (the error). This distinction can be made based on the pattern of methylation in newly replicated DNA. In *E. coli,* the A in GATC sequences is methylated subsequent to its synthesis. Thus an interval occurs during which the template strand is methylated and the newly synthesized strand is unmethylated. The mismatch repair system uses this difference in methylation state to excise the mismatched nucleotide in the nascent strand and replace it with the correct nucleotide by using the methylated parental strand of DNA as template (Figure 13.29).

Mismatch repair of DNA in *E. coli* requires the products of four genes, *mutH, mutL, mutS,* and *mutU* (= *uvrD*). The MutS protein recognizes mismatches and binds to them to initiate the repair process. MutH and MutL proteins then join the complex. MutH contains a **GATC-specific endonuclease activity** that cleaves the unmethylated strand at hemimethylated (that is, half methylated) GATC sites either 5' or 3' to

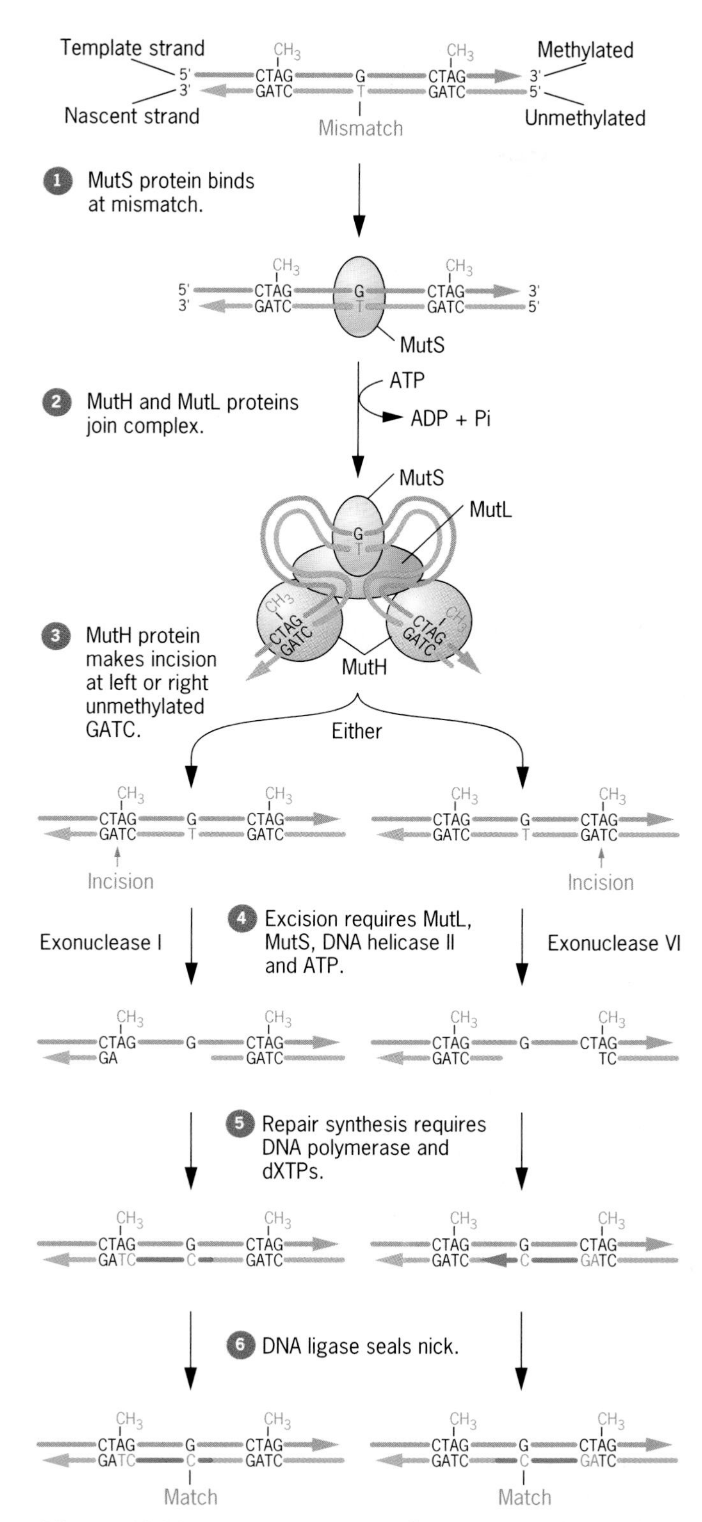

Figure 13.29 Mismatch repair of DNA in *E. coli*. The double helix has just replicated; the template strand is methylated, but the newly synthesized strand is still unmethylated. The MutH protein will cleave only umethylated GATC sites, assuring that the original (template strand) sequence is restored.

the mismatch. The incision sites may be 1000 nucleotide pairs or more from the mismatch. The subsequent excision process requires MutS, MutL, DNA helicase II (MutU), and an appropriate exonuclease. If the incision occurs at a GATC sequence 5' to the mismatch, a 5' → 3' exonuclease like *E. coli* exonuclease VII is required. If the incision occurs 3' to the mismatch, a 3' → 5' nuclease activity like that of *E. coli* exonuclease I is needed. After the excision process has removed the mismatched nucleotide from the unmethylated strand, DNA polymerase fills in the gap, and DNA ligase seals the nick.

Homologs of the *E. coli* MutS and MutL proteins have been identified in both *Saccharomyces cerevisiae* and humans—an indication that similar mismatch repair pathways occur in eukaryotes. In fact, mismatch excision has been demonstrated *in vitro* with nuclear extracts prepared from human cells. Thus mismatch repair is probably a universal or nearly universal mechanism for safeguarding the integrity of genetic information stored in double-stranded DNA.

Postreplication Repair

In *E. coli*, light-dependent repair, excision repair, and mismatch repair can be eliminated by mutations in the *phr* (**pho**to**r**eactivation), *uvr*, and *mut* genes, respectively. In multiple mutants deficient in these repair mechanisms, still another DNA repair system, called **postreplication repair**, is operative. When DNA polymerase III encounters a thymine dimer in a template strand, its progress is blocked. DNA polymerase restarts DNA synthesis at some position past the dimer, leaving a gap in the nascent strand opposite the dimer in the template strand. At this point, the original nucleotide sequence has been lost from both strands of this progeny double helix. The damaged DNA molecule is repaired by a recombination-dependent repair process (Figure 13.30) mediated by the *E. coli recA* gene product. The RecA protein, which is required for homologous recombination, stimulates the exchange of single strands between homologous double helices. During postreplication repair, the RecA protein binds to the single strand of DNA at the gap and mediates pairing with the homologous segment of the sister double helix. The gap opposite the dimer is filled with the homologous DNA strand from the sister DNA molecule. The resulting gap in the sister double helix is filled in by DNA polymerase, and the nick is sealed by DNA ligase. The thymine dimer remains in the template strand of the original progeny DNA molecule, but the complementary strand is now intact. If the thymine dimer is not removed by the nucleotide excision repair system, this postreplication re-

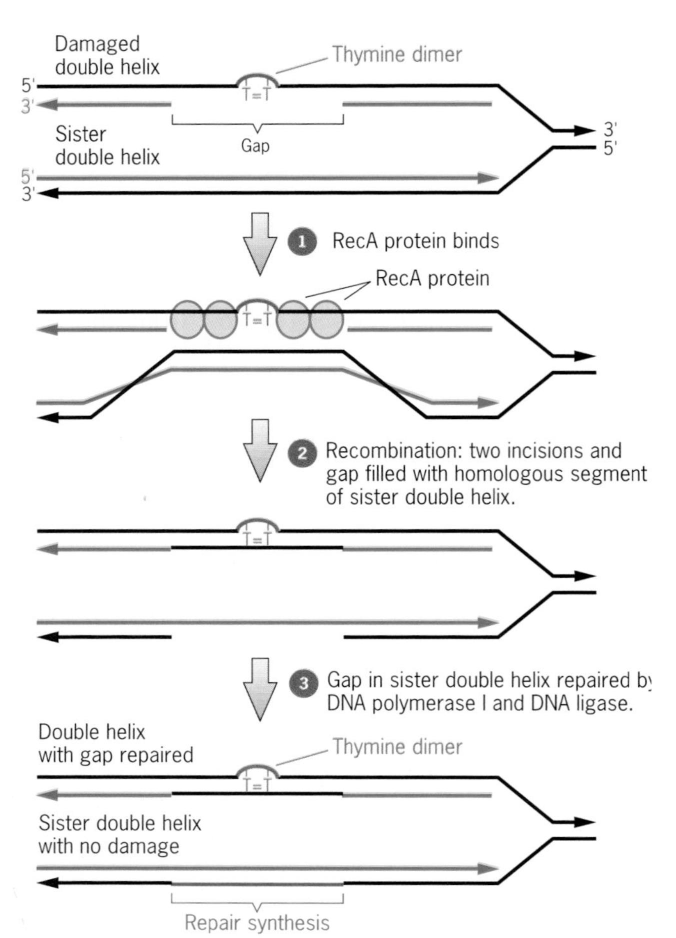

Figure 13.30 Postreplication repair of DNA in *E. coli*.

pair must be repeated after each round of DNA replication.

The Error-Prone Repair System

The DNA repair systems described so far are quite accurate. However, when the DNA of *E. coli* cells is heavily damaged by mutagenic agents such as UV light, the cells take some drastic steps in their attempt to survive. They go through a so-called **SOS response**, during which a whole battery of DNA repair, recombination, and replication proteins are synthesized. Two of these proteins, encoded by the *umuC* and *umuD* (**UV mu**table) genes, encode proteins that allow DNA replication to proceed across damaged segments of template strands, even though the nucleotide sequences in the damaged region can't be accurately replicated. This **error-prone repair** system eliminates gaps in the newly synthesized strands opposite damaged nucleotides in the template strands but, in so doing, sharply increases the frequency of replication errors. The SOS response appears to be a somewhat desperate and risky attempt to escape the lethal effects of heavily damaged DNA. When the error-prone repair system is operative, mutation rates increase sharply.

Key Points: **Multiple DNA repair systems have evolved to safeguard the integrity of genetic information in living organisms. Each repair pathway is designed to correct a certain type of damage in DNA.**

INHERITED HUMAN DISEASES WITH DEFECTS IN DNA REPAIR

As we discussed at the beginning of this chapter, individuals with xeroderma pigmentosum (XP) are extremely sensitive to sunlight. Exposure to sunlight results in a high frequency of skin cancer in XP patients. The cells of individuals with XP are deficient in the repair of UV-induced damage to DNA, such as thymine dimers. The XP syndrome can result from defects in any of at least nine different genes. The products of six of these genes, *XPA*, *XPB*, *XPC*, *XPD*, *XPF*, and *XPG*, are required for nucleotide excison repair. They have been purified and shown to be essential for excinuclease activity. Since excinuclease activity in humans requires at least 17 polypeptides, the list of *XP* genes will probably expand in the future. Two other human disorders, Cockayne syndrome and trichothiodystrophy, also result from defects in nucleotide excision repair. Individuals with Cockayne syndrome exhibit retarded growth and mental skills, but not increased rates of skin cancer. Patients with trichothiodystrophy are of short stature with brittle hair and scaly skin; they also have underdeveloped mental abilities. Individuals with either Cockayne syndrome or trichothiodystrophy are defective in a type of excision repair that is coupled to transcription. However, details of this transcription-coupled repair process are just starting to be worked out.

In addition to the damage to skin cells, some individuals with XP develop neurological abnormalities, which appear to result from premature death of nerve cells. This effect of defects in DNA repair on the very long-lived nerve cells has potentially interesting implications with respect to the causes of aging. One theory is that aging results from the accumulation of somatic mutations. If so, a defective repair system would be expected to speed up the aging process, and this appears to be the case with the nerve cells of XP patients. F. Macfarlane Burnet has emphasized that a low but significant level of spontaneous mutation must be maintained in living cells to provide sufficient flexibility for evolution to occur, and that aging may be an unavoidable consequence of this low level of mutation. Certainly, somatic mutations would be expected to contribute to the aging process. However, at present, we have little hard data that link somatic mutation to senescence.

Ataxia-telangiectasia, Fanconi anemia, and Bloom syndrome are three other inherited diseases in humans associated with known defects in DNA metabolism. All three disorders exhibit autosomal recessive patterns of inheritance, and all result in a high risk of malignancy, especially leukemia in the case of ataxia-telangiectasia and Fanconi anemia. Cells of patients with ataxia-telangiectasia exhibit an abnormal sensitivity to ionizing radiation, suggesting a defect in the repair of radiation-induced DNA damage. Cells of individuals with Fanconi anemia are impaired in the removal of DNA interstrand cross-links, such as those formed by the antibiotic mitomycin C. Individuals with Bloom syndrome exhibit a high frequency of chromosome breaks that result in chromosome aberrations (Chapter 6) and sister chromatid exchanges. While all three of these malignant-prone, inherited diseases are caused by defects in DNA metabolism, probably repair processes, the primary lesions are still unknown. The recent demonstration that the onset of hereditary nonpolyposis colorectal cancer is associated with a defect in the repair of mismatched base pairs in DNA hints that similar defects in other DNA repair pathways may be involved in the development of specific types of human cancer.

Key Points: **The importance of DNA repair pathways is documented convincingly by inherited human disorders that result from defects in DNA repair. Recent evidence indicates that the onset of certain types of cancer may be associated with defects in specific DNA repair pathways.**

DNA RECOMBINATION MECHANISMS

We discussed the main features of recombination between homologous chromosomes in Chapter 7, but we did not consider the molecular details of the process. Because many of the gene products that are involved in the repair of damaged DNA also are required for recombination between homologous chromosomes or crossing over, we will now examine some of the molecular aspects of this important process. Moreover, recombination usually, perhaps always, involves some DNA repair synthesis. Thus much of the information discussed in the preceding sections is relevant to the process of recombination.

In eukaryotes, crossing over is associated with the formation of the synaptinemal complex, which forms during prophase of the first meiotic division. This structure is composed primarily of proteins and RNA. For unknown reasons, crossing over occurs only rarely in male *Drosophila*. (Crossing over does occur in both sexes of most species; the near absence of crossing over in the heterogametic sex is unique to *Drosophila* and a few other species.) Of interest here is the fact that no synaptinemal complex is present during the first meiotic division of spermatogenesis in *Drosophila*. In addition, mutations in the *c3G* gene of *Drosophila* eliminate both crossing over and the formation of the synaptinemal complex in females. Thus, crossing over and the formation of the synaptinemal complex appear to be linked. A small amount of DNA synthesis occurs during the formation of the synaptinemal complex, and this DNA synthesis may be involved in synapsis or crossing over. Unfortunately, geneticists still know little about the functions of the components of the synaptinemal complex. Presumably, it plays a role in synapsis or crossing over, or both.

Recombination: Cleavage and Rejoining of DNA Molecules

In Chapter 7, we discussed the experiment of Creighton and McClintock, which indicated that crossing over occurs by breakage of parental chromosomes and rejoining of the parts in new combinations. Evidence demonstrating that recombination occurs by breakage and rejoining has also been obtained by autoradiography and other techniques. Indeed, the main features of the process of recombination are now well established, even though specific details remain to be elucidated.

Much of what we know about the molecular details of crossing over is based on the study of **recombination-deficient mutants** of *E. coli* and *S. cerevisiae*. Biochemical studies of these mutants have shown that they are deficient in various enzymes and other proteins required for recombination. Together, the results of genetic and biochemical studies have provided a fairly complete picture of recombination at the molecular level. Nevertheless, geneticists still do not understand all the details of crossing over in any species.

Many of the currently popular models of crossing over were derived from a model proposed by Robin Holliday in 1964. Holliday's model was one of the first that explained most of the genetic data available at the time by a breakage and reunion mechanism with associated repair synthesis. An updated version of the Holliday model is shown in Figure 13.31. This mechanism, like many others that have been invoked, begins when an endonuclease cleaves single strands of each of the two parental DNA molecules (breakage). Segments of the single strands on one side of each cut are then displaced from their complementary strands

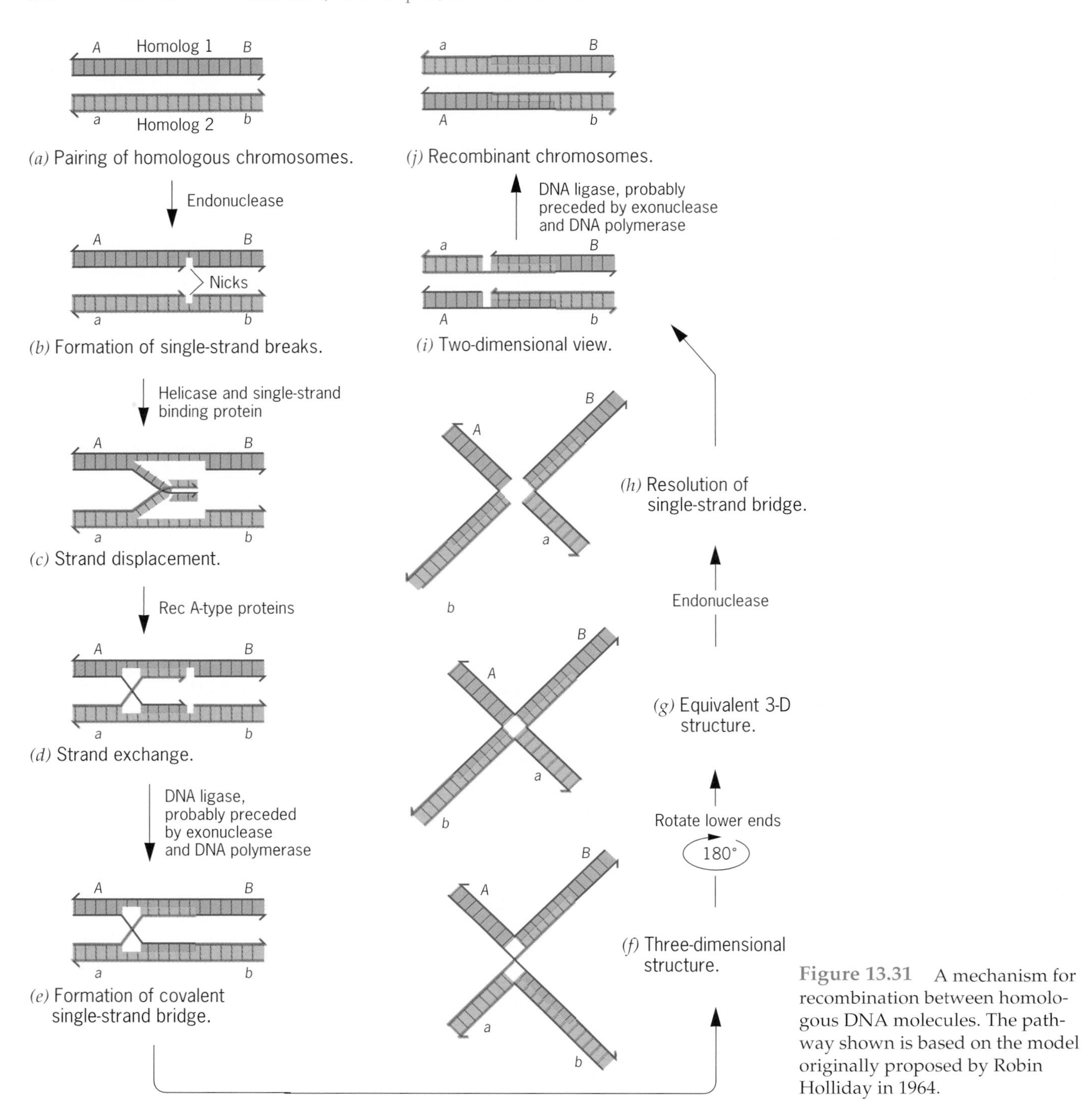

Figure 13.31 A mechanism for recombination between homologous DNA molecules. The pathway shown is based on the model originally proposed by Robin Holliday in 1964.

with the aid of DNA helicases and single-strand binding proteins. The helicases unwind the two strands of DNA in the region adjacent to single-strand incisions. In *E. coli*, the **RecBCD complex** contains both an endonuclease activity that makes single-strand breaks in DNA and a DNA helicase activity that unwinds the complementary strands of DNA in the region adjacent to each nick.

The displaced single strands then exchange pairing partners, base-pairing with the intact complementary strands of the homologous chromosomes. This process is stimulated by proteins like the *E. coli* **RecA protein**. RecA-type proteins have been characterized in many species, both prokaryotic and eukaryotic. RecA protein and its homologs stimulate **single-strand assimilation**, a process by which a single

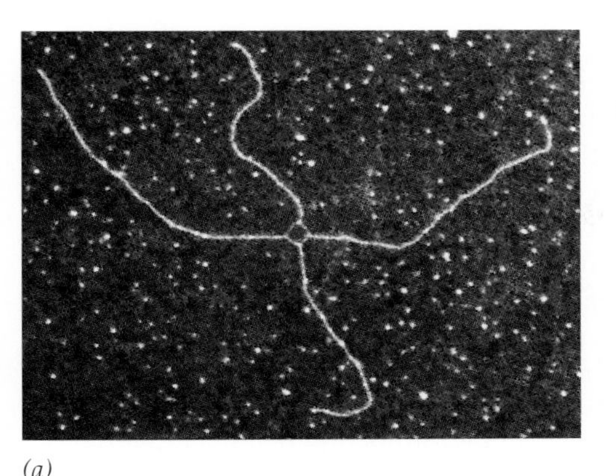
(a)

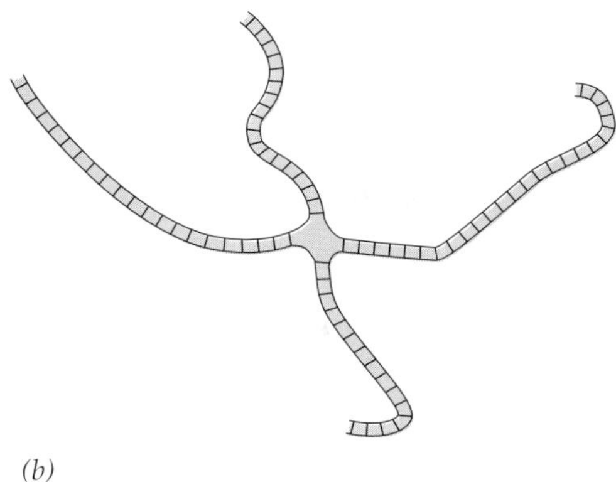
(b)

Figure 13.32 Electron micrograph (*a*) and diagram (*b*) of an X-shaped recombination intermediate or chi structure.

strand of DNA displaces its homolog in a DNA double helix. RecA-type proteins promote reciprocal exchanges of DNA single strands between two DNA double helices in two steps. In the first step, a single strand of one double helix is assimilated by a second, homologous double helix, displacing the identical or homologous strand and base-pairing with the complementary strand. In the second step, the displaced single strand is similarly assimilated by the first double helix. The RecA protein mediates these exchanges by binding to the unpaired strand of DNA, aiding in the search for a homologous DNA sequence, and, once a homologous double helix is found, promoting the replacement of one strand with the unpaired strand. If complementary sequences already exist as single strands, the presence of RecA protein increases the rate of renaturation by over 50-fold.

The cleaved strands are then covalently joined in new combinations (rejoining) by DNA ligase. If the original breaks in the two strands do not occur at exactly the same site in the two homologs, some tailoring will be required before DNA ligase can catalyze the reunion step. This tailoring usually involves the excision of nucleotides by an exonuclease and repair synthesis by a DNA polymerase. The sequence of events described so far will produce X-shaped recombination intermediates called **chi forms**, which have been observed by electron microscopy in several species (Figure 13.32). The chi forms are resolved by enzyme-catalyzed breakage and rejoining of the complementary DNA strands to produce two recombinant DNA molecules.

A substantial body of evidence indicates that homologous recombination occurs by more than one mechanism—probably by several different mechanisms. In *S. cerevisiae*, the ends of DNA molecules produced by double-strand breaks are highly recombinogenic. This fact and other evidence suggest that recombination in yeast often involves a double-strand break in one of the parental double helices. Thus, in 1983, Jack Szostak, Franklin Stahl, and colleagues proposed a **double-strand break model** of crossing over. According to their model, recombination involves a double-strand break in one of parental double helices, not just single-strand breaks as in the Holliday model. The initial breaks are then enlarged to gaps in both strands. The two single-stranded termini produced at the double-stranded gap of the broken double helix invade the intact double helix and displace segments of the homologous strand in this region. The gaps are then filled in by repair synthesis. This process yields two homologous chromosomes joined by two single-strand bridges. The bridges are resolved by endonucleolytic cleavage, just as in the Holliday model. Both the double-strand-break model and the Holliday model nicely explain the production of chromosomes that are recombinant for genetic markers flanking the region in which the crossover occurs.

Gene Conversion: DNA Repair Synthesis Associated with Recombination

Up to this point, we have discussed only recombination events that can be explained by breakage of homologous chromatids and the reciprocal exchange of parts. When crosses are done in Ascomycetes with genetic markers that are closely linked and all four products of meiosis are analyzed to obtain tetrad data, recombination is sometimes nonreciprocal. For example, if crosses are done between two closely linked mutations in *Neurospora*, and asci containing wild-type recombinants are analyzed, these asci frequently do not contain the reciprocal, double-mutant recombinant.

Consider a cross involving two closely linked mutations m_1 and m_2. In a cross of $m_1\ m_2^+$ with $m_1^+\ m_2$, asci of the following type are frequently observed:

Spore pair 1: $m_1^+\ m_2$
Spore pair 2: $m_1^+\ m_2^+$
Spore pair 3: $m_1\ m_2^+$
Spore pair 4: $m_1\ m_2^+$.

Wild-type m_1^+ m_2^+ spores are present, but the m_1 m_2 double-mutant spores are not present in the ascus. Reciprocal recombination would produce an m_1 m_2 chromosome whenever an m_1^+ m_2^+ chromosome was produced. In this ascus, the m_2^+: m_2 ratio is 3 : 1 rather than 2 : 2 as expected. One of the m_2 alleles appears to have been "converted" to the m_2^+ allelic form. Thus this type of nonreciprocal recombination was called **gene conversion**, and, despite its somewhat misleading connotation, the term has been used extensively for over three decades. We might assume that gene conversion results from mutation, except that it occurs at a higher frequency than the corresponding mutation events, always produces the allele present on the homologous chromosome, and is correlated about 50 percent of the time with reciprocal recombination of outside markers. The last observation strongly suggests that gene conversion results from events that occur during crossing over. Indeed, gene conversion is now believed to result from DNA repair synthesis associated with the breakage, excision, and rejoining events of crossing over.

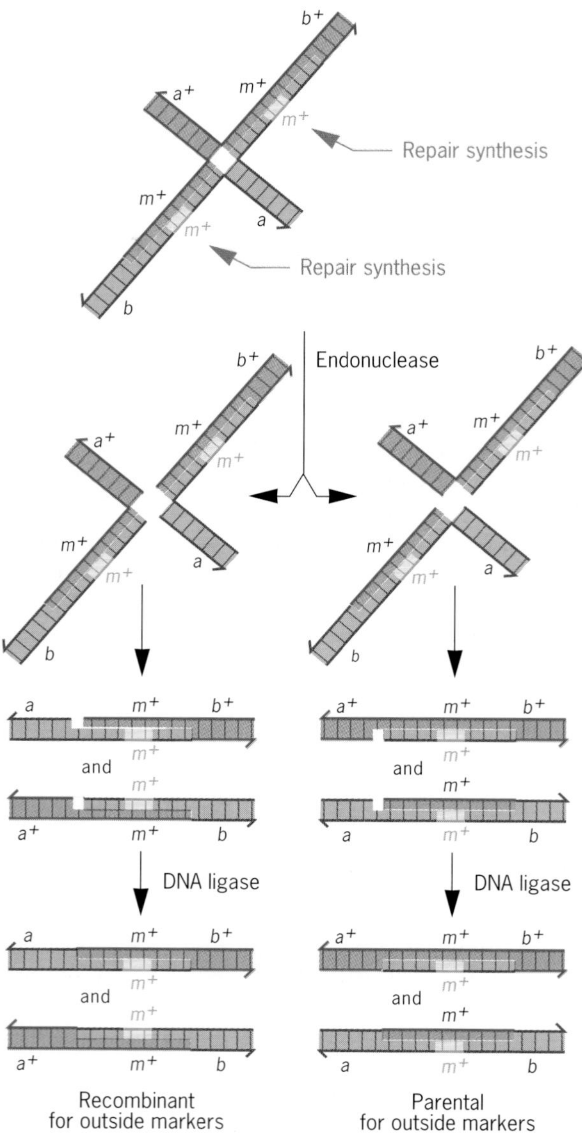

Figure 13.33 Formation of either the recombinant (bottom left) or parental combinations (bottom right) of outside markers in association with gene conversion. The recombination intermediate at the top is the equivalent to that illustrated in Figure 13.31*g*, but shows the mismatch-repaired chromatids of the tetrad diagrammed in the text. This tetrad produces a 3 m^+ to 1 m gene conversion ascus pattern. Cleavage of the single-strand bridge in the vertical plane (left) produces the parental (a^+ b^+ and a b) arrangement of outside markers, whereas cleavage in the horizontal plane yields the recombinant (a^+ b and a b^+) arrangement of the flanking markers.

With closely linked markers, gene conversion occurs more frequently than reciprocal recombination. In one study of the *his1* gene of yeast, 980 of 1081 asci containing his^+ recombinants exhibited gene conversion, whereas only 101 showed classical reciprocal recombination.

The most striking feature of gene conversion is that the input 1:1 allele ratio is not maintained. This can be explained easily if short segments of parental DNA are degraded and then resynthesized with template strands provided by DNA carrying the other allele. Given the mechanisms of excision repair discussed earlier in this chapter, the Holliday model of crossing over explains gene conversion for genetic markers located in the immediate vicinity of the crossover. In Figure 13.31 *d-i*, there is a segment of DNA between the *A* and *B* loci where complementary strands of DNA from the two homologous chromosomes are base-paired. If a third pair of alleles located within this segment were segregating in the cross, mismatches in the two double helices would be present. DNA molecules containing such mismatches, or different alleles in the two complementary strands of a double helix, are called **heteroduplexes.** Such heteroduplex molecules occur as intermediates in the process of recombination.

If Figure 13.31*e* were modified to include a third pair of alleles, and the other two chromatids were added, the tetrad would have the following composition:

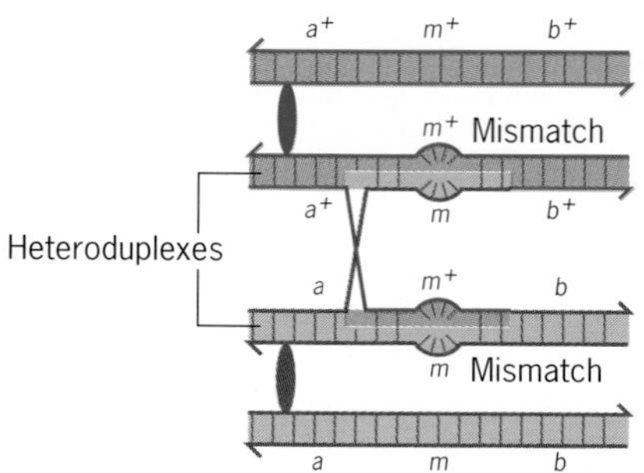

If the mismatches are resolved by excision repair (Figure 13.28), in which the *m* strands are excised and resynthesized with the complementary m^+ strands as templates, the following tetrad will result:

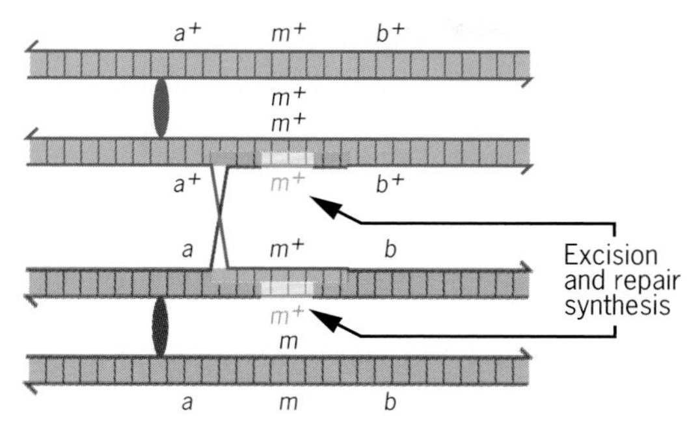

As a result of semiconservative DNA replication during the subsequent mitotic division, this tetrad will yield an ascus containing six m^+ ascospores and two *m* ascospores, the 3:1 gene conversion ratio.

Suppose that only one of the two mismatches in the tetrad just described is repaired prior to the mitotic division. In that case, the semiconservative replication of the remaining heteroduplex will yield one m^+ homoduplex and one *m* homoduplex, and the resulting ascus will contain a 5 m^+:3 *m* ratio of ascospores. Such 5:3 gene conversion ratios do occur. They result from postmeiotic (mitotic) segregation of unrepaired heteroduplexes.

Gene conversion is associated with the reciprocal recombination of outside markers approximately 50 percent of the time. This correlation is nicely explained by the Holliday model of recombination presented in Figure 13.31. If the two recombinant chromatids of the tetrad just diagrammed are drawn in a form equivalent to that shown in Figure 13.31*g*, the association of gene conversion with reciprocal recombination of outside markers can easily be explained (Figure 13.33). The single-strand bridge connecting the two chromatids must be resolved by endonucleolytic cleavage to complete the recombination process. This cleavage may occur either horizontally or vertically on the chi form drawn in Figure 13.33. Vertical cleavage will yield an ascus showing both gene conversion and reciprocal recombination of outside markers. Horizontal cleavage will yield an ascus showing gene conversion and the parental combination of flanking markers. Thus, if cleavage occurs in the vertical plane half of the time and in the horizontal plane half of the time, gene conversion will be associated with reciprocal recombination of outside markers about 50 percent of the time, as observed.

Key Points: **Crossing over involves the breakage of homologous DNA molecules and the rejoining of parts in new combinations. When genetic markers are closely linked, the four products of meiosis often contain three copies of one marker and one copy of the other marker. This gene conversion results from repair synthesis that occurs during the recombination process.**

TESTING YOUR KNOWLEDGE

1. Charles Yanofsky isolated a large number of auxotrophic mutants of *E. coli* that could grow only on medium containing the amino acid tryptophan. How can such mutants be identified? If a specific tryptophan auxotroph resulted from a nitrous acid-induced mutation, could it be induced to revert back to prototrophy by treatment with 5-bromouracil (5–BU)?

ANSWER

The culture of mutagenized bacteria must be grown in medium containing tryptophan so that the desired mutants can survive and reproduce. The bacteria should then be diluted, plated on agar medium containing tryptophan, and incubated until visible colonies are produced. The colonies are next replicated onto plates lacking tryptophan by the replica-plating technique developed by the Lederbergs (see Figure 13.3). The desired tryptophan auxotrophs will grow on the plates containing tryptophan, but not on the replica plates lacking tryptophan. Because nitrous acid and 5–BU produce transition mutations in both directions, A:T ⇔ G:C, any mutation induced with nitrous acid should be induced to back-mutate with 5–BU.

2. Assume that you recently discovered a new species of bacteria and named it *Escherichia mutaphilium*. During the last year, you have been studying the *mutA* gene and its polypeptide product, the enzyme trinucleotide mutagenase, in this bacterium. *E. mutaphilium* has been shown to use the established, nearly universal genetic code and to behave like *Escherichia coli* in all other respects relevant to molecular genetics.

The sixth amino acid from the amino terminus of the wild-type trinucleotide mutagenase is histidine, and the wild-type *mutA* gene has the triplet nucleotide-pair sequence

3'-G T A-5'
... .
5'-C A T-3'

at the position corresponding to the sixth amino acid of the gene product. Seven independently isolated mutants with single nucleotide-pair substitutions within this triplet have also been characterized. Moreover, the mutant trinucleotide mutagenases have all been purified and sequenced. All seven are different; they contain glutamine, tyrosine, asparagine, aspartic acid, arginine, proline, and leucine as the sixth amino acid from the amino terminus.

Mutants *mutA*1, *mutA*2, and *mutA*3 will not recombine with each other, but each will recombine with each of the other four mutants (*mutA*4, *mutA*5, *mutA*6, and *mutA*7) to yield true wild-type recombinants. Similarly, mutants *A*4, *A*5, and *A*6 will not recombine with each other but will each yield true wild-type recombinants in crosses with each of the other four mutants. Finally, crosses between *mutA*1 and *mutA*7 yield about twice as many true wild-type recombinants as do crosses between *mutA*6 and *mutA*7.

Mutants *A*1 and *A*6 are induced to back-mutate to wild-type by treatment with 5-bromouracil (5–BU), whereas mutants *A*2, *A*3, *A*4, *A*5, and *A*7 are not induced to back-mutate by treatment with 5–BU. Mutants *A*2 and *A*4 grow slowly on minimal medium, whereas mutants *A*3 and *A*5 carry null mutations (producing completely inactive gene products) and are incapable of growth on minimal medium. This difference has been used to select for mutation events from genotypes *mutA*3 and *mutA*5 to genotypes *mutA*2 and *mutA*4. Mutants *A*3 and *A*5 can be induced to mutate to *A*2 and *A*4, respectively, by treatment with 5-bromouracil or hydroxylamine. However, mutant *A*3 cannot be induced to mutate to *A*4, nor *A*5 to *A*2, by treatment with either mutagen.

Use the information given above and the nature of the genetic code (Table 12.2) to deduce which mutant allele specifies the mutant polypeptide with each of the seven different amino acid substitutions at position 6 of trinucleotide mutagenase, and describe the rationale behind each of your deductions.

ANSWER

The following deductions can be made from the information given.

(1) The wild-type His codon must be CAU based on the nucleotide-pair sequence of the gene.

(2) The codons for the seven amino acids found at position 6 in the mutant polypeptides must be connected to CAU by a single-base change because the mutants were all derived from wild-type by a single nucleotide-pair substitution. Thus the degeneracy of the genetic code is not a factor in deducing specific codon assignments.

(3) Because of the nature of the genetic code—specifically the degeneracy at the third (3') position in each codon, there are three possible amino acid substitutions due to single-base substitutions at each of the first two positions (the 5' base and the middle base), but only one possible amino acid change due to a single-base change at position 3 (the 3' base in the codon). For ease of discussion, the three nucleotide-pair positions in the triplet under consideration will be referred to as position 1 (corresponding to the 5' base in the codon), position 2 (the middle nucleotide pair), and position 3 (corresponding to the 3' base in the mRNA codon).

(4) Since *A*1, *A*2, and *A*3 do not recombine with each other, they must all result from base-pair substitutions at the same position in the triplet, at either position 1 or position 2. The same is true for *A*4, *A*5, and *A*6. Since *A*7 recombines with each of the other six mutant alleles, it must result from the single base-pair substitution at position 3 that leads to an amino acid change.

(5) The only amino acid with a codon connected to the His codon CAU by a single-base change at position 3 is Gln (codon CAA). Thus the *mutA*7 polypeptide must have glutamine as the sixth amino acid.

(6) Since *mutA*7 (the third position substitution) yields about twice as many wild-type recombinants in crosses with *mutA*1 as in crosses with *mutA*6, the *A*1 substitution must be at position 1 and the *mutA*6 subsititution must be at position 2. Combined with (4) above, this places the *A*2 and *A*3 substitutions at position 1 and the *A*4 and *A*5 substitutions at position 2.

(7) Since *mutA*1 and *mutA*6 are induced to revert to wild-type by 5-BU, they must be connected to the triplet of nucleotide pairs encoding His by transition mutations, that is

$$(mut\ A1)\ \frac{\text{ATA}}{\text{TAT}} \xleftrightarrow{\text{5-BU}} \frac{\text{GTA}}{\text{CAT}} \xleftrightarrow{\text{5-BU}} \frac{\text{GCA}}{\text{CGT}}\ (mut\ A6)$$

(8) Since *mutA*3 and *mutA*5 are induced to mutate to *mutA*2 and *mutA*4, resepectively, by hydroxylamine, *A*3 must be connected to *A*2 and *A*5 to *A*4 specifically by G:C → A:T transitions, that is,

$$(mut\ A3)\ \frac{\text{CTA}}{\text{GAT}} \xrightarrow{\text{HA}} \frac{\text{TTA}}{\text{AAT}}\ (mut\ A2)$$

and

$$(mut\ A5)\ \frac{\text{GGA}}{\text{CCT}} \xrightarrow{\text{HA}} \frac{\text{GAA}}{\text{CTT}}\ (mut\ A4)$$

Collectively, these deductions establish that the following relationships between the amino acids, codons, and nucleotide-pair triplets are present at the position of interest in the trinucleotide mutagenase polypeptides, mRNAs, and genes in the seven different mutants.

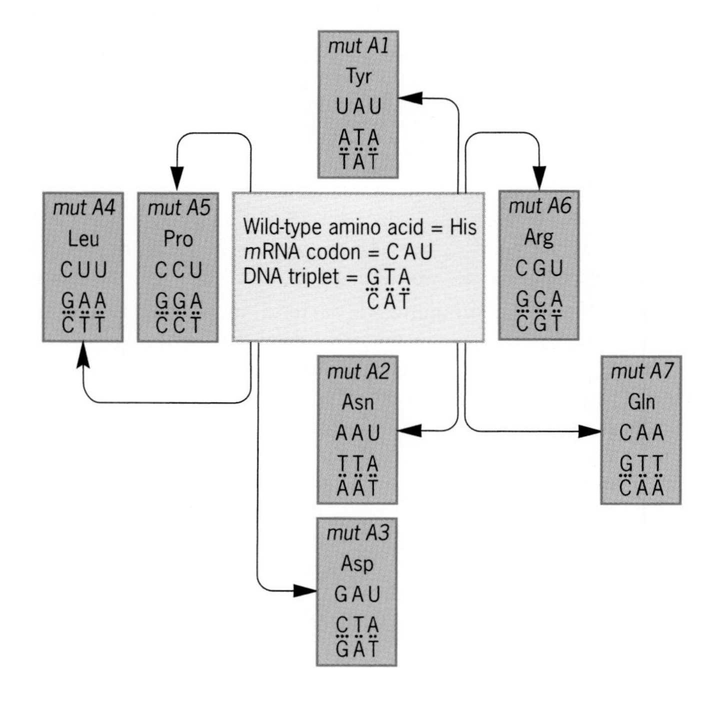

QUESTIONS AND PROBLEMS

13.1 Identify the following point mutations represented in DNA and in RNA as (1) transitions, (2) transversions, or (3) reading frameshifts. (a) A to G; (b) C to T; (c) C to G; (d) T to A; (e) UAU ACC UAU to UAU AAC CUA; (f) UUG CUA AUA to UUG CUG AUA.

13.2 Both lethal and visible mutations are expected to occur in fruit flies that are subjected to irradiation. Outline a method for detecting (a) sex-linked lethals and (b) sex-linked visible mutations in irradiated *Drosophila*.

13.3 How can mutations in bacteria causing resistance to a particular drug be detected? How can it be determined whether a particular drug causes mutations or merely identifies mutations already present in the organisms under investigation?

13.4 Published spontaneous mutation rates for humans are generally higher than those for bacteria. Does this indicate that individual genes of humans mutate more frequently than those of bacteria? Explain.

13.5 A precancerous condition (intestinal polyposis) in a particular human family group is determined by a single dominant gene. Among the descendants of one woman who died with cancer of the colon, 10 people have died with the same type of cancer and 6 now have intestinal polyposis. All other branches of the large kindred have been carefully examined, and no cases have been found. Suggest an explanation for the origin of the defective gene.

13.6 Juvenile muscular dystrophy in humans is dependent on a sex-linked recessive gene. In an intensive study, 33 cases were found in a population of some 800,000 people. The investigators were confident that they had found all cases that were well enough advanced to be detected at the time the study was made. The symptoms of the disease were expressed only in males. Most of those with the disease died at an early age, and none lived beyond 21 years of age. Usually, only one case was detected in a family, but sometimes two or three cases occurred in the same family. Suggest an explanation for the sporadic occurrence of the disease and the tendency for the gene to persist in the population.

13.7 Products resulting from somatic mutations, such as the navel orange and the Delicious apple, have become widespread in citrus groves and apple orchards. However, traits resulting from somatic mutations are seldom maintained in animals. Why?

13.8 If a single short-legged sheep should occur in a flock, suggest experiments to determine whether the short legs are the result of a mutation or an environmental effect. If due to a mutation, how can one determine whether the mutation is dominant or recessive?

13.9 How might enzymes such as DNA polymerase be involved in the mode of action of both mutator and antimutator genes (mutant genes that increase and decrease, respectively, mutation rates?

13.10 How could spontaneous mutation rates be optimized by natural selection?

13.11 A mutator gene *Dt* in maize increases the rate at which the gene for colorless aleurone (*a*) mutates to the dominant allele (*A*), which yields colored aleurone. When reciprocal crosses were made (i.e., seed parent *dt/dt, a/a* × *Dt/Dt, a/a* and seed parent *Dt/Dt, a/a* × *dt/dt, a/a*), the cross with *Dt/Dt* seed parents produced three times as many dots per kernel as the reciprocal cross. Explain these results.

13.12 A single mutation blocks the conversion of phenylalanine to tyrosine. (a) Is the mutant gene expected to be pleiotropic? (b) Explain.

13.13 How can normal hemoglobin (hemoglobin A) and hemoglobin S be distinguished?

13.14 If CTT is a DNA triplet (transcribed strand of DNA) specifying glutamic acid, what DNA and mRNA base triplet alterations could account for valine and lysine in position 6 of the β-globin chain?

13.15 Why is sickle-cell anemia called a molecular disease?

13.16 Assuming that the β-globin chain and the α-globin chain shared a common ancestor, what mechanisms might explain the differences that now exist in these two chains? What changes in DNA and mRNA codons would account for the differences that have resulted in unlike amino acids at corresponding positions?

13.17 In a given strain of bacteria, all of the cells are usually killed when a specific concentration of streptomycin is present in the medium. Mutations occur that confer resistance to streptomycin. The streptomycin-resistant mutants are of two types: some can live with or without streptomycin; others cannot survive unless this drug is present in the medium. Given a streptomycin-sensitive strain of this species, outline an experimental procedure by which streptomycin-resistant strains of the two types could be established.

13.18 One stock of fruit flies was treated with 1000 roentgens (r) of X rays. The X-ray treatment increased the mutation rate of a particular gene by 2 percent. What percentage increases in the mutation rate of this gene would be expected if this stock of flies was treated with X-ray doses of 1500 r, 2000 r, and 3000 r?

13.19 Why does the frequency of chromosome breaks induced by X rays vary with the total dosage and not with the rate at which it is delivered?

13.20 One person was in an accident and received 50 roentgens (r) of X rays at one time. Another person received 5 r in each of 20 treatments. Assuming no intensity effect, what proportionate number of mutations would be expected in each person?

13.21 How does ultraviolet light produce revertible mutations?

13.22 How does nitrous acid induce mutations? What specific end results might be expected in DNA and mRNA from the treatment of viruses with nitrous acid?

13.23 Are mutational changes induced by nitrous acid more likely to be transitions or transversions?

13.24 How does the action and mutagenic effect of 5-bromouracil differ from that of nitrous acid?

13.25 How do acridine-induced changes in DNA result in inactive proteins?

Use the known codon-amino acid assignments given in Chapter 12 to work the following problems.

13.26 Bacteriophage MS2 carries its genetic information in RNA. Its chromosome is analogous to a polygenic molecule of mRNA in organisms that store their genetic information in DNA. The MS2 minichromosome encodes 4 polypeptides (i.e., it has four genes). One of these four genes encodes the MS2 coat protein, a polypeptide 129 amino acids long. The entire nucleotide sequence in the RNA of MS2 is known. Codon 112 of the coat protein gene is CUA, which specifies the amino acid leucine. If you were to treat a replicating population of bacteriophage MS2 with the mutagen 5-bromouracil, what amino acid substitutions would you expect to be induced at position 112 of the MS2 coat protein (i.e., Leu → other amino acid)? (Note: Bacteriophage MS2 RNA replicates using a complementary strand of RNA and base-pairing like DNA.)

13.27 Would the different amino acid substitutions induced by 5-bromouracil at position 112 of the coat polypeptide that you indicated in Problem 13.26 be expected to occur with equal frequency? If so, why? If not, why not? Which one(s), if any, would occur more frequently?

13.28 Would such mutations occur if a nonreplicating suspension of MS2 phage was treated with 5-bromouracil?

13.29 Recalling that nitrous acid deaminates adenine, cytosine, and guanine (adenine → hypoxanthine, which base-pairs with cytosine; cytosine → uracil, which base-pairs with adenine; and guanine → xanthine, which base-pairs with cytosine), would you expect nitrous acid to induce any mutations that result in the substitution of another amino acid for a glycine residue in a wild-type polypeptide (i. e., glycine → another amino acid) if the mutagenesis was carried out on a suspension of mature (nonreplicating) T4 bacteriophages. (*Note:* After the mutagenic treatment of the phage suspension, the nitrous acid is removed. The treated phage are then allowed to infect *E. coli* cells to express any induced mutations.) If so, by what mechanism? If not, why not?

13.30 Keeping in mind the known nature of the genetic code, the information given about phage MS2 in Problem 13.26, and the information we have learned about nitrous acid in Problem 13.29, would you expect nitrous acid to induce any mutations that would result in amino acid substitutions of the type glycine → another amino acid if the mutagenesis were carried out on a suspension of mature (nonreplicating) MS2 bacteriophage? If so, by what mechanism? If not, why not?

13.31 Would you expect nitrous acid to induce a higher frequency of Tyr → Ser or Tyr → Cys substitutions? Why?

13.32 Which of the following amino acid substitutions should you expect to be induced by 5-bromouracil with the highest frequency? (a) Met → Leu; (b) Met → Thr; (c) Lys → Thr; (d) Lys → Gln; (e) Pro → Arg; or (f) Pro → Gln? Why?

13.33 Acridine dyes such as proflavin are known to induce primarily single base-pair additions and deletions. Suppose that the wild-type nucleotide sequence in the mRNA produced from a gene is

5' -AUGCCCUUUGGGAAAGGGUUUCCCUAA-3'

Also, assume that a mutation is induced within this gene by proflavin and, subsequently, a revertant of this mutation is similarly induced with proflavin and shown to result from a second-site suppressor mutation within the same gene. If the amino acid sequence of the polypeptide encoded by this gene in the revertant (double mutant) strain is

NH_2-Met-Pro-Phe-Gly-Glu-Arg-Phe-Pro-COOH

what would be the most likely nucleotide sequence in the mRNA of this gene in the revertant (double mutant)?

BIBLIOGRAPHY

AMES, B. N., J. MCCANN, AND E. YAMASAKI. 1975. "Methods for detecting carcinogens and mutagens with the *Salmonella*/mammalian-microsome mutagenicity test." *Mutation Research* 31:347–364.

DRAKE, J. W. 1970. *The Molecular Basis of Mutation*. Holden-Day, San Francisco.

DRESSLER, D., AND H. POTTER. 1982. "Molecular mechanisms in genetic recombination." *Annu. Rev. Biochem.* 51:727–761.

HANAWALT, P. C. 1994. "Transcription-coupled repair and human disease." *Science* 266: 1954–1956.

LEDERBERG, J., AND E. M. LEDERBERG. 1952. "Replica plating and indirect selection of bacterial mutants." *J. Bacteriology* 63: 399–406. (Reprinted in E. A. Adelberg, ed., 1966. *Papers on Bacterial Genetics*, 2nd ed., Little, Brown, Boston.)

MODRICH, P. 1994. "Mismatch repair, genetic stability, and cancer." *Science* 266: 1959–1960.

MULLER, H. J. 1927. "Artificial transmutation of the gene." *Science* 66:84–87. (Reprinted in J. A. Peters, ed., 1959. *Classical Papers in Genetics*. Prentice-Hall, Englewood Cliffs, NJ).

PRAKASH, L., S. PRAKASH, AND P. SUNG. 1993. "DNA repair genes and proteins of *Saccharomyces cerevisiae*." *Annu. Rev. Genet.* 27:33–70.

SANCAR, A. 1994. "Mechanisms of DNA excision repair." *Science* 266:1954–1956.

STAHL, F. W. 1979. *Genetic Recombination*. Freeman, San Francisco.

An albino giraffe with three normally pigmented companions.

14

Definitions of the Gene

CHAPTER OUTLINE

Sir Archibald Garrod and Human Inborn Errors of Metabolism

In 1902, just two years after the discovery of Mendel's work, Sir Archibald E. Garrod, a physician at the Hospital for Sick Children in London, published a paper entitled "The Incidence of Alkaptonuria: A Study in Chemical Individuality." Like Mendel's work, the concepts Garrod presented in this paper remained largely unknown by the scientists of the world for 40 years—until the concepts were independently formulated by George W. Beadle and Edward L. Tatum in 1941. Garrod's paper described the results of his studies of individuals with alkaptonuria, an innocuous, but easily detected, disorder. Because of the presence of a chemical called homogentisic acid (formerly called alkapton), the urine of affected individuals turns black when exposed to air. Garrod clearly recognized that alkaptonuria was inherited. In his 1902 paper, he wrote: "There are good reasons for thinking that alkaptonuria is not the manifestation of a disease but is rather the nature of an alternative course of metabolism, harmless and usually congenital and lifelong" (*Lancet* ii:1616).

In addition to alkaptonuria, Garrod studied albinism, cystinuria, pentosuria, and porphyrinurea in humans (See Human Genetics Sidelight: Human Inborn Errors of Metabolism). Garrod summarized the results of these studies in a treatise entitled "Inborn Errors of Metabolism," first presented as the Croonian Lectures to the Royal College of Physicians in London in 1908 and published in book form in 1909. In this book, Garrod clearly articulates the view that the metabolic defects observed in individuals with these disorders are caused by recessive mutant genes. Garrod probably did not develop his amazing insight into metabolic processes totally by himself. His father, Alfred B. Garrod, was also a physician and was the first to demonstrate the accumulation of the chemical uric acid in patients with gout, a painful inflammation of the joints in the hands and feet. In any case, Archibald E. Garrod was the first scientist to relate defects in genes to blocks in metabolic pathways. In this chapter, we focus on how Garrod's concept evolved into the current understanding of this basic unit of genetic information: the **GENE**.

The gene is to genetics what the atom is to chemistry. Thus throughout this text we have focused our attention on the gene and the alternate forms of a gene or alleles. In the preceding chapters, we have examined the patterns of transmission of independently assorting and linked genes, the chromosomal location of genes, the chemical composition of genes and chromosomes, the mechanism of replication of genes, mutational events in genes, and the mechanisms by which genes exert their effects on the phenotype of the organism. What is this unit of genetic information that we call the gene? As we will see, the concept of a gene is not static; it has evolved through several phases since Wilhelm Johannsen introduced the term in 1909, and it will undoubtedly evolve through additional refinements in the future.

The gene has been defined as the unit of genetic information that controls a specific aspect of the phenotype. Such a description, though accurate, does not provide a precise, unambiguous definition that can be used to identify a gene at the molecular level. At a more fundamental level, the gene has been defined as the unit of genetic information that specifies the synthesis of one polypeptide. However, it is not a very good operational definition. An **operational definition** spells out an operation or experiment that can be carried out to define or delimit something. One gene specifying one polypeptide is a poor operational definition because experimentally relating all the segments of DNA that represent genes with all the polypeptides is not feasible in a complex organism. Moreover, it is preferable to define the gene by using genetic approaches rather than biochemical experiments. Thus, in this chapter, we focus on the **complementation test** as an operational definition of the gene. We will also consider the limitations of the complementation test, along with the unique structural features of selected genes.

EVOLUTION OF THE CONCEPT OF THE GENE: SUMMARY

Before discussing evidence supporting the various concepts, let's summarize the important stages in the evolution of the gene concept. The gene theory of inheritance began with the publication of Mendel's classic paper in 1866 but did not become an accepted part of scientific knowledge until after the discovery of Mendel's work in 1900. Mendel's gene (not so-named) was the "character" or "constant factor" that controlled one specific phenotypic trait such as flower color in peas. At the time of the discovery of Mendel's work, the English physician Sir Archibald E. Garrod was studying several inherited diseases in humans. Garrod first recognized that homozygosity for recessive mutant alleles can cause defects in the normal processes of metabolism. His concept of the gene is probably stated most accurately as one mutant gene-one metabolic block, which over 30 years later was refined to the one gene-one enzyme concept enunciated by George W. Beadle and Edward L. Tatum. Since many enzymes contain two or more different polypeptides, each encoded by a separate gene, the one gene-one enzyme concept subsequently was modified to one gene-one polypeptide.

Prior to 1940, genes were considered analogous to beads on a string; recombination occurred between, but not within, genes. The gene was both the basic functional unit, which controlled one phenotypic trait, and the elementary structural unit, which could not be subdivided by recombination or mutation. Clarence Oliver's 1940 report that recombination had occurred within the *lozenge* gene of *Drosphila* stimulated both excitement and much debate about its significance. When the debate ended, the nucleotide pair had replaced the gene as the basic unit of structure, the unit of genetic material not subdivisible by mutation or recombination.

In the early 1940s, Edward B. Lewis developed the complementation, or *cis-trans*, test for functional al-

HUMAN GENETICS SIDELIGHT

Human Inborn Errors of Metabolism

As discussed in the text, Garrod's concept of one mutant gene–one metabolic block was based on his studies of a few inherited human disorders. Alkaptonuria, on which much of Garrod's information was based, is described in the text. However, Garrod also studied familial cases of cystinuria, pentosuria, porphyrinuria—inherited disorders characterized by elevated levels of the amino acid cystine, five-carbon sugars, and the iron-binding porphyrin component of hemoglobin, respectively, in urine. The most common disorder that Garrod studied was albinism, an autosomal recessive trait that occurs at a frequency of about 1 in 20,000 newborns in the United States. The red eyes and white skin and hair of albinos result from the absence of the black pigment melanin. The most common forms of albinism are caused by mutations that block the biosynthesis of melanin from the amino acid tyrosine. Of course, the pathway of melanin biosynthesis was unknown in 1909 when Garrod published his book *Inborn Errors of Metabolism*. Nevertheless, he clearly understood that mutations caused specific blocks in metabolic pathways.

Today, over 4000 inherited human disorders have been described, and the number of such hereditary abnormalities is constantly increasing. These disorders range from relatively innocuous disorders like alkaptonuria to those such as Tay-Sachs disease (characterized by rapid degeneration of the central nervous system) that are lethal in early childhood. Sickle-cell anemia and phenylketonuria (PKU) are perhaps the best known of the human inborn errors of metabolism. The severe anemia in individuals homozygous for the sickle-cell mutation is caused by a single amino acid substitution in their β-globin (Chapter 13). Children with PKU lack the enzyme phenylalanine hydroxylase. This enzyme deficiency results in the accumulation of phenylpyruvic acid, which is highly toxic to the central nervous system. If untreated, children with PKU develop severe mental retardation. However, if they are placed on a diet low in phenylalanine, they develop normal mental abilities (Chapter 20). Other inherited human diseases are discussed throughout the book.

In the United States alone, over 120,000 children with inherited defects are born each year. Long-term medical care of these birth defects is estimated to cost $10 to $20 billion. But the cost is even higher in terms of human suffering. For example, consider the tragic degeneration of the central nervous system that occurs in Huntington disease or the loss of body control resulting from progressive muscle degeneration associated with Duchenne muscular dystrophy (both discussed in Chapter 20) or the consequences of the loss of memory that occurs in individuals with Alzheimer's disease (Chapter 26). What are the causes of these inherited diseases? What can humans do, if anything, to keep the frequency of these disorders as low as possible? How many different human inborn errors of metabolism are there? Can the deleterious symptoms of these inherited disorders be eliminated by treatment? Can any of these inherited diseases be cured so that they won't be transmitted from parents to their children?

Human inherited diseases are caused by all of the types of mutations discussed in Chapter 13 and by changes in chromosome structure and number as discussed in Chapter 6. Because mutations are required for evolution, they can never be completely eliminated. However, humans can minimize the frequency of new mutations by avoiding contact with highly mutagenic agents such as irradiation and chemicals that damage DNA. All genes can mutate to nonfunctional states. Therefore, the number of human inborn errors of metabolism is almost certainly equal to the number of genes required to develop and maintain a normal human phenotype. Several inherited disorders such as PKU have proven treatable once their molecular bases are known. Others are currently being treated by somatic cell gene therapy (Chapter 20). In the future, some inherited human diseases may be cured by germ-line gene therapy; but serious ethical issues must be addressed before germ-line gene therapy can be performed on humans. Given the rapidly increasing number of known inherited human diseases and the identification and characterization of the genes responsible for these disorders (Chapter 20), the tools of molecular genetics will play an increasingly important role in the practice of medicine in the future.

lelism in *Drosophila*. This test subsequently was exploited by Seymour Benzer to define experimentally the gene in bacteriophage T4. The complementation test provides an operational definition of the gene. This test allows a geneticist to determine whether two independent mutations occurred in the same gene or in two different genes. Moreover, the gene as defined by the complementation test is a perfect fit to the one gene–one polypeptide concept.

In the 1960s, elegant experiments by Charles Yanofsky, Sydney Brenner, and their collaborators showed that the gene and its polypeptide product were colinear structures, with a direct correlation between the sequence of nucleotide pairs in the gene and the sequence of amino acids in the polypeptide. However, this simple concept of the gene as a continuous sequence of nucleotide pairs specifying a colinear sequence of amino acids in the polypeptide gene prod-

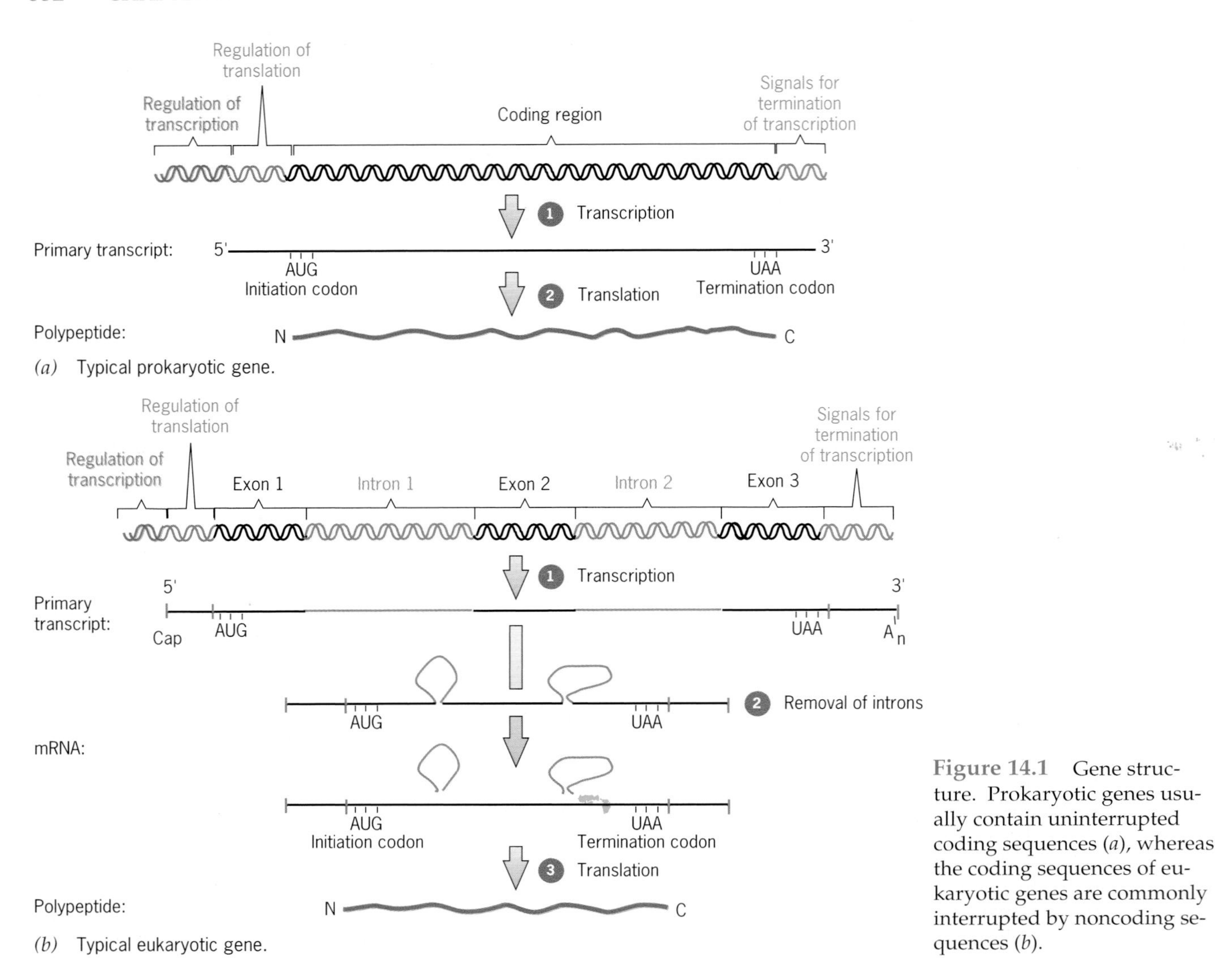

Figure 14.1 Gene structure. Prokaryotic genes usually contain uninterrupted coding sequences (*a*), whereas the coding sequences of eukaryotic genes are commonly interrupted by noncoding sequences (*b*).

uct was short-lived. Overlapping genes and genes-within-genes were discovered in the late 1960s, and the coding sequences of eukaryotic genes were shown to be interrupted by noncoding intron sequences in the late 1970s. Moreover, some genes, for example, genes encoding immunoglobulins, were shown to be stored in germ-line chromosomes as short "gene segments," which are assembled into mature, functional genes during development.

Thus the definition of the gene needs to remain somewhat pliable if it is to encompass all of the different structure/function relationships that occur in different organisms. Here, we define the gene as the unit of genetic information that controls the synthesis of one polypeptide or one structural RNA molecule. As just defined, the gene can be identified operationally by the complementation test. As such, the gene includes the 5' and 3' noncoding regions that are involved in regulating the transcription and translation of the gene and all noncoding sequences or introns within the gene (Figure 14.1). The structural gene refers to the portion that is transcribed to produce the RNA product. In the case of overlapping genes, this definition requires that some nucleotide-pair sequences be considered components of two or more genes. For those cases where exons are spliced together in various combinations to make related but different proteins, the gene may be defined as a DNA sequence that is a single unit of transcription and encodes a set of closely related polypeptides, sometimes called "protein isoforms." In germ-line chromosomes, the DNA sequences that encode segments of antibody chains probably should be called "gene segments," because this genetic information is not organized into units that fit any of the standard definitions of the gene (Chapter 24).

Key Points: **The concept of the gene has undergone many refinements since its discovery by Mendel in**

1866. Most genes encode one polypeptide and can be operationally defined by the complementation test.

EVOLUTION OF THE CONCEPT OF THE GENE: FUNCTION

In the preceding section, we briefly examined the evolution of the concept of the gene, the basic unit of genetic information. That summary included both functional and structural aspects of the gene. Now, let's take a closer look at the gene as a unit of function.

Mendel: Constant Factors Controlling Phenotypic Traits

The law of combination of different characters, which governs the development of the hybrids, finds therefore its foundation and explanation in the principle enunciated, that the hybrids produce egg cells and pollen cells which in equal numbers represent all constant forms which result from the combinations of the characters brought together in fertilisation. (Mendel, 1866; translation by William Bateson)

Mendel's characters or factors, which are now called genes, controlled specific phenotypic traits such as flower color, seed color, and seed shape. They were the basic units of function, the units of genetic information that governed one specific aspect of the phenotype. This definition of the gene as the basic unit of function is almost universally favored by present-day scientists. There has been no change in the concept of the gene as the basic unit of function since its discovery by Mendel in 1866. However, the discovery of the chemical nature of the genetic material raised questions about the structure of the gene, and the concept of the molecular structure of this basic unit of function has undergone several changes and refinements since the discovery of Mendel's work in 1900.

If we examine what is known about how genes control phenotypic traits, the need for a more precise definition of the gene will be obvious. The pathway by which a gene exerts its effect on the phenotype of an organism is often very complex (Figure 14.2). Several genes may have similar effects on the same phenotypic trait, making it difficult to sort out the effects of individual genes. All the genes of an organism are located in the same nuclei, and they do not all function independently. The phenotype of an organism is the product of the action of all the genes acting within the restrictions imposed by the environment. Each gene also has an effect on the population to which the organism carrying the gene belongs. Ultimately, each gene has a potential effect, small though it may be, on the cumulative phenotype of the biosphere, for each gene may affect the ability of the organism, or the population, or the species to compete for an ecological niche in the biosphere (Chapter 28).

Garrod: One Mutant Gene–One Metabolic Block

At the time of the discovery of Mendel's work in 1900, Sir Archibald Garrod was studying several congenital metabolic diseases in humans. One of these was the

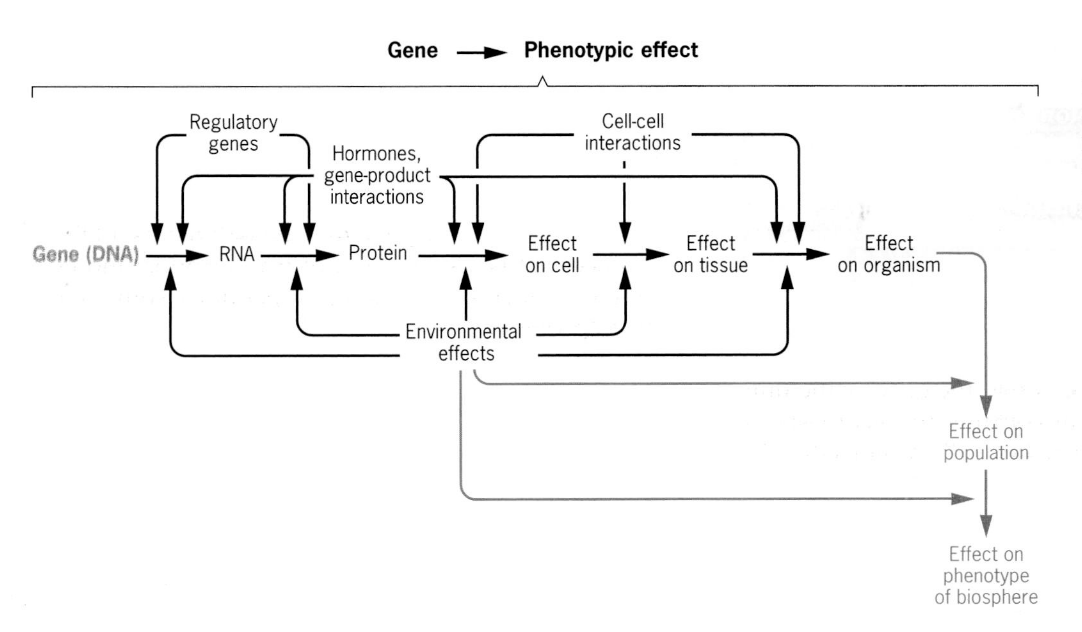

Figure 14.2 The complex pathway by which a gene exerts its effect on the phenotype of an organism, a population, or the biosphere.

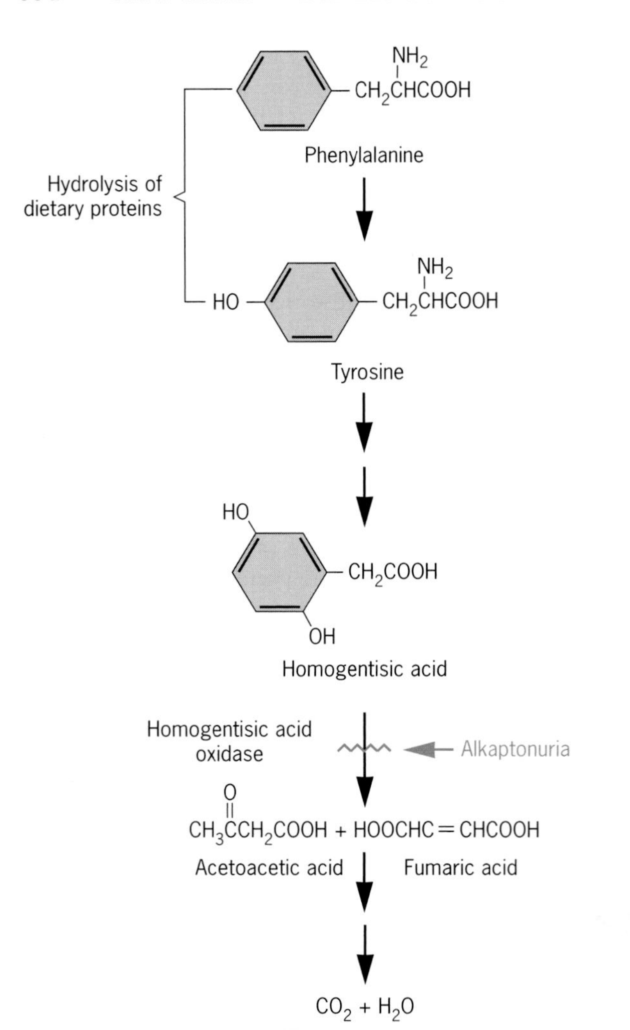

Figure 14.3 Alkaptonuria in humans results from a block in phenylalanine catabolism caused by a mutation in the gene encoding the enzyme homogentisic acid oxidase. When this enzyme is absent or inactive, its substrate, homogentisic acid, accumulates in tissues and in urine.

inherited disease **alkaptonuria**, which is easily detected because of the blackening of the urine upon exposure to air. The substance responsible for this blackening is alkapton (or homogentisic acid), an intermediate in the degradation of the aromatic amino acids tyrosine and phenylalanine (Figure 14.3). Garrod believed that the presence of homogentisic acid in the urine was due to a block in the normal pathway of metabolism of this compound. Moreover, on the basis of the family pedigree studies, Garrod proposed that alkaptonuria was inherited as a single recessive gene. The results of Garrod's studies of alkaptonuria and a few other congenital diseases in humans, such as albinism, were presented in detail in his book, *Inborn Errors of Metabolism*. Although the details of the biochemical pathway affected by the recessive mutations that cause alkaptonuria were not worked out until many years later, Garrod clearly understood the relationship between genes and metabolism. His concept might be best stated as **one mutant gene–one metabolic block**.

Early Evidence That Enzymes Are Controlled by Genes

In proposing that metabolic reactions were controlled by genes in 1902, Sir Archibald E. Garrod displayed great insight, because there was little direct evidence at the time to support his proposal. Some of the first evidence showing that genes control the enzyme-catalyzed reactions of metabolic pathways was obtained from studies of *Drosophila* eye color mutants. In 1935, George Beadle and Boris Ephrussi suggested that mutations in two genes controlling eye color resulted in blocks at two different steps in the biosynthesis of the brown eye pigment in *Drosophila*. Wild-type fruit flies have dark red eyes resulting from the presence of two eye pigments, one bright red and the other brown. The *vermilion* (*v*) and *cinnabar* (*cn*) mutant flies have bright red eyes owing to the absence of the brown pigment.

Like other insects, fruit flies undergo metamorphosis; fertilized eggs develop into larvae, then pupae, and finally adult flies (Chapter 2). The larval stages contain groups of embryonic cells called imaginal disks that will develop into the specific adult structures, such as eyes, wings, or legs, of the adult flies. Beadle and Ephrussi discovered that if they surgically transplanted eye imaginal disks from one larva into the abdomen of another larva, the transplanted disk would develop into a third eye, albeit nonfunctional, in the adult fly (Figure 14.4). When they transplanted eye disks from wild-type larvae into *vermilion* or *cinnabar* larvae, the transplanted disks developed into dark red wild-type eyes (Figure 14.4*a*). In these two cases, the genotype of the transplanted eye disk controlled its own phenotype; the phenotype of the third eye was not influenced by the genotype of the larval host. However, this was not always the case.

When Beadle and Ephrussi transplanted eye disks from *vermilion* or *cinnabar* larvae into the abdomens of wild-type larvae, the disks developed into dark red wild-type eyes (Figure 14.4*b*). Beadle and Ephrussi reasoned that the wild-type larval hosts must have provided some diffusible substance to the transplanted mutant disks that allowed them to bypass the metabolic block and synthesize the brown eye pigment. In these two cases, the development of the transplanted disks into adult eye structures was influenced by the genotype of the host organism.

Beadle and Ephrussi next transplanted eye disks from *vermilion* larvae into *cinnabar* hosts and eye disks from *cinnabar* larvae into *vermilion* hosts. The recipro-

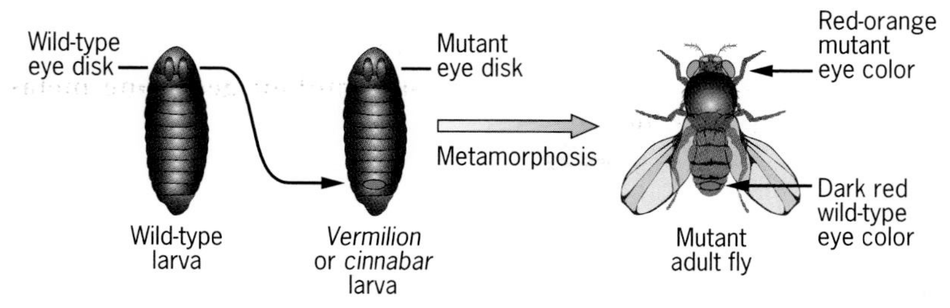

(*a*) Wild-type disks are transplanted into *vermilion* or *cinnabar* larvae.

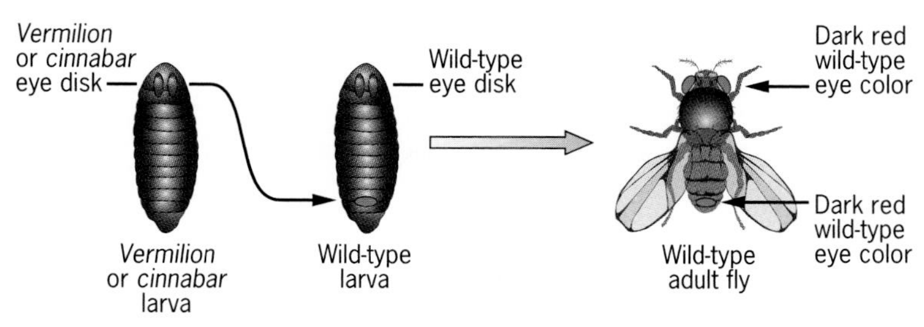

(*b*) *Vermilion* or *cinnabar* disks are transplanted into wild-type larvae.

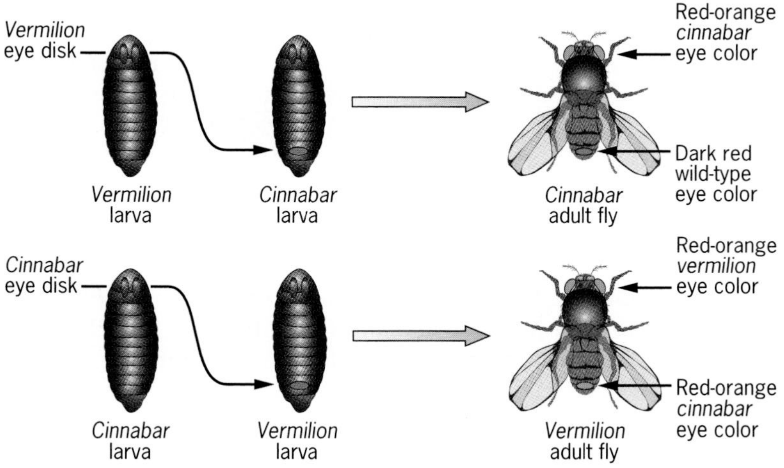

(*c*) *Vermilion* disks are transplanted into *cinnabar* larvae and *cinnabar* disks are transplanted into *vermilion* larvae.

Figure 14.4 Transplant experiments of *Drosophila* eye imaginal disks by Beadle and Ephrussi.

cal transplants yielded quite different results! The *vermilion* disks transplanted into *cinnabar* larvae developed into dark red wild-type eyes, whereas the *cinnabar* disks transplanted into *vermilion* larvae developed into bright red *cinnabar* eyes (Figure 14.4*c*). To explain these results, Beadle and Ephrussi proposed that the *cinnabar* hosts supplied a diffusible substance that the implanted *vermilion* disks converted to the brown eye pigment. They further suggested that the *vermilion* hosts did not produce a metabolite that could be converted to the brown pigment by the *cinnabar* disks.

In fact, the results of Beadle and Ephrussi's reciprocal transplant experiments can be easily explained if the brown eye pigment is synthesized by a series of enzyme-catalyzed reactions and the v^+ and cn^+ genes control different steps in the biosynthetic pathway (Figure 14.5). Beadle and Ephrussi proposed that (1) a precursor molecule X is converted to a diffusible intermediate substance Y by the product of the v^+ gene, and (2) intermediate Y is converted to the brown eye pigment by the product of the cn^+ gene. If their proposal was correct, the *cinnabar* larval hosts, which were v^+, would have provided diffusible substance Y to the transplanted *vermilion* disks. Because the *vermilion* disks were cn^+, they would have converted substance Y to the brown eye pigment and would have developed into dark red wild-type eyes. The *vermilion*

Figure 14.5 Beadle and Ephrussi's interpretation of the results of their reciprocal transplants of *vermilion* and *cinnabar* eye disks in *Drosophila*. They proposed that the v^+ and cn^+ genes control two different steps in the synthesis of the brown eye pigment and that the v^+ gene product acts before the cn^+ gene product.

larval hosts would not have provided anything comparable to substance Y to the transplanted *cinnabar* disks. Since the *cinnabar* disks were v^+, they would have converted precursor X to intermediate Y. However, in the absence of cn^+, intermediate Y could not be converted to the brown pigment. As a result, *cinnabar* disks transplanted into *vermilion* larvae developed into bright red mutant eyes.

During the 1940s, Beadle and Ephrussi's proposal was proven correct, albeit with the addition of a few steps to the pathway. The brown eye pigment, xanthommatin, was shown to be synthesized from the amino acid tryptophan (precursor X) by a sequence of enzyme-catalyzed reactions (Figure 14.6). The v^+ gene controls the synthesis of an enzyme called tryptophan pyrrolase, which converts tryptophan to *N*-formylkynurenine. Then, *N*-formylkynurenine is converted to kynurenine (intermediate Y) by the enzyme kynurenine formylase. The cn^+ gene controls the synthesis of the enzyme kynurenine hydroxylase, which converts kynurenine (substance Y) to 3-hydroxylkynurenine. Two additional enzyme-catalyzed reactions then convert 3-hydroxylkynurenine to the brown pigment xanthommatin.

These early studies of *Drosophila* eye color mutants and the pathways of eye pigment biosynthesis provided direct evidence for the genetic control of enzyme-catalyzed metabolic pathways. The results of subsequent investigations demonstrated the genetic control of metabolism in all types of living organisms, from viruses to humans.

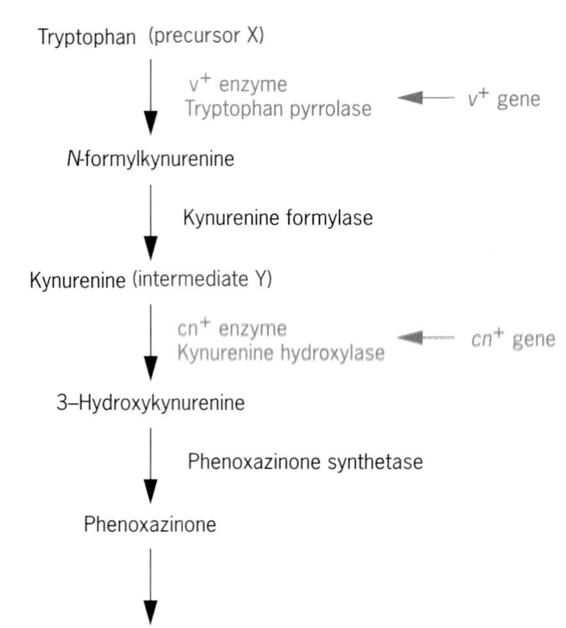

Figure 14.6 Biosynthesis of the brown eye pigment xanthommatin in *Drosphila*. The wild-type *vermilion* and *cinnabar* genes encode enzymes that catalyze two of the reactions in this pathway. Note the consistency of Beadle and Ephrussi's proposed sequence of reactions (Figure 14.5) with the actual biosynthetic pathway.

Beadle and Tatum: One Gene–One Enzyme

George Beadle's collaboration with Ephrussi in studies of the synthesis of the brown eye pigment in *Drosophila* led him to search for the ideal organism to use in extending this work. The pink bread mold *Neurospora crassa* can grow on medium containing only (1) inorganic salts, (2) a simple sugar, and (3) one vitamin, biotin. *Neurospora* growth medium containing only these components is called "minimal medium." George Beadle and Edward Tatum reasoned that *Neurospora* must be capable of synthesizing all the other essential metabolites, such as the purines, pyrimidines, amino acids, and other vitamins, *de novo*. Furthermore, they reasoned that the biosynthesis of these growth factors must be under genetic control. If so, mutations in genes whose products are involved in the biosynthesis of essential metabolites would be expected to produce mutant strains with additional growth-factor requirements.

Beadle and Tatum tested this prediction by irradiating asexual spores (conidia) of wild-type *Neurospora* with X rays or ultraviolet light, and screening the clones produced by the mutagenized spores for new growth-factor requirements (Figure 14.7). In order to select strains with a mutation in only one gene, they studied only mutant strains that yielded a 1:1 mutant to wild-type progeny ratio when crossed with wild-type. They identified mutants that grew on medium supplemented with all the amino acids, purines, pyrimidines, and vitamins (called "complete medium"), but could not grow on minimal medium. They analyzed the ability of these mutants to grow on medium supplemented with just amino acids, or just vitamins, and so on (Figure 14.7*a*). For example, Beadle

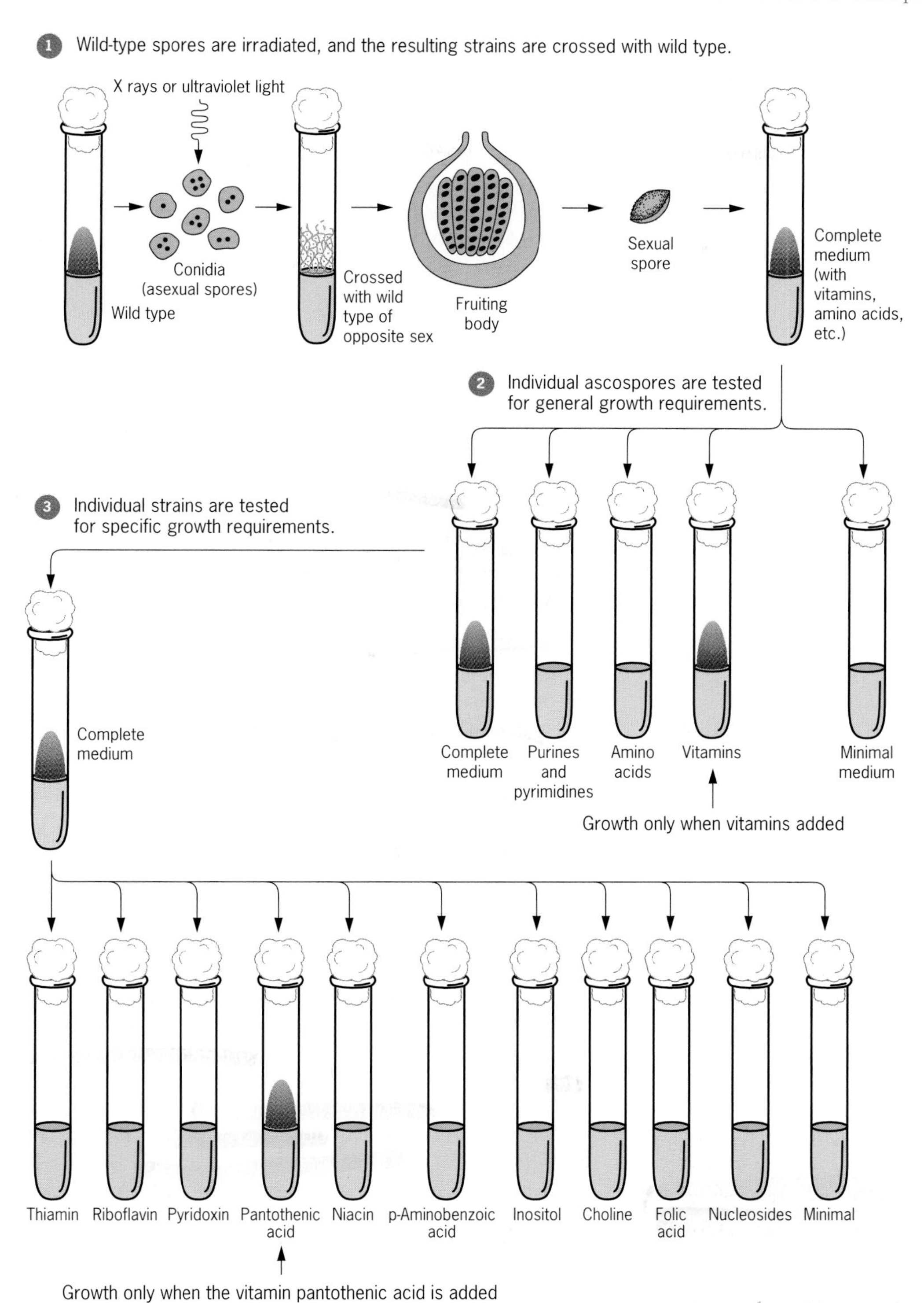

Figure 14.7 Diagram of Beadle and Tatum's experiment with *Neurospora* that led to the one gene–one enzyme hypothesis.

and Tatum identified mutant strains that grew in the presence of vitamins but could not grow in medium supplemented with amino acids or other growth factors. They next investigated the ability of these vitamin-requiring strains to grow on media supplemented with each of the vitamins separately (Figure 14.7*b*).

In this way, Beadle and Tatum demonstrated that each mutation resulted in a requirement for one growth factor. By correlating their genetic analyses with biochemical studies of the mutant strains, they demonstrated in several cases that one mutation resulted in the loss of one enzyme activity. This work, for which Beadle and Tatum received a Nobel Prize in 1958, was soon verified by similar studies of many other organisms in many laboratories. The **one gene–one enzyme** concept thus became a central tenet of molecular genetics.

Appropriately, in his Noble Prize acceptance speech, Beadle stated:

In this long, roundabout way, we had discovered what Garrod had seen so clearly so many years before. By now we knew of his work and were aware that we had added little if anything new in principle. . . . Thus we were able to demonstrate that what Garrod had shown for a few genes and a few chemical reactions in man was true for many genes and many reactions in Neurospora.

One Gene–One Polypeptide

Subsequent to the work of Beadle and Tatum, many enzymes and structural proteins were shown to be heteromultimeric, that is, to contain two or more different polypeptide chains, with each polypeptide encoded by a separate gene. For example, in *E. coli*, the enzyme tryptophan synthetase is a heterotetramer composed of two α polypeptides encoded by the *trpA* gene and two β polypeptides encoded by the *trpB* gene. Similarly, the hemoglobins, which transport oxygen from our lungs to all other tissues of our bodies, are tetrameric proteins that contain two α-globin chains and two β-globin chains, as well as four oxygen-binding heme groups (see Figure 12.4). In humans, the major form of adult hemoglobin contains two α-globin polypeptides encoded by the $Hb^{A}{}_{\alpha}$ gene on chromosome 16 and two β-globin polypeptides encoded by the $Hb^{A}{}_{\beta}$ gene on chromosome 11. Other enzymes, for example, *E. coli* DNA polymerase III (Chapter 10) and RNA polymerase II (Chapter 11), contain many different polypeptide subunits, each encoded by a separate gene. Thus the one gene–one enzyme concept was modified to **one gene–one polypeptide**.

Key Points: **The existence of a basic genetic element, the gene, that controlled a specific phenotypic trait was established by Mendel's work in 1866. Since the discovery of Mendel's results in 1900, the concept of the gene has evolved from the unit that can mutate to cause a specific block in metabolism, to the unit specifying one enzyme, to the sequence of nucleotide pairs in DNA encoding one polypeptide chain.**

EVOLUTION OF THE CONCEPT OF THE GENE: STRUCTURE

In the preceding section, we examined the evolution of the concept of the gene as the basic functional component of the genetic material. In this section, we examine the gene from a structural perspective. What is the structure of the gene? Do all genes have the same structure?

The Pre–1940 Beads-on-a-String Concept

Prior to 1940, the genes in a chromosome were considered analogous to beads on a string. Recombination was believed to occur only between the beads or genes, not within genes. The gene was believed to be indivisible. According to this beads-on-a-string concept, the gene was the basic unit of genetic information defined by three criteria: (1) function, (2) recombination, and (3) mutation. More specifically, the gene was

1. *The unit of function*, the unit of genetic material that controlled the inheritance of one "character" or one attribute of phenotype.
2. *The unit of structure*, operationally defined in two ways:
   a. *By recombination:* as the unit of genetic information not subdivisible by recombination.
   b. *By mutation:* as the smallest unit of genetic material capable of independent mutation.

Geneticists initially thought that all three criteria defined the same basic unit of inheritance, namely, the gene.

Geneticists now know that these criteria define two different units of inheritance. According to the current molecular concept, the gene is the unit of function, the unit of genetic information controlling the synthesis of one polypeptide chain or, in some cases, one RNA molecule. The unit of structure is simply the structural unit in DNA, the nucleotide pair. Because it clearly does not make sense to call each nucleotide pair a gene, geneticists have focused on the original definition of the gene as the unit of function and have discarded the beads-on-a-string view that the gene is not subdivisible by recombination or mutation. This is clearly appropriate since the emphasis in Mendel's work was on the *Merkmal* (or gene, as it is now called) controlling one phenotypic characteristic.

Discovery of Recombination Within the Gene

In 1940, Clarence P. Oliver published the first evidence indicating that recombination could occur within a gene. Oliver was studying mutations at the *lozenge* locus on the X chromosome of *Drosophila melanogaster*. Two mutations, lz^s ("spectacle" eye) and lz^g ("glassy" eye), were thought to be alleles, that is, different forms of the same gene. The data available prior to 1940 indicated that they mapped at the same position on the X chromosome. They had similar effects on the phenotype of the eye, and heterozygous lz^s/lz^g females had lozenge rather than wild-type eyes. However, when lz^s/lz^g females were crossed with either lz^s or lz^g males and large numbers of progeny were examined, wild-type progeny occurred with a frequency of about 0.2 percent.

These rare wild-type progeny could be explained by reversion of either the lz^s or the lz^g mutation. But there were two strong arguments against the reversion explanation. (1) The frequency of reversion of lz^s or lz^g to wild-type in hemizygous lozenge males was much lower than 0.2 percent. (2) When the lz^s/lz^g heterozygotes carried genetic markers bracketing the *lozenge* locus, the rare progeny with wild-type eyes always carried an X chromosome with lz^+ that was flanked by recombinant outside markers. Moreover, the same combination of outside markers always occurred, as though the sites of lz^s and lz^g were fixed relative to each other and crossing over was occurring between them. Different sets of outside markers were used and yielded the same result. If the lz^s/lz^g heterozygous female carried X chromosomes of the type

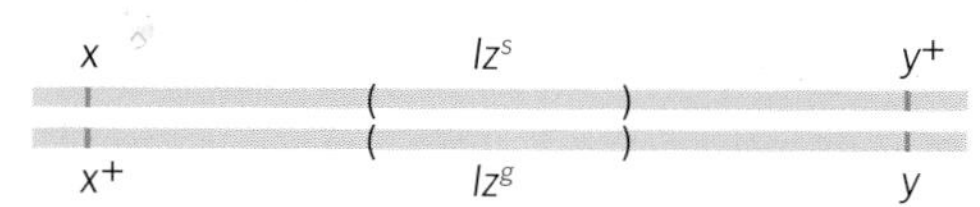

the rare progeny with wild-type eyes all (with one exception) contained an X chromosome with the following composition:

x lz+ y

Among progeny of these matings, the reciprocal combination of outside markers (x^+-y^+) never appeared in combination with lz^+. This result strongly suggested that the lz^s and lz^g mutations were located at distinct sites in the *lozenge* locus, and that the lz^+ chromosome was produced by crossing over between the two sites, as shown in the following diagram.

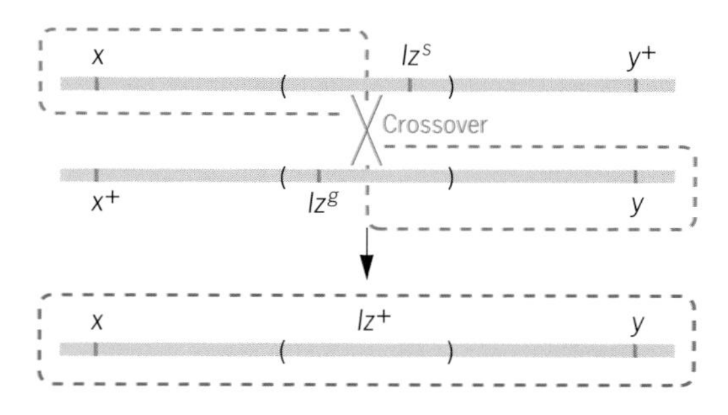

Definitive evidence for the involvement of recombination required the recovery and identification of the lz^s-lz^g double mutant with the reciprocal combination of outside markers—that is,

x+ lzg lzs y+

Oliver was not able to identify this double mutant because of the inability to distinguish it from the parental single-mutant phenotypes. The identification of both products, the wild-type and double mutant chromosomes, produced by crossing over within the *lozenge* gene, was first accomplished by Melvin M. Green, one of Oliver's students.

The results of these pioneering studies first indicated that the gene was more complex than a bead on a string. They showed that the gene was divisible, containing sites that were separable by crossing over. Oliver's and Green's results were the first step toward the present concept of the gene as a long sequence of nucleotide pairs, capable of mutating and recombining at many different sites along its length.

Recombination Between Adjacent Nucleotide Pairs

The results obtained by Oliver, Lewis, and Green in their studies of *Drosophila* genes all indicated that mutable sites that are separable by recombination can exist within a single gene. Seymour Benzer extended this picture of the gene by demonstrating the existence of 199 distinct sites of mutation that were separable by recombination within the *rIIA* gene of bacteriophage T4 (Chapter 15). Benzer's picture of the gene as a sequence of nucleotide pairs capable of mutating at many distinct sites was soon verified by the results of many researchers investigating gene structure in several different organisms, both prokaryotes and eukaryotes. Given this information about the structure of genes and the known structure of DNA, it followed that the smallest unit of genetic material capable of mutation might be the single nucleotide pair and that recombination might occur between adjacent nucleotide pairs, whether between or within genes. Recombination between adjacent nucleotide pairs of a gene was first demonstrated by Charles Yanofsky in his studies of the *trpA* gene encoding the α polypeptide of tryptophan synthetase in *E. coli.* This enzyme, a tetramer containing two α polypeptides and two β polypeptides, catalyzes the final step in the biosynthesis of the amino acid tryptophan.

Yanofsky and colleagues isolated and characterized a large number of tryptophan auxotrophs with mutations in the *trpA* gene. The wild-type *trpA* gene encodes an α polypeptide that is 268 amino acids long.

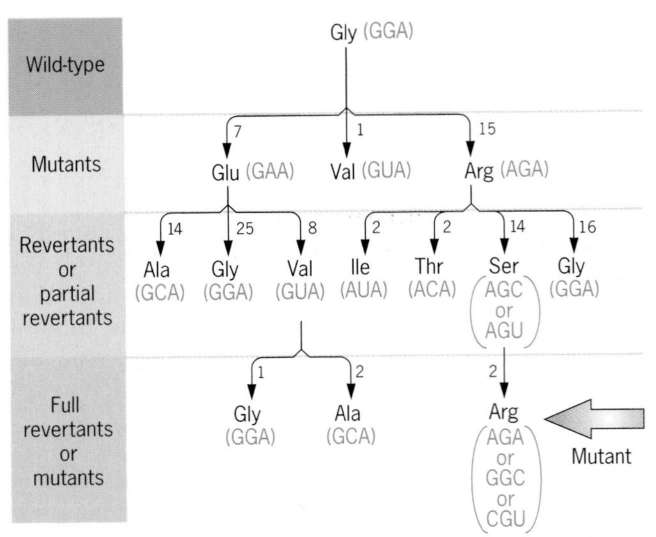

Figure 14.8 Pedigree of amino acid residue 211 (from the NH_2 terminus) of the α polypeptide of tryptophan synthetase of *E. coli*. Amino acid 211 is altered in *trpA*23 and *trpA*46 mutants (see Figure 14.11). The triplet codons shown in parentheses are the only codons specific to the indicated amino acids that will permit all of the observed amino acid replacements to occur by single base-pair substitutions. The number beside each arrow indicates the number of times that particular substitution was observed. These results indicate that the arginine and glutamic acid codons encoding amino acid 211 of the α polypeptides in *trpA*23 and *trpA*46 mutant cells are AGA and GAA, respectively. Thus these two mutations alter adjacent nucleotide pairs in the *trpA* gene.

Yanofsky and associates used the laborious techniques of protein sequencing to determine the complete amino acid sequence of the wild-type α polypeptide. They also determined the amino acid substitutions that had occurred in several mutant forms of the tryptophan synthetase α polypeptide. They mapped the mutations within the *trpA* gene by two- and three-factor crosses, and compared the map positions with the locations of the amino acid substitutions in the mutant polypeptides.

Yanofsky's correlated genetic and biochemical data for the *trpA* gene and the tryptophan synthetase α polypeptide showed that recombination can occur between mutations that alter the same amino acid. Mutations *trpA*23 and *trpA*46 both result in the substitution of another amino acid (arginine in the case of *A*23, glutamic acid in the case of *A*46) for the glycine present at position 211 of the wild-type tryptophan synthetase α polypeptide. However, these two mutations occur at different mutable sites; that is, the *A*23 and *A*46 sites are separable by recombination. Yanofsky and colleagues determined the amino acids present at position 211 of the α polypeptide in other mutants as well as in revertants and partial revertants of the *trpA*23 and *trpA*46 mutants. By using this information and the known codon assignments, they were able to determine which of the glycine, arginine, and glutamic acid codons were present in the *trpA* mRNA at the position encoding amino acid residue 211 of the α polypeptides present in trp^+, *trpA*23, and *trpA*46 cells, respectively (Figure 14.8).

Once the specific codons in mRNA are known, the corresponding base-pair sequences in the structural gene from which the mRNA is transcribed are also known. One strand of DNA will be complementary to the mRNA, and the second strand of DNA will be complementary to the first strand. Therefore, Yanofsky's data demonstrated that the mutational events that produced the *trpA*23 and *trpA*46 alleles were G:C to A:T transitions at adjacent nucleotide pairs. The trp^+ cells produced by recombination between chromosomes carrying mutations *A*23 and *A*46 demonstrated that recombination had occurred between adjacent nucleotide pairs in the *trpA* gene as shown in Figure 14.9. Yanofsky's results clearly showed that the

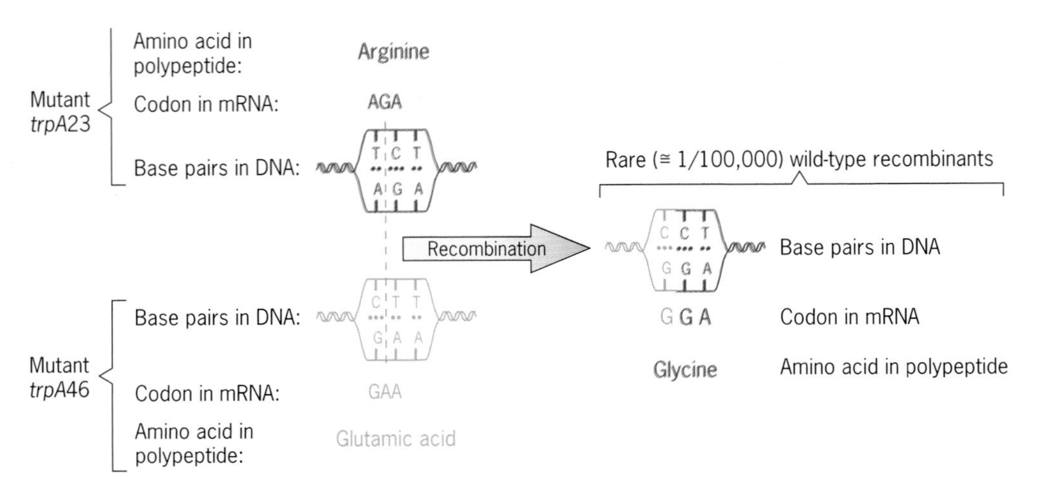

Figure 14.9 Recombination between mutations at adjacent nucleotide pairs in the *trpA* gene of *E. coli*. Mutations *A*23 and *A*46 both result in an amino acid substitution at position 211 of the trytophan synthetase α polypeptide. Wild-type *E. coli* has a glycine residue at this position of the α polypeptide. *A*23 causes a glycine to arginine substitution; *A*46 causes a glycine to glutamic acid substitution (see Figure 14.8).

unit of genetic material not divisible by recombination is the single nucleotide pair.

The pre–1940 concept of the gene as (1) the smallest unit of genetic material that could undergo mutation and (2) the unit of genetic information that could not be subdivided by recombination was clearly wrong. Each nucleotide pair of a gene can change or mutate independently, and recombination can occur between adjacent nucleotide pairs. The results of Oliver, Lewis, Green, Benzer, Yanofsky, and many others have compelled geneticists to focus on the gene as the unit of function, the sequence of nucleotide pairs controlling the synthesis and structure of one polypeptide or one RNA molecule.

Colinearity Between the Coding Sequence of a Gene and Its Polypeptide Product

The genetic information is stored in linear sequences of nucleotide pairs in DNA (or nucleotides in RNA, in some cases). Transcription and translation convert this genetic information into linear sequences of amino acids in polypeptides, which function as the key intermediaries in the genetic control of the phenotype.

It is now known that the nucleotide-pair sequences of the coding regions of the structural genes and the amino acid sequences of the polypeptides that they encode are **colinear.** That is, the first three base pairs of the coding sequence of a gene specify the first amino acid of the polypeptide, the next three base pairs (four to six) specify the second amino acid, and so on, in a colinear fashion (Figure 14.10*a*). It is also known that the coding regions of most of the genes in higher eukaryotes are interrupted by noncoding introns (Chapter 12). However, the presence of introns in genes does not invalidate the concept of colinearity. The presence of introns in genes simply means that there is no direct correlation in physical distances between the positions of base-pair coding triplets in a gene and the positions of amino acids in the polypeptide specified by that gene (Figure 14.10*b*).

The first strong evidence for colinearity between a gene and its polypeptide product resulted from stud-

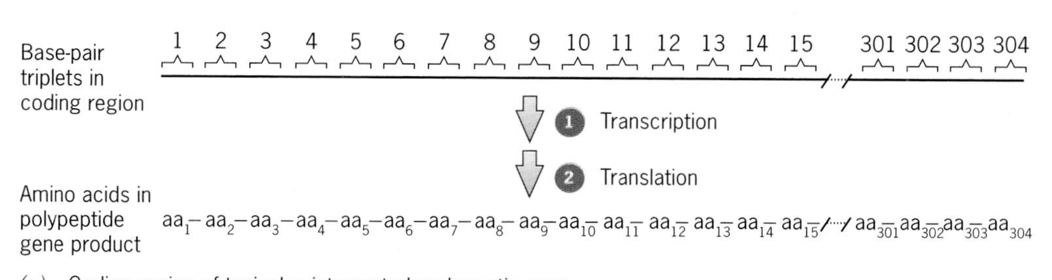

(*a*) Coding region of typical uninterrupted prokaryotic gene.

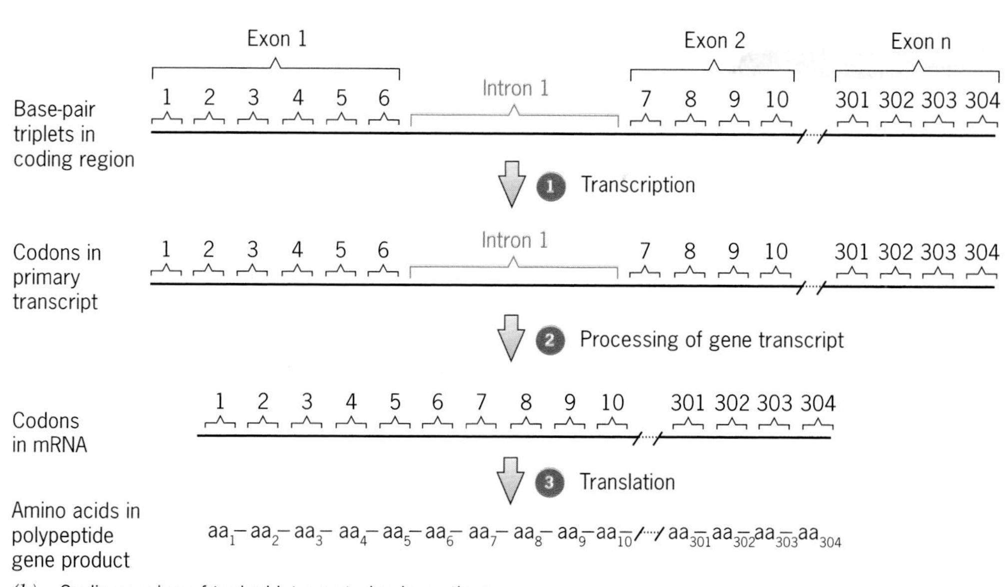

(*b*) Coding region of typical interrupted eukaryotic gene.

Figure 14.10 Colinearity between the coding regions of genes and their polypeptide products.

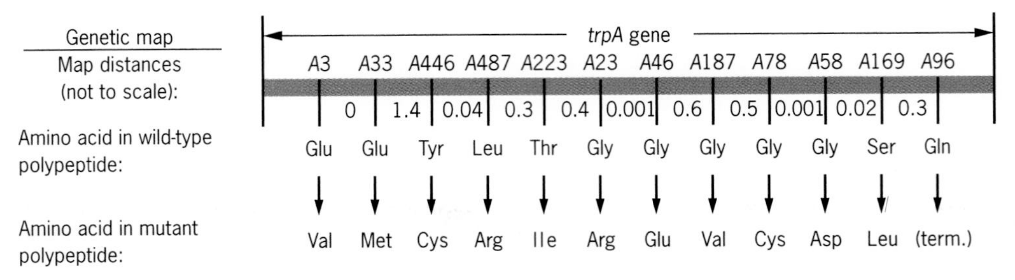

Figure 14.11 Colinearity between the *E. coli trpA* gene and its polypeptide product, the α polypeptide of tryptophan synthetase. The map positions of mutations in the *trpA* gene are shown at the top, and the locations of the amino acid substitutions produced by these mutations are shown below the map.

ies of Charles Yanofsky and colleagues on the *E. coli* gene encoding the α subunit of the enzyme tryptophan synthetase. Tryptophan synthetase catalyzes the final step in the biosynthesis of the amino acid tryptophan. As mentioned earlier, this enzyme contains two α polypeptides encoded by the *trpA* gene and two β polypeptides encoded by the *trpB* gene. Yanofsky and coworkers performed a detailed genetic analysis of mutations in the *trpA* gene and correlated the genetic data with biochemical data on the sequences of the wild-type and mutant tryptophan synthetase a polypeptides. They demonstrated that there was a direct correlation between the map positions of mutations in the *trpA* gene and the positions of the resultant amino acid substitutions in the tryptophan synthetase α polypeptide (Figure 14.11).

About the same time, Sydney Brenner and associates demonstrated a similar colinearity between the positions of mutations in the gene of bacteriophage T4 that encodes the major structural protein of the phage head and the positions in the polypeptide affected by these mutations. Brenner and colleagues studied *amber* (UAG chain-termination) mutations and demonstrated a direct correlation between the length of the polypeptide fragment produced and the position of the mutation within the gene.

In the yeast *S. cerevisiae*, Fred Sherman and colleagues demonstrated an uninterrupted colinear relationship between the map positions of mutations in the *CYCI* gene and the amino acid substitutions in the mutant forms of iso-1-cytochrome c, the polypeptide specified by the *CYC1* gene. More recently, colinearity has been documented for many interrupted genes in eukaryotes. The only difference observed with interrupted genes is that the linear sequence of nucleotide pairs encoding a colinear polypeptide is not one continuous sequence of nucleotide pairs. Instead, noncoding sequences (introns) intervene between coding sequences (exons).

Definitive evidence for colinearity has been provided by direct comparisons of the nucleotide sequences of genes and the amino acid sequences of their polypeptide products. One of the first cases where the amino acid sequence of a polypeptide and the nucleotide sequence of the gene encoding it were both determined experimentally involved the coat protein of bacteriophage MS2 and the gene that encodes it. This small virus has an RNA genome that encodes only four proteins, one being the coat protein that encapsulates the RNA. When the genetic code was used to compare the nucleotide sequence of the coat protein gene with the amino acid sequence of the coat polypeptide, the sequences exhibited perfect colinearity (Figure 14.12). Since then, similar results have established colinearity between many genes and their protein products in organisms ranging from viruses to humans. Thus colinearity between the coding regions of genes and the amino acid sequences of their polypeptide products is a universal or nearly universal feature of gene-protein relationships.

Key Points: **The concept of the gene has evolved from a bead on a string, not divisible by recombination or mutation, to a sequence of nucleotide pairs in DNA encoding one polypeptide chain. The unit of genetic material not divisible by recombination or mutation is the single nucleotide pair.**

A GENETIC DEFINITION OF THE GENE

With the emergence of the one gene–one polypeptide concept, scientists could define the gene biochemically, but they had no genetic tool to use in determining whether two mutations were in the same or different genes. This deficiency was resolved when Edward Lewis developed the complementation test for func-

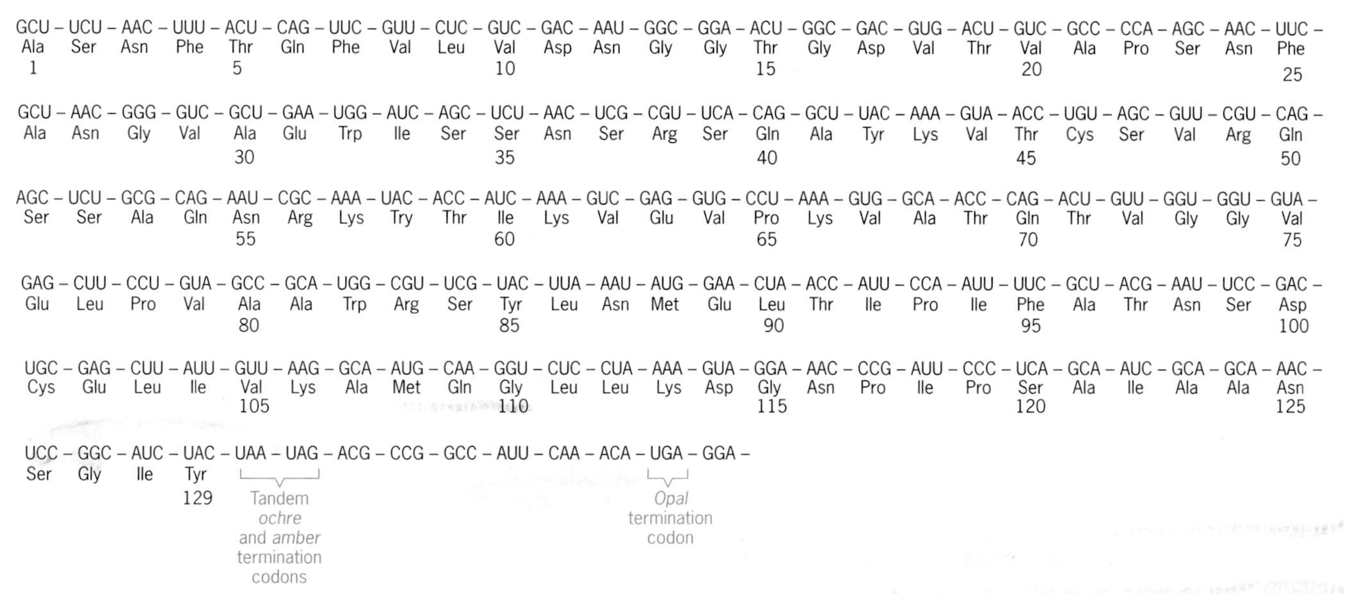

Figure 14.12 Colinearity between the nucleotide sequence of the bacteriophage MS2 coat protein gene and the amino acid sequence of the coat polypeptide that it encodes. Note that the amino acid sequence of this protein is precisely that predicted from the nucleotide sequence based on the genetic code. In addition, note that all three termination codons are present between the coat gene and the gene downstream from it on the MS2 chromosome.

tional allelism in 1942. Lewis was studying the *Star-asteroid* (small, rough eyes) locus in *Drosophila* and observed that flies of genotype $S\ ast^+/S^+\ ast$ had a more extreme mutant eye phenotype than flies of genotype $S\ ast/S^+\ ast^+$. His results were complicated by the partial dominance of the *Star* (*S*) mutation. Lewis subsequently performed similar experiments with two eye-color mutations, *white* (*w*) and *apricot* (*apr*), and obtained results that were easier to interpret.

The Complementation Test as an Operational Definition of the Gene

Fruit flies that are homozygous for the X-linked mutations *apr* (now w^a) and *w* have apricot-colored eyes and white eyes, respectively, in contrast to the red eyes of wild-type *Drosophila*.

Lewis reported that heterozygous *apr*/*w* females had light apricot-colored eyes and produced rare red-eyed progeny carrying recombinant apr^+w^+ chromosomes. In addition, Lewis was able to identify progeny flies that carried X chromosomes with the reciprocal *apr w* recombinant genoypte. The observed frequency of recombination between the *apr* and *w* mutations was 0.03 percent. Clearly, the *apr* and *w* mutations were separable by recombination, and *apr* and *w* were not alleles according to the pre–1940 concept of the gene. The two mutations appeared to be in the same unit of function but in two different units of structure. When Lewis produced flies of genotype $apr\ w/apr^+\ w^+$, they had red eyes just like those of wild-type flies. When he constructed flies of genotype $apr\ w^+/apr^+\ w$, they had light apricot-colored eyes. Both genotypes contain the same mutant and wild-type genetic information, but in different arrangements. The presence of different phenotypes in organisms that contain the same genetic markers, but with the markers present in different arrangements, is called a **position effect**, and the type of position effect observed by Lewis is referred to as a ***cis-trans*** **position effect**.

Before we analyze Lewis's results in greater detail, we need to define some terms. A double heterozygote, which carries two mutations and their wild-type alleles, that is, m_1 and m_1^+ plus m_2 and m_2^+, can exist in either of two arrangements (Figure 14.13). When the two mutations are on the same chromosome, the arrangement is called the **coupling** or ***cis*** **configuration**; a heterozygote with this genotype is called a ***cis*** **heterozygote** (Figure 14.13*a*). When the two mutations are on

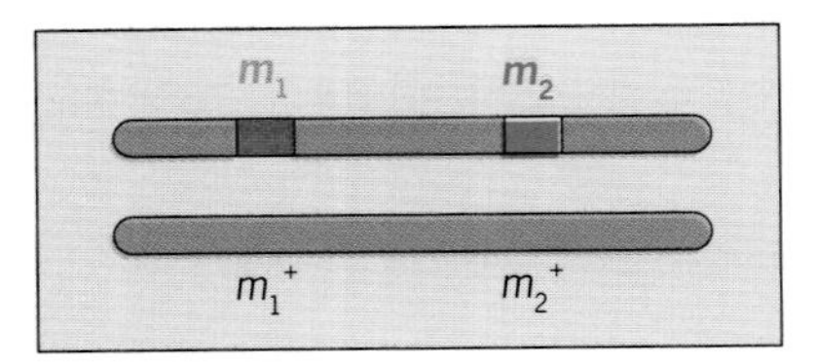

(a) *cis* heterozygote.

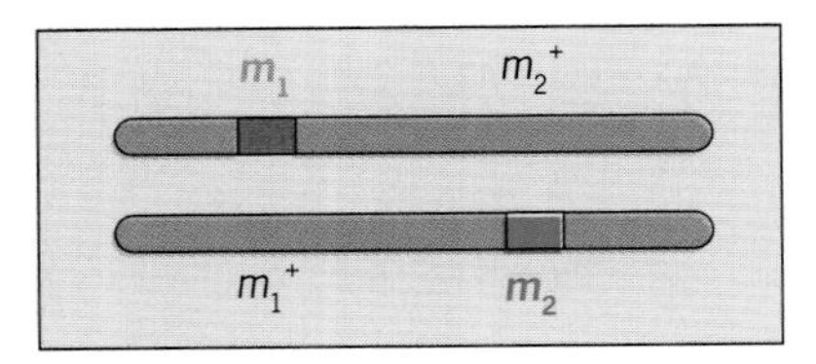

(b) *trans* heterozygote.

Figure 14.13 The arrangement of genetic markers in *cis* and *trans* heterozygotes.

different chromosomes, the arrangement is called the **repulsion** or ***trans* configuration**. An organism with this genotype is a ***trans* heterozygote** (Figure 14.13*b*).

Recall that *apr w*/*apr*$^+$ *w*$^+$ *cis* heterozygotes have wild-type red eyes, whereas *apr w*$^+$/*apr*$^+$ *w trans* heterozygotes have light apricot eyes. This is precisely the result that would be expected if *apr* and *w* are mutations at different sites in the same gene, the unit of genetic information encoding a single polypeptide (Figure 14.14). Thus *apr* and *w* are considered alleles of the same gene, and *apr* is now designated *w*^{a} in recognition of this relationship. If *apr* and *w* had been mutations in two different units of function, two different genes, both the *cis* and the *trans* heterozygotes should have expressed the wild-type phenotype, namely, red eyes. In the *trans* heterozygote, the *apr*$^+$ gene product would be produced by the *apr*$^+$ gene on the chromosome carrying the *w* mutation, and the *w*$^+$ gene product would be specified by the *w*$^+$ gene on the chromosome harboring the *apr* mutation .

Lewis's discovery of the *cis-trans* position effects with *Star-asteroid* and *w*^{a}-*w* led to the development of the **complementation test** or ***trans* test** for functional allelism. The complementation test allows geneticists to determine whether mutations that produce the same or similar phenotypes are in the same gene or in different genes. They must test mutations pairwise by determining the phenotypes of *trans* heterozygotes. That is, they must construct *trans* heterozygotes with each pair of mutations to be analyzed and determine whether these heterozygotes have mutant or wild-type phenotypes.

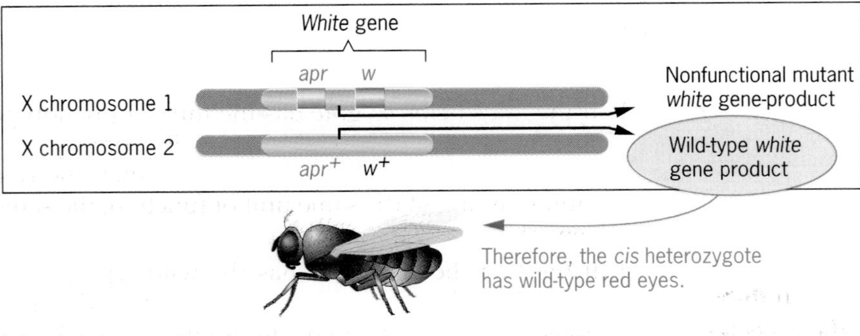

(a) *cis* heterozygote.

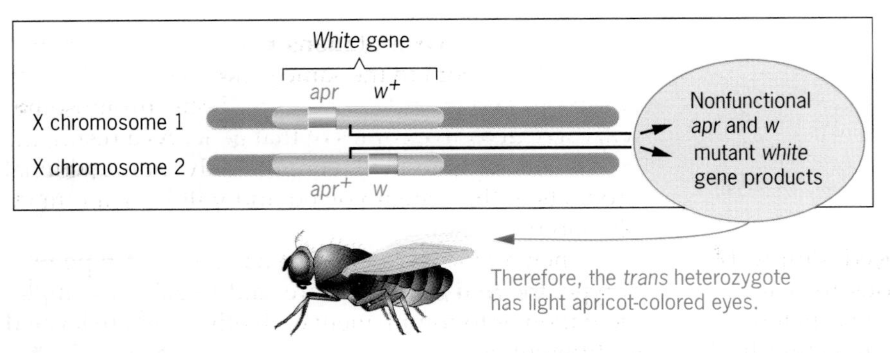

(b) *trans* heterozygote.

Figure 14.14 The *cis-trans* position effect observed by Edward Lewis with the *apr* and *w* mutations of *Drosophila*.

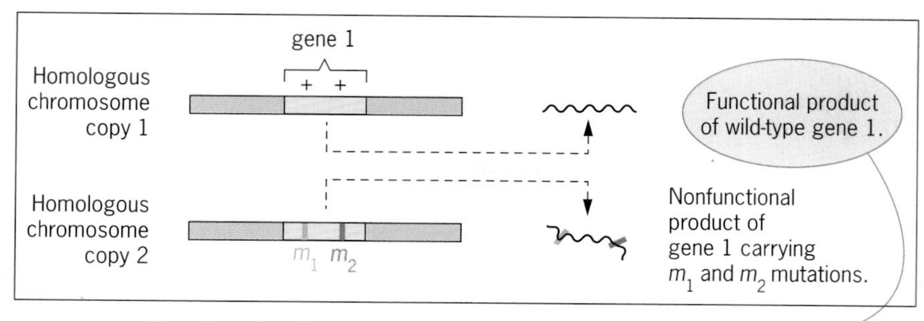

(*a*) *cis* heterozygote: mutations in one gene.

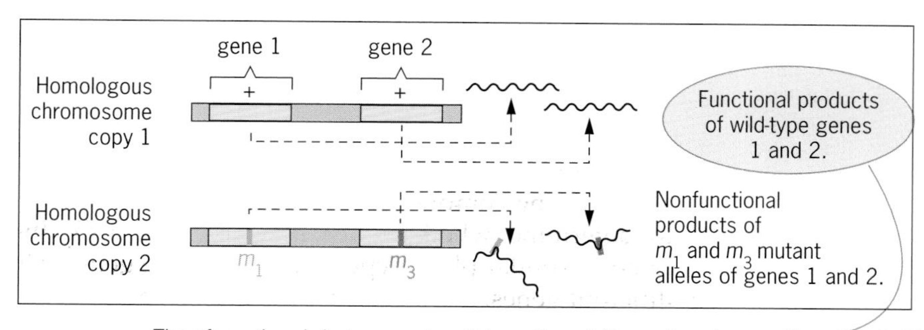

(*b*) *cis* heterozygote: mutations in two different genes.

Figure 14.15 The *cis* test. The *cis* heterozygote should have the wild-type phenotype whether the mutations are in the same gene (*a*) or in two different genes (*b*).

Ideally, the complementation or *trans* test should be done in conjunction with the ***cis* test**—a control that is often omitted. *Cis* tests are performed by constructing *cis* heterozygotes for each pair of mutations to be analyzed and determining whether they have mutant or wild-type phenotypes. Together, the complementation or *trans* test and the *cis* test are referred to as the ***cis-trans* test**. Each *cis* heterozygote, which contains one wild-type chromosome, should have the wild-type phenotype whether the mutations are in the same gene or in two different genes (Figure 14.15). Indeed, the *cis* heterozygote must have the wild-type phenotype for the results of the *trans* test to be valid. If the *cis* heterozygote has the mutant phenotype, the *trans* test cannot be used to determine whether the two mutations are in the same gene. Thus the *trans* test cannot be used with dominant mutations. Because *cis* heterozygotes contain one chromosome with wild-type copies of all relevant genes, these genes should specify functional products and a wild-type phenotype.

Whether two mutations are in the same gene or two different genes is determined by the results of the complementation or *trans* test. With diploid organisms, the *trans* heterozygote is produced simply by crossing organisms that are homozygous for each of the mutations of interest. With viruses, *trans* heterozygotes are produced by simultaneously infecting host cells with two different mutants. Regardless of how the two mutations are placed in a common protoplasm in the *trans* configuration, the results of the *trans* or complementation test provide the same information.

1. If the *trans* heterozygote has the mutant phenotype (the phenotype of organisms or cells homozygous for either one of the two mutations), then the two mutations are in the same unit of function, the same gene (Figure 14.16*a*).
2. If the *trans* heterozygote has the wild-type phenotype, then the two mutations are in two different units of function, two different genes (Figure 14.16*b*).

When the two mutations present in a *trans* heterozygote are both in the same gene, as shown for mutations m_1 and m_2 in Figure 14.16*a*, both chromosomes will carry defective copies of that gene. As a result, the *trans* heterozygote will contain only nonfunctional products of the gene involved and will have a mutant phenotype.

When a *trans* heterozygote has the wild-type phenotype, the two mutations are said to exhibit complementation or to complement each other and are located in different genes. In the example illustrated in Figure

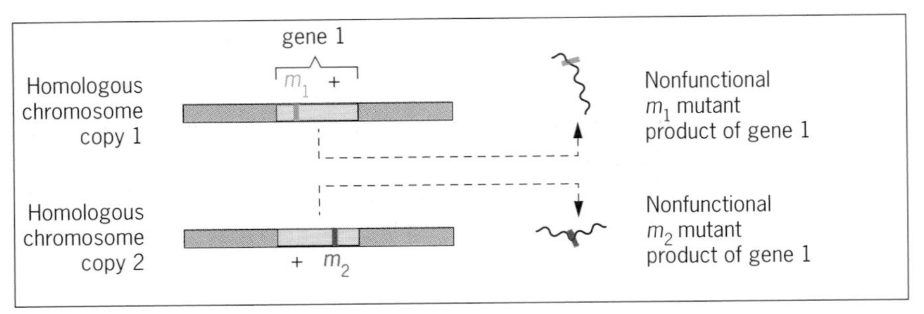

(*a*) *trans* heterozygote: mutations in one gene.

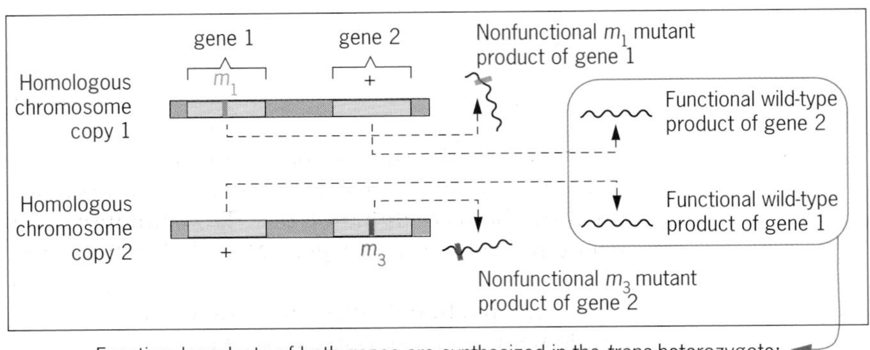

(*b*) *trans* heterozygote: mutations in two different genes.

Figure 14.16 The *trans* test. The *trans* heterozygote should have (*a*) the mutant phenotype if the two mutations are in the same gene, and (*b*) the wild-type phenotype if the mutations are in two different genes.

14.16*b*, the chromosome carrying mutation m_1 in gene 1 has a wild-type copy of gene 2, which specifies functional gene 2 product, and the chromosome carrying mutation m_3 in gene 2 has a wild-type copy of gene 1, which encodes functional gene 1 product. Thus the *trans* heterozygote shown contains functional products of both genes and has a wild-type phenotype.

Only the complementation test, or the *trans* part of the *cis-trans* test, is included in most genetic analyses. Constructing a chromosome that carries both mutations for the *cis* test is often difficult, especially with eukaryotes. Moreover, the *cis* heterozygotes almost always have wild-type phenotypes if the mutations being analyzed are recessive. Thus, in most instances, the results of complementation or *trans* tests can be interpreted correctly without carrying out the laborious *cis* tests.

Seymour Benzer introduced the term **cistron** to refer to the unit of function operationally defined by the *cis-trans* test. However, today, most geneticists consider the terms gene and cistron to be synonyms. Thus we will use gene, rather than cistron, throughout this text.

The gene is operationally defined as the unit of function by the complementation or *trans* test, which is used to determine whether mutations are in the same gene or different genes. The complementation test is one of the three basic tools of genetics. The other two genetic tools are recombination (Chapter 7) and mutation (Chapter 13).

The information provided by complementation tests is totally distinct from that obtained from recombination analyses.

The results of complementation tests indicate whether mutations are allelic, whereas the results of recombination analyses indicate whether mutations are linked and, if so, provide estimates of how far apart they are on a chromosome. Nevertheless, students sometimes confuse complementation and recombination. Thus we will contrast complementation and recombination by illustrating both phenomena with the same three mutations in bacteriophage T4. Phage T4 is very similar to phage T2, which we discussed in Chapter 10. Because of the simple structure of the virus and the direct relationship between specific gene products and phenotypes, the phage T4 system provides an excellent mnemonic visualization of the difference between complementation and recombination.

We discussed the *amber* mutations of bacteriophage T4 in Chapter 12 (see Figure 12.23) and the pathway of morphogenesis of phage T4 in Chapter 13 (see Figure 13.11) *Amber* mutations produce translation–termination triplets within the coding regions of genes. As a result, the products of the mutant genes are truncated polypeptides, which are almost always totally nonfunctional. Therefore, complementation tests performed with *amber* mutations are usually unambiguous. When *amber* mutations occur in essential genes, the mutant phenotype is lethality—that is, no progeny are produced when a restrictive host cell is infected; and the wild-type phenotype is a normal yield (about 300 phage per cell) of progeny phage in each infected restrictive host cell. In Chapter 13, we called such mutations conditional lethals and emphasized their utility in genetic analysis. With conditional lethals, the distinction between the mutant and wild-type phenotypes is maximal: lethality versus normal growth.

Two of the three *amber* mutations that we will consider (*am*B17 and *am*H32) are in gene *23,* which encodes the major structural protein of the phage head; the other mutation (*am*E18) is in gene *18,* which specifies the major structural protein of the phage tail. We can see from Figure 14.17 why complementation occurs between mutations *am*B17 (head gene) and *am*E18 (tail gene) and why complementation does not occur between mutations *am*B17 and *am*H32 (both in head gene). Complementation is the result of the interaction of the gene *products* specified by chromosomes carrying two different mutations when they are present in a common protoplasm. Complementation does not depend on recombination of the two chromosomes or involve any direct interaction between the chromosomes. *Complementation, or the lack of it, is assessed by the phenotype (wild-type or mutant) of each* trans *heterozygote.*

Figure 14.18 illustrates the occurrence of recombination between the *amber* mutations used to illustrate complementation in Figure 14.17. Although we haven't discussed viral genetics in detail (Chapter 15), we have covered the essential concepts of recombination (Chapter 7). Recombination of phage genes occurs by a process analogous to crossing over in eukaryotes, with linkage distances measured in map units, just as in eukaryotes. Recombination frequencies are measured by infecting permissive host cells with two mutants so that the mutant chromosomes can replicate and participate in crossing over. Then, the progeny are screened for wild-type recombinants by plating them on lawns of restrictive host cells (*E. coli* cells in which only the wild-type phage can grow). In the example shown in Figure 14.18, recombination is observed in both crosses: (1) *am*B17 × *am*E18, mutations in two different genes, and (2) *am*B17 × *am*H32, mutations in the same gene. The only difference is that more recombinants are produced in cross 1, which involves two *amber* mutations that are relatively far apart on the phage T4 chromosome, than in cross 2, which involves two mutations located near one another in the same gene. *Recombination involves direct interactions between the chromosomes carrying the mutations, the actual breakage of chromosomes, and reunion of parts to produce wild-type and double-mutant chromosomes.*

Complementation should occur in every *trans* heterozygote containing mutations in two different genes. Recombination is detected by examining the genotypes of the progeny of heterozygotes, not the phenotypes of the *trans* heterozygotes themselves. Moreover, a *trans* heterozygote will never produce more than 25 percent gametes (or progeny for haploids) with wild-type chromosomes. If the mutations are closely linked, like the *amber* mutations shown in Figure 14.18*b*, the frequency of wild-type recombinant chromosomes will be much lower than 25 percent.

Structural allelism is the occurrence of two or more different mutations at the same site and is determined by the recombination test. Two mutations that do not recombine are structurally allelic; the mutations either occur at the same site or overlap a common site. Functional allelism is determined by the complementation test as just described; two mutations that do not complement are in the same unit of function, the same gene. Mutations that are both structurally and functionally allelic are called **homoalleles**; they do not complement or recombine with each other. Mutant homoalleles have defects at the same site or overlap a common site in the same gene. Mutations that are functionally allelic, but structurally nonallelic, are called **heteroalleles**; they recombine with each other but do not complement one another. Mutant heteroalleles occur at different sites but within the same gene.

Intragenic Complementation

The results of complementation tests are usually unambiguous when mutations that result in the synthesis of no gene product, partial gene products, or totally defective gene products are used—for example, deletions of segments of genes, frameshift mutations, or polypeptide chain-terminating mutations. Of course, the mutations must be recessive. When mutations causing amino acid substitutions are used, the results of complementation tests are sometimes ambiguous because of the occurrence of a phenomenon called **intragenic complementation**.

The functional forms of some proteins are dimers or higher multimers consisting of two or more polypeptides. These polypeptides may be either homologous, the products of a single gene, or nonhomologous, the products of two or more distinct genes.

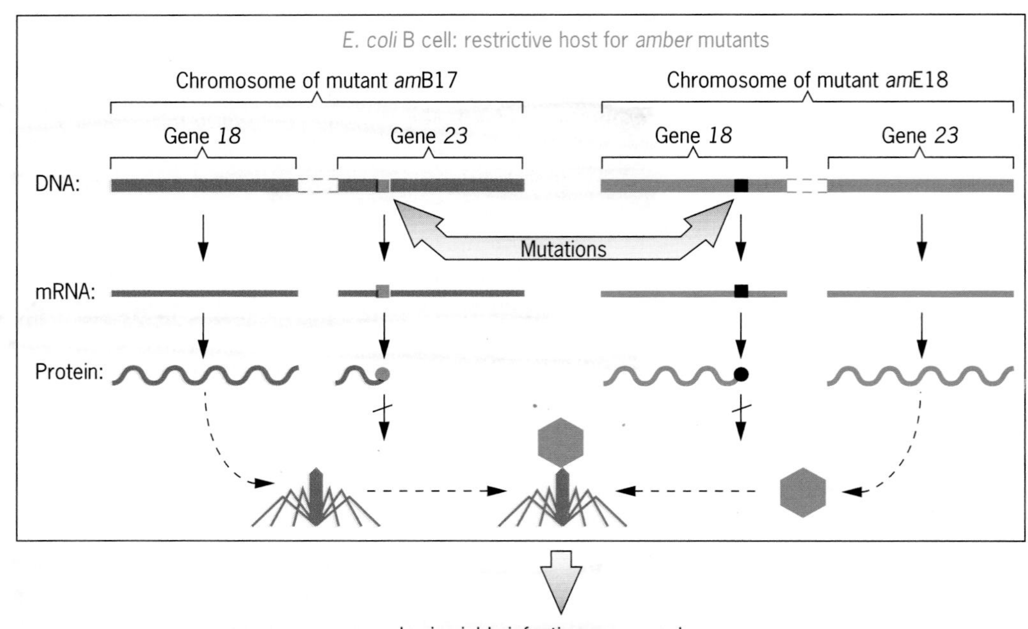

Lysis yields infective progeny phage; therefore, the *trans* heterozygote has the wild-type phenotype.

(*a*) Complementation between mutations *am*B17 and *am*E18.

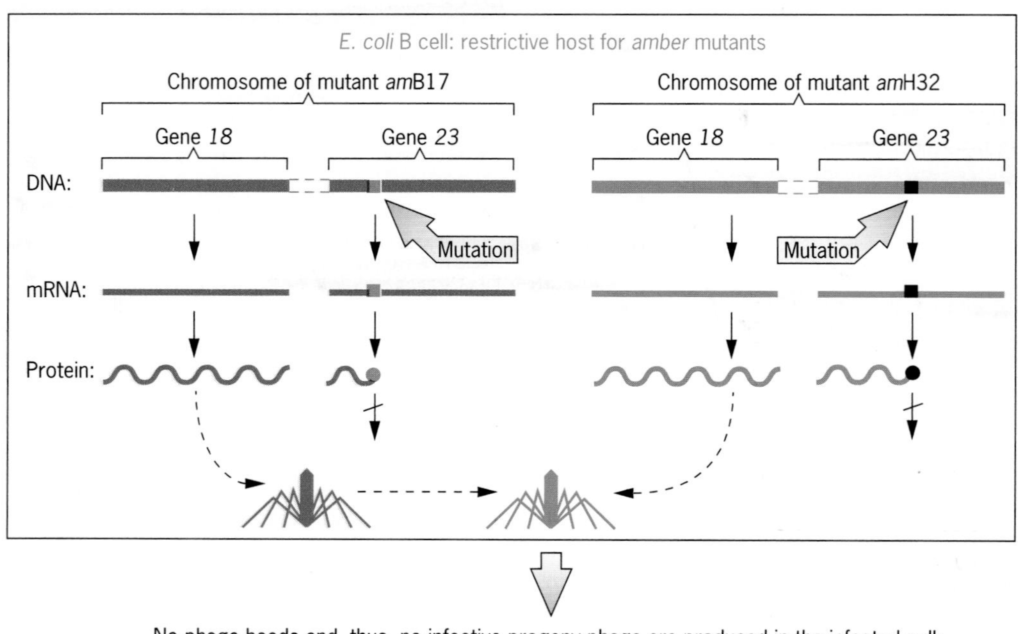

No phage heads and, thus, no infective progeny phage are produced in the infected cell; therefore, the *trans* heterozygote has the mutant phenotype.

(*b*) Lack of complementation between mutations *am*B17 and *am*H32.

Figure 14.17 Complementation and noncomplementation in *trans* heterozygotes. (*a*) Complementation between mutation *am*B17 in gene *23,* which encodes the major structural protein of the phage T4 head, and mutation *am*E18 in gene *18,* which encodes the major structural protein of the phage tail. Both heads and tails are synthesized in the cell, with the result that infective progeny phage are produced. (*b*) When the *trans* heterozygote contains two mutations (*am*B17 and *am*H32) in gene *23,* no heads are produced, and no infective progeny phage can be assembled. Compare with Figure 14.18.

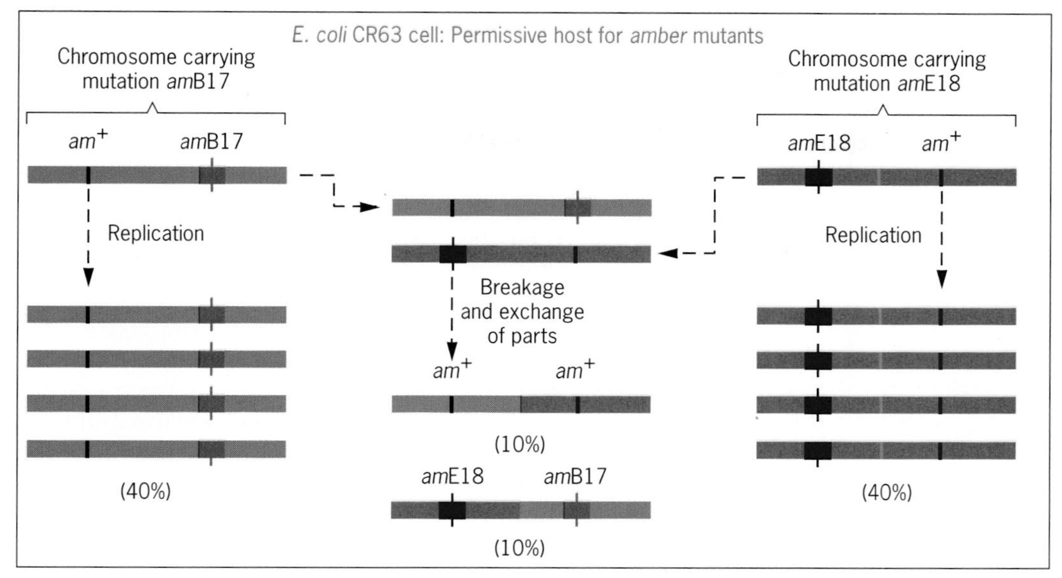

Lysis yields progeny phage of four genotypes:
Parental genotypes : ~ 40% *am*B17 and ~ 40% *am*E18;
Recombinant genotypes: ~ 10% wild-type (am^+) and ~ 10% double mutant (*am*E18-*am*B17).

(*a*) Recombination between phage T4 chromosomes carrying mutations *am*B17 and *am*E18.

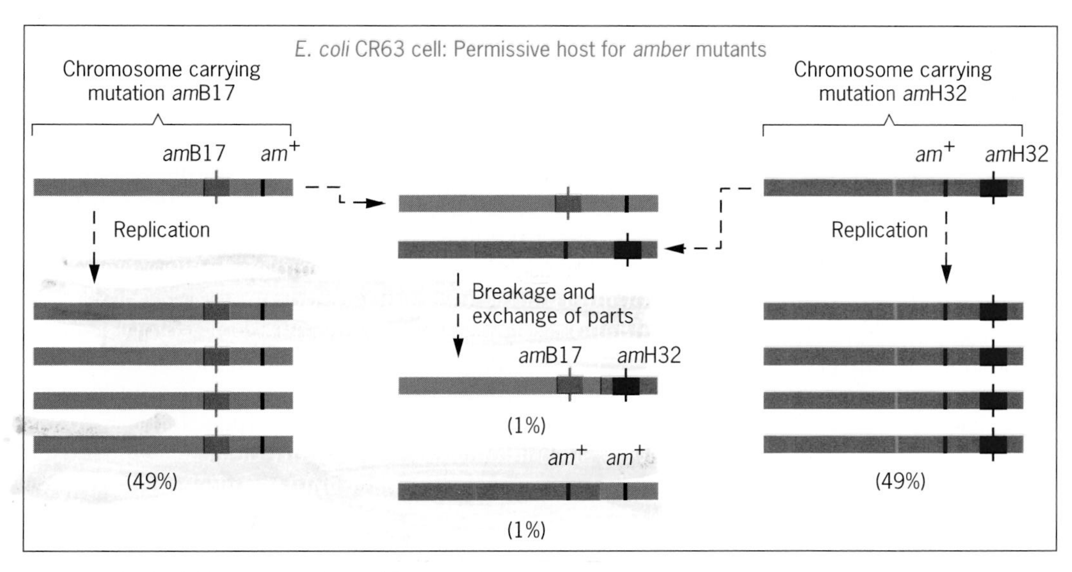

Lysis yields progeny phage of four genotypes:
Parental genotypes : ~ 49% *am*B17 and ~ 49% *am*H32;
Recombinant genotypes: ~ 1% wild-type (am^+) and ~ 1% double mutant (*am*B17-*am*H32).

(*b*) Recombination between phage T4 chromosomes carrying mutations *am*B17 and *am*H32.

Figure 14.18 Recombination between (*a*) the complementing mutations *am*B17 (gene *23*) and *am*E18 (gene *18*), and (*b*) the noncomplementing mutations *am*B17 and *am*H32 (both in gene *23*). Recombination occurs in both cases; however, fewer recombinants are produced in cells infected with *am*B17 and *am*H32 because the two mutations are located closer together on the phage T4 chromosome. Compare with Fig. 14.17.

When the active form of the protein contains two or more homologous polypeptides (it may or may not also contain nonhomologous polypeptides), intragenic complementation may occur. Intergenic complementation (discussed in the preceding section) and intragenic complementation (discribed below) are two distinct phenomena.

Let us consider an enzyme that functions as a homodimer, that is, a protein containing two copies of a specific gene product (Figure 14.19). In organisms

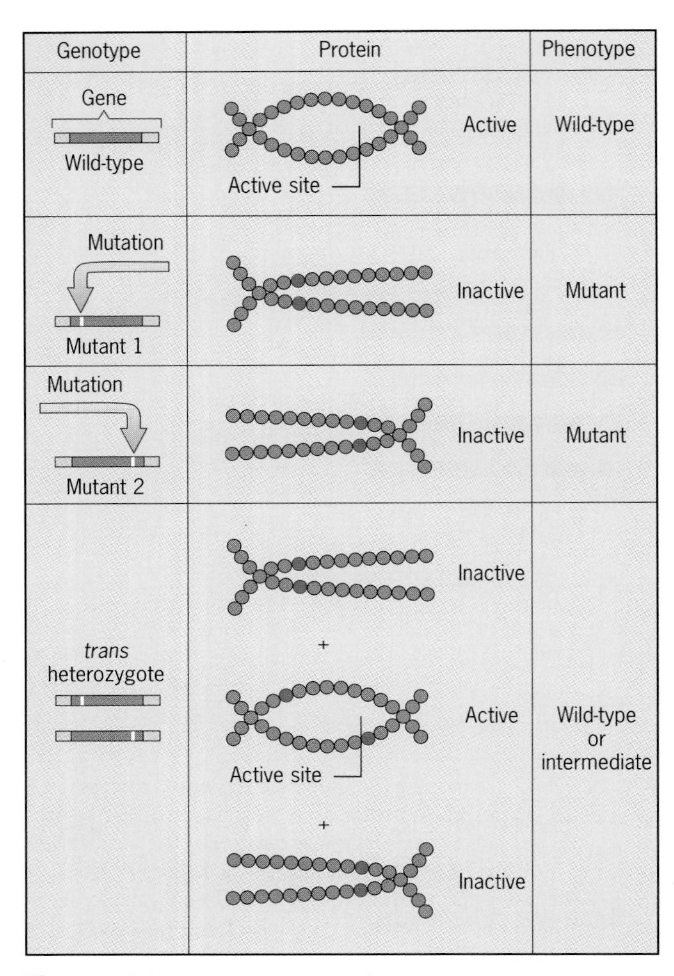

Figure 14.19 *Intra*genic complementation sometimes occurs when the active form of an enzyme or structural protein is a multimer that contains at least two copies of any one gene product. Here, the functional form of the enzyme is a dimer composed of two polypeptides encoded by one gene. The amino acids altered by the mutations are shown as red circles in the polypeptide chains.

that are homozygous for the wild-type allele of the gene, all the protein dimers will contain identical wild-type polypeptides. Similarly, organisms that are homozygous for any mutation in the gene will contain dimers with two mutant polypeptides. An organism that is heterozygous for two different mutations in the gene will produce some dimers that contain the two different mutant polypeptides. We call these heterodimers. Such heterodimers may have partial or complete (wild-type) function. If they do, intragenic complementation has occurred, and the *trans* heterozygote has a wild-type phenotype or a phenotype intermediate between mutant and wild-type (Figure 14.19, bottom). In the case of noncomplementing mutations in a gene encoding a multimeric protein, the heteromultimers are nonfunctional, just like the mutant homomultimers (protein multimers composed of two or more identical mutant polypeptides).

In several known cases of intragenic complementation, the active form of the protein in the heterozygote has been purified and shown to be a heterodimer or heterotetramer containing two different mutant polypeptides. Why such heteromultimers should be active when the two corresponding homomultimers are inactive is not clear. Apparently, the wild-type sequence of amino acids in the nonmutant segment of one mutant polypeptide somehow compensates for the mutant segment of the polypeptide encoded by the second mutant allele, and vice versa (Figure 14.19). However, most proteins have complex three-dimensional structures, and until the exact structures of a wild-type homomultimer, two mutant homomultimers, and an active heteromultimer composed of the two mutant polypeptides have been determined, the molecular basis of intragenic complementation will continue to be subject to speculation.

Limitations on the Use of the Complementation Test

The complementation test has been very useful in operationally delimiting genes. Usually, two or more mutations that produce the same phenotype can be assigned to one or more genes based on the results of complementation tests. However, in some cases, the results of complementation tests cannot be used to delimit genes. As previously mentioned, complementation tests are not informative in studies of dominant or codominant mutations or in cases where intragenic complementation occurs. In addition, complementation tests are sometimes uninformative because of epistatic interactions between the mutant gene products. If the *cis* test is done, such interactions are readily detected because the *cis* heterozygotes will have mutant phenotypes rather than the required wild-type phenotype.

Another limitation of the complementation test is encountered in working with so-called polar mutations. A **polar mutation** is a mutation that not only results in a defective product of the gene in which it is located, but also interferes with the expression of one or more adjacent genes. The adjacent genes are always located on one side of the gene carrying the mutation (thus the term *polar mutation*). Such polar mutations are frequently observed in prokaryotes in coordinately regulated sets of genes called operons (Chapter 21). They usually are mutations resulting in polypeptide chain-termination signals (nucleotide-pair triplets yielding UAA, UAG, and UGA codons in mRNA) within genes. These polar mutations interfere with the expression of genes located downstream (relative to the direction of transcription) of the mutant gene. As

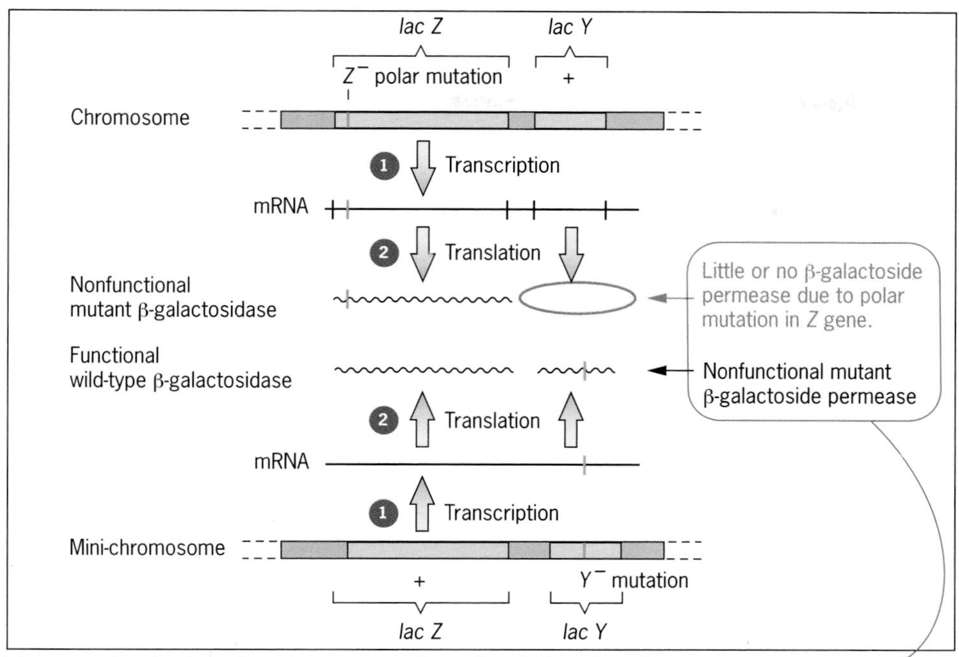

Little or no functional β-galactoside permease is produced. Therefore, the *trans* heterozygote has a mutant phenotype — that is, it is unable to utilize lactose as an energy source.

Figure 14.20 Lack of complementation between a polar mutation and a mutation in a downstream gene in the same transcription unit. The ability of *E. coli* cells to utilize lactose as an energy source depends on the products of two co-transcribed genes: *lacY*, which encodes β-galactoside permease, and *lacZ*, which encodes β-galactosidase. Transcription of the *lacY* and *lacZ* genes produces a multigenic mRNA, which is translated to provide the two proteins. A translation–termination mutation near the translation–start site in *lacZ* has a polar effect on translation of the *lacY* gene, reducing its translation efficiency to 1–2% of the normal level. Thus the polar mutation in *lacZ* will not complement a null mutation in *lacY*, even though the two point mutations are in two different genes.

a result, polar mutations fail to complement mutations in genes subject to the polar effect (Figure 14.20). Thus the results of complementation tests performed with polar mutations are often ambiguous.

Key Points: **The complementation or *trans* test provides an operational definition of the gene; it is used to determine whether mutations are in the same gene or different genes. Intragenic complementation may occur when a protein is a multimer containing at least two copies of one gene product.**

COMPLEX GENE–PROTEIN RELATIONSHIPS

Most prokaryotic genes consist of continuous sequences of nucleotide pairs, which specify colinear sequences of amino acids in the polypeptide gene products. As we discussed in Chapter 11, most eukaryotic genes are split into coding sequences (exons) and noncoding sequences (introns). However, because the spliceosomes usually excise introns from primary transcripts by *cis*-splicing mechanisms (processes that join exons from the same RNA molecule), the presence of introns in genes does not invalidate the complementation test as an operational definition of the gene. Nevertheless, in some cases, transcripts of split genes may undergo several different types of splicing, making the relationships between genes and proteins more complex than the usual one gene–one polypeptide. In other cases, expressed genes are assembled from "gene pieces" during the development of the specialized cells in which they are expressed.

Alternate Pathways of Transcript Splicing: Protein Isoforms

Many interrupted eukaryotic genes, such as the mammalian hemoglobin genes and the chicken ovalbumin

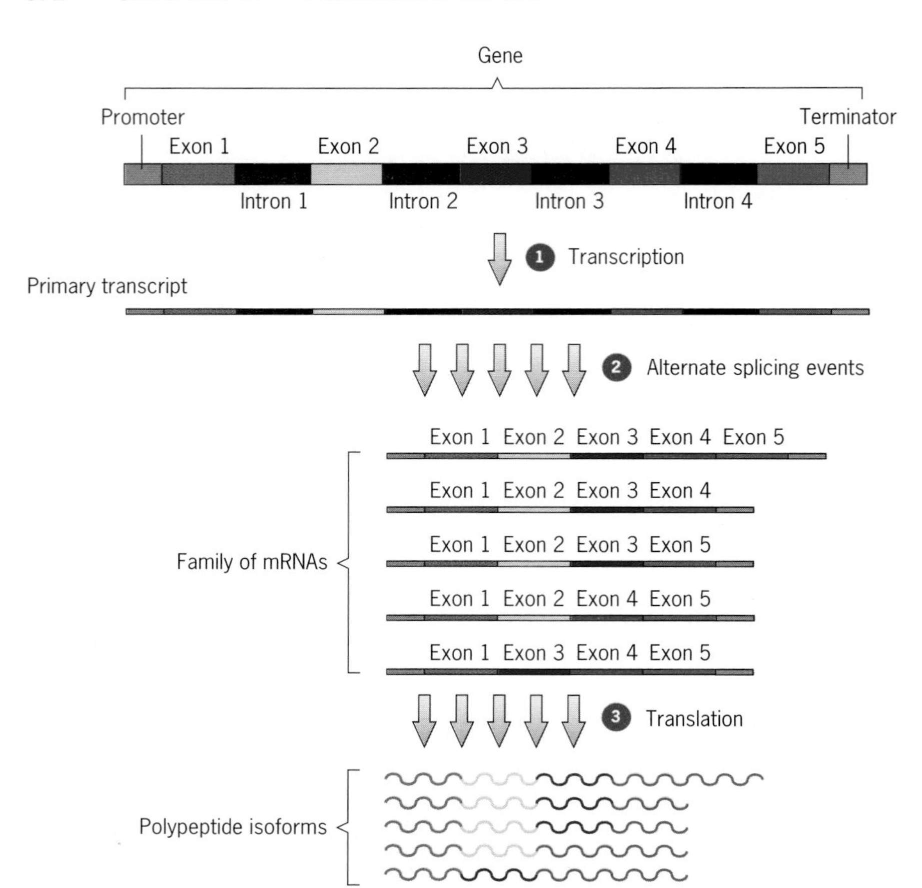

Figure 14.21 A single gene may produce a family of closely related polypeptides by using alternate pathways of exon splicing.

and 1α2 collagen genes discussed in Chapter 11, each encode a single polypeptide chain with a specific function. In these cases, the mRNA produced from a given gene contains all the exons of the gene joined together in the same order as they occur in the gene. However, the transcripts of some interrupted genes undergo alternate pathways of transcript splicing. That is, different exons of a gene may be joined to produce a related set of mRNAs encoding a small family of closely related polypeptides called **protein isoforms** (Figure 14.21). The alternate splicing pathways are often tissue-specific, producing related proteins that carry out similar, but not necessarily identical, functions in different types of cells. The mammalian tropomyosin genes provide striking examples of genes which each produce a family of protein isoforms. Tropomyosins are proteins involved in the regulation of muscle contraction in animals. Because the various organs of an animal contain different muscle types, all of which need to be regulated, the availability of a family of related tropomyosins might be beneficial. In any case, one mouse tropomyosin gene is known to produce at least 10 different tropomyosin polypeptides as a result of alternate pathways of transcript splicing. Genes of this type obviously do not fit the one gene–one polypeptide concept very well. For such genes, where alternate splicing pathways give rise to two or more different polypeptides, the gene can be defined as a DNA sequence that is a single unit of transcription and encodes a set of protein isoforms.

Assembly of Genes During Development: Human Antibody Chains

Genetic information is not always organized into genes of the type described in the preceding sections of this chapter. In rare cases, genes are assembled from a storehouse of **gene segments** during the development of an organism. The immune system of vertebrate animals depends on the synthesis of proteins called **antibodies** to provide protection against infections by viruses, bacteria, toxins, and other foreign substances. Each antibody contains four polypeptides, two identical heavy chains, and two identical light chains. The light chains are of two types: kappa and lambda. Each antibody chain contains a variable region, which exhibits extensive diversity from antibody to antibody, and a constant region, which is largely the same in all antibodies. In germ-line chromosomes,

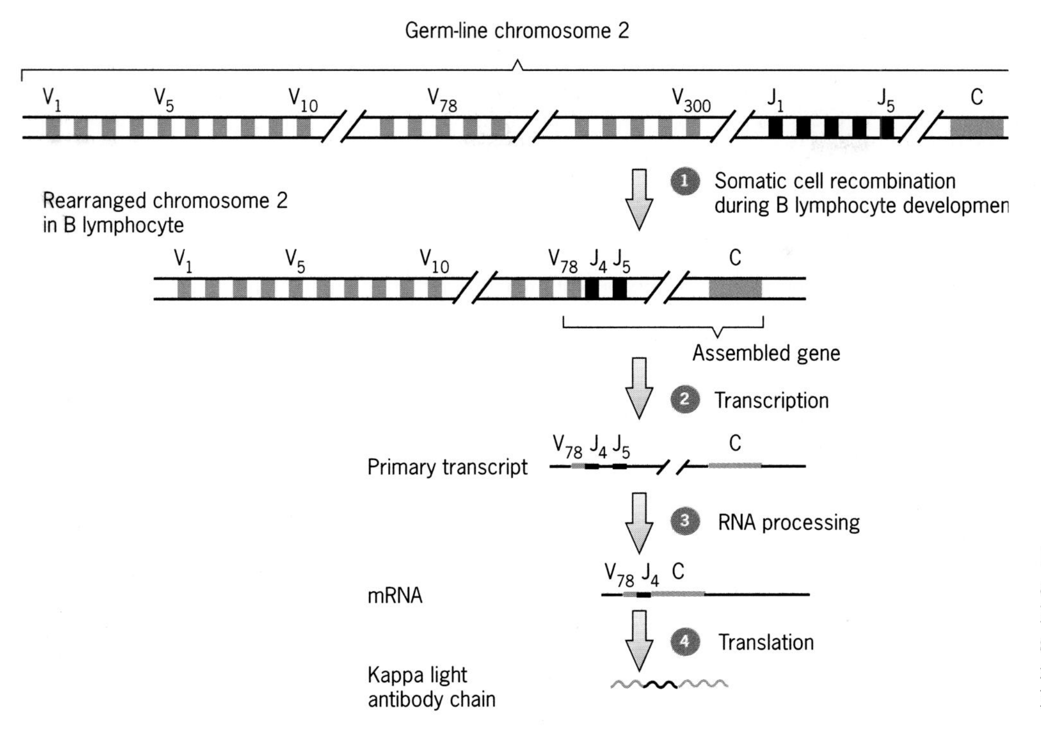

Figure 14.22 Assembly of a gene encoding an antibody kappa light chain from gene segments during the development of a B lymphocyte in humans.

the DNA sequences encoding these antibody chains are present in gene segments, and the gene segments are joined together to produce genes during the differentiation of the antibody-producing cells (B lymphocytes) from progenitor cells.

To illustrate this process of gene assembly during development, let us briefly consider the DNA sequences encoding kappa light chains in humans. (We discuss this topic in detail in Chapter 24.) A kappa light chain gene is assembled from three gene segments: V_k (**V** for **v**ariable region), J_k (**J** for **j**oining segment), and C_k (**C** for **c**onstant region), during B lymphocyte development. Together, the V_k and J_k gene segments encode the variable region of the kappa light chain, whereas the C_k gene segment encodes the constant region. No functional V_k-J_k-C_k kappa light chain gene is present in any human germ-line chromosome. Instead, human chromosome 2 contains a cluster of about 300 V_k gene segments, another cluster of five J_k gene segments, and a single C_k gene segment (Figure 14.22). During the differentiation of each B lymphocyte, recombination joins one of the V_k gene segments to one of the J_k gene segments. Any J_k segments remaining between the newly formed V_k-J_k exon and the C_k gene segment become part of an intron that is removed during the processing of the primary transcript. Similar somatic recombination events are responsible for the assembly of the genes encoding antibody heavy chains, lambda light chains, and T-lymphocyte receptor proteins (Chapter 24).

Key Points: **The transcripts of some genes undergo alternate pathways of splicing to produce mRNAs with different exons joined together. Translation of these mRNAs produces closely related polypeptides called protein isoforms. Other genes, such as those encoding antibody chains, are assembled from gene segments during development by regulated processes of somatic recombination.**

TESTING YOUR KNOWLEDGE

1. *In Drosophila, white, cherry,* and *vermilion* are all sex-linked mutations affecting eye color. All three mutations are recessive to their wild-type allele(s) for red eyes. A white-eyed female crossed with a vermilion-eyed male produces white-eyed male offspring and red-eyed (wild-type) female offspring. A white-eyed female crossed with a cherry-eyed male produces white-eyed sons and light cherry-eyed daughters. Do these results indicate whether or not any of the three mutations affecting eye color are located in the same gene? If so, which mutations?

ANSWER

The complementation test for allelism involves placing mutations pairwise in a common protoplasm in the *trans* configuration and determining whether the resulting *trans* heterozygotes have mutant or wild-type phenotypes. If the two mutations are in the same gene, both copies of the gene in the *trans* heterozygote will produce defective gene products, resulting in a mutant phenotype (see Figure 14.16*a*). However, if the two mutations are in different genes, the two mutations will complement each other, because the wild-type copies of each gene will produce functional gene products (see Figure 14.16*b*). When complementation occurs, the *trans* heterozygote will have the wild-type phenotype. Thus the complementation test allows one to determine whether any two recessive mutations are located in the same gene or in different genes.

If the *trans* heterozygote has the mutant phenotype, the two mutations are in the same gene. If the *trans* heterozygote has the wild-type phenotype, the two mutations are in two different genes. Because the mutations of interest are sex-linked, all the male progeny will have the same phenotype as the female parent. They are hemizygous, with one X chromosome obtained from their mother. In contrast, the female progeny are *trans* heterozygotes. In the cross between the white-eyed female and the vermilion-eyed male, the female progeny have red eyes, the wild-type phenotype. Thus the *white* and *vermilion* mutations are in different genes, as illustrated in the following diagram:

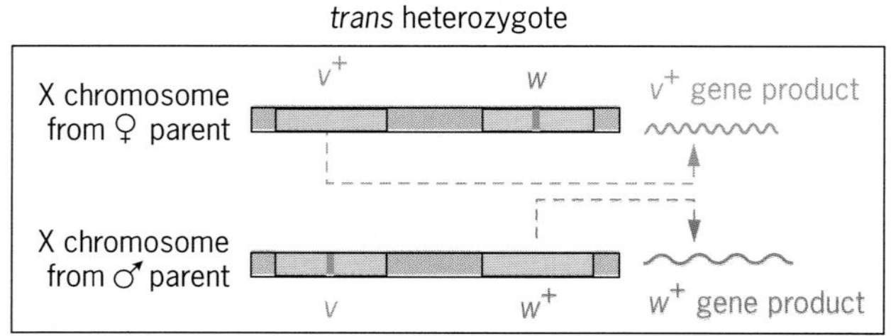

Complementation yields wild-type phenotype; both v^+ and w^+ gene products are produced in the *trans* heterozygote.

In the cross between a white-eyed female and a cherry-eyed male, the female progeny have light cherry-colored eyes (a mutant phenotype), not wild-type red eyes as in the first cross. Since the *trans* heterozygote has a mutant phenotype, the two mutations, *white* and *cherry*, are in the same gene:

trans heterozygote

X chromosome from ♀ parent — w

No active (w^+) gene product

X chromosome from ♂ parent — w^{ch}

No w^+ gene product; therefore, mutant phenotype.

2. Suppressor-sensitive (*sus*) mutants of bacteriophage ϕ29 can grow on *Bacillus subtilis* strain L15, but cannot grow (that is, are lethal) on *B. subtilis* strain 12A. Wild-type (sus^+) ϕ29 phage can reproduce on both strains, L15 and 12A. Thus the ϕ29 *sus* mutants are conditional lethal mutants like the *amber* mutants of bacteriophage T4 (see Figure 12.23). Seven different *sus* mutants of phage ϕ29 were analyzed for complementation by simultaneously infecting the restrictive host (*B. subtilis* strain 12A) with each possible pair of mutants. Single infections with each of the mutants and with wild-type ϕ29 were also done as controls. The results of these complementation or *trans* tests and the controls are given as progeny phage per infected cell in the accompanying table. Several infections performed with wild-type ϕ29 phage yielded 300 to 400 progeny phage per infected cell. The results of the *cis* controls are not given, but assume that all of the *cis* heterozygote controls yielded over 300 progeny phage per infected cell. Also assume that no intragenic complementation occurs between any of the *sus* mutants studied.

Phage ϕ29 Progeny per Infected Bacterium

Mutant:	1	2	3	4	5	6	7
7	365	384	344	371	347	333	0.01
6	341	301	351	369	329	0.1	
5	386	326	322	0.04	< 0.01		
4	327	398	374	0.06			
3	354	387	<0.01				
2	0.01	<0.01					
1	0.02						

(a) Based on these data, how many genes are identified by the seven *sus* mutants? (b) Which *sus* mutations are located in the same gene(s)?

ANSWER

The seven *sus* mutants yielded from <0.01 to 0.1 progeny phage per infected *B. subtilis* strain 12A cell; those data define the mutant phenotype (basically no progeny). Infections of strain 12A cells with wild-type ϕ29 produced 300 to 400 progeny phage per infected cell, defining the wild-type phenotype. We then examine the phenotypes of the *trans* heterozygotes to determine whether any of the mutations are located in the same gene(s). In each case, we must ask whether the *trans* heterozygote has the mutant or the wild-type phenotype. If a *trans* heterozygote has the mutant phenotype, the two *sus* mutations are in the same gene. If it has the wild-type phenotype, the two *sus* mutations are in different genes. If you are unsure of why this is true, review Figure 14.16. Of the 21 *trans* heterozygotes examined, 19 exhibited the wild-type phenotype, indicating that in each case the two mutations are in different genes. Two *trans* heterozygotes, (1) *sus*1 on one chromosome and *sus*2 on a second chromosome and (2) *sus*4 on one chromosome and *sus*5 on the another, had the mutant phenotype. Therefore, (*a*) the seven *sus* mutations are located in five different genes, with (*b*) mutations *sus*1 and *sus*2 in one gene and mutations *sus*4 and *sus*5 in another gene.

QUESTIONS AND PROBLEMS

14.1 In what ways does our present concept of the gene differ from the pre–1940 or classical concept of the gene?

14.2 What was the first evidence that indicated that the unit of function and the unit of structure of genetic material were not the same?

14.3 What is the currently accepted operational definition of the gene?

14.4 Of what value are conditional lethal mutations for genetic fine structure analysis?

14.5 Eight independently isolated mutants of *E. coli*, all of which are unable to grow in the absence of histidine (his⁻), were examined in all possible *cis* and *trans* heterozygotes (partial diploids). All of the *cis* heterozygotes were able to grow in the absence of histidine. The *trans* heterozygotes yielded two different responses: some of them grew in the absence of histidine; others did not. The experimental results, using + to indicate growth and 0 to indicate no growth, are given in the accompanying table. How many genes are defined by these eight mutations? Which mutant strains carry mutations in the same gene(s)?

Growth of Trans Heterozygotes (without Histidine)

Mutant:	1	2	3	4	5	6	7	8
8	0	0	0	0	0	0	+	0
7	+	+	+	+	+	+	0	
6	0	0	0	0	0	0		
5	0	0	0	0	0			
4	0	0	0	0				
3	0	0	0					
2	0	0						
1	0							

14.6 Assume that the mutants described in Problem 14.5 yielded the following results. How many genes would they have defined? Which mutations would have been in the same gene(s)?

Growth of Trans Heterozygotes (without Histidine)

Mutant:	1	2	3	4	5	6	7	8
8	+	+	+	+	+	+	0	0
7	+	+	+	+	+	+	0	
6	+	+	+	+	0	0		
5	+	+	+	+	0			
4	+	+	0	0				
3	+	+	0					
2	0	0						
1	0							

14.7 What determines the maximum number of different alleles that can exist for a given gene?

14.8 What is the difference between a pair of homoalleles and a pair of heteroalleles?

14.9 Two different inbred varieties of a particular plant species have white flowers. All other varieties of this species have red flowers. What experiments might be done to obtain evidence to determine whether the difference in flower color in these varieties is the result of different alleles of a single gene or the result of genetic variation in two or more genes?

14.10 The *amber* mutants of phage T4 are conditional lethal mutants. They grow on *E. coli* strain CR63 but are lethal on *E. coli* strain B. An *amber* mutant almost never exhibits *intragenic* complementation with any other *amber* mutant; for this problem, assume that no *intragenic* complementation occurs between any of the mutants involved. The following results were obtained when eight *amber* mutants were analyzed for complementation by infecting the restrictive host (*E. coli* strain B) with each possible pair of mutants. The results of mixed infections by pairs of mutants are shown as **0** if no progeny are produced and as **+** if progeny phage resulted from the infection with that particular pair of mutants.

Mutant:	1	2	3	4	5	6	7	8
8	+	+	+	+	+	+	0	0
7	+	+	+	+	+	+	0	
6	+	+	+	+	+	0		
5	0	+	0	+	0			
4	+	+	+	0				
3	0	+	0					
2	+	0						
1	0							

(a) These data indicate that the eight *amber* mutations are located in how many different genes?

(b) Which mutations are located in the same gene or genes?

14.11 Considering only base-pair substitutions, how many different mutant homoalleles can occur at one site in a gene?

14.12 Are the following statements concerning the genetic element referred to as the gene true or false?

(a) The classical (pre–1940) conception of the gene was that it was (1) a unit of physiological function or expression, (2) the smallest unit that could undergo mutation, and (3) a unit not subdivisible by recombination.

(b) In bacteria, the *cis-trans* test provides an operational definition by which we usually can identify a gene as the unit that specifies one mRNA molecule.

(c) Our present knowledge of the structure of the gene indicates that the units defined by criteria (2) and (3) in statement (a) above are both equivalent to a single nucleotide pair.

(d) Studies in the 1940s demonstrated the existence of heteroalleles, clearly indicating that many mutations that were allelic by the functional criterion could be separated by recombination, and thereby indicating that the units of function, mutation, and recombination are not equivalent.

(e) Homoalleles are functionally and structurally allelic; heteroalleles are functionally allelic but structurally nonallelic.

14.13 The *rosy* (*ry*) gene of *Drosophila* encodes the enzyme xanthine dehydrogenase; the active form of xanthine dehydrogenase is a dimer containing two copies of the *rosy* gene product. Mutations ry^2 and ry^{42} are both located within the region of the *rosy* gene that encodes the *rosy* polypeptide gene product. However, ry^2/ry^{42} *trans* heterozygotes have wild-type eye color. How can the observed complementation between ry^2 and ry^{42} be explained given that these two mutations are located in the same gene?

14.14 Both *temperature-sensitive* (*ts*) mutant alleles and *amber* (*am*) mutant alleles have been identified and studied for many of the genes of bacteriophage T4. Different *ts* mutations within the same gene are frequently found to complement each other, whereas different *am* mutations within the same gene practically never complement one another. Why is this difference to be expected?

14.15 Suppressor-sensitive (*sus*) mutants of bacteriophage lambda can grow on *E. coli* strain C600 but cannot grow (that is, are lethal) on *E. coli* strain W3350. In other words, *sus* mutants are conditional-lethal mutants. Seven *sus* mutants were analyzed for complementation by simultaneously infecting the restrictive host (*E. coli* strain W3350) with each possible pair of mutants. Single infections with each mutant and with wild-type lambda were also done as controls. The results of these complementation or *trans* tests and the controls are given as progeny per infected cell in the accompanying table. Several infections with wild-type lambda yielded 120 to 150 progeny phage per infected cell. The results of the *cis* heterozygote controls are not given, but assume that all of the *cis* heterozygotes yielded over 100 progeny phage per infected cell. Also assume that no intragenic complementation occurs between any of these *sus* mutants.

Lambda Progeny per Infected Cell

Mutant:	1	2	3	4	5	6	7
7	0.01	133	0.01	146	134	128	0.01
6	131	142	161	0.06	0.1	0.1	
5	120	126	134	0.05	<0.01		
4	147	129	134	0.06			
3	<0.01	147	<0.01				
2	170	<0.01					
1	0.02						

(a) Based on the above data, how many genes are defined by the seven *sus* mutants?

(b) Which *sus* muations are located in the same gene(s)?

14.16 Is the number of potential alleles of a gene directly related to the number of nucleotide pairs in the gene? Is such a relationship more likely to occur in prokaryotes or in eukaryotes? Why?

14.17 In *Drosophila*, *white*, *eosin*, and *carnation* are all sex-linked recessive mutations affecting eye color. A white-eyed female crossed with a carnation-eyed male produced white-eyed male progeny and red-eyed (wild-type) female offspring. A white-eyed female crossed with an eosin-eyed male produced white-eyed sons and light eosin-eyed daughters. Based on these data, which of the three mutations (*white*, *eosin*, and *carnation*), if any, are located in the same gene(s)?

14.18 Suppressor-sensitive (*sus*) mutants of bacteriophage ϕ29 can grow on *Bacillus subtilis* strain L15 but cannot grow (that is, are lethal) on *B. subtilis* strain 12A. Wild-type (sus^+) ϕ29 phage can reproduce on both strains, L15 and 12A. Thus the ϕ29 *sus* mutants are conditional lethal mutants like the *amber* mutants of bacteriophage T4. Seven different *sus* mutants of phage ϕ29 were analyzed for complementation by simultaneously infecting the restrictive host (*B. subtilis* strain 12A) with each possible pair of mutants. Single infections with each of the mutants and with wild-type ϕ29 were also done as controls. The results of these complementation or *trans* tests and the controls are given as progeny phage per infected cell in the accompanying table. Several infections performed with wild-type ϕ29 phage yielded 300 to 400 progeny phage per infected cell. The results of the *cis* controls are not given, but assume that all of the *cis* heterozygote controls yielded over 300 progeny phage per infected cell. Also assume that no intragenic complementation occurs between any of the *sus* mutants studied.

Phage ϕ29 Progeny per Infected Bacterium

Mutant	1	2	3	4	5	6	7
7	0.01	384	0.01	371	347	333	0.01
6	341	301	351	0.06	329	0.1	
5	386	326	322	367	<0.01		
4	327	398	374	0.06			
3	<0.01	387	<0.01				
2	354	<0.01					
1	0.02						

(a) Based on these data, how many genes are identified by the seven *sus* mutants?

(b) Which *sus* mutations are located in the same gene(s)?

14.19 Assume that the mutants described in Problem 14.18 had yielded the following results.

Phage ϕ29 Progeny per Infected Bacterium

Mutant	1	2	3	4	5	6	7
7	0.01	0.01	0.01	0.03	<0.01	0.01	0.01
6	0.08	0.09	0.05	0.06	0.1	0.1	
5	0.02	<0.01	<0.01	0.04	<0.01		
4	0.05	0.06	0.03	0.06			
3	<0.01	<0.01	<0.01				
2	0.01	<0.01					
1	0.02						

How many genes would they have defined? Which mutations would have been in the same gene(s)?

14.20 The recessive mutations *bl* (*black*) and *e* (*ebony*) in *Drosophila* both produce flies with black bodies rather than gray bodies like wild-type flies. Mapping studies showed that *bl* is located on chromosome 2, whereas *e* is on chromosome 3. When homozygous *bl/bl* flies are crossed with homozygous *e/e* flies, the heterozygous *bl/e* progeny have gray bodies. The observed complementation indicates that the two mutations are in two different genes. Was it necessary to perform a complementation test to conclude that the *bl* and *e* mutations were located in two different genes? If so, why? If not, why not?

14.21 Why was it necessary to modify Beadle and Tatum's one gene–one enzyme concept of the gene to one gene–one polypeptide?

14.22 In their analysis of gene function, Beadle and Tatum used *Neurospora* as an experimental organism, whereas Garrod had studied gene function in humans. What advantages does *Neurospora* have over humans for such studies?

14.23 Based on the information provided in Figure 14.11, (a) are mutations *trpA*3 and *trpA*33 heteroalleles or homoalleles? (b) Are mutations *trpA*78 and *trpA*58 heteroalleles or homoalleles?

14.24 Based on the information given in Figure 14.11, what is the maximum number of nucleotide pairs separating mutations *trpA*78 and *trpA*78?

14.25 Arthur Chovnick and colleagues have mapped a large number of recessive mutations that produce fruit flies with rose-colored eyes in the homozygous state. They also have performed complementation tests on these *ry* (*rosy*) mutations. Heterozygotes that carried mutations ry^{42} and ry^{406} in the *trans* configuration had wild-type eyes, whereas *trans* heterozygotes that harbored ry^{5} and ry^{41} had rose-colored eyes. The results of two- and three-factor crosses unambiguously demonstrated that mutations ry^{42} and ry^{406} both map between mutations ry^{5} and ry^{41}. How can these results be explained?

14.26 The sequences of nucleotide pairs that encode human antibody chains are usually referred to as gene segments rather than genes. Why?

14.27 Tropomyosins are proteins that mediate the interactions between actin and troponin and regulate muscle contractions. In *Drosophila*, six different tropomyosins that have some amino acid sequences in common, but differ in other sequences, are encoded by two tropomyosin genes (*TmI* and *TmII*). How can two genes encode six different polypeptides?

14.28 In *Drosophila*, *car* (*carnation*) and *g* (*garnet*) are sex-linked mutations that produce brown eyes, in contrast to the dark red eyes of wild-type flies. The *g* and *car* mutations map at positions 44.4 and 62.5, respectively, on the linkage map of the X chromosome. Is a complementation test needed to determine whether these two mutations are in the same gene or two different genes? If so, why? If not, why not?

14.29 The *loz* (*l*ethal *o*n *Z*) mutants of bacteriophage X are conditional lethal mutants that can grow on *E. coli* strain Y but cannot grow on *E. coli* strain Z. The results shown in the following table were obtained when seven *loz* mutants were analyzed for complementation by infecting *E. coli* strain Z with each possible pair of mutants. A + indicates that progeny phage were produced in the infected cells, and a 0 indicates that no progeny phage were produced. All possible *cis* tests were also done, and all cis heterozygotes produced wild-type yields of progeny phage.

Mutant	1	2	3	4	5	6	7
7	+	+	0	+	0	0	0
6	+	+	+	+	+	0	
5	+	+	0	+	0		
4	0	0	+	0			
3	+	+	0				
2	0	0					
1	0						

Given that intragenic complementation does not occur between any of the seven *loz* mutants analyzed here, (a) propose four plausible explanations for the apparently anomalous complementation behavior of *loz* mutant number 7. (b) What simple genetic experiments can be used to distinguish between the four possible explanations? (c) Explain why specific outcomes of the proposed experiments will distinguish between the four possible explanations.

BIBLIOGRAPHY

BEADLE, G. W., AND E. L. TATUM. 1942. Genetic control of biochemical reactions in *Neurospora*. *Proc. Natl. Acad. Sci. USA* 27:499–506.

CARLSON, E. A. 1966. *The Gene: A Critical History*. W. B. Saunders, Philadelphia.

FINCHAM, J.R.S. 1966. *Genetic Complementation*. Benjamin, Menlo Park, CA.

GARROD, A. E. 1909. *Inborn Errors of Metabolism*. Oxford University Press, New York. (Reprinted in H. Harris, 1963, *Garrod's Inborn Errors of Metabolism*, Oxford Monographs on Medical Genetics, Oxford University Press, London.)

OLIVER, C. P. 1940. A reversion to wild type associated with crossing over in *Drosophila melanogaster*. *Proc. Natl. Acad. Sci. USA* 26:452–454.

STURTEVANT, A. H. 1965. *A History of Genetics*. Harper & Row, New York.

YANOFSKY, C., AND V. HORN. 1972. "Tryptophan synthetase α chain positions affected by mutations near the ends of the genetic map of *trpA* of *Escherichia coli*." *J. Biol. Chem.* 247:4494–4498.

A CONVERSATION WITH
MARGARET KIDWELL

Tell us about your background and how you became interested in science.

I was born in a small, isolated village in England, and the time that I spent in formal education was rather minimal. I did a lot of learning by doing. I had only three years in primary school, seven years in secondary school, and three years as an undergraduate. I was actually the first of my family to go to college. While I was growing up in the countryside, I had a great curiosity about the diversity of living organisms, and I think that led to an interest in science. Furthermore, I did well in math and science in school, which served to emphasize my interest. As a child, I was particularly fascinated by animals and plants, probably because I lived on a farm.

What drew your attention to genetics?

My father was a pedigree breeder of poultry, and as a young girl I spent a lot of time working on the farm for the love of it rather than being a paid hand. I helped with recordkeeping and the selection of hens for the next generation. I kept bantam poultry and many other kinds of animals. I think that's how my basic interest in genetics got started. However, I didn't take a formal course in genetics until I went to graduate school in the United States.

Your research initially involved poultry, didn't it?

Yes. It developed in England while I was a poultry adviser in the civil service. After working for five years, I came to the United States to study poultry breeding at Iowa State University. I met my husband there. He was just completing a Guggenheim Fellowship and had switched his research from beef cattle to *Drosophila* and mice because they are much better experimental organisms. After we were married, we moved to Canada, and there weren't very many hens in the middle of Ottawa. Consequently, I also switched to *Drosophila.* I learned the *Drosophila* experimental techniques from my husband.

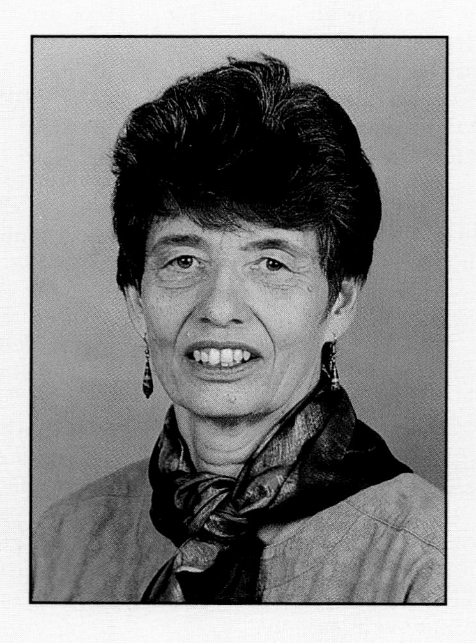

Even from a very young age, **Dr. Kidwell** was destined to become one of the most influential geneticists of her generation. She was born and raised in England, then emigrated to the United States for her higher degrees. She first began studying genetics at Iowa State University, where she earned her M.S. She switched to *Drosophila* as a Ph.D. candidate at Brown University. Dr. Kidwell's discovery of *P* transposable elements in *Drosophila* has enhanced our understanding of genetics enormously. Dr. Kidwell is a recipient of multiple honors and awards for her innovative research and outstanding contribution to the science community as a professor and distinguished lecturer. Dr. Kidwell is currently Regents' Professor and head of the Department of Ecology and Evolutionary Biology at the University of Arizona. In 1996 she was admitted to the National Academy of Sciences of the United States.

Tell us about the circumstances that led you to the discovery of transposable elements in *Drosophila*.

For my Ph.D. degree, I was working on a mutator line of *Drosophila*. The spontaneous mutation rate in this line was much higher than in lines that had been treated with a powerful mutagen. It was so high that eventually the line died out. As controls for this project, I started working with lines from natural populations. I discovered that when I crossed these lines in the laboratory, their offspring exhibited unusual properties, including, for example, a high mutation rate. During this work, I met John Sved, a geneticist from Australia. He had made similar observations, so we started a collaboration that eventually led to a paper describing these unusual phenomena. We called the phenomena "hybrid dysgenesis." A few years later, through a collaboration with two other geneticists, Gerald Rubin and Paul Bingham, I discovered that hybrid dysgenesis was caused by a mobile genetic element, one that we called the *P* element.

Has your work changed how we think about the genetic material?

We now think of the genome as being much more fluid and dynamic than we used to. In general, I think there are many properties of transposable elements that are consistent with the idea that they are genomic parasites. However, this idea is really too simplistic. Some of them may be mutualists for at least part of their life history. By that I mean that they may confer some benefit on their host, and the host may confer a benefit on them. Perhaps the general picture is one of serial parasitism combined with mutualism, with the parasite phase predominating.

Why is so little known about transposable elements in human beings?

It's important to realize that human beings are not so easy to work with genetically. We can't do experiments with them as we can with mice and *Drosophila*. When we can't do experiments, progress is less rapid. However, with the genome project we may be able to find out a lot more about transposable elements in human beings. I believe they are there, but that we just haven't been able to study them as readily as in other organisms.

You have been a researcher, a teacher, a wife, a mother, and also a department head. How have you been able to juggle all these roles?

That's a good question. I think especially in the early days I didn't try to do too many things at once. For example, I didn't go back to do my Ph.D. work until my children were in elementary school. I didn't get my Ph.D. degree until I was 40. By concentrating on one or two things at a time, I've managed. I think it's very difficult if you try to do too many things at once. There is plenty of time if you're patient. Also, I've had the support of many people during my life. In particular, a number of men have been wonderful mentors to me—my father, my husband, and then various bosses along the line. They were all extremely encouraging and helpful.

Do you have advice for students, especially those who might be interested in science as a career?

I would say that at an early stage you don't have to make up your mind exactly about what you're going to do later. Rather, be prepared to change as circumstances change, and let things go as they are. You don't need to worry if you're not too sure where you are going.

You've traveled extensively during your career and collaborated with researchers in many countries. Do you think that having an international perspective helps a person be a scientist?

Indeed, yes. I think that in science having a broad perspective and knowing many ways of dealing with problems is always an advantage. I've always felt that traveling and working with people from different backgrounds and cultures can be tremendously helpful. After all, that is how progress is made in science, as well as in other activities of life. Right now, I have collaborations with scientists in Russia, Brazil, Australia, Spain, and Austria. . . . So, it's still going on. It's something I enjoy very much.

15

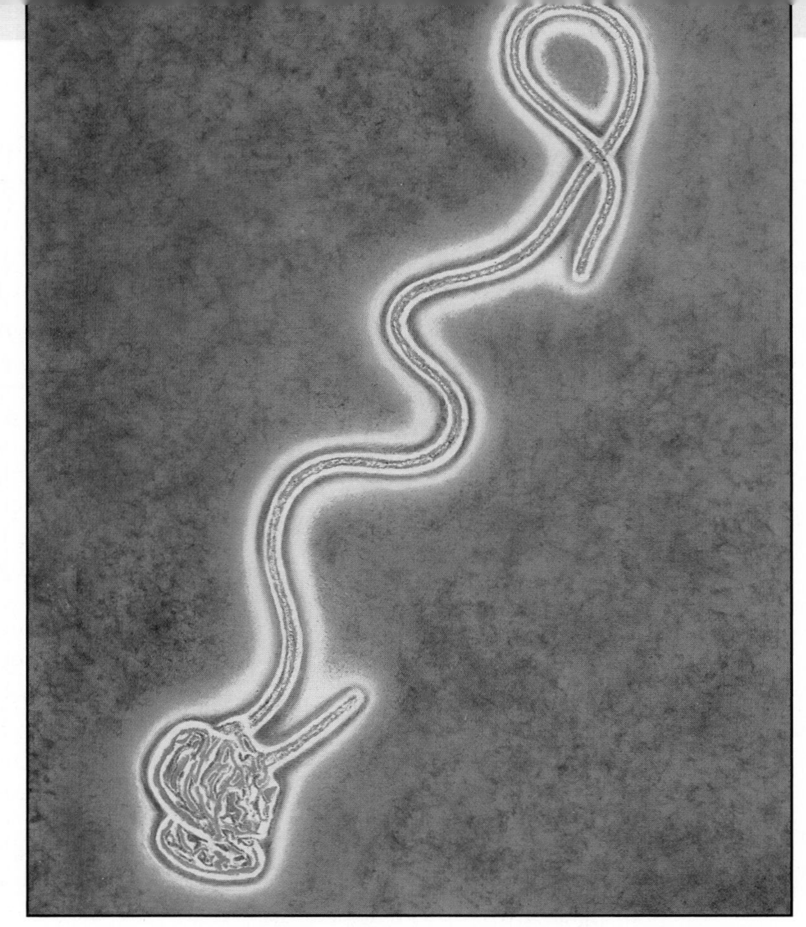

The Ebola *virus pictured here (x19,000) is one of a group of viruses that cause deadly hemorrhagic fevers. Wild vertebrates such as rats serve as reservoir hosts.*

The Genetics of Viruses

CHAPTER OUTLINE

A Killer Unleashed

In the spring of 1995, Joseph Kinfumu, a laboratory technician working in Kikwit, Zaïre, came into the emergency room of the local hospital complaining of dysentery and a fever. By themselves, these symptoms were not alarming, but the situation changed rapidly. He soon began to ooze blood from his eyes, nose, ears, mouth, and anus—virtually every orifice of his body. He vomited blood. Within four days, he was dead. An autopsy revealed that his internal organs appeared to liquify. But Mr. Kinfumu was not going to be the only person to die like this. Soon others were diagnosed with the same disease, and they too died. The disease spread through Kikwit and neighboring villages. Blood samples from the victims were rushed to the Centers for Disease Control in Atlanta, and a terrifying diagnosis was made: the cause was the deadly Ebola virus, the same virus that killed hundreds in Zaïre and Western Sudan in 1976 and 1986. Over 90 percent of people infected by Ebola die within nine days. This is truly a frightening disease.

The Ebola virus, like all viruses, requires a host cell in order to reproduce. Viral genetic material, which can be DNA or RNA (Ebola is RNA), is packaged inside a shell of proteins. Upon entering a host cell, the viral genetic material commandeers the cell's protein-synthesizing machinery to make copies of itself and in the process make proteins that, in the case of the Ebola virus infecting humans, make people so sick they usually die. The Ebola proteins cause a profound capillary leakage syndrome. Progeny viruses are released by infected cells, they infect other cells, and the disease progresses.

For obvious reasons, virus research has its roots deep in the field of medicine, but the simplicity of viruses makes them ideal for basic genetic analysis. By the 1930s, viruses had been implicated in cancer, in plant diseases, and in the destruction of bacteria. Because the simplest types of virus to analyze in the laboratory were bacterial viruses, researchers concentrated on them with the idea that the knowledge they gained would help them in their studies of the more complicated viruses. Early work with bacterial viruses was geared toward medical issues; however, as researchers came to realize that the bacterial virus was an ideal system to study the basic properties of the gene, they broadened their scope.

Viruses straddle the line between the living and the nonliving. For example, the tobacco mosaic virus (TMV), which causes tobacco mosaic disease and often creates a major economic problem for tobacco farmers, can be crystallized and stored on a shelf for years. In this state, it exhibits none of the properties normally associated with living systems: it does not reproduce; it does not grow or develop; it does not utilize energy; and it does not respond to environmental stimuli. However, if TMV is rubbed onto the leaf of a tobacco plant, the viruses infect the cells, reproduce, utilize energy supplied by the plant cell, and are responsive to cellular signals. They exhibit the properties of living systems.

It is their simplicity that has made viruses such an ideal research tool in modern genetic analysis. Questions that are difficult to answer using more complicated eukaryote systems can be addressed using viruses. These questions include the molecular nature of the genetic material, DNA replication, mechanisms of mutation, gene structure and function, and gene regulation.

In this chapter, we focus on how the study of viruses has illuminated important genetic concepts. We study how viruses transmit genetic information from generation to generation, how viral genes are mapped, how viral genes recombine, and how viral genomes are organized. Much of our discussion centers on bacterial viruses, or bacteriophage, because they were the main working tool in the researchers' quest to unravel the mystery of the gene. We also discuss the virus that infects eukaryotic cells causing acquired immune deficiency disease (AIDS).

THE DISCOVERY OF BACTERIAL VIRUSES

Today there is general agreement on the meaning of "virus": ultramicroscopic, obligate, intracellular parasites capable of autonomous replication inside a host cell. But the meaning of the term has shifted over time. Ancient Roman physicians called any "poison" of animal origin a virus and the diseases caused by such poisons "virulent." Thus any disease caused by or transmitted by an animal, even bubonic plague, later discovered to be caused by bacteria, was considered viral.

During the Middle Ages, the term *virus* was reserved for infectious agents that caused contagious diseases. Of course, many of these "infectious agents" were bacterial. Once bacteria were discovered, the definition of a virus became even more muddled.

It was not until the twentieth century that many researchers recognized viruses as a distinct class of organisms. They were defined as microbes that could pass through a filter specifically designed to trap bacteria. They could not be detected by a microscope, and they did not grow on any medium that supported the growth of bacteria. However, many researchers questioned whether these so-called viruses really merited a distinct classification. Perhaps they were just minute bacteria that could pass through a filter.

In 1915, the idea that viruses are indeed distinct from bacteria gained strength with the work of Frederick Twort. He was one of the few microbiologists who thought viruses were distinct from bacteria, and suggested that disease-causing viruses originated from nondisease-causing ancestors. Working on the supposition that these ancestors could grow on bacterial media, he innoculated plates containing bacterial medium with smallpox vaccine fluid. The only growth he observed came from bacteria in the fluid. But an interesting phenomenon occurred to these bacteria. Some of the bacterial colonies dissolved away, as though they were being killed. In his published observations, Twort suggested as one possible explanation that a virus had infected and killed the bacteria. Two years later, Felix D'Hérelle reported similar observations and suggested even more forcefully that this bacterial killing was due to viral infection. D'Hérelle called these bacteria-destroying viruses **bacteriophage**, or "eater of bacteria."

The primary motivation for continuing research into bacteriophages was medical, not genetic. D'Hérelle and others believed that these bacterial viruses could be used therapeutically to attack and destroy bacteria that cause such devastating diseases as cholera, thyphus, and scarlet fever. Unfortunately, they soon discovered that when these bacterial viruses are introduced into the body, they are destroyed by the body's immune or digestive systems.

Beginning in the 1930s, the use of bacteriophage, or simply phage, as a research tool became established. They attracted the attention of researchers who were intrigued more by their molecular simplicity than by any possible therapeutic function. Among these early investigators was Max Delbrück, who might rightly be referred to as the father of modern phage genetics. Trained as a physicist, Delbrück became interested in biological problems, especially in the nature of the gene. To him, the phage, because of its simplicity, was an ideal system with which to explore the nature of the gene. In 1935 he published a paper dealing with a physical model of the gene. Two years later, Delbrück and his colleague Emory Ellis described the key events in the bacteriophage life cycle: Within about 30 minutes after a phage infects a cell, the cell ruptures and releases over 100 progeny phage. Thus, by 1940 a major question had taken shape: How does a bacteriophage enter a cell, replicate, and produce so many progeny in such a short span of time?

Key Points: **Initially, viruses were thought to be a type of bacteria, but key observations by Twort and d'Hérelle suggested that viruses were distinct organisms that infected and then destroyed bacteria. Delbrück and Ellis recognized the genetic significance of viral simplicity and launched studies into the life cycle of viruses.**

THE STRUCTURE AND LIFE CYCLE OF A BACTERIAL VIRUS

Structurally, viruses are the simplest of all organisms. The bacterial virus, for example, is composed of a protein coat surrounding a nucleic acid core. Eukaryotic viruses are in general more complex structurally but less complex genetically than bacterial viruses. Viruses are classified according to three basic properties: (1) the type of nucleic acid they contain; (2) the nucleic acid strandedness; and (3) the presence or absence of a membranous envelope. The nucleic acid can be DNA or RNA and can be single-stranded or double-stranded (Figure 15.1*a*).

One of the groups of bacterial viruses most commonly used for research includes the T phages that infect *Escherichia coli*. Although there is some structural variation among members of this group, they are all tadpolelike in appearance (Figure 15.1*b*). The **head** is composed of various proteins and is shaped like an icosahedron, a polyhedron with 20 faces. Inside the head is a single molecule of double-stranded DNA. Attached to the head is a **tail** composed of two coaxial hollow tubes—an inner **needle** or **core**, and an outer **sheath**. At the end of the tail is a **tail plate** made up of **tail fibers** and **tail pins**. The tail plate is the point of attachment to the *E. coli* host cell. Upon attachment, the phage DNA passes through the needle and enters the host cell where it is replicated and subsequently packaged into progeny virus particles. The protein coat remains outside the cell, as Hershey and Chase showed (Chapter 9)

The life cycle of a T4 bacterial virus is in general representative of bacteriophage life cycles. At 37°C (the *E. coli* host grows at normal body temperature), the life cycle takes about 22 minutes to complete (Figure 15.2). The phage adsorbs to the bacterial cell wall and injects its DNA into the cell. Shortly after the entry of the phage DNA, the synthesis of bacterial DNA, RNA, and protein is shut down by **early phage proteins** that are translated from early phage mRNA. Some of these early proteins degrade host DNA into nucleotide subunits. Other early proteins begin replicating phage DNA. About half way through the life cycle, the **late proteins** that form the head and tail structures are synthesized. About 15 minutes after the initial infection, complete progeny phage accumulate inside the cell. At 22 minutes after the initial infection,

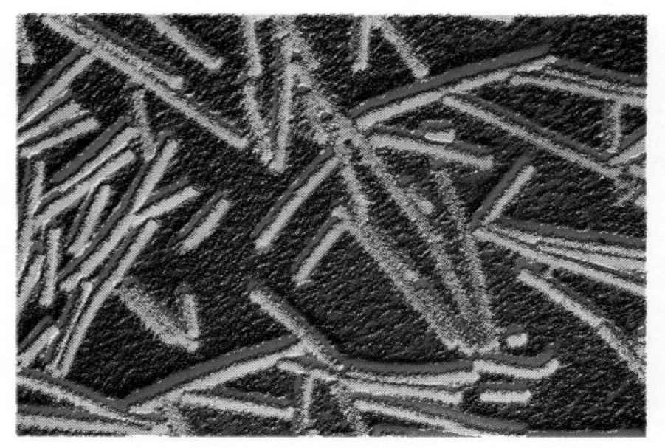

Tobacco Mosaic Virus (TMV)

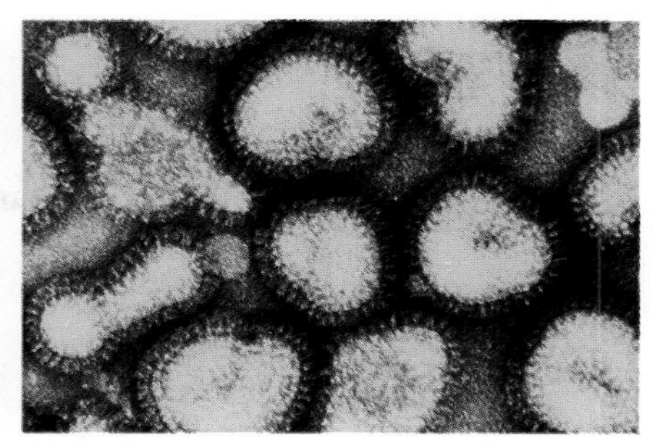

Influenza Virus

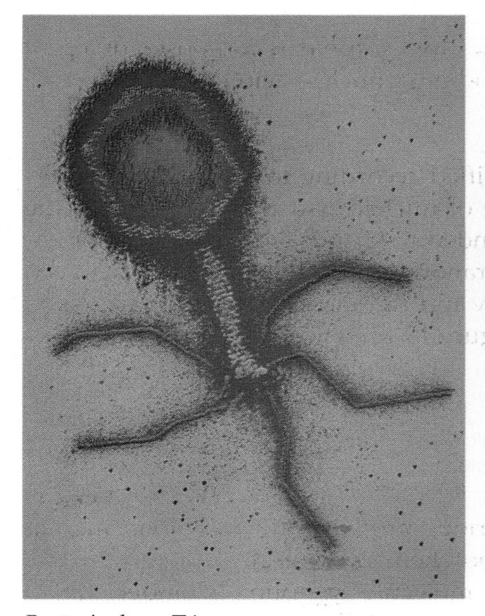

Bacteriophage T4

Bacteriophage λ

(*a*)

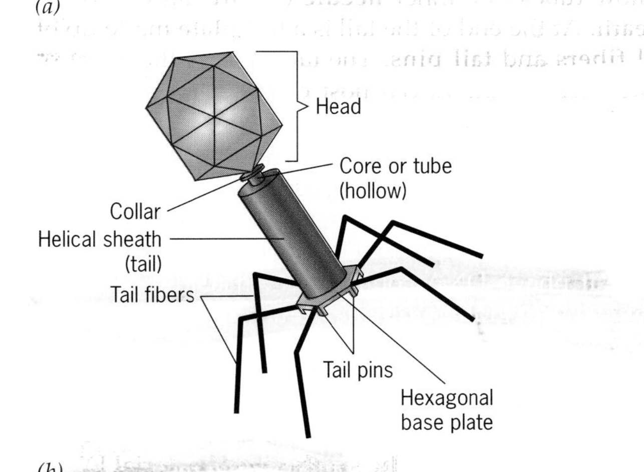

(*b*)

Figure 15.1 (*a*) Electron micrographs of a selection of viruses. TMV and influenza virus are single-stranded RNA viruses; λ and T4 are double-stranded DNA viruses. (*b*) A diagram of the structure of bacteriophage T4.

a phage-encoded enzyme called **lysozyme** acting in concert with other proteins causes the bacteria to lyse or rupture. Between 100 and 300 progeny phage are released. Let's examine some of the details of this life cycle.

Shortly after the life cycle begins, bacterial RNA polymerase is modified by complexing with early phage proteins so that it either does not recognize bacterial gene promoters or recognizes them poorly. This modification effectively shuts down all bacterial gene activity. The modified bacterial RNA polymerase does, however, recognize phage DNA. It transcribes early phage genes that have a specific type of promoter. The bacterial RNA polymerase is further modified by complexing with other phage proteins so that it recognizes promoters of phage genes that are activated later in the cycle. This modification process ensures a carefully orchestrated sequence of genes that are switched on and others that are switched off as the phage life cycle progresses.

Another important aspect of the T4 life cycle is the composition of the phage DNA. T4 DNA is unusual in that it contains 5-hydroxymethylcytosine (HMC) in-

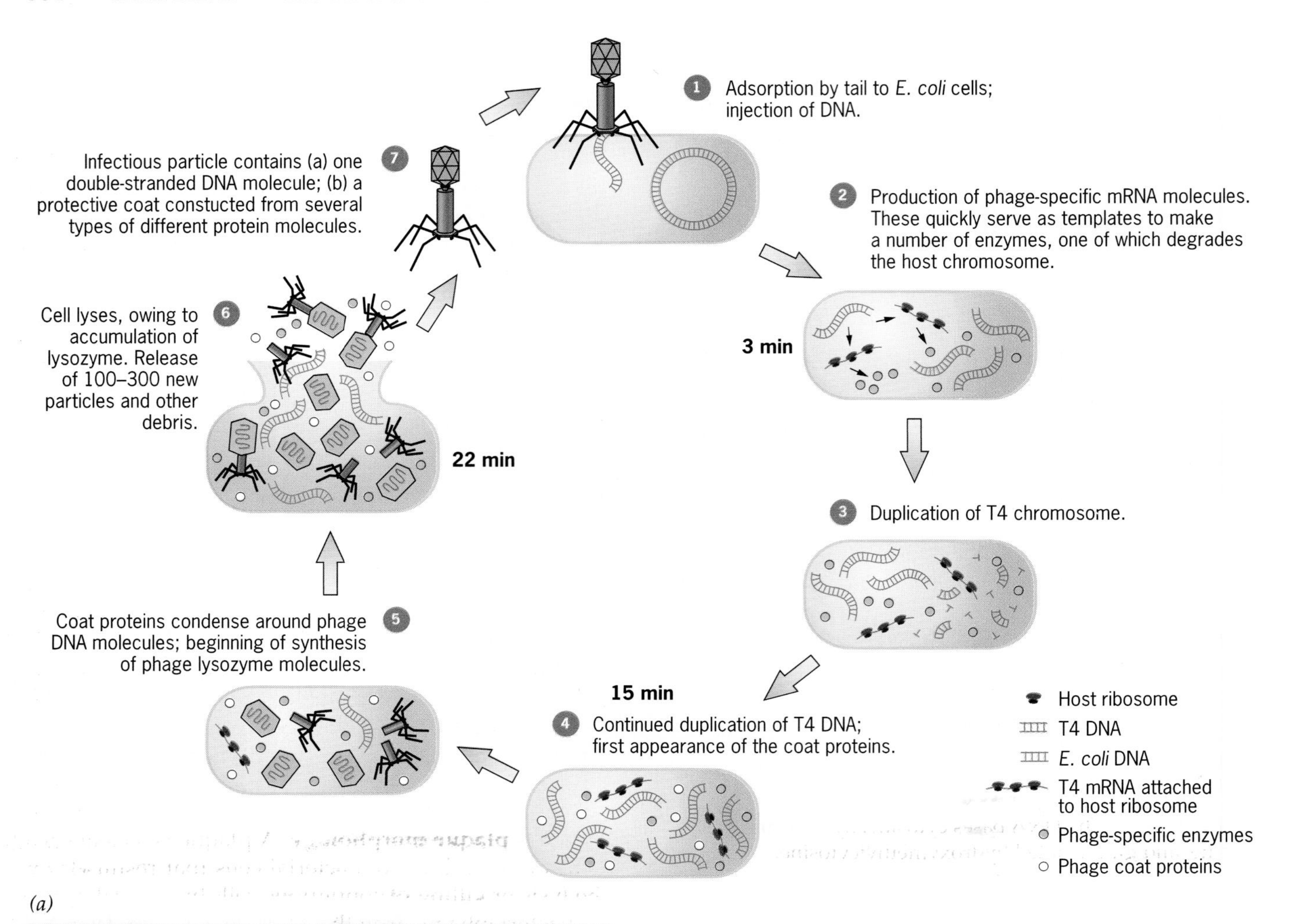

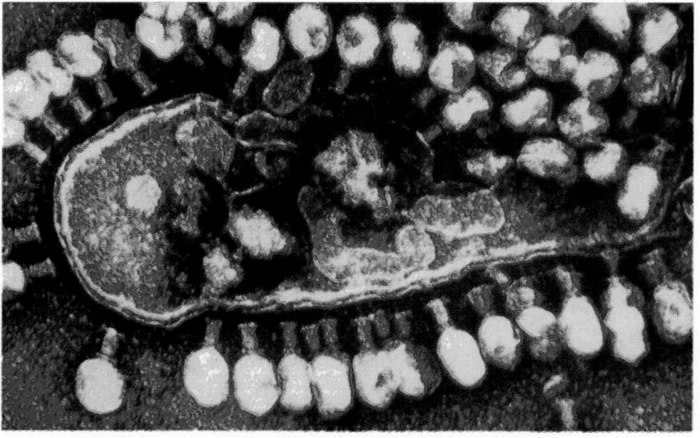

Figure 15.2 (*a*) The life cycle of bacteriophage T4. (*b*) An electron micrograph of an *E. coli* cell infected with T4.

stead of cytosine (Figure 15.3). In addition, glucose-like sugars are attached to the HMC. Where does this glucosylated HMC come from and what function does it serve?

Recall that once the phage DNA enters the cell, the bacterial DNA is degraded. Bacterial DNA degradation is carried out by early phage gene products that cleave cytosine-containing host DNA into mononucleotide subunits (dAMP, dCMP, dTMP, and dGMP). These sububits are used in the synthesis of new phage DNA. A phage-encoded enzyme converts C into HMC. Another phage-encoded enzyme blocks incorporation of C into phage DNA. This is important because if phage DNA contained C instead of HMC, it would be degraded by the same enzymes that degrade host cell DNA.

The addition of hydroxymethyl residues to C protects the phage DNA from its own DNA-degrading enzymes. But the bacterial cell has ways of protecting itself from foreign DNA . Bacteria have **restriction enzymes** that degrade HMC-containing DNA. Restriction enzymes protect bacteria from infection by bacterial viruses. Over time, viral genes have evolved that circumvent this bacterial defense. These viral genes code for enzymes that add glucose residues to HMC. The DNA-degrading bacterial enzymes are unable to recognize the phage DNA carrying glucosylated HMC DNA and thus do not degrade it. Phage carrying a mutated glucosylation gene (so that the HMC is not

5–OH–Methylcytosine (HMC)

5–OH–Methylcytosine with glucose attached

Cytosine (C)

Figure 15.3 The DNA bases cytosine, hydroxymethylcytosine, and glucosylated hydroxymethylcytosine.

glucosylated) do not survive in the host cell because their DNA is degraded by the host's restriction enzymes. If the host cell is mutant for the gene that degrades HMC-containing DNA, phage carrying the glucosylation mutation survive.

The life cycle of the T4 bacteriophage typifies the story of the coevolution of bacteria and the viruses that infect them. It is a fascinating story about the evolution of resistance in bacteria and the evolution of ways viruses have developed to overcome that resistance.

Key Points: **Bacteriophage require a bacterial host to complete their life cycle. The phage attach to the host cell and inject their DNA. The phage then take over the cellular protein-synthesizing machinery. Phage DNA directs the synthesis of progeny phage that are released from the cell and infect other cells. T4 viruses have glucosylated hydroxymethylcytosine instead of cytosine in their DNA. This protects them against degradation by phage-encoded nucleases and bacterial restriction enzymes.**

MAPPING THE BACTERIOPHAGE GENOME

Phage Phenotypes

Constructing a genetic map in any organism usually requires crossing organisms that carry different alleles of a gene. These alleles cause such phenotypic differences as red eyes versus white or wrinkled seeds versus smooth. But bacteriophage do not provide such obvious phenotypes. In fact, the virus itself can only be seen with an electron microscope. The phenotypes used to study bacteriophage genetics are usually based on the interaction between phage and host cell, or on variables in the phage life cycle.

A phage trait commonly used in basic phage research is **plaque morphology**. A plaque is a clear area in a confluent lawn of bacterial cells that results from the lysis or killing of contiguous cells by several cycles of bacteriophage growth. Plaques may be large or small, with sharp or fuzzy edges. Plaque morphology is a function of two factors: (1) the physiological interaction between phage and bacterium; and (2) the phage genotype. The most intensely studied plaque morphology mutant is the **rapid lysis**, or *r* mutant in the T phage. Wild-type T phage (r^+) produce a plaque that is small with fuzzy edges. The rapid lysis, or *r* mutant, produces a large plaque with sharp edges (Figure 15.4).

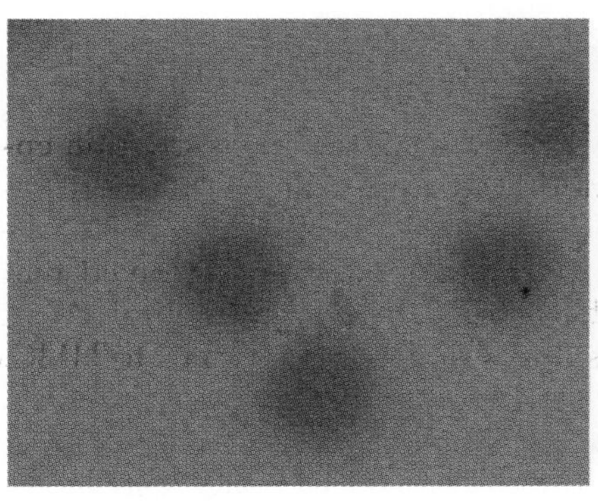

(*a*)

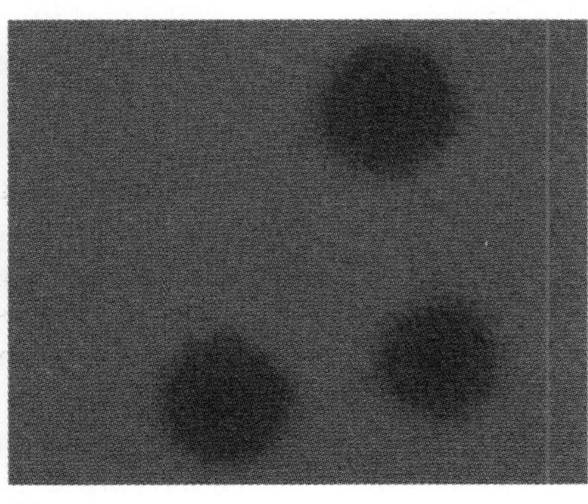

(*b*)

Figure 15.4 (*a*) Wild type (r^+) plaques. The plaques are clear, transparent spots. (*b*) *r* plaques.

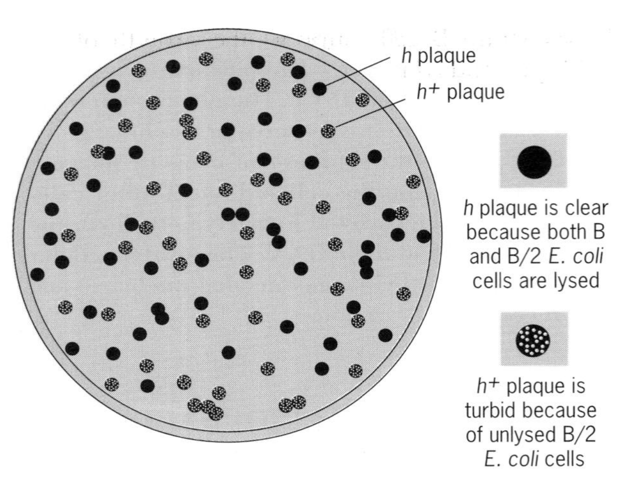

Figure 15.5 Plaques formed by T2 *h* and T2 h^+ phage growing on a mixed lawn of *E. coli* B and *E. coli* B/2 cells.

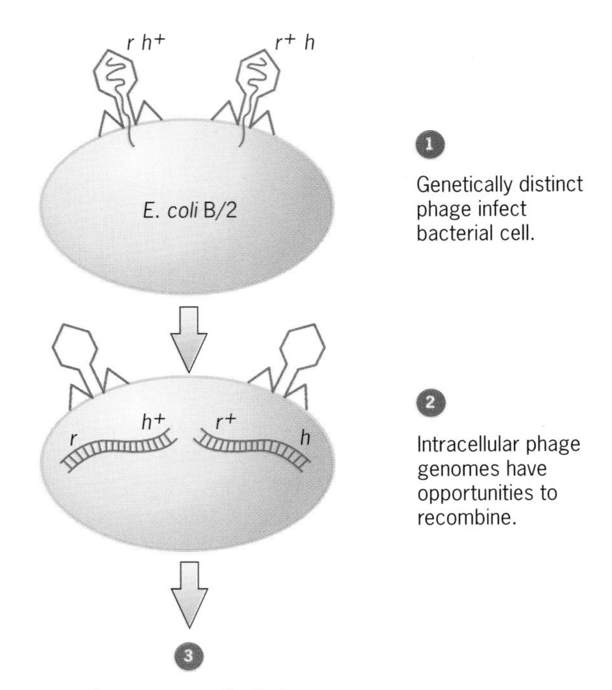

Figure 15.6 A cross between $r\ h^+$ and $r^+\ h$ bacteriophage.

Another commonly used phage trait is **host range**. Host range mutants are able to infect some cell strains but not others. For example, normally *E. coli* B cells can be infected by T2 bacteriophage. However, a strain of *E. coli* B, called B/2, carries a mutation making it resistant to T2 infection. A strain of T2, T2*h* , carries a host range mutation (*h*) that allows the virus to infect both *E. coli* B *and E. coli* B/2. This alternation of resistance and susceptibility may continue with bacterial mutations that make *E. coli* resistant to *both* T2 and T2*h*, and with viral mutations that allow infection of the new resistant strains. Host range mutants occur in several types of bacterial viruses.

T2 and T2*h* bacteriophage have the same plaque morphology, but the mutants can be distinguished from each other if the viruses are grown on a mixed lawn of *E. coli* B and *E. coli* B/2. The *h* mutants, which infect both B and B/2 cells, produce clear plaques because all cells in the plaque area are destroyed. T2h^+ viruses infect and destroy only the B cells, not the B/2, producing turbid plaques. The turbidity is caused by the B/2 cells remaining in the plaque area (Figure 15.5).

Genetic Recombination in Phage

In 1946, Max Delbrück and Alfred Hershey independently announced the discovery of genetic recombination in phage. In the wake of this discovery, Hershey and his colleague Raquel Rotman carried out the first detailed study of genetic recombination in phage. They infected *E. coli* with two different strains of T2 bacteriophage: $r\ h^+ \times r^+\ h$ (Figure 15.6). They collected the progeny from this mixed infection—the equivalent of a cross in phage research—and plated them on a mixture of B and B/2 cells. Each of the four possible genotypes produced a distinct plaque on these cells (Figure 15.7): the plaques were large and clear (*h r*), large and turbid ($h^+\ r$), small and clear ($h\ r^+$), or small and turbid ($h^+\ r^+$). Two of these genotypes were recombinant (*h r* and $h^+\ r^+$), and the other two nonre-

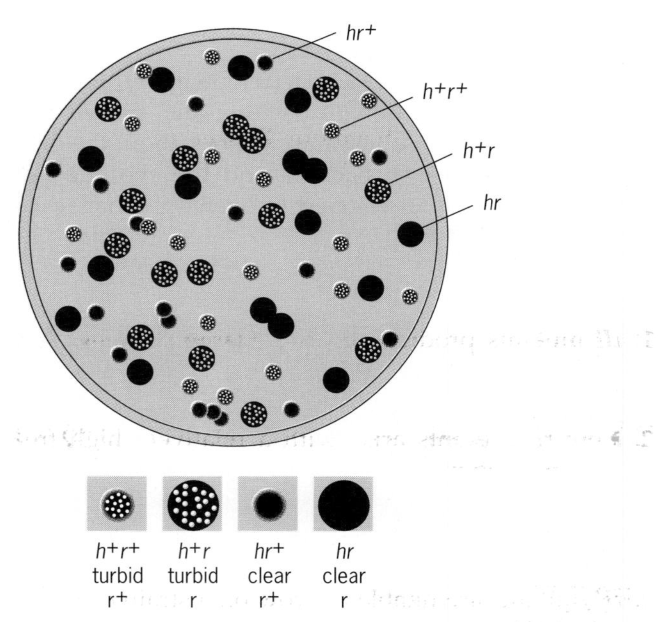

Figure 15.7 The types of plaques formed by T2 *r h*, T2 $r^+\ h^+$, T2 $r^+\ h$, and T2 $r\ h^+$ phage on a mixed lawn of *E. coli* B and *E. coli* B/2 cells.

combinant. In one cross, Hershey and Rotman found that about 2 percent of the progeny were recombinant and 98 percent were nonrecombinant. This exciting discovery allowed them to estimate that the distance between the *h* and *r* genes was about 2 map units.

Some important points emerged from these early phage mapping studies:

1. Recombination appeared to be a reciprocal event, as it is in meiosis. Thus, if one recombinant class appeared, the other appeared with approximately the same frequency. However, lysates from single cells did not contain equal numbers of reciprocal recombinants, suggesting that recombination is not reciprocal. Lysates from single cells do not usually show reciprocal recombination because of the peculiarities in the phage life cycle. Not all phage chromosomes are packaged into mature phage particles. When cells lyse, mature phage are released along with unpackaged DNA and other viral parts. If both members of a reciprocal event are not packaged into mature phage, the data will imply that recombination is not reciprocal.

2. Recombination in phage does not resemble meiotic recombination. In meiosis, which occurs only in sex cells, chromosomes synapse and cross over at a specific stage of cell division. In viruses, there is no process as complicated as meiosis. In fact, recombination can occur at any time during the phage life cycle as long as the DNA has not been packaged into protein capsids. Multiple rounds of genetic exchange can and do occur.

Genetic Fine Structure

Gene mapping took on an entirely new dimension in the mid-1950s when Seymour Benzer mapped over 2400 *rII* mutants to 308 sites in two contiguous genes in a small region of the bacteriophage T4 genome. Benzer's objective was to analyze the detailed organization of genes. He studied the *rII* region because of its special properties:

1. *rII* mutants produce distinctly large plaques with clear edges, compared to the smaller, fuzzy-edged wild-type plaques (Figure 15.4).
2. New *rII* mutants arise with a relatively high frequency of about 1 in every 100,000 r^+ progeny. This frequency can be enhanced considerably using mutagenic agents such as certain chemicals and ultraviolet radiation.
3. *rII* mutants are unable to grow on a strain of *E. coli* called K12. Although this strain of *E. coli* does not support the growth of *rII* mutants, it does support the growth of wild-type T4.
4. *E. coli* strain B cells support the growth of both wild-type and *rII* phage.

The system Benzer employed in his analysis of the *rII* region involved crossing two different, independently arisen *rII* mutants on *E. coli* B cells, then collecting and analyzing the progeny. If recombination occurred between the two mutants, a wild-type chromosome would result along with a double mutant (Figure 15.8). These would be extremely rare because the distance between the two mutant sites is so small. (Recombination frequencies could be as small as 10^{-6}.) The rarity of recombination events demanded a system that allowed easy detection of recombinants, and *E. coli* K12 provided that system. If progeny from the cross on B cells are plated on K12, the only phage capable of growth are the wild-type. Nonrecombinants do not grow on K12 because they are *r* mutants, and, of course, the double mutant does not grow. Thus a defined volume of progeny phage from B is plated

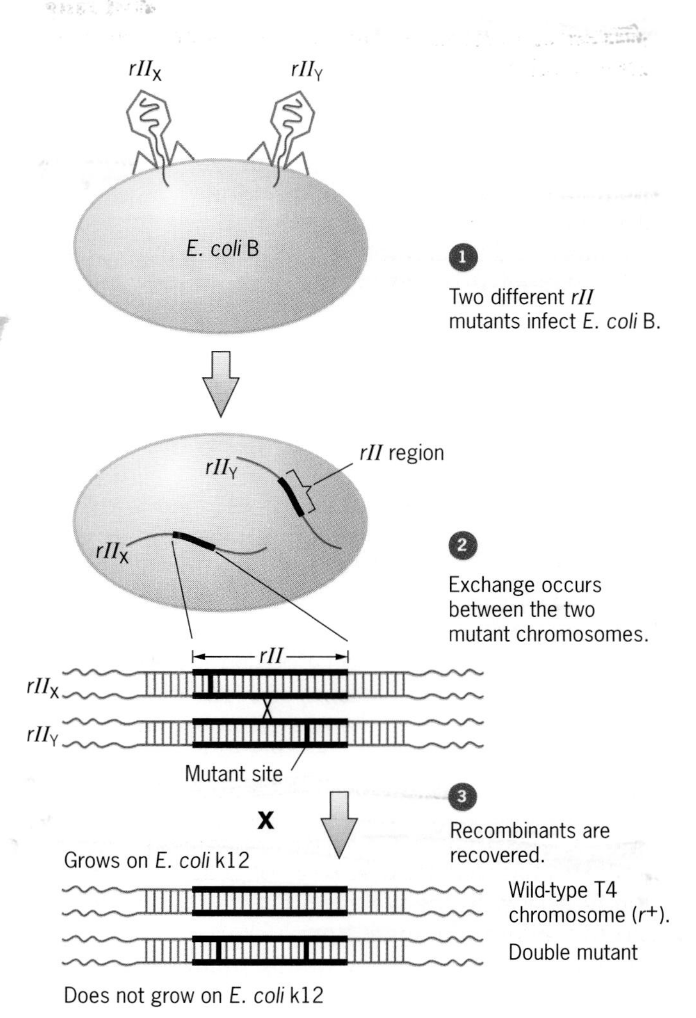

Figure 15.8 A genetic cross between two different *rII* mutants on *E. coli* B cells. The recombinant progeny are wild type (r^+) and double mutants.

out on a lawn of *E. coli* K12. If 2 wild-type (r^+) plaques appear, then there would be two wild-type recombinants. However, because there is one double mutant produced for each wild-type, there are a total of four recombinants.

There is one small flaw in this system. An *r* mutant may spontaneously revert back to r^+ by reverse mutation. The frequency of reverse mutation can be estimated for various *r* mutants so that recombination frequencies can be corrected for these reversion events.

To estimate the frequency of recombination, the total number of progeny produced per unit volume in the cross must be known. A sample of the progeny is diluted perhaps 10 million fold and plated on *E. coli B* cells. All genotypic classes of T4 grow on the B cells, including the single and double mutants. If 100 plaques are counted in a sample diluted 10^7 fold, then the original sample contained 10^9 phage per unit volume (100×10^7). Thus the recombination frequency between the two mutant sites would be approximately $4/10^9 = 0.000000004$.

All of Benzer's *r* mutants were mutants for one of two contiguous genes: *rIIA* or *rIIB*. He determined this by doing a **complementation test** (see Chapter 14). He infected cells with two different *rII* mutants. If the two *rII* mutants were mutant in the same gene (that is, both were *rIIA* mutants), they failed to grow on *E. coli* K12; if they were mutant in different genes (one was *rIIA* and the other was *rIIB*), they grew on *E. coli* K12.

Benzer mapped a number of *rII* mutants using this system. Although the system worked well, it was a labor-intensive procedure. After he had mapped about 60 mutants by this procedure, he developed a new, more efficient system to analyze the rest of the 2400 *rII* mutants in his collection.

The technique that Benzer developed was called **deletion mapping**. He found that a number of the 2400 *rII* mutants had deletions of part or all of the *rII* region. A phage carrying a deletion cannot revert back to wild-type, nor can it recombine with another phage that has a mutation in the region covered by the deletion. It cannot generate wild-type recombinants with a mutant that overlaps the deletion because neither phage has wild-type DNA in this region of the gene (Figure 15.9). Any *rII* mutant that produces wild-type recombinants when crossed with a deletion mutant must have a mutation that maps *outside* of the deletion, and any mutant that fails to generate wild-type recombinants must carry a mutation that maps *within* the area defined by the deletion.

To use deletions for mapping, Benzer first had to define them. He did this by crossing different deletions against each other (Figure 15.10). If the two deletions overlapped, no wild-type recombinants formed because both were missing the same segment of the *rII*

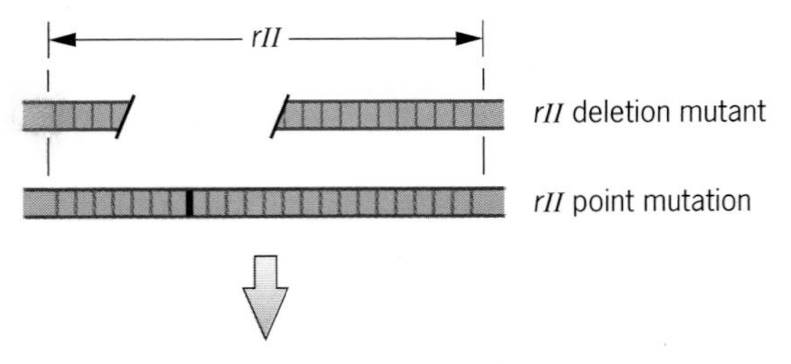

Figure 15.9 A cross between an *rII* point mutant and an rII deletion.

gene. The appearance of wild-type recombinant progeny indicated that the deletions did not overlap. Using this basic approach, Benzer estimated the extent of the deletions. He further refined the limits of each deletion by crossing them with the 50 plus mutants he had collected and mapped by the more tedious earlier methods.

Benzer identified seven large deletions (called the "big 7") that covered various large segments of the *rII* region, a region comprising the *A* and *B* genes. He also collected and defined a set of smaller secondary deletions (Figure 15.11). He then used these deletions to map the 2400 *rII* mutants that had been collected. Benzer first crossed an unknown *rII* mutant against the seven large deletions (Figure 15.12). If, for example, wild-type recombinants appeared with deletions 638, A105, and PB242 but not with the others, he concluded that the unknown mutant mapped to the left of the PB242 endpoint and to the right of the PT1 endpoint. (It does not overlap 638, A105, PB242). This area coincides with the A4 region.

Additional crosses with five smaller deletions in this region (EM66, 184, 221, 1368, and PB28) gave a more precise location (Figure 15.13). In this series of crosses, the only smaller deletion that failed to produce wild-type recombinants was EM66; this suggests

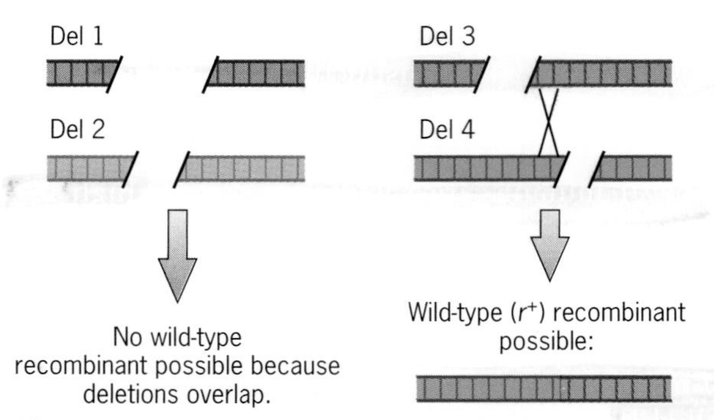

Figure 15.10 Crosses between pairs of deletion mutants. If the deletions overlap, no wild-type recombinants are produced; if they do not overlap, r^+ progeny are produced.

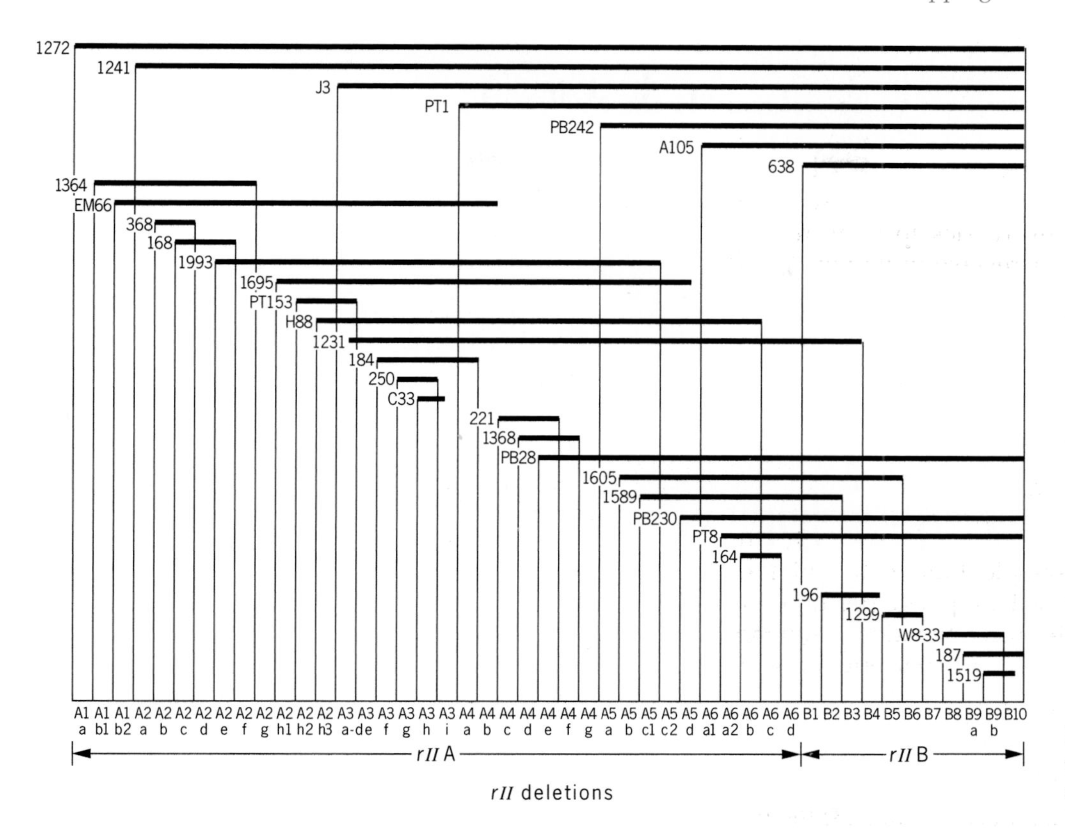

Figure 15.11 The deletions used by Benzer in his analysis of the T4*rII* region.

that the unknown mutant mapped in the region roughly defined by segment A4b.

With only 12 crosses, Benzer mapped a mutation to a small segment of the *rII* region. Once he had mapped each mutant to a particular *rII* segment, he used two-factor crosses to locate the mutant site more precisely within the segment. He thus produced a map of the T4 *rII* region (Figure 15.14) that was the most detailed genetic map at that time.

Four important conclusions emerged from Benzer's analysis of the *rII* region:

1. The 2400 *rII* mutants mapped to 308 different sites. However, the mutants were not randomly distributed over the 308 sites. Some *rII* sites mutated more

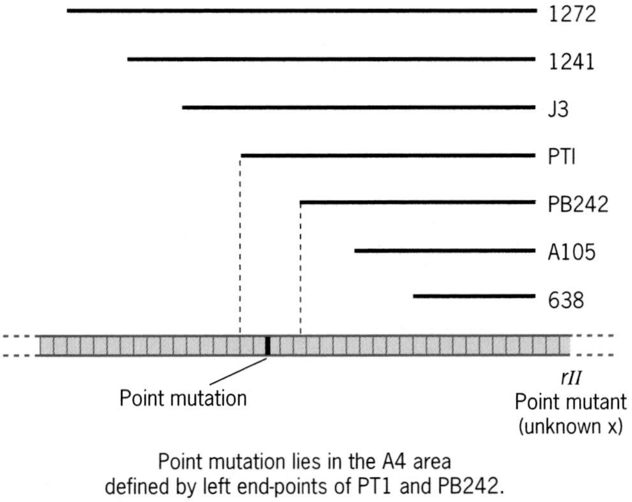

1272
1241
J3
PTI
PB242
A105
638
Point mutation
rII
Point mutant (unknown x)

Point mutation lies in the A4 area defined by left end-points of PT1 and PB242.

Figure 15.12 An unknown *rII* point mutant is crossed to the "big 7" deletions. r^+ recombinants appear only in crosses with PB242, A105, and 638. This places the unknown in the shaded area.

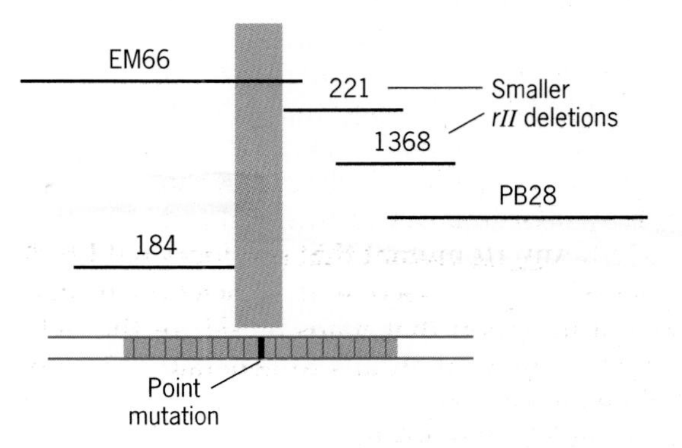

rII point mutation lies in the shaded area (A4b) because r^+ recombinants were produced by all crosses except for EM66 deletion.

Figure 15.13 The *rII* point mutant in Figure 15.12 crossed to smaller deletions. r^+ recombinants are found with all deletions except EM66.

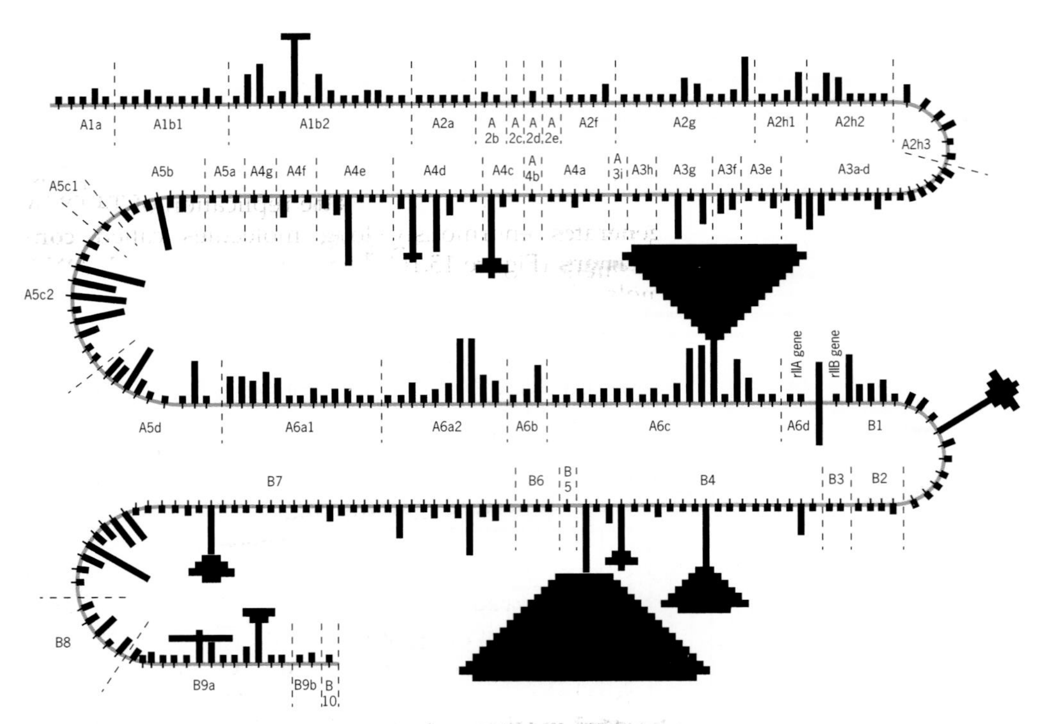

Figure 15.14 The genetic map of the *rIIA* and *rIIB* genes of phage T4. Each square corresponds to an independent occurrence of a mutation at a given site on the chromosome. Thus some regions ("hot spots") mutate much more frequently than others.

often than others. These sites are called mutational **hot spots**.

2. All 2400 *rII* mutants mapped to either the *rII*A gene or the *rII*B gene. Benzer suggested that the two genes encoded two different polypeptides, although at the time he had no evidence to support this idea. He has since been proven correct.
3. The smallest recombination frequency found by Benzer was 0.02 percent. Because the T4 map is about 1500 map units in length, this frequency represents 0.02/1500 = 0.00133 percent of the map. If recombinational events are equally likely over the entire T4 map, given that the T4 genome is about 173,000 base pairs in length, the smallest region in which recombination can occur is (0.0000133) × (173,000), or about 2.3 base pairs. Thus Benzer's research suggested that recombination can occur between any pair of nucleotides.
4. Benzer's work extended the results of earlier experiments with *Drosophila* (Chapter 14) which demonstrated that the idea that the gene was not divisible by mutation and recombination was incorrect.

Detailed knowledge of the T4 *rII* region has given researchers numerous opportunities to explore other complex issues. For example, the *rII* system has been successfully used to elucidate the mechanism of chemically induced mutational events. Knowledge of the triplet, nonoverlapping nature of the genetic code was made possible using *rII* mutations caused by the addition or deletion of individual base pairs. Researchers gained important insights into the mechanism of recombination using *rII* mutants. Benzer's analysis of the T4 *rII* system stands today as one of the greatest achievements in modern genetics.

Key Points: **Phage phenotypes, such as plaque morphology and host range, were utilized in early mapping studies. The process of recombination in phage is different than it is in sexually reproducing organisms. Benzer mapped over 2400 mutations in the *rII* region of T4 and demonstrated that the base pair was the smallest unit of mutation and that recombination probably occurred between adjacent base pairs. His results solidified the concept of the gene as a unit of function that is divisible by mutation and recombination.**

T4: A CIRCULAR GENETIC MAP BUT A LINEAR CHROMOSOME

One of the most widely used bacteriophages for genetic study is T4. Early results of gene mapping implied that the T4 genes might be distributed over as

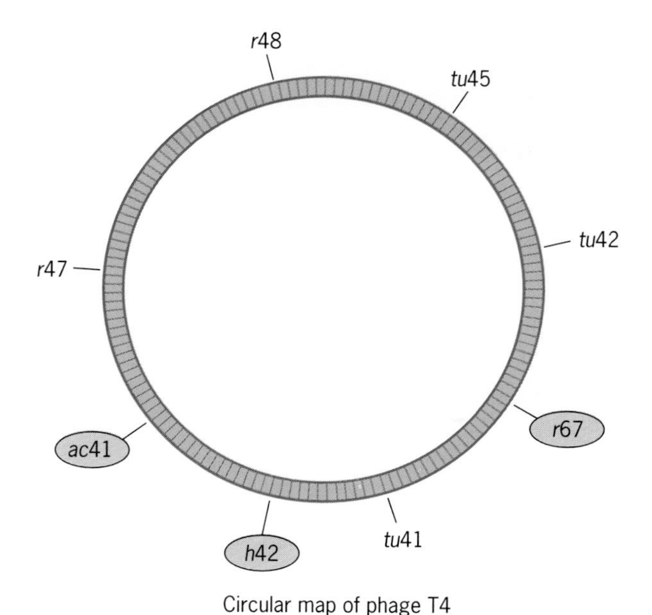

Figure 15.15 The circular map of phage T4.

many as seven different linkage groups. Physical studies showed, however, that T4 carried a single DNA molecule. All T4 genes were linked to this DNA molecule. T4 mapping studies uncovered some intriguing problems. For example, genetic crosses established the following linear gene order: *h*42—*ac*41—*r*67. When these three genes were used in three-point crosses, the rarest recombinant classes—presumably the double cross-overs—placed *h*42 in the middle. This contradiction of the map sequence created a dilemma.

The dilemma was resolved by suggesting that the T4 map was circular (Figure 15.15). With a circular map, *h*42 maps to the left and to the right of both *ac*41 and *r*67, depending on the direction taken around the circle. Other crosses confirmed this circular map. But is the T4 chromosome physically circular? Physical studies of T4 DNA showed that it was linear.

How is it possible that a linear chromosome generates a circular map? Based on a detailed study of the segregation pattern of r/r^+ heterozygotes by Gus Doermann, George Streisinger and Franklin Stahl suggested that the T4 chromosome was *terminally redundant* and *circularly permuted*: The ends of a T4 chromosome are the same—that is, they are redundant (*abcdefg........wxyzabc*). However, the endpoints for each chromosome are different—that is, one may be *abcdef .... xyzabc*, and another *defghi .... xyzabcdef*. Sometimes *d* is on the end, and sometimes it is in the middle. The combination of these two features results in a circular genetic map from a linear chromosome.

Remaining to be resolved was the enigma of how a single T4 virus with a single, linear, terminally redundant chromosome could produce progeny with a variety of terminally redundant, circularly permuted chromosomes. That is, how can a parent virus that is redundant for *ab* at its ends produce progeny that are redundant for *cd*, *de*, *ef*, and so forth? The key to the answer is the way that T4 DNA replicates and is packaged into the phage head. The replication of T4 DNA generates enormously long molecules called **concatamers** (Figure 15.16). They arise as the single DNA molecule replicates and the replicated molecules join end to end through a recombination process. T4 head proteins fold around these concatameric DNA molecules until the head is full. The amount of DNA required to fill a head is more than a single set of genes starting at *a* and ending at *z*. Because there is still room left in the head after *z*, genes *a*, *b*, and *c* are packed in again. The head is now full and the DNA is cut. This particular virus is redundant for genes *a*, *b*, and *c*. The next virus particle begins packaging DNA at *d*, gets to *c*, and then keeps going to *d*, *e*, and *f*. This virus particle is redundant for genes *d*, *e*, and *f*. The next virus begins packaging at *g* and is redundant for *g*, *h*, and *i*. This particular mode of DNA packaging, called the **headful mechanism** (Figure 15.17), explains both terminal redundancy and circular permutations: All the progeny viruses carry redundant DNA molecules, and the molecules are redundant for different genes.

Strong physical evidence supports the terminally redundant-circularly permuted model of the T4 chromosome. For example, terminal redundancy can be demonstrated when T4 DNA is treated with an exonuclease that digests DNA only from the 5' end of the molecule (Figure 15.18). This process creates single-stranded ends that are complementary to each other. With the creation of these conditions, the ends fold back onto themselves, hybridize, and form circular molecules. If the ends were not terminally redundant, there would be no circular molecules because the ends

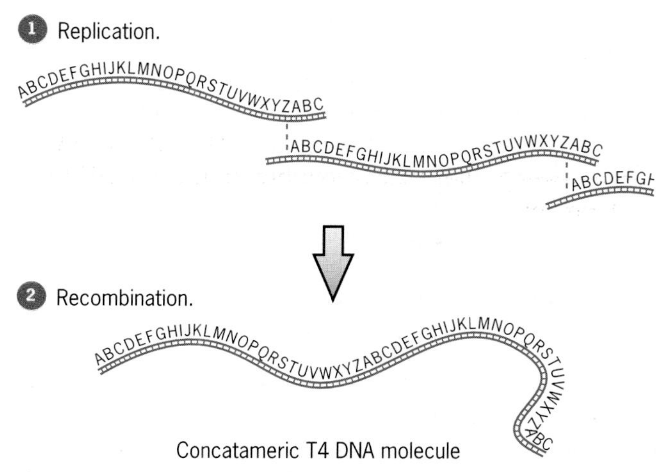

Figure 15.16 The formation of a concatameric T4 DNA molecule.

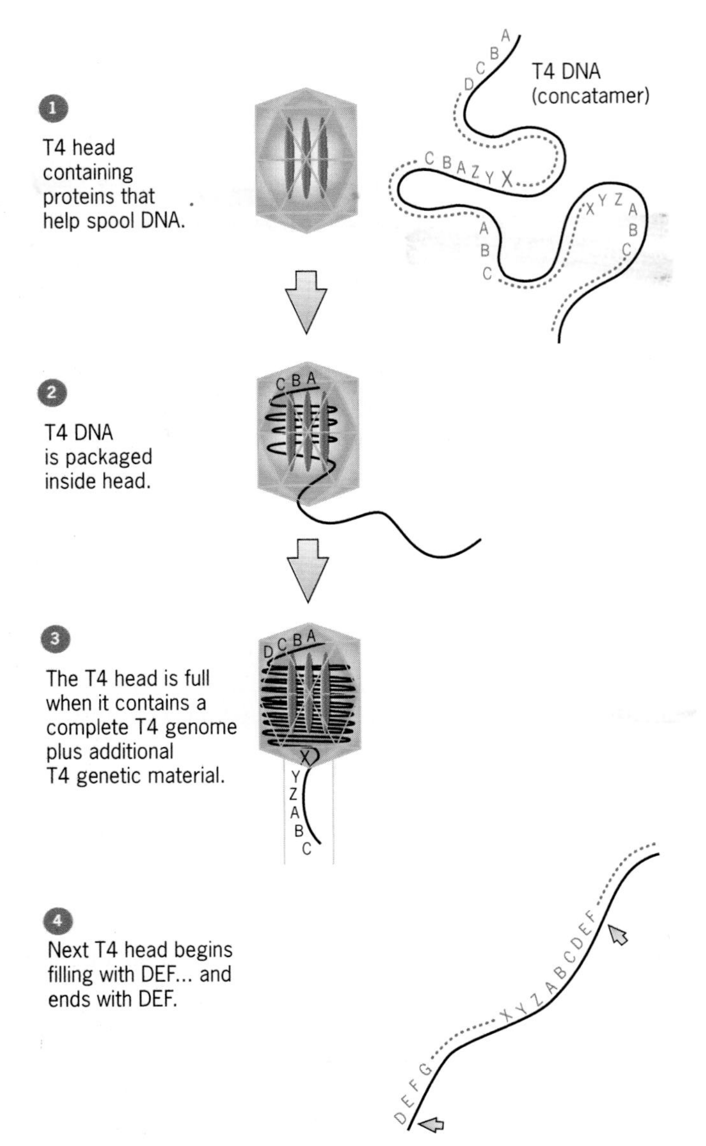

Figure 15.17 The headful mechanism.

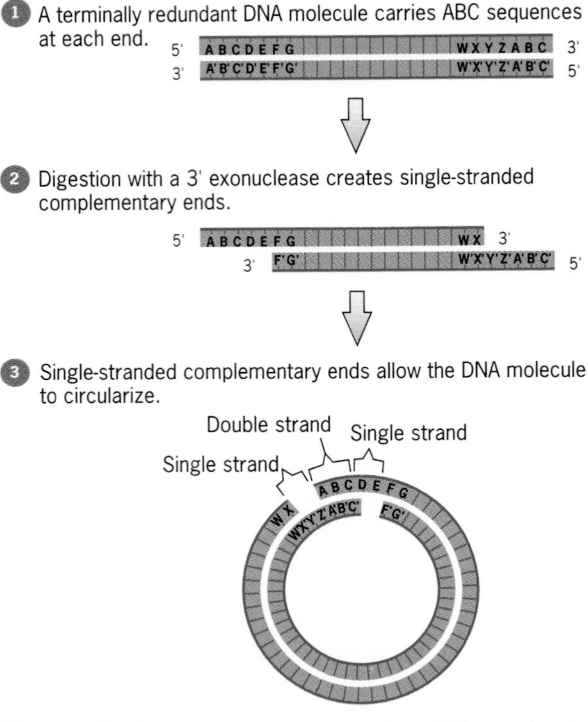

Figure 15.18 A terminally redundant DNA molecule and its identification by 3' exonuclease digestion and circularization. A nonredundant molecule would not circularize.

would not be complementary. That circular DNA molecules can be observed with the electron microscope when this procedure is followed supports the hypothesis that the DNA molecules are terminally redundant.

Evidence supporting circularly permuted chromosomes can be obtained if a mixture of different T4 chromosomes is denatured into single strands and then allowed to renature (Figure 15.19). During renaturation, single DNA strands from different molecules randomly come together and hybridize. These strands, however, have different redundancies; thus redundant ends of both strands cannot hybridize. These unhybridized single strands appear as tails in an otherwise circular molecule. If all the ends on all the DNA molecules were redundant for the same sequences, these tails would not appear. Electron micrographs of DNA circles with tails support this model.

Key Points: **Phage T4 has a circular genetic map, but the chromosome is physically linear. To resolve the paradox, experiments showed that the chromosome was terminally redundant and circularly permuted.**

GENES WITHIN GENES: BACTERIOPHAGE ΦX174

One of the most fascinating discoveries about gene organization in viruses has come from studies of the bacteriophage ΦX174. This phage has a circular, single-stranded DNA molecule, 5386 nucleotides in length. The complete nucleotide sequence of this DNA molecule has been determined. Early studies suggested that there might be six or seven genes, but later studies showed that the ΦX174 genome codes for 11 different proteins containing more than 2300 amino acids, and therein lies the puzzle. The genetic code is triplet: a sequence of three nucleotides codes for a single amino acid. If all 5386 nucleotides coded for amino acids, the maximum number of amino acids possible would be 5386/3 or 1795, more than 500 amino acid residues fewer than observed. How can this discrepancy be explained? Furthermore, since an average protein contains about 400 amino acids, the ΦX174 genome would encode only 4 or 5 proteins, not 11.

This puzzle was solved when it was shown that

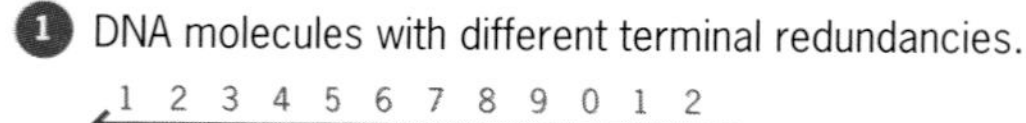

2 Denature DNA molecules then renature different single strands.

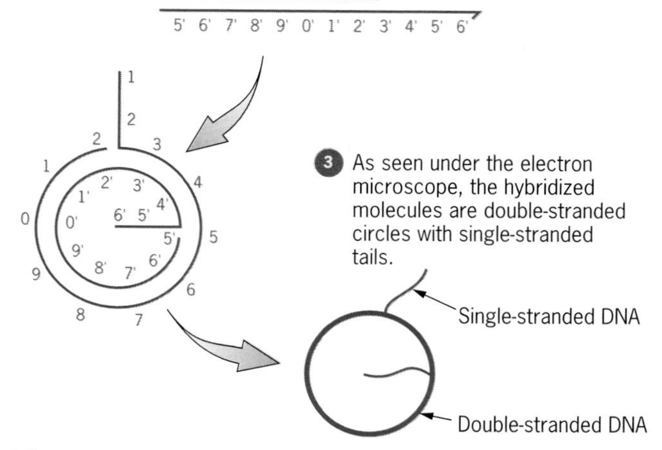

Figure 15.19 Circularly permuted DNA molecules and their identification by denaturation and renaturation. Renaturation of strands with different termini produces double-stranded circular molecules (heavy line) with short single-stranded branches (thin lines).

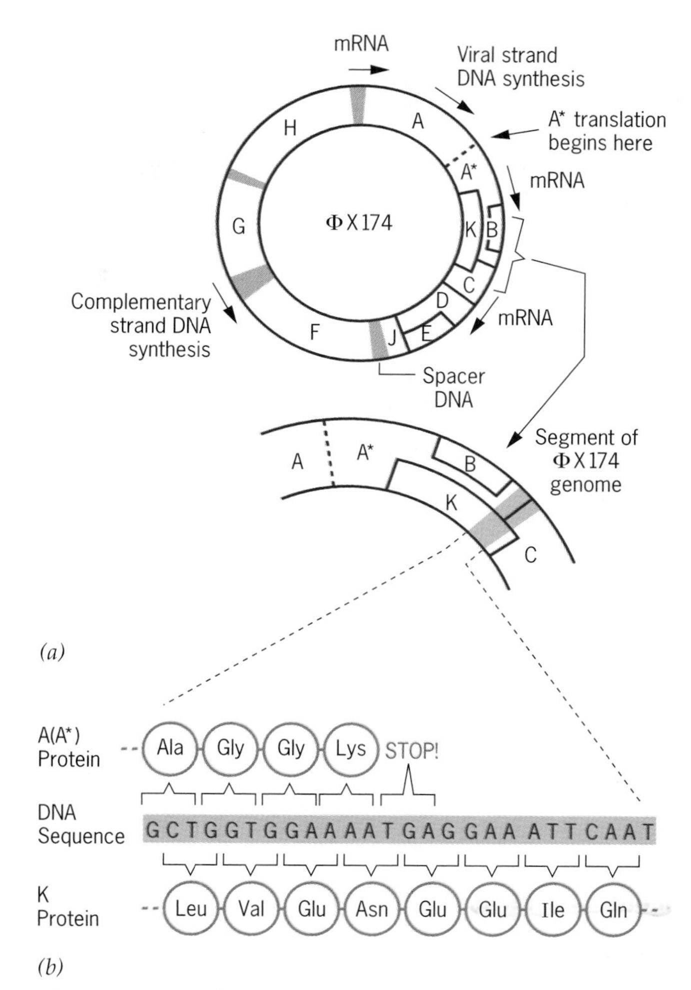

Figure 15.20 (*a*) The map of the phage ΦX174 showing the start and stop points for mRNA transcription and the boundaries of the individual protein products. The solid regions are spacers. (*b*) The DNA and amino acid sequence at the end of the *A* (*A**) gene and the *K* gene.

ΦX174 has **overlapping genes**: Different reading frames within the same DNA sequence code for different proteins (Figure 15.20). For example, protein K is encoded by a DNA sequence shared with the *A* and *A** gene sequence. (*A* and *A** are read in the same frame, but they start at different points.) However, the reading frames for *A* (*A**) and *K* are different. Thus, even though the same nucleotides are used, the amino acid sequences for *A* and *A** are different from *K*. The key feature of all the overlapping sequences is the presence of the initiation codon, AUG (TAC in the DNA molecule), within the gene sequence. This initiation codon, which codes for N-formylmethionine (F-met) signals the first amino acid in nearly all proteins (Chapter 12).

Overlapping genes allow the virus to make maximum usage of its 5386 nucleotides. A greater variety of proteins results from a relatively small amount of DNA. However, there is a tradeoff. A single mutation will affect the structure of more than one protein, thus slowing the evolutionary process because it is unlikely that a single mutational event would have a positive or neutral impact on two or three proteins. In spite of this constraint, overlapping genes have evolved in ΦX174 and other viruses.

Key Points: **The bacteriophage ΦX174 has enough genetic information in its single-stranded DNA chromosome to code for about four or five proteins. However, because its genes overlap, ΦX174 codes for 11 proteins.**

PHAGE HETEROZYGOSITY: A CLUE TO THE MECHANISM OF RECOMBINATION

Bacteriophage contain a single molecule of DNA (or sometimes RNA). Thus, except for terminally redundant gene sequences, a phage carries only one allele of each gene. In 1951, Alfred Hershey and Martha Chase suggested that under some circumstances a phage may appear as if it were actually heterozygous, a phenomenon normally observed only in diploid organisms. Their suggestion emerged from observations of

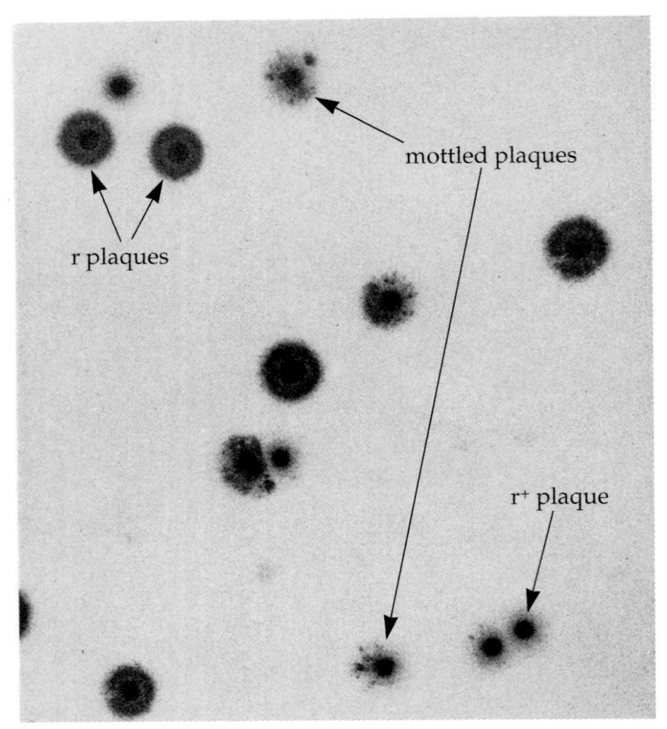

Figure 15.21 Mottled plaques appear after infection of *E. coli* cells by r^+ and *rII* phage.

Key Points: DNA molecules may break in such a way that they are left with single-stranded ends. If DNA fragments with complementary single-stranded ends come together, they may form a heteroduplex molecule: One DNA strand in a specific region carries genetic information for one allele, and the other DNA strand carries genetic information for a different allele. This discovery provided insight into the molecular mechanism of recombination.

mottled plaques (Figure 15.21) on *E. coli* lawns in which the progeny of T2 *r* phage crossed to wild-type (T2 r^+) were plated. The mottled plaques contain both T2 r^+ and T2 *r* phage; thus they have some *r* as well as some wild-type (r^+) qualities. Because each plaque is the result of a single virus particle, a mottled plaque indicates that a single parental phage carried both *r* and r^+ alleles. But how can a single virus particle carry two different alleles? Mutation seemed unlikely because the frequency of mottled plaques was too high to be accounted for by mutational events.

One explanation for the mottled plaques is the formation of **DNA heteroduplexes** (Figure 15.22). Heteroduplex molecules form as a result of recombination between *r* and r^+ molecules. Staggered breaks in the r regions of the two molecules create "sticky" ends that can hybridize because of their complementarity. A union of two DNA fragments creates a recombinant molecule carrying the *r* allele on one strand and an r^+ allele on the other. On replication of this heteroduplex structure, one daughter molecule will be r^+ and the other will be *r*. Thus a single DNA molecule gives rise to two different progeny classes.

The formation of heteroduplex molecules has been experimentally confirmed in a number of different systems. Heteroduplex molecules represent a common intermediate structure during the process of recombination.

HIV: A EUKARYOTIC VIRUS

In the summer of 1981, the Centers for Disease Control (CDC) reported the unexplained occurrence of *Pneumocyctis carinii* pneumonia in five previously healthy homosexual men in Los Angeles. The same year, the CDC reported a rare skin cancer called Kaposi's sarcoma in 26 previously healthy homosexual men in New York and Los Angeles. These two reports from the CDC represent the first time that **acquired immune deficiency syndrome** (**AIDS**) was recognized in the United States.

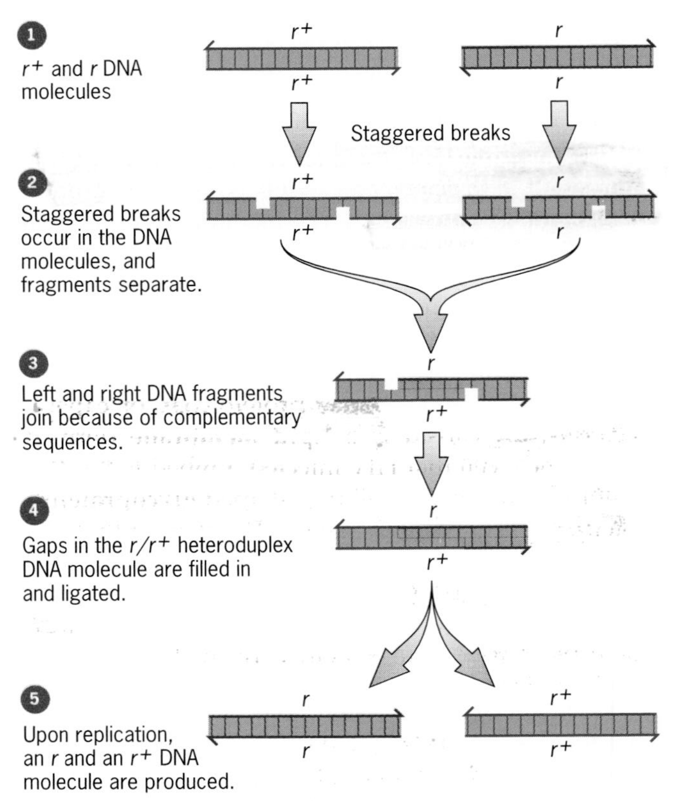

Figure 15.22 A model for the formation of a heteroduplex molecule and the generation of genetically different DNA molecules from a single parent molecule.

By 1985, researchers were referring to the 1980s as the decade of the HIV/AIDS pandemic. Many researchers thought that AIDS would be conquered before the 1990s began. They were wrong, for the 1990s mark the second decade of the AIDS pandemic. Today, nearly a million people in the world have been diagnosed with AIDS. In the United States alone, more than a million people are infected with the **human immunodeficiency virus** (**HIV**), the agent that causes this disease. These people are not yet showing symptoms of the disease, but they will unless a cure can be found.

What is AIDS?

The basic feature of AIDS is the slow and steady decline of the immune system. This deterioration is primarily the result of the decimation of one population of white blood cells. Because the immune system is so compromised, people with AIDS are vulnerable to a host of diseases that are rarely found in the healthy, non-HIV-infected population. These opportunistic infections occur because organisms normally unable to infect the body are able to do so. For example, *Pneumocyctis carinii*, a parasite that causes a rare type of pneumonia, is normally controlled by a healthy immune system. In an immune-compromised person, the parasite is devastating.

The Structure of HIV

A virus of any type can be viewed as a bit of bad news wrapped in a protein coat. Certainly, bacterial viruses are bad news for bacteria, and eukaryotic viruses are bad news for eukaryotes. The basic structure of HIV has much in common with other eukaryote viruses. Externally, HIV looks like a twenty-sided soccer ball (Figure 15.23). This shape is determined by the proteins that make up the outer protein coat, or **capsid**. Overlying the capsid is a **lipid membrane** derived from the host cell that HIV infected. Embedded in this membrane are various lollipop-shaped **glycoproteins**, proteins complexed with sugars. The stem of the "lollipop" has a molecular weight of 41 kilodaltons (kd), and so is called gp41; the top part of the lollipop has a molecular weight of 120 kd and is referred to as gp120. Inside the capsid is a **protein core** made of various protein molecules.

The genome of HIV is not like that of the T phages discussed earlier. HIV has RNA, not DNA, as its genetic material. Two identical molecules of this RNA are packaged inside the protein core, along with an enzyme called **reverse transcriptase** that transcribes the RNA into DNA.

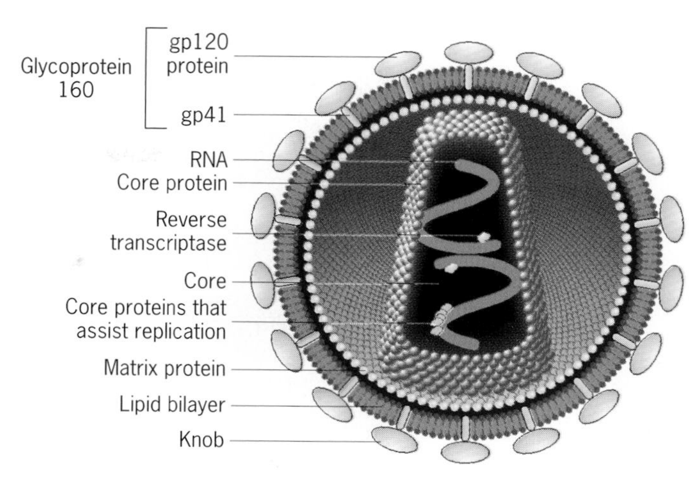

Figure 15.23 The structure of the human immunodeficiency virus (HIV). The core contains the viral genetic material and proteins required for replication. The protein capsid is surrounded by a membrane derived from the cell the virus infected. Viral proteins are embedded in the membrane.

The HIV Life Cycle

The hallmark of the HIV life cycle is the reverse transcription of genomic RNA into DNA by the viral enzyme, reverse transcriptase (Figure 15.24). The HIV life cycle begins with the binding of gp120 to the **CD4 receptor** on the membrane of a type of immune cell called the **helper T cell**. The CD4 receptor is also found on another type of immune cell called a **macrophage**. Other recently discovered HIV-binding receptors, such as fusin and CC-CKR-5, are also essential for HIV entry into cells. Once HIV has bound to the host cell, the HIV membrane fuses with the host cell membrane. Fusion is mediated in part by the gp41 protein. Inside the cell, the RNA is uncoated. Reverse transcriptase converts RNA into double-stranded DNA. The conversion of RNA into DNA is called reverse transcription, and for this reason, HIV belongs to a viral class called **retroviruses** (reverse viruses). (See Human Genetics Sidelight: Genomic Fossils.) The DNA migrates to the nucleus where it is integrated into the host cell genome by a virally encoded enzyme called **integrase**. The integrated DNA is a permanent part of the host cell genome. The integrated HIV genome may be inactive, or it may be transcriptionally active, producing progeny viral genomes.

Activation of the HIV genome requires virally encoded and host cell-encoded protein factors. If the infected cell is not activated, integration of the HIV genome into the host genome is inefficient. Furthermore, activation of the cell is required for the integrated HIV genome to be transcribed into either genomic or messenger RNA.

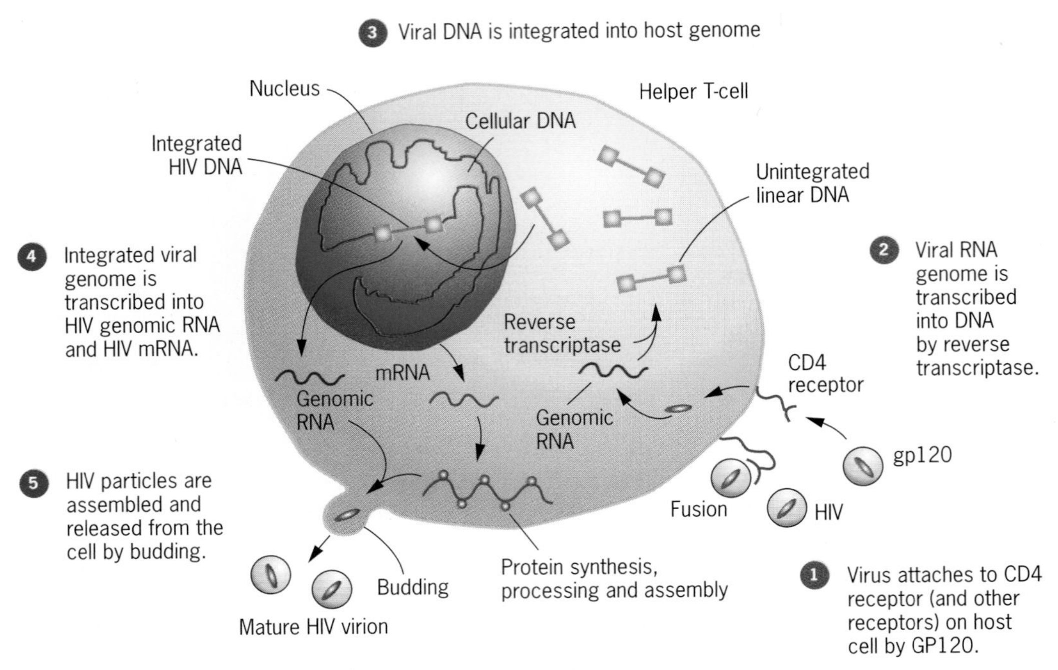

Figure 15.24 The life cycle of HIV.

Following transcription, HIV mRNA is translated into viral proteins. The viral core is formed at the cellular membrane. The core is composed of core proteins surrounding viral genomic RNA and enzymes. Budding of the progeny viruses occurs through the host cell membrane, where the core acquires its external envelope.

The HIV Genome

Genetic analysis of the HIV genome reveals that it is similar to that of other retroviruses (Figure 15.25). The *gag* gene encodes the proteins that form the core; the *pol* gene encodes reverse transcriptase and integrase; and *env* encodes the envelope glycoproteins. These are

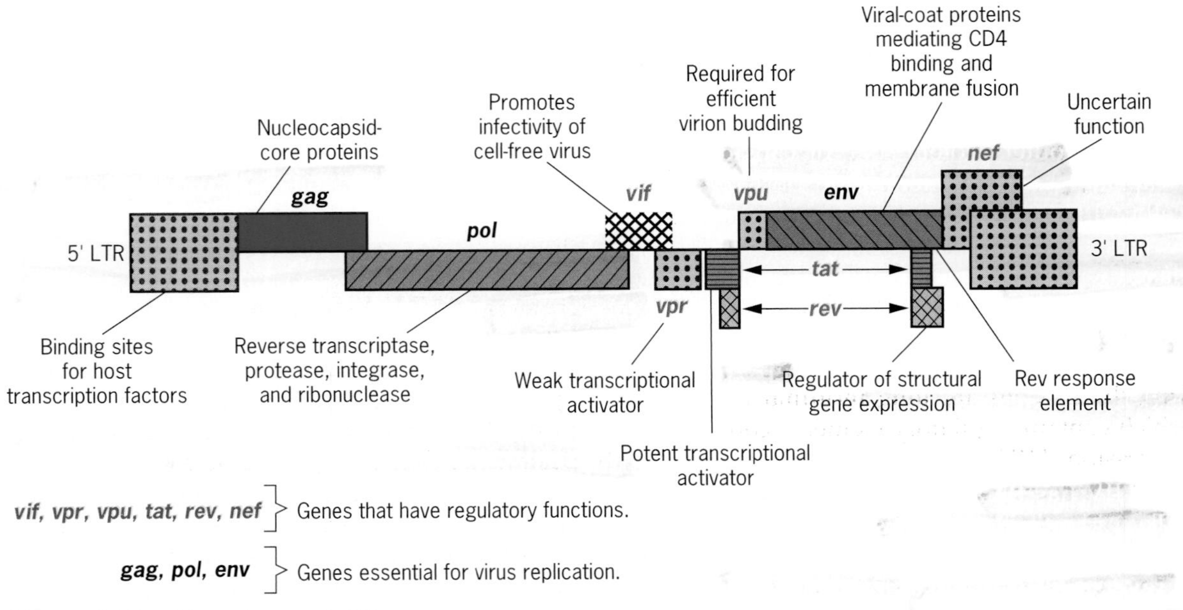

Figure 15.25 The integrated HIV genome.

HUMAN GENETICS SIDELIGHT

Genomic Fossils

Upon infecting a cell, eukaryotic RNA viruses employ reverse transcriptase to convert viral RNA into DNA. The DNA transcript of the viral genome then inserts into the host cell's genome. If the infecting retrovirus is HIV, the consequences are disastrous. But are all retroviral infections devastating? Perhaps not.

At least 0.1 percent of the human genome, approximately 3 million base pairs, is related to retroviral genomes. These remnants of retroviral genomes likely originated from ancient infections of embryonic cells. If the individual survived the infection and if virally infected cells became part of the germ tissue, the retroviral genome would be transmitted from generation to generation as if it were a Mendelian unit.

Some virus infections create a natural immunity to infection by other retroviruses. Powerful evidence supporting this claim comes from a study of a retroviral disease in a population of wild mice near Lake Casitas in southern California. The mice in this population were suffering from a virulent retroviral infection causing a fatal hind limb paralysis. About 20 percent of the mice in the Lake Casitas population were infected and died. When the same virus infected a laboratory strain of the same mouse species, over 90 percent of the infected mice died. Clearly, the Lake Casitas population was not as sensitive to the virus as the laboratory mouse population. What might cause this apparent immunity in the wild population?

Murray Gardner and his colleagues discovered that retroviral resistance was conferred in these mice by a gene called *Fv-4*. This gene blocked infection by the retrovirus in animals carrying it. Analysis of the *Fv-4* gene revealed a surprise: the gene was a remnant of the retroviral genome that caused the fatal hind-limb paralysis syndrome. This retroviral remnant was integrated into chromosome 12 of resistant mice. The remnant contained a portion of the retroviral *pol* gene (for reverse transcriptase), an incomplete copy of the *env* gene (for envelope proteins), and sequences in the 3' region that regulate gene expression. The incomplete *env* gene encodes an envelope glycoprotein that binds to and blocks the retroviral receptor protein, preventing the virus from infecting cells.

The Lake Casitas population apparently experienced an infection by the fatal retrovirus at a point in the distant past. In some infected individuals, a segment of the retroviral genome became integrated into the genome of cells destined to become germ cells and was transmitted to the offspring. These now resistant offspring had an advantage over mice lacking this retroviral insert. The number of resistant mice increased at the expense of the sensitive mice.

Expression of endogenous retroviral genomes in human cells is rare, and there is no evidence that endogenous retroviral gene expression confers resistance to any retroviral infection. There is the possibility, however, that nonhuman primates may have a natural immunity to a relative of human immunodeficiency virus (HIV), called the simian immunodeficiency virus (SIV). SIV is endogenous in certain species of African monkeys, such as the green monkey and the sooty mangabey. These monkeys do not develop any disease symptoms when infected by SIV. However, if Asian macaques are infected with SIV, they develop AIDS. One interpretation of SIV resistance in the African monkeys is that a previous and probably ancient SIV infection resulted in SIV resistance through integrated, truncated SIV genomes. SIV resistance was subsequently selected for. The integrated retroviral remnants have evolved since the original infection. The Asian primates were sensitive to SIV because they never encountered this virus before. They had no resistance.

In light of these instances of retroviral resistance, one cannot help but reflect on humans who are HIV positive, and have been for decades, but never developed AIDS. Is it possible that these individuals have acquired an immunity similar to the type found in the Lake Casitas mice or the SIV resistant African monkeys?

Gardner, M. B., M. B. Kozak, and S. J. O'Brien. 1991. The Lake Casitas wild mouse: Evolving genetic resistance to retroviral disease. *Trends in Genetics* 7:22–27.

the basic genes of all retroviruses, but HIV is much more complex. At least six other genes (*tat, rev, nef, vif, vpr,* and *vpu*) code for proteins that regulate HIV expression. Flanking these genes are **long terminal repeats (LTRs)**, which contain regulatory elements controlling gene expression. HIV contains genes within genes, a phenomenon observed in the bacteriophage, ΦX174 (see page 392). For example, *vif, pol,* and *vpr* have DNA sequences in common. Also, two HIV proteins, Tat and Rev, are produced by alternate RNA splicing.

The Course of HIV Infection

During the first two months following HIV infection, called the acute phase, the patients can experience flu-like symptoms: fatigue, fever, headache, and muscle aches. Scientists can detect large numbers of free HIV particles in the bloodstream during this time. However, toward the end of the acute phase, the number of free HIV particles in the blood declines and the virus becomes localized in the lymphatic system. During all stages of the infection, the immune system is actively fighting the virus. HIV-specific antibodies and killer T-cells are deployed in this battle.

Following the initial events of the acute phase, the flulike symptoms disappear and the infected person becomes asymptomatic. This asymptomatic phase lasts on average 8 to 10 years. During this period, there is a gradual decline in the number of helper T-cells. Some of the infected cells continue to produce HIV particles during this time, but the immune system still remains capable of managing the free viruses and the infected cells. Thus the individual remains healthy. The fact that infected people can remain healthy for 10 years suggests that the immune system has tremendous excess capacity.

Eventually, the immune system wears down. The population of helper T-cells diminishes, so that the load of HIV particles is not cleared from the system. Disease symptoms begin to appear. The normal function of helper T-cells is to direct the other cells of the immune system. When the helper T-cells are destroyed, the entire immune system fails and opportunistic infections occur.

The mechanism by which HIV kills the helper T-cells is not understood. Does the virus produce a protein that destroys the cell, or is cell death the result of some other type of viral activity? An uninfected individual normally has about 800 helper T-cells per microliter of blood. One way a person is diagnosed with AIDS is when the helper T-cell count falls below 200 cells per microliter of blood in a person who has HIV antibodies. This means that 75 percent of the helper T-cells have been lost. However, when these remaining helper T-cells are examined, only about 1 percent of them are HIV infected. How can so many cells die when so few are actually infected?

Some researchers have suggested that HIV infected cells have an indirect effect on uninfected cells, producing a product that kills them or a product that targets them for killing. One suggestion is that HIV infected cells secrete gp120 fragments from the HIV "lollipop" into the circulating blood. These gp120 proteins are picked up by helper T-cells and displayed on their surface. These cells now look as if they are infected. The gp120 fragments are targets for destruction by killer T-cells, even though the helper T-cells are not HIV-infected. Another suggestion is that gp120 on the surface of infected cells binds to the CD4 molecule on the surface of uninfected cells. This could lead to the development of a syncytium of uninfected and infected cells—a mass of cells all clumped together.

Some researchers are focusing on other HIV proteins. They suggest that Tat and Nef may be secreted by infected cells to regulate or even destroy the activity of infected but latent and uninfected helper T-cells. The strategy of research scientists is to target HIV genes that regulate the infection process and to look for ways to disrupt the viral life cycle. Unfortunately, HIV has evolved ways not only to evade the immune system, but also to destroy it. As the virus replicates within the host cell, variants emerge that are more efficient killers, more effective at infecting cells than their progenitors.

THE ORIGIN OF VIRUSES

All viruses, even the most complex ones, depend on the host cell, whether bacterial or eukaryotic, for their functions and their basic existence. Viruses do not code for ribosome components or for the enzymes that are involved with energy production. They are totally dependent on the host cell's metabolic machinery. This dependence strongly suggests that viruses evolved *after* cells appeared. They probably evolved from small segments of host cell DNA or RNA that had acquired the capacity to replicate independently of the regular host cell genome. If such nucleic acid segments acquired coding sequences that actually coded for an encapsulating protein enabling the nucleic acid segment to exit the cell and to infect another cell, then this encapsulated nucleic acid molecule could well represent the primitive viral form.

If viral genomes evolved from segments of host cell genomes, they had regions of sequence homology with this cellular DNA. This homology served as a basis for recombination between viral and host cell genomes and created opportunities for increasing the complexity of the viral genome. Interacting selection strategies between the viruses and the cells they infect have led to the abundance of viruses that parasitize or destroy cells today.

TESTING YOUR KNOWLEDGE

1. A T4 bacteriophage has the terminally redundant chromosome: ABCDEF .... QRSTUVWXYZABC. This phage experiences a deletion that removes EF. This deletion mutant then infects an *E. coli* cell and produces progeny phage. Which of the following progeny groups would not be expected:

(a) ABCDG....WXYZABC, BCDGHIJ....XYZABCD, GHIJK.... ABCDGHI
(b) ABCDG...ZABCDG, BCDGHI...YZABCDGH, GHIJK... BCDGHIJK
(c) ABCDGH...WXYZA, BCDGHI...WXYZAB, GHIJK... YZABCDG

ANSWER

The answer to this question depends on knowledge of the headful mechanism. Progeny group (b) is the only one that is consistent with the deletion of two genes from the T4 chromosome. Groups (a) and (c) are not expected. The redundant chromosome has 29 letters (26 + 3 extras). By deleting 2 letters, it means that the base of 26 letters + 3 extras now is 24 letters + 5 extras. Therefore, the redundancy will be 5 letters in length; (a) has 3 extras, and (c) has 1 extra.

2. Three T4 *rII* deletion mutants are intercrossed in pairwise fashion. When deletion X is crossed to deletion Y, no recombinants are found; When deletion X is crossed to deletion Z, no recombinants are found; when deletion Y is crossed to deletion Z, wild-type recombinants are found. What is the map order of these deletions?

ANSWER

The results of the various crosses tell us that X overlaps both Y and Z because no recombinants are found. Y and Z, however, do not overlap because recombinants are produced. The most reasonable map order would therefore be Y X Z:

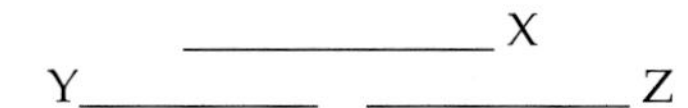

QUESTIONS AND PROBLEMS

15.1 A series of crosses is made using T4:

(a) $m^+ n\, r \times m\, n^+ r^+$
(b) $m\, n^+ r \times m^+ n\, r^+$
(c) $m\, n\, r^+ \times m^+ n^+ r$

Wild-type recombinants are much less frequent in cross (a) than in crosses (b) or (c). Using this information, determine the sequence of the genes?

15.2 Discuss the similarities and differences in recombination as it occurs in viruses on the one hand and eukaryotes on the other.

15.3 How has the study of bacteriophage genetics contributed to the chromosome theory of inheritance?

15.4 Andre Lwoff, in a poetic remark reminiscent of Gertrude Stein, stated that "a virus is a virus is a virus." If you disagree with Lwoff, how would you characterize the virus? If you agree with Lwoff's frustration, why can you not define the virus in any other way?

15.5 Two strains of T4 infect *E. coli* cells: strain 1 is $a\ b\ c$, and strain 2 is $a^+ b^+ c^+$. The recombination data are as follows:

Class	Frequency
$a\ b\ c$	0.34
$a^+ b^+ c^+$	0.36
$a\ b^+ c^+$	0.05
$a^+ b\ c$	0.05
$a\ b\ c^+$	0.08
$a^+ b^+ c$	0.09
$a\ b^+ c$	0.02
$a^+ b\ c^+$	0.02

Do these data support the hypothesis that the three genes are linked? If so, determine the sequence of the three genes and calculate the distances separating them. What is the coefficient of coincidence, and what does it mean?

15.6 Hershey and Chase performed mixed infections using three different T2 strains carrying alleles for three different genes (h, m, and ri) and obtained the following results:

	$h^+ m^+ ri^+$	$h^+ m^+ ri$	$h\ m^+ ri^+$	$h^+ m\ ri^+$	$h\ m^+ ri$	$h\ m\ ri^+$	$h^+ m\ ri$	$h\ m\ ri$
$h\ m^+ ri^+ \times$ $h^+ m\ ri^+ \times$ $h^+ m^+ ri$	25	17	18	16	7	8	6	3
$h\ m\ ri^+ \times$ $h^+ m\ ri \times$ $h\ m^+ ri$	3	4	7	10	16	22	15	23

What conclusions can you draw from this experiment pertaining to recombination?

15.7 The following T4 crosses are performed:

	$r\ h^+$	$r^+ h$	$r^+ h^+$	$r\ h$
$r_x h^+ \times r^+ h$	0.38	0.44	0.09	0.09
$r_y h^+ \times r^+ h$	0.43	0.48	0.04	0.05
$r_z h^+ \times r^+ h$	0.47	0.51	0.01	0.01

x, y, and z refer to different r genes.

(a) Construct a map for each cross.

(b) Construct the possible genetic map(s) for all four genes.

(c) The cross $r_y^+ r_z \times r_y r_z^+$ yields 13 percent recombinants. What does this tell you about the h - r_x - r_z order?

(d) From what you know about the T4 genetic map, can you resolve the remaining gene maps so that they are all compatible?

15.8 Study the following phage map:

x —— 1.2 —— m —— 0.9 —— n

From 80,000 progeny scored in the cross $x^+ m^+ n^+ \times x\ m\ n$, 76 are $x^+ m\ n^+$ or $x\ m^+ n$. What is the interference in this cross?

15.9 Three independently arising *rIIB* mutants are crossed in T4, and the following results are obtained:

	Percent Recombinants
$rIIB_1 \times rIIB_2$	0
$rIIB_1 \times rIIB_3$	0
$rIIB_2 \times rIIB_3$	0.9

Interpret these results.

15.10 How do you determine whether a particular mutation in phage is a deletion?

15.11 *E. coli* B cells are infected by two different T4 *rII* mutants (*p* and *q*). The progeny collected from this mating produce *r* plaques on a lawn of B cells. These same two mutants do not grow on *E. coli* K cells. If, however, the progeny from an *rIIp* × *rIIq* mating on *E. coli* B cells are plated on a lawn of *E. coli* K cells, a few wild-type plaques appear. Interpret these observations.

15.12 Compare and contrast the meanings of the term *heterozygous* as it is used in describing bacteriophage and *Drosophila*.

15.13 How can a physically linear chromosome produce a circular genetic map?

15.14 The phage lambda chromosome is linear except when it is replicating inside a host bacterium. Then it is circular. Yet even though it is circular and recombines as a circular molecule, it produces a linear genetic map. How would you explain this?

15.15 A newly discovered *rIIA* point mutant, when mapped against several deletion mutants, produced wild-type recombinants only when crossed with deletion 638 of the "big seven" deletions. It also produced wild-type recombinants with deletion H88 but not with 1231. Using Figure 15.11, indicate the position of this new mutant.

15.16 T4 DNA is cleaved into fragments about one-fourth the size of the intact T4 chromosome. These fragments are then treated with a 3' exonuclease to create single-stranded ends. Would you expect these fragments to circularize? Explain.

BIBLIOGRAPHY

CAIRNS, J., G. S. STENT, AND J. D. WATSON (eds). 1966. *Phage and the Origins of Molecular Biology*. Cold Spring Harbor, New York: Cold Spring Harbor Laboratory.

CANN, A. J. 1993. *Principles of Modern Virology*. Academic Press, San Diego.

FREIFELDER, D. 1987. *Microbial Genetics*. Jones and Bartlett, Boston.

HAYES, W. 1968. *The Genetics of Bacteria and Their Viruses*. 2nd ed. John Wiley and Sons, New York.

HOFFMAN, M. 1994. AIDS: Solving the molecular puzzle. *Amer. Sci.* (March/April):171–177.

LEVINE, A. J. *Viruses*. 1992. Scientific American Library, New York.

SEILLIER-MOISEIWITSCH, F., B. H. MARGOLIN, AND R. SWANSTROM. 1994. Genetic variability of the human immunodeficiency virus: Statistical and biological issues. *Annu. Rev. Genet.* 28:559–596.

STENT, G. S. 1963. *The Molecular Biology of Bacterial Viruses*. W. H. Freeman, San Francisco.

STENT, G. S. 1966. *Papers on Bacterial Viruses*. 2nd ed. Little, Brown, Boston.

STENT, G. S. AND R. CALENDAR. 1978. *Molecular Genetics: An Introductory Narrative*. 2nd ed. W. H. Freeman, San Francisco.

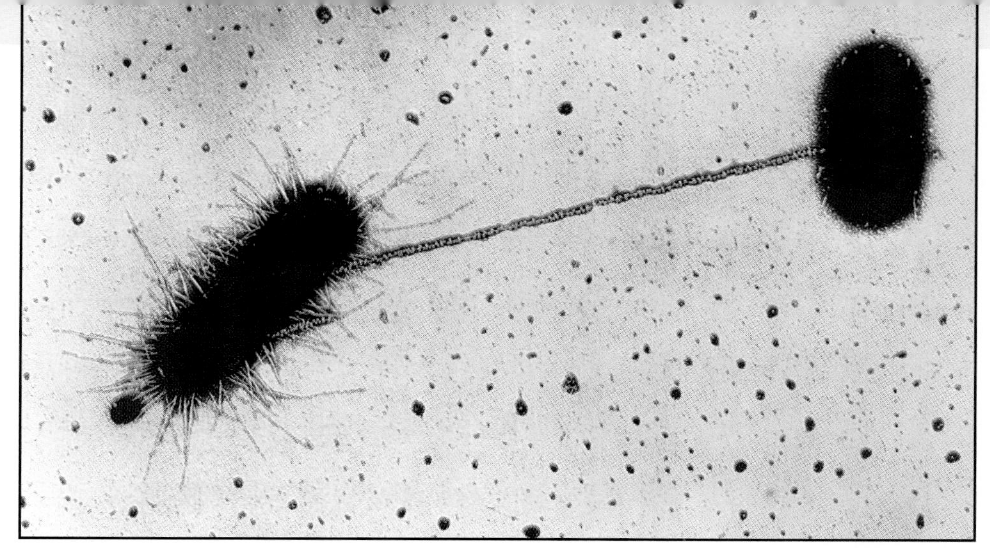

An elongated F^+ *Escherichia coli* cell (left), covered with thin hairs called fimbrae, is connected to a smaller F^- cell (right) by a single sex pilus during the process of conjugation.

16

The Genetics of Bacteria

CHAPTER OUTLINE

Drug Resistance and Sex in Bacteria

Bacteria reproduce asexually; that is, one cell divides by fission into two cells, two cells divide into four, and so on. Although we do not usually think of bacteria as having any kind of a sexual mode of reproduction, they do. In fact, it is one type of sexual reproduction that has created serious problems in hospitals and clinics all over the world. Today, the spread of drug-resistant pathogenic bacteria is one of the most serious threats to the successful treatment of microbial disease. Consider the case of Thomas Walker, middle-aged male with a heart murmur. He entered the hospital for surgery to replace his defective heart valve. This serious operation can sometimes have complications; however, it is seldom fatal. The surgery was successful, but three days later Walker developed a high temperature. Chest X rays showed fluid was building up in his lungs: he had contracted pneumonia. He was treated with two powerful antibiotics, but they failed to stop the infection. Despite the addition of a third antibiotic, he developed respiratory failure and died five days after surgery. Autopsy showed that he died from infection by an antibiotic-resistant bacterium in the genus *Pseudomonas*.

Why did drugs that had in the past been so successful at destroying these bacteria fail? The *Pseudomonas* strain in the hospital had acquired antibiotic resistance genes. Genes for drug resistance are present in the main bacterial chromosome and on plasmids, small, circular DNA molecules that exist separate from the main chromosome. Plasmids carrying resistance genes spread rapidly through a population. They are transmitted from bacterium to bacterium through normal gene exchange processes.

Plasmids carrying drug resistance genes are called R-plasmids (resistance plasmids). The plasmid resistance genes code for enzymes that destroy or modify drugs, rendering them ineffective. What makes R-plasmids so dangerous is the ease by which they are propagated through a population.

A single R-plasmid may carry genes for resistance to several drugs. Thus a patient being treated unsuccessfully with a single drug may also be simultaneously resistant to several others. Drug-resistant strains of pathogenic bacteria are creating disturbing scenarios in hospitals because physicians are running out of options for antibiotics to use in the treatment process, and new antibiotics are not being developed. Of special concern is the emergence of resistance to vancomycin, a powerful antibiotic that is the sole remaining weapon against some of the most deadly microbes. The resistant bacteria that killed Thomas Walker were introduced into his wounds by hospital personnel who did not properly wash their hands. Once in the wounds, the bacteria rapidly multiplied, unchecked by once lethal doses of antibiotics. Many medical researchers fear that the time is growing near when there will be no alternative antibiotic to turn to.

GENETIC EXCHANGE IN BACTERIA: AN OVERVIEW

The development of plasmid-mediated antibiotic resistance points out a mechanism for genetic exchange that has evolved in bacteria. It is not true sexual reproduction in which meiotically produced male and female gametes unite in fertilization, such as that observed in peas, fruit flies, and humans. Even though the process may differ, the end result is similar: The transmission of plasmid DNA from one organism to another creates new genotypes, in this case, drug-resistant genotypes.

Plasmid genes can be mapped, and certain types of plasmids can be used to map genes in the main bacterial chromosome. To map genes in bacteria, geneticists use the same basic experimental strategy: They cross genetically different strains and analyze the progeny for recombination events. They then estimate map distances between two genes by the frequency of exchange events between them. But there are important differences between crossing over in prokaryotes and eukaryotes. Crossing over in eukaryotes occurs only between paired chromosomes during the early stages of meiosis I and results in a reciprocal exchange of genetic material between two homologous chromosomes. Furthermore, both products of a single exchange can usually be recovered. The situation in prokaryotes is different: There is no meiotic process in prokaryotes; recombination is not necessarily reciprocal; and the products of a genetic exchange are not always recoverable. In the last chapter, we examined strategies for analyzing the viral genome. In this chapter, we examine the process of genetic recombination in bacteria and the approaches used for mapping genes.

A bacterium contains a single main circular chromosome carrying thousands of genes, all involved in various aspects of the bacterial life cycle, and one or more extrachromosomal elements called **plasmids**. Genetic exchanges in bacteria rarely involve the interaction between intact main chromosomes from different bacterial strains. Instead, recombination usually occurs between the main chromosome of one strain and a DNA fragment from another strain. The DNA fragment is contributed by a **donor strain** to the **recipient strain** with the intact main chromosome. Recombination occurs in the recipient strain. The transfer of genetic material is thus **unidirectional**.

Because the recipient chromosome is circular and the donor DNA fragment linear, recombination requires an even number of exchange events (Figure 16-1). Only one of the crossover products (the main chromosome) is usually recovered in this process. Recall that in eukaryotes, there can be both odd and even numbers of exchanges. Sometimes, a genetic exchange occurs between the main chromosome and a circular DNA molecule (such as a plasmid) that entered the recipient. In these cases, a single exchange results in the integration of the entire circular molecule into the main chromosome (Figure 16.2).

We discuss three main types of genetic transfer in this chapter. All three are **parasexual** processes because recombination is achieved by mechanisms other than meiosis:

Transformation is the process by which a donor DNA molecule is taken up from the external environment and incorporated into the genome of a recipient cell.

Transduction is the process by which DNA is transferred from one bacterial cell to another by a bacterial virus.

Conjugation is the process by which bacterial cells make effective contact with each other and DNA is transferred from one cell, the donor, to the other, the recipient cell.

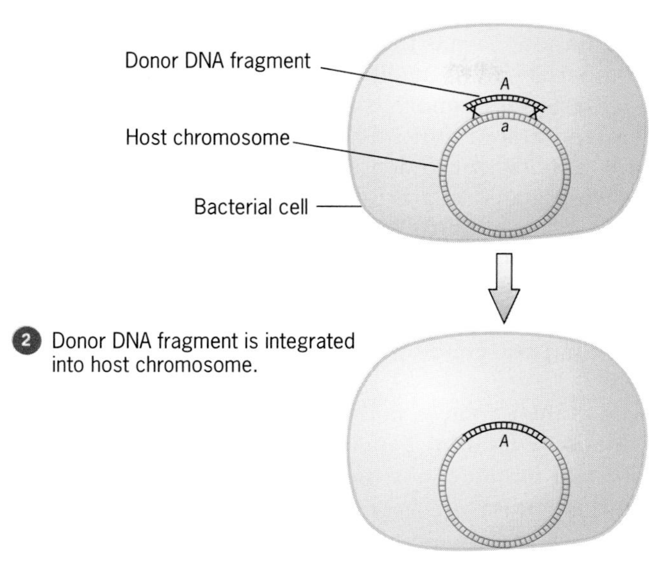

Figure 16.1 In order for successful recombination to occur, an even number of exchanges, two in this case, is required between the circular main chromosome and the linear fragment.

Recombinational events that occur as a consequence of these processes can be used to construct genetic maps of bacterial chromosomes. These maps are generally quite different from the type of genetic map produced by eukaryote crosses. For example, maps produced by conjugation analysis are measured in units of time, not frequencies of crossing over.

Key Points: **Three main mechanisms of genetic exchange in bacteria are transformation, conjugation, and transduction.**

MUTANT PHENOTYPES IN BACTERIA

Bacteria grow in a liquid medium or on the surface of a medium that has been solidified by agar. A single bacterium placed on an agar surface divides over and over again until it produces a visible cluster of cells called a **colony** (Figure 16.3). The number of bacteria growing in a liquid culture is estimated by diluting the bacterial suspension, spreading a known volume of the final dilution on an agar surface, and then counting the number of colonies that form. For example, if we observe 100 colonies on a plate—after we diluted the original bacterial suspension by a factor of 1 million (10^6), and then spread 0.1 ml of this final dilution on an agar surface—we estimate that the bacterial cell concentration per ml in the original suspension

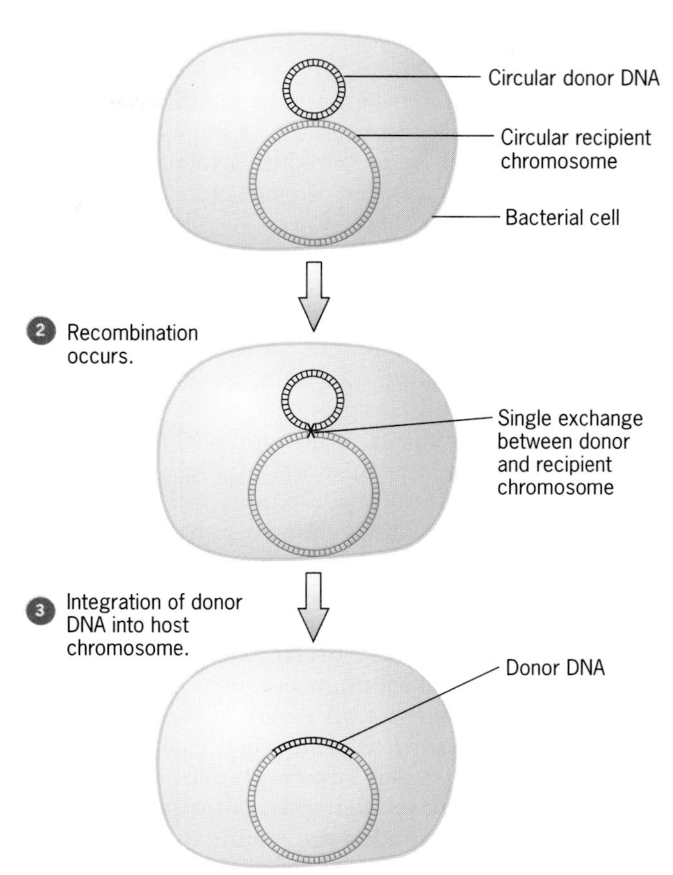

Figure 16.2 The integration of a circular donor DNA molecule into a circular recipient chromosome requires a single crossover event.

Figure 16.3 Colonies of *Staphylococcus aureus.*

was: 100 cells on the plate × 10 (the number of 0.1 ml aliquots per ml) × 1 million (the dilution factor) = 10^9 cells per ml.

The ability or inability to form colonies on a special medium can often be used to identify the genotype of the cell that produced the original colony. The genetic analysis of bacteria requires identifiable mutants. Of those mutants that are particularly useful, let us discuss three:

Antibiotic-resistant mutants. These mutants are able to grow and produce colonies on a medium containing an antibiotic such as streptomycin (Str) or tetracycline (Tet). For example, Str-sensitive cells (Str^s) fail to grow on medium containing streptomycin, but Str-resistant cells (Str^r) do grow.

Nutritional mutants. Wild-type bacteria are able to grow on a simple or **minimal medium** containing only a carbon source and some inorganic salts. From these materials, the cells synthesize all the proteins, nucleic acids, and lipids they require. Wild-type cells are said to be nutritionally independent, or **prototrophic**. Mutations may occur that result in the loss of enzyme function. These mutants are no longer able to grow on minimal medium because they have lost the ability to synthesize an essential nutrient. They grow only if that nutrient is supplied in the medium. These mutants are called **auxotrophic** for the nutrient they require as a medium supplement. For example, if a bacterium has a mutation preventing it from synthesizing the amino acid tryptophan, it is unable to grow on the minimal medium containing only a carbon source and some inorganic salts. However, the tryptophan auxotroph is able to grow on a minimal medium supplemented with tryptophan.

Carbon-source mutants. Some mutants are unable to use certain substances as sources of energy or carbon atoms. For example, lactose mutants are unable to grow and produce colonies on a minimal medium containing the sugar lactose as the sole carbon source.

Other bacterial mutant phenotypes that have also been useful in genetic analysis, include colony morphology, phage resistance, and temperature-sensitive mutants. These last mutants are especially useful for looking at mutations in essential function genes, such as DNA replication.

Bacterial genotypes can be determined by first plating the cells on a nutritionally complete medium containing all the necessary materials for growth. This is a **nonselective medium** because both mutant and wild-type cells grow. Samples of cells from the colonies on the complete medium are placed on various **selective media** in which one or more nutrients are missing or antibiotics have been added. Failure to grow and produce colonies on a selective medium missing a nutrient indicates that the colony from which the cell came on the complete medium was mutant. For example, a selective medium containing only lactose as the carbon source identifies lactose mutants. The ability to grow on a minimal medium containing an antibiotic means that the colony from which the cell originated had mutated to antibiotic-resistance.

Bacterial phenotypes are designated by three letters, the first of which is capitalized, followed by a + or − superscript to denote the presence or absence of the designated character, and r or s to denote resistance or sensitivity to a compound. Bacterial genotypes use the same three letters, all lower case and italicized, followed by +, −, r, or s superscripts. A mutant requiring arginine as a supplement is Arg^- phenotypically and arg^- genotypically.

Key Points: **Three main types of bacterial mutants commonly used for genetic analysis are antibiotic-resistant mutants, nutritional mutants, and carbon-source mutants.**

BASIC TEST FOR TRANSFORMATION, CONJUGATION, AND TRANSDUCTION

The three parasexual processes of transformation, conjugation, and transduction are distinguished by two simple criteria (Table 16.1): (1) sensitivity to deoxyribonuclease (DNase); and (2) dependence on cell–cell contact.

These two criteria are easily tested. The first criterion is tested by simply adding DNase to the medium containing the bacterial strains involved in recombination. If recombination occurs in the absence but not the presence of DNase, transformation must be occurring because the DNA, which is free in the medium, is vulnerable to the enzyme's activity.

In conjugation and transduction, DNA is not free in the medium, so it is not subject to degradation by DNase. Conjugation, which requires cell–cell contact, can be identified by the **U-tube experiment** in which bacteria of two different genotypes are placed in opposite arms of a U-shaped culture tube (Figure 16.4). The arms of the U-tube, and hence the two bacterial

TABLE 16.1
Criteria for the Three Parasexual Processes

	Criterion	
Recombination Process	*Cell Contact Required?*	*Sensitive to Dnase?*
Transformation	no	yes
Transduction	no	no
Conjugation	yes	no

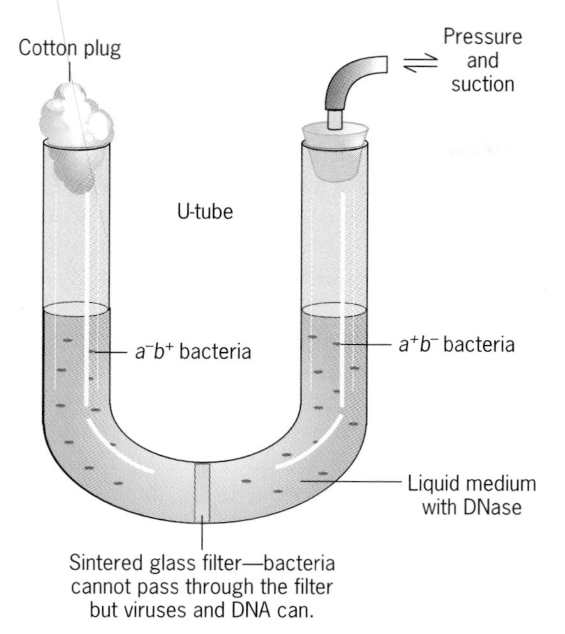

Figure 16.4 The U-tube experiment. Bacteria of different genotypes are in different arms of the tube, separated by a sintered glass filter that prevents cell–cell contact. This experiment tests for conjugation because cell–cell contact is required for the transfer of genetic material. DNase in the culture medium degrades free DNA, providing a test for transformation. If recombination occurs, it is likely taking place by means of transduction.

cultures, are separated by a sintered glass filter that allows DNA molecules and bacterial viruses but not bacterial cells to pass through. Conjugation cannot occur under these conditions because the filter prevents cell–cell contact. If DNase is present in the culture medium of the U-tube and if recombination occurs, it is likely taking place by the process of transduction.

Key Points: **To determine which of the three parasexual processes may be operating in a particular bacterial species, it is necessary to determine whether cell–cell contact is required and whether the process can be disrupted by the presence of DNase in the medium.**

TRANSFORMATION

Transformation was the earliest of the three mechanisms of genetic exchange in bacteria to be described. It was first discovered in 1928 in pathogenic strains of *Streptococcus pneumoniae* (also known as *Diplococcus pneumoniae*) by Frederick Griffith, although he knew nothing of the actual nature of the process (Chapter 9). Griffith's recorded observations were crucial to the eventual identification of DNA as the genetic material. In 1944 Avery, MacLeod, and McCarty demonstrated convincingly that the "transforming principle" described by Griffith was DNA. Their discovery that DNA was the genetic material was confirmed by the exprimental work of A. D. Hershey and M. Chase in 1952. Some details of the mechanism by which transformation occurs remain unclear, 70 years after Griffith's discovery. However, a reasonably complete picture of the overall process of transformation has been established.

The Process of Transformation

The movement of donor DNA molecules across the membrane and into the cytoplasm of recipient bacteria is an active, energy-requiring process. It does not involve the passive diffusion of DNA molecules through permeable cell walls and membranes. Transformation is not a naturally occurring process in all species of bacteria; rather, it takes place only in those species possessing the protein and enzymatic machinery required to bind free DNA molecules in the medium and transport them to the cytoplasm. Most transformation studies have used three species of bacteria, *S. pneumoniae*, *Bacillus subtilis*, and *Haemophilus influenzae*. Even in these species, all cells in a given population are not capable of being transformed. Only **competent cells**, which secrete a **competence factor**, a small protein that induces the synthesis of 8 to 10 new proteins required for transformation, are capable of serving as recipients in transformation. The proportion of bacteria in a culture that are physiologically competent to be transformed depends on the growth conditions. In most bacterial species, cells that are likely to be transformed are dividing at their maximal rate. These populations of cells are growing exponentially and are fast approaching the plateau phase where nutrients in the medium become a limiting factor in the continued growth of the population (Figure 16.5).

Since the mechanism of transformation has been studied extensively in *Streptococcus*, we will use that species as the basis for our discussion (Figure 16.6). A competent cell binds a large double-stranded fragment of DNA (at least about 500,000 daltons) at specific receptor sites on the surface of the bacterium. Because there are a limited number of these receptors, DNA molecules compete with each other for those sites. Following binding, DNA moves across the cell membrane. This is an active, energy-requiring process. One of the DNA strands is hydrolyzed by a membrane-bound exonuclease, providing energy to

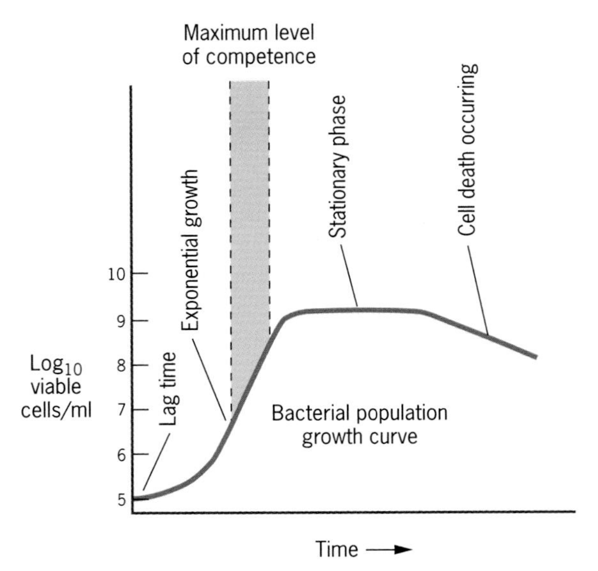

Figure 16.5 A bacterial growth curve. The maximum level of competence for transformation in most bacterial species occurs near the end of the exponential growth phase.

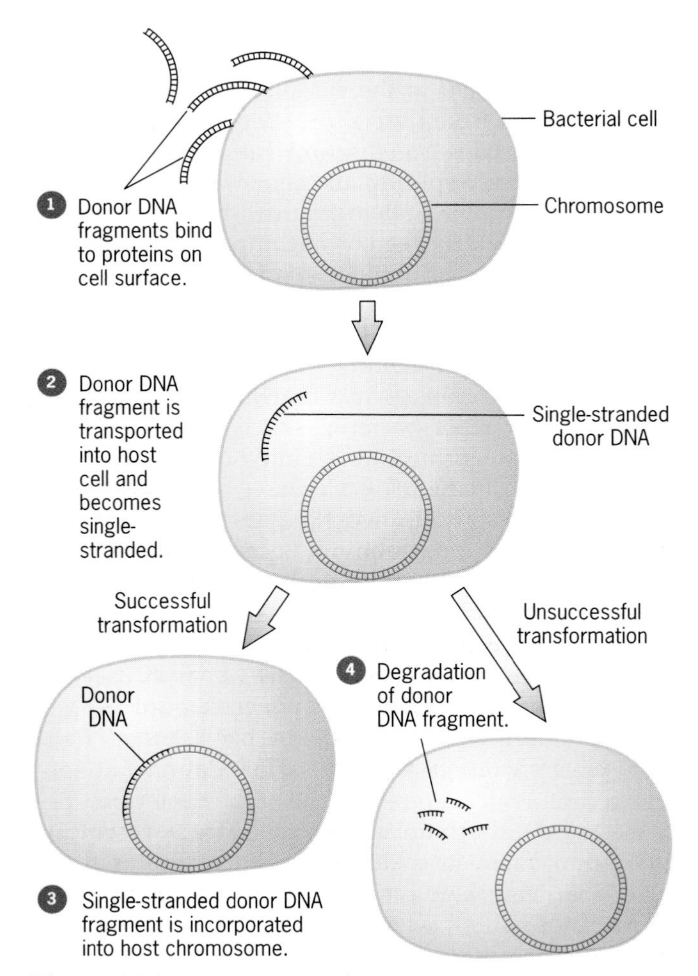

Figure 16.6 Bacterial transformation.

help fuel DNA transport across the membrane. The undegraded DNA strand associates with membrane proteins that were induced by the competence factor and is transported across the membrane. The single-stranded DNA fragment can then align with a homologous region of the host cell genome and be integrated in place of that region (Figure 16.7). This type of recombination is unidirectional and **nonreciprocal** because only the recipient cell becomes genetically altered.

In *Haemophilis*, the process of transformation is different. This bacterial species does not produce a competence factor. While *Streptococcus* can take up DNA from a variety of sources, *Haemophilus* can take up DNA from only closely related species. Double-stranded DNA is not degraded but rather is complexed with proteins and then taken into the cell by membrane vesicles. The specificity of *Haemophilis* transformation is due to a special 11 base-pair sequence (5'–AAGTGCGGGTCA–3') that is repeated about 600 times in the *Haemophilis* genome. For DNA to be transported successfully across the membrane, it must have this nucleotide sequence.

Transformation and Gene Mapping

Transformation is an important tool for use in gene mapping. When DNA is extracted or released from a cell, it fragments. DNA molecules extracted from *Streptococcus* bacteria typically break into about 50 fragments per genome depending upon the extraction procedure; thus each DNA fragment represents about 2 percent of the genome. In gene mapping experiments using transformation, two genes are said to be closely linked if they are frequently found on the same DNA fragment. This use of the term *linkage* is quite different from that in eukaryote gene mapping, where all genes on the same chromosome are said to be linked. Because bacteria have only one chromosome, all their genes are linked. If two genes are very close to each other, they will almost always be on the same DNA fragment and thus cells have a high probability of becoming **co-transformed** with these genes; if they are far apart, they will never be on the same fragment. The farther apart two genes are, the greater the chances that they will not be found on the same fragment. This is the basis of gene mapping experiments using transformation.

To determine whether two genes are closely linked, three different genetic classes of donor DNA molecules are used: a^+b^-, a^-b^+, and a^+b^+. The recipient cells are a^-b^-. Steadily decreasing concentrations of a^+b^- and a^-b^+ donor DNA are used in seperate experiments to measure the single-transformation frequencies of a^-b^- to a^+b^- and a^-b^+. Theoretically, the frequency of

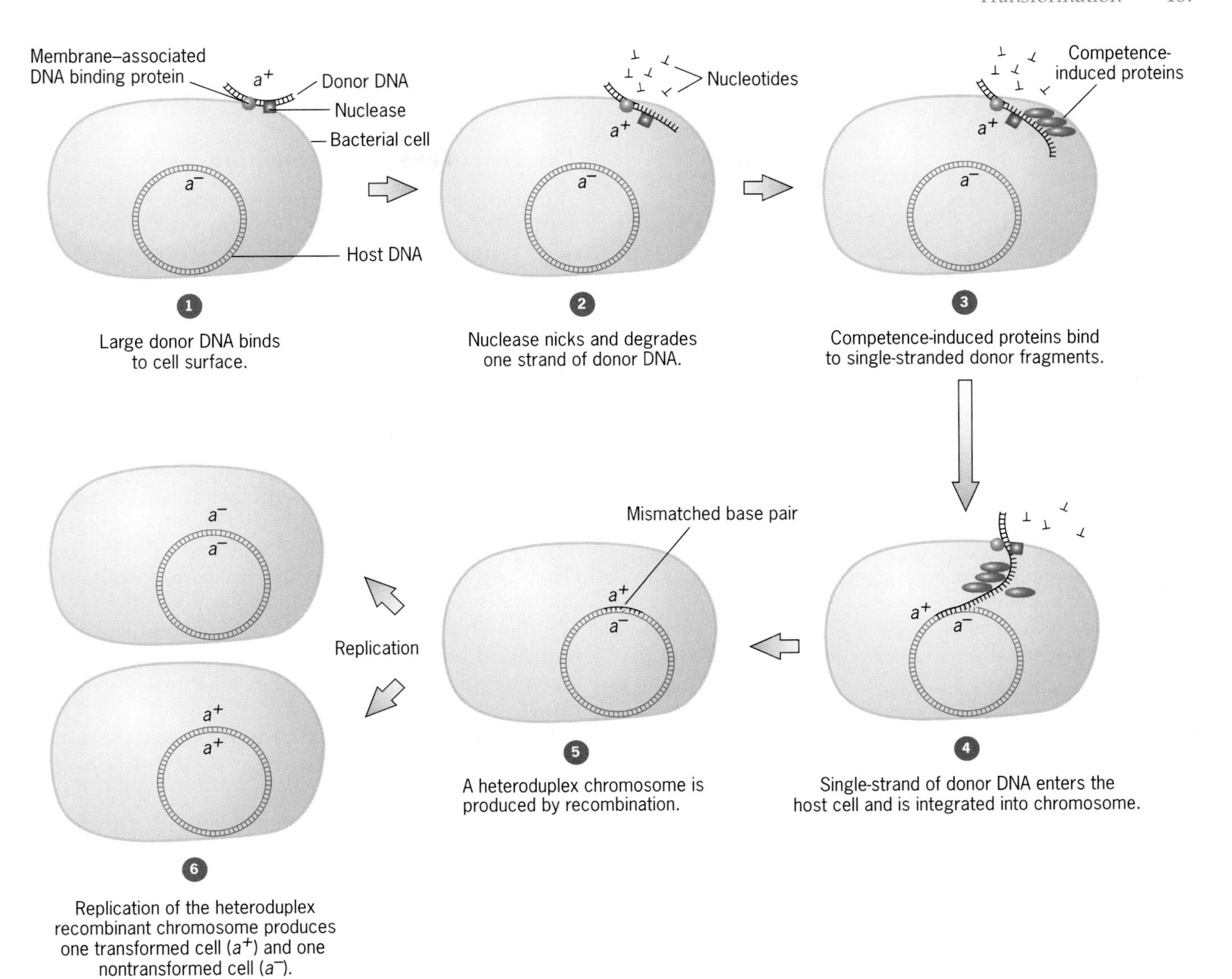

Figure 16.7 The mechanism of transformation.

single-transformants should decrease linearly as the concentration of donor DNA decreases (Figure 16.8*a*). Next the double-transformation frequency of a^-b^- to a^+b^+ using a^+b^+ donor DNA is determined as a function of decreasing DNA concentration. If the two genes are closely linked—that is, they are frequently found on the same DNA fragment—a curve with the same slope as single transformation is expected because double transformation, like single transformation, requires uptake and incorporation of a single DNA fragment (Figure 16.8*b*). If the two genes are far apart, double transformation can occur only if two different fragments are taken up by the recipient cells and two separate recombination events occur. Thus if the DNA concentration decreases ten-fold, the double-transformation frequency drops a hundred-fold: For example, when the DNA concentration decreases from 1.0 to 0.1 units per ml, single-transformation frequency decreases ten-fold from 1.0 to 0.1 cells per unit volume; the double-transformation frequency for two genes on two different fragments decreases to $0.1 \times 0.1 = 0.01$ cells per unit volume, a hundred-fold decrease. This exponential decline generates a curve with a much steeper slope (Figure 16.8*c*). Thus, by analyzing single- and double-transformation frequencies solely as a function of decreasing DNA concentration, we are able unambiguously to ascertain linkage.

Key Points: **In transformation of some bacterial species, a fragment of double-stranded DNA is transported across the cell membrane of competent cells, is converted into a single-stranded DNA, and then is physically integrated into the recipient cell's genome, replacing its homologous sequence. Linkage in trans-**

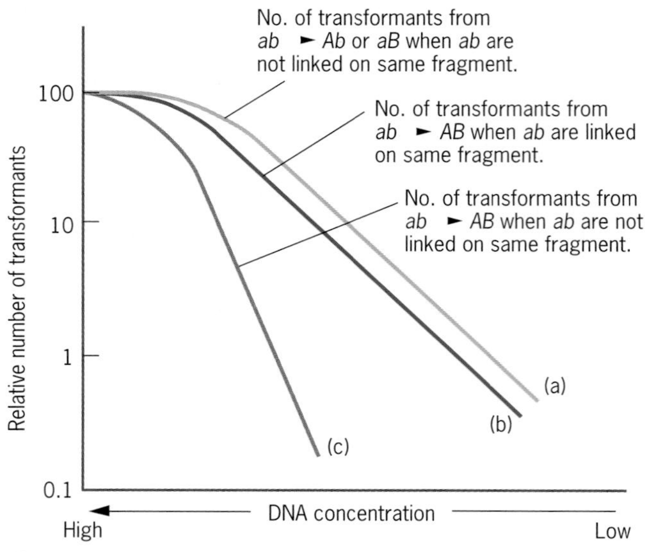

Figure 16.8 Single- and double-transformation frequencies as a function of decreasing DNA concentration. Double-transformation frequency requiring two events decreases exponentially as a function of decreasing DNA concentration; double-transformation frequency requiring one event decreases arithmetically.

formation experiments means that two genes are frequently found on the same DNA fragment. For a cell to be doubly transformed by genes far apart, two separate fragments must insert. If the two genes are close together on the same DNA fragment, only one fragment insertion is necessary for double transformation.

CONJUGATION

Escherichia coli is the bacterial species most widely used in genetic research, but it does not normally undergo transformation. Nevertheless, other processes have evolved in this species to generate recombinant genotypes. In 1946, Joshua Lederberg and Edward L. Tatum established for the first time that *E. coli* engages in a form of sexual reproduction involving a physical mating between sexually differentiated strains of bacteria. This mating process, called **conjugation**, accomplishes the transfer of genetic material from one strain to another unidirectionally, enabling bacteria of one strain to exchange genetic information with bacteria of another strain. During conjugation, chromosomal DNA may or may not be transferred to a recipient cell. Because genetic transfer is unidirectional and does not involve gamete formation or gamete fusion, conjugation is a parasexual process. Conjugation creates recombinant genotypes that have provided investigators the opportunity to analyze the *E. coli* genome with some rigor.

The Discovery of Conjugation

The discovery of a sexual process in bacteria by Lederberg and Tatum grew out of an effort to study the genetic control of metabolic processes. They analyzed various nutritional mutants in bacteria and in 1946 published a paper in the British journal *Nature* that began: "Analysis of mixed cultures of nutritional mutants has revealed the presence of new types which strongly suggest the occurrence of a sexual process in the bacterium *Escherichia coli.*" They had mixed together two auxotrophic strains of *E. coli*, each requiring two different nutritional supplements in the medium. One of the auxotrophic strains (A) required methionine and biotin for growth (Met^- Bio^-); a second strain (B) required threonine and leucine for growth (Thr^- Leu^-). After plating the mixture of cells on a minimal medium that would not support the growth of either strain, they discovered prototrophic colonies that could grow. These results suggested that genetic recombination had occurred, but what was the mechanism?

When Lederberg and Tatum mixed strain A ($met^-bio^-thr^+leu^+$) cells with strain B cells ($met^+bio^+thr^-leu^-$) cells on a nutrient medium for several hours and then plated cells on a minimal medium, genetically stable $met^+bio^+thr^+leu^+$ prototrophs appeared (Figure 16.9). Strains A and B alone did not yield prototropic colonies. Considering the relatively high frequency of prototrophic colonies from the mixed strains, they concluded that the prototrophs could not have arisen by reverse mutation ($- \Rightarrow +$) in the auxotrophic strains because two separate reversion events in each strain would be necessary to produce a prototroph. They suggested that either transformation was occurring or that the two mutant strains were somehow mating and exchanging genetic material.

Lederberg and Tatum eliminated transformation as a possibility by showing that DNA extracted from either auxotrophic strain could not transform the other strain. Transformation does not normally occur in *E. coli*. They did not prove that physical contact was required. Proof that cell–cell contact was required for gene exchange came in 1950 when Bernard Davis developed the U-tube (page 405). When strains A and B were separated by a sintered glass filter, gene transfer could not take place.

$F^+ \times F^-$ Mating

Two years after Davis proved that cell–cell contact was required for conjugation, William Hayes discov-

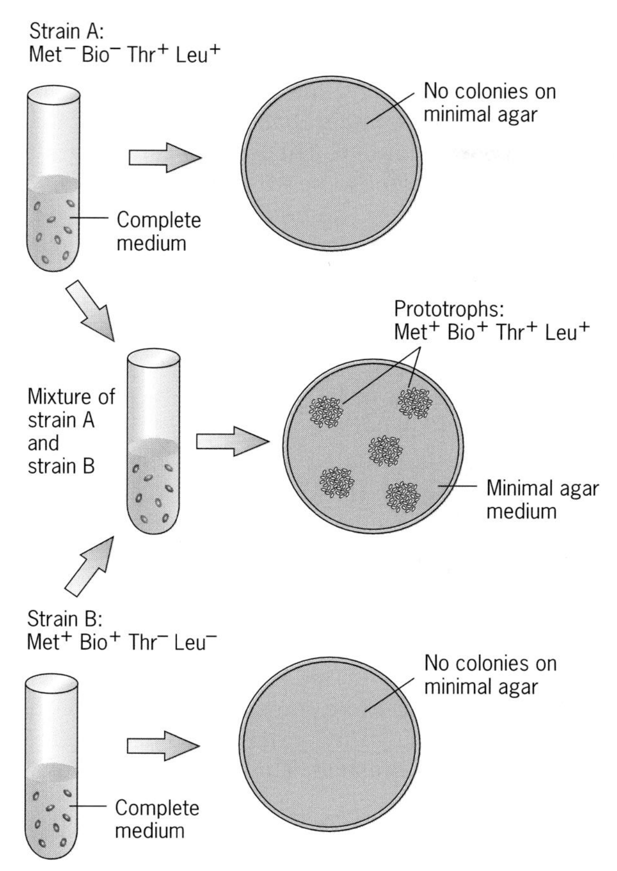

Figure 16.9 A summary of the classic Lederberg and Tatum experiment in which two auxotrophic strains of *E. coli* were mixed together, producing prototrophic genetic recombinants. Strains A and B alone did not produce prototrophic cells.

ered that gene transfer was unidirectional from a donor (F^+) strain to a recipient (F^-) strain. Hayes further discovered that in $F^+ \times F^-$ matings, genes on the main bacterial chromosome were rarely transferred but that F^- cells usually became F^+.

Donor or F^+ cells carry a plasmid called the fertility or **F factor**, a circular extrachromosomal molecule containing about 94,000 base pairs. Genes on the F factor are responsible for facilitating cell–cell contact and F-factor transfer from donor to recipient cell. Many F-factor genes direct the synthesis of **sex pili** (singular, pilus), filamentous appendages on the bacterial surface that attach the F^+ cell to an F^- cell.

The F factor and the main bacterial chromosome have **insertion sequences (IS)** that enable the F factor to insert into the chromosome by homologous recombination. Insertion sequences are short, transposable DNA sequences that contain a gene coding for an enzyme called transposase which facilitates the movement of the insertion sequence around the genome. Because the F factor can exist independently of the bacterial chromosome or be integrated into it, it is a special type of plasmid called an **episome**.

During $F^+ \times F^-$ conjugation, the F factor replicates by the rolling circle mechanism (Chapter 10), and a single-stranded copy of it moves into the F^- cell (Figure 16.10). It is not clear whether the DNA passes through the hollow sex pilus or through another conjugation channel. The single-stranded DNA entering the recipient becomes a double-stranded molecule. Be-

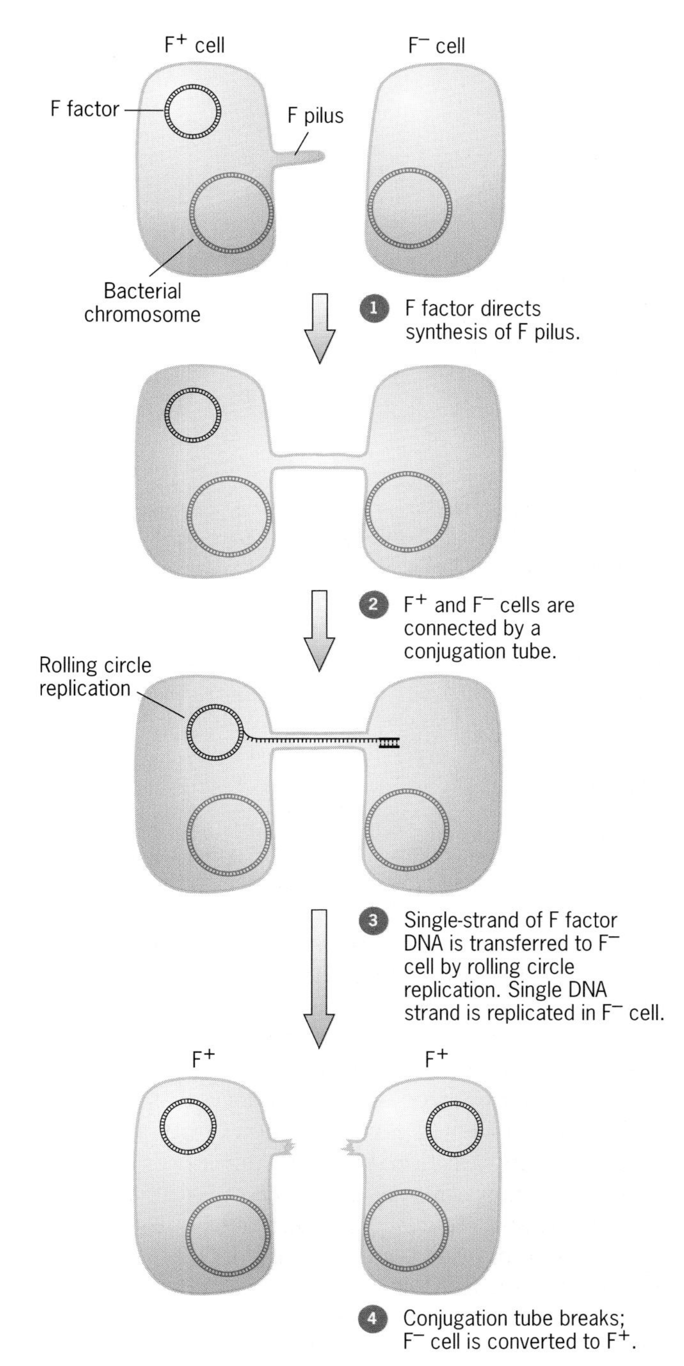

Figure 16.10 $F^+ \times F^-$ conjugation.

cause genes on the main bacterial chromosome are rarely transferred into the recipient with the independent F factor, the recombination frequency is low. It can only occur if the F factor integrates into the chromosome.

Hfr Conjugation

Some F+ strains transfer bacterial genes from the main chromosome into recipient cells with a high frequency, but they do not usually transfer the F factor. This is quite different from the F+ cell we have just considered. In these strains, the F factor does not exist independently of the main bacterial chromosome; rather, it is integrated into it. The F factor can integrate into the bacterial chromosome at several locations by recombination between homologous insertion sequences present on both the F factor and the chromosome. The integrated plasmid directs the synthesis of sex pili, undergoes rolling circle replication, and transfers genetic material to an F– cell. This type of donor cell is called an **Hfr cell** (high-frequency recombination) because of the high frequency of chromosomal gene transfer into F– cells.

DNA transfer is initiated at a point on the integrated F-factor (Figure 16.11). Rolling circle replication moves the donor strand from the main chromosome into the recipient cell where it becomes double-stranded. Because only a small part of the F-factor is transferred to the recipient, the recipient remains F– unless the entire bacterial chromosome is transferred. It takes about 100 minutes for the entire bacterial chromosome to be transferred, but the connection between conjugating cells is usually severed before the process is finished. Thus the recipient usually remains F–.

Hfr strains vary in the chromosomal location of F factor integration and in the orientation of the F factor, which determines whether the direction of gene trans-

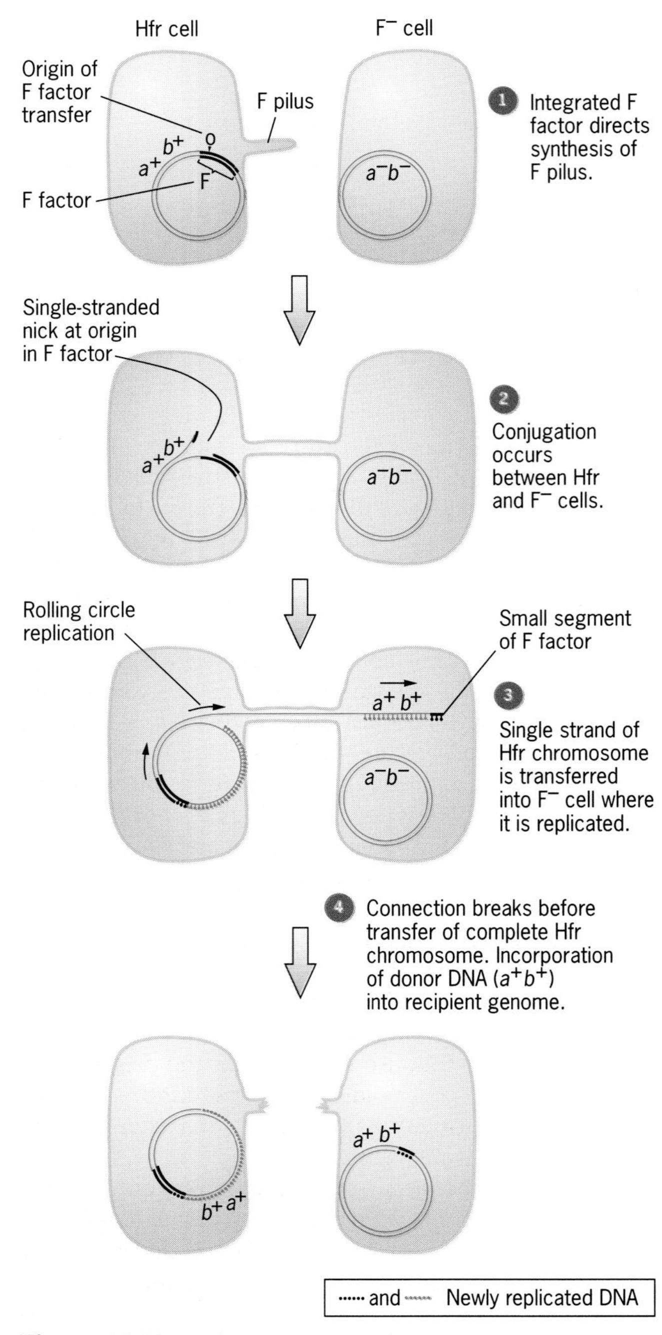

Figure 16.11 Hfr × F– conjugation.

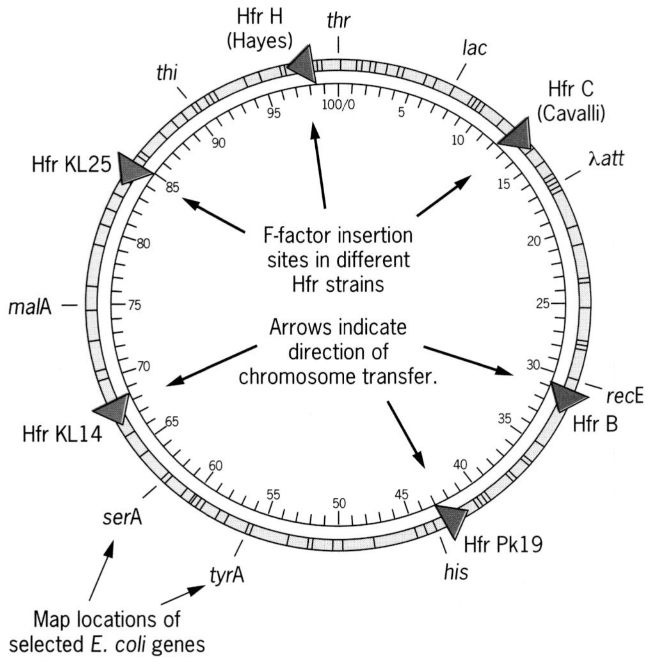

Figure 16.12 Some of the sites of F-factor integration in the *E. coli* chromosome. The arrows indicate the direction of transfer. A few gene loci are indicated as reference points.

fer is clockwise or counterclockwise around the circular chromosome (Figure 16.12). When the donor DNA enters the F^- cell, it may be degraded or it may recombine with the recipient chromosome. Hfr conjugation is the most efficient natural mechanism of gene transfer in bacteria.

F' Conjugation or Sexduction

The integrated F factor in an Hfr cell may exit the bacterial chromosome by reversing the steps that resulted in its integration (Figure 16.13). Sometimes there is an error in the excision of the F factor from the chromosome. This error results in the formation of an **F' factor**–one containing a segment of the chromosome. The mating of an F' cell with an F^- cell is basically the same as an F^+ times F^- mating, with one important difference: bacterial genes incorporated in the F' plasmid are transferred with very high frequency to recipient cells. These genes do not have to be integrated into the recipient cell's chromosome to be expressed. The recipient acquires the F' factor and is partially diploid for the genes that it carries. The transfer of bacterial genes by F' factors is called **sexduction**. Sexduction results in the rapid spread of bacterial genes from the main chromosome through a population.

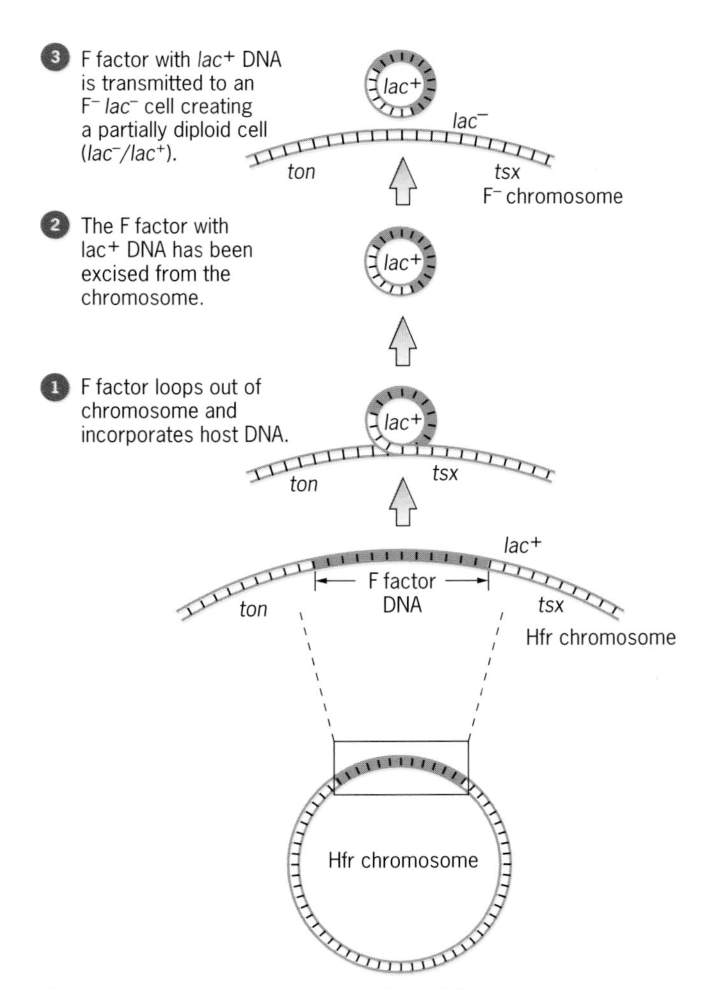

Figure 16.13 The formation of an F' factor.

Sexduction is a valuable tool for microbial geneticists. Because of the cell's partially diploid condition, it allows investigators to answer questions about dominance and recessiveness for alleles carried by the F' factor. Furthermore, if two genes are incorporated together into an F' factor, investigators conclude that the two genes are close together on the bacterial chromosome.

Conjugation and Gene Mapping

The discovery of Hfr cells opened the door to mapping genes in *E. coli*. The use of Hfr cells in gene mapping takes advantage of the fact that during conjugation, chromosomal DNA moves from donor to recipient at a constant rate. In an **interrupted mating experiment**, conjugating cells are placed in a blender and broken apart. The Hfr × F^- mating is stopped at various intervals after the start (Figure 16.14). The order and timing of gene transfer can be determined because the first gene to enter the recipient is closest to the origin of transfer, and genes farther from the origin enter the recipient later (Figure 16.15). In one Hfr × F^- conjugation experiment performed by Elie Wollman and François Jacob in the mid-1950s, an Hfr strain that carried the genetic markers *azi*^{s} *ton*^{s} *lac*$^+$ *gal*$^+$ was mated to an F^- strain that was *azi*^{r} *ton*^{r} *lac*$^-$ *gal*$^-$. (The genes determine sodium azide resistance or sensitivity, phage T1 resistance or sensitivity, and the ability to use lactose or galactose respectively.) The *azi* gene was the first to enter the recipient, at about nine minutes after the initiation of conjugation. The *gal* gene was the last to enter the recipient, about 26 minutes after conjugation began. The *ton* and *lac* genes entered about 10 and 17 minutes, respectively, after the initiation of conjugation.

The map generated by interrupted conjugation is circular and is marked off in units of time rather than crossover frequency. The technique can fairly precisely locate genes that are three or more minutes apart. The genes that are farther from the origin enter later and have a lower plateau because interruption of mating occurs spontaneously, and later markers are less likely to be transferred.

Because of the large size of the *E. coli* genome, it is not possible to map all the genes using a single Hfr strain. Maps are generated by using different Hfr strains with different F-factor integration sites and different orientations. These maps are then combined to produce a final map of the *E. coli* chromosome (Figure

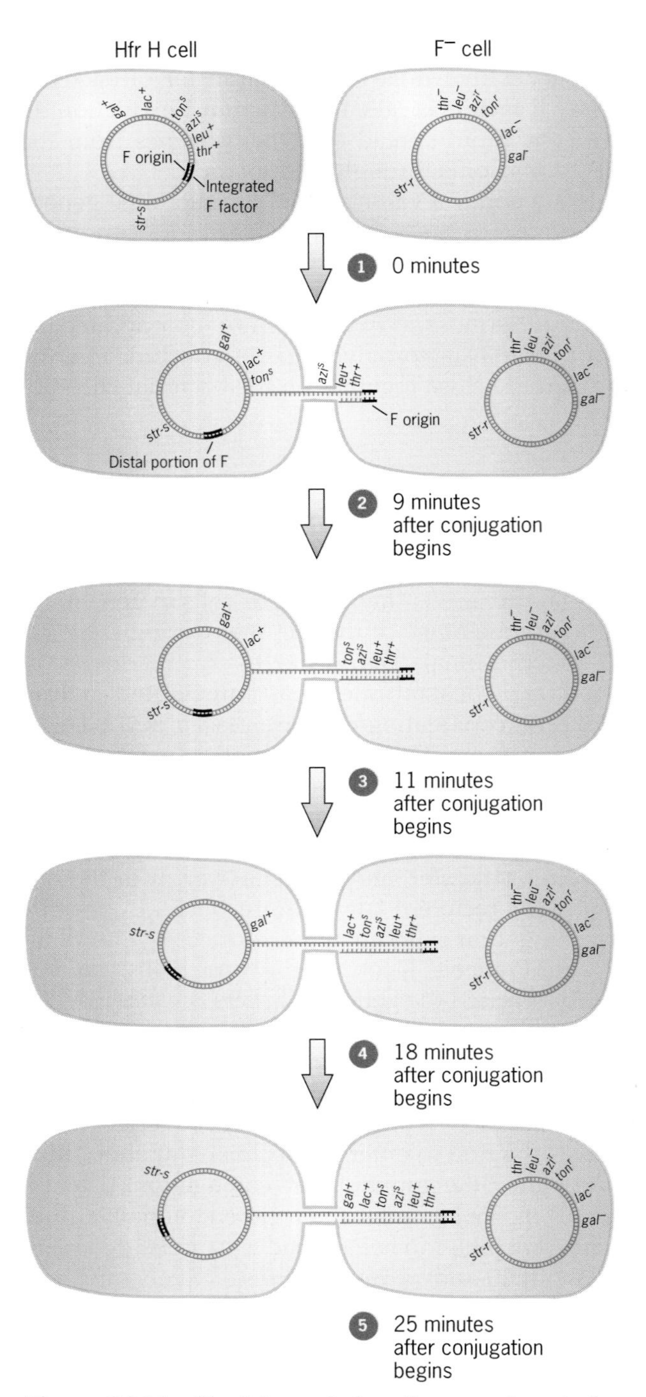

Figure 16.14 The interrupted mating experiment of Wollman and Jacob. The transfer of the Hfr chromosome to the F⁻ cell is linear, beginning at the origin point on the F factor.

16.16). The map is divided into 100 minutes, the time required to transfer a complete chromosome from donor to recipient. The 0 point is set at the *thr* locus because this is the gene closest to the point of insertion of the F factor first described by Hayes.

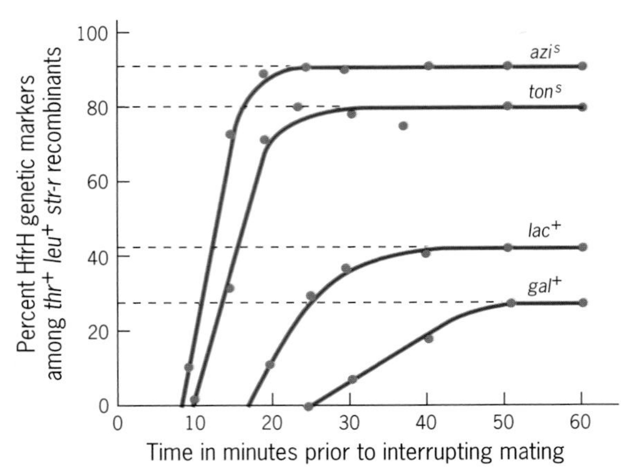

Figure 16.15 Data from the interrupted mating experiment. The frequencies of the unselected donor markers are shown on the ordinate and time of entry on the abscissa. The dashed lines indicate the plateau frequencies observed for the various donor markers.

Mapping Closely Linked Genes

Interrupted mating experiments provide approximate map distances for genes that are several map units apart, but they are usually uninformative for genes that are close together, say, one- or two-minute units apart. As in eukaryotes, mapping closely linked genes in bacteria usually necessitates three-point crosses. The rationale behind three-point crosses in bacteria is essentially the same as in eukaryotes: The double crossover class is the rarest, and this class identifies the middle gene.

Recombination in conjugating bacteria requires an interaction between the donor DNA fragment and the recipient cell's chromosome. The donor DNA segment must be incorporated into the recipient chromosome. Such an incorporation requires two crossovers or recombination events (Figure 16.1). A single exchange between donor and recipient DNA does not yield a structurally intact chromosome (Figure 16.17).

In three-point mapping experiments with eukaryotes, double crossover events are much less frequent than single crossover events. The rarity of double crossover events allows geneticists to determine the order of three genes (Chapter 7). Three-point crosses for closely linked bacterial genes employ the same basic principle, with one minor variation: recombination involving two exchanges is more frequent than recombination involving four exchanges.

To use conjugation to map closely linked genes, each three-point cross is performed in two ways: (1) a single mutant donor × a double mutant recipient; and (2) a double mutant donor × a single mutant recipient.

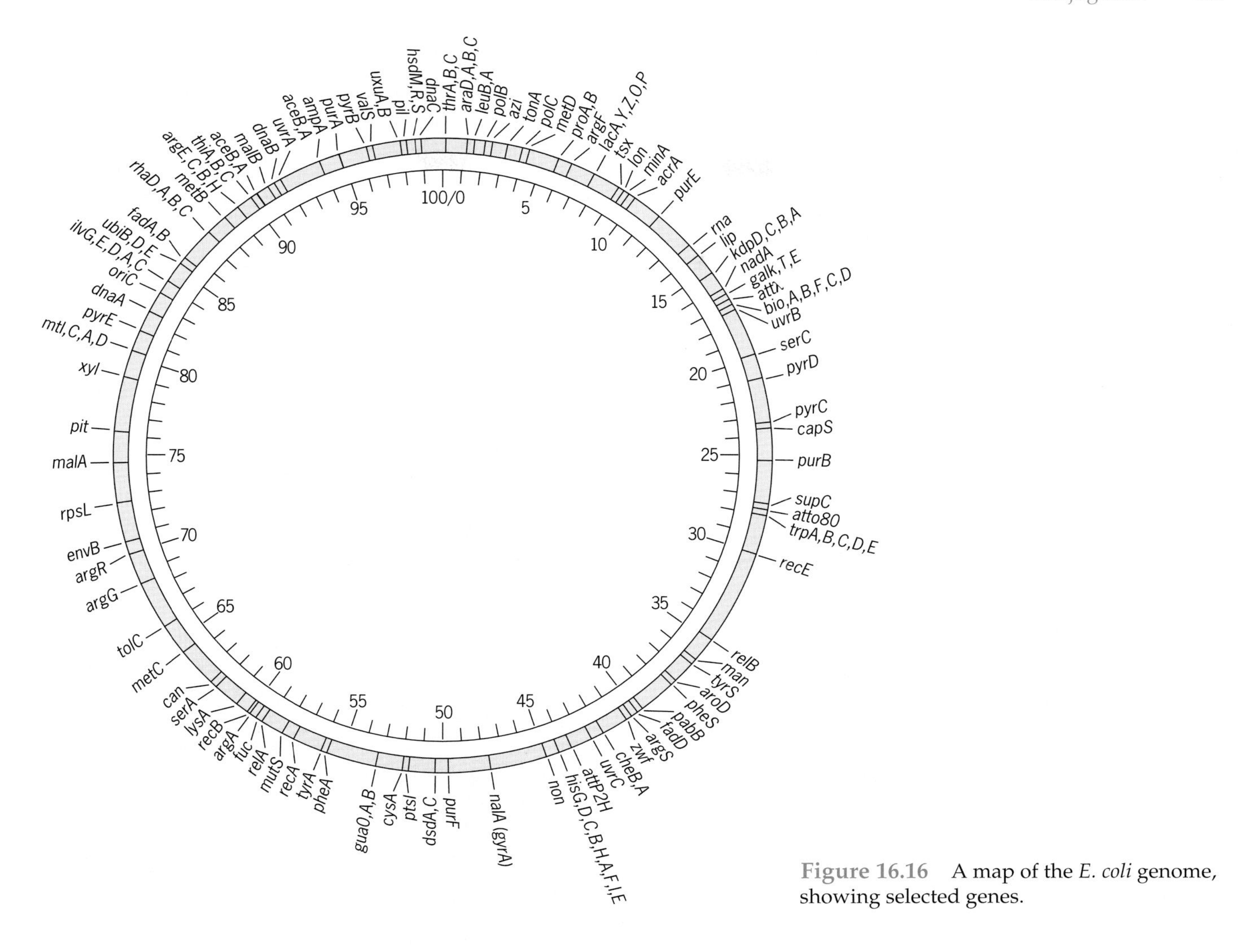

Figure 16.16 A map of the *E. coli* genome, showing selected genes.

Suppose, for example, that we want to order two closely linked genes (*b* and *c*) relative to a third (*a*) that is nearby. The two reciprocal crosses are:

	Donor		Recipient
(1)	$a^+ b^+ c^-$	×	$a^- b^- c^+$
(2)	$a^- b^- c^+$	×	$a^+ b^+ c^-$

Because the chromosome is circular, there are only two possible orders for these three genes: *a - b - c* or *a - c - b*.

By comparing the frequency of $a^+ b^+ c^+$ recombinants in the reciprocal crosses, we can discriminate between these two orders. If the order of the genes is *a-b-c*, then the $a^+ b^+ c^+$ recombinant frequencies in the reciprocal crosses will be about the same because two

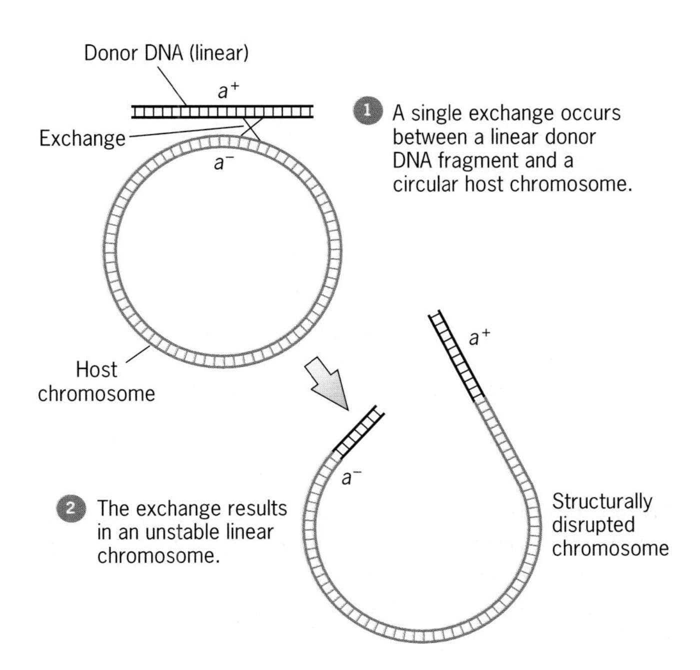

Figure 16.17 A single exchange between a linear donor DNA molecule and a circular recipient chromosome does not yield a structurally intact recombinant chromosome.

crossover events are required in both cases (Figure 16.18). If the order is *a-c-b*, the a^+ b^+ c^+ recombinant frequencies will differ dramatically in the reciprocal crosses because in one cross a double exchange will produce the a^+ b^+ c^+ recombinant, whereas in the reciprocal cross four exchanges are necessary to produce the a^+ b^+ c^+ recombinant. Thus the three-point cross can be used to establish the order of closely linked genes.

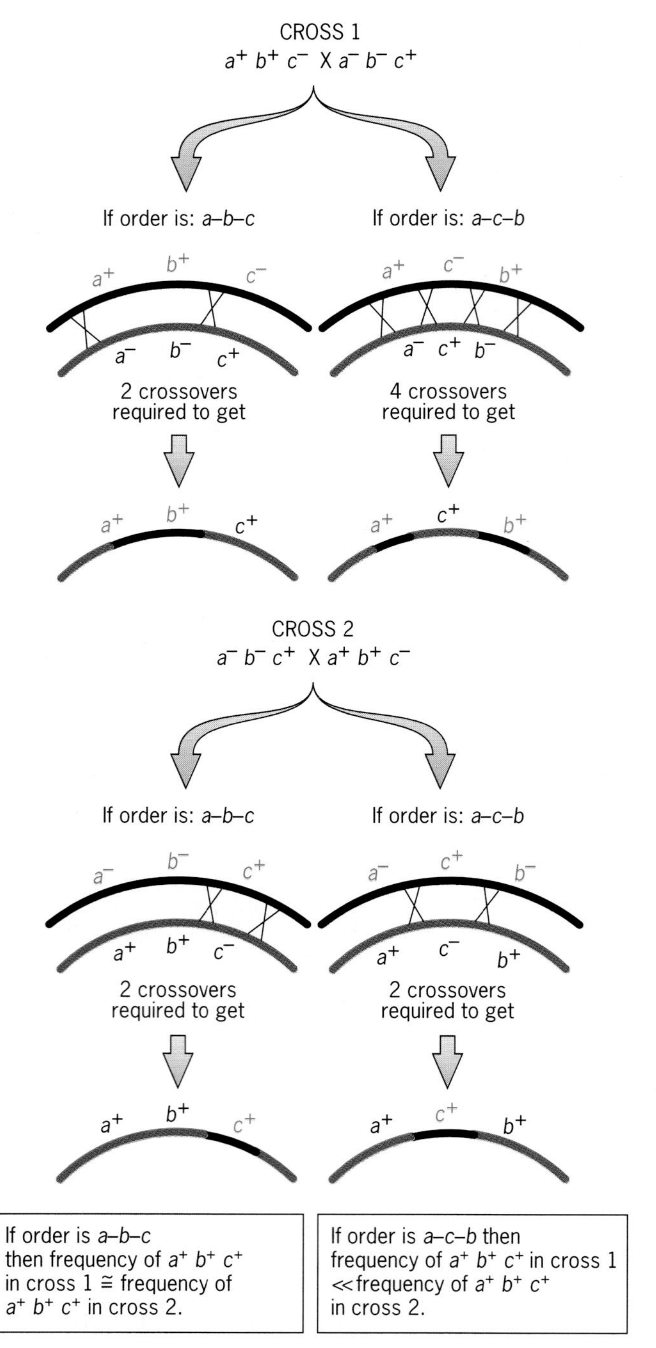

Figure 16.18 The rationale for ordering closely linked markers in a three-factor cross. Reciprocal crosses predict different a^+ b^+ c^+ frequencies, depending on the gene order.

On the Origin of Plasmids

Ever since their discovery, investigators have pondered the significance and origin of extrachromosomal elements, such as the F factor. Their origin remains an intriguing mystery. Many view them as descendants of phage DNA that have entered into symbiotic or mutually advantageous relationships with bacterial hosts. In 1963, P. Fredericq, referring to the origin of the F factor, stated: "A phage is a particle that can establish effective contact with a cell and inject into it genetic material. . . . What else is an F^+ cell?"

It is not at all unreasonable to suggest that plasmids and episomes are remnants of bacterial viruses. These remnants, instead of coding for viral proteins that package phage DNA, code for proteins that are expressed on the bacterial surface. These cell-surface viral proteins function like the protein coat of the bacteriophage, enabling one bacterial cell to physically attach to another and inject plasmid or episomal DNA.

Key Points: **Conjugation is a process in which bacteria make contact with each other and genetic material is transferred unidirectionally from donor to recipient cells. The F factor directs conjugation. Cells carrying an F factor are F^+, and those without it are F^-. A cell with the F factor integrated into the chromosome is called Hfr. An F factor that has incorporated some genes from the main chromosome is called an F' factor. Interrupted conjugation experiments showed that the Hfr chromosome is transferred unidirectionally and that the first genes to be transferred are located close to the point of F-factor integration. The *E. coli* chromosome is calibrated in minutes.**

TRANSDUCTION

Transduction, the third mechanism of gene transfer in bacteria, differs significantly from transformation and conjugation in that it is mediated by bacterial viruses. Bacterial genes are incorporated into a phage capsid because of errors that occur during the virus life cycle. The virus then injects these genes into another bacterium, transfering genetic material from one bacterium to another. Two different types of transduction have been identified: (1) generalized transduction, in which the phage can transfer any segment of the bacterial genome to another bacterium; and (2) specialized transduction, in which only restricted segments of the bacterial genome are transferred. Transduction is a common mechanism for gene exchange and recombination in bacteria and, like transformation and conjugation, has been extremely important in bacterial gene mapping.

The Discovery of Transduction

Generalized transduction was discovered in 1952 by Joshua Lederberg and his student, Norton Zinder. They were attempting to show that conjugation, which Lederberg and Tatum discovered in *E. coli* in 1946, could occur in other bacterial species. While studying *Salmonella typhimurium,* they found that when two multiply auxotrophic strains were incubated together (Phe⁻ Trp⁻ Tyr⁻ and Met⁻ His⁻) on an amino-acid-free minimal medium, prototrophs appeared with a frequency of about 1 per 100,000 cells. Because this frequency was too high to be accounted for by reverse mutation, they concluded that the bacteria were undergoing a process of genetic exchange. To test for transformation and conjugation, they carried out a U-tube experiment (page 404). The two strains were placed in the two arms of the U-tube, separated by a sintered glass filter, in a medium containing DNase. Prototrophs still appeared but in only one of the U-tube arms. These results could not be explained by transformation because DNase destroys free DNA. Conjugation could not explain the results because the filter prevented cell–cell contact. Lederberg and Zinder had discovered a new mechanism of bacterial gene transfer. Further analysis of this gene exchange process in *Salmonella* revealed that one of the bacterial strains harbored a virus called P22. When cells carrying these P22 viruses lysed, transducing particles were released. These particles passed through the filter and infected cells in the other arm of the U-tube, carrying with them genes from the virus-infected strain. These genes were subsequently incorporated into the genomes of the newly infected cells.

Generalized Transduction

In **generalized transduction**, a phage may carry any bacterial gene from one cell to another. Toward the end of the phage life cycle, when viral DNA is being packaged inside viral protein coats, fragments of the bacterial chromosome are mistakenly packaged into the phage head (Figure 16.19). This packaging error produces a phage particle that carries bacterial DNA and no phage DNA. The size of the bacterial DNA

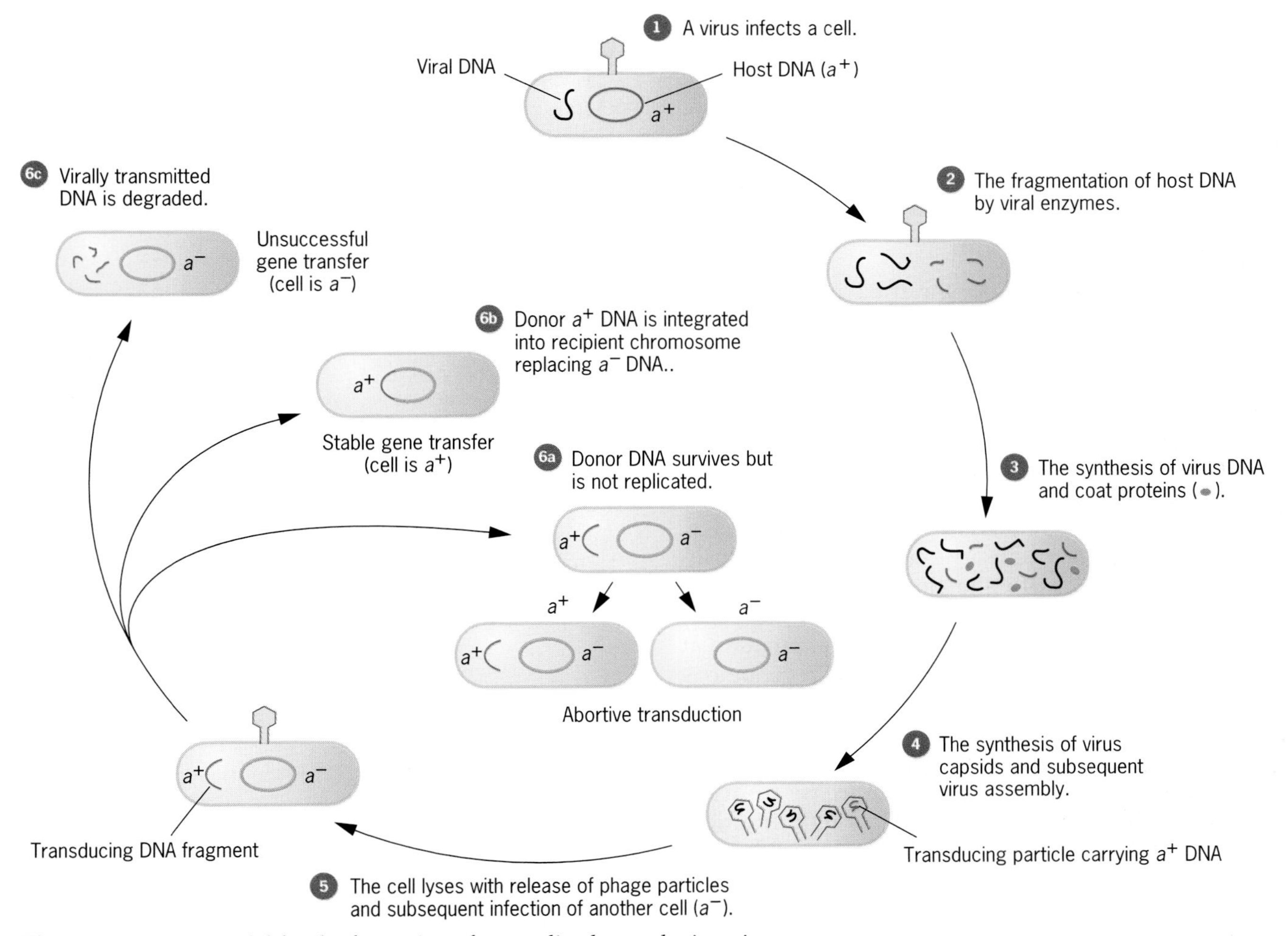

Figure 16.19 A model for the formation of generalized transducing viruses.

fragment packaged into the phage varies, but it is not unusual for it to consist of 1 or 2 percent of the bacterial genome. Thus, for a bacterium with 3000 genes, 30 to 60 genes may be packaged into a transducing virus particle.

A phage particle that carries donor bacterial DNA is released upon lysis of the cell. The phage protein coat functions as a vehicle for the transfer of bacterial DNA from one cell to another. The transducing particle attaches to a recipient cell and injects donor bacterial DNA into that cell. This transducing particle, which carries no phage DNA, is incapable of going through a life cycle. Once the DNA is injected, it must be incorporated into the host genome in order to be preserved.

The donor DNA is not always incorporated into the recipient's genome. Occasionally, the fragment of donor DNA survives and is expressed in the recipient's cytoplasm. This cell is partially diploid for the introduced genes. The donor DNA is not replicated because it does not have the proper recognition site for DNA polymerase. Thus, when the bacterium divides, only one of the two daughter cells receives the donor fragment and expresses the donor genotype. The other cell, having lost the donor fragment, reverts back to its original phenotype. This phenomenon is called **abortive transduction** and is the result of partially diploid bacteria containing nonintegrated, transduced DNA.

Specialized Transduction

In **specialized transduction**, the transducing particle carries only specific portions of the bacterial genome. In order to understand the nature of specialized transduction, it is first necessary to understand the life cycle of one of the most extensively studied and fascinating specialized transducing viruses, lambda (Figure 16.20). Lambda (λ), unlike the T phages, is a **temperate phage** because it has two options when it infects a cell: it can proceed through the lytic cycle, like the T phage, and destroy the host cell; or it can enter into a **lysogenic relationship** with the host, a relationship in which the λ chromosome circularizes and then integrates into the host chromosome by recombination at a specific site. In the integrated state, the λ genome is genetically inactive, except for a single gene, encoding a repressor, that maintains all other λ genes in a transcriptionally inactive state. However, if the repression breaks down, the λ chromosome can excise from the bacterial chromosome and commence a lytic infection. Thus, an integrated λ is still able to cause cell lysis at some time in the future. Details of the regulatory mechanisms that determine whether λ enters the lytic or lysogenic state are discussed in Chapter 21. In the lysogenic state, the λ DNA is replicated along with the host chromosome during the cell's replication cycle.

The integrated λ genome, called a **prophage,** can be induced by various means to exit the host chromosome. Excision of the λ prophage occurs when the λ DNA forms a loop and a site-specific recombination event mediated by λ-encoded enzymes releases a circular phage chromosome from the larger bacterial chromosome. Sometimes this recombination even oc-

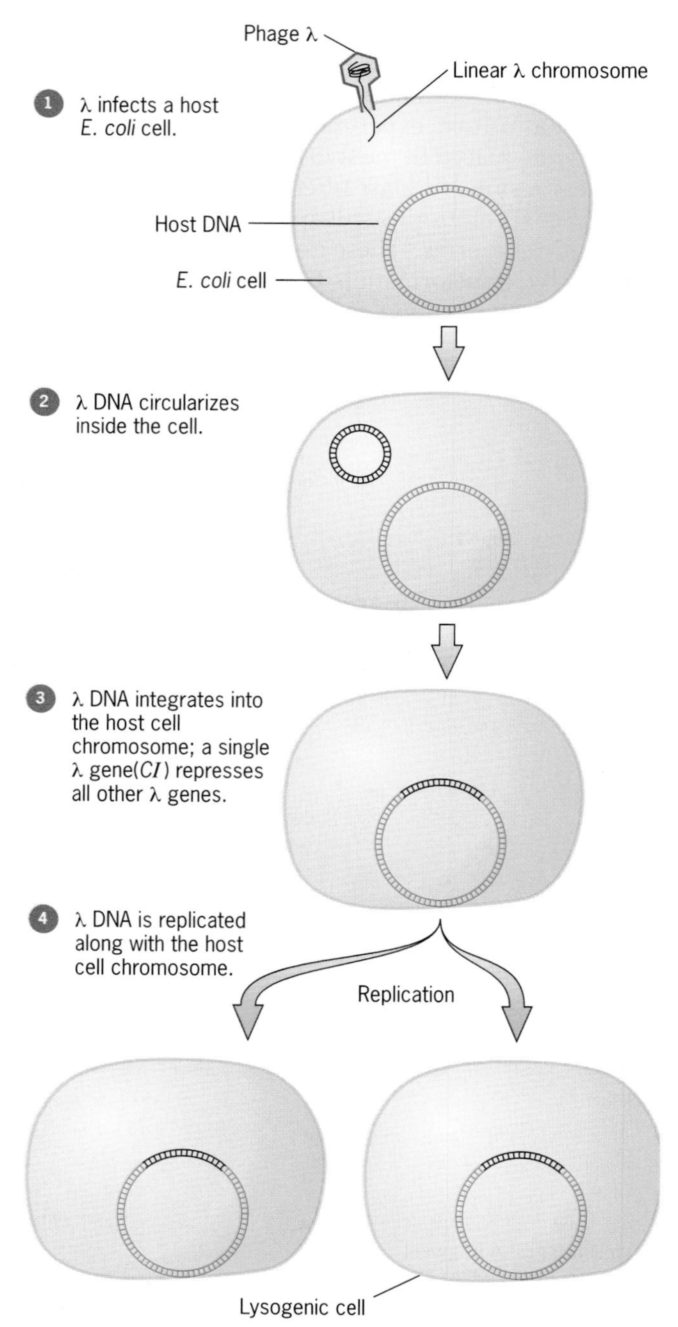

Figure 16.20 The establishment of lysogeny.

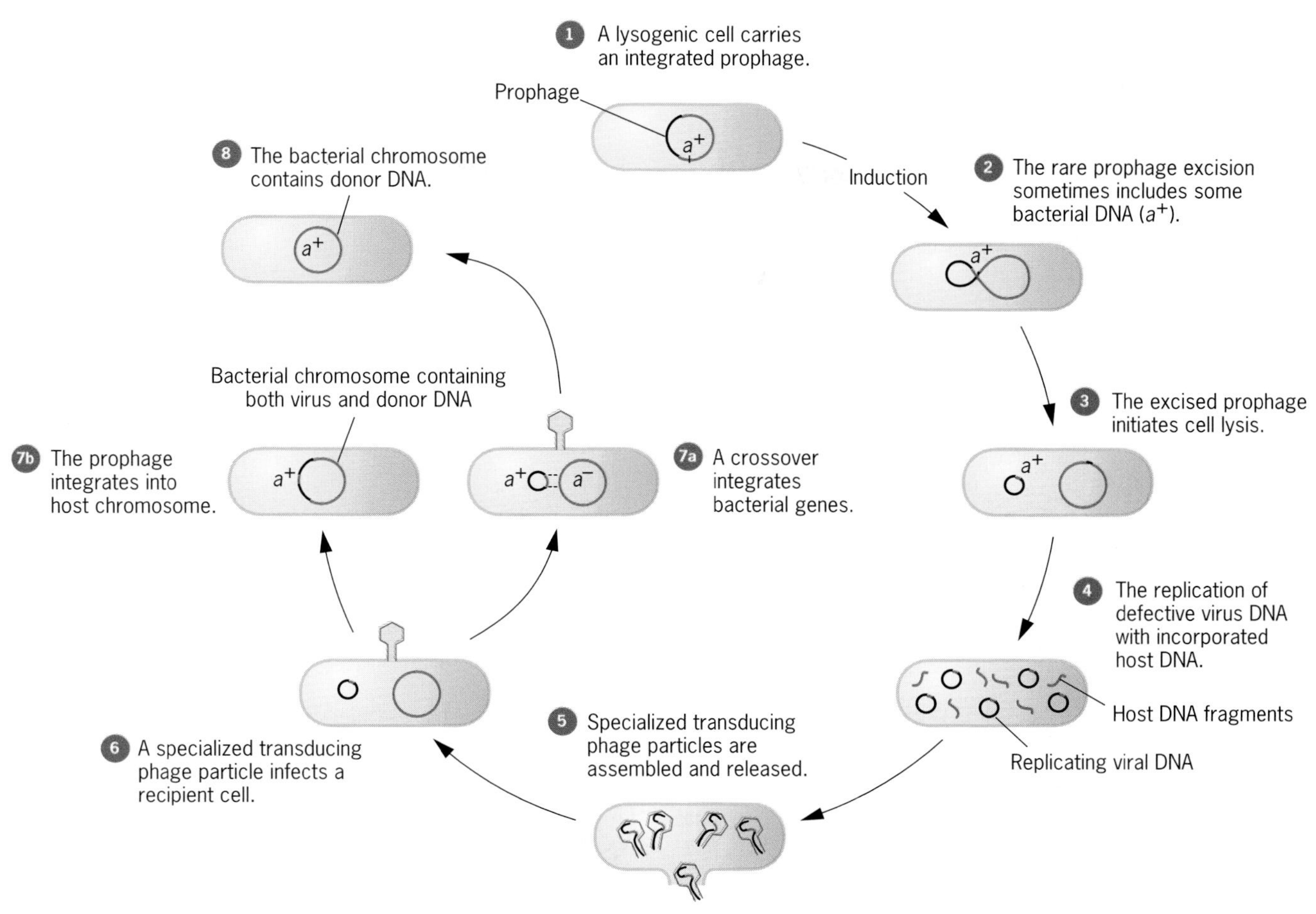

Figure 16.21 Specialized transduction.

curs aberrantly, releasing a phage genome that has had some of its own genetic material replaced by bacterial genetic material (Figure 16.21) This process is similar to the formation of an F' factor during conjugation. When it excises improperly, it can pick up only those bacterial genes that bracket its insertion site in the bacterial chromosome. A λ chromosome that is produced in this way is usually defective because it has left behind some of its own genetic material. When the improperly excised λ chromosome is packaged into a phage coat, it forms a specialized transducing particle. Unlike the generalized transducing particle discussed earlier (which carries only bacterial DNA), a specialized transducing particle carries a hybrid chromosome containing both bacterial and viral genetic material. Following lysis from the host cell, the specialized transducing particle injects its chromosome into another bacterium, thus transmitting bacterial genes from one cell to another. However, the defective chromosome of this particle is unable to direct the synthesis of progeny viruses. Instead, the bacterial genes carried by the transducing particle recombine with the genes of the newly infected cell and become incorporated into the recipient cell's genome.

Let's look more closely at λ's insertion site and how it becomes a transducing particle. Lambda inserts at a specific region of the *E. coli* chromosome called λ *att* (Figure 16.22). The *att* sites on the phage and bacterial chromosome are homologous. Pairing occurs, and crossing over integrates the λ chromosome into the bacterial chromosome. The λ *att* site on the bacterial chromosome lies between the *gal* and *bio* genes. When the λ prophage is improperly excised from the bacterial chromosome, it carries either galactose-utilization genes *(gal)* or biotin synthesis genes *(bio).* Lambda phage carrying *gal* genetic material are called lambda *dgal* because they are defective (*d*), having lost some of their own genes. Lambda *dbio* are defective and carry *bio* DNA.

When lysogenic cells lyse, most of the progeny are normal λ phage because excision is carried out properly. There are very few defective transducing particles in the lysate (about 1 in 10^5 or 10^6 particles); thus these lysates are called **low-frequency transduction**

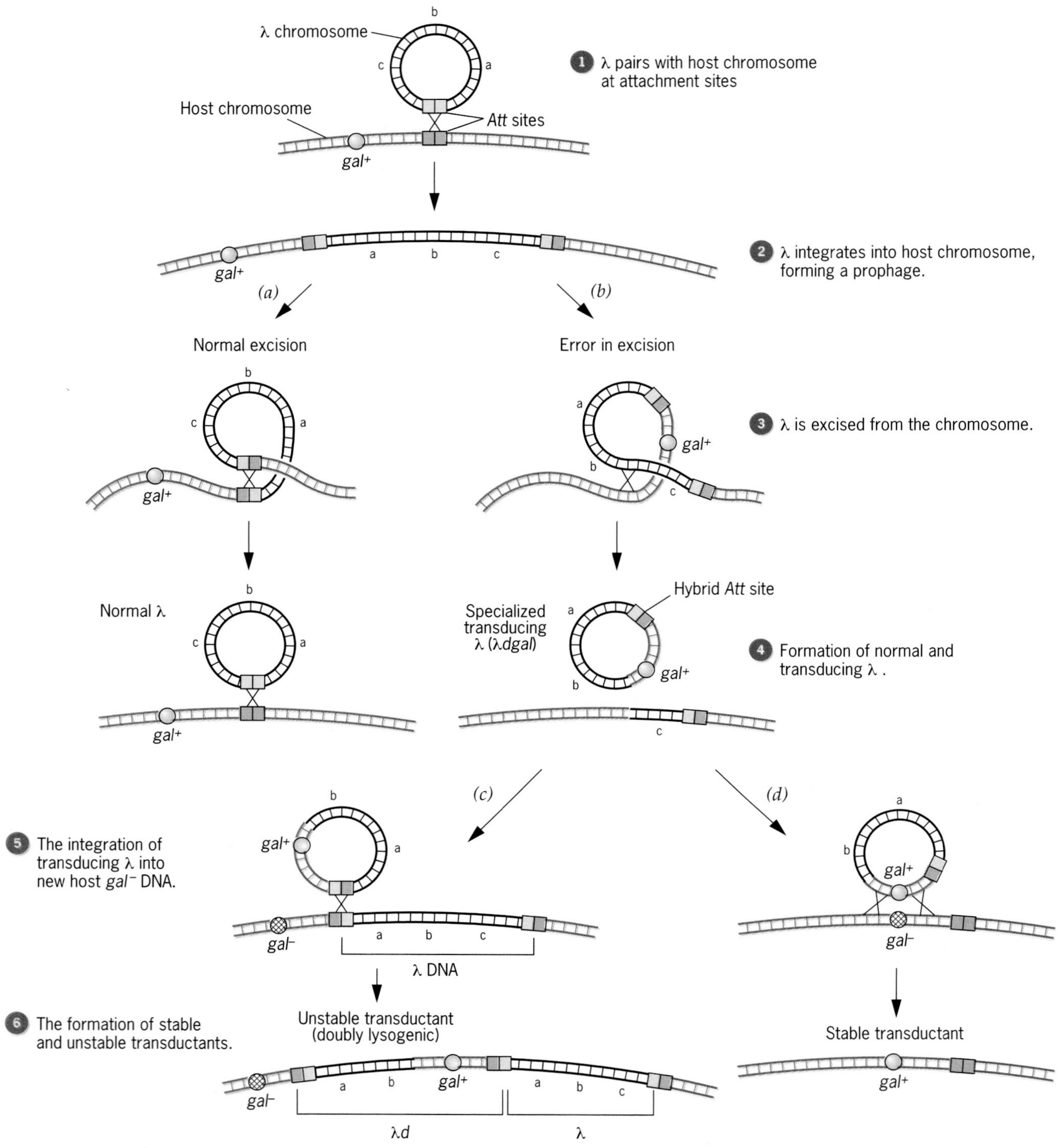

Figure 16.22 The mechanism of specialized transduction in λ bacteriophage. (*a*) Normal excision of λ from the *E. coli* chromosome; (*b*) inexact excision leading to the formation of a specialized transducing particle; (*c*) the formation of an unstable transductant; (*d*) the formation of a stable transductant.

lysates (LFT lysates). Normal λ phage have complete attachment sites, but defective transducing particles have a nonfunctional hybrid attachment site that is part bacterial and part phage in origin. Thus, defective phage chromosomes do not readily integrate into bacterial DNA.

Defective λd phage can integrate into the recipient chromosome if there is a normal λ in the same cell. The normal λ integrates, producing two bacterial/phage hybrid *att* sites where the defective λ can now insert (Figure 16.22*c*). Furthermore, the normal λ supplies genes that the defective λ is missing. This normal λ is called a **helper phage** because it allows the defective λ to integrate into the host chromosome and reproduce. The bacterial chromosome carrying the helper λ and the λd is an **unstable transductant** for two reasons:

1. The λd prophage can exit the chromosome and be lost, thus causing the cell to revert to its original phenotype.
2. The transductant can be induced to enter the lytic cycle by UV irradiation or some other means. The helper λ and λd are both excised from the chromosome and enter the lytic cycle. The lysate produced by this induction contains about 50 percent λd and 50 percent normal λ. Because it has such a high proportion of transducing particles, this lysate is called a **high-frequency transduction lysate** (**HFT lysate**).

A **stable transductant** may form when there is a crossover between the λd chromosome and the bacterial chromosome. The crossover involves an exchange on both sides of the transduced DNA (Figure 16.22*d*) and the replacement of one allele by another. The λ genome is not inserted into the host chromosome, so the host does not become lysogenic.

Transduction and Gene Mapping

Generalized transduction provides linkage information in much the same way as transformation. Linkages are usually expressed as cotransduction frequencies: the closer two genes are, the greater the chance they will be on a single DNA fragment packaged together into the same transducing particle. The *E. coli* phage P1 forms generalized transducing particles and is useful for mapping bacterial genes. In one mapping study, P1-infected bacteria carried wild-type alleles of three genes: leu^+ thr^+ and azi^r (*leu*= leucine, *thr* = threonine, and azi^r = sodium azide resistant). Progeny phage from this infection were collected and special techniques based on density differences were employed to separate transducing P1 phage from the regular P1 phage. The transducing phage were used to infect leu^- thr^- and azi^s bacteria. The newly infected cells were placed on a medium that selected for one or two of the genetic markers but not all three. For example, cells placed on a medium without azide but supplemented with threonine, leu^+ would be the selected marker because only leu^+ cells would grow on this medium. The *thr* and *azi* markers are unselected because both thr^+ and thr^- genotypes will grow on the threonine-supplemented medium, and in the absence of azide, both azi^r and azi^s cells will grow.

TABLE 16.2
Data from a Generalized Transduction Experiment Using Selected and Unselected Markers to Determine Gene Order

Experiment	*Selected Marker(s)*	*Unselected Markers(s)*	
1	leu^+	50% azi^r	2% thr^+
2	thr^+	3% leu^+	0% azi^r
3	leu^+, thr^+	0% azi^r	

For each of the three experiments with the selected marker, the frequency of the unselected marker(s) was determined (Table 16.2). For example, those cells found to be leu^+ in the first experiment were further tested to see if they were thr^+ or thr^-, and azi^r or azi^s.

Results from the first experiment indicated that the *leu* and *azi* loci were close together and that *leu* and *thr* were not close. The data showed that *leu* and *azi* were cotransduced 50 percent of the time, but in only 2 percent of the cases were *thr* and *leu* cotransduced. Thus there are two possible gene orders:

leu —- *azi* ——————————————- *thr*

or

azi —- *leu* ——————————————*thr*

The second experiment indicated that because *leu* and *thr* were cotransduced 3 percent of the time, and because *thr* and *azi* were never cotransduced, *leu* must be closer to *thr* than is *azi*. Thus the suggested gene order is

azi —- *leu* —————————————— *thr*

The data from the third experiment were consistent with this arrangement because the *thr leu* transducing fragment never carried the *azi* marker.

This study also revealed some interesting information about the maximum size of the transducing DNA fragment carried by P1 phage. The length of DNA between *thr* and *leu* must be close to the maximum size limit because they are cotransduced only 2 to 3 percent of the time. The *azi* locus, which is close to *leu*, is never cotransduced with *thr*. Thus the maximum size of the transducing fragment that P1 can carry extends from the *thr* locus to a point between *leu* and *azi*.

Specialized transduction can be used to map phage attachment sites and to analyze the genes that lie close to these sites. However, the attachment sites are usually studied using conjugation mapping. The genes closely linked to the phage attachment sites can be analyzed in detail by using specialized transducing particles that have incorporated segments of these genes. In general, however, specialized transducing particles are not as useful in gene mapping experiments as generalized transducing particles.

Key Points: **Transduction is the transport of bacterial genetic material from one bacterium to another by a phage. A generalized transducing phage can carry any segment of the bacterial genome. If the transduced DNA is not incorporated into the recipient genome, abortive transduction may result. Specialized transducing phage insert into a specific site on the bacterial chromosome. They are able to incorporate and transfer only those bacterial genes located around the insertion site. Generalized transducing particles are useful for mapping bacterial genes. The closer two genes are, the greater the likelihood that they will be cotransduced. Specialized transducing particles can be used to map phage attachment sites and to analyze the genes closely linked to these sites.**

THE EVOLUTIONARY SIGNIFICANCE OF SEXUALITY IN BACTERIA

Sexual reproduction maximizes the diversity of genotypes among progeny which is highly adaptive under unpredictable conditions. Bacteria do not reproduce sexually, but sexuality has evolved in these organisms. Genetic recombination is unquestionably as important to bacterial evolution as it is to eukaryote evolution. The ability to generate new genotypes in rapidly changing environments allows members of a bacterial population the opportunity to avoid potentially catastrophic environmental changes such as toxins, bacterial viruses, heavy metals, and antibiotics and to exploit new environments. Without parasexual processes, the only way that genetic variants would accumulate in bacterial populations would be through the relatively slow process of mutation.

TESTING YOUR KNOWLEDGE

1. By means of the interrupted mating technique, five different Hfr strains (**A, B, C, D, E**) were analyzed for the sequence in which they transmitted ten different genes (0, 1, 2, 3, 4, 5, 6, 7, 8, 9) to an F⁻ recipient. Each strain was found to transmit its genes in a unique sequence. The results of the experiment are presented in the following table (only the first six genes are recorded):

		Hfr Strain				
		A	B	C	D	E
Order	First	5	0	6	3	5
of	Second	7	2	7	4	8
Transmission	Third	6	1	5	6	9
	Fourth	4	3	8	7	0
	Fifth	3	4	9	5	2
	Sixth	1	6	0	8	1

What is the gene sequence in the original strain from which these five strains were derived?

ANSWER

In the **A** strain, the sequence must be: origin 576431. Strain **B** has a different F insertion point, and the sequence is origin 021346. The 1346 sequence is common to both strains, but in reverse order. The 02 genes are to the left of 1, and 75 are to the right, giving us a sequence of 02134675. In strain **C** the sequence is origin 675890, telling us that 890 map to the right of 5, giving us a sequence of 02134675890, a complete circle :

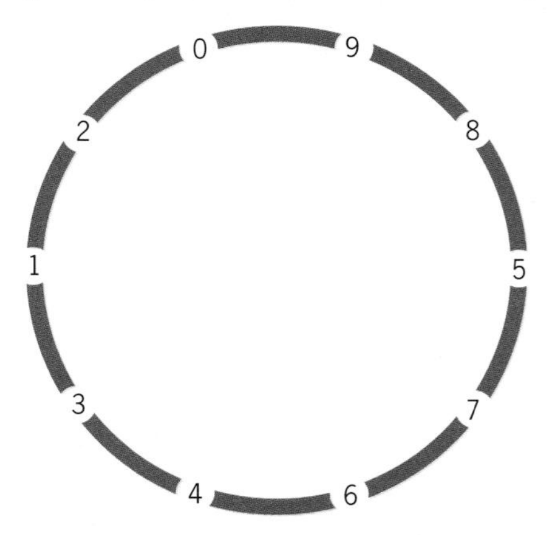

Strain **D** is 346758, which is consistent with this sequence; and strain **E** is 589021, which is also consistent. The F factors are inserted as follows: In **A** F is between 5 and 8, and the transmission is counterclockwise; in **B** it is between 0 and 9, and the transmission is clockwise; in **C** it is between 4 and 6, and the transmission direction is clockwise; in **D** it is between 1 and 3, and the transmission direction is clockwise; and in **E** it is between 5 and 7, and the direction is clockwise.

2. In a transformation experiment, a number of crosses were made between strains that differed in two genes:

Donor	Recipient	Frequency of Wild-Type Transformants (++)
$a^- b^+$	$a^+ b^-$	0.450
$c^- b^+$	$c^+ b^-$	0.190
$d^- b^+$	$d^+ b^-$	0.260

Assuming that the b gene is at one end, what is the sequence of these four genes?

ANSWER

In each of these crosses, two exchanges are required to insert the wild-type allele into the recipient. These are diagrammed as follows:

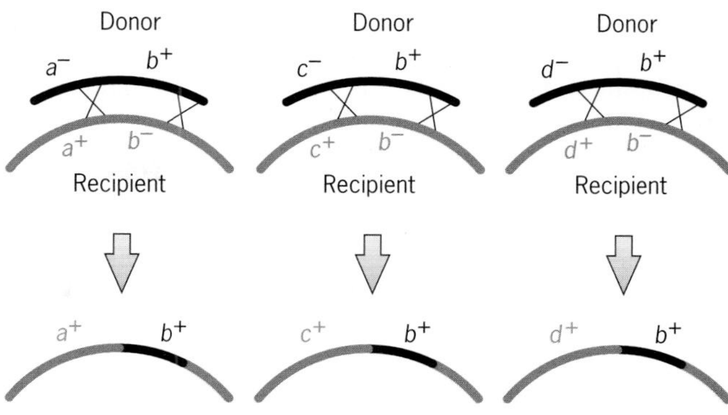

More exchanges occur between a and b than any of the other genes and b, so a must be farthest from b. The c gene is closer to b than d because it has a lower recombination frequency. Because the b and c genes are 19 map units apart, b and d are 26.3 map units apart, and a and b are 45 map units apart, the map would be:

a- - - - - -19- - - - - -*d*- - -7- - - *c*- - - - - -19- - - - - -*b*

QUESTIONS AND PROBLEMS

16.1 A nutritionally defective *E. coli* strain grows only on a medium containing thymine, whereas another nutritionally defective strain grows only on medium containing leucine. When these two strains are grown together, a few progeny are able to grow on a minimal medium with neither thymine or leucine. How can this result be explained?

16.2 Assume that you have just demonstrated genetic recombination in a previously undescribed species of bacteria (e.g., when a strain of genotype $a\ b^+$ is present with a strain of genotype $a^+\ b$, some recombinant genotypes, $a^+\ b^+$ and $a\ b$, are formed). Describe how you would go about (operationally) determining whether the observed recombination was the result of a process similar to transformation, a process similar to transduction, or a process similar to conjugation.

16.3 Compare, in table form, similarities and differences of the mechanisms through which (a) transduction, (b) sexduction, and (c) transformation may occur.

16.4 (a) What are the genotypic differences between F^- cells, F^+ cells, and Hfr cells? (b) What are the phenotypic differences? (c) By what mechanism are F^- cells converted to F^+ cells? F^+ cells to Hfr cells?

16.5 (a) Of what use are F' factors in genetic analysis? (b) How are F' factors formed? (c) By what mechanism does sexduction occur?

16.6 What are the basic differences between generalized transduction and specialized transduction?

16.7 What is the difference between a phage and a prophage?

16.8 How can genes be mapped by interrupted mating conjugation experiments?

16.9 What does the term *cotransduction* mean? How can cotransduction frequencies be used to map genetic markers?

16.10 An F^+ strain, marked at 10 loci, gives rise spontaneously to Hfr progeny whenever the F factor becomes incorporated into the chromosome of the F^+ strain. The F factor can integrate into the circular chromosome at many points, so that the various Hfr segregants transfer the genetic markers in different sequences. For any Hfr strain, the order of markers entering early can be determined by interrupted mating experiments. From the following data for several Hfr strains derived from the same F^+, determine the order of markers in the F^+ strain.

Hfr Strain	Markers Donated in Order
1	— Z-H-E-R →
2	— O-K-S-R →
3	— K-O-W-I →
4	— Z-T-I-W →
5	— H-Z-T-I →

16.11 In *E. coli,* the ability to utilize lactose as a carbon source requires the presence of the enzymes β-galactosidase and β-galactoside permease. These enzymes are coded for by two closely linked genes, *lacZ* and *lacY*, respectively. Another gene, *proC*, controls, in part, the ability of *E. coli* cells to synthesize the amino acid proline. The alleles str^r and str^s control resistance and sensitivity, respectively, to streptomycin. HfrH is known to transfer the two *lac* genes, *proC*, and *str*, in that order, during conjugation.

A cross was made between HfrH of genotype $lacZ^-\ lacY^+\ proC^+\ str^s$ and an F^- of genotype $lacZ^+\ lacY^-\ proC^-\ str^r$. After about two hours, the mixture was diluted and plated out on medium containing streptomycin but no proline. When the resulting $proC^+\ str^r$ recombinant colonies were checked for their ability to grow on medium containing lactose as the sole carbon source, very few of them were capable of fermenting lactose. When the cross HfrH $lacZ^+$ $lacY^-$ $proC^+$ str^s × F^- $lacZ^-$ $lacY^+$ $proC^-$ str^r was done, many of the $proC^+$ str^r recombinants were able to grow on medium containing lactose as the sole carbon source. What is the order of the *lacZ* and *lacY* genes relative to *proC*?

16.12 The data in the following table were obtained from three-point transduction tests made to determine the order of mutant sites in the *A* gene for tryptophan synthetase in *E. coli. Anth* is a linked, unselected marker. In each cross, trp^+ recombinants were selected and then scored for the *anth* marker ($anth^+$ or $anth^-$). What is the linear order of *anth* and the three mutant alleles of the *A* gene indicated by the data in the table?

Cross	Donor Markers	Recipient Markers	*anth* Allele in Recombinants	% $anth^+$
1	$anth^+$ *A34*	$anth^-$ *A223*	72 $anth^+$: 332 $anth^-$	18
2	$anth^+$ *A46*	$anth^-$ *A223*	196 $anth^+$: 180 $anth^-$	52
3	$anth^+$ *A223*	$anth^-$ *A34*	380 $anth^+$: 379 $anth^-$	50
4	$anth^+$ *A223*	$anth^-$ *A46*	60 $anth^+$: 280 $anth^-$	20

16.13 Two additional mutations in the *trp A* gene of *E. coli*, *trp A58* and *trp A487*, were ordered relative to *trp A223* and the outside marker *anth* by three-factor transduction crosses as described in Problem 16.12. The results of these crosses are summarized in the following table. What is the linear order of *anth* and the three mutant sites in the *trp A* gene?

Cross	Donor Markers	Recipient Markers	*anth* Allele in trp+ Recombinants	% $anth^-$
1	$anth^+$ *A487*	$anth^-$ *A223*	72 $anth^+$: 332 $anth^-$	82
2	$anth^+$ *A58*	$anth^-$ *A223*	196 $anth^+$: 180 $anth^-$	48
3	$anth^+$ *A223*	$anth^-$ *A487*	380 $anth^+$: 379 $anth^-$	50
4	$anth^+$ *A223*	$anth^-$ *A58*	60 $anth^+$: 280 $anth^-$	80

16.14 In an $a^+ b^+ c^- d^+ \times a^- b^- c^+ d^-$ cross using conjugating *E. coli* cells, the $b^+ c^+$ alleles were selected for among the recombinant offspring. The *a* and *d* alleles were not selected. When the $b^+ c^+$ recombinants were checked, most were $a^- d^-$. Which strain was the donor (Hfr)?

16.15 A three-point transformation mapping experiment is done in *Bacillus*, and the following results are obtained:

Donor: $a^+ b^+ c^+$
Recipient: $a^- b^- c^-$

	Class 1	2	3	4	5	6	7	Total
a:	–	–	+	–	+	+	+	
b:	–	+	–	+	–	+	+	
c:	+	–	–	+	+	–	+	
Number:	700	400	2,600	3,600	100	1,200	12,000	20,600

What is the gene order and the linkage distance between each pair of genes?

16.16 Two *Bacillus* genes, *a* and *b*, are studied in a transformation experiment. The results of the experiment are presented in the following table:

Experiment	Donor	Recipient	Transformant Classes *a*	*b*	Number
A	$a^+ b^-$	$a^- b^-$	+	–	232
	$a^- b^+$		–	+	341
			+	+	7
B	$a^+ b^+$	$a^- b^-$	+	–	130
			–	+	96
			+	+	247

(a) Are the two genes closely linked, according to the criteria for linkage in transformation experiments?

(b) If they are linked, what is the recombination frequency between them?

16.17 In a conjugation experiment, you find that there are two genes that lie so close together you are unable to determine a gene order by time-interval mapping. There are two possible gene orders to consider:

a————*b*–*c* or *a*————*c*–*b*

You set up reciprocal crosses so that the Hfr donor is + + *c* and the F^- recipient is *a b* + in one experiment (A) and the Hfr donor is *a b* + and the F^- recipient is + + *c* in another experiment (B). When you screen for the + + + recombinants and compare their frequencies in both experiments, you find that experiment A produces about 1/25th as many as experiment B. What does this suggest about the order of the three genes?

BIBLIOGRAPHY

ADELBERG, E. A. (ed). 1966. *Papers on Bacterial Genetics*, 2nd ed. Little, Brown, Boston.

BROCK, T. D. 1990. *The Emergence of Bacterial Genetics*. Cold Spring Harbor Laboratory Press, Cold Spring Harbor.

DALE, J. W. 1994. *Molecular Genetics of Bacteria*. 2nd ed. John Wiley and Sons, New York.

DAVIS, R. W., D. BOTSTEIN, AND J. R. ROTH. 1980. *Advanced Bacterial Genetics*. Cold Spring Harbor Laboratory Press, Cold Spring Harbor.

DONACHIE, W. D. 1993. The cell cycle of *Eschericia coli*. *Annu. Rev. Microbio.* 47:199–230.

HAYES, W. 1968. *The Genetics of Bacteria and Their Viruses*. 2nd ed. John Wiley and Sons, New York.

HOLLOWAY, B. W. 1993. Genetics for all bacteria. *Annu. Rev. Microbio.* 47:659–684.

INGRAHAM, J. L., MAALOE, AND F. C. NEIDHARDT. 1983. *Growth of the Bacterial Cell*. Sinauer, Sunderland, MA.

MALOY, S. R., E. J. CRONAN, AND D. FREIFELDER. 1994. *Microbial Genetics*. 2nd ed. Jones and Bartlett, Boston.

STENT, G. S. AND R. CALENDAR. 1978. *Molecular Genetics: An Introductory Narrative*. 2nd ed. W. H. Freeman, San Francisco.

Color variation among kernels of maize. Studies of the genetic basis of this variation led to the discovery of transposable elements.

17

Transposable Genetic Elements

Maize: From Colored Kernels to Transposable Elements

Maize is one of the world's most important crop plants. In the United States alone over 79 million acres are devoted to maize cultivation. In 1994 this crop yielded an average of 138.6 bushels of grain per acre, enough to feed tens of millions of farm animals and to produce a large assortment of foods for human consumption, including cereal, sugar, corn oil, and corn syrup. The cultivation of maize began at least 5000 years ago in Central America. By the time Christopher Columbus arrived in the New World, maize cultivation had spread north to Canada and south to Argentina.

The native peoples of North and South America developed many different varieties of maize, each adapted to particular conditions. Maize was cultivated in lowland clearings, in mountainous uplands, in the rain forests of Amazonia, and in the deserts of Arizona and New Mexico. This diversification of maize varieties occurred through a prescientific application of the practice of selective breeding. Ancient farmers cultivated plants with desirable characteristics—size, shape of the ears, number, texture and color of the kernels, dietary value, hardiness, and so on. After many generations of selective breeding, each locality developed a distinctive variety. This prescientific genetic manipulation was so extensive that today maize cannot grow without human assistance. It is a completely domesticated plant.

Maize cultivation has also had a cultural significance among these native peoples. They developed varieties that had colorful kernels—red, blue, yellow, white, and purple—and associated each color with a special aesthetic or religious value. To the peoples of the American Southwest, for example, blue maize is considered sacred, and each of the four cardinal directions of the compass is represented by a particular maize color. Some groups consider kernels with stripes and spots to be signs of strength and vigor.

The jewel-like patterns we see on maize also have an important scientific significance. Modern research has shown that the stripes and spots on maize kernels are the result of a genetic phenomenon called *transposition*. Within the maize genome—indeed, within the genomes of most organisms—geneticists have found a special class of DNA sequences, sequences that can move. These *transposable genetic elements*—or more simply, *transposons*—have the remarkable ability to jump from one site in the genome to another. During this process, a transposon may break chromosomes or mutate genes. In fact, these two phenomena are responsible for the beautiful mosaic kernels that fascinated the early American peoples and that continue to fascinate us today.

The research that revealed the existence of transposable elements was performed in the 1940s and 1950s by Barbara McClintock (Historical Sidelight: Barbara McClintock, The Discoverer of Transposable Elements), an American scientist who later received the Nobel Prize in Physiology or Medicine for her work. McClintock studied the genetic basis of kernel striping and spotting in maize, and found that it was due to the insertion and excision of mobile genetic elements. For many years, McClintock's work had no parallel, but in the 1960s and 1970s, as new techniques emerged to analyze genes at the molecular level, transposable elements were discovered in the genomes of assorted bacteria, fungi, plants, and animals; then, in the 1980s, they were found in human beings. Extensive research since the time of McClintock has indicated that these elements are present in the genomes of many—perhaps most—organisms.

In this chapter, we investigate the structure and behavior of different types of transposable elements, and we also discuss the experiments that led to their discovery. In addition, we explore the genetic and evolutionary significance of these mobile DNA sequences.

TRANSPOSABLE ELEMENTS IN BACTERIA

Bacterial transposons were the first to be studied at the molecular level. There are three main types: the Insertion Sequences, or IS elements, the composite transposons, and the Tn3 elements. These three types of transposons differ in size and structure. The IS elements are the simplest, containing only genes that encode proteins involved in transposition. The composite transposons and Tn3 elements are more complex, containing some genes that encode products unrelated to the transposition process.

IS Elements

The simplest bacterial transposons are the **Insertion Sequences**, or **IS elements**, so named because they can insert at many different sites in bacterial chromosomes and plasmids. IS elements were first detected in certain *lac*⁻ mutations of *E. coli*. These mutations had the unusual property of reverting to wild-type at a high rate. Molecular analyses eventually revealed that these unstable mutations possessed extra DNA in or near the *lac* genes. When DNA from the wild-type revertants of these mutations was compared with that from the mutations themselves, it was found that the extra DNA had been lost. Thus, these genetically unstable mutations were caused by DNA sequences that had inserted into *E. coli* genes and reversion to wild-type was caused by excision of these sequences. Similar insertion sequences have been found in many other bacterial species.

IS elements are compactly organized. Typically, they consist of less than 2500 nucleotide pairs and contain only genes whose products are involved in pro-

HISTORICAL SIDELIGHT

Barbara McClintock, the Discoverer of Transposable Elements

Scientific advances can come through the unflagging persistence of a single individual, someone who focuses on a question or problem and researches it for many years. Barbara McClintock was such an individual (Figure 1). Her long life was devoted to the study of maize genetics. Born in New England in 1902, she grew up in New York City and attended Cornell University, first as an undergraduate, then as a graduate student. After receiving her Ph.D., McClintock remained at Cornell for several years, collaborating with an illustrious group of maize geneticists, including George Beadle, R. A. Emerson, Charles Burnham, Marcus Rhoades, and Lowell Randolph. Together, these researchers developed the materials and methods of maize genetics into a rich intellectual discipline. McClintock played an important role in this work. As one who excelled in cytological analysis, she succeeded in identifying each of the 10 maize chromosomes and was able to connect them with linkage groups. She also did pioneering work on the mechanism of crossing over and on the origin of the nucleolus. However, McClintock's most notable achievement was to elucidate the genetic properties of transposable elements. Most of her research on this subject was done during the many years she spent in the Genetics Department of the Carnegie Institution at Cold Spring Harbor, New York. There she found the freedom to pursue her studies of transposon-mediated mutation and chromosome breakage.

McClintock published her first report about transposable elements in 1948. Several other reports followed, including a major paper in the 1951 *Cold Spring Harbor Symposium on Quantitative Biology*. For many reasons, the ideas that she espoused in these papers were not especially well received. The concept of transposition contradicted the established view that genes occupied fixed positions on chromosomes. McClintock's data were complex, and she had difficulty communicating them to her colleagues. In addition, transposition did not seem to be a general phenomenon. Although there was little doubt that it occurred in maize, no one had seen it in other organisms. This situation changed in the 1960s and 1970s, when transposition was discovered in bacteria and *Drosophila*. At last the scientific world awoke to the broad significance of McClintock's ideas.

Figure 1 Barbara McClintock.

McClintock was highly respected by her colleagues. In 1944, at the relatively young age of 42, she was admitted to the prestigious National Academy of Sciences of the United States. In 1945 she was elected president of the Genetics Society of America, and in 1970 she was awarded the National Medal of Science. Her Nobel Prize came in 1983, 35 years after her first publication on transposable elements. The day the Prize was announced, she was out in the woods collecting mushrooms. During her life, McClintock was a model of "adamant individuality and self-containment."[1] She died in 1992, a few months after her ninetieth birthday.

[1]Federoff, N. V., 1994. Barbara McClintock (June 16, 1902–September 2, 1992). *Genetics* 136:1–10.

moting or regulating transposition. Many distinct types of IS elements have been identified. The smallest, IS*1*, is 768 nucleotide pairs long. Each type of IS element is demarcated by short identical, or nearly identical, sequences at its ends (Figure 17.1). Because these terminal sequences are always in inverted orientation with respect to each other, they are called **inverted terminal repeats**. Their lengths range from 9 to 40 nucleotide pairs. Inverted terminal repeats are characteristic of most—but not all—types of transposons. When nucleotides in these repeats are mutated, the transposon usually loses its ability to move. These mutations therefore demonstrate that inverted terminal repeats play an important role in the transposition process.

At least some IS elements encode a protein that is needed for transposition. This protein, called **transposase**, seems to bind at or near the ends of part of the element, where it cuts both strands of the DNA. Cleavage of the DNA at these sites excises the element from the chromosome or plasmid, making it free to in-

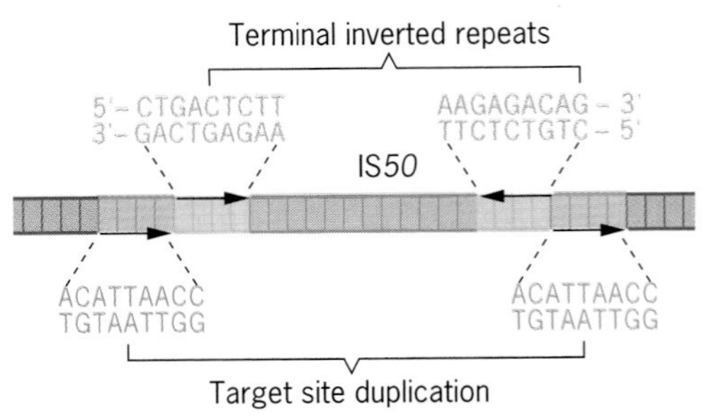

Figure 17.1 Structure of an inserted IS*50* element showing its terminal inverted repeats and target site duplication. The terminal inverted repeats are imperfect because the fourth nucleotide pair from each end is different.

sert at a new position in the same or a different DNA molecule. When IS elements insert into chromosomes or plasmids, they create a duplication of part of the DNA sequence at the site of the insertion. One copy of the duplication is located on each side of the element. These short (2 to 13 nucleotide pairs), directly repeated sequences, called **target site duplications**, are thought to arise from staggered cleavage of the double-stranded DNA molecule (Figure 17.2).

A bacterial chromosome may contain several copies of a particular type of IS element. For example, 6 to 10 copies of IS*1* are found in the *E. coli* chromosome. Plasmids may also contain IS elements. The F plasmid, for example, typically has at least two different IS elements, IS*2* and IS*3*. When a particular IS element resides in both a plasmid and a chromosome, it creates the opportunity for homologous recombination between different DNA molecules. Such recombination appears to be responsible for the integration of the F plasmid into the *E. coli* chromosome (Chapter 16).

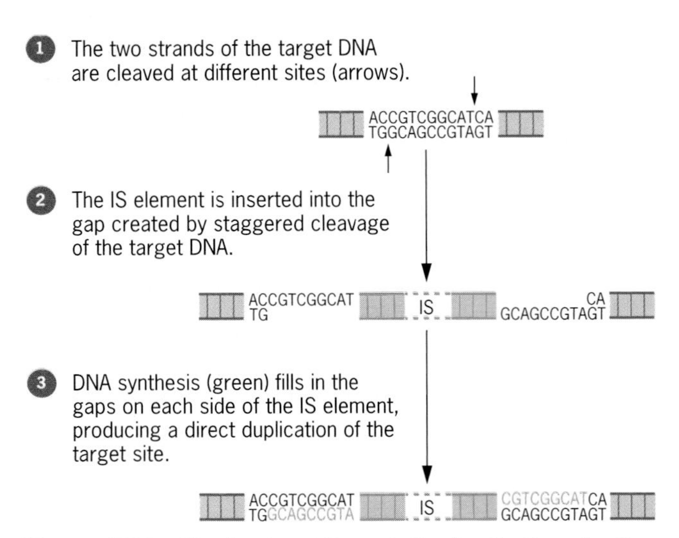

Figure 17.2 Production of target site duplications by the insertion of an IS element.

Both the *E. coli* chromosome and the F plasmid are circular DNA molecules. When the plasmid and the chromosome recombine in a region of homology—for example, in an IS element common to both of them—the smaller plasmid is integrated into the larger chromosome, creating a single circular molecule. Such integration events produce Hfr strains capable of transferring their chromosomes during conjugation. These strains vary in the integration site of the F plasmid because the IS elements that mediate recombination occupy different chromosomal positions in different *E. coli* strains—a result of their ability to transpose. Because of their role in forming Hfr strains, IS elements potentiate the exchange of chromosomal genes between different strains of bacteria. This exchange creates genetic variability, which, along with mutation, is the basis for evolutionary change in bacterial populations.

Composite Transposons

Composite transposons, denoted by the symbol Tn, are created when two IS elements insert near each other. The region between them can then be transposed by the joint action of the flanking elements. In effect, two IS elements "capture" a DNA sequence that is otherwise immobile and endow it with the ability to move. Figure 17.3 gives three examples. In Tn*9*, the flanking IS elements are in the same orientation with respect to each other, whereas in Tn*5* and Tn*10*, the orientation is inverted. The region between the IS elements in each of these transposons contains genes that have nothing to do with transposition. In fact, in all three transposons, the genes between the flanking IS elements confer resistance to antibiotics—chloramphenicol resistance in Tn*9*, kanamycin, bleomycin, and streptomycin resistance in Tn*5*, and tetracycline resistance in Tn*10*.

Sometimes the flanking IS elements in a composite transposon are not quite identical. For instance, in Tn*5*, the element on the right, called IS*50*R, is capable of producing a transposase to stimulate transposition, but the element on the left, called IS*50*L, is not. This difference is due to a change in a single nucleotide pair that prevents IS*50*L from specifying the active transposase.

Tn*5* also illustrates another feature of the composite transposons: their movement is regulated (Figure 17.4). When a bacterial cell is infected with a nonlytic bacteriophage that carries Tn*5* on its chromosome, the frequency of Tn*5* transposition is dramatically reduced if the infected cell already carries a copy of Tn*5*.

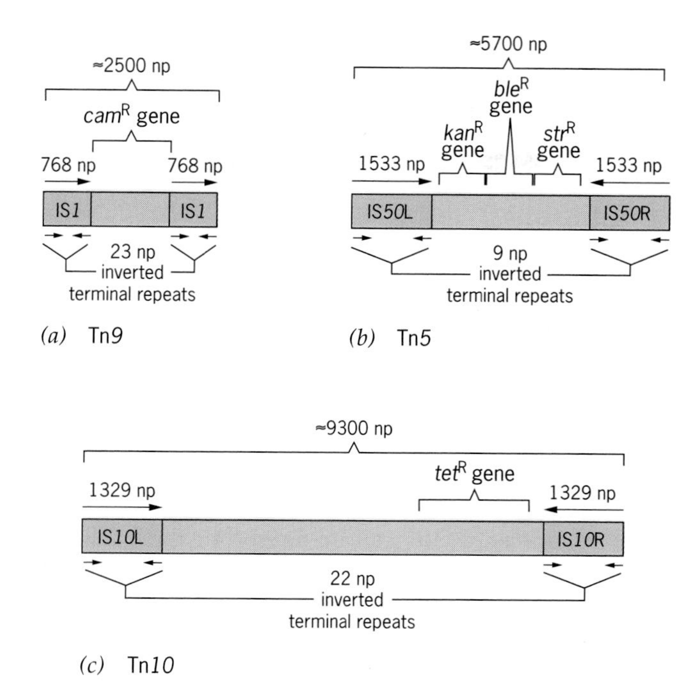

Figure 17.3 Genetic organization of composite transposons. The orientation and length (in nucleotide pairs, np) of the constituent sequences are indicated. (*a*) Tn*9* consists of two IS*1* elements flanking a gene for chloramphenicol resistance. (*b*) Tn*5* consists of two IS*50* elements flanking genes for kanamycin, bleomycin, and streptomycin resistance. (*c*) Tn*10* consists of two IS*10* elements flanking a gene for tetracycline resistance.

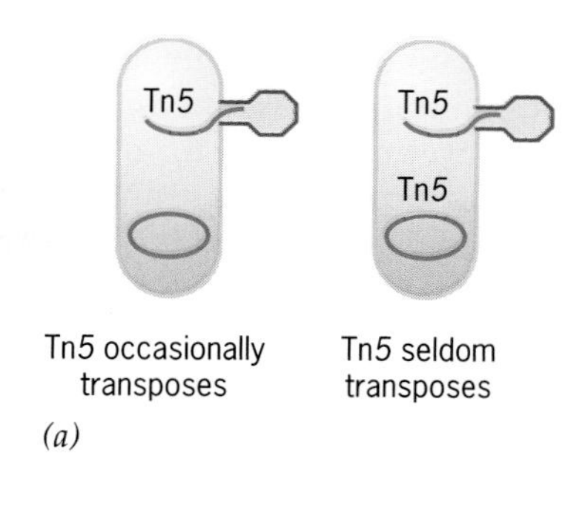

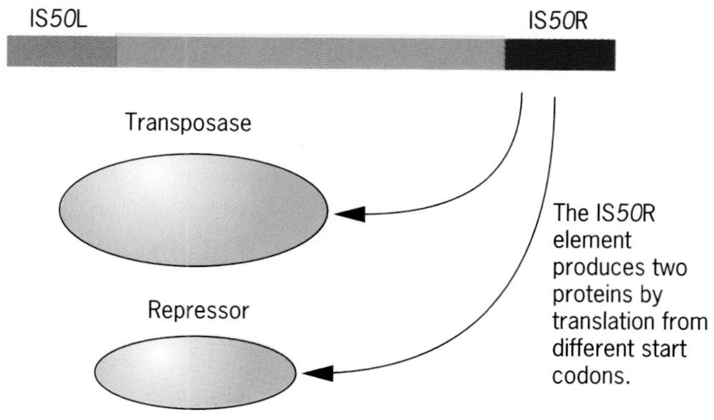

Figure 17.4 Regulation of Tn*5*. (*a*) Infection of *E. coli* cells with bacteriophages carrying Tn*5*. Cells that already possess a copy of Tn*5* repress transposition. (*b*) Genetic basis of Tn*5* regulation. One of the proteins produced by IS*50*R is a transposase that catalyzes transposition, but the other is a repressor that inhibits transposition. The effect of the repressor usually prevails.

This reduction implies that the resident transposon inhibits the transposition of an incoming transposon, possibly by synthesizing a repressor. Analyses by Michael Syvanen, William Reznikoff, and their colleagues have shown that this hypothesis is correct. The IS*50*R element of Tn*5* actually produces two proteins. One, the transposase, catalyzes transposition, whereas the other, a truncated transposase created by translation from a start codon within the transposase gene, prevents transposition. Because the shorter protein is the more abundant, Tn*5* transposition tends to be repressed.

Tn*3* Elements

The elements in this group of transposons are larger than the IS elements and usually contain genes that are not necessary for transposition. Like the IS elements, they have inverted repeat sequences at their termini (38 to 40 nucleotide pairs long), and they produce target site duplications upon insertion. The transposon known as Tn*3* is the most thoroughly studied example.

The genetic organization of Tn*3* is shown in Figure 17.5. There are three genes, *tnpA*, *tnpR*, and *bla*, encoding, respectively, a transposase, a resolvase/repressor, and an enzyme called beta lactamase. The beta lactamase confers resistance to the antibiotic ampicillin, and the other two proteins play important roles in transposition.

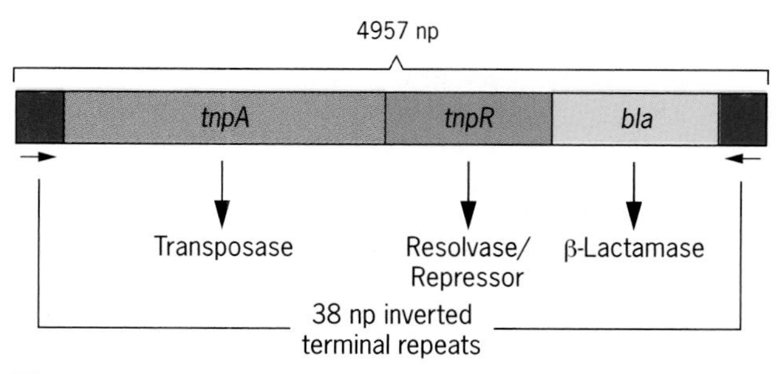

Figure 17.5 Genetic organization of Tn*3*. Lengths of DNA sequences are given in nucleotide pairs (np).

The transposition of Tn*3* occurs in two stages (Figure 17.6). First, the transposase mediates the fusion of two molecules, forming a structure called a **cointegrate**. During this process, the transposon is replicated, and one copy is inserted at each junction in the cointegrate; the two Tn*3* elements in the cointegrate are oriented in the same direction. In the second stage of transposition, the *tnpR*-encoded resolvase mediates a site-specific recombination event between the two Tn*3* elements. This event occurs at a sequence in Tn*3* called *res*, the *resolution site*, generating two molecules, each with a copy of the transposon.

The *tnpR* gene product has yet another function—to repress the synthesis of both the transposase and resolvase proteins. Repression occurs because the *res* site is located between the *tnpA* and *tnpR* genes. By binding to this site, the *tnpR* protein interferes with the transcription of both genes, leaving their products in chronic short supply. As a result, the Tn*3* element tends to remain immobile.

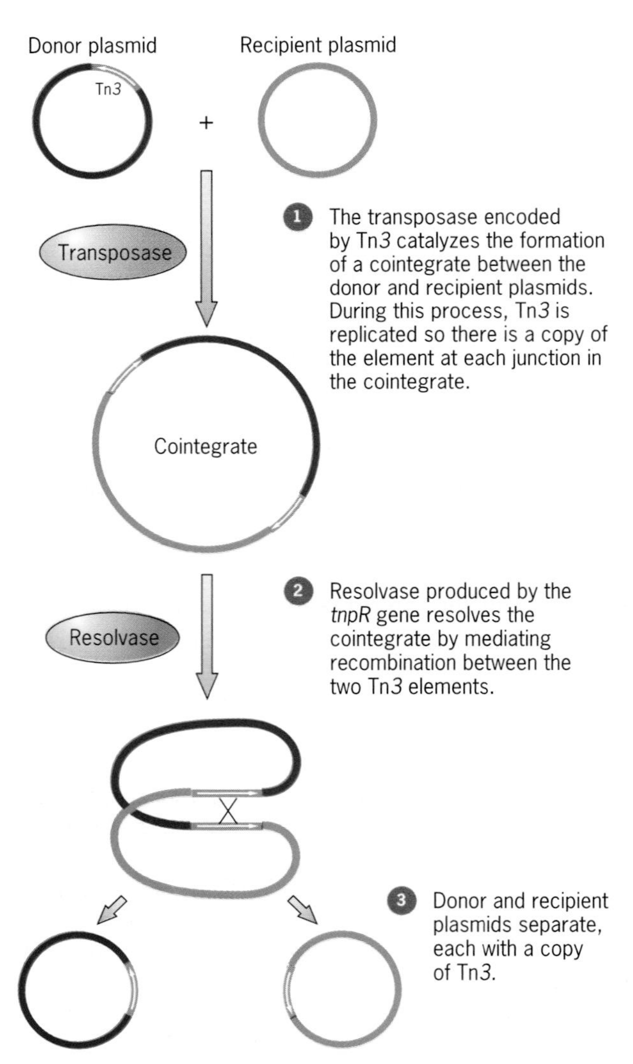

Figure 17.6 Transposition of Tn*3* via the formation of a cointegrate.

The Medical Significance of Bacterial Transposons

Many bacterial transposons carry genes for antibiotic resistance. Consequently, it is a relatively simple matter for these genes to move from one DNA molecule to another—for instance, from a chromosome to a plasmid. This genetic flux has a profound medical significance because many of the DNA molecules that acquire resistance genes can be passed on to other cells. Resistance to a particular antibiotic can therefore be spread horizontally as well as vertically in a bacterial population. Eventually, all or nearly all the bacterial cells become resistant, and the antibiotic is no longer useful in combating the organism.

This process has occurred in several species pathogenic to humans, including strains of *Staphlococcus*, *Enterococcus*, *Neisseria*, *Shigella*, and *Salmonella*. Today many bacterial infections causing diseases such as dysentery, tuberculosis, and gonorrhea are difficult to treat. In some cases, the pathogen has acquired resistance to several different antibiotics, making it particularly deadly.

The spread of multiple drug resistance in bacterial populations has been accelerated by the evolution of **conjugative R plasmids** that carry the resistance genes (Chapter 16). These plasmids have two components. One, called the **resistance transfer factor,** or **RTF**, contains the genes needed for conjugative transfer between cells; the other, called the **R-determinant**, contains the genes for antibiotic resistance. Often these resistance genes are carried by a transposon, or a set of transposons, that have been inserted into the plasmid (Figure 17.7). Conjugative R plasmids can be transferred rapidly between cells in a bacterial population. Surprisingly, this transfer is not limited to cells of the same species. Conjugative plasmids can, in fact, be passed from one species to another, even between quite dissimilar cell types—for example, between a *Coccus* and a *Bacillus*. Thus, once multiple drug resistance has evolved in a part of the microbial kingdom, it can spread to other parts with relative ease.

Key Points: **Bacterial transposons, including the IS elements, composite transposons made from them, and Tn*3*, have short inverted repeats at their termini. Upon insertion into a chromosome or plasmid, they create a short, direct duplication of DNA at the insertion site. Transposition is catalyzed by a protein, the transposase, produced by a gene within the transposon. Other transposon-encoded proteins may repress transposition. Composite transposons and Tn*3* ele-**

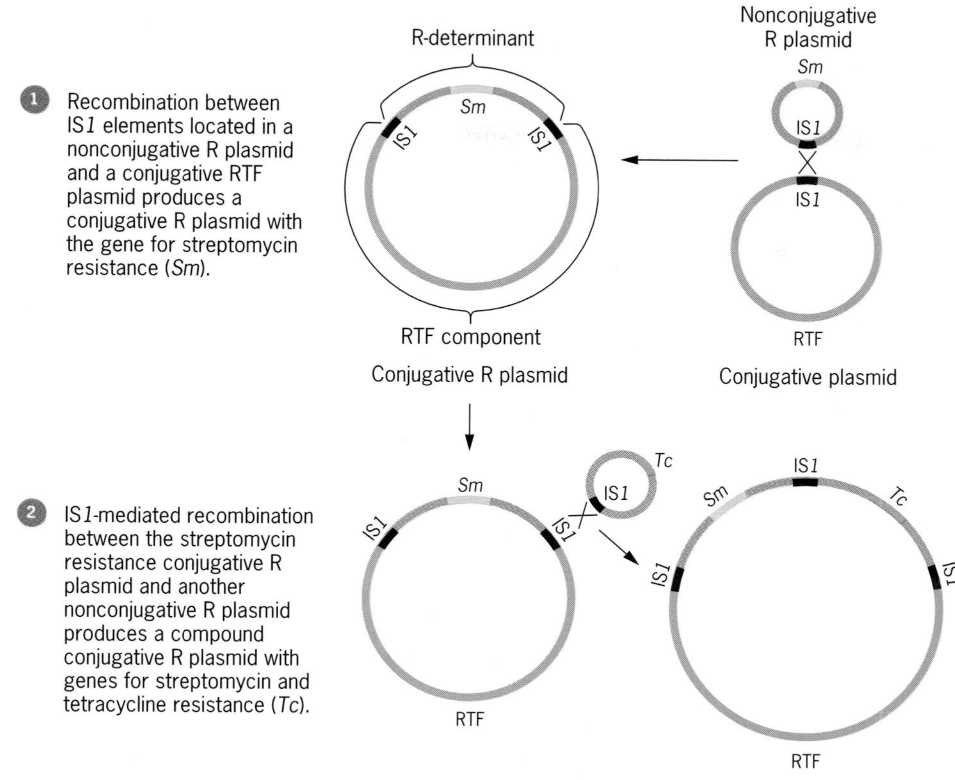

Figure 17.7 Evolution of conjugative plasmids carrying genes for antibiotic resistance.

ments carry genes that confer resistance to antibiotics.

TRANSPOSABLE ELEMENTS IN EUKARYOTES

Geneticists have found many different types of transposons in eukaryotes. These elements vary in size, structure, and behavior. Some are abundant in the genome, others rare. In the following sections, we discuss a few of the eukaryotic transposons that have been studied intensively. All these elements have inverted repeats at their termini and create target site duplications when they insert into DNA molecules. Some encode a transposase that catalyzes the movement of the element from one position to another.

Ac and *Ds* Elements in Maize

The *Ac* and *Ds* elements in maize were discovered through the pioneering work of Barbara McClintock. Through genetic analysis, McClintock showed that the activities of these elements are responsible for the striping and spotting of maize kernels. Many years later, Nina Federoff, Joachim Messing, Peter Starlinger and their colleagues isolated the elements and determined their molecular structure.

McClintock discovered the *Ac* and *Ds* elements by studying chromosome breakage. She used genetic markers that controlled the color of maize kernels to detect the breakage events. When a particular marker was lost, McClintock inferred that the chromosome segment on which it was located had also been lost, indicating that a breakage event had occurred. The loss of a marker was detected by a change in the color of the aleurone, the outermost layer of the triploid endosperm of maize kernels.

In one set of experiments, the genetic marker that McClintock followed was an allele of the *C* locus on the short arm of chromosome 9. Because this allele, C^I, is a dominant inhibitor of aleurone coloration, any kernel possessing it is colorless. McClintock fertilized *CC* ears with pollen from C^IC^I tassels, producing kernels in which the endosperm was C^ICC. (Recall from Chapter 2 that the triploid endosperm receives two alleles from the female parent and one from the male parent.) Although McClintock found that most of these kernels were colorless, as expected, some showed patches of brownish-purple pigment (Figure

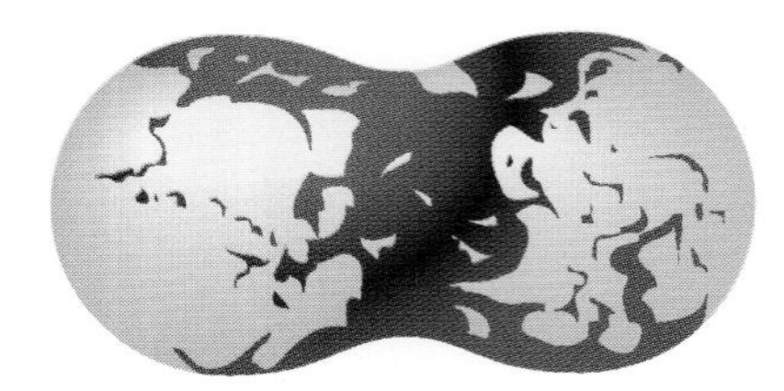

Figure 17.8 Maize kernel showing loss of the C^I allele for the inhibition of pigmentation in the aleurone. The pigmented patches are -*CC*, whereas the colorless patches are C^ICC.

17.8). McClintock guessed that in such mosaics, the inhibitory C^I allele had been lost sometime during endosperm development, leading to a clone of tissue that was able to make pigment. The genotype in such a clone would be -*CC*, where the dash indicates the missing C^I allele.

The mechanism that McClintock proposed to explain the loss of the C^I allele is diagrammed in Figure 17.9. A break at the site labeled by the arrow detaches a segment of the chromosome from its centromere, creating an acentric fragment. Such a fragment tends to be lost during cell division; thus all the descendants of this cell will lack part of the paternally derived chromosome. Because the lost fragment carries the C^I allele, none of the cells in this clone are inhibited from forming pigment. If any of them produces a part of the aleurone, a patch of color will appear, creating a mosaic kernel similar to the one shown in Figure 17.8.

McClintock found that the breakage responsible for these mosaic kernels occurred at a particular site on chromosome 9. She named the factor that produced these breaks ***Ds***, for **Dissociation**. However, by itself, this factor was unable to induce chromosome breakage. In fact, McClintock found that *Ds* had to be stimulated by another factor, called ***Ac***, for **Activator**. The *Ac* factor was present in some maize stocks but absent in others. When different stocks were crossed, *Ac* could be combined with *Ds* to create the condition that led to chromosome breakage.

This two-factor *Ac/Ds* system provided an explanation for the genetic instability that McClintock had observed on chromosome 9. Additional experiments demonstrated that this was only one of many instabilities present in the maize genome. McClintock found other instances of breakage at different sites on chromosome 9 and also on other chromosomes. Because breakage at these sites depended on activation by *Ac*, she concluded that *Ds* factors were also involved. To explain all these observations, McClintock proposed that *Ds* could exist at many different sites in the genome and that it could move from one site to another.

This explanation has been borne out by subsequent analyses. The *Ac* and *Ds* elements belong to a family of transposons. These elements are structurally related to each other and can insert at many different sites on the chromosomes. In fact, multiple copies of the *Ac* and *Ds* elements are often present in the maize genome. Through genetic analysis, McClintock demonstrated that both *Ac* and *Ds* can move. When one of these elements inserts in or near a gene, McClintock found that the gene's function is altered—sometimes completely abolished. Thus *Ac* and *Ds* can induce mutations by inserting into genes. To emphasize this effect on gene expression, McClintock called the *Ac* and *Ds* transposons **controlling elements**.

DNA sequencing has shown that *Ac* elements consist of 4563 nucleotide pairs bounded by inverted repeats that are 11 nucleotide pairs long (Figure 17.10*a*); these inverted terminal repeats are essential for transposition. Each *Ac* element is also flanked by direct re-

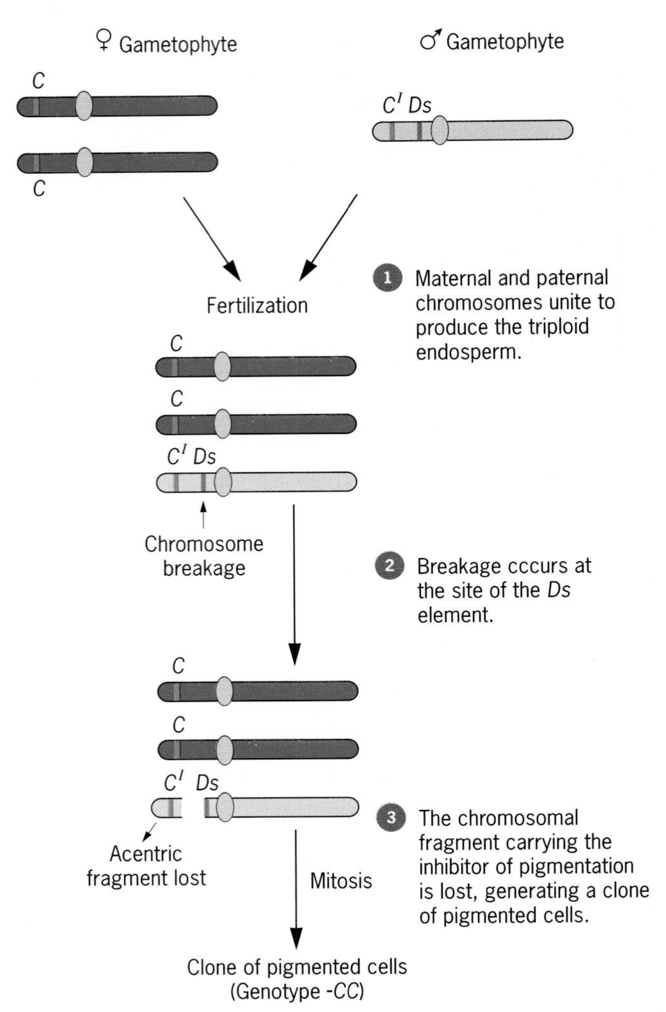

Figure 17.9 Chromosome breakage caused by the transposable element *Ds* in maize. The allele *C* on the short arm of chromosome 9 produces normal pigmentation in the aleurone; the allele C^I inhibits this pigmentation.

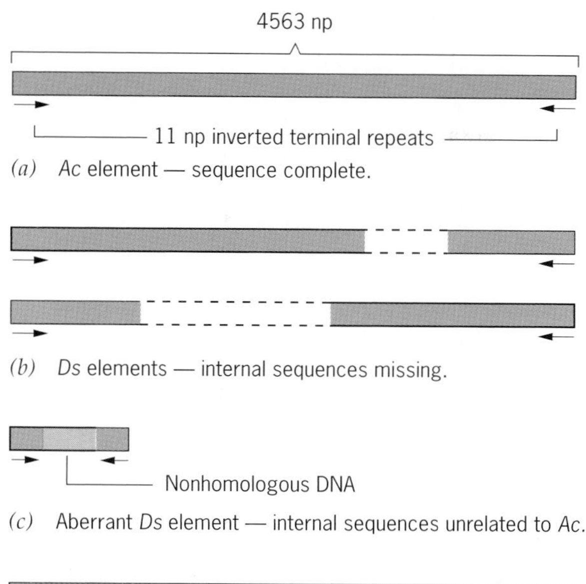

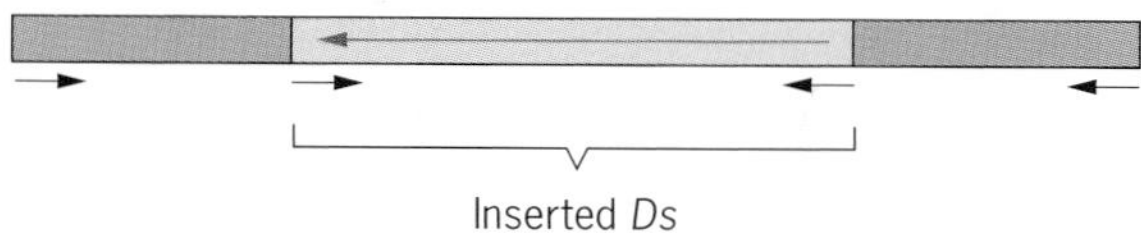

Figure 17.10 Structural organization of the members of the *Ac/Ds* family of transposable elements in maize. The terminal inverted repeats (short arrows underneath) and DNA sequence lengths (in nucleotide pairs, np) are indicated.

peats 8 nucleotide pairs long. Because the direct repeats are created at the time the element is inserted into the chromosome, they are target site duplications, not integral parts of the element.

Unlike *Ac*, *Ds* elements are structurally heterogeneous. They all possess the same inverted terminal repeats as *Ac* elements, demonstrating that they belong to the same transposon family, but their internal sequences vary. Some *Ds* elements appear to have been derived from *Ac* elements by the loss of internal sequences (Figure 17.10*b*). The deletions in these elements may have been caused by incomplete DNA synthesis during replication or transposition. Other *Ds* elements contain non-*Ac* DNA between their inverted terminal repeats (Figure 17.10*c*). These unusual members of the *Ac/Ds* family are called *aberrant Ds* elements. A third class of *Ds* elements is characterized by a peculiar piggybacking arrangement (Figure 17.10*d*); one *Ds* element is inserted into another but in an inverted orientation. These so-called *double Ds* elements are apparently the ones responsible for the chromosome breakage that McClintock observed in her experiments.

The activities of the *Ac/Ds* elements—excision and transposition, and all their genetic correlates, including mutation and chromosome breakage—are caused by a transposase encoded by the *Ac* elements. The *Ac* transposase apparently interacts with sequences at or near the ends of *Ac* and *Ds* elements, catalyzing their movement. Deletions or mutations in the gene that encodes the transposase abolish this catalytic function. Thus *Ds* elements, which have such lesions, cannot activate themselves. However, they can be activated if a transposase-producing *Ac* element is present somewhere in the genome. The transposase made by this element can diffuse through the nucleus, bind to a *Ds* element, and activate it. The *Ac* transposase is, therefore, a *trans*-acting protein.

P Elements and Hybrid Dysgenesis in *Drosophila*

Some of the most extensive research on transposable elements has focused on the *P* elements of *Drosophila melanogaster*. These transposons were identified through the cooperation of geneticists working in several different laboratories. In 1977 Margaret and James Kidwell, working in Rhode Island, and John Sved, working in Australia, discovered that crosses between certain strains of *Drosophila* produce hybrids with an assortment of aberrant traits, including frequent mutation, chromosome breakage, and sterility. The term **hybrid dysgenesis**, derived from Greek roots meaning "a deterioration in quality," was used to denote this syndrome of abnormalities.

Kidwell, Kidwell, and Sved found that they could classify *Drosophila* strains into two main types based on whether or not they produce dysgenic hybrids in testcrosses. The two types of strains are denoted M and P. Only crosses between M and P strains produce dysgenic hybrids, and they do so only if the male in the cross is from the P strain. Crosses between two different P strains, or between two different M strains, produce hybrids that are normal. We can summarize the phenotypes of the hybrid offspring from these different crosses in a simple table:

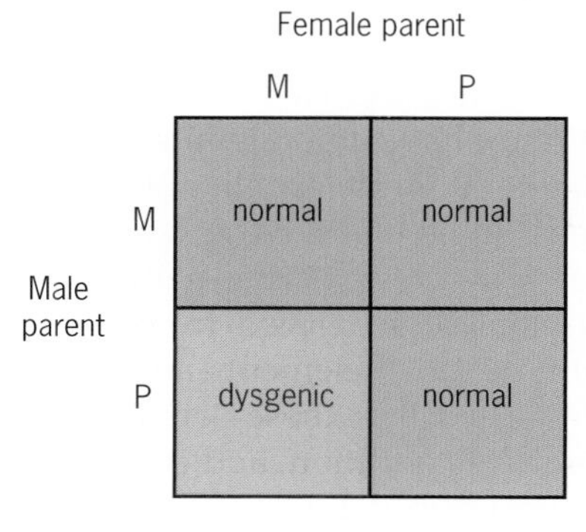

Male parent	Female parent M	Female parent P
M	normal	normal
P	dysgenic	normal

To Kidwell, Kidwell, and Sved, these findings suggested that the chromosomes of P strains carry genetic factors that are activated when they enter eggs made by M females, and that once activated, these factors induce mutations and chromosome breakage. Inspired by this work, William Engels, a graduate student at the University of Wisconsin, began to study mutations induced in dysgenic hybrids. In 1979 Engels found a mutation that reverted to wild-type at a high rate. This instability, which is reminiscent of the behavior of IS-induced mutations in *E. coli*, strongly suggested that a transposable element was involved.

The discovery by Michael Simmons and Johng Lim of dysgenesis-induced mutations in the *white* gene allowed the transposon hypothesis to be tested. In 1980, Simmons and Lim, working in Minnesota and Wisconsin, respectively, sent the newly discovered *white* mutations to Paul Bingham, a geneticist in North Carolina. Bingham and his collaborator, Gerald Rubin, a geneticist in Maryland, had just finished isolating DNA from the *white* gene. Using this DNA as a probe, Bingham and Rubin were able to isolate DNA from the mutant *white* alleles and compare it to the wild-type *white* DNA. In each mutation, they found that a small element had been inserted into the coding region of the *white* gene. Additional experiments demonstrated that these elements are present in multiple copies and at different locations in the genomes of P strains; however, they are completely absent from the genomes of M strains. Geneticists therefore began calling these P-strain-specific transposons ***P* elements**.

DNA sequence analysis has shown that *P* elements vary in size. The largest elements are 2907 nucleotide pairs long, including terminal inverted repeats of 31 nucleotide pairs. These *complete P* elements carry a gene that encodes a transposase. When the P transposase binds near the ends of a complete *P* element, it can move that element to a new location in the genome. *Incomplete P* elements (Figure 17.11) lack the ability to produce the transposase because some of their internal sequences are deleted; however, they do possess the terminal and subterminal sequences that bind the transposase. Consequently, these elements can be mobilized if a transposase-producing complete element is present somewhere in the genome.

Surveys of natural populations of *Drosophila* conducted by Dominique Anxolabéhère and his colleagues in France have demonstrated that there is considerable variation in the number of *P* elements in the chromosomes. Some flies have as many as 50, whereas others have only a few. Perhaps the most surprising discovery is that flies derived from strains captured before 1950 have no *P* elements at all. Margaret Kidwell has suggested that these "empty" strains represent the primitive condition, and that *P* elements have invaded natural populations of *Drosophila* during recent times. Curiously, the closest relatives of *D.*

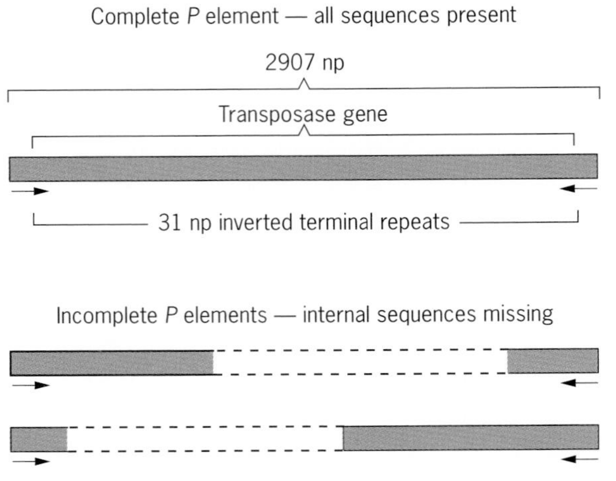

Figure 17.11 Structure of *P* elements in *Drosophila* showing orientations and lengths (in nucleotide pairs, np) of DNA sequences.

melanogaster have preserved the "empty" condition, but other, more distantly related species have acquired *P* elements. It is not possible to say how these species acquired their *P* elements, but one possibility is that the elements were carried into the genome by viruses that naturally infect *Drosophila*. Such a process would be analogous to the transduction of *E. coli* cells by a bacteriophage that carried an IS element.

Populations of *Drosophila* that possess *P* elements have evolved mechanisms to regulate their movement. In some strains, this regulation depends on **cytotype**, a cellular condition that is transmitted maternally through the egg cytoplasm. The P cytotype represses *P* element movement, and the M cytotype permits it. P cytotype is characteristic of strains that have *P* elements on their chromosomes, whereas M cytotype is characteristic of strains that lack *P* elements. When *P* elements are combined with the M cytotype by making appropriate crosses, they are induced to transpose. Although the regulatory role of cytotype has been studied by numerous researchers, its cellular nature is still unknown. One possibility is that the P cytotype is produced by polypeptides encoded by the *P* elements themselves and that the M cytotype is simply the absence of these *P*-encoded polypeptides.

The maternal transmission of cytotype can be seen in the offspring of reciprocal crosses between P and M cytotype strains (Figure 17.12). There are two crosses: (1) P cytotype female × M cytotype male, and (2) P cytotype male × M cytotype female. The offspring from both crosses inherit *P* elements on their chromosomes and are, in fact, genotypically identical; however, only those from the second cross allow *P* movement. This difference between the reciprocal crosses indicates

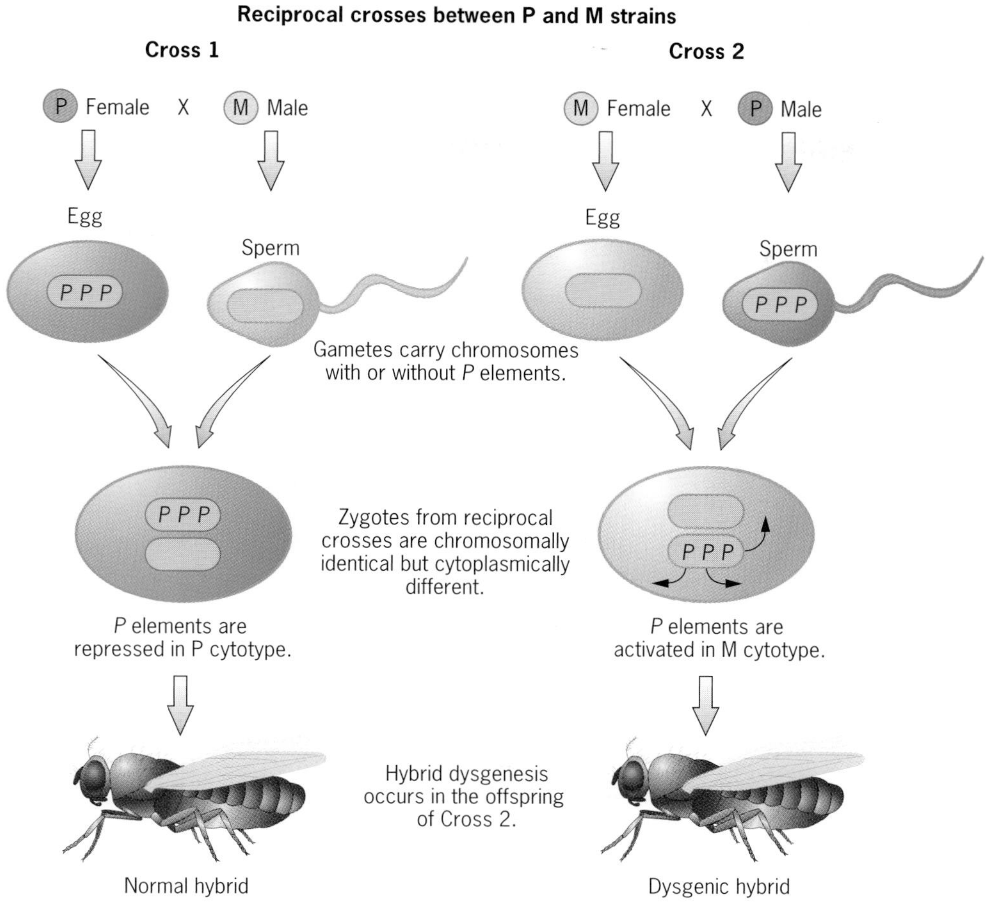

Figure 17.12 *P* element-mediated hybrid dysgenesis in *Drosophila*. Cytotype is inherited maternally. The P cytotype represses *P* element movement, whereas the M cytotype permits it. Dysgenesis occurs only in the M cytotype hybrids of Cross 2.

that the condition that permits or represses *P* movement is transmitted only by the female parent, presumably in the egg cytoplasm. In cross 1, the females transmit the P cytotype to their progeny, which then represses *P* movement, whereas in cross 2, the females transmit the M cytotype, which allows *P* movement.

Given the damage that can be caused by extensive *P* element movement, it may seem surprising that P male x M female crosses produce any viable progeny at all. Donald Rio, Frank Laski, and Gerald Rubin have shown that these progeny are healthy because *P* elements move only in the germ line. In the somatic tissues, where *P* element movement would cause serious problems, there is little, if any, transposition because the P transposase is not synthesized there. The metabolic block occurs at the level of RNA splicing; one of the introns remains in the transcript of the transposase gene, creating a stop codon that prematurely terminates translation. Thus, instead of making the transposase, somatic cells synthesize a shorter protein that does not have the transposase's catalytic function. Without this function, *P* elements are unable to move, and the fly is protected from massive damage in its somatic tissues.

Geneticists routinely use hybrid dysgenesis to obtain *P*-insertion mutations in the laboratory. Dysgenic hybrids produced by crossing P males with M females are mated to recover mutations that have occurred in their germ lines. These mutations are detected by appropriate techniques, such as Muller's *ClB* crossing scheme to identify X-linked lethals (Chapter 13). Hybrid dysgenesis has an advantage over traditional methods of inducing mutations because a gene that has been mutated by the insertion of a transposable element is "tagged" with a known DNA sequence. The transposon tag may subsequently be used to isolate the gene from a large, heterogeneous mixture of DNA. Mutagenesis by **transposon tagging** is therefore a standard technique in molecular genetics. The Technical Sidelight: Genetic Transformation of *Drosophila* with *P* Elements, discusses another important use of *P* elements in genetics research.

TECHNICAL SIDELIGHT

Genetic Transformation of *Drosophila* with *P* Elements

In the 1940s, Oswald Avery and colleagues discovered that bacteria could be genetically altered by treating them with isolated DNA. In this process, a DNA fragment enters a cell and is physically recombined into the chromosome. The re-

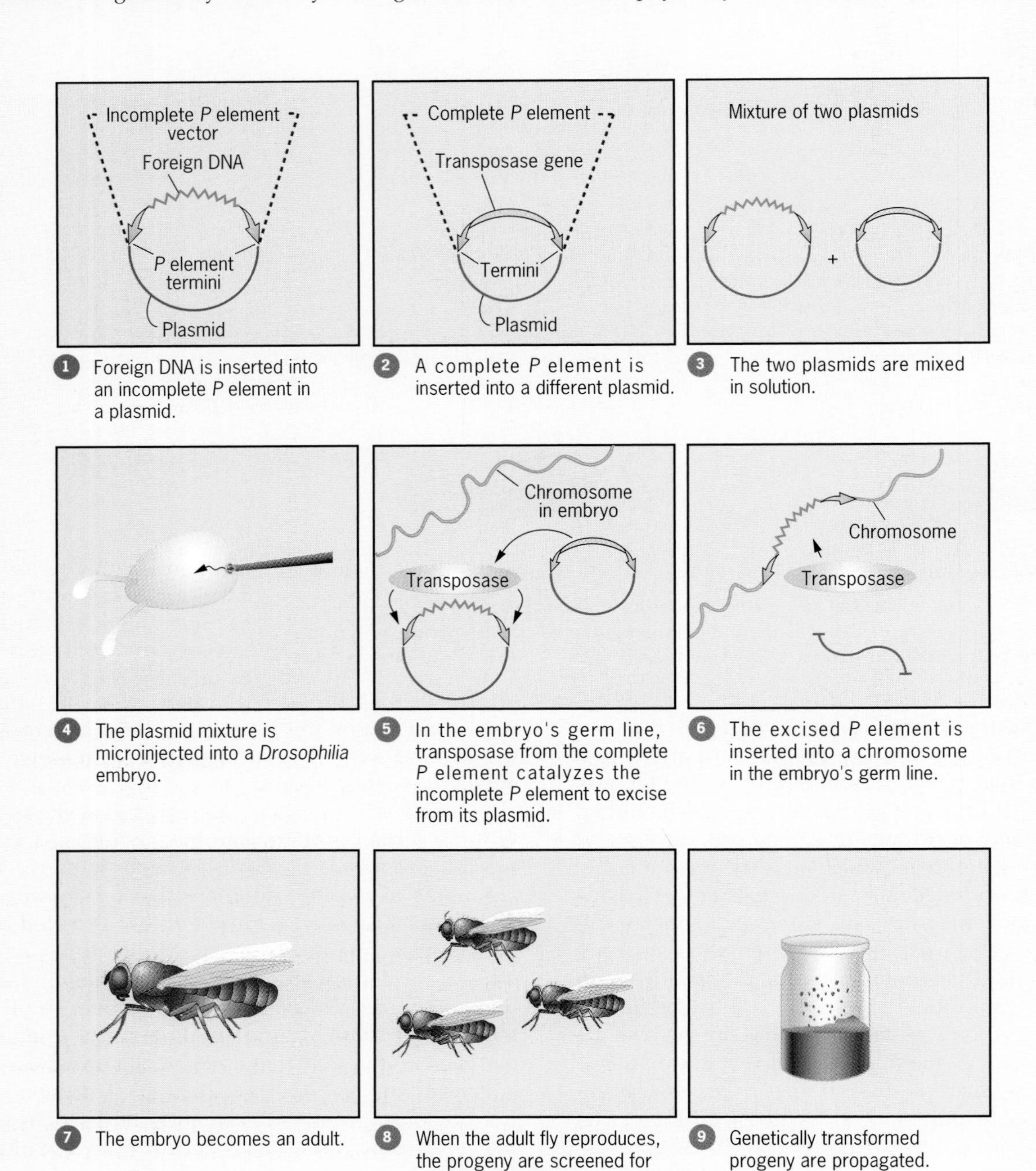

Figure 1 Genetic transformation of *Drosophila* using *P* element vectors. Foreign DNA inserted between *P* element termini is integrated into the genome through the action of a transposase encoded by the complete *P* element. Flies with this DNA in their genomes can be propagated in laboratory cultures.

combinant cell can then be cultured to produce a strain of genetically transformed organisms. Avery's discovery raised the hope that someday it would be possible to alter the genomes of eukaryotic organisms by inserting specific DNA fragments into them. When applied to human beings, such a procedure might provide a way of correcting genetic diseases.

For nearly 40 years, the experimental production of genetic transformants was limited to microorganisms. Many researchers attempted to transform higher eukaryotes, but none succeeded. This string of failures was broken in 1982, when Allan Spradling and Gerald Rubin produced the first genetically transformed *Drosophila*.

Spradling and Rubin used transposable *P* elements to insert purified DNA into living *Drosophila* embryos. First, they constructed two bacterial plasmids that contained *P* elements. One plasmid contained a complete *P* element capable of producing the *P* transposase *in vivo*. The other contained an incomplete *P* element into which a gene for wild-type eye color had been inserted. Next, Spradling and Rubin injected a mixture of the two plasmids into *Drosophila* embryos that were homozygous for a recessive mutation of the eye color gene. They hoped that transposase produced by the complete *P* element would catalyze the incomplete *P* element to jump from its plasmid into the chromosomes of the injected animals, carrying with it the eye color gene. When these animals matured, Spradling and Rubin mated them to flies homozygous for the eye color mutation and looked for progeny that had wild-type eyes. They found many, indicating that the wild-type gene carried by the incomplete *P* element had been successfully incorporated into the genomes of some of the injected embryos. In effect, Spradling and Rubin had corrected the mutant eye color phenotype by inserting a copy of the wild-type gene into the fly genome.

The technique that Spradling and Rubin developed is now routinely used to transform *Drosophila* with isolated DNA (Figure 1). An incomplete *P* element serves as the **transformation vector**, and a complete *P* element serves as the source of the transposase that is needed to insert the vector into the chromosomes of an injected embryo. The term *vector* comes from the Latin word for "carrier"; it is used in this context because the incomplete *P* element *carries* a fragment of DNA into the genome. Practically any DNA sequence can be placed into the vector and ultimately inserted into the animal. In fact, genes from organisms as diverse as bacteria and humans have successfully been incorporated into *Drosophila* chromosomes. It remains to be seen whether similar techniques will be developed for the genetic transformation of other organisms, including our own species.

mariner, an Ancient and Widespread Transposon

Two of the closest relatives of *Drosophila melanogaster*, *D. simulans* and *D. mauritiana*, possess a small transposon called *mariner* (1286 nucleotide pairs long with 28-nucleotide-pair inverted terminal repeats). Although *mariner* is not present in *D. melanogaster*, similar transposons are present in many other insects, in nematodes, fungi, and even humans. This widespread distribution suggests that *mariner* elements are ancient, dating from the earliest evolutionary times.

Sequence analysis of *mariners* from many different organisms has also suggested that these elements are occasionally transferred horizontally between species. Hugh Robertson has shown that *mariner*-like elements are present in several different orders of insects. Curiously, the *mariners* in distantly related species are sometimes more similar to each other than the *mariners* in closely related species. For example, the *mariner* elements in the Mediterranean fruit fly are very similar to those in the honeybee; yet these two species are separated by hundreds of millions of years of evolutionary time. A plausible explanation is that the *mariners* in these two species are actually recent invaders from some outside source, perhaps another insect. It is not known how these elements could have been transferred between species, but one possibility is that they hitchhiked in the genome of a virus with a wide host range. During infection in one host species, the virus may have acquired the transposon; then during infection in another species, the transposon may have jumped from the virus into the host genome.

DNA sequencing has also identified another group of transposons related to the *mariner* elements. These are slightly larger, 1.6 to 1.7 kilobases (kb; 1kb = 1000 base pairs) long, with inverted repeats at their termini. However, their repeat sequences typically range from 54 to 234 nucleotide pairs. The best studied transposon in this group, called *Tc1*, is found in the nematode worm *Caenorhabditis elegans*. Other *Tc1*-like elements have been found in insects. The *Tc1* and *mariner* elements therefore belong to a transposon superfamily of ancient origin.

Key Points: **The maize transposable element *Ds*, discovered through its ability to break chromosomes, is activated by another transposable element, *Ac*, which encodes a transposase. In *Drosophila*, transposable *P* elements are responsible for hybrid dysgenesis, a syndrome of germ-line abnormalities that occurs in the offspring of crosses between P and M strains. The ancient *mariner* transposable elements are present in many distantly related organisms and probably have spread horizontally between species during the course of evolution.**

RETROTRANSPOSONS

In addition to transposons such as *Ac*, *P*, and *mariner*, eukaryotic genomes contain transposable elements whose movement depends on the reverse transcription of RNA into DNA. This reversal in the flow of genetic information has led geneticists to call these elements **retrotransposons**, from a Latin prefix meaning "backwards." There are two main classes of retrotransposons: the retrovirus-like elements and the retroposons. The members of the first class resemble the chromosomes of a group of viruses that depend on reverse transcription for their propagation, and the members of the second class have a structure reminiscent of polyadenylated RNA.

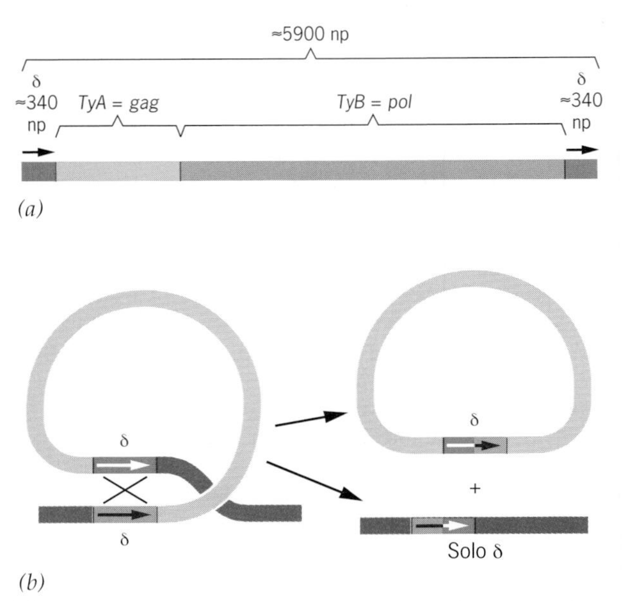

Figure 17.13(*a*) Genetic organization of yeast Ty element, showing the long terminal repeat sequences (LTRs, denoted by the Greek letter delta) and the two genes (*TyA* and *TyB*). Lengths of sequences are in nucleotide pairs (np). (*b*) Formation of a solo delta sequence by homologous recombination between the delta sequences at the ends of the element.

Retrovirus-like Elements

Retrovirus-like elements are found in many different organisms, including yeast, plants, and animals. Despite differences in size and nucleotide sequence, they all have the same basic structure: a central coding region flanked by **long terminal repeats,** or **LTRs**, which are oriented in the same direction. The repeated sequences are typically a few hundred nucleotide pairs long. Each LTR is, in turn, usually bounded by short, inverted repeats like those found in other types of transposons.

The coding region of a retrovirus-like element contains a small number of genes, usually only two. These are homologous to the *gag* and *pol* genes found in retroviruses (see Chapter 15); *gag* encodes a structural protein of the virus capsule, and *pol* encodes a reverse transcriptase/integrase protein. The retroviruses have a third gene, *env*, which encodes a protein component of the virus envelope. In the retrovirus-like elements, the gag and pol proteins play important roles in the transposition process.

One of the best-studied retrovirus-like elements is the **Ty transposon** from the yeast *Saccharomyces cerevisiae* (Figure 17.13*a*). This element is about 5.9 kilobase pairs long; its LTRs are about 340 base pairs long, and it creates a 5-bp target site duplication upon insertion into a chromosome. Most yeast strains have about 35 copies of the Ty element; sometimes they also contain LTRs that have been detached from Ty elements. These solo LTRs, or delta sequences as they are sometimes called, are apparently formed by recombination between the LTRs of complete Ty elements (Figure 17.13*b*). The recombination event puts the central coding region and a portion of each LTR onto a circular molecule. When the circle leaves the chromosome, the remaining portions of the LTRs fuse, creating the solo delta sequence.

Ty elements have only two genes, *TyA* and *TyB*, which are homologous to the *gag* and *pol* genes of the retroviruses. Biochemical studies have shown that the products of these two genes can form virus-like particles inside yeast cells. However, it is not known whether these particles are genuinely infectious. The transposition of Ty elements involves reverse transcription of RNA (Figure 17.14). After the RNA is synthesized from Ty DNA, a reverse transcriptase encoded by the *TyB* gene uses it as a template to make double-stranded DNA. Then the newly synthesized DNA is inserted somewhere in the genome, creating a new Ty element.

Retrovirus-like elements have also been found in *Drosophila*. One of the first that was identified is called *copia*, so named because it produces copious amounts of RNA. Another is called *gypsy*, because it seems to wander about the genome. Both of these transposons form virus-like particles inside *Drosophila* cells, and recent evidence indicates that the *gypsy* particles can occasionally move across cell membranes. Thus the *gypsy* element may be a genuine retrovirus. Many other families of retrovirus-like transposons have been found in *Drosophila*, but their activities are poorly understood.

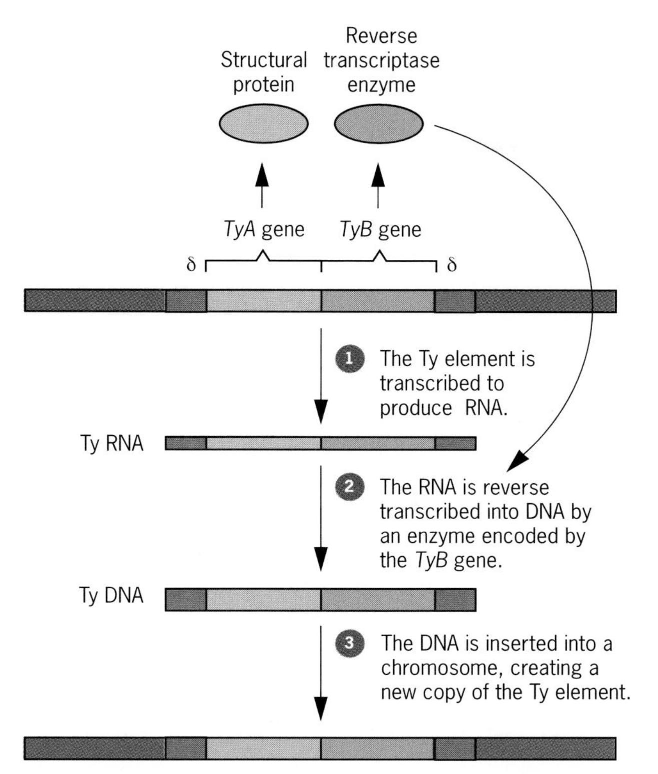

Figure 17.14 Transposition of the yeast Ty element.

Retroposons

The **retroposons** are a large and widely distributed class of retrotransposons, including the *F*, *G*, and *I* elements of *Drosophila* and the **Long Interspersed Nuclear Elements (LINEs)** of mammals. These elements move through an RNA molecule that is reverse transcribed into DNA, probably by a protein encoded by the elements themselves. Although they create a target site duplication when they insert into a chromosome, they do not have inverted or direct repeats as integral parts of their termini. Instead, they are distinguished by a homogeneous sequence of A:T base pairs at one end. This sequence is derived from reverse transcription of the poly-A tail that is added near the 3' end of the retroposon RNA during its maturation. Integrated retroposons therefore exhibit a vestige of their origin as reverse transcripts of polyadenylated RNAs.

L1, a Human Retroposon The ***LINE-1* retroposon**, also known as ***L1***, is the only transposable element known to be active in the human genome. (There may be others, but to date none has been found.) The first documented *L1* transposition event, reported by Haig Kazazian, Jr. and colleagues in 1988, involved an insertion into the X-linked gene for protein factor VIII, one of the clotting factors in the blood. The resulting mutation caused hemophilia, a failure of the blood to clot. The inserted *L1* element was apparently produced by reverse transcription of an RNA made by an *L1* element on chromosome 22. Curiously, the gorilla homolog of chromosome 22 also carries this *L1* element, suggesting that it was present before the human and gorilla lineages diverged—about 6 million years ago.

L1 elements are heterogeneous in size, and most are truncated at the 5' end—an indication of incomplete synthesis of the DNA by reverse transcription from an RNA template. These elements are present in 50,000 to 100,000 copies in the human genome, and account for about 5 percent of the DNA. Similar elements are also found in the genomes of other mammals.

Telomere-associated Retroposons: HeT-A *and* TART In *Drosophila*, retroposons are found at the ends (telomeres) of chromosomes, where they perform the critical function of replenishing DNA that is lost by incomplete chromosome replication. With each round of DNA replication, a chromosome becomes shorter. This happens because the DNA polymerase can only move in one direction, adding nucleotides to the 3' end of a primer (Chapter 10). Usually, the primer is RNA, and when it is removed, a single-stranded region is left at the end of the DNA duplex. In the next round of replication, the deficient strand produces a duplex that is shorter than the original. As this process continues, cycle after cycle, the chromosome loses material from its end.

To counterbalance this loss, *Drosophila* has evolved a curious mechanism involving at least two different retroposons, one called ***HeT-A*** and another called ***TART*** (for Telomere Associated Retrotransposon). Mary Lou Pardue, Robert Levis, Harold Biessmann, James Mason, and their colleagues have shown that these two elements transpose preferentially to the ends of chromosomes, extending them by several kilobases. Eventually, the transposed sequences are lost by incomplete DNA replication, but then a new transposition occurs to restore them. The *HeT-A* and *TART* retroposons therefore perform the important function of regenerating lost chromosome ends.

Key Points: **The movement of retrovirus-like elements and retroposons depends on reverse transcription of RNA into DNA. Retrovirus-like elements have Long Terminal Repeat (LTR) sequences at both ends and resemble the integrated chromosomes of retroviruses. Retroposons have a sequence of A:T base pairs at one end. The retroposon *L1* is the only transposable element known to be active in the human genome. The *HeT-A* and *TART* retroposons are associated with the ends of *Drosophila* chromosomes.**

THE GENETIC AND EVOLUTIONARY SIGNIFICANCE OF TRANSPOSABLE ELEMENTS

Transposable elements are widespread and probably ancient components of the genome. In some species, they constitute an appreciable fraction of the total DNA—in *Drosophila*, for example, 12 to 15 percent. Mobile DNA is therefore an important component of the genome. It also contributes significantly to the total mutation rate. In *Drosophila*, for example, perhaps half the mutations that occur spontaneously are caused by transposable element insertions. This ability to inflict mutational damage raises questions about the evolutionary status of transposable elements. Do they perform any useful function, or are they merely genetic parasites, causing mutations as they wander about the genome? Where did transposable elements come from? What mechanisms have evolved to control and limit their movement?

Transposons and Genome Organization

Certain chromosome regions are especially rich in transposon sequences. In *Drosophila*, where the most detailed studies have been carried out, transposons are concentrated in the centric heterochromatin and in the heterochromatin abutting the euchromatin of each chromosome arm. However, many of these transposons have mutated to the point where they cannot be mobilized; genetically, they are the equivalent of "dead." Heterochromatin therefore seems to be a kind of graveyard filled with degenerate transposable elements.

Transposable elements are also found in the euchromatin, where they are dispersed at many different sites. In *Drosophila*, approximately 40 distinct families of transposable elements have been identified. The number of members in each family ranges from a few to one or two hundred, but most families seem to have between 20 and 80 members.

There is some evidence, especially from cytological studies of *Drosophila* by Johng Lim, that transposable elements play a role in the evolution of chromosome structure. Several *Drosophila* transposons have been implicated in the formation of chromosome rearrangements, and a few seem to rearrange chromosomes at high frequencies. One possible mechanism is crossing over between homologous transposons located at different positions in a chromosome (Figure 17.15). If two transposons in opposite orientation pair and cross over, the segment between them will be inverted. If two transposons in the same orientation pair and cross over, the segment between them will be deleted. These events are examples of **ectopic intrachromosomal exchanges**—that is, exchanges between sequences at different sites within a single chromosome. **Ectopic interchromosomal exchanges** are also possible. In these types of events, sequences in two different chromosomes (either homologous or nonhomologous) pair and cross over, generating novel products. Figure 17.16 gives an example of an ectopic ex-

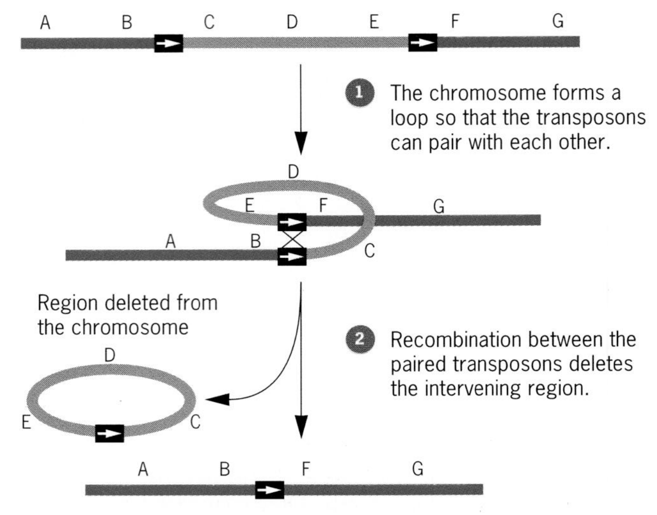

(a) Chromosome with two transposons oriented in the same direction.

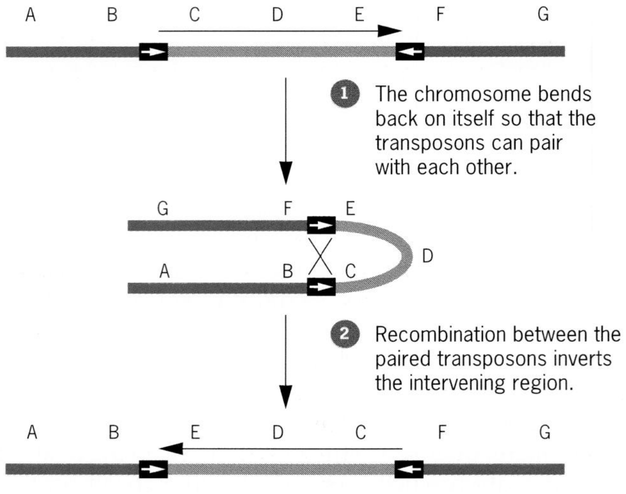

(b) Chromosome with two transposons oriented in opposite directions.

Figure 17.15 Transposon-mediated chromosome rearrangements. (*a*) Formation of a deletion by intrachromosomal recombination between two transposons in the same orientation. (*b*) Formation of an inversion by intrachromosomal recombination between two transposons in opposite orientations.

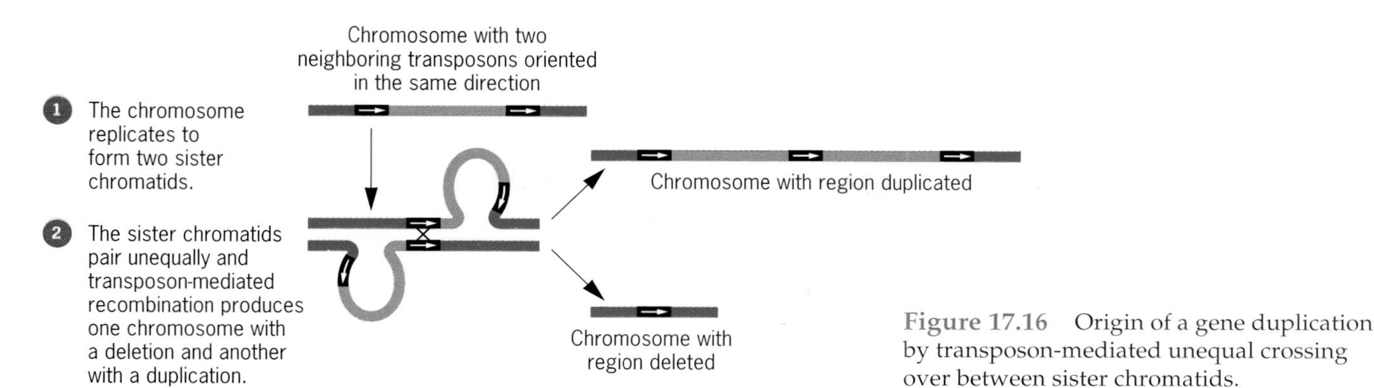

Figure 17.16 Origin of a gene duplication by transposon-mediated unequal crossing over between sister chromatids.

change between transposable elements in two sister chromatids. Notice that one product from this exchange lacks the segment between the two transposons, whereas the other contains a duplication of this segment. Transposable elements can therefore change chromosome structure by mediating ectopic exchanges.

Another example of ectopic exchange is the crossover event that inserts the F plasmid into the *E. coli* chromosome. Such events are mediated by IS elements located in both of these circular molecules.

Transposons and Mutation

Transposable elements are responsible for mutations in a wide variety of organisms. Most transposons have been discovered as insertions in various mutant genes. Figure 17.17 shows some of the transposon insertions that have been found in different mutant alleles of the *Drosophila white* gene. These include several types of elements—*P*, retrovirus-like elements, and retroposons. Some of these elements are inserted in exons, others in introns, and still others in regulatory DNA upstream of the actual gene. In fact, the very first mutant allele of *white*, w^1 discovered by T. H. Morgan, resulted from a transposon insertion.

Although transposon insertions are common in stocks of mutant organisms, the occurrence of new insertion mutations is a rare event. This suggests that the movement of many transposon families is regulated. When this regulation is upset, a burst of transposition may occur, causing many mutations simultaneously. This is apparently what happens when *P* elements are mobilized in dysgenic hybrids of *Drosophila*.

Evolutionary Issues Concerning Transposable Elements

The widespread distribution of transposable elements suggests that they have played a role in evolution. One hypothesis is that these elements are nature's tools for genetic engineering. Their ability to copy, transpose, and rearrange other DNA sequences, such as genes for antibiotic resistance, can be construed as a benefit for the organisms that carry them. Thus trans-

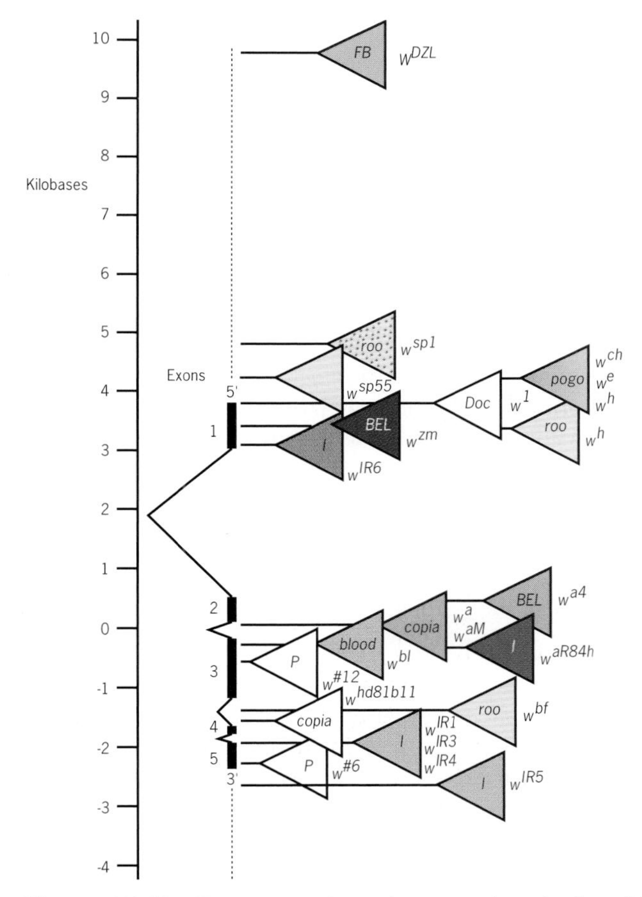

Figure 17.17 Transposon-insertion mutations in the *white* gene of *Drosophila*. Distance along the molecular map is given in kilobase pairs, with the zero coordinate arbitrarily positioned in the second intron of the gene at the site of the *copia* insertion in the w^a mutation. Each triangle represents a different transposon insertion in the *white* gene. Colors indicate the mutant eye colors.

posable elements may have spread because they confer a selective advantage. Another hypothesis is that transposable elements have spread simply because they have an ability to multiply independently of the normal replication machinery. According to this view, transposable elements are little more than genomic parasites—segments of DNA that replicate selfishly, possibly even to the detriment of their hosts.

How might the first transposons have evolved? Nancy Kleckner has suggested that a primordial transposon might arise by the modification of a gene that encodes an enzyme for the creation and repair of DNA breaks. All that would be needed is for the enzyme to develop a modest degree of specificity, perhaps by recognizing a particular DNA sequence of six or eight nucleotide pairs. Such a sequence might occur by chance in inverted orientation on either side of the gene, creating a situation in which the gene's product could interact with each of these flanking sequences. By "cutting and pasting" the DNA, this modified enzyme could then transpose the entire unit to a new position in the genome. Such a unit would behave as a primitive transposon.

Other questions concern the evolutionary relationship between retrovirus-like elements, such as Ty in yeast, and full-fledged retroviruses. Collectively, these entities have been referred to as **retroelements**. A. J. Kingsman and S. M. Kingsman have proposed that retroviruses have developed from the simpler retrotransposons by the addition of a gene (called *env*) that synthesizes a membrane protein. With this addition, the retroelement could produce a particle capable of escaping from one cell and entering another one. Such a particle would be infectious and would therefore provide the retroelement with the opportunity to transpose between genomes as well as within them. Of course, the situation could be reversed—a retrovirus could lose its ability to escape from a cell and become trapped inside. Such a mutant virus would be reduced to the status of a retrotransposon, capable of moving within cells but not between them. Using a musical metaphor, the Nobel Prize-winning virologist Howard Temin once described these contrasting scenarios as ascending and descending evolutionary scales. Retrotransposons can rise to the level of retroviruses, and retroviruses can fall to the level of retrotransposons.

Key Points: **Transposons constitute a significant fraction of the DNA in the genomes of some organisms. In nature, they may be a major cause of mutations and chromosome rearrangements. Some transposons may confer a selective advantage on their carriers; others may simply be genetic parasites.**

TESTING YOUR KNOWLEDGE

1. A copy of the wild-type *white* gene (w^+) from *Drosophila* was inserted in the middle of an incomplete *P* element contained within a plasmid. The plasmid was mixed with another plasmid that contained a complete *P* element, and the mixture was carefully injected into *Drosophila* embryos homozygous for a null mutation (w^-) of the *white* gene. The adults that developed from these injected embryos all had white eyes, but when they were mated to uninjected white flies, some of their progeny had red eyes. Explain the origin of these red-eyed progeny.

ANSWER

The complete *P* element in one of the plasmids would produce the *P* transposase, the enzyme that catalyzes *P* element transposition, in the germ lines of the injected embryos. The incomplete *P* element in the other plasmid would be a target for this transposase. If this incomplete *P* element were mobilized by the transposase to jump from its plasmid into the chromosomes of the injected embryo, the fly that developed from this embryo would carry a copy of the wild-type *white* gene in its germ line. (*P* element movement is limited to the germ line; therefore, the incomplete *P* element would not jump into the chromosomes of the somatic cells, such as those that eventually form the eye.) Such a genetically transformed fly would, in effect, have the germ-line genotype w^-/w^-; $P(w^+)$ or w^-/Y; $P(w^+)$, where $P(w^+)$ denotes the incomplete *P* element that contains the w^+ gene. This element could be inserted on any of the chromosomes. If the transformed fly were mated to an uninjected white fly, some of its offspring would inherit the $P(w^+)$ insertion, which, because it carries a wild-type white gene, would cause red eyes to develop. The red-eyed progeny are therefore the result of genetic transformation of a mutant white fly by the w^+ gene within the incomplete *P* element.

QUESTIONS AND PROBLEMS

17.1 It has been proposed that the *hobo* transposable elements in *Drosophila* mediate intrachromosomal recombination—that is, two *hobo* elements on the same chromosome pair and recombine with each other. What would such a recombination event produce if the *hobo* elements were oriented in the same direction on the chromosome? What if they were oriented in opposite directions?

17.2 The X-linked *singed* locus is one of several in *Drosophila* that controls the formation of bristles on the adult cuticle. Males that are hemizygous for a mutant *singed* allele have bent, twisted bristles that are often much reduced in size. Several *P* element insertion mutations of the *singed* locus have been characterized, and some have been shown to revert to the wild-type allele by excision of the inserted element. What conditions must be present to allow such reversions to occur?

17.3 In maize, the recessive allele *bz* produces a lighter color in the aleurone than does the dominant allele, *Bz*. Ears on a homozygous *bz/bz* plant were fertilized by pollen from a homozygous *Bz/Bz* plant. The resulting cobs contained kernels that were uniformly dark except for a few on which light spots occurred. Suggest an explanation.

17.4 Which of the following pairs of DNA sequences could qualify as the teminal repeats of a bacterial IS element: (a) 5'-GAATCCGCA-3' and 5'-ACGCCTAAG-3', (b) 5'-GAATCCGCA-3' and 5'-CTTAGGCGT-3', (c) 5'-GAATCCGCA-3' and 5'-GAATCCGCA-3', (d) 5'-GAATCCGCA-3' and 5'-TGCGGATTC-3'. Explain.

17.5 Sometimes solo copies of the LTR of a retrotransposon called *gypsy* are found in *Drosophila* chromosomes. How might these solo LTR's originate?

17.6 Which of the following pairs of DNA sequences could qualify as target site duplications at the point of an IS*50* insertion? (a) 5'-AATTCGCGT-3' and 5'AATTCGCGT-3', (b) 5'-AATTCGCGT-3' and 5'-TGCGCTTAA-3', (c) 5'-AATTCGCGT-3' and 5'-TTAAGCGCA-3', (d) 5'AATTCGCGT-3' and 5'-ACGCGAATT-3'. Explain.

17.7 One strain of *E. coli* is resistant to the antibiotic streptomycin, and another strain is resistant to the antibiotic ampicillin. The two strains were cultured together and then plated on selective medium containing streptomycin and ampicillin. Several colonies appeared, indicating that cells had acquired resistance to both antibiotics. Suggest a mechanism to explain the acquisition of double resistance.

17.8 In homozygous condition, a deletion mutation of the *c* locus, c^n, produces colorless (white) kernels in maize; the dominant wild-type allele, *C*, causes the kernels to be purple. A newly identified recessive mutation of the *c* locus, c^m, has the same phenotype as the deletion mutation (white kernels), but when c^mc^m and c^nc^n plants are crossed, they produce white kernels with purple stripes. If it is known that the c^nc^n plants harbor *Ac* elements, what is the most likely explanation for the c^m mutation?

17.9 Dysgenic hybrids in *Drosophila* have elevated mutation rates as a result of *P* element transposition. How could you take advantage of this situation to obtain *P* element-insertion mutations on the X chromosome?

17.10 Suggest a method to determine whether the *TART* retroposon is situated at the telomeres of each of the chromosomes in the *Drosophila* genome.

17.11 Approximately half of all spontaneous mutations in *Drosophila* are caused by transposable element insertions. In human beings, however, the accumulated evidence suggests that the vast majority of spontaneous mutations are *not* caused by transposon insertions. Propose an hypothesis to explain this difference.

17.12 What distinguishes IS and Tn*3* elements in bacteria?

17.13 The circular order of genes on the *E. coli* chromosome is *A B C D E F G H *, with the * indicating that the ends of the chromosome are attached to each other. Two copies of an IS element are located in this chromosome, one between genes C and D, and the other between genes D and E. A single copy of this element is also present in the F plasmid. Two Hfr strains were obtained by selecting for integration of the F plasmid into the chromosome. During conjugation, one strain transfers the chromosomal genes in the order D E F G H A B C, whereas the other transfers them in the order D C B A H G F E. Explain the origin of these two Hfr strains. Why do they transfer genes in different orders? Does the order of transfer reveal anything about the orientation of the IS elements in the *E. coli* chromosome?

17.14 By chance, an IS*1* element has inserted near an IS2 element in the *E. coli* chromosome. The gene between them, *sug*, confers the ability to metabolize certain sugars. Will the unit IS*1* *sug* IS2 behave as a composite transposon? Explain.

17.15 The composite transposon Tn*5* consists of two IS*50* elements, one on either side of a group of three genes for antibotic resistance. The entire unit IS*50*L *kan ble str* IS*50*R can transpose to a new location in the *E. coli* chromosome. However, of the two IS*50* elements in this transposon, only IS*50*R produces the catalytically active transposase. Would you expect IS*50*R to be able to excise from the Tn*5* composite transposon and insert elsewhere in the chromosome? Would you expect IS*50*L to be able to do this?

17.16 A researcher has found a new Tn*5* element with the structure IS*50*L *str ble kan* IS*50*L. What is the most likely origin of this element?

17.17 Would a Tn*3* element with a frameshift mutation early in the *tnpA* gene be able to form a cointegrate? Would a Tn*3* element with a frameshift mutation early in the *tnpR* gene be able to form a cointegrate?

17.18 In maize, the *O2* gene, located on chromosome 7, controls the texture of the endosperm, and the *C* gene, located on chromosome 9, controls its color. The gene on chromosome 7 has two alleles, a recessive, *o2*, which causes the endosperm to be soft, and a dominant, *O2*, which causes it to be hard. The gene on chromosome 9 also has two alleles, a recessive, *c*, which allows the endosperm to be colored, and a dominant, C^I, which inhibits coloration. In one homozygous C^I strain, a *Ds* element is inserted on chromosome 9 between the *C* gene and the centromere. This element can be activated by introducing an *Ac* element by appropriate crosses. Activation of *Ds* causes the C^I allele to be lost by chromosome breakage. In $C^I/c/c$ kernels, such loss produces patches of colored tissue in an otherwise colorless background (see Figures 17.8 and 17.9). A geneticist crosses a strain with the genotype *o2/o2*; C^I *Ds*/C^I *Ds* to a strain with the genotype *O2/o2*; *c/c*. The latter strain also carries an *Ac* element somewhere in the genome. Among the offspring, only those with hard endosperm show patches of colored tissue. What does this tell you about the location of the *Ac* element in the *O2/o2; c/c* strain?

17.19 If DNA from a *P* element-insertion mutation of the *Drosophila white* gene and DNA from a wild-type *white* gene were purified, denatured, mixed with each other, renatured, and then viewed with an electron microscope, what would the hybrid DNA molecules look like?

17.20 When complete *P* elements are injected into embryos from an M strain, they transpose into the chromosomes of the germ line, and progeny reared from these embryos can be used to establish new P strains. However, when complete *P* elements are injected into embryos from insects that lack these elements, such as mosquitoes, they do not transpose into the chromosomes of the germ line. What does this failure to insert in the chromosomes of other insects indicate about the nature of *P* element transposition?

17.21 What evidence suggests that some transposable elements are not simply genetic parasites?

17.22 Would you ever expect the genes in a retrotransposon to possess introns? Explain.

BIBLIOGRAPHY

DOMBROSKI, B. A., S. L. MATHIAS, E. NANTHAKUMAR, A. F. SCOTT, AND H. H. KAZAZIAN, JR. 1991. Isolation of an active human transposable element. *Science* 254: 1805-1810.

ENGELS, W. R. 1989. *P* elements in Drosophila. In D. E. Berg and M. M. Howe (Eds.), *Mobile DNA*. American Society for Microbiology.

FEDEROFF, N. V. 1989. About maize transposable elements and development. *Cell* 56:181–191.

GRINDLEY, N.D.F. 1983. Transposition of Tn3 and related transposons. *Cell* 32:3–5.

KIDWELL, M. G. 1994. The evolutionary history of the P family of transposable elements. *J. Heredity* 85: 339–346.

KINGSMAN, A. J. AND S. M. KINGSMAN. 1988. Ty: a retroelement moving forward. *Cell* 53:333–335.

MCCLINTOCK, B. 1956. Controlling elements and the gene. *Cold Spring Harbor Symp. Quant. Biol.* 21:197–216.

REZNIKOFF, W. S. 1982. Tn5 transposition and its regulation. *Cell* 31:307–308.

SHAPIRO, J. A. (Ed.) 1983. *Mobile genetic elements*. Academic Press, New York. (688 pp.).

A rain forest is a dramatic manifestation of the energy conversions that occur in chloroplasts and mitochondria.

18

The Genetics of Mitochondria and Chloroplasts

CHAPTER OUTLINE

Mitochondria, Chloroplasts, and the Biological Energy Wheel

The primitive earth was a very different world from the one we know today. The land was barren, devoid of any living organisms, and the air consisted primarily of methane, ammonia, and water vapor. The earth was probably much hotter than it is today, radiating great quantitites of heat from its molten core, and there were violent volcanic eruptions and intense electrical storms. Lethal ultraviolet radiation drenched the surface of the planet, energizing chemical reactions in the air and in shallow pools of water. In this hostile environment, life began, probably as a concretion of organic molecules that arose from the fusion of simpler compounds. Later, cellular organization evolved and diversified, leading eventually to a group of organisms that could think, speak, and conduct scientific research. Motivated by an intense curiosity, these organisms have attempted to look back over the long course of history and reconstruct the shadowy events of their own origins.

This research has revealed that the first true organisms were single cells that lived in water and fed on organic materials in the environment. These organisms sustained themselves by fermenting simple sugars, such as glucose, and by generating carbon dioxide (CO_2) gas as a waste product. The buildup of carbon dioxide provided a new resource on the primitive earth, and eventually organisms arose that could exploit it. These organisms evolved a mechanism to synthesize simple sugars from carbon dioxide and water by capturing energy from sunlight—the process called photosynthesis; the sugars created by photosynthesis

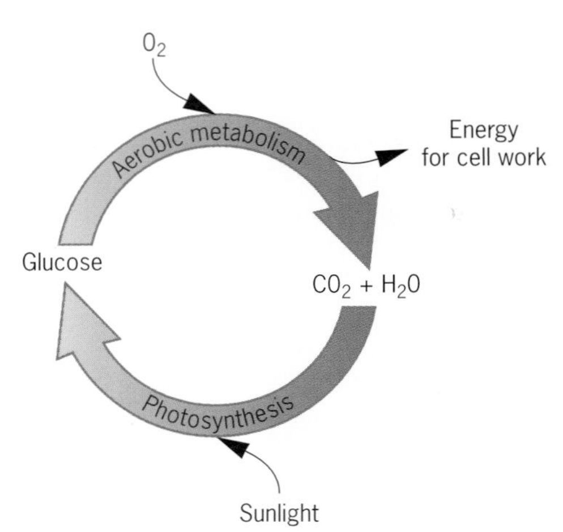

Figure 18.1 The biological energy wheel.

were then fermented to provide energy for life. As photosynthesis evolved and spread, it produced a new waste gas, oxygen (O_2), which then permitted a new group of organisms to evolve. These organisms were able to use the oxygen produced by photosynthesis to extract additional energy from sugars through a chemical process called oxidative, or aerobic, metabolism. Aerobic metabolism superseded the simpler anaerobic process of fermentation and led to the evolution of more complex, more active, and more diverse organisms.

Today, photosynthesis and aerobic metabolism are the complementary parts of a great energy cycle that encompasses the entire earth (Figure 18.1). Through photosynthesis, carbon dioxide and water are used to synthesize organic materials, and through oxidative metabolism, organic materials are decomposed into carbon dioxide and water. In eukaryotes, these processes are carried out in specialized subcellular organelles—photosynthesis in the **chloroplasts** and aerobic metabolism in the **mitochondria**. Both of these organelles appear to have arisen from simple, prokaryotic organisms that became established inside eukaryotic cells—a phenomenon called **endosymbiosis**—more than a billion years ago. Because these organisms brought along their genomes as well as their photosynthetic and oxidative powers, today all eukaryotic chloroplasts and mitochondria contain DNA. In this chapter, we explore the genetics and molecular biology of these important organelles.

THE CLASSICAL GENETICS OF ORGANELLES

The discovery of chloroplast and mitochondrial DNA was anticipated by studies that pointed to the existence of hereditary factors outside the nucleus. These studies date from the beginning of the twentieth century. In fact, the first of them was carried out by Carl Correns, one of the three botanists who rediscovered Mendel's principles.

Most of the early studies were performed with plants, which possess both mitochondria and chloroplasts. Consequently, it was not always possible to determine which of the two types of organelles was responsible for nonnuclear heredity. Later studies used yeast, for which the involvement of chloroplasts could be excluded. In the following sections, we review some of these classic studies. The common precept shared by all of them is that organelle heredity does not follow simple Mendelian rules. Instead, it is characterized by unequal contributions of the two parents and by an irregular segregation of alleles. Organelle heredity is therefore **non-Mendelian heredity**.

Leaf Variegation in Plants

The leaves in plants sometimes exhibit a striking pattern of **color variegation** (Figure 18.2). Some sectors are green, others pure white. Intermediate shades of pale or yellow green may also exist. These mosaic patterns are highly prized for their ornamental qualities, which partly explains why so many researchers have been interested in studying them.

Figure 18.2 Leaf variegation caused by the segregation of different types of chloroplasts.

Many cases of leaf variegation can be explained by the sorting of different types of chloroplasts during cell division (Figure 18.3). Cells may contain some chloroplasts that are capable of making the green pigment chlorophyll and others that are not. This mixture of two types of organelles within a cell is called **het-**

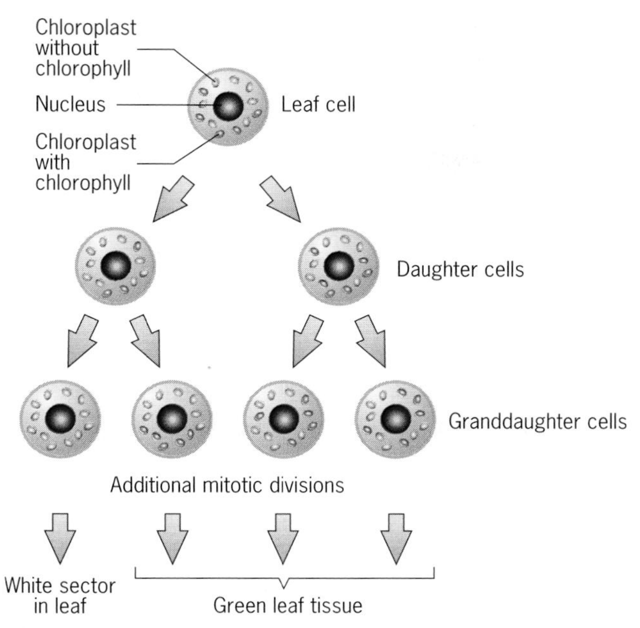

Figure 18.3 Chloroplast sorting during mitosis. The irregular distribution of chloroplasts during cell division may produce a cell that lacks chloroplasts capable of making chlorophyll. Through further divisions, such a cell will produce a white sector of tissue in an otherwise green leaf.

eroplasmy, from Greek words meaning "two different bodies." The presence of a single type of organelle within a cell is called **homoplasmy**. When a heteroplasmic cell divides, the two types of chloroplasts are distributed to the daughter cells in an irregular fashion. Over several divisions, this irregular distribution may produce a cell that totally lacks pigment-producing chloroplasts. Such a cell will then proliferate into a white sector in an otherwise green leaf. The white sectors produced by this process are irregular in size, shape, and position. However, in some plants, white sectors are formed in definite, predictable patterns in the leaves. These sectors are not due to chloroplast sorting. Rather, they are created by physiological systems that regulate the synthesis of pigment in a spatial and temporal framework within the leaves.

The inheritance of leaf variegation was first studied by the German botanists Carl Correns and Erwin Baur. Correns worked with variegating strains of the four o'clock, *Mirabilis jalapa*, a popular garden plant. He systematically made crosses between green and variegating strains. In these crosses it was possible to use flowers from a white sector that had developed on a variegating plant. Correns observed that the offspring of such crosses were always phenotypically identical to the tissue that produced the female gametes (Figure 18.4). Thus a cross made by fertilizing ovules from a green plant with pollen from a white sector on a variegating plant produced only green offspring. However, a cross made with ovules from a white sector on a variegating plant and pollen from a green plant produced only pure white offspring. This strict **maternal inheritance** could be explained if plant color was controlled by factors that were transmitted through the ovules but not through the pollen. Chloroplasts were the obvious candidates to contain these factors. In *Mirabilis*, the chloroplasts are transmitted to the offspring through the female reproductive cells, but are largely, if not entirely, excluded from the male reproductive cells. When a plant inherits a mixture of pigmented and unpigmented chloroplasts from the female parent, its tissues may variegate be-

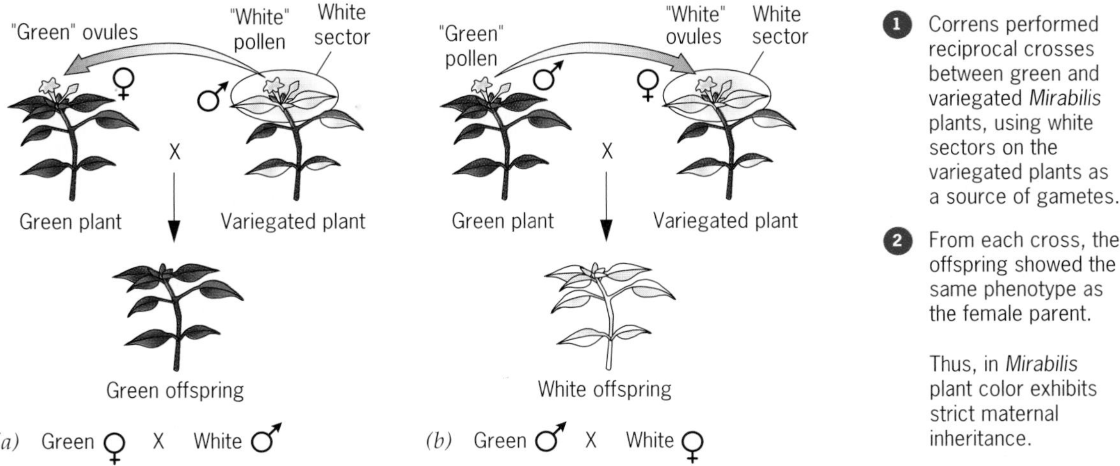

Figure 18.4 Correns' experiments on the inheritance of leaf variegation in *Mirabilis*. (*a*) Green female × white male (on a variegating plant). (*b*) Green male × white female (on a variegating plant).

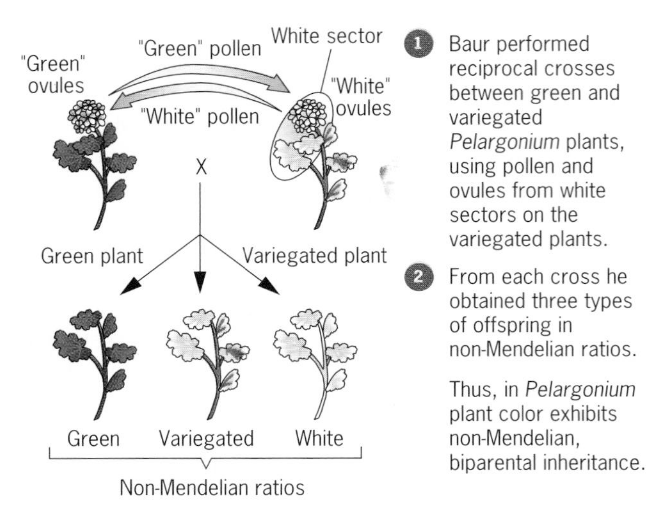

Figure 18.5 Baur's experiments on the inheritance of leaf variegation in *Pelargonium*. Both reciprocal crosses produced three types of progeny in non-Mendelian ratios.

cause the two types of chloroplasts sort themselves out during development.

Baur studied leaf variegation in another ornamental species, *Pelargonium zonale* (Figure 18.5). In this plant, crosses between green and variegating strains produce a mixture of offspring, some green, some variegating, and some pure white. However, these different phenotypes are not recovered in Mendelian proportions. This **non-Mendelian, biparental inheritance** indicates that the color of *Pelargonium* leaves is determined by a mixture of maternal and paternal factors located outside the nucleus, presumably in the chloroplasts. It therefore seems that in *Pelargonium* the chloroplasts are transmitted through the pollen as well as through the ovule.

The experiments of Correns and Baur demonstrated that leaf variegation is inherited as a non-Mendelian trait, one that is almost certainly controlled by factors located in the chloroplasts. The discovery that chloroplasts contain DNA has made this hypothesis very attractive. Yet, in spite of the work of many scientists, the molecular basis of leaf variegation is still unknown. It probably involves differences in the DNA molecules that are present within normal and pigment-deficient chloroplasts. More work is needed to determine the cause of this interesting botanical phenomenon.

Cytoplasmic Male Sterility in Maize

In maize, the male and female reproductive organs are produced on different parts of the plant. The female organs are in the ears, which develop at various positions along the stalk, and the male organs are in the tassels, which develop only at the top of the stalk. In some strains of maize the tassels fail to develop; thus these strains are male-sterile. However, because the ears develop and produce functional ovules, they can be crossed to other, male-fertile strains.

In the 1930s, Marcus Rhoades demonstrated that male sterility in maize can be caused by a maternally inherited factor (Figure 18.6). Rhoades pollinated the ears of a male-sterile strain with pollen from a male-fertile strain. All the offspring of this cross were male-sterile. Rhoades then backcrossed the male-sterile offspring to the male-fertile variety to see if male sterility was caused by a dominant allele inherited from the male-sterile strain. Because all the offspring were sterile, the dominant allele hypothesis was disproved. Rhoades continued to backcross the male-sterile offspring to male-fertile plants for several generations. At each step in the process, he found that the male-sterile phenotype persisted. This showed that male sterility was transmitted by a maternally inherited fac-

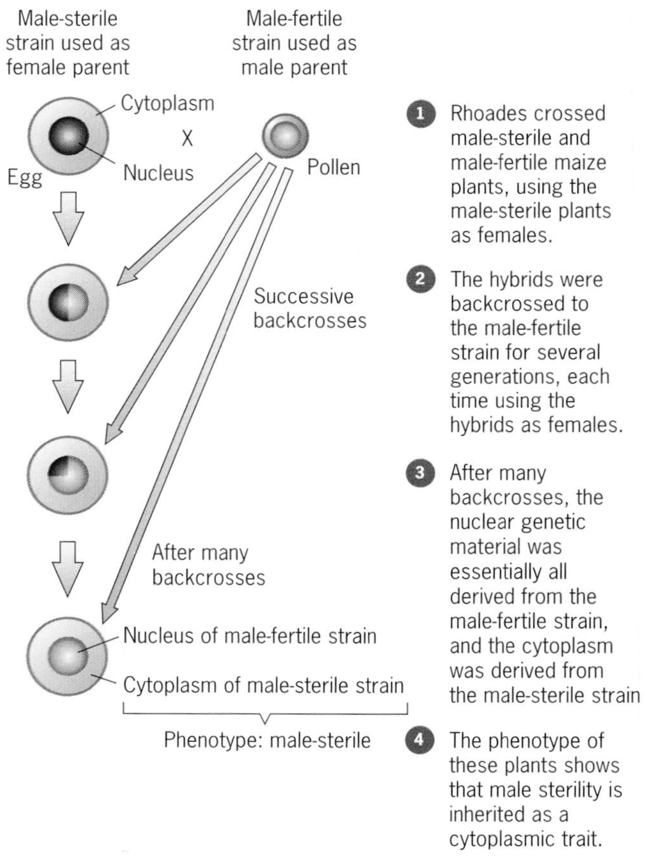

Figure 18.6 Cytoplasmic inheritance of male-sterility in maize. Successive backcrosses place the nucleus of a male-fertile strain into the cytoplasm of a male-sterile strain. Because the end result is male-sterile, the trait must be inherited through the cytoplasm.

tor, evidently something in the cytoplasm of the female reproductive cells. As the research progressed, Rhoades discovered that dominant mutations in several nuclear genes could suppress the effect of this cytoplasmic factor. A plant that had the cytoplasmic factor plus one of these dominant mutations could make functional pollen. The male-sterile phenotype therefore depended on an interaction between the cytoplasmic factor and the products of several different nuclear genes. With one combination of nuclear genes, the cytoplasmic factor could cause sterility, but with another combination, it could not.

Maize breeders have taken advantage of cytoplasmic male sterility to streamline the production of hybrid maize seed. However, the method has not been a panacea. In the early 1970s, overreliance on the sterilizing cytoplasm of a particular strain of maize had a disastrous effect on the United States' corn crop. Plants that carried this cytoplasm proved to be especially susceptible to infection by a mutant race of *Helminthosporium maydis*, a fungus that causes southern leaf blight. In some parts of the United States nearly 50 percent of the corn crop was destroyed. Fortunately, maize varieties that did not have the blight-susceptible cytoplasm had been preserved for use in breeding programs. These varieties were pressed into service, and the U.S. corn crop quickly rebounded.

What is responsible for the non-Mendelian inheritance of male sterility in maize? Maize plants possess both mitochondria and chloroplasts, so either organelle could be involved. A partial understanding of the basis of cytoplasmic male-sterility emerged in the 1970s and 1980s. During this time, the procedures for isolating and manipulating organelle DNA were developed. The application of these techniques to male-sterile maize revealed that some varieties contain two small accessory DNA molecules, called S1 and S2, in their mitochondria. These molecules appear to be involved in the male sterility phenomenon, but the mechanism by which they cause it is not known. In other strains of maize, male sterility seems to be caused by a mutation in the mitochondrial DNA itself. Strains with this mutation produce a small polypeptide that may disrupt the function of proteins on the inner mitochondrial membrane. The strains that make this polypeptide are also the ones that are susceptible to southern leaf blight. Thus, it is possible that blight susceptibility is mediated by the mutant polypeptide.

Antibiotic Resistance in *Chlamydomonas*

Photosynthetic green algae are an important part of the biosphere. They are abundant in the oceans, where they form the foundation of the food chain, but they are also found on the land. One terrestrial species,

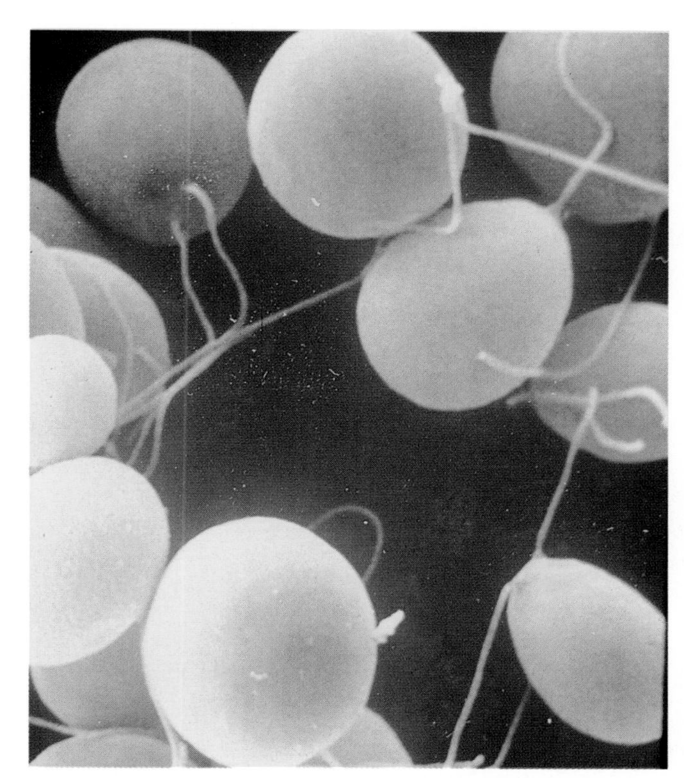

Figure 18.7 Cells of the unicellular alga *Chlamydomonas reinhardtii.*

Chlamydomonas reinhardtii (Figure 18.7), has been used extensively in genetics research. This unicellular, haploid organism exists in two different mating types, denoted plus and minus. Cells of opposite mating type can fuse to produce a diploid zygote. The zygote then undergoes meiosis to produce four haploid offspring, which divide mitotically to produce clones of vegetative cells. Analysis of these clones indicates that mating type is controlled by a nuclear gene with two alleles, mt^+ (specifying the plus mating type) and mt^- (specifying the minus mating type). Two of the four offspring from a *Chlamydomonas* zygote are therefore mt^+, and two are mt^-. Many other nuclear genes with a Mendelian pattern of segregation have been identified in *Chlamydomonas*.

Each *Chlamydomonas* cell contains a single large chloroplast and several mitochondria. When mt^+ and mt^- cells fuse, their organelles are combined. It is not altogether clear what happens to these organelles inside the zygote, but when the four haploid offspring are produced, each has one chloroplast and several mitochondria. The basic number of organelles is therefore preserved.

In 1954 Ruth Sager, an American geneticist, discovered that antibiotic resistance in *Chlamydomonas* is a non-Mendelian trait (Figure 18.8). Sager mated a mutant mt^+ strain that was resistant to the antibiotic streptomycin to an mt^- strain that was sensitive to this

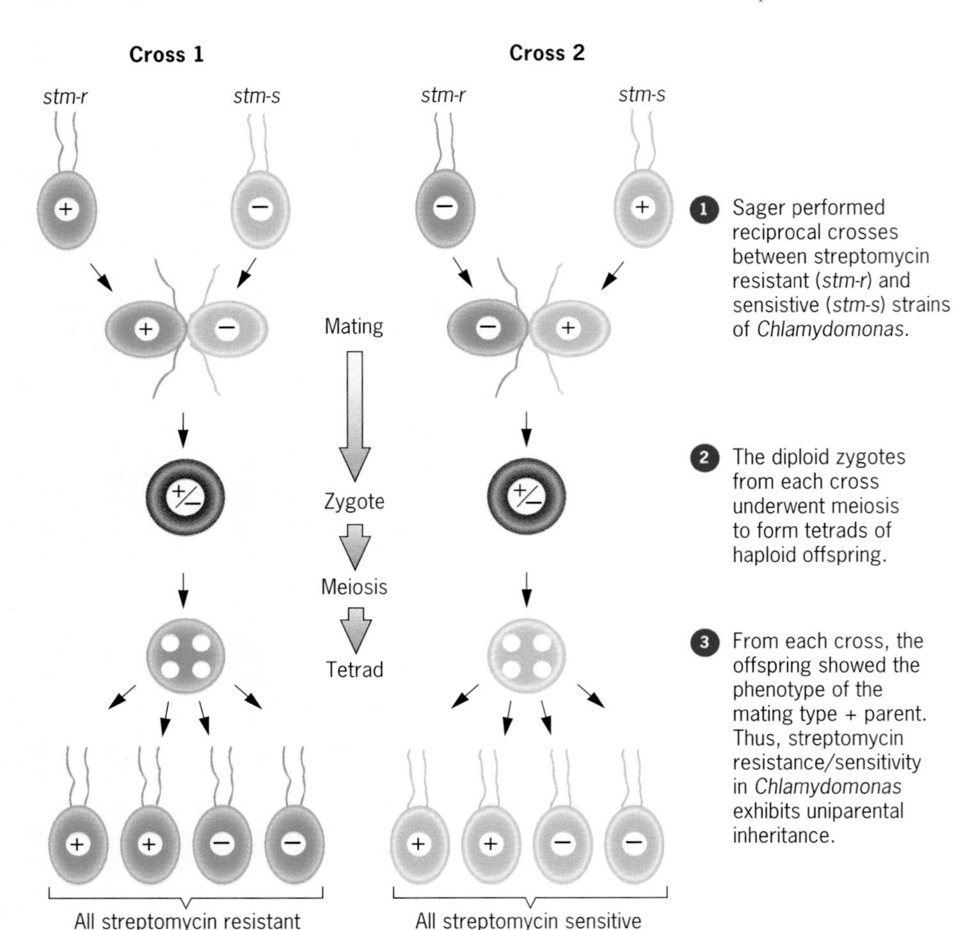

Figure 18.8 Sager's experiments showing uniparental inheritance of streptomycin resistance (*stm-r*) and sensitivity (*stm-s*) in *Chlamydomonas*. All the offspring of a cross have the same phenotype as the mt^+ parent.

antibiotic. All the progeny from this cross were streptomycin resistant, even though the mating type alleles segregated in a 1:1 fashion. Thus resistance to streptomycin in *Chlamydomonas* is controlled by a non-Mendelian factor. Sager then mated some of the resistant mt^- cells to sensitive mt^+ cells, and this time she found that all the progeny were sensitive. Resistance or sensitivity to streptomycin was therefore always inherited through the cytoplasm of the mt^+ cell. Sager soon found other traits that followed this same pattern of inheritance.

Sager and her associate Zenta Ramanis later discovered a way to alter the uniparental inheritance of these traits. This involved irradiating the mt^+ cells with UV light just prior to mating. When Sager and Ramanis mated irradiated mt^+ cells to unirradiated mt^- cells, they found that some of the offspring inherited the cytoplasmic trait of the mt^- parent instead of that of the mt^+ parent.

These phenomena were finally explained when DNA was discovered in the *Chlamydomonas* chloroplast. Both mt^+ and mt^- cells possess this DNA, but when mating occurs, the DNA contributed by the mt^- cell is degraded, probably by a nuclease that specifically recognizes it. The offspring from a cross therefore retain only the DNA inherited from the mt^+ chloroplast. As a result, the traits encoded by this DNA—and streptomycin resistance is one of them—are always inherited through the mt^+ parent. When mt^+ cells are irradiated prior to mating, the system that destroys the mt^- chlroplast DNA is partially incapacitated. Consequently, some of the mt^- chloroplast DNA survives and is transmitted to the offspring. This DNA may actually recombine with the mt^+ chloroplast DNA, generating offspring that inherit chloroplast genes from both parents. Sager and Ramanis used the frequency of such recombination to construct a map of the chloroplast "chromosome" (Figure 18.9). This map was the first detailed picture of how genes are organized on a nonnuclear DNA molecule.

Metabolic Defects in Yeast

Some mutant strains of yeast form tiny colonies when grown on a rich, glucose-containing medium. These strains are called **petite mutants**, from the French word for "tiny." Wild-type yeast strains form large, or **grande**, colonies on glucose medium. Additional tests with different food sources suggest that petite mu-

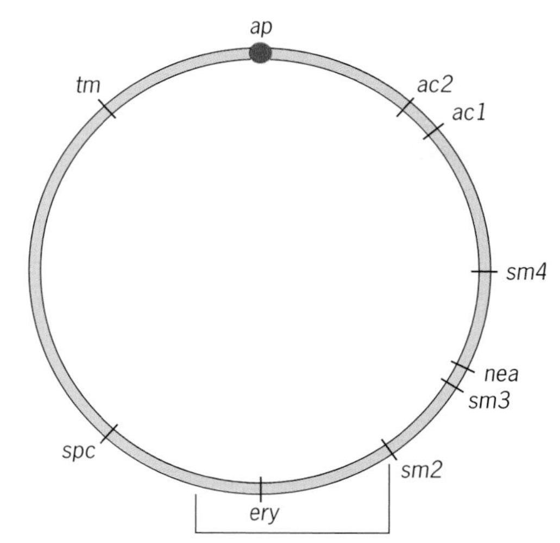

Figure 18.9 A genetic map of chloroplast DNA in *Chlamydomonas*, based on the work of Sager and Ramanis. Symbols: *ap*, attachment point; *ac1*, *ac2*, acetate requirement; *sm4*, streptomycin dependence; *nea*, neamine resistance; *sm3*, low-level streptomycin resistance; *sm2*, high-level streptomycin resistance; *ery*, erythromycin resistance; *spc*, spectinomycin resistance; *tm*, temperature sensitivity.

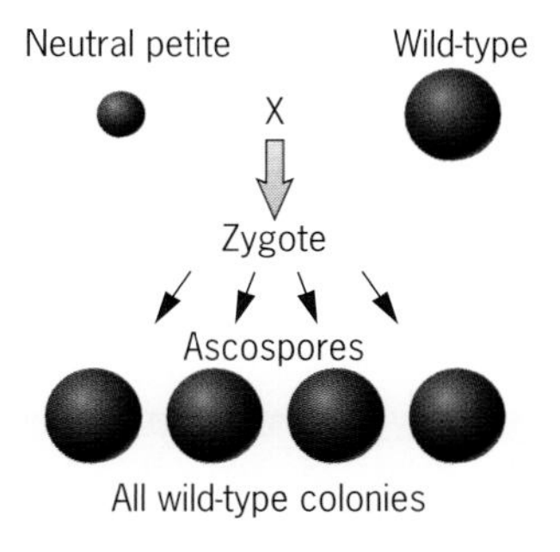

(*a*) A cross between a neutral petite strain and a wild-type strain produces only wild-type progeny.

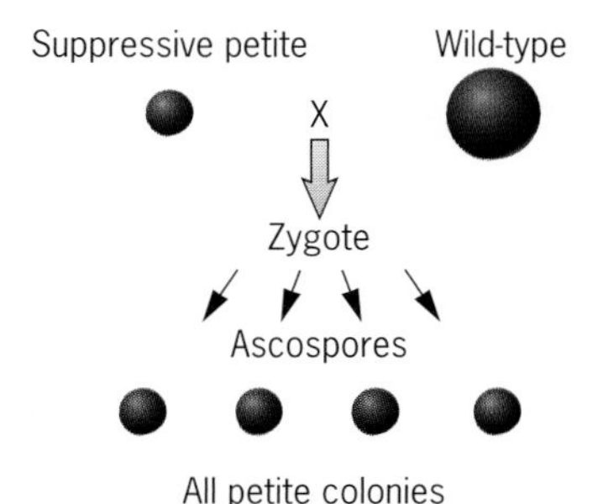

(*b*) A cross between a suppressive petite strain and a wild-type strain produces only petite progeny.

Figure 18.10 Non-Mendelian segregation of (*a*) neutral and (*b*) suppressive petite mutations in yeast.

tants suffer from a defect in glucose metabolism. This defect is, in fact, traceable to the mitochondria of the petite cells, which are ill-formed and lack many of the macromolecules found in wild-type mitochondria. These mitochondrial defects prevent petite cells from carrying out aerobic metabolism.

The first genetic analysis of the petite condition was carried out by the French researcher Boris Ephrussi in the 1940s and 1950s. His analysis revealed that there are two general classes of petite mutants, neutral petites and suppressive petites (Figure 18.10). The **neutral petite mutants** are characterized by an inability to transmit the petite phenotype to the offspring of crosses with wild-type strains. When such crosses are performed, all four haploid offspring from a pair of mated cells grow into grande colonies—suggesting that the petite mutation has been lost. **Suppressive petite mutants**, by contrast, are able to transmit the petite phenotype to all their progeny—suggesting that the wild-type condition has been lost. With both types of mutants, reciprocal crosses between the mutants and wild-type strains give identical results, demonstrating that the yeast mating types are not involved in the inheritance of the petite trait.

The molecular basis of the two types of petite mutants was determined in the 1970s and 1980s. Neutral petites lack any mitochondrial DNA, and suppressive petites have grossly mutated mitochondrial DNA (Figure 18.11). Thus both types of mutants are due to genetic abnormalities in the mitochondria. But what explains the different patterns of inheritance with these two types of mutants?

A cross between a neutral petite mutant and wild-type yields zygotes that inherit mitochondria from both parents; however, only the mitochondria from the wild-type parent contain any DNA. When these zygotes go through meiosis and form spores—the process called sporulation—the wild-type mitochondrial DNA is distributed to each of the four ascospores. As these grow and divide, healthy, functional mitochondria develop, allowing the cells to carry out aerobic metabolism. Thus, when colonies form, all of them have the grande phenotype.

A cross between a suppressive petite mutant and a wild-type strain yields zygotes with mitochondrial DNA from both parents. If these zygotes are sporulated immediately, the resulting ascospores inherit only the mutant mitochondrial DNA and grow into petite colonies. However, if the zygotes are first propagated mitotically in liquid culture and then sporulated, each ascospore inherits wild-type mitochondrial DNA and grows into a grande colony. These results suggest that the mutant mitochondrial DNA initially has some sort of advantage—it "suppresses" the wild-type DNA, perhaps because the wild-type DNA cannot replicate as quickly. However, during mitotic

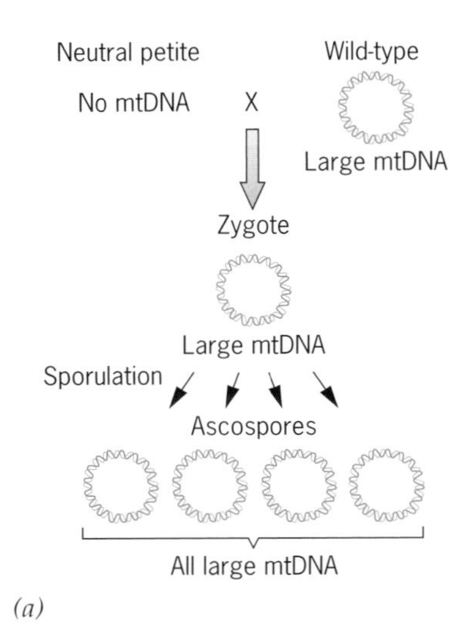

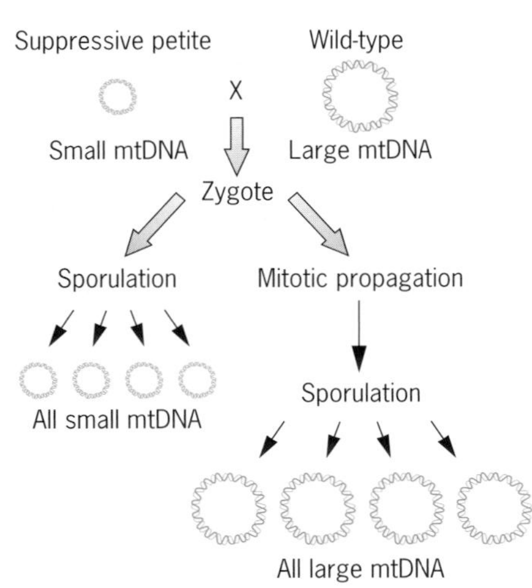

Figure 18.11 Inheritance of mitochondrial DNA in crosses between petite and wild-type strains of yeast. (*a*) Cross between neutral petite and wild-type strains. (*b*) Cross between suppressive petite and wild-type cells.

propagation, the replication of the wild-type DNA catches up with that of the mutant DNA and eventually surpasses it.

What is the nature of the mitochondrial DNA in suppressive petite mutants? Physical and chemical analysis has shown that it is a smaller DNA molecule than wild-type mitochondrial DNA and that it is also very A:T-rich. These two features may endow the suppressive petite mitochondrial DNA with a replicative advantage over wild-type. However, a full understanding of why the suppressive petite DNA is so effectively transmitted through crosses will require more research.

Key Points: **Organelle heredity is characterized by unequal contributions of the two parents and by irregular segregation of phenotypes. These non-Mendelian phenomena are due to the preferential transmission of chloroplasts or mitochondria through the gametes of one sex, which in higher eukaryotes is usually the female. Traits due to mitochondrial or chloroplast genes are therefore often inherited in a strictly maternal fashion.**

THE MOLECULAR GENETICS OF MITOCHONDRIA

Mitochondrial genetic systems consist of DNA plus the molecular machinery needed to replicate and express the genes contained in this DNA. This machinery includes the macromolecules needed for transcription and translation. (As remarkable as it may seem, mitochondria even possess their own ribosomes!) Many of these macromolecules are encoded by mitochondrial genes, but some are encoded by nuclear genes and are therefore imported from the cytosol. In the following sections, we explore the organization and function of the mitochondrial genetic system, and the ways in which it interacts with the nuclear genetic system.

Mitochondrial DNA

Mitochondrial DNA, or **mtDNA** as it is sometimes abbreviated, was discovered in the 1960s, initially through electron micrographs that revealed DNA-like

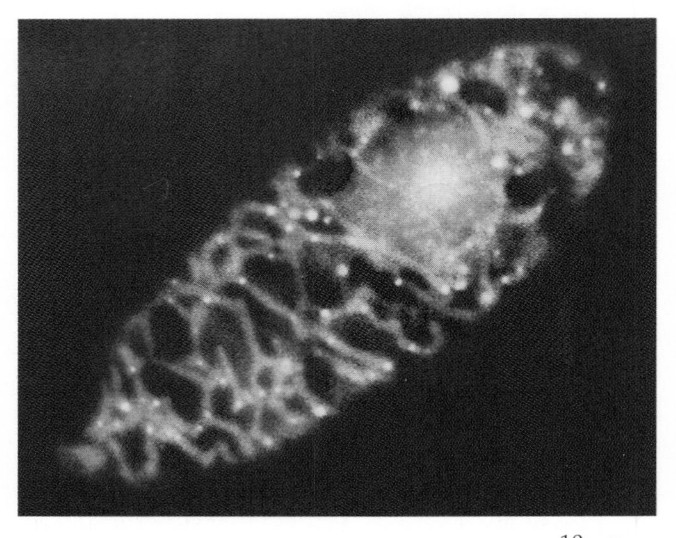

Figure 18.12 Mitochondrial DNA (yellow) in the unicellular organism *Euglena gracilis*. The nuclear DNA (red) is also visible.

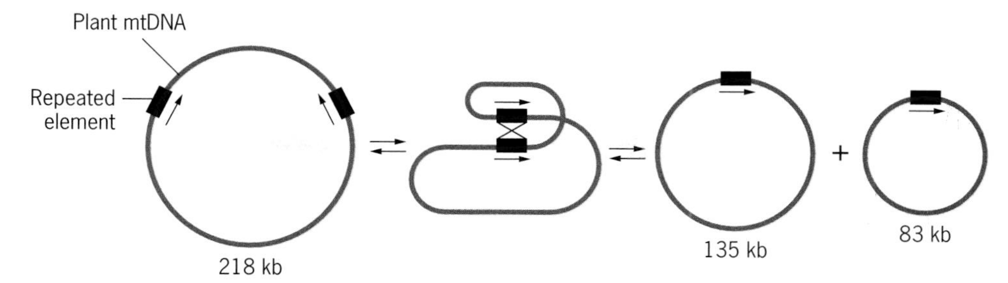

Figure 18.13 Intramolecular recombination in the mtDNA of the Chinese cabbage, *Brassica campestris*. Recombination between the repeated elements in the large circular DNA molecule partitions this molecule into two smaller ones. Alternatively, the repeated elements in the two small molecules may recombine with each other, thereby joining the molecules into a single large molecule.

fibers within the mitochondria. Later, these fibers were extracted and characterized by physical and chemical procedures. The advent of recombinant DNA techniques made it possible to analyze mtDNA in great detail. In fact, the complete nucleotide sequences of mtDNA molecules from several different species have now been determined.

Mitochondrial DNA molecules vary enormously in size, from about 16 to 17 kb in vertebrate animals to 2500 kb in some of the flowering plants. Each mitochondrion appears to contain several copies of the DNA, and because each cell usually has many mitochondria, the number of mtDNA molecules per cell can be very large (Figure 18.12). In a vertebrate oocyte, for example, it has been estimated that as many as 10^8 copies of the mtDNA are present. Somatic cells, however, have fewer copies, perhaps less than 1000.

Most mtDNA molecules are circular, but in some species, such as the alga *Chlamydomonas reinhardtii* and the ciliate *Paramecium aurelia*, they are linear. The circular mtDNA molecules, which have been studied the most thoroughly, appear to be organized in many different ways. The simplest arrangement is that seen in the vertebrates, where 37 distinct genes are packed into a 16- to 17-kb circle leaving little or no space between genes. The most complex arrangements exist in some of the flowering plants, where an unknown number of genes are dispersed over a very large circular DNA molecule hundreds or thousands of kilobases in circumference. In fact, in these plants the mitochondrial genes may become separated onto different circular molecules by a process of intramolecular recombination (Figure 18.13). This recombination is mediated by repetitive sequences located in the mtDNA. An exchange between two of the repetitive sequences can partition the "master" mtDNA circle into two smaller circles, a process that superficially resembles the excision of a lambda prophage from the *E. coli* chromosome. In some species, several DNA circles of different sizes are formed by recombination between pairs of repetitive sequences located at different positions around the master DNA circle. These molecules are difficult to study, and more research is needed to elucidate the mechanism that produces them.

The fine details of mtDNA organization have been studied by DNA sequencing. Animal mtDNA is small and compact (Figure 18.14). In human beings, for example, the mtDNA is 16,659 base pairs long and contains 37 genes, including two that encode ribosomal RNAs, 22 that encode transfer RNAs, and 13 that en-

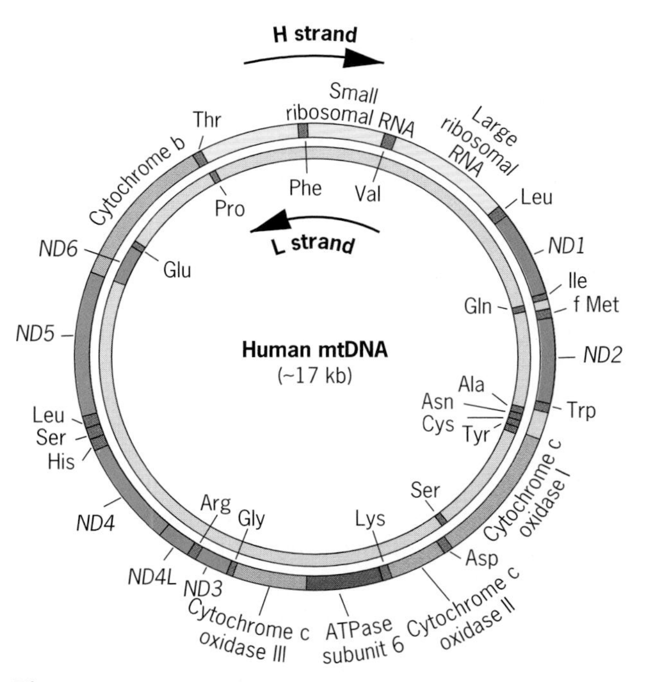

Figure 18.14 Map of human mtDNA showing the pattern of transcription. Genes on the inner circle are transcribed from the L strand of the DNA, whereas genes on the outer circle are transcribed from the H strand of the DNA. Arrows show the direction of transcription. *ND1-6* are genes encoding subunits of the enzyme NADH reductase; the tRNA genes in the mtDNA are indicated by abbreviations for the amino acids.

code polypeptides involved in oxidative phosphorylation. In mice, cattle, and frogs, the mtDNA is similar to that of human beings—an indication of a basic conservation of structure within the vertebrate subphyllum. Invertebrate mtDNA is about the same size as vertebrate mtDNA, but it has a somewhat different genetic organization. These differences seem to have been caused by structural rearrangements of the genes within the circular mtDNA molecule.

In fungi, the mtDNA is considerably larger than it is in animals. Yeast, for example, possesses circular mtDNA molecules 78 kb long. These molecules contain at least 33 genes, including two that encode ribosomal RNAs, 23 to 25 that encode transfer RNAs, one that encodes a ribosomal protein, and seven that encode different polypeptides involved in oxidative phosphorylation. The yeast mtDNA is larger than animal mtDNA because several of its genes contain introns and there are long noncoding sequences between some of the genes. Animal mtDNA does not contain introns.

Plant mtDNA is much larger than the mtDNA of other organisms. It is also more variable in structure. These conclusions come from crude physical and chemical analyses, and from limited DNA sequencing. At present, very few plant mtDNAs have been completely sequenced. One that has been sequenced is the mtDNA from the liverwort, *Marchantia polymorpha*. The mtDNA from this primitive, nonvascular plant is a 186-kb circular molecule with 94 substantial open reading frames (ORFs), some corresponding to known genes and others having still unassigned genetic functions. These latter ORFs are therefore called **URFs**, for unassigned reading frames. Thirty-two distinct introns have been found in the *Marchantia* mtDNA, accounting for about 20 percent of the entire molecule. In vascular plants, the mtDNA is larger than it is in *Marchantia*; for example, it is a 570-kb circle in maize and a 300-kb circle in the watermelon. Higher plant mtDNA molecules contain many noncoding sequences, including some that are duplicated. The actual number of genes is unknown, but it is probably greater than in the mtDNA of other organisms. Physical mapping of some of these genes has shown that they are located in different positions on the mtDNA circles of different species, even when the species are fairly closely related. This implies that the mtDNA of higher plants has undergone many genetic rearrangements during its evolution.

Expression of Mitochondrial Genes

The simple mtDNAs of vertebrates are organized into two large transcription units, each encoding the information of several genes. As an example, consider the human mtDNA shown in Figure 18.14. When the two strands of human mtDNA are separated by centrifugation in an alkaline solution, one proves to be denser than the other. This strand is referred to as the H (for heavy) strand, and its complement is referred to as the L (for light) strand. Both strands are transcribed in human mitochondria. The promoters for the H and L transcription units are situated just upstream of the phenlyalanine tRNA gene. The transcripts that initiate at these points are extended in opposite directions around the entire circumference of the mtDNA molecule. The transcript from the H strand encodes the two ribosomal RNAs, 14 tRNAs, and 12 polypeptides, and the transcript from the L strand encodes 8 tRNAs and one polypeptide. Each transcript is cleaved to separate the tRNAs from the ribosomal and mRNAs, and the mRNAs are polyadenylated. Each mRNA is then translated into polypeptides using the mitochondrial ribosomes and a combination of nuclear and ribosomal tRNAs.

Translation in the mitochondria proceeds much as it does on the ribosomes of the cytosol, except that some of the codons have a different meaning. AGA and AGG are termination codons in mammalian mitochondria, whereas in the cytosol they specify the incorporation of arginine into a polypeptide; UGA, which is a termination codon in the cytosol, is a tryptophan codon in the mitochondria and AUA, which encodes isoleucine in the cytosol, is the methionine initiation codon in the mitochondria. These and other mitochondrial codon variations indicate that the genetic code is not completely universal.

In fungi and plants, the mtDNA is organized into many separate transcription units, some containing the information for more than one gene. Little is known about the details of transcription, but in yeast, the mitochondrial RNA polymerase is a single polypeptide encoded by a nuclear gene. Thus, it is smaller than the RNA polymerase of *E. coli*. RNA processing separates plant mitochondrial transcripts into their constituent parts and also removes the introns, which are present in several plant mitochondrial genes. At present, the mechanics of these events are poorly understood.

Another peculiarity of plant mitochondrial gene expression is that many of the mtRNA transcripts undergo **editing**; that is, some of the nucleotides are changed after the transcript has been synthesized. The most frequent change is C to U, but occasionally, U is changed to C. Thus RNA editing alters the composition of codons in plant mitochondrial transcripts, including some that would otherwise signal the end of polypeptide synthesis. Editing corrects the information that is actually encoded in the mtDNA and allows functional polypeptides to be synthesized. Curiously, RNA editing is not found in the nonvascular plants (mosses and algae), even though all groups of higher

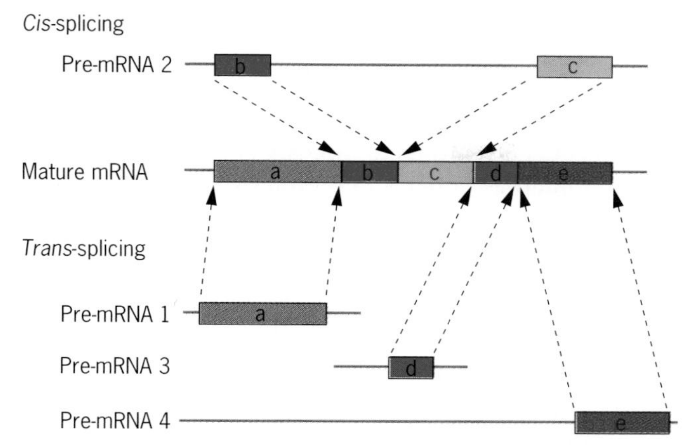

Figure 18.15 *Trans*-splicing in wheat mitochondria. Four different RNAs contribute to the final mRNA encoding a polypeptide of the enzyme NADH reductase.

plants (ferns, gymnosperms, and angiosperms) seem to have it. Thus the editing mechanism probably evolved sometime after plants had become established on the land. RNA editing also occurs in the mitochondria of protozoans, including the trypanosomes, where the mechanism has been studied in some detail. In these organisms, small RNA molecules that are partially complementary to the mtDNA transcripts serve as guides for the editing process (See Figure 11.22). They are, therefore, called guide RNAs (gRNAs). A similar guiding mechanism may operate in plants but the details are currently unknown.

Yet a third peculiarity of plant mitochondrial gene expression is that some mitochondrial messenger RNAs are formed by the process of ***trans*-splicing**. *Trans*-splicing occurs when segments of a gene are scattered over the mtDNA molecule. Each gene segment is transcribed independently, and the exons of the different transcripts are spliced together by interactions between the introns that flank them. For example, in wheat the *nad1* gene, which encodes a subunit of NADH reductase, a protein of oxidative phosphorylation, is partitioned into four segments in the mtDNA. Each of these segments is separately transcribed, and the resulting transcripts are then spliced together to form the mRNA (Figure 18.15). This process requires a single *cis*-splicing reaction and three *trans*-splicing reactions.

Interplay between Mitochondrial and Nuclear Gene Products

Most—perhaps all—mitochondrial gene products function solely within the mitochondrion. However, they do not function alone. Many nuclear gene products are imported to augment or facilitate their function. Mitochondrial ribosomes, for example, are constructed with ribosomal RNA transcribed from mitochondrial genes and with ribosomal proteins encoded by nuclear genes. The ribosomal proteins are synthesized in the cytosol and imported into the mitochondria for assembly into ribosomes.

Many of the polypeptides needed for aerobic metabolism are also synthesized in the cytosol. These include subunits of several proteins involved in oxidative phosphorylation, for example, the ATPase that is responsible for binding the energy of aerobic metabolism into ATP. However, because some of the subunits of this protein are synthesized in the mitochondria, the complete protein is actually a mixture of nuclear and mitochondrial gene products. This dual composition suggests that the nuclear and mitochondrial genetic systems are coordinated in some way so that equivalent amounts of their products are made; possible molecular mechanisms for this coordination are currently under investigation.

Key Points: **Mitochondrial DNA (mtDNA) molecules range from 16 kb to 2500 kb in size, and most appear to be circular. These molecules contain genes for some of the ribosomal RNAs, transfer RNAs, and polypeptides used within the mitochondrion. The structure, organization, and expression of these genes vary among species. In some groups of organisms, the transcripts of mitochondrial genes are edited after they are synthesized. Both mitochondrial and nuclear gene products are needed for proper mitochondrial function.**

MITOCHONDRIAL DNA AND HUMAN DISEASE

In yeast, petite mutants are associated with alterations in the structure of the mtDNA and in some cases, with the complete loss of this DNA. Are any phenotypes in higher eukaryotes traceable to mtDNA mutations?

Recent research has demonstrated that several human diseases are caused by mitochondrial defects, and in some cases, these defects are due to mutations in the mtDNA. One such disease is **Leber's hereditary optic neuropathy (LHON)**, a condition characterized by the sudden onset of blindness in adults. At a physiological level, this disease is associated with the death of the optic nerve, and at a molecular level, it is associated with mutations in any of several different mitochondrial genes. Each of these mutations changes an amino acid in one of the mitochondrial proteins, thereby reducing the efficiency of oxidative phosphorylation. The reduction is great enough to destroy the function of the optic nerve and cause total blindness. It is not known why this lethal effect is limited to the optic nerve. Perhaps the nerve cells are especially sensitive to disruptions in aerobic metabolism. As would be expected for a mitochondrial mutation, LHON is inherited strictly through the maternal line; there is never any transmission through males.

Another disorder caused by a mutation in the mtDNA is **Pearson marrow-pancreas syndrome**. This disease, characterized by a loss of bone-marrow cells during childhood, is frequently fatal. It is caused by a fairly large deletion in the mtDNA. People with the Pearson syndrome almost never have affected parents; thus the causative deletion probably occurs spontaneously during development in the child or during oogenesis in the mother. Individuals with the Pearson syndrome actually have a mixture of deleted and normal mtDNA—an example of mitochondrial heteroplasmy. Individuals who are homoplasmic for the deleted mtDNA of Pearson's syndrome have never been observed, probably because they die very early in development.

Key Point: **Mutations in mtDNA can cause human diseases such as Leber's hereditary optic neuropathy and Pearson marrow-pancreas syndrome.**

THE MOLECULAR GENETICS OF CHLOROPLASTS

Chloroplasts are specialized forms of a general class of plant organelles called **plastids**. Botanists distinguish among several kinds of plastids, including chromoplasts (plastids containing pigments), amyloplasts

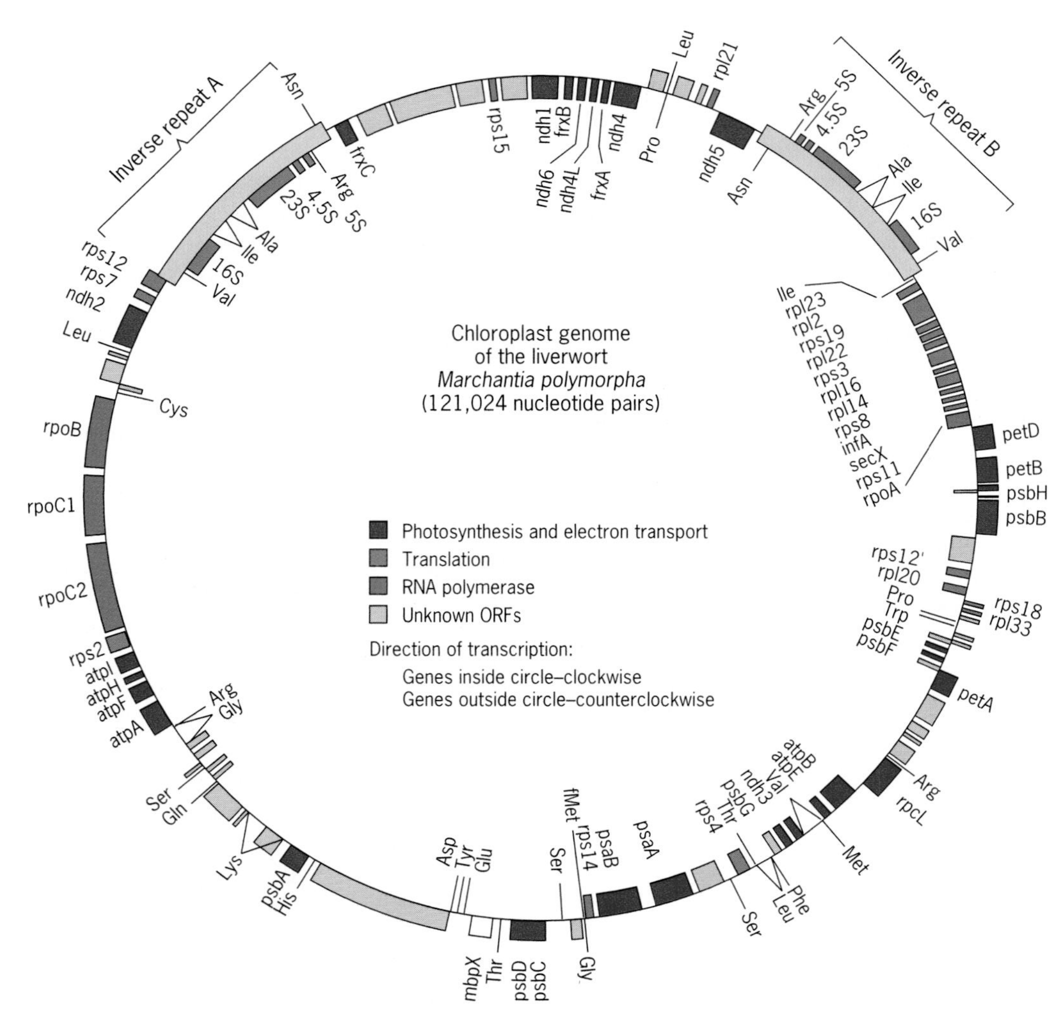

Figure 18.16 Genetic organization of the chloroplast DNA in the liverwort *Marchantia polymorpha*. Symbols: rpo, RNA polymerase; rps, ribosomal proteins of small subunit; rpl and secX, ribosomal proteins of large subunit; 4.5S, 5S, 16S, 23S, rRNAs of the indicated size; rbs, ribulose bisphosphate carboxylase; psa, photosystem I; psb, photosystem II; pet, cytochrome b/f complex; atp, ATP synthesis; infA, initiation factor A; frx, iron-sulfur proteins; ndh, NADH reductase; mpb, chloroplast permease (?); tRNA genes are indicated by abbreviations for the amino acids.

(plastids containing starch), and elaioplasts (plastids containing oil or lipid). All three types seem to develop from small membrane-bounded organelles called proplastids, and, within a particular plant species, all seem to contain the same DNA. This DNA is generally referred to as **chloroplast DNA**, abbreviated simply as **cpDNA**.

Chloroplast DNA

In higher plants, cpDNAs typically range from 120 to 160 kb in size, and in algae, from 85 to 292 kb. In a few species of green algae in the genus *Acetabularia*, the cpDNA is much larger, about 2000 kb. Among the 200 or so species of plants whose chloroplast DNA has been at least partially characterized, the cpDNA seems to be organized as a covalently closed circular molecule. However, in some species, especially those with large cpDNAs, a linear arrangement cannot be ruled out.

The number of cpDNA molecules in a cell depends on two factors: the number of chloroplasts and the number of cpDNA molecules within each chloroplast. For example, in the unicellular alga *Chlamydomonas reinhardtii* there is only one chloroplast per cell, and it contains about 100 copies of the cpDNA. In *Euglena gracilis*, another unicellular organism, there are about 15 chloroplasts per cell, and each contains about 40 copies of the cpDNA.

All cpDNA molecules carry basically the same set of genes, but in different species of plants these genes are arranged in different ways. The basic gene set includes genes for ribosomal RNAs, transfer RNAs, some ribosomal proteins, various polypeptide components of the photosystems that are involved in capturing solar energy, the catalytically active subunit of the enzyme ribulose 1,5-bisphosphate carboxylase, and four subunits of a chloroplast-specific RNA polymerase. Two cpDNA molecules have been sequenced in their entirety, one from the liverwort, *Marchantia polymorpha* (Figure 18.16), and the other from the tobacco plant, *Nicotiana tobacum*. The tobacco cpDNA is larger (155,844 bp) and probably contains about 150 genes. The best estimate for the gene number in the liverwort cpDNA (121,024 bp) is 136. Most cpDNAs have a pair of large inverted repeats that contain the genes for the ribosomal RNAs. These repeats range anywhere from 10 to 76 kb in length and are variously located in different cpDNA molecules.

Chloroplast Biogenesis

As mentioned earlier, all plastids develop from proplastids. In the case of chloroplasts, this development is stimulated by light and involves the transcription of

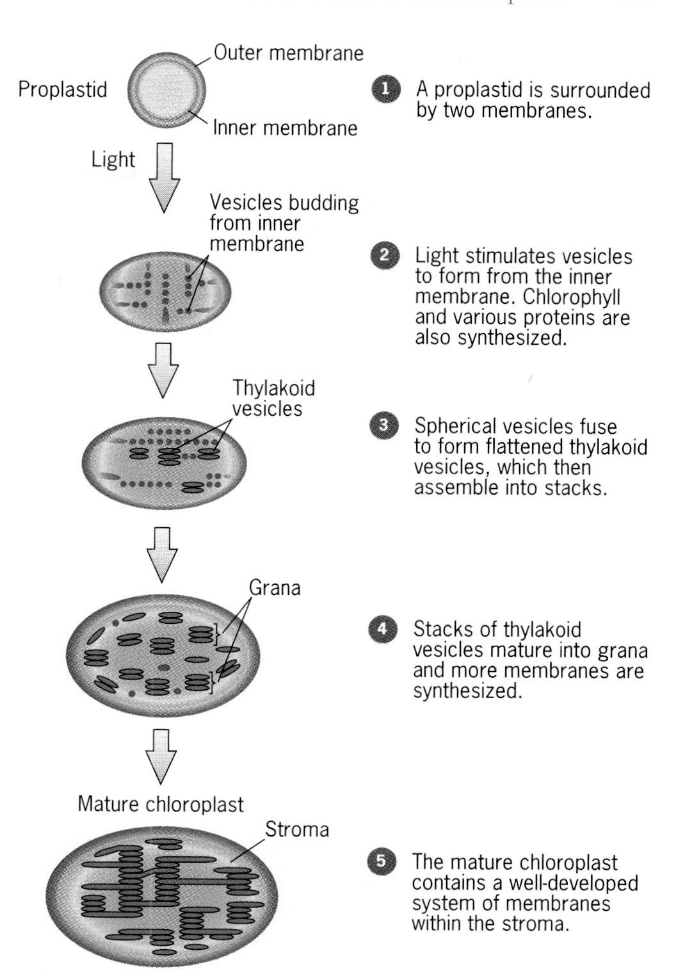

Figure 18.17 Chloroplast biogenesis. A mature chloroplast containing stacks of thylakoid membranes (grana) within its protoplasmic matrix (stroma) develops from a proplastid after exposure to light.

many genes, including some located in the nucleus. Illumination triggers a profound change in each proplastid. The organelle increases in size, and the inner of its two surrounding membranes expands to bud off vesicles that eventually arrange themselves in stacks forming structures called grana (Figure 18.17). All the proteins and chlorophyll pigments needed for photosynthesis are made and targeted to their appropriate locations within the emerging chloroplast. Some of these molecules are positioned in the thylakoid membranes that make up the grana, and others are localized in the stroma, which is the protoplasmic space surrounding the grana.

The formation of functional chloroplasts is a process referred to as **biogenesis**. Only some of the details are known. However, it is clear that light plays an important role. For example, the nuclear gene that encodes the small subunit of ribulose 1,5-bisphosphate carboxylase is vigorously transcribed when light is provided. A special class of pigmented proteins called **phytochromes** seems to mediate this and other responses to light. By absorbing light energy, these proteins acquire the ability to trigger other proteins to stimulate the transcription of genes involved in chloroplast biogenesis.

The formation of chloroplasts and the maintenance of their structure and function during the life of a plant depend on the coordinated expression of nuclear and chloroplast genes. Each genome produces a distinct set of mRNAs, which are translated in the cytosol and the chloroplast, respectively. Chloroplast biogenesis therefore involves considerable interplay between nuclear and chloroplast gene products.

Key Points: **Circular chloroplast DNA (cpDNA) molecules are typically 120 to 292 kb in size and may contain 150 or more genes. The organization of these genes varies among species of plants and algae. Other plastids also contain cpDNA molecules. Light induces chloroplasts to develop from unpigmented proplastids. This process, called biogenesis, involves the interplay of chloroplast and nuclear gene products.**

THE ORIGIN AND EVOLUTION OF MITOCHONDRIA AND CHLOROPLASTS

Life on earth originated more than 3 billion years ago, probably in shallow pools of nutrient-rich water. The first cellular organisms were prokaryotic; they lacked true nuclei and cytoplasmic organelles. Furthermore, these cells were heterotrophic; that is, they fed on nutrients present in the environment. Through the fermentation of nutrients, these primordial organisms recruited the energy they needed to survive. Later, as CO_2 accumulated in the atmosphere, photosynthetic organisms evolved, probably much like the cyanobacteria that are found in stagnant pools today. These organisms were also prokaryotic; their photosynthetic machinery was localized in membranes strewn within the cell—not organized within a special subcellular organelle. Photosynthesis generated substantial quantities of O_2, and as this gas built up in the atmosphere, aerobic metabolism evolved, probably 1 to 2 billion years ago. The first aerobic organisms were also prokaryotic.

Eukaryotic organisms evolved 1 to 1.5 billion years ago. The first known eukaryotes were filamentous green algae, with nuclei and an elaborate intracellular organization, including subcellular organelles. These organisms were capable of photosynthesis and aerobic metabolism, and probably contained both chloroplasts and mitochondria. Where did these cytoplasmic organelles come from?

Many scientists now believe that chloroplasts and mitochondria were derived from organisms that were internalized by primitive eukaryotic cells. This is not a new idea. In the nineteenth century, several microscopists suggested that chloroplasts originated as tiny algae, and in the early twentieth century, the American physician J. E. Wallin proposed that mitochondria were actually bacteria living inside cells. Today, the hypothesis that chloroplasts and mitochondria originated as separate organisms is widely accepted. These organisms evidently established a symbiotic relationship with the primitive eukaryotic cells that internalized them.

Eukaryotic Organelles as Endosymbionts

Symbiosis is a condition in which two or more organisms live together in close association for their mutual benefit. Lichens are a classic example. These widely distributed organisms are composed of an alga and a fungus that live together in an intimate, mutually beneficial association. **Endosymbiosis** is a special case in which one of the symbiotic partners lives inside the other. For example, cells of *Chlorella*, a unicellular alga, sometimes live inside the single-celled protozoan *Paramecium bursaria*. The chlorophyll in these endosymbiotic algae makes the *Paramecium* appear green.

Evidence from biochemical, genetic, and molecular studies suggests that mitochondria and chloroplasts originated as bacterial endosymbionts. Mitochondria and chloroplasts are both about the same size as some bacteria, and both contain circular DNA molecules much like circular bacterial chromosomes. Both mitochondria and chloroplasts contain their own distinctive ribosomes, and both utilize a distinctive RNA polymerase. In chloroplasts, this RNA polymerase is similar to the one found in *E. coli*. DNA sequencing studies have demonstrated that some genes in the mitochondrial and chloroplast DNAs are similar to bacterial genes. For example, the mitochondrial genes for ribosomal RNA are most closely related to the genes for ribosomal RNA in a group of purple photosynthetic bacteria, and the chloroplast genes for ribosomal proteins seem to be related to bacterial genes for these proteins. All these facts strongly suggest that mitochondria and chloroplasts originated when bacteria were incorporated into primitive eukaryotic cells, probably more than a billion years ago. The endosymbiotic bacteria benefited from a stable biological environment that provided food and an ability to reproduce, and the eukaryotic cells benefited from the oxidative and photosynthetic energy-recruiting powers of the endosymbiotic bacteria.

Since the formation of these relationships, each of the symbiotic partners has undergone significant changes, and considerable genetic shuffling has taken place among the mitochondrial, chloroplast, and nuclear DNAs. For example, the mtDNA in some higher plants contains a transfer RNA gene from cpDNA, and in one plant species, the mtDNA contains part of the chloroplast gene for the large subunit of ribulose 1,5-bisphosphate carboxylase. In most plant species, the small subunit of ribulose 1,5-bisphosphate carboxylase is encoded by a nuclear gene, but in some species of algae, it is encoded by a chloroplast gene. These differences indicate that some of the genes that were originally present in the DNA of the endosymbiotic bacteria have been moved to other cellular compartments, including many that have been transferred into the nucleus. Because of this gene shuffling, neither mitochondria nor chloroplasts are able to sustain themselves without materials specified by the nucleus. In fact, most of the materials that are needed for the assembly and function of these organelles are encoded by nuclear genes. Even major components of the genetic systems of the mitochondria and chloroplasts are derived from nuclear gene products. These include the DNA and RNA polymerases, some transfer RNAs, and many, if not all, of the ribosomal proteins. Over evolutionary time, the symbiotic partners have exchanged so many genes that they are now completely interdependent.

The Evolution of Mitochondria and Chloroplasts

Variation among the DNAs of mitochondria and chloroplasts indicates that the genomes of these organelles have undergone significant changes. It also suggests that these genomes may trace back to several different symbiotic partnerships that were established independently very long ago. Among organelle genomes that are clearly related, there are two kinds of variation.

First, there is structural variation, or variation in the order of genes within the DNA. For example, human and *Drosophila* mtDNAs contain the same set of genes, but these genes are arranged differently in the two molecules. The different arrangements suggest that regions of the mtDNA have been relocated since the vertebrate and invertebrate lineages diverged from a common ancestor 400 to 600 million years ago.

Second, there is variation in the nucleotide sequences of particular genes. For example, when we compare the genes for cytochrome b in human and *Drosophila* mtDNAs, we find many nucleotide differences, some of which alter the sequence of amino acids in the cytochrome b polypeptide. This kind of variation indicates that the sequences of these genes have evolved since they diverged from a common ancestor. Considerable effort has been expended to study the rate of mitochondrial gene evolution and to compare it with the rate of nuclear gene evolution. The surprising finding is that mitochondrial genes evolve faster, perhaps 5 to 10 times faster, than nuclear genes. The reasons for this rapid evolution are not clear. Perhaps mitochondrial gene products are more flexible and are therefore able to tolerate more amino acid changes than are nuclear gene products, or perhaps the mutation rate is simply greater in the mitochondria than it is in the nucleus.

An important feature of mitochondria and chloroplasts is that they are almost always transmitted through the female. There is some evidence for paternal transmission (see, for example, the earlier discussion of Baur's experiments on chloroplast inheritance in *Pelargonium*), but in many organisms it seems to be infrequent. Thus the genomes of these organelles evolve mainly by an asexual process. However, because recombination apparently does take place between organelle genomes within particular cells, genetic variants on two different molecules can, in principle, be combined together. The evolutionary significance of this possibility has not been thoroughly assessed.

Key Points: **Mitochondria and chloroplasts probably originated as bacteria that were incorporated into eukaryotic cells about a billion years ago. The mutually beneficial relationship that has existed between these cells and their internalized bacteria has evolved into a relationship of complete interdependence. In the course of this endosymbiosis, genes have been shuffled among the mitochondrial, chloroplast, and nuclear DNAs. For unknown reasons, mitochondrial genes seem to be evolving 5 to 10 times faster than nuclear genes.**

TESTING YOUR KNOWLEDGE

1. A mutant mt^+ strain of *Chlamydomonas* that was resistant to the antibiotic streptomycin (*stm-r*) and the herbicide *benomyl* (*ben-r*) was crossed to a wild-type mt^- strain that was sensitive to these two chemicals (*stm-s* and *ben-s*). Twenty tetrads of progeny were analyzed for mating type and resistance to streptomycin and benomyl. When the analysis was completed, the tetrads were classified into three types:

	Type 1	Type 2	Type 3
	mt^+ *stm-r ben-r*	mt^+ *stm-r ben-s*	mt^+ *stm-r ben-r*
	mt^+ *stm-r ben-r*	mt^+ *stm-r ben-s*	mt^+ *stm-r ben-s*
	mt^- *stm-r ben-s*	mt^- *stm-r ben-r*	mt^- *stm-r ben-r*
	mt^- *stm-r ben-s*	mt^- *stm-r ben-r*	mt^- *stm-r ben-s*
Number	8	9	3

Which of these traits seems to be due to a nonnuclear gene?

ANSWER

Streptomycin resistance/sensitivity. All the progeny from the cross are streptomycin-resistant, just like the mt^+ parent. This uniparental inheritance is diagnostic of a chloroplast gene. The other two traits, mating type and benomyl resistance/sensitivity, segregate 1:1 in the progeny, indicating that they are determined by nuclear genes. In fact, from the numbers of the three types of tetrads, we can infer that the *mt* and *ben* loci are on different chromosomes. This inference follows because the number of parental ditype tetrads (Type 1) is equal to the number of nonparental ditype tetrads (Type 2); see Chapter 8 for a full explanation of tetrad analysis for nuclear genes.

QUESTIONS AND PROBLEMS

18.1 When Correns studied the inheritance of leaf variegation in *Mirabilis*, he made crosses between green and variegating plants. Why didn't he make crosses between green and white plants?

18.2 Leaf coloration in variegating strains is inherited maternally in *Mirabilis* and biparentally in *Pelargonium*. Furthermore, in *Pelargonium* the pattern of inheritance is non-Mendelian. What do these facts suggest about the inheritance of chloroplasts in these species?

18.3 Explain how plants such as *Mirabilis* could have green, pale green, and white sectors in their leaves. If such sectors reached sexual maturity, what color characteristics would each type be expected to transmit through male and female gametes?

18.4 Reciprocal crosses with experimental animals or plants sometimes give different results in the F_1. Two possible explanations are (a) sex-linked inheritance and (b) organelle inheritance. How could an investigator determine which explanation is correct?

18.5 Oscar Renner performed reciprocal crosses between two types of evening primroses, *Oenothera hookeri* and *O. muricata*, known to have the same chromosome constitution. When the female parent was *O. hookeri*, the leaves of the progeny were yellow, but when the female parent was *O. muricata*, the leaves of the progeny were green. How might this difference in the results of reciprocal crosses be explained?

18.6 What is the practical significance of cytoplasmic male sterility in maize?

18.7 An mt^+ strain of *Chlamydomonas* that was resistant to the antibiotic spectinomycin was crossed with a sensitive mt^- strain. Half the progeny were mt^+ and half were mt^-, but all were resistant to spectinomycin. When the mt^- spectinomycin-resistant cells were crossed to mt^+ spectinomycin-sensitive cells, all the progeny were spectinomycin-sensitive. Explain the inheritance of spectinomycin resistance/sensitivity.

18.8 A mutant strain of yeast that produced tiny colonies when grown on glucose-rich medium was crossed to a wild-type strain. All the progeny were wild-type. What kind of mutation was responsible for the tiny colonies?

18.9 Another mutant yeast strain that produced tiny colonies was crossed to a wild-type strain. In this cross, however, all the progeny formed tiny colonies. What kind of mutation was responsible?

18.10 What are the main differences between plant and animal mtDNA molecules?

18.11 If a plant mitochondrial gene for NADH reductase were inserted into a plasmid and then placed into *E. coli* cells, would it produce catalytically active NADH reductase? Explain.

18.12 Can both men and women develop Leber's hereditary optic neuropathy (LHON)? Explain.

18.13 If chloroplast DNA were inherited exclusively through the female gamete, would it ever be possible to observe recombination between chloroplast genes?

18.14 An mt^+ strain of *Chlamydomonas* that was resistant to the antibiotic spectinomycin was crossed with an mt^- strain that was resistant to the antibiotic streptomycin. The genes for resistance to these two antibiotics are known to reside in cpDNA. Just prior to mating, the mt^+ cells were irradiated with ultraviolet light. Each mated pair of cells produced four haploid progeny. In some of these tetrads, all the progeny were resistant to spectinomycin and sensitive to streptomycin. In others, all the progeny were resistant to streptomycin and sensitive to spectinomycin. In still others, at least one of the progeny was resistant to both antibiotics. Explain these different results.

18.15 A male-sterile maize plant was crossed with a male-fertile plant that carried male-sterilizing cytoplasm and one copy of a dominant nuclear allele, *Rf*, that restores male fertility. What fraction of the offspring will be male-sterile? If the male-fertile offspring of this cross are intercrossed with each other, what fraction of their offspring will be male-sterile?

18.16 A male-sterile variety of maize was crossed with a male-fertile variety and all the offspring were male-sterile. These offspring were then crossed to a male-fertile variety that had normal (male-fertile) cytoplasm. What fraction of the offspring of this second cross are expected to be male-sterile?

18.17 A mammalian mitochondrial gene has been fused with a bacterial gene by using recombinant DNA techniques. The fusion has been made so that the 3' half of the mitochondrial gene (encoding 278 amino acids) is attached in frame to the 5' half of the bacterial gene (encoding 200 amino acids). If this fusion gene is introduced into bacterial cells by transformation, will it produce a fusion polypeptide 478 amino acids long? Explain.

18.18 What implication does *trans*-splicing of mitochondrial RNA have for the one gene–one polypeptide concept?

18.19 In plants, one subunit of the enzyme ribulose 1,5-bisphosphate carboxylase is encoded by a nuclear gene, and another is encoded by a chloroplast gene. Both subunits are needed for the enzyme to function in photosynthesis. What special problem does the separation of genes in different cellular compartments pose for the synthesis of this important protein?

18.20 What facts suggest that chloroplasts and mitochondria are derived from bacterial cells?

BIBLIOGRAPHY

AVISE, J. C. 1991. Ten unorthodox perspectives on evolution prompted by comparative population genetic findings on mitochondrial DNA. *Ann. Rev. Genet.* 25:45–69.

MARGULIS, L. 1971. Symbiosis and Evolution. *Sci. Amer.* 225(2):49–57.

PALMER, J. D. 1985. Comparative organization of chloroplast genomes. *Ann. Rev. Genet.* 19:325–354.

SAGER, R. 1972. *Cytoplasmic Genes and Organelles.* Academic Press, New York.

SCHUSTER, W., AND A. BRENNICKE. 1994. The plant mitochondrial genome: physical structure, information content, RNA editing and gene migration to the nucleus. *Ann. Rev. Plant Physiol. Plant Mol. Biol.* 45:61–78.

WALBOT, V. 1991. RNA editing fixes problems in plant mitochondrial transcripts. *Trends in Genetics.* 7:37–39.

WALLACE, D. C. 1993. Mitochondrial diseases: genotype versus phenotype. *Trends in Genetics* 9:128–133.

WISSINGER, B., A. BRENNICKE, AND W. SCHUSTER. 1992. Regenerating good sense: RNA editing and *trans* splicing in plant mitochondria. *Trends in Genetics* 8:322–328.

A CONVERSATION WITH

NANCY WEXLER

Dr. Wexler, we appreciate this opportunity to discuss the personal and scientific dimensions of Huntington's disease with you. What inspired you to enter the field of genetics, and specifically neurogenetics?

Well, I was sort of yanked into the field by my hair. I know that everybody has a choice, but it really surprised me that I would become as intimately involved in this field as I am. My mother actually was a geneticist and was one of the early workers in *Drosophila* genetics. She worked in Thomas Hunt Morgan's famous fly room at Columbia. She came from a very unusual family. Her father came to the United States from a small village in Russia in the early 1800s. He was a garment salesman but eventually had to stop working because he developed Huntington's disease. Mother's three older brothers put her through college and graduate school. She got a Master's degree in genetics. So, when I was growing up, she would try to educate my sister Alice and me in genetics, but you don't usually listen to your mother at that age. But then she brought me into genetics in a different way, a way that I didn't anticipate. We discovered that she had Huntington's disease, just like her father.

Nancy Wexler has made it her life's goal to find a treatment and cure for Huntington's disease (HD), a degenerative neurological disorder. She devotes her research to human genetics, specifically HD. Her personal experiences coupled with scientific discoveries have made Dr. Wexler the voice behind the Huntington's disease research. In the 1980s, her collaborative effort with other scientists led to the localization of the HD gene. One of the most phenomenal achievements is her work with several villages in Venezuela, where over 1,653 adults and 73 children carry a mutant form of the HD gene. Dr. Wexler and her team constructed a gigantic human pedigree of approximately 10,000 Venezuelan people. Currently, she is president of the Hereditary Disease Foundation and Higgin's professor at Columbia University.

Would you mind describing the circumstances of this shocking discovery for us?

Well, I knew all the time that I was growing up that there was something wrong with her, but I really didn't understand what it was. She was frequently depressed and became increasingly withdrawn. She had been, so I was told, a very lively, energetic person. But she became quiet and timid. We had moved from New York to California by this time, and she decided to go back to teaching high school. She needed to get certified, so she went to UCLA. For a while, she and I were hanging out at UCLA at the same time, she as a college student and I as a high school student participating in a special college program.

Things got worse all the time I was in college. I was taking a year off, I had a Fulbright Fellowship, and it was during this time she was diagnosed with Huntington's. But it was a very long kind of odyssey up to that point. I was 21 at the time; she was 53. She died ten years later. We later learned that all of her brothers died of Huntington's disease. All three of them were diagnosed with Huntington's within the same week. Imagine, within a one-week period, my mother learned that all three of her brothers were dying, that she had a 50–50 risk of getting Huntington's, and that her two daughters were at risk. It was a quintuple whammy in her life.

How do you cope with the knowledge that you and your sister Alice could have inherited the Huntington's disease allele?

Our first reaction upon learning that we were at risk for inheriting Huntington's disease was one of horror. It's not anything you would ever expect. If feels like a total invasion of everything that you are. It feels like it's taking away everything that you wanted to be in the future. When you are in high school or college, the thought of getting seriously ill or of dying . . . well, that's not on your radar. It's not what you think about. Suddenly, you feel as if you have been turned into a very old person. At this point, my sister and I decided not to have children.

Even though we both knew that this was a disease that tended to come on in your thirties and forties, we suddenly became extremely aware of every single little movement that we would make. I was at a Senate hearing on Huntington's disease, and I was really nervous. As I started the program, I spilled a glass of water all over everything, everybody. This was it. I was not only mortified and felt stupid, but in the back of my mind I was saying that now I have Huntington's and this is the first symptom and everyone knows it and is looking at me. So I become hyper and vigilant and very self-conscious. A person goes through a process of mourning. I know I did. I was depressed and irritable. But it was important to go through this phase. People just can't pretend that nothing has happened.

When people have a crisis in their life, if they don't allow themselves to mourn, they don't allow themselves to really recognize what has happened to them. It lasts much longer and it creeps out in other ways. For me, the turning-around point was when I started to fight it, started getting involved with it. At this point in time, there was not an awful lot going on in the science of Huntington's disease. I started meeting other families with Huntington's, and my father, who is a psychologist and really my role model in life, started the Hereditary Disease Foundation. So right away we started interacting with scientists. The science was so enthralling, so fascinating, so intriguing; and the people I met were and still are my closest friends; but this personal and scientific involvment in Huntington's disease was a life saver for me. It turned everything around.

What was your father's role in establishing the foundation?

My father came up with a brilliant idea that has been adopted by many other groups looking for cures for other diseases. The idea was an interdisciplinary workshop. At most scientific meetings, people stand up and present slides and everybody goes to sleep. You fall asleep because you spent the previous evening sitting at a bar, talking to people about the really interesting stuff in science. My father said, why not bring the bar to the workshop and forget the slide show. That is exactly what we did. We brought in 15 to 20 young people who were right in the midst of various research projects, graduate students, undergrads, post-docs, young instructors, and just got them thinking about the problem. It worked spectacularly. It has been an impetus for a tremendous amount of research on Huntington's disease, including the isolation of the gene.

The paper announcing the discovery of the Huntington's disease gene was authored by the Huntington's Collaborative Research Group. Were you pleased about this cooperative venture?

This was really a star experience. If I had to choose whether I'd rather be at risk for Huntington's disease and known those people and had that experience or not, I'm not sure that I could make that choice. But the fact that you can think about it as choice is really wonderful. The experience I've gotten and the fun that I've had, it's been incredibly joyous and fantastically interesting. There have been frustrations, of course, but that goes with the terrritory. This was an incredible group of people. I wish the story could get across more. So many people think that science is so competitive and cut-throat, but more scientists are realizing that collaborations are essential because they cannot do it all by themselves.

How do you feel about the future of science and your role in this future?

In a way, I wish I was starting all over again. This is such an exciting time. The intersection of genetics and medicine and biology is the future. What will we learn about human nature, how will this affect morality, what kind of world will open up to us?

We need to consider the challenges we face. What kind of people are we and what kind of a future do we want to create for ourselves? This is really an extraordinary time, a unique epoch all made possible by discoveries of DNA's structure and function. It's a growing time. It's a good time to go to the bar and learn more.

Photograph illustrating the effect of growth hormone on body size in humans. The individual on the left is over 8 feet tall; his companion, shown next to his miniature pony, is only 3 feet tall.

19

The Techniques of Molecular Genetics

CHAPTER OUTLINE

Treatment of Pituitary Dwarfism with Human Growth Hormone

Kathy was a typical child in most respects—happy, playful, a bit mischievous, and intelligent. Indeed, the only thing unusual about Kathy was her small stature. She was born with pituitary dwarfism, which results from a deficiency of human growth hormone (HGH). Kathy seemed destined to remain abnormally small throughout her life, like the midget clowns who performed with the traveling circuses of the past. Then, at age 10, Kathy began receiving treatments of HGH synthesized in bacteria. She grew five inches during her first year of treatments. By continuing to receive HGH during maturation, Kathy reached the short end of the normal height distribution for adults. Without these treatments, she would have remained abnormally small in stature.

The HGH that allowed Kathy to grow to near-normal size was one of the first products of genetic engineering, the use of designed or modified genes to synthesize desired products. HGH was initially produced in *E. coli* cells harboring a modified gene composed of the coding sequence for HGH fused to synthetic bacterial regulatory elements. This chimeric gene was constructed *in vitro* and introduced into *E. coli* by transformation. In

1985, HGH produced in *E. coli* became the second pharmaceutical product of genetic engineering to be approved for use in humans by the U.S. Food and Drug Administration. Human insulin, which was the first such product, was approved in 1982 (see Human Genetics Sidelight: Treatment of Diabetes Mellitus with Human Insulin Produced in Bacteria).

How do scientists construct a gene that will produce HGH or human insulin in *E. coli*? If they do engineer this gene *in vitro*, how can they introduce it into living cells? In this chapter, we focus on the powerful tools of molecular genetics that allow researchers to construct and express such genes.

Much of what is known about the structure of genes has been obtained by molecular analyses through the application of **recombinant DNA technologies**. Recombinant DNA approaches begin with the **cloning** of the gene(s) of interest. The cloning of a gene involves its isolation and insertion into a small self-replicating genetic element such as a plasmid or viral chromosome. This small self-replicating genetic element is referred to as the **cloning vector**.

Once a gene has been cloned, it can be subjected to a whole array of manipulations that allow investigations of gene structure–function relationships. Usually, a cloned gene is sequenced; that is, the nucleotide-pair sequence of the gene is determined. If the function of the gene is unknown, its nucleotide sequence can be compared with thousands of gene sequences stored in three large computer gene banks, one in Germany, a second in Japan, and a third in the United States. Sometimes the function of a gene can be deduced based on its similarity to other genes whose functions are known. Given the nucleotide sequence of a gene and knowledge of the genetic code, the amino acid sequence of the polypeptide encoded by the gene can be predicted. The predicted amino acid sequence of the polypeptide can then be searched for functional domains that may provide clues as to its function. Nucleic acid and protein sequence databases have become important resources for research in molecular genetics, and they will become increasingly important in both basic biological research and in diverse applications of this research.

BASIC TECHNIQUES USED TO CLONE GENES

The haploid genome of a mammal contains about 3×10^9 nucleotide pairs. If the average gene is 3000 nucleotide pairs long (many are larger), each gene will represent one of a million such sequences in the genome. Thus isolating any one gene is like searching for the proverbial needle in a haystack. Most techniques used in the analysis of genes and other DNA sequences require that the sequence be available in significant quantities in pure or essentially pure form. How can one identify the segment of a DNA molecule that carries a single gene and isolate enough of this sequence in pure form to permit molecular analyses of its structure and function?

The development of recombinant DNA and gene-cloning technologies has provided molecular geneticists with methods by which genes or other segments of large chromosomes can be isolated, replicated, and studied by nucleic acid sequencing techniques, electron microscopy, and other analytical techniques. The **gene-cloning** procedure involves two essential steps: (1) the incorporation of the gene of interest into a small self-replicating chromosome, and (2) the amplification of the recombinant minichromosome by its replication in an appropriate host cell. Step 1 involves the joining of two or more different DNA molecules *in vitro* to produce **recombinant DNA molecules**, for example, a human gene fused to an *E. coli* origin of replication. Step 2 is really the gene-cloning event, in which the recombinant DNA molecule is replicated or "cloned" to produce many identical copies for subsequent biochemical analysis. Thus, although the entire procedure is often referred to as the recombinant DNA or gene-cloning technique, these terms actually refer to two separate steps in the overall process.

The Discovery of Restriction Endonucleases

The ability to clone and sequence essentially any gene or other DNA sequence of interest from any species depends on a special class of enzymes called **restriction endonucleases** (from the Greek term *éndon* meaning "within"; endonucleases make internal cuts in DNA molecules). Many endonucleases make random cuts in DNA, but the restriction endonucleases are site-specific. They cleave DNA molecules only at specific nucleotide sequences called **restriction sites.** Dif-

HUMAN GENETICS SIDELIGHT

Treatment of Diabetes Mellitus with Human Insulin Produced in Bacteria

Diabetes mellitus is the third leading cause of death in the United States; only heart disease and cancer cause more deaths. Over 2 million individuals have severe cases of this disease in the United States alone. In diabetics, the polypeptide hormone insulin is either not produced in adequate amounts or, if produced, does not stimulate the appropriate responses in muscle, liver, and adipose (fat storage) cells. The genetics of diabetes is complex; many genes, most of which appear to be autosomal recessive, contribute to a predisposition to the disease. See Chapter 25 for a discussion of the different types of diabetes.

About 2 million diabetics in the United States require daily injections of insulin. Prior to 1982, this insulin was isolated from the pancreatic glands of cattle or pigs obtained from meat-packing plants. However, about 5 percent of the individuals with diabetes are allergic to these insulins, which are slightly different from human insulin. These individuals required injections of insulin isolated from other animals or from human cadavers—expensive treatments in either case. In 1982, human insulin produced in genetically engineered bacteria was approved for use in humans, eliminating the dependence of diabetics who were allergic to insulin from cattle and pigs on the limited supply of human insulin obtained from cadavers.

Insulin is a small peptide hormone composed of two chains joined by disulfide bonds:

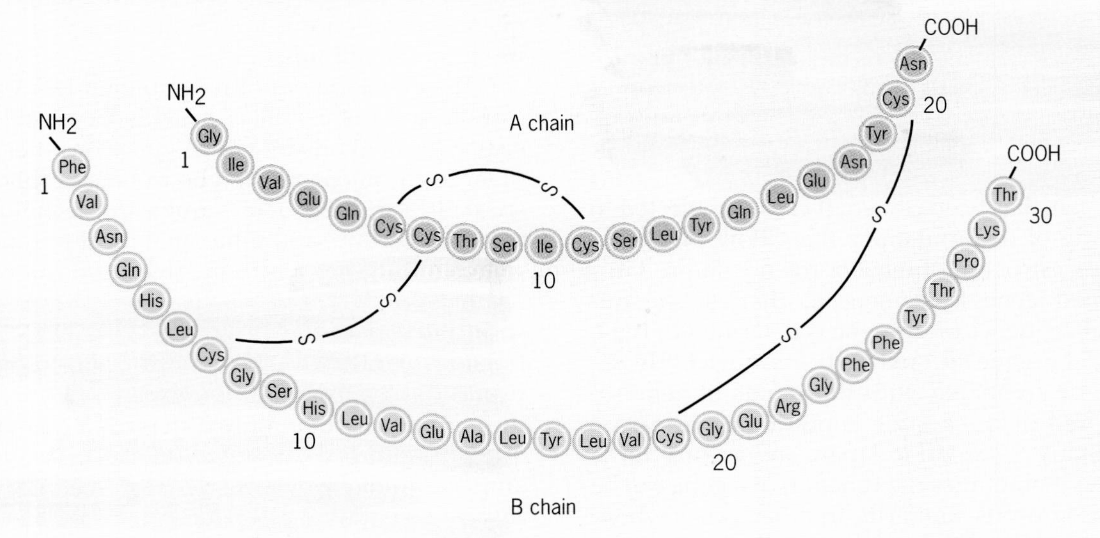

The A and B chains are 30 and 21 amino acids long, respectively. These two chains are synthesized as part of a larger gene product called preproinsulin, which contains an amino-terminal signal sequence that is excised during its secretion from β cells and a central region (the C chain) that is subsequently removed by proteolytic cleavage. The primary secretion product, proinsulin, is converted to insulin by the proteolytic excision of the C chain and the formation of three disulfide bonds as shown:

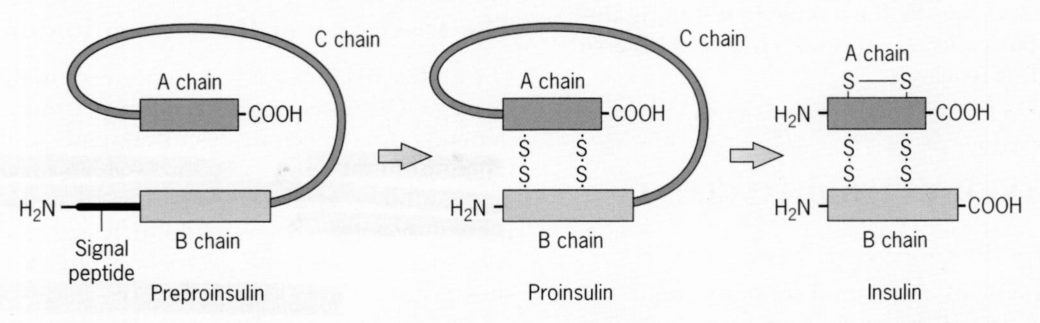

Human insulin was initially produced in *E. coli* cells by recombinant DNA techniques in 1979. The DNA sequences encoding the A and B chains of human insulin were chemically synthesized and joined to the *E. coli* gene encoding β-galactosidase. The chimeric genes were inserted into plasmid pBR322 (Figure 19.4) and introduced into *E. coli* cells by transformation. The fusion proteins specified by these chimeric genes contained a methionine residue between the bacterial β-galactosidase and the human insulin portions. This facilitated the removal of the insulin chain from the fusion protein by treatment with cyanogen bromide (which cleaves peptide chains on the carboxyl sides of methionine residues). This procedure is illustrated as follows.

1 Chemical synthesis of DNA molecules encoding human insulin A and B chains.

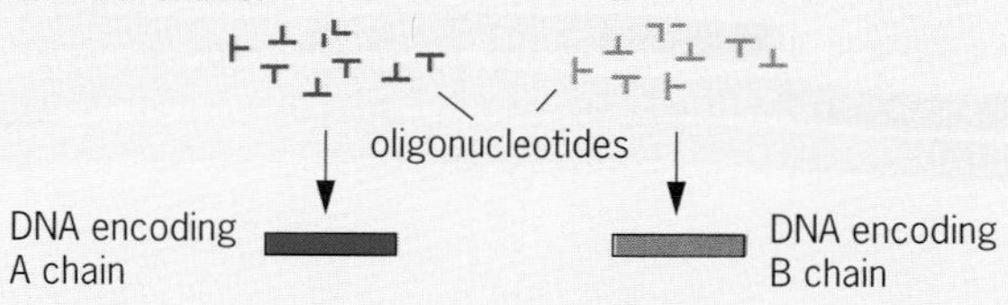

2 Fusion of these human insulin coding sequences to *E. coli* DNA encoding the enzyme β-galactosidase.

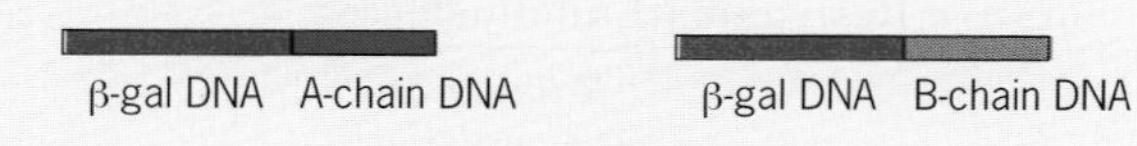

3 Insertion of chimeric genes into plasmid pBR322.

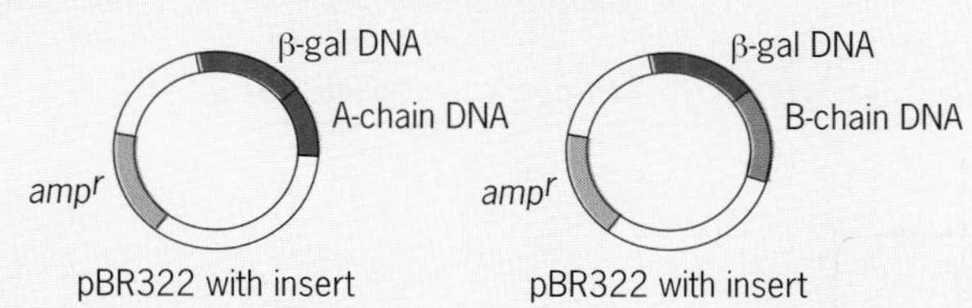

4 Introduce into and express in *E. coli*.

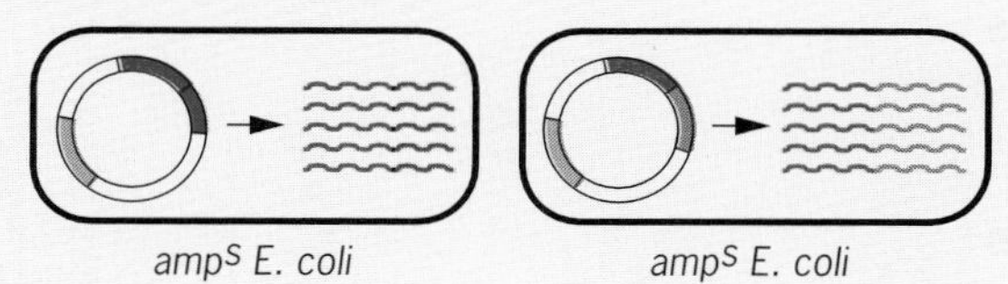

5 Extract and purify β-gal/insulin fusion proteins.

β-gal/A chain fusion protein β-gal/B chain fusion protein

6 Cleave with cyanogen bromide and purify insulin chains.

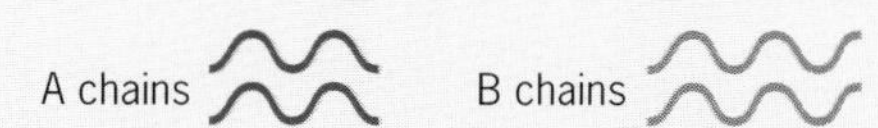

7 Mix A and B chains under appropriate conditions.

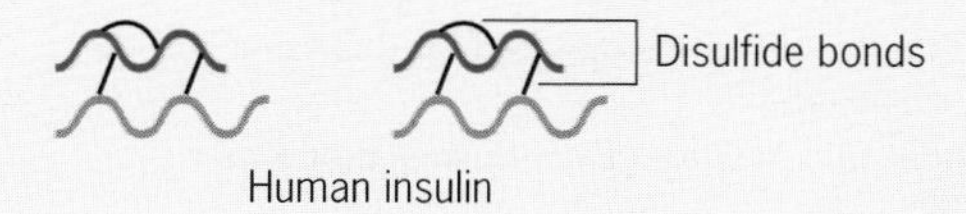

The β-galactosidase/human insulin fusion proteins encoded by the chimeric genes were extracted, purified, and cleaved with cyanogen bromide. The A and B chains of human insulin were then purified, mixed, reduced, and reoxidized to form the disulfide bonds present in native insulin. The human insulin produced by this procedure was identical to human insulin produced *in vivo*. Today, human insulin produced by recombinant DNA techniques is still widely used by diabetics, but this insulin is produced in animal cells (for example, Chinese hamster cells) growing in culture.

ferent restriction endonucleases are produced by different microorganisms and recognize different nucleotide sequences in DNA (Table 19.1). The restriction endonucleases are named by using the first letter of the genus and the first two letters of the species that produces the enzyme. If an enzyme is produced only by a specific strain, a letter designating the strain is appended to the name. The first restriction enzyme identified from a bacterial strain is designated I, the second II, and so on. Thus restriction endonuclease ***Eco*RI** is produced by *Escherichia coli* strain RY13. Over 150 different restriction enzymes have been characterized and purified; thus restriction endonucleases that cleave DNA molecules at many different DNA sequences are available.

Restriction endonucleases were discovered in 1970 by Hamilton Smith and Daniel Nathans. Smith and Nathans shared the 1986 Nobel Prize in physiology or medicine with Werner Arber, who carried out pioneering research that led to the discovery of restriction enzymes. In the 1960s, Arber and colleaques demonstrated that bacteriophage DNA is often restricted from replicating in one strain of *E. coli* if the DNA was synthesized and packaged in phage particles in a different strain of *E. coli*. It was later shown that the phage DNA is restricted in the new host because it is degraded by a restriction endonuclease produced by the bacterial cell. The biological function of restriction endonucleases is to protect the genetic material of bacteria from "invasion" by foreign DNAs, such as viral DNAs. As a result, restriction endonucleases are sometimes referred to as the immune systems of prokaryotes.

All cleavage sites in the DNA of an organism must be protected from cleavage by the organism's own restriction endonucleases; otherwise the organism would commit suicide by degrading its own DNA. In many cases, this protection of endogenous cleavage sites is accomplished by **methylation** of one or more nucleotides in each nucleotide sequence that is recog-

TABLE 19.1
Recognition Sequences and Cleavage Sites of Representative Restriction Endonucleases

Enzyme	*Source*	*Recognition Sequence*[a] *and Cleavage Sites*[b]	*Number of Recognition Sequences Per Chromosome of* Phage λ	*SV40 Virus*
*Eco*RI	*Escherichia coli*	G↓AA TTC / ● / CTT AA↑G	5	1
*Hin*dII	*Hemophilus influenzae*	GTPy↓PuAC[c] / ● / CAPu↑PyTG	34	7
*Hin*dIII	*Hemophilus influenzae*	A↓AG CTT / ● / TTC GA↑A	6	6
*Hpa*I	*Hemophilus parainfluenzae*	GTT↓AAC / ● / CAA↑TTG	11	4
*Hpa*II	*Hemophilus parainfluenzae*	C↓C GG / ● / GG C↑C	750	1

[a]The axis of symmetry in each palindromic recognition sequence is indicated by the dot.
[b]The position of each bond cleaved is indicated by an arrow. Note that the cuts are staggered (at different positions in the two complementary strands) with some restriction nucleases.
[c]Pu indicates that either purine (adenine or guanine) may be present at this position; Py indicates that either pyrimidine (thymine or cytosine) may be present.

nized by the organism's own restriction endonuclease. Methylation occurs rapidly after replication, catalyzed by site-specific methylases produced by the organism. Each restriction endonuclease will cleave a foreign DNA molecule (a DNA molecule from another species) into a fixed number of fragments, the number depending on the number of restriction sites in the particular DNA molecule (Table 19.1).

An interesting feature of restriction endonucleases is that they commonly recognize DNA sequences that are **palindromes**—that is, nucleotide-pair sequences that read the same forward or backward from a central axis of symmetry, as in the nonsense phrase

←——— ———→
AND MADAM DNA

In addition, a useful feature of many restriction nucleases is that they make staggered cuts; that is, they cleave the two strands of a double helix at different points. Because of the palindromic nature of the restriction sites, the staggered cuts produce segments of DNA with complementary single-stranded ends. For example, cleaving a DNA molecule of the following type:

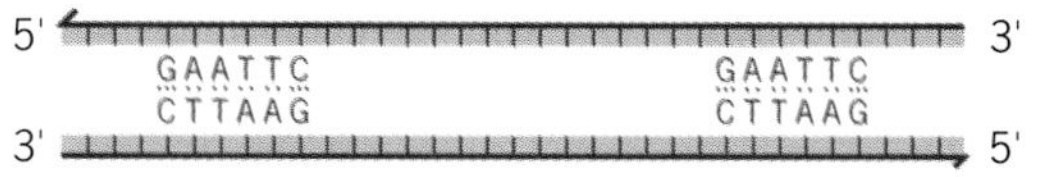

```
5' ───────────────────────────────────── 3'
      GAATTC                  GAATTC
      CTTAAG                  CTTAAG
3' ───────────────────────────────────── 5'
```

with the restriction endonuclease *Eco*RI will yield

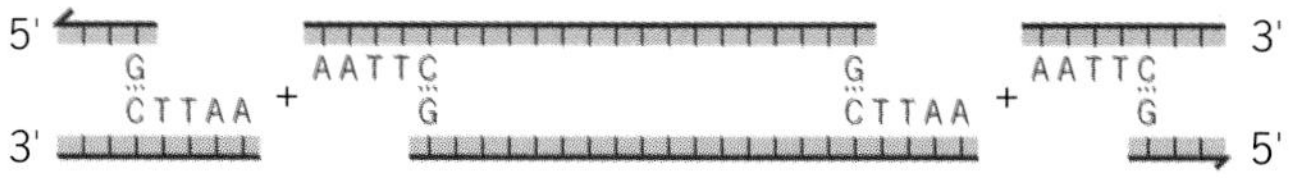

```
5' ────                  ──────────────                 ────── 3'
     G          AATTC         G            AATTC
     CTTAA  +       G         CTTAA    +       G
3' ────────          ────────────────               ──── 5'
```

Because all the resulting DNA fragments will have complementary single-stranded termini, they can be rejoined under the appropriate renaturation conditions by using the enzyme **DNA ligase** to reform the missing phosphodiester linkage in each strand. Thus DNA molecules can be cut into pieces, called **restriction fragments**, and the pieces can be joined together again with DNA ligase, almost at will.

The Production of Recombinant DNA Molecules *In Vitro*

A restriction endonuclease catalyzes the cleavage of a specific nucleotide-pair sequence regardless of the source of the DNA. It will cleave phage DNA, *E. coli* DNA, corn DNA, human DNA, or any other DNA, as long as the DNA contains the nucleotide sequence that it recognizes. Thus restriction endonuclease *Eco*RI will produce fragments with the same complementary single-stranded ends, AATT, regardless of the source of DNA, and two *Eco*RI fragments can be covalently fused regardless of their origin; that is, an *Eco*RI fragment from human DNA can be joined to an *Eco*RI fragment from *E. coli* DNA just as easily as two *Eco*RI fragments from *E. coli* DNA or two *Eco*RI fragments from human DNA can be joined. A DNA molecule of the type shown in Figure 19.1, containing DNA fragments from two different sources, is referred to as a recombinant DNA molecule. The ability of geneticists to construct such recombinant DNA molecules at will is the basis of the recombinant DNA technology that has revolutionized molecular biology in the last two decades.

The first recombinant DNA molecules were produced in Paul Berg's laboratory at Stanford University. Berg's research team constructed recombinant DNA molecules that contained phage lambda genes inserted into the small circular DNA molecule of simian virus 40 (SV40). In 1980, Berg was a co-recipient of the Nobel Prize in Chemistry as a result of this accomplishment. Shortly thereafter, Stanley Cohen and colleagues, also at Stanford, inserted an *Eco*RI restriction fragment from one DNA molecule into the cleaved, unique *Eco*RI restriction site of a circular, self-replicating plasmid. When this recombinant plasmid was introduced into *E. coli* cells by transformation, it exhibited normal autonomous replication, just like the parental plasmid.

Amplification of Recombinant DNA Molecules in Cloning Vectors

The various applications of recombinant DNA techniques require not only the construction of recombinant DNA molecules, as shown in Figure 19.1, but also the **amplification** of these recombinant molecules, that is, the production of many copies or **clones** of these molecules. This is accomplished by making sure that one of the parental DNAs incorporated into the recombinant DNA molecule is capable of self-replication. In practice, the gene or DNA sequence of interest is inserted into a specially chosen cloning vector. Most of the commonly used cloning vectors have been derived from viral chromosomes (Chapter 15) or plasmids (Chapter 16).

A cloning vector has three essential components: (1) an origin of replication, (2) a **dominant selectable marker gene**, usually a gene that confers drug resistance to the host cell, and (3) at least one **unique restriction endonuclease cleavage site**—a cleavage site that is present only once in the DNA of the vector (Figure 19.2). For maximum usefulness, a cloning vector should contain unique cleavage sites for several different restriction enzymes. The currently used cloning vectors contain a cluster of unique restriction sites called a **polylinker** or **polycloning site** (Figure 19.3).

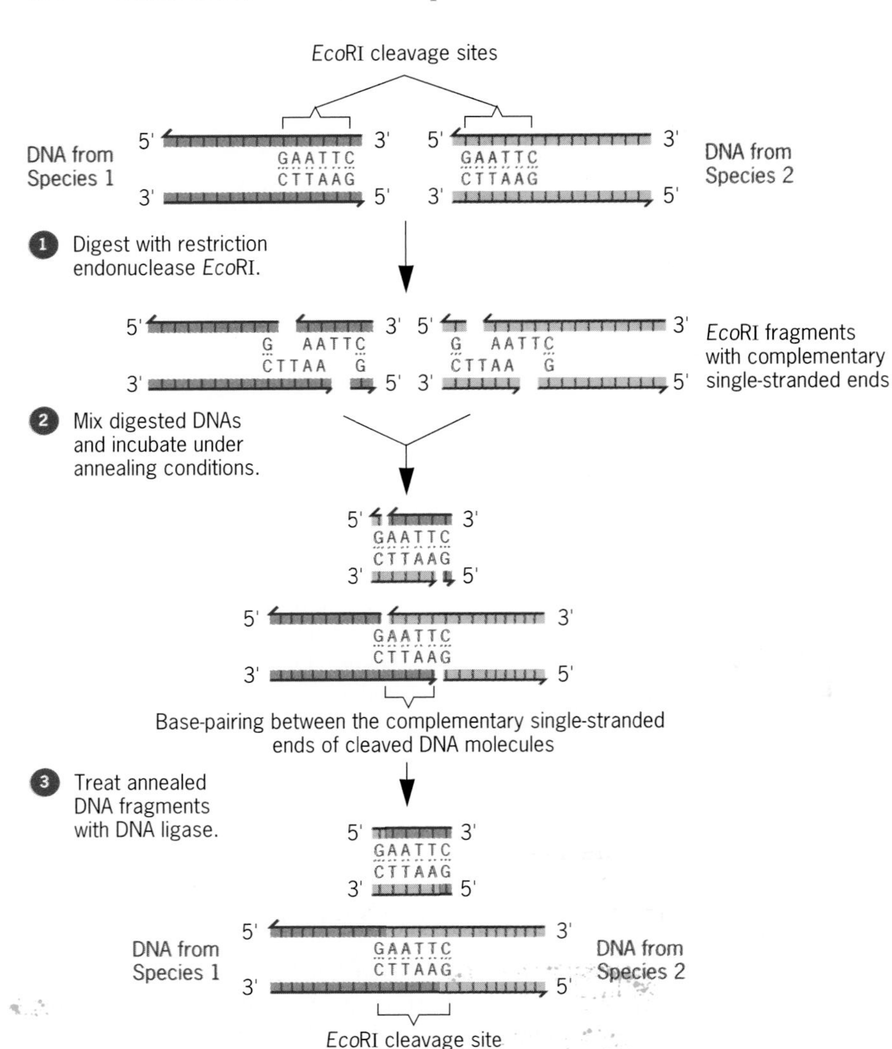

Figure 19.1 The construction of recombinant DNA molecules *in vitro*. DNA molecules isolated from two different species are cleaved with a restriction enzyme, mixed under annealing conditions, and covalently joined by treatment with DNA ligase. The DNA molecules can be obtained from any species—animal, plant, or microbe. The digestion of DNA with the restriction enzyme *Eco*RI produces the same complementary single-stranded AATT ends regardless of the source of the DNA.

Plasmid vectors: Plasmids are extrachromosomal, double-stranded circular molecules of DNA present in microorganisms, especially bacteria (Chapter 16). They range from about 1 kb (1 kilobase = 1000 base-pairs) to over 200 kb in size and replicate autonomously. Long before the recombinant DNA revolution began, some bacterial plasmids were known to carry antibiotic-resistance genes, which are ideal dominant selectable markers. Thus their use as cloning vectors was predictable. The first plasmid to be used as a cloning vector was *E. coli* plasmid pSC101, which contains a sin-

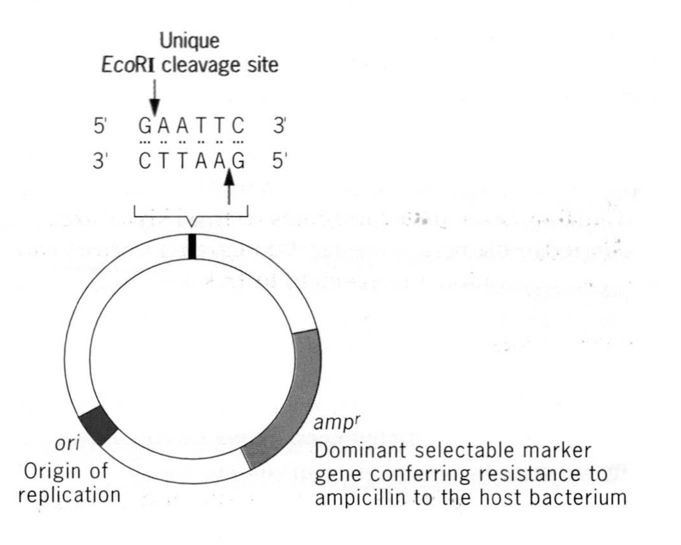

Figure 19.2 The essential features of a cloning vector. A unique *Eco*RI cleavage site is shown for illustrative purposes, but a unique cleavage site for any of a large number of different restriction endonucleases is just as appropriate.

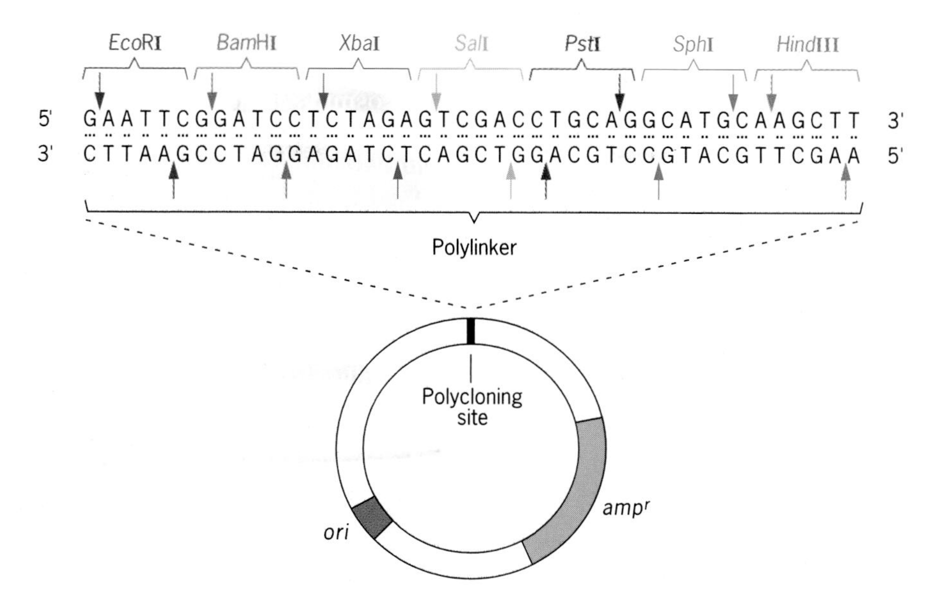

Figure 19.3 Structure of a polycloning site in a cloning vector. The phosphodiester bonds cleaved by the restriction endonucleases are indicated by arrows. The restriction endonucleases shown are produced by the following microorganisms: *Eco*RI, *Escherichia coli* strain RY13; *Bam*HI, *Bacillus amyloliquefaciens* strain H; *Xba*I, *Xanthomonas badrii; Sal*I, *Streptomyces albus; Pst*I, *Providencia stuartii; Sph*I, *Streptomyces phaeochromogenes;* and *Hin*dIII, *Haemophilus influenzae* strain R_d.

gle *Eco*RI cleavage site and a tetracycline-resistance (*tet^r*) gene. However, more versatile plasmid vectors were soon developed. Plasmid pBR322 was one of the first widely used cloning vectors; it contains both ampicillin- and tetracycline-resistance genes and a number of unique restriction enzyme cleavage sites (Figure 19.4). Many of the second- and third-generation cloning vectors in use today were derived, at least in part, from plasmid pBR322.

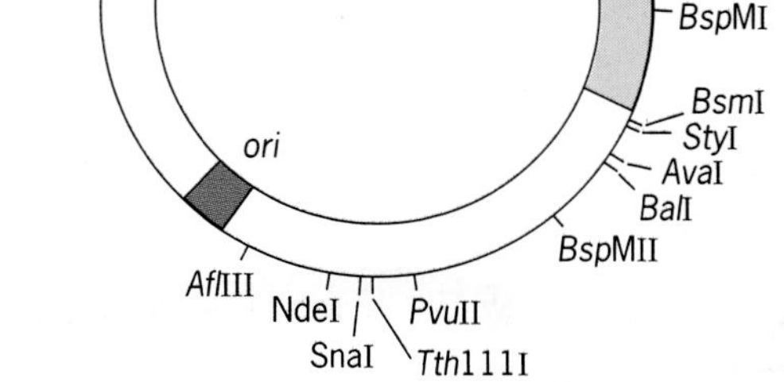

Figure 19.4 Structure of *E. coli* plasmid cloning vector pBR322, a circular DNA molecule 4.36 kb in size. The locations of unique restriction enzyme cleavage sites, the origin (*ori*) of replication, and the genes that confer resistance to the antibiotics ampicillin (*amp^r*) and tetracycline (*tet^r*) are shown on the map of the plasmid DNA molecule.

Bacteriophage vectors: Most bacteriophage cloning vectors have been constructed from the bacteriophage λ chromosome. The complete 48,502 nucleotide-pair sequence of the wild-type lambda genome is known, along with the functions of all of its genes. The central one-third of the λ chromosome contains genes that are required for lysogeny (Chapter 16), but not for lytic growth. Thus the central part (about 15 kb in length) of the λ chromosome can be excised with restriction enzymes and replaced with foreign DNA (Figure 19.5). The resulting recombinant DNA molecules can be packaged in phage heads *in vitro*. The phage particles can inject the recombinant DNA molecules into *E. coli* cells, where they will replicate and produce clones of the recombinant DNA molecules. DNA molecules that are too large or too small cannot be packaged in the lambda head; only molecules 45 to 50 kb in size are packaged. As a result, the lambda cloning vectors can only accommodate inserts of 10 to 15 kb.

Cosmid vectors: Some eukaryotic genes are larger than 15 kb in size and cannot be cloned intact in either plasmid or lambda cloning vectors. For this and other reasons, scientists have developed vectors that can accommodate larger DNA insertions. The first such vectors, called **cosmids** (for λ *cos* site and plasmid),

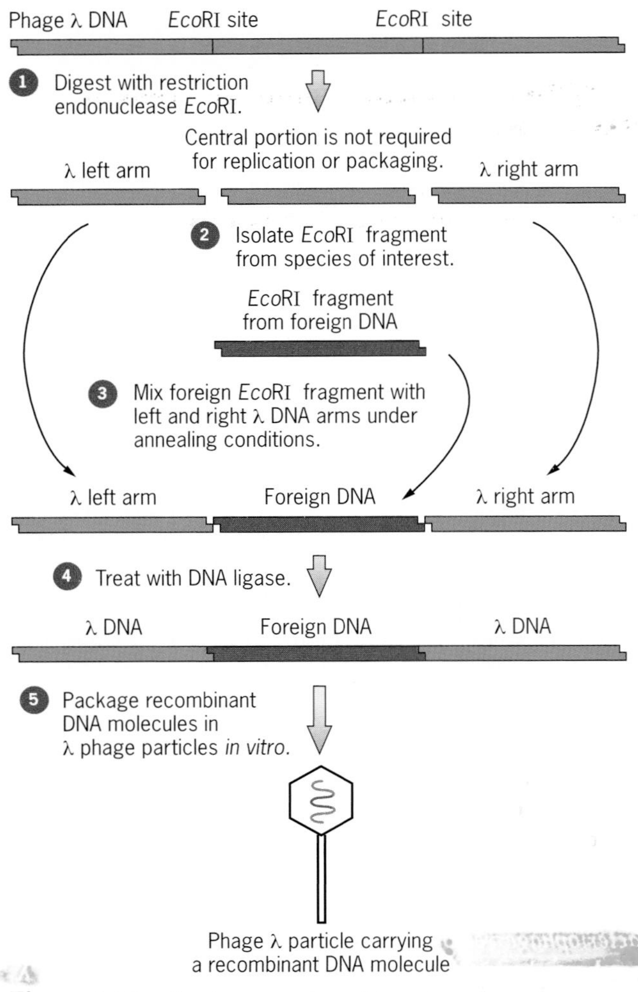

Figure 19.5 Strategy employed in using phage λ as a cloning vector.

were hybrids between plasmids and the phage λ chromosome. Cos stands for cohesive site, in reference to the 12-base complementary single-stranded termini on the mature λ chromosome (Chapter 10). The *cos* site is recognized by the phage λ DNA packaging apparatus, which makes staggered cuts at this site during packaging to produce the complementary cohesive ends of the mature lambda chromosome.

Cosmids combine the key advantages of plasmid and λ phage vectors; they possess (1) the plasmid's ability to replicate autonomously in *E. coli* cells and (2) the *in vitro* packaging capacity of the λ chromosome. The cosmid vectors (Figure 19.6) carry the origin of replication and one of the antibiotic-resistance genes of the parental plasmid and the λ *cos* site, which is needed to package DNA in λ heads, from the phage λ chromosome. The key advantage of cosmid vectors is their ability to accept foreign DNA inserts of 35 to 45 kb.

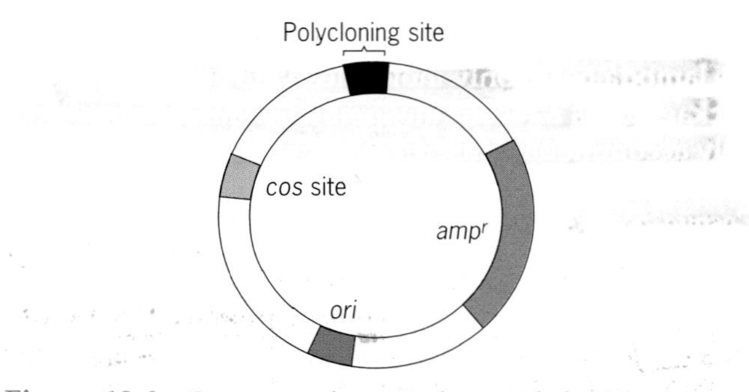

Figure 19.6 Structure of a typical cosmid cloning vector.

Eukaryotic and shuttle vectors: The plasmid, phage λ, and cosmid cloning vectors all replicate in *E. coli* cells. Because various taxonomic groups utilize distinct origins of replication and regulatory signals, different cloning vectors must be used in different species. Thus special cloning vectors have been developed to introduce recombinant DNA molecules into other prokaryotes and into eukaryotic organisms. Many unique cloning vectors are available for use in *S. cerevisiae*, *D. melanogaster*, mammals, plants, and other taxonomic groups.

E. coli is the host cell of choice for many of the manipulations performed on cloned DNAs. Thus, some of the most useful cloning vectors are **shuttle vectors** that can replicate in both *E. coli* and another species, such as a eukaryotic organism. Shuttle vectors designed for use in the yeast *S. cerevisiae* contain both *E. coli* and *S. cerevisiae* origins of replication and selectable marker genes, along with a polycloning site (Figure 19.7). Such shuttle vectors are extremely useful for genetic dissections. A yeast gene can be cloned in a shuttle vector, subjected to site-specific mutagenesis in an *E. coli* vector, and then moved back to yeast to examine the effects of the induced modifications in native host cells. Similar shuttle vectors are available for use in *E. coli* and various animal systems.

Artificial chromosomes: Some eukaryotic genes are very large. For example, the human dystrophin (an elongated protein that links filaments to membranes in muscle cells) gene is over 2000 kb in length. With the goal of cloning large segments of chromosomes, researchers have worked on the development of vectors that would accept DNA sequences longer than the 35 to 45 kb inserts of cosmid vectors. This research led to the development of **yeast artificial chromosomes (YACs)**, which can accommodate foreign DNA inserts of 200 to 500 kb. YAC vectors are genetically engineered yeast minichromosomes. They contain (1) a yeast origin of replication, (2) a yeast centromere, (3) two yeast telomeres, one at each end, (4) a selectable marker, and (5) a polycloning site (Figure 19.8). In yeast, origins of replication are called ***ARS*** (for Au-

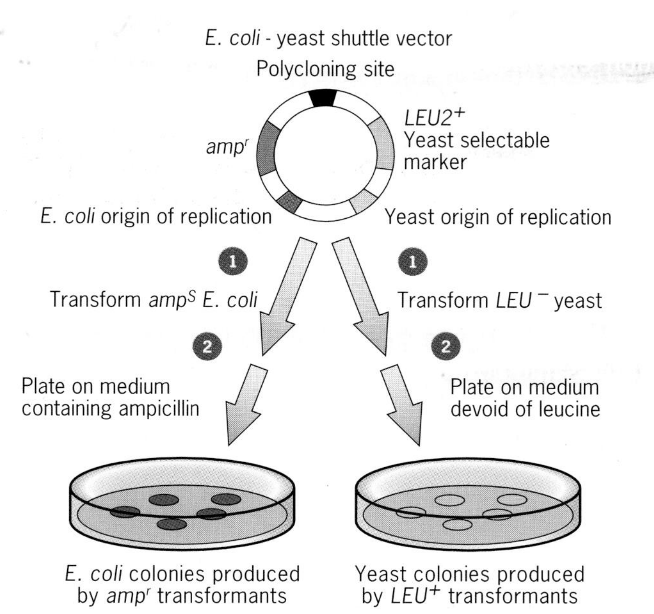

Figure 19.7 Basic structure and utility of an *E. coli*-yeast shuttle vector, which can replicate in both *E. coli* and *S. cerevisiae*. Shuttle vectors allow investigators to move genes back and forth between two organisms and study the structure or function of genes of interest in either host or in both hosts.

tonomously Replicating Sequence) elements. The selectable marker usually is a wild-type gene that confers prototrophy on the host cell. For example, *URA3*$^+$ can be used as the selectable marker in transformations of *URA3*$^-$ auxotrophs growing on medium lacking uracil.

YAC cloning vectors are especially valuable in investigations such as the Human Genome Project, which is designed to map, clone, and sequence large eukaryotic genomes (Chapter 20). The large DNA inserts make it much easier to identify and characterize clones covering an entire genome. If the average insert in cosmid vectors is 40 kb and the average insert in YAC vectors is 200 kb, five cosmid clones are needed to cover the region represented by one YAC clone. Because genome mapping projects require the isolation of overlapping clones, this difference in insert size translates into more than a fivefold difference in the effort required to obtain a complete physical map.

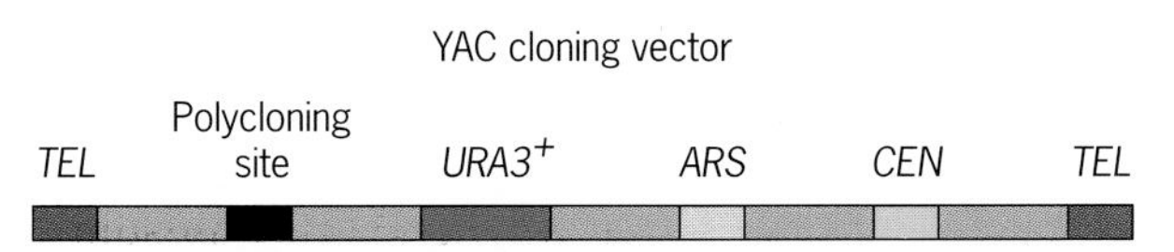

Figure 19.8 Structure of a YAC cloning vector. The components are: (1) *ARS*, autonomously replicating sequence (a yeast origin of replication), (2) *CEN*, a yeast centromere, (3) *TEL*, a yeast telomere, (4) *URA3*$^+$, a wild-type gene required for the biosynthesis of uracil, and (5) the polycloning site, containing the recognition and cleavage sites for a set of restriction endonucleases.

Recently, **bacterial artificial chromosomes** (**BACs**), which have many of the advantages of YACs, have been constructed from bacterial fertility (F) factors. Like YACs, BACs accept large inserts. However, BACs are less complex and, thus, easier to construct. In addition, BACs replicate in *E. coli* like plasmid, lambda, and cosmid vectors. Thus BAC vectors are beginning to replace YAC vectors in studies where large inserts are needed, such as in the construction of physical maps of entire chromosomes.

Key Points: **Recombinant DNA and gene-cloning techniques allow scientists to isolate and characterize virtually any gene or other DNA sequence from any organism. These techniques became possible with the discovery of restriction endonucleases, enzymes that recognize and cleave DNA in a sequence-specific manner. DNA sequences of interest are inserted into small, self-replicating DNA molecules called cloning vectors *in vitro*, and the resulting recombinant DNA molecules are amplified by replication *in vivo* after being introduced into cells, usually bacteria, by transformation. A variety of cloning vectors have been constructed, each with advantages for certain research purposes.**

CONSTRUCTION AND SCREENING OF DNA LIBRARIES

The first step in cloning a gene from an organism usually involves the construction of a **genomic DNA library**—a set of DNA clones collectively containing the entire genome. Sometimes, individual chromosomes of an organism are isolated by a procedure that sorts chromosomes based on size and DNA content. The DNAs from the isolated chromosomes are then used to construct **chromosome-specific DNA libraries**. The availability of chromosome-specific DNA libraries makes the search for a gene that is known to reside on a particular chromosome much easier, especially for organisms like humans with large genomes.

An alternative approach to gene cloning restricts the search for a gene to DNA sequences that are transcribed into mRNA copies. The RNA retroviruses (Chapter 15) encode an enzyme called reverse transcriptase, which catalyzes the synthesis of DNA molecules complementary to single-stranded RNA templates. The DNA molecules that are synthesized from RNA templates are called **complementary DNAs** (**cDNAs**). They can be converted to double-stranded cDNA molecules with DNA polymerases (Chapter

10), and the double-stranded cDNAs can be cloned in plasmid or phage λ vectors. By starting with mRNA, geneticists are able to construct a **cDNA library** that contains only the coding regions of the expressed genes of an organism.

Construction of Genomic Libraries

Genomic DNA libraries are usually prepared by isolating total DNA from an organism, digesting the DNA with a restriction endonuclease, and inserting the restriction fragments into an appropriate cloning vector. Two different procedures are used to insert the DNA fragments into the cloning vector. If the restriction enzyme that is used makes staggered cuts in DNA, producing complementary single-stranded ends, the restriction fragments can be ligated directly into vector DNA molecules cut with the same enzyme (Figure 19.9). An advantage of this procedure is that the foreign DNA inserts can be precisely excised from the vector DNA by cleavage with the restriction endonuclease used to prepare the genomic DNA fragments for cloning.

If the restriction enzyme cuts both strands of DNA at the same position, producing blunt ends, or if fragments are produced by shearing DNA molecules, complementary single-stranded ends must be added

1 Isolation of *E.coli* cosmid pJB8 DNA and mouse genomic DNA.

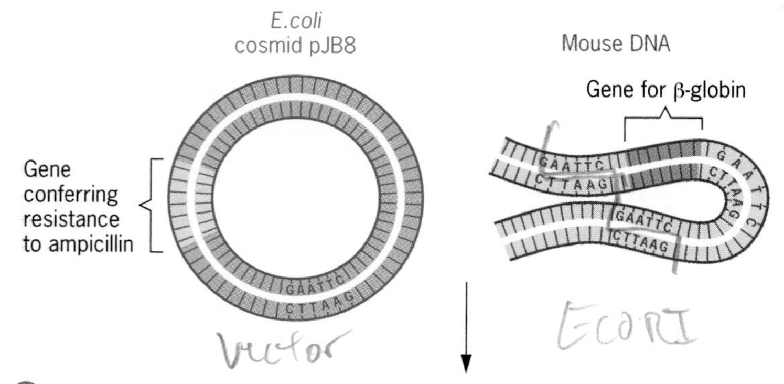

2 Cleave cosmid and mouse DNAs with restriction endonuclease *Eco*RI.

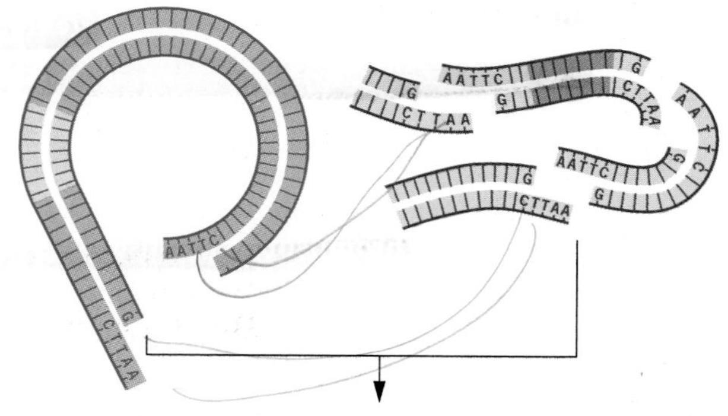

3 Mix cosmid and mouse DNAs under annealing conditions and treat with DNA ligase.

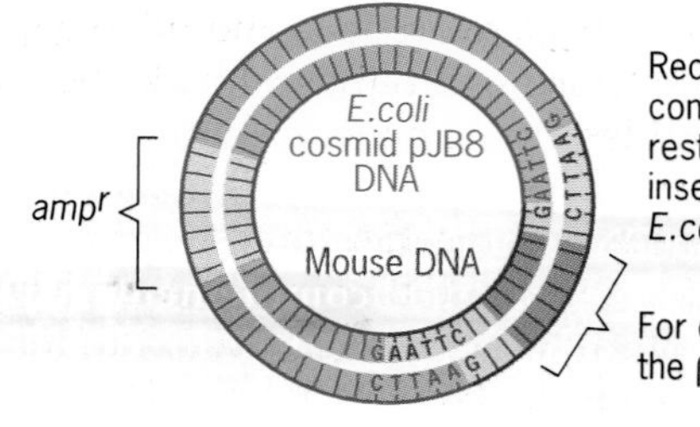

Figure 19.9 Procedure used to clone DNA restriction fragments with complementary single-stranded ends.

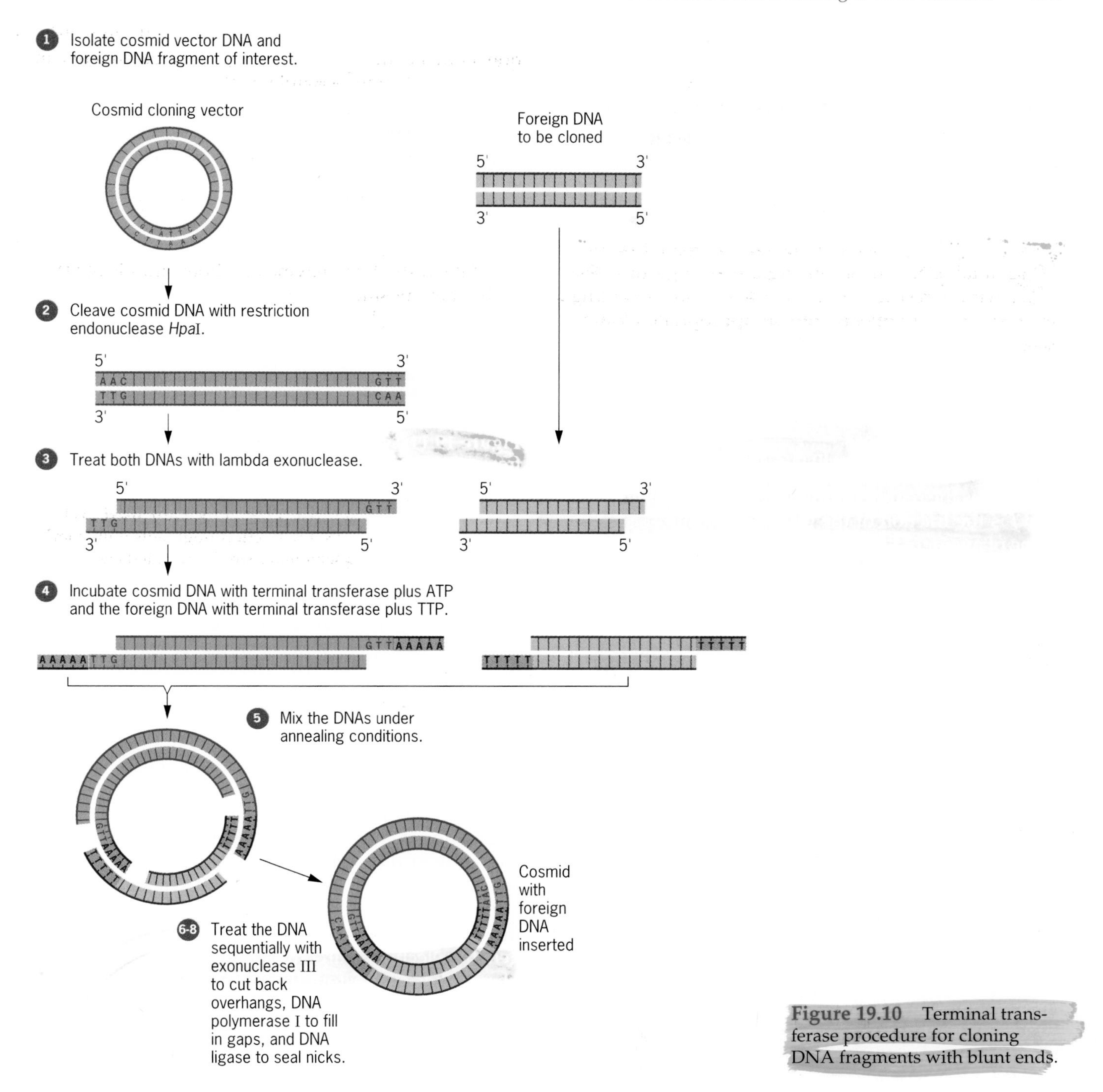

Figure 19.10 Terminal transferase procedure for cloning DNA fragments with blunt ends.

to the DNA fragments *in vitro*. This is accomplished by using the enzyme **terminal transferase** to add nucleotides to the 3' termini of the DNA strands. Usually, poly-A tails are added to the cleaved vector DNA, and poly-T tails are added to the genomic DNA fragments (Figure 19.10), or vice versa. Then, the T-tailed genomic DNA fragments are inserted into the A-tailed vector DNA molecules with DNA ligase. Since the T- and A-tails will not always be the same length, the *E. coli* enzymes exonuclease III and DNA polymerase I are used to cut back overhangs and fill in gaps, respectively. DNA ligase will only seal nicks between adjacent nucleotides; it will not add nucleotides if gaps are present.

Once the genomic DNA fragments are ligated into vector DNA, the recombinant DNA molecules must be introduced into host cells, usually *E. coli* cells, for amplification by replication *in vivo*. This step usually involves transforming antibiotic-sensitive recipient cells (Chapter 16). When *E. coli* is used, the bacteria must first be made permeable to DNA by treatment with a calcium salt or other chemicals. Transformed cells that harbor vector DNA are then selected by growing the cells under conditions where the selectable marker gene of the vector is essential for growth.

A good genomic DNA library contains essentially all of the DNA sequences in the genome of interest. For large genomes, complete libraries contain thousands to hundreds of thousands of different recombinant clones. If we assume that all DNA sequences have an equal chance of being inserted into a cloning vector, the number of clones required to produce a library that has a certain probability (P) of containing a given DNA sequence is given by the formula

$$N = \frac{\ln(1 - P)}{\ln(1 - f)}$$

where N is the required number of recombinant clones and f is the fraction of the genome present in a single clone of average size.

For example, suppose that you are constructing a cosmid library of the *Drosophila* genome, which is about 10^5 kb in size. If the average insert size is 40 kb, how many clones will be required to be 99 percent sure that a specific sequence is present in the library? The required number (N) of clones is 11,513, calculated by the equation

$$N = \frac{\ln[1 - 0.99]}{\ln[1 - (40/100{,}000)]}$$

Construction of cDNA Libraries

Most of the DNA sequences present in the large genomes of higher animals and plants are not expressed. Thus expressed DNA sequences can be identified more easily by working with complementary DNA (cDNA) libraries. Because most mRNA molecules contain 3' poly-A tails, poly-T oligomers can be used to prime the synthesis of complementary DNA strands by reverse transcriptase (Figure 19.11). Then, the RNA-DNA duplexes are converted to double-stranded DNA molecules by the combined activities of ribonuclease H, DNA polymerase I, and DNA ligase. Ribonuclease H degrades the RNA template strand, and short RNA fragments produced during degradation serve as primers for DNA synthesis. DNA polymerase I catalyzes the synthesis of the second DNA strand and replaces RNA primers with DNA strands, and DNA ligase seals the remaining single-strand breaks in the double-stranded DNA molecules. These double-stranded cDNAs can be inserted into plasmid or phage λ cloning vectors by the terminal transferase procedure shown in Figure 19.10.

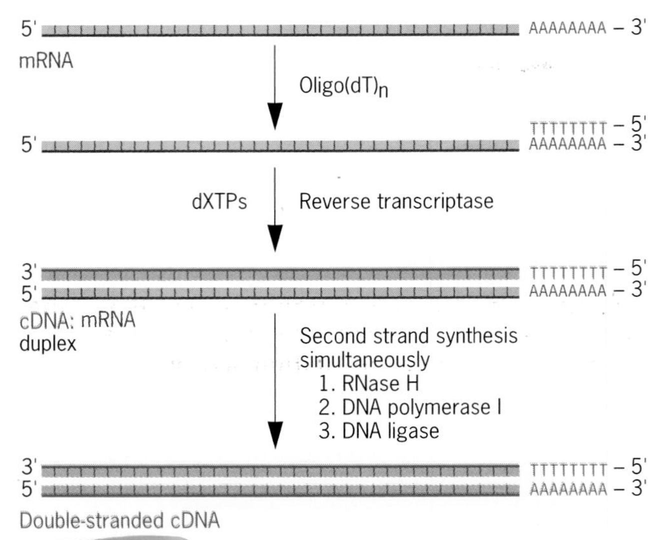

Figure 19.11 The synthesis of double-stranded cDNAs from mRNA molecules.

Screening DNA Libraries for Genes of Interest

The genomes of higher plants and animals are very large. For example, the human genome contains 3×10^9 nucleotide pairs. Thus searching genomic DNA or cDNA libraries of multicellular eukaryotes for a specific gene or other DNA sequence of interest requires the identification of a single DNA sequence in a library that contains a million or more different sequences. The most powerful screening procedure is genetic selection: searching for a DNA sequence in the library that can restore the wild-type phenotype to a mutant organism. When genetic selection cannot be employed, more laborious molecular screens must be carried out. Molecular screens usually involve the use of DNA or RNA sequences as hybridization probes or the use of antibodies to identify gene products synthesized by cDNA clones. When genetic selection is not available, the identification of the desired clone often is more difficult than the construction of the genomic or cDNA library.

Genetic selection: The simplest procedure for identifying a clone of interest is the use of **genetic selection**. For example, the *Salmonella typhimurium* gene that confers resistance to penicillin can be easily cloned. A genomic library is constructed from the DNA of a *pen*^{r} strain of *S. typhimurium*. Penicillin-sensitive *E. coli* cells are transformed with the recombinant DNA clones in the library and are plated on medium containing penicillin. Only the transformed cells harboring the *pen*^{r} gene will be able to grow in the presence of penicillin.

When mutations are available in the gene of interest, genetic selection can be based on the ability of the wild-type allele of a gene to restore the normal phenotype to a mutant organism. Although this type of selection is called **complementation screening**, it really depends on the dominance of wild-type alleles over mutant alleles that encode inactive products. For example, the genes of *S. cerevisiae* that encode histidine-biosynthetic enzymes were cloned by transforming *E. coli* histidine auxotrophs with yeast cDNA clones and selecting transformed cells that could grow on histidine-free medium. Indeed, many plant and animal genes have recently been identified based on their ability to complement mutations in *E. coli* or yeast.

One of the first genes of higher plants to be cloned by complementation screening encodes the enzyme glutamine synthetase in maize. Glutamine synthetase catalyzes the final step in the biosynthesis of the amino acid glutamine. Because glutamine is one of the 20 amino acids specified by the genetic code, it is essential for growth in all cellular organisms. Thus an *E. coli* cell requires glutamine for growth, but it does not care whether this glutamine is synthesized by a corn enzyme, a human enzyme, or an *E. coli* enzyme. Initially, glutamine synthetase cDNA clones from corn were identified on the basis of their ability to rescue glutamine synthetase-deficient *E. coli* mutants growing on medium lacking glutamine. The maize glutamine synthetase cDNA clone was then used to identify the corresponding genomic clone (the gene) from a genomic DNA library by the procedure outlined in the next section.

Complementation screening has some obvious limitations. Eukaryotic genes contain noncoding introns, which must be spliced out of gene transcripts prior to their translation. Because *E. coli* cells do not possess the machinery required to excise introns from eukaryotic genes, complementation screening of eukaryotic clones in *E. coli* is restricted to cDNAs, which have the intron sequences already spliced out. In addition, the complementation screening procedure depends on the correct transcription of the cloned gene in the new host. Eukaryotes have signals that regulate gene expression that are different from those in prokaryotes; therefore, the complementation approach is more likely to work with prokaryotic genes in prokaryotic organisms and eukaryotic genes in eukaryotic organisms. For this reason, researchers often use *S. cerevisiae* to screen eukaryotic DNA libraries by the complementation procedure.

Molecular hybridization: The first eukaryotic DNA sequences to be cloned were genes that are highly expressed in specialized cells. These genes included the mammalian α- and β-globin genes and the chicken ovalbumin gene. Red blood cells are highly specialized for the synthesis and storage of hemoglobin. Over 90 percent of the protein molecules sythesized in red blood cells during their period of maximal biosynthetic activity are globin chains. Similarly, ovalbumin is a major product of chicken oviduct cells. As a result, RNA transcripts of the globin and ovalbumin genes can be easily isolated from reticulocytes and oviduct cells, respectively. These RNA transcripts can be employed to synthesize radioactive cDNAs, which, in turn, can be used to screen genomic DNA libraries by ***in situ* colony** or **plaque hybridization** (Figure 19.12). Colony hybridization is used with libraries constructed in plasmid and cosmid vectors; plaque hybridization is used with libraries in phage lambda vectors. We shall focus on *in situ* colony hyridization here, but the two procedures are virtually identical.

The colony hybridization screening procedure involves replica plating the colonies formed by transformed cells onto nitrocellulose membranes, hybridization with a radioactive DNA or RNA probe, and autoradiography (Figure 19.12). If RNA transcripts of the gene of interest can be isolated from specialized cells, the RNAs can be used as templates for the *in vitro* synthesis of radioactive cDNAs by reverse transcriptase. Alternatively, if a cDNA clone can be isolated by genetic selection, as was done for the maize glutamine synthetase gene, the cDNA can be used to screen a genomic DNA library by colony hybridization. The labeled cDNA is employed as a probe for hybridization to denatured DNA from colonies grown on the nitrocellulose membranes. The DNA from the lysed cells is bound to the nitrocellulose membrane before hybridization so that it won't come off during subsequent steps in the procedure. After time is allowed for hybridization between complementary strands of DNA, the membranes are washed with buffered salt solutions to remove nonhybridized cDNA and are exposed to X-ray film to detect the presence of radioactivity on the membrane. Only colonies that contain DNA sequences complementary to the radioactive cDNA will yield radioactive spots on the autoradiographs (Figure 19.12). The locations of the radioactive spots are used to identify colonies that contain the desired sequence on the original replicated plates. These colonies are used to purify DNA clones harboring the gene or DNA sequence of interest.

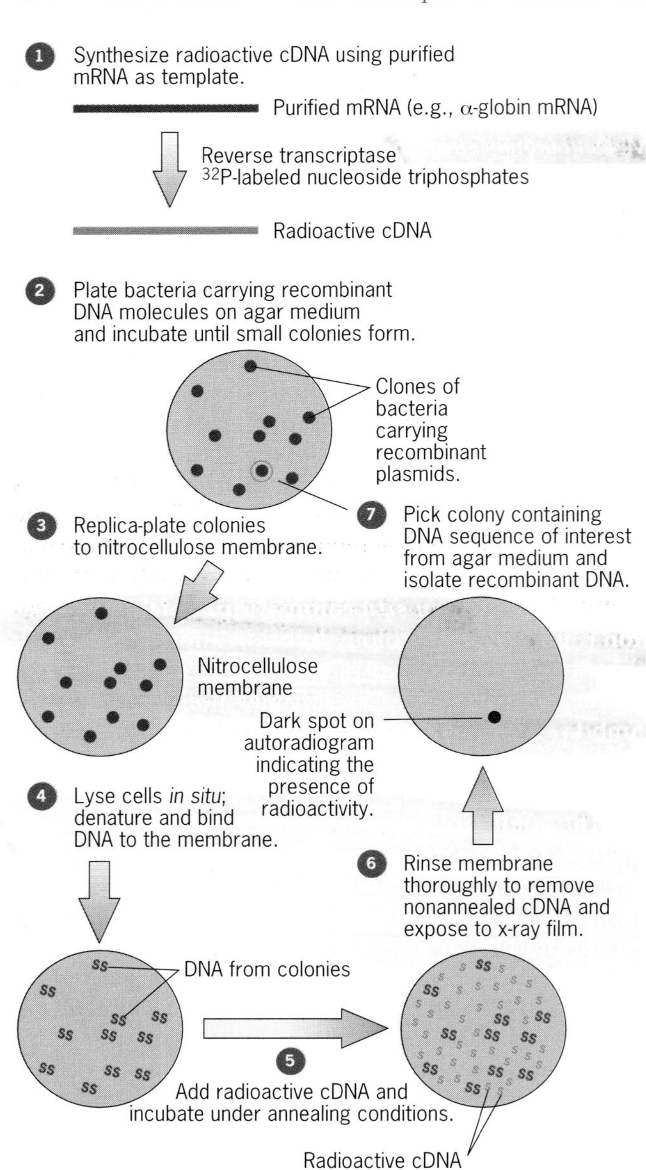

Figure 19.12 Screening DNA libraries by colony hybridization. Here, a radioactive cDNA is employed as a hybridization probe. Frequently, a purified restriction fragment is used as the hybridization probe. The colony hybridization procedure allows researchers to identify bacterial colonies harboring DNA sequences that hybridize to the radioactive DNA probe. These colonies are then isolated and used to prepare large quantities of the recombinant DNA molecules of interest.

Biological and Physical Containment of Recombinant DNA Molecules

As the methods for constructing recombinant DNA molecules and cloning were being developed, molecular biologists recognized that this technology would provide a powerful tool with which to study the structure and function of genetic material. They also wondered whether it might have potential hazards. For example, some retroviruses carry genes that participate in the transformation of animal cells to the malignant state. If these genes were cloned in *E. coli* plasmids, might the *E. coli* strains carrying the viral genes serve as vectors for the transmission of cancer-causing genes? With such possibilities in mind, several of the leading molecular biologists called for a voluntary moratorium on research of this type. Numerous conferences were held to discuss whether potential hazards existed, and, if so, what precautions should be taken to minimize the possibility of creating pathogenic recombinant organisms. A major concern of opponents of recombinant DNA work was that a pathogenic or otherwise harmful new organism might be introduced into the ecosystem. Proponents argued against the likelihood of this hazard and emphasized the benefits of recombinant DNA research.

After extensive discussions, the National Institutes of Health (NIH) of the U.S. Department of Health, Education, and Welfare (now the Department of Health and Human Services) established specific guidelines under which recombinant DNA research of various types would be done. These guidelines emphasized both physical and biological containment of the recombinant molecules constructed. **Physical containment** includes the use of sterile techniques, containment hoods, and specially designed laboratories to prevent vectors containing recombinant DNA molecules from escaping into natural ecosystems. **Biological containment** involves the use of organisms with specially constructed, weakened genotypes as vectors in cloning experiments. Ideally, these organisms should be unable to survive under conditions existing in any natural ecosystem.

As more work with recombinant DNA was performed, it became evident that bacteria and viruses that carry foreign genes are simply not very healthy. They have been found not to survive in competition with wild-type organisms under natural ecosystem conditions. Thus the NIH guidelines have been relaxed for the more routine types of gene-cloning experiments. Other kinds of experiments still require approval by an NIH panel and by a local institutional biosafety committee prior to their initiation.

Key Points: **DNA libraries can be constructed that contain complete sets of genomic DNA sequences or DNA copies (cDNAs) of mRNAs in an organism. Specific genes or other DNA sequences can be isolated from these libraries by genetic complementation and by hybridization to labeled nucleic acid probes containing DNA sequences of known function.**

THE MANIPULATION OF CLONED DNA SEQUENCES *IN VITRO*

Recombinant DNA technology has made huge contributions to biological and medical research. New information about the structures of eukaryotic genes and chromosomes has accumulated at an unprecedented rate since the development of these techniques. Practical applications also have been realized; human insulin and human growth hormone are now produced in genetically engineered bacteria. For diabetics who are allergic to nonhuman insulins or for individuals with defects in the synthesis of human growth hormone, these early achievements of scientists using recombinant DNA methodologies are of major significance.

As would be expected following the discovery of a major new research tool, technical advances have accumulated rapidly. Researchers can now clone and characterize any gene or other DNA sequence of interest from virtually any organism. These cloned genes can be modified *in vitro* and reintroduced into the same or different species for further study. A few of the most important methods for manipulating DNA are discussed in the following sections.

Phagemids: The Biological Purification of DNA Single Strands

Single-stranded DNA provides the optimal substrate for DNA sequencing experiments and is required for certain *in vitro* mutagenesis protocols. Several vectors have been constructed that provide a simple biological mechanism for purifying large amounts of single strands of cloned genes. The most useful of these are hybrid vectors called **phagemids**, which contain components from both phage chromosomes and plasmids. These vectors replicate in *E. coli* as normal double-stranded plasmids until a **helper phage** is provided. After addition of the helper phage, they switch to the phage mode of replication and package single strands of DNA in phage particles. The helper phage is a mutant that replicates its own DNA inefficiently, but provides viral replication enzymes and structural proteins for the production of phagemid DNA molecules that are packaged in phage coats.

Before discussing the phagemid vectors further, let's examine the life cycle of the filamentous phages that contain single-stranded DNA. The best-known of these phages are **M13, fl,** and **fd**, all of which have long threadlike morphologies and reproduce in *E. coli* cells. Their single-stranded DNA genomes replicate by the rolling-circle mechanism (Chapter 10). The filamentous single-stranded DNA phages infect cells by absorbing to and entering through F pili; thus these phages only infect F^+ or Hfr cells, not F^- cells. Furthermore, these phages do not lyse the host cells like phage T4 (Chapter 15). Instead, the progeny viruses are extruded through the cell membrane and cell wall without killing the host cell. Infected cells continue to grow and extrude thousands of progeny virus particles, each containing a single-stranded genome, into the medium. Because the virus particles are much smaller than the host cells, the bacteria can be removed by low-speed centrifugation. The virus particles can then be collected from the supernatant suspension by high-speed centrifugation, and their single-stranded DNA molecules can be isolated by simple phenol-chloroform extractions. The same DNA strand of the virus is always packaged; it is called the + strand, and its complement is the – strand. The packaged + strand has the same sense as mRNA; its nucleotide triplets correspond to the mRNA codons, but with T in place of U.

The major features of the M13 life cycle are shown in Figure 19.13. Note that *the packaging of single strands of phage DNA in progeny phage provides a neat biological*

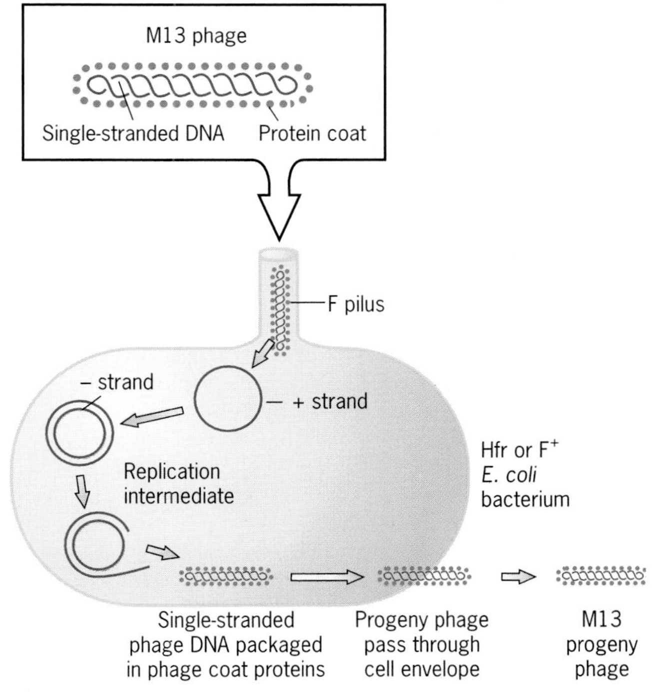

Figure 19.13 The life cycle of bacteriophage M13. The genetic information of the virus is stored in single-stranded DNA packaged within a long threadlike virion. The virus enters the cell through an F pilus. After the coat proteins are removed from the DNA, it replicates by the rolling-circle mechanism. Progeny single strands of DNA are packaged in new coats, and the progeny virions are extruded through the cell envelope without killing the host cell.

purification of single-stranded DNA. This is the case for a foreign gene cloned in the viral chromosome, just as it is for the phage genes themselves. The phagemid cloning vectors take advantage of this unique aspect of M13 reproduction.

The phagemid vectors **pUC118** and **pUC119** are virtually identical, but they contain the polycloning region in opposite orientations (turned end-for-end) relative to the rest of the genes of the vector (Figure 19.14). Thus, if a foreign DNA is inserted into a specific restriction site in both vectors, one vector will package one strand of the foreign DNA, and the other vector will package the complementary strand. Both strands of the cloned DNA can therefore be isolated, sequenced, subjected to site-specific mutagenesis, and so on. These vectors were designated **pUC** for plasmid and University of California, where they were constructed. Vectors pUC118 and pUC119 differ from earlier vectors in the pUC series by the addition of the origin of replication from phage M13. This distinction permits pUC118 and pUC119 to replicate either (1) as a double-stranded plasmid in the absence of helper phage or (2) as a single-stranded DNA, which is packaged in M13 phage coats and extruded from the cell, in the presence of helper phage. In the absence of helper phage, replication is controlled by the plasmid origin of replication. In the presence of helper phage, replication is directed by the phage M13 origin of replication.

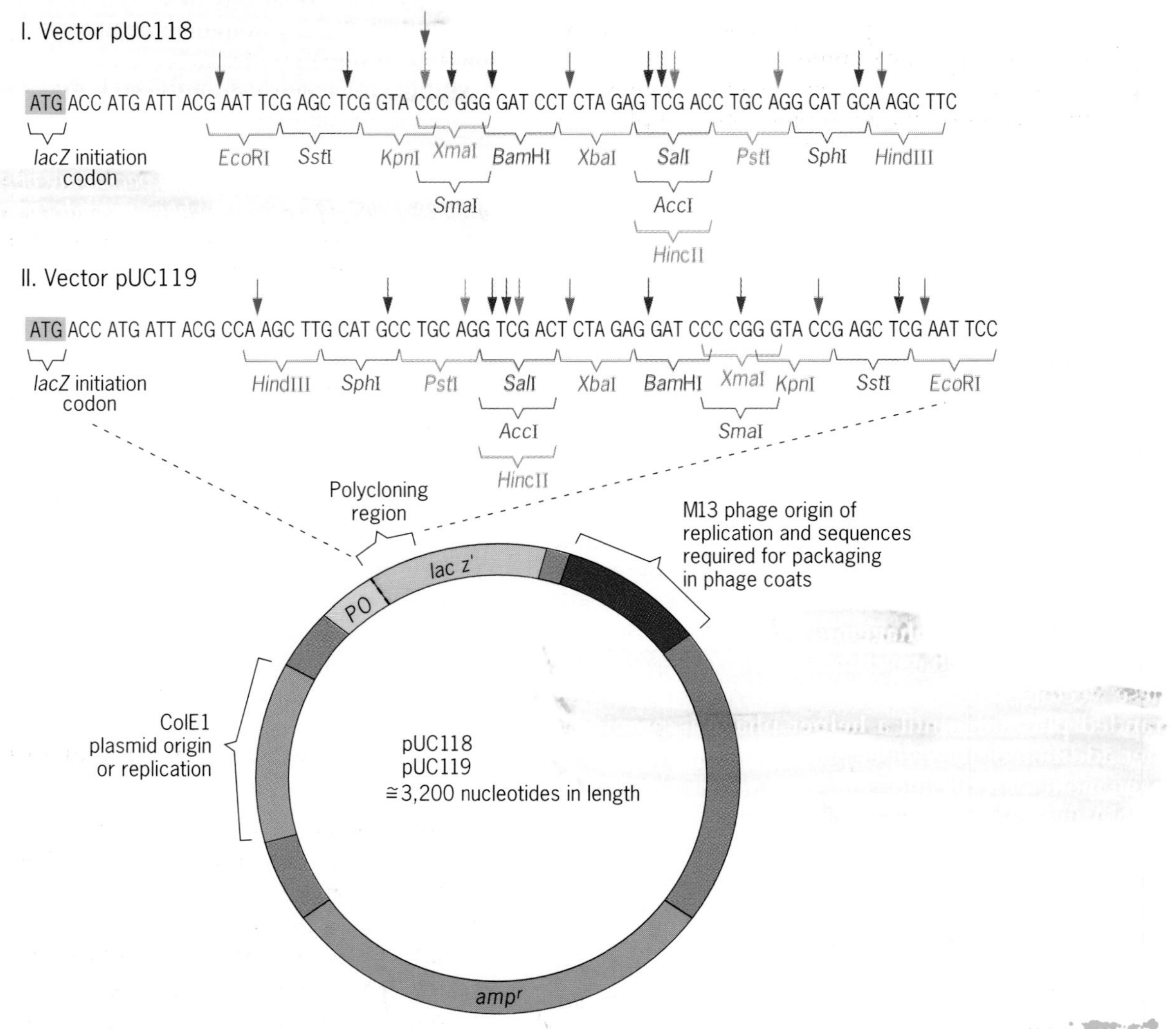

Figure 19.14 Important components of the phagemid vectors pUC118 and pUC119. *P* and *O* are the promoter (RNA polymerase binding site) and operator (repressor binding site) regions that regulate the transcription of the lactose biosynthetic enzymes (Chapter 21). Gene segment *lacZ'* encodes the amino terminal 147 amino acids of β-galactosidase. The polycloning site lies within *lacZ'* and contains a cluster of restriction enzyme cleavage sites (top). The ColE1 plasmid origin controls replication in the absence of helper phage, whereas the M13 origin controls replication in the presence of helper phage.

The utility of the pUC vectors is greatly enhanced by a simple color test that allows cells harboring plasmids with foreign DNA inserts to be distinguished from those harboring plasmids with no insert. The basis of this color indicator test is the functional inactivation of the 5' segment of the *lacZ* gene, which is present in the vector, by the insertion of foreign DNA into the polycloning region.

The *E. coli lacZ* gene encodes β-galactosidase, the enzyme that cleaves lactose into glucose and galactose. This is the first step in the catabolism of lactose in *E. coli* (Chapter 21). The presence of β-galactosidase in cells can be monitored on the basis of its ability to cleave the substrate 5-bromo-4-chloro-3-indolyl-β-D-galactoside (usually called Xgal) to galactose and 5-bromo-4-chloroindigo. Xgal is colorless; 5-bromo-4-chloroindigo is blue. Thus cells containing active β-galactosidase produce blue colonies on agar medium containing Xgal. Cells lacking β-galactosidase activity produce white colonies on Xgal plates.

The molecular basis of the β-galactosidase activity that provides the color indicator test for pUC vectors is somewhat more complex. The *lacZ* gene of *E. coli* is over 3000 nucleotide pairs long, and placing the entire gene on the plasmid would make the vector larger than desired. The pUC vectors contain only a small part of the *lacZ* gene. This *lacZ* gene segment encodes only the amino terminal portion of β-galactosidase. However, the presence of a functional copy of the *lacZ* gene segment can be detected because of a unique type of intragenic complementation. When a functional copy of the *lacZ* gene segment on the pUC plasmid is present in a cell that contains a particular *lacZ* mutant allele on the chromosome or on an F' plasmid, the two defective *lacZ* sequences yield polypeptide products that complement each other and produce β-galactosidase activity. The mutant allele, designated *lacZΔM15*, synthesizes a *lac* protein that lacks amino acids 11 through 14 from the amino terminus. The absence of these amino acids prevents the mutant polypeptides from interacting to produce the active tetrameric form of the enzyme.

By some mechanism that is still not completely understood, the presence of the amino terminal fragment (the first 147 amino acids) of the *lacZ* polypeptide encoded by the *lacZ* gene fragment on pUC plasmids facilitates tetramer formation by the ΔM15 deletion polypeptides. This yields active β-galactosidase, which can be detected by the Xgal color test. The unique type of intragenic complementation observed between the amino terminal fragment of β-galactosidase and the defective ΔM15 polypeptide is called **α complementation**. It conveniently permits the Xgal color test to be utilized without placing the entire *lacZ* gene in the pUC vectors. However, for this color test to work with these vectors, the host cell must contain a mutant *lacZΔM15* gene on the chromosome, or, more commonly, on an F' plasmid in a cell with a deletion of the chromosomal copy of the *lacZ* gene.

The standard protocol for cloning foreign DNAs into pUC vectors involves (1) ligation of the foreign DNA of interest into one of the restriction enzyme cleavage sites in the polycloning region of the vector, (2) transformation of cells of an appropriate *amp^s^ E. coli* strain carrying the *lacZΔM15* gene with the ligation products, and (3) plating the transformed cells in plates with a nutrient agar medium containing Xgal and ampicillin. Only transformed cells that contain a pUC plasmid will be able to grow in the presence of ampicillin. Colonies produced by cells harboring pUC plasmids without foreign DNA inserts will be blue. Colonies produced by bacteria harboring pUC vectors with foreign DNA inserts will be white because of the functional inactivation of the *lacZ* gene segment by the inserted DNA. Thus white colonies are picked and used for further characterization of foreign DNAs.

The ability to perform **forced cloning**, the insertion of a foreign DNA into the polycloning region in a predetermined orientation, is another important advantage of the pUC118-119 vector system. Consider a DNA sequence that has a *Sst*I site at one end and a *Pst*I site at the other end. If this DNA is cleaved with both enzymes, the resultingl *Sst*I-*Pst*I fragment can be inserted into either pUC118 DNA or pUC119 DNA that has been cut with *Sst*I and *Pst*I (Figure 19.14). This *Sst*I-*Pst*I fragment will be present in pUC118 and pUC119 in opposite orientations. Thus, if a gene or other DNA fragment is force cloned in both pUC118 and pUC119, one vector will package one strand of DNA and the other vector will package the complementary strand of DNA after the infection of host cells with helper phage (Figure 19.15). Thus, by using both vectors, each of the two complementary single strands of a cloned gene or other DNA sequence of interest can be isolated in large quantities.

In summary, the pUC118 and pUC119 cloning vectors (Figure 19.14) have several important features.

1. They contain *small, supercoiled, covalently closed circular DNA*; thus they are easily isolated and manipulated *in vitro*.
2. They carry the *amp^r^ gene as a selectable marker*; thus only bacteria harboring a plasmid will grow on medium containing ampicillin.
3. They are present in *high copy number*, up to 700 copies per bacterium; thus small cell cultures will yield large amounts of DNA.
4. They contain a *polycloning region* with several restriction enzyme cleavage sites; thus many different types of restriction fragments can be inserted.

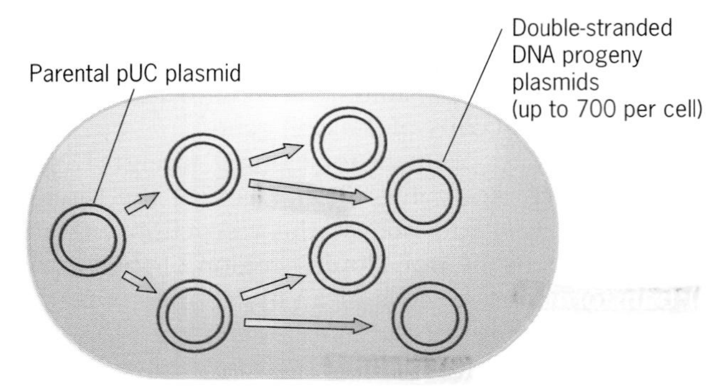

(*a*) Replication of pUC118 and pUC119 as double-stranded plasmid DNAs in the absence of helper phage.

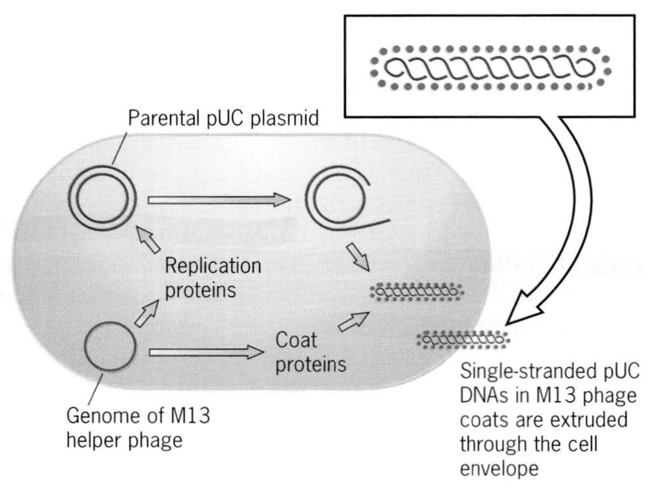

(*b*) Replication of pUC118 and pUC119 as single-stranded phage DNAs in the presence of helper phage.

Figure 19.15 The plasmid and phage modes of replication of the pUC118 and pUC119 phagemid vectors. (*a*) In the absence of M13 helper phage, replication is controlled by an origin of replication derived from plasmid ColE1. (*b*) In bacteria infected with an M13 helper phage, rolling-circle replication is controlled by the M13 origin of replication. The chromosome of the helper phage contains a modified origin of replication that results in inefficient replication; thus most of the extruded phage particles contain single-stranded pUC DNA molecules.

5. *The polycloning region interrupts the coding region of the 5' end of the E. coli lacZ gene,* which encodes β-galactosidase; thus colonies harboring plasmids with foreign DNA inserts can be distinguished from those carrying plasmids with no insert by a simple color test.
6. *The lacZ gene is under the control of the lac promoter;* thus genes inserted in frame (with codons in the proper reading frame) can be expressed to produce β-galactosidase-foreign protein fusion products.
7. They carry a *plasmid origin of replication;* thus replication in the absence of helper phage produces large amounts of double-stranded plasmid DNA.
8. They carry a *phage M13 origin of replication;* thus single-stranded DNA is produced and packaged in phage coats when helper phage is present.
9. *The polycloning regions are present in opposite orientations in pUC118 and pUC119;* thus if pUC118 packages one strand of a cloned gene, pUC119 will package the complementary strand.

Transcription Vectors: The Synthesis of RNA Transcripts *In Vitro*

Several bacteriophages, such as the *E. coli* phages T3 and T7 and *B. subtilis* phage SP6, encode their own RNA polymerases. The phage gene specifying the RNA polymerase is transcribed by the host cell's RNA polymerase, but then the viral "late proteins" or structural components of the progeny viruses are produced from genes that are transcribed by the respective phage RNA polymerase. These phage RNA polymerases transcribe only genes with unique phage-specific promoter sequences. They do not transcribe any genes of the host bacterium. Moreover, each phage RNA polymerase is highly specific for its own promoter sequence, in particular for a conserved sequence of 18 to 22 nucleotide pairs just upstream from the 5' GTP transcription start site (Chapter 11).

By adding T3, T7, or SP6 promoter sequences adjacent to the polycloning regions of vectors like pUC118 and pUC119, scientists have constructed vectors that allow foreign DNA inserts to be transcribed *in vitro* by the appropriate viral RNA polymerase. Because of the specificity of the viral RNA polymerase for its own promoter sequence, only the inserted gene or DNA of interest will be transcribed *in vitro.* No other plasmid genes will be adjacent to that particular phage promoter; thus none of the plasmid genes will be transcribed.

A number of *in vitro* transcription vectors have been constructed. Most commonly, they contain promoters for two different viral RNA polymerases on opposite sides of the polycloning region. These promoters are oriented so as to direct transcription of foreign DNAs located within the polycloning region. One set of transcription vectors, called pBS(+) and pBS(–)—one packages the + strand of DNA, the other the – strand—is essentially identical to pUC118 and pUC119 except for the addition of two phage promoter sequences. The general structure of a transcription vector such as pBS(+) is shown in Figure 19.16. Vector pBS(+) contains a T7 promoter on one side of the polycloning region and a T3 promoter in the opposite orientation on the other side of the polycloning re-

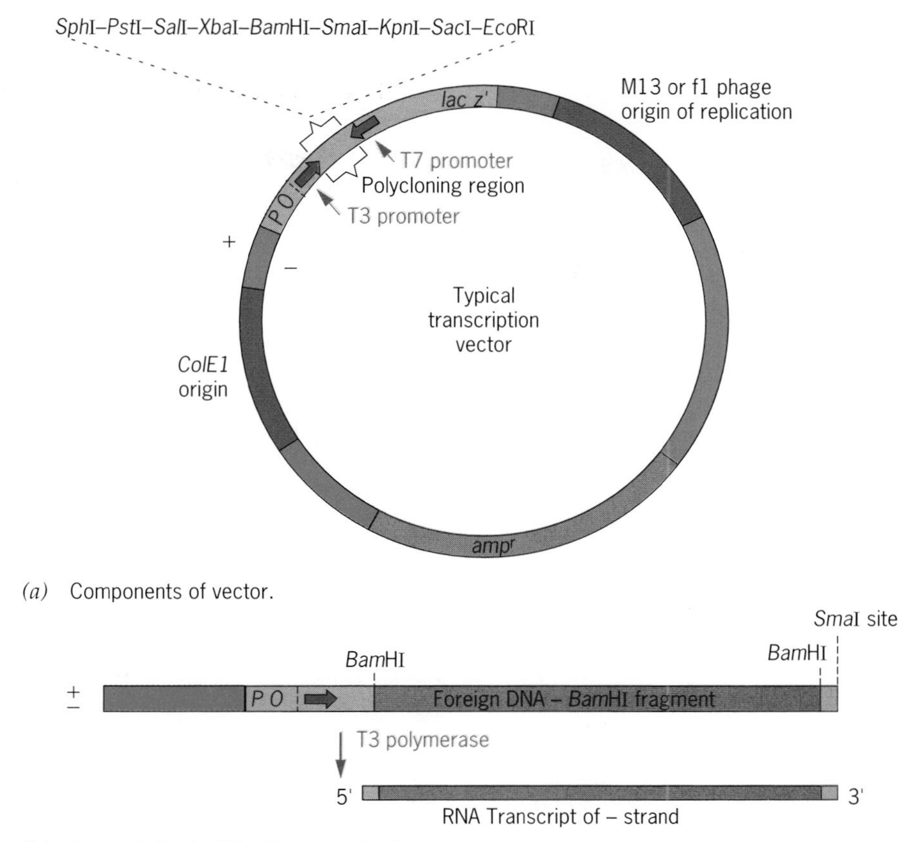

(*a*) Components of vector.

(*b*) Transcription by T3 polymerase *in vitro.*

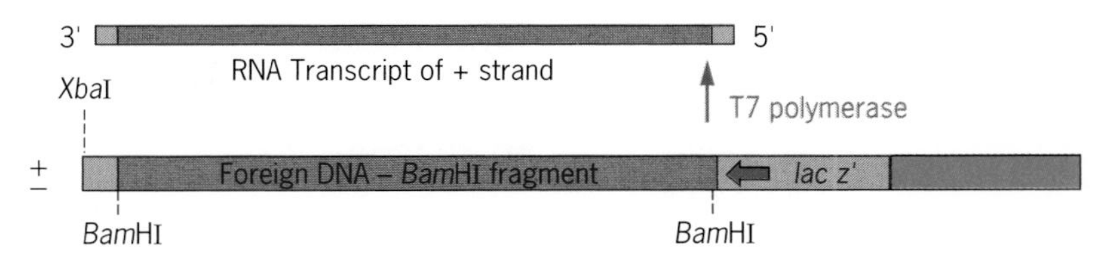

(*c*) Transcription by T7 polymerase *in vitro.*

Figure 19.16 Structure of a typical transcription vector (*a*) and *in vitro* transcription of the – strand of a foreign DNA by T3 RNA polymerase (*b*) or the + strand by T7 RNA polymerase (*c*). Components of the vector are essentially as described for the pUC118 and pUC119 vectors in Figure 19.14 except for the addition of T3 and T7 promoter sequences flanking the polycloning region. In the vector shown, T3 RNA polymerase will synthesize a transcript of the – strand after the vector has been linearized by cleavage with *Sma*I, whereas T7 RNA polymerase will synthesize a transcript of the + strand after the vector has been linearized by cleavage with *Xba*I.

gion. These promoters direct transcription through the polycloning region and any foreign DNA that is inserted in it. One promoter directs transcription with one strand of the DNA as template; the other directs transcription using the complementary strand as template. Because the foreign DNA insert may not contain a transcription termination signal, the insert-vector DNA is usually linearized by cleavage with a restriction enzyme that cuts the polycloning region just distal to the insert from the promoter being used (Figure 19.16 *b,c*). This procedure prevents transcription from continuing beyond the insert into vector sequences.

Transcription vectors are used to prepare radioactive RNA hybridization probes. These RNA probes are used in genomic and cDNA library screens, in various analyses of genome structure, and in studies of gene expression. The RNA transcripts can also be translated *in vitro*, facilitating the study of proteins encoded by the cloned genes. The resulting proteins are usually **fusion proteins** containing an amino terminal β-galactosidase peptide joined to the product of the cloned gene. Of course, proper translation requires that the coding sequence of the foreign gene be fused to the *lacZ'* gene segment in the correct reading frame.

For a randomly inserted DNA sequence, this will occur one-third of the time. In the other cases, the reading frame has to be modified by the addition or deletion of bases or by inserting linkers (see the following section). The present recombinant DNA technology is so advanced that DNA sequences and reading frames can be modified at will by a variety of procedures.

Joining DNAs with Linker and Adapter Molecules

The powerful tools of recombinant DNA and gene-cloning technologies are based almost exclusively on utilizing the sequence specificity of restriction endonucleases. In dissecting genes, chromosomes, and

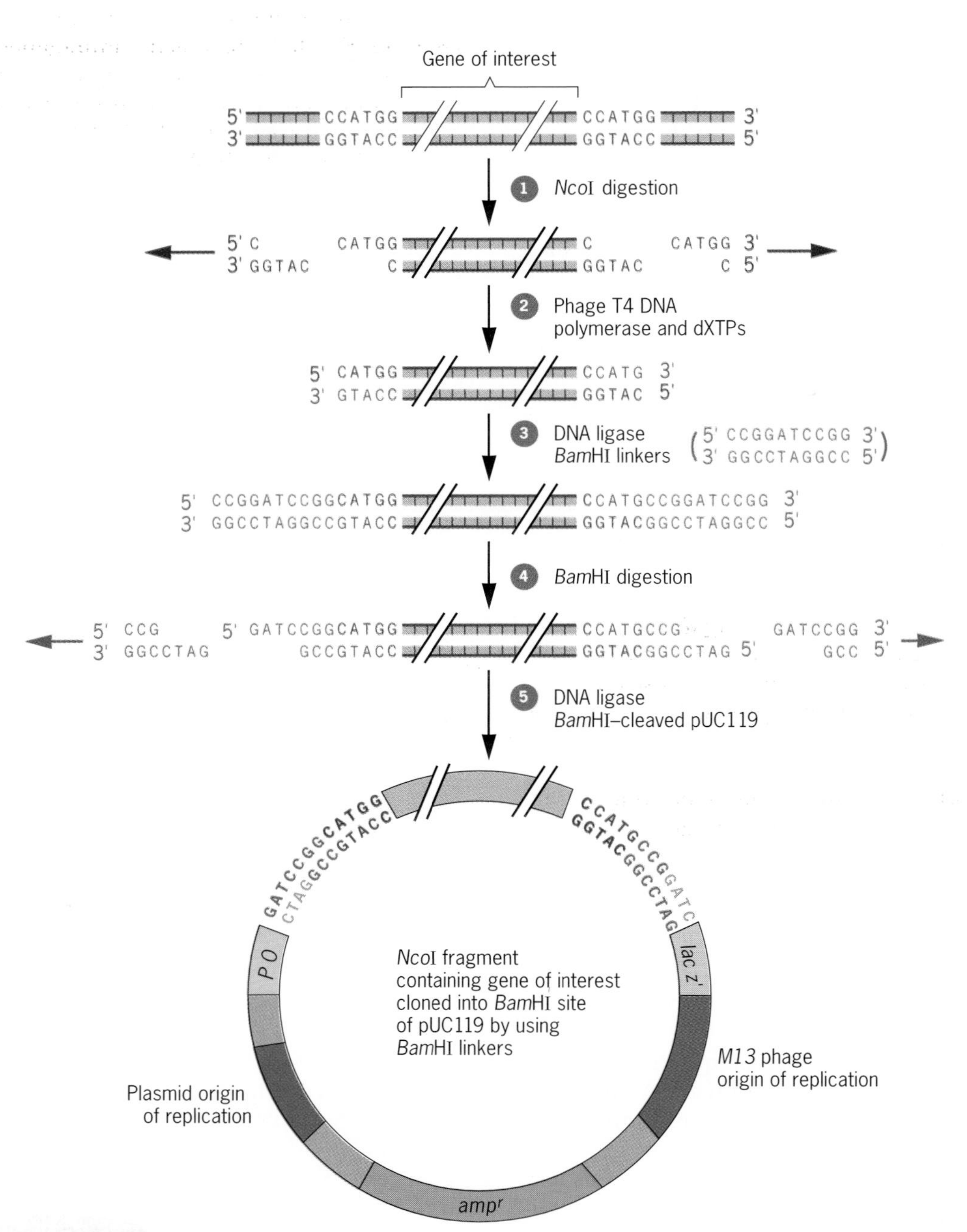

Figure 19.17 Use of *Bam*HI linkers to clone a gene on an *Nco*I restriction fragment into the *Bam*HI site in the polycloning region of pUC119. The phage T4 DNA polymerase I contains 5 —> 3' polymerase activity and 3' —> 5' proofreading exonuclease activity, but no 5 —> 3' exonuclease activity. Thus 5' ends of DNA strands remain intact, and the 3' ends are filled in by the polymerase until the template is exhausted. This process yields blunt-ended molecules that can be ligated to synthetic linkers.

genomes of various organisms, molecular biologists rely on their ability to cut DNA molecules at desired positions with restriction enzymes. Sometimes the cleavage sites of choice are not present at the desired locations. To circumvent this problem, researchers have taken two different approaches. One approach is to change the native DNA sequence at this position to the desired sequence—that is, to add the restriction enzyme cleavage site of choice. This is done by the procedure of site-specific mutagenesis, which is discussed in the next section of this chapter. The other approach is to add synthetic oligonucleotides containing the restriction sites of choice to the ends of DNA molecules produced by cleavage with a different enzyme.

Suppose that a gene is flanked by two *Nco*I cleavage sites and that we want to clone this gene into pUC119. Vector pUC119 does not contain an *Nco*I site. This lack of an *Nco*I site presents no major obstacle to cloning the gene in pUC119, in fact, into any one of several sites in the pUC119 polycloning region. Suppose that we want to clone this gene into the *Bam*HI site of pUC. We can do this by (1) excising the gene from its native site in the chromosome with *Nco*I, (2) isolating the *Nco*I restriction fragment, (3) filling in the 3' ends *in vitro* by using phage T4 DNA polymerase to produce blunt-ended molecules, (4) ligating *Bam*HI linkers to the blunt ends, (5) cleaving the *Bam*HI sites in the linkers by digestion with *Bam*HI, and (6) inserting the *Bam*HI fragment that has been produced into *Bam*HI-cut pUC119 (Figure 19.17). **Linkers** are short, double-stranded DNAs that contain restriction enzyme cleavage sites; they are synthesized from mononucleotides by simple protocols of organic chemistry.

Other manipulations of DNA ends are performed by using **adapters**, synthetic oligonucleotides that contain two restriction enzyme cleavage sites. By using adapters, the ends produced by cleavage of DNA with one restriction enzyme can be converted to cleavage sites for another restriction enzyme. For example, *Eco*RI-*Hin*dIII adapters will allow one to convert an *Eco*RI restriction fragment (a DNA fragment with *Eco*RI sites at its ends) to a *Hin*dIII restriction fragment (one with *Hin*dIII sites at the ends). Indeed, the present tools of molecular biology allow geneticists to modify DNA molecules at will—to change sequences, to add sequences, or to delete sequences.

In Vitro Site-Specific Mutagenesis

Recombinant DNA and gene-cloning technologies allow scientists to isolate and characterize any given gene from any living organism. In addition, these methodologies permit researchers to modify these genes *in vitro* in order to produce new restriction enzyme cleavage sites or to change a particular codon to a different codon or even a whole set of different codons. In short, scientists can now dissect genes at the nucleotide level by changing one nucleotide at a time and examining the effect of each change on the function of the gene or its product. Such nucleotide changes are accomplished by a procedure called **site-specific mutagenesis,** or, more precisely, **oligonucleotide-directed site-specific mutagenesis.**

Site-specific mutagenesis (Figure 19.18) can be carried out on genes cloned in vectors like pUC118 and pUC119 that provide for the isolation of DNA single strands (Figure 19.15). An oligonucleotide of 12 to 15

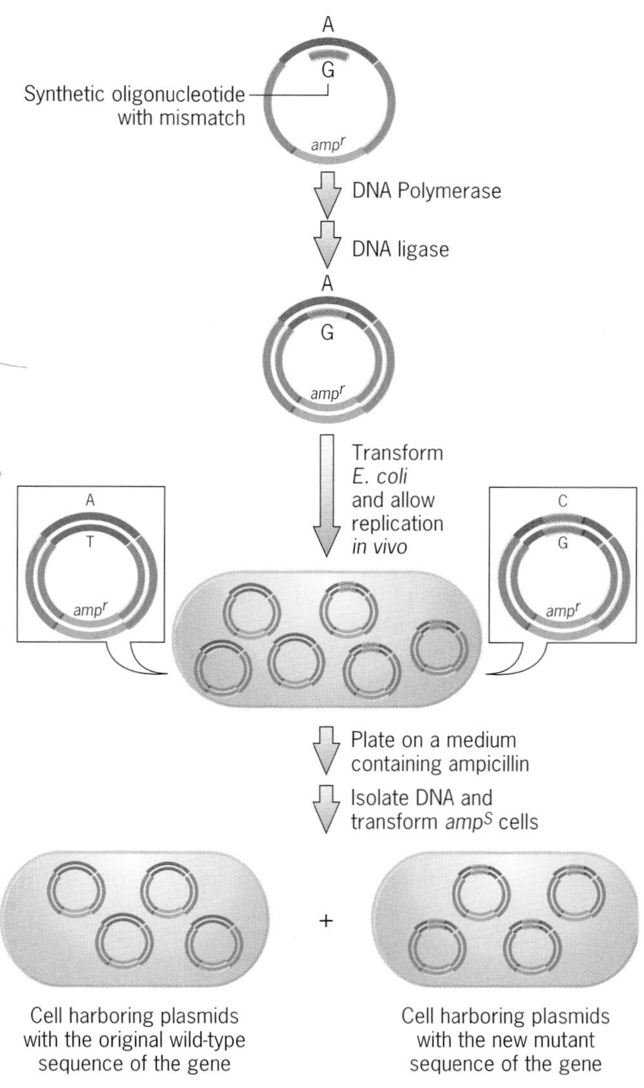

Figure 19.18 Oligonucleotide-directed site-specific mutagenesis. Single-stranded pUC119 DNA carrying the gene of interest is purified and annealed to a synthetic oligonucleotide that contains the desired change in sequence. Here, an A:T base pair is converted to a C:G base pair.

nucleotides in length, which is largely complementary to the purified single-stranded copy of the gene in the region of interest, but which contains one or more noncomplementary or "mismatched" bases, is synthesized. The mismatched bases will provide the desired mutant sequences. The synthetic oligonucleotide is annealed to the single strand of the gene in the vector, and a complementary strand is synthesized *in vitro* with a DNA polymerase. DNA ligase is then used to seal the single-strand break left by polymerase, and the double-stranded circular DNA products are introduced into *E. coli* cells by transformation. The semiconservative replication of the DNA molecules in *E. coli* will produce a population of progeny DNA molecules, half of which carry the original DNA sequence and half of which carry the new or mutant DNA sequence. The two types of molecules are separated by a second cycle of transformation, and the desired mutant genes are identified by hybridization to radioactively labeled synthetic oligonucleotides under conditions of high stringency (conditions that require perfect complementarity of the strands for duplexes to form) or by sequencing the modified region of the cloned gene.

Key Points: **Cloned genes and other DNA sequences can be manipulated almost at will in the test tube. Phagemids provide a powerful tool for purifying the single strands of a cloned DNA sequence. Transcription vectors allow scientists to express genes *in vitro* by using coupled transcription and translation systems. DNA fragments with different termini can be joined *in vitro* with synthetic oligonucleotides called linkers and adapters. The nucleotide sequences of cloned genes can be changed as desired by *in vitro* site-specific mutagenesis.**

THE MOLECULAR ANALYSIS OF DNA, RNA, AND PROTEIN

The development of recombinant DNA techniques has spawned a whole array of new approaches to the analysis of genes and gene products. Many questions that were totally unapproachable just 15 years ago can now be investigated with relative ease. Geneticists can isolate and characterize essentially any gene from any organism; however, the isolation of genes from large eukaryotic genomes is sometimes a long and laborious process (Chapter 20). Once a gene has been cloned, its expression can be investigated in even the most complex organisms such as humans.

Is a particular gene expressed in the kidney, the liver, bone cells, hair follicles, erythrocytes, or lymphocytes? Is this gene expressed throughout the development of the organism or only during certain stages of development? Is a mutant allele of this gene similarly expressed, spatially and temporally, during development? Or does the mutant allele have an altered pattern of expression? If the latter, is this altered pattern of expression responsible for an inherited syndrome or disease? These questions and many others can now be routinely investigated using well-established methodologies.

Clearly, a comprehensive discussion of all the techniques used to investigate gene structure and function is far beyond the scope of this text. However, let's consider some of the most important methods used to investigate the structure of genes (DNA), their transcripts (RNA), and their final products (usually proteins).

Analysis of DNAs by Southern Blot Hybridizations

Gel electrophoresis provides a powerful tool for the separation of macromolecules with different sizes and charges. DNA molecules have an essentially constant charge per unit mass; thus they separate in agarose and acrylamide gels almost entirely on the basis of size or conformation. Agarose or acrylamide gels act as molecular sieves, retarding the passage of large molecules more than small molecules. Agarose gels are better sieves for large molecules (larger than a few hundred nucleotides); acrylamide gels are better for separating small DNA molecules. Figure 19.19 describes the separation of DNA restriction fragments by agarose gel electrophoresis. The procedures used to separate RNA and protein molecules are largely the same in principle, but involve slightly different techniques because of the unique properties of each class of macromolecule.

In 1975, E. M. Southern published an important new procedure that allowed investigators to identify the locations of genes and other DNA sequences on restriction fragments separated by gel electrophoresis. The essential feature of this technique is the transfer of the DNA molecules that have been separated by gel electrophoresis onto a nitrocellulose or nylon membrane (Figure 19.20). Such transfers of DNA to membranes are called Southern blots after the scientist who developed the technique. The DNA is denatured either prior to or during transfer by placing the gel in an alkaline solution. After transfer is complete, the DNA is immobilized on the membrane by drying or UV irradiation. A radioactive DNA probe containing the sequence of interest is then hybridized or annealed (Chapter 9) with the immobilized DNA on the membrane. The probe will anneal only with DNA molecules on the membrane that contain a nucleotide se-

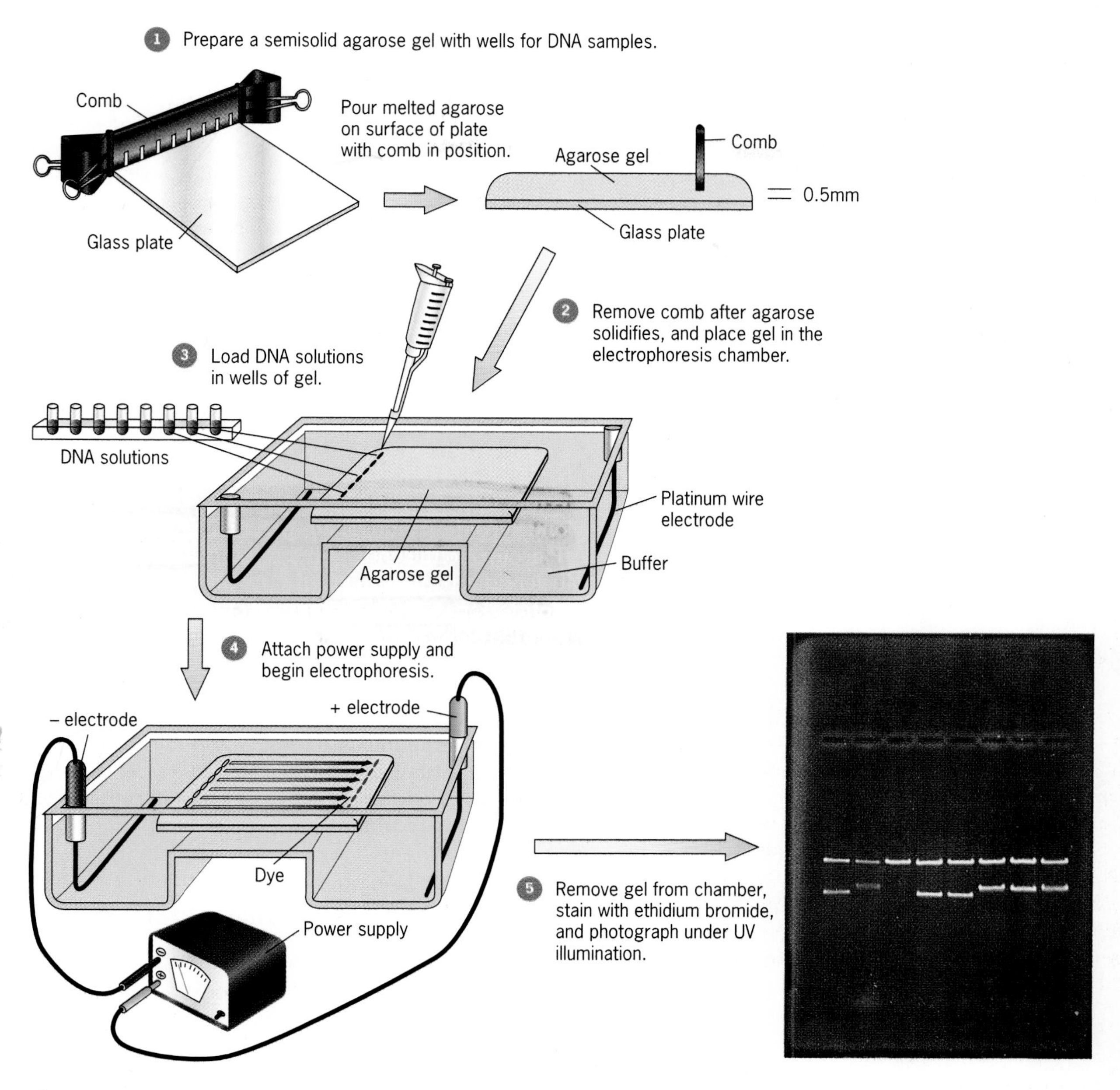

Figure 19.19 The separation of DNA molecules by agarose gel electrophoresis. The DNAs are dissolved in loading buffer with density greater than that of the electrophoresis buffer so that DNA samples settle to the bottoms of the wells, rather than diffusing into the electrophoresis buffer. The loading buffer also contains a dye to monitor the rate of migration of molecules through the gel. Ethidum bromide binds to DNA and fluoresces when illuminated with ultraviolet light. In the photograph shown, the third lane from the left contained *Eco*RI-cut pUC119 DNA; the other lanes contained *Eco*RI-cut pUC119 DNAs carrying maize glutamine synthetase cDNA inserts.

quence complementary to the sequence of the probe. Nonannealed probe is then washed off the membrane, and the washed membrane is exposed to X-ray film to detect the presence of the radioactivity in the bound probe. After the X ray is developed, the dark bands show the positions of DNA sequences that have hybridized with the probe (Figure 19.21).

The ability to transfer DNA restriction fragments or other DNA molecules that have been separated by gel electrophoresis to nitrocellulose or nylon membranes for hybridization studies and other types of analyses has proven to be extremely useful (see Technical Sidelight: Detection of a Mutant Gene Causing Cystic Fibrosis by Southern Blot Analysis). Radioactive probes for use in Southern blot hybridization experiments are prepared by a variety of procedures including (1) producing radioactive RNA probes by *in vitro* transcription of cloned sequences in transcription

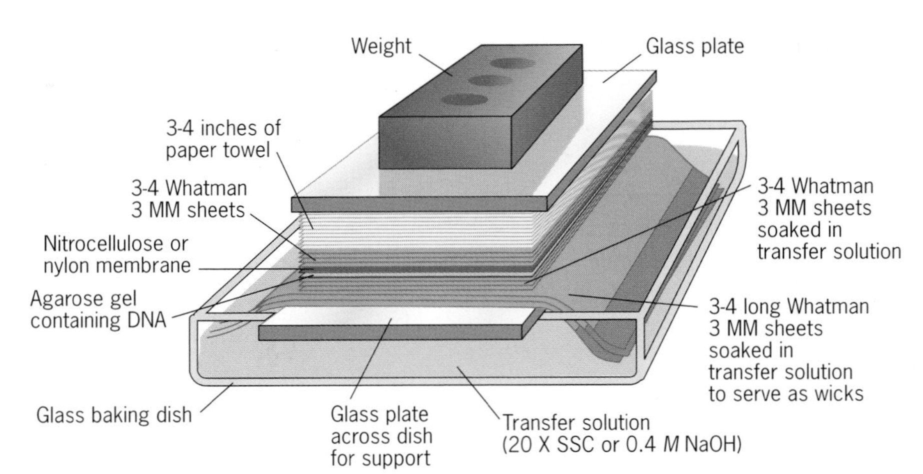

Figure 19.20 Procedure used to transfer DNAs separated by gel electrophoresis to nitrocellulose or nylon membranes. The transfer solution carries the DNA from the gel to the membrane as the dry paper towels on top draw the salt solution from the reservoir through the gel to the towels. The DNA binds to the membrane on contact. The membrane with the DNA bound to it is dried and baked under vacuum to firmly affix the DNA prior to hybridization.

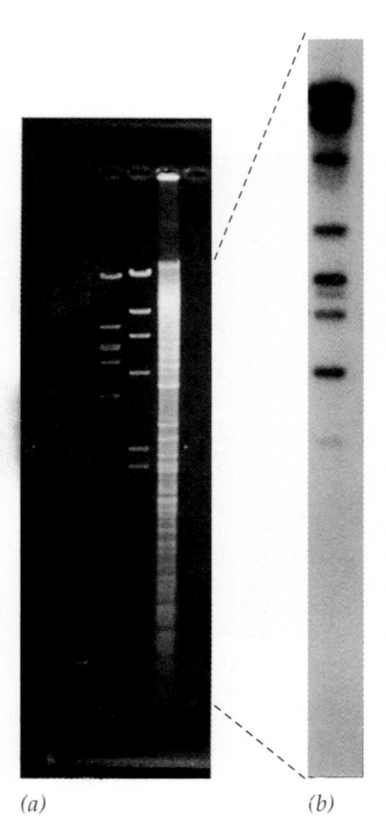

Figure 19.21 Identification of genomic restriction fragments harboring specific DNA sequences by the Southern blot hybridization procedure. (*a*) Photograph of an ethidium bromide-stained agarose gel containing (left lane) phage λ DNA disgested with *Eco*RI, (center lane) phage λ DNA digested with *Hin*dIII, and (right lane) *Arabidopsis thaliana* DNA digested with *Eco*RI. The λ DNA digests provide size markers. The *A. thaliana* DNA digest was transferred to a nylon membrane by the Southern procedure (Figure 19.20) and hybridized to a radioactive DNA fragment of a cloned β-tubulin gene. The resulting Southern blot is shown in (*b*); nine different *Eco*RI fragments hybridized with the β-tubulin probe.

vectors (Figure 19.16) with at least one radioactive ribonucleoside triphosphate precursor, (2) transferring ^{32}P groups to the ends of DNA molecules from labeled nucleotide triphosphates by the action of polynucleotide kinases, and (3) **nick-translation** of double-stranded DNA molecules. Nick-translation is performed by making single-strand cuts (nicks) in double helices with endonucleases, and then "translating" the nick along the DNA molecule by the concerted action of the 5' → 3' exonuclease activity and the 5' → 3' polymerase activity of *E. coli* DNA polymerase I (Chapter 9) in the presence of a labeled deoxyribonucleotide triphosphate precursor.

Analysis of RNAs by Northern Blot Hybridizations

If DNA molecules can be transferred from agarose gels to nitrocellulose or nylon membranes for hybridization studies, we might expect that RNA molecules separated by agarose gel electrophoresis could be similarly transferred and analyzed. Indeed, such RNA transfers are used routinely in molecular genetics laboratories. These RNA blots are called **northern blots** in recognition of the fact that the procedure is the mirror image of the Southern blotting technique. As we will discuss in the next section, this somewhat unusual terminology has been extended to the transfer of proteins from gels to membranes, a procedure called western blotting.

The northern blot procedure is essentially identical to that used for Southern blot transfers (Figure 19.20). However, RNA molecules are very sensitive to degradation by RNases. Thus care must be taken to prevent contamination of materials with these extremely stable enzymes. Furthermore, most RNA molecules contain considerable secondary structure. Thus RNA molecules must be kept denatured during electrophoresis in

TECHNICAL SIDELIGHT

Detection of a Mutant Gene Causing Cystic Fibrosis by Southern Blot Analysis

Southern blots are DNA gel blots prepared by the transfer of DNAs separated by gel electrophoresis to membranes and hybridized to labeled DNA or RNA probes (Figure 19.20). The Southern blotting procedure is one of the most important tools of molecular genetics. Southern blots are used to identify DNA molecules or restriction fragments that contain specific genes or DNA sequences (Figure 19.21). In some cases, Southern blots can be used to identify mutant alleles that are responsible for inherited diseases. Here, we examine how the Southern blotting procedure is used to identify the mutant allele responsible for the majority of the cases of cystic fibrosis (CF). This devastating disease, which is characterized by the accumulation of mucus in the lungs, pancreas, and liver, and the subsequent malfunction of these organs, is the most common inherited disease in humans of northern European descent. In Chapter 20, we discuss cystic fibrosis and the identification and characterization of the gene that causes it.

Approximately 70 percent of the cases of CF result from a specific mutant allele of the *CF* gene. This mutant allele, *CFΔF508*, contains a three-base deletion that eliminates a phenylalanine residue at position 508 in the polypeptide product. Because the nucleotide sequence of the *CF* gene is known and since the *CFΔF508* allele differs from the wild-type allele by three base pairs, it was possible to design oligonucleotide probes that hybridize specifically with the wild-type *CF* allele or the *ΔF508* allele under the appropriate conditions.

The wild-type *CF* gene and gene product have the following nucleotide and amino acid sequences in the region altered by the *ΔF508* mutation:

deleted in *ΔF508* (TTT)

bases in the
coding strand: 5'- AAA GAA AAT ATC ATC TTT GGT GTT-3'

amino acids
in product: NH_2-Leu Glu Asn Ile Ile Phe Gly Val-COOH,

amino acid 508 (Phe)

whereas the *ΔF508* allele and product have the sequences:

deletion

bases in the
coding strand: 5'- AAA GAA AAT ATC AT. . .T GGT GTT-3'

amino acids
in product: NH_2-Leu Glu Asn Ile Ile — Gly Val-COOH.

Phe absent

Based on these nucleotide sequences, Lap-Chee Tsui and colleagues synthesized oligonucleotides spanning this region of the mutant and wild-type alleles of the *CF* gene and tested their specificity. They demonstrated that at 37°C under a standard set of conditions, one oligonucleotide probe (oligo-N: 3'-CTTTTATAGTAGAAACCAC-5') hybridized only with the wild-type allele, whereas another oligonucleotide (oligo-ΔF: 3'-TTCTTTTATAGTA. . .ACCACAA-5') hybridized only with the *ΔF508* allele. Their results indicated that the oligo-ΔF probe could be used to detect the *ΔF508* allele in either the homozygous or heterozygous state by Southern blot hybridization experiments. When Tsui and coworkers used these allele-specific oligonucleotide probes to analyze CF patients and their parents for the presence of the *ΔF508* mutation, they found that many of the patients were homozygous for this mutation, whereas most of their parents were heterozygous as is expected. Some of their results are as shown at the bottom of this page.

The Southern blot hybridization test for the presence of the *ΔF508* mutation can be carried out with small amounts of DNA if this segment of the *CF* gene is first amplified by the polymerase chain reaction (Figure 19.25). Indeed, the Southern blot test for the *ΔF508* mutation can be carried out with DNA obtained from fetal cells by amniocentesis or even with DNA from a single cell of an eight-cell pre-embryo produced by *in vitro* fertilization (Chapter 20).

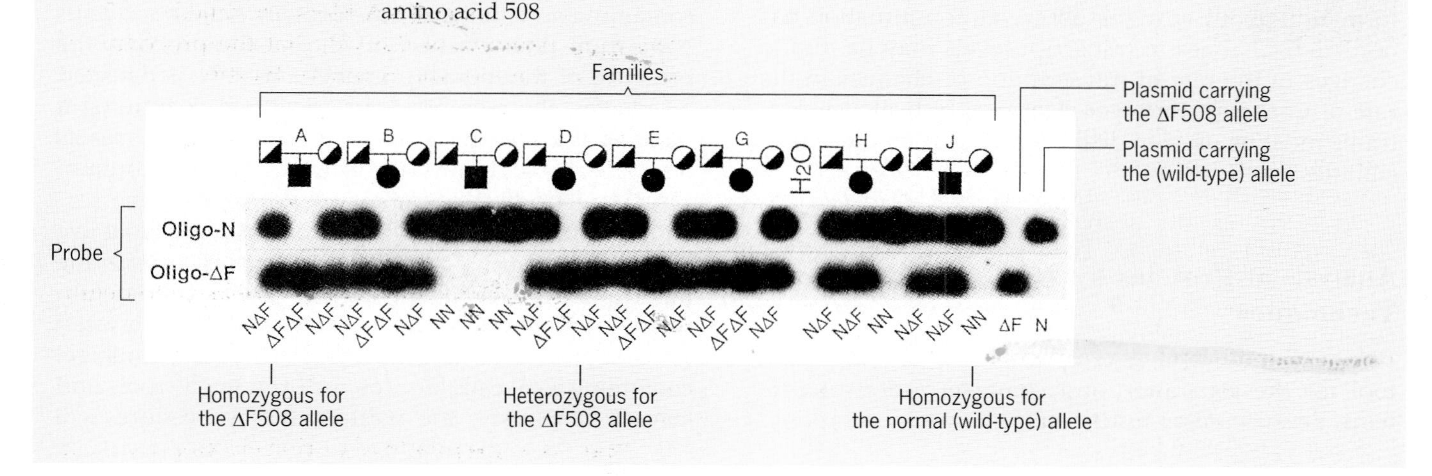

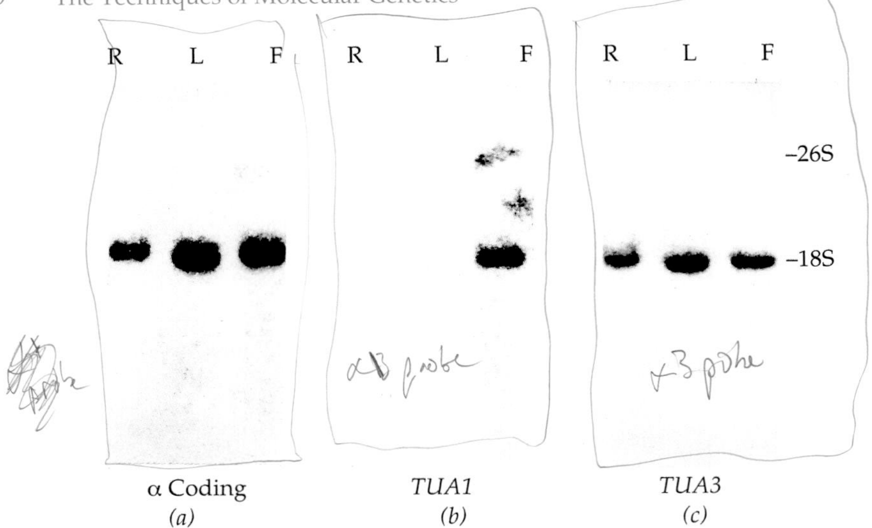

Figure 19.22 Typical northern blot hybridization data. Total RNAs were isolated from roots (R), leaves (L), and flowers (F) of *A. thaliana* plants, separated by agarose gel electrophoresis, and then transferred to nylon membranes. The autoradiogram shown in (*a*) is of a blot that was hybridized to a radioactive probe containing an α-tubulin coding sequence. This probe hybridizes to the transcripts of all six α-tubulin genes in *A. thaliana*. The autoradiograms shown in (*b*) and (*c*) are of RNA blots that were hybridized to DNA probes specific for the α1- and α3-tubulin genes, respectively. The results show that the α3-tubulin transcript is present in all organs analyzed, whereas the α1-tubulin transcript is present only in flowers. The 18S and 26S ribosomal RNAs provide size markers.

order to separate them on the basis of size. Denaturation is accomplished by adding formaldehyde or some other chemical denaturant to the buffer used for electrophoresis. After transfer to an appropriate membrane, the RNA blot is hybridized to either RNA or DNA probes just as with a Southern blot.

Northern blot hybridizations (Figure 19.22) are extremely helpful in studies of gene expression. They can be used to determine whether a particular gene is transcribed in all tissues of an organism or only in certain tissues. They can also be used to study the temporal pattern of expression of individual genes during growth and development. However, we must remember that northern blot hybridizations measure only the accumulation of RNA transcripts. They provide no information about why the observed accumulation has occurred. Changes in transcript levels may be due to changes in the rate of transcription or changes in the rate of transcript turnover. More sophisticated procedures must be used to distinguish between these possibilities.

Analysis of Proteins by Western Blot Techniques

Polyacrylamide gel electrophoresis is an important tool for the separation and characterization of proteins. Because many functional proteins are composed of two or more subunits, individual polypeptides are separated by electrophoresis in the presence of the detergent sodium dodecyl sulfate (SDS), which denatures the proteins. After electrophoresis, the proteins are detected by staining with Coomassie blue or silver stain. However, the separated polypeptides also can be transferred from the gel to a nitrocellulose membrane, and individual proteins can be detected by using specific antibodies. This transfer of proteins from acrylamide gels to nitrocellulose membranes, called **western blotting**, is performed by using an electric current to move the proteins from the gel to the surface of the membrane (Figure 19.23). After transfer, a specific protein of interest is identified by placing the membrane with the immobilized proteins in a solution containing an antibody to the protein. Nonbound antibodies are then washed off the membrane, and the presence of the initial (primary) antibody is detected by placing the membrane in a solution containing a secondary antibody. This secondary antibody reacts with immunoglobulins (the group of proteins containing all antibodies) in general (Chapter 24). The secondary antibody is conjugated to either a radioactive isotope (permitting autoradiography) or an enzyme that produces a visible product when the proper substrate is added. Figure 19.24 shows the use of a western blot to detect a single protein in an acrylamide gel containing total cellular proteins from maize roots and leaves. Obviously, the western blot procedure is a

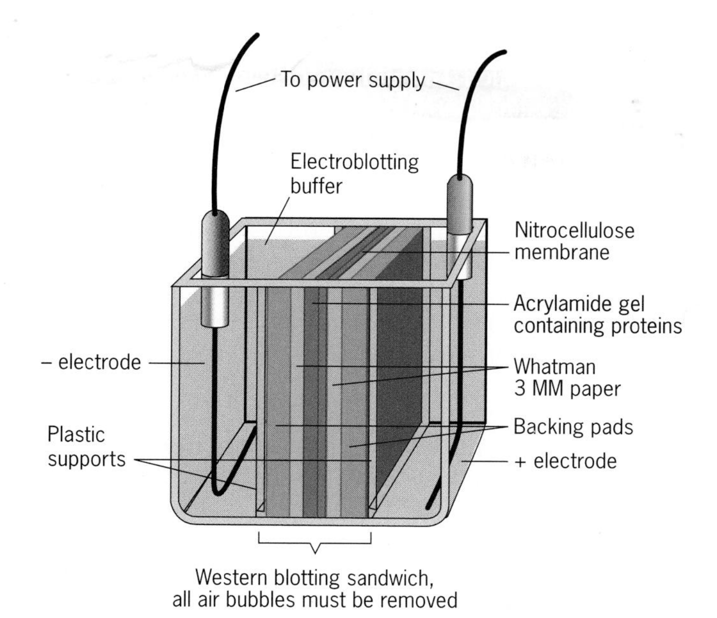

Figure 19.23 A typical western blotting or electroblotting apparatus. An electric current is used to transfer the proteins from a gel to a nitrocellulose membrane placed next to it in the blotting sandwich. All other components of the sandwich function to provide gentle but firm support; tight contact between the gel and the membrane is essential for good transfer.

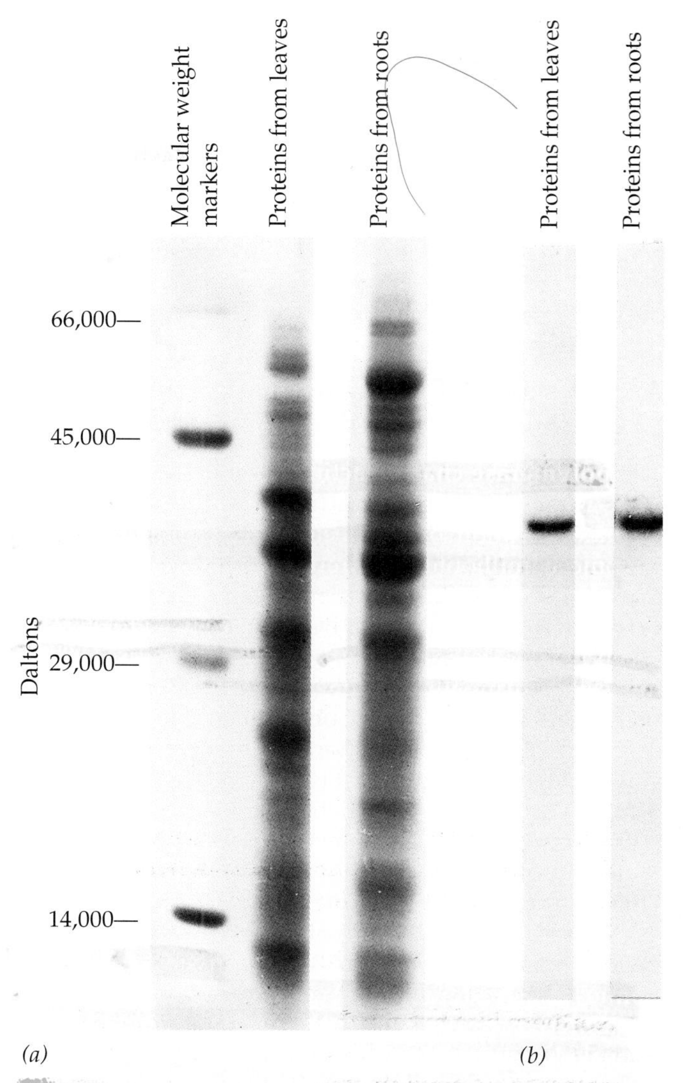

Figure 19.24 The use of western blots to identify individual proteins separated by polyacrylamide gel electrophoresis. (*a*) Proteins isolated from roots or leaves of maize were separated by polyacrylamide gel electrophoresis and stained with Coomassie blue. (*b*) The chloroplastic form of glutamine synthetase was identified by western blot analysis of the gel shown in (*a*).

powerful tool for identifying and characterizing specific gene products.

Key Points: DNA restriction fragments and other small DNA molecules can be separated by agarose and acrylamide gel electrophoresis and transferred to solid supports, usually nitrocellulose or nylon membranes, to produce DNA gel blots called Southern blots. The DNAs on these blots can be hybridized to labeled DNA probes to detect sequences of interest by autoradiography. The same procedure can be applied to RNA molecules separated by gel electrophoresis to produce RNA gel blots called northern blots. When proteins are transferred from gels to membranes and detected with antibodies, the products are called western blots.

THE MOLECULAR ANALYSIS OF GENES AND CHROMOSOMES

Recombinant DNA techniques allow geneticists to determine the structure of genes, chromosomes, and entire genomes. Indeed, molecular geneticists are constructing detailed physical maps of the genomes of many organisms. Complete physical maps are now available for most of the chromosomes of the worm *C. elegans* and for two human chromosomes, and major genome mapping projects are underway for *Drosophila*, *Arabidopsis thaliana* (a popular research plant), several important agricultural plants, and humans.

The ultimate physical map of a genetic element is its nucleotide sequence, and the complete nucleotide sequences of the genomes of many viruses, of mitochondria, and of two bacteria, *Hemophilus influenzae* and *Mycoplasma genitalium*, have already been determined. As is discussed in Chapter 20, the goal of the Human Genome Project is to determine the complete nucleotide sequence of the human genome within the next 10 years. In the following sections, we discuss the amplification of specific segments of DNA, the construction of restriction enzyme cleavage site maps of genes and chromosomes, and the determination of DNA sequences.

Amplification of DNAs by the Polymerase Chain Reaction (PCR)

The **polymerase chain reaction**, usually referred to as **PCR**, is an extremely powerful procedure that allows the amplification of a selected DNA sequence in a genome a millionfold or more. The PCR procedure was devised by Kary Mullis, who received a Nobel Prize in 1993 for this work. PCR can be used to clone a given DNA sequence *in vitro*—without the use of living cells during the cloning process. However, the procedure can be applied only when the nucleotide sequence of at least one short DNA segment on each side of the region of interest is known. The PCR procedure involves using synthetic oligonucleotides complementary to these known sequences to prime enzymatic amplification of the intervening segment of DNA in the test tube.

The PCR procedure involves three steps, each repeated many times (Figure 19.25). In step 1, the genomic DNA containing the sequence to be amplified is denatured by heating. In step 2, the denatured DNA is annealed to an excess of the synthetic oligonucleotide primers. In step 3, DNA polymerase is used to replicate the DNA segment between the sites complementary to the oligonucleotide primers. The primer provides the free 3'-OH required for covalent extension, and the denatured genomic DNA provides the required template function (Chapter 9). The products of the first cycle of replication are then denatured, annealed to oligonucleotide primers, and replicated again with DNA polymerase. The cycle is then repeated many times until the desired level of amplication is achieved. Note that *amplification occurs exponentially*. One DNA double helix will yield 2 double helices after one cycle of replication, 4 after two cycles, 8 after three cycles, 16 after four cycles, 1024 after ten cycles, and so on.

Initially, PCR was performed with DNA polymerase I of *E. coli* as the replicase. Because this enzyme is heat-inactivated during the denaturation step, new enzyme had to be added at step 3 of each cycle. A major improvement in PCR amplification of DNA came with the discovery of a heat-stable DNA polymerase in the thermophilic bacterium, *Thermus aquaticus*. This polymerase, called **Taq polymerase** (*T. aquaticus* polymerase), remains active during the heat denaturation step. As a result, polymerase does not have to be added after each cycle of denaturation. Instead, excess *Taq* polymerase and oligonucleotide primers can be added at the start of the PCR process, and amplification cycles can be carried out by sequential alterations in temperature. PCR machines are now available that cycle the temperature automatically, making the PCR amplification of a specific DNA sequence a simple task.

PCR technologies provide shortcuts for many cloning and sequencing applications. These procedures permit scientists to obtain definitive structural data on genes and DNA sequences when very small amounts of DNA are available. One important application occurs in the diagnosis of inherited human diseases, especially in cases of prenatal diagnosis, where limited amounts of fetal DNA are available. A second major application occurs in forensic cases involving the identification of individuals by using DNA isolated from very small tissue samples. No criteria can provide more definitive evidence of identity than DNA sequences. After all, other phenotypic traits are controlled by the expression of these DNA sequences. By using PCR amplification, DNA sequences can be obtained from minute amounts of DNA isolated from small blood samples, sperm, or even individual human hairs. Thus **PCR DNA fingerprinting** experiments play important roles in legal cases involving disputed identity. These applications of PCR are discussed in Chapter 20.

Physical Maps of DNA Molecules Based on Restriction Enzyme Cleavage Sites

Most restriction endonucleases cleave DNA molecules in a site-specific manner (see Table 19.1). As a result, they can be used to generate **physical maps** of chromosomes that are of tremendous value in assisting researchers in isolating segments of DNA carrying genes or other DNA sequences of interest. The sizes of the restriction fragments can be determined by polyacrylamide or agarose gel electrophoresis (Figure 19.19). Because of the nucleotide subunit structure of DNA, with one phosphate group per nucleotide, DNA has an essentially constant charge per unit of mass. Thus the rates of migration of DNA fragments during electrophoresis provide accurate estimates of their lengths; the rate of migration is inversely proportional to length.

A simple diagram of the procedure that is used to map the restriction enzyme cleavage sites on a DNA

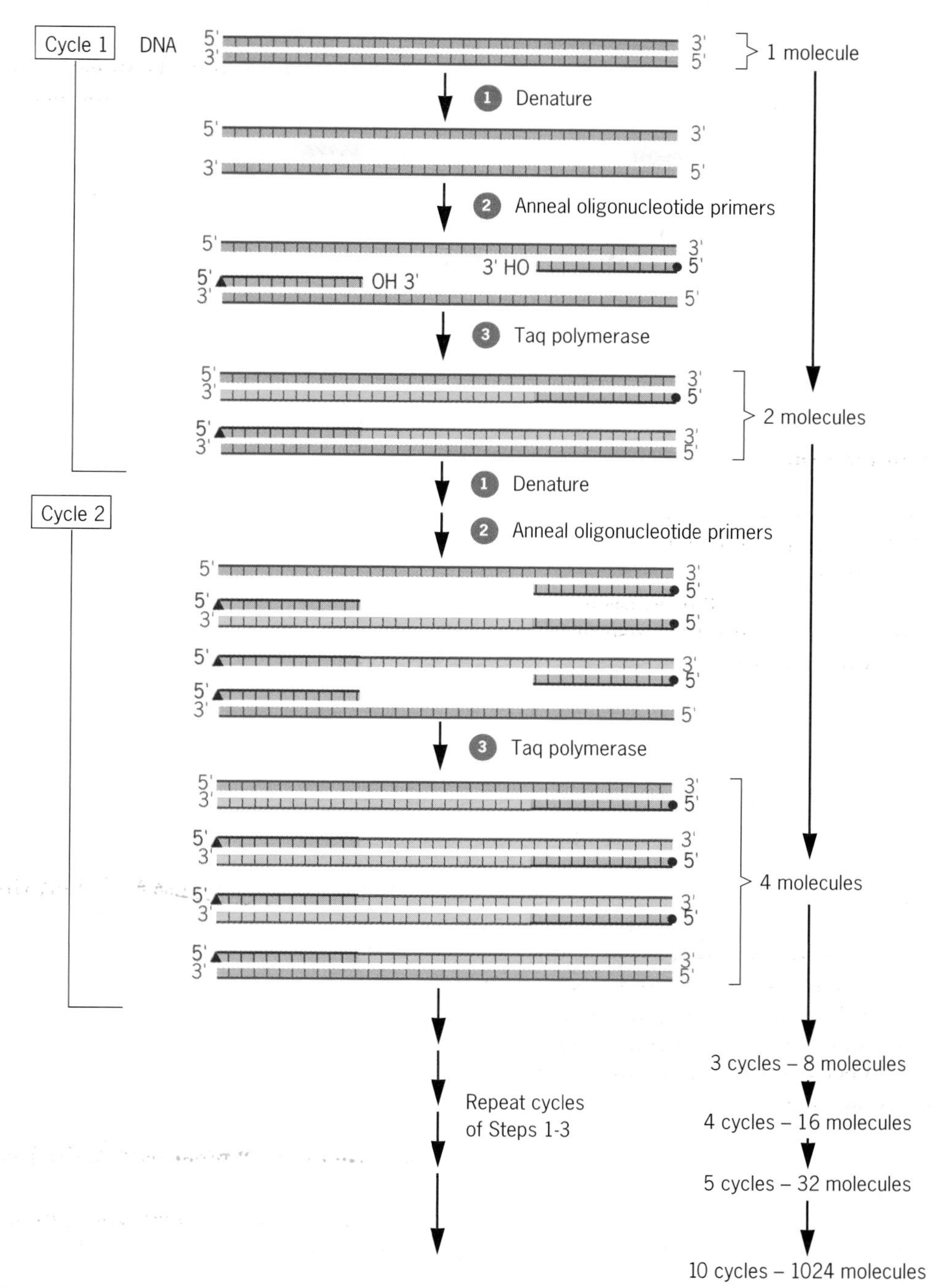

Figure 19.25 The use of PCR to amplify DNA molecules *in vitro*. Each cycle of amplification involves three steps: (1) denaturation of the genomic DNA being analyzed, (2) annealing of the denatured DNA to chemically synthesized oligonucleotide primers with sequences complementary to sites on opposite sides of the DNA region of interest, and (3) enzymatic replication of the region of interest by *Taq* polymerase.

molecule is illustrated in Figure 19.26. The sizes of DNA restriction fragments are estimated by using a set of DNA markers of known size. In Figure 19.26, a set of DNA molecules that differ in length by 1000 nucleotide pairs are used as size markers. Consider a DNA molecule approximately 6000 nucleotide pairs (6

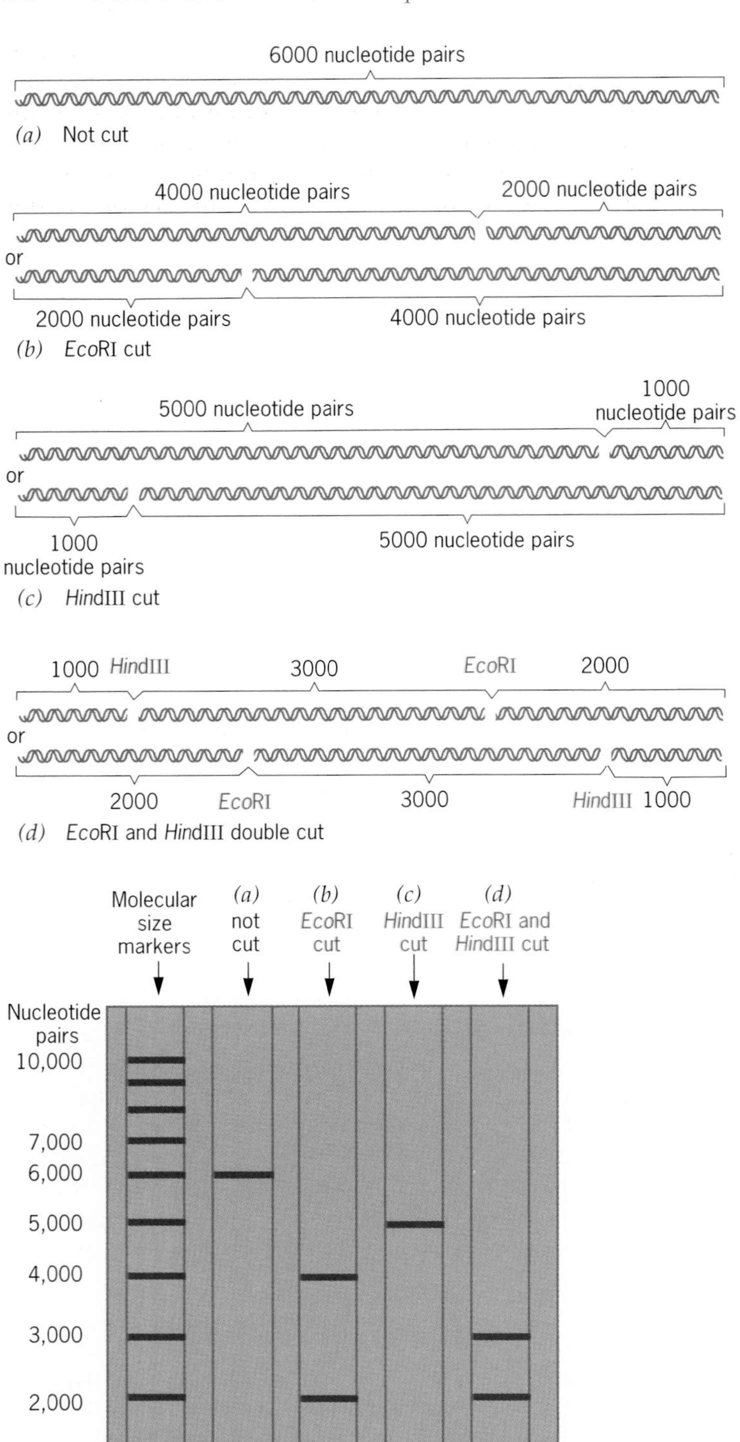

Figure 19.26 Procedure used to map restriction enzyme cleavage sites in DNA molecules. (*a-d*) Structures of the DNA molecule or of restriction fragments of the molecule either (*a*) uncut or cut with (*b*) *Eco*RI, (*c*) *Hin*dIII, or (*d*) *Eco*RI and *Hin*dIII. (*e*) The separation of these DNA molecules and fragments by agarose gel electrophoresis. The left lane on the gel contains a set of molecular size markers, a set of DNA molecules of size 1000 nucleotide pairs and multiples thereof.

kb) in length. When the 6-kb DNA molecule is cut with *Eco*RI, two fragments of sizes 4000 and 2000 nucleotide pairs are produced. The possible positions of the single *Eco*RI cleavage site in the molecule are shown in Figure 19.26*b*. When the same DNA molecule is cleaved with *Hind*III, two fragments of sizes 5000 and 1000 nucleotide pairs result.

The possible locations of the single *Hind*III cleavage site are shown in Figure 19.26*c*. Note that at this stage of the analysis no deductions can be made about the relative positions of the *Eco*RI and *Hind*III cleavage sites. The *Hind*III cleavage site may be located in either of the two *Eco*RI restriction fragments. The molecule is then simultaneously digested with both *Eco*RI and *Hind*III, and three fragments of sizes 3000, 2000, and 1000 nucleotide pairs are produced. This result establishes the positions of the two cleavage sites relative to one another on the molecule. Since the 2000 nucleotide-pair *Eco*RI restriction fragment is still present (not cut by *Hind*III), the *Hind*III cleavage site must be at the opposite end of the molecule from the *Eco*RI cleavage site (Figure 19.26*d*). By extending this type of analysis to include the use of several different restriction enzymes, more extensive maps of restriction sites can be constructed. When large numbers of restriction enzymes are employed, detailed maps of entire chromosomes can be constructed. An important aspect of restriction maps is that, unlike genetic maps (Chapter 7), restriction maps reflect true physical distances along the DNA molecule.

During their early work on the characterization of restriction endonucleases, Hamilton Smith and Daniel Nathans used the DNA chromosome of simian virus 40 (SV40) for their studies. SV40 is an animal virus that can transform cells to the cancerous state. Its circular chromosome is only 5226 nucleotide pairs in length, making it well suited for restriction enzyme studies. As a result of their work, Smith and Nathans' were able to construct a **restriction enzyme cleavage site map** or more simply **restriction map**, of the SV40 chromosome.

The restriction enzyme *Eco*RI cleaves the SV40 chromosome at only one site (Figure 19.27*a*). This site has been arbitrarily set as position 0 on the SV40 restriction map. Restriction enzymes *Hpa*I and *Hind*III cleave the SV40 chromosome at four and six sites, respectively (Figure 19.27*b* and *c*). When the SV40 chromosome is digested with all three of these enzymes, 11 distinct restriction fragments are produced (Figure 19.27*d*). To prepare a restriction map of the SV40 chromosome, Smith and Nathans had to determine the order of these restriction fragments in the chromosome.

The three major procedures for ordering restriction fragments all utilize polyacrylamide or agarose gel electrophoresis to obtain accurate estimates of fragment size. They are: (1) sequential digestion of chromosomes with two or more different restriction enzymes, (2) partial digestion of chromosomes after labeling the ends with a radioactive isotope, and (3) determination of whether the strands of different restriction fragments can hybridize with each other after denaturation. Smith and Hamilton used the first procedure to map the SV40 chromosome.

If the SV40 chromosome is cleaved with *Hind*III (Figure 19.27*c*), a large fragment *E* is produced. This same fragment is still present after subsequent digestion with *Hpa*I (Figure 19.27*d*), indicating that there is no *Hpa*I cleavage site within the *Hind*III *E* fragment. By contrast, when *Hpa*I restriction fragment *B* + *C* (Figure 19.27*b*) is subsequently digested with HindIII, it is cleaved into two fragments *B* and *C* (Figure 19.27*d*). Similarly, subsequent digestion of *Hind*III fragment *C* + *D* (Figure 19.27*c*) with *Hpa*I yields fragments *C* and *D* (Figure 19.27*d*). These data, taken together, show that the common sequence (or overlap sequence) in the *Hpa*I fragment *B* + *C* and *Hind*III fragment *C* + *D* is fragment *C* and establishes the order *B-C-D* (or *D-C-B*). By continuing to analyze the fragments produced by sequential digestions with these and other restriction enzymes, the order of the fragments usually can be determined.

By combining computer-assisted restriction mapping with other molecular techniques, it is possible to construct physical maps of entire genomes. The first multicellular eukaryote for which this was accomplished is *Caenorhabditis elegans*, a worm that is important for studies of the genetic control of development (Chapter 23). Moreover, the physical map of the *C. elegans* genome has been correlated with its genetic map. Thus, when an interesting new mutation is identified in *C. elegans*, its position on the genetic map often can be used to obtain clones of the wild-type gene from a large international *C. elegans* clone bank. One of the goals of the Human Genome Project is to construct a physical map of this type for the entire human genome (Chapter 20).

Nucleotide Sequences: The Ultimate Fine Structure Maps

The ultimate fine structure map of a gene or a chromosome is its nucleotide-pair sequence, complete with a chart of all nucleotide-pair changes that alter the function of that gene or chromosome. Prior to 1975, the thought of trying to sequence entire chromosomes was barely conceivable—at best, a laborious task requiring years of work. By late 1976, however, the entire 5386 nucleotide-long chromosome of phage ϕX174 had been sequenced. Today, sequencing is a routine

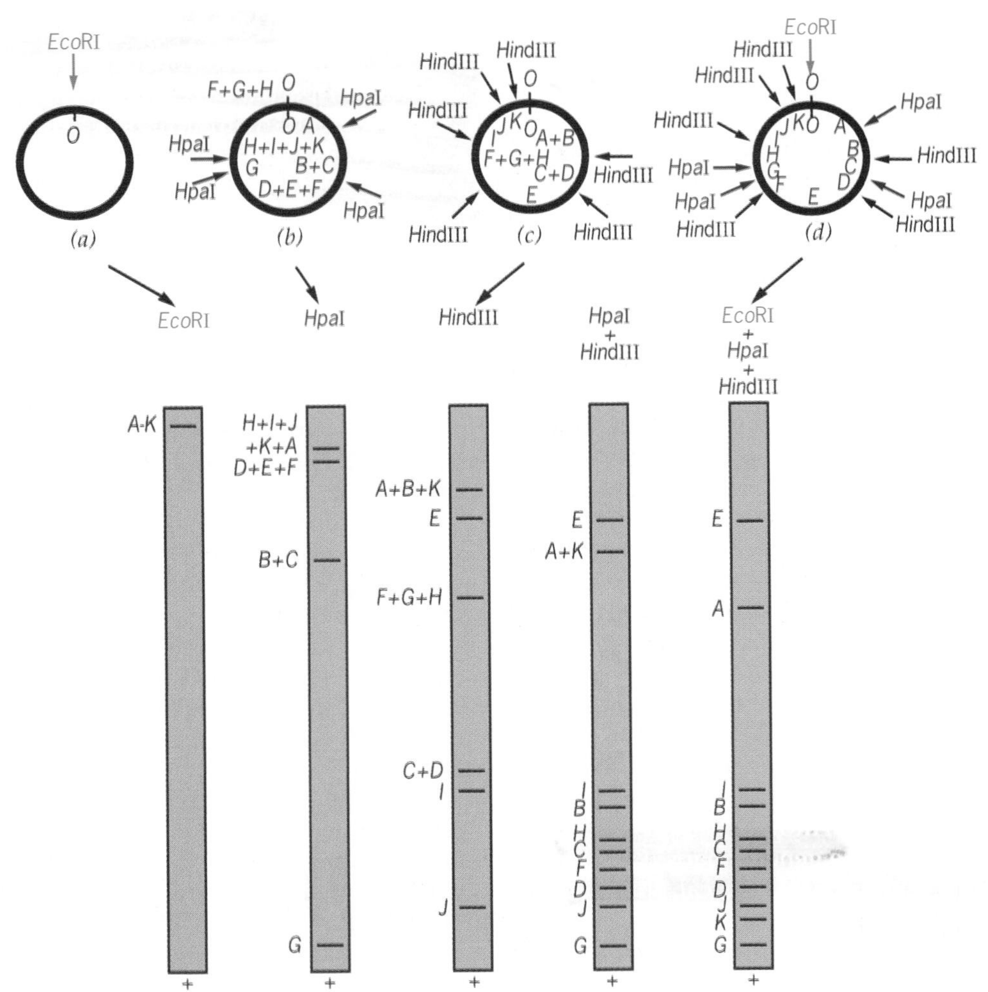

Figure 19.27 Restriction enzyme cleavage site maps of the simian virus 40 (SV40) chromosome (top) and the separation of SV40 restriction fragments by polyacrylamide gel electrophoresis (bottom). The restriction enzymes *Eco*RI (*a*), *Hpa*I (*b*), and *Hind*III (*c*) cleave the SV40 chromosome at 1, 4, and 6 sites, respectively. The 11 fragments produced when the SV40 chromosome is cut with all three restriction enzymes are shown in (*d*); they are arbitrarily labeled alphabetically starting at the unique *Eco*RI cleavage site and progressing in the clockwise direction.

laboratory procedure. The entire chromosomes of several viruses and two bacteria, along with large segments of eukaryotic chromosomes, have been sequenced. Within the next few decades, the sequences of entire eukaryotic genomes will be determined and stored in computer databanks for subsequent reference.

The present ability to sequence essentially any DNA molecule is the result of four major developments. The most important breakthrough was the discovery of restriction enzymes and their use in preparing homogeneous samples of specific segments of chromosomes. A second major advance was the improvement of gel electrophoresis procedures to the point where DNA fragments that differ in length by a single nucleotide can be resolved. The availability of gene-cloning techniques greatly facilitate the preparation of large quantities of a particular DNA molecule. Finally, researchers invented two different procedures by which the nucleotide sequences of DNA molecules can be determined.

Both DNA sequencing procedures depend on the generation of a population of DNA fragments that all have one end in common (all end at exactly the same nucleotide) and terminate at all possible positions (every consecutive nucleotide) at the other end. These fragments are then separated on the basis of chain length by polyacrylamide gel electrophoresis. In both cases, four separate biochemical reactions are carried out simultaneously, each of which generates a set of fragments terminating at one of the four bases (A, G, C, or T) in DNA.

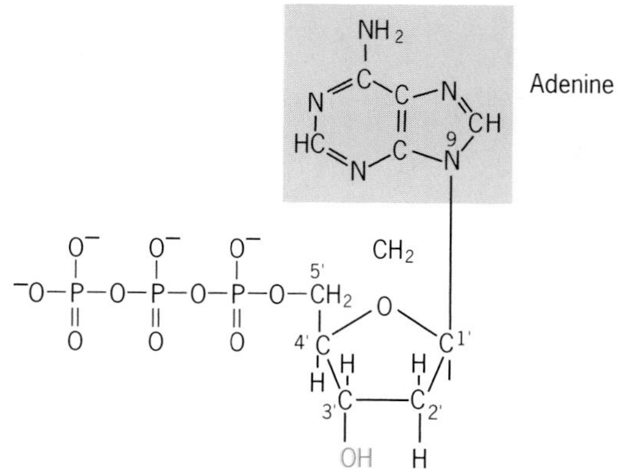

Adenine

The normal DNA precursor
2'-deoxyadenosine triphosphate (dATP)

Adenine

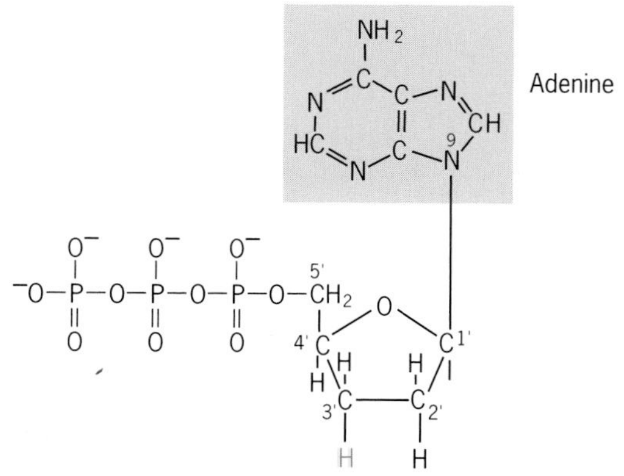

The chain-termination precursor
2', 3' -dideoxyadenosine triphosphate (ddATP)

Figure 19.28 Comparison of the structures of the normal DNA precursor 2'-deoxyadenosine triphosphate (dATP) and the chain-terminator 2', 3'-dideoxyadenosine triphosphate (ddATP) used in DNA sequencing reactions. Note the absence of the 3'-OH on ddATP.

The first procedure, called the Maxam and Gilbert procedure after Allan Maxam and Walter Gilbert who invented it, uses four different chemical reactions to cleave DNA chains specifically at As, Gs, Cs, or Cs + Ts. The second procedure, developed by Fred Sanger and colleagues, uses *in vitro* DNA systhesis in the presence of radioactive nucleotides and specific chain-terminators to generate four populations of radioactively labeled fragments that end at As, Gs, Cs, and Ts, respectively. We will discuss the Sanger procedure.

2',3'-Dideoxyribonucleoside triphosphates (Figure 19.28) are the chain-terminators most frequently used in the Sanger sequencing procedure. Recall that DNA polymerases have an absolute requirement for a free 3'-OH on the DNA primer strand (Chapter 10). If a 2',3'-dideoxynucleotide is added to the end of a chain, it will block subsequent extension of that chain since the 2', 3'- dideoxynucleotides have no 3'-OH. By using (1) 2',3'-dideoxythymidine triphosphate (ddTTP), (2) 2', 3'-dideoxycytidine triphosphate (ddCTP), (3) 2', 3'-dideoxyadenosine triphosphate (ddATP), and (4) 2', 3'-dideoxyguanosine triphosphate (ddGTP) as chain-terminators in four separate DNA synthesis reactions, four populations of fragments can be generated, and each population will contain chains that all terminate with the same base (T, C, A, or G) (Figure 19.29).

In a given reaction, the ratio of dXTP:ddXTP (where X can be any one of the four bases) is kept at approximately 100:1, so that the probability of termination at a given X in the nascent chain is about 1/100. This yields a population of fragments terminating at all potential (X) termination sites within a distance of a few hundred nucleotides from the original primer terminus.

After the DNA fragments generated in the four parallel reactions are released from the template strands by denaturation, they can be separated by polyacrylamide gel electrophoresis and their positions in the gel can be detected by autoradiography. The bands on the autoradiograms will correspond to radioactive chains of different lengths; they will produce a "ladder" defining the nucleotide sequence of the longest chain that has been synthesized (Figures 19.29 and 19.30).

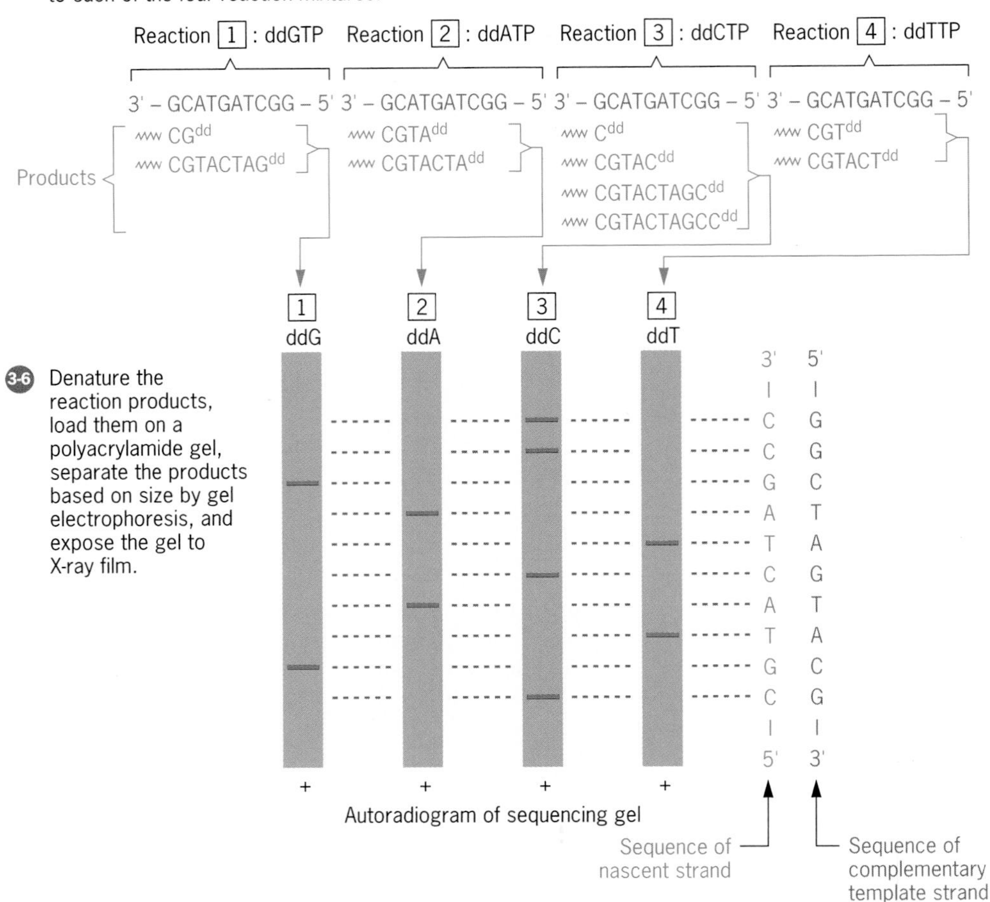

Figure 19.29 Sequencing DNA by the 2',3'-dideoxynucleoside triphosphate chain-termination procedure. Four reactions are carried out in parallel, each of which contains one of the four 2', 3'-dideoxy chain-terminators: ddGTP, ddATP, ddCTP, and ddTTP. All four reaction mixtures contain the components required for DNA synthesis *in vitro*. One radioactive DNA precursor (^{32}P-dCTP here) is present in each reaction so that the products can be detected by autoradiography. The products of the four reactions are separated by polyacrylamide gel electrophoresis, and the positions of the nascent DNA chains in the gel are determined by autoradiography. Because the shortest chain migrates the greatest distance, the nucleotide sequence of the nascent chains (shown in red at the right of the autoradiogram) is obtained by reading the gel from the bottom (anode) to the top (cathode).

The shortest fragment will migrate the greatest distance and give rise to the band nearest the anode (the positive electrode). Each successive band will contain chains that are one nucleotide longer than the chains in the preceding band of the ladder. The 3'-terminal nucleotide of the chain in each band will be the dideoxynucleotide chain-terminator present in the reaction mixture (1, 2, 3, or 4) in which that specific chain was produced (see Figure 19.29). By reading the ladder produced by autoradiography of the polyacrylamide gels used to separate the fragments generated in each of the four parallel reactions, the complete nucleotide sequence of a DNA chain can be determined. This is illustrated in Figure 19.29 for a hypothetical nucleotide sequence. An autoradiogram of an actual dideoxynucleotide chain-terminator sequencing gel is shown in Figure 19.30. Under optimal conditions, long sequences of several hundred nucleotides can be determined from a single sequencing gel.

Key Points: **The polymerase chain reaction can be used to amplify specific DNA sequences *in vitro* a millionfold or more. Detailed physical maps of DNA molecules can be prepared by identifying the sites that are cleaved by various restriction endonucleases. The nucleotide sequences of DNA molecules can be determined by the Maxam and Gilbert or Sanger procedures; these nucleotide sequences provide the ultimate physical maps of genes and chromosomes.**

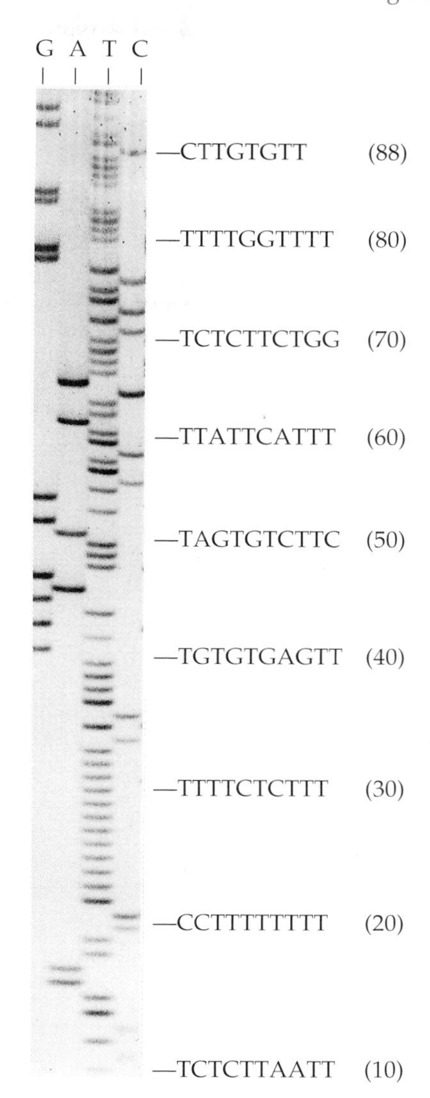

Figure 19.30 Photograph of an autoradiograph of a 2',3'-dideoxynucleotide chain-terminator sequencing gel. The sequence shown is that of a segment of one strand of a glutamine synthetase cDNA of maize.

TESTING YOUR KNOWLEDGE

1. The human genome (haploid) contains about 3×10^9 nucleotide pairs of DNA. If you digest a preparation of human DNA with *Not*I, a restriction endonuclease that recognizes and cleaves the octameric sequence 5'-GCGGCCGC-3', how many different restriction fragments would you expect to produce? Assume that the four bases (G, C, A, and T) are equally prevalent and randomly distributed in the human genome.

ANSWER

Assuming that the four bases are present in equal amounts and are randomly distributed, the chance of a specific nucleotide occurring at a given site is 1/4. The chance of a specific dinucleotide sequence (e.g., AG) occurring is $1/4 \times 1/4 = (1/4)^2$, and the probability of a specific octanucleotide sequence is $(1/4)^8$ or 1/65,536. Therefore, *Not*I will cleave such DNA molecules an average of once in every 65,536 nucleotide pairs. If a DNA molecule is cleaved at n sites, $n + 1$ fragments will result. A genome of 3×10^9 nucleotide pairs should contain about 45,776 ($3 \times 10^9/65{,}536$) *Not* I cleavage sites. If the entire human genome consisted of a single molecule of DNA, *Not*I would cleave it into 45,776 + 1 fragments. Given that these cleavage sites are distributed on 23 different chromosomes, complete digestion of the human genome with *Not*I should yield about 45,776 + 23 restriction fragments.

2. The maize gene (*gln2*) that encodes the chloroplastic form of the enzyme glutamine synthetase is known to contain a single cleavage site for *Hin*dIII, but no cleavage site for *Eco*RI. You are given an *E. coli* plasmid cloning vector that contains a unique *Hin*dIII cleavage site within the gene (amp^r) that confers resistance to the antibiotic ampicillin on the host cell and a unique *Eco*RI cleavage site within a second plasmid gene (tet^r) that makes the host cell resistant to

the antibiotic tetracycline. You are also given an *E. coli* strain that is sensitive to both ampicillin and tetracycline (*amp*^{s} *tet*^{s}). How would you go about constructing a maize genomic DNA library that includes clones carrying a complete *gln2* gene?

ANSWER

Maize genomic DNA should be purified and digested with *Eco*RI. Vector DNA should be similarly purified and digested with *Eco*RI. The maize *Eco*RI restriction fragments and the *Eco*RI-cut plasmid DNA molecules will now have complementary single-stranded ends (5'-AATT-3'). The maize restriction fragments should next be mixed with the *Eco*RI-cut plasmid molecules and covalently inserted into the linearized vector molecules in an ATP-dependent reaction catalyzed by DNA ligase. The ligation reaction will produce circular recombinant plasmids, some of which will contain maize *Eco*RI fragment inserts. Insertion of maize DNA fragments into the *Eco*RI site of the plasmid disrupts the *tet*^{r} gene so that the resulting recombinant plasmids will no longer confer tetracycline resistance to host cells. *Amp*^{s} *tet*^{s} *E. coli* cells should then be transformed with the recombinant plasmid DNAs, and the cells should be plated on medium containing ampicillin to select for transformed cells harboring plasmids. The majority of the cells will not be transformed and, thus, will not grow in the presence of ampicillin. The cells that grow on ampicillin-containing medium should be pooled and frozen at -80°C in 20 percent glycerol. This collection of cells harboring different *Eco*RI fragments of the maize genome represents a clone library that should contain clones with an intact *gln2* gene since it contains no *Eco*RI cleavage site. Note that the *Hin*dIII site of the vector could be used to construct a similar maize genomic *Hin*dIII fragment library, but such a library would not contain intact *gln2* genes because of the *Hin*dIII cleavage site in *gln2*.

QUESTIONS AND PROBLEMS

19.1 (a) In what ways is the introduction of recombinant DNA molecules into host cells similar to mutation? (b) In what ways is it different?

19.2 In what way(s) do restriction endonucleases differ from other endonucleases?

19.3 Of what value are recombinant DNA and gene-cloning technologies to geneticists?

19.4 What determines the sites at which DNA molecules will be cleaved by a restriction endonuclease?

19.5 Restriction endonucleases are invaluable tools for biologists. However, genes encoding restriction enzymes obviously did not evolve to provide tools for scientists. Of what possible value are restriction endonucleases to the microorganisms that produce them?

19.6 Why is the DNA of a microorganism not degraded by a restriction endonuclease that it produces, even though its DNA contains recognition sequences normally cleaved by the endonuclease?

19.7 One of the procedures for cloning foreign DNA segments takes advantage of restriction endonucleases like *Hin*dIII (see Table 19.1) that produce complementary single-stranded ends. These enzymes produce identical complementary ends on cleaved foreign DNAs and on the vector DNAs into which the foreign DNAs are inserted. What major advantage does this cloning strategy have over procedures that use terminal transferase to synthesize complementary single-stranded ends on foreign DNAs and vector DNAs *in vitro*?

19.8 You are working as part of a research team studying the structure and function of a particular gene. Your job is to clone the gene. A restriction map is available for the region of the chromosome in which the gene is located; the map is as follows.

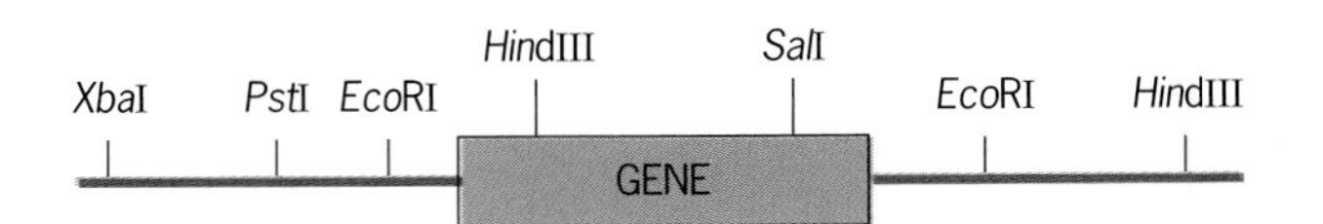

Your first task is to prepare a genomic DNA library that contains clones carrying the entire gene. Describe how you would prepare such a library in plasmid vector pUC118 (see Figure 19.14), indicating which restriction enzymes, media, and host cells you would use.

19.9 Compare the nucleotide-pair sequences of genomic DNA clones and cDNA clones of specific genes of higher plants and animals. What is the most frequent difference that you would observe?

19.10 Most of the genes of higher plants and animals that were cloned soon after the development of recombinant DNA and gene-cloning technologies were genes encoding products that are synthesized in large quantities in specialized cells or tissues. For example, about 90 percent of the protein synthesized in mature red blood cells of mammals consists of α- and β-globin chains, and the globin genes were among the first mammalian genes cloned. Why were genes of this type so prevalent among the first higher eukaryotic genes that were cloned?

19.11 Genomic clones of the chloroplastic glutamine synthetase gene (*gln2*) of maize are cleaved into two fragments by digestion with restriction endonuclease *Hin*dIII, whereas full-length maize *gln2* cDNA clones are not cut by *Hin*dIII. Explain these results.

19.12 You are studying a gene in *E. coli* that is expressed at 45°C but not at 37°C. You have shown that the regulation of this gene occurs at the level of transcription. In addition, you have isolated a protein required for induction of transcription of the gene at high temperature and have demonstrated that it binds to a specific octameric nucleotide-pair sequence upstream (5') from the gene. You have cloned the complete gene plus the upstream regulatory sequence in a plasmid, and you have available an *E. coli* mutant carrying a deletion of the entire gene. You now want to determine which of the nucleotide pairs in the 5' octameric protein binding site are involved in the interaction with (binding of) the regulatory protein. How could you determine this experimentally?

19.13 What major advantage does the polymerase chain reaction (PCR) technique have over all other methods for analyzing nucleic acid structure and function?

19.14 Certain types of molecular analyses are facilitated by the availability of large quantities of pure single-stranded DNA from a normally double-stranded DNA molecule. However, the separation and purification of the two single strands of a given double helix are usually very difficult to accomplish by standard biochemical techniques. How have molecular biologists taken advantage of a natural biological mechanism for the purification of DNA single strands?

19.15 Almost all the sophisticated cloning vectors—phages, plasmids, or phage-plasmid hybrids—in use today have one component (in addition to some type of origin of replication) in common. What is this component, and what is its function?

19.16 What advantages do transcription vectors have over other types of cloning vectors?

19.17 Suppose that a gene that you wish to clone is located on a *Hpa*I restriction fragment (that is, the gene is flanked by restriction endonuclease *Hpa*I cleavage sites). Further suppose that you want to clone this *Hpa*I fragment containing the gene of interest into the *Hin*dIII site of pUC119. What is the simplest way to obtain the desired clone? (The *Hpa*I cleavage site is shown in Table 19.1.)

19.18 An important tool of molecular biologists involves the transfer of proteins that have been separated by gel electrophoresis to nitrocellulose membranes and the detection of specific proteins on the membranes by using antibodies and coupled radioactive labels or coupled enzymatic reactions. When this procedure is used, the resulting display of the visualized protein bands is called a "western blot." What is the significance of the name western blot?

19.19 Cystic fibrosis (CF) is an autosomal recessive disorder of humans characterized by unusually viscous mucus that interferes with the normal functioning of several exocrine glands, including those in the skin (sweat glands), lungs (mucus glands), liver, and pancreas. The available evidence suggests that the CF gene product is a transmembrane protein involved in the regulation of ion transport across epithelial cell membranes, but the exact function of the CF gene product remains unknown. The *CF* gene (location: chromosome 7, region q31) has been cloned and sequenced, and studies of CF patients have shown that about 70 percent of them are homozygous for a mutant *CF* allele that has a specific three nucleotide-pair deletion (equivalent to one codon). This deletion results in the loss of a phenylalanine residue at position 508 in the predicted CF gene product. Assume that you are a genetic counselor responsible for advising families with CF in their pedigrees regarding the risk of CF among their offspring. How might you screen putative CF patients and their parents and relatives for the presence of the *CFΔ508* mutant gene? What would the detection of this mutant gene in a family allow you to say about the chances that CF will occur again in the family?

19.20 Cereal grains are major food sources for humans and other animals in many regions of the world. However, most cereal grains contain inadequate supplies of certain of the amino acids that are essential for monogastric animals such as humans. For example, corn contains insufficient amounts of lysine, tryptophan, and threonine. Thus a major goal of plant geneticists has been and is to produce corn varieties with increased kernel lysine content. As a prerequisite to the engineering of high-lysine corn, molecular biologists need more basic information about the regulation of the biosynthesis and the activity of the enzymes involved in the synthesis of lysine. The first step in the anabolic pathway unique to the biosynthesis of lysine is catalyzed by the enzyme dihydrodipicolinate synthase. Assume (1) that you have recently been hired by a major U.S. plant research institute, and (2) that the job you have been assigned is to isolate a clone of the nucleic acid sequence encoding dihydrodipicolinate synthase in maize. Briefly describe four different approaches that you might take in attempting to isolate such a clone and include at least one genetic approach.

19.21 (a) What common experimental procedure is carried out in Southern, northern, and western blot analyses? (b) What is the major difference between Southern, northern, and western blot analyses?

19.22 You have isolated a cDNA clone encoding a protein of interest in a higher eukaryote. This cDNA clone is *not* cleaved by restriction endonuclease *Eco*RI. When this cDNA is used as a radioactive probe for blot hybridization analysis of *Eco*RI-digested genomic DNA, three radioactive bands are seen on the resulting Southern blot. Does this result indicate that the genome of the eukaryote in question contains three copies of the gene encoding the protein of interest?

19.23 You are studying a circular plasmid DNA molecule of size 10.5 kilobase pairs (kb). When you digest this plasmid with restriction endonucleases *Bam*HI, *Eco*RI, and *Hin*dIII, singly and in all possible combinations, you obtain linear restriction fragments of the following sizes.

Enzyme(s)	Fragment Size(s) in kb
*Bam*HI	7.3, 3.2
*Eco*RI	10.5
*Hin*dIII	5.1, 3.4, 2.0
*Bam*HI + *Eco*RI	6.7, 3.2, 0.6
*Bam*HI + *Hin*dIII	4.6, 2.7, 2.0, 0.7, 0.5
*Eco*RI + *Hin*dIII	4.0, 3.4, 2.0, 1.1
*Bam*HI + *Eco*RI + *Hin*dIII	4.0, 2.7, 2.0, 0.7, 0.6, 0.5

Draw a restriction map (map of the restriction enzyme cleavage sites) for the plasmid that fits your data.

19.24 Ten micrograms of a decanucleotide-pair *Hpa*I restriction fragment were isolated from the double-stranded DNA chromosome of a small virus. Octanucleotide poly-A tails were then added to the 3'-ends of both strands using terminal transferase and dATP, that is,

5'-X X X X X X X X X X X-3'
3'-X'X'X'X'X'X'X'X'X'X'X-5'
↓ terminal transferase, dATP
5'-X X X X X X X X X X X A A A A A A A A-3'
3'-A A A A A A A A X'X'X'X'X'X'X'X'X'X'X'-5'

where X and X' can be any of the four standard nucleotides, but X' is always complementary to X.

The two complementary strands (Watson strand and Crick strand) were then separated and sequenced by the 2', 3'-dideoxyribonucleoside triphosphate chain-termination method. The reactions were all primed using a radioactive (^{32}P-labeled) synthetic poly-T octamer, that is,

Watson strand

3'-A A A A A A A A X'X'X'X'X'X'X'X'X'X'X'-5'
5'-^{32}P-T T T T T T T TOH

Crick strand

5'-X X X X X X X X X X X A A A A A A A A-3'
HO T T T T T T T T T-^{32}P-5'

The usual four parallel reactions: (1) ddTTP, (2) ddCTP, (3) ddATP, and (4) ddGTP (plus DNA polymerase and all other substrates and required components) were carried out for both strands. Each reaction mixture was applied to a lane in a polyacrylamide gel, fractionated by electrophoresis, and autoradiographed. The autoradiogram of the sequencing gel for one of the strands is shown on the left of the following diagram. Draw the banding pattern that would be expected on the autoradiogram of the gel for the complementary strand on the right in the diagram.

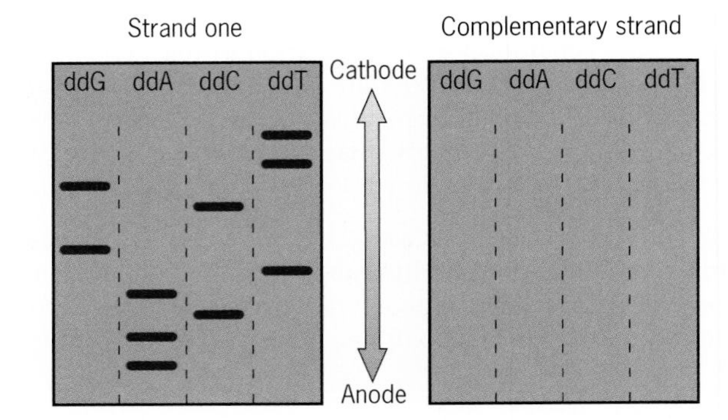

BIBLIOGRAPHY

AUSUBEL, F. M., R. BRENT, R. E. KINGSTON, D. D. MOORE, J. D. SEIDMAN, J. A. SMITH, AND K. STRUHL. 1988. *Current Protocols in Molecular Biology*. Green Publishing Associates/John Wiley & Sons, New York.

COHEN, S. N. 1975. "The manipulation of genes." *Sci. Amer.* 233(1):24–33.

INNIS, M. A., D. H. GELFAND, J. J. SNINSKY, AND T. J. WHITE. 1989. *PCR Protocols, a Guide to Methods and Applications*. Academic Press, San Diego.

MULLIS, K. B. 1990. "The unusual origin of the polymerase chain reaction." *Sci. Amer.* 262(4):56–65.

SAMBROOK, J., E. F. FRITSCH, AND T. MANIATIS. 1989. *Molecular Cloning, A Laboratory Manual*, 2nd edition. Cold Spring Harbor Laboratory Press, Cold Spring Harbor, New York.

SANGER, F., S. NICKLEN, AND A. R. COULSON. 1977. "DNA sequencing with chain-terminating inhibitors." *Proc. Natl. Acad. Sci., U.S.* 74:5463–5467.

SOUTHERN, E. M. 1975. "Detection of specific sequences among DNA fragments separated by gel electrophoresis." *J. Mol. Biol.* 98: 503–517.

VIEIRA, J., AND J. MESSING. 1987. "Production of single-stranded plasmid DNA." *Methods in Enzymology* 153: 3–11.

WATSON, J. D., M. GILMAN, J. WITKOWSKI, AND M. ZOLLER. 1992. *Recombinant DNA*, 2nd ed. Scientific American Books, New York.

WATSON, J. D., AND J. TOOZE. 1981. *The DNA Story*. W. H. Freeman, San Francisco.

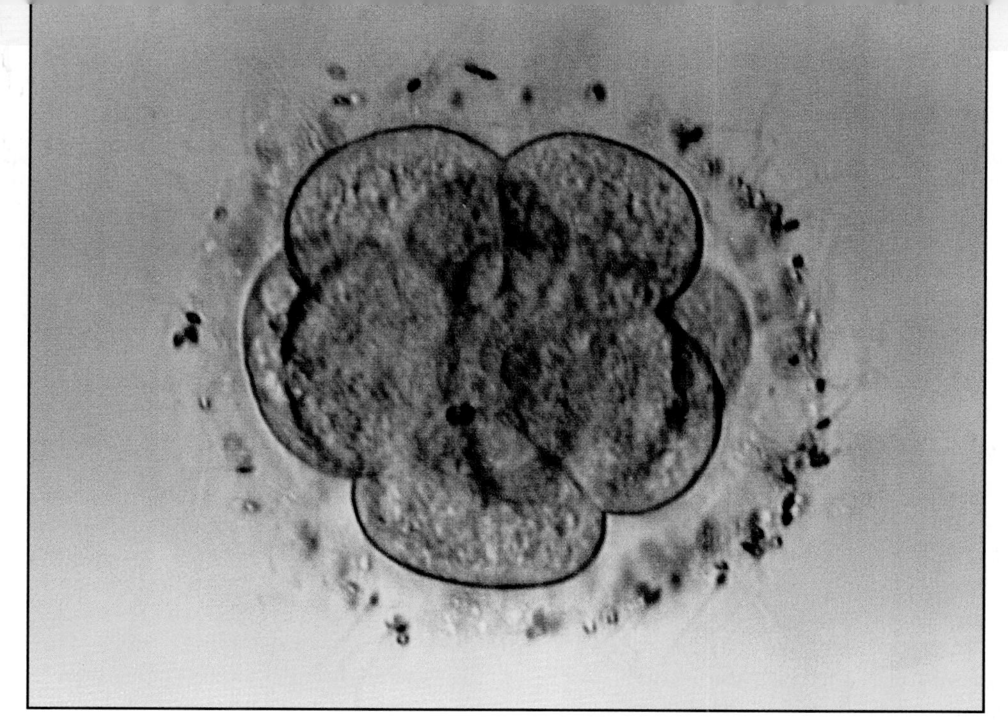

An eight-cell human pre-embryo.

20

Molecular Analysis of Genes and Gene Products

CHAPTER OUTLINE

Detection of the Tay-Sachs Mutation in Eight-Cell Pre-embryos

Brittany was conceived *in vitro* from ova and sperm obtained from her mother and father. Three days after conception, one of the eight cells of the pre-embryo from which Brittany developed was used to conduct DNA tests for a mutant gene that causes Tay-Sachs disease. The results of the tests were negative, and the pre-embryo was implanted in the womb of Brittany's mother. Nine months later, Brittany was delivered with no concern that she might someday die from Tay-Sachs disease as her older sister had.

Tay-Sachs disease (see Human Genetics Sidelight in Chapter 13) is a lethal autosomal recessive disorder. Infants with Tay-Sachs are normal at birth, but undergo rapid neurological degeneration leading to blindness, paralysis, mental retardation, and death at three to four years of age. The degeneration results from the absence of an enzyme called hexosaminidase A, which catalyzes the first step in the breakdown of a complex lipid named ganglioside GM_2. In the absence of hexosaminidase A, ganglioside GM_2 accumulates in neurons and causes progressive degeneration of the central nervous system.

Both of Brittany's parents are heterozygous for the mutant gene that causes Tay-Sachs disease. They knew that if they conceived another child, it would have a 1/4th chance of suffering from Tay-Sachs disease, as did their first daughter. Although they longed for a child, Brittany's parents were not willing to risk having another infant with Tay-Sachs disease. Tay-Sachs disease can be detected prenatally; however, because of religious convictions, the couple had ruled out the abortion of an affected fetus. *In vitro* fertilization and a DNA test for the mutant hexosaminidase A gene that both parents carried allowed confirmation that Brittany was not homozygous for the mutant gene before implantation at the pre-embryo stage. To her parents, who suffered through the degeneration and death of their first child, Brittany is a priceless treasure.

The DNA tests for Tay-Sachs and other human disorders became possible with the discovery of the genetic defects that cause these diseases. These defects have been identified by searching the human genome using sophisticated chromosome mapping and recombinant DNA techniques. In this chapter, we will consider a few of the many applications of the recombinant DNA and gene-cloning technologies described in the preceding chapter. These technologies are, without doubt, the most powerful tools ever developed in the field of biology. The ability to carry out carefully designed genetic engineering of living cells has many potential applications. Several of these applications have already reached fruition. If used wisely, genetic engineering promises to enhance the quality of human life. On the other hand, if used carelessly or in an unethical manner, genetic engineering could have a negative impact on our quality of life. As with any powerful tool, the impact of recombinant DNA techniques on our lives will depend on the wisdom of the humans who use them.

Just as geneticists now know the complete pathway of morphogenesis of bacteriophage T4 (Chapter 13), in the future they will know the complete pathway of morphogenesis of a yeast cell, a fruit fly, an *Arabidopsis* plant, or, indeed, even a human being. Moreover, at some point, biologists will understand the molecular basis of learning and memory and will know what molecular events underlie the aging process. Most important, they will understand the complex mechanisms that regulate cell division in humans and should be able to use this knowledge to prevent or cure at least some types of human cancer and life-treatening viral infections? Clearly, much important work remains to be done by the next generation of biologists, and the powerful tools of molecular genetics will play central roles in expanding our understanding of the world we live in. In this chapter, we discuss some of these tools.

MAP POSITION-BASED CLONING OF GENES

The first eukaryotic genes to be cloned were genes that are expressed at very high levels in specialized tissues or cells. For example, about 90 percent of the protein synthesized in mammalian reticulocytes is hemoglobin. Thus α- and β-globin mRNAs could be easily isolated from reticulocytes and used to prepare radioactive cDNA probes for genomic library screens. However, as most genes are not expressed at such high levels in specialized cells, how are genes that are expressed at moderate or low levels cloned? One important approach is to map the gene rather precisely and to search for a clone of the gene by using procedures that depend on its location in the genome. This approach, called **positional cloning**, can be used to identify any gene, given an adequate map of the region of the chromosome in which it is located. Indeed, as we shall discuss later in this chapter, the human genes responsible for such inherited disorders as Huntington's disease and cystic fibrosis have been identified by using the positional cloning approach.

Because the utility of positional cloning depends on the availability of a detailed map of the region of the chromosome where the gene of interest resides, major efforts have focused on the development of detailed maps of the human genome and the genomes of important model systems such as *D. melanogaster*, *C. elegans*, and *A. thaliana*. The goal of this research is to construct correlated genetic and physical maps with markers distributed at relatively short intervals throughout the genome. In the case of the human and *Drosophila* genomes, the genetic and physical maps can also be correlated with cytogenetic maps (banding patterns) of the chromosomes.

Recall that genetic maps are constructed from recombination frequencies, with 1 centiMorgan (cM) equal to the distance that yields an average frequency of recombination of 1 percent (Chapter 7). In contrast, physical maps, such as the restriction maps discussed in Chapter 19, are based on molecular distances—base pairs (bp), kilobases (kb, 1000 bp), and megabases (mb, 1 million bp). Physical distances do not correlate directly with genetic map distances because recombination frequencies are not always proportional to molecular distances. However, the two are often reasonably well correlated in euchromatic regions of chromosomes. In humans, 1 cM is equivalent, on average, to about 1 mb of DNA.

Physical maps can be constructed in several different ways. In addition to restriction maps, overlapping genomic clones can be used to build contig (contiguous clone) maps, and short unique DNA sequences, called sequence tagged sites (STSs), can then be used to relate clones to known chromosomal sites. Of course, the ultimate physical map of a chromosome is its complete nucleotide sequence. A laborious, but important, task is to link the genetic, cytogenetic, and physical maps of a chromosome to each other. We will discuss some of the techniques that are used to construct and interrelate genetic and physical maps of chromosomes in the following sections of this chapter.

Restriction Fragment-Length Polymorphism Maps

When mutations change the nucleotide sequences in restriction enzyme cleavage sites, the enzymes no longer recognize them (Figure 20.1*a*). Other mutations may create new restriction sites. These mutations result in variations in the lengths of the DNA fragments produced by digestion with various restriction enzymes (Figure 20.1*b*). Such restriction fragment-length polymorphisms, or RFLPs, have proven invaluable in constructing detailed genetic maps for use in positional cloning. The RFLPs are mapped just like other

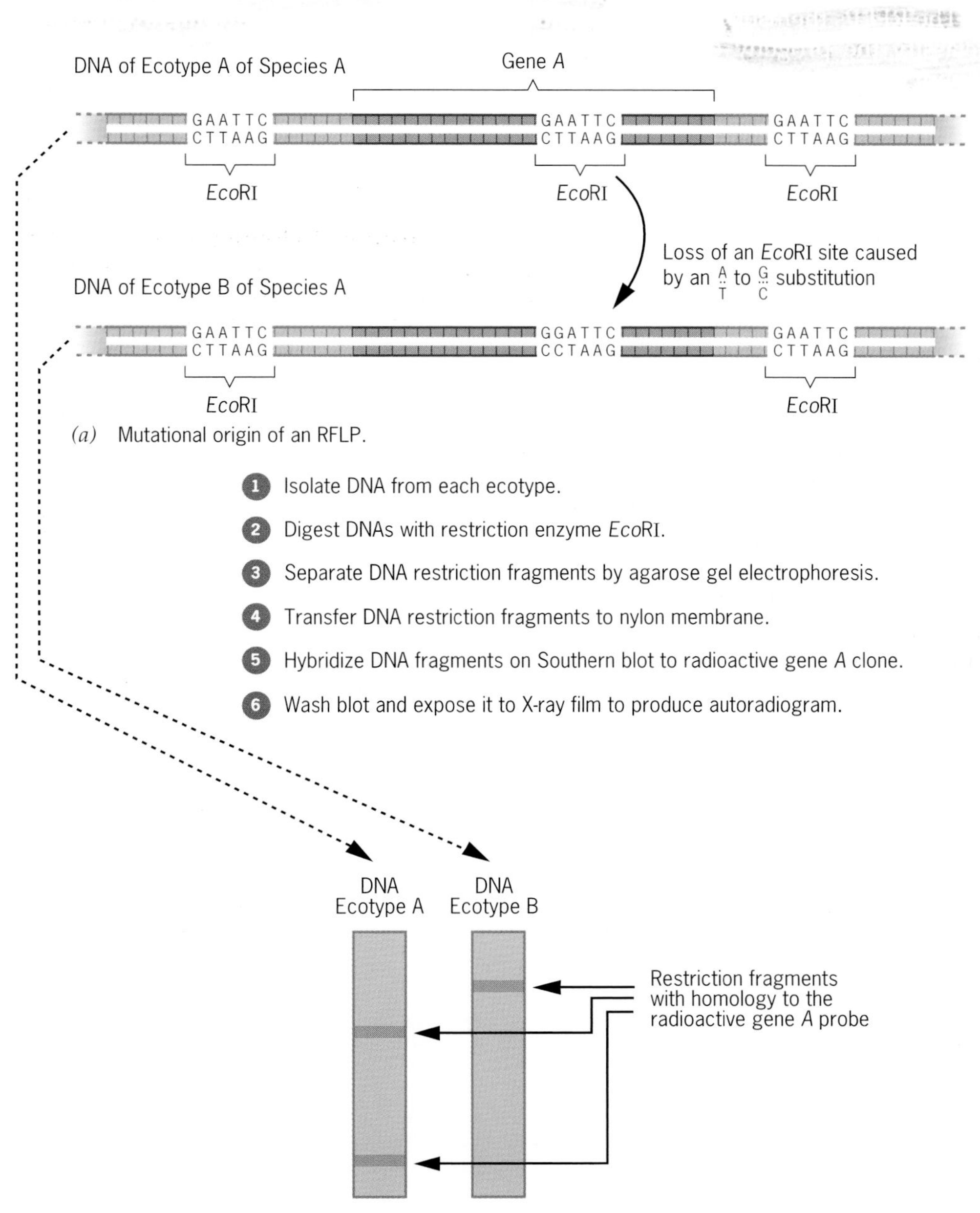

Figure 20.1 The mutational origin (*a*) and detection (*b*) of RFLPs in different ecotypes of a species. In the example shown, an A:T → G:C base-pair substitution results in the loss of the central *Eco*RI recognition sequence present in gene *A* of the DNA of ecotype A. This mutation could have occurred in a common progenitor of the two ecotypes or in an ecotype B ancestor during the early stages of its divergence from ecotype A. In ecotype A, gene *A* sequences are present on two *Eco*RI restriction fragments, whereas in ecotype B, all gene *A* sequences are present on one large *Eco*RI restriction fragment.

genetic markers; they segregate in crosses as codominant genetic markers.

The DNAs of different geographical isolates, different ecotypes (strains adapted to different environmental conditions), and different inbred lines of a species contain many RFLPs that can be used to construct detailed genetic maps. Indeed, the DNAs of different individuals—even relatives—often exhibit RFLPs. The RFLPs can be visualized directly when the fragments in DNA digests are separated by agarose gel electrophoresis, stained with ethidium bromide, and viewed under ultraviolet illumination. Individual RFLPs can be detected by using specific cDNA or genomic clones as radioactive hybridization probes on genomic Southern blots (Figure 20.1*b*). We discussed these techniques in Chapter 19. The RFLPs themselves are the phenotypes used to classify the progeny of crosses as parental or recombinant. Figure 20.2 illustrates the use of RFLPs as genetic markers in mapping experiments in the model plant *Arabidopsis*. An RFLP map of chromosome 1 of *Arabidopsis* is shown in Figure 20.3. Because its genome is small (about 10^8 nucleotide pairs), *Arabidopsis* is an excellent system in which to clone genes based on their locations.

RFLP markers have proven especially valuable in mapping the chromosomes of humans, where researchers must rely on the segregation of spontaneously occurring mutant alleles in families to estimate map distances. In humans, the most useful polymorphisms involve short sequences that are present as tandem repeats. The number of copies of each sequence present at a given site on a chromosome is highly variable. Thus these sites, called **variable number tandem repeats**, or **VNTRs**, are highly polymorphic. VNTRs vary in fragment length not because of differences in the positions of restriction enzyme

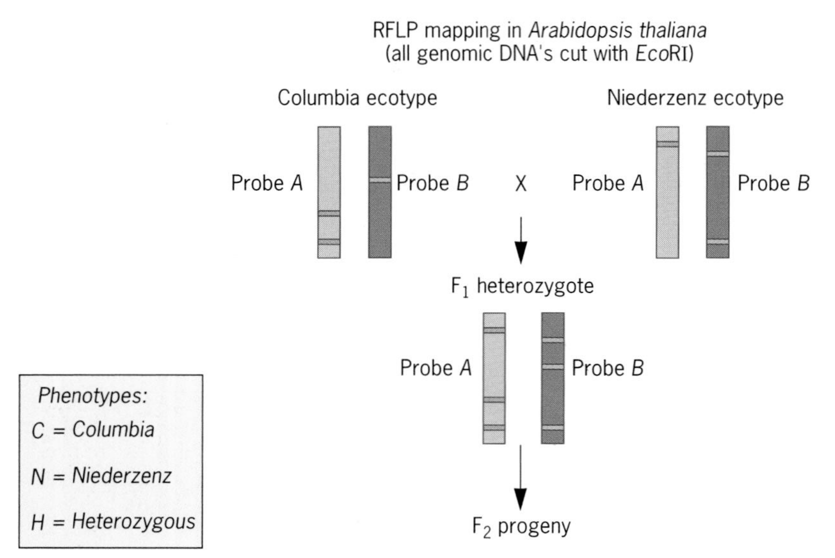

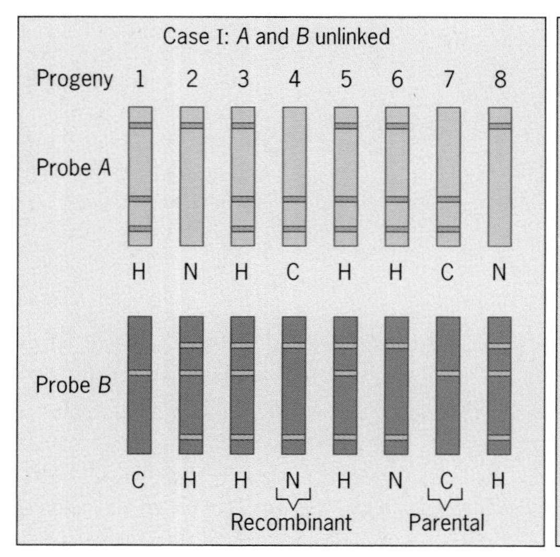

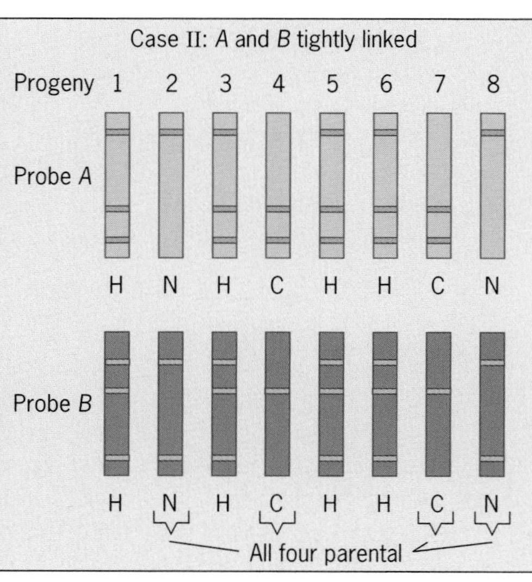

Figure 20.2 Mapping RFLPs in *Arabidopsis*. All Southern blots are hybridized sequentially to probe *A* (shown in blue) and probe *B* (shown in green). The F_1 progeny are heterozygous, containing the restriction fragments of both parents as expected. When such F_1 plants are self-pollinated, a 1:2:1 (C homozygote:heterozygote:N homozygote) segregation ratio is expected for each RFLP. If the two RFLPs assort independently (*left*), parental and recombinant combinations should occur with equal frequency among the F2 progeny. If they are tightly linked (*right*), recombinants should be rare. The frequency of recombinants can be used to calculate the map distance between RFLPs, just as for other genetic markers.

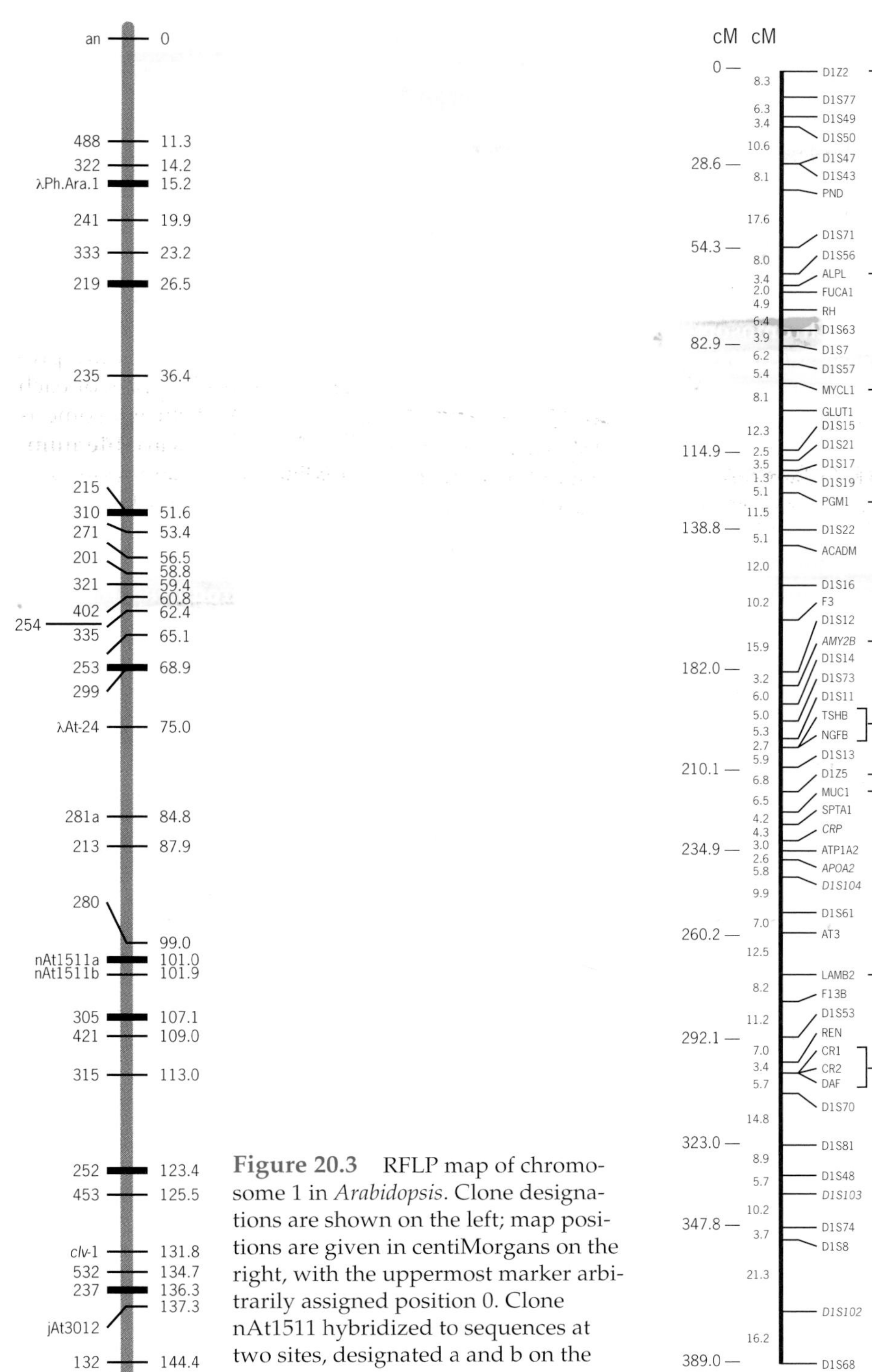

Figure 20.3 RFLP map of chromosome 1 in *Arabidopsis*. Clone designations are shown on the left; map positions are given in centiMorgans on the right, with the uppermost marker arbitrarily assigned position 0. Clone nAt1511 hybridized to sequences at two sites, designated a and b on the map.

Figure 20.4 Correlation of the RFLP map (left) and the cytogenetic map (right) of human chromosome 1. Distances are in centiMorgans (cM), with the uppermost marker set at position 0 on the left and distances between adjacent markers shown in the center.

cleavage sites, but because of differences in the number of copies of the repeated sequence between the restriction sites.

We discussed the use of data from family pedigrees to map genes by the Lod score method in Chapter 8. The Lod score compares the probabilities that the genetic markers segregating in the pedigree are unlinked or linked by various map distances (Table 8.1). Geneticists have used the Lod score method to map about 2000 RFLPs to sites on the 24 human chromosomes. Figure 20.4 shows the RFLP map of human chromosome 1 and its correlation with the cytogenetic map of this chromosome. These RFLP maps have made it possible to identify and characterize mutant genes that are responsible for several human diseases such as cystic fibrosis and Duchenne's muscular dystrophy.

Chromosome Walks and Jumps

Positional cloning is accomplished by mapping the gene of interest, identifying an RFLP marker near the gene, and then "walking" or "jumping" along the chromosome until the gene is reached. **Chromosome walking** is very difficult in species with large genomes (the walk is usually too far) and an abundance of dispersed repetitive DNA (each repeated sequence is a potential roadblock). Chromosome walking is easier in organisms such as *Arabidopsis* and *C. elegans*, which have small genomes and little dispersed repetitive DNA.

Chromosome walks are initiated by the selection of a molecular marker (RFLP or known gene clone) close to the gene of interest and the use of this clone as a hybridization probe to screen a genomic library for overlapping sequences. Restriction maps are constructed for the overlapping clones identified in the library screen, and the restriction fragment farthest from the original probe is used in a second screen of the genomic library. Repeating this procedure several times and isolating a series of overlapping genomic clones allow a researcher to walk the required distance down the chromosome to the gene of interest (Figure 20.5). Without information about the orientation of the starting clone on the linkage map, the initial walk will have to proceed in both directions until another RFLP is identified and it is determined whether the new RFLP is closer to or farther away from the gene of interest than the starting RFLP.

Verification that a clone of the gene of interest has been isolated is accomplished in various ways. In experimental organisms such as *Drosophila* and *Arabidopsis*, verification is achieved by introducing the wild-

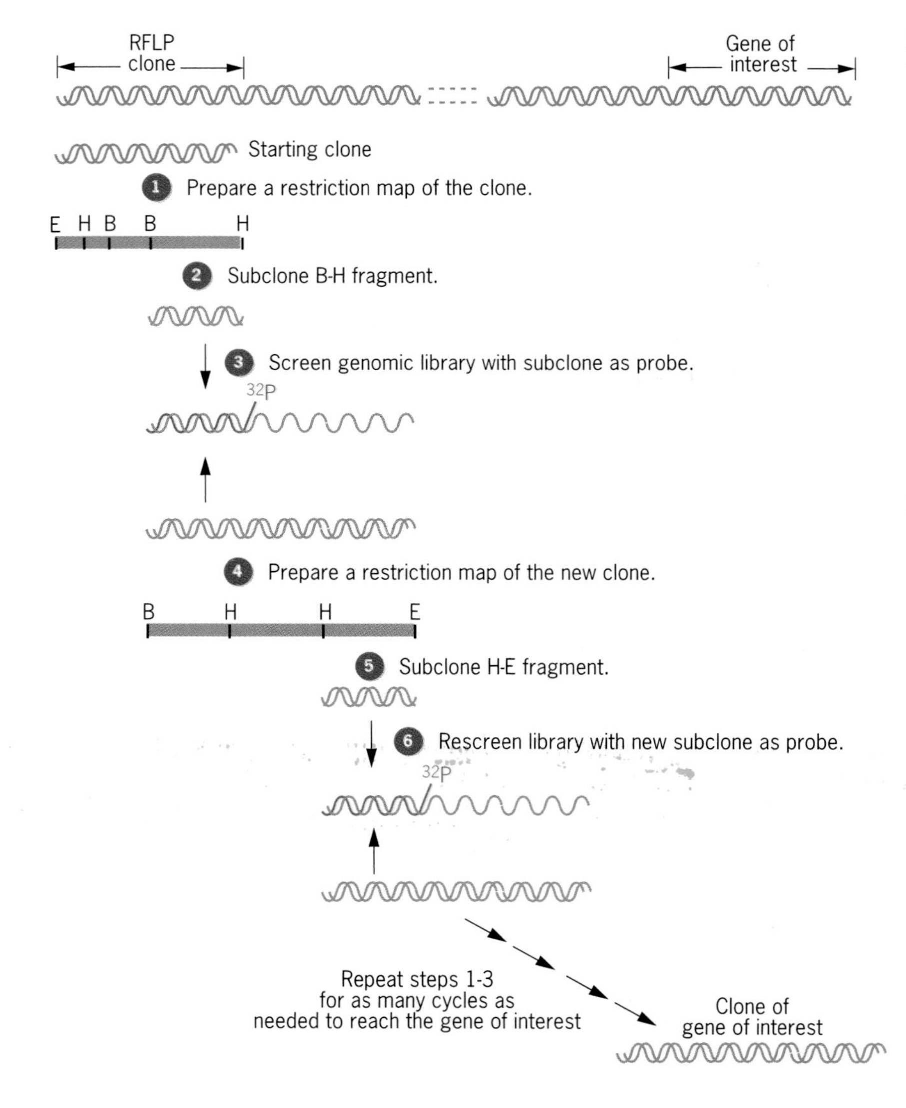

Figure 20.5 Positional cloning of a gene by chromosome walking. A walk starts with the identification of the molecular marker—such as the RFLP shown—close to the gene of interest. A restriction map is prepared for the initial RFLP clone, and the restriction fragment closest to the gene of interest is used as a probe to screen a genomic library for overlapping clones. Restriction maps are then prepared for the new genomic clones (for simplicity, only one is shown), and, again, the restriction fragment proximal to the gene of interest is subcloned and used as a probe to isolate a second set of overlapping genomic clones. Restriction maps of the new clones are prepared, and the process is repeated until the walk reaches the gene of interest.

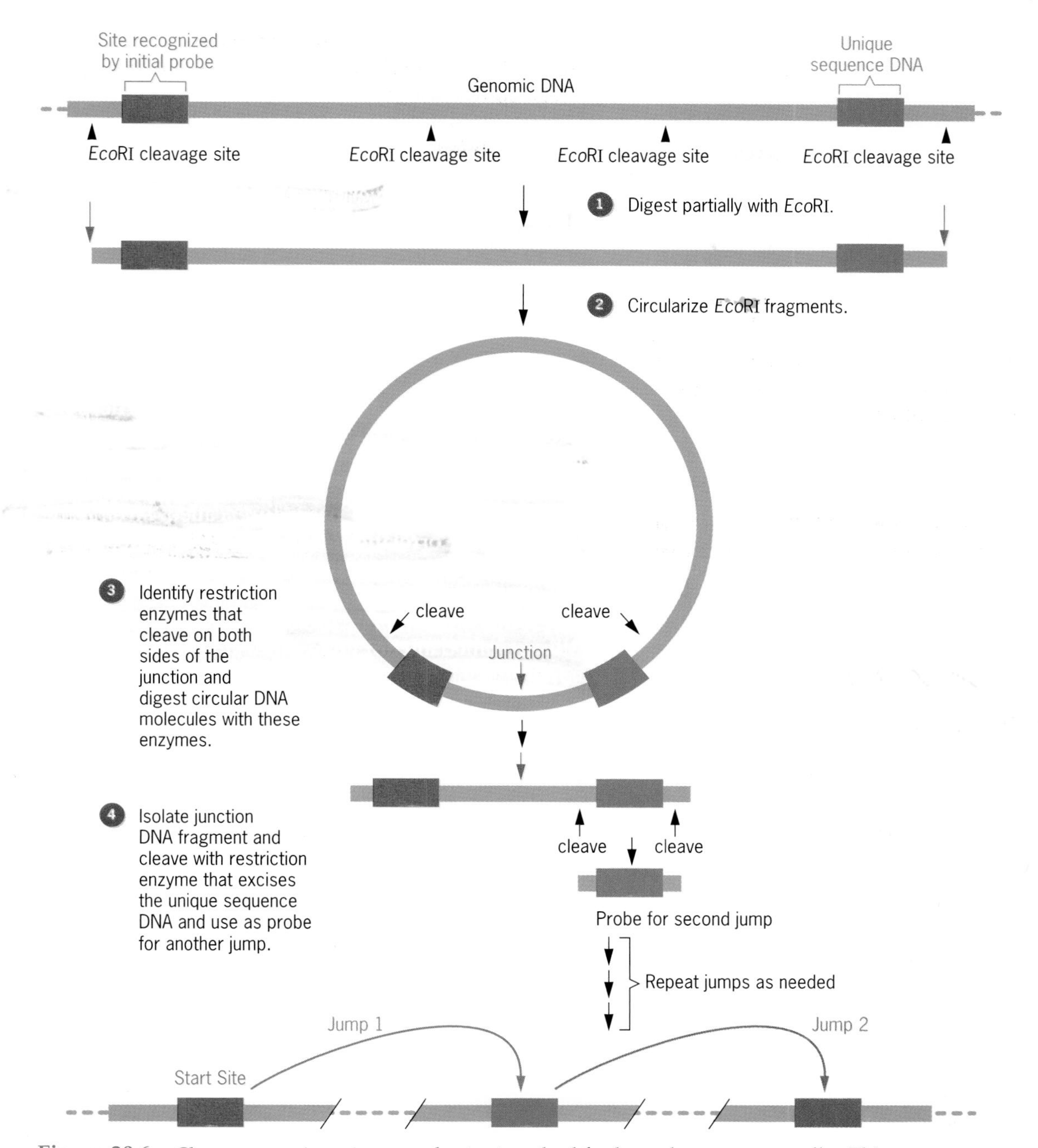

Figure 20.6 Chromosome jumping as a shortcut method for long chromosome walks. This procedure can also be used to jump over repetitive DNA sequences that block chromosome walks.

type allele of the gene into a mutant organism and showing that it restores the wild-type phenotype. In *Drosophila*, this is done by transforming mutant embryos with transposable element vectors (see Technical Sidelight in Chapter 17) that carry the wild-type allele of the gene. In humans, verification usually involves determining the nucleotide sequences of the wild-type gene and several mutant alleles and showing that the coding sequences of the mutant genes are defective and unable to produce functional gene-products.

When the distance from the closest molecular marker to the gene of interest is large, a technique called **chromosome jumping** can be used to speed up an otherwise long walk. Each jump may cover a distance of 100 kb or more. The chromosome jumping procedure is illustrated in Figure 20.6. Like a walk, a jump is initiated by using a molecular probe such as an RFLP as a start point. However, with chromosome jumps, large DNA fragments are prepared by partial digestion of genomic DNA with a restriction endonuclease. The large genomic fragments are then circularized with DNA ligase. A second restriction endonuclease is used to excise the junction fragment from the circular molecule. This junction fragment will contain both ends of the long fragment; it can be identified by hybridizing the DNA fragments on Southern blots to the initial molecular probe. A restriction map of the junction fragment is prepared, and a restriction fragment corresponding to the distal end of the long genomic fragment is cloned and used to initiate a chromosome walk or a second chromosome jump. Chromosome jumping has proven especially useful in work with large genomes such as the human genome. Chromosome jumps played a key role in identifying the gene responsible for cystic fibrosis, which we will discuss in a later section.

Physical Maps and Clone Banks

The RFLP mapping procedure has been used to construct detailed genetic maps of chromosomes, which, in turn, have made positional cloning feasible. These genetic maps are now being supplemented with physical maps of chromosomes. By isolating and preparing restriction maps of large numbers of genomic clones, overlapping clones can be identified and used to construct physical maps of entire chromosomes and even entire genomes. In principle, this procedure is simple (Figure 20.7). However, in practice, it is a formidable task, especially for large genomes. The restriction maps of the genomic clones are analyzed by computer and organized in overlapping sets of clones called **contigs**. As more data are added, adjacent contigs are joined; when the physical map of a genome is complete, each chromosome will be represented by a single contig map.

Once a physical map of a chromosome has been constructed, it needs to be correlated with the genetic and cytogenetic maps (Figure 20.8) for maximum utility. This is accomplished in several ways. The positions of cloned genes on the cytogenetic map can be determined by their hybridization (Chapter 9) with denatured DNA in chromosomes prepared for microscopic examination—a procedure called *in situ* hybridization. Correlations between the genetic and physical maps are established by locating clones of genetically mapped genes or RFLPs on the physical map. Genes that are mapped both genetically and physically are called **anchor genes**; they anchor the physical map to the genetic map and vice versa. Physical maps

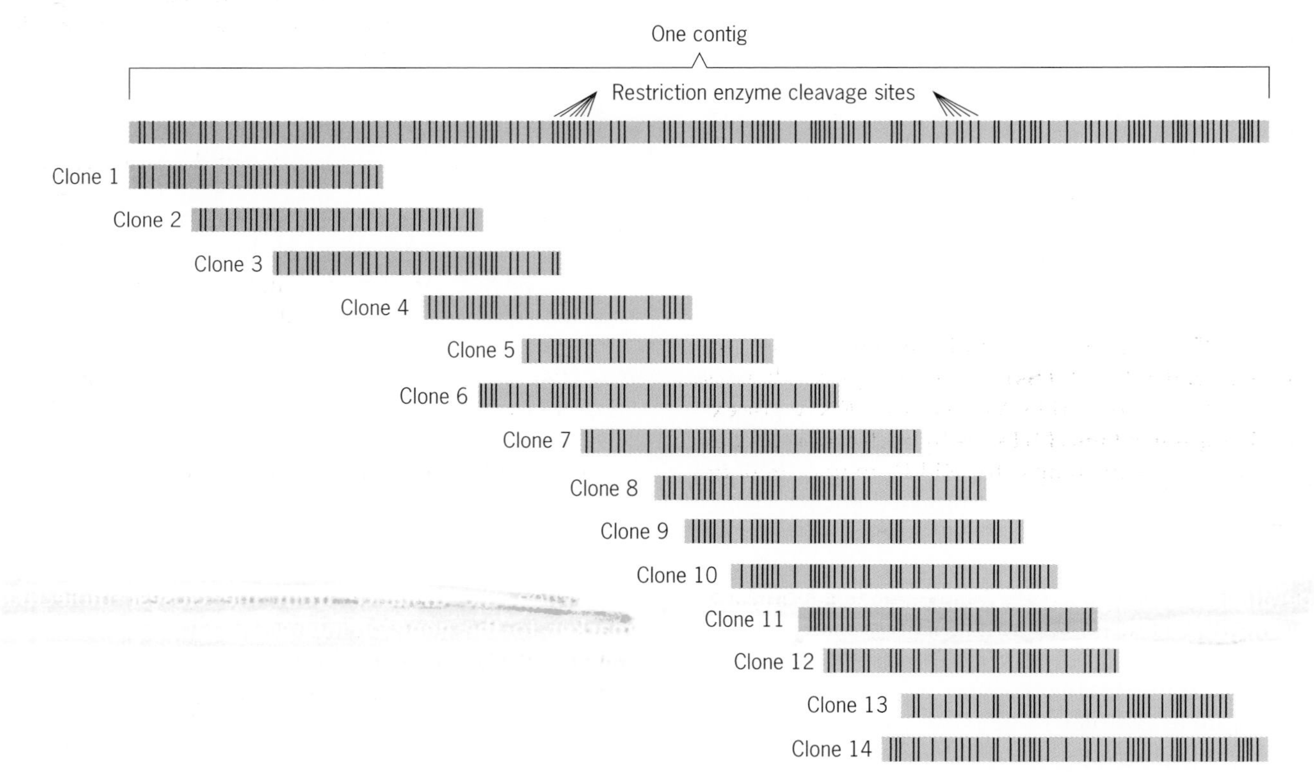

Figure 20.7 A contig map produced from overlapping genomic clones. Large—200 to 500 kb—genomic clones, such as those present in YAC or BAC vectors (Chapter 19), are used to construct contig maps. Restriction maps of individual clones are prepared and searched by computer for overlaps. Overlapping clones are then organized into contig maps like the one shown here. When the physical map of a genome is complete, each chromosome will be represented by a single contig map.

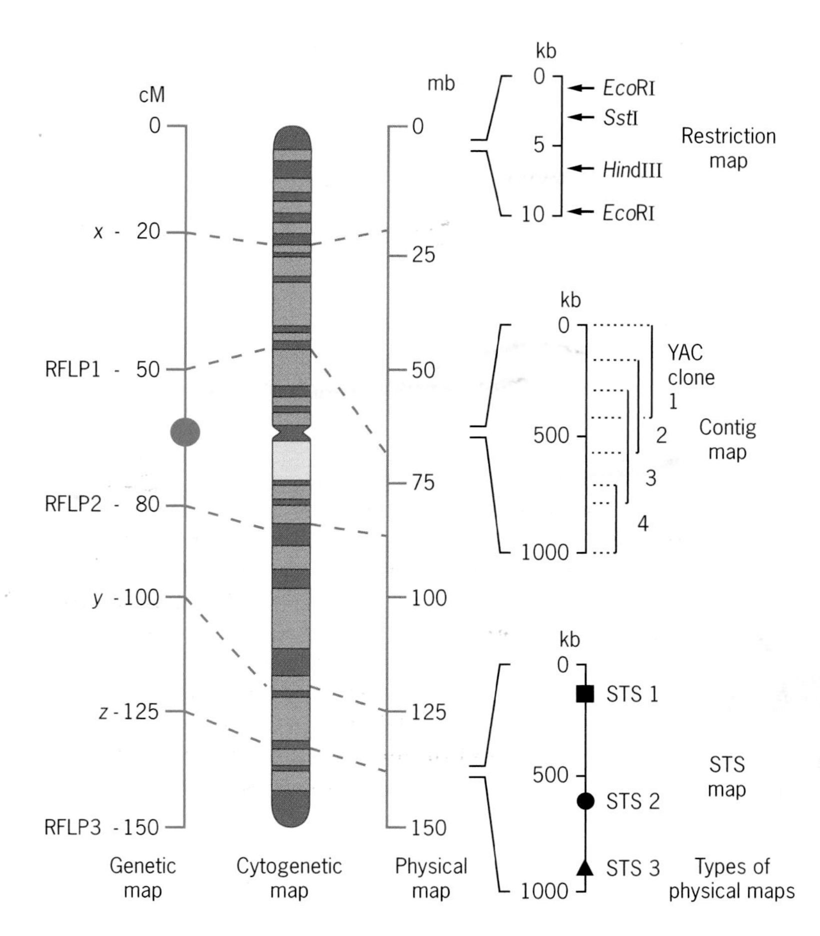

Figure 20.8 Correlation of the genetic, cytogenetic, and physical maps of a chromosome. Genetic map distances are based on crossing over frequencies and are measured in percentage recombination or centiMorgans (cM), whereas physical distances are measured in kilobase pairs (kb) or megabase pairs (mb). Restriction maps, contig maps, and STS (sequence-tagged site) maps are described in the text.

of chromosomes are currently being extended and correlated with genetic and cytogenetic maps by two additional methods. One procedure utilizes PCR (see Figure 19.25) to amplify short—usually 200 to 500 bp—unique genomic DNA sequences, Southern blots to relate these sequences to overlapping clones, and *in situ* hybridization to determine their chromosomal locations. These sequences provide anchors called **sequence-tagged sites (STSs)**. Another approach uses short cDNA sequences (DNA copies of mRNAs) or **expressed-sequence tags (ESTs)** as hybridization probes to anchor physical maps to RFLP maps (genetic maps).

The construction of physical maps of entire genomes requires that vast amounts of data be searched for overlaps. Nevertheless, detailed physical maps are available for major portions of the genomes of *C. elegans* and *A. thaliana*. In addition, complete physical maps are now available for 2 of the 24 human chromosomes (21 and Y). These physical maps are being used to prepare clone banks that contain catalogued clones collectively spanning entire chromosomes. Thus, if a researcher needs a clone of a particular gene or segment of a chromosome, that clone may already have been catalogued in the clone bank and be available on request. Obviously, the availability of such clone banks and the correlated physical maps of entire genomes will dramatically accelerate genetic research. Indeed, searching for a specific gene with and without the aid of a physical map is like searching for a book in a huge library with and without a card catalog or computer index giving the locations of the books in the library.

Key Points: **Detailed genetic and physical maps of genomes allow genes to be identified, cloned, and characterized based on their chromosomal locations. A closely linked molecular marker is used to initiate walks or jumps along the chromosome to the gene of interest. Positional cloning of genes can be done without laborious walks or jumps once complete physical maps and clone banks of chromosomes are available.**

USE OF RECOMBINANT DNA TECHNOLOGY TO IDENTIFY HUMAN GENES

Recombinant DNA techniques have revolutionized the search for defective genes that cause human diseases. Indeed, several major "disease genes" have already been identified by positional cloning. In addition, the mutations responsible for the diseases have been determined by comparing the nucleotide sequences of wild-type and mutant alleles of the genes. The coding sequences of the wild-type alleles were then translated by computer to predict the amino acid sequences of the gene products. Oligopeptides were synthesized based on the predicted amino acid sequences and used to produce antibodies, which, in turn, were used to localize the gene products and to investigate their functions *in vivo*. In some cases, the resulting information about gene function has suggested possible approaches to treating the diseases. The results of these studies will allow future treatment of some of these diseases by gene therapy.

Huntington's Disease

Huntington's disease (HD) is an insidious disorder caused by an autosomal dominant mutation, which occurs in about one of 10,000 individuals of European descent. Individuals with HD undergo progressive degeneration of the central nervous system, usually beginning at age 30 to 50 years and terminating in death 10 to 15 years later. To date, HD is untreatable. However, the identification of the gene and the mutational defect responsible for HD has kindled hope for an effective treatment in the future. Because of the late age of onset of the disease, most HD patients already have children before the disease symptoms appear. Since the disorder is caused by a dominant mutation, each child of a heterozygous HD patient has a 50 percent chance of being afflicted with the disease. Thus these children observe the degeneration and death of their HD parent, knowing that they have a 50:50 chance of suffering the same fate.

The HD gene was one of the first human genes shown to be tightly linked to an RFLP. In 1983, James Gusella, Nancy Wexler, and coworkers demonstated that the HD gene cosegregated with an RFLP that mapped near the end of the short arm of chromosome 4. This was accomplished largely based on data from studies of two large families, one in Venezuela and one in the United States. Subsequent research showed that the linkage was about 96 percent complete; 4 percent of the offspring of HD heterozygotes were recombinant for the RFLP and the HD allele. Given this early localization of the HD gene to a relatively short segment of chromosome 4, some geneticists predicted that the HD gene would soon be cloned and characterized. However, the task was more difficult than anticipated and took a full 10 years to accomplish. Unlike many other inherited disorders, no case of Hunting-

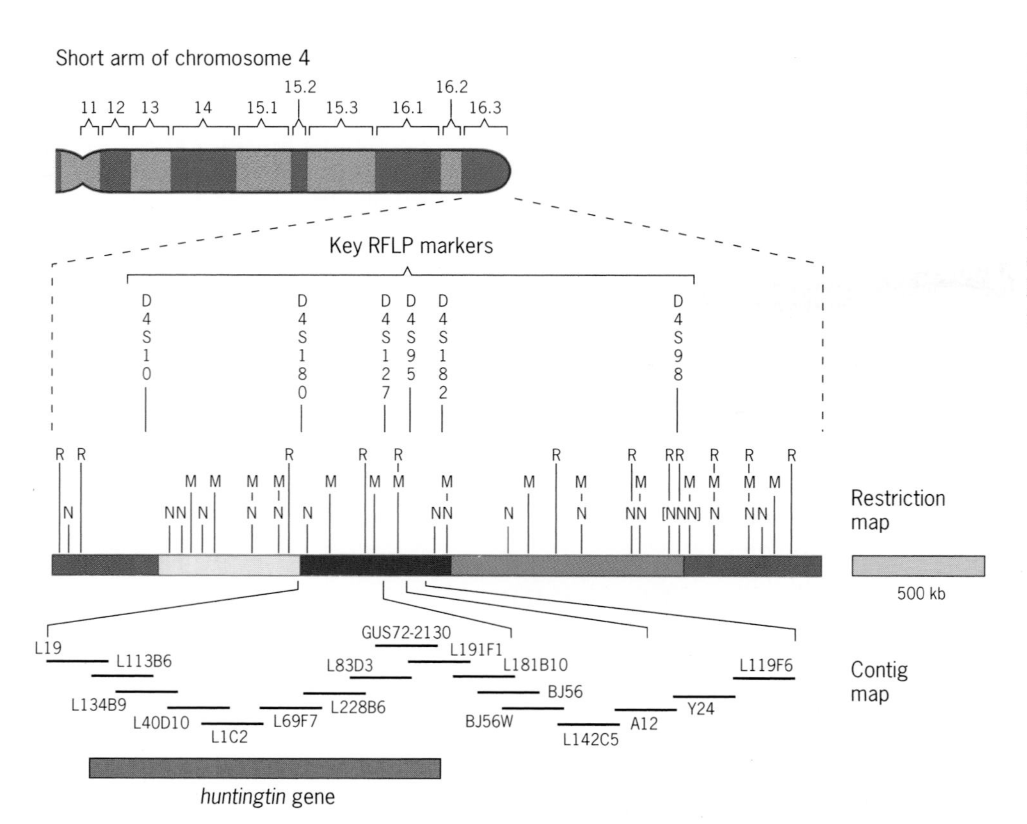

Figure 20.9 Identification of the gene responsible for Huntington's disease by positional cloning. The cytogenetic map of the short arm of chromosome 4 is shown at the top. The RFLP markers, restriction map, and contig map used to locate the *huntingtin* gene are shown below the cytogenetic map.

ton's disease is caused by a chromosome rearrangement with a breakpoint that could be used to pinpoint the location of the gene. Therefore, even though the RFLP cosegregation data focused the search on a 500 kb region of chromosome 4, researchers still had to identify all of the genes in this region and determine which one was responsible for HD. Detailed restriction and contig maps were prepared of the 500 kb region (Figure 20.9). Then, a procedure called **exon amplification** was used to identify coding sequences of genes within the region of interest.

Exon amplification (Figure 20.10) is used to identify coding sequences that are flanked by 5' and 3' intron splice sites. In principle, the procedure is simple. Genomic DNA fragments are inserted into a cloning site within an intron from a mammalian virus. The viral intron contains 5' and 3' splice sites flanking the cloning site. Therefore, if an exon flanked by 3' and 5' intron splice sites is inserted into the cloning site of the vector, the exon will lie between two introns that should be spliced out *in vivo*. This intron-cloning site is present in a viral cloning vector, which can be intro-

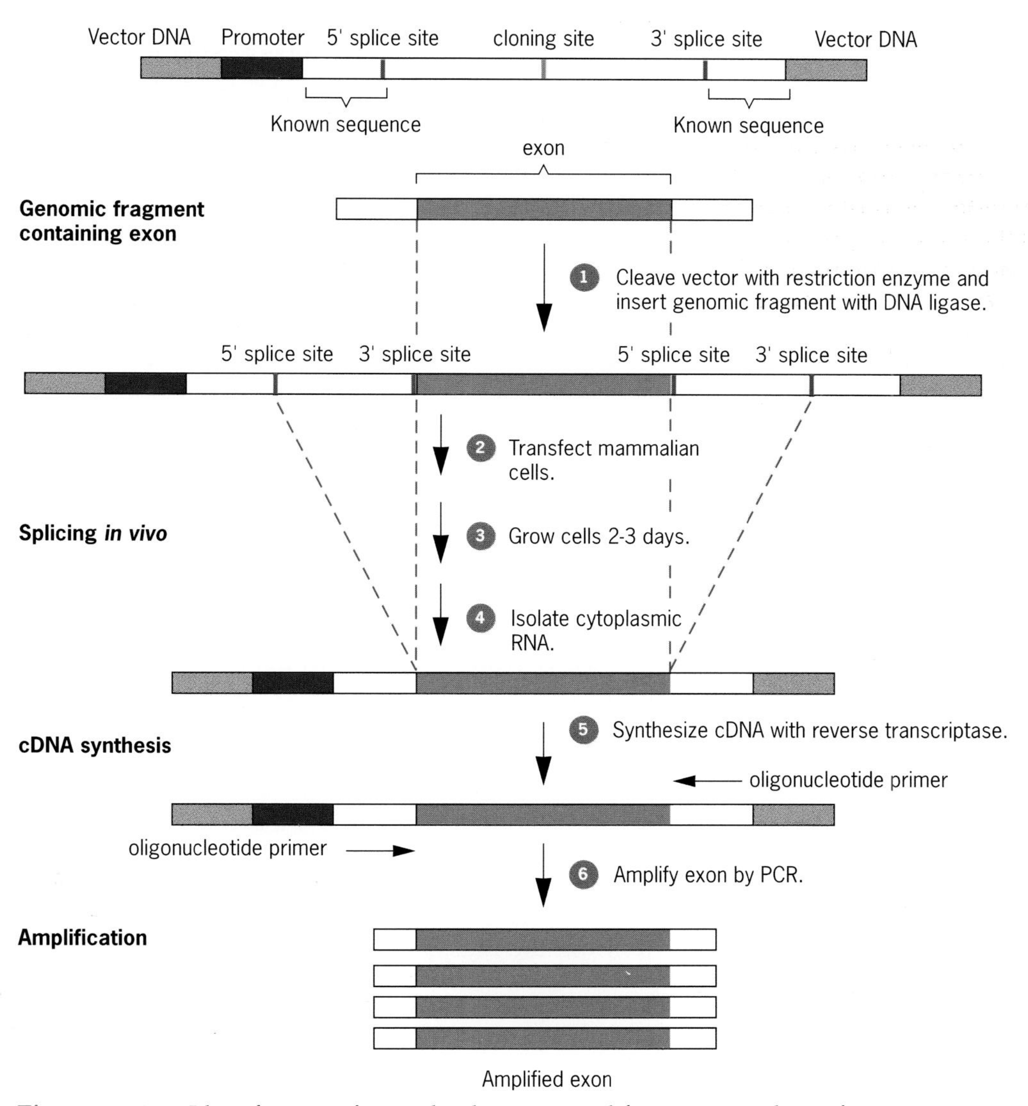

Figure 20.10 Identification of exons by the exon amplification procedure. If a genomic DNA fragment contains an exon with intact 3' and 5' splice sites, the adjoining intron sequences will be spliced out *in vivo* after its insertion into a viral splicing vector. The splicing events will move the oligonucleotide primer sites adjacent to the exon so that it is amplified during the subsequent PCR step. If 3' and 5' splice sites are not present in the fragment, splicing will not occur, and the exon will not be amplified by this procedure.

duced into host cells by transfection. The vector contains a viral origin of replication and a strong promoter adjacent to the cassette so that the entire region is transcribed efficiently. After transfection, the mammalian culture cells are grown for a few days to allow transcription and RNA splicing to occur. Then, cytoplasmic RNAs are isolated, cDNAs are synthesized with reverse transcriptase (Figure 19.11), and the exon is amplified by PCR (Figure 19.25) using primers complementary to flanking vector DNA sequences. The amplified exon sequences are then used as probes to screen cDNA libraries by colony or plaque hybridization (Figure 19.12). The cDNA clones, in turn, are used to identify complementary genomic DNA clones. The nucleotide sequences of the cDNAs and genomic DNAs are compared in order to define the introns and exons in the genes. These procedures allow identification of all genes within a given region of a chromosome. Then, the gene responsible for a given disorder must be identified by comparing the sequences of the alleles in affected and normal individuals.

Using all of the above procedures, Gusella, Wexler, and coworkers identified a gene, first called IT15 (for Interesting Transcript number 15) and subsequently named *huntingtin*, that spans about 210 kb near the end of the short arm of chromosome 4 (Figure 20.9). This gene contains a trinucleotide repeat, $(CAG)_n$, which is present in from 11 to 34 copies on each chromosome 4 of normal individuals. In HD patients, the chromosome carrying the HD mutation contains from 42 to 100 or more copies of the CAG repeat in this gene. Moreover, the age of onset of HD is inversely correlated with the number of copies of the trinucleotide repeat. Rare juvenile onset of the disease occurs in children with an unusually high repeat copy number. The trinucleotide repeat regions of HD chromosomes are unstable, with repeat number often expanding and sometimes contracting between generations. Gusella, Wexler, and collaborators detected expanded CAG repeat regions in chromosomes from 72 different families with HD, leaving little doubt that they had identified the correct gene.

The *huntingtin* gene is expressed in many different cell types, producing a large 10–11 kb mRNA. The coding region of the *huntingtin* mRNA predicts a protein 3144 amino acids in length. Unfortunately, the predicted amino acid sequence of the huntingtin protein has provided no hint as to its function. The amino acid sequence is not closely related to that of any known protein. The dominance of the HD mutation indicates that the mutant protein has some new function that causes the disease, but what this function might be is unknown.

HD was the fourth human disease to be associated with an unstable trinucleotide repeat. Fragile X syndrome (the most common form of mental retardation in humans), and myotonic dystrophy and spino-bulbar muscular atrophy (both diseases associated with loss of muscle control), had previously been shown to result from expanded trinucleotide repeats. More recently, two other neurodegenerative diseases, spinocerebellar ataxia type 1 and dentatorubro-pallidoluysian atrophy, have been shown to result from similar enlarged trinucleotide repeat regions. These results indicate that the expansion of such repeat regions may be a common mutational event in humans.

Although the identification of the genetic defect, the expanded trinucleotide repeat in the *huntingtin* gene, has contributed little insight into the molecular and cellular basis of the neurological degeneration that occurs in HD patients, it has provided a simple and accurate DNA test for the HD mutation (Figure 20.11). Once the nucleotide sequences of the *huntingtin* gene on either side of the trinucleotide repeat region were known, oligonucleotide primers could be synthesized and used to amplify the region by PCR, and the number of CAG repeats could be determined by polyacrylamide gel electrophoresis. Thus individuals at risk of carrying the mutant *huntingtin* gene can easily be tested for its presence. Because the PCR procedure requires very little DNA, the test for HD also can be performed prenatally on fetal cells obtained by amniocentesis or chorionic biopsy (Human Genetics Sidelight in Chapter 6).

Given the availability of the DNA test for the HD mutation, individuals who are at risk of transmitting the defective gene to their children can determine whether they carry it before starting a family. However, the potential psychological effects of a positive identification of the HD mutation can be traumatic. Given a 50:50 chance of a diagnosis of HD:no HD, would you want to know early that you carry the mutation, or, instead, hold on to the hope for as long as possible that you do not carry it? For those individuals who test negative for the HD mutation, a heavy burden is lifted. For those who test positive, will knowing help them prepare emotionally and financially for the onset of the disease? Clearly, the DNA tests for HD must be performed with great care and only in conjunction with proper counseling. Still, for the young men and women at risk for HD and not wanting to transmit the mutation to their children, the DNA test offers hope. Each person with a heterozygous parent has a 50 percent chance of not carrying the defective gene. If the test is negative, she or he can begin a family with no concern about transmitting the mutation.

If the test is positive, the fetus can be tested prenatally, or the couple can consider *in vitro* fertilization, as did the parents we discussed at the beginning of this chapter. If the eight-cell pre-embryo tests negative for the HD mutation, it can be implanted in the mother's uterus with the knowledge that it carries two normal

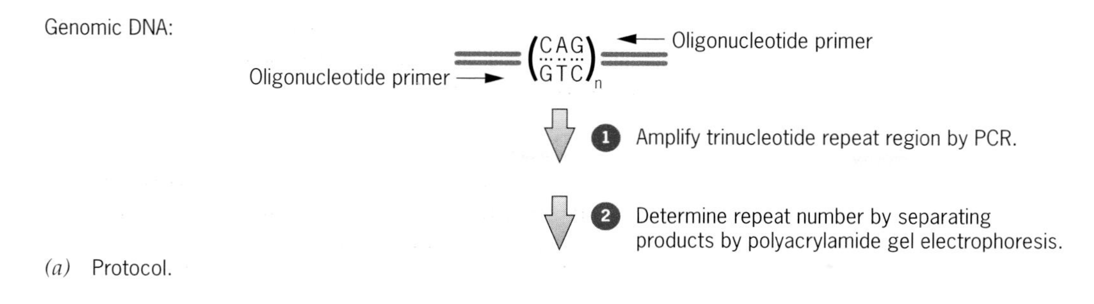

(*a*) Protocol.

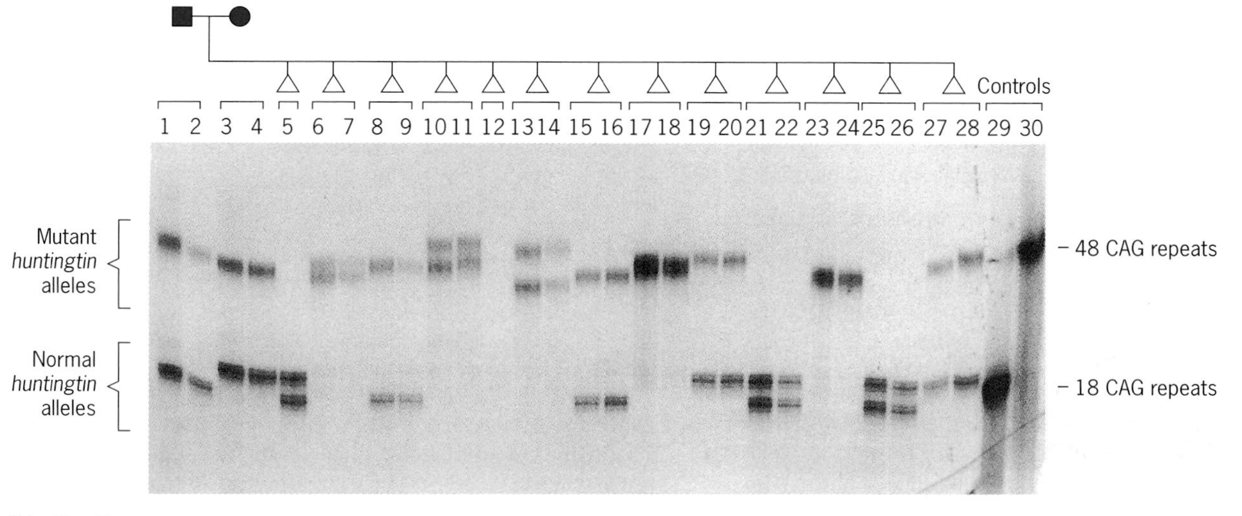

(*b*) Results.

Figure 20.11 Testing for the expanded trinucleotide repeat regions in the *huntingtin* gene that are responsible for Huntington's disease by PCR. The results shown are from a Venezuelan family in which the parents are heterozygous for the same mutant *huntingtin* allele. The order of birth of the children has been changed, and their sex is not given to assure anonymity. All of the family members but two (DNA samples 5 and 12) were tested twice using DNA from cells obtained at least one full year apart.

copies of the *huntingtin* gene. If used conscientiously, the DNA test for the HD mutation should diminish human suffering from this dreaded disease.

Cystic Fibrosis

Cystic fibrosis (**CF**) is one of the most common inherited diseases in humans, affecting 1 in 2,000 newborns of Northern European heritage. CF is inherited as an autosomal recessive mutation, and the frequency of heterozygotes is estimated to be about 1 in 25 in Caucasian populations. In the United States alone, over 30,000 people suffer from this devastating disease. One easily diagnosed symptom of CF is excessively salty sweat, a largely benign effect of the mutant gene. Other symptoms are anything but benign. The lungs, pancreas, and liver become clogged with a thick mucus, which results in chronic infections and the eventual malfunction of these vital organs. In addition, mucus often builds up in the digestive tract, causing individuals to be malnourished no matter how much they eat. Lung infections are recurrent, and patients often die from pneumonia or related infections of the respiratory system. In 1940, the average life expectancy for a newborn with CF was less than two years. With improved methods of treatment, life expectancy has gradually increased. Today, the life expectancy for someone with CF is about 30 years, but the quality of life is poor.

The identification of the *CF* gene is one of the major successes of positional cloning. Biochemical analyses of cells from CF patients had failed to identify any specific metabolic defect or mutant gene product. Then, in 1989, Francis Collins and Lap-Chee Tsui and their coworkers identified the *CF* gene and characterized some of the mutations that cause this tragic disease. The cloning and sequencing of the *CF* gene quickly led to the identification of its product, which, in turn, has suggested approaches to clinical treatment

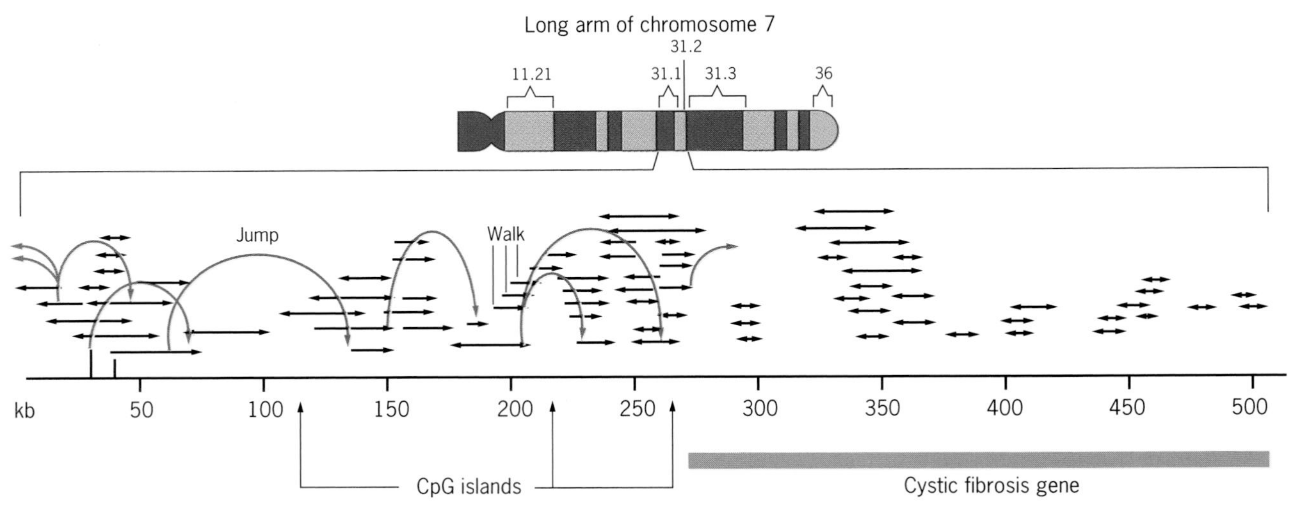

Figure 20.12 The sequence of chromosome walks and jumps used to locate and characterize the cystic fibrosis gene. The positions of CpG islands used as landmarks in locating the 5' end of the gene are also shown.

of the disease and hope for successful gene therapy for CF patients in the future.

The *CF* gene was first mapped to the long arm of chromosome 7 by its cosegregation with specific RFLPs. Further RFLP mapping indicated that the gene was located within a 500-kb region of chromosome 7. The two RFLP markers closest to the *CF* gene were then used to initiate chromosome walks and jumps and to begin construction of a detailed physical map of the region (Figure 20.12). Three kinds of information were used to narrow the search for the *CF* gene.

1. Human genes are often preceded by clusters of cytosines and guanines called **CpG islands**. Three such clusters are present just upstream from the *CF* gene (Figure 20.12).
2. Important coding sequences usually are conserved in related species. When exon sequences from the *CF* gene were used to probe Southern blots containing restriction fragments from human, mouse, hamster, and bovine genomic DNAs (often called **zoo blots**), the exons were found to be highly conserved.
3. As previously mentioned, CF is known to be associated with abnormal mucus in the lungs, pancreas, and sweat glands. A cDNA library was prepared from mRNA isolated from sweat gland cells growing in culture and screened by colony hybridization using exon probes from the *CF* gene (candidate *CF* gene at the time).

Use of the sweat gland cDNA library proved to be critical in identifing the *CF* gene, because northern blot experiments subsequently showed that this gene is expressed only in epithelial cells of the lungs, pancreas, salivary glands, intestine, and reproductive tract. Thus cDNA clones of the *CF* gene would not have been identified using cDNA libraries prepared from other tissues and organs. The northern blot results also showed that the putative *CF* gene is expressed in the appropriate tissues.

The identification of a candidate gene as a disease gene hinges on comparisons of normal and mutant alleles from several different families. CF is unusual in that 70 percent of the mutant alleles contain the same three-base deletion, Δ*F508*, which eliminates the phenylalanine residue at position 508 in the *CF* gene product. Unlike the *huntingtin* gene, the nucleotide sequence of the *CF* gene proved very informative. The gene is huge, spanning 250 kb and containing 24 exons (Figure 20.13). The *CF* mRNA is about 6.5 kb in length and encodes a protein of 1480 amino acids. A computer search of the protein data banks quickly showed that the *CF* gene product is similar to several ion channel proteins, which form pores between cells through which ions pass. The *CF* gene product, called the **cystic fibrosis transmembrane conductance regulator**, or **CFTR protein**, forms ion channels (Figure 20.13) through the membranes of cells that line the respiratory tract, pancreas, sweat glands, intestine, and other organs and regulates the flow of salts in and out of these cells. A large body of evidence indicates that the CFTR ion channel is activated by a cyclic ATP-dependent protein kinase, although the exact mechanism of activation remains to be worked out. Because the mutant CFTR protein does not function properly in CF patients, salt accumulates in epithelial cells and mucus builds up on the surfaces of these cells.

Although 70 percent of the cases of CF are due to the Δ*F508* trinucleotide deletion, over 170 different *CF* mutations have been identified (Figure 20.14). About

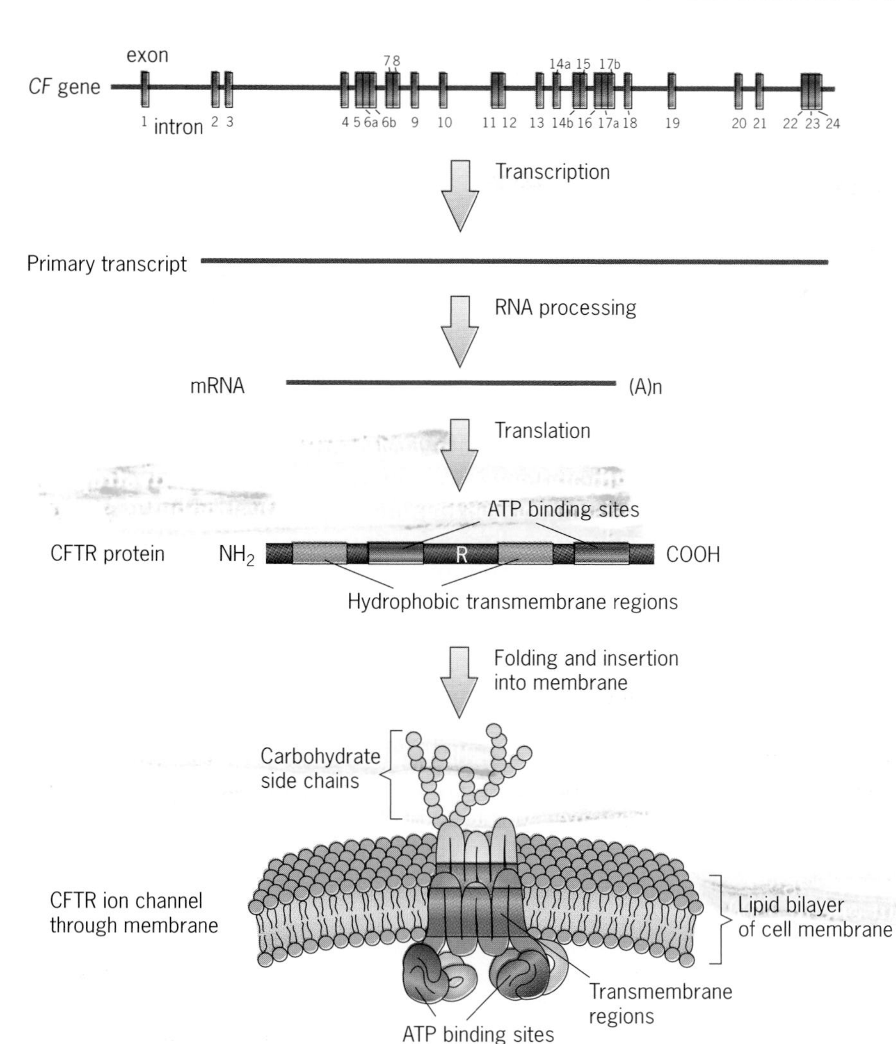

Figure 20.13 The structure of the *CF* gene and its product, the CFTR protein. The CFTR protein forms ion channels through the membranes of epithelial cells of the lungs, intestine, pancreas, sweat glands, and some other organs.

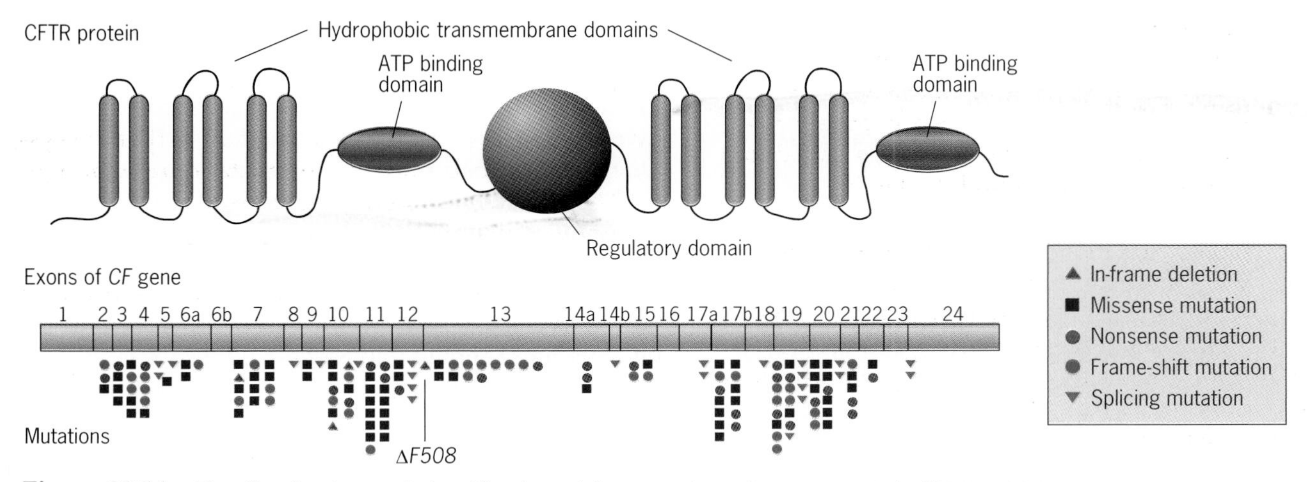

Figure 20.14 The distribution and classification of the mutations that cause cystic fibrosis are shown below the exons of the *CF* gene. A schematic diagram of the CFTR protein is shown above the exon map to illustrate the domains of the protein that are altered by the mutations. About 70 percent of all cases of CF result from mutation *ΔF508*, which deletes the phenylalanine present at position 508 of the normal CFTR protein.

20 of these mutations are quite common, others are rare, and many have been identified in only one individual. Several of these mutations can be detected by DNA screens such as the test for the Δ*F508* deletion illustrated in the Technical Sidelight in Chapter 19. These tests can be performed on fetal cells obtained by amniocentesis or chorionic biopsy. They have also been done successfully on eight-cell pre-implantation embryos produced by *in vitro* fertilization.

The rapid accumulation of new information about the CFTR protein has dramatically increased prospects for the successful treatment of CF by gene therapy. The ion channel defect has been corrected in CF culture cells by the introduction of a functional *CF* gene, and many strategies for gene therapy treatment of the disease are currently being evaluated. One possibility involves the delivery of the wild-type *CF* gene to lung epithelial cells with one of the viruses that causes the common cold. Given the wealth of new information about the function of the CFTR protein at the molecular and cellular level, we can at least be optimistic that CF, once a deadly childhood killer, may be successfully treated in the not-too-distant future.

Duchenne Muscular Dystrophy

Duchenne muscular dystrophy (DMD) is a tragic X-linked recessive disease that afflicts about one in 3300 newborn males. Individuals with DMD undergo progressive muscle degeneration beginning within the first few years of life. They usually are confined to a wheelchair by age twelve and die in the late teens or early twenties, often from cardiac or respiratory failure. Because of the early onset of DMD, males seldom transmit the mutant allele. Thus DMD is extremely rare in females. Transmission of the defective gene is almost exclusively from heterozygous women to half of their sons. A less severe form of the disorder, called Becker muscular dystrophy (BMD), occurs in about 1 of 30,000 newborn males. Muscle degeneration is less severe, onset is later, and survival is longer in individuals with BMD. Both diseases are caused by mutations in the same gene; DMD results from a total loss of gene-product function, whereas BMD occurs because of a partial loss of function.

The *DMD* gene was identified by positional cloning involving chromosome walks like those used to isolate the *huntingtin* and *CF* genes. However, locating the *DMD* gene was easier because of numerous deletions and a key translocation breakpoint within the gene. The translocation breakpoint indicated that the *DMD* gene was located in region 21 on the short arm of the X chromosome. This translocation moved rRNA genes from the tip of chromosome 21 to a position adjacent to the *DMD* gene on the X chromosome, allowing researchers to begin walking to the *DMD* gene with rRNA probes. The identification of the *DMD* gene was greatly assisted by the existence of numerous mutant alleles containing deletions (Figure 20.15). Indeed, in one study of 160 individuals with DMD, 98 (61 percent) contained a deletion of all or part of the *DMD* gene.

In contrast to its identification, the characterization of the *DMD* gene proved to be a major challenge because of its enormous size. The *DMD* gene is almost ten times larger than any other human gene sequenced to date. It contains 79 exons and spans more than 2500 kb. The *DMD* mRNA is about 14 kb in length and encodes a huge protein, called **dystrophin**, consisting of 3685 amino acids. Dystrophin is synthesized primarily in muscle cells as would be expected given the symptoms of DMD. Lower levels of *DMD* gene expression occur in the brain. Although the function of dystrophin is uncertain, it has been associated with a complex of proteins on the intracellular membranes of muscle cells and is thought to link actin filaments in muscle to these membranes. In addition to dystrophin, smaller proteins, called **apo-dystrophins**, are encoded by the distal portion of the *DMD* gene. They are synthesized in nonmuscle cells from small transcripts controlled by a promoter located between exons 62 and 63 in the *DMD* gene. The functions of the apo-dystrophins are still unknown, but their existence and tissue-specific synthesis suggest that the *DMD* gene may have different functions in different tissues.

DNA tests have been developed for many of the *DMD* deletions; these tests can be performed on fetal cells collected by amniocentesis or chorionic biopsy. *DMD* deletions have also been detected using DNA from a single cell of eight-cell pre-embryos produced by *in vitro* fertilization. Thus both prenatal and pre-implantation screening are available to women who are heterozygous for a *DMD* mutation. In addition, the implantation of normal muscle progenitor cells called myoblasts may offer some relief for DMD patients. One problem encountered in using gene therapy to treat DMD is that the dystrophin gene is too large to fit in any of the standard viral cloning vectors. However, this has not eliminated hope for successful gene therapy treatment of DMD in the future. In fact, scientists have constructed a functional *DMD* mini-gene by deleting unessential regions of the wild-type gene; this mini-gene does fit into retroviral cloning vectors. Thus DMD may be treated by gene therapy in the future.

Key Points: **The human genes that cause Huntington's disease, cystic fibrosis, Duchenne muscular dystropy, and several other disorders have been identified by positional cloning. The nucleotide sequences**

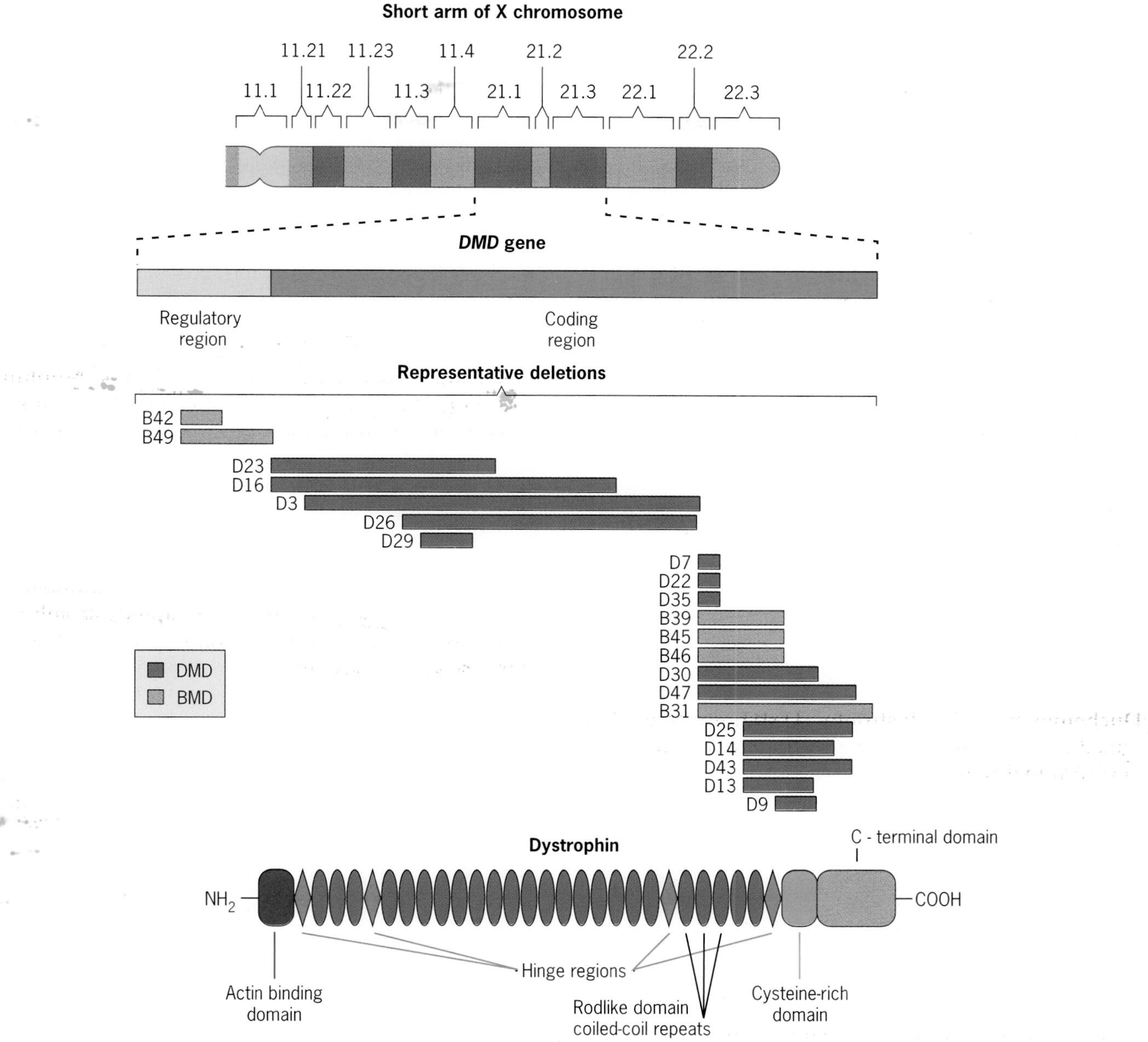

Figure 20.15 Structure of the human Duchenne muscular dystrophy (*DMD*) gene and its product, dystrophin. The locations of a few of the many deletions in the *DMD* gene also are shown. The four major domains in the huge dystrophin polypeptide are shown at the bottom. Proline-rich hinge regions separate the actin-binding and cysteine-rich domains from the central rodlike domain, and two other hinge regions interrupt the rodlike domain. The C-terminal domain is thought to anchor dystrophin to a complex of membrane-bound proteins. DMD = Duchenne muscular dystrophy; BMD = Becker muscular dystrophy.

of these genes were used to predict the amino acid sequences of their polypeptide products and to obtain valuable information about the functions of the gene products. The identification of these genes led directly to new methods of treating the diseases, possible approaches to gene therapy, and DNA tests for the mutations that cause the diseases.

MOLECULAR DIAGNOSIS OF HUMAN DISEASES

Once the gene responsible for a human disease has been cloned and sequenced and the nature of the mutations that cause the disorder is known, a molecular test for the mutant alleles usually can be designed.

These tests can be performed on small amounts of DNA by using PCR to amplify the DNA segment of interest (Figure 19.25). Thus they can be performed prenatally on fetal cells obtained by amniocentesis or chorionic biopsy, or even on a single cell from a pre-embryo produced by *in vitro* fertilization.

Some molecular diagnoses involve simply testing for the presence or absence of a specific restriction enzyme cleavage site in DNA. For example, the mutation that causes sickle-cell anemia (Chapter 14) removes a cleavage site for the restriction enzyme *Mst*II (Figure 20.16). The $Hb_{\beta}{}^{s}$ (sickle-cell) allele can be distinguished from the normal β-globin allele ($Hb_{\beta}{}^{a}$) by amplifying part of the β-globin gene by PCR, cutting the amplified DNA with *Mst*II, separating the resulting restriction fragments by agarose gel electrophoresis, preparing a Southern blot of the separated fragments, and hybridizing the DNA on the blot to a probe spanning the site of the mutation. The probe will hybridize with two small fragments from the normal β-globin gene, but with only one fragment from the sickle-cell β-globin gene (Figure 20.16). Thus this test allows the detection of heterozygotes as well as homozygotes.

For inherited disorders such as Huntington's disease and fragile X syndrome, which result from expanded trinucleotide repeat regions in genes, PCR and Southern blots can be used to detect the mutant alleles. The DNA test for the *huntingtin* gene is illustrated in Figure 20.11. Other types of mutations can be detected by using allele-specific oligonucleotides to probe genomic Southern blots. This procedure is illustrated for the *ΔF508* mutation in the *CF* gene—the most frequent cause of cystic fibrosis—in the Technical Sidelight in Chapter 19. Indeed, once the mutations responsible for a disease have been characterized, the development of DNA tests to detect them is usually routine. Clearly, the availability of diagnostic tests for mutations that cause human diseases has contributed greatly to the field of genetic counseling, providing families in which the genetic defects occur with invaluable information.

Key Points: **The mutant genes responsible for several inherited human diseases can be accurately diagnosed by screening genomic DNAs for the genetic defect. The results of these tests provide information that allows genetic counselors to inform families of the risk of an affected child.**

THE HUMAN GENOME PROJECT

As the recombinant DNA, gene cloning, and DNA sequencing technologies improved in the 1970s and

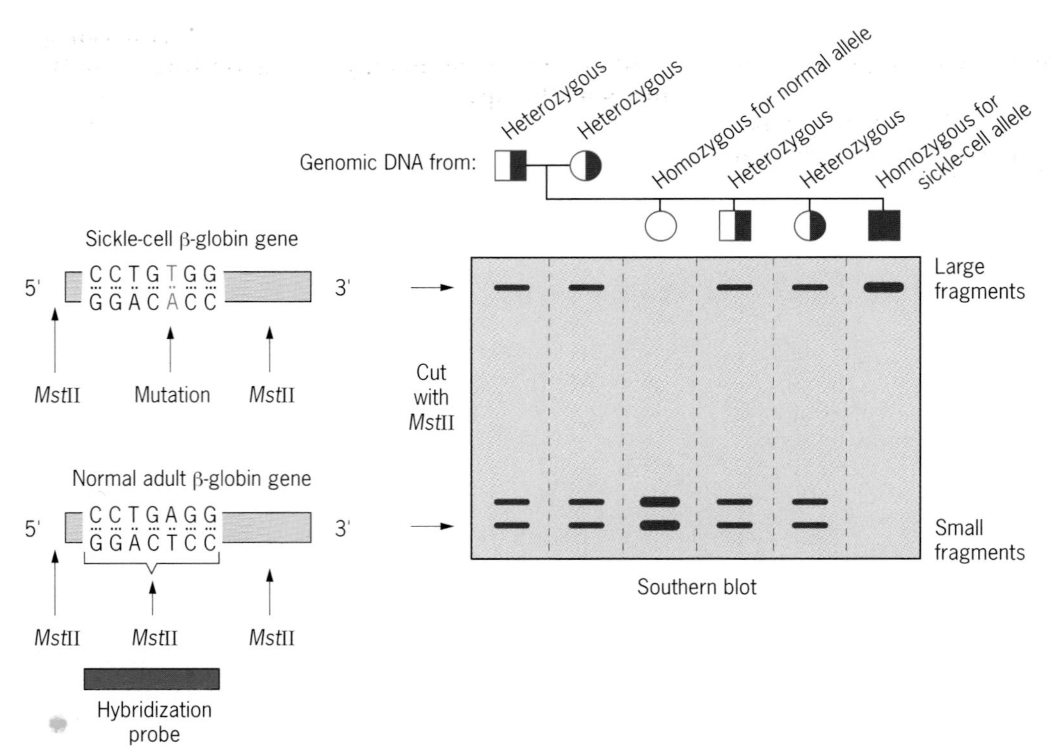

Figure 20.16 Detection of the sickle-cell hemoglobin mutation by Southern blot analysis of genomic DNAs cut with restriction enzyme *Mst*II.

early 1980s, scientists began discussing the possibility of sequencing all 3×10^9 nucleotide pairs in the human genome. These discussions led to the launching of the **Human Genome Project** in 1990. The goals of the Human Genome Project are (1) to map all of the 50,000 to 100,000 human genes, (2) to construct a physical map of the entire human genome, and (3) to determine the nucleotide sequences of all 24 human chromosomes. James Watson, who, with Francis Crick, discovered the double-helix structure of DNA, was the first director of this ambitious project, which was expected to take two decades to complete and to cost in excess of $3 billion. Scientists soon realized that this huge undertaking should be a worldwide effort. Therefore, an international **Human Genome Organization (HUGO)** was organized to coordinate the efforts of human geneticists around the world.

Indeed, progress in mapping the human genome has been quite spectacular. Complete physical maps of chromosomes Y and 21 and detailed RFLP maps of the X chromosome and all 22 autosomes were published in 1992. These maps have already proven invaluable to researchers cloning genes by their locations in the genome. In 1993, Francis Collins, who, with Lap-Chee Tsui, led the research teams that identified the cystic fibrosis gene, replaced Watson as director of the Human Genome Project. At present, the gene mapping work is ahead of schedule, but progress toward sequencing the entire genome is well behind schedule.

Whereas knowing the complete sequence of the human genome will help identfy genes responsible for human diseases and might lead to successful gene therapies for these diseases, it will not tell us what these genes do or how they control biological processes. Indeed, by itself, the nucleotide sequence of a gene, a chromosome, or an entire genome is uninformative. Only when supplemented with information about their functions do sequences become truly meaningful. Thus information about the functions of nucleotide sequences must still be obtained by traditional genetic studies and by molecular analyses. If geneticists want to understand the genetic control of the growth and development of a mature human from a single fertilized egg (Chapter 23), they will need to know much more than the sequence of the human genome. But the availability of the ultimate map of the human genome, its nucleotide sequence, will certainly accelerate progress toward understanding the programs of gene expression that control morphogenesis.

Key Points: **Important goals of the Human Genome Project are to map all 50,000 to 100,000 human genes and to determine the complete nucleotide sequence of all 24 human chromosomes.**

HUMAN GENE THERAPY

Of the over 4000 inherited human diseases catalogued to date, only a few are currently treatable. For many of these diseases, the missing or defective gene product cannot be supplied exogenously, as insulin is supplied to diabetics. Most enzymes are unstable and cannot be delivered in functional form to their sites of action in the body, at least not in a form that provides for long-term activity. Cell membranes are impermeable to large macromolecules such as proteins; thus enzymes must be synthesized in the cells where they are needed. Therefore, treatment of inherited diseases is largely restricted to those cases where the missing metabolite is a small molecule that can be distributed to the appropriate tissues of the body through the circulatory system. For many other inherited diseases, **gene therapy** offers the most promising approach to successful treatment. Gene therapy involves adding a normal (wild-type) copy of a gene to the genome of an individual carrying defective copies of the gene. A gene that has been introduced into a cell or organism is called a **transgene** (for transferred gene) to distinguish it from endogenous genes, and the organism carrying the introduced gene is said to be **transgenic**. If gene therapy is successful, the transgene will synthesize the missing gene-product and restore the normal phenotype.

Before considering specific examples, we need to discuss two types of gene therapy: **somatic-cell** or **nonheritable gene therapy**, and **germ-line** or **heritable gene therapy**. In higher animals such as humans, the reproductive or germ-line cells are produced by a cell lineage separate from all somatic cell lineages. Thus somatic-cell gene therapy will treat the disease symptoms of the individual but will not cure the disease. That is, the defective gene(s) will still be present in the germ-line cells of the patient after somatic-cell gene therapy and may be transmitted to his or her children. All of the gene-therapy treatments of human diseases that we will discuss here are somatic-cell gene therapies. Germ-line gene therapy is being performed on mice and other animals, but not on humans.

The distinction between somatic-cell and germ-line gene therapy is important when we discuss humans. The frequently expressed concerns about humankind's "tinkering with nature" or "playing God" apply to germ-line gene transfers, not to somatic-cell gene therapy. Major moral and ethical considerations are involved in any decision to perform germ-line modifications of human genes. In contrast, somatic-cell gene therapy is no different from enzyme (gene-product) therapy or cell, tissue, and organ transplants. In transplants, entire organs, with all the foreign genes present in the genome of every cell in the organ, are implanted in patients. In current somatic-cell gene therapies, some of the patient's own cells are removed, repaired, and reimplanted in the patient. Thus somatic-cell gene therapy is a less extreme manipulation of an individual than an organ transplant.

Human gene therapy is performed under strict guidelines developed by the National Institutes of Health (NIH). Each proposed gene-therapy procedure is scrutinized by review committees at both the local (institution or medical center) and national (NIH) levels. Several requirements must be fulfilled before a gene-therapy procedure will be approved:

1. The gene must be cloned and well characterized; that is, it must be available in pure form.
2. An effective method must be available for delivering the gene into the desired tissue(s) or cells.
3. The risks of gene therapy to the patient must have been carefully evaluated and shown to be minimal.
4. The disease must not be treatable by other strategies.
5. Data must be available from preliminary experiments with animal models or human cells and must indicate that the proposed gene therapy should be effective.

A gene-therapy proposal will not be approved by the local and national review committees until they are convinced that all of the above conditions have been fulfilled.

The first use of gene therapy in humans occurred in 1990, when a four-year-old girl with **adenosine deaminase-deficient severe combined immunodeficiency disease** (**ADA⁻SCID**) received her first transgene treatment. Several other ADA⁻SCID patients have subsequently been treated, along with a few individuals with other inherited disorders. SCID is a rare autosomal disease of the immune system (Chapter 24). Individuals with SCID have essentially no immune system, so that even minor infections can become serious and sometimes fatal. Some SCID patients lack an enzyme called adenosine deaminase (ADA). In the absence of this enzyme, toxic levels of the phosphorylated form of its substrate, deoxyadenosine, accumulate in T lymphocytes (white blood cells essential to an immune response) and kill them. T lymphocytes stimulate cells called B lymphocytes to develop into antibody-producing plasma cells. Thus, in the absence of T lymphocytes, no immune response is possible, and newborns with ADA⁻SCID seldom live more than a few years.

Four factors made ADA⁻SCID a good candidate for somatic-cell gene therapy. First, the *ADA* gene was one of the first human disease genes to be cloned and characterized. Second, white blood cells can easily be obtained from ADA⁻SCID patients and reintroduced after functional copies of the *ADA* gene are added. Third, even a small amount of functional ADA will restore partial immune function. Fourth, the overproduction of ADA does not appear to have toxic effects on the patient.

As a prerequisite to gene therapy, a retroviral gene transfer vector was constructed with the human *ADA* gene under the control of a strong viral promoter (Figure 20.17), and expression of the chimeric gene was demonstrated in white blood cells infected with the retroviral vector. The actual treatment of ADA⁻SCID by gene therapy involves four sequential steps: (1) isolation of white blood cells from the patient, (2) the introduction of functional copies of the *ADA* gene into these cells, (3) a demonstration of transgene expression in these cells, and (4) the infusion of the transgenic cells back into the patient (Figure 20.18).

Although the number of ADA⁻SCID patients who have been treated by gene therapy is still small and

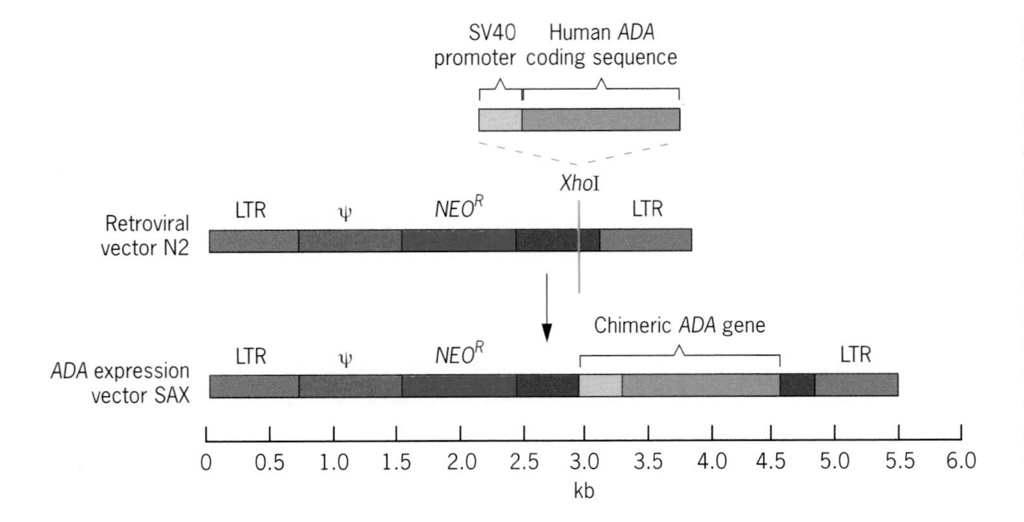

Figure 20.17 Construction of the human adenosine deaminase (*ADA*) gene-transfer vector SAX (for **s**imian virus 40, ***A****DA*, ***X****ho*I site). In this vector, expression of the human *ADA* coding sequence is controlled by a strong promoter obtained from simian virus 40. The NEO^R gene makes cells resistant to the antibiotics neomycin, kanamycin, and G418. LTR refers to the long terminal repeats from the Moloney murine leukemia virus; ψ is a DNA sequence required to package vector DNA in virions.

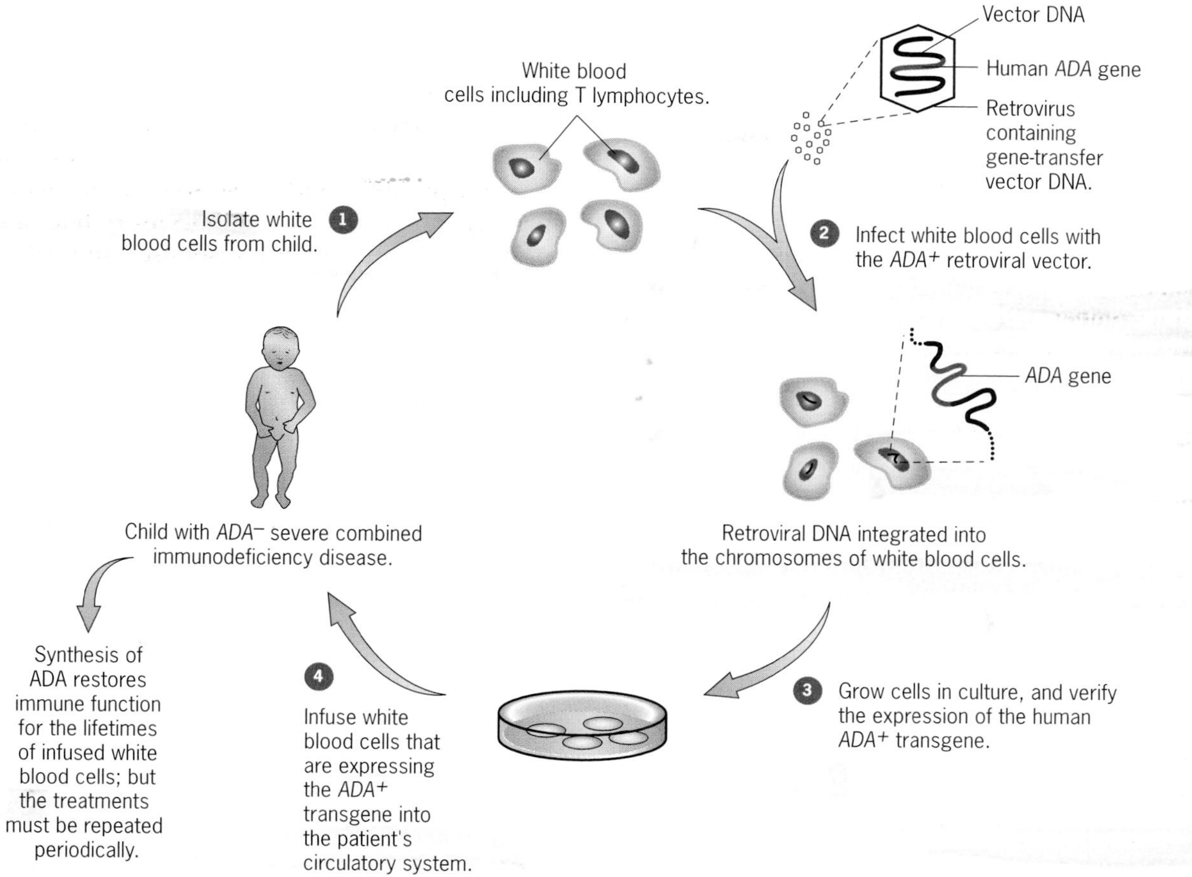

Figure 20.18 Treatment of adenosine deaminase-deficient severe combined immunodeficiency disease (ADA^-SCID) by somatic-cell gene therapy. ADA^-SCID results from the loss or lack of activity of the enzyme adenosine deaminase (ADA). Gene therapy is performed by isolating white blood cells from the patient, introducing a wild-type copy of the *ADA* gene into these cells with a retroviral vector, verifying the expression of the transgene in cultured cells, and infusing the transformed white blood cells back into the patient.

these patients have continued to receive treatments with purified ADA, the therapy has resulted in improved immune function in these patients. Most are attending normal schools, and some have survived infectious diseases such as chicken pox. It was anticipated that the effects of gene therapy using white blood cells would be short-lived due to the limited lifespan of these cells. Thus repeated infusions of white blood cells carrying functional ADA genes have been performed on the patients. The major problem encountered in attempts to treat ADA^-SCID patients by gene therapy has not been the short lifespan of the white blood cells, but the transient expression of the introduced *ADA* transgenes. Within a few weeks after treatment, transcription of the *ADA* transgene has subsided. The activity of the viral promoters that control this transcription has apparently been arrested by an unknown mechanism. These results indicate that additional research is needed to more fully understand the fate of genes introduced into mammalian cells with retroviral vectors.

To avoid the limitations resulting from the short lifespan of white blood cells, the bone marrow stem cells that give rise to white blood cells (Chapter 24) could be used to treat immune disorders such as ADA^-SCID. The modified stem cells should continually produce T lymphocytes with the *ADA* transgene and could provide a permanent or long-term treatment of the disease. Indeed, stem-cell gene therapy was first used to treat two infants with ADA^-SCID in 1993, and this procedure will undoubtedly be the method of choice in the future.

Although the initial somatic-cell gene therapy experiments have been less effective than scientists and physicians had hoped, this approach to the treatment of human diseases holds great promise for the future. With the recent identification and cloning of several genes that cause serious human diseases, gene therapy is clearly on the horizon for individuals suffering from several of these disorders. We previously discussed the possible treatment of cystic fibrosis and Duchenne muscular dystropy by gene therapy. Other candidates for somatic-cell gene therapy include phenylketonuria (Chapter 26), the thalassemias (hemoglobin deficiencies), sickle-cell anemia (Chapter 13), citrullinemia (a life-threatening hyperammonemia caused by a deficiency of the urea-cycle enzyme argininosuccinate synthetase), Lesch-Nyhan syndrome (caused by a defect in purine metabolism; see Chapter 26), and Hurler syndrome (caused by a defect in polysaccharide metabolism).

Current somatic-cell gene therapy protocols are **gene-addition** procedures; they simply add functional copies of the gene that is defective in the patient to the genomes of recipient cells. They do not replace the defective gene with a functional gene. In fact, the introduced genes are inserted at random or nearly random sites in the chromosomes of the host cells. The ideal gene-therapy protocol would replace the defective gene with a functional gene. **Gene replacements** would be mediated by homologous recombination and would place the introduced gene at its normal location in the host genome. In humans, gene replacements are usually referred to as targeted gene transfers. Oliver Smithies and coworkers first used homologous recombination to target DNA sequences to the β-globin locus of human tissue-culture cells in 1985. However, the frequency of the targeted gene transfer was very low (about 10^{-5}). Since then, Smithies, Mario Capecchi, and others have developed improved gene targeting vectors and selection strategies. As a result, more efficient targeted gene replacements are possible, and cells with the desired gene replacement can be identified more easily. In the future, targeted gene replacements will probably become the method of choice for somatic-cell gene therapy treatment of human diseases.

Key Points: **Gene therapy involves the addition of a normal (wild-type) copy of a gene to the genome of an individual who carries defective copies of the gene. Physicians are currently testing the effectiveness of somatic-cell gene therapy in the treatment of patients with adenosine deaminase-deficient severe combined immunodeficiency disease. Although many technical details must still be worked out, somatic-cell gene therapy holds great promise for the treatment of many inherited human diseases. In the future, somatic-cell gene therapy will probably involve gene replacements, rather than gene additions.**

DNA FINGERPRINTS

Fingerprints have played a central role in human identity cases for decades. Indeed, fingerprints often provide the key evidence that places a suspect at the crime scene. The use of fingerprints in forensic cases is based on the premise that no two individuals will have identical prints. Similarly, no two individuals, except for identical twins, will have genomes with the same nucleotide sequences. The human genome contains 3×10^9 nucleotide pairs; each site is occupied by one of the four base pairs in DNA. Many base-pair substitutions are silent; they occur in unessential noncoding sequences or at genomic positions corresponding to the third base of codons and do not change the amino acid sequences of the gene products because of the degeneracy in the code. Therefore, such nucleotide-pair substitutions accumulate in genomes during the course of evolution. In addition, duplications and deletions of DNA sequences and other genome rearrangements contribute to the evolutionary divergence of genomes. Indeed, recent evidence has demonstrated that the human genome contains large families of DNA polymorphisms of many different types, polymorphisms that can provide valuable evidence in cases of uncertain human identity. These polymorphisms can be used to produce **DNA fingerprints**, specific banding patterns on Southern blots of genomic DNA cleaved with a specific restriction enzyme and hybridized to an appropriate DNA probe.

The power and utility of the DNA fingerprinting procedure in personal identity cases are obvious to anyone familiar with molecular genetics and the techniques utilized in the production of DNA prints. The controversies regarding the use of DNA fingerprints in forensic cases relate to the competency of the research laboratories involved, the probability of human error in producing prints, and the methods for calculating the probability that two individuals have identical fingerprints. To make accurate estimates of the likelihood of identical prints, researchers must have reliable information about the frequency of the polymorphisms in the population in question. For example, if inbreeding (matings between related individuals) is common in the population, the probability of identical fingerprints will increase. Thus accurate estimates of the probability that two individuals will have matching fingerprints requires reliable information about the frequencies of the polymorphisms in the relevant population. Data obtained from one population should never be extrapolated to another population, especially if the two populations include individuals with different ethnic backgrounds.

DNA fingerprinting provides an extremely powerful forensic tool if used properly. The DNA prints can be prepared from extremely minute amounts of

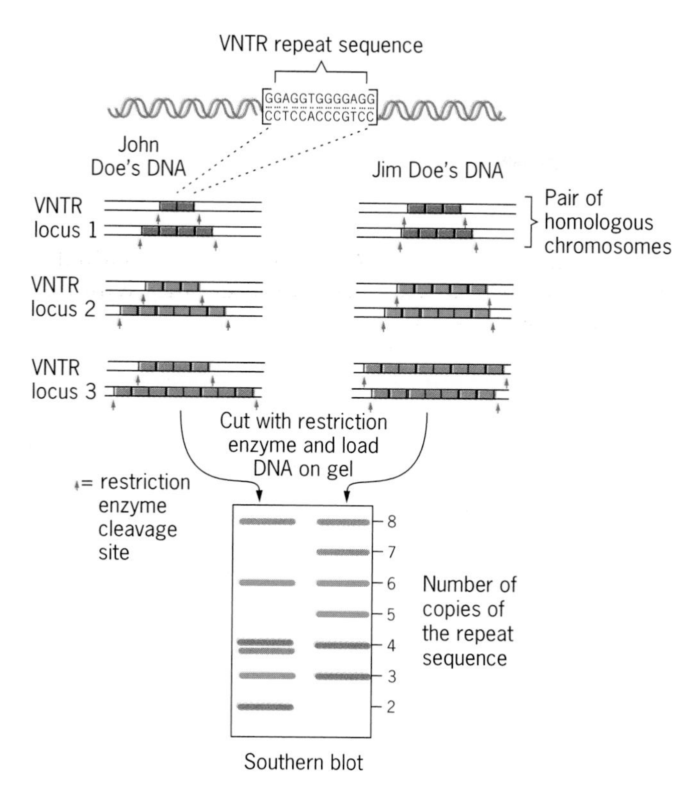

Figure 20.19 Simplified diagram of the use of variable number tandem repeats in preparing DNA fingerprints.

blood, semen, hair bulbs, or other cells of the body. The DNA is extracted from these cells, amplified by PCR, and analyzed with carefully chosen DNA probes by the Southern blot procedure (Chapter 19). Indeed,fingerprinting can sometimes be done with DNA isolated from preserved tissues of individuals long after their deaths (Technical Sidelight: DNA Tests and the Mystery of the Duchess Anastasia). As previously mentioned, the human genome contains numerous short DNA sequences that are present as tandem repeats of varied lengths at several chromosomal locations. These variable number tandem repeats, or VNTRs, are important components of DNA fingerprints (Figure 20.19). Although DNA prints are applicable in all cases of questionable identity, they have proven especially useful in paternity and forensic cases.

Paternity Tests

In the past, cases of uncertain paternity often have been decided by comparing the blood types of the child, the mother, and possible fathers. Blood-type data can be used to prove that men with particular blood types could not have fathered the child. Unfortunately, these blood-type comparisons contribute little toward a positive identification of the father. In contrast, DNA fingerprints not only exclude misidentified fathers, but also come close to providing a positive identification of the true father. DNA samples are obtained from cells of the child, the mother, and possible fathers, and DNA fingerprints are prepared as described in Figure 20.19. When the fingerprints are compared, all the bands in the child's DNA print should be present in the DNA prints of the parents. For each pair of homologous chromosomes, the child will have received one from each parent. Thus approximately half of the bands in the child's DNA print will result from DNA sequences inherited from the mother, and the other half from DNA sequences inherited from the father.

Figure 20.20 shows the DNA fingerprints of a child, the mother, and two men suspected of being the child's father. In this case, the DNA prints indicate

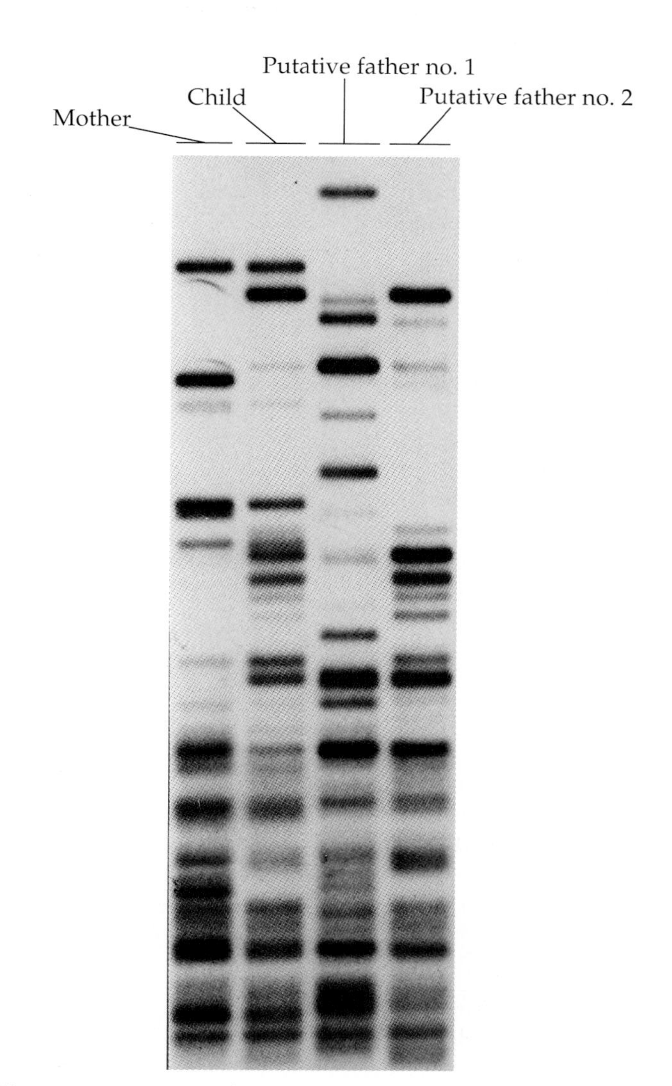

Figure 20.20 DNA fingerprints of a mother, her child, and two men, each of whom claimed to be the child's father.

TECHNICAL SIDELIGHT

DNA Tests and the Mystery of the Duchess Anastasia

According to historical records, the Russian royal family—Tsar Nicholas II, Tsarina Alexandra, and their five children: Alexis, Olga, Tatiana, Marie, and Anastasia—were executed on July 16, 1918, by a revolutionary Bolshevik firing squad and buried in a single grave in the Ural Mountains. However, in 1920, an unknown woman, "Fraulein Unbekannt," who was pulled from a canal in Berlin in a state of hypothermia, claimed that she was the Duchess Anastasia. Although she did not speak Russian, Fraulein Unbekannt, or Anna Anderson Manahan, as she was subsequently known, was amazingly well-informed about details of life in the Imperial Russian Court. Her claim to be Anastasia was vigorously rejected by the surviving relatives of the Russian Royal Family. The Grand Duke of Hesse even hired a private detective to investigate Anna's heritage. The detective concluded that Anna was really Franzisca Schanzkowska, but the dispute continued. Although little is known about Franzisca, she was born in the northern part of Germany, lived in Berlin during World War I, and was severely injured by an explosion while working in a munitions factory. She was subsequently admitted to two mental hospitals for treatment. She disappeared in 1920, at about the same time that Anna Anderson Manahan was rescued from the Berlin canal and claimed to be Anastasia.

When Princess Irene of Prussia, Anastasia's aunt, was persuaded to meet with the women who claimed to be her niece, Anna ran and hid in her room. Anna's bizarre behavior made her claim to be Anastasia difficult to evaluate, and the controversy over the identity of Anna Anderson Manahan continued for over 70 years. Was Anna really Anastasia? Her supporters were steadfast in their belief that she was indeed the Duchess. Disbelievers were equally adamant that she was not Anastasia.

In 1979, a Russian geologist discovered a shallow grave believed to contain the remains of the Royal Family. Because of the political climate in the Soviet Union at the time, the geologist reburied the bodies. Twelve years later, when the political climate was more favorable, the bodies were exhumed, and their authenticity was established by comparing DNA from the skeletons with DNA from surviving relatives. However, the controversy about the identity of Anna was rekindled by the absence of two bodies, those of Anastasia and her brother Alexis. Had they somehow escaped execution? Although there is still no definitive answer to this question, the results of recent DNA tests indicate that Anna Anderson Manahan was not the Duchess Anastasia.

Anna Anderson Manahan died in 1984 at the age of 83. However, during surgery performed in 1979 at the Martha Jefferson Hospital in Charlottesville, Virginia, intestinal tissues were removed, fixed in formaldehyde, and preserved in paraffin blocks. In addition, a few of Anna's hair follicles were preserved. DNA tests—VNTR (variable number tandem repeat) prints and nucleotide sequences of specific noncoding regions of mitochondrial DNA—were performed on these preserved tissues and on relatives of Franzisca Schanzkowska and of the Royal Family. These tests were performed independently in three different laboratories: (1) the Armed Forces DNA Identification Laboratory in the United States, (2) the Forensic Science Service in the United Kingdom, and (3) the Department of Anthropology at Pennsylvania State University. The results obtained by the three laboratories all indicate that Anna Anderson Manahan was not Anastasia. Indeed, the results strongly suggest that Anna was Franzisca Schanzkowska.

Figure 1 The children of Tsar Nicholas II: (left to right) Marie, Tatiana, Anastasia, Olga, and Alexis.

Of five different VNTRs examined, four were inconsistent with the possibility that Anna was the daughter of Tsar Nicholas II and Tsarina Alexandra. DNA sequence comparisons also argued that Anna was not related to the Royal Family. Instead, the nucleotide sequence data indicated that Anna was Ms. Schanzkowska. At the six variable positions shown below, Anna's mitochondrial DNA contained the same nucleotides as the DNA from Carl Maucher, Franzisca Schanzkowska's grand nephew, and differed from those in the DNA of the Duke of Edinburgh, the grand nephew of Tsarina Alexandra.

	Variable Nucleotides in Mitochondrial DNA					
Position:	1	2	3	4	5	6
Anna Anderson Manahan	C	C	T	T	C	T
Carl Maucher (Grand nephew of Franzisca)	C	C	T	T	C	T
Duke of Edinburgh (Grand nephew of Alexandra)	T	T	C	C	T	C

When these mitochondrial DNA sequences were compared with over 300 published and unpublished DNA sequences in data banks, no match to Carl Maucher and Anna Anderson Manahan's sequence was found. These results indicate that the Maucher-Manahan sequence is rare. If the data base sequences are representative of Caucasians of Northern European origin, the chance of matching DNA profiles is less than 1 in 300 if Carl and Anna are unrelated. The results clearly support the hypothesis that Anna was Franzisca Schanzkowska rather than the Duchess Anastasia, and finally put to rest the controversy about the identity of Anna Anderson Manahan.

that the second father candidate is probably the child's biological father. The accuracy of DNA fingerprints in identifying child-parent relationships can be enhanced by increasing the number of hybridization probes used in the analysis. With the use of more probes, more polymorphisms can be surveyed and a larger proportion of the genomes of the child and parents can be compared; the result is a more reliable identification.

Forensic Applications

DNA fingerprints were first used as evidence in a criminal case in 1988. In 1987, a Florida judge denied the prosecutor's request to present statistical interpretations of DNA evidence against an accused rapist. After a mistrial, the suspect was released. Three months later, he was again in court, accused of another rape. This time the judge allowed the prosecutor to present a statistical analysis of the data, based on appropriate population surveys. The analysis showed that the DNA fingerprint prepared from semen recovered from the victim had a probability of about one in 10 billion of matching the DNA fingerprint of the suspect purely by chance. This time the suspect was convicted. There can be no question about the value of DNA prints in forensic cases of this type. If performed carefully by trained scientists and interpreted conservatively using valid population-based data on the distributions of the polymorphisms involved, DNA fingerprints can provide a much needed and powerful tool in the ongoing fight against crime.

Figure 20.21 illustrates one type of DNA fingerprints used in forensic cases, namely, VNTR prints. Even a novice can see that the DNA fingerprint prepared from the bloodstain at the crime scene matches the DNA print from suspect 1, but not the prints from the other two suspects. Of course, these matching DNA prints by themselves do not prove that suspect 1 committed the crime, but, if combined with additional DNA prints and supporting evidence, they can provide strong evidence that suspect 1 was at the scene of the crime. In addition, these prints clearly show that the blood cells in the stain were not from either of the other two suspects.

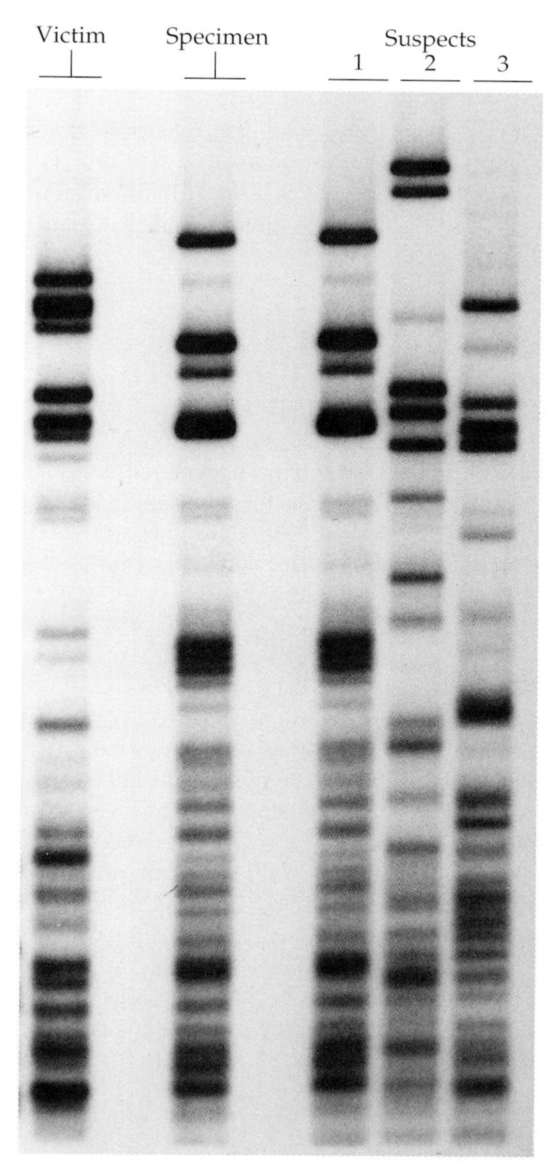

Figure 20.21 DNA fingerprints prepared from DNA isolated from a bloodstain at the site of the crime and from blood obtained from three individuals suspected of commiting the crime.

By combining VNTR fingerprints with prints produced using other types of DNA probes, the possibility that DNA fingerprints from two individuals will match just by chance can virtually be eliminated. The rationale behind using DNA fingerprints to identify individuals is that each person's DNA has a unique nucleotide sequence. Despite the human population explosion, there are far more possible combinations of the four base pairs in a human genome of 3×10^9 base pairs than there are humans on planet Earth. Thus, except for identical twins, no two humans should have identical genomes. DNA fingerprints provide a tool by which these differences can be detected and recorded, just as fingerprints have been recorded for decades.

Key Points: **DNA fingerprints detect and record polymorphisms in the genomes of individuals. DNA prints provide strong evidence of individual identity, evidence that may be extremely valuable in paternity and forensic cases.**

PRODUCTION OF EUKARYOTIC PROTEINS IN BACTERIA

For decades, microorganisms have been used to produce important products for humans. We are all aware of the impact of the penicillin antibiotics on human health; fewer of us are aware of the economic importance of penicillins. The wholesale market value of the penicillins in the United States in 1979 was $221 million. The wholesale market value of all antibiotics in the United States in 1979 was over 1 billion. Microbes also play important roles in the production of many other materials, for example, antifungal drugs, amino acids, and vitamins. Today, because of genetic engineering, bacteria are being used in the production of important eukaryotic proteins such as human insulin, human growth hormone, and the entire family of human interferons. In addition, genetically engineered microbes are being used to synthesize valuable enzymes and other organic molecules and to provide metabolic machinery for the detoxification of pollutants and the conversion of biomass to combustible compounds.

Human Growth Hormone

In 1982, human insulin became the first commercial success of the new recombinant DNA technologies in the field of pharmaceuticals. Since then, several other human proteins with medicinal value have been synthesized in bacteria. Some of the first human proteins to be produced in microorganisms were blood-clotting factor VIII (lacking in individuals with one type of hemophilia), plasminogen activator (a protein that disperses blood clots), and human growth hormone (a protein deficient in certain types of dwarfism). As an example, let's examine the synthesis of **human growth hormone (HGH)** in *E. coli*. HGH, which is required for normal growth, is a single polypeptide chain 191 amino acids in length. In contrast to insulin, porcine and bovine pituitary growth hormones do not work in humans. Only growth hormones from humans or from closely related primates function effectively in humans. Thus, prior to 1985, the major source of growth hormone suitable for treatment of humans was from human cadavers.

To obtain expression in *E. coli*, the HGH coding sequence must be placed under the control of *E. coli* regulatory elements. Therefore, the HGH coding sequence was joined to the promoter and ribosome-binding sequences of the *E. coli lac* operon (a set of genes encoding proteins required for growth on the sugar lactose; see Chapter 21). To accomplish this, a *Hae*III cleavage site in the nucleotide-pair triplet specifying codon 24 of HGH was used to fuse a synthetic DNA sequence encoding amino acids 1–23 to a partial cDNA sequence encoding amino acids 24-191. This unit was then inserted into a plasmid carrying the *lac* regulatory signals and introduced into *E. coli* by transformation. The structure of the first plasmid used to produce HGH in *E. coli* is shown in Figure 20.22.

The HGH produced in *E. coli* in these first experiments contained methionine at the amino terminus (the methionine specified by the ATG initiator codon). Native HGH has an amino-terminal phenylalanine; the initial methionine is enzymatically removed after synthesis. *E. coli* also removes many amino-terminal methionine residues post-translationally. However, the excision of the terminal methionine is sequence dependent, and *E. coli* cells do not excise the amino-terminal methionine residue from HGH. Nevertheless, the HGH synthesized in *E. coli* was found to be fully active in humans despite the presence of the extra amino acid. More recently, a DNA sequence encoding a signal peptide (the amino acid sequence required for transport of proteins across membranes) has been added to a HGH gene construct similar to the one shown in Figure 20.22. With the signal sequence added, HGH is both secreted and correctly processed. That is, the methionine residue is removed with the rest of the signal peptide during the transport of the primary translation product across the membrane. This product is identical to native HGH. In 1985, HGH became the second genetically engineered pharmaceutical to be approved for use in humans by the U.S. Food and Drug Administration. Human insulin produced in *E. coli* was approved for use by diabetics in 1982.

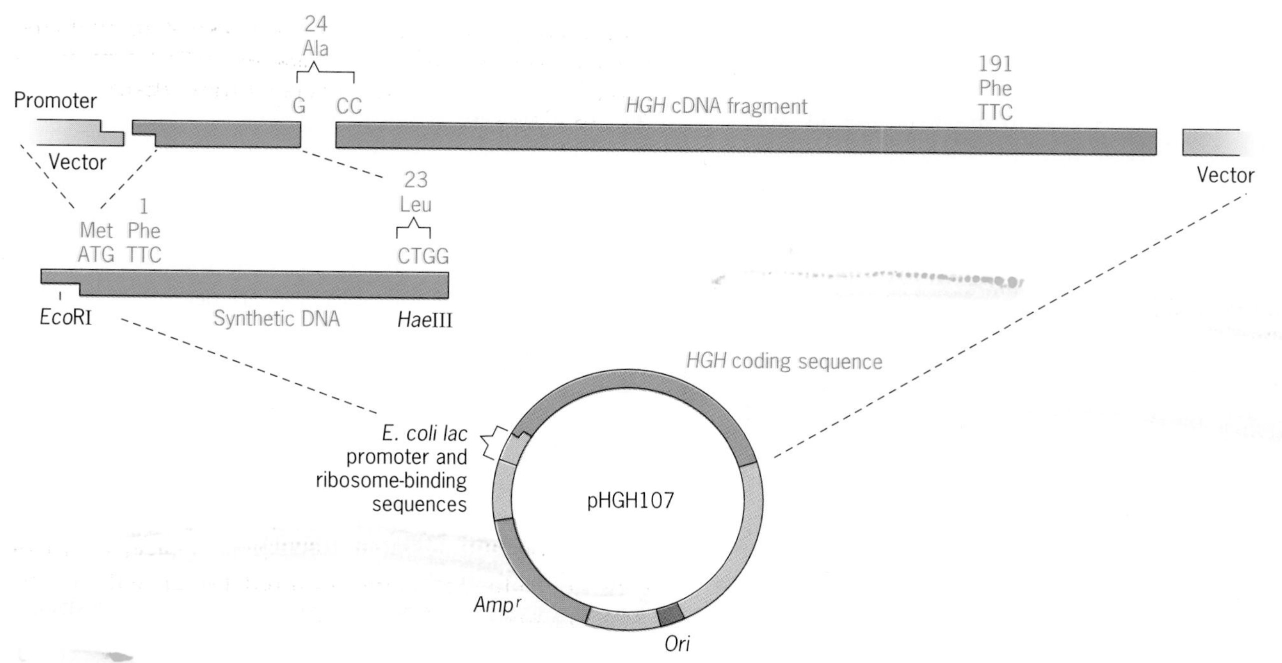

Figure 20.22 Structure of the first vector used to produce human growth hormone (HGH) in *E. coli*. The *Amp*^{r} gene provides resistance to ampicillin; *Ori* is the plasmid's origin of replication.

Proteins with Industrial Applications

Some enzymes with important industrial applications have been manufactured for many years by using microorganisms to carry out their synthesis. For example, proteases have been produced from *Bacillus licheniformis* and other bacteria. These proteases have been employed extensively as cleaning aids in detergents and in smaller amounts as meat tenderizers and as digestive aids in animal feeds. Amylases have been widely used to break down complex carbohydrates such as starch to glucose. The glucose is then converted to fructose with the enzyme glucose isomerase, and this fructose is used as a food sweetener. The amylases and glucose isomerase are all manufactured by microbiological processes. In 1980, worldwide production of α-amylase and glucose isomerase totaled 320 and 70 tons, respectively. Clearly, these are major industrial products; they account for millions of dollars in sales each year. The protein rennin is used in making cheeses; 26 tons were used in 1980 (wholesale value $64 million). Prior to the advent of genetic engineering, rennin was extracted from the fourth stomach of cattle. Genetically engineered bacteria are now used for the commercial production of rennin. These examples are all proteins that have had important industrial applications for some time. In the future, we can expect many additional enzymes to be manufactured and used in industrial applications because of the ease of producing these proteins by means of recombinant microorganisms.

In fact, the major tools of molecular biology today are themselves enzymes—namely, the vast array of restriction endonucleases, ligases, polymerases, and reverse transcriptases that are used to manipulate DNA molecules *in vitro*. These enzymes are now commonly produced by cloning the respective genes and expressing these genes on high-copy-number plasmids in bacteria. Thus the development of the recombinant DNA and gene-cloning technologies has spawned a whole new industry, a group of companies that manufacture and sell the enzymes, vectors, and so on that are used by genetic engineers. New applications in this field are inevitable, as it is now possible to produce essentially any enzyme or other protein of interest by cloning the gene that encodes this protein and expressing that gene at very high levels in microorganisms.

Key Points: **Valuable proteins that could be isolated from eukaryotes only in small amounts and at great expense can now be produced in large quantities in genetically engineered bacteria. Proteins such as human insulin and human growth hormone are valuable pharmaceuticals used to treat diabetes and pituitary dwarfism, respectively. Other proteins have important industrial applications.**

TRANSGENIC PLANTS AND ANIMALS

Although the first recombinant DNA molecules were constructed and expressed in microorganisms, methods are now available that allow foreign or synthetic genes to be introduced into and expressed in higher plants and animals. To date, the application of genetic engineering technologies to plants and animals has focused on three major goals: (1) the introduction or enhancement of desirable traits in agronomic plants and domestic animals, (2) the production of valuable products by transgenic plants and animals, and (3) the use of transgenic animals and plants to investigate basic biological processes such as gene expression. Although a complete discussion of the methods used to produce transgenic plants and animals is beyond the scope of this book, let us discuss a few commonly used procedures and some of the initial applications of recombinant DNA technologies in plant and animal breeding.

Transgenic Animals: Microinjection of DNA into Fertilized Eggs

Two methods are predominantly used to produce transgenic animals: (1) microinjection of DNA into the pronuclei of fertilized eggs, and (2) infection of preimplantation embryos with retroviral vectors. Most of the transgenic animals studied to date were produced by microinjection of DNA into fertilized eggs. Prior to microinjection, the eggs are surgically removed from the female parent and are fertilized *in vitro*. The DNA is then microinjected into the male pronucleus (the haploid nucleus contributed by the sperm, prior to nuclear fusion) of the fertilized egg through a very fine-tipped glass needle (Figure 20.23). The injected DNA can be either linear or circular. However, linear DNA integrates with a higher (about fivefold) efficiency than circular DNA. Although the reason for this difference is not known, it probably results from the high frequency of recombination at the ends of DNA molecules owing to their ability to invade homologous double helices. Usually, several hundred to several thousand copies of the gene of interest are injected into each egg, and multiple integrations often occur. Surprisingly, when multiple copies do integrate into the genome, they usually do so as tandem, head-to-tail arrays at a single chromosomal site. These DNA molecules carrying multiple tandem copies of the transgene are believed to form prior to their integration into the host genome, by recombination events between injected DNA molecules. The integration of injected DNA molecules appears to occur at random sites in the genome.

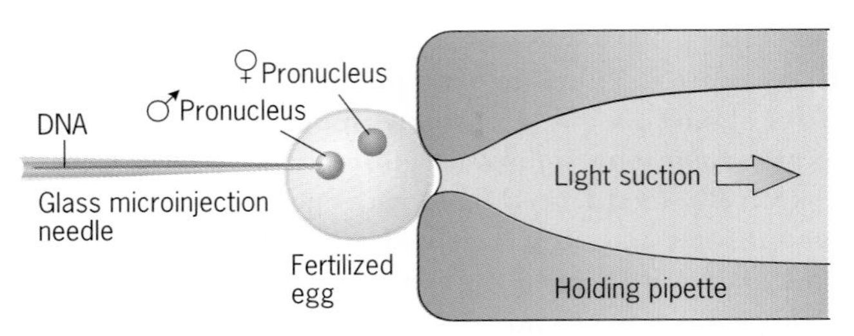

Figure 20.23 The production of transgenic animals by microinjection of DNA into fertilized eggs.

Because the DNA is injected so early during development, even before fusion of the haploid nuclei to form the diploid nucleus of the zygote, integration of the injected DNA molecules usually occurs early during embryonic development. As a result, some germ-line cells usually carry the transgene. As would be expected, the animals that develop from the injected eggs—called the **G_0 generation**—are almost always genetic mosaics, with some somatic cells carrying the transgene and others not carrying it. The initial (G_0) transgenic animals must be mated and G_1 progeny produced to obtain animals in which all cells carry the transgene. In most of the cases where progeny studies have been performed, the transgenes were transmitted to progeny. In a few cases, transgenes have undergone rearrangements in progeny generations. In mammals, injected DNA molecules are not maintained by autonomous replication as they are in frogs, sea urchins, and *Caenorhabditis elegans*.

Transgenic mice are produced routinely in laboratories throughout the world. They provide valuable tools for the study of gene expression in mammals and an excellent model system with which to test various gene-transfer vectors and methodologies for possible use in humans. Hundreds of different transgenes have been studied in mice, and, in most cases, the transgenes show normal patterns of inheritance, indicating that they have been integrated into the host genome. We discuss some of the applications of transgenic mice in studies of development in Chapter 23.

One of the early dramatic effects of a transgene was the increased growth rate that occurred when rat, bovine, or human growth hormone genes were expressed in mice (Figure 20.24). This prompted animal breeders to ask whether the introduction of either (1) extra copies of the homologous (same-species) growth hormone gene or (2) copies of heterologous growth hormone genes from related species might result in domestic animals with enhanced growth rates. Transgenic pigs were produced with the hope that enhanced growth hormone levels might result in leaner pigs with improved meat quality and with faster growth. Other scientists introduced growth hormone transgenes into fish and chickens with similar objectives. The experiments with transgenic pigs indicated

Figure 20.24 The transgenic mouse on the left, which carries a chimeric human growth hormone gene, is about twice the size of the control mouse on the right.

that growth rates were not increased on standard diets, but were increased on high-protein diets. However, the transgenic pigs were found to be leaner than the controls, as was expected, because growth hormone favors the synthesis of proteins instead of fats.

Unfortunately, the transgenic pigs also exhibited several undesirable side effects of the higher growth hormone levels. Most notably, the female transgenic pigs were sterile. In addition, transgenic animals of both sexes were lethargic with weak muscles and were highly susceptible to arthritis and ulcers. Although scientists are hopeful that ways can be found to overcome these side effects, the initial results suggest that attempts to improve growth rates and enhance meat quality in domestic animals by increasing growth hormone levels with transgenes may not be very effective.

Clearly, the genetic control of the native levels of growth hormone represents the end product of many generations of natural selection for optimal fitness in these animals, and any changes in hormone levels induced by transgenes or other manipulations are likely to decrease fitness by upsetting the normal metabolic balance. Thus deleterious side effects are to be expected. This does not mean that increased growth rates or improved meat quality cannot be achieved without unacceptable decreases in fitness of the transgenic animals. Moderate decreases in vigor may well be offset by highly desirable changes in meat quality (for example, lower cholesterol levels) or by increases in the efficiency of conversion of feed into meat products.

Other transgenic animals are being tested for resistance to viral infections. Avian leukosis virus (ALV) is a major viral pathogen of chickens, causing losses to the poultry industry of $50 million to $100 million per year. Obviously, the availability of an ALV-resistant strain of chickens would be of major commercial value. Therefore, researchers have produced transgenic chickens that carry a defective ALV genome. These chickens produce viral RNA and the viral envelope protein, but no progeny viruses. Most importantly, they are resistant to infection by ALV. The synthesis of large amounts of the retroviral envelope protein somehow blocks the reproductive cycle of intact, pathogenic ALV viruses. That ALV resistance has been transmitted to several generations of progeny indicates that the trait is stable. These encouraging results suggest that the introduction of defective viral genomes into domestic animals may be a useful tool in the production of virus-resistant genotypes.

Another potentially important use of transgenic animals is for the production and secretion of valuable proteins in milk. Many native human proteins contain carbohydrate or lipid side groups that are added post-translationally. Bacteria do not contain the enzymes that catalyze the addition of these moieties to nascent proteins. In such cases, recombinant bacteria cannot be used to synthesize the final product; they will synthesize the polypeptide only in its unmodified form. For this reason, some researchers have begun to explore alternative methods for producing valuable human proteins, especially glycoproteins and lipoproteins.

One promising approach involves the use of transgenic animals that secrete human proteins in milk. In some of the first experiments of this kind, transgenic sheep were produced that secrete either of two human proteins, the blood-clotting protein factor IX or the elastase inhibitor α1-antitrypsin, in their milk. A deficiency of factor IX occurs in one type of hemophilia in humans, and a lack of α1-antitrypsin sometimes contributes to the development of emphysema (degenerative lung disease) in humans. Thus both proteins have important medicinal applications. Researchers inserted the coding sequences for these proteins into the β-lactoglobulin (a milk protein) gene of sheep and microinjected the chimeric constructs into fertilized eggs. The injected eggs were then implanted into foster mothers to produce transgenic lambs. Although the concentrations of human factor IX and human α1-antitrypsin were low in the milk of these sheep, researchers believe that the concentrations can be increased by appropriately manipulating the regulatory sequences of the chimeric genes. Importantly, the transgenic sheep exhibited no apparent side effects from the production of either human protein in their milk. Nevertheless, further research must be done before the potential applicability of this approach can be adequately evaluated.

Transgenic Plants: The Ti Plasmid of *Agrobacterium tumefaciens*

Plants have been genetically manipulated by plant breeders for decades. Today, however, plant breeders can directly modify the DNA of plants and can quickly add genes from other species to plant genomes by recombinant DNA techniques. Indeed, transgenic plants can be produced by several different procedures. For example, one widely used procedure, called **microprojectile bombardment**, involves shooting DNA-coated tungsten or gold particles into plant cells. Another procedure, called **electroporation**, uses a short burst of electricity to get the DNA into the cell. However, more transgenic plants are produced by ***Agrobacterium tumefaciens*-mediated transformation** than by the other procedures. *Agrobacterium tumefaciens* is a soil bacterium that has evolved a natural genetic engineering system; it contains a segment of DNA that is transferred from the bacterium to plant cells.

An important feature of plant cells is their **totipotency**—that is, the ability of a single cell to produce all the differentiated cells of the mature plant. Many differentiated plant cells are able to dedifferentiate to the embryonic state and subsequently to redifferentiate to new cell types. Thus there is no separation of germline cells from somatic cells as in higher animals. When excised tissues from mature plants are placed in the appropriate sterile tissue culture conditions (notably in the presence of the plant hormone 2,4-dichlorophenoxyacetic acid, 2,4-D), cells in these tissue explants will dedifferentiate and grow into highly unorganized cell masses called **calli** (or calluses; singular = callus). If these undifferentiated callus cell-clumps are subsequently transferred to growth medium favoring differentiation (medium lacking 2,4-D, but containing growth hormones such as kinetin), plantlets will often regenerate. This totipotency of plant cells is a major advantage for genetic engineering because it permits the regeneration of entire plants from individual modified somatic cells.

Agrobacterium tumefaciens is the causative agent of crown gall disease of dicotyledonous plants. The name refers to the galls or tumors that often form at the crown (junction between the root and the stem) of infected plants (Figure 20.25*a*). Because the crown of the plant is usually located at the soil surface, it is here that a plant is most likely to be wounded (for example, from a soil abrasion as it blows in a strong wind) and infected by a soil bacterium such as *A. tumefaciens*. However, *A. tumefaciens* can infect a plant and induce

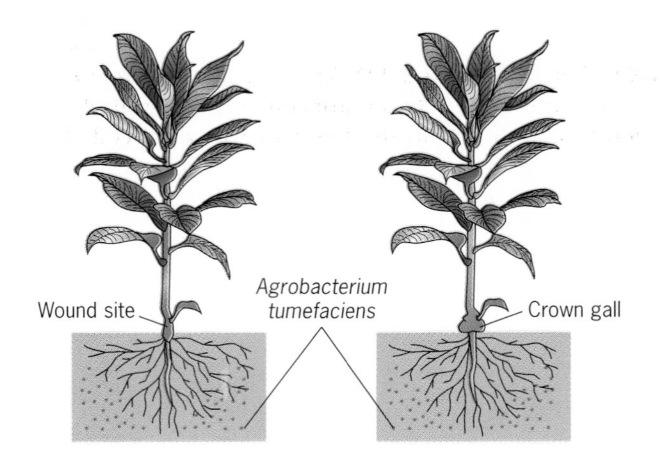

(*a*) Formation of a crown gall at the soil surface.

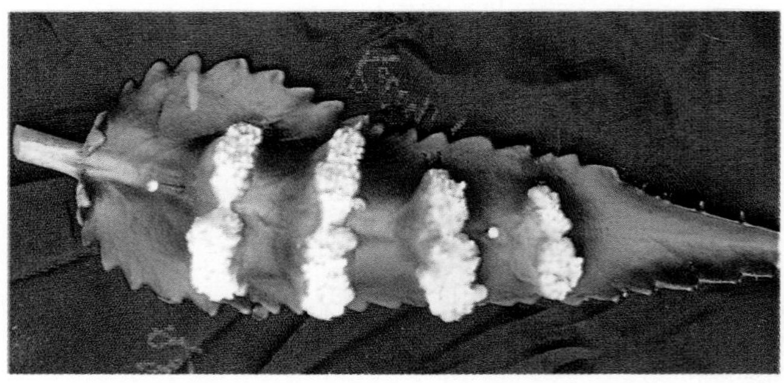

(*b*) Formation of tumors or galls at wound sites on a leaf.

Figure 20.25 The formation of tumors or "galls" on plants at wound sites infected by *Agrobacterium tumefaciens*.

a tumor at any wound site (Figure 20.25*b*). During the infection of a wound site by *A. tumefaciens*, two key events occur: (1) the plant cells begin to proliferate and form tumors, and (2) they begin to synthesize an arginine derivative called an opine. The opine synthesized is usually either nopaline or octopine depending on the strain of *A. tumefaciens*. These opines are catabolized and used as energy sources by the infecting bacteria. *Agrobacterium tumefaciens* strains that induce the synthesis of nopaline can grow on nopaline, but not on octopine, and vice versa. Clearly, an interesting interrelationship has evolved between *A. tumefaciens* strains and their plant hosts. *Agrobacterium tumefaciens* is able to divert the metabolic resources of the host plant to the synthesis of opines, which are of no apparent benefit to the plant but which provide sustenance to the bacterium.

The ability of *A. tumefaciens* to induce crown galls in plants is controlled by genetic information carried on a large (about 200,000 nucleotide pairs) plasmid called the **Ti plasmid** for its Tumor-inducing capacity. Two components, the **T-DNA** and the ***vir* region**, of the Ti plasmid are essential for the transformation of plant cells. During the transformation process, the T-DNA (for Transferred DNA) is excised from the Ti plasmid, transferred to a plant cell, and integrated (covalently inserted) into the DNA of the plant cell. The available data indicate that integration of the T-DNA occurs at random chromosomal sites; moreover, in some cases, multiple T-DNA integration events occur in the same cell. In nopaline-type Ti plasmids, the T-DNA is a 23,000-nucleotide-pair segment that carries 13 known genes. In octopine-type Ti plasmids, there are two separate T-DNA segments. For the sake of brevity, we consider only nopaline-type Ti plasmids in the subsequent discussion.

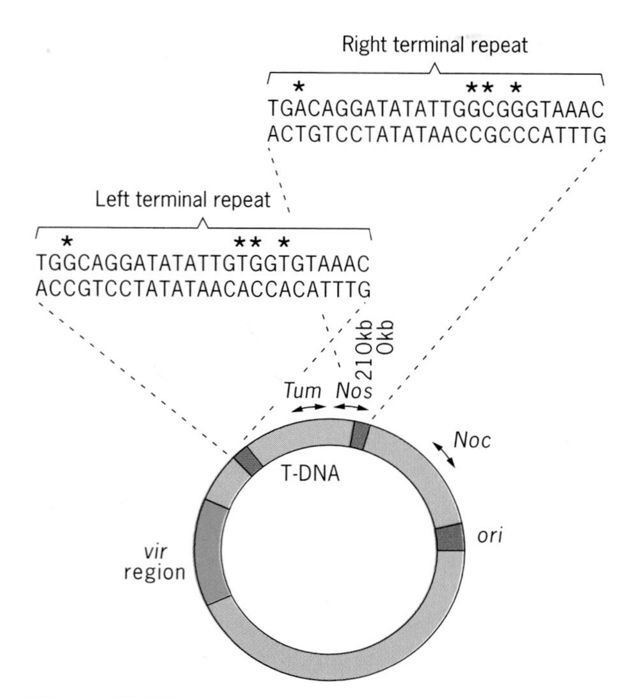

Figure 20.26 Structure of the nopaline Ti plasmid pTi C58, showing only components required for DNA transfer from *Agrobacterium* to plant cells. Symbols used are: *ori*, origin of replication; *Tum*, genes responsible for tumor formation; *Nos*, genes involved in nopaline biosynthesis; *Noc*, genes involved in the catabolism of nopaline; *vir*, virulence genes required for T-DNA transfer. The nucleotide-pair sequences of the left and right terminal repeats are shown at the top; the asterisks mark the four base pairs that differ in the two border sequences.

The structure of a typical nopaline Ti plasmid is shown in Figure 20.26. Some of the genes on the T-DNA segment of the Ti plasmid encode enzymes that catalyze the synthesis of phytohormones (the auxin indoleacetic acid and the cytokinin isopentenyl adenosine). These phytohormones are responsible for the tumorous growth of cells in crown galls. The T-DNA region is bordered by 25-nucleotide-pair imperfect repeats, which are required in *cis* for T-DNA excision and transfer. The deletion of either border sequence completely blocks the transfer of T-DNA to plant cells.

The *vir* (for virulence) region of the Ti plasmid contains the genes required for the T-DNA transfer process. These genes encode the DNA processing enzymes required for excision, transfer, and integration of the T-DNA segment during the transformation process. The *vir* genes can supply the functions needed for T-DNA transfer when located either *cis* or *trans* to the T-DNA. They are expressed at very low levels in *A. tumefaciens* cells growing in soil. However, exposure of the bacteria to wounded plant cells or exudates from plant cells induces enhanced levels of expression of the *vir* genes. Surprisingly, this induction process is very slow for bacteria, taking 10 to 15 hours to reach maximum levels of expression. Phenolic compounds such as acetosyringone act as inducers of the *vir* genes, and transformation rates can often be increased by adding acetosyringone to plant cells inoculated with *Agrobacterium*. The transformation of plant cells by the Ti plasmid of *A. tumefaciens* occurs as illustrated in Figure 20.27.

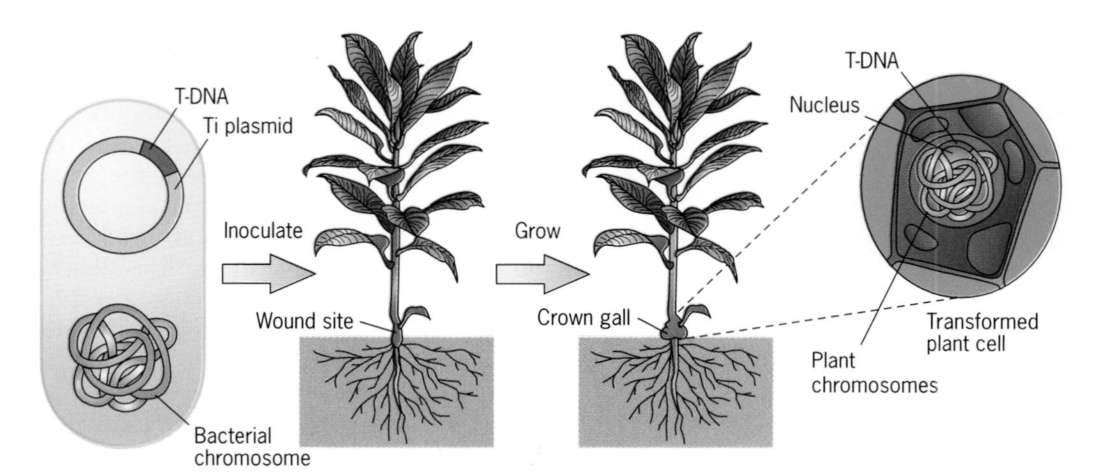

Figure 20.27 Transformation of plant cells by *Agrobacterium tumefaciens* harboring a wild-type Ti plasmid. Plant cells in the tumor contain the T-DNA segment of the Ti plasmid integrated into chromosomal DNA.

Once it had been established that the T-DNA region of the Ti plasmid of *A. tumefaciens* is transferred to plant cells and becomes integrated in plant chromosomes, the potential use of *Agrobacterium* in plant genetic engineering was obvious. Foreign genes could be inserted into the T-DNA with the hope that these genes would be transferred to the plant with the rest of the T-DNA segment. In fact, this works very well. The problem is that the transformed plant cells carrying wild-type T-DNAs lose their normal control of cell division and form tumors. This feature of T-DNA renders wild-type Ti plasmids incompatible with the goals of most gene-transfer experiments. Fortunately, the solution to this problem came early with the identification of the genes in the T-DNA that were responsible for tumor formation (Figure 20.26). The deletion of one or more of these genes produces a disarmed Ti plasmid. Unfortunately, the deletion of the tumor-causing genes also makes it extremely difficult to identify plant cells that have received the disarmed T-DNA. With wild-type Ti plasmids, the recipient plant cells form tumors and are easily identified by the tumor phenotype; with disarmed Ti plasmids, the recipient plant cells continue to grow just like their neighbors that do not harbor the T-DNA. Thus a way to identify plant cells transformed with disarmed Ti plasmids—ideally, a good selectable marker gene located within the T-DNA region of the disarmed Ti plasmid—is needed.

A good selectable marker gene is one that will provide resistance to a drug, an antibiotic, or another agent that arrests the growth of normal plant cells. The selective agent should inhibit the growth of plant cells or kill them slowly. Agents that kill cells rapidly result in the release of phenolic compounds and other substances that are toxic to the growth of the remaining, otherwise resistant cells. To date, three selectable marker genes have been used extensively in plant systems; they provide resistance to the antibiotics (1) kanamycin and the related aminoglycoside G418 and (2) hygromycin, and to the drug (3) methotrexate. Of these, kanamycin (and G418) has been the most widely used selective agent in plants.

The *Kan*^{r} gene from the *E. coli* transposon Tn*5* has been used extensively as a selectable marker in plants; it encodes an enzyme called neomycin phosphotransferase type II (NPTII). NPTII is one of several prokaryotic enzymes that detoxify the kanamycin family of aminoglycoside antibiotics by phosphorylating them. Because the promoter sequences and transcription-termination signals are different in bacteria and plants, the native Tn*5 Kan*^{r} gene cannot be used in plants. Instead, the NPTII coding sequence must be provided with a plant promoter (5′ to the coding sequence) and plant termination and polyadenylation signals (3′ to the coding sequence). Such constructions with prokaryotic coding sequences flanked by eukaryotic regulatory sequences are called **chimeric selectable marker genes**.

Regulatory sequences from several different plant genes have been used to construct chimeric marker genes. The two most frequently used regulatory sequences are (1) the promoter of the nopaline synthase (*nos*) gene of the Ti plasmid and (2) the promoter that controls the synthesis of the 35S transcript of cauliflower mosaic virus (CaMV). The most frequently used termination and polyadenylation sequence is that from the nopaline synthase gene of the Ti plasmid. One widely used chimeric selectable marker gene has the structure CaMV 35S promoter/NPTII coding sequence/Ti *nos* termination sequence; this chimeric gene is usually symbolized 35S/NPTII/*nos*. The Ti vectors used to transfer genes into plants have the tumor-inducing genes of the plasmid replaced with a chimeric selectable marker gene such as 35S/NPTII/*nos*. A large number of sophisticated Ti plasmid gene-transfer vectors are now used routinely to transfer genes into plants.

Agrobacterium tumefaciens-mediated transformation is by far the most frequently used method for generating transgenic plants. However, *A. tumefaciens*-mediated gene transfers are no longer performed by inoculating whole plants with the bacterium. Instead, tissue explants are co-cultivated with *A. tumefaciens* cells harboring the Ti plasmid of choice, and plants are regenerated from transformed callus cells formed on the cut surfaces of the tissue explant.

Herbicide-Resistant Plants

The use of genetic engineering to develop herbicide-tolerant varieties of agronomically important plants such as corn, soybeans, and the cereals promises to have a major impact on agriculture, both economically and on production practices. Weeds compete with crops for soil nutrients and routinely lead to significant losses in yield. Modern agriculture makes use of herbicides to control weeds and minimize the losses. Unfortunately, the available herbicides seldom provide the degree of specificity that is desired, and most herbicides will control only certain classes of weeds. Broad-spectrum herbicides may give good weed control but, in so doing, usually have deleterious effects on the growth of the crop plant as well. As a result, scientists are now evaluating alternate approaches to weed control. The most promising of these approaches is the development of herbicide-tolerant plant varieties for use with broad-spectrum herbicides that are less toxic to the environment. Obviously, the potential economic value of herbicide-tolerant plant varieties is substantial; thus large amounts of money

Figure 20.28 Glyphosate-tolerant petunia plants (top) carrying a chimeric 35S/EPSP synthase/*nos* gene. The dead plants at the bottom were wild-type controls. The plants were sprayed with glyphosate.

and effort have been invested in the production of herbicide-tolerant plants.

Herbicides are simply chemical compounds that kill or inhibit the growth of plants without deleterious effects on animals. Herbicides usually inhibit processes that are unique to plants, for example, photosynthesis. Most frequently, herbicides act as inhibitors of essential enzyme reactions. Thus anything that diminishes the level of inhibition will provide increased herbicide tolerance. The two most common sources of herbicide tolerance are overproduction of the target enzyme and mutations resulting in enzymes that are less sensitive to the inhibitor.

Glyphosate is one of the most potent broad-spectrum herbicides in use today; it is marketed under the trade name Roundup. Glyphosate acts by inhibiting the enzyme 5-enolpyruvylshikimate 3-phosphate synthase (EPSP synthase), an essential enzyme in the biosynthesis of the aromatic amino acids tyrosine, phenylalanine, and tryptophan. These aromatic amino acids are essential components in the diets of higher animals since the enzymes that catalyze the biosynthesis of these amino acids are not present in higher animals. Therefore, since higher animals contain no EPSP synthase, glyphosate has no toxic effects on animal systems. In this respect, glyphosate is an ideal herbicide.

Glyphosate does inhibit the EPSP synthases of microorganisms as well as those of plants. By selecting for growth in the presence of glyphosate concentrations that inhibit the growth of wild-type bacteria, researchers have been able to isolate glyphosate-tolerant mutants of *Salmonella typhimurium, Aerobacter aerogenes*, and *Escherichia coli*. In bacteria, EPSP synthase is encoded by the *aroA* gene. When the mutant bacterial *aroA* genes were fused to plant promoters and polyadenylation signals and were introduced into plants, the transgenic plants exhibited increased tolerance to glyphosate (Figure 20.28). Glyphosate-tolerant varieties of several agronomically important plant species are currently being evaluated. In addition, plant varieties with tolerance to several other herbicides have been produced by genetic engineering. Herbicide-tolerant plants will undoubtedly play a significant role in feeding the world's population in the future.

Use of Antisense RNAs to Block Gene Expression

Another new tool that became possible with the advent of recombinant DNA and gene-cloning technologies is the use of **antisense RNA** to regulate gene expression. The antisense RNA method involves the synthesis of RNA molecules that are complementary to the mRNA molecules produced by transcription of a given gene. The normal mRNA of a gene is said to be sense because it carries the codons that are read during translation to produce the specified sequence of amino acids in the polypeptide gene product. Normally, the complement to the mRNA sense strand will not contain a sequence of codons that can be translated to produce a functional protein; thus this complementary strand is called antisense RNA. In addition, an antisense RNA usually will not contain the regulatory sequences required for translation.

When antisense RNA molecules are present in the same cytosol with sense (mRNA) molecules from a gene, the antisense RNA and mRNA molecules will anneal to form duplex RNA molecules. These duplex RNA molecules cannot be translated; thus the presence of antisense RNA will block translation of the mRNA of the affected gene.

The antisense RNA approach has proven useful in dissecting pathways of gene expression. It has also been a commercial success. The first genetically engineered plant product approved for human consumption, the FlavrSavr™ tomato, was introduced in supermarkets in California and Illinois in 1994. The FlavrSavr™ tomato, which remains firm longer dur-

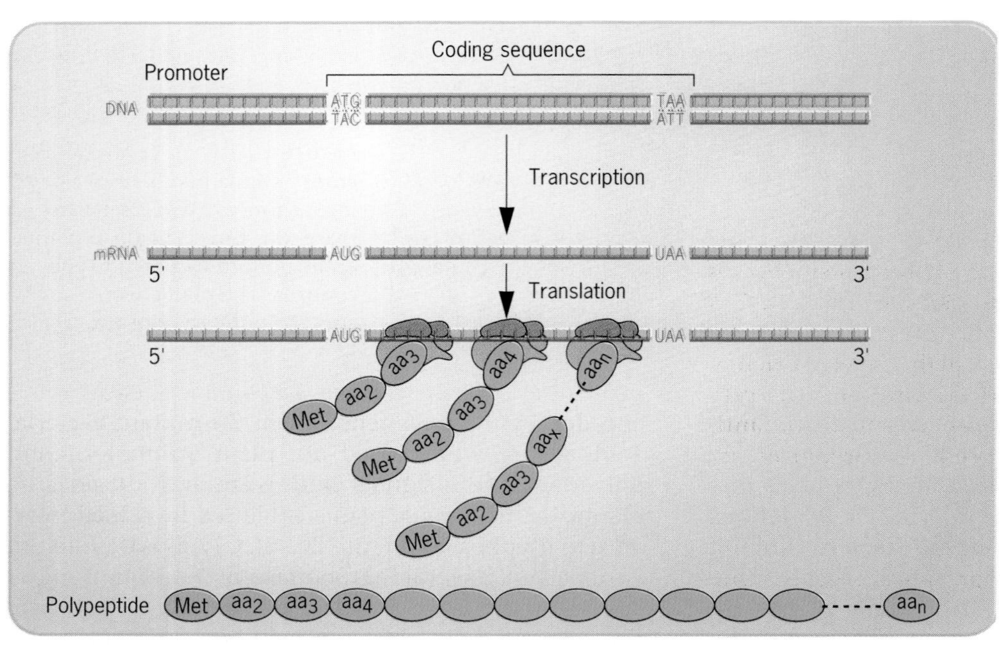

(a) Normal cell: only "sense RNA" (=mRNA).

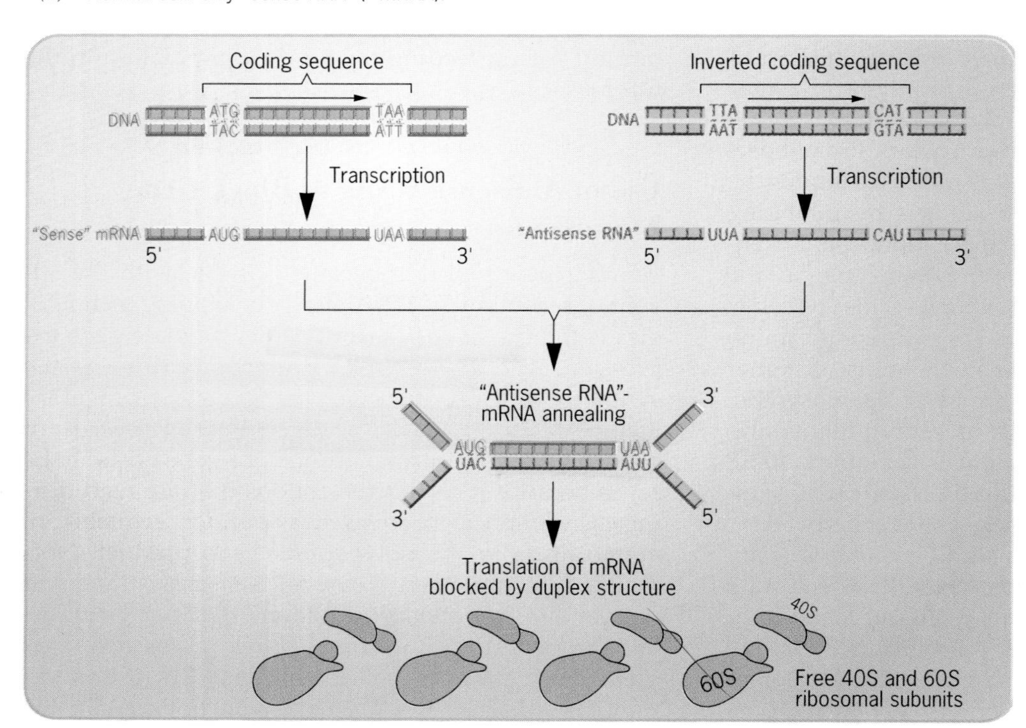

(b) Recombinant cell: "sense RNA" + "antisense RNA".

Figure 20.29 The antisense RNA procedure for blocking or reducing the level of expression of a specific gene.

ing the ripening process, was produced by using antisense RNA to decrease the rate of expression of an endogenous gene encoding an enzyme called polygalacturonase to 10 percent of the normal level. Polygalacturonase is an enzyme that breaks down cell walls and causes softening in tomatoes as they ripen.

As most of us know from first-hand experience, vine-ripened tomatoes are much more flavorful than those that are picked green and allowed to ripen en route to the marketplace. However, vine-ripened tomatoes are too soft to survive the handling required during transport; they bruise too easily. The Flavr-Savr™ tomato has partially solved this problem. Flavr-Savr™ tomatoes remain firm further into the ripening process, allowing them to remain on the vines longer before they are picked and shipped to market.

The simplest way to produce antisense RNA of a gene within a cell or an organism is to (1) clone the gene of interest, (2) separate the coding sequence of the gene from its promoter by cleavage of the DNA with an appropriate restriction enzyme, (3) ligate the coding sequence to its promoter in the inverse orientation (flipped end-for-end) and (4) introduce the promoter/inverted coding sequence construct or **antisense gene** into the host cell or organism by transformation. This procedure is illustrated in Figure 20.29. The net result of this procedure is that transcription of the antisense gene will produce antisense RNA transcripts. These antisense RNAs will hybridize with mRNA (sense RNA) molecules in cells and block translation of the mRNAs. As a result, little or no protein gene product will be synthesized in these cells.

Another potential application of the antisense RNA technology is the treatment or prevention of viral diseases. In principle, this approach is simple. A biologist identifies a gene that is completely essential for the reproduction of a virus and blocks its expression with antisense RNA. The difficult step in this approach is delivering the antisense gene to the right cells in the desired tissue(s) in order to intercept the reproduction of the target virus. Nevertheless, some scientists believe that the antisense RNA strategy may be the best hope for treating and/or preventing AIDS in the future. Whether antisense RNA will prove useful in combating pathogenic viruses remains uncertain. Currently, geneticists know that antisense RNAs shut off some genes and do not shut off other genes, but they do not know why.

Key Points: **Synthetic, modified, or other foreign genes can now be introduced into most plant and animal species. The resulting transgenic organisms provide valuable systems in which to study the functions of the introduced genes. The Ti plasmid of *Agrobacterium tumefaciens* is an important tool for transferring genes into plants. Antisense RNA, which is complementary to sense RNA (mRNA), can be used to shut off or reduce the expression of individual genes.**

TESTING YOUR KNOWLEDGE

1. Spinocerebellar ataxia (type 1) is a progressive neurological disease with onset typically occurring between ages 30 and 50. The neurodegeneration results from the selective loss of specific neurons. Although it is not understood why selective neuronal death occurs, it is known that the disease is caused by the expansion of a CAG trinucleotide repeat, with normal alleles containing about 28 copies and mutant alleles harboring 43 to 81 copies of the trinucleotide. Given the nucleotide sequences on either side of the repeat region, how would you test for the presence of the expanded trinucleotide repeat region responsible for type 1 spinocerebellar ataxia?

ANSWER

The DNA test for spinocerebellar ataxia (type 1) would be similar to the test for the *huntingtin* allele described in Figure 20.11. You would first make PCR primers corresponding to DNA sequences on either side of the CAG repeat region. These primers would be used to amplify the desired CAG repeat region from genomic DNA of the individual being tested by PCR. Then, the sizes of the trinucleotide repeat regions would be determined by measuring the sizes of the PCR products by gel electrophoresis. Any gene with less than 30 copies of the CAG repeat would be considered a normal allele, whereas the presence of a gene with 40 or more copies of the trinucleotide would be diagnostic of the mutant alleles that cause spinocerebellar ataxia.

2. Assume that you have just performed the DNA test for spinocerebellar ataxia on a 25-year-old woman whose mother died from the disease. The results came back positive for the ataxia mutation. The woman and her husband long for their own biological children, but do not want to risk transmitting the defective gene to any of these children. What are their options?

ANSWER

Their options will depend on their religious and moral convictions. One possibility involves the use of amniocentesis or chorionic biopsy to obtain fetal cells early in pregnancy, performing the DNA test for the expanded trinucleotide region responsible for spinocerebellar ataxia on the fetal cells, and allowing the pregnancy to continue only if the defective gene is not present. Another possibility is the use of *in vitro* fertilization. The ataxia DNA test is then performed on a cell from an eight-cell pre-embryo, and the pre-embryo is implanted only if the test for the defective ataxia gene is negative. A third option may be available in the future, namely, an effective method of treating the disease prior to the onset of neurodegeneration, perhaps by gene-replacement therapy.

QUESTIONS AND PROBLEMS

20.1 Distinguish between a genetic map, a cytogenetic map, and a physical map. How can each of these types of maps be used to identify a gene by positional cloning?

20.2 What is the difference between chromosome walking and chromosome jumping? Why must chromosome jumps sometimes be used to identify a gene of interest rather than just chromosome walks?

20.3 What is a contig? an RFLP? a VNTR? an STS? an EST? How are each of these used in the construction of chromosome maps?

20.4 What are CpG islands? Of what value are CpG islands in positional cloning of human genes?

20.5 Why is the mutant gene that causes Huntington's disease called *huntingtin*? Why might this gene be renamed in the future?

20.6 How was dystrophin discovered? How was the nucleotide sequence of the *CF* gene used to obtain information about the structure and function of its gene product?

20.7 How might the characterization of the *CF* gene and its product lead to the treatment of cystic fibrosis by somatic-cell gene therapy? What obstacles must be overcome before cystic fibrosis can be treated successfully by gene therapy?

20.8 Myotonic dystrophy (MD), occurring in about 1 of 8000 individuals, is the most common form of muscular dystrophy in adults. The disease, which is characterized by progressive muscle degeneration, is caused by a dominant mutant gene that contains an expanded CAG repeat region. Wild-type alleles of the *MD* gene contain from 5 to 30 copies of the trinucleotide. Mutant MD alleles contain from 50 to over 2000 copies of the CAG repeat. The complete nucleotide sequence of the MD gene is available. Design a diagnostic test for the mutant gene responsible for myotonic dystrophy that can be carried out using genomic DNA from newborns, fetal cells obtained by amniocentesis, and single cells from eight-cell pre-embryos produced by *in vitro* fertilization.

20.9 In humans, the absence of an enzyme called purine nucleoside phosphorylase (PNP) results in a severe T-cell immunodeficiency similar to that of severe combined immunodeficiency disease (SCID). PNP deficiency exhibits an autosomal recessive pattern of inheritance, and the gene encoding human PNP has been cloned and sequenced. Would PNP deficiency be a good candidate for treatment by gene therapy? Design a procedure for the treatment of PNP deficiency by somatic-cell gene therapy.

20.10 What are the goals of the Human Genome Project? What impact will achieving these goals have on the practice of medicine in the future?

20.11 Human proteins can now be produced in bacteria such as *E. coli*. However, one cannot simply introduce a human gene into *E. coli* and expect it to be expressed. What steps must be taken to construct an *E. coli* strain that will produce a mammalian protein such as human growth hormone?

20.12 You have constructed a synthetic gene that encodes an enzyme that degrades the herbicide glyphosate. You wish to introduce your synthetic gene into *Arabidopsis* plants and test the transgenic plants for resistance to glyphosate. How could you produce a transgenic *Arabidopsis* plant harboring your synthetic gene by *A. tumefaciens*-mediated transformation?

20.13 DNA fingerprints have played central roles in many recent rape and murder trials. What is a DNA fingerprint? What roles do DNA prints play in these forensic cases? In some cases, geneticists have been concerned that DNA fingerprint data were being used improperly. What were some of their concerns, and how can these concerns be properly addressed?

20.14 The DNA fingerprints shown below were prepared using genomic DNA from blood cells obtained from a woman, her daughter, and three men who all claim to be the girl's father. Based on the DNA prints, what can be determined about paternity in this case?

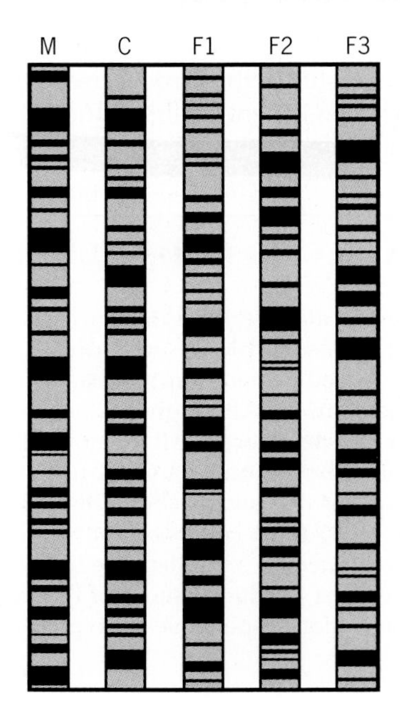

20.15 The Ti plasmid contains a region referred to as T-DNA. Why is this region called T-DNA, and what is its significance?

20.16 Disarmed retroviral vectors can be used to introduce genes into higher animals, including humans. What advantages do retroviral vectors have over other kinds of gene-transfer vectors? What disadvantages?

20.17 Transgenic mice are now routinely produced and studied in research laboratories throughout the world. How are transgenic mice produced? What kinds of information can be obtained from studies performed on transgenic mice? Does this information have any importance to the practice of medicine? If so, what?

20.18 What is antisense RNA? Of what use is antisense RNA in genetic research? in agriculture? How can a cell or an organism that produces a specific antisense RNA be produced?

20.19 Many valuable human proteins contain carbohydrate or lipid components that are added post-translationally. Bacteria do not contain the enzymes needed to add these components to primary translation products. How might these proteins be produced using transgenic animals?

20.20 Richard Meagher and coworkers have cloned a family of 10 genes that encode actins (a major component of the cytoskeleton) in *Arabidopsis thaliana*. The 10 actin gene products are similar, often differing by just a few amino acids. Thus the coding sequences of the 10 genes are also very similar, so that the coding region of one gene will cross-hybridize with the coding regions of the other nine genes. In contrast, the noncoding regions of the 10 genes are quite divergent. Meagher has hypothesized that the 10 actin genes exhibit quite different temporal and spatial patterns of expression. You have been hired by Meagher to test this hypothesis. Design experiments that will allow you to determine the temporal and spatial pattern of expression of each of the 10 actin genes in *Arabidopsis*.

BIBLIOGRAPHY

Blaese, R. M., K. W. Culver, A. D. Miller, C. S. Carter, T. Fleisher, M. Clerici, G. Shearer, L. Chang, Y. Chiang, P. Tolstoshev, J. J. Greenblatt, S. A. Rosenberg, H. Klein, M. Berger, C. A. Mullen, W. J. Ramsey, L. Muul, R. A. Morgan, and W. F. Anderson. 1995. T lymphocyte-directed gene therapy for ADA$^-$ SCID: Initial trial results after 4 years. *Science* 270: 475–480.

Bordignon, C., L. D. Notarangelo, N. Nobili, G. Ferrari, G. Casorati, P. Panina, E. Mazzolari, D. Maggioni, C. Rossi, P. Servida, A. G. Ugazio, and F. Mavilio. 1995. Gene therapy in peripheral blood lymphocytes and bone marrow for ADA$^-$immunodeficient patients. Science 270: 470–475.

Huntington's Disease Collaborative Research Group. 1993. A novel gene containing a trinucleotide repeat that is expanded and unstable on Huntington's disease chromosomes. *Cell* 72:971–983.

Kristjansson, K., S. S. Chong, I. B. Van den Veyver, S. Subramarian, M. C. Snabes, and M. R. Hughes. 1994. Preimplantation single cell analyses of dystrophin gene deletions using whole genome amplification. *Nature Genetics* 6:19–23.

Lewis, R. 1994. *Human Genetics, Concepts and Applications*. Wm. C. Brown, Dubuque, Iowa.

Mol, J.N.M., R. van Blokland, P. de Lange, M. Stam, and J. M. Kooter. 1994. Post-translational inhibition of gene expression: sense and antisense genes. In *Homologous Recombination and Gene Silencing in Plants*, pp. 309–334, J. Paszkowski, ed. Kluwer, Dordrecht, The Netherlands.

NIH/CEPH Collaborative Mapping Group. 1992. A comprehensive genetic linkage map of the human genome. *Science* 258:67–86.

Rennie, J. 1994. Grading the gene tests. *Sci. Amer.* 270(6):88–97.

Rommens, J. M., M. C. Iannuzzi, B. Kerem, M. L. Drumm, G. Melmer, M. Dean, R. Rozmahel, J. L. Cole, D. Kennedy, N. Hidaka, M. Zsiga, M. Buchwald, J. R. Riordan, L-C. Tsui, and F. S. Collins. 1989. Identification of the cystic fibrosis gene: chromosome walking and jumping. *Science* 245:1059–1065.

Tennyson, C. N., H. J. Klamut, and R. G. Worton. 1995. The human dystrophin gene requires 16 hours to be transcribed and is cotranscriptionally spliced. *Nature Genetics* 9:184–190.

Weintraub, H. M. 1990. Antisense RNA and DNA. *Sci. Amer.* 262(1):40–46.

Willems, R. J. 1994. Dynamic mutations hit double figures. *Nature Genetics* 8:213–215.

A CONVERSATION WITH

EDWARD B. LEWIS

What inspired you to become a biologist? Did you have any mentors or role models who were especially inspiring to you?

An interest in animals, especially toads, turtles, and snakes, probably inspired me to become a biologist. In high school I was most influenced by a book, *The Biological Basis of Human Nature*, by H.S. Jennings in which there were figures of the results of *Drosophila* matings that demonstrated sex linkage of the white eye gene. I had no mentors in high school, but a fellow student, Edward Novitski, (now Emeritus Professor of Genetics at the University of Oregon) and I teamed up in high school to carry out experiments with *Drosophila*. Novitski corresponded with Calvin Bridges of Caltech, who provided us with cultures and instructions. As an undergraduate at the University of Minnesota, I was greatly indebted to Professor C. P. Oliver (a student of H. J. Muller) who gave me a desk in his laboratory so I could carry on my *Drosophila* experiments. He was studying lozenge eye mutants which he showed undergo mutation associated with crossing over—a result that was a forerunner of pseudoallelism.

Edward B. Lewis is currently Professor Emeritus at the California Institute of Technology in Pasadena. He received his Ph.D. at the California Institute of Technology in 1942. He served as Captain in the United States Army Air Force from 1942 to 1945, as a meteorologist and oceanographer. He received an honorary Doctor of Science degree from the University of Umea in Sweden in 1982, and an honorary Doctor of Science degree from the University of Minnesota in 1993. His research has focused on developmental genetics and the somatic effects of radiation. He has received numerous awards and prizes including the Nobel Prize in Physiology or Medicine in 1995. Dr. Lewis, Christiane Nüsslein-Volhard, and Eric Wieschaus were awarded this Nobel Prize for their research on the genetic control of development in *Drosophila*. He was a member of the National Advisory Committee on Radiation from 1958 to 1961 and served on several Advisory Committees of the National Academy of Sciences on the Biological Effects of Ionizing Radiation.

You concentrate most of your research on *Drosophila* genetics. What is so enticing about *Drosophila* genetics?

Drosophila genetics combines virtually all of the requirements needed to deduce the laws of inheritance for higher organisms: a short life cycle of 10 days at room temperature; high fertility; only 4 pairs of chromosomes; giant salivary gland chromosomes. For a long time it did not seem to be a good organism for studying embryology; but new tools and a host of mutant genes now make *Drosophila* a leading organism for studying development.

Do studies of *Drosophila* genes help us to understand how human genes function?

Now there are many examples in which genes first found in *Drosophila* turn out to have homologues in human beings; for example, the homeotic genes that control the head, thorax and abdomen of the fly and of the human being are highly conserved between these organisms in the homebox portion of these genes.

In your career as a scientist, what was one of your greatest challenges, and how did you overcome it?

One of my greatest challenges as a scientist was how to alert the public to the potential dangers of fallout from atom-bomb testing by the United States and other countries. This led to a publication [E.B. Lewis, Leukemia and Ionizing Radiation, *Science* 125:965-972 (1957)] in which I analyzed several groups of persons who had been exposed to ionizing radiation (including A-bomb survivors) and showed that the risk of leukemia in these groups was compatible with a linear relation between radiation dose and number of cases of leukemia with no indication of a threshold dose below which no cases would be expected. By extrapolation, the leukemia risk was estimated as one or two cases per million persons per year (0.1 rad being the approximate annual dose to the body from natural background radiation). A threshold dose of many rads for cancer induction had been assumed by those in charge of testing atomic weapons. Two years later I pointed out ["Thyroid Radiation Doses from Fallout," E. B. Lewis. *Proc. Natl. Acad. Sci.* 45: 894-897 (1959)] that above-ground testing of such weapons in Nevada produced levels of radioiodine in the thyroid of infants consuming fresh milk that probably averaged 0.1 rad per year for several years (with some regions of the United States receiving doses that could have averaged one rad). Infants receive a dose ten to twenty times that which the adult thyroid receives for the same amount taken up by the gland, owing to the small size of the infant thyroid, and this had been overlooked by those in charge of testing. The public had been assured that fallout levels would result in exposures that were well below the natural background level of 0.1 rad. I also carried out an epidemiological study [E. B. Lewis, Leukemia, multiple myeloma and aplastic anemia in American radiologists. *Science* 142: 1492–1494.] that showed that American radiologists, occupationally exposed to radiation, experienced during the years 1948 to 1961, inclusive, a statistically highly significant increased death rate from leukemia, multiple myeloma, and aplastic anemia.

That cancers result from somatic mutation had been proposed by H. J. Muller [*Science* 66:844 (1927)] in his famous paper showing that x-rays can induce genetic mutations in *Drosophila*. The absence of a detectable threshold for certain cancers as well as for germ line mutations after exposure to ionizing radiation lent further support to Muller's hypothesis. Molecular evidence for such mutations is now well established for certain human cancers.

What do you see as the most challenging and exciting problems to be explored in genetics today?

The most exciting challenges are to determine how genes are able to control the development and behavior of organisms, including human beings, and to determine how gene duplication has been able to accomplish the evolution of higher organisms. Computer analysis of DNA sequence data will be increasingly helpful in identifying increasing numbers of genes involved in producing birth defects and other heritable diseases as well as those involved in cancer.

What advice would you give to undergraduate students interested in the study of genetics?

One, obtain a firm foundation in all branches of genetics and in the related disciplines of cytology, statistics, biochemistry, and molecular biology and two, retain a keen interest in the biology of living organisms, whether they be plants or animals.

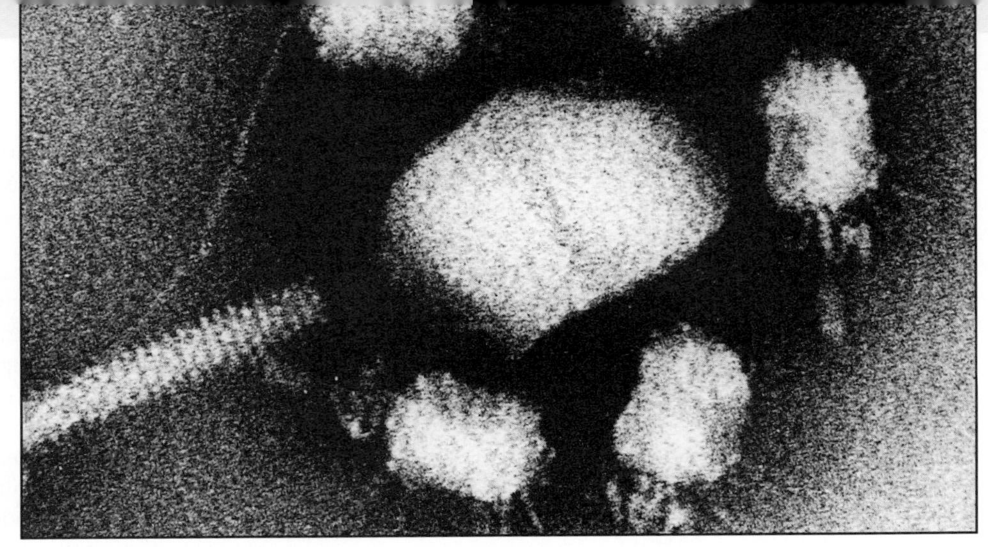

21

A quintet of Bacillus subtilis *bacteriophage φ29 particles surrounding an Escherichia coli T2 bacteriophage.*

Regulation of Gene Expression in Prokaryotes

D'Hérelle's Dream of Treating Dysentery in Humans by Phage Therapy

In 1910, the French-Canadian microbiologist Felix d'Hérelle was in Mexico investigating a bacterial disease that was killing entire populations of locusts. The infected locusts developed severe diarrhea, excreting almost pure suspensions of bacilli prior to death. When he isolated and studied the fecal bacteria, d'Hérelle observed circular clear spots in the bacterial cultures grown on agar. However, when he examined the material in the clear spots microscopically, he could not see anything. In 1915, d'Hérelle returned to the Pasteur Institute in Paris, where he studied an epidemic of bacterial dysentery that was raging through army units stationed in France. He once again observed clear spots in lawns of bacteria. In addition, he demonstrated that whatever was killing the *Shigella* could pass through a porcelain filter that retained all known bacteria. In 1917, d'Hérelle published his results and named the submicroscopic bacteriocidal agents bacteriophages (from the Greek word for "bacteria-devouring"). About the same time, an English medical bacteriologist, Frederick W. Twort, discovered a similar submicroscopic agent that killed micrococci. Twort's results were published in 1915 but remained largely unknown until after the publication of d'Hérelle's work. Unfortunately, Twort's research was soon interrupted by a call to serve in the Royal Army Medical Corps in World War I.

Meanwhile, d'Hérelle continued to study his submicroscopic agents that killed the bacteria responsible for diarrhea in locusts and the *Shigella* that cause dysentery in humans. He provided the following first-hand account of one of his experiments: "in a flash I had understood: what caused my clear spots was in fact an invisible microbe, a filtrable virus, but a virus parasitic on bacteria. . . . 'If this is true, the same thing has probably occurred during the night in the sick man, who yesterday was in serious condition. In his intestine, as in my test-tube, the dysentery baccilli will have dissolved away under the action of their parasite. He should now be cured.' I dashed to the hospital. In fact, during the night, his condition had greatly improved and convalescence was beginning" (d'Hérelle, 1949, "The Bacteriophage," *Science News*, No. 14, Penguin, Harmondsworth).

Indeed, d'Hérelle became obsessed with his belief that human diseases caused by bacteria could be treated, perhaps even eradicated, by bacteriophage therapy. Unfortunately, it was soon proven that bacteriophage therapy is not effective in treating bacterial infections because, too frequently, the bacteria mutate to phage-resistant forms. Nevertheless, d'Hérelle's work set the stage for research that would eventually produce a whole new field—microbial genetics, lead to the use of viral chromosomes as gene-transfer vectors (Technical Sidelight: Proviruses as Vectors for Human Gene Therapy), and yield insights into the mechanisms by which gene expression is regulated. In this chapter, we examine some of these mechanisms.

Microorganisms exhibit remarkable capacities to adapt to diverse environmental conditions. This adaptability depends in part on their ability to turn on and turn off the expression of specific sets of genes in response to changes in the environment. Most microorganisms exhibit a striking ability to regulate the expression of specific genes in response to environmental signals. The expression of particular genes is turned on when the products of these genes are needed for growth. Their expression is turned off when the gene products are no longer needed. Clearly, the ability of an organism to regulate gene expression in this way increases its ability to reproduce under a variety of environmental conditions. The synthesis of gene transcripts and translation products requires the expenditure of considerable energy. By turning off the expression of genes when their products are not needed, an organism can avoid wasting energy and can utilize the conserved energy to synthesize products that maximize growth rate. What, then, are the mechanisms by which microorganisms regulate gene expression in response to changes in the environment? Is there a single mechanism by which the expression of different genes is regulated? Or, are different genes controlled by different mechanisms?

Gene expression in prokaryotes is regulated at several different levels: transcription, mRNA processing, mRNA turnover, translation, and enzyme function (Figure 21.1). However, the regulatory mechanisms with the largest effects on phenotype act at the level of transcription.

Based on what is known about the regulation of transcription, the various regulatory mechanisms seem to fit into two general categories:

1. *Mechanisms that involve the rapid turn-on and turn-off of gene expression in response to environmental changes.* Regulatory mechanisms of this type are important in microorganisms because of the frequent exposure of these organisms to sudden changes in environment. They provide microorganisms with considerable "plasticity," an ability to adjust their metabolic processes rapidly in order to achieve maximal growth and reproduction under a wide

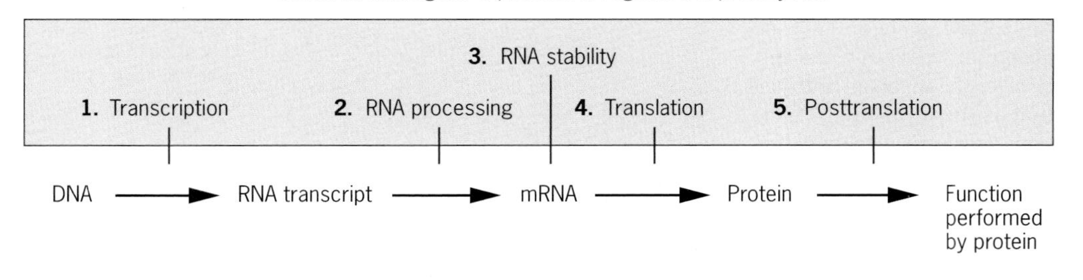

Figure 21.1 An abbreviated pathway of gene expression, showing five important levels of regulation in prokaryotes.

TECHNICAL SIDELIGHT

Proviruses as Vectors for Human Gene Therapy

The treatment of inherited human diseases by gene therapy has progressed from fantasy to reality during the last decade. Partial immune responses have been documented in children with Severe Combined Immunodeficiency Disease (SCID) who were treated by somatic-cell gene therapy (Chapter 20). With over 4000 inherited human disorders already described and 50,000 to 100,000 mutable genes in the human genome, the potential future impact of human gene therapy is enormous.

The treatment of inherited human diseases by gene therapy involves the addition of wild-type genes to cells of individuals carrying defective copies of the gene. How can genes be transferred into human cells? Will the introduced genes be transmitted to progeny cells? Ideally, human geneticists would like to insert the wild-type genes into the chromosomes of the mutant cells to maximize the likelihood that these genes will be inherited by progeny cells. How might this be accomplished? To date, the most promising human gene-transfer vectors are the proviruses of RNA tumor viruses or retroviruses; these proviruses become covalently inserted into the chromosomal DNA of the infected cells (Chapters 15 and 22). If a wild-type human gene is present in a provirus, the gene will be incorporated into the DNA of the host cell along with the proviral DNA (Figure 22.23).

Before examining the use of proviruses as vectors for human gene therapy, let's briefly consider the origin of the provirus paradigm. The idea that the chromosome of a virus might be able to insert itself into the chromosome of the host cell evolved from studies of lysogenic bacteria. In 1950, André Lwoff proposed that the genetic information of a bacteriophage could exist in a bacterium in a dormant state, which he called the prophage. However, it was not until 1962 that Allan Campbell correctly proposed that phage λ lysogeny was established in *E. coli* by a site-specific recombination event that covalently inserted the λ chromosome into the *E. coli* chromosome, and the accuracy of this proposal was not firmly established until 1965. The pioneering research on the λ prophage discussed in this chapter clearly provided the paradigm that led to the concept of proviruses in eukaryotes.

That RNA tumor viruses might exist in a proviral state in eukaryotic chromosomes was first proposed in 1960 by Howard Temin, co-recipient of the 1975 Nobel Prize in Physiology or Medicine. In 1964, Temin proposed that proviruses are DNA copies of the RNA genomes of these viruses, and the authenticity of his DNA provirus concept was established by the results of research conducted in several laboratories during the early 1970s. Subsequent research demonstrated that the proviruses are covalently inserted into the DNA of the host chromosomes, just as the λ prophage is inserted into the *E. coli* chromosome. The only major difference between the two systems is that the λ prophage is inserted at a specific site in the *E. coli* chromosome, whereas the proviruses of RNA tumor viruses are incorporated at random or nearly random sites in eukaryotic genomes.

To date, all the retroviral vectors used to transfer genes into human cells have been derived from the Moloney Murine Leukemia Virus (MMLV). Gene-transfer vectors are constructed by replacing viral genes with wild-type human genes, such as the adenosine deaminase gene that is being used in gene therapy of ADA⁻SCID patients (Chapter 20), and appropriate regulatory sequences. Together, the wild-type human gene and the regulatory element(s) are called an expression cassette, and the desired cassette is inserted into the retroviral genome in place of viral genes required for infection and replication of the virus, as shown in the following:

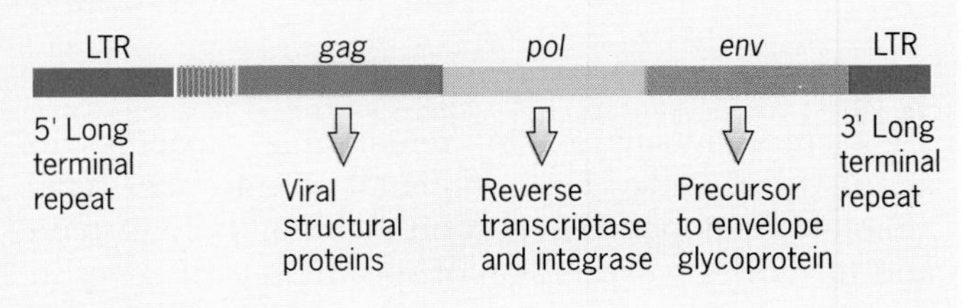

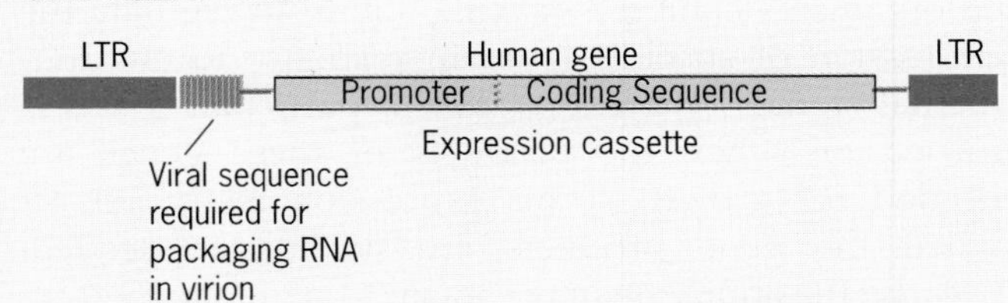

Because viral genes required for replication have been replaced with the expression cassette, the retroviral gene-transfer vector cannot reproduce by itself. The replication-deficient vector must be amplified and packaged in virions in a special helper or packaging cell line that produces the viral *gag*, *pol*, and *env* gene products, but no viral genomes, as shown in the following illustration:

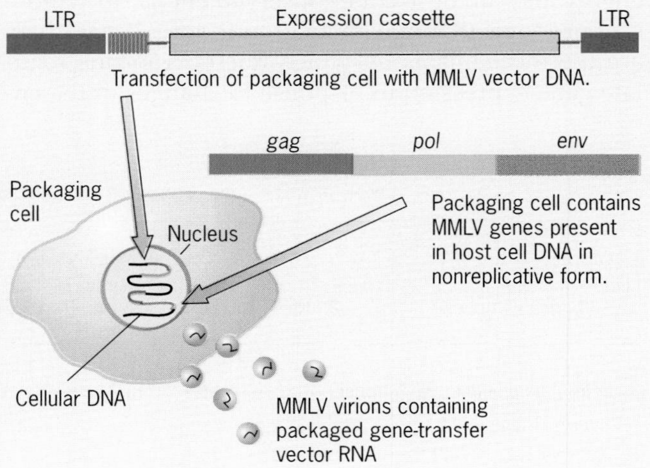

The virions containing the expression cassette are then used to deliver the desired gene to mutant cells from individuals lacking the respective gene-product activity. After the vector RNA enters a mutant cell, it is converted to DNA by reverse transcriptase, and the resulting proviral DNA—including the expression cassette—is integrated into the DNA of the mutant cell by recombination, as shown in the following diagram:

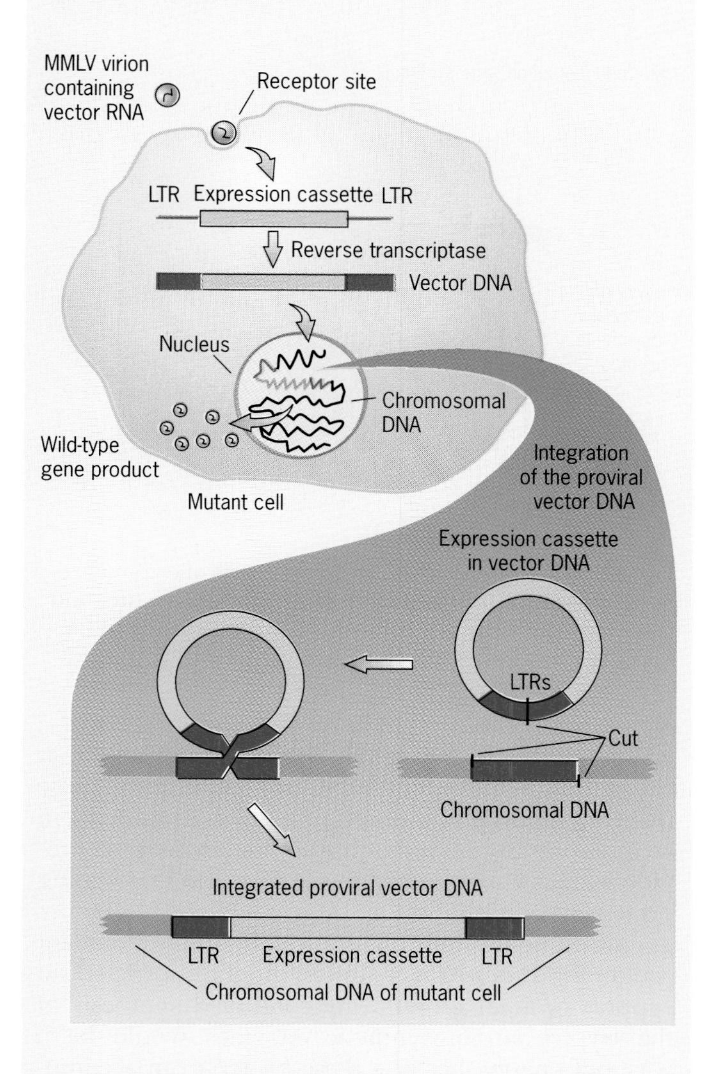

Transcription and translation of the expression cassette carried by the integrated provirus should produce the enzyme or structural protein that had been missing in the mutant cells. The genetically modified cells are tested for appropriate expression of the introduced gene before they are re-implanted in the patient (Figure 20.18), where the presence of the wild-type gene product should alleviate the symptoms of the disease. Although many technical problems remain to be worked out before somatic gene therapy can be used extensively in humans, none of these problems appears to be insurmountable. Thus it seems safe to predict that somatic-cell gene therapy will play an increasingly important role in the practice of medicine in the future.

range of environmental conditions. These rapidly responding on-off switches seem to be less important in multicellular eukaryotes, perhaps because the circulatory systems of higher animals buffer cells against many of the sudden environmental changes that microorganisms encounter.

2. *Mechanisms referred to as preprogrammed circuits or cascades of gene expression*. In these cases, some event triggers the expression of one set of genes. The product (or products) of one (or more) of these genes functions by turning off the transcription of the first set of genes or turning on the transcription of a second set of genes. In turn, one or more of the products of the second set acts by turning on a third set, and so on. In these cases, the sequential expression of genes is genetically preprogrammed, and the genes cannot usually be turned on out of sequence. Such preprogrammed sequences of gene expression are well documented in both prokaryotes and eukaryotes (Chapter 22). When a lytic bacteriophage infects a bacterium, the viral genes are expressed in a predetermined sequence, and this sequence is directly correlated with the temporal sequence of gene-product involvement in the reproduction and morphogenesis of the virus. In most of the known examples of preprogrammed gene expression, the circuitry is cyclical. For example, during viral infections, some event associated with the packaging of the viral DNA or RNA in the protein coat resets the genetic program so that the proper sequence of gene expression occurs once again when a progeny virus infects a new host cell.

CONSTITUTIVE, INDUCIBLE, AND REPRESSIBLE GENE EXPRESSION

Certain gene products, such as tRNA molecules, rRNA molecules, ribosomal proteins, RNA polymerase subunits, and enzymes catalyzing metabolic processes that are frequently referred to as cellular "housekeeping" functions are essential components of almost all living cells. Genes that specify products of this type are continually being expressed in most cells. Such genes are said to be expressed constitutively and are referred to as **constitutive genes**.

Other gene products are needed for cell growth only under certain environmental conditions. Constitutive synthesis of such gene products would be wasteful, using energy that could otherwise be utilized for more rapid growth and reproduction. The evolution of regulatory mechanisms that provide for the synthesis of such gene products only when and where they are needed would clearly endow the organisms that possess these regulatory mechanisms with a selective advantage over organisms that lack them. This undoubtedly explains why currently existing organisms, including the "primitive" bacteria and viruses, exhibit highly developed and efficient mechanisms for the control of gene expression.

Escherichia coli and most other bacteria are capable of growth using any one of several carbohydrates, for example, glucose, sucrose, galactose, arabinose, and lactose, as an energy source. If glucose is present in the environment, it will be preferentially metabolized by *E. coli* cells. However, in the absence of glucose, *E. coli* cells can grow very well on other carbohydrates. Cells growing in medium containing the sugar lactose, for example, as the sole carbon source synthesize two enzymes, β-galactosidase and β-galactoside permease, which are uniquely required for the catabolism of lactose. A third enzyme, β-galactoside transacetylase, is also synthesized, but its function is unknown. β-Galactoside permease pumps lactose into the cell, where β-galactosidase cleaves it into glucose and galactose. Neither of these enzymes is of any use to *E. coli* cells if no lactose is available to them. The synthesis of these two enzymes requires the utilization of considerable energy (in the form of ATP and GTP; see Chapters 11 and 12). Thus, *E. coli* cells have evolved a regulatory mechanism by which the synthesis of these lactose-catabolizing enzymes is turned on in the presence of lactose and turned off in its absence.

In natural environments (intestinal tracts and sewers), *E. coli* cells probably encounter an absence of glucose and the presence of lactose relatively infrequently. Therefore, the *E. coli* genes encoding the enzymes involved in lactose utilization are probably turned off most of the time. If cells growing on a carbohydrate other than lactose are transferred to medium containing lactose as the only carbon source, they quickly begin to synthesize the enzymes required for lactose utilization (Figure 21.2*a*). This process of turning on the expression of genes in response to a substance in the environment is called **induction.** Genes whose expression is regulated in this manner are called **inducible genes;** their products, if enzymes, are called **inducible enzymes.**

Enzymes that are involved in **catabolic** (degradative) **pathways,** such as in lactose, galactose, or arabinose utilization, are characteristically inducible. As we discuss below, induction occurs at the level of transcription. Induction alters the rate of synthesis of enzymes, not the activity of existing enzyme molecules. Induction should not be confused with enzyme activation, which occurs when the binding of a small molecule to an enzyme increases the activity of the enzyme, but does not affect its rate of synthesis.

Bacteria can synthesize most of the organic molecules, such as amino acids, purines, pyrimidines, and vitamins, required for growth. For example, the *E. coli* genome contains five genes encoding enzymes that catalyze steps in the biosynthesis of tryptophan. These five genes must be expressed in *E. coli* cells growing in an environment devoid of tryptophan in order to provide adequate amounts of this amino acid for ongoing protein synthesis.

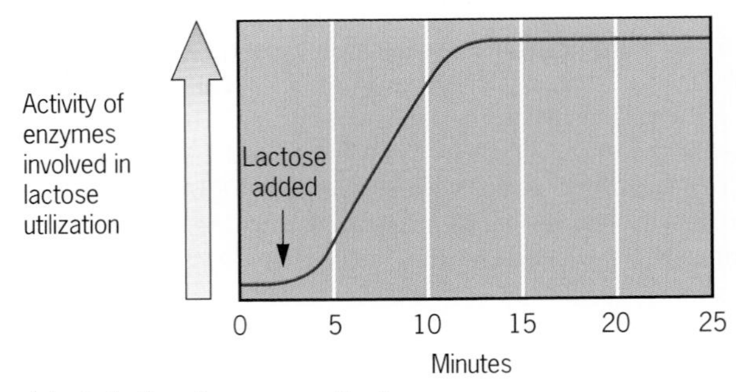

(*a*) Induction of enzyme synthesis.

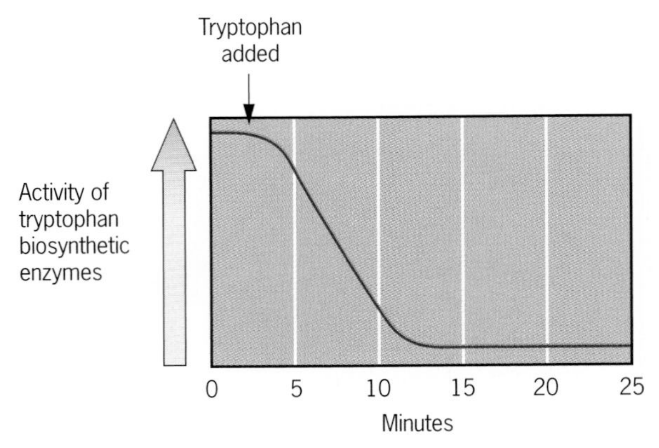

(*b*) Repression of enzyme synthesis.

Figure 21.2 (*a*) Induction of the synthesis of enzymes required for the utilization of lactose as an energy source and (*b*) repression of the synthesis of the enzymes required for the biosynthesis of tryptophan, both in *E. coli*.

When *E. coli* cells are present in an environment containing concentrations of tryptophan sufficient to support optimal growth, the continued synthesis of the tryptophan biosynthetic enzymes would be a waste of energy, because these bacteria can accumulate and utilize tryptophan from the environment. Thus a regulatory mechanism has evolved in *E. coli* that turns off the synthesis of the tryptophan biosynthetic enzymes when external tryptophan is available (Fig. 21.2*b*). This process of turning off the expression of genes in response to an environmental signal is called **repression.** A gene whose expression has been turned off in this way is said to repressed; when its expression is turned on, a gene of this type is said to be derepressed. Such a response is called **derepression.**

Enzymes that are components of **anabolic** (biosynthetic) **pathways** often are repressible. Repression, like induction, occurs at the level of transcription. Repression should not be confused with feedback inhibition, which occurs when the product of a biosynthetic pathway binds to and inhibits the activity of the first enzyme in the pathway, but does not affect the synthesis of the enzyme.

Key Points: **Genes that specify housekeeping functions such as rRNAs, tRNAs, ribosomal proteins, and DNA and RNA polymerase subunits are expressed constitutively. Other genes usually are expressed only when their products are needed. Genes that encode enzymes involved in catabolic pathways are often expressed only in the presence of the substrates of the enzymes; their expression is inducible. Genes that encode enzymes involved in anabolic pathways usually are turned off or repressed in the presence of the end product of the pathway; their expression is repressible. Although gene expression can be regulated at many levels, transcriptional regulation is the most common.**

OPERONS: COORDINATELY REGULATED UNITS OF GENE EXPRESSION

Induction and repression of gene expression can be accomplished by the same mechanism, with one relatively minor modification. This mechanism, known as the **operon model**, was developed in 1961 by François Jacob and Jacques Monod to explain the regulation of genes encoding the enzymes required for lactose utilization in *E. coli.* Jacob and Monod, 1965 Nobel Prize recipients, proposed that the transcription of one or a set of contiguous structural genes (genes specifying the amino acid sequences of enzymes or structural proteins) is regulated by two controlling elements (Figure 21.3*a*). One of these elements, the **regulator** or **repressor gene**, encodes a protein called the **repressor.** Under the appropriate conditions, the repressor binds to the second element, the **operator**. The operator is always contiguous with the structural gene or genes whose expression it regulates.

When the repressor is bound to the operator, transcription of the structural genes cannot occur. Transcription is initiated at RNA polymerase-binding sites called **promoters** located just upstream (5′) from the coding regions of structural genes (Chapter 11). Promoters were unknown at the time of Jacob and Monod's proposal, but they have since been shown to be essential components of operons. When repressor is bound to the operator, it sterically prevents RNA polymerase from transcribing the structural genes in the operon. Operator regions are always contiguous with promoter regions; sometimes operators and promoters even overlap, sharing a short DNA sequence. Operator regions are usually located between the promoters and the structural genes that they regulate. The complete contiguous unit, including the structural gene or genes, the operator, and the promoter, is called an **operon** (Figure 21.3*a*).

Whether the repressor will bind to the operator and turn off the transcription of the structural genes in an operon is determined by the presence or absence of **effector molecules** in the environment. These effector molecules are usually small molecules such as amino acids, sugars, and similar metabolites. In the case of inducible operons, the effector molecules are **inducers.** Those active on repressible operons are called **co-repressors.** These effector molecules, inducers and co-repressors, are bound by repressors and cause changes in the three-dimensional structures of the repressors. Conformational changes in protein structure resulting from the binding of small molecules are called **allosteric transitions**. Conformational changes in proteins frequently result in alterations in their activity. In the case of repressors, allosteric transitions caused by the binding of effector molecules usually alter their ability to bind to operator regions.

Inducible operons and repressible operons can be distinguished from one another by determining whether the naked repressor or the repressor-effector molecule complex is active in binding to the operator.

1. *In the case of an inducible operon, the free repressor binds to the operator, turning off transcription* (Figure 21.3*b*). When the effector molecule (the inducer) is present, it is bound by the repressor, causing the repressor to be released from the operator; that is, the **repressor-inducer complex** cannot bind to the operator. Thus the addition of inducer turns on or induces the transcription of the structural genes in the operon (Figure 21.3*b*).

2. *In the case of a repressible operon, the situation is just reversed. The free repressor cannot bind to the operator. Only the repressor-effector molecule (co-repressor) complex is active in binding to the operator* (Figure 21.3*c*). Thus transcription of the structural genes in a repressible operon is turned on in the absence of and turned off in the presence of the effector molecule (the co-repressor).

Except for this difference in the operator-binding behavior of the free repressor and the repressor-effector molecule complex, inducible and repressible operons are identical. They operate by exactly the same mechanism.

A single mRNA transcript carries the coding information of an entire operon. Thus the mRNAs of operons consisting of more than one structural gene are multigenic. For example, the tryptophan operon mRNA of *E. coli* is a huge macromolecule containing the coding sequences of five different genes. Because

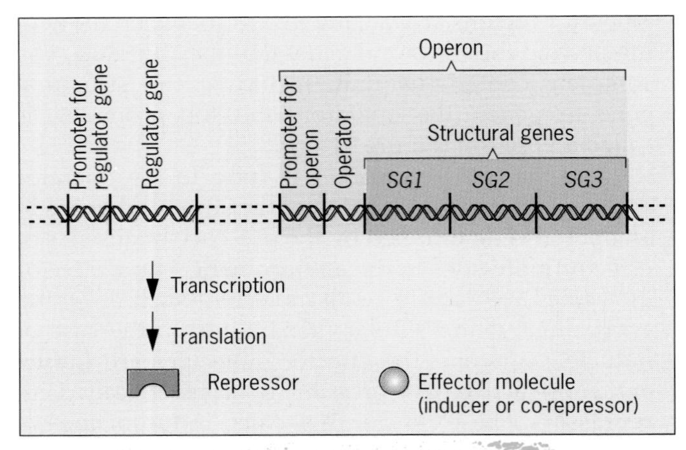

(a) The operon: components.

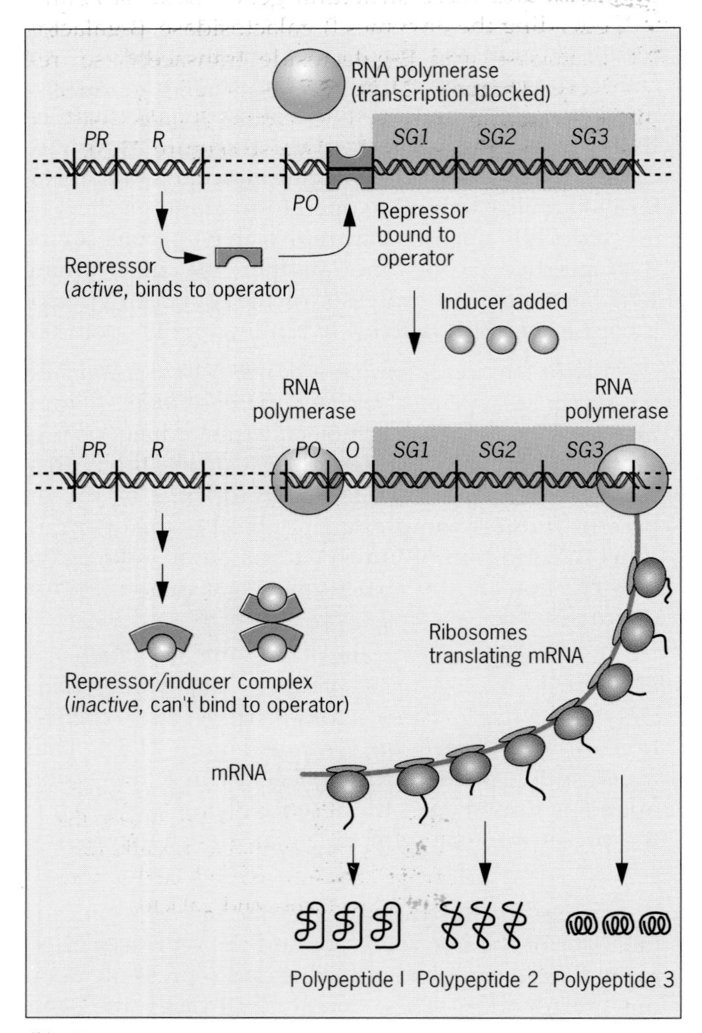

(b) The operon: induction.

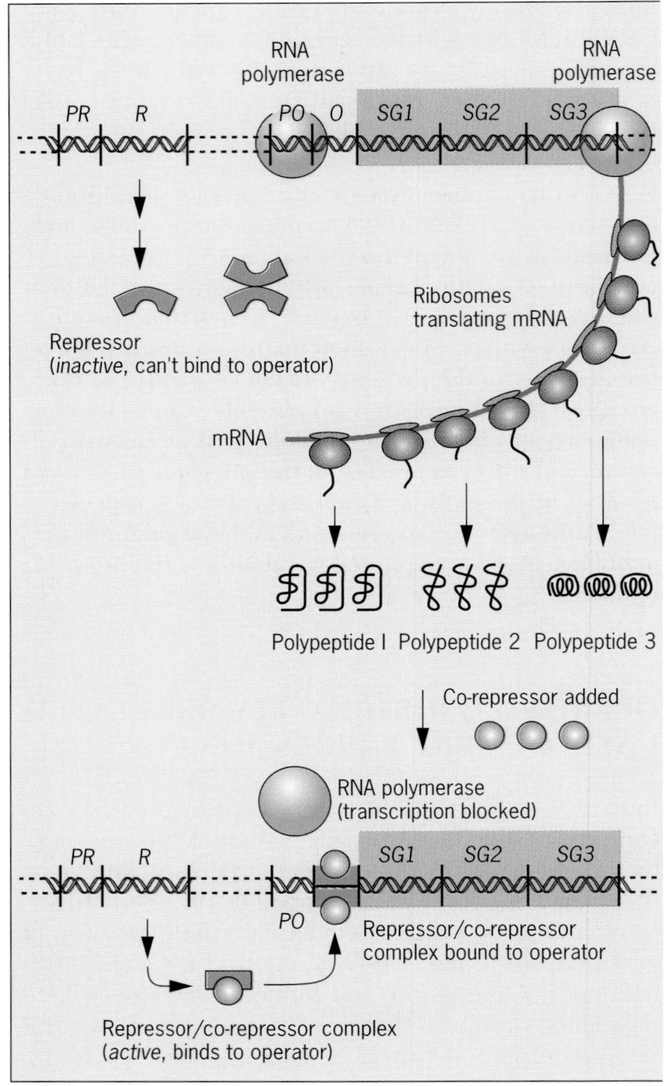

(c) The operon: repression.

Figure 21.3 Regulation of gene expression by the operon mechanism. (*a*) Essential components of an operon: one or more structural genes (three, *SG1*, *SG2*, and *SG3*, are shown) and the adjoining operator and promoter sequences. The transcription of the regulator gene (*R*) is initiated by RNA polymerase, which binds to its promoter (*PR*). When the repressor is bound to the operator, it sterically prevents RNA polymerase from initiating transcription of the structural genes. The difference between an inducible operon (*b*) and a repressible operon (*c*) is that free repressor binds to the operator of an inducible operon, whereas the repressor-effector molecule complex binds to the operator of a repressible operon. Thus an inducible operon is turned off in the absence of the effector (inducer) molecule, and a repressible operon is turned on in the absence of the effector (co-repressor) molecule.

they are co-transcribed, all structural genes in an operon are coordinately expressed. Whereas the molar quantities of the different gene products need not be the same (due to different efficiencies of initiation of translation), the relative amounts of the different polypeptides specified by genes in an operon remain the same, regardless of the state of induction or repression.

Because the product of the regulator gene, the repressor, acts by shutting off the transcription of structural genes, the operon mechanism is referred to as a **negative control system**. In a **positive control system**, the products of regulator genes are required to turn on transcription. We will discuss examples of both negative and positive control mechanisms in this chapter.

Key Points: **In bacteria, genes with related functions frequently occur in coordinately regulated units called operons. Each operon contains a set of contiguous structural genes, a promoter (the binding site for RNA polymerase) and an operator (the binding site for a regulatory protein called a repressor). When the repressor is bound to the operator, RNA polymerase cannot transcribe the structural genes in the operon. When the operator is free of repressor, RNA polymerase can transcribe the operon.**

THE LACTOSE OPERON IN *E. COLI:* INDUCTION AND CATABOLITE REPRESSION

Jacob and Monod proposed the operon model largely as a result of their studies of the lactose (*lac*) operon in *E. coli.* More is known about this operon than any other operon. The *lac* operon contains a promoter (*P*), an operator (*O*), and three structural genes, *lacZ, lacY,* and *lacA,* encoding the enzymes β-galactosidase, β-galactoside permease, and β-galactoside transacetylase, respectively (Figure 21.4). β-Galactoside permease "pumps" lactose into the cell, where β-galactosidase cleaves it into glucose and galactose (Figure 21.5). The biological role of the transacetylase is not known.

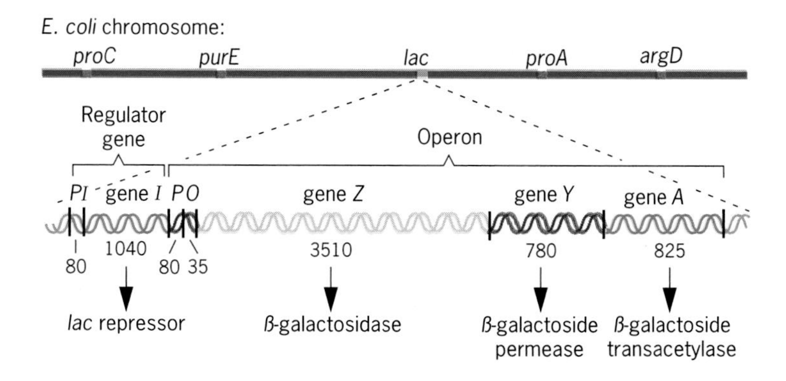

Figure 21.4 The *lac* operon of *E. coli.* The *lac* operon consists of three structural genes, *Z, Y,* and *A,* plus the promoter (*P*) and operator (*O*) regions adjoining the *Z* gene. The regulator gene (*I*) is contiguous with the operon in the case of *lac.* The regulator gene has its own promoter [*PI*]. The numbers below the genes indicate their sizes in nucleotide pairs.

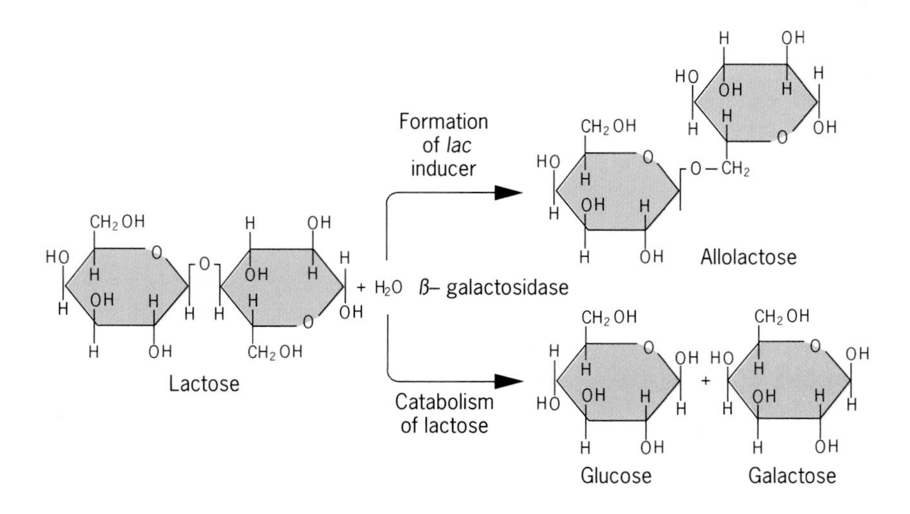

Figure 21.5 Two physiologically important reactions catalyzed by β-galactosidase: (1) conversion of lactose to the *lac* operon inducer allolactose, and (2) cleavage of lactose to produce the monosaccharides glucose and galactose.

Induction

The lac operon is an inducible operon; the *lacZ, lacY,* and *lacA* genes are expressed only in the presence of lactose. The *lac* regulator gene, designated the *I* gene, encodes a repressor that is 360 amino acids long. However, the active form of the *lac* repressor is a tetramer containing four copies of the *I* gene product. In the absence of inducer, the repressor binds to the *lac* operator sequence, which, in turn, prevents RNA polymerase from catalyzing the transcription of the three structural genes (see Figure 21.3*b*). A few molecules of the *lacZ, lacY,* and *lacA* gene products are synthesized in the uninduced state, providing a low background level of enzyme activity. This background activity is essential for induction of the *lac* operon, because the inducer of the operon, allolactose, is derived from lactose in a reaction catalyzed by β-galactosidase (Figure 21.5). Once formed, allolactose is bound by the repressor, causing the release of the repressor from the operator. In this way, allolactose induces the transcription of the *lacZ, lacY,* and *lacA* structural genes (Figure 21.3*b*).

The *lacI* gene, the *lac* operator, and the *lac* promoter were all initially identified genetically by the isolation of mutant strains that exhibited altered expression of the *lac* operon genes. Mutations in the *I* gene and the operator frequently result in constitutive synthesis of the *lac* gene products. These mutations are designated I^- and O^c, respectively. The I^- and O^c constitutive mutations can be distinguished not only by map position, but also by their behavior in partial diploids (Chapter 16) in which they are located in *cis*- and *trans*-configurations relative to mutations in *lac* structural genes.

Like haploid wild-type ($I^+P^+O^+Z^+Y^+A^+$) cells, partial diploids of genotype F′ $I^+P^+O^+Z^+A^+/I^+P^+O^+Z^-Y^-A^-$ or of genotype F′ $I^+P^+O^+Z^-Y^-A^-/I^+P^+O^+Z^+Y^+A^+$ are inducible for the utilization of lactose as a carbon source. The wild-type alleles (Z^+, Y^+, and A^+) of the three structural genes are dominant to their mutant alleles (Z^-, Y^-, and A^-). This dominance is to be expected because the wild-type alleles produce functional enzymes, whereas the mutant alleles produce no enzymes or defective (inactive) enzymes. Partial diploids of genotype $I^+P^+O^+Z^+Y^+A^+/I^-P^+O^+Z^+Y^+A^+$ are also inducible for the synthesis of the three enzymes specified by the *lac* operon. Thus I^+ is dominant to I^- as expected, since I^+ encodes a functional repressor molecule and its I^- allele specifies an inactive repressor. The dominance of I^+ over I^- also indicates that the repressor is diffusible, because the repressor produced by the $lacI^+$ allele on one chromosome can turn off the lac structural genes on both chromosomes in the cell (Figure 21.6*a*).

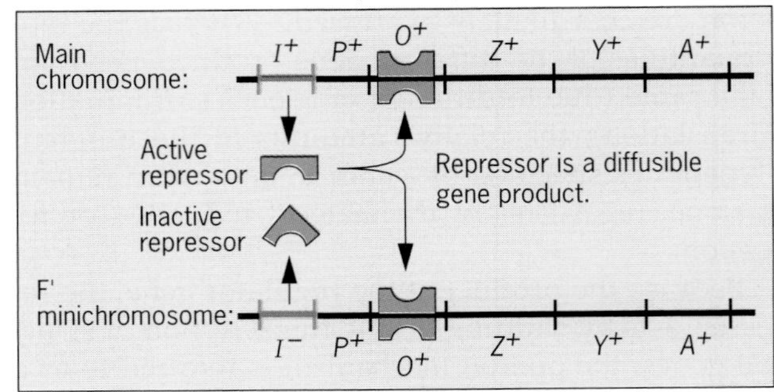

Inducible synthesis of *lac* operon gene products because the wild-type ($lacI^+$) repressor binds to the *lac* operators on both chromosomes

(*a*) Dominance of $lacI^+$ over $lacI^-$.

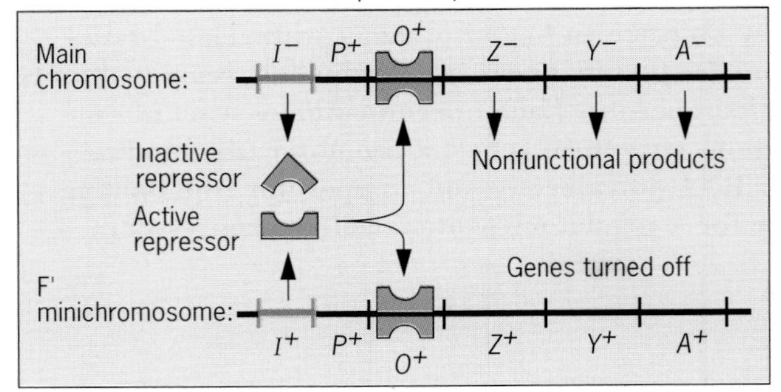

Inducible synthesis of the *lac* operon gene products

(*b*) *Cis*–dominance of $lacI^+$: I^+ located *cis* to Z^+, Y^+ and A^+.

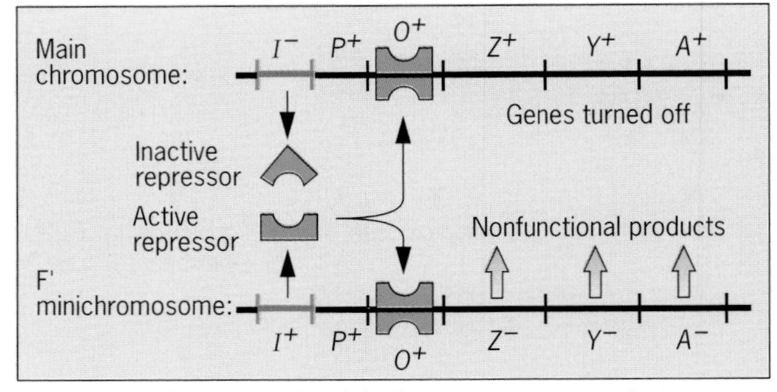

Inducible synthesis of the *lac* operon gene products

(*c*) *Trans*–dominance of $lacI^+$: I^+ located *trans* to Z^+, Y^+ and A^+.

Figure 21.6 The $lacI^+$ gene is dominant to $lacI^-$ alleles (*a*) and acts on *lac* operators located either *cis* (*b*) or *trans* (*c*) to itself. These effects demonstrate that the *lacI* gene encodes a diffusible product.

Like wild-type cells, partial diploids of genotype F′ $I^+P^+O^+Z^+Y^+A^+/I^-P^+O^+Z^-Y^-A^-$ or genotype F′ $I^+P^+O^+Z^-Y^-A^-/I^-P^+O^+Z^+Y^+A^+$ are inducible for β-galactosidase, β-galactoside permease, and β-galactoside transacetylase. The inducibility of these genotypes demonstrates that the *lac* repressor (I^+ gene

E. coli partial diploid

Inducer absent; no functional *lac* operon gene products are synthesized

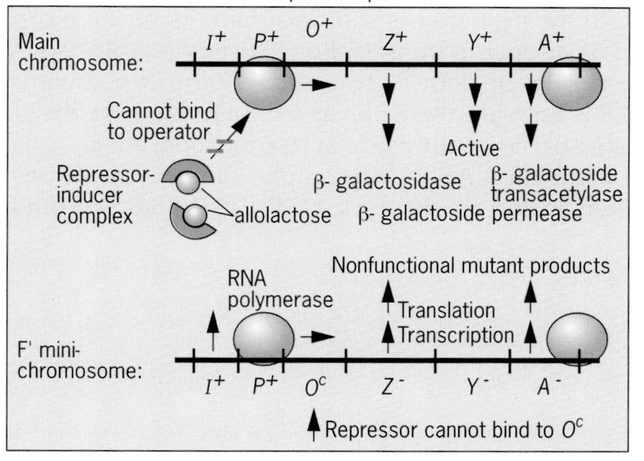

Inducer present; functional *lac* operon gene products are synthesized

(*a*) Inducible synthesis of the *lac* operon gene products in an F' $I^+ P^+ O^c Z^- Y^- A^- / I^+ P^+ O^+ Z^+ Y^+ A^+$ bacterium.

E. coli partial diploid

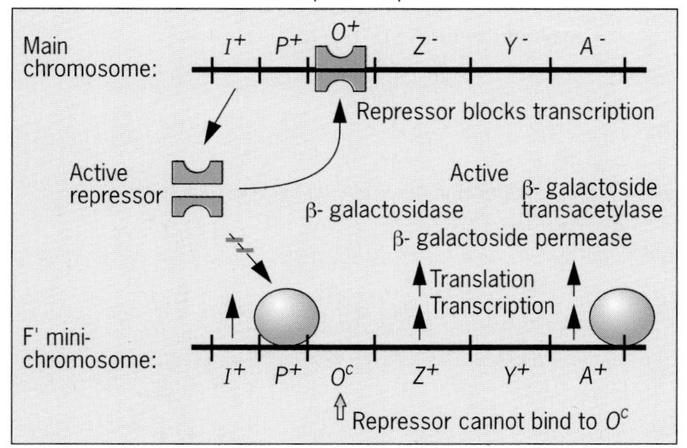

Inducer absent; functional *lac* operon gene products are synthesized

(*b*) Constitutive synthesis of the *lac* operon gene products in an F' $I^+ P^+ O^c Z^+ Y^+ A^+ / I^+ P^+ O^+ Z^- Y^- A^-$ bacterium.

product) controls the expression of structural genes located either *cis* (Figure 21.6 *b*) or *trans* (Figure 21.6*c*) to the *lacI*$^+$ allele.

Figure 21.7 The operator acts only in the *cis* configuration. The synthesis of functional β-galactosidase, β-galactoside permease, and β-galactoside transacetylase is (*a*) inducible in an *E. coli* partial diploid of genotype F' $I^+P^+O^cZ^-Y^-A^-/I^+P^+O^+Z^+Y^+A^+$ and (*b*) constitutive in an *E. coli* partial diploid of genotype F' $I^+P^+O^cZ^+Y^+A^+/I^+P^+O^+Z^-Y^-A^-$. These results demonstrate that the operator (*O*) is *cis*-acting; that is, it only regulates structural genes located on the same chromosome.

The operator constitutive (O^c) mutations act only in ***cis***. That is, O^c mutations affect the expression of only those structural genes located on the same chromosome. The *cis*-acting nature of O^+ mutations is logical given the function of the operator. O^+ mutations should not act in *trans* if the operator is the binding site for the repressor; as such, the operator does not encode any product, diffusible or otherwise. A regulator gene should act in *trans* only if it specifies a diffusible product. Therefore, a partial diploid of genotype F' $I^+P^+O^cZ^-Y^-A^-/I^+P^+O^+Z^+Y^+A^+$ is inducible for the three enzymes specified by the structural genes of the *lac* operon (Figure 21.7*a*), whereas a partial diploid of genotype F' $I^+P^+O^cZ^+Y^+A^+/I^+P^+O^+Z^-Y^-A^-$ synthesizes these enzymes constitutively (Figure 21.7*b*).

Some of the *I* gene mutations, those designated I^{-d}, are dominant to the wild-type allele (I^+). This dominance results from the inability of heteromultimers (recall that the *lac* repressor functions as a tetramer) that contain both wild-type and mutant polypeptides to bind to the operator. Other *I* gene mutations, those designated I^{-s} (*s* for superrepressed), cause the *lac* operon to be uninducible. In strains carrying these I^{-s} mutations, the *lac* structural genes can usually be induced to some degree with high concentrations of inducer, but they are not induced at normal concentrations of inducer. When studied *in vitro*, the mutant I^{-s} polypeptides form tetramers that bind to *lac* operator DNA. However, they either do not bind inducer or exhibit a low affinity for inducer. Thus the I^{-s} mutations alter the inducer binding site of the *lac* repressor.

Promoter mutations do not change the inducibility of the *lac* operon. Instead, they modify the levels of gene expression in the induced and uninduced state by changing the frequency of initiation of *lac* operon transcription—that is, the efficiency of RNA polymerase binding.

The *lac* promoter actually contains two separate components: (1) the RNA polymerase binding site, and (2) a binding site for another protein called *Catabolite Activator Protein* (abbreviated *CAP*) that prevents the *lac* operon from being induced in the presence of glucose. This second control circuit assures the preferential utilization of glucose as an energy source when it is available. We consider this second regulatory circuit next.

Catabolite Repression

Jacob and Monod proposed the operon model to explain the induction of the biosynthesis of the enzymes involved in lactose utilization when this sugar is added to the medium in which *E. coli* cells are growing. However, the presence of glucose has long been known to prevent the induction of the *lac* operon, as well as other operons controlling enzymes involved in carbohydrate catabolism. This phenomenon, called **catabolite repression** (or the **glucose effect**), assures that glucose is metabolized when present, in preference to other, less efficient, energy sources.

The catabolite repression of the *lac* operon and several other operons is mediated by a regulatory protein called **CAP** (for **C**atabolite **A**ctivator **P**rotein) and a small effector molecule called **cyclic AMP** (adenosine-3',5'-phosphate; abbreviated cAMP) (Figure 21.8). Because CAP binds cAMP when this mononucleotide is present at sufficient concentrations, it is sometimes called the cyclic AMP receptor protein.

We know that the *lac* promoter contains two separate binding sites: (1) one for RNA polymerase and (2) one for the CAP-cAMP complex (Figure 21.9). The CAP-cAMP complex must be present in its binding site in the *lac* promoter in order for the operon to be induced. The CAP-cAMP complex thus exerts positive control over the transcription of the *lac* operon. It has an effect exactly opposite to that of repressor binding to an operator. Although the precise mechanism by which CAP-cAMP stimulates RNA polymerase binding to the promoter is still uncertain, its positive control of *lac* operon transcription is firmly established by the results of both *in vivo* and *in vitro* experiments. CAP is known to function as a dimer; thus, like the *lac* repressor, it is multimeric in its functional state.

Only the CAP-cAMP complex binds to the *lac* promoter; in the absence of cAMP, CAP does not bind.

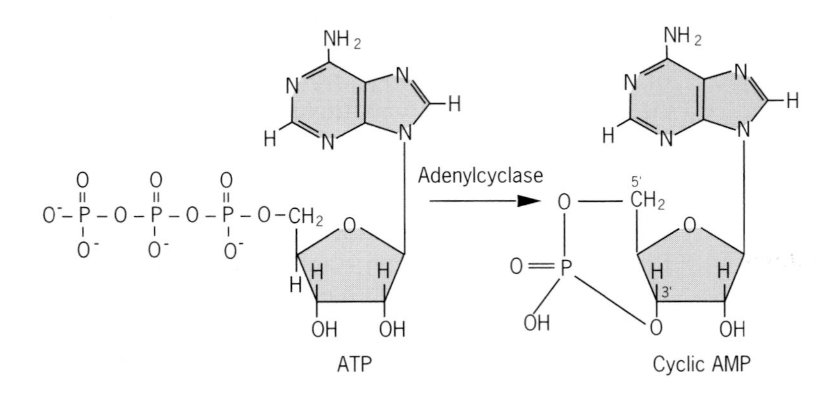

Figure 21.8 The adenylcyclase-catalyzed synthesis of cyclic AMP (cAMP) from ATP.

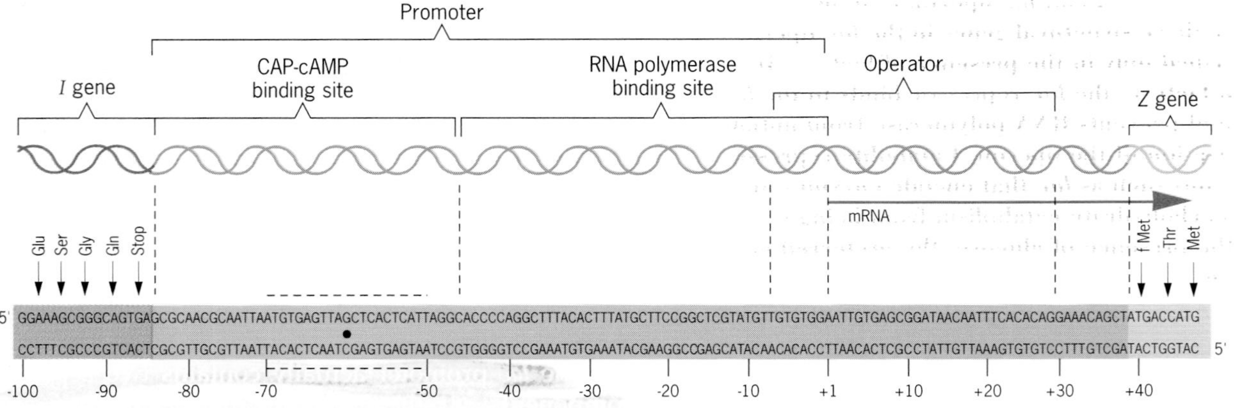

Figure 21.9 Organization of the promoter-operator region of the *lac* operon. The promoter consists of two components: (1) the site that binds the CAP-cAMP complex and (2) the RNA polymerase binding site. The adjacent segments of the *lacI* (repressor) and *lacZ* (β-galactosidase) structural genes are also shown. The horizontal line labeled mRNA shows the position at which transcription of the operon begins (the 5'-end of the *lac* mRNA). The numbers at the bottom give distances in nucleotide pairs from the site of transcript initiation (position +1). The dot between the two nucleotide strands indicates the center of symmetry of an imperfect palindrome.

Thus cAMP acts as the effector molecule, determining the effect of CAP on *lac* operon transcription. The intracellular cAMP concentration is sensitive to the presence or absence of glucose. High concentrations of glucose cause sharp decreases in the intracellular concentration of cAMP. How glucose controls the cAMP concentration is not clear. Perhaps glucose, or some metabolite that forms in the presence of sufficient concentrations of glucose, inhibits the activity of adenylcyclase, the enzyme that catalyzes the formation of cAMP from ATP. Whatever the mechanism, the presence of glucose results in a decrease in the intracellular concentration of cAMP. In the presence of a low concentration of cAMP, CAP cannot bind to the *lac* operon promoter. In turn, RNA polymerase cannot bind efficiently to the *lac* promoter in the absence of bound CAP-cAMP. The overall result of the positive control of transcription of the *lac* operon by the CAP-cAMP complex is that in the presence of glucose *lac* operon transcription never exceeds 2 percent of the induced rate observed in the absence of glucose. By similar mechanisms, CAP and cAMP keep the arabinose (*ara*) and galactose (*gal*) operons of *E. coli* from being induced in the presence of glucose.

The complete nucleotide-pair sequence of the *lac* operon regulatory region is shown in Figure 21.9. Comparative nucleotide-sequence studies of mutant and wild-type promoters and operators plus *in vitro* CAP-cAMP, RNA polymerase, and repressor binding studies are providing detailed information about the nature of these important **sequence-specific protein-nucleic acid interactions.** The results of these studies provide an excellent example of the advantages of integrating genetic and biochemical approaches in attempting to understand biological phenomena.

Key Points: **The *E. coli lac* operon is an inducible system; the three structural genes in the *lac* operon are transcribed only in the presence of lactose. In the absence of lactose, the *lac* repressor binds to the *lac* operator and prevents RNA polymerase from initiating transcription of the operon. Catabolite repression keeps operons such as *lac* that encode enzymes involved in carbohydrate catabolism from being induced in the presence of glucose, the preferred energy source.**

THE TRYPTOPHAN OPERON IN *E. COLI:* REPRESSION AND ATTENUATION

The *trp* operon of *E. coli* controls the synthesis of the enzymes that catalyze the biosynthesis of the amino acid tryptophan. The functions of the five structural genes and the adjacent regulatory sequences of the *trp* operon have been analyzed in detail by Charles Yanofsky and colleagues. The five structural genes encode enzymes that convert chorismic acid to tryptophan. The expression of the *trp* operon is regulated at two levels: repression, which controls the initiation of transcription, and attenuation, which governs the frequency of premature transcript termination. We will discuss these regulatory mechanisms in the following two sections.

Repression

The *trp* operon of *E. coli* is probably the best known repressible operon. The organization of the *trp* operon and the pathway of biosynthesis of tryptophan are shown in Figure 21.10. The *trpR* gene, which encodes the *trp* repressor, is not closely linked to the *trp* operon. The operator (*O*) region of the *trp* operon lies within the primary promoter (P_1) region. There is also a weak promoter (P_2) at the operator-distal end of the *trpD* gene. The P_2 promoter results in a somewhat increased basal level of transcription of the *trpC, trpB,* and *trpA* genes. Two transcription termination sequences (*t* and *t'*) are located downstream from *trpA.* The *trpL* region specifies a 162-nucleotide-long mRNA leader sequence.

The regulation of transcription of the *trp* operon is as diagrammed in Figure 21.3*c*. In the absence of tryptophan (the co-repressor), RNA polymerase binds to the promoter region and transcribes the structural genes of the operon. In the presence of tryptophan, the co-repressor/repressor complex binds to the operator region and prevents RNA polymerase from initiating transcription of the genes in the operon.

The rate of transcription of the *trp* operon in the derepressed state (absence of tryptophan) is 70 times the rate that occurs in the repressed state (presence of tryptophan). In *trpR* mutants that lack functional repressor, the rate of synthesis of the tryptophan biosynthetic enzymes (the products of the structural genes of the *trp* operon) is still reduced about tenfold by the addition of tryptophan to the medium. This additional reduction in *trp* operon expression is caused by attenuation, which is discussed next.

Attenuation

Deletions that remove part of the *trpL* region (Figure 21.10) result in increased rates of expression of the *trp* operon. However, these deletions have no effect on the repressibility of the *trp* operon; that is, repression and derepression occur just as in $trpL^+$ strains. These results

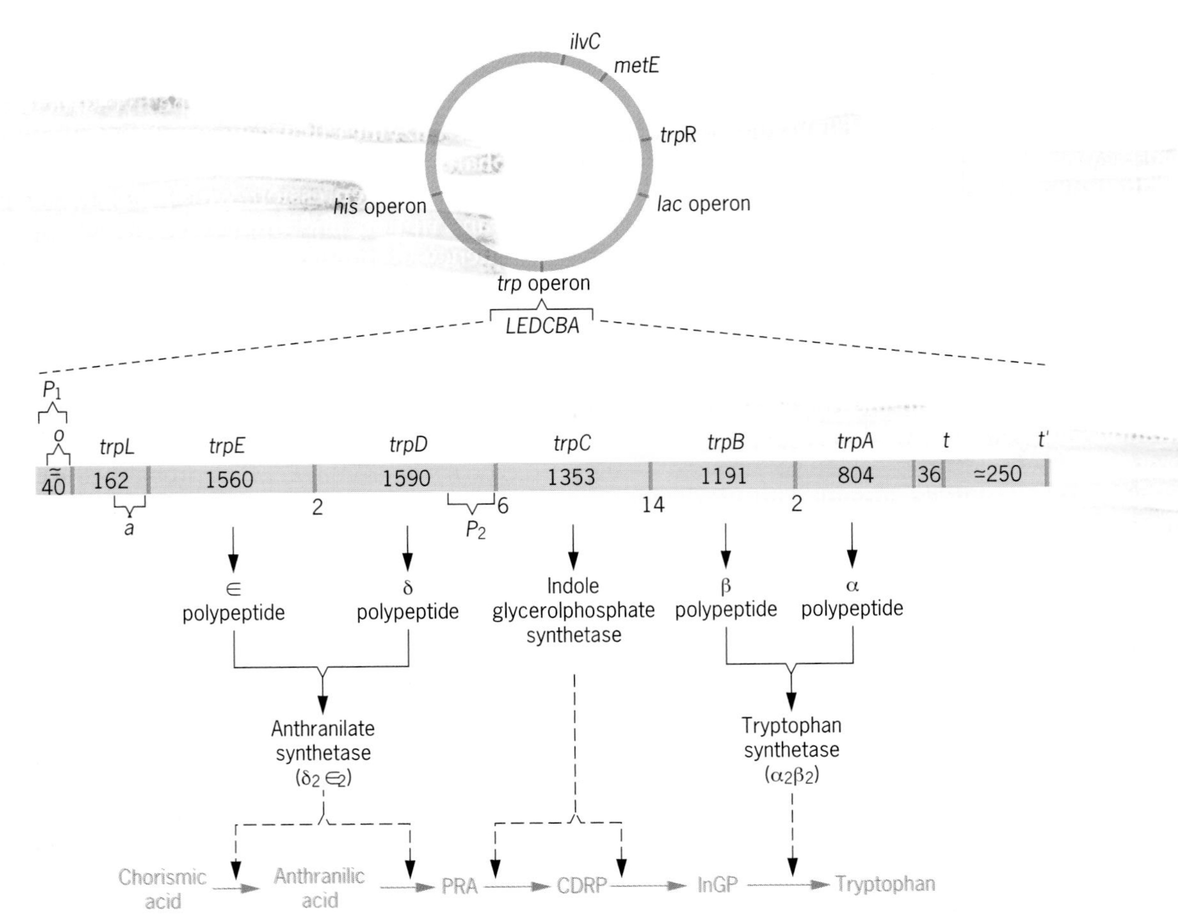

Figure 21.10 Organization of the *trp* (tryptophan) operon in *E. coli*. The *trp* operon contains five structural genes that encode enzymes involved in the biosynthesis of tryptophan, as shown at the bottom. The length of each gene or region is given in nucleotide pairs; the intergenic distances are shown below the gene sequence. Key: PRA, phosphoribosyl anthranilate; CDRP, carboxyphenylamino-deoxyribulose phosphate; InGP, indole-glycerol phosphate.

indicate that the synthesis of the tryptophan biosynthetic enzymes is regulated at a second level by a mechanism that is independent of the state of repression/derepression of the *trp* operon and requires nucleotide sequences present in the *trpL* region of the *trp* operon.

This second level of regulation of the *trp* operon is called **attenuation,** and the sequence within *trpL* that controls this phenomenon is called the **attenuator** (Figure 21.11). Attenuation occurs by control of the termination of transcription at a site near the end of the mRNA leader sequence. This "premature" termination of *trp* operon transcription occurs only in the presence of tryptophan-charged $tRNA^{Trp}$. When this premature termination or attenuation occurs, a truncated (140 nucleotides) *trp* transcript is produced.

The attenuator region has a nucleotide-pair sequence essentially identical to the **transcription-termination signals** found at the ends of most bacterial operons (including the *trp* operon). These termination signals contain a G:C-rich palindrome followed by several A:T base pairs. Transcription of these termination signals yields a nascent RNA with the potential to form a hydrogen-bonded hairpin structure followed by several Us. When a nascent transcript forms this hairpin structure, it is believed to cause a conformational change in the associated RNA polymerase, resulting in termination of transcription within the following, more weakly hydrogen-bonded $[(A:U)_n]$ region of DNA-RNA base-pairing.

The nucleotide sequence of the attenuator therefore explains its ability to terminate *trp* operon transcription prematurely. But how can this be regulated by the presence or absence of tryptophan?

First, recall that transcription and translation are coupled in prokaryotes; that is, ribosomes begin translating mRNAs while they are still being produced by transcription. Thus events that occur during translation may also affect transcription.

Second, note that the 162 nucleotide-long leader sequence of the *trp* operon mRNA contains sequences that can base-pair to form alternate stem–and–loop or

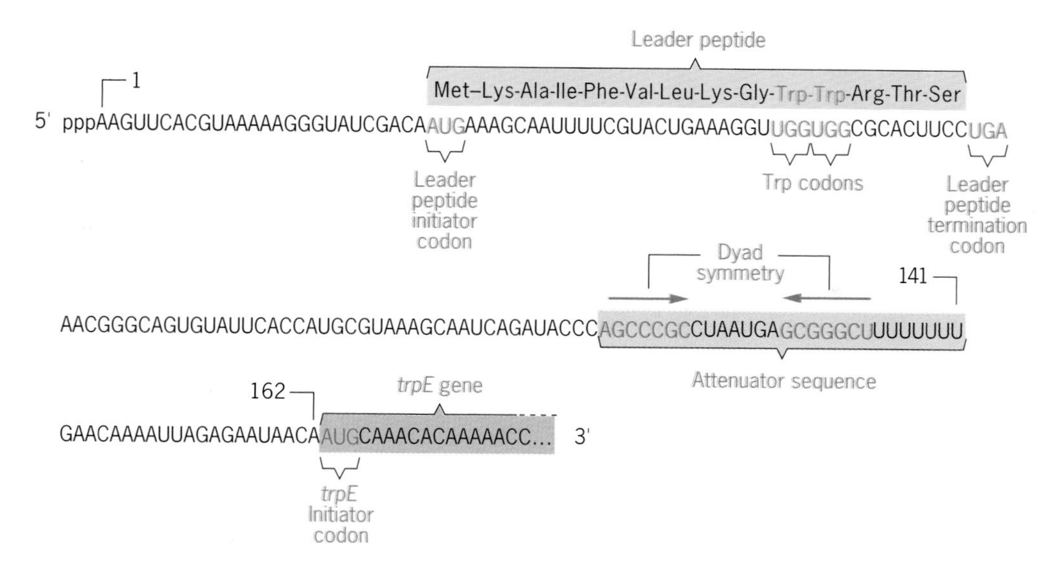

Figure 21.11 Nucleotide sequence of the leader of the *trp* operon mRNA. The region of dyad symmetry in the attenuator forms a transcription-termination hairpin (Figure 21.13*c*). The two tandem tryptophan codons in the sequence encoding the leader peptide are responsible for the control of attenuation by tryptophan.

hairpin structures (Figure 21.12). The four leader regions that can base-pair to form these structures are: (1) nucleotides 59–67, (2) nucleotides 71–79, (3) nucleotides 110–121, and (4) nucleotides 126–134. The nucleotide sequences of these four regions are such that region 1 can base-pair with region 2, region 2 can

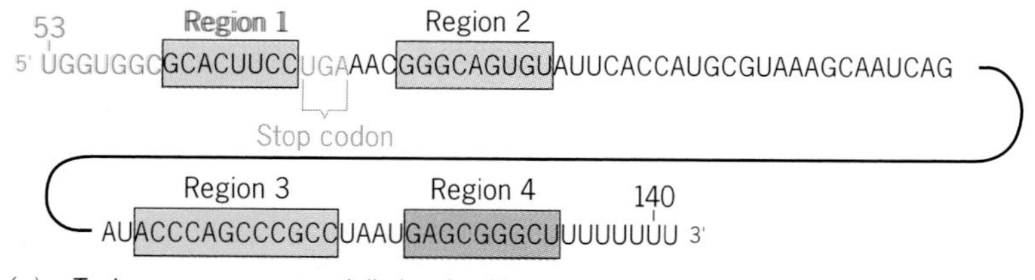

(*a*) *TrpL* sequences potentially involved in secondary structures.

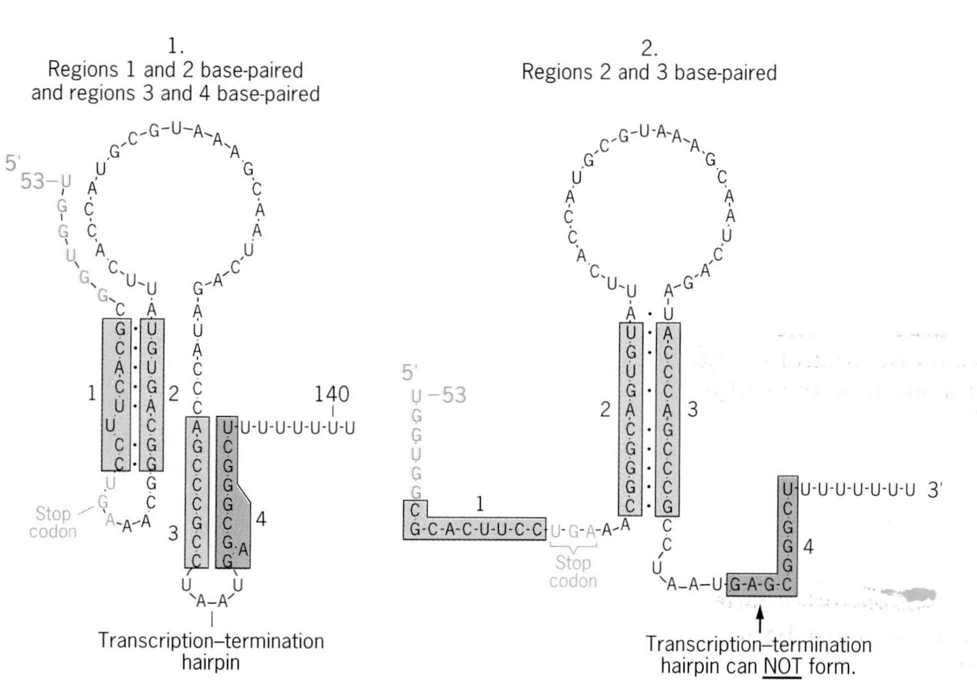

(*b*) Alternate secondary structures.

Figure 21.12 Regions of the *trp* mRNA leader sequence involved in regulation of *trp* operon expression by attenuation. The four shaded regions, labeled 1, 2, 3, and 4 (*a*), can form stem–and–loop or hairpin structures by imperfect base-pairing, as shown in (*b*). Either (1) region 1 will pair with region 2 and region 3 with region 4, forming a transcription-termination hairpin, or (2) region 2 will base-pair with region 3, preventing region 3 from pairing with region 4. The presence or absence of tryptophan determines which of these structures will form during the transcription of the *trp* operon.

pair with region 3, and region 3 can base-pair with region 4. Region 2 can base-pair with either region 1 or region 3, but, obviously, it can pair with only one of these regions at any given time. Thus there are two possible secondary structures for the *trp* leader sequence: (1) region 1 paired with region 2 and region 3 paired with region 4 or (2) region 2 paired with region 3, leaving regions 1 and 4 unpaired. The pairing of regions 3 and 4 produces the previously mentioned transcription-termination hairpin. If region 3 is base-

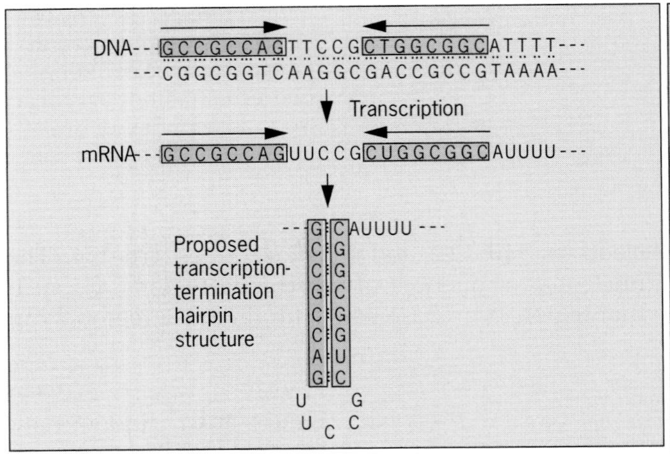

(a) Structure of *trp* operon transcription-termination sequence *t* and formation of the transcription-termination hairpin.

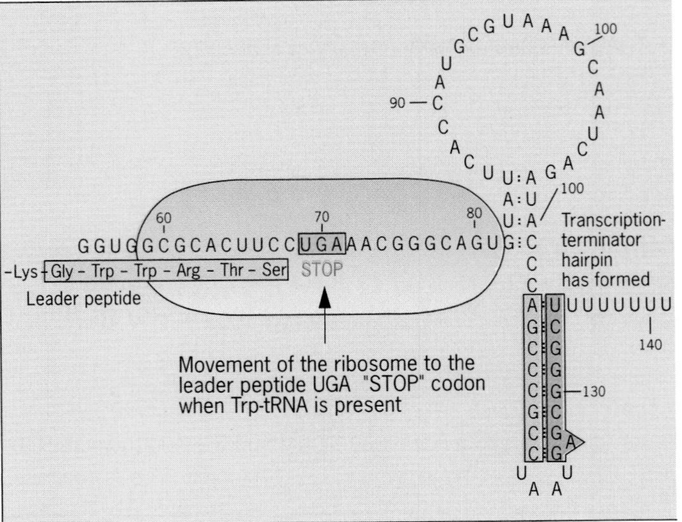

(c) In the presence of tryptophan, translation proceeds past the Trp codons to the termination codon and disrupts the base-pairing between leader regions 2 and 3. This process leaves region 3 free to pair with region 4 to form the transcription-termination hairpin, which stops transcription at the attenuator sequence.

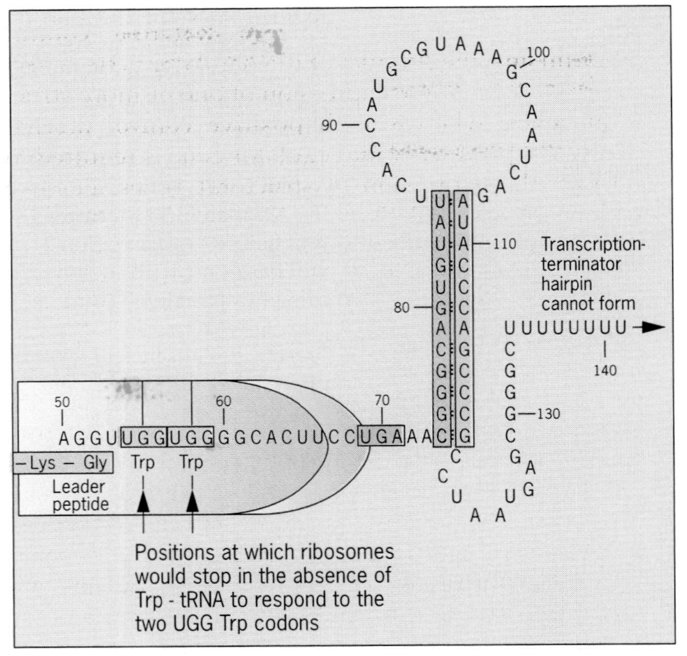

(b) In the absence of tryptophan, translation of the leader sequence stalls at one of the Trp codons. This stalling allows leader regions 2 and 3 to pair, which prevents region 3 from pairing with region 4 to form the transcription-termination hairpin. Thus transcription proceeds through the entire *trp* operon.

Figure 21.13 Control of the *trp* operon by attenuation. (*a*) The transcription-termination signal in *E. coli* contains a region of dyad symmetry that results in mRNA sequences that can form hairpin structures. (*b*) In the absence of tryptophan, transcription proceeds past the attenuator sequence through the entire *trp* operon. (c) In the presence of tryptophan, transcription terminates at the attenuator sequence.

paired with region 2, it cannot pair with region 4, and the transcription-termination hairpin cannot form. As you have probably guessed by now, the presence or absence of tryptophan determines which of these alternative structures will form.

Third, note that the leader sequence contains an AUG translation-initiation codon, followed by 13 codons for amino acids, followed, in turn, by a UGA translation-termination codon (Figure 21.11). In addition, the *trp* leader sequence contains an efficient ribosome-binding site located in the appropriate position for the initiation of translation at the leader AUG initiation codon. All the available evidence indicates that a 14 amino acid "leader peptide" is synthesized as diagrammed in Figure 21.11. The synthesis of this putative leader peptide *in vivo* has not been verified directly. However, because short peptides of this type are very unstable in cells, the inability to detect it is not surprising.

The formation of the normal *trp* operon transcription-termination hairpin is shown in Figure 21.13*a*, and the proposed mechanism of attenuation of *trp* operon transcription is diagrammed in Figure 21.13*b* and *c*. The leader peptide contains two contiguous tryptophan residues. The two Trp codons are positioned such that in the absence of tryptophan (and thus the absence of Trp-tRNA^{Trp}), the ribosome will stall before it encounters the base-paired structure formed by leader regions 2 and 3 (Figure 21.13*b*). Because the pairing of regions 2 and 3 precludes the formation of the transcription-termination hairpin by the base-pairing of regions 3 and 4, transcription will continue past the attenuator into the *trpE* gene in the absence of tryptophan.

In the presence of tryptophan, the ribosome can translate past the Trp codons to the leader-peptide termination codon. In the process, it will disrupt the base-pairing between leader regions 2 and 3. This disruption leaves region 3 free to pair with region 4, forming the transcription-termination hairpin (Figure 21.13*c*). Thus, in the presence of tryptophan, transcription frequently (about 90 percent of the time) terminates at the attenuator, reducing the amount of mRNA for the *trp* structural genes.

The transcription of the *trp* operon can be regulated over a range of almost 700-fold by the combined effects of repression (up to 70-fold) and attenuation (up to tenfold).

Regulation of transcription by attenuation is not unique to the *trp* operon. Five other operons (*thr, ilv, leu, phe,* and *his*) are known to be regulated by attenuation. Of these six operons, *trp* and possibly *phe* are also regulated by repression. The *his* operon, which was thought to be repressible for many years, is now believed to be regulated entirely by attenuation. Although minor details of the attenuation mechanism vary from operon to operon, the main features of attenuation are the same for all six operons.

Key Points: **The *E. coli trp* operon is a repressible system; transcription of the five structural genes in the *trp* operon is repressed in the presence of tryptophan. Operons such as *trp* that encode enzymes involved in amino acid biosynthetic pathways often are controlled by a second regulatory mechanism called attenuation. The level of expression of these operons is reduced by the premature termination of transcription at an attenuator site located in the mRNA leader sequence when the amino acid produced by the pathway is present.**

THE ARABINOSE OPERON IN *E. COLI:* POSITIVE AND NEGATIVE CONTROLS

The mechanisms by which the *lac* and *trp* operons are regulated are documented by an extensive body of experimental data. Other operons, such as the *ara* (arabinose) operon of *E. coli*, are controlled by more complex regulatory mechanisms. In the *lac* and *trp* operons, the product of the regulator gene, the repressor, functions in a negative manner, turning off the transcription of the operon. In contrast, the catabolite activator protein (CAP) exerts a positive control over the *lac* operon by stimulating transcription. With **negative control mechanisms**, the product of the regulator gene is necessary to shut off the expression of one or more structural genes, whereas with **positive control mechanisms**, the product of the regulator gene is required to turn on the expression of structural genes. Positive and negative regulation are illustrated for both inducible and repressible systems in Figure 21.14.

The major regulatory protein of the *ara* operon exhibits both negative and positive regulatory effects on the transcription of the structural genes of the operon depending on the environmental conditions. Moreover, several *cis*- and *trans*-acting elements are involved in controlling transcription of the *ara* operon. One *cis*-acting element controls a promoter that is located over 200 nucleotide pairs away. Although not all the details of the mechanism by which the *ara* operon is regulated are established unambiguously, the main features are well documented by experimental evidence. The *ara* operon provides a good example of the complex circuitry controlling some bacterial operons.

The *ara* operon of *E. coli* contains three structural genes, *araB, araA,* and *araD,* which encode the three enzymes involved in the catabolism of arabinose (Figure 21.15*a*). Transcription of these three structural

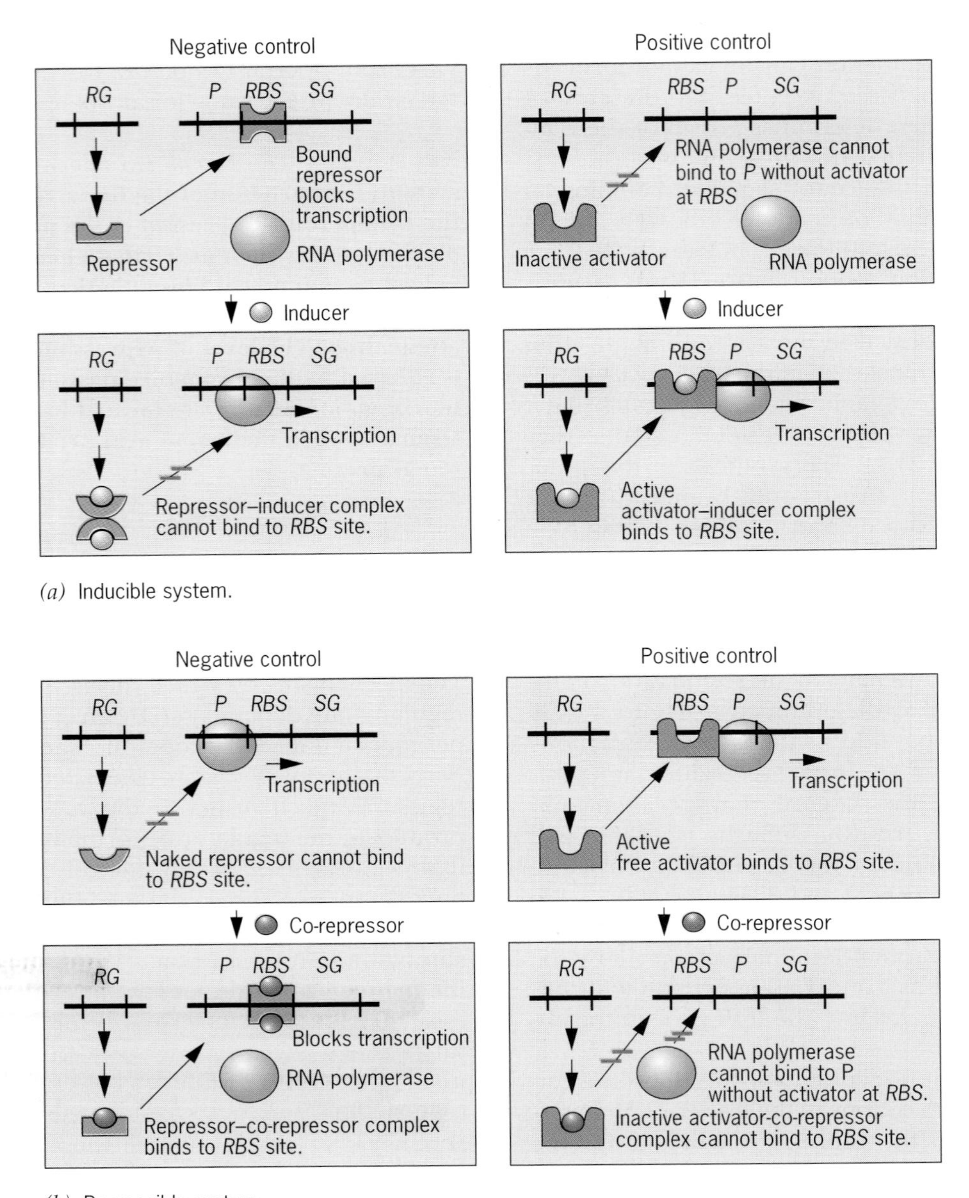

Figure 21.14 Negative and positive control of inducible and repressible gene expression. Key: *RG*, regulator gene; *P*, promoter; *RBS*, regulator protein binding site, whether activator (positive control) or repressor (negative control); and *SG*, structural gene or genes. The regulator gene product is required to turn on gene expression in positive control circuits and to turn off gene expression in negative control systems.

genes is initiated at a promoter called P_{BAD} and yields a single mRNA that carries all three coding sequences. (Active transport of arabinose into cells is carried out by the products of genes *araE, araF,* and *araG*. These genes are located at sites far from the *araBAD* operon and will not be discussed further.)

The major regulatory gene product, the AraC protein, of the *ara* operon is produced from a transcript that is initiated at a promoter called P_C. The P_C promoter is about 100 nucleotide pairs away from P_{BAD}, but the two promoters initiate transcription in opposite directions (Figure 21.15*b*). The **AraC protein** acts as a negative regulator (a repressor) of transcription of the *araB, araA,* and *araD* structural genes from the P_{BAD} promoter in the absence of arabinose and cyclic AMP (cAMP). It acts as a positive regulator (an activator) of transcription of these genes from the P_{BAD} promoter when arabinose and cAMP are present. Thus, depending on whether the effector molecules arabinose and cAMP are present or absent, the *araC* regulatory gene product may exert either a positive or a negative effect on transcription of the *araB, araA,* and *araD* structural genes.

Since, like the *lac* operon, the *ara* operon is subject to catabolite repression and thus to positive control by CAP and cAMP, induction of the *ara* operon depends on the positive regulatory effects of two proteins, the AraC protein and CAP. The binding sites for these two proteins and for RNA polymerase all are located in a region of the *ara* operon historically called *araI* (I for induction), located between the three structural genes of the operon and the *araC* regulator gene (Figure 21.15*a*).

Scientists studying the regulation of the *ara* operon initially thought that all the binding sites for the AraC regulatory protein and the cAMP-CAP complex were in the *araI* region. More recent evidence has shown that repression of the *ara* operon results from the binding of AraC protein to a site called $araO_2$ (O for operator, 2 because it was the second *ara* operator identified) located 211 nucleotide pairs upstream (relative to the direction of transcription from P_{BAD}) from the AraC protein-binding site in *araI* (Figure 21.15*b*). Operator $araO_1$, the first *ara* operator to be identified, controls the transcription of the *araC* regulator gene, which is initiated at promoter P_C.

Repression of the *ara* operon requires dimers of AraC protein to bind at both the *araI* site and the $araO_2$ site. The AraC protein dimers at these two sites then interact so that the intervening DNA forms a loop (Figure 21.15*c*). If five nucleotide pairs are inserted or deleted in the region between *araI* and $araO_2$, normal repression of the operon cannot occur. Such an insertion or deletion will rotate one AraC protein-binding site halfway around the double helix (to the opposite face) relative to the other AraC protein-binding site. This rotation would prevent the AraC dimers bound at the *araI* and $araO_2$ sites from interacting and forming the DNA loop required for repression. The DNA loop (Figure 21.15*c*) is believed to prevent or interfere with the binding of RNA polymerase at the P_{BAD} promoter.

The *ara* operon is induced when arabinose and cAMP are present (Figure 21.15*d*). Under these conditions, the AraC protein becomes an activator of transcription from the P_{BAD} promoter. Presumably, arabinose stimulates an allosteric transition in the AraC protein that converts it to a positive regulator of transcription of the operon. The arabinose-AraC and the cAMP-CAP complexes must somehow open the loop by binding at their *araI* sites. This, in turn, must stimulate the binding of RNA polymerase to promoter P_{BAD} and the transcription of the *araB*, *araA*, and *araD* structural genes.

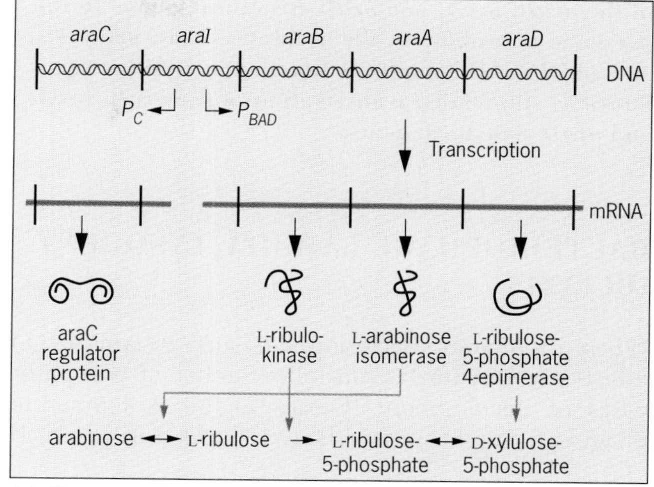

(*a*) Structure and expression of the *ara* operon of *E. coli.*

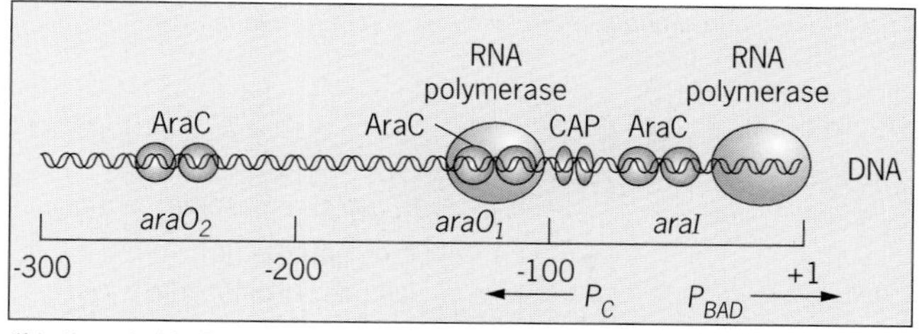

(*b*) Protein binding sites in the regulatory region of the *ara* operon.

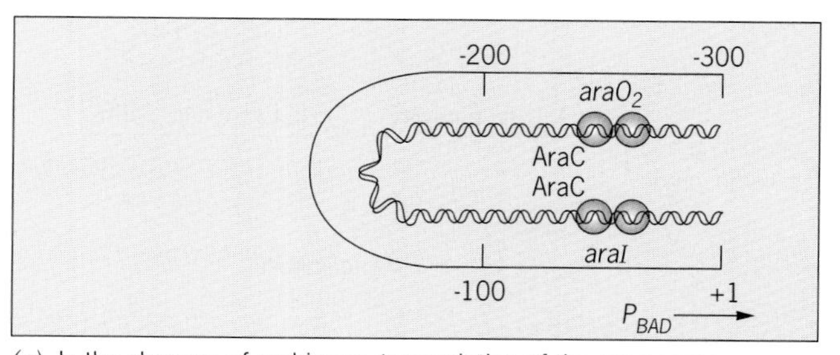

(*c*) In the absence of arabinose, transcription of the *ara* operon is repressed by AraC protein.

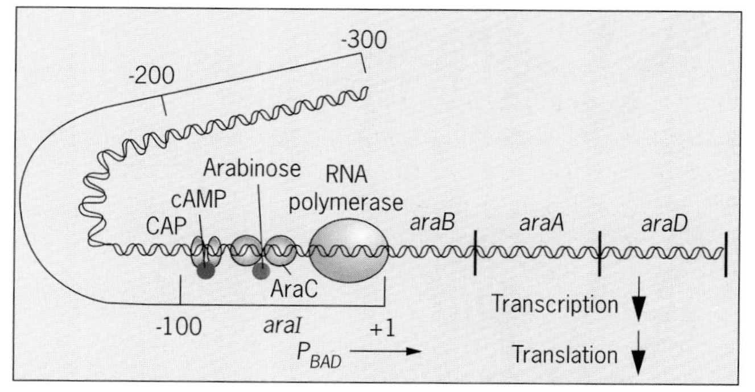

(*d*) In the presence of arabinose, transcription of the *ara* operon is induced by arabinose–AraC protein and cAMP–CAP.

Figure 21.15 Structure and regulation of the arabinose (*ara*) operon of *E. coli*. In the absence of arabinose, AraC protein acts as a negative regulator of transcription of the operon. In the presence of arabinose, the arabinose-AraC protein complex acts as a positive regulator of transcription of the operon. In the absence of glucose, the cAMP-CAP complex also acts as a positive regulator of transcription of the *ara* operon.

Key Points: Transcription of the arabinose operon of *E. coli* is controlled by both positive and negative regulatory mechanisms. In the absence of arabinose, the AraC protein acts as a negative regulator by binding to the *araO*$_2$ operator and blocking transcription of the *araB*, *araA*, and *araD* structural genes. In the presence of arabinose, the arabinose-AraC protein and cAMP-CAP complexes function as positive regulators by stimulating transcription of the *araB*, *araA*, and *araD* structural genes.

BACTERIOPHAGE LAMBDA: LYSOGENY OR LYSIS

When a temperate bacteriophage such as lambda (λ) infects a bacterium, it can follow either of two pathways of development (Figure 21.16). A temperate phage can either (1) enter the lytic cycle, during which it reproduces and lyses the host cell, just like a virulent phage (Chapter 15), or (2) enter the lysogenic pathway (Chapter 16), during which its chromosome is inserted into the chromosome of the host and thereafter is replicated like any other segment of that chromosome. When integrated in the chromosome of the host cell, the temperate phage chromosome is called a prophage. In a lysogenic bacterium, the genes of the prophage that encode products involved in the lytic pathway must not be expressed. The prophage genes specifying enzymes involved in the replication of phage DNA, structural proteins required for phage morphogenesis, and the lysozyme that catalyzes cell lysis must be kept turned off to maintain a stable lysogenic state. Indeed, André Lwoff shared the 1965 Nobel Prize in Physiology or Medicine with Jacob and Monod for his discovery and characterization of the prophage state of temperate bacteriophages.

The lytic pathway genes of the prophage are kept turned off in lysogenic cells by a simple repressor-oper-

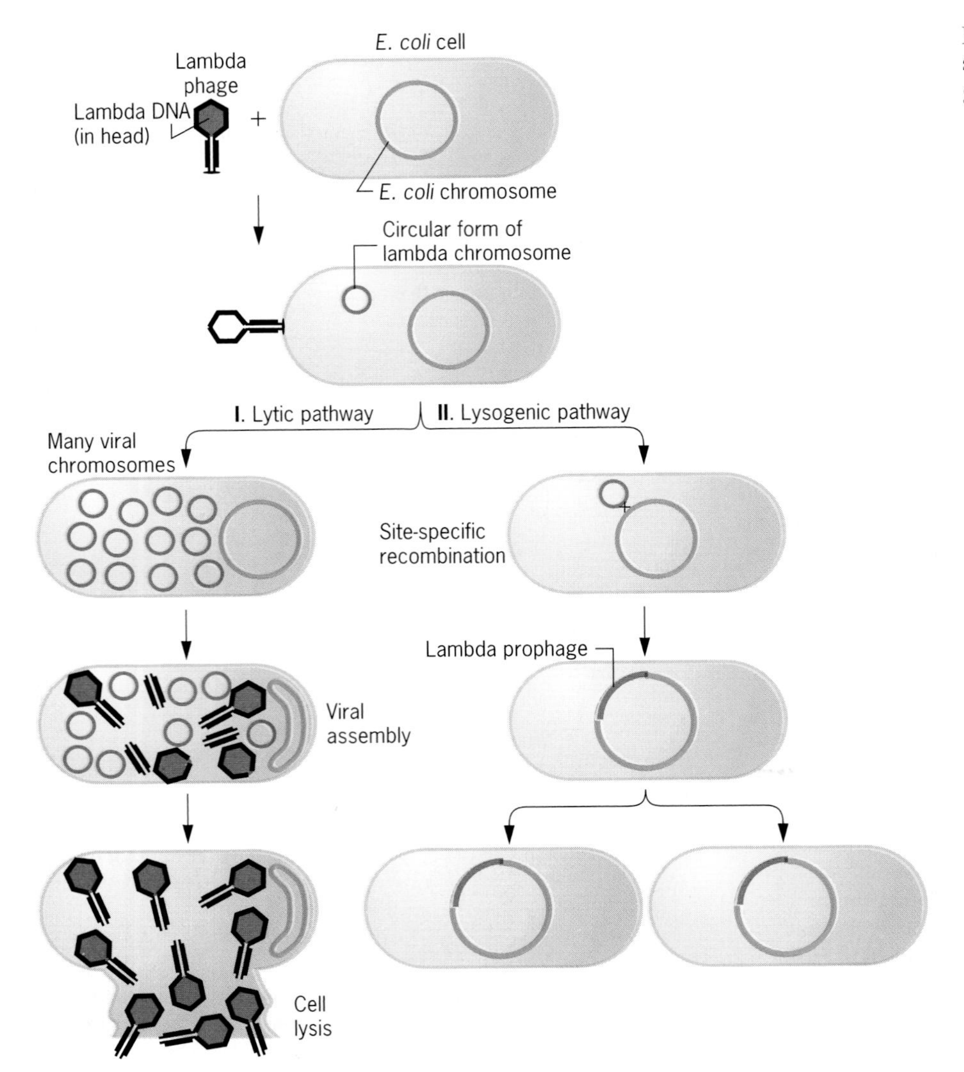

Figure 21.16 The two intracellular states of bacteriophage lambda: lytic growth and lysogeny.

ator-promoter circuit, much like the negative regulatory circuits of bacterial operons. However, the really interesting aspect of phage lambda is how the virus makes the initial choice between the lytic pathway and the lysogenic pathway. This decision involves an elegant genetic switch that sends the infecting phage chromosome down one of the two pathways: (1) the lytic regulatory cascade or (2) the autogenously maintained repression circuit of lysogeny. The interplay between these two regulatory networks is one of the most fascinating stories in the field of molecular genetics.

Repression of Lambda Lytic Pathway Genes During Lysogeny

The phage lambda genes that encode functions involved in lytic development are maintained in the off state by a negative regulatory mechanism analogous to those controlling the *lac* and *trp* operons of *E. coli*. The C_I gene of phage λ encodes a repressor, a well-characterized protein with a molecular weight of 27,000. The C_I repressor, as a dimer, binds to two operator regions that control transcription of the λ genes involved in lytic growth (Figure 21.17). These two operator regions, designated O_L (for transcription in a leftward direction) and O_R (for transcription in the rightward direction), overlap with promoter sequences at which RNA polymerase binds to initiate transcription of the genes controlling lytic development. With repressor bound to the two operators, RNA polymerase cannot bind to the two promoters and cannot initiate transcription. In this way, the lambda lytic genes continue to be repressed, allowing the dormant prophage to be transmitted from parental host cells to progeny cells generation after generation.

When the phage lambda operator and promoter regions were sequenced, each operator was found to

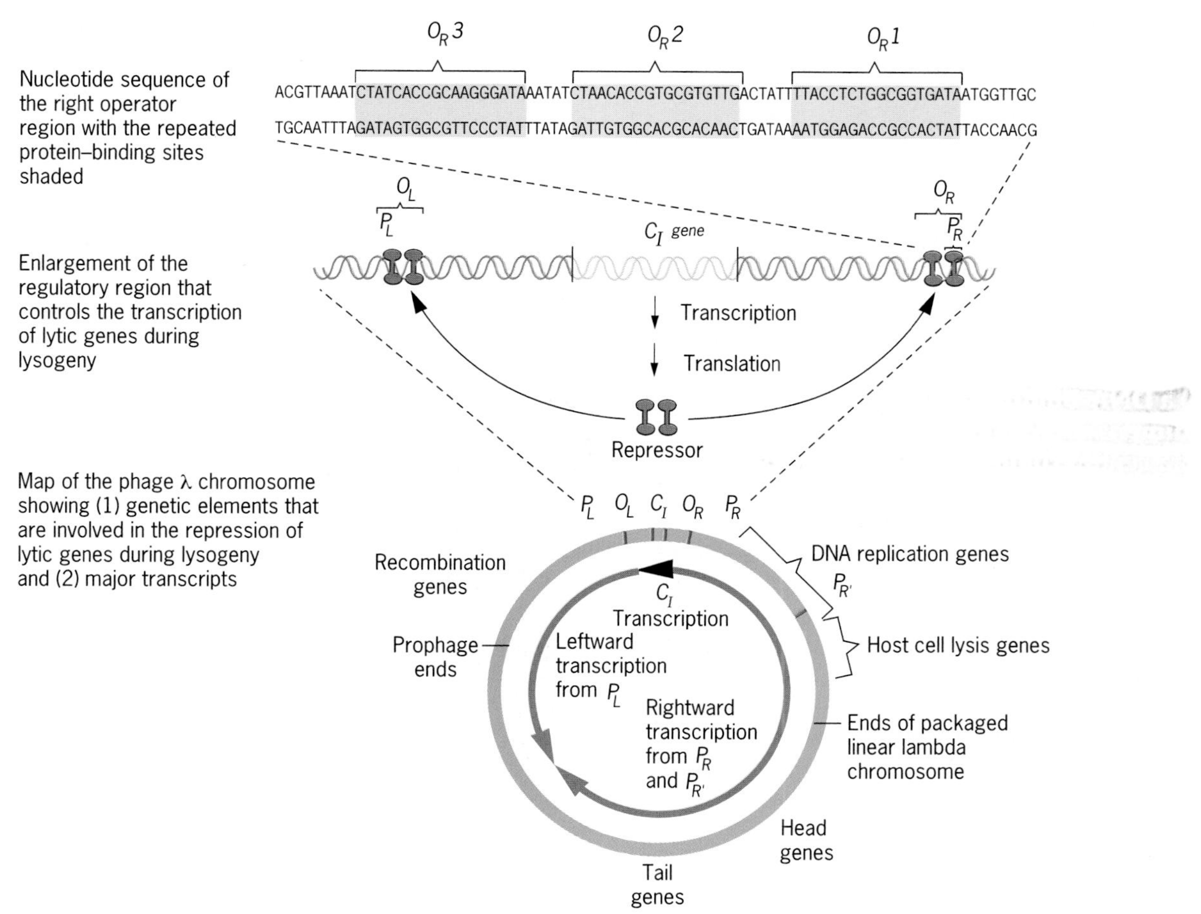

Figure 21.17 Repression of the lambda lytic genes in a lysogenic *E. coli* cell. Transcription of the lytic genes is repressed by the binding of the C_I gene product (the lambda repressor) to two operator sequences (O_L and O_R), which regulate leftward and rightward transcription of the lambda chromosome (bottom). The lambda repressor is encoded by the C_I gene (center) and represses the synthesis of the major transcripts by binding to triplicate protein binding sites in O_L and O_R. The arrows show the relative sizes and directions of synthesis of the major lambda transcripts.

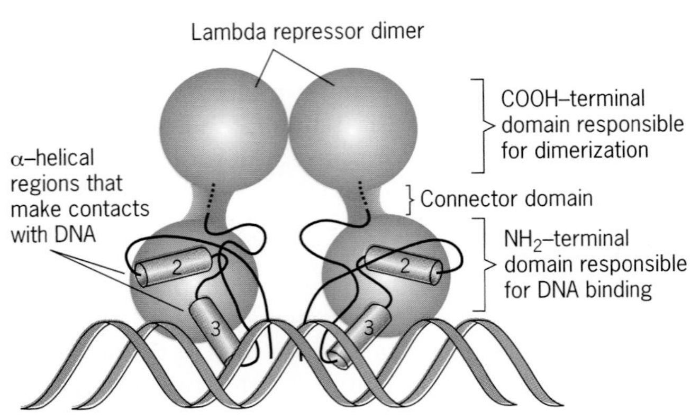

Figure 21.18 The lambda repressor dimer contacts its binding sites in the lambda operator (O_L and O_R) regions primarily through two α-helical regions (green cylinders) in the amino-terminal DNA-binding domain. Whereas the exact conformation of the carboxyl-terminal domain is unknown, the three-dimensional structure (black line and green cylinders) of the amino-terminal domain has been determined. This DNA-binding domain contains five α-helical regions, two of which (labeled 2 and 3 on the diagram) are primarily responsible for the specific binding of the repressor to the lambda operator regions.

contain three repressor binding sites, each 17 nucleotide pairs long, with similar, but not identical, nucleotide sequences. Each repressor binding site has partial twofold symmetry around the central base pair. This symmetry plays an important role in the interaction of repressor dimers with these operator binding sites (Figure 21.18).

The lambda repressor has been studied in great detail. The repressor monomer contains three regions: (1) a **DNA-binding domain** at the amino terminus, (2) a **dimerization domain** at the carboxyl terminus, and (3) a central **connector region**. The three-dimensional structure is known only for the DNA-binding domain. This domain, 92 amino acids in length, contains five regions that form α-helical structures; the α-helices are numbered 1 through 5, starting from the NH_3 terminus. Two of the α-helical regions, numbers 2 and 3, are largely responsible for the DNA-binding specificity of the repressor. When a repressor dimer makes contact with its DNA binding site (Figure 21.18), the two region 3 helices lie almost entirely within adjacent major grooves on one face of the λ operator DNA. Each monomer contacts one side (half-site) of the palindromic nucleotide-pair sequence in each repressor binding site. The two region 2 helices lie across the major grooves occupied by the region 3 helices.

Because the lambda repressor has the highest affinity for sites O_{L1} and O_{R1}, the first repressor dimers usually bind at these two sites. The λ repressor exhibits cooperative binding to (1) sites O_{L1} and O_{L2}, and (2) sites O_{R1} and O_{R2}. Thus the presence of a dimer at site O_{L1} or O_{R1} increases the affinity of repressor for sites O_{L2} or O_{R2}, respectively. This cooperativity does not extend to O_{L3} or O_{R3}; at normal intracellular concentrations of repressor, only sites O_{L1}, O_{L2}, O_{R1}, and O_{R2} are occupied by repressor. When repressor dimers are bound to O_{R1}-O_{R2} and O_{L1}-O_{L2}, RNA polymerase cannot bind to P_R and P_L and, therefore, cannot initiate transcription. Thus the λ genes encoding functions involved in lytic development are maintained in a repressed state.

Lambda lysogeny is quite stable. The lambda lytic pathway genes are tightly repressed in a lysogenic cell; thus lytic-function gene products are not produced by transcription of prophage genes. In addition, spontaneous switches from lysogeny to lytic development are rare under normal growth conditions. Populations of lysogenic cells can be induced to enter the lytic pathway only by drastic treatments, such as by irradiation with ultraviolet light (UV). UV irradiation of lambda lysogens activates a host cell protease that cleaves the connector region of the λ repressor and renders it nonfunctional.

A major reason for the stability of the λ lysogenic state is that the synthesis of the repressor is autogenously regulated; that is, the presence of λ repressor stimulates the synthesis of more repressor. By what mechanism is the synthesis of the λ repressor self regulated?

The λ C_I gene, which encodes the repressor, is transcribed from a promoter designated P_{RM} (for promoter for repressor maintenance), located between the C_I gene and O_{R3} (Figure 21.19). Promoter P_{RM} overlaps operator O_{R3}, much as P_R overlaps O_{R1} (Figure 21.17). However, transcription initiated at P_{RM} occurs in the opposite direction (to the left as drawn in Figure 21.17) from transcription initiated at P_R. The key factor in the maintenance of the lysogenic state is that repressor bound at O_{R2} acts as a positive regulator of transcription from P_{RM}. In the absence of repressor at O_{R2} (lytic growth), RNA polymerase cannot bind at P_{RM} to initiate transcription of the C_I gene. In the presence of repressor, the O_{R1} and O_{R2} sites are occupied by repressor dimers (Figure 21.19). Repressor bound at O_{R2} stimulates the binding of RNA polymerase at P_{RM}, leading to the synthesis of more repressor (positive self-regulated synthesis).

Although the repressor has lower affinity for O_{R3} than for O_{R2} or O_{R1}, it will bind at O_{R3} and block transcription from P_{RM} if present at high concentration. This block will lead to a lower concentration of repressor, which, in turn, will yield an open O_{R3} and renewed transcription from P_{RM}. The control of repressor synthesis by this autogenous regulatory circuit enhances the stability of the lambda lysogenic state. It

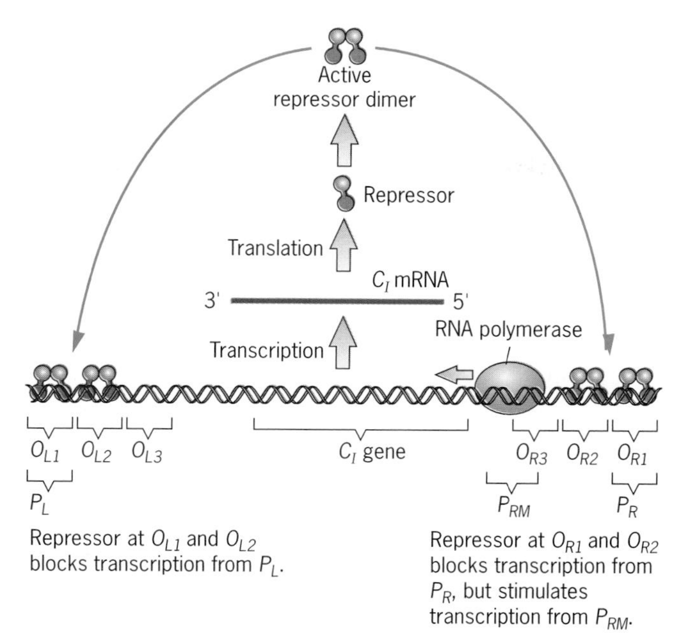

Figure 21.19 Autogenous regulation of phage lambda repressor synthesis. The presence of λ repressor stimulates the synthesis or more repressor by acting as a positive regulator of transcription of the C_I gene from promoter P_{RM}. At the same time, the repressor functions as a negative regulator of genes involved in lytic development. Repressor dimers bound at O_{L1}–O_{L2} and O_{R1}–O_{R2} prevent transcription of lytic function genes from promoters P_L and P_R, respectively.

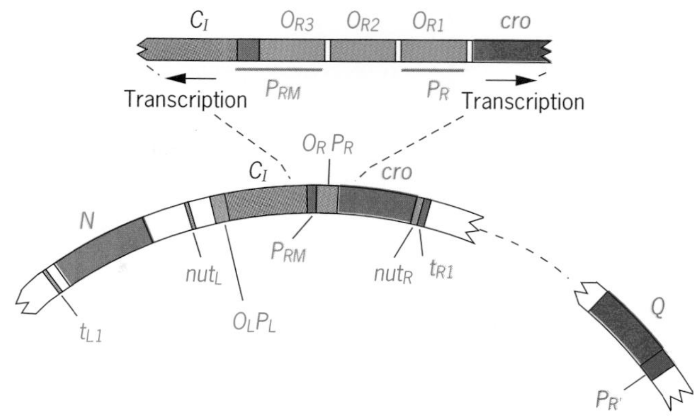

Figure 21.20 Genes and recognition sites involved in the lambda lytic regulatory cascade. Genes *cro*, *N*, and *Q* encode regulatory proteins required for lytic development. Operators O_L and O_R control leftward and rightward transcription from promoters P_L and P_R, respectively. O_{R1}, O_{R2}, and O_{R3} are the three repressor binding sites within O_R. Sequences nut_L and nut_R are the sites at which the *N* gene product must act to prevent transcription termination at sites t_L and t_{R1}, respectively.

also helps explain why repressor is not synthesized during the lytic growth of this virus. If no repressor is present in the cell, the O_{R2} site will be free, and repressor will not be synthesized from transcripts initiated at P_{RM}. As a result, once lambda has started down the lytic pathway, P_{RM} will not be utilized to produce repressor, and the absence of repressor will allow lytic development to proceed unabated.

The Lambda Lytic Regulatory Cascade

The lambda lytic regulatory cascade is straightforward, once a commitment has been made to enter the lytic pathway. The λ genes can be placed in three groups based on when they are expressed during the lytic cycle: (1) immediate early genes, (2) delayed early genes, and (3) late genes. Lambda has only two immediate early genes, *cro* and *N*, both encoding regulatory proteins required for lytic development. The λ chromosome contains a dozen delayed early genes, which specify products required for DNA replication, recombination, and additional regulation. The remaining 23 genes of phage lambda encode late-function proteins involved in head and tail morphogenesis and in lysis of the host cell. One of the immediate early gene products, **N protein**, is required to express the delayed early genes. In turn, one of the delayed early gene products, **Q protein**, must be present to express the λ late genes. Let's look at the functions of the key regulatory genes, *cro*, *N*, and *Q*, involved in the lytic growth of phage λ (Figure 21.20).

In the absence of the lambda repressor, the C_I gene product described earlier, RNA polymerase initiates transcription at promoters P_L and P_R. The first gene to be transcribed starting at P_R is *cro* (for control of repressor and other gene products). **Cro protein** is a repressor with a DNA-binding region very similar to that of the λ repressor. Cro binds to the same DNA sites in O_L and O_R as the λ repressor; but Cro has a higher affinity for O_{R3} than for O_{R1} and O_{R2}. Thus, when Cro is first synthesized, it binds to O_{R3} and keeps the synthesis of repressor turned off by blocking the initiation of transcription at P_{RM} (Figure 21.21*a*). Later, when more Cro protein has accumulated in the cell, it also binds to O_{R1}, O_{R2}, O_{L1}, and O_{L2}, which suppresses transcription from P_R and P_L, decreasing the rate of synthesis of early gene products (Figure 21.21*b*).

Synthesis of the delayed early gene products requires the product of the second immediate early gene, *N*, which is the first gene transcribed from promoter P_L. The gene *N* product functions to prevent the termination of transcription at two sites, one (t_{R1}) located adjacent to *cro* and the other (t_{L1}) just downstream from

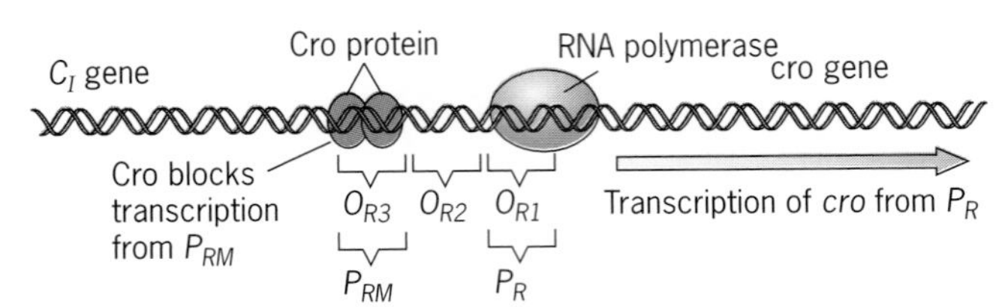

(*a*) Early: low concentration of Cro protein.

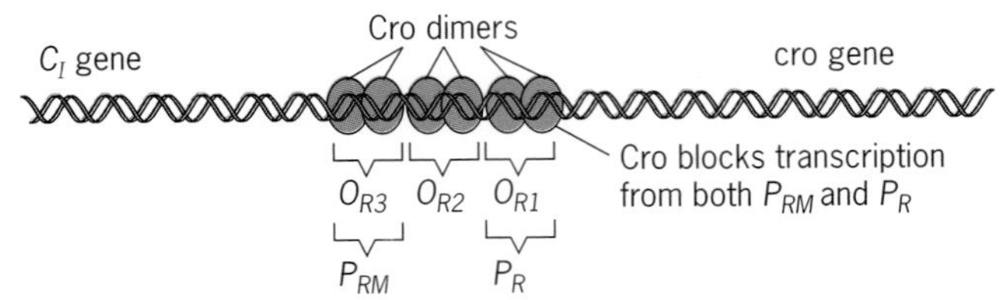

(*b*) Late: high concentration of Cro protein.

Figure 21.21 Functions of the *cro* gene product at early (*a*) and late (*b*) stages of lambda lytic growth. Cro protein has higher affinity for O_{R3} than O_{R1} or O_{R2}, resulting in its presence only at O_{R3} at low concentrations and at all three O_R sites (and O_L sites) at high concentrations.

gene *N* itself (Figure 21.20). Thus N protein acts as a **transcriptional antiterminator**; it permits RNA polymerase to continue transcription past termination signals t_{L1} and t_{R1} and on through the adjacent delayed early genes. N protein carries out its antitermination function only in the presence of DNA sequences called *nut* (for N utilization) sites located upstream from the termination sites (Figure 21.22). N protein binds to the *nut* sites and, with the aid of *E. coli* proteins called Nus (for N utilization substance) factors and ribosomal protein S10, modifies the specificity of RNA polymerase as it passes the *nut* site so that termination does not occur when the RNA polymerase complex reaches termination signals t_{L1} and t_{R1}.

The antiterminator role of N protein is illustrated in Figure 21.22. In the absence of functional N protein, transcription terminates at t_{R1} and t_{L1}. In the presence of active N protein, it binds to the *nut* sites, and, along with proteins NusA, NusB, and S10, modifies RNA polymerase so that termination of transcription does not occur at t_{R1} and t_{L1}. As a result, transcription continues into the adjacent delayed early genes.

One of the delayed early genes, *Q*, encodes another transcription antiterminator that is required for late gene expression. Q protein functions much like N protein but recognizes a different DNA sequence, called *qut* (for Q utilization). Q protein prevents the arrest of transcription initiated at the constitutive promoter $P_{R'}$ at termination site t_{R3}, allowing transcription to continue through the late genes.

Nondefective lambda prophages in lysogenic *E. coli* cells can be induced to enter the lytic pathway by exposure to ultraviolet light (UV). UV irradiation of cells causes damage to DNA, which results in a number of physiological changes collectively called the SOS response (Chapter 13). One component of the SOS response in *E. coli* is the conversion of the RecA protein (an important recombination protein; see Chapter 13) to a protease that cleaves the connector region of the λ repressor. This cleavage event prevents the dimerization of the DNA-binding domain of the repressor. In the absence of active repressor dimers, Cro and N protein are synthesized. Cro binds to O_{R3} and prevents the synthesis of repressor. With Cro and N protein, but no repressor, present in the cell, lytic development is almost inevitable.

The Lambda Switch: Lytic Development or Lysogeny

When phage lambda injects its DNA into an *E. coli* cell, that cell contains neither repressor nor Cro protein.

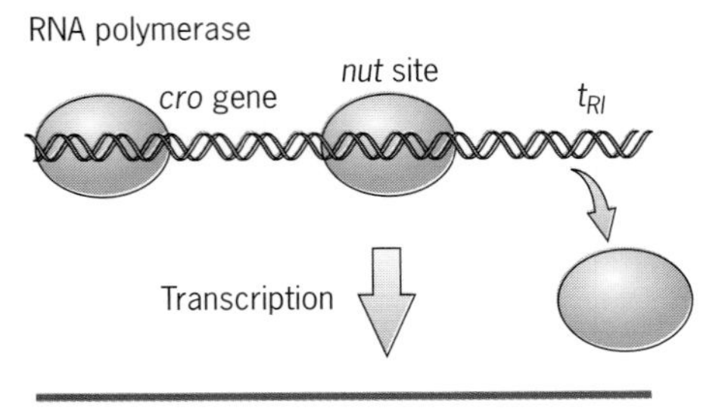

(*a*) Transcription is terminated at t_{R1} in the absence of N protein.

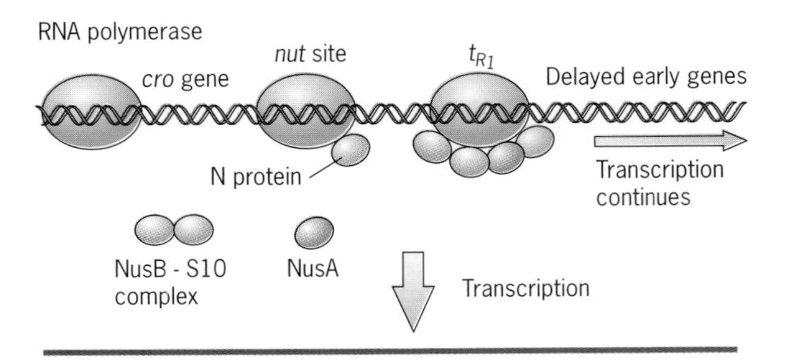

(*b*) Transcription continues through t_{R1} in the presence of N protein.

Figure 21.22 The phage lambda N protein functions as an antiterminator of transcription at sites t_{R1} and t_{L1} (not shown). (*a*) In the absence of N protein, transcription of the early gene *cro*, which is initiated at P_R, terminates at t_{R1}. (*b*) In the presence of N protein, transcription initiated at P_R continues through t_{R1} into the delayed early genes. At the N protein utilization site, *nut*, RNA polymerase is modified by the association of N protein and *E. coli* proteins NusA, NusB, and S10. The modified RNA polymerase complex reads through t_{R1} and transcribes the adjacent delayed early genes.

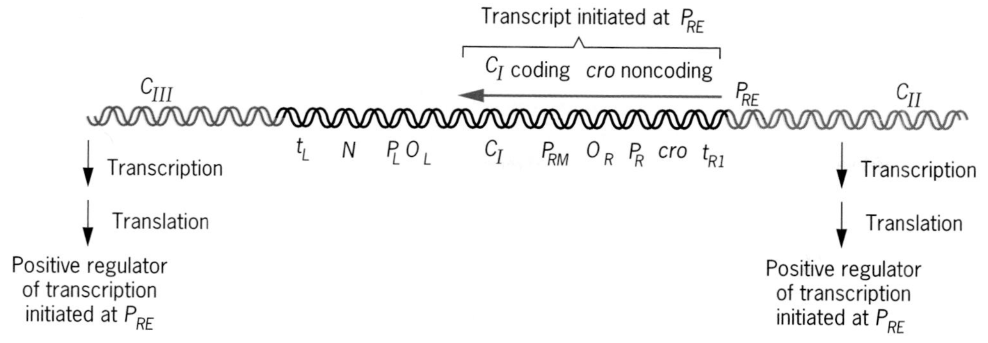

Figure 21.23 Regulatory elements involved in setting the lambda genetic switch to lysogeny or to lytic growth. The C_{II} and C_{III} genes encode proteins that are required for the initiation of transcription at promoter P_{RE}.

What then determines whether the injected lambda chromosome will (1) enter the lysogenic pathway and insert itself into the host chromosome, or (2) enter the lytic pathway and produce progeny phage at the expense of the host cell? Having examined the lambda regulatory circuits involved in the lysogenic and lytic pathways, let's consider the switch that controls whether lysogeny or lysis follows an infection of *E. coli* by a lambda phage. At the outset, we should realize that no one can accurately predict which response will occur in a specific cell that has been infected with a wild-type λ phage.

The decision between lysogeny and lytic growth hinges on a delicate balance between the λ repressor and Cro protein. Which protein will occupy the three binding sites in operators O_L and O_R and, therefore, control transcription initiation at P_L, P_R, and P_{RM}? The decision is finalized when one of these proteins gains occupancy of the O_L and O_R sites. If λ repressor occupies these operators, lysogeny will result. If Cro protein occupies O_L and O_R, lytic growth will occur.

The genetic switch that sends lambda down either the lysogenic or the lytic pathway involves two key regulatory genes, C_{II} and C_{III}, and one promoter, P_{RE} (promoter for repressor establishment). P_{RE} is located on the opposite side of the *cro* gene from P_{RM} (Figure 21.23). The transcript initiated at P_{RE} contains the noncoding or antisense sequence of the *cro* gene and the normal mRNA coding or sense sequence of the C_I gene. Both C_{II} and C_{III} are delayed early genes; therefore, their expression is dependent on N protein. The C_{II} protein is required for the initiation of transcription at P_{RE}; RNA polymerase can initiate transcription at P_{RE} only when C_{II} protein is present (Figure 21.24). However, C_{II} protein is very unstable; it is rapidly degraded by a host cell protease. The function of the C_{III} protein is to stabilize the C_{II} protein, allowing RNA polymerase to initiate transcription at P_{RE}. Actually, the effect of C_{III} on the stability of C_{II} is a function of the growth conditions. Under poor growth conditions, C_{III} protects C_{II} from degradation by proteases more completely, favoring lysogeny, whereas under optimal growth conditions, C_{III} is less active and, thus, C_{II} is less stable, favoring lytic growth.

The transcript initiated at P_{RE} favors establishment of the lysogenic pathway in two ways. (1) Lambda repressor is produced by translation of the C_I coding sequence. (2) The 5' portion of this transcript contains the noncoding or antisense sequence of the *cro* gene. This *cro* antisense sequence is complementary to the *cro* coding or sense sequence in authentic *cro* mRNA synthesized by transcription initiated at P_R.

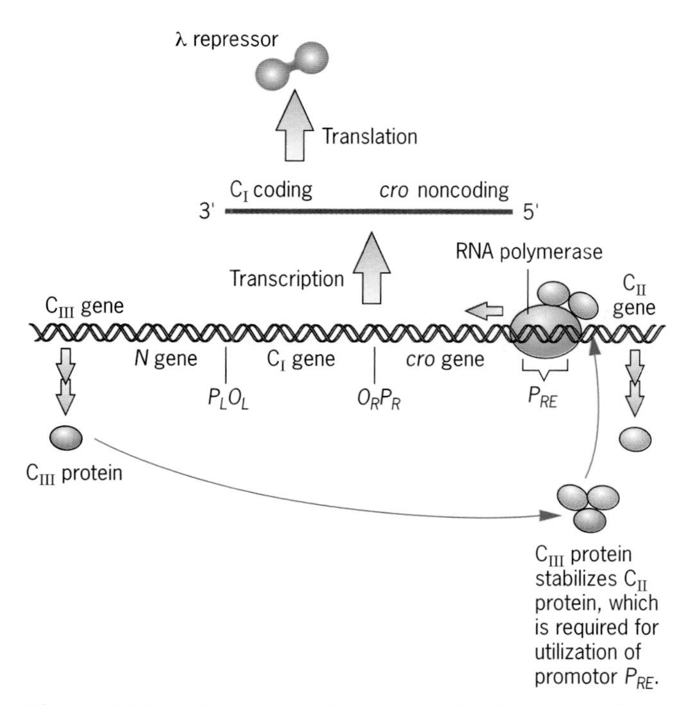

Figure 21.24 Diagram of the phage λ C_{II}-C_{III}-P_{RE} regulatory circuit involved in the decision between lytic and lysogenic development.

Thus these *cro* antisense and sense RNAs hybridize (form double helices) with each other, which blocks translation of the *cro* mRNA.

Once repressor has been synthesized by use of the C_{II}-C_{III}-P_{RE} circuit, it binds at O_{R1} and O_{R2}. The presence of repressor at O_{R1} and O_{R2} inhibits the expression of lytic function genes by blocking the initiation of transcription at P_R and establishes the autogenous repressor maintenance circuit mediated by P_{RM}. Figure 21.25 summarizes the regulatory genes and *cis*-acting elements involved in setting the lysogeny-lytic switch and in the lysogenic and lytic pathways of development after the switch has been set.

In summary, in each infected cell, the genetic switch for either lysogeny or lytic development is set by the race between Cro protein and λ repressor for occupancy of the O_L and O_R sites. If Cro protein occupies these operators, lytic development ensues. If repressor occupies these operators, lysogeny is established. The C_{II} and C_{III} regulatory proteins play important roles in deciding the outcome of this race between Cro protein and repressor. Environmental influences on the outcome of the competition between Cro and repressor act through C_{III} and its effect on the stability of C_{II} protein. Increased stability of C_{II} favors lysogeny; decreased stability of C_{II} favors lytic growth.

Key Points: **Temperate bacteriophages can follow either of two pathways: (1) lytic growth, during which they reproduce and kill the host cells, and (2) lysogeny, during which their chromosomes exist as a dormant prophage covalently inserted in the chromosomes of the host bacteria. During lysogeny, the lytic genes of the prophage are kept turned off by a repressor-operator-promoter circuit similar to those of bacterial operons. Whether a lambda phage will enter the lysogenic state or undergo lytic development is determined by which of two regulatory proteins, λ repressor or Cro protein, occupies key operator sites that control transcription of the λ genome. The lysogenic state is maintained by the autogenous control of λ repressor, whereas lytic development is controlled by a regulatory cascade in which transcriptional antiterminators play a major role.**

TEMPORAL SEQUENCES OF GENE EXPRESSION DURING PHAGE INFECTION

Regulation of gene expression during the life cycles of virulent bacteriophages is quite different from the reversible on-off switches characteristic of bacterial operons. In phage-infected bacteria, the viral genes are expressed in genetically preprogrammed sequences or cascades. Although different bacterial viruses exhibit variations of the specific mechanisms involved, a common picture emerges. One set of phage genes, usually called early genes, is expressed immediately after infection. The product(s) of one or more of the early genes is (are) responsible for turning on the expression of the next set of genes and turning off the expression of the early genes, and so on. Two to four sets of genes are usually involved, depending on the virus. In all cases studied so far, the regulation of sequential gene expression during phage infection occurs primarily at the level of transcription.

In three of the most extensively studied bacterial viruses, *E. coli* phages T7 and T4 and *Bacillus subtilis* phages SP01, the sequential gene expression is controlled by modifying the specificity of RNA polymerase for different promoter sequences, either by the synthesis of a new RNA polymerase (T7) or by phage-induced alterations of the host cell's RNA polymerase (T4 and SP01).

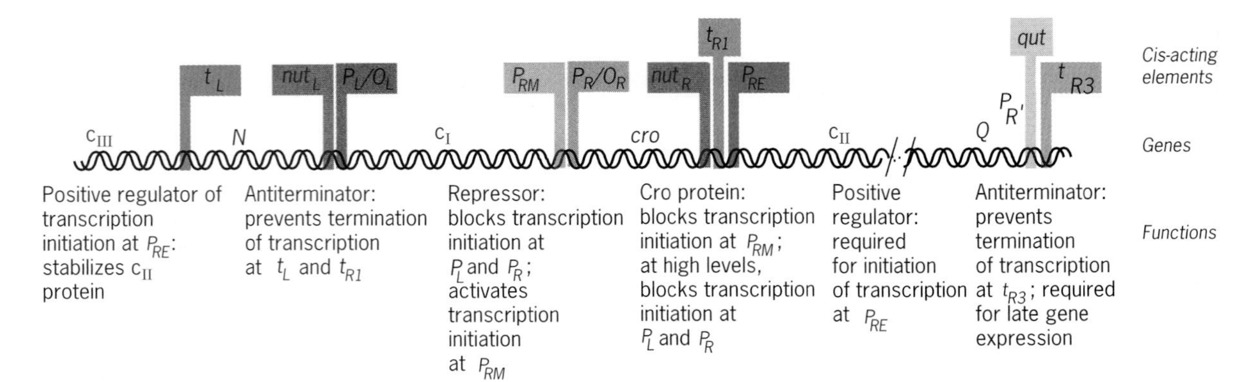

Figure 21.25 Summary of the lambda regulatory elements that control lysogenic and lytic development. *Cis*-acting elements are shown at the top, genes encoding *trans*-acting factors are given directly above the DNA molecule, and the functions of the *trans*-acting regulators are described at the bottom.

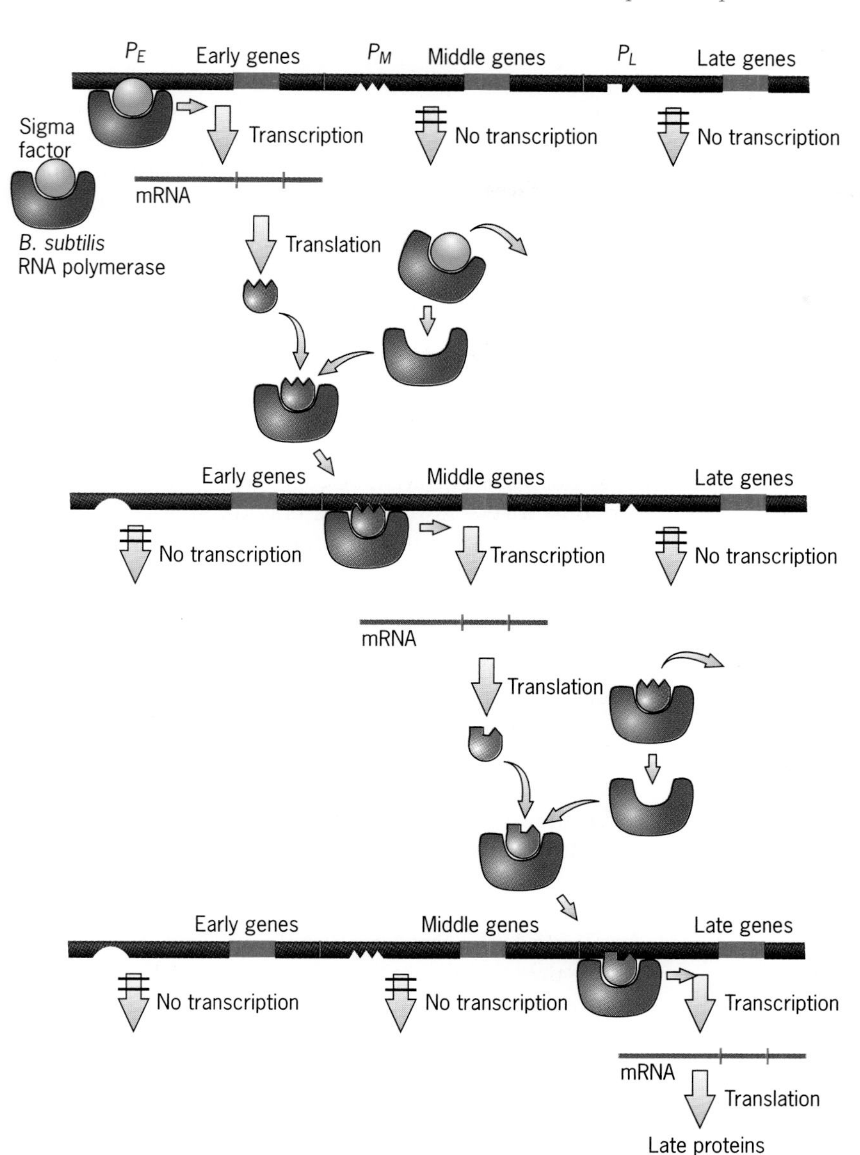

Figure 21.26 The *Bacillus subtilis* phage SP01 regulatory cascade. Early genes are transcribed by the RNA polymerase of the host cell. Middle and late genes are transcribed by RNA polymerases in which the host σ factor has been replaced by one or two SP01 proteins, respectively. P_E, P_M, and P_L are the promoters for early, middle, and late transcripts, respectively.

In phage T7-infected cells, the early genes are transcribed by the *E. coli* RNA polymerase. One of the early genes encodes the T7 RNA polymerase, which then transcribes all of the late genes encoding the T7 structural proteins and lysozyme. *Bacillus subtilis* phage SP01 exhibits a slightly more complex pathway of sequential gene expression, including three sets of genes—the early, middle, and late genes, in reference to their time of expression during the phage reproductive cycle (Figure 21.26). The phage SP01 early genes are transcribed by the *B. subtilis* RNA polymerase. One of the early gene products is a polypeptide that binds to the host cell's RNA polymerase, changing its specificity so that the modified RNA polymerase transcribes the middle genes of SP01. Two of the products of the middle genes are, in turn, polypeptides that associate with the *B. subtilis* RNA polymerase, further changing its specificity so that it then transcribes the SP01 late genes.

Phage T4 exhibits an even more complex pattern of sequential gene expression, involving several different modifications of the host cell's RNA polymerase. Thus, in the case of these bacterial viruses, the control of the observed sequential gene expression occurs primarily at the level of transcription and is mediated by modifications of the specificity of RNA polymerase for different promoter sequences.

Key Points: **Preprogrammed temporal sequences of viral gene expression occur in bacteriophage-infected cells. The first viral genes expressed in an infected cell are transcribed by unmodified bacterial RNA**

polymerase. Subsequent sets of expressed viral genes are transcribed either by an RNA polymerase encoded by the phage genome or by bacterial RNA polymerase modified by the addition of viral protein(s).

TRANSLATIONAL CONTROL OF GENE EXPRESSION

In the preceding sections of this chapter, we have focused on regulatory mechanisms that act at the level of transcription. Transcriptional regulation of gene expression is paramount in prokaryotes. Nevertheless, gene expression is often fine-tuned by regulatory mechanisms that act at the translation level.

In prokaryotes, mRNA molecules are frequently multigenic, carrying the coding sequences of several genes. For example, the *E. coli lac* operon mRNA harbors nucleotide sequences encoding β-galactosidase, β-galactoside permease, and β-galactoside transacetylase. Thus the three genes encoding these proteins must be turned on and turned off together at the transcription level, because the genes are co-transcribed. Nevertheless, the three gene products are not synthesized in equal amounts. An *E. coli* cell that is growing on rich medium with lactose as the sole carbon source contains about 3000 molecules of β-galactosidase, 1500 molecules of β-galactoside permease, and 600 molecules of β-galactoside transacetylase. Clearly, the different molar quantities of these proteins per cell must be controlled post-transcriptionally.

Remember that transcription, translation, and mRNA degradation are coupled in prokaryotes; an mRNA molecule usually is involved in all three processes at any given time. Thus gene products may be produced in different amounts from the same transcript by several mechanisms.

1. **Unequal efficiencies of translational initiation** are known to occur at the ATG start codons of different genes.
2. **Altered efficiencies of ribosome movement** through intergenic regions of a transcript are quite common. Decreased translation rates often result from hairpins or other forms of secondary structure that impede ribosome migration along the mRNA molecule.
3. **Differential rates of degradation** of specific regions of mRNA molecules also occur.

The synthesis of the ribosomal proteins of *E. coli* provides several well-documented examples of translational regulation of gene expression. *E. coli* cells that are growing rapidly under optimal conditions need more ribosomes for protein synthesis than those that are growing slowly under adverse conditions. Thus, under optimal growth conditions, *E. coli* cells devote a larger share of their energy to the production of ribosomes than cells growing under less favorable conditions. Recall (Chapter 12) that the *E. coli* ribosome contains three RNA molecules and 52 polypeptides. The syntheses of these structural components must be coordinated to assure proper stoichiometry for ribosome assembly. The synthesis of one or more components in excess would be a waste of energy. Thus regulatory mechanisms have evolved that assure that ribosomal RNAs and proteins are synthesized in the appropriate amounts, and some of these regulatory mechanisms act at the level of translation.

Most of the *E. coli* genes encoding ribosomal proteins are located in clusters, and the genes in each cluster are co-transcribed. All but one of the ribosomal proteins are utilized in equimolar amounts during ribosome assembly. Moreover, the synthesis of the ribosomal proteins must be coordinated with the synthesis of the three ribosomal RNAs. This coordination between ribosomal protein synthesis and ribosomal RNA synthesis occurs at the level of translation.

The ribosomal protein *S10* (for small subunit protein number 10) gene cluster provides a good illustration of how this regulation works. The *S10* transcriptional unit contains 10 coordinately regulated genes, all encoding ribosomal proteins (Figure 21.27*a*). In the case of the *S10* transcriptional unit and at least five other such units containing ribosomal protein genes, the product of one of the genes inhibits the translation of the multigenic transcript. Thus the gene encoding the regulatory protein is itself one of the regulated genes, so that the regulatory gene is negatively self-regulated. In the *S10* transcriptional unit, the regulatory gene is *rplD* (for ribosomal protein large subunit gene D), which encodes protein L4 (large subunit protein number 4). When free rRNAs are present in the cell, nascent L4 protein binds to the RNA (Figure 21.27*b*) and is assembled into ribosomes. In the absence of rRNA, the L4 protein binds to the 5′ end of the *S10* transcriptional unit mRNA and inhibits its translation (Figure 21.27*c*). This prevents the synthesis of ribosomal proteins that cannot be used by the cell.

The inhibition of translation of an mRNA molecule by one of the products that it encodes is common in both prokaryotes and eukaryotes. This mechanism is referred to as **negative self-regulation** or **negative autogenous regulation**. When the gene product is a structural component of the cell or of some organelle within the cell, the autoregulation often is carried out by the free monomers present in the cell.

Key Points: **Regulatory fine-tuning often occurs at the level of translation by modulation of the rate of**

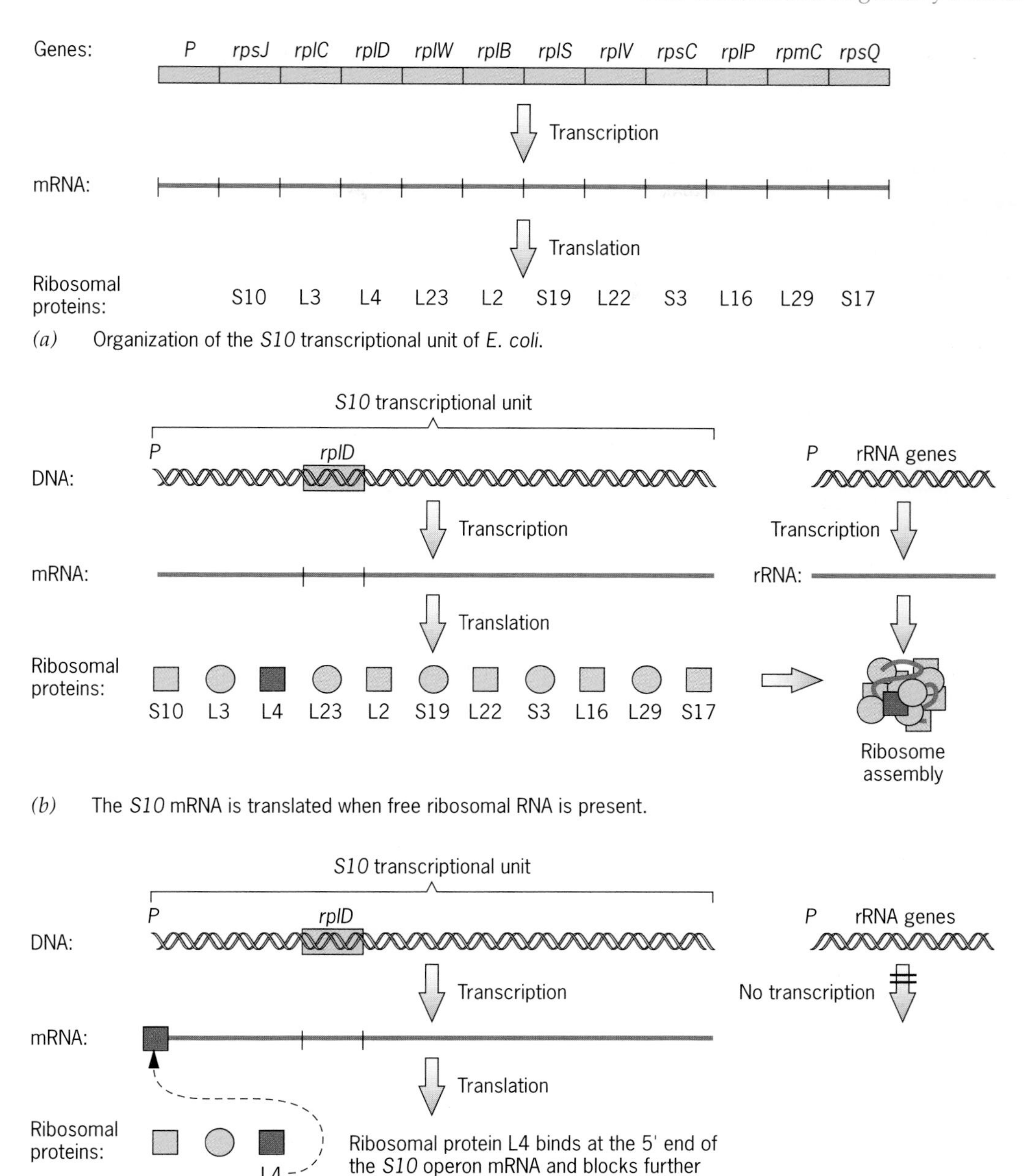

Figure 21.27 Organization (*a*) and translational regulation (*b* and *c*) of the *E. coli S10* transcriptional unit, which contains 11 genes encoding ribosomal proteins. Translation of the mRNA of the *S10* transcriptional unit is regulated by ribosomal protein L4, which binds to a nucleotide sequence near the 5' end of the *S10* transcript. (*b*) In the presence of ribosomal RNA, the ribosomal proteins interact with the rRNA in the assembly of ribosomes. (*c*) In the absence of ribosomal RNA, ribosomal protein L4 binds near the 5' end of the *S10* transcript and blocks its translation.

either polypeptide chain initiation or chain elongation. Sometimes regulation occurs by the differential degradation of specific regions of polygenic mRNAs. The inhibition of translation of a specific mRNA by a protein that it encodes is called negative autogenous regulation.

POST-TRANSLATIONAL REGULATORY MECHANISMS

Earlier in this chapter, we discussed the mechanism by which the transcription of bacterial genes encoding enzymes in a biosynthetic pathway is repressed when the product of the pathway is present in the medium in which the cells are growing. A second, and more rapid, regulatory fine-tuning of metabolism often occurs at the level of enzyme activity. The presence of a sufficient concentration of the end product of a biosynthetic pathway frequently results in the inhibition of the first enzyme in the pathway (Figure 21.28). This phenomenon is called **feedback inhibition** or **end-product inhibition**. Feedback inhibition results in an almost instantaneous arrest of the synthesis of the end product when it is added to the medium.

Feedback inhibition-sensitive enzymes contain an end-product binding site (or sites) in addition to the substrate binding site (or sites). In the case of multimeric enzymes, the **end-product or regulatory binding site** often is on a subunit (polypeptide) different from that of the substrate site. Upon binding the end product, such enzymes undergo allosteric transitions that reduce their affinity for the substrates. Proteins that undergo such conformational changes are referred to as allosteric proteins. Many, perhaps most, enzymes undergo allosteric transitions of some kind.

Allosteric transitions also appear to be responsible for enzyme activation, which often occurs when an enzyme binds one or more of its substrates or some other small molecule. Some enzymes exhibit a broad spectrum of activation and inhibition by many different effector molecules. The enzyme glutamine synthetase catalyzes the final step in the biosynthesis of the amino acid glutamine. Glutamine synthetase is a complex multimeric enzyme in both prokaryotes and eukaryotes. The glutamine synthetase of *E. coli* has been shown to respond, either by activation or inhibition, to

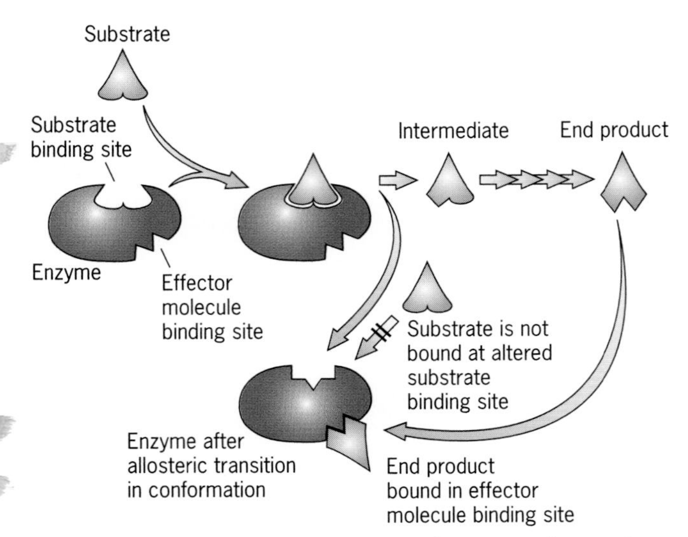

Figure 21.28 Feedback inhibition of gene-product activity. The end product of a biosynthetic pathway often inhibits the activity of the first enzyme in the pathway, rapidly stopping synthesis of the end product.

16 different metabolites, presumably through allosteric transitions.

Key Points: **The end product of a biosynthetic pathway often inhibits the activity of the first enzyme in the pathway, rapidly shutting off the synthesis of the product. This regulatory mechanism is called feedback inhibition. Enzyme activation occurs when a substrate or other effector molecule enhances the activity of an enzyme.**

TESTING YOUR KNOWLEDGE

1. The operon model for the regulation of enzyme synthesis concerned in lactose utilization by *E. coli* includes a regulator gene (*I*), an operator region (*O*), a structural gene (*Z*) for the enzyme β-galactosidase, and another structural gene (*Y*) for β-galactoside permease. β-Galactoside permease transports lactose into the bacterium, where β-galactosidase cleaves it into galactose and glucose. Mutations in the *lac* operon have the following effects: Z^- *and* Y^- mutant strains are unable to make functional β-galactosidase and β-galactoside permease, respectively, whereas I^- *and* O^C mutant strains synthesize the *lac* operon gene products constitutively. The following diagram is of a partially diploid strain of *E. coli* that carries two copies of the *lac* operon. On the diagram, fill in a genotype that will result in the constitutive synthesis of β-galactosidase and the inducible synthesis of β-galactoside permease by this partial diploid.

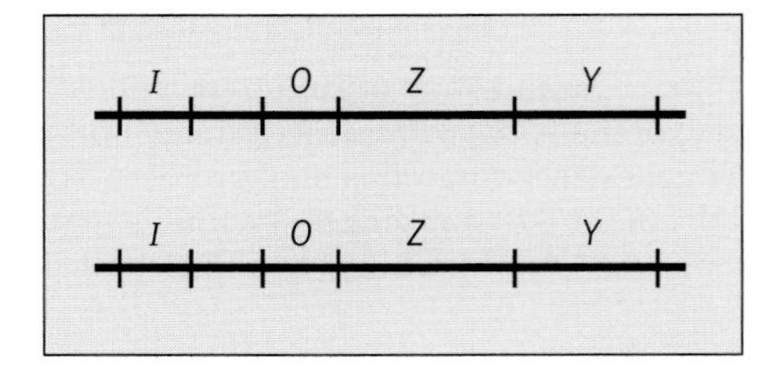

ANSWER

Several different genotypes will produce β-galactosidase constitutively and β-galactoside permease inducibly. They must meet two key requirements: (1) the cell must contain at least one copy of the I^+ gene, which encodes the repressor, and (2) the Z^+ gene and an O^C mutation must be on the same chromosome, because the operator acts only in *cis;* that is, it only affects the expression of genes on the same chromosome. In contrast, the cell can be either homozygous or heterozygous for the I^+ gene, and, if heterozygous, I^+ may be on either chromosome, because I^+ is dominant to I^- and I^+ acts in both the *cis* and *trans* arrangement. One possible genotype is given in the following diagram. How many other genotypes can you devise that will synthesize β-galactosidase constitutively and β-galactoside permease inducibly?

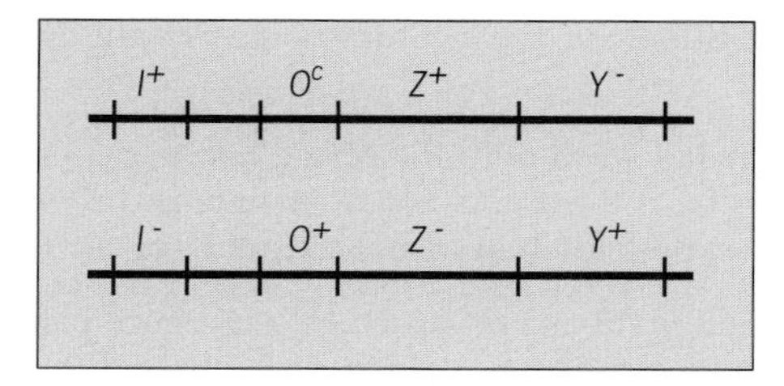

2. The arabinose (*ara*) operon of *E. coli* contains three genes that encode enzymes required for the catabolism of arabinose. One of these genes, *araA*, encodes an enzyme called arabinose isomerase. Transcription of the *ara* operon is controlled by a regulatory protein called AraC, encoded by the *araC* gene. You are given *E. coli* strains with the following genotypes: (a) $araA^-$ $araC^-$ / F' $araA^+$ $araC^-$, (b) $araA^-$ $araC^-$ / F' $araA^-$ $araC^+$, (c) $araA^+$ $araC^+$ / F' $araA^-$ $araC^-$, and (d) $araA^+$ $araC^-$ / F' $araA^-$ $araC^+$. Can these strains synthesize arabinose isomerase? Will the synthesis be constitutive or inducible?

ANSWER

All of the genotypes except that shown in (b) can synthesize arabinose isomerase. The (b) strain is homozygous for an $araA^-$ mutation and cannot produce a functional isomerase. The genotype shown in (a) will synthesize a low level (about 30 molecules per cell) of isomerase constitutively, whereas the genotypes shown in (c) and (d) will exhibit inducible synthesis of arabinose isomerase (about 12,000 molecules per cell when fully induced). The (a) strain is homozygous for an $araC^-$ mutation and cannot synthesize the AraC regulatory protein, which is required for induction of the *ara* operon. Thus transcription of the *ara* operon is not induced by arabinose in $araC^-$ strains. The genotypes shown in (c) and (d) both produce isomerase inducibly (only when arabinose is present in the medium). Both strains are heterozygous for $araC^+$ and $araC^-$, with $araC^+$ dominant to $araC^-$. The difference between them is that the $araC^+$ allele is located *cis* to $araA^+$ in strain (c) and *trans* to $araA^+$ in strain (d). However, recall that $araC^+$ encodes a diffusible product that binds to the two *ara* operator regions. Thus the $araC^+$ regulatory gene is active whether located *cis* or *trans* to the *ara* operon.

QUESTIONS AND PROBLEMS

21.1 How can inducible and repressible enzymes of microorganisms be distinguished?

21.2 Distinguish between (a) repression and (b) feedback inhibition caused by the end product of a biosynthetic pathway. How do these two regulatory phenomena complement each other to provide for the efficient regulation of metabolism?

21.3 In the lactose operon of *E. coli*, what is the function of each of the following genes or sites: (a) regulator, (b) operator, (c) promoter, (d) structural gene *Z* and (e) structural gene *Y*?

21.4 What would be the result of inactivation by mutation of the following genes or sites in the *E. coli* lactose operon: (a) regulator, (b) operator, (c) promoter, (d) structural gene *Z*, and (e) structural gene *Y*?

21.5 Groups of alleles associated with the lactose operon are as follows (in order of dominance for each allelic series): repressor, I^s (superrepressor), I^+ (inducible), and I^- (constitutive); operator, O^c (constitutive, *cis*-dominant) and O^+ (inducible, *cis*-dominant); structural, Z^+ and Y^+. (a) Which of the following genotypes will produce β-galactosidase and β-galactoside permease if lactose is present: (1) $I^+O^+Z^+Y^+$, (2) $I^-O^cZ^+Y^+$, (3) $I^sO^cZ^+Y^+$, (4) $I^sO^+Z^+Y^+$, and (5) $I^-O^+Z^+Y^+$? (b) Which of the above genotypes will produce β-galactosidase and β-galactoside permease if lactose is absent? Why?

21.6 For each of the following partial diploids indicate whether enzyme formation is constitutive or inducible (see Problem 21.5 for dominance relationships): (a) $I^+O^+Z^+Y^+/I^+O^+Z^+Y^+$, (b) $I^+O^+Z^+Y^+/I^+O^CZ^+Y^+$, (c) $I^+O^CZ^+Y^+/I^+O^CZ^+Y^+$, (d) $I^+O^+Z^+Y^+/I^-O^+Z^+Y^+$, (e) $I^-O^+Z^+Y^+/I^-O^+Z^+Y^+$. Why?

21.7 Write the partial diploid genotype for a strain that will (a) produce β-galactosidase constitutively and permease inducibly and (b) produce β-galactosidase constitutively but not permease either constitutively or inducibly, even though a Y^+ gene is known to be present.

21.8 Constitutive mutations produce elevated enzyme levels at all times; they may be of two types: O^c or I^-. Assume that all other DNA present is wild-type. Outline how the two constitutive mutants can be distinguished with respect to (a) map position, (b) regulation of enzyme levels in O^c/O^+ versus I^-/I^+ partial diploids, and (c) the position of the structural genes affected by an O^c mutation versus the genes affected by an I^- mutation in a partial diploid.

21.9 How could the tryptophan operon in *S. typhimurium* have developed and been maintained by evolution?

21.10 Of what biological significance is the phenomenon of catabolite repression?

21.11 How might the concentration of glucose in the medium in which an *E. coli* cell is growing regulate the intracellular level of cyclic AMP?

21.12 Is the CAP-cAMP effect on the transcription of the *lac* operon an example of positive or negative regulation? Why?

21.13 Would it be possible to isolate *E. coli* mutants in which the transcription of the *lac* operon is not sensitive to catabolite repression? If so, in what genes might the mutations be located?

21.14 Using examples, distinguish between negative regulatory mechanisms and positive regulatory mechanisms.

21.15 A deletion of the regulator *(I)* gene of the *lac* operon in *E. coli* is expected to result in the constitutive synthesis of the *lac* operon enzymes. What will the expected phenotype be for a mutant strain of *E. coli* with a deletion of the regulator gene *(araC)* of the *ara* operon? Why?

21.16 The following table gives the relative activities of the enzymes β-galactosidase and β-galactoside permease in cells with different genotypes at the *lac* locus in *E. coli*. The level of activity of each enzyme in wild-type *E. coli* not carrying F's was arbitrarily set at 100; all other values are relative to the observed levels of activity in these wild-type bacteria. Based on the data given in the table for genotypes 1 through 4, fill in the levels of enzyme activity that would be expected for the fifth genotype.

	β-galactosidase		β-galactoside permease	
Genotype	– Inducer	+ Inducer	– Inducer	+ Inducer
1. $I^+O^+Z^+Y^+$	0.1	100	0.1	100
2. $I^-O^+Z^+Y^+$	100	100	100	100
3. $I^+O^CZ^+Y^+$	25	100	25	100
4. $I^-O^+Z^+Y^-/F'\ I^-O^+Z^+Y^+$	200	200	100	100
5. $I^-O^CZ^-Y^+/F'\ I^+O^+Z^+Y^+$	—	—	—	—

21.17 What effect, if any, will deletion of the $araO_2$ site have on the expression of the *araBAD* operon?

21.18 The rate of transcription of the *trp* operon in *E. coli* is controlled by both (1) repression/derepression and (2) attenuation. By what mechanisms do these two regulatory processes modulate *trp* operon transcript levels?

21.19 What effect will deletion of the *trpL* region of the *trp* operon have on the rates of synthesis of the enzymes encoded by the five genes in the *trp* operon in *E. coli* cells growing in the presence of tryptophan?

21.20 By what mechanism does the presence of tryptophan in the medium in which *E. coli* cells are growing result in premature termination or attenuation of transcription of the *trp* operon?

21.21 Suppose that you used site-specific mutagenesis to modify the *trpL* sequence such that the two UGG Trp codons at positions 54–56 and 57–60 (see Figure 21.11) in the mRNA leader sequence were changed to GGG Gly codons. Will attenuation of the *trp* operon still be regulated by the presence or absence of tryptophan in the medium in which the *E. coli* cells are growing?

21.22 What is a prophage? In what ways does the phage λ prophage differ from the λ chromosome present during lytic infections?

21.23 In what ways does the phage λ lytic regulatory cascade differ from the autogenously maintained repression of viral gene expression that occurs during lysogeny?

21.24 What is the major structural difference between the operator region in the *lac* operon of *E. coli* and the phage λ operators O_L and O_R?

21.25 The λ lysogenic state is very stable, in part because the synthesis of the λ repressor is autogenously regulated. What does this mean? How does autogenous regulation of repressor enhance the stability of lysogeny?

21.26 During lytic growth, the phage λ genes exhibit a rather precise pattern of temporal expression. The immediate early genes are expressed first, then the delayed early genes, and lastly the late genes. How does this regulatory cascade work? Can the late genes be expressed without the delayed early genes being expressed first? If not, why not? What role does transcriptional termination play in the λ lytic regulatory cascade?

21.27 The N protein of phage λ is a transcriptional antiterminator. What role does it play in the lambda regulatory cascade?

21.28 A lambda phage has a UAG chain-termination mutation in the middle of the C_I gene. What phenotype will result from this mutation?

21.29 When phage λ infects an *E. coli* cell, it can either undergo lytic growth and lyse the host cell or enter the lysogenic pathway. What determines which response will occur in any given infected cell? Will lytic or lysogenic development occur in a cell infected with a λ phage with a deletion of the C_I gene? Which will occur in a cell infected with a λ phage with a deletion of the *Cro* gene?

21.30 Would attenuation of the type that regulates the level of *trp* transcripts in *E. coli* be likely to occur in eukaryotic organisms?

BIBLIOGRAPHY

BECKWITH, J. R., AND D. ZIPSER. 1970. *The Lactose Operon*. Cold Spring Harbor Laboratory Press, Cold Spring Harbor, New York.

DAS, A. 1993. Control of transcription termination by RNA-binding proteins. *Ann. Rev. Biochem.* 62:893–930.

JACOB, F., AND J. MONOD. 1961. Genetic regulatory mechanisms in the synthesis of proteins. *J. Mol. Biol.* 3:318–356.

KOLB, A., S. BUSBY, H. BUC, S. GARGES, AND S. ADHYA. 1993. Transcriptional regulation by cAMP and its receptor. *Ann. Rev. Biochem.* 62:749–795.

LEE, D.-H., AND R. F. SCHLEIF. 1989. *In vivo* DNA loops in

araCBAD: size limits and helical repeat. *Proc. Natl. Acad. Sci., U. S.* 86:476–480.

NEIDHARDT, F. C., J. L. INGRAHAM, K. B. LOW, B. MAGASANIK, M. SCHAECHTER, AND H. E. UMBARGER. (eds.) 1987. Escherichia coli *and* Salmonella typhimurium, *Cellular and Molecular Biology*. American Society for MIcrobiology, Washington, DC. (See in particular Vol. 2, Part IV, entitled "Regulation of gene expression," pp. 1231–1526.)

PTASHNE, M. 1987. *A Genetic Switch: Gene Control and Phage* λ. Cell Press, Cambridge, MA.

PTASHNE, M. 1989. How gene activators work. *Sci. Amer.* 260(1):40–47.

PTASHNE, M., A. D. JOHNSON, AND C. O. PABO. 1982. A genetic switch in a bacterial virus. *Sci. Amer.* 247(5):128–140.

YANOFSKY, C. 1981. Attenuation in the control of expression of bacterial operons. *Nature* 289:751–758.

22

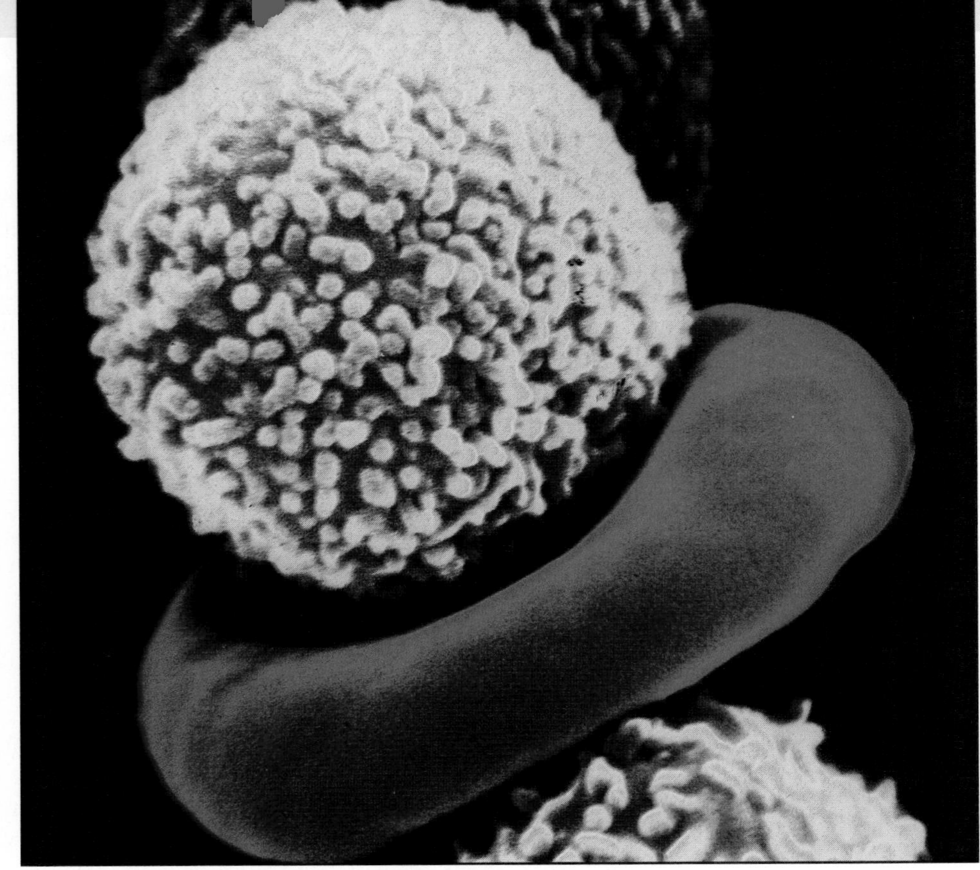

False-color scanning electron micrograph of abnormal white blood cells (B lymphocytes) along with one normal red blood cell (center) from a patient with chronic lymphocytic leukemia. These malignant B lymphocytes are abnormally small and are covered with tiny fingerlike projections. They are not able to carry out their usual functions, leaving the individual more susceptible to infections.

Gene Regulation in Eukaryotes and the Genetic Basis of Cancer

CHAPTER OUTLINE

Cancer: Evidence for the Importance of Regulated Gene Expression

Lawrence, a biology professor at a large American university, had trouble finishing one of his lectures. His thoughts were labored and unfocused, and he felt tired and disoriented. After dismissing his class, Lawrence sought medical attention. The preliminary diagnosis was that Lawrence's mental and motor abilities had been impaired by a slight stroke, the result of a rupture in a tiny blood vessel in the brain. However, further examination revealed a more serious problem. Lawrence's brain had developed a tumor, probably as an offshoot of another tumor located on one of his kidneys. These two masses of cancerous tissue had grown to the point where they were beginning to interfere with important biological functions. Lawrence had surgery to remove the kidney tumor and was given radiation therapy to kill the brain tumor. Unfortunately, the growth of the brain tumor could not be stopped, and 18 months after teaching his last class, Lawrence died.

What causes tumors to form and what causes some of them to spread? Why do some types of tumors tend to run in families? Is cancer an inherited disease? In recent years, these and other questions have stimulated an enormous amount of research on the basic biology of cancer. Although many details are still unclear, the fundamental finding is that cancers result from genetic malfunctions. The mis-expression of critical genes can cause cellular processes to go awry and lead to unregulated growth. Each case of cancer is therefore sobering testimony to the importance of regulated gene expression.

In this chapter, we consider the mechanisms that regulate gene expression in eukaryotes. These mechanisms are more elaborate than those in prokaryotes because many eukaryotic organisms are multicellular, and within each such organism there may be many different types of cells. The formation of these cell types requires that groups of genes be expressed in precise spatial and temporal patterns within the organism. We will see that this spatial and temporal regulation of gene expression is accomplished by a diverse array of molecular mechanisms.

SPATIAL AND TEMPORAL REGULATION OF EUKARYOTIC GENES

The thousands of genes that are present in the genomes of multicellular eukaryotes are normally expressed in a controlled fashion. One dimension of this control is spatial. Not every gene product is needed in every tissue. Some genes are expressed in nerve cells, others in blood cells, and still others in reproductive cells. In fact, the complexity of multicellular eukaryotes is partly due to the tissue-specific expression of many different genes. A second dimension of eukaryotic gene regulation is temporal. Different genes are expressed at different times, some in response to biological signals such as hormones, and others in response to environmental stimuli. Temporal specificity is most dramatically seen during development, when a fertilized egg grows into a multicellular organism. As this organism forms, batteries of genes are expressed in an orderly sequence to direct the formation of tissues and organs. The temporal and spatial regulation of genes is therefore an important aspect of eukaryotic biology.

Spatial Regulation of Tubulin Genes in Plants

The genes for the tubulin polypeptides provide a dramatic example of expression that is spatially regulated. These polypeptides are the building blocks of microtubules (Chapter 2). There are two general types of tubulin polypeptides, α and β, and one molecule of each type aggregates to form a dimer. These dimers then assemble in parallel rows to form hollow, cylindrical microtubules about 24 nm in diameter. Several microtubules may aggregate with each other to create more complex structures, such as cilia and flagella. Within cells, microtubules are found in many places—in the cytoplasm just below the plasma membrane, around the nuclear membrane, and in a special region called the microtubule-organizing center. Microtubules play an important role in cell movement. In cilia and flagella, their wavelike bending helps to move a cell from one position to another, and inside cells, they are responsible for moving chromosomes during mitosis.

The α and β tubulins are encoded by distinct sets of genes. Typically, several different genes for each type of tubulin are present in the genome. For example, the weedy plant *Arabidopsis thaliana* has six α tubulin genes and nine β tubulin genes. Each of these genes has been isolated, and some of them have been studied for expression in whole plants. The results of this work show that each gene is expressed in a specific spatial pattern. For example, the *TUA1* gene, which encodes a variant of the α tubulin polypeptide, is expressed preferentially in pollen grains (Figure 22.1); there is also some expression of *TUA1* in young anthers, but little or no expression of it in leaves, stems, or roots. In contrast, *TUB1*, a β tubulin gene, is expressed only in a section of the root immediately above the rapidly dividing root tip. Another β tubulin

Figure 22.1 Flowers of *Arabidopsis thaliana* stained to detect expression from the promoter of the *TUA1* alpha tubulin gene. The *TUA1* promoter has been fused to the coding sequence of the β-glucouronidase gene from *Escherichia coli*, which acts as a reporter of expression (blue color).

gene, *TUB8*, is preferentially expressed in the vascular tissues of the plant. The spatially specific patterns of expression of these and other tubulin genes suggest that slightly different types of microtubules are needed in different parts of *Arabidopsis* plants.

Temporal Regulation of Globin Genes in Animals

Perhaps one of the most dramatic examples of temporally regulated gene expression comes from the study of hemoglobin, the protein that is responsible for transporting oxygen in the blood of vertebrate animals. In higher vertebrates, this protein is a tetramer of polypeptides called globins; in each tetramer there are two α and two β globin chains. Molecules of an iron-containing compound called heme are loosely joined to each of these polypeptides, forming pockets that can bind molecular oxygen. In human beings, multiple genes for the α and β globins are located in two separate sites in the genome. The α globin genes occupy a 28-kb region on chromosome 16, and the β globin genes occupy a 45-kb region on chromosome 11 (Figure 22.2). Because the genes within each cluster are duplicates of an ancestral globin gene, they form a small multigene family. Over evolutionary time, the members of these families have diverged from each other by random mutation so that today, each one encodes a slightly different polypeptide. In some of these duplicate genes, a frameshifting or chain-terminating mutation has abolished the ability to make a polypeptide; such non-coding genes are called **pseudogenes** and are denoted by the Greek letter psi (ψ).

A remarkable feature of both the α and β gene clusters is that their members are expressed at different times during development. The genes on one side in the clusters are expressed only in the embryo, those in the middle are expressed only in the developing fetus, and those on the other side are expressed only after birth. This sequential activation of genes from one side to the other in a cluster is apparently related to the need to produce slightly different kinds of hemoglobin during the course of human development. Embryo, fetus, and infant have different oxygen requirements, different circulatory systems, and different physical environments. The temporal switching in globin gene expression is apparently an adaptation to this changing array of conditions.

Key Points: **In eukaryotes, gene expression is spatially and temporally regulated. Some genes, such as those encoding the α and β tubulins in *Arabidopsis*, are expressed in a tissue-specific manner. Other genes, such as those encoding the α and β globins in vertebrates, are expressed in a specific temporal pattern during development.**

WAYS OF REGULATING EUKARYOTIC GENE EXPRESSION

As in prokaryotes, the expression of genes in eukaryotes involves the transcription of DNA into RNA and the subsequent translation of that RNA into polypep-

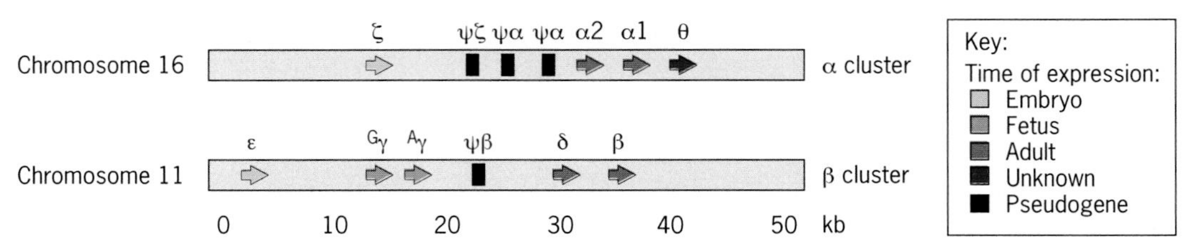

Figure 22.2 Organization and temporally specific expression of the human α and β globin genes.

tides. However, prior to translation, most eukaryotic RNA is "processed." During processing, the RNA is capped at its 5' end, polyadenylated at its 3' end, and altered internally by losing its noncoding intron sequences (see Chapter 11). Prokaryotic RNAs typically do not undergo these terminal and internal modifications. Gene expression is therefore more complicated in eukaryotes than it is in prokaryotes.

There is greater complexity of gene expression in eukaryotes because eukaryotic cells are compartmentalized by an elaborate system of membranes. This compartmentalization subdivides the cells into separate organelles, the most conspicuous one being the nucleus; eukaryotic cells also possess mitochondria, chloroplasts (if they are plant cells), and an endoplasmic reticulum. Each of these organelles performs a different function. The nucleus stores the genetic material, the mitochondria and chloroplasts recruit energy, and the reticulum transports materials within the cell.

The subdivision of eukaryotic cells into organelles physically separates the events of gene expression. The primary event, transcription of DNA into RNA, occurs in the nucleus. RNA transcripts are also modified in the nucleus by capping, polyadenylation, and the removal of introns. The resulting messenger RNAs are then exported to the cytoplasm where they become associated with ribosomes, many of which are located on the membranes of the endoplasmic reticulum. Once associated with ribosomes, these mRNAs are translated into polypeptides. This physical separation of the events of gene expression makes it possible for regulation to occur in different places (Figure 22.3). Regulation can occur in the nucleus at either the DNA or RNA level, or in the cytoplasm at either the RNA or polypeptide level.

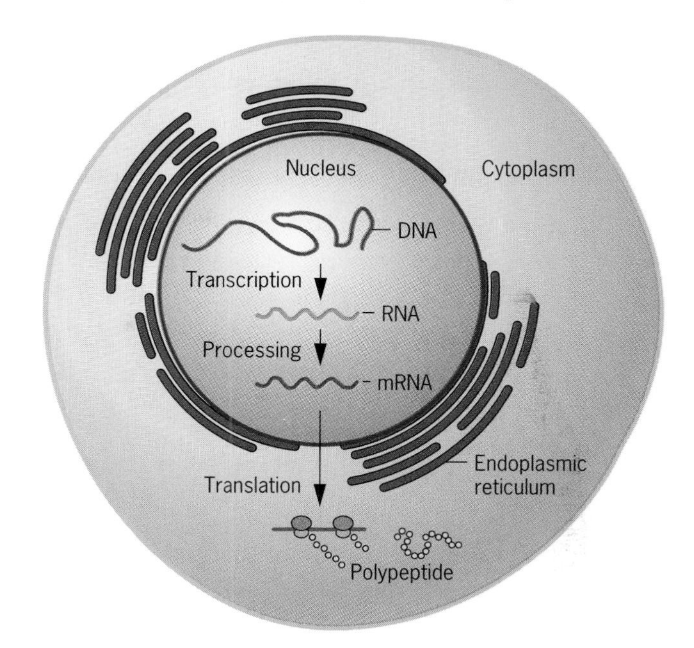

Figure 22.3 Eukaryotic gene expression showing the stages at which expression can be regulated: transcription, processing, and translation.

Controlled Transcription of DNA

In prokaryotes, gene expression is regulated mainly by controlling the transcription of DNA into RNA. A gene that is not transcribed is simply not expressed. Transcription occurs in prokaryotes when negative regulatory molecules such as the *lac* repressor protein have been removed from the vicinity of a gene and positive regulatory molecules such as the cyclic AMP-catabolite activator protein (CAP) complex have bound to it (Chapter 21). These protein-DNA interactions control whether or not a gene is accessible to RNA polymerase. Furthermore, because unicellular prokaryotes are at the mercy of their environments, the mechanisms that have evolved to control transcription in these organisms must respond quickly to environmental changes. As we discussed in Chapter 21, this hair-trigger control is an efficient strategy for prokaryotic survival.

The control of transcription is more complex in eukaryotes than it is in prokaryotes. One reason is that genes are sequestered in the nucleus. Before they can have any effect on the level of transcription, environmental signals must be transmitted from the cell surface, where they are usually received, through the cytoplasm and the nuclear membrane, and onto the chromosomes. Eukaryotic cells therefore need internal signaling systems to control the transcription of DNA. Another complicating factor is that many eukaryotes are multicellular. Environmental cues may have to pass through layers of cells in order to have an impact on the transcription of genes in a particular tissue. Intercellular communication is therefore an important aspect of eukaryotic transcriptional regulation.

As in prokaryotes, eukaryotic transcriptional regulation is mediated by protein-DNA interactions. Positive and negative regulator proteins bind to specific regions of the DNA and stimulate or inhibit transcription. As a group, these proteins are called **transcription factors**. Many different types have been identified, and most seem to have a characteristic domain that allows them to interact with DNA. The structure of these proteins, and the nature of their interactions with DNA, will be discussed in a later section.

Alternate Splicing of RNA

Most eukaryotic genes possess introns, the noncoding sequences embedded within the sequence that speci-

Exons in rat troponin T gene

5' 1 2 3 4 5 6 7 8 9 10 11 12 13 14 15 16 17 18 3'

Alternate splicing of exons produces 64 different mRNAs.

Exons 1-3, 9-15, and 18 exons present in all mRNAs.

Exons 4-8 present in various combinations in mRNAs.

Exons 16 or 17, but not both, present in all mRNAs.

Examples of mRNAs

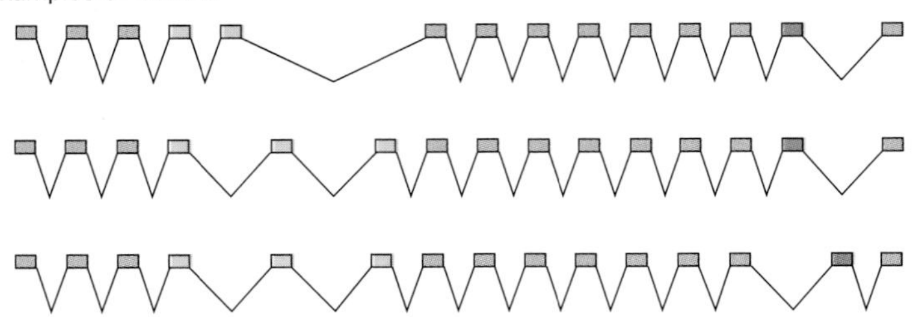

Figure 22.4 Alternate splicing of transcripts from the rat troponin T gene. Only 3 of 64 different mRNAs are shown.

fies the incorporation of amino acids into a polypeptide. Each intron must be removed from the RNA transcript of a gene in order for the coding sequence to be expressed properly. As we discussed in Chapter 11, this process involves the precise joining of the coding sequences, or exons, into a messenger RNA. The formation of the mRNA is mediated by tiny nuclear organelles called spliceosomes.

Genes with multiple introns present a curious problem to the RNA splicing machinery. These introns can be removed separately or in combination, depending on how the splicing machinery interacts with the RNA. If two successive introns are removed together, the exon between them will also be removed. Thus, the splicing machinery has the opportunity to modify the coding sequence of an RNA by deleting some of its exons. This phenomenon of splicing an RNA transcript in different ways is apparently a way of economizing on genetic information. Instead of duplicating genes, or pieces of genes, the **alternate splicing** of transcripts makes it possible for a single gene to encode different polypeptides.

One example of alternate splicing occurs during the expression of the gene for troponin T, a protein found in the skeletal muscles of vertebrates; the size of this protein ranges from about 150 to 250 amino acids. In the rat, the troponin T gene is more than 16 kb long and contains 18 different exons (Figure 22.4). Transcripts of this gene are spliced in different ways to create a large array of mRNAs. When these are translated, many different troponin T polypeptides are produced. All these polypeptides share amino acids from exons 1–3, 9–15, and 18. However, the regions encoded by exons 4–8 may be present or absent, depending on the splicing pattern, and apparently in any combination. Additional variation is provided by the presence or absence of regions encoded by exons 16 and 17; if 16 is present, 17 is not, and vice versa. These different forms of troponin T presumably function in slightly different ways within the muscles, contributing to the variability of muscle cell action.

Another example of alternate splicing occurs during the expression of genes involved in sex determination in *Drosophila*. The master regulator of the sex-determination process is the X-linked *Sex-lethal (Sxl)* gene. In chromosomal females, the transcript of this gene is spliced to produce an mRNA that encodes a regulatory protein. In chromosomal males, the *Sxl* transcript is alternately spliced to include an exon with a stop codon; thus, when this RNA is translated, it generates a short polypeptide without regulatory function (Figure 22.5). In XX embryos, where the full-length *Sxl* protein is present, a particular set of genes is expressed that causes these embryos to develop as females. In XY embryos, where the full-length *Sxl* protein is absent, a different set of genes is expressed that causes these embryos to develop as males. Female-specific or male-specific splicing is ultimately controlled by the ratio of X chromosomes to autosomes. If the X:A ratio is 1.0, the female-specific pattern of splicing occurs; if the X:A ratio is 0.5, the male-specific pattern occurs. Thus alternate splicing of the *Sxl* RNA is responsible for sexual differention in *Drosophila.*

Cytoplasmic Control of Messenger RNA Stability

Messenger RNAs are exported from the nucleus to the cytoplasm where they serve as templates for polypeptide synthesis. Once in the cytoplasm, a particular mRNA can be translated by several ribosomes that

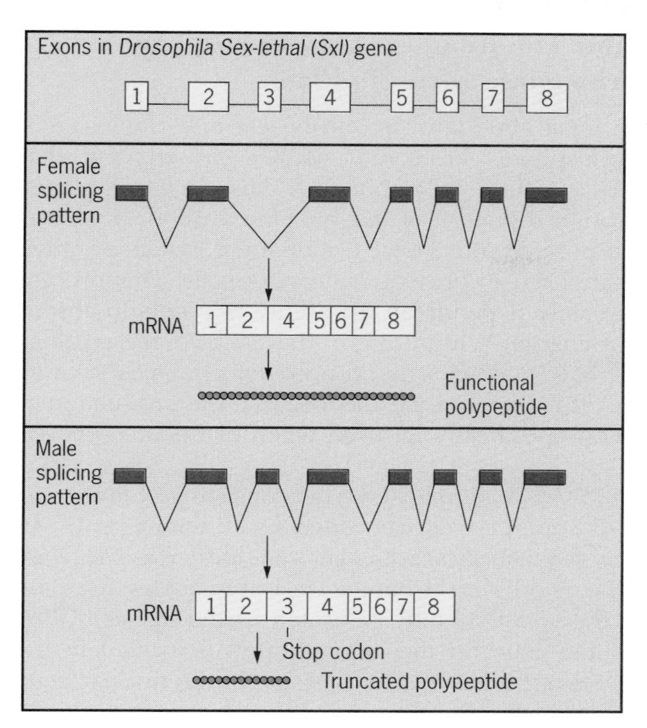

Figure 22.5 Alternate splicing of transcripts from the *Sex-lethal* gene in male and female *Drosophila.*

move along it in sequential fashion. This translational assembly line continues until the mRNA is degraded. Messenger RNA degradation is therefore another control point in the overall process of gene expression. Long-lived mRNAs can support multiple rounds of polypeptide synthesis, whereas short-lived mRNAs cannot.

An mRNA that is rapidly degraded must be replenished by additional transcription; otherwise, the polypeptide it encodes will cease to be synthesized. This cessation of polypeptide synthesis may, of course, be part of a developmental program. Once the polypeptide has had its effect, it may no longer be needed; in fact, its continued synthesis may be harmful. In such cases, rapid degradation of the mRNA would be a reasonable way of preventing undesired polypeptide synthesis.

Although the longevity of mRNAs has been difficult to study, a few insights have been obtained by monitoring radioactively labeled RNAs over time. The results of these labeling experiments indicate that mRNA longevity can be influenced by several factors, including the length of the poly-A tail, the structure of the 3' untranslated region preceding the tail, and the metabolic state of the cell.

Poly-A tails seem to stabilize mRNAs. Histone mRNAs, for example, lack poly-A tails and are extremely short-lived. The sequence of the 3' untranslated region (3'UTR) also seems to affect mRNA stability. Several short-lived mRNAs have the sequence AUUUA repeated several times in their 3' untranslated regions. When this sequence is artificially transferred to the 3' untranslated region of more stable mRNAs, they too become unstable. Other studies have suggested that the stability of mRNAs can be influenced by chemical factors such as hormones. In the toad *Xenopus laevis*, the vitellogenin gene is transcriptionally activitated by the steroid hormone estrogen. However, in addition to inducing transcription of this gene, estrogen also increases the longevity of its mRNA. All these findings demonstrate that by controlling the rate of mRNA degradation, cells have additional ways of modulating the overall process of gene expression.

Key Points: **Eukaryotic gene expression can be regulated at the transcriptional, processing, or translational levels. Transcriptional regulation involves protein-DNA interactions, processing regulation involves the alternate splicing of primary gene transcripts, and translational regulation involves mRNA stability.**

INDUCTION OF TRANSCRIPTIONAL ACTIVITY BY ENVIRONMENTAL AND BIOLOGICAL FACTORS

In their study of the *lactose* operon in *E. coli*, Jacob and Monod discovered that the genes for lactose metabolism were specifically transcribed when lactose was given to the cells. Thus they demonstrated that lactose was an **inducer** of gene transcription. Following in the footsteps of Jacob and Monod, many researchers have attempted to identify specific inducers of eukaryotic gene transcription. Although these efforts have met with considerable success, the overall extent to which eukaryotic genes are induced by environmental and nutritional factors seems to be less than it is in prokaryotes. Here we will consider three examples of inducible gene expression in eukaryotes. The first two involve induction by environmental factors—temperature and light—and the third involves induction by that special group of signaling molecules called hormones.

Temperature: The Heat-shock Genes

When organisms are subjected to the stress of high temperature, they respond by synthesizing a group of proteins that help to stabilize the internal cellular environment. These **heat-shock proteins,** found in both prokaryotes and eukaryotes, are among the most con-

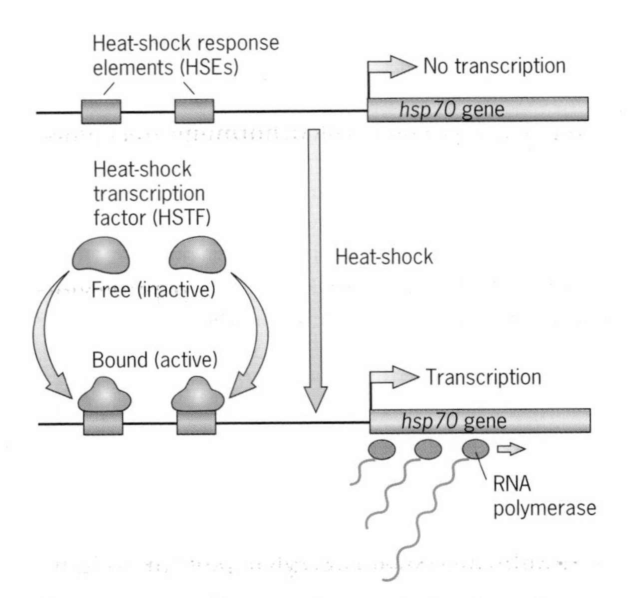

Figure 22.6 Induction of transcription from the *Drosophila hsp70* gene by heat shock. The HSEs are located between 40 and 90 base pairs upstream of the transcription initiation site (bent arrow).

served polypeptides known. Comparisons of the amino acid sequences of heat-shock proteins from organisms as diverse as *E. coli* and *Drosophila* show that they are 40 to 50 percent identical—a remarkable finding considering the length of evolutionary time separating these organisms.

The expression of the heat-shock proteins is regulated at the transcriptional level; that is, heat stress specifically induces the transcription of the genes encoding these proteins (Figure 22.6). In *Drosophila*, for example, one of the heat-shock proteins called **HSP70** (for heat-shock protein, molecular weight 70 kilodaltons) is encoded by a family of genes located in two nearby clusters on one of the autosomes. Altogether, there are five to six copies of these *hsp70* genes in the two clusters. When the temperature exceeds 33°C, as it does on hot summer days, each of the genes is transcribed into RNA, which is then processed and translated to produce HSP70 polypeptides. This heat-induced transcription of the *hsp70* genes is mediated by a polypeptide called the **Heat-Shock Transcription Factor**, or **HSTF**, which is present in the nuclei of *Drosophila* cells. When *Drosophila* are heat stressed, the HSTF is chemically altered by phosphorylation. In this altered state, it binds specifically to nucleotide sequences upstream of the *hsp70* genes and makes the genes more accessible to RNA polymerase II, the enzyme that transcribes most protein-encoding genes. The transcription of the *hsp70* genes is then vigorously stimulated. The sequences to which the phosphorylated HSTF binds are called **heat-shock response elements (HSEs)**.

Light: The Ribulose 1,5-bisphosphate Carboxylase Genes in Plants

The most abundant protein on earth is ribulose 1,5-bisphosphate carboxylase (RBC), an enzyme that plays a critical role in photosynthesis in green plants. Through the work of this enzyme, carbon dioxide is incorporated into molecules of sugar, which are then metabolized to provide energy for cells. This process ultimately depends on the ability of plants to absorb light energy. Without light, the entire process comes to a halt, and there is no need for enzymes such as RBC. It is therefore no surprise that the production of RBC is specifically induced when plants are exposed to light.

RBC is a complex enzyme consisting of large and small subunits, each encoded by different genes. In some plant species, both genes are located in the DNA of the chloroplast (Chapter 18), but in others, the gene for the small subunit is located in the nuclear DNA and the gene for the large subunit is located in the chloroplast DNA. The expression of the nuclear gene for the small subunit (denoted *rbcS)* has been analyzed in several different plant species. These studies have shown that *rbcS* is vigorously transcribed after plants have been exposed to light (Figure 22.7).

The light-inducible transcription of *rbcS* is only partly understood. One important event is the absorbtion of light by a cytoplasmic protein called phytochrome (Chapter 18). A light-absorbing molecule called a chromophore is attached to each molecule of phytochrome. The absorption of light by the chromophore apparently causes a conformational change in the phytochrome polypeptide, which then triggers changes in other proteins. Although the details of subsequent events are sketchy, it seems that eventually some of these proteins bind to regions upstream of the *rbcS* gene and stimulate its transcription. This response is quick and vigorous; a substantial amount of *rbcS* RNA is generated after exposure to light, providing plenty of templates for the synthesis of the small

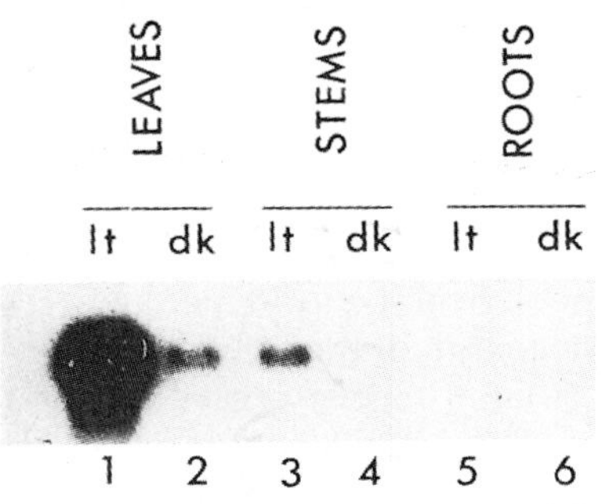

Figure 22.7 Light-induced transcription of the gene for the small subunit of ribulose 1,5-bisphosphate carboxylase.

subunit of RBC. A similar process may be involved in the production of the large subunit. Exposure to light therefore induces the production of one of the key enzymes needed for photosynthesis.

Signal Molecules: Genes That Respond to Hormones

In multicellular eukaryotes, one type of cell can signal another by secreting a **hormone**. Hormones circulate through the body, make contact with their target cells, and then initiate a series of events that regulate the expression of particular genes. In animals there are two general classes of hormones. The first class, the **steroid hormones**, are small, lipid-soluble molecules derived from cholesterol. Because of their lipid nature, they have little or no trouble passing through cell membranes. Examples are estrogen and progesterone, which play important roles in female reproductive cycles, testosterone, a hormone of male differentiation, the glucocorticoids, which are involved in regulating blood sugar levels, and ecdysone, a hormone that controls developmental events in insects. Once these types of hormones have entered a cell, they interact with cytoplasmic proteins called **hormone receptors**. The receptor-hormone complex that is formed then enters the nucleus where it acts as a transcription factor to regulate the expression of certain genes (Figure 22.8).

The second class of hormones, the **peptide hormones**, are linear chains of amino acids. Like all other polypeptides, these molecules are encoded by genes. Examples are insulin, which regulates blood sugar levels, somatotropin, which is a growth hormone, and prolactin, which targets tissues in the breasts of female mammals. Because peptide hormones are typically too large to pass freely through cell membranes, the signals they convey must be transmitted to the interior of cells by **membrane-bound receptor proteins** (Figure 22.9). When a peptide hormone interacts with its receptor, it causes a conformational change in the receptor that eventually leads to changes in other proteins inside the cell. Through a cascade of such changes, the

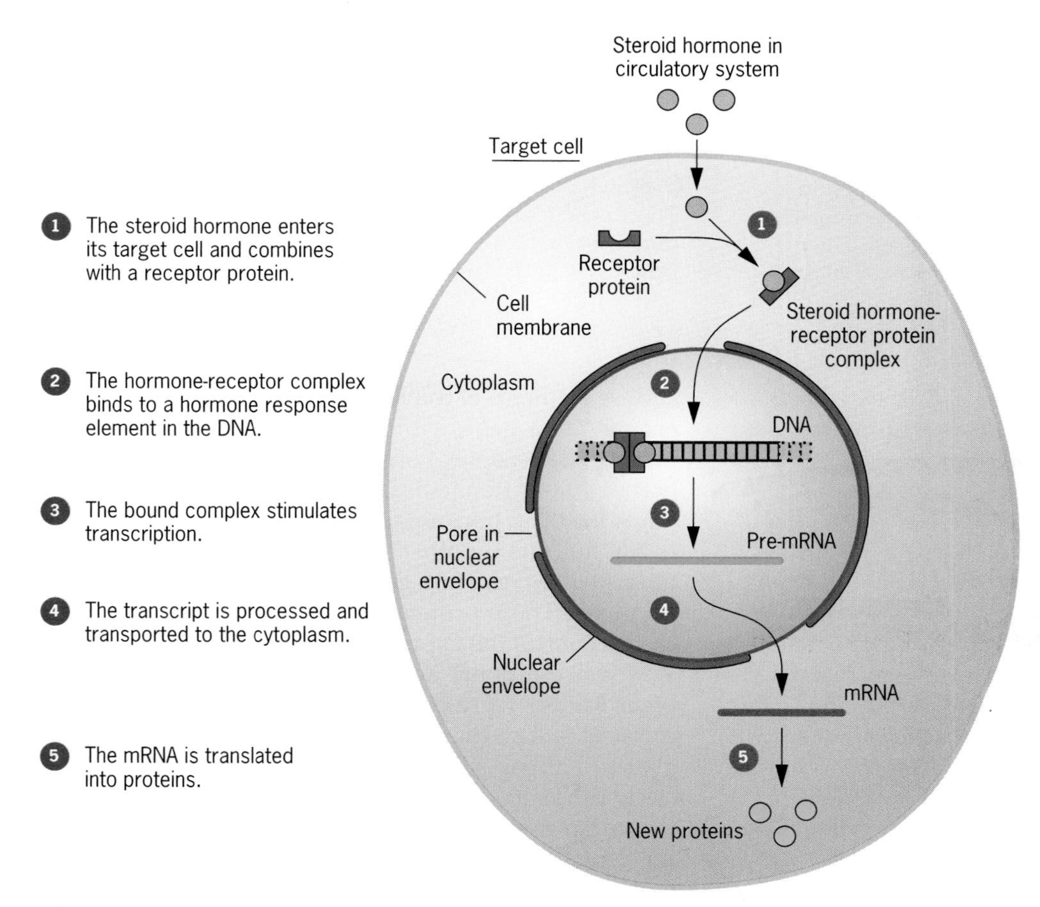

Figure 22.8 Regulation of gene expression by steroid hormones. The hormone interacts with a receptor inside its target cell, and the resulting complex moves into the nucleus where it activates the transcription of particular genes.

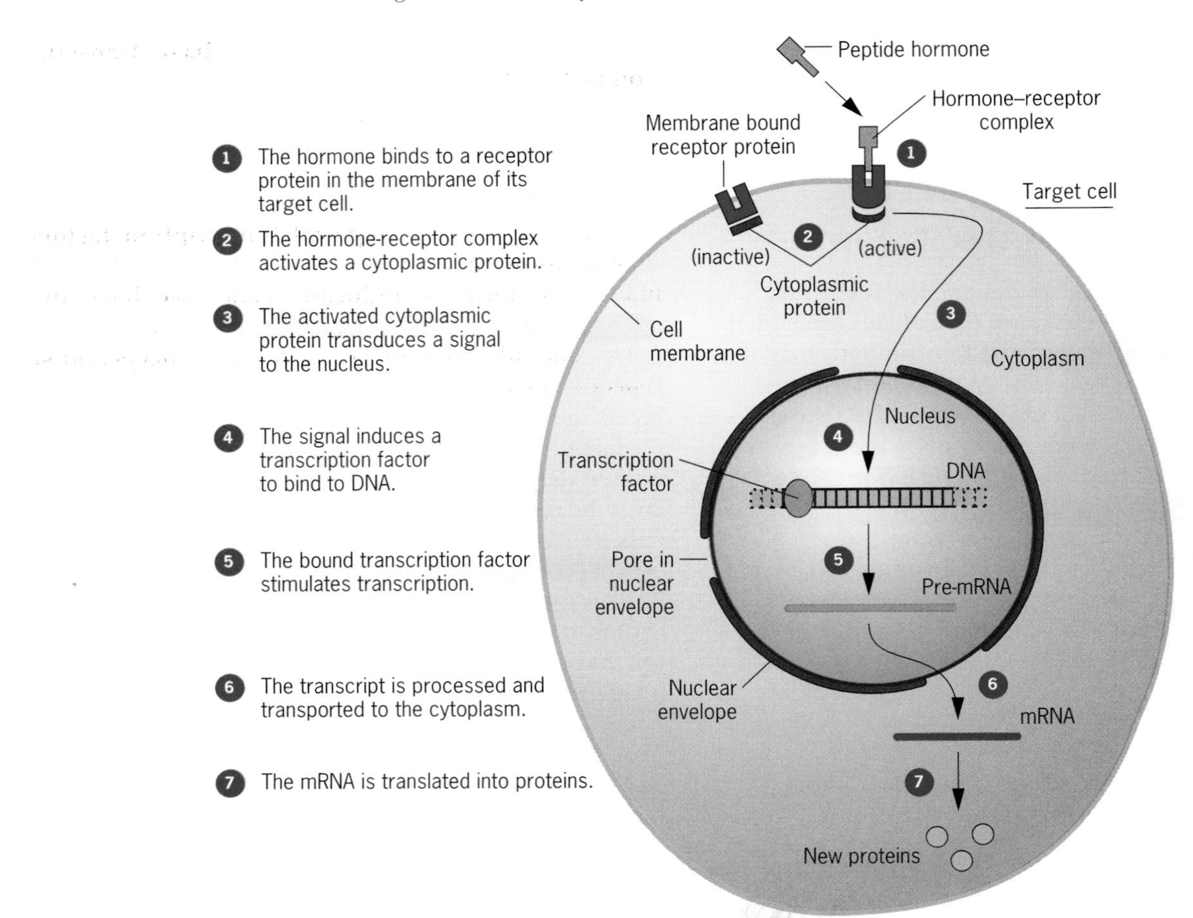

Figure 22.9 Regulation of gene expression by peptide hormones. The hormone (an extracellular signal) interacts with a receptor in the membrane of its target cell. The resulting hormone-receptor complex activates a cytoplasmic protein that triggers a cascade of intracellular changes. These changes transmit the signal into the nucleus, where a transcription factor stimulates the expression of particular genes.

hormonal signal is transmitted through the cytoplasm of the cell and into the nucleus, where it ultimately has the effect of regulating the expression of specific genes. This process of transmitting the hormonal signal through the cell and into the nucleus is called **signal transduction**.

Hormone-induced gene expression is mediated by specific sequences in the DNA. These sequences, called **hormone response elements (HREs)**, are analogous to the heat-shock response elements discussed earlier. They are situated near the genes they regulate and serve to bind specific proteins, which then act as transcription factors. With steroid hormones such as estrogen, the HREs are bound by the hormone-receptor complex, which then stimulates transcription. The vigor of this transcriptional response depends on the number of HREs present. When there are multiple response elements, hormone-receptor complexes bind cooperatively with each other, significantly increasing the rate of transcription; that is, a gene with two response elements is transcribed more than twice as vigorously as a gene with only one. With peptide hormones, the receptor usually remains in the cell membrane, even after it has formed a complex with the hormone. The hormonal signal is therefore conveyed to the nucleus by other proteins, some of which bind to sequences near the genes that are regulated by the hormone. These proteins then act as transcription factors to control the expression of the genes.

Transcriptional activity can be induced by many other kinds of proteins that are not hormones in the classical sense, that is, not produced by a particular gland or organ. These include a variety of secreted, circulating molecules such as nerve growth factor, epidermal growth factor and platelet-derived growth factor, and other noncirculating molecules associated with cell surfaces or with the matrix between cells. Although each of these proteins has its own peculiarities, the general mechanism whereby they induce transcription resembles that of the peptide hormones. An interaction between the signaling protein and a membrane-bound receptor initiates a chain of events inside the cell that ultimately results in specific transcription factors binding to particular genes, which are then transcribed.

Key Points: **Eukaryotic gene expression can be induced by environmental factors such as heat and light, and by signaling molecules such as hormones and growth factors. Hormone-induced gene expression is mediated by proteins that interact with the hormones. Some of these hormone receptors act directly as transcription factors by binding to DNA sequences in the vicinity of a gene; others control transcription indirectly through a signal transduction pathway that targets transcription factors to a gene.**

MOLECULAR CONTROL OF TRANSCRIPTION IN EUKARYOTES

Much of the current research on eukaryotic gene expression is focusing on the factors that control transcription. This heavy emphasis on transcriptional control is partly due to the development of experimental techniques that have permitted this aspect of gene regulation to be analyzed in great detail. However, it is also due to the appeal of ideas that emerged from the study of prokaryotic genes. In both prokaryotes and eukaryotes, transcription is the primary event in gene expression; it is therefore the most fundamental level at which gene expression can be controlled.

DNA Sequences Involved in the Control of Transcription: Enhancers and Silencers

Transcription is initiated in the promoter of a gene, the region recognized by the RNA polymerase. However, as we discussed in Chapter 11, the accurate initiation of transcription from eukaryotic gene promoters requires several accessory proteins, or **basal transcription factors**. Each of these proteins binds to a sequence within the promoter to facilitate the proper alignment of the RNA polymerase on the template strand of the DNA.

The transcription of eukaryotic genes is also controlled by a variety of **special transcription factors**, such as those involved in the regulation of the heat, light, and hormone inducible genes we have discussed. These factors bind to response elements, or, more generally, to sequences called **enhancers** and **silencers**. The term *enhancer* is used when the binding of the factor stimulates transcription, and the term *silencer* is used when the binding represses transcription. Because enhancers seem to be much more common than silencers, we will focus our discussion on them.

Enhancers exhibit three fairly general properties: (1) they act over relatively large distances—up to several thousand base pairs from their regulated gene(s); (2) their influence on gene expression is independent of orientation—they function equally well in either the normal or inverted orientation within the DNA; and (3) their effects are independent of position—they can be located upstream, downstream, or within an intron of a gene and still have a profound effect on the gene's expression. These three characteristics distinguish enhancers from promoters, which are typically located immediately upstream of the gene and which function only in one orientation.

Enhancers can be relatively large, up to several hundred base pairs long. They sometimes contain repeated sequences that have partial regulatory activity by themselves. Most enhancers function in a tissue-specific manner; that is, they stimulate transcription only in certain tissues; in other tissues they are simply ignored. A clear example of this tissue specificity comes from the study of the *yellow* gene in *Drosophila* (Figure 22.10). This gene is responsible for pigmentation in many parts of the body—in the wings, legs, thorax, and abdomen. Wild-type flies show a dark brownish-black pigment in all these structures,

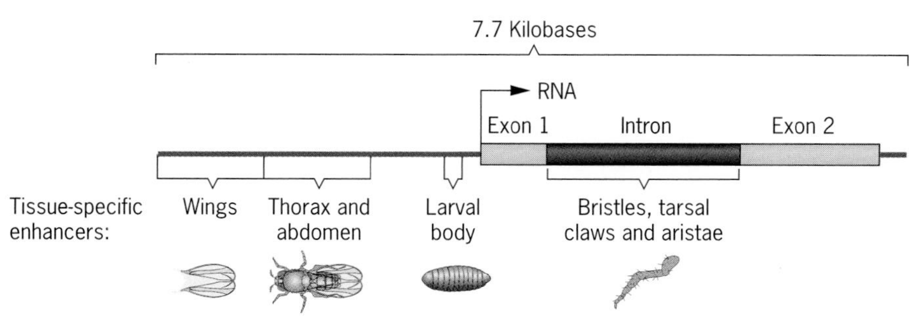

Figure 22.10 The tissue-specific enhancers of the *Drosophila yellow* gene.

whereas mutant flies show a lighter yellowish-brown pigment. However, in some mutants, there is a mosaic pattern of pigmentation, brownish-black in some tissues and yellowish-brown in others. These mosaic patterns are due to mutations that alter the transcription of the *yellow* gene in some tissues but not in others. Pamela Geyer and Victor Corces have shown that the *yellow* gene is regulated by several enhancers, some of which are located within an intron, and that each enhancer activates transcription in a different tissue. If, for example, the enhancer for expression in the wing is mutated, the bristles on the wings are yellowish-brown instead of brownish-black. The battery of enhancers associated with the *yellow* gene allows its expression to be controlled in a tissue-specific way.

One of the first enhancers to be studied extensively is located in the chromosome of the eukaryotic virus SV40 (Chapter 19). This virus infects monkey cells and has been widely used in biological research. Its 5.2-kb circular chromosome contains a single prominent enhancer about 220 base pairs long (Figure 22.11). The enhancer regulates the transcription of two groups of genes on the virus chromosome. One group, situated to the right of the enhancer, is transcribed early during infection, and the other group, situated to the left, is transcribed later. The SV40 enhancer contains two 72-bp direct repeats, either of which is sufficient for enhancer function. It can be inverted or moved to different locations on the SV40 chromosome and still retain its regulatory ability. Furthermore, if it is inserted upstream or downstream from other eukaryotic genes, it stimulates their transcription. These effects are presumably mediated by proteins that bind to the enhancer. Curiously, when SV40 chromosomes are examined with the electron microscope, the enhancer region is not wrapped around nucleosomes (Figure 22.11*c*). A plausible interpretation is that enhancer-binding transcription factors prevent nucleosome formation.

How do enhancers influence the transcription of genes? Although a detailed answer is not known, it appears that the proteins that bind to enhancers influ-

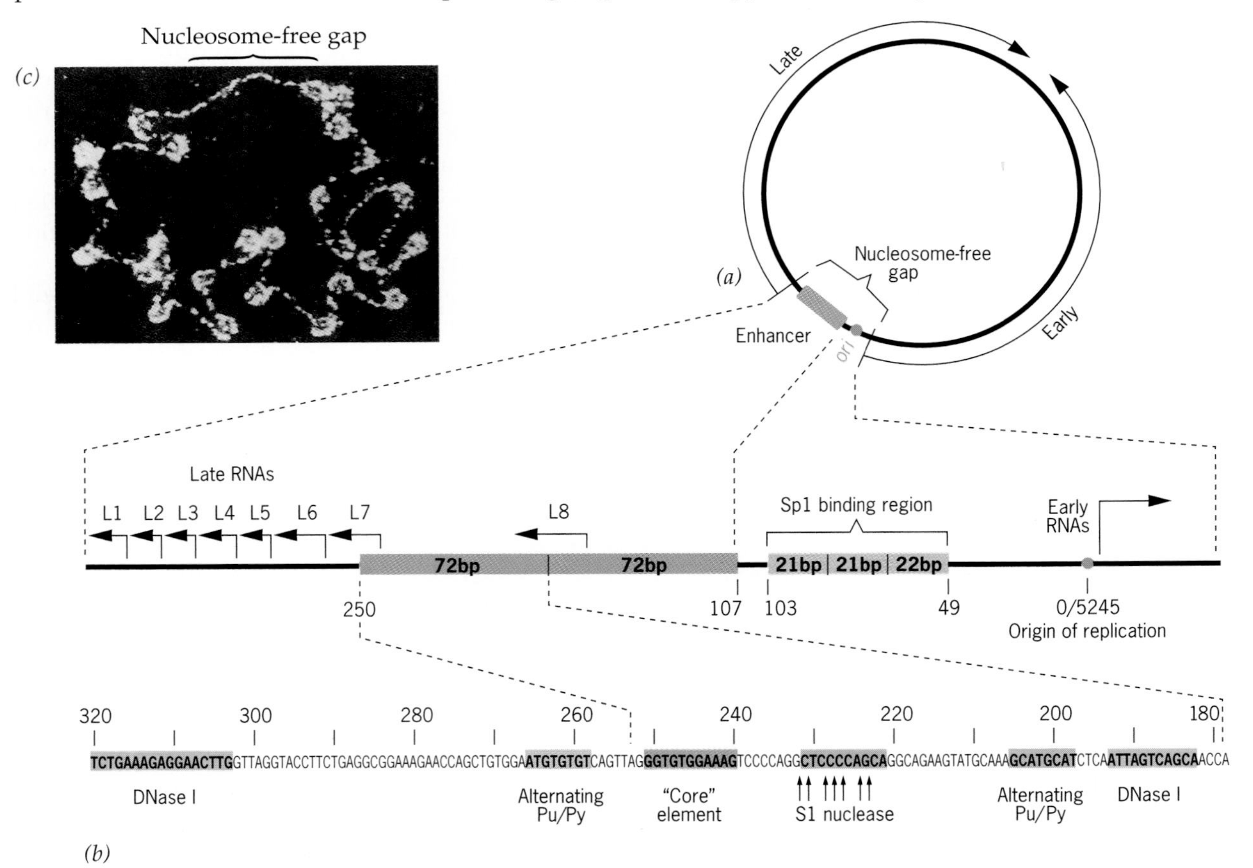

Figure 22.11 Structure of the enhancer of the simian virus 40 (SV40). (*a*) Diagram of the SV40 chromosome showing the location of the enhancer and the origin of replication (*ori*). (*b*) Diagram showing the components of the enhancer. Each of the two 72-base-pair repeats contains regions that are sensitive to DNase I and S1 nuclease digestion, regions with alternating purines (Pu) and pyrimidines (Py) and a core-enhancer element. A 64-base-pair element that binds the transcription factor SP1 is adjacent to one of the repeats. The origin or replication is at coordinate 0/5245 on the circular chromosome. (*c*) Electron micrograph of an SV40 chromosome showing nucleosomes everywhere except in the vicinity of the enhancer.

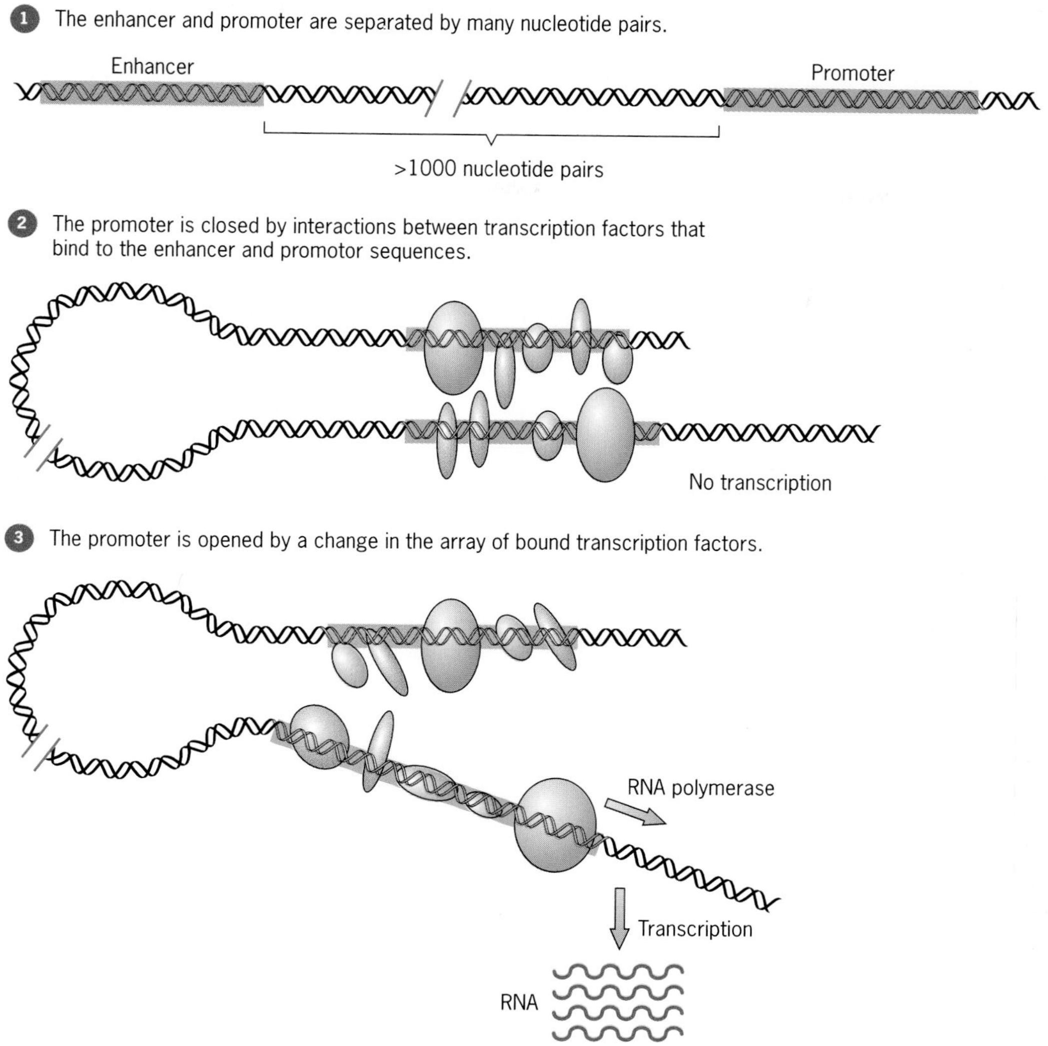

1 The enhancer and promoter are separated by many nucleotide pairs.

Enhancer

Promoter

>1000 nucleotide pairs

2 The promoter is closed by interactions between transcription factors that bind to the enhancer and promotor sequences.

No transcription

3 The promoter is opened by a change in the array of bound transcription factors.

RNA polymerase

Transcription

RNA

Figure 22.12 Model for the control of transcription by proteins bound to an enhancer and a promoter.

ence the activity of the proteins that bind to promoters (Figure 22.12). This finding suggests that the two groups of proteins are in physical contact, perhaps because the enhancer and promoter elements are brought together by some sort of DNA folding. Models of enhancer action therefore resemble the model for the regulation of the *arabinose* operon in *E. coli* (Chapter 21), except that more regulatory proteins are involved.

Proteins Involved in the Control of Transcription: Transcription Factors

Recent research has identified a large number of eukaryotic proteins that stimulate transcription. Many of these proteins appear to have at least two important chemical domains: a DNA-binding domain and a transcriptional activation domain. These domains may occupy separate parts of the molecule, or they may be overlapping. In the GAL4 transcription factor from yeast (see Technical Sidelight: GAL4, a Transcription Factor that Regulates the Genes Involved in Galactose Metabolism in Yeast), for example, the DNA-binding domain is situated near the amino terminus of the polypeptide; two transcriptional activation domains are present in this polypeptide, one more or less in the middle and one near the carboxy terminus. In the steroid hormone receptor proteins, which are transcription factors in animals, the DNA-binding domain is centrally located and seems to overlap a transcriptional activation domain that extends toward the amino terminus; steroid hormone receptors also have a third domain that specifically binds the steroid hormone.

TECHNICAL SIDELIGHT

GAL4, a Transcription Factor That Regulates the Genes Involved in Galactose Metabolism in Yeast

Yeast cells growing in a medium that contains galactose but not glucose specifically express several genes whose products are involved in galactose metabolism. Two of these genes, *GAL1* and *GAL10*, are less than a kilobase apart on one of the yeast chromosomes (Figure 1). The transcription of these genes is controlled by a sequence located between them. This sequence, called the Upstream Activating Sequence, or UAS, is the binding site for a protein that activates transcription from the *GAL1* and *GAL10* promoters. Because these promoters are oriented in opposite directions, the *GAL1* and *GAL10* genes are divergently transcribed.

The protein that binds to the UAS between *GAL1* and *GAL10* is encoded by an unlinked gene, *GAL4*. This gene is constitutively transcribed in yeast cells. In the presence of galactose, the GAL4 protein is able to activate the transcription of *GAL1* and *GAL10*, but in the absence of galactose, the GAL4 protein cannot activate the transcription of these genes. Thus galactose can be considered an inducer of *GAL1* and *GAL10* gene expression—through an effect of the GAL4 protein.

The GAL4 protein is 881 amino acids long. Experiments with truncated GAL4 proteins have shown that the first 73 amino acids are sufficient for binding to the UAS. This portion of the protein contains a zinc finger motif, a structural feature frequently associated with the ability to bind DNA. Once the GAL4 protein has bound to DNA, it can activate the transcription of nearby genes. Studies with portions of the GAL4 protein have shown that this activating property resides in two separate domains. One spans amino acids 148–196, and the other spans amino acids 768–881. Transcriptional activation is thought to involve contact between these parts of the GAL4 protein and other proteins that have bound to the gene's promoter, including, possibly, RNA polymerase II.

The UAS located between the *GAL1* and *GAL10* genes resembles the enhancers found in higher eukaryotes. It functions in either orientation and at some distance from the genes it regulates; however, unlike an enhancer, it cannot function downstream of a gene's transcription initiation site. Molecular analysis has revealed that the UAS contains four GAL4 binding sites. Each is an imperfect palindrome of eight base pairs centered on an A:T base pair.

When the UAS is artificially inserted upstream of a gene, that gene comes under the control of the GAL4 protein. This phenomenon was first demonstrated by Hitoshi Kakidani and Mark Ptashne in experiments with Chinese hamster cells that were transfected with two plasmids, one carrying a version of the *GAL4* gene and the other carrying a "reporter" gene located downstream of the UAS. The reporter gene encoded an enzyme (chloramphenicol acetyltransferase, CAT) for which a simple biochemical assay was available. Kakidani and Ptashne set out to determine whether the GAL4 protein expressed from one of the plasmids could stimulate the expression of the *CAT* reporter gene on the other plasmid by binding to the UAS near it. Their results (Figure 2) showed that plasmids carrying an intact *GAL4* gene, or a *GAL4* gene encoding the UAS-binding domain and at least one of the transcriptional activation domains of the protein, were able to stimulate expression of the *CAT* reporter gene. However, a plasmid with a *GAL4* gene encoding only the UAS-binding domain could not stimulate *CAT* gene expression. Thus both the UAS-binding and transcriptional activation domains of the GAL4 protein are necessary for biological function.

The discovery that the GAL4 protein can regulate the expression of a gene near a UAS has become the basis of a procedure to study the effects of eukaryotic gene expression *in vivo*. In this procedure, a UAS is inserted upstream of a

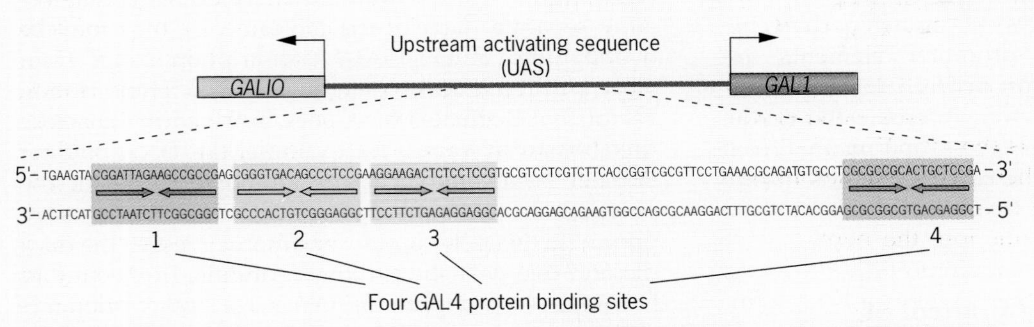

Figure 1 The *GAL1* and *GAL10* genes in yeast. The GAL4 protein coordinately regulates these divergently transcribed genes by binding to four sites within the Upstream Activating Sequence (UAS) between them.

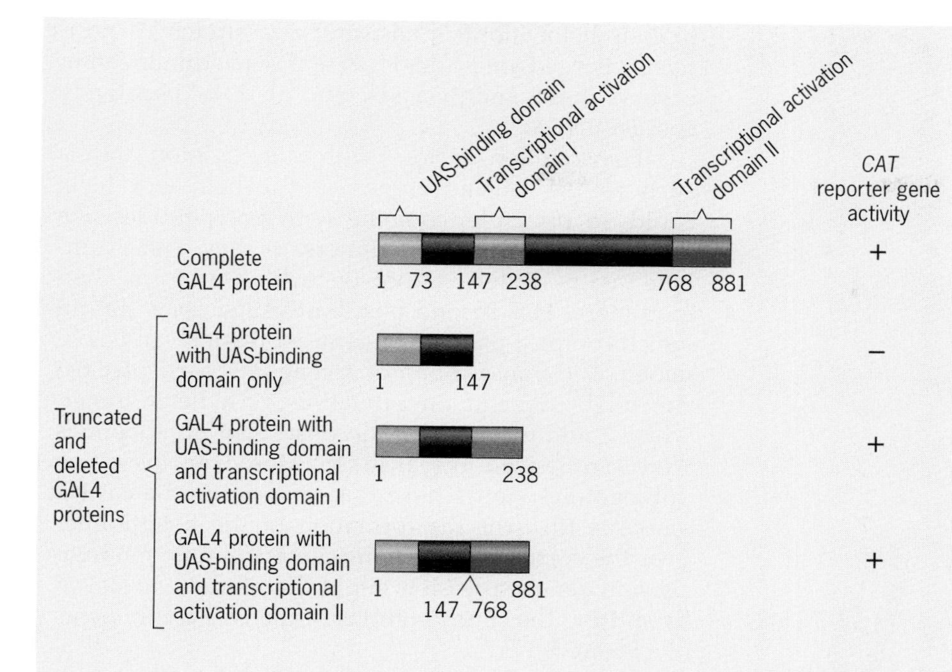

Figure 2 Results of the experiments of Kakidani and Ptashne on GAL4 regulation of a *CAT* reporter gene fused to a UAS. The complete GAL4 protein and two deleted proteins are capable of stimulating expression of the reporter gene. All three of these GAL4 proteins contain the UAS-binding domain and at least one of the transcriptional activation domains. A truncated GAL4 protein containing only the UAS-binding domain does not stimulate reporter gene expression.

gene's transcription initiation site, and the resulting fusion gene is introduced into the cells of an organism—either transiently by transfection or permanently by transformation. A functional *GAL4* gene is then introduced into the cells to stimulate the expression of the fusion gene. This procedure provides a way to induce the expression of the fusion gene and thereby to ascertain the significance of that gene's function in a particular population of cells.

Although the detailed mechanism of transcriptional activation is not fully understood, it appears to involve physical interactions between proteins. A transcription factor that has bound to an enhancer may make contact with one or more proteins at other enhancers, or it may interact directly with proteins that have bound in the promoter region. Through these contacts and interactions, the transcriptional activation domain of the factor may then induce conformational changes in the assembled proteins, paving the way for the RNA polymerase to initiate transcription.

Many eukaryotic transcription factors have characteristic structural motifs that result from associations between amino acids within their polypeptide chains. One of these motifs is the **zinc finger**, a short peptide loop that forms when two cysteines in one part of the polypeptide and two histidines in another part nearby jointly bind a zinc ion; the peptide segment between the two pairs of amino acids then juts out from the main body of the protein as a kind of finger (Figure 22.13*a*). Mutational analysis has demonstrated that these fingers play important roles in DNA binding.

A second motif in many transcription factors is the **helix-turn-helix**, a stretch of three short helices of amino acids separated from each other by turns (Figure 22.13*b*). Genetic and biochemical analyses have shown that the helical segment closest to the carboxy terminus is required for DNA binding; the other helices seem to be involved in the formation of protein dimers. In many transcription factors, the helix-turn-helix motif coincides with a highly conserved region of approximately 60 amino acids called the **homeodomain**, so named because it occurs in proteins encoded by the homeotic genes of *Drosophila*. Classical analysis has demonstrated that mutations in these genes alter the developmental fates of groups of cells (Chapter 23). Thus, for example, mutations in the *Antennapedia* gene can cause antennae to develop as legs. This bizarre phenotype is an example of a homeotic transformation—the substitution of one body part for another during the developmental process. Molecular analyses of the homeotic genes in *Drosophila* have demonstrated that each encodes a protein with a homeodomain and that these proteins can bind to DNA. The homeodomain proteins stimulate the transcription of particular genes in a spatially and temporally specific manner during development. Homeodomain proteins have also been identified in other organisms, including human beings, where they may play important roles as transcription factors.

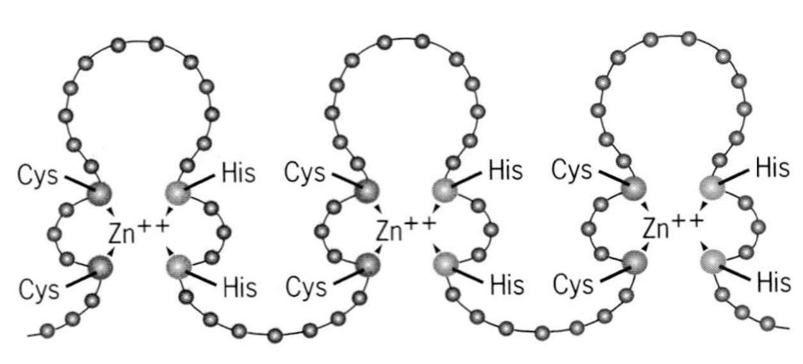

(a) Zinc finger motif.

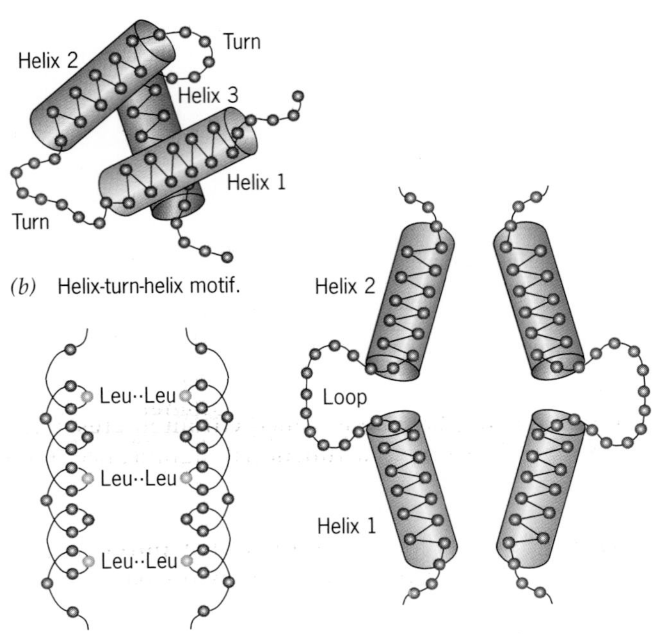

(b) Helix-turn-helix motif.

(c) Leucine zipper motif. (d) Helix-loop-helix motif.

Figure 22.13 Structural motifs within different types of transcription factors. (*a*) Zinc-finger motifs in the mammalian transcription factor SP1. (*b*) Helix-turn-helix motif in a homeodomain transcription factor. (*c*) A leucine zipper motif that allows two polypeptides to dimerize and then bind to DNA. (*d*) A helix-loop-helix motif that allows two polypeptides to dimerize and then bind to DNA.

A third structural motif found in transcription factors is the **leucine zipper,** a stretch of amino acids with a leucine at every seventh position (Figure 22.13*c*). Polypeptides with this feature can form dimers by interactions between the leucines in each of their zipper regions. Usually, the zipper sequence is adjacent to a positively charged stretch of amino acids. When two zippers interact, these charged regions splay out in opposite directions, forming a surface that can bind to negatively charged DNA.

A fourth structural motif found in some transcription factors is the **helix-loop-helix**, a stretch of two helical regions of amino acids separated by a non-helical loop (Figure 22.13*d*). The helical regions permit dimerization between two polypeptides. Sometimes the helix-loop-helix motif is adjacent to a stretch of positively charged amino acids, so that when dimerization occurs, these amino acids can bind to negatively charged DNA.

Transcription factors with dimerization motifs such as the leucine zipper or the helix-loop-helix could, in principle, combine with polypeptides like themselves to form homodimers, or they could combine with different polypeptides to form heterodimers. This second possibility suggests a way in which complex patterns of gene expression can be obtained. The transcription of a gene in a particular tissue might depend on activation by a heterodimer, which could form only if its constituent polypeptides were synthesized in that tissue. Moreover, these two polypeptides would have to be present in the correct amounts to favor the formation of the heterodimer over the corresponding homodimers. Subtle modulations in gene expression might therefore be achieved by shifting the concentrations of the two components of a heterodimer.

Key Points: **Enhancers and silencers act in an orientation-independent manner over considerable distances to regulate transcription from a gene's promoter. This regulation is mediated by different types of transcription factor proteins that bind to these DNA sequences. Different motifs have been identified in the amino acid sequences of these proteins.**

GENE EXPRESSION AND CHROMOSOME ORGANIZATION

The primary event in gene expression is transcription of DNA into RNA. For this to occur, the DNA must be accessible to RNA polymerase and an assortment of other proteins that help to initiate transcription. If the DNA is tightly bound by histones or other kinds of "packaging" proteins, it may be too condensed to allow transcription. Chromosomal position may also affect the transcription of genes because genes that have been transposed to different sites within a chromosome often show altered expression. These alterations may be caused by enhancers at the new chromosomal site or by grosser aspects of chromosome structure. All these phenomena indicate that gene expression is influenced by chromosome organization.

Transcription in Lampbrush Chromosome Loops

Few if any of the genes in highly condensed chromosomes, such as those found at metaphase of mitosis or

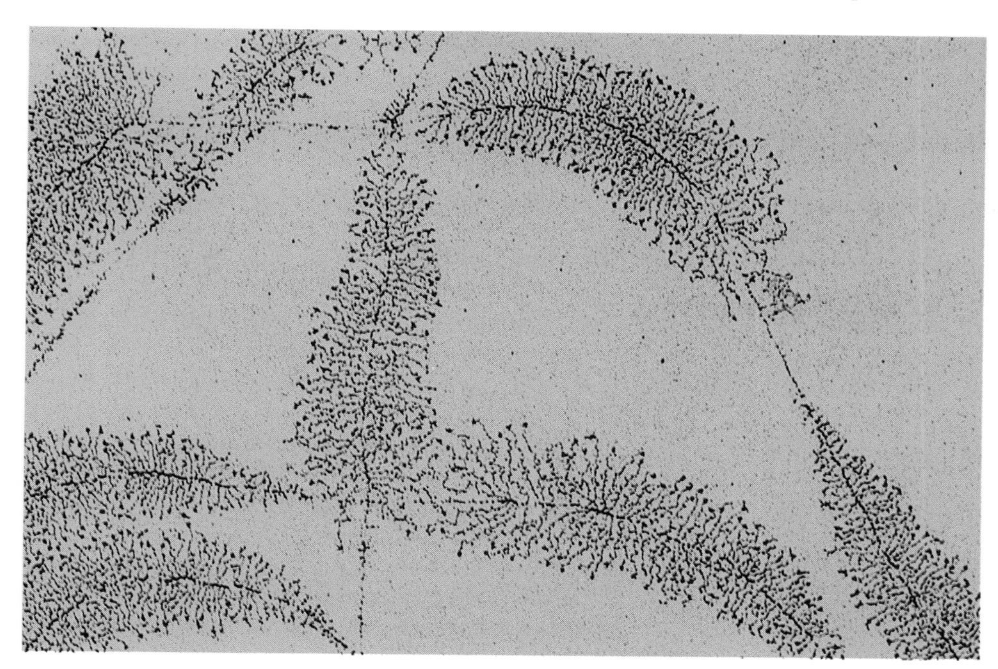

Figure 22.14 Photomicrograph showing transcription from the DNA loops of amphibian lampbrush chromosomes.

in mature sperm, are transcribed. In order to be transcribed, the chromatin must be "open" to the transcriptional apparatus. Some of the first evidence supporting this idea came from the cytological localization of transcription in the meiotic chromosomes of amphibian oocytes. These very large (400 to 800 μm long), duplicated chromosomes—called lampbrush chromosomes—consist of a highly condensed axis surrounded by numerous pairs of lateral loops (Chapter 9).

The lateral loops in lampbrush chromosomes are regions of intense transcriptional activity (Figure 22.14). This has been demonstrated by pulse-labeling amphibian oocytes with ^{3}H-uridine and then examining the chromosomes by autoradiography. The ^{3}H-uridine is incorporated into newly synthesized RNA. Autoradiography of pulse-labeled oocytes reveals that the radioactive uridine is localized around the lateral loops of the lampbrush chromosomes rather than around the condensed axes; thus the more loosely organized loops are actively involved in RNA synthesis.

Transcription in Polytene Chromosome Puffs

Additional cytological evidence for transcription in "open" chromosome regions comes from the study of polytene chromosomes in *Drosophila* and other Dipteran insects. As discussed in Chapter 6, these chromosomes consist of hundreds of sister chromatids aligned side by side, creating a large, thick cable that is longitudinaly differentiated into light and dark bands. For many years, the bands, called **chromomeres**, have been thought to have a functional significance. One reason for this speculation is that during the course of development, particular bands expand into diffuse, less densely staining structures called **puffs** (Figure 22.15). *In situ* hybridization experiments have shown that puffs contain genes that are actively transcribed. In these experiments, a radioactively–labeled RNA or DNA that contains a specific sequence is denatured and hybridized with denatured RNA or DNA in tissues that have been prepared for cytological analysis. The labeled RNA or DNA probe binds to its complementary sequence in the tissues, that is, it binds *in situ*; the location of the bound probe is then revealed by

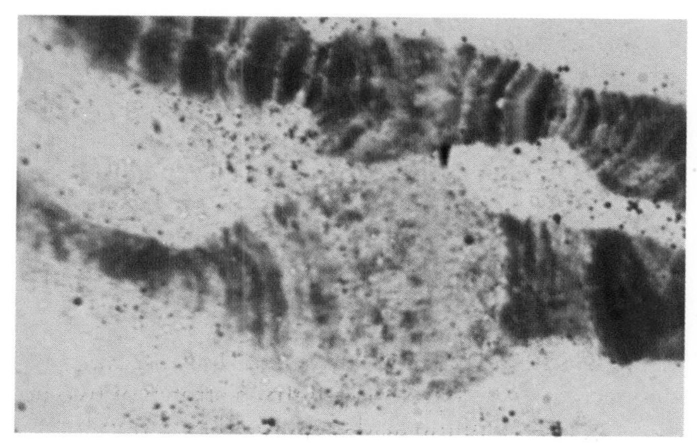

Figure 22.15 Puffs in the polytene chromosomes of *Drosophila*.

autoradiography. Application of this technique shows that puffs in polytene chromosomes contain DNA sequences that are actively transcribed into RNA. Over time, particular puffs regress and new ones form at different sites in the chromosomes. This temporal sequence of puffing is controlled by **ecdysone**, a steroid hormone that appears in pulses at different times during development. A plausible interpretation is that ecdysone induces transcription in puffed bands. Puffing may also be induced by other factors, for example, heat shock. When *Drosophila* larvae are incubated at temperatures greater than 33°C, several new puffs appear in the polytene chromosomes. These new puffs correspond to the loci for the heat-shock genes. These cytological studies with *Drosophila* therefore show that diffuse, expanded chromosomal regions are sites of intense RNA synthesis.

Molecular Organization of Transcriptionally Active DNA

The findings with lampbrush and polytene chromosomes raise questions about the molecular organization of transcriptionally active DNA. Is this DNA packaged in nucleosomes? If it is, what structural changes occur in the nucleosomes during transcription? Are the nucleosomes "opened" and "closed" as the RNA polymerase passes along the DNA template? Are there sites in the DNA that define transcriptionally active domains such as lampbrush chromosome loops and polytene chromosome puffs?

Various techniques have been used to answer these questions, including electron microscopy, enzymatic digestion of DNA, and biochemical analysis of chromosomal proteins. The application of these techniques has shown that DNA that is being transcribed is indeed packaged into nucleosomes, and that it has the same frequency and spacing of core particles as nontranscribed DNA. However, the fact that transcribed DNA is considerably more sensitive to degradation by enzymes such as pancreatic deoxyribonuclease I (DNase I) suggests that it is more "open" than nontranscribed DNA.

This increased sensitivity to DNase was demonstrated in 1976 by M. Groudine and H. Weintraub in a comparative study of transcribed and nontranscribed genes in chicken red blood cells. Groudine and Weintraub extracted chromatin from these cells and partially digested it with DNase I. They then probed the residual material for sequences of two genes, β-globin, which is actively transcribed in red blood cells, and ovalbumin, which is not. They found that over 50 percent of the β-globin DNA had been digested by the DNase I enzyme, compared with only 10 percent of the ovalbumin DNA. The results strongly implied that the actively transcribed gene was more "open" to nuclease attack. Subsequent research has shown that the nuclease sensitivity of transcriptionally active genes depends on at least two small nonhistone proteins, HMG14 and HMG17 (HMG for high mobility group, because they have high mobility during gel electrophoresis). When these proteins are removed from active chromatin, nuclease sensitivity is lost; when they are again added, it is restored.

The treatment of isolated chromatin with a very low concentration of DNase I causes the DNA to be cleaved at a few specific sites, appropriately called **DNase I hypersensitive sites**. Some of these sites have been shown to lie upstream of transcriptionally active genes, either in promoter or enhancer regions. The functional significance of these hypersensitive sites is still unclear, but some evidence suggests that they may mark regions in which the DNA is locally unwound perhaps because transcription has begun.

Are chromosomes organized into functional domains by special sites in the DNA? Some evidence from the study of *Drosophila* heat-shock genes suggests that the answer is yes. Paul Schedl, Rebecca Kellum, and colleagues have identified short elements in the DNA on either side of a pair of *hsp70* genes situated in polytene chromosome band 87A7 (Figure 22.16). These elements, called **specialized chromatin structures** (denoted **scs** and **scs'**), are themselves relatively impervious to DNase I digestion, but each is flanked by DNase I hypersensitive sites. The scs and scs' elements apparently insulate the *hsp70* genes within the 87A7 chromomere from the surrounding chromatin. This insulating property was demonstrated when these elements were artificially placed on either side of another gene, and the resulting scs–gene–scs' complex was tested for function in different genomic positions. As hypothesized, the "sandwiched" gene was expressed uniformly in a wide variety of chromosomal sites; thus it could function independently of the surrounding chromatin. However, when either of the bordering elements was eliminated, the gene's function became subject to the effects of chromosomal position. These results therefore suggest that chromosomes may be organized into functional domains by short DNA elements.

Euchromatin and Heterochromatin

In most species, the chromosomes of interphase cells cannot readily be identified. However, variation in the density of chromatin within the nuclei of these cells causes certain regions to stain deeply when dyes such as the Feulgen reagent are applied (Chapter 6). This deeply staining material is called **heterochromatin**, and its lightly staining counterpart is called **euchro-**

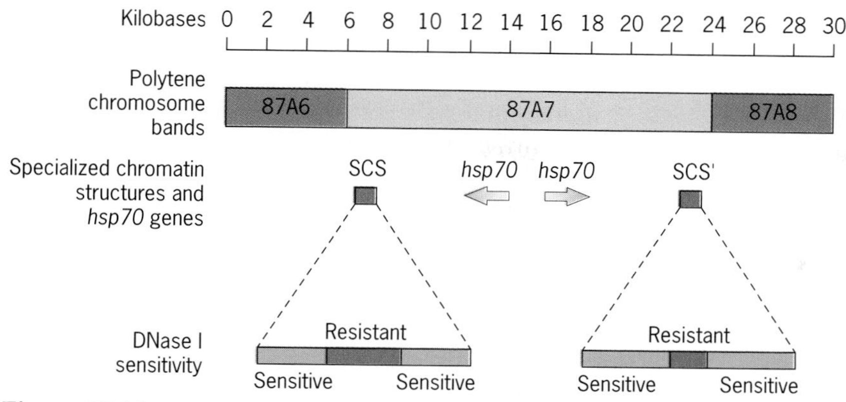

Figure 22.16 Location of the specialized chromatin structures (scs and scs') in the vicinity of the *hsp70* genes in band 87A7 of chromosome 3 in *Drosophila*. The sites of DNase I sensitivity that flank the structures are indicated.

matin. What, if any, is the functional significance of this difference in chromatin staining?

A combination of genetic and molecular analyses has shown that the vast majority of eukaryotic genes are located in euchromatin. Moreover, when euchromatic genes are artificially transposed to a heterochromatic environment, they tend to function abnormally, and in some cases, not to function at all. This impaired ability to function can create a mixture of normal and mutant characteristics in the same individual, a condition referred to as **position-effect variegation**. This term is used because the variability in the phenotype is caused by changing the position of the euchromatic gene, specifically by relocating it to the heterochromatin. Many examples of position–effect variegation have been discovered in *Drosophila*, usually in association with inversions or translocations that move a euchromatic gene into the heterochromatin. The *white* mottled mutation discussed in Chapter 6 is a good example. In this case, a wild-type allele of the *white* gene has been relocated by an inversion, with one break near the *white* locus and the other in the basal heterochromatin of the X chromosome. This rearrangement interferes with the normal expression of the *white* gene and causes a mottled-eye phenotype. Apparently, the euchromatic *white* gene cannot function well in a heterochromatic environment. This and other examples have led to the view that heterochromatin represses gene function, perhaps because it is condensed into a form that is not accessible to the transcriptional machinery. Ongoing research is attempting to identify the proteins that might be involved in this condensation process.

Gene Amplification

Sometimes, gene expression is facilitated by an increase in gene number. This process, called **gene amplification**, is clearly designed to increase the number of DNA templates for RNA synthesis. Perhaps the most dramatic example of expression-related gene amplification involves the ribosomal RNA genes in amphibian oocytes. These genes are needed to produce structural components of the ribosomes. In eukaryotes, there are three main types of rRNA genes: the 5S gene, which encodes a 120-base rRNA; the 18S gene, which encodes a 1.8-kb rRNA; and the 28 S gene, which encodes a 4.7-kb rRNA. In the genome of the toad *Xenopus laevis*, thousands of 5S genes are scattered over all the chromosomes, but there are far fewer copies of the 18S and 28S genes, perhaps only 800 to 1000 per diploid cell. These latter genes are all concentrated at a single site called the **nucleolar organizer**, so named because the nucleolus attaches to it. Molecular mapping has shown that the 18S and 28S rRNA genes belong to a single transcriptional unit that is tandemly repeated within the nucleolar organizer.

All three types of ribosomal RNA genes must be transcribed to produce RNAs for *Xenopus* ribosomes. In oocytes, this transcription must be especially vigorous because the egg needs to accumulate enough ribosomes to support the early development of the embryo after fertilization. Perhaps as many as 10^{12} ribosomes must be synthesized in each oocyte. Because each ribosome contains exactly one molecule of each type of rRNA, 10^{12} molecules of each of the 5S, 18S, and 28S rRNAs must be synthesized. This re-

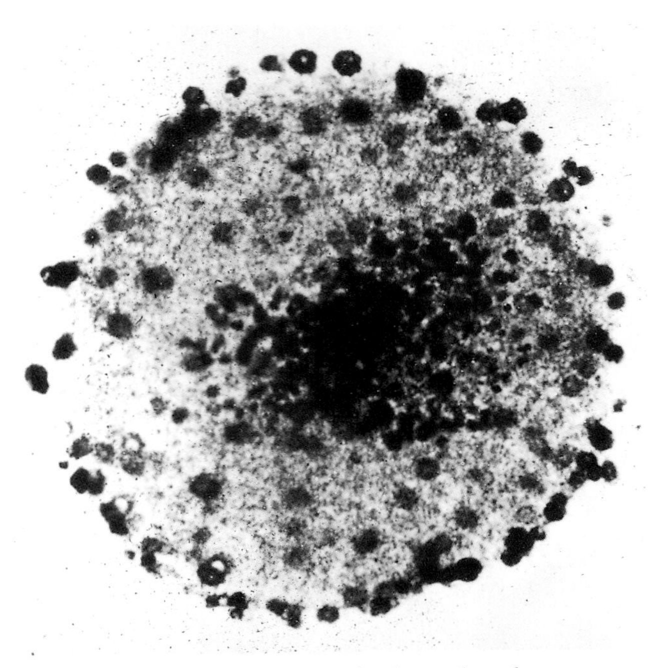

Figure 22.17 Photomicrograph of a nucleus from an oocyte of the toad *Xenopus laevis* showing supernumerary nucleoli filled with ribosomal DNA molecules.

quirement is fulfilled in two ways. First, *Xenopus* has about 24,000 5S genes, including many that are specifically activated in oocytes. This enormous set of genes is therefore able to generate the 5S rRNAs that are needed for ribosome production in the egg. Second, the smaller number of 18S and 28S genes is specifically amplified in oocytes by the creation of extrachromosomal copies of these genes. Small, covalently closed circular DNAs carrying the 18S and 28S genes are formed. These replicate by a rolling-circle mechanism (Chapter 10) to produce many copies, which are gathered into supernumerary nucleoli within the oocyte (Figure 22.17). Transcription from these circular DNAs provides much of the 18S and 28S rRNA that is needed for ribosome assembly in the egg. The mechanism that generates these extrachromosomal DNA molecules is not known. However, it might involve some sort of intrachromosomal recombination between repetitive units within the nucleolar organizer.

Researchers have identified other cases of selective gene amplification in eukaryotes. During oogenesis in *Drosophila*, for example, a group of X-linked genes encoding eggshell proteins is amplified within the follicle cells of the ovary by localized replication of a small part of the X chromosome. This process forms a branching network of DNA duplexes within the single DNA duplex of the chromosome. Transcription from each of the replicated molecules increases the overall production of mRNA from this group of genes, which, in turn, leads to vigorous synthesis of the eggshell proteins. Large quantities of these proteins are needed to form protective shells around *Drosophila* eggs.

In mammals, the gene for dihydrofolate reductase (DHFR), an enzyme required for DNA synthesis, is sometimes amplified when cells are exposed to methotrexate, a chemotherapeutic drug that binds to dihydrofolate reductase and inhibits its activity. This amplification is accomplished either by multiplying the gene within the chromosome to form a tandem array or by creating numerous extrachromosomal copies. The first mechanism visibly increases the size of the chromosome, forming a region within the chromosome that stains uniformly during mitotic metaphase—a **homogeneously staining region (HSR)** (Figure 22.18*a*). The second mechanism generates

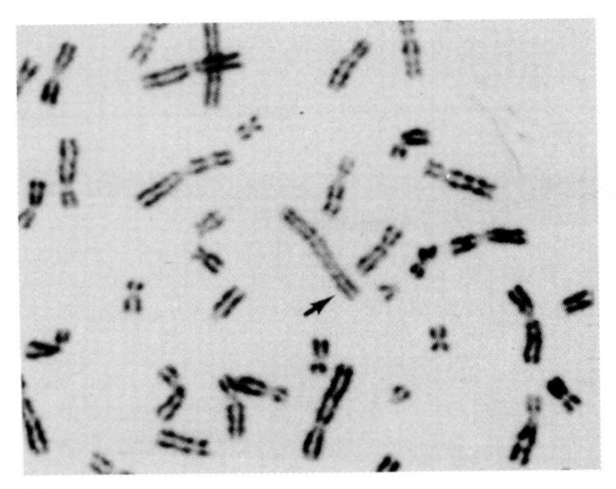

(*a*)

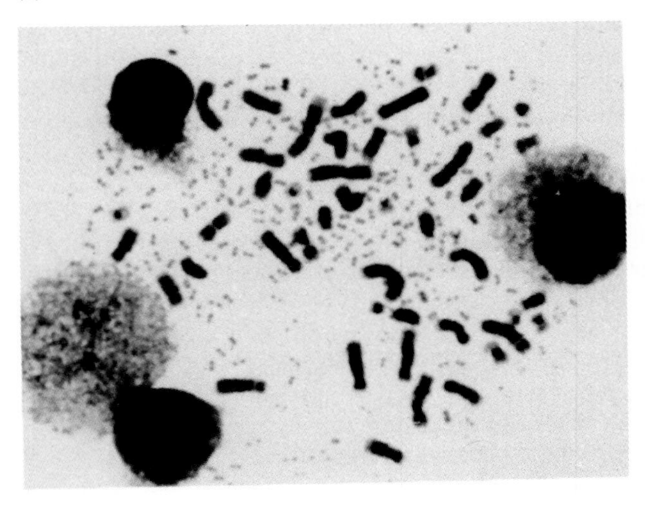

(*b*)

Figure 22.18 Amplification of chromosome segments in human cancer cells. (*a*) Amplified segment of a chromosome (arrow) in a cell line derived from a colon carcinoma. (*b*) Double-minute chromosomes (tiny dots) in a cell from an abdominal solid tumor.

many tiny chromatin fragments that segregate irregularly during cell division—dotlike bodies called **double minutes (DMs)** (Figure 22.18*b*). Amplification of the DHFR gene by either of these mechanisms causes dihydrofolate reductase to be overproduced. Because some of the excess enzyme remains unbound by methotrexate, cells in which the DHFR gene has been amplified are able to replicate their DNA and divide. Thus amplification of the DHFR gene can thwart a physician's attempts to inhibit the spread of a cancer by using methotrexate as a chemotherapeutic drug.

Key Points: **Transcription occurs preferentially in loosely organized chromosome regions exemplified by the loops of lampbrush chromosomes and the puffs of polytene chromosomes. Transcriptionally active DNA tends to be more sensitive to digestion with DNase I and may be delimited by specialized chromatin structures that insulate it from neighboring DNA. Heterochromatin is associated with the repression of transcription. Increased gene expression may be achieved by amplifying the DNA, either within chromosomes or on extrachromosomal DNA molecules.**

ACTIVATION AND INACTIVATION OF WHOLE CHROMOSOMES

Organisms with an XX/XY or XX/XO sex-determination system face the problem of equalizing the activity of X-linked genes in the two sexes. In mammals, this problem is solved by randomly inactivating one of the two X chromosomes in females; each female therefore has the same number of transcriptionally active X-linked genes as a male. In *Drosophila*, neither of the two X chromosomes in a female is inactivated; instead, the genes on the single X chromosome in a male are transcribed more vigorously to bring their output in line with that of the genes on the two X chromosomes in a female. Still another solution to the problem of unequal numbers of X-linked genes has been found in the nematode worm *Caenorhabiditis elegans*. In this organism, XX individuals are hermaphrodites (they function as both male and female) and XO individuals are males. X-linked transcriptional activity is equalized in these two genotypes by partial repression of the genes on both of the X chromosomes in the hermaphrodites. Therefore, mammals, flies, and worms have solved the problem of X-linked gene dosage in different ways (Figure 22.19). In mammals, one of the X chromosomes in females is inactivated, in *Drosophila*, the single X chromosome in males is hyperactivated, and in *C. elegans*, both of the X chromosomes in hermaphrodites are hypoactivated.

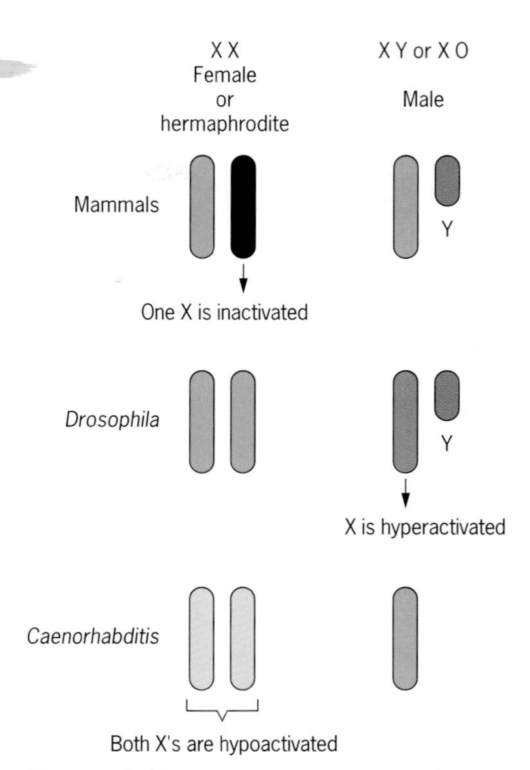

Figure 22.19 Three mechanisms of dosage compensation for X-linked genes: inactivation, hyperactivation, and hypoactivation.

These three different mechanisms of **dosage compensation—inactivation**, **hyperactivation**, and **hypoactivation**—have an important feature in common: many different genes are coordinately regulated because they are on the same chromosome. This chromosomewide regulation is superimposed on all other regulatory mechanisms involved in the spatial and temporal expression of these genes. What might be responsible for such a global regulatory system? For decades, geneticists have been trying to elucidate the molecular basis of dosage compensation. The working hypothesis has been that some factor or factors bind specifically to the X chromosome and alter its transcriptional activities. Recent discoveries indicate that this idea is correct.

Inactivation of X Chromosomes in Mammals

In mammals, X chromosome inactivation begins at a particular site called the **X inactivation center (XIC)** and then spreads in opposite directions toward the ends of the chromosome (Chapter 5). Curiously, not all genes on an inactivated X chromosome are transcriptionally silent. One that remains active is called *XIST* (for X inactive specific transcript); this gene is located within the XIC (Figure 22.20). Recent research

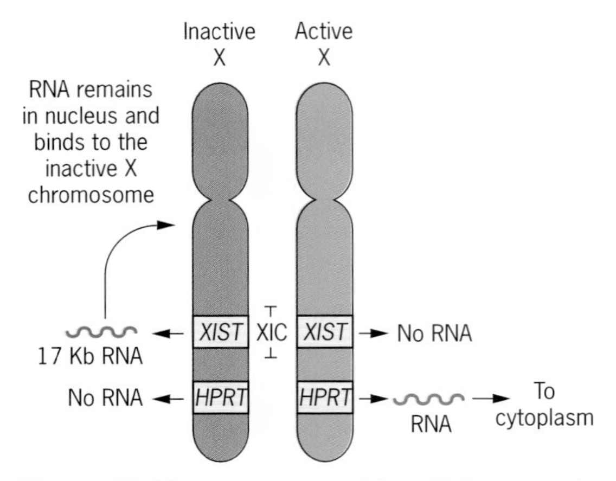

Figure 22.20 Expression of the *XIST* gene in the inactive X chromosome of human females. For comparison, the expression of the *HPRT* gene on the active X chromosome is shown. This gene encodes hypoxanthine phosphoribosyl transferase, an enzyme that plays a role in the metabolism of purines.

has shown that in human beings, the *XIST* gene encodes a 17-kb transcript devoid of any significant open reading frames. It therefore seems unlikely that the *XIST* gene codes for a protein. Instead, the RNA itself is probably the functional product of the *XIST* gene. This RNA is restricted to the nucleus and has been specifically localized to inactivated X chromosomes; it does not appear to be associated with active X chromosomes in either males or females. These observations suggest (but certainly do not prove) that X chromosome inactivation in mammals is caused by the transcription of the *XIST* gene. Perhaps these transcripts remain associated with the inactive X chromosome, repressing the transcription of other genes. In this view, the X chromosome that is not inactivated is the one that represses the transcription of the *XIST* gene. Any X chromosome that does not repress *XIST* would make the *XIST* RNA and inactivate itself.

Inactive X chromosomes are readily identified in mammalian cells. During interphase, they condense into a darkly staining mass associated with the nuclear membrane. This mass, the Barr body (Chapter 5), is an example of **facultative heterochromatin**; that is, heterochromatin that appears and disappears during the cell cycle. In contrast, **constitutive heterochromatin**, such as that found around the centromeres of chromosomes, is present throughout the cell cycle. During S-phase, the Barr body decondenses to allow the inactive X chromosome to be replicated; however, because this process takes some time, the inactive X replicates later than the rest of the chromosomes. Inactive X chromosomes must therefore have a very different chromatin structure than that of other chromosomes. Recent evidence suggests that this difference is partly determined by the kinds of histones associated with the DNA. One of the four core histones, H4, can be chemically modified by the addition of acetyl groups to any of several lysines in the polypeptide chain. Acetylated H4 is associated with all the chromosomes in the human genome. However, on the inactive X it seems to be restricted to three fairly narrow bands, each corresponding to a region that contains some active genes. Acetylated H4 is also depleted in areas of constitutive heterochromatin on the other chromosomes. These findings suggest that acetylated H4 is involved in the maintenance of transcriptional activity.

Hyperactivation of X Chromosomes in *Drosophila*

In *Drosophila*, dosage compensation requires the products of at least four different genes. Null mutations in these genes result in male-specific lethality because the single X chromosome is not hyperactivated. Mutant males usually die during the late larval or early pupal stages. These dosage compensation genes are therefore called male-specific lethal (*msl*) loci. The two that have been cloned encode proteins that bind specifically to the X chromosome in males (Figure 22.21). One of these proteins, the product of the *msl* gene called *maleless* (*mle*), is homologous to DNA and RNA helicases, which are enzymes capable of unwinding nucleic acids. The putative helicase function of the maleless protein is consistent with the idea that the X chromosome in a *Drosophila* male must be "opened" for vigorous transcription. It is also consistent with the observation that polytene X chromosomes in males appear bloated and have a diffuse banding pattern, perhaps because they are hyperactivated. Recent research has also shown that a particular acetylated version of histone H4 is exclusively associated with the X chromosome in males. It therefore appears that in *Drosophila*, dosage compensation depends on the products of the *msl* genes, and that these form some sort of protein complex that hyperactivates X-linked genes by altering chromatin structure.

Hypoactivation of X Chromosomes in *Caenorhabditis*

In *C. elegans*, dosage compensation involves the partial repression of X-linked genes in hermaphrodites. The mechanism is not fully understood, but the product of one gene called *dumpy-27* (*dpy-27*) plays a key role. Like the *msl* proteins in *Drosophila*, the *dpy-27* protein binds specifically to the X chromosome. However, unlike the situation in *Drosophila*, the *dpy-27* protein binds only when two X chromosomes are present. It does not bind to the single X chromosome in XO

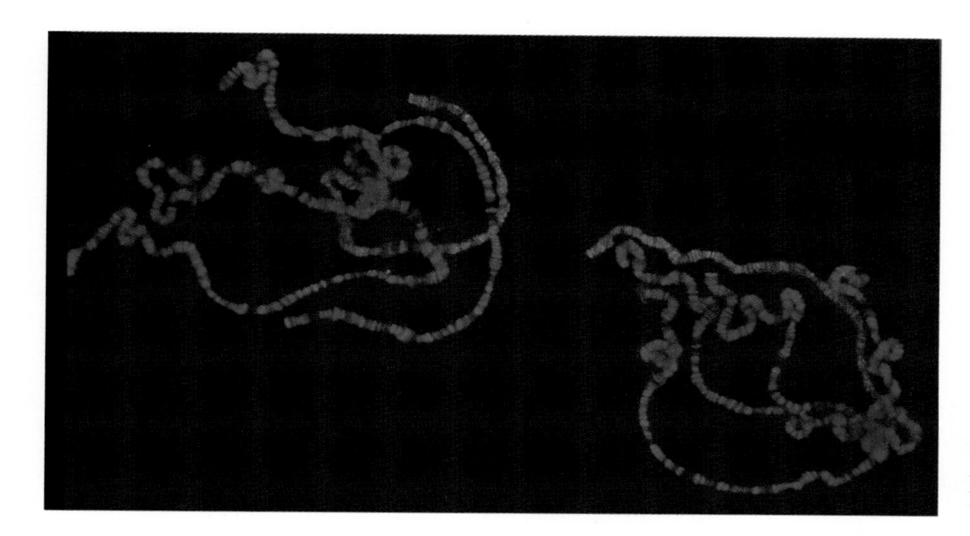

Figure 22.21 Binding of the protein product of the *Drosophila mle* gene to the single X chromosome in males.

males, nor does it bind to any of the autosomes in either males or hermaphrodites. Dosage compensation in *C. elegans* therefore seems to involve a mechanism exactly opposite to the one in *Drosophila*. A protein complex binds to the X chromosomes and represses rather than enhances transcription.

Key Points: **In mammals, dosage compensation is achieved by inactivating one of the X chromosomes in XX females, in *Drosophila* by hyperactivating the single X chromosome in XY males, and in *Caenorhabditis*, by hypoactivating the two X chromosomes in XX hermaphrodites.**

GENE EXPRESSION AND CANCER

Cancer is a disease in which cell growth and division are unregulated. Without regulation, the cells divide ceaselessly, piling up on top of each other to form tumors (Figure 22.22). When cells detach from a tumor and invade the surrounding tissues, the tumor is **malignant**. When the cells do not invade the surrounding tissues, the tumor is **benign**. Malignant tumors may spread to other locations in the body, forming secondary tumors. This process is called **metastasis**, from Greek words meaning to "change state." In both benign and malignant tumors, something has gone wrong with the systems that control cell division. Two decades of research have now firmly established that this loss of control is due to abnormal gene expression.

Tumor-inducing Retroviruses and Viral Oncogenes

Fundamental insights into the genetic basis of cancer have come from the study of tumor-inducing viruses. Among these viruses, many have a genome composed of RNA instead of DNA. After entering a cell, the viral RNA is used as a template to synthesize complementary DNA, which is then inserted at one or more positions in the cell's chromosomes. The synthesis of DNA from RNA is catalyzed by the viral enzyme reverse transcriptase. This reversal of the normal flow of genetic information from DNA to RNA has prompted biologists to call these pathogens retroviruses (Chapter 15).

The first tumor-inducing virus was discovered in 1910 by Peyton Rous; it caused a special kind of tumor, or sarcoma, in the connective tissue of chickens and has since been called the **Rous sarcoma virus** (Figure 22.23). Modern research has shown that the RNA

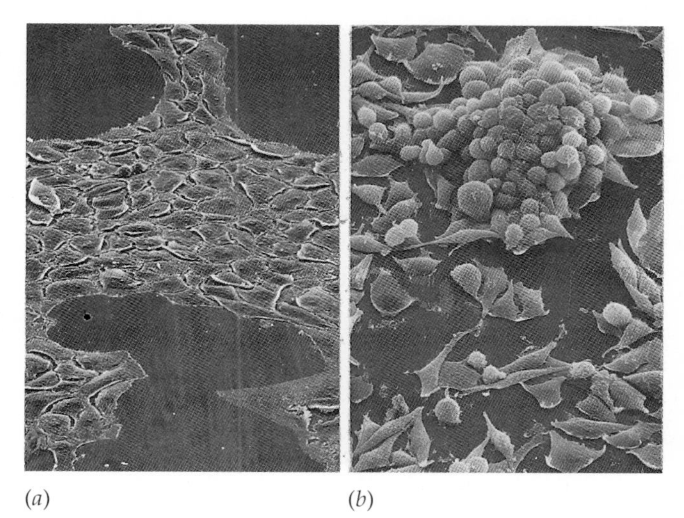

(*a*) (*b*)

Figure 22.22 Scanning electron micrographs showing the features of normal (*a*) and cancerous (*b*) rat kidney cells growing in culture. The normal cells adhere to the surface of the culture dish, forming monolayers of flat cells. The cancerous cells overgrow each other, forming clumps. Magnification: (*a*) 200X; (*b*) 280X.

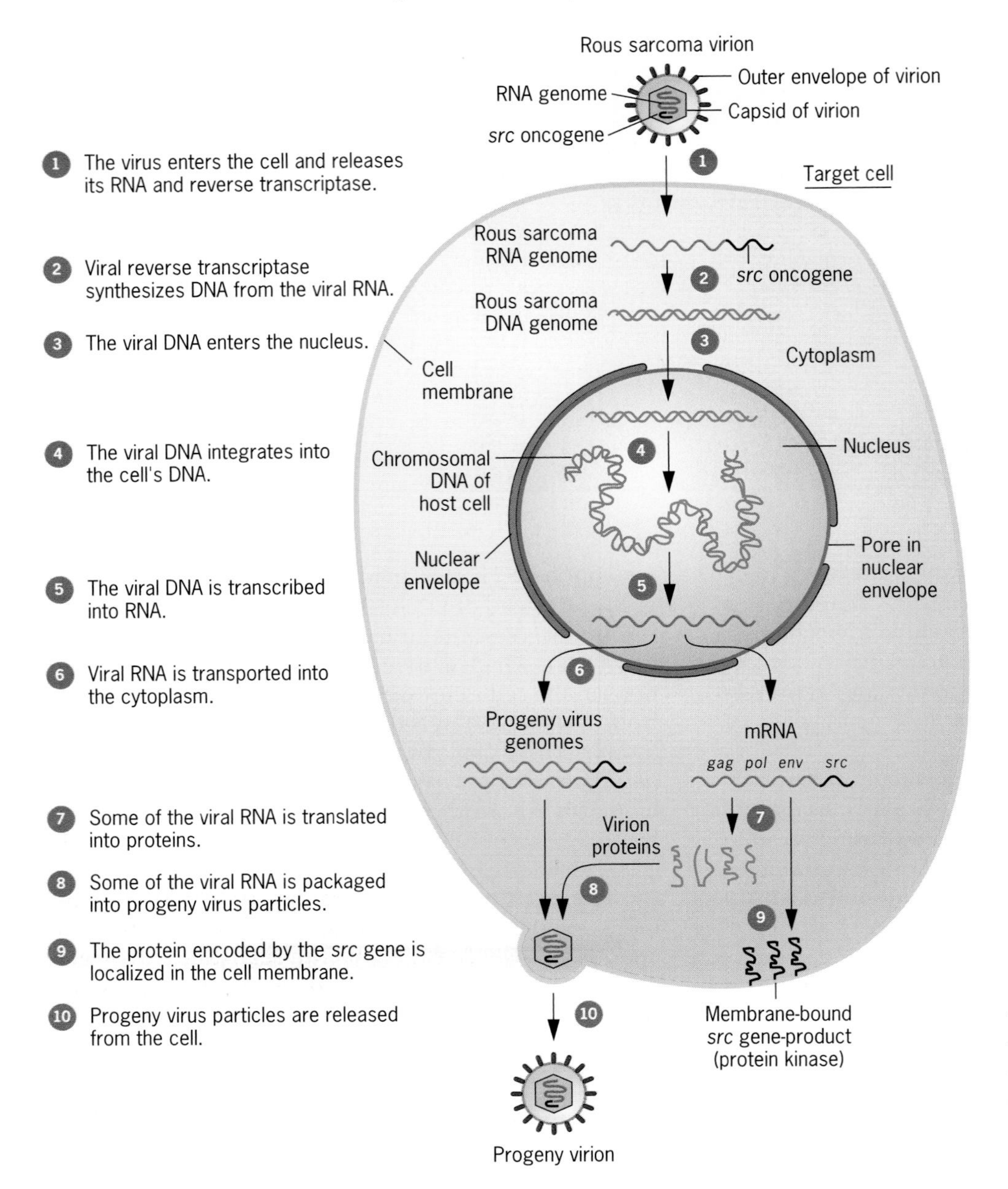

Figure 22.23 Life cycle of the Rous sarcoma RNA tumor virus. The ability of the virus to cause cancer resides in the *src* oncogene (black), which encodes a membrane-bound protein kinase.

genome of this retrovirus contains exactly four genes: *gag*, which encodes the capsid protein of the virion; *pol*, which encodes the reverse transcriptase; *env*, which encodes a protein of the viral envelope; and *v-src*, which encodes a protein kinase that attaches to the plasma membranes of infected cells. The distinguishing feature of a kinase is that it can phosphorylate other proteins. Of these four genes, only one, the *v-src* gene, is responsible for the virus' ability to form tumors. A virus in which the *v-src* gene has been deleted is infectious but unable to induce tumors. Cancer-causing genes such as *v-src* are called **oncogenes** (from the Greek word for "tumor").

Studies with other tumor-inducing retroviruses have uncovered at least 20 different viral oncogenes (usually denoted *v-onc*) (Table 22.1). Some of these are related to genes that encode growth factors. For example, *v-sis*, an oncogene from the simian sarcoma virus, codes for a version of the platelet-derived growth factor (PDGF). PDGF is normally produced by platelet cells to promote wound healing; it does this by stimulating the growth of cells at the wound site. Simian sarcoma viruses carrying the *v-sis* gene induce tumors in monkeys. They also transform cultured cells to a cancerous state, presumably by producing large amounts of the *v-sis* version of PDGF, which then causes uncontrolled cell growth.

Other viral oncogenes encode proteins that are similar to growth-factor and hormone receptors. For example, the *v-erbB* gene from the avian erythroblasto-

TABLE 22.1
Retroviral Oncogenes

Oncogene	*Virus*	*Host Species*	*Function of Gene Product*
abl	Abelson murine leukemia virus	Mouse	Tyrosine-specific protein kinase
erbA	Avian erythroblastosis virus	Chicken	Analog of thyroid hormone receptor
erbB	Avian erythroblastosis virus	Chicken	Truncated version of epidermal growth factor (EGF) receptor
fes	ST feline sarcoma virus	Cat	Tyrosine-specific protein kinase
fgr	Gardner-Rasheed feline sarcoma virus	Cat	Tyrosine-specific protein kinase
fms	McDonough feline sarcoma virus	Cat	Analog of colony stimulating growth factor (CSF-1) receptor
fos	FJB osteosarcoma virus	Mouse	Transcriptional activator protein
fps	Fuginami sarcoma virus	Chicken	Tyrosine-specific protein kinase
jun	Avian sarcoma virus 17	Chicken	Transcriptional activator protein
mil (*mht*)	MH2 virus	Chicken	Serine/threonine protein kinase
mos	Moloney sarcoma virus	Mouse	Serine/threonine protein kinase
myb	Avian myeloblastosis virus	Chicken	Transcription factor
myc	MC29 myelocytomatosis virus	Chicken	Transcription factor
raf	3611 murine sarcoma virus	Mouse	Serine/threonine protein kinase
H-ras	Harvey murine sarcoma virus	Rat	GTP-binding protein
K-ras	Kirsten murine sarcoma virus	Rat	GTP-binding protein
rel	Reticuloendotheliosis virus	Turkey	Transcription factor
ros	URII avian sarcoma virus	Chicken	Tyrosine-specific protein kinase
sis	Simian sarcoma virus	Monkey	Analog of platelet-derived growth factor (PDGF)
src	Rous sarcoma virus	Chicken	Tyrosine-specific protein kinase
yes	Y73 sarcoma virus	Chicken	Tyrosine-specific protein kinase

sis virus encodes a protein much like the cellular receptor for epidermal growth factor, and the *v-fms* gene from the feline sarcoma virus encodes a protein much like the receptor for the cellular CSF-1 growth factor. Both of these growth-factor receptors are transmembrane proteins with a growth-factor-binding domain on the outside of the cell and a protein kinase domain on the inside. This domain on the inside enables the protein to phosphorylate certain amino acids, usually tyrosine, in other proteins that interact with it.

Many viral oncogenes, including *v-src*, encode tyrosine kinases that do not span the plasma membrane. Instead, these proteins are situated on the inner face of the plasma membrane, where they perform their phosphorylating function. Some of the best-studied of the membrane-associated tyrosine kinases are encoded by the *v-ras* oncogenes. These proteins bind GTP and exhibit GTPase activity, much like the cellular G proteins, which play an important role in regulating the level of cyclic AMP.

Another group of viral oncogenes encode proteins that apparently function as transcription factors. These include the *v-jun*, *v-fos*, *v-erbA*, and *v-myc* genes, each carried by a different retrovirus. These proteins are homologous to cellular proteins that bind to DNA and regulate transcription.

Each type of viral oncogene therefore appears to encode a protein that could theoretically play a key role in regulating the expression of cellular genes. Some of these proteins may act as signals to stimulate certain types of cellular activity, for instance, division; others may act as receptors to pick up certain signals, or as intracellular agents to convey those signals from the plasma membrane to the nucleus; yet another category of viral oncogene proteins may act as transcription factors to stimulate gene expression.

Cellular Homologs of Viral Oncognes: The Proto-oncogenes

The proteins encoded by viral oncogenes are similar to cellular proteins with important regulatory functions. Many of these cellular proteins were actually identified by first isolating the cellular homolog of the viral oncogene. For example, the cellular homolog of the *v-src* gene was obtained by screening a genomic DNA library made from uninfected chicken cells. For this screening, the *v-src* gene was used as a hybridization probe to detect recombinant DNA clones that could base-pair with it. Analysis of these clones established that the chicken cells contained a gene that was similar to *v-src*—indeed, that was related to it in an evolutionary sense. However, this gene was not associated with an integrated sarcoma provirus, and it differed from the *v-src* gene in a very important respect: it contained introns (Figure 22.24). There are, in fact, 11 introns in the chicken homolog of *v-src*, compared to zero in the

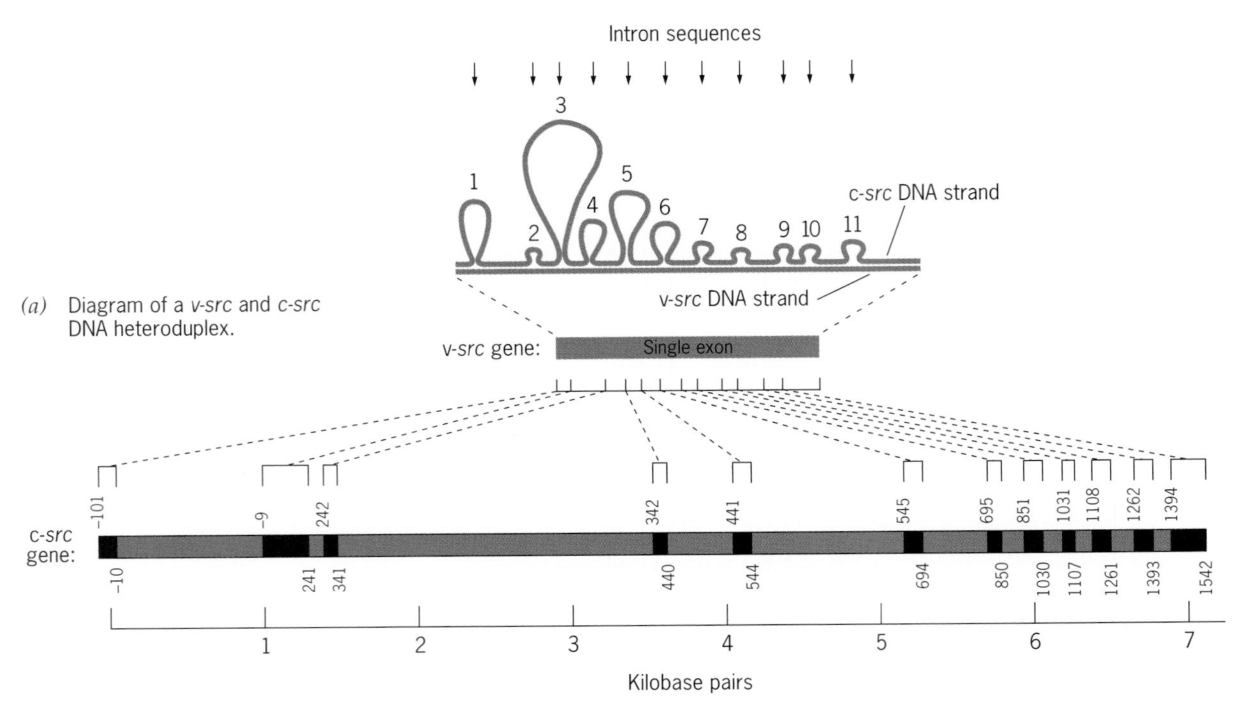

Figure 22.24 Structures of the *v-src* and *c-src* genes. (*a*) Diagram of the predicted DNA heteroduplex formed by hybridization of a strand from the *c-src* gene (top) and a partially complementary strand from the *v-src* gene (bottom). The introns (numbered 1–11) form single-stranded loops. (*b*) Schematic comparison of these two genes, with exons shown in black. The coordinate system for the exons in the *c-src gene* is based on the first nucleotide in the coding sequence (position 1). The first exon (position –101 to –10) is in the 5' leader region of the gene.

v-src gene itself. This startling discovery suggested that perhaps *v-src* had evolved from a normal cellular gene and that, concomitantly, it had lost its introns.

The cellular homologs of viral oncogenes are called **proto-oncogenes**, or more often, **normal cellular oncogenes**, denoted ***c-onc***. The cellular homolog of *v-src* is therefore *c-src*. The coding sequences of these two genes are very similar, differing in only 18 nucleotides; *v-src* encodes a protein of 526 amino acids, and *c-src* encodes a protein of 533 amino acids. Using *v-onc* genes as probes, other *c-onc* genes have been isolated from many different organisms. As a rule, these cellular oncogenes show considerable conservation in structure. *Drosophila*, for example, carries close homologs of the vertebrate cellular oncogenes *c-abl*, *c-erbB*, *c-fps*, *c-raf*, *c-ras*, and *c-myb*. The similarity of oncogenes from different species strongly suggests that the proteins they encode are involved in important cellular functions.

Why do *c-oncs* have introns whereas *v-oncs* do not? The most plausible answer is that *v-oncs* were derived from *c-oncs* by the insertion of a fully processed *c-onc* RNA into the genome of a retrovirus. A virion that packaged such a recombinant molecule would then be able to transduce the *c-onc* gene whenever it infected another cell. During infection, the recombinant RNA would be reverse-transcribed into DNA and then integrated into the cell's chromosomes. What could be of greater value to a virus than to have a new gene that stimulates increased growth of its host, while its integrated genome goes along for the ride? In many cases, the retroviral acquisition of an oncogene has been accompanied by the loss of viral genetic material. Because this lost material is needed for viral replication, these oncogenic viruses are able to reproduce only if a helper virus is present. In this respect, they resemble the defective transducing bacteriophages we discussed in Chapter 16.

Why do *v-oncs* induce tumors, whereas normal *c-oncs* do not? In general, the answer is not known. However, a strong possibility is that the viral oncogenes produce much more protein than their cellular counterparts, perhaps because they are transcriptionally activated by enhancers in the viral genome. In chicken tumor cells, for example, the *v-src* gene produces 100 times as much tyrosine kinase as the *c-src* gene. This vast oversupply of the kinase evidently upsets the delicate signaling mechanisms that control cell division, causing unregulated growth. A second possibility is that a *v-onc,* under the control of viral regulator sequences, is expressed at inappropriate times during the cell cycle. A third possibility is that the *v-onc,* being a mutant version of a *c-onc,* acquires a new function that transforms a normal cell into a cancer cell.

Key Points: **Cancer-causing genes, or oncogenes, were initially discovered in the genomes of retroviruses. These genes encode a variety of different proteins, including growth factors, growth-factor receptors, tyrosine kinases, and transcription factors. Later, these genes were found to have normal cellular homologs, but unlike their viral counterparts, these cellular oncogenes possess introns.**

GENETIC BASIS OF HUMAN CANCERS

In the human population, cancer is responsible for no small amount of misery. Approximately one-third of all people in the United States will eventually die of some form of cancer. Of course, cancer is not a single disease, but rather, a collection of diseases characterized by uncontrolled cell growth. Cancers can originate in many different tissues of the body. Some grow aggressively, others more slowly. In the last 15 years, medical research has made great progress in elucidating the molecular and cellular basis of several types of cancer. This research has shown that cancer may be caused by mutant cellular oncogenes, or by the inappropriate expression of normal cellular oncogenes that have been relocated in the genome by a chromosome rearrangement. It may also be caused by the loss or inactivation of genes that suppress tumor formation.

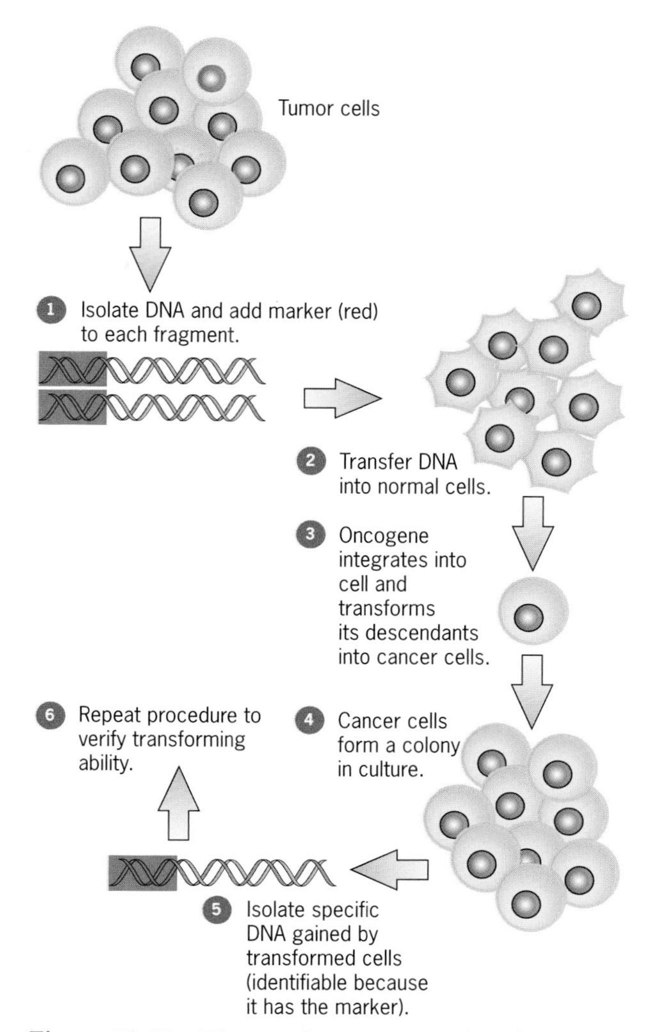

Figure 22.25 The transfection test to identify DNA sequences (oncogenes) capable of transforming normal cells into cancer cells.

Mutant Cellular Oncogenes and Cancer

The products of the *c-oncs* play key roles in regulating cellular activities. Clearly, a mutation in one of these genes could disrupt regulation and cause cancerous growth. Unfortunately, this unappealing prospect has been realized in a large number of documented cases. Many human cancers are caused by spontaneous mutations in cellular oncogenes.

The first evidence linking cancer to a mutant *c-onc* came from the study of a human bladder cancer. The mutation responsible for this bladder cancer was isolated by Robert Weinberg and colleagues using a **transfection test** (Figure 22.25). DNA was extracted from the cancerous tissue and fragmented into small pieces; then each of these pieces was joined to a segment of bacterial DNA, which served as a molecular marker. The marked DNA fragments were then introduced, or transfected, into cells growing in culture to see if any of them could transform the cells into a cancerous state. This state could be recognized by the tendency of the cancer cells to form small clumps, or foci, when grown on soft agar plates. The DNA from such cells was extracted and screened to see if it carried the molecular marker that was linked to the original transfecting fragments. If it did, this DNA was retested for its ability to induce the cancerous state. After several tests, Weinberg's research team identified a DNA fragment from the original bladder cancer that reproducibly transformed cultured cells into cancer cells. This fragment carried an allele of the *c-H-ras* oncogene, a homolog of an oncogene in the Harvey strain of the rat sarcoma virus. DNA sequence analysis subsequently showed that a nucleotide in codon 12 of this allele had been mutated, substituting a valine for the glycine normally found at this position in the *c-H-ras* protein.

Geneticists now have some understanding of how this mutation causes cells to become cancerous. Unlike viral oncogenes, the mutant *c-H-ras* gene does not synthesize abnormally large amounts of protein. Instead, the valine-for-glycine substitution at position 12 impairs the ability of the mutant *c-H-ras* protein to hydrolyze one of its substrates, GTP. Because of this impairment, the mutant protein is kept in an active signaling mode, transmitting information that ultimately stimulates the cells to divide in an uncontrolled way (Figure 22.26).

Mutant versions of *c-ras* oncogenes have now been found in a large number of different human tumors, including lung, colon, mammary, and bladder tumors, as well as neuroblastomas (nerve cell cancers), fibrosarcomas (cancers of the connective tissues), and teratocarcinomas (cancers that contain different embryonic cell types). In all cases, the mutations involve amino acid changes in one of three positions—12, 59, or 61. Each of these amino acid changes impairs the ability of the mutant *ras* protein to switch out of its active signaling mode. These types of mutations are

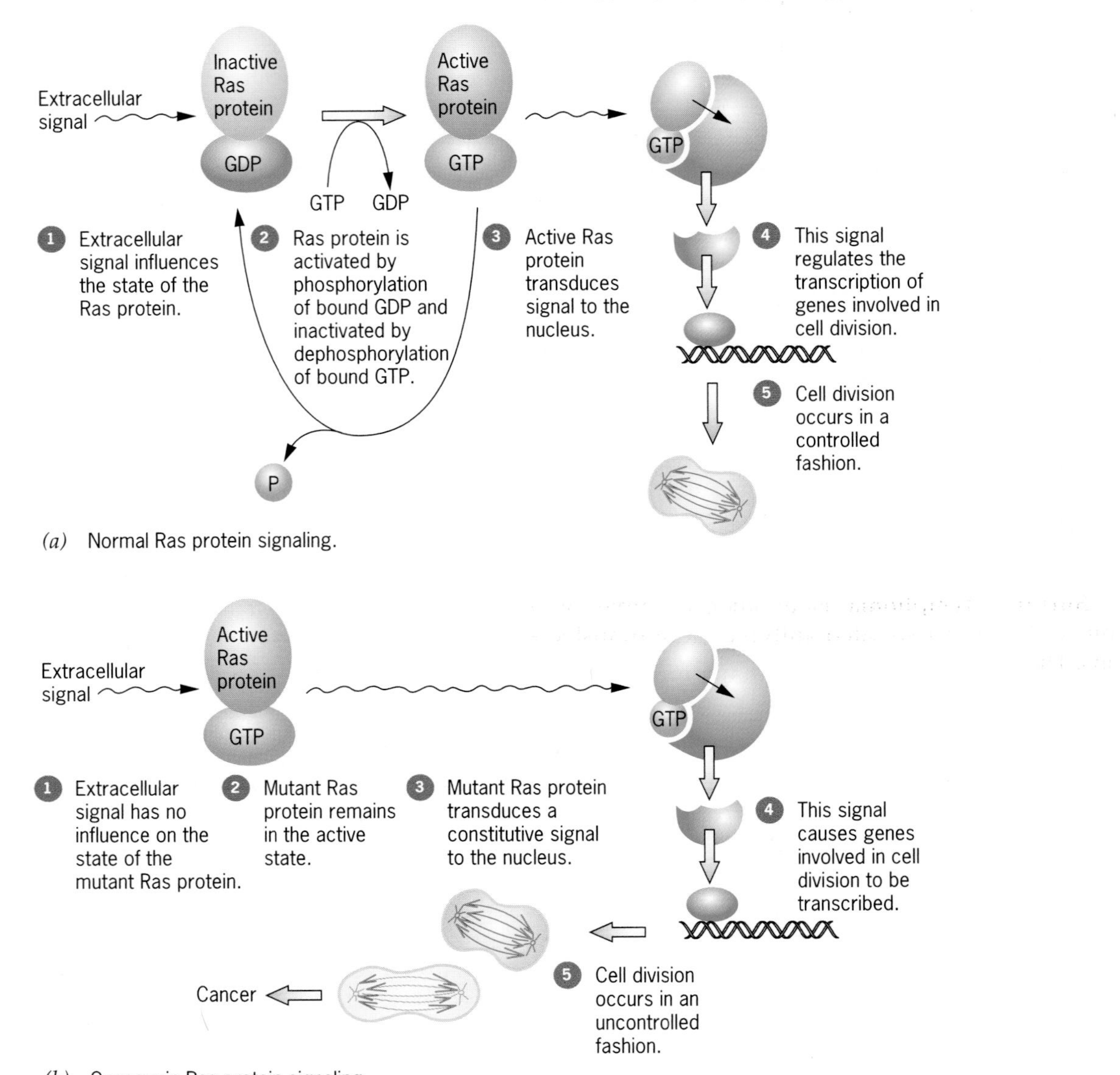

Figure 22.26 Ras protein signaling and cancer. (*a*) The normal protein product of the *ras* gene alternates between inactive and active states, depending on whether it is bound to GDP or GTP. Extracellular signals, such as growth factors, stimulate the conversion of inactive Ras to active Ras. Through active Ras, these signals are transmitted to other proteins and eventually to the nucleus, where they induce the expression genes involved in cell division. Because this signaling is intermittent and regulated, cell division occurs in a controlled manner. (*b*) Mutant Ras proteins exist mainly in the active state. These proteins transmit their signals more or less constantly, leading to uncontrolled cell division, the hallmark of cancer.

therefore **dominant activators** of uncontrolled cell growth.

Chromosome Rearrangements and Cancer

For many years, it has been known that certain types of cancer are associated with chromosome rearrangements. For example, chronic myelogenous leukemia (CML) in human beings is associated with an aberration of chromosome 22. This abnormal chromosome was originally discovered in the city of Philadelphia, and thus is called the **Philadelphia chromosome**. Initially, it was thought to have a simple deletion in its long arm; however, subsequent analysis using molecular techniques has shown that the Philadelphia chromosome is actually involved in a reciprocal translocation with chromosome 9. (For a general discussion of translocations, see Chapter 6.) In this translocation, the tip of the long arm of chromosome 9 has been joined to the body of chromosome 22, and the distal portion of the long arm of chromosome 22 has been joined to the body of chromosome 9 (Figure 22.27). The translocation breakpoint on chromosome 9 is in the *c-abl* oncogene, and the breakpoint on chromosome 22 is in a gene called *bcr*, which encodes a tyrosine kinase. Through the translocation, the *bcr* and *c-abl* genes have been physically joined, creating a fusion gene whose polypeptide product has the amino terminus of the *bcr* protein and the carboxy terminus of the *abl* protein. Although it is not understood precisely why, this fusion polypeptide causes white blood cells to become cancerous.

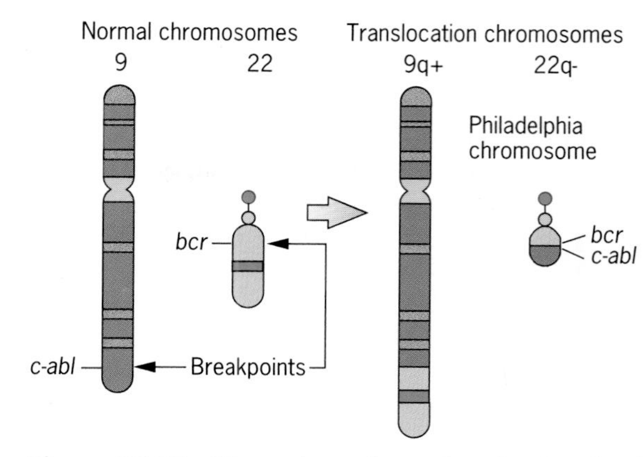

Figure 22.27 The reciprocal translocation involved in the Philadelphia chromosome associated with chronic myelogenous leukemia.

Burkitt's lymphoma is another example of a white cell cancer associated with reciprocal translocations. These translocations invariably involve chromosome 8 and one of the three chromosomes (2, 14, and 22) that carry genes encoding immunoglobulins (also known as antibodies; see Chapter 24). Translocations involving chromsomes 8 and 14 are the most common (Figure 22.28). In these translocations, the *c-myc* oncogene on chromosome 8 is juxtaposed to the genes for the immunoglobulin heavy chains on chromosome 14. As a result of this rearrangement, the *c-myc* gene is overexpressed in cells that produce immunoglobulin heavy chains, that is, in B lymphocytes, and this overexpression causes those cells to become cancerous.

Tumor-Suppressor Genes and Cancer

The normal alleles of genes such as *c-ras* and *c-myc* produce proteins that regulate the cell cycle. When these genes are overexpressed, or when they produce proteins that function as dominant activators, the cell cycle becomes unregulated, leading to the uncontrolled division that is characteristic of cancer. However, not all cancers are caused by the overexpression of cellular oncogenes or by mutations in those genes that produce dominant activator proteins. In fact, some cancers are caused by the loss of particular gene functions.

An example is retinoblastoma, a cancer of the eye. All the available evidence indicates that this cancer is caused by loss-of-function mutations in a gene called *RB*, which is located in band q14 on the long arm of

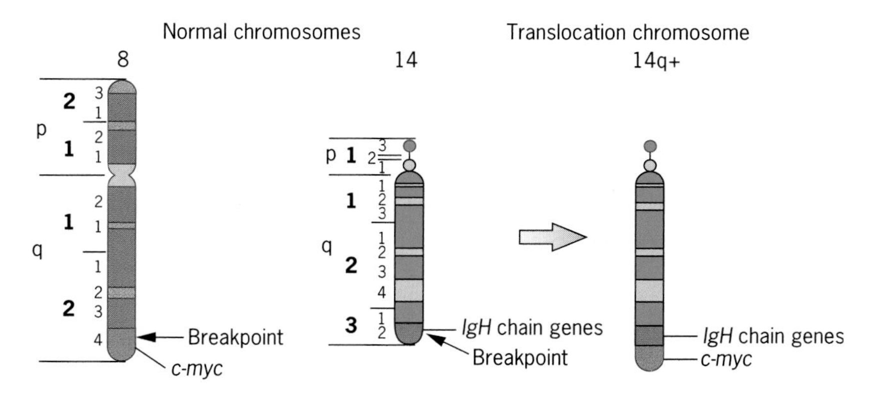

Figure 22.28 A reciprocal translocation involved in Burkitt's lymphoma. Only the translocation chromosome (14q+) that carries both the *c-myc* oncogene and the immunoglobulin heavy chain genes (*IgH*) is shown.

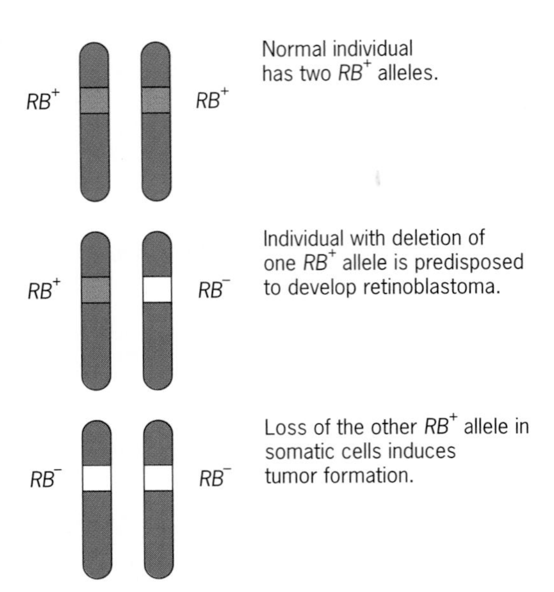

Figure 22.29 The genetic basis of retinoblastoma. The *RB* gene is located in band q14 in the long arm of chromosome 13.

chromosome 13. In fact, many retinoblastoma patients have inherited a deletion of this band in one of their chromosomes 13. In the cancerous eye tissue of these patients, the *RB* gene on the other chromosome is also missing or nonfunctional, probably because of a somatic mutation (Figure 22.29). Thus, these individuals have lost the function of both copies of their *RB* gene. Molecular and biochemical studies have indicated that the *RB* gene encodes a protein of 928 amino acids, and that this protein plays a role in regulating the cell cycle. In its absence, cell growth is abnormal, and tumors develop in the retina of the eye. The normal function of the *RB* protein is therefore needed to prevent, or suppress, tumors from forming. For this reason, the *RB* gene is called a **tumor-suppressor gene**. Other tumor-suppressor genes have been identified in the human genome. For example, the *BRCA1* gene on chromosome 17 is involved in suppressing breast and ovarian tumors (see Human Genetics Sidelight: *BRCA1*, a Gene that Predisposes Women to Develop Breast Cancer).

Is Cancer an Inherited Disease?

Cancers such as retinoblastoma have a clear hereditary component. Someone who has inherited a mutation or a deletion in the *RB* gene has a much higher probablity of developing retinoblastoma than someone who has not. Such people are said to have a predisposition to develop the cancer. This predisposition falls short of an absolute certainty because a single mutant *RB* gene is insufficient to cause cancer. Both copies of the gene must be mutant in a single cell in order to eliminate the *RB* gene product and disrupt cellular metabolism. This idea, called the two-hit hypothesis, was initially formulated by Alfred Knudson long before there was any molecular evidence to support it. Knudson based his hypothesis on the relatively frequent occurrence of retinoblastoma in certain families. He conjectured that these families were segregating a primary retinoblastoma mutation, which was inherited as a simple Mendelian factor, and that, by chance, some family members acquired a second mutation in their somatic cells. This combination of an inherited *RB* mutation with a somatic mutation in the other *RB* gene caused retinoblastoma to develop. Knudson proposed that nonfamilial or sporadic retinoblastoma occurs when a single somatic cell experiences two mutational events that knock out the function of that cell's *RB* genes.

Knudson's hypothesis has been extended to other types of hereditary cancer, including forms of colon, kidney, and breast cancer. Some of these cancers have been linked to specific genes: colon cancer to the *hMSH2* gene, which encodes a protein involved in DNA repair, and breast cancer to the *BRCA1* gene, which encodes a putative DNA-binding protein. However, the mechanism by which mutations in these genes cause cancer is not fully understood.

Only a small fraction of all cancers involve an inherited predisposition. Most develop sporadically in the population, without any obvious familial pattern. Thus, for example, 60 percent of retinoblastomas and 95 percent of breast cancers are not associated with an inherited predisposition. These and other sporadic cancers occur in an unpredictable fashion, probably because multiple somatic mutations are required in genes that regulate cell growth. They may be associated with risk factors, such as smoking, exposure to mutagenic chemicals and radiation, and particular types of diets. Scientists believe that many of these cancers are the result of a stepwise process of accumulating mutations in several genes. A mutation in one gene might increase the rate of cell division, leading to the formation of a benign tumor. A mutation in another gene might cause this tumor to grow larger, and mutations in still other genes might endow it with the ability to invade other tissues, and eventually, to metastasize throughout the body. Thus most cancers may prove to be caused by a series of mutations rather than by just one or two. In the future, the identification of the genes that have mutated may help physicians devise more effective therapies to treat their cancer patients.

Key Points: **Some cancers are caused by mutant proteins that function as dominant activators of cell**

HUMAN GENETICS SIDELIGHT

BRCA1, a Gene that Predisposes Women to Develop Breast Cancer

Each year in the United States, 180,000 new cases of breast cancer are diagnosed. Most of these cases involve women who had no prior warning that they were at risk to develop this disease. In the general population, the odds that a woman will develop breast cancer sometime during her lifetime are about one in eight. However, for some women, the risk is much greater, perhaps as high as 85 percent. These women are heterozygous for mutations in the *BRCA1* gene, a gene that may account for 5 percent of all cases of breast cancer.

The *BRCA1* gene was discovered by analyzing pedigrees with a family history of breast and ovarian cancer. In these pedigrees, the predispositon to develop cancer follows an autosomal dominant inheritance pattern. Coarse linkage studies had shown that the predisposition is associated with molecular markers on the long arm of chromosome 17, and more refined linkage tests had indicated that the cancer-causing locus is situated within a 600-kb region between bands 17q12 and 17q21 (see Figure 1.3 in Chapter 1). This localization allowed a research group led by Mark H. Skolnick of Myriad Genetics and the University of Utah to identify the cancer-causing gene.

Skolnick and his colleagues focused their attention on transcripts produced by genes in the 600-kb interval. These transcripts were characterized by size, sequence, and expression pattern. Eventually, a transcription unit comprising 22 exons in the genomic DNA was identified. This large gene produces a 7.8-kb mRNA capable of encoding a polypeptide of 1863 amino acids. It is expressed in breasts and ovaries, as well as in testes and the thymus. By sequencing this gene from PCR-amplified DNA, Skolnick's research team showed that it was mutant in each of several individuals who carried molecular markers linked to the predisposition for breast and ovarian cancer. However, these mutations were not found in relatives who lacked the molecular markers; nor were they found in the general population. Thus these mutations appeared to be responsible for the predisposition to develop breast and ovarian cancer. Skolnick and colleagues concluded that they had identified a gene responsible for the inherited form of these cancers. They named it *BRCA1* (for Breast Cancer 1).

Subsequent studies have shown that *BRCA1* is a tumor suppressor gene that encodes a zinc-finger type protein—possibly a transcription factor—that is localized in the nucleus, and that this protein is mislocalized in the cytoplasm of cells derived from breast and ovarian cancers. The normal *BRCA1* protein may therefore be required to regulate the expression of genes involved in breast and ovarian cell growth.

Mutations in *BRCA1* account for some, but not all, inherited cases of breast cancer. Another gene, *BRCA2*, located on the long arm of chromosome 13, appears to be responsible for as many cases of breast cancer as *BRCA1*. Part of the *BRCA2* gene was identified at the end of 1995 by an international research team led by Richard Wooster and Michael Stratton of the British Institute of Cancer Research and Andrew Futreal of Duke University in the United States. More research is needed to determine the nature of the *BRCA2* gene product.

Curiously, sporadic cases of breast cancer—that is, those without a family history—are not associated with mutations in *BRCA1*. These cases constitute the vast majority of all breast cancers that are diagnosed. The noninvolvement of *BRCA1* in these cancers suggests that other genes may be responsible for the suppression of breast and ovarian tumors, and that when one or more of these genes is mutated, a cancer may develop.

The identification of genes involved in breast cancer will facilitate the diagnosis, treatment, and prevention of this potentially deadly disease. Already Skolnick's company, Myriad Genetics, has developed a test to detect mutations in the *BRCA1* gene. Women who have a family history of breast cancer and whose test results are positive, have an 85 percent chance of developing the disease. By contrast, women whose test results are negative have a much lower lifetime risk, only about 10 percent. Thus the *BRCA1* test can help a woman with a family history of breast cancer decide whether she should have a preventive mastectomy—an operation to remove breast tissue likely to become cancerous. The cloning of *BRCA2* and other genes involved in breast cancer will almost certainly lead to more genetic tests and, ultimately, to better medical approaches to this disease.

division. Mutations that produce these proteins have been identified by transfecting cultured cells with DNA isolated from tumor cells. Other cancers, such as Burkitt's lymphoma and chronic myelogenous leukemia, are associated with chromosome translocations that alter the expression of a particular cellular oncogene. Still other cancers, such as retinoblastoma, are associated with deletions of genes that suppress tumor formation. Such a deletion may predispose an individual to develop a cancer if a spontaneous somatic mutation eliminates the other copy of the gene.

TESTING YOUR KNOWLEDGE

1. The bacterial *lacZ* gene for β-galactosidase was inserted into a transposable *P* element from *Drosophila* (Chapter 17) so that it could be transcribed from the *P* element promoter. This fusion gene was then injected into the germline of a *Drosophila* embryo along with an enzyme that catalyzes the transposition of *P* elements. During development, the modified *P* element became inserted into the chromosomes of some of the germ-line cells. Progeny from this injected animal were then individually mated to flies from a standard laboratory stock to establish strains that carried the *P/lacZ* fusion gene in their genomes. Three of these strains were analyzed for *lacZ* expression by staining dissected tissues from adult flies with X-gal, a chromogenic substrate that turns blue in the presence of β-galactosidase. In the first strain, only the eyes stained blue, in the second, only the intestines stained blue, and in the third, all the tissues stained blue. How do you explain these results?

ANSWER

The three strains evidently carried different insertions of the *P/lacZ* fusion gene. In each strain, the expression of the *P/lacZ* fusion gene must have come under the influence of a different regulatory sequence, or enhancer, capable of interacting with the *P* promoter and initiating transcription into the *lacZ* gene. In the first strain, the modified *P* element must have inserted near an eye-specific enhancer, which would drive transcription only in eye tissue. In the second strain, it must have inserted near an enhancer that drives transcription in the intestinal cells, and in the third strain, it must have inserted near an enhancer that drives transcription in all, or nearly all, cells, regardless of tissue affiliation. Presumably each of these different enhancers lies near a gene that would normally be expressed under its control. For example, the eye-specific enhancer would be near a gene needed for some aspect of eye function or development. These results show that random insertions of the *P/lacZ* fusion gene can be used to identify different types of enhancers and, through them, the genes they control. These fusion gene insertions are therefore often called enhancer traps.

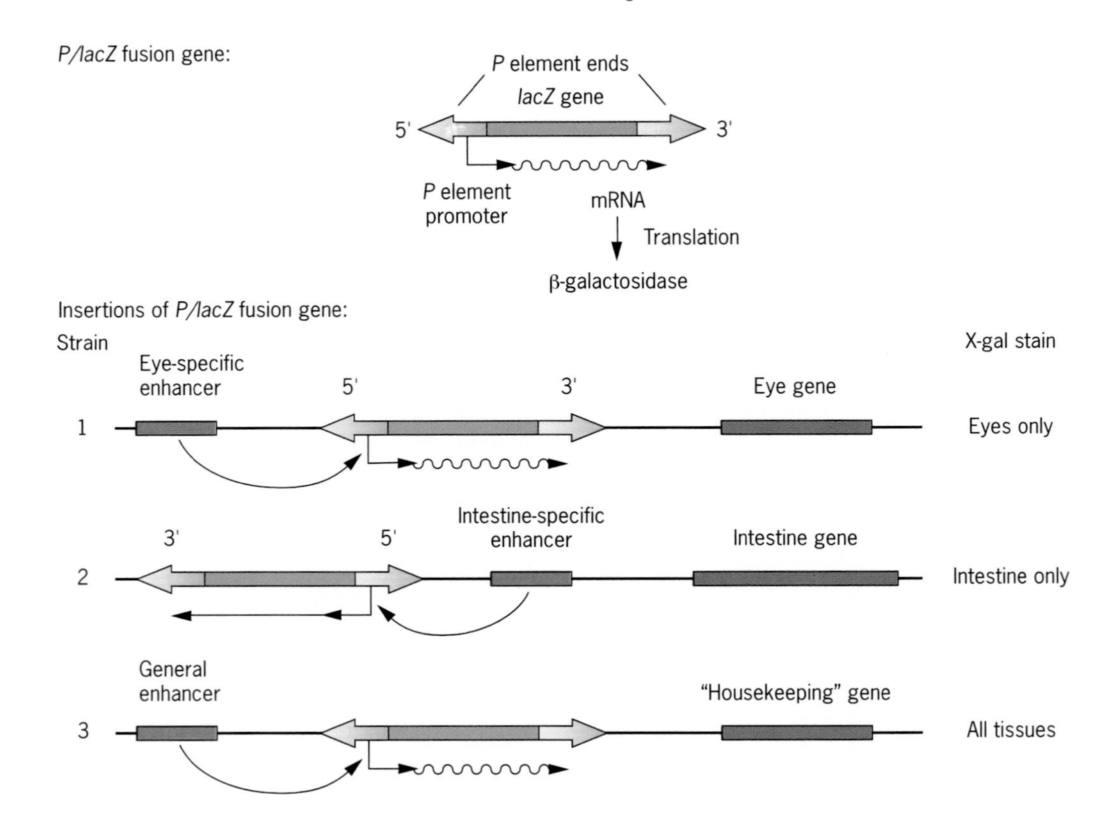

2. An oncogene within the genome of a retrovirus has a high probability of causing cancer, but an oncogene in its normal chromosomal position does not. If these two oncogenes encode exactly the same polypeptide, how can we explain their different properties?

ANSWER

There are at least three possibilities. One is that the virus simply adds extra copies of the oncogene to the cell and that collectively these produce too much of the polypeptide. An excess of polypeptide might cause uncontrolled cell division, that is, cancer. Another possibility is that the viral oncogene is expressed inappropriately under the control of enhancers in the viral DNA. These enhancers might trigger the oncogene to be expressed at the wrong time or to be overexpressed constitutively. In either case, the polypeptide would be inappropriately produced and might thereby upset the normal controls on cell division. A third possibility is that integration of the virus into the chromosomes of the infected cell might put the viral oncogene in the vicinity of an enhancer in the chromosomal DNA and that this enhancer might elicit inappropriate expression. All three explanations stress the idea that the expression of an oncogene must be correctly regulated. Misexpression or overexpression could lead to uncontrolled cell division.

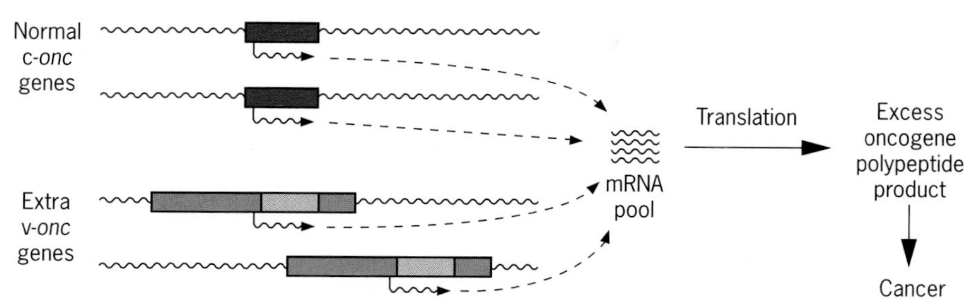

(*a*) The virus adds extra copies of the oncogene to the cell.

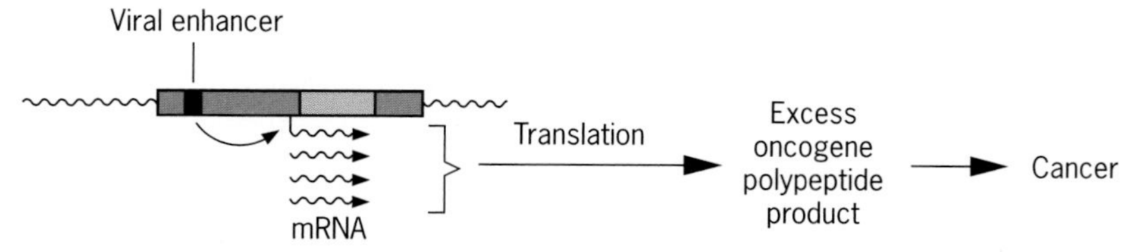

(*b*) The viral oncogene is expressed inappropriately under the control of a viral enhancer.

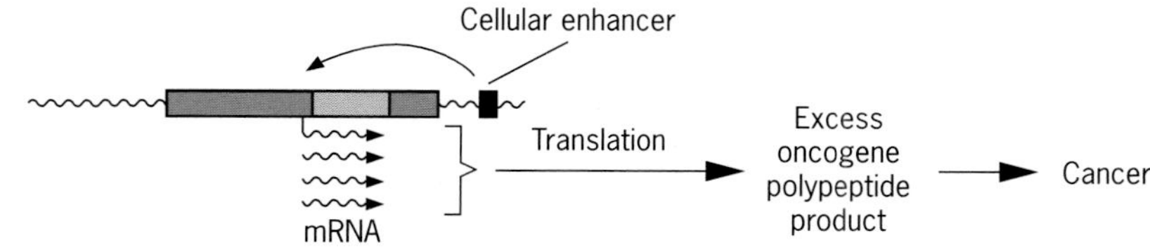

(*c*) The viral oncogene is expressed inappropriately under the control of a cellular enhancer.

QUESTIONS AND PROBLEMS

22.1 Operons are common in bacteria but not in eukaryotes. Suggest a reason why.

22.2 What are the similarities and differences in the regulation of eukaryotic gene expression by steroid and peptide hormones?

22.3 How can you use the polytene chromosomes of Dipteran insects to study the regulation of transcription?

22.4 How would you distinguish between an enhancer and a promoter?

22.5 Tropomyosins are proteins that mediate the interaction of actin and troponin, two proteins involved in muscle contractions. In higher animals, tropomyosins exist as a family of closely related proteins that share some amino acid sequences but differ in others. Explain how these proteins could be created from the transcript of a single gene.

22.6 In *Arabidopsis*, one β tubulin gene (*TUB1*) is expressed in the roots and another β tubulin gene (*TUB8*) is expressed in the vascular tissues. Suggest a mechanism to account for this tissue-specific expression.

22.7 What techniques could be used to show that a plant gene is transcribed when the plant is illuminated with light?

22.8 Does a pseudogene specify an RNA? Does it specify a functional polypeptide?

22.9 Using the techniques of genetic engineering, a researcher has constructed a fusion gene containing the heat-shock response elements from a *Drosophila hsp70* gene and the coding region of a jellyfish gene for green fluorescent protein (*gfp*). This fusion gene has been inserted into the chromosomes of living *Drosophila* by the technique of transposon-mediated transformation (Chapter 17). Under what conditions will the green fluorescent protein be synthesized in these genetically transformed flies? Explain.

22.10 Suppose that the segment of the *hsp70* gene that was used to make the *hsp70/gfp* fusion in the preceding problem had mutations in each of its heat-shock response elements. Would the green fluorescent protein encoded by this fusion gene be synthesized in genetically transformed flies?

22.11 Why does it make sense for the photosynthetic enzyme ribulose1,5-bisphophate carboxylase (RBC) to be synthesized specifically when plants are exposed to light?

22.12 Distinguish between the zinc finger and leucine zipper motifs found in some eukaryotic transcription factors.

22.13 The protein product of the *Drosophila Sex-lethal* gene is needed in females but not in males. Predict the phenotype of XY *Drosophila* that are hemizygous for a null mutation in the X-linked *Sex-lethal* gene.

22.14 The alternately spliced forms of the RNA from the *Drosophila doublesex* gene encode proteins that are needed to block the development of one or the other set of sexual characteristics. The protein that is made in female animals blocks the development of male characteristics, and the protein that is made in male animals blocks the development of female characteristics. Predict the phenotype of XX and XY animals homozygous for a null mutation in the *doublesex* gene.

22.15 In *Drosophila*, expression of the *yellow* gene is needed for the formation of dark pigment in many different tissues; without this expression, a tissue appears yellow in color. In the wings, the expression of the *yellow* gene is controlled by an enhancer located upstream of the gene's transcription initiation site. In the tarsal claws, expression is controlled by an enhancer located withn the gene's only intron. Suppose that by genetic engineering, the wing enhancer is placed within the intron and the claw enhancer is placed upstream of the transcription initiation site. Would a fly that carried this modified *yellow* gene in place of its natural *yellow* gene have darkly pigmented wings and claws? Explain.

22.16 Describe the cytological evidence that RNA is preferentially transcribed from "loosely" organized segments of chromosomes.

22.17 Would a *Drosophila* gene that was flanked by specialized chromatin structures be susceptible to position–effect variegation?

22.18 In *Drosophila* larvae, the single X chromosome in males appears diffuse and bloated in the polytene cells of the salivary gland. Is this observation compatible with the idea that X-linked genes are hyperactivated in *Drosophila* males?

22.19 What chemical modification of histone H4 is associated with transcriptionally active genes?

22.20 Why do null mutations in the *mle* gene in *Drosophila* have no effect in females?

22.21 How do we know that normal cellular oncogenes are not simply integrated retroviral oncogenes that have acquired the proper regulation?

22.22 When cellular oncogenes are isolated from different animals and compared, the amino acid sequences of the polypeptides they encode are found to be very similar. What does this suggest about the functions of these polypeptides?

22.23 The majority of the *c-ras* oncogenes obtained from cancerous tissues have mutations in codon 12, 59, or 61 in the coding sequence. Suggest an explanation.

22.24 How might normal cellular oncogenes be mutated into cancer-causing oncogenes?

22.25 A mutation in the *ras* cellular oncogene can cause cancer when it is in heterozygous condition, but a mutation in the *RB* cellular oncogene can cause cancer only when it is in homozygous condition. What does this difference between dominant and recessive mutations imply about the roles that the *ras* and *RB* gene products play in normal cellular metabolism?

BIBLIOGRAPHY

BAKER, B. S. 1989. Sex in flies: the splice of life. *Nature* 340: 521–524.

BROWN, C. J., B. D. HENDRICH, J. L. RUPERT, R. G. LAFRENIERE, Y. XING, J. LAWRENCE, AND H. F. WILLARD. 1992. The human XIST gene: Analysis of a 17 kb inactive X-specific RNA that contains conserved repeats and is highly localized within the nucleus. *Cell* 71:527–542.

CAVENEE, W. K. AND R. L. WHITE. 1995. The genetic basis of cancer. *Scientific American* 272 (March): 72–79.

DARNELL, J., H. LODISH, AND D. BALTIMORE. 1990. *Molecular Cell Biology*. W. H. Freeman, New York.

EVANS, R. M. 1988. The steroid and thyroid hormone receptor superfamily. *Science* 240:889–895.

GORMAN, M., AND B. S. BAKER. 1994. How flies make one equal two: dosage compensation in *Drosophila*. *Trends in Genet.* 10:376–380.

KARP, J. E. AND S. BRODER. 1995. Molecular foundations of cancer: new targets for intervention. *Nature Medicine* 1:309–320.

PARKHURST, S. M., AND P. M. MENEELY. 1994. Sex determination and dosage compensation: Lessons from flies and worms. *Science* 264:924–932.

These modifications of insect wings are the result of differences in the way that sets of genes control development.

23

The Genetic Control of Animal Development

Insect Wings: Evidence for the Genetic Control of Development

No one knows how many different species of animals are currently on the earth, but the vast majority of them are insects. These small creatures originated about 400 million years ago when life spread from the seas onto the land. Many different groups of insects were established in those ancient times; today, their descendants are the most abundant eukaryotes on the planet. All insects have segmented bodies and three pairs of legs, but beyond that, their features are amazingly diverse. Consider, for example, insect wings. Some groups of insects, such as butterflies, moths, and dragonflies, have two pairs of membranous wings, but others, such as grasshoppers, beetles, and flies, have only one. The wings of butterflies and moths are delicate and colorful, whereas those of dragonflies are sturdy and plain. In grasshoppers, what would have been an anterior pair of wings has been modified into protective sheaths, and in beetles, this modification has been extended even further, forming tough, shell-like structures called elytra. In flies, what would have been a posterior pair of wings has been transformed into short, knoblike structures called halteres, which help the animal to balance during flight.

Figure 23.1 A fruit fly with four wings instead of the normal two. This unusual phenotype is caused by a trio of tightly linked autosomal mutations.

All these differences demonstrate that one of the basic features of the insect body plan—two pairs of flight-related appendages—has been extensively modified during evolution. These modifications are, of course, due to genetic differences among the various groups of insects. Why does a fly have halteres instead of a second pair of wings? Because its genes make it that way. Geneticists know this because mutations in certain fly genes cause wings to develop instead of halteres (Figure 23.1). This remarkable transformation produces a fly with two pairs of wings, like a dragonfly, and convincingly shows that the development of specific body parts is under genetic control.

In this chapter, we explore the genetic basis of animal development—how genes shape the basic body plan and all its anatomical details. This is one of the most exciting areas in modern biology, one that is growing at a rapid pace because of intensive research, much of it focused on a few carefully chosen organisms. The goal of all this effort is to obtain a general understanding of how the body of a multicellular animal—particularly that of humans—develops from a fertilized egg.

THE PROCESS OF DEVELOPMENT IN ANIMALS

Animal development actually begins before fertilization of the egg. Critical substances are synthesized by the female and sequestered in the egg to nourish and control the development of the zygote into a multicellular organism. After fertilization, a rapid series of mitotic divisions increases the number of cells in the developing animal, usually without much growth in overall size (Figure 23.2). The resulting mass of cells, the **blastula**, is relatively nondescript, but even at this stage, some of its cells have been earmarked to form specific tissues in the animal's body. Such cells have been fated, or determined, to develop in a certain way. As development continues, the blastula is reorganized by a series of cell migrations, collectively called **gastrulation**, to produce an embryo with three distinct cell layers—**ectoderm** (on the outside), **mesoderm** (in the middle), and **endoderm** (on the inside). These three layers give rise to specific tissues, which are then assembled to form organs. From gastrulation onward, cells become different from each other, a process called **differentiation**. This is a key feature of animal development, providing the many specialized functions that are needed for life.

Oogenesis and Fertilization

All the developmental potential of an animal is contained in a fertilized egg. When sperm and egg unite, they bring together the paternal and maternal sets of genes, which provide the detailed instructions for development. However, development is also profoundly

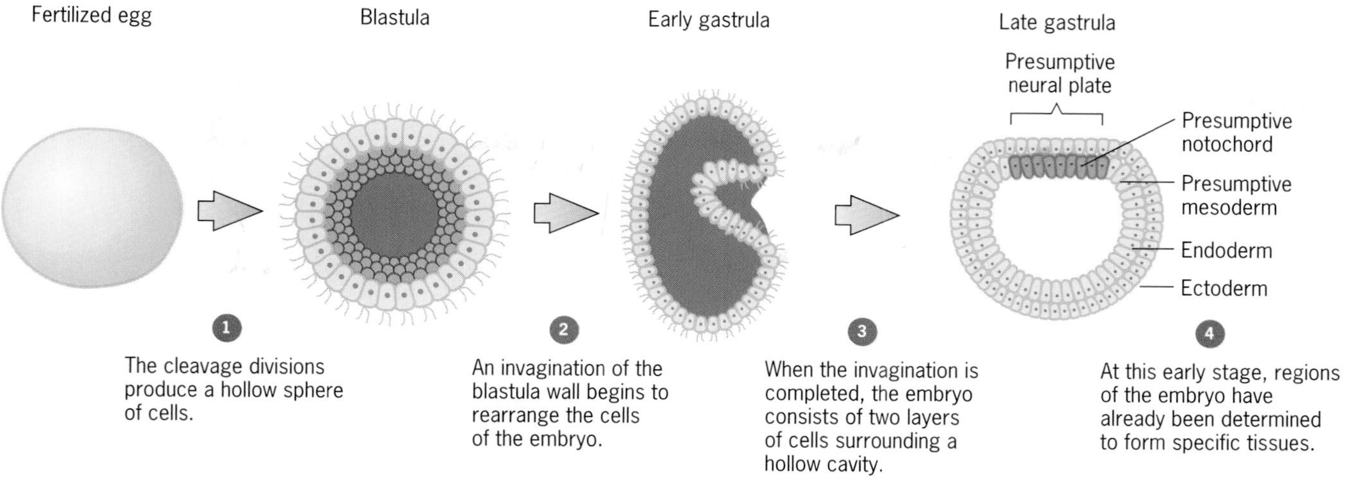

(*a*) Early stages of development in *Amphioxus*.

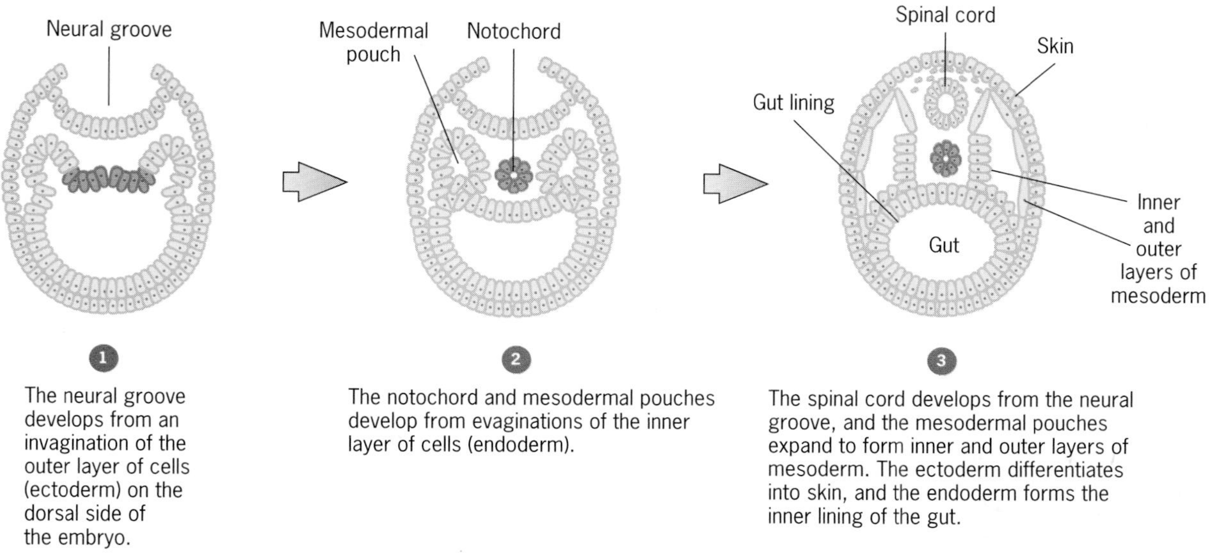

(*b*) Morphogenesis in *Amphioxus*.

Figure 23.2 Embryonic development of the marine animal *Amphioxus*, a protochordate. (*a*) Early stages of embryonic development. (*b*) Morphogenesis.

affected by cytoplasmic materials present in the egg. Some of these materials, such as yolk, are nutritive and serve as food for the developing animal. However, other materials are determinative in that they help to lay out the basic pattern of development itself. Both the nutritive and determinative components of the egg cytoplasm are manufactured during oogenesis, the process of egg formation. This involves a complex series of events, including, of course, the two meiotic divisions, which reduce the chromosome number from diploid to haploid; curiously, the second meiotic division is sometimes not completed until after the egg has been fertilized.

In female animals, the meiotic divisions produce cells of unequal size (Chapter 2). The first division separates a large cell, the secondary oocyte, from a small cell, the first polar body. The secondary oocyte goes on to develop into the egg, whereas the polar body degenerates. This unequal division of cells is clearly a way of mustering materials into the egg cytoplasm. Very little cytoplasm is allotted to the polar body; instead, most of it is reserved for the oocyte. The second meiotic division is also unequal in that most of the cytoplasm is reserved for the egg; only a tiny bit is pinched off to form the second polar body.

Neighboring cells also enrich the cytoplasm of the egg. In many organisms, these cells synthesize materials that are transported into the oocyte, performing a nourishing or nursing function. Surrounding cells may also provide materials to envelop the oocyte. In birds, for example, a hard, calcium-rich shell is formed by secretions in the oviduct, a part of the female reproductive tract through which the egg passes on its way out the body.

Fertilization occurs when an egg is penetrated by a sperm. This is a complex process, involving interactions between different proteins on the surfaces of the two gametes. When the sperm makes contact with the egg, it triggers a response in the egg that causes the cytoplasm near the surface of the egg to mound up around the sperm and internalize it. After the sperm has entered the egg, the egg completes meiosis (if it has not already done so) and the sperm and egg nuclei fuse to form the nucleus of the zygote. In some organisms, for example, *Drosophila*, the sperm and egg nuclei actually divide once mitotically *before* fusing.

The Embryonic Cleavage Divisions and Blastula Formation

A newly formed zygote quickly divides into two cells, each of which divides into two more, and so on, until many cells are present in the developing embryo. These initial divisions are called **cleavage divisions**. Sometimes they occur asymmetrically so that the daughter cells receive different amounts of cytoplasm (Figure 23.3). However, except for a few curious cases, each cell seems to get the same genetic material. Thus their subsequent differentiation into specific cell types is not due to the segregation of genes or chromosomes, but rather, to the activation or repression of specific sets of genes.

The differentiation of cell types is preceded by **determination**, a process in which cells in the blastula are assigned developmental fates. This has been confirmed by experiments in which a piece of the blastula is surgically excised and moved to a different position, or even to a different organism. In some cases, these transplants survive and differentiate just as they would have if they had not been moved. Thus, even before there is any visible evidence of specialization, some of the cells in the blastula are determined to develop in a specific way. By the time of blastula formation, the basic developmental plan of the animal has already been laid down.

Gastrulation and Morphogenesis

The cells of the blastula are reorganized by gastrulation, a series of movements in which groups of cells migrate to new positions in the embryo. This converts the nondescript blastula, which is essentially a sphere of cells, into an organism with distinct cell layers, often called the **primary germ layers** (not to be confused with the germ cells). These layers form the primitive tissues of the animal—ectoderm, mesoderm, and endoderm—

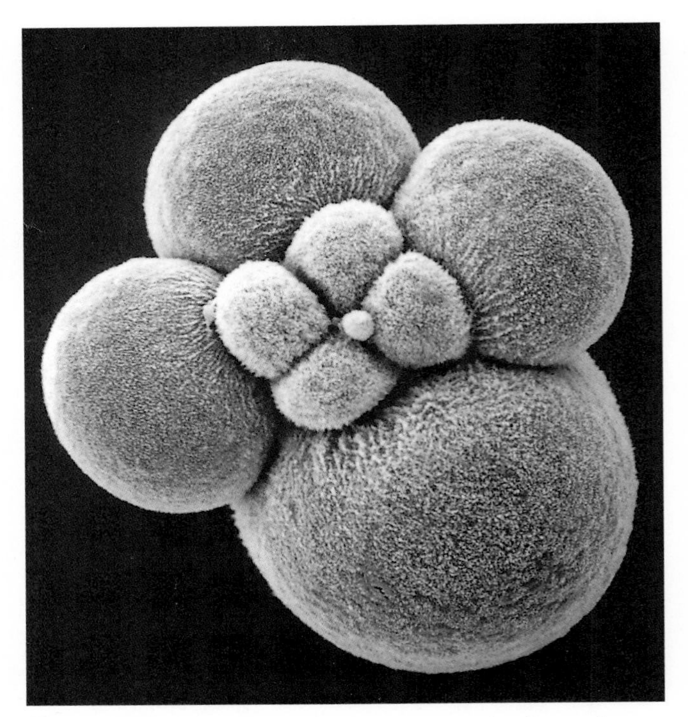

Figure 23.3 Scanning electron micrograph of an embryo of the marine snail *Ilyanassa obsoleta* at the eight-cell stage. The tiny cell in the center is a polar body.

and eventually differentiate into specific tissues—nerve, epithelium, muscle, blood, bone, hair, glands, viscera, and so forth. It is therefore through a combination of cell movement and differentiation that the body of an animal takes shape. In the course of this process, often called **morphogenesis**, individual organs such as brain, heart, intestine, and liver are formed.

Key Points: **The development of a fertilized egg into a multicellular animal involves cell division, the assignment of fates to individual cells (determination), and the subsequent realization of those fates (differentiation). Key materials for the early phases of this process are transferred into the egg during oogenesis. After fertilization, the egg proceeds through several cleavage divisions to form a sphere of cells (the blastula), which is subsequently reorganized by cell movements (gastrulation) to form the primary germ layers (ectoderm, endoderm, and mesoderm). These primitive tissues eventually produce mature tissues and organs.**

GENETIC ANALYSIS OF DEVELOPMENT IN MODEL ORGANISMS

Modern developmental biology depends heavily on genetic analysis, which must, of course, begin with the identification of relevant genes. This is usually accomplished by collecting mutations with diagnostic phenotypes; for example, if we wish to study the development of an insect's wings, we would begin by collecting mutations that alter or prevent wing development. We would then map these mutations and test them for allelism with each other to define and position the relevant genetic loci. Once we have identified these loci, we would combine representative mutations from each in pairwise fashion to determine whether some are epistatic over others. Such epistasis testing can sometimes provide valuable insights into how different genes contribute to a developmental process (Chapter 4). Further analysis might involve studying the nature of gene expression by creating animals that are composed of a mixture of genotypes; these so-called genetic mosaics permit us to study the cell lineages in which gene expression is required. Finally, to investigate the molecular basis of gene action, we would have to clone individual genes involved in the developmental process and study them with the full panoply of techniques now available, including, of course, DNA sequencing.

Not many animals are suitable for such an aggressive research program. Consequently, developmental geneticists have concentrated their efforts on a few model species in which it is possible to do intensive genetic analysis *and* high-resolution molecular biology. Currently, the best species are the fruit fly *Drosophila melanogaster* and the roundworm *Caenorhabditis elegans*.

Drosophila as a Model Organism

Drosophila was introduced to the world of genetics in 1909 through the laboratory of T. H. Morgan. Since then, intensive research has made it one of the best understood eukaryotes. Adult *Drosophila* develop from ellipsoidal eggs about 1 mm long and 0.5 mm wide at their maximum diameter (Figure 23.4). Each egg is surrounded by a **chorion**, a tough shell-like structure that is made of materials synthesized by somatic cells in the ovary. The anterior end of the egg is distinguished by two filaments that help to bring oxygen

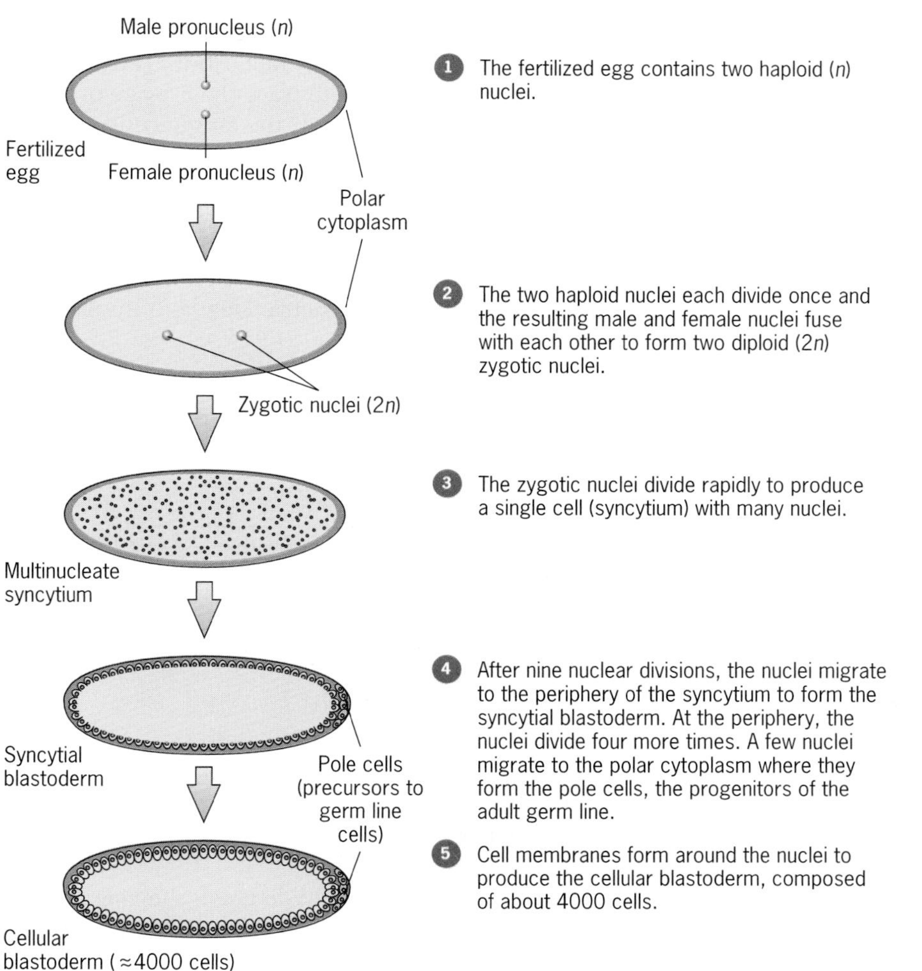

Figure 23.4 Early embryonic development in *Drosophila*.

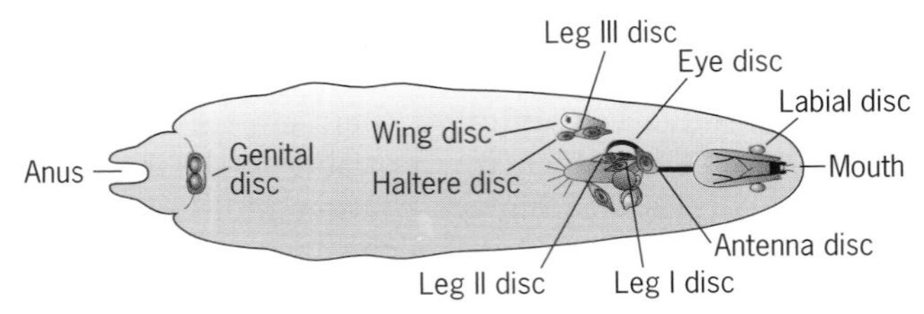

Figure 23.5 The location of the imaginal discs in *Drosophila* larvae.

into the interior. Sperm enter the egg through another anterior structure, the **micropyle**. The divisions that follow fertilization are rapid—so rapid, in fact, that there is no time for membranes to form between daughter cells. Consequently, the early *Drosophila* embryo is actually a single cell with many identical nuclei; such a cell is called a **syncytium**. After division cycle 9 within the syncytium, the 512 nuclei that have been created migrate to the cytoplasmic membrane on the periphery of the embryo, where they continue to divide four more times. In addition, a few of the nuclei migrate to the posterior pole of the embyro. At division cycle 13, all the nuclei in the syncytium become separated from each other by cell membranes, creating a single layer of cells on the embryo's surface. This single layer, called the **cellular blastoderm**, is equivalent to the blastula of other organisms; it will give rise to all the somatic tissues of the animal. Cellularization of the nuclei at the posterior pole creates the **pole cells**, which give rise to the adult germ line. Thus, at this very early stage of development, the somatic- and germ-cell lineages of the future adult have already been separated.

It takes about a day for the *Drosophila* embryo to develop into a wormlike **larva**. This larva hatches by chewing its way through the egg shell and begins feeding voraciously. It sheds its skin twice to accommodate increases in size, and then, about five days after hatching, becomes immobile and hardens its skin to form a **pupa**. During the next four days, many of the larval tissues are destroyed, and flat packets of cells that were sequestered during the larval stages expand and differentiate into adult structures such as antennae, eyes, wings, and legs. Because an adult insect is called an **imago**, these packets are referred to as **imaginal discs** (Figure 23.5). When this anatomical reorganization is completed, a radically different animal emerges from the pupal casing—an animal that can fly and reproduce!

Caenorhabditis as a Model Organism

Caenorhabditis elegans (Figure 23.6) became a part of the genetics scene in the 1960s, long after *Drosophila* had been firmly established as an experimental organism. Sydney Brenner, who initially worked with bacteriophages, chose this small, free-living soil nematode as a model for studying animal development. Adult *C. elegans* are about 1 mm long, roughly the same size as a *Drosophila* egg. They reproduce quickly and prolifically and can be reared easily on agar plates that have been seeded with *E. coli* bacteria as food. Under optimal conditions, the life cycle is completed in a mere three days.

C. elegans is a hermaphroditic species, that is, one in which individual organisms are capable of producing both sperm and eggs. Because there is no system of self-incompatibility, sperm can fertilize eggs from the same animal. To the researcher, this self-fertilization is a convenient way of making recessive mutations homozygous. Hermaphrodites have two X chromosomes and five pairs of autosomes. Occasionally, animals with a single X chromosome and five pairs of autosomes are produced by meiotic nondisjunction; these animals are males, capable of making sperm but not eggs. XO males can be crossed with XX hermaphrodites to carry out standard genetic procedures such as recombination mapping and complementation testing.

C. elegans is a transparent animal, affording researchers the opportunity to observe each cell in the course of development. In fact, John Sulston and coworkers have been able to trace the lineage of all the cells in the adult worm from the single-celled zygote (Figure 23.7). This work has shown that the events of *C. elegans* development are essentially invariant. Each adult hermaphrodite consists of 959 somatic nuclei (some cells are multinucleate) as well as an indeterminant number of germ cells. A *C. elegans* zygote goes through a series of asymmetric cleavage divisions to

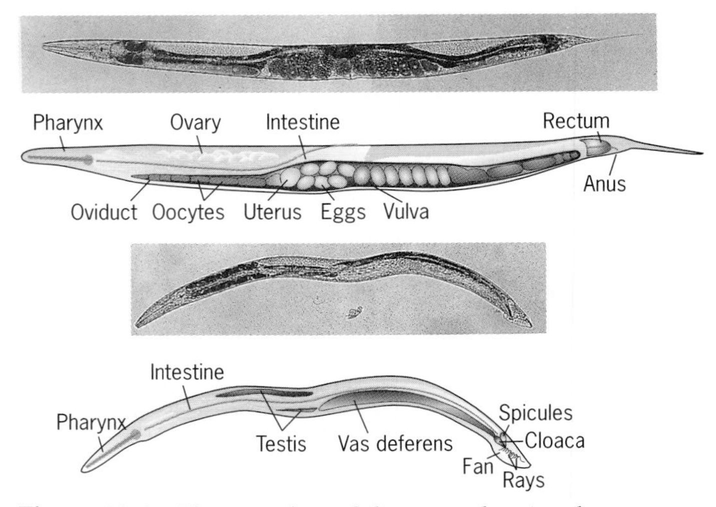

Figure 23.6 Photographs and diagrams showing the phenotypes of a *Caenorhabditis* hermaphrodite (top) and male (bottom).

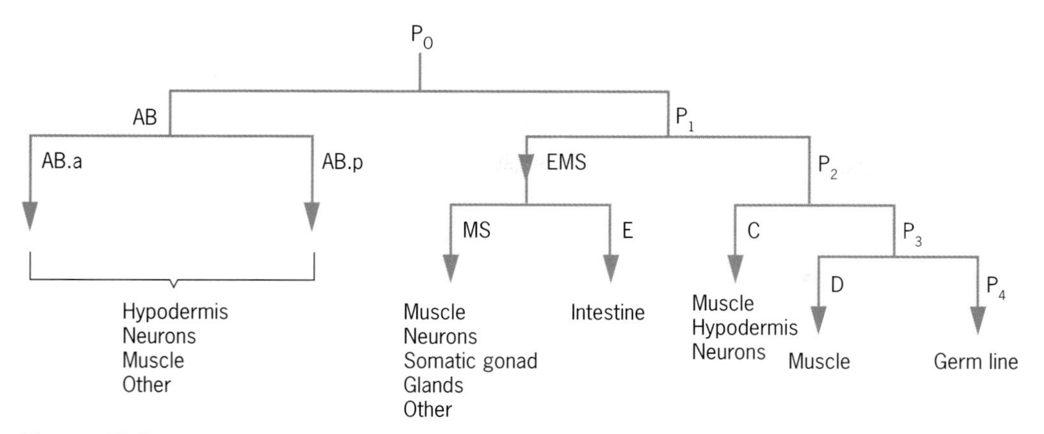

Figure 23.7 A portion of the cell lineage of the *C. elegans* hermaphrodite. P_0, the fertilized egg, divides to produce two cells denoted AB and P_1. Each of these then divides to produce two pairs of cells, one pair denoted AB.a and AB.p and the other denoted EMS and P_2. Further divisions of these cells and their descendants produce all the cells of the adult hermaphrodite.

produce six "founder" cells. One of these eventually gives rise to the entire germ line, one to the intestine and one to the body wall muscles. The other three founder cells produce mainly nerve and muscle cells. The **invariant cell lineages** that form the tissues of the adult make *C. elegans* an excellent organism for the study of development.

Key Points: **The fruit fly *Drosophila melanogaster* and the hermaphroditic nematode *Caenorhabditis elegans* are model organisms with features that make them ideal for studies of the genetic control of development.**

GENETIC ANALYSIS OF DEVELOPMENTAL PATHWAYS

The notion that phenotypes result from a series of steps in a pathway and that genes control each of these steps emerged in the 1930s and 1940s from a combination of genetic and biochemical investigations. By this time, it had already been established that cells carry out a multitude of biochemical reactions, each catalyzed by an enzyme, and that several reactions can be linked to form a pathway. Genetics provided the key insight that each enzyme in a pathway is encoded by a gene. A mutation in one of these genes could therefore inactivate an essential enzyme, blocking the entire pathway and causing a mutant phenotype. Geneticists quickly realized that by studying such mutations, they could identify each step in the pathway. Furthermore, by analyzing pairs of mutations, they could sometimes arrange the steps in the correct temporal order. This genetic dissection of biochemical pathways was soon extended to other biological processes, including development.

A developmental pathway consists of the events involved in the differentiation of tissues and organs. Different gene products participate in these events, including some that are signal molecules, some that are signal receptors, others that are signal transducers, and still others that are transcription factors. Other kinds of regulatory proteins may also be involved. Ultimately, a pathway generates the components that form the structures of particular tissues and organs. It therefore creates a phenotype.

The components of a pathway are causally ordered by the way in which they bring about the phenotype. Gene *A* may produce a secreted protein that acts as a signal to stimulate the transcription of gene *X*, whose product is a component of a differentiated cell. This stimulation may be mediated by other gene products, including, for example, a membrane-bound receptor for the A protein, intracellular proteins that are activated by this receptor, and one or more transcription factors that respond to this activation by binding to enhancers near gene *X* to induce its expression. The overall structure of the pathway is

Gene *A* →	Gene *R* →	Gene *C* →	Gene *T* →	Gene *X*
↓	↓	↓	↓	↓
secreted signal protein A	membrane-bound receptor protein	cytoplasmic protein	transcription factor	protein X in a differentiated cell

In this developmental pathway, the different proteins are ordered by a causal chain that ultimately produces a protein that is characteristic of a particular differentiated cell. The arrows between the genes in the pathway indicate the order in which the gene products act to bring about a differentiated state.

In *Drosophila* and *Caenorhabditis*, the developmental pathways that have been most thoroughly studied are responsible for the phenotypes of the two sexes. These pathways control the differentiation of the somatic tissues into male or female structures. Other, less well understood pathways govern the development of the tissues in the male and female germ-lines. The *Drosophila* and *Caenorhabditis* somatic sex-determination pathways actually involve entirely different molecular mechanisms. In *Drosophila*, the key genes encode proteins that regulate RNA splicing, whereas in *Caenorhabditis,* they encode transcription factors, signaling molecules and their receptors. In both animals, however, the sex-determinaton pathway responds to the same fundamental signal, which is the ratio of X chromosomes to autosomes. When this is 1.0 or greater, the pathway produces a female phenotype, whereas when it is 0.5 or less, it produces a male phenotype.

Sex Determination in *Drosophila*

The sex-determination pathway in *Drosophila* has three components: (1) a system to ascertain the X:A ratio very early in the embryo; (2) a system to convert this ratio into a developmental signal; and (3) a system to respond to this signal by producing either male or female structures. Intensive research has led to a detailed understanding of the first two components; however, the last component is still something of a mystery.

The system to ascertain the X:A ratio involves interactions between maternally synthesized proteins that have been deposited in the egg cytoplasm, and embryonically synthesized proteins that are encoded by several X-linked genes (Figure 23.8). These latter proteins are twice as abundant in XX embryos as in XY

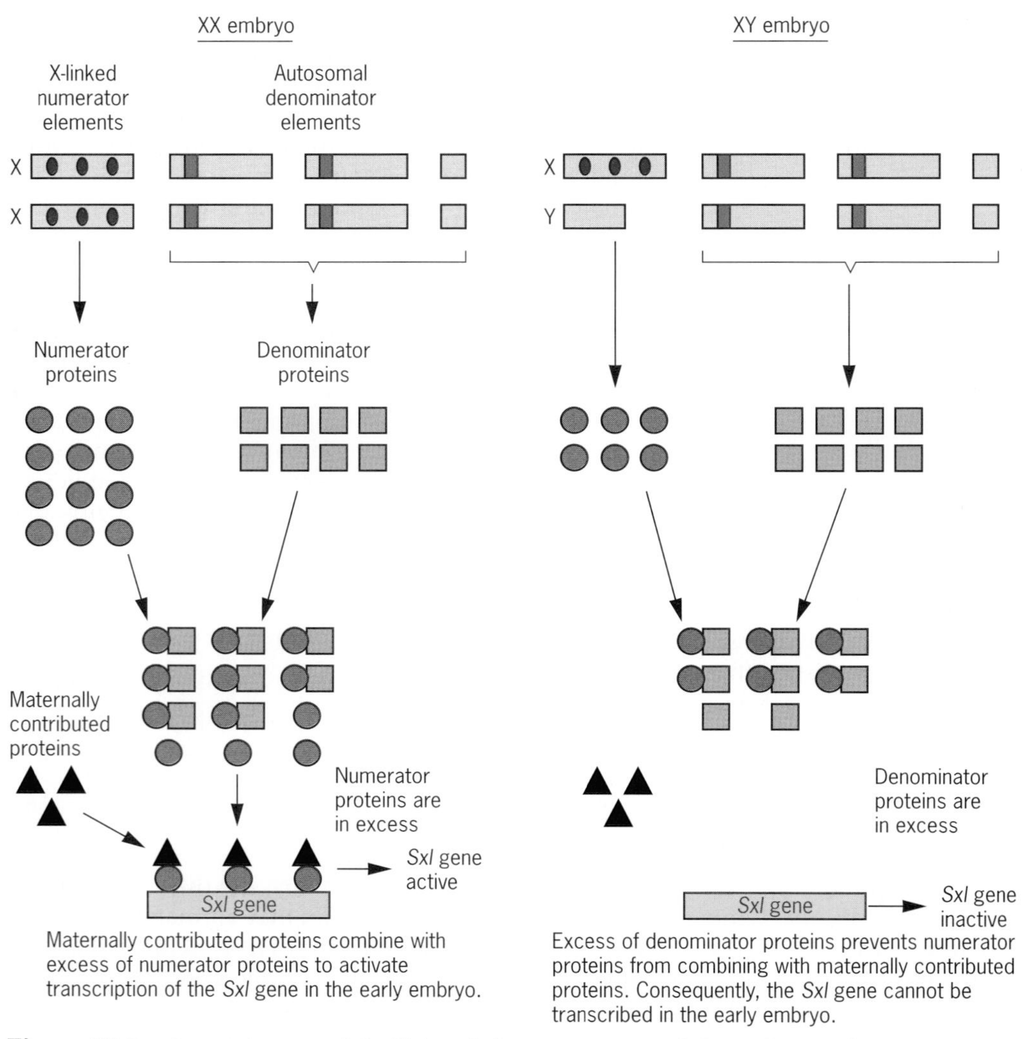

Figure 23.8 Ascertainment of the X:A ratio by numerator and denominator elements in *Drosophila*. The ratio of X chromosomes to sets of autosomes is established by interactions between the protein products of these genes.

embryos, and therefore provide a means of "counting" the number of X chromosomes present. Because the genes that encode these proteins affect the numerator of the X:A ratio, they are called **numerator elements**. Other genes located on the autosomes affect the denominator of the X:A ratio. These so-called **denominator elements** encode proteins that antagonize the products of the numerator elements. Consequently, as the dosage of denominator elements is increased, the "perceived" dosage of numerator elements is decreased, and the number of X chromosomes present in the genotype is underestimated. This process occurs in *Drosophila* with two X chromosomes and three pairs of autosomes (genotype XX; AAA); such flies develop as intersexes rather than as females. The system for ascertaining the X:A ratio in *Drosophila* is therefore based on antagonism between X-linked (numerator) and autosomal (denominator) gene products.

Once the X:A ratio has been ascertained, it is converted into a molecular signal that controls the expression of the X-linked *Sex-lethal* (*Sxl*) gene, the master regulator of the sex-determination pathway (Figure 23.9). Early in development, this signal activates transcription of the *Sxl* gene, but only in XX embryos. These early transcripts are processed and translated to produce functional *Sex-lethal* protein (denoted SXL). Later in development, the *Sxl* gene is also turned on in XY embryos. However, in these embryos, transcription is initiated at a different promoter, and the resulting transcripts are alternately spliced to include an exon with a stop codon. Thus, when *Sxl* mRNA is translated in XY embryos, it generates a short polypeptide without regulatory function. This alternate splicing effectively prevents functional SXL protein from being made in an embyro that is destined to become a male. In an XX embryo—where SXL was initially made in response to the X:A signal—*Sxl* transcripts are spliced to encode more SXL protein. This splicing pattern is maintained because the SXL protein is itself a regulator of the splicing process. When the SXL protein is present, it causes *Sxl* transcripts to be spliced in such a way that they encode the full-length SXL protein. In XX embryos, this protein is therefore a positive regulator of its own synthesis. This curious feedback mechanism maintains the expression of the SXL protein in XX embryos and prevents it in XY embryos.

The SXL protein also regulates the splicing of transcripts from another gene in the sex-determination pathway, *transformer* (*tra*) (Figure 23.10). These tran-

1 Transcription:

In XX embryos, a molecular signal based on the X:A ratio initiates transcription of the *Sxl* gene from promoter P_E.

In XY embryos, transcription is eventually initiated at promoter P_M.

2 Splicing:

In XX embryos, the *Sxl* transcript is spliced to contain all the exons except exon 3.

In XY embryos, the *Sxl* transcript is spliced to contain all the exons, including exon 3.

3 Translation:

In XX embryos, the *Sxl* mRNA is translated into a polypeptide (SXL) that acts as a regulator of splicing, including the splicing of *Sxl* transcripts.

In XY embryos, a stop codon in exon 3 prevents the *Sxl* mRNA from being translated into a functional polypeptide.

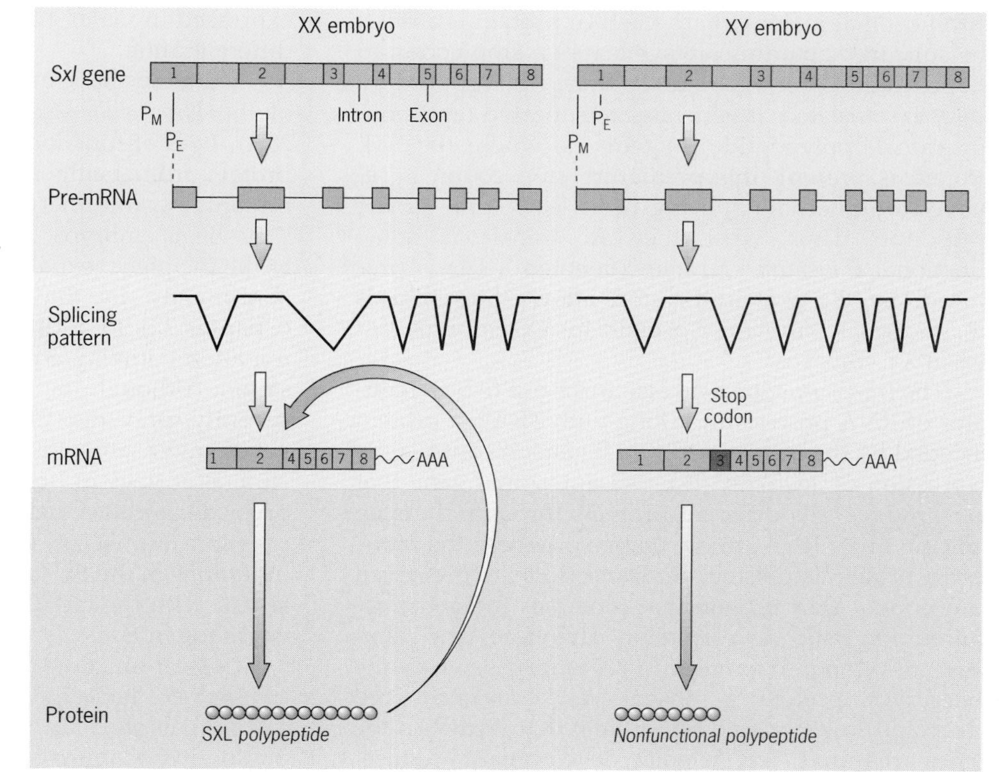

Figure 23.9 Sex-specific expression of the *Sex-lethal* (*Sxl*) gene in *Drosophila*. Although this gene is transcribed in both XX and XY embryos, alternate splicing of its RNA limits the synthesis of the SXL protein to XX embryos, which develop as females. The absence of SXL protein in XY embryos causes them to develop as males.

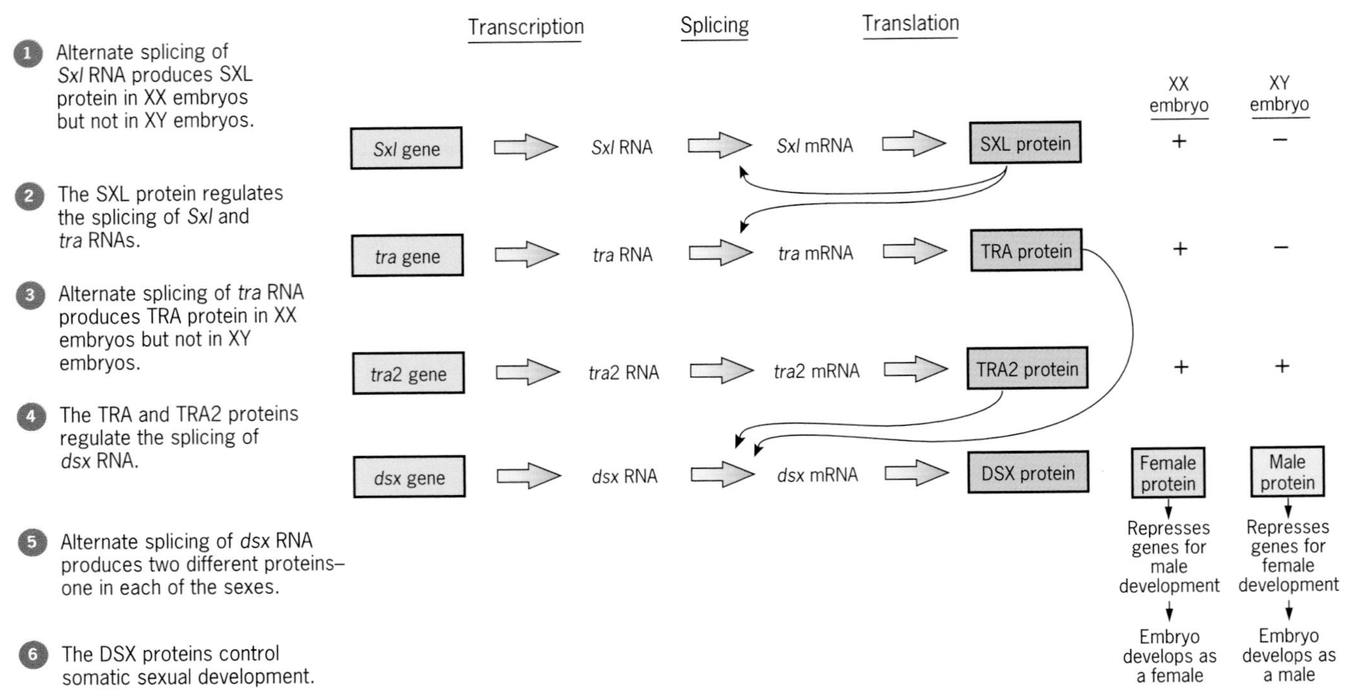

Figure 23.10 Regulation of sex determination in *Drosophila* by the *Sex-lethal* (*Sxl*) gene. The *Sxl* gene regulates the expression of the *transformer* (*tra*) gene, which, in turn, regulates the expression of the *doublesex* (*dsx*) gene; the *transformer2* (*tra2*) gene also participates in the regulation of *dsx*.

scripts can be processed in two different ways. In chromosomal males, where the SXL protein is absent, the splicing apparatus always leaves a stop codon in the second exon of the *tra* RNA; thus, when spliced *tra* RNA is translated, it generates a truncated (and nonfunctional) polypeptide. In females, where the SXL protein is present, this premature stop codon is removed by alternate splicing in at least some of the transcripts. Thus, when they are translated, some functional *transformer* protein (denoted TRA) is produced.The SXL protein therefore allows the synthesis of functional *transformer* protein in XX embryos but not in XY embryos.

The *transformer* protein also turns out to be a regulator of RNA processing. Along with TRA2, a protein encoded by the *transformer2* (*tra2*) gene, it controls the expression of *doublesex* (*dsx*), an autosomal gene that can produce two different proteins through alternate splicing of its RNA. In XX embryos, where the *transformer* protein is present, *dsx* transcripts are processed to encode a DSX protein that represses the genes required for male development. Therefore, such embryos develop into females. In XY embryos, where the *transformer* protein is absent, *dsx* transcripts are processed to encode a DSX protein that represses the genes required for female development. Consequently, such embryos develop into males. The *dsx* gene is therefore the switch point at which a male or female developmental pathway is explicitly chosen. From this point, different sets of genes are specifically expressed in males and females to bring about sexual differentiation.

Mutations have been obtained in each of the genes of the *Drosophila* sex-determination pathway (Table 23.1). Loss-of-function mutations in *Sxl* prevent SXL protein from being made in females. Homozygous mutants would therefore develop into males—but they die as embryos. This embryonic death is not due to the incipient sexual transformation but rather to an abnormality in the dosage compensation system (Chapter 22). In addition to regulating the sex-determination pathway, *Sxl* also regulates dosage compensation. Although the mechanism is not known, it apparently prevents the hyperactivation of X-linked genes in XX animals. When this hyperactivation occurs, as it does in homozygous *Sxl* mutants, XX embryos die because there is too much X-linked gene expression. However, in XY animals, loss-of-function mutations in the *Sxl* gene have no effect, which is consistent with the fact that the SXL protein is normally not made in males.

Loss-of-function mutations in *transformer* and *transformer2* have the same phenotype: both XX and XY animals develop into males. The sexual transformation in XX animals demonstrates that both the tra^+ and $tra2^+$ genes are needed for female development; however, they are perfectly dispensable for male development. Loss-of-function mutations in the *dsx* gene

TABLE 23.1
Phenotypes of Loss-of-Function Mutations in Sex-Determination Genes in *Drosophila melanogaster* and *Caenorhabditis elegans*.

Gene	*XX mutant phenotype*	*XY (or XO) mutant phenotype*
Drosophila melanogaster		
Numerator gene	lethal	no effect
Denominator gene	no effect	reduced viability
Sxl	lethal	no effect
tra	male	no effect
tra2	male	sterile male
dsx	sterile intersex	sterile intersex
Caenorhabditis elegans		
xol-1	no effect	lethal
sdc-1	masculinized	no effect
sdc-2	masculinized	no effect
sdc-3	no sex-determination effect	no effect
her-1	no effect	fertile hermaphrodite
tra-2	male	no effect
tra-3	male	no effect
fem-1	female	female
fem-2	female	female
fem-3	female	female
tra-1	male	minor effects in gonad

Source: S. M. Parkhurst and P. M. Meneely, *Science* 264:924–932, 1994.

cause both XX and XY embryos to develop into intersexes. The intersexual phenotype appears because *both* the male and female developmental pathways are activated in *dsx* mutants.

Sex Determination in *Caenorhabditis*

The somatic sex-determination pathway in *Caenorhabditis* involves at least 10 different genes (Figure 23.11). As in *Drosophila*, mutations in these genes alter sexual development (Table 23.1). For example, loss-of-function mutations in the two *transformer* genes *tra-1* and *tra-2* (not to be confused with the *tra* and *tra2* genes of *Drosophila*) cause XX animals to develop into males, and loss-of-function mutations in the *hermaphrodite her-1* gene cause XO animals to develop into hermaphrodites. These phenotypic alterations show that the *tra-1* and *tra-2* gene products are needed for normal hermaphrodite development and that the *her-1* gene product is needed for normal male development.

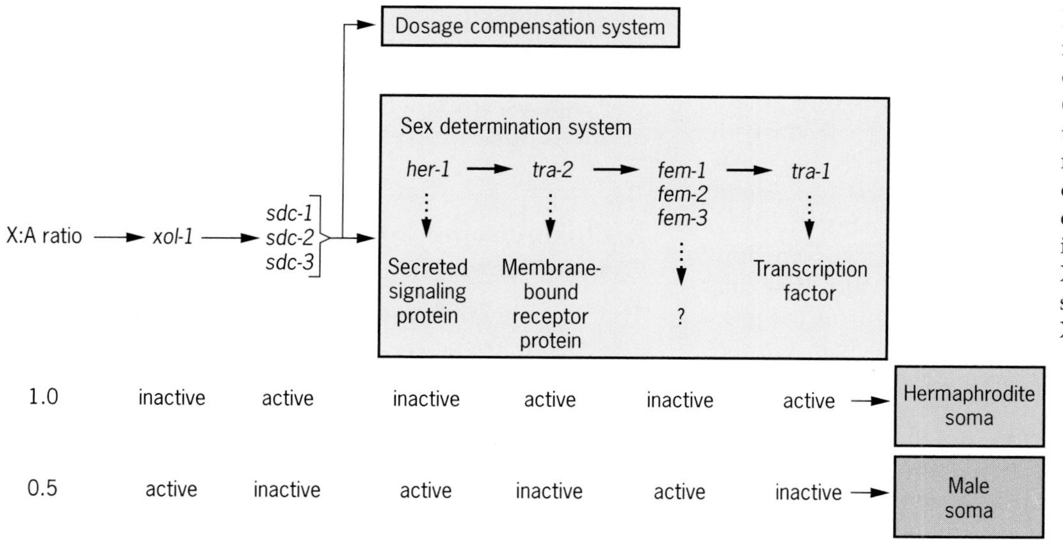

Figure 23.11 Developmental pathway for sex determination in *Caenorhabditis*. Differentation of the animal as a male or as a hermaphrodite depends on a cascade of gene expression, which, in turn, depends on the X:A ratio. Dosage compensation also depends on the X:A ratio.

Loss-of-function mutations in three other genes, *fem-1*, *fem-2*, and *fem-3* (for feminization), cause XO animals to develop into spermless females. This feminized phenotype shows that the *fem* gene products are needed for normal male development.

The phenotypes of double mutant combinations have established an epistasis hierarchy for these six sex-determination genes:

tra-1 >> *fem-1*, *fem-2*, *fem-3* >> *tra-2* >> *her-1*

For example, a *tra-1*, *her-1* double mutant develops as a male, whether it is XO or XX. The masculinizing *tra-1* mutation is therefore epistatic over the feminizing *her-1* mutation. Using these epistatic relationships and additional information about how the genes might function, researchers have been able to arrange these six genes in a developmental pathway.

The first gene, *her-1*, encodes a secreted protein that is likely to be a signaling molecule. The next gene, *tra-2*, encodes a membrane-bound protein, which may function as a receptor for the *her-1* signaling protein. The products of the *fem* genes are not known, but the last gene in the pathway, *tra-1*, apparently encodes a zinc-finger type transcription factor, which may regulate the genes involved in sexual differentiation.

In *Caenorhabditis*, the sex-determination pathway seems to involve a series of negative regulators of gene expression. In XO animals, the secreted *her-1* gene product apparently interacts with the *tra-2* gene product, causing it to become inactive. This interaction allows the three *fem* gene products to be activated, and they collectively inactivate the *tra-1* gene product, which is a positive regulator of female differentiation. Because the animal cannot develop as a hermaphrodite without active *tra-1* protein, it develops as a male. In XX animals, the *her-1* protein is either not made or not secreted; thus, its putative receptor, the *tra-2* protein, remains active. Active *tra-2* protein causes the *fem* gene products to be inactivated, which in turn allows the *tra-1* protein to stimulate female differentiation. The animal therefore develops as a hermaphrodite.

Sexual development in *Caenorhabditis* fundamentally depends on the X:A ratio, just as it does in *Drosophila*. However, in *Caenorhabitis* it is not clear how this ratio is ascertained. There may be a system of numerator and denominator elements, but to date, none has been identified. The X:A ratio is somehow converted into a molecular signal that controls sexual differentiation. This same signal also controls the phenomenon of dosage compensation, which in *Caenorhabditis* involves hypoactivation of the two X chromosomes in the hermaphrodite (Chapter 22). The signal from the X:A ratio is funneled into the sex-determination and dosage-compensation pathways through a short pathway involving at least four genes. One of these, *xol-1*, is required in males but not in hermaphrodites. Loss-of-function mutations in *xol-1* cause XO animals to die, hence the name of the gene, *XO-lethal*. Three other genes, *sdc-1*, *sdc-2*, and *sdc-3* (for sex determination and dosage compensation), are negatively regulated by *xol-1*. Loss-of-function mutations in these genes either kill XX animals or turn them into males. Thus, the *sdc* genes are needed in hermaphrodites but not in males.

Key Points: **The genetic dissection of a developmental pathway consists of analyzing genes whose products are involved in the differentiation of specific tissues and organs. In *Drosophila*, for example, the pathway that controls sexual differentiation into male or female involves some genes that ascertain the X:A ratio, some that convert this ratio into a developmental signal, and others that respond to the signal by producing either male or female structures. In *Caenorhabditis*, the sexual-differentiation pathway involves genes that encode signaling proteins, their receptors, and transcription factors.**

THE USE OF GENETIC MOSAICS TO STUDY DEVELOPMENT

Because all the cells in a multicellular organism are derived mitotically from a single fertilized egg, we would expect every one of them to have the same genotype. For the most part, this seems to be the case. The phenotypic differences that emerge among cells during development are caused by differential gene expression rather than by differences in gene content. However, abnormalities in mitosis sometimes produce daughter cells that have different genotypes. Replication of these daughter cells within the developing animal can give rise to genetically different clones. An animal with such clones is called a **genetic mosaic**. The analysis of genetic mosaics has allowed researchers to trace cell lineages during development as well as to ascertain whether genes control intracellular or intercellular processes.

X Chromosome Loss in *Drosophila*: Gynandromorphs

The first genetic mosaics to be studied in *Drosophila* arose from loss of an X chromosome very early in development (Figure 23.12). These mosaics occur spontaneously at a low rate and are composed of two types of cells, one containing two X chromosomes and the other containing only one. Such XX/XO mosaics develop into flies that are part female and part male, called gynandromorphs (Chapter 6). The XX tissues in a gynan-

Figure 23.12 A *Drosophila* gynandromorph. Both front legs are male (sex-combs present). The left side of the head and thorax is wild-type and female; the right side shows two X-linked mutant phenotypes, eosin eye and forked bristles, and is male.

dromorph differentiate into female structures, and the XO tissues differentiate into male structures. The perfect coincidence between the sexual phenotype of a tissue and its chromosomal genotype demonstrates that in *Drosophila*, sex determination occurs wholly within cells. Communication between cells—for example, by hormones—does not seem to play a role. Instead, every cell develops its sexual phenotype independently of every other cell. Thus sexual development in *Drosophila* is said to be a **cell-autonomous process**.

Other aspects of *Drosophila* development are also cell-autonomous—for example, the formation and pigmentation of the sensory bristles and hairs that cover much of the adult body. Two genes controlling these processes are located on the X chromosome. The *singed* (*sn*) gene controls bristle formation and the *yellow* (*y*) gene controls bristle pigmentation. Loss-of-function mutations in the *singed* gene cause the bristles to be gnarled and bent, instead of straight; loss-of-function mutations in the *yellow* gene cause the bristles to be brownish-yellow instead of black. If a zygote that is heterozygous for these recessive mutations (genotype *y sn*/+ +) loses the wild-type X chromosome during one of the early cleavage divisions, it will develop into a gynandromorph with brownish-yellow and singed bristles in the male tissue and wild-type bristles in the female tissue. The male tissue will be *y sn*/O and the female tissue will be *y sn*/+ +. Within the male tissue, the bristles will manifest both of the mutant phenotypes, even if they are adjacent to female tissue on the adult body. Careful examination shows that each bristle develops from a single cell. Thus the formation and pigmentation of bristles in *Drosophila* is a cell-autonomous process.

Somatic Recombination in *Drosophila*: Twin Spots

Genetic mosaics can also be produced by somatic recombination between homologous chromosomes. This is a rare event in *Drosophila*, even though homologous chromosomes pair in the somatic cells. However, its frequency can be increased by X-irradiation during development. Radiation-induced breaks in the DNA apparently initiate recombination events between paired chromosomes. As an example, let's consider an animal that is a repulsion heterozygote for mutations in the *yellow* and *singed* genes (genotype *y* +/+ *sn*). Both of these genes are far away from the centromere of the X chromosome. Consequently, it is possible to induce a recombination event between the centromere and the segment of the X that carries the two genes.

Figure 23.13 shows such a recombination event in a tetrad of chromatids produced by chromosome replication. When these chromatids disjoin during the following mitosis, each daughter cell may wind up homozygous for both of the loci. One cell will be ho-

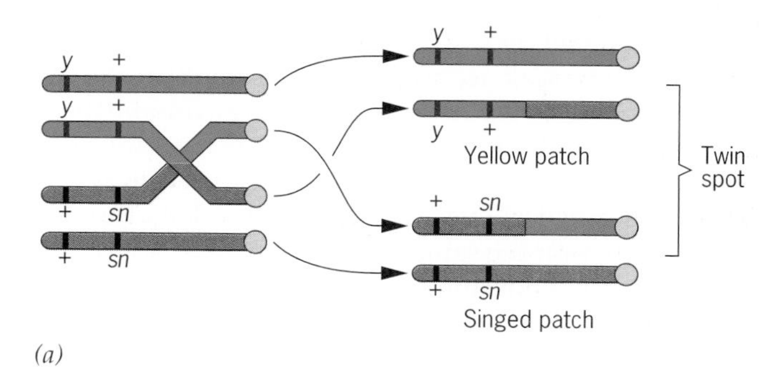

(a)

(b)

Figure 23.13 Production of twin spots in *Drosophila* by somatic recombination. (a) Recombination between the genes *yellow* (*y*) and *singed* (*sn*) and the centromere produces genetically different daughter cells, one homozygous for the *yellow* mutation and the other homozygous for the *singed* mutation. (b) A yellow/singed twin spot on the thorax of a fly.

mozygous for the mutant allele of *yellow* and the wild-type allele of *singed* (genotype *y* +/*y* +), and the other cell will have the reciprocal genotype (+ *sn*/+ *sn*). Of course, this is not the only possible outcome. The chromatids may disjoin so that both cells remain heterozygous for both of the loci. However, if disjunction produces cells that are homozygous for the different mutants, and if these cells generate clones in the bristle-forming tissues, the adult will show a patch of yellow bristles adjacent to a patch of singed bristles in an otherwise wild-type background. This condition is referred to as a **twin spot**.

Somatic recombination can be induced at any time during development. When the recombination is induced early, the resulting mosaic patches are large; when it is induced late, they are small. The size of a mosaic patch therefore indicates how early in development the causative recombination event occurred. Because all the cells in a patch are descendants of a common ancestor, the shape of a patch can be used to trace cell lineages during development.

Cell-autonomous genes such as *yellow* and *singed* can be used as markers to determine whether other genes linked to them are also cell autonomous (Figure 23.14). Let's suppose, for example, that we are interested in a gene located between *yellow* and *singed* on the X chromosome that is essential for life. Loss-of-function mutations in this gene would be recessive lethals (denoted by *l*). To determine whether the gene is cell-autonomous, we could induce somatic recombination in *y l* +/+ + *sn* heterozygotes. We would expect that exchanges between the *singed* locus and the centromere would produce twin spots on the adult body, as we discussed above. However, if the lethal mutation, which is on the same X chromosome as the *yellow* mutation, functions autonomously, the homozygous *y l* +/*y l* + cells produced by such exchanges will die, leaving their + + *sn*/+ + *sn* sister cells to grow into solitary patches of singed tissue on the adult. The appearance of these solitary singed patches would therefore be an indication that the lethal mutation is cell-autonomous. If it were nonautonomous, homozygous *y l* +/*y l* + cells could be rescued by the surrounding heterozygous tissue, and yellow/singed twin spots would appear. This technique therefore allows us to determine whether a gene controls intracellular (autonomous) or intercellular (nonautonomous) functions.

Key Points: **Genetic mosaics caused by mitotic chromosome loss or somatic recombination can be used to trace cell lineages during development and to ascertain whether a gene functions autonomously within cells. In *Drosophila*, such mosaics can be created by X chromosome loss or by radiation-induced somatic recombination.**

MOLECULAR ANALYSIS OF GENES INVOLVED IN DEVELOPMENT

The detailed analysis of developmentally significant genes requires that they be cloned and sequenced. For both *Drosophila* and *Caenorhabditis*, this is a fairly straighforward process involving the techniques we discussed in Chapters 19 and 20. Genomic or cDNA libraries can be constructed from either species and screened for particular genes. Screening is facilitated by the relatively small size of the genome (170,000 kb in *Drosophila* and 80,000 kb in *Caenorhabditis*) and by the wealth of genetic, cytological, and molecular data already collected from these organisms. In *Drosophila*, for example, cloned DNA sequences can be localized

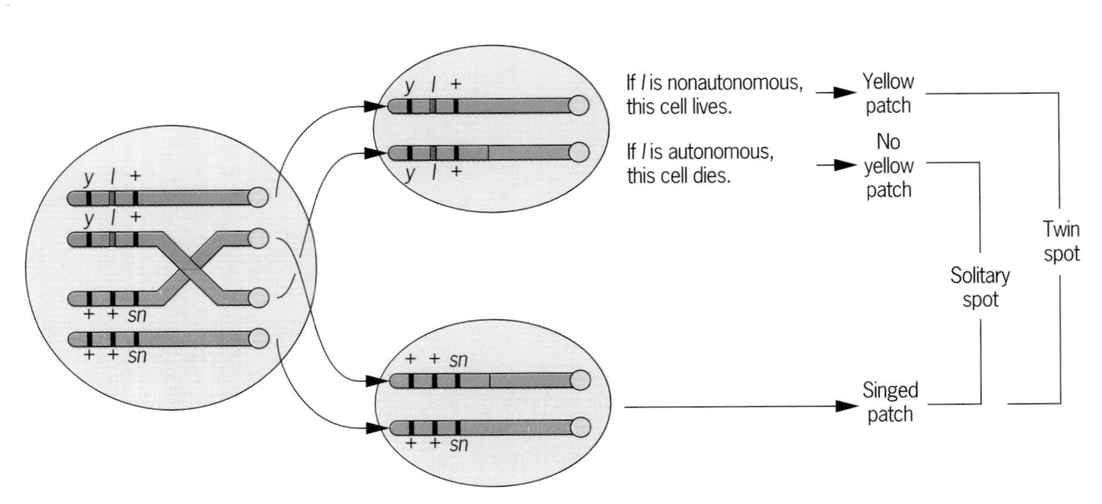

Figure 23.14 Somatic recombination in *Drosophila* with markers to determine if a lethal mutation (*l*) is autonomous or nonautonomous.

cytologically by *in situ* hybridization to polytene chromosome squashes. In *Caenorhabditis*, clones can be localized on a detailed physical map of the genome by hybridization with a comprehensive array of tester clones fixed to nylon membranes. Transposon tagging, a procedure whereby genes are mutated by the insertion of mobile genetic elements (Chapter 17), is also possible in these organisms. Once tagged, a mutant gene can be cloned by using a transposon probe to screen a recombinant DNA library made from mutant animals.

Cloning a gene makes it possible to conduct experiments that can help scientists understand the gene's developmental significance. Cloned sequences can be used to make probes to detect transcripts from the gene in different tissues and at different times during development. They can also be used to generate antibodies to the protein encoded by the gene. Once antibodies have been made, they can be used to detect the protein in specific tissues at different developmental stages. All these procedures help researchers understand where and when a gene is expressed in a developing animal.

For both *Drosophila* and *Caenorhabditis*, a cloned gene can also be reintroduced into the animal, perhaps after altering it *in vitro*, to determine how the gene functions *in vivo*. The techniques for genetic transformation with cloned DNA were first developed for *Drosophila* using transposable elements as vectors to insert the cloned DNA into the genome (Chapter 17). These techniques involve injecting the DNA into the primordial germ line of *Drosophila* embryos prior to blastoderm formation. At this stage, it is possible for the injected DNA to be integrated into the chromosomes of the future germ line. If integration occurs and the injected animal survives to become an adult, the inserted DNA may be passed on to the animal's offspring. This procedure of germ-line transformation can therefore create a stock of flies in which cloned DNA has been inserted into the genome. The inserted DNA is called a transgene, and the flies are called transgenic animals. Although the most common procedure involves transforming *Drosophila* with *Drosophila* DNA, it is also possible to transform them with DNA from other organisms—bacteria, other invertebrates, even vertebrates. In this way, *Drosophila* can be used as a living laboratory to investigate the functions of genes from many different species, including human beings.

Cloned DNA can also be introduced into *Caenorhabditis* by injecting it into a syncytial region in the gonad of a hermaphrodite. In this case, however, the DNA usually does not integrate into the chromosomes but rather, remains unincorporated, probably as a tandem array of linked molecules. By chance, some of the DNA may be packaged into sperm that are formed in this part of the gonad and then may be subsequently transmitted to a zygote. During development, this extrachromosomal DNA segregates irregularly during cell division, creating mosaic animals that may or may not pass the DNA on to their offspring.

Key Points: **Cloning a gene is fundamental to determining its role in development. Genetically transformed, or transgenic, animals can provide information on how a cloned gene functions *in vivo*.**

MATERNAL GENE ACTIVITY IN DEVELOPMENT

Important events occur in animal development even before an egg is fertilized. At this time, nutritive and determinative materials are transported into the egg from surrounding cells, laying up food stores and organizing the egg for its subsequent development—the molecular equivalent of a mother's love. These materials are generated by the expression of genes in the female reproductive system, some being expressed in somatic reproductive tissues and others only in germline tissues. Collectively, these genes help to form eggs that can develop into embryos after fertilization. In some species, these maternal gene products lay out the basic body plan of the embryo, distinguishing head from tail and back from belly. These maternally supplied materials therefore establish a molecular coordinate system to guide an embryo's development.

Maternal-Effect Genes

Mutations in genes that contribute to the formation of healthy eggs may have no effect on the viability or appearance of the female making those eggs. Instead, their effects may be seen only in the next generation. Such mutations are called **maternal-effect mutations** because the mutant phenotype in the offspring is caused by a mutant genotype in its mother.

Genes identified by such mutations are called **maternal-effect genes**. The *dorsal* (*dl*) gene in *Drosophila* is a good example (Figure 23.15). Matings between flies homozygous for recessive mutations in this gene produce inviable progeny. This lethal effect is strictly maternal. A cross between homozygous mutant females and homozygous wild-type males produces inviable progeny, but the reciprocal cross (homozygous mutant males x homozygous wild-type females) produces viable progeny. The lethal effect of the *dorsal* mutation is therefore manifested only if females are homozygous for it. The male genotype is irrelevant.

Molecular characterization of the *dorsal* gene has revealed the basis for this maternal effect. The *dorsal*

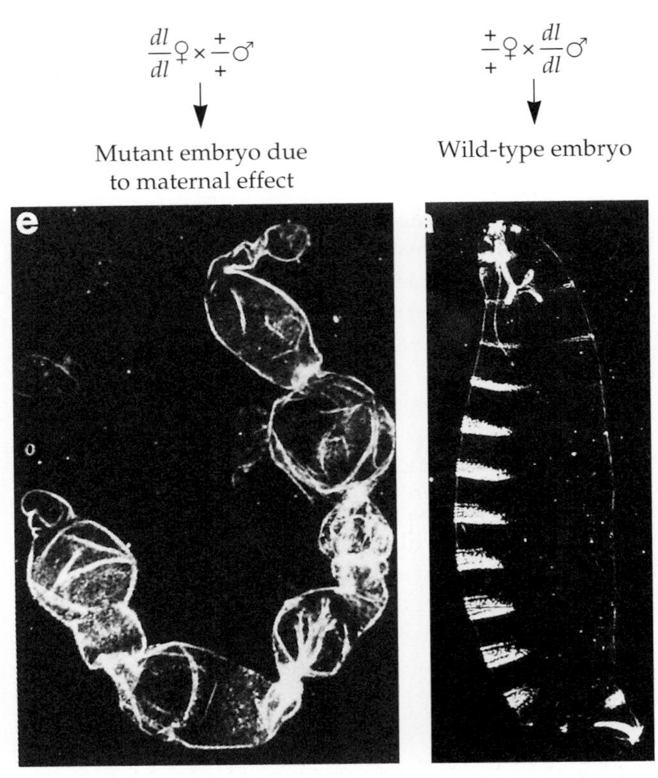

Figure 23.15 The maternal effect of a mutation in the *dorsal* (*dl*) gene of *Drosophila*. The mutant phenotype is an embryo that lacks ventral tissues, that is, it is dorsalized.

gene encodes a transcription factor that is produced during oogenesis and stored in the egg. Early in development, this transcription factor plays an important role in the differentiation of the dorsal and ventral parts of the embryo. When it is missing, the ventral parts incorrectly differentiate as if they were on the dorsal side, creating an embryo with two dorsal surfaces. This lethal condition cannot be prevented by a wild-type *dorsal* allele inherited from the father because *dorsal* is not transcribed in the embryo. Rather, because its expression is limited to the female germ line, mutations in *dorsal* are strict maternal-effect lethals.

Determination of the Dorsal-Ventral and Anterior-Posterior Axes in *Drosophila* Embryos

Animals with bilateral symmetry have two primary body axes, one distinguishing back from belly (dorsal from ventral) and the other distinguishing head from tail (anterior from posterior). Both of these axes are established very early in development, in some species even before fertilization. In *Drosophila*, the processes of axis formation have been dissected genetically by collecting mutations that affect early embryonic development.

In the 1970s and 1980s, massive searches for such mutations were carried out by Christiane Nüsslein-Volhard, Eric Weischaus, Trudi Schüpbach, Gerd Jurgens, and others. These researchers used chemical mutagens to induce mutations in each of the *Drosophila* chromosomes. Many mutations were identified, including maternal-effect lethals in genes such as *dorsal*. Molecular and genetic analyses of these mutations have provided considerable insight into the events of early *Drosophila* development.

In *Drosophila* the formation of the dorsal-ventral axis hinges on the action of the transcription factor encoded by the *dorsal* gene (Figure 23.16). This protein is synthesized maternally and stored in the cytoplasm of egg. At the time of blastoderm formation, the *dorsal* protein enters the nuclei on the ventral side of the embryo, inducing the transcription of two genes called *twist* and *snail* (whimsically named for their mutant phenotypes). In these same nuclei, it represses the genes *zerknullt* (from the German for "crumpled") and *decapentaplegic* (from the Greek words for "15" and "stroke"). This selective induction and repression of genes causes the ventral cells to differentiate into em-

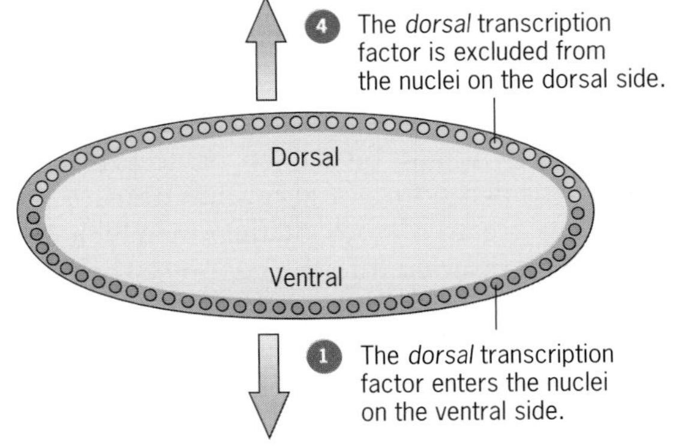

Figure 23.16 Determination of the dorsal-ventral axis in *Drosophila* by the *dorsal* protein, a transcription factor that acts only in the nuclei on the ventral side of the embryo. The genes *twist*, *snail*, *zerknullt*, and *decapentaplegic* are regulated by *dorsal* protein.

bryonic mesoderm. On the opposite side of the embryo, where the *dorsal* protein is excluded from the nuclei, *twist* and *snail* are not induced and *zerknullt* and *decapentaplegic* are not repressed. Consequently, these cells differentiate into embryonic epidermis. The entrance of the *dorsal* transcription factor into the ventral nuclei and its excusion from the dorsal nuclei therefore create the dorsal-ventral axis.

But what triggers the *dorsal* protein to move into the nuclei on only one side of the embryo? The answer lies in a complex exchange of signals between the tissues of the female germ line, which consist of the oocyte and its associated nurse cells, and the surrounding somatic follicle cells (Figure 23.17). During oogenesis, several gene products synthesized in the germ line transmit a signal to the follicle cells, inducing them to differentiate into dorsal and ventral types. Gene products from the ventral follicle cells then transmit a signal to the ventral side of the oocyte. This signal is received by a membrane-bound protein encoded by a gene called *Toll* (from the German for "tuft"). After the *Toll* receptor protein has been activated by this signal, it induces the transcription factor encoded by the *dorsal* gene to enter the nuclei on the ventral side of the embryo, thereby creating the dorsal-ventral axis.

1 The germ line (nurse cells and oocyte) signals the somatic follicle cells to differentiate into dorsal and ventral types.

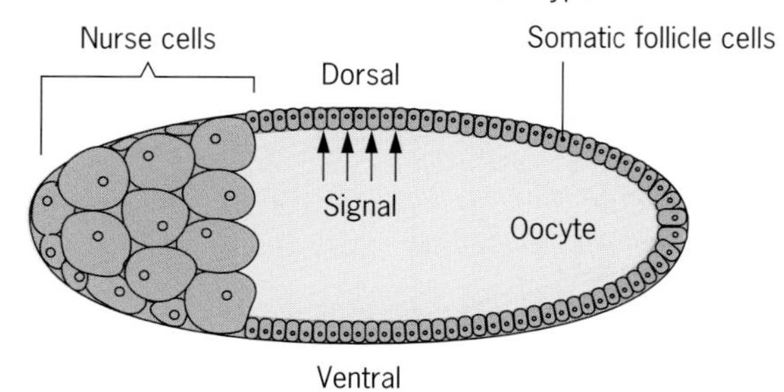

2 The ventral follicle cells signal the ventral surface of the oocyte to activate the *Toll* receptor protein on the oocyte's ventral surface.

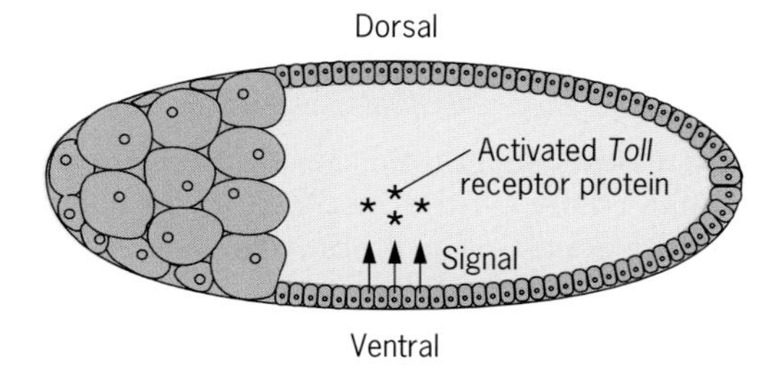

3 Eventually the activated *Toll* receptor protein stimulates the *dorsal* transcription factor to enter the nuclei on the ventral side of the embryo.

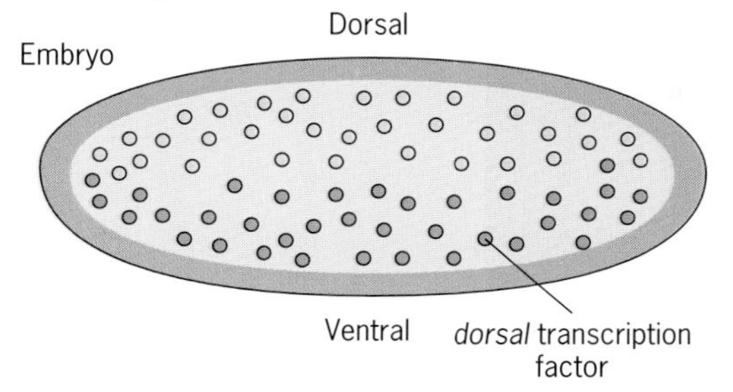

Figure 23.17 Signaling between the maternal germ line and the soma eventually determines the dorsal ventral axis of a *Drosophila* embryo.

The anterior-posterior axis in *Drosophila* (Figure 23.18) is created by the asymmetric synthesis of a transcription factor encoded by the *hunchback* gene (named for its embyronic mutant phenotype). This gene is transcribed in the maternal germ line during oogenesis, and its RNA becomes distributed uniformly throughout the young oocyte. However, *hunchback* RNA is translated only in the anterior half of the embryo; in the posterior half, it is degraded. This asymmetric synthesis of *hunchback* protein is the foundation of the anterior-posterior axis.

What limits the synthesis of *hunchback* protein to the anterior parts of the embryo? It turns out that two maternally supplied RNAs are involved. One is transcribed from the *bicoid* gene in the nurse cells of the germ line. This RNA is transported into the oocyte, where it becomes anchored at the anterior end. After fertilization, the *bicoid* RNA is translated into a protein that diffuses in a posterior direction, forming a concentration gradient over the anterior two-thirds of the embryo (Figure 23.19). The *bicoid* protein is a transcription factor capable of activating other genes, including *hunchback*. When *bicoid* activates *hunchback* in the anterior of the embryo, it augments the supply of hunchback RNA already there, leading to the synthesis of more *hunchback* protein. This protein then activates the genes that control the formation of anterior structures. It is, therefore, an anterior determinant.

As *bicoid*-regulated expression of *hunchback* defines the anterior of the embryo, repression of *hunchback* by another maternally supplied RNA defines the posterior. This RNA is transcribed from the *nanos* gene in the nurse cells of the ovary and is then transported to the posterior pole of the oocyte. Translation of this RNA creates a concentration gradient of *nanos* protein extending toward the anterior end of the embryo. Although the details are still somewhat unclear, *nanos* protein apparently functions to prevent the translation of *hunchback* RNA, possibly by binding to its 3' UTR and causing its degradation. This keeps *hunchback* protein, the anterior determinant, out of the posterior region of the embryo, which then develops its own characteristic structures.

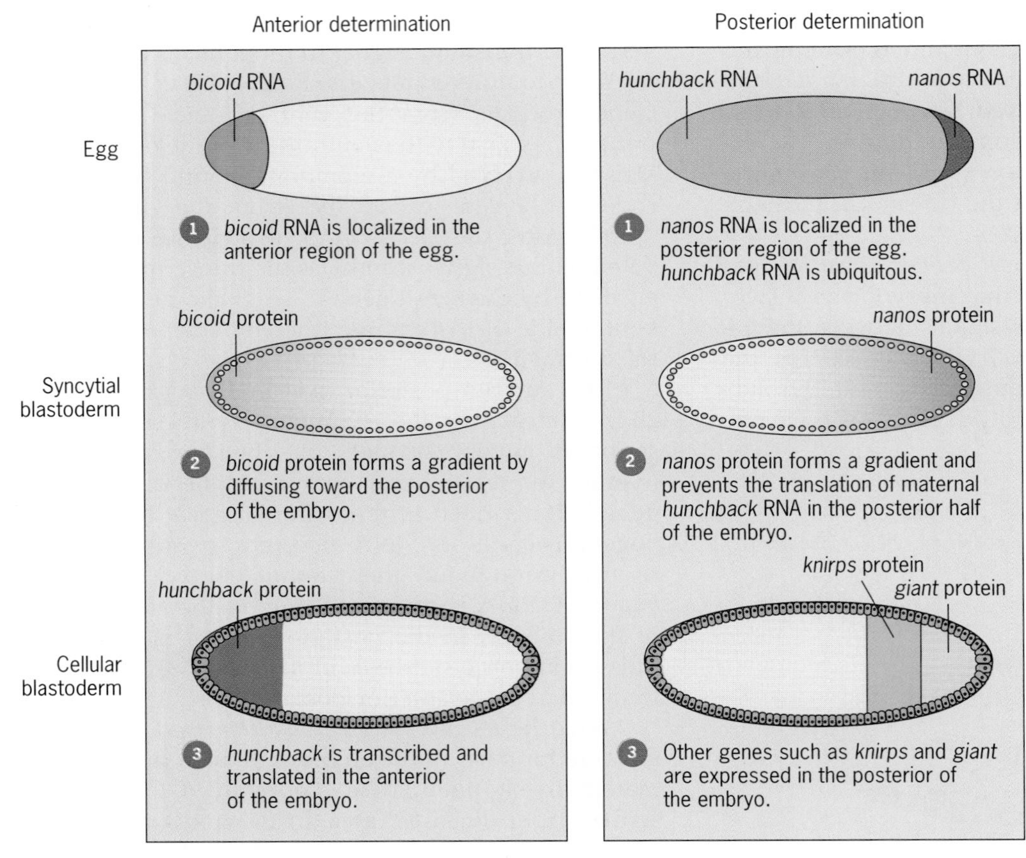

Figure 23.18 Determination of the anterior-posterior axis in *Drosophila* by maternally supplied RNA from the *hunchback*, *bicoid*, and *nanos* genes.

Key Points: **Materials transported into the egg during oogenesis play a major role in its development. These are the products of maternal-effect genes such as *dorsal*, *bicoid*, and *nanos* in *Drosophila*, all of which function in the determination of the embryonic axes. Recessive mutations in maternal effect genes are expressed only in embryos produced by homozygous females.**

ZYGOTIC GENE ACTIVITY IN DEVELOPMENT

The earliest events in animal development are controlled by maternally synthesized factors. However, at some point, the genes in the embryo are selectively activated, and new materials are made. This process is referred to as **zygotic gene expression**, because it oc-

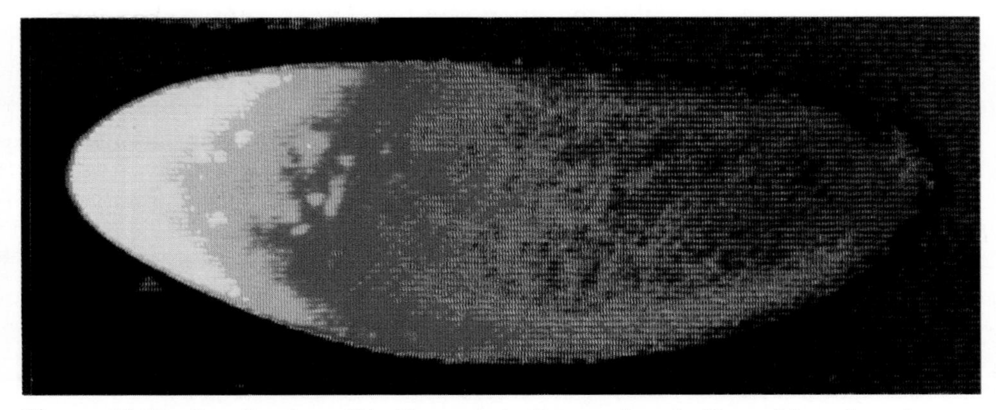

Figure 23.19 Localization of *bicoid* protein in the anterior of a *Drosophila* embryo by antibody-staining.

curs after the egg has been fertilized. The initial wave of zygotic gene expression is a response to maternally synthesized factors. In *Drosophila*, for example, the maternally supplied *dorsal* transcription factor activates the zygotic genes *twist* and *snail*. As development proceeds, the activation of other zygotic genes leads to complex cascades of gene expression. We shall now examine how these zygotic genes carry the process of development forward.

Body Segmentation

In many invertebrates the body consists of an array of adjoining units called **segments**. An adult *Drosophila*, for example, has a head, three distinct thoracic segments, and eight abdominal segments. Within the thorax and abdomen, each segment can be identified by coloration, bristle pattern, and the kinds of appendages attached to it. These segments can also be identified in the embryo and the larva (Figure 23.20). In vertebrates, a segmental pattern is not so evident in the adult, but it can be recognized in the embryo from the way that nerve fibers grow from the central nervous system, from the formation of branchial arches in the head, and from the organization of muscle masses along the anterior-posterior axis. Later in development, these features are modified, and the original segmental pattern becomes obscured. Nonetheless, in both vertebrates and many invertebrates, segmentation is a key aspect of the overall body plan.

Homeotic Genes Interest in the genetic control of segmentation began with the discovery of mutations that transform one segment into another. The first such mutation was found in *Drosophila* in 1915 by Calvin Bridges. He named it *bithorax* (*bx*) because it affected two thoracic segments; in this mutant, the third thoracic segment was transformed, albeit weakly, into the second, creating a fly with a small pair of rudimentary wings in place of the halteres (Figure 23.21). Later, other segment-transforming mutations were found in *Drosophila*, for example, *Antennapedia* (*Antp*), a mutant that partially transforms the antennae on the head into legs, which characteristically grow from the thorax. These mutations have come to be called **homeotic mutations** because they cause one body part to look like another. The word homeotic comes from William Bateson, who coined the term **homeosis** to refer to cases in which "something has been changed into the likeness of something else." Like so many other words Bateson coined, this one has become a standard term in the modern genetics vocabulary.

The bithorax and Antennapedia phenotypes result from mutations in homeotic genes. Several such genes have now been identified in *Drosophila*, where they form two large clusters on one of the autosomes (Figure 23.22). The **Bithorax Complex**, usually denoted **BX-C**, consists of three genes, *Ultrabithorax* (*Ubx*), *Ab-*

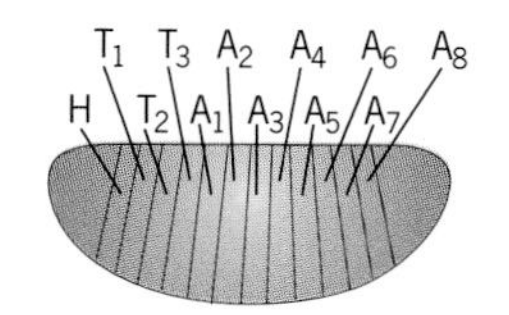

(*a*) Blastoderm

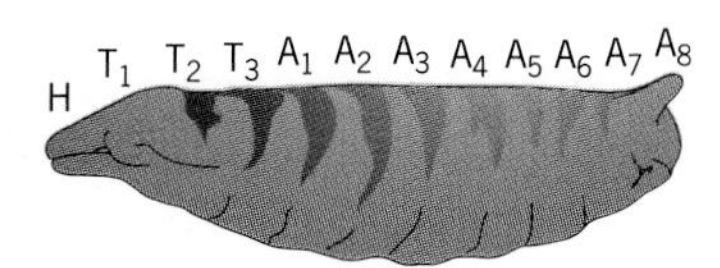

(*b*) Larva

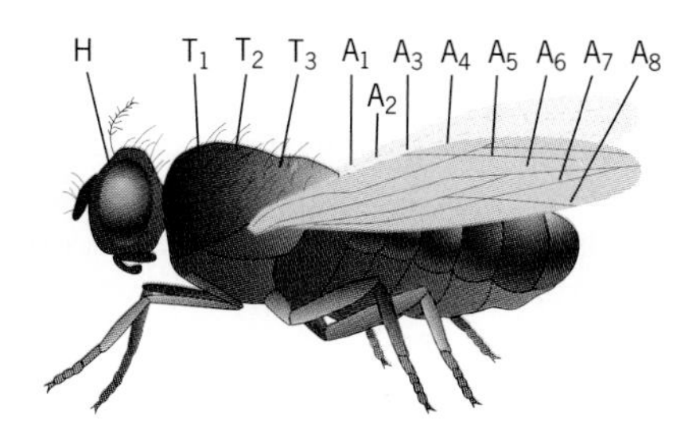

(*c*) Adult

Figure 23.20 Segmentation in *Drosophila* at the (*a*) blastoderm, (*b*) larval, and (*c*) adult stages of development. Although segments are not visible in the blastoderm, its cells are already committed to form segments as shown; H, head segment; T, thoracic segment; A, abdominal segment.

Figure 23.21 The phenotype of a *bithorax* mutation in *Drosophila*.

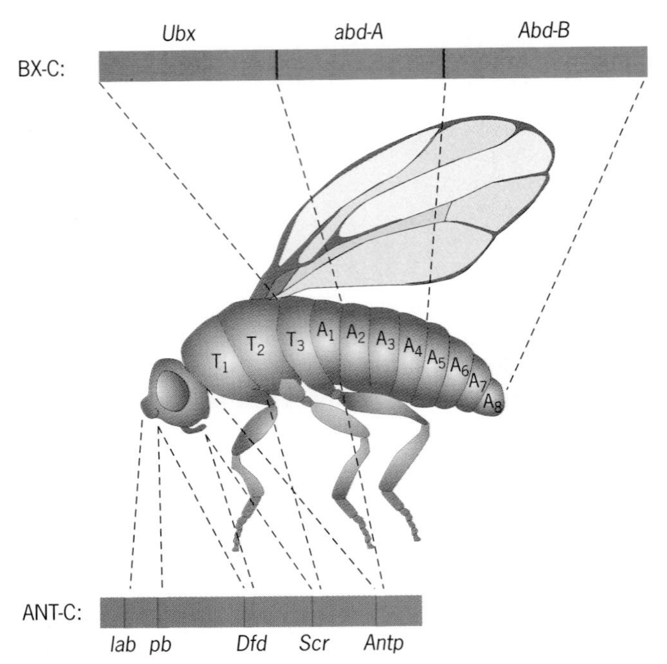

Figure 23.22 The homeotic genes in the Bithorax Complex (BX-C) and Antennapedia Complex (ANT-C) of *Drosophila*. The body regions in which each is expressed are indicated.

dominal-A (*Abd-A*), and *Abdominal-B* (*Abd-B*); the **Antennapedia Complex**, denoted **ANT-C**, consists of five genes, *labial (lab), proboscipedia (pb), Deformed (Dfd), Sex combs reduced (Scr),* and *Antennapedia (Antp).* Molecular analysis of these genes has demonstrated that they all encode helix-turn-helix transcription factors with a conserved region of 60 amino acids. This region, called the homeodomain, is involved in DNA binding.

The BX-C was the first of the two homeotic gene complexes to be dissected genetically. Analysis of this complex began in the late 1940s with the work of Edward Lewis. By studying mutations in the BX-C, Lewis showed that the wild-type function of each part of the complex is restricted to a specific region in the developing animal. Molecular analyses later reinforced and refined this conclusion. Study of the ANT-C began in the 1970s, principally through the work of Thomas Kaufman, Matthew Scott, and their collaborators. Through a combination of genetic and molecular analyses, these investigators showed that the genes of the ANT-C are also expressed in a regionally specific fashion. However, the ANT-C genes are expressed more anteriorly than the BX-C genes. Curiously, the pattern of expression of the ANT-C and BX-C genes along the anterior-posterior axis corresponds exactly to the order of the genes along the chromosome (Figure 23.22); it is not yet clear why this is so. The analysis of genetic mosaics has shown that these homeotic genes control the identities of individual segments in a cell-autonomous way. In fact, the developmental pathway that each cell takes seems to depend simply on the set of homeotic genes that are expressed within it. Because the homeotic genes play such a key role in selecting the segmental identities of individual cells, they are often called **selector genes**.

The proteins encoded by the homeotic genes are homeodomain transcription factors. These proteins bind to regulatory sequences in the DNA, including some within the Bithorax and Antennapedia Complexes themselves. For example, the UBX and ANTP proteins bind to a sequence within the promoter of the *Ubx* gene—a suggestion that the homeotic genes can regulate themselves and each other. Other gene targets of the homeodomain transcription factors have been identified, including some that encode other types of transcription factors. The homeotic genes therefore seem to control a regulatory cascade of target genes, which in turn act to determine the segmental identities of individual cells. However, the homeotic genes do not stand at the top of this regulatory cascade. Their activities are controlled by another group of genes expressed earlier in development—the segmentation genes.

Segmentation Genes Most of the homeotic genes were identified by mutations that alter the phenotype of the adult fly. However, these same mutations also have phenotypic effects in the embryonic and larval stages. This finding suggests that other genes involved in segmentation might be discovered by screening for mutations that cause embryonic and larval defects. In the 1970s and 1980s, Christiane Nüsslein-Volhard and Eric Wieschaus carried out such screens. They found a whole new set of genes required for segmentation along the anterior-posterior axis. Nüsslein-Volhard and Wieschaus classified these **segmentation genes** into three groups based on embryonic mutant phenotypes.

1. ***Gap Genes.*** These genes define segmental regions in the embryo. Mutations in the gap genes cause an entire set of contiguous body segments to be missing; that is, they create an anatomical gap along the anterior-posterior axis. Four gap genes have been well characterized: *Krüppel* (from the German for "cripple"), *giant, hunchback,* and *knirps* (from the German for "dwarf"). Each is expressed in characteristic regions in the early embryo under the control of the maternal effect genes *bicoid* and *nanos.* The gap genes encode transcription factors.
2. ***Pair-rule Genes.*** These genes define a pattern of segments within the embryo. The pair-rule genes

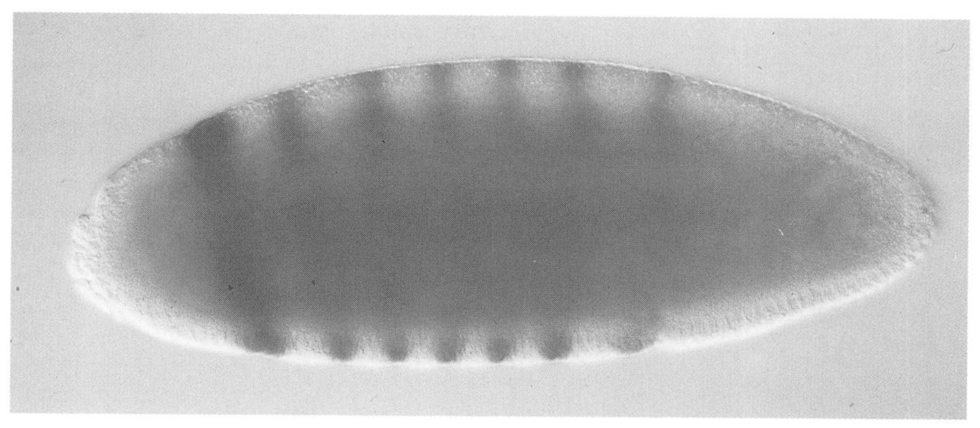

Figure 23.23 The seven-stripe pattern of RNA expression of the pair-rule gene *fushi tarazu* (*ftz*) in a *Drosophila* blastoderm embryo. The RNA was detected by *in situ* hybridization with a *ftz*-specific probe. Anterior is at the left; dorsal is at the top.

are regulated by the gap genes and are expressed in seven alternating bands, or stripes, along the anterior-posterior axis, in effect dividing the embryo into 14 distinct zones, or **parasegments** (Figure 23.23). Mutations in each of the several pair-rule genes produce embryos with only half as many parasegments as wild-type. In each mutant, every other parasegment is missing, although the missing parasegments are not the same in different pair-rule mutants. Examples of pair-rule genes are *fushi tarazu* (from the Japanese for "something missing") and *even-skipped*. In *fushi tarazu* mutants, each of the odd-numbered parasegments is missing; in *even-skipped* mutants, each of the even-numbered parasegments is missing. Most of the pair-rule genes encode transcription factors.

3. ***Segment-Polarity Genes***. These genes define the anterior and posterior compartments of individual segments along the anterior-posterior axis. Mutations in segment-polarity genes cause part of each segment to be replaced by a mirror-image copy of an adjoining half-segment. For example, mutations in the segment-polarity gene *gooseberry* cause the posterior half of each segment to be replaced by a mirror-image copy of the adjacent anterior half-segment. Many of the segment-polarity genes are expressed in 14 narrow bands along the anterior-posterior axis. Thus they refine the segmental pattern established by the pair-rule genes. Two of the best-studied segment-polarity genes are *engrailed* and *wingless*; *engrailed* encodes a transcription factor, and *wingless* encodes a signaling molecule.

These three groups of genes form a regulatory hierarchy (Figure 23.24). The gap genes, which are regionally activated by the maternal effect genes, regulate the expression of the pair-rule genes, which in turn, regulate the expression of the segment-polarity genes. Concurrent with this process, the homeotic genes are activated under the control of the gap and pair-rule genes to give unique identities to the segments that form along the anterior-posterior axis. Interactions among the products of all these genes then refine and stabilize the segmental boundaries. In this way, the *Drosophila* embryo is progressively subdivided into smaller and smaller developmental units. For an examination of developmental events in another group of insects, see Technical Sidelight: The Development of Butterfly Wings: Insights from *Drosophila* Genetics.

Specification of Cell Types

The genetic control of segmentation shows how sets of genes work in a regulatory cascade to determine the identities of groups of cells in different regions of the embryo. However, within these regions, what causes differentiation into specific cell types? What causes some cells to become neurons and others to become neuronal support cells? The answer to this kind of question is only now coming into scientific grasp, largely through the study of very simple situations involving a few distinct cell types. One such situation occurs in the development of the *Drosophila* eye (Figure 23.25).

Each of the large compound eyes in *Drosophila* originates as a flat sheet of cells in one of the imaginal discs. Initially, all the cells in this epithelial sheet look the same, but late during the larval stage, a furrow forms near the posterior margin of the disc. As this furrow moves in the anterior direction across the disc,

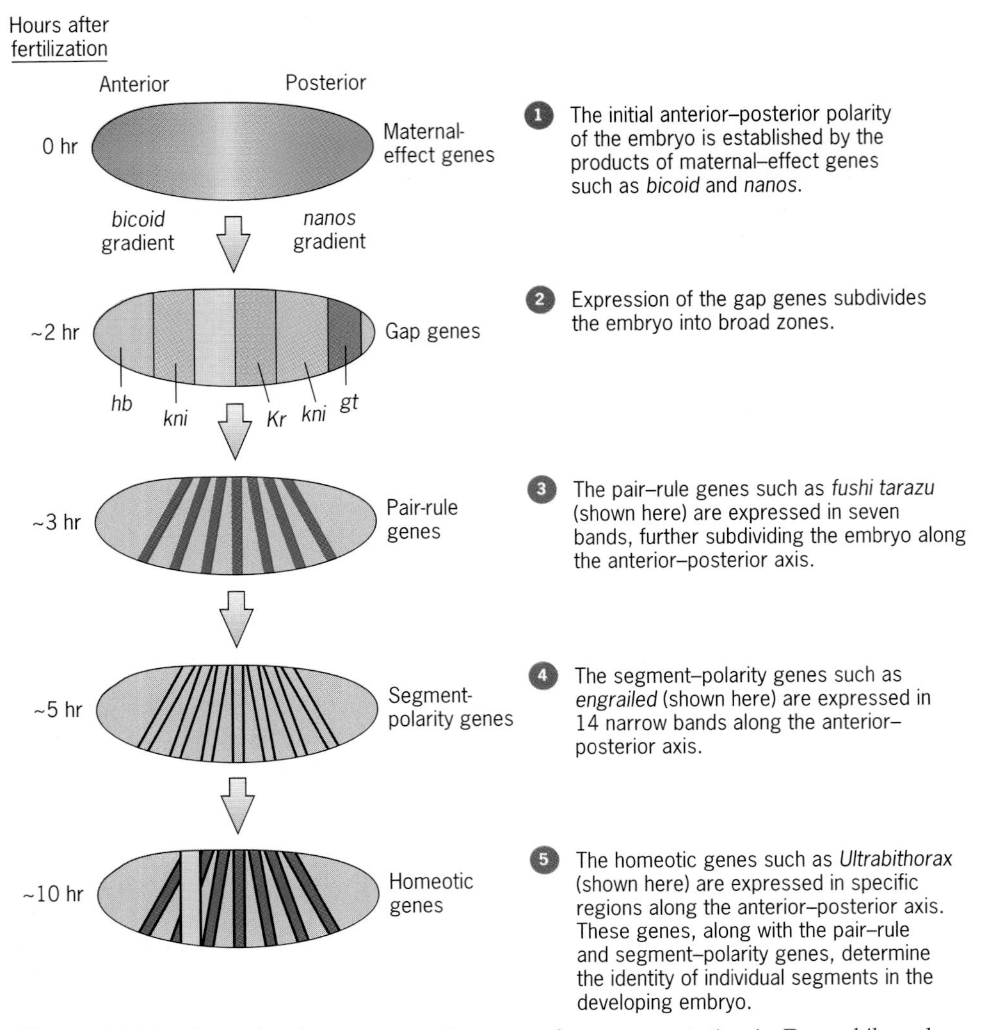

Figure 23.24 Cascade of gene expression to produce segmentation in *Drosophila* embryos.

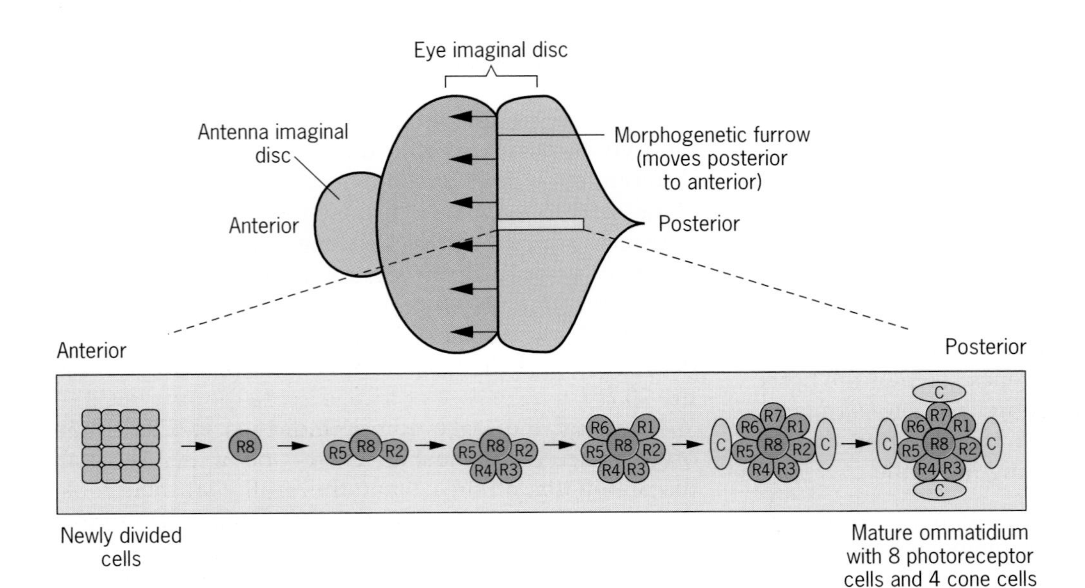

Figure 23.25 Development of the *Drosophila* eye. As the morphogenetic furrow moves toward the anterior of the eye-antenna imaginal disc, a wave of cell division follows in its wake. The newly divided cells then begin to differentiate into specific types. The insert shows the differentiation of the photoreceptor (R1-R8) and cone cells (C) that form each ommatidium of the compound eye.

it triggers a wave of cell division in its wake. The newly divided cells then differentiate into specific cell types to form the 800 individual facets of the adult eye. Each facet consists of 20 cells. Eight are photoreceptor neurons designed to absorb light; four are cone cells that secrete a lens to focus light into the photoreceptors; six are sheath cells to provide insulation and support; and the two remaining cells of a facet form sensory hairs on the eye's surface. Thus a highly patterned array of intricately differentiated facets develops in what had been a flat sheet of identical cells. What brings about this array?

Gerald Rubin and his collaborators have attempted to answer this question by collecting mutations that disrupt eye development. Their research has led to the idea that the specification of cell types within each facet depends on a series of cell–cell interactions. This is illustrated in the differentiation of the eight photoreceptor cells, denoted R1, R2, . . . R8 (Figure 23.26). In a fully formed facet, six of the photoreceptors (R1-R6) are arranged in a circle around the other two (R7, R8). One of the central cells, R8, is the first to differentiate in the developing facet. Its appearance is followed by the differentiation of the peripheral cells R2 and R5, then by R3 and R4, and R1 and R6; finally, the second central cell, R7, differentiates into a photoreceptor.

This last event has been studied in great detail. Rubin and his colleagues have shown that the differentiation of the R7 cell depends on reception of a signal from the already differentiated R8 cell. To receive this signal, the R7 cell must synthesize a specific receptor, a membrane-bound protein encoded by a gene called *sevenless* (*sev*). Mutations in this gene abolish the function of the receptor and prevent the R7 cell from differentiating as a neuron; instead, it differentiates as a cone cell. The signal for the R7 receptor is produced by a gene called *bride of sevenless* (*boss*) and is specifically expressed on the surface of the R8 cell. Contact between the differentiated R8 cell and the undifferentiated R7 cell allows the R8 signal, or **ligand** as it is technically called, to interact with the R7 receptor and activate it. This activation induces a cascade of changes within the the R7 cell that ultimately prompt it to differentiate as a light-receiving neuron. This differentiation is presumably mediated by one or more transcription factors acting on genes within the R7 nucleus. Thus the signal from the R8 cell is "transduced" into the R7 nucleus, where it alters the pattern of gene expression. The analysis of eye development in *Drosophila* therefore shows that **induction**, the process of determining the fate of an undifferentiated cell by a signal from a differentiated cell, can play an important role in the specification of cell types.

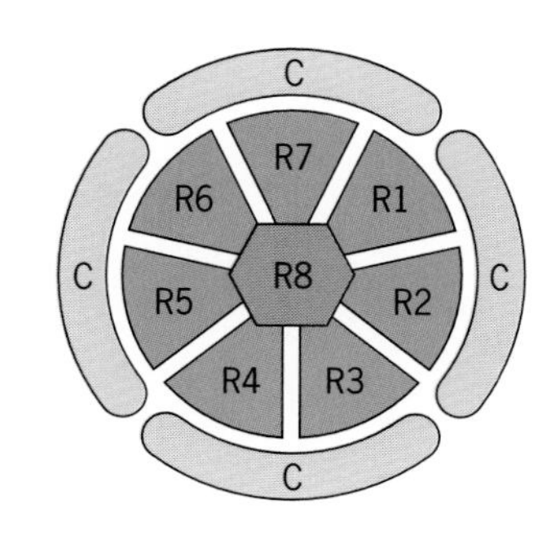

(*a*)

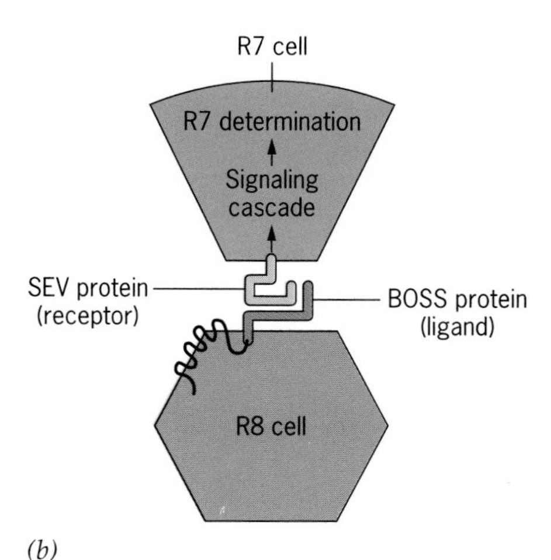

(*b*)

Figure 23.26 Determination of the R7 photoreceptor of an ommatidium in the *Drosophila* compound eye. (*a*) Arrangement of the eight photoreceptors (1-8) and four cone cells (C) in an ommatidium. (*b*) Signaling between the differentiated R8 cell and the presumptive R7 cell. The *bride of sevenless* protein (BOSS) on the R8 cell is the ligand for the *sevenless* receptor protein (SEV) on the surface of the R7 cell. Activation of this receptor initiates a signaling cascade within the R7 cell that induces it to differentiate.

Organ Formation

When many different types of cells are organized for a specific purpose, they form an organ. The heart, stomach, kidney, liver, and eye are all examples of organs. One of the remarkable features of an organ is that it forms in a specific part of the body. The development of a heart in the head, or an eye in the thorax would be extremely abnormal, and we would wonder what had gone wrong. Anatomically correct organ formation is obviously under tight genetic control.

Geneticists have obtained insights into the nature of this control from the study of another gene in *Drosophila*. This gene is called *eyeless* after the pheno-

TECHNICAL SIDELIGHT

The Development of Butterfly Wings: Insights from *Drosophila* Genetics

More than 17,000 species of butterflies have been identified, including many with beautiful patterns on their wings. These insects have long been the object of collectors and recently have caught the eyes of geneticists. What genes are involved in the formation of butterfly wings? What molecular events are responsible for producing their colorful patterns? How might the system for wing development have evolved in different groups of insects?

Sean Carroll, a molecular geneticist at the University of Wisconsin, is attempting to answer these questions by applying knowledge gained from the study of *Drosophila* wing development. The wings of butterflies and fruit flies both develop from imaginal discs, which are packets of undifferentiated cells sequestered in the larva. Initially, each disc is organized as a relatively small number of cells within the thousands of cells that are found in a particular body segment. During larval growth, the disc cells proliferate, and regions of the disc become determined to produce characteristic adult structures. For example, a pair of discs that form in one of the thoracic segments of *Drosophila* larvae will eventually differentiate into membranous wings that consist of two cell layers—dorsal and ventral; these discs also produce the thoracic tissue that anchors the wings to the adult's body. Carroll's research has shown that the developmental events that produce a butterfly's wing are similar to those that occur in *Drosophila*.

A wing disc in third instar *Drosophila* larvae consists of a single layer of cells (Figure 1). In one portion of the disc, the

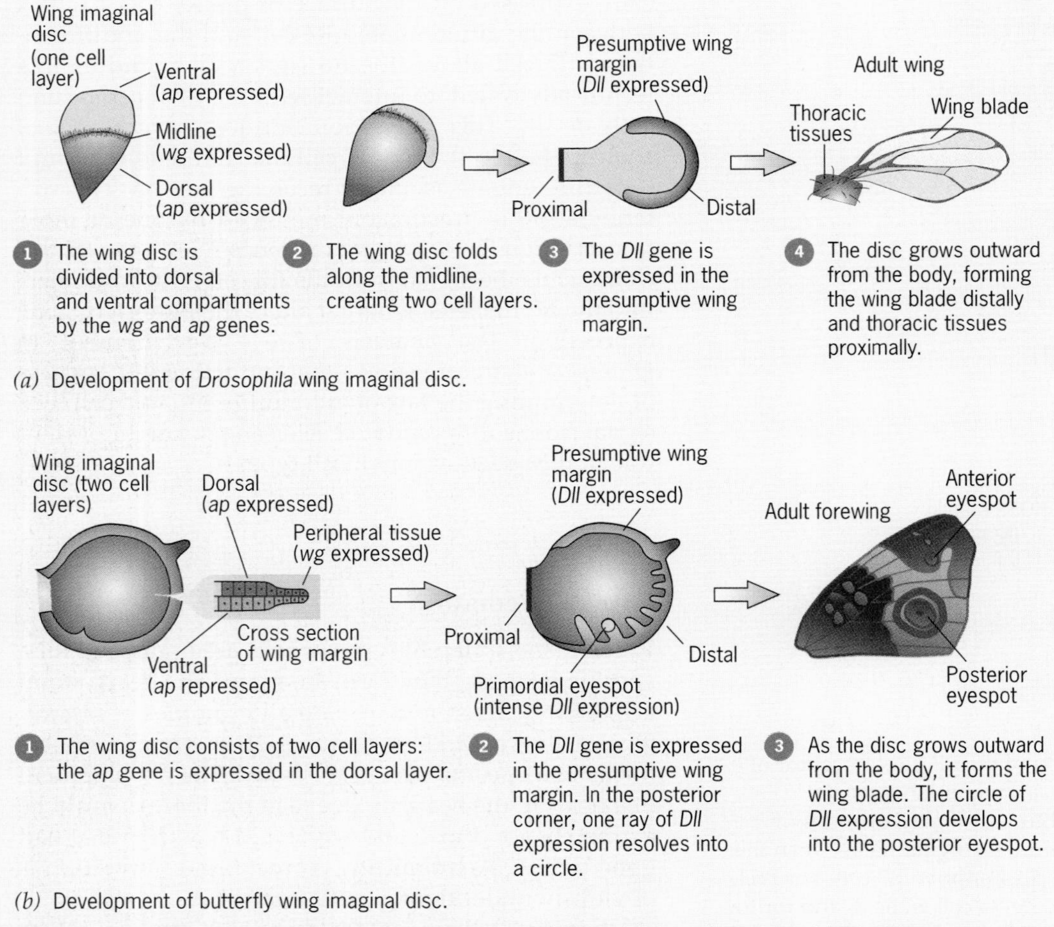

Figure 1 Comparison of development in *Drosophila* and butterfly wings. (*a*) Development of the *Drosophila* wing imaginal disc. (*b*) Development of a butterfly wing imaginal disc.

cells are determined to become the dorsal side of the adult wing; in another portion, they are determined to become the ventral side. This determination is associated with the expression of a gene called *apterous* (*ap*), which encodes a transcription factor. The *ap* gene is expressed in the presumptive dorsal portion of the disc; in the presumptive ventral portion, it is repressed. Another gene, *wingless* (*wg*), is expressed along the midline that divides the dorsal and ventral portions of the disc. This gene encodes a signaling molecule that organizes the cells destined to form the distal margin of the adult wing. During metamorphosis, the wing disc folds along its midline, creating dorsal and ventral wing surfaces. After folding, another gene, *Distal-less* (*Dll*), produces a transcription factor in the wing margin. As the wing develops, it grows outward—that is, distal—from the body. The expression of the *Dll* gene therefore helps to define the proximal-distal axis for wing development.

In butterflies, the wing discs consist of two layers of cells, one destined to form the dorsal surface of the adult wing and the other the ventral surface (Figure 1). Carroll studied the development of these two layers by monitoring the expression of the butterfly homologs of *Drosophila* genes. As in *Drosophila*, the presumptive dorsal layer is marked by the expression of a gene that is homologous to *ap*, and the boundary between the presumptive dorsal and ventral layers is marked by the expression of a gene that is homologous to *wg*. Later in development, a homolog of the *Dll* gene is expressed in the presumptive margin of the butterfly wing. Thus these three genes—*ap*, *wg* and *Dll*—are expressed in similar patterns in the wing discs of both types of insects. However, in the developing butterfly wing, the expression of *Dll* has a feature not seen in *Drosophila*. In butterflies, the zone of *Dll* expression extends inward from the wing margin in a series of rays. One of these rays eventually resolves itself into a circle of intense expression. Carroll and his colleagues discovered that this circle ultimately forms one of the pigmented eyespots on the adult butterfly wing. Thus the *Dll* transcription factor appears to be responsible for producing a striking element in the pattern of a butterfly's wings—one that probably has a significant adaptive value.

Carroll's research shows that homologous genes in butterflies and fruit flies have preserved some of their basic functions, even though these organisms diverged from a common ancestor about 200 million years ago. However, these genetic functions are not absolutely conserved, as the difference in *Dll* expression indicates. Additional comparative studies of the type that Carroll has initiated may eventually reveal how variations in the expression of particular genes—such as *Dll*—alter the form and function of wings in many different types of insects. Thus these studies may provide an understanding of how insect wings evolved.

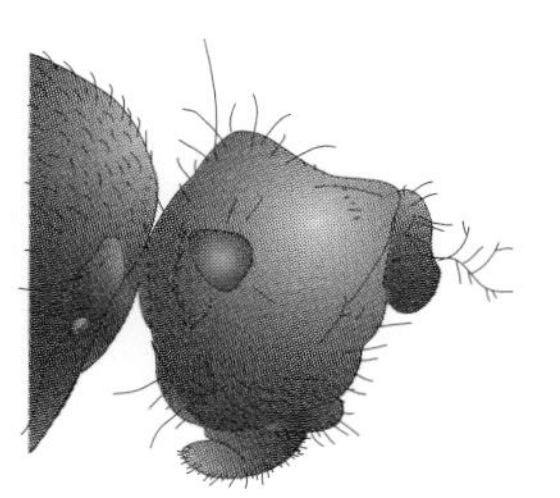

Figure 23.27 The phenotype of an *eyeless* mutation in *Drosophila*.

type of flies that are mutant for it (Figure 23.27). The wild-type *eyeless* gene encodes a homeodomain transcription factor whose action switches on a developmental pathway that involves several thousand genes. Initially, several subordinate regulatory genes are activated. Their products then trigger a cascade of events that create specific cell types within the developing eye.

The role of the *eyeless* gene has been demonstrated by expressing it in tissues that normally do not form eyes (Figure 23.28). Walter Gehring and colleagues did this by creating transgenic flies in which the *eyeless* gene was fused to a promoter that could be activated in specific tissues. Activation of this promoter caused transcription of the *eyeless* gene outside its normal domain of expression. This, in turn, caused eyes to form in unorthodox places such as wings, legs, and anten-

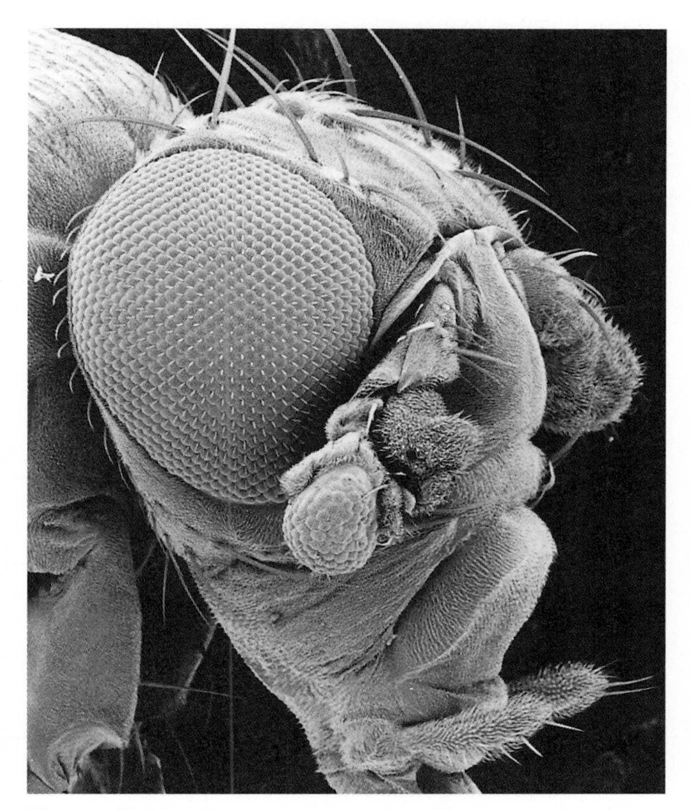

Figure 23.28 An extra eye produced by expressing the wild-type *Drosophila eyeless* gene in the antenna of a fly.

nae. These extra (or ectopic) eyes were anatomically well developed and functional; in fact, their photoreceptors responded to light.

An even more remarkable finding is that a mammalian homolog of the *eyeless* gene also produces these extra eyes when it is inserted into *Drosophila* chromosomes. Gehring and coworkers used the mouse homolog of *eyeless* to transform *Drosophila* and got the same result as they did with the *eyeless* gene itself. This showed that the mouse gene, which also encodes a homeodomain protein, is functionally equivalent to the *Drosophila* gene; that is, it regulates the pathway for eye development. However, when the mouse gene is put into *Drosophila*, it produces *Drosophila* eyes, not mouse eyes. *Drosophila* eyes develop because the genes that respond to the regulatory command of the inserted mouse gene are normal *Drosophila* genes, which must, of course, specify the formation of a *Drosophila* eye. In mice, mutations in the homolog of the *eyeless* gene reduce the size of the eyes. For that reason, the gene is called *Small eye*. A homolog of *eyeless* and *Small eye* has also been found in human beings. Mutations in this gene cause a syndrome of eye defects called *anridia* (also the name of the gene).

The discovery of homologous genes that control eye development in different organisms has profound evolutionary implications. It suggests that the function of these genes is very ancient, dating back to the common ancestor of flies and mammals. Perhaps the eyes in this ancestral organism were nothing more than a cluster of light-sensitive cells organized through the regulatory effects of a primitive *eyeless* gene. Over evolutionary time, this gene continued to regulate the increasingly more complicated process of eye development, so that today, eyes as different as those in insects and mammals are still formed under its control.

Key Points: **The zygotic genes are activated after fertilization in response to maternal gene products. In *Drosophila*, a cascade of zygotic gene activity brings about the differentiation of the dorsal-ventral and anterior-posterior dimensions of the embyro. Ultimately, this activity subdivides the embryo into a series of segments. The identity of each segment is determined by the homeotic genes of the Bithorax and Antennapedia Complexes, acting in response to a regulatory hierarchy of segmentation genes. Once segmental identities have been established, specific cell types differentiate. The formation of an entire organ may depend on the product of a master regulatory gene, such as *eyeless* in *Drosophila*.**

GENETIC ANALYSIS OF DEVELOPMENT IN VERTEBRATES

Much of the knowledge about the genetic control of development comes from the study of *Drosophila* and *Caenorhabditis*, two model invertebrates. Geneticists would like to apply and extend this knowledge to other groups of animals, in particular, to vertebrates; the ultimate goal, of course, would be to learn about the genetic control of development in our own species.

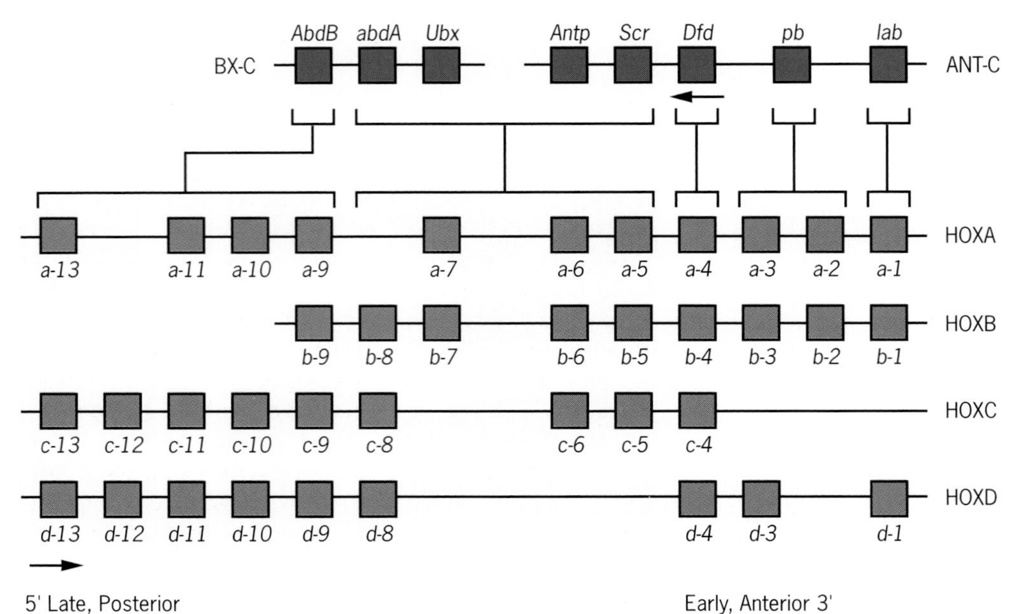

Figure 23.29 Organization and expression of the mammalian *Hox* genes homologous to the *Drosophila* genes in the BX-C and ANT-C. Homologies are indicated by brackets. All the genes except *Dfd* are transcribed from left to right. The time and anatomical location of expression are indicated.

One strategy for achieving this goal is to use invertebrate genes as probes to isolate and clone their vertebrate homologs. Another is to study model vertebrate species with techniques similar to those that are being used in *Drosophila* and *Caenorhabditis*.

Vertebrate Homologs of Invertebrate Genes

Once a gene has been cloned, it can be used as a probe to clone homologous genes from other organisms. If the gene's sequences have been reasonably well conserved over evolutionary time, this procedure works even for distantly related species. Thus it has been possible to clone genes from various vertebrate species by using probes made from *Drosophila* and *Caenorhabditis* genes. The cloning of a vertebrate gene then makes many kinds of experimental analyses possible, including assays for the gene's expression at both the RNA and protein levels.

One of the most dramatic applications of this approach has shown that vertebrates contain homologs of the homeotic genes of *Drosophila*. These so-called *Hox* genes were initially identified by probing Southern blots of mouse and human genomic DNA with segments of the *Drosophila* homeotic genes. Subsequently, the cross-hybridizing DNA fragments were cloned, mapped with restriction enzymes, and sequenced. The results of all this analysis have established that mice, humans, and all other vertebrates so far examined have 38 *Hox* genes in their genomes (Figure 23.29). These genes are organized in four clusters (a, b, c, and d), each about 120 kb long; in mice and humans each cluster is located on a different chromosome. It seems that the four *Hox* gene clusters were created by the quadruplication of a primordial cluster very early in the evolution of the vertebrates, probably 500 to 600 million years ago.

The molecular analysis of the vertebrate *Hox* genes has revealed striking structural and functional similarities to the homeotic genes of the Bithorax and Antennapedia Complexes in *Drosophila*. These latter genes actually seem to have been members of a larger cluster, called HOM-C, which was split by a chromosome rearrangement during the evolution of the flies. In more primitive insects such as the flour beetle, *Tribolium castaneum*, the Bithorax and Antennapedia complexes are united in a single cluster. Comparison between vertebrates and invertebrates shows that the basic organization of the *Hom/Hox* genes has been preserved during evolution. The structural homologs of the ANT-C genes are at one end of each vertebrate *Hox* gene cluster and the structural homologs of the BX-C genes are at the other end. Moreover, within each cluster, the physical order of the *Hox* genes corresponds to the position of their expression along the anterior-posterior axis of the embryo, just as it does for the homeotic genes in *Drosophila* (Figure 23.30). With one exception (the *Deformed* gene in *Drosophila*), all the HOM and *Hox* genes are transcribed in the same direction, with expression proceeding from the 3′ end of a cluster to the 5′ end, both spatially (anterior to posterior in the embryo) and temporally (early to late in development). This phenomenon, called **colinearity**, suggests that the HOM and *Hox* genes provide a common molecular mechanism for establishing the identities of specific regions in many different types of embryos. However, even in *Drosophila*, researchers have only begun to investigate how this mechanism operates.

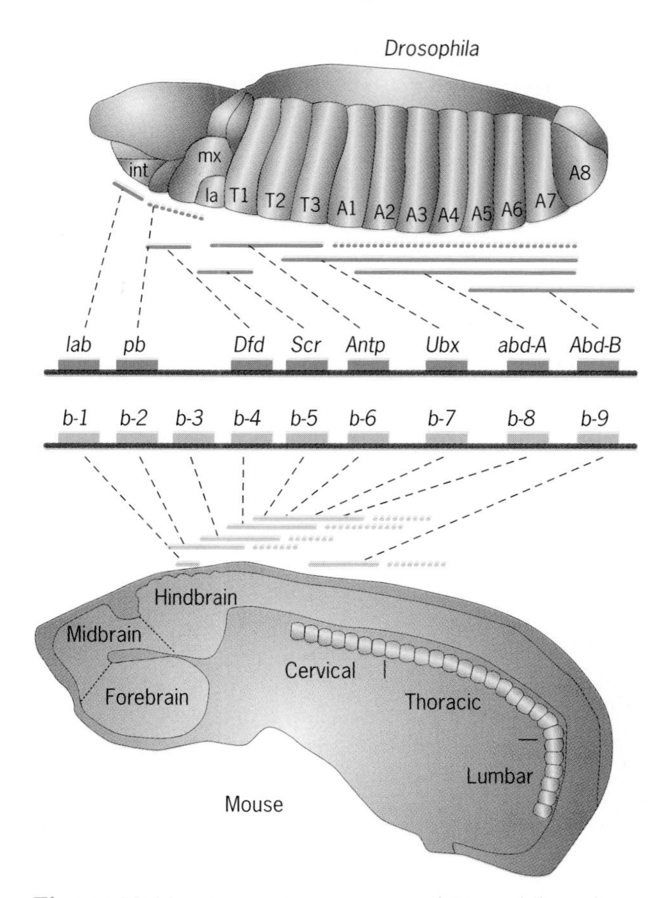

Figure 23.30 Expression patterns of *Drosophila* and mouse homeotic genes. Top: A 10-hour-old *Drosophila* embryo showing the approximate expression of the homeotic genes in the epidermis. In the head, int, mx, and la designate intercalary, maxillary, and labial segments, respectively. Thoracic and abdominal segments are indicated as T1-T3, and A1-A8. Bottom: A 12-day-old mouse embryo showing the approximate expression of the *Hoxb* genes in the central nervous system. The dotted lines indicate extension of the expression toward the posterior.

The Mouse: Insertional Mutagenesis, Transgenics, and Knockout Mutations

The genetic control of development cannot be studied in vertebrates with the same thoroughness as it can in invertebrates such as *Drosophila*. Obviously, there are technical and logistical constraints. Vertebrates have comparatively long life cycles, their husbandry is expensive, and typically few mutant strains, especially those with a developmental significance, are available. In spite of these shortcomings, geneticists have been able to make some headway in the genetic analysis of development in one complex vertebrate—the mouse—and efforts are currently underway to analyze development in a simpler vertebrate—the zebrafish. Over 500 loci responsible for genetic diseases have been identified in the mouse, and some are involved in developmental processes. Most of these loci were discovered through ongoing projects to collect spontaneous mutations. Such work requires that very large numbers of mice—hundreds of thousands, if not millions—be reared and examined for phenotypic differences, and that whatever differences are found be tested for genetic transmission. This is painstaking, costly work that can be supported only at a few facilities in the entire world—one in England and two in the United States. Once a mutation is detected, it can be mapped and tested for allelism and epistatic interactions with other mutations. However, cloning the mutant gene may prove to be as difficult and costly as finding the original mutant mouse. Fortunately, the discovery of new techniques has expedited this process.

These techniques are based on the introduction of particular DNA molecules into the mouse genome. This DNA may cause mutations by inserting into genes in the mouse chromosomes. Such insertional mutations are much easier to clone than spontaneous mutations because they have been tagged by the inserted DNA (Figure 23.31). A probe made from this DNA can be used to isolate the mutated gene from a recombinant DNA library constructed from the mutant animal. Insertional mutagenesis in the mouse is therefore analogous to transposon tagging in *Drosophila* and *Caenorhabditis*. However, the DNA used for insertional mutagenesis in the mouse is not a *bona fide* transposon but some other sequence capable of integrating more or less randomly into the chromosomes. For example, replication-defective retroviruses have been used for this purpose.

As we discussed in an earlier section, a "foreign" DNA sequence that has been inserted into the genome is called a transgene. There are two general methods of introducing transgenes into mouse chromosomes. One relies on the injection of DNA into fertilized eggs or embryos (Figure 23.32), and the other relies on the injection or transfection of DNA into large populations of cultured cells that were derived from very young mouse embryos (Figure 23.33). These **embryonic stem cells** (or **ES cells**) come from the inner cell mass, a group of cells found in the blastula stage of mouse embryos. Such cells can be cultured *in vitro*, transfected or injected with DNA, and then introduced into other developing mouse embryos. By chance, some of the introduced ES cells may contribute to the formation of adult tissues, so that when the mouse is born, it may consist of a mixture of two

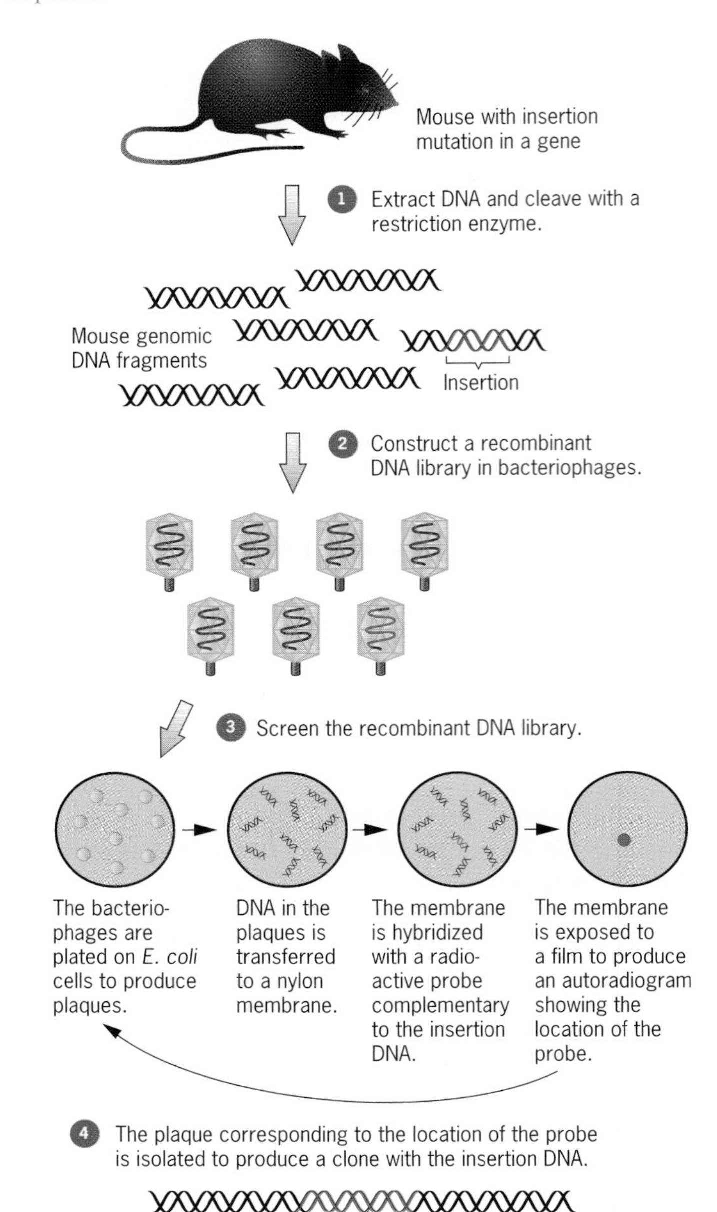

Figure 23.31 Cloning an insertion mutation from a mouse.

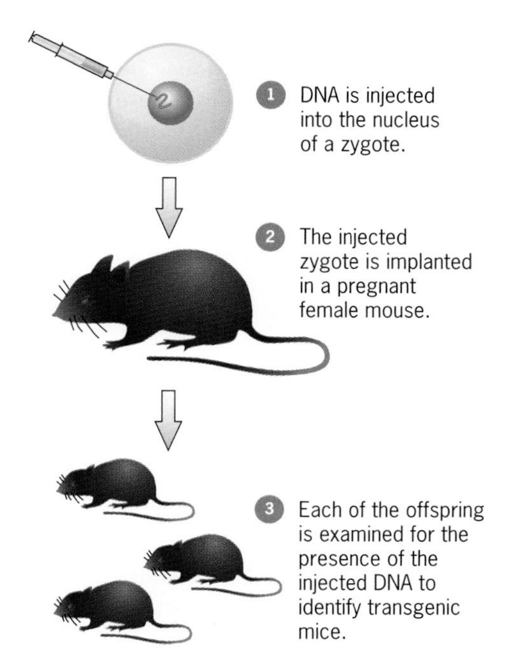

Figure 23.32 Producing transgenic mice by injecting eggs and implanting them into females to complete their development.

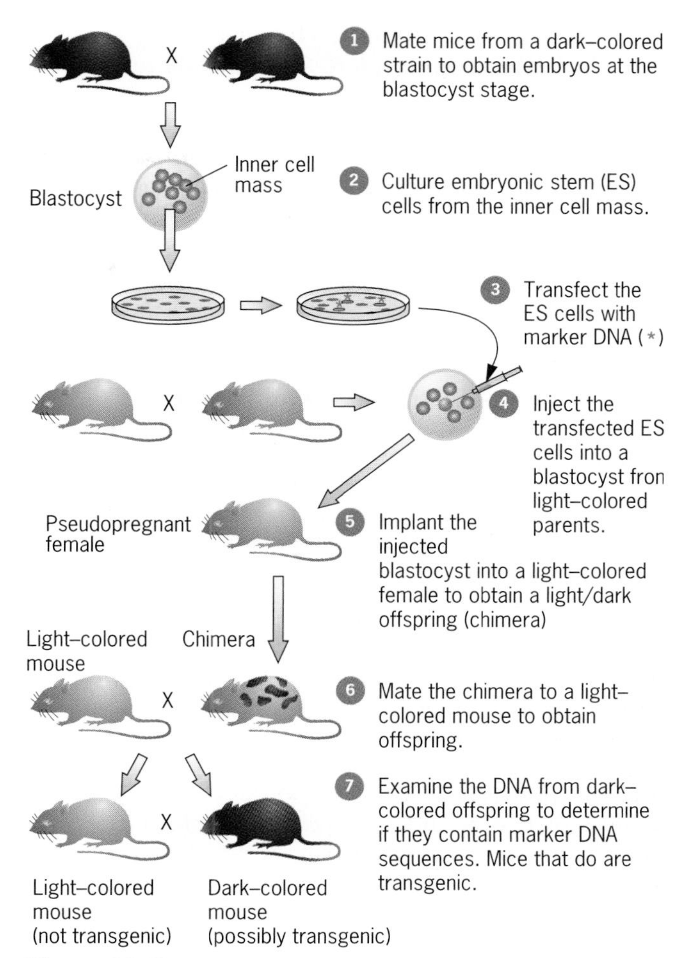

Figure 23.33 Producing transgenic mice by embryonic stem (ES) cell technology.

types of cells, its own and those derived from the cultured (and potentially mutagenized) ES cells. Such mice are called **chimeras**. If the ES cells happened to contribute to the chimera's germ line, a mutation that was induced in them by the insertion of foreign DNA has a chance of being transmitted to the next generation. Breeding a chimeric mouse may therefore establish a mutant transgenic strain. Transgenic strains can also be produced by breeding mice derived from embryos that were directly injected with DNA. To date, by using either of these methods, thousands of transgenic strains have been created, and among them, many developmentally significant mutations have been detected.

A special application of transgenic mutagenesis is possible with genes that have already been cloned. For such genes, it is possible to create mutations that knockout the gene's function. A **"knockout" mutation** can help a reseacher determine what role the normal gene plays during development. To create a knock-out mutation, the sequence of the cloned gene must be altered *in vitro* and then introduced into cultured ES cells. At a low frequency, the mutated transgene will replace its normal allele on the chromosome by homologous recombination, a process called **targeting**. ES cells that contain a targeted knockout mutation can be used to create chimeras, which can then be bred to produce transgenic strains that carry the knockout mutation. It is then a simple matter of intercrossing two heterozygotes to make the knockout mutation homozygous and determine what effect it has on development. As an example, let's consider the effect of a knockout mutation in the *Hoxc-8* gene. Mice that are homozygous for this mutation develop an extra pair of ribs posterior to the normal sets of ribs; they also have clenched toes on their forepaws. The extra-rib phenotype in these mutant mice is reminiscent of the segmental transformations that are seen with homeotic mutations in *Drosophila*.

The genetic analysis of development in mice is providing clues about development in our own species. For example, mutations in at least two different mouse genes mimic the development of abnormal left-right asymmetries in human beings. Normally, humans, mice, and other vertebrates exhibit structures that are asymmetric along the left-right body axis. The heart tube always loops to the right, and the liver, stomach and other viscera are shifted either to the left or right away from the body's midline. In mutant individuals, these characteristic asymmetries are not seen, perhaps because of a defect in the mechanisms that establish the basic body plan. Studying these types of mutants in the mouse may therefore help to elucidate how the organs are positioned in human beings.

Key Points: **Many vertebrate genes have been identified by homology to genes isolated from model organisms such as *Drosophila* and *Caenorhabditis*. For example, the mammalian *Hox* genes are structurally and functionally similar to the homeotic genes of *Drosophila*. Among vertebrates, the mouse provides opportunities to study mutations that affect development. Transgenic mice can be created by injecting DNA into eggs or embryos, or by inserting transfected embryonic stem cells into developing embryos. These mice can be a source of insertional mutations, including knockout mutations, in genes that have a developmental significance.**

TESTING YOUR KNOWLEDGE

1. The protein product of the *dorsal* (*dl*) gene in *Drosophila* has been called a ventral morphogen—that is, a substance that brings about the formation of ventral structures in the embryo by virtue of its high concentration in the nuclei on the ventral side of the blastoderm. However, the *dorsal* protein can only enter these ventral nuclei if a receptor on the embryo's ventral surface has been activated. This receptor is encoded by the *Toll* (*Tl*) gene. The extracellular ligand for the *Toll* receptor is most likely encoded by the *spatzle* (*spz*) gene (German for "little dumpling"). However, this ligand can exist in two states, "native" and "modified," and the modified state is needed for the activation of the *Toll* receptor. The products of three genes, *snake* (*snk*), *easter* (*ea*, discovered on Easter Sunday) and *gastrulation defective* (*gd*), are required to convert the native ligand into the modified ligand. All three of these gene products are serine proteases, proteins capable of cleaving other proteins at certain serines in the polypeptide chain. Using these facts, diagram the developmental pathway that ultimately causes the *dorsal* protein to induce the formation of ventral structures in the *Drosophila* embryo.

ANSWER

Here is one representation.

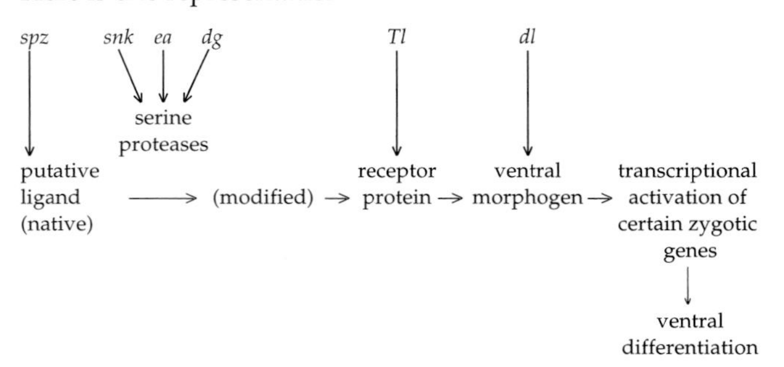

The protein product of the *spz* gene is modified by the serine proteases made by the *snk*, *ea*, and *gd* genes. In its modified form, this ligand is able to activate the *Toll* receptor protein, but the activation is restricted to the ventral side of the embryo. (The reasons for this localized activation are still unclear.) When the *Toll* receptor has been activiated (presumably, by binding the modified *spatzle* ligand), it transduces a signal into the cytoplasm of the embryo. This signal ultimately causes the *dorsal* protein to move into the nuclei on the ventral side of the embryo, where it acts as a transcription factor to regulate the expression of the zygotic genes involved in the differentiation of ventral fates.

2. Considering the pathway described above, what would be the phenotypes of recessive loss-of-function mutations in the *spz* and *Tl* genes?

ANSWER

For reference, we should note that loss-of-function mutations in *dl* are maternal effect lethals; that is, embryos from *dl/dl* mothers die during development. When these dying embryos are examined, they are found to lack ventral structures. Geneticists say that they are "dorsalized." This peculiar phenotype is due to the failure of the *dorsal* transcription factor to induce appropriate development in the ventral nuclei of the embryo. In the absence of this induction, the ventral cells differentiate as if they were on the dorsal side of the embryo. Mutations in *spz* and *Tl* might be expected to have the same phenotypic effect because they would block steps in the pathway that ultimately causes the *dorsal* protein to induce ventral differentiation. Recessive mutations in *spz* and *Tl* are therefore maternal effect lethals. Females homozygous for these mutations produce dorsalized embryos that die during development.

QUESTIONS AND PROBLEMS

23.1 What events form the blastula during animal development? What events form the gastrula?

23.2 During oogenesis, what mechanisms enrich the cytoplasm of animal eggs with nutritive and determinative materials?

23.3 Distinguish between "determination" and "differentiation."

23.4 Are the "primary germ layers" components of the germ line? Explain.

23.5 Outline the main steps in the genetic analysis of development in a model organism such as *Drosophila* or *Caenorhabditis*.

23.6 Why is the early *Drosophila* embryo a syncytium?

23.7 In *Drosophila*, what larval tissues produce the external organs of the adult?

23.8 In *Caenorhabditis*, homozygosity for a *dumpy* (*dpy*) mutation causes the animal to be shorter than wild-type. If a hermaphrodite heterozygous for such a mutation were self-fertilized, what fraction of its progeny would be dumpy?

23.9 A researcher wants to perform a complementation test between two independently isolated *dumpy* mutations in *Caenorhabditis*. Both mutations are autosomal recessives. In addition to being homozygous for the *dumpy* mutation, one strain is homozygous for an X-linked recessive mutation (*unc*) that causes the worms to be uncoordinated. The other strain is homozygous for the *dumpy* mutation alone. How should the complementation test be performed?

23.10 Loss-of-function mutations in the *Drosophila Sxl* gene cause females to die, but have no effect in males. What phenotype would a gain-of-function mutation most likely have?

23.11 A *Drosophila* researcher has discovered an autosomal recessive mutation that causes XX animals to develop into sterile males. Propose a scheme to find out if the mutation is an allele of either the *tra* or *tra2* genes.

23.12 Diagram the sex-determination pathway in *Drosophila*.

23.13 What phenotype would you expect in an XX *Drosophila* that was homozygous for loss-of-function mutations in both the *tra* and *dsx* genes? Explain.

23.14 Triploid *Drosophila* with two X chromosomes develop as intersexes. Suggest a mechanism to explain this phenotype.

23.15 Predict the phenotype of XO and XX *Caenorhabiditis* that were homozygous for loss-of-function mutations in both the *tra-1* and *fem-1* genes.

23.16 Loss-of-function mutations in the *Caenorhabditis xol-1* gene cause males to die, whereas loss-of-function mutations in the *Drosophila Sxl* gene cause females to die. What is the reason for these opposite effects?

23.17 A *Drosophila* researcher has observed a peculiar fly. The forelegs have sex combs, which are characteristic of males, and the abdomen has an ovipositor, which is characteristic of females. Furthermore, the bristles on the forelegs and part of the thorax are yellow whereas those on the rest of the body are black. Suggest an explanation for the unusual phenotype of this fly.

23.18 *Drosophila* larvae that were heterozygous for a null mutation of the X-linked *white* gene (*w*) and its wild-type allele (w^+) were irradiated with X rays and then reared to adulthood. When the adults emerged from the pupal cases, a few had white patches in their otherwise red eyes. What caused these patches to develop? Do the patches indicate whether the *white* gene functions autonomously or nonautonomously? Explain.

23.19 A *Drosophila* developmental geneticist has identified a recessive lethal mutation (*l*) on an X chromosome that is marked with a recessive allele of the *singed* gene (*sn*). Through recombination analysis, this lethal has been mapped between *sn* and *yellow* (*y*), which is at the tip of the X chromosome far away from the centromere (denoted by a dot). The geneticist has produced *y* + + . / + *l sn* . heterozygous larvae and has irradiated them with X rays to induce recombination between *sn* and the centromere. Some of the adults that have developed from these larvae have solitary patches of yellow tissue on an otherwise black body; however, none of the tissue (either yellow or black) is singed. What does this pattern of pigmentation reveal about the function of the lethal mutation?

23.20 A researcher is trying to clone the *dpy-3* gene from *Caenorhabditis*. Recessive mutations in this gene cause worms to be shorter than normal, a phenotype referred to as "dumpy." The researcher has mapped the *dpy-3* gene relative to other genes on one of the *Caenorhabditis* chromosomes and, using the map data, has obtained cosmid clones with wild-type DNA from the vicinity of the *dpy-3* gene. To determine which clone contains the *dyp-3* gene, the researcher has injected each of them into the gonads of hermaphrodites homozygous for a recessive mutation in *dpy-3*. She has then scored the transgenic progeny derived from these injected hermaphrodites for the dumpy phenotype. From the results shown below, where on the molecular map should the researcher place the *dpy-3* gene?

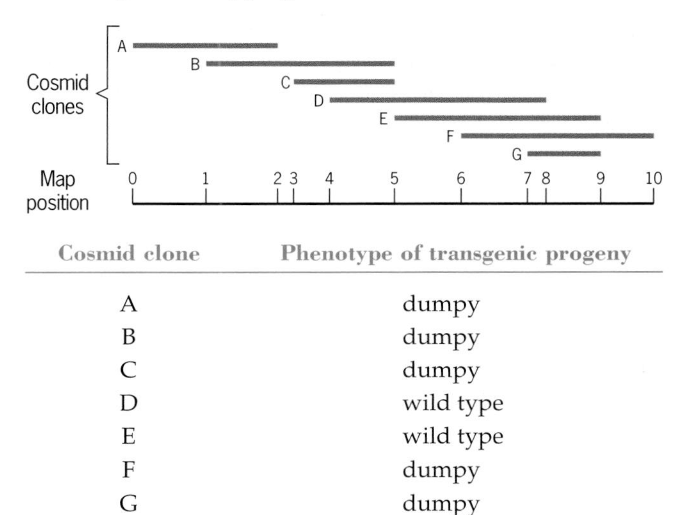

Cosmid clone	Phenotype of transgenic progeny
A	dumpy
B	dumpy
C	dumpy
D	wild type
E	wild type
F	dumpy
G	dumpy

23.21 Like *dorsal*, *bicoid* is a strict maternal-effect gene in *Drosophila*; that is, it has no zygotic expression. Recessive mutations in *bicoid* (*bcd*) cause embryonic death by preventing the formation of anterior structures. Predict the phenotypes of (a) *bcd/bcd* animals produced by mating heterozygous males and females; (b) *bcd/bcd* animals produced by mating *bcd/bcd* females with *bcd*/+ males; (c) *bcd*/+ animals produced by mating *bcd/bcd* females with *bcd*/+ males; (d) *bcd/bcd* animals produced by mating *bcd*/+ females with *bcd/bcd* males; (e) *bcd*/+ animals produced by mating *bcd*/+ females with *bcd/bcd* males.

23.22 In *Drosophila*, recessive mutations in the dorsal-ventral axis gene *dorsal* (*dl*) cause a dorsalized phenotype in embryos produced by *dl/dl* mothers; that is, no ventral structures develop. Predict the phenotype of embryos produced by females homozygous for a recessive mutation in the anterior-posterior axis gene *nanos*.

23.23 A researcher is planning to collect mutations in maternal-effect genes that control the earliest events in *Drosophila* development. What phenotype should the researcher look for in this search for maternal-effect mutations?

23.24 A researcher is planning to collect mutations in the gap genes, which control the first steps in the segmentation of *Drosophila* embryos. What phenotype should the researcher look for in this search for gap gene mutations?

23.25 Diagram a pathway that shows the contributions of the *sevenless* (*sev*) and *bride of sevenless* (*boss*) genes in the differentiation of the R7 photoreceptor in the ommatidia of *Drosophila* eyes. Where would *eyeless* (*ey*) fit in this pathway?

23.26 When the mouse gene *Small eye*, which is homologous to the *Drosophila* gene *eyeless*, is expressed in *Drosophila*, it produces extra compound eyes with ommatidia, just like normal *Drosophila* eyes. If the *Drosophila eyeless* gene were introduced into mice and expressed there, what effect would you expect? Explain.

23.27 How might you show that two mouse *Hox* genes are expressed in different tissues and at different times during development?

BIBLIOGRAPHY

BEDDINGTON, R. 1992. Transgenic mutagenesis in the mouse. *Trends in Genet*. 8:345–347.

CLINE, T. W. 1993. The *Drosophila* sex determination signal: how do flies count to two? *Trends in Genet.* 9:385–390.

GOVIND, S., AND R. STEWARD. 1991. Dorsoventral pattern formation in *Drosophila*. *Trends in Genet*. 7:119–125.

GREENWALD, I., AND G. M. RUBIN. 1992. Making a difference: the role of cell-cell interactions in establishing separate identities for equivalent cells. *Cell* 68:271–281.

GURDON, J. B. 1992. The generation of diversity and pattern in animal development. *Cell* 68:185–199.

HALDER, G., P. CALLAERTS, AND W. J. GEHRING. 1995. Induction of ectopic eyes by targeted expression of the *eyeless* gene in *Drosophila*. *Science* 267:1788–1792.

KENYON, C. 1988. The nematode *Caenorhabditis elegans*. *Science* 240:1448–1453.

LEWIS, E. B. 1994. Homeosis: the first 100 years. *Trends in Genet*. 10:341–343.

MCGINNIS, W., AND R. KRUMLAUF. 1992. Homeobox genes and axial patterning. *Cell* 68:283–302.

PARKHURST, S. M., AND P. M. MENEELY. 1994. Sex determination and dosage compensation: lessons from flies and worms. *Science* 264:924–932.

RUBIN, G. M. 1988. *Drosophila melanogaster* as an experimental organism. *Science* 240:1453–1459.

ST JOHNSTON, D., AND C. NÜSSLEIN-VOLHARD. 1992. The origin of pattern and polarity in the Drosophila embryo. *Cell* 68:201–219.

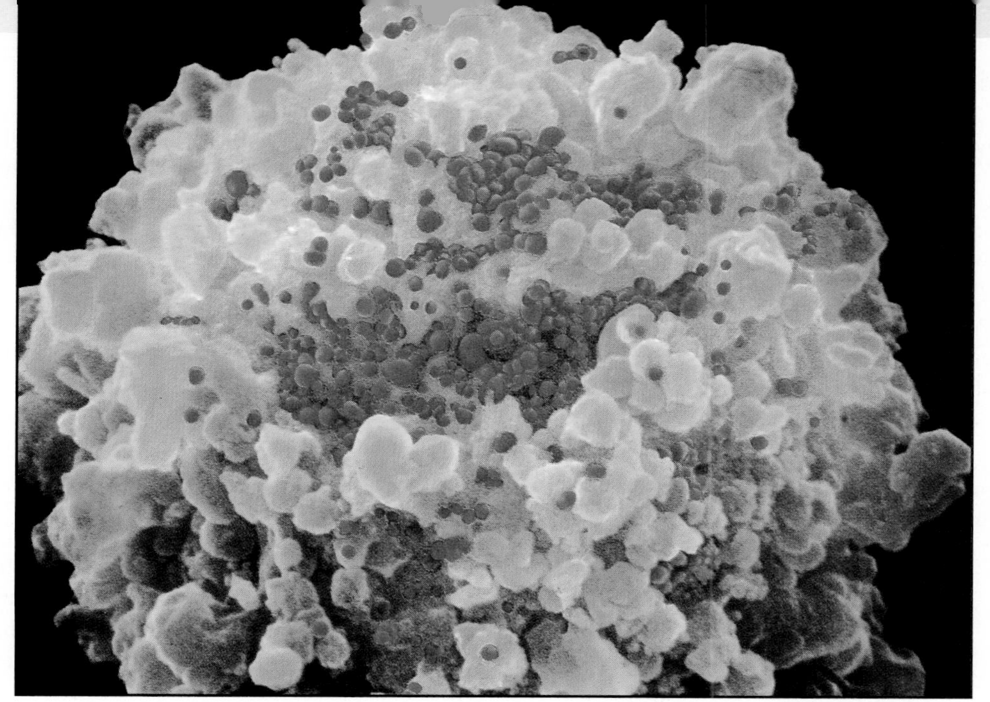

A human T lymphocyte (green) that is infected with human immunodeficiency viruses (red). Individuals infected with HIV usually develop acquired immune deficiency syndrome (AIDS), the plague of the twentieth century.

24

The Genetic Control of the Vertebrate Immune System

CHAPTER OUTLINE

AIDS Emphasizes the Vital Role of the Immune System

In 1971, when Brandon was just a toddler, he fell on the front step of his house and cut his knee. When the bleeding would not stop, his mother took him to the emergency room of the local hospital. Blood tests revealed that Brandon had hemophilia A, an inherited deficiency of blood clotting factor VIII (Chapter 5). Individuals with this X-linked recessive disorder exhibit excessive bleeding from even small cuts and lacerations. However, the bleeding can be arrested by treatment with the missing clotting factor. In 1982, when Brandon was injured in a car accident, he received blood transfusions and treatment with clotting factor VIII. He recovered from his injuries and continued to live a normal, active life. Eight years later, Brandon began to experience fatique, recurrent fever, and weight loss. He seemed to become infected with every virus that he encountered. Shortly thereafter, the frightening diagnosis was made: Brandon had acquired immune deficiency syndrome (AIDS), the fatal disease caused by the human immunodeficiency virus (HIV). A few months later, Brandon contracted a type of pneumonia caused by *Pneumocystis carinii.* This protozoan is an opportunistic pathogen that infects individuals, such as AIDS patients, with weakened immune systems. HIV destroys the immune system of its hosts by killing a type of white blood cell that is required for a normal immune response (Chapter 15). Brandon died less than a year later.

Brandon became infected with HIV in contaminated blood used in the transfusion after his car accident. Prior to 1985, blood serum used in transfusions was not tested for HIV, which was identified in 1983. Today, all donated blood is tested for HIV before being deposited in blood banks. Thus the risk of contracting AIDS from a blood transfusion is extremely low—approximately one case in 10^5 transfusions.

Until about a decade ago, when AIDS burst into our lives, most of us took our immune systems for granted. As children, we received vaccinations to protect us from diseases such as measles, diphtheria, and whooping cough; we gave little thought, however, to why we remained, for the most part, healthy. Today, many of us have watched the health of a relative, friend, or neighbor with AIDS deteriorate before our eyes. If not, we have observed the horrible effects of AIDS on the human body through the news media. The pain and suffering caused by this dreaded disease are truly shocking. We have seen what happens when the immune system can no longer do its job. Most of us will never again take our immune systems for granted.

AIDS emphasizes the vital role that the immune response plays in our survival. On a nearly daily basis, our immune systems save us from certain death due to infections by a vast array of pathogenic viruses, bacteria, fungi, and other agents. Thus the recent elucidation of the genetic mechanisms that govern the immune response has proven exciting to all students of biology, regardless of their area of special interest. Investigations of the genetic control of the immune system have provided an amazing and elegant picture of this product of the evolutionary process. Scientists have discovered that the genes that encode the key proteins of the immune response, the antibodies and T cell receptors, are assembled from sets of gene segments during the differentiation of the cells that produce these proteins.

All vertebrate animals are protected by immune systems. However, for the sake of brevity and focus, we will restrict our discussion in this chapter to the immune systems of mammals, with special emphasis on humans.

COMPONENTS OF THE MAMMALIAN IMMUNE SYSTEM

The environment in which we live contains an enormous number of different viruses, bacteria, fungi, and other agents. Moreover, these organisms—many of which are infectious—are constantly evolving and producing new strains. Viruses, in particular, accumulate new mutations and evolve very rapidly (Chapter 15). How, then, can our immune systems protect us from all of these different pathogens, some of which are newly evolved and are confronting cells of the immune system for the first time?

The most striking feature of the immune system is its **specificity**. This specificity is provided by (1) a team of highly specialized cells, each performing its job in a highly coordinated fashion; (2) two armies of proteins, antibodies and T cell receptors, each with seemingly unlimited ability to recognize foreign substances; and (3) a specialized set of proteins, the histocompatibility antigens, that facilitate communication between cells and allow cells of the immune system to recognize foreign substances, and to distinguish foreign cells (nonself) from an individual's own cells (self). The most important of these immune system cells and proteins are listed in Table 24.1. In total, the immune system of an adult human consists of something like a trillion (10^{12}) white blood cells and 10^{15} or more special protein molecules, all acting in concert to eliminate foreign substances.

When a foreign agent such as a virus or bacterium invades the human body, it is recognized by cells of the immune system and triggers the production of the two sets of proteins, antibodies and T cell receptors, that specifically recognize and bind to the foreign substance. A substance that is bound by an antibody or T

TABLE 24.1
Components of the Mammalian Immune System

Component	Description and Function
I. Proteins	
Antigens	Substances that trigger an immune response
Antibodies	Proteins produced by the immune system that bind to antigens and assist in their destruction
T cell receptors	Proteins produced by the immune system in response to antigens; they become anchored on the surface of killer T cells and bind antigens with the help of major histocompatibility antigens
Major histocompatibility complex (MHC) antigens	Cell surface proteins that (1) allow immune system cells to distinguish foreign substances from self and (2) facilitate communication between cells
II. Cells	
Stem cells	Undifferentiated bone marrow cells that give rise to the various cells of the immune system
Phagocytes	Large cells that capture, ingest, and destroy invading foreign agents such as viruses, bacteria, and fungi
Macrophages	Phagocytic cells that ingest antigens and display them on the cell surface for interactions with other cells of the immune system
B lymphocytes	Cells that differentiate (in bone marrow) into antibody-producing plasma cells and memory B cells
Plasma cells	Antibody-producing white blood cells derived from B lymphocytes
Memory B cells	B cells that facilitate the rapid production of a given antibody in response to a second and subsequent encounters with an antigen
T lymphocytes	Cells that undergo important differentiation events in the thymus gland and further differentiate into the various types of T cells
Helper T cells	T cells that respond to the display of an antigen by a macrophage by stimulating B lymphocytes to produce antibodies and T lymphocytes to produce T cell receptors
Suppressor T cells	T cells that suppress the production of antibodies and T cell receptors by B cells and T cells, respectively
Cytotoxic or killer T cells	T cells that carry T cell receptors and kill cells displaying the recognized antigens
Memory T cells	T cells that facilitate the rapid production of a given T cell receptor in second and subsequent encounters with an antigen

cell receptor is called an **antigen**, and an antigen that elicits an immune response is called an **immunogen**. Virtually all biological macromolecules, including polypeptides, polysaccharides, and nucleic acids, can act as antigens, and most can function as immunogens.

The mammalian immune response has two major components: the antibody-mediated or humoral response and the T cell-mediated or cellular response. The humoral immune response involves the production and secretion of antibodies into the circulatory system, where these antibodies bind to and sequester their respective antigens. The antibody-antigen complexes are then ingested and destroyed by a special class of white blood cells. The humoral immune system is primarily reponsible for protecting mammals from foreign cells (for example, bacteria and fungi) and from free viruses before they infect cells of the host organism. The cellular immune response involves the production of T cell receptors that coat the surfaces of special white blood cells, and allow these cells to recognize and kill infected cells of the host organism. Thus the cellular immune system is largely reponsible for protecting mammals from viral infections. The cells and macromolecules that participate in the antibody-mediated and T cell-mediated immune responses of mammals are described in the following sections.

Specialized Cells Mediate the Immune Response

The immune response to a foreign substance relies on the coordinated activities of a variety of highly specialized cells. All of these cells are derived from undifferentiated precursor cells, called **stem cells**, present in bone marrow. One major component of the immune response is carried out primarily by **B lymphocytes**, called **B cells** because, in mammals, they mature in bone marrow; the other major component of the immune response is performed largely by **T lymphocytes**, called **T cells** because they undergo key differentiation events in the thymus gland. The Greek origins for the word "lymphocyte" mean colorless cell, an appropriate name for these small white blood cells. The nearly colorless phenotype of these lymphocytes is a striking contrast to the highly pigmented phenotype of their hemoglobin-containing neighbors, the red blood cells. Another group of immune system cells are the **phagocytes** (from the Greek for "eating cells"). One type of phagocyte, the macrophage (from the Greek for "big eater"), plays a central role in initiating both components of the immune response.

Upon receiving the appropriate signals, B lymphocytes differentiate into **plasma cells**, which produce antigen-binding proteins called antibodies, and **memory B cells**, which facilitate a more rapid production of antibodies during a later exposure to the same antigen. The T lymphocytes differentiate into four important types of T cells.

1. **Cytotoxic** or **killer T cells** carry antigen-binding proteins called T cell receptors on their cell surfaces, and, as the name implies, kill cells displaying the appropriate antigens.
2. **Helper T cells** stimulate B and T lymphocytes to differentiate into antibody-producing plasma cells and killer T cells, respectively.
3. **Suppressor T cells** assist in "suppressing" or down-regulating the activity of plasma cells and killer T cells.
4. **Memory T cells** "remember" the antigen and provide for a rapid production of killer T cells during subsequent encounters with the same antigen.

Figure 24.1 summarizes the origins and functions of the cells involved in important aspects of the immune response of mammals.

Specialized Proteins Provide Immunological Specificity

Immunological specificity is provided by two classes of proteins, antibodies and T cell receptors, which mediate the two major components—the antibody-mediated or humoral response and the T cell-mediated or cellular response—of the mammalian immune system, respectively. Each of these classes of proteins displays a seemingly unlimited variety of antigen-binding sites, and the mechanisms by which this variability is generated is most fascinating.

Antibodies belong to the class of proteins called **immunoglobulins.** Each antibody is a tetramer composed of four polypeptides, two identical **light chains**

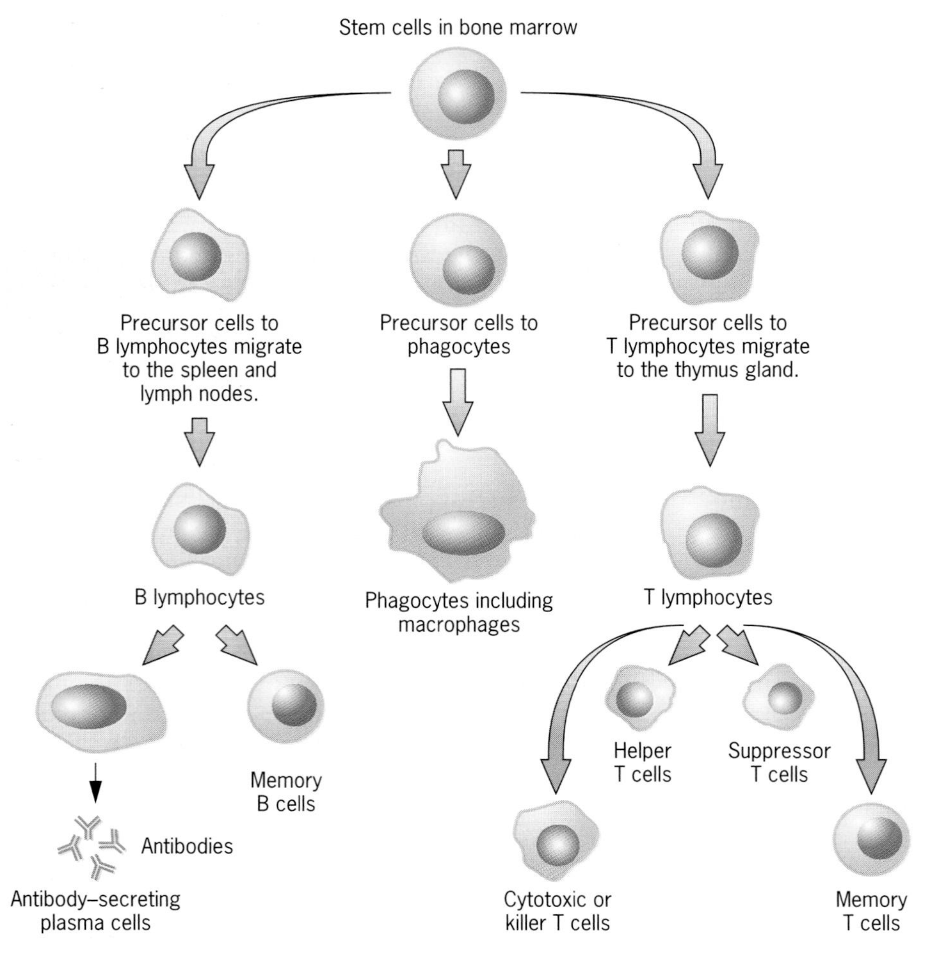

Figure 24.1 The most important cells of the immune system are derived from bone marrow stem cells by three separate cell lineages. They give rise to (1) B cells and antibody-producing plasma cells, (2) the various types of T cells, and (3) the phagocytes, including macrophages.

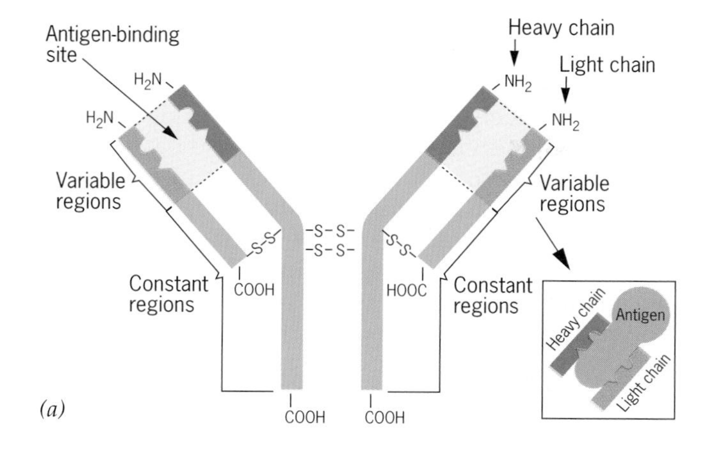

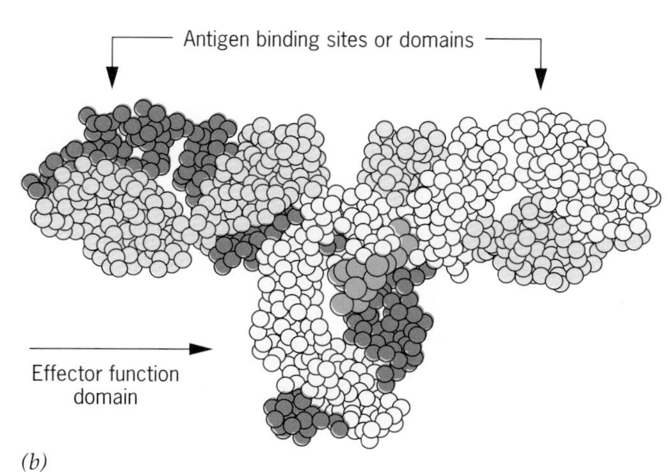

Figure 24.2 (*a*) Diagram and (*b*) space-filling model of antibody structure. Each antibody is a tetramer composed of four polypeptide chains: two identical light chains and two identical heavy chains. (*a*) Each chain consists of a variable region (blue) and a constant region (green), and each antibody has two antigen-binding sites, formed by heavy and light chain variable regions. (*b*) The two heavy chains are shown in white and blue, respectively. The two light chains are shown in yellow. The green-colored components are the associated carbohydrate moieties; their location and structure will vary depending on the immunoglobulin (Ig) class. The structure shown is for a human IgG molecule.

and two identical **heavy chains,** joined by disulfide bonds (Figure 24.2). The light chains are about 220 amino acids long, and the heavy chains are about 440 to 450 amino acids long. Every chain, heavy and light, has an amino-terminal **variable region,** within which the amino acid sequence varies among antibodies specific for different antigens, and a carboxyl-terminal **constant region,** within which the amino acid sequence is the same for all antibodies of a given immunoglobulin (Ig) class, regardless of antigen-binding specificity. The variable regions of all antibody chains are about 110 amino acids long.

Regions of proteins that carry out particular functions are called **domains.** Each antibody has two **antigen-binding sites** or domains, each of which is formed by the variable regions of one light chain and one heavy chain (Figure 24.2). These variable region domains form binding sites that interact with the antigen in a lock-and-key fashion (Figure 24.3). In addition, the constant regions of the two heavy chains interact to form a third domain, called the **effector function domain,** which is responsible for the proper interaction of the antibody with other components of the immune system.

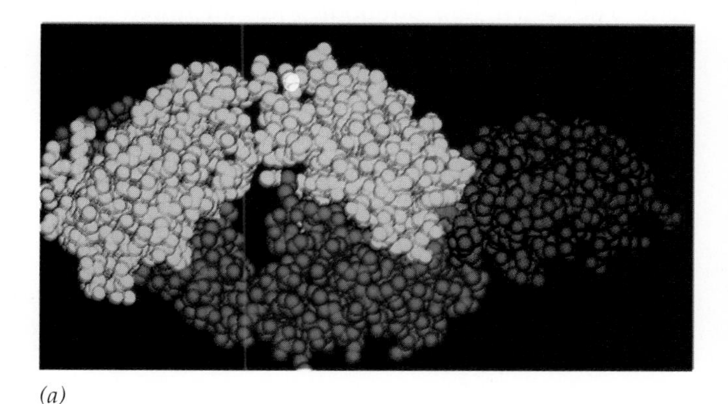

(*a*)

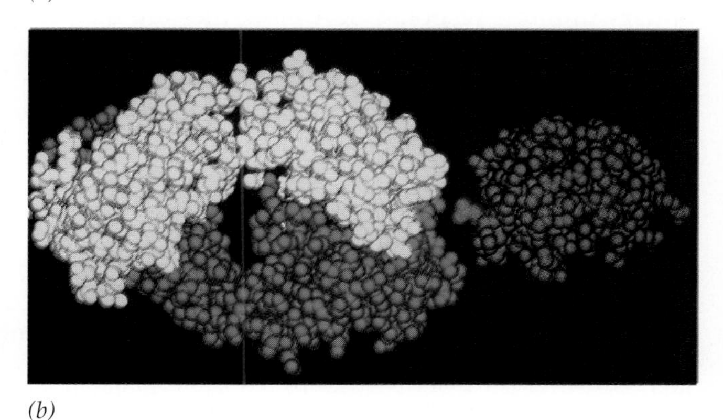

(*b*)

Figure 24.3 Three-dimensional structure of an antigen-antibody complex. Only one of the two antigen-binding sites of a typical antibody is shown. The antigen (green) is the enzyme lysozyme. The antigen-binding site of the antibody is formed by the amino-terminal portions of a light chain (yellow) and a heavy chain (blue). A glutamine residue that protrudes from lysozyme at the epitopic site is shown in red. The structures shown are based on X-ray diffraction data. (*a*) The fully engaged antigen-antibody complex. (*b*) The antigen and antibody have been pulled apart to show their complementary lock-and-key structures.

The light chains of antibodies are of two types, **kappa chains** and **lambda chains**, with type being determined by the structure of the constant region. As we will see, antibodies may have the same antigen-binding specificity, as determined by the variable regions of the four chains, but different immunological functions, as determined by the constant regions of the two heavy chains.

There are five classes of antibodies: **IgA**, **IgD**, **IgE**, **IgG**, and **IgM** (Table 24.2). The class to which an antibody belongs, and thus the function that it carries out, is determined by the structure of its heavy chain constant region (i.e., the structure of its effector function domain). IgA, IgD, IgE, IgG, and IgM antibodies have α, δ, ϵ, γ, and μ heavy chains, respectively. In humans, there are two subclasses of IgA antibodies and four subclasses of IgG antibodies; each subclass has different, but closely related, heavy chains. The first class of antibody produced in a developing B cell is always IgM. The initial IgM antibodies are anchored in the membrane of the cell, providing it with cell-surface antigen receptors. In a primary immune response (the first encounter with a given antigen), a B cell subsequently will begin producing **secreted IgM antibodies**, which are pentameric molecules with ten antigen-binding sites. Later, the B cell may switch to the production of other classes of antibodies, a process called **class switching**.

IgG antibodies account for about 80 percent of the immunoglobulin molecules in human blood. They are produced in large amounts during the second and subsequent encounters with a given antigen, the secondary immune response. IgA molecules are the predominant immunoglobulins in secretions such as saliva, tears, and milk, whereas IgE molecules are found as membrane-bound antigen receptors on special cells called mast cells in various tissues. The binding of antigens by these mast cell surface receptors triggers the secretion of histamines and other amines, which are largely responsible for the allergic reactions observed in individuals with asthma and hay fever.

If we were to examine the structure of antibodies, we would see that their diversity resides almost entirely within the variable regions of the molecules. If these polypeptides were synthesized from colinear nucleotide-pair sequences of genes, one gene per polypeptide chain, the genome would have to contain a vast array of genes with highly variable sequences at one end and essentially identical sequences at the other end. However, this is not the case. Recombinant DNA techniques have made it possible to isolate and sequence many of the segments of chromosomal DNA of mice and humans encoding antibody chains. The results of these studies have provided an elegant explanation for the generation of proteins with great diversity in certain regions and constancy in other regions.

The cellular immune response is highly antigen-specific, just like the antibody-mediated immune response, and is equally important in defending mammals against infections. In the cell-mediated immune response, the antigen specificity is provided by **T cell receptors**. Each T cell receptor is a dimer composed of an α polypeptide and a β polypeptide (Figure 24.4). Both polypeptides have amino-terminal variable regions and carboxyl-terminal constant regions, like antibody chains. However, T cell receptors have only one antigen-binding site, rather than two, as in antibodies. As the name implies, these receptors are present on the surfaces of T cells, anchored in the cell membrane by their constant regions. T cell receptors are present on both cytotoxic or killer T cells and on

TABLE 24.2
Classes of Immunoglobulins

Type	*Heavy Chain*	*Structure*	*Proportion*	*Location*	*Function*
IgA	α	α_2L_2	14%	Secretions: milk, saliva, tears	Protection against microbes at potential entry sites
IgD	δ	δ_2L_2	1%	Blood; B cells	Uncertain; may stimulate B cells to make other classes of antibodies
IgE	ϵ	ϵ_2L_2	<1%	Tissues; mast cells[a]	Receptors for antigens that lead to the secretion of histamines by mast cells
IgG	γ	γ_2L_2	80%	Blood; macrophages, plasma cells	Activation of complement during a secondary immune response[b]
IgM	μ	μ_2L_2	5%	Blood; B cells	Activation of complement during a primary immune response[b]

[a]Cells that are widely distributed in various tissues and release histamines during the nonspecific inflammatory response to an infection.

[b]Complement is a system of approximately 20 soluble proteins that circulate in the bloodstream and form large protein complexes in response to antibody-antigen complexes. The complement proteins kill cells by a variety of mechanisms, such as punching holes in the cell membrane. This system of proteins is called "complement" because it complements the activity of antibodies in fighting pathogenic foreign cells.

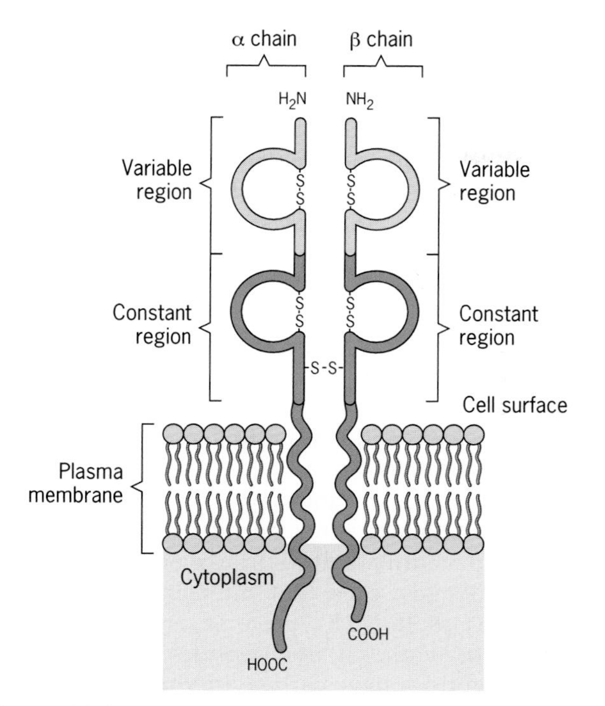

Figure 24.4 Structure of a T cell receptor anchored in the cell membrane.

helper T cells, where, with the help of major histocompatibility antigens, they bind antigens and participate in the activities of these important immune cells.

Major Histocompatibility Antigens: Distinguishing Self from Nonself

Organ transplants and skin grafts between unrelated individuals usually result in immune rejections of the transplanted tissues. In contrast, organ transplants between identical twins usually are accepted. Also, skin can be transplanted from one location to another on the same individual without immune rejection. These results indicate that the immune system can distinguish foreign or "nonself" cells from "self" cells. Carefully controlled experiments have shown that the observed rejections of nonself tissue grafts are controlled by a complex set of cell-surface macromolecules called **histocompatibility antigens**. The most important of these histocompatibility antigens are the **major histocompatibility complex** (**MHC**) proteins. Many of the proteins encoded by the MHC are deposited on cell surfaces where they are highly antigenic to other mammals as well as "nonidentical" individuals of the same species. In humans, the MHC proteins are called the **HLA antigens** (human-leukocyte-associated antigens) because they were discovered on the surfaces of leukocytes (white blood cells). The HLA antigens are encoded by a large cluster of genes, the ***HLA* locus**, on chromosome 6 (Figure 24.5). The *HLA* gene cluster is over 2×10^6 nucleotide pairs in length.

The genes of the *HLA* locus are highly polymorphic, with some genes having up to 100 or more differ-

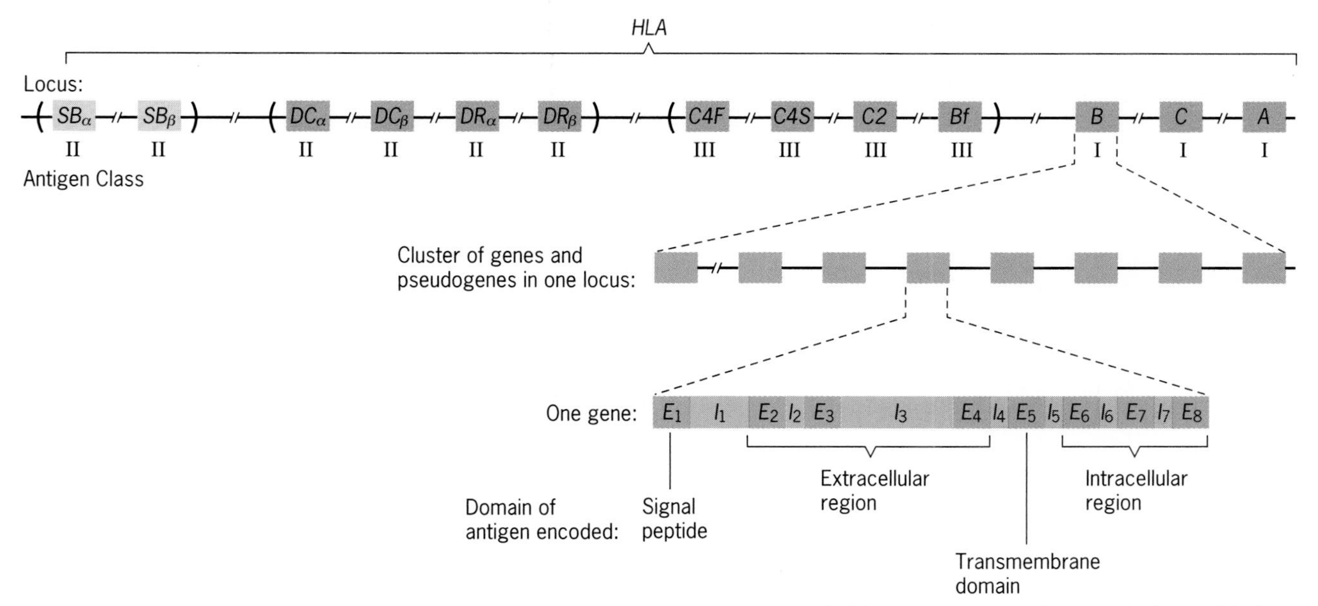

Figure 24.5 Organization of the major histocompatibility complex (*HLA*) on human chromosome 6. The relative positions of mapped loci within this huge gene complex are shown at the top. The order of loci enclosed in parentheses is uncertain. The entire *HLA* complex is over 2×10^6 nucleotide pairs in length. The class of histocompatibility antigen encoded by genes in each locus is indicated below the map. Note that each of the loci within the *HLA* complex is itself a complex locus containing several genes and pseudogenes (*center*). The structure of a typical class I gene is shown at the bottom. Note that different exons encode different functional domains of the polypeptide gene product.

ent alleles. Thus the probability that two unrelated individuals will carry the same alleles of all the *HLA* genes is very low. As a result, finding organ donors with HLA antigens that precisely match those of a recipient is very difficult.

The *HLA* genes encode three different classes of MHC proteins, with each class of molecules involved in a different aspect of the immune response. The ***HLA* class I genes** encode proteins that are called the **transplantation antigens** because of their role in tissue rejections during organ and tissue transplants. The class I proteins exist as glycoproteins anchored as integral membrane proteins with the antigenic determinants exposed on the outside of cells. They are present on virtually all cells of an organism and provide killer T cells with a mechanism for distinguishing "foreign" from "self." Each *HLA* class I gene encodes an α polypeptide with a carboxyl-terminal transmembrane domain and three extracellular globular domains (Figure 24.6*a*). Each class I MHC antigen contains an α polypeptide complexed with a small extracellular protein called α-microglobulin. The two polymorphic amino-terminal domains of the α polypeptide form a fairly nonspecific antigen-binding pocket. Thus each class I MHC protein can bind a rather broad spectrum of antigens and present them to killer T cells.

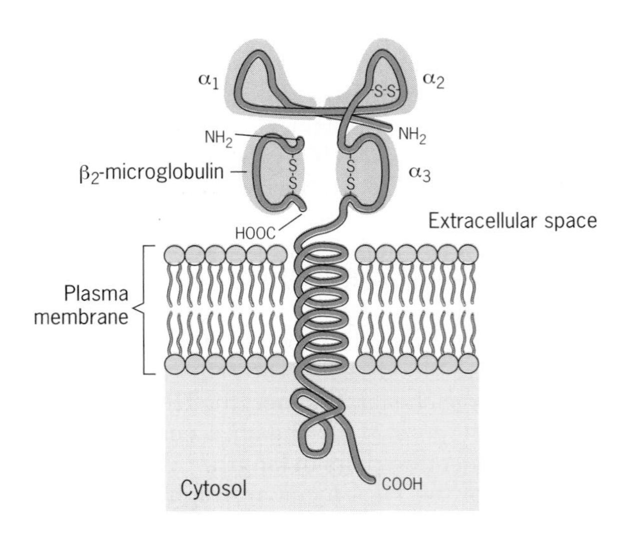

(*a*) MHC class I protein.

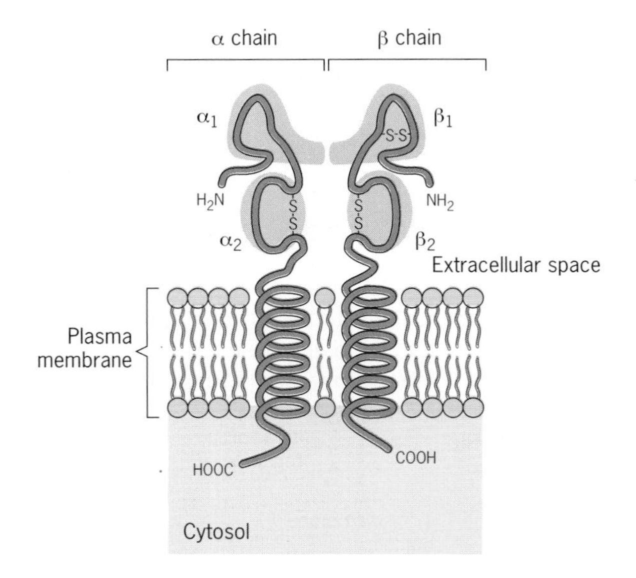

(*b*) MHC class II protein.

Figure 24.6 Structures of the MHC class I (*a*) and class II (*b*) proteins. The extracellular domains of both class I and class II proteins form antigen-binding pockets that bind antigens, usually fragments of partially degraded proteins, and present them to the appropriate T cells.

The ***HLA* class II genes** encode polypeptides that are located primarily on the surfaces of B lymphocytes and macrophages, cells that interact with helper T cells. The class II MHC proteins are heterodimers composed of α and β polypeptides (Figure 24.6*b*). Both polypeptides have carboxyl-terminal transmembrane domains and amino-terminal extracellular domains. Together, the polymorphic amino-terminal domains of the two polypeptides form an antigen-binding groove that binds a wide range of related peptides and displays them to helper T cells. Finally, the ***HLA* class III genes** encode **complement proteins** that interact with antibody-antigen complexes and help destroy them by proteolysis. Complement proteins also kill infected cells by disrupting the integrity of cell membranes.

Key Points: **The immune response of mammals involves the coordinated activities of several specialized white blood cells. After exposure to a foreign substance—an antigen—B lymphocytes differentiate into plasma cells that produce antigen-binding proteins called antibodies. Similarly, T lymphocytes develop into killer T cells that carry T cell antigen receptors on their surfaces and kill cells displaying the recognized antigens. Highly polymorphic major histocompatibility antigens on cells allow killer T cells to distinguish foreign cells from "self" cells, the cells of the individual mounting the immune response.**

THE IMMUNE RESPONSE IN MAMMALS

When a mammal is infected by a pathogenic virus, bacterium, fungus, or other invading cell, the body mounts a complex immune response. Each immune response involves three essential steps:

1. *Recognition* of the foreign entity.

2. *Communication* of this recognition to the appropriate cells.
3. *Elimination* of the invading entity.

Some components of the immune response are directed against infectious agents in general and are called **nonspecific immune responses**. They include inflammation of the infected tissues resulting in increased blood flow and the recruitment of phagocytes (Figure 24.7) that ingest and destroy viruses, bacteria, and fungi in a nonspecific manner. However, the most important responses are antigen-specific and thus are called **specific immune responses**. For the rest of this chapter, we focus on the two major specific immune responses: the synthesis of antibodies and production of killer T cells. However, these two responses do not occur independently; rather, the cells involved in the two responses must communicate with each other for either response to be effective. An overview of the two components of the immune response in mammals and of the communication that occurs between immune system cells is presented in Figure 24.8.

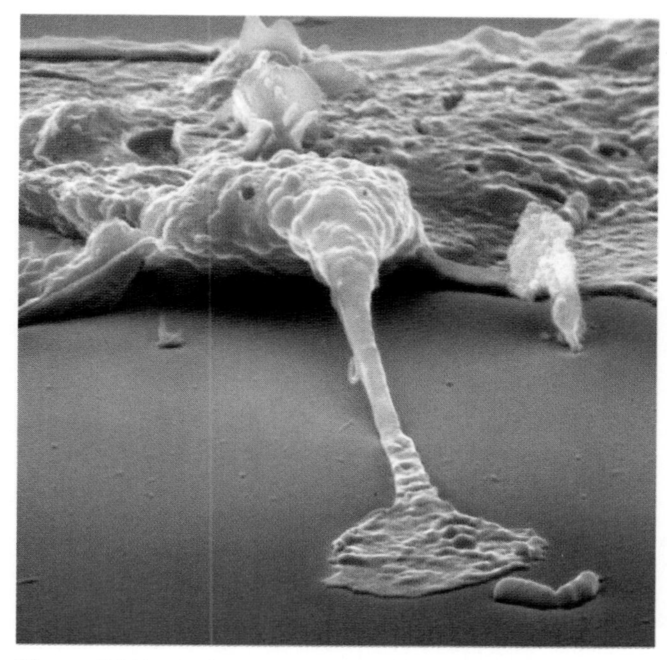

Figure 24.7 A phagocyte stalking a rod-shaped bacterium.

The Antibody-Mediated (Humoral) Immune Response

When a foreign antigen, for example, a glycoprotein on the surface of a bacterium, enters the bloodstream of a mammal, it triggers a defense mechanism, the **antibody-mediated** or **humoral immune response,** which results in the synthesis of antibodies. These antibodies bind to the antigen with exceptional specificity, thus facilitating the removal of the invading bacterium from the circulatory system. The antibody-mediated immune response is primarily responsible for protecting mammals from bacterial and fungal infections, and for the removal of viruses before they infect host cells. It involves several different classes of immune cells and leads to the production of antibodies specific to the invading antigen by plasma cells derived from B lymphocytes (Figure 24.8, *left*).

The antibody-mediated immune response is initiated when a macrophage encounters a foreign substance such as a bacterium and engulfs and destroys it. Antigens from the partially degraded bacterium are anchored on the surface of the macrophage, where they are recognized by helper T cells. Once presented with antigens by macrophages, the activated helper T cells stimulate B lymphocytes to differentiate into plasma cells, which synthesize and secrete large quantities of antibodies, and memory B cells, which facilitate a faster immune response during subsequent encounters with the same antigen.

Actually, one specific structural feature of an antigen, called an **epitope**, stimulates the production of antibodies, and the resulting antibodies specifically recognize and bind to that particular structural component of the antigen. For proteins, an epitope may consist of only a short peptide, sometimes containing as few as six amino acids.

Antibodies bind specifically to the antigens eliciting the immune response and help inactivate them or remove them from the body in several ways. However, antibody-antigen complexes are eliminated by two major pathways (Figure 24.9). (1) Many antibody-antigen complexes are ingested and degraded by phagocytes. (2) IgG and IgM antibodies recruit the complement proteins, which degrade viruses and other proteinaceous substances by proteolysis and kill bacteria and fungi by cell lysis. In other instances, the binding of the antibody to the foreign substance may inactivate it. For example, viruses contain specific surface proteins that are used to attach to and infect host cells. When antibodies bind to these attachment proteins, the viruses can no longer infect host cells. Similarly, toxins and related agents usually are inactivated when agglutinated by antibodies.

As B lymphocytes begin to differentiate, each B cell synthesizes one specific type of antibody. That is, the antibodies synthesized by any one B cell are all identical, with exactly the same antigen-binding specificity. How, then, does an individual produce large amounts of a particular antibody in order to combat an infection by an invading pathogen? Remember that the circulatory system of a mammal contains millions of B cells. Thus the entire population of B cells in an individual will be producing millions of different anti-

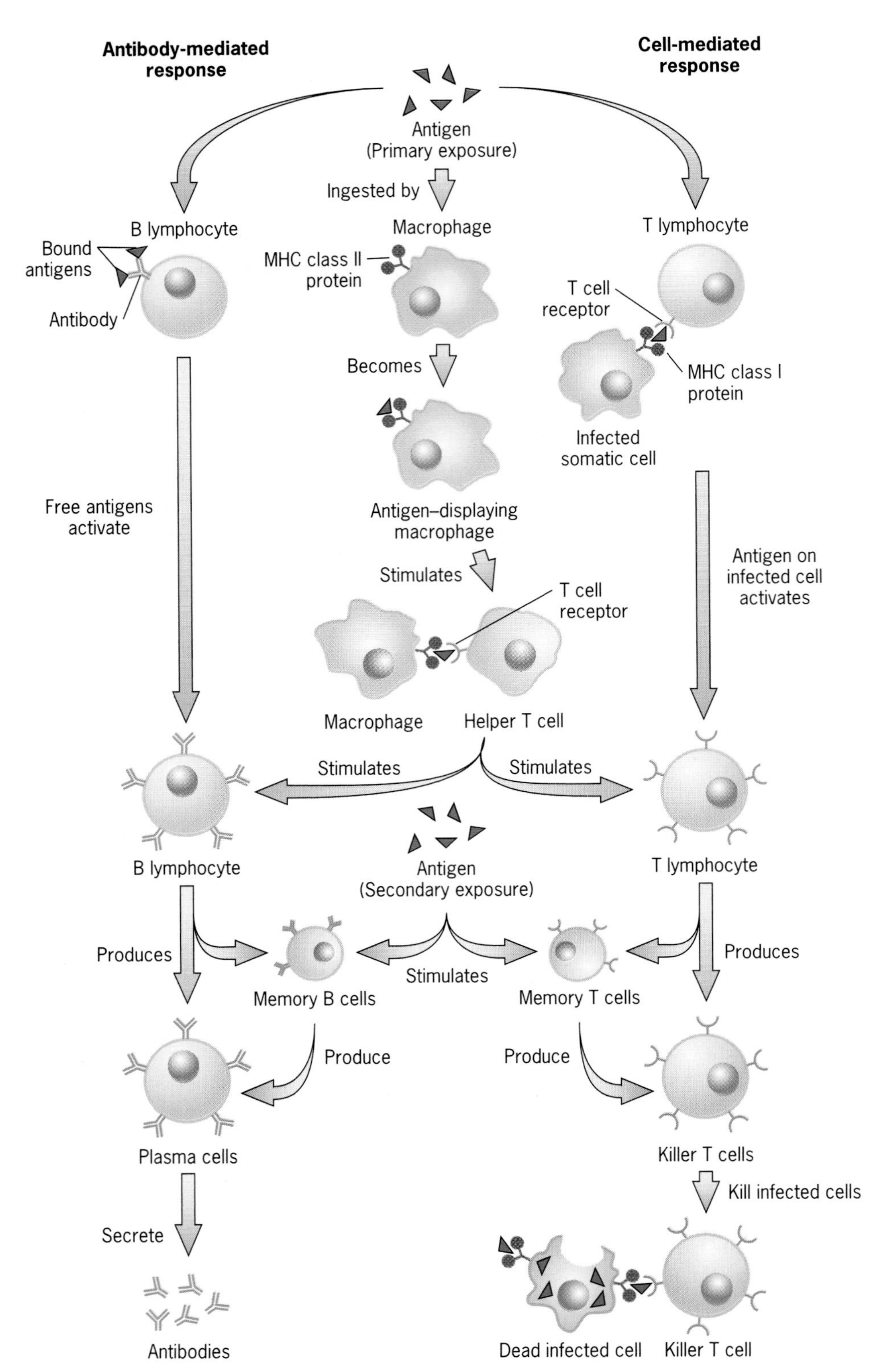

Figure 24.8 Summary of the immune response in mammals. The antibody-mediated (humoral) immune response, which is activated by free antigens, is carried out by B lymphocytes, and the cell-mediated (cellular) immune response, which is activated by antigens displayed on infected cells, is performed by T lymphocytes. Both responses are stimulated by activated helper T cells. The rapid immune responses that occur after a second or subsequent exposure to an antigen are mediated by memory B and T lymphocytes that rapidly produce antibody-producing plasma cells and killer T cells, respectively.

bodies. If a B cell is producing an antibody and is exposed to an antigen bound by that antibody, the B cell is stimulated to multiply and produce a population of plasma cells all producing the same antibody (Figure 24.10). Because this process results in a **clone** of identical plasma cells, all synthesizing a given antibody, it is called **clonal selection**. Because a mature plasma cell can synthesize and secrete 2,000 to 20,000 antibodies per second, an entire clone of such cells can unleash an enormous molecular army to combat an invading

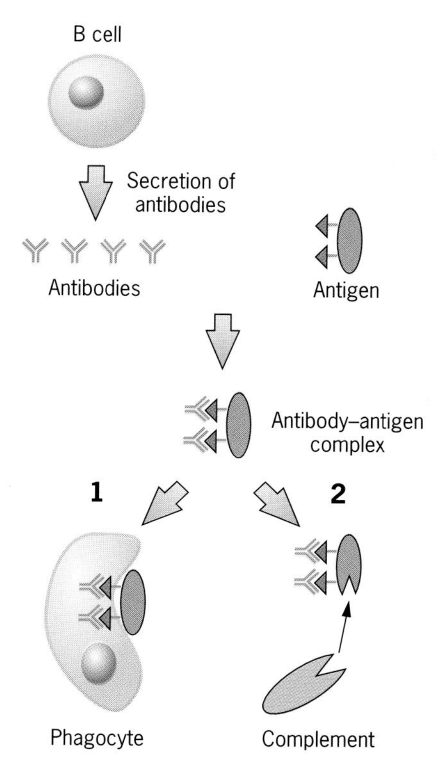

Figure 24.9 Antibody-antigen complexes are eliminated by two major pathways: (1) ingestion and degradation by phagocytes and (2) the proteolytic activity of complement proteins, which degrade the protein coats of viruses and lyse cellular pathogens.

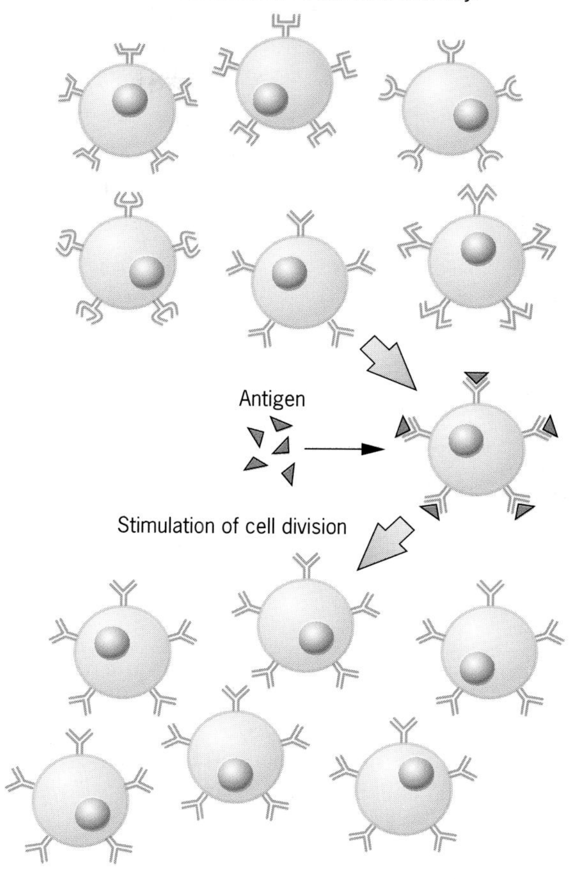

Figure 24.10 The role of clonal selection in the antibody-mediated immune response. The process by which a B lymphocyte is stimulated to divide and produce a clone of plasma cells all synthesizing the same antibody is complex. Some of the steps are described in the text and illustrated in Figure 24.8.

pathogen. Once the foreign antigen has been eliminated from the body, the suppressor T cells somehow signal the antibody-producing cells to turn off the production of that particular antibody, at least until a subsequent encounter with the offending antigen.

The T Cell-Mediated (Cellular) Immune Response

In addition to the antibody-mediated immune response, mammals possess a second antigen-specific defense mechanism called the **T cell-mediated** or **cellular immune reponse** (Figure 24.8, *right*). As the name implies, this defense mechanism is mediated directly by cells, specifically by cytotoxic or killer T cells, which attack and kill cells displaying foreign antigens on their surfaces. Killer T cells seek out and destroy cells infected with viruses, bacteria, and other pathogens, as well as some types of cancer cells (Figure 24.11). The killer T cells have antibody-like T cell receptors on their surfaces and use these receptors to seek out infected cells carrying the appropriate antigen. The antigens that are recognized by T cell receptors are usually fragments of partially degraded viral or bacterial proteins; these antigens are displayed on the surfaces of infected cells by proteins encoded by the major histocompatibility complex.

The T cell-mediated immune response begins just like the antibody-mediated response when a macrophage ingests and partially degrades a foreign agent. Antigens produced by partial degradation of the foreign substance are then displayed on the surface of the macrophage by the major histocompatibility complex proteins. The displayed antigens are recognized by developing killer T cells and by helper T cells. Having been activated by antigen-bearing macrophages, helper T cells stimulate the development of immature B and T lymphocytes by secreting classes of signal molecules collectively called cytokines, lymphokines,

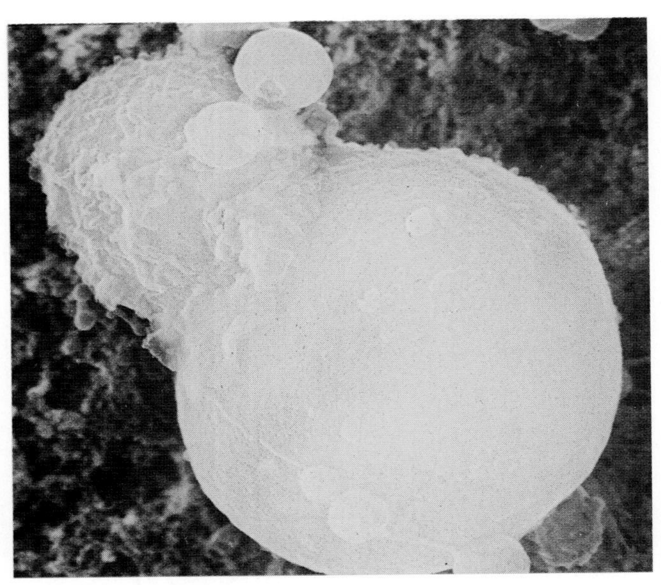

Figure 24.11 Lysis of a target cell displaying a foreign antigen by a killer T cell.

and interleukins. Once activated, the immature killer T cells develop into mature killer T cells and memory T cells. As with B lymphocytes, a population of killer T cells produces an enormous number of different antigen-specific T cell receptors. Moreover, effective populations of killer T cells armed with receptors for a given antigen are produced by clonal selection, just as they are for B cells (Figure 24.10).

Mature killer T cells attach to target cells displaying the appropriate antigens and kill them by a special form of cell lysis. The killer cells secrete a protein called perforin, which inserts itself into the membranes of target cells, producing pores. The cytoplasm of the target cells leaks out through these pores, causing cell death.

Immunological Memory: Primary and Secondary Immune Responses

When a human is first exposed to a particular foreign antigen, it takes a week to 10 days for the body to produce a significant concentration of antibodies specific to the antigen and two to three weeks to reach maximal antibody production. The first encounter with a foreign antigen results in a relatively slow immune response, called the **primary immune response**. A subsequent encounter with the same antigen produces a more rapid response, called the **secondary immune response**. Not only is a secondary immune response faster, but it also results in a higher concentration of antibodies. Similar slow primary responses and rapid secondary responses are observed for T cell receptors. Thus the immune system has memory; it can remember previously encountered antigens, which is why vaccinations against pathogens are often effective.

The cellular basis of immunological memory is known, but the molecular mechanisms responsible for this memory must still be determined. B and T lymphocytes that have not encountered a foreign antigen are referred to as **virgin** or **naive cells**. After exposure to an antigen, the virgin cells differentiate into two types of cells: **memory cells** and so-called **activated B or T cells** (Figure 24.8). The activated cells will multiply and differentiate into antibody-producing plasma cells or receptor-producing T cells, respectively.

In contrast to the short lifespan (usually days to a week) of plasma cells and killer T cells, memory cells are long-lived, with life spans of months or even years. Moreover, memory B and T cells are already in an activated state and thus can develop into antibody-producing plasma cells and receptor-producing killer T cells, respectively, more rapidly than can virgin cells. During a secondary encounter with an antigen, the memory B cells and memory T cells are stimulated to multiply and differentiate into plasma cells and killer T cells. Therefore, secondary immune responses are faster and result in larger populations of plasma cells and killer T cells than primary immune responses.

Key Points: **The immune response in mammals involves three steps: (1) recognition of the foreign substance, (2) communication of this recognition to the responding cells, and (3) elimination of the invading agent. Nonspecific responses include the recruitment of phagocytes to ingest and destroy viruses and microorganisms. Antigen-specific responses include antibody production and the activation of killer T cells. The production of large amounts of antibodies specific for an invading antigen results from clonal selection, the stimulation of a B cell producing an antibody that recognizes the antigen to multiply and produce a population of plasma cells all producing the same antibody. Long-lived memory cells facilitate a faster immune response during a second encounter with a foreign substance.**

GENETIC CONTROL OF VAST ANTIBODY DIVERSITY

The most remarkable aspect of the immune response from a genetics standpoint is the seemingly infinite variety of antibodies that can be synthesized in response to antigens that the animal has not previously encountered. How can an organism have prepared to synthesize an antibody designed to bind very specifically to a particular antigen without ever having made contact with the antigen? Moreover, how can an organism store enough genetic information to encode the amino acid sequences of a virtually unlimited variety of antibodies? This question had puzzled geneticists for many years. Then, in 1976, Nobumichi Hozumi and Susumu Tonegawa discovered that an expressed kappa light chain gene in the mouse contained DNA sequences that were present at two different locations on chromosome 6 in cells that were not involved in the production of antibodies. These DNA sequences had been joined together to produce a functional kappa light chain gene in antibody-producing plasma cells. In the following years, functional antibody genes were shown to be assembled during B lymphocyte differentiation from a storehouse of gene segments clustered at different sites in germ-line chromosomes.

Geneticists do not know how many different antibodies a mouse or a human can produce; they do know, however, that the number is very large, almost certainly in the millions. This huge number has presented a paradox. The mammalian genome contains about 3×10^9 nucleotide pairs. If all this DNA were in the form of uninterrupted coding sequences of genes each about 1000 nucleotide pairs long, the genome would contain about 3 million genes. Of course, many of these genes encode various RNA molecules, enzymes, and structural proteins. Also, geneticists know that many of these genes contain long noncoding introns. In addition, they know that the highly repetitive DNA sequences in the mouse and human genomes are not transcribed. The paradox is obvious: the mammalian genome does not appear to be large enough to encode the observed plethora of different antibodies and all the other essential gene products.

Early attempts to explain the genetic control of antibody diversity can be grouped into three different hypotheses.

1. The germ-line hypothesis, which stated that there is a separate germ-line gene for each antibody. This concept fit with our early knowledge about protein synthesis but presented the paradox of not enough DNA.
2. The somatic mutation hypothesis, which stated that only one or a few germ-line genes specify each major class of antibodies and that the diversity is generated by a high frequency of somatic mutation—mutation that occurs in the antibody-producing somatic cells or in cell lineages leading to antibody-producing cells. There was no precedent for a high frequency of mutation occurring in only certain genes and in only certain types of cells. By what mechanism could this occur, and how could it be regulated?
3. The mini-gene hypothesis, which stated that the diversity is generated by the shuffling of many small segments of a few genes into a multitude of possible combinations. The shuffling would occur by recombination processes in somatic cells. This possibility required totally novel mechanisms for rearranging segments of DNA.

All three hypotheses were correct in certain respects. The mini-gene hypothesis explains a great deal of the observed antibody variability, but somatic mutation contributes additional diversity. In addition, the constant region of each antibody chain is specified by a gene segment that is present in germ-line DNA in only a few copies. Details of the mechanisms by which this vast antibody diversity is generated are discussed in the following sections of this chapter.

Key Points: **Humans and other vertebrates can produce a large array of antibodies specific to antigens not previously encountered. The germ line, somatic mutation, and mini-gene hypotheses all contributed to an understanding of the genetic control of antibody diversity.**

GENOME REARRANGEMENTS DURING B LYMPHOCYTE DIFFERENTIATION

How can the human genome store enough genetic information to encode a seemingly infinite number of different antibodies? The answer to that question is simple if we overlook the molecular details. The genetic information encoding antibody chains is stored in bits and pieces, and these bits and pieces are put together in the appropriate sequences by genome rearrangements occurring during the development of the antibody-producing cells of the body. Each antibody chain is synthesized using information stored as several different gene segments in germ-line chromosomes. The antibody genes that are expressed in B lymphocytes and plasma cells are spliced together from these gene segments by recombination events that occur during the differentiation of these cells from stem cells in bone marrow. During these genomic rearrangements, the appropriate gene segments are joined by sequence-specific recombination processes that delete the regions of DNA between these gene segments. Lambda and kappa light chain genes and heavy chain genes are spliced together by very similar mechanisms. The only major difference is that lambda light chain genes, kappa light chain genes, and heavy chain genes are assembled from two, three, and four different gene segments, respectively. However, note that the gene segments encoding lambda, kappa, and heavy chains are located on three different human chromosomes (chromosomes 22, 2, and 14, respectively). Thus the assembly of human lambda, kappa, and heavy-chain genes involves three distinct somatic recombination processes.

Lambda Light Chain Genes Assembled from Two Gene Segments

In germ-line chromosomes, the DNA sequences that encode lambda light chains are present in two distinct types of gene segments (Figure 24.12). One type of gene segment, designated $\boldsymbol{L_\lambda V_\lambda}$ (for leader peptide and variable region), encodes (1) the amino-terminal hydrophobic leader peptide that is excised from the antibody chain after it directs the nascent polypeptide through the membrane of the rough endoplasmic reticulum and (2) the NH_2-terminal 97 amino acids of the **variable region** of the lambda light chain. The other type of gene segment, called $\boldsymbol{J_\lambda C_\lambda}$ (for joining segment and constant region), contains the so-called **joining sequence** and the region encoding the COOH-terminal **constant region** of the lambda light chain. The joining sequence encodes the last 13 to 15 amino acids of the variable region of lambda light chains. A complete lambda light chain gene is assembled by joining an $L_\lambda V_\lambda$ gene segment with a $J_\lambda C_\lambda$ gene segment, with the deletion of the intervening DNA, during B lymphocyte differentiation (Figure 24.12). The gene assembly process is controlled by sequence-specific recombination events; the sequences that regulate

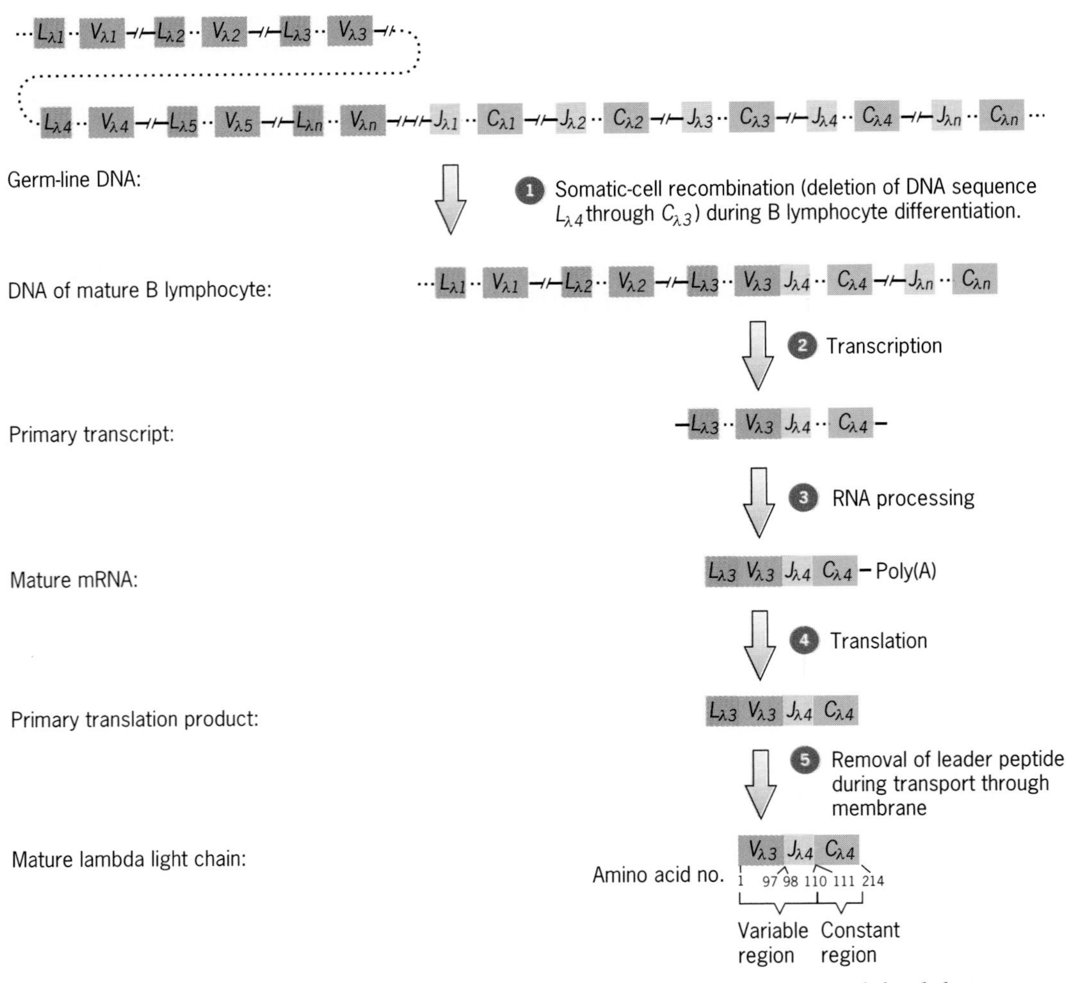

Figure 24.12 The genetic control of human antibody lambda light chains. Each lambda light chain gene is assembled from two pools of gene segments present on germ-line chromosome 22 during B lymphocyte differentiation.

this process are discussed in a subsequent section of this chapter.

In humans, there are about 300 different $L_\lambda V_\lambda$ gene segments clustered near the centromere on chromosome 22 and nine different $J_\lambda C_\lambda$ gene segments—at least six of which encode functional gene products—located somewhat farther from the centromere on chromosome 22. During the differentiation of B lymphocytes, one $L_\lambda V_\lambda$ gene segment is joined to one $J_\lambda C_\lambda$ gene segment, with the deletion of the intervening DNA (Figure 24.12). Any $L_\lambda V_\lambda$ gene segment may be joined to any $J_\lambda C_\lambda$ gene segment. The newly assembled $L_\lambda V_\lambda J_\lambda C_\lambda$ gene contains two introns, one 93 nucleotide pairs long between the L_λ and V_λ exons, and one about 1200 nucleotide pairs long between the J_λ and C_λ exons. These two introns are spliced out of the primary transcript during RNA processing just like the introns of other genes.

In contrast to the over 300 $L_\lambda V_\lambda$ gene segments in the human genome, the mouse genome contains only two $L_\lambda V_\lambda$ gene segments. Thus it is not surprising that only 5 percent of the antibodies of mice are of the lambda type, whereas 40 percent of the antibodies of humans have lambda light chains.

Kappa Light Chain Genes Assembled from Three Gene Segments

Synthesis of kappa light chains is controlled by three different gene segments: (1) an $\boldsymbol{L_K V_K}$ gene segment, which encodes the leader peptide and the NH_2-terminal 95 amino acids of the variable region, (2) a $\boldsymbol{J_K}$ gene segment, which encodes the last 13 amino acids of the variable region, and (3) a $\boldsymbol{C_K}$ gene segment, which encodes the COOH-terminal constant region. The arrangement of the kappa chain gene segment in germ-line cells and in all other cells *not* involved in antibody production is shown in Figure 24.13. In humans, chromosome 2 contains about 300 $L_K V_K$ gene

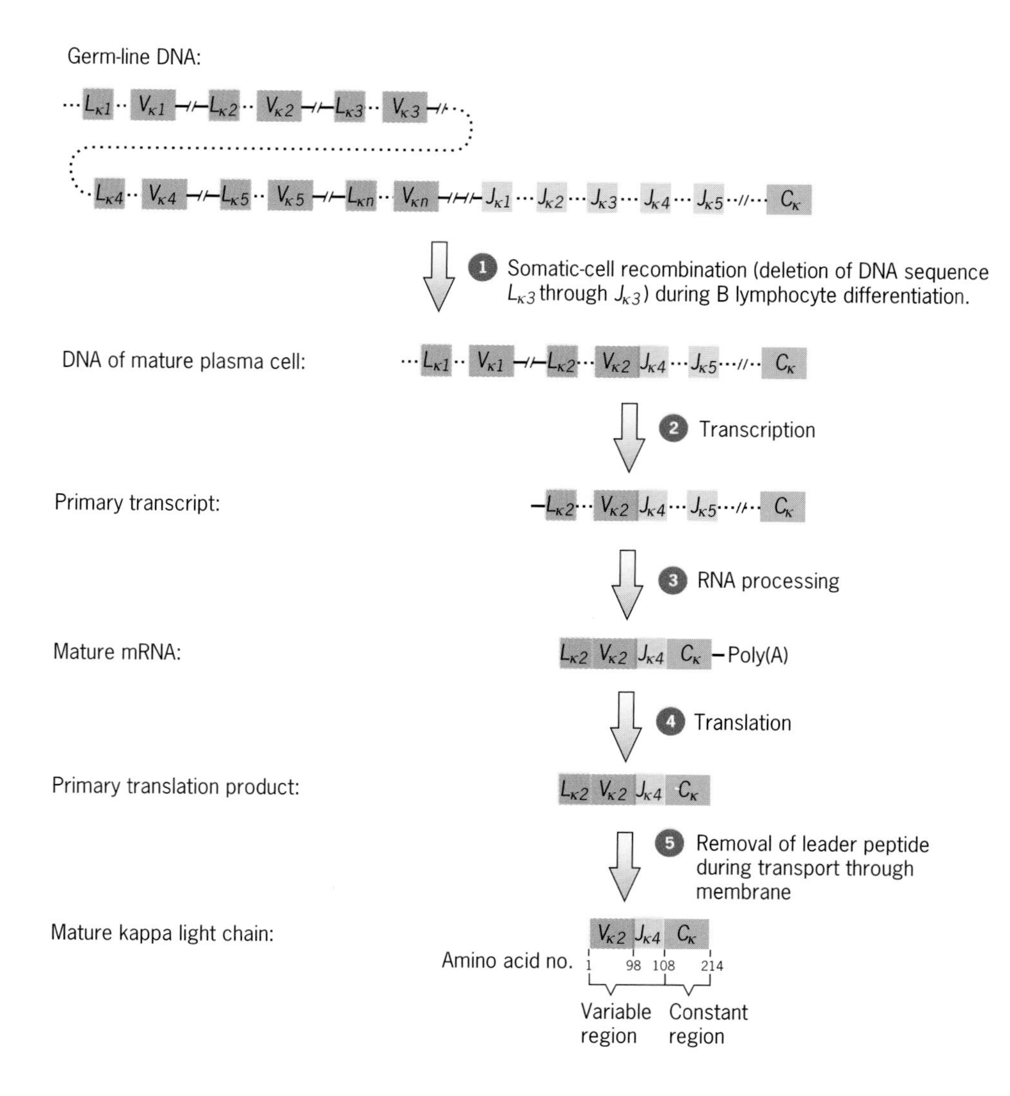

Figure 24.13 The genetic control of human antibody kappa light chains. Each kappa light chain gene is assembled from three different gene segments located on chromosome 2 during the differentiation of B lymphocytes into antibody-producing plasma cells.

segments, five J_K gene segments, and a single C_K gene segment. The J_K gene segments are located between the $L_K V_K$ gene segments and the C_K gene segment.

In germ-like cells, the five J_K segments are separated from the $L_K V_K$ segments by a long noncoding sequence and from the C_K segment by an approximately 2000 nucleotide-pair-long noncoding sequence. During the development of a B lymphocyte, the particular kappa light chain gene that will be expressed in that cell is assembled from one $L_K V_K$ segment, one J_K segment, and the single C_K segment by a process of somatic recombination (Figure 24.13, step 1). During this process, recombination can join any one of the approximately 300 $L_K V_K$ gene segments with any one of the five J_K segments, with the deletion of all intervening DNA (Figure 24.13). It produces an $L_K V_K J_K$ fusion that encodes the entire variable region of the kappa light chain. The noncoding sequence between the cluster of J_K gene segments and the C_K gene segment along with any nondeleted C_K-proximal J_K gene segments are retained (between the new $L_K V_K J_K$ fusion and the C_K segment) in the DNA of plasma cells (Figure 24.13). This entire DNA sequence ($L_K V_K J_K$-noncoding-C_K) is transcribed (Figure 24.13, step 2), and the noncoding sequences are removed during RNA processing (Figure 24.13, step 3), just as are the introns of any other eukaryotic gene. The resulting mRNA is translated (Figure 24.13, step 4), and the leader peptide is excised from the nascent kappa polypeptide during its transport through the membrane of the rough endoplasmic reticulum (Figure 24.13, step 5).

Heavy Chain Genes Assembled from Four Gene Segments

The genetic information encoding antibody heavy chains is organized into $L_H V_H$, J_H, and C_H gene segments analogous to those for kappa light chains; but there is one additional gene segment, called ***D*** for diversity, that encodes 2 to 13 amino acids of the variable region. The variable region of the heavy chain is thus specified by three separate gene segments, $L_H V_H$, J_H, and D, which must be joined during B-lymphocyte development. In addition, there are from one to four C_H gene segments for each Ig class.

In the mouse, there are a total of eight C_H gene segments, all functional, arranged on chromosome 12 in the sequence $C_{H\mu}$, $C_{H\delta}$, $C_{H\gamma3}$, $C_{H\gamma1}$, $C_{H\gamma2b}$, $C_{H\gamma2a}$, $C_{H\epsilon}$, $C_{H\alpha}$, (Figure 24.14). $C_{H\mu}$, $C_{H\delta}$, $C_{H\epsilon}$ and $C_{H\alpha}$ encode the

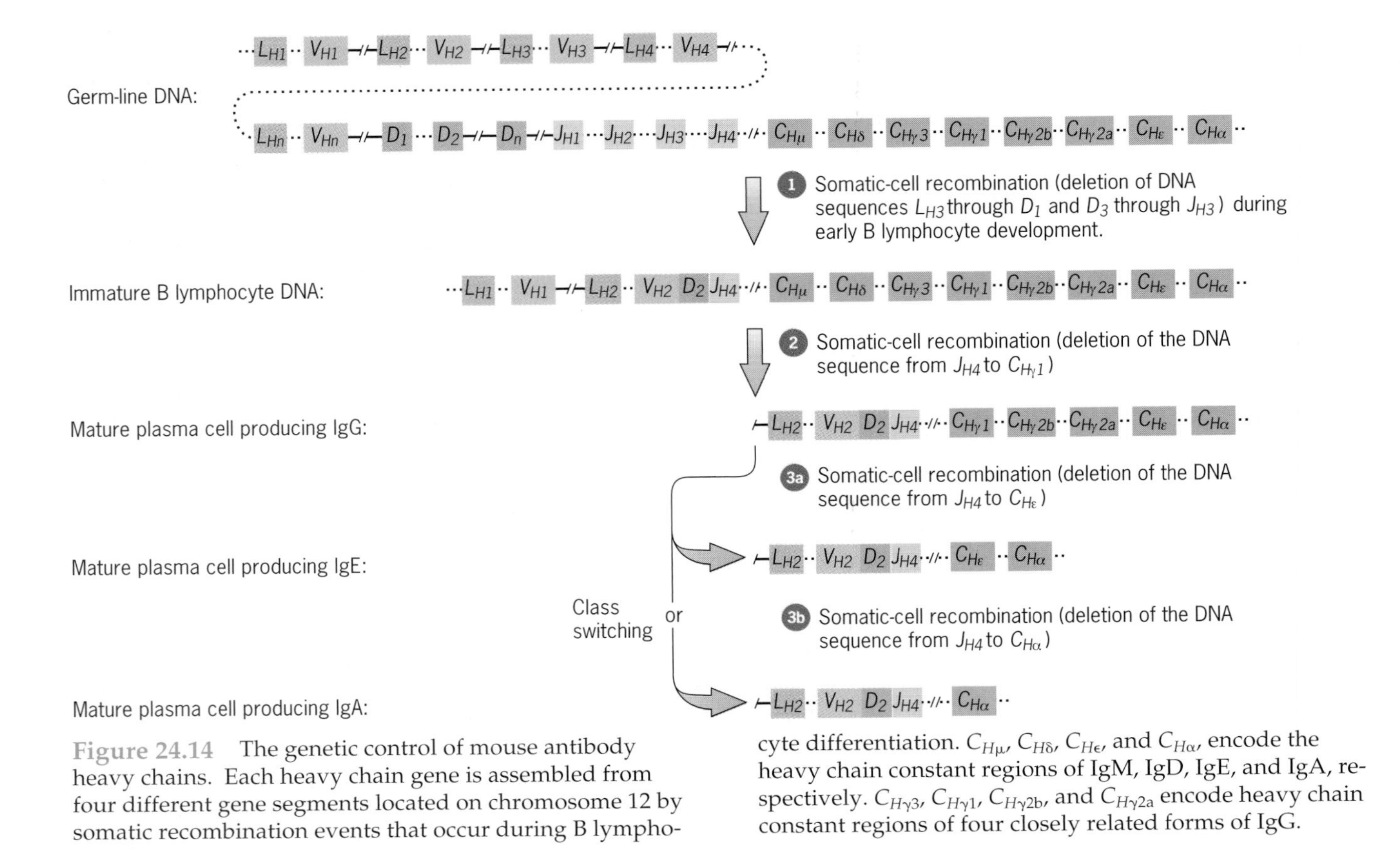

Figure 24.14 The genetic control of mouse antibody heavy chains. Each heavy chain gene is assembled from four different gene segments located on chromosome 12 by somatic recombination events that occur during B lymphocyte differentiation. $C_{H\mu}$, $C_{H\delta}$, $C_{H\epsilon}$, and $C_{H\alpha}$, encode the heavy chain constant regions of IgM, IgD, IgE, and IgA, respectively. $C_{H\gamma3}$, $C_{H\gamma1}$, $C_{H\gamma2b}$, and $C_{H\gamma2a}$ encode heavy chain constant regions of four closely related forms of IgG.

heavy chain constant regions of IgM, IgD, IgE, and IgA, respectively. Four gene segments, $C_{H\gamma3}$, $C_{H\gamma1}$, $C_{H\gamma2b}$, and $C_{H\gamma2a}$, encode IgG heavy chain constant regions. In humans, there are 9 or 10 functional C_H gene segments: $C_{H\mu}$, $C_{H\delta}$, $C_{H\gamma1}$, $C_{H\gamma2}$, $C_{H\gamma3}$, $C_{H\gamma4}$, $C_{H\epsilon1}$, probably $C_{H\epsilon2}$, $C_{H\alpha1}$, and $C_{H\alpha2}$. The human C_H gene cluster also contains two pseudogenes—sequences that are very similar to genes, but that contain in-frame stop codons and are thus nonfunctional.

In mouse germ-line cells, there are about 300 L_HV_H gene segments, from 10 to 50 D gene segments, 4 J_H gene segments, and 8 C_H gene segments, arranged on chromosome 12 in the order given above (Figure 24.14). During B lymphocyte differentiation, somatic recombination events join one L_HV_H gene segment with one D gene segment and one J_H gene segment, deleting the two intervening sequences of DNA, to form one continuous DNA sequence, $L_HV_HDJ_H$, which encodes the entire heavy chain variable region (Figures 24.14, step 1).

Signal Sequences Control Somatic Recombination Events

How are the genome rearrangements that occur during B lymphocyte development regulated? What controls the somatic recombination events such that a *V* gene segment is joined to a *J* segment and not to another *V* segment or directly to a *C* segment? Several long segments of chromosomal DNA carrying clusters of *V* gene segments, *D* gene segments, and *J* gene segments of both mice and humans have been sequenced, and the resulting nucleotide-pair sequences suggest the presence of specific *V-J*, *V-D*, and *D-J* joining signals. The same signal sequences are found adjacent to all *V* gene segments. Similarly, all *J* gene segments have identical signal sequences located adjacent to their coding sequences; however, their signal sequence is different from that adjacent to *V* gene segments. *D* and *C* gene segments also have their own signal sequences.

The signal sequences controlling *V-J*, *V-D*, and *D-J* joining contain 7-base-pair-long (heptamer) and 9-base-pair-long (nonamer) sequences separated by spacers of different, but specific lengths. For L_kV_k-J_k joining, the spacer in the L_kV_K signal sequence is 12 nucleotide pairs long, whereas that in the J_k signal sequence is 22 nucleotide pairs long. The heptamer and nonamer sequences located 3' to or "after" (to the right as drawn in Figure 24.13) the L_kV_k gene segments are complementary (with the exception of one base pair) to those 5' to or "preceding" (to the left as drawn in Figure 24.13) the J_k gene segments. These signal sequences have the potential to form "stem and loop" structures (Figure 24.15), thus bringing the L_kV_k and J_k gene segments into juxtaposition for joining. Apparently, joining will occur only when one signal sequence contains a 12-base-pair spacer and the other contains a 22-base-pair spacer. This requirement presumably is enforced by the proteins that mediate the DNA splicing processes. Very similar signal sequences appear to control L_HV_H-D and D-J_H joining.

Additional Diversity: Variable Joining Sites and Somatic Hypermutation

A vast amount of antibody diversity is generated by the combinatorial joining of antibody gene segments as previously described. For example, consider the number of different kappa light chains possible in humans. If there are 300 L_KV_K gene segments and 5 J_k segments, then 1500 fused $L_kV_kJ_k$ gene segments can be produced by joining the L_kV_k and J_k gene segments in all possible combinations. Similarly, given 300 $L_\lambda V_\lambda$ and 6 $J_\lambda C_\lambda$ gene segments, 1800 different lambda light chains can be produced by combinatorial joining. The heavy chain variable region provides even greater diversity because of the multiple *D* gene segments. If there are 300 V_H gene segments, 25 *D* gene segments, and 6 J_H gene segments in human germ line cells, 45,000 different heavy chain variable regions can be assembled. If there are 3300 (1500 + 1800) different light chains and 45,000 different heavy chains, then 3300 × 45,000 or 148,500,000 different antibodies can be produced by combinatorial joining alone. Clearly, the combinatorial joining of antibody gene segments produces a vast amount of antigen-binding diversity. However, molecular analyses of antibodies and antibody genes have shown that additional diversity is generated by other mechanisms.

A comparison of the diversity of amino acid sequences present in antibody molecules with that predicted from the sequences of gene segments that encode these antibodies reveals that there is more variation in amino acid sequences at the *V-J* junctions than is predicted by the nucleotide sequences. Some of this additional diversity has been shown to result from the use of alternate sites of recombination during the joining events that are involved in the assembly of mature antibody genes. An example of the use of alternate sites of joining of V_K and J_K gene segments in the mouse is illustrated in Figure 24.16. During the joining of gene segments V_{K41} and J_5, recombination events have been shown to occur between four adjacent nucleotide positions at the junction sites. As shown in Figure 24.16*d*, these recombination events produce four different nucleotide sequences that encode three distinct amino acids at position 96 in the mouse kappa light chain. Since amino acid 96 occurs

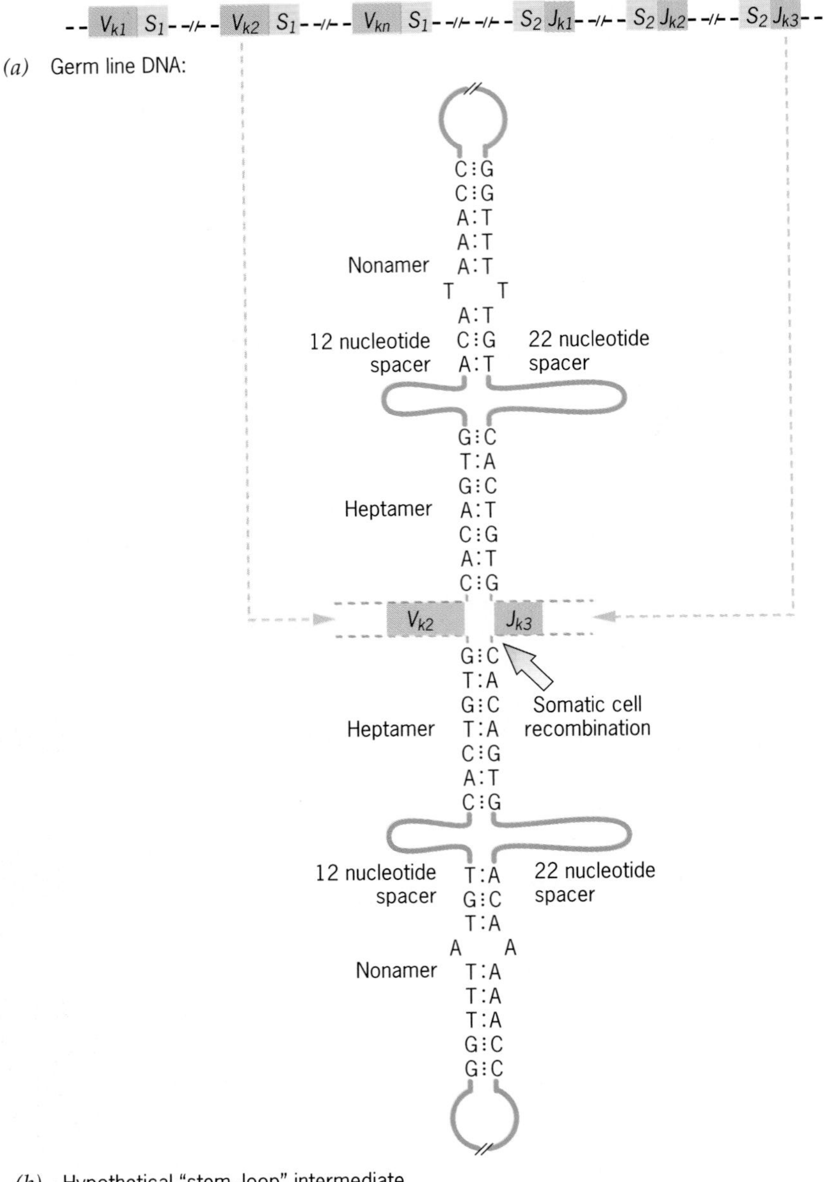

Figure 24.15 Signal sequences direct $V_\kappa J_\kappa$ joining during the assembly of kappa light chain genes. The germ-line and plasma cell DNA arrangements are shown in (*a*) and (*c*), respectively. The heptameric and nonameric signal sequences and the spacer regions that control the DNA splicing events are shown in (*b*). The "stem and loop" structure shown in (*b*) is a hypothetical intermediate.

in a region of the antibody chain that is involved in antigen binding, alternate *V-J* joining events of this type contribute significantly to the great diversity of antibody specificity that is observed in mammals. Similar alternate joining events have been documented for $V\lambda$-$J\lambda$ and V_H-D-J_H joining reactions.

Despite the vast array of antibody diversity produced by (1) the joining of large families of *V*, *D*, and *J* gene segments, and (2) the use of alternate positions of recombination during the joining reactions, considerable data demonstrate that still another mechanism must be involved in the generation of antibody diversity. This has been established by comparing (1) the nucleotide-pair sequences of expressed genes with the sequences of germ-line gene segments and (2) the actual amino acid sequences of antibody chains with the

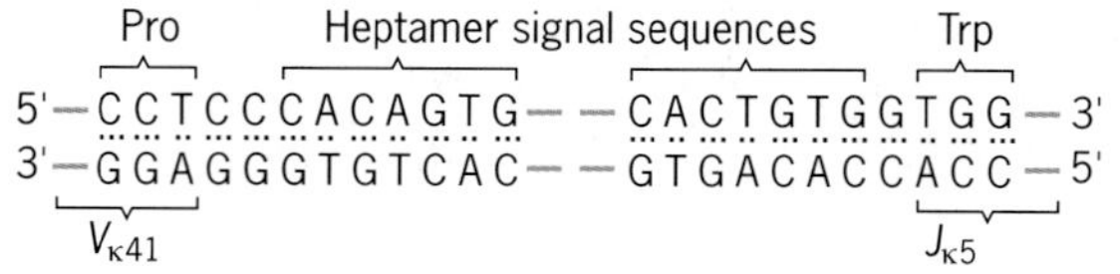

Pro Heptamer signal sequences Trp

5'—CCTCCCACAGTG— —CACTGTGGTGG—3'
3'—GGAGGGTGTCAC— —GTGACACCACC—5'

$V_{\kappa 41}$ $J_{\kappa 5}$

(*a*) Nucleotide sequences of the V_{K41} and J_5 gene segments in the mouse.

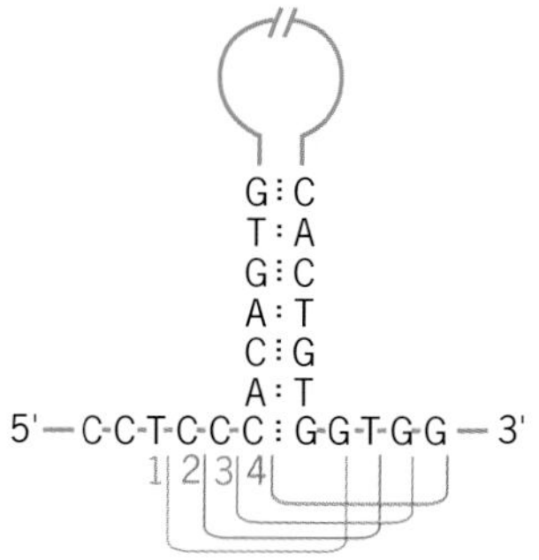

G:C
T:A
G:C
A:T
C:G
A:T
5'—C-C-T-C-C-C : G-G-T-G-G—3'
1 2 3 4

(*b*) Putative role of the signal sequences in the juxtaposition of the gene segments.

$V_{\kappa 41}$ CCTCCC
1 2 3 4
GTGG $J_{\kappa 5}$

(*c*) Alternate sites of somatic recombination.

1 5'—CCTTGG—3'
Pro Trp

2 5'—CCTCGG—3'
Pro Arg

3 5'—CCTCCG—3'
Pro Pro

4 5'—CCTCCC—3'
Pro Pro

Amino acid number 95 96

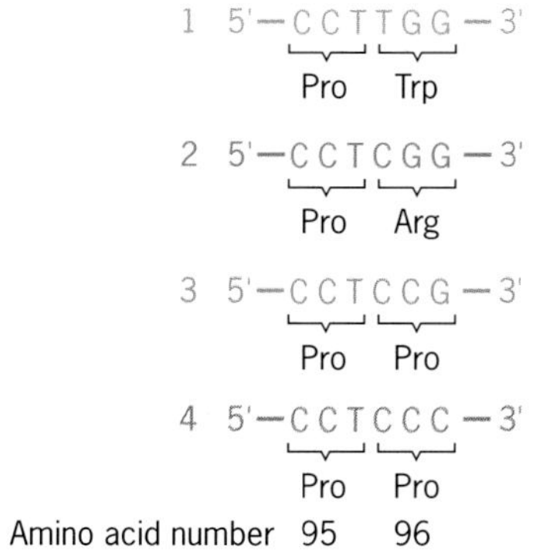

(*d*) The resulting nucleotide sequences and the dipeptides that they encode.

Figure 24.16 Antibody diversity at the V_kJ_k junction is produced by variation in the exact position of the joining reaction.

amino acid sequences predicted from the nucleotide sequences of the genes. For example, when the actual amino acid sequences of different mouse λ_1 chains were compared with the amino acid sequences predicted from the nucleotide-pair sequences of λ_1 gene segments, differences were found at sites in the variable regions other than the junction sites. Similar observations have been made in studies of heavy chain variable regions. In essentially all cases, the changes have resulted from single nucleotide-pair substitutions. Such substitutions may represent up to 2 percent of the nucleotide pairs of the gene segments encoding the variable regions of antibodies. These nucleotide-pair substitutions must occur by some mechanism of **somatic mutation** that is restricted to the DNA sequences encoding the variable regions of antibody chains. Because these changes in the variable segments of antibody genes occur at such a high frequency, the process by which they occur is called **somatic hypermutation.** The mechanism by which somatic hypermutation occurs is unknown.

Somatic hypermutation of regions of antibody genes that encode antigen-binding sites may be of great value to the organism. Without this mechanism for generating antibody diversity, the range of available antibody specificity would be fixed in terms of the sequences present in the genome at birth and the combinations that could be produced by the various levels of gene-segment joining reactions. Viruses and other pathogens are constantly evolving and producing new variants with new antigenic determinants. To provide an adequate defense against the changing antigenic composition of these viruses and other components of the environment, the immune system must also be capable of responding rapidly to these changes. What better way to provide this safeguard than to endow antibody genes with their own mechanism—somatic hypermutation—for rapid adaptation to new antigens that might evolve in the future.

Key Points: **The enormous variety of antibodies produced by mammals results from the combinatorial joining of antibody gene segments by somatic recombination during the differentiation of B lymphocytes into antibody-producing plasma cells. Signal sequences flanking antibody gene segments control these somatic recombination events. Additional antibody diversity is produced by variability in the sites of gene-segment joining and by a high frequency of mutation—somatic hypermutation—in the DNA sequences that encode antigen-binding sites.**

ANTIBODY CLASS SWITCHING

At the time that antibody synthesis begins in a developing B lymphocyte, all the C_H gene segments are still present, separated from the newly formed $L_HV_HDJ_H$ gene fusion by a short noncoding sequence (Figure 24.14). At this stage, all antibodies synthesized have IgM heavy chains ($C_{H\mu}$ gene products). However, some plasma cells may soon begin to produce other classes of antibodies (with IgA, IgD, IgE, or IgG heavy chains). This process is called **class switching.** The first type of class switching occurs when certain plasma cells begin to produce both IgM and IgD antibodies. These antibodies differ only in their effector function domains; they have identical antigen-binding domains, specified by the same fused $V_k J_k$ or $V_\lambda J_\lambda$ and $V_H D J_H$ gene segments. In these cells, a primary transcript that extends through both the $C_{H\mu}$ and $C_{H\delta}$ gene segments is synthesized. During processing, the $V_H D J_H$ transcript sequence may be spliced to either the $C_{H\mu}$ sequence or the $C_{H\delta}$ sequence. As a result, both types of heavy chain are synthesized in the same cell (Figure 24.17).

Other B cells will switch from producing antibodies of class IgM to producing IgA, IgD, IgE, or IgG antibodies. This type of class switching involves further genome rearrangements during which the C_H gene segments closest to the previously joined $L_H V_H D J_H$ gene segments are deleted (Figure 24.14, step 3). The class of antibodies produced after class switching depends on which C_H coding region remains juxtaposed to the previously fused $L_H V_H D J_H$ coding sequence.

A further complexity observed in the antibody-mediated immune response is the sequential production of membrane-bound and secreted forms of a given antibody. The first antibodies to appear in developing B lymphocytes are membrane-bound IgM molecules. Subsequently, these cells switch to the production of a secreted form of IgM. These two forms of IgM differ only in the COOH-terminal portions of their heavy chains. The heavy chain of the membrane-bound form is 21 amino acids longer than that of the secreted form. The membrane-bound heavy chain has a 41 amino acid-long hydrophobic sequence at the carboxyl terminus that anchors it to the cell surface. This hydrophobic sequence is replaced by a 20-amino-acid-long hydrophilic sequence in the secreted form. The distinct heavy chains of the membrane-bound and secreted forms of IgM are produced by alternate pathways of transcript splicing.

Key Points: **Developing B lymphocytes can switch from the production of IgM antibodies to other classes of antibodies by genome rearrangements or by alternate pathways of transcript splicing.**

ASSEMBLY OF T CELL RECEPTOR GENES BY SOMATIC RECOMBINATION

Several types of T lymphocytes play central roles in both the antibody-mediated and and T cell-mediated immune responses (Table 24.1). The various T cells carry out their specific functions in part by responding to antigens bound to T cell receptors anchored on their cell surfaces (Figure 24.4). Cytotoxic or killer T cells recognize antigens on the surface of cells and kill the cells carrying these antigens. Like the antibodies produced by plasma cells, T cells can recognize and destroy cells carrying an amazing variety of antigens. Thus the T cell response also exhibits a phenomenal degree of specificity. How is this specificity produced? The answer is that the diversity of T cell receptor specificity is produced by genome rearrangements analogous to those involved in antibody production. But how do killer T cells avoid interacting with free antigens, so that they don't simply duplicate the function of B cells? The answer is that T cells must simultaneously recognize both the offending antigen on the cell surface and the major histocompatibility complex (MHC) protein that displays the antigen to the T cell (Figure 24.8). The cell-surface MHC protein that the T cell receptor recognizes is the product of one of many genes at the *MHC* locus of mammals (the *HLA* locus in humans, Figure 24.5). Killer T cells are able to recognize and destroy cells that are producing a given antigen, for example, the coat protein of a virus, and an MHC protein in essentially any tissue of the body.

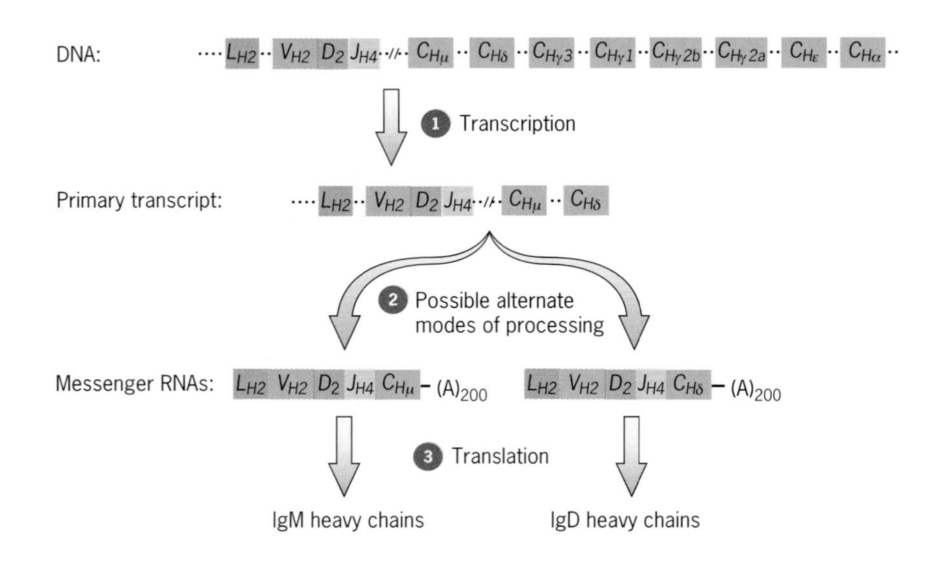

Figure 24.17 Antibody class switching may occur by alternate pathways of transcript splicing. The two splicing pathways shown result in the production of both IgM and IgD heavy chains by the same plasma cell.

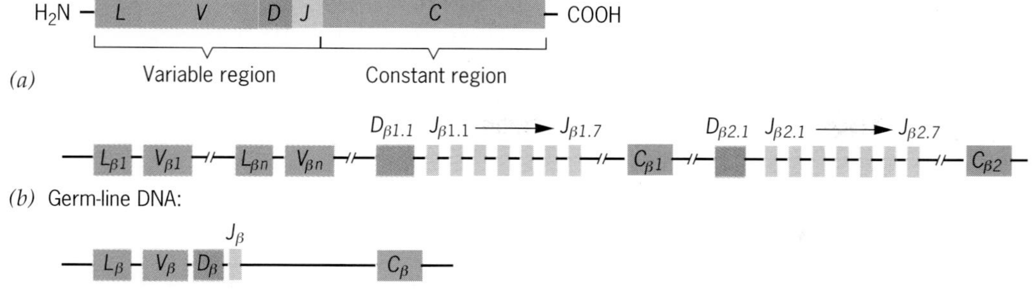

Figure 24.18 Diagrams showing (*a*) the regions of the T cell receptor α and β proteins that are encoded by the *L-V*, *D*, *J*, and *C* gene segments, respectively, (*b*) the germ-line organization of the clusters of gene segments that encode the β polypeptides of T cell receptors in the mouse, and (*c*) the structure of an assembled gene encoding a T cell receptor β chain. The C_{β} gene segments each contain three introns, which are not shown in order to minimize complexity. The large intron between the *J* segment and the *C* segment of the assembled gene is removed during processing of the primary transcript, as are the three introns within the *C* segment of the gene.

Like antibody heavy chains, the two polypeptide chains of T cell receptors are encoded by *L-V*, *D*, *J*, and *C* gene segments. Figure 24.18*a* shows the relationship of these gene segments to the regions of the T cell receptor that they encode. The variable regions of the T cell receptor proteins are encoded by multiple *L-V*, *D*, and *J* gene segments; the constant regions are encoded by a small number of *C* gene segments. The T cell receptor genes are assembled by genomic rearrangements that occur during the differentiation of T lymphocytes from stem cells just as in the case of antibody genes in developing B lymphocytes. The α and β receptor proteins are encoded by gene segments that are arranged in clusters much like those of gene segments encoding antibody chains. In humans, the α and β gene-segment clusters are located on chromosomes 14 and 7, respectively.

The germ-line organization of the gene segments that encode T cell receptor β-polypeptides in the mouse is shown in Figure 24.18*b*. The arrangement of the receptor β chain locus in humans is very similar to that in the mouse; however, only 12 of the 14 *J* gene segments are functional in humans. The organization of an assembled β-polypeptide gene is shown in Figure 24.18*c*. Heptameric and nonameric signal sequences similar to those controlling antibody gene rearrangements appear to guide the DNA splicing process. T cell receptor variable regions are encoded by only about 30 *LV* gene segments, in contrast to the approximately 300 *LV* gene segments that encode the variable regions of the three types of antibody chains in humans. However, there are more *J* gene segments in the T cell receptor gene clusters. For example, there are 12 functional *J* gene segments that encode amino acid sequences in β receptor polypeptides (Figure 24.18*b*). The T cell receptor α polypeptide gene cluster has a similar organization; however, the α polypeptide gene cluster contains 18 *J* gene segments and only a single C sequence. Thus, T cell receptor diversity, like antibody diversity, is generated by the combinatorial joining of pools of gene segments during T lymphocyte differentiation.

Key Points: T cell receptor genes are assembled from gene segments during T lymphocyte differentiation by genome rearrangements analogous to those involved in antibody production.

REGULATION OF IMMUNOGLOBIN GENE EXPRESSION

Antibody gene segments are not transcribed or are transcribed at very low levels in germ-line cells. In plasma cells, 10 to 20 percent of the mRNA molecules are antibody gene transcripts. What activates the transcription of antibody genes during or immediately after their assembly? Why is only one type of antibody produced by a given plasma cell? What prevents the synthesis of more than one type of antibody by a plasma cell?

Allelic Exclusion: Only One Functional Rearrangement per Cell

Each plasma cell produces only one type of antibody. Why? Mammalian cells are diploid; they carry two sets of genetic information encoding each of the antibody chains. However, only one productive genome

TECHNICAL SIDELIGHT

Monoclonal Antibodies: Powerful Probes for Detecting Proteins and Other Antigens

In addition to defending the body against viruses, bacteria, fungi, and other foreign substances, antibodies provide scientists with powerful research tools that can be used to identify specific proteins or other antigenic molecules. For many years, the problem with using antibodies as molecular probes was the difficult task of purifying those specific to the antigen of interest. In 1975, Cesar Milstein and Georges Köhler developed a procedure that solved this problem. They discovered how to produce immortal, antibody-secreting cell culture clones. They fused B lymphocytes with cancer cells called myelomas to produce hybrid somatic cells. The resulting hybrid cells, called **hybridomas**, were immortal; moreover, they could be cloned and screened for the production of the antibodies of interest. Once identified, the desired hybridoma cell lines could be used to produce essentially unlimited quantities of pure **monoclonal antibodies**, antibodies all recognizing the same antigenic epitope.

The procedure used to produce monoclonal antibodies is shown in Figure 1. The first step is to inject a mouse with the antigen of interest. This will stimulate the immune system of the mouse to begin producing antibodies specific to the antigen. The second step is to remove the spleen from the injected mouse and to place the spleen cells, many of which are B lymphocytes, in cell culture medium. The third step involves fusion of the B cells with rapidly growing myeloma cells to produce hybridoma cells. This is the key step in the production of monoclonal antibodies. Each B cell will produce a single type of antibody due to allelic exclusion. However, one cannot simply clone B cells because they will live for only a few days whether *in vitro* or *in vivo*. Cancer cells, such as the myelomas, are immortal in culture; they divide rapidly and continuously, exhibiting totally uncontrolled growth and division. But myeloma cells do not produce antibodies. Fortunately, the hybrid cells acquire the immortality of the parental myeloma cells and the capacity to produce antibodies from the parental B lymphocytes. Individual hybrid cells are separated and cultured to produce clones of antibody-producing hybridomas. Each clone is assayed for the production of the antibody of interest. Once identified, the hybridomas that are producing the desired antibodies can be grown continuously in culture or stored frozen in liquid nitrogen for later use. Thus the monoclonal antibody procedure allows one to produce an essentially unlimited quantity of an important antibody and to store the antibody-producing cells for use many years later.

Because of their phenomenal specificity, monoclonal antibodies are invaluable research tools. They can be used to identify individual polypeptides that have been separated by gel electrophoresis (Chapter 20) or to determine the location of macromolecules in tissues and cells by combining microscopy with immunolocalization techniques. Monoclonal antibodies also can be used in the purification of proteins by immunoprecipitation. In recent years, monoclonal antibodies have begun to be used in the diagnosis of inherited and acquired diseases. Although monoclonal antibodies have not provided the hoped-for cure of cancer, they have proven to be invaluable tools in biological research and medical diagnosis.

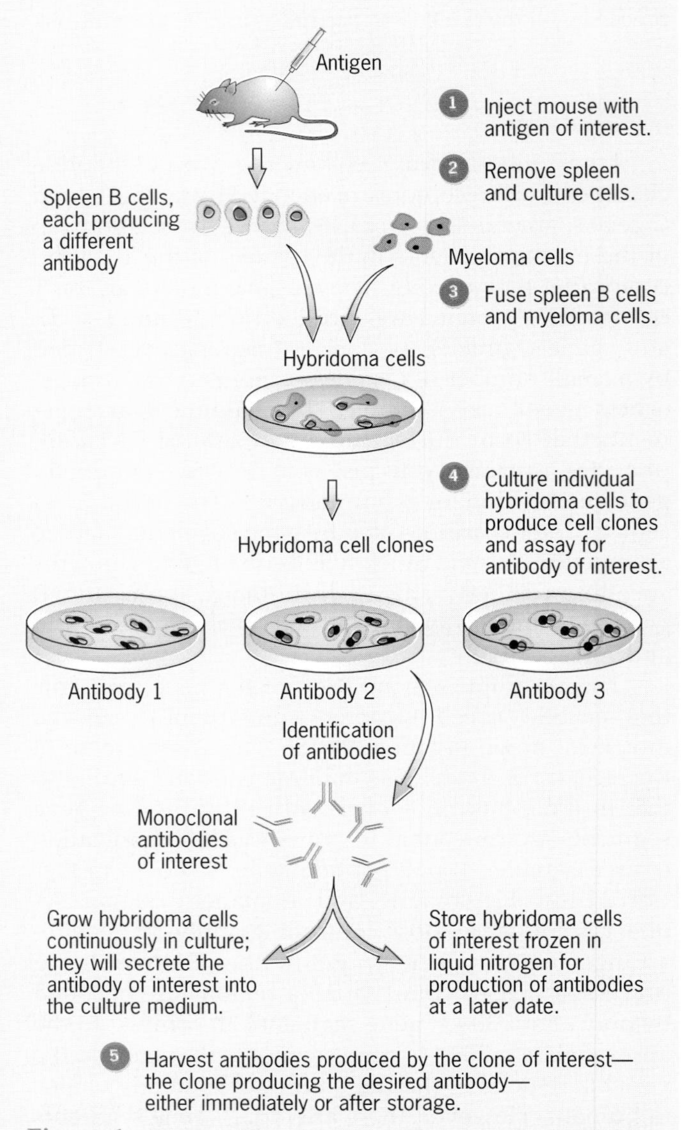

Figure 1 The production of monoclonal antibodies by hybridoma cell clones.

rearrangement of light chain coding sequences and one productive genome rearrangement of heavy chain coding sequences occur during the differentiation of each B lymphocyte! This phenomenon is called **allelic exclusion** because one of the "alleles" is excluded from being expressed. How? Why? At present, geneticists don't know. They do know that both "alleles" are sometimes rearranged in the same cell, but, in these cases, only one "allele" has undergone a productive rearrangement to a functional gene. The other "allele" is always only partially assembled or contains deletions of essential coding sequences.

Clearly, there must be some type of feedback mechanism that arrests the antibody gene-assembly process after one functional gene has been constructed and the cell has started to synthesize antibodies. The simplest mechanism would involve inhibition of the assembly process by the antibody itself. However, further work will be required to establish the mechanism by which allelic exclusion is enforced. It is known that the allelic exclusion mechanism works independently for the heavy chain gene segments and the light chain gene segments. Only one heavy chain and one light chain are synthesized in a given plasma cell. Regardless of the mechanism, allelic exclusion, which assures the production of a single type of antibody by each plasma cell, has one invaluable practical consequence: it permits researchers to obtain large amounts of an antibody that binds a single epitope, a so-called **monoclonal antibody** (Technical Sidelight: Monoclonal Antibodies: Powerful Probes for Detecting Proteins and Other Antigens), without using laborious protein purification methodologies.

Heavy Chain Gene Transcription: A Tissue-Specific Enhancer

By what mechanism are antibody genes activated during their assembly from gene segments? In the case of heavy chain genes, the answer is that the assembly process brings the promoters located upstream of L_HV_H gene segments into the range of influence of a strong enhancer element located in the intron between the J_H gene segments and the $C_{H\mu}$ gene segment (Figure 24.19). Each L_HV_H gene segment contains an upstream promoter. Prior to the genomic rearrangement events that lead to heavy chain synthesis, the closest L_HV_H promoter is over 100,000 nucleotide pairs away from the enhancer (Figure 24.19, *top*), and enhancers have little effect on transcription initiated at promoters located that far away. During B cell differentiation, the heavy chain gene-assembly process (Figure 24.14) brings the promoter of the closest L_HV_H gene segment and the enhancer to within 2000 nucleotide pairs of each other (Figure 24.19, *bottom*). The enhancer now activates transcription from the promoter located upstream from the L_HV_H gene segment. This enhancer is tissue specific; it activates transcription only in B cells, having no effect in other types of cells. Presumably, the activation process requires the presence of a transcription factor that is present in B lymphocytes but not in other types of cells.

A similar enhancer element is present in the intron between the light chain J_k gene segment cluster and the C_k coding sequence. Thus it seems likely that the movement of antibody gene promoters into the range of influence of tissue-specific enhancers may be a general mechanism of activation of antibody genes during B lymphocyte differentiation.

Key Points: Antibody gene segments are not transcribed at significant levels in stem cells. Their transcription is activated by the genome rearrangements that occur during the differentiation of B lymphocytes. Only one productive rearrangement occurs per cell. Transcription of assembled heavy chain genes is activated by transfer of their promoters to positions where they are controlled by a strong tissue-specific enhancer.

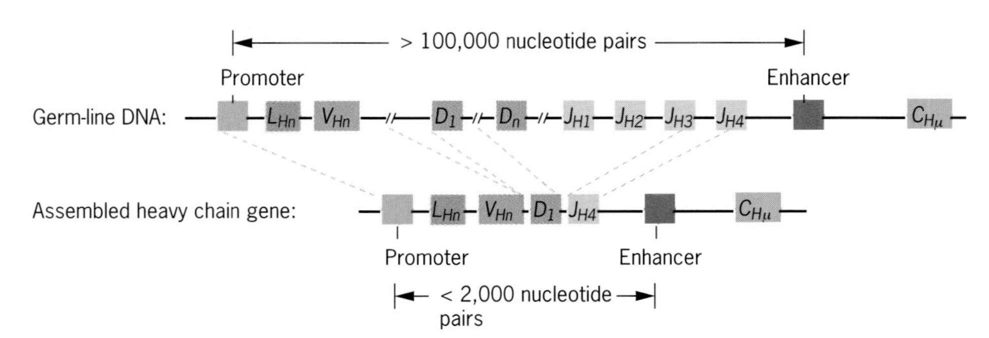

Figure 24.19 Activation of heavy chain gene transcription by movement of a heavy chain gene promoter (orange box) into proximity with a tissue-specific enhancer (red box) during gene assembly in B lymphocytes.

DISORDERS OF THE HUMAN IMMUNE SYSTEM

The biosphere in which we live is teeming with potentially pathogenic microorganisms. Nevertheless, we stay healthy most of the time because our immune systems protect us from serious infections by these pathogens. But what happens when an individual's immune system malfunctions, either because of a mutation or environmentally induced damage to some component of the system? Unfortunately, humans and other vertebrates seldom survive for very long without functional antibody-mediated and T cell-mediated immune systems.

Inherited Defects of the Immune System: Autoimmune Diseases, SCID and Agammaglobulinemia

Given the over 4000 documented inborn errors of metabolism in humans, it is not surprising that many inherited diseases involve defects in the immune system. These defects range from an almost complete loss of both the antibody-mediated and T cell-mediated immune responses to the loss or malfunction of one specific type of immune cell to the inability to synthesize a particular class of immunoglobulins. A brief discussion of a few of the more severe inherited immunodeficiencies—autoimmune diseases, severe combined immunodeficency syndrome, and agammaglobulinemia—will emphasize the importance of a functional immune system to good health and to our very survival.

In the **autoimmune diseases**, the affected individuals produce antibodies against "self" antigens. These antibodies may cause damage to specific tissues or organs or general damage throughout the body. In most autoimmune diseases, this damage results in severe pain to the affected individuals. **Juvenile diabetes**, **hemolytic anemia**, **myasthenia gravis**, **scleroderma**, and **systemic lupus erythamatosus** are examples of autoimmune diseases in which individuals produce antibodies that attack self-antigens on pancreatic beta cells, red blood cells, skeletal muscles, connective tissue cells, and neurons plus blood cells, respectively. In juvenile diabetes, these antibodies attack and destroy the insulin-producing beta cells in the pancreas, which results in the dependence of diabetics on insulin obtained from animals or human insulin produced in bacteria or cultured animal cells by recombinant DNA techniques (see Human Genetics Sidelight in Chapter 19). Without diagnosis and treatment with insulin, juvenile diabetes results in eventual blindness, kidney malfunction, heart disease, and premature death.

Diseases collectively referred to as **severe combined immunodeficiency syndrome (SCID)** are the most extreme of the inherited diseases of the immune system. Individuals with SCID are unable to mount an immune response of any kind, either antibody-mediated or T cell-mediated. These individuals have no functional B or T lymphocytes; in most cases, no B or T cells are even present. There are multiple forms of SCID, with about half of the cases caused by X-linked recessive mutations and the other half by autosomal recessive defects. Among the autosomally inherited cases of SCID, about two-thirds are due to autosomal recessive mutations in a gene or genes of unknown function. The other one-third are caused by mutant alleles of the gene encoding an enzyme called adenosine deaminase (ADA).

If untreated, autosomally inherited ADA-deficient SCID also usually results in death from severe infections early in childhood. Adenosine deaminase catalyzes the first step in the degradation of adenosine and deoxyadenosine to uric acid, which is excreted in urine. In the absence of ADA, toxic concentrations of deoxyATP accumulate via the phosphorylation of deoxyadenosine by kinases. Helper T cells are poisoned by the toxic concentrations of deoxyATP, and without helper T cells, B and T lymphocytes are not stimulated to differentiate into antibody-producing plasma cells or killer T cells, respectively. Thus ADA^{-}SCID patients are unable to mount either cellular or humoral immune responses.

In 1990, ADA-deficient SCID became the first human disease to be treated by somatic-cell gene therapy (Chapter 20). A four-year-old girl with ADA^{-}SCID was infused with her own white blood cells after they had been modified by the insertion of a functional *ADA* gene. Two other ADA-deficient children received similar gene-therapy treatments shortly thereafter. However, all these ADA^{-}SCID patients continued to receive treatment with adenosine deaminase to minimize the chance of life-threatening infections. To date, these pioneering gene-therapy treatments have not proven as effective as scientists had hoped (Chapter 20). The results indicate that further research on methods of introducing genes into somatic cells is needed before gene therapy can be effectively used to treat human diseases. Nevertheless, somatic-cell gene therapy is a promising tool; it will undoubtedly play a significant role in the treatment of human diseases in the future.

Individuals with **X-linked agammaglobulinemia** are unable to synthesize antibodies of any kind. They totally lack any antibody-mediated immune response. Because of a defect in B cell development, B lymphocytes and plasma cells are usually absent in affected individuals. In some cases, immature B lymphocytes are present, but they never develop into functional antibody-producing plasma cells. Because the genetic defect is X-linked, the vast majority of individuals with this form of agammaglobulinemia are hemizygous males. As expected, these individuals are highly susceptible to bacterial and fungal infections. However, they have a normal complement of T cells and exhibit a normal T cell-mediated immune response. Thus they are usually able to combat viral infections effectively. Prior to the development of successful bone marrow transplant techniques, X-linked agammaglobulinemia patients usually died early in life because of severe and recurrent infections. Today, this immune disorder can be effectively treated by bone marrow transplants if a matched donor is available. The infusion of bone marrow from a healthy donor provides the patient with stem cells that produce functional B cells and restore antibody-mediated immunity. Thus X-linked agammaglobulinemia is an example of a disease for which an understanding of the biology of the immune system has led to the development of an effective method of treating the disease.

Acquired Immunodeficiencies: Nongenetic DiGeorge Syndrome and AIDS

Not all immunodeficiences are inherited; some result from accidents during fetal development or childhood, and others are the result of damage to components of the immune system by toxic chemicals or pathogenic microorganisms. Because T lymphocytes undergo important differentiation events in the thymus gland, any damage to this gland may lead to immunodeficiencies. Individuals with **DiGeorge syndrome** have no functional thymus or parathyroid glands. Nongenetic forms of DiGeorge syndrome are usually caused by injuries that occur during fetal development and block the normal development of the thymus gland. As a result, individuals with this disorder have no T cells and cannot mount an effective immune response. Without bone marrow transplants or other treatments, infants with DiGeorge syndrome usually succumb to recurrent infections soon after birth.

Of the various immunodeficiencies, **acquired immunodeficiency syndrome** or **AIDS** is unique in its widespread and disastrous effects on humankind. As the AIDS epidemic spreads throughout the world, there is still no effective treatment of this fatal disease, which is sometimes called the "twentieth-century plague" (Chapter 15). This disease is caused by the **human immunodeficiency virus** (**HIV**), which is really not one virus but a group of closely related and rapidly evolving viruses. The major effect of HIV is to destroy the immune system of the infected individual. HIV infects and kills helper T cells, preventing them from stimulating T and B lymphocytes to differentiate into killer T cells and antibody-producing plasma cells, respectively (Figure 24.20). HIV enters helper T

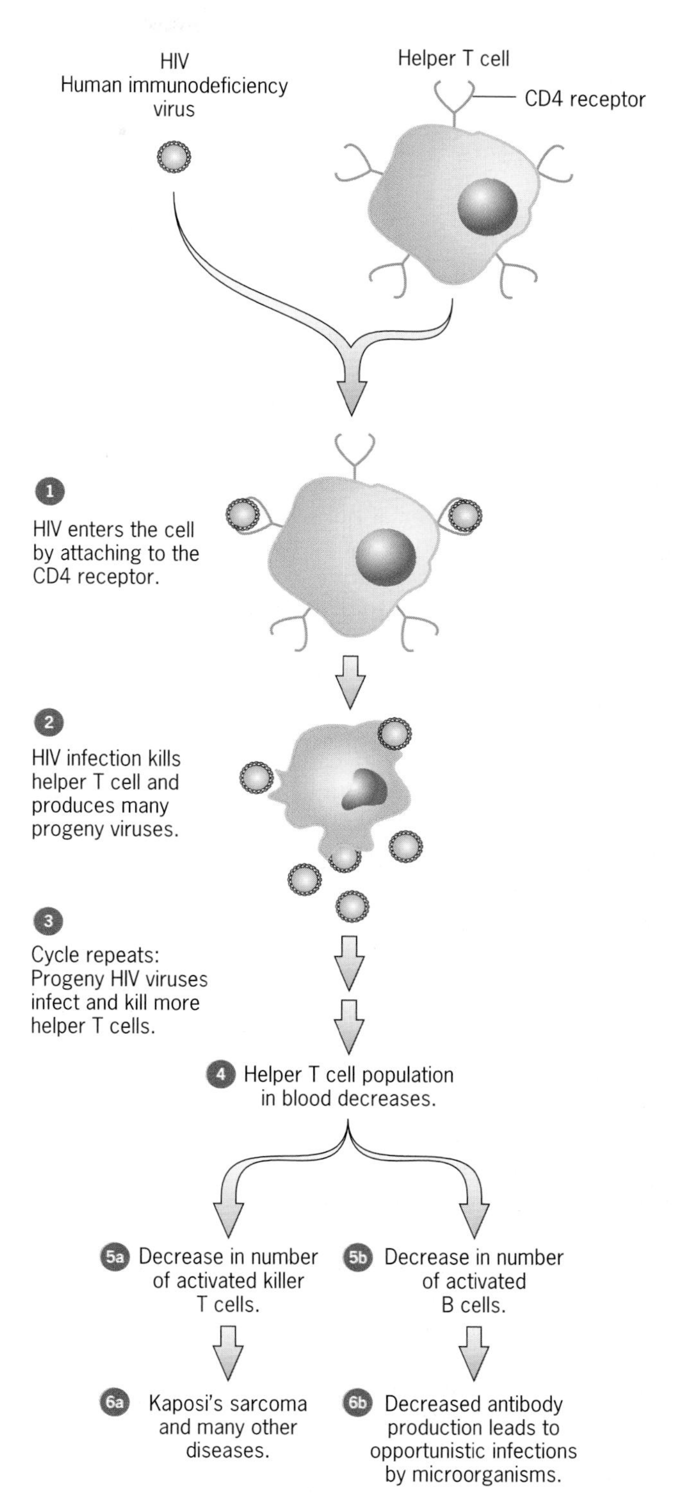

Figure 24.20 The human immunodeficiency virus (HIV) destroys the bodies immune system by infecting and killing helper T cells.

cells by adsorbing to the CD4 receptors on the surface of these cells. The CD4 receptor plays an important role in the interaction of helper T cells with other cells of the immune system; it helps stabilize the binding of helper T cells to other T and B lymphocytes and, in so doing, facilitates intercellular communication.

HIV is a retrovirus (Chapter 15); its genetic information is stored in RNA. Upon infecting a helper T cell, the viral reverse transcriptase copies the genetic information of the virus into double-stranded DNA form. The double-stranded DNA of HIV then inserts itself into the DNA of the host cell. The integrated viral DNA, now called a provirus, may remain dormant in the host cell's genome for many years, or it may be activated and begin producing viral RNA and proteins soon after the initial infection. We still do not understand what factors trigger the transcription of the HIV provirus.

The initial infection by HIV is suppressed rapidly by the immune system of the affected individual. Thereafter, the number of helper T cells in the circulatory system of the infected individual gradually decreases, whereas the relative concentration of HIV particles increases. Eventually, there is an almost total loss of helper T cells. At this point, the HIV-infected individual begins to exhibit the symptoms of AIDS, with recurrent infections by opportunistic viruses, bacteria, fungi, and protozoa. AIDS patients also frequently develop a form of cancer, Kaposi's sarcoma, which afflicts individuals with weakened immune systems. AIDS patients may suffer damage to the brain and central nervous system as well, but death is usually directly attributable to the immunodeficiency, the inability to combat microbial infections.

In 1983, when AIDS was shown to be caused by a virus, many scientists were optimistic that a vaccine could be developed to prevent the spread of the virus. Unfortunately, this optimism has proven ill-founded. The RNA genome of HIV mutates at an extremely high rate, in part, because the reverse transcriptase responsible for its replication contains no proofreading activity (Chapter 10). Thus mutations accumulate rapidly during the replication of the HIV genome. In fact, any HIV-infected individual contains a heterogeneous population of HIV particles, with many variants produced by new mutations that have occurred after the initial infection. The immune system of an individual initially combats HIV infections quite well, but never completely eliminates the virus, probably because of the presence of mutant forms that are not affected by the initial anti-HIV immunoglobulin.

To date, there is no effective treatment of AIDS; the mortality rate is very close to 100 percent. Some antiviral drugs help extend the lives of AIDS patients but do not prevent eventual death. Many scientists believe that the best chance of combating HIV is to learn how to stimulate the immune system to more effectively combat HIV; indeed, scientists have recently reported the identification of chemicals that block the growth of HIV in cultured white blood cells. These chemicals are produced by a specific type of T lymphocyte, but their mode of action is still unknown. At present, there is no information regarding the potential effect(s) of these chemicals on the growth of HIV in AIDS patients or in preventing the transmission of HIV.

At present, the best hope for slowing the worldwide AIDS epidemic is to educate people about how to prevent the spread of HIV. HIV is transmitted through blood and semen, not by casual contact with infected individuals. The spread of HIV can be largely prevented if uninfected individuals will avoid sharing needles or sexual activities with HIV-infected individuals. The use of condoms during sexual encounters decreases, but does not eliminate, the risk of transmitting HIV to an uninfected participant.

Key Points: **Survival of humans and other vertebrates depends on a functional immune system. Severe immunodefiencies, either inherited or acquired, usually are fatal.**

TESTING YOUR KNOWLEDGE

1. If the genome of a particular primate contains 300 V_k, 5 J_k, 1 C_k, 150 V_λ, 5 J_λ–J_λ, 300 V_H, 4 J_H, 50 D, and 10 C_H gene segments, and these can be joined in all possible appropriate combinations during the genome arrangements that occur in the differentiation of B lymphocytes, how many different antibodies could this primate produce?

ANSWER

Over a billion different antibodies (1.35 billion) could be assembled by combinatorial joining of the various gene segments and the different light and heavy antibody chains. There would be 2250 different light chains, consisting of 1500 kappa light chains (300 V_k gene segments $\times$ 5 J_k segments $\times$ 1 C_k segment) and 750 lambda light chains (150 V_λ gene segments $\times$ 5 J_λ–C_λ segments), and 600,000 heavy chains (300 V_H gene segments $\times$ 4 J_H segments $\times$ 50 D segments $\times$ 10 C_H gene segments). The number of combinations of light and heavy chains would be 2250 light chains $\times$ 600,000 heavy chains, or 1,350,000,000 different antibodies.

2. The human immunodeficiency virus (HIV) causes many of the symptoms of AIDS (acquired immunodeficiency syndrome) by infecting and killing specific types of T lympho-

cytes. HIV gains entry to T cells by binding to a specific cell-surface antigen called CD4. The CD4 antigen is found primarily on T lymphocytes called helper T cells, and the depletion of helper T cells is a major factor in the deterioration of the immune system in individuals with AIDS. How might this knowledge of the process by which HIV infects cells be used to devise a method to prevent or at least retard the effects of HIV?

ANSWER

If HIV could be prevented from infecting T lymphocytes, many of the symptoms of AIDS could be eliminated. Since HIV must first bind to the CD4 surface antigen of T cells, one could prevent entry of the virus by blocking this binding step. One possible way to block this binding step would be to introduce an altered, soluble form of CD4 into the circulatory systems of HIV-positive individuals. This soluble CD4 would bind to the virus and block its interaction with authentic CD4 on the surface of T cells. Alternatively, one might implant transgenic white blood cells carrying a modified gene that produced a secreted, soluble form of CD4. However, a potential problem with these approaches is that that the altered CD4 might cause serious side effects by interfering with the normal function of the CD4 antigen. If the virus binds to a region in CD4 that is not essential to its normal function, one might be able to use an oligopeptide that contains the HIV-binding domain, but not other regions of the CD4 protein required for its normal function(s). Alternatively, one might be able to produce a monoclonal antibody that binds to the HIV-binding domain of CD4 without interfering with normal CD4 function(s). Unfortunately, there are many unknowns in these potential stategies for treating and preventing AIDS that must be investigated before their feasibility can be evaluated. Nevertheless, one must continue to hope that successful methods for combating HIV will be forthcoming in the future.

QUESTIONS AND PROBLEMS

24.1 Is the genetic information specifying antibody chains stored in germ-line cells in the same format as that specifying most other polypeptides?

24.2 How many polypeptide chains are present in each antibody molecule? How many antigen-binding sites are present per antibody? How many different antibodies are produced in each plasma cell?

24.3 What are three different sources of antibody variability?

24.4 Antibodies are of five different classes. What determines the class of an antibody?

24.5 Does class switching during B lymphocyte differentiation occur at the DNA level or the RNA level? By what mechanism does it occur?

24.6 (a) Name three different types of white blood cells that play important roles in the immune response of vertebrates. (b) What function(s) does each of these types of cells perform in the immune response?

24.7 Of what importance is the *MHC* locus of humans in medical practice?

24.8 The genes of the *MHC* locus of humans are said to be highly polymorphic. (a) What does this mean? (b) Of what significance is this in the field of medicine?

24.9 Plasma cells make up a population of cells that collectively produce a vast array of different antibodies. Yet, each plasma cell produces only a single type of antibody (sometimes slightly different antibodies, but all specific for a single antigen). After exposure of a human or other vertebrate to a foreign antigen such as the coat protein of a virus, the individual rapidly initiates the production of large quantities of antibodies specific to the foreign antigen to which she or he was exposed. By what mechanism does this response occur?

24.10 Antibodies are encoded by families of gene segments that are assembled into the genes that direct their synthesis during the development of the plasma cells that produce them. Are any other vertebrate proteins encoded by gene segments that are assembled in a similar manner during cell differentiation?

24.11 In what ways are the structures of antibodies and T cell antigen receptors similar?

24.12 Why are secondary immune responses faster and stronger than primary immune responses?

24.13 Microbial infections of mammals trigger several different responses, some specific to the invading agent, others nonspecific. What are the major nonspecific immune responses? What are the major specific immune responses?

24.14 In what way do memory cells differ from most other cells of the immune system?

24.15 In what ways are SCID and AIDS similar? different?

24.16 Monoclonal antibodies have proven to be very powerful tools for studies of biological processes. (a) How are monoclonal antibodies produced? (b) How are they used in biological research?

24.17 The AIDS pandemic is now in its second decade with no cure or effective treatment in sight. (a) Why have HIV infections of humans proven so difficult to combat? (b) What is the most effective means of preventing the spread of HIV in the human population?

24.18 Why do AIDS patients often succumb to infections by microorganisms that are not normally dangerous human pathogens?

24.19 Prior to 1990, severe combined immunodeficiency disease (SCID) was fatal during early childhood. Today, the prognosis for children born with this immunodeficiency disease is much better. Why?

24.20 How many different antibody heavy chains are usually produced by a given plasma cell? What role does allelic exclusion play in antibody production by plasma cells?

24.21 Helper T cells are required to mount both antibody-mediated and T cell-mediated immune responses. Why?

24.22 What are the roles of phagocytes and complement proteins, respectively, in the immune responses of mammals? In what ways are the activities of phagocytes and complement proteins dependent on the production of antibodies?

24.23 What are autoimmune diseases, and why are they so-named?

24.24 If the genome of a particular mammal contains 200 V_k, 4 J_k, 1 C_k, 150 V_λ, 6 J_λ–C_λ, 200 V_H, 4 J_H, 100 D, and 8 C_H gene segments, and these can be joined in all possible appropriate combinations during the genome rearrangements that occur in the differentiation of B lymphocytes, how many different antibodies could this mammal produce?

24.25 In humans, the absence of an enzyme called purine nucleoside phosphorylase (PNP) results in a severe deficiency of T cells. Deoxyguanosine, the PNP substrate, accumulates in the absence of the enzyme and is especially toxic to T lymphocytes. What effect would you expect this PNP deficiency to have on the cellular and humoral immune systems of individuals with the disorder?

BIBLIOGRAPHY

ALBERTS, B., D. BRAY, J. LEWIS, M. RAFF, K. ROBERTS, AND J. D. WATSON. 1994. *Molecular biology of the cell.* 3rd ed. (Chapter 23, "The immune system"). Garland Publishing, New York.

CHEN, J., AND F. ALT. 1993. Gene rearrangements and B-cell development. *Curr. Opin. Immunol.* 5:194–200.

CRABTREE, G. R., AND N. A. CLIPSTONE. 1994. Signal transmission between the plasma membrane and the nucleus of T lymphocytes. *Annu. Rev. Biochem.* 63:1045–1083.

GOLDE, D. 1991. The stem cell. *Sci. Amer.* 265(12):86–93.

HUNKAPILLER, T., AND L. HOOD. 1986. The growing immunoglobulin gene superfamily. *Nature* 323:15–16.

LEDER, P. 1982. The genetics of antibody diversity. *Sci. Amer.* 246(5):102–115.

LIEBER, M. R. 1992. The mechanism of V(D)J recombination: A balance of diversity, specificity, and stability. *Cell* 70:873–876.

TONEGAWA, S. 1985. The molecules of the immune system. *Sci. Amer.* 253(10):122–131.

YOUNG, J. D-E., AND Z. A. COHN. 1988. How killer cells kill. *Sci. Amer.* 258(1):38–44.

A CONVERSATION WITH

THOMAS J. BOUCHARD, JR.

When did you first become interested in the genetics of human behavior, and what sparked this interest?

One of my first teaching assignments—in 1969—was a course called "Human Intelligence." That was the year that Jensen's very controversial paper in the "Harvard Educational Review" was published. Having to teach a course on this topic, I had to read Jensen's paper in order to be able to answer students' questions on the subject. To become informed on the topic, I looked up and studied almost all the 150 or so references listed in the paper. These papers sparked my interest in the subject. A few years later, after I had moved to Minnesota, my teaching responsibilities included a course on "Human Individual Differences," and it included material on behavioral genetics. So, I thought that I really ought to do research on the genetics of human behavior. My colleague David Lykken was already studying twins; but there was little ordinary family data to use as a base for comparisons with the twin data. Thus my students and I launched a couple of large-scale family studies. A few years later, a student in my course on "Industrial Organizational Psychology" asked me if I had read an article in the Minneapolis newspaper about some identical twins reared apart. This article really started me thinking about the uniqueness of such data and led to the initiation of our research on twins. I said to myself, "Every psychologist should do a clinical case study—like a clinician working up a disease. I've never done that, and these identical twins reared apart would be a wonderful clinical case study."

Thomas J. Bouchard, Jr. is currently professor of psychology at the University of Minnesota. He earned both his B.A. and Ph.D. from the University of California, Berkeley. Dr. Bouchard's influence extends beyond genetic research to psychological consulting and writing for numerous journals in psychology, biology, and genetics. He has a special interest in the role of genetics in human behavior. Dr. Bouchard is renowned for his studies of twins. These studies have helped us gain insights into the ways in which genes and environmental factors influence human behavior. He has received many distinguished awards throughout his 30 year career as a scientist, psychologist, and professor.

From what you've said, your students and colleagues appear to have played an important role in the development of your work in human behavior genetics.

Yes. At the University of Minnesota, I have many colleagues with the expertise to carry out all the necessary measurements and analyses. I was able to persuade them to collaborate so that we could do a really good clinical case study. Initially, the difficult task was contacting the twins and funding the project. Once, in a discussion with an AP reporter, I was asked how I was going to obtain funding for the research. I remember telling her: "I will beg, borrow, or steal, but I will get it." A few days later, there was an article in the *New York Times* that stated: "Psychologist will beg, borrow, or steal to study twins." Anyway, we did obtain funding, and we did find the required twins. But it took a long time to build up a large pool of twins. The study grew slowly over the years. We've been at it for 16 years and now have 130 pairs of twins.

Do you consider yourself primarily a psychologist or a geneticist?

I'm primarily a psychologist, and I collaborate with lots of other people. For example, some of my own students are far more versed in genetics than I am. Virtually all of our research on twins is collaborative, because you need the specialized knowledge and talents of many people. That is one of the wonderful things about working at a large university.

The results of your studies of twins reared apart led to appearances on the "Phil Donahue Show" and similar shows. Did that kind of exposure have an impact on your scientific career?

Yes, no question about that! It never got us any grants, but it was a major factor in finding twins during the initial years of our study. After about five years, I discovered that there are large national organizations of adoptees in the United States, Great Britain, New Zealand, and Australia. These groups assist adoptees that are searching for their biological families, and they have been of great help in identifying twins reared apart.

Was there a mentor or role model who inspired you to study psychology?

No. There really wasn't. The individual who came the closest was my father. He had only an elementary school education, but he kept taking college courses throughout his adult years. I do remember him saying: "This psychology stuff is really interesting." Originally, I had no intention of going into psychology. When I started college, my plan was to major in chemical engineering. But I enrolled in some psychology courses, and, just like my father, I found them fascinating. Also, I've always had a strong interest in biology.

During your many years of research on the behavior of identical twins reared apart, what result surprised you the most and why?

Well, that is an easy question to answer, because it was pretty clear to me—and wrongly clear—that some things are obviously influenced by genetic factors and others are not. People know that certain traits run in families; the observations are clear. But I'd also taken many courses in learning and conditioning, and I knew that human beings are capable of learning, and that behavior can be shaped by learning. Therefore, I was of the opinion that we would find certain classes of behavioral traits that would be heavily influenced by genetic factors and other traits that would be shaped largely by the family and the environment. Well, even after the first 10 or so pairs of twins—very small samples—when we analyzed the data, there seemed to be a genetic influence on literally everything that we looked at. This was surprising; this was not at all what we had expected. As the samples got larger, this continued to be the case, and it continues to be the case today. Over the 16 years that we have been doing these twin studies, there have been many other behavior geneticists doing similar studies, and, by and large, this tends to be the result across the board. So, that really has been the biggest surprise.

How will the Human Genome Project help us understand human behavior?

All the evidence that behavior geneticists like myself bring to bear on this question is very indirect. Fundamental science advances by specifying mechanisms. Fundamental knowledge has to involve specifying the step-by-step mechanisms. In that sense, we've not explained a great deal yet. We are setting the groundwork that justifies looking at other things, so that eventually all the steps can be filled in. If we couldn't show that there were genetic effects on behavior, then others wouldn't look for those steps. The Human Genome Project is going to make it possible to demonstrate the existence of allelic differences between people that score high and low on these behavior scales, and that there are physical correlates of psychological traits. Much work remains to be done before we will understand the molecular mechanisms that affect behavior.

What are your wildest and noblest expectations for the Human Genome Project?

One of my philosophical beliefs is that there is enormous diversity in terms of personalities, abilities, attitudes, and the like, in humans. In my opinion, diversity is one of the salient characteristics of the human species. Indeed, diversity is what makes humans so interesting, and I believe that the Human Genome Project will demonstrate that this diversity is influenced to a large degree by our genes.

What advice would you give to undergraduate students interested in the study of human genetics or human behavior?

Take lots of mathematics and courses in human evolution. I think evolution is the most fascinating of all intellectual subjects. I don't think anything makes sense without evolution, and I strongly recommend that students take as many evolution courses as possible. Also, read widely. The greatest fun that I can recall was roaming the library at Berkeley, pulling books off the shelves, and reading them.

The Club-footed Boy, *1642, by Jusepe de Ribera.*

25

Complex Inheritance Patterns

CHAPTER OUTLINE

The Club-footed Boy

On a recent trip to Paris, a geneticist spent a few wonderful hours at the Louvre. One painting there resonated with his interests in genetics. The painting, *The Club-footed Boy,* was done by the Spanish artist Jusepe de Ribera in 1642 and depicted a young boy with club feet and what appeared to be a mild paralysis of one side of his body. He carried a sign that translated from the Latin read "For the love of God give me alms." Club foot occurs in about 1 percent of live births. About half of the cases are related to the position of the fetus in the uterus and are nongenetic in nature. However, the other half have a genetic component.

Club foot is one of many traits that are clearly familial but do not follow any simple Mendelian inheritance pattern. Some traits, such as variation in skin color or height, are common. Others, such as club foot or cleft lip and palate, are less common or even rare. For many inherited traits, such as spina bifida and other neural tube disorders, diabetes, and high blood pressure, it is difficult to dissect out the segregating genetic elements. For these complex traits, geneticists recognize that:

1. The genotype contributes to the phenotype, although it is not usually known how many genes are involved, their function, or their location.
2. Nongenetic factors also play an important role in determining the traits.

In this chapter, we shall compare continuous and discontinuous patterns of variation in a Mendelian context, and examine various strategies for analyzing quantitative traits with complex inheritance patterns. We also discuss examples of complex human disorders controlled by genetic and nongenetic factors.

CONTINUOUS AND DISCONTINUOUS VARIATION

Darwinism and Mendelism: Two Views of Variation

One of the greatest challenges to biologists in 1900, the year that Mendel's work was discovered, was the reconciliation of the different views of variation held by Darwinians and Mendelians. Extensive observations of the natural world led Charles Darwin to conclude that evolutionary change was gradual. Natural selection acted on small, incremental changes in inherited variation to produce gradual changes in populations. Sudden or dramatic inherited changes in phenotype, such as dwarf sheep or extra digits (Darwin referred to such mutational events as "sports"), were not evolutionarily significant because they would be quickly eliminated by natural selection. Organisms can tolerate only small changes. Thus Darwin and others sought principles to explain gradual, continuous patterns of variation (Figure 25.1).

Darwin's cousin, Francis Galton, had interesting ideas about the inheritance of traits showing continuous variation. He agreed with Darwin that traits were determined by genetic elements called "gemmules," but Galton's gemmule concept was more complex than Darwin's. Galton suggested that the more gemmules involved in determining a trait, the more variation there would be for that trait. For example, if three gemmules controlled a trait, and if a gemmule could, with equal probability, be either black or white, four

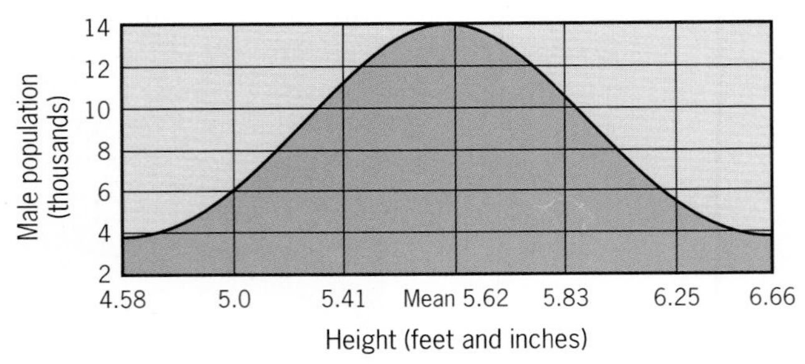

Figure 25.1 Height distribution in a population of 91,163 English males. The average height was found to be 5.62 feet. This is an example of a trait that shows continuous variation.

different combinations and shades would be possible: 3 black (black), 2 black 1 white (dark gray), 1 black 2 white (light gray), or 3 white (white). If six gemmules controlled a trait, seven phenotypic classes would occur: black, white, and five shades of gray in a ratio of 1:6:15:20:15:6:1 (Figure 25.2).This type of phenotypic distribution begins to approximate a continuous curve. Galton, who rejected the Lamarckian view of acquired characteristics, further suggested that each gemmule could mutate or vary, providing an additional source of variation. It was precisely this type of scheme that Darwin was seeking in order to explain inherited continuous variation. Mendelian principles would not at first glance seem to provide a basis for continuous variation. Mendel's laws explained the inheritance patterns of contrasting forms of a trait. The traits Mendel studied showed discontinuous variation: plants were either tall or short; flowers were either purple or white, axial or terminal; and seeds were wrinkled or smooth.

When Mendel's principles of inheritance were rediscovered in 1900, there was intense, sometimes rancorous debate about the applicability of these laws to the evolutionary process. Could Mendel's laws governing the inheritance of discrete phenotypic classes be applied to inherited quantitative traits showing a continuous pattern of variation? The answer to this

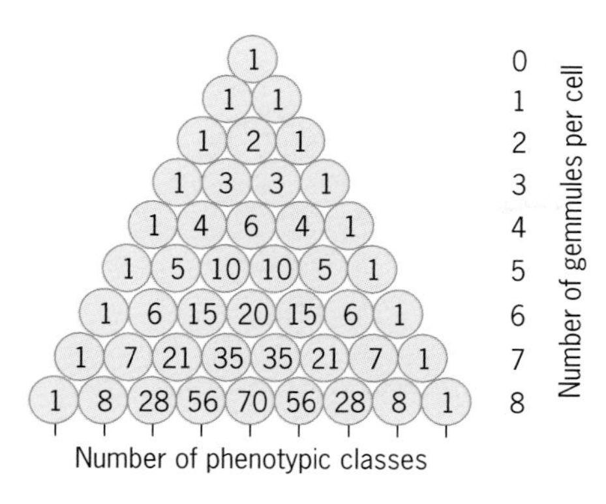

Figure 25.2 A model of continuous variation proposed by Francis Galton. According to Galton, variation was controlled by gemmules, and the more gemmules controlling a trait, the more variation in that trait. If six gemmules controlled color (black, white, or in between), there would be seven possible phenotypic classes: one black, one white, and five shades of gray. The ratio of these phenotypic classes would be 1:6:15:20:15:6:1. This phenotypic distribution approximates a continuous distribution.

question turned out to be a resounding yes, and the person who was most responsible for resolving this dilemma was Sir Ronald A. Fisher, a British statistician and geneticist. Fisher constructed a model for quantitative traits that takes into account the relative contributions of genes and environment, the number of genes affecting a trait, the nature of individual gene effects, and Mendelian segregation. Developed early in this century, Fisher's model provides a foundation for modern quantitative genetics by fusing the newly emerging Mendelian theory with the more biometrical approach to the study of heredity. Fisher's approach united the Mendelian and the more statistically based biometrical points of view. In the sections that follow, we examine components of Fisher's ideas.

Multiple Alleles and Continuous Variation

Continuous variation in a trait may occur when there are two or more forms of an enzyme, and each form has a slightly different pattern of enzymatic activity. For example, there are three major variants of acid phosphatase in human red blood cells. This monomeric enzyme removes or transfers phosphate groups in a wide variety of tissues. Each variant has slightly different properties, and each is encoded by a different allele of the *acid phosphatase* gene (*ACP*). The three main alleles are ACP^A, ACP^B, and ACP^C. Three alleles provide for six possible genotypes: ACP^A/ACP^A, ACP^B/ACP^B, ACP^C/ACP^C, ACP^A/ACP^B, ACP^A/ACP^C, and ACP^B/ACP^C. The A, B, and C variants of acid phosphatase are electrophoretically distinct from each other, which indicates that even though they are all functionally similar to each other, they differ from one another by one or more amino acids. Variant enzymatic polypeptides produced by allelic forms of the same gene are called **allozymes**. Each of the six genotypes produces a different pattern of electrophoretic variants (Figure 25.3*a*). Each of the three homozygous classes produces two distinct bands because the same identical polypeptide assumes two different conformations and each conformation has a different mobility in the electrophoresis system.

Red cell acid phosphatase activity when measured in a population generates a continuous curve (Figure 25.3*b*). When genotypes are determined for all the individuals in a population, it is found that for each genotype there is a range of enzymatic activity, and the range for each genotype is a continuous curve. The activity patterns for different genotypic classes often overlap. The sum of the ACP activity for the individuals in a population (in this case, a sample from the

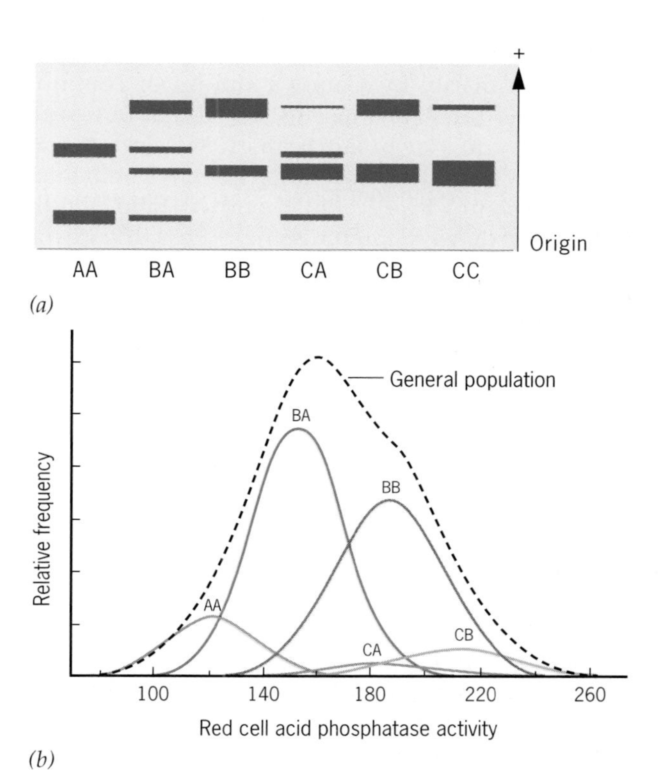

Figure 25.3 (*a*) Electrophoretic patterns of different red cell acid phosphatase phenotypes in a phosphate buffer system. There are two bands for each of the homozygous classes because the same protein assumes two different conformations in the buffer system. (*b*) The distribution of red cell phosphatase activities in the general English population (broken line) and in the separate genotypes.

population of Great Britain), *without regard for genotype*, generates a continuous curve.

What do the results of this acid phosphatase study mean? They mean that if the ACP activity of a population is measured, the phenotypic variation would be continuous. Yet underlying this continuous pattern of variation is a relatively simple Mendelian system of multiple alleles: a single gene with three major alleles produces a continuous pattern of phenotypic variation. By further analysis of this population using gel electrophoresis, the separate genetic elements can be identified.

Multiple Loci and Environmental Factors

The debate that began in 1900 about whether Mendel's principles could explain continuous patterns of variation did not last much longer than a decade. Important studies carried out over the next 10 to 15 years proved that an inherited trait showing continuous variation was determined by two factors: (1) two or more pairs of alleles segregating and assorting independently; and (2) the environment.

In 1909, Wilhelm Johannsen drew a clear distinction between genotype, phenotype, and environment. Motivated primarily by Galton's studies of continuously varying traits, Johannsen showed that in a pure line (homozygous) of broad beans (*Phaseolus*), variation in bean size is attributed largely to environmental influences (Figure 25.4). Under normal field conditions where extensive intercrossing occurs, the beans of *Phaseolus* hybrids varied in weight from 150 to 750 mg, ranging along a quantitative continuum. If either large or small beans were selected and inbred for several generations, pure lines were established. Yet these pure lines still varied, and further selection from within them failed to alter the phenotypic curve, whether the largest beans from the small line or the smallest beans from the large line were selected. Johannsen clearly showed that the environment influenced the expression of a genetically controlled trait because homozygous lines still varied. In its broader application, Johannsen's work led to the realization that a phenotype is controlled by two components—one genotypic, the other environmental. In fact, Johannsen coined the terms *phenotype*, *genotype*, and *gene* as a result of his studies.

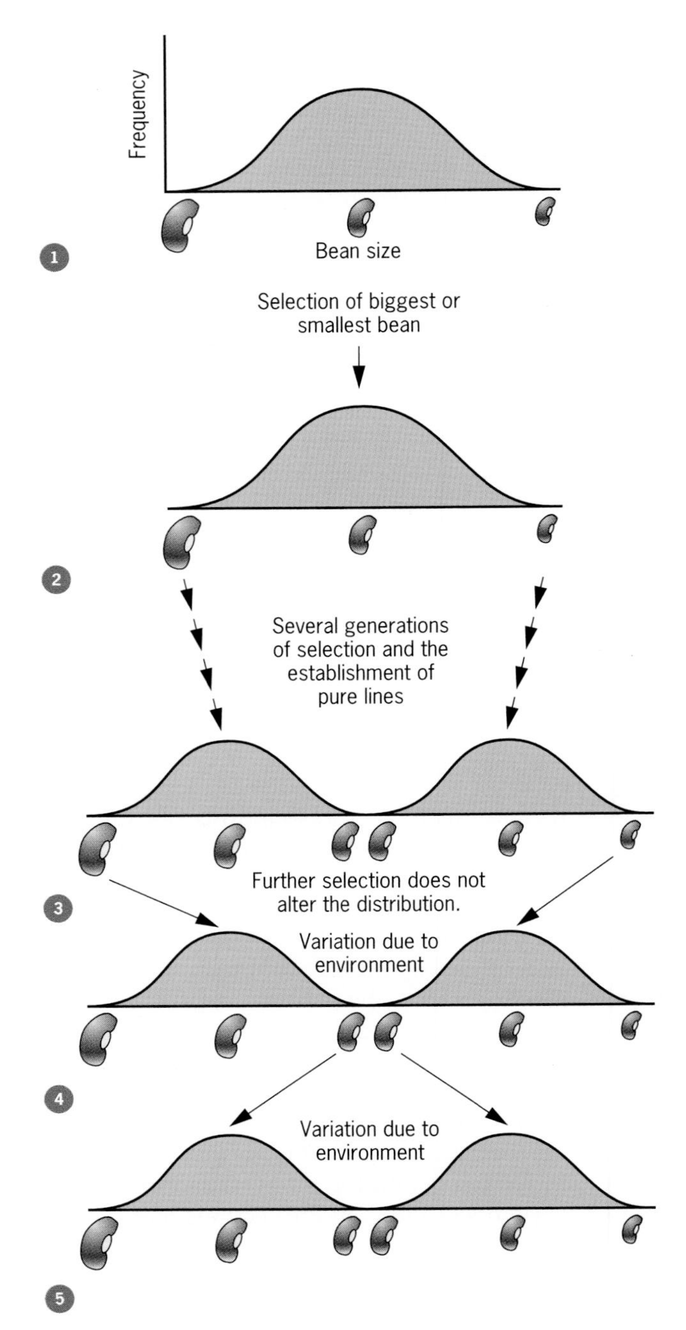

Figure 25.4 Johannsen's selection experiments with the broad bean, *Phaseolus*, showed that he could establish pure lines for bean size. However, within the pure lines, there was still variation caused by environmental variables.

Over the next few years, genetic studies convincingly showed that more than one gene can control the expression of a single trait. In 1909, Herman Nilsson-Ehle, working with wheat, showed that Mendelian principles explained continuous patterns of variation. Nilsson-Ehle proposed that three independently assorting pairs of alleles determined grain color in wheat. When he crossed a white-grained variety of wheat to a dark red-grained variety, the F_1 had an intermediate red phenotype (Figure 25.5). Self-fertilization of the F_1 produced an F_2 with a distinct phenotypic ratio of 1 dark red, 6 moderately dark red, 15 red, 20 intermediate red, 15 light red, 6 very light red, and 1 white. The sum of these numbers is 64, the basic denominator of a trihybrid ratio. The most reasonable

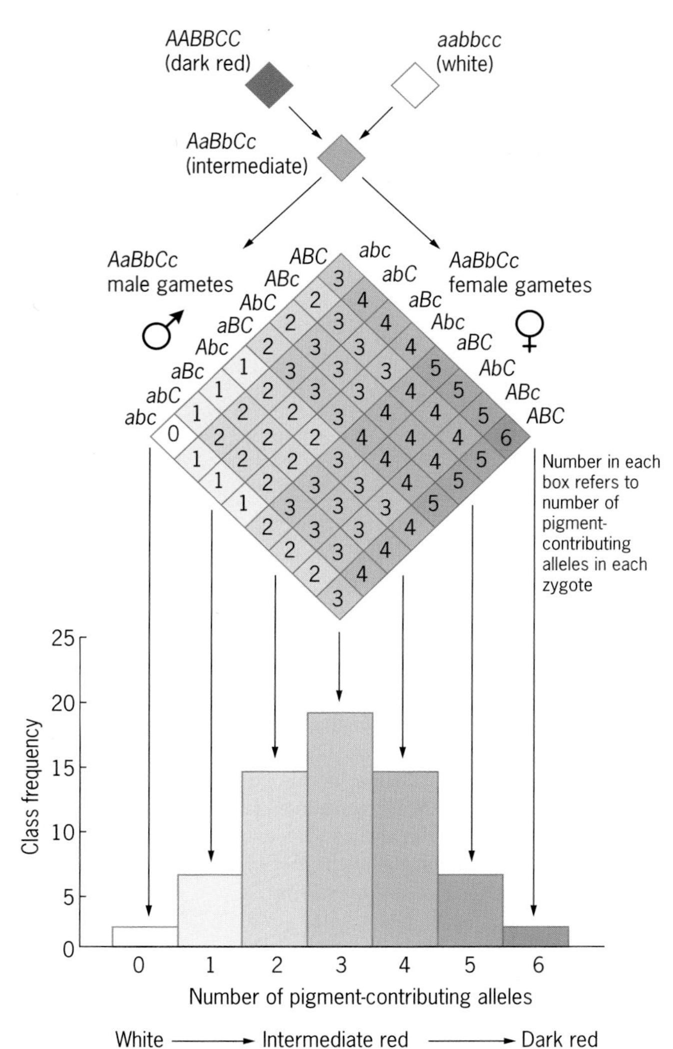

Figure 25.5 Nilsson-Ehle's experiment with kernel color in wheat indicated that three pairs of independently assorting alleles and incomplete dominance accounted for the kernel color variation.

way to interpret the results of this cross in a Mendelian context is to suggest that three independently assorting and segregating pairs of incompletely dominant alleles control this trait. The intensity of the red pigment is a function of how many pigment-contributing alleles are present. The darkest red has 6 such alleles, and white has none. All the intermediate reds have between 5 and 1 dominant alleles.

In 1916, Edward M. East extended Nilsson-Ehle's studies to a trait that did not show Mendelian ratios and that furthermore was subject to environmental influences. East crossed pure lines of tobacco that differed in the length of the flower's corolla—41 cm versus 93 cm (Figure 25.6). Within each pure line, East noted that there was a certain amount of phenotypic variation. Because each line is essentially homozygous, the intrastrain variation must be the result of environmental influences. The interstrain differences are the result of genetic differences. The F_1 from this cross was genetically uniform and intermediate in corolla length. Thus the F_1 phenotypic variation was due to environmental differences.

Intercrossing the F_1 produced an F_2 with a wide range of phenotypes. There are two reasons for the F_2 pattern of variation: (1) the segregation and independent assortment of different pairs of alleles; and (2) varying environmental influences on the different genotypes. East inbred some of the F_2 classes and observed in the F_3 less variation within the different lines due to the segregation of fewer pairs of alleles. Each F_3 peak corresponds closely to the value of the trait in the F_2 individual that was self-fertilized. This is a powerful argument for the genetic basis of the trait and further suggests that the F_3 differences between families are genetic. Within a family, the differences are due to genetic and environmental differences.

It is sometimes possible to estimate the number of independently assorting genes controlling a trait by determining the proportion of F_2 individuals with parental phenotypes. For example, if 1/64th of the F_2 progeny have one of the parental phenotypes, we could infer that three pairs of independently assorting alleles controlled this trait. (In a trihybrid cross, the two parental classes are each 1/64th of the total progeny.) This was the situation in Nilsson-Ehle's experiment with wheat grain color. East raised 444 F_2 plants without finding a single parental phenotype. If four independently assorting pairs of alleles controlled the trait, we might expect to find 1/256 of the total in each parental class; with five pairs of alleles, we expect 1/1024 of the total to be in each parental class. We can thus suggest that the number of pairs of alleles controlling corolla length is at least 5. Since East's study was first published, additional work has indicated that there are nine independently assorting pairs of alleles influencing corolla length in this tobacco species.

Caution must be exercised in trying to estimate the number of genes influencing a trait. First, all the different loci may not be contributing equally to the phenotype. Second, several genes influencing a trait may be tightly linked and behave as a single segregating unit. Third, the effect of the environment may be more profound on some genotypes compared to others. Nevertheless, this method is useful as an approximation.

DDT Resistance: A Trait Controlled by Several Genes

Insects can become resistant to DDT, and this resistance is a genetically determined trait. In one study, DDT resistance in an Illinois housefly population in-

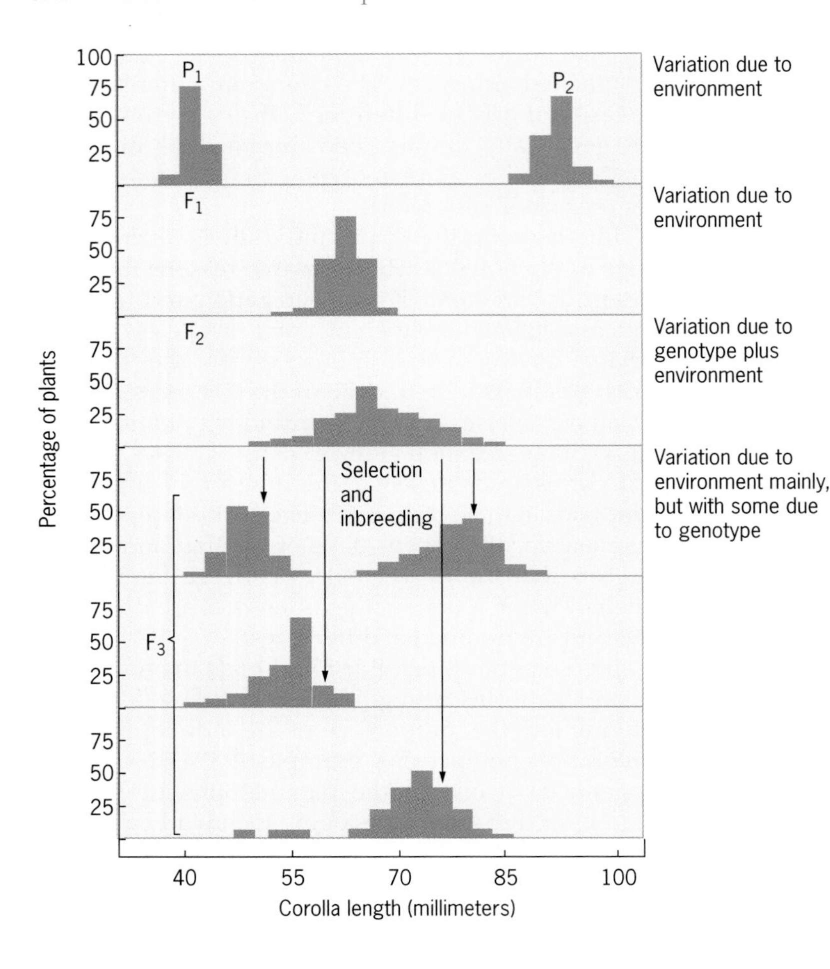

Figure 25.6 E. M. East studied the inheritance of corolla length in tobacco (*Nicotiana longiflora*). The figure shows the percentage frequencies with which individuals fall into various corolla-length classes. The P_1 and P_2 classes are pure lines; the F_1 is uniformly heterozygous. Any variation in the P and F_1 classes is due to the environment. The F_2 is genetically heterogeneous. Variation in this generation is due to genetic and environmental factors. The means of the four F_3 populations are correlated with the corolla-length of the F_2 plants from which they were obtained by selfing.

creased from less than 5 percent of the population to almost 50 percent over a period of six years following regular exposure to DDT.

No single physiological mechanism accounts for DDT resistance. Resistance may occur by several mechanisms, including

1. An increase in cellular lipid content enabling the lipid-soluble DDT to be more diluted and sequestered from other parts of the organism.
2. Variant enzymes that more effectively inactivate the poison.
3. Genetically determined changes in the neuron membranes making nerve cells less susceptible to the toxic effects of DDT.
4. Genetic changes in the insect's chitinous exoskeleton reducing the ability of DDT to gain access to the body's tissues.

Each one of these events is influenced by the genotype. DDT resistance is complex because there are so many ways for resistance to develop. Many genes are involved.

James Crow (see Epilogue) analyzed the contribution that each *Drosophila* chromosome makes to the development of DDT resistance. Crow selected strains of *Drosophila* that were either highly resistant or highly sensitive to DDT. These two strains had chromosomes that carried different dominant genetic markers, enabling him to follow individual chromosomes. By intercrossing the two strains and keeping track of the chromosomes from each, he showed that genes responsible for DDT resistance are located on each of the major *Drosophila* chromosomes (Figure 25.7). Thus each gene in a polygenic system (that is, multiple genes) may have small and qualitatively different incremental effects on the DDT resistance phenotype. Crow also found that very high insecticide resistance in houseflies and other species is usually monogenic.

The Threshold Model: Complex Inheritance with Simple Alternative Phenotypes

DDT resistance, corolla length, kernel color, and bean size are continuously varying traits. However, some discontinuous traits (expressed or not expressed) are controlled by multiple factors, both genetic and nongenetic. An example of such a trait is cleft lip with or without cleft palate, a developmental error in which there is a failure of midline fusion in early embryogen-

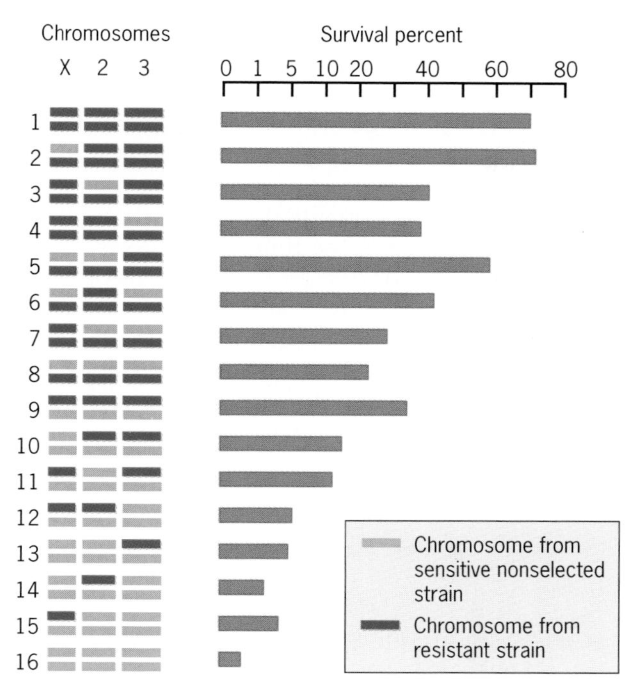

Figure 25.7 Percent survival of 16 different types of *Drosophila melanogaster* flies exposed to a uniform dose of DDT. Each fly carries a marked set of chromosomes derived from DDT-resistant and DDT-sensitive strains.

esis. Twin and family studies of this condition suggest that multiple genetic and environmental factors influence the expression of this trait.

Cleft lip with or without cleft palate is a prime example of a **threshold trait** because individuals either express the trait or they do not. Although the trait is discontinuous, the genetic variation underlying the trait is complex and falls along a continuum. Only those genotypes at the tail of the continuum are at risk to develop the phenotype (Figure 25.8). Some proportion of the individuals above a certain threshold of genetic liability will express a threshold trait, but only if they are exposed to appropriate environmental factors.

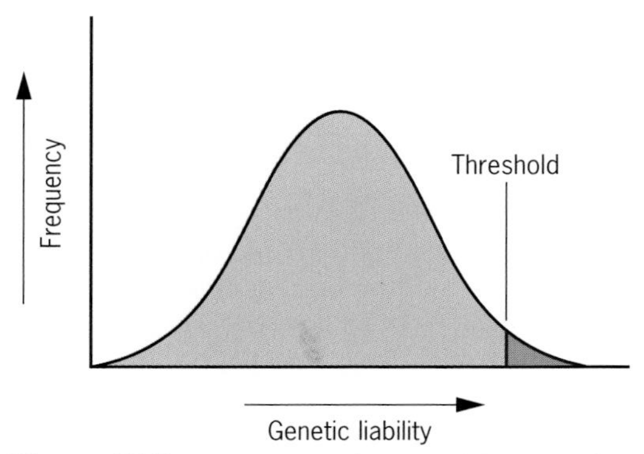

Figure 25.8 The threshold model for multifactorial inheritance. Those genotypes to the right of the dotted line have the potential to express the trait, whereas those to the left do not.

How do geneticists know that cleft lip with or without cleft palate is inherited? There are two main lines of evidence suggesting that genetic factors are important but that the environment plays a significant role as well. The first line of evidence comes from twin studies (Table 25.1*a*). When monozygotic twins (who are genetically identical) were examined, it was found that in 40 percent of the cases where one twin expresses the trait the other twin also expresses the trait (we say the concordance rate is 40 percent). For dizygotic or fraternal twins, who have 50 percent of their genes in common, the concordance rate was 4 percent. This suggests that genes predispose an individual to express the trait, but that environmental factors trigger its ultimate development. If only the genotype controlled the expression of this trait, then in monozygotic twin studies, the concordance rate would be 100 percent. The table indicates concordance rates for other complex human traits.

The second line of evidence supporting a genetic basis for cleft lip with or without cleft palate comes from family studies (Table 25.1*b*). First-degree rela-

TABLE 25.1*a*

Concordance Among Monozygotic and Dizygotic Twins for Common Malformations and Diseases

	Concordance	
Trait	*MZ*	*DZ*
	%	
Cleft lip ± cleft palate	40	4
Club foot	30	2
Pyloric stenosis	22	2
Congenital dislocation of hip	33	3
Schizophrenia	60	10
Insulin-dependent diabetes mellitus	50	10

TABLE 25.1*b*

Family Studies of Cleft Lip, With or Without Cleft Palate[a]

Relatives	*Percentage of Relatives Affected*	*Incidence Relative to General Population*
First degree		
Sibs	4.1	× 40
Children	3.5	× 35
Second degree		
Aunts and uncles	0.7	× 7
Nephews and nieces	0.8	× 8
Third degree		
First cousins	0.3	× 3

[a]Carter, C.O. 1969. Genetics of common disorders. Br. Med. Bull. 25:52–57.

tives of a person (siblings and children) with the trait are far more likely to express the trait than are second- and third-degree relatives. This makes sense, for first-degree relatives share about 50 percent of their genes, whereas second- and third-degree relatives share only about 25 and 12.5 percent of their genes, respectively. Risk decreases as genetic relationships become more distant.

Two points should be emphasized about threshold traits in general: (1) environmental factors play a significant role in the expression of most threshold traits; and (2) the nature of the genetic factors involved in these traits is poorly understood.

Key Points: **A phenotype can exhibit a continuous pattern of variation and be controlled by a single gene with several alleles. The sum of the various genotypes produces a continuous pattern of variation. Experiments with wheat demonstrated that a continuously varying trait was determined by three independently assorting pairs of alleles showing incomplete dominance. Experiments with tobacco showed that a continuously varying trait was determined by both genetic and environmental factors. A threshold trait is one with a continuously distributed liability or risk in which individuals with a liability greater than a critical value (threshold) exhibit the phenotype.**

ANALYSIS OF QUANTITATIVE TRAITS

In natural populations, variation in most traits forms a continuum rather than discrete phenotypic classes. The variation is quantitative, not qualitative. Because it is difficult to discern the Mendelian units that influence these quantitative traits, it is necessary to use statistical techniques to analyze and describe them. A major objective in quantitative genetics is to determine how the genotype and environment interact to produce a trait with a continuous distribution.

The first step in the analysis of a complex trait is to describe it quantitatively. This is normally done by taking random samples from a population, analyzing them, and then extrapolating to the larger population. Values calculated from samples of data are called **statistics** and are essential to the study of quantitative genetics. Let's examine some of the ways that quantitative traits are described.

Frequency Distribution

Data for a quantitative trait can be presented graphically as a **frequency distribution**. The horizontal axis, the x-axis, measures values of the trait obtained in a sample from the population. The x-axis is divided into regular intervals that represent various expressions of the trait, such as a range of heights. The vertical axis, the y-axis, measures the frequency of the observations at each interval.

An example of a quantitative trait that can be graphically displayed is the time wheat takes to mature. In one study, four different populations of wheat were analyzed for time to maturation: two parental strains, an F_1, and an F_2 population (Figure 25.9). All four populations were grown under conditions that minimized environmental differences. A sample from each population was monitored for maturation time.

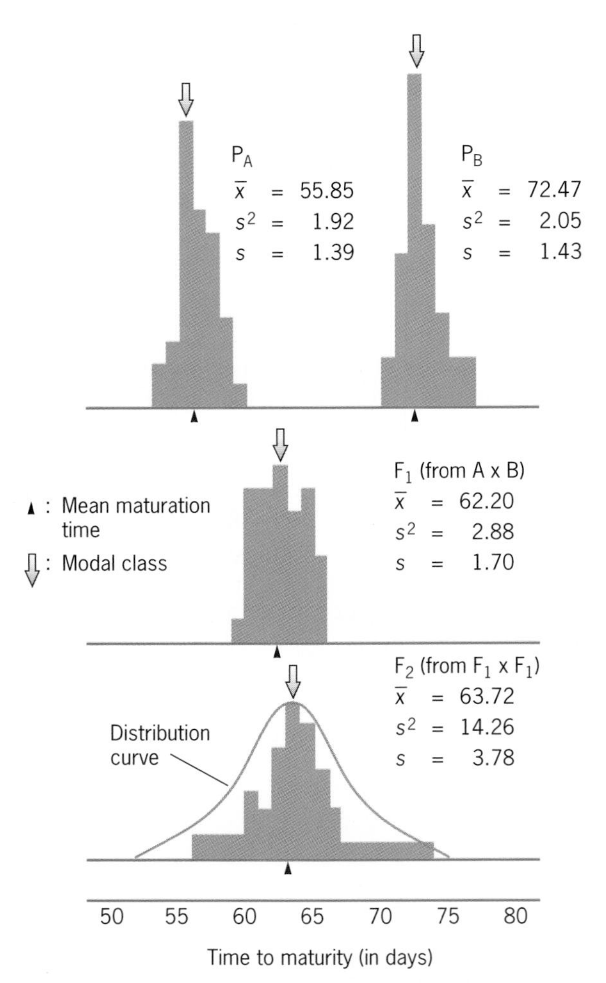

Figure 25.9 Frequency distributions and descriptive statistics of time to maturity in four populations of wheat. P_A and P_B are inbred strains that were crossed to produce F_1 hybrids. The F_1 plants were intercrossed to produce an F_2. Seed from all four populations was planted in the same season to determine the time to maturity. In each case, data were obtained from 40 plants. The mean, mode, variance, and standard deviation are given.

The time to maturity for each plant in the sample was recorded in days. Parental strains A and B were highly inbred varieties and thus essentially homozygous. Strain A matured quickly and B more slowly. There was no phenotypic overlap in the two strains, and environmental factors caused some variation within each strain.

The F_1, produced by intercrossing strains A and B, was intermediate between the two parental strains. The F_1 distribution was also somewhat broader and less peaked than the distributions of the inbred parental strains because of environmental factors but also because of small genetic differences among the F_1. Genetic differences arise when the parental strains are not homozygous at all loci.

Interbreeding the F_1 produced an F_2 in which the variability in maturation time among the plants sampled was considerably greater than it was in the inbred or F_1 samples. The range of F_2 maturation times overlapped the parental phenotypes and included all times in between.

This type of analysis accomplishes two main objectives: (1) It describes the trait (maturation time) quantitatively; and (2) it tells us that there are genetic and environmental factors affecting the expression of the trait. We are now in a position to describe other features of the trait.

The Mean and Modal Class

The essential characteristics of a frequency distribution can often be represented by simple descriptive statistics. One of these is called the **mean** or **average**. It gives us the "center of the distribution" curve. We calculate the **sample mean** (X) by summing all the data (ΣX_i) and dividing by the total number of observations in the sample (n). Thus the mean is:

$$X = \Sigma X_i / n$$

In this formula, the Greek letter Σ is a mathematical shorthand for the sum of all individual measurements ($X_1 + X_2 + X_3 + \ldots X_n$). X_i represents all the individual data points, and n represents the sample size. In Figure 25.9, for example, the means are indicated by the triangles; the numerical values are given on the right. The means of the F_1 and F_2 samples are 62.20 and 63.72 days, respectively; both are a little less than the average of the means of the two inbred parental strains (64.16 days).

The **modal class** is that class containing more individuals than any other in a frequency distribution. In Figure 25.9, the modal classes are indicated by small arrows.

The Variance and Standard Deviation

A sample mean tells us nothing about how data points are scattered about the mean—that is, how they are dispersed around this central point. The scatter could be broad or it could be narrow. The **variance** measures the scatter of the data points about the mean. Data that are widely dispersed produce a large variance, whereas data tightly clustered around the mean have a small variance. The sample variance, s^2, is calculated from the formula

$$s^2 = \Sigma(X_i - X)^2/(n-1)$$

In this formula, $(X_i - X)^2$ is the squared difference between the ith observation and the sample mean (often called the **squared deviation from the mean**), and the Greek letter Σ indicates that all such squared deviations are summed. The sum of the squared deviations is averaged by dividing by $n - 1$. (For technical reasons, the divisor is one less than the sample size.) Because the variance is a squared statistic, it is always positive. Although this feature has useful mathematical properties, it sometimes makes the variance difficult to interpret. One problem is that the units of measurement are squared (for example, $s^2 = 2.88$ days^2). Consequently, another statistic, called the **standard deviation**, is often used to describe the relationship between the mean and the variance. The standard deviation (s) is expressed in the same units as the original measurements. The standard deviation is simply the square root of the sample variance

$$s = \sqrt{s^2}$$

Variances and standard deviations for the four wheat populations are given in Figure 25.9. The F_2 population clearly has the greatest variance and standard deviation values, the result of both genetic and environmental differences. The smallest variance and standard deviation values are found in the two parental strains that have very little genetic variation owing to their homozygosity. Most of the variability in the parental strains results from environmental differences, although some can be ascribed to genetic differences if they are not pure homozygotes.

Many quantitative traits exhibit a symmetrical, bell-shaped distribution about the mean, similar to the one superimposed over the F_2 data in Figure 25.9. This type of distribution is called a **normal distribution**. Knowing the mean and standard deviation of a trait, we can specify a theoretical normal distribution: 66 percent of all the individuals have values within one standard deviation of the mean; 95 percent of the individuals fall within two standard deviations of the mean; and 99 percent of all individuals fall within three standard deviations of the mean (Figure 25.10).

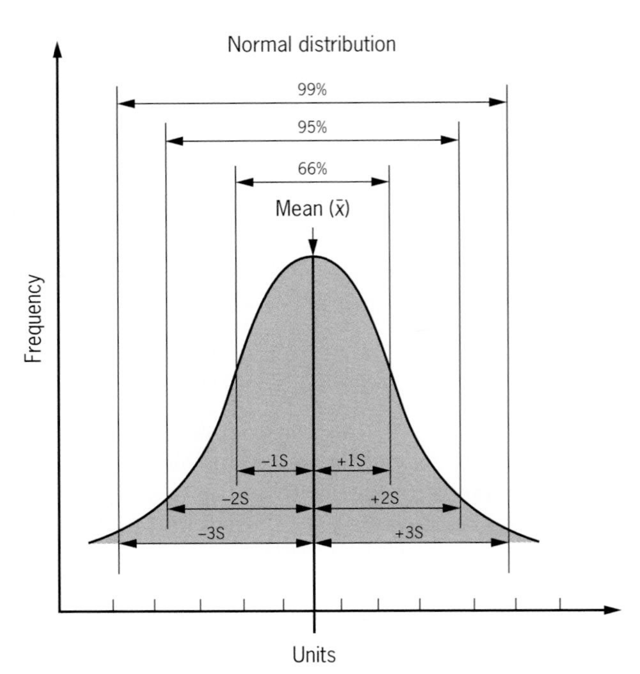

Figure 25.10 A normal distribution curve showing the proportions of the data in the distribution included within 1, 2, and 3 standard deviations from the mean.

Correlation Coefficients

Sometimes one set of measurements is related to another. For example, researchers can examine the relationship between two population variables, such as height and weight, or they can examine the relationship between a single variable measured in pairs of genetically related individuals (siblings, parent–offspring, monozygotic twins, fraternal twins, and so forth). In such circumstances, it is sometimes useful to calculate correlations between the matched data. For example, is there any relationship between the heights of fathers and the heights of sons? The **correlation coefficient** (r) addresses this question of relationships between variables and is calculated as follows:

$$r = \Sigma[(X_i - \overline{X})(Y_i - \overline{Y})] / ((n-1)s_X\, s_Y)$$

Here, X_i and Y_i are the data from the ith pair of observations, $\overline{X}$ and $\overline{Y}$ are the sample means of the two sets of measurements, and s_X and s_Y are their standard deviations. Again, n is the sample size. The correlation coefficient ranges from –1 to +1; –1 indicates a perfect negative correlation between X and Y (taller fathers have consistently shorter sons) and +1 indicates a positive correlation (taller fathers have consistently taller sons). A 0 indicates that there is no correlation (Figure 25.11). Positive and negative correlations may suggest that there is a causal link between two traits. For example, in humans, height and weight are positively correlated because they are both related to growth processes. However, in poultry, egg mass and egg number are negatively correlated because a hen can either lay a few large eggs or many small ones. A correlation coefficient of 0 indicates that there is no connection between the two variables. For example, IQ scores and height are in no way connected and show a correlation coefficient of 0.

Fingerprints are an example of a trait that is set early in embryonic life and that correlates positively with genetic relatedness: The closer the genetic relationship, the more similar the fingerprint pattern. The trait is quantitative, with virtually no environmental influence. One feature of fingerprints called total dermal ridge count correlates positively with the degree

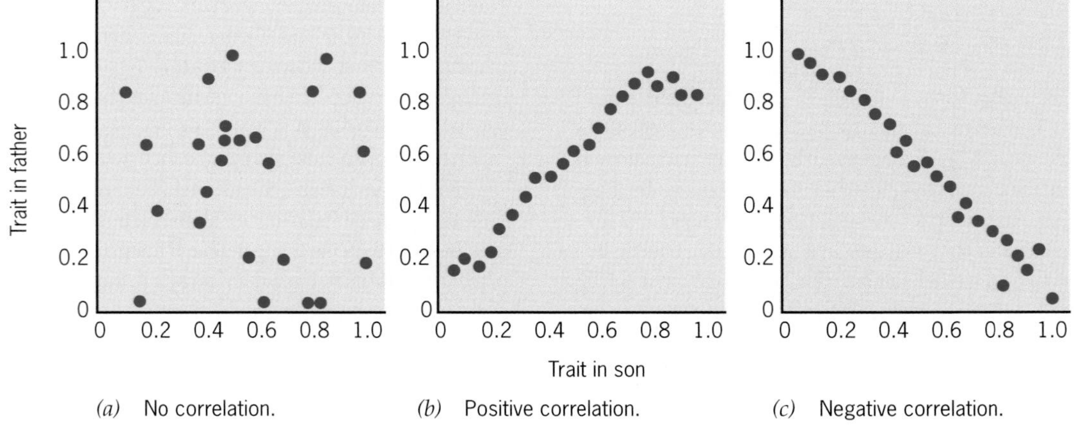

Figure 25.11 Statistical associations and the correlation coefficient. In (*a*) there is no correlation between the two variables; if one goes up, the other can go down or up. In (*b*) there is a positive correlation between the two variables; if one goes up (or down), the other goes up (or down). In (*c*) there is a negative correlation between the two variables; if one goes up the other goes down, and vice versa.

Correlation coefficients for various pairs of relatives for total dermal ridge count

Relationship	No. of Pairs	Observed Correlation Coefficient	Theoretical Correlation Coefficient
Parent–child	810	0.48	0.50
Mother–child	405	0.48 ± 0.04	0.50
Father–child	405	0.49 ± 0.04	0.50
Father–mother	200	0.05 ± 0.07	0.00
Midparent–child	405	0.66 ± 0.03	0.71
Sib–sib	642	0.50 ± 0.04	0.50
Monozygotic twins	80	0.95 ± 0.01	1.00
Dizygotic twins	92	0.49 ± 0.08	0.50

Figure 25.12 This table shows correlations between relatives for total dermal ridge count.

of genetic relationship (Figure 25.12). The correlation coefficient between unrelated individuals is 0.05, a value suggesting that the fingerprint pattern for one individual of a pair is virtually unrelated to the pattern of the other individual. Because unrelated individuals have few alleles in common, this inference makes sense. For monozygotic twins the correlation coefficient is 0.95; that is, the pattern for one individual in the pair is almost identical to the pattern of the other individual. This is expected because monozygotic twins are genetically identical.

In the above formula for the correlation coefficient, the term

$$\Sigma[(X_i - \overline{X})(Y_i - \overline{Y})]/(n-1)$$

is called the **covariance** between X and Y, and is symbolized $Cov(X,Y)$ or Cov_{XY}. Covariance measures the statistical association between two sets of data; it gives the correlation coefficient its sign. When two sets of data are independently distributed, the covariance between them is 0.

A cautionary note is necessary about the interpretation of correlation coefficients. That two variables are correlated does not mean that there is a causal connection between them. For example, there is a positive correlation between the number of school teachers and the amount of milk consumed in cities over 50,000. There is no causal link between school teachers and the amount of milk consumed in these cities. Rather, in larger cities there will be more milk consumed and there will be more school teachers.

Regression

The heights of fathers and the heights of their sons are positively correlated: taller fathers have taller sons. Correlation coefficients, however, have little predictive value. They cannot be used to predict the height of the son of a man who is 6 feet 3 inches tall. To obtain this type of information, regression analysis is used. **Regression** provides quantitative information about the dynamics of the relationship between two variables. In other words, it states how much a change in one variable is associated with a change in another variable. The relationship between two variables is expressed as a **regression line**. This mathematically calculated line represents the best fit of a line to the points. In other words, the squared vertical distance from the points to the regression line is minimized. The regression line is described by the equation

$$y = bx + a$$

where x and y represent the values of the two variables, b is the slope or regression coefficient, and a is the y intercept. For example, the locomotor activity of parents versus offspring in *Drosophila* is plotted in Figure 25.13. This figure shows the regression of offspring on the average phenotype of the two parents (sometimes called the midparent value). The equation of the line is

$$y = 0.51x + 15.56$$

Thus, with $b = 0.51$, there is a positive association between midparent and offspring. Offspring activity depends in some way on the parental values. This example illustrates a way to estimate how much the change in one variable is caused by a change in the other variable.

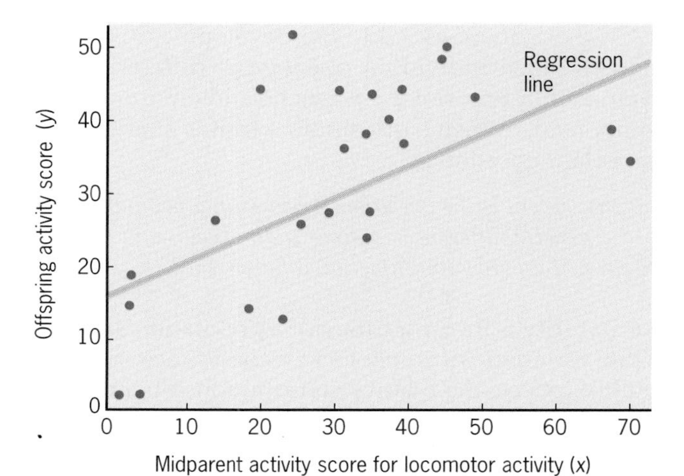

Figure 25.13 Locomotor activity scores in *Drosophila melanogaster*. Regression of offspring scores on midparent scores (average of the two parental scores).

The slope (b) is calculated from the covariance of x and y in the following manner:

$$\text{slope} = b = \text{Cov}_{xy}/s^2_x$$

The slope indicates how much an increase in the variable on the y-axis is associated with a unit increase in the variable on the x-axis. For example, if the slope is 0.4 for the regression of mother and daughter head circumference, it means that for each unit of increase in head circumference in the mother, the predicted increase would be 0.4 units in the daughter.

Key Points: **A frequency distribution arranges phenotypic measurements as a graph showing either the relative or absolute incidence of classes in a population. The mean is the sum of all measurements or values in a sample divided by the sample size. The mode is the single class in a statistical distribution having the greatest frequency. Variance and its square root, the standard deviation, are measures of variation in a population. The correlation coefficient is a measure of the extent to which variations in one variable are related to variations in another. Regression analysis assesses the quantitative relationship between two variables.**

HERITABILITY

R. A. Fisher, whose ideas reconciled the views of the biometricians with those of the Mendelian school, suggested that quantitative traits are **multifactorial**: they are controlled by several gene loci, multiple alleles at each locus, and environmental factors that modify genotype expression. This combination of multiple environmental and genetic factors determines the individual's phenotype.

Fisher's theories led to the development of a system in which individual phenotypic differences are evaluated in terms of their genetic and environmental components. When a quantitative trait is analyzed, an important question is:

To what extent is the variation observed in a population due to the genetic differences among individuals, and to what extent is it due to environmental differences?

Heritability is the proportion of a population's phenotypic variation attributable to genetic factors. Heritability reflects the relative contributions of genes and environment to variation in a specific trait.

Why is heritability so important to understand? In agricultural programs, many multifactorial traits—body weight in cattle, the number of eggs laid by chickens, or the amount of milk produced by dairy cows—are influenced by the genotype. Knowledge of heritability for these traits allows agriculturalists to develop goal-oriented selective breeding programs. Because many ecologically important traits such as body size, fertility, and disease resistance are also multifactorial, knowledge of heritability for these traits is important for understanding how natural populations evolve and how they can be best managed in conservation programs. Finally, human traits such as blood pressure and plasma cholesterol levels are multifactorial. Knowing the extent to which genes affect the variation in these and other traits may be useful in health care programs.

Partitioning the Variance

To estimate heritability, researchers must determine the total variance in a trait (V_t) and then partition this variance into the variance due to genetic differences (V_g) and the variance due to environmental differences (V_e). Thus, $V_t = V_e + V_g$. A simple model with three different scenarios illustrates how this is done.

The activity of enzyme A is determined for different genotypes in different environments. In the first scenario, we examine four individuals with two genotypes (A and B) in two different environments (1 and 2):

Individual	Genotype	Environment	Enzyme Activity
1	A	1	32
2	A	2	28
3	B	1	24
4	B	2	20

The mean enzyme activity in this sample is 26, and the variance is 26.66. (Verify these values using the formulas given earlier.) The 26.66 value represents the total phenotypic variance, or V_t, and it is due to both genetic and environmental differences.

In the second scenario, we eliminate all variance due to environmental differences (V_e) such that $V_e = 0$. We can thus solve for the variance due to genetic differences (V_g):

Individual	Genotype	Environment	Enzyme Activity
1	A	1	32
2	A	1	32
3	B	1	24
4	B	1	24

The mean in this population is 28, and the variance is 21.33. Because there is no environmental variance ($V_e = 0$), all variance is due to genetic differences: $V_g = V_t = 21.33$.

In the third scenario, the genotypes are identical ($V_g = 0$), but the environment varies:

Individual	Genotype	Environment	Enzyme Activity
1	A	1	32
2	A	1	32
3	A	2	28
4	A	2	28

The mean in this sample is 30; the variance is 5.33. All the variation in this case is due to environmental differences; therefore, $V_e = V_t = 5.33$.

Using this simple example, we can partition the total variance into variance due to genetic differences and variance due to environmental differences:

$$V_t = V_e + V_g$$
$$26.66 = 5.33 + 21.33$$

In this procedure of variance partitioning, simplifying assumptions were made; for example, genotype A was better than B in all environments, and by the same amount. This assumption rarely, if ever, is true in the natural world. Genotypes and the environment interact with each other so that sometimes a genotype is superior in one environment and inferior in another. We will explore this issue shortly. A second simplifying assumption was that environments and genotypes could be rigorously controlled; however, in actual experiments, researchers achieve only a limited degree of control.

Broad-Sense Heritability

Broad-sense heritability (***H^2***) is that fraction of the total variance (V_t) due to genetic differences (V_g). Thus, $H^2 = V_g/V_t$. The symbol H^2 is used to remind us that the broad-sense heritability is calculated from variances, which are squared quantities. In our enzyme activity example, heritability is

$$H^2 = 21.33/26.66 = 0.80$$

Broad-sense heritability ranges from 0 to 1. A 0 means that none of the phenotypic variability is due to genetic differences; a 1 means that all the phenotypic variability is due to genetic differences.

Let us return to the earlier discussion of corolla length in tobacco flowers to estimate broad-sense heritability and use East's data (Figure 25.6) to estimate broad-sense heritability (H^2). The P_1, P_2, and F_1 are all genetically homogeneous, or nearly so ($V_g = 0$): the two parental classes are highly inbred and homozygous; the F_1 are all identically heterozygous. Because there is essentially no genetic variation in these three populations, all of the observed variation must be due to environmental variation (V_e). We can thus estimate V_e for this trait by calculating the mean of the variances for these three samples (using East's published data):

$$V_e\,(\text{estimate}) = (V_{P1} + V_{P2} + V_{F1})/3 = 7$$

The F_2 generation is genetically and environmentally heterogeneous. The total variance in this population, 43, is due to genetic and environmental differences. Thus the estimate of heritability for corolla length is

$$V_t = V_e + V_g$$
$$43 = 7 + V_g$$
$$V_g = 36$$
$$H^2 = 36/43 = 0.84$$

Broad-sense heritability includes all types of genetic variation. It includes variance due to dominant alleles, alleles that are epistatic to alleles at other loci, and alleles that have additive effects (alleles that lack dominance so that each contributes incrementally to the phenotype). Because of these genetic complexities, it is often difficult to make any predictions about the offspring of parents expressing a quantitative trait with a well-characterized broad-sense heritability value. This is a problem in agricultural breeding progams.

Narrow-Sense Heritability

Estimates of genetic variance are important in predicting the phenotypes of offspring. However, the precision of these predictions depends on the amount of genetic variation that is due to the additive effects of alleles. Genetic variation that is due to the effects of dominance and epistasis has little predictive power.

To see how dominance limits the ability to make predictions, consider the ABO blood types in humans (Table 4.1 in Chapter 4). This trait is determined strictly by the genotype; environmental variation has essentially no effect on the phenotype. However, because of dominance, two individuals with the same phenotype can have different genotypes. For example, a person with type A blood could be either I^AI^A or I^AI^O. If two people with type A blood produce a child, we cannot predict precisely what phenotype the child will have. It could be either type A or type O, depending on the genotypes of the parents; however, we know that it will not have type B or type AB blood. Thus, although we can make some kind of prediction about the child's phenotype, dominance prevents us from making a precise prediction.

Our ability to make predictions about an offspring's phenotype is improved in situations where the genotypes are not confused by dominance. Consider, for example, the inheritance of flower color in the snapdragon, *Antirrhinum majus*. Flowers in this plant are white, red, or pink, depending on the genotype (Figure 4.1 in Chapter 4). As with the ABO blood types, variation in flower color has essentially no environmental component; all the variance is the result of genetic differences. However, for the flower color trait, the phenotype of an individual is determined without the complications of dominance. A plant with two *w* alleles is white, a plant with one *w* allele and one *W* allele is pink, and a plant with two *W* alleles is red. Thus, in this system, the phenotype depends simply on the number of *W* alleles present; each *W* allele intensifies the color by a fixed amount. Thus, we can say that the color-determining alleles contribute to the phenotype in a strictly additive fashion; dominance is not involved. This lack of dominance improves our ability to make predictions in crosses between different plants. A red x red mating predictably produces all red offspring; a white x white mating produces all white offspring; and a red x white mating produces all pink offspring. The only uncertainty is in a cross between two heterozygotes, and in this case the uncertainty is due to Mendelian segregation, not to dominance.

Quantitative geneticists distinguish between genetic variance that is due to alleles that act additively (such as those in the flower color example just discussed) and genetic variance that is due to dominance. These different variance components are symbolized as:

V_a = additive genetic variance
V_d = dominance variance

and together, they constitute the total genetic variance:

$$V_g = V_a + V_d$$

The **narrow-sense heritability (h^2)** is based on the additive genetic variance. Specifically, h^2 is the proportion of the total phenotypic variance that is additive genetic variance:

$$h^2 = V_a/V_t$$

Like the broad-sense heritability, h^2 lies between 0 and 1. The closer it is to one, the greater the proportion of the total phenotypic variance is additive genetic variance, and the greater is our ability to predict an offspring's phenotype.

Narrow-sense heritability is estimated in a variety of ways. One way is to compare the phenotypes of a selected group of parents with those of their offspring. This is an empirical method that determines how well the phenotypes of the parents can be used to predict the phenotypes of their offspring; that is, it assesses the extent to which a trait is heritable. Let's suppose, for example, that some trait is measured in a population and that the resulting data have been compiled into a frequency distribution (Figure 25.14). The mean of the distribution is 40 units. The parents selected for the next generation have a mean of 46 units. If the variation in the trait is due almost entirely to additive genetic differences, then the offspring from these selected parents will have a mean that is close to that of the selected patents (46 units). If the variation is due largely to environmental factors, the mean of the offspring will be close to the mean of the original population (40 units). If the variation is due to a combinantion of genetic and environmental factors, the mean of the offspring will be somewhere between that of the original population and that of the parents. Thus, the predictability of an offspring's phenotype depends on the extent to which the variance in the trait is caused by additive allelic effects. In other words, it depends on the magnitude of the narrow-sense heritability.

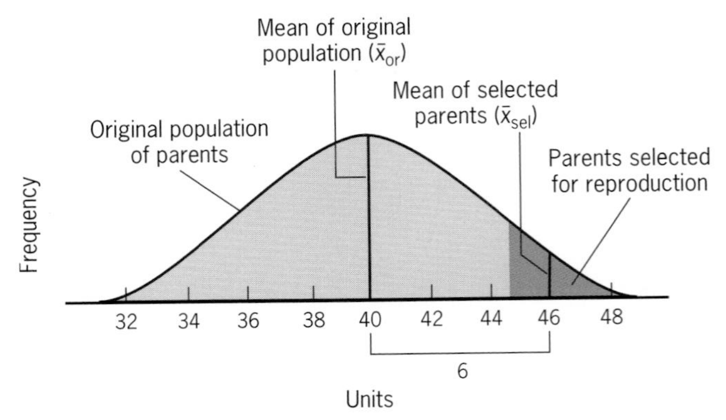

Figure 25.14 A population showing continuous variation for some trait. The population mean (X_{or}) and the mean of the individuals selected for reproduction (X_{sel}) are also shown. The more highly heritable the trait, the closer the mean of the offspring will be to the mean of the selected parents. Three possible outcomes are seen in Figure 25.15.

To illustrate these scenarios more quantitatively, let's suppose that the mean of the offspring is 45 (Figure 25.15*a*). Then the narrow-sense heritability is calculated as:

46 – 40 = 6 units = the expected phenotypic difference if all variance was due to the additive effects of alleles.

45 – 40 = 5 = the actual phenotypic difference.

5/6 = 0.83 = estimated narrow-sense heritability.

If the mean of the offspring is 43 (Figure 25.15*b*), the estimate of narrow-sense heritability is

$$(43 - 40)/(46 - 40) = 3/6 = 0.5$$

If the mean of the offspring is 40.5 (Figure 25.15c), the estimate of narrow-sense heritability is

$$(40.5 - 40)/(46 - 40) = 0.5/6 = 0.08$$

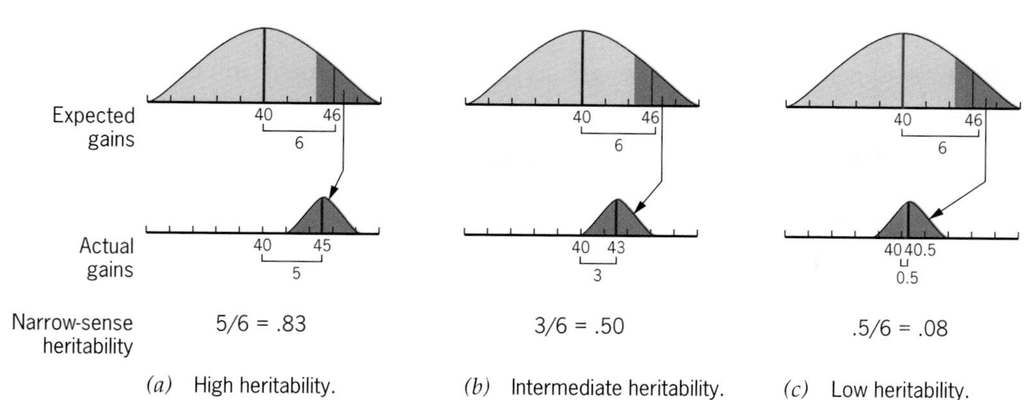

Figure 25.15 Heritability (narrow-sense) calculated from the outcome of selection experiments (see Figure 25.14). Narrow-sense heritability is the ratio between ($X_{offspring}$ - X_{or}) - response to selection and (X_{sel} - X_{or}) - selection differential. The offspring on the left show high heritability because the mean of the offspring is very similar to that of the selected parents. The offspring on the right show low heritability because the mean of the offspring is very close to the mean of the original population. The offspring in the middle show an intermediate heritability.

Artificial Selection

The narrow-sense heritability is used to predict the outcome of selective breeding programs with crop plants and livestock. We can see the usefulness by examining a selection experiment that was performed by Frank Enfield. In this experiment, Enfield selected for increased body size (pupal weight) in the common flour beetle, *Tribolium castaneum*, for 125 generations. Each generation, the largest pupae were selected to complete development and to serve as parents for the next generation. At the start of the experiment, the weight of the individual pupa ranged from 1800 to 3000 μg, the mean was 2400 μg, and the variance was 40,000 μg^2. After 125 generations of selection, the mean pupa weight increased to 5800 μg, more than twice the mean of the starting population (Figure 25.16*a*). Moreover, none of the individuals in the selected population was as small as the largest individuals in the original starting population (Figure 25.16*b*). This complete lack of overlap in the frequency distributions indicates that the genetic makeup of the population had been radically altered.

To achieve this stunning result, Enfield used a selection differential of 200 μg in each generation, the difference between the population mean and the parental mean. Initially, the narrow-sense heritability for pupa weight was estimated to be about 0.3 (actual difference/expected difference); thus the predicted re-

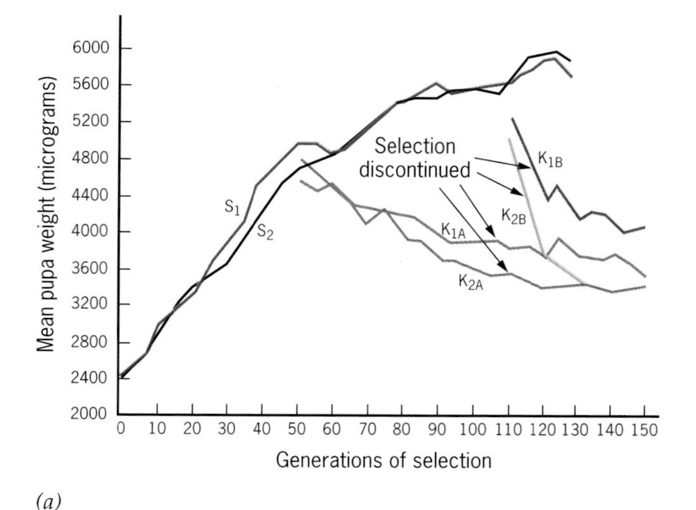

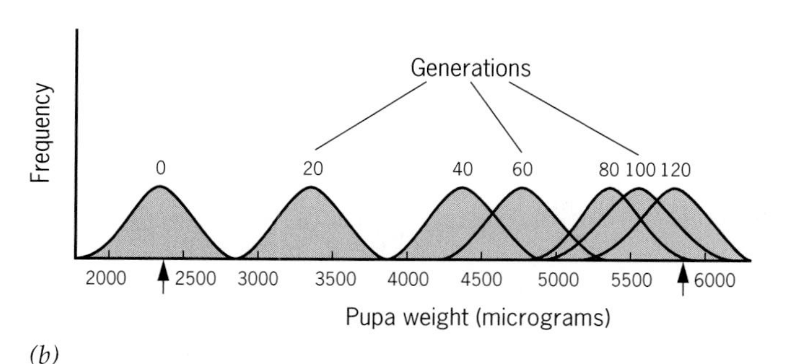

Figure 25.16 (*a*) Artificial selection for increased size in the flour beetle, *Tribolium*. Size was measured by pupa weight. The curves S_1 and S_2 show the response to selection in two replicate populations. The curves K_{1A}, K_{2A}, K_{1B}, and K_{2B} show what happened when artificial selection was discontinued in subpopulations that were established from the selected populations. (*b*) Frequency distributions of pupa weight in *Tribolium* populations selected for increased size. The shape of the distributions is only approximate. The means at generations 0 and 120 are indicated by arrows.

sponse to selection was 0.3 × 200 μg = 60 μg per generation (see Figure 25.15). For the first 40 generations, this was approximately what Enfield observed. However, the cumulative response during this time was 2000 μg, a little less than the 2400 μg that was expected (60 μg/generation × 40 generations). This discrepancy was due to factors that reduced the selection efficiency, including such things as infertility among the selected individuals. Thus, although the narrow-sense heritability is a reasonably good predictor of the response to selection over a few generations, in the long term it tends to overestimate this response.

The later generations of Enfield's project dramatically demonstrate this point. Between generations 40 and 125, the cumulative response was 1400 μg, which, although impressive, is much less than the expected response of 5100 μg (60 μg/generation × 85 generations). Enfield checked narrow-sense heritability in these later generations to see if any of the additive genetic variance had been lost during the long selective process. To his surprise, he found that h^2 was only slightly changed, indicating that the population still retained selectable genetic variability. A more detailed analysis demonstrated that during these generations, the efficiency of selection was severely reduced by a negative correlation between size and reproductive ability. (After a certain point, the larger the beetle, the less reproductively successful it is.) This reduced the effective selection differential and made it difficult to select for further increases in size. In fact, when selection was relaxed in generation 50 and again in generation 110, mean pupa weight began dropping back to a nearly normal value. This result suggests that a force of **natural selection** strongly opposed the artificial selection that was practiced in the main experiment. Enfield's attempts to increase pupa weight beyond 5800 μg failed, not for any lack of genetic variability, but simply because artificial selection had exceeded the natural limits.

The Interaction Between the Genotype and Environment

One further point concerning the estimation of heritability must be made. In the simple example discussed earlier of using enzyme activity to illustrate how the variance is partitioned, genotype A always was superior to genotype B and by the same amount, no matter what the environment. Furthermore, environment 1 always prompted more enzyme activity than did environment 2. Such consistent relationships may not occur in nature. For example, a genotype may have identical expressions in two different environments; or perhaps genotype A has a greater level of expression in environment 1 while genotype B has a greater expression in environment 2. In other words, the genotype and environment interact with each other in complex ways. This interaction must be considered in any estimate of heritability.

A study by R. Cooper and J. P. Zupek clearly illustrates the complexity of genotype and environment interactions. Under standard laboratory conditions, rats were selected for their ability to maneuver through a maze. Those making the journey the fastest and with the fewest errors were placed in the "bright" category and selected as parents for the next generation. Those taking the most time to journey through the maze and making the most errors were classified as "dull." They too were selected out and interbred to produce the next generation. Carried out over several generations, this type of artificial selection experiment led to the development of two lines: a "maze bright" and a "maze dull" line. The fact that the researchers were able to select for the trait (maze brightness or dullness) indicates that genetic factors are involved. But do these phenotypic differences exist in all environments?

Placing maze bright rats and maze dull rats into different environments caused the phenotypic differences to disappear. When both strains were placed into an enriched, stimulating environment, they did equally well; the phenotypic differences disappeared. When the two strains were placed into a stressful, deprived environment, they both did equally poorly. Thus there was a genotypic difference between the two lines, but it could be seen phenotypically only in certain environments. The importance of this genotype–environment interaction (V_{ge}) will be discussed further, along with other aspects of heritability, in the next chapter.

Key Points: **The phenotypic variance is equal to the variance due to genetic differences *plus* the variance due to environmental differences. Heritability is the proportion of phenotypic variation attributable to genetic differences. Broad-sense heritability includes the effects of dominant, additive, and epistatic alleles. Narrow-sense heritability includes only the additive effects of alleles. Narrow-sense heritability for a quantitative trait can be used to predict the outcome of a selective breeding program. Some of the variance in a quantitative trait may be caused by genotype-environment interactions.**

INBREEDING DEPRESSION AND HETEROSIS

The hills of eastern Kentucky are home to some families who trace their ancestry to a single man, Martin Fugate, a Frenchman who settled in Troublesome

Creek, Kentucky in 1820. What makes Martin Fugate interesting to geneticists is the fact that he had a rare but relatively benign recessive disorder that caused his skin to take on a blue, almost purplish hue. Fugate was missing an enzyme called diaphorase, and as a consequence he was unable to properly convert methemoglobin back to hemoglobin. Methemoglobin is the oxidized form of hemoglobin, containing iron in the ferric (Fe^{3+}) rather than the ferrous (Fe^{2+}) state. Methemoglobin has a reduced capacity to bind oxygen and is bluish in color, which is why the skin is blue. Normally, hemoglobin is slowly converted to methemoglobin. If this conversion continued unabated, all the body's hemoglobin would become functionless; but diaphorase (also called NADH-methemoglobin reductase) converts methemoglobin back to hemoglobin. Fugate and his descendants accumulated methemoglobin owing to a missing enzyme and thus had a bluish appearance.

The blue skin allele was transmitted to Martin Fugate's children and to his grandchildren. Over several generations the allele for this trait spread through the Troublesome Creek population. Because people in this region of Kentucky are relatively isolated, marriages often involve neighbors and neighbors are often related to each other. The blue skin trait became common in this region because frequently both parents were carriers of this otherwise rare allele.

Any time there is a departure from random mating in a population, such as the case of the families in Troublesome Creek, rare recessive traits may appear with a frequency that is much higher than in the general population. An extreme example of a departure from random mating is self-fertilization, seen commonly in some animal species such as worms and snails and in many plant species, including Mendel's peas. If a heterozygote, *Aa*, self-fertilizes, it will produce three kinds of progeny: *AA*, *Aa*, and *aa* in a 1:2:1 ratio. The heterozygote frequency is 0.5. However, if self-fertilization continues for another generation, the homozygotes (*AA* and *aa*) will breed true, but the heterozygotes will again segregate, producing three genotypic classes and reducing the heterozygote frequency to 0.25. With every succeeding generation of self-fertilization, the heterozygote frequency drops by 50 percent, reaching 0.008 by generation 7 and 0.001 by generation 10. At this point, the population is 99.9 percent homozygous.

Self-fertilization is an extreme form of nonrandom mating called **inbreeding,** mating between close relatives. Other examples of inbreeding include sib × sib mating, first cousin mating, parent–offspring mating, and aunt–nephew mating. They all have the same consequence: They increase the homozygote frequency and decrease the heterozygote frequency. In populations where inbreeding is common, there are elevated levels of homozygosis and thus a higher than expected frequency of rare recessive traits.

Inbreeding affects not only basic Mendelian traits, but also quantitative traits such as body size, vigor, fertility, and yield in agricultural crops. Detailed studies have shown that traits such as vigor and fertility decline more or less linearly with the intensity of inbreeding. The more inbred a population is, the less vigorous and fertile it is. The loss of vigor and fertility associated with inbreeding is called **inbreeding depression.** One explanation for this depression is that deleterious recessive alleles are made homozygous by inbreeding.

Outbreeding occurs when matings occur preferentially between unrelated individuals. Some of the most striking examples of genetically enforced outbreeding are found in plants that have developed a system of sexual incompatibility genes ensuring outbreeding. Incompatibility is based on a genetically complex locus (*S*) with multiple alleles. In species with **gametophytic incompatibility**, pollen cannot fertilize an ovule if the pollen grain carries an allele that is present in the maternal reproductive tissue. For example, pollen with the S_1 allele cannot fertilize an ovule produced by an S_1S_2 plant; however, it can fertilize an ovule produced by an S_2S_3 plant. Clearly, in such a system, homozygotes can never be formed, thus the *S* locus is usually a highly variable one that promotes heterozygosity.

Outbreeding is also a standard feature of crop production and animal husbandry. For example, seed corn is commercially produced by crossing different inbred lines, each of which has been bred for certain traits such as disease resistance and sugar content. Although the inbred strains themselves have a high degree of homozygosity and are not usually vigorous, their hybrid progeny are quite vigorous and often yield an abundant crop. This **hybrid vigor**, or **heterosis**, is probably due to increased heterozygosity.

Key Points: **Inbreeding is the preferential mating between close relatives. It leads to an increase in homozygote frequency and a decrease in heterozygote frequency. Outbreeding is the preferential mating between unrelated individuals. Although inbreeding often leads to inbreeding depression, outbreeding can lead to hybrid vigor.**

MOLECULAR TECHNIQUES IN THE ANALYSIS OF QUANTITATIVE TRAITS

Gene loci responsible for quantitative traits are called **quantitative trait loci (QTLs)**. Mapping QTLs in plants and animals is important and has many applications (see Human Genetics Sidelight: The Epilepsies

HUMAN GENETICS SIDELIGHT

The Epilepsies are Complex and Often Dramatic

Time and time again, studies of rare Mendelian disorders provide important insights into basic biological processes. Such is the case for epilepsy, a heterogeneous group of disorders in which there is a predisposition to recurrent seizures or fits. Seizures represent an occasional, excessive, and disorderly discharge of neurons. This discharge often produces dramatic convulsions or involuntary contractions of body muscles. There are many types of epileptic syndromes, but most fall into one of two categories: **partial** or **focal epilepsies**, in which seizures originate from a specific area of the cerebral cortex; and **generalized epilepsies** in which seizures originate from the activation of the entire cerebral cortex. Generalized epilepsies are complex, having genetic and environmental components. The concordance rate is higher in monozygotic twins than in dizygotic twins, but the genetics is poorly understood.

The partial or focal epilepsies are yielding to genetic analysis. These epilepsies are usually associated with lesions in the cerebral cortex, lesions caused by tumors, injuries, abnormal brain development, or circulatory problems. If a partial epilepsy is expressed but no lesion is evident, it has usually been assumed that the lesion is too small to be detected. However, several lines of research now suggest that certain types of partial epilepsy may have a genetic basis.

Ruth Ottman and her colleagues studied a family containing 11 individuals with partial epilepsy (Figure 1).[1] The trait in this family exhibited an autosomal dominant pattern of inheritance with reduced penetrance. They used molecular markers to locate the epilepsy susceptibility gene in this family to a region on the long arm of chromosome 10.

J. C. Mulley, H. A. Phillips, and their colleagues identified a gene on the long arm of chromosome 20 that causes another rare type of partial epilepsy called autosomal dominant, nocturnal, frontal lobe epilepsy (ADNFLE).[2] The Mulley study is especially intriguing because it suggests that another rare type of partial epilepsy called benign neonatal epilepsy (EBN) may be an allelic variant of ADNFLE. Researchers have located an EBN gene on the long arm of chromosome 20 (*EBN1*) that accounts for most families with EBN. A second gene on the long arm of chromosome 8 (*EBN2*) accounts for EBN in at least one family, so the disorder appears to be genetically heterogeneous. However, research suggests that there are two phenotypically distinct epilepsies caused by two alleles of the same gene.

What is the genetic mechanism for partial epilepsy? One possibility being explored is that the gene is normally expressed in only a limited region of the cerbral cortex. The expression or lack of expression of the mutant allele in this region leads to epileptic seizures. Another possibility is that the gene is expressed widely in the brain, but only certain cortical cells are sensitive to the variant gene products.

The observation that seizures or fits run in families dates back to Hippocrates, but probably goes back even further. Today researchers are beginning to unravel the elusive genetic mechanisms underlying epilepsy.

[1] Ottman, R., *et al.* 1995. Localization of a gene for partial epilepsy to chromosome 10q. *Nature Genet.* 10:56–60.

[2] Phillips, H. A., *et al.* 1995. Localization of a gene for autosomal dominant nocturnal frontal lobe epilepsy to chromosome 20q13.2. *Nature Genet.* 10:117–118.

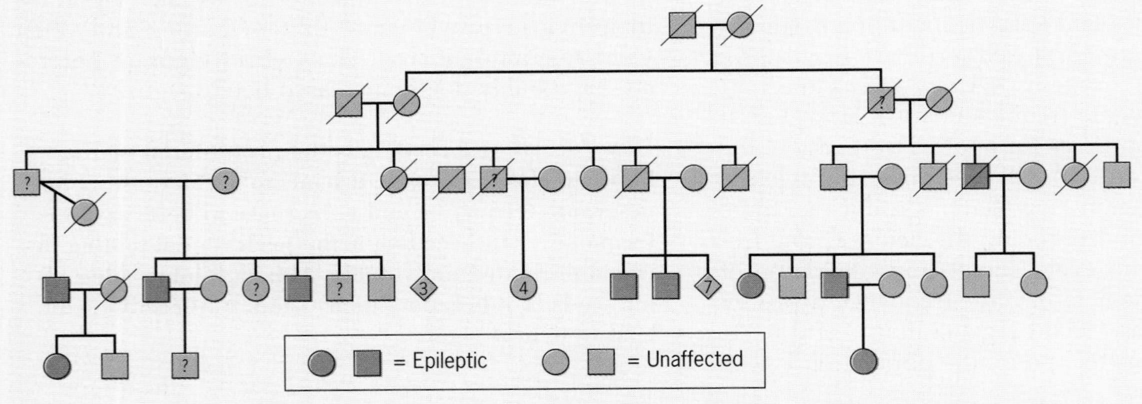

Figure 1 A pedigree of a family with partial (Focal) epilepsy. "?" denotes individuals are possible but unconfirmed epileptics.

Are Complex and Often Dramatic). For example, genes influencing the development of diseases such as hypertension, colon cancer, and diabetes have been analyzed in humans and in nonhuman animal models such as the mouse and rat. Furthermore, mapping QTLs that influence agriculturally important traits

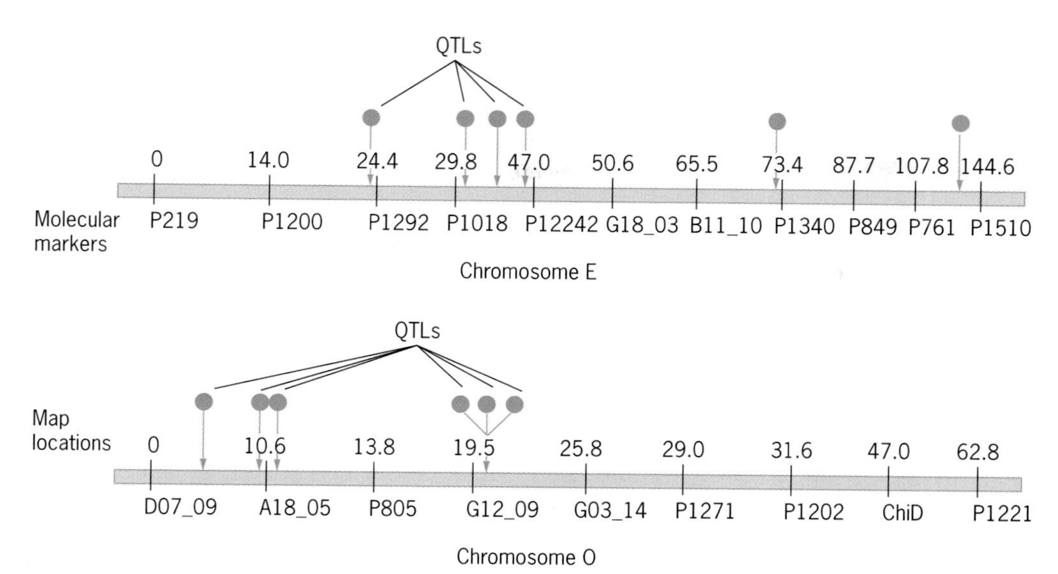

Figure 25.17 QTL clusters associated with stem growth on two chromosomes in *Populus*, a genus of trees.

such as tomato fruit mass, grain yield in maize, milk production, and fatness and growth in pigs can lead to a better understanding of these traits and to novel improvement programs.

Mapping QTLs has become more feasible with the advent of techniques to detect molecular variation. An enormous amount of genetic variation has been revealed by techniques such as DNA sequencing, restriction site polymorphism, and protein electrophoresis. Detailed genetic maps have been constructed for hundreds of molecular genetic markers evenly spaced throughout the genome. Powerful computer programs have been developed to analyze the complex data. Mapping strategies, which are beyond the scope of this text, depend upon backcrosses or intercrosses using organisms where the entire genome has been densely mapped with molecular markers. The objective of QTL mapping is to associate a QTL with a molecularly marked region of the genome.

QTL mapping has been successfully applied to many traits, including traits important in forestry programs. For example, H. D. Bradshaw and R. F. Stettler recently analyzed poplars, cottonwoods, and aspens, common forest trees in the genus *Populus*, for QTLs associated with stem growth. They found several gene loci clustered on chromosomes E and O in this genus (Figure 25.17). Armed with this type of knowledge, Forest Service researchers will be able to breed and select for trees with more desirable genetic qualities.

Key Points: **Quantitative trait loci (QTLs) are loci that influence quantitative traits. QTLs can be mapped using molecular markers distributed densely but uniformly over the genome.**

DIABETES: A MULTIFACTORIAL TRAIT

Geneticists have been successful in identifying and isolating the genes involved in monogenic human disorders, such as Duchenne's muscular dystrophy and cystic fibrosis, but success has been more elusive in identifying genes associated with more complex multifactorial genetic disorders, such as diabetes. However, new molecular techniques are being applied with increasing success to the analysis of multifactorial disorders. These techniques, when applied to large families with several affected members, have enabled investigators to focus on candidate regions of the genome that house genes for disease susceptibility. They are also gaining insight into environmental factors that may trigger the expression of the disease. One objective of this research effort is to improve genetic counseling to families with multifactorial disorders (see Human Genetics Sidelight: Genetic Counseling and Multifactorial Traits).

Diabetes mellitus is a major public health problem that afflicts more than 10 million people in the United States. It is characterized by hyperglycemia (elevated blood glucose levels) owing to defects in the transport and utilization of glucose. The long-term complications of diabetes involve deteriorating function of the kidneys, eyes, nerves, and blood vessels. The common denominator in all forms of diabetes is **insulin**, a pancreatic hormone secreted by the beta cells of the islets of Langerhans. This hormone is essential for the oxidation and utilization of glucose. All forms of diabetes involve absolute or relative insulin deficiency or insulin resistance, or both. There are two primary categories of idiopathic diabetes: Insulin-dependent diabetes melli-

HUMAN GENETICS SIDELIGHT

Genetic Counseling and Multifactorial Traits

For traits that do not follow single-gene modes of inheritance, genetic counseling can be difficult. Nevertheless, there are ways to test for polygenic inheritance and ways to counsel families that express multifactorial traits. C. O. Carter suggests five features of multifactorial traits that are important in counseling situations.[1]

1. For polygenic traits, the risk of recurrence is greater in relatives of an affected person than the incidence in the general population. The magnitude of the risk difference decreases as the incidence of the trait in the general population increases. For example, cleft lip/palate has an incidence of about 0.001 in the general population, but first-degree relatives of a person with cleft lip/palate have a 50 times higher risk. Spina bifida, on the other hand, has a higher general population incidence of about 0.005, but in first-degree relatives of an affected person, the risk is increased only about sevenfold.
2. For polygenic traits, the risk of recurrence increases with increasing severity of the malformation. For example, if a couple has a child with mild cleft lip but no cleft palate, the risk that their next child will have the trait is 2.5 percent. However, if the first child is severely affected with cleft lip and palate, the risk that their next child will have the trait more than doubles to about 6 percent.
3. In single-gene inheritance, the risks do not increase with the number of affected family members. For example, in cystic fibrosis, if a couple has an affected child, the risk is 25 percent for each of their subsequent children. That risk will not change no matter how many affected children they have. For multifactorial traits, however, the risk of recurrence increases with the number of affected family members. For example, if a couple has a child with spina bifida, the risk for the next child is about 5 percent. However, if the couple has two children with spina bifida, the risk for their next child doubles to about 10 percent.
4. For polygenic traits, if the trait is expressed more in one sex than in the other, the risk of recurrence is greater in the offspring of the affected person belonging to the least frequently affected sex. For example, pyloric stenosis (narrowing of the pylorus, the band of muscle that functions like a valve at the junction of the stomach and small intestine) is five times more common in males than in females. The risk to a child is greater if the mother had pyloric stenosis than if the father had it.
5. The incidence of multifactorially determined traits is higher in the offspring from matings between relatives than in matings between unrelated persons.

In addition to these five features of polygenic inheritance, genetic counselors must be absolutely certain that they are in fact dealing with a multifactorial trait. Although many conditions are multifactorial, some are the result of a pleiotropic effect of a single gene or a chromosome abnormality. For example, cleft palate without cleft lip is most likely the result of a single autosomal dominant allele. Cleft lip alone or cleft lip with cleft palate is usually multifactorial.

[1]Carter, C. O. 1969. Genetics of common disorders. *Br. Med. Bull.* 25:52–57.

tus (IDDM, type I), and noninsulin-dependent diabetes mellitus (NIDDM, type II).

The genetic and nongenetic factors contributing to diabetes are poorly understood, although there are intriguing clues. The complexity of the disease phenotype and its multifactorial nature has prompted some to refer to diabetes as a geneticist's nightmare.

Insulin-Dependent Diabetes Mellitus (Type I, IDDM)

IDDM is predominantly, if not exclusively, an autoimmune disorder in which the body makes antibodies against the insulin-producing cells of the pancreas (the islet β cells). Patients with this disorder are entirely dependent on exogenous insulin therapy. This form of diabetes usually, but not always, develops prior to early adulthood.

A major gene (*IDDM*1) involved in IDDM was discovered to be tightly linked to the HLA region on chromosome 6 (see Chapter 24). The HLA genetic system is responsible for distinguishing between self and nonself. A second gene, *IDDM*2, on the short arm of chromosome 11 near the genes for insulin and insulin-like growth factor gene (*IGF*) has also been implicated in predisposing individuals to IDDM.

IDDM is found predominantly in individuals with the *HLA* alleles *DR4* and *DR3*. About 45 percent of Caucasians in the United States carry *DR3* or *DR4* alleles, but over 95 percent of patients with IDDM have at least one of these *DR* alleles. Focus has recently shifted to the *HLA-DQ* locus where there is an even stronger association between particular *HLA-DQ* alleles and susceptibility or resistance to IDDM.

How do the *HLA-DQ* or *DR* alleles influence susceptibility to IDDM? Probably by having a greater affinity for the diabetogenic peptides that are to be presented to the immune system than other *HLA* alleles. The strong presentation of diabetogenic peptides stimulates a powerful immune response.

The fact that the discordance rate in monozygotic twins (one twin has IDDM, the other does not) is about 50 percent suggests that nongenetic factors are required for expression of IDDM. The environmental factor in many cases of IDDM is thought to be a virus that infects the pancreatic β cells. Predisposing *HLA* alleles would present the viral antigen for an immune response, but the immune response would be so powerful that β cell destruction ensues.

Recent reports that exposure to cow's milk early in life predisposes susceptible individuals to IDDM have generated much interest. The environmental trigger in this case is a protein called bovine albumin, and it operates by a mechanism called **molecular mimicry**. For genetically susceptible individuals, antibodies made to bovine albumin cross react with a protein on the surface of β cells and destroy these cells.

Recent genome-wide searches for IDDM suscepibility genes (QTLs) have revealed over 20 independent chromosome regions associated with predisposition to the disease. Research also suggests that a mitochondrial gene mutation is associated with both IDDM and NIDDM. Mitochondria play an important role in diabetes susceptibility because oxidative phosporylation is a crucial process in the secretion of insulin by β cells in response to glucose and other nutrients. Patients with diabetes caused by this mitochondrial mutation also commonly have sensory hearing loss. This research emphasizes the genetic and environmental complexity of this disease (Figure 25.18).

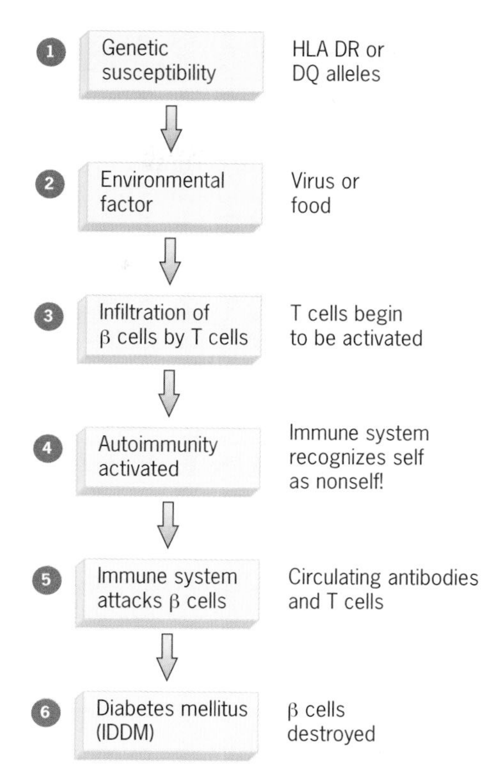

Figure 25.18 Steps in the development of diabetes mellitus (IDDM).

Noninsulin-Dependent Diabetes Mellitus (Type II, NIDDM)

Approximately 85 percent of all diabetes cases are NIDDM. Treatment of NIDDM and its secondary complications is estimated to account for up to 10 percent of total health care expenditures. Although NIDDM runs in families, modes of inheritance are not known, except for the dominantly inherited variant known as **maturity-onset diabetes of the young** (**MODY**). NIDDM commonly develops later in life than IDDM and is not usually associated with autoimmunity. Patients with NIDDM usually secrete insulin, often at elevated levels, but are insulin resistant.

MODY follows a classic dominant mode of inheritance, and serves as a model for understanding the genetics of type II diabetes. Through the use of polymorphic molecular markers in large pedigrees in which MODY is segregating, it has been discovered that at least two separate genes are involved, one of which is on the short arm of chromosome 7. This gene codes for glucokinase, which catalyzes the conversion of glucose to glucose-6-phosphate in the pancreatic β cells. Several mutations in this gene affect the enzyme's binding site, suggesting that MODY is caused by a reduced efficiency for binding glucose. This means that glucose levels would be necessarily higher in order for phosphorylation to occur and for insulin to be released. The glucose thermostat is thus set at a higher level in this type of diabetes.

Glucokinase mutations have also been implicated in the more classical forms of type II diabetes. In certain ethnic groups, such as Americans of African descent, there is a strong correlation between the frequency of type II diabetes and the frequency of certain mutant glucokinase alleles.

Type II diabetes has also recently been associated with mutations in the glycogen synthetase gene. This enzyme catalyzes the synthesis of glycogen from glucose subunits. Defects in this function cause glucose levels to rise. Other mutations are also associated with NIDDM.

Whatever the genetic mechanisms controlling NIDDM, they are powerful. The concordance rate of NIDDM in monozygotic twins is nearly 100 percent. About 40 percent of the siblings of a NIDDM patient also develop the disease, and about a third of the offspring of NIDDM patients develop NIDDM.

Diabetes is only one of many multifactorial disorders. Hypertension, or elevated arterial blood pressure, was once thought to be determined entirely by environmental factors, such as smoking, high sodium intake, and lack of exercise. These are certainly important factors in the development of the disease, but they are not the only causes. Using QTL analysis, researchers are identifying genes that play an important role in the development of hypertension. The same is true for obesity. Twin and adoption studies dramatically point out the connection between genes and body mass. And recent studies identifying specific genes controlling obesity further emphasize this point. Several genes controlling multiple sclerosis (ms) an inflammatory diseases of the central nervous system, have recently been identified. Modern techniques of genetic analysis will soon make the analysis of complex traits easier and more rewarding.

Key Points: **Diabetes is a complex multifactorial disorder. Insulin-dependent diabetes (IDDM) is an autoimmune disease. Noninsulin-dependent diabetes (NIDDM) is due to several genes involved in glucose metabolism.**

TESTING YOUR KNOWLEDGE

1. Two inbred strains of *Drosophila* differ from each other in the number of abdominal bristles they have. These two strains are intercrossed, and the F_1 and F_2 are analyzed. The mean number of bristles in the F_1 is 25, with a standard deviation of 2. The F_2 mean is also 25, but the standard deviation is 3. Calculate V_g, V_e, and H^2 for this quantitative trait.

ANSWER

The F_1 is genetically homogeneous and all heterozygous because the parental generation is inbred. Thus V_g for the F_1 is 0. The total variance (V_t) in the F_1 is s^2 or 4 because s is the square root of the variance. Since V_g is 0, all variance must be due to V_e, thus $V_e = 4$.

The F_2 is genetically heterogeneous because of independent assortment and segregation of the alleles affecting abdominal bristle number. The V_t in the F_2 is 3^2 or 9, and because we have already estimated that V_e is 4, V_g must be 5. H^2 is therefore $5/9 = 0.56$.

2. In a cross between two pure strains of broad beans, the standard deviation for plant height is 2.2 in the F_1 and 4.44 in the F_2. Estimate H^2 for plant height in the broad bean.

ANSWER

The standard deviation is the square root of the variance; thus the F_1 and F_2 variances are 4.84 and 19.71, respectively. The F_1 is uniformly heterozygous, so $V_g = 0$ and $V_t = V_e = 4.84$. The F_2 V_t is due to both genetic and environmental differences; therefore, $V_t = 19.71 = V_g + 4.84$, and $V_g = 14.87$. H^2 is therefore $14.87/19.71 = 0.75$.

QUESTIONS AND PROBLEMS

25.1 A wheat variety with red kernels (genotype $A'A'B'B'$) was crossed with a variety with white kernels (genotype $AABB$). The F_1 were intercrossed to produce an F_2. If each primed allele increases the amount of pigment in the kernel by an equal amount, what phenotypes will be expected in the F_2? Assuming that the A and B loci are unlinked, what will the phenotypic frequencies be?

25.2 How would you distinguish between a phenotype controlled by multifactorial inheritance and one controlled by a single gene with two alleles and variable expressivity?

25.3 Nilsson-Ehle studied the inheritance of kernel color in wheat. In one cross, he crossed a true-breeding red variety to a true-breeding white variety. The F_1 was red as was the F_2. When he self-fertilized the 78 F_2 plants, he obtained the following results:

Number of F_2 Plants	Progeny from Self-Fertilized Plants
50	all red
15	15 red:1 white
8	3 red:1 white
5	63 red:1 white

How many genes are segregating in this series of crosses? What are the parental and F_2 phenotypes?

25.4 A sample of 20 plants from a population was measured in inches as follows: 18, 21, 20, 23, 20, 21, 20, 22, 19, 20, 17, 21, 20, 22, 20, 21, 20, 22, 19 and 23. Calculate (a) the mean, (b) the variance, and (c) the standard deviation.

25.5 Quantitative geneticists use the variance as a measure of scatter in a sample of data; they calculate this statistic by averaging the squared deviations between each measurement and the sample mean. Why don't they simply measure the scatter by computing the average of the deviations without bothering to square them?

25.6 Two inbred strains of corn were crossed to produce an F_1; this was then intercrossed to produce an F_2. Data on ear length from a sample of F_1 and F_2 individuals gave phenotypic variances of 15.2 cm^2 and 27.6 cm^2, respectively. Why was the phenotypic variance for the F_2 greater than that for the F_1?

25.7 A study of quantitative variation for abdominal bristle number in female *Drosophila* yielded estimates of $V_t = 6.08$, $V_g = 3.17$, and $V_e = 2.91$. What was the broad-sense heritability?

25.8 The following table presents data on the variances of two phenotypic traits in sparrows (wing span and beak length):

Wing Span	Beak Length
$V_t = 271.4$	$V_t = 627.8$
$V_e = 71.2$	$V_e = 107.3$

Estimate the the broad-sense heritability for each trait.

25.9 Measurements on ear length were obtained from three populations of corn—two inbred varieties, and a randomly pollinated population derived from a cross between the two inbred strains. The phenotypic variances were 9.2 cm^2 and 9.6 cm^2 for the two inbred varieties and 26.4 cm^2 for the randomly pollinated population. Estimate the broad-sense heritability of ear length for these populations.

25.10 The broad-sense heritability can be estimated by computing the correlation coefficient between identical twins (t and t'). The formula is $r = Cov(t, t')/V_t = V_g/V_t = H^2$. Why is $Cov(t,t')$ equal to V_g?

25.11 A quantitative geneticist claims that the narrow-sense heritability for body mass in human beings is 0.7 while the broad-sense heritability is only 0.3. Why must there be an error?

25.12 The mean value of a trait is 100 units, and the narrow-sense heritability is 0.4. A male and a female measuring 124 and 126 units, respectively, mate and produce a large number of offspring, which are reared in an average environment. What is the expected value of the trait among these offspring?

25.13 The narrow-sense heritability for abdominal bristle number in a population of *Drosophila* is 0.3. The mean bristle number is 12. A male with 10 bristles is mated to a female with 20 bristles, and a large number of progeny are scored for bristle number. What is the expected number of bristles among these progeny?

25.14 A breeder is trying to decrease the maturation time in a population of sunflowers. In this population, the mean time to flowering is 100 days. Plants with a mean flowering time of only 90 days were used to produce the next generation. If the narrow-sense heritability for flowering time is 0.2, what will the average time to flowering be in the next generation?

25.15 A fish breeder wishes to increase the rate of growth in a stock by selecting for increased length at six weeks after hatching. The mean length of six-week-old fingerlings is currently 10 cm. Adult fish that had a mean length of 15 cm at six weeks of age were used to produce a new generation of fingerlings. Among these, the mean length was 12.5 cm. Estimate the narrow-sense heritability of fingerling length at six weeks of age and advise the breeder about the feasibility of the plan to increase growth rate.

25.16 A selection differential of 40 μg per generation was used in an experiment to select for increased pupa weight in *Tribolium*. The narrow-sense heritability for pupa weight was estimated to be 0.3. If the mean pupa weight was initially 2000 μg and selection was practiced for 10 generations, what was the mean pupa weight expected to become?

25.17 Suppose the narrow-sense heritability for antler size in mule deer is estimated to be 0.96 and the heritability for body size is estimated to be about 0.89. The genetic correlation between antler size and body size is 0.90. What effect would selection for larger antler size have on body size?

BIBLIOGRAPHY

Atkinson, M. A., and N. K. Maclaren. 1994. The pathogenesis of insulin-dependent diabetes mellitus. *N. Engl. J. Med.* 331:1428–1436.

Bell, J. I. and G. M. Lathrop. 1996. Multiple loci for multiple sclerosis. *Nature Genetics* 13:377–378.

Bodmer, W. F., and L. L. Cavalli-Sforza. 1976. *Genetics, Evolution, and Man.* W. H. Freeman, San Francisco.

East, E. M. 1916. Studies on size inheritance in *Nicotiana*. Genetics 1:164–176.

Falconer, D. 1989. *Introduction to Quantitative Genetics.* 3rd ed. Longman Scientific and Technical Publications, London.

Lander, E. S., and D. Botstein. 1989. Mapping Mendelian factors underlying quantitative traits using RFLP linkage maps. *Genetics* 121:185–199.

McDevitt, H. O. 1995. Autoimmune diabetes and its antigenic triggers. *Hosp. Practice* 30: 55–62.

Marx, Jean. 1990. Dissecting the complex diseases. *Science* 247:1540–1542.

Naggert, J. K., et al. 1995. Hyperproinsulinemia in obese fat/fat mice associated with a carboxypeptidase E mutation which reduces enzyme activity. *Nature Genet.* 10:135–142.

A set of monozygotic twin sisters.

26

The Genetic Control of Behavior

Twins and Personality

Oskar Stohr and Jack Yufe are monozygotic, or identical, twins who were separated shortly after their birth. Oskar was taken by his mother to Germany where he was raised by his grandmother as a Christian and a Nazi. Jack was raised by his father in the Caribbean as a Jew, spending part of his youth on an Israeli kibbutz. At the age of 47, these two brothers were reunited. The similarities in their personalities and behavior were remarkable. Both men sported mustaches, wore two-pocket shirts with epaulets, and used wire-rimmed glasses. Each loved spicy foods and sweet liqueurs and tended to fall asleep in front of the television. Each thought it amusing to sneeze in a room full of strangers. Both flushed the toilet before using it, stored rubber bands on their wrists, read magazines from back to front, and dipped buttered toast in their morning coffee. They were raised very differently, yet it was clear to all observers that they shared astounding similarities in their mannerisms.

Thomas Bouchard and his colleagues at the University of Minnesota are studying Oskar and Jack and many other sets of identical twins who had been separated shortly after birth in order to determine the role the genotype plays in influencing behavior. As we will discuss later in this chapter, they are finding that more aspects of human behavior are genetically determined or influenced than ever imagined.

Bouchard is one of a long list of investigators who have used twins to try to understand the biological basis of human behavior. Francis Galton, Charles Darwin's cousin (see Chapter 25), introduced the twin-study method of distinguishing between the effects of nature and nurture in human behavior. Although there were many flaws in his methodology and interpretations, Galton's idea that identical twins were ideal natural experiments with which to study the nature/nurture issue has proved to be a powerful and persistent one.

Twins, however, are not the only subjects used by researchers to analyze the genetic control of behavior. They can often study the effect of single genes on a behavior, or the influence of a particular chromosome on a behavior. To what extent do genes influence a behavior? What is the relationship between a gene product and a behavior? We explore these and other questions in this chapter.

WHAT IS A BEHAVIOR?

The term **behavior** refers to a coordinated neuromuscular response to changes or signals in the external and internal environment. This response requires the integration of sensory, neural, and hormonal factors.

Many gene products play a role directly and indirectly in the structure and function of a nervous system. If one of these gene products is altered in any way, either qualitatively or quantitatively, there is every reason to expect some altered behavior. The relationship between gene products and a specific behavior is usually complex. Genes code for products that influence cellular physiology so that under certain conditions a particular behavior appears (Figure 26.1). There is no single gene product that when present in altered form causes a person to hoist glasses of whiskey and by so doing become addicted to alcohol. There are, however, genes that control ethanol metabolism, and alleles of some of these genes may make a person more sensitive to the effects of alcohol. This sensitivity, coupled with certain personality traits, certain neuron properties, and environmental stresses may produce the disease called alcoholism. (We discuss this topic in more detail in a later section of this chapter.)

Genes do influence behavior. In some cases, a genotype may actually determine a behavioral trait. For example, if left untreated, the autosomal recessive metabolic disorder phenylketonuria (PKU) results in serious mental retardation. It is important to remember, however, that the relationship between the genotype and behavior is not usually a simple one. Genetically influenced behaviors are usually multifactorial, and it is the complexity of the interaction between the genotype and environment that makes the study of behavior so difficult. Sometimes researchers are able

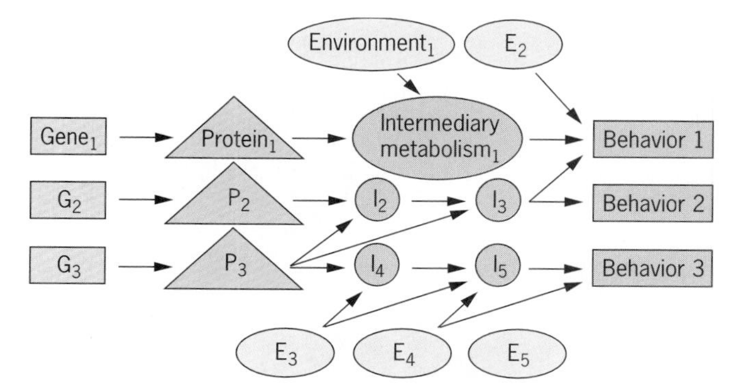

Figure 26.1 Behavior is influenced by the genotype (G) interacting with the environment (E). Genes code for proteins (P) whose functions are influenced by multiple environmental factors.

to dissect a behavior and evaluate the roles of the genotype and environment. We consider examples of this in invertebrates, mammals, and humans.

Key Points: **A behavior is a coordinated neuromuscular response to environmental signals. Genes and the environment influence many behaviors.**

GENETIC ANALYSIS OF BEHAVIOR IN INVERTEBRATES

Insects and vertebrates (fish, amphibians, reptiles, and mammals) have complex nervous systems and complex behaviors. Most behaviors are multifactorial, controlled by environmental and genetic elements. Sometimes, single genes that affect a complex behavior can be identified and analyzed. Invertebrates such as bees and fruit flies are ideal to use for behavior genetic studies because they have well-defined ritualized behaviors and because identifiable mutations affect those behaviors. The fruit fly is especially ideal to study because many mutations in this species influence behavior. Some of the *Drosophila* behaviors that have been studied include response to light or gravity, courtship behavior, ability to learn, and biological rhythms synchronized with the day–night cycle.

Nest-Cleaning Behavior in Honeybees

Honeybee larvae are sometimes killed by a bacterial infection that causes a disease known as American foulbrood. Honeybee larvae mature in chambers or cells in the hive, and these cells are maintained by the worker bees. To maintain a "hygienic" environment within the hive, larvae infected by American foulbrood are removed by the workers. This removal process requires two genetically controlled behaviors: first, workers uncap the cell containing the diseased larvae; second, they remove them. Some bees are "unhygienic" because they do not follow this behavior pattern of removing diseased larvae.

In 1964 N. Rothenbuhler discovered that nest-cleaning behavior is inherited and follows a basic Mendelian dihybrid pattern. When unhygienic bees are crossed to hygienic bees, all the offspring are unhygienic—a finding suggesting that unhygienic behavior is dominant. When F_1 hybrids are backcrossed to the parental hygienic strains, four classes of offspring are produced in roughly a 1:1:1:1 ratio:

1. One-fourth of the offspring are hygienic.
2. One-fourth of the offspring uncap the cells containing diseased larvae but do not remove them.
3. One-fourth do not uncap the cells but remove the diseased larvae if an experimenter first uncaps the cell.
4. One-fourth of the offspring can neither uncap the cells nor remove the diseased larvae.

This type of genetic analysis suggests that two pairs of independently assorting alleles operate to control hygienic behavior: one pair controlling the uncapping procedure and one pair controlling larval removal (Figure 26.2).

(*a*)

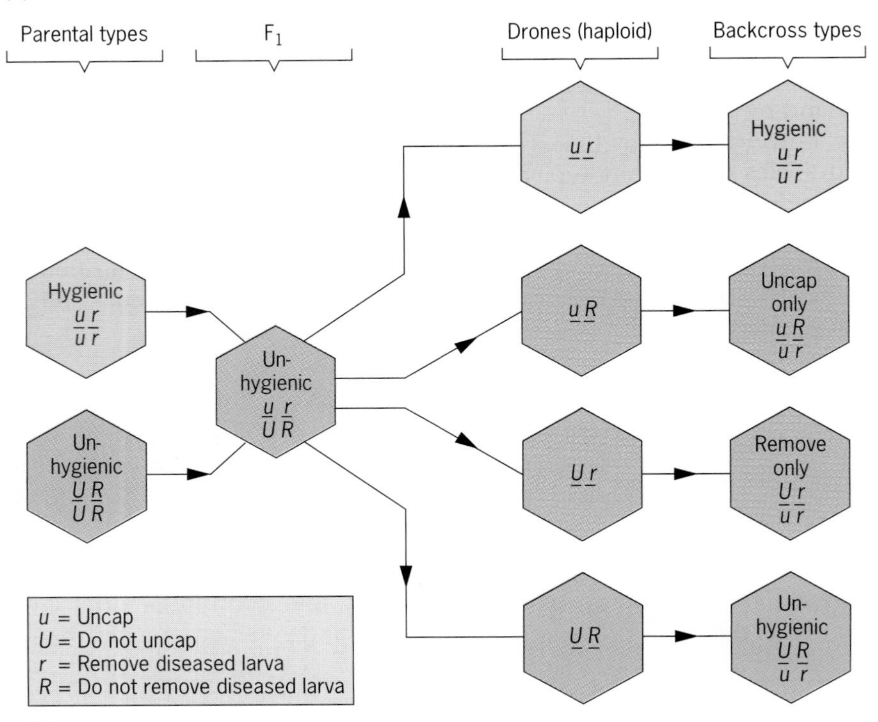

(*b*)

Figure 26.2 (*a*) A honeybee comb. The empty cells have been uncapped, and the infected larvae have been removed by hygienic bees. (*b*) Hygienic behavior in honeybees follows a classic Mendelian dihybrid inheritance pattern. Two independently assorting genes, each with two alleles (*U* and *u*; *R* and *r*), control uncapping and larva removal.

The gene products controlling this remarkable behavior and their functions are unknown. Perhaps a chemical signal given off by infected cells attracts the hygienic workers but not the unhygienic ones. The capped cells containing the diseased larvae may give off low levels of this chemical signal, levels so low they do not induce the uncapping mode of behavior in the uncapping-deficient strain. Once the cap is removed, the chemical signal may now be strong enough to induce expression of the gene-determining larval removal behavior. For workers that uncap but do not remove the infected larvae, the signal may be sufficient to trigger the uncapping behavior but, in the uncapped cell, be so strong that it inhibits the larval removal behavior. Some evidence suggests that the strength of stimulus is a factor in these behaviors, but the model is speculative. Whatever the mechanism, "hygienic behavior" is clearly controlled by genes that affect parts of the bee nervous system.

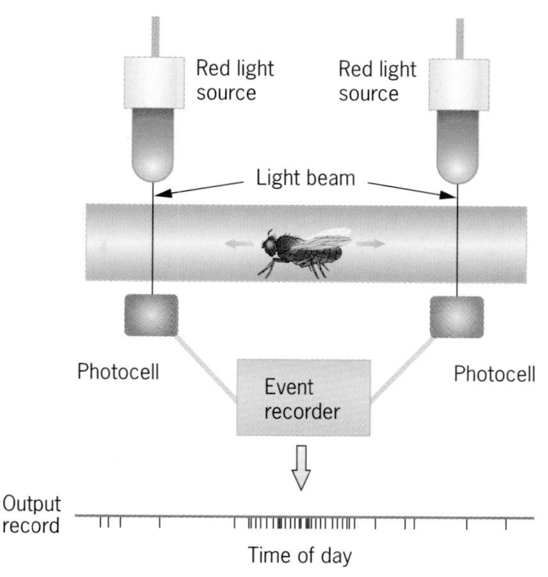

Figure 26.3 Locomotor activity in *Drosophila* is circadian, synchronized with the 24-hour day/night cycle. Activity is monitored by placing the fly in a glass tube that is illuminated with a red light at each end. The fly cannot see red light. When the fly breaks the beam of light at either end of the tube, the event is recorded. Each vertical line in the output record is an activity event caused by the movement of the fly to one end of the glass tube.

The Genetics of *Drosophila* Rhythms

Rhythms in the plant and animal world have been observed and recorded for centuries. The most obvious of these rhythms are those that are synchronized to daily cycles of light and darkness. Recently, investigators discovered that many rhythmic behavior patterns are under the control of the genotype, even in the absense of environmental timing cues. Timekeeping systems are of profound importance to the earth's life forms because they allow organisms to adapt to the daily and seasonal cycles of their environment, cycles that are in turn generated by the earth's rotation and its journey around the sun.

Plants, animals, fungi, even bacteria, exhibit rhythmic behaviors. For example, morning glory flowers open in the early morning and then close for the rest of the day. The flowers of the baobob tree close during the day but open at night, a rhythm that works to their advantage since they are pollinated by nocturnal bats. In the filamentous fungus *Neurospora* (see Chapters 1 and 8), asexual spore production is rhythmic and controlled by the *frq* (*frequency*) gene. In the tiny laboratory plant *Arabidopsis*, the *CAB* gene (chlorophyll-a/c-binding protein) is switched on during the day and switched off at night. *Arabidopsis* maintains this rhythm—directed by its internal clock—even when the plant is kept in constant darkness.

The *per* Gene in *Drosophila*

A functional equivalent to *Neurospora*'s *frq* gene and one of the most intensively studied genes influencing biological rhythms is the *period* or *per* gene located on the X chromosome in *Drosophila*. *Drosophila* exhibits rhythmic behavior patterns synchronized with the 24-hour cycle of day and night. The term **circadian** (from the Latin words *circa* meaning "about" and *dies* meaning "day") describes endogenous rhythms that are synchronized with the 24-hour cycle of day and night. The *per* gene influences the 24-hour pattern of activity, the time the adult fly emerges from its pupal case, the fly's courtship pattern, as well as other circadian rhythms.

Drosophila is most active around dawn and just after dusk. The fly's activity is measured by placing it in a glass tube in the dark and shining a narrow beam of red light (the fly cannot see red light) across the tube (Figure 26.3). Every time the fly moves across the beam and disrupts it, a photoreceptor cell records the event. The more active a fly is, the more it interrupts the light beam. A typical pattern of *Drosophila* activity for a wild-type fly (per^+) is seen in Figure 26.4. Various *per* alleles alter this activity pattern. For example, the per^0 allele causes the activity pattern to be arrhythmic, the per^S (short) allele causes the activity pattern to cycle through 19-hour intervals, and the per^L (long) allele causes the activity pattern to cycle through 29-hour intervals.

The *per* locus also controls the time of emergence of adult flies from their pupal cases. Normally, per^+ flies emerge from their pupal cases in the morning. This emergence time is advantageous because it gives

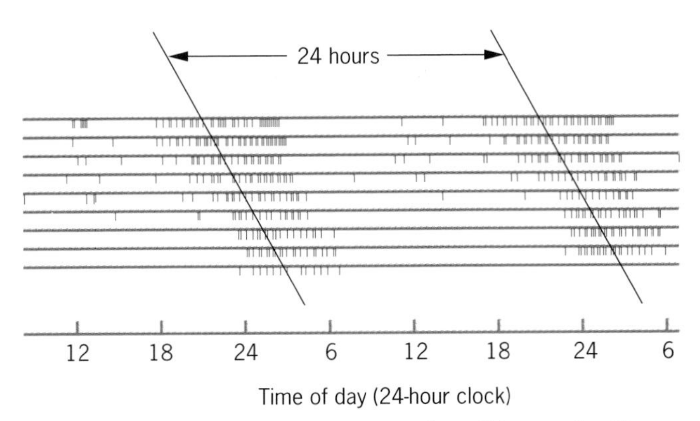

Figure 26.4 The activity pattern of a wild-type fly. The activity is maximal where the vertical lines are the densest. The activity pattern has a 24-hour periodicity.

the immature flies an opportunity to unfold and dry their wings so they can fly off. This emergence pattern is synchronized to a 24-cycle. For the *per*^{0} mutant, which is arrhythmic and shows no circadian rhythym, emergence may occur at any time of the day or night. The *per*^{S} mutant shows emergence patterns that run on 19-hour cycles, and the *per*^{L} mutant runs on a 29-hour cycle (Figure 26.5).

One of the most intriguing aspects of *per* activity is the effect it has on *Drosophila* behavior that is *not* circadian, such as courtship behavior. *Drosophila* courtship is a highly ritualized series of male behaviors. If properly executed, these behaviors result in successful copulation. Each *Drosophila* species has its own distinctive courtship ritual. In general, the male approaches the female from the front at an angle and then moves toward her rear. He extends his wing at a 90° angle and vibrates it, generating a song that stimulates the female and makes her more receptive to his advances. He moves to the rear of the female and touches her genitalia with his proboscis. He then attempts to copulate. If she accepts him, mating is completed; if she rejects him, which she often does, he will go through the entire sequence again, and continue to do so until either he is successful or unsuccessful.

The male song generated by rapidly vibrating his wing is characteristic of the species. A typical *Drosophila* song pattern, as recorded by a special recording device, is a "hum" followed by a series of "pulses" (Figure 26.6). The intervals between pulses vary in length from 15 to 100 milliseconds. The characteristic interval pattern usually starts short, becomes longer, and then becomes short again (Figure 26.7*a*). When the intervals between each songburst are plotted over a 60-second time span, a pattern similar to the one in Figure 26.7*b* is obtained. Over a 60-second span, there is a clear rhythmicity to the songburst pattern, going from short to long and back to short.

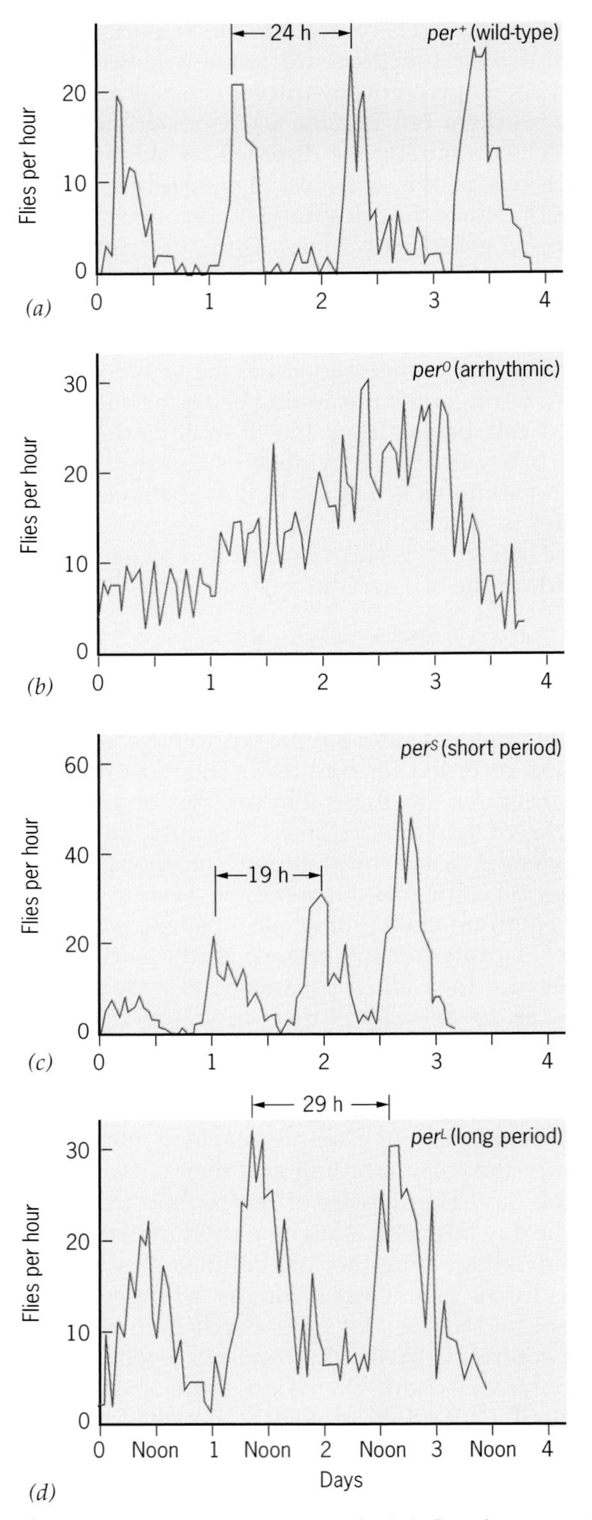

Figure 26.5 The emergence of adult flies from pupal cases follows a circadian rhythm. Each graph represents a pattern for various *per* alleles. The flies are kept in total darkness after pupation. (*a*) The wild-type adults tend to emerge in the early morning hours. (*b*) The *per*^{0} mutants show no rhythmicity. (*c*) The *per*^{S} mutant runs on a 19-hour cycle. (*d*) The *per*^{L} mutant runs on a 29-hour cycle.

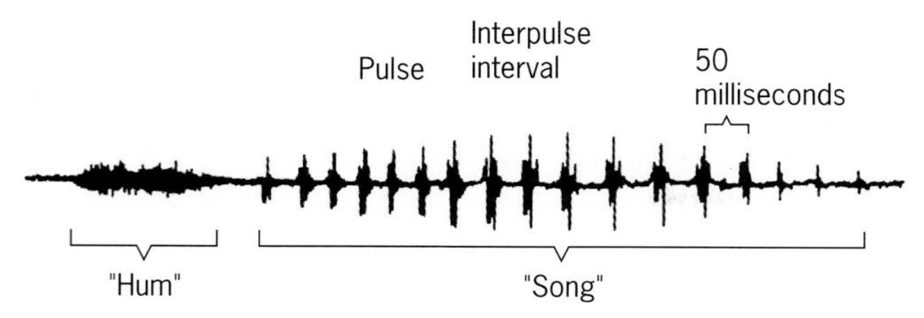

Figure 26.6 Typical songburst pattern in *Drosophila* courtship. The burst is preceded by a "hum," followed by a series of discrete pulses. The time between two pulses is an interpulse interval.

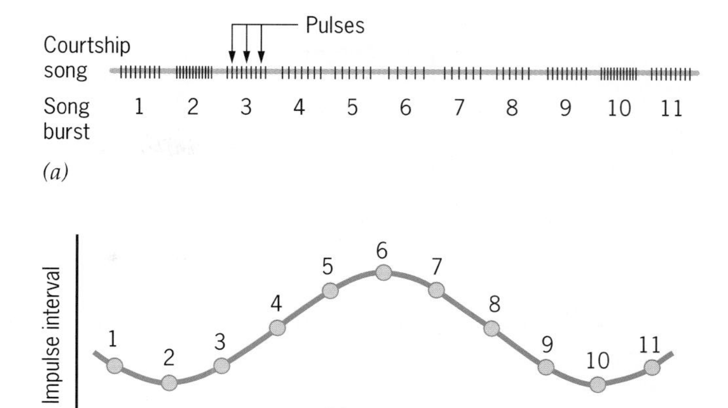

Figure 26.7 The *Drosophila* courtship song, composed of wing vibrations, is rhythmic. (*a*) A pattern of 11 separate songbursts. Notice that the interpulse intervals within the songbursts expand and contract. (*b*) A plot of the interpulse intervals as a function of time. There is a periodicity of about 60 seconds.

The various *per* alleles alter the periodicity of the courtship song. The per^S allele shortens the periodicity to a 40-second cycle; the per^L allele lengthens it to about 90 seconds (Figure 26.8).

What evidence supports the conclusion that the *per* gene controls these songburst pattern differences? The most compelling evidence comes from experiments where alleles of the *per* gene are transferred from one species to another using recombinant DNA techniques. For example, in *Drosophila melanogaster* the song-burst pattern follows a 60-second cycle. In *Drosophila simulans*, it normally follows a 35-second interval. If the wild-type *per* gene from *D. simulans* is transferred into *D. melanogaster* males, the *D. melanogaster* males express a 35-second song-burst pattern instead of a 60-second pattern.

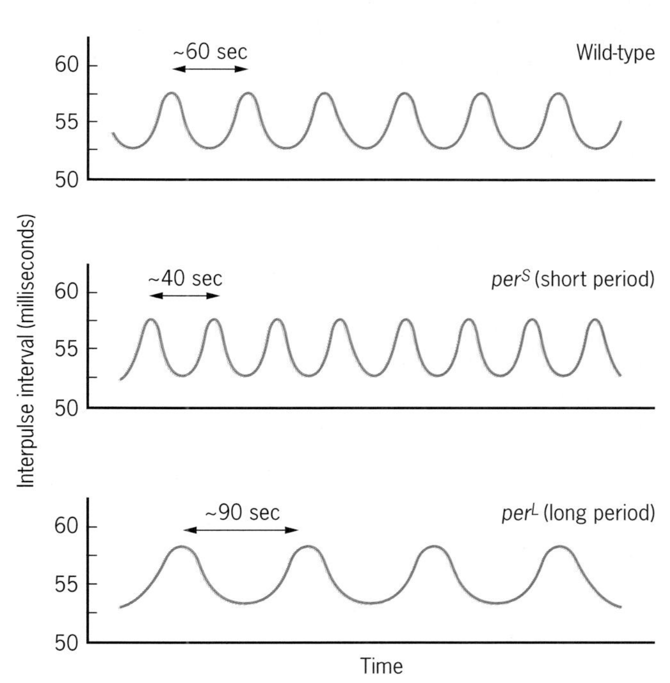

Figure 26.8 Different alleles of *per* affect the *Drosophila* courtship song. (*a*) The wild-type *Drosophila* courtship song has a 60-second period (see Figure 26.7). (*b*) The per^S mutant has a 40-second period. (*c*) The per^L mutant has a 90-second period.

How Does *Per* Control Biological Rhythms?

The *per* gene is involved in timekeeping functions, but how is this accomplished and in what cells is the *per* gene product expressed? Geneticists have isolated the *per* gene and identified its protein product (PER). Using fluorescently tagged antibodies to PER to identify those cells that are synthesizing it, they found that PER is expressed throughout the brain, in glial cells (the supportive structure of nervous tissue), and in some other nerve cells (Figure 26.9). PER is also expressed in tissues other than nerve tissue.

Investigators are trying to identify the specific cells responsible for controlling circadian rhythms. They are focusing on a small set of lateral neurons found in the brain that show circadian fluctuations in the PER protein. That PER is concentrated in cell nuclei suggests that the protein, either by itself or in combination with other proteins, somehow regulates gene activity at the level of transcription. Researchers have also found that the PER protein and *per* mRNA have their own rhythms: they are synthesized during the day. The per^0 mutant, which is arrythmic, shows no mRNA rhythms.

One way that *per* may be able to control circadian rhythms is through a negative feedback loop. Analysis of PER shows that it does not have any DNA binding domains. However, it does have a domain that facili-

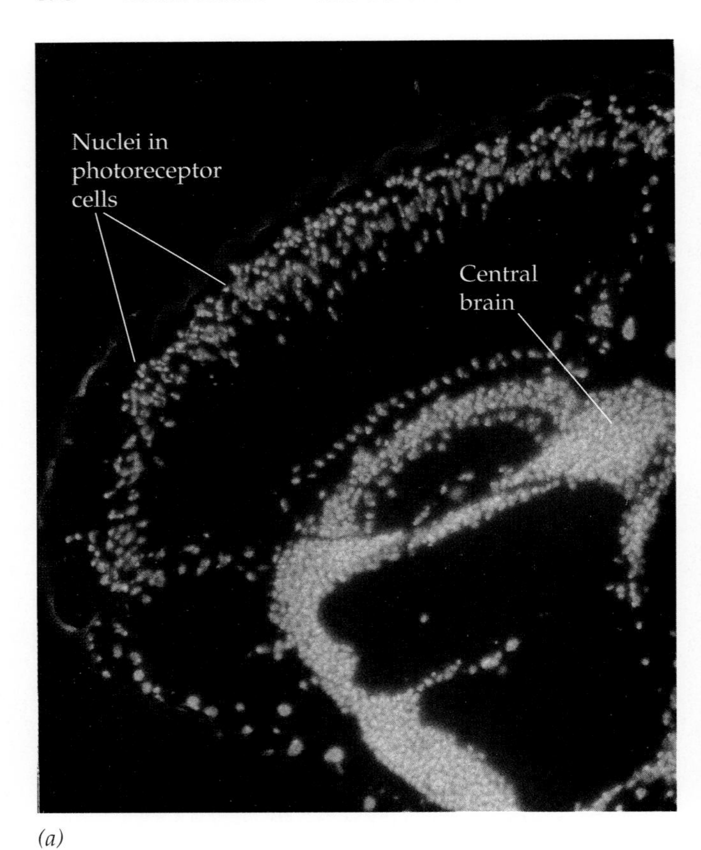

(a)

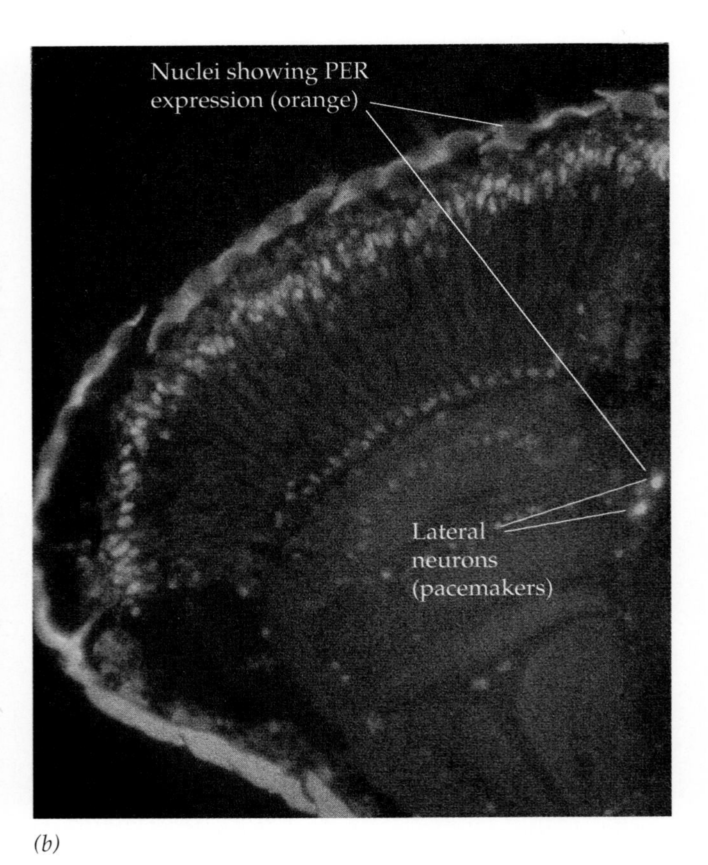

(b)

Figure 26.9 PER expression in *Drosophila* nerve cells. The photos show the same cross section through the head of a fly with different stains detected using different filters. In (*a*), the DNA in the nuclei has been stained fluorescent blue. In (*b*), monoclonal antibodies to PER were complexed with Cy3 fluorochrome. PER expression is in orange. PER is expressed ubiquitously, but its expression in the lateral neurons conveys rhythmicity in adults.

tates protein–protein interactions. This suggests that PER may interact or complex with itself, with another protein or with various proteins to repress the transcription of *per* and cause *per* mRNA levels to decline. *Per* mRNA synthesis decreases at night, which means that less PER protein is made. As the PER levels decline, the gene becomes derepressed and mRNA transcription increases toward dawn. This cycle of repression and derepression runs on a 24-hour cycle.

Recent discoveries have provided important insights into how PER works. PER interacts with a second clock protein called timeless (TIM). Both the *per* and the *tim* genes are turned on in the morning, and mRNAs from these genes accumulate during the day (Figure 26.10). As the day progresses, the PER and TIM levels increase. By dusk, PER and TIM have such high levels that the two proteins complex with each other and move to the nucleus where they stop the transcription of their own genes. The *per* and *tim* mRNA levels decline during the night, followed by the decline of the PER and TIM protein levels. By dawn, the PER and TIM levels are so low that they no longer form a complex to repress the *per* and *tim* genes. Transcription of these genes begins again, and PER and TIM levels rise.

An important question about this cycling pattern of gene activity is the role that light plays. It has now been discovered that the TIM protein is inactivated by light. In the presence of light (e.g., when dawn breaks), TIM breaks down and so does the TIM–PER complex. The *per* and *tim* genes are now active. As light levels decrease (dusk to night), TIM levels rise, allowing the shutdown of the *per* and *tim* genes and the subsequent drop in PER and TIM proteins. Experimentally administered flashes of light during the night causes the destruction of TIM and a resetting of the clock. Researchers do not yet know how the light destroys TIM.

Despite great success in analyzing *per*, much work remains to be done. For example, how does light destroy TIM? What is the relationship between the PER and TIM levels to the periodicity of the male courtship song? Rhythms in humans are regulated by cells in the hypothalamus, specifically in the superchiasmatic nucleus. Are there functional *per* and *tim* homologs in humans, perhaps being expressed exclusively in this

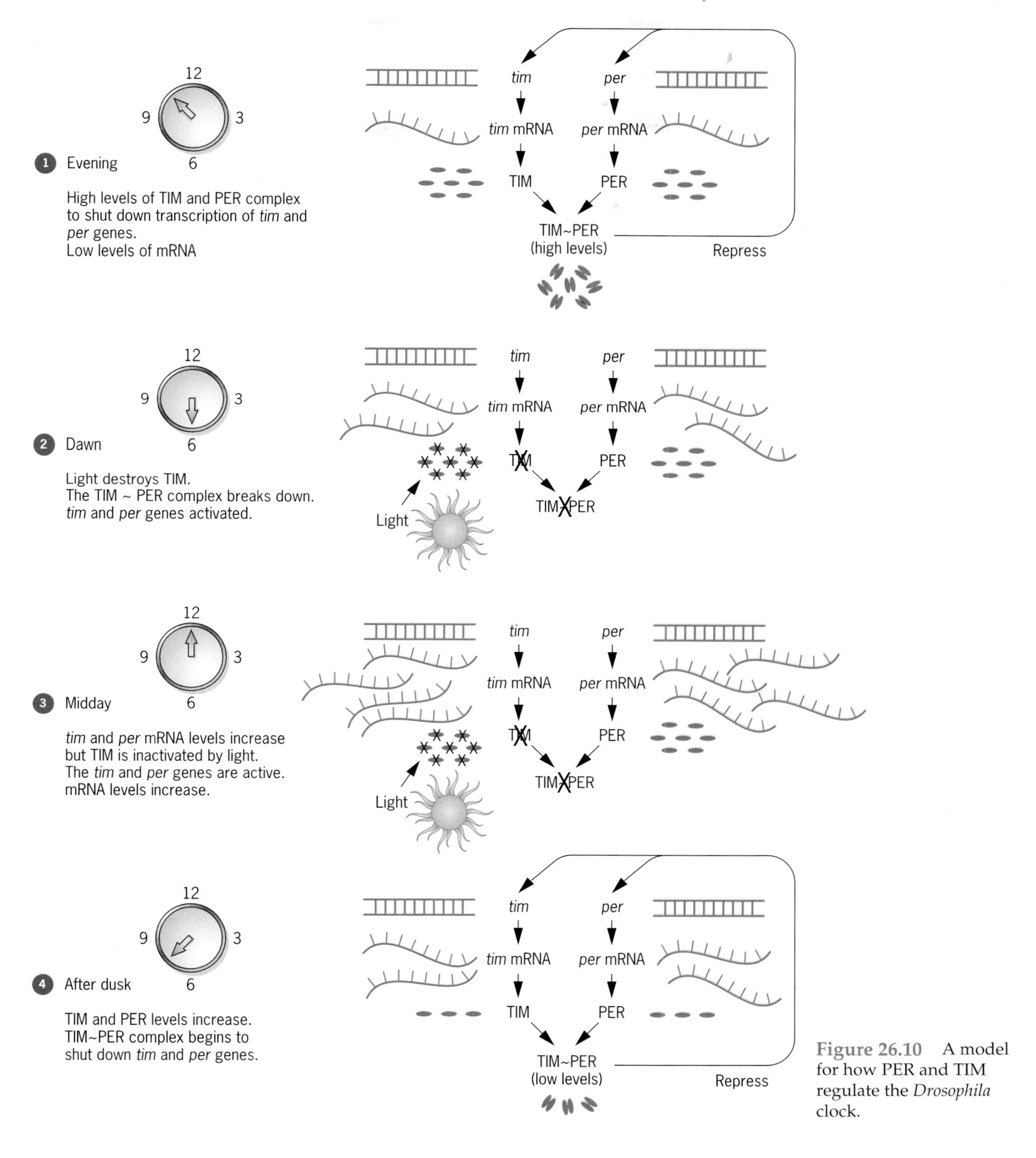

Figure 26.10 A model for how PER and TIM regulate the *Drosophila* clock.

area of the brain? Researchers continue to explore these and other questions.

Key Points: **Nest-cleaning behavior in bees is controlled by two independently assorting genes. Circadian rhythms are behaviors or functions that are synchronized with the 24-hour day/night cycle. *Drosophila's* circadian activity pattern is controlled by two genes, *per* and *tim*; *per* alleles lengthen or shorten the cycle or cause it to be arrhythmic. They**

also affect other *Drosophila* behaviors. PER and TIM reach their highest levels during the day and form a complex that represses transcription of the *per* and *tim* genes at night. TIM is destroyed by light.

THE GENETIC ANALYSIS OF BEHAVIOR IN DOGS AND MICE

Mammalian behavior, though often complex, can be successfully analyzed through the use of inbred lines and through controlled matings. In humans, distinct behaviors can be analyzed in families by pedigree analysis and by twin studies. In this section we analyze specific behavior patterns in dogs and mice, and in the next section we examine the genetic basis of human behavior.

The Genetic Basis of Dog Behavior

The distinctive and complex behavior of different dog breeds is strongly influenced by the genotype. This was shown by J. P. Scott and J. L. Fuller who studied behavior in five pure dog breeds: American cocker spaniel, fox terrier, African basenji, Shetland sheep dog, and beagle. The cocker spaniel (Figure 26.11*a*), which descended from bird dogs, is now primarily a house pet. They are very people-oriented and nonaggressive. Spaniels originated in Spain where they were used in the sport of hawking and in net-hunting of birds. The dog was trained to crouch close to the ground near birds that it detected. A net was then thrown over both the dog and the birds it had discovered. With the advent of shotguns, the dog was trained simply to stop and point to the birds. Springer spaniels actually jump into bushes (hence the name "springer") and flush the birds, whereas the tendency to crouch near the birds is found to exist only in the cocker spaniel.

Terriers are considerably more agressive and scrappy. They were bred to creep into burrows and drive out small animals. Shetland sheep dogs have been bred to perform complex functions under the direction of their controller. They are not hunters.

The African basenji (Figure 26.11*b*) is used by Pygmies and other African tribes for general hunting. Basenjis are wary animals that rarely bark (although they do howl on occasion, usually at night). The basenji puppy is wild compared to the friendly and tame cocker spaniel. They do not like to be handled and will fight any kind of restraint. However, they tame rapidly with handling and human contact.

Scott and Fuller (1974) used controlled matings between basenjis and cocker spaniels to study the genetic basis of their behavior. When a basenji is crossed to a cocker spaniel, the behavior of the hybrid puppies is much like that of the basenji parent, suggesting that the "wild" behavior pattern of the basenji is dominant. ("Wild" behavior includes avoidance and vocalization in reaction to handling, and struggle against restraint.) Backcrossing the F_1 hybrid to the parental types indicates that a single major gene and several modifier genes may determine the "wild" behavior. Playful aggressiveness at 13 to 15 weeks of age, which is high in basenjis and low in cockers, appears to be the result of two genes with incomplete dominance. Barking characteristics and sexual behavior also appear to be largely under the control of one or two genes.

The Scott and Fuller study of dog behavior is important because it shows that many aspects of complex mammalian behaviors are strongly influenced by the genotype. The study also shows that specific behaviors can be selected for successfully.

Single Genes That Affect Circadian Rhythms and Nurturing in Mice

Circadian behavior in mice is precise and easily quantitated, just as it is in *Drosophila*. When tested in constant darkness, a normal mouse exhibits a robust circadian rhythm of wheel-running activity. The period is a bit over 23 hours. A recently isolated mouse mu-

(*a*)

(*b*)

Figure 26.11 Photos of cocker spaniel and Basenji dogs.

(*a*)

(*b*)

Figure 26.12 (*a*) Typical nursing posture of wild type female shortly after giving birth. (*b*) Typical behavior of *fosB* mutant female shortly after giving birth. Her pups were scattered and neglected.

tant called *Clock,* when homozygous, has a periodicity of 27 to 28 hours. Heterozygotes for the *Clock* mutation have a periodicity of about 25 hours. After about two weeks in constant darkness, homozygous *Clock* mutants lose circadian rhythmicity. This observation tells us two important things about the *Clock* gene: (1) It regulates the intrinsic circadian rhythm; and (2) it regulates the persistence of circadian rhythmicity. *Clock* is an essential gene for this behavior. It has no other morphological or developmental effects, so in this sense it is a pure behavioral mutation. Using inbred mouse strains and molecular techniques, the *Clock* gene has been mapped to mouse chromosome 5 in a region of the chromosome that bears molecular similarity to human chromosome 4. It remains to be seen whether there is any molecular similarity to the *Drosophila per* gene. The human homolog of *Clock* has not been identified.

Recently, Michael Greenberg and his colleagues identified a gene that has a profound effect on nurturing behavior in mice. *fosB* mutant mice do not nurture their pups (Figure 26.12) but are normal with respect to other cognitive and sensory functions. The *fosB* gene encodes a transcription factor. The absence of this transcription factor in the preoptic region of the hypothalamus results in the loss of normal nurturing behavior. The FosB protein controls a complex behavior by regulating a specific neuronal circuit.

Key Points: **In carefully conducted breeding experiments using different dog breeds, it has been shown that many aspects of dog behavior are genetically controlled. The *Clock* gene in mice controls circadian rhythms and *fosB* controls nurturing behavior.**

ANEUPLOIDY AND HUMAN BEHAVIOR

The genetic analysis of behavior is most complicated in humans. The behaviors are complex, and the analysis of those behaviors is difficult. Nevertheless, important insights into the genetic basis of human behavior have emerged from the effects of aneuploidic conditions on human behavior, from single gene defects that affect human behavior, and from family studies of complex, multifactorial behavorial traits.

Mistakes occurring during meiosis (Chapter 6) may result in numerical changes in chromosome number that usually have a profound effect on the phenotype, including behavior. For example, one of the most common causes of mental retardation in human populations is trisomy 21 or Down syndrome (Figure 6.16). Individuals with this disorder have three copies of chromosome 21 and are usually diagnosed at birth or shortly thereafter. Down syndrome children typically have IQs that range from 25 to 50. If they survive beyond the age of 30, and many do because of improved medical care, they invariably develop Alzheimer's disease. Down syndrome results from having three copies of chromosome 21 genes rather than having a specific mutant allele. It is a gene dosage syndrome (see Technical Sidelight: A Mouse Model for Down Syndrome).

The presence of extra chromosomes can also have a more subtle effect on a phenotype. For example, males with Klinefelter syndrome have two X chromosomes and a Y. The XXY males are not usually perceived to be phenotypically distinct until after puberty. At that time, the male secondary sexual characteristics develop poorly. The XXY male may have behavioral problems in school, tends to have a lower than normal verbal IQ, is usually less aggressive and less active than his XY peers, and is more prone to social stresses.

The XYY male has been the subject of intense study and debate. How different is the XYY male phenotype from the XY phenotype? About 3 percent of the males in prisons and mental institutions are XYY. In the male population over six feet tall in these institutions, the XYY percent jumps to over 20 percent. Among all male live births, the XYY frequency is

TECHNICAL SIDELIGHT

A Mouse Model for Down Syndrome

Down syndrome (trisomy 21) is a leading cause of mental retardation. It places a tremendous burden on individuals with the disorder, on their families, and on society. A major objective of Down syndrome research is to correlate dosage imbalance of specific chromosome 21 genes with different aspects of the Down syndrome phenotype. One way this is accomplished is by studying individuals with segmental trisomy 21 arising from translocations or duplications resulting in partial duplication of a subset of chromosome 21 genes. There are two problems with this approach: first, the sample size of well-characterized cases of duplication or translocation Down syndrome is small; second, the phenotypes of these partial trisomic individuals are variable. In the absence of human subjects, researchers are turning to animal models of Down syndrome.

Human chromosome 21 shares a large region of genetic homology with mouse chromosome 16, which led to the creation of trisomic 16 mice as models for Down syndrome. This mouse model has problems, however. First, trisomy 16 mice die *in utero*, so they cannot be studied postnatally. Second, some human chromosome 21 gene homologs are found on mouse chromosomes 17 and 10. Third, some mouse chromosome 16 genes have human homologs on chromosomes other than 21. In spite of these problems, researchers have constructed a mouse strain that is informative about the consequences of gene dosage effects for some human chromosome 21 genes. Roger Reeves and his colleagues created a reciprocal translocation between mouse chromosomes 16 and 17.[1] One of the translocation products contains most of chromosome 16 and nearly all of 17 (16^{17}). The reciprocal product (17^{16}) contains the distal segment of chromosome 16 translocated very close to the centromere of 17 (Figure 1). Females carrying this reciprocal translocation, when mated to normal males, produced some offspring carrying the 17^{16} translocated chromosome. These offspring had segmental trisomy for chromosome 16. This segment of chromosome 16 carries genes homologous to those found in region q21-22.1 of human chromosome 21 (Figure 2).

Reeves and his colleagues compared the phenotypes of the partial trisomy 16 mice with comparable partial and complete trisomy 21 humans, and mouse trisomy 16. Mice with full trisomy 16 and some trisomy 21 humans have characteristic heart defects that are not observed in the partial trisomy 16 mice. This finding suggests that the specific cardiac defects observed in trisomy 16 mice and human trisomy 21 are due to genes outside the region of human chromosome 21 homology. Humans with trisomy 21 invariably develop Alzheimer's disease, but mice with partial trisomy 16 do not develop any Alzheimer's pathology. This finding suggests that the development of early-onset Alzheimer's disease in patients with Down syndrome may be due in part to chromosome 21 genes not carried on the chromosome 16 segment. The brains of the partial trisomic mice displayed mini-

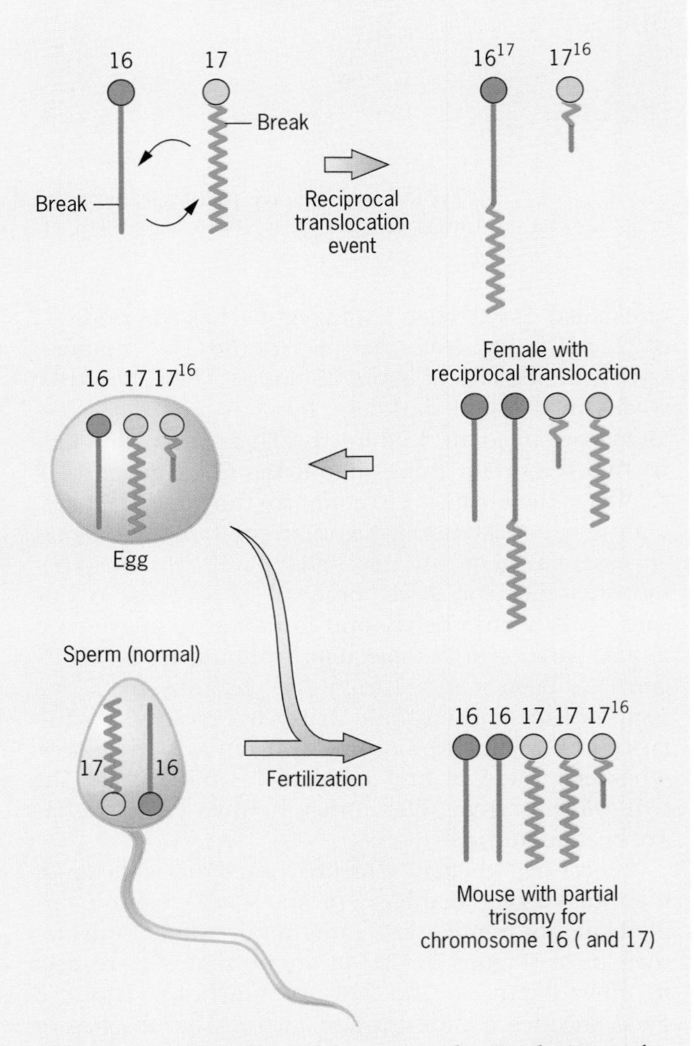

Figure 1 The creation of a mouse with partial trisomy for chromosome 16.

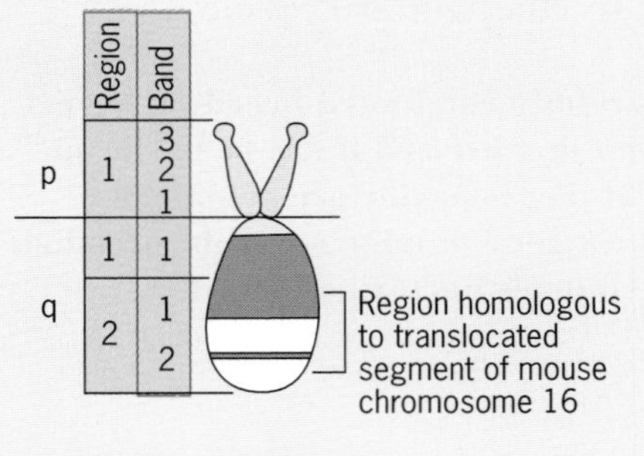

Figure 2 Human chromosome 21 showing the region of genetic homology to the partial trisomy 16 mouse.

mal gross pathological changes, when compared to the brains of Down syndrome patients. However, the partial trisomic mice exhibited serious learning and behavioral disabilities. Therefore, human homologs of chromosome 16 genes carried by the partial trisomic mice may contribute to cognitive deficits when a dosage imbalance exists. Of course, these mice also carry an imbalance of chromosome 17 genes, which may contribute to the observed phenotype. However, the segment of chromosome 17 that is trisomic is very small, and careful studies of chromosome 17 markers indicate that the contribution of chromosome 17 to the Down syndrome phenotype is minimal.

The mouse with partial trisomy for chromosome 16 is an important model for studying the development of Down syndrome. The genes carried by this segment of chromosome 16, as well as the genes it does not carry, provide insight into the relationship between the phenotypic characteristics associated with Down syndrome and specific regions of chromosome 21.

[1]Reeves, Roger H., et al. 1995. A mouse model for Down syndrome exhibits learning and behaviour deficits. *Nature Genetics* 11:177–184.

about 1 in a 1000. Does the high frequency of XYY males in prisons and mental institutions imply that an extra Y chromosome prompts aggressive behavior? The linkage between the extra Y and criminal behavior is not well established. However, the XYY male tends to have some educational difficulties, has a more volatile temper, tends to be more moody, and seems to be more hyperactive than his XY peers.

The XXX female does have a more distinct behavioral phenotype than her XX counterparts. XXX females are phenotypically normal, though they tend to be taller than their XX counterparts. They have problems in speech development, their verbal IQ scores are usually lower, and they have poor interpersonal skill development. The XO female (Turner syndrome; Figure 6.18) has mild central nervous system defects that include hearing impairment (50 percent of the cases). Although their intelligence is average or above average, they often have problems with spatial perception and fine motor skills. As a consequence, their nonverbal IQ scores tend to be lower than their verbal IQ scores. It is clear that alteration in the number of chromosomes not only affects the physical phenotype, but also can have important effects on behavior.

Key Points: **Various aneuploidic conditions have effects on behavior. Trisomy 21 individuals are mentally retarded; XXY and XXX individuals are abnormal in certain aspects of their mental functions; XYY males may express some aggressive tendencies; and XO females tend to have hearing problems and problems with spatial perception.**

SINGLE-GENE DEFECTS THAT AFFECT HUMAN BEHAVIOR

Many single-gene defects have a dramatic effect on human behavior. Some of these genes code for enzymes and are inherited as recessive traits; others are dominant and control the synthesis of membrane-bound receptor proteins. Some genes control neurotransmitters or encode signal transducers that convey signals from outside the cell to its interior (see Human Genetics Sidelight: A Genetic Basis for Aggression). Often, even though a trait follows a simple Mendelian inheritance pattern and a gene product has actually been identified, the function of the gene product is unknown.

Phenylketonuria, Formerly a Major Cause of Mental Retardation

Phenylketonuria (PKU), one of the more common metabolic disorders, affects about 1 in every 10,000 Caucasians and about 1 in every 16,500 Asians. Classic PKU is an autosomal recessive trait, the result of a defective liver enzyme, phenylalanine hydroxylase (PAH). This enzyme converts the amino acid phenylalanine into tyrosine (Figure 26.13). If the enzyme is defective, the conversion does not occur. Phenylalanine accumulates in the plasma; abnormal metabolites produced by alternative pathways for phenylalanine metabolism accumulate and are excreted in the urine, and insufficient tyrosine is produced. The accumulation in the plasma of high phenylalanine levels as well

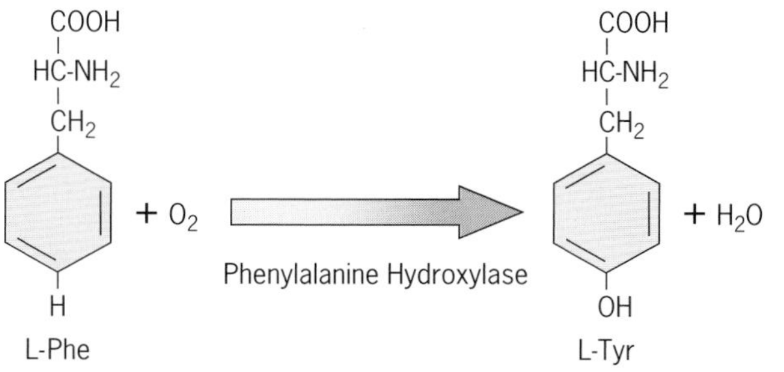

Figure 26.13 Phenylalanine is hydroxylated to tyrosine by phenylalanine hydroxylase. Classic phenylketonuria (PKU) is the result of a deficiency in this enzyme.

HUMAN GENETICS SIDELIGHT

A Genetic Basis for Aggression

A large Dutch family included several members with serious behavioral problems. Many of the males in this family were borderline mentally retarded. These same males were also behaviorally abnormal: they exhibited impulsive aggression, they were exhibitionists, they attempted rape, and they were arsonists. The pedigree of this family suggested that these behavioral problems were inherited in an X-linked recessive manner (Figure 1).

H. G. Brunner and his colleagues noted that aggressive behavior in animals and humans may be associated with the altered metabolism of serotonin, and to a lesser extent with that of dopamine and noradrenaline.[1] They thus looked for a link between aggressive behavior in the males of this family and the metabolism of these neurotransmitters. Theirs is the first report linking a specific mutation in humans to a strictly behavioral phenotype.

Brunner and his colleagues analyzed urine samples from the aggressive males and found markedly disturbed catecholamine metabolism. The catecholamines (dopamine, norepinephrine, and epinephrine) are neurotransmitters in a number of areas of the brain that relate to motor control, cognition, emotion, positive reinforcement, and visceral regulation functions. The principle degradative enzyme for the catecholamines is monoamine oxidase (MAO), which is mitochondrially located but encoded by nuclear genes. The two MAO genes, *MAO-A* and *MAO-B*, are closely linked on the X chromosomes and encode proteins that show over 70 percent amino acid sequence similarity. The males who expressed the abnormal behavior in this family were deficient for MAO-A. They all carried the same mutation, a point mutation in the eighth exon changing a glutamine codon to a stop codon. This results in a truncated, nonfunctional protein.

O. Cases and his colleagues recently created a mouse strain that was deficient for MAO-A and found that the MAO-A deficient males showed increased aggressiveness.[2] The MAO-A deficient males had greatly elevated levels of the catecholamines. It remains to be seen whether similar mutations are found in other pedigrees with similar behavior patterns.

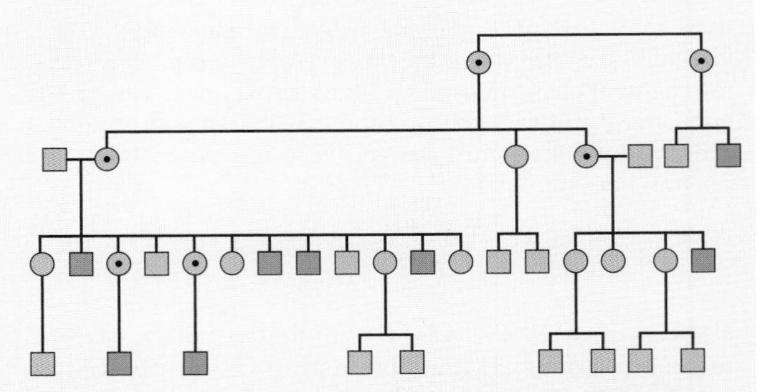

Figure 1 A pedigree of a family with X-linked borderline mental retardation and prominent behavioral disturbance.

It is not clear that all of the biochemical alterations caused by MAO-A deficiency are necessary for development of the abnormal behavior. MAO inhibition in adults does not lead to aggressive behavior. However, it may be that only certain stages of life are sensitive to MAO-A deficiency.

[1]Brunner, H. G., et al. 1993. Abnormal behavior associated with a point mutation in the structural gene for monoamine oxidase A. *Science* 262:578–580.

[2]Cases, O., et al. 1995. Aggressive behavior and altered amounts of brain serotonin and norepinephrine in mice lacking MAO-A. *Science* 268:1763–1766.

as other phenylalanine-derived metabolites causes abnormal myelination of nerve cells and the failure of normal brain development. The consequence of these metabolic problems is mental retardation. In addition, a lower than normal tyrosine level impairs the synthesis of melanin, which is derived from tyrosine; this could account for the fact that patients with PKU have less pigmentation than expected.

The range of PKU phenotypes in untreated individuals is broad—from mildly elevated levels of phenylalanine and nearly normal mental functions to very high levels of plasma phenylalanine and drastically reduced IQ scores. This phenotypic heterogeneity may be a consequence of different *PAH* alleles. Alleles coding for a totally defective form of the enzyme produce the most severe form of PKU, and alleles that code for enzyme forms that retain partial function produce less severe cases of PKU.

A defective *PAH* gene causes mental retardation indirectly because PAH-deficient liver cells malfunction, thus causing a buildup in the plasma of a metabolite that results in neuron dysfunction. If this metabolic defect is not treated by dietary restriction of phenylalanine intake, mental retardation develops.

However, PKU, once a devastating disorder, is now successfully treated by restricting the dietary intake of phenylalanine. By keeping phenylalanine intake to an appropriate level, the mentally impairing toxic buildup of phenylalanine and its metabolites is avoided.

Lesch-Nyhan Syndrome and Self-Mutilation

Like PKU, Lesch-Nyhan syndrome (LNS) is also an inborn error of metabolism due to a defective enzyme. LNS is an X-linked trait caused by a defect in hypoxanthine-guanine phosphoribosyltransferase (HPRT). Because LNS is an X-linked recessive trait, it appears almost exclusively in males. LNS is a devastating disorder characterized by mental retardation and ceaseless and involuntary slow, sinuous, writhing movements, especially in the hands. The most intriguing and bizarre symptom of LNS is a compulsive self-mutilating behavior (Figure 26.14). Patients usually begin by biting their lips and the inside of their mouth. Then they move to the fingers and hands. To protect the patient from this self-destructive behavior, restraints are often used, and the patient's teeth are removed.

HPRT is an enzyme involved in purine metabolism. The enzyme is normally expressed at higher levels in the brain, suggesting that the neurological abnormalities result from changes in brain purine levels.

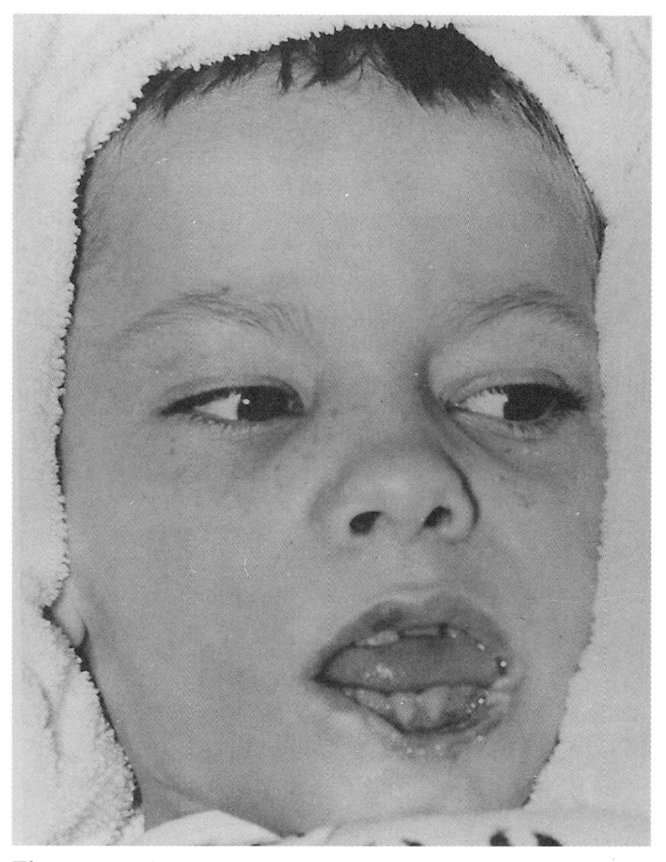

Figure 26.14 The consequences of the self-mutilation behavior in Lesch-Nyhan syndrome.

The Tragedy of Huntington's Disease

Huntington's disease (HD) is a fatal neurodegenerative disorder inherited in an autosomal dominant fashion. The consequence of inheriting the mutant *HD* allele is the gradual loss of neurons, mainly in the caudate nucleus and putamen located in the central region of the brain (Figure 26.15). These areas of the brain are critically important in the control of coordinated movements. Other nerve cells in the brain are also affected by the *HD* allele, but it is unclear whether these are primary or secondary effects. There is a steady loss of neurons as the disease progresses.

Not all nerve cells in the caudate nucleus are equally affected by the *HD* allele. Some neurons are severly affected early on, whereas others appear to be unaffected. It is possible that the type of neurotransmitter (a molecule that transmits signals from one neuron to another) used by a neuron may determine how sensitive it is to the effects of the *HD* allele. Those neurons that are relatively unaffected tend to use neurotransmitters that are different from those used by the more severely affected neurons. Thus neurotransmitter receptors may hold a key to understanding the mechanism of HD.

The *HD* gene was isolated in 1993, and the protein it encodes was identified not long after (see Chapter 20). At this writing, the function of the protein, huntingtin, remains largely unknown. The trait is a true dominant; that is, one copy of the *HD* allele produces the same phenotype as two copies. This finding suggests some type of pathologic gain-of-function: the defective protein has acquired a new or an enhanced function, with devastating consequences. *HD* mRNA is found in all tissues of the brain and in other tissues as well. That some areas of the brain express higher levels of *HD* mRNA than others may be a clue to how

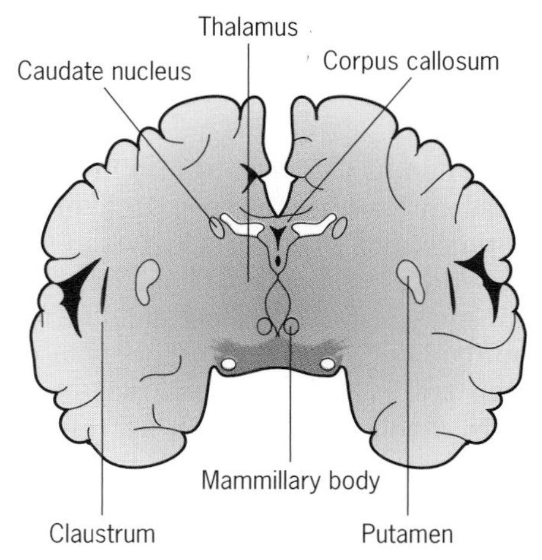

Figure 26.15 Drawing of a frontal section of the brain showing the caudate nucleus and putamen, areas of the brain that are severely affected by Huntington's disease.

huntingtin acts. Another intriguing observation is the presence of two different sizes of *HD* mRNA; this finding suggests that *HD* RNA may be processed differently in different cell types, producing proteins with distinct functions. Recent research has found that huntingtin binds tightly to certain cellular proteins involved in cell death. Mutant forms of huntingtin are more readily cleaved by these proteins, suggesting that HD may be a disorder of inappropriate brain cell death.

Huntingtin plays an essential role in normal neuronal function. HD is a direct and primary consequence of the expression of a mutant *HD* allele in a subset of neurons. PKU and Lesch-Nyhan syndrome differ from HD in that enzymatic defects alter metabolite levels that secondarily alter neuronal functions.

Many other disorders of the nervous system can be traced to defects in single genes. In each instance, there exists a gene product that has either a direct (HD) or an indirect (PKU and LNS) effect on the normal functioning of the nervous system.

Key Points: **PKU is a metabolic disorder caused by an enzyme deficiency that results in the accumulation of phenylalanine and phenylalanine-derived metabolites. Elevated levels of these substances cause central nervous system dysfunction. Lesch-Nyhan syndrome is caused by an enzymatic defect that disrupts purine metabolism and leads to mental retardation and self-mutilation. Huntington's disease is a fatal neurodegenerative disease.**

THE ANALYSIS OF PERSONALITY

Most traits, whether physical or behavioral, do not exhibit simple Mendelian inheritance patterns. They are multifactorial, influenced by genetic factors as well as nongenetic factors. In this section and those that follow, we examine some human traits that are multifactorial.

In 1979, Thomas J. Bouchard and his colleagues at the University of Minnesota (see Conversation with Bouchard preceding Chapter 25) began a study of twins who were reared apart. Their objective was to describe the genetic and environmental factors that influence human behavior. From the results of their study, published in 1990, two remarkable conclusions emerged: (1) genetic factors exert a pronounced and pervasive influence on behavior; (2) the effect of being raised in the same home environment is negligible for most psychological traits. Their study suggests that individual differences in ability, personality, interests, and even social attitudes are strongly influenced by the genotype.

Bouchard and his colleagues examined over 100 sets of twins from around the world who had been reared apart , including the pair described in the opening of this chapter. All the twins had been separated very early in life, reared apart during their formative years, and then reunited as adults. Extensive testing was done to determine whether the twins were monozygotic or dizygotic. Each person completed over 50 hours of carefully controlled medical and psychological assessment.

The Minnesota twin study concluded that the broad-sense heritability (H^2) of intelligence (as defined by performance on IQ tests) was about 0.70. In other words, about 70 percent of the variance in IQ in this population was found to be associated with genetic variation. This was not surprising because earlier studies arrived at similar heritability estimates. Later in this chapter, we discuss heritability and IQ in more detail. However, the Bouchard group also made some rather surprising discoveries. On various measures of personality, temperament, occupational and leisure-time interests, and social attitudes, monozygotic twins reared apart are about as similar as monozygotic twins raised together (Table 26.1). The effect of being raised together is negligible for many psychological traits. Other studies support these findings. A reasonable conclusion that can be drawn from these studies is that genetic factors influence personality and behavior.

The Minnesota twin study has several important implications. The first is that general intelligence, as measured by various IQ tests, is strongly affected by genetic factors. The study in no way suggests that innate general intelligence cannot be enhanced by enriched environments or blunted by deprived conditions; indeed, it concludes just the opposite. The Minnesota study states simply that in the current environments of industrialized societies, about 70 percent of the observed variance in IQ can be attributed to genetic variation.

A second implication is that the institutions and practices of modern Western society do not greatly constrain the development of individual differences in psychological traits. Culture is an important issue in any discussion of heritability of psychological traits. In a culturally uniform environment (V_e is small), heritability will tend to be large because most of the observed differences will be the result of genetic differences. As V_e increases, heritability decreases (see Chapter 25 for a review of heritability). People in Western societies vary tremendously in personality traits, interests, and attitudes, yet heritabilities for these traits are high. This suggests that the cultural environment is not as determinative in molding psychological traits as psychologists and psychiatrists once assumed.

The home environment, of course, has a profound impact on psychological development. If parents are

TABLE 26.1
Minnesota Twin Study Data for Several Variables in MZ Twins Reared Apart (MZA) or Together (MZT)

Variables	*MZAs*		*MZTs*	
	Correlation coefficients	*Pairs (no.)*	*Correlation coefficients*	*Pairs (no.)*
Anthropometric variables				
Fingerprint ridge count	0.97	54	0.96	274
Height	0.86	56	0.93	274
Weight	0.73	56	0.83	274
Electroencephalographic (brainwave) variables				
Amount of 8- to 12-Hz (alpha) activity	0.80	35	0.81	42
Midfrequency of alpha activity	0.80	35	0.82	42
Psychophysiologic variables				
Systolic blood pressure	0.64	56	0.70	34
Heart rate	0.49	49	0.54	160
Electrodermal response (EDR) amplitude				
Males	0.82	20	0.70	17
Females	0.30	23	0.54	19
Trials to habituation EDR	0.43	43	0.42	36
Information processing ability factors				
Speed of response	0.56	40	0.73	50
Acquisition speed	0.20	40	NA	NA
Speed of spatial processing	0.36	40	NA	NA
Mental ability—general factor				
WAIS IQ—full scale	0.69	48	0.88	40
WAIS IQ—verbal	0.64	48	0.88	40
WAIS IQ—performance	0.71	48	0.79	40
Raven, Mill-Hill composite	0.78	42	0.76	37
First principal component of special mental abilities	0.78	43	NA	NA
Special mental abilities				
Mean of 15 Hawaii-battery scales	0.45	45	NA	NA
Mean of 13 Comprehensive Ability Battery Scales	0.48	41	NA	NA
Personality variables				
Mean of 11 Multidimensional Personality Questionnaire (MPQ) scales	0.50	44	0.49	217
Mean of 18 California Psychological Inventory (CPI) scales	0.48	38	0.49	99
Psychological interests				
Mean of 23 Strong Campbell Interest Inventory scales (SCII)	0.39	52	0.48	116
Mean of 34 Jackson Vocational Interest Survey scales (JVIS)	0.43	45	NA	NA
Mean of 17 Minnesota Occupational Interest scales	0.40	40	0.49	376
Social attitudes				
Mean of 2 religiosity scales	0.49	31	0.51	458
Mean of 14 nonreligious social attitude items	0.34	42	0.28	421
MPQ traditionalism scale	0.53	44	0.50	217

Source: Bouchard, T. J. et al. 1990. *Science* 250: 223–228.

especially stimulating, supportive, and encouraging, or if they are not, the children will often reflect that rearing environment. The Minnesota twin study does not argue that the extraordinary similarity in social attitudes in monozygotic twins raised apart cannot be influenced by the parents, but it shows that this parental influence does not tend to occur in most families.

The third important implication of the Minnesota twin study is that monozygotic twins raised apart are similar psychologically because they seek out similar environments. Research has shown that MZ twins elicit, select, seek out, or create very similar effective environments. Thus the genotype–environment interaction guides the developing human through environmental options. The genotype influences personality by influencing the character, selection, and impact of experiences during development.

Despite the impressive methodology of Bouchard's twin study, it has not escaped criticism. The method of twin selection, for example, may be biased. The researchers relied heavily on media coverage to recruit twins, who then came to Minnesota for study amid wide publicity. This modus operandi may select only those twins who are interested in the publicity and fame, reducing V_e and artificially elevating heritability estimates. Also, it now appears that some of the twin pairs, such as Jack Yufe and Oskar Stohr, actually met before the reunion in Minnesota. Although the Minnesota researchers have acknowledged these facts and even factored them into their conclusions, they are quick to point out that while their work remains valid, it should not be overinterpreted.

Key Points: **The Minnesota study of MZ twins reared apart presented strong evidence that personality and behavioral traits are heavily influenced by the genotype.**

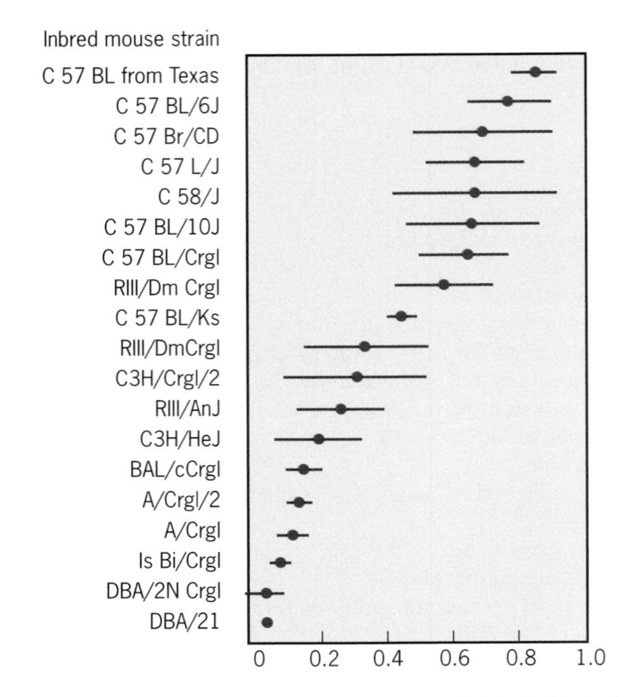

Figure 26.16 Differences in alcohol preferences in some inbred strains of mice. The C 57 BL from Texas had the highest alcohol preference at over 80 percent.

THE GENETIC BASIS OF ALCOHOLISM

Approximately 90 million people in the United States use alcohol. About 18 million of these are problem drinkers who have difficulties with their personal and social lives, and about 10 million of these 18 million have severe alcohol dependence with associated medical, psychological, and social problems. It is difficult to determine how many are true alcoholics as opposed to heavy drinkers or alcohol abusers because of variations in the clinical definition of alcoholism. One estimate places the number of alcoholics at 3 to 5 percent of all adult males and 0.1 to 1 percent of all adult females.

It has long been recognized that alcoholism tends to run in families, but this fact could mean that alcoholism is due to genetic factors, environmental factors, or both. What evidence is there that alcoholism has an underlying genetic basis? Some evidence comes from studies of mice. Various strains of mice differ in their preference for alcohol (Figure 26.16). When mice are given their free choice of water or 10 percent alcohol (about the same alcoholic content of wine), some strains will consume 80 percent or more of their daily fluid from the alcohol solution, and some strains avoid alcohol almost entirely.

Not only are there strain-specific differences in alcohol preference, but also there are strain-specific differences in the effects of alcohol. In general, strains with the highest alcohol preference are the least susceptible to alcohol's effects. For example, alcohol-induced sleep is shortest in strains with the highest alcohol preference. The strain differences apparently are due to both the brain's sensitivity to alcohol and the liver's ability to metabolize the alcohol via the enzyme alcohol dehydrogenase (ADH).

A growing body of research on humans suggests that the genotype plays a role in alcoholism. Studies have shown, for example, that:

1. There is nearly a 50 percent lifetime risk of alcoholism in sons and brothers of alcoholic fathers.
2. The concordance rate for alcoholism is 55 percent in monozygotic twins, but only 28 percent in dizygotic twins.
3. Adopted sons of alcoholic men show a rate of alcoholism more like that of their biological fathers than that of their adoptive fathers.

Alcoholics show metabolic differences that may be linked to the development of the disorder. For example, after a moderate dose of alcohol, male offspring of alcoholic parents have higher levels of acetaldehyde (a breakdown product of ethanol) than do control males. Other metabolic differences suggest that people who are at risk for becoming alcoholics metabolize alcohol differently from people who have no risk factors for becoming alcoholics.

TABLE 26.2
DRD2 A1 Allele Frequencies Reported for Alcoholics and Controls

	Alcoholics					Controls					
Investigator	*Total*	*A1A1*	*A1A2*	*A2A2*	*f(A1)*	*Total*	*A1A1*	*A1A2*	*A2A2*	*f(A1)*	*Control Type*
Blum, *et al.* 1990	35	2	22	11	0.37	35	2	5	28	0.13	Autopsy screened
Blum, *et al.* 1991	89	3	39	47	0.25	31	0	6	25	0.10	Screened
Total	**124**	**5**	**61**	**58**	**0.29**	**66**	**2**	**11**	**53**	**0.12**	

Source: JAMA (1993). 269:1674.

There are two distinct general categories of heritable alcoholism. In **type I alcoholism**, the disorder begins after the age of 25, occurs equally in both males and females, is associated with a rapid loss of control (dependency), and is characterized by alternating binges and abstinence. Type I alcoholics tend to be worriers, are deliberate, cautious, friendly, empathic, and compassionate, and exhibit social dependency. In **type II alcoholism**, the disorder generally has a teenage onset, occurs predominantly in males, and is associated with aggressive, antisocial behavior. Type II alcoholics are commonly unable to stop drinking and frequently engage in binge drinking. They tend to be impulsive, risk-taking, aloof, antisocial, and vengeful. They also tend to be adventurers and to express social hostility. The personality traits in both types of alcoholism have been shown to be heritable, but type II alcoholism seems to have a higher heritability than does type I.

Although it may be difficult to imagine the kind of genes involved in susceptibility to overindulgence in alcohol, some candidates have been proposed. In 1990, Kenneth Blum and his colleagues announced the discovery of a genetic marker that was associated with alcoholism: the *A1* allele of the D2 dopamine receptor gene (*DRD2*). They examined the brains of 35 deceased alcoholics and 35 controls and found that the *A1* allele was present in 69% of alcoholics and 20% of the controls. A year later they published more data supporting their conclusion (Table 26.2). This was an exciting discovery because dopamine is the chief neurotransmitter in the brain's "pleasure center," and the dopaminergic system is important in alcohol-related and in other substance abuse behaviors.

The association of the *DRD2 A1* allele with alcoholism is controversial. Although some studies support Blum's findings, others have been unable to replicate his results. Indeed, if the Blum data are disregarded, the sum of the other studies show no link between *A1* and alcoholism (Table 26.3). Part of the discrepancy may be due to differences in the definition and diagnosis of alcoholism. Another may be attributed to the wide variation in the *A1* frequency in different ethnic groups. For example, the frequency of the *A1* allele is about 0.50 in Japan, but the alcoholism rate is low in this population. Blum points out that the alcoholism rate is much lower than expected in the Japanese population because they have a high frequency of an allele of the aldehyde dehydrogenase gene that prevents them from metabolizing alcohol, making them much more sensitive to alcohol's effects. Thus the controversy is far from resolved.

Part of the problem with the *DRD2 A1* allele studies may be the heterogeneity of the study samples. The

TABLE 26.3
DRD2 A1 Allele Frequencies Reported for Alcoholics and Controls (Blum Data Excluded)

	Alcoholics					Controls					
Investigator	*Total*	*A1A1*	*A1A2*	*A2A2*	*f(A1)*	*Total*	*A1A1*	*A1A2*	*A2A2*	*f(A1)*	*Control Type*
Bolos, *et al.*	40	3	12	25	0.22	127	8	30	89	0.18	Unscreened
Comings, *et al.*	104	3	41	60	0.23	39	0	6	33	0.08	Screened
Cook, *et al.*	20	0	6	14	0.15	20	1	4	15	0.15	Screened
Gelernter, *et al.*	44	1	18	25	0.23	68	3	21	44	0.20	Unscreened
Goldman, *et al.*	46	0	14	32	0.15	36	2	11	23	0.21	Screened
Grandy, *et al.*	...	...	...	...	...	43	5	11	27	0.24	Unscreened
Parsian, *et al.*	32	0	13	19	0.20	25	0	3	22	0.06	Screened
Schwab, *et al.*	45	0	11	34	0.12	69	8	14	47	0.22	Unscreened
Turner, *et al.*	47	0	9	38	0.10	...	...	...	...	...	...
Total	**378**	**7**	**124**	**247**	**0.18**	**427**	**27**	**100**	**300**	**0.18**	

Source: Gelernter, J., D. Goldman, and N. Risch. 1993. JAMA 269:1673–1677.

samples were not usually classified as type I or type II alcoholics. If the alcoholics are classified by type, the association between the *DRD2 A1* allele and alcoholism may be clarified. The *DRD2 A1* allele story does indeed require clarification, and until the controversies are resolved, researchers are analyzing other candidate alcoholism genes, of which there are many, including other dopamine receptor genes.

A recent study by J. Tiihonen and his colleagues (1995) reported that dopamine transporter densities are drastically lower in type I (nonviolent) alcoholics than in controls and type II (violent) alcoholics. In fact, dopamine transporter densities in type II alcoholics are even higher than age- and sex-matched controls. This finding is significant because increased dopamine activity is associated with aggressive behavior, which is typical of type II alcoholics. Tiihonen's results showing drastically lower dopamine transporter densities for type I alcoholics are the opposite of those for type II alcoholics. The dopamine transporter gene (*DAT1*) may be the next logical place to look for a genetic basis of alcoholism.

There is little doubt that the genotype influences alcoholism, but alcohol dependence is a very complex phenotype and most likely is genetically complex as well. Even if a link is established between dopamine receptor and transporter alleles and alcoholism, it will be necessary to explain how these alleles cause alcoholism.

Key Points: **Alcoholism is strongly influenced by the genotype. One way the genotype may increase the risk for developing alcoholism is in the way it controls ethanol metabolism. Two genes involved with dopamine functions have been implicated in alcoholism. The association of the *A1* allele of the *DRD2* gene with alcoholism remains controversial. The association of the *DAT1* gene with alcoholism is a potentially important one.**

INSIGHT INTO ALZHEIMER'S DISEASE

As people age, there is a normal deterioration of mental functions. However, this gradual, normal pattern of deterioration pales in comparison to the serious memory deficits, gross lack of judgment, and deterioration in temperament and behavior that form the clinical basis of **dementia**. There are many causes of dementia. We have already discussed dementia associated with Huntington's disease. Another form of dementia is Alzheimer's disease (AD), which affects about 5 percent of individuals over 65 and nearly 25 percent of those over 80. The brain of an AD sufferer shows a marked loss of neurons and the accumulation of **senile plaques**, which are thickened nerve cell processes (axons and dendrites) surrounding an **amyloid β protein** deposit. Also visible in the AD brain are curious structures called **neurofibrillary tangles**, intraneuronal accumulations of filamentous material in the form of loops, coils, or tangled masses (Figure 26.17).

Although some forms of AD do not appear to be inherited, others do. One important piece of evidence implicating genes in AD comes from several observations that people with Down syndrome (trisomy 21)

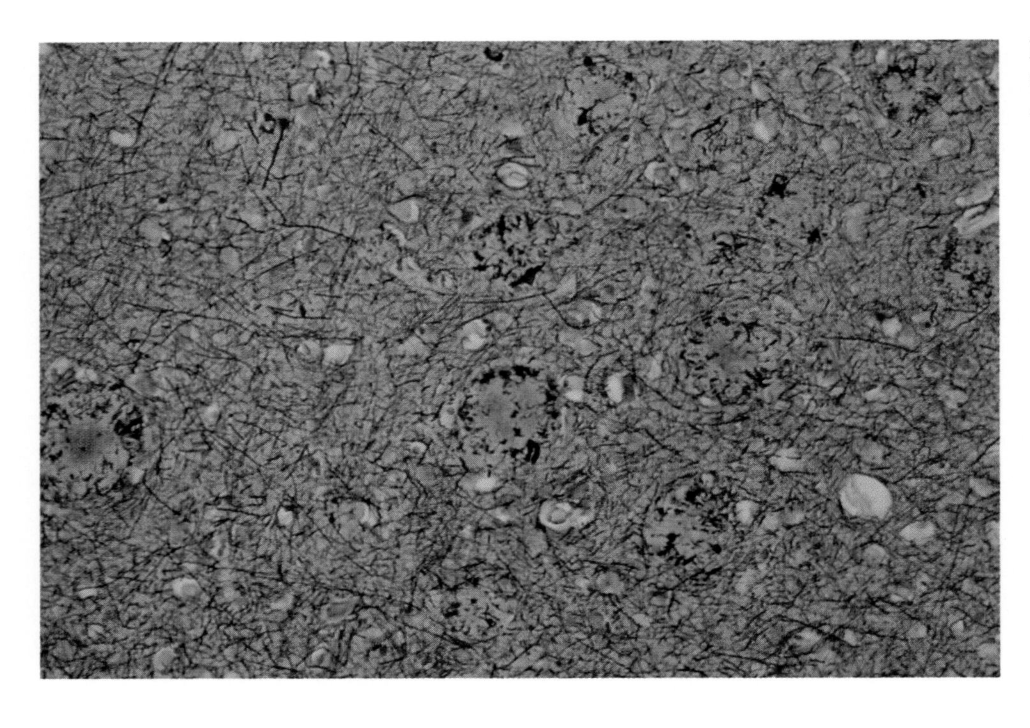

Figure 26.17 Neurofibrillary tangles and senile plaques in Alzheimer's brain tissue.

commonly develop early AD. This suggests that the extra dose of a gene or genes on chromosome 21 causes AD to develop.

A second observation further implicating a chromosome 21 gene comes from studies of the composition of the senile plaques seen in AD patients. These plaques are made up in part of a small protein called amyloid β protein, which contains about 40 amino acids. Amyloid β protein is derived by cleavage from a larger precursor protein called **amyloid precursor protein** (APP) that contains approximately 700 amino acids. The gene coding for APP has been mapped to chromosome 21. Researchers have shown that mutations in the *APP* gene lead to a dominant form of AD called **early-onset Familial Alzheimer's Disease** (FAD). This rare form of AD accounts for only about 3 percent of all familial cases of AD, but it is important because of the insight it gives us into the mechanism of Alzheimer's disease.

A major challenge in Alzheimer's research has been to understand exactly how the amyloid β protein contributes to the disease. Researchers are now beginning to understand the relationship between AD and amyloid β protein. APP is inserted into cell membranes, with about two-thirds of it protruding out of the cell into the extracellular matrix. In some cases, the APP is enzymatically cleaved into smaller fragments, and in other cases the APP is packaged into lysosomes (see Chapter 2)—small intracellular vesicles containing proteolytic enzymes—and degraded. The lysosomes release amyloid β proteins into the extracellular matrix (Figure 26.18). In AD patients, there is a six- to eightfold increase in amyloid β protein secretion compared to controls.

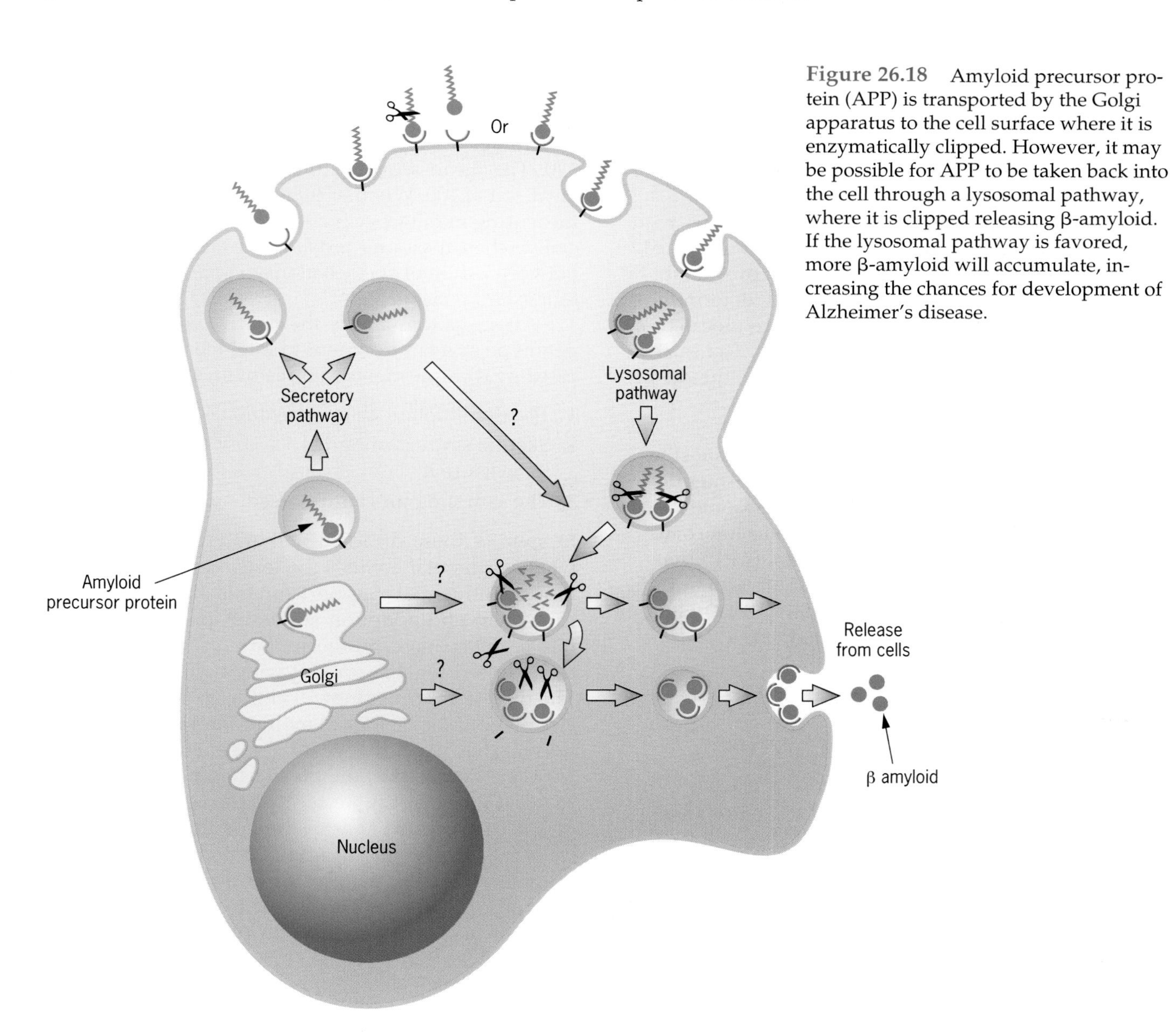

Figure 26.18 Amyloid precursor protein (APP) is transported by the Golgi apparatus to the cell surface where it is enzymatically clipped. However, it may be possible for APP to be taken back into the cell through a lysosomal pathway, where it is clipped releasing β-amyloid. If the lysosomal pathway is favored, more β-amyloid will accumulate, increasing the chances for development of Alzheimer's disease.

Researchers have identified a dominant gene family on chromosome 1 that may hold the key to understanding some forms of familial AD. This gene family codes for membrane-bound proteins, called the presenilins. Mutations in presenilin genes account for 70 to 80 percent of all familial cases of AD. Although the function of presenilin has not yet been precisely determined, its amino acid sequence suggests that it may be involved in protein transport. If correct, then the disease may be triggered by a change in the way APP travels through the cell. A change in the structure of APP or a change in the transport mechanism may lead to increased production of extracellular amyloid β protein and the development of AD.

At least four different genes are associated with Alzheimer's disease. In addition to the *APP* and presenilin genes, a third gene has been mapped to chromosome 19. This gene codes for a protein called ApoE that is involved in transporting cholesterol into the brain. An allele of this gene called *ApoE4* increases the risk for developing AD. A fourth gene for familial AD is known but has not yet been mapped. It has been excluded from chromosomes 1, 19, and 21. This autosomal dominant gene accounts for about 20 percent of familial AD.

Researchers are employing linkage analysis with known molecular markers to try to locate other AD genes in other families where AD is prevalent. It seems likely, however, that the genes determining most cases of familial AD have either been located or are undergoing continuing investigation. Genes that increase the risk of developing AD, such as the *ApoE4* gene, are also being investigated.

Key Points: **People with Down syndrome almost invariably develop Alzheimer's disease, suggesting that the overexpression of a gene(s) on chromosome 21 is one cause of AD. A candidate gene on chromosome 21, encoding amyloid precursor protein (APP), has been identified. Amyloid ß protein, which accumulates in senile plaques, is cleaved from APP. Other genes have been implicated in familial AD.**

INTELLIGENCE, A MULTIFACTORIAL TRAIT

Intelligence is a complex trait influenced both by the genotype and by environmental factors. It is a constellation of related mental abilities that include

1. Defining and understanding words.
2. Rapidly thinking of words.
3. Analyzing mathematical relationships.
4. Analyzing spatial relationships.
5. Memorizing and recalling information.
6. Perceiving similarities and differences among objects.
7. Formulating rules, principles, or concepts for solving problems or understanding situations.

An extreme view of intelligence is that it is determined almost entirely by the genotype. Such a deterministic view is not widely accepted. Another extreme view of intelligence is that the genotype plays no role. The Minnesota twin study that we discussed earlier emphasizes the importance of both genotype and environment in intelligence.

General reasoning ability is influenced by both genetic and environmental factors. Intelligence tests, and there are many of them, generate a single number, the **intelligence quotient** or **IQ**, and this number is considered an indication of a person's innate intelligence. The IQ is derived by dividing the person's mental age (as determined by the tests) by his or her actual age and then multiplying by 100. Thus a 10-year-old whose test scores are equivalent to those achieved by an average 12-year-old has an IQ of $12/10 \times 100 = 120$. The mean score for the U.S. population is standardized at 100. Whether IQ tests are valid devices for measuring intelligence is a topic of much debate, especially when the same tests are used for comparisons between different socioeconomic classes and different ethnic and racial groups.

Attempts to quantify the relative contributions of genotype and environment to intelligence are complicated by disagreements in three main areas:

1. The definition or characterization of intelligence.
2. Skepticism regarding the validity of tests designed to measure it.
3. The sampling procedures used.

In spite of these difficulties, there is little doubt that whatever mental abilities the tests measure, the genotype is influential. Support for this statement comes from twin and adoption studies. Monozygotic twins raised apart have more similar IQs than dizygotic twins raised together or dizygotic twins raised apart. The IQ scores of adopted children correlate more closely with the biological mother than with the adoptive mother. And the closer the genetic relationship between two people, the greater the correlation in IQ scores (Figure 26.19). Recall that the Minnesota study of MZ twins raised apart estimated that the heritability of IQ was about 0.70, a finding that strengthens the argument that the genotype has an important influence on intelligence. This is not to say that the environment has a minimal role to play, for it does not. Mental skills can develop fully only if the environment supports it, as we will see shortly.

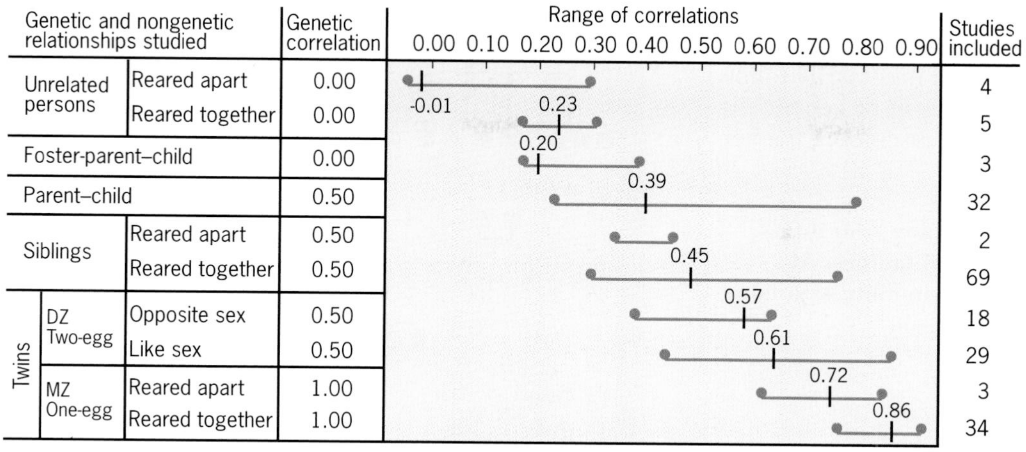

Genetic and nongenetic relationships studied		Genetic correlation	Studies included
Unrelated persons	Reared apart	0.00	4
	Reared together	0.00	5
Foster-parent–child		0.00	3
Parent–child		0.50	32
Siblings	Reared apart	0.50	2
	Reared together	0.50	69
Twins: DZ Two-egg	Opposite sex	0.50	18
	Like sex	0.50	29
Twins: MZ One-egg	Reared apart	1.00	3
	Reared together	1.00	34

Figure 26.19 Correlation scores on intelligence tests taken by pairs of people with varying degrees of hereditary and environmental similarity. The vertical lines are the average correlation in each study group. Individuals who are most alike genetically (MZ twins) have a correlation coefficient of almost 1.0, and this is true whether they were raised apart or together. Unrelated persons have a very low correlation coefficient.

Are there inherent differences in the intelligence of peoples of different races? In a controversial study in 1969, psychologist Arthur Jensen claimed that there are inherent differences in the average intelligence of different racial groups. He examined several studies showing that blacks as a group score about 15 points lower than whites on IQ tests. He further estimated that the heritability of IQ was about 0.8. After standardizing his samples, he concluded that the difference in test scores of blacks and whites is due largely to genetic factors.

Jensen's ideas and conclusions have been intensely criticized. Some of the data that Jensen cited in his study have since been shown to be fraudulent. In addition, researchers point out problems with definitions of a race, the IQ tests themselves, bias in sampling procedures, and perhaps most importantly, use of the term *heritability* (see Chapter 25). Let us examine the issue of how heritability can be misused.

Consider a hypothetical situation in which genetically identical individuals ($V_g = 0$) are placed into extremely heterogeneous environments. We determine the IQ of each individual. Because the genotypes are identical, there is no genetic variation. All variation in IQ scores would be due to variation in the environments, and the heritability under these circumstances would be 0.0 ($H^2 = V_g/V_t = 0/V_t = 0.0$). Next, several genetically unrelated individuals are placed into identical environments and their IQs determined. The environmental variation (V_e) is 0; therefore, all the variation in IQ scores would be the result of genetic variation. Heritability in this case would be 1.0 ($H^2 = V_g/(V_g + V_e) = V_g/(V_g + 0) = 1.0$). In this latter situation, the IQ scores would be higher if the environment was challenging and stimulating, and the scores would be lower if the environment lacked stimulation. Only the IQ scores, not the heritability estimate, would vary. In a third situation, unrelated individuals are placed into heterogeneous environments. The IQ of each individual is determined. Because both environmental and genetic factors vary, the heritability value will fall somewhere between 0.0 and 1.0. Three important points emerge from these hypothetical examples:

1. Heritability does not express the extent to which a trait is determined by genes. It expresses the proportion of phenotypic variation that is the result of genetic variation.
2. Heritability measures only a specific population in a specific environment at a specific time. In the hypothetical model it was shown that heritability values ranged from 0 to 1 for the same trait. Heritability varied because the conditions were manipulated. Thus, we cannot extrapolate directly from one population to another.
3. Just because heritability is high, say 1.0, it does not follow that observed variation between two populations must be the result of genetic differences. High heritability estimates were obtained from the two situations in which genetically unrelated individuals were placed into enriched or deprived environments; yet it is obvious that the IQ differences between the two groups were due to environmental, not genetic, differences.

These examples illustrate a major problem with the use of heritability and its application to group differences. Some traits, such as IQ, have a high heritability value, and there are documentable intergroup differences in IQ scores. But it does not necessarily follow that differences in scores are the result of genetic differences. Care must be taken when discussing the heritability of any trait.

Key Points: **Intelligence is a complex trait influenced by genetic and environmental factors. Although IQ scores are heritable, heritability estimates must be used cautiously.**

A GENETIC BASIS FOR SEXUAL ORIENTATION?

One of the most controversial areas in behavioral genetics research is the genetic basis of sexual orientation. Several recent twin and adoption studies as well as brain anatomy studies have indicated that there is a significant genetic or biological component in homosexual behavior. If so, then there must be a genotype(s) that predisposes an individual toward homosexuality.

In a 1991 study, Simon LeVay studied a small region of the hypothalamus in the brain known to control sexual response. He looked at the brains of homosexual men, heterosexual men, and heterosexual women. He found that this region of the hypothalamus is almost twice as large in the heterosexual males as in the homosexual males or heterosexual females. There is a morphological difference, but is this difference genetically based or is it caused by some other nongenetic factor?

That same year, another study strengthened a genetic link to homosexuality. J. Michael and R. Pillard studied 161 homosexual men, each of whom had at least one identical (MZ) or fraternal (DZ) twin or an adopted brother. They found that 52 percent of the MZ twins were both homosexual, 22 percent of the DZ twins were both homosexual, and 11 percent of the adopted brothers of homosexuals were homosexual. Similar results were found for homosexual women. The concordance rate for monozygotic twins compared to dizygotic twins and to genetically unrelated persons raised in the same household suggests that the genotype is somehow involved in the development of homosexual behavior. A genetic influence on homosexual orientation is also suggested by a few cases of identical twins concordant for homosexuality who were separated early in life and reared apart.

In 1993, researchers at the National Cancer Institute studying homosexual behavior identified a candidate region of the X chromosome and suggested that this region carried a gene that influenced sexual be-

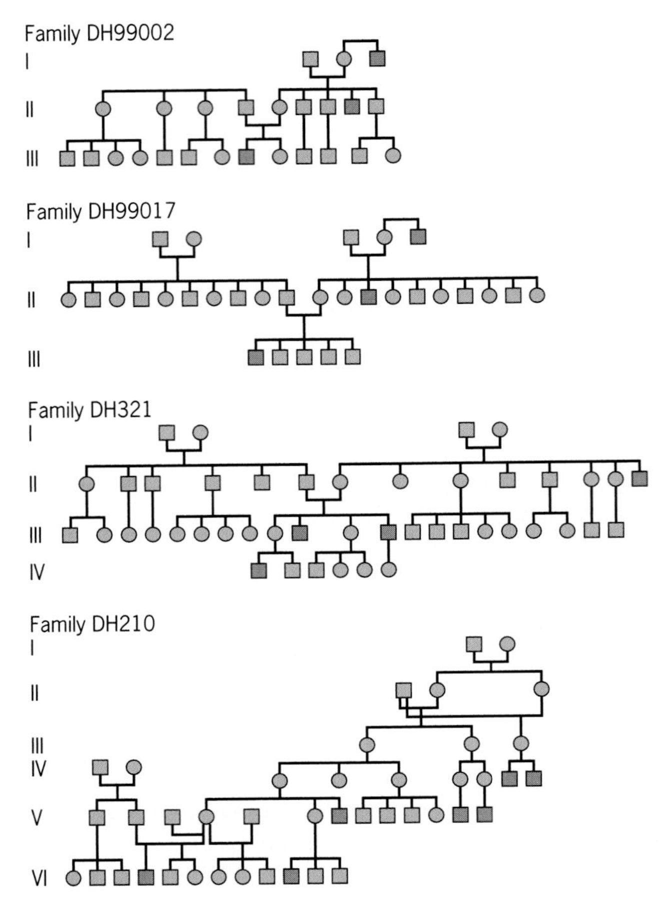

Figure 26.20 Family pedigrees displaying apparent maternal transmission of male homosexuality. Red symbol: homosexual; blue symbol: nonhomosexual.

havior. Dean Hamer and his colleagues identified 76 homosexual men and constructed pedigrees for each. After identifying individuals in each family that were homosexual, they found that 13.5 percent of the men's brothers were homosexual, compared to 2 percent of the men in the general population. Their results argued for a genetic link, as had the earlier studies. There was another interesting feature of the pedigrees: There were more homosexual relatives on the maternal side than on the paternal side (Figure 26.20). Homosexuality in these pedigrees seemed to follow an X-linked inheritance pattern.

Hamer and his colleagues began to look for regions of the X chromosome that might carry a gene that influences sexual behavior. Their approach was a simple one. A pair of brothers have about half of their X chromosome DNA in common. If they are both homosexual, then whatever genetic factors might be involved in homosexuality would be carried on that part of the DNA they have in common. The research team screened all homosexual brothers for DNA sequences

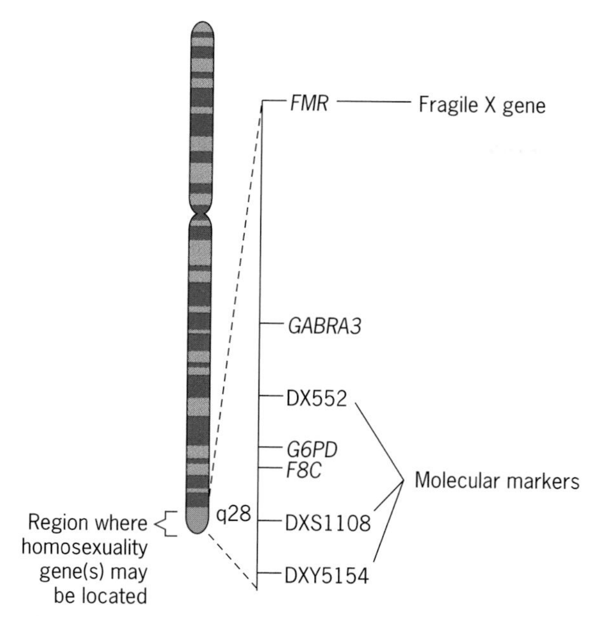

Figure 26.21 The X chromosome, showing the markers that appear linked to a gene that predisposes males to homosexual behavior. The possible gene site is Xq28.

they had in common. They found that of 40 pairs of homosexual brothers, 33 pairs shared five identical molecular markers located near the tip (Xq28) of the X chromosome (Figure 26.21).

Hamer's research findings were greeted with much publicity and some skepticism from other researchers. His findings have now been replicated in a study of another group of homosexuals in Colorado. A significant majority of 33 pairs of homosexual brothers shared the same Xq28 markers. Some of the homosexual men had heterosexual brothers who failed to share the markers with their homosexual brothers. The researchers also found that the Xq28 markers were not shared to a significant extent within 36 pairs of homosexual sisters, suggesting that different genes influence homosexual behavior in males and females.

Efforts are underway to identify candidate genes in the Xq28 region and elsewhere in the genome that may influence homosexuality. Researchers are also searching for environmental factors that may influence homosexual behavior. The search for environmental factors is concentrating on agents that control prenatal hormone surges.

There are many questions about candidate genes that might influence homosexual behavior. What proteins do these genes encode? What do the proteins do? What are the differences between the proteins in homosexual and heterosexual men? What do the genes do in females? There are also questions about the role of environmental factors. What are they and what do they do?

Key Points: **Homosexual behavior is now thought to be influenced by the genotype. A candidate region on the tip of the X chromosome is currently being studied.**

GENETICS AND BEHAVIOR IN PERSPECTIVE

It is a disturbing thought that our behavior, mental abilities, personality, and mental well-being can be destroyed by small alterations in DNA. However, although researchers continue to establish links between genes and behavior, they are also revealing the enormous complexity of the interaction between the genetic information being expressed and the environment in which it is being expressed.

Researchers have been tremendously successful in identifying single-gene mutations that are responsible for important neurological disorders. But complex neurological diseases, such as schizophrenia and manic depression, are likely the result of several genes and multiple environmental factors. Schizophrenia and manic depression have a well-established genetic basis, but it is important to recognize that these disorders may be expressed in genetically predisposed individuals only under special environmental circumstances, such as extreme stress.

It is important to continue studying complex mental disorders because they will give us insights into the biochemical and cellular basis of neurological functions. Brain functions result from nature and nurture, and research must continue to analyze the genetic and nongenetic factors that influence these functions.

Key Points: **Much of human behavior is influenced by the genotype interacting with the environment. This interaction is complex.**

TESTING YOUR KNOWLEDGE

1. You discover a *Drosophila* mutant that exhibits abnormal behavior. Not only is the male's courtship song different, but also its pattern of locomotor activity runs on a cycle that is longer than 24 hours. The new mutant, which is recessive, may be a mutant *per* allele. How would you test this?

ANSWER

Both genetic and molecular tests should be employed in this analysis. Because *per* is X-linked, crosses should be done to see if the new mutant is X-linked. If it is, then mapping experiments must be conducted to see if the gene maps to the same chromosome region as *per*. If it does, the next step is to do a complementaion analysis of the new mutant against *per*. By crossing the new mutant against per^0/per^0 females, you would get per^0/"*new mutant*" heterozygote females. The phenotype of the heterozygote would indicate allelism: If the phenotype is normal, the new mutant is not a *per* allele because the female would be heterozygous at two loci; if the phenotype is mutant, then it is probably a *per* allele because the female would carry two mutant *per* alleles. (Review complementation analysis in Chapter 14.) If all indications are that this is a *per* allele, DNA sequence analysis should be done to compare the wild-type *per* allele with the new mutant and presumptive *per* allele. The mutant allele should have a different DNA sequence. If sequences are different, it would then be necessary to demonstrate that the specific altered DNA sequence was responsible for the observed behavioral change.

2. Two highly inbred strains of mice differ in their ability to traverse a maze. When motivated by standard mouse food at the end of the maze, one of the strains called "maze bright" gets through the maze much faster than the other strain, which is called "maze dull." However, if the end of the maze has different food items in a dark, damp environment, the "maze bright" and "maze dull" strains arrive at the food at roughly the same time. What does this observation tell us about learning and genetics?

ANSWER

If maze brightness and dullness were selected in the strains to begin with, it tells us that there is a genetic basis for this behavior. The fact that the interstrain differences disappear when the conditions change argues that the genotype and environment interact with each other. This has important human implications. Just because two populations differ for a heritable trait, it does not mean that those differences will persist when environmental conditions change.

QUESTIONS AND PROBLEMS

26.1 Can two nonhygienic bees produce hygienic offspring? Explain.

26.2 Two mice expressing a behavioral aberrancy called zigzag were crossed. They produced 84 zigzag offspring and 58 normal offspring. What can you conclude about the inheritance of zigzag from this cross? Explain.

26.3 Discuss the problems associated with the use of heritability in IQ studies.

26.4 You discover a new *Drosophila* mutant that is extremely sensitive to loud noises. When you bang on the table sharply, the fly stops, quivers for a while, and then falls over. It recovers after a few minutes. You call this mutant "sensitive" (*sen*). In order to assess a possible genetic basis for the trait, you cross it to a wild-type fly from the same small population from which *sen* originated. Half of the offspring are wild-type and half express *sen*. When the *sen* mutants are crossed among themselves, all the offspring are *sen*; when wild-type flies from the small population that produced *sen* are intercrossed, some produce all wild-type offspring, but some of the wild-type × wild-type matings produce *sen* offspring. On the basis of this information, is it likely that *sen* is due to a dominant or a recessive allele?

26.5 If you have a strain of mice that is highly inbred, would you expect to be able to select for maze bright and maze dull lines? Explain.

26.6 In what ways are phenylketonuria and Lesch-Nyhan syndromes pleiotropic disorders?

26.7 How do genes indirectly affect behavior? Give examples.

26.8 Why is the *per* gene considered to be a pure "behavior gene"?

26.9 What evidence supports the hypothesis that *per* controls *Drosophila* songburst patterns?

26.10 PER does not appear to bind to DNA, yet it is implicated in transcription control. How would you explain this?

26.11 What is the evidence that aneuploidy causes behavioral problems?

26.12 What are the similarities and differences between PKU and Lesch-Nyhan on the one hand, and Huntington's disease on the other?

26.13 What two important conclusions emerged from the Minnesota twin study?

26.14 Discuss the implications of the Minnesota twin study.

26.15 Evaluate the evidence that alcoholism has a genetic basis.

26.16 What makes the Blum study showing a link between the *DRD2 A1* allele and alcoholism so controversial?

26.17 What is the connection between amyloid precursor protein (APP) and Alzheimer's disease?

26.18 What is the significance of the observation that patients with Down syndrome invariably develop Alzheimer's disease?

26.19 Discuss the evidence that homosexual behavior might have a genetic basis?

BIBLIOGRAPHY

BARINAGA, M. 1996. Researchers find the reset button for the fruit fly clock (Research News). *Science* 271:1671–1672.

BOUCHARD, T. J. 1994. Genes, environment, and personality. *Science* 264:1700–1701.

BROWN, J. R., et al. 1996. A defect in nurturing in mice lacking the immediate early gene *fosB*. *Cell* 86:297–309.

CLONINGER, R. C. 1995. The psychological regulation of social cooperation. *Nature Medicine* 1:623–625.

CRABBE, J. C., J. K. BELKNAP, AND K. J. BUCK. 1994. Genetic animal models of alcohol and drug abuse. *Science* 264:1715–1723.

FRIEDMAN, R. C. AND J. I. DOWNEY. 1994. Homosexuality. *New England J. Med.* 331:923–930.

GOLDMAN, D. 1995. Dopamine transporter, alcoholism, and other diseases. *Nature Medicine* 1:624–625.

HALL, J. C. 1994. The mating of a fly. *Science* 264:1702–1714.

HAMER, D. H., ET AL. 1993. A linkage between DNA markers on the X chromosome and male sexual orientation. *Science* 261:321–327.

LEWONTIN, R. C. 1975. Genetic aspects of intelligence. *Ann. Rev. Genet.* 9:387–405.

PLOMIN, R., M. J. OWEN, AND P. MCGUFFIN. 1994. The genetic basis of complex human behaviors. Science 264:1733–1739.

ROSES, A. D. 1996. From genes to mechanisms to therapies: Lessons to be learned from neurological disorders. *Nature Medicine* 2:267–269.

ROTHENBUHLER, W. C., J. M. KULINCEVIC, AND W. E. KERR. 1968. Bee genetics. *Ann. Rev. Genet.* 2:413–438.

SCOTT, J. P. AND J. L. FULLER (eds). 1974. *Dog Behavior: The Genetic Basis.* University of Chicago Press, Chicago.

TIIHONEN, J., ET AL. 1995. Altered striatal dopamine re-uptake site densities in habitually violent and non-violent alcoholics. *Nature Med.* 1:654–657.

YANKNER, B. A. 1996. New Clues to Alzheimer's disease: Unraveling the roles of amyloid and tau. *Nature Medicine* 2:850–852.

A CONVERSATION WITH

DEBORAH AND BRIAN CHARLESWORTH

How did you become interested in genetics and specifically in evolutionary genetics?

D.C.: Actually, I have been interested in genetics since I was a teenager, though I was interested in biology much earlier than that. When I was in high school, my biology teacher suggested that I should listen to some programs on the radio which were talks by Julian Huxley, a renowned evolutionary biologist and a member of the famous Huxley family of biologists. I found the talks fascinating and later managed to get hold of Huxley's book, *Evolution: The Modern Synthesis*. From that point on, I was hooked. It was an exciting book to read. It really introduced me to genetics, which was not taught in our high school, and it introduced me to evolutionary genetics. So I got both of them right away at the beginning. It was an eye opener to me that biology could have such generality and be so beautiful. Brian, I believe you had a similar type of experience.

B.C.: Well, I can be brief because my experience was indeed highly similar, except that it was not a particular book that ignited my interest. As a child, I was always very interested in natural history. In high school I had a very good biology teacher who taught us some genetics. I got excited about biology and genetics, and how genetics connected with evolution; it just went on from there.

Brian Charlesworth and **Deborah Charlesworth** are an amazing team; they collaborate as scientists and have authored many of their scientific papers together. They were born in England and received their Ph.D. degrees in genetics from the University of Cambridge. In 1985, they emigrated to the United States. Their work in population genetics and evolutionary biology have won them international recognition. Brian Charlesworth is currently George Wells Beadle Distinguished Service Professor at the University of Chicago. Deborah Charlesworth is currently professor of ecology and evolution at the University of Chicago.

In July, 1997, they will return to the UK, where Brian Charlesworth will be a Royal Society Research Professor, and Deborah Charlesworth will be a Senior Research Fellow.

You both have published extensively on a wide variety of topics such as transposable genetic elements, mating systems, mutational load, and chromosome evolution. How do you decide which research topics to pursue?

B.C.: Well, I don't think I actually plan ahead very carefully. Basically, you think about one thing, and it occurs to you that it might relate to something else, and I think you go on from day to day building on ideas and making connections. What do you think, Deborah?

D.C.: Yes, I think one thing leads to another, quite unexpectedly sometimes. For example, our work on the evolution of separate sexes was, oddly enough, the result of previous research we were doing on mimicry in butterflies and on the close linkage of genes involved in controlling the different mimetic patterns. So we got interested in looking at other evolutionary or genetic situations where you knew that several genes were involved but they were so closely linked they behaved as a block. This block of tightly linked genes looked like a single locus with different alleles. The control of sex by the sex chromosomes is a classic case of the same phenomenon, and that led me further into plant mating systems and breeding systems in general.

This line of investigation led us later into studies of deleterious mutations because we were looking at how inbreeding leads to a loss of vigor. We were thinking about possible reasons why organisms might have systems to ensure outbreeding. Our interest in deleterious mutations has recently led us into studying the effect of deleterious mutations in shaping molecular evolution. It's quite strange really how things that at first might seem quite disparate do relate to one another. Our research is actually a very nice interconnected system.

I think that this is one of the wonderful things about genetics. It's trite to say that it integrates so much of biology, but that is one of the attractions of studying and working in genetics. You learn about all different types of biological systems, and you don't have to specialize too narrowly on a single system. From reading the literature you find useful information from a different organism or a different system that relates to your own system, and this is always fascinating.

Let's explore Darwinism for a moment. There continues to be resistance to Darwinism, the most recent example being the attempt by the Tennessee State Legislature to pass a law requiring teachers to refer to evolution as a theory and not fact. Why do you think that there is this continuing resistance to evolution and what can we do as biologists to counter it?

B.C.: To some extent, this is more of a problem in the United States than in Europe, and I think ultimately the answer to why people are hostile to Darwinism is religious prejudice. Opponents of evolution feel that the idea that humans and other organisms are the product of natural forces extending over billions of years overthrows their fundamental beliefs about the nature of their relation to the universe. It is not something that most professional biologists have a problem with, because we are obviously trained to think of the world in terms of scientific evidence, not something which is written down in a book and taken for granted. I think that it is the same kind of problem that Galileo encountered with the Catholic Church 300 years ago. People do not want to see themselves as cosmically insignificant, so to speak.

What we can do about it is a more difficult question. I think there should certainly be a push for much better science education in schools at all levels, and that people should be introduced to science in a way which is not just *here are the facts, learn them*. Students need to understand the ongoing process of exploring nature. They should be taught how to relate ideas and concepts to data and to learn how to test ideas against facts very early on. One should be absolutely hard-nosed about the fact that the evidence of evolution is based on the same kind of scientific evidence as any other theory in science. If you begin to question evolution, you might as well give up as far as the whole of science is concerned. I think it is not really widely appreciated that the creationist attitude is fundamentally irrational, and is essentially a threat to our entire enterprise of rational investigation of the natural world.

We should be concerned about efforts by the fundamentalists to interject religion into science. They are politically astute in getting control of school boards and this kind of thing, and they will undoubtedly use that power as a vehicle for propagating creationist views. Anti-evolutionary views could be a problem at the national level if they affect the funding for basic research, although I don't see any evidence for this at the moment.

D.C.: I think Brian put it very clearly, and I agree with everything he says. I do think that the main thing we can do against this type of irrational thinking is educational. The idea of evolution is really simple in outline, and I think that many people do not accept it because they haven't been taught it in a clear way. The idea that we are animals and the product of evolution along with other animals comes very naturally of course, once you see the evidence and follow the reasoning.

I certainly think that one of the defects of a lot of science education is that people are just told to learn the facts. Science teaching ought to include more of a debate component. People need to understand that the scientific process involves a lot of debate. I remember as a high school student arguing with teachers. Often I was being ridiculous, of course, and I was trying to defend indefensible positions, but in doing that sort of debate you get an idea of what the support really is for ideas, and where it is lacking. You quickly realize the weakness of your position when you are arguing for something that is unsupported by data. Presenting only what is accepted in textbooks is a difficulty for young people studying science. They don't want to swallow something merely because it's written in a book. Science does not work that way.

27

A male lion with a cub he sired.

Population Genetics and Evolution

In the Jungle

When a young male lion assumes control of a harem, one of the first things he does is kill all the lion cubs in the pride. The behavior is difficult to watch and until recently has been difficult to understand. However, in 1964 W. D. Hamilton developed a theory that helped put this behavior into a context that made biological sense. The basic idea behind this theory, called kin selection, is simple and obvious: Individuals fight for the transmission of their genes from one generation to the next. The lion who inherited this harem will probably control it for a very short period of time. During this time, he will want to reproduce as often as possible so that his genes are well represented in the next generation. Lionesses who are nursing cubs sired by a different male do not ovulate and cannot become pregnant. Therefore, the new controlling male kills the young lion cubs, who are carrying genes from the deposed male, so that the females can be impregnated with his sperm and the offspring will carry his genes. This is Darwinian evolution in all its glory. Organisms that are reproductively more successful will have their genes better represented in future generations.

The theory of population genetics enables researchers to evaluate the mechanisms of evolution. In this chapter and the next we examine some of those mechanisms. In Chapter 27, we trace the development of evolutionary thought and focus on what we might call *microevolutionary changes*: those changes that occur in populations that may eventually lead to the formation of new species. In Chapter 28, we examine more closely the processes of speciation, as well as evolutionary events above the species level.

THE EMERGENCE OF EVOLUTIONARY THEORY

Darwin's theory of evolution by natural selection, as proposed in 1859, precipitated one of the most fundamental, complex, and profound of all intellectual revolutions that have occurred in human history. It affected not only our view of the natural world, it also affected our social and religious views.

Any explanatory system developed to explain biodiversity has to address three of its general features.

1. Living organisms exhibit great structural complexity.
2. Living organisms possess adaptive features that appear to have been specially designed.
3. Tremendous diversity exists in the living world.

These general observations in turn lead to a series of questions:

1. How have complex organisms and structures come into existence?
2. What forces have operated to mold adaptive characteristics?
3. How did diversity originate in the living world, and how is this diversity maintained?

The theory of evolution, proposed by Charles Darwin, answers these questions.

Early Evolutionary Theories

The idea of evolution—of descent with modification—had existed for more than 100 years before Darwin's *Origin of Species* was first published in 1859, but prominent biologists and geologists largely rejected the notion. One reason why evolution was not accepted was the enormous power of the theologically based idea of **special creation**, in which species have separate origins and do not change or change only within narrow limits.

Two other factors stymied evolutionary thinking. First, the religiously rooted doctrine of **catastrophism** attempted to explain fossil discoveries in the context of the Great Flood described in the Old Testament. A second factor was **typological thinking** or **essentialism**. Essentialists viewed the living world as consisting of basic organismal types, such as dog or cat. Although limited variation within each type (different breeds of dogs and cats) was acknowledged, essentialists saw no evolutionary link between the basic types. In other words, dogs and cats did not share a common ancestor. Both catastrophism and essentialism retarded the development of evolutionary thinking.

Fifty years before Darwin, Jean Baptiste Lamarck had proposed a theory for the origin of species which stated that one species was gradually transformed into another through the accumulation of acquired characteristics. This idea was similar to Darwinian evolution in its suggestion that an organic link existed between different species, and it stood in marked contrast to typological thinking. Lamarck's theory denied extinction, however. Lineages of species were thought to be transformed into new species via "internal forces" and acquired characteristics, but they did not become extinct. They were modified according to environmental dictates.

Charles Lyell, a nineteenth century geologist, made enormously important contributions to evolutionary theory. In his 1830 book *Principles of Geology* he adhered to the then radical position that natural forces operating on the earth have been fairly constant throughout the earth's history. These forces, he stated, operating over millions of years, gradually changed the surface of the earth, creating mountains, canyons, deserts, grasslands, and other features of the earth's outer crust. Darwin read and admired Lyell's *Principles of Geology* and was profoundly influenced by it.

The Voyage of the *Beagle*

What led Darwin to his theory of evolution? The most important influence on his thinking on evolution was his voyage around the world on the *H.M.S. Beagle* (Figure 27.1). Observations he made on that five-year journey (1831–1836) led him to conclude that there was a natural explanation for the origin of adapted species. Although Darwin collected volumes of data on this voyage, three sets of observations stand out:

1. In South America, Darwin observed fossils that bore a marked similarity to currently existing species (Figure 27.2). He puzzled over this similarity. If the fossils were remnants of extinct species wiped out by a great flood, and if a new creation had followed the flood, why were entirely new organisms not created? Were they similar because the extinct organisms were somehow related to those currently occupying the region?
2. In Chile, Darwin observed the full effects of a massive earthquake and noted the dramatic alterations of the landscape that resulted. Land had moved and been rearranged. What had once been sea floor became dry land. Could this type of land movement explain why fossil sea creatures are found 10,000 feet above sea level in the Andes? Perhaps what were once sea beds had been thrust up over millions of years to become mountain tops. The Chilean

Figure 27.1 The H.M.S. *Beagle* in the Straits of Magellan.

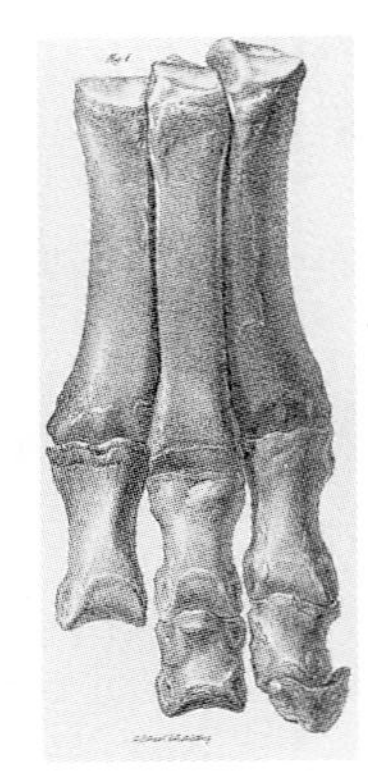

(*a*) Patagonian Guanaco

Macrauchenia Forefoot

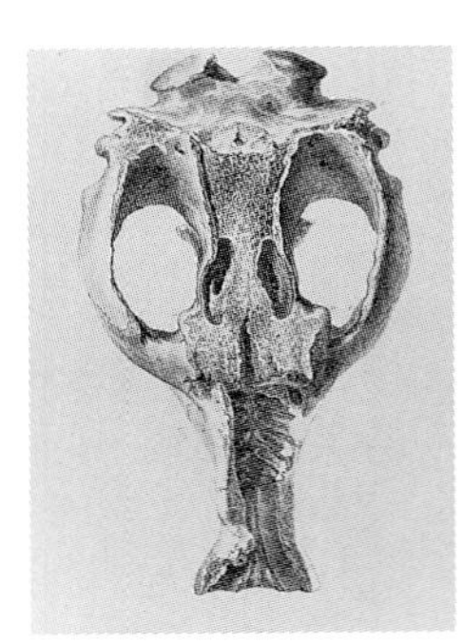

(*b*) Capybaras

Toxodon Skull

Figure 27.2 The Patagonian guanaco is structurally similar to the extinct Macrauchenia (*a*). The capybara is structurally similar to the extinct Toxodon (*b*).

earthquake and other observations convinced Darwin that Charles Lyell's theory of a continuously and gradually changing earth was correct.

3. In the Galapagos Islands, Darwin observed 13 rather dull-looking species of finches (Figure 27.3), which were similar to finches that lived on the west coast of South America. All 13 differed from each other in body size, and in beak morphology and function. They were morphologically similar to each other, and yet they differed as a result of occupying different ecological niches. Darwin could reasonably assume that a single finch or perhaps a few finches somehow made their way to the Galapagos Islands and founded a population that subsequently diversified throughout the island chain, adapting to different ecological niches on the individual islands.

Upon returning to England, Darwin enjoyed celebrity status, for the reports he published while serving as the *Beagle's* naturalist had won him wide recognition in the scientific community. His challenge now was to confront and interpret the enormous amount of data he had collected. Although he concluded that species evolved naturally and that natural forces had guided this evolutionary process, he had not yet identified a mechanism to explain it. He knew that species changed over time, but how did these changes occur?

Fifteen months after his return to England and just after reading Thomas Malthus's essay on the principle of population, Darwin had his mechanism. Malthus's essay provided the key: members of a population compete for limited resources. Darwin's theory of evolution states that populations tend to increase exponentially but do not do so because resources are limited. Limited resources result in competition. Specifically, those individuals in the population with favorable inherited variations are more likely to survive and reproduce. Over time the genetic composition of the population changes, with favorable variants replacing less favorable variants. Because the ecological conditions of the earth change, populations vary in terms of their adaptive features from the an-

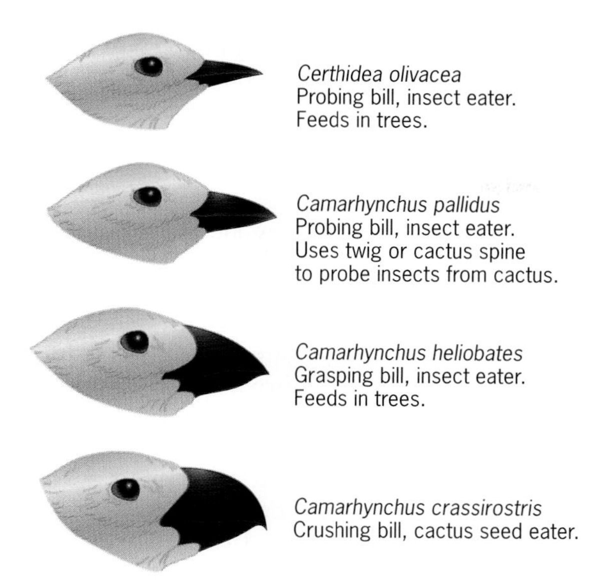

Figure 27.3 Examples of beak shape variation in different species of Galapagos finches, as correlated with feeding habits.

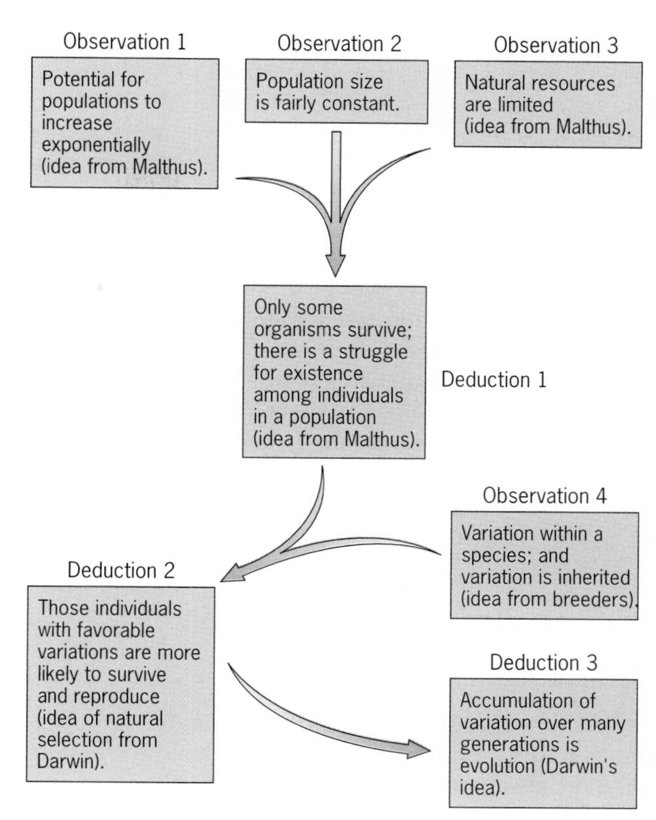

Figure 27.4 The observations and deductions made by Darwin in the development of his evolutionary theory.

cestral population. The observations and deductions that culminated in Darwin's theory of evolution are outlined in Figure 27.4.

The Darwinian Revolution

The Darwinian revolution continued for years after Darwin published the *Origin of Species* in 1859; indeed, it goes on to this day. It challenged the human-centered view of the world by showing that the same natural forces that shaped other species also helped shape the human species. Darwinian theory has affected nearly all of the metaphysical and ethical concepts of Western society, and it has had an enormous impact on the history and literature of most of the world. The eminent zoologist, Ernst Mayr, characterized the Darwinian revolution as consisting of

1. The recognition of the vast age of the earth (not just 4000 to 6000 years, as suggested by some biblical scholars).
2. The rejection of catastrophism.
3. The rejection of a striving for "perfection," or a goal-oriented process.
4. The rejection of special creation.
5. The replacement of typological thinking with populational thinking.
6. The abolition of anthropocentrism—the view that the human species is the center of the universe.

The complexity of these elements explains why the Darwinian concept of evolution was slow in gaining a foothold and why opposition to it continues today.

Darwin's theory did have a major flaw, however. It lacked a sound mechanism that could explain how genetic variation originated, how the variation was transmitted, and how the genetic composition of populations changed over time. Darwin eventually suggested a mechanism of variation that was rooted in the widely accepted concept of acquired characteristics, an idea commonly associated with Lamarck, but an idea that actually predated him. The mechanism Darwin suggested was called pangenesis. Units of inheritance called gemmules filtered through the body where they were packaged into gametes and transmitted to the next generation. According to Darwin these gemmules were subject to directed mutation and thus could change according to environmental dictates.

Mendelism eventually provided a basis for understanding variation and its inheritance. Before the Darwinian evolution could be fully understood, however, the fusion of Darwinian with Mendelian principles required a new dimension: Principles had to be formulated that would allow geneticists to evaluate gene frequency changes in natural populations of sexually reproducing organisms. In other words, they had to learn how to monitor the flow of genetic material in a freely interbreeding population and to assess the forces that act to alter the frequencies of different alleles—forces that over time could produce new species.

Key Points: **Darwin's theory of evolution states that populations tend to increase exponentially but do not do so because resources are limited. Limited resources result in competition. Individuals with favorable inherited variations are more likely to survive and reproduce. Over time the genetic composition of the population changes, with favorable variants replacing less favorable variants.**

THE THEORY OF POPULATION GENETICS

Two major schools of thought led to modern **population genetics**, the discipline that studies the frequencies of alleles in populations. The first—the biometrical school of thought—began with Francis Galton and ended with H. Nilsson-Ehle (Chapter 25). It demonstrated multifactorial inheritance: several loci and the environment interact to influence the expression of a trait. This concept was important because it provided a basis for understanding the inheritance of small phenotypic differences in traits that lie along a continuum. The second school of thought began with Mendel and progressed to the analysis of inbreeding and crossbreeding in sexually reproducing populations. This second approach, coupled with the principles of quantitative genetics, led to the establishment of population genetics and, later, to the formulation of mechanisms that account for the processes of organic evolution.

The Hardy–Weinberg Principle of Genetic Equilibrium

Attempts to apply Mendelian principles to freely interbreeding populations of sexually reproducing individuals began shortly after Mendel's work was rediscovered. Udny Yule pointed out in 1902 that when members of an F_2 population, segregating for a single pair of alleles (*A* and *a*), interbreed at random, the three possible genotypes (*AA*, *Aa*, and *aa*) are represented in the same proportions in the F_3 and all succeeding generations. This idea was expanded in 1903 by William Castle, who showed that the genotypic ratios would change in each generation if the *aa* class were selected out at each generation. (They were physically removed or prevented from mating.)

Yule, who had shown how Mendelism applied to F_2 populations, had difficulty understanding the case of brachydactyly, an autosomal dominant human trait characterized by shortened fingers and toes. Yule said that over the course of time brachydactyly should appear in the population in a classic 3:1 phenotypic ratio. Since it does not, he contended that Mendelian principles may not apply to human populations. However, G. H. Hardy, a mathematician, suggested to Yule in a 1908 letter to the editor in the journal *Science* that non-mathematically inclined biologists tend to make erroneous extrapolations of Mendel's work and its applicability. Hardy pointed out that in order for a 3:1 F_2 phenotypic ratio to occur, the frequency of the dominant and recessive alleles each had to be 0.5, as they are in a monohybrid cross between *AA* and *aa* parents. At allele frequencies other than 0.5, the phenotypic ratio would deviate from 3:1. Yule failed to realize that the phenotypic frequencies are a function of the allele frequencies.

That same year (1908) Wilhelm Weinberg, a German physician, applied Mendelian principles to human populations. Working independently, Hardy and Weinberg developed the basic principle that relates the frequencies of genotypes to the frequencies of alleles in populations. This principle, now known as the Hardy–Weinberg Principle, makes one important assumption: that the members of the population mate at random to produce the next generation. Hardy and Weinberg discovered that under this assumption of random mating, the frequencies of the alleles in the population can be used to predict the frequencies of the genotypes in that population.

Let's suppose that in a population a gene is segregating two alleles, *A* and *a*, and that the frequency of *A* is p and the frequency of *a* is q. If we assume that the members of the population mate randomly, the diploid genotypes of the next generation will be formed by the random union of haploid sperm and eggs. The probability that a sperm (or egg) carries *A* is p and the probability that it carries *a* is q. Thus, the probability of producing an *AA* zygote in this population is simply $p \times p = p^2$, and the probability of producing an *aa* zygote is $q \times q = q^2$. For the *Aa* heterozygotes, there are two possibilities: An *A* sperm can unite with an *a* egg, or vice versa. Each event has a probability of $p \times q$, and because both events are equally likely, the total probability of forming an *Aa* zygote in the population is $2pq$. Thus, on the assumption of random mating, the frequencies of the three genotypes *AA*, *Aa*, and *aa* can be predicted from the allele frequencies p and q. The predicted genotype frequencies, p^2, $2pq$, and q^2, are simply the terms in the expansion of the binomial expression $(p + q)^2 = p^2 + 2pq + q^2$. This simple relationship between genotype frequencies and allele frequencies will persist as long as the population mates randomly. Thus, the Hardy–Weinberg genotype frequencies p^2, $2pq$ and q^2 are sometimes called the *equilibrium* genotype frequencies.

To illustrate the Hardy–Weinberg Principle, let's consider a hypothetical population composed of 1000 individuals, 490 *AA*, 420 *Aa* and 90 *aa*. First we must calculate the frequencies of the two alleles, *A* and *a*. This is done by counting the two types of alleles and dividing each count by the total number of alleles in the population.

$$\text{Frequency of } A = (2 \times 490 + 420)/2000 = 0.70 = p$$
$$\text{Frequency of } a = (2 \times 90 + 420)/2000 = 0.30 = q$$

We also note that $p + q = 0.70 + 0.30 = 1$.

Next, we use the allele frequencies and the Hardy–Weinberg Principle to predict the genotype frequencies in the next generation, assuming that the members of the population mate randomly:

$$AA\text{: } p^2 = 0.49$$
$$Aa\text{: } 2pq = 0.42$$
$$aa\text{: } q^2 = 0.09$$

Thus, in a population of 1000 progeny, we predict 490 *AA*, 420 *Aa* and 90 *aa*. These are the same numbers we began with, illustrating that once the Hardy–Weinberg genotype frequencies have been obtained, they will persist in the population indefinitely. Of course, this persistence is conditioned on the continuation of random mating, and on the absence of evolutionary forces such as mutation and natural selection. Such forces would change the underlying allele frequencies or alter the frequencies of the genotypes after the zygotes had been formed by mating.

In real populations the genotypes are often found in Hardy–Weinberg frequencies, or in frequencies very close to them. For example, let's consider data from a human population that was studied for the MN blood types. These blood types are determined by two codominant alleles, L^M and L^N (Chapter 5):

Genotype	L^ML^M	L^ML^N	L^NL^N	Total
Observed numbers	1787	3039	1303	6,129

$$\text{Frequency of } L^M = (2 \times 1787 + 3039)/12{,}258 = 0.5395 = p$$
$$\text{Frequency of } L^N = (2 \times 1303 + 3039)/12{,}258 = 0.4605 = q$$

Hardy–Weinberg genotype frequencies	$p^2 =$ 0.2911	$2pq =$ 0.4968	$q^2 =$ 0.2121
Hardy–Weinberg genotype numbers	1784.2	3044.8	1300

In this example the predicted number for each genotype is very close to the observed number. Thus, the population appears to be in Hardy–Weinberg equilibrium for the gene that controls the MN blood types.

Now let's consider data from a study of chromosome rearrangements in the larvae of *Drosophila tropicalis*, a relative of *D. melanogaster*. There are two rearrangements, *A* and *D*, in one of the autosomes. Because these rearrangements do not recombine with each other, they can be treated as if they were alleles:

Genotype	*AA*	*AD*	*DD*	Total
Observed numbers	3	134	3	140

$$\text{Frequency of } A = (2 \times 3 + 134)/280 = 0.5 = p$$
$$\text{Frequency of } D = (2 \times 3 + 134)/280 = 0.5 = q$$

Hardy–Weinberg genotype frequencies	$p^2 =$ 0.25	$2pq =$ 0.50	q^2 0.25
Hardy–Weinberg genotype numbers	35	70	35

In this example, the genotypes are obviously not in Hardy–Weinberg frequencies. There is an excess of heterozygotes and a corresponding deficiency in both classes of homozygotes in the sample of flies that was studied. One possible explanation is that the heterozygotes have a much greater chance of surviving to the larval stage, when the chromosome rearrangements were identified. In this population, natural selection may therefore be altering the frequencies of the genotypes during zygotic development.

Applications of the Hardy–Weinberg Principle

The Hardy–Weinberg principle has a number of important applications. For example, it can be used to estimate the frequency of individuals in a population heterozygous for a deleterious recessive allele. Consider the allele that causes phenylketonuria (PKU), a metabolic disorder that results in severe mental retardation if not properly treated. In some human populations, the PKU frequency is 1/10,000, an *aa* frequency of 0.0001. Since the *AA* and *Aa* individuals are phenotypically identical, allele frequencies cannot be calculated directly by counting alleles. Instead, we use the Hardy–Weinberg principle in reverse to estimate the allele frequencies. Because $aa = q^2 = 0.0001$, the *a* frequency is estimated to be the square root of 0.0001, or 0.01. The *A* frequency is $1.0 - 0.01 = 0.99$. These allele frequencies are then used to estimate the genotype frequencies in the population:

$$AA\text{: } p^2 = (0.99)^2 = 0.9801$$
$$Aa\text{: } 2pq = 2\,(0.99)(0.01) = 0.0198$$
$$aa\text{: } q^2 = (0.01)^2 = 0.0001$$

The vast majority of the deleterious recessive PKU alleles are sequestered in the heterozygotes. If there were a plan to try to decrease the frequency of this allele in the population by somehow preventing *aa* individuals from reproducing, the plan would fail because it would target only a small fraction of the PKU alleles in the population.

A second application of the Hardy–Weinberg principle is in the analysis of multiple alleles. Consider the example of the locus that controls the ABO human blood groups. The ABO system is a multiple allelic system consisting of three different major alleles: I^A, I^B, and I^O, or simply *A*, *B*, and *O*. *O* is recessive to *A* and *B*; and *A* and *B* are codominant alleles, meaning that *AB* individuals are phenotypically A and B. Given this background, there are four phenotypic classes (A, B, O, and AB) and six genotypic classes (*AA*, *AO*, *BB*, *BO*, *AB*, *OO*). The *AA* and *AO* classes are phenotypically indistinguishable, as are the *BB* and *BO* classes.

A sample of 500 individuals displayed the following blood types:

A: 195 B: 70 AB: 25 O: 210

To estimate allele frequencies, we assume random mating (that is, pairs do not form on the basis of blood type). Because there are three alleles, another symbol, *r*, is necessary to extend the Hardy–Weinberg principle:

$$p = A \text{ frequency}$$
$$q = B \text{ frequency}$$
$$r = O \text{ frequency}$$

The genotype frequencies are represented by the expansion of the trinomial expression $(p + q + r)^2$:

	$p^2 + 2pr + r^2$		$+ \quad 2pq$	$+ \quad 2qr + q^2 = 1$
Genotypes	*AA AO*	*OO*	*AB*	*BO BB*
Phenotypes	A	O	AB	B
Phenotype frequencies	195/500 = 0.39	210/500 = 0.42	25/500 = 0.05	70/500 = 0.14

To estimate allele frequencies, we begin with the *O* allele:

$$r^2 = 0.42 = OO \text{ frequency}$$
$$r = \sqrt{0.42} = 0.65 = O \text{ frequency}$$

The combined frequencies of the A and O phenotypes are given by

$$p^2 + 2pr + r^2 = (p + r)^2$$
$$(p + r)^2 = 0.39 + 0.42 = 0.81$$

Taking the square root of both sides, we see that $p + r = 0.9$. Because $r = 0.65$, $p = 0.25$. Because $p + q + r = 1$, $q = 0.1$. The Hardy–Weinberg genotype frequencies can be calculated from these estimated allele frequencies:

$$\left.\begin{aligned} AA&: \ p^2 = (0.25^2)(500) = 30\ AA \\ AO&: \ 2pr = (2)(0.25)(0.65)(500) = 165\ AO \end{aligned}\right\} 195 \text{ A}$$
$$AB: \ 2pq = (2)(0.25)(0.10)(500) = 25\ AB$$
$$OO: \ r^2 = (0.65^2)(500) = 210\ OO$$
$$\left.\begin{aligned} BO&: \ 2qr = (2)(0.10)(0.65)(500) = 65\ BO \\ BB&: \ q^2 = (0.10^2)(500) = 5\ BB \end{aligned}\right\} 70 \text{ B}$$

Because observed genotype numbers agree with expected numbers, the population appears to be in Hardy–Weinberg equilibrium.

Another application of the Hardy–Weinberg principle is in the analysis of X-linked alleles. Estimations of allele frequencies for X-linked alleles must take into account the fact that females carry twice as many X-linked alleles as males because they have two X chromosomes compared to only one in males. At equilibrium, therefore, the genotype frequencies are

$p + q$ for males (two genotypic classes)

$p^2 + 2pq + q^2$ for females (three genotypic classes)

Color blindness is a trait caused by an X-linked recessive allele. In a sample of 2000 individuals (50 percent male and 50 percent female), 90 males and 3 females were color blind. In the males, the allele frequencies are estimated to be:

$$q = \text{freq. } a \text{, allele for colorblindness} = 90/1000 = 0.09$$
$$p = \text{freq. } A \text{, normal allele} = 1 - q = 0.91$$

With $p = 0.91$ and $q = 0.09$, and assuming identical allele frequencies in females and random mating in the population, we can estimate the frequencies of *aa*, *Aa*, and *AA* females:

$$AA: (0.91)^2\,(1000) = 828$$
$$Aa: (2)(0.91)(0.09)(1000) = 164$$
$$aa: (0.09)^2\,(1000) = 8$$

Thus, we expect over 99 percent of the women to be phenotypically normal, compared with 91 percent for males.

Key Points: **The Hardy–Weinberg principle of genetic equilibrium describes the relationship between allele frequencies and genotype frequencies under the assumption of random mating. This principle applies to loci with two alleles, loci with multiple alleles, and X-linked loci.**

THE THEORY OF NATURAL SELECTION

Natural selection is the principal force that alters allele frequencies and brings about evolutionary change. Darwin's premise, interpreted in a modern context, was that the genetic composition of populations changes as the environment changes. He argued that because organisms produce more offspring than the environment can support, not all of them survive. Because genetic differences exist among all these progeny and because genetic differences account for phenotypic differences, some progeny are better able to survive and reproduce than others and so leave more offspring than the others. Their genotypes increase in frequency at the expense of the less successful progeny. This is the essence of **natural selection**: the differential and nonrandom perpetuation of dissimilar genotypes to the next generation.

An important component of natural selection is **fitness**, a difficult term to define precisely. Herbert Spencer, an English philosopher and founder of the modern science of sociology, popularized the term *evolution* (Darwin rarely used the term) and invented the phrase "survival of the fittest" to describe natural selection. This phrase is circular in that the fittest are those that survive and those that survive are the fittest. For our purposes here, we define fitness as the capacity of a phenotype to donate genes to the next generation. The greater the fitness, the greater the reproductive capacity.

Natural selection changes allele frequencies in a population. A simple model illustrates how this phenomenon occurs and how fitness and selection relate to each other. Consider three genotypes: *AA*, *Aa*, and *aa*, where *A* is dominant over *a* and the *aa* genotype is inviable or sterile. What happens to the *A* and *a* allele frequencies and the *AA*, *Aa*, *aa* genotype frequencies over time?

Let *s* represent the *selection pressure* against a genotype:

$s_{A-} = 0$ against *AA* and *Aa* (no selection)

$s_{aa} = 1.0$ against *aa* because it is inviable or sterile

The selection pressure (s) has a range from 0 to 1. Fitness, symbolized by w, also has a range from 0 to 1. Fitness and selection have a complementary relationship:

$$w = 1 - s$$

Because selection against *aa* is 1.0, the fitness of *aa* is 0; because there is no selection against *AA* or *Aa*, these genotypes have a fitness of 1.0. Each genotype contributes to the next generation in proportion to its fitness. Thus multiplying each genotype frequency by its fitness allows us to predict allele and genotype frequencies for the next generation. Starting with a population in which the frequencies of *A* and *a* are each 0.5, and assuming Hardy–Weinberg genotype proportions, we have the following fitness relationships:

$$\tfrac{1}{4}AA\ (1.0) : \tfrac{1}{2}Aa\ (1.0) : \tfrac{1}{4}aa\ (0.0)$$

fitness values

The only members of the population that contribute genes to the next generation are *AA* and *Aa*. Two-thirds of these individuals are *Aa* and one-third *AA*. Thus the genotype frequencies of the parents producing the next generation are

$$AA : \tfrac{1}{3} \qquad Aa : \tfrac{2}{3}$$

and the allele frequencies among these reproducing individuals are

$$A : \tfrac{1}{3} + (\tfrac{1}{2})(\tfrac{2}{3}) = \tfrac{2}{3}$$
$$a : (\tfrac{1}{2})(\tfrac{2}{3}) = \tfrac{1}{3}$$

With random mating, the genotype frequencies in the next generation will be

$$AA : (\tfrac{2}{3})^2 = \tfrac{4}{9}$$
$$Aa : 2(\tfrac{2}{3})(\tfrac{1}{3}) = \tfrac{4}{9}$$
$$aa : (\tfrac{1}{3})^2 = \tfrac{1}{9}$$

After one generation, the *A* frequency increased from $\tfrac{1}{2}$ to $\tfrac{2}{3}$, and the *a* frequency has declined from $\tfrac{1}{2}$ to $\tfrac{1}{3}$. In the formation of the next generation, we continue to observe allele frequency changes:

$$\tfrac{4}{9}AA\ (1.0) : \tfrac{4}{9}Aa\ (1.0) : \tfrac{1}{9}aa\ (0.0)$$

fitness values

The only genotypes that contribute genetic material to the next generation are *AA* and *Aa*. The *AA* genotype accounts for $\frac{\frac{4}{9}}{\frac{8}{9}} = \frac{4}{8}$; *Aa* represents $\frac{\frac{4}{9}}{\frac{8}{9}} = \frac{4}{8}$. Thus the allele frequencies in the next generation are

$$A : \tfrac{1}{2} + \tfrac{1}{2}(\tfrac{1}{2}) = \tfrac{3}{4} = 0.75$$
$$a : \tfrac{1}{2}(\tfrac{1}{2}) = \tfrac{1}{4} = 0.25$$

and the genotype frequencies are

$$AA : (\tfrac{3}{4})^2 = \tfrac{9}{16} = 0.56$$
$$Aa : 2(\tfrac{3}{4})(\tfrac{1}{4}) = \tfrac{6}{16} = 0.38$$
$$aa : (\tfrac{1}{4})^2 = \tfrac{1}{16} = 0.06$$

After two generations, the *A* frequency has increased from 0.50 to 0.75, and the *a* frequency has dropped from 0.50 to 0.25.

Under this type of extreme selection pressure, *a* continues to decrease in frequency; given enough time, it would disappear from the population were it not replaced by new mutations and migration. This is a dramatic case, but genotypes with a fitness value of 0 do occur. In humans, Tay-Sachs disease, Lesch Nyhan syndrome, and Duchenne muscular dystrophy are genetic conditions in which selection against the trait is 1 (fitness is 0) because individuals with these disorders do not survive long enough to reproduce. The same is true for nonlethal genotypes that prevent reproduction through sterility or through the inability to find a mate.

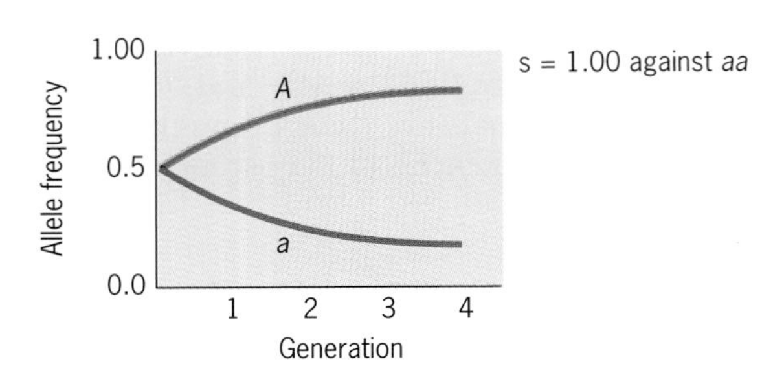

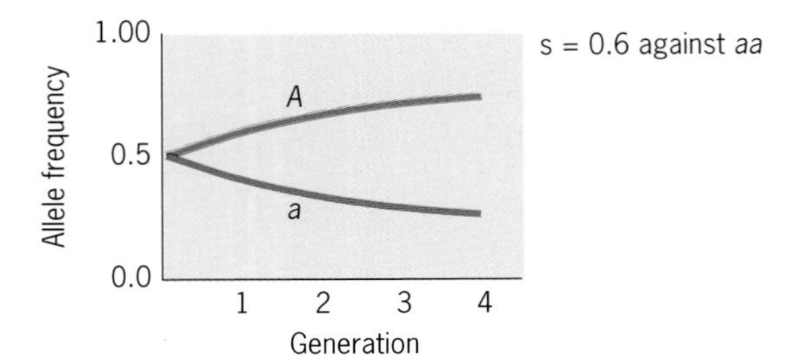

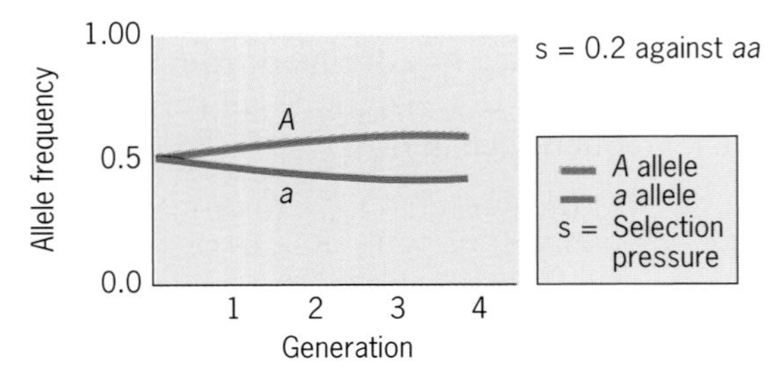

Figure 27.5 The effects of three different selection pressures on the frequencies of a pair of alleles (*A* and *a*) over four generations. At generation 0, $A = a = 0.5$.

As discussed earlier, selection pressures vary from 0 to 1. The stronger the selection pressure, the more rapid is the rate of loss of a deleterious allele from a population (Figure 27.5).

Selection operates on entire phenotypes and therefore on entire genotypes. Gene complexes or linkage groups are often a unit of selection. Thus, although an allele may by itself reduce fitness, it may be preserved and actually increase in frequency in a population because it is part of a linkage complex that, when taken as a whole, confers some selective advantage. For example, gene loci contained within an inversion are held together as a block because crossing over within an inversion produces aneuploid gametes (Chapter 7). Thus all the genes in an inversion heterozygote function as a "supergene" and are acted upon as a unit.

A possible example of a deleterious allele that is not eliminated by selection is the recessive cystic fibrosis allele (Chapter 20). Although this is a serious, usually fatal metabolic disorder, one particular *CF* allele carrying the *ΔF508* mutation appears in the heterozygous state in the Caucasian population at a frequency higher than expected based on its homozygote frequency and its low fitness. One reason for this higher frequency may be something called the "hitch hiker effect": This *CF* allele may be tightly linked to an allele of another gene that is favored by selection. Thus, the *ΔF508 CF* allele is "carried along for the ride."

Two Forms of Natural Selection

Darwin recognized two different forms of natural selection. In one form, the organism struggles directly with the physical constraints of the environment to survive and reproduce. For example, a desert plant struggles to conserve water and survive drought conditions. The fitness of the desert plant does not depend on whether it is common or rare in the population because it is an individual struggle with the physical environment that determines survival, not a competitive struggle with other members of the population. This form of fitness is **frequency-independent.**

The second form of selection involves competition between organisms for a limited resource. In this form, the relative frequencies of the competing phenotypes determine the fitness values, and these fitness values change as the composition of the population changes. This form of fitness is called **frequency-dependent fitness.** A well-documented example of frequency-dependent selection is mimicry in butterflies. Some butterfly species, such as the monarch with its vivid orange and black color pattern, are distasteful to birds (Figure 27.6). Birds quickly learn to avoid the monarch. A palatable butterfly species, such as the viceroy, which has evolved a color pattern similar to the monarch, is protected from predation because potential predators perceive it as unpalatable. The viceroy mimics the monarch and is thus afforded protection. However, as the frequency of the mimic increases in the population, predators learn that butterflies with the monarch pattern are not always distasteful. The protective value of the monarch pattern diminishes, and predation pressure increases on the mimics. The selective advantage of mimicry is lost. In frequency-dependent selection, the less frequent the mimic the greater its fitness.

Selection in Natural Populations

Natural selection can be demonstrated in the laboratory, but it is more difficult to demonstrate in nature. One type of selection operating in nature is **heterozygote superiority**. In this type of selection, different genotypes are maintained in the population through the superior fitness of the heterozygote. E. B. Ford re-

(a) (b)

Figure 27.6 A blue jay eating a monarch butterfly (*a*), which induces vomiting in the jay (*b*). This experience causes the jay to avoid monarchs or palatable butterflies that mimic them.

ferred to this phenomenon as **balanced polymorphism** because it provided a way to maintain, by natural selection, genetic variability in a population through the selective advantage of heterozygotes. A gene is considered to be polymorphic if at least two alleles of the gene are maintained in a population with a frequency of at least 1 percent for the second most frequent allele. A well-known and widely cited example of balanced polymorphism is the sickle-cell allele in humans (Figure 27.7). Heterozygotes for this allele and the normal β globin allele (Hb^A/Hb^S) survive and reproduce better than either normal (Hb^A/Hb^A) or sickle-cell homozygotes (Hb^S/Hb^S) in regions where the malarial parasite is present.

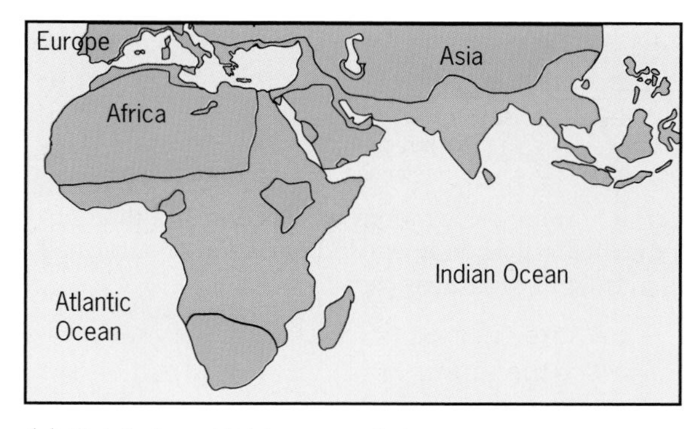

(*a*) Distribution of Falciparum malaria.

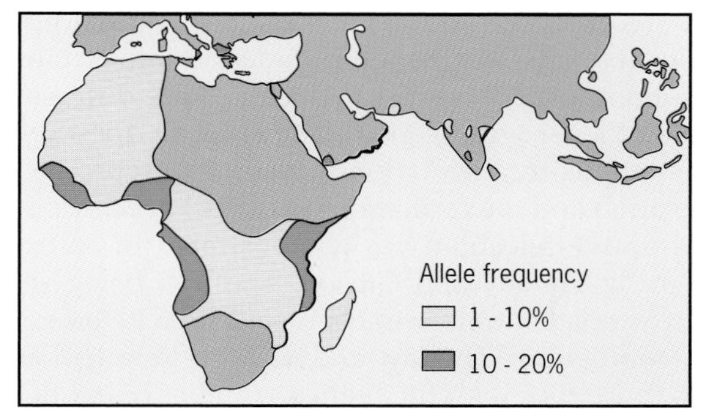

(*b*) Distribution of sickle-cell anemia allele (Hb^S).

Figure 27.7 Distribution of Falciparum malaria (*a*) and the distribution of the sickle-cell anemia allele (Hb^S) (*b*) in the Old World.

Another well-known example of selection in a polymorphic genetic system is **industrial melanism** in the moth *Biston betularia.* Over the past hundred years or so, several moth populations have changed color from whitish-gray to charcoal gray. This change has accompanied industrial expansion in regions of Germany and England where factory smoke and soot have discolored the trees and buildings.

The dark color in moths is determined by a dominant allele *C*, whereas the whitish-gray color is determined by a recessive allele *c*. The two alleles are maintained at different frequencies in different environments: In polluted environments, the *C* allele is present in higher frequency than *c*; in less polluted regions the *c* allele is present in higher frequency. These observations indicate that the change in color from whitish-gray to charcoal gray in polluted environments is the result of the dark variety's increased fitness in the polluted environment. Sitting on a tree darkened by pollution, the dark variety is less conspicuous to bird predators than the light-colored variety. In effect it is camouflaged. In the countryside, the situation is reversed: the light variety is now less conspicuous to predators against the lighter background (Figure 27.8). The most reasonable interpretation of these observations is that selection favors moths that blend best with their environment. Is there experimental evidence to support this hypothesis?

H.B.D. Kettlewell demonstrated that selection is, indeed, responsible for the color polymorphism in these moth populations when he showed that in industrial regions, predator birds eat darker moths less frequently than lighter moths. In the countryside, which is far less polluted, the opposite is true.

(*a*) (*b*)

Figure 27.8 Melanic form of moth *Biston betularia* resting on an unpolluted, lichen-covered tree trunk (*a*). A typical form of the moth resting on a polluted, soot-covered tree trunk (*b*).

Industrial melanism in *Biston betularia* also illustrates frequency-dependent fitness. As the frequency of the camouflaged variety increases, its fitness decreases because the predators learn to identify it more efficiently.

Modes of Natural Selection

Selection operates in populations in different ways and with different consequences. In this section, we consider three different modes of selection: directional selection, stabilizing selection, and disruptive selection.

Stabilizing Selection Sometimes selection favors an optimal phenotype and works against the extremes, a regime called **stabilizing selection** (Figure 27.9*a*). Human birth weight is an example of stabilizing selection. This is a heritable trait, and babies born around 8 pounds have the best survival chance, whereas babies less than 7 pounds and greater than 9 have reduced survival chances.

Stabilizing selection also occurs when selection favors the heterozygote over the two homozygote classes. For example, in a malarial environment, heterozygotes for the sickle-cell allele (Hb^A/Hb^S) are adaptively superior to either homozygotes for the mutant allele or homozygous normal individuals (Hb^SHb^S and Hb^A/Hb^A; Chapter 12).

Directional Selection Selection that favors an extreme phenotype, thus shifting the population mean in one or the other direction, is called **directional selection** (Figure 27.9*b*). Directional selection regimes are commonly observed in agricultural situations where breeders select for the more economically advantageous expressions of traits such as increased egg production, increased yield per acre, increased milk production, or increased fruit size. Directional selection occurs when the environment is steadily changing.

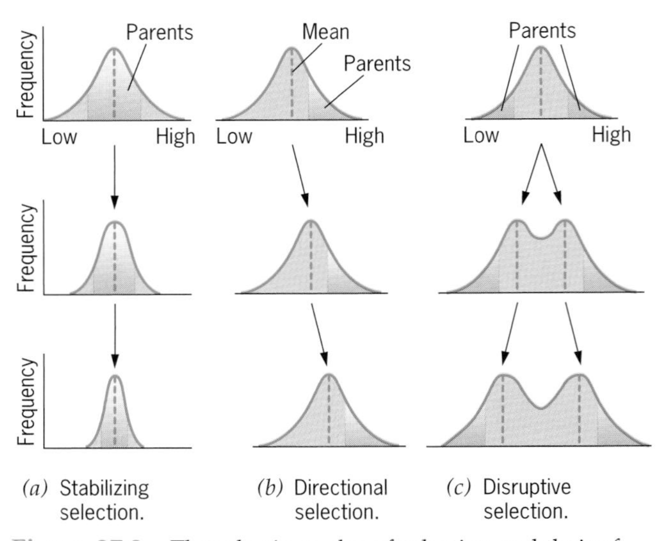

(*a*) Stabilizing selection. (*b*) Directional selection. (*c*) Disruptive selection.

Figure 27.9 Three basic modes of selection and their effects on the mean (dashed lines) and variation of a normally distributed quantitative character.

Disruptive Selection **Disruptive selection** favors the more extreme variants and selects against the intermediates (Figure 27.9*c*). A major consequence of disruptive selection is to fragment a population into two or more subpopulations, each of which is quite different from the other(s) with respect to the trait in question. Over time, this fragmentation may lead to reproduction isolation and the formation of distinct species.

Disruptive selection was demonstrated in an experiment by Thoday and Gibson. The trait being selected was bristle number in *Drosophila*. In a heterogeneous population, Thoday and Gibson selected as parents those flies with the highest number of bristles and those with the lowest number. Those flies with an average number did not contribute genetic information to the next generation. The investigators counteracted this selection of flies with high and low bristle numbers by placing all the parent flies in a single bottle so that they had opportunities to intermate. After about 24 hours, they discarded the males and separated the females into "high" and "low" bristle number lines. They again selected the progeny from these matings for "high" and "low" bristle numbers, placed them in a common bottle, and allowed them to mate. The males were subsequently discarded and the females were again separated into "high" and "low" lines in separate bottles. This cycle continued for several generations.

The results of Thoday and Gibson's experiment showed that the population diverged into "high" and "low" lines, in spite of the crossbreeding opportunities. The "highs" preferred to mate with other "highs" and the "lows" with other "lows," even though they could have interbred (Figure 27.10). The populations were becoming reproductively isolated from each other.

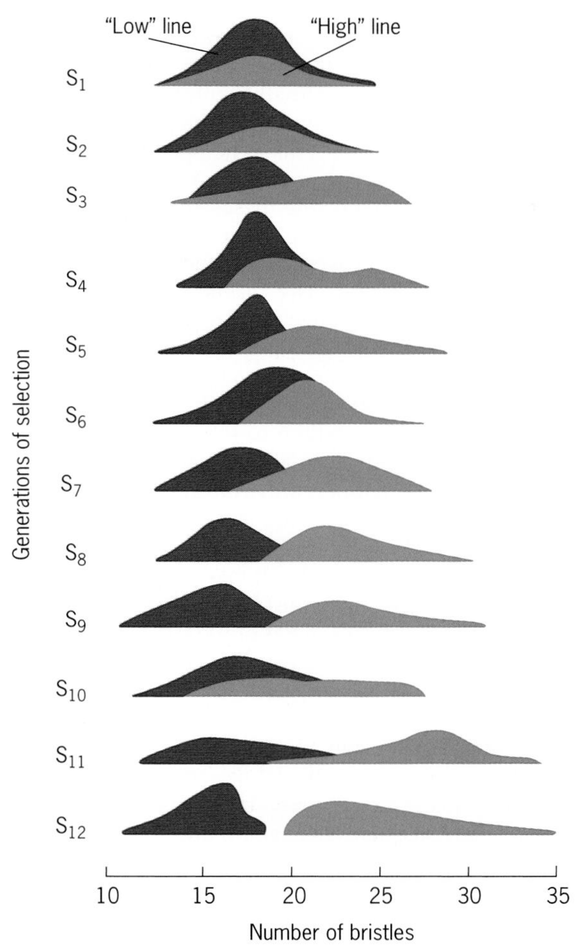

Figure 27.10 Divergence of "high" and "low" lines of *Drosophila melanogaster* that was obtained by disruptive selection despite the opportunity for random mating to occur. Selected males and females from "high" and "low" lines were placed together in a single mating chamber.

Key Points: **Fitness (w) is the relative reproductive ability of a genotype. Selection (s) refers to the intrinsic differences in the ability of genotypes to survive and reproduce. Quantitatively, $w = 1 - s$. The fitness of a phenotype may depend on its frequency in the population.**

When the heterozygote is the most fit genotype, a balanced polymorphism results. Directional selection favors an extreme phenotype. Stabilizing selection favors the heterozygote or intermediate type over the extremes. Disruptive selection acts against the intermediate type.

CHANGES IN THE GENETIC STRUCTURE OF POPULATIONS

Deviation from Random Mating: Inbreeding

Any departure from random mating naturally leads to complications in the relationships between allele frequencies and genotype frequencies. The process of repeated self-fertilization discussed in Chapter 25 illustrates the general effects of a form of nonrandom mating called **inbreeding**. During inbreeding, the allele frequencies remain the same, but the genotype frequencies change. Inbreeding, or **consanguineous mating**, occurs whenever mates are genetically related. Self-fertilization is the most extreme example of inbreeding, but other examples (Figure 27.11) such as the mating of full siblings, first cousins, parents and offspring, uncle and niece or aunt and nephew, and second cousins have the same general effect—namely, to increase the frequency of homozygotes and decrease the frequency of heterozygotes. As a result, Hardy–Weinberg genotypic frequencies are not obtained.

Inbreeding allows rare alleles to become homozygous, and it greatly increases the likelihood that rare recessive disorders will appear in the offspring of related parents. For example, in xeroderma pigmentosum, a rare autosomal recessive condition of DNA repair, more than 20 percent of the cases reported come from marriages between first cousins. Inbreeding also explains the high frequency of albinism in the Hopi Indians of northeast Arizona (see Human Genetics Sidelight: A Cultural Effect on the Frequencies of Alleles).

The general genetic risk to the offspring of marriages between related people is not as great as is sometimes imagined. The risk to the offspring of a first cousin marriage is less than twice the overall risk for marriages between unrelated persons. The risk to

HUMAN GENETICS SIDELIGHT

A Cultural Effect on the Frequencies of Alleles

Human populations sometimes produce allele frequencies that seem to defy explanation by the usual rules. A fascinating case involves albinism among certain Native Americans.

The frequency of albinism varies widely among different human populations. In Norway it affects about 1 in 10,000 people; in Southern Europe the frequency is more on the order of 1 in 30,000. One in 20,000 is a usual figure for albinism among Caucasians in the United States. But among certain Native American tribes, notably the Cuna of Panama, the Hopi of Arizona, and the Jemez and Zuni of New Mexico, frequencies of about 1 albino in every 200 people have been observed repeatedly. What accounts for this incredibly high frequency of albinism?

The most detailed studies to date deal with the Hopi. Their high frequency of albinism has been recognized for at least a century, but it took research by geneticists from Arizona State University to give the first analytical account of the genetic structure of the population.[1]

Hundreds of years ago, the tribe sought refuge from its traditional enemies, the Navajo and the Utes, by building homes on the tops of mesas in northeastern Arizona. Indeed, one of their towns, Oraibi, is recognized as the oldest continuously inhabited community in the continental United States, having been established about 850 years ago. During historic times the tribe has never numbered more than a few thousand, and probably has never been large. The population, then, is a small one.

Can one find a means of selection, of reproductive advantage based on some *biologic* superiority, that has favored the high frequency of albinos? All evidence indicates that albinos are not biologically superior. They are known to have higher rates of skin cancer and, often, poor eyesight (myopia and lateral nystagmus). Their extreme sensitivity to sunlight necessitates constant guarding against sunburn in the hot American Southwest. Natural selection acting on some element of biological advantage, then, does not seem to offer any answer.

Could genetic drift be the explanation? The Hopi are divided into three major enclaves of approximately equal size, one on each of the three tips of Black Mesa. The three major communities (called First Mesa, Second Mesa, and Third Mesa) are located on a generally east-west line. First Mesa lies more than 11 miles to the east of Second Mesa; Third Mesa lies westerly of Second Mesa at a distance of about 10 miles. Historically, the distances and the rugged terrain have kept the three populations rather separate; they are definitely not the "randomly mating population" needed for a simple treatment of community structure. In the Arizona State study, the 26 albinos identified were either from Second Mesa or Third Mesa and about equally divided between the two. Such a pattern demonstrates that the populations of the three communities are not randomly intermixed. Would genetic drift account for the high frequency of albinism in two *independent* mesas? It does not seem likely (and the cases of albinism in the Jemez, Zuni, and Cuna tribes further argue against genetic drift as the causal mechanism).

The construction of detailed pedigrees revealed considerable inbreeding within the Hopi communities. The Hopi of First Mesa were the least inbred as a population, having had considerable intermixing with a nearby population of Tewa. The Second Mesa population was the most inbred and did indeed have just slightly more albinos than Third Mesa. Inbreeding, then, seems a plausible explanation for the high frequency of albinos in the Second and Third Mesa populations: remember, it tends to bring recessive alleles into homozygous condition. It is also recognized that the tribe likely derives from a rather small group of migrants into the area in the distant past; thus, founder effects must also be considered. We can do little more than recognize their influence, however, because we know virtually nothing about the genetic makeup of the original founders of the tribe.

Yet another mechanism seems to be operating, one that derives directly from the nature of albinism and the Hopi life style. Because the tops of the mesas contain little good soil, the Hopi have had to locate their farms in the broad lowlands to the south and west of the mesas, where sufficient water could be found. The farms, then, were often many miles from the mesas. They were worked primarily by the men and boys of the tribe, who would customarily rise well before sunrise, walk to the fields, labor throughout the day, and return to the mesas each evening.

Albinos, being very sensitive to the sun, were not expected to work in the fields; thus albino males stayed behind on the mesas with the women and children taking up domestic occupations such as weaving in order to contribute to the tribe's economy. They spent their long days essentially alone with the women of the tribe. It takes little imagination to understand that the albino males had a marked reproductive advantage over the other males of the tribe.

Possible founder effects, inbreeding, and a reproductive advantage generated purely by the culture of the people clarify why the Hopi have such a high frequency of albinism. In this instance, humans have added a significant new dimension—a form of cultural selection—to the usual rules governing allele frequencies. There are many human genes whose allele frequencies have no immediate explanation based on the usual rules. As we learn more about natural selection and its operation in human societies, we will undoubtedly gain insights into some of these unusual cases. Among those insights may well rest a number of uniquely human stories similar to that of the Hopi.

[1]Woolf, C. M., and F. C. Dukepoo. 1969. Hopi Indians, inbreeding, and albinism. *Science* 164:30–37.

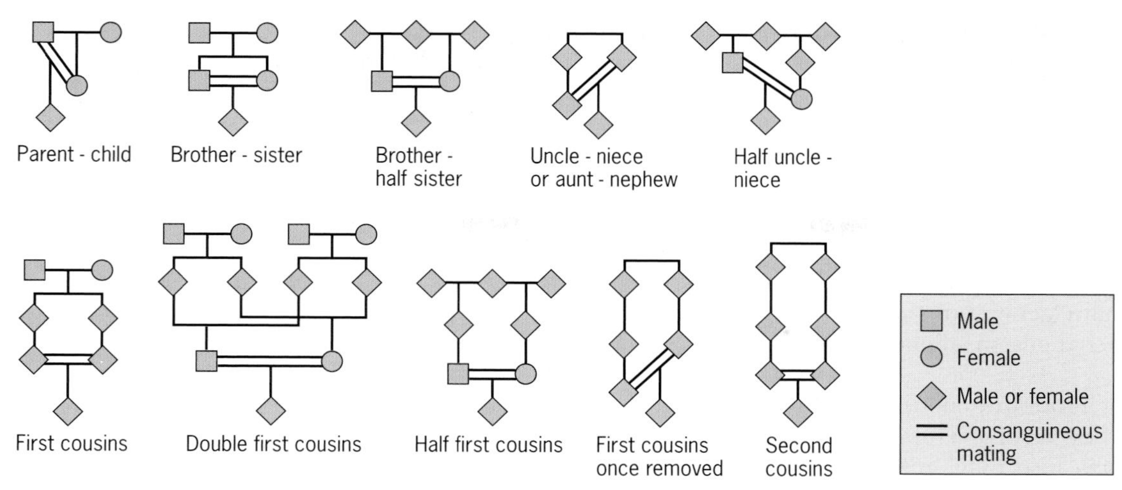

Figure 27.11 Various types of consanguineous matings.

the offspring of third cousin matings is basically the same as that for unrelated individuals.

Genetic Drift: Random Changes in Small Populations

During reproduction, allele frequencies may change as a consequence of random events. Consider, for example, a pool of gametes, 50 percent of which are white and 50 percent red (Figure 27.12). A random sample of 10 gametes produces 5 individuals who form the first generation. Six of the gametes are white and 4 are red. The gamete pool (allele frequencies) in the first generation is now 60 percent white and 40 percent red. A sample of 10 gametes from the first generation produces 5 individuals who form the sec-

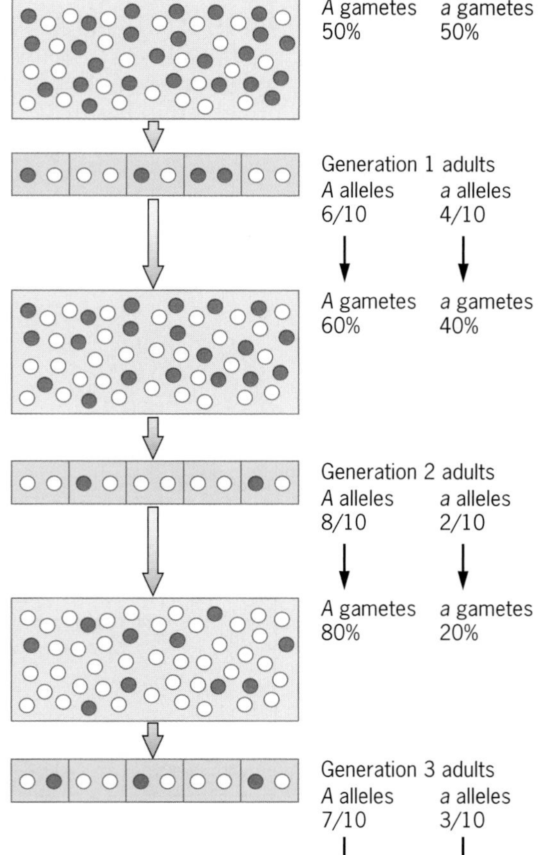

Figure 27.12 Genetic drift operating in a small population. The *a* allele frequency changes from 0.5 to 0.3 as a consequence of sampling errors. There is a chance that the *a* allele could become extinct.

ond generation. Eight of the gametes are white and 2 are red. Our starting allele frequencies were 50 percent red and 50 percent white, but after two generations, the frequencies shifted to 80 percent white and 20 percent red. The frequencies changed as a result of chance fluctuations in small samples.

Such random changes in allele frequencies may occur in small populations. The cumulative effect of these statistical fluctuations is a process called **random genetic drift**. Over many generations, genetic drift can produce significant changes in allele frequencies in small populations.

All the effects of random genetic drift have one feature in common: they involve a loss of genetic variability. As populations lose their genetic variability, homozygosity increases and heterozygosity decreases. Alleles may become lost or fixed in a population because of genetic drift. Warwick Kerr and Sewall Wright showed how genetic drift operates in small populations of *Drosophila*. They started a *Drosophila* population with eight flies: four females and four males. The females were of three genotypes: +/+, +/*f* and *f/f*, where *f* (forked bristles) is an X-linked recessive allele, in a 1:2:1 ratio. Two of the males were +/Y and two were *f*/Y. The *f* and + allele frequencies were 0.5 in both sexes. At each generation, four males and four females were randomly selected to be parents for the next generation. If all eight parental flies were phenotypically wild-type, they were interbred; if the offspring were all wild-type, then Kerr and Wright concluded that the *f* allele was absent from the parents and the + allele had become fixed. If all eight parental flies were forked, then the + allele was absent and the *f* had become fixed. If the eight parent flies were both *f* and + phenotypically, or if they were wild-type but produced forked offspring, there was no allele fixation and mating continued.

Using this protocol, Kerr and Wright set up 96 lines and recorded at each generation whether a line was fixed for *f*, fixed for +, or unfixed (Table 27.1). In the early stages of the experiment, little allele fixation took place. By generation 7, alleles were becoming fixed at a greater frequency. By generation 16, nearly 73 percent of the lines were fixed for either the + or the *f* allele. That more lines were fixed for + than for *f* (41 versus 29) suggested natural selection was acting against the forked phenotype. Genetic drift causes fluctuations in allele and genotype frequencies, but natural selection continues to operate throughout the process.

Genetic drift also occurs in human populations. Bentley Glass studied a population of "Dunkers" (German Baptist Brethren), numbering about 350, in south-central Pennsylvania. He analyzed genetic traits that were either adaptively neutral or had extremely low selection pressures operating on them (the ABO and MN blood groups). The Dunkers are a very small population, which makes them ideal for the operation of genetic drift. Furthermore, they rarely interbreed with members of the surrounding non-Dunker population. In addition, the Dunkers' parent population and the population that surrounds the Dunkers are genetically well defined with respect to the traits in question. A model of the population and its analysis are given in Figure 27.13.

TABLE 27.1
Genetic Drift in Small Populations of *D. melanogaster*

Generation	*Wild (+)*	*Unfixed*	*Forked*
0	0	96	0
1	1	94	1
2	1	92	3
3	2	87	7
4	7	79	10
5	10	70	16
6	11	66	19
7	16	59	21
8	17	56	23
9	20	52	24
10	24	47	25
11	29	39	28
12	31	37	28
13	34	34	28
14	37	30	29
15	38	29	29
16	41	26	29

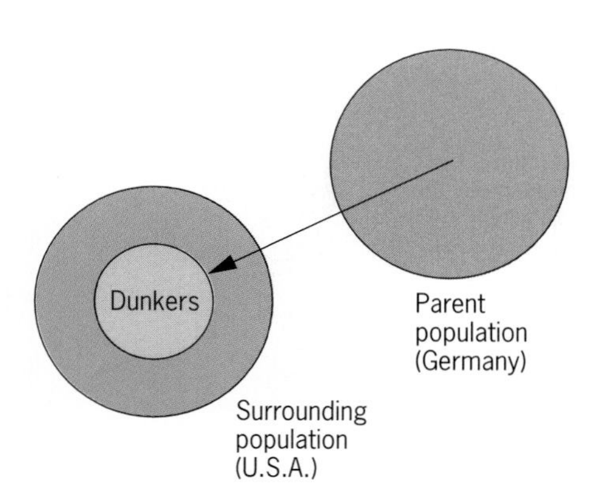

Blood type	Frequency		
	U.S.A.	Dunker	Germany
A	0.40	0.60	0.45
B or AB	0.15	0.05	0.15
M	0.30	0.45	0.30
MN	0.50	0.41	0.50
N	0.20	0.14	0.20

Figure 27.13 A model of the Dunker community, comparing some of its genotype frequencies with the frequencies in the parent and surrounding populations.

The Dunkers migrated from Germany early in the eighteenth century, and their communities have in large part remained reproductively isolated from each other and from the surrounding U.S. population. Marriages between Dunkers and individuals from other communities have been monitored so that compensations in allele frequencies can be made. The frequency of blood type A is 0.6 in the Dunkers but only 0.40 in the United States and 0.45 in the parent German population. The B and AB blood types are almost entirely absent from the Dunkers (0.05), but they comprise 0.15 of the U.S. and German populations. This same type of allelic fluctuation was found for a number of other traits as well. In all cases, the alleles in the Dunkers have frequencies unlike those of either the Germans or the Americans surrounding them, and unlike anything in between. Instead, the frequencies fluctuated widely from one extreme to the other, strongly suggesting that genetic drift has been at work.

Some of the most dramatic examples of the power of genetic drift can be found in island fauna where populations might have been started by a few migrant individuals. Nowhere are the effects of genetic drift more evident than in the Hawaiian Island *Drosophila* species. The Hawaiian Islands are remote and geologically well characterized.

At some time millions of years ago, one or a very small number of founder "picture-winged" *Drosophila* species from Asia managed to get to the island of Kauai. Because their numbers were so small, perhaps only a single pregnant female, these founders (or founder) carried alleles and chromosome rearrangements in frequencies that did not reflect the parent population. This was followed by the expansion of the founder(s) into a large population that diversified into the various niches on the island. Additional founding events took place when small numbers of flies moved from Kauai to the younger islands in the Hawaiian chain (Figure 27.14). There were further founding events from older to younger islands, so that today the islands have more than 100 species of picture-winged *Drosophila* species.

Hampton Carson has related all the picture-winged species to each other and to a presumptive founder from the Asian mainland by studying patterns of chromosome inversions in the various species. For example, Carson determined that the 26 species on the youngest island, Hawaii, are all derived from 19 founders: 15 from the Maui complex, 3 from Oahu, and 1 from Kauai. Genetic drift characterized each founding event, which was followed by population expansion and diversification. Genetic drift, operating through this **founder principle**, has been instrumental in forming these *Drosophila* species.

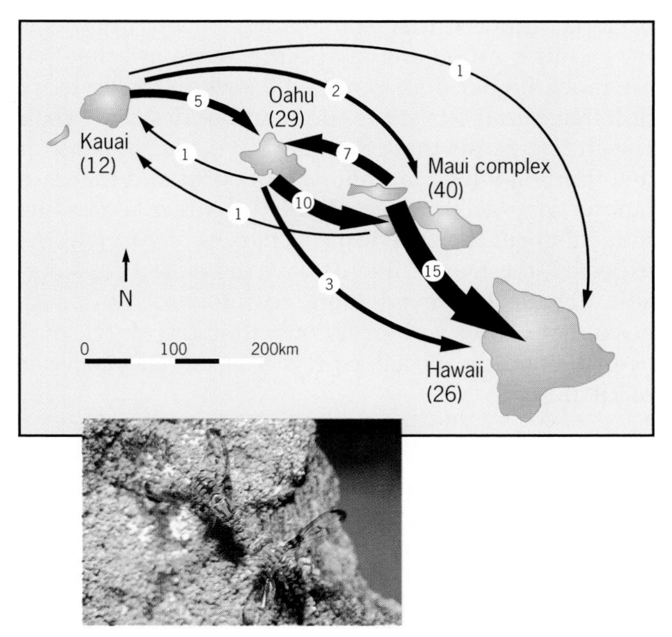

Figure 27.14 A summary of the proposed founder events invoked to explain the origin of the picture-winged *Drosophila* species (inset) on the Hawaiian Islands. The width of the arrows is proportional to the number of proposed founders in each case. The number of species on each island is given in parentheses.

Migration Changes Allele Frequencies

One population may modify the genetic makeup of another through migration. Figure 27.15 shows a simple case in which a large continental population donates individuals (and therefore genes) to a small island population. The rate of migration is m, which is equal to the fraction of genes on the island that are replaced by genes from the continent each generation. If q_i is the frequency of a particular allele, a, on the island and q_c is the corresponding frequency on the continent, then after migration, the frequency on the island will be

$$q_i' = (1 - m)q_i + mq_c$$

This follows because a fraction $(1 - m)$ of the island's genes are not replaced, and among them the fre-

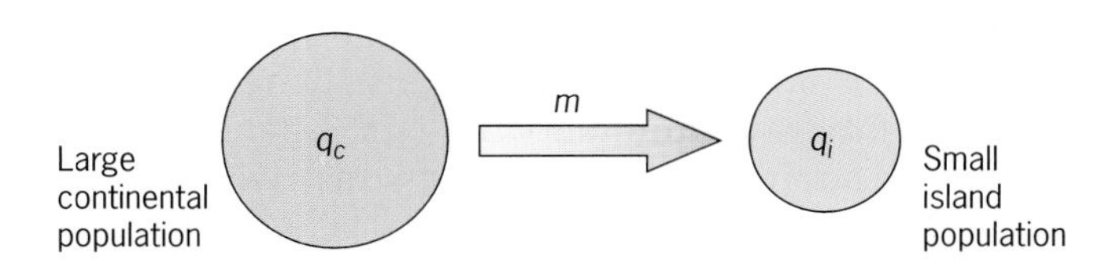

Figure 27.15 A model of migration between a large continental population and a small island.

quency of a is q_i, whereas a fraction m of the island's genes are replaced, and among the replacements the frequency of a is q_c. The change in the frequency of a (Δq_i) in one generation will therefore be

$$\Delta q_i = q_i' - q_i = m\,(q_c - q_i)$$

The change in the frequency of a will become zero either when migration stops ($m = 0$) or when the frequency of a on the island equals the frequency of a on the continent ($q_i = q_c$).

This example shows that migration makes populations genetically similar and also allows for the exchange of new genetic information between populations. More complicated cases arise when migration goes in both directions or when there are several populations capable of exchanging genes. Whether or not a population maintains its own genetic identity depends on how many migrants it receives. Theoretical studies have indicated that even a few migrants per generation are sufficient to eliminate the differences among geographically separated populations. Thus migration can be a powerful homogenizing force in population genetics.

Mutations: A Source of New Variation

In the broadest sense, a mutation is any change in the genetic material, and this includes chromosome rearrangements, deletions, duplications, as well as point mutations. In this section, we restrict our discussion to point mutations: qualitative changes within a gene (Chapter 13). Mutation is important in population genetics, and therefore in evolution, because it provides the raw material on which natural selection acts.

From the perspective of population genetics and evolution, mutations are classified as beneficial, harmful, or neutral, depending on whether they increase, decrease, or do not change the viability or fertility of their carriers. In other words, mutations either increase, decrease, or do not alter fitness. Natural selection and genetic drift determine whether a mutation persists in a population or is eliminated.

Strains of some organisms produce mutations at abnormally high rates. This high mutability may be due to error-prone DNA replication, defective DNA repair systems, or transposable genetic elements. Mutator strains have been used to study the relationship between fitness and mutation rate. In some instances, a mutator strain is more fit than a strain with a more normal mutation rate (10^{-5} to 10^{-6} per gene per generation); this finding suggests that under some circumstances a higher mutation rate is advantageous. However, the mutation rate cannot be too high because the frequency of harmful mutations would be too great for the population to survive. A mutation rate of zero is also undesirable, because mutation is the only source of new genetic variation in a constantly changing environment. In nature, selection favors a balance so that mutation rates produce enough, but not too much, genetic variability for the population.

Are Mutations Truly Random?

Lamarck, Darwin, and others before them suggested that specific mutations arise in response to a specific environmental stimulus. Does the environment direct mutations to occur in specific directions, or are mutations spontaneous and random? Although there is no clear proof for the occurrence of **postadaptive mutations**, directed adaptive mutations that occur *after* exposure to a selective environment, there are many recorded instances of **preadaptive mutations**: undirected, spontaneous mutations that were subsequently selected. One of the most elegant experiments showing that mutations are preadaptive was Joshua and Esther Lederberg's **replica-plating experiment** (see Chapter 13 for a discussion of replica plating) showing that drug-resistant mutations existed *before* bacteria were actually exposed to the drugs. Other experiments have confirmed the preadaptive nature of mutations, but recently experiments by John Cairns and his colleagues have challenged the idea that all mutations are preadaptive. Their experiments with *E. coli* bacteria indicate that some selective environments may induce mutations in specific genes at rates that are much higher than expected based on the assumption that mutations are random and undirected. Their research suggests that cells may have mechanisms for determining which mutations will occur and that mutations may arise nonrandomly, yielding a product that enhances the individual's chances of survival. As expected, these experiments have generated considerable debate. However, more recent work now suggests that Cairns's observations, though correct, may be explained by special aspects of bacterial metabolic regulation.

Key Points: **Inbreeding, the mating between genetically related individuals, increases homozygosity and decreases heterozygosity. Genetic drift is the random fluctuation in allele frequency from generation to generation resulting from small population size.**

One population may alter the genetic makeup of another population through migration. The net result of migration is to homogenize genetic differences between populations.

Some mutations are harmful, some are beneficial, and some are neutral or nearly so. Experiments

have shown that most, if not all mutations are preadaptive, not postadaptive.

A BALANCE BETWEEN SELECTION AND MUTATION

Deleterious alleles are introduced into a population by mutation from wild-type alleles. If the mutant allele is recessive, it will be eliminated by selection against the recessive homozygotes or by genetic drift. However, if the mutant allele is dominant, it will be eliminated by selection against the heterozygote. In either case, mutation-selection equilibrium is achieved when the rate of introduction of new alleles is balanced by the rate of elimination by selection (Figure 27.16).

The equilibrium frequency of a harmful recessive allele is obtained by noting that selection acts only against the recessive homozygotes. The rate of elimination by selection is approximately sq^2, where s is the selection pressure and q^2 is the recessive homozygote frequency (Figure 27.17). If the mutation rate is u, the rate of introduction of the recessive allele is up, which is essentially u since p is very close to 1. Thus, at equilibrium, rate of introduction = rate of elimination implies that $u = sq^2$ and $q^2 = u/s$. Thus the equilibrium frequency of the recessive allele will be $q = \sqrt{u/s}$. For a dominant allele, a similar argument leads to the conclusion that the equilibrium frequency is approximately u/s.

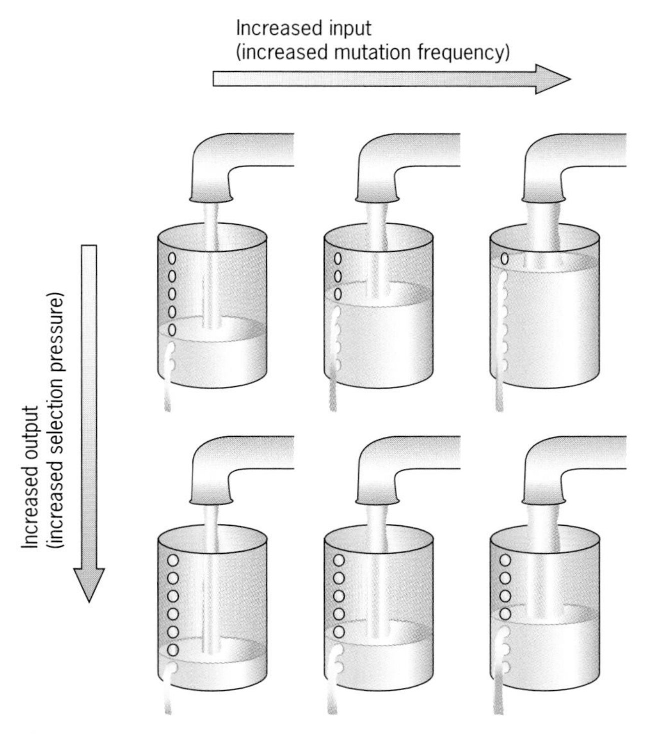

Figure 27.16 A model of mutation-selection equilibrium. Water level indicates the equilibrium frequency of an allele. Water input indicates mutation frequency. Hole size indicates selection pressure. When the holes are small (small selection pressure), the water level (equilibrium level) is higher than when the holes are larger (greater selection pressure).

Let's consider a case of equilibrium reached in the human population in which mutation pressure increases the frequency of a dominant allele and selection intensity decreases its frequency. **Achondroplasia** is a form of dwarfism caused by a dominant allele that is lethal in the homozygous state. About 1 in every 25,000 births is an achondroplastic, and about 80 percent of these births are to normal parents with no family history of the trait. Because this is a fully penetrant trait, these 80 percent are due to new mutations. This information allows us to estimate the mutation rate for the achondroplasia allele:

1. 1/25,000 births means 1/50,000 gametes carry a mutation.
2. 0.8/50,000 of these mutations are new.
3. The mutation rate = 1.6/100,000 = 1.6×10^{-5}.

Selection operates on achondroplasia in two ways. The homozygous dominant is lethal, and the heterozygote has a fitness value (w) of about 0.2 (selection pressure (s) against the achondroplasia allele is about 0.8). We thus estimate the equilibrium frequency for the achondroplasia allele to be

$$u/s = 1.6 \times 10^{-5}/0.8 = 2 \times 10^{-5}$$

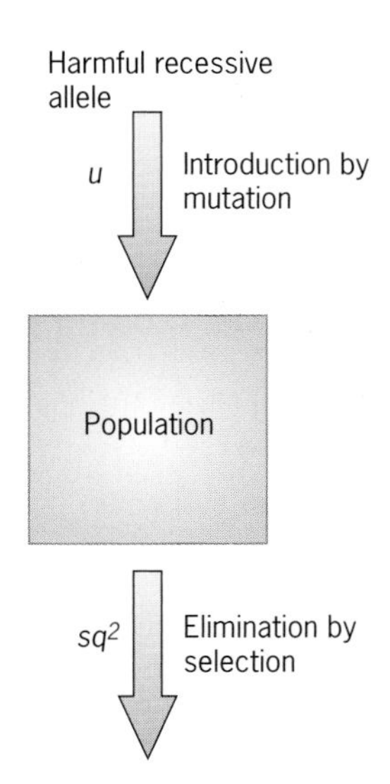

Figure 27.17 Mutation-selection balance for a recessive allele. Genetic equilibrium is reached when the introduction of the harmful allele into the population by mutation is offset by the elimination of the harmful allele by selection (s) acting on aa (frequency q^2).

Key Points: **An equilibrium between mutation and selection will be achieved when the rate of introduction of new alleles is balanced by the rate of elimination by selection.**

THE ROLE OF POPULATION GENETICS IN GENETIC COUNSELING

Knowledge of allele frequencies in specific populations is extremely important in genetic counseling. Consider, for example, a family that has a history of cystic fibrosis, an autosomal recessive disorder (Figure 27.18). A woman (III-1) whose uncle (II-1) has cystic fibrosis (CF) wants to know what the chances are that she will have an affected baby. First, we must determine the probability that the woman is a carrier. Her grandparents (I-1 and I-2) must be heterozygotes because they have an affected child. The woman's mother (II-3) is unaffected, so she obviously cannot be *aa*. Thus she is either homozygous normal (*AA*) or heterozygous (*Aa*). The probability that she is *Aa* is 2/3 because in a classic 1:2:1 Mendelian ratio where 1/4 of the possible outcomes are excluded (*aa* in this case), 2/3 of the remaining possibilities are *Aa*. If she is a carrier, then the probability that her daughter (III-1) inherited a CF allele from her is 1/2. Thus the woman's chance of being a carrier is 2/3 (the probability that II-1 is *Aa*) × 1/2 (the probability that III-1 received the CF allele from II-1) = 1/3. The woman's husband is of Northern European ancestry, and the frequency of CF heterozygotes in that population is about 1/25. Thus the chance that the woman and her husband are both carriers is 1/3 × 1/25 = 1/75. If they are both carriers, the probability that the couple will have an affected child is 1/4. Thus the overall probability that this couple will have an affected child is 1/3 × 1/25 × 1/4 = 1/300. If the woman's husband came from a population where the frequency of the CF allele was very low, then the chances that they would have a CF child would be essentially 0. Other genetic traits, such as Tay-Sachs disease and sickle-cell anemia, have strong ethnic associations. Knowledge of allele frequencies in specific populations and ethnic groups aids genetic counselors in risk assessment.

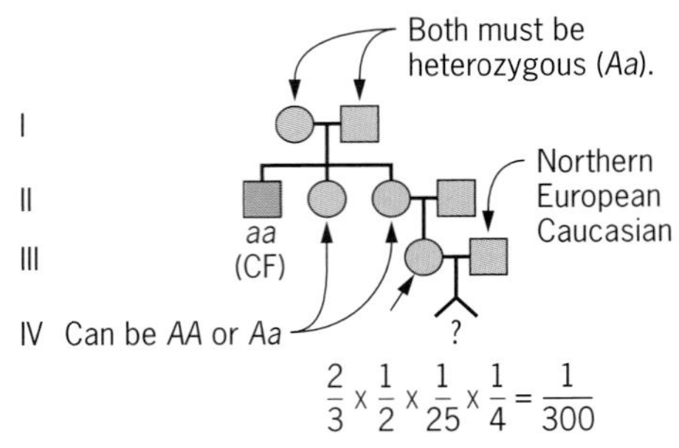

Figure 27.18 About 4 percent of the European Caucasian population is heterozygous for the cystic fibrosis allele (CF), a fact that is important in genetic counseling situations. In this family, the risk that the couple will have a CF child is about 1 in 300 because III-2 has a 1 in 25 chance of being a carrier. If III-2 came from a population where CF was extremely rare, the risk would approach 0.

Key Points: **Knowledge of allele frequencies in populations or ethnic groups helps genetic counselors more accurately assess the risk of inheriting a genetic disorder.**

TESTING YOUR KNOWLEDGE

1. Among a population of 700 South Dakota Sioux, the following blood types are observed:

O	A	B	AB
338	326	20	16

(a) Estimate the A, B, and O allele frequencies.

(b) How well does the observed distribution of phenotypes agree with those expected based on the calculated allele frequencies?

ANSWER

(a) Let p = *A* frequency, q = *B* frequency, and r = *O* frequency; furthermore $(p + q + r)^2 = p^2 + 2pq + q^2 + 2qr + 2pr + r^2 = 1$ for genotype frequencies *AA*, *AB*, *BB*, *BO*, *AO*, and *OO*, respectively. Since type O is genetically *OO*, r^2 = 338/700 = 0.483. Taking the square root of 0.483, we get r = 0.69, the frequency of *O*.

Solving for p and q, we get: $p^2 + 2pr + r^2 = AA + AO + OO$ = 326/700 + 338/700 = 0.949; $(p + r)^2 = 0.949$; taking the square root, we get $p + r = 0.974$; $p = 0.974 - r = 0.284$, the frequency of the *A* allele. Because $p + q + r = 1$, $q = 1 - (p + r) = 0.026$, the frequency of the *B* allele.

(b) To determine whether the phenotype frequencies agree with those expected based on our estimates of the *A*, *B*, and *O* frequencies, we plug the allele frequencies into the genotype frequency equation:

Type A = $AA + AO = p^2 + 2pr = (0.284)^2 + 2(0.284)(0.69) = 0.47$
Type B = $BB + BO = q^2 + 2qr = (0.026)^2 + 2(0.026)(0.69) = 0.04$
Type AB = $2pq = 2(0.284)(0.026) = 0.01$
Type O = $r^2 = (0.69)^2 = 0.48$

These are our expected phenotype frequencies; the observed frequencies were

Type A = 326/700 = 0.47
Type B = 20/700 = 0.03
Type AB = 16/700 = 0.02
Type O = 338/700 = 0.48

The observed and expected frequencies match very closely; the population is in Hardy–Weinberg equilibrium.

2. Hemophilia is caused by an X-linked recessive allele. In a population, you find a frequency of 1 hemophiliac per 4000 males. Assuming that this population is in Hardy–Weinberg equilibrium, how many hemophiliac females would you expect to find?

ANSWER

Males are hemizygous, so the frequency of the hemophilic allele (h) is $1/4000 = 0.00025$ in the male population. If the population is in equilibrium, the male and female allele frequencies are the same. Thus the expected frequency of hemophiliac females (hh) would be $(0.00025)^2$ or 6.2×10^{-8}.

3. Pattern baldness in humans is an autosomal, sex-influenced trait: it is dominant in males and recessive in females. A male with at least one copy of the pattern baldness allele (PB) is bald; for a female to be bald, she must be homozygous for the pattern baldness allele (PB/PB). The frequency of the pattern baldness allele in a population is 0.3. Assuming random mating, what is the expected frequency of pattern bald males and pattern bald females in a population of 500 males and 500 females?

ANSWER

If the frequency of PB is 0.3, the frequency of PB^+ is 0.7. With random mating, we expect the following genotype frequencies:

$$p^2(PB^+/PB^+) + 2pq(PB^+/PB) + q^2(PB/PB) = 1$$
$$p^2 = (0.7)^2 = 0.49$$
$$2pq = 2(0.7)(0.3) = 0.42$$
$$q^2 = (0.3)^2 = 0.09$$

Based on these genotype frequencies, we expect to find $(0.42 + 0.09)(500) = 255$ pattern bald males and $(0.09)(500) = 45$ pattern bald females.

QUESTIONS AND PROBLEMS

27.1 The frequency of an allele in a large randomly mating population is 0.2. What is the frequency of heterozygous carriers?

27.2 The incidence of recessive albinism is 0.0004 in a human population. What is the frequency of the recessive allele? (Assume random mating.)

27.3 The frequency of newborn infants homozygous for a recessive lethal allele is about 1 in 25,000. What is the expected frequency of the carriers of this allele in the population?

27.4 The following data for the MN blood types were obtained from Indian villages in North and Central America.

Group	Population Size	M	MN	N
Central American	86	53	29	4
North American	278	78	61	139

Calculate the frequencies of the L^M and L^N alleles for the two groups.

27.5 In a sample from an African population, the frequencies of the L^M and L^N alleles were 0.78 and 0.22, respectively. Assuming random mating what are the expected frequencies of the M, MN, and N phenotypes?

27.6 A locus has three alleles, A_1, A_2, and A_3, with frequencies 0.6, 0.3, and 0.1, respectively. Assuming random mating, what is the expected frequency of all the heterozygotes in the population?

27.7 Human beings carrying the dominant allele T can taste the substance phenylthiocarbamide (PTC). In a population in which the frequency of this allele is 0.4, what is the probability that a particular taster is homozygous ?

27.8 In a wild prairie grass, the genotypes TT and Tt produce plants that are 100 cm tall. In contrast, the genotype tt produces plants that are only 50 cm tall. In a randomly mating population, the frequency of the t allele is 0.6. What is the average height of plants in this population?

27.9 In estimating the number of colorblind females expected in a population of 1000, we obtained the number 8 (see page 724). What factors could account for the fact that we observed only 3 colorblind females?

27.10 Comment on the following statement: The frequency of a mutant allele in a population is an inverse function of the selection pressure against it.

27.11 Using population genetics theory, how would you account for the lack of eyes in some cave-dwelling fish?

27.12 What are the A and a allele frequencies in a human population in which the Aa heterozygote frequency is 0.50? (Assume the population is in equilibrium).

27.13 A population is composed of AA, Aa, and aa genotypes in a 1:1:1 ratio. The offspring produced by these genotypes number 996, 996, and 224, respectively. What are the selection pressures on the three genotypes?

27.14 Bruce Wallace studied the elimination of an autosomal recessive lethal mutation (a) from an experimental population of *Drosophila melanogaster*. He obtained the data shown in the following table:

Generation	Population size	Observed a frequency	Expected a frequency
0	—	0.500	—
1	454	0.284	0.333
2	194	0.232	0.250
3	212	0.189	0.200
4	260	0.188	0.167
5	290	0.090	0.143
6	398	0.085	0.125
7	366	0.082	0.111
8	382	0.065	0.100
9	388	0.054	0.091
10	394	0.041	0.083

Are these data perfectly consistent with the hypothesis that selection against *AA* and *Aa* is 0 and that selection against *aa* is 1.0? Explain.

27.15 How will allele frequencies change in three generations if selection against *AA* and *aa* is 0, and selection against *Aa* is 1.0, with initial allele frequencies of $A = a = 0.5$ (assume random mating)?

27.16 How will the allele frequencies in Problem 27.15 change if the initial allele frequencies are $A = 0.4$ and $a = 0.6$?

27.17 C. Gordon (*Amer. Nat.* 69:381, 1935) released a large number of *Drosophila* heterozygous for *e* (ebony body), an autosomal recessive allele, in a part of England that is normally devoid of this species. After six generations, the frequency of the *e* allele was 0.11. How would you interpret this finding?

27.18 In a natural population of field mice, you find the following distribution of genotypes for the X-linked alleles *s* and s^+ (*s* = striped). What are the *s* and s^+ frequencies, and is this population in Hardy–Weinberg equilibrium?

	ss	ss^+	s^+s^+	sY	s^+Y
male	—	—	—	32	40
female	17	22	26	—	—

27.19 What frequencies of *A* and *a* in a population produce the greatest frequency of heterozygotes?

27.20 Consider two large isolated populations, each carrying a *different* recessive lethal allele of a particular locus. In population 1, the recessive lethal (*a*) has a frequency of 0.06. In population 2 the recessive lethal allele (*b*) has a frequency of 0.03. The *b* allele is not found in population 1, and the *a* allele is not found in population 2. Following an environmental change, the two populations merge and the members of the combined population mate randomly. What is the frequency of the lethal phenotype in population 1, population 2, and the new population (once it reaches equilibrium)?

27.21 A study of 108 achondroplastic dwarfs showed that they produced 27 offspring. These dwarfs had 457 normal siblings who produced 582 children. Estimate the fitness value of the dwarf genotype.

27.22 Of 94,075 children born to normal (nonachondroplastic dwarf) parents, 8 were achondroplastic dwarfs. These would all be heterozygous (homozygotes are lethal) and the result of new mutations. Use this information to estimate the mutation rate at this locus.

27.23 In a class of 25 students, 14 were found to be phenylthiocarbamide nontasters (they were *tt*). The other 11 were all tasters (*TT* or *Tt*). How many were heterozygous?

27.24 T. Dobzhansky and O. Pavlovsky studied two autosomal chromosomal inversions (*A* and *D*) in *Drosophila tropicalis*. Individuals can be either *AA*, *AD*, or *DD* for the two inversions. In a sample of 200 flies, they found 20 *AA*, 150 *AD*, and 30 DD. On the basis of Hardy–Weinberg principle, is this population in equilibrium for the three genotypes? If you decide the answer is no, how would you interpret these data?

27.25 In a large randomly mating population, 0.84 of the individuals express the phenotype of the dominant allele (*A*) and 0.16 express the phenotype of the recessive allele (*aa*). (a) What is the frequency of the dominant allele? (b) If the *aa* individuals are 5 percent less fit than the *A-* individuals, what will the frequency of *A* be in the next generation?

27.26 Mice with the genotype *Hh* are twice as fit as either of the homozygotes. Assuming random mating, what is the expected frequency of the *h* allele?

27.27 Can selection operate in conjunction with random genetic drift? Explain.

BIBLIOGRAPHY

AYALA, F. 1978. The mechanisms of evolution. *Scientific American*, September.

BODMER, W. F. AND L. L. CAVALLI–SFORZA. 1976. *Genetics, Evolution, and Man.* W. H. Freeman, New York.

CAVALLI–SFORZA, L. L. 1974. The genetics of human populations. *Scientific American*, September.

CLARKE, C. A. AND P. M. Sheppard. 1966. A local survey of the distribution of industrial melanic forms of the moth *Biston betularia* and estimates of the selective values of these in an industrial environment. *Proc. R. Soc. Lond. (Biol.)* 165:424–439.

CROW, J. F. 1986. *Basic Concepts in Population, Quantitative, and Evolutionary Genetics.* W. H. Freeman, New York.

DARWIN, C. 1860. *On the Origin of Species by Means of Natural Selection, or the Preservation of Favoured Races in the Struggle for Life.* Appleton, New York.

DOBZHANSKY, T. 1970. *Genetics of the Evolutionary Process.* Columbia University Press, New York.

HARDY, G. H. 1908. Mendelian proportions in a mixed population. *Science* 28:49–50.

HARTL, D. L. 1988. *A Primer of Population Genetics*, 2nd ed. Sinauer, Sunderland, MA.

LEWONTIN, R. C. 1985. Population genetics. *Annu. Rev. Genet.* 19:81–102.

STRICKBERGER, M. W. 1996. *Evolution,* 2nd Edition. Jones and Bartlett, Boston.

Grunion laying eggs on a Southern Califormia beach.

28

Genetics and Speciation

A Dance on the Beach

Legend has it that when the Spaniards arrived on the Pacific coast of California in the mid-eighteenth century, the Indians told them a story of fish who came out of the ocean and danced on the beach when the moon was full. As we now know, there is truth in this legend. A local fish species known as a grunion comes out of the water at night, laying its eggs in the sand above the normal high tide mark. The females bury themselves tail first in the sand. The males wiggle or dance around the half-buried females and fertilize the eggs as they are deposited. The full moon is an important part of this ritual because it produces larger than normal high tides and allows the eggs time to incubate undisturbed in the sand for nearly a month. The fish must time their reproduction perfectly so that after they lay their eggs they get picked up by an incoming wave and carried back out to sea. The next high tide will wash the newly hatched fish out to sea where they will complete their development. Over millions of years, the forces of evolution have acted on the grunion, producing a species that lives in sychrony with the rhythms of the tides and the moon.

The natural world is full of these remarkable adaptations which testify to the power of natural selection. They are constant reminders of the earth's biodiversity and the evolutionary forces that produced it.

Evolution means descent with modification. In its most fundamental genetic sense, it is a change in allele frequencies and chromosome combinations in populations over time. Populations evolve largely as a result of natural selection, the force that shapes or molds randomly generated genetic variation. With time, the genetic characteristics of a population change, and sometimes those changes are of such a magnitude that new species emerge.

There are two main forms of speciation. In the first form, selection produces changes in a species over time so that it evolves or is transformed into a new species. In this process, the number of species remains the same, and the change consists only in the emergence of a new species from within a single lineage. This is **phyletic speciation**, and we discuss it only briefly in this chapter because it does not increase biodiversity. The second form, a branching type of speciation in which two or more species evolve from a single parent species, is responsible for the earth's biodiversity. We concentrate on **branching speciation** in this chapter.

The key to speciation in sexually reproducing organisms is the accumulation of genetic differences in populations, which ultimately result in reproductive isolation. These genetic differences produce phenotypic differences that under natural conditions prevent individuals from the two populations from mating and producing fertile offspring. To better understand species formation, we need to understand how barriers to interbreeding evolve and how the accumulation of genetic changes produce reproductive isolation.

(a)

(b)

Figure 28.1 (*a*) Red and orange color in the salmonberry (*Rubus spectabilis*), and (*b*) the blue and white color of the snow goose (*Chen hyperborea*) are polymorphisms controlled by single genes with two alleles.

GENETIC VARIATION IN NATURAL POPULATIONS

Almost all populations exhibit phenotypic diversity. For most traits, there are differences in phenotypic expression among the individuals of a population. Humans, for example, exhibit variation in height, weight, skin color, hair color, eye color, hair texture, and many other traits controlled or strongly influenced by genes. Our concern in this chapter is with variation due to genetic differences (see Chapter 25). Because genetic variation is so fundamental to the evolutionary process, it is important to know how much genetic variation actually exists in a population.

Phenotypic Variation

Individuals in a natural population often exhibit genetically based morphological variation. The variation is sometimes continuous and sometimes discontinuous. Discontinuous variation is seen in salmonberry (*Rubus spectabilis*) fruit color and snow goose (*Chen hyperborea*) plumage color. In salmonberry, the fruits are either red or orange (Figure 28.1*a*), a trait controlled by a single gene with two alleles. The blue and white snow goose colors (Figure 28.1*b*) are also controlled by two alleles of a single gene. Plumage color strongly influences snow goose mating patterns; because of behavioral imprinting early in their life, white geese tend to mate only with white and blue only with blue.

Continuous, not discontinuous, variation is the norm in nature. For example, the Asiatic ladybeetle, *Harmonia axyridis*, exhibits a wide variety of genetically controlled pattern variants (Figure 28.2*a*). Color variation in the starfish (*Pisaster ochraceus*) is genetically determined and ranges from purple to orange (Figure 28.2*b*).

Polymorphism of Chromosome Structure

As a general rule, each species has a distinctive karyotype—that is, all members of a species have the same chromosome number and the same basic chromosome

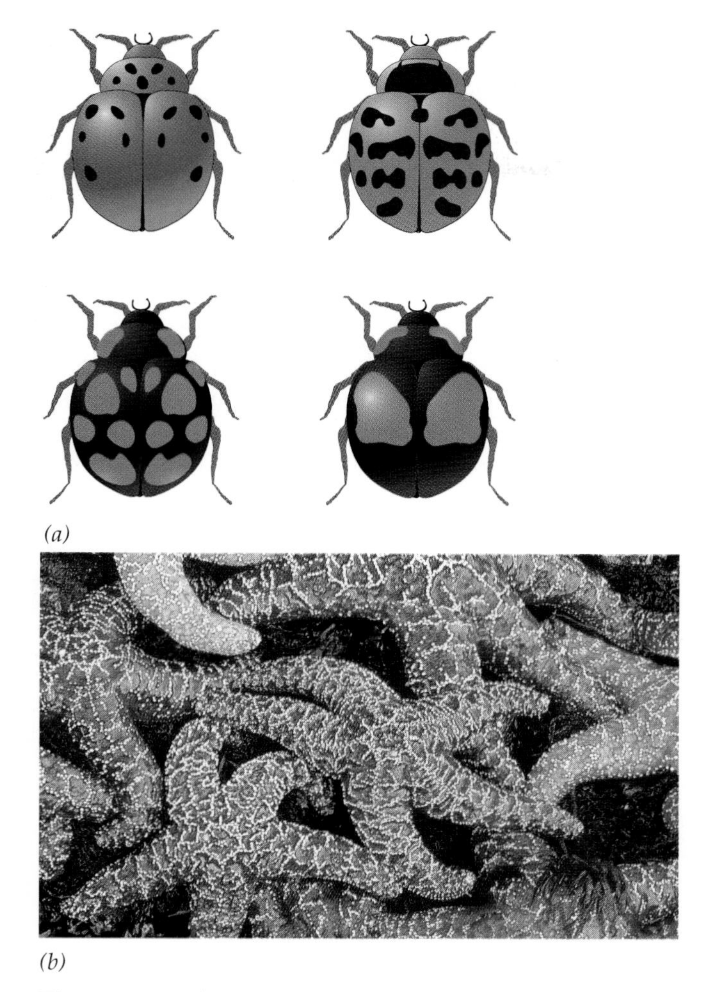

(a)

(b)

Figure 28.2 (*a*) Pattern morphs in the Asiatic ladybeetle (*Harmonia axyridis*) and (*b*) color morphs in the starfish (*Pisaster ochraceus*).

structure. However, variation often exists in certain structural features of chromosomes. For example, in populations of *Drosophila pseudoobscura* in the Mount San Jacinto area of California, three common third chromosome inversion patterns have been identified: Standard (ST), Arrowhead (AR), and Chiricahua (CH). At first, these and other inversions were thought to be adaptively neutral. However, when populations from the same locale were examined at different times of the year, it was discovered that the frequency of ST decreases and that of CH increases from March to June; the reverse takes place from June to August, the hot season (Figure 28.3). Although the flies are outwardly identical in phenotype, these inversions must be associated with subtle physiological differences such that strong selection favors one inversion type over another at different times of the year.

Laboratory studies support the field studies of *Drosophila pseudoobscura* (Figure 28.4). Laboratory populations composed of 10 percent ST and 90 percent CH were established and sampled over a period of several months. By the fourth month (about four generations in this species), ST had increased in frequency to about 0.43, and CH had decreased to about 0.57. After about a year, under standard laboratory conditions, the population stabilized with ST at about 0.7 and CH at about 0.3. When fitness estimates were calculated for the three possible genotypes (ST/ST, ST/CH, and CH/CH), it was determined that ST/CH was the most fit; it was thus assigned a fitness value of 1.0. (ST/ST had a relative fitness of 0.89, and CH/CH had a relative fitness of 0.41.) Thus variation in chromosome structure has an important impact on the fitness of a species. We return to inversions later in this chapter when we discuss the role of chromosome rearrangements in primate evolution.

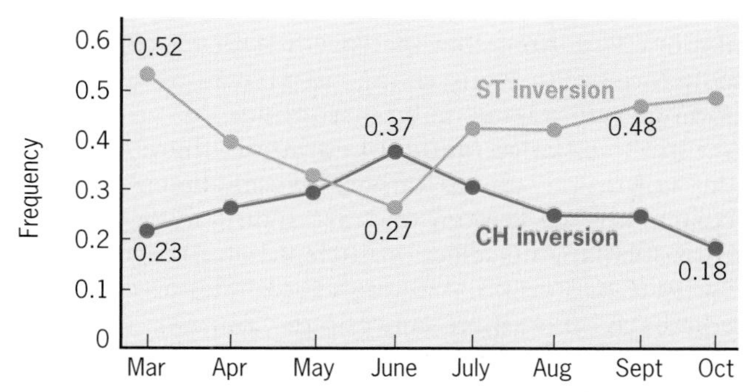

Figure 28.3 Seasonal changes in the frequencies of chromosome 3 inversions, ST and CH, in a population of *Drosophila pseudoobscura*.

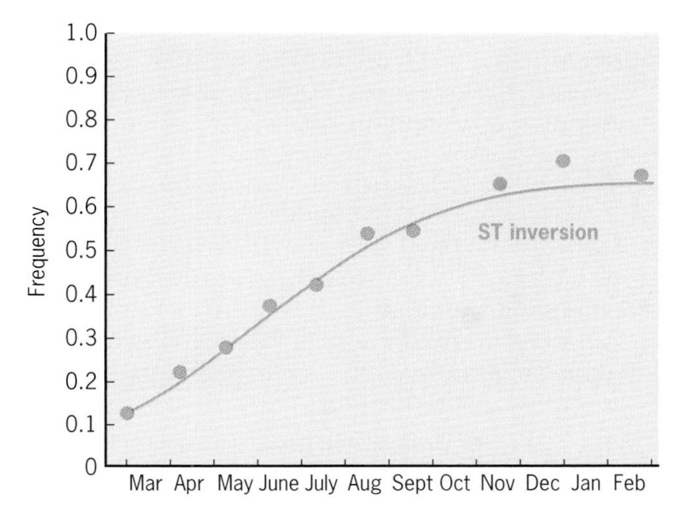

Figure 28.4 Frequency changes in the ST inversion of chromosome 3 in an experimental population of *D. pseudoobscura* in which ST competed with the CH inversion.

The Classical Versus Balance Models of Variation

Field studies show that chromosome rearrangements vary in a population. How much gene variation actually exists in a population? The two main models of the genetic structure of populations are the classical and the balance models (Figure 28.5). According to the **classical model**, a "wild-type" allele exists for each gene, with a frequency approaching 1.0. Mutant alleles are detrimental and exist at very low frequencies. A typical individual is homozygous for the wild-type allele at most loci, and heterozygous for a rare mutant allele at very few loci. Mutant alleles introduced into the population are usually deleterious and are eliminated by selection. However, some mutations may have such small effects on fitness that they are essentially neutral (see Chapter 13 and the Sidelight, Molecular Evolution and the Neutral Theory). These mutations may increase in frequency through random genetic drift and actually become common in the population. Mutations that increase fitness are considered to be rare because any change in a harmoniously functioning system tends to be disruptive.

In the **balance model** of variation, there is no single "wild-type" allele. Rather, several alleles of a gene exist in the gene pool and are maintained by some form of balancing selection such as heterozygote superiority (Chapter 27). Although the alleles are different from each other, there is no single "normal" allele; instead, there are several alleles, each of which may function successfully in the range of environments the population normally encounters.

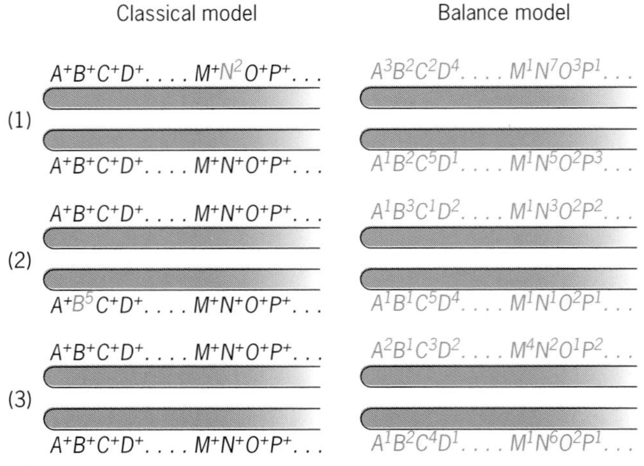

Figure 28.5 The classical and balance models of the genetic structure of populations. The hypothetical genotypes of three individuals are represented according to each model. The capital letters represent genes. The "+" represents the wild-type allele; each number represents a different allele.

Genetic Variation at the Molecular Level

Observable variation in natural populations is widespread, but it does not always reflect the amount of genetic variation that exists in a population. To determine this amount, researchers screen for allelic variants of randomly selected individual genes. They can carry out this task by analyzing polypeptides. Variation in the amino acid sequence of a polypeptide reflects variation in the gene encoding it because the nucleotide sequence determines the amino acid sequence. A simple way to detect amino acid variation in a polypeptide is to use gel electrophoresis to iden-

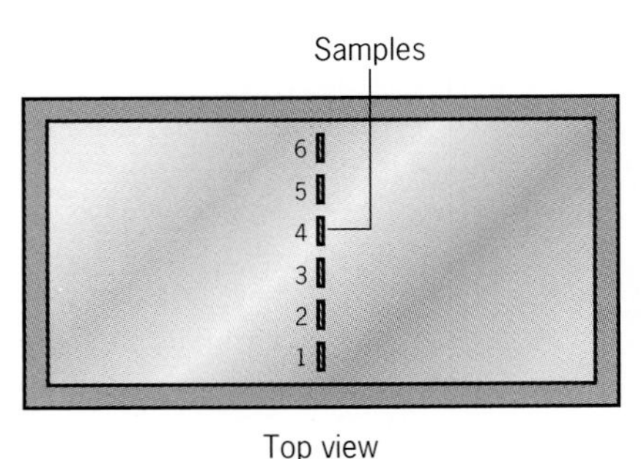

(*a*) Tissue homogenates are placed in wells in a gel providing a homogeneous matrix.

(*b*) The gel with the samples in it is placed in an electric field for several hours.

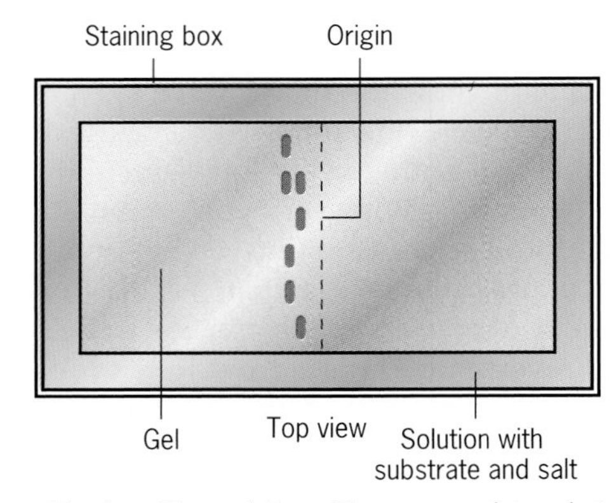

(*c*) The gel is placed in a solution with an appropriate substrate for the enzyme to be assayed and a salt that produces a colored band when it reacts with the product of the enzyme's activity.

Figure 28.6 A schematic representation of gel electrophoresis.

tify charge differences between protein molecules. In this procedure, tissue samples are homogenized to release proteins, including enzymes. Samples are then placed in a gel and subjected to an electric current (Figure 28.6). The proteins migrate in the gel toward the positive or negative pole, depending on their net charge; the rate of migration is a function of a polypeptide's size and charge. The gel is then stained to locate specific proteins or enzymes. For example, 11 *Drosophila pseudoobscura* were analyzed for variation in the *Pgm*-1 gene (phosphoglucomutase). There are two alleles of this gene, *100* and *104*. The proteins they encode are distinguished by their electrophoretic mobilities (Figure 28.7). There are no known functional differences in the proteins. Individuals can be homozygous for either allele (for example, individuals 1 and 2), or they can be heterozygous (individuals 5 and 10).

Two alleles of *Pgm*-1 are detectable by gel electrophoresis, but there may actually be more. Some allelic variants may go undetected for a variety of reasons. For example, a DNA base-pair change may not result in an amino acid change (e.g., both UUA and UUG code for leucine); or a DNA base-pair change may result in the substitution of one amino acid by another amino acid of the same charge (e.g., valine to alanine). Finally, variation may occur in a region of the gene that does not encode any amino acid sequence, such as an intron, a promoter, or a downstream sequence. Thus electrophoretic variation in proteins is an underestimate of the actual genetic variation. Even so, Richard Lewontin and his colleagues detected a large amount of genetic variation in natural populations of *Drosophila willistoni* using electrophoretic techniques (Table 28.1). Harry Harris and his colleagues analyzed human enzyme polymorphism and found a similarly high degree of polymorphism (Table 28.2).

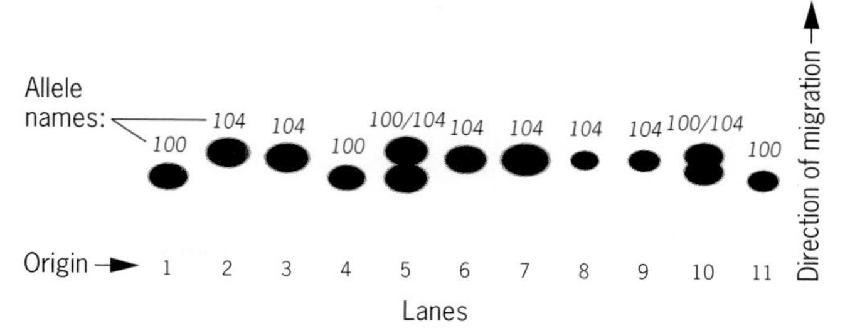

Figure 28.7 Genetic variation at the *Pgm-1* locus (encoding phosphoglucomutase) in a sample of 11 *D. pseudoobscura* females from a natural population. Homozygous flies exhibit only one band; heterozygous flies exhibit two bands.

Variation in DNA sequences is more extensive than variation in amino acid sequences for the reasons just discussed. Consider, for example, the variation

TABLE 28.1
Variation at Three Gene Loci for Enzymes in Five Natural Populations of *Drosophila willistoni*

		Localities				
Gene Locus	*Alleles*	*Puerto Rico*	*Dominican Republic*	*Tame, Colombia*	*Santarem, Brazil*	*São Paulo, Brazil*
Lap-5	a	0.00	0.005	0.01	0.02	0.01
(leucine	b	0.04	0.03	0.12	0.14	0.07
(aminopeptidase)	c	0.60	0.74	0.28	0.39	0.25
	d	0.34	0.22	0.55	0.43	0.57
	e	0.02	0.005	0.04	0.02	0.09
	*H**	0.527	0.408	0.600	0.649	0.618
	Genomes sampled:	320	620	192	492	1,806
Est-5	a	0.00	0.00	0.00	0.01	0.01
(esterase)	b	0.13	0.18	0.02	0.06	0.04
	c	0.86	0.81	0.95	0.91	0.93
	d	0.01	0.006	0.04	0.01	0.02
	*H**	0.252	0.307	0.102	0.159	0.136
	Genomes sampled:	636	320	190	82	1,916
To	a	0.01	0.00	0.00	0.00	0.000
(tetrazolium	b	0.00	0.00	0.01	0.00	0.004
oxidase)	c	0.99	1.00	0.97	0.98	0.995
	d	0.00	0.00	0.01	0.02	0.001
	*H**	0.012	0.000	0.025	0.032	0.010
	Genomes sampled:	508	260	161	244	1,038

*H**: Heterozygote frequency

TABLE 28.2
Incidence of Electrophoretic Polymorphisms in Different Types of Human Enzymes

Enzyme Type	*No. of Loci Studied*	*Polymorphic Loci*	*Percent Polymorphic Loci*
1. Oxidoreductases	24	7	29
2. Transferases	29	10	34
3. Hydrolases	38	13	34
4. Lyases	10	3	30
5. Isomerases	3	–	–
Totals	104	33	

represented by restriction fragment-length polymorphisms, or RFLPs (Chapter 19). Restriction enzymes cleave DNA molecules at specific DNA sequences causing the molecules to fall apart into fragments of specific length. These fragments are separated according to size by gel electrophoresis. If there is any variation in the cleavage sites, the size of the fragments will be altered.

The *D4S10* locus—function unknown—maps close to the Huntington's disease gene on the tip of chromosome 4 (Figure 28.8). *D4S10* is polymorphic with respect to *Hind*III restriction sites. Variation in the restriction sites produces four restriction fragment patterns; each pattern is referred to as a haplotype, a term similar to "allele" in meaning (Figure 28.9). Each individual has two copies of chromosome 4; therefore, each has two haplotypes. Southern blotting (Chapter 19) determines the haplotypes of individuals in a population. With four haplotypes (A, B, C, and D), there are 10 possible combinations or genotypes: AA, BB, CC, DD, AB, AC, AD, and so forth. In the absence of a selective advantage, the 10 genotypes are expected to be in Hardy–Weinberg proportions, based on the haplotype frequencies. Similar variation can be observed at other restriction enzyme sites, revealing that the nucleotide sequence of the *D4S10* locus varies within the population.

There is also variation in the Huntington's disease (*HD*) gene. Huntington's disease is a fatal neurodegenerative disease caused by the excessive amplification of the number of CAG trinucleotide repeats near the 5' end of the *HD* gene (see Chapter 20). About 17 alleles of the nonmutant gene are known with from 11 to 34 copies of the CAG repeat. The disease-producing

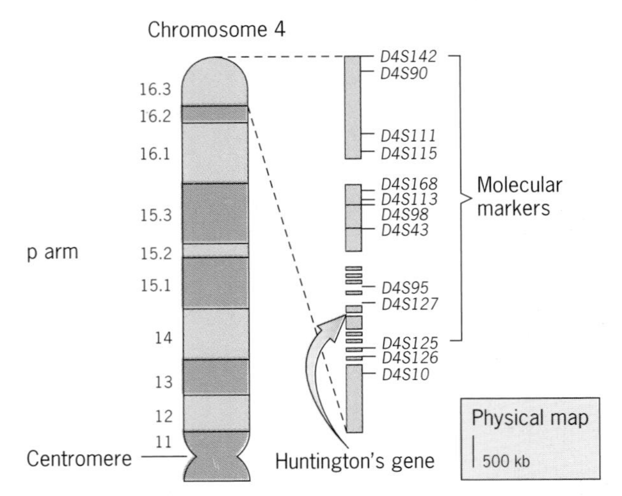

Figure 28.8 The location of the Huntington's disease gene and molecular markers near the tip of chromosome 4.

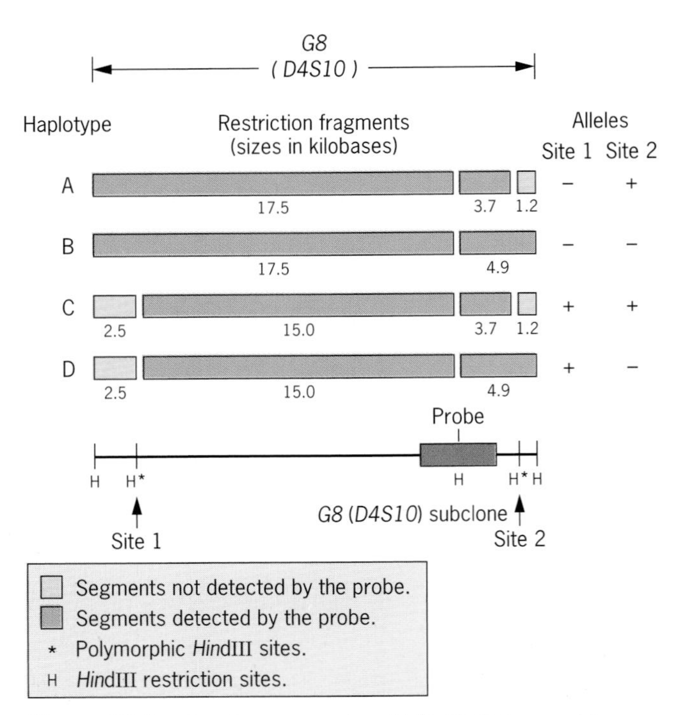

Figure 28.9 Haplotypes at the *G8* (*D4S10*) locus. At *Hind*III sites 1 and 2, there are two possible alleles (+ and −), producing four possible haplotypes.

HD alleles have from 42 to over 100 copies of this trinucleotide repeat. Like the *D4S10* locus, the *HD* gene (and thousands of others scattered all over the human genome) exhibit a staggering amount of sequence variation. Combining data on amino acid and DNA sequence variation leads to the conclusion that populations carry a prodigious amount of both concealed and visible variation. Some of this variation is probably maintained by balancing selection, and some is probably due to recurrent mutation that produces neutral or slightly deleterious alleles.

Key Points: **Members of a population have considerable genotypic variation. Some populations are polymorphic for chromosome rearrangements which can be associated with effects on fitness. Many populations are polymorphic for electrophoretic protein variants and for DNA sequence variants. This variation is probably maintained by a combination of recurrent mutation and balancing selection.**

THE SPECIES CONCEPT

The forces of natural selection shape genetic variation in such a way as to produce reproductively successful organisms suitably, though not necessarily perfectly adapted to their surroundings. Over time, a population may split into two or more subpopulations that eventually become isolated from each other so that they cannot interbreed. The attainment of such reproductive isolation is the key event in the formation of a species.

Charles Darwin considered species to be dynamic and constantly evolving. Some species within the same genus, such as the leopard frog, *Rana pipiens*, and the bull frog, *Rana catesbiana*, are distinct from one another in appearance (Figure 28.10), whereas other congeneric species such as *Drosophila persimilis* and *Drosophila pseudoobscura* are very similar to each other outwardly. In both situations, however, the organisms undoubtedly belong to single and distinct species. Neither the two *Rana* species nor the two *Drosophila* species interbreed. Thus these examples illustrate that both phenotypic and reproductive criteria contribute to the species concept.

Defining or characterizing a species poses a great challenge. The difficulty in finding universal agreement on what a species is illustrates the dynamic and fluid nature of species. With species constantly evolving a rigid definition of a species is not easily achieved.

The **phenotypic** or **phenetic species concept** characterizes a species in terms of phenotypic features. Supporters of this view argue that species are groups of individuals who share certain phenotypic similarities. Unfortunately, this concept lacks a solid and consistent philosophic foundation, arbitrarily selecting certain traits for the purpose of classification. According to one set of characteristics, then, species may be grouped one way; but according to a different set, they may be grouped another way. The inability to resolve these differences suggests that the phenotypic species concept is problematical. Nevertheless, this is an important basis for species determination in the fossil record because it is not possible to apply other criteria.

The **biological species concept**, which is widely accepted, defines species as a group of interbreeding populations reproductively isolated from other such groups. The key component of this species characterization, "population," suggests that biological species are subject to the principles of population genetics. Gene flow may be observed between members of the same population and between populations of the same species, but there is normally no gene flow between populations of different species. The phenotypic species concept implies that morphological similarities reflect genetic similarities. While in many cases this may be true, in others it may not. The biological species concept, on the other hand, argues that members of the same species resemble each other because they share a common gene pool whose composition

Figure 28.10 *Rana pipiens* (left) and *Rana catesbiana* (right).

differs from that of other species (see Chapter 27). This concept establishes a more solid and less arbitrary basis for the similar morphologies expressed by members of the same species than does the phenotypic species concept.

Phenotypic and reproductive units may differ. For example, *Drosophila pseudoobscura* and *Drosophila persimilis* are so similar phenotypically that they are indistinguishable to all but an expert. Nevertheless, the two species are reproductively isolated from each other. Advocates of the phenotypic species concept might consider these two biological species to be members of a single species based on phenotypic criteria. However, advocates of the biological species concept would regard them as **sibling species** that have only recently become reproductively isolated from each other through the accumulation of genetic differences. Apparently there has been insufficient time for these two species to become phenotypically distinct.

Key Points: **The phenotypic species concept groups organisms according to phenotypic similarities. The biological species concept is based on reproductive isolation.**

REPRODUCTIVE ISOLATION: THE KEY TO THE SPECIES CONCEPT

D. pseudoobscura and *D. persimilis* are phenotypically almost identical, and they occupy the same habitat, yet they do not interbreed in nature. The two species are reproductively isolated from each other. Various mechanisms have evolved to prevent different species from successfully exchanging genetic information. These mechanisms fall into two general categories: (1) **premating** or **prezygotic isolating mechanisms**, which prevent mating and therefore the formation of hybrid offspring from individuals belonging to different species; and (2) **postmating** or **postzygotic isolating mechanisms**, which reduce the viability or fertility of any hybrid offspring produced in the event that interspecific mating does occur. In most cases, a combination of prezygotic and postzygotic mechanisms may operate together to reinforce reproductive isolation. By various combinations of prezygotic and postzygotic mechanisms, species achieve and maintain reproductive isolation.

Theodosius Dobzhansky proposed different categories of prezygotic and postzygotic isolating mechanisms (Table 28.3). For example, species may be reproductively isolated from each other because they live in different habitats (ecological isolation), because they have different courtship patterns (behavioral isolation), because their gametes may not fuse (gametic isolation), or because hybrids between them are inviable. One or more or these mechanisms may operate to keep species reproductively isolated. For example, two closely related species of toads (*Bufo fowleri* and *Bufo americanus*) are reproductively isolated because of geography and because they have different breeding seasons (temporal isolation). In regions where there is a geographical overlap, they do not interbreed because their mating seasons and mating calls are different.

Key Points: **Prezygotic isolating barriers to reproduction prevent the union of male and female gametes. In the event of a successful fertilization between two species, postzygotic isolating barriers to reproduction assure that the hybrid will not successfully reproduce.**

TABLE 28.3
A Classification of Prezygotic and Postzygotic Isolating Mechanisms

1. *PREZYGOTIC MECHANISMS prevent the formation of hybrid zygotes.*
   (a) **Ecological Isolation.** Populations occupy different habitats in the same general region.
   (b) **Temporal Isolation.** Mating or flowering times occur at different seasons.
   (c) **Behavioral Isolation.** Mutual attraction between the sexes of different species is weak.
   (d) **Mechanical Isolation.** Physical noncorrespondence of the genitalia or flower parts prevents copulation or pollen transfer.
   (e) **Isolation by Different Pollinators.** Related flowering plant species may be specialized to attract different insects as pollinators.
   (f) **Gametic Isolation.** Gametes may not attract each other, or gametes may be inviable in the reproductive tract of the opposite sex.
2. *POSTZYGOTIC MECHANISMS reduce the viability or fertility of hybrid zygotes.*
   (g) **Hybrid Inviability.** Hybrid zygotes are inviable or have reduced viability.
   (h) **Hybrid Sterility.** Hybrids of one or both sexes fail to produce functional gametes.
   (i) **Hybrid Breakdown.** The F_2 or backcross hybrids have reduced fertility or viability.

MODES OF SPECIATION

There are several modes by which interbreeding populations become separated into two or more reproductively isolated groups or species. We shall consider here only some of the major ones and evidence that supports them. These include the classic allopatric mode of speciation, the more controversial sympatric mode, parapatric speciation, and quantum speciation.

The Allopatric Mode of Speciation

The classic model of speciation, the allopatric or geographical model (Figure 28.11), portrays a population diversifying over its ecological range. Different subpopulations become adapted through natural selection to the different environments within this range and acquire their own genetic identities. Mutations accumulate independently in each of the subpopulations. Gene flow occurs between the different subpopulations but is somewhat restricted. These subpopulations are called **subspecies** or **races**. Races of a species can interbreed, but each has its own unique allele frequency differences. Race is a relative, not an absolute term.

Races may eventually become split into two or more separate or allopatric populations by a physical barrier, such as a canyon or a river, which prevents the flow of genetic material between the two populations. If the populations are somehow reunited, they will still be able to exchange genetic information with each other. As genetic differences accumulate, races evolve into *semispecies*.

Over time, the populations adapt to the prevailing ecological conditions and independently continue to

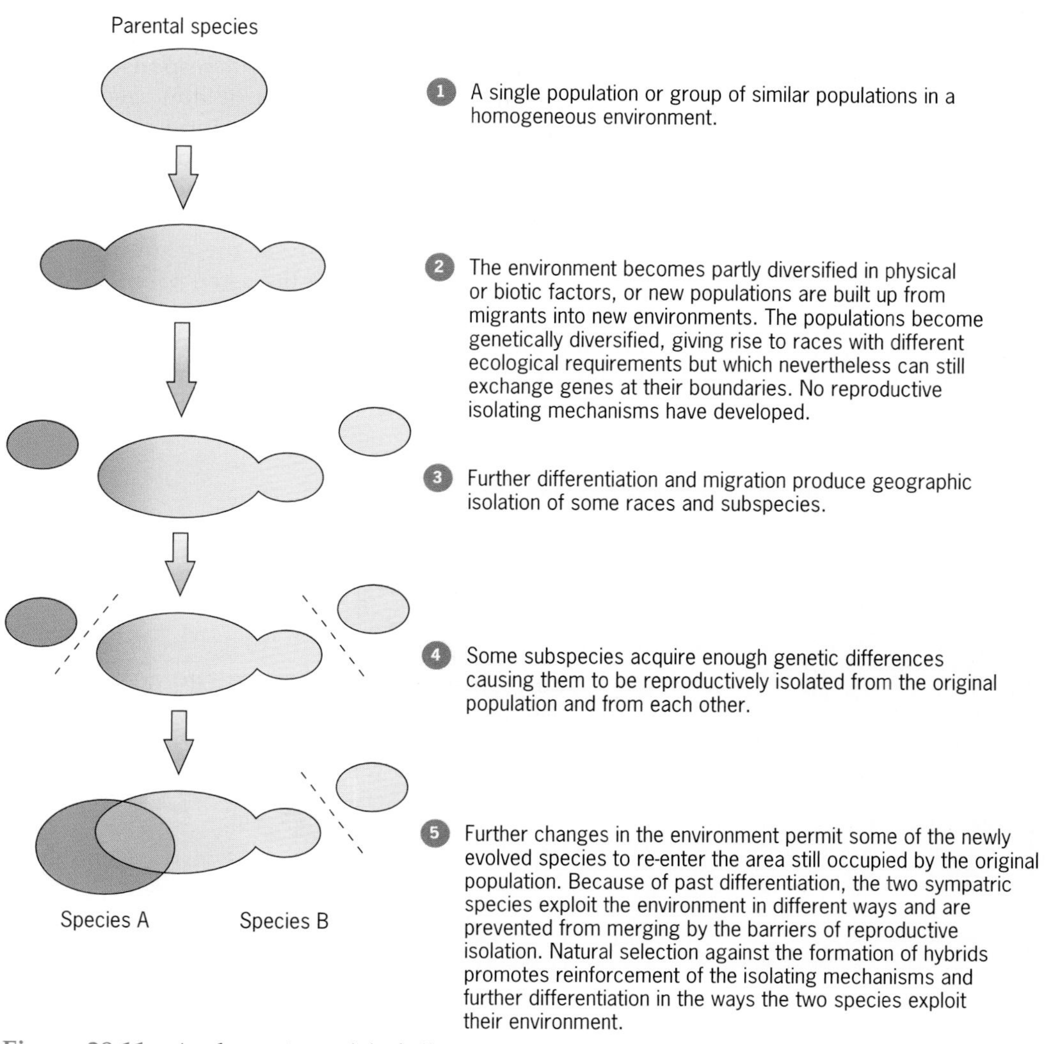

Figure 28.11 A schematic model of allopatric speciation.

accumulate their own genetic mutations, both genic and chromosomal. Natural selection is the most influential evolutionary force operating on these populations, although genetic drift may further accentuate the differences. As a result of the genetic differences that accumulate in the isolated populations, reproductive isolation develops between them. Once reproductive isolation is complete, speciation has occurred. Even if the geographical barriers that separated the populations disappear, the two populations/species remain reproductively isolated through a variety of pre- and postzygotic mechanisms.

Species that have been separate and subsequently rejoined often compete with each other. The consequence of this competition is more intense selection that creates greater character differences. Because of intense selection pressure, the two species may diverge faster from each other *after* they resume contact than they did when they were separated. This process in which there is selection for accentuated differences in species' characters under conditions of interspecific competition is called **character displacement.** The finches of the Galapagos Islands which Darwin studied illustrate character displacement. In this case, the beak size of the finch species that feed primarily on seeds has become significantly displaced. When small and medium-sized finch species share the same habitat on the larger islands, beak sizes differ widely, averaging 8.4 mm on the small finch to 13.3 mm on the larger finch. On the smaller islands, where only one of the two finch species exists, the beaks are intermediate in size, averaging about 9.7 mm.

Evidence from both laboratory and field studies supports the hypothesis that divergent selection in two geographically separate populations can lead to reproductive isolation. In a laboratory study using the domestic house fly, *Musca domestica*, L. E. Hurd and R. M. Eisenberg demonstrated that at least partial prezygotic reproductive isolation occurred as a consequence of selection for different responses to gravity. In one line of flies, these researchers selected for flies that moved against the force of gravity (*negative geotaxis*); in another line, they selected for flies that moved with the force of gravity (*positive geotaxis*). The two populations of flies were kept geographically isolated from each other for 16 generations while this selection regime was carried out. After 16 generations of selection, flies from the two selected lines were brought together to determine whether there was any evidence of prezygotic isolation. Hurd and Eisenberg found that 80 percent of the matings in the merged population were between individuals from the same selected line. Geotactic response is a genetically influenced trait; thus an association, either pleiotropy or linkage, between that and mating preference can be reasonably assumed. Perhaps the genes controlling geotactic response have a pleiotropic effect on mating preference; or perhaps genes influencing mating preference are closely linked to geotactic response genes.

A frequently cited example of how geographic isolation leads to reproductive isolation in a natural population is the evolution of two populations of tuft-eared squirrels (*Sciurus aberti*) on the north and south rims of the Grand Canyon (Figure 28.12). The phenotypic and molecular similarity of these squirrels suggests that they evolved from the same parent population. The formidable canyon separating the two populations reduces gene flow between them to close to zero. Over time, the two populations have developed their own genetic identities. For example, the

Figure 28.12 The Grand Canyon looking to the East. The North Rim, home to the Kaibab population, is to the left and the South Rim, home to the Abert population, is to the right. The Canyon serves as a powerful geographic barrier to gene flow.

northern population is darker and has whiter tails than the southern populations. Although many mammalogists had considered the two populations to be different species, increasingly the two are now believed to belong to the same species: *S. aberti aberti* on the south rim and *S. aberti kaibabensis* on the north rim. The canyon is an effective physical barrier, but it is not perfect. A very few organisms may breach the barrier and mate with members of the other population and produce viable hybrids. Reproductive isolation is not yet complete.

Sympatric Speciation: A Controversial Mode

Sympatric speciation refers to a process by which reproductive isolation occurs among groups of individuals within a continuous interbreeding population (Figure 28.13). This population may or may not exhibit a gradient of different genotype frequencies in different locales. Allele frequency differences that do exist may reflect adaptations to subtle differences existing in the various locales. In this theory speciation occurs *without* geographical isolation, but how can reproductive isolation become established in a contiguous population sharing in a common gene pool? How can a group or subpopulation accumulate enough genetic variation so that it becomes reproductively isolated from other members of the contiguous population? Unique mutational events would be expected to spread through the population through interbreeding. For this reason many investigators think that sympatric speciation is highly improbable: Some kind of isolation is required so that sufficient genetic differences can accumulate between populations to establish reproductive isolation. Others, however, think that sympatric speciation is possible under certain circumstances.

A study of two species of green lacewing flies by C. A. Tauber and M. J. Tauber suggests that sympatric speciation may have occurred. The two species, *Chrysoperla carnea* and *C. downesi*, are sympatric in North America. Their subtle coloration differences are associated with habitat preferences: *C. carnea* is light green in the spring and early summer and changes to brown in the fall. This species lives in grassy areas in the early part of the year and moves to deciduous trees in the fall. Its coloration pattern is adaptive to this habitat. *C. downesi* is dark green all year and lives on evergreens that do not change color much during the year. No evidence exists that the two species were ever physically separated from each other. They are, however, reproductively isolated from each other by seasonal mechanisms and habitat preference. *C. carnea* breeds in the winter and summer, and *C. downesi* breeds only in the spring. The breeding time is controlled by light/dark cycles. If the two species are brought into the laboratory and the light/dark cycles manipulated, the two species can be induced to intermate. The hybrid offspring are fertile. Thus, under natural conditions, the species are reproductively isolated by habitat and seasonal mechanisms.

The genetic control of coloration and mating time in these species is not complicated. A single gene with two alleles controls color: G_1G_1 produces light green that changes to brown; G_2G_2 produces the darker green; and G_1G_2 produces an intermediate phenotype that is never observed in nature. The fact that the intermediate is not observed in nature further suggests that the two species do not normally interbreed. The breeding cycle is controlled by two genes. A dominant allele at either of the two loci causes a winter/summer breeding cycle, whereas recessive alleles at both loci result in a spring breeding pattern.

The Taubers hypothesize that polymorphism at the color locus caused the first divergence that re-

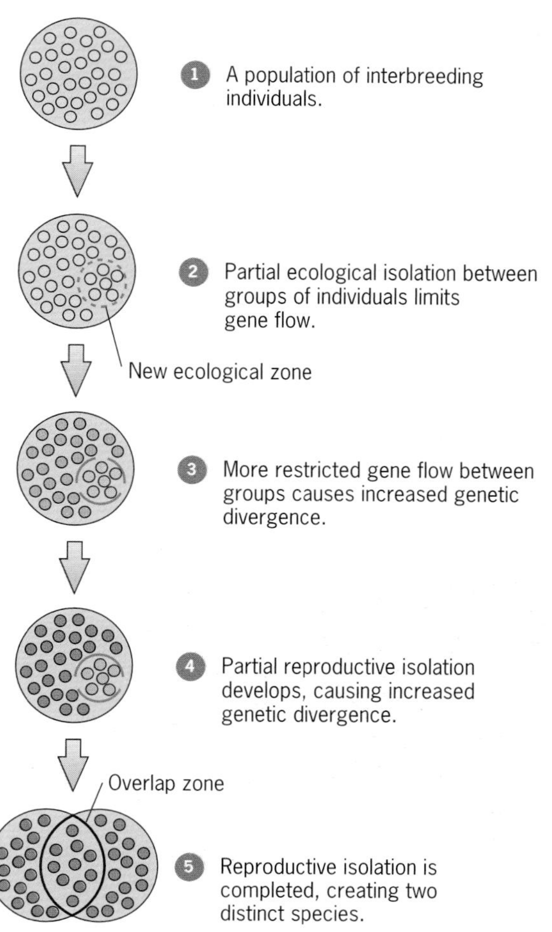

Figure 28.13 A schematic model of sympatric speciation.

stricted gene flow between the two populations. One homozygote was better adapted to the grass/deciduous tree habitat; the other homozygote was better adapted to the conifers. The heterozygote was not well adapted to either. This disruptive selection favored the development of a mating pattern in which like mated with like. Once this geographical separation occurred, other genetic changes accumulated that strengthened mating between like types only. Some changes that reinforced isolation between the two populations include (1) polymorphism for breeding cycles; a mutation altering the breeding cycle would have reinforced the isolation between the two populations, and (2) genetically based differences in courtship behavior. Unfortunately, it cannot be proved that the populations were sympatric prior to reproductive isolation. It is possible that they were geographically isolated.

Although this study and others indicate that sympatric speciation may have occurred, the idea remains controversial. It is difficult to exclude an initial geographical separation event preceding reproductive isolation.

Parapatric Speciation

A variation of the sympatric speciation model is **parapatric speciation,** a rapid process requiring only a few individuals of adjacent populations whose distributions do not overlap but are in contact with each other at one or more points. Reproductive isolation develops when structural rearrangements or other genetic changes occur in a relatively small number of individuals, often on the periphery of a population's range. Those individuals that carry the new rearrangements become isolated from the other members of the population because of problems associated with chromosome pairing and meiosis.

An example of parapatric speciation is the Old World mole rat, *Spalax ehrenbergi* (Figure 28.14). The mole rat lives its life in a small area around a burrow. The mole rats remain in the burrow during the heat of the day and emerge at night to feed; they do not stray far from the burrow opening. There are four chromosomal subspecies of *Spalax* in and around Israel: two in the north by the Golan Heights, one in central Israel, and one in the south around Jerusalem and the Sinai Desert. Each subspecies has a characteristic chromosome number: The two northern subspecies have 52 and 54 chromosomes; the central Israeli subspecies has 58; and the southern subspecies has 60. At the molecular level, the proteins examined in all four subspecies are virtually identical.

Interspecific hybrids can be produced under laboratory conditions, but they all have reduced viability.

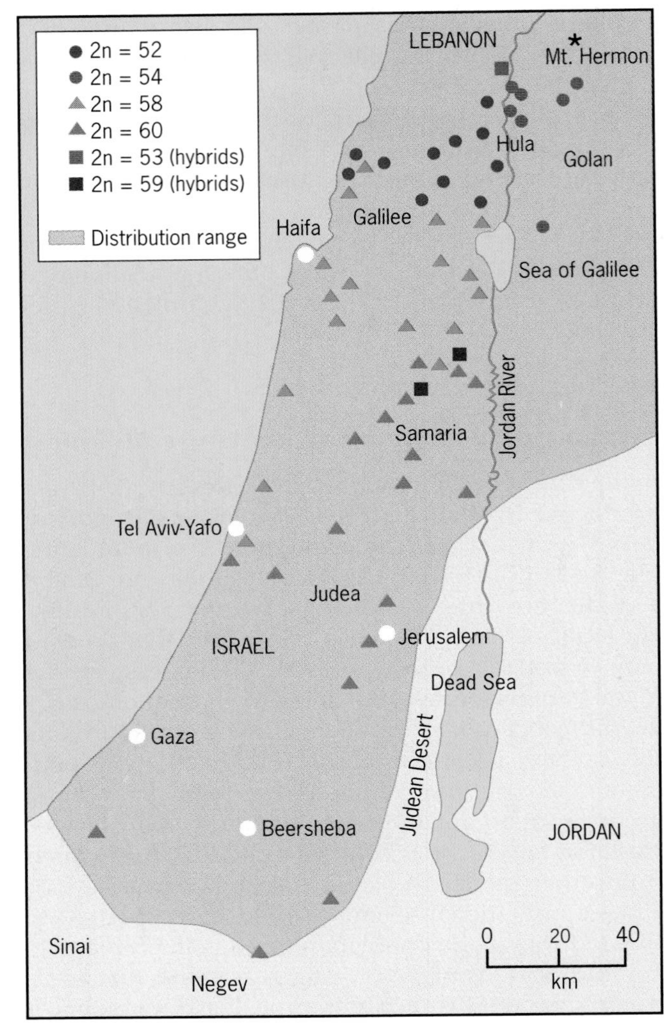

Figure 28.14 The distribution in Israel of the four forms of the mole rat, *Spalax ehrenbergi*.

In nature, hybrids rarely occur, and when they do, fertility is reduced. It is entirely reasonable to consider these four subspecies as distinct species. Fossil evidence indicates that all four subspecies (or species) are descended from an ancestral species, *Spalax mimtus*, which lived in the area about 500,000 years ago.

It is hypothesized that subpopulations of the ancestral species migrated into different areas. Chromosome changes occurred in one or a few individuals. It has been speculated that these changes were adaptively neutral, but that they served to isolate small groups from the ancestral population. Other genetic changes reinforced reproductive isolation and led to further divergence. The burrowing life-style allowed little opportunity for intermating, even though the ranges were contiguous. Today the four subspecies are effectively isolated from each other, with narrow zones separating them.

Quantum Speciation

Quantum speciation (Figure 28.15) is the budding off of a new and different daughter species from a semi-isolated peripheral population of the cross-breeding ancestral species. The idea was originally proposed by George Gaylord Simpson in 1944 and was recently elaborated upon by Verne Grant. The best documented example of quantum speciation is the evolution of *Clarkia lingulata,* an angiosperm that grows in the central Sierra Nevada Mountains of California (Figure 28.16*a*). *C. lingulata* evolved rapidly via chromosome rearrangements from *C. biloba,* the ancestral species (Figure 28.16*b*).

We can understand the relationship between *C. biloba* and *C. lingulata* by looking at meiotic chromosome pairing in the F_1 hybrid (Figure 28.16*c*). *C. biloba* has eight pairs of chromosomes, and *C. lingulata* has nine. Thus the F_1 hybrid has 17 chromosomes. Meiotic chromosome pairing in this hybrid reveals the close relationship between these two species: four pairs of chromosomes synapse and are completely homologous; there is a ring of four chromosomes; and there is a chain of five chromosomes. The formation of rings and chains—characteristic of chromosomes that have experienced reciprocal translocations (see Chapter 6)—suggests that *C. lingulata* diverged from *C. biloba* as a result of chromosome rearrangements.

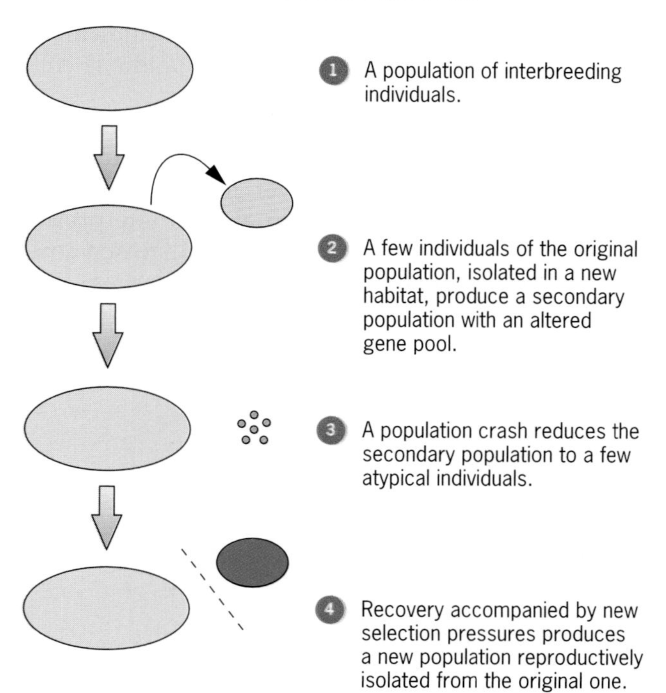

Figure 28.15 A schematic model of quantum speciation.

This theory of *Clarkia*'s evolutionary history is strengthened by analyzing *C. lingulata* 's geographical location (Figure 28.16*d*). There are only two populations of this species, and both of them are found in one small region at the junction of the Merced River with the South Fork River near the town of El Portal, just west of Yosemite National Park. The two populations are surrounded by *C. biloba,* which has a much wider range, suggesting that *C. lingulata* diverged from *C. biloba* through chromosome repatterning.

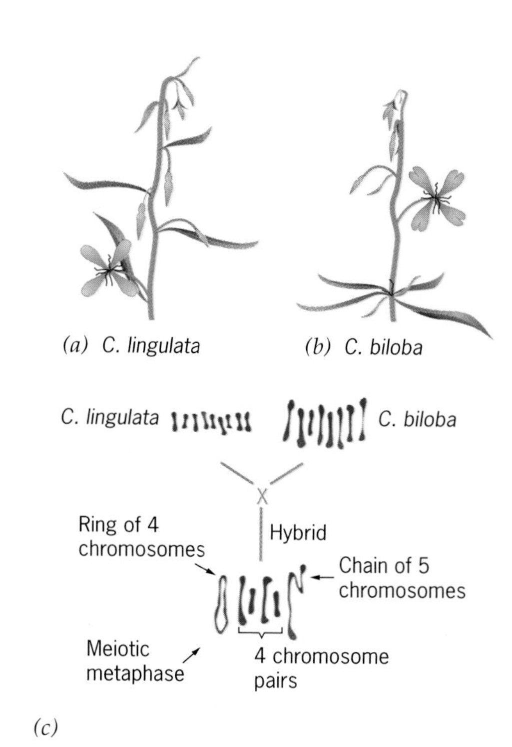

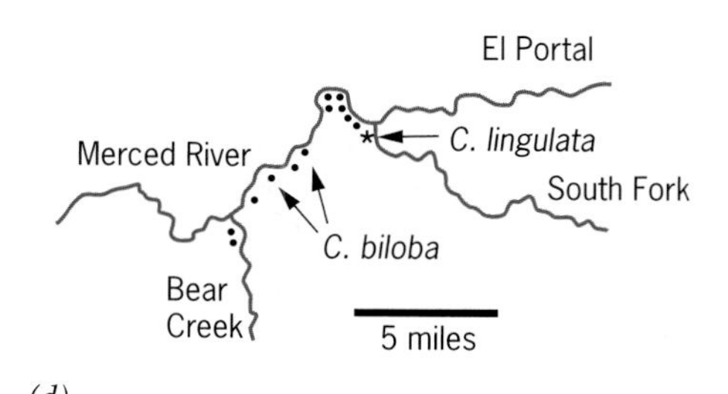

Figure 28.16 *Clarkia lingulata (a); Clarkia biloba (b);* a cross between *C. biloba* and *C. lingulata* with eight and nine pairs of chromosomes, respectively, produces a hybrid with 17 chromosomes. Paired chromosomes at meiotic metaphase form four pairs, a ring of four, and a chain of five, indicating that the parental species differ from each other with respect to at least two reciprocal translocations of chromosome segments *(c).* A map of the region around the Merced River Canyon in the Sierra Nevada of California *(d).* The black dots indicate populations of *C. biloba,* and the asterisk indicates the two known populations of *C. lingulata.*

Chromosomal rearrangements are not always a necessary prerequisiite for quantum speciation. In the Hawaiian *Drosophila* species discussed in Chapter 27, rapid speciation occurred in many cases as a consequence of dramatic shifts in allele frequencies in small founder populations owing to genetic drift and in other cases to chromosomal inversions.

Unlike allopatric speciation, quantum speciation is relatively rapid, requiring only a few generations. The true ancestors of the new species in quantum speciation are very small in number and may be genetically different from the parent population. Population crashes whereby the number of individuals in the population plummets precipitously followed by population expansion accentuate genetic differences between the new and ancestral populations. These crashes may also occur during allopatric speciation, but they are not required. In allopatric speciation, natural selection is the dominant evolutionary force; in quantum speciation, chance events are dominant.

Key Points: Allopatric speciation is a gradual process of speciation via geographical isolation. Sympatric speciation occurs without geographic separation. Parapatric speciation is a rapid process that does not require geographic isolation. Quantum speciation is the budding off of a new species from a semi-isolated peripheral population of the ancestral species.

PUNCTUATED EQUILIBRIUM: THE RATE OF EVOLUTIONARY CHANGE IS NOT ALWAYS GRADUAL

In 1972, graduate students Stephen J. Gould and Niles Eldridge began a study of invertebrate fossils, expecting to find support for a gradual evolutionary process, a view held by most Darwinians. Instead, their findings led them to challenge the widely held neo-Darwinist view that speciation is a gradual process taking millions of years to occur. The gradualistic view of evolution had posed a problem for Darwin because the fossil record appeared to suggest that changes were sudden. Darwin concluded that the fossil record was incomplete, with transition forms either not fossilized or not yet discovered (Figure 28.17). Studies of the fossil record by paleontologists continued to turn up data that show sudden changes in the fossil record, suggesting that speciation occurs episodically rather than gradually.

Gould and Eldridge maintain that the fossil record accurately reflects speciation events. They refer to this pattern of evolution as **punctuated equilibrium**: speciation events occur in bursts after long periods of stasis (no change). Such events occur in small isolated populations along the periphery of larger populations. Because these populations are small and the sites are peripheral, a fossil record of the event would be unlikely.

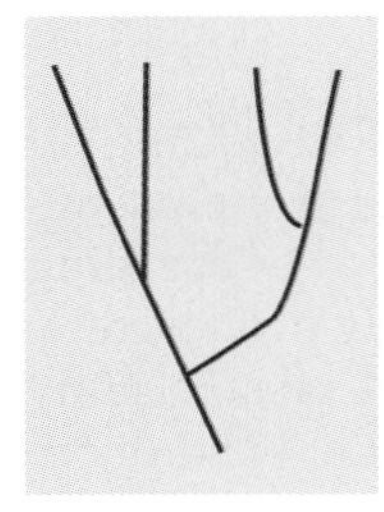

(a) Real pattern of evolution.

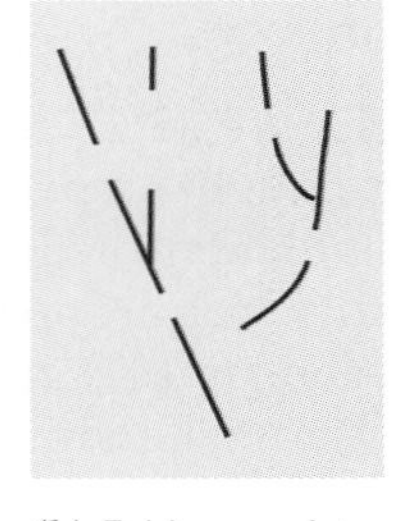

(b) Fairly complete record; some of the gradual change can still be seen.

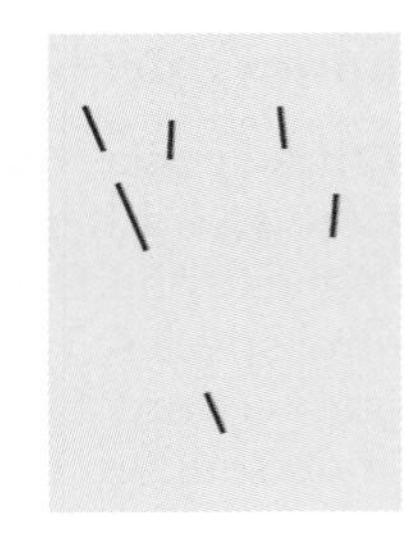

(c) Incomplete fossil record; evidence of gradual change is lost.

Figure 28.17 The record of gradual evolution is disrupted if the fossil record is not complete.

The key difference between the punctuated view and the gradualistic view centers on the relative rates of evolution during and between speciation events. According to the gradualists, the rate of change remains fairly constant throughout both periods, whereas the punctuationalists argue that the rate of change is much more rapid during speciation events than between them (Figure 28.18). Between the periods of speciation bursts are periods of stasis characterized by little or no evolutionary change. The fossil record has been interpreted as being consistent with either punctuated equilibrium or gradualism. In fact, it is likely that both processes have occurred. Disagreement focuses not on which of these two viewpoints is correct, but rather on their relative importance.

A study of fossil snails in the sediments of Lake Turkana in Kenya illustrates the essential features of punctuated equilibrium. (However, it should be noted that the results of this study can also be interpreted within a gradualist framework.) P. G. Williamson analyzed specimens from several species lineages, and concluded that the snails showed no obvious morpho-

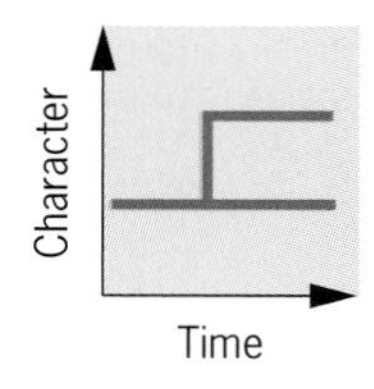

(a) Punctuated equilibrium.

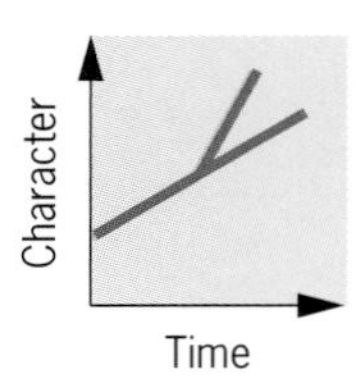

(b) Gradual change.

Figure 28.18 The main difference between punctuated equilibrium and gradualism is the rate of evolutionary change at, and between, speciation events.

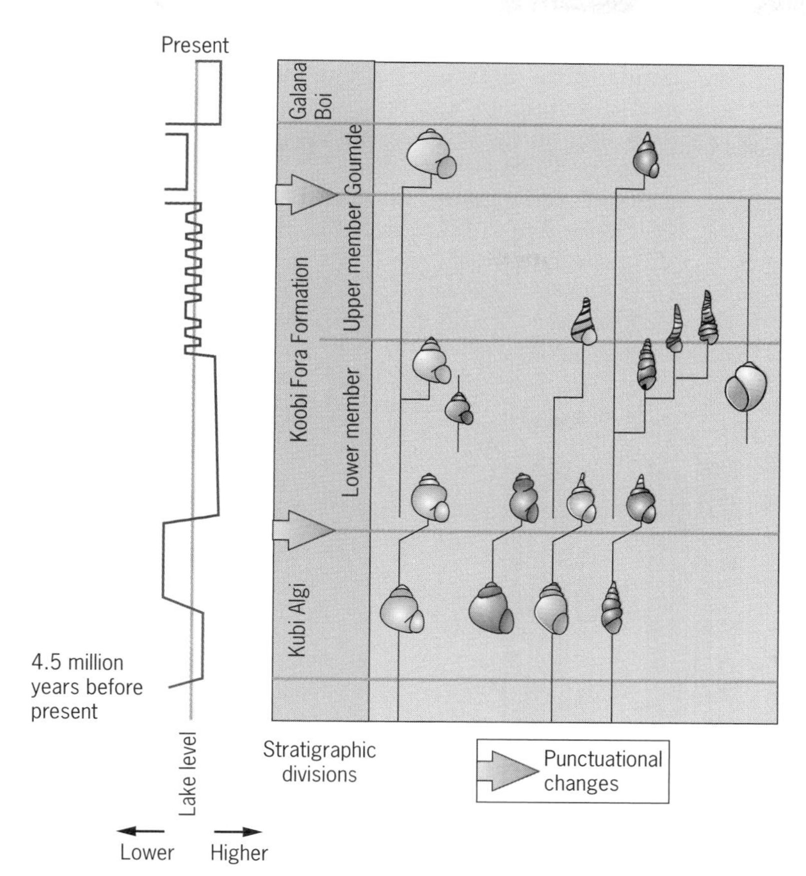

Figure 28.19 Williamson's study of Lake Turkana snails showing a punctuated equilibrium pattern of evolution. Changes in the lake level are shown on the left. Speciation events often accompany changes in the lake level.

logical change for extended periods of time and then exhibited bursts of change (Figure 28.19). These bursts of change coincided with changes in the water level. Williamson hypothesized that when the water level dropped, the snail populations became isolated on small islands; such conditions would be ideal for rapid speciation events.

Williamson's interpretations have been criticized by the gradualists. It has been suggested, for example, that phenotypic changes in the snails may have been triggered by environmental stress. The morphological differences observed in the fossil record may not represent new species at all, but rather morphological variations within the same species. Evidence found in other species supports this possibility. For example, rearing snail eggs of a single species under different conditions produces different adult phenotypes. Thus Williamson's species may not be different species at all, but rather phenotypic variants of the same species.

Another complication is that during the period when little or no morphological change appeared to be taking place, speciation events may have been occurring. Recall that two phenotypically similar *Drosophila* species have different courtship behaviors, habitats, chromosome structures, and so on. These types of species differences do not fossilize and yet they represent key differences between species. Williamson's analysis of the snails is consistent with the concept of punctuated equilibrium, but other interpretations of the same data must be considered.

One of the strongest studies supporting the theory of punctuated equilibrium was published by Jeremy Jackson and Alan Cheetham. These researchers analyzed the morphology of fossilized bryozoans, small coral-like marine invertebrates. To test the validity of their system for classifying fossil bryozoans based on morphology, they applied the same system to living bryozoan species. The system worked. The 46 morphological features of the bryozoan skeleton that defined fossil species also defined living species. They found that over a span of 15 million years these species remained virtually unchanged for periods of 2 to 6 million years at a time, and then, in bursts of less than 160,000 years, split off new species that continued to coexist with the ancestral species. The methodology of Jackson and Cheetham has been extended to several other species, including snails. These studies suggest that punctuated equilibrium is the dominant evolutionary pattern.

Key Points: **Punctuated equilibrium refers to an evolutionary process in which populations exhibit little change for millions of years, punctuated by periods of rapid speciation.**

PRIMATE EVOLUTION AS SEEN IN THE CHROMOSOMES

Studies of modern primates show the importance of chromosome repatterning in hominid evolution. Detailed analysis of G-banded metaphase chromosomes (see Chapter 6) in the human, chimpanzee, gorilla, and orangutan reveal a general homology of chromosome bands in the four species. The banding patterns suggest a common origin for chimpanzee, gorilla, orangutan, and humans. Comparisons of G-banded chromosomes in humans, apes, and Old World monkeys have enabled investigators to work backward and to suggest karyotypes of the ancestors of humans and apes.

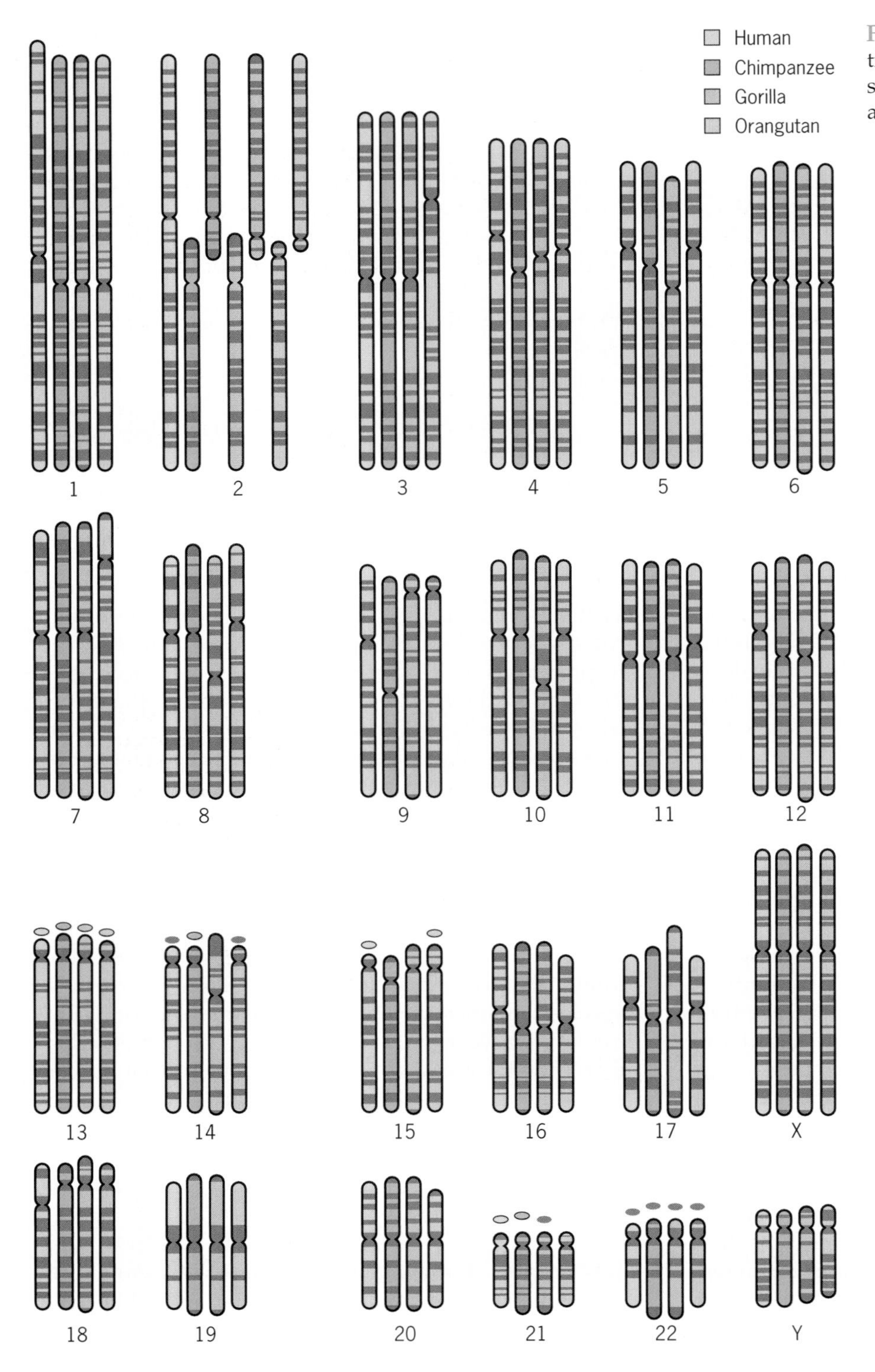

Figure 28.20 Schematic representation of late prophase banded chromosomes of a human, chimpanzee, gorilla, and orangutan.

Three general conclusions can be drawn from these analyses: (1) there was a precursor to orangutans and the hominoid ancestor of gorillas, chimpanzees, and humans; (2) a hominoid ancestor emerged after the divergence of the orangutan from the hominoid-orang precursor; and (3) a progenitor of chimpanzee and human emerged after the divergence of the gorilla from the human-chimpanzee progenitor.

The banding pattern in all four species is remarkably similar (Figure 28.20). Most of the differences are due to inversions that are pericentric (the inverted segment includes the centromere), the result of which is often a change of centromere position. In addition to inversions, some chromosomes have undergone reciprocal translocations (for example, 5; 17 in the gorilla), insertions (20p13 to 8q11.2 in the orangutan), and telomeric fusions (2p and 2q, with inactivation of the 2q centromere in humans). There are other differences too.

Chromosome banding analysis suggests the evolutionary history of these primates (Figure 28.21). A precursor of the ancestral hominoids and orangutan had the same chromosomes as the hominoid ancestor except for five chromosomes: 3, 7, 10, and Y are similar to those of orangutan, and chromosome 17 is like that of the rhesus monkey and the baboon. It appears that the orangutan diverged from this ancestor, with alterations in the structure of chromosomes 2q, 4, 8, 11, 17, and 20. The hominoid ancestor's chromosomes were similar to those of the human, chimpanzee, and gorilla. Before the divergence of humans and chimpanzees, their ancestor shared similar chromosomes 2p, 7, and 9. Human divergence from the chimpanzee is marked by the fusion of chromosomes 2p and 2q into chromosome 2, and by inversions in chromosomes 1 and 18. By contrast, the gorilla experienced nine chromosomal changes and the chimpanzee seven.

Chromosome analysis supports other studies, mainly molecular, suggesting that humans and the great apes could be placed in one family, the Hominidae, with the orangutan in the subfamily Ponginae and the gorilla, chimpanzee, and human in the subfamily Homininae. Evidence at the chromosomal and molecular levels also indicates that the great apes and humans had three ancestors from which first the orangutan, then the gorilla, and finally the chimanzee and human diverged. Chromosome repatterning reflects the evolution of the primates, as it has in other species.

Key Points: **A comparison of the banding patterns of human, chimpanzee, gorilla, and orangutan chromosomes shows clear evolutionary relationships. The evolution of the primates is characterized by chromosome rearrangements, including inversions, and translocations.**

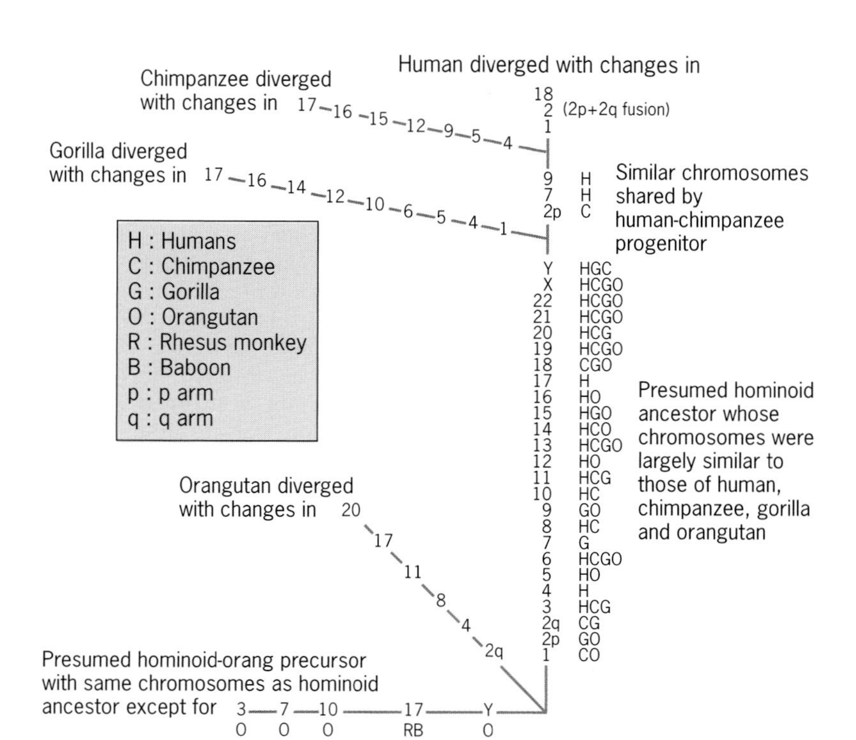

Figure 28.21 Presumed ancestor of human, chimpanzee, gorilla, and orangutan based on chromosome similarities. The chromosomes indicated on the branches are those chromosomes that have changed. For example, before the split between humans and chimpanzees took place, their ancestor shared similar chromosomes 2p, 7, and 9. Human divergence is marked by the fusion of 2p and 2q, and by small inversions in chromosomes 1 and 18.

TECHNICAL SIDELIGHT

Molecular Evolution and the Neutral Theory

At a fundamental level, evolution is change. DNA sequences change by base-pair substitution, insertions, or deletions. If these mutational events occur in a protein-encoding region of DNA, an amino acid substitution may result. Given enough time, such changes can accumulate, leading to a molecule that is very different from its progenitor. Recent advances in molecular biology have made it possible to determine the nucleotide sequences of DNA and the amino acid sequences of polypeptides. By comparing related sequences, the molecular details of evolution can be analyzed.

Amino acid sequence analysis when combined with fossil data provides considerable insight into evolutionary events. For example, the analysis of the α-globin protein in various vertebrates allows researchers to estimate the evolutionary rate of the α-globin gene (Figure 1). All of the organisms in Figure 1 are derived from an evolutionary line that appeared about 440 million years ago. Since that time, this line has split into many different branches, each giving rise to a different vertebrate class—the cartilaginous fishes, the bony fishes, the amphibians, the marsupials, and the placental mammals. Figure 1 shows the number of amino acids that are different when human α-globin is compared to each of the other α-globin molecules. Altogether, there are 141 amino acids in the α-globin polypeptide. The shark α-globin is the most unlike the human α-globin, followed by carp, newt, kangaroo, and cow α-globins, in that order. The degree of difference depends on the time that has elapsed since the evolutionary lines diverged from a common ancestor: the greater the time, the greater the amino acid differences.

The relationship between the degree of difference (D) and the evolutionary time (T) can be used to estimate the rate at which α-globin has evolved. Consider, for example, the human and carp evolutionary lines. The common ancestor of these two species existed about 400 million years ago, so the total time that has elapsed since the two lines diverged is 800 million years (400 million years for each line). During this time, human and carp α-globins have accumulated differences in 68 amino acid sites. The degree of difference is therefore $68/141 = 0.482$ differences per amino acid site. Therefore, that α-globin has evolved at a rate of $D/T = 0.482$ amino acid substitutions per site over 800 million years, or 0.6×10^{-9} substitutions at each site per year.

Actually, this value underestimates the true evolutionary rate because some amino acid sites may have changed more than once and there is no way of knowing which ones, or how many times they may have changed. Fortunately, it is possible to make adjustments for these unseen changes and to obtain a more realistic value for the evolutionary rate of α-globin. The corrected rate is about one substitution per amino acid site every 1 billion years.

Similar analyses have been performed with other polypeptides, and the results indicate that each polypeptide has its own evolutionary rate. For instance, the fibrinopeptides, which are components of blood clots, are evolving about eight times faster than α-globin, whereas the histones, which bind to eukaryotic DNA, are evolving about 1000 times slower.

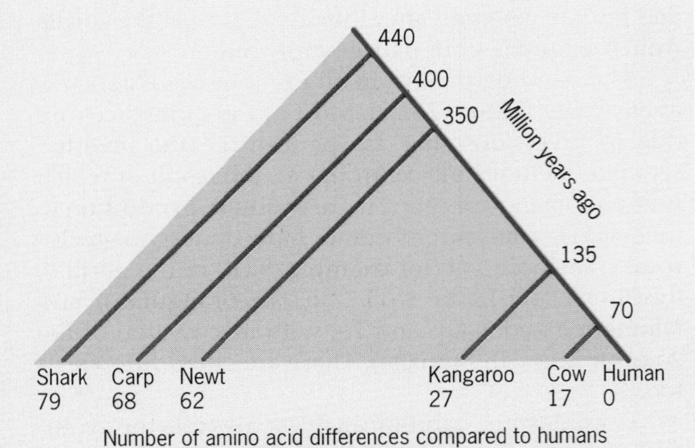

Figure 1 The evolution of α-globin in several vertebrate groups.

How can these observations be explained? The late Motoo Kimura proposed that the great majority of these molecular changes are caused by the random fixation of selectively neutral, or nearly neutral, mutations. In this neutral theory of molecular evolution, mutation and random genetic drift play the critical roles. The probability of random fixation of a selectively neutral mutation is equal to the mutation rate. Thus, in the neutral theory, the rate of evolution is completely determined by the force of mutation. Because mutation pressure is more or less constant, this would explain why polypeptides like α-globin evolve at a fairly uniform rate in different taxonomic groups. However, it does not explain why there are different rates for different polypeptides.

To account for such differences, it is necessary to invoke selection pressure. Kimura and others have proposed that some polypeptides are functionally constrained. This means that some sites in a polypeptide cannot be changed without impairing the polypeptide's function. For example, the active site of an enzyme would not easily tolerate a change while other areas of the enzyme molecule might. The more constrained a polypeptide, the lower the probability of a neutral amino acid substitution and the lower the evolutionary rate. In this view, polypeptides such as the histones are highly constrained, whereas those such as the fibrinopeptides are not.

This theory of molecular evolution has been controversial because it places so much emphasis on random processes. Although the neutral theory admits a role for selection as a purifying agent—that is, as a force that eliminates harmful mutations—it has little or no place for selection in the positive Darwinian sense. According to the neutral theory, species do not get better by fixing beneficial mutations; they simply do not get worse by fixing only neutral or nearly neutral mutations. For this reason, the neutral theory is sometimes called the non-Darwinian theory of evolution. Most evolutionary geneticists would agree that it has been an important influence on our thinking about the evolutionary process.

Gillespie, J. H. 1992. *The Causes of Molecular Evolution.* Oxford University Press, New York.

HUMANS AND CHIMPANZEES: EVOLUTION AT TWO LEVELS

Based on chromosome structure, the chimpanzee is the human's closest relative. Just how closely humans and chimpanzees are related is clear from a molecular analysis.

In a landmark study of humans (*Homo sapiens*) and chimpanzees (*Pan troglodytes*), M-C. King and A. C. Wilson showed that the molecular differences between these two species are very small in comparison to the considerable organismal differences (anatomy, physiology, behavior, and ecology). King and Wilson used a variety of biochemical techniques to estimate the genetic distance between the two species: electrophoresis, immunology, and amino acid sequencing to study proteins; and annealing techniques to study nucleic acids. In particular, they studied the amino acid sequences of several proteins and found them to be almost identical. King and Wilson determined the average degree of difference between human and chimpanzee proteins was 7.2 amino acids per 1000. In other words, the amino acid sequences of human and chimpanzee polypeptides are more than 99 percent identical!

King and Wilson also compared human and chimpanzee nucleic acids. Using DNA denaturation techniques, they estimated that on average there are 33 nucleotide-pair differences per 3000 DNA base pairs (representing 1000 amino acids) in the two species. DNA variation is greater than the amino acid variation for three reasons: (1) some base-pair changes do not produce amino acid changes due to the genetic code's degeneracy; (2) base-pair changes in the upstream and downstream regulatory sequences would not be reflected in amino acid differences; and (3) genomic DNA includes noncoding introns, which vary more in their sequence than the coding DNA regions or exons. (At the time of King and Wilson's publication, introns and exons had not yet been discovered.) (See Technical Sidelight: Molecular Evolution and the Neutral Theory)

King and Wilson used this molecular information to estimate the genetic distance between humans and chimpanzees. They estimated an average of 0.62 electrophoretically detectable codon differences per gene between homologous human and chimpanzee proteins. This distance is 25 to 60 times greater than the estimated genetic distance between different human races (Figure 28.22). The genetic distance between humans and chimpanzees is even less than that for recently diverged sibling species (Figure 28.23). The molecular data clearly suggest that humans and chimpanzees are very closely related, yet according to or-

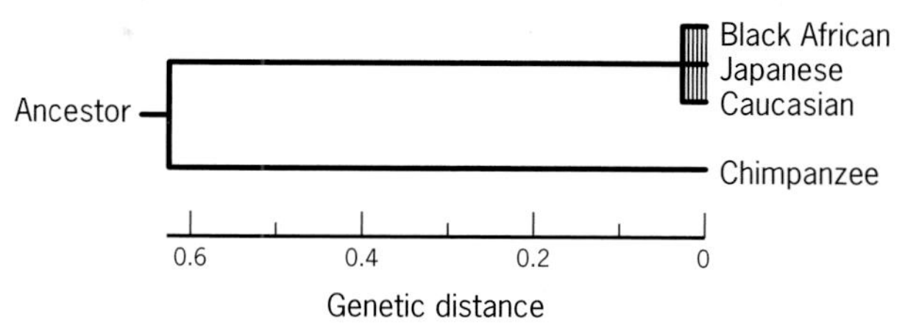

Figure 28.22 The phylogenetic relationship between human populations and chimpanzees. The genetic distances are based on electrophoretic comparison of proteins. The vertically hatched area between the three human lineages indicates that the populations are not really separate, owing to gene flow.

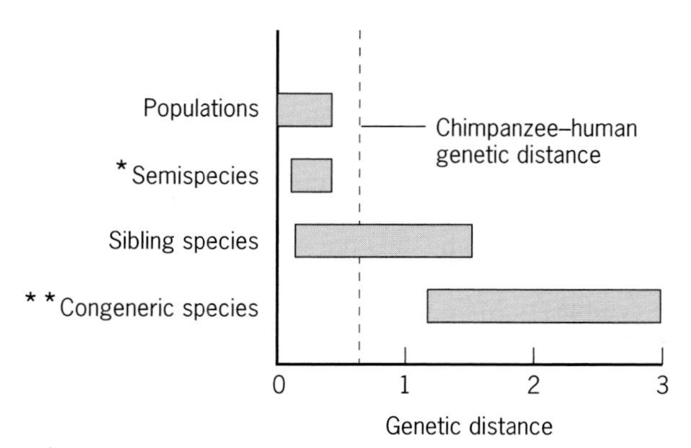

Figure 28.23 The genetic distance between humans and chimpanzees (dashed line) compared to the genetic distances between other taxonomic groups. These data are based on amino acid differences.

HUMAN GENETICS SIDELIGHT

Mitochondrial Eve

All evidence indicates that chimpanzees are the closest living relatives of humans. Fossil evidence indicates that evolution within the genus *Homo* involved many speciation events. But where did modern humans, *Homo sapiens*, come from? This question has generated intense debate in recent decades. Some suggest that modern human populations throughout the world descended from different populations of *Homo erectus* that migrated out of Africa about 1 million years ago: Asian populations of *Homo erectus* became Asian *Homo sapiens*; European *Homo erectus* became European *Homo sapiens*, and so on. This is phyletic speciation whereby one species evolves over time into another species without splitting.

Another suggestion is that populations of *Homo erectus* migrated out of Africa about a million years ago and established populations throughout the Old World. However, these archaic populations were replaced by more recent migrations (about 150,000 years ago) out of Africa of anatomically modern humans. Recent molecular data offer strong support for this viewpoint. The molecular evidence is based on studies of mitochondrial DNA from populations all over the world. These studies suggest that there was a *"Mitochondrial Eve"* that was the progenitor of all modern humans. It is, needless to say, a controversial but important idea. Let's examine the Mitochondrial Eve hypothesis more closely.

Mitochondria are small, membrane-bound cellular organelles responsible for aerobic energy metabolism (Chapters 2 and 18). They have their own genome made up of a circular double-stranded molecule of DNA containing 37 genes and 16,569 base pairs. To an evolutionist, mitochondrial DNA is like a molecular clock: it accumulates mutations at a reasonably constant and rapid rate so that by comparing nucleotide differences between two populations, it is possible to make certain inferences about their genetic similarities. Estimates can be made about the amount of time that has transpired since these populations diverged from a common ancestral population. Mitochondrial DNA is better than nuclear DNA for studying more recent evolutionary events because it mutates about 10 times faster than nuclear DNA.

Mitochondrial DNA has an important second advantage: it is transmitted only through the maternal germ line, and it does not recombine. It is not normally transmitted to the zygote through the male germ line because sperm have so little cytoplasm. Male mitochondria coalesce at the base of the sperm tail to provide energy for the tail. Nuclear genes are a combination of maternal and paternal genes, which makes the assessment of genetic histories difficult. The analysis of mitochondrial DNA is ideal for analyzing the history of modern human populations.

In 1983, Douglas Wallace and his colleagues studied mitochondrial DNA (mtDNA) from several human populations from around the world. They looked specifically at restriction enzyme sites in the DNA. If the mitochondrial DNA from all people had the same identical sequence, a specific restriction enzyme would cut all samples of mtDNA into the same pattern of fragments. Any variation in the fragment patterns reflects variation in restriction sites. The greater the variation, the greater the degree of genetic differences.

Wallace and his team made three important discoveries in their study of mitochondrial DNA. (1) There was little variation in mtDNA sampled from various modern human groups, which suggests that all modern humans originated relatively recently, about 200,000 years ago. (2) Africans had by far the greatest amount of genetic variation. If the rate of change of mtDNA is the same in all populations then the African populations must be the oldest and Africa must be the origin of modern humans. (3) Geographic populations carried specific types of mtDNA changes that produce specific fragment patterns. Wallace and his colleagues constructed a genealogical tree based on these mtDNA data. These conclusions remain valid to this day.

In 1987, Allan Wilson and his colleagues published a followup study confirming Wallace's work. They looked at

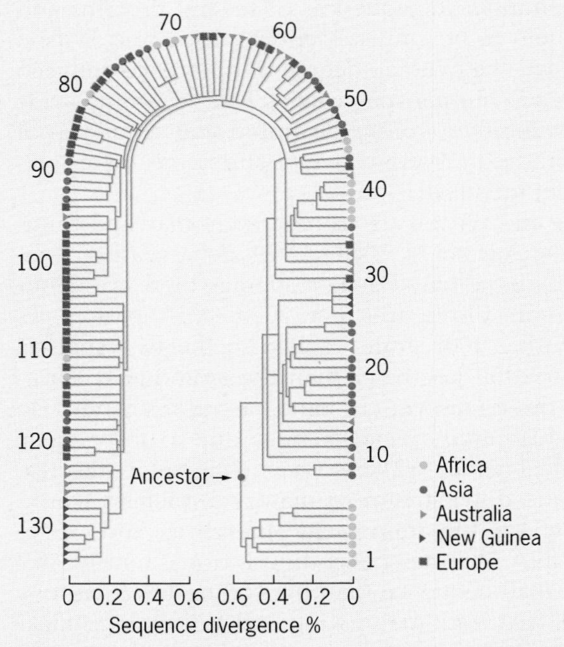

Figure 1 The genealogy of different geographic human populations based on 134 types of human mitochondrial DNA.

mtDNA restriction sites in 147 people representing human groups from Africa, Asia, Australila, Europe, and New Guinea. Wilson found 133 different restriction fragment patterns among the 147 people. Some of these patterns suggested that only a single restriction site had been altered; other patterns indicated more complex changes. Wilson analyzed these patterns and related them to each other, using an analytical strategy that looks for the smallest number of changes required to link the different patterns and eventually trace these patterns back to a common ancestral sequence. This analysis resulted in the production of the most reasonable genealogical tree from the given data. From this analysis, Wilson constructed a genealogical tree of the 134 different patterns based on their mtDNA sequences (Figure 1). He concluded that, "All these mitochondrial DNAs stem from one woman who is postulated to have lived about 200,000 years ago, probably in Africa." The press quickly picked this up, and thus was born the term "Mitochondrial Eve."

Additional studies of mtDNA are consistent with the idea that modern humans have a common mitochondrial DNA ancestor who lived in Africa about 150,000 years ago. Although there is some disagreement over data interpretation, some concepts are not in dispute. First, genetic variation in mtDNA is small, suggesting that modern humans have a recent origin. Second, the African population has the greatest amount of mtDNA sequence variation, suggesting that the human populations on this continent are the oldest. Third, molecular studies, coupled with fossil evidence, is providing us with rich insights into our distant past.

Wallace and Wilson and their colleagues never meant to imply that a single woman alone gave rise to all modern humans. Rather their argument is more of a statistical argument based on the idea that the inheritance of mtDNA is subject to statistical uncertainties. Suppose, for example, that there were 100 females and each had two children. About 25 of these women would have two male children, so their mtDNA line would become extinct inasmuch as males do not pass on mtDNA to their offspring. At each generation, a certain number of females would see their mtDNA line become extinct for the same reason. Given enough time, only a single mtDNA line will exist. Thus, if an ancestral population consisted of 10,000 mating pairs, after about 10,000 generations, the mtDNA from only a single female of that population would be represented in modern humans. Because of segregation, independent assortment, and crossing over, the inheritance of nuclear DNA is more complicated and cannot be traced to a single individual.

Mitochondrial DNA analysis suggests that modern humans originated in Africa about 150,000 years ago. Fossil evidence indicates that about 100,000 years ago groups began migrating into other parts of the world. They reached Southeast Asia and Australia about 50,000 years ago, Europe about 40,000 years ago, and Northern Asia about 20 to 35,000 years ago. Modern humans emerged out of Africa and have become the most dominant (and perhaps most destructive) species on Earth today.

Wilson, A. C. and R. L. Cann. 1992. The recent African genesis of humans. *Scientific American*, pp. 68–73 (April).

ganismal criteria they are placed in different families—pongids (the great apes) and hominids.

How can we reconcile molecular and organismal evolution in these species (Figure 28.24)? Although genomes increase in size and complexity through mechanisms such as DNA duplication, other mechanisms also play a role in speciation and divergence. In the case of the hominids and pongids, researchers now think that divergence has largely been a consequence of changes in gene regulation. Chromosomal inversions and translocations have moved genes into locations that have altered their activity patterns. However, the genes themselves have not changed much, nor have the amino acid sequences of the proteins they encode. Regulatory changes may therefore account for the tremendous organismal differences between humans and chimpanzees. Human evolutionary history is discussed from the perspective of mitochondrial DNA in the Human Genetics Sidelight: Mitochondrial Eve.

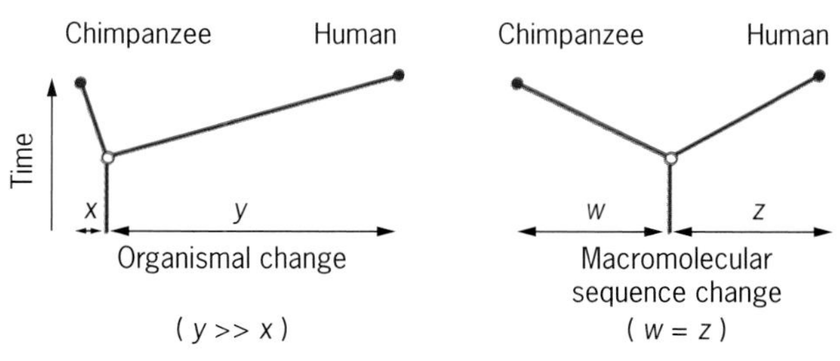

Figure 28.24 The contrast between biological and molecular evolution since the divergence of the human and chimpanzee lineages from a common ancestor. Far more biological change (left) has taken place in the human lineage *(y)* than in the chimpanzee lineage *(y >> x)*. Both protein and nucleic acid evidence indicate that as much change has occurred in chimpanzee genes *(w)* as in human genes *(z)*.

Key Points: **The molecular distance between humans and chimpanzees is less than that for sibling species, yet the two species belong to different taxonomic families. The organismal differences between the two species are quite large, which justifies the classification scheme. The organismal differences may be due to differences in gene regulation.**

TESTING YOUR KNOWLEDGE

1. You discover four different isolated populations of *Drosophila pseudoobscura*. They are phenotypically identical. The only major distinguishing feature in these populations is cytological. Each population carries a different second chromosome inversion:

Population 1 has chromosome sequence A B ▪ D C F G H
Population 2 has chromosome sequence A C G F B ▪ D H
Population 3 has chromosome sequence A C D ▪ B F G H
Population 4 has chromosome sequence A B ▪ D G F C H

centromere

Your colleague suggests that based on these sequences, population 4 is ancestral to the other three. Do you agree with this conclusion?

ANSWER

In approaching this type of problem, you want to construct an evolutionary tree requiring the fewest chromosome changes. Your colleague is incorrect. The most likely ancestral population is population 3. A single inversion in this population would produce population 1; and a single inversion in population 1 would produce population 4. A different inversion in population 3 would produce population 2:

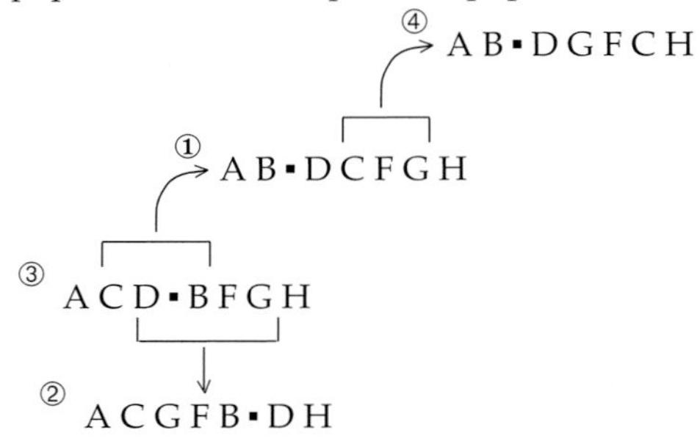

2. Match each of the following descriptions of isolation events (1-5) with the appropriate isolation mechanism (a-f):

1. *Tradescantia canaliculata* and *T. subaspera* are sympatric species but their flowers bloom at different times.

2. *Drosophila melanogaster* and *D. simulans* are sibling species. The males have different courtship behaviors.

3. In some *Drosophila* species, an insemination reaction occurs in the vagina of the female, causing it to swell thus preventing fertilization.

4. A cross between two frog species produced a zygote, but there was no cleavage of the zygote.

5. A male hybrid produced by a cross between *Drosophila pseudoobscura* and *D. persimilis* had immotile sperm.

a. Gametic Incompatibility
b. Behavioral Isolation
c. Hybrid Inviability
d. Seasonal Isolation
e. Mechanical Isolation
f. Hybrid Sterility

ANSWER

The match-ups are: 1d, 2b, 3a or 3e, 4c, and 5f.

QUESTIONS AND PROBLEMS

28.1 You discover two geographically separate populations of *Drosophila*. How would you ascertain whether the two populations are members of the same species or of different species?

28.2 Compare and contrast allopatric speciation, sympatric speciation, parapatric speciation, phyletic speciation, and quantum speciation.

28.3 Why is sympatric speciation a controversial model?

28.4 Why would you criticize the contention of a person who claims that he or she can look at an organism and tell what race it belongs to?

28.5 Comment on the statement that race or subspecies is not a valid taxonomic category.

28.6 You cross two presumably different species in the laboratory and obtain hybrids that are partially or completely fertile. Would you now reject the classification scheme that places the organisms into two distinct species?

28.7 Can a single mutational event give rise to a new species? Explain.

28.8 Working at a fossil dig site in Wyoming, a paleontologist uncovers the following sequence of fossils:

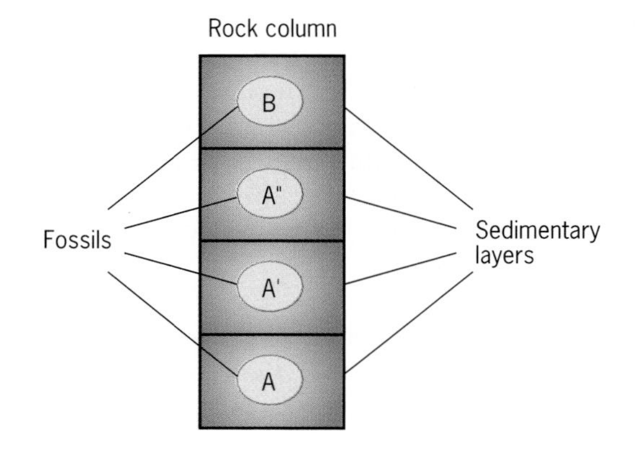

The fossils at the bottom represent species A. At the next (more recent) level, species A' is only slightly different. The next level is A'', which is slightly different from A'. Above the A'' level is a more radically different species called B. It is very different from any of the A species but is still clearly related. The paleontologist considers the fossil evidence here to be consistent with the gradual mode of speciation. He suggests that A evolved gradually into A', which evolved gradually into A''. Between A'' and B, there was a long period when sediments were not laid down, so that fossil events were rare. When conditions were right and fossils once again formed, evolution had progressed gradually to B.

Offer an alternative interpretation.

28.9 The human species is genetically diverse, yet we still consider it a single species. Explain.

28.10 What is a sexually reproducing population, and why is it so important to our concept of evolution?

28.11 On a trip to a zoo, your friend sees a gorilla and remarks, "you mean we descended from that animal?" How would you respond to your friend's remark about the relationship between gorilla and *Homo sapiens?*

28.12 You observe two morphologically similar populations that are geographically isolated from each other. How would you determine whether they are different races or distinct species?

28.13 Five populations of *Drosophila* differ in a third chromosome inversion pattern. The patterns are: 178956234, 128956734, 176598234, 172659834, and 128376594. Which of these arrangements is ancestral to the others? Construct the sequence of events that produced the five populations.

28.14 How do Darwin's finches epitomize the concept of character displacement?

28.15 What isolating mechanisms tend to occur early in the separation of species?

28.16 Speciation by polyploidy is much more common in plants than in animals. Propose an explanation for this phenomenon.

28.17 Discuss the possibilitiy that hybrids between two diploid species could develop into a new, third species.

BIBLIOGRAPHY

DOBZHANSKY, T. 1970. *Genetics and the Origin of Species.* Columbia University Press, New York.

FUTUYMA, D. J. 1986. *Evolutionary Biology,* Second Edition. Sinauer, Sunderland, MA.

GOULD, S. J. 1982. The meaning of punctuated equilibrium and its role in validating a hierarchical approach to macroevolution. In: R. Milkman, Ed. *Perspectives in Evolution,* pp. 83–104. Sinauer, Sunderland, MA.

GRANT, V. 1981. *Plant Speciation,* 2nd ed. Columbia University Press, New York.

HURD, L. E., and R. M. EISENBERG. 1975. Divergent selection for geotactic response and evolution of reproductive isolation in sympatric and allopatric populations of houseflies. *Amer. Natur.* 109:353–358.

JACKSON, J.B.C., and A. H. CHEETHAM. 1994. Phylogeny reconstruction and the tempo of speciation in Cheilostome bryozoa. *Paleobiolo.* 20:407–423.

KING, M-C., and A. C. WILSON. 1975. Evolution at two levels in humans and chimpanzees. *Science* 188:107–116.

LEWONTIN, R. C. 1974. *The Genetic Basis of Evolutionary Change.* Columbia University Press, New York.

MAYR, E. 1988. The why and how of species. *Biol. Phil.* 3:431–441.

PRICE, P. W. 1996. *Biological Evolution.* Saunders, Philadelphia.

RIDLEY, MARK. 1993. *Evolution.* Blackwell Scientific Publications, Boston.

TAUBER, C. A., and M. J. TAUBER. 1989. Sympatric speciation in insects: perception and perspective. In : D. Otte, and J. A. Endler, Eds. *Speciation and Its Consequences.* Sinauer, Sunderland, MA.

WILLIAMSON, P. G. 1981. Paleontological documentation of speciation in Cenozoic molluscs from Turkana Basin. *Nature* 293:437–443.

YUNIS, J. J., and O. PRAKASH. 1982. The origin of man: A chromosomal pictorial legacy. *Science* 212:1525–1530.

A CONVERSATION WITH

JAMES F. CROW

When did you first become interested in genetics, and what stimulated that interest?

I was fascinated by Mendelian ratios in my first college biology course and enjoyed the problems. Chemistry was also appealing, and the outcome was a biology-chemistry double major. I applied for graduate study in both fields and accepted an assistantship at the University of Texas to study genetics. Later, an offer came in biochemistry. If these had come in the other order, I would be a biochemist.

You have been a role model to many of my colleagues. Did you have a role model?

I never had a role model in the sense of wanting to mimic such a person or of feeling the need for one. But I certainly did have heroes, and still do.

During graduate school days at the University of Texas, the person whose work I admired was Sewall Wright. I found his papers interesting, but bordering on the occult, and hoped for a postdoctoral fellowship to study with him. Since I graduated in 1941, and war was imminent, a fellowship was out of the question. But I spent a great deal of my spare time reading his papers.

My other hero at the time was H. J. Muller. During World War II my job was teaching naval trainees at Dartmouth College. Muller was then at Amherst, and I visited him several times. After the first visit, I was so excited by new ideas that I didn't sleep at all that night—the only time this has ever happened.

Dr. James F. Crow, Professor Emeritus at the University of Wisconsin, Madison, has the credentials to write about genetics yesterday, today, and tomorrow. He received a Ph.D. in 1941 from the University of Texas, taught at Dartmouth College for a while, and then joined the faculty of the University of Wisconsin, where he served as Chairman of the Genetics Department and Acting Dean of the Medical School. Dr. Crow's expertise is in population and *Drosophila* genetics. He is the author of many scholarly articles and three textbooks, and has been active in numerous scientific organizations, including the Genetics Society of America and the National Academy of Sciences. For many years he has edited the Perspectives series of articles in *Genetics*, and has written many of the articles himself. Dr. Crow's contributions to science and education have been widely recognized. He received Distinguished Teaching and Service Awards from the University of Wisconsin, and an honorary doctorate from the University of Chicago. His avocation is playing the viola in the Madison Civic Symphony. Though officially "retired," Dr. Crow still gives lectures in genetics courses at the University of Wisconsin.

Later, I became well acquainted with both men, not as teachers, but as colleagues. Since then I've added R. A. Fisher, J. B. S. Haldane, and A. H. Sturtevant to the list of heroes that I was fortunate in knowing personally.

What advice would you give to undergraduate students interested in the study of genetics?

First, get a good background in the physical sciences and math; it's probably unnecessary to mention computer literacy. I am sure that geneticists in the future will be increasingly dependent on such knowledge. There is a risk that the beautifully neat, clean results of molecular genetics will produce a generation of super-technicians. If you are to be an innovator, not a recipe-follower, you need the underlying principles.

Let me urge you also to get a broad general education. I hope you will continue to learn throughout your lives, but this is the last time in which learning is your main business. Knowledge of literature and the arts may not make you a better geneticist, but it will make you a better person and your life a more satisfying one.

In addition to your career as a geneticist, you are known for your avocation as a musician. How did you find the time to pursue this interest?

As a child, I inveigled my parents into giving me violin lessons, and I started out to be a musician. I gradually came to realize that I lacked the talent and drive to perform professionally. The final nail in my musician's coffin was harmony, which didn't come easily. I was recently asked by a group of aspiring high-school students if I had ever had a teacher who inspired me to be a scientist. I answered: Yes, my harmony teacher. I have not given up music, however, and have continued to play viola with undiminished enthusiasm. I managed to squeeze out time to play in the Madison Symphony for 47 years and to see it grow from a loose coterie of light-hearted amateurs into a band of serious professionals, now setting a standard too high for my arthritic fingers.

You have known most of the famous geneticists of the twentieth century. What is your favorite story about one of them?

My favorite story about Wright—everybody's favorite story—is his absent-mindedly erasing the blackboard with a guinea pig; but it is probably apocryphal. Here is a true story. In the early 1950s Newton Morton and I took a train to Chicago to ask Wright about some ideas on effective population number. To each of our questions, he simply said that he didn't know the answer, and the conference was over in a few minutes. On the way back, Morton, who then as now was not given to understating his views, said that if Wright was the world's greatest population geneticist, the subject was in bad shape. Then, a few days later I got a 15-page letter from Wright with every question answered. Later, I had many confirmations of this Wrightian trait; he would not guess or speculate, but worked out each idea carefully before speaking about it.

During your many years of teaching genetics, do any classroom incidents stand out?

One of the most satisfying teaching experiences is eliciting an unexpectedly deep insight from a student. After one lecture, a student corrected an erroneous formula for predicting yield of some corn crosses. I had unthinkingly accepted a wrong formula in a technical paper. I tried to seduce this student into genetics, but he was committed to organic chemistry.

Is there an "artistic" element in science?

There is certainly an aesthetic element. Some of my greatest intellectual satisfaction has come from neat, clever mathematical derivations, often of old results previously arrived at in a clumsy way. R. A. Fisher's genetical and statistical work epitomizes elegance, as does Malecot's formulation of the inbreeding coefficient. Some of the work that Motoo Kimura and I did together has given me aesthetic pleasure not unlike that of an artfully crafted Mozart String Quartet.

Can you single out any one research result as the most exciting of your career?

Most of my experimental research has involved long-time experiments planned well in advance, so the results were more solid than exciting. One purely accidental discovery, however, brought me genuine excitement. In 1957, a student, Yuichiro Hiraizumi, discovered a gene in a natural *Drosophila* population that violated Mendel's laws by segregating, not in the regulation 1:1 ratio, but more like 98:1. At first I didn't believe the result and thought there was a mistake somewhere. But when it became clear, I realized that we now had, in a species where a thorough genetic analysis was possible, a beautiful example of meiotic drive. The late Larry Sandler was joining me the next year as a postdoctoral fellow, and I was pleased to tell him that we had something great for him to work on—an instance of the phenomenon that he had predicted and called meiotic drive. The study of Segregation Distortion is still going on, with recent new breakthroughs from molecular technology.

In your long life as a scientist you have spent a great deal of time teaching and on administrative and committee work, both local and national. How do you feel about it now?

I have indeed spent a great deal of time teaching, and I don't regret a minute of it. Teaching was fun, but I thought it important and took it very seriously. As for committee and administrative work, this was less satisfying, but on the whole I am glad to have done it—to have had some influence on University and national policy. Of course, if I had not done all this, I would have published more papers. But I doubt that my scientific contribution would have amounted to any more; all the ideas that I regard as important have found their way into print.

EPILOGUE

Genetics Yesterday, Today, and Tomorrow
A Personal View

James F. Crow

Because genetics has such a short past, the authors of this book thought that a history of the subject, as seen by one who has lived through a large part of that history, might be appropriate. I am delighted to oblige. This review is spotty and personal, but I hope it conveys some of the excitement that came with the rapid succession of often-surprising discoveries. I also hope that it conveys some of the uncertainties I feel about the future.

You have now finished the book and have perforce acquired a great deal of genetic knowledge. I trust you have found the experience interesting, even exciting. I would hope that you are just a bit curious as to how all this came about. So, I invite you to view the subject through the eyes of one who was around when much of what you have learned was being discovered.

Genetics is a new science, a twentieth-century phenomenon. The subject is unusual in that its origin can be pinpointed to a single event, the publication of Gregor Mendel's paper in 1866. Or perhaps I could better say it started in 1900 when Mendel's work, ignored for 34 years, was finally rediscovered.

I find it clarifying to divide genetics into three periods. The first period dates from the rediscovery of Mendel's laws in 1900 to the beginning of World War II. During this period the subject was dominated by breeding experiments and cytology. The main emphasis, and the main success, was transmission genetics. The second period starts at the end of the war and extends to the present. It was dominated by the genetics of microorganisms and molecular techniques. The emphasis was on molecular understanding, and it is still going on. Thanks to new tools, development and neurobiology have come to the front. The third period, overlapping the second, crept in gradually just one or two decades ago. Still in its early stages, this period marks the beginning of a major impact of molecular genetics on agriculture and medicine.

I started college in 1933 and completed my graduate work in 1941, just at the end of the first period. It was a different time. What did we know then? And, perhaps more interesting, what did we *not* know?

THE FIRST PERIOD, 1900–1941

The year 1900 brought a genetic spectacular. Of course, the main event was the rediscovery of Mendel's laws by three men, working in three different countries. But there were significant sideshows. The year also saw Karl Pearson's invention of the χ^2 test and Karl Landsteiner's discovery of blood groups, which gave genetics its first useful human polymorphism and marked the beginning of immunogenetics.

In the years before 1900, a number of discoveries prepared the way for the instant acceptance of Mendel's work. By the turn of the century, the details of mitosis and meiosis had been worked out; thus, once Mendel's laws were announced, it was immediately obvious that his particles were carried by the chromosomes. The key papers pointing this out were written by T. H. Montgomery and W. S. Sutton in the United States and T. Boveri in Germany. Undoubtedly, the idea occurred to many others, but they didn't write about it. Sutton, incidentally, was still a student and later went into medicine. The other two are well-known scientists.

Another nineteenth-century discovery turned out to be premature. In the 1870s F. Miescher discovered a mixture of proteins and nucleic acids in pus and sperm cells and called the substance nuclein. In 1895 the cytologist E. B. Wilson wrote:

> *Now, chromatin is well known to be similar to, if not identical with, a substance known as nuclein. . . . And thus we reach the remarkable conclusion that inheritance may, perhaps, be effected by the physical transmission of a particular chemical compound from parent to offspring.*

A year later Wilson suggested that nucleic acid was the component responsible for genetic transmission. He was a half-century ahead of his time.

I'll mention one other nineteenth-century finding: the realization that a single egg and sperm (or pollen grain) unite to produce a zygote. Without knowing this fact, Mendel could hardly have arrived at his pathbreaking laws.

With Mendel's discovery of particulate inheritance, the identification of chromosomes as the vehicles of inheritance, and the suggestion that nucleoproteins constitute the chemical basis of inheritance, the stage was set for the subject to take off. And take off it did. But the successes and the emphasis were mainly on the rules of transmission, not on the chemical basis of heredity. The time wasn't ripe for this; chemical techniques were not up to the job.

Before 1900 the understanding of heredity was vague, and the subject seemed hopelessly complicated. There were no general rules, only countless anecdotes. Then Mendel's elegant rules appeared, bringing simplicity and order out of chaos. Thus came the realization that the elements of heredity—Mendel's *Merkmale,* or in contemporary English, genes—are particulate and are carried by the chromosomes. What could be neater?

The X and Y chromosomes were soon discovered, bringing a solution to the mystery of sex determination. And along with X chromosomes came X-linked genes, the classical example being the white-eyed *Drosophila* mutant in T. H. Morgan's Columbia University laboratory. Actually, the first sex-linked trait had been discovered earlier, in *Abraxis,* a moth. The genetic data did not fit the cytological observations, however, and you can imagine the confusion before it was realized that in moths the female is XY and the male is XX.

The next big step was the use of linkage to map the position of genes on the chromosome. The idea that the frequency of crossing over between pairs of genes could be used as a measure of their distance apart occurred to A. H. Sturtevant, an undergraduate student in Morgan's laboratory. He took the data home one evening, and the next morning, having neglected his regular homework, produced the first chromosome map (see Chapter 7). It is remarkably similar to the contemporary one shown in Figure 7.1.

Mutation as a phenomenon was realized early in the century. It is interesting to note that the original "mutations" found by Hugo deVries in the evening primrose, *Oenothera,* were in fact not mutations at all, but segregation from complex translocation heterozygotes. But the idea was on target; this was not the only time in science that a good hypothesis has come from misinterpreted data. The kinetics of mutation (for example, its independent occurrence in different genes and its temperature dependence) were laboriously worked out, mainly by the *Drosophila* group, who developed techniques for measuring mutation rates accurately. The climax came in 1927 when H. J. Muller, a former Morgan student, discovered that the mutation rate could be greatly enhanced by ionizing radiation.

Population genetics had its beginning with the Hardy–Weinberg rule. G. H. Hardy, who was England's most distinguished mathematician, regarded this simple application of the binomial theorem as utterly trivial. Wilhelm Weinberg was a busy physician responsible for delivering hundreds of babies; yet he found time to invent all sorts of clever ways to bypass the difficulties caused by the fact that humans have small families and do not choose to mate so as to facilitate genetic analysis. Subsequent developments came almost exclusively from the great trinity, J. B. S. Haldane and R. A. Fisher in Britain and Sewall Wright in the United States. Wright invented the inbreeding coefficient in 1922. By the end of the first period, population genetics was solidly established, and a mathematical theory of evolution had emerged. It still stands.

Meanwhile, much of the pleasing simplicity disappeared. The particulate nature of the gene and its chromosomal location remained intact, but all sorts of complications were tacked on. Various exceptions to Mendel's rules turned up, at first puzzling, then interesting, and eventually revealing and sometimes useful. For example, many plants reproduce asexually. The dandelion is one. Although the plant produces pollen and goes through the motions of sex, meiosis is bypassed and the seed is a genetic replica of the mother. As an instance of the occasional perversity of nature, the hawkweed reproduces this way. This explains the great disappointment that Mendel experienced when he tried to duplicate his pea results in this species. Genetics might have gotten off to an earlier start and its history might have been quite different had the hawkweed behaved in a more orthodox way.

Sex determination was not always found to be standard XX-XY. Sometimes the Y was missing. Sometimes there were multiple X's. As mentioned earlier, the system is reversed in moths and butterflies. And the mechanism in bees and wasps turned out to depend on haploidy versus diploidy. In still other organisms, sex determination was not chromosomal at all but environmental. Alligators and some turtles develop into males or females according to the temperature at a crucial period. (Curiously, this discovery, which might have been made at any time, didn't come until later.)

Many plants and some animals turned out to be

polyploid. At first, the inheritance pattern in these species seemed utterly bewildering, but it was immediately clarified when the chromosomal basis was straightened out. Plant breeders quickly put polyploidy to use, for example, using allopolyploidy to produce useful, fertile hybrids. The process was eased by the discovery that colchicine induces polyploidy. Breeders also discovered that conveniently seedless fruits, like bananas, had odd chromosome numbers and hence could not have a normal meiosis. Hitoshi Kihara in Japan used triploidy (produced by crossing a diploid with a tetraploid) to engineer seedless watermelons.

There were many studies of developmental and physiological genetics. The creative use of mutant genes permitted a fine dissection of developmental processes, much finer than the surgical techniques of the time permitted. Early in the century, Garrod discovered that mutant genes, usually recessive, led to metabolic diseases and that the diseased state was the consequence of enzymatic failure. So biochemical genetics had its origins long ago. The great mathematical theorists, J. B. S. Haldane and Sewall Wright, applied the theory of enzyme kinetics to genetic problems with considerable success. But there was no deep knowledge of the chemical details such as we have today. And embryology lacked the tools for major advances.

The theory of quantitative genetics developed rapidly during this first period. R. A. Fisher, while teaching high school in 1915, wrote a paper giving the basis for statistical analysis of quantitative traits. The paper was rejected by the referees and was not published until 1918, and then with the help of a subsidy from Charles Darwin's son Leonard. Even the greatest can have trouble getting their work published. This must surely be one of the deepest analyses ever made by a schoolteacher, working in his spare time. Our whole theory of efficient breeding for performance in plants and livestock had its origins in this paper. Fisher, along with Sewall Wright in the United States, made us realize that what an individual inherits from its parents is best described as a deviation from the population average, and the very useful concepts of additive genetic variance and heritability emerged.

So, to view this period with the advantage of your much deeper knowledge, what did we know and what did we not know at the beginning of World War II? In particular, how much did we know about the deepest genetic question, the nature of the gene? So, let's look at what was understood in 1941, through the hindsight of your vastly superior knowledge, through a retrospectroscope, say. What *did* geneticists know?

We knew a great deal about the gene. We knew from Mendel that it is particulate. We knew from cytology and chromosome mapping that the genes are arranged in linear order along the chromosome. We knew from developmental and chemical studies that the genes often exert their influence by producing proteins, enzymes in particular. We knew a great deal about the kinetics of mutation. The chief idea-man, Muller, had clearly delineated what the gene had to do: (1) It had to carry information, enormous amounts of it; (2) it had to replicate itself every time the cell divides; (3) it had to mutate, meaning that it had to copy errors; and (4) it had to control developmental and physiological processes.

But the most interesting point is what we *didn't* know. We didn't know *what* a gene is. Needless to say, geneticists speculated endlessly about this, *the* central question. To quote a reviewer of the 1932 International Genetics Congress: "Oceans of words were spilled in formal and informal gatherings to discuss the vital question: 'What is the gene?' but that important entity is still elusive." My fellow graduate students in the late 1930s, usually over glasses of beer on Saturday night, opined—parroting the views of our teachers—that we would not live long enough to see the day when we would know what a gene is. So we had better work on other things. How wrong we were! Although the advice to work on other things might have been good.

INTERLUDE: WORLD WAR II

In most genetics laboratories, especially in Europe and Japan, the lights were out. All the stocks of *Drosophila ananasse,* a species extensively studied in Japan, were lost. Research went on in other parts of the world but at a considerably reduced pace. Yet some key findings emerged during this dark period. One was the discovery of biochemical mutations in *Neurospora* by George Beadle and Edward L. Tatum. Not only was this the beginning of modern biochemical genetics, but also it was the first step toward applying genetic methods to microorganisms.

A second great discovery during this period was Oswald Avery and colleagues' demonstration that the transforming principle in *Pneumococcus* was DNA. This came as a big surprise to most geneticists, who were betting on protein. But recall that Wilson had made the right guess in the 1890s.

The third big breakthrough during the war was the discovery of chemical mutagens by Charlotte Auerbach in Edinburgh. This discovery led to increasingly sophisticated mutagenesis studies and ultimately to a rational chemical study of the mutation process. That mustard gas might cause mutations was suggested by the similarity of mustard and X-ray injuries, as Auerbach painfully discovered for herself. Although her discoveries were made during the war, they were not published until afterward. Because mustard was a war gas, all studies of it were military secrets and her results could not be disclosed. (Actually they were not completely secret, for during the war Muller told me that Auerbach had found something very exciting. He clearly deduced what she had done and would like to have discussed it with someone else privy to the information; but since I was not, he was a good citizen and didn't tell me. I had to wait until the war was over.)

All three of these discoveries were forerunners of the great molecular advances soon to follow. In 1945, Muller wrote about "the coming chemical attack on the gene," and the attack came almost immediately. Within a few years studies of ever finer resolution by geneticists and of ever larger molecules by chemists had converged.

THE SECOND PERIOD

Shortly after the discovery of *Neurospora* mutants with specific vitamin or amino acid requirements, Tatum extended these methods to *E. coli.* This opened the door to a major breakthrough. Joshua Lederberg was a medical student at Columbia University who had reason (based on the permutational patterns of antigen types) to think there might be a scrambling process in bacteria. He also thought that such a sexual process would be rare, so some selective mechanism would be needed to isolate recombinants. The idea seemed promising enough that he took leave from his medical studies to work with Tatum at Yale. There they mixed strains with different nutritional requirements chosen so that only recombinants would grow in minimal medium. They were lucky; the first experiment worked, and a medical career ended. At last, the stage was set for the genetic analysis of a prokaryote. At about the same time virologists, led by Max Delbrück, found recombination in bacteriophage. Much of genetic progress, such as the study of mutations and fine-structure mapping, had already depended on devising tricks to study very rare events. Microorganisms extended this power by several orders of magnitude. Genetic resolution was carried to its logical extreme by Seymour Benzer, with recombination mapping down to the nucleotide level in phage T4.

Avery's *Pneumococcus* transforming experiments argued very convincingly for DNA as the genetic material, but curiously the less definitive experiments of Al Hershey and Martha Chase in phage had a larger impact on the genetics community. By this time, 1952, DNA was becoming increasingly accepted as the key substance. A year later James Watson and Francis Crick proposed their model of DNA, and genetics was forever changed.

Seldom in science does a single finding bring immediate order out of confusion, but it's wonderful when it does. The Watson–Crick model was unexpectedly simple and surprisingly elegant. The very structure itself shouted the ways, quickly confirmed, in which Muller's questions could be answered. The information lay in the base sequences, a simple copying mechanism insured the continuity of the sequence, and mutation was explained as the copying of errors by blindly trusting enzymes of replication. Furthermore, the linearity of the information suggested an answer to Muller's forth point: some sort of translating mechanism transforms the linear DNA information into one-dimensional polypeptides, which fold into three-dimensional proteins.

In this period, genetics was again beautifully simple, as it had been in the teens. Messenger and transfer RNA were soon discovered. The simple three-letter code was solved. The *lac* operon provided a particularly neat means of developmental control. It appeared that biology would have some of the elegant simplicity of physics. Just as the period after 1900 brought a great generality and simplification to genetics, so did the DNA revolution and the beginnings of molecular biology.

Then the complications started, complications of which you, as the reader of this book, are well aware, perhaps painfully so. The deepest generalities remained, but questions of replication, transcription, translation, and regulation became more and more complex and, still more distressing, different from one species to another. The eukaryotic genome turned out to be surprisingly complicated. Genetics since the 1960s has been very much like the genetics of the 1930s. The basic principles, particulate inheritance in the early period and the structure of DNA in the later, are intact. But there are innumerable detailed ways in which the system operates, and, as before, much of the interest in biology lies in the diverse ways that different species do a job, often the same job. But for students, this means learning a large number of facts rather than deducing them from a few general principles.

One of the most remarkable of the new findings is error correction. DNA replication is much too accurate to be operating on the simple principles of physical chemistry. We now know about such things as excision repair and proofreading that collectively bring DNA replication to its astonishingly low error rate. And be thankful, for the last thing we need is a higher mutation rate.

Who would have expected Okazaki fragments, leading and trailing strands, and the need for an RNA primer in DNA replication? Everyone thought that there would be both 5' and 3' polymerases. An engineer would surely have arranged it that way. Many other surprises and complications I will simply mention. They are all in the book, along with many others. Some of these are overlapping genes, introns, reverse transcriptase, Z-DNA, antisense RNA, junk DNA, lines, sines, different codes in mitiochondria and some microorganisms, imprinting, pseudogenes, attenuation, microsatellites, chaperones, and transposons.

One of the most striking developments of recent years has been the rapid growth of gene mapping in higher plants and animals, including humans. The discovery of an almost unlimited number of molecular markers has totally changed human gene mapping. A few years ago only half a dozen linkages were known. Cytological studies had localized a few genes to specific chromosomes, or chromosome regions, but the process was slow. Molecular polymorphisms, most of which have no known function, can be used as markers. Through positional cloning and other methods, genes can be identified and then mapped by standard methods. The human linkage map has jumped from a dozen or so to thousands of genes in less than a decade. And now the program is underway to determine the entire DNA sequence. At this moment, the largest organism whose total sequence is known is a bacterium, but it won't be long until the list is extended to yeast, to nematodes, to various plants, to flies, and to mice and men.

Some of the most striking advances have been in the genetic control of development. Again, the details differ from species to species. At first, it looked as if the *lac*-operon would be *the* method of gene regulation. Now it turns out to be only one of many.

DNA chemistry has broken a major barrier in evolution studies. Previously, these studies were confined to animals or plants that could be crossed, and this usually meant only closely related species. There was even doubt as to whether differences between more remotely related forms were due to genes. Now, it is commonplace to compare genes throughout the animal and plant kingdoms. You have to have lived through an earlier period to realize how remarkable this is.

Molecular biology has created the totally new field of molecular evolution. This has permitted an accurate record of evolutionary changes at the DNA level. Although nothing in molecular studies has undermined the centrality of Darwin's theory of evolution by natural selection, Motoo Kimura in Japan has made us conscious of the large role that random changes play in molecular evolution.

One byproduct of molecular evolution studies has been more accurate phylogenetic trees. It is remarkable that some genes, such as homeobox genes, have been maintained through a whole evolutionary history. Molecular studies have also shown how species can become more complicated by adding new genes. But the new genes are not really new, they are adaptations of old genes recently duplicated and redundant, and therefore ready to take on a new function while the old one is retained. By gene duplication, evolution really *can* have its cake and eat it.

Now here's a curious question. Why does genetics, and especially developmental genetics, have to be so complicated—so needlessly complicated, it seems? And why do many processes that might be done in a uniform way differ so greatly from species to species? Every species seems to want to do things its own way. As I have said, the most basic rules are general. But detailed implementation of these rules is anything but general; it is remarkably disorderly. The key to understanding this, I think, is history. Remember that each contemporary species is at the end of a long evolutionary tree. As changes occurred, they altered something that was already there, patching up and tacking on. As a result, development does not proceed in the orderly way that it might if it were being designed from scratch using engineering principles of simplicity and efficiency. Rather, it is a series of changes grafted on other changes, and often the process would make sense only if we could reconstruct the history. Stated another way, this seeming disorder reflects the opportunism of evolution. Evolution incorporates whatever change is best at the moment, and it may not fit into any rational long-range plan. So, biology is complicated and varied, and often illogical; history has made it so. I know you find it bewildering, but I hope you also find it challenging and interesting.

Let's pause a moment before going into the next section. What, among the many unknowns in genetics, would you most like to understand? When I was a student, almost all geneticists would have given the same answer: What is the gene? What is it made of? What is its structure? How does it carry information? How does it replicate? How does it mutate? How does it control development and physiology?

Now things are different. There doesn't seem to be a single, central question. Would you opt for an understanding of aging? Or how life originated, on earth or elsewhere? Or what is mind? Or might you prefer less profound questions, such as what is the use of introns? Or are transposons useful to the organism, or simply DNA's way of doing what it does best, replicate? Let me suggest that, before reading on, you pause a moment and ask yourself this question. You might even write down your answer, and stash it away. You will enjoy looking at it 10, 25, or 50 years hence, and perhaps your question will have been answered.

THE THIRD PHASE: THE FUTURE IS NOW

Deep new scientific discoveries usually bring important new applications. The profound and revolutionary understanding that came with the Watson–Crick model and the development of molecular genetics ought to have practical consequences. Yet, these were slow in coming. The earliest applications involved the use of microorganisms: for example, developing strains to produce specific chemicals.

The third phase of genetics, the practical application of molecular genetics to agriculture and medicine, began more recently. The beginning was gradual, but the acceleration phase is here. About 15 years ago, it first became feasible to clone human genes and their mouse homologs. The ability to detect genes at the DNA level means that geneticists can now begin to perform in plants and animals, including humans, the kinds of tricks previously possible only in microbes.

In agriculture, molecular methods have made it feasible to introduce useful genes into farm crops. An example is a tomato that carries an introduced gene delaying the normal softening process, so that the tomatoes can be shipped with less spoilage. Similarly, genes for disease resistance can be introduced, as they have recently been in the important food crop, rice. The old process of introgressing useful genes from the wild relatives of cultivars involved several generations of backcrossing to remove unwanted hitchhikers that came in with the desired gene. Molecular methods can introduce only the genes that are wanted, without unwelcome linked interlopers.

Improvement of quantitative traits still proceeds by selection, and the methods used by animal and plant breeders are as efficient as sophisticated statistical methods can make them. But it would be better if some of the individual gene components of performance could be identified. To some extent this has happened as quantitative trait loci (QTLs) have been discovered. These discoveries can partially, but surely not totally, replace the slower process of crossing and selection to get improved performance.

The greatest successes in human genetics have been aids to diagnosis and genetic prognosis. These advances can be valuable in warning future parents of a possible diseased child. But improved diagnosis is often of little comfort if there is no treatment. Except for a few very rare diseases, human gene therapy has so far been a disappointment. But the successes, at least partial, in mammals argue that there will be successes in humans as well. But how soon? We shall have to wait to see.

WHAT OF THE FUTURE?

The human genome project has an intermediate goal of physically mapping 30,000 sites by 1998. This is one marker sequence every 100,000 bases. As I write this epilogue, the halfway point has been reached. The ultimate goal is the complete nucleotide sequence by 2005. Whether this is too optimistic remains to be seen, but surely a substantial part will be completed and will include virtually all the functional regions.

Then what? Knowing the complete sequence and finding all the genes won't tell us a lot of things we should like to know about development. There are problems of regulation, and who knows how complicated the interactions will be? But the sequence will provide an enormous technical boost as the process of development, complex as it is, is further dissected, with both intellectual satisfaction and practical uses.

We can look forward to a time in which the great bulk of severe monogenic diseases will be preventable. The main exception will be new mutations, since it is not likely that screening all fetuses for a large number of diseases will be feasible for some time to come. But we can count on better, cheaper, and less invasive methods of detecting fetal abnormalities, especially morphological ones. Almost all parents will be assured of a "perfect" baby—that is, one with no overt abnormalities. We can also count on increasing ethical dilemmas (e.g., privacy) as increasingly detailed knowledge of ourselves, and our neighbors, becomes available.

We can safely predict that chemical understanding, partly from genetics and partly from other sources, will bring mental disease into the realm of chemical intervention. We can expect simpler and better treatments, with untold reduction in human despair. The methods of discovery will be more rational and less empirical. Future advances in genetics, if intelligently applied, can surely decrease human suffering and anguish. But, as with any revolutionary change, there will be problems.

Gene therapy, as it develops, will almost certainly not be perfect, especially at first. And it is likely to be expensive. So we can expect an increase in medical costs. Diseases, genetic and nongenetic, that once caused early death will be treated. But everyone dies eventually, and a frequent consequence of better treatment will be trading an early, quick death for a protracted, expensive one later on. For this, and many other reasons, we can expect an increasingly aged and infirm population. Natural selection has no interest in people past the reproductive and child-caring ages and has made no provisions for the welfare of those euphemistically referred to as "seniors" or "mature" (postmature would be better). Are we really interested in prolonging life of low quality? Who wants to spend years dying? Will society soon condone assisted suicide for the terminally ill? Or perhaps go farther and institute some sort of euthanasia, as some societies that we like to call "primitive" have long done? The ethical dilemmas are clear: balancing societal needs and personal desires against cherished traditions and deeply held beliefs.

One problem that is not likely to have an easy solution is mutation. Molecular methods can deal with major mutations. But if *Drosophila* tells us anything, it is that the great bulk of mutations are those that cause small, nonspecific, harmful effects. This can only mean a gradual weakening of the population, unless this is compensated by environmental improvements or opposed by selection. But natural selection has been and we hope will continue to be less and less effective as each zygote has a high probability of surviving and being able to reproduce. Will future societies try to do something about this? Fortunately, this is a problem that will arise gradually over dozens of generations. We have time to learn more before acting precipitously.

We can be sure that human evolution will not stop, but can we predict the course of future genetic changes? Some things, like mutation accumulation, we can foresee. But for changes in frequencies of genes for health, intelligence, special talents, emotional stability, and other matters that we care about, there is no predicting. The reason is obvious: the genetic future depends on who produces more children than someone else. And that depends on human decisions, and human decisions are notoriously unpredictable. So you will get no predictions from me, only a feeling of uneasiness.

This raises the question of eugenics. Will human society be content to let its genetic future be determined by individual human decisions? How far can we count on molecular intervention and improved environments to substitute for selection? In Chapter 1 you learned of the crudity and naiveté of some of the early eugenicists in this country. Yet the idea, as formulated by Galton and many followers, had a large element of idealism and respect for individual liberties. The subject was dealt a death blow by Nazism. But genetic problems won't go away. Can we hope for a society that is sufficiently sophisticated, far-sighted, and compassionate to use its power wisely? Or will our successors decide not to step in where angels fear to tread?

And there are other problems. Try global warming. But to stay with biology, our greatest biological problem is not genetic; it is demographic. Unless the birthrate declines, and declines rapidly, this already overcrowded planet will become unlivable for most of humanity. If we don't solve the numbers problem, we won't have the luxury of worrying about mutation accumulation, eugenics, and ethical dilemmas arising from genetic knowledge. Yours may be the last generation in which reproduction is regarded as a right rather than a rationed privilege; this is already happening in parts of China. It is hard to be optimistic about a political solution to the world population problem. Maybe there will be a technical fix (a cheap, candy-coated, long-lasting contraceptive pill, perhaps?). The alternative to population control is an ever-widening extension of the poverty, disease, and starvation that we already see in too many parts of the globe.

Well, you have read one person's mixture of optimism and pessimism, of excitement and apprehension. One thing you can be sure of: the years ahead won't be dull. The road is full of potholes and land mines. But it promises the thrill of new insights, and the hope for a better world. I would like to be traveling it with you, but I'm afraid that is not in the cards.

GLOSSARY

This glossary provides an introduction to some basic and recurring terms in the text. Names of chemical compounds, definitions of specialized terms, and variants of basic names have been omitted from the glossary but are given in the index. Please locate terms that are not in the glossary by referring to the index.

Abscissa. The horizontal scale on a graph.

Acentric chromosome. Chromosome fragment lacking a centromere.

Acridine dyes. A class of positively charged polycyclic molecules that intercalate into DNA and induce frameshift mutations.

Acrocentric. A modifying term for a chromosome or chromatid that has its centromere near the end.

Acrosome. An apical organelle in the head of the sperm.

Adaptation. Adjustment of an organism or a population to an environment.

Adaptor. A synthetic oligonucleotide that contains two or more restriction enzyme cleavage sites.

Additive allelic effects. Genetic factors that raise or lower the value of a phenotype on a linear scale of measurement.

Adenine. A purine base found in RNA and DNA.

A-DNA. A right-handed DNA double helix that has 11 base pairs per turn. DNA exists in this form when partially dehydrated.

***Agrobacterium tumefaciens*-mediated transformation**. A naturally occurring process of DNA transfer from the bacterium *A. tumefaciens* to plants.

AIDS (acquired immunodeficiency syndrome). The usually fatal human disease in which the immune system is destroyed by the human immunodeficiency virus (HIV).

Albinism. Absence of pigment in skin, hair, and eyes of an animal. Absence of chlorophyll in plants.

Aleurone. The outermost layer of the endosperm in a seed.

Alkaptonuria. An inherited metabolic disorder. Alkaptonurics excrete excessive amounts of homogentisic acid (alkapton) in the urine.

Alkylating agents. Chemicals that transfer alkyl (methyl, ethyl, and so on) groups to the bases in DNA.

Allelic exclusion. A phenomenon whereby only one functional allele of an antibody gene can be assembled in a given B lymphocyte. The "allele" on the other homologous chromosome in a diploid mammalian cell cannot undergo a functional rearrangement, which would result in the production of two different antibodies by a single plasma cell.

Allele (allelomorph; *adj.*, allelic, allelomorphic). One of a pair, or series, of alternative forms of a gene that occur at a given locus in a chromosome. Alleles are symbolized with the same basic symbol (for example, *D* for tall peas and *d* for dwarf). (See **Multiple alleles.**)

Allele frequency. The proportion of one allele relative to all alleles at a locus in a population.

Allopatric speciation. Speciation occurring at least in part because of geographic isolation.

Allopolyploid. A polyploid having chromosome sets from different species; a polyploid containing genetically different chromosome sets derived from two or more species.

Allosteric transition. A reversible interaction of a small molecule with a protein molecule that causes a change in the shape of the protein and a consequent alteration of the interaction of that protein with a third molecule.

Allotetraploid. An organism with four genomes derived from hybridization of different species. Usually, in forms that become established, two of the four genomes are from one species and two are from another species.

Allozygote. A diploid individual in which the two genes of a particular locus are not identical by descent from a common ancestor.

Amino acid. Any one of a class of organic compounds containing an amino (NH_2) group and a carboxyl (COOH) group. Amino acids are the building blocks of proteins. Alanine, proline, threonine, histidine, lysine, glutamine, phenylalanine, tryptophan, valine, arginine, tyrosine, and leucine are among the common amino acids.

Aminoacyl (A) site. The ribosome binding site that contains the incoming aminoacyl-tRNA.

Aminoacyl-tRNA synthetases. Enzymes that catalyze the formation of high energy bonds between amino acids and tRNA molecules.

Amniocentesis. A procedure for obtaining amniotic fluid from a pregnant woman. Chemical contents of the fluid are studied directly for the diagnosis of some diseases. Cells are cultured, and metaphase chromosomes are examined for irregularities (for example, Down syndrome).

Amnion. The thin membrane that lines the fluid-filled sac in which the embryo develops in higher vertebrates.

Amniotic fluid. Liquid contents of the amniotic sac of higher vertebrates containing cells of the embryo (not of the mother). Both fluid and cells are used for diagnosis of genetic abnormalities of the embryo or fetus.

Amorph. A mutation that obliterates gene function; a null mutation.

Amphidiploid. A species or type of plant derived from doubling the chromosomes in the F_1 hybrid of two species; an allopolyploid. In an amphidiploid the two species are known, whereas in other allopolyploids they may not be known.

Anabolic pathway. A pathway by which a metabolite is synthesized; a biosynthetic pathway.

Anaphase. The stage of mitosis or meiosis during which the daughter chromosomes pass from the equatorial plate to opposite poles of the cell (toward the ends of the spindle). Anaphase follows metaphase and precedes telophase.

Anchor gene. A gene that has been positioned on both the physical map and the genetic map of a chromosome.

Androgen. A male hormone that controls sexual activity in vertebrate animals.

Anemia. Abnormal condition characterized by pallor, weakness, and breathlessness, resulting from a deficiency of hemoglobin or a reduced number of red blood cells.

Aneuploid. An organism or cell having a chromosome number that is not an exact multiple of the monoploid (n) with one genome, that is, hyperploid, higher (for example, $2n + 1$); or hypoploid, lower (for example, $2n - 1$). Also applied to cases where part of a chromosome is duplicated or deficient.

Antibody. Substance in a tissue or fluid of the body that acts in antagonism to a foreign substance (antigen).

Antibody class. In mammals, there are five classes of antibodies—IgA, IgD, IgE, IgG, and IgM; the class to which an antibody belongs depends on the type of heavy chain present.

Antibody-mediated (humoral) immune response. The synthesis of antibodies by plasma cells in response to an encounter of the cells of the immune system with a foreign immunogen.

Anticodon. Three bases in a transfer RNA molecule that are complementary to the three bases of a specific codon in messenger RNA.

Antigen. A substance, usually a protein, that is bound by an antibody or a T cell receptor when introduced into a vertebrate organism (cf. **Immunogen**).

Antihemophilic globulin. Blood globulin that reduces the clotting time of hemophilic blood.

Antisense gene. A gene that produces a transcript that is complementary to pre-mRNA or mRNA of a normal gene (usually constructed by inverting the coding region relative to promoter).

Antisense RNA. RNA that is complementary to the pre-mRNA or mRNA produced from a gene.

Artificial selection. The practice of choosing individuals from a population for reproduction, usually because these individuals possess one or more desirable traits.

Ascospore. One of the spores contained in the ascus of certain fungi such as *Neurospora*.

Ascus (*pl.*, **asci**). Reproductive sac in the sexual stage of a type of fungi (Ascomycetes) in which ascospores are produced.

Asexual reproduction. Any process of reproduction that does not involve the formation and union of gametes from the different sexes or mating types.

Assortative mating. Mating in which the partners are chosen because they are phenotypically similar.

Asynapsis. The failure or partial failure in the pairing of homologous chromosomes during the meiotic prophase.

ATP. Adenosine triphosphate: an energy-rich compound that promotes certain activities in the cell.

Attenuation. A mechanism for controlling gene expression in prokaryotes that involves premature termination of transcription.

Attenuator. A nucleotide sequence in the 5' region of a prokaryotic gene (or in its RNA) that causes premature termination of transcription, possibly by forming a secondary structure.

Autocatalytic reaction. A reaction catalyzed by a substrate without the involvement of any other catalytic agent.

Autoimmune diseases. Disorders in which the immune systems of affected individuals produce antibodies against self antigens—antigens synthesized in their own cells.

Autonomous. A term applied to any biological unit that can function on its own, that is, without the help of another unit. For example, a transposable element that encodes an enzyme for its own transposition (cf. **Nonautonomous**).

Autopolyploid. A polyploid that has multiple and identical or nearly identical sets of chromosomes (genomes). A polyploid species with genomes derived from the same original species.

Autoradiograph. A record or photograph prepared by labeling a substance such as DNA with a radioactive material such as tritiated thymidine and allowing the image produced by decay radiations to develop on a film over a period of time.

Autosome. Any chromosome that is not a sex chromosome.

Auxotroph. A mutant microorganism (for example, bacterium or yeast) that will not grow on a minimal medium but that requires the addition of some compound such as an amino acid or a vitamin.

Backcross. The cross of an F_1 hybrid to one of the parental types. The offspring of such a cross are referred to as the backcross generation or backcross progeny. (See **Testcross.**)

Back mutation. A second mutation at the same site in a gene as the original mutation, which restores the wild-type nucleotide sequence.

BACs (bacterial artificial chromosomes). Cloning vectors constructed from bacterial fertility (F) factors; like YAC vectors, they accept large inserts of size 200 to 500 kb.

Bacteriophage. A virus that attacks bacteria. Such viruses are called bacteriophages because they destroy their bacterial hosts.

Balanced lethal. Lethal mutations in different genes on the same pair of chromosomes that remain in repulsion because of close linkage or crossover suppression. In a closed population, only the *trans*-heterozygotes ($l_1 +/+ l_2$) for the lethal mutations survive.

Balanced polymorphism. Two or more types of individuals maintained in the same breeding population.

Barr body. A condensed mass of chromatin found in the nuclei of placental mammals that contains one or more X chromosomes; named for its discoverer, Murray Barr.

Basal body. Small granule to which a cilium or flagellum is attached.

Base analogs. Unnatural purine or pyrimidine bases that differ slightly from the normal bases and can be incorporated into nucleic acids. They are often mutagenic.

Base substitution. A single base change in a DNA molecule. (See **Transition; Transversion.**)

Binomial expansion. Exponential multiplication of an expression consisting of two terms connected by a plus (+) or minus (–) sign, such as $(a + b)^n$.

Biometry. Application of statistical methods to the study of biological problems.

Bivalent. A pair of synapsed or associated homologous chromosomes that have undergone the duplication process to form a group of four chromatids.

Blastomere. Any one of the cells formed from the first few cleavages in animal embryology.

Blastula. In animals, an early embryo form that follows the morula stage; typically, a single-layered sheet or ball of cells.

B lymphocytes (B cells). An important class of cells that mature in bone marrow and are largely responsible for the antibody-mediated or humoral immune response; they give rise to the antibody-producing plasma cells and some other cells of the immune system.

Breeding value. In quantitative genetics, the part of the deviation of an individual phenotype from the population mean that is due to the additive effects of alleles.

Broad-sense heritability. In quantitative genetics, the proportion of the total phenotypic variance that is the genotypic variance.

CAAT box. A conserved nucleotide sequence in eukaryotic promoters involved in the initiation of transcription.

Carcinogen. A substance capable of inducing cancer in an organism.

Carrier. An individual who carries a recessive allele that is not expressed (that is, is obscured by a dominant allele).

Catabolic pathway. A pathway by which an organic molecule is degraded in order to obtain energy for growth and other cellular processes; degradative pathway.

Catabolite repression. Glucose-mediated reduction in the rates of transcription of operons that specify enzymes involved in catabolic pathways (such as the *lac* operon).

cDNA (complementary DNA). A DNA molecule synthesized *in vitro* from an RNA template.

cDNA library. A collection of cDNA clones containing copies of the RNAs isolated from an organism or a specific tissue or cell type of an organism.

Cell cycle. The cyclical events that occur during the divisions of mitotic cells. The cell cycle oscillates between mitosis and the interphase, which is divided into G_1, S, and G_2.

Cellular immune response. See **T cell-mediated immune response.**

Centimorgan. See **Crossover unit.**

Centriole. An organelle in many animal cells that appears to be involved in the formation of the spindle during mitosis.

Centromere. Spindle-fiber attachment region of a chromosome.

Character (*contraction of the word* **characteristic**). One of the many details of structure, form, substance, or function that make up an individual organism.

Chemotaxis. Attraction or repulsion of organisms by a diffusing substance.

Chiasma (*pl.*, **Chiasmata**). A visible change of partners or a crossover in two of a group of four chromatids during the first meiotic prophase. In the diplotene stage of meiosis, the four chromatids of a bivalent are associated in pairs, but in such a way that one part of two chromatids is exchanged. This point of "change of partner" is the chiasma.

Chimera (animal). Individual derived from two embryos by experimental intervention.

Chimera (plant). Part of a plant with a genetically different constitution as compared with other parts of the same plant. It may result from different zygotes that grow together or from artificial fusion (grafting); it may either be pernical, with parallel layers of genetically different tissues, or sectorial.

Chimeric selectable marker gene. A gene that is constructed from parts of two or more different genes and allows the host cell to survive under conditions where it would otherwise die.

Chi-square. A statistic used to test the goodness of fit of data to the predictions of an hypothesis.

Chloroplastid. A green organelle in the cytoplasm of plants that contains chlorophyll and in which starch is synthesized. A mode of cytoplasmic inheritance, independent of nuclear genes, has been associated with these cytoplasmic organelles.

Chromatid. In mitosis or meiosis, one of the two identical strands resulting from self-duplication of a chromosome.

Chromatin. The deoxyribonucleohistone in chromosomes; originally named because of the readiness with which it stains with certain dyes (chromaticity).

Chromatin fibers. A basic organizational unit of eukaryotic chromosomes that consists of DNA and associated proteins assembled into strands of average diameter 30 nm.

Chromatography. A method for separating and identifying the components from mixtures of molecules having similar chemical and physical properties.

Chromocenter. Body produced by fusion of the heterochromatic regions of the chromosomes in the polytene tissues (for example, the salivary glands) of certain *Diptera.*

Chromomeres. Small bodies, described by J. Belling, that are identified by their characteristic size and linear arrangement along a chromosome.

Chromonema (*pl.*, **chromonemata**). An optically single thread forming an axial structure within each chromosome.

Chromosome aberration. Abnormal structure or number of chromosomes; includes deficiency, duplication, inversion, translocation, aneuploidy, polyploidy, or any other change from the normal pattern.

Chromosome banding. Staining of chromosomes in such a way that light and dark areas occur along the length of the chromosomes. Lateral comparisons identify pairs. Each human chromosome can be identified by its banding pattern.

Chromosome jumping. A procedure that uses large DNA fragments to "jump" along a chromosome from one site to another site. (See **Positional cloning.**)

Chromosomes. Darkly staining nucleoprotein bodies that are observed in cells during division. Each chromosome carries a linear array of genes.

Chromosome theory of inheritance. The theory that chromosomes carry the genetic information and that their behavior during meiosis provides the physical basis for segregation and independent assortment.

Chromosome walking. A procedure that uses overlapping clones to "walk" down a chromosome from one site to another site. (See **Positional cloning.**)

Cilium (*pl.*, **cilia;** *adj.*, **ciliate**). Hairlike locomotor structure on certain cells; a locomotor structure on a ciliate protozoan.

***cis*-acting sequence.** A nucleotide sequence that only affects the expression of genes located on the same chromosome, that is, *cis* to itself.

***cis* configuration.** See **Coupling.**

***cis* heterozygote.** A heterozygote that contains two mutations arranged in the *cis* configuration—for example, $a^+ b^+ / a\ b$)

Cistron. See **Gene.**

Class switching. The process during which a plasma cell stops producing antibodies of one class and begins producing antibodies of another class.

Clonal selection. The production of a population of plasma cells all producing the same antibody in reponse to the interaction between a B lymphocyte producing that specific antibody and the antigen bound by that antibody.

Clone. All the individuals derived by vegetative propagation from a single original individual. In molecular biology, a population of identical DNA molecules all carrying a particular DNA sequence from another organism.

Cloning vector. A small, self-replicating DNA molecule—usually a plasmid or viral chromosome—into which foreign DNAs are inserted in the process of cloning genes or other DNA sequences of interest.

Codominant alleles. Alleles that produce independent effects when heterozygous.

Codon. A set of three adjacent nucleotides in an mRNA molecule that specifies the incorporation of an amino acid into a polypeptide chain or that signals the end of polypeptide synthesis. Codons with the latter function are called termination codons.

Coefficient. A number expressing the amount of some change or effect under certain conditions (for example, the coefficient of inbreeding).

Coenzyme. A substance necessary for the activity of an enzyme.

Coincidence. The ratio of the observed frequency of double crossovers to the expected frequency, where the expected frequency is calculated by assuming that the two crossover events occur independently of each other.

Cointegrate. A DNA molecule formed by the fusion of two different DNA molecules, usually mediated by a transposable element.

Colchicine. An alkaloid derived from the autumn crocus that is used as an agent to arrest spindle formation and interrupt mitosis.

Colinearity. A relationship in which the units in one molecule occur in the same sequence as the units in another molecule which they specify; for example, the nucleotides in a gene are colinear with the amino acids in the polypeptide encoded by that gene.

Competence. Ability of a bacterial cell to incorporate DNA and become genetically transformed.

Complementarity. The relationship between the two strands of a double helix of DNA. Thymine in one strand pairs with adenine in the other strand, and cytosine in one strand pairs with guanine in the other strand.

Complementation test (*trans* test). Introduction of two recessive mutations into the same cell to determine whether they are alleles of the same gene, that is, whether they affect the same genetic function. If the mutations are allelic, the genotype $m_1 +/+ m_2$ will exhibit a mutant phenotype, whereas if they are nonallelic, it will exhibit the wild phenotype.

Complement proteins. Proteins (encoded by the *HLA* class III genes in humans) that bind to antibody-antigen complexes and help degrade the complexes by proteolysis.

Composite transposon. A transposable element formed when two identical or nearly identical transposons insert on either side of a nontransposable segment of DNA—for example, the bacterial transposon Tn5.

Compound chromosome. A chromosome formed by the union of two separate chromosomes, as in attached-X chromosomes or attached X-Y chromosomes.

Concordance. Identity of matched pairs or groups for a given trait—for example, identical twins both expressing the same genetic syndrome.

Conditional lethal mutation. A mutation that is lethal under one set of environmental conditions—the restrictive conditions—but is viable under another set of environmental conditions—the permissive conditions.

Conidium (*pl.*, conidia). An asexual spore produced by a specialized hypha in certain fungi.

Conjugation. Union of sex cells (gametes) or unicellular organisms during fertilization; in *Escherichia coli,* a one-way transfer of genetic material from a donor ("male" cell) to a recipient ("female" cell).

Consanguinity. Related by descent from a common ancestor.

Consensus sequence. The nucleotide sequence that is present in the majority of genetic signals or elements that perform a specific function.

Constitutive enzyme. An enzyme that is synthesized continually regardless of growth conditions (cf. **Inducible enzyme** and **Repressible enzyme**).

Constitutive gene. A gene that is continually expressed in all cells of an organism.

Contig. A set of overlapping clones that provide a physical map of a portion of a chromosome.

Continuous variation. Variation not represented by distinct classes. Individuals grade into each other, and measurement data are required for analysis (cf. **Discontinuous variation**). Multiple genes are usually responsible for this type of variation.

Controlling element. In maize, a transposable element such as *Ac* or *Ds* that is capable of influencing the expression of a nearby gene.

Coordinate repression. Correlated regulation of the structural genes in an operon by a molecule that interacts with the operator sequence.

Copolymers. Mixtures consisting of more than one monomer; for example, polymers of two kinds of organic bases such as uracil and cytosine (poly-UC) have been combined for studies of the genetic code.

Co-repressor. An effector molecule that forms a complex with a repressor and turns off the expression of a gene or set of genes.

Correlation. A statistical association between variables.

Cosmids. Cloning vectors that are hybrids between phage λ chromosomes and plasmids; they contain λ *cos* sites and plasmid origins of replication.

Coupling (*cis* configuration). The condition in which a double heterozygote has received two linked mutations from one parent and their wild-type alleles from the other parent (for example, *a b/a b* × + +/+ + produces *a b*/+ + (cf. **Repulsion**).

Covalent bond. A bond in which an electron pair is equally shared by protons in two adjacent atoms.

Covariance. A measure of the statistical association between variables.

cpDNA. The DNA of plant plastids, including chloroplasts.

CpG islands. Clusters of cytosines and guanines that often occur upstream of human genes.

Crossbreeding. Mating between members of different races or species.

Crossing over. A process in which chromosomes exchange material through the breakage and reunion of their DNA molecules. (See **Recombination.**)

Crossover unit. A measure of distance on genetic maps that is based on the average number of crossing-over events that take place during meiosis. A map interval that is one crossover unit in length (sometimes called a centimorgan) implies that only one in every hundred chromatids recovered from meiosis will have undergone a crossing-over event in this interval.

Cytogenetics. Area of biology concerned with chromosomes and their implications in genetics.

Cytokinesis. Cytoplasmic division and other changes exclusive of nuclear division that are a part of mitosis or meiosis.

Cytology. The study of the structure and function of cells.

Cytoplasm. The protoplasm of a cell outside the nucleus in which cell organelles (mitochondria, plastids, and the like) reside; all living parts of the cell except the nucleus.

Cytoplasmic inheritance. Hereditary transmission dependent on the cytoplasm or structures in the cytoplasm rather than the nuclear genes; extrachromosomal inheritance. Example: Plastid characteristics in plants may be inherited by a mechanism independent of nuclear genes.

Cytosine. A pyrimidine base found in RNA and DNA.

Cytotype. A maternally inherited cellular condition in *Drosophila* that regulates the activity of transposable P elements.

Dalton. The mass of a hydrogen atom.

Deficiency (deletion). Absence of a segment of a chromosome, reducing the number of loci.

Degeneracy (of the genetic code). The specification of an amino acid by more than one codon.

Deme. A local population of organisms.

Denaturation. Loss of native configuration of a macromolecule, usually accompanied by loss of biological activity. Denatured proteins often unfold their polypeptide chains and express changed properties of solubility.

***de novo*.** Arising anew, afresh, once more.

Deoxyribonuclease (DNase). Any enzyme that hydrolyzes DNA.

Deoxyribonucleic acid. See **DNA.**

Derepression. The process of turning on the expression of a gene or set of genes whose expression has been repressed (turned off).

Determination. Process by which undifferentiated cells in an embryo become committed to develop into specific cell types, such as neuron, fibroblast, and muscle cell.

Deviation. As used in statistics, a departure from an expected value.

Diakinesis. A stage of meiosis just before metaphase I in which the bivalents are shortened and thickened.

Dicentric chromosome. One chromosome having two centromeres.

Dicot. A plant with two cotyledons, or seed leaves.

Differentiation. A process in which unspecialized cells develop characteristic structures and functions.

Dihybrid. An individual that is heterozygous for two pairs of alleles; the progeny of a cross between homozygous parents differing in two respects.

Dimer. A compound having the same percentage composition as another but twice the molecular weight; one formed by polymerization.

Dimorphism. Two different forms in a group as determined by such characteristics as sex, size, or coloration.

Diploid. An organism or cell with two sets of chromosomes ($2n$) or two genomes. Somatic tissues of higher plants and animals are ordinarily diploid in chromosome constitution in contrast with the haploid (monoploid) gametes.

Diplonema (*adj.*, **diplotene**). That stage in prophase of meiosis I following the pachytene stage, but preceding diakinesis, in which the chromosomes of bivalents separate from each other at and around their centromeres.

Discontinuous variation. Phenotypic variability involving distinct classes such as red versus white, tall versus dwarf (cf. **Continuous variation**).

Discordant. Members of a pair showing different, rather than similar, characteristics.

Disjunction. Separation of homologous chromosomes during anaphase of mitotic or meiotic divisions. (See **Nondisjunction.**)

Disome. See **Monosomic.**

Ditype. In fungi, a tetrad that contains two kinds of meiotic products (spores) (for example, $2AB$ and $2ab$).

Dizygotic twins. Two-egg or fraternal twins.

DNA. Deoxyribonucleic acid; the information-carrying genetic material that comprises the genes. DNA is a macromolecule composed of a long chain of deoxyribonucleotides joined by phosphodiester linkages. Each deoxyribonucleotide contains a phosphate group, the five-carbon sugar 2-deoxribose, and a nitrogen-containing base.

DNA fingerprint. The banding pattern on a Southern blot of an individual's genomic DNA that has been cleaved with a restriction enzyme(s) and hybridized to an appropriate nucleic acid probe(s).

DNA gyrase. An enzyme in bacteria that catalyzes the formation of negative supercoils in DNA.

DNA helicase. An enzyme that catalyzes the unwinding of the complementary strands of a DNA double helix.

DNA ligase. An enzyme that catalyzes covalent closure of nicks in DNA double helices.

DNA polymerase. An enzyme that catalyzes the synthesis of DNA.

DNA primase. An enzyme that catalyzes the synthesis of short strands of RNA that initiate the synthesis of DNA strands.

DNA repair enzymes. Enzymes that catalyze the repair of damaged DNA.

DNA topoisomerase. An enzyme that catalyzes the introduction or removal of supercoils from DNA.

Dominance. A condition in which one member of an allele pair is manifested to the exclusion of the other.

Dominant selectable marker gene. A gene that allows the host cell to survive under conditions where it would otherwise die.

Dosage compensation. A phenomenon in which the activity of a gene is increased or decreased according to the number of copies of that gene in the cell.

Drift. See **Random genetic drift.**

Duplication. The occurrence of a segment more than once in the same chromosome or genome; also, the multiplication of cells.

Ecdysone. A hormone that influences development in insects.

Eclosion. Emergence of an adult insect from the pupal stage.

Ecotype. A population or strain of organisms that is adapted to a particular habitat.

Effector molecule. A molecule that influences the behavior of a regulatory molecule, such as a repressor protein, thereby influencing gene expression.

Egg (ovum). A germ cell produced by a female organism.

Electrophoresis. The migration of suspended particles in an electric field.

Electroporation. A process whereby cell membranes are made permeable to DNA by applying an intense electric current.

Elongation factors. Soluble proteins that are required for polypeptide chain elongation.

Embryo. An organism in the early stages of development; in humans, the first two months in the uterus.

Embryo sac. A large thin-walled space within the ovule of the seed plant in which the egg and, after fertilization, the embryo develop; the mature female gametophyte in higher plants.

Endogenote. The part of the bacterial chromosome that is homologous to a genome fragment (exogenote) transferred from the donor to the recipient cell in the formation of a merozygote.

Endomitosis. Duplication of chromosomes without division of the nucleus, resulting in increased chromosome number within a cell. Chromosomes strands separate, but the cell does not divide.

Endonuclease. An enzyme that breaks strands of DNA at internal positions; some are involved in recombination of DNA.

Endoplasmic reticulum. Network of membranes in the cytoplasm to which ribosomes adhere.

Endopolyploidy. A state in which the cells of a diploid organism contain multiples of the diploid chromosome number (that is, $4n$, $8n$, and so on).

Endosperm. Nutritive tissue that develops in the embryo sac of most angiosperms. It usually forms after the fertilization of the two fused primary endosperm nuclei of the embryo sac with one of the two male gamete nuclei. In most diploid plants, the endosperm is triploid ($3n$), but in some (for example, the lily) it is $5n$.

End-product inhibition. See **Feedback inhibition.**

Enhancer. A substance or object that increases a chemical activity or a physiological process; a major or modifier gene that increases a physiological process; a DNA sequence that influences transcription of a nearby gene.

Environment. The aggregate of all the external conditions and influences affecting the life and development of an organism.

Enzyme. A protein that accelerates a specific chemical reaction in a living system.

Epigenetic. A term referring to the nongenetic causes of a phenotype.

Episome. A genetic element that may be present or absent in different cells and that may be inserted in a chromosome or independent in the cytoplasm (for example, the fertility factor (F) in *Escherichia coli*).

Epistasis. Interactions between products of nonallelic genes. Genes suppressed are said to be hypostatic. Dominance is associated with members of allelic pairs, whereas epistasis is interaction among products of nonalleles.

Epitope. A specific structural feature of an antigen that stimulates the production of antibodies.

Equational division. Mitotic-type division that is usually the second division in the meiotic sequence; somatic mitosis and the nonreductional division of meiosis.

Equatorial plate. The figure formed by the chromosomes in the center (equatorial plane) of the spindle in mitosis.

Equilibrium. A state of dynamical systems in which there is no net change.

Equilibrium density gradient centrifugation. A procedure used to separate macromolecules based on their density (mass per unit volume).

Estrogen. Female hormone or estrus-producing compound.

ESTs (expressed sequence tags). Short cDNA sequences that are used to link physical maps and genetic (RFLP) maps.

Euchromatin. Genetic material that is not stained so intensely by certain dyes during interphase and that comprises many different kinds of genes (cf. **Heterochromatin**).

Eugenics. The application of the principles of genetics to the improvement of humankind.

Eukaryote. A member of the large group of organisms that have nuclei enclosed by a membrane within their cells (cf. **Prokaryote**).

Euploid. An organism or cell having a chromosome number that is an exact multiple of the monoploid (n) or haploid number. Terms used to identify different levels in an euploid series are diploid, triploid, tetraploid, and so on (cf. **Aneuploid**).

Excinuclease. The endonuclease-containing protein complex that excises a segment of damaged DNA during excision repair.

Excision repair. DNA repair processes that involve the removal of the damaged segment of DNA and its replacement by the synthesis of a new strand using the complementary strand of DNA as template.

Exit (E) site. The ribosome binding site that contains the free tRNA prior to its release.

Exogenote. Chromosomal fragment homologous to an endogenote and donated to a merozygote.

Exon amplification. A procedure that is used to identify coding regions (exons) that are flanked by 5' and 3' intron splice sites.

Exons. The segments of a eukaryotic gene that correspond to the sequences in the final processed RNA transcript of that gene.

Exonuclease. An enzyme that digests DNA or RNA, beginning at the ends of strands.

Expressivity. Degree of expression of a trait controlled by a gene. A particular gene may produce different degrees of expression in different individuals.

Extrachromosomal. Structures that are not part of the chromosomes; DNA units in the cytoplasm that control cytoplasmic inheritance.

F_1. The first filial generation; the first generation of descent from a given mating.

F_2. The second filial generation produced by crossing *inter se* or by self-pollinating the F_1. The inbred "grandchildren" of a given mating, but in controlled genetic experimentation, self-fertilization of the F_1 (or equivalent) is implied.

F factor. A bacterial episome that confers the ability to function as a genetic donor ("male") in conjugation; the fertility factor in bacteria.

Feedback inhibition. The accumulated end product of a biochemical pathway stops synthesis of that product. A late metabolite of a synthetic pathway regulates synthesis at an earlier step of the pathway.

Fertilization. The fusion of a male gamete (sperm) with a female gamete (egg) to form a zygote.

Fetus. Prenatal stage of a viviparous animal between the embryonic stage and the time of birth; in humans, the final seven months before birth.

Filial. See F_1 and F_2.

Fitness. The number of offspring left by an individual, often compared with the average of the population or with some other standard, such as the number left by a particular genotype.

Fixation. An event that occurs when all the alleles at a locus except one are eliminated from a population. The remaining allele, with frequency 100 percent, is said to have been fixed.

Flagellum (*pl.*, **flagella;** *adj.* **flagellate**). A whiplike organelle of locomotion in certain cells; locomotor structures in flagellate protozoa.

Folded genome. The condensed intracellular state of the DNA in the nucleoid of a bacterium. The DNA is segregated into domains, and each domain is independently negatively supercoiled.

Forced cloning. The insertion of a foreign DNA into a cloning vector in a predetermined orientation.

Founder principle. The possibility that a new, small, isolated population may diverge genetically because the founding individuals are a random sample from a large, main population.

Frameshift mutation. A mutation that changes the reading frame of an mRNA, either by inserting or deleting nucleotides.

Frequency distribution. A graph showing either the relative or absolute incidence of classes in a population. The classes may be defined by either a discrete or a continuous variable; in the latter case, each class represents a different interval on the scale of measurement.

Fusion protein. A polypeptide made from a recombinant gene that contains portions of two or more different genes. The different genes are joined so that their coding sequences are in the same reading frame.

Gall. A tumorous growth in plants.

Gamete. A mature male or female reproductive cell (sperm or egg).

Gametogenesis. The formation of gametes.

Gametophyte. That phase of the plant life cycle that bears the gametes; the cells have n chromosomes.

Gametophytic incompatibility. A botanical phenomenon controlled by the complex S locus in which a pollen grain cannot fertilize an ovule produced by a plant that carries the same S allele as the pollen grain. For example, S_1 pollen cannot fertilize an ovule made by an S_1/S_2 plant.

Gap gene. A gene that controls the formation of adjacent segments in the body of *Drosophila.*

Gastrula. An early animal embryo consisting of two layers of cells; an embryological stage following the blastula.

Gene. A hereditary determinant of a specific biological function; a unit of inheritance (DNA) located in a fixed position on a chromosome; a segment of DNA encoding one polypeptide and defined operationally by the *cis-trans* or complementation test.

Gene addition. The addition of a functional copy of a gene to the genome of an organism.

Gene cloning. The incorporation of a gene of interest into a self-replicating DNA molecule and the amplification of the resulting recombinant DNA molecule in an appropriate host cell.

Gene conversion. A process, often associated with recombination, during which one allele is replicated at the expense of another, leading to non-Mendelian segregation ratios. In whole tetrads, for example, the ratio may be 6:2 or 5:3 instead of the expected 4:4.

Gene expression. The process by which genes produce RNAs and proteins and exert their effects on the phenotype of an organism.

Gene flow. The spread of genes from one breeding population to another by migration, possibly leading to allele frequency changes.

Gene pool. The sum total of all different alleles in the breeding members of a population at a given time.

Gene replacement. The incorporation of a transgene into a chromosome at its normal location by homologous recombination, thus replacing the copy of the gene originally present at the locus.

Gene therapy. The treatment of inherited diseases by introducing wild-type copies of the defective gene causing the disorder into the cells of affected individuals. If reproductive cells are modified, the procedure is called *germ-line* or *heritable gene therapy.* If cells other than reproductive cells are modified, the procedure is called *somatic-cell* or *noninheritable gene therapy.*

Genetic code. The set of 64 nucleotide triplets that specify the 20 amino acids and polypeptide chain initiation and termination.

Genetic drift. See **Random genetic drift.**

Genetic equilibrium. Condition in a group of interbreeding organisms in which the allele frequencies remain constant over time.

Genetics. The science of heredity and variation.

Genome. A complete set (n) of chromosomes (hence, of genes) inherited as a unit from one parent.

Genomic DNA library. A collection of clones containing the genomic DNA sequences of an organism.

Genotype. The genetic constitution (gene makeup) of an organism (cf. **Phenotype**).

Germ cell. A reproductive cell capable when mature of being fertilized and reproducing an entire organism (cf. **Somatic cell**).

Germ plasm. The hereditary material transmitted to the offspring through the germ cells.

Globulins. Common proteins in the blood that are insoluble in water and soluble in salt solutions. Alpha, beta, and gamma globulins can be distinguished in human blood serum. Gamma globulins are important in developing immunity to diseases.

Glucocorticoid. A steroid hormone that regulates gene expression in higher animals.

Gonad. A sexual organ (that is, ovary or testis) that produces gametes.

Guanine. A purine base found in DNA and RNA.

Guide RNAs. RNA molecules that contain sequences that function as templates during RNA editing.

Gynandromorph. An individual in which one part of the body is female and another part is male; a sex mosaic.

Haploid (monoploid). An organism or cell having only one complete set (n) of chromosomes or one genome.

Haptoglobin. A serum protein, alpha globulin, in the blood.

Hardy-Weinberg equilibrium. Mathematical relationship that allows the frequencies of genotypes in a population to be predicted from their constituent allele frequencies; a consequence of random mating.

Helix. Any structure with a spiral shape. The Watson and Crick model of DNA is in the form of a double helix.

Helper T cells. T cells that respond to an antigen displayed by a macrophage by stimulating B and T lymphocytes to develop into antibody-producing plasma cells and killer T cells, respectively.

Hemizygous. The condition in which only one allele of a pair is present, as in sex linkage or as a result of deletion.

Hemoglobin. Conjugated protein compound containing iron, located in erythrocytes of vertebrates; important in the transportation of oxygen to the cells of the body.

Hemolymph. The mixture of blood and other fluids in the body cavity of an invertebrate.

Hemophilia. A bleeder's disease; tendency to bleed freely from even a slight wound; hereditary condition dependent on a sex-linked recessive gene.

Heredity. Resemblance among individuals related by descent; transmission of traits from parents to offspring.

Heritability. Degree to which a given trait is controlled by inheritance. (See **Broad-sense heritability** and **Narrow-sense heritability**.)

Hermaphrodite. An individual with both male and female reproductive organs.

Heteroalleles. Mutations that are functionally allelic but structurally nonallelic; mutations at different sites in a gene.

Heterochromatin. Chromatin staining darkly even during interphase, often containing repetitive DNA with few genes.

Heteroduplex. A double-stranded nucleic acid containing one or more mismatched (noncomplementary) base pairs.

Heterogametic sex. Producing unlike gametes with regard to the sex chromosomes. In humans, the XY male is heterogametic, and the XX female is homogametic.

Heterogeneous nuclear RNA (hnRNA). The population of primary transcripts in the nucleus of a eukaryotic cell.

Heterokaryon. A cell containing two or more different nuclei.

Heteroplasmy. A cellular condition in which two genetically different types of an organelle are present (cf. **Homoplasmy**).

Heteropyknosis (*adj.*, **heteropyknotic**). Property of certain chromosomes, or of their parts, to remain more dense and to stain more intensely than other chromosomes or parts during the cell cycle.

Heterosis. Superiority of heterozygous genotypes in respect to one or more traits in comparison with corresponding homozygotes.

Heterozygote (*adj.*, **heterozygous**). An organism with unlike members of any given pair or series of alleles that consequently produces unlike gametes.

Hfr. High-frequency recombination strain of *Escherichia coli;* in such strains, the F episome is integrated into the bacterial chromosome.

Histones. Group of proteins rich in basic amino acids. They function in the coiling of DNA in chromosomes and in the regulation of gene activity.

HIV (human immunodeficiency virus). The retrovirus that causes AIDS in humans.

HLA antigens. See **Major histocompatibility antigens.**

***HLA* locus.** See **Major histocompatibility complex.**

Holoenzyme. The form of a multimeric enzyme in which all of the component polypeptides are present.

Homeobox. A DNA sequence found in several genes that are involved in the specification of organs in different body parts in aminals; characteristic of genes that influence segmentation in animals. The homeobox corresponds to an amino acid sequence in the polypeptide encoded by these genes; this sequence is called the homeodomain.

Homeotic mutation. A mutation that causes a body part to develop in an inappropriate position in an organism; for example, a mutation in *Drosophila* that causes legs to develop on the head in the place of antennae.

Homoalleles. Mutations that are both functionally and structurally allelic; mutations at the same site in the same gene.

Homogametic sex. Producing like gametes with regard to the sex chromosomes (cf. **Heterogametic sex**).

Homologous chromosomes. Chromosomes that occur in pairs and are generally similar in size and shape, one having come from the male parent and the other from the female parent. Such chromosomes contain the same array of genes.

Homoplasmy. A cellular condition in which all copies of an organelle are genetically identical (cf. **Heteroplasmy**).

Hormone. An organic product of cells of one part of the body that is transported by the body fluids to another part where it influences activity or serves as a coordinating agent.

Human Genome Project. A huge international effort to map and sequence the entire human genome.

Humoral immune response. See **Antibody-mediated immune response.**

Hybrid. An offspring of homozygous parents differing in one or more genes; more generally, an offspring of a cross between unrelated strains.

Hybrid dysgenesis. In *Drosophila,* a syndrome of abnormal germ-line traits, including mutation, chromosome breakage, and sterility, which results from transposable element activity.

Hybridization. Interbreeding of species, races, varieties, and so on, among plants or animals; a process of forming a hybrid by cross pollination of plants or by mating animals of different types.

Hybrid vigor (heterosis). Unusual growth, strength, and health of heterozygous hybrids from two less vigorous homozygous parents.

Hydrogen bonds. Weak interactions between electronegative atoms and hydrogen atoms (electropositive) that are linked to other electronegative atoms.

Hydrophobic interactions. Association of nonpolar groups with each other when present in aqueous solutions because of their insolubility in water.

Hyperploid. A genetic condition in which a chromosome or a segment of a chromosome is overrepresented in the genotype (cf. **Hypoploid**).

Hypersensitive sites. Regions in the DNA that are highly susceptible to digestion with endonucleases.

Hypomorph. A mutation that reduces but does not completely abolish gene expression.

Hypoploid. A genetic condition in which a chromosome or segment of a chromosome is underrepresented in the genotype (cf. **Hyperploid**).

Hypostasis. See **Epistasis.**

Imaginal disc. A mass of cells in the larvae of *Drosophila* and other holometabolous insects that gives rise to particular adult organs such as antennae, eyes, and wings.

Immunogen. A substance, usually a protein, that elicits an immune response when introduced into a vertebrate.

Immunoglobulin. See **Globulin.**

Inbreeding. Matings between related individuals.

Incomplete dominance. Expression of two alleles in a heterozygote that allows the heterozygote to be distinguished from either of its homozygous parents.

Independent assortment. The random distribution of alleles to the gametes that occurs when genes are located in different chromosomes. The distribution of one pair of alleles is independent of other genes located in nonhomologous chromosomes.

Inducer. A substance of low molecular weight that is bound by a repressor to produce a complex that can no longer bind to the operator; thus, the presence of the inducer turns on the expression of the gene(s) controlled by the operator.

Inducible enzyme. An enzyme that is synthesized only in the presence of the substrate that acts as an inducer.

Inducible gene. A gene that is expressed only in the presence of a specific metabolite, the inducer.

Induction. The process of turning on the expression of a gene or set of genes by an inducer.

Inhibitor. Any substance or object that retards a chemical reaction; a major or modifier gene that interferes with a reaction.

Initiation factors. Soluble proteins required for the initiation of translation.

In situ. From the Latin, meaning in the natural place; refers to experimental treatments performed on cells or tissue rather than on extracts from them.

***In situ* colony or plaque hybridization.** A procedure for screening colonies or plaques growing on plates or membranes for the presence of specific DNA sequences by the hybridization of nucleic acid probes to the DNA molecules present in these colonies or plaques.

Interaction. In statistics, an effect that cannot be explained by the additive action of contributing factors; a departure from strict additivity.

Intercalating agent. A chemical capable of inserting between adjacent base pairs in a DNA molecule.

Interference. Crossing over at one point that reduces the chance of another crossover nearby; detected by studying the pattern of crossing over with three or more linked genes.

Interphase. The stage in the cell cycle when the cell is not dividing; the metabolic stage during which DNA replication occurs; the stage following telophase of one division and extending to the beginning of prophase in the next division.

Intersex. An organism displaying secondary sexual characters intermediate between male and female; a type that shows some phenotypic characteristics of both males and females.

Intragenic complementation. Complementation that occurs between two mutant alleles of a gene; common only when the product of the gene functions as a homomultimer.

Introns. Intervening sequences of DNA bases within eukaryotic genes that are not represented in the mature RNA transcript because they are spliced out of the primary RNA transcript.

Invariant. Constant, unchanging, usually referring to the portion of a molecule that is the same across species.

Inversion. A rearrangement that reverses the order of a linear array of genes in a chromosome.

Inverted repeat. A sequence present twice in a DNA molecule but in reverse orientation.

in vitro. From the Latin meaning "within glass"; biological processes made to occur experimentally outside the organism in a test tube or other container.

in vivo. From the Latin meaning "within the living organism."

Ionic bonds. Attractions between oppositely charged chemical groups.

Ionizing radiation. The portion of the electromagnetic spectrum that results in the production of positive and negative charges (ion pairs) in molecules. X rays and gamma rays are examples of ionizing radiation.

IS element (insertion sequence). A short (800–1400 nucleotide pairs) DNA sequence found in bacteria that is capable of transposing to a new genomic location; other DNA sequences that are bounded by IS elements may also be transposed.

Isoalleles. Different forms of a gene that produce the same phenotype or very similar phenotypes.

Isochromosome. A chromosome with two identical arms and identical genes. The arms are mirror images of each other.

Isoform. A member of a family of closely related proteins—proteins that have some amino acid sequences in common and some different.

Isogenic stocks. Strains of organisms that are genetically uniform; completely homozygous.

Kappa chain. One of two classes of antibody light chains (cf. **Lambda chain**).

Karyotype. The chromosome constitution of a cell or an individual; chromosomes arranged in order of length and according to position of centromere; also, the abbreviated formula for the chromosome constitution, such as 47, XX + 21 for human trisomy-21.

Killer T cells (cytotoxic T cells). T cells that carry T-cell receptors and kill cells displaying the recognized antigens.

Kinetics. A dynamic process involving motion.

Kinetosome. Granular body at the base of a flagellum or a cilium.

Lagging strand. The strand of DNA that is synthesized discontinuously during replication.

Lambda chain. One of two classes of antibody light chains (cf. **Kappa chain**).

Lamella. A double-membrane structure, plate, or vesicle that is formed by two membranes lying parallel to each other.

Lampbrush chromosomes. Large diplotene chromosomes present in oocyte nuclei, particularly conspicuous in amphibians. These chromosomes have extended regions called loops, which are active sites of transcription.

Leader sequence. The segment of an mRNA molecule from the 5' terminus to the translation initiation codon.

Leading strand. The strand of DNA that is synthesized continuously during replication.

Leptonema (*adj.*, leptotene). Stage in meiosis immediately preceding synapsis in which the chromosomes appear as single, fine, threadlike structures (but they are really double because DNA replication has already taken place).

Lethal mutation. A mutation that renders an organism or a cell possessing it inviable.

Ligand. A molecule that can bind to another molecule in or on cells.

Ligase. An enzyme that joins the ends of two strands of nucleic acid.

Ligation. The joining of two or more DNA molecules by covalent bonds.

LINEs (long interspersed nuclear elements). Families of long (average length = 6500 bp) moderately repetitive transposable elements in eukaryotes.

Linkage. A relationship among genes in the same chromosome. Such genes tend to be inherited together.

Linkage equilibrium. A state in which the alleles of linked loci are randomized with respect to each other on the chromosomes of a population.

Linkage map. A linear or circular diagram that shows the relative positions of genes on a chromosome as determined by genetic analysis.

Linker. A short, double-stranded DNA that contains one or more restriction enzyme cleavage sites.

Locus (*pl.,* loci). A fixed position on a chromosome that is occupied by a given gene or one of its alleles.

Lod score. The logarithm of the ratio of odds for pedigree data, where the odds are calculated under the assumption of linkage with a specified frequency of recombination and under the assumption of no linkage, that is, 50 percent recombination.

LTR (Long terminal repeat). A DNA sequence present at each end of a retrotransposon.

Lymphocyte. A general class of white blood cells that are important components of the immune system of vertebrate animals.

Lysis. Bursting of a cell by the destruction of the cell membrane following infection by a virus.

Lysogenic bacteria. Those harboring temperate bacteriophages.

Macromolecule. A large molecule; term used to identify molecules of proteins and nucleic acids.

Macrophages. Large, white blood cells that ingest foreign substances and display antigens produced from them on their surfaces to be recognized by other cells of the immune system.

Major histocompatibility antigens. A complex set of cell-surface macromolecules that allow the immune system to distinguish foreign or "nonself" cells from "self" cells. These are the antigens that must be matched between donors and recipients during organ and tissue transplants to prevent rejections. Called **HLA (human-leukocyte-associated) antigens** in humans.

Major histocompatibility complex. The large cluster of genes that encode the major histocompatibility antigens in mamals. Called the **HLA locus** in humans.

Map unit. See **Crossover unit.**

Mass selection. As practiced in plant and animal breeding, the choosing of individuals for reproduction from the entire population on the basis of the individuals' phenotypes rather than on the phenotypes of their relatives.

Maternal effect. Trait controlled by a gene of the mother but expressed in the progeny.

Maternal inheritance. Inheritance controlled by extrachromosomal (that is, cytoplasmic) factors that are transmitted through the egg.

Maturation. The formation of gametes or spores.

Mean. The arithmetic average; the sum of all measurements or values in a sample divided by the sample size.

Median. In a set of measurements, the central value above and below which there are an equal number of measurements.

Meiosis. The process by which the chromosome number of a reproductive cell becomes reduced to half the diploid ($2n$) or somatic number; results in the formation of gametes in animals or of spores in plants; important source of variability through recombination.

Meiotic drive. Any mechanism that causes alleles to be recovered unequally in the gametes of a heterozygote.

Melanin. Brown or black pigment.

Memory cells. Long-lived B and T cells that mediate rapid secondary immune responses to a previously encountered antigen.

Mendelian population. A natural interbreeding unit of sexually reproducing plants or animals sharing a common gene pool.

Merozygote. Partial zygote produced by a process of partial genetic exchange, such as transformation in bacteria. An exogenote may be introduced into a bacterial cell in the formation of a merozygote.

Mesoderm. The middle germ layer that forms in the early animal embryo and gives rise to such parts as bone and connective tissue.

Messenger RNA (mRNA). RNA that carries information necessary for protein synthesis from the DNA to the ribosomes.

Metabolic cell. A cell that is not dividing.

Metabolism. Sum total of all chemical process in living cells by which energy is provided and used.

Metacentric chromosome. A chromosome with the centromere near the middle and two arms of about equal length.

Metafemale (superfemale). In *Drosophila,* abnormal female, usually sterile, with an excess of X chromosomes compared with sets of autosomes (for example, XXX; AA).

Metaphase. That stage of cell division in which the chromosomes are most discrete and arranged in an equatorial plate; stage following prophase and preceding anaphase.

Metastasis. The spread of cancer cells to previously unaffected organs.

Microprojectile bombardment. A procedure for transforming plant cells by shooting DNA-coated tungsten or gold particles into the cells.

Microtubules. Hollow filaments in the cytoplasm making up a part of the locomotor apparatus of a motile cell; component of the mitotic spindle.

Midparent value. In quantitative genetics, the average of the phenotypes of two mates.

Mismatch repair. DNA repair processes that correct base pairs that are not properly hydrogen-bonded.

Missense mutation. A mutation that changes a codon specifying an amino acid to a codon specifying a different amino acid.

Mitochondria. Organelles in the cytoplasm of plant and animal cells where oxidative phosphorylation takes place to produce ATP.

Mitosis. Disjunction of duplicated chromosomes and division of the cytoplasm to produce two genetically identical daughter cells.

Modal class. In a frequency distribution, the class having the greatest frequency.

Model. A mathematical description of a biological phenomenon.

Modifier (modifying gene). A gene that affects the expression of some other gene.

Monoclonal antibody. An antibody produced by a population (clone) of identical hybridoma cells growing in culture.

Monohybrid. An offspring of two homozygous parents that differ from one another by the alleles present at only one gene locus.

Monohybrid cross. A cross between parents differing in only one trait or in which only one trait is being considered.

Monomer. A single molecular entity that may combine with others to form more complex structures.

Monoploid (haploid). Organism or cell having a single set of chromosomes or one genome (chromosome number n).

Monosomic. A diploid organism lacking one chromosome of its proper complement ($2n - 1$); an aneuploid. Monosome refers to the single chromosome, disome to two chromosomes of a kind, and trisome to three chromosomes of a kind.

Monozygotic twins. One-egg or identical twins.

Morphogen. A substance that stimulates the development of form or structure in an organism.

Morphology. Study of the form of an organism; developmental history of visible structures and the comparative relation of similar structures in different organisms.

Mosaic. An organism or part of an organism that is composed of cells of different genotypes.

mtDNA. The DNA of mitochondria.

Multigene family. A group of genes that are similar in nucleotide sequence or that produce polypeptides with similar amino acid sequences.

Multiple alleles. A condition in which a particular gene occurs in three or more allelic forms in a population of organisms.

Mutable genes. Genes with an unusually high mutation rate.

Mutagen. An environmental agent, either physical or chemical, that is capable of inducing mutations.

Mutant. A cell or individual organism that shows a change brought about by a mutation; a changed gene.

Mutation. A change in the DNA at a particular locus in an organism. The term is used loosely to include point mutations involving a single gene change as well as a chromosomal change.

Mutation pressure. A constant mutation rate that adds mutant genes to a population; repeated occurrences of mutations in a population.

Mycelium (*pl.*, mycelia). Threadlike filament making up the vegetative portion of thallus fungi.

Narrow-sense heritability. In quantitative genetics, the proportion of the phenotypic variance that is due to the additive effects of alleles.

Natural selection. Differential survival and reproduction in nature that favors individuals that are better adapted to their environment; elimination of less fit organisms.

Negative autogenous regulation (negative self-regulation). Inhibition of the expression of a gene or set of coordinately regulated genes by the product of the gene or the product of one of the genes.

Negative control system. A mechanism in which the regulatory protein(s) is required to turn off gene expression.

Neutral mutation. A mutation that changes the nucleotide sequence of a gene but has no effect on the fitness of the organism.

Neutral theory. The theory that the evolution of traits with little or no effect on fitness is a random process involving mutation and genetic drift.

Nick-translation. A procedure for labeling DNA by nicking it with an endonuclease and "translating" the nick along the DNA molecule in the presence of labeled deoxyribonucleoside triphosphates by the concerted action of the 5' —> 3' exonuclease and 5' —> 3' polymerase activities of *E. coli* DNA polymerase I.

Nonautonomous. A term referring to biological units that cannot function by themselves; such units require the assistance of another unit, or "helper" (cf. **Autonomous**).

Nondisjunction. Failure of disjunction or separation of homologous chromosomes in mitosis or meiosis, resulting in too many chromosomes in some daughter cells and too few in others. Examples: In meiosis, both members of a pair of chromosomes go to one pole so that the other pole does not receive either of them; in mitosis, both sister chromatids go to the same pole.

Nonhistone chromosomal proteins. All of the proteins in chromosomes except the histones.

Nonsense mutation. A mutation that changes a codon specifying an amino acid to a termination codon.

Nontemplate strand. In transcription, the nontranscribed strand of DNA. It will have the same sequence as the RNA transcript, except that T is present at positions where U is present in the RNA transcript.

Northern blot. The transfer of RNA molecules from an electrophoretic gel to a cellulose or nylon membrane by capillary action.

Nuclease. An enzyme that catalyzes the degradation of nucleic acids.

Nucleic acid. A macromolecule composed of phosphoric acid, pentose sugar, and organic bases; DNA and RNA.

Nucleolar Organizer (NO). A chromosomal segment containing genes that control the synthesis of ribosomal RNA, located at the secondary constriction of some chromosomes.

Nucleolus. An RNA-rich, spherical sack in the nucleus of metabolic cells; associated with the nucleolar organizer; storage place for ribosomes and ribosome precursors.

Nucleoprotein. Conjugated protein composed of nucleic acid and protein; the material of which the chromosomes are made.

Nucleoside. An organic compound consisting of a base covalently linked to ribose or deoxyribose.

Nucleosome. Spherical subunits of eukaryotic chromatin that are composed of a core particle consisting of an octamer of histones and 146 nucleotide pairs.

Nucleotide. A subunit of DNA and RNA molecules containing a phosphate group, a sugar, and a nitrogen-containing organic base.

Nucleus. The part of a eukaryotic cell that contains the chromosomes; separated from the cytoplasm by a membrane.

Nullisomic. An otherwise diploid cell or organism lacking both members of a chromosome pair (chromosome formula $2n - 2$).

Null mutation. A mutation that abolishes the expression of a gene (See **Amorph**).

Octoploid. Cell or organism with eight genomes or sets of chromosomes (chromosome number $8n$).

Oligonucleotide-directed site-specific mutagenesis. A procedure by which a specific nucleotide sequence can be changed to another predetermined sequence.

Oncogene. A gene that can cause neoplastic transformation in animal cells growing in culture and tumor formation in animals themselves.

Oocyte. The egg-mother cell; the cell that undergoes two meiotic divisions (oogenesis) to form the egg cell. Primary oocyte—before completion of the first meiotic division; secondary oocyte—after completion of the first meiotic division.

Oogenesis. The formation of the egg or ovum in animals.

Oogonium (*pl.*, oogonia). A germ cell of the female animal before meiosis begins.

Open reading frame. A sequence of nucleotide triplets that lacks a termination codon.

Operational definition. An operation or procedure that can be carried out to define or delimit something.

Operator. A part of an operon that controls the expression of one or more structural genes by serving as the binding site for one or more regulatory proteins.

Operon. A group of genes making up a regulatory or control unit. The unit includes an operator, a promoter, and structural genes.

Ordinate. The vertical axis in a graph.

Organelle. Specialized part of a cell with a particular function or functions (for example, the cilium of a protozoan).

Organizer. An inductor; a chemical substance in a living system that determines the fate in development of certain cells or groups of cells.

Otocephaly. Abnormal development of the head of a mammalian fetus.

Outbreeding. Mating of unrelated individuals.

Ovary. The swollen part of the pistil of a plant flower that contains the ovules; the female reproductive organ or gonad in animals.

Overdominance. A condition in which heterozygotes are superior (on some scale of measurement) to either of the associated homozygotes.

Ovule. The macrosporangium of a flowering plant that becomes the seed. It includes the nucellus and the integuments.

P. Symbol for the parental generation or parents of a given individual.

Pachynema (*adj.*, **pachytene**). A mid-prophase stage in meiosis immediately following zygonema and preceding diplonema. In favorable microscopic preparations, the chromosomes are visible as long, paired threads. Rarely, four chromatids are detectable.

Pair-rule gene. A gene that influences the formation of body segements in *Drosophila.*

Palindrome. A segment of DNA in which the base-pair sequence reads the same in both directions from a central point of symmetry.

Panmictic population. A population in which mating occurs at random.

Panmixis. Random mating in a population.

Paracentric inversion. An inversion that is entirely within one arm of a chromosome and does not include the centromere.

Parameter. A value or constant based on an entire population (cf. **Statistic**).

Parthenogenesis. The development of a new individual from an egg without fertilization.

Paternal. Pertaining to the father.

Pathogen. An organism that causes a disease.

PCR. See **Polymerase chain reaction.**

Pedigree. A table, chart, or diagram representing the ancestry of an individual.

Penetrance. The percentage of individuals that show a particular phenotype among those capable of showing it.

Peptide. A compound containing amino acids; a breakdown or buildup unit in protein metabolism.

Peptide bond. A chemical bond holding amino acid subunits together in proteins.

Peptidyl (P) site. The ribosome binding site that contains the tRNA to which the growing polypeptide chain is attached.

Peptidyl transferase. An enzyme activity—built into the large subunit of the ribosome—that catalyzes the formation of peptide bonds between amino acids during translation.

Pericentric inversion. An inversion including the centromere, hence involving both arms of a chromosome.

Petite mutant. A respiration-deficient yeast mutant that produces small colonies when grown on glucose-containing medium.

Phage. See **Bacteriophage.**

Phagemids. Cloning vectors that contain components derived from both phage chromosomes and plasmids.

Phagocytes. Immune system cells that ingest and destroy viruses, bacteria, fungi, and other foreign substances or cells.

Phenocopy. An organism whose phenotype (but not genotype) has been changed by the environment to resemble the phenotype of a different (mutant) organism.

Phenotype. The observable characteristics of an organism.

Phenylalanine. See **Amino acid.**

Phenylketonuria. Metabolic disorder resulting in mental retardation; transmitted as a Mendelian recessive and treated in early childhood by special diet.

Photoreactivation. A DNA repair process that is light-dependent.

Phylogeny. Evolutionary history of populations of related organisms.

Plaque. Clear area on an otherwise opaque culture plate of bacteria where the bacteria have been killed by a virus.

Plasma cells. Antibody-producing white blood cells devived from B lymphocytes.

Plasmid. An extrachromosomal hereditary determinant that exists in an autonomous state and is transferred independently of chromosomes.

Plastid. A cytoplasmic body found in the cells of plants and some protozoa. Chloroplastids, for example, produce chlorophyll that is involved in photosynthesis.

Pleiotropy (*adj.*, **Pleiotropic**). Condition in which a single gene influences more than one trait.

Point mutations. Changes that occur at specific sites in genes. They include nucleotide-pair substitutions and the insertion or deletion of one or a few nucleotide pairs.

Polar bodies. In female animals, the smaller cells produced at meiosis that do not develop into egg cells. The first polar body is produced at division I and may not go through division II. The second polar body is produced at division II.

Polar mutation. A mutation that influences the functioning of genes that are downstream in the same transcription unit.

Pole cells. A group of cells in the posterior of *Drosophila* embryos that are precursors to the adult germ line.

Polyadenylation. The addition of poly(A) tails to eukaryotic gene transcripts (RNAs).

Poly(A) polymerase. An enzyme that adds the poly(A) tails to the 3′ termini of eukaryotic gene transcripts (RNAs).

Polycloning site. See **Polylinker.**

Polydactyly. The occurrence of more than the usual number of fingers or toes.

Polygene (*adj.*, **polygenic**). One of many genes involved in quantitative inheritance.

Polylinker (polycloning site). A segment of DNA that contains a set of unique restriction enzyme cleavage sites.

Polymer. A compound composed of many smaller subunits; results from the process of polymerization.

Polymerase. An enzyme that catalyzes the formation of DNA or RNA.

Polymerase chain reaction (PCR). A procedure involving multiple cycles of denaturation, hybridization to oligonucleotide primers, and polynucleotide synthesis that amplifies a particular DNA sequence.

Polymerization. Chemical union of two or more molecules of the same kind to form a new compound having the same elements in the same proportions but a higher molecular weight and different physical properties.

Polymorphism. Two or more kinds of individuals maintained in a breeding population.

Polynucleotide. A linear sequence of joined nucleotides in DNA or RNA.

Polypeptide. A linear molecule with two or more amino acids and one or more peptide groups. They are called dipeptides, tripeptides, and so on, according to the number of amino acids present.

Polyploid. An organism with more than two sets of chromosomes ($2n$ diploid) or genomes—for example, triploid ($3n$), tetraploid ($4n$), pentaploid ($5n$), hexaploid ($6n$), heptaploid ($7n$), octoploid ($8n$).

Polysaccharide capsules. Carbohydrate coverings with antigenic specificity that are present on some types of bacteria.

Polytene chromosomes. Giant chromosomes produced by interphase replication without division and consisting of many identical chromatids arranged side by side in a cablelike pattern.

Population. Entire group of organisms of one kind; an interbreeding group of plants or animals; the extensive group from which a sample might be taken.

Population (effective). Breeding members of the population.

Population genetics. The branch of genetics that deals with frequencies of alleles and genotypes in breeding populations.

Positional cloning. The isolation of a clone of a gene or other DNA sequence based on its map position in the genome.

Position effect. A difference in phenotype that is dependent on the position of a gene or group of genes, often caused by heterochromatin that is nearby.

Positive control system. A mechanism in which the regulatory protein(s) is required to turn on gene expression.

Postreplication repair. A recombination-dependent mechanism for repairing damaged DNA.

Primary immune response. The immune response that occurs during the first encounter of a mammal with a given antigen (cf. **Secondary immune response**).

Primary transcript. The RNA molecule produced by transcription prior to any post-transcriptional modifications; also called a pre-mRNA in eukaryotes.

Primer. A short nucleotide sequence with a reactive 3' OH that can initiate DNA synthesis along a template.

Primosome. A protein replication complex that catalyzes the initiation of Okazaki fragments during discontinuous synthesis. It contains DNA primase and DNA helicase activities.

Probability. The frequency of occurrence of an event.

Proband. The individual in a family in whom an inherited trait is first identified.

Progeny testing. The practice of ascertaining the genotype of an individual by mating it to an individual of known genotype and examining the progeny.

Progerias. Inherited diseases characterized by premature aging.

Prokaryote. A member of a large group of organisms (including bacteria and blue-green algae) that lack true nuclei in their cells and that do not undergo meiosis.

Promoter. A nucleotide sequence to which RNA polymerase binds and initiates transcription; also, a chemical substance that enhances the transformation of benign cells into cancerous cells.

Proofreading. The enzymatic scanning of DNA for structural defects such as mismatched base pairs.

Prophage (provirus). The genome of a temperate bacteriophage integrated into the chromosome of a lysogenic bacterium and replicated along with the host chromosome.

Prophase. The stage of mitosis between interphase and metaphase. During this phase, the centriole divides and the two daughter centrioles move apart. Each sister DNA strand from interphase replication becomes coiled, and the chromosome is longitudinally double except in the region of the centromere. Each partially separated chromosome is called a chromatid. The two chromatids of a chromosome are sister chromatids.

Protamines. Small basic proteins that replace the histones in the chromosomes of some sperm cells.

Protease. Any enzyme that hydrolyzes proteins.

Protein. A macromolecule composed of one to several polypeptides. Each polypeptide consists of a chain of amino acids linked together by peptide bonds.

Proto-oncogene. A normal cellular gene that can be changed to an oncogene by mutation.

Protoplast. A plant or bacterial cell from which the wall has been removed.

Prototroph. An organism such as a bacterium that will grow on a minimal medium.

Provirus. A viral chromosome that has integrated into a host—either prokaryotic or eukaryotic—genome (cf. **Prophage**).

Pseudoautosomal gene. A gene located on both the X and Y chromosomes.

Pseudogene. An inactive but stable component of a genome resembling a gene; apparently derived from active genes by mutation.

Pulsed-field gel electrophoresis. A procedure used to separate very large DNA molecules by alternating the direction of electric currents across a semisolid gel in a pulsed manner.

Punctuated equilibrium. The occurrence of speciation events in bursts separated by long intervals of species stability.

Purine. A double-ring nitrogen-containing base present in nucleic acids; adenine and guanine are the two purines present in most DNA and RNA molecules.

Pyrimidine. A single-ring nitrogen-containing base present in nucleic acids; cytosine and thymine are commonly present in DNA, whereas uracil usually replaces thymine in RNA.

Quantitative inheritance. Inheritance of measurable traits (height, weight, color intensity) that depend on the cumulative action of many genes, each producing a small effect on the phenotype.

Quantitative trait loci (QTL). Two or more genes that affect a single quantitative trait.

Quantum speciation. The formation of a new species in one or a few generations by selection and genetic drift.

Race. A distinguishable group of organisms of a particular species.

Radioactive isotope. An unstable isotope (form of an atom) that emits ionizing radiation.

Random genetic drift. Changes in allele frequency in small breeding populations due to chance fluctuations.

Reading frame. The series of nucleotide triplets that are sequentially positioned in the *A* site of the ribosome during translation of an mRNA; also, the sequence of nucleotide-pair triplets in DNA that correspond to these codons in mRNA.

Receptor. A molecule that can accept the binding of a ligand.

Recessive. A term applied to one member of an allelic pair lacking the ability to manifest itself when the other or dominant member is present.

Reciprocal crosses. Crosses between different strains with the sexes reversed; for example, female A × male B and male A × female B are reciprocal crosses.

Recombinant DNA molecule. A DNA molecule constructed *in vitro* by joining all or parts of two different DNA molecules.

Recombination. The production of gene combinations not found in the parents by the assortment of nonhomologous chromosomes and crossing over between homologous chromosomes during meiosis. For linked genes, the frequency of recombination can be used to estimate the genetic map distance; however, high frequencies (approaching 50%) do not yield accurate estimates.

Reduction division. Phase of meiosis in which the maternal and paternal chromosomes of the bivalent separate (cf. **Equational division**).

Regulator gene. A gene that controls the rate of expression of another gene or genes. Example: The *lac I* gene produces a protein that controls the expression of the structural genes of the *lac* operon in *Escherichia coli.*

Release factors. Soluble proteins that recognize termination codons in mRNAs and terminate translation in response to these codons.

Renaturation. The restoration of a molecule to its native form. In nucleic acid biochemistry, this term usually refers to the formation of a double-stranded helix from complementary single-stranded molecules.

Repetitive DNA. DNA sequences that are present in a genome in multiple copies—sometimes a million times or more.

Replica plating. A procedure for duplicating the bacterial colonies growing on agar medium in one petri plate to agar medium in another petri plate.

Replication. A duplication process that is accomplished by copying from a template (for example, reproduction at the level of DNA).

Replicon. A unit of replication. In bacteria, replicons are associated with segments of the cell membrane that control replication and coordinate it with cell division.

Replisome. The complete replication apparatus—present at a replication fork—that carries out the semiconservative replication of DNA.

Repressible enzyme. An enzyme whose synthesis is diminished by a regulatory molecule.

Repression. The process of turning off the expression of a gene or set of genes in response to an environmental signal.

Repressor. A protein that binds to DNA and turns off gene expression.

Repressor gene. A gene that encodes a repressor.

Repulsion (*trans* configuration). The condition in which a double heterozygote has received a mutant and a wild-type allele from each parent; for example, $a+/a+ \times +b/+b$ produces $a+/+b$ (cf. **Coupling.**)

Resistance factor A plasmid that confers antibiotic resistance to a bacterium.

Restitution nucleus. A nucleus with unreduced or doubled chromosome number that results from the failure of a meiotic or mitotic division.

Restriction enzyme. An endonuclease that recognizes a specific short sequence in DNA and cleaves the DNA molecule at or near that site.

Restriction fragment. A fragment of DNA produced by cleaving a DNA molecule with one or more restriction endonucleases.

Restriction map. A linear or circular physical map of a DNA molecule showing the sites that are cleaved by different restriction enzymes.

Restriction site. A DNA sequence that is cleaved by a restriction enzyme.

Reticulocyte. A young red blood cell.

Retroelement. Any of the integrated retroviruses or the transposable elements that resemble them.

Retroposon. A transposable element that moves via reverse transcription of RNA into DNA but lacks the long terminal repeat sequences.

Retrotransposon. A transposable element that resembles the integrated form of a retrovirus.

Retrovirus. A virus that stores its genetic information in RNA and replicates by using reverse transcriptase to synthesize a DNA copy of its RNA genome.

Reverse transcriptase. An enzyme that catalyzes the synthesis of DNA using an RNA template.

Reversion (reverse mutation). Restitution of a mutant gene to the wild-type condition, or at least to a form that gives the wild phenotype; more generally, the appearance of a trait expressed by a remote ancestor.

RFLP (Restriction Fragment Length Polymorphism). A genetic difference among individuals that is detected by comparing DNA fragments released by digestion with one or more restriction enzymes.

Rh factor. Antigen in the red blood corpuscles of certain people. A pregnant Rh negative woman carrying an Rh positive child may develop antibodies, causing the child to develop a hemolytic disease.

Ribonuclease. Any enzyme that hydrolyzes RNA.

Ribonucleic acid. See **RNA.**

Ribosomal RNAs (rRNAs). The RNA molecules that are structural components of ribosomes.

Ribosome. Cytoplasmic organelle on which proteins are synthesized.

R-loops. Single-stranded DNA regions in RNA-DNA hybrids formed *in vitro* under conditions where RNA-DNA duplexes are more stable than DNA-DNA duplexes.

RNA. Ribonucleic acid; the information-carrying material in some viruses; more generally, a molecule derived from DNA by transcription that may carry information (messenger or mRNA), provide subcellular structure (ribosomal or rRNA), transport amino acids (transfer or tRNA) or facilitate the biochemical modification of itself or other RNA molecules.

RNA editing. Post-trancriptional processes that alter the information encoded in gene transcripts (RNAs).

RNA polymerase. An enzyme that catalyzes the synthesis of RNA.

Roentgen (r). Unit of ionizing radiation.

Satellite DNA. A component of the genome that can be isolated from the rest of the DNA by density gradient centrifugation. Usually, it consists of short, highly repetitious sequences.

Scaffold. The central core structure of condensed chromosomes. The scaffold is composed of nonhistone chromosomal proteins.

SCID (severe combined immunodeficiency syndrome). A group of diseases characterized by the inability to mount an immune response, either humoral or cellular.

Secondary immune response. The rapid immune response that occurs during the second and subsequent encounters of the immune system of a mammal with a specific antigen (cf. **Primary immune response**).

Secondary oocyte. See **Oocyte.**

Secondary spermatocyte. See **Spermatocyte.**

Segment-polarity gene. A gene that functions to define the anterior and posterior components of body segments in *Drosophila.*

Segregation. The separation of paternal and maternal chromosomes from each other at meiosis; the separation of alleles from each other in heterozygotes; the occurrence of different phenotypes among offspring, resulting from chromosome or allele separation in their heterozygous parents; Mendel's first principle of inheritance.

Selection. Differential survival and reproduction among genotypes; the most important of the factors that change allele frequencies in large populations.

Selection coefficient. A number that measures the fitness of a genotype relative to a standard.

Selection differential. In plant and animal breeding, the difference between the mean of the individuals selected to be parents and the mean of the overall population.

Selection pressure. Effectiveness of differential survival and reproduction in changing the frequency of alleles in a population.

Selection response. In plant and animal breeding, the difference between the mean of the individuals selected to be parents and the mean of their offspring.

Selector gene. A gene that influences the development of specific body segments in *Drosophila;* a homeotic gene.

Self-fertilization. The process by which pollen of a given plant fertilizes the ovules of the same plant. Plants fertilized in this way are said to have been selfed. An analogous process occurs in some animals, such as nematodes and molluscs.

Semisterility. A condition of only partial fertility in plant zygotes (for example, maize); usually associated with translocations.

Sense RNA. A primary transcript or mRNA that contains a coding region (contiguous sequence of codons) that is translated to produce a polypeptide.

Serology. (*adj.,* **serological**). The study of interactions between antigens and antibodies.

Sex chromosomes. Chromosomes that are connected with the determination of sex.

Sexduction. The incorporation of bacterial genes into F factors and their subsequent transfer by conjugation to a recipient cell.

Sex factor. A bacterial episome (for example, the F plasmid in *E. coli*) that enables the cell to be a donor of genetic material. The sex factor may be propagated in the cytoplasm, or it may be integrated into the bacterial chromosome.

Sex-influenced dominance. A dominant expression that depends on the sex of the individual. For example, horns in some breeds of sheep are dominant in males and recessive in females.

Sex-limited. Expression of a trait in only one sex. Examples: milk production in mammals; horns in Rambouillet sheep; egg production in chickens.

Sex linkage. Association or linkage of a hereditary trait with sex; the gene is in a sex chromosome, usually the X; often used synonymously with X-linkage.

Sex mosaic. See **Gynandromorph.**

Sexual reproduction. Reproduction involving the formation of mature germ cells (that is, eggs and sperm).

Shine-Dalgarno sequence. A conserved sequence in prokaryotic mRNAs that is complementary to a sequence near the 5′ terminus of the 16S ribosomal RNA and is involved in the initiation of translation.

Shuttle vector. A plasmid capable of replicating in two different organisms, such as yeast and *E. coli.*

Sib-mating (crossing of siblings). Matings involving two individuals of the same parentage; brother-sister matings.

Sigma factor. The subunit of prokaryotic RNA polymerases that is responsible for the initiation of transcription at specific initiation sequences.

Silencer. A DNA sequence that helps to reduce or shut off the expression of a nearby gene.

SINEs (short interspersed nuclear elements). Families of short (150 to 300 bp), moderately repetitive transposable elements of eukaryotes. The best known SINE family is the Alu family in humans.

Single-strand DNA-binding protein. A protein that coats DNA single strands, keeping them in an extended state.

Site-specific mutagenesis. See **Oligonucleotide-directed site-specific mutagenesis.**

Small nuclear ribonucleoproteins (snRNPs). RNA-protein complexes that are components of spliceosomes.

Small nuclear RNAs (snRNAs). Small RNA molecules that are located in the nuclei of eukaryotic cells; most snRNAs are components of the spliceosomes that excise introns from pre-mRNAs.

Somatic cell. A cell that is a component of the body, in contrast with a germ cell that is capable, when fertilized, of reproducing the organism.

Somatic hypermutation. A high frequency of mutation that occurs in the gene segments encoding the variable regions of antibodies during the differentiation of B lymphocytes into antibody-producing plasma cells.

SOS response. The synthesis of a whole set of DNA repair, recombination, and replication proteins in bacteria containing severely damaged DNA (for example, following exposure to UV light).

Southern blot. The transfer of DNA fragments from an electrophoretic gel to a cellulose or nylon membrane by capillary action.

Species. Interbreeding, natural populations that are reproductively isolated from other such groups.

Sperm (abbreviation of spermatozoon, *pl.*, spermatozoa). A mature male germ cell.

Spermatids. The four cells formed by the meiotic divisions in spermatogenesis. Spermatids become mature spermatozoa or sperm.

Spermatocyte (sperm mother cell). The cell that undergoes two meiotic divisions (spermatogenesis) to form four spermatids; the *primary* spermatocyte before completion of the first meiotic division; the *secondary* spermatocyte after completion of the first meiotic division.

Spermatogenesis. The process by which maturation of the gametes (sperm) of the male takes place.

Spermatogonium (*pl.*, spermatogonia). Primordial male germ cell that may divide by mitosis to produce more spermatogonia. A spermatogonium may enter a growth phase and give rise to a primary spermatocyte.

Spermiogenesis. Formation of sperm from spermatids; the part of spermatogenesis that follows the meiotic divisions of spermatocytes.

Spheroplast. A plant or bacterial cell from which the wall has been removed. See **Protoplast.**

Splicing. The process that covalently joins exon sequences of RNA and eliminates the intervening intron sequences.

Sporophyte. The diploid generation in the life cycle of a plant that produces haploid spores by meiosis.

Standard deviation. A measure of variability in a set of data; the square root of the variance.

Standard error. A measure of variation among a population of means.

Statistic. A value based on a sample or samples of a population from which estimates of a population value or parameter may be obtained.

Stem cell. A mitotically active somatic cell from which other cells arise by differentiation.

Sterility. Inability to produce offspring.

Structural gene. A gene that specifies the synthesis of a polypeptide.

STSs (sequence-tagged sites). Short, unique DNA sequences (usually 200 to 500 bp) that are amplified by PCR and used to link physical maps and genetic maps.

Subspecies. One of two or more morphologically or geographically distinct but interbreeding populations of a species.

Supercoil. A DNA molecule that contains extra twists as a result of overwinding (positive supercoils) or underwinding (negative supercoils).

Suppressor mutation. A mutation that partially or completely cancels the phenotypic effect of another mutation.

Suppressor-sensitive mutant. An organism that can grow when a second genetic factor—a suppressor—is present, but not in the absence of this factor.

Symbiont. An organism living in intimate association with another, dissimilar organism.

Sympatric speciation. The formation of new species by populations that inhabit the same or overlapping geographic regions.

Synapsis. The pairing of homologous chromosomes in the meiotic prophase.

Synaptinemal complex. A ribbonlike structure formed between synapsed homologs at the end of the first meiotic prophase, binding the chromatids along their length and facilitating chromatid exchange.

Syndrome. A group of symptoms that occur together and represent a particular disease.

Synkaryon. A nucleus formed by the fusion of nuclei from two different somatic cells during somatic-cell hybridization.

Synteny. The occurrence of two loci on the same chromosome, without regard to the distance between them.

Taq polymerase. A heat-stable DNA polymerase isolated from the thermophilic bacterium *Thermus aquaticus*.

Target site duplication. A sequence of DNA that is duplicated when a transposable element inserts; usually found at each end of the insertion.

TATA box. A conserved promoter sequence that determines the transcription start site.

Tautomeric shift. The transfer of a hydrogen atom from one position in an organic molecule to another position.

T cell-mediated (cellular) immune response. The synthesis of antigen-specific T cell receptors and the development of killer T cells in response to an encounter of immune system cells with a foreign immunogen.

T cell receptor. An antigen-binding protein that is located on the surfaces of killer T cells and mediates the cellular immune response of mammals. The genes that encode T cell antigens are assembled from gene segments by somatic recombination processes that occur during T lymphocyte differentiation.

T-DNA. The segment of DNA in the Ti plasmid of *Agrobacterium tumefaciens* that is transferred to plant cells and inserted into the chromosomes of the plant.

Telomerase. An enzyme that adds telomere sequences to the ends of eukaryotic chromosomes.

Telomere. The unique structure found at the end of eukaryotic chromosomes.

Telophase. The last stage in each mitotic or meiotic division in which the chromosomes are assembled at the poles of the division spindle.

Temperate phage. A phage (virus) that invades but may not destroy (lyse) the host (bacterial cell) (cf. **Virulent phage**). However, it may subsequently enter the lytic cycle.

Temperature-sensitive mutant. An organism that can grow at one temperature but not at another.

Template. A pattern or mold. DNA stores coded information and acts as a model or template from which information is copied into complementary strands of DNA or transcribed into messenger RNA.

Template strand. In transcription, the DNA strand that is copied to produce a complementary strand of RNA.

Terminalization. Repelling movement of the centromeres of bivalents in the diplotene stages of the meiotic prophase that tends to move the visible chiasmata toward the ends of the bivalents.

Terminal transferase. An enzyme that adds nucleotides to the 3′ termini of DNA molecules.

Termination signal. In transcription, a nucleotide sequence that specifies RNA chain termination.

Testcross. Backcross to the recessive parental type, or a cross between genetically unknown individuals with a fully recessive tester to determine whether an individual in question is heterozygous or homozygous for a certain allele. It is also used as a test for linkage.

Tetrad. The four cells arising from the second meiotic division in plants (pollen tetrads) or fungi (ascospores). The term is also used to identify the quadruple group of chromatids that is formed by the association of duplicated homologous chromosomes during meiosis.

Tetraploid. An organism whose cells contain four haploid ($4n$) sets of chromosomes or genomes.

Tetrasomic (*noun,* tetrasome). Pertaining to a nucleus or an organism with four members of one of its chromosomes whereas the remainder of its chromosome complement is diploid. (Chromosome formula: $2n + 2$).

Tetratype. In fungi, a tetrad of spores that contains four different types; for example, *AB, aB, Ab* and *ab.*

Thymine. A pyrimidine base found in DNA. The other three organic bases—adenine, cytosine, and guanine—are found in both RNA and DNA, but in RNA, thymine is replaced by uracil.

Ti plasmid. The large plasmid in *Agrobacterium tumefaciens.* It is responsible for the induction of tumors in plants with crown gall disease and is an important vector for transferring genes into plants, especially dicots.

T lymphocytes (T cells). Cells that differentiate in the thymus gland and are primarily responsible for the T cell-mediated or cellular immune response.

Topoisomerase. An enzyme that introduces or removes supercoils from DNA.

Totipotent cell (or nucleus). An undifferentiated cell (or nucleus) such as a blastomere that when isolated or suitably transplanted can develop into a complete embryo.

***trans*-acting.** A term describing substances that are diffusable and that can affect spatially separated entities within cells.

***trans* configuration.** See **Repulsion.**

Transcription. Process through which RNA is formed along a DNA template. The enzyme RNA polymerase catalyzes the formation of RNA from ribonucleoside triphosphates.

Transcriptional antiterminator. A protein that prevents RNA polymerase from terminating transcription at specfic transcription-termination sequences.

Transcription factor. A protein that regulates the transcription of genes.

Transcription unit. A segment of DNA that contains transcription initiation and termination signals and is transcribed into one RNA molecule.

Transcription vector. A cloning vector that allows the foreign gene or DNA sequence to be transcribed *in vitro.*

Transduction (t). Genetic recombination in bacteria mediated by bacteriophage. Abortive t: Bacterial DNA is injected by a phage into a bacterium, but it does not replicate. Generalized t: Any bacterial gene may be transferred by a phage to a recipient bacterium. Restricted t: Transfer of bacterial DNA by a temperate phage is restricted to only one site on the bacterial chromosome.

Transfection. The uptake of DNA by a eukaryotic cell, followed by the incorporation of genetic markers present in the DNA into the cell's genome.

Transfer RNAs (tRNAs). RNAs that transport amino acids to the ribosomes, where the amino acids are assembled into proteins.

Transformation (bacteria). Genetic alteration of bacteria brought about by the incorporation of foreign DNA in the bacterial cells.

Transformation (eukaryotic cells). The conversion of eukaryotic cells growing in culture to a state of uncontrolled cell growth (similar to tumor cell growth).

Transgene. A foreign or modified gene that has been introduced into an organism.

Transgenic. A term applied to organisms that have been altered by introducing DNA molecules into them.

Transgressive variation. The appearance in the F_2 (or later) generation of individuals showing more extreme development of a trait than either of the original parents.

***trans* heterozygote.** A heterozygote that contains two mutations arranged in the *trans* configuration—for example, $a\ b^+ / a^+\ b$).

Transition. A mutation caused by the substitution of one purine by another purine or one pyrimidine by another pyrimidine in DNA or RNA.

Translation. Protein (polypeptide) synthesis directed by a specific messenger RNA; occurs on ribosomes.

Translocation. Change in position of a segment of a chromosome to another part of the same chromosome or to a different chromosome.

Transposable genetic element. A DNA element that can move from one location in the genome to another.

Transposase. An enzyme that catalyzes the movement of a DNA sequence to a different site in a DNA molecule.

Transposons. DNA elements that can move ("transpose themselves") from one position in a DNA molecule to another.

Transposon tagging. The insertion of a transposable element into or nearby a gene, thereby marking that gene with a known DNA sequence.

Transversion. A mutation caused by the substitution of a purine for a pyrimidine or a pyrimidine for a purine in DNA or RNA.

Trihybrid. The offspring from homozygous parents differing in three pairs of genes.

Trinucleotide repeats. Tandem repeats of three nucleotides that are present in many human genes. In several cases, these trinucleotide repeats have undergone expansions in copy number that have resulted in inherited diseases.

Trisomic. An otherwise diploid cell or organism that has an extra chromosome of one pair (chromosome formula: $2n + 1$).

Tubulin. The major protein component of the microtubules of eukaryotic cells.

Ultraviolet (UV) radiation. The portion of the electromagnetic spectrum—wavelengths from about 1 to 350 nm—between ionizing radiation and visible light. UV is absorbed by DNA and is highly mutagenic to unicellular organisms and to the epidermal cells of multicellular organisms.

Unequal crossing over. Crossing over between repeated DNA sequences that have paired out of register, creating duplicated and deficient products.

Univalent. An unpaired chromosome at meiosis.

Universality (of the genetic code). The codons have the same meaning, with minor exceptions, in virtually all species.

Uracil. A pyrimidine base found in RNA but not in DNA. In DNA, uracil is replaced by thymine.

Van der Waals interactions. Weak attractions between atoms placed in close proximity.

Variance. A measure of variation in a population; the square of the standard deviation.

Variation. In biology, the occurrence of differences among individuals.

Vector. A plasmid or viral chromosome that may be used to construct recombinant DNA molecules for introduction into living cells.

Velocity density gradient centrifugation. A procedure used to separate macromolecules based on their rate of movement through a density gradient.

Viability. The capability to live and develop normally.

Virulent phage. A phage (virus) that destroys the host (bacterial) cell (cf. **Temperate phage**).

Viscoelastometry. A method to study the physical properties of molecules in solution.

VNTR (variable number tandem repeat). A short DNA sequence that is present as tandem repeats and in highly variable copy number.

Western blot. The transfer of proteins from an electrophoretic gel to a cellulose or nylon membrane by means of an electric force.

Wild type. The customary phenotype or standard for comparison.

Wobble hypothesis. Hypothesis to explain how one tRNA may recognize two codons. The first two bases of the mRNA codon and anticodon pair properly, but the third base in the anticodon has some play (or wobble) that permits it to pair with more than one base.

X chromosome. A chromosome associated with sex determination. In most animals, the female has two, and the male has one X chromosome.

YACs (yeast artificial chromosomes). Linear cloning vectors constructed from essential elements of yeast chromosomes. They can accommodate foreign DNA inserts of 200 to 500 kb in size.

Y chromosome. The partner of the X chromosome in the male of many animal species.

Z-DNA. A left-handed double helix that forms in GC-rich DNA molecules. The Z refers to the zig-zagged paths of the sugar-phosphate backbones in this form of DNA.

Zygonema (*adj.*, zygotene). Stage in meiosis during which synapsis occurs; after the leptotene stage and before the pachytene stage in the meiotic prophase.

Zygote. The cell produced by the union of two mature sex cells (gametes) in reproduction; also used in genetics to designate the individual developing from such a cell.

Answers to Odd Numbered Questions and Problems

CHAPTER 2

2.1 (a) 0; (b) 0; (c) 0 (d) 0; (e) 0; (f) +; (g) +.
2.3 (a) 200; (b) 50.
2-5 See Figure 2.13 (oogenesis). At the start of oogenesis, the woman has a pair of chromosomes, one carrying *M* and one carrying *m*; at the completion of oogenesis, a mature egg carrying the *M* allele, when fertilized by an *m* carrying sperm, produces a myopic child (*M*//*m*).
2.7 Meiosis includes a pairing (synapsis) of homologous maternal and paternal chromosomes. In the cell division that follows, the chromosomes that have previously paired separate. This results in a reduction of chromosome number from $2n$ (diploid) to n (haploid).
2.9 An egg nucleus and two polar nuclei are developed in the ovule. Two haploid nuclei are introduced into the ovule by the pollen tube. One nucleus fuses with the egg nucleus to produce the $2n$ (diploid) zygote and the other with the two polar nuclei to produce the $3n$ (triploid) endosperm nucleus. The zygote develops into the embryo, and the endosperm forms the nutrient material that supports the developing embryo.
2.11 See Figure 2.13 (oogenesis). The primary oocyte is *a*//*a H*//*h*; following meiosis I and II, 1/2 of the mature eggs will carry *a* and *H* chromosomes.
2.13 The man is *A*//*a B*//*b* .
2.15 (a) Early primary oocyte; (b) prophase of meiosis I; (c) suspended prophase I; (d) the first meiotic division is completed just before ovulation of each egg.
2.17 Four.
2.19 Haploid cells differentiate into gametes that fuse to produce a diploid zygote. The diploid zygote then undergoes meiosis to produce four haploid cells, each of which divides mitotically to produce the haploid individual. A haploid individual cannot undergo meiosis directly because its chromosomes do not exist as homologous pairs.

CHAPTER 3

3.1 (a) All tall; (b) 3/4 tall, 1/4 dwarf; (c) all tall; (d) 1/2 tall, 1/2 dwarf
3.3 The data suggest that coat color is controlled by a single gene with two alleles, *C* (gray) and *c* (albino), and that *C* is dominant over *c*. On this hypothesis, the crosses are: Gray (*CC*) × albino (*cc*) → F_1 gray (*Cc*); $F_1 \times F_1$ → 3/4 gray (2 *CC*: 1 *Cc*), 1/4 albino (*cc*). The expected results in the F_2 are 203 gray, 67 albino. To compare the observed and expected results, compute χ^2 with one degree of freedom: $(198–203)^2/203 + (67–72)^2/72 = 0.470$, which is not significant at the 5% level. Thus, the results are consistent with the hypothesis.
3.5 (a) Checkered, red (*CC BB*) × plain, brown (*cc bb*) → F_1 all checkered, red (*Cc Bb*); (b) F_2 progeny: 9/16 checkered, red (*C- B-*), 3/16 plain, red (*cc B-*), 3/16 checkered, brown (*C- bb*), 1/16 plain, brown (*cc bb*)
3.7 Among the F_2 progeny with long, black fur, the genotypic ratio is 1 *BB RR*: 2 *BB Rr*: 2 *Bb RR*: 4 *Bb Rr*; thus 1/9 of the rabbits with long, black fur are homozygous for both genes.
3.9 Half the children from *Aa* × *aa* matings would be albino. In a family of three children, the chance that one will be normal and two albino is $3 \times (1/2)^1 \times (1/2)^2 = 3/8$.
3.11 Man (*Cc ff*) × woman (*cc Ff*). (a) *cc ff*, (1/2) × (1/2) = 1/4; (b) *Cc ff*, (1/2) × (1/2) = 1/4; (c) *cc Ff*, (1/2) × (1/2) = 1/4; (d) *Cc Ff*, (1/2) × (1/2) = 1/4
3.13

	F_1 gametes	F_2 genotypes	F_2 phenotypes
(a)	2	3	2
(b)	2 × 2 = 4	3 × 3 = 9	2 × 2 = 4
(c)	2 × 2 × 2 = 8	3 × 3 × 3 = 27	2 × 2 × 2 = 8
(d)	2^n	3^n	2^n, where n is the number of genes

3.15 $(1/2)^3 = 1/8$
3.17 (20/64) + (10/64) + (5/64) + (1/64) = 36/64
3.19 (a) (1/2) × (1/4) = 1/8; (b) (1/2) × (1/2) × (1/4) = 1/16; (c) (2/3) × (2/3) × (1/4) = 1/6; (d) (2/3) × (1/2) × (1/2) × (1/4) = 1/24
3.21 For III-1 × III-2, the chance of an affected child is 1/2. For IV-2 × IV-3, the chance is zero.
3.23 (a) $(1/4)^3 = 1/64$; (b) $(1/2)^3 = 1/8$; (c) $3 \times (1/2)^1 \times (1/2)^2 = 3/8$; (d) 1 - probability that the offspring is not homozygous for the recessive allele of any gene = $1 - (3/4)^3 = 37/64$; (e) $(3/4)^3 = 27/64$

CHAPTER 4

4.1 M and MN
4.3

	Parents	Offspring
(a)	Yellow × yellow	2 yellow: 1 light belly
(b)	Yellow × light belly	2 yellow: 1 light belly: 1 black and tan
(c)	Black and tan × yellow	2 yellow: 1 black and tan: 1 black
(d)	Light belly × light belly	all light belly
(e)	Light belly × yellow	1 yellow: 1 light belly
(f)	Agouti × black and tan	1 agouti: 1 black and tan
(g)	Black and tan × black	1 black and tan: 1 black
(h)	Yellow × agouti	1 yellow: 1 light belly
(i)	Yellow × yellow	2 yellow: 1 light belly

4.5 (a) all AB; (b) 1 A: 1 B; (c) 1 A: 1 B: 1 AB: 1 O; (d) 1 A: 1 O
4.7 No.The woman is I^AI^B. One man could be either I^AI^A or I^AI^O; the other could be either I^BI^B or I^BI^O. Given the uncertainty in the genotype of each man, either could be the father of the child.
4.9 Cross homozygous *waltzing* with homozygous *tango*. If the mutations are alleles, all the offspring will have an uncoordinated gait; if they are not alleles, all the offspring will be wild-type. If the two mutations are alleles, they could be denoted with the symbols v (*waltzing*) and v^t (*tango*).
4.11 9/16 dark red, 7/16 brownish-purple
4.13 The allele for yellow fur is homozygous lethal.
4.15 Dominant. The condition appears in every generation, and nearly every affected individual has an affected parent. The exception, IV-2, had a father who carried the ataxia allele but did not manifest the trait—an example of incomplete penetrance.
4.17 *Rr pp* × *Rr Pp*
4.19 13/16 white, 3/16 colored
4.21 9 black: 3 gray: 52 white

CHAPTER 5

5.1 The male-determining sperm carries a Y chromosome; the female-determining sperm carries an X chromosome.
5.3 All the daughters will be green, and all the sons will be rosy.
5.5 XX is female, XY is male, XXX is female (but barely viable), XO is male (but sterile).
5.7 No. Defective color vision is caused by an X-linked mutation. The son's X chromosome came from his mother, not his father.
5.9 The risk for the child is P(mother is *C*/*c*) × P(mother transmits *c*) × P(child is male) = (1/2) × (1/2) × (1/2) = 1/8; if the couple has already had a child with color blindness, P(mother is *C*/*c*) = 1, and the risk for each subsequent child is 1/4.

5.11 Each of the rare white-eyed daughters must have resulted from the union of an X(*w*) X (*w*) egg with a Y-bearing sperm. The rare diplo-X eggs must have originated through nondisjunction of the X chromosomes during the second meiotic division in the mother.
5.13 Three-fourths will be phenotypically female (genotypically *tfm*/*Tfm*, *Tfm*/*Tfm*, or *tfm*/Y). Among the females, 2/3 (*tfm*/*Tfm* and *Tfm*/*Tfm*) will be fertile; the *tfm*/Y females will be sterile.
5.15 (a) Female; (b) intersex; (c) intersex; (d) male; (e) female; (f) male
5.17 Yes. The gene for feather patterning is on the Z chromosome. If we denote the allele for barred feathers as *B* and the allele for nonbarred feathers as *b*, the crosses are: *B*/*B* (barred) male × *b*/W (nonbarred) female → F_1: *B*/*b* (barred) males and *B*/W (barred) females. Intercrossing the F_1 produces *B*/*B* (barred) males, *B*/*b* (barred) males, *B*/W (barred) females, and *b*/W (nonbarred) females, all in equal proportions.
5.19 (a) Zero; (b) one; (c) one; (d) two; (e) three; (f) zero

CHAPTER 6

6.1 Use one of the banding techniques.
6.3 46, XX, 22q- or 46, XY, 22q-, depending on the sex chromosome constitution
6.5 In allotetraploids, each member of the different sets of chromosomes can pair with a homologous partner during prophase I and then disjoin during anaphase I. In triploids, disjunction is irregular because homologous chromosomes associate during prophase I by forming either bivalents and univalents or by forming trivalents.
6.7 The F_1 hybrid had 5 chromosomes from species X and 7 from species Y, for a total of 12. When this hybrid was backcrossed to species Y, the few progeny that were produced had 5 + 7 = 12 chromosomes from the hybrid and 7 from species Y, for a total of 19. This hybrid was therefore a triploid. Upon self-fertilization, a few F_2 plants were formed, each with 24 chromosomes. Presumably, the chromosomes in these plants consisted of 2 × 5 = 10 from species X and 2 × 7 = 14 from species Y. These vigorous and fertile F_2 plants were therefore allotetraploids.
6.9 1/2
6.11 Approximately half the progeny should be disomic *ey*/+ and half should be monosomic *ey*/O. The disomic progeny will be wild-type, and the monosomic progeny will be eyeless.
6.13 XYY men would produce more children with sex chromosome abnormalities because their three sex chromosomes will disjoin irregularly during meiosis. This irregular disjunction will produce a variety of aneuploid gametes, including the XY, YY, XYY, and nullo sex chromosome constitutions.
6.15 The animal is heterozygous for an inversion:

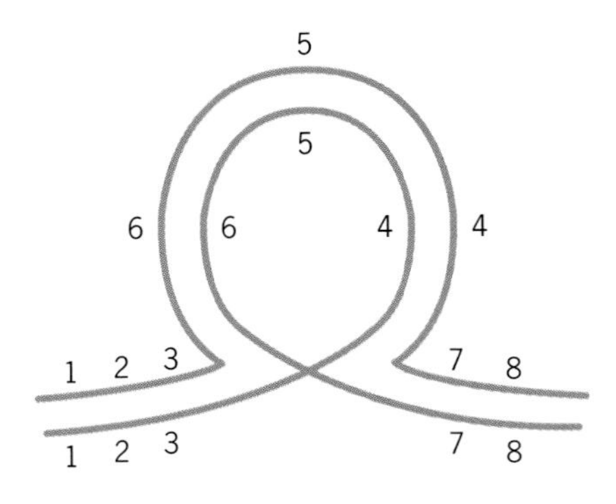

6.17

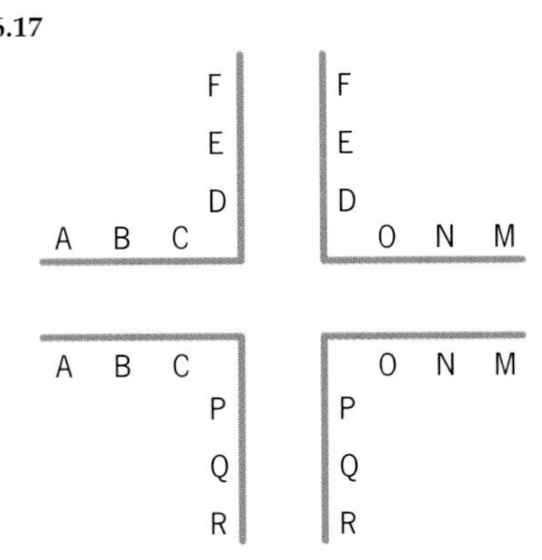

6.19 The boy carries a translocation between chromosome 21 and another chromosome, say No. 14. He also carries a normal chromosome 21 and a normal chromosome 14. The boy's sister carries the translocation, one normal chromosome 14, and two normal copies of chromosome 21.
6.21 All the daughters will be yellow-bodied, and all the sons will be white-eyed.
6.23 The three populations are related by a series of inversions:

P1 1 2 3 4 5 6 7 8 9 10

P2 1 2 3 9 8 7 6 5 4 10

P3 1 2 3 9 8 5 6 7 4 10

6.25 The phenotype in the female offspring is mosaic because one of the X chromosomes is inactivated in each of their cells. If the translocated X is inactivated, the autosome attached to it could also be partially inactivated by a spreading of the inactivation process across the translocation breakpoint. This spreading could therefore inactivate the color-determining gene on the translocated autosome and cause patches of tissue to be phenotypically mutant.
6.27 XX zygotes will develop into males because one of their X chromosomes carries the *TDF* gene that was translocated from the Y chromosome. XY zygotes will develop into females because their Y chromosome has lost the *TDF* gene.

CHAPTER 7

7.1 The class represented by 351 offspring indicates that at least two of the three genes are linked.
7.3 A two-strand double crossover must have occurred.
7.5 $(7!)/7^7 = 0.0061$
7.7 (a) Cross: $a^+ b^+/a^+ b^+ \times a\,b/a\,b$
Gametes: $a^+ b^+$ from one parent, $a\,b$ from the other
F_1: $a^+ b^+/a\,b$
(b) 40% $a^+ b^+$, 40% $a\,b$, 10% $a^+ b$, 10% $a\,b^+$
(c) F_2 from testcross: 40% $a^+ b^+/a\,b$, 40% $a\,b/a\,b$, 10% $a^+ b/a\,b$, 10% $a\,b^+/a\,b$
(d) Coupling linkage phase
(e) F_2 from intercross:

		Sperm			
		40% $a^+ b^+$	40% $a\,b$	10% $a^+ b$	10% $a\,b^+$
Eggs	40% $a^+ b^+$	16% $a^+ b^+/a^+ b^+$	16% $a^+ b^+/a\,b$	4% $a^+ b^+/a^+ b$	4% $a^+ b^+/a\,b^+$
	40% $a\,b$	16% $a\,b/a^+ b^+$	16% $a\,b/a\,b$	4% $a\,b/a^+ b$	4% $a\,b/a\,b^+$
	10% $a^+ b$	4% $a^+ b/a^+ b^+$	4% $a^+ b/a\,b$	1% $a^+ b/a^+ b$	1% $a^+ b/a\,b^+$
	10% $a\,b^+$	4% $a\,b^+/a^+ b^+$	4% $a\,b^+/a\,b$	1% $a\,b^+/a^+ b$	1% $a\,b^+/a\,b^+$

Summary of phenotypes:

a^+ and b^+	66%
a^+ and b	9%
a and b^+	9%
a and b	16%

7.9 Coupling heterozygotes $a^+ b^+/a\,b$ would produce the following gametes: 30% $a^+ b^+$, 30% $a\,b$, 20% $a^+ b$, 20% $a\,b^+$; repulsion heterozygotes $a^+ b/a\,b^+$ would produce the following gametes: 30% $a^+ b$, 30% $a\,b^+$, 20% $a^+ b^+$, 20% $a\,b$. In each case, the frequencies of the testcross progeny would correspond to the frequencies of the gametes.
7.11 Yes. Recombination frequency = (24 + 26)/(126 + 24 + 26 + 124) = 0.167. Cross:

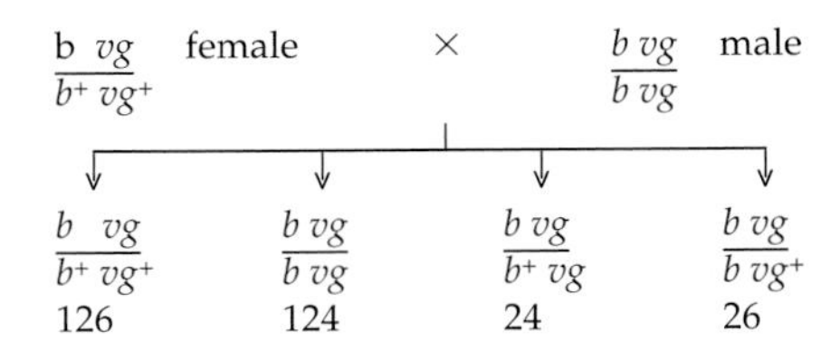

7.13 Yes. Recombination frequency is estimated by the frequency of black offspring among the colored offspring: 34/(66 + 34) = 0.34. Cross:

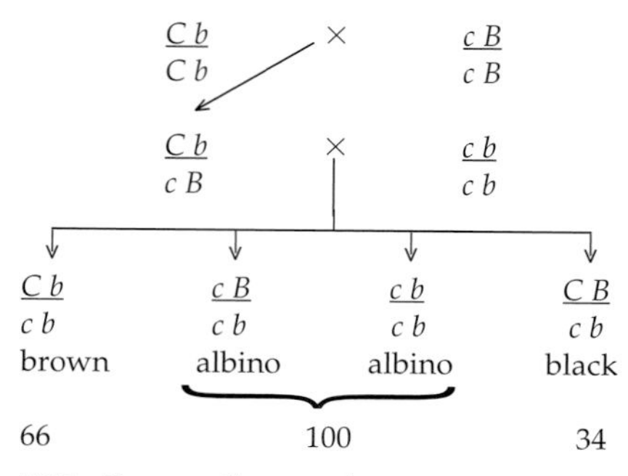

7.15 Because the two chromosomes assort independently, the genetic makeup of the gametes (and, therefore, of the backcross progeny) can be obtained from the following table.

		Chromosome 3 in gametes			
		$c\ d$ 0.425	$c^+ d^+$ 0.425	$c\ d^+$ 0.075	$c^+ d$ 0.075
Chromosome 2 in gametes	$a\ b$ 0.40	$a\ b\ c\ d$ 0.17	$a\ b\ c^+ d^+$ 0.17	$a\ b\ c\ d^+$ 0.03	$a\ b\ c^+ d$ 0.03
	$a^+ b^+$ 0.40	$a^+ b^+ c\ d$ 0.17	$a^+ b^+ c^+ d^+$ 0.17	$a^+ b^+ c\ d^+$ 0.03	$a^+ b^+ c^+ d$ 0.03
	$a^+ b$ 0.10	$a^+ b\ c\ d$ 0.0425	$a^+ b\ c^+ d^+$ 0.0425	$a^+ b\ c\ d^+$ 0.0075	$a^+ b\ c^+ d$ 0.0075
	$a\ b^+$ 0.10	$a\ b^+ c\ d$ 0.0425	$a\ b^+ c^+ d^+$ 0.0425	$a\ b^+ c\ d^+$ 0.0075	$a\ b^+ c^+ d$ 0.0075

7.17 (a) The F_1 females, which are $cn\ vg^+/cn^+\ vg$, produce four types of gametes: 45% $cn\ vg^+$, 45% $cn^+\ vg$, 5% $cn^+\ vg^+$, 5% $cn\ vg$. (b) 45% cinnabar eyes, normal wings; 45% reddish-brown eyes, vestigial wings; 5% reddish-brown eyes, normal wings; 5% cinnabar eyes, vestigial wings.

7.19 In the enumeration below, classes 1 and 2 are parental types, classes 3 and 4 result from a single crossover between *Pl* and *Sm*, classes 5 and 6 result from a single crossover between *Sm* and *Py*, and classes 7 and 8 result from a double crossover, with one of the exchanges between *Pl* and *Sm* and the other between *Sm* and *Py*.

Class	Phenotypes	(a) Frequency with no interference	(b) Frequency with complete interference
1	purple, salmon, pigmy	0.405	0.40
2	green, yellow, normal	0.405	0.40
3	purple, yellow, normal	0.045	0.05
4	green, salmon, pigmy	0.045	0.05
5	purple, salmon, normal	0.045	0.05
6	green, yellow, pigmy	0.045	0.05
7	purple, yellow, pigmy	0.005	0
8	green, salmon, normal	0.005	0

7.21 The double crossover classes, which are the two that were not observed, establish that the gene order is *y—w—ec*. Thus, the F_1 females had the genotype *y w ec*/+ + +. The distance between *y* and *w* is estimated by the frequency of recombination between these two genes: (8 + 7)/1000 = 0.015; similarly, the distance between *w* and *ec* is (18 + 23)/1000 = 0.041. Thus, the genetic map for this segment of the X chromosome is *y*—1.5 cM—*w*—4.1 cM—*ec*.

7.23 (a) Two of the classes (the parental types) vastly outnumber the other six classes (recombinant types); (b) *st* + +/+ *ss e*; (c) *st—ss—e*; (d) [(145 + 122) × 1 + (18) × 2]/1000 = 30.3 cM; (e) (122 + 18)/1000 = 14.0 cM; (f) (0.018)/(0.163 × 0.140) = 0.789; (g) *st* ++/+ *ss e* females × *st ss e* / *st ss e* males → 2 parental classes and 6 recombinant classes.

7.25 Ignore the female progeny and base the map on the male progeny. The parental types are ++ *z* and *x y* +. The two missing classes (+ *y* + and *x* + *z*) must represent double crossovers; thus, the gene order is *y—x—z*. The distance between *y* and *x* is (32 + 27)/1000 = 5.9 cM and that between *x* and *z* is (31 + 39)/1000 = 7.0 cM. Thus, the map is *y*—5.9 cM—*x*—7.0 cM—*z*. The coefficient of coincidence is zero.

7.27 $(P/2)^2$

7.29 From the parental classes, + + *c* and *a b* +, the heterozygous females must have had the genotype + + *c*/*a b* +. The missing classes, + *b* + and a + *c*, which would represent double crossovers, establish that the gene order is *b—a—c*. The distance between *b* and *a* is (96 + 110)/1000 = 20.6 cM and that between *a* and *c* is (65 + 75)/1000 = 14.0 cM. Thus, the genetic map is *b*—20.6 cM—*a*—14.0 cM—*c*.

7.31 5.4 chiasmata

7.33 M2 carries an inversion that suppresses recombination in the chromosome.

CHAPTER 8

8.1 PD >> NPD, so the genes are linked; the distance between the genes is estimated as [(1/2) × 23 + 3]/48 = 30 cM.

8.3 The distance is half the frequency of second division segregation asci: (1/2) × (84/200) = 21 cM.

8.5 The *arg* and *thi* loci are unlinked; however, the *arg* and *leu* loci are linked. The distance between *arg* and *leu* is [(1/2) × 44]/200 = 11 cM; the distance between *arg* and its centromere is (1/2) × (1/300) = 0.17 cM; thus the genetic map for the chromosome that carries *arg* and *leu* is centromere—0.17 cM—*arg*—11 cM—*leu*. The *thi* gene is very tightly linked to its centromere.

8.7 The exceptional females result from crossing over between the genes and the centromere. The *y* locus is farther away from the centromere than the *sn* locus.

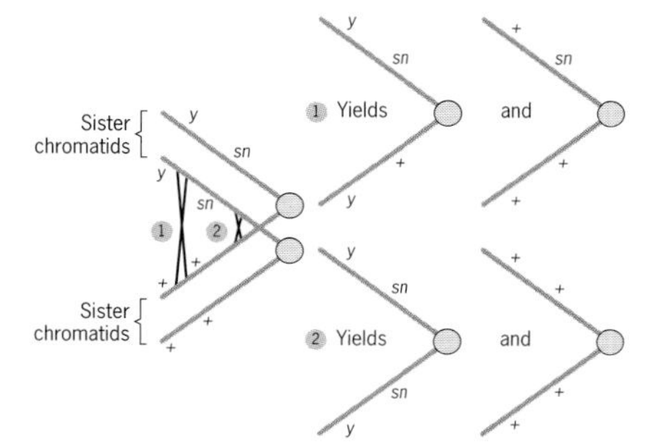

8.9 *C H*/*c h* (*c* = color blind, *C* = normal vision; *h* = hemophilia, *H* = normal)

8.11 40% *A b*, 40% *a B*, 10% *A B*, 10% *a b*

8.13 Chromosome 18

8.15 *ACP* is in 2p distal to the translocation breakpoint.

8.17 I) *A*—10 cM—*B*—6 cM—*C*; II) *B*—10 cM—*A*—16 cM—*C*

8.19 Chromosome 10

CHAPTER 9

9.1 (a) Griffith's *in vivo* experiments demonstrated the occurrence of transformation in pneumococcus. They provided no indication as to the molecular basis of the transformation phenomenon. Avery and colleagues carried out *in vitro* experiments, employing biochemical analyses to demonstrate that transformation was mediated by DNA. (b) Griffith showed that a transforming substance existed; Avery et al. defined it as DNA. (c) Griffith's experiments did not include any attempt to characterize the substance responsible for transformation. Avery *et al.* isolated DNA in "pure" form and demonstrated that it could mediate transformation.

9.3 Purified DNA from Type III cells was shown to be sufficient to transform Type II cells. This occurred in the absence of any dead Type III cells.

9.5 DNA contains phosphorus (normally ^{31}P) but no sulfur; it can be labeled with ^{32}P. Proteins contain sulfur (normally ^{32}S) but usually no phosphorus; they can be labeled with ^{35}S.

9.7 (a) The ladderlike pattern was known from X-ray diffraction studies. Chemical analyses had shown that a 1:1 relationship existed between the organic bases adenine and thymine and between cytosine and guanine. Physical data concerning the length of each spiral and the stacking of bases were also available. (b) Watson and Crick developed the model of a double helix, with the rigid strands of sugar and phosphorus forming spirals around an axis, and hydrogen bonds connecting the complementary bases in nucleotide pairs.

9.9 400,000; (b) 20,000; (c) 400,000; (d) 68,000 nm

9.11 (a) DNA has one atom less of oxygen than RNA in the sugar part of the molecule. In DNA, thymine replaces the uracil that is present in RNA. (In certain bacteriophages, DNA-containing uracil is present.) DNA is most frequently double-stranded, but bacteriophages such as ϕX174 contain single-stranded DNA. RNA is most frequently single-stranded. Some viruses, such as the Reoviruses, however, contain double-stranded RNA chromosomes.

9.13 No. TMV RNA is single-stranded.

Thus the base-pair stoichiometry of DNA does not apply.

9.15 (1) The nucleosome level; the core containing an octamer of histones plus 146 nucleotide pairs of DNA arranged as 1 3/4 turns of a supercoil (see Figure 9.25), yielding an approximately 10 nm diameter spherical body; or juxtaposed, a roughly 10 nm diameter fiber. (2) The 30 nm fiber observed in condensed mitotic and meiotic chromosomes; it appears to be formed by coiling or folding the 10-nm nucleosome fiber. (3) The highly condensed mitotic and meiotic chromosomes (for example, metaphase chromosomes); the tight folding or coiling maintained by a "scaffold" composed of nonhistone chromosomal proteins (see Figure 9.28).

9.17 The satellite DNA fragments would renature much more rapidly than the main-band DNA fragments. In *D. virilus* satellite DNAs, all three have repeating heptanucleotide-pair sequences. Thus essentially every 40 nucleotide-long (average) single-stranded fragment from one strand will have a sequence complementary (in part) with every single-stranded fragment from the complementary strand. Many of the nucleotide-pair sequences in main-band DNA will be unique sequences (present only once in the genome).

9.19 (a) (1) Euchromatin; (2) euchromatin; (3) heterochromatin. (b) (1) Yes; (2) no. (c)(1) Most of the single-copy DNA sequences are believed to be structural genes encoding proteins: structural proteins and the vast repertoire of enzymes employed by living organisms; (2) Essentially nothing is known regarding the functions of the highly repetitive DNA sequences—your hypotheses are probably as valid as anyone else's. (d) Some moderately repetitive DNA sequences specify products such as ribosomal RNA molecules that are required by cells in large quantities. Others are believed to be binding sites for proteins that regulate gene expression and replication of the multiple replicons in the giant DNA molecules of eukaryotic chromosomes. Some moderately repetitive sequences probably play structural roles in chromosomes, especially during the condensations of mitosis and meiosis. Others undoubtedly carry out functions that are still unknown.

9.21 (a) (1) Centromeres function as spindle-fiber attachment sites on chromosomes; they are required for the separation of homologous chromosomes to opposite poles of the spindle during anaphase I of meiosis and for the separation of sister chromatids during anaphase of mitosis and anaphase II of meiosis. (2) Telomeres provide at least three important functions: (i) prevention of exonucleolytic degradation of the ends of the linear DNA molecules in eukaryotic chromosomes, (ii) prevention of the fusion of ends of DNA molecules of different chromosomes, and (iii) provision of a mechanism for replication of the distal tips of linear DNA molecules in eukaryotic chromosomes. (b) Yes. Most telomeres studied to date contain DNA sequence repeat units (for example, TTAGGG in human chromosomes), and, at least in some species, telomeres terminate with single-stranded 3' overhangs that form "hairpin" structures. The bases in these hairpins exhibit unique patterns of methylation that presumably contribute to the structure and stability of telomeres. (c) Telomerase adds the terminal DNA sequences or telomeres to the linear chromosomes in eukaryotes. (d) The broken ends resulting from irradiation will not contain telomeres; as a result, the free ends of the DNA molecules are apparently subject to the activities of enzymes such as exonucleases, ligases, and the like, which modify the ends. They can regain stability by fusing to broken ends of other DNA molecules that contain terminal telomere sequences.

9.23 (a) Two. The axial region of a "lampbrush" chromosome contains the two chromatids of one homologous chromosome (postreplication). (b) One. Each lateral loop of a "lampbrush" chromosome is a segment of a single chromatid.

9.25 (a) Histones have been highly conserved throughout the evolution of eukaryotes. A major function of histones is to package DNA into nucleosomes and chromatin fibers. Since DNA is composed of the same four nucleotides and has the same basic structure in all eukaryotes, one might expect that the proteins that play a structural role in packaging this DNA would be similarly conserved. (b) The nonhistone chromosomal proteins exhibit the greater heterogeneity in chromatin from different tissues and cell types of an organism. The histone composition is largely the same in all cell types within a given species—consistent with the role of histones in packaging DNA into nucleosomes. The nonhistone chromosomal proteins include proteins that regulate gene expression. Because different sets of genes are transcribed in different cell types, one would expect heterogeneity in some of the nonhistone chromosomal proteins of different tissues.

9.27 The 6 percent of the human DNA that is already renatured at $t = 0$ in standard DNA renaturation experiments results from the presence of single strands that themselves contain complementary sequences with opposite chemical polarity (one sequence reading 5' to 3' complementary to another sequence reading 3' to 5'). Single strands containing such complementary sequences form double-stranded "hairpin" or "foldback" structures. Such reactions are concentration independent, because collisions between two molecules are not required for the renaturation events to occur. Thus they occur very fast (too fast to be measured in the standard renaturation experiments) with unimolecular reaction kinetics. Some DNA sequences that exhibit unimolecular renaturation kinetics are called palindromes; such DNAs contain sequences that are the same when read in opposite direction starting from a central point of symmetry (Chapter 19).

CHAPTER 10

10.1 (a) (i) One-half of the DNA molecules with ^{15}N in both strands and 1/2 with ^{14}N in both strands; (ii) all DNA molecules with one strand containing ^{15}N and the complementary strand containing ^{14}N; (iii) all DNA molecules with both strands containing roughly equal amounts of ^{15}N and ^{14}N. (b) (i) 1/4 of the DNA molecules with ^{15}N in both strands and 3/4 with ^{14}N in both strands; (ii) 1/2 of the DNA molecules with one strand containing ^{15}N and the complementary strand containing ^{14}N and the other 1/2 with ^{14}N in both strands; (iii) all DNA molecules with both strands containing about 1/4 ^{15}N and 3/4 ^{14}N. See Figure 10.2.

10.3 (a) Both 3' → 5' and 5' → 3' exonuclease activities. (b) The 3' → 5' exonuclease "proofreads" the nascent DNA strand during its synthesis. If a mismatched base pair occurs at the 3'-OH end of the primer, the 3' → 5' exonuclease removes the incorrect terminal nucleotide before polymerization proceeds again. The 5' → 3' exonuclease is responsible for the removal of RNA primers during DNA replication and functions in pathways involved in the repair of damaged DNA (see Chapter 13). (c) Yes, both exonuclease activities appear to be very important. Without the 3' → 5' proofreading activity during replication, an intolerable mutation frequency would occur. The 5' → 3' exonuclease activity is essential to the survival of the cell. Conditional mutations that alter the 5' → 3' exonuclease activity of DNA polymerase I are lethal to the cell under conditions where the exonuclease is nonfunctional.

10.5 If nascent DNA is labeled by exposure to ^3H-thymidine for very short periods of time, continuous replication predicts that the label would be incorporated into chromosome-sized DNA molecules, whereas discontinuous replication predicts that the label would first appear in small pieces of nascent DNA (prior to covalent joining, catalyzed by polynucleotide ligase).

10.7
Two 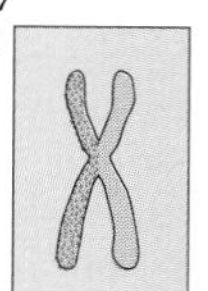 Plus two 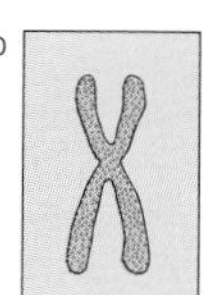 For both the large and small chromosomes

10.9 That DNA replication was unidirectional rather than bidirectional. As the intracellular pools of radioactive ^{3}H-thymidine are gradually diluted after transfer to nonradioactive medium, less and less ^{3}H-thymidine will be incorporated into DNA at each replicating fork. This will produce autoradiograms with tails of decreasing grain density at each growing point. Since such tails appear at only one end of each track, replication must be unidirectional. Bidirectional replication would produce such tails at both ends of an autoradiographic track (see Figure 10.29).
10.11 DNA polymerases α, β, δ, and ϵ are located in the nuclei of cells; polymerase γ is located in mitochondria and chloroplasts. Current evidence suggests that polymerases α and δ are both required for the replication of nuclear DNA. Polymerase δ is thought to catalyze the continuous synthesis of the leading strand, and polymerase α is believed to catalyze discontinuous synthesis of the lagging strand because it contains the primase activity required for the repeated initiation of "Okazaki fragments." Polymerase γ presumably catalyzes replication of organellar chromosomes. Polymerases β and ϵ function in DNA repair pathways like DNA polymerase I of *E. coli*.
10.13 Sucrose velocity density gradient centrifugation is the standard technique for separating DNA molecules in this size range. Pulsed-field gel electrophoresis (Chapter 9) could also be used.
10.15 (a) DNA gyrase; (b) primase; (c) the 5' → 3' exonuclease activity of DNA polymerase I, (d) the 5' → 3' polymerase activity of DNA polymerase III, and (e) the 3' → 5' exonuclease activity of DNA polymerase III.
10.17 The 5' → 3' exonuclease activity of DNA polymerase I is essential to the survival of the bacterium, whereas the 5' → 3' polymerase activity of the enzyme is not essential.
10.19 Because A:T base pairs are held together by only two hydrogen bonds instead of the three hydrogen bonds present in G:C base pairs, the two strands of A:T-rich regions of double helices are separated more easily, providing the single-stranded template regions required for DNA replication.
10.21 DNA polymerase III does not have a 5' → 3' exonuclease activity that acts on double-stranded nucleic acids. Thus it cannot excise RNA primer strands from replicating DNA molecules. DNA polymerase I is present in cells at much higher concentrations and functions as a monomer. Thus DNA polymerase I is able to catalyze the removal of RNA primers from the vast number of Okazaki fragments formed during the discontinuous replication of the lagging strand.
10.23 DNA polymerase I is a single polypeptide of molecular weight 109,000, whereas DNA polymerase III is a complex multimeric protein. The DNA polymerase holoenzyme has a molecular mass of about 900,000 daltons and is composed of at least 20 different polypeptides. The *dnaN* gene product, the β subunit of DNA polymerase III, forms a dimeric clamp that encircles the DNA molecule and prevents the enzyme from dissociating from the template DNA during replication.
10.25 The primosome is a protein complex that initiates the synthesis of Okazaki fragments during lagging strand synthesis. The major components of the *E. coli* DNA primosome are DNA primase and DNA helicase. Geneticists have been able to show that both DNA primase and DNA helicase are required for DNA replication by demonstrating that mutations in the genes encoding these enzymes result in the arrest of DNA synthesis in mutant cells under conditions where the altered proteins are inactive.
10.27 Grow *E. coli* cells for a few seconds in medium containing ^{3}H-thymidine, isolate total DNA from these cells, and determine the sizes of the radioactive DNA molecules by sucrose velocity density gradient centrifugation. If replication is continuous on one strand and discontinuous on the other strand, 50 percent of the radioactivity will be present in Okazaki fragments that are 1000 to 2000 nucleotides long and the other 50 percent will be present in large (chromosome-size) DNA molecules. If replication is discontinuous on both strands, all of the radioactivity will be present in Okazaki fragments.
10.29 (1) DNA replication usually occurs continuously in rapidly growing prokaryotic cells but is restricted to the S phase of the cell cycle in eukaryotes. (2) Most eukaryotic chromosomes contain multiple origins of replication, whereas most prokaryotic chromosomes contain a single origin of replication. (3) Prokaryotes utilize two catalytic complexes that contain the same DNA polymerase to replicate the leading and lagging strands, whereas eukaryotes utilize two distinct DNA polymerases for leading and lagging strand synthesis. (4) Replication of eukaryotic chromosomes requires the partial disassembly and reassembly of nucleosomes as replisomes move along parental DNA molecules. In prokaryotes, replication probably involves a similar partial disassembly/reassembly of nucleosome-like structures. (5) Most prokaryotic chromosomes are circular and thus have no ends. Most eukaryotic chromosomes are linear and have unique termini called telomeres that are added to replicating DNA molecules by a unique, RNA-containing enzyme called telomerase.
10.31 The chromosomes of haploid yeast cells that carry the *est1* mutation become shorter during each cell division. Eventually, chromosome instability results from the complete loss of telomeres, and cell death occurs because of the deletion of essential genes near the ends of chromosomes.

CHAPTER 11

11.1 (a) RNA contains the sugar ribose, which has an hydroxyl (OH) group on the 2-carbon; DNA contains the sugar 2-deoxyribose, with only hydrogens on the 2-carbon. RNA usually contains the base uracil at positions where thymine is present in DNA. However, some DNAs contain uracil, and some RNAs contain thymine. DNA exists most frequently as a double helix (double-stranded molecule); RNA exists more frequently as a single-stranded molecule; but some DNAs are single-stranded and some RNAs are double-stranded. (b) The main function of DNA is to store genetic information and to transmit that information from cell to cell and from generation to generation. RNA stores and transmits genetic information in some viruses that contain no DNA. In cells with both DNA and RNA, (1) mRNA acts as in intermediary in protein synthesis, carrying the information from DNA in the chromosomes to the ribosomes (sites at which proteins are synthesized), (2) tRNAs carry amino acids to the ribosomes and function in codon recognition during the synthesis of polypeptides, and (3) rRNA molecules are essential components of the ribosomes. (c) DNA is located primarily in the chromosomes (with some in cytoplasmic organelles, such as mitochondria and chloroplasts), whereas RNA is located throughout cells.
11.3 3'—GACTA—5'
11.5 Protein synthesis occurs on ribosomes. In eukaryotes, most of the ribosomes are located in the cytoplasm and are attached to the extensive membranous network of endoplasmic reticulum. Some protein synthesis also occurs in cytoplasmic organelles such as chloroplasts and mitochondria.
11.7 The entire nucleotide-pair sequences—including the introns—of the genes are transcribed by RNA polymerase to produce primary transcripts that still contain the intron sequences. The intron sequences are then spliced out of the primary transcripts to produce the mature, functional RNA molecules. In the case of protein-encoding nuclear genes of higher eu-

karyotes, the introns are spliced out by complex macromolecular structures called spliceosomes (see Figure 11.31).

11.9 Spliceosomes excise intron sequences from nuclear gene transcripts to produce the mature mRNA molecules that are translated on ribosomes in the cytoplasm. Spliceosomes are complex macromolecular structures composed of snRNA and protein molecules (see Figure 11.32).

11.11 (a) Sequence 5. It contains the conserved intron sequences: a 5' GU, a 3' AG, and a UACUAAC internal sequence providing a potential bonding site for intron excision. Sequence 4 has a 5' GU and a 3' AG, but contains no internal A for the bonding site during intron excision. (b) 5'—UAGUCUCAA—3'; the putative intron from the 5' GU through the 3' AG has been removed.

11.13

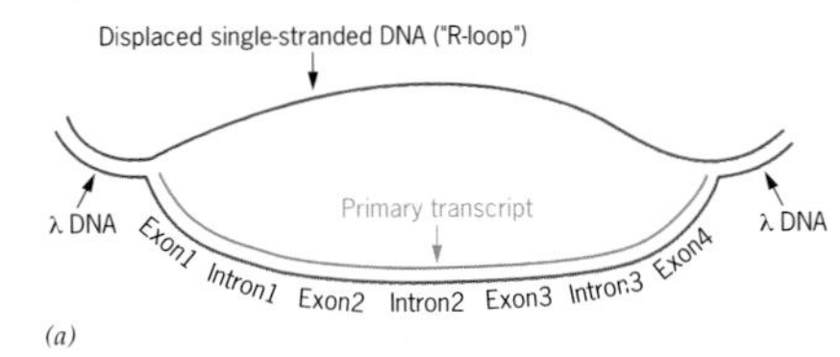

(a)

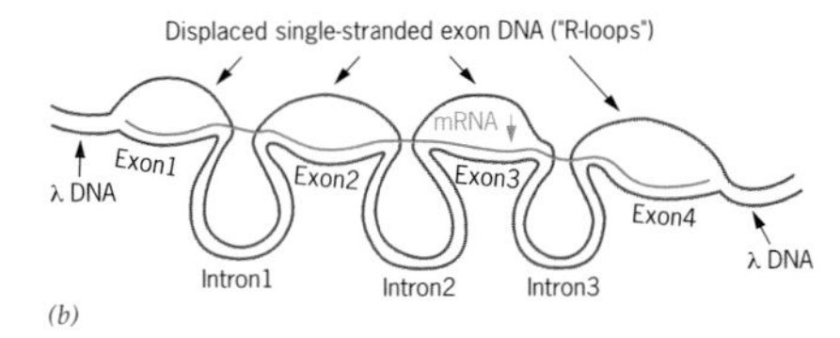

(b)

11.15 Assuming that there is a -35 sequence upstream from the consensus -10 sequence in this segment of the DNA molecule, the nucleotide sequence of the transcript will be 5'- ACCCGACAUAGCUACGAUGACGAUAAGCGACAUAGC-3'.

11.17 Assuming that there is a CAAT box located upstream from the TATA box shown in this segment of DNA, the nucleotide sequence of the transcript will be 5'-ACCCGACAUAGCUACGAUGACGAUA-3'.

11.19 Although in theory it would be possible to produce six different enzyme activities by alternative splicing of one gene transcript or by combining the polypeptide products of two or three genes into different combinations with distinct enzyme activities, in reality these possibilities are unlikely. In living organisms, six enzymes are usually specified by at least six genes—with each gene encoding a single polypeptide. However, many enzymes are composed of two or more distinct polypeptides. Thus the synthesis of six enzymes may require more than six genes.

11.21 DNA, RNA, and protein synthesis all involve the synthesis of long chains of repeating subunits. All three processes can be divided into three stages: chain initiation, chain elongation, and chain termination.

11.23 The primary transcripts of eukaryotes undergo more extensive post-transcriptional processing than those of prokaryotes. Thus the largest differences between mRNAs and primary transcripts occur in eukaryotes. Transcript processing is usually restricted to the excision of terminal sequences in prokaryotes. In contrast, eukaryotic transcripts are usually modified by (1) the excision of intron sequences, (2) the addition of 7-methyl guanosine caps to the 5' termini, and (3) the addition of poly(A) tails to the 3' termini. In addition, the sequences of some eukaryotic transcripts are modified by RNA editing processes.

11.25 The gene that contains a single exon will be transcribed in the least time. The rate of RNA chain extension is the same—about 30 nucleotides per second—for both intron and exon sequences. Thus the synthesis of the long intron sequence will take about 23 minutes. On the other hand, the time required to translate the two mRNAs will be the same because intron sequences are spliced out of transcripts prior to their translation.

11.27 A simple pulse- and pulse/chase-labeling experiment will demonstrate that RNA is synthesized in the nucleus and is subsequently transported to the cytoplasm. This experiment has two parts. (1) Pulse label eukaryotic culture cells by growing them in [^{3}H]uridine for a few minutes, and localize the incorporated radioactivity by autoradiography. (2) Repeat the experiment, but this time add a large excess of nonradioactive uridine to the medium in which the cells are growing after the labeling period, and allow the cells to grow in the nonradioactive medium for about an hour. Then localize the incorporated radioactivity by autoradiography. The expected results are shown in Figure 11.6

11.29 During DNA synthesis, (1) both strands of DNA are used as templates for the synthesis of complementary nascent strands, (2) the precursors are deoxribonucleoside triphosphates, (3) new chains are initiated with RNA primers, (4) the parental DNA strands are completely separated, and (5) chain extension occurs at a rate of about 500 nucleotides per second (prokaryotes). During RNA synthesis, (1) only one strand of DNA is used as a template for the synthesis of a complementary nascent strand, (2) the precursors are ribonucleoside triphosphates, (3) new chains are initiated *de novo*, (4) synthesis occurs within a localized region of strand separation, and (5) chain extension occurs at a rate of 40 to 50 nucleotides per second (prokaryotes).

11.31 The first preparation of RNA polymerase is probably lacking the sigma subunit and, as a result, initiates the synthesis of RNA chains at random sites along both strands of the *argH* DNA. The second preparation probably contains the sigma subunit and initiates RNA chains only at the site used *in vivo*, which is governed by the position of the -10 and -35 sequences of the promoter.

11.33 The simplest procedure for determining which of the three RNA polymerases catalyzes the transcription of the gene is to measure the sensitivity of its transcription to α-amanitin. Since the gene is expressed in cells growing in culture, you can simply add α-amanitin to the culture medium. If the gene is transcribed by RNA polymerase I, α-amanitin will have no effect on its transcription. If the gene is transcribed by RNA polymerase II, its transcription will be completely blocked by the presence of α-amanitin. If the gene is transcribed by RNA polymerase III, the rate of transcription will be reduced, but not blocked, by α-amanitin.

11.35 (1) Intron sequences are spliced out of gene transcripts to provide contiguous coding sequences for translation. (2) The 7-methyl guanosine caps added to the 5' termini of most eukaryotic mRNAs help protect them from degradation by nucleases and are recognized by proteins involved in the initiation of translation. (3) The poly(A) tails at the 3' termini of mRNAs play an important role in their transport from the nucleus to the cytoplasm and enhance their stability.

11.37 The introns of tRNA precursors, *Tetrahymena* rRNA precursors, and nuclear pre-mRNAs are excised by completely different mechanisms. (1) Introns in tRNA precursors are excised by cleavage and joining events catalyzed by splicing nucleases and ligases, respectively. (2) Introns in *Tetrahymena* rRNA precursors are excised autocatalytically. (3) Introns of nuclear pre-mRNAs are excised by spliceosomes. snRNAs are involved in nuclear pre-mRNA splicing as structural components of spliceosomes. In addition, snRNA U1 is required for the cleavage events at the 5' termini of introns; U1 is thought to base-pair with a partially complementary consensus sequence at this position in pre-mRNAs.

11.39 This individual will probably be nonviable because the gene product is essential and the elimination of the 5' splice site will almost certainly result in the production of a nonfunctional gene product.

CHAPTER 12

12.1 Proteins are long chainlike molecules made up of amino acids linked together by peptide bonds. Proteins are composed of carbon, hydrogen, nitrogen, oxygen, and usually sulfur. They provide the enzymatic capacity and much of the structure of living organisms. DNA is composed of phosphate, the pentose sugar 2-deoxyribose, and four nitrogen-containing organic bases (adenine, cytosine, guanine, and thymine). DNA stores and transmits the genetic information in most living organisms. Protein synthesis is of particular interest to geneticists because proteins are the primary gene products—the key intermediates through which genes control the phenotypes of living organisms.

12.3 Ribosomes are from 10 to 20 nm in diameter. They are located primarily in the cytoplasm of cells. In bacteria, they are largely free in the cytoplasm. In eukaryotes, many of the ribosomes are attached to the endoplasmic reticulum. Ribosomes are complex structures composed of over 50 different polypeptides and three to five different RNA molecules.

12.5 Messenger RNA molecules carry genetic information from the chromosomes (where the information is stored) to the ribosomes in the cytoplasm (where the information is expressed during protein synthesis). The linear sequence of triplet codons in an mRNA molecule specifies the linear sequence of amino acids in the polypeptide(s) produced during translation of that mRNA. Transfer RNA molecules are small (about 80 nucleotides long) molecules that carry amino acids to the ribosomes and provide the codon-recognition specificity during translation. Ribosomal RNA molecules provide part of the structure and function of ribosomes; they represent an important part of the machinery required for the synthesis of polypeptides.

12.7 A specific aminoacyl-tRNA synthetase catalyzes the formation of an amino acid-AMP complex from the appropriate amino acid and ATP (with the release of pyrophosphate). The same enzyme then catalyzes the formation of the aminoacyl-tRNA complex, with the release of AMP. The amino acid-AMP and aminoacyl-tRNA linkages are both high-energy phosphate bonds.

12.9 (a) The genetic code is degenerate in that all but 2 of the 20 amino acids are specified by two or more codons. Some amino acids are specified by six different codons. The degeneracy occurs largely at the third or 3′ base of the codons. "Partial degeneracy" occurs where the third base of the codon may be either of the two purines or either of the two pyrimidines and the codon still specifies the same amino acid. "Complete degeneracy" occurs where the third base of the codon may be any one of the four bases and the codon still specifies the same amino acid. (b) The code is ordered in the sense that related codons (codons that differ by a single base change) specify chemically similar amino acids. For example, the codons CUU, AUU, and GUU specify the sructurally related amino acids, leucine, isoleucine, and valine, respectively. (c) The code appears to be almost completely universal. Known exceptions to universality include strains carrying suppressor mutations that alter the reading of certain codons (with low efficiencies in most cases) and the use of UGA as a tryptophan codon in yeast and human mitochondria.

12.11 (a) Met → Val. This substitution occurs as a result of a transition. All other amino acid substitutions listed would require transversions.

12.13 (a) By a complex reaction involving mRNA, ribosomes, initiation factors (IF-1, IF-2, and IF-3), GTP, the initiator codon AUG, and a special initiator tRNA ($tRNA_f^{Met}$). It also appears to involve a base-pairing interaction between a base sequence near the 3'-end of the 16S rRNA and a base sequence in the "leader sequence" of the mRNA. (b) By recognition of one or more of the chain-termination codons (UAG, UAA, and UGA) by the appropriate protein release factor (RF-1 or RF-2).

12.15 Crick's wobble hypothesis explains how the anticodon of a given tRNA can base-pair with two or three different mRNA codons. Crick proposed that the base-pairing between the 5' base of the anticodon in tRNA and the 3' base of the codon in mRNA was less stringent than normal and thus allowed some "wobble" at this site. As a result, a single tRNA often recognizes two or three of the related codons specifying a given amino acid (see Table 12.3).

12.17 (a) Singlet and doublet codes provide a maximum of 4 and $(4)^2$ or 16 codons, respectively. Thus neither code would be able to specify all 20 amino acids. (b) 20. (c) $(20)^{146}$.

12.19 (a) Attachment of an amino acid to the correct tRNA. (b) Recognition of termination codons UAA and UAG and release of the nascent polypeptide from the tRNA in the *P* site of the ribosome. (c) Formation of a peptide bond between the amino group of the aminoacyl-tRNA in the *A* site and the carboxyl group of the growing polypeptide on the tRNA in the *P* site. (d) Formation of the initiation complex required for translation; all steps leading up to peptide bond formation. (e) Translocation of the peptidyl-tRNA from the *A* site on the ribosome to the *P* site.

12.21 A nonsense mutation changes a codon specifying an amino acid to a chain-termination codon, whereas a missense mutation changes a codon specifying one amino acid to a codon specifying a different amino acid. (b) Missense mutations are more frequent. (c) Of the 64 codons, only three specify chain termination. Thus the number of possible missense mutations is much larger than the number of possible nonsense mutations. Moreover, nonsense mutations almost always produce nonfunctional gene products. As a result, nonsense mutations in essential genes are usually lethal in the homozygous state.

12.23 (a) The incoming aminoacyl-tRNA enters the *A* site of the ribosome, whereas the nascent polypeptide-tRNA occupies the *P* site. (b) In order for peptide bond formation to occur, the amino group of an aminoacyl-tRNA must be placed in juxtaposition to the carboxyl group of a peptidyl-tRNA. For this to occur, ribosomes must contain binding sites for at least two tRNAs.

12.25 (a) The Shine–Dalgarno sequence is a conserved polypurine tract, consensus AGGAGG, that is located about seven nucleotides upstream from the AUG initiation codon in mRNAs of prokaryotes. It is complementary to, and is believed to base-pair with, a sequence near the 5' terminus of the 16S ribosomal RNA. (b) Prokaryotic mRNAs with the Shine–Dalgarno sequence deleted are either not translated or are translated inefficiently.

12.27 Incorporation of alanine into polypeptide chains.

12.29 NH_2-Met-Ala-Ile-Cys-Leu-Phe-Gln-Ser-Leu-Ala-Ala-Gln-Asp-Arg-Pro-Gly-COOH.

CHAPTER 13

13.1 (a) Transition; (b) transition; (c) transversion; (d) transversion; (e) frameshift; (f) transition.

13.3 Bacteria treated with a mutagen or expected to carry mutations may be introduced into media with particular drugs in appropriate concentrations. Colonies that appear have originated from cells carrying preexisting mutations for resistance. This may be verified by the replica-plating technique (see Figure 13.3). The frequency of mutations of wild-type (drug-sensitive) cells to drug resistance can be measured in the presence or absence of the drug.

13.5 A dominant mutation presumably occurred in the woman in whom the condition was first known.

13.7 Plants can be propagated vegetatively, but no such methods are available for widespread use in animals.

13.9 Enzymes may discriminate among the different nucleotides that are being incorporated. Mutator enzymes may utilize a higher proportion of incorrect nucleotides, whereas antimutator enzymes may select fewer incorrect bases in DNA replication. In the case of the phage T4 DNA polymerase, the relative efficiencies of polymerization and proofreading by the polymerase's 3' → 5' exonuclease activity play key roles in determining the mutation rate.

13.11 *Dt* is a mutator gene that induces somatic mutations in developing kernels.
13.13 These hemoglobins can be distinguished by mobility of molecules in an electric field (electrophoretic mobility) and by the amino acid sequences of their β polypeptides.
13.15 The label "molecular disease" became common in speaking of sickle-cell anemia because its molecular basis (the substitution of a valine residue for the glutamic acid residue at amino acid position number 6 in the β chain) was recognized quite early during the emergence of the science of molecular biology. Actually, most if not all inherited diseases probably have very similar molecular bases. We just don't know what the molecular defects are in most instances.
13.17 Irradiate the nonresistant strain and plate the irradiated organisms on a medium containing streptomycin. Those that survive and produce colonies are resistant. They could then be replicated to a medium without streptomycin. Those that survive would be of the first type; those that can live with streptomycin but not without it would be the second type.
13.19 Each quantum of energy from the X rays that is absorbed in a cell has a certain probability of hitting and breaking a chromosome. Hence, the greater the number of quanta of energy or dosage, the more likely breaks are to occur. The rate at which this dosage is delivered does not change the probability of each quantum inducing a break.
13.21 During the replicating process, ultraviolet light produces mispairing alterations mostly in pyrimidines (for example, cytosine to thymine transitions). Thymine may be altered to cytosine (or a modified pyrimidine with the base-pairing potential of cytosine), which pairs with guanine. A reverse mutation may occur when cytosine is changed to thymine (or a derivative of cytosine with the hydrogen-bonding potential of thymine), which pairs with adenine. The T-A base pair may thus be changed to a C-G, and the reverse mutation may occur from C-G to T-A.
13.23 Transitions.
13.25 Mutations induced by acridine dyes are primarily insertions or deletions of single base-pairs. Such mutations alter the reading frame (the in-phase triplets specifying mRNA codons) for that portion of the gene distal (relative to the direction of transcription and translation) to the mutation (see Figure 13.15b). This would be expected to totally change the amino acid sequences of polypeptides distal to the mutation site and produce inactive polypeptides. In addition, such frameshift mutations frequently produce in-frame termination codons that result in truncated proteins.
13.27 No. Leucine → proline would occur more frequently. Leu (CUA) $\xrightarrow{5\text{-BU}}$ Pro (CGA) occurs by a single base-pair transition, whereas Leu (**CU**A) $\xrightarrow{5\text{-BU}}$ Ser (**UC**A) requires two base-pair transitions. Recall that 5-bromouracil (5-BU) induces only transitions (see Figure 13.17).
13.29 Yes:

DNA:	←GGX— / —CGX'→	$\xrightarrow{HNO_2}$	←GGX— / —UCX'→
mRNA:	GGX		AGX
Polypeptide:	Gly		Ser or Arg (depending on X)

or

DNA:	←GGX— / —CGX'→	$\xrightarrow{HNO_2}$	←GGX— / —CUX'→
mRNA:	GGX		GAX
Polypeptide:	Gly		Asp or Glu (depending on X)

or

DNA:	←GGX— / —CGX'→	$\xrightarrow{HNO_2}$	←GGX— / —UUX'→
mRNA:	GGX		AAX
Polypeptide:	Gly		Asn or Lys (depending on X)

Note: The X at the third position in each codon in mRNA and in each triplet of base pairs in DNA refers to the fact that there is complete degeneracy at the third base in the glycine codon. Any base may be present in the codon, and it will still specify glycine.
13.31 Tyr → Cys substitutions; Tyr to Cys requires a transition, which is induced by nitrous acid. Tyr to Ser would require a transversion, and nitrous acid is not expected to induce transversions.
13.33 5'-AUGCCCUUUGGG**GAAAGG**UUUCCCUAA-3'

CHAPTER 14

14.1 Prior to 1940, the gene was considered a "bead-on-a-string," not subdivisible by recombination or mutation. Today, the gene is considered to be the unit of genetic material that codes for one polypeptide. The unit of structure, not subdivisible by recombination or mutation, is known to be the single nucleotide pair.
14.3 The *cis-trans* test, which defines the unit of genetic material specifying the amino acid sequence of one polypeptide.
14.5 Two genes; mutations 1, 2, 3, 4, 5, 6, and 8 are in one gene; mutation 7 is in a second gene.
14.7 The size of the gene (assuming that all nucleotide pairs in the gene are capable of undergoing base-pair substitutions, as seems highly probable). Dominant lethal alleles and recessive lethal alleles in haploids will (under normal conditions) exist only transiently, of course.
14.9 (1) Cross the two white-flowered varieties. The F_1 plants will be *trans*-heterozygotes. If the F_1 plants have white flowers, the two varieties probably carry mutations in the same gene, causing white flowers. (2) Cross white-flowered varieties with red-flowered varieties and self-pollinate or intercross the F_1 plants. If alleles of a single gene are involved, monohybrid F_2 ratios should be observed in all cases.
14.11 A maximum of three mutant homoalleles in addition to the wild-type base pair at any one site.
14.13 The observed complementation between ry^2 and ry^{42} is *intra*genic complementation. Xanthine dehydrogenase is a dimeric protein, and dimers that contain one polypeptide encoded by the ry^2 allele and one polypeptide encoded by the ry^{42} allele are partially active. Presumably, the wild-type segment of the ry^2 polypeptide somehow stabilizes the mutant segment of the ry^{42} polypeptide, and vice versa, yielding a dimer with enzymatic activity.
14.15 (a) The seven *sus* mutants are located in three different genes. (b) Mutant strains 1, 3, and 7 contain mutations in one gene; 4, 5, and 6 carry mutations in a second gene; 2 has a mutation in a third gene.
14.17 *White* and *eosin* are located in the same gene; *carnation* is located in a different gene.
14.19 One gene; all seven mutations are in the same gene.
14.21 Several enzymes were shown to contain two or more different polypeptides, and these polypeptides were sometimes controlled by genes that mapped to different chromosomes. Thus the mutations clearly were not in the same gene.
14.23 (a) Homoalleles; (b) heteroalleles.
14.25 All four mutations are in the same gene. However, ry^{42} and ry^{406} exhibit intragenic complementation with each other, whereas ry^5 and ry^{41} do not.
14.27 Several different polypeptides can be produced from a single gene by alternate pathways of transcript splicing. In the case of the tropomyosins, the exons of the transcripts are known to be spliced together in different combinations to produce overlapping, but distinct polypeptides.
14.29 Mutations 1, 2, and 4 do not complement one another and thus appear to be located in the same gene, as do mutations 3 and 5. The anomaly is that mutation 7 does not complement mutations 3, 5, and 6, even though mutation 6 does complement mutations 3 and 5. (a) There are four simple explanations of the seemingly anomalous complementation behavior of mutation 7. (1) It is a deletion spanning all or parts of two genes. (2) It is a double mutation with defects in two genes. (3) It is a polar mutation in the promoter-proximal gene of a multigenic transcription unit. (4) It exhibits rare intergenic noncomplementation with either mutations 3 and 5 or mutation 6 be-

cause it is present in a gene that encodes a product that interacts with the product of the other gene. (b) Three simple genetic operations will distinguish between these four possibilities. (1) Reversion. Plate a large number of mutant 7 phage on *E. coli* strain Z and look for wild-type revertants. (2) Backcross mutant 7 to wild-type phage and test the mutant progeny for the ability to complement mutations 3 and 6. (3) Introduce F's carrying tRNA nonsense suppressor genes into *E. coli* strain Z and determine whether any of them suppress the *loz7* mutation. (c) If mutation 7 is a deletion, it will not revert, and, if it is a double mutation, the reversion rate will probably be below the level of detection in your experiment. On the other hand, if it is a polar nonsense mutation or a noncomplementing missense mutation, loz^+ revertants will be obtained. If mutation 7 is a deletion, no new genotypes will be produced in the backcross to wild-type. However, if it is a double mutation, some recombinant single-mutant progeny will be produced in the backcross to wild-type, and these single mutations will complement either mutation 3 or mutation 6. If mutation 7 is a polar nonsense mutation, it should be suppressed by one or more of the tRNA suppressor genes introduced into *E. coli* strain Z. If mutation 7 (1) reverts to loz^+, (2) does not yield any single-mutant progeny in the backcross to wild-type, and (3) is not suppressed by any of the suppressor tRNA genes, then rare intergenic noncomplementation is probably responsible for its unusual behavior.

CHAPTER 15

15.1 Because wild-type recombinants are rarest in cross (a), this must be the double crossover class, and *m* must be the middle gene. In crosses (b) and (c), wild-type recombinants can be generated by single exchanges.

15.3 The chromosome theory of inheritance is the theory that genes are located on chromosomes. The gene is DNA, and Hershey and Chase's studies on bacterial viruses (Chapter 9) demonstrated that DNA was the genetic material. Studies on the mechanism of recombination have correlated genetic recombination with the physical exchange of DNA molecule segments.

15.5 The evidence suggests that the genes are linked and that the order is *a - b - c*. The rarest classes ($a^+ b c^+$ and $a b^+ c$) when compared to the noncrossover classes (*abc* and $a^+ b^+ c^+$) establish *b* as the middle gene. The distance between *a* and *b* is 0.05 + 0.05 + 0.02 + 0.02 = 0.14, or 14 map units; the distance between *b* and *c* is 0.08 + 0.09 + 0.02 + 0.02 = 0.21, or 21 map units. The coefficient of coincidence (observed double crossovers/expected double crossovers) is 0.040/0.036 = 1.11. There are slightly more double exchanges than expected, probably reflecting the multiple exchanges occurring in the population of DNA molecules.

15.7 (a) The data suggest that *x* and *h* are 18 map units apart; *y* and *h* are 9 map units apart; and *z* and *h* are 2 map units apart. (b) The possible maps for these four genes are:

1. *x* - - - - -9- - - - -*y*- - - - 7- - -*z*- -2- -*h*
2. *x* - - - - - -16- - - - - -*z*- -2- -*h*- - - - -9- - - - -*y*
3. *x* - - - - -9- - - - -*y*- - - - -9- - - - -*h*- -2- -*z*
4. *x* - - - - - - -18- - - - - - -*h*- -2- -*z*- - - -7- - - -*y*

(c) With 13 map units between *y* and *z*, orders 2 and/or 3 are the most likely. (d) A way to resolve these data is to construct a circular genetic map.

15.9 One possibility is that $rIIB_2$ and $rIIB_3$ are point mutations 0.9 map unit apart; since $rIIB_1$ does not recombine with either of them, it may be a deletion that spans the B_2 and B_3 sites. Additional tests would need to be done to confirm that B_1 is a deletion, such as reverse mutation analysis.

15.11 Because the two mutants fail to grow on K cells, they do not complement each other, so they are different mutants of the same gene. The few wild-type progeny that appear after a cross on B cells indicate that recombination has occurred between the mutant sites, producing wild-type progeny. (The double mutant recombinants do not grow on K cells.)

15.13 Linear chromosomes that are terminally redundant and circularly permuted produce a circular genetic map.

15.15 The new mutant maps in the A6c-A6d region.

CHAPTER 16

16.1 Recombination has occurred between the two strains, producing wild-type (prototrophic) bacteria.

16.3

Recombination Process	Agent Mediating DNA Transfer	Size of DNA units Transferred	State of Donor DNA in Recombinant Cell
Transformation	Active uptake of free DNA	Small (about 1/200 to 1/100 of a chromosome	Single-stranded; integrated
Transduction	Bacteriophage	Small (about 1/100 to 1/50 of a chromosome	Double-stranded; integrated into host chromosome (except in abortive transduction)
Sexduction	F factor	Variable	Initially added to the host cell as separate plasmid; may undergo recombination with host chromosome to yield stable transductant

16.5 (a) F' factors are useful for genetic mapping, for complementation analysis (see Chapter 14), and for studies of dominance relationships. (b) F' factors are formed by errors during excision of F factors from Hfr chromosomes. (c) Sexduction occurs by the conjugative transfer of an F' factor from a donor to a recipient (F^-) cell.

16.7 A prophage is a phage chromosome that has become integrated into the host chromosome. The prophage is dormant in the sense that the phage genes involved in lytic development (replication and maturation) are repressed. The prophage is replicated during host chromosome replication just as any other segment of that chromosome.

16.9 Cotransduction refers to the simultaneous transduction of two different genetic markers into a single recipient cell. Since bacteriophage particles can package a maximum of 1 to 2 percent of the total bacterial chromosome, only markers that are relatively closely linked can be cotransduced. The frequency of cotransduction of any two markers will be an inverse function of the distance between them on the chromosome. As such, this frequency can be used as an estimate of the linkage distance. Specific cotransduction-linkage relationships must be prepared for each phage-host system studied.

16.11 *lac*Y - *lac*Z - *pro*C.

16.13 *anth - A487 - A223 - A58.*

16.15 The data indicate that the sequence is *a-b-c* because the rarest class ($a^+b^-c^+$) is a double crossover where the middle gene has switched position.

The distance between a and b is calculated as the number of $a^+ b^-$ and $a^- b^+$ divided by the number of individuals who are either $a^+ b^-$, $a^- b^+$, or $a^+ b^+$:
400 + 2600 + 3600 + 100 = 6700/ 19900 = 0.34 (34 map units).
The distance between *b* and *c* is calculated in a similar fashion:
700 + 400 + 100 + 1200 = 2400/18000 = 0.13 (13 map units).
The distance between *a* and *c* is calculated to be:
700 + 2600 + 3600 + 1200 + = 8100/20200 = 0.40 (40 map units).
The map is thus:

a ---- 34 ---- *b* ---- 13 ---- *c*
← ------- 40 -------------- →

16.17 If the sequence is *a-b-c*, the reciprocal crosses (A and B) should give the same frequency of + + + recombinants because in both crosses, two exchanges are required:

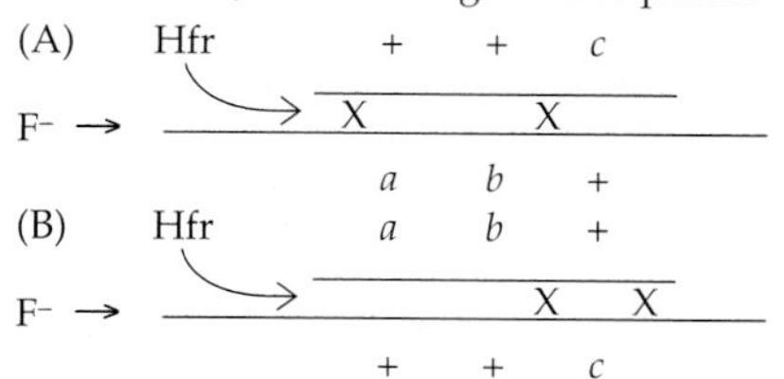

If the sequence is *a-c-b*, the reciprocal crosses give different frequencies of + + +

recombinants because four exchanges are required in (A) and would be much less frequent than (B), where two are required:

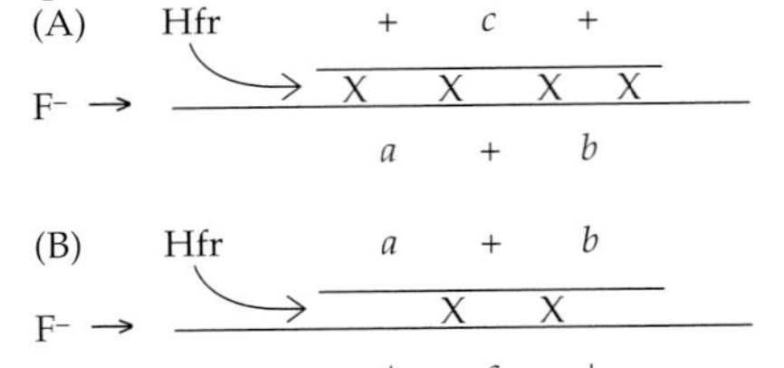

Thus the sequence is: *a-c-b*

CHAPTER 17

17.1 Same orientation: a deletion; opposite orientation: an inversion.
17.3 The paternally inherited *Bz* allele was inactivated by a transposable element insertion.
17.5 Through crossing over between the LTRs of a *gypsy* element.
17.7 Resistance for the second antibiotic was acquired by conjugative gene transfer between the two types of cells.
17.9 Cross dysgenic (highly mutable) males carrying a wild-type X chromosome to females homozygous for a balancer X chromosome; then cross the heterozygous F_1 daughters individually to their brothers and screen the F_2 males that lack the balancer chromosome for mutant phenotypes, including failure to survive (lethality). Mutations identified in this screen are probably due to *P* element insertions in X-linked genes.
17.11 The transposition rate in humans may be very much less than it is in *Drosophila*.
17.13 In the first strain, the F factor integrated into the chromosome by recombination with the IS element between genes *C* and *D*. In the second strain, it integrated by recombination with the IS element between genes *D* and *E*. The two strains transfer their genes in different orders because the two chromosomal IS elements are in opposite orientations.
17.15 Both IS*50* elements should be able to excise from the transposon and insert elsewhere in the chromosome.
17.17 The *tnpA* mutation: no; the *tnpR* mutation: yes.
17.19

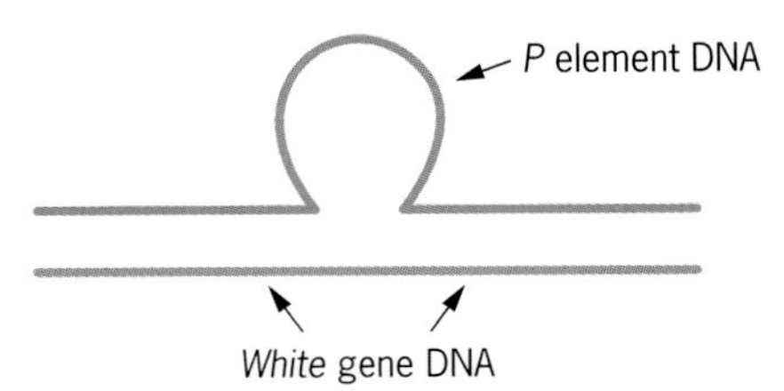

17.21 *TART* and *HeT-A* replenish the ends of *Drosophila* chromosomes.

CHAPTER 18

18.1 The white phenotype is lethal in whole plants.
18.3 Color depends on the proportion of wild-type and mutant chloroplasts in the tissue. Female gametes from green sectors should transmit the green color, female gametes from white sectors should transmit the white color, and female gametes from pale green sectors should transmit all three colors. Male gametes should have no effect on the trait.
18.5 If *O. hookeri* had yellow plastids, and *O. muricata* had green plastids.
18.7 Spectinomycin resistance is encoded by a chloroplast gene.
18.9 Suppressive petite
18.11 No. The genetic code for mitochondrial genes is not the same as that for *E. coli* genes.
18.13 Very rarely. Mutations would have to occur in two different cpDNA molecules within the same cell, and then these would have to recombine to produce a molecule that carried both mutations—an extremely unlikely series of events.
18.15 Half the offspring will be male-sterile. After the intercross, one-fourth of the offspring will be male-sterile.
18.17 Probably not. The polypeptide could be shorter or longer because of differences between the mitochondrial and bacterial genetic codes.
18.19 The expression of genes in two different genetic systems must be coordinated.

CHAPTER 19

19.1 (a) Both introduce new genetic variability into the cell. In both cases, only one gene or a small segment of DNA representing a small fraction of the total genome is changed or added to the genome. The vast majority of the genes of the organism remain the same. (b) The introduction of recombinant DNA molecules, if they come from a very different species, is more likely to result in a novel, functional gene product in the cell, if the introduced gene (or genes) is capable of being expressed in the foreign protoplasm. The introduction of recombinant DNA molecules is more analogous to duplication mutations (see Chapter 6) than to other types of mutations.
19.3 Recombinant DNA and gene-cloning techniques allow geneticists to isolate essentially any gene or DNA sequence of interest and to characterize it structurally and functionally. Large quantities of a given gene can be obtained in pure form, which permits one to determine its nucleotide-pair sequence (to "sequence it" in common lab jargon). From the nucleotide sequence and our knowledge of the genetic code, geneticists can predict the amino acid sequence of any polypeptide encoded by the gene. By using an appropriate subclone of the gene as a hybridization probe in northern blot analyses, geneticists can identify the tissues in which the gene is expressed. Based on the predicted amino acid sequence of a polypeptide encoded by a gene, geneticists can synthesize oligopeptides and use these to raise antibodies that, in turn, can be used to identify the actual product of the gene and localize it within cells or tissues of the organism. Thus recombinant DNA and gene-cloning technologies provide very powerful tools with which to study the genetic control of essentially all biological processes. These tools have played major roles in the explosive progress in the field of biology during the last two decades.
19.5 Restriction endonucleases are believed to provide a kind of primitive immune system to the microorganisms that produce them — protecting their genetic material from "invasion" by foreign DNAs from viruses or other pathogens or just DNA in the environment that might be taken up by the microorganism. Obviously, these microorganisms do not have a sophisticated immune system like that of higher animals (Chapter 24).
19.7 A foreign DNA cloned using an enzyme that produces single-stranded complementary ends can always be excised from the cloning vector by cleavage with the same restriction enzyme that was originally used to clone it. For example, if a *Hin*dIII fragment from the human genome is cloned into *Hin*dIII-cleaved pUC119, the human *Hin*dIII fragment can be excised from a plasmid DNA preparation of this clone by cleavage with restriction endonuclease *Hin*dIII. The human *Hin*dIII fragment will be flanked in the recombinant plasmid DNA clone by *Hin*dIII cleavage sites. When terminal transferase is used to add complementary single-stranded ends during cloning, the original restriction endonuclease cleavage sites are destroyed. Thus, the restriction enzyme used to generate the fragment for cloning cannot be used to excise the original fragment from the cloning vector.
19.9 Most genes of higher plants and animals contain noncoding intron sequences. These intron sequences will be present in genomic clones, but not in cDNA clones since cDNAs are synthesized using mRNA templates and intron sequences are removed during the processing of the primary transcripts to produce mature mRNAs.

19.11 The maize *gln2* gene contains many introns, and one of the introns contains a *Hin*dIII cleavage site. The intron sequences (and thus the *Hin*dIII cleavage site) are not present in mRNA sequences and thus are also not present in full-length *gln2* cDNA clones.

19.13 The PCR technique has much greater sensitivity than any other method available for analyzing nucleic acids. Thus PCR procedures permits analysis of nucleic acid structure given extremely minute amounts of starting material. DNA sequences can be amplified and structurally analyzed from very small amounts of tissue like blood or sperm in assault and rape cases. In addition, PCR methods permit investigators to detect the presence of rare gene transcripts (for example, in specific types of cells) that could not be detected by less sensitive procedures such as northern blot analyses or *in situ* hybridization studies.

19.15 All modern cloning vectors contain a "polycloning site"—a cluster of cleavage sites for a number of different restriction endonucleases in a nonessential region of the vector into which foreign DNAs can be inserted. In general, the greater the complexity of the polycloning site—that is, the more restriction endonuclease cleavage sites that are present—the greater the utility of the vector for cloning a wide variety of different restriction fragments. For example, see the polycloning site present in pUC118 and pUC119 shown in Figure 19.14.

19.17 Restriction endonuclease *Hpa*I produces fragments with blunt ends. The simplest way to clone the desired *Hpa*I fragment into the *Hin*dIII site of pUC119 would be to use *Hin*dIII linkers. To obtain the desired clone, the *Hpa*I fragment of interest would be isolated by agarose gel electrophoresis, extracted from the agarose in the slice cut out of the gel, ligated to *Hin*dIII linkers, digested with *Hin*dIII restriction enzyme, separated from the remaining linker fragments by agarose gel electrophoresis and reextraction, and ligated into *Hin*dIII-cut pUC119. *Amp*^{s} *E. coli* cells would then be transformed with the ligation products and plated on medium containing ampicillin and X-gal (5-bromo-4-chloro-3-indolyl-β-D-galactoside). Only bacteria containing pUC119 plasmids will be able to grow on the ampicillin-containing medium. Two kinds of colonies will be present: blue colonies that harbor pUC119 plasmids with no foreign DNA inserts and white colonies that harbor plasmids with inserts. A white colony should be used to inoculate rich broth medium containing ampicillin to grow a culture of cells containing the desired clone. Plasmid DNA should then be isolated from the resulting cell culture and used to verify the presence of the *Hpa*I fragment of interest. The entire procedure can be carried out in two or three days.

19.19 Because the nucleotide-pair sequences of both the normal *CF* gene and the *CF*Δ*508* mutant gene are known, labeled oligonucleotides can be synthesized and used as hybridization probes to detect the presence of each allele (normal and Δ*508*). Under high-stringency hybridization conditions, each probe will hybridize only with the *CF* allele that exhibits perfect complementarity to itself. Since the sequences of the *CF* gene flanking the Δ*508* site are known, oligonucleotide PCR primers can be synthesized and used to amplify this segment of the DNA obtained from small tissue explants of putative CF patients and their relatives by PCR. The amplified DNAs can then be separated by agarose gel electrophoresis, transferred to nylon membranes, and hybridized to the respective labeled oligonucleotide probes, and the presence of each *CF* allele detected by autoradiography. For a demonstration of the utility of this procedure, see B. Kerem et al. "Identification of the cystic fibrosis gene: Genetic analysis." *Science* 245: 1073–1080, 1989. Kerem and coworkers used two synthetic oligonucleotide probes (oligo-N = 3'-CTTTTATAGTAGAAACCAC-5' and oligo-ΔF = 3'-TTCTTTTATAGTA- - -ACCACAA-5'; the dashes indicate the deleted nucleotides in the *CF*Δ*508* mutant allele) to analyze the DNA of CF patients and their parents. For confirmed CF families, the results of these Southern blot hybridizations with the oligo-N (normal) and oligo-ΔF (*CF*Δ*508*) labeled probes were often as follows:

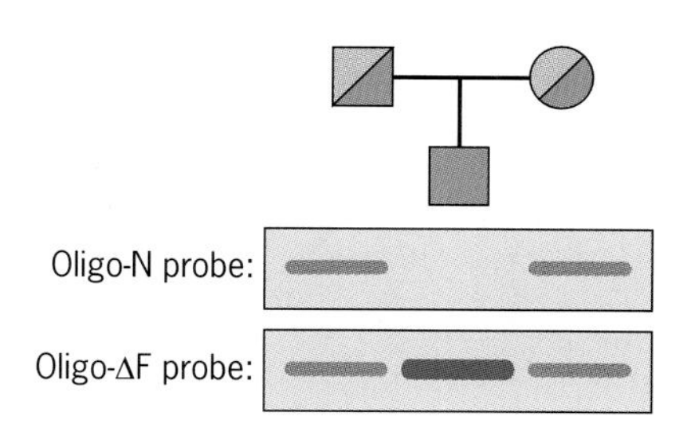

Both parents were heterozygous for the normal *CF* allele and the mutant *CF*Δ*508* allele as would be expected for a rare recessive trait, and the CF patient was homozygous for the *CF*Δ*508* allele. In such families, one-fourth of the children would be expected to be homozygous for the Δ*508* mutant allele and exhibit the symptoms of CF, whereas three-fourths would be normal (not have CF). However, two-thirds of these normal children would be expected to be heterozygous and transmit the allele to their children. Only one-fourth of the children of this family would be homozygous for the normal *CF* allele and have no chance of transmitting the mutant *CF* gene to their offspring. Note that the screening procedure described here can be used to determine which of the normal children are carriers of the *CF*Δ*508* allele: that is, the mutant gene can be detected in heterozygotes as well as homozygotes.

19.21 (a) Southern, northern, and western blot procedures all share one common step, namely, the transfer of macromolecules (DNAs, RNAs, and proteins, respectively) that have been separated by gel electrophoresis to a solid support—usually a nitrocellulose or nylon membrane—for further analysis. (b) The major difference between these techniques is the class of macromolecules that are separated during the electrophoresis step: DNA for Southern blots, RNA for northern blots, and protein for western blots.

19.23

E B H 0.6 0.5 4.0 2.7 0.7 B 2.0 H H

Restriction enzyme cleavage sites for *Bam*HI, *Eco*RI, and *Hin*dIII are denoted by B, E, and H, respectively. The numbers give distances in kilobase pairs.

CHAPTER 20

20.1 Genetic map distances are determined by crossover frequencies. Cytogenetic maps are based on chromosome morphology or physical features of chromosomes. Physical maps are based on actual physical distances—the number of nucleotide pairs (0.34 nm per bp)—separating genetic markers. If a gene or other DNA sequence of interest is shown to be located near a mutant gene, a specific knob on a chromosome, or a particular DNA restriction fragment, that genetic or physical marker (mutation, knob, or restriction fragment) can be used to initiate a chromosome walk (see Figure 20.5).

20.3 A contig (contiguous clones) is a physical map of a chromosome or part of a chromosome prepared from a set of overlapping genomic DNA clones. An RFLP (restriction fragment length polymorphism) is a variation in the length of a specific restriction fragment excised from a chromosome by digestion with one or more restriction endonucleases. A VNTR (variable number tandem repeat) is a short DNA sequence that is present in the genome as tandem repeats and in highly variable copy number. An STS (sequence tagged site) is a unique DNA sequence that has been mapped to a specific site on a chromosome. An EST (expressed sequence tag) is a cDNA sequence—a genomic sequence that is transcribed. Contig maps permit researchers to obtain clones harboring genes of interest directly from DNA Stock Centers—to "clone by phone." RFLPs are used to construct the high-density genetic maps that are needed for positional cloning. VNTRs are especially valuable RFLPs that are used to identify multiple sites in genomes. STSs and ESTs provide molecular probes that can be used to initiate chromosome walks to nearby genes of interest.

20.5 The gene was named *huntingtin* after the disease that it causes when defective. The gene will probably be renamed after the function of its gene product has been determined.

20.7 Once the function of the CF gene product has been established, scientists should be able to develop procedures for introducing wild-type copies of the *CF* gene into the appropriate cells of cystic fibrosis patients to alleviate the devastating effects of the mutant gene. A major obstacle to somatic-cell gene-therapy treatment of cystic fibrosis is the size of the *CF* gene—about 250 kb, which is too large to fit in the standard gene transfer vectors. Perhaps a shortened version of the gene constructed from the *CF* cDNA—about 6.5 kb—can be used in place of the wild-type gene. A second major obstacle is getting the transgene into enough of the target cells of the cystic fibrosis patient to alleviate the symptoms of the disease. A third challenge is to develop an expression vector containing the gene that will result in long-term expression of the introduced gene in transgenic cells. Another concern is how to avoid possible side effects caused by overexpression or inappropriate expression of the transgene in cystic fibrosis patients. Despite these obstacles, many scientists are optimistic that cystic fibrosis will be effectively treated by somatic-cell gene therapy in the future.

20.9 Yes. A somatic-cell gene therapy procedure similar to that used for ADA-SCID (see Figure 20.18) might be effective in treating purine nucleoside phosphorylase (PNP) deficiency. White blood cells could be isolated from the patient, transfected with a vector carrying a wild-type *PNP* gene, grown in culture and assayed for the expression of the *PNP* transgene, and then infused back into the patient after the expression of the transgene had been verified.

20.11 The transcription initiation and termination and translation initiation signals of eukaryotes differ from those of prokaryotes such as *E. coli*. Therefore, to produce a human protein in *E. coli*, the coding sequence of the human gene must be joined to appropriate *E. coli* regulatory signals—promoter, transcription terminator, and translation initiator sequences. Moreover, if the gene contains introns, they must be removed or the coding sequence of a cDNA must be used, because *E. coli* does not possess the spliceosomes required for the excision of introns from nuclear gene transcripts. In addition, many eukaryotic proteins undergo post-translational processing events that are not carried out in prokaryotic cells. Such proteins are more easily produced in transgenic eukaryotic cells growing in culture.

20.13 DNA fingerprints are the specific patterns of bands present on Southern blots of genomic DNAs that have been digested with particular restriction enzymes and hybridized to appropriate DNA probes such as VNTR sequences. DNA fingerprints, like epidermal fingerprints, are used as evidence for identity or nonidentity in forensic cases. Geneticists have expressed concerns about the statistical uses of DNA fingerprint data. In particular, they have questioned some of the methods used to calculate the probability that DNA from someone other than the suspect could have produced the observed DNA fingerprints. This concern is based in part on the lack of adequate databases for various human subpopulations and the lack of precise information about the amount of variability in DNA fingerprints for individuals of different ethnic backgrounds. These concerns can be best addressed by the acquisition of data on fingerprint variability in different subpopulations and ethnic groups.

20.15 The T in T-DNA is an abbreviation for "transferred." The T-DNA region of the Ti plasmid is the segment that is transferred from the Ti plasmid of the bacterium to the chromosomes of the plant cells during *Agrobacterium tumefaciens*-mediated transformation.

20.17 Transgenic mice are usually produced by microinjecting the genes of interest into pronuclei of fertilized eggs or by infecting pre-implantation embryos with retroviral vectors containing the genes of interest. Transgenic mice provide invaluable tools for studies of gene expression, mammalian development, and the immune system of mammals. Transgenic mice are of major importance in medicine; they provide the model system most closely related to humans. They have been, and undoubtedly will continue to be, of great value in developing the tools and technology that will be used for human gene therapy in the future.

20.19 Post-translationally modified proteins can be produced in transgenic eukaryotic cells growing in culture or in transgenic plants and animals. Indeed, transgenic sheep have been produced that secrete human blood-clotting factor IX and α1-antitrypsin in their milk. These sheep were produced by fusing the coding sequences of the respective genes to a DNA sequence that encodes the signal peptide required for secretion and introducing this chimeric gene into fertilized eggs that were then implanted and allowed to develop into transgenic animals. In principle, this approach could be used to produce any protein of interest.

CHAPTER 21

21.1 By studying the synthesis or lack of synthesis of the enzyme in cells grown on chemically defined media. If the enzyme is synthesized only in the presence of a certain metabolite or a particular set of metabolites, it is probably inducible. If it is synthesized in the absence but not in the presence of a particular metabolite or group of metabolites, it is probably repressible.

21.3

Gene or Regulatory Element	Function
(a) Regulator gene	Codes for repressor
(b) Operator	Binding site of repressor
(c) Promoter	Binding site of RNA polymerase and CAP-cAMP complex
(d) Structural gene *Z*	Encodes β-galactosidase
(e) Structural gene *Y*	Encodes β-galactoside permease

21.5 (a) 1, 2, 3*, and 5; (b) 2, 3*, and 5; *3 may be either noninducible or constitutive, depending on whether the specific O^c mutation eliminates binding of the I^s "superrepressor."

21.7 (a) $\dfrac{I^+ O^c Z^+ Y}{I^+ O^+ Z^- Y^+}$ (b) $\dfrac{I^s O^c Z^+ Y^-}{I^s O^+ Z^- Y^+}$

21.9 The system could have developed from a series of tandem duplications of a single ancestral gene. Mutational changes that make the system more efficient and, therefore, favored in selection could have brought the system to its present level of efficiency.

21.11 Possibly by directly or indirectly inhibiting the enzyme adenylcyclase, which catalyzes the synthesis of cyclic AMP from ATP.

21.13 Yes; in the gene coding for CAP. Some mutations in this gene might result in a CAP that binds to the promoter in the absence of cAMP. Also, mutations in the gene (or genes) coding for the protein (or proteins) that regulate the cAMP level as a function of glucose concentration.

21.15 Uninducible, but with a higher level of baseline synthesis of the arabinose enzymes; or, stated differently, a low level of constitutive synthesis of the enzymes. The product of the arabinose regulator gene is required for induction; in its absence, induction could not occur. However, in the absence of arabinose, the regulator gene product represses the level of *ara* operon transcription. This effect would also be eliminated in such a deletion mutant, resulting in a higher baseline level of synthesis (or a low level of constitutive synthesis).

21.17. Constitutive synthesis of the *araB*, *araA*, and *araD* proteins would occur at basal or noninduced levels because the *araC* protein could no longer bind at $araO_2$ and form the loop structure (see Figure 21.15*c*).

21.19 Deletion of the *trpL* region would result in the levels of the tryptophan biosynthetic enzymes in cells growing in the presence of tryptophan being increased about tenfold because attenuation would no longer occur if this region were absent.

21.21 No. Attenuation of the *trp* operon would now be controlled by the presence or absence of Gly-$tRNA^{Gly}$.

21.23 The autogenously maintained repression of λ lytic genes during lysogeny occurs at the level of transcriptional initiation, whereas the λ lytic cascade is regulated at the level of transcriptional termination. In the first case, a repressor protein binds to the promoter and blocks the transcription of the λ lytic genes. In the second case, two key antiterminator proteins act to prevent the termination of transcription at sites distal to the immediate early and delayed early coding sequences. These antitermination events result in the transcription of delayed early and late λ genes.

21.25 Autogenous means self-generated. The λ repressor regulates its own synthesis; the presence of λ repressor stimulates the synthesis of more repressor. This autoregulatory mechanism enhances the stability of the lysogenic state by virtually assuring that once repressor has been synthesized in an infected bacterium, it will continue to be synthesized in amounts that are sufficient to keep the λ lytic genes repressed.

21.27 The N protein, the product of the first gene transcribed from promoter P_L, prevents the termination of transcription at two sites (t_{R1} and t_{L1}), which results in the transcription of the delayed early genes located downstream from these sites. N protein binds to *nut* sites just upstream from the termination sites, and, together with the Nus protein and ribosomal protein S10, modifies the specificity of RNA polymerase so that termination does not occur at sites t_{R1} and t_{L1} (see Figure 21.22).

21.29 Whether a λ phage will undergo lytic growth or enter the lysogenic pathway when it infects an *E. coli* cell is determined by a genetic switch controlled by two regulatory proteins—λ repressor and Cro protein—that bind to the λ operators and govern transcription of the λ genome. If Cro protein occupies these sites, lytic development occurs. If λ repressor occupies these sites, lysogeny results. The C_I gene encodes the λ repressor. Thus lytic development occurs in cells infected with λ phage that carry a deletion of the C_I gene. In contrast, lysogeny will occur in bacteria infected with λ phage that harbor a deletion of the *cro* gene.

CHAPTER 22

22.1 In multicellular eukaryotes, the environment of an individual cell is relatively stable. There is no need to respond quickly to changes in the external environment. In addition, the development of a multicellular organism involves complex regulatory hierarchies composed of hundreds of different genes. The expression of these genes is regulated spatially and temporally, often through intricate intercellular signaling processes.

22.3 By monitoring puffs in response to environmental signals, such as heat shock, or to hormonal signals.

22.5 By alternate splicing of the transcript.

22.7 Northern blotting of RNA extracted from plants grown with and without light, or PCR amplification of cDNA made by reverse transcribing these same RNA extracts.

22.9 The green fluorescent protein will be made after the flies are heat shocked.

22.11 This enzyme plays an important role in photosynthesis, a light-dependent process. Thus it makes sense that its production should be triggered by exposure to light.

22.13 The flies would be phenotypically normal males.

22.15 Yes. Enhancers are able to function in different positions in and around a gene.

22.17 No.

22.19 H4 is acetylated.

22.21 They possess introns.

22.23 Mutations in these codons cause amino acids changes that activate the Ras protein.

22.25 Ras protein is an activator of cell metabolism; RB protein is a suppressor of cell metabolism.

CHAPTER 23

23.1 Cell division forms the blastula; cell movement forms the gastrula.

23.3 Determination establishes cell fates; differentiation realizes those fates.

23.5 Collect mutations with diagnostic phenotypes; map the mutations and test them for allelism with one another; perform epistasis tests with mutations in different genes; clone individual genes and analyze their function at the molecular level.

23.7 Imaginal discs.

23.9 Mate homozygous *unc* hermaphrodites that carry one of the *dumpy* mutations with males that carry the other *dumpy* mutation and score the non-uncoordinated (*unc*/+) F_1 hermaphrodites for the dumpy phenotype. If these worms are dumpy, the two mutations are allelic; if they are wild-type, the two mutations are not allelic.

23.11 Cross heterozygous carriers separately with *tra*/+ and *tra2*/+ flies. If any of the XX progeny of these matings are transformed into sterile males, the new mutation is an allele of the *tra* or *tra2* tester.

23.13 Intersex. The *dsx* mutation is epistatic to the *tra* mutation.

23.15 XX, male phenotype; XO, female phenotype.

23.17 The fly is a gynandromorph. It probably arose through the loss of a wild-type X chromosome early in the development of a *y*/+ zygote.

23.19 The lethal is autonomous.

23.21 (a) Wild-type; (b) embryonic lethal; (c) embryonic lethal; (d) wild-type; (e) wild-type.

23.23 Female sterility.

23.25 *ey* —> *boss* —> *sev* —> R7 differentiation

23.27 Northern blotting of RNA extracted from the tissues at different times during development. Hybridize the blot with gene-specific probes.

CHAPTER 24

24.1 The genetic information specifying antibody chains is stored in sequences of nucleotide pairs encoding sequences of amino acids, just like the genetic information specifying other polypeptides. However, the information specifying an antibody chain is stored in bits and pieces that are assembled into functional genes encoding antibody chains by genome rearrangements (somatic-cell recombination events) occurring during the development of the B lymphocytes (the antibody-producing cells). See Figures 24.12–24.14.

24.3 (1) The joining of different *V*, *D*, and *J* gene segments by somatic recombination during B lymphocyte development (see Figures 24.12–24.14); (2) variability in the exact location of the joining reaction during *V-D-J* joining events (see Figure 24.16); and (3) somatic mutation.

24.5 At the DNA level; class switching occurs by somatic recombination during B lymphocyte differentiation (see Figure 24.14*c–e*).

24.7 The *MHC* locus encodes the transplantation antigens that play a major role in the rejection of foreign tissues after organ and tissue transplant operations.

24.9 This response occurs by a process called clonal selection during which antibodies on the surface of the B lymphocyte producing them bind to the antigen. This in turn stimulates the cell to divide and produce a population (clone) of plasma cells all producing the same antibody.

24.11 Both contain polypeptides with variable regions that form the antigen-binding domains and constant regions that facilitate interactions with cell membranes and other components of the immune response.

24.13 Nonspecific immune responses include inflammation of infected tissues resulting in increased blood flow to the tissues and recruitment of phagocytes to ingest and destroy the infecting agent(s). The two major specific immune responses are the production of antigen-specific antibodies and killer T cells.

24.15 In both cases—SCID and AIDS—the disease symptoms and eventually death are the result of a nonfunctional immune system. SCID is inherited; AIDS is caused by a virus (HIV).

24.17 (a) HIV mutates at an extremely high rate. As a result, infected individuals contain a heterogeneous population of viruses, which helps them escape the immune response of their hosts. (b) Because HIV is spread by sexual contact, blood transfusions, or other forms of blood transfer, the spread of AIDS can best be prevented by practicing safe sexual behavior (abstinence, sexual encounters only with uninfected individuals, and the use of condoms), by not sharing the use of hypodermic needles, and by testing all blood used in transfusions to make certain that it is free of HIV.

24.19 Some forms of SCID can now be treated by bone marrow transplants or by infusions of functional T lymphocytes. In addition, many scientists are optimistic that certain types of SCID will be effectively treated by somatic-cell gene therapy in the near future.

24.21 When helper T cells encounter antigens on the surfaces of macrophages, they secrete cytokines, lymphokines, and interleukins that stimulate B and T lymphocytes to differentiate into antibody-producing plasma cells and killer T cells, respectively. Thus helper T cells are involved in both the antibody-mediated and T cell-mediated immune responses.

24.23 Autoimmune diseases are disorders in which the immune system of an individual attacks specific tissues or cells of her or his own body ("self" cells). The combining term "auto" means self; they are called autoimmune diseases because of the immune system's attack on the patient's own cells.

24.25 Because helper T cells are required to stimulate the development of both antibody-producing plasma cells and killer T cells, the toxicity of deoxyguanosine—the substrate of purine nucleoside phosphorylase—to T lymphocytes would be expected to interfere with both the humoral and cellular immune reponses in an individual lacking this enzyme.

CHAPTER 25

25.1 The phenotypes are determined by the number of primed alleles in the genotypes. This is a classic dihybrid cross with no dominance. 1/16 of the F_2 will have 0 primed alleles; 4/16 will have 1 primed allele; 6/16 will have 2 primed alleles; 4/16 will have 3 primed alleles; and 1/16 will have 4 primed alleles.

25.3 This is a trihybrid cross because of the 1/64th total in one of the classes. The parental genotypes were AABBCC × A'A'B'B'C'C', and the F_1 was A'AB'BC'C. We expect 1/64 of the F_2 progeny to be white and to produce all white offspring, but in a sample of 78, white is so low in frequency (1/64 × 78 = 1.2) that it may not have been produced.

25.5 Because $\Sigma(x_i - \text{mean}) = 0$.

25.7 3.17/6.08 = 0.52.

25.9 V_e can be estimated by the average of the variances of the inbreds; V_g is obtained as the difference between the variances of the randomly pollinated population and the inbreds; thus the broad-sense heritability is (26.4 – 9.4)/26.4 = 0.64.

25.11 Broad-sense heritability must be greater than narrow-sense heritibility. $H^2 = Vg/Ve$; $h^2 = Va/Vt$, and $Va \leq Vg$.

25.13 (15 – 12)(0.3) + 12 = 12.9

25.15 (12.5 – 10)/(15 – 10) = 0.5; selection for increased growth rate should be effective.

25.17 The heritabilities for body size and antler size in the mule deer are high, which means that both traits will respond strongly to selection. The strong positive correlation means that if selection is for larger antler size, the body size will increase too, and vice versa.

CHAPTER 26

26.1 Yes; in a mating of *Uu Rr* × *Uu Rr*, 1/16 of the offspring will be *uu rr*.

26.3 Certainly among the problems associated with heritability and IQ is the testing procedure used to determine IQ. Other problems include making heritability determinations, quantifying the interaction between the genotype and the environment, estimating additive genetic variation, applying heritability studies on one population at one point in time to other populations at different times, and using heritability studies in educational programs.

26.5 Highly inbred lines are homozygous for most alleles. The greater the degree of homozygosity, the less successful the selection will be. Thus selection for maze-bright and maze-dull lines in a highly inbred line would not be effective.

26.7 Examples of genes that affect behavior indirectly are described in 26.6. Both PKU and Lesch-Nyhan syndrome are metabolic disorders that affect the metabolic activity of several tissues and organ systems. The metabolic abnormalities associated with these disorders cause brain cells to malfunction, which in turn produces behavioral abnormalities.

26.9 Different *per* alleles alter the periodicity of the *Drosophila* song-burst pattern. When *per* alleles are transferred from one *Drosophila* species to another, the recipient species exhibits the song pattern of the donor species.

26.11 Down syndrome, Klinefelter syndrome, Turner syndrome, and other aneuploid conditions all have an impact on neurological functions, some more dramatically than others.

26.13 Two important conclusions are: (1) genetic factors exert a profound influence on behavioral variability; (2) the effect of being raised in the same home environment is negligible for most psychological traits.

26.15 Studies with mice indicate that preference for alcohol varies among different inbred lines as does susceptibility to the effects of alcohol. Twin and adoption studies in humans also suggest a genetic link. Research has recently suggested that variation in dopamine receptors or dopamine transport are related to alcoholism.

26.17 Amyloid β protein, which accumulates to abnormal levels in the brains of Alzheimer's patients, is derived from APP through a process of cleavage.

26.19 Twin studies suggest a genetic link, but Hamer's study goes much further, identifying a region of the X chromosome that appears to be related to homosexual behavior.

CHAPTER 27

27.1 (2)(.2)(.8) = 0.32 = heterozygote frequency

27.3 The recessive mutant allele frequency is the square root of 1/25,000 = 0.006; the normal allele frequency is 1 - 0.006 = 0.994; the carrier frequency is estimated to be 2 (0.006)(0.994) = 0.012.

27.5 M = $(0.78)^2$ = 0.61, MN = 2(0.78)(0.22) = 0.34, N = $(0.22)^2$ = 0.05

27.7 The frequency of homozygous tasters is $(0.4)^2$ = 0.16, and the frequency of heterozygous tasters is 2(0.4)(0.6) = 0.48. Thus

0.16/(0.16 + 0.48) = 0.25 = the probability that a taster is homozygous.

27.9 An expectation of 8 per 1000 females is quite small, so finding 3 per 1000 may simply be the result of sampling error. There is another way to view this problem. Nearly all color blind females will come from matings between carrier mothers and color blind fathers. The probability of such a mating is $2pq \times q = 2pq^2 = 2(0.91)(0.09)^2 = 0.015$. Among the daughters of this type of mating, half will be color blind. Thus, the frequency of color blind individuals among all females in the population will be essentially (0.5)(0.015) = 0.0075, a little less than 8 per 1000. Of course, color blind females can be produced by matings between color blind women and color blind men. However, the probability of this type of mating is $q^2 \times q = q^3 = (0.09)^3 = 0.0007$, which is insignificant compared to the probability of a mating between a carrier woman and a color blind man.

27.11 The evolution of the eye is complex, but remains consistent with Darwinian principles. Darwin argued that all known organs could have indeed evolved in small steps. The eye is a case in point. At its most primitive level, the eye is a simple spot of pigmented cells. At its most complex level, the eye has a cornea, a lens, a retina, an iris, and other structures. There are all stages of eye evolution in between. Many of the genes involved in eye development control developmental processes that are reversible. In total darkness, the eye becomes useless and may even have a negative adaptive value if it increases the risk of infection or injury. Random mutations in developmental genes causing eye reduction could be selected for under these conditions. In small populations, these mutations could be fixed by a combination of drift and selection.

27.13 On the assumption that *AA* is indistinguishable from *Aa*:

Genotype:	*AA*	*Aa*	*aa*
Fitness (w):	1	1	0.22
Selection pressure (s):	0	0	0.78

27.15 They will not change.

27.17 Initially, the *e* frequency was 0.5 but decreased to 0.11 after six generations. Either *ee* was effectively lethal or selection acted against *ee* and *Ee*.

27.19 $A = a = 0.5$

27.21 (27/108)/(582/457) = 0.197 = approximate fitness

27.23 The frequency of *t* is $\sqrt{14/25} = 0.75$, and the frequency of T is 1 – 0.75 = 0.25. Thus, because the frequency of heterozygotes should be 2(0.25)(0.75) = 0.38, 9 to 10 (on average 9.46) of the 11 tasters are expected to be heterozygous.

27.25 (a) The frequency of $aa = 0.16$; the frequency of a = square root of 0.16 = 0.4; therefore the frequency of $A = 0.6$. (b) With no selection, we expect $AA = 0.36$; $Aa = 0.48$; and $aa = 0.16$. If the fitness of *aa* is 0.95, then the selection coefficient is 0.05 and the effective contribution of *aa* to the next generation decreases to 0.15 ($0.95 \times 0.16 = 0.15$). The contributions of *AA* and *Aa* increase accordingly. Thus the *a* frequency decreases to about 0.39, and the *A* frequency increases to about 0.61.

27.27 The Kerr and Wright experiment with small populations of *Drosophila* shows how drift and selection operate together on a population.

CHAPTER 28

28.1 At one level, morphological similarities should be evaluated. They should also be tested for reproductive isolation by trying to interbreed them. If they do interbreed, do they produce viable offspring? Are the offspring fertile?

28.3 It is controversial because (except under conditions of self-fertilization or polyploidy) it is difficult to demonstrate that reproductive isolation develops without geographic separation. In a freely interbreeding population, new genetic combinations and new mutations would get "swamped" out through outcrossing. In spite of the controversial nature of sympatric speciation, there may be examples in nature where it has occurred.

28.5 Race or subspecies can be considered a valid taxonomic category characterized by special genotypic and phenotypic characters and often by localization in a given geographic region; however, because of the large amount of genetic and phenotypic variability in most species, the number of racial distinctions that can be made is often arbitrary. Different races of a species can still interbreed successfully with the remainder of the species.

28.7 If one considers the formation of a polyploid to be a single event, then the answer is yes.

28.9 All members of the human species can successfully interbreed.

28.11 The human and the gorilla share a common ancestor, but each species has undergone extensive evolutionary divergence since that ancestor existed.

28.13 The ancestral sequence is 176598234. This ancestral sequence evolved in one direction as follows (underlined markers are the inverted sequences): $17\underline{6598}234 \longrightarrow 17\underline{89562}34 \longrightarrow 172659834$; The ancestral sequence evolved in another direction as follows: $1\underline{765982}34 \longrightarrow 128\underline{95673}4 \longrightarrow 128376594$.

28.15 Most commonly, the first event in the establishment of reproductive isolation is geographic separation of two populations. This would be followed by one or more of the other prezygotic isolating mechanisms, although postzygotic mechanisms could come into play here as well.

28.17 If a hybrid between two species is viable, it may not be fertile because the chromosomes do not have homologous partners and so do not go through a normal meiosis. However, if the hybrid undergoes chromosome duplication, it would have homologous chromosome pairs and be reproductively isolated from the parental species because it has a new chromosome number. The evolution of wheat is an example of this phenomenon.

PHOTO CREDITS

Conversation 1
Opener: Courtesy of James V. Neel

Chapter 1
Opener: Courtesy of the American Museum of Natural History. Figure 1.1: John Jenkins. Figure 1.2: From Peter Royce & Beat Steinman, *Connective Tissue and Its Disorders: Molecular, Genetic, and Medical Breakthroughs,* © Copyright 1993, Wiley-Liss, Inc. Photo by Dr. B. L. Hazleman, Rheumatology Research Unit, Addenbrooke's Hospital, Cambridge, UK. Figure 1.5 a): UPI/Corbis-Bettmann. Figure 1.5 b): Courtesy of Dr. David Rayle, Department of Biology, San Diego State University. Figure 1.6: Leslye Borden/PhotoEdit/PNI. Figure 1.7: Larry Lefever/Grant Heilman Photography. Figure 1.8 (top left): Andy Sacks/Tony Stone Images/New York, Inc. Figure 1.8 (top center): Renee Lynn/Tony Stone Images/New York, Inc. Figure 1.8 (top right): Sean Arbabi/Tony Stone Images/New York, Inc. Figure 1.8 (bottom left and center): Grant Heilman/Grant Heilman Photography. Figure 1.8 (bottom right): Photo supplied by SHORTHORN COUNTRY. Figure 1.9: Photo courtesy of Steven Sims & David Fischhoff, Monsanto Company. Page 12: UPI/Corbis-Bettmann.

Chapter 2
Opener: Richard Green/Photo Researchers. Figure 2.3 a): K. R. Porter/Photo Researchers. Figure 2.3 b): Newcomb & Wergin/Tony Stone Images/New York, Inc. Figure 2.4: Gunther F. Bahr/AFIP/Tony Stone Images/New York, Inc. Figure 2.9 a): David M. Phillips/Visuals Unlimited. Figure 2.9 b): Cabisco/Visuals Unlimited. Figure 2.11: © Science Software Systems, Educational Images Ltd. Figure 2.19: Carolina Biological Supply/Phototake. Figure 2.20: © Lennart Nilsson, *A CHILD IS BORN*, Dell Publishing Company.

Conversation 2
Opener: Courtesy of Johng Lim.

Chapter 3
Opener: Courtesy of the New York Public Library and the American Museum of Natural History. Figure 3.10 a): Larry Lefever/Grant Heilman Photography. Figure 3.15 a): UPI/Corbis-Bettmann. Figure 3.16 a): Courtesy of McKay Kunz and Charles Kunz.

Chapter 4
Opener: West Light. Figure 4.10: Verjik Martin/Amstock. Figure 4.11 a): Courtesy Lester V. Bergman & Associates, Inc. Page 71 (left) and (center left): The Granger Collection. Page 71 (center right): Isabella Stewart Gardner Museum, Boston. Page 71 (right): Roger Viollet/Gamma Liaison. Figure 4.13 a): Courtesy of Ralph Somes. Figures 4.15 and 4.16: NYPL.

Chapter 5
Opener: Courtesy of Stanley J. P. Iyadurai. Page 84: Columbia University, Columbiana Collection. Figure 5.10 a): Corbis-Bettmann. Figure 5.12 a): From Richards and Sutherland, *Trends in Genetics*, vol. 8 (7), p. 249, 1992. Photo courtesy of Grant Sutherland. Figure 5.21 b): Grant Heilman/Grant Heilman Photography. Figure 5.23: From N. Ason/DeHaan, Fig. 9.8, *Biological World*, 1973.

Chapter 6
Opener: Kunio Owaki/The Stock Market. Figure 6.2: From S. E. Schlarbaum and T. Tsuchiya, "Chromosomes of incense cedar." *J. of Heredity* 66:41–42, (1975) Fig. 1A. Figure 6.3: From C. G. Vosa, 1971. "The quinacrine-flourescence patterns of the chromosomes of Albium carinatum," *Chromosoma* 33:382–385, Fig. 1. Figure 6.4 a): From R. M. Patterson & J. C. Petricciani, 1973. "A comparison of prophase & metaphase G-bands in the muntjak." *J. of Heredity* 64 (2): 80–82. Fig. 2A. Figure 6.4 b): From R. S. Verma, H. Dosik, and H. A. Lubs, Jr., 1977. "Demonstration of color & size polymorphisms in human acrocentric chromosomes by acridine orange reverse banding." *J. of Heredity* 68:262–263, Fig. 1. Figure 6.4 c): From A.N. Bruére, H. M. Chapman, P. M. Jaine, & R. M. Morris, 1976. "Origin and significance of centric fusions in domestic sheep." *J. of Heredity* 67: 149–154, Fig. 1. Figure 6.5: From "The Normal Human Karotype." Originally reproduced from "An International System for Human Cytogenetic Nomenclature (1978)," in *Birth Defects: Original Article Series*, XIV: 8, 1978, The National Foundation, New York, and *Cytogenetics and Cell Genetics* 21: 6, 1978, S. Karger AG, Basel Switzerland, reproduced with permission. Figure 6.6: From R. L. Phillips, "Molecular Cytogenetics of the Nucleolus Organizer Region," pp. 711–741, *Maize Breeding and Genetics*, D. B. Waldon, ed., copyright © by John Wiley & Sons. Original photograph courtesy of R. L. Phillips. Figure 6.7 (left): C. G. Maxwell/National Audubon Society/Photo Researchers. Figure 6.7 (top center): Eric Crichton/Bruce Coleman, Inc. Figure 6.7 (bottom center): Ken Wagner/Phototake. Figure 6.7 (right): M. Timothy O'Keefe/Bruce Coleman, Inc. Figure 6.13: Peter J. Bryant/Tony Stone Images/New York, Inc. Figure 6.16 a): Steve Dunwell/The Image Bank. Figure 6.16 b): Courtesy of R. M. Fineman. Page 115: IMS Creative/Custom Medical Stock Photo. Figure 6.19 a): Courtesy of Irene A. Uchida, Department of Pediatrics, McMaster University. Figure 6.19 b): Courtesy of University of Utah Cytogenetics Laboratory. Figures 6.21 b) and 6.25b): Courtesy of J. K. Lim. Page 122: David Jackson/A–Z Botanical Collection Ltd.

Chapter 7
Opener: B. John Cabisco/Visuals Unlimited. Figure 7.2: UPI/Corbis-Bettmann. Figure 7.8 (Insert): J. Forsdyke/Gene Cox/Science Photo Library/Photo Researchers. Figure 7.11: Photograph courtesy of R. N. Shoffner, Cytogenetics Laboratory, Department of Animal Science, University of Minnesota.

Chapter 8
Opener: Biophoto Associates/Photo Researchers. Figure 8.6 a): From *AN INTRODUCTION TO GENETIC ANALYSIS*, 5/e by Griffiths, et al. Copyright © 1993 by W. H. Freeman and Company. Used with permission. Figure 8.8: Science

Source/Photo Researchers. Page 162: From B. H. Judd, M. W. Shen, and T. C. Kaufman, *Genetics* 71:139–152, 1972. Figure 8.21 a): Courtesy of Dr. Beverly Emanuel, Children's Hospital, Philadelphia, PA.

Conversation 3
Opener: Courtesy of Mary Lou Pardue, Biology Department, Massachusetts Institute of Technology.

Chapter 9
Opener: Dr. Gopal Murti/Science Photo Library/Photo Researchers. Figure 9.8: Ted Thai/Time Magazine. Figure 9.9: Professor M. H. F. Wilkins, Biophysics Dept., King's College, London. Figure 9.14: Courtesy of Dr. Svend Freytag. Figure 9.20 a): Omikron/Science Source/Photo Researchers. Figure 9.21: from Vollrath & Davis, *Nucleic Acid Res.* 15:7876 (1987) Oxford University Press. Photo courtesy of Dr. Douglas Vollrath. Figure 9.22: From R. Kavenoff, L. C. Klotz, & B. H. Zimm, *Cold Spring Harbor Symp. Quant. Biol.* 38:1–8, 1973. Copyright © 1973 by Cold Spring Harbor Laboratory. Original photo courtesy of R. Kavenoff. Figure 9.23: Courtesy of Dr. F. Thoma. Figure 9.26: From E. DuPraw, *DNA & Chromosomes*, Holt, Rinehart, & Winston, New York, 1970. Original photo courtesy of E. DuPraw. Figure 9.28: From J. R. Paulson & U. K. Laemmli, *Cell* 12: 817–828, 1977. Copyright © 1977, MIT; published by MIT Press. Original photo courtesy of U. K. Laemmli.

Chapter 10
Opener: Courtesy of Dr. Karl Fredga, Department of Genetics, Uppsala University, Sweden. Figure 10.4: From J. H. Taylor, "The Replication and Organization of DNA in Chromosomes," *Molecular Genetics*, Part I, J. H. Taylor (ed.), Academic Press, New York, 1963. Figure 10.5 a): J. Cairns, Imperial Cancer Research Fund, London, England. Reproduced with permission from *Cold Spring Harbor Sympos. Quant. Biol.* 28:43, 1963. Original photo courtesy of John Cairns. Figure 10.8 c): Photo reproduced with permission from M. Schnös and R. B. Inman, *J. Mol. Biol.* 51: 61–73, 1970. Copyright ©1970 by Academic Press, Inc. (London), Ltd. Figure 10.10: From J. Wolfson, D. Dressler, and M. Magazin, *Proc. Natl. Acad. Sci.*, U.S. 69: 499, 1972. Original micrographs courtesy of D. Dressler. Figure 10.14 a): From X.P. Kong et al., *Cell* 69:425–437, 1992. © Cell Press. Original photograph courtesy of Dr. John Kuriyan, Howard Hughes Medical Institute, Rockefeller University. Figure 10.28: From K. Koths and D. Dressler, Harvard Medical School. Figure 10.29: Reproduced with permission from J. A. Huberman and A. D. Riggs, *J. Mol. Biol.* 32:327–341, 1968. Copyright © 1968 by Academic Press. Original photographs courtesy of J. A. Huberman. Figure 10.30: From D. R. Wolstenholme, *Chromosoma*, 43:1, 1973. Original photograph courtesy of D. R. Wolstenholme. Figure 10.32 a): Courtesy of Steven L. McKnight and Oscar L. Miller, Jr. Figure 10.34: AP/Wide World Photos.

Chapter 11
Opener: Reproduced with permission from R. Reed, J. Griffith, and T. Maniatis, *Cell* 53: 949–961, 1988. Copyright © 1988 by Cell Press; published by Cell Press. Original micrograph courtesy of J. D. Griffith, Department of Microbiology and Immunology, University of North Carolina Medical School, Chapel Hill. Figure 11.6: From D. M. Prescott, "Cellular Sites of RNA Synthesis," *Prog. Nucleic Acid Res. Mol. Biol.* 3:33–57, 1964. Figure 11.14: From O.L. Miller, Jr., B.A. Hamkalo, and C.A. Thomas, Jr., *Science* 169:392–395, 1970. Copyright © 1970 by the American Association for the Advancement of Science. Original micrograph courtesy of O. L. Miller, Jr. Figure 11.18: Courtesy of Dr. Seth Darst. Figure 11.25: From Shirley M. Tilghman, et al., *Proc. Natl. Acad. Sci.* U.S.A. 75:1309–1312, 1978. Figure 11.27 a): Courtesy of Pierre Chambon. Figure 11.31: Courtesy of Jack Griffith, Lineberger Comprehensive Cancer Center, University of North Carolina at Chapel Hill.

Chapter 12
Opener: Stanley Flegler/Visuals Unlimited. Figure 12.8: Courtesy of E. G. Jordon, Molecular Biology, King's College, London. Figure 12.9: O. L. Miller, B. R.Beatty, D. W. Fawcett/Visuals Unlimited. Figure 12.12 a): From S. H. Kim, F. L. Suddath, G. J. Quiqley, A. McPherson, J. L. Sussman, A. H. J. Wang, N. C. Seeman, and A. Rich, *Science* 185: 435–440, 1974 by the American Association for the Advancement of Science. Original photo courtesy of S. H. Kim. Figure 12.18: From S. L. McKnight, N. L. Sullivan, and O. L. Miller, Jr., *Prog. Nucleic Acid Res. Mol. Biol.* 19:313–318, 1976. Micrograph courtesy of S. L. McKnight and O. L. Miller, Jr., University of Virginia.

Chapter 13
Opener: David Scharf/Peter Arnold, Inc. Figure 13.1: Photograph by P. E. Polani, Guy's Hospital, London. Reprinted with permission of Macmillan Publishing Co., Inc., from E. Novitski, *Human Genetics.* Copyright © 1977 by Edward Novitski. Figure 13.2 a): Frank P. Rossotto/The Stock Market. Figure 13.2 b): Kenneth W. Fink/Bruce Coleman, Inc. Figure 13.25: From B. N. Ames, J. McCann, and E. Yamasaki, *Mutat. Res.* 31:347, 1975. Photograph courtesy of B. N. Ames. Figure 13.32 a): Courtesy of H. Potter and D. Dressler, Harvard Medical School.

Chapter 14
Opener: Reinhard Kunkel/Peter Arnold, Inc.

Conversation 4
Opener: Courtesy of Margaret Kidwell, Department of Ecology and Evolutionary Biology, University of Arizona.

Chapter 15
Opener: Barry Dowsett/Science Photo Library/Photo Researchers. Figure 15.1 a) (top left): Science Source/Photo Researchers. Figure 15.1 a) (top right): Omikron/Photo Researchers. Figure 15.1 a) (bottom left): Lee D. Simon/Science Source/Photo Researchers. Figure 15.1 a) (bottom right): Courtesy of Harold Fisher, University of Rhode Island. Figure 15.2 b): Lee D. Simon/Photo Researchers. Figure 15.4: Carolina Biological Supply/Phototake. Figure 15.21: From:

Hayes, *Genetics of Bacteria and their Viruses,* 2/e, p. 488. Copyright © 1968 by John Wiley & Sons, Inc.

Chapter 16
Opener: Courtesy of Charles C. Brinton, Jr. and Judith Carnahan. Figure 16.3: Copyright © Boehringer Ingelheim International GmbH, photo Lennart Nilsson/Bonnier Alba AB.

Chapter 17
Opener: Tom Bean/Tony Stone Images/Seattle. Page 425: Used with permission of Marjorie M. Bhavnani, photo from Cold Spring Harbor Laboratory Archives.

Chapter 18
Opener: Frans Lanting/Minden Pictures, Inc. Figure 18.2: Courtesy of Dr. Scott Poethig, University of Pennsylvania, Department of Biology. Figure 18.7: R. Kessel & C. Y. Shih/Visuals Unlimited. Figure 18.12: From Y. Hayashi and K. Ueda, *Journal of Cell Science* 93, 565, 1989. Company of Biologists Ltd. Original Photograph supplied by K. Ueda.

Conversation 5
Opener: Roy Gumpel.

Chapter 19
Opener: Corbis-Bettmann. Figure 19.19: Photograph by D. P. Snustad. Figure 19.21: Photographs courtesy of D. G. Oppenheimer and D. P. Snustad, Department of Genetics and Cell Biology, University of Minnesota. Figure 19.22: Photograph courtesy of S. R. Ludwig and D. P. Snustad, Department of Genetics and Cell Biology, University of Minnesota. Figure 19.24: Photographs courtesy of M. G. Li and D. P. Snustad, Department of Genetics and Cell Biology, University of Minnesota. Page: 487: From Kerem, et al. (1989), *Science* 245:1073–1080. (Fig. 2). Figure 19.30: Courtesy of D. P. Snustad, Department of Genetics and Cell Biology, University of Minnesota.

Chapter 20
Opener: Courtesy of Susan Lanzendorf, Ph.D., Jones Institute for Reproductive Medicine/Eastern Virginia Medical School. Figure 20.11b): From the Huntington's Disease Collaborative Research Group, "The RFLP, restriction Map and Contig Map," *Cell*, Vol 72:971–983, Copyright © 1993 by Cell Press. Figure 20.20: Courtesy of Cellmark Diagnostics, Germantown, Maryland. Page 524: Hulton Deutsch Collection. Figure 20.21: Courtesy of Cellmark Diagnostics, Germantown, Maryland. Figure 20.24: Dr. R. L. Brinster, School of Veterinary Medicine, University of Pennsylvania. Figure 20.25 b): Photograph courtesy of G. Panzour, A. Das, Department of Biochemistry, University of Minnesota. Figure 20.28: Photograph courtesy of D. M. Shah, et al, *Science* 233:478–481, 1986. Copyright © 1986 by the American Association for the Advancement of Science.

Chapter 21
Opener: Courtesy of Dwight Anderson.

Conversation 6
Opener: Courtesy of Edward B. Lewis.

Chapter 22
Opener: Professor Aaron Polliack/Science Photo Library/Photo Researchers. Figure 22.1: D. Peter Snustad. Figure 22.7: From Gloria Coruzzi, Richard Broglie, Carol Edwards, and Nam-Hai Chua, "Tissue-specific and light-regulated expression of a pea nuclear gene encoding the small subunit of ribulose-1, 5–bisphosphate carboxylase," *EMBO J.* 3: 1671–1679, 1984. Figure 22.11 c): Reproduced with permission from E. Serfling, M. Jasin, and W. Schaffner, *Trends in Genet.* 1:224–230, 1985. Figure 22.14: O. L. Miller, B. R. Beatty, and D. W. Fawcett/Visuals Unlimited. Figure 22.15 a): Jack M. Bostrack/Visuals Unlimited. Figure 22.17: From D. W. Brown and I. B. Dawid, *Science* 160: 272–280, 1968. Copyright © 1968 by the American Association for the Advancement of Science. Figure 22.18 a): Photograph courtesy of B. Streifel and D. Arthur, Dept. of Laboratory Medicine and Pathology, University of Minnesota. Figure 22.18 b): Photograph courtesy of Jaroslav Cervenka, Division of Cytogenetics and Cell Genetics, Health Sciences, University of Minnesota. Figure 22.21: From M. I. Kuroda, et al., *Cell* 66:935–947, 1991, Fig. 6. Photograph courtesy of Dr. Mitzi Kuroda. Figure 22.22: Photographs courtesy of Mike Atkinson and Chris Frethem, Department of Cell Biology and Neuroanatomy, University of Minnesota.

Chapter 23
Opener (top left): Ross Hamilton/Tony Stone Images/New York, Inc. Opener (top right): Rod Planck/Tony Stone Images/New York, Inc. Opener (bottom left): James H. Carmichael, Jr./The Image Bank. Opener (bottom right): Stephen Dalton/Photo Researchers. Figure 23.1: Courtesy of Edward B. Lewis, California Institute of Technology. Figure 23.3: Courtesy of Michael M. Craig, Biomedical Sciences Dept., Southwest Missouri State University. Figure 23.6: From J. E. Sulston and H. R. Horvitz, *Develop. Biol.* 56: 110–156, 1977. Photograph courtesy of Dr. John Sulston. Figure 23.15: Daniel St. Johnston and Christiane Nüsslein-Volhard, *Cell* 68:201–219, 1992. Photograph courtesy of Christiane Nüsslein-Volhard. Figure 23.19: From W. Driever and C. Nüsslein-Volhard, "A Gradient of Biocoid Protein in Drosophila Embryos." *Cell* 54: 83–93, 1988. Reproduced with permission from the cover of *Cell,* July 1, 1988. Copyright © 1988 by Cell Press, Cambridge, MA. Figure 23.23: Courtesy of Matthew Scott, Howard Hughes Medical Institute. Figure 23.28: Courtesy of Walter Gehring, Universitat Basel, Switzerland.

Chapter 24
Opener: NIBSC/Science Photo Library/Photo Researchers. Figure 24.3 a): From Amit, et al., *Science* 233: 747, Copyright © 1986 the American Association for the Advancement of Science. Photograph courtesy of R. J. Poljak. Figure 24.3 b): From Amit, et al, *Science* 233: 747, Copyright © 1986 American Association for the Advancement of Science. Figure

24.7: Copyright © Boehringer Ingelheim International GmbH, photo Lennart Nilsson/Bonnier Alba AB. Figure 24.11: Dr. A. Liepins/Science Photo Library/Photo Researchers.

Conversation 7
Opener: Courtesy of Thomas Bouchard, Jr., Dept. of Psychology, University of Minnesota.

Chapter 25
Opener: Erich Lessing/Art Resource, NY. Ribera, Jusepe de. *The Club-footed Boy*, 1642. Louvre, Paris. France.

Chapter 26
Opener: Richard Wiess/Peter Arnold, Inc. Figure 26.2 a): Treat Davidson/Photo Researchers. Figure 26.9: Courtesy of Kathleen K. Siwicki, Biology Dept., Swarthmore College. Figure 26.11(left): Jean-Paul Ferrero/Jacana/Photo Researchers. Figure 26.11 (right): Mero/Jacana/Photo Researchers. Figure 26.12: From Jennifer R. Brown, 1996, "Nurturing and Not Nurturing in Mice," *Cell*, 86: 297–309, Fig. 2 c) and d). Copyright © Cell Press. Photo courtesy of Jennifer R. Brown, Children's Hospital Neurobiology Lab, Harvard Medical School. Figure 26.14: Courtesy of Professor William L. Nyhan, M.D., Ph.D., Department of Pediatrics, University of California, San Diego. Figure 26.17: Martin Rotker/Phototake.

Conversation 8
Opener: Courtesy of Brian & Deborah Charlesworth.

Chapter 27
Opener: Renee Lynn/Tony Stone Images/ New York, Inc. Figure 27.1: Granger Collection. Figure 27.2 a): Courtesy of the New York Public Library. Figure 27.2 b) (left): Jany Sauvanet/Photo Researchers. Figure 27.2 b) (right): Courtesy of the New York Public Library. Figure 27.6: Lincoln P. Brower. Figure 27.8: Courtesy of Professor Lawrence Cook, University of Manchester. Figure 27.14 (inset): Courtesy of Kenneth Kaneshiro, University of Hawai'i at Manoa, Photo by William Mull.

Chapter 28
Opener: Tom McHugh/Photo Researchers. Figure 28.1 a/left): Alexander Lowry. Figure 28.1a/right): Ron Sanford/Photo Researchers. Figure 28.1 b): Jim Brandenburg/Minden Pictures, Inc. Figure 28.2b): Darrell Gulin/Tony Stone Images/New York, Inc. Figure 28.10a): Barry L. Runk/Grant Heilman Photography. Figure 28.10b): Leonard Lee Rue III/Animals Animals. Figure 28.12 (left): Pat & Tom Leeson/Photo Researchers. Figure 28.12 (center): Carr Cliffton. Figure 28.12 (right): C. K. Lorenz/Photo Researchers.

Epilogue
Opener: Courtesy of James Crow.

ILLUSTRATION CREDITS

Chapter 1

Figure 1.4: From A. Baer, *Heredity and Society,* 2ed. Reprinted by permission of Prentice Hall, 1977.

Chapter 4

Figure 4.11b: From *Principles of Human Genetics,* 3ed., by C. Stern. Copyright (c) 1973 by Curt Stern. Used with permission of W.H. Freeman and Company.

Chapter 6

Technical Sidelight, Chapter 6, Figure 2: Burnham, C., *Discussions in Cytogenetics,* Copyright 1962. Reprinted by permission of Burgess International Group, Inc. Figure 6.20 a, b: From *Principles of Human Genetics,* 3ed., by Stern. Copyright (c) 1973 by Curt Stern. Used with permission of W.H. Freeman and Company.

Chapter 8

Technical Sidelight, Chapter 8, Figure 1b: From B.H. Judd, M.W. Shen and T.C. Kaufman, *Genetics*, 71:139, 1972. Reprinted by permission.

Chapter 10

Figure 10.25: From *DNA Replication,* 2ed., by Kornberg and Baker. Copyright (c) 1992 W.H. Freeman and Company. Used with permission.

Chapter 11

Figure 11.30: From N.J. Zang, P.J. Grabowski, and T.R. Cech. Reprinted with permission from *Nature*, 301:578–583. Copyright 1983 Macmillan Magazines Limited. Figure 11.32: From D.A. Brow and C. Guthrie. Reprinted with permission from *Nature,* 337:14–15. Copyright 1989 by Macmillan Magazines Limited.

Chapter 12

Figures 12.3, 12.4a,b: From Bruce Alberts, Dennis Bray, Julian Lewis, Martin Raff, Keith Roberts, and James D. Watson, *Molecular Biology of the Cell,* 3ed., page 114, Copyright 1994. Reprinted by permission of Garland Publishing, Inc. Figure 12.5: From Claude A. Villee, *Biology,* 2ed. Copyright 1989 by Saunders College Publishing, reproduced by permission of the publisher. Figure 12.11: From R.W. Holley, et al., in *Science*, Vol 147:1462–1465. Copyright 1965, American Association for the Advancement of Science. Reprinted by permission.

Chapter 15

Figures 15.11, 15.14: From S. Benzer, *Proceedings of the National Academy of Sciences,* 47:403 and 410, 1961. Reprinted by permission of S. Benzer. Figure 15.24: From Fauci in *Science*, 239:617. Copyright 1988. American Association for the Advancement of Science. Reprinted by permission. Figure 15.25: From W.C. Greene in *The New England Journal of Medicine*, 324:308, 1991. Reprinted by permission of The New England Journal of Medicine and permission of Dr. Warner C. Greene, Gladstone Institute of Virology and Immunology at the University of California, San Francisco.

Chapter 17

Figure 17.15: From J.K. Lim and M.J. Simmons, *Bioessays*, 1994 Vol 16:269–275. Reprinted by permission. Figure 17.17: From D.L. Lindsley and G. Zimm, *The Genome of Drosophila Melanogaster,* 1992. Reprinted by permission of Academic Press, Inc.

Chapter 18

Figures 18.8, 9: After Ruth Sager, *Cytoplasmic Genes and Organelles*, 1972, Academic Press, Inc. Reprinted by permission. Figures 18.13, 18.17: From *Molecular Cell Biology* by Darnell, Lodish, and Baltimore. Copyright (c) 1990 by Scientific American Books, Inc. Used with permission of W.H. Freeman and Company. Figure 18.15: From B. Wissinger, A. Brennicke, and W. Schuster, "Regenerating Good Senses RNA Editing and Trans Splicing in Plant Mitochondria." Vol 8:322–328, 1992. Reprinted by permission of *Trends in Genetics.*

Chapter 20

Figure 20.4: Based on data from NIH/CEPH Collaboration Mapping Group, *Science,* Vol 258:67–86. Copyright 1992, American Association for the Advancement of Science. Reprinted by permission. Figures 20.9: From the Huntington's Disease Collaborative Research Group, "The RFLP, restriction Map and Contig Map," *Cell*, Vol 72:971–983, Copyright 1993 by Cell Press. Reprinted by permission. Figure 20.12: From J.L. Marx, *Science*, 245:923–925, 1989. Reprinted by permission of Dr.Lap-Chee Tsui, The Hospital for Sick Children, Toronto, Canada. Figures 20.13,20.14: From F.S. Collins, *Science*, Vol 256:774–779. Copyright 1992, American Association for the Advancement of Science. Reprinted by permission. Figure 20.15: From Blake, et al., "Dystrophin Domain Structure." Reprinted by permission of *Trends in Cell Biology*, Vol 4:19–23, 1994.

Chapter 21

Figures 21.18, 21.25: From Benjamin Lewin, *Genes V*. Published by Oxford University Press, 1994. Reprinted by permission. Figure 21.27: From Jinks-Robertson and M. Nomura, "In Escherishia Coli and Salmonella Typhimurium," *Cellular and Molecular Biology*, Vol 2:1358–1385. Copyright 1987. Reprinted by permission of American Society for Microbiology.

Chapter 22

Figure 22.6: From *Molecular Cell Biology,* 2ed by Darnell, Lodish, and Baltimore. Copyright (c) 1990 by Scientific American Books, Inc. Used with permission of W.H. Freeman and Company.

Chapter 23

Figure 23.29: From Duboule and Morata in *Trends in Genetics*, Vol 10:358–364, 1994. Reprinted by permission. Figure 23.30: From McGinnis and Krumlauf in *Cell,* Vol 68:283–302. Copyright 1992 by Cell Press. Reprinted by permission.

Chapter 24
Figure 24.2b: From E.W. Silverton, M.A. Navia, and D.R. Davies, *Proceedings of the National Academy of Sciences*, Vol 74:5140. Copyright 1977. Reprinted by permission. Figure 24.6: From Alberts, et al, *Molecular Biology of The Cell,* 3ed., Copyright 1994. Reprinted by permission of Garland Publishing, Inc.

Chapter 25
Figure 25.3b: From H. Harris, *Principles of Human Biochemical Genetics,* 3ed., *Revised*. Copyright 1980 by Elsevier Science, B.V. Reprinted by permission of the Author. Figure 25.4: From Elof Carlson, *Gene: A Critical History*. Copyright 1966. Reprinted by permission of W.B. Saunders Company.

Chapter 26
Figure 26.3: From Daniel Hartl, *Genetics,* 3ed., Copyright 1994 Boston: Jones and Bartlett Publishers. Reprinted with permission. Figure 26.4: From Y. Citri, H.V. Colot, et al., "Locomotor Activity in Drosophila." Reprinted with permission of *Nature* 346:82, Copyright 1987, Macmillan Magazines Limited. Figure 26.5: From R.J. Konopka and S. Benzer, *Proceedings of the National Academy of Science*, 1971. Reprinted by permission of S.Benzer. Figures 26.6, 26.7a,b: From C.P. Kryiacou, et al., *Molecular Genetics of Biological Rhythms*, Marcel Dekkar, Inc. New York. Reprinted by permission.

Chapter 27
Figure 27.12: From *Genetics, Evolution, and Man* by Bodmer and Cavalli-Sforza. Copyright (c) 1976 by W.H. Freeman and Company. Used with permission. Figure 27.14: From H.L. Carson, *Genetics* 103:465, 1983. Reprinted by permission of Genetics Society of America.

Chapter 28
Human Genetics Sidelight, Chapter 28, Figure 1: From Rebecca L. Cann, et al., *Nature* 325:31, Copyright 1987. Reprinted with permission of *Nature,* Macmillan Magazines Limited. Figures 28.3, 28.4: From T. Dobzhansky, *Genetics of the Evolutionary Process*, Copyright (c) 1972 by Columbia University Press. Reprinted with permission of the publisher. Figures 28.6, 28.11, 28.15, 28.1 (table): From *Evolution* by Dobzhansky, et al., Copyright (c) 1977 by W.H. Freeman and Company. Used with permission. Figure 28.14: From *Modes of Speciation* by White. Copyright (c) 1978 by W.H. Freeman and Company. Used with permission. Figure 28.20: From Prakash, Yunis, *Science,* 215:1525–29. Reprinted with Permission from American Association for the Advancement of Science. Copyright 1982. Figures 28.23, 28.24: From King and Wilson, *Science*, 188:107–116. Reprinted with permission from American Association for the Advancement of Science. Copyright 1982.

INDEX

NOTE: An *f* after a page number denotes a figure; *t* denotes a table; *C* denotes a conversation; *HS* denotes a Historical Sidelight; *K* denotes a key point; *TS* denotes a Technical Sidelight; *HG* denotes a Human Genetics Sidelight.

E

F

O

P

S

T

Y

Z